Wolpert's Principles of Development

Wolpert's
Principles of Development

Seventh Edition

Cheryll Tickle

Alfonso Martinez Arias

Marysia Placzek

Lewis Wolpert

Mary Baylies | James Locke | Peter Lawrence

OXFORD
UNIVERSITY PRESS

OXFORD
UNIVERSITY PRESS

Great Clarendon Street, Oxford, OX2 6DP,
United Kingdom

Oxford University Press is a department of the University of Oxford.
It furthers the University's objective of excellence in research, scholarship,
and education by publishing worldwide. Oxford is a registered trade mark of
Oxford University Press in the UK and in certain other countries

© Oxford University Press 2025

The moral rights of the authors have been asserted

Fourth Edition 2011
Fifth Edition 2015
Sixth Edition 2019

All rights reserved. No part of this publication may be reproduced, stored in a retrieval system, transmitted, used for text and data mining, or used for training artificial intelligence, in any form or by any means, without the prior permission in writing of Oxford University Press, or as expressly permitted by law, by licence or under terms agreed with the appropriate reprographics rights organization. Enquiries concerning reproduction outside the scope of the above should be sent to the Rights Department, Oxford University Press, at the address above.

You must not circulate this work in any other form
and you must impose this same condition on any acquirer

Published in the United States of America by Oxford University Press
198 Madison Avenue, New York, NY 10016, United States of America

British Library Cataloguing in Publication Data
Data available

Library of Congress Control Number: 2024942421

ISBN 978-0-19-289661-2

Printed in the UK by
Bell & Bain Ltd., Glasgow

The manufacturer's authorised representative in the EU for product safety is
Oxford University Press España S.A. of El Parque Empresarial San Fernando de Henares,
Avenida de Castilla, 2 – 28830 Madrid (www.oup.es/en or product.safety@oup.com).
OUP España S.A. also acts as importer into Spain of products made by the manufacturer.

Links to third party websites are provided by Oxford in good faith and
for information only. Oxford disclaims any responsibility for the materials
contained in any third party website referenced in this work.

Lewis Wolpert
(1929–2021)

Lewis Wolpert, the originator of this textbook, was an eminent developmental biologist whose ideas have had a profound and lasting impact on the way we think about development. His philosophy forms the basis of the textbook, first published in 1998. He believed that, even though the development of different organisms seems overwhelmingly complex, there are general principles. Lewis suggested there are four main developmental processes—pattern formation, cell differentiation, morphogenesis, and growth—and that these are accomplished by a limited repertoire of cell activities. His view was that 'development can be best understood by understanding how genes control cell behavior'. This short tribute outlines his career and how his ideas developed.

Lewis did not set out to be a developmental biologist. He was born in South Africa and studied engineering at the University of Witwatersrand. After working for several years as a soil mechanic, he embarked on a PhD in cell biology (aged 25) at King's College London. This change in direction occurred because his friend, Wilfred Stein, chanced on a paper about the mechanics of cell division and thought that, with his background in mechanics, Lewis might be interested in this topic. To obtain a PhD in biology, Lewis not only had to carry out his thesis research but also take a final-year course on vertebrates—and pass the exams!

During his PhD, Lewis spent his summers at the Kristineberg Marine Laboratory in Sweden, collecting and fertilizing sea urchin eggs to study the mechanics of their cell division. Here he was introduced to developmental biology. He met the developmental biologist Trygve Gustafson who was making films of gastrulating sea urchin embryos. Lewis is famous for the typically witty comment attributed to him, and still often quoted by developmental biologists: 'The most important event in your life is not birth, marriage, or death, but gastrulation.'

A key question in development—both in morphogenesis and pattern formation—is where and when cells undergo different activities. Lewis realized that any explanation had to take account of regulative development and made a vivid analogy with the French flag, which has the same pattern of blue, white, and red stripes, irrespective of its size (see Section 1.15). To explain how patterns form and regulate, he put forward the concept of positional information, in which cells are first informed of their position and then interpret this information to behave appropriately, published in a landmark paper in 1969.

Lewis first applied his ideas to regeneration in *Hydra* and then, when he moved to the Middlesex Hospital Medical School in London in the late 1960s, to the developing chick wing. Experiments carried out by his students and postdocs provided evidence that cell position in the chick wing is specified by the combination of morphogen gradients and timing. Thus, the chick wing, a seemingly esoteric topic, became a universal model for pattern formation. But in the 1970s, this was all theoretical. For example, no morphogens were known. It was only in the 1980s that developmentally important genes, including genes that encode morphogens, were discovered. Lewis's ideas provided the framework for understanding the molecular basis of pattern formation and developmental biologists continue to build on his foundational work.

Preface

How a fertilized egg gives rise to an organism is a source of amazement and wonder, but is also one of the most challenging problems in biology. Lewis Wolpert designed this textbook to set out the general principles that govern embryonic development, showing how they are used in laying down the body plan, in the development of selected organs, and in regeneration in animals and plants. To recognize Lewis as the founding author, and his work as inspiration for the textbook, this new seventh edition is titled *Wolpert's Principles of Development*. The general principles have not changed since the first edition but we are now able to answer fundamental questions in new ways thanks to technical advances, and so gain greater insight into how these core principles are established and employed.

The principles governing development emerged through the study of model organisms but the ultimate aim of developmental biology has always been to understand the development of human embryos. Since the sixth edition, significant advances in establishing new models of mammalian development using cultured stem cells from both mice and humans have opened up avenues to tackle this most ambitious aim. This exciting area is discussed in a new chapter—Chapter 7—in this edition. In addition, the CRISPR-Cas9 technique for gene editing has become a standard tool for genetic manipulations of developing embryos, and our description of the technique is now in Chapter 1. Another technique that has become more widely used is single-cell RNA sequencing and examples of its use will be found throughout.

We concentrate on the development of vertebrates and *Drosophila* but include other animals, such as the nematode, the sea urchin, and *Hydra*, where they best illustrate a particular concept. As in previous editions, there is also a chapter on plant development, which illustrates similarities and differences to animal development. Chapter 1 provides a brief history of embryology and an introduction to the main principles and processes involved. Chapter 2 considers the process of pattern formation in laying down the body plan in *Drosophila*. This small fly has played, and still plays, a central role in elucidating developmental mechanisms. Chapters 3, 4, and 5 in the previous edition, which covered the process of pattern formation in laying down the body plan in our vertebrate model organisms—*Xenopus*, zebrafish, chick, and mouse—have been reorganized into two chapters—Chapter 3 and Chapter 4, with many new figures. The new Chapter 3 begins by outlining the main stages in vertebrate development together with the methods used to study them. This is followed by a description of the stages in development of *Xenopus* and how the body axes and the organizer are established in this organism. We then consider these aspects of development in the other model organisms. Chapter 3 ends with a description of the laying down of the body plan in human embryos and the development of the placenta. The developmental events described in Chapter 3 overlap in time with those described in Chapter 4, where we consider the patterning events that take place during gastrulation and the initiation of development of the nervous system. We go on to look at aspects of how the body plan is completed, focusing mainly on chick and mouse, and end by considering how left–right asymmetry is established.

Chapter 5 (Chapter 7 in the sixth edition) focuses on the fundamental developmental process of morphogenesis, including details of how cell movements are coordinated during gastrulation and the early development of the nervous system. The discussion on differentiation (Chapter 8 in the sixth edition) has been reorganized into two chapters. Chapter 6 covers cell differentiation in general, the technique of single-cell RNA analysis and its application to early mouse and human embryos, the differentiation of selected tissues, and the role of stem cells in the development of tissues that are renewed throughout life. Discussion of embryonic stem cells, organoids, and potential clinical uses of stem cells has been moved to a new chapter—Chapter 7—which has been expanded to cover the use of embryonic stem cells as a model for developmental biology. It presents entirely new work on blastoids, gastruloids, and stem cell embryos—three-dimensional structures derived from stem cells that mimic early stages of mammalian development. Chapter 8 deals with germ cells, fertilization, and sex determination. Organogenesis (Chapter 9) and the development of the nervous system (Chapter 10) are huge topics, so, as previously, we have had to be very selective in our coverage. Chapter 10 has been extensively revised and updated, with many new figures. Growth and regeneration are considered in Chapter 11 and the sections on regeneration, a very active area of research, have been updated. Chapter 12 deals with plant development and includes new work on the patterning of leaves, roots, and flowers, and on the mechanism of vernalization. Chapter 13 deals with development in relation to evolution, and highlights the impact of genomics and the many insights it has brought to this aspect of developmental biology.

Wolpert's Principles of Development is designed for undergraduates, and we have tried to make these principles as clear as possible and to provide numerous summaries, in both words and pictures. The illustrations are a special feature and have been carefully designed and chosen to illuminate both experiments and mechanisms. New diagrams and photographs are included throughout the book, together with information about their sources.

We have assumed that students have some familiarity with basic cell and molecular biology and genetics, but all key concepts, such as the control of gene activity, are explained in the text. There is an extensive Glossary, which means that the book is self-contained. In providing further reading, our prime concern has been to guide the student to particularly helpful papers and reviews, rather than to give credit to all the scientists who have made major contributions: to those whom we have neglected, we apologize.

The main authors from the previous edition, Cheryll Tickle and Alfonso Martinez Arias, have been joined by Marysia Placzek who revised the chapter on the nervous system and helped to reorganize Chapters 3 and 4. The long-standing co-author, Peter Lawrence, and co-author James Locke from the previous edition have been joined by a new co-author, Mary Baylies, who reviewed Chapter 2 and helped to update Chapters 6 and 7. Each chapter has also been reviewed by experts (see page xxvi), and we thank them for their input. The authors made the initial revisions, which were then edited and incorporated by our editor, Eleanor Lawrence. Eleanor is more familiar with the textbook than anyone else and her invaluable expertise and influence pervades the book. We also thank Matthew McClements for creating the new illustrations for this edition. Both Eleanor and Matthew have been involved since the first edition and their commitment over more than 25 years has ensured continuity. Finally, we are indebted to Jonathan Crowe at Oxford University Press for his help and guidance over the past five editions and to Judith Lorton at the Press who is responsible for preparing the e-version.

C.T.

Bath, August 2023

A.M.A.

Barcelona, August 2023

M.P.

Sheffield, August 2023

Learning from this book

Wolpert's Principles of Development includes a number of features to help make it easy to use, and to make your learning as effective as possible.

At the start of each chapter

At the start of each chapter we provide a bullet-pointed list of the sections of the chapter, a summary of the chapter content, and the chapter introduction. This all helps to give you an overview of the content you're about to read.

Throughout the chapter

Experimental Boxes discuss both classic and current experimental research, demonstrating 'how we know what we know'.

Cell Biology Boxes equip you with a robust conceptual framework on which to add further detail from the vast amount of scientific information available to us today.

Medical Boxes illustrate the direct relevance of developmental biology to medicine and health-related issues.

Special Interest Boxes highlight topics of interest such as 'Identical twins' in Chapter 1, 'Putting it together: the development of a simple neural circuit' in Chapter 10 and the 'Origins of morphological diversity in dogs' in Chapter 13.

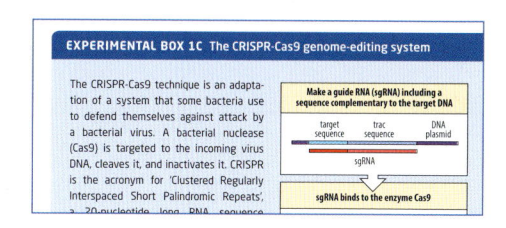

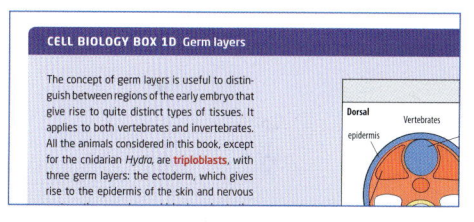

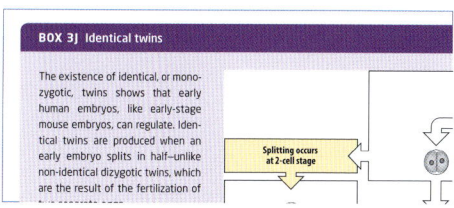

Learning from this book

Summary Boxes provide a brief overview of each main section, which we hope you will find particularly useful when revising for examinations.

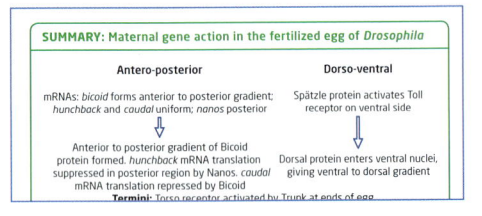

Summary diagrams provide brief visual overviews of key concepts in that section.

At the end of each chapter

Summary
A bullet-point review of the chapter's major concepts at the end of each chapter.

End of chapter questions
These are concept questions that require longer answers than the shorter questions you'll have come across throughout the chapter.

About the authors

Cheryll Tickle, Emeritus Professor, University of Bath. Her research focused on how cells become organized in embryos using the developing chick limb as her main experimental model. She has lectured on developmental biology to biology undergraduates, also medical and dental students, and to scientists from a wide range of different backgrounds, including industry and veterinarians on continuing professional development courses.

Alfonso Martinez Arias, ICREA Research Professor at the Universitat Pompeu Fabra (UPF). His research makes use of pluripotent stem cells to develop *in vitro* models of early mammalian development with a focus on the process of gastrulation. Before moving to the UPF he spent 35 years at the University of Cambridge (UK) where he did research and taught developmental biology in graduate and undergraduate courses.

Marysia Placzek, Professor at the University of Sheffield. Marysia is a neurobiologist who studies development of the vertebrate neural tube, with a particular focus on the hypothalamus. For the last 30 years she has combined research with undergraduate and postgraduate teaching, coordinating modules in basic and translational developmental biology.

Lewis Wolpert was Emeritus Professor of Biology as Applied to Medicine at University College London. He was an eminent developmental biologist whose ideas have had a profound and lasting impact on the way we think about development. He believed that, even though the development of different organisms seems to be overwhelmingly complex, there are general principles. His philosophy forms the basis of *Wolpert's Principles of Development*, which was first published in 1998. Lewis also authored several books aimed at the general public including *The Triumph of the Embryo* and an influential book on depression, *Malignant Sadness*, based partly on his own experiences.

Mary Baylies, Member, Sloan Kettering Institute, Memorial Sloan Kettering Cancer Center and Professor, Weill Cornell Graduate School of Biomedical Sciences, Cornell Medical School.

James Locke, Professor, Sainsbury Laboratory, University of Cambridge.

Peter Lawrence, Emeritus Scientist, MRC Laboratory of Molecular Biology, Cambridge.

Eleanor Lawrence, Freelance Science Writer and Editor.

Matthew McClements, Science Illustrator.

Summary of contents

Preface		vii
Learning from this book		ix
About the authors		xi
List of boxes		xxv
Reviewer acknowledgments		xxvii

Chapter 1	History and basic concepts		1
Chapter 2	Development of the *Drosophila* body plan		43
Chapter 3	Vertebrate development I: establishing the body plan		101
Chapter 4	Vertebrate development II: completing the body plan		187
Chapter 5	Morphogenesis: change in form in the early embryo		247
Chapter 6	Cell differentiation and stem cells		311
Chapter 7	Embryonic stem cells, embryo models, and regenerative medicine		367
Chapter 8	Germ cells, fertilization, and sex determination		401
Chapter 9	Organogenesis		441
Chapter 10	Development of the nervous system		517
Chapter 11	Growth, post-embryonic development, and regeneration		571
Chapter 12	Plant development		629
Chapter 13	Evolution and development		673
	Glossary		727
	Index		751

Contents

Preface	vii
Learning from this book	ix
About the authors	xi
List of boxes	xxv
Reviewer acknowledgments	xxvii

Chapter 1 History and basic concepts — 1

Introduction — 1

■ Box 1A Basic stages of *Xenopus laevis* development — 3

The origins of developmental biology — 4

1.1 Aristotle first defined the problem of epigenesis versus preformation — 4

1.2 Cell theory changed how people thought about embryonic development and heredity — 5

■ Cell Biology Box 1B The mitotic cell cycle — 6

1.3 Two main types of development were originally proposed — 6

1.4 The discovery of induction showed that one group of cells could determine the development of neighboring cells — 8

1.5 Developmental biology emerged from the coming together of genetics and embryology — 8

1.6 Development is studied through selected model organisms — 10

1.7 The first developmental genes were identified as spontaneous mutations — 12

■ Experimental Box 1C The CRISPR-Cas9 genome-editing system — 15

Summary — 16

A conceptual tool kit — 17

1.8 Development involves the emergence of pattern, change in form, cell differentiation, and growth — 17

■ Cell Biology Box 1D Germ layers — 19

1.9 Cell behavior provides the link between gene action and developmental processes — 20

1.10 Genes control cell behavior by specifying which proteins are made — 21

1.11 The expression of developmental genes is under tight control — 23

■ Experimental Box 1E Visualizing gene expression in embryos — 24

■ Cell Biology Box 1F Gene regulatory networks and cell-fate decisions — 26

1.12 Development is progressive and the fates of cells become determined at different times — 27

1.13 Inductive interactions make cells different from each other — 29

■ Cell Biology Box 1G Signal transduction and intracellular signaling pathways — 31

■ Medical Box 1H When development goes awry — 32

1.14 The response to inductive signals depends on the state of the cell — 33

1.15 Patterning can involve the interpretation of positional information — 33

1.16 Lateral inhibition can generate spacing patterns — 35

1.17 Localization of cytoplasmic determinants and asymmetric cell division can make daughter cells different from each other — 36

1.18 The embryo contains a generative rather than a descriptive program — 37

1.19 The reliability of development is achieved by various means — 38

1.20 The complexity of embryonic development is due to the complexity of cells themselves — 38

1.21 Development is a central element in evolution — 39

Summary — 40

Summary to Chapter 1 — 40

Chapter 2 Development of the *Drosophila* body plan — 43

Introduction — 43

***Drosophila* life cycle and outline of development** — 44

2.1 The early *Drosophila* embryo is a multinucleate syncytium — 45

2.2 Cellularization is followed by gastrulation and segmentation — 46

2.3 After hatching, the *Drosophila* larva develops through several larval stages, pupates, and then undergoes metamorphosis to become an adult — 48

2.4 Many developmental genes were identified in *Drosophila* through large-scale genetic screening for induced mutations — 49

■ Experimental Box 2A Mutagenesis and genetic screening strategy for identifying developmental mutants in *Drosophila* — 50

Summary — 51

Setting up the body axes — 51

2.5 The body axes are set up while the *Drosophila* embryo is still a syncytium — 51

2.6 Maternal factors set up the body axes and direct the early stage of *Drosophila* development — 53

2.7 Three classes of maternal genes specify the antero-posterior axis — 54

2.8 Bicoid protein provides an antero-posterior gradient of a morphogen — 55

2.9 The posterior pattern is controlled by the gradients of Nanos and Caudal proteins — 56

2.10 The anterior and posterior extremities of the embryo are specified by activation of a cell-surface receptor — 58

2.11 The dorso-ventral polarity of the embryo is specified by localization of maternal proteins in the egg vitelline envelope — 58

2.12 Positional information along the dorso-ventral axis is provided by the Dorsal protein — 59

■ **Cell Biology Box 2B** The Toll signaling pathway: a multifunctional pathway — 60

Summary — 61

Localization of maternal determinants during oogenesis — 61

2.13 The antero-posterior axis of the *Drosophila* egg is specified by signals from the preceding egg chamber and by interactions of the oocyte with follicle cells — 62

■ **Cell Biology Box 2C** The JAK-STAT signaling pathway — 64

2.14 Localization of maternal mRNAs to either end of the egg depends on the reorganization of the oocyte cytoskeleton — 65

2.15 The dorso-ventral axis of the egg is specified by movement of the oocyte nucleus followed by signaling between oocyte and follicle cells — 67

Summary — 67

Patterning the early embryo — 68

2.16 The expression of zygotic genes along the dorso-ventral axis is controlled by Dorsal protein — 68

2.17 The Decapentaplegic protein acts as a morphogen to pattern the dorsal region — 71

2.18 The antero-posterior axis is divided up into broad regions by gap gene expression — 73

2.19 Bicoid protein provides a positional signal for the anterior expression of zygotic *hunchback* — 74

2.20 The gradient in Hunchback protein activates and represses other gap genes — 75

■ **Experimental Box 2D** Targeted gene expression and misexpression screening — 76

Summary — 77

Activation of the pair-rule genes and the establishment of parasegments — 78

2.21 Parasegments are delimited by expression of pair-rule genes in a periodic pattern — 78

2.22 Gap gene activity positions stripes of pair-rule gene expression — 78

Summary — 82

Segmentation genes and segment patterning — 82

2.23 Expression of the *engrailed* gene defines the boundary of a parasegment, which is also a boundary of cell-lineage restriction — 82

2.24 Segmentation genes stabilize parasegment boundaries — 84

2.25 Signals generated at the parasegment boundary delimit and pattern the future segments — 84

■ **Cell Biology Box 2E** The Hedgehog signaling pathway — 87

■ **Experimental Box 2F** Mutants in denticle pattern provided clues to the logic of segment patterning — 87

Summary — 89

Specification of segment identity — 90

2.26 Segment identity in *Drosophila* is specified by Hox genes — 90

2.27 Homeotic selector genes of the bithorax complex are responsible for diversification of the posterior segments — 91

2.28 The Antennapedia complex controls specification of anterior regions — 93

2.29 The order of Hox gene expression corresponds to the order of genes along the chromosome — 93

2.30 The *Drosophila* head region is specified by genes other than the Hox genes — 94

Summary — 94

Summary to Chapter 2 — 95

Chapter 3 Vertebrate development I: establishing the body plan — 101

Introduction — 102

Outline of vertebrate development — 103

3.1 All vertebrate embryos pass through a broadly similar set of developmental stages — 103

Experimental approaches to studying vertebrate development — 107

3.2 Normal development can be visualized by various types of imaging, by detecting patterns of gene and protein expression, and by making fate maps — 108

3.3 Disrupting the developmental process in specific ways helps to discover underlying mechanisms — 109

■ **Experimental Box 3A** Fate mapping and lineage tracing reveal which cells in which parts of the early *Xenopus* embryo give rise to particular structures in the tadpole — 110

3.4 Not all techniques are equally applicable to all vertebrates — 112

■ **Experimental Box 3B** Ways of making transgenic mice — 115

■ **Experimental Box 3C** Large-scale mutagenesis screens for recessive mutations in zebrafish — 116

Xenopus development and setting up the body plan — 118

3.5 The frog *Xenopus laevis* is the model amphibian for studying development of the body plan — 118

3.6 The earliest stages in *Xenopus* development are controlled by maternal factors — 122

■ **Cell Biology Box 3D** Intercellular protein signals in vertebrate development — 124

3.7 Cortical rotation specifies the future dorsal side of the *Xenopus* embryo — 125

■ **Cell Biology Box 3E** The Wnt/β-catenin signaling pathway — 128

3.8 The Nieuwkoop center, a signaling center, develops on the dorsal side of the blastula — 129

3.9 Control of development is transferred from the maternal to the zygotic genome at the mid-blastula transition — 130

3.10 The axes of the blastula are transformed by gastrulation into the body axes of the tailbud embryo — 130

Summary — 131

The origins and specification of the germ layers and formation of the organizer in *Xenopus* — 132

3.11 Mesoderm is induced and its patterning is initiated by signals from vegetal and dorsal regions of the blastula — 132

■ **Cell Biology Box 3F** Signaling by members of the TGF-β family of growth factors — 133

■ **Experimental Box 3G** Timing and cell-cell interactions in mesoderm specification — 135

3.12 Members of the TGF-β family have been identified as mesoderm inducers and induce formation of the organizer — 137

■ **Experimental Box 3H** Investigating receptor function using dominant-negative proteins — 138

3.13 Threshold responses to gradients of signal proteins are likely to pattern the mesoderm — 139

3.14 Regulation is still possible at the blastula stage — 140

Summary — 141

Zebrafish development and setting up the body plan — 142

3.15 The zebrafish embryo develops around a large yolk cell — 142

3.16 Genetic studies show that the axes of the zebrafish blastoderm are established by maternal determinants — 145

3.17 The germ layers are specified in the zebrafish blastoderm by signals similar to those in *Xenopus* — 146

Summary — 148

Chick development and setting up the body plan — 148

3.18 The chick is a classic embryological model, especially from gastrulation to organogenesis stages — 149

3.19 In the chick, the main reference axis for development of the body plan is the antero-posterior axis — 154

3.20 Mesodermal transcription factors are upregulated during primitive streak formation — 156

3.21 A fate map for the chick can be made after primitive streak formation — 156

Summary — 157

Mouse development and setting up the body plan — 157

3.22 The mouse is the major model for mammalian development — 158

3.23 Blastomeres in the morula are specified as trophectoderm or inner cell mass depending on their position — 162

3.24 The inner cell mass becomes differentiated into primitive endoderm and cells that will form the embryo proper — 164

3.25 The first asymmetry in mouse development appears in the blastocyst — 165

3.26 The antero-posterior body axis of the mouse embryo is predicted by the movement of the anterior visceral endoderm — 166

3.27 Apart from its topology the fate map of the mouse is similar to that of the chick — 168

Summary — 168

Development of the human body plan — 169

3.28 The early development of a human embryo is generally similar to that of the mouse but the topology is more like that of the chick — 169

■ **Medical Box 3I** Pre-implantation genetic diagnosis — 172

3.29 The timing of formation and the anatomy of the human placenta differ from that in the mouse — 173

3.30 Some studies of human development are possible but are subject to strict laws — 175

■ **Box 3J** Identical twins — 176

Summary — 178

Summary to Chapter 3 — 178

Chapter 4   Vertebrate development II: completing the body plan — **187**

Introduction — 187

The *Xenopus* Spemann organizer patterns the dorso-ventral and antero-posterior body axes — 189

4.1 Secreted proteins from the organizer pattern the dorso-ventral axis by antagonizing the effects of ventral signals — 189

4.2 Secreted proteins from the organizer pattern the antero-posterior axis by antagonizing the effects of posterior signals — 191

Summary — 193

The *Xenopus* Spemann organizer induces neural tissue and initiates neural patterning — 193

4.3 The neural plate is induced from the ectoderm through the action of BMP antagonists — 194

■ **Cell Biology Box 4A** The fibroblast growth factor signaling pathway — 197

4.4 Factors that pattern the antero-posterior body axis also pattern the antero-posterior axis of the nervous system — 197

Contents

4.5 The final *Xenopus* body plan emerges by the end of gastrulation and neurulation — 199

Summary — 200

Organizer activity in the zebrafish, chick, and mouse — 200

4.6 The shield in zebrafish has organizing activity — 200

■ **Experimental Box 4B** A zebrafish gene regulatory network — 203

4.7 Hensen's node in chick is the embryonic organizer — 203

4.8 Mesoderm patterning in mouse is similar to that in other model organisms, but antero-posterior organizer functions are spatially and temporally separated — 206

■ **Cell Biology Box 4C** Fine-tuning Nodal signaling — 207

■ **Experimental Box 4D** The Cre/*loxP* system: a strategy for making gene knock-outs in mice — 208

Summary — 211

Elongation of the body and the origin of posterior tissues — 211

4.9 Axial structures in chick and mouse are generated from self-renewing cell populations — 212

■ **Cell Biology Box 4E** Retinoic acid: a small-molecule intercellular signal — 214

Summary — 216

Somite formation and the Hox code in antero-posterior patterning — 216

4.10 Somites are formed in a well-defined order along the antero-posterior axis — 217

■ **Cell Biology Box 4F** The Notch signaling pathway — 219

4.11 Identity of somites along the antero-posterior axis is specified by Hox gene expression — 221

■ **Box 4G** The Hox genes — 223

4.12 Deletion or overexpression of Hox genes causes changes in axial patterning — 226

4.13 Hox gene expression is activated in an anterior to posterior pattern — 228

4.14 The fate of somite cells is determined by signals from the adjacent tissues — 229

Summary — 231

The origin, behavior, and patterning of neural crest — 232

4.15 Neural crest cells arise from neural plate border cells and migrate to give rise to a wide range of different cell types — 232

4.16 Neural crest cells migrate from the hindbrain to populate the branchial arches — 235

Summary — 236

Determination of left-right asymmetry — 236

4.17 The bilateral symmetry of the early embryo is broken to produce left-right asymmetry of internal organs — 237

Summary — 240

Summary to Chapter 4 — 240

Chapter 5 Morphogenesis: change in form in the early embryo — 247

Introduction — 247

Cell adhesion — 249

■ **Cell Biology Box 5A** Cell-adhesion molecules and cell junctions — 250

5.1 Sorting out of dissociated cells demonstrates differences in cell adhesiveness in different tissues — 251

5.2 Cadherins can provide adhesive specificity — 252

5.3 The activity of the cytoskeleton regulates the mechanical properties of cells and their interactions with each other — 254

■ **Cell Biology Box 5B** The cytoskeleton, cell-shape change, and cell movement — 254

5.4 Transitions of tissues from an epithelial to a mesenchymal state, and vice versa, involve changes in adhesive junctions — 255

Summary — 256

Cleavage and formation of the blastula — 257

5.5 The orientation of the mitotic spindle determines the plane of cleavage at cell division — 258

5.6 The positioning of the spindle within the cell also determines whether daughter cells will be the same or different sizes — 260

5.7 Cells become polarized in the sea urchin blastula and the mouse morula — 261

5.8 Fluid accumulation as a result of tight-junction formation, ion transport, and hydraulic fracturing forms the blastocoel of the mammalian blastocyst — 264

Summary — 264

Gastrulation movements — 266

5.9 Gastrulation in the sea urchin involves an epithelial-to-mesenchymal transition, cell migration, and invagination of the blastula wall — 266

5.10 Mesoderm invagination in *Drosophila* is due to changes in cell shape controlled by genes that pattern the dorso-ventral axis — 270

5.11 Germ-band extension in *Drosophila* involves myosin-dependent remodeling of cell junctions and cell intercalation — 271

5.12 Planar cell polarity confers directionality on a tissue — 273

5.13 Gastrulation in amphibians and fish involves involution, epiboly, and convergent extension — 276

■ **Box 5C** Convergent extension — 278

5.14 *Xenopus* notochord development illustrates the dependence of medio-lateral cell elongation and cell intercalation on a pre-existing antero-posterior polarity — 281

5.15 Gastrulation in chick and mouse embryos involves the separation of individual cells from the epiblast and their ingression through the primitive streak — 282

5.16 Differences in gastrulation between vertebrate species result from the global control of conserved mechanical features of cell populations — 286

Summary — 287
Neural tube formation — 288

5.17 Neural tube formation is driven by changes in cell shape and convergent extension — 289

■ **Cell Biology Box 5D** Eph receptors and their ephrin ligands — 292

■ **Medical Box 5E** Neural tube defects — 293

Summary — 294
Formation of tubes and branching morphogenesis — 294

5.18 The *Drosophila* tracheal system is a prime example of branching morphogenesis — 295

5.19 The vertebrate vascular system develops by vasculogenesis followed by sprouting angiogenesis — 296

5.20 New blood vessels are formed from pre-existing vessels in angiogenesis — 297

Summary — 298
Cell migration — 298

5.21 Embryonic neural crest gives rise to a wide range of different cell types — 298

5.22 Neural crest migration is controlled by environmental cues — 299

5.23 The formation of the lateral-line primordium in fishes is an example of collective cell migration — 301

5.24 Body wall closure occurs in *Drosophila*, *Caenorhabditis*, mammals, and chick — 302

Summary — 304
Summary to Chapter 5 — 304

Chapter 6 Cell differentiation and stem cells — 311

Introduction — 312

■ **Box 6A** Conrad Waddington's 'epigenetic landscape' provides a framework for thinking about how cells differentiate — 313

The control of gene expression — 315

6.1 Control of transcription involves both general and tissue-specific transcriptional regulators — 316

6.2 Gene expression is also controlled by epigenetic chemical modifications to DNA and histone proteins that alter chromatin structure — 319

■ **Cell Biology Box 6B** Epigenetic control of gene expression by chromatin modification — 323

6.3 Patterns of gene activity can be inherited by persistence of gene-regulatory proteins or by maintenance of chromatin modifications — 326

6.4 Changes in patterns of gene activity during differentiation can be triggered by extracellular signals — 327

Summary — 328
Cell differentiation and stem cells — 329

6.5 The early stages of cell differentiation during gastrulation can be detected by single-cell transcriptional analysis — 329

■ **Experimental Box 6C** Single-cell analysis of cell-fate decisions — 330

6.6 Muscle differentiation is determined by the MyoD family of transcription factors — 334

6.7 The differentiation of muscle cells involves withdrawal from the cell cycle, but is reversible — 337

6.8 All blood cells are derived from multipotent stem cells — 338

6.9 Intrinsic and extrinsic changes control differentiation of the hematopoietic lineages — 341

6.10 Developmentally regulated globin gene expression is controlled by regulatory sequences far distant from the coding regions — 343

■ **Medical Box 6D** Towards a therapy for β-globin gene defects — 346

6.11 The epidermis of adult mammalian skin is continually being replaced by derivatives of stem cells — 346

■ **Medical Box 6E** Treatment of junctional epidermolysis bullosa with skin grown from genetically corrected stem cells — 349

6.12 Stem cells use different modes of division to maintain tissues — 350

6.13 The lining of the gut is another epithelial tissue that requires continuous renewal — 351

6.14 Skeletal muscle and neural cells can be renewed from stem cells in adults — 353

Summary — 355
The plasticity of the differentiated state — 355

6.15 Patterns of gene activity in differentiated cells can be changed by cell fusion — 356

6.16 The differentiated state of a cell can change by transdifferentiation — 357

6.17 Nuclei of differentiated cells can support development — 359

Summary — 361
Summary to Chapter 6 — 361

Chapter 7 Embryonic stem cells, embryo models, and regenerative medicine — 367

Introduction — 368
Embryonic stem cells and induced pluripotent stem cells — 368

7.1 Embryonic stem cells are pluripotent cells derived from mammalian blastocysts and can contribute to normal development *in vivo* — 368

7.2 Human ES cells differ from mouse ES cells in several of their properties — 370

7.3 Adult differentiated cells can be reprogrammed to form pluripotent stem cells — 372

■ **Experimental Box 7A** Induced pluripotent stem cells — 373

7.4 Pluripotent stem cells are used as models of cell differentiation — 375

Summary — 377
Embryo models — 377

7.5 ES cells can be used to provide models of early development — 377

- Box 7B Self-organization and self-assembly — 379
- 7.6 Pluripotent stem cells can model the process of gastrulation — 381
- Box 7C Somitogenesis in Matrigelled gastruloids — 383
- Summary — 387

Medical applications of stem cells — 387

- 7.7 Stem cells could be a key to regenerative medicine — 388
- 7.8 Various approaches could be used to generate differentiated cells for cell-replacement therapies — 389
- 7.9 ES cells can be used to generate organoids and tissues — 391
- Experimental Box 7D Intestinal organoids can be produced from adult stem cells or ES cells — 392
- 7.10 Stem cells might be used to grow human organs in other animals for transplantation — 394
- Summary — 395
- Summary to Chapter 7 — 396

Chapter 8 Germ cells, fertilization, and sex determination — 401

Introduction — 401

The development of germ cells — 402

- 8.1 Germ cell fate is specified in some embryos by a distinct germplasm in the egg — 403
- 8.2 In mammals germ cells are induced by cell–cell signaling during development — 406
- 8.3 Germ cells migrate from their site of origin to the gonad — 407
- 8.4 Germ cells are guided to their destination by chemical signals — 408
- 8.5 Germ cell differentiation involves a halving of chromosome number by meiosis — 408
- Box 8A Polar bodies — 410
- 8.6 Oocyte development can involve gene amplification and contributions from other cells — 412
- 8.7 Factors in the cytoplasm maintain the totipotency of the egg — 413
- 8.8 In mammals some genes controlling embryonic growth are 'imprinted' — 413
- Summary — 416

Fertilization — 417

- 8.9 Fertilization involves cell-surface interactions between egg and sperm — 418
- 8.10 Changes in the egg plasma membrane and enveloping layers at fertilization block polyspermy — 420
- 8.11 Sperm–egg fusion causes a calcium wave that results in egg activation — 422
- Summary — 423

Determination of the sexual phenotype — 424

- 8.12 The primary sex-determining gene in mammals is on the Y chromosome — 424
- 8.13 Mammalian sexual phenotype is regulated by gonadal hormones — 426
- 8.14 The primary sex-determining factor in *Drosophila* is the number of X chromosomes and is cell autonomous — 427
- 8.15 Somatic sexual development in *Caenorhabditis* is determined by the number of X chromosomes — 429
- 8.16 Determination of germ cell sex depends on both genetic constitution and intercellular signals — 430
- 8.17 Various strategies are used for dosage compensation of X-linked genes — 433
- Summary — 435
- Summary to Chapter 8 — 436

Chapter 9 Organogenesis — 441

Introduction — 441

The insect wing and leg — 442

- 9.1 Imaginal discs arise from the ectoderm in the early *Drosophila* embryo — 442
- 9.2 Imaginal discs arise across parasegment boundaries and are patterned by signaling at compartment boundaries — 444
- 9.3 The adult wing emerges at metamorphosis after folding and evagination of the wing imaginal disc — 445
- 9.4 A signaling center at the boundary between anterior and posterior compartments patterns the *Drosophila* wing disc along the antero-posterior axis — 446
- Box 9A Positional information and morphogen gradients — 449
- 9.5 A signaling center at the boundary between dorsal and ventral compartments patterns the *Drosophila* wing along the dorso-ventral axis — 451
- 9.6 Vestigial is a key regulator of wing development that acts to specify wing identity and control wing growth — 451
- 9.7 The *Drosophila* wing disc is also patterned along the proximo-distal axis — 453
- 9.8 The leg disc is patterned in a similar manner to the wing disc, except for the proximo-distal axis — 454
- 9.9 Different imaginal discs can have the same positional values — 456
- Summary — 456

The vertebrate limb — 458

- 9.10 The chick wing is a model for limb development — 458
- 9.11 Genes expressed in the lateral plate mesoderm induce limb development and specify limb polarity and identity — 461
- 9.12 The apical ectodermal ridge is required for limb bud outgrowth and the formation of structures along the proximo-distal axis of the limb — 463
- 9.13 Formation and outgrowth of the limb bud involves oriented cell behavior — 464
- 9.14 Positional values along the proximo-distal axis of the limb are specified by a combination of graded signaling and a timing mechanism — 466

9.15 The polarizing region specifies position along the limb's antero-posterior axis	469
9.16 Sonic hedgehog is a good candidate for the polarizing region morphogen	471
■ **Medical Box 9B** Too many fingers: mutations that affect antero-posterior patterning can cause polydactyly	472
■ **Cell Biology Box 9C** Sonic hedgehog signaling and the primary cilium	473
9.17 The dorso-ventral pattern of the limb is controlled by the ectoderm	475
■ **Medical Box 9D** External agents and the consequences of damage to the developing embryo	476
9.18 Development of the limb is integrated by interactions between signaling centers	478
9.19 Hox genes control regional identity in the limb	479
9.20 Self-organization might be involved in the development of the limb	482
9.21 Limb muscle is patterned by the connective tissue	483
■ **Box 9E** Reaction–diffusion mechanisms	484
9.22 The initial development of cartilage, muscles, and tendons is autonomous	485
9.23 Joint formation involves secreted signals and mechanical stimuli	486
9.24 Separation of the digits is the result of programmed cell death	487
Summary	487
Teeth	489
9.25 Tooth development involves epithelial–mesenchymal interactions and a homeobox gene code specifies tooth identity	489
Summary	492
Branching organs: the vertebrate lung and the mammalian mammary gland	492
9.26 The vertebrate lung develops from buds of endoderm	492
■ **Medical Box 9F** What developmental biology can tell us about cancer	493
9.27 Morphogenesis of the lung involves three modes of branching	495
9.28 The mammary glands develop from buds of ectoderm	496
9.29 The branching pattern of the mammary gland epithelium is determined by the mesenchyme	498
Summary	499
The vertebrate heart	499
9.30 The development of the vertebrate heart involves morphogenesis and patterning of a mesodermal tube	499
9.31 Several different cell populations contribute to making a heart	500
9.32 Key transcription factors specify development of the heart and cardiac differentiation	502

Summary	503
The vertebrate eye	503
9.33 Development of the vertebrate eye involves interactions between an extension of the forebrain and the ectoderm of the head	504
9.34 Eye-specifying transcription factors are conserved	507
Summary	508
Summary to Chapter 9	509
Chapter 10 Development of the nervous system	**517**
Introduction	517
Specification of cell identity in the nervous system	519
10.1 The vertebrate brain is organized into three main regions	520
10.2 The vertebrate brain is regionalized by local signaling centers	521
■ **Medical Box 10A** Alcohol exposure and brain development	522
10.3 Parts of the nervous system develop through a process of self-organization	523
10.4 Specialized boundaries formed in the developing brain delimit areas of cell-lineage restriction	524
10.5 Evolutionarily conserved transcription factors provide positional information along the antero-posterior axis of the developing brain	526
10.6 Opposing morphogen gradients pattern the spinal cord along the dorso-ventral axis	526
10.7 Stable progenitor domains in the ventral spinal cord are established over time	528
■ **Medical Box 10B** Directed differentiation of induced pluripotent stem cells towards functional neuronal types	530
10.8 Hox genes specify motor neuron subtypes in the spinal cord	530
10.9 Temporal mechanisms contribute to the generation of neural cell diversity	531
Summary	532
The formation and migration of neurons	532
10.10 Neurons in *Drosophila* arise from proneural clusters	532
10.11 The development of neurons in *Drosophila* involves asymmetric cell divisions and timed changes in gene expression	536
10.12 In the vertebrate neural tube, neurons are formed in the proliferative zone and migrate outwards	537
■ **Cell Biology Box 10C** Glial cells in the nervous system	538
10.13 Specialized features in the cortex support high levels of neurogenesis	539
10.14 Inhibitory interneurons undergo tangential migration into the cortex	540
■ **Experimental Box 10D** Timing the birth of cortical neurons	541
10.15 Delta-Notch signaling and lateral inhibition govern vertebrate neurogenesis	541

Summary 544
Axon navigation and mapping 544
10.16 The growth cone's response to guidance cues controls the path taken by a growing axon 545
■ **Box 10E** Putting it together: the development of a simple neuronal circuit 547
10.17 Axon guidance is a dynamic process 549
10.18 Axons that cross the ventral midline provide a model system for axon guidance 549
10.19 Motor neuron innervation of the chick limb is controlled by patterns of gene expression 551
10.20 The retino-tectal system is an example of a topographic map 552
Summary 555
Synapse formation and map refinement 556
10.21 Synapse formation involves reciprocal interactions 556
10.22 Many neurons die through apoptosis during normal development 559
■ **Medical Box 10F** Autism: a developmental condition that involves synapse dysfunction 560
10.23 Neuronal cell death and survival involve both intrinsic and extrinsic factors 561
10.24 Neural activity refines synapses and topographic maps 562
10.25 The mammalian brain has a high degree of plasticity, even in adults 564
Summary 564
Summary to Chapter 10 565

Chapter 11 Growth, post-embryonic development, and regeneration 571
Introduction 571
Growth 572
11.1 Tissues can grow by cell proliferation, cell enlargement, or accretion 573
11.2 Cell proliferation is controlled by regulating entry into the cell cycle 574
11.3 Cell division in early development can be controlled by an intrinsic developmental program 575
11.4 Extrinsic signals coordinate cell division, cell growth, and cell death in the developing *Drosophila* wing 576
■ **Cell Biology Box 11A** The core Hippo signaling pathways in *Drosophila* and mammals 577
11.5 Cancer can result from mutations in genes that control cell proliferation 578
11.6 The relative contributions of intrinsic and extrinsic factors in controlling size differ in different mammalian organs 580

11.7 Overall body size depends on the extent and the duration of growth 582
11.8 Hormones and growth factors coordinate the growth of different tissues and organs and contribute to determining overall body size 584
11.9 Elongation of the long bones illustrates how growth can be determined by a combination of an intrinsic growth program and extracellular factors 585
■ **Box 11B** Digit length ratio is determined in the embryo 586
11.10 The amount of nourishment an embryo receives can have profound effects in later life 589
Summary 590
Molting and metamorphosis 590
11.11 Arthropods must molt in order to grow 591
11.12 Insect body size is determined by the rate and duration of larval growth 591
11.13 Metamorphosis in amphibians is under hormonal control 594
Summary 595
Regeneration 596
11.14 Regeneration involves repatterning of existing tissues and/or growth of new tissues 597
11.15 Regeneration in *Hydra* involves repatterning of existing tissues 597
■ **Box 11C** Planarian regeneration 598
11.16 Amphibian limb regeneration involves cell dedifferentiation and new growth 601
11.17 Limb regeneration in amphibians depends on the presence of nerves 605
11.18 The limb blastema gives rise to structures with positional values distal to the site of amputation 606
11.19 Retinoic acid can change proximo-distal positional values in regenerating limbs 608
11.20 Mammals can regenerate the tips of the digits 610
■ **Box 11D** Why can't we regenerate our limbs? 611
11.21 Insect limbs intercalate positional values by both proximo-distal and circumferential growth 612
11.22 Heart regeneration in zebrafish involves the resumption of cell division by cardiomyocytes 614
Summary 616
Aging and senescence 617
11.23 Genes can alter the timing of senescence 618
11.24 Cell senescence blocks cell proliferation 619
11.25 The ability of adult salamanders to regenerate tissues does not diminish with age 621
Summary 621
Summary to Chapter 11 622

Chapter 12 Plant development — 629

Introduction — 629

Introduction to *Arabidopsis thaliana* — 631

12.1 The model plant *Arabidopsis thaliana* has a short life cycle and a small diploid genome — 631

Embryonic development — 632

12.2 Plant embryos develop through several distinct stages — 632

■ Box 12A Angiosperm embryogenesis — 633

12.3 Gradients of the signal molecule auxin establish the embryonic apical-basal axis — 636

12.4 Plant somatic cells can give rise to embryos and seedlings — 637

12.5 Cell enlargement is a major process in plant growth and morphogenesis — 639

■ Experimental Box 12B Plant transformation and genome editing — 640

Summary — 641

Meristems — 642

12.6 A meristem contains a small central zone of self-renewing stem cells — 642

12.7 The size of the stem-cell area in the meristem is kept constant by a feedback loop to the organizing center — 643

12.8 The fate of cells from different meristem layers can be changed by changing their position — 644

12.9 A fate map for the embryonic shoot meristem can be deduced using clonal analysis — 645

12.10 Meristem development is dependent on signals from other parts of the plant — 646

12.11 Gene activity patterns the proximo-distal and adaxial-abaxial axes of leaves developing from the shoot meristem — 648

12.12 The regular arrangement of leaves on a stem is generated by regulated auxin transport — 650

12.13 The outgrowth of secondary shoots is under hormonal control — 651

12.14 Root tissues are produced from *Arabidopsis* root apical meristems by a highly stereotyped pattern of cell divisions — 653

12.15 Root hairs are specified by a combination of positional information and lateral inhibition — 656

Summary — 656

Flower development and control of flowering — 657

12.16 Homeotic genes control organ identity in the flower — 658

■ Box 12C The basic model for the patterning of the *Arabidopsis* flower — 660

12.17 The *Antirrhinum* flower is patterned dorso-ventrally as well as radially — 662

12.18 The internal meristem layer can specify floral meristem patterning — 662

12.19 The transition of a shoot meristem to a floral meristem is under environmental and genetic control — 663

■ Box 12D The circadian clock coordinates plant growth and development — 664

12.20 Vernalization reflects the epigenetic memory of winter — 665

12.21 Most flowering plants are hermaphrodites, but some produce unisexual flowers — 667

Summary — 668

Summary to Chapter 12 — 668

Chapter 13 Evolution and development — 673

Introduction — 673

■ Box 13A Darwin's finches — 676

The evolution of development — 677

13.1 Multicellular organisms evolved from single-celled ancestors — 677

13.2 Genomic evidence is throwing light on the evolution of animals — 679

■ Box 13B The metazoan family tree — 680

13.3 How gastrulation evolved is not known — 681

13.4 More general characteristics of the body plan develop earlier than specializations — 682

13.5 Embryonic structures have acquired new functions during evolution — 683

13.6 Evolution of different types of eyes in different animal groups is an example of parallel evolution — 685

Summary — 686

The diversification of body plans — 687

13.7 Hox gene complexes have evolved through gene duplication — 687

13.8 Differences in Hox gene expression determine the variation in position and type of paired appendages in arthropods — 689

13.9 Changes in Hox gene expression and their target genes contributed to the evolution of the vertebrate axial skeleton — 693

13.10 The basic body plan of arthropods and vertebrates is similar but the dorso-ventral axis is inverted — 694

Summary — 695

Evolutionary modification of specialized characters — 696

13.11 Limbs evolved from fins — 696

13.12 Studies of zebrafish fin development give insights into the origins of tetrapod limbs — 697

■ Experimental Box 13C Transgenic tests of the function of Hox digit enhancers — 699

13.13 Limbs have evolved to fulfill different specialized functions — 701

13.14 The evolution of limblessness in snakes is associated with changes in axial gene expression and mutations in a limb-specific enhancer — 702

Contents

13.15 Butterfly wing markings have evolved by redeployment of genes previously used for other functions — 703

■ **Experimental Box 13D** Using CRISPR-Cas9 genome-editing techniques to test the functioning of the snake ZRS — 704

13.16 Adaptive evolution within the same species provides a way of studying the developmental basis for evolutionary change — 707

■ **Experimental Box 13E** Pelvic reduction in sticklebacks is based on mutations in a gene control region — 709

Summary — 710

Changes in the timing of developmental processes — 710

13.17 Changes in growth rates can modify the basic body plan — 710

■ **Box 13F** Origins of morphological diversity in dogs — 712

13.18 Evolution can be due to changes in the timing of developmental events — 713

13.19 The evolution of life histories has implications for development — 716

■ **Box 13G** Long- and short-germ development in insects — 717

Summary — 719

Summary to Chapter 13 — 720

Glossary — 727

Index — 751

List of boxes

Special Interest Boxes

Box 1A Basic stages of *Xenopus laevis* development	3
Box 3J Identical twins	176
Box 4G The Hox genes	223
Box 5C Convergent extension	278
Box 6A Conrad Waddington's 'epigenetic landscape' provides a framework for thinking about how cells differentiate	313
Box 7B Self-organization and self-assembly	379
Box 7C Somitogenesis in Matrigelled gastruloids	383
Box 8A Polar bodies	410
Box 9A Positional information and morphogen gradients	449
Box 9E Reaction–diffusion mechanisms	484
Box 10E Putting it together: the development of a simple neuronal circuit	547
Box 11B Digit length ratio is determined in the embryo	586
Box 11C Planarian regeneration	598
Box 11D Why can't we regenerate our limbs?	611
Box 12A Angiosperm embryogenesis	633
Box 12C The basic model for the patterning of the *Arabidopsis* flower	660
Box 12D The circadian clock coordinates plant growth and development	664
Box 13A Darwin's finches	676
Box 13B The metazoan family tree	680
Box 13F Origins of morphological diversity in dogs	712
Box 13G Long- and short-germ development in insects	717

Cell Biology Boxes

Box 1B The mitotic cell cycle	6
Box 1D Germ layers	19
Box 1F Gene regulatory networks and cell-fate decisions	26
Box 1G Signal transduction and intracellular signaling pathways	31
Box 2B The Toll signaling pathway: a multifunctional pathway	60
Box 2C The JAK–STAT signaling pathway	64
Box 2E The Hedgehog signaling pathway	87
Box 3D Intercellular protein signals in vertebrate development	124
Box 3E The Wnt/β-catenin signaling pathway	128
Box 3F Signaling by members of the TGF-β family of growth factors	133
Box 4A The fibroblast growth factor signaling pathway	197
Box 4C Fine-tuning Nodal signaling	207
Box 4E Retinoic acid: a small-molecule intercellular signal	214
Box 4F The Notch signaling pathway	219
Box 5A Cell-adhesion molecules and cell junctions	250
Box 5B The cytoskeleton, cell-shape change, and cell movement	254
Box 5D Eph receptors and their ephrin ligands	292
Box 6B Epigenetic control of gene expression by chromatin modification	323
Box 9C Sonic hedgehog signaling and the primary cilium	473
Box 10C Glial cells in the nervous system	538
Box 11A The core Hippo signaling pathways in *Drosophila* and mammals	577

Experimental Boxes

Box 1C The CRISPR-Cas9 genome-editing system	15
Box 1E Visualizing gene expression in embryos	24
Box 2A Mutagenesis and genetic screening strategy for identifying developmental mutants in *Drosophila*	50
Box 2D Targeted gene expression and misexpression screening	76
Box 2F Mutants in denticle pattern provided clues to the logic of segment patterning	87
Box 3A Fate mapping and lineage tracing reveal which cells in which parts of the early *Xenopus* embryo give rise to particular structures in the tadpole	110
Box 3B Ways of making transgenic mice	115
Box 3C Large-scale mutagenesis screens for recessive mutations in zebrafish	116
Box 3G Timing and cell–cell interactions in mesoderm specification	135
Box 3H Investigating receptor function using dominant-negative proteins	138
Box 4B A zebrafish gene regulatory network	203
Box 4D The Cre/*loxP* system: a strategy for making gene knock-outs in mice	208
Box 6C Single-cell analysis of cell-fate decisions	330
Box 7A Induced pluripotent stem cells	373
Box 7D Intestinal organoids can be produced from adult stem cells or ES cells	392
Box 10D Timing the birth of cortical neurons	541
Box 12B Plant transformation and genome editing	640
Box 13C Transgenic tests of the function of Hox digit enhancers	699
Box 13D Using CRISPR-Cas9 genome-editing techniques to test the functioning of the snake ZRS	704
Box 13E Pelvic reduction in sticklebacks is based on mutations in a gene control region	709

Medical Boxes

Box 1H When development goes awry — 32
Box 3I Pre-implantation genetic diagnosis — 172
Box 5E Neural tube defects — 293
Box 6D Towards a therapy for β-globin gene defects — 346
Box 6E Treatment of junctional epidermolysis bullosa with skin grown from genetically corrected stem cells — 349
Box 9B Too many fingers: mutations that affect antero-posterior patterning can cause polydactyly — 472
Box 9D External agents and the consequences of damage to the developing embryo — 476
Box 9F What developmental biology can tell us about cancer — 493
Box 10A Alcohol exposure and brain development — 522
Box 10B Directed differentiation of induced pluripotent stem cells towards functional neuronal types — 530
Box 10F Autism: a developmental condition that involves synapse dysfunction — 560

Reviewer acknowledgments

We would like to reiterate our thanks to the people who contributed to the first five editions. Many people gave their advice based on the sixth edition, and others reviewed the draft chapters for the edition as they emerged. We wish to express our gratitude to the following colleagues.

Stefan Baumgartner, Exp. Medical Sciences, Lund University

Josephine Bowles, School of Biomedical Sciences, University of Queensland

James Briscoe, The Francis Crick Institute, London

Brendan Davis, School of Biology, University of Leeds

Philippa Francis-West, Centre for Craniofacial and Regenerative Biology, King's College London

Brigid Hogan, Department of Cell Biology, Duke University

Maike Kittelmann, Biological and Medical Sciences, Oxford Brookes University

Kenneth Poss, Department of Cell Biology, Duke University

Nikolina Radulovich, Department of Chemistry and Biology, Ryerson University

Rebecca Rolfe, Zoology, University of Dublin, Trinity College Dublin

Christina Schilde, School of Life Sciences, University of Dundee

Guojun Sheng, Graduate School of Medical Sciences, Kumamoto University

Henrietta Standley, School of Biosciences, Cardiff University

Ben Steventon, Department of Genetics, University of Cambridge

Shahragim Tajbakhsh, Institut Pasteur, Paris

Mikiko Tanaka, Department of Life Science and Technology, Institute of Science, Tokyo

Prof. dr. Gert Jan C. Veenstra, Molecular Developmental Biology, Radboud University

Fiona Wardle, Randall Centre of Cell and Molecular Biophysics, King's College London

Alison Woollard, Department of Biochemistry, University of Oxford

1

History and basic concepts

- The origins of developmental biology
- A conceptual tool kit

How does a single cell—the fertilized egg—give rise to a multicellular organism, in which a multiplicity of different cell types is organized into tissues and organs to make up a three-dimensional body? This is the big question that developmental biologists aim to answer. The question can be studied from many different viewpoints: which genes are expressed, and when and where; how cells communicate with each other; how cells proliferate and differentiate into specialized cell types; and how major changes in body shape are produced. All the information for embryonic development (embryogenesis) is contained within the fertilized egg. We shall see that an organism's development is ultimately driven by the regulated expression of its genes in space and time, dictating which proteins are present in which cells and when. In turn, proteins largely determine how a cell behaves. The genes provide a generative program for development in which their actions are translated into developmental outcomes through cellular behavior such as intercellular signaling, cell proliferation, cell differentiation, and cell movement that ultimately shape the recognizable and functional body.

Introduction

The development of a multicellular organism from a single cell—the fertilized egg—is a brilliant triumph of evolution. The fertilized egg divides to give rise to many millions of cells, which form structures as complex and varied as eyes, arms, heart, and brain. This amazing achievement raises many questions. How do the cells arising from division of a single cell, the fertilized egg, become different from each other? How do they become organized into structures such as limbs and brains? What controls the behavior of individual cells so that such highly organized patterns emerge? How are the organizing principles of development embedded within the egg, and, in particular, within the genetic material, DNA? Much of the excitement in developmental biology today comes from our growing understanding of how genes direct these developmental processes, and genetic control is one of the main themes of this book. Thousands of genes are involved in controlling development, but we will focus only on those that have key roles and illustrate general principles.

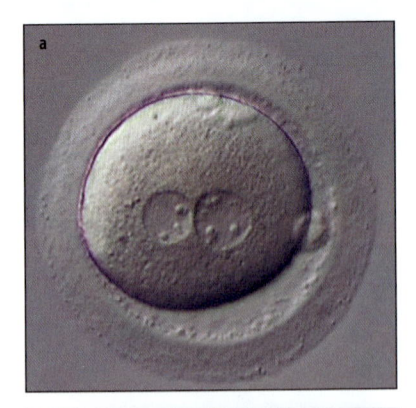

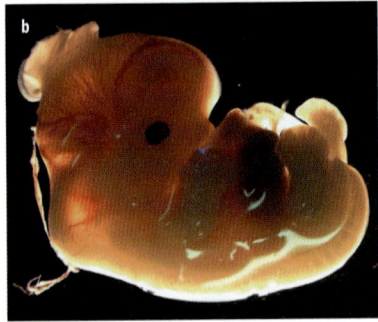

Fig. 1.1 Human fertilized egg and embryo. (a) Human fertilized egg. The nuclear membranes of the sperm and egg nuclei (pronuclei) have not yet broken down to allow the parental chromosomes to mingle. (b) Human embryo at around 51 days' gestation (Carnegie stage 20), which is equivalent to a mouse embryo at 13.5 days post-fertilization. A human embryo at this stage is about 21–23 mm long.

(a) Courtesy of Alpesh Doshi, CRGH, London. (b) Reproduced courtesy of the MRC/Wellcome-funded Human Developmental Biology Resource.

Fig. 1.2 The South African claw-toed frog, *Xenopus laevis*. Scale bar = 1 cm.

Photograph courtesy of J. Smith.

Understanding how embryos develop is a huge intellectual challenge, and one of the ultimate aims of the science of developmental biology is to understand how we humans develop (Fig. 1.1). We need to understand human development for several reasons. We need to understand why it sometimes goes wrong and why a **fetus** may fail to be born or a baby be born with a congenital condition. The link between developmental biology and genetics is very close, as mutations in genes can lead to differences in an embryo's developmental program; environmental factors, such as drugs and infections, can affect it too. Another area of medical research related to developmental biology is regenerative medicine—finding out how to use cells to repair damaged tissues and organs. The focus of regenerative medicine is currently on **stem cells**, a special class of cells that can proliferate without differentiating, and act as a reservoir for the development of the different tissues of the body. These cells can be grown in culture and this offers a new way of studying developmental events. In addition, such cells may, in the future, also provide a source of tissues and organs for repair. Cancer cells also display some properties of embryonic cells, such as the ability to divide indefinitely. Also, many of the genes that control embryonic development are involved in cancer, and so the study of embryonic cells and their behavior could lead to new and better treatments for cancer.

Many different animals have been studied during the modern history of developmental biology, from the fruit fly *Drosophila melanogaster* and the nematode worm *Caenorhabditis elegans* through other invertebrates such as the sea urchin and the cnidarian *Hydra* to a range of vertebrates including the frog *Xenopus laevis*, the zebrafish, the chick, and the mouse. Is it necessary to cover so many different organisms to understand the basic features of development? The answer is yes. Developmental biologists do, indeed, believe that there are general principles of development that apply to all animals, but life is too wonderfully diverse to find all the answers in a single organism. As it is, developmental biologists have tended to focus their efforts on a relatively small number of animals, chosen because they are convenient to study and amenable to experimental manipulation or genetic analysis. This is why some creatures, such as the frog *Xenopus laevis* (Fig. 1.2) and the fruit fly *Drosophila*, have such a dominant place in developmental biology. Similarly, work with the thale cress, *Arabidopsis thaliana*, has uncovered many features of plant development.

One of the most exciting and satisfying aspects of developmental biology is that understanding a developmental process in one organism can help to illuminate developmental processes elsewhere—for example, giving insights into how humans develop. Nothing illustrates this more dramatically than the influence that our understanding of *Drosophila* development, and especially of its genetic basis, has had throughout developmental biology. The identification of genes controlling **embryogenesis**—the development of an embryo from a fertilized egg—in *Drosophila* has led to the discovery of related genes used in similar ways in the development of mammals and other vertebrates. Such discoveries encourage us to believe in the existence of general developmental principles.

Amphibians have long been favorite organisms for studying the early development of animals because their eggs are large and their embryos are easy to grow in a simple culture medium and relatively easy to experiment on. Box 1A illustrates embryogenesis in the South African frog *Xenopus*.

In the rest of this chapter we first look briefly at the history of **embryology**—as the study of developmental biology used to be called. The term **developmental biology** itself is of much more recent origin and reflects the appreciation that development is not restricted to the embryo alone. An important difference between the two disciplines refers to the fact that embryology is largely a descriptive discipline that accounts for the orderly emergence of the organism during its development. By contrast, developmental biology has a mechanistic aspect that aims to capture the molecular and cellular processes underlying the changes described by embryology

BOX 1A Basic stages of *Xenopus laevis* development

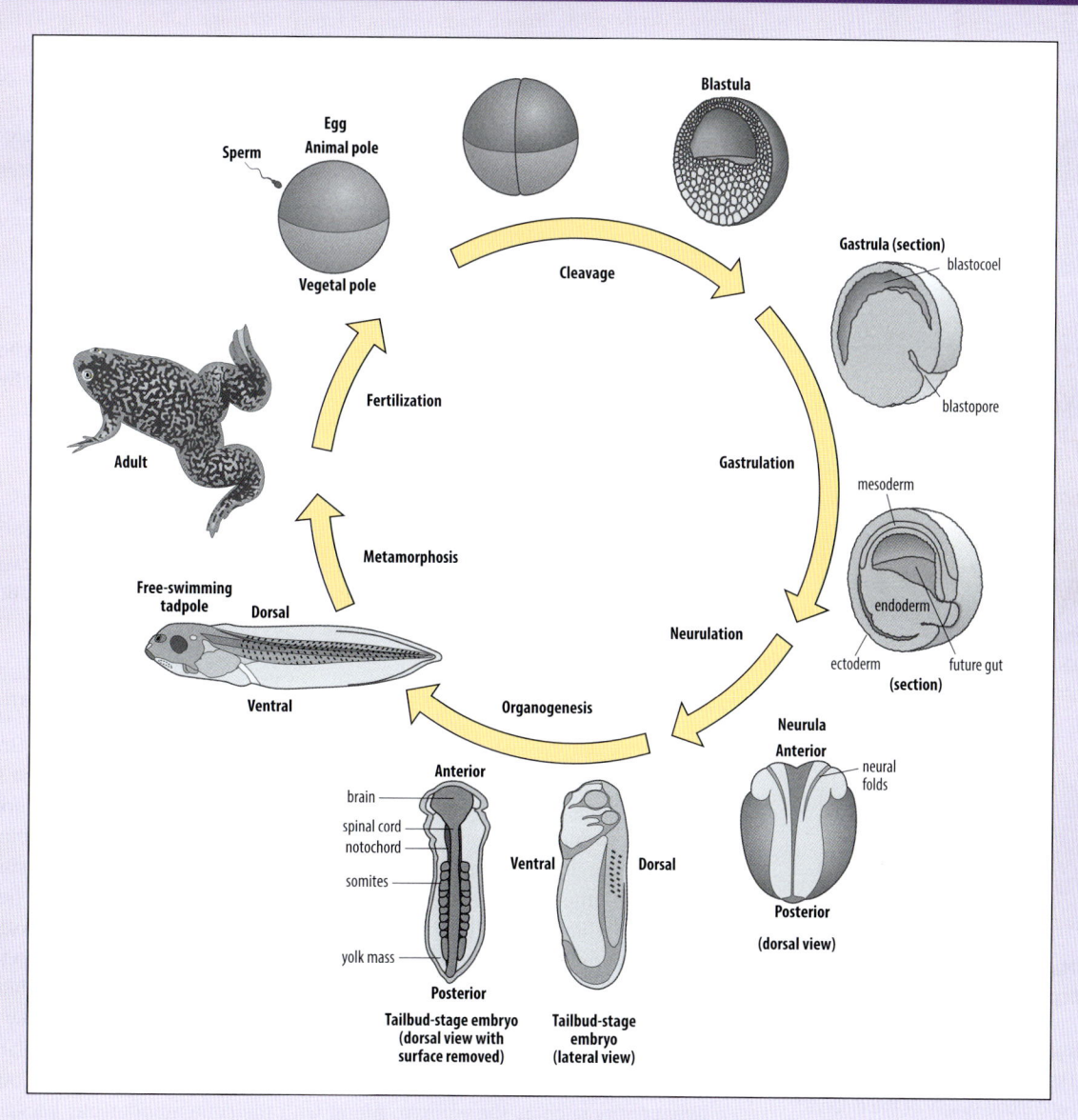

Fig. 1

Although vertebrate development is varied, there are basic stages that can be illustrated by following the development of the frog *Xenopus laevis* (Fig. 1). These basic stages include fertilization, cleavage, gastrulation, neurulation, and organogenesis.

In *Xenopus*, the unfertilized egg is a large cell. It has a pigmented upper surface (the **animal pole**) and a lower region (the **vegetal pole**) characterized by an accumulation of yolk granules. After **fertilization** of the egg by a **sperm**, and the fusion of male and female pronuclei, **cleavage** begins. Cleavages are mitotic divisions in which cells do not grow between each division, and so with successive cleavages the cells become smaller. After about 12 cleavage divisions, the embryo, now known as a **blastula**, consists of a sheet of many small cells surrounding a fluid-filled cavity (the **blastocoel**) above the larger yolky cells. Already, changes have occurred within the cells and they have interacted with each other so that the three **germ layers**—**mesoderm**, **endoderm**, and **ectoderm**—are specified (see Box 1D). The **animal region** gives rise to ectoderm, which forms both the epidermis of the skin and the nervous system. The **vegetal region** and the **equatorial region** give rise to the endoderm and mesoderm respectively, which will form internal organs. At this stage, these cells are still on the surface of the embryo. During the next stage—**gastrulation**—there is a dramatic rearrangement of cells; the endoderm and mesoderm move inside, and the basic body plan of the tadpole is established. Internally, the mesoderm gives rise to a rod-like structure (including the notochord), which runs from the head to the tail, and lies centrally beneath the future nervous system. On either side of the notochord are segmented blocks of mesoderm called somites, which will give rise to the muscles and vertebral column, as well as the dermis of the skin (somites can be seen in the cutaway view of the later tailbud-stage embryo).

Shortly after gastrulation, the ectoderm above the axial mesoderm folds to form a tube (the **neural tube**), which gives rise to the brain and spinal cord—a process known as **neurulation**. By this time, other organs, such as limbs, eyes, and gills, are specified at their future locations, but only develop a little later, during **organogenesis**. During organogenesis, specialized cells such as muscle, cartilage, and neurons differentiate. By 4 days after fertilization, the embryo has become a free-swimming tadpole with typical vertebrate features.

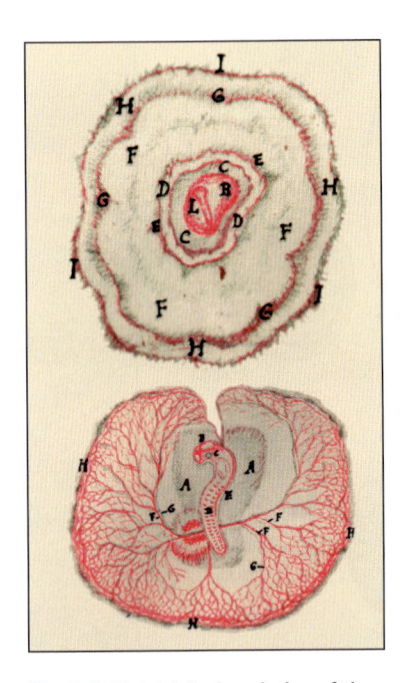

Fig. 1.3 Malpighi's description of the chick embryo. The figure shows Malpighi's drawings, made in 1673, depicting the early embryo (top) and at 2 days' incubation (bottom). His drawings accurately illustrate the shape and blood supply of the embryo.

Reprinted by permission of the President and Council of the Royal Society.

and in other moments of the life of the organism. In the second part of the chapter we will introduce some key concepts that are used over and over again in studying and understanding development.

The origins of developmental biology

Many questions in embryology were first posed hundreds, and in some cases thousands, of years ago. Appreciating the history of these ideas helps us to understand why we approach developmental problems in the way that we do today.

1.1 Aristotle first defined the problem of epigenesis versus preformation

The study of embryology in Europe started with the ancient Greeks. The Greek philosopher Aristotle formulated a question that was to dominate much of the thinking about development until the end of the nineteenth century. How do the different parts of the embryo form? He considered two possibilities: one was that everything in the embryo was preformed from the very beginning and simply got bigger during development; the other was that new structures arose progressively, a process he termed **epigenesis** (which means 'upon formation') and that he likened metaphorically to the 'knitting of a net'. Aristotle favored epigenesis and his conjecture was correct. Aristotle's influence on European thought was enormous and his ideas remained dominant well into the seventeenth century. The contrary view to epigenesis, namely that the embryo was preformed from the beginning, was championed anew in the late seventeenth century. Many could not believe that physical or chemical forces could mold a living entity such as the embryo. Along with the contemporaneous background of belief in the divine creation of the world and all living things was the belief that all embryos had existed from the beginning of the world, and that the first embryo of a species must contain all future embryos.

Even the brilliant seventeenth-century Italian embryologist Marcello Malpighi could not free himself from preformationist ideas. While he provided a remarkably accurate description of the development of the chick embryo, he remained convinced, against the evidence of his own observations, that the fully formed embryo was present from the beginning (Fig. 1.3). He argued that at very early stages the parts were so small that they could not be seen, even with his best microscope. Other preformationists believed that the sperm contained the embryo, and some even claimed to see a tiny human—a homunculus—in the head of each human sperm (Fig. 1.4).

The preformation/epigenesis issue was vigorously debated throughout the eighteenth century. But the problem could not be resolved until one of the great advances in biology had taken place—the recognition that living things, including embryos, were composed of cells.

1.2 Cell theory changed how people thought about embryonic development and heredity

The invention of the microscope around 1600 was essential for the discovery of cells, but the 'cell theory' of life was only developed between 1820 and 1880 by, among others, the German botanist, Matthias Schleiden, and the physiologist, Theodor Schwann. It recognized that all living organisms consist of cells, that these are the basic units of life, and that new cells can only be formed by the division of pre-existing cells. The cell theory was one of the most illuminating advances in biology and had an enormous impact. Multicellular organisms, such as animals and plants, could now be viewed as communities of cells. Development could not, therefore, be based on preformation, but must be by epigenesis, because during development many new cells are generated by division from the egg, and new types of cells are formed. A crucial step forward in understanding development was the recognition, in the 1840s, that the egg itself is but a single, albeit specialized, cell.

An important advance in embryology was the proposal by the nineteenth-century German biologist, August Weismann, that an offspring does not inherit its characteristics from the body (the soma) of the parent but only from the **germ cells**—egg and sperm. Weismann drew a fundamental distinction between germ cells and the body cells or **somatic cells** (Fig. 1.5). Characteristics acquired by the body during an animal's life cannot be transmitted to the germline. As far as heredity is concerned, the body is merely a carrier of germ cells. As the English novelist and essayist Samuel Butler put it: 'A hen is only an egg's way of making another egg.'

Work on sea urchin eggs showed that after fertilization the egg contains two nuclei, which eventually fuse; one of these nuclei belongs to the egg while the other comes from the sperm. Fertilization therefore results in a single cell—the **zygote**—carrying a nucleus with contributions from both parents, and it was concluded that the cell nucleus must contain the physical basis of heredity. The climax of this line of research was the demonstration, towards the end of the nineteenth century, that the chromosomes within the nucleus of the zygote are derived in equal numbers from the two parental nuclei, and the recognition that this provided a physical basis for the transmission of genetic characters according to the laws developed by the Austrian botanist and monk, Gregor Mendel. The number of chromosomes is kept constant from generation to generation by a specialized type of cell division that produces the germ cells, called **meiosis**, which halves the chromosome number; the full complement of chromosomes is then restored at fertilization. The zygote and the somatic cells that arise from it divide by the process of **mitosis**, which maintains chromosome number (Box 1B). Germ cells contain a single copy of each chromosome and are called **haploid**, whereas germ cell precursor cells and the other somatic cells of the body contain two copies and are called **diploid**.

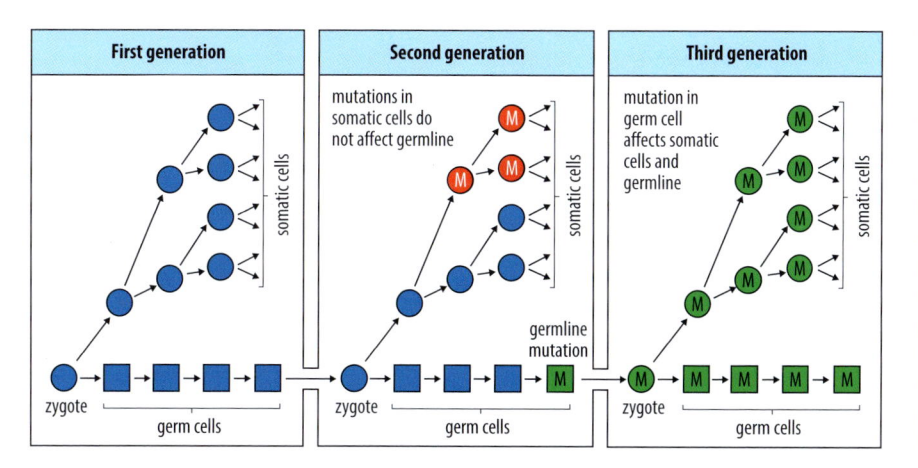

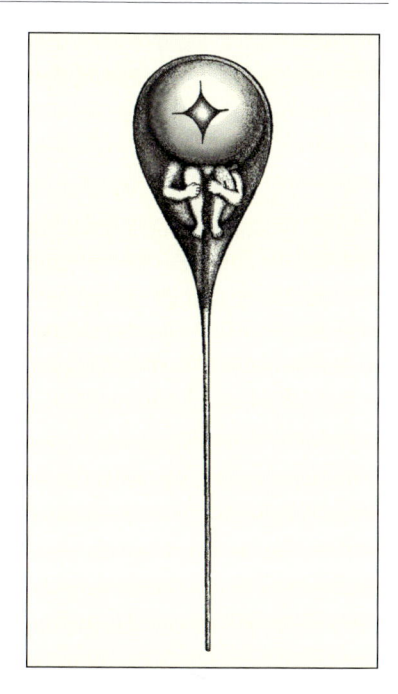

Fig. 1.4 Some preformationists believed that a homunculus was curled up in the head of each sperm.
An imaginative drawing, after Nicholas Harspeler (1694).

Fig. 1.5 The distinction between germ cells and somatic cells. In each generation a germ cell contributes to the zygote, which gives rise to both somatic cells and germ cells, but inheritance is through the germ cells only (first panel). Changes that occur due to a mutation (red) in a somatic cell can be passed on to its daughter cells but do not affect the germline, as shown in the second panel. In contrast, a mutation in the germline (green) in the second generation will be present in every cell in the body of the new organism to which that cell contributes and will also be passed on to the third and future generations through the germline, as shown in the third panel.

CELL BIOLOGY BOX 1B The mitotic cell cycle

When a eukaryotic cell duplicates itself it goes through a fixed sequence of events called the **cell cycle**. The cell grows in size, the DNA is replicated, and the replicated chromosomes then undergo mitosis and become segregated into two daughter nuclei. Only then can the cell divide to form two daughter cells, which can go through the whole sequence again.

The standard eukaryotic mitotic cell cycle is divided into well-marked phases (Fig. 1). At the M phase, mitosis and cell cleavage give rise to two new cells. The rest of the cell cycle, between one M phase and the next, is called interphase. Replication of DNA occurs during a defined period in interphase, the S phase (the S stands for synthesis of DNA). Preceding S phase is a period known as G_1 (the G stands for gap), and after it another interval known as G_2, after which the cells enter mitosis (see Fig. 1). G_1, S phase, and G_2 collectively make up interphase, the part of the cell cycle during which cells synthesize proteins and grow, as well as replicate their DNA. When somatic cells are not proliferating they are usually in a state known as G_0, into which they withdraw after mitosis. The decision to enter G_0 or to proceed through G_1 may be controlled by both intracellular state and extracellular signals such as growth factors. Growth factors enable the cell to proceed out of G_0 and progress through the cell cycle. Cells such as neurons and skeletal muscle cells, which do not divide after differentiation, are permanently in G_0.

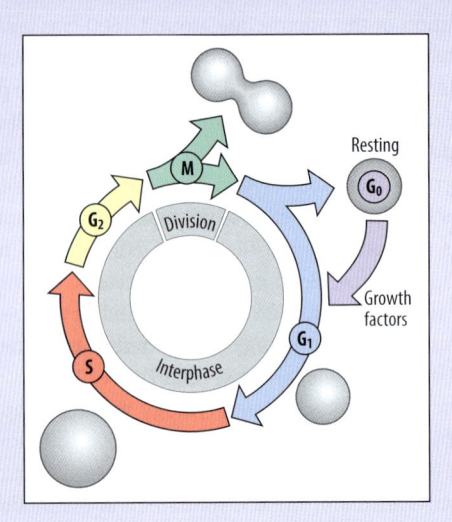

Fig. 1

Particular phases of the cell cycle are absent in some cells: during cleavage of the fertilized *Xenopus* egg G_1 and G_2 are virtually absent, and cells get smaller at each division. In *Drosophila* salivary glands there is no M phase, as the DNA replicates repeatedly without mitosis or cell division, leading to the formation of giant **polytene chromosomes**.

1.3 Two main types of development were originally proposed

The next big question was how cells become different from one another during embryonic development. With the increasing emphasis on the role of the nucleus, in the 1880s Weismann put forward a model of development in which the nucleus of the zygote contained special factors, or **determinants** (Fig. 1.6). He proposed that while the fertilized egg underwent the rapid cycles of cell division known as **cleavage** (see Box 1A), these nuclear determinants would be distributed unequally to the daughter cells and so would control the cells' future development. The fate of each cell was therefore predetermined in the egg by the factors it would receive during cleavage.

Fig. 1.6 Weismann's theory of nuclear determination. Weismann assumed that there were factors in the nucleus that were distributed asymmetrically to daughter cells during cleavage and directed their future development.

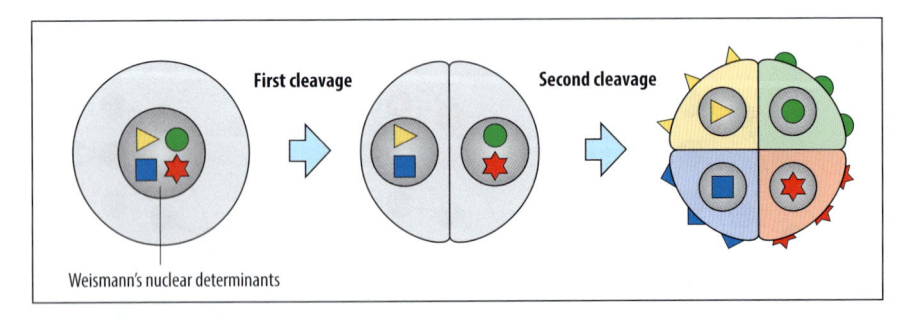

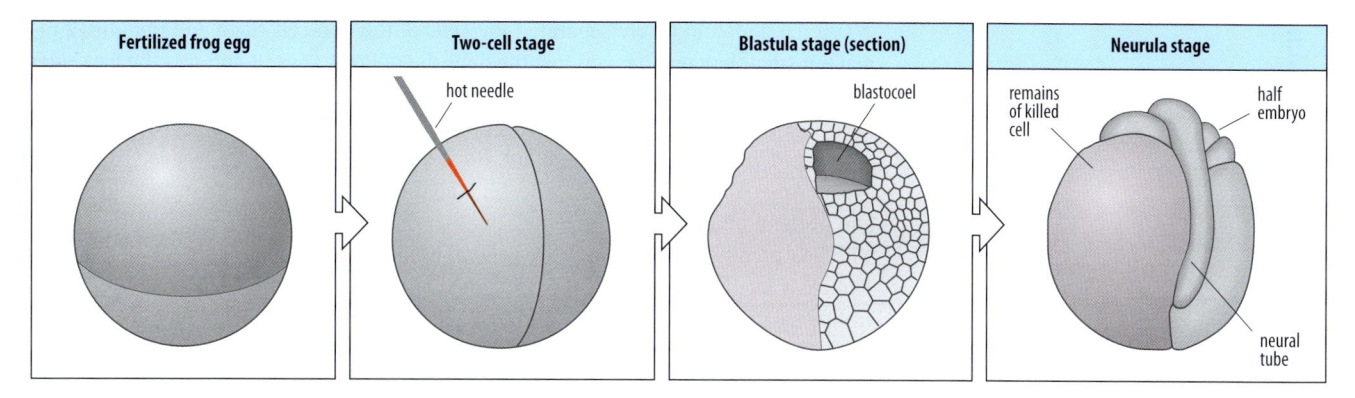

This type of model was termed 'mosaic', as the egg could be considered to be a mosaic of discrete localized determinants. Central to Weismann's theory was the assumption that early cell divisions must make the daughter cells quite different from each other because of unequal distribution of nuclear components.

In the late 1880s, initial support for Weismann's ideas came from experiments carried out independently by the German embryologist, Wilhelm Roux, who experimented with frog embryos. Having allowed the first cleavage of a fertilized frog egg, Roux destroyed one of the two cells with a hot needle and found that the remaining cell developed into a half-larva (Fig. 1.7). He concluded that the 'development of the frog is based on a mosaic mechanism, the cells having their character and fate determined at each cleavage'.

But when Roux's fellow countryman, Hans Driesch, repeated the experiment on sea urchin eggs, he obtained quite a different result (Fig. 1.8). He wrote later: 'But things turned out as they were bound to do and not as I expected; there was, typically, a whole gastrula on my dish the next morning, differing only by its small size from a normal one; and this small but whole gastrula developed into a whole and typical larva.'

Fig. 1.7 Roux's experiment to investigate Weismann's theory of mosaic development. After the first cleavage of a frog embryo, one of the two cells is killed by pricking it with a hot needle; the other remains undamaged. At the blastula stage the undamaged cell can be seen to have divided as normal into many cells that fill half of the embryo. The development of the blastocoel, a small fluid-filled space in the center of the blastula, is also restricted to the undamaged half. In the damaged half of the embryo, no cells appear to have formed. At the neurula stage, the undamaged cell has developed into something resembling half a normal embryo.

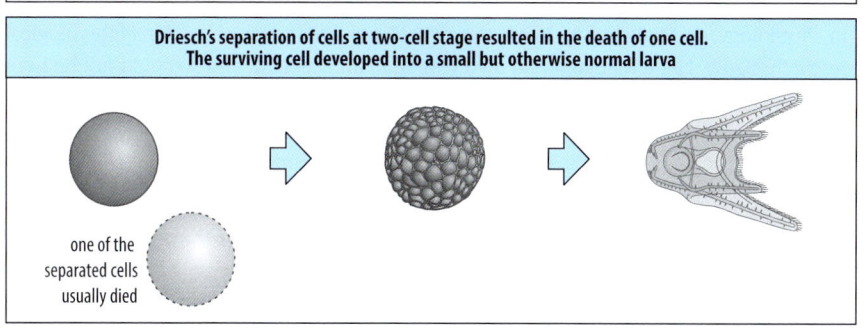

Fig. 1.8 The outcome of Driesch's experiment on sea urchin embryos, which first demonstrated the phenomenon of regulation. After separation of cells at the two-cell stage, a separated cell could develop into a small, but whole, normal larva. This is the opposite of Roux's earlier finding that when one of the cells of a two-cell frog embryo is damaged, the remaining cell develops into a half-embryo only (see Fig. 1.7).

Driesch had completely separated the cells at the two-cell stage and obtained normal but small larvae. That was just the opposite of Roux's result and was the first clear demonstration of the developmental process known as **regulation**, that is, the ability to restore normal development, even if some portions of the embryo are removed or rearranged very early in development. Roux's experiment on frogs was later repeated by the American T.H. Morgan, who separated the two blastomeres instead of killing one of them and leaving it attached, and he obtained the same result as Driesch did with sea urchins. This showed the general ability of vertebrate embryos to regulate. The basis for this phenomenon is explained later in the chapter. The extent to which embryos can regulate differs in different species and we shall see many examples of regulation throughout the book. The existence of regulation does not mean, however, that the unequal distribution of determinants that make two daughter cells different from each other is not important during development. But Weismann was wrong in one crucial respect, in that such determinants are not nuclear, but are located in the cell cytoplasm. We shall see many examples of developmentally important proteins and RNAs that act in this way as **cytoplasmic determinants**.

1.4 The discovery of induction showed that one group of cells could determine the development of neighboring cells

The fact that embryos can regulate implies that cells must communicate and interact with each other, but the central importance of **cell–cell interactions** in embryonic development was not really established until the discovery of the phenomenon of **induction**. This is where one cell, or tissue, directs the development of another, neighboring, cell or tissue.

The importance of induction and other cell–cell interactions in development was proved dramatically in 1924 when Hans Spemann and his student Hilde Mangold carried out a now famous transplantation experiment in amphibian embryos. They showed that a partial second embryo could be induced by grafting one small region of an early newt embryo onto another at the same stage (Fig. 1.9). The grafted tissue was taken from the dorsal lip of the **blastopore**—the slit-like invagination that forms where gastrulation begins on the dorsal surface of the amphibian embryo (see Box 1A). The origin of the tissues in the partial second embryo was ascertained by grafting the dorsal lip of the blastopore from an unpigmented species of newt onto the embryo of another species of newt that is pigmented. This showed that the development of the second body axis was directed by the graft, because most of its structures—neural tube, somites, embryonic kidneys, epidermis—were largely made up of host tissue. The graft gave rise to the notochord in the second axis and only made small contributions to somites and neural tube. Spemann and Mangold called the small region of the embryo that had this dramatic effect the **organizer**, as it seemed to be ultimately responsible for controlling the organization of a complete embryonic body. It is now known as the **Spemann–Mangold organizer**, or just the **Spemann organizer**. For their discovery, Spemann received the Nobel Prize in Physiology or Medicine in 1935, the first Nobel Prize ever given for embryological research. Sadly, Hilde Mangold had died earlier, in an accident, and so could not be honored. Spemann may have been influenced by the earlier work of Ethel Browne. She had performed similar experiments in the cnidarian *Hydra*, a simpler organism, identifying a region with the ability to induce a new axis in this organism.

1.5 Developmental biology emerged from the coming together of genetics and embryology

When Mendel's laws were rediscovered in 1900 there was a great surge of interest in mechanisms of inheritance, particularly in relation to evolution, but less so in relation to embryology. Genetics was seen as the study of the transmission of hereditary

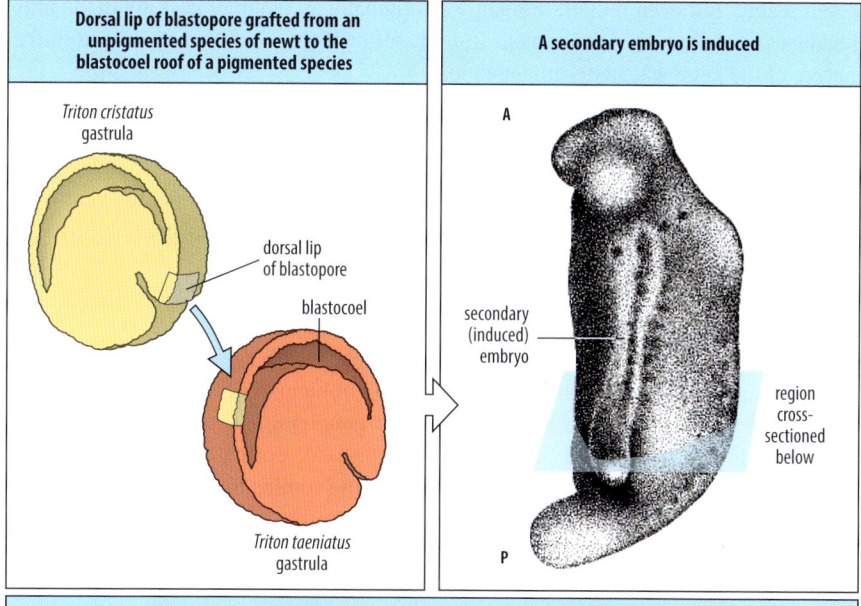

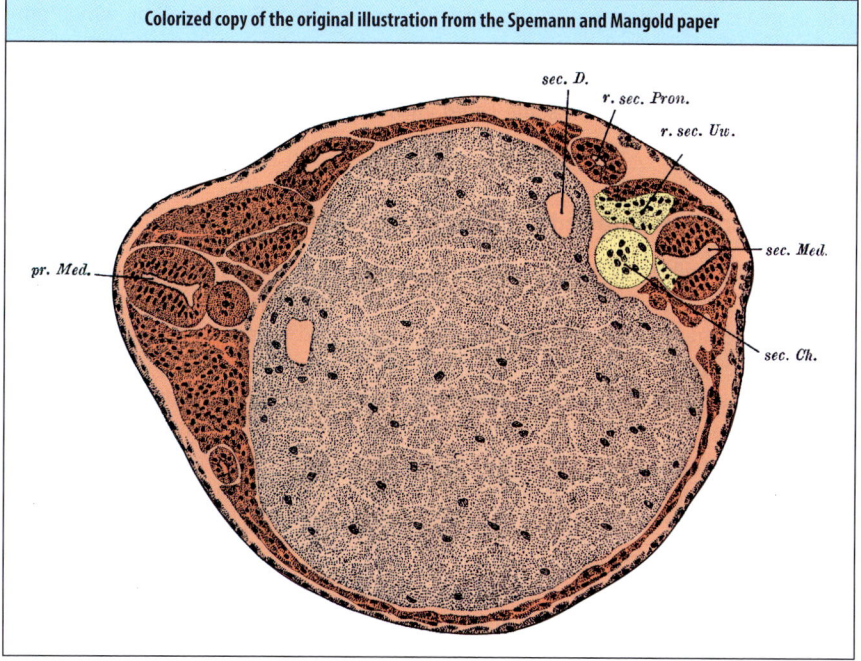

Fig. 1.9 The dramatic demonstration by Spemann and Mangold of the induction of a new main body axis by the organizer region in the early amphibian gastrula. Top left panel: a piece of tissue (yellow) from the dorsal lip of the blastopore of a newt (*Triton cristatus*) gastrula was grafted to the opposite side of a gastrula of another, pigmented, newt species (*Triton taeniatus*, red). Top right panel: original illustration from their paper showing that the grafted tissue induced a new body axis. The embryo is viewed from the left side and shows a surface view of the secondary embryo, with tailbud, neural tube, and somites visible. Anterior is to the top. Bottom panel: a colorized copy of their original illustration of a section through the secondary axis of the embryo shown above, at the level indicated. The primary axis structures are to the left of the section and the secondary axis structures to the right. The unpigmented graft tissue (colored yellow here) formed a notochord at its new site and also contributed to the somites, but most of the neural tube and the other structures of the new axis have been induced from the pigmented host tissue (red). The organizer region discovered by Spemann and Mangold is known as the Spemann organizer. Spemann's original labels are in German: *pr. Med*, primary neural tube; *sec. Ch*, secondary notochord; *r. sec. Uw.*, right secondary somite; *r. sec. Pron.*, right secondary pronephric duct; *sec. Med.*, secondary neural tube.

Top right and bottom panels are copied from Spemann, H., Mangold, H.: **Induction of embryonic primordia by implantation of organizers from a different species.** *Int. J. Dev. Biol. 2001,* **45***: 13-38 [reprinted translation of their 1924 paper from the original German].*

elements from generation to generation, whereas embryology was the study of how an individual organism develops and, in particular, how cells in the early embryo became different from each other. Genetics seemed, in this respect, to be irrelevant to development.

The fledgling science of genetics was put on a firm conceptual and experimental footing in the first quarter of the twentieth century by T.H. Morgan. Morgan chose the fruit fly *Drosophila melanogaster* as his experimental organism. He noticed a fly with white eyes rather than the usual red eyes, and by careful cross-breeding he showed that inheritance of this mutant trait was linked to the sex of the fly. He found three other sex-linked traits and worked out that they were each determined by three distinct 'genetic loci', which occupied different positions on the same chromosome, the fly's X chromosome. The rather abstract hereditary 'factors' of Mendel had been

Fig. 1.10 The difference between genotype and phenotype. These identical twins have the same genotype because one fertilized egg split into two during development. Their slight difference in appearance is due to non-genetic factors, such as environmental influences.

Photograph courtesy of Robert and Lewis McCaffrey.

given reality. But even though Morgan was originally an embryologist, he made little headway in explaining development in terms of genetics. That had to wait until the nature of the gene was better understood.

An important concept in understanding how genes influence physical and physiological traits is the distinction between **genotype** and **phenotype**. This was first put forward by the Danish botanist, Wilhelm Johannsen, in 1909. The genetic endowment of an organism—the genetic information it inherits from its parents—is the genotype. The organism's visible appearance, internal structure, and biochemistry comprise the phenotype. While the genotype certainly controls development, environmental factors interacting with the genotype influence the phenotype. Despite having identical genotypes, identical twins can develop differences in their phenotypes as they grow up (Fig. 1.10), and these tend to become more evident with age.

Following Morgan's discoveries in genetics, the problem of development could now be posed in terms of the relationship between genotype and phenotype: how the genetic endowment becomes 'translated' or 'expressed' during development to give rise to a functioning organism, its phenotype. But the coming together of genetics and embryology was slow and tortuous. The discovery in the 1940s that genes are made of DNA and encode proteins was a major turning point. It was already clear that the properties of a cell are determined by the proteins it contains, and so the fundamental role of genes in development could at last be appreciated. By controlling which proteins were made in a cell, and when, genes could control the changes in cell properties and behavior that occurred during development. A further major advance in the 1960s was the discovery that some genes encode proteins that control the activity of other genes.

1.6 Development is studied through selected model organisms

Although the embryology of many different species has been studied at one time or another, a relatively small number of organisms provide most of our knowledge about developmental mechanisms. We can thus regard them as 'models' for understanding the processes involved, and they are often called **model organisms**. Sea urchins and amphibians were the main animals used for the first experimental investigations because their developing embryos are easy to obtain and, in the case of amphibians, relatively easy to manipulate experimentally, even at quite late stages. Among vertebrates, the frog, *Xenopus laevis*, the mouse, *Mus musculus*, the chicken, *Gallus gallus*, and the zebrafish, *Danio rerio*, are the main model organisms now studied. Among invertebrates, the fruit fly *Drosophila melanogaster* and the nematode *Caenorhabditis elegans* have been the focus of most attention, because a great deal is known about their developmental genetics and they can be easily genetically modified. With the advent of modern methods of genetic analysis, there has also been a resurgence of interest in the sea urchin *Strongylocentrotus purpuratus*. For plant developmental biology, the thale cress *Arabidopsis thaliana* serves as the main model organism. The life cycles and background details for these model organisms are given in the relevant chapters later in the book. The evolutionary relationships of these organisms, and their relationship to humans, are shown in Figure 1.11.

The reasons for these choices are partly historical—once a certain amount of research has been done on one animal it is more efficient to continue to study it rather than start at the beginning again with another species—and partly a question of ease of study and biological interest. Each species has its advantages and disadvantages as a developmental model. *Drosophila* develops fast and produces very large numbers of embryos that, as we shall see, can be screened for mutations. The chick embryo, however, has long been studied as a model for vertebrate development because fertile eggs are easily available and the embryo withstands experimental microsurgical manipulation very well. A disadvantage, however, was that until very recently little was known about the chick's

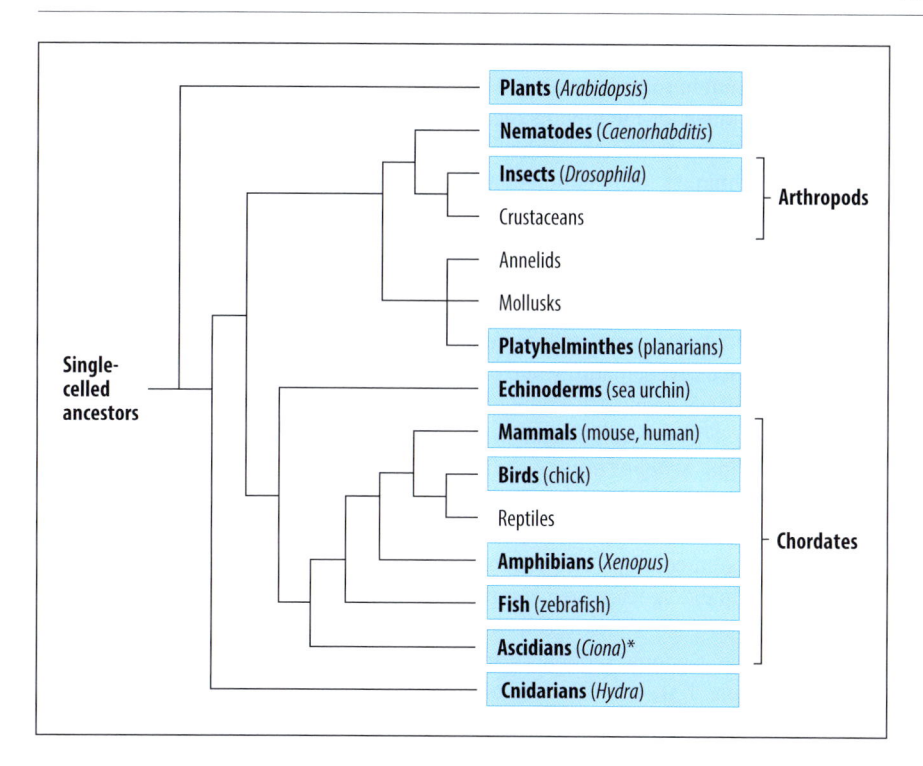

Fig. 1.11 Phylogenetic tree showing the placement of the main developmental model organisms. The organisms discussed in this book are highlighted in blue.

genetics. By contrast, we know a great deal about the genetics of the mouse, although the mouse is more difficult to study in some ways, as development normally takes place entirely within the mother. It is, however, possible to fertilize mouse eggs in culture, let them develop for a short time outside the uterus, when they can be observed, and then re-implant the early embryos for them to complete their development. Many developmental mutations have been identified in the mouse, and it is also amenable to genetic modification by **transgenic** techniques, by which genes can be introduced, deleted, or modified in living organisms. It is the best experimental model we have for studying mammalian development. The zebrafish is a more recent addition to the select list of vertebrate model organisms; it is easy to breed in large numbers, the embryos are transparent and so cell divisions and tissue movements can be followed visually, and it has great potential for genetic investigations.

However, as stated earlier, the ultimate goal of developmental biology is to understand how we humans develop, with the idea of using this knowledge to understand and treat disease. Until recently, our knowledge about our own development has been derived from the observation of medical collections of normal embryos and those in which development has gone wrong. Although we have a good description of human development (see Chapter 3), we have had to rely on the study of model organisms to provide clues to the genes and mechanisms involved. This has been very successful. For example, alterations in the human counterparts of developmental genes first discovered in *Drosophila* were uncovered by clinical geneticists studying people with developmental differences. It is now possible, however, to fertilize human eggs in culture and to study directly the very earliest stages in human development (discussed in Chapter 3). Over the past few years, as well, the discovery and use of human stem cells to study development has opened up many possibilities to learn about our development in an ethical manner (see Chapter 7).

A major goal of developmental biology is to understand how genes control embryonic development, and to do this one must first identify which genes, out of the many thousands in the organism, are critically and specifically involved in controlling development. This task can be approached in various ways, depending on the organism involved, but the general starting point is to identify gene mutations that alter development in some specific and informative way, as described in the following section. Techniques for identifying such developmental genes and for detecting and manipulating their expression in the organism are described throughout the book, along with techniques for manipulating the genes themselves.

Some of our model organisms are more amenable to conventional genetic analysis than others. Despite its importance in developmental biology, little conventional genetics has been done on *Xenopus laevis*, which has the disadvantage of being **tetraploid** (it has four sets of chromosomes in its somatic cells, as opposed to the two sets carried by the cells of diploid organisms, such as humans and mice) and a relatively long period of 1–2 years to reach sexual maturity and breed. Using modern genetic and bioinformatics techniques, however, many developmental genes have been identified in *Xenopus laevis* by direct DNA sequence comparison with known genes in *Drosophila* and mice. The closely related frog *Xenopus tropicalis* is a more attractive organism for genetic analysis; it is diploid and can also be genetically manipulated to produce transgenic organisms.

The hereditary information of a particular organism, comprising its complete set of genes and non-coding DNA sequences, is known as its **genome**. In sexually reproducing diploid species, two complete copies of the genome are carried by the chromosomes in each somatic cell—one copy inherited from the father and one from the mother. Genome sequences are available for all our model organisms highlighted in Figure 1.11, even for the tetraploid *Xenopus laevis*. In the zebrafish, a genome-duplication event in its evolutionary history means that there are duplicates of at least 2900 out of the estimated 23,000 protein-coding genes in its genome. Thus, the zebrafish genome contains over 26,000 protein-coding genes, whereas the mouse genome has only 22,000. The genomes of many different people have now also been sequenced. Having the complete DNA sequences of the genomes of our model organisms helps enormously in identifying developmental genes and other developmentally important DNA sequences in humans as well. These developmentally important sequences include non-coding control regions that regulate the expression of the protein-coding genes.

In general, when an important developmental gene has been identified in one animal, it has proved very rewarding to see whether a corresponding gene is present and is acting in a developmental capacity in other animals. Such genes are often identified by a sufficient degree of nucleotide sequence similarity to indicate descent from a common ancestral gene. Genes that meet this criterion are known as **homologous genes**. As we shall see in Chapter 4, this approach identified a hitherto unsuspected class of vertebrate genes that control the regular segmented pattern from head to tail, which is represented by the different types of vertebrae at different positions. These genes were identified by their **homology** with genes that determine the identities of the different body segments in *Drosophila* (described in Chapter 2). The DNA sequences that act as gene control regions can also often be identified by homology.

1.7 The first developmental genes were identified as spontaneous mutations

Most of the organisms dealt with in this book are sexually reproducing diploid organisms: their somatic cells contain two copies of each gene, with the exception of those on the sex chromosomes. In a diploid species, one copy, or **allele**, of each gene is contributed by the male parent and the other by the female. For many genes there are several different alleles present in the population, which leads to the variation in phenotype one normally sees within any sexually reproducing species. Occasionally, however, a mutation will occur spontaneously in a gene and there will be a marked change, usually deleterious, in the phenotype of the organism.

Many of the genes that affect development have been identified by spontaneous or induced mutations that disrupt their function and produce an abnormal phenotype.

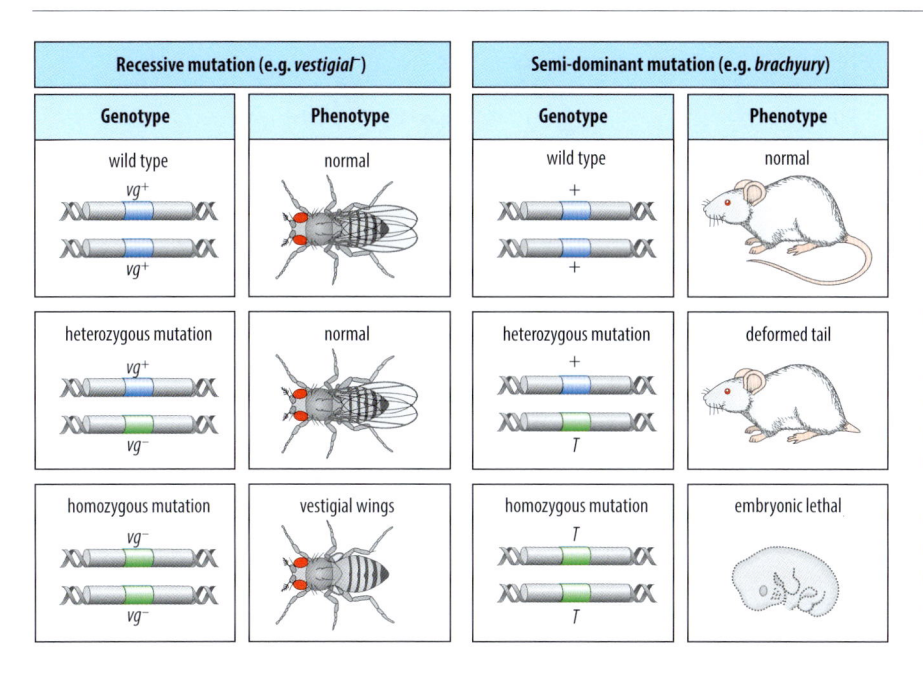

Fig. 1.12 Types of mutations. Left: a mutation is recessive when it only has an effect in the homozygous state; that is, when both copies of the gene carry the mutation. The *vestigial* mutation in *Drosophila* leads to the absence of wings when homozygous. A plus sign (+) denotes the 'normal' allele (wild type), and a minus sign (−) the recessive allele. Right: by contrast, a dominant or semi-dominant mutation produces an effect on the phenotype in the heterozygous state; that is, when just one copy of the gene carries the mutation. The Brachyury mutation in the mouse has an obvious but non-lethal effect in the heterozygous state, but is lethal in the homozygous state. + denotes the normal allele, and *T* denotes the dominant mutant allele of the *brachyury* gene (now called *TbxT*).

Mutations are classified broadly according to whether they are dominant or recessive (Fig. 1.12). **Dominant** and **semi-dominant** mutations are those that produce a distinctive phenotype when the mutation is present in just one allele of a pair; that is, it exerts an effect in the **heterozygous** state. By contrast, **recessive** mutations alter the phenotype only when both alleles are of the mutant type; this is known as the **homozygous** state. In recessive mutations, the function of the one wild-type gene is sufficient to confer a normal phenotype in heterozygotes, whereas in dominant or semi-dominant mutations, the effect of the mutant allele overrides that of the wild-type gene in the heterozygote or one copy of the normal gene is not enough to provide function.

In general, dominant mutations are more easily recognized, particularly if the phenotype involves alterations in gross anatomy or coloration, provided that they do not cause the early death of the embryo in the heterozygous state. True dominant mutations are rare. However, if one functional allele is not enough to provide the full function of a gene, a heterozygous mutant will have an abnormal phenotype. An example of this is seen in mice that have one mutant allele of the *TbxT* gene, denoted by *T*; heterozygous mice (*T*/+) have short tails. Because of this phenotype, the gene affected in the mutants was originally called *brachyury* (from the Greek *brachy*, short, and *oura*, tail). When the mutation is homozygous (*T/T*), it has a much greater effect, and embryos die at an early stage, with very short bodies, indicating that the *TbxT* gene is required for normal embryonic development (Fig. 1.13). Once breeding studies had confirmed that a single gene was involved, the mutation could be mapped to a location on a particular chromosome by classic gene-mapping techniques, and it was found that the *T* allele represents a deletion of the *TbxT* gene. In this case, the dominant mutation is due to an insufficiency of one copy of the gene to provide function. In other cases, for example the *Drosophila* mutation Antennapedia (*Antp*/+) in which the fly has legs on the head instead of antennae (see Fig. 9.2), the reason is an overactivity of one of the alleles. Both are cases of dominant mutations but the molecular mechanism is different.

Identifying genes affected by recessive mutations is more laborious, as the heterozygote has a phenotype identical to a normal wild-type animal, and a carefully worked-out breeding program is needed to obtain homozygotes. In mammals, identifying potentially lethal recessive developmental mutations requires careful observation and analysis, as the homozygotes may die unnoticed inside the mother.

Fig. 1.13 Genetics of the semi-dominant *T* mutation, which deletes the *TbxT* gene in the mouse. A male heterozygote carrying the *T* mutant allele of *TbxT* merely has a short tail. When mated with a female homozygous for the normal, or wild-type, *TbxT* gene (+/+), some of the offspring will also be heterozygotes and have short tails. Mating two heterozygotes together will result in some of the offspring being homozygous (*T*/*T*) for the mutation, resulting in a lethal developmental abnormality in which the body is very short because posterior mesoderm does not develop.

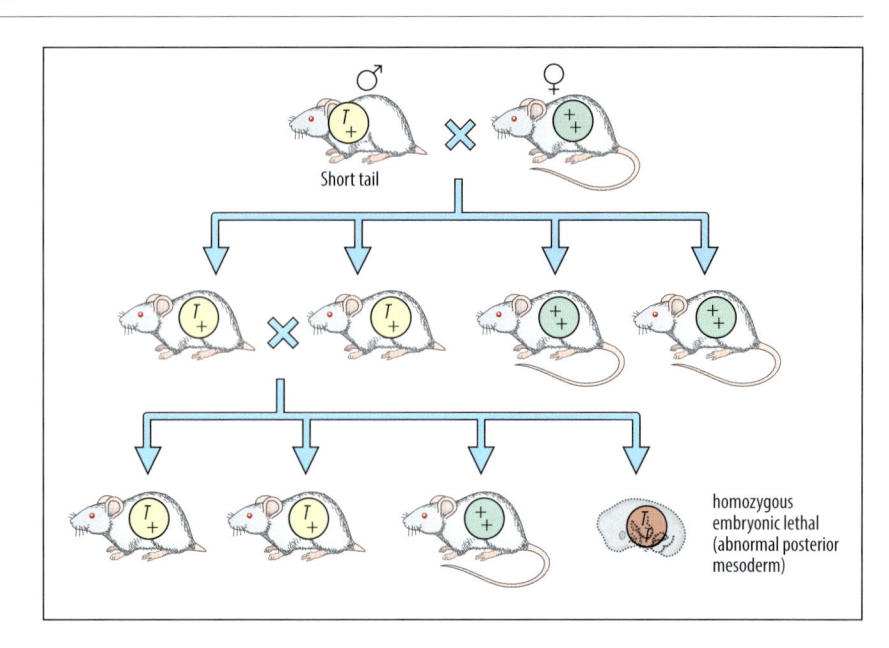

Many mutations in invertebrates have been identified as **conditional mutations**. The effects of these mutations only show up when the animal is kept in a particular condition—most commonly at a higher temperature, in which case the mutation is known as a **temperature-sensitive mutation**. At the usual ambient temperature, the animal appears normal. Temperature sensitivity is usually due to the encoded mutant protein being able to fold into a functional structure at the normal temperature, but being less stable at the higher temperature.

Very rigorous criteria must be applied to identify mutations that are affecting a genuine developmental process and not just affecting some vital but routine 'housekeeping' function without which the animal cannot survive. One simple criterion for a developmental mutation is that it leads to the death of the embryo, but mutations in genes involved in vital housekeeping functions can also have this effect. Mutations that produce observable abnormal embryonic development are more promising candidates for true developmental mutations. In later chapters we shall see how large-scale screening for mutations after mutagenesis by chemicals or X-rays has identified many more developmental genes than would ever have been picked up just by looking for rare spontaneous mutations.

In recent years, new techniques for identifying developmental genes have appeared. These depend on knowing the existence and sequence of a gene (e.g. from genomic sequences) and working backwards to determine its function in the animal, usually by removing the gene or blocking its function. Such methods are known as **reverse genetics**, as opposed to the traditional methods of **forward genetics**, outlined above, which start with a mutant phenotype and then identify the gene responsible. Examples of reverse genetics include **gene knock-out**, in which the gene is effectively deleted from the animal's genome by transgenic techniques (discussed in Box 3B), and **gene knockdown** or **gene silencing**, in which gene expression is prevented by techniques such as RNA interference or antisense RNA. More recently, a new DNA-editing technology, **CRISPR-Cas9**, has revolutionized the production of gene knock-outs and other genetic approaches to development (Box 1C). This technique allows an efficient and clean production of mutations in any organism in which the sequence of the gene to be targeted is known and the editing components can be effectively introduced into cells. Mutations can be introduced in both protein-coding and regulatory DNA sequences.

EXPERIMENTAL BOX 1C The CRISPR-Cas9 genome-editing system

The CRISPR-Cas9 technique is an adaptation of a system that some bacteria use to defend themselves against attack by a bacterial virus. A bacterial nuclease (Cas9) is targeted to the incoming virus DNA, cleaves it, and inactivates it. CRISPR is the acronym for 'Clustered Regularly Interspaced Short Palindromic Repeats', a 20-nucleotide long RNA sequence encoded in the bacterial genome that matches sequences in the viral DNA and targets the nuclease to it. The CRISPR RNA (crRNA) binds to Cas9 and guides the enzyme to the target; the nuclease then makes a double-strand break in the virus DNA. As well as the Cas9 nuclease and the crRNA, the bacterial system includes another essential piece of RNA called transactivating CRISPR RNA (tracrRNA). The crRNA and tracrRNA form the complex that positions the Cas9 nuclease at the target DNA sequence. Successful targeting also requires a short nucleotide motif in the target DNA, which is usually N(any base)GG.

Once the mechanism of action of the CRISPR-Cas9 system was discovered, researchers realized that it could be adapted to introduce modifications into eukaryotic genomes. By replacing the virus-specific CRISPR sequence with another sequence, CRISPR-Cas9 could be engineered to target any DNA sequence—whether a protein-coding sequence or a regulatory DNA element such as a promoter or an enhancer. Retooling of the bacterial system has created a simple and effective means of genome editing in eukaryotes, with a wide range of applications.

The basic CRISPR system used to edit the genomes of mammalian and other eukaryotic cells consists of two components: the Cas9 nuclease and the crRNA and the tracrRNA fused into a single unit, with the crRNA sequence designed to match the desired target. The RNA ensemble is called the single guide RNA (sgRNA). The cells to be edited are transfected with DNA plasmids that encode and express the desired sgRNA and the Cas9 nuclease, although in some applications the protein and RNAs themselves are introduced. The sgRNA-Cas9 complex then finds the target sequence and positions the nuclease to introduce a double-strand break (Fig. 1).

The DNA repair systems of the cell now intervene and act in either of two ways (see Fig. 1). If there is no DNA template to guide repair, a pathway called non-homologous end joining (NHEJ) will attempt to repair the break, and in the process will introduce small deletions or insertions into the DNA that, in general, will create loss-of-function mutations. Alternatively, if there is a repair template, the homology-directed repair (HDR) system will use the process of homologous recombination to repair the break according to the template DNA sequence. This makes it possible to change target sequences in a precise manner—to repair existing mutations, for example, or introduce desired ones—by providing an appropriate template together with the Cas9 and sgRNA. HDR can also be used to introduce quite extensive modifications, such as the insertion of DNA sequences that encode reporters for gene expression or tags for proteins. The editing system leaves no potentially harmful residue, because the CRISPR and Cas9 sequences are not incorporated into the genome and so, as the edited cells divide, they are diluted and lost. CRISPR-Cas9 is now the most common method for creating gene knock-outs.

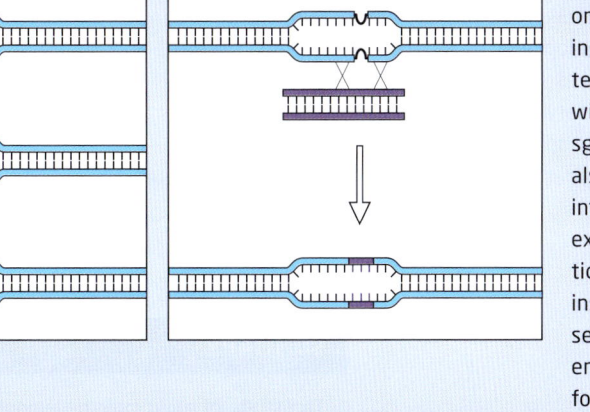

Fig. 1

The CRISPR-Cas9 system has been modified in various ways for different applications. Novel Cas9-like proteins have been found in bacteria that are more efficient and easier to deliver. Bacterial enzymes that produce single base changes in targeted genes—known as **base editing**—without producing a double-strand break in the DNA have been developed, making the editing process even more precise. These enzymes can be targeted to the required gene by linking them to a CRISPR-Cas9 system in which the Cas9 nuclease is enzymatically inactive. Altogether, these innovations are creating an ever-growing toolkit for genome modification. Another development also involves Cas9 proteins without nuclease activity, which when fused to transcription factors or chromatin modifiers can deliver these activities to specific DNA sequences.

CRISPR-Cas9 technology cuts down the time it takes to breed a line of organisms with a desired mutation. In mice, for example, the enzyme and the RNA can be injected directly into the fertilized egg, thus producing modifications in one generation, and because several genes can be targeted at once, it is possible to generate double or triple mutants (Fig. 2). The same is true for the fertilized eggs of other vertebrates, such as zebrafish.

CRISPR-Cas9 has been adapted to a large number of applications, ranging from the generation of reporters for any given gene to the execution of mutant screens by using collections of CRISPR sequences. The simplicity and versatility of the system has enabled the genetic modification of organisms that do not lend themselves to conventional genetic analysis, such as crustaceans and cnidarians. It can also be applied to plants. CRISPR-Cas9 is thus having a huge effect on our understanding of developmental processes and their universality.

One important caveat that needs to be considered, particularly in light of future potential therapeutic applications, is 'off-target' effects: these are due to the occasional lack of specificity which, in addition to the desired genomic change, introduces additional uncontrolled ones elsewhere in the genome. The development of increasingly specific sgRNAs and Cas9 derivatives is minimizing these unwanted effects, but it is still necessary to sequence the genome of the altered cell or organism completely to detect whether an off-target effect has occurred.

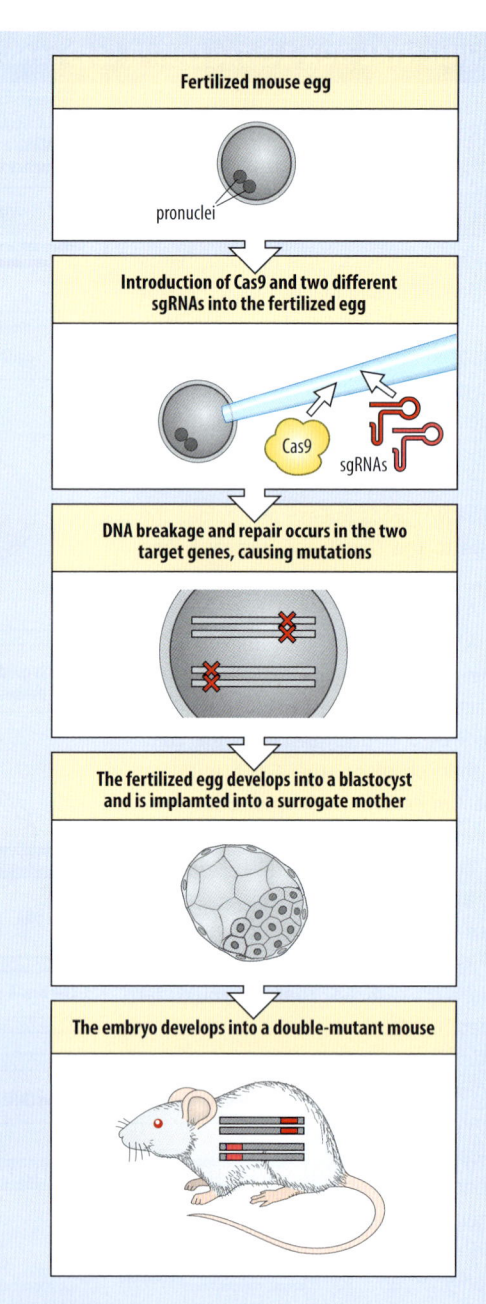

Fig. 2

SUMMARY

The study of embryonic development in Europe started with the Greeks more than 2000 years ago. Aristotle put forward the idea that embryos were not contained completely preformed in miniature within the egg, but that form and structure emerged gradually, as the embryo developed (a process he called epigenesis). This idea was challenged in the seventeenth and eighteenth centuries by those who believed in preformation, the idea that all embryos that had been, or ever would be, had existed from the beginning of the world. The emergence of the cell theory in the nineteenth century finally settled the issue in favor of epigenesis, and it was realized that the sperm and egg are single, albeit highly specialized, cells. Some of the earliest experiments showed that very early sea urchin embryos can regulate—that is, develop normally even if cells are removed or killed. This established the important principle that development must depend, at least in part, on communication between the cells of the embryo. Direct evidence for the importance of cell–cell interactions came from the organizer graft experiments carried out by Spemann and Mangold in 1924, showing that the cells of the amphibian organizer region could induce a new partial secondary embryo from host tissue when transplanted into another embryo. The role of genes in controlling development, by determining which proteins are made, has only been fully appreciated in the past 50 years and the study of the genetic basis of development has been made much easier in recent times by the techniques of molecular biology and the availability of the DNA sequences of whole genomes.

A conceptual tool kit

Development into a multicellular organism is the most complicated fate a single living cell can undergo; in this lies both the fascination and the challenge of developmental biology. Yet only a few basic principles are needed to start to make sense of developmental processes. The rest of this chapter is devoted to introducing these key concepts. These principles are encountered repeatedly throughout the book, as we look at different organisms and developmental systems, and should be regarded as a conceptual tool kit, essential for embarking on a study of development.

Genes control development by determining where and when proteins are synthesized. Gene activity sets up intracellular networks of interactions between proteins and genes, and between proteins and proteins, that give cells their particular properties. One of these properties is the ability to communicate with, and respond to, other cells. It is these cell–cell interactions that determine how the embryo develops; no developmental process can therefore be attributed to the function of a single gene or single protein. The amount of genetic and molecular information on developmental processes is now enormous. In this book, we will be highly selective and describe only those molecular details that give an insight into the mechanisms of development and illustrate general principles.

1.8 Development involves the emergence of pattern, change in form, cell differentiation, and growth

In most embryos, fertilization is immediately followed by cell division. This stage is known as **cleavage** and divides the fertilized egg into a number of smaller cells (Fig. 1.14). In the embryos of some animals, such as *Xenopus*, there is no increase in cell size between each cleavage division; the cleavage cell cycles consist simply of phases of DNA replication, mitosis, and cell division (see Box 1B). The early embryos are therefore no bigger than the zygote, and there is little increase in size during embryogenesis. By contrast, bird and mammalian embryos increase in size during early development. This is related to their access to a substantial source of material they can use to grow—for example, the large yolk available to the chick embryo and the nutrients provided to mammalian embryos through the placenta.

Development is, in essence, the emergence of organized structures from an initially very simple group of cells. It is convenient to distinguish four main developmental processes, which occur in roughly sequential order in development, although, in reality,

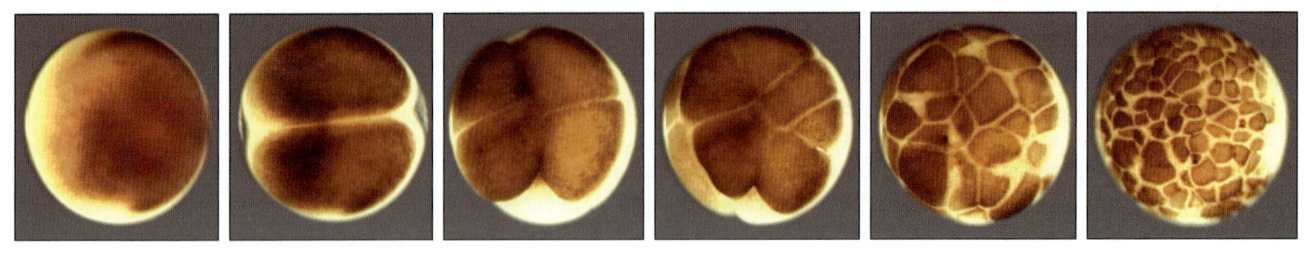

Fig. 1.14 Light micrographs of cleaving *Xenopus* eggs.

Images courtesy of Dr H. Williams.

they overlap with, and influence, each other considerably. They are **pattern formation**, **morphogenesis** (change in form), **cell differentiation**, and **growth**. Pattern formation is the process by which cell identity and activity is organized in space and time so that a well-ordered structure develops within the embryo. It is of fundamental importance in the early embryo and also later in the formation of organs. In the developing arm, for example, pattern formation enables cells to 'know' whether to make an upper arm or fingers, which finger to make, and where the muscles should form and attach. There is no universal strategy of patterning; rather, it is achieved by various cellular and molecular mechanisms in different organisms and at different stages of development.

Early in development, pattern formation involves laying down the overall **body plan**—defining the main body axes of the embryo that run from head (**anterior**) to tail (**posterior**), and from back (**dorsal**) to underside (**ventral**) as well as a midline that will create bilateral symmetry. Most of the animals in this book have a head at one end and a tail at the other, with the left and right sides of the body being outwardly bilaterally symmetrical across the midline—that is, a mirror image of each other. In such animals, the main body axis is the **antero-posterior axis**, which runs from the head to the tail. Bilaterally symmetrical animals also have a **dorso-ventral axis**, running from the back to the belly. A striking feature of these axes is that they are almost always at right angles to one another and so can be thought of as making up a system of coordinates on which any position in the body could be specified (Fig. 1.15). Internally, animals have distinct differences between the left and right sides in the disposition of their internal organs—the human heart, for example, is on the left side. How the basic body plans of our four model vertebrates are laid down is covered in Chapters 3 and 4. In plants, the main body axis runs from the growing tip (the apex) to the roots and is known as the **apical–basal axis**. Plants also have radial symmetry, with a **radial axis** running from the center of the stem outwards. Plant development is discussed in Chapter 12.

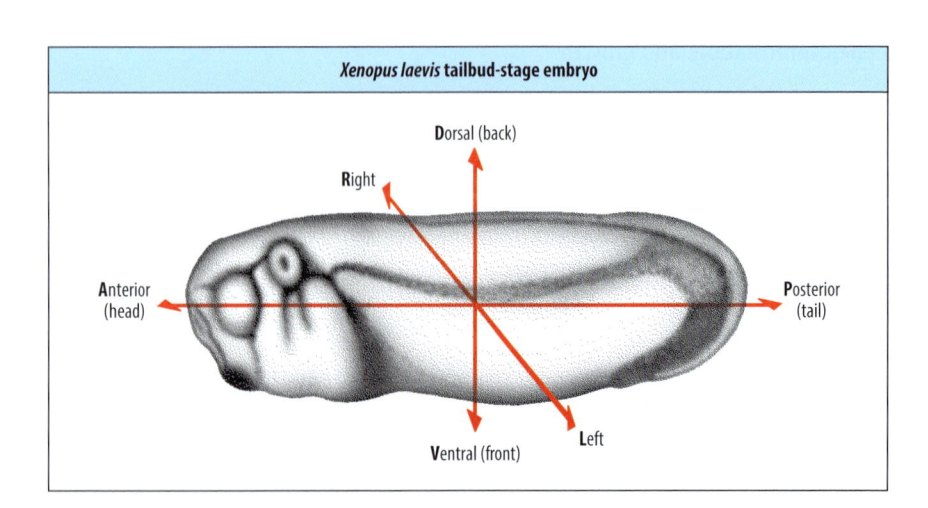

Fig. 1.15 The main axes of a developing embryo. The antero-posterior axis and the dorso-ventral axis are at right angles to one another, as in a Cartesian coordinate system.

Even before the body axes become clear, eggs and embryos often show a distinct **polarity**, which in this context means that they have an intrinsic orientation, with one end different from the other. Polarity can be specified by a gradient in a protein or other molecule, as the slope of the gradient provides direction. Individual cells within the embryo can also have their own polarity. One type of polarity is called **apico-basal polarity**, and is seen in simple **epithelia**, single-layered sheets of cells in which the two sides of the sheet (the apical and basal sides) are different, and which are building blocks of tissues and organs (discussed in detail in Chapter 5). A second axis of polarity in cells can be discerned along the plane of the tissue and this provides a larger-scale coordinate system for the tissue. This type of individual cell polarity is known as **planar cell polarity** and is described in Chapter 5. Planar cell polarity is sometimes easily visible in structures—the hairs produced by epidermal cells on the edge of the *Drosophila* wing all point in the same direction.

At the same time that the axes of the embryo are being formed, cells are being allocated to the different **germ layers**—**ectoderm**, **mesoderm**, and **endoderm** (Box 1D showing stylized body sections). During further pattern formation, cells of these germ layers acquire different identities so that organized spatial patterns of differentiated cells finally emerge, such as the arrangement of skin, muscle, and cartilage in developing limbs, and the arrangement of neurons in the nervous system.

The second important developmental process is change in form, or morphogenesis (discussed in Chapter 5). Embryos undergo remarkable changes in three-dimensional form—you need only think about the complicated anatomy of structures such as your hands and feet. At certain stages in development, there are characteristic and dramatic changes in form, of which **gastrulation** is the most striking. All but the very simplest multicellular animals undergo gastrulation, during which the main body plan emerges. During gastrulation, some cells from the outside of the embryo move

---

CELL BIOLOGY BOX 1D Germ layers

The concept of germ layers is useful to distinguish between regions of the early embryo that give rise to quite distinct types of tissues. It applies to both vertebrates and invertebrates. All the animals considered in this book, except for the cnidarian *Hydra*, are **triploblasts**, with three germ layers: the ectoderm, which gives rise to the epidermis of the skin and nervous system; the mesoderm, which gives rise to the skeleto-muscular system, connective tissues, and other internal organs, such as the kidney and heart; and the endoderm, which gives rise to the gut and its derivatives, such as the liver and lungs in vertebrates (Fig. 1). These are specified early in development. There are, however, some exceptions to the types of tissues that arise from a particular germ layer. The neural crest in vertebrates, for example, is ectodermal in origin, but gives rise not only to neural tissue, but also to some skeletal elements that generally arise from mesoderm.

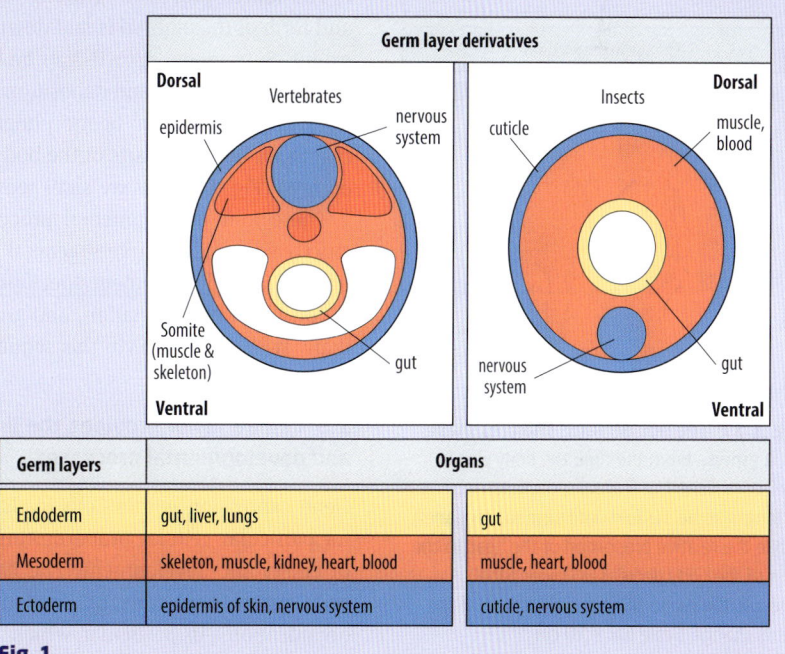

Germ layers	Organs	
Endoderm	gut, liver, lungs	gut
Mesoderm	skeleton, muscle, kidney, heart, blood	muscle, heart, blood
Ectoderm	epidermis of skin, nervous system	cuticle, nervous system

Fig. 1

Fig. 1.16 Gastrulation in the sea urchin. Gastrulation transforms the spherical blastula into a structure with a tube—the gut—through the middle. One side of the embryo has been removed.

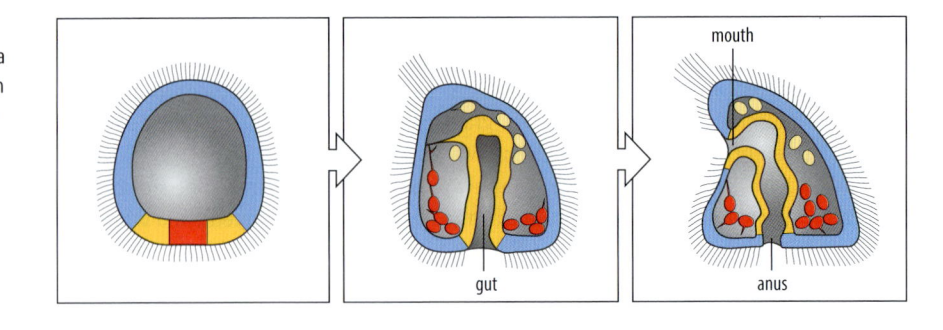

inwards and transform a hollow spherical structure into a complex three-dimensional structure with a tube through the middle—the gut (Fig. 1.16). Morphogenesis in animal embryos can also involve extensive cell migration. Most of the cells of the human face, for example, are derived from cells that migrated from a tissue called the neural crest, which originates on the dorsal side of the embryo. Morphogenesis can also involve developmentally programmed cell death, called **apoptosis**, which is responsible, for example, for the separation of our fingers and toes from a solid plate of tissue.

The third developmental process we must consider here is **cell differentiation**, in which cells become structurally and functionally different from each other, ending up as distinct cell types, such as blood, muscle, or skin cells. Differentiation is a gradual process, with cells often going through several divisions between the time at which they start differentiating and the time they are fully differentiated (when many cell types stop dividing altogether) and is discussed in Chapter 6. In humans, the fertilized egg gives rise to hundreds of clearly distinguishable types of cells.

Pattern formation and cell differentiation are very closely interrelated, as we can see by considering the difference between human arms and legs. Both contain the same types of cells—muscle, cartilage, bone, skin, and so on—yet the pattern in which they are arranged is clearly different. It is essentially pattern formation that makes us look different from elephants and chimpanzees.

The fourth process is **growth**—the increase in size. In general, the basic pattern and form of the embryo is laid down on a small scale, always less than a millimeter in extent. Subsequent growth can be brought about in various ways: cell proliferation, increase in cell size, and deposition of extracellular materials, such as, for example, in bone. Growth can also be morphogenetic, in that differences in growth rates between organs, or between parts of the body, can generate changes in the overall shape of the embryo (Fig. 1.17), as we shall see in more detail in Chapter 11.

These four developmental processes are neither independent of each other nor strictly sequential. In very general terms, however, one can think of pattern formation in early development specifying differences between cells that lead to changes in form, cell differentiation, and growth. But in any real developing system there are many twists and turns in this sequence of events.

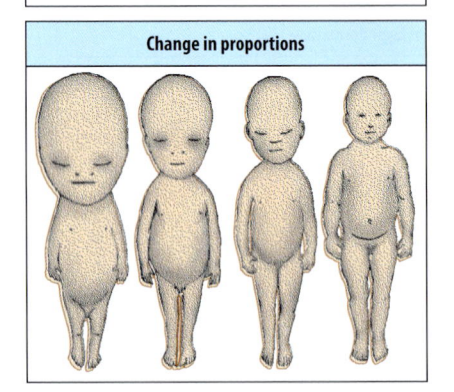

Fig. 1.17 The human fetus changes shape as it grows. From the time the body plan is well established, at 8 weeks, the fetus increases in length some 10-fold until birth (upper panel), while the relative proportion of the head to the rest of the body decreases (lower panel). As a result, the proportions and the shape of the fetus change. Scale bar = 10 cm.

After Moore, K.L.: The Developing Human. *Philadelphia, PA: Saunders, 1983.*

1.9 Cell behavior provides the link between gene action and developmental processes

Gene expression within cells leads to the synthesis of proteins that are responsible for particular cellular properties and behavior, which, in turn, determine the course of embryonic development. The past and current patterns of gene activity confer a certain state, or identity, on a cell at any given time, which is reflected in its molecular organization—in particular which proteins are present. As we shall see, embryonic cells and their progeny undergo many changes in state as development progresses. Other categories of cell behavior that will concern us are intercellular communication, also known as **cell–cell signaling** or **cell–cell interactions**, changes in cell shape and cell movement, cell proliferation, and cell death.

Changing patterns of gene activity during early development are essential for pattern formation. They give cells identities that determine their future behavior and lead eventually to their final differentiation. And, as we saw in the example of induction by the Spemann organizer, the capacity of cells to influence each other's fate by producing and responding to signals is crucial for development. By their response to signals for cell movement or a change in shape, for example, cells generate the physical forces that bring about morphogenesis (Fig. 1.18). The curvature of a sheet of cells into a tube, as happens in *Xenopus* and other vertebrates during formation of the neural tube (see Box 1A), is the result of contractile forces generated by cells changing their shape at certain positions within the cell sheet. An important feature of cell surfaces is the presence of adhesive proteins known as cell-adhesion molecules, which serve various functions: they hold cells together in tissues; they enable cells to sense the nature of the surrounding extracellular matrix; and they serve to guide migratory cells such as the neural crest cells of vertebrates, which leave the neural tube to form structures elsewhere in the body.

We can therefore describe and explain developmental processes in terms of how individual cells and groups of cells behave. Because the final structures generated by development are themselves composed of cells, explanations and descriptions at the cellular level can provide an account of how these adult structures are formed.

Because development can be understood at the cellular level, we can pose the question of how genes control development in a more precise form. We can now ask how genes are controlling cell behavior. The many possible ways in which a cell can behave therefore provide the link between gene activity and the morphology of the adult animal—the outcome of development. Cell biology provides the means by which the genotype becomes translated into the phenotype.

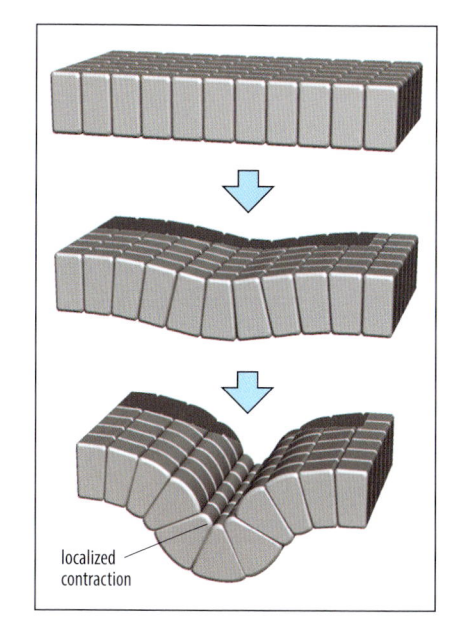

Fig. 1.18 Localized contraction of particular cells in a sheet of cells can cause the whole sheet to fold. Contraction of a line of cells at their apices due to changes in the cells' internal structure causes a furrow to form in a sheet of epithelium.

1.10 Genes control cell behavior by specifying which proteins are made

What a cell can do is determined very largely by the proteins it makes. The hemoglobin in red blood cells enables them to transport oxygen; the cells lining the gut secrete specialized digestive enzymes; and skeletal muscle cells can contract because they contain contractile structures composed of the proteins myosin, actin, and tropomyosin, as well as other muscle-specific proteins needed to make muscle function correctly. All these are specialized proteins and are not involved in the 'housekeeping' activities that are common to all cells and keep them alive and functioning. Housekeeping activities include the production of energy and the metabolic pathways involved in the breakdown and synthesis of molecules necessary for the life of the cell. Although there are qualitative and quantitative variations in housekeeping proteins in different cells, they are not important players in development. In development we are concerned primarily with those proteins that make cells different from one another and which are often known as **tissue-specific** proteins.

Genes control development mainly by specifying which proteins are made in which cells and when. Proteins are the agents that directly determine cell behavior, which includes determining which genes are expressed. To produce a particular protein, its gene must be switched on and transcribed (**transcription**) into **messenger RNA (mRNA)**. This process is known as **gene expression**. The protein encoded by the mRNA is then synthesized in the process of **translation**. The **initiation of transcription**, or switching on of a gene, involves combinations of specialized **gene-regulatory proteins**, which include **transcription factors** and **RNA polymerase**, binding to the **control region** of the gene. The control region is made up of *cis*-**regulatory regions** in the DNA (the *cis* simply refers to the fact that the regulatory region is on the same DNA molecule as the gene it controls). These *cis*-regulatory regions are often modular, and in many developmental genes the time of gene expression and the tissue in which the gene is expressed are controlled by different *cis*-**regulatory modules**.

Both transcription and translation are subject to several layers of control, and translation does not automatically follow transcription. Figure 1.19 shows the main stages in gene expression at which the production of a protein can be controlled. Even after a gene has been transcribed, for example, the mRNA might be degraded before it can be exported from the nucleus. Even if an mRNA reaches the cytoplasm, its translation can be prevented or delayed. In the eggs of many animals, preformed mRNA is prevented from being translated until after fertilization. mRNAs may also be targeted by specific **microRNAs (miRNAs)**, short **non-coding RNAs** that can render the mRNA inactive and so block translation. Some microRNAs are known to be involved in gene regulation in development. Another process that determines which proteins are produced is **RNA processing**. In eukaryotic organisms—all organisms other than bacteria and archaea—the initial RNA transcripts of most protein-coding genes are cut and rejoined in a processing step called **RNA splicing** to give the functional mRNA. In some genes, the transcripts can be cut and spliced in different ways, called **alternative splicing**, to give rise to two or more different mRNAs; in this way, different proteins with different properties can be produced from a single gene.

Even if a gene has been transcribed and the mRNA translated, the protein may still not be able to function immediately. Many newly synthesized proteins require further **post-translational modification** before they acquire biological activity. One very common permanent modification that occurs to proteins destined for the cell membrane or for secretion is the addition of carbohydrate side chains, or **glycosylation**. Reversible post-translational modifications, such as phosphorylation, can also significantly alter protein function. Together, alternative RNA splicing and post-translational modification mean that the number of functionally different proteins that can be produced is considerably greater than the number of protein-coding genes would indicate. For many proteins, their subsequent localization in a particular part of the cell, for example the nucleus, is essential for them to carry out their function.

An intriguing question is how many genes out of the total genome are **developmental genes**—that is, genes specifically required for embryonic development. This is not easy to estimate. In the early development of *Drosophila*, at least 60 genes are directly involved in pattern formation up to the time that the embryo becomes divided into segments. In *Caenorhabditis*, at least 50 genes are needed to specify a small reproductive structure known as the vulva. These are quite small

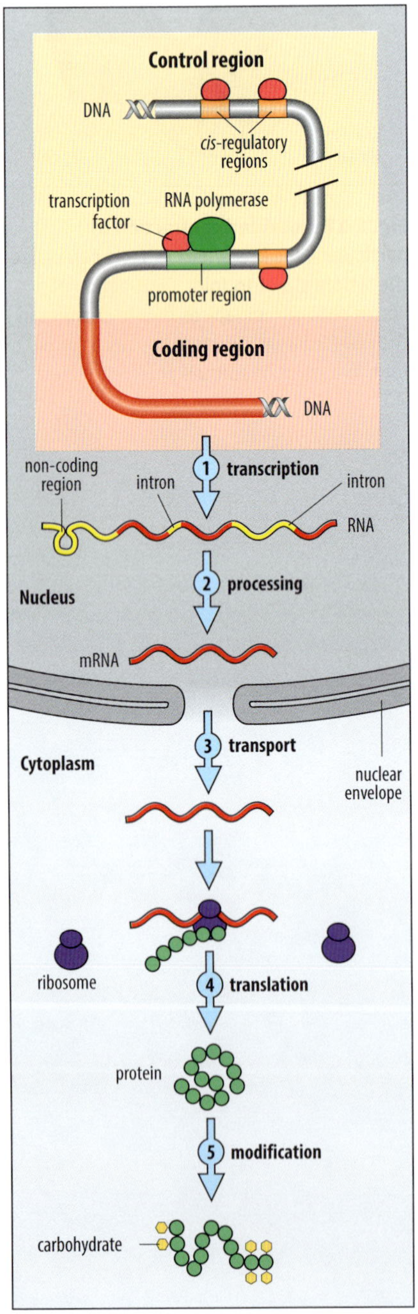

Fig. 1.19 Gene expression and protein synthesis. A protein-coding gene comprises a **coding region**—a stretch of DNA that contains the instructions for making the protein—and adjacent DNA sequences that act as a control region. The control region includes the promoter, to which general transcription factors and the enzyme RNA polymerase bind to start transcription, and a *cis*-regulatory region, consisting of one or more modules, at which other transcription factors bind to switch the gene on or off. This latter control region may be thousands of base pairs away from the promoter. When the gene is switched on, the DNA sequence is transcribed to produce RNA (1). The RNA formed by transcription is spliced to remove introns and the non-coding region at the start of the gene (yellow) and processed within the nucleus (2) to produce mRNA. This is exported from the nucleus to the cytoplasm (3) for translation into protein at the ribosomes (4). Control of gene expression and protein synthesis occurs mainly at the level of transcription but can also occur at later stages. For example, mRNA may be degraded before it can be translated, or if it is not translated immediately it may be stored in inactive form in the cytoplasm for translation at a later stage. mRNAs may also be targeted by specific microRNAs so that translation is blocked. Some proteins require a further step of post-translational modification (5) to become biologically active. A very common post-translational modification is the addition of carbohydrate side chains (glycosylation), as shown here.

numbers compared to the thousands of genes that are active at the same time; some of those genes are essential in that they are necessary for maintaining life, but they provide no or little information that influences the course of development. Many genes change their activity in a systematic way during development, suggesting that they could be developmental genes. The total number of genes in *Caenorhabditis* and *Drosophila* is around 19,000 and 14,000, respectively, and the number of genes involved in development is likely to be in the thousands. A systematic analysis of most of the *Caenorhabditis* genome suggests that approximately 9% of the genes (1722 out of 19,000) are involved in development.

Developmental genes typically code for proteins involved in the regulation of cell behavior: receptors; intercellular cell–cell signal proteins such as growth factors; intracellular signaling proteins; and gene-regulatory proteins. Many of these genes, especially those for receptors and signaling proteins, are used throughout an organism's life, but others are active only during embryonic development. Some of the techniques used to find out where and when a gene is active are described in Box 1E and Section 3.2. There are techniques that can document all the genes expressed in the embryo, or in a particular tissue, or even in a single cell at a particular time in development.

It is also possible to prevent the expression of a given gene and so prevent its protein from being made. This can be done using **antisense RNAs**, which are small RNA molecules of about 25 nucleotides complementary in sequence to part of a gene or mRNA. Such antisense RNAs can block gene function by binding to the gene or its mRNA. Artificial RNA molecules called morpholinos are used, which are more stable than normal RNA. Another technique for blocking gene expression is **RNA interference (RNAi)**, which operates in a similar way, but uses a different type of small RNA called small interfering RNAs (siRNAs) to exploit cells' own metabolic pathways for degrading mRNAs.

1.11 The expression of developmental genes is under tight control

All the somatic cells in an embryo are derived from the fertilized egg by successive rounds of mitotic cell division. Thus, with rare exceptions, they all contain identical genetic information, the same as that in the zygote. The fact that genetic information is not lost during differentiation was first shown in a landmark series of experiments by John Gurdon working with *Xenopus*. In these experiments, nuclei from differentiated cells were placed into unfertilized eggs in which the nucleus had been destroyed to test the ability of these nuclei to support normal development. In many cases tadpoles were obtained, and in a small number of cases even adult frogs, showing that differentiated cells still contain all the genetic information required to make a complete animal (Fig. 1.20). This is the principle of **genetic equivalence**. The differences between cells must therefore be generated by differences in gene activity that lead to the synthesis of different proteins. This is the principle of **differential gene expression**. John Gurdon was awarded the Nobel Prize in Physiology or Medicine in 2012 for this work, together with the Japanese biologist Shinya Yamanaka for his work on mouse and human stem cells (discussed in Chapters 6 and 7).

Turning the correct genes on or off in the right cells at the right time therefore becomes the central issue in development (Box 1E). As we shall see throughout the book, genes do not provide a blueprint for development but a set of instructions to be used by cells. Key elements in regulating the readout of these instructions are the *cis*-regulatory regions associated with developmental genes and with genes for specialized proteins (see Fig. 1.19). These are acted on by gene-regulatory proteins, which switch genes on or off by activating or repressing transcription. Some gene-regulatory proteins, called transcription factors, act by binding directly to the

EXPERIMENTAL BOX 1E Visualizing gene expression in embryos

To understand how gene expression is guiding development, it is essential to know exactly where and when particular genes are active. Genes are switched on and off during development and patterns of gene expression are continually changing. Several powerful techniques can show where a gene is being expressed in the embryo.

One set of techniques uses *in situ* **nucleic acid hybridization** to detect the mRNA that is being transcribed from a gene. If a DNA or RNA probe is complementary in sequence to an mRNA being transcribed in the cell, it will base-pair precisely (hybridize), with the mRNA (Fig. 1). A DNA probe can therefore be used to locate its complementary mRNA in a tissue slice or a whole embryo. The probe may be labeled in various ways—with a radioactive isotope, a fluorescent tag (fluorescence *in situ* hybridization (FISH)), or an enzyme—to enable it to be detected. Radioactively labeled probes are detected by autoradiography, whereas probes labeled with fluorescent dyes are observed by fluorescence microscopy. Enzyme-labeled probes are detected by the enzyme's conversion of a colorless substrate into colored product. Labeling with fluorescent dyes or enzymes has the advantage over radioactive labeling in that the expression of several different genes can be detected simultaneously, by labeling the different probes with different colored detectors. It also means that researchers do not have to work with radioactive chemicals.

Proteins can also be detected *in situ* by immunostaining with antibodies specific for the protein. The antibodies may be tagged directly with a colored dye or an enzyme, or, more usually, the protein-specific antibody is detected by a second round of staining with an anti-immunoglobulin antibody tagged either with a fluorescent dye (as pictured here) or an enzyme (see Fig. 2.34). Figure 2 shows a section through the neural tube of a chick embryo at stage 16 (around 2–2.75 days after laying) which has been immunostained for expression of transcription factors that pattern the neural tube dorso-ventrally. This image was generated by immunostaining adjacent sections separately for each of the five transcription factors and then merging these images. The transcription factor was first detected *in situ* using an antibody specific for the protein and then visualized by applying a secondary anti-immunoglobulin antibody labeled with a fluorescent dye (purple, Foxa2; red, Nkx2.2; yellow, Olig2; green, Pax6; blue, Pax7). These *in situ* labeling techniques, whether for RNA or protein, can only be applied to fixed embryos and tissue sections.

The pattern and timing of gene expression can also be followed by inserting, using transgenic techniques, a 'reporter' gene into an animal. The reporter gene codes for some easily detectable protein and is placed under the control of an appropriate promoter that will switch on the expression of the protein at a particular time and place. One commonly used reporter gene is the *lacZ* sequence coding for the bacterial enzyme β-galactosidase. The presence of the enzyme can be detected by treating the specimen after fixation with a modified substrate, which the enzyme reacts with to give a blue-colored product (see Fig. 4.35), or by using a fluorescent-tagged antibody specific for β-galactoside.

Other very commonly used reporter genes are those for small fluorescent proteins, such as green fluorescent protein (GFP), which fluoresce in various colors and were isolated from jellyfish and other marine organisms. These proteins are extremely useful as they are harmless to living cells and are readily made visible by illuminating cultured cells, embryos, or even whole animals with light of a suitable wavelength. Gene expression can therefore be visualized and followed in living embryos and cells (see Fig. 5.47c). The three scientists who discovered and developed GFP and other fluorescent proteins for use in live imaging were awarded the 2008 Nobel Prize in Chemistry.

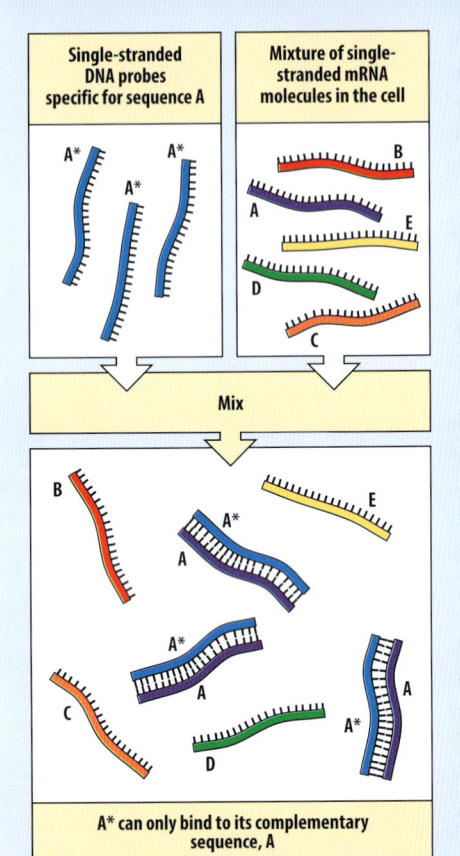

Fig. 1

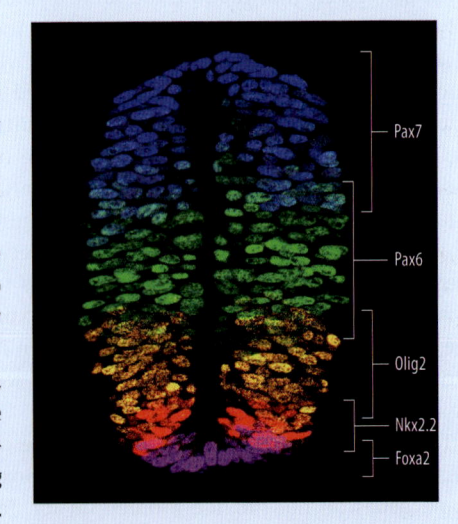

Fig. 2

Photograph courtesy of G. Le Dréau and E. Martí. From Le Dréau, G., Marti, E.: Dev. Neurobiol. *2012,* **72**: *1471–1481.*

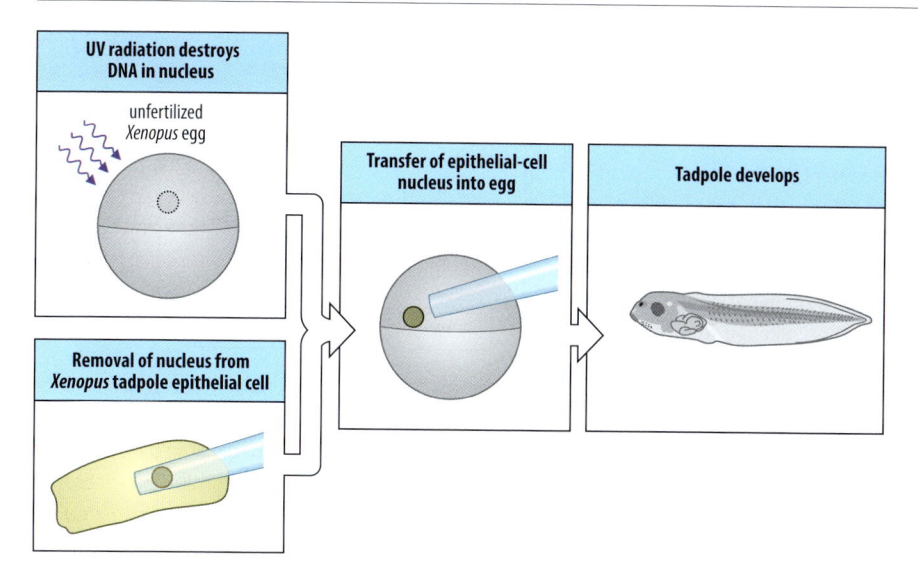

Fig. 1.20 Differentiated cells still contain all the genetic information required to make a complete animal. The haploid nucleus of an unfertilized *Xenopus* egg is rendered non-functional by treatment with ultraviolet radiation. A diploid nucleus taken from an adult skin cell is transferred into the enucleated egg and can support the development of an embryo until at least the tadpole stage.

DNA of the control regions (see Fig. 1.19), whereas others interact with transcription factors already bound to the DNA, and act as **co-activators** or **co-repressors**. Whether a gene is active or inactive therefore depends on the presence of transcription factors that can activate or repress its activity. There are also higher levels of gene control that depend on the state of the **chromatin**—the complex of DNA and protein of which the chromosomes are made. The physical state of the chromatin—whether it is tightly wound or unwound—and the particular chemical modifications to the DNA and the chromatin proteins determine the accessibility of the gene's DNA to transcription factors and other gene-regulatory proteins. This type of gene control is called **epigenetic** ('above the gene') because it can be propagated through cell division independently of the activities and the factors that initiated it, and can thus produce long-term changes in gene expression without a change in the DNA sequence of the gene. 'Epigenetic' gene control is quite different from the 'epigenesis' mentioned earlier, which refers to the progressive emergence of structures during development. Epigenetic control of gene expression is of great importance during development and is discussed in Chapter 6.

Developmental genes are highly regulated to ensure they are switched on only at the right time and place in development. This is a fundamental feature of development. To achieve this, developmental genes usually have extensive and complex *cis*-regulatory regions composed of one or more modules. Each module contains multiple binding sites for different transcription factors, and it is the precise combination of factors that binds that determines whether the gene is on or off. On average, a module will have binding sites for 4–8 different transcription factors, and these factors will, in turn, often bind other co-activator or co-repressor proteins.

The modular nature of the regulatory region means that each module can function somewhat independently. This means that a gene with more than one regulatory module is usually able to respond quite differently to different combinations of inputs, and so can be expressed at different times and in different places within the embryo in response to developmental signals. Different genes can have the same regulatory module, which usually means that they will be expressed together, or different genes may have modules that contain some, but not all, of the binding sites in common, introducing subtle differences in the timing or location of expression. Thus, an organism's genes are linked in complex interdependent networks of expression through the modules of their *cis*-regulatory regions and the proteins that bind to them. Common examples of such regulation

CELL BIOLOGY BOX 1F Gene regulatory networks and cell-fate decisions

Gene regulatory networks (GRNs) underlie many developmental processes and are crucial features of development. They are ensembles of genes, extracellular signals, and transcription factors that interact with each other in an orderly manner to control the expression of particular genes in space and time. A powerful approach to investigating development relies on a combination of bioinformatics and experimental molecular and cell biology to identify interactions between transcription factors and their target genes, and so describe these networks.

Transcription factors act in combinations to define cell types, but they also mediate the way cells make choices about their fates over time and how these states are maintained. They do that by organizing themselves into GRNs. In this age of genomics, some GRNs simply represent collections of transcription-factor genes involved in a certain developmental process, such as mesoderm specification, whose expression is coordinated by the same extracellular signals, and then the genes that these transcription factors regulate, and so on. However, it is sometimes possible to identify small networks that exhibit recognizable behaviors associated with particular cell-fate decisions. For example, Figure 1 shows two simple networks that create temporal patterns of gene expression that influence cell fate. Both are the same type of feed-forward loop (a variant of the one shown in Fig 1.21), but they operate in different developmental contexts and reflect different types of players. The first (Fig. 1 (left)) is from *Drosophila* and shows the sequence of gene/protein expression leading to the initial specification of the second stripe of *even-skipped* expression (eve stripe2) in the embryo (see Fig. 2.35). Together

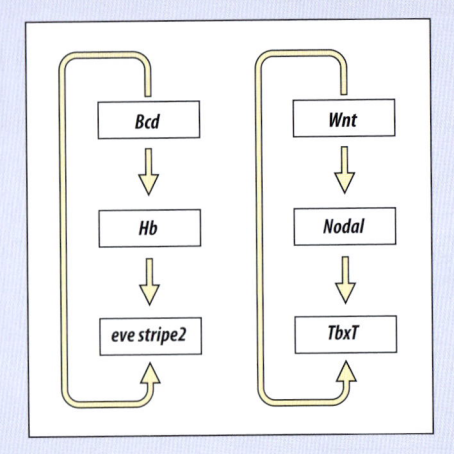

Fig. 1

with other genes, *eve* specifies a particular segment in the insect. Bicoid (Bcd) and Hunchback (Hb) are transcription factors and the expression of *even-skipped* requires the direct inputs of both of them. In addition, the expression of the *hunchback* gene requires Bcd. The second feed-forward loop (Fig. 1 (right)) operates early in mammalian development, during gastrulation. In this case Wnt is an extracellular signal protein, whose activity is required to turn on the *Nodal* gene in the responding cells (note that most signaling pathways end in specific gene activation, see Box 1G). Nodal is also a signal protein and both Nodal and Wnt signals are required to turn on the gene *TbxT*, which is a very early and enduring marker of mesoderm.

are **positive-feedback** and **negative-feedback** loops, in which a transcription factor promotes or represses, respectively, the expression of a gene whose product maintains that gene's expression (Fig. 1.21). Indeed, extensive networks of gene interactions, known as **gene regulatory networks** (Box 1F), have been worked out for various stages of development for animals such as the zebrafish (illustrated in Chapter 4), *Xenopus*, and the sea urchin and now in mice.

One might be tempted to think of development simply in terms of mechanisms for controlling gene expression. But that would be a mistake. Genes direct the synthesis of proteins, and gene expression is only the first step in a cascade of cellular processes that lead, via protein activity, to changes in cell behavior—cell proliferation, changes in cell shape, cell signaling, cell–cell contacts, cell migration, and cell death—and so direct the course of embryonic development. Proteins are the machines that drive development. Crucial aspects of cell biology, such as changes in cell shape, are often initiated, via proteins, at several steps removed from gene activity. In fact, there are very few cases where the complete sequence of events from gene expression to altered cell behavior has been worked out. The route leading from gene activity to a structure such as the five-fingered hand is tortuous, but progress in following it has been made, as discussed in Chapter 9.

1.12 Development is progressive and the fates of cells become determined at different times

As embryonic development proceeds, the organizational complexity of the embryo becomes vastly increased over that of the fertilized egg, as first recognized by Aristotle's 'epigenesis' principle outlined earlier in the chapter. Many different cell types are formed, spatial patterns emerge, and there are major changes in shape. All this occurs gradually, depending on the particular organism. But, in general, the embryo is first divided up into a few broad regions, such as the future germ layers (mesoderm, ectoderm, and endoderm). Subsequently, the cells within these regions have their fates more and more finely determined. **Determination** implies a stable change in the internal state of a cell, and an alteration in the pattern of gene activity is assumed to be the initial step, leading to a change in the proteins produced in the cell. Different mesoderm cells, for example, eventually become **determined** as muscle, cartilage, bone, the fibroblasts of connective tissue, and the cells of the dermis of the skin.

An important concept in the process of cell determination is that of cell **lineage**, the direct line of descent of a particular cell. Once a cell is determined, this stable change is inherited by its descendants. Thus, the lineage of a cell has important consequences in terms of what type of cell it can eventually become. For example, once determined as ectoderm, cells will not subsequently give rise to endodermal tissues, or vice versa. In some organisms, such as *Caenorhabditis*, the entire body is derived from the zygote according to very strict lineages, but in most organisms, the fate of cells is not determined according to a fixed, invariant lineage.

It is important to understand clearly the distinction between the normal fate of a cell at any particular stage, and its state of determination. The **fate** of a group of cells merely describes what they will normally develop into. By marking cells of the early embryo one can find out, for example, which ectodermal cells will normally give rise to the nervous system, and of those, which, in particular, will give rise to the neural retina of the eye. In these types of experiments one obtains a **fate map** of the embryo (see Box 3A). However, the map in no way implies that those cells in the early embryo can only develop into a retina or are already determined, or committed, to do so.

A group of cells is called **specified** if, when isolated and cultured in the neutral environment of a simple culture medium away from the embryo, they develop more or less according to their normal fate (Fig. 1.22). For example, cells at the animal pole of the amphibian blastula (see Box 1A) are specified to form ectoderm, and will form epidermis when isolated. Cells that are 'specified' in this technical sense need not yet be 'determined', for influences from other cells can change their normal fate. If tissue from the animal pole is put in contact with cells from the vegetal pole, some animal pole tissue will form mesoderm instead of epidermis. At a later stage of development, however, the cells in the animal region have become determined as ectoderm and their fate is unlikely to be altered. Tests for specification rely on culturing the tissue in a neutral environment lacking any inducing signals, and this is often difficult to achieve, whereas it is much easier to test for determination.

The state of determination of cells at any developmental stage can be demonstrated by transplantation experiments. At the gastrula stage of the amphibian embryo, one can graft the ectodermal cells that give rise to the eye into the side of the body and show that the cells develop according to their new position; in this case into mesodermal cells. At this early stage, their potential for development is much greater than their normal fate. However, if the same operation is done at a later stage, then the cells

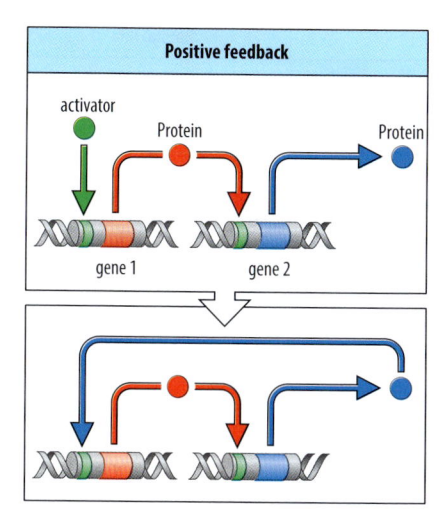

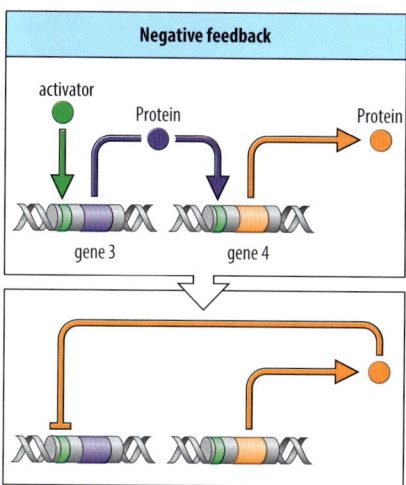

Fig. 1.21 Simple genetic feedback loops. Top: gene 1 is turned on by an activating transcription factor (green); the protein it produces (red) activates gene 2. The protein product of gene 2 (blue) not only acts on further targets, but also activates gene 1, forming a positive-feedback loop that will keep genes 1 and 2 switched on even in the absence of the original activator. Bottom: a negative-feedback loop is formed when the product of the gene at the end of a pathway (gene 4) can inhibit the first gene (gene 3), thus shutting down transcription of both genes. Arrows indicate activation; a barred line indicates inhibition.

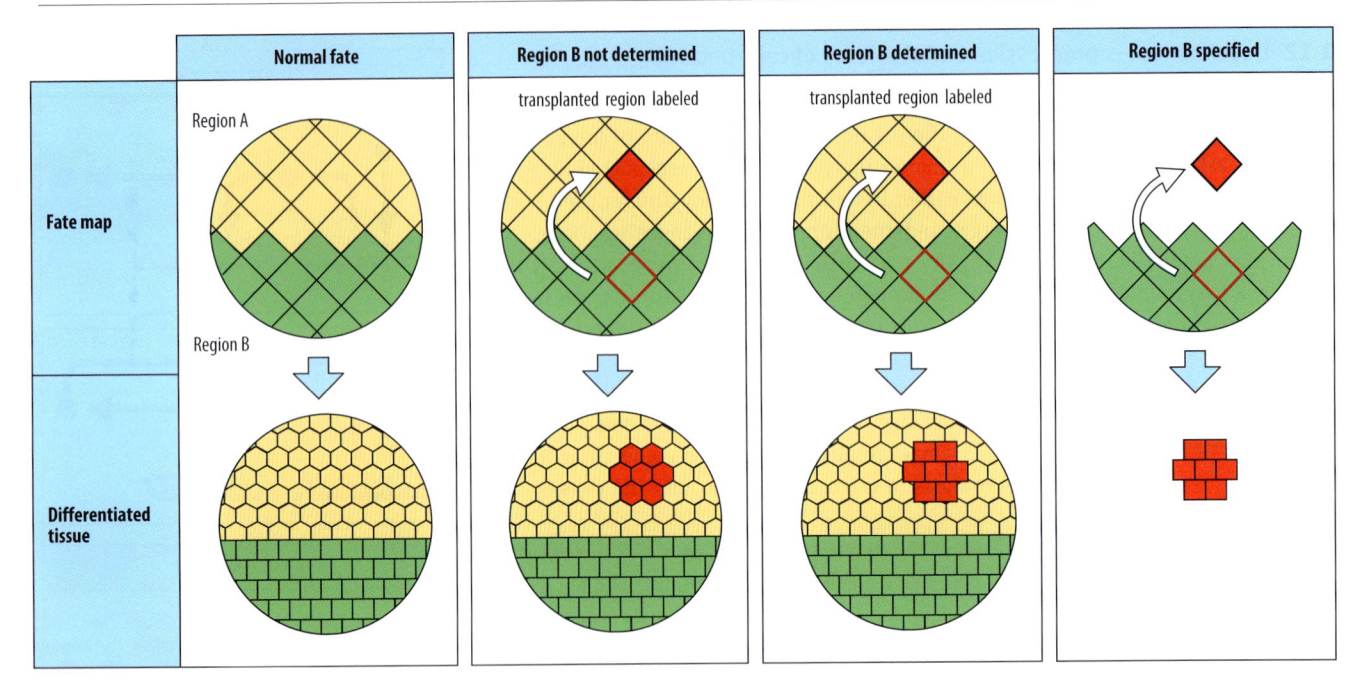

Fig. 1.22 The distinction between cell fate, determination, and specification. In this idealized system, regions A and B differentiate into two different sorts of cell, depicted as hexagons and squares. The fate map (first panel) shows how they would normally develop. If cells from region B are grafted into region A and now develop as A-type cells, the fate of region B has not yet been determined (second panel). By contrast, if region B cells are already determined when they are grafted to region A, they will develop as B cells (third panel). Even if B cells are not determined, they may be specified, in that they will form B cells when cultured in isolation from the rest of the embryo (fourth panel).

of the future eye region will form structures typical of an eye (Fig. 1.23). At the earlier stage the cells were not yet determined as prospective eye cells, whereas later they had become so, and so the graft development was **autonomous**—that is, it developed as eye irrespective of its new location.

It is a general feature of development that cells in the early embryo are less narrowly determined than those at later stages; with time, cells become more and more restricted in their developmental potential. We assume that determination involves a change in the genes that are expressed by the cell, and that this change fixes or restricts the cell's fate, thus reducing its developmental options.

We have already seen how, even at the two-cell stage, the cells of the sea urchin embryo do not seem to be determined. Each has the potential to generate a whole new larva (see Section 1.3). Embryos like these, where the developmental potential of cells is much greater than that indicated by their normal fate, are said to be capable of **regulation** and are described as **regulative**. Vertebrate embryos are among those capable of considerable regulation. By contrast, those embryos where, from a very early stage, the cells can develop only according to their early fate are termed **mosaic**. The term 'mosaic' has a long history (see Section 1.3). It is now used to describe eggs and embryos that develop as if their pattern of future development is laid down very early, even in the egg, as a spatial mosaic of different molecules in the egg cytoplasm. These cytoplasmic determinants get distributed in a regular fashion into different cells during cleavage, thus determining the fate of that cell lineage at an early stage. The different parts of the embryo then develop quite independently of each other. *Caenorhabditis elegans* is an example of an embryo with significant amounts of mosaic development. In such embryos, cell–cell interactions may be quite limited. But the demarcation between regulative and mosaic strategies of development is not always sharp, and in part reflects the time at which determination occurs; it occurs much earlier in mosaic systems.

The difference between regulative and mosaic embryos also reflects the relative importance of cell–cell interactions in each system. The occurrence of regulation absolutely requires interactions between cells, for how else could normal development take place and how could deficiencies be recognized and restored? A truly mosaic embryo, however, would, in principle, not require such interactions. No purely mosaic embryos are known to exist.

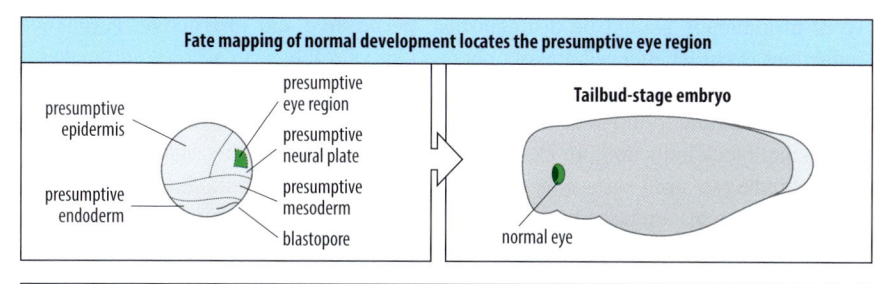

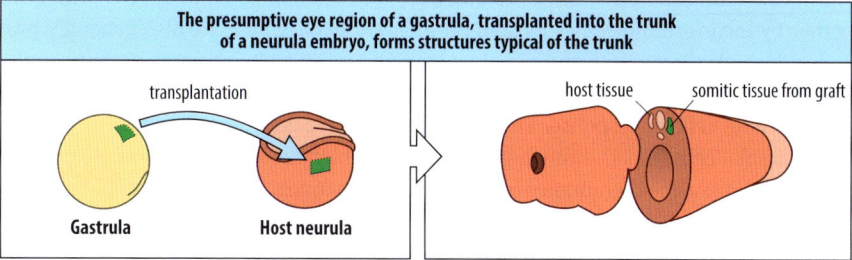

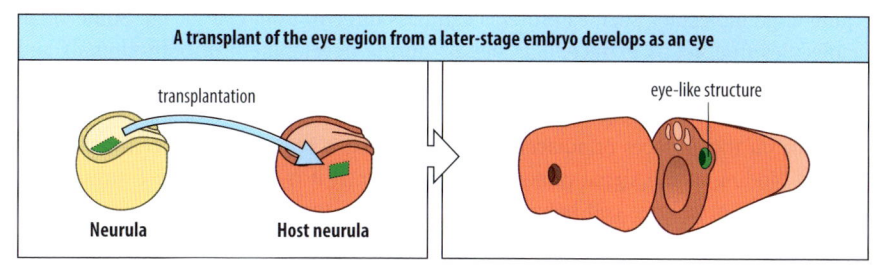

Fig. 1.23 Determination of the eye region with time in amphibian development. If the region of the gastrula that will normally give rise to an eye is grafted into the trunk region of a neurula (middle panel), the graft forms structures typical of its new location, such as mesodermal notochord and somites. If, however, the eye region from a neurula is grafted into the same site (bottom panel), it develops as an eye-like structure because at this later stage it has become determined.

Underlying the phenomena of regulative and mosaic-type development is the question of whether the effects of a given gene being expressed are restricted purely to the expressing cell or whether they are influencing and are influenced by other cells not expressing that gene. Effects restricted to the gene-expressing cell are called **cell-autonomous**, whereas gene expression in one cell that has an effect on other cells is called **non-cell-autonomous**. **Non-autonomous** action is typically due to a gene product such as an intercellular signal protein that is secreted by one cell and acts on others, whereas cell-autonomous effects are due to proteins such as transcription factors that act only within the cell that makes them.

Cell-autonomous versus non-autonomous genetic effects can be tested in mice produced by fusing together two very early embryos with different genetic constitutions (Fig. 1.24). The resulting mouse will be a **chimera**, a mosaic of cells with different genetic constitutions. If early embryos from a brown-haired and a white-haired mouse are fused, the resulting mouse has distinct patches of brown and white hair, reflecting patches of skin cells derived from the brown and white embryos, respectively. The brown cells still produce brown pigment when put into a white mouse, and they do not have any effect on the white cells; pigmentation is therefore a cell-autonomous character.

1.13 Inductive interactions make cells different from each other

Making cells different from one another is central to development. There are many times in development when a signal from one group of cells influences the development of an adjacent group of cells. This is known as induction, and the classic example is the action of the Spemann organizer in amphibians (see Section 1.4). Inducing signals can be propagated over a few, or even many, cells, or be highly localized. The inducing signals from the amphibian organizer affect many cells, whereas other inducing signals might be transmitted from a single cell only to its immediate neighbor. Two different

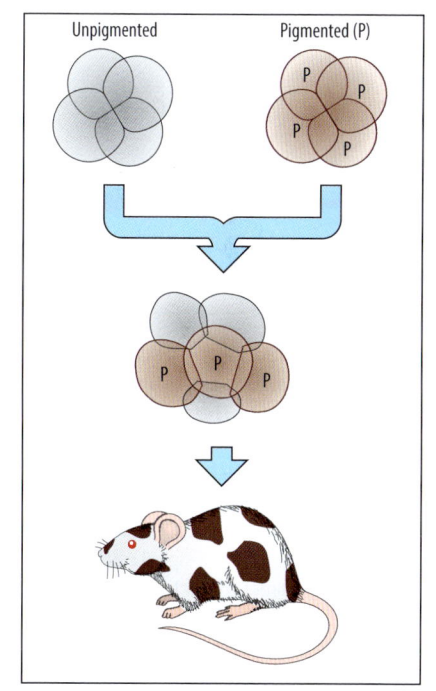

Fig. 1.24 Fusion of mouse embryos gives rise to a chimera. If a four-cell stage embryo of an unpigmented strain of mouse is fused with a similar embryo of a pigmented strain, the resulting embryo will give rise to a chimeric animal, with a mixture of pigmented and unpigmented cells. The distribution of the different cells in the skin gives this chimera a patchy coat.

types of induction should be distinguished: **permissive** and **instructive**. Permissive inductions occur when a cell makes only one kind of response to a signal and makes it when a given level of signal is reached. By contrast, in an instructive induction, the cells respond differently to different concentrations of the signal. 'Antagonistic' signal molecules that block induction, for example by preventing the inducing signal reaching the cell or binding to a cell-surface receptor, are also important in controlling development.

Inducing signals are transmitted between cells in three main ways (Fig. 1.25). First, the signal can be transmitted through the extracellular space in the form of a secreted diffusible molecule, generally a protein. Second, cells may interact directly with each other by means of molecules located on their surfaces, which are also generally proteins. In both these cases, the signal is generally received by receptor proteins in the cell membrane and is subsequently relayed through intracellular signaling systems to produce the cellular response. Third, the signal may pass from cell to cell directly. In most animal cells, this is through gap junctions, which are specialized protein pores in the apposed plasma membranes providing direct channels of communication between the cytoplasms of adjacent cells through which small molecules can pass. Plant cells are connected by strands of cytoplasm called plasmodesmata, through which even quite large molecules such as proteins can pass directly from cell to cell. Some small molecules that act as extracellular signals during development of both animals and plants can pass through the plasma membrane.

In the case of signaling by a diffusible protein, or by direct contact between two proteins on the surfaces of adjacent cells, the signal is received at the cell membrane by the binding of the signal protein to its receptor. If it is to alter gene expression in the nucleus, or change cell behavior in other ways, that information must be transmitted from the membrane to the cell's interior. This process is known generally as **signal transduction** or **intracellular signaling** and is carried out by relays of intracellular signaling molecules that are activated when the extracellular signal protein binds to its receptor. Highly simplified outlines of different types of intracellular signaling pathways, representing those initiated by commonly used intercellular signal proteins in development, are given in Box 1G. More detailed diagrams of individual pathways will be found throughout the book. We shall encounter these signals and intracellular signaling pathways again and again in many different animals and developmental situations.

Within the cell, signaling proteins interact with one another to transmit the signal onward in the cell. Activation and deactivation of pathway proteins by phosphorylation is an important feature of most signaling pathways. In development, most signaling leads to the activation or repression of genes by switching transcription on or off. Another important target of extracellular signals in development is the **cytoskeleton**, the internal network of protein fibers whose rearrangement enables cells to change shape, to move, and to divide. Signaling pathways are also used to alter enzyme activity and metabolic activity within cells temporarily and to initiate nerve impulses in neurons. The various signals a cell may be receiving at any one time are integrated by cross-talk between the different intracellular signaling pathways to produce an appropriate response. An example of what can happen when an important developmental signaling pathway is defective is illustrated in Box 1H. In general, organizing regions function by producing extracellular signal molecules or factors that influence intracellular signaling pathways in neighboring cells to determine their fate. These distinct signal-producing regions are also known as **signaling centers**.

An important feature of induction is whether or not a cell is able to respond to the inducing signal. The ability to respond, or **competence**, may depend, for example, on the presence of the appropriate receptor and transducing mechanism, or on the presence of particular transcription factors needed for gene activation. A cell's competence for a particular response can change with time; for example, the Spemann organizer can induce changes in the cells it affects only during a restricted window of time.

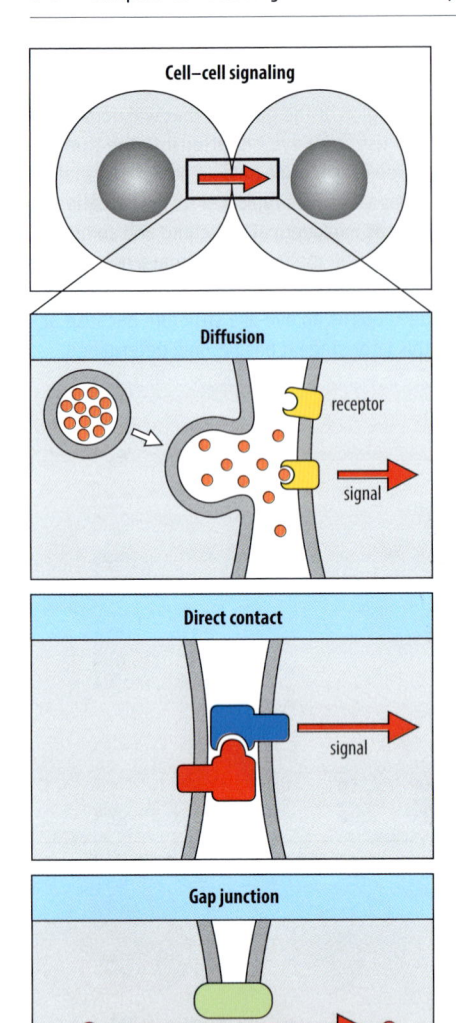

Fig. 1.25 An inducing signal can be transmitted from one cell to another in three main ways. The signal can be a secreted diffusible molecule, generally a protein, which interacts with a receptor on the target cell surface, activating the signaling pathway in the responding cell (second panel), or the signal can be transduced by direct contact between two complementary proteins at the cell surfaces (third panel). If the signal involves a small molecule it may pass directly from cell to cell through gap junctions in the plasma membrane (fourth panel).

CELL BIOLOGY BOX 1G Signal transduction and intracellular signaling pathways

Extracellular signal proteins act on a target cell via receptors located in the cell's plasma membrane. These receptors are transmembrane proteins and are specific for different signal proteins. This means that the receptors a cell expresses determine which signals it can respond to. The signal protein itself does not enter the cell. Instead, when it binds to the extracellular face of its cell-surface receptor this activates a relay of intracellular signaling molecules that results in a cellular response.

For many developmental signals, the end of the pathway is a change in the pattern of gene expression in the responding cell. The information that a signal molecule has bound to a receptor on the cell surface has to be transmitted from the receptor through the cytoplasm to the nucleus. Receptors that do this are generally known as signaling receptors, in contrast to receptors that, for example, bind to a molecule and internalize it. Intracellular signaling from the receptor is achieved by a pathway composed of a series of interacting proteins; different signaling pathways are employed by different receptors. Just as the same families of signal molecules are used by different animal species, the general features of the corresponding signaling pathways in different species are also similar.

There are various general types of intracellular signaling pathways (Fig. 1). In the absence of a signal, the receptor is not occupied and many of the intracellular proteins involved in the signaling event are in an inactive state (Fig. 1, first panel). When a signal binds its receptor, it can trigger a cascade of intracellular phosphorylation events (Fig. 1, second panel). The information that the signal protein has bound to its receptor is transmitted through the cytoplasm by a protein relay that involves a series of protein phosphorylations (compare the first and second panels of Fig. 1). Protein phosphorylation, which changes the activity of proteins and their interactions with other proteins, is ubiquitous in intracellular signaling pathways as one means of transmitting the signal onwards and is carried out by enzymes called protein kinases. A large group of receptors signal in this way. The first step in the pathway involves phosphorylation of the intracellular domain of the receptor by a protein kinase, and the pathway culminates in phosphorylation and activation of a protein kinase that enters the nucleus to phosphorylate and activate a specific transcription factor, thus selectively controlling gene expression. Fibroblast growth factor (see Box 4A)—a secreted signal protein used in many different contexts throughout development—epidermal growth factor, and many other extracellular signals use pathways of this type.

For some other extracellular signals, activation of the intracellular pathway leads to the movement of a transcription factor or other gene-regulatory protein (such as a co-activator or co-repressor) from the cytoplasm into the nucleus (Fig. 1, third panel). Extracellular signal proteins that use this type of pathway are the Wnt

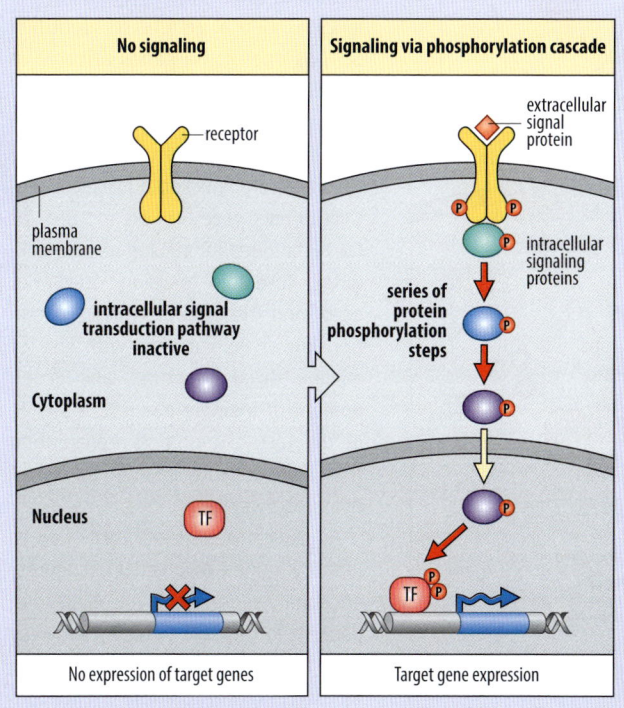

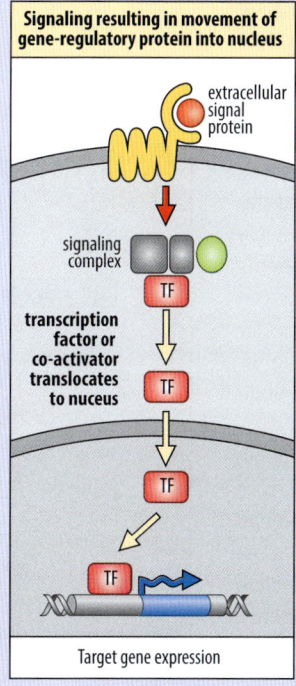

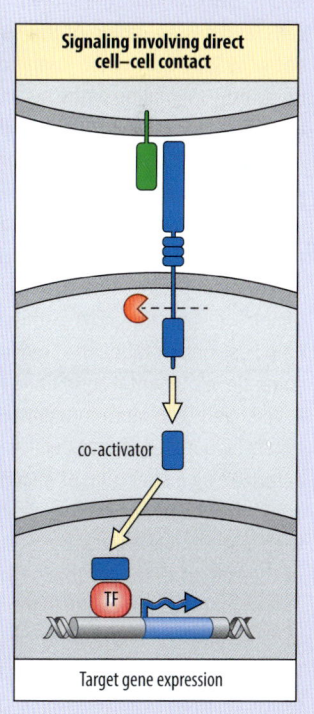

Fig. 1

proteins (see Box 3E), the Hedgehog protein family (see Box 2E and Box 9C), and proteins of the transforming growth factor-β family (see Box 3F), which we shall meet throughout the book.

The receptor Notch controls yet another type of signaling pathway, which involves direct cell–cell contact (Fig. 1, fourth panel). This transmembrane protein is activated by direct contact with its ligands (which are also transmembrane proteins) on another cell (**ligand** is the general term for a molecule that binds specifically to another biomolecule). In this case, binding of the transmembrane ligand to Notch results in cleavage of the intracellular portion of the receptor, which is then free to translocate to the nucleus, where it acts as a co-activator (see Fig. 1, fourth panel). The Notch pathway (see Box 4F) is commonly used to help determine cell fate.

It is important to ensure that intracellular signaling pathways remain inactive when the cell is not receiving a signal. For example, in the absence of a signal, transcription factors and other gene-regulatory proteins may be bound to other proteins in an inactive complex in the cytoplasm or be targeted for degradation. Signaling pathways can also contain feedback loops that modulate their own activity. In addition, cells can be receiving several signals at the same time and there can be cross-talk between the intracellular signaling pathways to ensure the appropriate cellular response.

Some small molecules that act as extracellular signals during development (e.g. retinoic acid (see Box 4E), steroid hormones, and some plant hormones, such as ethylene) can pass through the cell membrane unaided. They interact with specific receptors present in the cytoplasm or in the nucleus.

MEDICAL BOX 1H When development goes awry

The signaling pathway stimulated by the secreted signal protein Sonic hedgehog (SHH, named after the Hedgehog protein in *Drosophila*, which was the first member of this family to be discovered) regulates many developmental processes in humans and other vertebrates from very early in embryogenesis onwards. In adults, SHH controls the growth of stem cells, and misregulation of this signaling pathway leads to tumor formation. Like many other signaling pathways in development (discussed in Box 1G), the end results of SHH signaling are changes in gene expression.

SHH signaling in human brain and craniofacial development is a good example where knowledge of the components of intracellular signaling pathways can help us understand the reasons for some developmental conditions (Fig. 1). A relatively common congenital condition is holoprosencephaly, which is characterized by a spectrum of midline brain and facial differences. The severity of the condition varies greatly: at one end of the spectrum the forebrain (the prosencephalon) fails to divide in the midline to form right and left brain hemispheres, there is complete absence of the midline of the face, and the fetuses do not survive; at the other end, the division of the

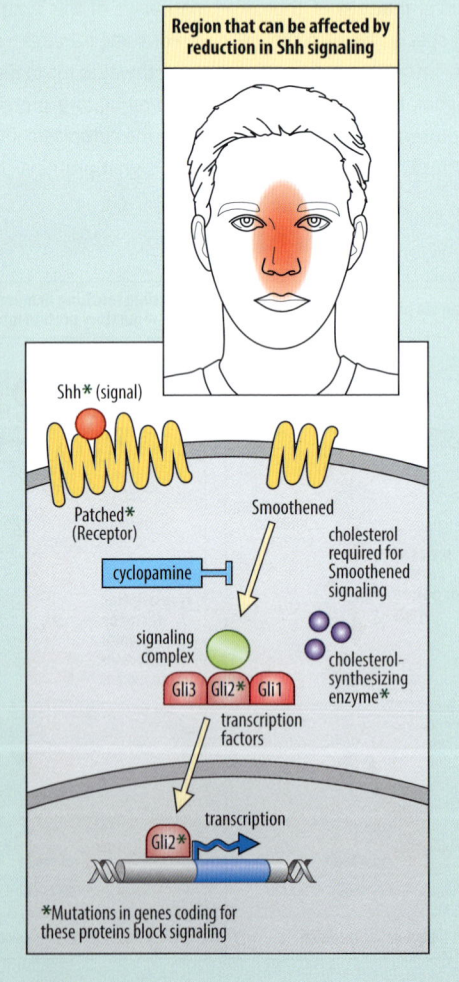

Fig. 1

forebrain is only mildly affected and the only observable difference might be a cleft lip or the presence of one incisor tooth rather than two.

The first indications that Shh was involved in brain development came in the early 1990s, when *Shh* was discovered to be expressed in the ventral midline of the nervous system in mice. In 1996, clinical geneticists identified *SHH* as one of the genes affected in people with inherited holoprosencephaly. At around the same time, it was found that when the *Shh* gene is completely knocked out experimentally in mice (see Box 1C and Box 3B for techniques for making gene knock-outs), forebrain division does not occur and the embryos have a condition known as cyclopia, with loss of midline facial structures and development of a single eye in the middle of the face—a condition representing the most severe manifestation of holoprosencephaly.

As clinical geneticists continued to identify genes involved in holoprosencephaly, more genes encoding components in the SHH signaling pathway were implicated. In all cases, the cause of the condition could be ascribed to a deficiency in SHH signaling, but different genes were affected in different cases. In some people, the condition was caused by mutations in the *SHH* gene itself, but in others the affected genes encoded either the receptor for SHH (Patched) or one of the transcription factors (GLI2) that moves into the nucleus to control gene expression. Yet other people had mutations in the gene for an enzyme required for cholesterol synthesis. Cholesterol promotes the activity of a transmembrane protein called Smoothened, which acts together with Patched to control the SHH signaling pathway. Figure 1 shows a simplified version of the SHH signaling pathway and how mutations in various components of the pathway (marked with asterisks) will block the pathway, resulting in a reduction or loss of Shh signaling (the full SHH signaling pathway is illustrated in Box 9C). The region of the face that is affected by reduction in SHH signaling is indicated. SHH signaling in vertebrates takes place on cellular structures called primary cilia, which are immotile cilia present on most cell types, and some people with genetic conditions in which formation of these structures is defective have distinctive facial features. Thus, different mutations that produce similar conditions in humans can highlight genes that encode proteins in a common developmental pathway and also genes that encode proteins that impinge on the function of pathway components but are not directly involved.

Facial differences like those produced by mutations in genes for SHH signaling pathway proteins can be caused by environmental factors. Pregnant ewes grazed in pastures in which the California False Hellebore (*Veratrum californicum*) is growing may give birth to lambs with severe facial abnormalities, such as cyclopia. The chemical in the plant that causes these defects is the alkaloid cyclopamine, and it too disrupts SHH signaling. Cyclopamine interferes with the activity of Smoothened and reduces Shh signaling (see Fig. 1). This property of cyclopamine is now being explored in a therapeutic context to develop drugs to treat cancers that display abnormal SHH signaling.

In embryos, it seems that small is generally beautiful where inductive signals direct signaling and pattern formation. Most patterns are specified on a scale involving just tens of cells and distances of only 100–500 μm. The final organism may be very big, but this is almost entirely due to growth after the basic pattern has been formed.

1.14 The response to inductive signals depends on the state of the cell

Although an inductive signal can be viewed as an instruction to cells about how to behave, it is important to realize that responses to such a signal are entirely dependent on the current state of the cells, which depends on their developmental history. Not only does the individual cell have to be competent to respond, but the number of possible responses is usually very limited. An inductive signal can only select one response from a small number of possible cellular responses. Thus, instructive inducing signals would be more accurately called selective signals. A truly instructive signal would be one that provided the cell with entirely new information and capabilities, by providing it with, for example, new genes, which is not thought to occur during development.

The fact that, at any given time, an inductive signal selects one out of several possible responses has several important implications for biological economy. On the one hand, it means that different signals can activate a particular gene at different stages of development: the same gene is often turned on and off repeatedly during development. On the other hand, the same signal can be used to elicit different responses in different cells. A particular signal molecule, for example, can act on several types of cells, evoking a characteristic and different response from each, depending on their current state. As we will see in future chapters, evolution has been lazy with respect to this aspect of development, and a small number of intercellular signal molecules are used over and over again for different purposes.

1.15 Patterning can involve the interpretation of positional information

One general mode of pattern formation can be illustrated by considering the patterning of a simple non-biological model—the French flag (Fig. 1.26). The French flag has a simple pattern: one-third blue, one-third white, and one-third red, along just one

Fig. 1.26 The French flag.

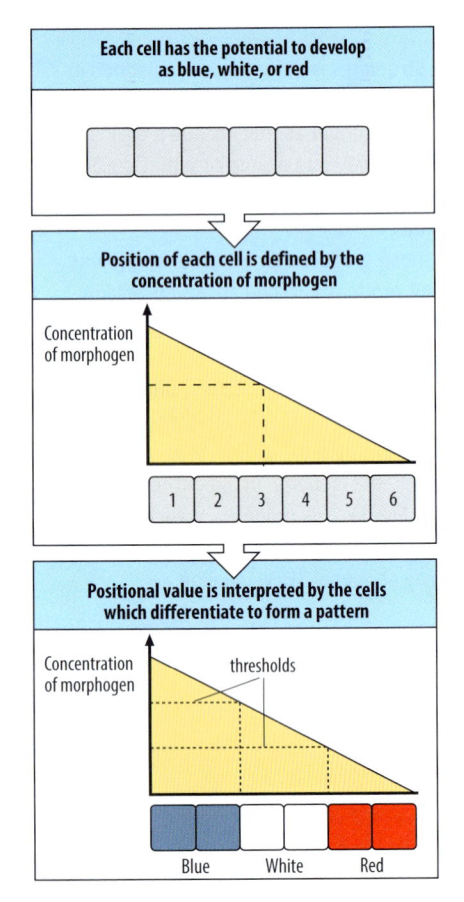

Fig. 1.27 The French flag model of pattern formation. Each cell in a line of cells has the potential to develop as blue, white, or red. The line of cells is exposed to a concentration gradient of some substance and each cell acquires a positional value defined by the concentration at that point. Each cell then interprets the positional value it has acquired and differentiates into blue, white, or red, according to a predetermined genetic program, thus forming the French flag pattern. Substances that can direct the development of cells in this way are known as morphogens. The basic requirements of such a system are that the concentration of substance at either end of the gradient must remain different from each other but constant, thus fixing boundaries to the system. Each cell must also contain the necessary information to interpret its positional value into a developmental outcome. Interpretation of the positional value is based on different threshold responses to different concentrations of morphogen.

axis. Moreover, the flag comes in many sizes but always the same pattern, and thus can be thought of as mimicking the capacity of an embryo to regulate. Given a line of cells, any one of which can be blue, white, or red, and given also that the line of cells can be of variable length, what sort of mechanism is required for the line to develop the pattern of a French flag?

One solution to the French flag problem, in which the pattern is invariant no matter what its size, is for the group of cells to use **positional information**—that is, information on their position along the line with respect to boundaries at either end. Each cell thus acquires its own **positional value**. After they have acquired their positional values, the cells interpret this information by differentiating according to their genetic program. Those in the left-hand third of the line will become blue, those in the middle-third white, and so on. The way in which regeneration occurs, for example in *Hydra* and in amphibian and insect limbs, in which the missing regions are precisely replaced, can be thought of in terms of the French flag model (discussed in Chapter 11).

Pattern formation using positional values implies at least two distinct stages: first the positional value has to be related to some boundary, and then it has to be interpreted by the cell. The separation of these two processes has an important implication: it means that there does not need to be a set relation between the positional values and how they are interpreted. In different circumstances, the same positional information and set of positional values could be used to generate the Italian flag (the same format as the French flag but different colors) or to generate another pattern. How positional values will be interpreted depends on the particular genetic instructions active in the group of cells at the time and thus the cells' developmental history.

Cells could have their positional values specified by various mechanisms. The simplest is based on a gradient of some substance. If the concentration of some chemical decreases from one end of a line of cells to the other, then the concentration in any cell along the line provides positional information that effectively specifies the position of the cell with respect to the boundary (Fig. 1.27). A chemical whose concentration varies and which is essential for pattern formation is called a **morphogen**. In the case of the French flag, we assume a source of morphogen at one end and a sink at the other, and that the concentrations of morphogen at both ends are kept constant but are different from each other. Then, as the morphogen diffuses down the line, its concentration at any point effectively provides positional information. If the cells can respond to **threshold concentrations** of the morphogen—for example, above a particular concentration the cells develop as blue, whereas below this concentration they become white, and at yet another, lower, concentration they become red—the line of cells will develop as a French flag (see Fig. 1.27). An exponential gradient will be produced if the morphogen is removed throughout the line of cells rather than by a localized sink, and such a gradient can similarly provide positional information with cells responding to threshold concentrations of morphogen. Thresholds can represent the amount of morphogen that must bind to receptors to activate an intracellular signaling system, or concentrations of transcription factors required to activate particular genes. The use of threshold concentrations of transcription factors to specify position is most beautifully illustrated in the early development of *Drosophila*, as we shall see in Chapter 2. Positional information provided by gradients of signal proteins is involved in patterning the limb (Chapter 9) and the developing neural tube (Chapter 10). Other ways of providing positional information and specifying positional values are by direct intercellular interactions and by timing mechanisms.

The French flag model illustrates two important features of development in the real world. The first is that, even if the length of the line varies, the system regulates and the pattern will still form correctly, given that the boundaries of the system are properly defined by keeping the different concentrations of morphogen constant at either end. The second is that the system could also regenerate the complete original pattern if it were cut in half, provided that the boundary concentrations were re-established.

It is therefore truly regulative. Here we have discussed making the pattern of the French flag as a one-dimensional patterning problem, but the model can easily be extended to provide patterning in two dimensions (Fig. 1.28).

There has been much debate about whether morphogen diffusion would be reliable enough to provide positional information. However, in a very well-studied example in *Drosophila*, it has been shown that diffusion aided by particular proteins is sufficient to produce normal development of the wing (see Box 9A). But in a more general view, there are many things we still do not know about positional information. We do not know how fine-grained the differences in position are. For example, does every cell in the early embryo have a distinct positional value of its own, and is it able to make use of this difference?

Another way in which a pattern can be generated in a group of cells is when cells that have differentiated in a salt-and-pepper fashion then 'sort out' so that they occupy a particular position (Fig. 1.29). Cells in the group might initially differentiate in a random spatial arrangement because of differences in sensitivity to a global signal or in response to differences in timing. Sorting out has been demonstrated in various experimental systems in which two or more different tissues are disaggregated into single cells and the cells are then mixed together so that they reaggregate. In many of these mixed reaggregates, the same type of cells tend to associate preferentially. This can result in the formation of separate aggregates, or in cells of one type adopting a particular position within a mixed aggregate. The basis for sorting out is cells having different cell-adhesion molecules on their surfaces or differing amounts of the same cell-adhesion molecules. As we will see, although positional information is the general mode of pattern formation in embryos, there are some instances where this alternative strategy is used—for example, in the generation of mesoderm and endoderm in zebrafish embryos and extra-embryonic and embryonic tissue in early mouse embryos. Sorting out has also been invoked as a mechanism for creating sharp boundaries between different cell types (e.g. see Section 5.1 and Chapter 10).

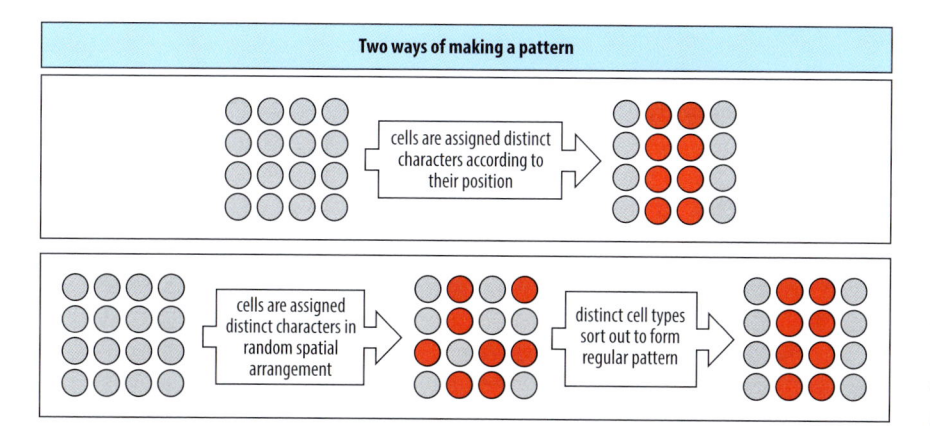

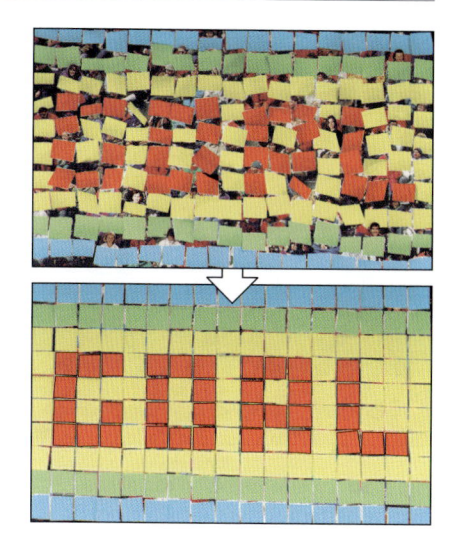

Fig. 1.28 Positional information can be used to generate patterns. A good example, as shown here, is where the people seated in a stadium each have a positional value defined by their row and seat numbers. Each position has an instruction about which colored card to hold up, and this makes the pattern. If the instructions were changed, a different pattern would be formed, and so positional information can be used to produce an enormous variety of patterns.

1.16 Lateral inhibition can generate spacing patterns

Many structures, such as the feathers on a bird's skin, are found to be more or less regularly spaced with respect to one another. One mechanism that gives rise to such spacing is called **lateral inhibition** (Fig. 1.30). Given a group of cells that all have the potential to differentiate in a particular way, for example as feathers, it is possible to regularly space the cells that will form feathers by a mechanism in which the first cells to begin to form feathers (which occurs essentially at random) inhibit adjacent cells from doing so. This is reminiscent of the regular spacing of trees in a forest by

Fig. 1.29 Two ways in which a pattern can be produced from a group of cells. The upper panel illustrates how a red stripe can be generated by positional information. The lower panel shows how the same red stripe can be generated by cells in the group differentiating randomly and then associating preferentially with cells of the same type.

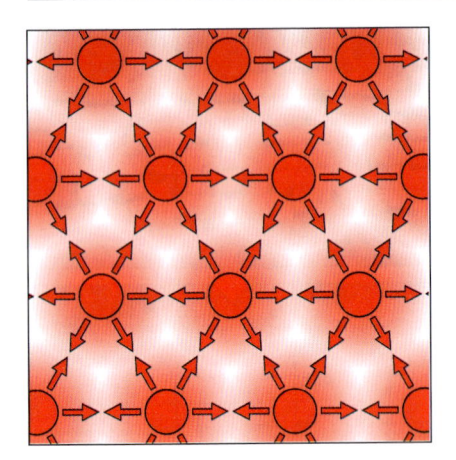

Fig. 1.30 Lateral inhibition can give a spacing pattern. Lateral inhibition occurs when developing structures produce an inhibitor that prevents the formation of any similar structures in the area adjacent to them, and the structures become evenly spaced as a result.

competition for sunlight and nutrients. In embryos, lateral inhibition is often the result of the differentiating cell producing an inhibitory molecule that acts locally on the nearest neighbors to prevent them developing in a similar way.

1.17 Localization of cytoplasmic determinants and asymmetric cell division can make daughter cells different from each other

Positional information is just one way in which cells can be given a particular identity. A separate mechanism is based on **cytoplasmic localization** of factors and **asymmetric cell division** (Fig. 1.31). Asymmetric divisions are so called because they result in daughter cells having properties different from each other, independently of any environmental influence. The properties of such cells therefore depend on their lineage, and not on environmental cues. Although some asymmetric cell divisions are also unequal divisions, in that they produce cells of different sizes, this is not usually the most important feature in animals; it is the unequal distribution of cytoplasmic factors that makes the division asymmetric. An alternative way of making the French flag pattern from the egg would be to have chemical differences (representing blue, white, and red) distributed in the egg in the form of determinants that foreshadowed the French flag. When the egg underwent cleavage, these cytoplasmic determinants would become distributed among the cells in a particular way and a French flag would develop. This would require no interactions between the cells, which would have their fates determined from the beginning.

Although such extreme examples of mosaic development (see Section 1.12) are not known in nature, there are well-known cases where eggs or cells divide so that some cytoplasmic determinant becomes unequally distributed between the two daughter cells and they develop differently. This happens at the first cleavage of the nematode egg, for example, and defines the antero-posterior axis of the embryo. The germ cells of *Drosophila* are also specified by cytoplasmic determinants, in this case contained in the cytoplasm located at the posterior end of the egg. And in *Xenopus*, a key cytoplasmic determinant in the first stages of embryonic development is a protein known as VegT, which is localized in the vegetal region of the fertilized egg. In general, however, as development proceeds, daughter cells become different from each other because of signals from other cells or from their extracellular environment, rather than because of the unequal distribution of cytoplasmic determinants.

A very particular type of asymmetric division is seen in stem cells. In one mode of division, these self-renewing cells can produce one new stem cell and another daughter cell that will differentiate into one or more cell types (Fig. 1.32). Stem cells that are capable of giving rise to all the cell types in the body are present in very early embryos and are known as **embryonic stem cells (ES cells)**, whereas in adults, stem cells of more limited potential are responsible for the continual renewal of tissues such as

Fig. 1.31 Cell division with asymmetric distribution of cytoplasmic determinants. If a particular molecule is distributed unevenly in the parent cell, cell division will result in its being shared unequally between the cytoplasm of the two daughter cells. The more localized the cytoplasmic determinant is in the parental cell, the more likely it will be that one daughter cell will receive all of it and the other none, thus producing a distinct difference between them.

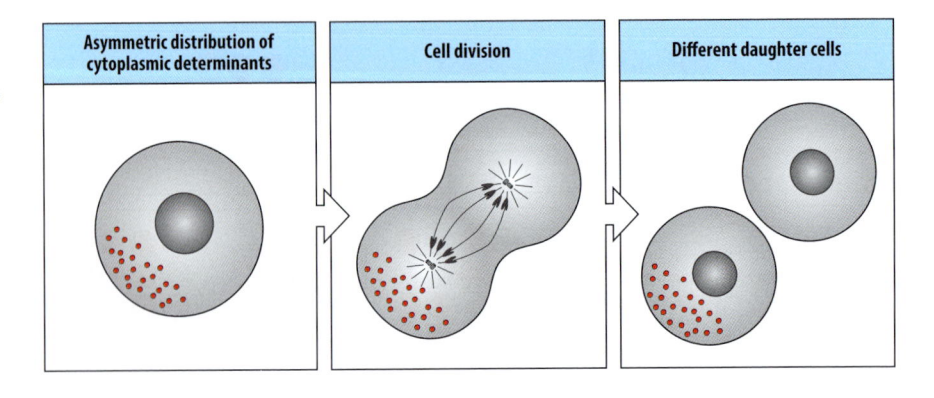

blood, skin epidermis, and gut epithelium, and for the replacement of tissues such as muscle when required. The difference in the behavior of the daughter cells produced by division of a stem cell can be due either to asymmetric distribution of cytoplasmic determinants or to the effects of extracellular signals. The generation of neurons in *Drosophila*, for example, depends on the asymmetric distribution of cytoplasmic determinants in the neuronal stem cells, whereas the production of the different types of blood cells seems to be regulated to a large extent by extracellular signals.

ES cells, which can give rise to all the types of cells in the body, are called **pluripotent**, whereas stem cells that give rise to a more limited number of cell types are known as **multipotent**. The hematopoietic stem cells in bone marrow, which give rise to all the different types of blood cells, are an example of multipotent stem cells. Both pluripotent and multipotent stem cells are of great interest in relation to regenerative medicine, as they could provide a means of repairing or replacing damaged organs (see Chapter 7). For most animals, the only cell that can give rise to a complete new organism is the fertilized egg, and this is described as **totipotent**.

Induced pluripotent stem cells (iPS cells) can be made by inducing adult somatic cells such as fibroblasts to revert to a pluripotent state by the introduction of a small number of transcription factors that induce pluripotency. iPS cells can be made from mouse or human terminally differentiated cells and can be derived from a child's or adult's own cells, removing any ethical concerns about using ES cells (see Chapter 7). This means they could be used as a personalized way of studying disease or testing treatments.

ES cells and iPS cells cultured *in vitro* can be directed to differentiate into all the different cell types in the body. ES cells, iPS cells, or stem cells from adult tissues can be grown in culture to form **organoids**, miniature three-dimensional structures with many of the features of a particular tissue or organ which can be used to study disease and test treatments. It is now even possible to coax mouse or human ES cells or iPS cells grown *in vitro* to form structures called **gastruloids** that have many features of post-gastrulation embryos, including the development of the positional coordinate system that governs the body plan. Gastruloids lack many features of a 'real' embryo—they do not develop an anterior 'brain' for example, or extra-embryonic tissues, and do not develop very far, but they are proving very useful models for studying aspects of early embryogenesis, especially of humans—for example, the influence of extracellular chemical and mechanical constraints on development. Organoids, gastruloids, and other embryo models are described in Chapter 7.

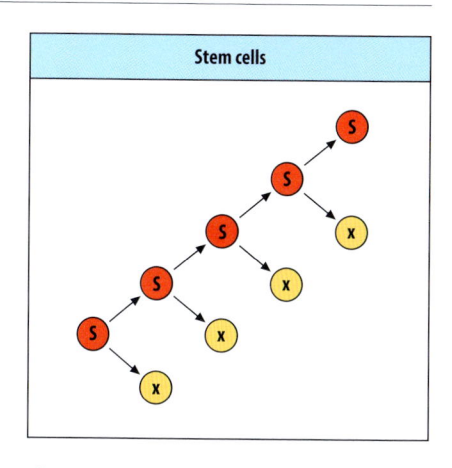

Fig. 1.32 Stem cells. Stem cells (S) are cells that both renew themselves and give rise to differentiated cell types. Thus, a daughter cell arising from stem-cell division can either develop into another stem cell or give rise to a different type of cell (X).

1.18 The embryo contains a generative rather than a descriptive program

All the information for embryonic development is contained within the fertilized egg. So how is this information interpreted to give rise to an embryo? One possibility is that the structure of the organism is somehow encoded as a descriptive program in the genome. Does the DNA contain a full description of the organism to which it will give rise: is it a blueprint for the organism? The answer is no. Instead, the genome contains a program of instructions for making the organism—a **generative program**—that determines where and when different proteins are synthesized and thus controls how cells behave.

A descriptive program, such as a blueprint or a plan, describes an object in some detail, whereas a generative program describes how to make an object. For the same object the programs are very different. Consider origami, the art of paper folding. By folding a piece of paper in various directions it is quite easy to make a paper hat or a bird from a single sheet. To describe in any detail the final form of the paper with the complex relationships between its parts is really very difficult, and not of much help in explaining how to achieve it. Much more useful and easier to formulate are instructions on how to fold the paper. The reason for this is that simple instructions about folding have complex spatial consequences. In development, gene action similarly sets in motion a sequence of events that can bring about profound changes in the embryo. One can thus think of the genetic information in the fertilized egg as equivalent to the folding instructions in origami; both contain a generative program for making a particular structure.

1.19 The reliability of development is achieved by various means

Development is remarkably consistent and reliable; you need only look at the similar lengths of your legs, each of which develops quite independently of the other for some 15 years (see Chapter 11). Embryonic development thus needs to be reliable so that the adult organism can function properly. For example, both wings of a bird must be very similar in size and shape if it is to be able to fly satisfactorily. How such reliability is achieved is an important problem in development.

Development needs to be reliable in the face of fluctuations that occur either within the embryo or in the external environment. Internal fluctuations include small changes in the concentrations of molecules, and also changes due, for example, to mutations in genes not directly linked to the development of the organ in question. External factors that could perturb development include temperature and environmental chemicals. However, the mechanisms underlying the developmental processes withstand large changes of these kinds and therefore are said to be **robust**.

One of the central problems of reliability relates to pattern formation. How are cells instructed to behave in a particular way in a particular position in a robust manner? Two ways in which reliability is assured during development are apparent redundancy of mechanism and negative feedback. **Redundancy** is when there are two or more ways of carrying out a particular process; if one fails for any reason another will still function. It is like having two batteries in your car instead of just one. True redundancy—such as having two identical genes in a haploid genome with identical functions—is rare except for cases such as rRNA, where there are often hundreds of copies of similar genes. Apparent redundancy, however, where a given process can be specified by several different mechanisms, is probably one of the ways that embryos can achieve such precise and robust results. It is like being able to draw a straight line with either a ruler or a piece of taut string. This type of situation is not true redundancy but may give the impression of being so if one mechanism is removed and the outcome is still apparently normal. Multiple ways of carrying out a process make it more robust and resistant to environmental or genetic perturbation.

Negative feedback also has a role in ensuring consistency; here, the end product of a process inhibits an earlier stage and thus keeps the level of product constant (see, e.g., Fig. 1.21). The classic example is in metabolism, where the end product of a biochemical pathway inhibits an enzyme that acts early in the pathway. Yet another reliability mechanism relates to the complexity of the networks of gene activity (GRNs) that operate in development. There is evidence that these networks, which involve multiple pathways, are robust and relatively insensitive to small changes in, for example, the rates of the individual processes involved.

1.20 The complexity of embryonic development is due to the complexity of cells themselves

Cells are, in a way, more complex than the embryo itself. In other words, the network of interactions between proteins and DNA within any individual cell contains many more components and is very much more complex than the interactions between the cells of the developing embryo. However clever you think cells are, they are almost always far cleverer. Each of the basic cell activities involved in development, such as cell division, response to cell–cell signals, and cell movements, are the result of interactions within a population of many different intracellular proteins whose composition varies over time and between different locations in the cell. Cell division, for example, is a complex cell-biological program that takes place over a period of time, in a set order of stages, and requires the construction and precise organization of specialized intracellular structures at mitosis.

Any given cell is expressing thousands of different genes at any given time. Much of this gene expression may reflect an intrinsic program of activity, independent of external signals. It is this complexity that determines how cells respond to the signals they receive; how a cell responds to a particular signal depends on its internal state. This state reflects the cell's developmental history—cells have good memories—and so different cells can respond to the same signal in very different ways. We shall see many examples of the same signals being used over and over again by different cells at different stages of embryonic development, with different biological outcomes.

We have at present only a fragmentary picture of how all the genes and proteins in a cell, let alone a developing embryo, interact with one another. But new technologies are now making it possible to detect the simultaneous activity of hundreds of genes in a given tissue. The discipline of systems biology is also beginning to develop techniques to reconstruct the highly complex signaling networks used by cells. How to interpret this information, and make biological sense of the patterns of gene activity revealed, is a massive task for the future.

1.21 Development is a central element in evolution

The similarity in developmental mechanisms and genes in such very different animals as flies and vertebrates is the result of the evolutionary process. All animals have evolved from a single multicellular ancestor, and it is therefore inevitable that some developmental mechanisms will be held in common by many different species, whereas new ones have arisen in different animal groups as evolution proceeded. The evolution of development is discussed in more detail in Chapter 13.

The basic Darwinian theory of evolution by natural selection states that changes in genes alter how an organism develops, and that such changes in development in turn determine how the adult will interact with their environment. If a developmental change gives rise to adults better adapted to survive and reproduce in the prevailing environment, the underlying genetic change will be selected to be retained in the population. Thus, changes in development due to changes in the genes are fundamental to evolution.

Very clear examples of the workings of evolution in development can be seen in the limbs of vertebrates. In Chapter 13 we will see how analysis of fossil evidence shows that the limbs of land vertebrates originally evolved from the fins of lobed-finned fish, and how the genetic and developmental basis of this evolution is being reconstructed. We will also see how the basic pattern of the five-digit limb then evolved to give limbs as diverse as those of bats, horses, and humans. In bats, the digits of the forelimb grow extremely long and support a leathery wing membrane, whereas in horses, one digit of the 'hand' in the forelimb and the 'foot' in the hindlimb has become a long bone that forms the lower part of the leg and the hoof, and some other digits have been lost. The fossil record gives many other examples of evolutionary changes.

While some major steps in evolution can be identified, the genes involved are not easy to identify. In the few examples so far in which the mutations underlying evolutionary changes have been identified, it has been found that the mutations are not in the protein-coding sequences themselves but in their control regions. Thus, there is increasing focus on the importance of regulatory sequences as the instruments of evolution. Recent advances in whole-genome sequencing have made it easier to look for the differences between members of different species that can give information about their evolutionary history. There are also questions as to how big the genetic steps to new forms were, and there are questions about how completely new structures evolve. But there is no doubt that it is all down to changes in embryonic development.

SUMMARY

Development results from the coordinated behavior of cells. The major processes involved in development are pattern formation, morphogenesis or change in form, cell differentiation, and growth. These involve cell activities such as intercellular signaling and turning genes on and off. Genes control cell behavior by controlling where and when proteins are synthesized, and thus cell biology provides the link between gene action and developmental processes. During development, cells undergo changes in the genes they express, in their shape, in the signals they produce and respond to, in their rate of proliferation, and in their migratory behavior. All these aspects of cell behavior are controlled largely by the presence of specific proteins; gene activity controls which proteins are made. As the somatic cells in the embryo all generally contain the same genetic information, the changes that occur in development are controlled by the differential activity of selected sets of genes in different groups of cells. The genes' control regions are fundamental to this process. Development is progressive and the fate of cells becomes determined at different times. The potential for development of cells in the early embryo is usually much greater than their normal fate, but this potential becomes more restricted as development proceeds. Inductive interactions, involving signals from one tissue or cell to another, are one of the main ways of changing cell fate and directing development. Asymmetric cell divisions, in which cytoplasmic components are unequally distributed to daughter cells, can also make cells different. One widespread means of pattern generation is through positional information; cells first acquire a positional value with respect to boundaries and then interpret their positional values by behaving in different ways. Another strategy for generating a pattern involves cells becoming different and then sorting out. Developmental signals are more selective than instructive, choosing one or other of the developmental pathways open to the cell at that time. The embryo contains a generative, not a descriptive, program—it is more like the instructions for making a structure by paper folding than a blueprint. Various mechanisms, including apparent redundancy and negative feedback, are involved in making development remarkably reliable and robust. The complexity of development lies within the cells. Evolution is intimately linked to development because heritable changes in the adult organism must inevitably be the result of changes in the genetic program of embryonic development.

Summary to Chapter 1

- All the information for embryonic development is contained within a single cell—the fertilized egg. Once the sperm and egg nuclei have fused, this cell is called a zygote and is diploid. The genome of the zygote contains a program of instructions for making the organism. The cytoplasmic constituents of the egg, and of the cells it gives rise to, are essential players in working out this developmental program along with the genes.
- Strictly regulated gene activity, by controlling which proteins are synthesized, and where and when, directs a sequence of cellular activities that brings about the profound changes that occur in the embryo during development.
- The major processes involved in development are pattern formation, morphogenesis, cell differentiation, and growth, which are controlled by communication between the cells of the embryo and changes in gene expression.
- Development is progressive, with the fate of cells becoming more precisely determined as it proceeds.
- Cells in the early embryo usually have a much greater potential for development than is evident from their normal fate, and this enables embryos, particularly vertebrate

embryos, to develop normally even if cells are removed, added, or transplanted to different positions.
- Cells' developmental potential becomes much more restricted as development proceeds.
- Many genes are involved in controlling the complex interactions that occur during development, and reliability is achieved in various ways.
- Changes in embryonic development are the basis for the evolution of multicellular organisms.
- A full understanding of development is far from being achieved. While basic principles and certain developmental systems are quite well understood, there are still many gaps.

■ End of chapter questions

Check your understanding of the content of this chapter by attempting the following long-answer concept questions.

1. The ultimate goal of the science of developmental biology is to understand human development. In your own words, discuss three areas of medicine where the study of developmental biology has an impact.

2. The Greek philosopher Aristotle considered two opposing theories for development: preformation and epigenesis. Describe what is meant by these two ideas; refer to the concept of the 'homunculus' in describing preformation, and to the art of origami in describing epigenesis. Which better describes our current view of development?

3. Weismann's concept of 'determinants' was supported by Roux's experiments demonstrating 'mosaic development' in the frog. How would determinants lead to mosaic development? Why does Driesch's concept of 'regulation' during development better explain human twinning?

4. Pattern formation is a central process in the study of development. Describe three examples of pattern formation and explain what 'patterns' are generated.

5. Briefly define the following cell behaviors: intercellular signaling, cell proliferation, cell differentiation, cell movement, changes in cell shape, gene expression, and cell death.

6. Contrast the role of housekeeping genes and proteins during development with the role of tissue-specific genes and proteins. Which class of genes and proteins is more important during development—housekeeping or tissue-specific? Give examples of the two types.

7. All cells in an organism contain the same genes; however, the organism contains different types of cells. How is this paradox of 'genetic equivalence' resolved by the concept of 'differential gene expression'?

8. How do transcription factors work? What is their relationship to control regions in DNA?

9. Suppose you wanted to inactivate a gene involved in gastrulation in a mouse and decided to use the CRISPR-Cas9 system. How would you do it? Describe the technique, as well as the test you would apply to know that the mutation has been created.

10. What is the difference between cell fate, cell specification, cell determination, and cell differentiation? What kinds of experiments allow the distinction between specification and determination to be made?

11. What are three ways in which inducing signals can be transmitted in developing embryos? In what circumstance would an inducing signal be a 'morphogen'?

12. What is the relationship between positional information and pattern formation? Give an example.

13. Why do you think feathers are spaced in the skin rather than every cell producing a feather?

■ Section further reading

The origins of developmental biology

De Robertis, E.M.: **Spemann's organizer and the self-regulation of embryonic fields**. *Mech. Dev.* 2009, **126**: 925–941.

Hamburger, V.: *The Heritage of Experimental Embryology: Hans Spemann and the Organizer*. New York: Oxford University Press, 1988.

Harris, H.: *The Birth of the Cell*. New Haven, CT: Yale University Press, 1999.

Harrison, M.M., Jenkins, B.V., O'Connor-Giles, K.M., Wildonger J.: **A CRISPR view of development**. *Genes Dev.* 2014, **28**: 1859–1872.

Lenhoff, H.M.: **Ethel Browne, Hans Spemann, and the discovery of the organizer phenomenon**. *Biol. Bull.* 1991, **181**: 72–80.

Milestones in Development [https://www.nature.com/collections/pmwvdkzjld] (date accessed 27 July 2024)

Needham, J.: *A History of Embryology*. Cambridge: Cambridge University Press, 1959.

Peng, Y., Clark, K.J., Campbell, J.M., Panetta, M.R., Guo, Y., Ekker, S.C.: **Making designer mutants in model organisms**. *Development* 2014, **141**: 4042–4054.

Sander, K.: **'Mosaic work' and 'assimilating effects' in embryogenesis: Wilhelm Roux's conclusions after disabling frog blastomeres**. *Roux's Arch. Dev. Biol.* 1991, **200**: 237–239.

Sander, K.: **Shaking a concept: Hans Driesch and the varied fates of sea urchin blastomeres**. *Roux's Arch. Dev. Biol.* 1992, **201**: 265–267.

Spemann, H., Mangold, H.: **Induction of embryonic primordia by implantation of organizers from a different species**. *Int. J. Dev. Biol.* 2001, **45**: 13–38 [reprinted translation of their 1924 paper from the original German].

A conceptual tool kit

Alberts, B., Heald, R., Hopkin, K., Johnson, A., Morgan, D., Roberts, K., Walter, P.: *Essential Cell Biology: An Introduction to the Molecular Biology of the Cell.* 6th edn. New York: W.W. Norton & Company, 2023.

Ashe, H.L., Briscoe, J.: **The interpretation of morphogen gradients**. *Development* 2006, **133**: 385-394.

Barolo, S., Posakony, J.: **Three habits of highly effective signalling pathways: principle of transcriptional control by developmental cell signalling**. *Genes Dev.* 2002, **16**: 1167-1181.

Bird, A.: **Perceptions of epigenetics**. *Nature* 2007, **447**: 396-398.

Jordan, J.D., Landau, E.M., Iyengar, R.: **Signaling networks: the origins of cellular multitasking**. *Cell* 2000, **103**: 193-200.

Levine, M., Cattoglio, C., Tjian, R.: **Looping back to leap forward: transcription enters a new era**. *Cell* 2014, **157**: 13-25.

Levine, M., Tjian, R.: **Transcriptional regulation and animal diversity**. *Nature* 2003, **424**: 147-151.

Martindale, M.Q.: **The evolution of metazoan axial properties**. *Nat. Rev. Genet.* 2005, **6**: 917-927.

Martinez Arias A., Nicholls, J., Schröter, C.: **A molecular basis for developmental plasticity in early mammalian embryos**. *Development* 2013, **140**: 3499-3510.

Nelson, W.J.: **Adaptation of core mechanisms to generate polarity**. *Nature* 2003, **422**: 766-774.

Papin, J.A., Hunter, T., Palsson, B.O., Subramanian, S.: **Reconstruction of cellular signalling networks and analysis of their properties**. *Nat. Rev. Mol. Biol.* 2005, **6**: 99-111.

Peter, I.S., Davidson, E.H.: **Assessing regulatory information in developmental gene regulatory networks**. *Proc. Natl Acad. Sci. USA* 2017, **114**: 5862-5869.

Rudel, D., Sommer, R.J.: **The evolution of developmental mechanisms**. *Dev. Biol.* 2003, **264**: 15-37.

Volff, J.N. (Ed.): *Vertebrate Genomes* (Genome Dynamics Series). Basel: Karger, 2006.

Wolpert, L.: **Do we understand development?** *Science* 1994, **266**: 571-572.

Wolpert, L.: **One hundred years of positional information**. *Trends Genet.* 1996, **12**: 359-364.

Wolpert, L.: **Positional information and pattern formation**. *Curr. Top. Dev. Biol.* 2016, **117**: 597-608.

Xing, Y., Lee, C.: **Relating alternative splicing to proteome complexity and genome evolution**. *Adv. Exp. Med. Biol.* 2007, **623**: 36-49.

Box 1G Signal transduction and intracellular signaling pathways

Alberts, B., Heald, R., Johnson, A., Morgan, D., Raff, M., Roberts, K., Walter, P., Wilson, J., Hunt, T.: *Molecular Biology of the Cell.* 7th edn. New York: W.W. Norton & Company, 2022.

Box 1H When development goes awry

Chen, J.K.: **I only have eye for ewe: the discovery of cyclopamine and development of Hedgehog pathway-targeting drugs**. *Nat. Prod. Rep.* 2016, **33**: 595-601.

Cohen, M.M. Jr.: **Holoprosencephaly: clinical, anatomic, and molecular dimensions**. *Birth Defects Res. A Clin. Mol. Teratol.* 2006, **76**: 658-673.

Geng, X., Oliver, G.: **Pathogenesis of holoprosencephaly**. *J. Clin. Invest.* 2009, **119**: 1403-1413.

Roessler, E., Muenke, M.: **How a Hedgehog might see holoprosencephaly**. *Hum. Mol. Genet.* 2003, **12**: R15-R25.

Zaghloul, N.A., Brugmann, S.A.: **The emerging face of primary cilia**. *Genesis* 2011, **49**: 231-246.

Development of the *Drosophila* body plan

- *Drosophila* life cycle and outline of development
- Setting up the body axes
- Localization of maternal determinants during oogenesis
- Patterning the early embryo
- Activation of the pair-rule genes and the establishment of parasegments
- Segmentation genes and segment patterning
- Specification of segment identity

The early development of the fruit fly Drosophila melanogaster *is better understood than that of any other animal of similar or greater morphological complexity. Its early development clearly illustrates many of the principles introduced in Chapter 1. The genetic basis of development is particularly well understood in the fly, and many developmental genes in other organisms, particularly vertebrates, have been identified by their homology with genes in* Drosophila, *in which they were first identified. Like many other animals,* Drosophila's *earliest development is directed by maternal gene products laid down in the egg, and these specify the main axes and regions of the body. We will see how gradients of these maternal morphogens act on the zygote's own genes to specify the antero-posterior and dorso-ventral pattern of the early embryo, to trigger the assignment of cells to particular germ layers, and to start the partitioning of the embryo into segments. Another set of zygotic genes—the Hox genes—acting along the antero-posterior axis—then confers on each segment its unique character, as displayed by the appendages the segment eventually bears in the adult, such as wings, legs, or antennae.*

Introduction

We are much more like flies in our development than you might think. Astonishing discoveries in developmental biology over the past 30 years or so have revealed that many of the genes that control the development of the fruit fly *Drosophila melanogaster* are like those controlling development in vertebrates and, indeed, in many other animals. It seems that once evolution finds a satisfactory way of patterning animal bodies, it tends to use the same strategies, mechanisms, and molecules over and over again with, of course, important organism-specific modifications. So, although insect and vertebrate development might seem very different, much of what we have learned from *Drosophila* can be applied to vertebrate development. Many crucial developmental signaling pathways and their associated transcription factors are very similar in insects and vertebrates, as we shall see later in this chapter, and *Drosophila* has also spearheaded our understanding of how the activity of cells contributes to the building of tissues and organs. Furthermore, many genes involved in human disease are present in *Drosophila*, including genes involved in Alzheimer's disease, Parkinson's disease, and some muscular dystrophies, thus making the fly a useful model system to understand the cellular bases of these pathologies. The pre-eminent place of *Drosophila* in modern developmental biology was recognized by the award of the 1995 Nobel Prize in Physiology or Medicine for work that led to a fundamental understanding of how genes control development in the fly embryo, only the second time that the prize had been awarded for work in developmental biology.

Drosophila has over 14,000 protein-coding genes, as estimated from its genome sequence. This is only twice the number of genes in unicellular yeast, and fewer than the nearly 20,000 genes in the morphologically simpler nematode, *Caenorhabditis elegans*. However, a large-scale analysis of the RNAs transcribed in developing and adult *Drosophila* has revealed a high incidence of alternative RNA splicing (see Section 1.10), which increases the number of different proteins that could be made. In addition, the fly genome contains around 1100 genes encoding functional RNAs other than messenger RNAs (mRNAs), such as transfer RNAs (tRNAs) and microRNAs (miRNAs). Thus, the complete arsenal of genetic information is more than sufficient to specify the large number of different cell types and more complex behavior of the fruit fly compared with the nematode.

Over the past few years, other functional genome components, called small open reading frames (smORFs), have been identified in *Drosophila*, and subsequently in other animals. The *Drosophila* genome contains tens of thousands of these short DNA sequences, which encode small peptides. Some of them act as hormones and some have a role in modulating, and sometimes changing, the function of other proteins with which they interact.

The first part of this chapter describes the life cycle of *Drosophila* and an outline of its development. The remaining parts examine how the basic **body plan**—the overall organization of the *Drosophila* larva, for example, the relative positions of the 'head' and 'tail' ends—is established. The embryo becomes divided into segments, and we will see how the segments are patterned and acquire their unique identities. More about *Drosophila* gastrulation, germ cell development and sex-determination mechanisms, adult organ development, neural development, growth and metamorphosis, and evolution can be found in Chapters 5, 8, 9, 10, 11, and 13, respectively.

Drosophila life cycle and outline of development

The fruit fly *Drosophila melanogaster* is a small dipteran insect (the Diptera have two wings and two much smaller structures called halteres on their thorax), about 3 mm long as an adult. It undergoes embryonic development inside an egg and hatches from that egg as a larva. This goes through two more larval stages, growing bigger each time. Eventually, it becomes a pupa, which undergoes metamorphosis to produce the adult. The life cycle of *Drosophila* is shown in Figure 2.1.

Drosophila life cycle and outline of development   **45**

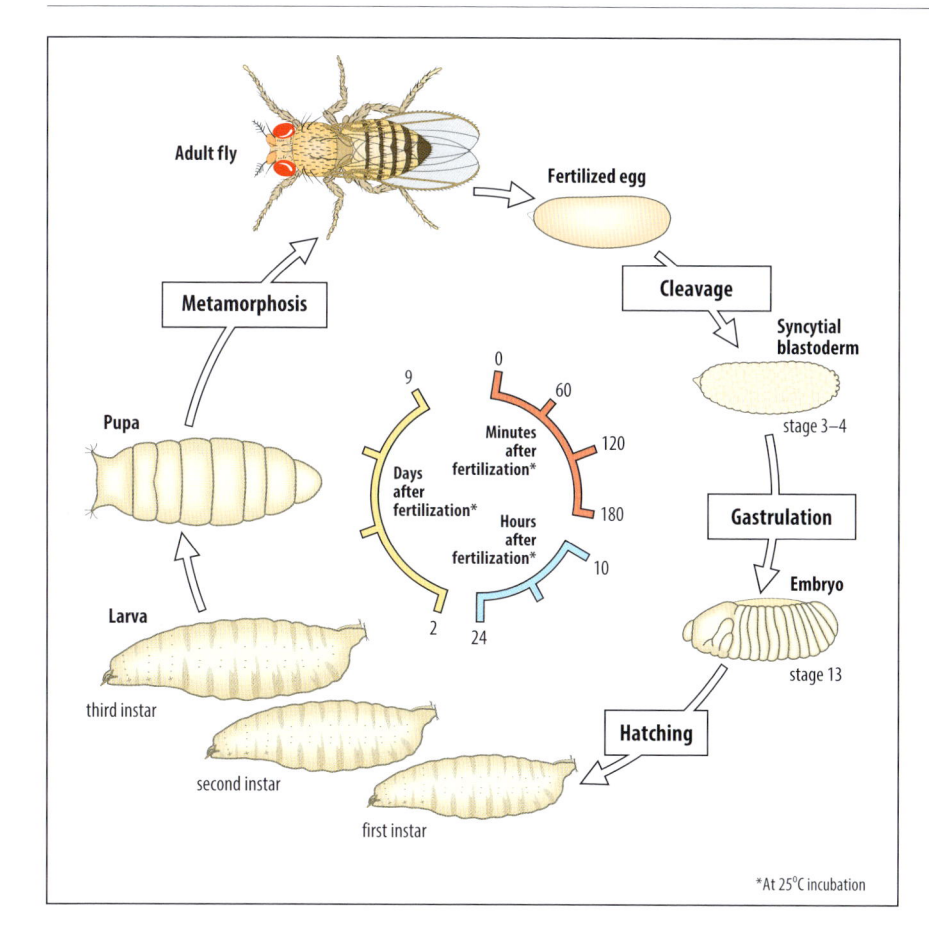

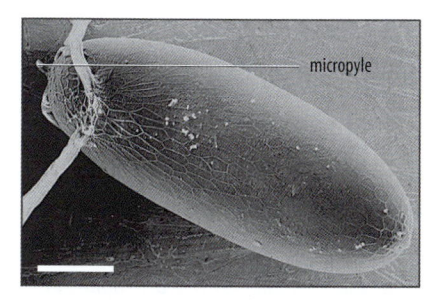

Fig. 2.1 Life cycle of *Drosophila melanogaster*. After cleavage and gastrulation, the embryo becomes segmented and hatches out as a feeding larva. The larva grows and goes through two more rounds of growth and molting (instars), eventually forming a pupa that will metamorphose into the adult fly. The photographs show scanning electron micrographs. Top: a *Drosophila* egg before fertilization. The sperm enters through the micropyle. The dorsal filaments (white ribbons) are extra-embryonic structures. Center: a first-instar larva. Bottom: a pupa. Anterior is to the left, and larva and pupa are shown in dorsal view. Scale bars = 0.1 mm.

Top photograph reproduced with permission from Turner, F.R., Mahowald, A.P.: Scanning electron microscopy of Drosophila *embryogenesis. I. The structure of the egg envelopes and the formation of the cellular blastoderm. Dev. Biol. 1976, **50**: 95–108. Center photograph reproduced with permission from Turner, F.R., Mahowald, A.P.: Scanning electron microscopy of* Drosophila melanogaster *embryogenesis. III. Formation of the head and caudal segments. Dev. Biol. 1979, **68**: 96–109.*

2.1 The early *Drosophila* embryo is a multinucleate syncytium

The *Drosophila* egg is oblong and about 0.5 mm in length. Its future anterior end is easily recognizable by the micropyle, a nipple-shaped structure in the tough external coat surrounding the egg. Sperm enter the anterior end of the egg through the micropyle. After fertilization and fusion of the sperm and egg nuclei, the zygote nucleus undergoes a series of rapid mitotic divisions, one about every 9 minutes. Unlike most other embryos, however, there is initially no cleavage of the cytoplasm and no formation of cell membranes to separate the nuclei. After 12 nuclear divisions, the result is a **syncytium** in which around 6000 nuclei are present in a common cytoplasm (Fig. 2.2); the embryo essentially remains a single cell during its early development. After nine divisions the nuclei move to the periphery to form the **syncytial blastoderm**. This comprises a superficial layer of nuclei and cytoplasm, which surrounds a central mass of yolky cytoplasm. The syncytial blastoderm is equivalent to the blastula stage in embryos such as *Xenopus* (see, e.g., Box 1A). Shortly thereafter, membranes grow in

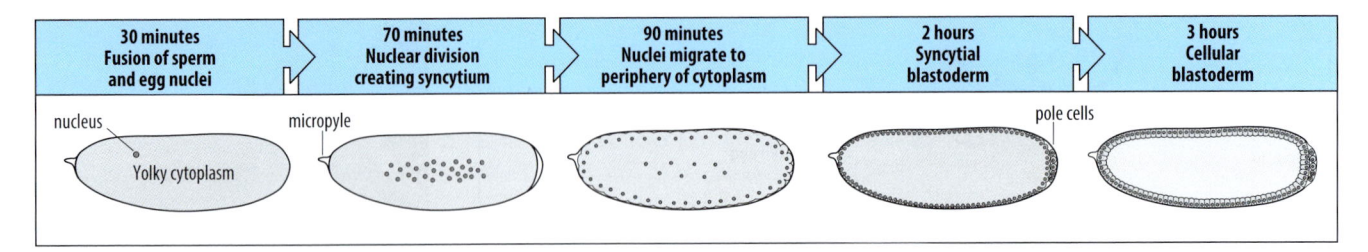

Fig. 2.2 Cleavage of the *Drosophila* embryo. After fusion of the sperm and egg nuclei, rapid nuclear division occurs, but no cell membranes are formed around the nuclei. This results in a syncytium of many nuclei in a common yolky cytoplasm. After the ninth division the nuclei move to the periphery to form the syncytial blastoderm, but some are delayed. About 3 hours after egg laying, cell membranes develop, giving rise to the cellular blastoderm. About 15 pole cells, which will give rise to germ cells, form a separate group at the posterior end of the embryo. Times given are for incubation at 25°C.

from the surface to enclose the nuclei and form cells, and the blastoderm becomes truly cellular after 14 mitoses. Because of the syncytium, even large molecules such as proteins can diffuse between nuclei during the first 3 hours of development and, as we shall see, this is of great importance to early *Drosophila* development.

At the syncytial stage, a small number of nuclei move to the posterior end of the embryo and become surrounded by cell membranes to form the **pole cells**, which end up on the outer surface of the blastoderm (see Fig. 2.2). The pole cells eventually give rise to the **germ cells** from which the gametes—sperm and eggs—develop in the adult insect, whereas the blastoderm gives rise to the somatic cells of the embryo. This setting aside of the germline from the rest of the embryo at a very early stage is a common strategy in animal development and is thought to help protect the germ cells from the inductive signals that drive the somatic developmental program in the embryo.

In this chapter we will be concerned with the development of the embryo to the stage at which it hatches out of the egg as a larva. Embryonic development naturally takes place inside the opaque egg case, and to study it embryos are freed from the egg case by treating with bleach and then placed in a special oil for observation. This can be done at any stage and they will continue to develop and even hatch.

2.2 Cellularization is followed by gastrulation and segmentation

All the future tissues, except for the germline cells, are derived from the single epithelial layer of the cellular **blastoderm**. This gives rise to the three germ layers—**ectoderm**, **mesoderm**, and **endoderm** (see Box 1D). In *Drosophila*, the prospective mesoderm is in the most ventral region, whereas the future midgut derives from two regions of prospective endoderm, one at the anterior and the other at the posterior end of the embryo. Endodermal and mesodermal tissues move to their future positions inside the embryo during **gastrulation**, leaving ectoderm as the outer layer (Fig. 2.3). Gastrulation starts at about 3 hours after fertilization, when the future mesoderm in the ventral region invaginates to form a furrow along the ventral midline. The mesoderm cells are initially internalized by formation of a tube of mesoderm, which will be described in more detail in Section 5.10. The mesoderm cells then separate from the surface layer of the tube and migrate under the ectoderm to internal locations, where they later give rise to muscle and other connective tissues. Each of these germ layers becomes a unit of lineage; that is, once a cell has been specified to one of these fates—mesoderm, ectoderm, or endoderm—all its descendants will adopt that fate. The mesoderm subsequently becomes subdivided into the visceral mesoderm, which gives rise to connective tissues associated with the gut and internal organs, and the somatic mesoderm, which gives rise to the muscles of the body wall that enable the larva to hatch out, feed, and move.

In insects, as in all arthropods, the main nerve cord lies ventrally, rather than dorsally as in vertebrates. Shortly after the mesoderm has invaginated, ectodermal cells

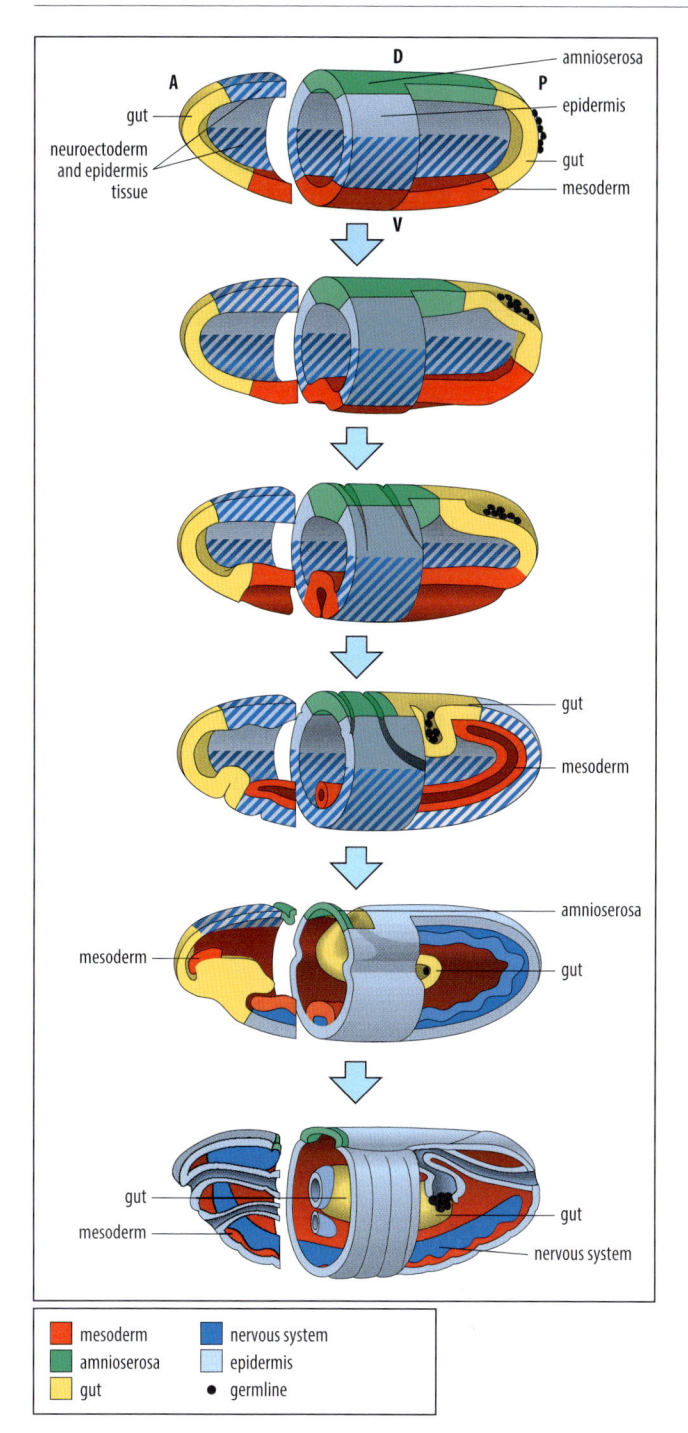

Fig. 2.3 Gastrulation in *Drosophila*. Gastrulation begins when the future mesoderm (red) invaginates in the ventral (V) region, first forming a furrow and then an internalized tube (red). The mesoderm cells then leave the tube and migrate internally under the ectoderm (see last panel). The nervous system comes from cells that leave the surface of the ventral blastoderm and form a layer between the remaining ventral ectoderm and mesoderm. Regions striped light blue and dark blue represent ectodermal tissue that will give rise to both nervous system (neuroectoderm, dark blue) and epidermis (light blue). The anterior (A) striped area gives rise to part of the brain and the epidermis of the head. The gut forms from two invaginations at the anterior and posterior (P) end that will fuse in the middle (yellow). The midgut region is derived from endoderm, whereas the foregut and hindgut are of ectodermal origin. The amnioserosa (green), an extra-embryonic membrane, is discussed in Section 2.5. D, dorsal.

of the ventral region, which will give rise to the nervous system, leave the surface individually and form a layer of prospective neural cells (neuroblasts) between the mesoderm and the remaining outer ectoderm (see Fig. 2.3, last panel). At the same time, two tube-like invaginations develop at the sites of the future anterior and posterior midgut. These grow inward towards each other and eventually fuse to form the endoderm of the midgut, while ectoderm is dragged inward behind them at each end to form the foregut and the hindgut. The outer ectoderm layer develops into the epidermis. The ectoderm and mesoderm continue to divide during gastrulation. Once the epidermal cells stop dividing, they secrete a thin cuticle composed largely of protein and the polysaccharide chitin.

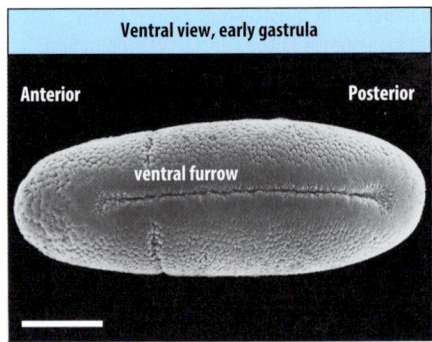

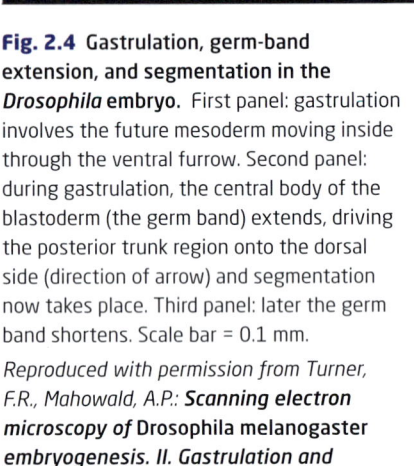

Fig. 2.4 Gastrulation, germ-band extension, and segmentation in the *Drosophila* embryo. First panel: gastrulation involves the future mesoderm moving inside through the ventral furrow. Second panel: during gastrulation, the central body of the blastoderm (the germ band) extends, driving the posterior trunk region onto the dorsal side (direction of arrow) and segmentation now takes place. Third panel: later the germ band shortens. Scale bar = 0.1 mm.

Reproduced with permission from Turner, F.R., Mahowald, A.P.: **Scanning electron microscopy of** *Drosophila melanogaster* **embryogenesis. II. Gastrulation and segmentation.** *Dev. Biol. 1977,* **57***: 403-416.*

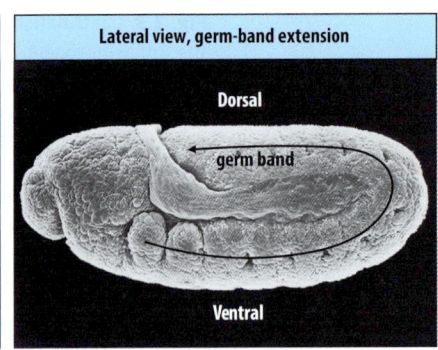

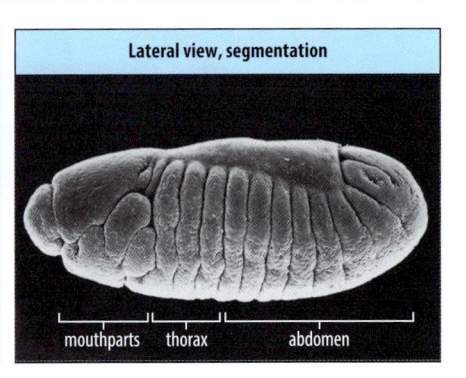

Also during gastrulation, the central body of the blastoderm, or **germ band**, which comprises the main trunk region of the embryo, undergoes a process of extension along the antero-posterior axis (Fig. 2.4). **Germ-band extension** drives the posterior trunk regions round the posterior end and onto what was the dorsal side, as described in Section 5.11. The germ band later retracts as embryonic development is completed. At the time of germ-band extension, the first external signs of **segmentation** can be seen. A series of evenly spaced grooves form more or less simultaneously, and these demarcate developmental regions called **parasegments**. As we shall see later, the segments of the larva and adult do not correspond exactly to the embryonic parasegments: parasegments act as the developmental basis for segments, which result from the fusion of the posterior and anterior halves of adjacent parasegments. There are 14 parasegments: three contribute to the head; three to the thoracic region; and eight to the abdomen.

2.3 After hatching, the *Drosophila* larva develops through several larval stages, pupates, and then undergoes metamorphosis to become an adult

The larva (Fig. 2.5) hatches about 24 hours after fertilization, but the different regions of the larval body are well defined several hours before that. The head is a complex structure, largely hidden from view before the larva hatches. A specialized structure associated with the most anterior region of the head is the acron. Another specialized structure, called the telson, marks the posterior end of the larva. Between the head and telson, three thoracic segments and eight abdominal segments can be distinguished by specializations in the cuticle. On the ventral side of each segment are belts of small tooth-like outgrowths from the epidermis called **denticles**, which are used for traction during movement, and other cuticular structures characteristic of each segment. As the larva feeds and grows, it molts, shedding its cuticle. This process occurs twice, each stage being called an **instar**. Over the 4-day larval period, the larva increases in size by an amazing 200-fold. After the third instar, the larva becomes a **pupa**, inside which **metamorphosis** into the adult fly occurs, and there is a major transformation of overall form.

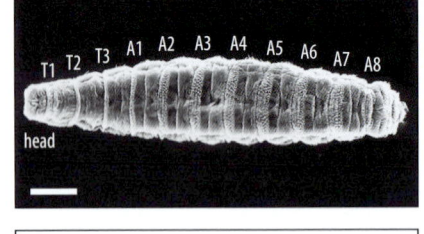

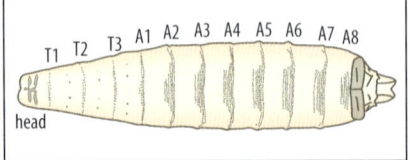

The *Drosophila* larva has neither wings nor legs; these and other organs characteristic of the adult are formed as the larva grows (discussed in Chapter 9). The primordia for these structures are set aside in the embryo and are present in the larva as **imaginal discs**. At the time that they are formed, each disc comprises a group of about 20–40 prospective epidermal cells. The discs grow by cell proliferation throughout larval life and form sacs of epithelia that fold to accommodate their increase in size. There are imaginal discs for each of the six legs, two wings, and the two halteres (balancing organs), and for the genital apparatus, eyes, antennae, and other adult head structures (Fig. 2.6). Associated with several discs are the adepithelial cells, which are progenitors for the adult muscles. At metamorphosis, the discs and associated adepithelial cells differentiate and develop into the adult organs. The adult abdominal epidermis develops from small groups of cells called **histoblasts**, which remain dormant during larval life. They proliferate in the pupal stage and become organized to form the epidermis of the

Fig. 2.5 The *Drosophila* first-instar larva. Ventral view. T1–T3 are the thoracic segments, and A1–A8 the abdominal segments. The characteristic pattern of denticles can be seen in the anterior region of each abdominal segment. Scale bar = 0.1 mm.

Photograph courtesy of F.R. Turner.

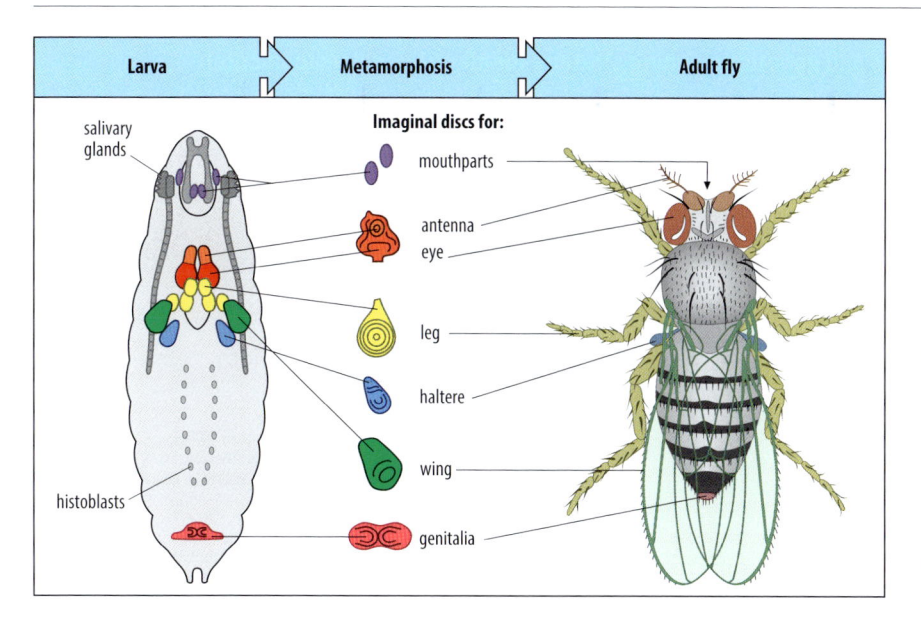

Fig. 2.6 Imaginal discs give rise to adult structures at metamorphosis. The imaginal discs in the *Drosophila* larva are small sheets of epithelial cells; at metamorphosis they give rise to a variety of adult structures. The abdominal epidermis of the adult is derived from groups of tissue-forming cells (histoblasts) located in each larval abdominal segment.

adult abdomen during metamorphosis. The imaginal discs provide continuity between the pattern of the larval body and that of the adult, even though metamorphosis intervenes. Their development is discussed in Chapter 9.

2.4 Many developmental genes were identified in *Drosophila* through large-scale genetic screening for induced mutations

Although spontaneous mutations have been valuable in the study of development, suitable developmentally informative mutations are rare. Most of the known developmental genes were identified by inducing random mutations in many individuals through chemical treatments or irradiation with X-rays, and then screening their offspring for developmental defects associated with mutations in individual genes. This **mutagenesis screening** approach is best used in organisms such as *Drosophila* that breed rapidly, so that large numbers of individuals can be obtained and treated conveniently. The overall aim of projects such as these is to treat a large enough population over many experiments so that at least one informative mutation is induced in every gene in the genome. In practice, the aim is to produce a family of mutations in each gene so that its function and nature can be worked out.

Many of the developmental mutations that led to our present understanding of early *Drosophila* development came from a brilliantly successful screening program that searched the *Drosophila* genome systematically for mutations affecting the patterning of the early embryo. Its success was recognized with the award of the Nobel Prize in Physiology or Medicine to Edward Lewis, Christiane Nüsslein-Volhard, and Eric Wieschaus in 1995.

In this screening program, thousands of flies were treated with a chemical mutagen, were subsequently bred, and their offspring screened according to the strategy described in Box 2A. Given the number of progeny involved, it was important to devise a strategy that would reduce the number of flies that had to be examined to find a mutant. So the search was restricted to mutations on just one chromosome at a time. As described in the box, the program incorporated a means of identifying flies homozygous for the male-derived chromosomes carrying the induced mutations and included various ways in which flies that did not carry a mutated chromosome, or that had mutations that did not affect early embryonic patterning, could be easily excluded.

The screen took advantage of the precise and reproducible pattern of the stereotyped pattern of denticles in the cuticle formed at the end of embryogenesis (see Fig. 2.5). The screens identified changes in the number of segmental structures as well as in the arrangement of denticles (see Box 2A). In this way, the key genes involved in

EXPERIMENTAL BOX 2A Mutagenesis and genetic screening strategy for identifying developmental mutants in *Drosophila*

The mutagen ethyl methane sulfonate (EMS) was applied to large numbers of male flies homozygous for a recessive but viable mutation (e.g. white eyes instead of the normal red eyes, see Fig. 1.12) on the selected chromosome, 'a' in Figure 1.

The treated males, which now produce sperm with various induced mutations on this chromosome ($a*$), were crossed with untreated females that carried different mutations (*DTS* and *b*) on their two 'a' chromosomes, but were otherwise wild type. These mutations track the untreated female-derived chromosomes and automatically eliminate all embryos carrying two female-derived chromosomes in subsequent generations. *DTS* is a dominant temperature-sensitive mutation that kills flies raised at 29°C, whereas *b* is an embryonic lethal recessive mutation without a visible phenotype that allows the elimination of embryos not bearing the $a*$ chromosome. The female flies also carried a special balancer chromosome (not shown) that prevented the recombination of male and female chromosomes.

To identify the new recessive mutations ($a*$) caused by the chemical treatment, a large number of the heterozygous males arising from this first cross were outcrossed to *DTS/b* females. Of the offspring of each cross, only $a*/b$ flies survive when placed at 29°C; all other combinations die. The surviving siblings of each cross bear the same $a*$ mutation. They were intercrossed and the offspring screened for patterning mutants. There are three possible outcomes: flies homozygous for the induced mutation $a*$; heterozygous $a*/b$ flies; and homozygous *b* flies (which die as normal-looking embryos).

$a*$ patterning mutations of interest are embryonic lethal, and so any culture tubes containing adult white-eyed flies can be discarded immediately, as their $a*$ mutations must have allowed them to develop to adulthood even when homozygous. If no white-eyed flies are present in a tube, then the homozygous $a*/a*$ embryos might have died or been arrested in their development because of abnormal development, making the mutation potentially interesting. The embryos or larvae from this cross can then be examined for pattern defects, as in the examples illustrated here (Fig. 2). The phenotypically wild-type adults in this tube are heterozygous for $a*$ and are used as breeding stock to study the mutant further. This whole program was repeated for each of the four chromosome pairs of *Drosophila*.

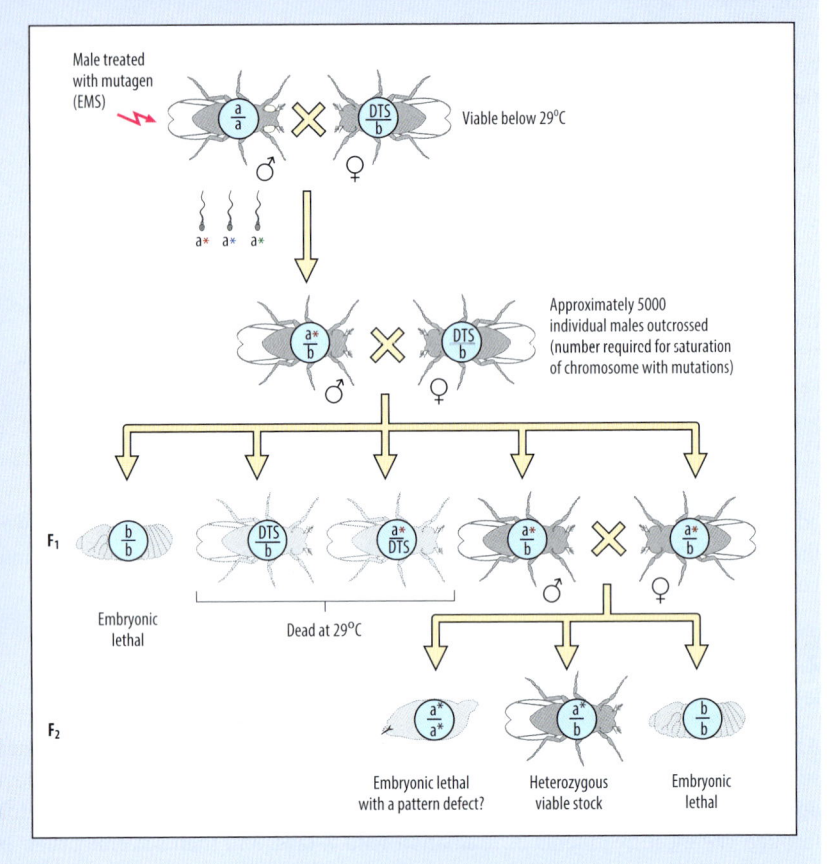

Fig. 1

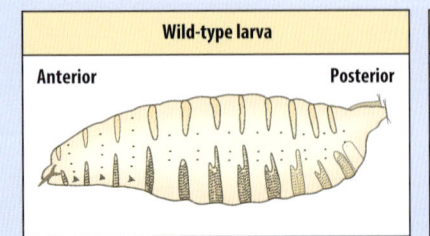

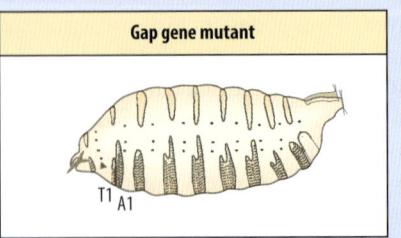

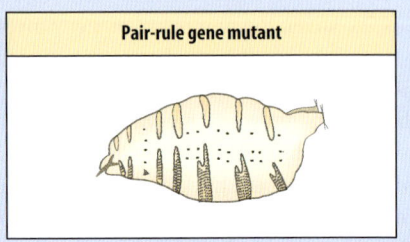

Fig. 2

patterning the early *Drosophila* embryo were first identified. The screens also yielded genes involved in the specification of the germ layers, genes affecting mitosis and the cell cycle, and many others involved in determining the structure and physiology of cells. These genes are conserved throughout the animal kingdom—that is, they are very similar in every organism. For this reason, characterizing their functions has helped us to understand the fundamental relationships between the molecular activities inside cells and the processes that regulate all animal development.

Some mutations that affect *Drosophila* embryonic development act by way of the mother and need a somewhat different type of screen to detect them. These mutations are called **maternal-effect mutations**, and the genes affected by the mutations are called **maternal-effect genes** (or sometimes just **maternal genes** in this context). When present in the mother, such mutations do not affect her appearance or physiology, but have profound effects on the development of her offspring, independently of the father's contribution. In other words, the paternal genome in the embryo cannot 'rescue' the defect. As we shall see later in the chapter, some maternal-effect genes exert their effects by being expressed in the follicle cells of the ovary, which form a bag that contains the germline-derived **oocyte** (the immature egg) and nurse cells. Others are expressed in the nurse cells, which eventually contribute their cytoplasm to the oocyte, or in the oocyte itself, producing developmentally important mRNAs and proteins that are deposited in the oocyte in a particular spatial pattern. This initial patterning process is crucial for correct development.

SUMMARY

The embryo of the fruit fly *Drosophila* develops within the egg case and hatches out as a larva, which goes through several cycles of growth and molting before pupating and undergoing metamorphosis to produce the adult fly. Maternal proteins and mRNAs laid down in the egg guide the early stages of development, and we shall look next at how these maternal gene products specify the antero-posterior and dorso-ventral axes of the embryo, and begin to pattern them.

Setting up the body axes

The insect body, like that of vertebrates and most of the other organisms described in this book, is bilaterally symmetrical. Like all animals with bilateral symmetry, the *Drosophila* larva has two distinct and largely independent axes: the antero-posterior and dorso-ventral axes, which are at right angles to each other. These axes are already partly set up in the *Drosophila* egg, and we will see here how they become established and start to be patterned in the very early embryo.

2.5 The body axes are set up while the *Drosophila* embryo is still a syncytium

The antero-posterior and dorso-ventral axes become fully established and start to become patterned while the embryo is still in the syncytial blastoderm stage. Along the antero-posterior axis the embryo becomes divided into several broad regions, which will become the **head**, **thorax**, and **abdomen** of the larva (Fig. 2.7). The thorax and abdomen become divided into segments as the embryo develops, while two regions of endoderm at each end of the embryo invaginate at gastrulation to form the gut (see Fig. 2.3). Each segment, and the head, has its own unique character in the larva, as revealed by both its external cuticular structures and its internal organization.

Fig. 2.7 Patterning of the *Drosophila* embryo. The body plan is patterned along two distinct axes. The antero-posterior and dorso-ventral axes are at right angles to each other and are laid down in the egg. In the early embryo, the dorso-ventral axis is divided into four regions: mesoderm (red), ventral ectoderm (neuroectoderm) (dark blue), dorsal ectoderm (light blue), and amnioserosa (an extra-embryonic membrane; green). The ventral ectoderm gives rise to both ventral epidermis and neural tissue, the dorsal ectoderm to epidermis. The antero-posterior axis becomes divided into different regions that later give rise to the head, thorax, and abdomen. After the initial division into broad body regions, segmentation begins. The future segments can be visualized as transverse stripes by staining for specific gene activity; these stripes demarcate 14 parasegments, 10 of which are indicated in this diagram. The embryo develops into a segmented larva. By the time the first-instar larva hatches out, the 14 parasegments have been converted into three thoracic (T1–T3) and eight abdominal (A1–A8) segments, with each segment being made up of the posterior half of one parasegment and the anterior half of the next. The rest of the parasegments contribute to structures in the head. Different segments are distinguished by the patterns of hairs and denticles on the cuticle. Specialized structures, the acron and telson, develop at the head and tail ends of the embryo, respectively.

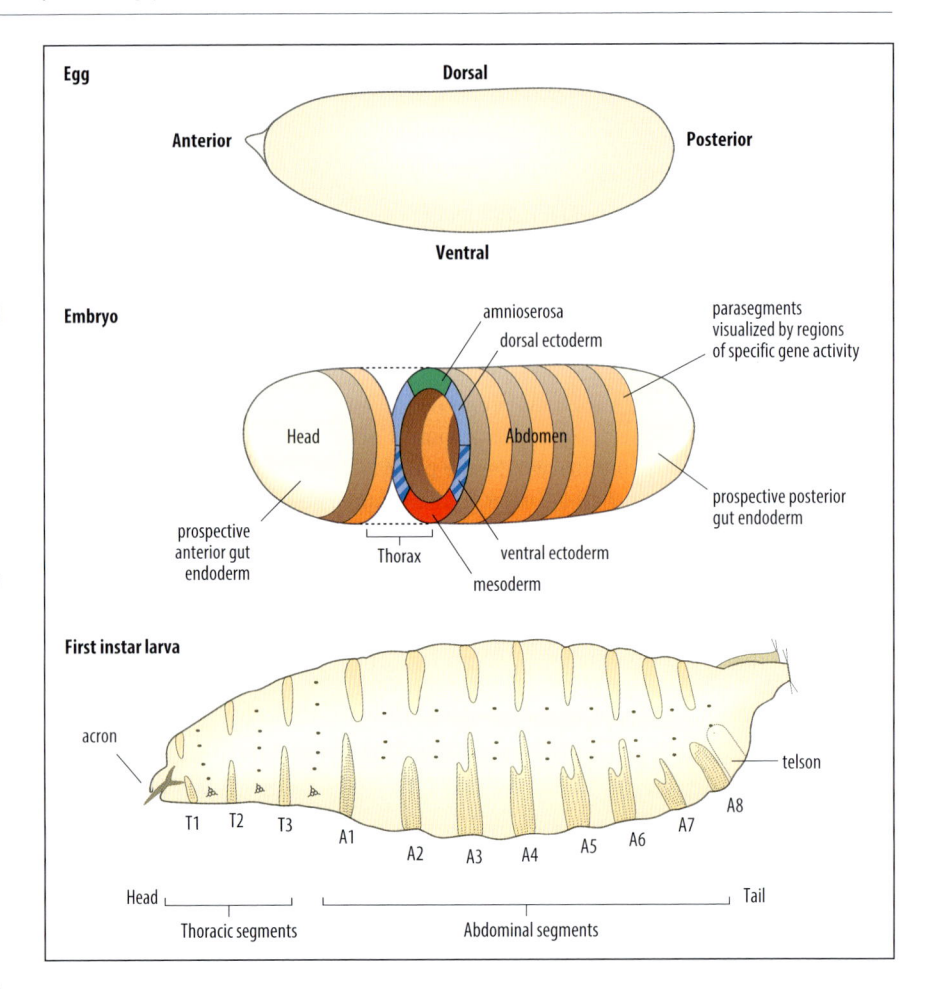

The dorso-ventral axis of the embryo becomes divided into four regions early in embryogenesis: these are, from ventral to dorsal, the **mesoderm**, which will form muscles and other internal connective tissues; the **ventral ectoderm** or **neuroectoderm**, which gives rise to the larval nervous system and epidermis; the **dorsal ectoderm**, which gives rise to larval epidermis; and the **amnioserosa**, which gives rise to an extra-embryonic membrane on the dorsal side of the embryo (see Figs 2.3 and 2.7). Organization along the antero-posterior and dorso-ventral axes of the early embryo develops more or less simultaneously, but each axis is specified by independent mechanisms and by different sets of genes.

The early development of *Drosophila* is peculiar to certain types of insects, as patterning occurs within a multinucleate syncytial blastoderm (see Fig. 2.2). Figure 2.8 shows an embryo at the syncytial blastoderm stage, with the nuclei stained for two maternal transcription factors. Only after the beginning of segmentation does the embryo become truly multicellular. At the syncytial stage, many proteins, including transcription factors that are not normally secreted from cells, can diffuse throughout the blastoderm and enter other nuclei. This allows the formation of concentration gradients of transcription factors along the length of the embryo. These gradients provide **positional information** for the nuclei to interpret (see Section 1.15).

Early development is essentially two-dimensional, because patterning occurs mainly in the blastoderm, which is initially a single layer of nuclei, that later becomes a single layer of cells. But the larva itself is a three-dimensional object, with internal structures coordinately organized along the axis. This third dimension develops later, at gastrulation, when parts of the surface layer move into the interior to form the gut, the mesodermal structures that will give rise to muscle, and the ectodermally derived nervous system (see Fig. 2.3).

2.6 Maternal factors set up the body axes and direct the early stage of *Drosophila* development

A key general principle of development is that an embryo is patterned in a series of steps, and *Drosophila* illustrates this particularly clearly. Broad regional differences are established first, and within these domains, the expression of specific genes and their interactions produce a larger number of smaller developmental domains, each characterized by a unique profile of gene activity. Developmental genes act in a strict temporal sequence. They form a hierarchy of gene activity, in which the action of one set of genes is essential for another set of genes to be activated, and thus for the next stage of development to occur.

The earliest stage of *Drosophila* embryonic development is guided by **maternal factors**—mRNAs and proteins that are synthesized and laid down in the egg by the mother. Genetic analysis has identified about 50 maternal genes involved in setting up the two axes and a basic molecular framework of positional information, which is then interpreted by the embryo's own genetic program. In contrast to the maternal-effect genes, the genes expressed in the embryo's own nuclei during development are known as **zygotic genes**. All later patterning, which involves the regulated expression of the zygotic genes, is built on the framework of maternal gene

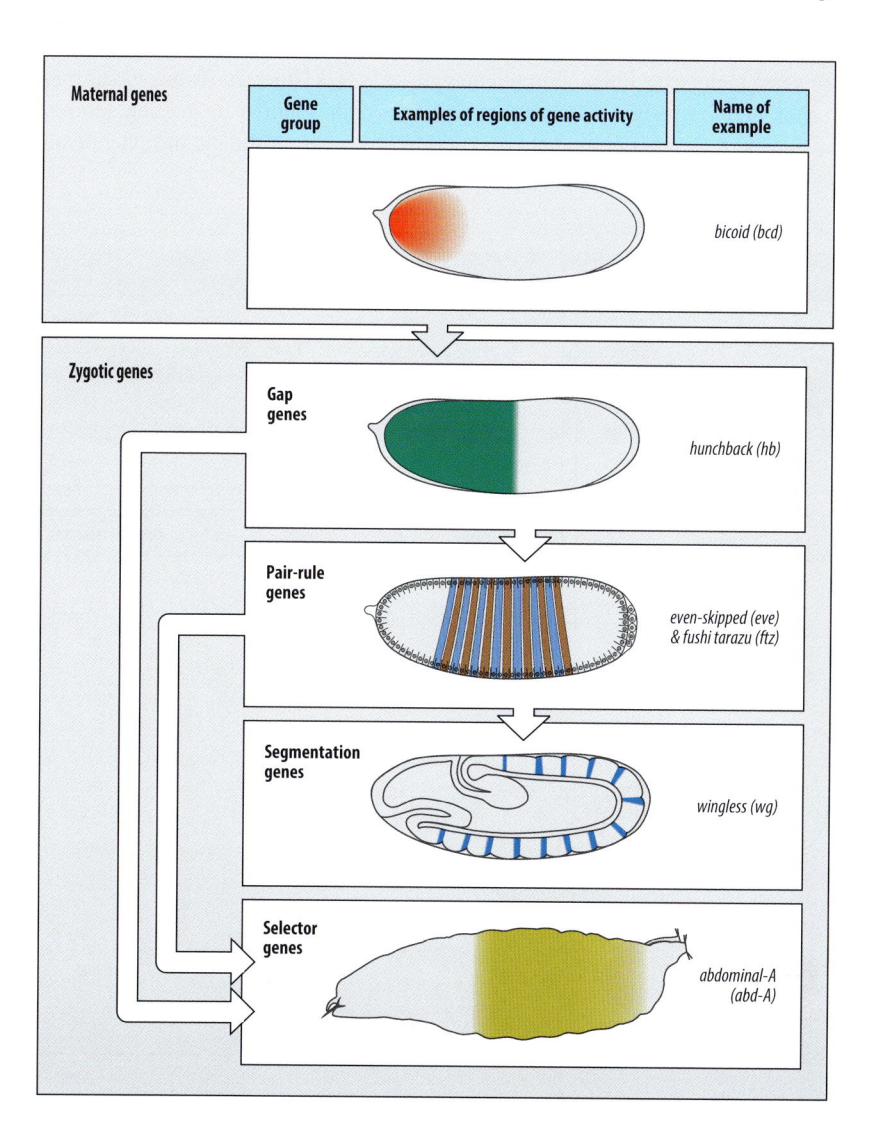

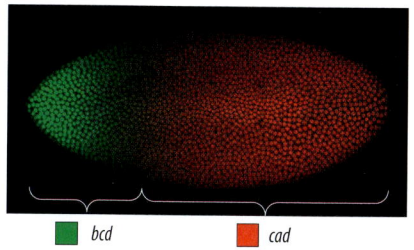

Fig. 2.8 Initial patterning of *Drosophila* occurs at the syncytial blastoderm stage. The photograph shows an embryo about 2 hours after egg laying, in which two translated maternal proteins, the transcription factors Bicoid (*bcd*, green) and Caudal (*cad*, red), have entered the blastoderm nuclei. These proteins are involved in patterning the antero-posterior axis of the embryo. Bicoid protein prevents the synthesis of Caudal protein at the anterior end of the embryo, thus restricting Caudal to the posterior end. The embryo has been fixed and the proteins stained using fluorescent antibodies.

Reproduced with permission from Surkova, S., et al.: Characterization of the Drosophila segment determination morphome. Dev. Biol. 2008, 313: 844-862.

Fig. 2.9 The sequential expression of different sets of genes establishes the body plan along the antero-posterior axis. Upper panel: after fertilization, maternal gene products laid down in the egg, such as *bicoid* mRNA (red), are translated. When translated, the Bicoid protein diffuses and its concentration along the main axis provides positional information that activates the zygotic genes (the Bicoid protein gradient can be seen in Figs 2.8 and 2.12). Lower panel: the four main classes of zygotic genes acting along the antero-posterior axis are the gap genes, the pair-rule genes, the segmentation genes, and the homeotic selector genes. The gap genes define regional differences, which result in the expression of a periodic pattern of gene activity by the pair-rule genes, which define the parasegments and foreshadow segmentation. The segmentation genes elaborate patterning within the parasegments, and thus the future segments, and the homeotic selector genes determine the identity of segments (whether a segment is thoracic or abdominal, for example, and which specific segment it is). The functions of each of these classes of genes are discussed in this chapter.

products (Fig. 2.9). After development begins, the maternal RNAs are translated, and the resulting proteins, many of which are transcription factors, act on the embryo's nuclei to activate zygotic genes in a specific spatial pattern along each axis, thus setting the scene for the next round of patterning. The transition between maternally deposited RNAs and proteins and activation of the embryo's or zygote's genome is called the **maternal–zygotic transition (MZT)**. MZT is found in many other organisms. In *Drosophila*, the MZT completes just before cellularization.

2.7 Three classes of maternal genes specify the antero-posterior axis

We will first look at how maternal gene products specify the antero-posterior axis of the embryo. The expression of maternal genes during egg formation in the mother creates regional differences in the egg along the antero-posterior axis even before it is fertilized. These differences already distinguish the future anterior and posterior ends of the larva. The roles of individual maternal genes can be deduced from the effects of maternal-effect mutations (see Section 2.4) on the embryo. These fall into three classes: those that affect only anterior regions; those that affect only posterior regions; and those that affect both terminal regions (Fig. 2.10). Mutations of genes in the anterior class, such as *bicoid*, lead to a reduction or loss of head and thoracic structures in the larva, and in some cases their replacement with posterior structures. Posterior-group mutations, such as *nanos*, cause the loss of abdominal regions, leading to a smaller than normal larva, while those of the terminal class, such as *torso*, affect the acron and telson.

The idiosyncratic naming of genes in *Drosophila* usually reflects the attempts by their discoverers to describe the mutant phenotype—*nanos* is Greek for dwarf, *bicoid* means two ends, and *torso* reflects the fact that both ends of the embryo are missing. Moreover, gene names in *Drosophila* are italicized and in lower case if the mutant phenotype of

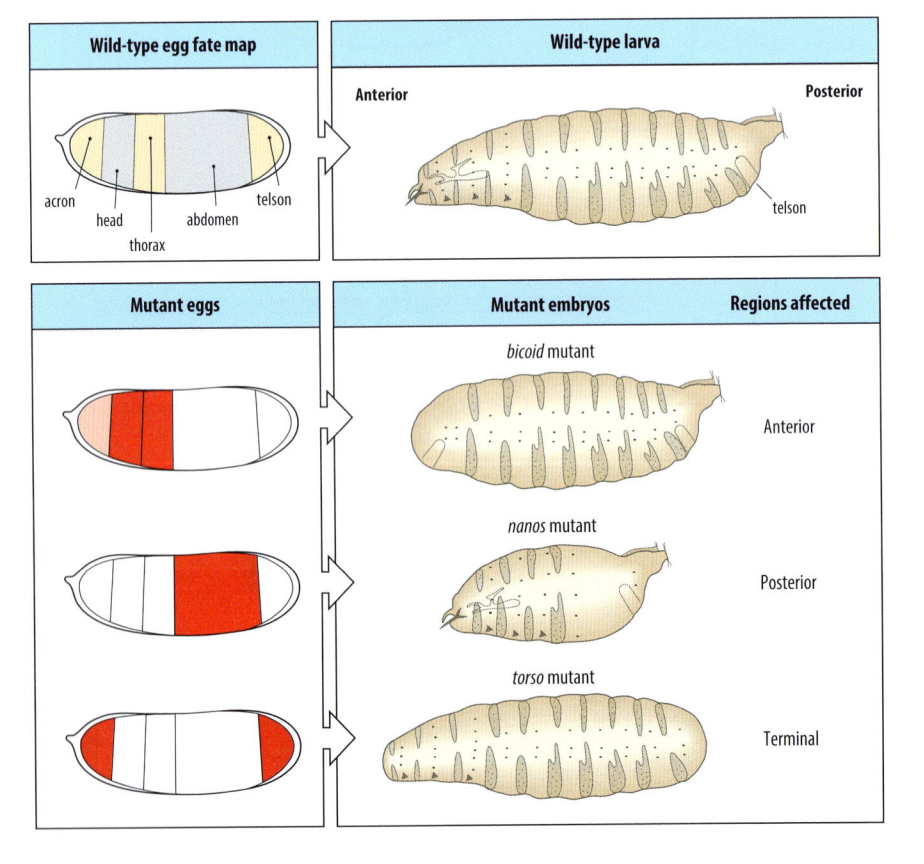

Fig. 2.10 The effects of mutations in the maternal gene system. Mutations in maternal genes lead to deletions and abnormalities in anterior, posterior, or terminal structures. The wild-type fate map shows which regions of the egg give rise to particular regions and structures in the larva. Regions that are affected in mutant eggs, and which lead to lost or altered structures in the larva, are shaded in red. In *bicoid* mutants there is a partial loss of anterior structures and the appearance of a posterior structure—the telson—at the anterior end. *nanos* mutants lack a large part of the posterior region. *torso* mutants lack both acron and telson.

that gene is recessive (e.g. *nanos* or *bicoid*). If the mutant phenotype is dominant, the gene name is still italicized, but it begins with a capital letter (e.g. *Notch*). All *Drosophila* proteins start with a capital letter and are in Roman font (e.g. Nanos, Notch).

2.8 Bicoid protein provides an antero-posterior gradient of a morphogen

Maternal *bicoid* mRNA is localized at the anterior end of the unfertilized egg during oogenesis. After fertilization it is translated into Bicoid protein, which diffuses from the anterior end to form a concentration gradient along the antero-posterior axis. Historically, the Bicoid protein gradient provided the first concrete evidence for the existence of the morphogen gradients that have been postulated to control pattern formation (see Section 1.15). **Morphogens** are chemicals (often, but not always, proteins) that are distributed non-uniformly within a tissue, and influence cells to adopt different fates at different concentrations, thus patterning the tissue.

The role of the *bicoid* gene was first elucidated by a combination of genetic and classic embryological experiments on the *Drosophila* embryo. Female flies that do not express *bicoid* produce embryos that have no proper head and thorax (see Fig. 2.10). In a separate line of investigation into the role of localized cytoplasmic factors in anterior development, normal eggs were pricked at the anterior ends and some cytoplasm allowed to leak out. The embryos that developed bore a striking resemblance to *bicoid* mutant embryos. Together, these observations suggested that normal eggs have some factor(s) in the cytoplasm at their anterior end that is absent in *bicoid* mutant eggs. This was confirmed by showing that anterior cytoplasm from wild-type embryos could partially 'rescue' the *bicoid* mutant embryos—in the sense they developed more normally—if it was injected into their anterior regions (Fig. 2.11). Moreover, if normal

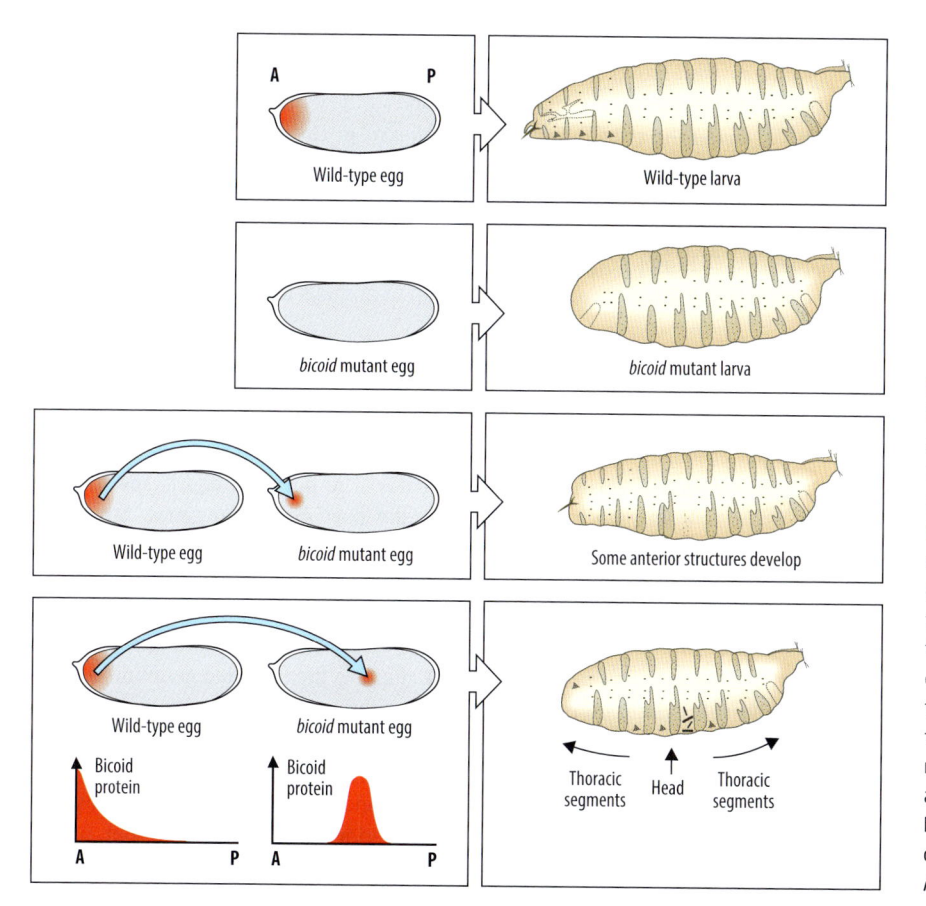

Fig. 2.11 The *bicoid* gene is necessary for the development of anterior structures. Embryos whose mothers lack the *bicoid* gene lack anterior regions (second row). Transfer of anterior cytoplasm from wild-type embryos to *bicoid* mutant embryos causes some anterior structures to develop at the injection site (third row). If wild-type anterior cytoplasm is transplanted to the middle of a *bicoid* mutant egg or early embryo, head structures develop at the injection site, flanked on both sides by thoracic-type segments (fourth row). These results can be interpreted in terms of the anterior cytoplasm setting up a gradient of Bicoid protein with the high point at the site of injection (see graphs, bottom left panel). A, anterior; P, posterior.

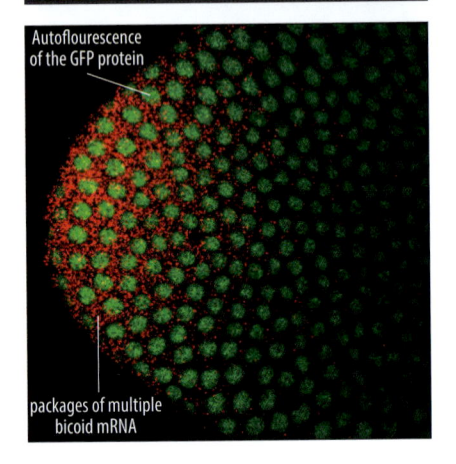

56 Chapter 2 Development of the *Drosophila* body plan

Fig. 2.12 Bicoid protein enters the nuclei and forms a nuclear gradient from anterior to posterior. The top photograph shows the nuclear Bicoid protein gradient in a fixed transgenic *Drosophila* embryo carrying a *bicoid-GFP* fusion gene. In these embryos the concentration of Bicoid protein could be accurately measured either by the intensity of the autofluorescence of the green fluorescent protein (GFP) itself (not shown) or by the intensity of fluorescence of an anti-GFP antibody tagged with a fluorescent dye, as shown in a surface view of the embryo here. The lower photograph shows the anterior tip of a fixed transgenic *Drosophila* embryo carrying a *bicoid-GFP* fusion gene labeled with fluorescently tagged cDNA probes that bind *bicoid* mRNA molecules. Green color shows autofluorescence of the GFP protein in the nuclei; red color shows packages of multiple *bicoid* mRNA molecules in the anterior cytoplasm.

Upper photograph courtesy of J.O. Dubuis and T. Gregor; lower photograph courtesy of S.C. Little and T. Gregor.

anterior cytoplasm was injected into the middle of a fertilized *bicoid* mutant egg, head structures developed at the site of injection and the adjacent segments became thoracic segments, setting up a mirror-image body pattern at the site of injection. The simplest interpretation of these experiments is that the *bicoid* gene is necessary for the establishment of the anterior structures because it establishes a gradient of Bicoid protein, whose source and highest level is at the anterior end.

bicoid mRNA was originally shown to be tightly localized to the anterior-most region of the unfertilized egg by the technique of *in situ* hybridization (see Box 1E). Staining with an antibody raised against Bicoid protein then showed that this is absent in the unfertilized egg; after fertilization, however, *bicoid* mRNA is translated into protein at the anterior pole. The protein then diffuses and forms a gradient with the highest point at the anterior end. Bicoid is a transcription factor, and so must enter the embryo's nuclei to carry out its function of switching on zygotic genes. By the time the syncytial blastoderm is formed, there is a clear gradient of intranuclear Bicoid protein concentration along the antero-posterior axis, with its high point at the anterior end.

With advances in imaging technology, the dynamics of Bicoid gradient formation have been observed in living embryos by measuring the intranuclear concentrations of Bicoid protein genetically tagged with green fluorescent protein (Bicoid–GFP) (Fig. 2.12). The measurements indicate that the concentration of Bicoid protein inside a nucleus at a given position along the gradient is somehow maintained at a constant level, despite the increase in the number of nuclei at each mitotic cycle and the outflow of Bicoid into the cytoplasm from nuclei when nuclear membranes break down at mitosis.

Bicoid protein acts as a morphogen that patterns the anterior portion of the embryo. As described in more detail later in the chapter, it switches on specific zygotic genes at different **threshold concentrations**, thereby initiating new patterns of gene expression along the axis (the concept of thresholds is discussed in Section 1.15). Thus, *bicoid* is a key maternal gene in early *Drosophila* development. The other maternal genes of the anterior group are mainly involved in the localization of *bicoid* mRNA to the anterior end of the egg during oogenesis and in the control of its translation after fertilization. Given the crucial importance of *bicoid* in *Drosophila* development, it is worth noting that the *bicoid* gene is only present in a small group of the most recently evolved dipterans—such as fruit flies and blowflies. Different mechanisms of early development have evolved in other groups of insects and some of these are discussed in Chapter 13. It is not surprising that many different developmental mechanisms have evolved in such a large and diverse group of animals as the insects.

2.9 The posterior pattern is controlled by the gradients of Nanos and Caudal proteins

For proper patterning along an axis, both ends need to be specified, and Bicoid defines only the anterior end of the antero-posterior axis. The posterior end is specified by the actions of at least nine maternal genes—the posterior-group genes. Mutations in posterior-group genes, such as *nanos*, result in larvae that are shorter than normal because there is no abdomen (see Fig. 2.10). One function of the products of other posterior-group genes (e.g. *oskar*) is to localize *nanos* mRNA at the extreme posterior pole of the unfertilized egg. Another function is to assemble the posterior **germplasm** in the egg. This is cytoplasm that contains the so-called germline factors; in the embryo this cytoplasm is incorporated into the pole cells (see Section 2.1) and will give rise to eggs or sperm.

nanos mRNA is localized to the posterior end of the egg and, like *bicoid* mRNA, it is translated only after fertilization. It produces a concentration gradient of Nanos protein, in this case with the highest level at the posterior end of the embryo. But, unlike Bicoid, Nanos is not a transcription factor and does not act directly as a morphogen to specify the abdominal pattern; it has a quite different action. Its function is to suppress translation of the maternal mRNA of another gene, *hunchback*, in the posterior region of the embryo.

Maternal *hunchback* mRNA is distributed throughout the embryo and starts to be translated after fertilization. However, *hunchback* is also one of the zygotic genes activated by the Bicoid protein as part of its patterning activity in the anterior half of the embryo. Restriction of Hunchback protein to the anterior region is crucial for correct patterning along the antero-posterior axis. To avoid interference by maternal Hunchback protein in the posterior region, and the production of anterior structures there, its translation is prevented in the posterior. This is the sole task of the Nanos protein (Fig. 2.13): it prevents *hunchback* mRNA translation by binding to a complex of the mRNA and Pumilio protein, which is encoded by another of the posterior-group genes. If one uses a maternal-effect mutation in *hunchback*, one can eliminate all maternal Hunchback from an embryo, and then it can be shown that there is no need for Nanos. In this situation, even if *nanos* is mutated and non-functional, the embryo will develop normally right through to an adult fly.

Evolution can only work on what is already there: it cannot 'look' at the whole system and redesign it economically. So, if a gene causes a difficulty by being expressed in the 'wrong' place, the problem may be solved not by reorganizing the gene-expression pattern, but by bringing in a new function, such as *nanos*, to remove the unwanted protein.

Another maternal product crucial to establishing the posterior end of the axis is *caudal* mRNA. Again, this is only translated after fertilization. *caudal* mRNA itself is uniformly distributed throughout the egg, but after fertilization, a posterior-to-anterior gradient of Caudal protein is established by the specific inhibition of Caudal protein synthesis by Bicoid protein, which binds to a site in the 3′ untranslated region of the *caudal* mRNA. This function is quite separate from Bicoid's action as a transcription factor. Because the concentration of Bicoid is low at the posterior end of the embryo, Caudal protein concentration is highest there (see Fig. 2.8). Mutations in the *caudal* gene result in the abnormal development of abdominal segments.

Soon after fertilization, therefore, several gradients of maternal proteins have been established along the antero-posterior axis. Two gradients—Bicoid and Hunchback proteins—run in an anterior-to-posterior direction, whereas Caudal protein is graded posterior to anterior. We next look at the quite different mechanism that specifies the two termini of the embryo.

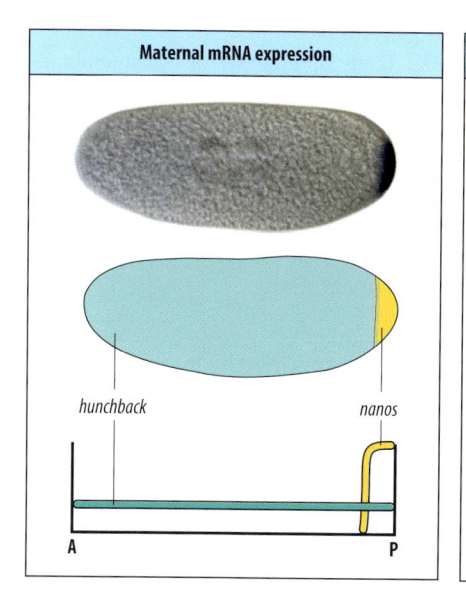

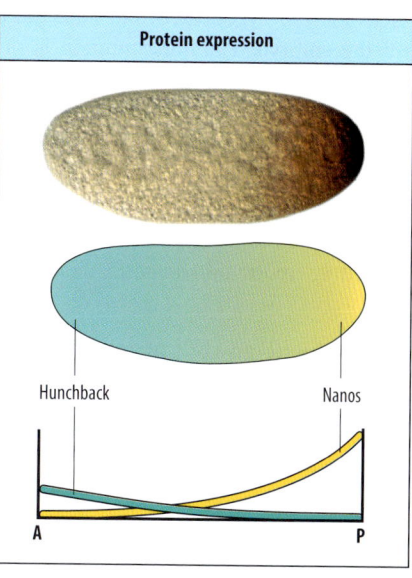

Fig. 2.13 Establishment of a maternal gradient in Hunchback protein. Left panel: in the unfertilized egg, maternal *hunchback* mRNA (turquoise) is present at a relatively low level throughout the egg, whereas *nanos* mRNA (yellow) is located posteriorly. The photograph is of an *in situ* hybridization showing the location of *nanos* mRNA (black). Right panel: after fertilization, *nanos* mRNA is translated and Nanos protein blocks translation of *hunchback* mRNA in the posterior regions, giving rise to a shallow antero-posterior gradient in maternal Hunchback protein. The photograph shows the graded distribution of Nanos, detected with a labeled antibody.

Photographs courtesy of R. Lehmann, from Griffiths, et al.: An Introduction to Genetic Analysis. *6th edn. New York: W.H. Freeman & Co., 1996.*

2.10 The anterior and posterior extremities of the embryo are specified by activation of a cell-surface receptor

A third group of maternal genes specifies the structures at the extreme ends of the antero-posterior axis—the acron and the head region at the anterior end, and the telson and the most posterior abdominal segments at the posterior end. A key gene in this group is *torso*; mutations in *torso* can result in embryos developing neither acron nor telson (see Fig. 2.10). This indicates that the two terminal regions, despite their topographical separation, are not specified independently but use the same pathway.

The terminal regions are specified by an interesting mechanism that involves the localized activation of a receptor protein that is itself present throughout the plasma membrane of the fertilized egg. The activated receptor sends a signal to the adjacent cytoplasm that specifies it as terminal. The receptor is known as Torso, as mutations in the maternal *torso* gene produce embryos lacking terminal regions. After fertilization, the maternal *torso* mRNA is translated, and the Torso protein is uniformly distributed throughout the fertilized egg plasma membrane. Torso is, however, activated only at the ends of the fertilized egg because the protein ligand that stimulates it is only present there.

This ligand for Torso is thought to be a fragment of a protein known as Trunk: *trunk* mRNA is deposited in the egg by the nurse cells, and Trunk protein is secreted into the **perivitelline space** during oocyte development. The perivitelline space is the space between the oocyte plasma membrane and a protective membrane of extracellular matrix called the **vitelline membrane** or **vitelline envelope**, which surrounds the oocyte. The Trunk protein is thought to be present throughout the perivitelline space, but the Trunk fragment that acts as a ligand for Torso is generated only at the poles of the egg, because the processing activity that produces it from Trunk is present only in these two regions. The activity of a protein called Torso-like is crucial to the localization of the Trunk fragment. Torso-like is produced exclusively by follicle cells adjacent to the two poles of the egg and is present in the vitelline envelope at the poles of the fertilized egg. By the time development begins after fertilization, small quantities of the Trunk fragment have been produced at the poles and are present in the perivitelline space, where they can bind Torso. Because the Trunk fragment is only present in small quantities, most of it becomes bound to Torso at the poles, with little left to diffuse further away. In this way, a localized area of receptor activation is set up at each pole (Fig. 2.14).

Stimulation of Torso by its ligand produces a signal that is transmitted across the plasma membrane to the interior of the developing embryo. This signal directs the activation of zygotic genes in nuclei at both poles, thus defining the two extremities of the embryo. The Torso protein is one of a large superfamily of transmembrane receptors, known as receptor tyrosine kinases, whose cytoplasmic portions have tyrosine protein kinase activity. The kinase is activated when the extracellular part of the receptor binds its ligand, and the cytoplasmic part of the receptor transmits the signal onwards by phosphorylating cytoplasmic proteins.

This ingenious mechanism for setting up a localized area of receptor activation is not confined to determination of the terminal regions of the embryo, but is also used in setting up the dorso-ventral axis, which we consider next.

2.11 The dorso-ventral polarity of the embryo is specified by localization of maternal proteins in the egg vitelline envelope

The dorso-ventral axis is specified by a different set of maternal genes from those that specify the antero-posterior axis. The basic mechanism is very similar to that described in Section 2.10 for the ends of the embryo. The receptor involved in dorso-ventral axis organization is a maternal protein called Toll, which is present throughout the plasma membrane of the fertilized egg. The ventral end of the axis is determined

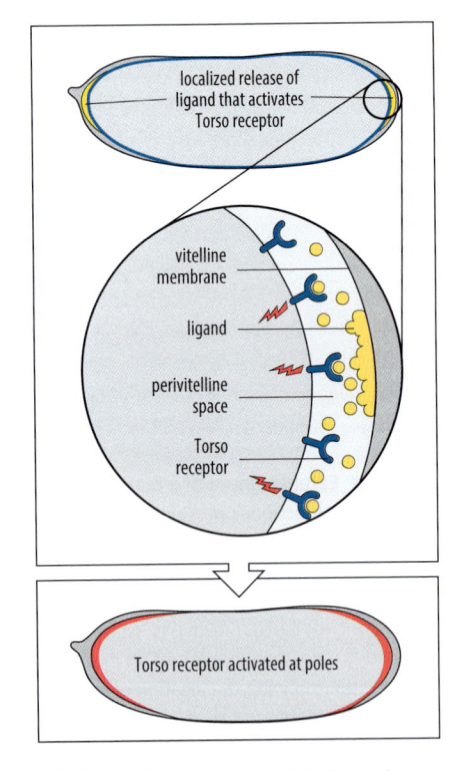

Fig. 2.14 The receptor protein Torso is involved in specifying the terminal regions of the embryo. The receptor protein encoded by the gene *torso* is present throughout the egg plasma membrane. Its ligand is laid down in the vitelline membrane at each end of the egg during oogenesis. After fertilization, the ligand is released and diffuses across the perivitelline space to activate the Torso protein at the ends of the embryo only.

by the localized production of the ligand for this receptor in the ventral vitelline envelope. The ligand for Toll is a protein fragment produced by the proteolytic processing of a maternal protein called Spätzle. After fertilization, Spätzle itself is uniformly distributed throughout the extra-embryonic perivitelline space. The localized processing of Spätzle is controlled by a small set of maternal genes that are expressed only in the follicle cells adjoining the future ventral region of the developing egg—about a third of the total surface area of the egg. The key gene here is *pipe*, which codes for an enzyme—a heparan sulfate sulfotransferase—that is secreted into the oocyte vitelline envelope by these cells. The activity of the Pipe enzyme triggers a protease cascade that leads to the localized processing of Spätzle, so that the active Spätzle fragment is only present in the ventral perivitelline space (see Box 2B).

Toll mRNA is laid down in the oocyte and is probably not translated until after fertilization. Although Toll is present throughout the membrane of the fertilized egg, it is only activated in the future ventral region of the embryo, owing to the localized production of its ligand, the Spätzle fragment. Toll activation is greatest where the concentration of its ligand is highest, and falls off rapidly on all sides, probably as a result of the limited amount of ligand being rapidly mopped up by Toll. Activation of Toll sends a signal to the adjacent cytoplasm of the embryo. At this stage, the embryo is still a syncytial blastoderm, and this signal causes a maternal gene product in the cytoplasm—the Dorsal protein—to enter nearby nuclei (Fig. 2.15). This protein is a transcription factor with a vital role in organizing the dorso-ventral axis.

2.12 Positional information along the dorso-ventral axis is provided by the Dorsal protein

The initial dorso-ventral organization of the embryo is established at right angles to the antero-posterior axis at about the same time that this axis is being divided into terminal, anterior, and posterior regions. The embryo initially becomes divided into four regions along the dorso-ventral axis (see Section 2.5 and Fig. 2.7), and this patterning is controlled by the distribution of the maternal protein Dorsal.

Unlike Bicoid, Dorsal protein is uniformly distributed in the egg. Initially it is restricted to the cytoplasm, but under the influence of signals from the ventrally activated Toll proteins, it enters nuclei in a graded fashion, with the highest concentration in ventral nuclei where Spätzle is active and the concentration progressively decreasing in a dorsal direction, as the Toll signal becomes weaker (see Fig. 2.15). Thus, there is little or no Dorsal in nuclei in the dorsal regions of the embryo.

The role of Toll was first established by the observation that mutant embryos lacking it are strongly **dorsalized**—that is, no ventral structures develop. In these embryos, Dorsal protein does not enter nuclei but remains uniformly distributed in the cytoplasm. Transfer of wild-type cytoplasm into *Toll*-mutant embryos results in specification of a new dorso-ventral axis, the ventral region always corresponding to the site of injection. This occurs because in the absence of Toll, the Spätzle fragments produced on the original ventral side diffuse throughout the perivitelline space because there is no Toll protein to bind them. When the wild-type cytoplasm is injected, the Toll proteins it contains enter the membrane at the site of the injection. Spätzle fragments then bind to these receptors, setting in motion the chain of events that defines the ventral region at the site of injection.

In the absence of a signal from Toll protein, Dorsal protein is prevented from entering nuclei by being bound in the cytoplasm by another maternal gene product, the Cactus protein. As a result of Toll activation, Cactus is degraded and no longer binds Dorsal, which is then free to enter the nuclei. The pathway leading from Toll to the activation of Dorsal is shown in Box 2B. In embryos lacking Cactus protein, almost all of the Dorsal protein is found in the nuclei; there is a very poor concentration gradient and the embryos are **ventralized**—that is, no dorsal structures develop.

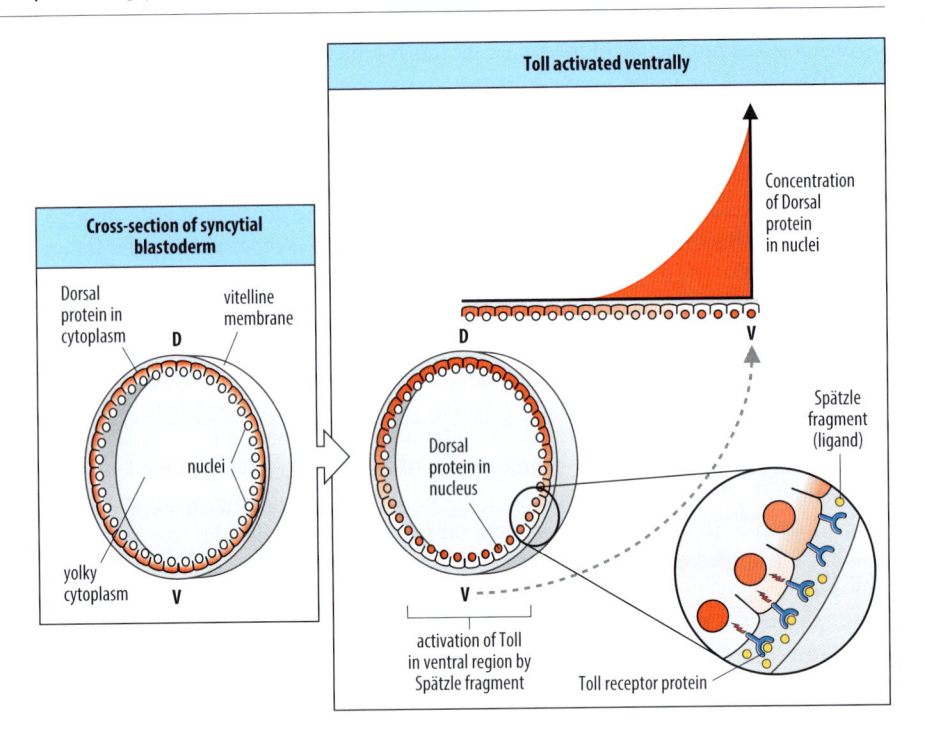

60 Chapter 2 Development of the *Drosophila* body plan

Fig. 2.15 Toll protein activation results in a gradient of intranuclear Dorsal protein along the dorso-ventral axis. Before Toll protein is activated, the Dorsal protein (red) is distributed throughout the peripheral band of cytoplasm. The Toll protein is a receptor that is only activated in the ventral (V) region, by a maternally derived ligand (the Spätzle fragment), which is processed in the perivitelline space after fertilization. The localized activation of Toll results in the entry of Dorsal protein into nearby nuclei. The intranuclear concentration of Dorsal protein is greatest in ventral nuclei, resulting in a ventral-to-dorsal (D) gradient.

CELL BIOLOGY BOX 2B The Toll signaling pathway: a multifunctional pathway

The interaction between the Dorsal and Cactus proteins in the Toll signaling pathway of *Drosophila* is of more than local interest: Dorsal protein is a transcription factor with homology to the Rel/nuclear factor kappa B (NFκB) family of vertebrate transcription factors, which are involved in regulation of gene expression in vertebrate immune responses, and the Toll signaling pathway is also used in the adult fly in defense against infection. The discovery of these innate immune defense pathways in the fly and the recognition of the importance of Toll-like receptors and the NFκB pathway in human immunity was recognized by the award of the Nobel Prize in Physiology or Medicine in 2011 to a researcher in *Drosophila* innate immunity alongside workers in human innate immunity. So what might seem at first sight a rather specialized mechanism for confining transcription factors to the cytoplasm until it is time for them to enter the nucleus during embryonic development is likely to be widely used for controlling gene expression and cell differentiation.

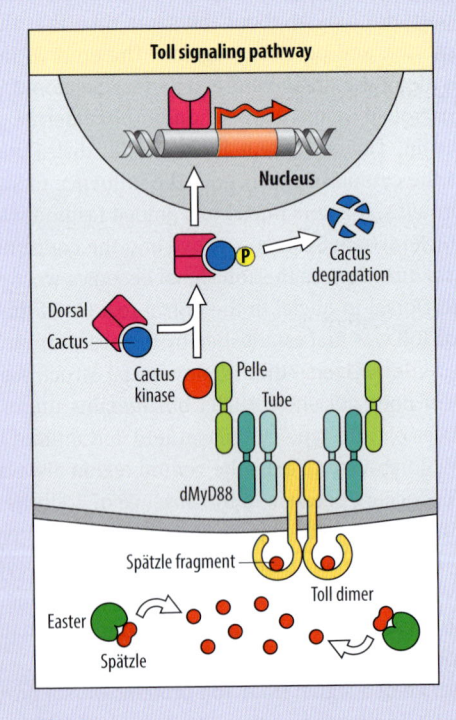

The Toll signaling pathway is a good example of a conserved intracellular signaling pathway that is used by multicellular organisms in different contexts, in this case ranging from embryonic development to defense against disease. All members of the Rel/NFκB family are typically held inactive in the cytoplasm until the cell is stimulated through an appropriate receptor. This leads to degradation of the inhibitory protein, which releases the transcription factor. This then enters the nucleus and activates gene transcription (Fig. 1). In the *Drosophila* embryonic Toll pathway, Dorsal is held inactive in the syncytium cytoplasm by the protein Cactus. The ligand for Toll is a fragment of the protein Spätzle, which is generated in the ventral vitelline space by the protease Easter, the last member in a protease cascade triggered by the activity of the heparan sulfate sulfotransferase enzyme Pipe, which is expressed only in the ventral follicle cells. When Toll is activated by binding the Spätzle fragment, its cytoplasmic domain binds a complex of the adaptor proteins dMyD88 and Tube. Tube interacts with, and activates, the protein kinase Pelle. Activation of Pelle leads to the phosphorylation and degradation of Cactus. This releases Dorsal, which is then free to enter a nucleus.

In adult *Drosophila*, the receptor protein Toll is stimulated by fungal and bacterial infection, and the signaling pathway results in the production of antimicrobial peptides. Toll is one of a family of nine Toll receptors encoded in the *Drosophila* genome. Toll (Toll-1) acts both in embryonic development and adult immunity; Toll-2, Toll-6, and Toll-8 are involved in germ-band extension in the embryo (discussed in Section 5.11); Toll-5 has a role in antifungal immunity in the adult; and Toll-3 and Toll-4 are expressed in the testis, but removal of each has mild, if any, effect on male fertility or viability; Toll-6 and Toll-7 act as receptors for the peptide growth factor neurotrophin in the central nervous system. In humans, Toll-like receptors acting by essentially the same pathway are also involved in innate immunity to microbial infection. IRAK and IκB are the mammalian homologs of Pelle and Cactus, respectively, and have the same roles in this pathway. In vertebrates, NFκB is also activated in response to signaling through receptors other than Toll.

SUMMARY

Maternal genes act in the ovary of the mother fly to set up differences in the egg in the form of localized deposits of mRNAs and proteins. After fertilization, maternal mRNAs are translated and provide the embryonic nuclei with positional information in the form of protein gradients or localized protein. Along the antero-posterior axis there is an anterior-to-posterior gradient of maternal Bicoid protein, which controls patterning of the anterior region. For normal development it is essential that maternal Hunchback protein is absent from the posterior region and its suppression is the function of the posterior-to-anterior gradient of Nanos. The extremities of the embryo are specified by localized activation of the receptor protein Torso at the poles. The dorso-ventral axis is established by intranuclear localization of the Dorsal protein in a graded manner (ventral to dorsal), due to ventrally localized activation of the receptor protein Toll by a fragment of the protein Spätzle.

SUMMARY: Maternal gene action in the fertilized egg of *Drosophila*

Antero-posterior	Dorso-ventral
mRNAs: *bicoid* forms anterior to posterior gradient; *hunchback* and *caudal* uniform; *nanos* posterior	Spätzle protein activates Toll receptor on ventral side
⇩	⇩
Anterior to posterior gradient of Bicoid protein formed. *hunchback* mRNA translation suppressed in posterior region by Nanos. *caudal* mRNA translation repressed by Bicoid	Dorsal protein enters ventral nuclei, giving ventral to dorsal gradient
Termini: Torso receptor activated by Trunk at ends of egg	

Localization of maternal determinants during oogenesis

Having seen how localized maternal gene products in the egg set up the basic framework for development, we now look at how they come to be localized so precisely. When the *Drosophila* egg is released from the ovary it already has a well-defined organization: *bicoid* mRNA is located at the anterior end and *nanos* and *oskar* mRNAs at the opposite end. Torso-like protein is present in the vitelline envelope at both poles, and other maternal proteins are localized in the ventral vitelline envelope. Numerous other maternal mRNAs, such as those for *caudal*, *hunchback*, *Toll*, *torso*, *dorsal*, and *cactus*, are distributed uniformly. How do these maternal mRNAs and proteins get laid down in the egg during **oogenesis**—its period of development in the ovary—and how are they localized in the correct places?

The development of an egg in the *Drosophila* ovary is shown in Figure 2.16. A diploid germline stem cell in the **germarium** divides asymmetrically to produce another stem cell and a cell called the cystoblast, which undergoes a further four mitotic divisions to give 16 cells with cytoplasmic bridges between them; this group of cells is known as the **germline cyst**. One of these 16 cells will become the oocyte; the other 15 will develop into **nurse cells**, which produce large quantities of proteins and RNAs that are exported into the oocyte through the cytoplasmic bridges. Somatic ovarian cells make up a sheath of **follicle cells** around the nurse cells and oocyte to form the **egg chamber**. Successively produced egg chambers are joined to each other by 'stalks' derived from the follicle cells. This gives the string of egg chambers (the **ovariole**) a distinct architectural polarity, which is translated into oocyte polarity by signaling between the oocyte and its surrounding follicle cells. In the diagram in Figure 2.16, the future anterior end of each egg chamber is to the left.

The follicle cells have a key role in patterning the egg's axes. During oogenesis, they become subdivided into functionally different populations at various locations around the oocyte; these subpopulations express different genes and thus have differing effects on the parts of the oocyte adjacent to them. Follicle cells also secrete the materials of the vitelline envelope and egg case that surround the mature egg. The oocyte eventually fills the whole of the egg chamber, the stalk breaks down, and the mature egg is released from the ovary and laid. During most of the stages of development discussed here the oocyte is arrested in the first prophase stage of meiosis (see Fig. 8.8). Meiosis is completed after fertilization.

2.13 The antero-posterior axis of the *Drosophila* egg is specified by signals from the preceding egg chamber and by interactions of the oocyte with follicle cells

The antero-posterior axis is the first axis to be established in the oocyte. The first visible sign of symmetry breaking is the movement of the oocyte from a central position surrounded by nurse cells to the posterior end of the developing egg chamber, where it is in direct contact with follicle cells. This rearrangement occurs while the egg

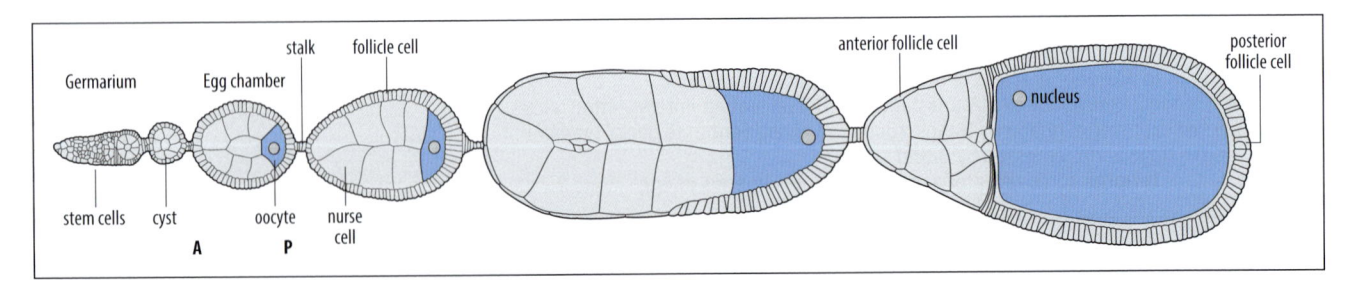

Fig. 2.16 Egg development in *Drosophila*. Oocyte development begins in a germarium, with stem cells at one end. One stem cell will divide four times to give 16 cells with cytoplasmic connections between each other, forming a structure called a cyst. One of the cells that is connected to four others will become the oocyte, the others will become nurse cells. The nurse cells and oocyte become surrounded by follicle cells and the resulting structure buds off from the germarium as an egg chamber. Successively produced egg chambers are still attached to each other at the poles. The oocyte grows as the nurse cells provide material through the cytoplasmic bridges. The resulting chain has an intrinsic polarity, which is reflected in that of the oocyte, with its anterior end being the side in contact with the nurse cells and closer to the germarium. The follicle cells have a key role in patterning the oocyte.

Localization of maternal determinants during oogenesis | **63**

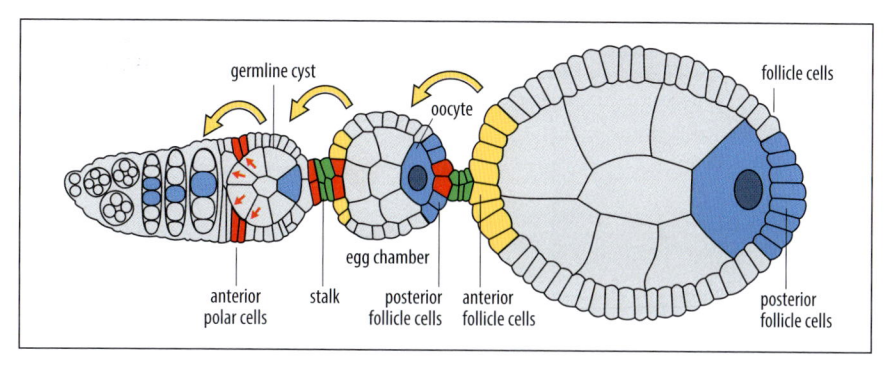

Fig. 2.17 Signals from older to younger egg chambers initially polarize the *Drosophila* oocyte. As a germline cyst buds from the germarium, it signals through the Delta–Notch pathway (small red arrows) to induce the formation of anterior polar cells (red). These, in turn, signal to the adjacent cells anterior to them and induce them to become stalk (green). Signals from the stalk induce adjacent follicle cells in the younger cyst to become posterior polar cells (red), and induce the younger cyst to round up and the oocyte to become positioned at the posterior of the egg chamber. Signals from anterior and posterior polar cells then specify adjacent follicle cells as posterior and anterior follicle cells, respectively (shown in more detail in Fig. 2.18). The yellow arrows indicate the overall direction of signaling from older to younger egg chambers.

chamber is becoming separated from the germarium, and is caused by preferential adhesion between the future posterior end of the oocyte and the adjacent posterior follicle cells. Oocyte antero-posterior polarity is the result of signaling from the anterior of the older egg chamber to the posterior of the younger chamber (Fig. 2.17).

The older germline cyst first signals to its anterior follicle cells through a widely used signaling pathway in animal development, the Delta–Notch pathway, which will be described in detail later in the book (see Box 4F). The germline cells produce the ligand Delta, which interacts with the receptor Notch on the follicle cells. This signaling event specifies several of the anterior follicle cells to become specialized **polar follicle cells**. The polar cells, in turn, secrete a signal molecule called Unpaired, which acts on receptors on adjacent follicle cells. On binding, it stimulates the JAK–STAT signaling pathway (Box 2C), which induces the follicle cells to form a stalk between the two egg chambers. The signals that induce stalk formation and the specification of anterior and posterior follicle cells are shown in more detail in Figure 2.18. The stalk cells upregulate production of the adhesion molecule cadherin, and adhesive interactions between the oocyte in the younger cyst and the stalk

Fig. 2.18 Stalk formation and specification of anterior and posterior follicle cells. First panel: in the older egg chamber, Delta (red arrows) signals from the germline cells to specify anterior polar cells (red). Second panel: these produce the signal molecule Unpaired, which signals adjacent pre-stalk cells to become stalk cells (green). Adhesive interactions anchor the oocyte to the posterior of the younger egg chamber and maternal mRNAs move to the posterior of the oocyte. Third panel: the most posterior follicle cells of the younger cyst are specified as polar cells and secrete Unpaired. mRNA in the oocyte encoding the protein Gurken (orange, see text) is translated and Gurken signals to the adjacent follicle cells. A second round of Delta signaling from the germline initiates differentiation of the follicle cells. Those exposed to both Unpaired and Gurken signals differentiate into posterior follicle cells; those exposed to Unpaired in the absence of Gurken differentiate into anterior follicle cells.

After Figure 4 from Roth, S. and Lynch, J.A.: **Symmetry breaking during** *Drosophila* **oogenesis.** *Cold Spring Harb. Perspect. Biol. 2009, 1: a001891.*

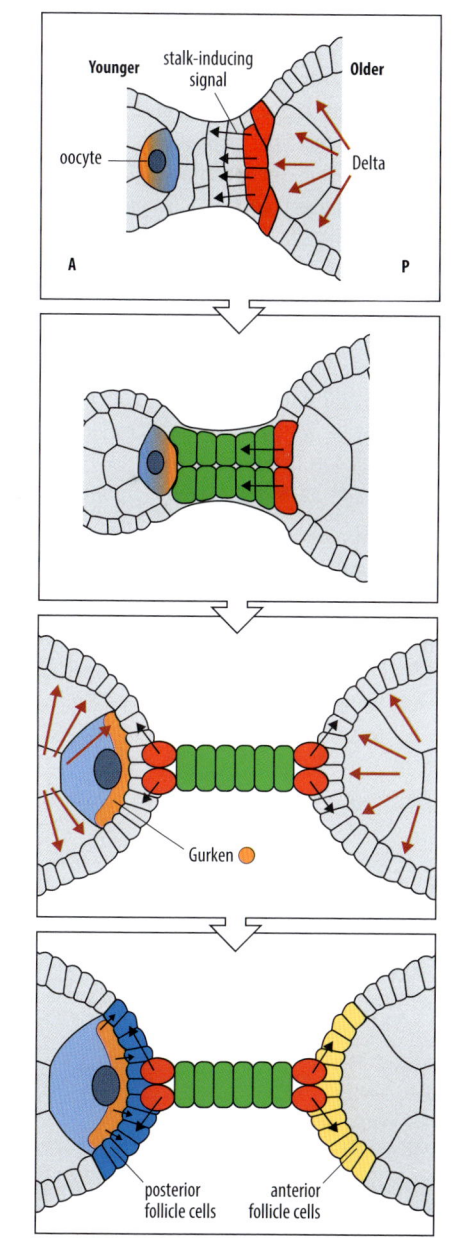

64 Chapter 2 Development of the *Drosophila* body plan

> **CELL BIOLOGY BOX 2C** The JAK-STAT signaling pathway
>
> This relatively simple signaling pathway was first identified in mammals as an element in the cellular response to cytokines, including the interferons and many other cytokines crucial to the development and function of the immune system. In *Drosophila*, the pathway was identified genetically in the context of a variety of functions, including segmentation of the embryo, growth and pattern formation of the imaginal discs, oogenesis (see Section 2.13), and in antimicrobial responses in adult flies. The central component of the JAK-STAT pathway is a cytoplasmic tyrosine protein kinase of the Janus kinase (JAK) family. This protein family takes its name from the mythical Roman two-headed god Janus, as JAKs have two kinase domains. JAKs are associated with the cytoplasmic tails of the cytokine receptors, and act as a bridge between the receptor and the transcriptional effectors of the pathway, the STAT proteins (signal transducers and activators of transcription; Fig. 1). *Drosophila* has a single receptor for the JAK-STAT pathway called Domeless (Dome), a single JAK called Hopscotch, and a single STAT. The ligands for Dome are the Unpaired proteins. Mammals have multiple receptors, ligands, JAKs, and STATs.
>
> Like the mammalian cytokine receptors, Dome consists of two transmembrane protein chains, each associated with a Hopscotch protein. When Unpaired binds to the receptor, the JAKs are activated, and phosphorylate each other and the receptor tails at selected tyrosine residues. This creates binding sites for STAT proteins, which in the absence of ligand binding shuttle between cytosol and nucleus. The phosphorylated STATs dimerize and translocate to the nucleus to effect transcription of a variety of target genes. In *Drosophila*, the pathway has a negative feedback in the form of the protein SOCS, whose gene is a target of STAT and which interferes with the activation between Hopscotch and STAT.
>
> **Fig. 1**

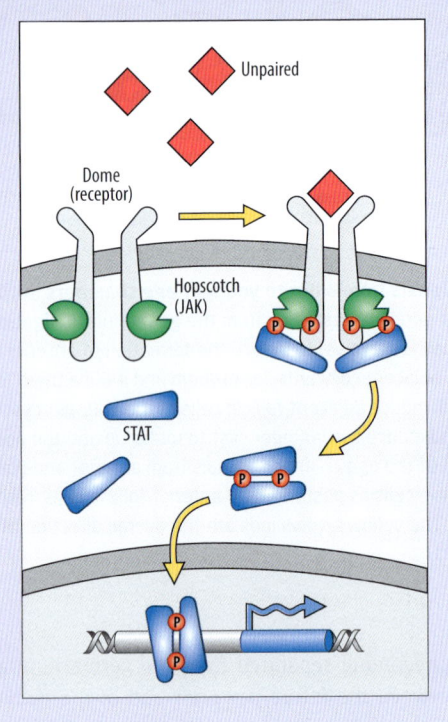

cells immediately adjacent to it position the oocyte firmly at the posterior end of its egg chamber. (The way in which cadherins and other adhesion molecules work is explained in Box 5A.) Thus, an antero-posterior polarity is propagated from one egg chamber to the next one to be formed.

The posterior localization of the oocyte is accompanied by a movement of maternal mRNAs from the future anterior end of the oocyte, where they have been deposited by nurse cells, to its future posterior end (see Fig. 2.18). This movement depends on a reorientation of the oocyte microtubule cytoskeleton, on which the maternal mRNAs are transported. At this stage of oocyte development, the nucleus is also positioned at the posterior end of the oocyte. Reorientation of the oocyte microtubule cytoskeleton depends on proteins known as PAR proteins, for 'partitioning defective'. Among other functions, PAR proteins are involved in determining the antero-posterior polarity of cells in many different situations in animal development. Once the oocyte has become located posteriorly in its egg chamber, the next steps in its antero-posterior polarization are mediated by a protein called Gurken, which is a member of the transforming growth factor (TGF)-α family. Examples of the most commonly used growth factor families in *Drosophila* are given in Figure 2.19; we shall meet many of these again in vertebrates and other animals. At this stage of oocyte development, maternal *gurken* mRNA is located posteriorly.

Cells at the anterior end of the stalk have now also become specified as polar cells and secrete the ligand Unpaired, which specifies about 200 adjacent follicle cells at the posterior end of the younger egg chamber as 'terminal cells' through the JAK–STAT pathway (see Fig. 2.18). At the same time, *gurken* mRNA is translated at the posterior end of the oocyte close to the nucleus. This produces a local posterior concentration of Gurken protein, which is secreted across the oocyte membrane. Gurken induces the adjacent terminal follicle cells to adopt a posterior fate by locally stimulating the receptor protein Torpedo, which is present on the surface of follicle cells. Torpedo is a receptor tyrosine kinase and is the *Drosophila* equivalent of the epidermal growth factor (EGF) receptor of mammals. This receptor signals by the same type of pathway as the fibroblast growth factor receptor (see Box 4A). In response to Gurken signaling through Torpedo, the posterior follicle cells produce an as-yet-unidentified signal that induces another reorientation of the oocyte's microtubule cytoskeleton, so that about midway through oogenesis, the minus ends of most microtubules are directed towards the anterior, and the plus ends towards the posterior of the oocyte.

2.14 Localization of maternal mRNAs to either end of the egg depends on the reorganization of the oocyte cytoskeleton

Microtubule reorientation is essential for the final localization of maternal mRNAs, such as *bicoid* and *oskar*, at either end of the egg. *bicoid* mRNA is originally made by nurse cells located next to the anterior end of the developing oocyte, and is transferred

Fig. 2.19 Common intercellular signals in *Drosophila*. These signaling pathways are universally used by all animals and their configurations were initially worked out largely through genetic screens in *Drosophila*.

Common intercellular signals used in *Drosophila*			
Signaling pathway and signal proteins (ligands)	**Receptors**	**Nuclear effector**	**Examples of roles in *Drosophila* development**
Hedgehog (see Box 2E)			
Hedgehog	Patched	Cubitus interruptus (Ci)	Patterning of insect segments (this chapter), positional signaling in imaginal discs (Chapter 10)
Wingless (Wnt) (see Box 4B)			
Wingless and five other Wnt proteins	Frizzled	β-catenin/TCF complex	Patterning of segments (this chapter), patterning of imaginal discs (Chapter 10), development of nervous and muscle systems
Delta/Notch (see Box 5D)			
Delta, Serrate	Notch	Nintra (cleaved intracellular domain of the Notch receptor)/Suppressor of hairless (Su(H))	Lateral inhibition in the development of the nervous (Chapter 11) and muscle systems, multiple roles in tissue patterning and cell-type specification
Transforming growth factor (TGF-α)/epidermal growth factor (EGF) receptor			
Gurken, Spitz, Vein	Torpedo (EGF receptor)	Pointed	Polarization of oocyte (this chapter), eye development, wing vein differentiation
Bone morphogenetic protein (BMP) and transforming growth factor-β (TGF-β) (Box 4C)			
Decapentaplegic, screw, glass bottom boat (BMP homologs)	Type I (e.g. Thickveins) and type II (e.g. Punt) subunits form heterodimeric receptor serine-threonine kinases	SMADs (Mad, Medea)	Patterning of the dorso-ventral axis (this chapter), patterning of the antero-posterior axis and growth of imaginal discs (Chapter 10)
Fibroblast growth factor (FGF) (Box 4E)			
Branchless, Pyramus, Thisbe	Breathless, Heartless	Not known if pathway activates transcription in *Drosophila*	Migration of tracheal cells (Chapter 7) and of mesoderm
JAK-STAT (Box 2C)			
Unpaired	Domeless (associates with cytoplasmic serine-threonine kinase, Hopscotch)	STAT	Polarization of oocyte (this chapter), segmentation, patterning of imaginal discs

Fig. 2.20 *bicoid* and *oskar* mRNAs are localized to the anterior and posterior ends of the oocyte, respectively. Maternal mRNAs delivered to the oocyte by the nurse cells are localized by transport along microtubules. Left panel: the motor protein dynein transports *bicoid* mRNA (red) towards the minus ends of microtubules. Right panel: *oskar* mRNA (green) is transported towards the plus ends of microtubules by the motor protein kinesin.

Illustration after St Johnston, D.: **Moving messages: the intracellular localization of mRNAs.** *Nat. Rev. Mol. Cell Biol. 2005, **6**: 363–375.*

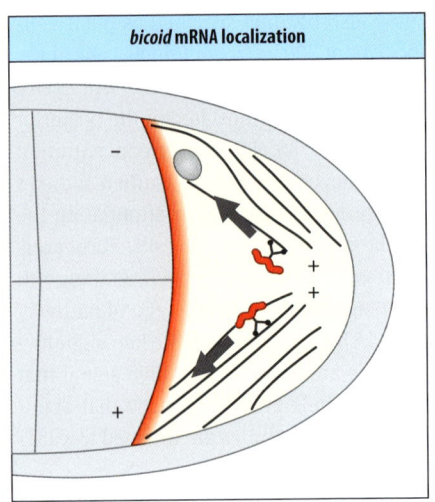

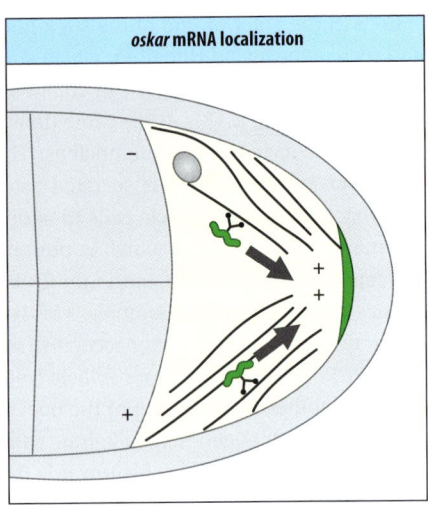

from them to the oocyte. It is subsequently transported to the posterior as a result of the microtubule reorganization that accompanies posterior positioning of the oocyte. After a second microtubule reorganization (described in Section 2.13), *bicoid* mRNA is transported along the reoriented microtubules, probably by the motor protein dynein, to its final location at the anterior-most end of the egg (Fig. 2.20, left panel). *oskar* mRNA is delivered into the oocyte by nurse cells and transported towards the posterior end of the oocyte by another motor protein, kinesin, which transports its cargo along microtubules in the opposite direction to dynein (Fig. 2.20, right panel). Both these localizations require the RNA-binding protein Staufen, which can itself be observed to localize to the posterior and anterior ends of the oocyte (Fig. 2.21).

Observations of *oskar* mRNA particles showed that they move in all directions but have a bias towards the posterior end that is sufficient to send the bulk to the posterior end. One role of *oskar* mRNA and protein is to nucleate the assembly of the germ-plasm at the posterior end of the egg. In the embryo this cytoplasm is incorporated into the pole cells (see Section 2.1) and directs them to form primordial germ cells.

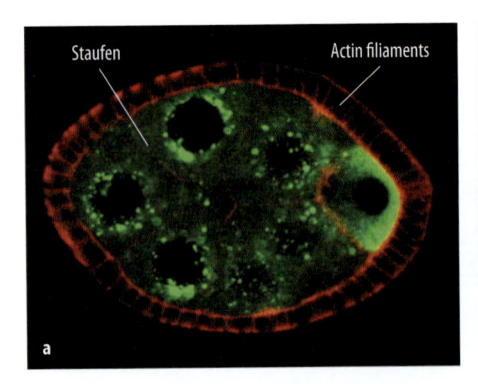

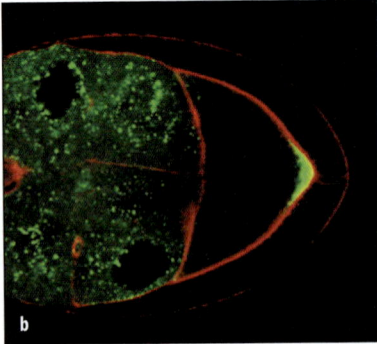

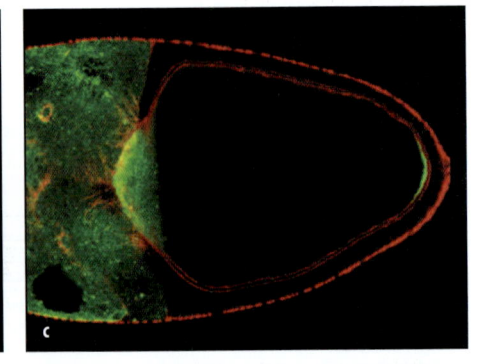

Fig. 2.21 The RNA-binding protein Staufen localizes to each pole of the *Drosophila* oocyte. The location of the Staufen protein at different stages of oocyte development is made visible by linking its gene to the coding sequence of green fluorescent protein and making flies transgenic for this construct. (a) In a stage-6 egg chamber, Staufen (green) accumulates in the oocyte. The fixed egg chambers were counterstained with rhodamine phalloidin, which labels actin filaments (red). (b) In a stage-9 egg chamber, Staufen localizes to the posterior of the oocyte. (c) In a stage-10b egg chamber, Staufen is localized to both the anterior and posterior poles of the oocyte.

From Martin, S.G., et al.: **The identification of novel genes required for Drosophila anteroposterior axis formation in a germline clone screen using GFP-Staufen.** *Development 2003, **130**: 4201–4215.*

Localization of maternal determinants during oogenesis 67

Fig. 2.22 **The dorso-ventral axis of the egg chamber and oocyte are specified by further interactions between the oocyte and follicle cells.** Signaling from the newly specified posterior (P) follicle cells results in the oocyte nucleus moving to a position on the anterior (A) margin of the oocyte. *gurken* mRNA coming into the oocyte from the nurse cells is now transported on microtubules to the region surrounding the nucleus. *gurken* mRNA is translated and the local release of Gurken protein from the oocyte specifies the adjacent follicle cells as dorsal (D) follicle cells and that side of the oocyte as the future dorsal side. The photograph shows *gurken* mRNA at its new location. V, ventral.
*Illustration after González-Reyes, A., et al.: **Polarization of both major body axes in** Drosophila **by gurken-torpedo signalling.** Nature 1995, 375: 654-658. Photograph courtesy of Prof. D. St Johnston, Gurdon Institute, Wellcome Trust.*

The posterior localization of *oskar* mRNA and the production of Oskar protein there is also required for the subsequent localization of *nanos* mRNA to the posterior end (see Section 2.9), as *nanos* RNA is recruited to wherever Oskar protein is made.

2.15 The dorso-ventral axis of the egg is specified by movement of the oocyte nucleus followed by signaling between oocyte and follicle cells

The setting-up of the egg's dorso-ventral axis involves a further set of oocyte–follicle cell interactions, which occur after the posterior end of the oocyte has been specified. In response to a signal from the posterior follicle cells, the oocyte nucleus is released from its location at the posterior of the oocyte and moves to a site on the anterior margin (Fig. 2.22). This movement results from a pushing force generated by the growth of astral microtubules—microtubules that anchor the cellular structures called centrosomes to the cortex. Gurken protein is translated at this new site from mRNA that has been transported on microtubules from its original posterior location.

The locally produced Gurken protein acts as a signal to the adjacent follicle cells, specifying them as dorsal follicle cells; the side away from the nucleus thus becomes the ventral region by default. Figure 2.23 shows the definition of the dorsal side in terms of specific gene expression in the dorsal anterior region only. The ventral follicle cells produce proteins, such as Pipe (see Section 2.11), that are deposited in the ventral vitelline envelope of the oocyte and are instrumental in establishing the ventral side of the axis.

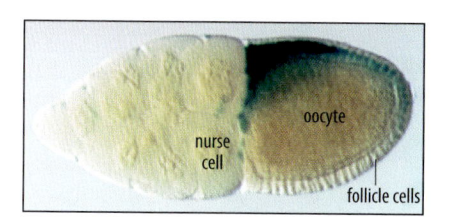

Fig. 2.23 *Drosophila* **oocyte development.** A developing *Drosophila* oocyte is shown attached to its 15 nurse cells and surrounded by a monolayer of 700 follicle cells. The oocyte and follicle layer are cooperating at this time to define the future dorso-ventral axis of the egg and embryo, as indicated by the expression of a gene only in the follicle cells overlying the dorsal anterior region of the oocyte (blue staining).
Photograph courtesy of A. Spradling.

SUMMARY

Drosophila oocytes develop inside individual egg chambers that are successively produced from a germarium, which contains germline stem cells that give rise to the oocyte and nurse cells, and stem cells that give rise to the somatic follicle cells that surround the oocyte. The nurse cells provide the oocyte with large amounts of mRNAs and proteins, some of which become localized at particular sites. As a result of signals from the adjacent older egg chamber, an oocyte becomes localized posteriorly in its egg chamber due to differential cell adhesion to posterior follicle cells. Subsequently, the oocyte sends a signal to these follicle cells, which respond with a signal that causes a reorganization of the oocyte cytoskeleton that localizes *bicoid* mRNA to the anterior end and other mRNAs to the posterior end of the oocyte. Together, these set up the beginnings of the embryonic antero-posterior axis. The dorso-ventral axis of the oocyte is also initiated by a local signal from the oocyte to follicle cells on the future dorsal side of the egg, thus specifying them as dorsal follicle cells. Thus, directly, or indirectly, follicle cells on the opposite side of the oocyte specify the ventral side of the oocyte by deposition of maternal proteins in the ventral vitelline envelope. Follicle cells at either end of the oocyte similarly specify the termini by localized deposition of maternal protein in the vitelline envelope.

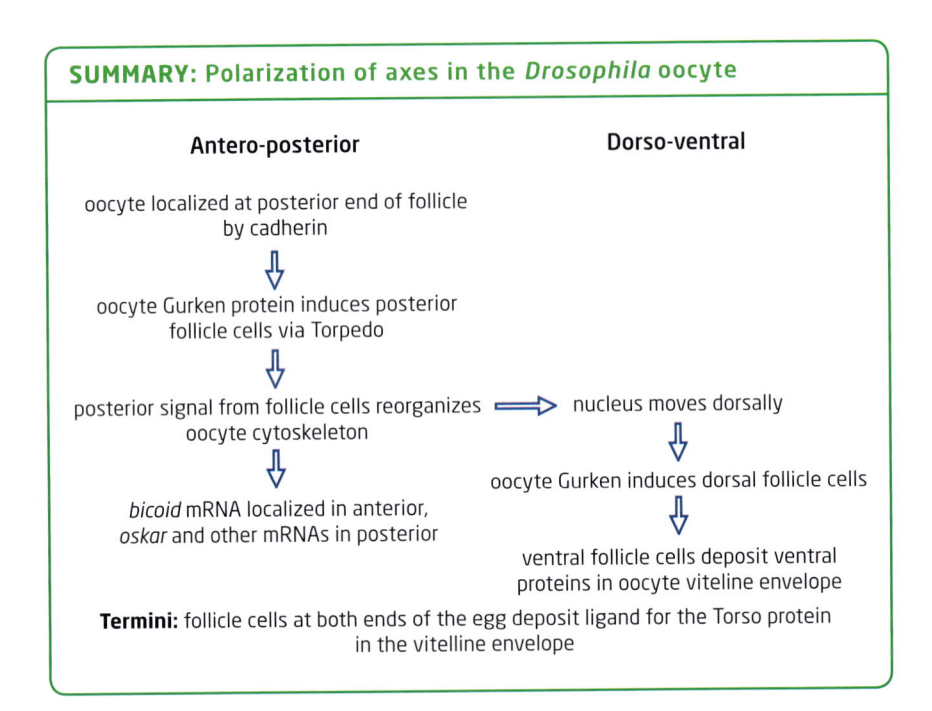

SUMMARY: Polarization of axes in the *Drosophila* oocyte

Antero-posterior

oocyte localized at posterior end of follicle by cadherin
⇩
oocyte Gurken protein induces posterior follicle cells via Torpedo
⇩
posterior signal from follicle cells reorganizes oocyte cytoskeleton ⟹
⇩
bicoid mRNA localized in anterior, *oskar* and other mRNAs in posterior

Dorso-ventral

nucleus moves dorsally
⇩
oocyte Gurken induces dorsal follicle cells
⇩
ventral follicle cells deposit ventral proteins in oocyte viteline envelope

Termini: follicle cells at both ends of the egg deposit ligand for the Torso protein in the vitelline envelope

Patterning the early embryo

We have seen how gradients of Bicoid, Hunchback, and Caudal proteins are established along the antero-posterior axis, and how intranuclear Dorsal protein is graded along the ventral-to-dorsal axis. This maternally derived framework of positional information is, in essence, the start of a transcriptional cascade that is interpreted and elaborated on through the regulatory regions of the zygotic genes, many of which encode transcription factors, to give each region of the embryo an identity. In this part of the chapter, we describe how the embryo is patterned along the antero-posterior and dorso-ventral axes. This patterning occurs simultaneously along both axes but involves different sets of transcription factors and signaling proteins. We first consider patterning along the dorso-ventral axis, which involves specifying the mesoderm and ectoderm, and the subdivision of the ectoderm to specify the neuroectoderm, which gives rise to the central nervous system.

2.16 The expression of zygotic genes along the dorso-ventral axis is controlled by Dorsal protein

Dorsal protein entering the nuclei of the syncytial blastoderm forms an intranuclear concentration gradient from ventral to dorsal (see Section 2.12). This gradient lasts for about 2 hours, during which the blastoderm becomes cellular. Dorsal is a transcription factor and its gradient divides the dorso-ventral axis into at least four well-defined regions by the expression of key target genes in response to different concentrations of Dorsal (Fig. 2.24). This is also the time at which the germ layers are defined, and Dorsal acts to specify the boundaries of these. Going from ventral to dorsal, the main regions are mesoderm, ventral ectoderm, dorsal ectoderm, and prospective amnioserosa. The mesoderm gives rise to internal soft tissues such as muscle and connective tissue; the ventral ectoderm becomes the neuroectoderm, which gives rise to the nervous system, as well as ventral epidermis; and the dorsal ectoderm gives rise to epidermis. The third germ layer, the endoderm, is located at either end of the embryo and we do not consider it here; it gives rise to the midgut (see Fig. 2.3).

Patterning along the axes poses a problem like that of patterning the French flag (see Section 1.15). It is estimated that 50 genes are direct targets of the Dorsal protein gradient, and it is remarkable that the graded expression of Dorsal results in a precise and reproducible pattern of expression of these genes. How this is achieved is now understood, and provides a good example of how a gradient of a transcription factor can directly impart positional information to a field of nuclei that are initially equivalent. Intranuclear Dorsal protein falls off rapidly in the dorsal half of the embryo and as a result, little Dorsal protein is found in nuclei above the equator. Dorsal can act as either an activator or a **repressor** of gene expression and, because of its actions, specific genes are expressed in particular regions of the dorso-ventral axis.

The Dorsal protein gradient is one of the best and simplest examples known of how the activity of a single transcription factor can generate different patterns of gene expression, and thus different developmental fates, in the cells in which it acts. Control sites in the target genes for Dorsal along the axis have differences in binding strength, or **affinity**, for the Dorsal protein, and also differ in whether the binding of Dorsal tends to activate or repress the gene, depending on its interactions with adjacent factors. In addition, the control regions of the Dorsal target genes have binding sites for other sets of transcription factors, which act in combination with Dorsal to regulate target-gene expression.

The most ventral region (a strip 12–14 cells wide) is defined by the highest intranuclear concentrations of Dorsal and its interactions with genes that have low-affinity binding sites for Dorsal in their regulatory regions. Such genes will only be expressed where Dorsal is at high concentration. Expression of the genes *snail* and *twist*, both of which encode transcription factors, characterizes this region (see Fig. 2.24). Genes expressed in the ventral region also have binding sites for the Twist protein, a transcriptional regulator that reinforces and maintains their expression.

The neuroectoderm is induced by intermediate levels of Dorsal through the recognition of high-affinity binding sites on target genes in combination with other transcription factors that interact with Dorsal to maintain it on the DNA. A boundary one cell wide between mesoderm and neuroectoderm is characterized by the expression of the gene *single-minded*, whose expression is controlled by a high-affinity site for Dorsal protein in conjunction with sites for other transcription factors. This boundary region is prospective **mesectoderm**, which will give rise to specialized glial cells of the nervous system. The neuroectoderm is defined by expression of the gene *rhomboid*, which encodes a membrane protein involved in regulating cell–cell signaling by the EGF receptor pathway. Because *single-minded* and *rhomboid* are activated by Dorsal, they could, in principle, be expressed in the most ventral domain, but their expression is prevented there by the products of the more ventral genes induced by Dorsal, such as *twist* and *snail*. For example, the regulatory regions of the *rhomboid* gene contain binding sites for both Dorsal and Snail proteins; in this gene, Dorsal activates, whereas Snail represses (Fig. 2.25). Thus, *rhomboid* can only be expressed in regions where there is no Snail but where there are sufficient levels of Dorsal, which restricts it to the ventrolateral domain that gives rise to the neuroectoderm.

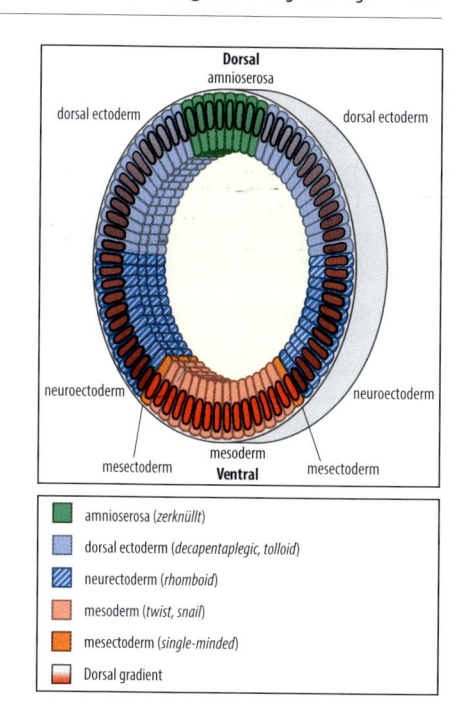

Fig. 2.24 The subdivision of the dorso-ventral axis into different regions by the interactions between Dorsal and its target genes. In the dorsal regions, where nuclear Dorsal protein is absent, the genes *tolloid*, *zerknullt*, and *decapentaplegic*, for which Dorsal acts as a repressor, can be expressed. In the most ventral region, the high concentration of Dorsal protein activates the genes *twist* and *snail*, which have low-affinity binding sites for Dorsal. The prospective neuroectoderm is characterized by expression of the gene *rhomboid*. The Twist protein helps maintain expression of both its own gene and *snail*, while the Snail protein helps to form the boundary between mesoderm and neuroectoderm by repressing *rhomboid* in the ventral region. The boundary is characterized by the expression of the gene *single-minded*, whose expression is controlled by a high-affinity site for Dorsal protein in conjunction with sites for other transcription factors.

In the more dorsal regions of the embryo, there is virtually no Dorsal protein in the nuclei. Genes that have control sites at which Dorsal protein acts as a repressor can therefore be expressed in these regions (see Fig. 2.25). Key genes expressed here are *decapentaplegic* (*dpp*), *tolloid*, and *zerknüllt* (*zen*). *zen* is expressed most dorsally and specifies the amnioserosa. *dpp* and *tolloid* are expressed throughout a broader dorsal region (see Fig. 2.24). Decapentaplegic protein is a secreted signaling protein with a key role in the specification of pattern in the dorsal part of the *Drosophila* dorso-ventral axis and is considered in detail in Section 2.17.

Mutations in the maternal genes that regulate the formation and readout of the Dorsal gradient can cause dorsalization or ventralization of the embryo (see Section 2.12). In dorsalized embryos, Dorsal protein is excluded uniformly from the nuclei. This has several effects, one of which is that the *dpp* gene is no longer repressed and is expressed everywhere. In contrast, *twist* and *snail* are not expressed at all in dorsalized embryos, as they need high intranuclear levels of Dorsal protein to be activated. The opposite result is obtained in mutant embryos in which the Dorsal protein is present at high concentration in all the nuclei. The embryos are ventralized, *twist* and *snail* are expressed throughout, and *dpp* is not expressed at all (Fig. 2.26).

An approximately twofold difference in the level of Dorsal determines whether an unspecified embryonic cell forms mesoderm or neuroectoderm. For simplicity, we have concentrated here on the broad division into mesoderm and neuroectoderm,

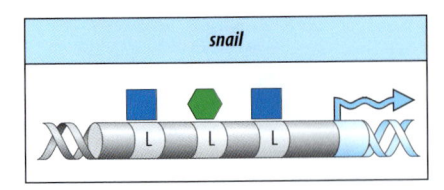

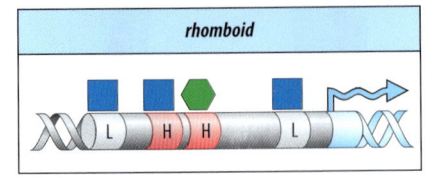

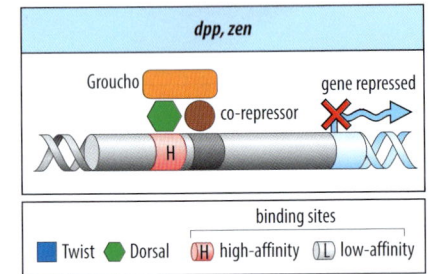

Fig. 2.25 Binding sites in Dorsal target genes. Dorsal protein can act as an activator or a repressor depending on its concentration and its partners on the DNA. High levels lead to the activation of genes such as *twist* and *snail*, whose products then reinforce their own expression by binding to the same DNA control regions as Dorsal. Twist and Snail also act as repressors of genes such as *rhomboid*, which means that these genes, whose expression also depends on Dorsal, are repressed where Twist and Snail are co-expressed. For some genes, such as *decapentaplegic* (*dpp*) and *zerknüllt* (*zen*), Dorsal acts as a repressor by interacting with co-repressors and recruiting repressors such as Groucho. These genes can only be expressed in the absence of Dorsal, and this is why their expression is restricted to the most dorsal region of the embryo.

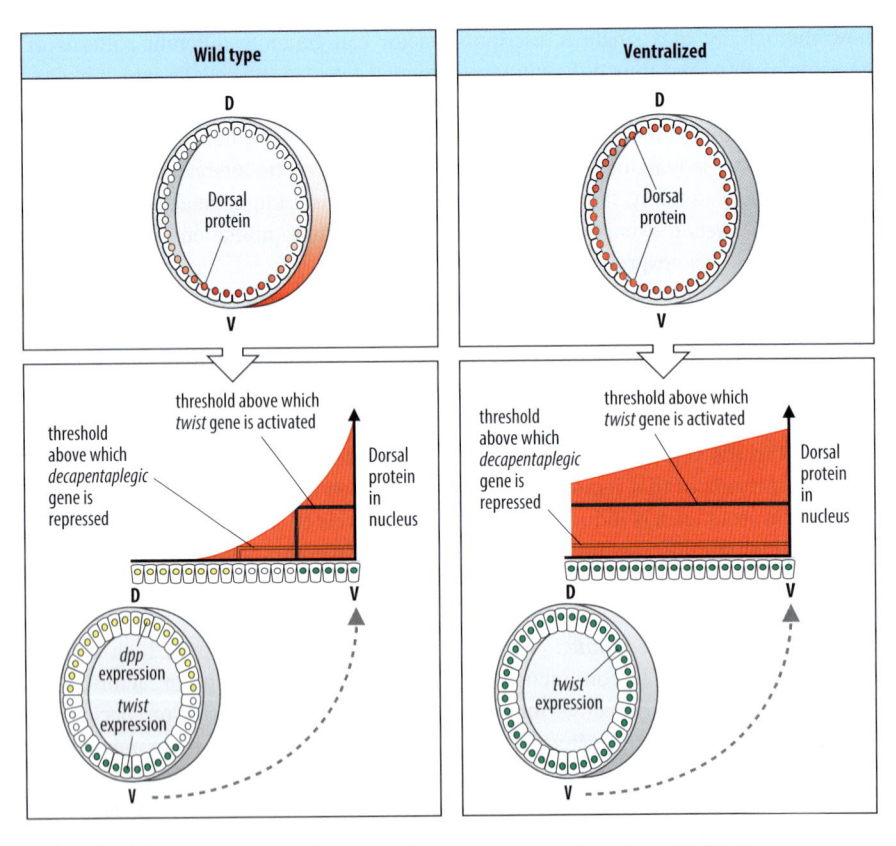

Fig. 2.26 Changes in expression of *twist* and *decapentaplegic* in ventralized embryos. Left panels: in normal embryos, the *twist* gene is activated above a certain threshold concentration (green line) of Dorsal protein, whereas above a lower threshold (yellow line), the *decapentaplegic* (*dpp*) gene is repressed. Right panels: in embryos ventralized due to mutations in maternal genes, the Dorsal protein is present in all nuclei; *twist* is now also expressed everywhere, whereas *dpp* is not expressed at all, because Dorsal protein is above the threshold level required to repress it everywhere.

but within this region, a number of different thresholds in Dorsal concentration pattern the future ventral midline and the neuroectoderm. The neuroectoderm, for example, is subdivided into three regions along the dorso-ventral axis, which will form three distinct columns of neurons in the future nerve cord (we shall return to this topic in Chapter 10). This subdivision is primarily the result of activation of genes for three different transcription factors at distinct thresholds of the Dorsal gradient.

The gradient of Dorsal protein is therefore effectively acting as a morphogen gradient along the dorso-ventral axis, activating specific genes at different threshold concentrations, and so defining a dorso-ventral pattern. The regulatory sequences in these genes can be thought of as developmental switches, which when thrown by the binding of transcription factors activate new genes and set cells off along new developmental pathways. The Dorsal protein gradient is one solution to the French flag problem. But it is not the whole story—yet another gradient is also involved.

2.17 The Decapentaplegic protein acts as a morphogen to pattern the dorsal region

The gradient of Dorsal protein, with its high point in the ventral-most region, specifies the initial pattern of zygotic gene activity along the dorso-ventral axis. It also further patterns the mesoderm and neuroectoderm by activating other target genes, such as those specifying different parts of the central nervous system, at different thresholds (see Section 2.16). The further patterning of the dorsal region is not simply specified by a low level of Dorsal, however, but by a gradient in the activity of another morphogen, the Decapentaplegic protein (Dpp). Thus, the combination of low levels of Dorsal and the activity of Dpp specifies the dorsal ectoderm and the most dorsal region, the amnioserosa. As we shall see in Chapter 3, Dpp is a homolog of bone morphogenetic protein (BMP)-4, a TGF-β growth factor in vertebrates that is also involved in patterning the dorso-ventral axis (the general signaling pathway for these proteins is shown in Box 3F).

Soon after the gradient of intranuclear Dorsal protein has become established, the embryo becomes cellular, and transcription factors can no longer diffuse between nuclei. Secreted or transmembrane proteins and their corresponding receptors must now be used to transmit developmental signals between cells. Dpp is one such secreted signaling protein (see Fig. 2.19). Dpp is also involved in many other developmental processes throughout *Drosophila* development, including the patterning of the wing imaginal discs (discussed in Chapter 9).

The *dpp* gene is expressed throughout the dorsal region where Dorsal protein is not present in the nuclei. Dpp protein then diffuses ventrally away from this source, forming a gradient of Dpp activity with its high point in the dorsal region. The cells have receptors that allow them to assess accurately how much Dpp is present and to respond by activating transcription of the appropriate genes, thus dividing up the dorsal region into different sub-regions characterized by different patterns of gene expression.

Evidence that a Dpp gradient specifies dorsal pattern comes from experiments in which *dpp* mRNA is introduced into an early wild-type embryo. As more mRNA is introduced and the concentration of Dpp protein increases above the normal level, the cells along the dorso-ventral axis adopt a more dorsal fate than they would normally. Ventral ectoderm becomes dorsal ectoderm and, at very high concentrations of *dpp* mRNA, all the ectoderm develops as amnioserosa. A most important target of Dpp signaling is the gene *zen*, which is repressed by Dorsal (see Fig. 2.25). Initially, *dpp* and *zen* are expressed in similar domains, but as the gradient of Dpp protein activity emerges, *zen* gene expression becomes restricted to the region with higher levels of Dpp signaling and defines the amnioserosa.

Dpp protein is initially produced uniformly throughout the dorsal region just as cellularization is beginning, but within less than an hour, its activity becomes restricted to a dorsal strip of some 5–7 cells and is much lower in the adjacent prospective dorsal ectoderm (Fig. 2.27). This sharp peak of Dpp concentration is not a matter of simple diffusion. Instead, it illustrates how a gradient in activity can be derived from a fairly uniform initial distribution of the protein by interactions with other proteins. In addition, Dpp activity is also thought to be modified by interactions with different forms of its receptors, further refining the activity gradient. It also interacts with collagen in the extracellular matrix, which restricts its movement, thereby restricting its signaling range. Dpp is a good example of how the formation of a signal gradient is usually more complicated than simple diffusion.

The sharp transition between the most dorsal ectoderm cells and those just lateral to them involves the proteins Short gastrulation (Sog), Twisted gastrulation (Tsg), and Tolloid. Sog and Tsg are BMP-related proteins that can bind Dpp and prevent it binding to its receptors, thus inhibiting its action. Tolloid is a metalloproteinase that degrades Sog when it is bound to Dpp, thus releasing Dpp.

The gene *sog* is expressed throughout the prospective neuroectoderm. Binding of Dpp by the Sog protein prevents Dpp activity from spreading into this region. Sog protein diffusing into the dorsal region is, in turn, degraded by Tolloid, which is expressed throughout the dorsal region; this sets up a gradient in Sog with a high point in the neuroectoderm and a low point at the dorsal midline and ensures that Sog, along with its bound Dpp, will be shuttled towards the dorsal region. Degradation of Sog or Tsg

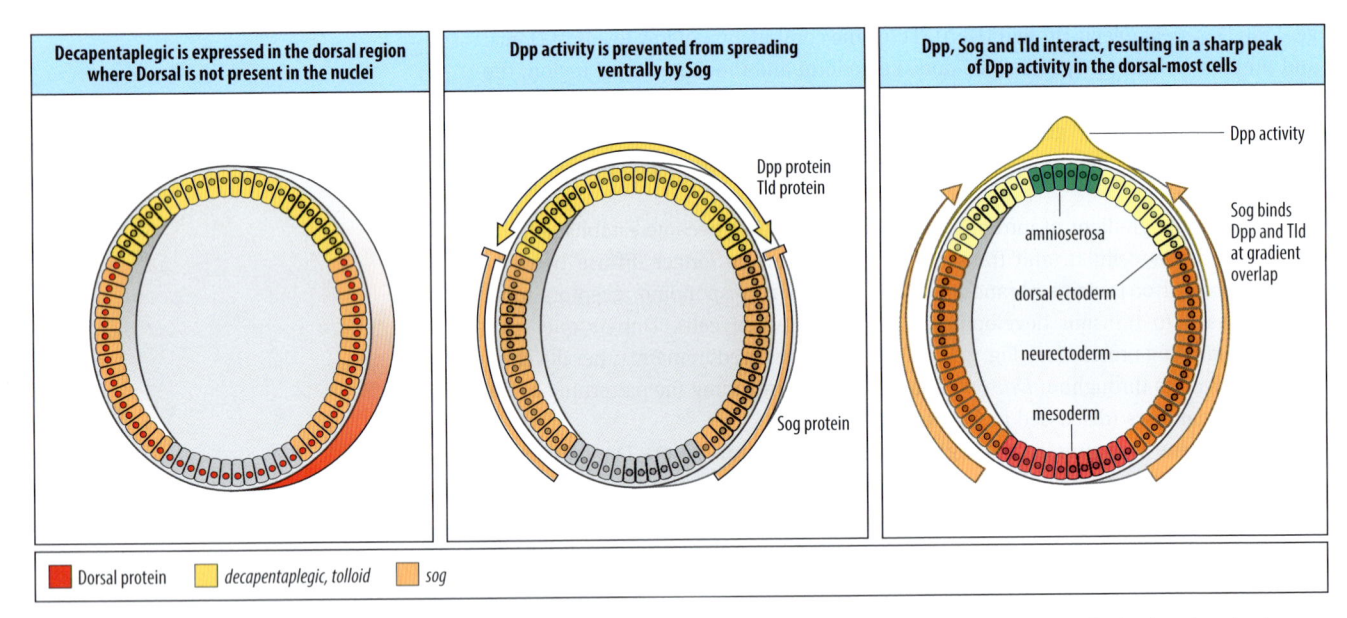

Fig. 2.27 Decapentaplegic protein (Dpp) activity is restricted to the dorsal-most region of the embryo by the antagonistic activity of the Short gastrulation protein (Sog). Left panel: the gradient of intranuclear Dorsal protein leads to the expression of the gene *short gastrulation* (*sog*) throughout the prospective neuroectoderm. *decapentaplegic* (*dpp*) and *tolloid* are expressed throughout the dorsal ectoderm. Center panel: *sog* encodes a secreted protein, Sog, that forms a ventral-to-dorsal gradient, while the secreted Dpp protein is initially present throughout the dorsal region. The protease Tolloid (Tld) is also expressed in the same region as Dpp. Where Sog meets Dpp, it binds it and thus prevents Dpp from interacting with its receptors and Dpp signaling from spreading ventrally into the neuroectoderm. Right panel: Dpp bound to Sog is also carried towards the dorsal region by the developing Sog gradient, thus helping to concentrate its activity in a sharp peak in the dorsal-most region. Tld is involved in creating the final sharp dorsal peak of Dpp activity. It binds to the Sog–Dpp complex and cleaves Sog. This releases Dpp, which can then bind to its receptors. The dorsal region therefore becomes subdivided by Sog activity into a region of high Dpp signaling, which will become the amnioserosa, and a zone of lower Dpp signaling, which is the dorsal ectoderm.

After Ashe, H.L., Levine, M.: **Local inhibition and long-range enhancement of Dpp signal transduction by Sog.** Nature 1999, **398**: 427-431.

by Tolloid releases free Dpp, which helps to generate a gradient of Dpp activity with its high point in the dorsal-most region (see Fig. 2.27).

Another factor proposed to sharpen the gradient is the rapid internalization and degradation of Dpp when it binds to its receptors. Dpp activity is also affected by its synergistic action with another member of the TGF-β family, Screw, which forms heterodimers with Dpp that signal more strongly than homodimers of Dpp or Screw alone; these are likely to be responsible for most of the Dpp activity. Experimental observations and mathematical modeling suggest that heterodimers are formed preferentially in the dorsal-most region, which helps to explain the high Dpp activity here. Homodimers of Dpp and of Screw are responsible for lower levels of signaling elsewhere in the prospective dorsal ectoderm.

The Dpp/Sog pairing has a counterpart in vertebrates, the BMP-4/Chordin partnership (Chordin is the vertebrate homolog of Sog), which is also involved in patterning the dorso-ventral axis. The nerve cord runs dorsally in vertebrates, however, not ventrally as in insects, and thus the dorso-ventral axis is reversed in vertebrates compared with insects. We shall see in Chapter 4 how this reversal in anatomical position is reflected in a reversal of the pattern of activity of BMP-4 and Chordin during early vertebrate development. We now return to the *Drosophila* antero-posterior axis.

2.18 The antero-posterior axis is divided up into broad regions by gap gene expression

Patterning along the antero-posterior axis starts, like the patterning of the dorso-ventral axis, when the *Drosophila* embryo is still acellular. The **gap genes** are the first zygotic genes to be expressed along the antero-posterior axis (see Fig. 2.9), and all code for transcription factors. Gap genes were initially recognized by their mutant phenotypes, in which quite large continuous sections of the body pattern along the antero-posterior axis are missing (see Fig. 1 in Box 2A). The mutant phenotype of a gap gene usually shows a gap in the antero-posterior pattern in the region in which the gene is normally expressed, but there are also more wide-ranging effects. This is because gap gene expression is also essential for later development along the axis.

Gap gene expression is initiated by the antero-posterior gradient of Bicoid protein while the embryo is still essentially a single multinucleate cell. Bicoid primarily activates anterior expression of the gap gene *hunchback*, which, in turn, is instrumental in switching on the expression of the other gap genes, among which are *giant*, *Krüppel*, and *knirps*, which are expressed in this sequence along the antero-posterior axis (Fig. 2.28) (*giant* is, in fact, expressed in two bands, one anterior and one posterior, but its posterior expression does not concern us here).

The overall mechanism of patterning the antero-posterior axis through transcriptional regulation by *bicoid* and the various gap genes is very similar to that we have just seen for Dorsal: transient gradients of transcription factors are formed; these transcription factors have different affinities for different target genes; and gap gene proteins act in a combinatorial manner with other transcription factors. Because the blastoderm is still acellular at this stage, the gap gene proteins can diffuse away from their site of synthesis. They are short-lived proteins with half-lives of minutes. Their distribution, therefore, extends only slightly beyond the region in which the gene is expressed, and this typically gives a bell-shaped profile of protein concentration. The overlaps at these boundaries are used to sharpen the domains of the different regions of the embryo through cross-regulatory interactions. Zygotic Hunchback protein is an exception, as it is expressed fairly uniformly over a broad anterior region with a steep drop in concentration at the posterior boundary of its expression. The control of zygotic *hunchback* expression by Bicoid protein is best understood and will be considered first.

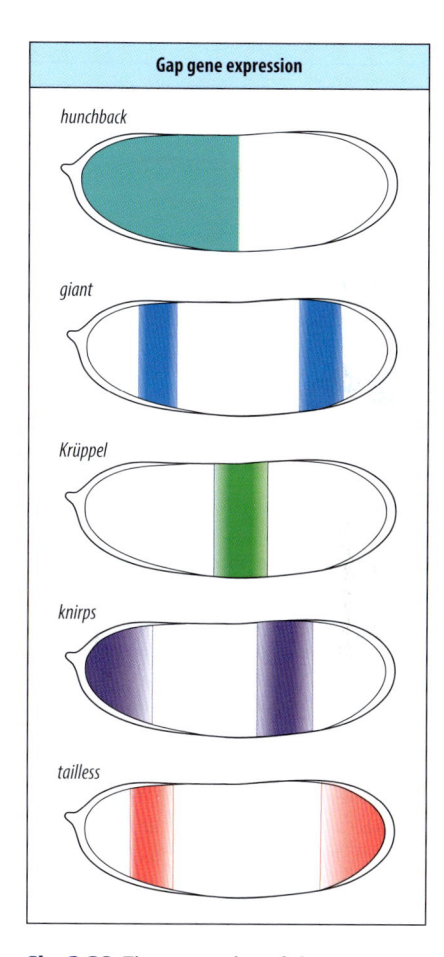

Fig. 2.28 The expression of the gap genes *hunchback*, *Krüppel*, *giant*, *knirps*, and *tailless* in the early *Drosophila* embryo. Gap gene expression at different points along the antero-posterior axis is controlled by the concentration of Bicoid and Hunchback proteins, together with interactions between the gap genes themselves. The expression pattern of the gap genes provides an aperiodic pattern of transcription factors along the antero-posterior axis, which delimits broad body regions.

2.19 Bicoid protein provides a positional signal for the anterior expression of zygotic *hunchback*

The Bicoid protein induces expression of the zygotic *hunchback* genes over most of the anterior half of the embryo. This zygotic expression is superimposed on low levels of *hunchback* mRNA, whose translation is suppressed posteriorly by Nanos (see Section 2.9).

The localized anterior expression of Hunchback is an interpretation of the positional information provided by the Bicoid protein gradient. The *hunchback* gene is switched on only when Bicoid, a transcription factor, is present above a certain threshold concentration. This level is attained only in the anterior half of the embryo, close to the site of Bicoid synthesis, which restricts *hunchback* expression to this region.

The relationship between Bicoid concentration and *hunchback* gene expression can be illustrated by looking at how *hunchback* expression changes when the Bicoid concentration gradient is changed by increasing the maternal dosage of the *bicoid* gene (Fig. 2.29). Expression of *hunchback* then extends more posteriorly, because the region in which the concentration of Bicoid is above the threshold for *hunchback* activation is also extended in this direction.

A smooth gradient in Bicoid protein is translated into a sharp boundary of *hunchback* expression about halfway along the embryo. Neighboring nuclei along the antero-posterior axis experience Bicoid concentrations that are very similar, differing by only about 10% from one cell to the next. This can be seen by measuring the gradient of Bicoid protein and the pattern of *hunchback* expression in many different embryos. It is then possible to see the very tight correlation between a value of Bicoid and the onset of *hunchback* expression (Fig. 2.30). How do nuclei distinguish such small differences and form a sharp boundary of gene expression against a background of biological 'noise' (i.e. fluctuations in the concentration of transcription factors at the region of the threshold) such as that resulting from small random variations in the dynamics of Bicoid behavior? This question has been addressed by using expression of a Bicoid–GFP construct (see Section 2.8) to directly observe and measure the distribution of Bicoid protein and its concentrations in nuclei at different points along the antero-posterior axis. Bicoid distribution, intranuclear concentrations at different positions, and the sharp boundary of *hunchback* expression all turn out to be highly reproducible over a large number of embryos, which suggests that background noise is, in fact, kept low by precise controls in the embryo. These might, for example, include communication between nuclei, but the exact mechanism for such precision remains unknown.

Bicoid activates the *hunchback* gene by binding to regulatory sites within the promoter region. Direct evidence of *hunchback* activation by Bicoid was originally obtained by experiments in which a fusion gene constructed from the *hunchback* promoter regions and a bacterial reporter gene, *lacZ*, was introduced into the fly genome using transgenic techniques such as P-element-mediated transformation. The large promoter region required for completely normal *bicoid* gene expression can be whittled down to an essential sequence of 263 base pairs that will still give almost normal activation of *hunchback*. This sequence has several sites at which Bicoid can bind, and it seems likely that the threshold response involves **cooperativity** between the different binding sites: that is, binding of Bicoid at one site makes binding of Bicoid at a nearby site easier, and so facilitates further binding.

Some other techniques for visualizing gene expression and forcing new patterns of gene expression in *Drosophila* are described in Box 2D. In addition, the CRISPR-Cas9 technology for precise gene editing has revolutionized our ability to tinker with gene expression in a wide variety of species, including *Drosophila*, in a minimally invasive manner (see Box 1C).

The regulatory region of a gene such as *hunchback* is yet another example of a developmental switch that directs nuclei along a new developmental pathway. Transcriptional switches such as this are a universal feature of embryonic development.

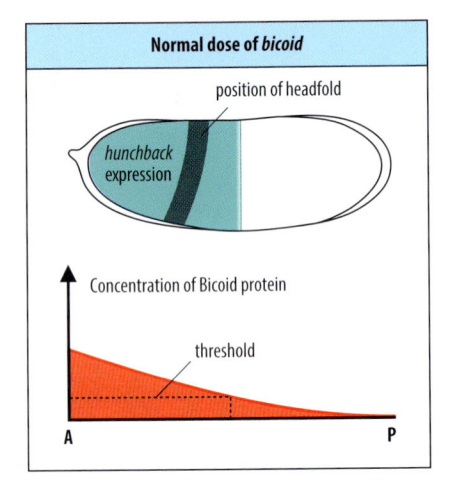

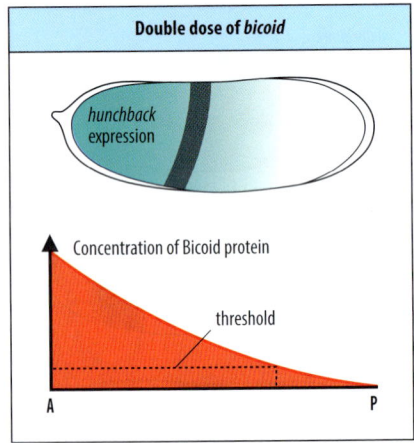

Fig. 2.29 Maternal Bicoid protein controls zygotic *hunchback* expression. If the dose of maternal *bicoid* is increased twofold, the extent of the Bicoid gradient also increases. The activity of the *hunchback* gene is determined by the threshold concentration of Bicoid, so at the higher dose its region of expression is extended towards the posterior end of the embryo because the region in which Bicoid concentration exceeds the threshold level also extends more posteriorly (see graph, bottom panel). The position of the head furrow, shown, is also shifted posteriorly at increasing doses of *bicoid*, showing the disruption to the body plan. A, anterior; P, posterior.

Patterning the early embryo   **75**

Fig. 2.30 Quantitative relation of Bicoid and Hunchback (Hb) expression. The photographs show the gradient of maternal Bicoid protein (top) in nuclei and the pattern of zygotic Hb protein (bottom) in nuclei in a late syncytial blastoderm embryo. To distinguish the overlapping patterns of expression, the embryo is stained for both proteins using specific antibodies that are visualized by secondary antibodies carrying different fluorescent tags. The fluorescence intensity is then recorded separately for each protein. These images have been false-colored. The graph plots the concentration of Bicoid against that of Hb along the length of the embryo. From the anterior end up to about 45% of embryo length, Bicoid concentration is at or above the threshold level required to activate the *hb* gene. Note the sharp cut-off of Hb expression as the Bicoid concentration in nuclei reduces to below the threshold level. (The *hb* gene has a second domain of expression at the most posterior region of the embryo that is independent of Bicoid.)
Original scans courtesy of J. Reinitz, M. Samsonova, and A. Pisarev. Reproduced with permission from Reinitz, J.: ***A ten per cent solution.*** *Nature 2007,* ***448****: 420-421.*

2.20 The gradient in Hunchback protein activates and represses other gap genes

The Hunchback protein is a transcription factor and acts as a morphogen to which the other gap genes respond. The other gap genes are expressed in transverse stripes across the antero-posterior axis (see Fig. 2.28). The stripes are delimited by mechanisms that depend on the control regions of these genes being sensitive to different concentrations of Hunchback protein and to other proteins, including Bicoid, which now contribute to the sharpening of the pattern. Expression of the *Krüppel* gene, for example, is activated by low levels of Hunchback, but is repressed at high concentrations. Within this concentration window *Krüppel* remains activated (Fig. 2.31, top panel). But below a lower threshold concentration of Hunchback, *Krüppel* is not activated. In this way, the gradient in Hunchback protein locates a band of *Krüppel* gene activity near the center of the embryo (Fig. 2.32). Refinement of this spatial localization is brought about by repression of *Krüppel* by other gap gene proteins. Such relationships were worked out by altering the concentration profile of Hunchback protein systematically, while all other known influences were eliminated or held constant. Increasing the dose of Hunchback protein, for example, results in a posterior shift in its concentration profile, and this results in a posterior shift in the posterior boundary of *Krüppel* expression. In another set of experiments using embryos lacking Bicoid protein (so that only the maternal Hunchback protein gradient is present), the level of Hunchback is so low that the *Krüppel* gene is even activated at the anterior end of the embryo (Fig. 2.31, bottom panel).

Hunchback protein is also involved in specifying the anterior borders of the bands of expression of the gap genes *knirps* and *giant*, again by a mechanism involving thresholds for repression and activation of these genes. At high concentrations of Hunchback, *knirps* is repressed, and this specifies its anterior margin of expression. The posterior margin of the *knirps* band is specified by a similar type of interaction with the product of another gap gene, *tailless*. Where the regions of expression of the gap genes overlap, there is extensive cross-inhibition between them, their proteins all being transcription factors. These interactions are essential to sharpen and stabilize the pattern of gap gene expression. For example, the anterior border of *Krüppel* expression lies 4-5 nuclei posterior to nuclei that express *giant*, and is set by low levels of Giant protein.

Fig. 2.31 *Krüppel* **gene activity is specified by Hunchback protein.** Top panel: above a threshold concentration of Hunchback protein, the *Krüppel* gene is repressed; at a lower concentration, above another threshold value, it is activated. Bottom panel: in mutants lacking the *bicoid* gene, and thus also lacking zygotic *hunchback* gene expression, only maternal Hunchback protein is present, which is located at the anterior end of the embryo at a relatively low level. In these mutants, *Krüppel* is activated at the anterior end of the embryo, giving an abnormal pattern. A, anterior; P, posterior.

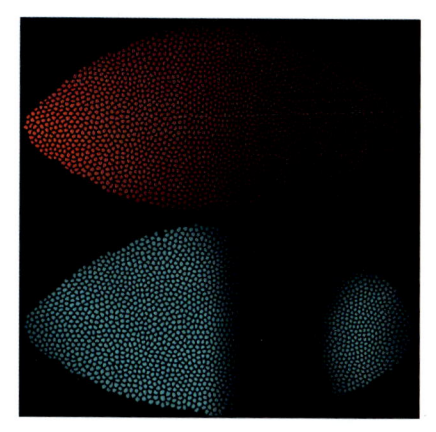

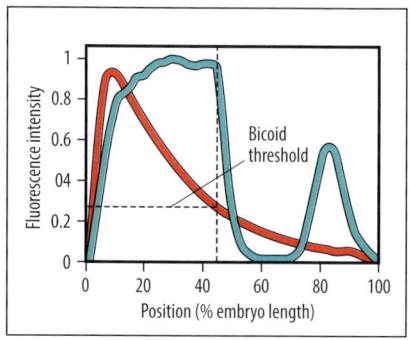

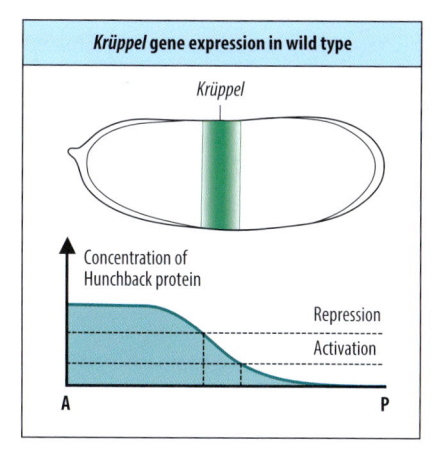

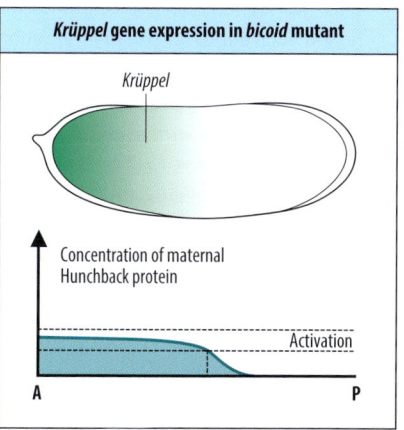

EXPERIMENTAL BOX 2D Targeted gene expression and misexpression screening

The ability to turn on the expression of a gene in a particular place and time during development is very useful for analyzing its role in development. This is called **targeted gene expression** and can be achieved in several different ways in *Drosophila*. One is to give the selected gene a heat-shock promoter using standard genetic techniques. This enables the gene to be switched on by a sudden rise in the temperature at which the embryos are being kept. By adjusting the temperature, the timing of expression of genes attached to this promoter can be controlled; the effects of expressing a gene at different stages of development can be studied in this way.

Another approach to targeted gene expression uses the transcription factor Gal4 from yeast in the system called Gal4–UAS. Gal4 can activate transcription of any gene whose promoter has a Gal4-binding site (the UAS or upstream activation sequence). In *Drosophila*, genes with Gal4-responsive promoters can be created by inserting the Gal4-binding site. To turn on the target gene, Gal4 itself has to be produced in the embryo. In one approach, Gal4 can be produced in a designated region or at a particular time in development by introducing a P-element transgene in which the Gal4-coding region is attached to a *Drosophila* regulatory region known to be activated in that situation (Fig. 1). Another approach is to cross a fly with an inserted Gal4 sequence linked to a known *Drosophila* gene or regulatory sequence with a fly carrying the UAS linked to a target gene.

A more versatile approach is based on the so-called **enhancer-trap** technique. The Gal4-coding sequence is attached to a vector that integrates randomly into the *Drosophila* genome. The Gal4 gene will come under the control of the promoter and enhancer region adjacent to its site of integration, and so Gal4 protein will be produced where or when that gene is normally expressed. To activate a target gene in a particular tissue, for example, flies that express Gal4 in that tissue are crossed with flies in which the target gene has Gal4-binding sites in its regulatory region. Many *Drosophila* lines with Gal4 expression in different tissues or cells, along with many lines with Gal4-binding sites in different genes' regulatory regions, are now available to researchers.

A novel pattern of gene expression can also be obtained using the Gal4–UAS system. For example, the pair-rule gene

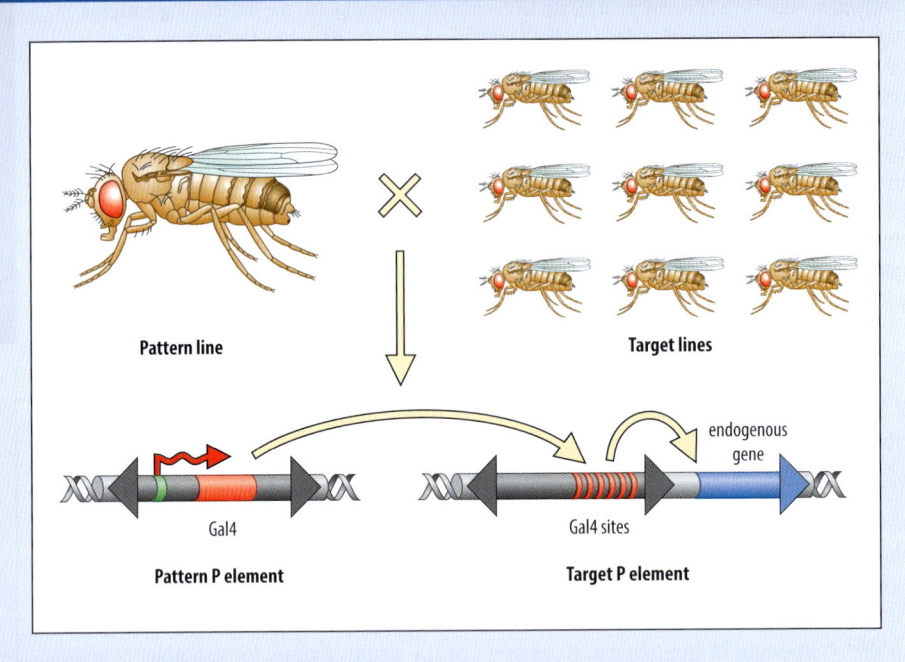

Fig. 1

even-skipped was expressed in even-numbered rather than odd-numbered parasegments, and this led to changes in the pattern of denticles on the cuticle. The Gal4–UAS system was invented in the 1980s and revolutionized fly genetics.

There are now many more similar bipartite expression systems used in *Drosophila* to control gene expression in a desired tissue or cell type, each with their particular advantages. For example, the Gal80 ts system (using a temperature-sensitive allele of the transcription factor Gal80) enables earlier stages of development to be bypassed and gene expression only in later stages to show up. This is useful as many genes are expressed at different stages in development.

The Gal4–UAS approach has also been adapted to large-scale screens that look systematically for genes whose overexpression or misexpression in a particular tissue causes a mutant phenotype. This is known as **misexpression screening**. In this case, flies expressing Gal4 in the tissue of interest are crossed with large numbers of flies carrying random insertions of the Gal4-binding site, and the progeny screened for a mutant phenotype. This approach is a useful complement to the more conventional genetic screens described in Box 2A, which generally detect loss-of-function mutations. If the target flies also carry a mutation in a known gene, misexpression screening can be used to identify genes whose overexpression enhances or suppresses the mutation. This approach can identify genes whose products interact directly or indirectly, or are part of the same pathway.

The antero-posterior axis thus becomes divided into a number of unique regions on the basis of the overlapping and graded distributions of different transcription factors. This method of delimiting regions can only work in an embryo such as the acellular syncytial blastoderm of *Drosophila*, where the transcription factors are able to diffuse freely throughout the embryo. But, as we shall see in other organisms, the general principle of first dividing the early embryo into broad regions, which are then further subdivided and patterned, is universal, even though the mechanisms for doing it differ.

In *Drosophila*, the regional distribution of the gap-gene products provides the starting point for the next stage in development along the antero-posterior axis—the activation of the pair-rule genes, cellularization, and the beginning of segmentation.

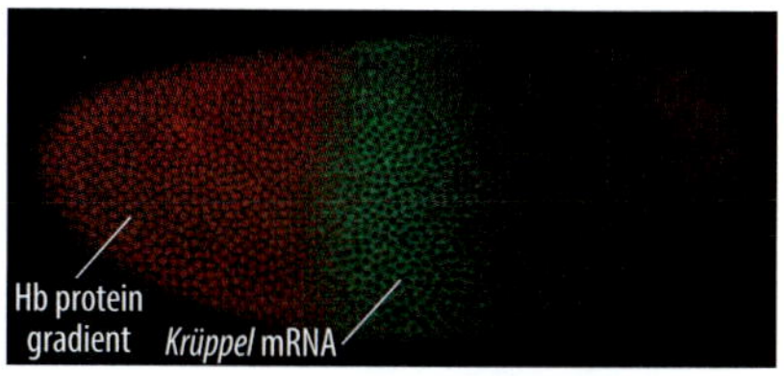

Fig. 2.32 Localization of *Krüppel* expression by the Hunchback protein gradient. An embryo stained with fluorescent antibodies to simultaneously detect the Hb protein gradient (red) and the expression of *Krüppel* mRNA (green) early in nuclear cycle 14.

From Yu, D., Small, S.: **Precise registration of gene expression boundaries by a repressive morphogen in *Drosophila*.** Curr. Biol. 2008, **18**: 868–876.

SUMMARY

Gradients of maternally derived transcription factors along the dorso-ventral and antero-posterior axes provide positional information that activates zygotic genes at specific locations along these axes. Along the antero-posterior axis the maternal gradient in Bicoid protein initiates the activation of zygotic gap genes to specify general body regions. Interactions between the gap genes and their products, all of which code for transcription factors, help to define their borders of expression. The dorso-ventral axis becomes divided into four regions: ventral mesoderm, ventral ectoderm (neuroectoderm), dorsal ectoderm (dorsal epidermis), and amnioserosa. A ventral-to-dorsal gradient of maternal Dorsal protein both specifies the ventral mesoderm and defines the dorsal region; a second gradient, of the Decapentaplegic protein, patterns the dorsal ectoderm. Patterning along the dorso-ventral and antero-posterior axes divides the embryo into discrete regions, each characterized by a unique pattern of zygotic gene activity.

SUMMARY: Early expression of zygotic genes

Antero-posterior

Bicoid protein gradient switches *hunchback* on at high concentration
⇩
Hunchback activates and represses gap genes such as *Krüppel, knirps, giant*
⇩
gap-gene products and gap genes interact to sharpen expression boundaries
⇩
axis is divided into unique domains containing different combinations of transcription factors

Dorso-ventral

ventro-dorsal gradient of intranuclear Dorsal protein forms
⇩
ventral activation of *twist, snail*, and repression of *decapentaplegic*
⇩
Decapentaplegic expressed dorsally
⇩
gradient of Decapentaplegic activity patterns dorsal region
⇩
dorso-ventral axis divided into prospective mesoderm, neuroectoderm, epidermis, and amnioserosa

Activation of the pair-rule genes and the establishment of parasegments

The most obvious feature of a *Drosophila* larva is the regular segmentation of the larval cuticle along the antero-posterior axis, each segment carrying cuticular structures that define it as, for example, thorax or abdomen. This segmental cuticle pattern reflects the patterning of the underlying epidermis, which produces the cuticle and its structures. More generally, it reflects the segmentally repeating organization of the larval body, in which elements such as the tracheal breathing system, the larval nervous system, and the muscles of the body wall are organized in a modular fashion. A similar organization of repeated units that are diversified in a regional manner is also characteristic of vertebrates and suggests that this is a central principle of animal development.

In *Drosophila*, each segment first acquires a unique identity in the embryo, and the pattern and identity of segments is carried over into the adult at metamorphosis. Adult appendages such as wings, halteres, and legs are attached to specific segments in the form of imaginal discs (see Fig. 2.6), but the discernible segments of the newly hatched larva are not in fact the first units of segmentation along this axis. The basic developmental modules, whose definition we will follow in some detail, are the parasegments, which are specified in the embryo and from which the segments derive.

2.21 Parasegments are delimited by expression of pair-rule genes in a periodic pattern

The first visible signs of segmentation are transient grooves that appear on the surface of the embryo after gastrulation. These grooves define the 14 parasegments. Once each parasegment is defined, it behaves as an independent developmental unit, and in this sense at least the embryo can be thought of as being modular in construction. The parasegments are initially like each other in their developmental potential, but each will soon acquire a unique identity. The parasegments are out of register with the definitive segments of the late embryo and hatched larva by about half a segment; this means that each segment is made up of the posterior region of one parasegment and the anterior region of the next. The relation between parasegments, larval segments, and segment identity in the adult fly is illustrated in Fig. 2.33. In the anterior head region, the segmental arrangement is lost when some of the anterior parasegments fuse.

The parasegments in the thorax and abdomen are delimited by the action of the **pair-rule genes**, each of which is expressed in a series of seven transverse stripes along the embryo, each stripe corresponding to every second parasegment. When pair-rule gene expression is visualized by staining for the pair-rule proteins, a striking zebra-striped embryo is revealed (Fig. 2.34).

The positions of the stripes of pair-rule gene expression are determined by the pattern of gap gene expression—a non-repeating pattern of gap gene activity is converted into repeating stripes of pair-rule gene expression. We now consider how this is achieved.

2.22 Gap gene activity positions stripes of pair-rule gene expression

Pair-rule genes are expressed in stripes, with a periodicity corresponding to alternate parasegments. Mutations in these genes thus affect alternate segments, and the pair-rule genes were identified originally through mutations that resulted in a discontinuous and periodic loss of tissue in the larva (see Box 2A). Some pair-rule genes

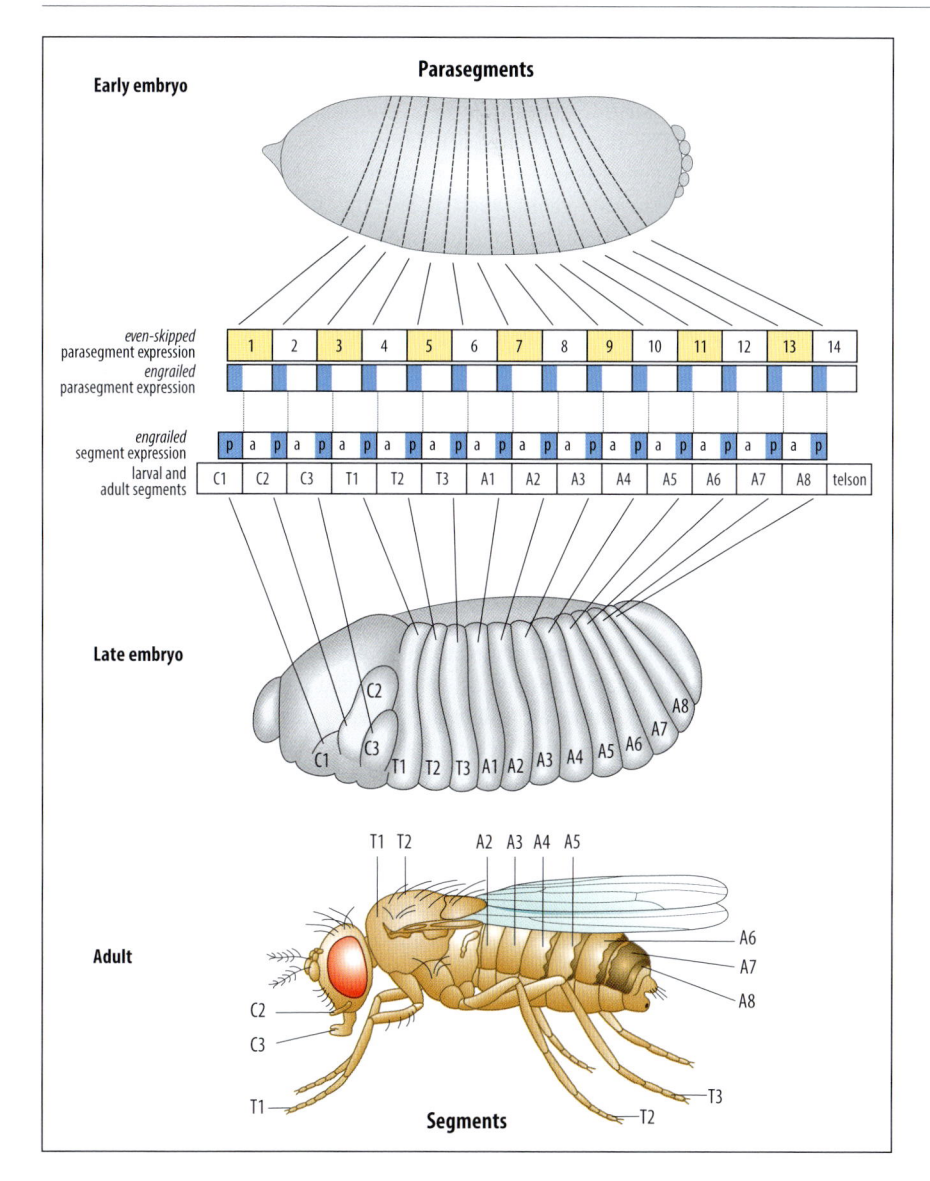

Activation of the pair-rule genes and the establishment of parasegments 79

Fig. 2.33 The relationship between parasegments and segments in the early embryo, late embryo, and adult fly. Pair-rule gene expression specifies the parasegments in the future thorax and abdomen, and some head parasegments in the embryo. *even-skipped* (yellow), for example, specifies the odd-numbered parasegments. The selector gene *engrailed* (blue), is expressed in the anterior region of every parasegment, and delimits the anterior margin of each parasegment. When segmentation occurs, the anterior region of a parasegment becomes the posterior portion of a segment. Each larval segment is therefore composed of the posterior region of one parasegment and the anterior region of the next, and segments are thus offset from the original parasegments by about half a segment. *engrailed* continues to be expressed in the posterior region of each segment throughout all larval stages and in the adult. In this figure, 'a' and 'p' refer to the anterior and posterior regions of the final segments. The segment specification is carried over into the adult and results in particular appendages, such as legs, wings, and mouthparts, developing on specific segments only. C1, mandibular segment; C2, maxillar segment; C3, labial segment. The adult mandibular segment lacks an appendage. T, thoracic segments; A, abdominal segments.

Illustration after Lawrence, P.: The Making of a Fly. Oxford: Blackwell Scientific Publications, 1992.

Fig. 2.34 The striped patterns of activity of pair-rule genes in the *Drosophila* embryo just before cellularization. Parasegments are delimited by pair-rule gene expression, each pair-rule gene being expressed in alternate parasegments. Expression of the pair-rule genes *even-skipped* (blue) and *fushi tarazu* (brown) is visualized by staining with antibody for their protein products. *even-skipped* is expressed in odd-numbered parasegments and *fushi tarazu* in even-numbered parasegments. Scale bar = 0.1 mm.

From Lawrence, P.: The Making of a Fly. Oxford: Blackwell Scientific Publications, 1992.

(e.g. *even-skipped* (*eve*)), define odd-numbered parasegments, whereas others (e.g. *fushi tarazu*) define even-numbered parasegments (see Fig. 2.34). The striped pattern of expression of pair-rule genes is present even before cells are formed, while the embryo is still a syncytium, although cellularization occurs soon after expression begins. Each pair-rule gene is expressed in seven stripes, each of which is only a few cells wide. For some pair-rule genes, such as *eve*, the anterior margin of the stripe corresponds to the anterior boundary of a parasegment; the domains of expression of other pair-rule genes, however, cross parasegment boundaries.

The striped expression pattern appears gradually: the stripes of the *eve* gene are initially fuzzy but eventually acquire a sharp anterior margin. At first sight this type of patterning would seem to require some underlying periodic process, such as the setting up of a wave-like concentration of a morphogen, with each stripe forming at the crest of a wave. It was surprising, therefore, to discover that each stripe is specified independently.

As an example of how the pair-rule stripes are generated, we will look in detail at the expression of the second *eve* stripe (Fig. 2.35). The appearance of this stripe depends on the expression of Bicoid protein and three gap genes—*hunchback*, *Krüppel*, and *giant* (only the anterior band of *giant* expression is involved in specifying the second *eve* stripe; see Fig. 2.28). Bicoid and Hunchback proteins are required to activate the *eve* gene, but they do not define the boundaries of the stripe. These are defined by Krüppel and Giant proteins, by a mechanism based on repression of *eve*. When concentrations of Krüppel and Giant are above certain threshold levels, *eve* is repressed, even if Bicoid and Hunchback are present. The anterior edge of the stripe is localized at the point of the threshold concentration of Giant protein, whereas the posterior border is similarly specified by the Krüppel protein.

By contrast, the positioning of the third and fourth *eve* stripes depends on regulatory regions that are repressed by the high concentration of Hunchback in the anterior region. The third stripe is expressed about halfway along the embryo, where Hunchback concentration starts to drop sharply, while the fourth *eve* stripe is expressed

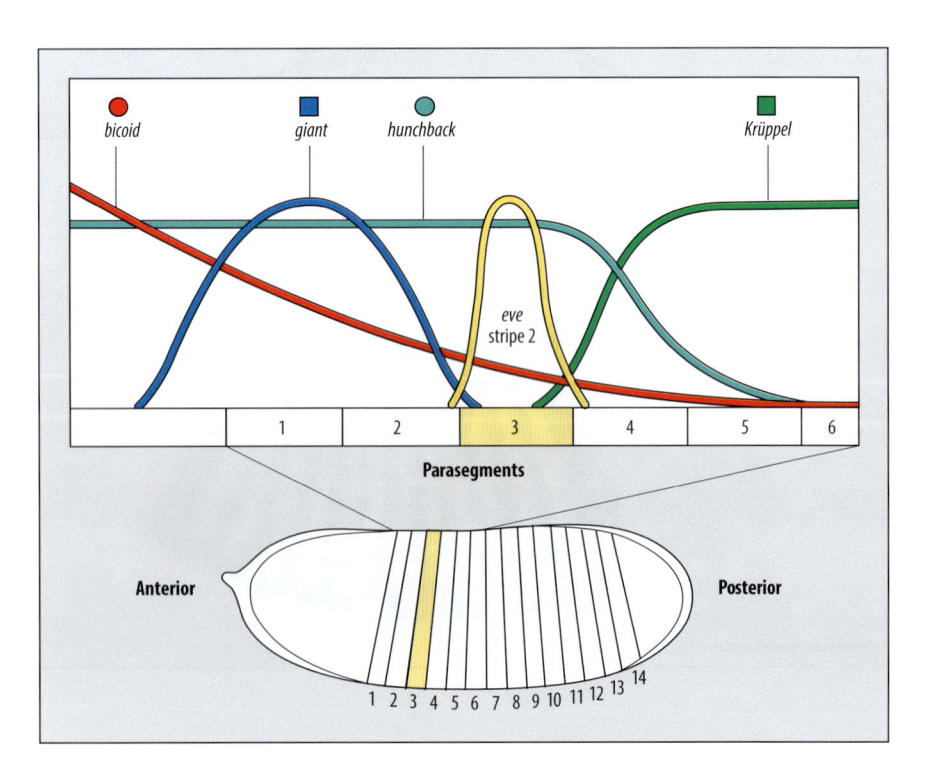

Fig. 2.35 The specification of the second *even-skipped* (*eve*) stripe by gap gene proteins. The different concentrations of transcription factors encoded by the gap genes *hunchback*, *giant*, and *Krüppel* localize *even-skipped*—expressed in a narrow stripe at a particular point along their gradients—in parasegment 3. Bicoid and Hunchback proteins activate *eve* in a broad domain, and the anterior and posterior borders are formed through repression of the gene by Giant and Krüppel proteins, respectively.

more posteriorly, where Hunchback is at an even lower level (Fig. 2.36). The posterior boundary of the third *eve* stripe is delimited by repression by the gap protein Knirps.

The independent localization of each of the stripes by the gap gene transcription factors requires that, in each stripe, the pair-rule gene responds to different concentrations and combinations of the gap gene transcription factors. The pair-rule genes thus require complex *cis*-regulatory control regions with multiple binding sites for each of the different factors. We have already seen how this can lead to the patterning of distinct regions in the case of the Dorsal protein and the dorso-ventral axis. Examination of the control regions of the *eve* gene reveals five separate modules, each of which controls the localization of one or two stripes. Control regions of around 500 base pairs have been isolated, each of which determines the expression of a single stripe (Fig. 2.37). The *eve* gene is an excellent example of modular *cis*-regulatory regions controlling the expression of a gene at different locations.

The presence of control regions that, when activated, lead to gene expression at a specific position in the embryo is a fundamental feature of developmental genes. Other examples of this type of control that we have seen in *Drosophila* are the localized expression of the gap genes, and the expression of genes in distinct regions along the dorso-ventral axis. Each regulatory region of such a gene contains binding sites for different transcription factors, some that activate the gene and some that repress it. In this way, the combinatorial activity of the gap gene proteins regulates pair-rule gene expression in each parasegment; the interactions of gap genes and pair-rule genes are an example of a **gene regulatory network** (see Box 1F).

Some pair-rule genes, such as *fushi tarazu*, may not be regulated by the gap genes directly, but may depend on the prior expression of so-called primary pair-rule genes, such as *eve* and *hairy*. With the initiation of pair-rule gene expression, the future segmentation of the embryo has been initiated. When the patterning along the dorso-ventral axis is also considered, the embryo is now divided into many unique regions, which each have a different developmental fate. These regions are characterized mainly by the combinations of transcription factors being expressed in each. These include proteins encoded by the gap genes, the pair-rule genes, and the genes expressed along the dorso-ventral axis.

The pair-rule transcription factors set up the spatial framework for the next round of patterning. Their combined activities will result in the expression of segmentation genes encoding transcription factors and signaling molecules on either side of the

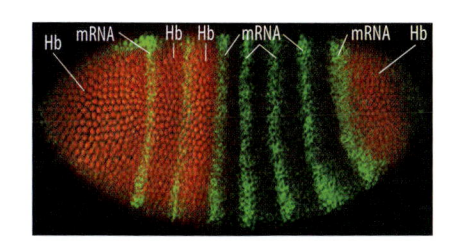

Fig. 2.36 Positioning of the third and fourth *eve* stripes by the Hunchback gradient. Embryo stained with fluorescent antibodies to simultaneously detect the Hb protein (red) and the mRNA of its target gene *eve* (green). The position of the third and fourth *eve* stripes is due directly to lack of repression of *eve* by Hunchback as its gradient drops off sharply.

*From Yu, D., Small, S.: **Precise registration of gene expression boundaries by a repressive morphogen in Drosophila**. Curr. Biol. 2008, **18**: 868–876.*

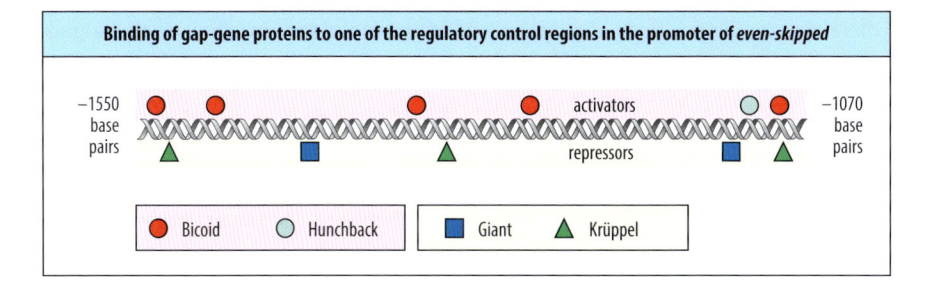

Fig. 2.37 Sites of action of activating and repressing transcription factors in the *eve* control region involved in formation of the second *eve* stripe. A control region of around 500 base pairs, located between 1070 and 1550 base pairs upstream of the transcription start site in the *eve* gene, directs formation of the second *eve* stripe. Gene expression can occur when the Bicoid and Hunchback transcription factors, acting as activators, are present above a given threshold concentration. Even in the presence of Bicoid and Hunchback, however, the gene is repressed when the Giant and Krüppel proteins, acting as repressors, are above threshold levels. They therefore set the boundaries to the stripe (see Fig. 2.35). The repressors may act by preventing binding of activators.

boundary of each parasegment. The products of these genes will act to delineate and pattern the parasegments, and thus the future segments. Also, together with the gap gene proteins, the pair-rule genes provide regional identities that will create functional specializations. Other strategies of specifying segmentation have evolved in insects and we shall compare some of these with *Drosophila* in Chapter 13.

> **SUMMARY**
>
> The activation of the pair-rule genes by the gap genes results in the transformation of the embryonic pattern along the antero-posterior axis from a variable regionalization to a periodic one. The pair-rule genes define 14 parasegments. Each parasegment is defined by narrow stripes of pair-rule gene activity. These stripes are uniquely defined by the local concentration of gap gene transcription factors acting on the regulatory regions of the pair-rule genes. Each pair-rule gene is expressed in alternate parasegments—some in odd-numbered, others in even-numbered ones. Most pair-rule genes code for transcription factors.

> **SUMMARY: Pair-rule genes and parasegments in *Drosophila***
>
> production of local combinations of gap-gene transcription factors
> ⇩
> activation of each pair-rule gene in seven transverse stripes along the antero-posterior axis
> ⇩
> pair-rule gene expression defines 14 parasegments, each pair-rule gene being expressed in alternate parasegments

Segmentation genes and segment patterning

The expression of the pair-rule genes defines the anterior and posterior boundaries of each parasegment but, like the gap genes, their activity is only temporary. How, therefore, are the positions of parasegment boundaries fixed, and how do the boundaries of the segments that are visible in the larval epidermis become established? This is the role of the **segmentation genes**. Segmentation genes are activated in response to pair-rule gene expression. During pair-rule gene expression the blastoderm becomes cellularized, so the segmentation genes are acting in a cellular rather than a syncytial environment. Unlike the gap genes and pair-rule genes, which act in the syncytial environment and all encode transcription factors, the segmentation genes are a diverse group of genes that encode signaling proteins, their receptors, and signaling pathway components, as well as transcription factors, and thus enable intercellular communication.

2.23 Expression of the *engrailed* gene defines the boundary of a parasegment, which is also a boundary of cell-lineage restriction

A key gene in establishing parasegment boundaries is the *engrailed* gene, which encodes a transcription factor. Unlike the gap genes and pair-rule genes, whose activity is transitory, *engrailed* is expressed throughout the life of the fly. *engrailed* activity first appears at the time of blastoderm cellularization as a series of 14 transverse stripes. It is initially expressed in lines of cells coincident with high levels of expression of the pair-rule genes *fushi tarazu* and *eve*, and delineates the anterior margin of each parasegment, which is only about four cells wide at this time (Fig. 2.38). Figure 2.39 shows *engrailed* expression in relation to the transient parasegment grooves at the later extended germ-band stage, when part of the ventral blastoderm (the germ band) has extended over the dorsal side of the embryo.

Evidence that the pair-rule genes control *engrailed* expression is provided, for example, by embryos carrying mutations in *fushi tarazu*, in which *engrailed* expression is absent only in even-numbered parasegments (in which *fushi tarazu* is normally expressed).

The anterior boundaries of the parasegments defined by *engrailed* are also boundaries of **cell-lineage restriction**. In other words, cells and their descendants in one parasegment never move into adjacent ones. Such domains of lineage restriction are known as **compartments**. A compartment can be defined as a distinct region that contains all the descendants of the cells present when the compartment is set up, and no others. It therefore has the potential to operate as a cellular unit, in which the cells express some key gene or genes that help to give the compartment an identity and distinguish it from other compartments. Compartments were originally defined in *Drosophila* epidermal structures, but they have since been discovered in the developing limbs and the brain in vertebrates (discussed in Chapters 9 and 10, respectively).

The existence of compartments in *Drosophila* can be detected by cell-lineage studies in which a single cell is marked in such a way that the mark is passed to all its descendants (which thus form a **clone** of marked cells) that can be followed into specific structures through many generations. One technique is to inject the egg with a harmless fluorescent compound that is incorporated into all the cells of the embryo. In the early embryo, a fine beam of ultraviolet light directed onto a single cell activates the fluorescent compound. All the descendants of that cell will be fluorescent and can therefore be identified. Examination of these clones, together with an analysis of *engrailed* expression, shows that cells at the anterior margin of the parasegment have no descendants that lie on the other side of that margin. The anterior margin is therefore a boundary of lineage restriction; that is, cells and their descendants that are on one or other side of the boundary when it is formed cannot cross it during subsequent development (Fig. 2.40).

engrailed expression marks the anterior boundary of the compartment. Unlike the pair-rule and gap genes, whose activity is transitory, *engrailed* needs to be expressed continuously throughout the life of the fly to maintain the character of the posterior compartment of the segment. As well as conferring a posterior segment identity on the cells, *engrailed* expression is required to change their surface properties so that they cannot mix with the cells adjacent to them. Furthermore, unlike other genes

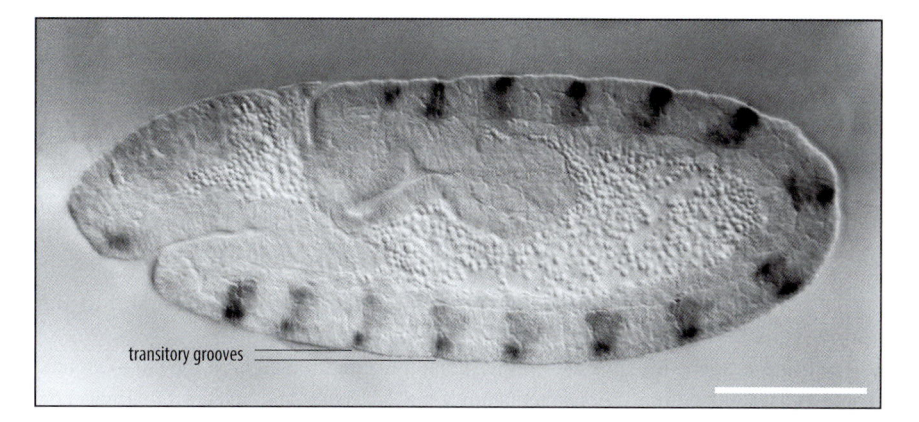

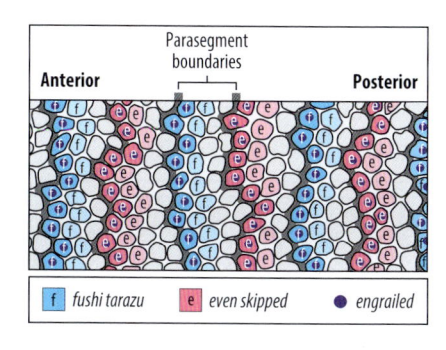

Fig. 2.38 The expression of the pair-rule genes *fushi tarazu* (blue), *even-skipped* (pink), and *engrailed* (purple dots) in parasegments. *engrailed* is expressed at the anterior margin of each stripe and delimits the anterior border of each parasegment. The boundaries of the parasegments become sharper and straighter later on.

After Lawrence, P.: The Making of a Fly. Oxford: Blackwell Scientific Publications, 1992.

Fig. 2.39 The expression of the *engrailed* gene in a late (stage 11) *Drosophila* embryo. The gene is expressed in the anterior region of each parasegment and the transient grooves between parasegments can be seen. At this stage of development, the germ band has temporarily extended and is curved over the back of the embryo. Scale bar = 0.1 mm.

From Lawrence, P.: The Making of a Fly. Oxford: Blackwell Scientific Publications, 1992.

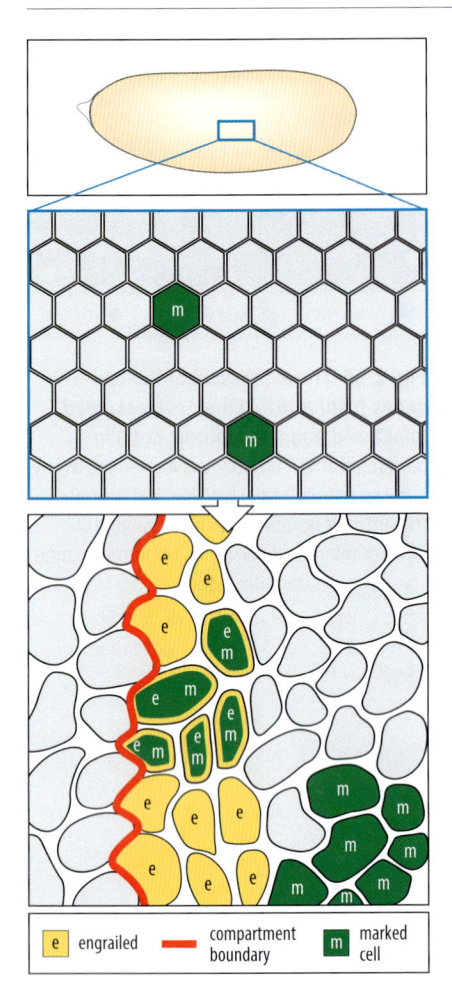

Fig. 2.40 Demonstration that the parasegment boundary is a boundary of lineage restriction. Individual cells are marked (green) in the blastoderm, when all cells are the same and the embryo is unpatterned. The cells divide as the embryo grows and give rise to small clones. When the clones are mapped with respect to the expression of *engrailed* (yellow), it is possible to visualize the position of the clones with respect to the anterior boundary of *engrailed* expression, corresponding to the parasegment boundary. It was found that no clones crossed this boundary, indicating that the boundary is a boundary of lineage restriction and delimits a compartment.

After Vincent, J.-P., O'Farrell, P.H.: **The state of engrailed expression is not clonally transmitted during early** *Drosophila* **development**. Cell 1992, **68**: 923-931.

involved in segmentation, it is expressed in the same pattern throughout development, and always in association with *hedgehog*. *engrailed* is thus an example of a **selector gene**—a gene whose activity is sufficient to cause cells to adopt a particular fate. Selector genes can control the development of a region such as a compartment or a segment, and, by controlling the activity of other genes, give the region a particular identity.

The compartment boundaries, as well as the expression of *engrailed* and *hedgehog*, are therefore carried over into the imaginal discs that derive from specific embryonic segments (see Fig. 2.6), and into the adult structures that derive from the discs, such as wings and legs. As discussed in Chapter 9, compartments can be more clearly visualized in these structures, where each compartment is composed of thousands of cells.

The definitive segments of the late embryo and larva are offset from the parasegments, with the anterior boundary of each segment developing immediately posterior to the *engrailed*-expressing cells. This means that each segment is divided into anterior and posterior compartments, with *engrailed* expression defining the posterior compartment (see Fig. 2.33).

2.24 Segmentation genes stabilize parasegment boundaries

Establishment of the parasegment boundary depends on an intercellular signaling circuit being set up between adjacent cells, which delimits the boundary between them, and which is discussed in detail in Section 2.26. As well as *engrailed*, this circuit primarily involves the segmentation genes *wingless* and *hedgehog*, which are also expressed in restricted domains within the parasegment in response to the pair-rule proteins (Fig. 2.41). In contrast to *engrailed*, which encodes a transcription factor, *wingless* and *hedgehog* encode secreted signal proteins, Wingless and Hedgehog, respectively. These act via receptor proteins on cell surfaces to activate intracellular signaling pathways that alter gene expression. *wingless* is named after the effect of its loss-of-function mutation on the adult fly; *hedgehog* was so named because the embryo is much shorter than normal and the ventral surface is completely covered in denticles, so that the embryo somewhat resembles a hedgehog. **Wingless** is one of the prototype members of the so-called **Wnt family** of signal proteins (see Fig. 2.19), which are key players in development in vertebrates, as well as invertebrates, and are involved in determining cell fate and cell differentiation in many different aspects of development. **Hedgehog**, similarly, has homologs in other animals, such as Sonic hedgehog in vertebrates, and is involved in similar processes.

The result of the initial activity of the pair-rule genes is the establishment, in the blastoderm, of one-cell-wide stripes of *engrailed* and *hedgehog* expression that will delimit the anterior boundary of each parasegment, and similar stripes of *wingless* expression posterior to them (see Fig. 2.41). Thus, cells expressing *wingless* abut cells expressing *engrailed* and *hedgehog* at the parasegment boundary.

2.25 Signals generated at the parasegment boundary delimit and pattern the future segments

By the late embryo stage, the continuous epithelium of the thorax and abdomen has become divided up into repeating units—the future segments—each composed of anterior and posterior compartments delimited by the original parasegment boundaries (see Fig. 2.41). We shall now see how the signals generated on either side of the parasegment boundary delimit and pattern the parasegments, and thus the future segments.

The *engrailed* gene is expressed in cells along the anterior margin of the parasegment. These cells also express the segmentation gene *hedgehog* and secrete the Hedgehog

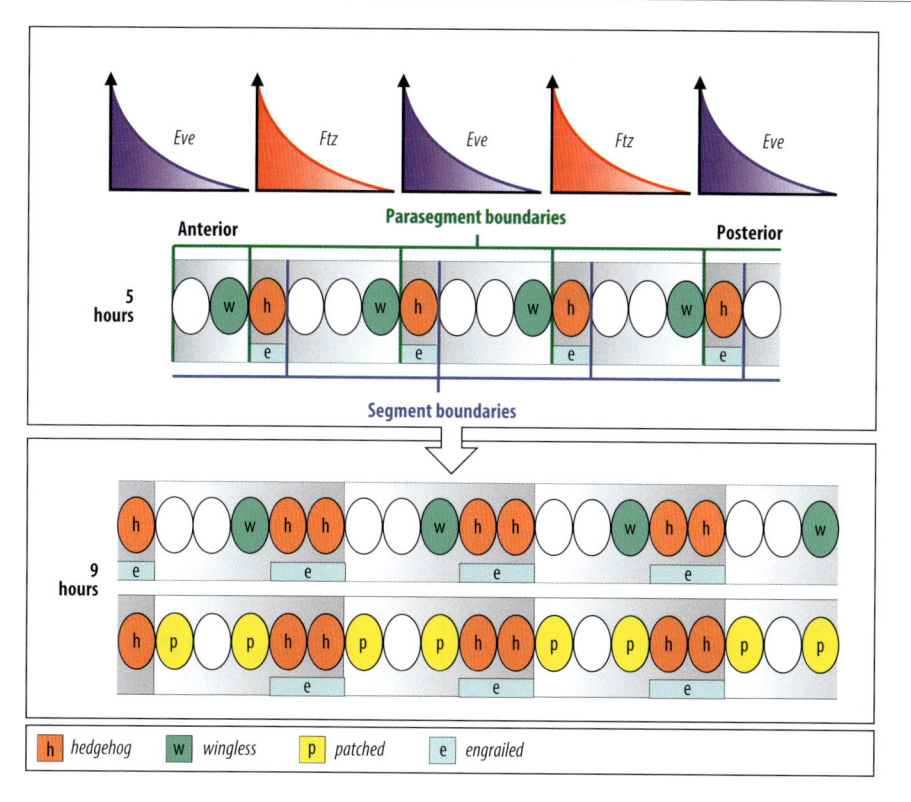

Fig. 2.41 The domains of expression of some key segmentation genes. Upper panel: expression of the pair-rule proteins Ftz and Eve in narrow stripes along the embryo delimit the parasegments. They activate expression of the *engrailed* and *hedgehog* genes in a strip one cell wide at the anterior margin of each parasegment, while the gene *wingless*, which is repressed by Ftz and Eve, is expressed in a strip one cell wide at the posterior of the parasegment as pair-rule gene expression fades away. Lower panel: as the embryo begins to grow, *hedgehog* is being expressed in two cells at the anterior margin of the parasegment, while *wingless* is expressed in an adjacent cell on the other side of the boundary (top row). At this time *patched* (which encodes the receptor for the Hedgehog protein) is induced by the secreted Hedgehog protein and is therefore expressed on either side of the domain of *engrailed/hedgehog* expression (bottom row). All the cells express Frizzled, the receptor for Wingless (not shown). The definitive segments of the late embryo and larva (shaded gray in the bottom two rows) are offset from the parasegments such that the original parasegment boundary now delimits the posterior compartment of the segment. This posterior compartment continues to be marked by *engrailed* expression throughout the fly's life.

signaling protein. Adjacent cells on either side express Patched, the receptor for Hedgehog, and the *patched* gene is itself a target of Hedgehog signaling (see Fig. 2.41). Hedgehog protein acting on these cells initiates an intracellular signaling pathway that results in the activation and maintenance of expression of *wingless* in the cells. The Wingless protein is secreted and feeds back on the *engrailed*-expressing cells to initiate a signaling event that maintains expression of the *engrailed* and *hedgehog* genes. All the cells express Frizzled, the receptor for Wingless, but only those close to the source of Wingless signal respond to it. These interactions stabilize and maintain the parasegment boundary (Fig. 2.42), and establish it as a **signaling center** that produces signals (Hedgehog and Wingless) that pattern the segment. As noted earlier, the parasegment boundary is the anterior boundary of the posterior compartment in the future segment. This means that in the segment itself, *engrailed* and *hedgehog* are expressed in the posterior compartment, and *wingless* in cells immediately anterior to this boundary (see Fig. 2.42).

The signaling pathway by which *Drosophila* Hedgehog acts is shown in Box 2E. At parasegment boundaries, and in many of their other developmental functions, Wingless and the vertebrate Wnt proteins act through a conserved signaling pathway illustrated in Box 3E, which is often called the **canonical Wnt/β-catenin pathway**. **β-catenin** is a potential transcriptional co-activator that in the absence of Wingless signaling is degraded in the cytoplasm by a complex of proteins. When Wingless binds to its receptor, Frizzled, this protein complex is inhibited, and β-catenin can then enter the nucleus and switch on the Wingless target genes. Computer modeling of the Wingless–Hedgehog–Engrailed circuitry shows that it is remarkably robust and resistant to variation, such as variations in the levels of expression of the genes governing the circuit's behavior.

About 3 hours after the parasegment boundary has been established, a deep groove develops at the posterior edge of each *engrailed* stripe and this marks the anterior boundary of each segment. A segmented late embryo is shown in Fig. 2.4 (third panel) and Fig. 2.43.

86 Chapter 2 Development of the *Drosophila* body plan

Fig. 2.42 Interactions between *hedgehog*, *wingless*, and *engrailed* genes and proteins establish the parasegment boundaries and control denticle pattern. Top left panel: the segmentation gene *hedgehog* is expressed in cells along the anterior margin of the parasegment, which is also delineated by expression of the gene *engrailed*. Hedgehog protein is secreted and its action on adjacent cells results in activation and maintenance of *wingless* gene expression. Secreted Wingless protein signals back to the cells expressing *engrailed* and *hedgehog* and maintains the expression of these genes. These interactions stabilize and maintain the boundary. Top right panel: in mutants where the *wingless* gene is inactivated and the Wingless protein is absent, neither *hedgehog* nor *engrailed* genes are expressed. This leads to loss of the parasegment boundary and conversion of the whole segment into 'anterior' type. Bottom left panel: the denticle bands on the ventral cuticle of the abdominal segments are useful markers for segment pattern. In the wild-type larva, the denticle bands are confined to the anterior part of the segment, whereas the posterior cuticle is naked, and this pattern is dependent on the activity of the *hedgehog* and *wingless* genes. Bottom right panel: in the *wingless* mutant, the normally well-defined pattern of denticles within each abdominal segment is lost. Denticles are present across the whole ventral surface of the segment in what looks like a mirror-image repeat of the anterior segment pattern.

Illustration after Lawrence, P.: The Making of a Fly. Oxford: Blackwell Scientific Publications, 1992.

Fig. 2.43 Denticle pattern of a *Drosophila* late embryo just before hatching. Each segment has a characteristic pattern of denticles on its ventral surface. Denticles are confined to the anterior of each segment. This view shows the ventral surface of the embryo.

*From Goodman, R.M., et al.: **Sprinter: a novel transmembrane protein required for Wg secretion and signalling.** Development 2006, **133**: 4901–4911.*

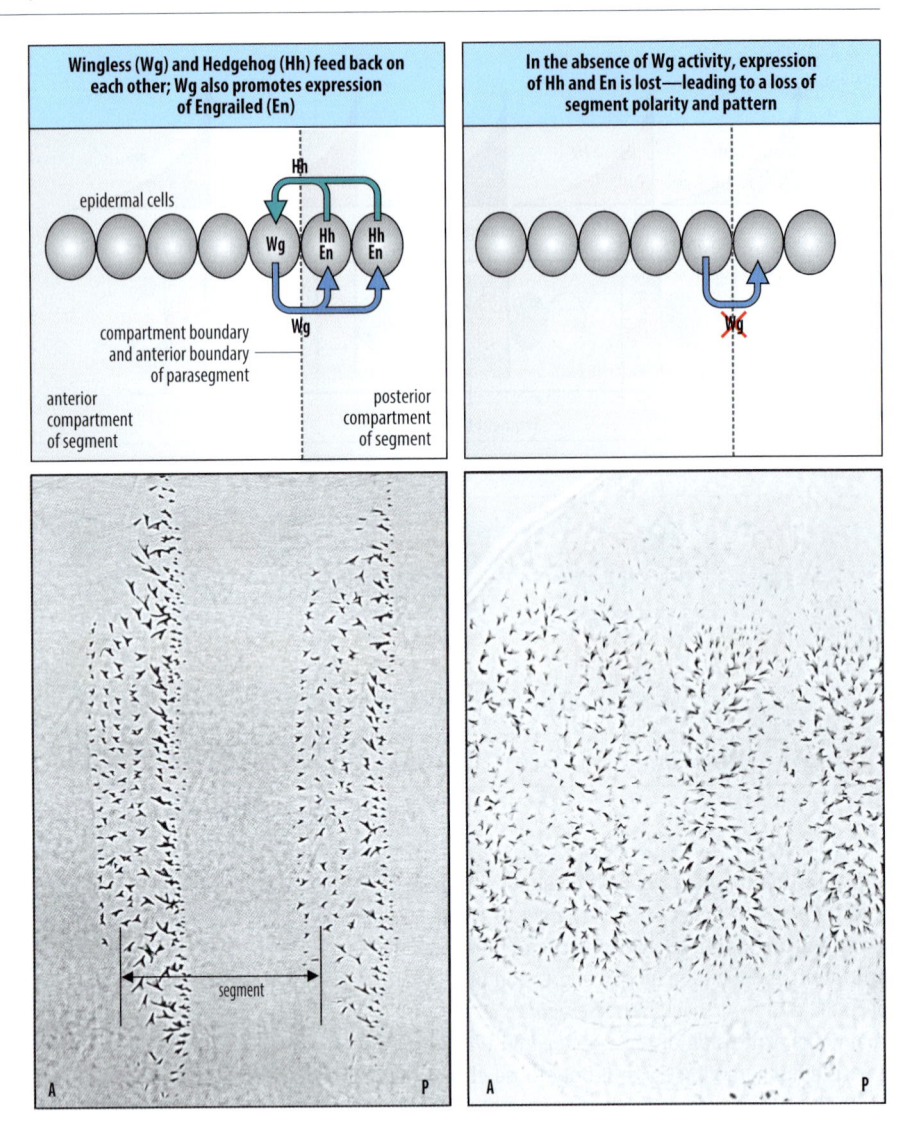

Once the parasegment boundary is consolidated, the signals localized on either side set up the pattern of each future segment, leading eventually to the differentiation of the epidermal cells to produce the cuticle pattern evident on the late embryo and larva. Each segment in the newly hatched larva has a well-defined antero-posterior pattern, which is most easily seen on the ventral epidermis of the abdomen: the anterior region of each segment bears denticles, whereas in most segments the posterior region is naked. The denticles appear in the late stages of embryonic development (see Fig. 2.43) and reflect the prior patterning of the embryonic epidermis by the actions of the segmentation genes. The rows of denticles make a distinct pattern on each segment and mutations in segmentation genes often alter the pattern in a systematic way—this is how these genes were first discovered (Box 2F).

The pattern of denticle belts is specified in response to the signals provided by Wingless and Hedgehog. At this later stage of embryonic development, Hedgehog and Wingless expression are no longer dependent on each other and can be analyzed separately. Wingless protein moves posteriorly over the parasegment boundary and also moves anteriorly to pattern that part of the segment immediately anterior to the cells in which it is expressed. Wingless moves a shorter distance posteriorly than

Segmentation genes and segment patterning

CELL BIOLOGY BOX 2E The Hedgehog signaling pathway

In the absence of Hedgehog (Fig. 1, left panel), the membrane protein Patched, which is the receptor for Hedgehog, inhibits the membrane protein Smoothened. In the absence of Smoothened activity, the transcription factor Cubitus interruptus (Ci) is held in the cytoplasm in two protein complexes—one associated with Smoothened and another with the protein Suppressor of fused (Su(fu)). In the absence of Hedgehog, Ci in the Smoothened complex is also phosphorylated by several protein kinases (glycogen synthase kinase (GSK-3), protein kinase A (PKA), and casein kinase 1 (CK1)), which results in the proteolytic cleavage of Ci and formation of the truncated protein CiRep. This enters the nucleus and acts as a repressor of Hedgehog target genes. When Hedgehog is present, it binds to Patched (Fig. 1, right panel) and this lifts the inhibition on Smoothened and blocks production of CiRep. Ci is released from both its complexes in the cytoplasm, enters the cell nucleus, and acts as a gene activator. Genes activated in response to Hedgehog signaling include *wingless* (*wg*), *decapentaplegic* (*dpp*), and *engrailed* (*en*).

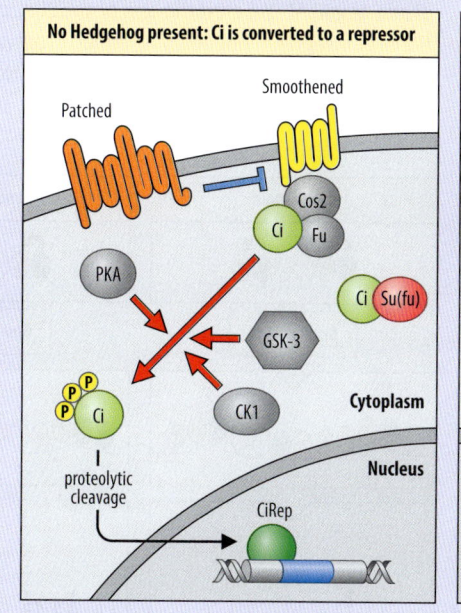

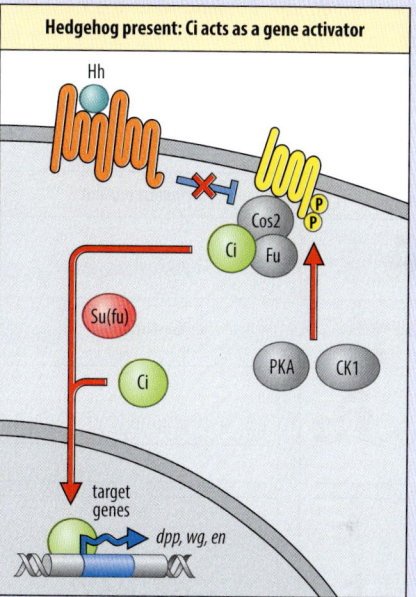

Fig. 1

EXPERIMENTAL BOX 2F Mutants in denticle pattern provided clues to the logic of segment patterning

Mutations that disrupted denticle pattern were discovered long before the signaling pathways underlying segment patterning were worked out. Although in theory there are many ways in which denticle pattern might be disrupted, mutations in denticle pattern all belonged to one of three classes of pattern defects (Fig. 1). The diagram shows the position of cells expressing the genes *wingless* and *engrailed* in a parasegment in the embryo before and after gastrulation. The photos beneath show the pattern of denticles on the larval cuticle.

The existence of these classes meant that there must be a logic to the construction of the epidermal patterning, and that the products of genes with a similar mutant phenotype must be functionally related. This was, indeed, the case. These three mutant classes identified Wingless and Hedgehog proteins as the main regulators of denticle pattern, and were instrumental in helping to identify the elements of their signaling pathways and the order in which they function. We now know that cells expressing Engrailed also express the protein Hedgehog, and that Hedgehog is needed to activate and maintain *wingless* gene expression, while Wingless protein is needed to maintain expression of *hedgehog* and *engrailed* (see Fig. 2.42).

One class is exemplified by the loss-of-function mutation in the *wingless* gene. In these mutants, the anterior (A) region of each segment has been duplicated in mirror-image polarity, and the smooth cuticle in the posterior (P) region is lost. This suggested that *wingless* expression might normally prevent denticle formation in the region of the segment nearest to the *wingless*-expressing cell. We now know that *wingless* encodes a diffusing signal protein, Wingless, whose action prevents denticle formation. The *hedgehog* mutation is also a member of this class, which suggested a functional relationship between the *hedgehog* and *wingless* genes. As we know now, Hedgehog is also a signal protein and the similarity in pattern reflects the fact that in the absence of Hedgehog signaling, *wingless* expression is lost.

A second class of mutation is exemplified by a loss-of-function mutation in a gene called *axin*, in which the anterior region of every segment is transformed into posterior and the denticles are substituted by smooth cuticle. This is the opposite effect to the *wingless* mutation, which suggested that *axin* might encode a negative regulator of the Wingless intracellular signaling pathway—that is, a protein that normally inhibits the pathway in cells not exposed to the Wingless signal. The Axin protein indeed turned out to be part of a protein complex that keeps β-catenin (the end product of the Wingless signaling pathway) inactive in the absence of a Wingless signal (see Box 3E). Its loss results in β-catenin being active in cells in which it is normally inactive. Mutations in other negative regulators of the Wingless signaling pathway have the same phenotype, which is also produced when the *wingless* gene is expressed throughout the whole embryo.

The third class is exemplified by mutations in *patched*, which encodes a receptor for Hedgehog. This produces a rather complex pattern of denticles on the cuticle in which the second one-third of the segment is a mirror image of the first one-third. An additional parasegment boundary is formed in between the two original ones. Mutations in other elements of the Hedgehog signaling pathway have a similar phenotype.

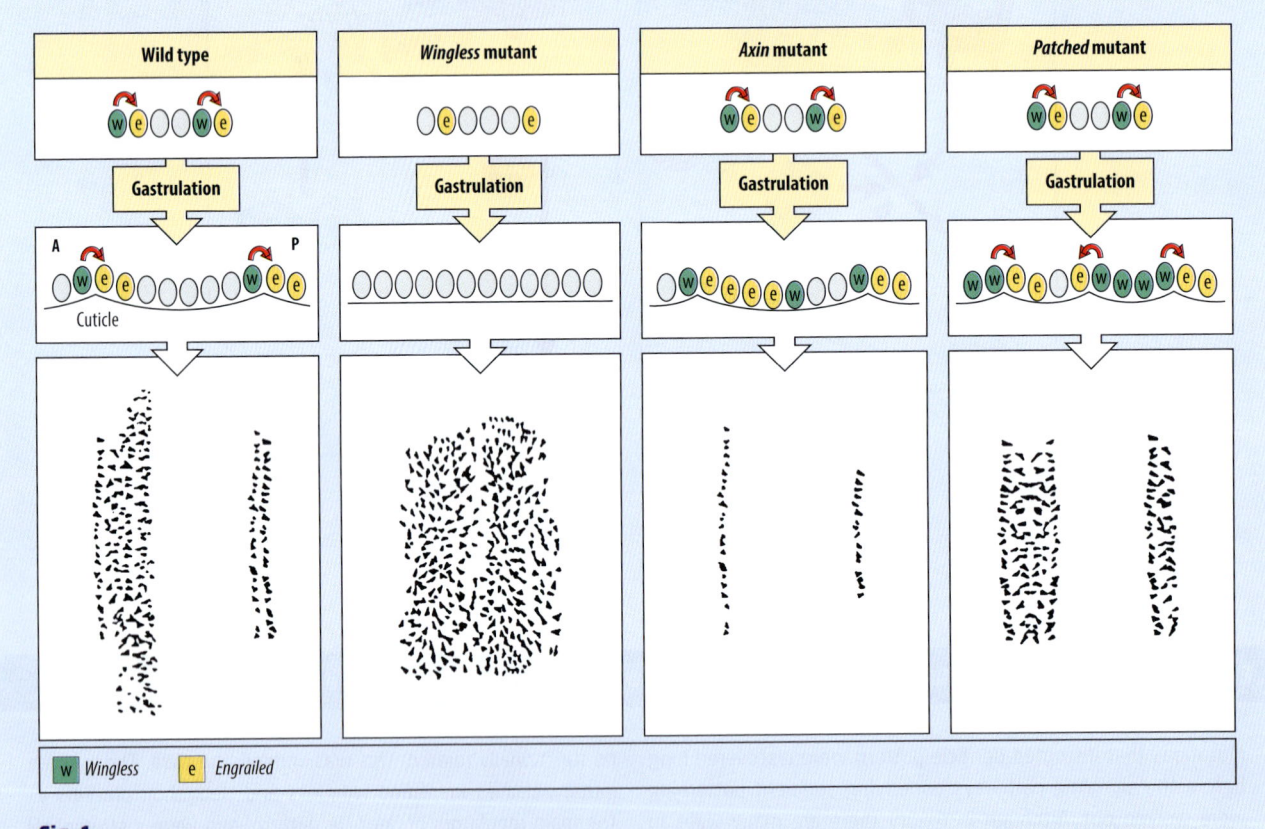

Fig. 1

anteriorly, as it is more rapidly degraded in the posterior part of the segment, and so its effects on patterning extend over a longer range in the anterior region (Fig. 2.44).

Wingless represses genes required for denticle formation, and so the cells that lie immediately anterior to the Wingless-expressing line of cells are specified as epidermal cells that will produce smooth cuticle, while the anterior extent of Wingless signaling defines the posterior edge of the denticle belts in that segment. In mutants that lack *wingless* function, denticles are produced in the normally smooth region (see Fig. 2.42).

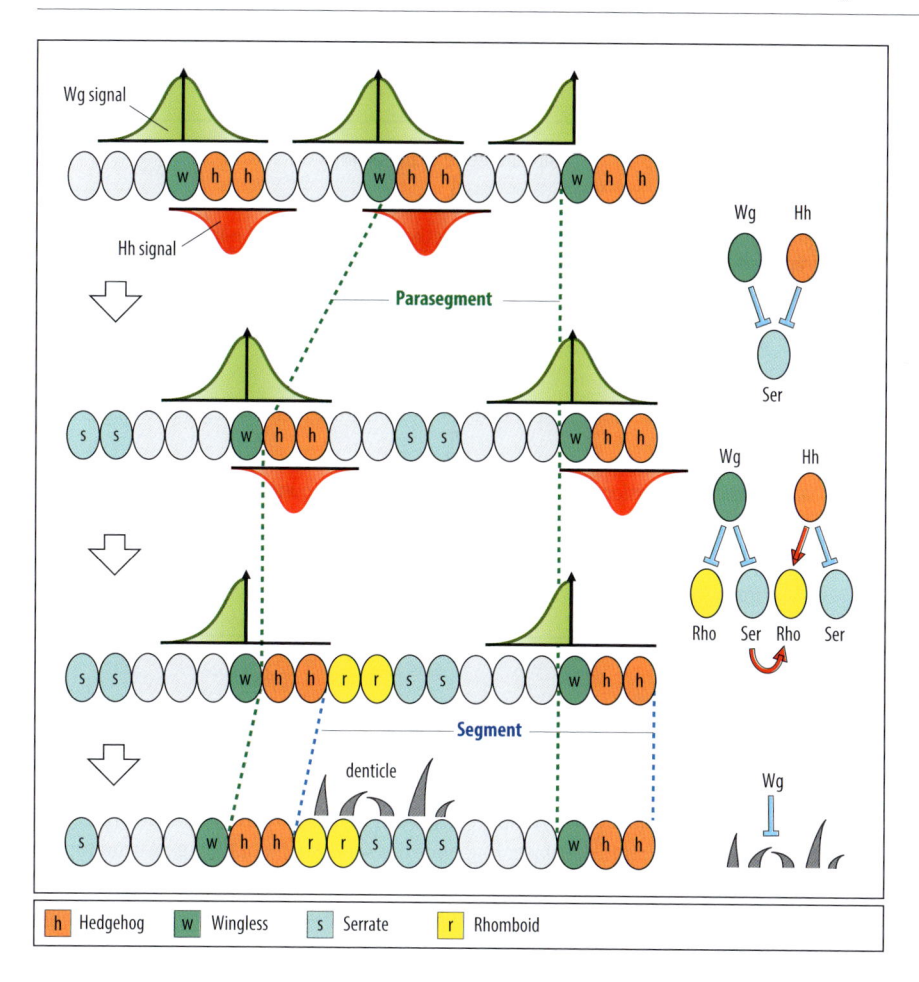

Fig. 2.44 Segment patterning restricts the production of denticles to specific cells. After gastrulation, the diffusion of Wingless and Hedgehog covers the whole of the segment. As the embryo grows, domains within each segment lose the influence of Hedgehog and Wingless and genes that are repressed by these signaling pathways can be expressed. One of these genes encodes Serrate (Ser), which is a ligand for Notch. Serrate then sends a signal to adjacent cells that leads to the activation of the gene *rhomboid*, which encodes a protein involved in regulating epidermal growth factor receptor signaling and which is required for denticle formation. At the same time, the diffusion of Wingless becomes polarized, and only cells anterior to its expression domain are exposed to Wingless protein. This sequence of interactions and gene expression creates a complex landscape, which results in the detailed patterning of the segment. Wingless suppresses the production of denticles, which results in the domain of cells that 'see' Wingless becoming naked cuticle, whereas in the rest of the segment, the fine-grained pattern of gene expression created as the embryo grows and matures results in the emergence of different kinds of denticles on each row of cells, as indicated. The denticle pattern in the second abdominal segment is shown here. Engrailed continues to be expressed in the Hedgehog-expressing cells (not shown for simplicity).

The Hedgehog signals move anteriorly to maintain Wingless function and also more posteriorly to help pattern the anterior region of the next segment. The signals provided by Hedgehog and Wingless lead, by a complex set of interactions, to the expression of a number of genes in non-overlapping narrow stripes in the late embryo that specify the larval segmental pattern of smooth cuticle or specific rows of denticles (see Fig. 2.44).

SUMMARY

The segmentation genes are involved in subdividing the length of the embryo into repeated units—the parasegments—and creating molecular signposts for their patterning through the actions of the *wingless* and *hedgehog* genes, which encode ligands that activate major cell-cell signaling pathways and regulate each other's expression. After the subdivision into 14 identical units, the activity of the gap and pair-rule genes positions Wingless and Hedgehog protein activity at the posterior (Wingless) and anterior (Hedgehog) sides of each parasegment boundary, which thus serve as a source of patterning information. The *engrailed* gene is activated at the same time and is expressed in the same cells as *hedgehog*. Regions of cell-lineage restriction, called compartments, were first discovered in *Drosophila*. The parasegment boundary is a compartment boundary, and this becomes the boundary between the anterior and posterior compartments of a segment in larval and adult tissue. The expression of Engrailed and Hedgehog, but not that of Wingless, is preserved in the adult, and these proteins also play a part in the patterning of the adult.

> **SUMMARY: Gene expression and intercellular signaling define segment compartments**
>
> pair-rule gene expression
> ⇩
> segmentation and selector gene *engrailed* activated in anterior of each parasegment, defining anterior compartment of parasegment and posterior compartment of segment
> ⇩
> *engrailed*-expressing cells also express segmentation gene *hedgehog*
> ⇩
> cells on other side of compartment boundary express segmentation gene *wingless*
> ⇩
> *engrailed* expression maintained and compartment boundary stabilized by Wingless and Hedgehog proteins
> ⇩
> compartment boundary provides signaling center from which segment is patterned

Specification of segment identity

Each segment has a unique identity, most easily seen in the larva in the characteristic pattern of denticles on the ventral surface. But since the same segmentation genes are turned on in each segment, what makes the segments different from each other? Their identity is specified by a class of genes known as **homeotic selector genes**, which set the future developmental pathway that each segment will follow and their individual structure and pattern. Homologs of the *Drosophila* homeotic selector genes that control segment identity have subsequently been discovered in virtually all other animals, where they broadly control identity along the antero-posterior axis in a similar manner, as we shall see in Chapter 4 for vertebrates, and in Chapter 13 in an evolutionary context.

2.26 Segment identity in *Drosophila* is specified by Hox genes

The first evidence for the existence of genes that specify segment identity came from unusual and striking mutations in *Drosophila* that produced **homeotic transformations**—the conversion of one segment into another. Eventually, the genes that produced these mutations were identified and the intricate way in which they specified segment identity was worked out. The discovery of these genes, now called the **Hox genes**, had an enormous impact on developmental biology. Homologous genes were found in other animals, where they acted in a similar way, essentially specifying identity along the antero-posterior axis. Hox genes are now considered as one of the fundamental defining features of animals. In 1995, the American geneticist Edward Lewis shared the Nobel Prize in Physiology or Medicine for his pioneering work on the *Drosophila* homeotic gene complexes and how they function. In *Drosophila*, the Hox genes are organized into two **gene clusters** or **gene complexes**, which together make up the **HOM-C complex** (Fig. 2.45; see also Box 4G). The Hox genes all encode transcription factors and take their name from the distinctive **homeobox** DNA motif that encodes part of the DNA-binding site of these proteins.

The two homeotic gene clusters in *Drosophila* are called the **bithorax complex** and the **Antennapedia complex** and are named after the mutations that first revealed

Specification of segment identity   91

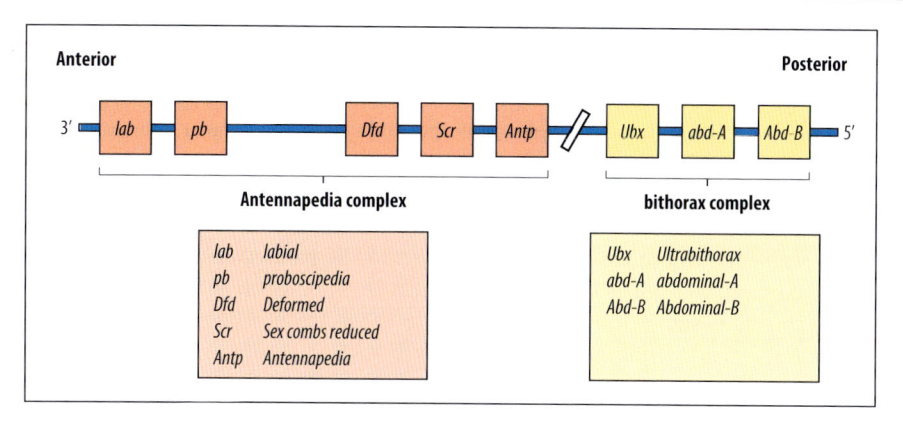

Fig. 2.45 The Antennapedia and bithorax homeotic selector gene complexes. The order of the genes from 3′ to 5′ in each gene complex reflects both the order of their spatial expression (anterior to posterior) and the timing of expression (3′ earliest).

their existence. Flies with the *bithorax* mutation have part of the haltere (the balancing organ on the third thoracic segment) transformed into part of a wing (Fig. 2.46), whereas flies with the dominant *Antennapedia* mutation have their antennae transformed into legs. Genes identified by such mutations are called **homeotic genes** because when mutated they result in **homeosis**—the transformation of a whole segment or structure into another related one, as in the transformation of antenna to leg. These bizarre transformations arise out of the homeotic genes' key role as identity specifiers. They control the activity of other genes in the segments, thus determining, for example, that a particular imaginal disc will develop as wing or haltere. The bithorax complex controls the development of parasegments 5–14, whereas the Antennapedia complex controls the identity of the more anterior parasegments. The action of the bithorax complex is the best understood and will be discussed first.

2.27 Homeotic selector genes of the bithorax complex are responsible for diversification of the posterior segments

The bithorax complex of *Drosophila* comprises three **homeobox genes**: *Ultrabithorax* (*Ubx*), *abdominal-A* (*abd-A*), and *Abdominal-B* (*Abd-B*). These genes are expressed in parasegmental units in a combinatorial manner (Fig. 2.47, top panel). *Ubx* is expressed in all parasegments from 5 to 12, *abd-A* is expressed more posteriorly in parasegments 7–13, and *Abd-B* more posteriorly still from parasegment 10 onwards. Because the genes are active to varying extents in different parasegments, their combined activities

Fig. 2.46 Homeotic transformation of the wing and haltere by mutations in the bithorax complex. Top panel: in the normal adult, both wing and haltere are divided into an anterior (A) and a posterior (P) compartment. Center panel: the *bithorax* mutation transforms the anterior compartment of the haltere into an anterior wing region. The mutation *postbithorax* acts similarly on the posterior compartment, converting it into posterior wing. Bottom panel: if both mutations are present together, the effect is additive and the haltere is transformed into a complete wing, producing a four-winged fly. Another mutation, *Haltere mimic*, causes a homeotic transformation in the opposite direction: the wing is transformed into a haltere (not shown).
*Illustration after Lawrence, P.: The Making of a Fly. Oxford: Blackwell Scientific Publications, 1992. Photograph reproduced with permission from Bender, W., et al.: **Molecular genetics of the bithorax complex in Drosophila melanogaster.** Science 1983, **221**: 23–29 (image on front cover).*

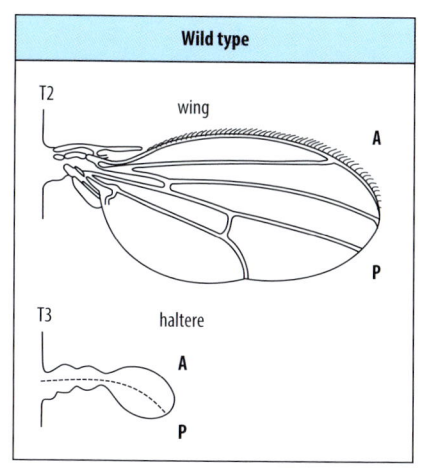

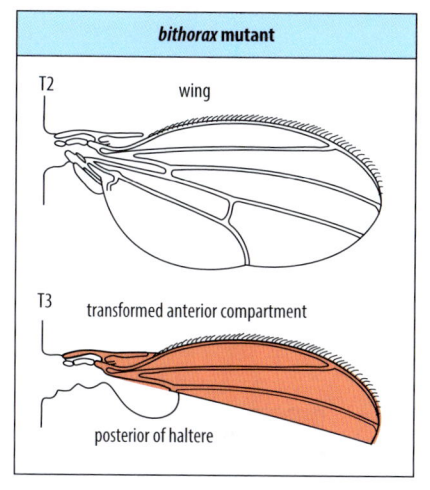

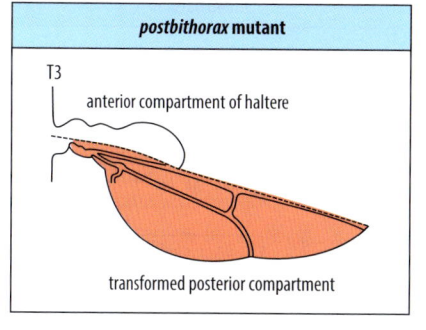

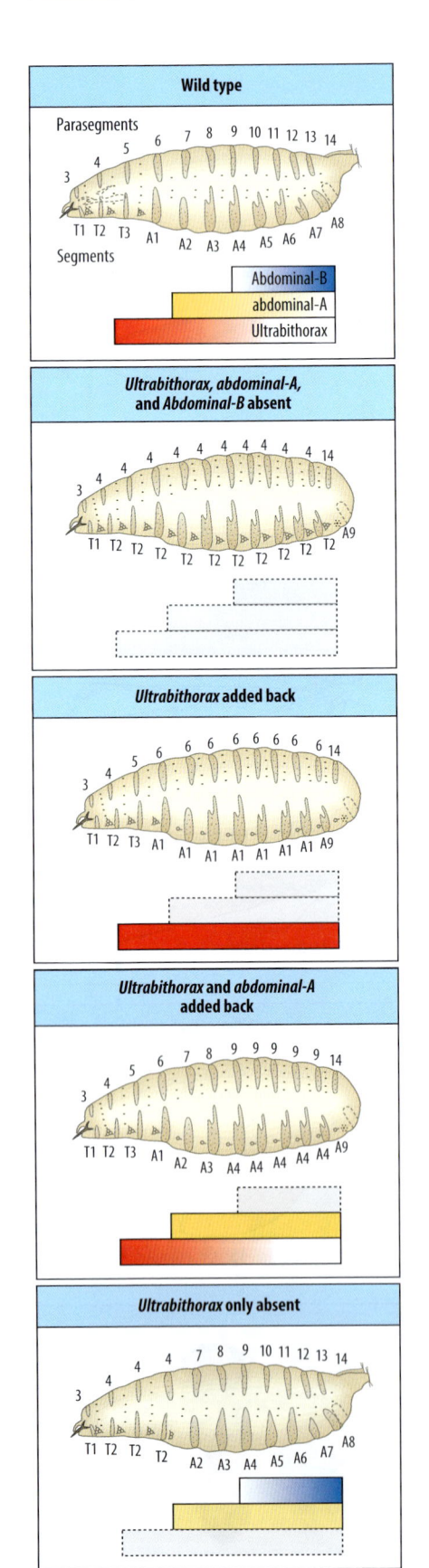

define the character of each parasegment. *Abd-B* also suppresses *Ubx*, such that *Ubx* expression is very low by parasegment 14, as *Abd-B* expression increases. The pattern of activity of the bithorax complex genes is determined by gap and pair-rule genes.

The role of the bithorax complex was first indicated by classical genetic experiments. In larvae lacking the whole of the bithorax complex (see Fig. 2.47, second panel), every parasegment from 5 to 13 develops in the same way and resembles parasegment 4. The bithorax complex is therefore essential for the diversification of these segments, whose basic pattern is represented by parasegment 4. This parasegment can be considered as being a type of 'default' state, which is modified in all the parasegments posterior to it by the proteins encoded by the bithorax complex. It is because the genes of the bithorax complex can superimpose a new identity on the default state that they are called selector genes.

An indication of the role of each gene of the bithorax complex can be deduced by looking at embryos constructed so that the genes are introduced, one at a time, into embryos lacking the whole gene complex (see Fig. 2.47, bottom three panels). If only the *Ubx* gene is present, the resulting larva has one parasegment 4, one parasegment 5, and eight parasegments 6. Clearly, *Ubx* has some effect on all parasegments from 5 backward and can specify parasegments 5 and 6. If *abd-A* and *Ubx* are put into the embryo, then the larva has parasegments 4, 5, 6, 7, and 8, followed by five parasegments 9. So *abd-A* affects parasegments from 7 backward, and in combination, *Ubx* and *abd-A* can specify the character of parasegments 7, 8, and 9. Similar principles apply to *Abd-B*, whose domain of influence extends from parasegment 10 backward, and which is expressed most strongly in parasegment 14. Hence, differences between individual segments reflect differences in the spatial and temporal pattern of Hox gene expression.

These results illustrate an important principle, namely that the character of the parasegments is specified by the genes of the bithorax complex acting in a combinatorial manner. This probably reflects the fact that the combinations of Hox proteins can regulate genes that are not affected by individual Hox proteins. Their combinatorial effect can also be seen by taking the genes away one at a time from the wild type. In the absence of *Ubx*, for example, parasegments 5 and 6 become converted to parasegment 4 (see Fig. 2.47, bottom panel). There is a further effect on the cuticle pattern in parasegments 7–14. Structures characteristic of the thorax are now present in the abdomen, showing that *Ubx* exerts an effect in all these segments. Such abnormalities may result from expression of 'nonsense' combinations of the bithorax genes. For example, in such a mutant, Abd-A protein is present in parasegments 7–9 without Ubx protein, and this is a combination never found normally. Which segments bear appendages, such as legs, is determined by the Hox genes. For example, expression of the genes *Antp* and *Ubx* specifies the segments from which the second and third pair of legs, respectively, will emerge. The evolution of different patterns of arthropod appendages in relation to Hox gene expression is discussed in Chapter 13. The downstream targets of Hox proteins are beginning to be identified, and there may be a very

Fig. 2.47 The spatial pattern of expression of genes of the bithorax complex characterizes each parasegment. In the wild-type embryo (top panel) the expression of the genes *Ultrabithorax, abdominal-A,* and *Abdominal-B* is required to confer an identity on each parasegment. Mutations in the bithorax complex result in homeotic transformations in the character of the parasegments and the segments derived from them. When the bithorax complex is completely absent (second panel), parasegments 5–13 are converted into nine parasegments 4 (corresponding to segment T2 in the larva), as shown by the denticle and bristle patterns on the cuticle. The three lower panels show the transformations caused by the absence of different combinations of genes. When the *Ultrabithorax* gene alone is absent (bottom panel), parasegments 5 and 6 are converted into parasegment 4. In each case the spatial extent of gene expression is detected by *in situ* hybridization (see Box 1E). Note that the specification of parasegment 14 is relatively unaffected by the bithorax complex.

large number of genes whose expression is affected in the mutants, which highlights the fact that the morphological identity of a segment results from the coordinated activity of many genes.

The gap and pair-rule proteins initially control the pattern of Hox gene expression, but these proteins disappear within about 4 hours. The continued correct expression of the Hox genes involves two other groups of genes—the Polycomb and Trithorax groups. The proteins of the Polycomb group bind to chromatin and maintain the long-term silencing of Hox genes that have been shut down in early development, whereas the chromatin-binding Trithorax-group proteins maintain the long-term expression of Hox genes that have been turned on. Members of these two groups of proteins were originally identified in *Drosophila* and are present in all animals and in plants. They comprise proteins with different functions that control gene expression by a variety of epigenetic mechanisms, including histone modification and chromatin remodeling, which determine the accessibility of DNA to transcriptional regulators (see Section 1.11). These mechanisms of epigenetic control are discussed in Box 6B.

2.28 The Antennapedia complex controls specification of anterior regions

The Antennapedia complex comprises five homeobox genes (see Fig. 2.45), which control the behavior of the parasegments anterior to parasegment 5 in a manner like the bithorax complex. Because there are no novel principles involved, we will not go into much detail. Several genes within the complex are critically involved in specifying particular parasegments. Mutations in the gene *Deformed* affect the ectodermally derived structures of parasegments 0 and 1, those in *Sex combs reduced* affect parasegments 2 and 3, and those in the *Antp* gene affect parasegments 4 and 5, the parasegments that produce the leg imaginal discs. The *Antennapedia* mutation that transforms antennae into legs in the adult fly (see Fig. 9.2) causes the misexpression of the *Antp* gene in anterior segments in which it is not normally expressed.

2.29 The order of Hox gene expression corresponds to the order of genes along the chromosome

The bithorax and Antennapedia complexes possess some striking features of gene organization. Within both, the order of the genes in the complex is the same as the spatial and temporal order in which they are expressed along the antero-posterior axis during development. *Ubx*, for example, is on the 3′ side of *abd-A* on the chromosome, is more anterior in its pattern of expression, and is activated earlier. As we will see, the related Hox gene complexes of vertebrates, whose ancestors diverged from those of arthropods hundreds of millions of years ago, show the same correspondence between gene order and timing of expression. This highly conserved **temporal** and **spatial co-linearity** must be related to the mechanisms that control the expression of these genes.

The complex, yet subtle, regulation of the bithorax complex genes is revealed in experiments in which production of Ultrabithorax protein is forced in all segments. This targeted gene misexpression was initially achieved by linking the Ultrabithorax protein-coding sequence to a heat-shock promoter—a promoter that is activated at 29°C—and introducing the novel DNA construct into the fly genome using a P element. When this transgenic embryo was given a heat shock for a few minutes, the extra *Ubx* gene was transcribed, and the protein was made at high levels in all cells. This had no effect on posterior parasegments (in which Ubx protein is normally present), except for parasegment 5 which, for unknown reasons, is transformed into parasegment 6 (this may reflect a quantitative effect of the Ubx protein). However, all parasegments anterior to 5 are also transformed into parasegment 6. While this seems a reasonably straightforward and expected result, consider what happens to parasegment 13.

Transcription of *Ubx* is normally suppressed in parasegment 13 in wild-type embryos, yet when the protein itself is produced in the parasegment as a result of heat shock, it has no effect. It is thought that the Ubx protein is rendered inactive in this parasegment by the actions of Abd-B. This phenomenon is quite common in the specification of the parasegments and is known as 'phenotypic suppression' or 'posterior dominance': it means that Hox gene products normally expressed in anterior regions are suppressed by more posterior products.

Although the roles of the bithorax and Antennapedia complexes in controlling segment identity are well established, we know little about how they interact with the genes that act next in the developmental pathway. These are the genes that actually specify the structures that give the segments their unique identities. What, for example, is the pathway that leads to a thoracic rather than an abdominal segment structure? Does each Hox gene activate just a few or many target genes? Do target genes with different functions act together to form a particular structure? Answers to these questions will help us understand how changes in a single gene can cause a dramatic homeotic transformation such as antenna into leg. The role of the Hox genes in the evolution of different body patterns and in the evolution of vertebrate limbs from fins is considered in Chapter 13.

2.30 The *Drosophila* head region is specified by genes other than the Hox genes

As mentioned earlier in the chapter, the *Drosophila* larval head anterior to the mandibles is formed by the three most-anterior parasegments. The segmental structure is also evident in the embryonic brain, which is composed of three segments (neuromeres). The specification of the anterior head and embryonic nervous system is not, however, under the control of either the pair-rule genes or the Hox genes. Instead, it appears to be determined at the blastoderm stage by the overlapping expression of the genes *orthodenticle*, *empty spiracles*, and *buttonhead*, which all encode gene-regulatory proteins and resemble the gap genes in their effect on phenotype; they are sometimes referred to as the 'head gap genes'. Unlike the gap genes regionalizing the trunk and abdomen, however, they do not regulate each other's activity. It is not known whether they work in a combinatorial way like the Hox genes to also confer specific identities on the head segments. *orthodenticle* and *empty spiracles* have homologs that carry out similar functions in vertebrates, and this head-specification system, like the Hox genes, is of ancient evolutionary origin in animals.

The genes *orthodenticle* and *empty spiracles* are examples of **homeodomain** transcription factors that lie outside the Hox gene clusters. The DNA-binding homeodomain is not confined to the proteins encoded by the Hox gene clusters, and similar domains are found in many other transcription factors involved in development in both *Drosophila* and vertebrates.

> **SUMMARY**
>
> Segment identity is conferred by the action of homeotic selector genes, which direct the development of different parasegments so that each acquires a unique identity. Two clusters of homeotic selector genes, known generally as the Hox genes, are involved in specification of segment identity in *Drosophila*: the Antennapedia complex, which controls parasegment identity in the head and first thoracic segment; and the bithorax complex, which acts on the remaining parasegments. Segment identity is determined by the combination of Hox genes active in the segment. The spatial pattern of expression of Hox genes along the antero-posterior axis of the embryo

is initially determined by the preceding gap gene activity, but Hox genes need to be switched on, or kept shut down, continuously throughout development to maintain the required phenotype. This long-term control is accomplished by epigenetic modification of Hox gene chromatin by proteins of the Polycomb group (which keep the genes shut down) and the Trithorax group (which keep the genes active). Mutations in genes of the Antennapedia and bithorax complexes can bring about homeotic transformations, in which one segment or structure is changed into a related one—for example, an antenna into a leg. The Antennapedia and bithorax complexes are remarkable, in that gene order on the chromosome corresponds to the spatial and temporal order of gene expression along the body.

SUMMARY: Homeotic genes and segment identity

gap and pair-rule gene expression delimits parasegments
⇩
selector genes of the HOM-C complex are expressed along the antero-posterior axis in an order co-linear with the order of genes on the chromosome
↙ ↘

Antennapedia complex
lab, pb, Dfd, Scr, Antp
⇩
specify segment identity in part of head and first thoracic segment

bithorax complex
Ubx, abd-A, Abd-B
⇩
specify segment identity in second and third thoracic segments and abdominal segments

---

Summary to Chapter 2

- During oogenesis, maternal proteins and mRNAs are deposited in the egg in a particular spatial pattern. This pattern will determine the main body axes and lay down a framework of positional information.
- The earliest stages of development in *Drosophila* take place when the embryo is a multinucleate syncytium. After fertilization, maternally supplied mRNAs are translated, and gradients of transcription factors are formed that broadly regionalize the body along the antero-posterior and dorso-ventral axes.
- These transcription factors enter the nuclei and turn on the embryo's own genes, activating a cascade of zygotic gene activity that further patterns the body.
- Along the dorso-ventral axis, zygotic gene expression defines several regions, including the future mesoderm and future neural tissue.
- The first zygotic genes to be activated along the antero-posterior axis are the gap genes, all encoding transcription factors, whose pattern of expression subdivides the embryo into a number of regions.
- The transition to a segmented body organization starts with the activity of the pair-rule genes, whose sites of expression are specified by the gap gene proteins and which divide the antero-posterior axis into 14 parasegments.
- At the time of pair-rule gene expression, the blastoderm becomes cellularized.
- The pair-rule genes initially define the spatial expression of segmentation genes, which pattern the parasegments. Once the embryo becomes cellular, patterning depends on intercellular communication via signaling proteins and their receptors as well as on transcription factors.

- After gastrulation, the definitive final segments of the late embryo and larva are delimited, offset from the original parasegments by about half a segment.
- The parasegment boundaries delimit regions of cell-lineage restriction in the epidermis, known as compartments. These compartment boundaries persist in the segments, where they delimit the anterior and posterior regions of each segment, and in the imaginal discs and the adult structures, such as wings and legs, that derive from the imaginal discs.
- Segment and compartment boundaries are involved in patterning and polarizing segments.
- The identity of each segment is determined by two gene complexes that contain homeotic selector genes, known generally as the Hox genes. The spatial pattern of expression of these genes is largely determined by gap gene activity.

■ End of chapter questions

Check your understanding of the content of this chapter by attempting the following long-answer concept questions.

1. As we will see in future chapters, cleavage divides the eggs of vertebrates and many other animals into a cluster of separate cells. By contrast, the equivalent stage in the *Drosophila* embryo generates a syncytium. What is a syncytium, and how does this influence the early development of *Drosophila*?

2. Much attention has been paid to the process of segmentation in the *Drosophila* embryo. What are segments, and what is their significance to the strategy used by *Drosophila* to carry out pattern formation? Do humans have segments? Give examples.

3. Briefly summarize the mutation screen devised by Nüsslein-Volhard and Wieschaus to find developmental mutants. What role did denticle patterns play in their search for developmental mutants?

4. Developmental genes act hierarchically during pattern formation, first defining broad regions, which are secondarily refined to form a larger number of smaller regions (see Section 2.6). Summarize how this general principle is illustrated by early *Drosophila* embryogenesis.

5. The anterior end of the embryo is established by a locally high level of activity of the transcription factor Bicoid; the ventral side of the embryo is established by a locally high level of activity of the transcription factor Dorsal. Contrast the mechanisms by which these activities come to be localized in the required places.

6. Bicoid was the first morphogen whose mechanism of action was molecularly characterized; an important outcome of this work was the understanding of how a smoothly declining gradient of Bicoid can lead to a sharply delimited boundary of Hunchback expression. Use the Bicoid–Hunchback system to illustrate how an activation threshold can translate a smooth gradient into a sharp developmental boundary. Be sure to incorporate the role of cooperative binding of Bicoid to the *hunchback* promoter in your answer.

7. The dorso-ventral axis is patterned by a gradient of nuclear localization of the transcription factor Dorsal. Outline the steps that lead to this gradient of nuclear localization: include the Pipe enzyme, the Spätzle ligand, the Toll receptor, the Cactus protein, and, finally, Dorsal.

8. The specification of the 14 parasegments along the antero-posterior axis depends on precise spatial expression of the pair-rule genes. Pair-rule gene expression depends in turn on the gap genes. Using the second stripe of *even-skipped* expression as a model, summarize the interactions that lead to the correct expression of Bicoid, Hunchback, Krüppel, and Giant. In particular, describe how the anterior border of Krüppel is set, and how this boundary sets the posterior border of *eve* stripe 2 expression.

9. What is the special role of *engrailed* in the patterning of the *Drosophila* body plan?

10. What is meant by a homeotic transformation? Give an example.

11. Experiments demonstrate that the role of the bithorax complex is to drive the identity of the parasegments in the posterior portion of the animal away from the default identity of parasegment 4. Describe these experiments. How do these experiments illustrate the importance of the combinations of gene products present in specifying segment identity?

■ General further reading

Adams, M.D., *et al.*: **The genome sequence of *Drosophila melanogaster***. *Science* 2000, **267**: 2185–2195.

Bate, M., Martinez Arias, A. (Eds): *The Development of* Drosophila melanogaster. Cold Spring Harbor, NY: Cold Spring Harbor Press, 1993.

FlyBase [https://flybase.org] (subscription only).

Greenspan, R.: *Fly Pushing: The Theory and Practice of* Drosophila *Genetics*. Cold Spring Harbor, NY: Cold Spring Harbor Press, 2004.

Lawrence, P.A.: *The Making of a Fly*. Oxford: Blackwell Scientific Publications, 1992.

Mohr, S.E., Hu, Y., Kim, K., Housden, B.E., Perrimon, N.: **Resources for functional genomics studies in *Drosophila melanogaster***. *Genetics* 2014, **197**: 1–18.

Perrimon, N., Pitsouli, C., Shilo, B.Z.: **Signaling mechanisms controlling cell fate and embryonic patterning**. *Cold Spring Harb. Perspect. Biol.* 2012, **4**: a005975.

Stolc, V., Stolc, V., Gauhar, Z., Mason, C., Halasz, G., van Batenburg, M.F., Rifkin, S.A., Hua, S., Herreman, T., Tongprasit, W., Barbano, P. E., Bussemaker, H. J., White, K.P.: **A gene** expression map for the euchromatic genome of *Drosophila melanogaster*. *Science* 2004, **306**: 655–660.

The Interactive Fly [http://www.sdbonline.org/sites/fly/aimain/1aahome.htm] (date accessed 18 October 2022).

■ Section further reading

2.2 Cellularization is followed by gastrulation and segmentation

Foe, V.E., Alberts, B.M.: **Studies of nuclear and cytoplasmic behaviour during the five mitotic cycles that precede gastrulation in *Drosophila* embryogenesis**. *J. Cell Sci.* 1983, **61**: 31–70.

Leptin, M.: **Gastrulation in *Drosophila*: the logic and the cellular mechanisms**. *EMBO J.* 1999, **18**: 3187–3192.

Martin, A.C. **The physical mechanisms of *Drosophila* gastrulation: mesoderm and endoderm invagination**. *Genetics* 2020, **214**: 543–560.

Stathopoulos, A., Levine, M.: **Whole-genome analysis of *Drosophila* gastrulation**. *Curr. Opin. Genet. Dev.* 2004, **14**: 477–484.

2.4 Many developmental genes were identified in *Drosophila* through large-scale genetic screening for induced mutations

Nüsslein-Volhard, C., Wieschaus, E.: **Mutations affecting segment number and polarity in *Drosophila***. *Nature* 1980, **287**: 795–801.

Box 2A Mutagenesis and genetic screening strategy for identifying developmental mutants in *Drosophila*

Greenspan, R.: *Fly Pushing: The Theory and Practice of* Drosophila *Genetics*. Cold Spring Harbor, NY: Cold Spring Harbor Press, 2004.

St Johnston, D. **The art and design of genetic screens: *Drosophila melanogaster***. *Nat. Rev. Genet.* 2002, **3**: 176–188.

2.7 Three classes of maternal genes specify the antero-posterior axis

St Johnston, D., Nüsslein-Volhard, C.: **The origin of pattern and polarity in the *Drosophila* embryo**. *Cell* 1992, **68**: 201–219.

2.8 Bicoid protein provides an antero-posterior gradient of a morphogen

Driever, W., Nüsslein-Volhard, C.: **The bicoid protein determines position in the *Drosophila* embryo in a concentration dependent manner**. *Cell* 1988, **54**: 95–104.

Ephrussi, A., St Johnston, D.: **Seeing is believing: the bicoid morphogen gradient matures**. *Cell* 2004, **116**: 143–152.

Gibson, M.C.: **Bicoid by the numbers: quantifying a morphogen gradient**. *Cell* 2007, **130**: 14–16.

Gregor, T., Wieschaus, E.F., McGregor, A.P., Bialek, W., Tank, D.W.: **Stability and nuclear dynamics of the bicoid morphogen gradient**. *Cell* 2007, **130**: 141–152.

Morrison, A.H., Scheeler, M., Dubuis, J., Gregor, T.: **Quantifying the bicoid morphogen gradient in living fly embryos**. *Cold Spring Harb. Protoc.* 2012, **2012**: 398–406.

Porcher, A., Dostatni, N.: **The bicoid morphogen system**. *Curr. Biol.* 2010, **20**: R249–R254.

2.9 The posterior pattern is controlled by the gradients of Nanos and Caudal proteins

Irish, V., Lehmann, R., Akam, M.: **The *Drosophila* posterior-group gene *nanos* functions by repressing *hunchback* activity**. *Nature* 1989, **338**: 646–648.

Murafta, Y., Wharton, R.P.: **Binding of pumilio to maternal *hunchback* mRNA is required for posterior patterning in *Drosophila* embryos**. *Cell* 1995, **80**: 747–756.

Rivera-Pomar, R., Lu, X., Perrimon, N., Taubert, H., Jackle, H.: **Activation of posterior gap gene expression in the *Drosophila* blastoderm**. *Nature* 1995, **376**: 253–256.

Struhl, G.: **Differing strategies for organizing anterior and posterior body pattern in *Drosophila* embryos**. *Nature* 1989, **338**: 741–744.

2.10 The anterior and posterior extremities of the embryo are specified by activation of a cell-surface receptor

Casali, A., Casanova, J.: **The spatial control of Torso RTK activation: a C-terminal fragment of the Trunk protein acts as a signal for Torso receptor in the *Drosophila* embryo**. *Development* 2001, **128**: 1709–1715.

Casanova, J., Struhl, G.: **Localized surface activity of torso, a receptor tyrosine kinase, specifies terminal body pattern in *Drosophila***. *Genes Dev.* 1989, **3**: 2025–2038.

Coppey, M., Boettiger, A.N., Berezhkovskii, A.M., Shvartsman, S.Y.: **Nuclear trapping shapes the terminal gradient in the *Drosophila* embryo**. *Curr. Biol.* 2008, **18**: 915–919.

Furriols, M., Casanova, J.: **In and out of Torso RTK signalling**. *EMBO J.* 2003, **22**: 1947–1952.

Furriols, M., Ventura, G., Casanova, J.: **Two distinct but convergent groups of cells trigger Torso receptor tyrosine kinase activation by independently expressing torso-like**. *Proc. Natl Acad. Sci. USA* 2007, **104**: 11660–11665.

Li, W.X.: **Functions and mechanisms of receptor tyrosine kinase Torso signaling: lessons from *Drosophila* embryonic terminal development**. *Dev. Dyn.* 2005, **232**: 656–672.

2.11 The dorso-ventral polarity of the embryo is specified by localization of maternal proteins in the egg vitelline envelope

Anderson, K.V.: **Pinning down positional information: dorsal–ventral polarity in the *Drosophila* embryo**. *Cell* 1998, **95**: 439–442.

Sen, J., Goltz, J.S., Stevens, L., Stein, D.: **Spatially restricted expression of *pipe* in the *Drosophila* egg chamber defines embryonic dorsal-ventral polarity**. *Cell* 1998, **95**: 471–481.

Shilo, B.Z., Haskel-Ittah, M., Ben-Zvi, D., Schejter, E.D., Barkai, N.: **Creating gradients by morphogen shuttling**. *Trends Genet.* 2013, **29**: 339–347.

Box 2B The Toll signaling pathway: a multifunctional pathway

Ferrandon, D., Imler, J.L., Hoffmann, J.A.: **Sensing infection in *Drosophila*: Toll and beyond**. *Semin. Immunol.* 2004, **16**: 43–53.

Imler, J.L., Hoffmann, J.A.: **Toll receptors in *Drosophila*: a family of molecules regulating development and immunity**. *Curr. Top. Microbiol. Immunol.* 2002, **270**: 63–79.

Levin, T.C., Malik, H.S. **Rapidly evolving *Toll-3/4* genes encode male-specific Toll-like receptors in *Drosophila*.** *Mol. Biol. Evol.* 2017, **34**: 2307–2323.

Valanne, S., Wang, J.H., Rämet, M.: **The *Drosophila* Toll signaling pathway.** *J. Immunol.* 2011, **186**: 649–656.

2.12 Positional information along the dorso-ventral axis is provided by the Dorsal protein

Moussian, B., Roth, S.: **Dorsoventral axis formation in the *Drosophila* embryo—shaping and transducing a morphogen gradient.** *Curr. Biol.* 2005, **15**: R887–R899.

Roth, S., Stein, D., Nüsslein-Volhard, C.: **A gradient of nuclear localization of the dorsal protein determines dorso-ventral pattern in the *Drosophila* embryo.** *Cell* 1989, **59**: 1189–1202.

Steward, R., Govind, R.: **Dorsal-ventral polarity in the *Drosophila* embryo.** *Curr. Opin. Genet. Dev.* 1993, **3**: 556–561.

2.13 The antero-posterior axis of the *Drosophila* egg is specified by signals from the preceding egg chamber and by interactions of the oocyte with follicle cells

González-Reyes, A., St Johnston, D.: **The *Drosophila* AP axis is polarised by the cadherin-mediated positioning of the oocyte.** *Development* 1998, **125**: 3635–3644.

Huynh, J.R., St Johnston, D.: **The origin of asymmetry: early polarization of the *Drosophila* germline cyst and oocyte.** *Curr. Biol.* 2004, **14**: R438–R449.

Riechmann, V., Ephrussi, A.: **Axis formation during *Drosophila* oogenesis.** *Curr. Opin. Genet. Dev.* 2001, **11**: 374–383.

Roth, S., Lynch, J.A.: **Symmetry breaking during *Drosophila* oogenesis.** *Cold Spring Harb. Perspect. Biol.* 2009, **1**: a001891.

Torres, I.L., Lopez-Schier, H., St Johnston, D.: **A Notch/Delta-dependent relay mechanism establishes anterior-posterior polarity in *Drosophila*.** *Dev. Cell* 2005, **5**: 547–558.

Zimyanin, V.L., Belaya, K., Pecreaux, J., Gilchrist, M.J., Clark, A., Davis, I., St Johnston, D.: **In vivo imaging of *oskar* mRNA transport reveals the mechanism of posterior localization.** *Cell* 2008, **134**: 843–853.

Box 2C The JAK-STAT signaling pathway

Arbouzova, N.I., Zeidler, M.P.: **JAK/STAT signalling in *Drosophila*: insights into conserved regulatory and cellular functions.** *Development* 2006, **133**: 2605–2616.

Hombría, J.C., Sotillos, S.: **JAK-STAT pathway in *Drosophila* morphogenesis: from organ selector to cell behavior regulator.** *JAKSTAT* 2013, **2**: e26089.

2.14 Localization of maternal mRNAs to either end of the egg depends on the reorganization of the oocyte cytoskeleton

Becalska, A.N., Gavis, E.R.: **Lighting up mRNA localization in *Drosophila* oogenesis.** *Development* 2009, **136**: 2493–2503.

Lasko, P.: **mRNA localization and translational control in *Drosophila* oogenesis.** *Cold Spring Harb. Perspect. Biol.* 2012, **4**: a012294.

Martin, S.G., Leclerc, V., Smith-Litière, K., St Johnston, D.: **The identification of novel genes required for *Drosophila* anteroposterior axis formation in a germline clone screen using GFP-Staufen.** *Development* 2003, **130**: 4201–4215.

McIlroy, G., Foldi, I., Aurikko, J., Wentzell, J.S., Lim, M.A., Fenton, J.C., Gay, N.J., Hidalgo, A. **Toll-6 and Toll-7 function as neurotrophin receptors in the *Drosophila melanogaster* CNS.** *Nat. Neurosci.* 2013, **16**: 1248–1256.

2.15 The dorso-ventral axis of the egg is specified by movement of the oocyte nucleus followed by signaling between oocyte and follicle cells

González-Reyes, A., Elliott, H., St Johnston, D.: **Polarization of both major body axes in *Drosophila* by gurken-torpedo signaling.** *Nature* 1995, **375**: 654–658.

Jordan, K.C., Clegg, N.J., Blasi, J.A., Morimoto, A.M., Sen, J., Stein, D., McNeill, H., Deng, W.M., Tworoger, M., Ruohola-Baker, H.: **The homeobox gene *mirror* links EGF signalling to embryonic dorso-ventral axis formation through Notch activation.** *Nat. Genet.* 2000, **24**: 429–433.

Roth, S., Neuman-Silberberg, F.S., Barcelo, G., Schupbach, T.: **Cornichon and the EGF receptor signaling process are necessary for both anterior-posterior and dorsal-ventral pattern formation in *Drosophila*.** *Cell* 1995, **81**: 967–978.

van Eeden, F., St Johnston, D.: **The polarisation of the anterior-posterior and dorsal-ventral axes during *Drosophila* oogenesis.** *Curr. Opin. Genet. Dev.* 1999, **9**: 396–404.

2.16 The expression of zygotic genes along the dorso-ventral axis is controlled by Dorsal protein

Cowden, J., Levine, M.: **Ventral dominance governs segmental patterns of gene expression across the dorsal-ventral axis of the neurectoderm in the *Drosophila* embryo.** *Dev. Biol.* 2003, **262**: 335–349.

Harland, R.M.: **A twist on embryonic signalling.** *Nature* 2001, **410**: 423–424.

Markstein, M., Zinzen, R., Markstein, P., Yee, K.P., Erives, A., Stathopoulos, A., Levine, M.: **A regulatory code for neurogenic gene expression in the *Drosophila* embryo.** *Development* 2004, **131**: 2387–2394.

Reeves, G.T., Stathopoulos, A.: **Graded dorsal and differential gene regulation in the *Drosophila* embryo.** *Cold Spring Harb. Perspect. Biol.* 2009, **1**: a000836.

Reeves, G.T., Trisnadi, N., Truong, T.V., Nahmad, M, Katz, S., Stathopoulos, A.: **Dorsal-ventral gene expression in the *Drosophila* embryo reflects the dynamics and precision of the dorsal nuclear gradient.** *Dev. Cell* 2012, **22**: 544–557.

Rusch, J., Levine, M.: **Threshold responses to the dorsal regulatory gradient and the subdivision of primary tissue territories in the *Drosophila* embryo.** *Curr. Opin. Genet. Dev.* 1996, **6**: 416–423.

Stathopoulos, A., Levine, M.: **Dorsal gradient networks in the *Drosophila* embryo.** *Dev. Biol.* 2002, **246**: 57–67.

2.17 The Decapentaplegic protein acts as a morphogen to pattern the dorsal region

Affolter, M., Basler, K.: **The Decapentaplegic morphogen gradient: from pattern formation to growth regulation.** *Nat. Rev. Genet.* 2007, **8**: 663–674.

Ashe, H.L., Levine, M.: **Local inhibition and long-range enhancement of Dpp signal transduction by Sog.** *Nature* 1999, **398**: 427–431.

Mizutani, C.M., Nie, Q., Wan, F.Y., Zhang, Y.T., Vilmos, P., Sousa-Neves, R., Bier, E., Marsh, J.L., Lander, A.D.: **Formation of the**

BMP activity gradient in the *Drosophila* embryo. *Dev. Cell* 2005, **8**: 915–924.

Shimmi, O., Umulis, D., Othmer, H., O'Connor, M.B.: **Facilitated transport of a Dpp/Scw heterodimer by Sog/Tsg leads to robust patterning of the *Drosophila* blastoderm embryo.** *Cell* 2005, **120**: 873–886.

Srinivasan, S., Rashka, K.E., Bier, E.: **Creation of a Sog morphogen gradient in the *Drosophila* embryo.** *Dev. Cell* 2002, **2**: 91–101.

Wang, Y.-C., Ferguson, E.L.: **Spatial bistability of Dpp-receptor interactions driving *Drosophila* dorsal-ventral patterning.** *Nature* 2005, **434**: 229–234.

Wharton, K.A., Ray, R.P., Gelbart, W.M.: **An activity gradient of decapentaplegic is necessary for the specification of dorsal pattern elements in the *Drosophila* embryo.** *Development* 1993, **117**: 807–822.

2.18 The antero-posterior axis is divided up into broad regions by gap gene expression

Rivera-Pomar, R., Jäckle, H.: **From gradients to stripes in *Drosophila* embryogenesis: filling in the gaps.** *Trends Genet.* 1996, **12**: 478–483.

2.19 Bicoid protein provides a positional signal for the anterior expression of zygotic *hunchback*

Brand, A.H., Perrimon, N.: **Targeted gene expression as a means of altering cell fates and generating dominant phenotypes.** *Development* 1993, **118**: 401–415.

Gibson, M.C.: **Bicoid by the numbers: quantifying a morphogen gradient.** *Cell* 2007, **130**: 14–16.

Gregor, T., Tank, D.W., Wieschaus, E.F., Bialek W.: **Probing the limits to positional information.** *Cell* 2007, **130**: 153–164.

Jaeger, J., Martinez Arias, A.: **Getting the measure of positional information.** *PLoS Biol.* 2009, **7**: e81.

Ochoa-Espinosa, A., Yucel, G., Kaplan, L., Pare, A., Pura, N., Oberstein, A., Papatsenko, D., Small, S.: **The role of binding site cluster strength in Bicoid-dependent patterning in *Drosophila*.** *Proc. Natl Acad. Sci. USA* 2005, **102**: 4960–4965.

Simpson-Brose, M., Treisman, J., Desplan, C.: **Synergy between the Hunchback and Bicoid morphogens is required for anterior patterning in *Drosophila*.** *Cell* 1994, **78**: 855–865.

Struhl, G., Struhl, K., Macdonald, P.M.: **The gradient morphogen Bicoid is a concentration-dependent transcriptional activator.** *Cell* 1989, **57**: 1259–1273.

2.20 The gradient in Hunchback protein activates and represses other gap genes

Struhl, G., Johnston, P., Lawrence, P.A.: **Control of *Drosophila* body pattern by the hunchback morphogen gradient.** *Cell* 1992, **69**: 237–249.

Wu, X., Vakani, R., Small, S.: **Two distinct mechanisms for differential positioning of gene expression borders involving the *Drosophila* gap protein giant.** *Development* 1998, **125**: 3765–3774.

Box 2D Targeted gene expression and misexpression screening

Duffy, J.B.: **GAL4 system in *Drosophila*: a fly geneticist's Swiss knife.** *Genesis* 2002, **34**: 1–15.

Zhong, J., Yedvobnick, B.: **Targeted gain-of-function screening in *Drosophila* using GAL4-UAS and random transposon insertions.** *Genet. Res. (Camb.)* 2009, **91**: 243–258.

2.21 Parasegments are delimited by expression of pair-rule genes in a periodic pattern

Clyde, D.E., Corado, M.S., Wu, X., Paré, A., Papatsenko, D., Small, S.: **A self-organizing system of repressor gradients establishes segmental complexity in *Drosophila*.** *Nature* 2003, **426**: 849–853.

Hughes, S.C., Krause, H.M.: **Establishment and maintenance of parasegmental compartments.** *Development* 2001, **128**: 1109–1118.

Jaynes, J.B., Fujioka, M.: **Drawing lines in the sand: even skipped et al. and parasegment boundaries.** *Dev. Biol.* 2004, **269**: 609–622.

Yu, D., Small, D.: **Precise registration of gene expression boundaries by a repressive morphogen in *Drosophila*.** *Curr. Biol.* 2008, **18**: 888–876.

2.22 Gap gene activity positions stripes of pair-rule gene expression

Small, S., Levine, M.: **The initiation of pair-rule stripes in the *Drosophila* blastoderm.** *Curr. Opin. Genet. Dev.* 1991, **1**: 255–260.

2.23 Expression of the *engrailed* gene defines the boundary of a parasegment, which is also a boundary of cell-lineage restriction and 2.24 Segmentation genes stabilize parasegment boundaries and 2.25 Signals generated at the parasegment boundary delimit and pattern the future segments

Alexandre, C., Lecourtois, M., Vincent, J.-P.: **Wingless and hedgehog pattern *Drosophila* denticle belts by regulating the production of short-range signals.** *Development* 1999, **126**: 5689–5698.

Chanut-Delalande, H., Fernandes, I., Roch, F., Payre, F., Plaza, S.: **Shavenbaby couples patterning to epidermal cell shape control.** *PLoS Biol.* 2006, **4**: e290.

Dahmann, C., Basler, K.: **Compartment boundaries: at the edge of development.** *Trends Genet.* 1999, **15**: 320–326.

Hatini, V., DiNardo, S.: **Divide and conquer: pattern formation in *Drosophila* embryonic epidermis.** *Trends Genet.* 2001, **17**: 574–579.

Hooper, J.E., Scott, M.P.: **Communicating with Hedgehogs.** *Nat. Rev. Mol. Cell Biol.* 2005, **6**: 306–317.

Larsen, C.W., Hirst, E., Alexandre, C., Vincent, J.-P.: **Segment boundary formation in *Drosophila* embryos.** *Development* 2003, **130**: 5625–5635.

Lawrence, P.A., Casal, J., Struhl, G.: **Hedgehog and engrailed: pattern formation and polarity in the *Drosophila* abdomen.** *Development* 1999, **126**: 2431–2439.

Sanson, B.: **Generating patterns from fields of cells. Examples from *Drosophila* segmentation.** *EMBO Rep.* 2001, **2**: 1083–1088.

Tolwinski, N.S., Wieschaus, E.: **Rethinking Wnt signaling.** *Trends Genet.* 2004, **20**: 177–181.

Vincent, J.P., O'Farrell, P.H.: **The state of *engrailed* expression is not clonally transmitted during early *Drosophila* development.** *Cell* 1992, **68**: 923–931.

von Dassow, G., Meir, E., Munro, E.M., Odell, G.M.: **The segment polarity network is a robust developmental module.** *Nature* 2000, **406**: 188–192.

Wiese, K., Nusse, R, van Amerongen, R.: **Wnt signaling: conquering complexity.** *Development* 2018, **145**: dev165902.

Box 2E The Hedgehog signaling pathway

Briscoe, J., Thérond, P.P.: **The mechanisms of Hedgehog signalling and its roles in development and disease.** *Nat. Rev. Mol. Cell Biol.* 2013, **14**: 416–429.

Box 2F Mutants in denticle pattern provided clues to the logic of segment patterning

Martinez Arias, A.: **A cellular basis for pattern formation in the insect epidermis.** *Trends Genet.* 1989, **5**: 262–267.

Martinez Arias, A., Baker, N.E., Ingham, P.W.: **Role of segment polarity genes in the definition and maintenance of cell states in the *Drosophila* embryo.** *Development* 1988, **103**: 157–170.

2.26 Segment identity in *Drosophila* is specified by Hox genes

Hueber, S.D., Lohmann, I.: **Shaping segments: Hox gene function in the genomic age.** *BioEssays* 2008, **30**: 965–979.

Lewis E.B.: **A gene complex controlling segmentation in *Drosophila*.** *Nature* 1978, **276**: 565–570.

Mallo, M., Alonso, C.R.: **The regulation of Hox gene expression during animal development.** *Development* 2013, **140**: 3951–3963.

2.27 Homeotic selector genes of the bithorax complex are responsible for diversification of the posterior segments and

2.28 The Antennapedia complex controls specification of anterior regions

Castelli-Gair, J., Akam, M.: **How the Hox gene *Ultrabithorax* specifies two different segments: the significance of spatial and temporal regulation within metameres.** *Development* 1995, **121**: 2973–2982.

Lawrence, P.A, Morata, G.: **Homeobox genes: their function in *Drosophila* segmentation and pattern formation.** *Cell* 1994, **78**: 181–189.

Mann, R.S., Morata, G.: **The developmental and molecular biology of genes that subdivide the body of *Drosophila*.** *Annu. Rev. Cell Dev. Biol.* 2000, **16**: 143–271.

Mannervik, M.: **Target genes of homeodomain proteins.** *BioEssays* 1999, **4**: 267–270.

Simon, J.: **Locking in stable states of gene expression: transcriptional control during *Drosophila* development.** *Curr. Opin. Cell Biol.* 1995, **7**: 376–385.

2.29 The order of Hox gene expression corresponds to the order of genes along the chromosome

Morata, G.: **Homeotic genes of *Drosophila*.** *Curr. Opin. Genet. Dev.* 1993, **3**: 606–614.

2.30 The *Drosophila* head region is specified by genes other than the Hox genes

Rogers B.T., Kaufman, T.C.: **Structure of the insect head in ontogeny and phylogeny: a view from *Drosophila*.** *Int. Rev. Cytol.* 1997, **174**: 1–84.

Vertebrate development I: establishing the body plan

- Outline of vertebrate development
- Experimental approaches to studying vertebrate development
- *Xenopus* development and setting up the body plan
- The origins and specification of the germ layers and formation of the organizer in *Xenopus*
- Zebrafish development and setting up the body plan
- Chick development and setting up the body plan
- Mouse development and setting up the body plan
- Development of the human body plan

In this and the next chapter we consider how the body plan is laid down in vertebrate embryos. The development of the body plan is a continuous process from the fertilized egg to a stage at which the clearly recognizable antero-posterior and dorso-ventral axes of the body have been established and the nervous system is developing. In this chapter we focus on the initial establishment of the body axes. We describe the development of the body axes in particular detail in the African clawed frog Xenopus laevis, introducing the classic experiments that revealed the processes and key cell populations—in particular the embryonic organizer (the Spemann organizer)—that establish the characteristic body plan. We then introduce our other model organisms—the zebrafish Danio rerio, the chick, and the mouse—and explain the steps that lead to axis formation and organizer development, pointing out key similarities and differences in the different species. Finally, we look at the evidence that the human body plan is set up according to the same principles that have been discovered in other vertebrates. The events described in this chapter overlap, in time, with those described in Chapter 4, where we consider the role of the organizer in the elaboration of the dorso-ventral and antero-posterior axes and the establishment of the nervous system.

Chapter 3 Vertebrate development I: establishing the body plan

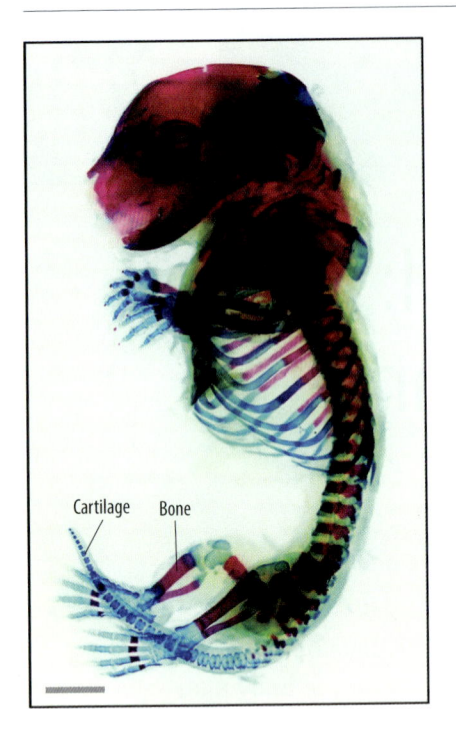

Fig. 3.1 The skeleton of a 17.5-day (E17.5) mouse embryo illustrates the vertebrate body plan. The skeletal elements in this embryo have been stained with Alcian blue (which stains cartilage) and Alizarin red (which stains bone). The brain is enclosed in a bony skull. The vertebral column runs along the dorsal side (back) of the body and is divided up into cervical (neck), thoracic (chest), lumbar (lower back), and sacral (hip and lower) regions. The paired limbs can also be seen. The rib cage extends from the dorsal to the ventral side of the thoracic region. Scale bar = 1 mm.

Photograph courtesy of M. Maden.

Introduction

In *Drosophila*, mRNAs and proteins laid down in the egg by the mother (maternal factors) interact with each other to specify broadly different regions of the body (see Chapter 2). This initial plan is then elaborated on by the embryo's own genes. We now look at how the task of establishing the body plan is achieved in vertebrate development. All vertebrates, despite their many outward differences, have a similar basic body plan. The defining structures are a segmented backbone or **vertebral column** surrounding the spinal cord, ahead of which is the brain, enclosed in a bony or cartilaginous skull (Fig. 3.1). These prominent structures mark the **antero-posterior axis**, the main body axis of vertebrates. The head is at the anterior end of this axis, followed by the trunk with its paired appendages—limbs in terrestrial vertebrates (with the exception of snakes and limbless lizards) and fins in fish—and in many vertebrates this axis terminates at the posterior end in a post-anal tail. The vertebrate body also has a distinct **dorso-ventral axis** running from the back to the belly, with the brain and spinal cord running along the dorsal side in the midline, and the mouth and gut on the ventral side. The antero-posterior and dorso-ventral axes together define the left and right sides of the animal. Vertebrates have a general **bilateral symmetry** around the midline so that outwardly the right and left sides are mirror images of each other. Some internal organs, such as lungs, kidneys, and gonads, are present as symmetrically paired structures, although their detailed anatomy or position can differ on the right and left sides of the body. Single organs such as the heart and liver are arranged asymmetrically with respect to the midline, with the heart on the left and the liver on the right. The overall similarity of the body plan in all vertebrates suggests that the developmental processes that establish it are broadly similar in the different animals. This is, indeed, largely the case, although there are some notable differences in development, especially at the earliest stages.

Our understanding of the general principles of vertebrate development comes from studying several different model organisms. In this chapter, we first introduce our model vertebrates. We outline the key stages in vertebrate development that result in an embryo with well-defined antero-posterior and dorso-ventral body axes. At the same time as the axes of the embryo are being formed, cells are also being allocated to the different germ layers—ectoderm, mesoderm, and endoderm (see Box 1D). The ectoderm gives rise to the epidermis of the skin and the nervous system; the mesoderm gives rise to the skeleton, muscle, heart, blood, kidneys, and some other internal organs and tissues; and the endoderm gives rise to the gut and associated glands and organs such as the liver and pancreas. We then consider the experimental techniques for studying vertebrate development and the advantages and disadvantages of the different models. After this, we describe each of the model organisms in turn and consider the early stages of development from the fertilized egg up to the time that the embryo undergoes gastrulation—the dynamic process that results in germ layers in their final locations and that makes the body axes of the larva or adult evident.

We focus in particular on early development in *Xenopus laevis*. This is because the classic experiments that provided insight into the mechanisms that establish the vertebrate body plan were carried out on amphibian embryos. We cover the stages from fertilization to gastrulation, and describe the mechanisms behind germ-layer formation and the onset of mesoderm patterning—including development of the embryonic organizer. We then examine the mechanisms that lead to germ-layer formation, the establishment of the organizer, and gastrulation in the zebrafish, the chick, and the mouse, pointing out key similarities and differences in the way in which the early embryo develops and sets up the body plan. Finally, we touch on early human development and outline the advances that have enabled the earliest stages in human embryos to be studied.

In Chapter 4, we look at how the organizer directs the ongoing patterning of the germ layers, including an all-important function—the induction of the future nervous system—and expand on how these events elaborate the body axes. We go on to consider how the body plan is completed, through the further patterning of the mesoderm and the development of the peripheral nervous system, using the chick and the mouse as our main models. Finally, we touch on how left–right asymmetry is established. The development of individual structures and organs, such as limbs, eyes, heart, and the nervous system, is covered in Chapters 9 and 10. In Chapter 5 we take a more detailed look at the mechanics of vertebrate early development and gastrulation, and the underlying cell biology that allows such extensive remodeling of tissues.

Outline of vertebrate development

All vertebrates pass through the same basic stages in early development, including fertilization, cleavage, gastrulation, neurulation, and organogenesis, illustrated in the simplified version of the life cycle of *Xenopus* in Box 1A. Early development has been particularly well studied in four vertebrates—*Xenopus*, zebrafish, chick, and mouse. The reason we need to consider all these different models is that some provide more information than others about certain aspects of development and are more amenable to particular experimental approaches. And even though the earliest stages in human development can now be studied directly, some model organisms are particularly informative in helping us to understand how the general principles involved in laying down the vertebrate body plan apply to humans.

3.1 All vertebrate embryos pass through a broadly similar set of developmental stages

Figure 3.2 compares the embryos of the four model vertebrates and a human embryo at equivalent stages of development. The details of the developmental stages for each model organism and the human embryo are shown in life-cycle diagrams later in the chapter.

The eggs of frog (*Xenopus*), zebrafish, chicken, and mouse are very different in size; human and mouse eggs are similar in size (see Fig. 3.2, top row). These size differences are in part due to the amount of the yolk in the egg, although the very large chicken egg also contains albumin (egg white). Yolk provides all the nutrients for fish, amphibian, reptile, and bird embryonic development, and for the few egg-laying mammals such as the platypus. The eggs of most mammals, in contrast, are small and non-yolky, and mammalian embryos are nourished for the first few days by fluids in the **oviduct**. After that, the embryos implant in the uterine wall. The embryo is then nourished by the mother and extra-embryonic tissues develop that contribute to the formation of the **placenta**.

After fertilization, the **zygote** undergoes **cleavage**—a series of rapid cell divisions that divide the embryo into a number of smaller cells, often called **blastomeres**. Cleavage is affected by the amount of yolk in the egg. In *Xenopus* eggs, which contain a relatively small amount of yolk, cleavage divides the fertilized egg completely into smaller cells and results in a spherical ball of cells. The same is true for mouse and human. In zebrafish and chicken eggs, which contain a substantial amount of yolk, cleavage occurs incompletely throughout the fertilized egg and this results in a number of smaller cells lying on top of the yolk. So, by the end of cleavage, the embryos of the different vertebrates look very different.

Cleavage is followed by a stage in which the cells in the embryo become arranged in cell sheets. In *Xenopus*, this is accomplished by the development of an internal cavity, around which forms a continuous spherical sheet, several cells deep, that will

104 Chapter 3 Vertebrate development I: establishing the body plan

| Xenopus | Zebrafish | Chick | Mouse | Human |

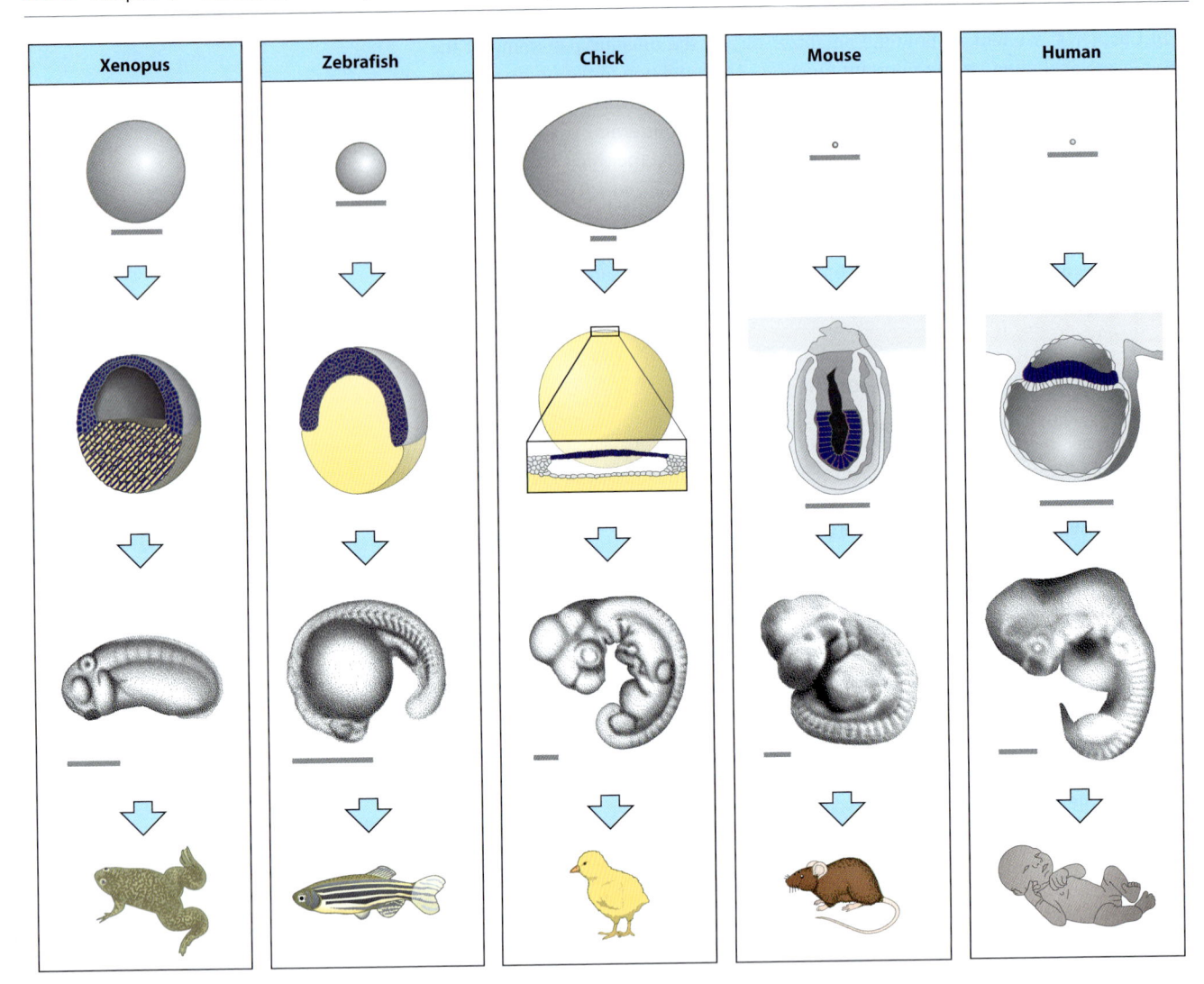

Fig. 3.2 Vertebrate embryos show considerable differences in form before gastrulation but subsequently all go through a stage at which they look similar. Top row: relative sizes of the eggs of frog (*Xenopus*), zebrafish, chicken, mouse, and human. Scale bars in this row represent 1 mm, except for the chicken egg, where the scale bar represents 10 mm. Second row: at a stage just before gastrulation, the early development of all these species differs. All embryos are shown in cross-section. The yolk is shown in yellow and the cells at this stage that will form the tissues of the embryo in zebrafish, chicken, mouse, and human in purple. The *Xenopus* embryo is a ball of cells with an internal cavity and, as seen in cross-section, basically consists of a continuous sheet of cells surrounding the cavity. The developing zebrafish is a mound of cells lying on top of the yolk. The embryo arises from the deep layer of cells that lies under the outer enveloping layer. The deep layer thins out into a sheet several cells thick as the blastoderm spreads over the yolk. The developing chick is a disc of cells seen in the high-power cross-section to comprise two epithelial sheets lying on top of the yolk; the upper epithelium (called the **epiblast**) gives rise to the embryo. The mouse and human embryos at an equivalent stage have implanted into the uterine wall and have already developed some extra-embryonic membranes required for formation of the placenta. The developing mouse is the lower half of the cup-shaped structure (the egg cylinder) at the center, seen in cross-section to comprise U-shaped epithelial sheets; the inner epithelium (the epiblast) gives rise to the embryo. The developing human is similar to the chick—a disc of cells, seen in cross-section as comprising two epithelial sheets; the upper epithelium (the epiblast) gives rise to the embryo. Scale bars for mouse and human embryos represent 100 μm. Third row: after gastrulation and formation of the neural tube, vertebrate embryos pass through an embryonic stage in which they all look rather similar. The body plan has been laid down, and notochord, somites, and neural tube are present. This is known as the phylotypic stage and is when organogenesis begins. Scale bars represent 1 mm. Bottom row: after the phylotypic stage, the development of different vertebrates diverges again. Paired appendages, for example, develop into fins in fish, wings and legs in the chick, and arms and legs in humans.

give rise to the embryo (see Fig. 3.2, second row, for *Xenopus* blastula and equivalent stages in zebrafish, chick, mouse, and human). The *Xenopus* embryo at this stage is known as a **blastula** (see Box 1A). At the equivalent stage in zebrafish, called the **blastoderm** or **sphere stage**, the embryo consists of a mound of cells lying on top of the yolk. The **deep layer**—the rounded cells under the outer enveloping layer—thin out to form a sheet several cells thick as the blastoderm spreads over the yolk, and these cells give rise to the embryo. The equivalent stage in chick is known as the **blastoderm** or **blastodisc**. The blastoderm is a disc of cells lying on top of the yolk, shown in cross-section to comprise two epithelial sheets; the upper epithelium gives rise to the embryo. The lower epithelium will give rise to extra-embryonic membranes that are used to obtain nutrients from the yolk, to exchange oxygen and carbon dioxide through the permeable eggshell, and to dispose of waste.

In mouse and human embryos, cleavage divisions give rise to a ball of cells, within which a cavity forms, leaving a small group of cells internally at one end—the **inner cell mass**—which will give rise mainly to the embryo (Fig. 3.3). This stage is known as the **blastocyst**. Cells from the inner cell mass of the blastocyst can be cultured as **embryonic stem cells (ES cells)**, which are pluripotent and can differentiate into all cell types *in vitro* (discussed in Chapter 7). The blastocyst implants into the uterus and the layer of cells on the outside of the blastocyst, the **trophectoderm**, gives rise to extra-embryonic tissues that contribute to the placenta and membranes that protect the developing embryo. The equivalent stage in mouse to the *Xenopus* blastula comes later, after implantation, at a time when the mouse embryo and associated structures have elongated and are known as the **egg cylinder** (see Fig. 3.2). The egg cylinder is cup-shaped, shown in cross-section to comprise two U-shaped epithelial cell sheets; the inner epithelium of the lower half of the egg cylinder gives rise to the embryo while the outer epithelium forms extra-embryonic membranes. The equivalent stage in human embryos is like that in the chick. It consists of a disc of cells, shown in cross-section to comprise two epithelial sheets, and is known as the **bilaminar germ disc**; the upper epithelium gives rise to the embryo while the lower epithelium contributes to extra-embryonic membranes. Mammals and birds, both of which form an extra-embryonic membrane called an amnion, are known as **amniotes**, whereas amphibians and fish, which do not form this extra-embryonic membrane, are known as **anamniotes**.

The next stage in development is **gastrulation**. The embryo during gastrulation is known as a **gastrula** (see Box 1A). Gastrulation is orchestrated by the **embryonic organizer**, a specialized group of cells whose properties will be introduced in this chapter and expanded on in Chapter 4. Gastrulation is a set of highly coordinated cell movements during which the germ layers assume their correct position within the body plan. The endoderm and mesoderm cells move from the outer surface of the embryo to the inside where they give rise to internal organs. The cells remaining on the outside of the embryo are ectoderm cells and give rise to the epidermis and the nervous system (see Box 1D).

Gastrulation is the stage at which the main antero-posterior and dorso-ventral body axes become evident. A rod-like structure develops by rearrangement of the cells in and around the organizer and runs from head to tail along the antero-posterior axis. At its anterior end is the **prechordal mesoderm**; the long posterior region is the **notochord**. The prechordal mesoderm and the notochord are collectively known as **axial mesoderm**.

The axial mesoderm defines the dorsal midline. Mesodermal regions that lie progressively more 'ventrally' to the notochord form, respectively, **somites**, **intermediate mesoderm**, and **lateral plate mesoderm**, each arranged with bilateral symmetry around the midline notochord (Fig. 3.4). The somites are blocks of mesoderm that give rise to the vertebral column and contribute to the trunk and limb musculature. The intermediate mesoderm develops into the kidneys. The lateral plate

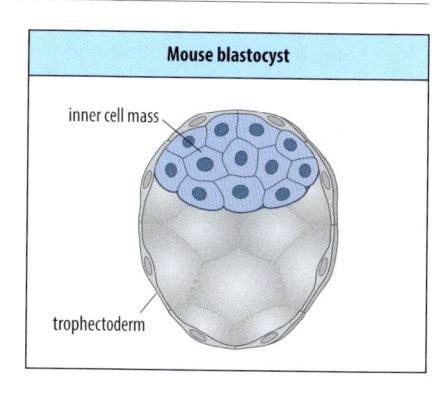

Fig. 3.3 Mouse blastocyst. The fertilized egg has undergone rapid cleavage divisions and has formed a hollow blastocyst with a small group of internal cells, the inner cell mass (blue), at one end of the cavity. The embryo will develop from the inner cell mass. The layer of cells around the outside of the blastocyst, the trophectoderm, will develop into extra-embryonic structures. A human blastocyst has the same anatomy.

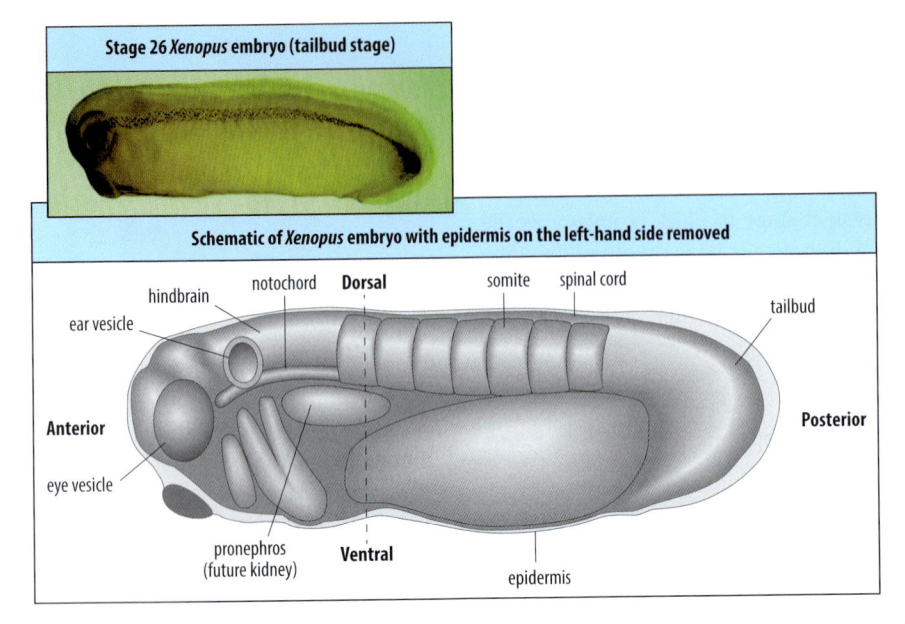

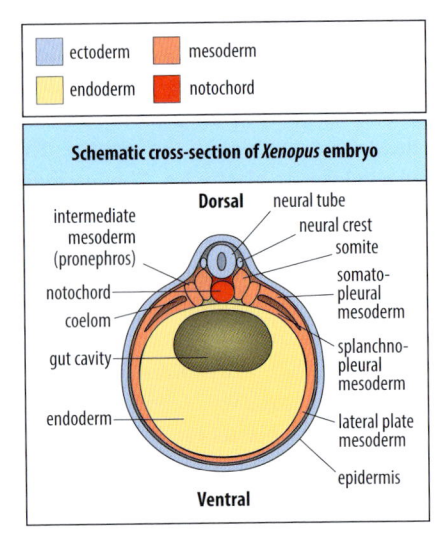

Fig. 3.4 A *Xenopus* embryo at the phylotypic stage. The side view in the photograph at the top shows the notochord (stained brown), the signature structure of the chordates, running along the dorsal midline of the embryo from anterior to posterior. Scale bar = 1 mm. Left panel: diagram of the side view of the embryo with epidermis removed on the left side. At the anterior end of the embryo, there is a well-developed head with prominent sense organs—eye and ear vesicles. The anterior part of the neural tube has formed the brain. More posteriorly, the neural tube forms the hindbrain and spinal cord; the notochord can be seen lying under the developing hindbrain. In the mesoderm, a succession of somites lies on either side of the notochord, and the embryonic kidney (pronephros) is beginning to form. At the posterior end of the embryo is the tailbud, which will give rise to the tail of the tadpole by forming a continuation of somites, neural tube, and notochord. Right panel: a cross-section taken across the trunk of the embryo at the position of the dotted line shows an internal view. The neural tube that will form the spinal cord is on the dorsal side of the embryo. The neural crest is shown on either side of the neural tube. The mesoderm has formed notochord, somites, intermediate mesoderm, and lateral plate mesoderm. Ventral to the notochord and somites is the gut, which terminates in the anus. The coelom has developed in the mesoderm, splitting the lateral plate mesoderm into an outer somatopleural layer and an inner splanchnopleural layer.

Photograph courtesy of B. Herrmann.

mesoderm extends ventrally to form a continuous layer lying between ectoderm and endoderm. Mesodermal regions along the antero-posterior axis form different structures. For example, the anterior somites form cervical vertebrae whereas posterior somites form lumbar vertebrae. Likewise, the anterior lateral plate mesoderm forms the heart, whereas the limbs arise from more posterior lateral plate mesoderm at specific positions along the antero-posterior axis.

Gastrulation is followed by the stage known as **neurulation** (see Box 1A), in which a morphologically distinct neural tube forms. This neural tube will develop into the **central nervous system** (discussed in Chapters 5 and 10). The embryo at this stage is known as a **neurula**. The earliest visible signs of neurulation are the neural folds, which form on the edge of the neural plate, an area of thickened ectoderm overlying the axial mesoderm and adjacent mesoderm. The folds rise up, fold towards the midline, and fuse together to form the **neural tube**, which then sinks beneath the epidermis (see Box 1A). This stage is called the **phylotypic stage** and a *Xenopus* embryo at this stage is shown in Figure 3.4. The anterior neural tube above the prechordal mesoderm gives rise to the forebrain; further back, the neural tube overlying the notochord develops into the hindbrain and spinal cord. Neural crest cells arise at the borders of the neural plate in both the head and the trunk, and migrate throughout the body to form an enormous range of different tissues and cell types, including cells of the peripheral nervous system and pigment cells of the skin (discussed in Chapters 4 and 5).

During the neurula stage, a cavity develops in the lateral plate mesoderm. This cavity becomes the **coelom**, the fluid-filled cavity in vertebrates that later surrounds the internal organs and is lined by a mesoderm-derived epithelium. The coelom splits the lateral plate mesoderm into an outer **somatopleural** layer and an inner **splanchnopleural** layer (see Fig. 3.4). The somatopleural mesoderm contributes to the connective tissues of the body wall and the limbs, whereas the splanchnopleural layer gives rise to the heart, blood vessels, and the connective tissues of the gut wall. By this stage, a gut tube lined by the endoderm has formed in the ventral part of the embryo, running along the antero-posterior axis. The anterior region of the tube forms the foregut, which also gives rise to organs such as the lungs. More posterior regions form midgut and hindgut.

The final developmental stage is **organogenesis**, in which the various organs form, including the paired appendages (fins or limbs), sense organs, and internal organs. Early in organogenesis, all vertebrate embryos pass through a stage at which they resemble each other more closely (see Fig. 3.2, third row) and show embryonic features characteristic of the chordates—the phylum of animals to which vertebrates belong.

The signature structure of the chordates is the notochord. The notochord is a transitory structure, and its cells eventually become incorporated into the vertebral column that forms the spine (see Fig. 3.1). After the phylotypic stage, the morphology of different vertebrate embryos again diverges as organogenesis proceeds and features unique to different vertebrate groups, such as beaks, wings, and fins, develop. After organogenesis, there may be a period of continued growth and maturation that can be extensive, as in human development.

The time taken from fertilization to each defined developmental stage varies in the different embryos. The blastula stage or its equivalent (shown in Fig. 3.2) is reached after about 6 hours in *Xenopus* and zebrafish embryos, after about 20 hours in chick, after 5 or 6 days in mouse, and after 14 days in the human embryo. The longer times taken by mammals can partly be explained by the need to generate the extra-embryonic tissues that will contribute to the placenta. Nevertheless, even taking this into account, the tempo of development is still slower in mammalian embryos, as it takes the mouse embryo 3–4 days to complete the cleavage stages and the human embryo 5–6 days. The differences in timing persist at later stages, with the phylotypic stage being reached about 18 hours after the blastula stage in *Xenopus* and the blastoderm stage in zebrafish, but almost 3 days after the equivalent stages in chick, 4 days in mouse, and 26 days in human embryos.

The timing of development is likely to involve both intrinsic and extrinsic factors and regulation of the speed of the cell cycle. In amphibians, for example, temperature affects the rate of development. Because such environmental factors can affect the rate of development, developmental biologists divide the normal embryonic development of each species into a series of numbered stages, which are identified by their main morphological features—for example, the number of somites—rather than by time after fertilization. Links to websites that describe all these numbered stages for our four model vertebrates are given in the 'General further reading' section.

Experimental approaches to studying vertebrate development

Developmental biologists harness a wide range of experimental techniques to study the events that underlie formation of the embryo. Here we focus on experimental approaches that are specific to or particularly useful to developmental biology. Details of some other general techniques that are widely used in developmental biology are noted elsewhere in the book.

3.2 Normal development can be visualized by various types of imaging, by detecting patterns of gene and protein expression, and by making fate maps

The very first studies of vertebrate embryonic development involved careful and detailed analysis of whole and dissected embryos under the microscope, and simply watching, describing, and drawing the embryos. As development is a dynamic process, time-lapse filming became an important tool to monitor changes in the shape of the embryo during development and to investigate the behavior of individual cells. Technical advances in imaging have benefited developmental biology over the past 20 years, with improvements in computer-aided microscopic imaging techniques, the development of fluorescent labels in a vast range of colors that allow the imaging of living embryos (see Box 1E), and the introduction of new forms of imaging, such as **optical projection tomography (OPT)**, **magnetic resonance imaging (MRI)**, X-ray microtomography, and **light-sheet fluorescence microscopy (LSFM)**. OPT generates three-dimensional digitized images of fixed embryos, whereas the latter three techniques can be used to image live embryos and film their development. LSFM allows individual cell behaviors in living embryos to be filmed at high resolution. In this technique, the repeated exposure of the embryo to the harmful effects of light is mitigated by illuminating consecutive thin 'slices' through the embryo and using computer programs to assemble the images in three dimensions and analyze the data.

As cells in the early embryo begin to develop into tissues and organs, they are characterized both through their position in the embryo and by the expression of cell-type-associated genes. Anatomical and cellular descriptions of normal development are therefore supplemented with information about where and when genes are expressed. The spatial patterns of expression of developmentally important genes are visualized by *in situ* **nucleic acid hybridization**. This technique uses labeled RNA or DNA probes complementary to an RNA of interest to detect gene transcripts, and so can visualize gene expression in cells and tissues (see Box 1E). A range of different colors of fluorescent dyes allows the expression of more than one target RNA to be mapped in the same tissue preparation or whole-mount embryo (Fig. 3.5); this is particularly important because a cell type is defined by a combination of expressed genes, and not simply by one or a few.

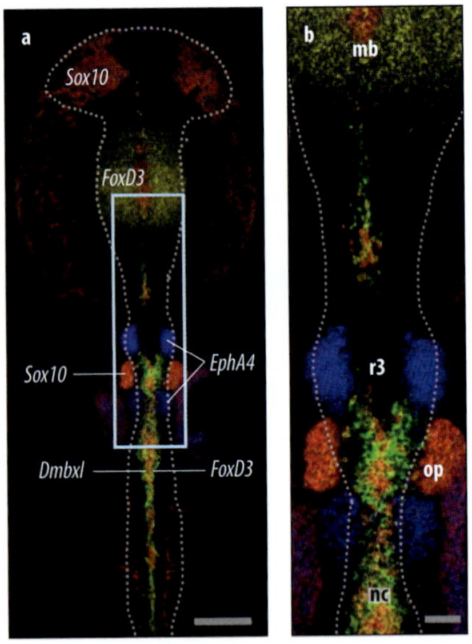

The most recent development of this approach is the technique of *in situ* **DNA hybridization chain reaction**. In this technique, the DNA probes are engineered such that once bound to their target they initiate a chain reaction in which large numbers of fluorescently tagged DNAs become tethered to the target RNA, thus amplifying the signal. This technique has the advantage over previous hybridization techniques in that multiple target RNAs can be detected simultaneously in the same specimen (see Fig. 3.5), and furthermore, the same specimen can be repeatedly re-probed. Routinely, therefore, 12–15 target RNAs can be examined in a single specimen.

Gene expression at the protein level can be visualized using **immunohistochemistry**. The proteins expressed from gene transcripts can be detected by specific antibodies labeled with fluorescent dyes (see Box 1E). Transcription factors act in the cell in which they are made, and so the distribution of the protein will be the same as the distribution of gene transcripts. But secreted proteins can be more widely distributed and act on cells some distance away from their source. So, to understand the operation of the many important secreted signal proteins in development, nucleic acid hybridization is needed to locate the transcripts, and immunohistochemistry to locate where the proteins are acting. We can also identify the location of the cells responding to the signal proteins by monitoring activation of the signal transduction pathway or expression of known target genes.

The entire set of genes being expressed in a particular tissue or whole embryo at a particular stage in development can be documented by techniques such as

Fig. 3.5 *In situ* hybridization chain reaction imaged in whole-mount chick embryo at Hamilton and Hamburger stage 10. (a) Four mRNAs expressed in different regions of the brain and neural crest have been targeted. Four-channel confocal micrograph with all four probes: *FoxD3* (green), *EphA4* (blue), *Sox10* (red), and *Dmbxl* (yellow). *FoxD3* labels pre-migrating and migrating neural crest at all axial levels; *EphA4* labels hindbrain rhombomere 3 and cells that have migrated to the otic placode (scattered blue cells in closer view); *Sox10* labels migrating neural crest, head mesenchyme, and otic placodes; *Dmbx1* labels midbrain. (b) Closer view of the boxed region in panel (a). White dots show the outline of neural tube. mb, midbrain; nc, neural crest; op, otic placode; r3, rhombomere 3. Scale bars = 200 μm (a) and 50 μm (b).

From Choi, H.M.T. et al.: *Third-generation* in situ *hybridization chain reaction: multiplexed, quantitative, sensitive, versatile, robust.* Development 2018, **145**: dev165753.

DNA microarray analysis and **RNA seq**. These techniques rely on genomic information to identify the mRNAs being transcribed. This field has progressed rapidly, and it is now commonplace to perform **single-cell RNA seq (scRNA-seq)** to analyze the total gene expression (the transcriptome) in a single cell (discussed in Box 6C). The field of **single-cell transcriptional analysis** has exploded in the past few years, and we now have unparalleled insight into changes in gene expression that occur at key stages of embryonic development, as well as changes that occur during the development of specific tissues and cells. Until recently, scRNA-seq analysis has relied on disaggregating cells from regions of interest dissected from the embryo, as in the scRNA analysis of gene expression in cells of both gastrulating mouse and human embryos (discussed in Section 6.5). Now, techniques are being developed that can integrate the single-cell transcriptomic data with spatial information on the location of the cells within the developing tissues, a technique termed **spatial transcriptomics**. For example, this approach has been used to follow the patterning of the midbrain–hindbrain boundary and the developing gut tube in the mouse embryo.

Another technique that depends on knowing the genome sequence is chromatin immunoprecipitation followed by DNA microarray analysis (**ChIP-on-chip** or **ChIP-chip**) or DNA sequencing (**ChIP-seq**). Specific antibodies are used to pull down a target protein such as a transcription factor, together with the DNA to which it is bound, from a sample. The DNA is then identified by its sequence. This technique can be used to identify the regulatory regions in the target genes that a given transcription factor is bound to *in vivo*. This information is used to help describe the gene regulatory networks that underlie many developmental processes (see Box 1F).

Other important information about normal development comes from making **fate maps**, which show what a particular region of the embryo or cell population will give rise to at a later time (see Section 1.12). Cells in particular regions can be labeled physically by injecting a non-toxic marker such as a fluorescent dye, or can be labeled genetically. Genetic fate mapping, often called **genetic lineage analysis**, involves the creation of a transgenic 'reporter' animal, in which, commonly, a cell-specific promoter is fused to a gene for a reporter such as green fluorescent protein (GFP). The marking of the cell involves a permanent change in its DNA that allows the cell to be identified from its unmarked neighbors. The genetic nature of the labeling ensures the transmission of the marker to all the cell's progeny, circumventing the problem of dilution of a dye marker during cell division. By choosing the appropriate promoter, a population of cells of interest can be precisely marked genetically at any time point during development and its progeny then followed in real time.

Fate maps can only indicate what the cells will normally give rise to if left in place, and not whether their fate is already determined or is still plastic. In other words, we know what the cells will give rise to—but the embryo does not. To find out whether the fate of cells has been **specified**, they are placed in culture and observed to see what tissues they give rise to; cells that are specified as a particular cell type or tissue can, however, still have their fates changed. To find out whether cells have lost this plasticity and become **determined**, the classic experiment is to transplant them to a different position on another embryo and see whether they develop in line with their original fate or in line with the fate of their new site (see Section 1.12). Techniques for making fate maps and their application to the early *Xenopus* embryo are explained in Box 3A.

3.3 Disrupting the developmental process in specific ways helps to discover underlying mechanisms

Descriptions on their own cannot unravel the mechanisms underlying developmental processes, and the only way to find out more about these is to disturb the developmental process through 'gain-of-function' or 'loss-of-function' experiments and see what

EXPERIMENTAL BOX 3A Fate mapping and lineage tracing reveal which cells in which parts of the early *Xenopus* embryo give rise to particular structures in the tadpole

For both biological and technical reasons, fate mapping in the early embryo is easiest in *Xenopus*. Even at the blastula stage, cells that will eventually form the three germ layers are arranged in distinct regions and are accessible from the surface of the embryo. Single blastomeres can be easily labeled by injecting a non-toxic marker such as fluorescein-linked dextran amine or by expression of a fluorescent protein such as GFP from injected RNA. Fluorescein-labeled dextran amine is a stable, high-molecular-weight molecule, which cannot pass through cell membranes and so is restricted to the injected cell and its progeny; as fluorescein fluoresces green when excited by ultraviolet (UV) light, the fluorescein-labeled dextran can be easily detected under a fluorescence microscope. GFP produces a green light when excited by UV light and stable forms are now widely used because the intense fluorescence can be viewed in living embryos and allows long-term lineage tracing, a general technique known as genetic lineage analysis. Figure 1 shows a fate map for one of the cells at the 32-cell stage in the *Xenopus* embryo. In this case the cell was labeled with fluorescein-dextran amine.

Figure 1 (left panel) shows the labeling of a single cell (C4) in the embryo by injection of fluorescein-dextran amine, which fluoresces green under UV light. Figure 1 (right panel) shows a cross-section of the embryo after it has developed to the tailbud stage, which reveals that the labeled cell has given rise to lateral plate mesoderm and part of a ventral somite on one side of the embryo.

By following the fate of individual cells or groups of cells in *Xenopus* embryos, we can build up a map on the blastula surface showing the regions that will give rise to, for example, notochord, somites, nervous system, or gut. Figure 2 shows a fate map for *Xenopus* in which the cells were labeled at the late blastula stage. The fate map shows that dorsal and ventral regions of ectoderm (shades of blue) give rise to the nervous system and epidermis, respectively. Along the dorso-ventral axis, the mesoderm (red and peach) gives rise to axial mesoderm (prechordal mesoderm and notochord), somites, heart, kidneys, and blood. The endoderm (yellow) gives rise to the gut.

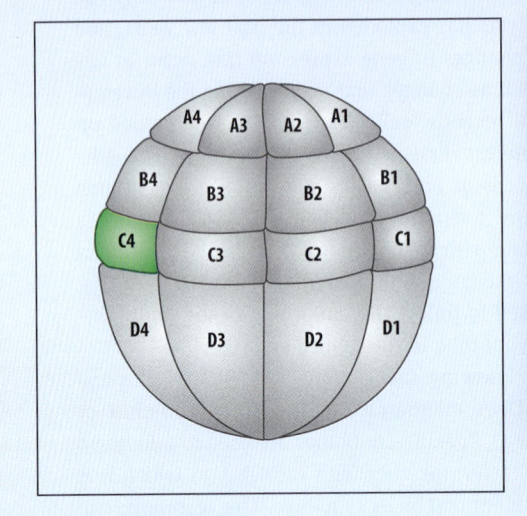

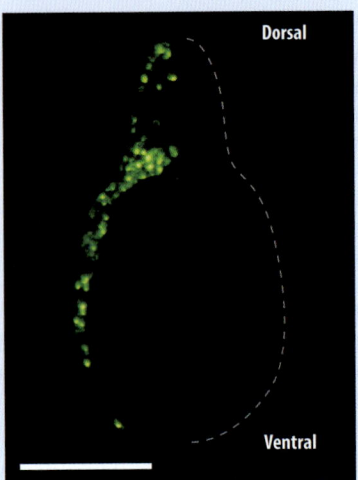

Fig. 1 Scale bar = 0.5 mm.
Photograph courtesy of L. Dale.

The fate map shows where the tissues of each germ layer normally come from, but it indicates neither the full potential of each region for development nor to what extent its fate is already specified in the blastula. Vertebrate embryos have considerable capacity for regulation when pieces are removed or are transplanted to a different part of the same embryo (see Section 1.12). This implies the existence and the persistence of considerable developmental plasticity, and that the actual fate of cells is heavily dependent on the signals they receive from their neighbors.

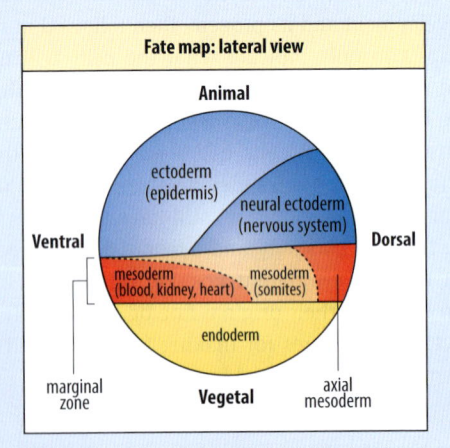

Fig. 2

happens. Such experiments can take a classic experimental embryological approach, for instance, assessing whether cells are determined by physically manipulating embryos—for example, by transplanting blocks of cells from one embryo to another (see Section 1.12) or by adding or removing cells from developing embryos. Alternatively, such experiments can take a genetic approach: the expression of developmentally important genes can be disturbed by mutation or gene silencing—removing expression of the gene at the time and/or the place where it is normally expressed—or can be disturbed by overexpression, or by misexpression—expressing the gene at a time or place where it is not normally expressed.

Developmentally important genes can be disrupted either by spontaneous mutations or by making new mutations experimentally. The mutant organism is recognized by its unusual phenotype, and genetic experiments are then carried out to find the gene responsible. This approach to identifying developmental genes is known as **forward genetics** (see Section 1.7). Rare spontaneous mutations were for a long time the only way that genetic disturbances of development could be studied and were the starting point for a forward-genetic approach. Many more developmental genes can be identified by inducing random mutations in large numbers of experimental organisms by chemical treatments or irradiation by X-rays, and then screening their progeny for mutants with unusual phenotypes of developmental interest. The aim of such an approach, known as **mutagenesis screening**, is to treat large enough populations so that, overall, a mutation is induced in every gene in the genome. So, it is best used in organisms that breed rapidly and that can conveniently be obtained and treated in very large numbers. We have already seen the value of this approach in discovering genes that govern development in *Drosophila* (see Box 2A).

The other main approach to disturb expression of developmentally important genes is **reverse genetics**. This starts with a gene whose existence and sequence are known and then analyzes the phenotype produced when that gene is overexpressed (**gain of function**) or deleted (**loss of function**). A particular gene may be chosen for investigation either because it is associated with a disease or, often, because it has been identified as important for development in an invertebrate, usually *Drosophila*, and we want to know its function in vertebrates. The gene can be altered using **transgenesis**, which involves introducing an altered or additional piece of DNA (a **transgene**) into the genome of the organism, which is then inherited in a normal Mendelian way—that is, it is present in every cell in the body. Animals into which a transgene has been introduced into the germline genome are known as **transgenic** animals (see Box 3B for ways of making transgenic animals). The transgene can be constructed to target a specific gene and make a mutation in a manner that permanently inactivates the gene. This is called a **gene knock-out**. Other transgenes are made to replace a gene with an altered but still functional copy, and this is called a **gene knock-in**.

A more recent reverse-genetic technique is **genome editing**, in which specific mutations in a target gene are made by introducing specialized DNA nucleases that act in combination with the cell's DNA-repair systems. One approach is to engineer the required specificity into DNA sequences that encode nucleases that cleave DNA—**zinc-finger nucleases (ZFNs)** and **TALENS** (transcription activator-like effector nucleases). Another approach is to engineer specificity into DNA sequences that encode RNAs that guide a nuclease to the desired gene (the **CRISPR-Cas9 system**). CRISPR-Cas9 genome editing, in particular, has revolutionized the ability to generate specific mutations, and because it is more efficient and more accurate than other methods, has now become the method of choice for this purpose (see Box 1C).

Gene expression can also be disturbed by **transient transgenesis** (also known as **somatic transgenesis**) in which an mRNA or DNA construct is introduced directly into somatic cells to test the effect of overexpression or misexpression of that gene. **Gene silencing** (also known as **gene knockdown**) is a commonly used technique in which a gene is not mutated or removed from the genome, but its expression is blocked by targeting the mRNA and preventing its translation. Genes can be silenced by the injection of **morpholino antisense RNAs**, which are stable RNAs designed to be complementary to a specific mRNA. They pair with the target mRNA, thus preventing its translation. The technique of **RNA interference (RNAi)** is also used in experimental developmental biology. In this case, **short interfering RNAs (siRNAs)** of defined sequence act as specific guide RNAs to target a nuclease to the relevant mRNA, which is then degraded.

A potential problem with all these gene-manipulation techniques, including transgenesis and genome editing, is the possibility of off-target effects. This highlights the importance of testing gene function using various approaches.

3.4 Not all techniques are equally applicable to all vertebrates

You will soon notice as you read further in this book that the different model organisms are used preferentially for different types of experiment. For instance, studies involving the microsurgical manipulation of vertebrate embryos largely feature *Xenopus* and the chick, whereas traditionally, zebrafish and mouse feature more prominently in forward- and reverse-genetic approaches to disturb the expression of developmentally important genes. In this section we briefly compare the advantages and disadvantages of each model organism.

The suitability of our four vertebrate models for imaging, fate mapping, and experimental embryology depends on being able to obtain living embryos and to follow their development (Fig. 3.6). *Xenopus* and zebrafish eggs are fertilized externally, and the development of the embryos can be followed using modern imaging techniques. Fertilized zebrafish eggs, injected with mRNA for a nuclear-localized fluorescent protein that acts as a marker for cell position, have been filmed using light-sheet fluorescence microscopy over the entire 2 days it takes to complete development. Light-sheet

Fig. 3.6 The suitability of the different model vertebrates for classic experimental embryological manipulation.

The suitability of embryos from the different model vertebrates for classic experimental embryological manipulation and imaging				
	Frog (*Xenopus laevis*)	**Zebrafish**	**Chick**	**Mouse**
Are living embryos easily available and observable at all stages of their development?	Yes. Eggs can be obtained and externally fertilized; embryos develop to swimming tadpole stage in dechlorinated tap water. Large batches of eggs can be fertilized at the same time and develop synchronously to give large numbers of embryos at the same developmental stage.	Yes. As for *Xenopus*. The embryo is transparent, which helps in observing development.	For most of it. Eggs are fertilized internally and so the very early stages of development—to blastoderm stage—occur within the mother hen. Laid eggs are easily obtainable, can be incubated in the laboratory, and the development of the embryo observed through a window in the eggshell at any time up to hatching. Early embryos can also be cultured out of the egg up to the stage at which the embryo has about 10 somites (early neurula).	Only very early stages. *In vitro* fertilized eggs can be cultured up to egg cylinder stage. Fertilized eggs cultured up to the blastocyst stage can be replaced into a surrogate mother to continue development. Isolated post-implantation gastrula stage embryos can be cultured for up to 2 days.
Is real-time imaging of development easy?	Yes. Light-sheet fluorescent microscopy (LSFM) to film cell behavior during gastrulation is possible. Tissue movements in embryos at this stage have also been filmed using MRI and time-lapse X-ray microtomography.	Yes. LSFM first successfully used to film zebrafish development from start to finish.	Yes. LSFM has been used for filming the behavior of cells during gastrulation in cultured embryos. MRI used to image live embryos developing inside the egg at later stages in organogenesis.	Yes. LSFM used to film cell behavior in cultured post-implantation embryos during gastrulation and early organogenesis.
Is the embryo amenable to experimental manipulation, e.g. blastomere removal or addition, or tissue grafting?	Embryo can be manipulated surgically up to the neurula stage (see Chapter 4). Embryos are comparatively large, and are highly resistant to infection after microsurgery.	Surgical manipulation up to the neurula stage and to some extent in later stages is possible but is technically challenging.	Manipulations possible in cultured blastoderm. Can also be manipulated surgically through a window in the shell up to organogenesis stages, including stages in limb development (see Chapters 3, 4, and 10).	Only very early embryos up to blastocyst stage can be manipulated if they are to be replaced in a surrogate mother for further development. Cultured post-implantation embryos can be experimentally manipulated, although this is technically very challenging.

Fig. 3.7 Chicks transgenic for green fluorescent protein (GFP). Transgenic chickens can be obtained by injecting DNA carried by a self-inactivating lentiviral vector beneath the blastodisc in newly laid eggs. The chicks illustrated are second-generation offspring from an original transgenic bird that carried the GFP transgene in the germline. The green color is present in all the cells but is only visible here in the naked skin. Elsewhere it is masked by the feathers. The bird in the center is not transgenic.

Photograph from McGrew, M.J., et al.: **Efficient production of germline transgenic chickens using lentiviral vectors.** *EMBO Rep. 2004,* **5:** *728-733.*

microscopy has also been used to follow development in a living mouse embryo. Tissue movements during gastrulation have also been filmed in *Xenopus* embryos using MRI and time-lapse X-ray microtomography.

The accessibility of *Xenopus* embryos and the relatively large size of the blastomeres make the frog particularly suited for producing fate maps at cleavage and blastula stages. Individual blastomeres can be injected with non-toxic markers or RNA encoding a fluorescent protein (such as GFP) that is then expressed in that cell and its progeny. *Xenopus* embryos are also tough and can be manipulated surgically up to the neurula stage. Fate maps can similarly be made of cleavage and blastoderm stages in zebrafish, but the embryos are relatively delicate, and microsurgery is technically very challenging.

Chick and mouse eggs are fertilized internally. In the chick, cleavage takes place as the fertilized egg passes down the hen's oviduct and the embryo is at the blastoderm stage when the egg is laid. The embryo can be observed through a window in the shell and development followed through to organogenesis. When the blastoderm is placed in culture, its development can be studied up to neurula stages. Individual behaviors of all the cells in early embryos have been followed for extended periods of time by filming cultured embryos of transgenic chickens expressing a membrane-localized fluorescent reporter.

The accessibility of the chicken embryo in the egg makes fate mapping and microsurgery relatively easy over a range of stages. Fate maps can be made by marking small groups of cells with lipophilic dyes, such as DiI, that stain cell membranes, by electroporating GFP to mark cells and their descendants, or by making **chimeras**, organisms composed of cells of two different genotypes. In chick–quail chimeras, cells in a specific region of a chick embryo are replaced by cells from the same region of a quail embryo. Because the quail cells can be distinguished histologically, this provides a way of finding out what the cells in that region normally give rise to. Another way of distinguishing quail cells is by carrying out immunohistochemistry using an anti-quail antibody. Chick chimeras can also be made by grafting tissue from embryos of a transgenic line of chickens that express GFP in every cell (Fig. 3.7).

All development in mice normally occurs within the mother, but thanks to the technique of ***in vitro* fertilization (IVF)**, mouse eggs can be fertilized outside the mother and cultured *in vitro* up to the egg cylinder stage (see Section 3.1). Fertilized eggs cultured through cleavage stages up to the blastocyst stage *in vitro* can be implanted in surrogate mothers and successfully complete development. Fate mapping at cleavage stages can be carried out by making chimeras (see Section 1.12) and experimental manipulations of the blastomeres are possible. **Post-implantation** mouse embryos isolated at gastrula stages can be cultured for up to 2 days to early organogenesis. Individual cell movements in such cultured embryos have been filmed using light-sheet fluorescence microscopy and analyzed to generate **dynamic fate maps**, that trace over time what each cell gives rise to. Experimental manipulations of cultured embryos are possible but extremely difficult. For fate mapping later stages of development,

	Frog (Xenopus laevis)	**Zebrafish**	**Chick**	**Mouse**
Genome sequenced	Yes. The diploid *X. tropicalis* genome was sequenced first.	Yes	Yes	Yes
Spontaneous mutations	No	Yes	Yes	Yes
Induced mutations (forward mutational genetic screens)	No, but some screens have been carried out in *X. tropicalis* which has a generation time of 4 months compared with the generation time of 1–2 years for *X. laevis*.	Yes, very suitable because of short generation time of 90 days (see Box 3C for details of how to detect recessive mutations).	No	Yes, suitable because of relatively short generation time of 63 days, although detection of recessive mutations is laborious.
Reverse loss-of-function genetic screens	No, although screens using genome editing in *X. tropicalis* may be possible in future.	Yes	No, genome editing is carried out at blastoderm stage and long generation time of 6 months makes screening impractical.	Yes, targeted gene knock-outs using ES cells, more recently using genome editing (see Box 3B).
Transgenic animals and gain-of-function mutations	Yes, now accomplished in both species by introducing DNA transgene into sperm nuclei and then transplanting these nuclei into unfertilized eggs. Reporter lines of animals available for imaging.	Yes, inject DNA transgene into the one-cell embryo. Gain-of-function mutations can be made and reporter lines of animals are available.	Yes, injection of lentiviral vectors containing the transgene under the blastoderm of newly laid egg. Reporter lines of animals available for imaging and fate mapping.	Yes, very routine by injection of transgene into fertilized egg. Gain-of-function mutations can be made and many reporter lines of animals available (see Box 3B).
Gene silencing or gene knock-out in somatic cells	Gene silencing by morpholino antisense RNAs (MOs) (Box 5B).	Gene silencing by MOs and RNA interference (RNAi) (Box 5B).		Gene knock-out by transgenic technique that restricts gene inactivation to specific cells/tissues, (Box 4D).
Gene overexpression or misexpression in time and space	Injection of RNAs into the fertilized egg and early embryo. Expression can be spatially restricted to some extent by injection into specific blastomeres.	Yes, as for *Xenopus*, but more challenging as blastomeres much smaller.	Yes, in somatic tissue, genes delivered by electroporation or in replication-competent viruses and expression can be spatially restricted.	Transgenic mice with inducible or cell-type specific expression of a transgene can be made.

Fig. 3.8 The suitability of the different model vertebrates for genetics-based study of development.

including organogenesis, transgenic techniques have been developed which depend on being able to induce expression of a transgene encoding an indelible marker—which is then transmitted to daughter cells—at a specific time and/or in a particular tissue *in vivo*.

The suitability of our four model organisms for genetics-based study of development is summarized in Fig. 3.8. The genomes of all four have been sequenced, so genomic techniques such as RNA-seq and CHIP-seq can be carried out on all of them.

Zebrafish and mice are most suited for forward- and reverse-genetic screens because they are diploid organisms and breed rapidly. The main forward-genetic approach is chemical mutagenesis, which can generate both dominant and recessive mutations. Dominant mutations can be recognized in offspring in the F_1 generation. More complicated breeding programs have to be used to reveal recessive mutations, which only alter the phenotype when both alleles of the gene are mutated, as shown later in this chapter. Some mutational screens have been carried out in *Xenopus tropicalis*, which is diploid and has a shorter generation time than *Xenopus laevis* (see Fig. 3.8).

The main reverse-genetic approach until quite recently was based on transgenesis and has been most highly developed in mice (see Box 3B). Transgenic mice that overexpress a given gene in particular cells or at particular times during development can be produced, and the effect of misexpressing that gene in a tissue, or at a time, where it is not usually active can be investigated. But it is more useful to test the function of a given gene by analyzing the phenotype of mice that have loss-of-function mutations in that gene. Until recently, mice were the only one of our model vertebrates in which genes could be specifically targeted for gene knock-out (see Box 3B). Targeting could be accomplished by genetically manipulating ES cells in culture and then introducing the cells into a mouse blastocyst, where they can contribute to all the tissues in the mouse that develops.

EXPERIMENTAL BOX 3B Ways of making transgenic mice

The development of methods for making transgenic mice depended on being able to fertilize mouse eggs *in vitro*, culturing them to the blastocyst stage, and then implanting the blastocyst into surrogate mothers to complete development. One technique is simply to inject a transgene composed of DNA encoding the required gene and any necessary control regions directly into a fertilized egg. If the transgene becomes incorporated into the genome, it will be present in all the cells of the embryo, including the germline, and will be expressed according to the promoter and other control regions it contains. In this way, mice that overexpress a given gene in particular cells or at particular times during development can be produced. Transgenes that encode reporters such as fluorescent proteins can also be injected to produce mice that can be used, for example, in light-sheet fluorescence microscopy or for producing fate maps.

Another technique for producing transgenic mice uses **embryonic stem cells (ES cells)**. ES cells will be discussed in Chapter 7. Until recently, this technique has been the one most commonly used in mice for producing specific gene knock-outs. ES cells can be isolated from the inner cell mass of the mouse blastocyst (see Fig. 3.3) and maintained in culture. They can then be injected back into another blastocyst. If the blastocyst is implanted in a surrogate mother to continue its development, the ES cells are incorporated into the embryo and can contribute to all the tissues of the mouse that develops, even giving rise to germ cells and gametes. For example, if ES cells from a brown-pigmented mouse are introduced into the blastocyst of a white mouse, the mouse that develops from this embryo will be a chimera made up of cells of the ES cell genotype and the blastocyst cell genotype. In the skin, this mosaicism is visible as patches of brown and white hairs (Fig. 1).

The animals produced by ES cell transfer are chimeras composed of both genetically modified and normal cells. For this reason, they may show few, if any, effects of the mutation. If the ES cells with the mutant gene have entered the germline, however, strains of transgenic mice heterozygous for the altered gene can be intercrossed to produce homozygotes, which will show the effects on development when the gene is mutated.

To generate transgenic mice with a particular mutation, the ES cells are genetically manipulated while growing in culture (Fig. 2) before being injected into the early blastocyst. Mutations are targeted to a particular gene by the technique of **homologous recombination**, in which specially tailored DNA constructs are introduced into the cell by **transfection**. A transfected DNA molecule will usually insert randomly into the genome but by including some sequence that is homologous with the desired target gene, it is possible to make it insert at a predetermined site (Fig. 2). Homologous recombination between the transfected DNA and the target gene in the ES cells results in an insertion that renders the gene non-functional. The introduced DNA usually carries a drug-resistance gene so that ES cells containing the insertion can be selected by adding the drug, which kills all the other unmodified cells.

Obtaining a mutation in a desired gene in mice by targeted gene knock-out using ES cells is laborious, however. CRISPR-Cas9 genome editing is now the method of choice for manipulating genes in mice. The editing components are injected into the fertilized egg (see Box 1C) and all the cells of the mouse that develops will have the same modified genome, so that a knock-out mouse can be produced in one generation.

A significant number of knock-outs of a single gene result in mice developing without any obvious abnormality, or with fewer and less severe abnormalities than might be expected from the normal pattern of gene activity. A striking example is that of *Myod*, a key gene in muscle differentiation. In *Myod* knock-outs, the mice are anatomically normal, although they do have a reduced survival rate. The most likely explanation for this is that the process of

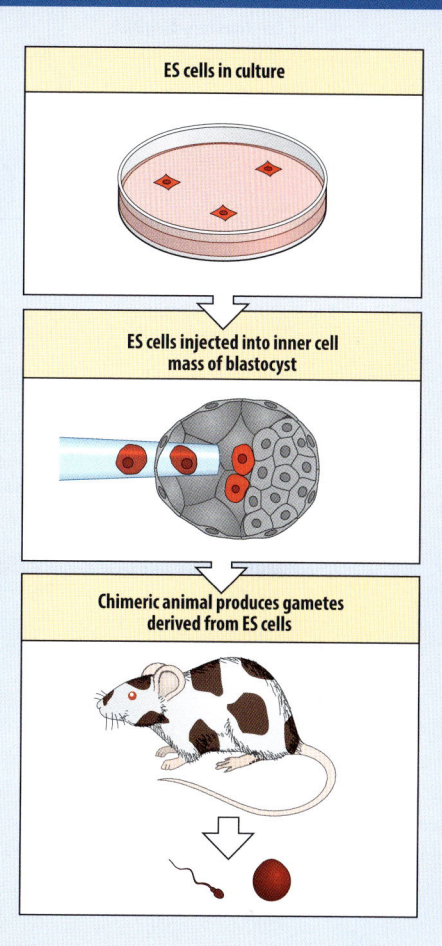

Fig. 1

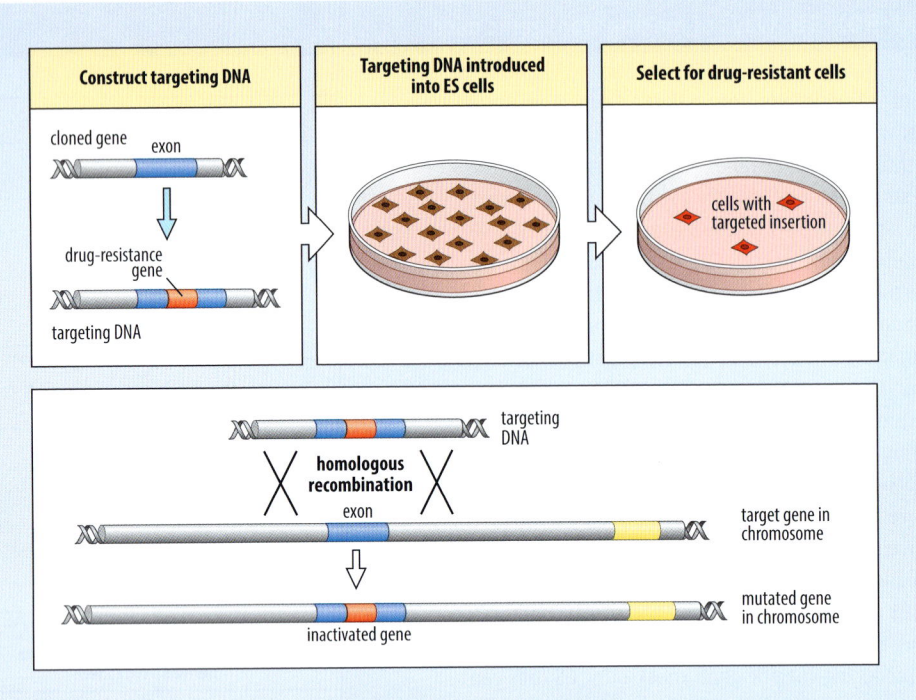

Fig. 2

muscle differentiation contains a certain amount of redundancy and that other genes can substitute for some of the functions of *Myod*. For more on muscle differentiation see Section 6.6.

However, it is unlikely that any gene is without any value at all to an animal. It is much more likely that there is an altered phenotype in these apparently normal animals that is too subtle to be detected under the artificial conditions of life in a laboratory. Redundancy is thus probably apparent rather than real. A further complication is the possibility that, under such circumstances, related genes with similar functions may increase their activity to compensate for the mutated gene.

A large international reverse-genetic screen (the International Mouse Phenotyping Consortium) was set up in 2011 to produce and document mutations and their phenotypes in each of the 20,000 protein-coding genes in the mouse genome (see 'General further reading' for URL). Reverse-genetic screens have also been carried out in zebrafish. A screen covering 60% of the zebrafish protein-coding genes using the **TILLING** technique was completed in 2016. The TILLING technique combines reverse genetics with large-scale mutagenesis screens. Random mutagenesis is used to generate families of mutant organisms. Fish with the desired mutation are first identified by DNA analysis of the sequence of interest and then their phenotype is analyzed to characterize any changes. The technique is explained in Box 3C.

EXPERIMENTAL BOX 3C Large-scale mutagenesis screens for recessive mutations in zebrafish

A screening program for recessive mutations in zebrafish involves breeding for three generations (see Fig. 1). Male fish treated with a chemical mutagen (ethylnitrosourea; ENU) are crossed with wild-type (+/+) females to generate F_1 offspring, each of which is likely to be heterozygous for a point mutation (m*) in a different gene. Each F_1 male offspring is crossed again with a wild-type female to generate different F_2 families of fish. If the parent F_1 fish carries a mutation (say, m1*), then 50% of the fish in the

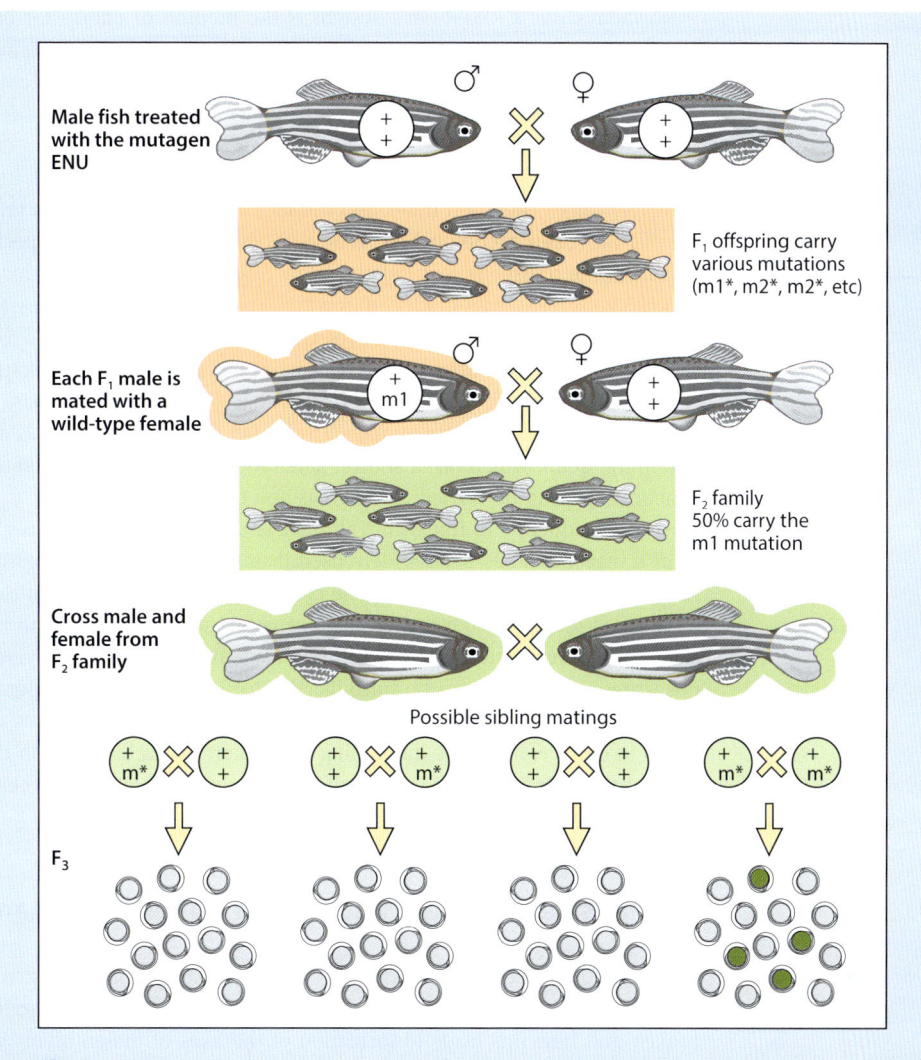

Fig. 1

m1 F_2 family will be heterozygous (m1*/+) for that mutation, so that when the female and male siblings from this F_2 family are themselves crossed, 25% of these matings will be between two heterozygotes and 25% of their offspring (the F_3 generation, shown here as fertilized eggs) will be homozygous for the mutation. The final step is to identify the genetic lesion that has led to the mutant phenotype. Zebrafish can also be made to develop as haploids by fertilizing the egg with sperm that have been heavily irradiated with UV light. This allows one to detect early-acting recessive mutations without having to breed the fish to obtain homozygous embryos.

Offspring of male fish treated with a chemical mutagen, which will have mutations in different genes, can be screened to identify an individual fish with a mutation in a specific gene using an approach known as TILLING (targeting-induced local lesions in genomes). This technique was first developed for plants, but has since been applied to *Drosophila*, *Caenorhabditis*, and the rat, as well as to zebrafish. All that is necessary is knowledge of the DNA sequence of the gene. Male fish treated with ENU are crossed with wild-type females. A sample of tissue from each F_1 male offspring, which will have a point mutation in a different gene, is then collected and, at the same time, its sperm are frozen and stored. To identify the individual fish that has a mutation in the gene of interest, the polymerase chain reaction (PCR) is used to generate large amounts of the DNA from the relevant genomic region from the tissue samples of all F_1 male offspring. This DNA is then hybridized with unmutated DNA from the gene in question. Mismatched bases will reveal the site of a mutation and the DNA is then sequenced to identify the precise change. The individual male fish with the desired mutation and/or its frozen sperm is then used to generate families of mutant fish, which will reveal any phenotypic changes associated with the mutant gene.

Nowadays, the CRISPR-Cas9 system is the method of choice for creating loss-of-function mutations in specific genes in mice by injecting the genome-editing components into the fertilized egg (see Box 1C). The CRISPR-Cas9 system has also been adapted for making targeted gene knock-outs in zebrafish and has been used in reverse-genetic screens.

Xenopus and chickens are not suited to reverse-genetic screens because of their relatively long generation times and the fact that *Xenopus laevis* is tetraploid. Transgenesis has been accomplished in both model organisms using special methods to introduce the transgene (see details in Fig. 3.8) and lines of transgenic *Xenopus* frogs and chickens have been produced in which reporter genes are expressed. An example is the line of chickens transgenic for GFP shown in Figure 3.7. ES cells have been isolated from chick embryos, but they have not been used successfully to produce chickens with targeted mutations. CRISPR-Cas9 genome editing has, however, now been accomplished in both animals. The use of this technology in *Xenopus tropicalis* and in the quail, which has a shorter generation time than the chicken, might enable genetic screens to be carried out in frogs and birds.

As the production of transgenic animals is not routine in *Xenopus* and chick, transient transgenesis has been widely used instead to produce changes in gene expression. In early *Xenopus* embryos, gene constructs and morpholino antisense RNAs can be injected directly into cleavage-stage cells for gain-of-function and loss-of-function experiments, respectively. In chick embryos, gene constructs for overexpression and misexpression studies can be introduced into the cells by electroporation or in retroviral vectors. In addition, the effects of signal proteins themselves can be directly tested in chick embryos by implanting small inert beads loaded with the protein into the embryo. The beads act as controlled-release carriers and the protein released acts locally. RNAi has also been used in chick embryos.

Xenopus development and setting up the body plan

We now look at the key processes that occur in *Xenopus* before gastrulation and that set up the body plan. We consider how the radial symmetry of the fertilized egg is broken and how this leads to formation of the axes of the embryo. We discuss how the germ layers are specified—an event that, in *Xenopus*, takes place before gastrulation. We discuss the initial patterning of the mesoderm, paying special attention to the origin of the mesodermal embryonic organizer, and how its identity and position in the early embryo are specified. Cells of the organizer will begin to differentiate, and as they do so, will move inside the embryo during gastrulation. The cell movements of gastrulation will bring the germ layers into their correct position and establish the pattern of structures along the antero-posterior and dorso-ventral axes of the body. This highlights one of the key principles of developmental biology—the importance of cell rearrangements as a driver of morphological change, and we will discuss this in more detail in Chapter 5. We first set the scene by describing the morphological changes and cell rearrangements that occur over the period of time from fertilization to gastrulation in the *Xenopus* life cycle, then describe the mechanisms that underlie these changes.

3.5 The frog *Xenopus laevis* is the model amphibian for studying development of the body plan

The *Xenopus* life cycle and developmental stages are illustrated in Figure 3.9. Even before fertilization the *Xenopus* egg has a distinct polarity, with a dark, pigmented **animal region** and a pale, yolky, and heavier **vegetal region** (Fig. 3.9, top photograph). The pigmented animal region comes to lie uppermost after being released

Xenopus development and setting up the body plan   **119**

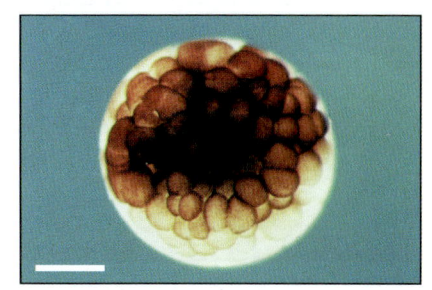

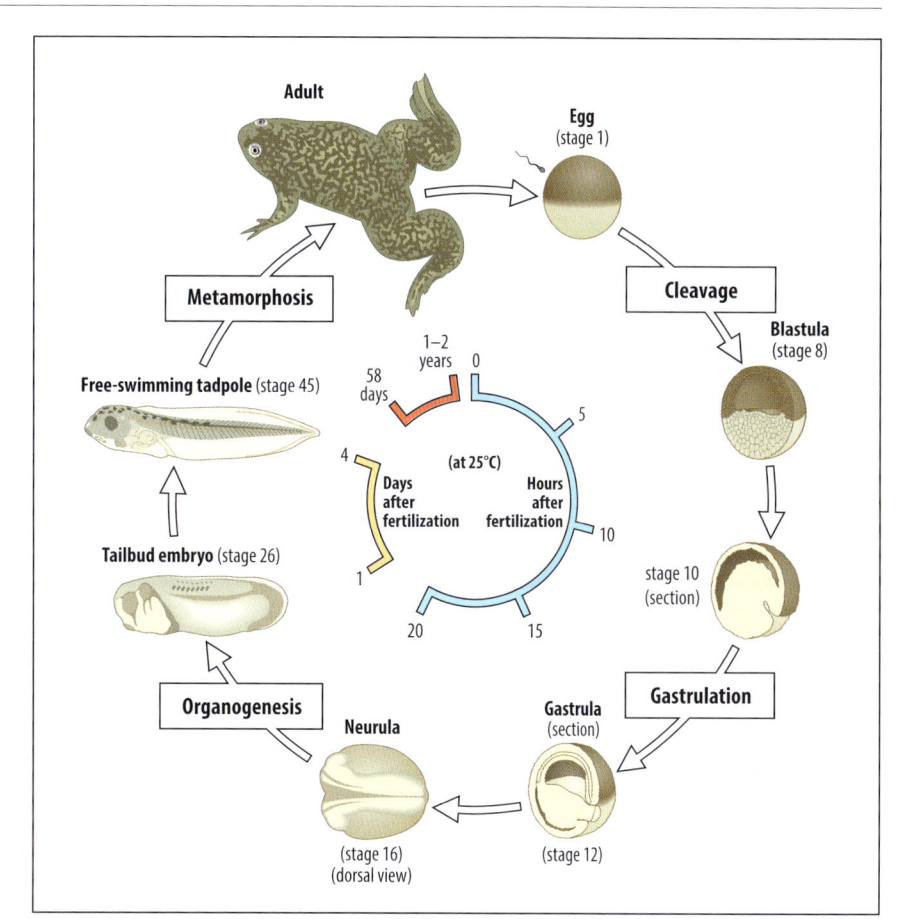

Fig. 3.9 Life cycle of the African clawed frog *Xenopus laevis*. The schematic shows key developmental stages in the life cycle. The numbered stages refer to standardized stages of *Xenopus* development. An illustrated list of all stages can be found on the Xenbase website listed in 'General further reading'. The photographs show an egg at stage 1 with its distinct polarity (top), an embryo at the blastula stage (middle), and a tadpole at stage 41 (bottom). Scale bars = 1 mm.
Top and bottom photographs courtesy of J. Smith. Middle photograph courtesy of J. Slack.

into the water. Most of the yolk, however, becomes located in the vegetal region under the influence of gravity. The axis running from the **animal pole** to the **vegetal pole** is known as the **animal–vegetal axis**. The egg is enclosed in a protective **vitelline membrane**, which is embedded in a gelatinous coat. Meiosis is not yet complete. The first meiotic division has resulted in a small cell—a **polar body**—forming at the animal pole (Fig. 3.10, top row). The second meiotic division is only completed after fertilization, when the second polar body also forms at the animal pole (see Box 8A). At fertilization, one sperm enters the egg, always penetrating the animal region. The egg completes meiosis and the female pronucleus (from the egg) and the male pronucleus (from the sperm) fuse to form the diploid zygote nucleus.

About 90 minutes after fertilization, cleavage begins. The planes of the early cleavages are related to the animal–vegetal axis. The first cleavage division of the fertilized egg proceeds along the plane of the animal–vegetal axis, dividing the embryo into equal left and right halves (Fig. 3.10, top and bottom rows). Further cleavages follow rapidly at intervals of about 30 minutes. The second cleavage is also along the animal–vegetal axis but at right angles to the first. The third cleavage is equatorial, and divides the embryo into four animal cells and four larger vegetal cells. There is no cell growth between cell divisions at this early stage and so continued cleavage results in the formation of smaller and smaller cells. Cleavage initially occurs synchronously,

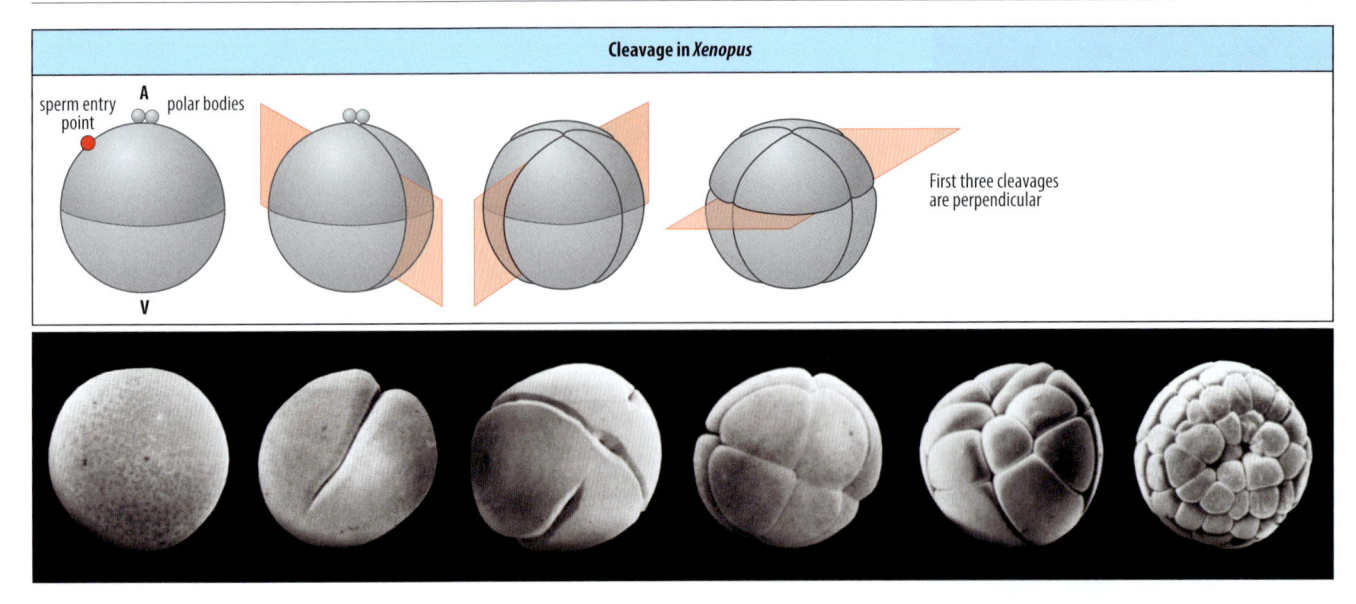

Fig. 3.10 External views of cleavage in the *Xenopus* embryo. Top row: diagrams showing the fertilized egg and the first three cleavage divisions seen from the side. Polar bodies shown in the fertilized egg and the two-cell stage are attached to the animal pole. Bottom row: photographs showing the cleaving *Xenopus* egg taken from various angles (see also Fig. 1.14). A, animal pole; V, vegetal pole.

Photographs reproduced with permission from Kessel, R.G., Shih, C.Y.: Scanning Electron Microscopy in Biology: A Student's Atlas of Biological Organization. London: Springer-Verlag, 1974. © 1974 Springer-Verlag GmbH & Co.

but because the yolk impedes the cleavage process, division is slowed down in the yolky vegetal half of the embryo, resulting in cells in this region being larger than those in the animal half.

Around 5 hours after fertilization a fluid-filled cavity—the **blastocoel**—develops in the animal region, marking the blastula stage. Figure 3.11 (top row, first panel) shows a cross-section of the *Xenopus* embryo so you can see what is happening inside the blastula. The embryo consists of a continuous sheet of cells surrounding the blastocoel. Although not visibly obvious, the germ layers have now been specified. Cells in the animal region are ectoderm; cells in the vegetal region are endoderm; in between these, in a ring-like equatorial region called the **marginal zone**, cells are mesoderm. At the late blastula stage the embryo has gone through about 12 cell divisions and is made up of several thousand cells.

Around 9 hours after fertilization, a small slit-like infolding—the **blastopore**—forms on the surface of the blastula, always on the side opposite the site of sperm entry (see Fig. 3.11, top row, second image and bottom row, first image). This is the first obvious sign that the embryo is beginning to undergo gastrulation. One edge of the blastopore is known as the **dorsal lip** of the blastopore (see Fig. 3.11, top row, second panel). The slit extends into a circle, which becomes smaller during gastrulation as the embryo elongates and the body axes become apparent (see Fig. 3.11, bottom row). The edge of the circular blastopore opposite the dorsal lip is sometimes known as the **ventral lip** (see Fig. 3.11 top row, third panel).

The mesoderm cells at the dorsal lip (also known as dorsal mesoderm cells) are the embryonic organizer, known as the **Spemann organizer** in amphibians. Immediately after they form, these cells start to rearrange and move inside the gastrula by rolling inwards as a coherent sheet of cells—a movement known as **involution** (see Fig. 3.11, top row, third panel). The involuting mesodermal cells take with them the overlying endoderm. As mesoderm and endoderm move inside, they converge towards the dorsal midline, elongate beneath the dorsal ectoderm, and the mesoderm differentiates into the axial mesoderm. As this occurs, the entire embryo lengthens and the

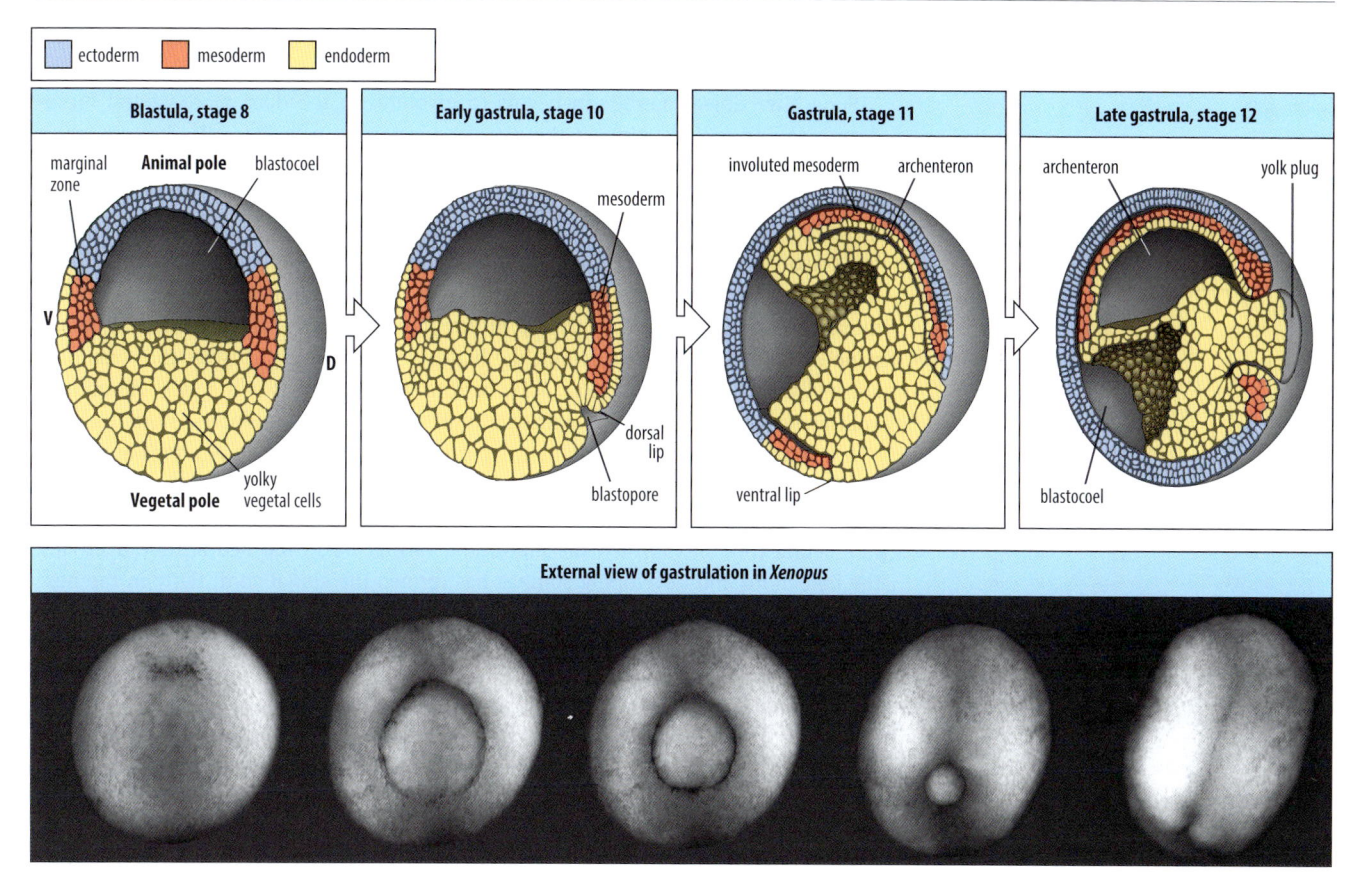

Fig. 3.11 Gastrulation in *Xenopus*. The schematic diagrams show cross-sectional views of what is happening during gastrulation. The blastula has been cut in half along the animal-vegetal axis, in a plane that passes through the site of blastopore formation. First panel: the blastula contains several thousand cells. The three germ layers have formed and the dorsal (D) and ventral (V) sides of the blastula have been determined, indicating the dorso-ventral axis of the pre-gastrula embryo. The vegetal region is made up of endoderm, the marginal zone is made up of mesoderm, and the animal region is made up of ectoderm. A fluid-filled cavity, the blastocoel, lies beneath the cells at the animal pole. Second panel: gastrulation begins at the dorsal lip of the blastopore. Future dorsal mesoderm (orange), together with the overlying endoderm (yellow), moves inside through the dorsal lip of the blastopore. Third panel: gastrulation continues, with the mesoderm elongating and ending up sandwiched between endoderm and ectoderm in the animal region. Fourth panel: future endoderm also moves inside through the ventral lip of the blastopore. The tissue movements create a new internal cavity—the archenteron—that will become the gut cavity, completely lined with endoderm. At the end of gastrulation, the blastocoel is considerably reduced in size. During gastrulation the ectoderm extends downwards (third and fourth panels) and will eventually cover the whole embryo. Bottom row: an external view of blastopore formation and how the blastopore changes in shape during gastrulation. The images show successively: the formation of the slit-like blastopore; the blastopore spreading laterally and ventrally; the blastopore spreading to form a complete circle encompassing cells from the vegetal region; as the tissues move inside and converge, the dorsal midline becomes apparent and the embryo elongates; by the end of gastrulation the blastopore becomes reduced to a small slit, which will form the anus.

Diagrams after Balinsky, B.I.: An Introduction to Embryology. 4th edn. Philadelphia, PA: W.B. Saunders, 1975. Photographs of blastopore formation from a video kindly provided by David R. Shook.

antero-posterior axis becomes obvious (see Fig. 3.11, bottom row). The details of how neural tissue forms from the ectoderm—an event that occurs at the same time as gastrulation—are described in Chapter 4. The famous experiment by Hans Spemann and Hilde Mangold that led to the discovery of the embryonic organizer in amphibian embryos and showed how it rearranges and differentiates is described in Figure 1.9.

The involuting sheet of endoderm adheres closely to the mesoderm, and the space between it and the yolky vegetal cells is known as the **archenteron** (see Fig. 3.11, top row, third panel). This is the precursor of the gut cavity. As the gut cavity forms, the blastocoel is displaced, shrinks, and eventually disappears. At the same time as the mesoderm and endoderm move inside, the ectoderm spreads downward to cover the whole embryo by a process known as **epiboly** (see Fig. 3.11, top row, third and fourth panels).

Soon after dorsal mesoderm begins to involute, so too does more-ventral mesoderm, leading to formation of the circular blastopore (see Fig. 3.11, bottom row). As progressively more-ventral mesoderm cells move inside they become positioned lateral to the notochord—and form somites, intermediate mesoderm, and lateral plate mesoderm on each side of it. In this way, the pattern of mesoderm structures along the dorso-ventral axis of the body is laid down (in Chapter 4 we describe the further development of these mesodermal structures). At the same time, more endoderm moves inside through the ventral lip of the blastopore (see Fig. 3.11, top row fourth panel). By the end of gastrulation, the blastopore has almost closed—the remaining slit will form the anus (see Fig. 3.11, bottom row). There is still a large amount of yolk present, which provides nutrients until the larva—the tadpole—starts feeding. The cell and tissue movements that occur during gastrulation in *Xenopus* and other animals are discussed in detail in Chapter 5.

By 20 hours after fertilization, the embryo has reached the neurula stage. The anterior neural tube has formed the brain, and further back, the neural tube overlying the notochord and developing somites has formed the spinal cord. Early in the next stage—organogenesis—the embryo passes through the phylotypic stage, which, in *Xenopus*, is the tailbud stage (see Fig. 3.4). At this stage the body plan has been laid down and we can recognize the main vertebrate features. The post-anal tail of the tadpole is formed last. It develops from the tailbud, which gives rise to a continuation of notochord, somites, and neural tube in the tail. After organ formation is complete, the tadpole hatches out of its jelly covering and begins to swim and feed. Later, the tadpole larva will undergo metamorphosis to produce the adult frog; the tail regresses and the limbs grow (metamorphosis is discussed in Chapter 11).

We now turn to the mechanisms that underlie the development of *Xenopus* from egg to gastrula.

3.6 The earliest stages in *Xenopus* development are controlled by maternal factors

The very earliest stages in amphibian development are exclusively under the control of maternal factors present in the egg. Maternal factors are the mRNA and protein products of genes that are expressed by the mother during **oogenesis** (egg formation) and laid down in the egg. The embryo's own genes—the **zygotic genes**—are not switched on until later, during the blastula stage. The maternal factors are localized in different regions of the unfertilized egg and, as we shall see, some are then re-localized in the first cell cycle after fertilization. This means that when the fertilized egg undergoes its stereotypical pattern of cleavage, blastomeres in different positions in the embryo inherit different maternal factors. The maternal factors act as **cell-fate determinants**, making cells different and fixing their subsequent development.

The polarization of the unfertilized *Xenopus* egg presages a first embryonic axis, the animal–vegetal axis (see Fig. 3.10, top row). The egg contains large amounts of maternal mRNAs. Some of these are stored in a translationally inactive state until later in embryogenesis when their encoded proteins are needed. The egg also contains large amounts of stored proteins; there is, for example, enough histone protein for the assembly of more than 10,000 nuclei, quite enough to see the embryo through the first 12 cleavages until its own histone genes begin to be expressed.

The pigment itself has no role in development but is a useful marker for real differences that exist between animal and vegetal regions of the egg. A specialized cohort of stored mRNAs are transported into the vegetal pole of the developing egg (the **oocyte**) and anchored to the cortex—an outer cytoplasmic layer just under the cell membrane. Another group of maternal mRNAs is concentrated in the animal region.

These localized mRNAs encode proteins that help to establish the future germ layers and future axes of the embryo. If the vegetal-most cytoplasm is removed prior to the first cell cycle leading up to the first cleavage division, the embryo never develops beyond a radially symmetric mass of tissue that lacks defined axes.

Various mechanisms contribute to the appropriate transport and localization of maternal mRNAs in the oocyte. Early in oogenesis, some mRNAs are concentrated in the vegetal cortex by transport involving a cytoplasmic region known as the message transport organizer region (METRO), which is associated with the mitochondrial cloud—a region of cytoplasm in the oocyte where mitochondria are being generated. Later in oogenesis, other mRNAs are transported to the vegetal cortex along microtubules powered by kinesin motor proteins, similar to the mRNA-transport mechanisms described in the *Drosophila* oocyte (see Fig. 2.20). mRNA localization by either route seems to depend on sequences in the 3'-untranslated region of the mRNAs that are recognized by specific *trans*-acting localization factors. Figure 3.12 shows the localization of a late-pathway mRNA in the vegetal region during oocyte development. Initially, it is distributed throughout the oocyte cytoplasm, then becomes localized in a wedge-shaped region beneath the oocyte nucleus, and eventually becomes localized to the cortex throughout the vegetal region.

An important mRNA localized at the animal pole of the fertilized egg is that for the E3 ubiquitin-ligase Ectodermin, which is a strong candidate for a maternal determinant of ectoderm, as we shall see later. Cells that develop from the animal region inherit Ectodermin mRNA. Developmentally important mRNAs located at the vegetal pole include the late-pathway mRNAs for the transcription factor VegT, a key maternal determinant of endoderm, and the signal protein GDF-1 (formerly known as Vg-1), which is involved in mesoderm formation. Cells that develop from the vegetal region inherit VegT, which promotes expression of zygotic genes encoding transcription factors that are needed for the cells' development as endoderm and their segregation from mesoderm cells. As we shall see later, VegT is also essential, in a cell non-autonomous manner, for the formation of mesoderm.

Maternal mRNAs for Wnt signal proteins and some components of the intracellular Wnt signaling pathway are also localized to the vegetal pole and are translated soon after fertilization. We will see later in the chapter that the Wnt signaling pathway is involved in a critical symmetry-breaking event, the specification of the dorsal side of the blastula. The main families of secreted signal proteins used in vertebrate early development are listed in Box 3D.

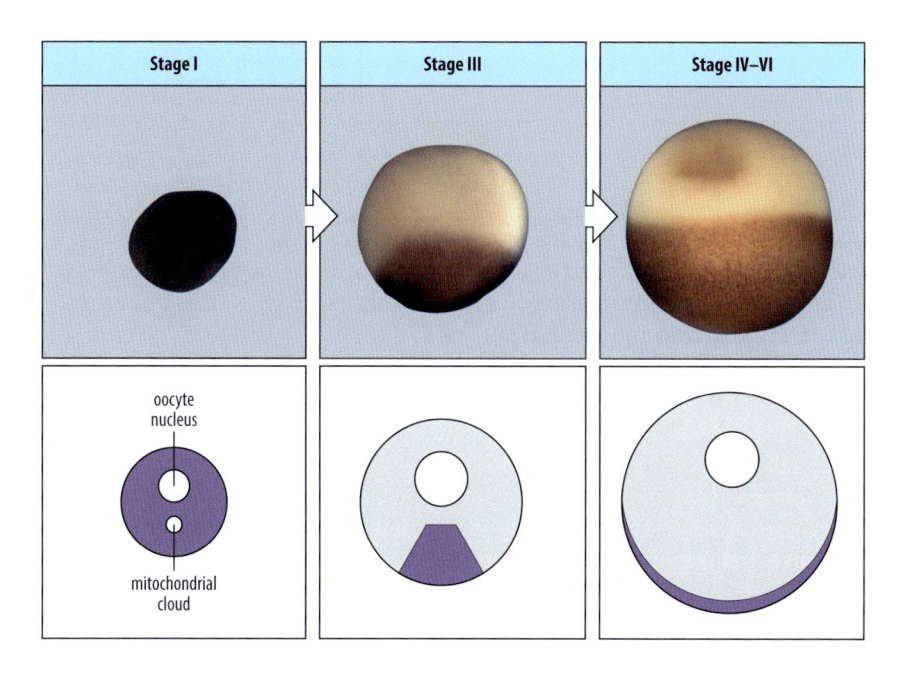

Fig. 3.12 Localization of maternal mRNAs in the *Xenopus* oocyte. Left panels: late-pathway RNAs such as that for the transcription factor VegT and the signal protein GDF-1 (purple) are ubiquitously distributed in the cytoplasm of stage I oocytes. Center panels: with the end of stage II/beginning of stage III, RNAs are enriched in a wedge-shaped region below the oocyte nucleus and start to accumulate at the vegetal cortex. Right panels: late-pathway RNAs localize to the entire cortex of the vegetal region (oocytes at stages IV-VI). The photographs show localization of GDF-1 RNA detected by *in situ* hybridization in albino oocytes.

From Claussen, M., Pieler, T.: **Identification of vegetal RNA-localization elements in Xenopus oocytes.** Methods 2010 **51**: 146-151.

CELL BIOLOGY BOX 3D Intercellular protein signals in vertebrate development

Proteins that are known to act as signals between cells during development belong to seven main families. Members of some of these families, such as some of the fibroblast growth factors (FGFs), were originally identified because they were essential for the survival and proliferation of mammalian cells in tissue culture, and so have become known as growth factors, even though they have many other developmental functions. The members of the seven families of signal proteins are either secreted or span the plasma membrane, and they provide intercellular signals that mediate cell-cell interactions in many tissues at many stages of development. Other signal proteins, such as the insulin-like growth factors, and the neurotrophins, are used by embryos for particular tasks, but the seven families listed in the table (Fig. 1) are the main families involved throughout development in vertebrates. Homologs of these proteins are also important in embryonic development in non-vertebrates (see Chapter 2, and examples throughout the book).

The signal molecules listed in the table all act by binding to membrane-spanning receptors on the surface of a target cell. Each type of signal protein has a corresponding receptor or set of receptors, and only cells with the appropriate receptors on their surface can respond. In this way the signal is passed, or transduced, across the cell membrane, by the receptor, to the intracellular biochemical signaling pathways. Thus, a developmental signal secreted from a 'signaling cell', stimulates the 'intracellular signaling pathway' (also called the 'signal transduction pathway') in a second, responding, cell. The intracellular signaling pathways can be complex and involve numerous different proteins. We have already introduced the general principles employed in intracellular signaling pathways in Box 1G. The Wnt/β-catenin pathway is illustrated in Box 3E, the transforming growth factor (TGF)-β pathways are illustrated in Box 3F, and the FGF pathway in Box 4A. Figures illustrating the other signaling pathways are cross-referenced in Fig. 1. In the developmental context, the important responses of cells to a signal are either a change in gene expression, with specific genes being switched on or off, or alterations in the cytoskeleton that lead to changes in cell behavior, such as changes in cell shape and motility (see Box 5B).

Some signal proteins, such as those of the TGF-β family, act as dimers; two molecules form a covalently linked complex that binds to a receptor that consists of two protein subunits (see Box 3F). In some cases, the active form of the protein signal is a heterodimer, made up of two different members of the same family. Secreted signals can diffuse or be transported short distances through tissue and set up concentration gradients. Some signal molecules, however, remain bound to the cell surface, and thus

Common intercellular signaling proteins in vertebrate embryonic development

Signal protein family	Receptors	Examples of roles in development
Fibroblast growth factor (FGF)		
Twenty-five members of the FGF family have been identified in vertebrates, not all of which are present in all vertebrates	FGF receptors (receptor tyrosine kinases)	Roles at all stages in development. Maintenance of mesoderm formation (this chapter). Induction of spinal cord (see this chapter and Box 4A for signaling pathway). Signal from apical ectodermal ridge in limb bud (Chapter 10)
Epidermal growth factor (EGF)		
Epidermal growth factor	EGF receptors (receptor tyrosine kinases)	Cell proliferation and differentiation. *Drosophila* limb patterning (Chapter 10)
Transforming growth factor-β (TGF-β)		
A large protein family that includes Activin, GDF-1, bone morphogenetic proteins (BMPs), Nodal, and Nodal-related proteins	Type I and Type II receptor subunits (receptor serine-threonine kinases) Receptors act as heterodimers	Roles at all stages in development. Signaling pathways in Box 3F. Mesoderm induction and patterning (this chapter)
Hedgehog		
Sonic hedgehog and other members	Patched	Positional signaling in limb and neural tube (Chapters 10 and 11). Signaling pathway in Box 10C. Involved in determination of left–right asymmetry in chick embryos (Chapter 4)
Wingless (Wnt)		
Nineteen members of the Wnt family have been identified in vertebrates, not all of which are present in all vertebrates	Frizzled family of receptors (seven-span transmembrane proteins)	Roles at all stages in development. For canonical and non-canonical signaling pathways see Box 3E and Box 6C. Dorso-ventral axis specification in *Xenopus* (this chapter). Limb development (Chapter 10). Stem-cell regulation (Chapters 7 and 8)
Delta and Serrate		
Transmembrane signaling proteins	Notch	Roles at all stages in development. Left–right asymmetry (Chapter 4). Somite development (see Chapter 4, signaling pathway in Box 4G)
Ephrins		
Transmembrane signaling proteins	Eph receptors (receptor tyrosine kinases)	Formation of rhombomeres in embryonic hindbrain (Chapter 11). Guidance of developing blood vessels (Chapter 6). Guidance of axon navigation in the nervous system (Chapter 11)

Fig. 1

can only interact with receptors on a cell that is in direct contact. One such is the signal protein Delta, which is membrane-bound and interacts with the receptor protein Notch on an adjacent cell (see Box 4F for the Notch signaling pathway).

For consistency in this book we have hyphenated the abbreviations for all secreted signal proteins where the name allows—for example, BMP-4, Wnt-11, GDF-1 (formerly known as Vg-1) and TGF-β.

3.7 Cortical rotation specifies the future dorsal side of the *Xenopus* embryo

Fertilization leads to an asymmetry in the developing embryo, which will predict the future dorso-ventral axis. Sperm entry, which can occur anywhere in the animal region of the egg, initiates a series of events that result in the dorsal side forming more or less opposite the sperm's entry point. During the elongated first cell cycle, the *Xenopus* egg undergoes a cytoplasmic rearrangement in which the cortex and the immediately underlying layer of cytoplasm in the vegetal region physically rotate with respect to the inner yolky mass to a position opposite the point of sperm entry (Fig. 3.13). This process is referred to as **cortical rotation** and creates a new molecular asymmetry in the zygote in the horizontal dimension by re-localizing some of the maternal factors that had been laid down in the egg—initiating the dorso-ventral axis of the blastula that is approximately perpendicular to the animal–vegetal axis.

Cortical rotation is driven by an array of oriented microtubules that forms immediately beneath the cortex in the vegetal region. The formation of these microtubules and direction of rotation is influenced by the **sperm aster**, which is produced from the sperm centrosome (the sperm centrosome enters the egg together with the sperm nucleus at fertilization). The sperm centrosome initially remains in an eccentric position in the animal region near the point of sperm entry and some of the microtubule 'rays' of the aster extend down into the vegetal region. Microtubules form rapidly as the rays meet the vegetal cortex and become oriented in the same direction as the

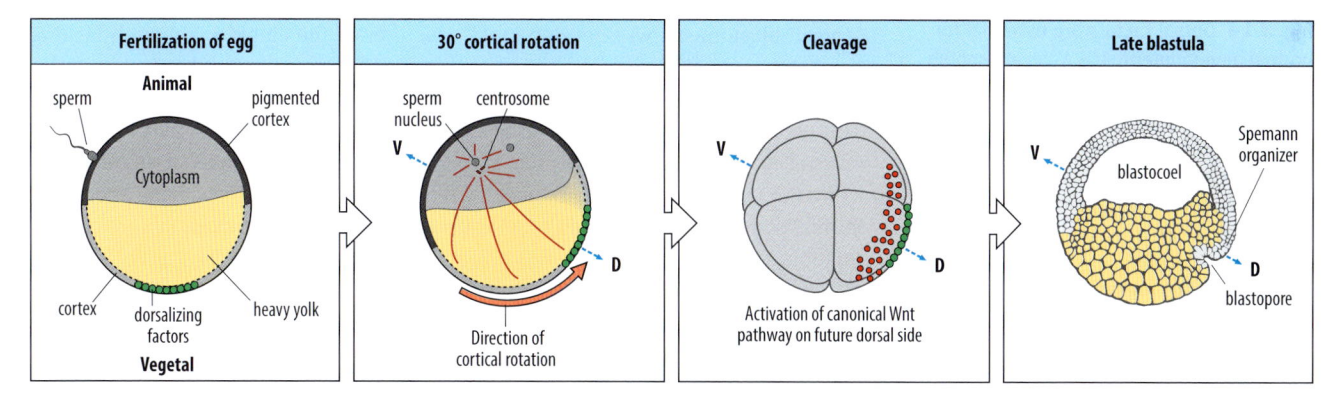

Fig. 3.13 The future dorsal side of the amphibian embryo develops opposite the site of sperm entry as a result of the redistribution of maternal dorsalizing factors by cortical rotation. After fertilization (first panel), the outer cortex, just under the cell membrane, rotates by 30° towards the future dorsal region (second panel). This results in dorsalizing factors such as *wnt11* mRNA and Dishevelled (Dsh) protein—indicated by green circles—being relocated from their initial position at the vegetal pole to a position approximately opposite the site of sperm entry. The factors are transported along aligned microtubules (not shown here). The formation of these microtubules and direction of rotation are influenced by the microtubule 'rays' of the sperm aster (red), some of which extend down into the vegetal region. Microtubules form rapidly as the rays meet the vegetal cortex and become oriented in the same direction as the rays, with their plus ends pointing away from the site of sperm entry. In cells that inherit these dorsalizing factors during cleavage, the Wnt pathway is activated (third panel). Red circles indicate activation of the Wnt signaling pathway and an accumulation of β-catenin in nuclei. This results in activation of Wnt signaling in cells on the future dorsal side of the blastula (not shown here). In the late blastula (fourth panel), the Spemann organizer is formed in this dorsal region, and is where gastrulation will start. V, ventral; D, dorsal.

rays in the aster, with their plus ends pointing away from the site of sperm entry. This array of subcortical microtubules acts as a directional transport system. The cortex and immediately underlying cytoplasm rotate from their initial position at the vegetal pole taking with them maternal factors so that these become distributed in the egg opposite the site of sperm entry—the future dorsal side of the blastula. Cells that inherit these maternal factors during the following cleavage divisions will express 'dorsal genes', and this will result in subsequent formation of the Spemann organizer and dorsal structures; for this reason these factors are often called **dorsalizing factors** (see Fig. 3.13). Thus, additional crucial decisions about the body plan have already been made during this first cell cycle and involve the asymmetric re-localization of maternal factors.

Cortical rotation can be prevented by irradiating the ventral side of the egg with ultraviolet (UV) light, which disrupts the microtubules responsible for transporting the dorsalizing factors. Embryos developing from irradiated eggs are ventralized; that is, they are deficient in structures normally formed on the dorsal side, and instead develop excessive amounts of blood-forming mesoderm, a tissue normally only present on the ventral side of the embryo. With very high doses of radiation, both dorsal and anterior structures are lost, and the ventralized embryo appears as little more than a small distorted cylinder. The effects of UV irradiation can, however, be 'rescued' by tipping the egg. Because of gravity, the dense yolk slides against the cortex, thus effectively restoring rotation and the localization of maternal factors.

At the first cleavage division, the fertilized egg is divided into two blastomeres that both inherit some dorsalizing factors but with subsequent divisions, only some blastomeres inherit them. Because the plane of the first cleavage division is usually aligned with the site of sperm entry, it essentially divides the embryo into right and left halves. This explains why, when they are separated, each can give rise to a near-perfect embryo, as described in Section 1.3. The second cleavage division at right angles to the first results in two 'dorsal' blastomeres containing dorsalizing factors, and two 'ventral' blastomeres, which contain no dorsalizing factors. When the embryo is divided into two at this four-cell stage, so that one half comprises the two 'dorsal' blastomeres and the other the two 'ventral' blastomeres, the 'dorsal' blastomeres will develop most structures of the embryo, although lacking a gut. By contrast, the 'ventral' blastomeres give rise to embryos that lack anterior and dorsal structures, in some cases being radially symmetrical 'belly' pieces (Fig. 3.14). This experiment shows that dorsal and ventral regions are specified by the four-cell stage. The ventralized embryos that develop from the two 'ventral' blastomeres are similar to those that can be produced by removal of the vegetal-most yolk soon after fertilization of the egg, before the relocalization of the dorsalizing factors by cortical rotation.

Fig. 3.14 Dorsal factors are essential for normal development. If a *Xenopus* embryo is divided into a dorsal half and a ventral half at the four-cell stage, the half containing dorsal factors develops as a dorsalized embryo lacking a gut (compare with normal embryo shown at the bottom), while the ventral half, which does not contain any dorsal factors, is ventralized and can lack both dorsal and anterior structures.

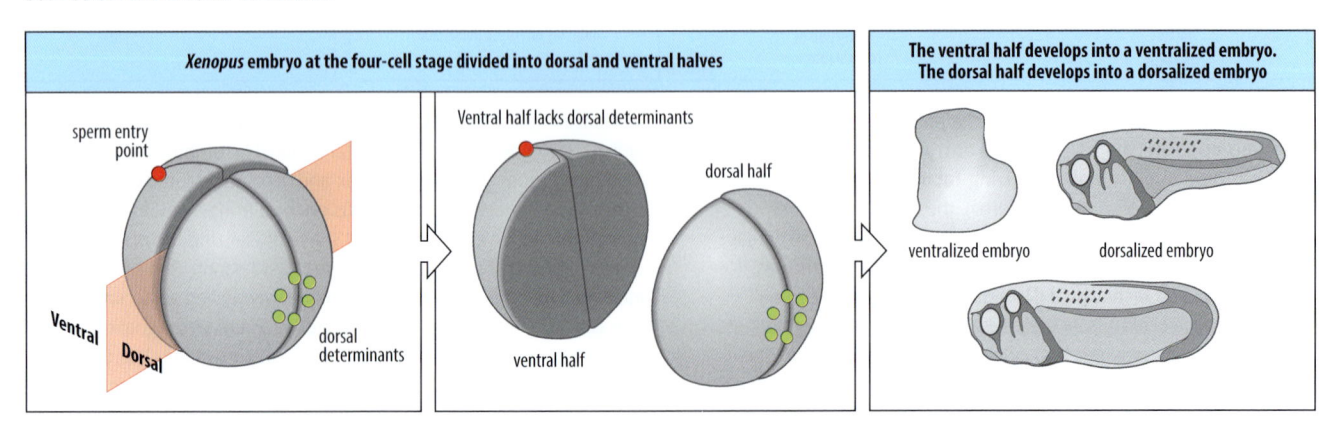

Fig. 3.15 Transplantation of the Spemann organizer can induce a new axis in *Xenopus*. When the Spemann organizer from a *Xenopus* gastrula is grafted into the ventral region of another gastrula, a partially twinned embryo results, with the second body axis induced by the transplanted organizer. The second axis has a complete dorso-ventral axis and a distinct head. The organizer therefore produces signals that not only result in patterning the mesoderm dorso-ventrally, but also induce anterior structures and neural tissue. Scale bar = 1 mm.
Photograph courtesy of J. Smith.

Establishment of a dorso-ventral axis is a key event in *Xenopus* development, as it sets up the conditions necessary for the formation of the embryonic organizer, the Spemann organizer, at the dorsal blastopore lip of the gastrula. As we saw in Chapter 1, the Spemann embryonic organizer was originally discovered by grafting the dorsal lip of the blastopore of one newt embryo into the blastocoel of another newt embryo and finding that it could induce the formation of a second embryonic axis. When the dorsal blastopore lip from an early *Xenopus* gastrula is grafted to the opposite side of another early *Xenopus* gastrula, a twinned embryo also develops. A second main body axis is induced, having its own head at the anterior, and a complete dorso-ventral axis (Fig. 3.15).

Among the maternal dorsalizing factors that move along the oriented microtubules to accumulate opposite the point of sperm entry are components of the Wnt pathway. These factors include *wnt11* mRNA, *wnt5a* mRNA, and Dishevelled (Dsh) protein—the last a member of the Wnt intracellular signaling pathway. After translation, the Wnt-11 and Wnt-5a signal proteins form a complex that activates the **Wnt/β-catenin pathway** (Box 3E) in 'dorsal' cells. Maternal β-catenin is initially present throughout the vegetal region. The activation of the canonical Wnt signaling pathway by the Wnt signal proteins prevents the degradation of β-catenin on the dorsal side of the blastula. The Dsh protein stops formation of the destruction complex required for β-catenin degradation. The destruction complex contains the enzyme GSK-3β, and treatment of early *Xenopus* embryos with lithium chloride, which blocks GSK-3β activity, dorsalizes them. By the two-cell stage, β-catenin has started to accumulate in the cytoplasm on one side of the two cells, and by the second or third cleavages, it has entered the nuclei of the dorsal cells, where it later helps to switch on zygotic genes, including those required to establish the Spemann organizer (see Fig. 3.13).

The crucial role of activation of the Wnt signaling pathway in specifying dorsal cells is demonstrated by both gain-of-function and loss-of-function experiments. For example, injection of oocytes with *Wnt11* mRNA leads to dorsalization, whereas injection of *Wnt11* antisense morpholino oligonucleotides leads to ventralization. Similarly, injection of oocytes with antisense morpholino oligonucleotides that inhibit maternal expression of one of the Wnt receptors, Frizzled 7 (Fz7), also produces ventralized embryos. This is because in the absence of the receptor, the Wnt/β-catenin pathway cannot be activated by an extracellular Wnt signal. Experiments manipulating levels of β-catenin have also been carried out. When β-catenin mRNA is injected into a ventral vegetal cell of a 32-cell stage *Xenopus* embryo, twinned embryos develop as a result of the unusual presence of active β-catenin in the ventral cells. This leads to expression of 'dorsal' genes in these cells and to a series of events that culminates in the formation of a second organizer.

Wnt signaling is also limited to the dorsal region of the blastula by a maternally provided antagonist of Wnt signaling, the protein Dickkopf1 (Dkk1), which is present in the ventral region. Dkk1 is a secreted protein that binds to the Wnt co-receptor LRP5/6 (see Box 3E), blocking its function and inhibiting activation of the pathway. As we shall see throughout this chapter, inhibition of the activity of secreted signal proteins is an important developmental strategy for restricting signaling to specific regions.

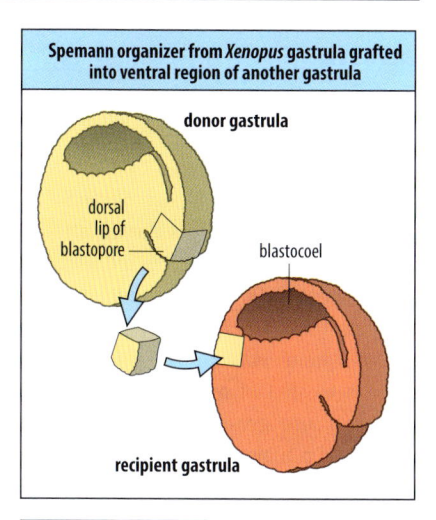

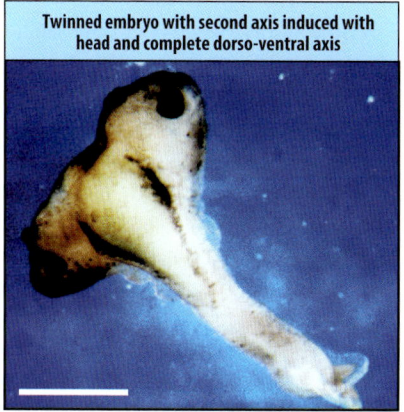

CELL BIOLOGY BOX 3E The Wnt/β-catenin signaling pathway

The important developmental signaling pathway initiated by the Wnt family of intercellular signal proteins (pronounced 'Wints') is named after the proteins encoded by the *wingless* gene in *Drosophila* and the *Int-1* gene (now *Wnt1*) in mammals. The *Int-1* gene was so called because it is located at one of the sites in the mouse genome where integration of a tumor virus causes cancer. Its essential role in development in mammals and other vertebrates was discovered as a result of its close similarity to *Drosophila wingless*, whose role in development was already known. Wnt proteins are involved in specifying cell fate in a multiplicity of developmental contexts in all multicellular animals studied so far. We have already met functions of Wingless in *Drosophila* and will meet many more of the functions of the Wnt family throughout the book. In adult humans, Wnt is required to maintain stem cells in their self-renewing state (discussed in Fig. 6.30), and incorrect control of the Wnt intracellular signaling pathway is implicated in common cancers such as colon cancer.

Wnt proteins can activate several different intracellular signaling pathways: the one we illustrate here is involved in specifying cell fates in the early development of all animals. It is known as the **canonical Wnt/β-catenin pathway**, or simply the **Wnt/β-catenin pathway**, as it was the first to be discovered, and it leads to the accumulation of the protein β-catenin and its action as a transcriptional co-activator. (β-catenin can also act as a protein that links cell-adhesion molecules to the cytoskeleton (see Box 5A); that function is quite distinct from its role as a transcriptional co-activator.) Other Wnt pathways will be described when they are encountered in the book. A simplified version of the canonical Wnt/β-catenin pathway is shown in Figure 1. As with many developmental signaling pathways, the initial Wnt signal results in changes in gene expression in the nucleus of the responding cell. In this case, the Wnt signal prevents the degradation of the protein β-catenin in the cytoplasm, and by doing so enables β-catenin to accumulate and to enter the nucleus, where it binds to and activates transcription factors of the T-cell-specific factor (TCF) family and so activates the expression of target genes.

In the absence of a Wnt signal, β-catenin (pale green circle) is bound by a 'destruction complex' of proteins in the cytoplasm (left-hand side of Fig. 1). This includes protein kinases (CK1γ and GSK-3β) that phosphorylate β-catenin, thus targeting it for ubiquitination and degradation in the proteasome. In the absence of β-catenin in the nucleus, transcriptional co-repressors (CTBP (C-terminal binding protein), histone deacetylase (HDAC), and Groucho) bind to the TCF transcription factors and prevent gene expression. Like most extracellular signals, Wnt acts on its target cells by binding to a specific transmembrane receptor protein (Frizzled) at the cell surface (right-hand side of Fig. 1). Frizzled and an associated co-receptor, LRP5/6, then transmit the Wnt signaling function across the membrane: Wnt itself does not enter the cell. Activation of these receptors causes the same protein kinases that phosphorylated β-catenin in the destruction complex to now become associated with the membrane and phosphorylate the cytoplasmic tail of the activated LRP5/6. Protein phosphorylation, which changes the activity of proteins or their interactions with other proteins, is ubiquitous in intracellular signaling pathways as a means of transmitting the signal onwards. The intracellular signaling protein Dishevelled (Dsh) and the protein Axin (which is a component of the destruction complex) are recruited to the tails of LRP5/6 and Frizzled at the membrane. This prevents the destruction complex from forming and leaves β-catenin free to accumulate and to enter the nucleus. Inside the nucleus, β-catenin binding to TCF displaces co-repressors, lifts the repression, and enables the target genes to be expressed.

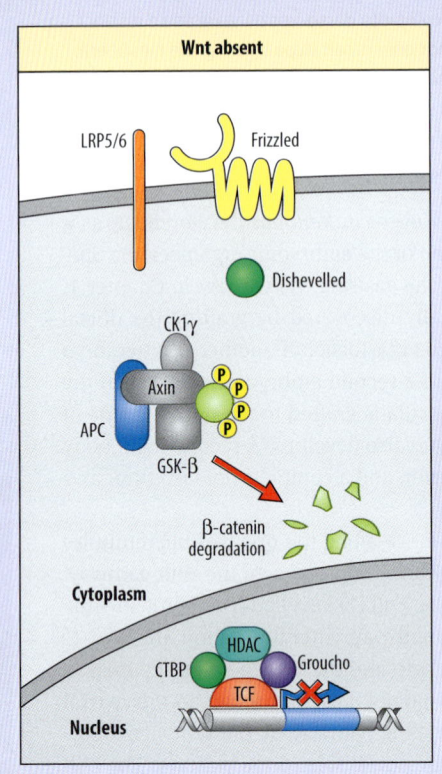

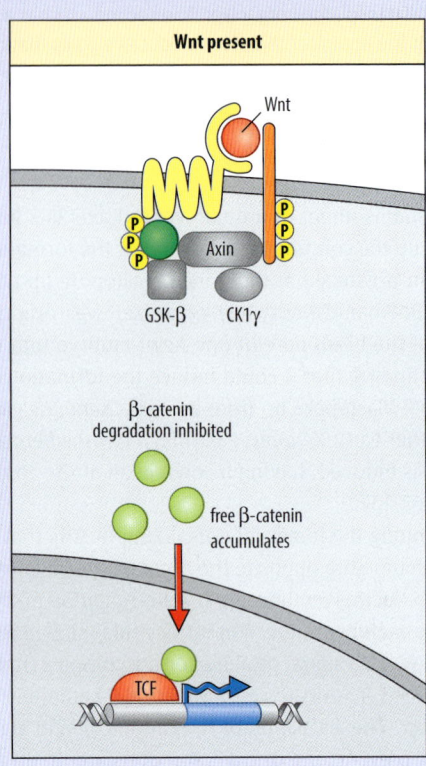

Fig. 1

Fig. 3.16 Establishment of the Nieuwkoop center. In *Xenopus*, cortical rotation moves dorsal factors (green) towards the future dorsal side of the embryo, creating a large region where β-catenin moves from the cytoplasm into nuclei (red). Cells in the region of the blastula in which β-catenin is intranuclear, and where VegT is also present, express the transcription factor Siamois (blue shading), which marks the Nieuwkoop center in the dorsal vegetal region of the early blastula.

3.8 The Nieuwkoop center, a signaling center, develops on the dorsal side of the blastula

The establishment of the dorso-ventral axis sets up the conditions for the formation of the Spemann organizer. This occurs because the combined actions of Wnt-11, Wnt-5a, β-catenin, and other maternal factors present in the dorsal side of the vegetal region lead to the formation of a signaling center known as the **Nieuwkoop center** in this region (Fig. 3.16). The Nieuwkoop center expresses particular transcription factors including Siamois, and consequently produces signals that induce the Spemann embryonic organizer in the adjacent dorsal marginal region.

The signaling activity of the Nieuwkoop center and its role in specifying the Spemann organizer can be demonstrated by grafting the cells from the dorsal vegetal region of a 32-cell *Xenopus* embryo into the ventral side of another embryo. This gives rise to a twinned embryo with two dorsal sides (Fig. 3.17). Cells from the graft itself contribute to endodermal tissues, but not to the mesodermal and neural tissues of the new axis, which instead are formed from the host embryo. This shows that the grafted cells act as a signaling center and induce these fates in the cells of the host embryo. By contrast, grafting ventral vegetal cells to the dorsal side has no effect.

The Nieuwkoop center is named after the Dutch embryologist Pieter Nieuwkoop, who discovered its ability to induce the Spemann organizer cells through experiments in which explants of tissue from the animal region of blastulas were combined with explants of tissue from the vegetal region (discussed in detail in Section 3.12).

The key maternal factor that works in combination with Wnts to specify the Nieuwkoop center is the VegT transcription factor, which was localized to the vegetal pole in the oocyte (see Section 3.6) but, unlike the Wnts, was not moved dorsally during cortical rotation. Instead, VegT remains in the vegetal region and is inherited by all vegetal blastomeres. Wnt signals and VegT functionally interact, and together

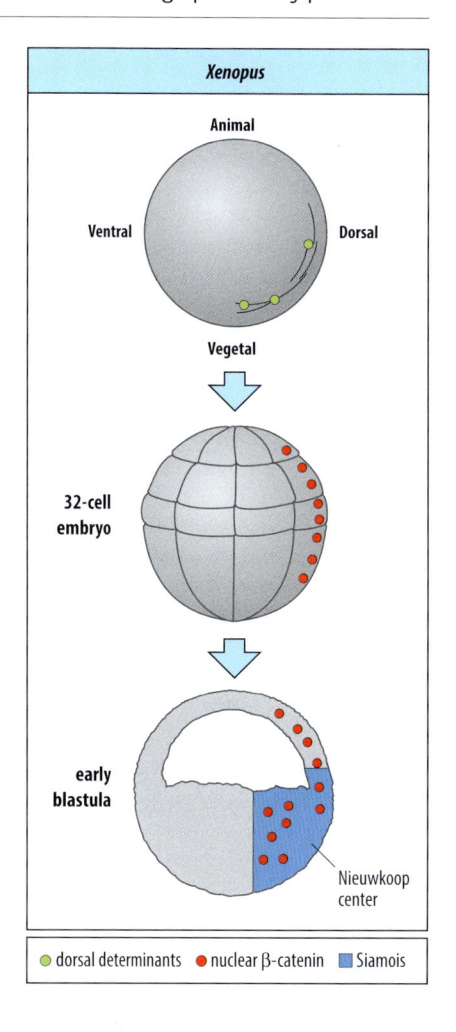

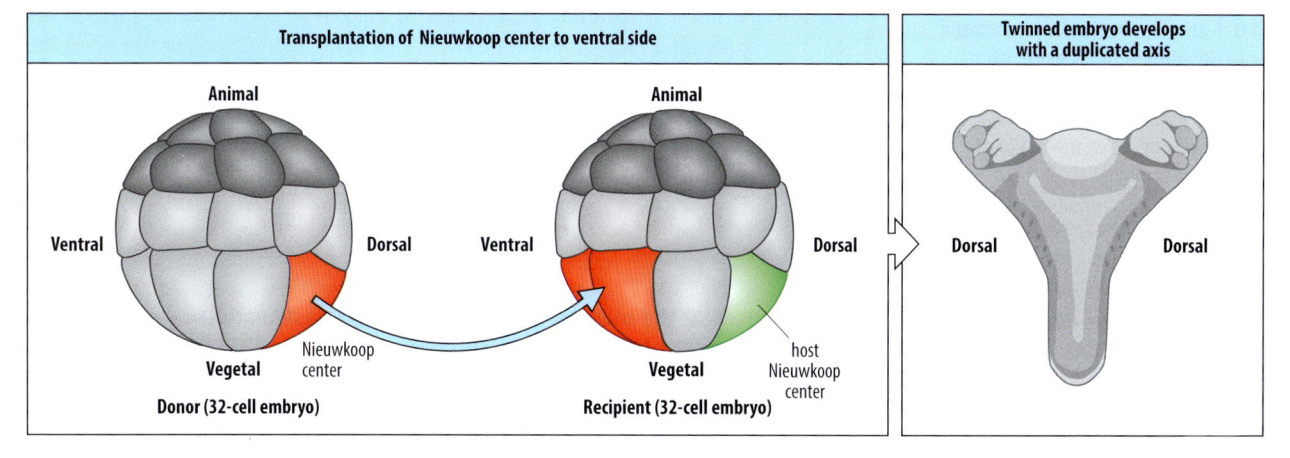

Fig. 3.17 The Nieuwkoop center can specify a new dorsal side. When vegetal cells containing the Nieuwkoop center from the dorsal side of a 32-cell *Xenopus* blastula are grafted to the ventral side of another blastula, a second axis forms and a twinned embryo develops. The grafted cells contribute to endodermal tissues but not to the mesodermal and neural tissues of the second axis.

promote the expression of genes that define the Nieuwkoop center, including *siamois*. *siamois* is among the earliest zygotic genes (genes encoded by the embryo's own DNA) expressed—as early as the 256-cell stage (cleavage cycle 8)—as a result of **zygotic gene** activation. In *Xenopus*, most zygotic genes required for development are activated later at what is known as the **mid-blastula transition** and we will briefly consider this next.

3.9 Control of development is transferred from the maternal to the zygotic genome at the mid-blastula transition

As in *Drosophila*, early development in *Xenopus* is under the control of maternal factors, but subsequent development is then orchestrated by the embryo's own genes. The transfer of the control of development from the maternal genome to the zygotic genome involves eliminating maternal mRNAs and proteins and initiating zygotic gene activation. In *Xenopus*, the major wave of activation of zygotic genes and synthesis of new mRNA occurs during the blastula stage, after 12 cleavages have taken place and the embryo contains 4096 cells. At the same time, the cell cycles also lengthen and become asynchronous. This event is known as the **mid-blastula transition**, although it does, in fact, occur in the late blastula, just before gastrulation. As we shall see, the timing of the activation of zygotic genes occurs at different stages in other vertebrates, as does the timing of the changes in cell-cycle dynamics.

How is the timing of the mid-blastula transition controlled? Suppression of cleavage, but not DNA synthesis, does not alter the timing of transcriptional activation, and so the timing is not linked directly to cell division. Nor are cell–cell interactions involved, because dissociated blastomeres undergo the transition at the same time as intact embryos. One key factor in the timing of the activation of zygotic gene expression and the increase in length of the cell cycle seems to be the ratio of DNA to cytoplasm—the quantity of DNA present per unit mass of cytoplasm.

Direct evidence for this comes from increasing the amount of DNA artificially by allowing more than one sperm to enter the egg, or by injecting extra DNA into the egg. In both cases, transcription of zygotic genes occurs prematurely. Conversely, in other experiments using hybrid *Xenopus* embryos in which the amount of DNA was decreased, the transcription of zygotic genes and the increase in the length of the cell cycle were delayed. These results suggest a model in which there is some fixed amount of a general repressor of transcription initially present in the egg cytoplasm. As the egg cleaves and the amount of DNA increases, the amount of repressor in relation to DNA gets smaller and smaller because the amount of cytoplasm does not change. When there is insufficient repressor to bind to all the available sites on the DNA, the repression is lifted, and transcription is activated. Thus, the timing of the mid-blastula transition seems to fit an hourglass egg-timer model (Fig. 3.18). In this model, the concentration of available repressor decreases as the amount of DNA increases, until a threshold is reached. The threshold is determined by the initial concentration of the repressor in the cytoplasm, whose volume does not increase. Good candidates for this repressor are maternal histone proteins that were stored in the egg.

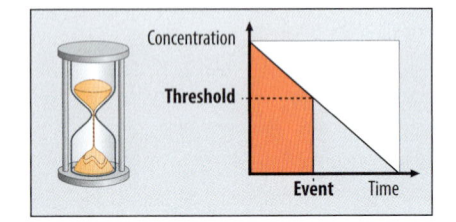

Fig. 3.18 A timing mechanism that could operate in development. A mechanism based on an analogy to an egg timer could measure time to the mid-blastula transition. The decrease in the available concentration of some molecule, such as a transcriptional repressor, could occur with time, and the transition could occur when the repressor reaches a critically low threshold concentration. This would be equivalent to all the sand running into the bottom of the egg timer. In the embryo, a reduction in repressor activity per nucleus would, in fact, occur, because repressor concentration in the embryo as a whole remains constant but the number of nuclei increases as a result of cell division. The amount of repressor per nucleus thus gets progressively smaller over time.

3.10 The axes of the blastula are transformed by gastrulation into the body axes of the tailbud embryo

The dorso-ventral and the antero-posterior axis of the final body plan are predicted by the dorso-ventral axis of the blastula. At the same time as the dorso-ventral axis of the blastula is being set up, cells are being allocated to the three germ layers. The fate map of a late *Xenopus* blastula shows that the vegetal region gives rise to the endoderm, the animal region to ectoderm, and the marginal zone to mesoderm

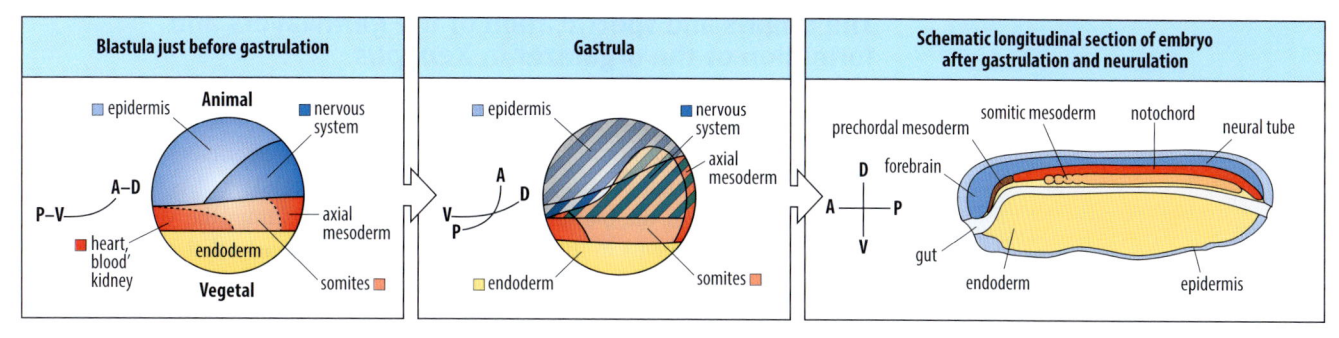

(Fig. 3.19, first panel). During gastrulation, the mesoderm and endoderm move inside and rearrange. The rearrangement of cells during gastrulation changes the relative positions of tissues, resulting in the emergence of the body axes and an embryo with a recognizable vertebrate body plan (Fig. 3.19, second and third panels).

A look at the details of the *Xenopus* fate map shows that distinct regions of mesoderm along the dorso-ventral axis of the blastula have different fates going from dorsal to ventral—axial mesoderm (prechordal mesoderm and notochord), somites, heart, kidney, and blood-forming tissues (see Box 3A, Fig. 2). The dorsal ectoderm gives rise to the nervous system, the more-ventral ectoderm to the epidermis. At this stage, the future dorsal and anterior ends of the body axes are in the same region, as are the future posterior and ventral ends of the body axes.

During gastrulation, the first mesodermal cells to move inside are the most-dorsal mesoderm cells. These form the prechordal mesoderm in the most dorsal and anterior position in the body plan and the overlying ectoderm forms the brain. Next, the adjacent mesoderm cells move inside to form notochord, which lies posterior to the prechordal mesoderm; cells from progressively more ventral regions move in later and end up in more ventral and posterior positions in the body plan, ultimately forming the very posterior end of the tailbud embryo. In this way, dorso-ventral positional information embedded in the blastula is revealed when the dorso-ventral and antero-posterior axes of the post-gastrula embryo emerge.

Fig. 3.19 Unfolding of the principal body axes during gastrulation in *Xenopus*. Left panel: fate map of the *Xenopus* late blastula embryo. Colors depict the regions that will give rise to mesoderm (red, peach) and ectoderm (dark and light blue) at different positions along the dorso-ventral axis and the region fated to give rise to the endoderm (yellow). At this stage, the future anterior (A) and dorsal (D) ends of the body axes are in the same region, as are the future posterior (P) and ventral (V) ends of the body axes. Center panel: during gastrulation, dorsal mesoderm moves inside to take up a position at an anterior and dorsal position in the body plan (shown here is notochord in red) and the antero-posterior body axis begins to emerge; more-ventral mesoderm (peach) also moves inside to take up its appropriate position. Some endoderm has also moved inside (not shown). Right panel: D-V and A-P axes have been unfolded by gastrulation and are now in their final positions in the body plan and the fate of the cells in different regions of the late blastula is realized.

*Based partly on Martinez Arias, A., Steventon, B.: **On the nature and function of organizers.** Development 2018, **145**: dev159525.*

SUMMARY

In *Xenopus*, an animal-vegetal axis is determined through the differential localization of maternal factors. The dorso-ventral axis of the blastula is specified by the site of sperm entry and the resulting cortical rotation. This relocates a set of maternal factors to a point opposite the site of sperm entry, which is thus specified as the dorsal side. These 'dorsalizing' maternal factors include components of the Wnt signaling pathway. Their actions in concert with the vegetally located maternal VegT transcription factor lead to the local establishment of a signaling center—the Nieuwkoop center—in the vegetal region on the dorsal side of the blastula. Signals from the Nieuwkoop center induce the formation of the Spemann embryonic organizer in the adjacent dorsal equatorial region of the late blastula and when the Nieuwkoop center is grafted to the ventral side of another blastula, a second organizer is induced, and a twinned embryo develops. During the blastula stage, in a period called the mid-blastula transition, the embryo's own genes begin to be transcribed, and thereafter, development is under their control. The dorso-ventral axis of the blastula predicts the dorso-ventral and antero-posterior axes in the final body plan.

The origins and specification of the germ layers and formation of the organizer in *Xenopus*

From the blastula fate map, we saw that mesoderm arises in the equatorial region of the blastula. In this part of the chapter we will look at the experiments that showed that mesoderm has not yet been specified in the early blastula, but is induced from the animal region in the blastula-stage embryo by signals from the vegetal region. Before those experiments in the late 1960s, it had been thought that mesoderm was intrinsically specified in a particular cytoplasmic region in the amphibian egg, as are endoderm and ectoderm. The experiments showed instead that mesoderm could be induced in animal cap cells that were cultured with vegetal region cells. We describe how mesoderm is specified through the actions of proteins encoded by zygotic genes, and how it is initially patterned, focusing on the specification of dorsal mesoderm (the Spemann organizer) versus ventral mesoderm. The further elaboration of germ-layer patterning in *Xenopus* will be considered in Chapter 4.

3.11 Mesoderm is induced and its patterning is initiated by signals from vegetal and dorsal regions of the blastula

The fate map of the blastula (see Fig. 3.19) shows which cells will give rise to which parts of the embryo during normal development but does not indicate when the fate of particular cell types is specified. A classic experiment to test for specification in *Xenopus* embryos is to dissect out and culture small fragments of tissue, called **explants**, from very specific regions of the embryo (see Section 1.12). When explants from different regions of an early *Xenopus* blastula are cultured, tissue from the region near the animal pole, usually referred to as the **animal cap**, forms ectoderm—specifically epidermis—whereas explants from the vegetal region form endoderm (Fig. 3.20). These results are in line with the normal fates of these regions. They also show that the developmental instructions that specify ectoderm and endoderm are already present, and spatially distinct, in the early blastula.

A strong candidate for a maternal determinant of ectoderm is the mRNA for the E3 ubiquitin-ligase Ectodermin, which is located in the animal region of the fertilized egg. Ectodermin mRNA is inherited by cells that develop from this region and Ectodermin protein becomes localized in the nuclei of cells throughout the animal region of the blastula (Fig. 3.21). Ectodermin is an enzyme that acts in the nucleus by adding ubiquitin to Smad4, a gene-regulatory protein involved in TGF-β type signaling by bone morphogenetic protein (BMP) and Nodal proteins (Box 3F). Ubiquitination targets Smad4 for degradation and thereby reduces signaling through both these pathways. In *Xenopus*, Nodal-type signaling is mediated by proteins called **Nodal-related proteins**,

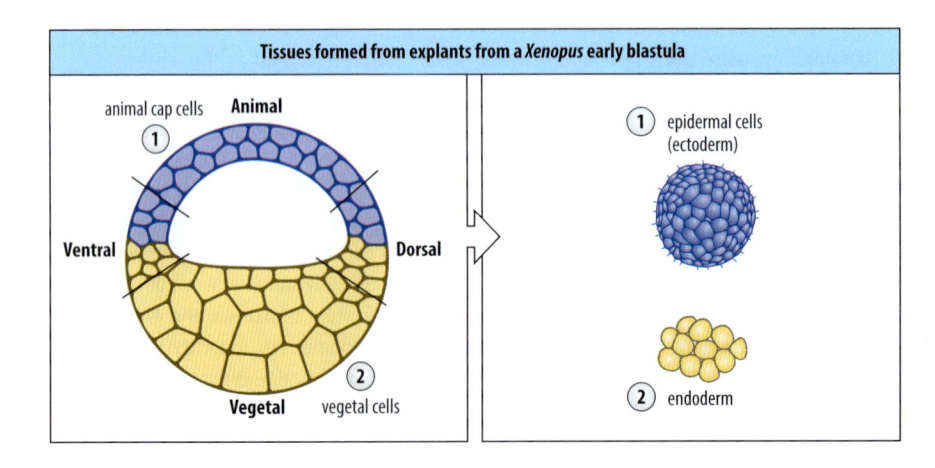

Fig. 3.20 Development of explants of animal cells and vegetal cells from a *Xenopus* early blastula. Animal cap cells (blue, 1) explanted on their own form epidermis, whereas vegetal cell explants (yellow, 2) form endoderm.

Fig. 3.21 Factors for germ-layer specification in *Xenopus*. In the vegetal region of the blastula, the maternal factor VegT is essential for both endoderm specification and for the production of Nodal-related proteins (Nodals), which are one of the main signals that induce mesoderm from the cells of the animal cap and are also involved in endoderm formation. In the animal region of the blastula, the maternal factor Ectodermin counteracts potential mesoderm-inducing signals, and limits mesoderm formation to the equatorial zone. The zygotic transcription factor FoxI1e is expressed in the animal region and is necessary for the specification of the animal cells as ectoderm and for preventing their mixing with more vegetal cells and developing as mesoderm or endoderm. In turn, signaling by the Nodal-related proteins inhibits the expression of FoxI1e in the mesoderm.

after the mouse signal protein Nodal, which was discovered first. As we shall see, Nodal-related proteins produced by the vegetal region are involved in mesoderm induction, and the main function of Ectodermin is to limit mesoderm formation in the blastula to the marginal zone. Another, later, determinant of ectoderm is the zygotic transcription factor FoxI1e, which is expressed in the cells of the animal region of the

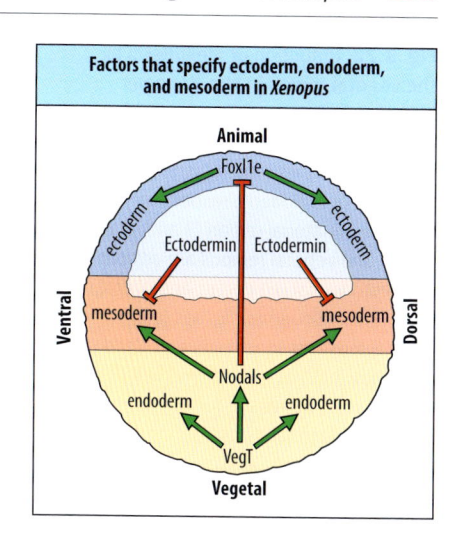

CELL BIOLOGY BOX 3F Signaling by members of the TGF-β family of growth factors

Signaling by the TGF-β family uses different combinations of receptor subunits and transcriptional regulatory proteins to activate different sets of target genes. Members of this family involved in early vertebrate development, such as Nodal and the Nodal-related proteins, bone morphogenetic proteins (BMPs), and Activin (also known as Inhibin), are dimeric ligands that act at cell-surface receptors. The members of the TGF-β family can function as heterodimers, for example, heterodimers of two different Nodal-related proteins or of a Nodal-related protein and GDF-1 (formerly known as Vg-1). The receptors are heterodimers of two different subunits—type I and type II—with intracellular serine/threonine kinase domains. There are several different forms of type I and type II subunits and these combine to form distinctive receptors for different TGF-β family members. Binding of ligand causes subunit II to phosphorylate receptor subunit I, which, in turn, phosphorylates intracellular proteins called Smads (Fig. 1). These are named for the Sma and Mad proteins in *Drosophila*, in which they were discovered. The phosphorylated Smads themselves bind to another type of Smad, sometimes called co-Smads (Smad4 is a representative example), forming a transcriptional regulatory complex that enters the nucleus to either activate or repress target genes. Different receptors use different Smads, thus leading to the activation of different sets of target genes. The identity of the type I subunit determines which Smads get activated. The biological response depends on the combination of the activated target genes and the particular cellular environment.

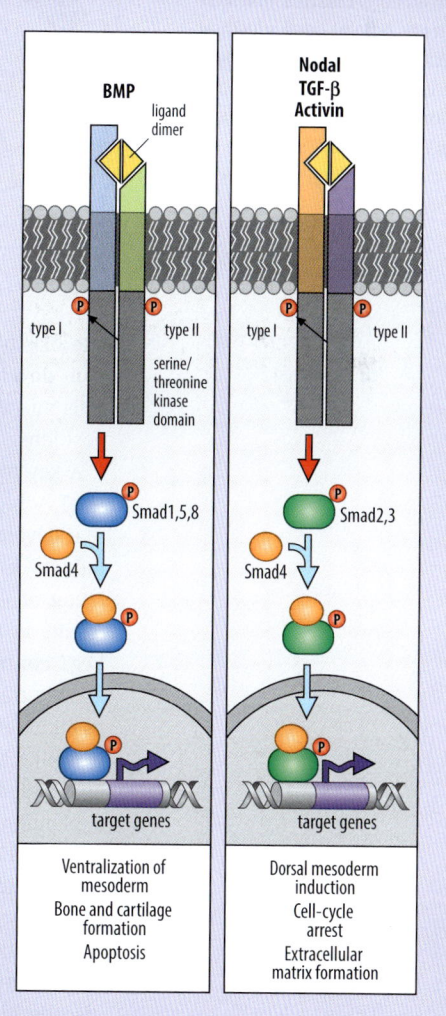

Fig. 1

Fig. 3.22 Induction of mesoderm by the vegetal region in the *Xenopus* blastula. Upper panels: explants of animal cap cells (blue, 1) or vegetal cells (yellow, 2) on their own from a late blastula form only ectoderm or endoderm, respectively. Explants from the equatorial region (3 and 4), where animal and vegetal regions are adjacent, form both ectodermal (epidermis and neural tube, blue) and mesodermal tissues (e.g. notochord, muscle, loose mesenchyme, and blood cells, red), showing that mesoderm induction has taken place. The reason for the differences in the mesodermal tissues formed by ventral and dorsal explants is explained in the text. Lower panels: the standard experiment for studying mesoderm induction is to take explants of animal cap cells and vegetal cells that would normally form only ectoderm and endoderm, respectively, and culture them together for 3 days. When such explants are taken from an early blastula, in which the mesoderm has not yet been specified, and are combined and cultured, mesodermal tissues (red) including notochord, muscle, loose mesenchyme, and blood are induced from the animal cap tissue.

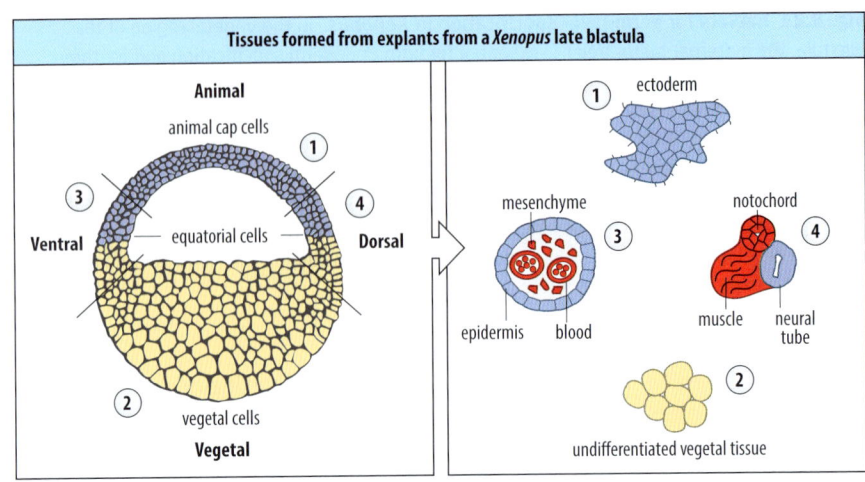

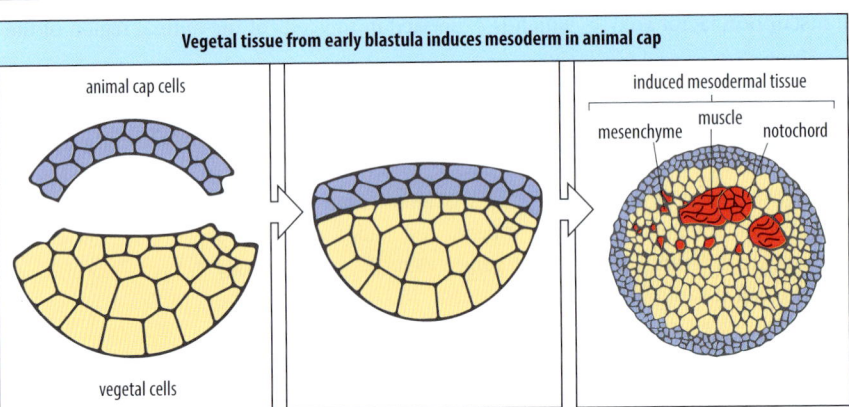

late blastula. FoxI1e is required to maintain the regional identity of these cells and their development as ectoderm (see Fig. 3.21). In the absence of FoxI1e, the animal cells mix with cells of the other germ layers and differentiate according to their new positions.

A key maternal determinant of endoderm is the mRNA for the transcription factor VegT, which is located in the vegetal region of the egg and is inherited by the cells that develop from this region. Injection of *vegt* mRNA into cells in the animal region induces expression of endoderm-specific markers. Conversely, blocking the translation of *vegt* mRNA in the vegetal region, by injecting antisense oligonucleotides, results in a loss of endoderm and instead, ectoderm forms throughout the embryo. VegT controls endoderm formation by promoting the expression of zygotic genes that encode transcription factors in vegetal cells that are necessary for their development as endoderm and promote their segregation from mesoderm cells.

Unlike the ectoderm and endoderm, no mesoderm is formed from explants from any region of early-blastula stage embryos, but mesoderm cells do develop from explants taken from the marginal zone of late-blastula stage embryos (Fig. 3.22, top panels). The results of these explant experiments suggest that mesoderm specification occurs after the onset of zygotic gene activation.

In early studies of such explanted tissues, mesoderm was distinguished by its histology because, after a few days' culture, it can differentiate into tissues such as notochord, muscle, loose mesenchyme (connective tissue), and blood. Mesoderm could also be identified by the typical proteins that cells of mesodermal origin produce, such as muscle-specific actin. Nowadays, however, mesoderm is distinguished by its expression of specific genes, in particular those encoding transcription factors.

The gene that encodes the T-box transcription factor TbxT (formerly known as Brachyury) is one of the earliest genes expressed in cells that are adopting a mesodermal identity. It is expressed throughout the presumptive mesoderm in the late blastula/early gastrula and can readily be detected in cultured explants of the marginal zone of late-blastula stage embryos.

Explant experiments showed that there is a period of about 7 hours during the blastula stage when animal cap cells are **competent** (i.e. able) to respond to a mesoderm-inducing signal; an exposure of about 2 hours to the inducing signal is sufficient for the induction of at least some mesoderm (Box 3G).

EXPERIMENTAL BOX 3G Timing and cell–cell interactions in mesoderm specification

We might expect that the timing of expression of mesoderm-specific genes, such as those of muscle, would be closely coupled to the time at which the mesoderm is induced—but it is not. There seems to be an intrinsic timing mechanism for mesodermal gene expression that is independent of the time of mesoderm induction. Explant experiments showed that animal cap cells isolated from a *Xenopus* early blastula are competent to respond to mesoderm-inducing signals only for a period of about 7 hours, between 4 and 11 hours after fertilization. For expression of target genes to occur, exposure to the inducer must be for at least 2 hours within this period. Irrespective of when the induction occurs within this competence period, muscle-gene activation occurs at the same time: 16 hours after fertilization, in the mid-gastrula stage (Fig. 1).

Differentiation of mesoderm into muscle seems to depend on a so-called **community effect** in the responding cells. A few animal cap cells placed in contact with vegetal tissue will not be induced to become mesodermal cells and will not express muscle-specific genes. By contrast, larger aggregates of animal cap cells respond by strongly expressing muscle-specific genes. A minimum number of 100–200 animal cap cells must be present for induction of muscle differentiation to occur (Fig. 2). The community effect is due to positive feedback in the animal cap cells that results in levels of a local signal being kept above the threshold required for differentiation. A certain minimum number of cells is needed for this to be achieved.

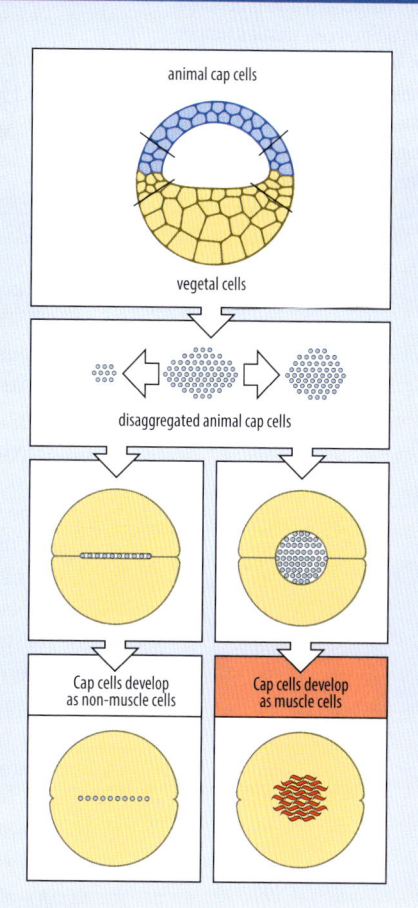

Fig. 2

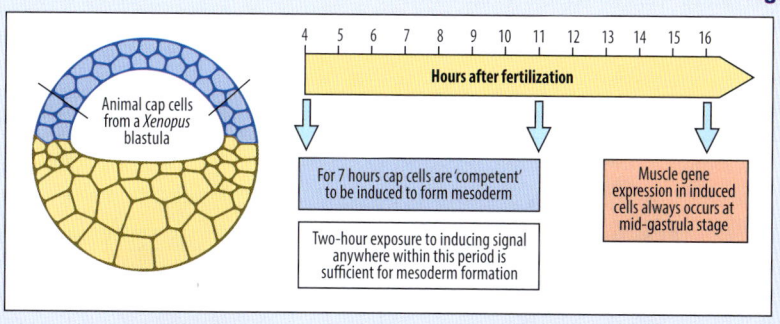

Fig. 1

136 Chapter 3 Vertebrate development I: establishing the body plan

Fig. 3.23 Differences in mesoderm induction by dorsal and ventral vegetal regions. The dorsal vegetal region of the *Xenopus* blastula, which contains the Nieuwkoop center, induces dorsal mesodermal structures, including organizer cells (which form notochord and prechordal mesoderm) from animal cap tissues (upper panel), whereas ventral vegetal cells induce ventral mesoderm such as blood and associated tissues (lower panel). Note also that muscle develops from animal cap tissue cultured with the dorsal vegetal region. The reason for this is discussed in Chapter 4.

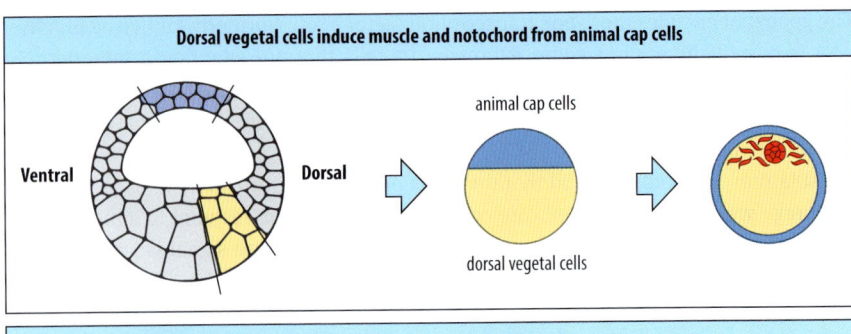

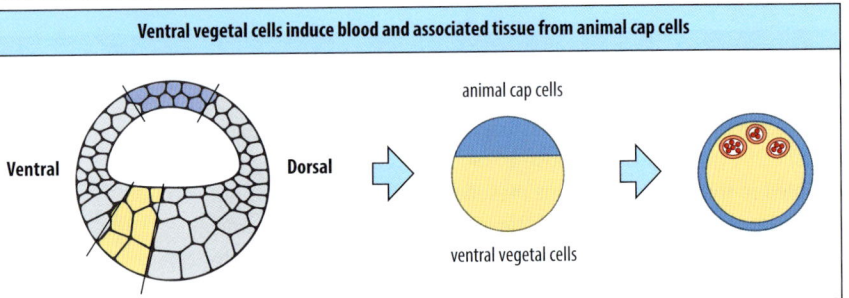

The key experiment showing that mesoderm is induced by interactions between animal and vegetal cells in the blastula involved culturing the animal cap in contact with cells taken from the vegetal region (see Fig. 3.22, lower panels). Mesodermal tissues are formed, and this explant culture system has become a standard system for studying mesoderm induction. Confirmation that it is the animal cap cells, and not the vegetal cells, that are forming mesoderm was obtained by pre-labeling the animal region of the blastula with a cell-lineage marker and showing that the labeled cells form the mesoderm. Clearly, the vegetal region of the blastula is producing a signal or signals that can induce mesoderm.

A series of experiments in which explants of different regions of the blastula were combined showed that signals produced by different parts of the vegetal region of the blastula initiate patterning of the mesoderm. Dorsal vegetal tissue induces dorsal mesoderm (including notochord) from animal cap cells, whereas ventral vegetal tissue induces ventral mesoderm (mainly blood-forming tissue; Fig. 3.23). The dorsal vegetal region is where the Nieuwkoop center is located (see Section 3.8) and the formation of the notochord indicates that the organizer has been induced. These results suggested that mesoderm is first patterned along the dorso-ventral axis due to differences in signaling activity and strength along the dorso-ventral axis of the vegetal region. In this way, specification of dorsal mesoderm (Spemann organizer) occurs above the Nieuwkoop center and specification of ventral mesoderm occurs above ventral vegetal regions (Fig. 3.24).

Fig. 3.24 Signals involved in mesoderm induction originate from the vegetal region. First panel: signals from the vegetal region of the blastula initially induce the mesoderm from the animal region. Second panel: the organizer tissue is specified on the dorsal-most side where the inducing signals are strongest and of longest duration.

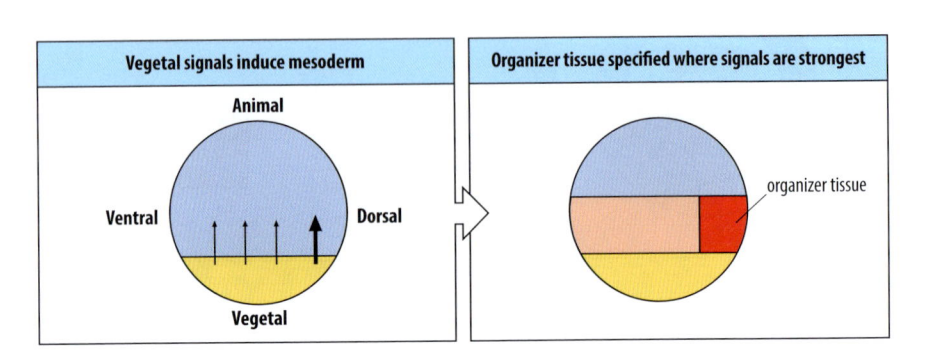

3.12 Members of the TGF-β family have been identified as mesoderm inducers and induce formation of the organizer

The explant experiments discussed in the previous section suggested that the vegetal region of the blastula induces mesoderm, and that different parts of the vegetal region induce different types of mesoderm. If the animal and vegetal explants are separated by a filter with pores too small to allow cell–cell contacts to form, mesoderm induction still takes place. This suggests that the mesoderm-inducing and patterning signals are secreted molecules that diffuse across the extracellular space, and do not pass directly from cell to cell. So what is the nature of these signals?

Two main approaches were used to try and identify these signals. One was to apply candidate proteins directly to isolated animal caps in culture. The other was to inject mRNA encoding suspected inducers into animal cells of the early blastula and then culture the cells. By itself, however, the ability to induce mesoderm in culture does not prove that the protein is a natural inducer in the embryo. Rigorous criteria must be met before such a conclusion can be reached. These include the presence of the protein signal and its receptor in the right concentration, place, and time in the embryo; the demonstration that the appropriate cells can respond to it; and the demonstration that blocking the response prevents induction. On all these criteria, growth factors of the TGF-β family that act as signal proteins (see Box 3F) have been identified as the main mesoderm inducers. Members of this family involved in early *Xenopus* mesoderm induction include the Nodal-related proteins Xnr-1 and Xnr-2, the protein Derrière, now known as growth differentiation factor 3 (GDF-3), GDF-1 (formerly known as Vg-1), and Activin (also known as Inhibin). The frog Nodal-related proteins are named after the mouse protein Nodal, with which they are homologous, and members of the Nodal protein family are key signal proteins in mesoderm induction in all vertebrates.

The expression of at least some Nodal-related proteins is under the control of maternal VegT (see Section 3.11). Because VegT is distributed throughout the vegetal region, Nodal-related proteins are likewise present throughout this region. But the presence of nuclear β-catenin on the dorsal side further stimulates *Xnr* transcription. This results in a gradient of Nodal-related proteins from dorsal to ventral, with their high point in the Nieuwkoop center (Fig. 3.25). Therefore, although mesoderm arises throughout the equatorial region of the blastula, the high levels of Nodals secreted from the Nieuwkoop center induce the specialized type of mesoderm cells at the dorsal margin that form the Spemann organizer.

A key concept in vertebrate developmental biology is that tissues develop in an integrated way: ultimately the body has to function as a single unit, and this can

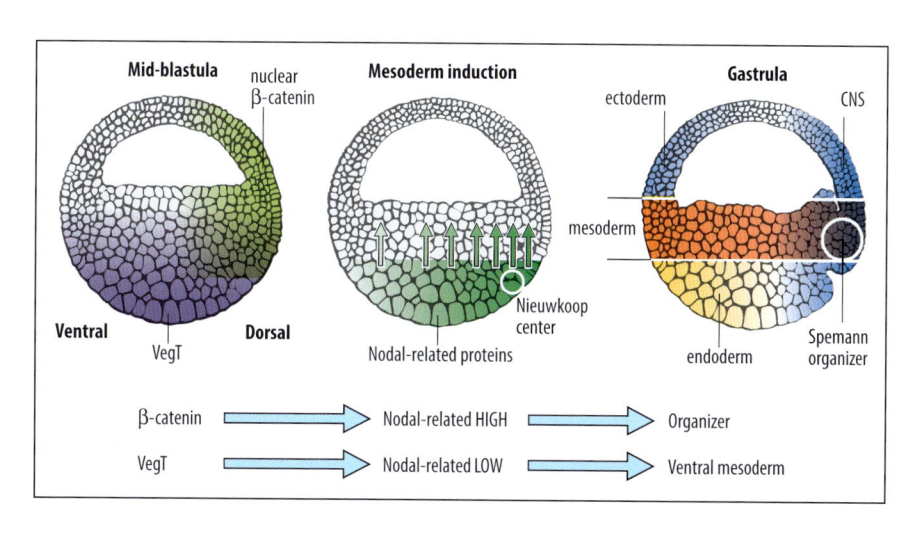

Fig. 3.25 A gradient in Nodal-related proteins provides the initial signals in mesoderm induction. Maternal VegT in the vegetal region initially activates the transcription of *nodal-related* genes. The presence of nuclear β-catenin further stimulates *nodal-related* transcription on the dorsal side, resulting in a dorsal-to-ventral gradient of Nodal-related proteins. These induce mesoderm and, at high doses, specify the Spemann organizer on the dorsal side. This picture is a simplification, because other signal proteins, such as GDF-1, Derrière, and Activin, also play a part in mesoderm formation. CNS, central nervous system.

be achieved if different parts of the body develop in a coordinated way. The role of VegT in both endoderm and mesoderm specification nicely illustrates how this can be achieved through the action of a single transcription factor. VegT acts directly in the cells in which it is expressed to specify them as endoderm (i.e. acting in a cell-autonomous or cell-intrinsic manner to direct endoderm fate), but acts indirectly, by turning on expression of a secreted signal, to specify adjacent cells as mesoderm (i.e. acting in a cell non-autonomous or cell-extrinsic manner to direct mesoderm). In this way, endoderm and mesoderm-derived structures are coordinated as they develop.

The signaling pathways stimulated by TGF-β growth factors are illustrated in Box 3F. Confirmation of the role of TGF-β growth factors in mesoderm induction came initially from experiments that blocked the activation of the receptors for these proteins (Box 3H). The Activin type II receptor is a receptor for several TGF-β family growth factors and is uniformly distributed throughout the early *Xenopus* blastula. When mRNA for a truncated receptor subunit is injected into the early embryo and blocks TGF-β family signaling, mesoderm formation is prevented. Because different TGF-β growth factors can act through the same receptor, these experiments were not able to pinpoint the individual proteins responsible.

It is now possible to study mesoderm induction using more specific inhibitors of Nodal family proteins, and also to detect the presence of signaling by using labeled

---

EXPERIMENTAL BOX 3H Investigating receptor function using dominant-negative proteins

The receptors for proteins of the TGF-β family function as dimers of type I and type II subunits, with the type II subunit participating only in binding and the type I subunit activating the signal transduction pathway (see Box 3F). One way of investigating the requirement for a particular receptor in a developmental process is to block receptor function (Fig. 1). In the case of dimeric receptors, receptor function in a cell can be blocked by introducing mRNA encoding a truncated receptor subunit that lacks most of the cytoplasmic domain and so cannot function. This receptor subunit can bind ligand and forms heterodimers with normal receptor subunits, but the complex cannot activate the pathway. The subunit lacking the cytoplasmic domain thus acts in a dominant-negative fashion, preventing receptor function. When mRNA encoding a dominant-negative truncated Activin receptor subunit is injected into the cells of the two-cell *Xenopus* embryo, subsequent mesoderm formation is blocked. No mesoderm or axial structures are formed.

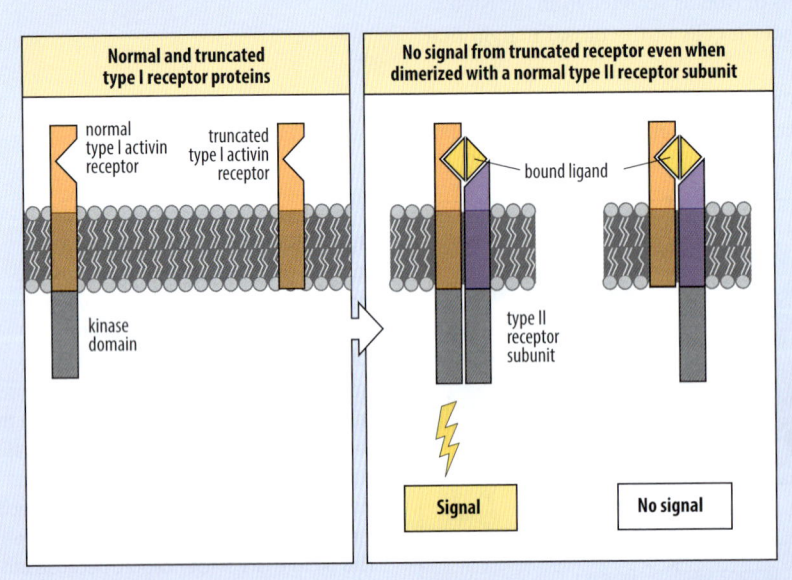

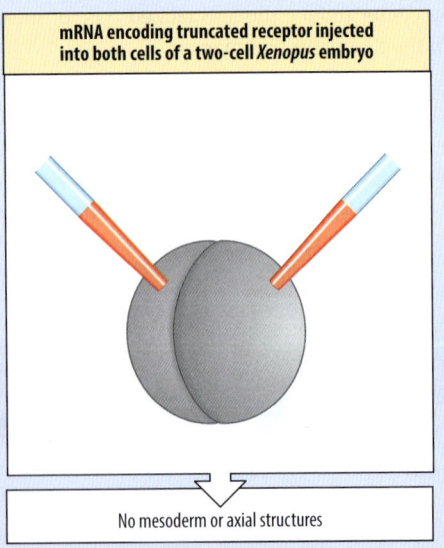

Fig. 1

antibodies to detect phosphorylated Smads (the transcription factors that are activated as a result of TGF-β signaling) and to distinguish their type. This latter technique enables the spatial patterns and dynamics of normal growth factor signaling to be visualized directly, and it is now widely used to detect signaling by TGF-β family members. For example, evidence that Nodal-related proteins act in a graded manner has come through the direct observation of the spatial pattern of phosphorylated Smad2 or 3 (see Box 3F).

3.13 Threshold responses to gradients of signal proteins are likely to pattern the mesoderm

The mesoderm-inducing signals described in previous sections exert their developmental effects by inducing the expression of specific genes that are required for the further development of the mesoderm. The *TbxT* gene is one of the earliest genes expressed in mesoderm, and TbxT protein is a key transcription factor in mesoderm specification and patterning. *TbxT* is expressed throughout the presumptive mesoderm in the late blastula and early gastrula (Fig. 3.26), later becoming confined to the notochord and to the tailbud (posterior mesoderm). Various members of the fibroblast growth factor (FGF) family of growth factors are also expressed in the mesoderm in the late blastula and early gastrula (the FGF signaling pathway is shown in Box 4A). One of the functions of FGF signaling in the mesoderm at these stages is to maintain expression of *TbxT*; in turn, TbxT protein maintains expression of the genes encoding the FGFs. This positive-feedback loop (see Fig. 1.21) ensures that the mesoderm continues robustly to express *TbxT*.

One of the first zygotic genes to be expressed specifically in the dorsal-most mesoderm that will become the Spemann organizer is *goosecoid*, which encodes a transcription factor somewhat similar to both the Gooseberry and Bicoid proteins of *Drosophila*—hence the name. In line with its presence in the organizer, microinjection of *goosecoid* mRNA into the ventral region of a *Xenopus* blastula mimics to some extent the transplantation of the Spemann organizer (see Fig. 3.15), resulting in formation of a secondary axis. Genes for other transcription factors are also specifically expressed in the organizer (Fig. 3.27).

It is still not clear how the secreted signal proteins present in the mesoderm turn on genes such as *goosecoid* and *TbxT* in the right place. For example, *TbxT* can be experimentally switched on in *Xenopus* animal caps by Activin, but it is more likely to be induced in the presumptive mesoderm *in vivo* by Nodal-related protein signaling, which also induces expression of FGFs. As we have seen, there is a gradient of activity of secreted signal molecules throughout the mesoderm that could provide positional information for switching on developmental genes.

Experiments with Activin provide a beautiful example of how a diffusible protein could pattern a tissue by turning on particular genes at specific threshold concentrations. Activin is one of the proteins that plays a part in mesoderm patterning *in vivo*

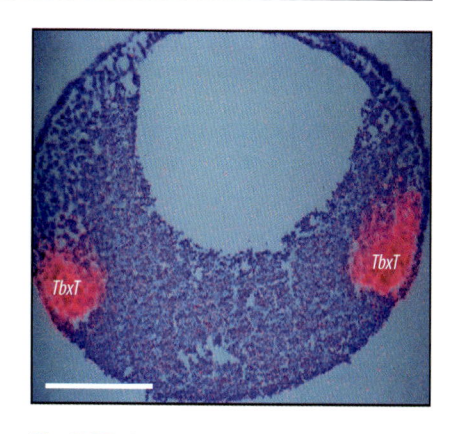

Fig. 3.26 Expression of *TbxT* (*brachyury*) in a *Xenopus* late blastula. *TbxT* (red) is expressed in the prospective mesoderm. This is a cross-section through the middle of the embryo along the animal–vegetal axis, so the entire ring of mesoderm is not apparent. Scale bar = 0.5 mm.

Photograph courtesy of M. Sargent and L. Essex.

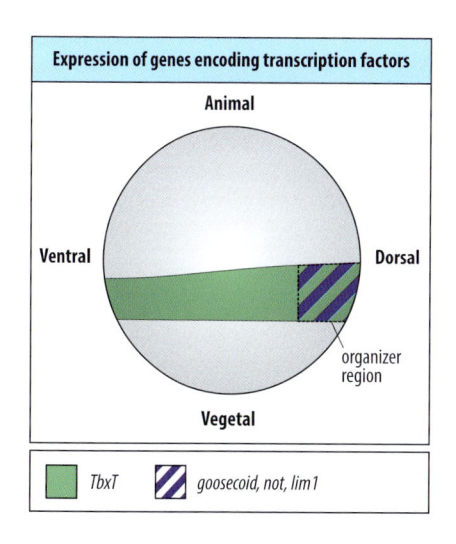

Fig. 3.27 Expression of transcription factor genes in a *Xenopus* late blastula/early gastrula. The expression domains of a number of zygotic genes that code for transcription factors reflect quite well the initial patterning of the mesoderm as judged by the distribution of their mRNAs. *TbxT* is expressed in a ring around the embryo corresponding closely to the future mesoderm (see also Fig. 5.33). A number of transcription factors are specifically expressed in the dorsal-most region of mesoderm that will give rise to the Spemann organizer and are required for its function. Not (Notochord homeobox protein) seems to have a role in the specification of notochord, while Goosecoid and Lim1 (LIM homeobox 1 protein) seem to have a role in the specification of prechordal mesoderm. These two proteins together are required for a transplanted Spemann organizer to be able to induce the formation of a new head.

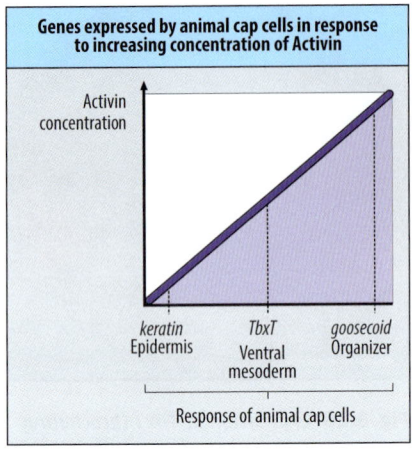

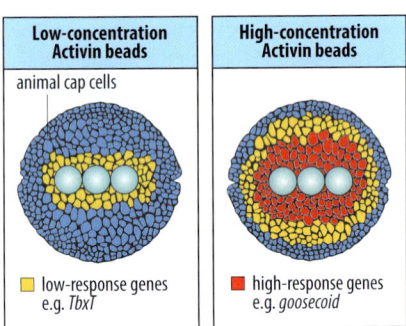

Fig. 3.28 Graded responses of *Xenopus* animal cap cells to increasing concentrations of Activin. When animal cap cells are treated with increasing concentrations of Activin, particular genes are activated at specific concentrations, as shown in the top panel. As the concentration of Activin begins to be increased, *TbxT* is induced, whereas *goosecoid*, which is characteristic of the organizer region, is induced only at higher concentrations. If beads releasing a low concentration of Activin are placed in the center of a mass of animal cap cells (lower left panel), expression of low-response genes such as *TbxT* is induced immediately around the beads. With beads releasing a high concentration of Activin (lower right panel), *goosecoid* and other high-response genes are now expressed around the beads and the low-response genes farther away.

and acts through the same receptor as the Nodal-related proteins. Explanted animal cap cells from a *Xenopus* blastula in culture respond to increasing doses of Activin by activation of different mesodermal genes at different threshold concentrations. In this explant system, increasing concentrations of Activin specify ventral and then dorsal mesoderm. At the lowest concentrations of Activin, only epidermal genes are expressed and no mesoderm is induced. Then, as the concentration increases and exceeds a particular threshold, *TbxT* is expressed. With a further increase in Activin concentration, *goosecoid* is expressed, corresponding to the dorsal-most region of the mesoderm—which will form the Spemann organizer (Fig. 3.28). One can see, therefore, how graded growth-factor signaling along the dorso-ventral axis of the blastula could, in principle, activate transcription factors in particular regions and thus pattern the mesoderm. Just how gradients are set up *in vivo* with the necessary precision is not clear. Although simple diffusion of morphogen molecules might play a part, more complex processes are likely to be involved in producing a graded distribution. Some common cellular mechanisms that could be involved in gradient formation are discussed in Box 9A.

How do cells distinguish between different concentrations of growth factor? Occupation of just 100 Activin receptors per cell is required to activate expression of *TbxT*, whereas 300 receptors must be occupied for *goosecoid* to be expressed. The link between the strength of the signal and gene expression may not be so simple, however. There are additional layers of intracellular regulation that act later to refine distinct regions of gene expression. Cells expressing *goosecoid* at high Activin concentrations repress *TbxT* expression, and this involves the action of the Goosecoid protein itself, together with other proteins. In turn, TbxT protein inhibits *goosecoid* expression, although this is not direct. The mutual repression between *TbxT* and *goosecoid* enables the graded Activin signal to be converted into two discrete domains of gene expression with a sharp boundary between them.

3.14 Regulation is still possible at the blastula stage

By the end of the blastula stage in *Xenopus* embryos, a coordinate system has been set up that ultimately will be transformed into the two main body axes during gastrulation. The three germ layers have been specified and the mesoderm has been broadly patterned into two domains, one of which is the organizer. Remarkably, however, despite this extensive pattern formation, regulation can still occur when the blastula is bisected, as long as each half contains some of the Spemann organizer (Fig. 3.29). This shows that at the blastula stage many cells are not yet determined and have considerable developmental plasticity (see Section 1.12). In other words, the developmental potential of cells at these quite late stages in laying down the body plan is greater than their position on the fate map suggests.

Transplantation experiments confirm that cells from blastula-stage embryos are not yet determined. For example, the progeny of single cells from the vegetal region of

Fig. 3.29 Both halves of a bisected *Xenopus* blastula can give rise to a complete tadpole. Upper panel: the bisection of the *Xenopus* blastula. Lower panel: formation of identical twins from each half (lower embryos) shows that regulation can take place at quite a late stage of development. Twins are only formed if each half contains some of the Spemann organizer. A normal embryo is shown above for comparison. A, animal pole; V, vegetal pole.

*Reproduced with permission from De Robertis, E.M.: **Spemann's organizer and self-regulation in amphibian embryos.** Nature Rev. Mol. Cell Biol. 2006, **7**: 296-302.*

early blastula embryos when introduced into the blastocoel of another embryo can contribute to a wide range of other tissues, such as muscle or nervous tissue. Similarly, the progeny of single animal-region cells from an early blastula, whose normal fate is epidermis or nervous tissue, can form endoderm or mesoderm tissues. With time, cells gradually become determined, so that cells taken from later blastulas and early gastrulas increasingly tend to develop according to their fate at the time of transplantation.

The recent advent of large-scale transcriptomic datasets of developing tissues provides support for the idea that at early stages, cells are biased to particular identities but retain a plasticity until late stages in their differentiation program. For instance, in human **pre-implantation** embryos, the fate markers *IFI16* and *GATA4* gradually become mutually exclusive when epiblast and primitive endoderm, respectively, become determined.

Having considered the process of axis formation in detail in *Xenopus*, we will now cover the same ground in other model organisms, and in humans. Much of the process of laying down the body plan in fish, avian, and mammalian embryos follows the same principles as in amphibians; fate maps at equivalent stages show strong similarities. However, there are some striking differences in early development between birds, mammals, and amphibians, in particular the timing of germ layer formation and organizer development relative to axis formation, and we shall focus on these.

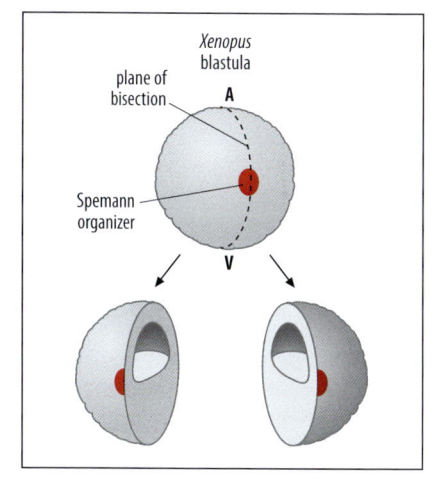

SUMMARY

The endoderm and ectoderm are maternally specified in *Xenopus*, but the mesoderm is induced in the animal region by signals from the vegetal region. This was shown by experiments with explanted fragments of the vegetal and animal regions of the blastula—on their own or in combination. In the embryo, the mesoderm forms a band around the equator of the blastula. In dorsal regions of the embryo, signals from the Nieuwkoop center induce a specialized type of mesoderm termed the Spemann organizer. Members of the TGF-β family have been identified as mesoderm inducers, including inducers of the Spemann organizer. These vegetal signals form a gradient from dorsal to ventral, and two distinct types of mesoderm are induced in response to different threshold concentrations, with highest levels of signals inducing Spemann organizer cells. Mesoderm-inducing signals act on responding animal cells to transcriptionally activate genes that are then required for further development of the mesoderm. However, the *Xenopus* embryo can still undergo considerable regulation at the blastula stage. This implies that interactions between cells, rather than intrinsic factors, have a central role in early amphibian development.

SUMMARY: MESODERM INDUCTION IN *XENOPUS*

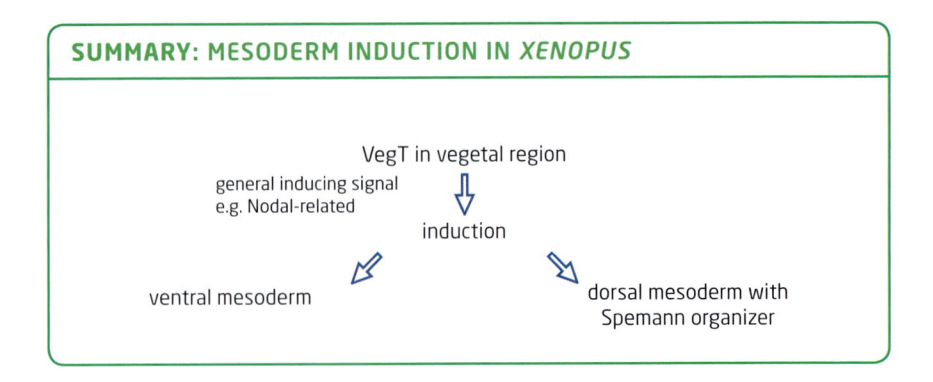

```
                        VegT in vegetal region
    general inducing signal      ⇓
       e.g. Nodal-related
                              induction
                      ⇙               ⇘
    ventral mesoderm              dorsal mesoderm with
                                    Spemann organizer
```

Zebrafish development and setting up the body plan

The way in which the body plan is established in zebrafish (*Danio rerio*) is similar to that in *Xenopus*. Mutagenesis screens for mutant fish with defects in the body plan (see Box 3C) indicate that, in general, the same families of transcription factors and signal proteins are involved in laying down the body plan in both *Xenopus* and zebrafish. Indeed, the study of zebrafish has provided outstanding genetic evidence for the roles of Wnt and Nodal signaling pathways in establishment of the vertebrate body plan. Here we will concentrate on general principles, highlighting where the study of zebrafish has brought insights, and where it is significantly different from *Xenopus*. One key difference is the way the endoderm forms—a consequence of there being much more yolk in the zebrafish egg compared with the *Xenopus* egg, so that cleavage divisions do not extend into it.

3.15 The zebrafish embryo develops around a large yolk cell

The life cycle of the zebrafish is shown in Figure 3.30. The zebrafish egg has a clear animal–vegetal axis: the cytoplasm and nucleus at the animal pole sit upon a large mass of yolk. After fertilization, the zygote undergoes cleavage, but cleavage does not extend into the yolk and results in a mound of blastomeres perched above a large yolk cell. The first five cleavages are all vertical, and the first horizontal cleavage gives rise to the 64-cell stage about 2 hours after fertilization (Fig. 3.31).

Further cleavage leads to the blastoderm (sphere) stage, in which the embryo is now in the form of a blastoderm of around 1000 cells lying on top of the yolk cell. The hemispherical blastoderm has an outer layer of flattened cells, one cell thick, known as the outer enveloping layer: this is largely protective and is eventually lost. The embryo develops from a deep layer of more rounded cells (Fig. 3.32). During the early blastoderm stage, blastomeres at the margin of the blastoderm collapse into the yolk cell, forming a continuous layer of multinucleated non-yolky cytoplasm underlying the blastoderm. This is called the **yolk syncytial layer**. The blastoderm, together with the yolk syncytial layer, spreads in a vegetal direction by epiboly and eventually covers the yolk cell.

Although the fish blastoderm and the amphibian blastula are different in shape, they have corresponding stages of development (see Fig. 3.2). In the fish, the endoderm is derived from the deep cells at the blastoderm margin. This narrow marginal layer is often known as **mesendoderm** because the descendants of a single cell in this region can give rise to either endoderm or mesoderm. Deep cells located between

Zebrafish development and setting up the body plan   **143**

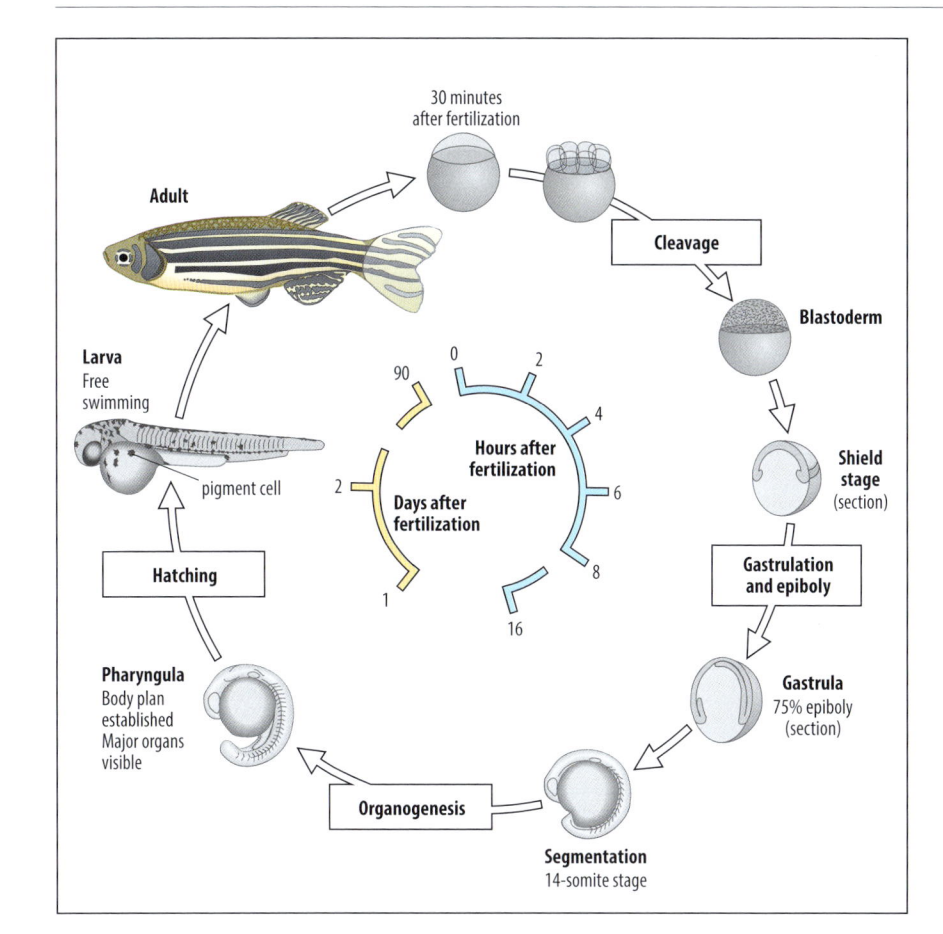

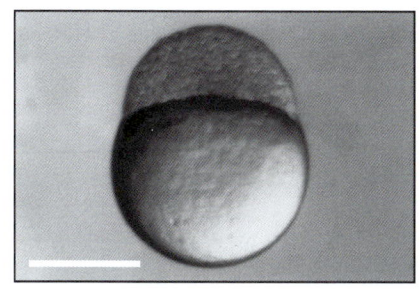

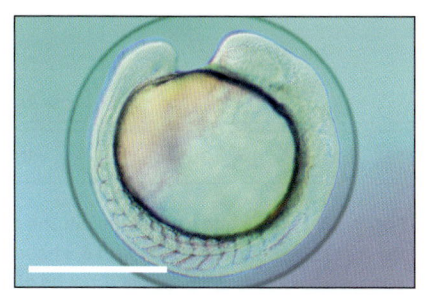

Fig. 3.30 Life cycle of the zebrafish. The schematic shows key stages in the life cycle. The zebrafish embryo develops as a mound-shaped blastoderm sitting on top of a large yolk cell. It develops rapidly and by 2 days after fertilization the tiny fish, still attached to the remains of its yolk, hatches out of the egg. The top photograph shows a zebrafish embryo at the blastoderm (sphere) stage of development, with the embryo sitting on top of the large yolk cell (scale bar = 0.5 mm). The middle photograph shows an embryo at the 14-somite stage, showing the segmental trunk muscle that has developed. Its transparency is useful for observing cell behavior (scale bar = 0.5 mm). The bottom photograph shows an adult zebrafish (scale bar = 1 cm). An illustrated list of the stages in zebrafish development can be found on the website listed in the 'General further reading' section.

Photographs courtesy of C. Kimmel (top, from Kimmel, C.B., et al.: **Stages of embryonic development of the zebrafish.** *Dev. Dyn. 1995, 203: 253–310), N. Holder (middle), and M. Westerfield (bottom, the Zebrafish Information Network, ZFIN.org).*

four and six cell diameters away from the margin give rise exclusively to mesoderm. This overlap between prospective endoderm and mesoderm in the zebrafish differs somewhat from the arrangement in *Xenopus*, in which the endoderm and mesoderm occupy more distinct locations. The ectoderm in the fish embryo, as in *Xenopus*, comes from cells in the animal region of the blastoderm.

Fig. 3.31 Cleavage of the zebrafish embryo is initially confined to the animal half of the embryo.

From Kessel, R.G., Shih, C.Y.: Scanning Electron Microscopy in Biology: A Student's Atlas of Biological Organization. London: Springer-Verlag, 1974. © 1974 Springer-Verlag GmbH & Co.

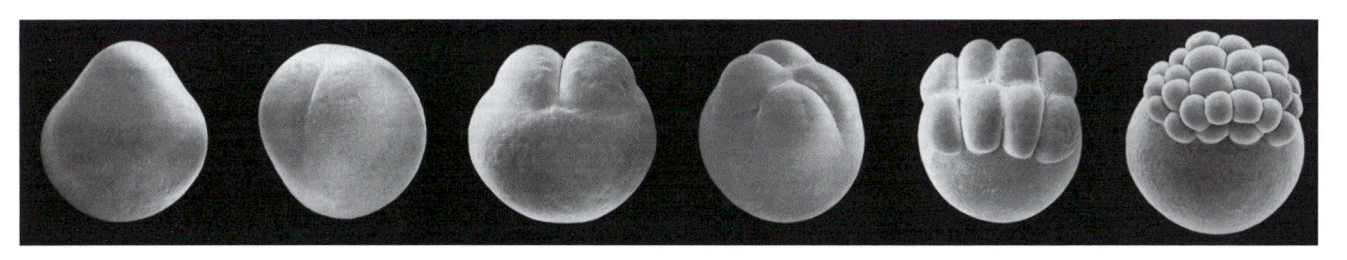

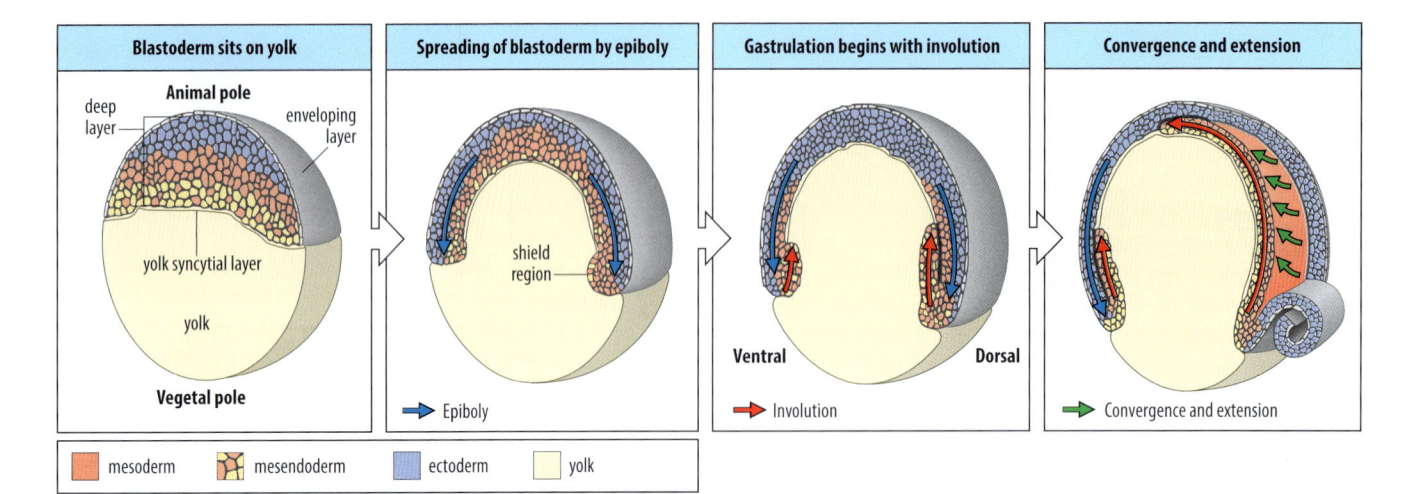

Fig. 3.32 Epiboly and gastrulation in the zebrafish. At the end of the first stages of cleavage, around 4.3 hours after fertilization, the zebrafish embryo is composed of a mound of blastomeres sitting on top of the yolk, separated from the yolk by a multinucleate layer of cytoplasm called the yolk syncytial layer (first panel). With further cleavage and spreading out of the layers of cells (epiboly), the upper half of the yolk becomes covered by a blastoderm with a thickened edge, the germ ring, and a shield-shaped region is visible in the blastoderm on the dorsal side (second panel). Gastrulation occurs by involution of cells in a ring around the edge of the blastoderm (third panel). The involuting cells converge on the dorsal midline to form the body of the embryo encircling the yolk (fourth panel).

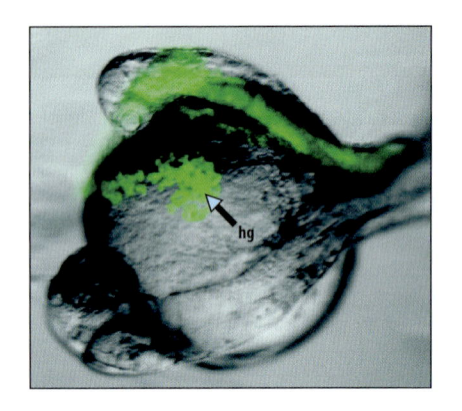

Fig. 3.33 Transplantation of the shield can induce a new axis in the zebrafish. The shield region from the dorsal region of a gastrula-stage embryo in which all the cells had been labeled by fluorescein-linked dextran amine was transplanted to the ventral region of an embryo at the same stage. The transplant induces a complete second axis with head, brain, and heart from the host tissue: the transplanted cells give rise to prechordal plate and notochord in the second axis, and also to the hatching gland (hg).

From Saude, L., et al.: Axis-inducing activities and cell fates of the zebrafish organizer. Development 2000, 127: 3407-3417.

By about 5.5 hours after fertilization, the blastoderm has spread halfway to the vegetal pole, and the deep-layer cells accumulate to form a thickening around the blastoderm edge known as the **embryonic germ ring** (see Fig. 3.32, second panel). At the same time, deep-layer cells within the **germ ring** converge towards the dorsal side of the embryo, eventually forming a compact, thickened shield-shaped region in the germ ring on the dorsal side. The shield becomes visible at about 6 hours after fertilization, and this stage is known as the **shield stage**. The **shield region** is analogous to the Spemann organizer of *Xenopus*: the cells of the shield can induce a complete body axis if transplanted to another embryo at an appropriate stage and reorganize and differentiate to give rise to the notochord and **prechordal plate**—the term given to tissue that is composed of prechordal mesoderm and underlying endoderm (prechordal plate is also known as **prechordal mesendoderm**). Like the *Xenopus* organizer, then, the shield is crucial to development of the dorso-ventral and antero-posterior body axes (Fig. 3.33).

Gastrulation ensues, and the mesendodermal cells and mesoderm cells roll under the blastoderm margin by involution and move into the interior under the ectoderm. The inward movement of cells occurs all around the periphery of the blastoderm at about the same time, with the earliest cells that involute—predominantly from the dorsal and lateral margins of the blastoderm—forming the endoderm. As the mesendoderm is becoming internalized, the ectoderm continues to undergo epiboly, spreading in the vegetal direction until it covers the whole embryo, including the yolk (discussed in more detail in Chapter 5).

Once internalized into the gastrulating embryo, the mesendodermal cells migrate under the ectoderm towards the animal pole, the tissue converging to form the main axis of the embryo and extending and elongating the embryo in an antero-posterior direction, as in *Xenopus*. The future mesodermal and endodermal cells are now beneath the ectoderm, and by the time that the blastoderm has spread about three-quarters of the way to the vegetal pole, an inner layer of endoderm cells has formed in the embryo next to the yolk, with the more superficial cells becoming mesoderm. By 9 hours, the notochord and prechordal plate have become distinct in the dorsal midline of the embryo, and involution of cells around the blastoderm margin is complete by 10 hours. Somite formation, neurulation, and migration of neural crest cells then follow.

Over the next 12 hours the embryo elongates further, and the first signs of the organ systems become recognizable. Neurulation in zebrafish begins, as in *Xenopus*, towards the end of gastrulation with the formation of the neural plate, an area of thickened ectoderm overlying the notochord. But unlike in *Xenopus*, the whole of the neural plate in the zebrafish first forms a solid rod of cells that later becomes hollowed out internally to form the neural tube. The nervous system develops rapidly. Optic vesicles, which give rise to the eyes, can be distinguished at 12 hours as bulges from the brain, and by 18 hours the body starts to twitch. At 48 hours the embryo hatches, and the young fish begins to swim and feed.

3.16 Genetic studies show that the axes of the zebrafish blastoderm are established by maternal determinants

The very earliest stages in development of the zebrafish, as in *Xenopus*, are exclusively under the control of maternal factors laid down in the egg. These are inherited by different blastomeres during cleavage of the fertilized egg and determine the cells' subsequent fate. In the zebrafish egg, maternal factors are localized along the animal–vegetal axis and key factors for axis formation are present in the vegetal region. The importance of such localized factors is shown by the fact that removal of the vegetal-most yolk early in the first cell cycle of the fertilized egg results in radially symmetric embryos that lack both dorsal and anterior structures, as in *Xenopus*. And, as in *Xenopus*, the dorso-ventral axis is also specified by activation of Wnt signaling and nuclear localization of β-catenin in cells of the future dorsal side, and this determines where the organizer—known as the shield in zebrafish—will develop.

The zebrafish has provided genetic evidence for the importance of the Wnt/β-catenin pathway in dorsal specification. Mutants have been isolated in which expression of Wnt/β-catenin pathway components is affected. For instance, in the *ichabod* mutant, in which the mutation maps near the zebrafish *beta-catenin2* gene, expression of β-catenin2 protein is downregulated and, instead of developing dorsal structures, the mutant embryos are radially symmetric. In zebrafish, the maternal dorsalizing factor is *wnt8a* mRNA rather than the *wnt11* and *wnt5a* mRNAs in *Xenopus*. The *wnt8a* mRNA and other maternal dorsalizing factors are initially located in the vegetal region of the oocyte, in the yolk cytoplasm, and thus after fertilization have to be transported into the developing blastoderm. They are transported to the future dorsal side on arrays of microtubules that assemble shortly after fertilization, in a movement similar to cortical rotation in *Xenopus*. Unlike *Xenopus*, however, where we saw that the site of sperm entry determines the direction of cortical rotation, how the microtubules are oriented in zebrafish is unknown. Wnt-8a protein activates the β-catenin pathway in the dorsal yolk syncytial layer and dorsal margin blastomeres, and β-catenin accumulates in nuclei there (Fig. 3.34). In zebrafish, maternally provided Wnt antagonists also limit Wnt signaling to the future dorsal side.

The dorsal yolk syncytial layer can be thought of as being equivalent to the *Xenopus* Nieuwkoop center, and the overlying marginal blastomeres as being equivalent to the Spemann organizer. Zebrafish embryos, like *Xenopus*, go through a mid-blastula transition at which transcription of zygotic genes begins. It occurs at the 512-cell stage, coincidentally with the formation of the syncytial layer at the margin between the blastoderm and yolk. Shortly after this, β-catenin helps to activate expression of a zygotic gene encoding a transcription factor called Dharma (encoded by the locus *bozozok*); it is a transcriptional repressor whose expression defines a region that is functionally equivalent to the Nieuwkoop center. Dharma/bozozok activity works together with Nodal-related proteins to induce the organizer of the zebrafish (the shield) and promote the development of axial mesoderm and head and trunk regions.

146 Chapter 3 Vertebrate development I: establishing the body plan

Fig. 3.34 A comparison of the establishment of the organizer in *Xenopus* and zebrafish embryos. Left panels: in *Xenopus*, cortical rotation moves the dorsal determinants such as *wnt11* and *wnt5a* mRNAs (green) towards the future dorsal side of the embryo, creating a large region where, starting at the 32-cell stage, β-catenin moves from the cytoplasm into nuclei (red). Cells in the region in which β-catenin is intranuclear and where VegT is also present express the transcription factor Siamois, which defines the Nieuwkoop center in the dorsal vegetal region of the blastula (blue shading). The Spemann organizer will arise in the equatorial region above the Nieuwkoop center in the early gastrula. Right panels: in zebrafish, the dorsal determinant *wnt8a* mRNA is transported to the future dorsal side of the embryo by oriented parallel arrays of microtubules and is inherited by the cells of the syncytial layer in that region. At the mid-blastula transition, β-catenin moves into nuclei of the cells of the dorsal yolk syncytial layer where it induces the expression of a zygotic gene encoding the transcription factor Dharma/bozozok (blue shading). The shield region, the zebrafish organizer, will arise in the dorsal region from the overlying marginal blastomeres.

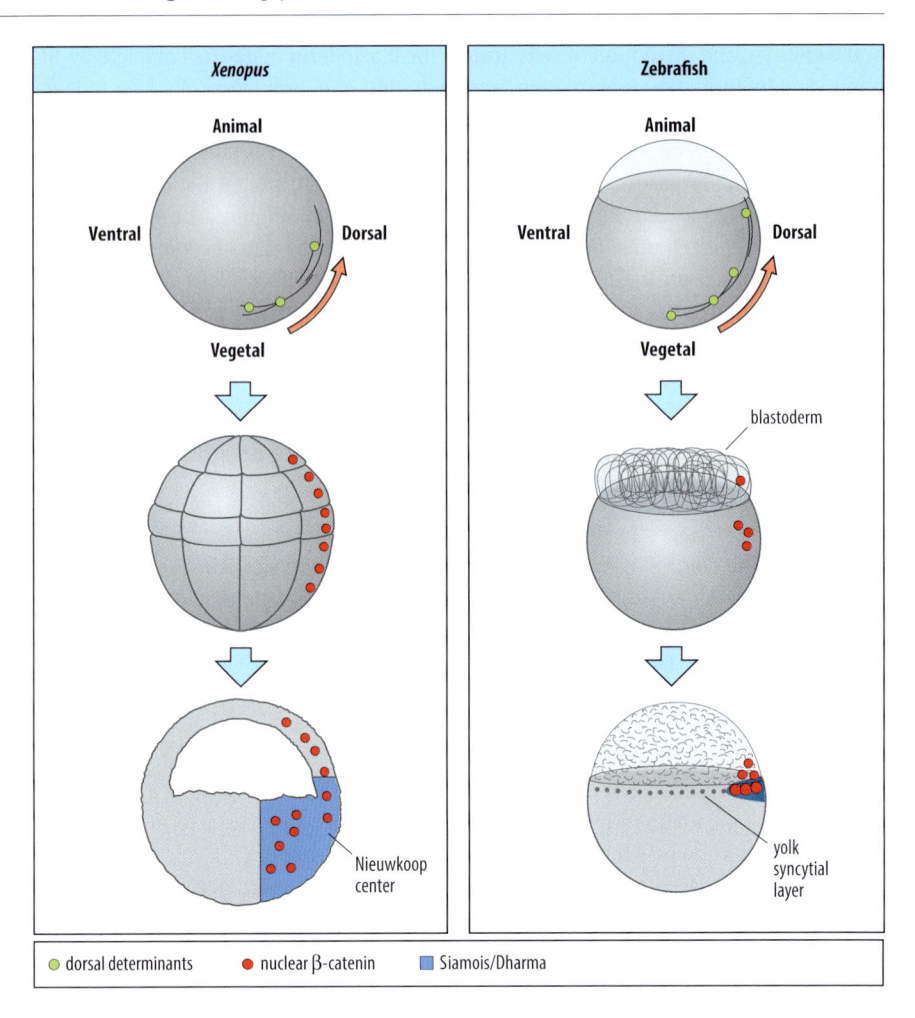

3.17 The germ layers are specified in the zebrafish blastoderm by signals similar to those in *Xenopus*

There is extensive cell mixing during the transition from blastula to gastrula in the zebrafish embryo, and so it is not possible to construct a reproducible fate map at cleavage stages, but only from the early gastrula stage. At the beginning of gastrulation, the fate of cells in the deep layer, from which all the cells of the embryo will come, is related to their position with respect to the animal pole (Fig. 3.35). Cells nearest to the animal pole give rise to the ectoderm, whereas cells beneath them slightly farther away from the animal pole give rise to mesoderm, and cells at the dorsal and lateral margins of the blastoderm form a layer of mesendoderm sitting on top of the yolk. The mesendoderm produces both mesoderm and endoderm cells that sort out as they differentiate. In general terms, the fate map of the zebrafish is rather

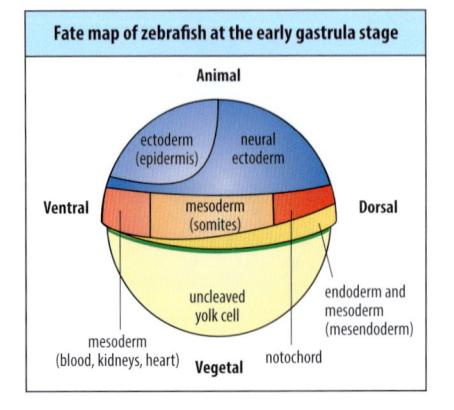

Fig. 3.35 Fate map of zebrafish at the early gastrula stage. The three germ layers come from the blastoderm, which sits on the lower hemisphere composed of an uncleaved yolk cell. The future endoderm comes from the dorsal and lateral margins of the blastoderm, which also give rise to mesoderm (mesendoderm), and some has already moved inside. The cells in the equatorial region give rise to mesoderm, and the ectoderm, including the neural ectoderm, comes from the cells nearest the animal pole. The green line indicates the location of tissue involution at gastrulation.

similar to that of *Xenopus*, if one imagines the vegetal region of the *Xenopus* late blastula being replaced by one large yolk cell.

As in *Xenopus*, the mesodermal germ layer in the zebrafish embryo is not pre-specified by maternal factors localized in the egg, and one of the key events during early development is the induction of the mesoderm. Mutagenesis screens have identified genes in zebrafish similar to those involved in mesoderm induction and dorso-ventral patterning in *Xenopus*, and they confirm the general outline of germ-layer specification. Studies in zebrafish have illuminated, in particular, the mechanisms through which Nodal gradients are established, which, as in *Xenopus*, provide positional information. Maternal mRNAs, including the transcription factor Nanog, are required to turn on the expression of zygotic *nodal-related* (*ndr*) genes, including *ndr1*, in the yolk syncytial layer and the cells immediately overlying it; here, as first suggested in *Xenopus*, they specify the future endoderm and induce dorsal and ventral mesoderm.

Ndr proteins act by forming heterodimers with a maternally provided member of the TGF-β family, GDF-1. Ndr–GDF-1 heterodimers activate the receptor and predominate in signaling (see Box 3F). Ndr-1–GDF-1 heterodimers secreted from cells at the blastoderm margin diffuse into the overlying blastoderm to set up a gradient of Nodal-type signaling in the blastoderm cells. High levels of signal near the blastoderm margin specify mesendoderm, from which endoderm and mesoderm will form, whereas lower levels of signal slightly farther away specify only mesoderm. As in *Xenopus*, the endoderm arises from the *ndr*-expressing cells. At the same time, a long-range antagonist of Nodal signaling, Lefty, is produced by cells at the blastoderm margin. This prevents Ndr–GDF-1 from binding to the receptor and so prevents Nodal-type signaling in the region of the blastoderm that will form the ectoderm (Fig. 3.36). (Lefty is so called because the *lefty* gene was first identified in mice as a gene that is expressed very strongly on the left side of the developing embryo.)

Nodal-related proteins signal via Smad2 (see Box 3F), and the evidence for a Nodal signaling gradient in zebrafish comes through analysis of nuclear accumulation of Smad2, visualized in living transgenic zebrafish embryos carrying a *smad2* gene fused with a sequence encoding a fluorescent tag. This experiment revealed a graded distribution of Nodal-type signaling, most probably due to Ndr-1–GDF-1 heterodimers, in the blastoderm of these embryos, with highest activity at the margin of the blastoderm and grading off towards the animal pole. This confirms a gradient of Ndr signaling that could provide cells in the blastoderm with positional information. Ndr signaling also seems to be strongest and of longest duration in the dorsal marginal region of the blastoderm, suggesting that it might also help to pattern the mesoderm dorso-ventrally. Injection of *ndr1* mRNA into a single cell of an early zebrafish embryo, which is expressing *gdf1*, results in high-threshold Ndr-1 target genes being activated in adjacent cells, whereas low-threshold genes are activated in more distant cells. This

Fig. 3.36 Specification of endoderm and mesoderm in the zebrafish blastoderm. Nodal-related (Ndr) signal proteins are secreted by cells at the blastoderm margin and diffuse into the overlying blastoderm to set up a gradient of Nodal-type signaling in the blastoderm cells, probably mediated by heterodimers of Ndr and GDF-1, which is also expressed in this region (not shown for simplicity). High levels of Nodal signals near the blastoderm margin specify mesendoderm, from which endoderm and mesoderm will form, whereas lower levels of Nodal signals slightly farther away specify mesoderm. Ndr-1 signaling acts over a long range, whereas signaling involving Ndr-2 acts at short range. An inhibitor of Nodal signaling, Lefty, is also secreted by cells at the blastoderm margin and diffuses into the blastoderm. Lefty has long-range activity and limits the range of Nodal signaling, ensuring that ectoderm cells are not exposed to Nodal signaling.

After Schier, A.F.: **Nodal morphogens.** Cold Spring Harb. Perspect. Biol. *2009;* **1**: a003459.

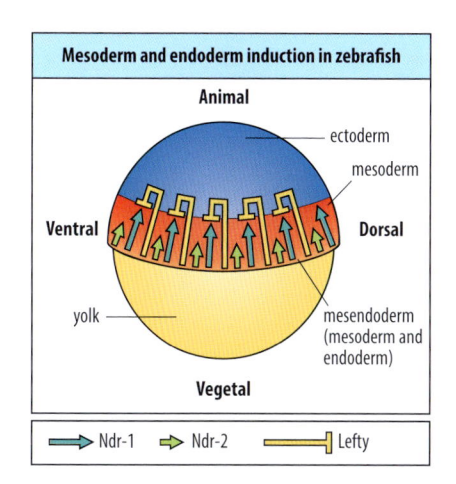

is just what would be expected if Ndr-1 acts as a morphogen, turning on different mesodermal genes at different threshold concentrations.

Analysis of mutants indicates that different Nodal-related proteins work together. *squint* mutant embryos, which lack functional Ndr-1, have defects in dorsal mesendoderm formation, but when combined with *cyclops* mutants, which have defects in the gene encoding Ndr-2 protein, they show a stronger phenotype: double mutants of *squint* and *cyclops* have no endoderm and almost no mesoderm. This indicates that together with Ndr-2, Ndr-1 participates in specifying the marginal blastomeres as future endoderm and mesoderm.

SUMMARY

Establishment of the zebrafish body plan is very similar to that in *Xenopus*, although the morphology of the early embryo is quite different. The difference is largely the result of the zebrafish embryo developing around a large yolk cell. Cleavage divisions do not go into this cell and so the embryo develops from a blastoderm of cells sitting on top. As in *Xenopus*, the mesoderm is not pre-specified by maternal factors but is induced from cells in the animal region. The embryo undergoes gastrulation by the involution of cells all around the margin of the blastoderm and elongates so that the head and tail become evident. The area equivalent to the organizer in *Xenopus* is called the shield region in the zebrafish and is similarly situated in the most-dorsal mesoderm. Because the zebrafish is much more suitable for genetic studies than *Xenopus* it has shed light on details of how the Wnt signaling pathway specifies the dorsal embryonic axis, and how Nodal signals can provide positional information to specify mesendoderm and mesoderm and induce the two distinct types of mesoderm that include organizer tissue.

Chick development and setting up the body plan

Before we describe the course of early development in chick embryos, it is worth emphasizing four features of their development that differ from those of *Xenopus* and zebrafish. The first is the shape of the early embryo. The avian structure that corresponds to the spherical amphibian blastula just before gastrulation is not a hollow blastula, but a layer of epithelium called an epiblast (see Fig. 3.2).

The second difference is that the timing of the specification of the germ layers is somewhat different from that in *Xenopus*, because there are no distinct regions within the epiblast that correspond to the future ectoderm, endoderm, and mesoderm. Instead of being specified before gastrulation, mesoderm and endoderm cells are specified during gastrulation. There is also considerable cell proliferation in the epiblast in chick embryos before and during gastrulation, which causes cell mixing at this stage. By contrast, there is little cell proliferation during gastrulation in *Xenopus*.

The third difference is that the gastrulation process that leads to internalization of endoderm and mesoderm and the organization of the germ layers is rather different in appearance from that in *Xenopus*. Internalization of cells occurs along a straight furrow, the **primitive streak**, rather than through a circular blastopore.

A fourth difference is that in the chick, unlike in *Xenopus* and the zebrafish, the antero-posterior axis is determined very early, preceding gastrulation.

We will now outline the development of the chick and look in more detail at the early stages.

3.18 The chick is a classic embryological model, especially from gastrulation to organogenesis stages

Once fertilized, the very large yolky egg cell begins to undergo cleavage while still in the hen's oviduct. Because of the massive yolk, cleavage is confined to a small patch of cytoplasm several millimeters in diameter, which comes to lie on top of the yolk. The early cleavage furrows extend downward from the surface of the cytoplasm but do not completely separate the cells, whose ventral faces initially remain open to the yolk. Further cleavages result in the formation of a disc of cells called the blastoderm or blastodisc, which is analogous to the early amphibian blastula. The blastoderm is composed of some 32,000–60,000 cells and shows features that resemble those in zebrafish, including the formation of a yolk syncytial layer. The egg is laid some 20 hours after fertilization. During its passage down the oviduct, the egg also becomes surrounded by extracellular albumen (egg white), the shell membranes, and the shell (Fig. 3.37). The chicken life cycle is shown in Figure 3.38.

At the time of laying, then, the embryo is a disc-shaped cellular blastoderm lying on top of a massive yolk (see Fig. 3.37). The central region of the blastoderm, which is separated from the yolk by a cavity called the **subgerminal space**, is translucent and is known as the **area pellucida**. The area pellucida is surrounded by the **area opaca** (Fig. 3.39). The blastoderm consists of two sheets of cells. The upper sheet of cells in the area pellucida—the epiblast—forms the embryo. The lower sheet of cells in the area pellucida is the **hypoblast**, and this together with the area opaca eventually gives rise to extra-embryonic structures such as those that connect the embryo to its source of nutrients in the yolk.

The first morphological structure that presages the antero-posterior axis of the chick is a crescent-shaped ridge of small cells called **Koller's sickle** (see Fig 3.39). This arises at the boundary between the area opaca and area pellucida in a region known as the **posterior marginal zone**. Koller's sickle defines the position in which the primitive streak will start to form at gastrulation. The streak appears as a cell-dense region that gradually extends forwards, creating a furrow in the midline of the epiblast (see Fig. 3.39, bottom panel). The primitive streak determines the future antero-posterior body axis: the start of the streak marks the posterior end of the embryo and the streak elongates in a direction that predicts the future anterior end of the embryo.

Like the amphibian blastopore, the primitive streak reveals, morphologically, how gastrulation occurs. The streak forms when epiblast cells converge towards the midline, pass through it as individual cells in a process known as **ingression**, and then generate mesoderm and endoderm. Unlike gastrulation in *Xenopus*, cell proliferation occurs throughout chick gastrulation, and the embryo elongates as the primitive streak extends. This proliferation, and the individual movement of cells, contrasts with the gastrulation movements of amphibians, where cells move as a coherent sheet and do not proliferate. After they ingress through the primitive streak, cells spread out anteriorly and laterally beneath the upper layer, forming a layer of loosely connected cells, or **mesenchyme**, between the epiblast and the endoblast (see Fig. 3.39, fifth and sixth rows). The mesenchyme cells give rise to a lower layer of endoderm and a middle layer of mesoderm, whereas the epiblast cells that remain on the surface give rise to the ectoderm. The endoderm cells displace those of the endoblast to give a layer of endoderm. In this way, the germ layers of the post-gastrula stage form during gastrulation and take up their relative positions in the body plan, with ectoderm on the outside, endoderm on the inside, and mesoderm sandwiched in between.

The primitive streak extends forwards, to about two-thirds of the way across the area pellucida. After it has extended to its maximum, a condensation of cells called the **node** (often called **Hensen's node**) forms at the streak's anterior-most end. Hensen's node is functionally equivalent to the Spemann organizer/zebrafish shield:

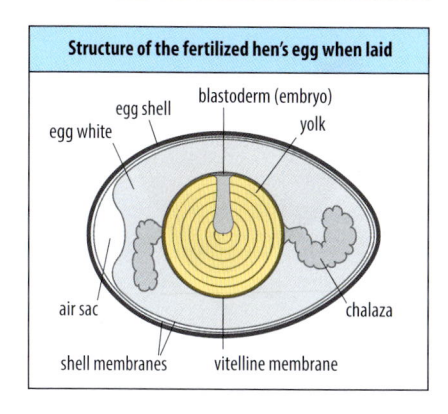

Fig. 3.37 The structure of a hen's egg at the time of laying. Cleavage begins after fertilization while the egg is still in the oviduct. The albumen (egg white) and shell are added during the egg's passage down the oviduct. At the time of laying, the embryo is a disc-shaped cellular blastoderm lying on top of a massive yolk, surrounded by the egg white and shell. The twisted chalazae are thought to act as balancers to support the yolk.

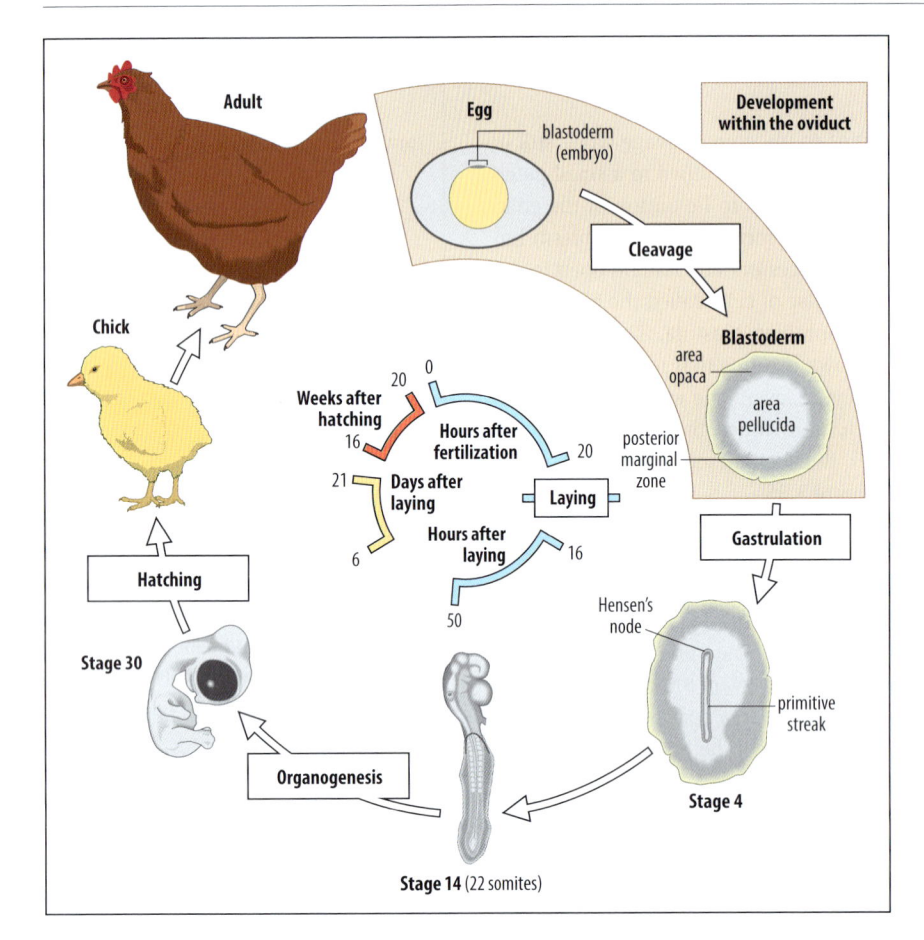

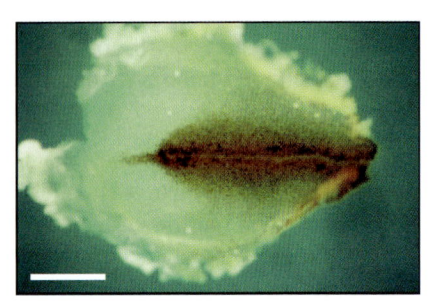

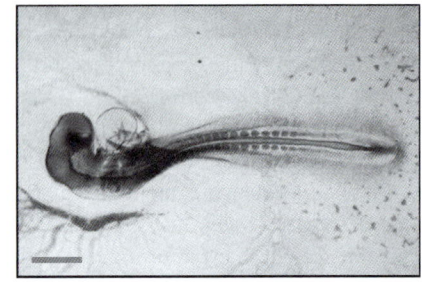

Fig. 3.38 Life cycle of the chicken. The egg is fertilized in the hen and by the time it is laid cleavage is complete and a cellular blastoderm lies on the yolk. The start of gastrulation is marked by the appearance of the primitive streak at the posterior marginal zone of the area pellucida. Hensen's node (the equivalent of the *Xenopus* Spemann organizer) develops at the anterior end of the streak, which then starts to regress, accompanied by the appearance of the main antero-posterior body axis and somite formation. The photographs show top: the fully extended primitive streak, and thinner head process (both stained brown with antibody against TbxT protein), surrounded by the area pellucida (scale bar = 1 mm); middle: a stage-14 embryo (50-53 hours after laying) with 22 somites (the head region is well defined and the transparent organ adjacent to it is the ventricular loop of the heart; scale bar = 1 mm); bottom: a stage-35 embryo, about 8.5-9 days after laying, with a well-developed eye and beak (scale bar = 10 mm). A descriptive list of the numbered stages (Hamilton and Hamburger (HH) stages) in chick development can be found on the website listed in the 'General further reading' section.

*Top photograph courtesy of B. Herrmann. Reproduced with permission from Kispert, A., et al.:***The chick Brachyury gene: developmental expression pattern and response to axial induction by localized activin.** *Dev. Biol. 1995, **168**: 406-415.*

when grafted to a second host embryo, it is capable of inducing a complete body axis, including a second neural plate (Fig. 3.40). And similarly to the situation at the *Xenopus* dorsal blastopore lip, a set of epiblast cells that ingress through Hensen's node migrate in an anterior direction under the midline of the epiblast and give rise to axial mesoderm cells—that is, to prechordal mesoderm and anterior ('head') notochord that eventually underlies the hindbrain. In chick these tissues are collectively known as the **head process**.

The primitive streak then begins to regress, and, as it does so, trunk notochord is laid down in its wake along the dorsal midline of the embryo, contributing to elongation of the antero-posterior axis (Fig. 3.41). This will be discussed in detail in Chapter 4. As the trunk notochord is laid down, epiblast cells that have moved in at more posterior regions of the streak take up positions lateral to the notochord. The mesoderm immediately on each side of the notochord forms the somites. These are flanked by intermediate mesoderm and then lateral plate mesoderm, which becomes

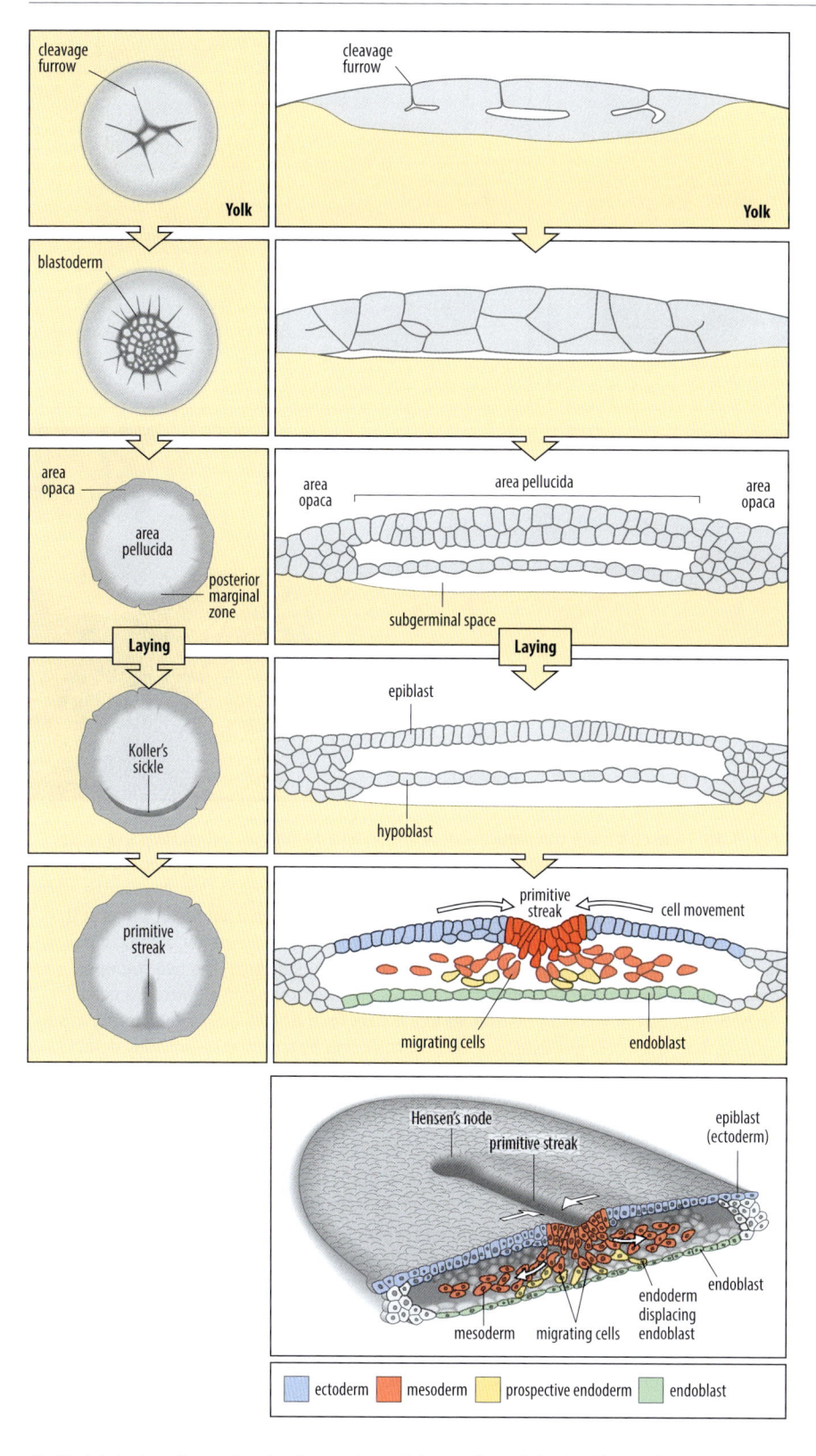

Chick development and setting up the body plan | **151**

Fig. 3.39 Cleavage and epiblast formation in the chick embryo. The panels on the left show views of the embryo from above; the panels on the right show cross-sections through the embryo. After fertilization, the first cleavage furrows extend downward from the surface of the egg cytoplasm and initially do not completely separate the cells from the yolk. Further cleavage results in the formation of a disc of cells called the blastoderm. In this cellular blastoderm, the central area overlying the subgerminal space is called the area pellucida and the surrounding region is called the area opaca. By the time the egg is laid, the blastoderm in the area pellucida consists of two layers of cells. The upper layer, the epiblast, is an epithelial cell sheet and gives rise to the embryo. The lower layer, the hypoblast, has developed below the epiblast. The hypoblast together with the area opaca will give rise to extra-embryonic structures. A crescent-shaped ridge of cells, known as Koller's sickle, arises at the boundary between the area opaca and the area pellucida in a region of the blastoderm known as the posterior marginal zone. Koller's sickle defines the position at which the primitive streak will start to develop. The primitive streak starts to form when the hypoblast becomes displaced forward from the posterior marginal zone by a new layer of cells called the endoblast, which grows out from the zone. Cells of the epiblast move toward the center to form the primitive streak (arrows), move through it, and then outward again underneath the surface to give rise to the mesoderm and endoderm internally, the latter displacing the endoblast. The cells that remain in the epiblast give rise to ectoderm.

divided into two layers by the formation of the coelom (Fig. 3.42). At the same time, the neural plate, a thickened region of ectoderm, develops above the axial mesoderm and adjacent mesoderm, and fuses in the dorsal midline to form the neural tube. The anterior region of the neural tube will give rise to the brain. As the embryo elongates, the trunk spinal cord together with the medial regions of the somites are progressively

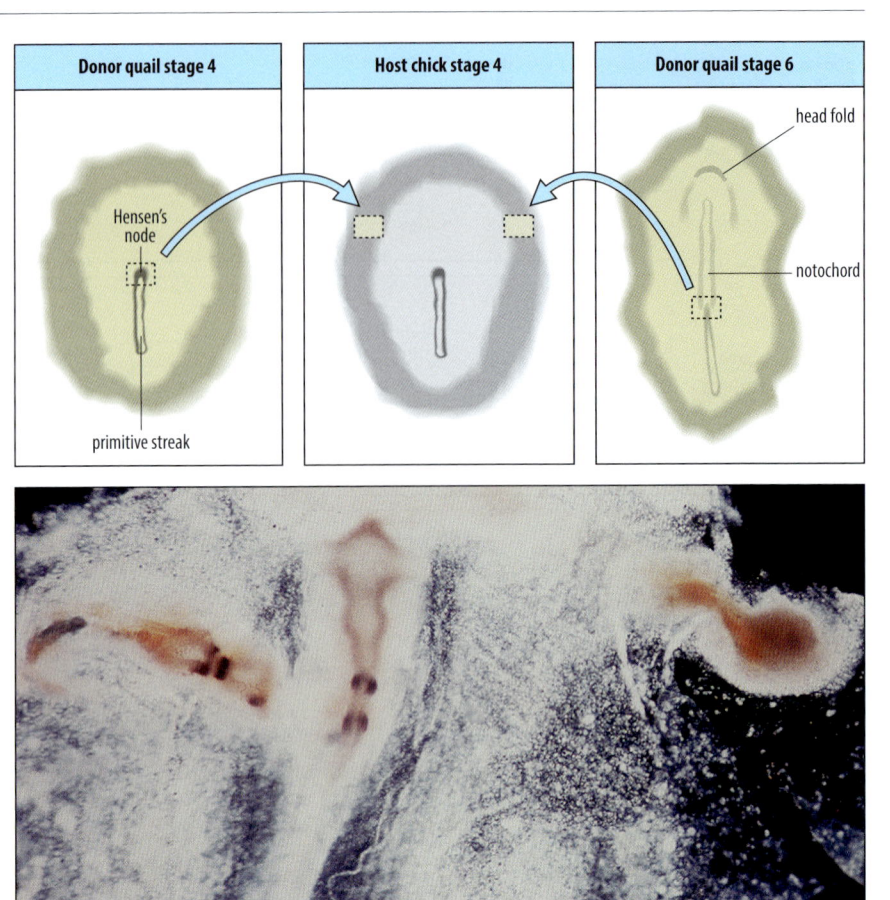

Fig. 3.40 Hensen's node can induce a new axis in avian embryos. When Hensen's node from a stage-4 quail embryo (left) is grafted beneath the extra-embryonic area opaca (dark gray), just outside the embryonic area pellucida (light gray), of a host chick embryo at the same stage of development (center), a near-complete new axis, including brain tissue, forms at the site of transplantation (photo, embryo on the left). The embryo in the center of the photograph has formed from the original primitive streak of the host chick embryo. The two stripes of dark purple staining in the left and center embryos indicate expression of *Krox20*, a gene encoding a transcription factor that serves as a marker for the hindbrain, as detected by *in situ* hybridization. Quail cells are distinguished from chick cells by an anti-quail antibody detected by immunohistochemistry staining reddish brown. Although some of the tissue of the new axis (embryo on left) is formed from the quail graft, most has been induced from chick tissue that does not normally form an embryo. A node from a quail embryo at the head-fold stage of development (stage 6, right) grafted to a stage-4 chick embryo produces a new axis with only trunk neural tissue (photo, right embryo) and most of the new tissue is derived from the quail graft, as indicated by the reddish-brown staining throughout.

From Storey, K.G., et al.: **Neural induction and regionalisation in the chick embryo.** *Development 1992,* **114***: 729-741; and Stern, C.D.:* **Neural induction: old problem, new findings, yet more questions.** *Development 2005,* **132***: 2007-2021.*

laid down by cells produced by the **stem zone**—a region of the epiblast on either side of the streak posterior to the node (discussed in Chapter 4).

In the frog, the equatorial region of the spherical blastula predicts both dorso-ventral and antero-posterior axes of the body plan (see Fig. 3.19). In the chick, however, as described above, gastrulation gives rise to a flat embryo with the axial mesoderm marking the dorsal midline along the entire antero-posterior axis and the pattern of structures along the future dorso-ventral axis mapped out on either side in the germ layers.

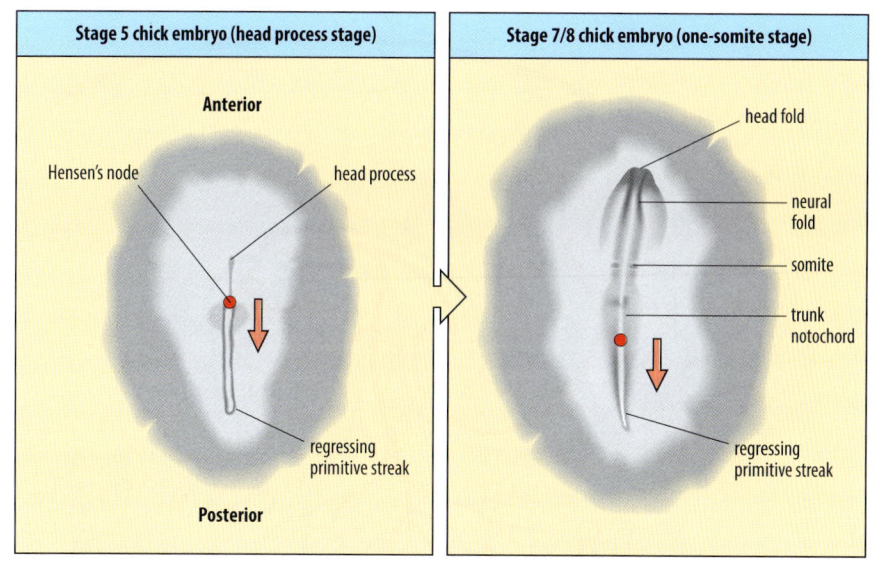

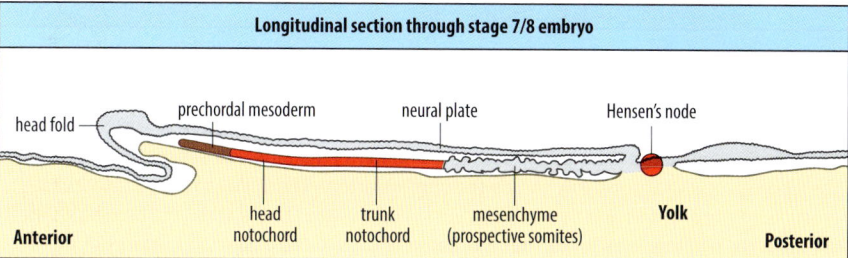

Fig. 3.41 Notochord and head-fold formation in the chick embryo. Upper panels: after the streak extends to its full length, the head process (prechordal mesoderm and head notochord) is formed by cells that are the first to emerge from the anterior tip of the streak and move forward (left upper panel). The head fold and neural plate form in the epiblast anterior to the furthest extent of the primitive streak. Right upper panel: as the primitive streak begins to regress posteriorly, trunk notochord is laid down in its wake, the edges of the neural plate begin to rise up to form the neural folds, and the first somite appears. Lower panel: schematic diagram of a sagittal section through the chick embryo around the stage of head-fold formation as Hensen's node is regressing. The prechordal mesoderm and notochord in the head region (the head process) are continuous with trunk notochord. The undifferentiated mesenchyme on either side of the notochord will form somites. The anterior somitic mesoderm (in which a somite has already formed, see upper right panel) has been cut away to show the trunk notochord.

The description of the dorso-ventral axis as the **medio-lateral axis** at this stage, medial being alongside the midline, is more useful in respect to the chick. The flat embryo then becomes converted into the three-dimensional form of the body by all three germ layers folding underneath. At the head end, this folding generates the **head fold**, where the pharynx and foregut arise (see Fig. 3.41); at the tail end it forms the hindgut; and the rest of the embryo folds laterally to internalize the rest of the gut and form the ventral body wall. This key morphogenetic event is known as **ventral closure**. As this lateral folding takes place, the two heart primordia that start out on either side come together at the midline to form one organ lying ventral to the gut (discussed in Chapter 9). By just over 2 days after laying, the heart and blood vessels have formed (see Fig. 3.38) and the embryo starts to turn on its side so that the right eye comes to lie uppermost towards the shell. At 4 days after laying, extra-embryonic membranes have developed through which the embryo gets its nourishment from the yolk and which also provide protection (Fig. 3.43). The **amnion** surrounds a fluid-filled amniotic sac in which the embryo lies and which provides mechanical protection; the **chorion** lies outside the amnion just beneath the shell; the **allantois** receives excretory waste products and provides the site of oxygen and carbon dioxide exchange; and the **yolk sac** surrounds the yolk. By about 9 days after laying, the embryo is well developed (see Fig. 3.38, bottom photograph). In the remaining time before hatching, the embryo grows considerably in size, the internal organs become fully developed, and down feathers grow on the body. The chick hatches out 21 days after the egg is laid.

The chick embryo is particularly suited to studying later development, such as neural tube and limb development, as the embryo can be surgically or genetically manipulated within the egg through a window cut in the eggshell, which can then be sealed to allow further development to be viewed. Early embryos can also be cultured out of the egg up to the stage at which the embryo has about 10 somites.

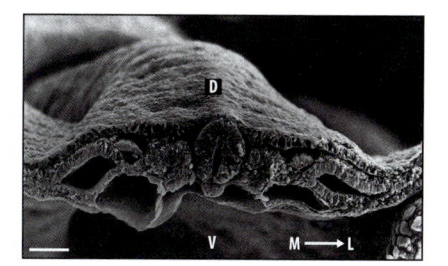

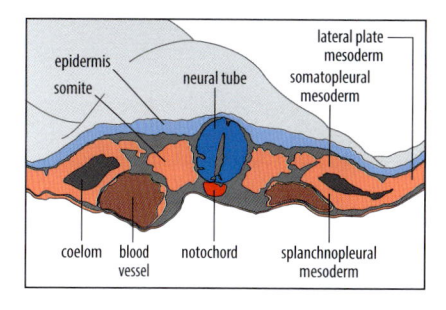

Fig. 3.42 Scanning electron micrograph of chick showing early somites and neural tube. Somites can be seen adjacent to the neural tube, with the notochord lying beneath it. The lateral plate mesoderm flanks the somites and has become divided into two layers by the formation of the coelom. D, dorsal; V, ventral; M, medial; L, lateral. Scale bar = 0.1 mm.

Photograph courtesy of J. Wilting.

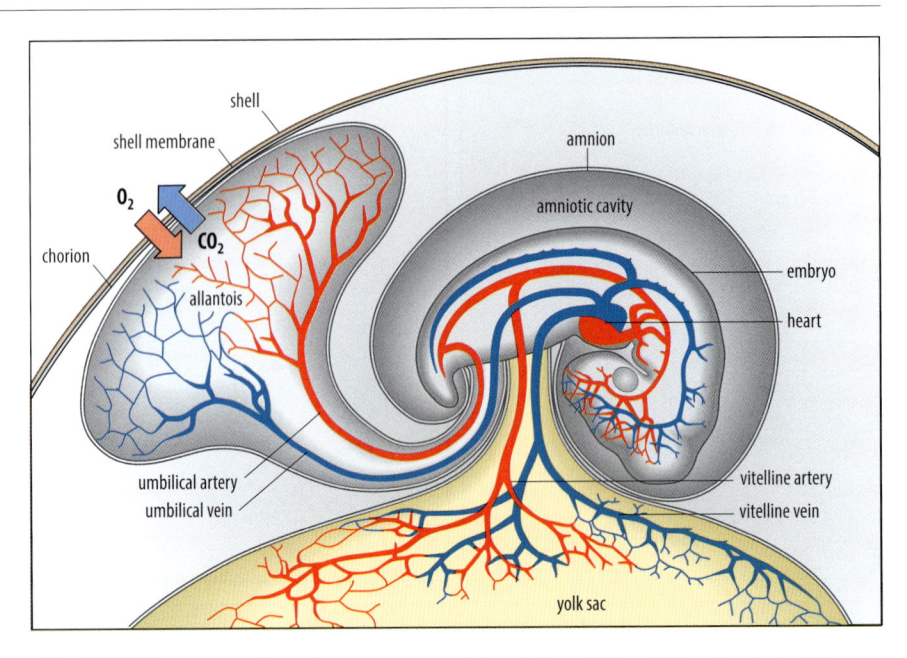

Fig. 3.43 The extra-embryonic structures and circulation of the chick embryo. A chick embryo at 4 days of incubation is depicted. The embryo lies within the fluid-filled amniotic cavity enclosed by the amnion, which provides a protective chamber. The yolk is surrounded by the yolk sac membrane. The vitelline vein takes nutrients from the yolk sac to the embryo and the blood is returned to the yolk sac via the vitelline artery. The umbilical artery takes waste products to the allantois and the umbilical vein brings oxygen to the embryo. The arteries are shown in red and the veins in blue but this does not denote the oxygenation status of the blood. As the embryo grows, the amniotic cavity enlarges; the allantois also increases in size and its outer layer fuses with the chorion (the membrane under the shell), while at the same time the yolk sac shrinks. Note that in this diagram the allantois has been enlarged so that the umbilical vessels can be shown clearly.

3.19 In the chick, the main reference axis for development of the body plan is the antero-posterior axis

The future dorso-ventral body axis is determined by the polarity of the blastoderm relative to the subgerminal space. But how is the radial symmetry of the chick blastoderm broken and the location of the posterior marginal zone determined? We saw in *Xenopus* that the symmetry of the blastula is broken by the asymmetric redistribution of maternal dorsal determinants shortly after fertilization. In the chick embryo, cleavage stages are similarly under the control of maternal factors laid down in the egg—but it is not clear that any factors are inherited asymmetrically by cells and determine the embryonic axes. The major wave of zygotic gene expression occurs between cleavage cycles 7 and 9, making the timing similar to that in *Xenopus* and zebrafish.

The antero-posterior axis seems to be specified later, at the blastoderm stage, as a result of rotational forces as the egg passes through the hen's oviduct and uterus (also known as the shell gland). As the egg moves down the shell gland, it rotates slowly around its long axis. The blastoderm is held in a tilted position and the posterior marginal zone develops at the uppermost edge of the blastoderm (Fig. 3.44). It is not clear whether there are molecules in the yolk that might be redistributed by gravity and specify the antero-posterior axis of the blastoderm, or whether mechanical forces are involved. Furthermore, the axis is not irreversibly determined by these events. The chick blastoderm is highly regulative. Even at the stage when the blastoderm contains

Fig. 3.44 Gravity defines the antero-posterior axis of the chick. Top panel: the egg rotates as it passes down the shell gland of the hen. Center panel: this results in the blastoderm (crescent of gray cells) being tilted in the direction of rotation, although it tends to remain uppermost. Gravitational forces reorient the yolk contents, as indicated by blue shading. Bottom panel: the posterior marginal zone (P) develops at the side of the blastoderm that was uppermost and initiates the primitive streak. A, anterior. It has been known since the early nineteenth century that the head-to-tail polarity of the chick embryo is related to the axis of the egg, and this is known as van Baer's rule. When the egg is placed on its side with the blunt end to the left, the head of the embryo will develop in the direction away from the observer; the rule applies to about 70% of embryos.

several tens of thousands of cells, it can be cut into many fragments, each of which can develop a complete embryonic axis.

The posterior marginal zone is a signaling center analogous to the Nieuwkoop signaling center in *Xenopus*. When the posterior marginal zone from one blastoderm is grafted to a second blastoderm of the same age it can induce a second primitive streak at this new position, and hence a new body axis. Among the earliest signal proteins involved in streak initiation in the chick are the TGF-β family members GDF-1, which is expressed in the epiblast of the posterior marginal zone above Koller's sickle, and Nodal, which is expressed in the epiblast cells adjacent to the posterior marginal zone (Fig. 3.45). Wnt-8C is present throughout the marginal zone with a high point of concentration at the posterior marginal zone. These signal proteins are the same type that operate in *Xenopus* and the zebrafish, where Wnts are associated with the formation of dorsal signaling centers in the blastula and GDF-1 is involved in mesoderm induction and endoderm specification. When cells of a fibroblast cell line expressing GDF-1 are grafted to another part of the marginal zone in a chick blastoderm they can induce a complete extra new streak. This is an example of the same proteins being used in different vertebrates for slightly different functions within a generally conserved process.

Studies in the chick have provided insight into why only one primitive streak develops in an embryo. Often, after grafting a second posterior marginal zone, only one axis develops—either the host's normal axis or the one induced by the graft. It is thought that the more advanced of the two signaling centers inhibits streak formation elsewhere. It has been shown that GDF-1 induces an inhibitory signal that can travel across the embryo (a distance of around 3 mm) in about 6 hours, to prevent another streak forming. GDF-1 is thus required for initiating streak formation, but at the same time is involved in inhibiting the formation of a second streak.

GDF-1 from the posterior marginal zone acts with other signals to restrict primitive streak development. At laying, epiblast cells—from which the streak will develop—are underlain by a continuous layer of hypoblast cells (see Fig. 3.39). The primitive streak starts to form in the epiblast only when the hypoblast becomes displaced away from the posterior marginal region by another layer of cells known as the **endoblast**, which grows out from the posterior marginal zone. The inhibition of streak formation by the hypoblast can be shown by experiments in which its removal leads to the formation of multiple streaks.

The inhibition of streak formation is due to the hypoblast producing the protein Cerberus, which antagonizes the signal protein Nodal. *Nodal* is expressed in epiblast cells that lie adjacent to the posterior marginal zone and is induced through Wnt-8C and GDF-1 signals. Nodal signaling by the epiblast, together with fibroblast growth factor (FGF) signaling from Koller's sickle, is required for streak formation (see Fig. 3.45), and so the primitive streak only starts to form as hypoblast becomes displaced. The capacity of the chick embryo to regulate persists right up to the time at which the primitive streak starts to form. When the embryo is cut into pieces, a new site of GDF-1 expression appears in each piece.

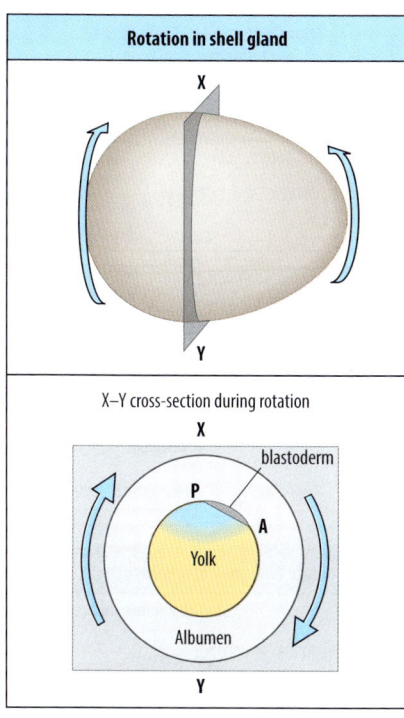

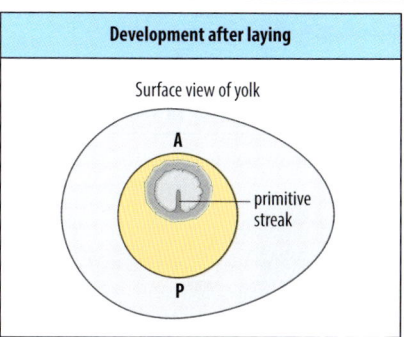

Fig. 3.45 Signals at the posterior marginal zone of the chick epiblast initiate primitive streak formation. Epiblast cells in the posterior marginal zone overlying Koller's sickle secrete GDF-1 and Wnts. These signals induce expression of the gene nodal in neighboring epiblast cells, but Nodal protein function is blocked by Cerberus, which is produced by the hypoblast. When the hypoblast has been displaced by endoblast, Nodal signals in the epiblast and FGF signals from Koller's sickle induce the internalization of epiblast cells and formation of the primitive streak.

3.20 Mesodermal transcription factors are upregulated during primitive streak formation

In the chick, mesoderm induction and patterning begins as cells pass through the primitive streak. Chick epiblast isolated before streak formation will form ventral mesoderm such as blood cells, but will not form dorsal mesodermal structures, such as notochord. Like the experiments in *Xenopus* with animal cap cells (see Section 3.13), treatment of the isolated epiblast with the Nodal-like signal protein Activin results in the appearance of notochord. This shows that TGF-β family members act as mesoderm-inducing signals in chick embryos as they do in *Xenopus*.

As in *Xenopus* and zebrafish, a strong candidate for the chick mesoderm-inducing signal is the TGF-β family member Nodal, which is produced first by the posterior marginal zone and then by the streak. Nodal might work together with FGF, produced by Koller's sickle, to form mesoderm. TbxT, a transcription factor that defines mesoderm in *Xenopus* and zebrafish, is detected in the chick primitive streak, in Hensen's node, and in the head-process mesoderm (see Fig. 3.38, top photograph). Likewise, the transcription factor Goosecoid (Gsc), which defines the *Xenopus* Spemann organizer, is detected in Hensen's node and the prechordal mesoderm. This type of conservation indicates the importance of these genes in mesoderm specification across vertebrate species.

3.21 A fate map for the chick can be made after primitive streak formation

It is difficult to make reproducible fate maps of chick embryos at the stages that correspond to the blastula stage in *Xenopus*. This is because the cells that will contribute to the different germ layers are mixed up in the epiblast and there are extensive cell movements both before and during primitive streak formation. The picture becomes clearer once the primitive streak has formed, and cells have started to move inside and become determined as endoderm and mesoderm.

The fate map shown in Figure 3.46 is of a chick embryo when the primitive streak has fully formed. At this stage the embryo has a three-layered structure. We are viewing the upper surface, which is the epiblast. Some cells from the epiblast have already ingressed through the primitive streak into the interior and formed mesodermal and endodermal layers—the fate of these layers is not represented in this diagram. The mesoderm cells that will form the heart have ingressed and migrated in an antero-lateral direction on each side and will give rise to the bilateral first heart fields (discussed in Section 9.31). Most of the cells in the anterior epiblast are now ectoderm, and will form neural tube and epidermis, but there are still cells in regions of the epiblast that will move through the streak and give rise to mesoderm. Cells that lie in and around Hensen's node and the anterior end of the streak are mesoderm. Fine-detailed fate mapping shows that cells that ingress through Hensen's node when the streak is maximally extended migrate in an anterior direction and give rise to the prechordal mesoderm and head notochord that underlies the hindbrain. By contrast, cells that are in the node as it starts to regress are left in its wake as more posterior, or trunk notochord, or will migrate laterally to contribute to the

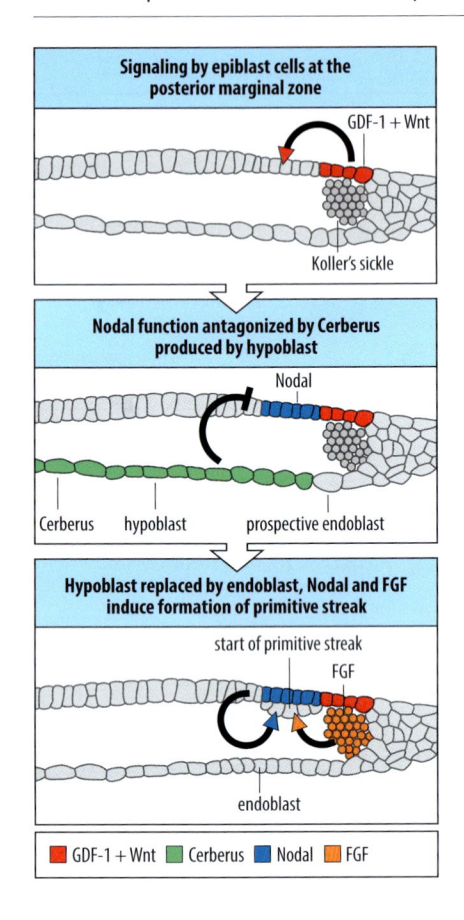

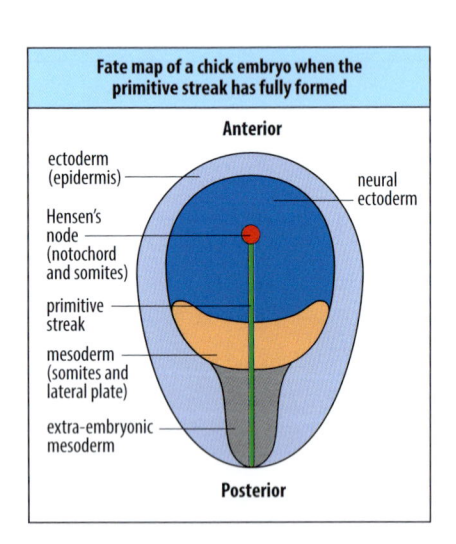

Fig. 3.46 Fate map of a chick embryo when the primitive streak has fully formed. The diagram shows a view of the upper surface of the embryo. Almost all the endoderm has already moved through the streak to form a lower layer, so is not represented.

somites, the lateral plate mesoderm, and organs such as the kidney. The most posterior region of the streak becomes incorporated into the tail bud and also forms extra-embryonic mesoderm.

Having seen how the chick develops in its early stages, and the differences from *Xenopus* and zebrafish in the topology of gastrulation and the timing of germ-layer specification, we will now look at our fourth model organism—the mouse. The mouse is a mammal, and so its embryo differs from the other three in developing completely inside the mother and being dependent on the mother for its nutrition.

> **SUMMARY**
>
> The chick develops from a flat, roughly circular blastoderm lying on top of the yolk. The blastoderm develops into the epiblast. A furrow, called the primitive streak, forms in the posterior midline of the epiblast, and then extends forward. The primitive streak marks the site of gastrulation and so is analogous to the blastopore lip in *Xenopus*. The primitive streak forms when epiblast cells move to the midline, ingress, and then migrate laterally beneath remaining epiblast cells. The chick germ layers are established with reference to the primitive streak, with ingressing cells induced to become mesoderm and endoderm, and remaining epiblast cells giving rise to ectoderm. The chick axes are also established with reference to the primitive streak. The dorso-ventral axis is formed as cells ingress and migrate laterally and give rise to medio-lateral tissues which then fold. The antero-posterior axis likely is specified by rotational forces during passage of the egg down the oviduct prior to laying and the direction of streak elongation predicts the future anterior end of the embryo. A structure called Hensen's node forms in anterior-most streak cells and gives rise to axial mesoderm that defines the antero-posterior axis. Hensen's node acts as an organizer, similar to the Spemann organizer. As in other vertebrates, Wnts initiate the events that lead to chick gastrulation and TGF-β family members act as mesoderm-inducing signals, and experiments in chick provide details of how their interactions set up a robust single gastrula. Likewise, similar transcription factors define the two distinct types of mesoderm that include organizer tissue in chick, *Xenopus*, and zebrafish.

Mouse development and setting up the body plan

A major difference between the development of mammals and the development of *Xenopus*, zebrafish, and chick is that mammalian eggs contain no yolk. The first stages in development of the mouse are concerned with producing the **placenta** through which the embryo will be nourished by the mother, and with establishing three initial cell lineages: the inner cell mass from which the embryo develops, and the primitive endoderm and trophectoderm lineages which give rise to **extra-embryonic** structures. In *Xenopus* and zebrafish, early development and specification of the axes depends on maternal factors but in mice, zygotic genes are activated very early and the axes are specified only later in development. In mouse, as in chick, the starting point for gastrulation is a single-cell-layered epiblast, and there are similarities in the way that gastrulation proceeds in the two species. But in the mouse, there are considerable differences in the overall form of the embryo at gastrulation: the epiblast is cup-shaped rather than flat. The mouse life cycle is shown in Figure 3.47.

In this part of the chapter we first review the development of the mouse and then focus on the earliest decisions in the embryo that determine whether cells will form the embryo or the placenta. We then discuss how the antero-posterior and dorso-ventral axes are defined and how the three germ layers are formed.

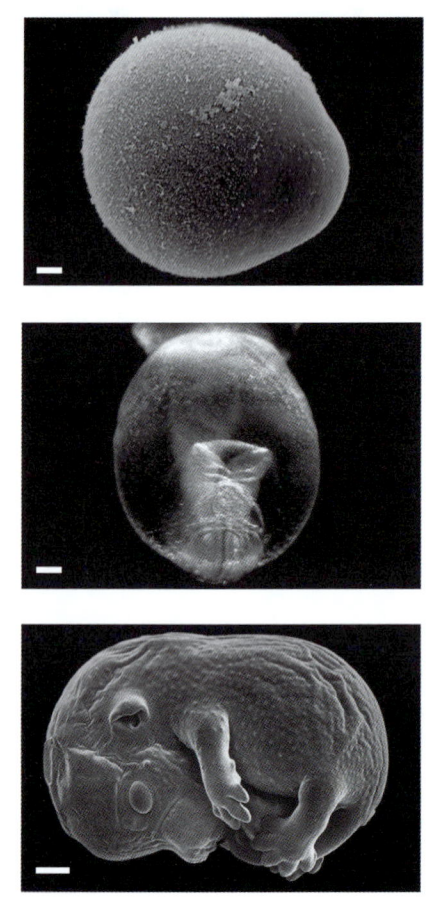

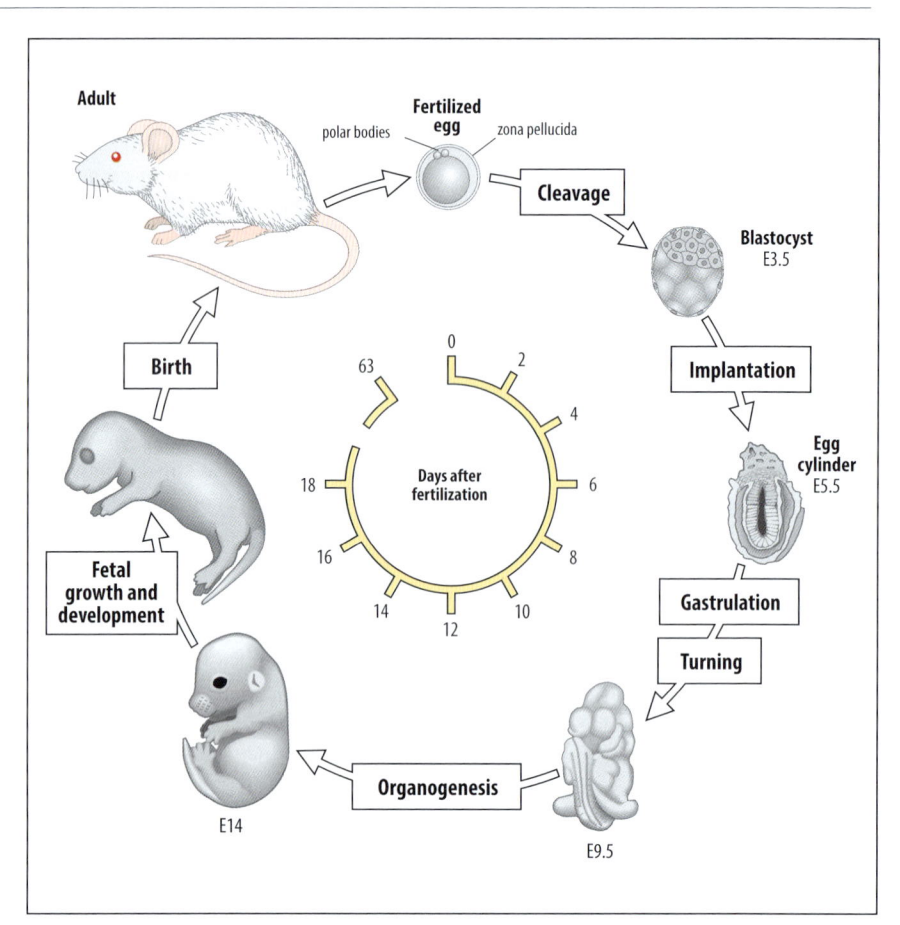

Fig. 3.47 The life cycle of the mouse. The egg is fertilized in the oviduct, where cleavage also takes place before the blastocyst implants into the uterine wall at about 4.5 days after fertilization. Gastrulation and organogenesis then take place over a period of about 7 days, and the remaining 6 or so days before birth are largely a time of overall growth. After gastrulation, the mouse embryo undergoes a complicated movement known as 'turning', in which it becomes surrounded by its extra-embryonic membranes (not shown here). The photographs show (from top): a fertilized mouse egg just before the first cleavage (scale bar = 10 μm); the anterior view of a mouse embryo at E8 (scale bar = 0.1 mm) enclosed in the parietal yolk sac with the heart and developing neural folds visible; and a mouse embryo at E14 (scale bar = 1 mm). An illustrated list of the stages in mouse development can be found on the website listed in 'General further reading'. Stages in mouse development are designated by 'embryonic day' (E), which is the estimated time in days after fertilization.

Top photograph reproduced with permission from Bloom, T.L.: **The effects of phorbol ester on mouse blastomeres: a role for protein kinase C in compaction?** *Development 1989,* **106***: 159-171. Middle photograph courtesy of N. Brown. Bottom photograph courtesy of J. Wilting.*

3.22 The mouse is the major model for mammalian development

Mouse eggs are much smaller than those of either chick or *Xenopus*, about 80–100 μm in diameter, and they contain no yolk. The unfertilized egg is shed from the ovary into the oviduct and is surrounded by a protective external coat, the **zona pellucida**, which is composed of mucopolysaccharides and glycoproteins. Fertilization takes place internally in the oviduct. Meiosis is then completed and the second polar body forms (see Box 8A). Cleavage starts while the fertilized egg is still in the oviduct. Unlike *Xenopus* and zebrafish, the early cleavage divisions are asynchronous, are very slow and do not follow a stereotypical pattern. The first cleavage occurs about 24 hours after fertilization, the second about 20 hours later, and subsequent cleavages at about 12-hour intervals. However, after implantation, the cells of the epiblast proliferate rapidly.

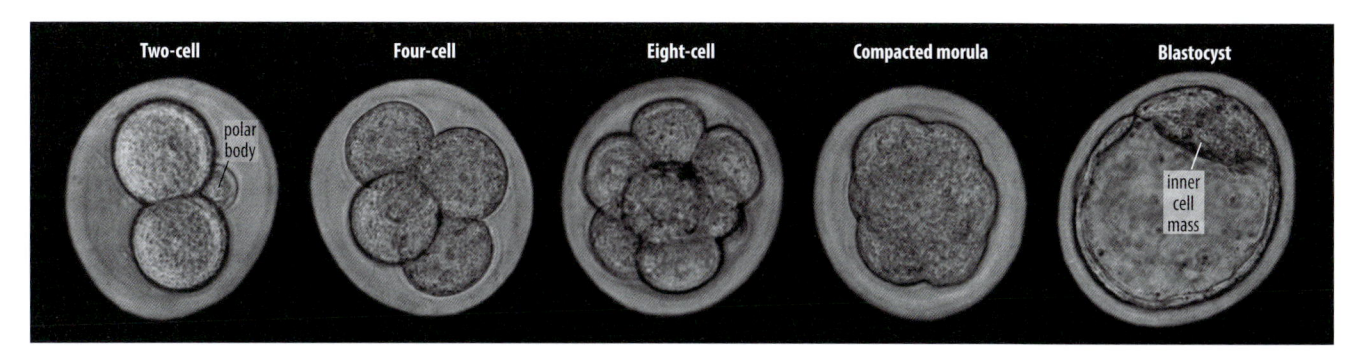

| Two-cell | Four-cell | Eight-cell | Compacted morula | Blastocyst |

Fig. 3.48 Cleavage in the mouse embryo. The photographs show the cleavage of a fertilized mouse egg from the two-cell stage through to the formation of the blastocyst. All these stages take place while the embryo is enclosed in the zona pellucida. After the eight-cell stage, compaction occurs, forming a solid ball of cells called the morula, in which individual cell outlines, while present, can no longer be discerned. The morula develops into the blastocyst, a hollow fluid-filled structure. The outer cells of the morula give rise to the trophectoderm—the outer epithelium of the blastocyst—while the internal cells give rise to the inner cell mass, which is the compact clump at one end of the blastocyst. It is from this that the embryo proper forms. The trophectoderm gives rise to extra-embryonic structures.

Photographs courtesy of T. Fleming.

Cleavage produces a solid ball of blastomeres, called a **morula** (Fig. 3.48). At the eight-cell stage, the blastomeres increase the area of cell surface in contact with each other in a process called **compaction**. After compaction, the blastomeres are polarized; their apical (exterior) surfaces carry microvilli and are characterized by localization of specific proteins, whereas their inner surfaces are smooth. Further cleavages are somewhat variable and are both radial and tangential, so that by the 16-cell stage, some cells end up on the inside while others remain on the outside. The outside cells are polarized but the inside cells are surrounded by other cells and are not polarized.

By the 32-cell stage (E3.5), the morula has developed into the blastocyst, a hollow, fluid-filled sphere of epithelium containing a clump of some 10–15 cells—the inner cell mass—attached to the epithelium at one end. The outer epithelium of the blastocyst, derived from the outside cells of the morula, is the trophectoderm, which gives rise to extra-embryonic structures. The blastocoel is formed by the trophectoderm pumping fluid into the interior of the blastocyst. The inner cell mass arises from the internal cells of the morula and gives rise to the embryo. The cells of the inner cell mass are pluripotent, meaning that they can give rise to all the cell types in the embryo (see Section 1.17). Cells from the inner cell mass can be cultured and are known as embryonic stem cells (ES cells), which can differentiate into all cell types *in vitro* (discussed in Chapter 7). Between E3.5 and E4.5 the inner cell mass becomes divided into two regions (Fig. 3.49). The surface layer that forms in contact with the blastocoel is the **primitive endoderm**, and will contribute to extra-embryonic membranes, whereas the remainder of the inner cell mass is the pluripotent epiblast from which the embryo develops. The blastocyst is then released from the zona pellucida and starts to implant in the uterus at E4.5. By this late stage, the position of the inner cell mass has become tilted with respect to the long axis of the blastocyst. The **mural trophectoderm** at the opposite end of the blastocyst to the inner cell mass adheres to the uterine epithelium, which is then stimulated to envelop the developing embryo.

By E5.0, the implanted blastocyst has grown considerably and is elongated in shape. The mural trophectoderm cells replicate their DNA without cell division, giving rise to primary **trophoblast giant cells** that invade the uterus wall and surround most of the implanted embryo, forming an interface with the maternal tissue.

The **polar trophectoderm** cells in contact with the epiblast continue to divide, forming the **ectoplacental cone** and the **extra-embryonic ectoderm**, both of which contribute to the placenta. The inner cells of the ectoplacental cone differentiate into cells that contribute to making up the chorion, the outermost membrane surrounding

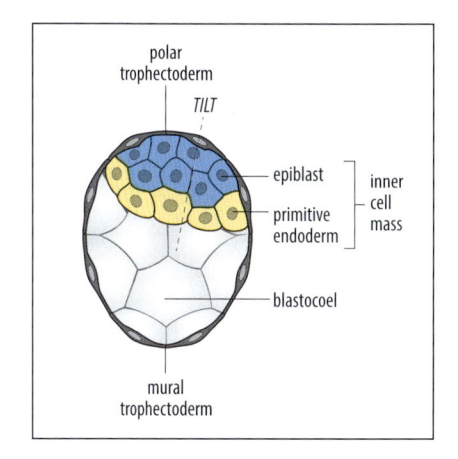

Fig. 3.49 The late blastocyst before implantation. The inner cell mass has become divided into a surface layer of primitive endoderm, which will form extra-embryonic tissues, and a pluripotent epiblast from which the embryo develops.

160 Chapter 3 Vertebrate development I: establishing the body plan

the embryo. They also form the **placental labyrinth**, the layer of the placenta that is the transport interface for nutrients and toxic waste. The outer cells of the ectoplacental cone differentiate into secondary trophoblast giant cells and other trophoblast cell types that attach the placenta to the uterine wall and remodel maternal arteries to increase blood flow into the labyrinth.

The extra-embryonic ectoderm produced by the ectoplacental cone pushes the epiblast across the blastocoel (Fig. 3.50, first panel). A cavity forms within the epiblast making it into a cup shape. Some cells from the primitive endoderm migrate to cover the whole inner surface of the mural trophectoderm. They become the **parietal endoderm**, which eventually becomes Reichert's membrane, a sticky

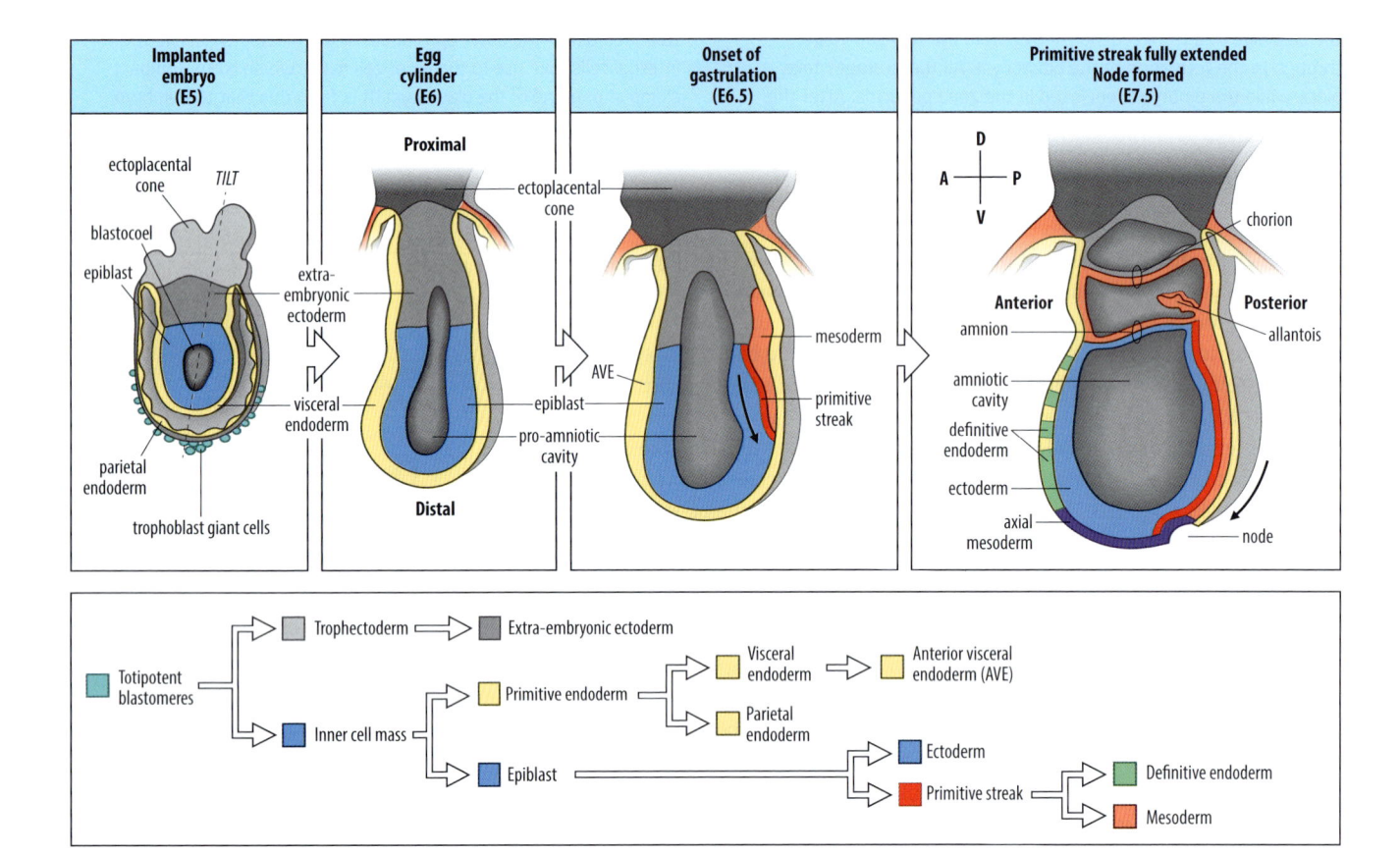

Fig. 3.50 Early post-implantation development of the mouse embryo. Top row: successive stages in development from implantation to the development of the primitive streak and mesoderm formation. First panel: the mural trophectoderm gives rise to trophoblast giant cells which invade the uterine wall, helping to anchor the blastocyst to it. The blastocyst becomes surrounded by the uterine wall (not shown). Some cells of the primitive endoderm migrate to cover the inner surface of the mural trophectoderm and become the parietal endoderm; the remaining endoderm cells become visceral endoderm covering the epiblast. The polar trophectoderm (in contact with the epiblast) proliferates and forms extra-embryonic tissues—the extra-embryonic ectoderm and ectoplacental cone—which contribute to the placenta. An internal cavity develops in the epiblast, giving it a cup-shaped form. Second panel: the embryo elongates and the internal cavity fuses with a cavity that has developed in the extra-embryonic ectoderm to form the pro-amniotic cavity. (The outer extra-embryonic tissues such as the parietal endoderm are not shown here for simplicity.) Third panel: the beginning of gastrulation is marked by the appearance of the primitive streak (red) and the formation of mesoderm (pink) at the posterior of the epiblast. Fourth panel: the primitive streak extends anteriorly (arrow) towards the bottom of the cup and the node develops at its anterior end. Axial mesoderm formed at the node moves in an anterior direction to underlie the ectoderm. Cells from the primitive streak enter the visceral endoderm to form the definitive endoderm (green). Bottom row: the lineage diagram shows the origin of the tissues shown in this figure and in Fig. 3.58. A, anterior; P, posterior; D, dorsal; V, ventral.

Upper panels from Thowfeequ, S., Srinivas, S.: **Mammalian embryo: establishment of the embryonic axes.** eLS 2021, **2**: 1-15. Lower panel from Nahaboo, W., Migeotte, I.: **Cleavage and gastrulation in the mouse embryo.** eLS (John Wiley, 2018) [https://doi.org/10.1002/9780470015902.a0001068.pub3].

layer of cells and extracellular matrix that has a barrier function. The rest of the primitive endoderm covering the epiblast and extra-embryonic ectoderm forms the **visceral endoderm**. The mouse epiblast is now a single layer of epithelium formed into a cup shape, U-shaped in section, with a layer of visceral endoderm on the outside.

By E6, the embryo has elongated further, and forms a cylindrical structure—the egg cylinder—containing the epiblast and the extra-embryonic ectoderm covered by visceral endoderm (see Fig. 3.50, second panel). The cavity within the epiblast has fused with a cavity that forms in the extra-embryonic ectoderm to give the pro-amniotic cavity. The egg cylinder has a proximo-distal polarity in relation to the site of implantation, with the proximal end closest to the ectoplacental cone and the distal end furthest away. The egg cylinder is peculiar to rodents; other mammals, including humans, develop from a flat disc.

The antero-posterior axis and the germ layers of the mouse embryo become apparent in gastrulation. The onset of gastrulation at E6.5 is marked by the formation of the primitive streak (see Fig. 3.50, third panel). The initial development of the primitive streak is like that in the chick and starts as a localized thickening in the epiblast. The streak forms at the edge of the cup on one side, at the junction with the extra-embryonic ectoderm. This side will correspond to the future posterior end of the embryo. Over the next 12–24 hours the primitive streak elongates until it reaches the bottom of the cup. Here, a condensation of cells—the **node**—becomes distinguishable at the anterior end of the extended streak (see Fig. 3.50, fourth panel). The node has similar properties to the Spemann organizer: when grafted to a second host embryo, it can induce a partial second body axis. But unlike in the frog, this second body axis lacks anterior structures (discussed in Chapter 4). To make primitive streak formation in the mouse easier to compare with that of the chick, imagine the epiblast cup spread out flat (see Fig. 3.39).

In mouse gastrulation, as in chick, epiblast cells converge on the primitive streak, and proliferating cells migrate through it to spread out laterally and anteriorly between the ectoderm and the visceral endoderm to form a mesodermal layer (Fig. 3.51). In addition to forming the mesoderm of the embryo, epiblast cells that pass through the posterior streak form extra-embryonic mesoderm that contributes to the amnion, allantois, and chorion. Some epiblast-derived cells pass through the streak

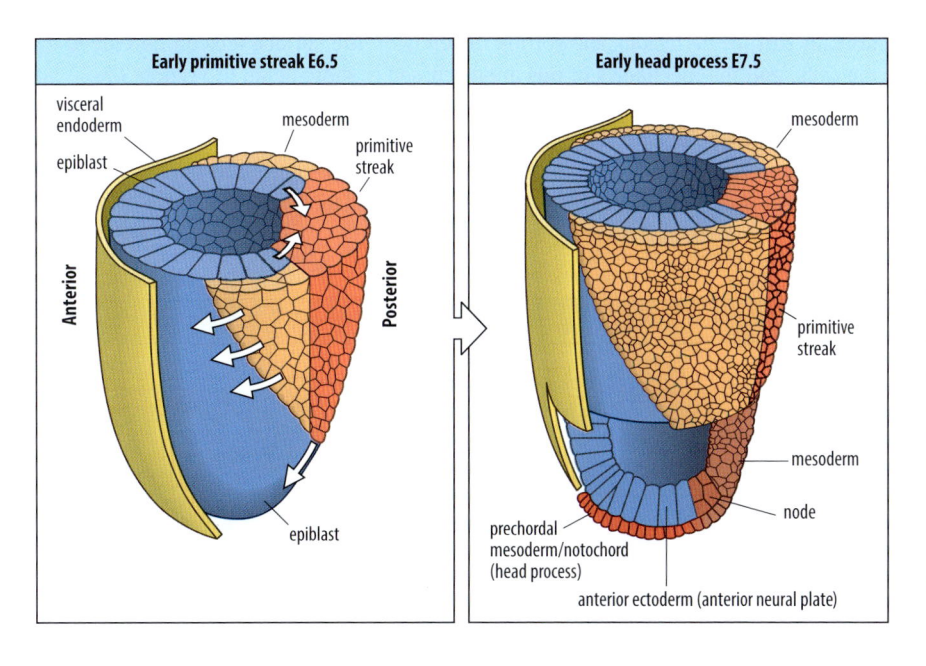

Fig. 3.51 Gastrulation in the mouse embryo. Left panel: as in the chick, gastrulation in the mouse embryo begins when epiblast cells converge on the posterior of the epiblast and move under the surface, forming the denser primitive streak (brown) where the cells are becoming internalized. Once inside, the proliferating cells spread out laterally between the epiblast and the visceral endoderm to give a layer of prospective mesoderm (light brown). Some of the internalized cells will eventually replace the visceral endoderm to give definitive endoderm (not shown on these diagrams for simplicity), which will form the lining of the gut. Right panel: as gastrulation proceeds, the primitive streak lengthens and reaches the bottom of the cup, with the node at the anterior end. Cells from the node move anteriorly to give rise to the prechordal mesoderm and notochord in the head region, which together are known as the head process. Part of the visceral endoderm and the mesoderm has been cut away in this diagram to show the node and notochord. Note that, given the topology of the mouse embryo at this stage, the germ layers appear inverted (ectoderm on the inner surface of the cup, endoderm (not shown) on the outer) if compared with the frog gastrula.

*Illustration adapted, with permission, from McMahon, A.P.: **Mouse development. Winged-helix in axial patterning.** Curr. Biol. 1994, **4**: 903–906.*

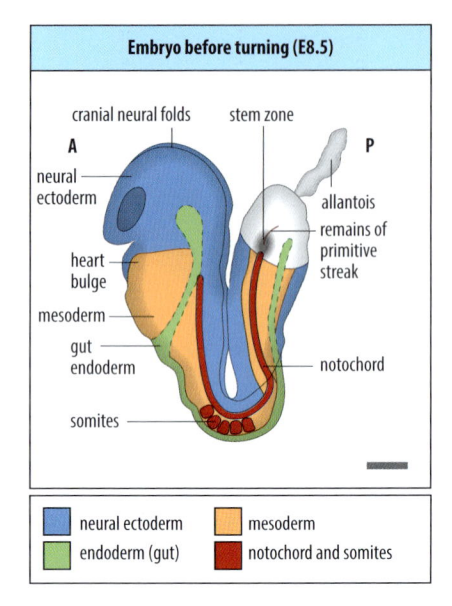

162 Chapter 3 Vertebrate development I: establishing the body plan

Fig. 3.52 Mouse embryo after ventral closure and before turning. By E8.5 there has been further growth of the embryo anterior to the node, the head is distinct, the neural folds have formed, the foregut and hindgut have closed, and somites are beginning to form on either side of the notochord. The embryo is covered with a layer of ectoderm that will form the epidermis (not shown). Scale bar = 100 μm.

Fig. 3.53 Turning in the mouse embryo. Between E8.5 and E9.5, the mouse embryo becomes entirely enclosed by the protective amnion and lies in the amniotic cavity containing amniotic fluid. The visceral yolk sac, a major source of nutrition, surrounds the amnion and the allantois connects the embryo to the placenta.
Illustration after Kaufman, M.H.: The Atlas of Mouse Development. London: Academic Press, 1992.

and intercalate into the visceral endoderm to form the **definitive endoderm** on the outside of the cup (see Fig. 3.50, fourth panel). This will form the lining of the gut. The epiblast cells that remain on the inside of the cup will form ectoderm. Once the node has formed, axial mesoderm cells migrate anteriorly from it and give rise to the head process (the prechordal mesoderm and notochord in the head region) underlying the ectoderm that will form the anterior neural plate. As in the chick, the node regresses, and trunk notochord is progressively laid down in its wake, with the spinal cord developing above it from a stem zone, and somites forming on either side (described in Chapter 4).

By E8.5, the embryo anterior to the node has grown rapidly, the head is apparent, and neural folds have started to form. The embryonic endoderm—initially exposed on the outside of the embryo—becomes internalized to form foregut and hindgut. Lateral folding then internalizes the gut completely and forms the ventral body wall, a process involving ventral closure as in the chick (Fig. 3.52). A complicated 'turning' process occurs at around E9.0. The cup-shaped embryo turns inside out so that the back of the embryo is now convex instead of concave and the embryo becomes enclosed by the amnion and lies in the amniotic cavity. The umbilical cord that connects the embryo to the placenta and incorporates the allantois now arises from the underside of the embryo (Fig. 3.53). (Turning is another developmental quirk peculiar to rodents; human embryos are surrounded by their extra-embryonic membranes from the beginning.) Organogenesis in the mouse proceeds very much as in the chick. From fertilization to birth takes around 18–21 days, depending on the strain of mouse.

3.23 Blastomeres in the morula are specified as trophectoderm or inner cell mass depending on their position

The first cell-fate decisions that specify the mouse body plan are made at the morula stage, and the position of a blastomere—whether it is on the outside or on the inside—determines whether it will give rise to the trophectoderm or to the inner cell mass, respectively. Up to the eight-cell stage, each blastomere seems to be equivalent in its ability to give rise to either tissue. At this stage, compaction occurs and the blastomeres become polarized. Assignment of cells to be either outside cells or inside cells depends on whether a cell generated by subsequent divisions inherits an apical domain, and thus becomes an outside cell, or whether it does not, and becomes an inside cell. Any cells on the outside of the embryo that happen not to inherit an apical domain sort to the inside. The sorting depends on the surface tension developed by the cells, and the mechanics of this are discussed in more detail in Section 5.7.

The most direct evidence for the effect of position comes from making chimeric embryos (see Section 3.4). Individual blastomeres disaggregated from eight-cell embryos are labeled and then combined in different positions with unlabeled blastomeres from other embryos. If the labeled cells are placed on the outside of a group

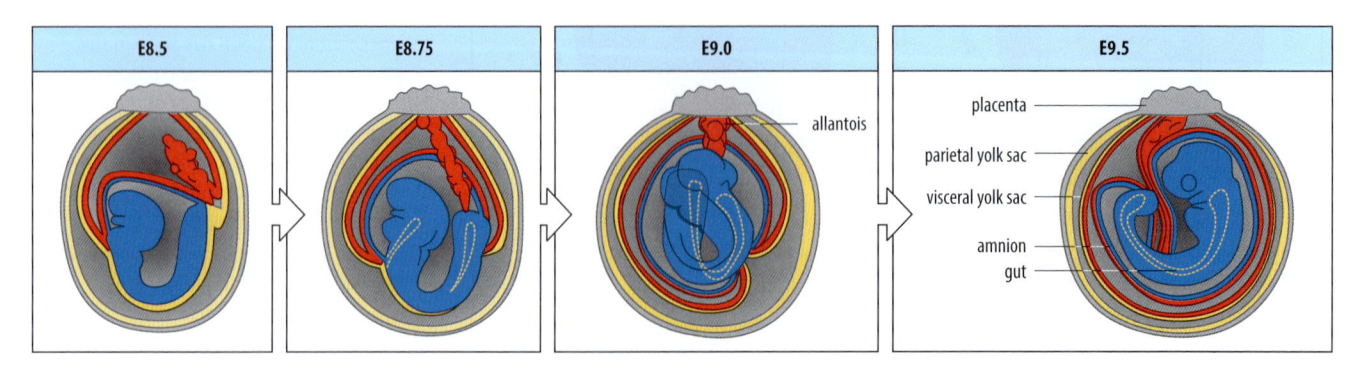

Fig. 3.54 The specification of the blastomeres that will form the inner cell mass of a mouse embryo depends on their position. The specification of blastomeres as inner cell mass depends on whether they are on the outside or the inside of the embryo. Left panels: following cleavage, the mouse embryo is a solid ball of cells (a morula). This develops into the hollow blastocyst, which consists of an inner cell mass surrounded by a layer of trophectoderm. Right panels: to investigate whether the position of a blastomere in the morula determines whether it will develop into trophectoderm or inner cell mass, labeled blastomeres (blue) from an eight-cell mouse embryo are separated and combined with unlabeled blastomeres (gray) from another embryo. By following the fate of the descendants of the labeled cells, it can be seen that blastomeres on the outside of the aggregate more often give rise to trophectoderm, 97% of the labeled cells ending up in that layer. The reverse is true for the origin of the inner cell mass, which derives predominantly from blastomeres inside the aggregate.

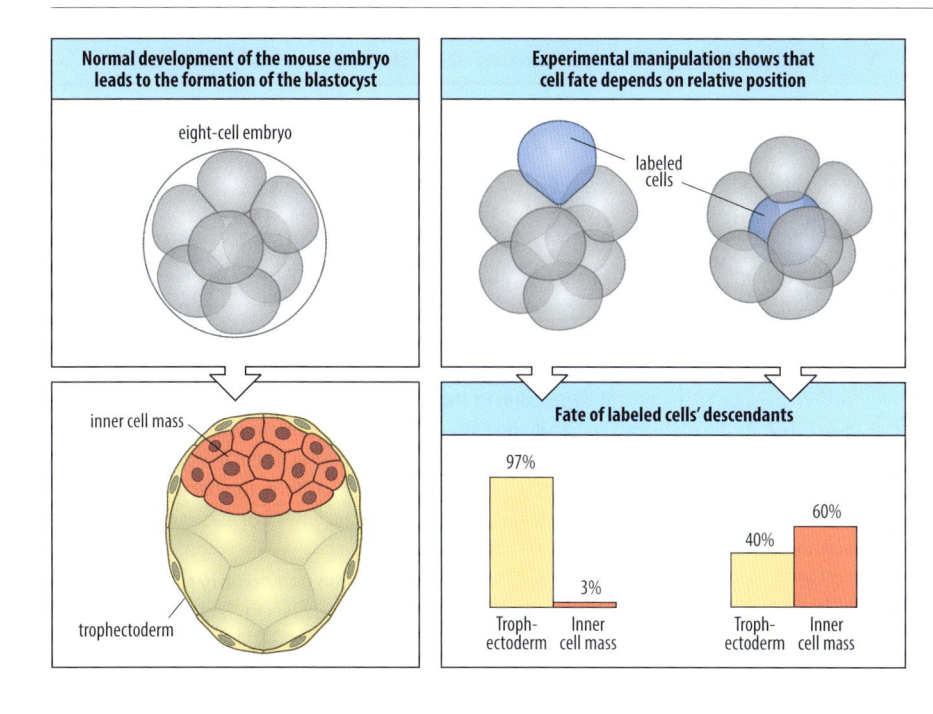

of unlabeled cells they usually give rise to trophectoderm; if they are placed inside, so that they are surrounded by unlabeled cells, they more often give rise to inner cell mass (Fig. 3.54, right panels). Aggregates composed entirely of either 'outside' or 'inside' cells of early embryos can develop into normal blastocysts, showing that there is no specification of these cells, other than by position, at this stage. Another way in which the effect of position has been demonstrated is by using chimeric mice created by combining blastomeres from strains of mice that have different coat colors (see Fig. 1.24). In this case, coat color acts as an indelible label. When cells from an eight-cell embryo from a white strain of mouse are placed on the outside of a group of cells from an eight-cell embryo from a black strain of mouse, the resulting chimeric mouse will have a black coat, showing that it has been formed by the 'inside' cells.

So how do the cells in the morula know whether they are on the outside or the inside? This information is conveyed by the Hippo signaling pathway (described in Box 11A). The Hippo pathway is known to act as a sensor of a cell's environment and its interactions with other cells. The pathway is differentially activated in outside and inside cells (see Section 5.7). In the outside cells, in which Hippo signaling is inactive, the transcriptional co-activator Yap accumulates in the nucleus, whereas in the non-polarized inside cells, in which Hippo signaling is active, Yap is sequestered in the cytoplasm. These differences in the activity of the pathway result in high levels of expression of the transcription factor Cdx2 in outside cells and low levels of expression in inside cells. Cdx2 is required for the differentiation of trophectoderm. If Cdx2 function is absent, trophectoderm initially forms in the blastocyst but does not differentiate and is not maintained.

While the transcription factor Cdx2 is involved in differentiation of the trophectoderm, other transcription factors are involved in the differentiation of the inner cell mass. Three transcription factors—Oct4, Nanog, and Sox2—keep the cells of the inner cell mass locked in a pluripotent state. The *Oct4* gene (known as *Pou5f1*) is expressed from the oocyte onwards and by the eight-cell stage all blastomeres express both *Oct4* and *Cdx2* (Fig. 3.55). As the morula compacts and the 16-cell stage is reached, levels of Cdx2 protein become slightly higher in the outside cells as a result of Hippo signaling suppressing *Cdx2* expression in inside cells. This difference in

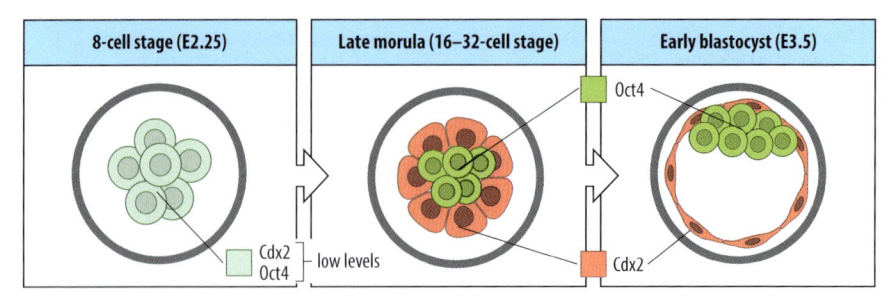

Fig. 3.55 Establishing the first two cell lineages in the mouse embryo. The first two cell types to be specified in the mouse embryo are the inner cell mass and the trophectoderm. This involves progressive restriction of expression of the genes that code for transcription factors Oct4 and Cdx2. At the eight-cell stage, all the cells express both genes at low levels, but by the late morula/early blastocyst stage, Cdx2 expression has increased in, and eventually becomes restricted to, the cells on the outside of the embryo, and these will form trophectoderm. Expression of the gene encoding Oct4 becomes restricted to the inner cell mass cells at early blastocyst stage, and these will form the embryo.

*After Stephenson, R.O., et al.: **Intercellular interactions, position and polarity in establishing blastocyst cell lineages and embryonic axes.** Cold Spring Harb. Perspect. Biol. 2012, **4**: a008235.*

expression of the two transcription factors in outer and inner cells is reinforced by **reciprocal negative feedback** in which the Cdx2 protein represses expression of the *Oct4* gene and vice versa—*Cdx2* expression increases and *Oct4* expression decreases in the outer cells, whereas *Oct4* expression increases and *Cdx2* expression decreases in the inner cells. By the blastocyst stage, *Cdx2* is expressed only in the trophectoderm and *Oct4* only in the cells of the inner cell mass. The reciprocal negative feedback of two transcription factors that results in the stabilization of two adjacent cell fates is a common principle used throughout development.

Cdx2 gene expression can be demonstrated to correlate with the commitment of cells to the trophectoderm lineage by following visually the behavior of the inside and outside cells using transgenic embryos expressing a gene for enhanced green fluorescent protein (eGFP) fused with the *Cdx2* gene. At the 32-cell stage the embryos are disaggregated into single cells and individual cells expressing different levels of Cdx2 (as shown by different levels of eGFP) are isolated. Each cell is then reaggregated with a morula from a normal mouse to make a chimera. Donor cells expressing low levels of eGFP move inside the morula and eventually become incorporated into the inner cell mass, whereas donor cells expressing high levels of eGFP usually remain on the surface of the morula and contribute to the trophectoderm.

At these early stages in mouse development, there is a remarkable degree of regulation. Even a single cell of a 32-cell mouse embryo can generate an entire embryo when aggregated with tetraploid cells, which will form the trophectoderm of the chimeric embryo and support its development. At the other end of the scale, giant embryos formed by the aggregation of several embryos in early cleavage stages can achieve normal size within about 6 days by reducing cell proliferation.

3.24 The inner cell mass becomes differentiated into primitive endoderm and cells that will form the embryo proper

The second cell-fate decision comes at the early blastocyst stage, when cells of the inner cell mass differentiate into either epiblast, from which the embryo develops,

Fig. 3.56 Two cell lineages arise from the inner cell mass. Within the inner cell mass, two separate cell types arise: the epiblast cells, from which the embryo will develop; and the primitive endoderm, which will form extra-embryonic structures. This involves a progressive restriction of expression of the genes for the transcription factors Nanog and Gata6. At the early blastula stage, some of the inner cell mass cells apparently at random start to express higher levels of Nanog, whereas the other cells express higher levels of Gata6. By the late blastula, the cells expressing the two different transcription factors have sorted out within the inner cell mass so that cells expressing high levels of Gata6 form an outer epithelial layer of primitive endoderm enclosing the epiblast cells expressing high levels of Nanog.
*After Stephenson, R.O., et al.: **Intercellular interactions, position and polarity in establishing blastocyst cell lineages and embryonic axes.** Cold Spring Harb. Perspect. Biol. 2012, 4: a008235.*

or primitive endoderm, which gives rise to extra-embryonic tissues. The differentiation of two separate lineages is driven by the expression of two transcription factors, Nanog and Gata6. Cells of the morula express low levels of both factors. As the mouse blastocyst matures, cells of the inner cell mass emerge that express higher levels of Nanog and low levels of Gata6, and vice versa, and, as a consequence, two cell populations become distinguished in a 'salt-and-pepper' arrangement. The two cell types then sort out, with *Nanog*-expressing cells giving rise to the epiblast, and *Gata6*-expressing cells giving rise to the primitive endoderm (see Fig. 3.56). FGF signaling is important in this separation. Nanog-expressing cells make FGF, which promotes *Gata6* expression in surrounding cells, which express FGF receptors, enhancing the probability that they will become primitive endoderm.

The two different cell types—epiblast and primitive endoderm—are specified in the inner cell mass, possibly according to time, in a more-or-less random arrangement and then sort out to take up their appropriate positions (see Fig. 3.56). Cell death might also play a part by eliminating cells in the wrong position. The initial salt-and-pepper expression of Nanog and Gata6 in the inner cell mass seems to be conserved in all mammalian embryos, including human embryos.

3.25 The first asymmetry in mouse development appears in the blastocyst

Early-stage development in the other model organisms we have considered is controlled by maternal factors, but this is less so in the mouse. There are signs of polarity in the mouse egg, but the major wave of activation of zygotic transcription begins at the late two-cell stage and is essential for development beyond this stage. All the blastomeres generated by early cleavages have the same developmental potential and can contribute to inner cell mass or trophectoderm. In addition, normal development still occurs even after cells are removed or added to a blastocyst before it implants into the uterus wall. Nevertheless, despite this developmental plasticity, there are indications that blastomeres of the two-cell mouse embryo are not identical. When the two blastomeres are separated, twin mice are produced only in a minority of cases. But these are difficult experiments and separation of the blastomeres might damage them. There is conflicting evidence as to whether early blastomeres are already biased by the four-cell stage in mice, with evidence for and against.

Before the blastocyst stage, the mouse embryo is a symmetric spheroidal ball of cells. The first asymmetry appears in the blastocyst, where the blastocoel develops asymmetrically, leaving the inner cell mass attached to one part of the trophectoderm (Fig. 3.57, left panel). The blastocyst now has a distinct axis running from the site where the inner cell mass is attached (the **embryonic pole**) to the opposite (**abembryonic pole**) site, with the blastocoel occupying most of the abembryonic half. This axis is known as the **embryonic–abembryonic axis**.

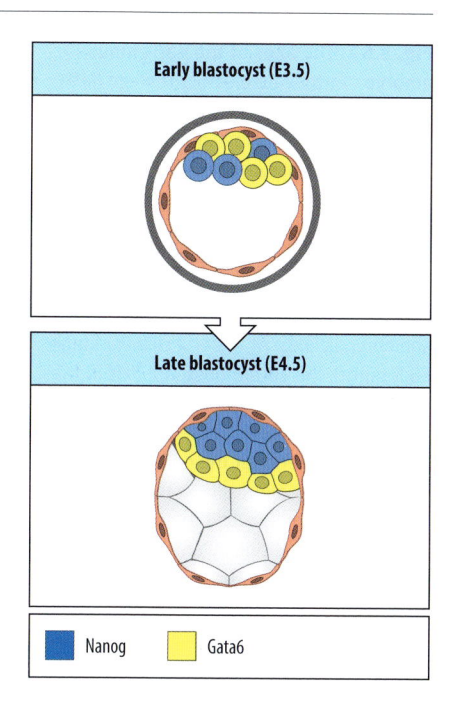

Fig. 3.57 The axes of the early mouse embryo. At the late blastocyst stage (left), at about E4.5, the inner cell mass is confined to the embryonic region and this defines an embryonic–abembryonic axis (which relates geometrically, although not in terms of cell fate, to the dorso-ventral axis of the future epiblast). The blastocyst is oval and has an axis of bilateral symmetry. The inner cell mass is tilted at an angle to the embryonic–abembryonic axis. At the egg cylinder stage, E6.5 (right), the antero-posterior axis in the epiblast becomes visible, with the formation of the primitive streak at the posterior (P) end. The interior of the epiblast cup corresponds geometrically to the dorsal side of the future embryo, and the outer side to the future ventral side. A, anterior.

Adapted from Thowfeequ, S., Srinivas, S.: *Mammalian embryo: establishment of the embryonic axes.* eLS 2021, **2**: 1-15.

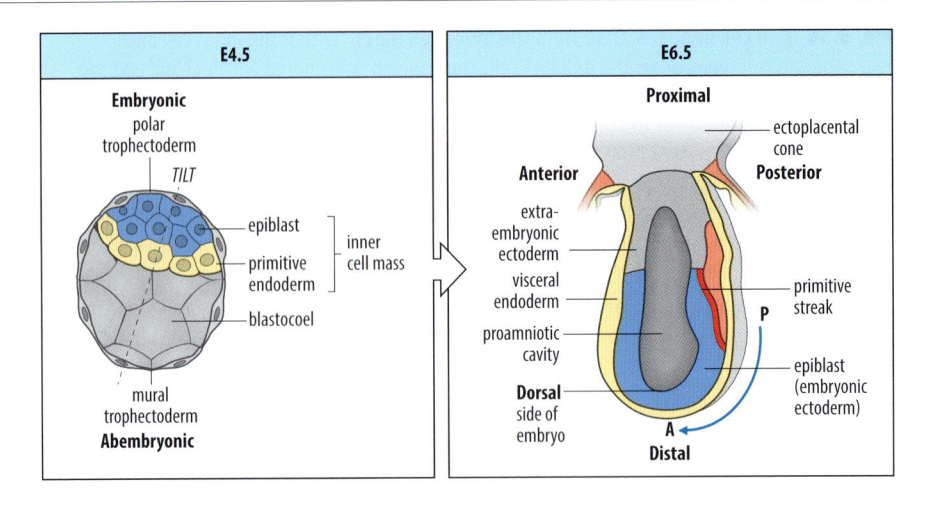

The embryonic–abembryonic axis of the blastocyst forms the proximo-distal axis of the egg cylinder of the post-implantation embryo (Fig. 3.57, right panel). The proximo-distal axis of the egg cylinder can be considered to prefigure the future dorso-ventral axis of the body in that the dorsal neural tube characteristic of the vertebrate body plan is derived from the epiblast and the visceral endoderm contributes to the development of the ventral gut. The antero-posterior axis of the epiblast is marked by the formation of the primitive streak at one point on the rim of the epiblast cup, which will be the future posterior end of the body. Just before the blastocyst implants, the inner cell mass becomes tilted with respect to the embryonic–abembryonic axis, which might be the first indications of the overt asymmetry that establishes the antero-posterior axis in the embryo.

3.26 The antero-posterior body axis of the mouse embryo is predicted by the movement of the anterior visceral endoderm

The crucial issue for specifying the antero-posterior body axis in the mouse embryo is the position of the primitive streak, as where it starts marks the posterior end of the embryo. The location of the primitive streak depends on signaling cues from a small region of visceral endoderm at the distal-most end of the epiblast cup. This sub-population of cells is known as the **distal visceral endoderm (DVE)** (Fig. 3.58). The distal visceral endoderm cells undergo unidirectional migration to the rim on one side of the cup and produce signals that restrict primitive streak formation to the opposite side. The side of the cup occupied by the distal endoderm cells defines the future anterior of the embryo, and hence the migrated distal visceral endoderm is now known as the **anterior visceral endoderm (AVE)**. In this way, the symmetry of the epiblast is broken and an antero-posterior axis is predicted by the AVE at one end and the primitive streak at the other.

Nodal and Wnt signals control the formation of the primitive streak. Immediately post-implantation, Nodal and Wnt signaling occurs throughout the epiblast. Expression of *Nodal* and *Wnt3* are induced by BMP-4 signaling from the extra-embryonic ectoderm proximal to the epiblast (see Fig. 3.58, first panel), and a gradient of Nodal signaling is established in the epiblast running from proximal to distal. The primitive streak could potentially form anywhere around the rim of the epiblast cup where Nodal and Wnt signaling is high. The DVE forms where levels of Nodal signaling are low. The DVE and resulting AVE express genes for secreted Wnt and Nodal antagonists, including the Wnt antagonist Dickkopf-1, and the Nodal antagonists Cerberus-1 and Lefty-1. The name 'Lefty' comes from the fact that this

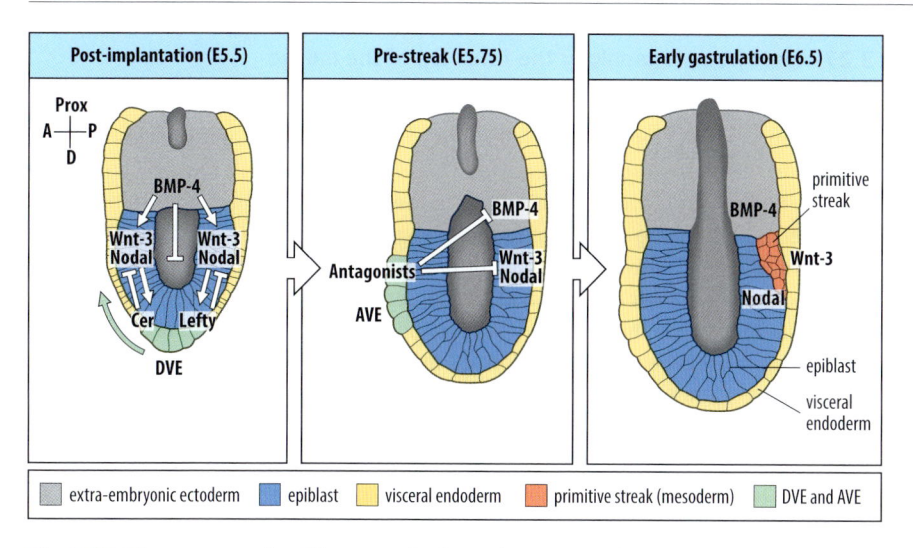

Fig. 3.58 The symmetry-breaking event in the early mouse embryo is the movement of the anterior visceral endoderm to one side of the epiblast cup. First panel: at around E5.5, before primitive streak formation, BMP-4 signaling by the extra-embryonic ectoderm at the boundary with the epiblast induces Nodal and Wnt expression in the adjacent epiblast. Visceral endoderm at the distal-most end of the cup, where Nodal signaling is lowest, expresses Lefty-1 (Lefty) and Cerberus-1 (Cer) and forms distal visceral endoderm (DVE; green). The rest of the visceral endoderm is shown in yellow. The direction in which the DVE will migrate is shown by the green arrow. Second panel: at around E5.75, the DVE has recruited additional visceral endoderm cells and migrated to one side of the cup and is now known as the anterior visceral endoderm (AVE). Antagonists produced by the AVE restrict Nodal and Wnt signaling to the cells at the opposite side of the cup. Third panel: at E6.5, the primitive streak (pink) has formed in the epiblast on the rim of the cup opposite to the AVE, thus marking the posterior end of the axis, and then begins to elongate distally. A, anterior; P, posterior; Prox., proximal; D, distal.

From Bardot, E.S., Hadjantonakis, A.-K.: **Mouse gastrulation: coordination of tissue patterning, specification and diversification of cell fate.** Mech. Dev. 2020, **163**: 103617.

antagonist was first identified because it is expressed very strongly on the left side of the embryo at later stages (the various ways in which Nodal signaling can be fine-tuned are shown in Box 4C). As a result of the secretion of the antagonists by the AVE, Nodal and Wnt signaling in the epiblast become restricted to a group of cells at the opposite side of the rim of the epiblast cup (see Fig. 3.58, second and third panels).

The DVE might originate earlier in the primitive endoderm of the blastocyst, where cells characterized by the expression of the gene for Lefty-1 are located asymmetrically on one side. These cells are thought to later recruit adjacent visceral endoderm to make up the group of cells that migrates to form the AVE. The direction of migration may similarly be related to the tilt in the proximal–distal axis in the pre-implantation blastocyst.

Despite the differences in topology, this process is developmentally similar to that described earlier for the chick (see Section 3.18): the AVE is functionally equivalent to the chick hypoblast. Like the hypoblast, the AVE prevents primitive streak formation. In both cases, this influence is due to these cells producing proteins that counteract Nodal signaling: Cerberus in chick; Lefty-1 and Cerberus-1 in mouse. In addition, the signals that initiate formation of the primitive streak are Wnt and Nodal—the same signals that are involved in initiating the streak at the posterior marginal zone of the chick epiblast (see Fig. 3.45).

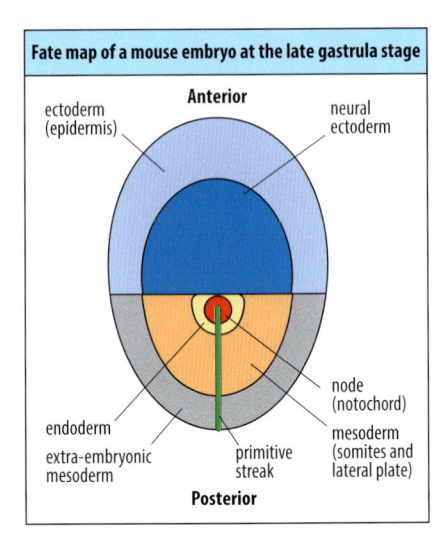

Fig. 3.59 Fate map of a mouse embryo at E7.5. The embryo is depicted as if the 'cup' has been flattened and is viewed from the inside surface. At this stage the primitive streak (green) is at its full length.

3.27 Apart from its topology the fate map of the mouse is similar to that of the chick

It is difficult to make reproducible fate maps of mouse embryos at the stages that correspond to the blastula stage in *Xenopus*. This is because the cells that will contribute to the different germ layers are mixed up in the epiblast and there are extensive cell movements both before and during primitive streak formation. The picture becomes clearer once the primitive streak has formed, and cells have started to move inside and become determined as endoderm and mesoderm.

The fate map of the mouse embryo at around E7.5 is shown in Figure 3.59. At this stage the epiblast is becoming transformed into the three germ layers by gastrulation. The fate map shows the mouse embryo as if the cup were opened out and flattened and viewed from the surface, which was on the inside of the cup. This shows that the fate map of the mouse at the primitive-streak stage is basically similar to that of the chick. It may be helpful to look back at Figure 3.51, which shows the cup-shaped mouse epiblast, to help interpret its fate map.

As in the fate map of the chick epiblast, the anterior region of the mouse epiblast will give rise to the ectoderm, and the node will give rise to the notochord. The cells that will form the heart have already ingressed and moved forward of the node, as have some cells that will give rise to the definitive gut endoderm, prechordal mesendoderm, and/or prechordal mesoderm (the small size of the mouse embryo has made it difficult to distinguish mesoderm from mesendoderm). The extreme anterior end of the primitive streak will give rise to more endoderm, whereas the middle part of the streak gives rise mainly to somites and lateral plate mesoderm. The posterior part of the streak will become the tailbud, and also provides the extra-embryonic mesoderm. The mouse embryo retains considerable capacity to regulate until late in gastrulation. Even at this primitive streak stage, up to 80% of the cells of the epiblast can be destroyed by treatment with the cytotoxic drug mitomycin C, and the embryo can still recover and develop with relatively minor abnormalities. This shows the continuing importance of cell–cell interactions.

> **SUMMARY**
>
> The fertilized mouse egg undergoes cleavage to form a solid ball of cells called a morula. The morula develops a fluid-filled cavity and is now called a blastocyst, composed of an outside layer of trophectoderm cells and an internal cluster of cells at one end, called the inner cell mass. The trophectoderm will form the extra-embryonic tissues of the placenta, while all the cells of the embryo's body are derived from the inner cell mass. Pluripotent stem cells—embryonic stem cells—can be isolated from the inner cell mass and grown in culture. There is no evidence in the mouse embryo that cytoplasmic localization of maternal factors makes cells different, and zygotic gene expression starts at the two-cell stage. The first cell-fate decision in the mouse embryo—the assignment of morula cells to become either trophectoderm or inner cell mass—is determined by their position in the morula (outer or inner cells) and this results in progressive restriction of gene expression to the two different cell types. A second cell-fate decision occurs at the blastocyst stage and is the assignment of inner cell mass cells to become either epiblast, from which the embryo develops, or primitive endoderm which gives rise to extra-embryonic tissues. The cells in the inner cell mass differentiate into the two cell types and then sort out. The epiblast becomes cup-shaped but as in the chick, gastrulation occurs through the primitive streak, with a node forming at the anterior end. Also like the chick, specification of the germ layers occurs during gastrulation and requires cell–cell interactions. As in *Xenopus*, zebrafish, and the chick, localized Wnt signaling specifies the position where the primitive streak will form in mouse embryos and Nodal is also involved.

> **SUMMARY: Vertebrate axis determination**
>
> | | Dorso-ventral axis | Antero-posterior axis |
> |---|---|---|
> | **Xenopus** | sperm entry point and cortical rotation lead to specification of dorsal side opposite the point of sperm entry | specified in organizer |
> | **Zebrafish** | initial symmetry-breaking event unknown. Localization of maternal Ndr1 and nuclear β-catenin species dorsal side and position of organizer | specified in organizer |
> | **Chick** | posterior marginal zone specifies dorsal end of dorso-ventral axis | gravity determines position of posterior marginal zone and thus the posterior end of the A-P axis |
> | **Mouse** | interaction between inner cell mass and trophectoderm | specification and movement of dorsal visceral endoderm |

Development of the human body plan

Studies of the embryonic development of our four main model vertebrates have provided general principles that help us to understand how a human embryo develops. The human life cycle is shown in Figure 3.60. The length of time from fertilization to birth is 38 weeks, much longer than in any of our model vertebrates. A landmark advance in the 1970s was the devising of methods for fertilizing human eggs *in vitro* and then culturing them as they undergo the earliest stages in development. Not only could the earliest stages in human development now be observed for the first time, but there was also a major and lasting impact on the treatment of infertility. This advance also opened up new clinical treatments, for example **pre-implantation genetic diagnosis** (Box 3I). It has also become possible, building on technical advances pioneered in mice, to study the events that occur just after implantation, by removing the zona pellucida from human blastocysts and then culturing them attached to a dish. Our knowledge of the later stages of human development comes from descriptions of collections of clinical material.

3.28 The early development of a human embryo is generally similar to that of the mouse but the topology is more like that of the chick

Unfertilized human eggs are about the same size as mouse eggs, about 100 μm in diameter. Fertilization normally takes place in the oviduct (in humans also known as the Fallopian or uterine tube) and the earliest stages of development then take place while the fertilized egg travels towards the uterus. These stages are similar to those in mice—cleavage divisions, compaction, and blastocyst formation (Fig. 3.61). The major wave of activation of zygotic gene transcription occurs at the 4–8-cell stage transition, slightly later than in the mouse. Implantation usually occurs at around day 7 of development, when about 256 cells are present. As in mouse blastocysts, cells from the inner cell mass of a human blastocyst can be cultured as pluripotent ES cells, and these can provide a model system for studying processes in human development such as cell differentiation (see Chapter 7).

The human embryo just before gastrulation looks more like the flat chick embryo than the cup-shaped mouse embryo (Fig. 3.62, upper panel). Most other non-rodent

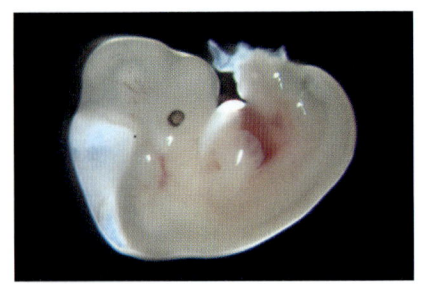

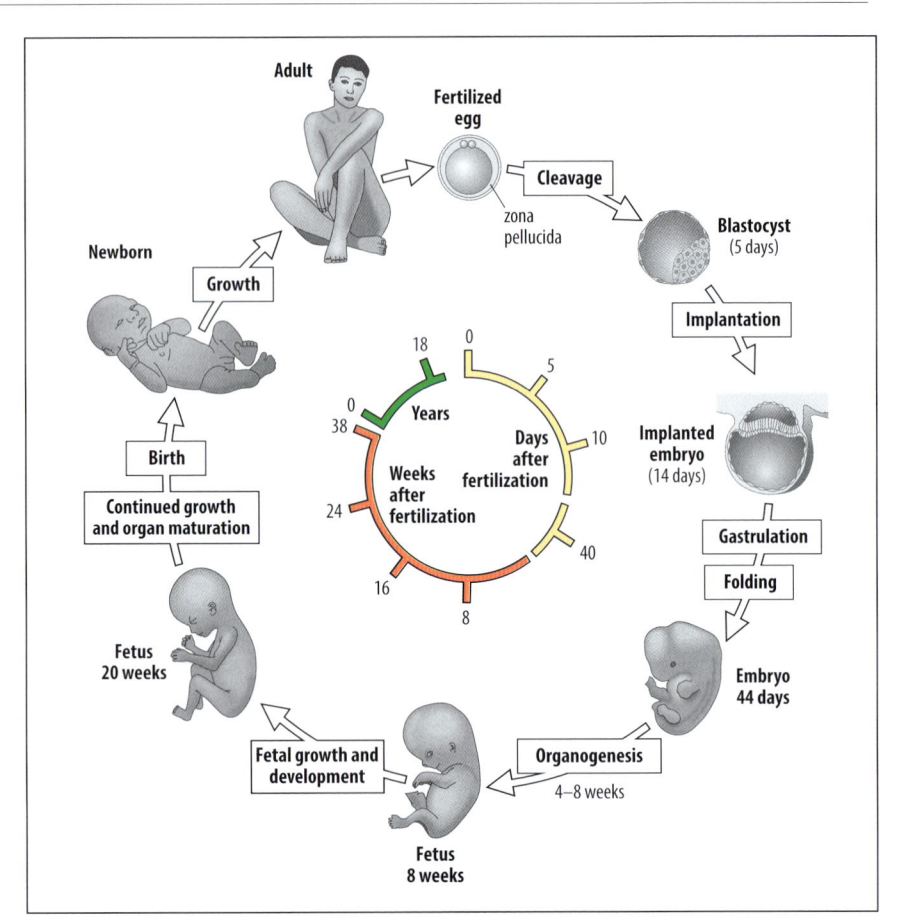

Fig. 3.60 Life cycle of a human being. The egg is fertilized in the oviduct. By the time it reaches the uterus—about 5 days after fertilization—it has developed into a blastocyst; this is released from the zona pellucida and implants in the uterus wall. During the next week, the embryo proper emerges as the epiblast—the upper layer of epithelium in a bilaminar disc of cells, with the amniotic cavity above and the yolk sac cavity below. The other changes that take place at this time relate to the development of the placenta and the formation of protective membranes. Gastrulation occurs during the third week of development, giving rise to the three germ layers. This is followed by the rapid folding of the flat embryo into a three-dimensional body. Organogenesis takes place between 4 and 8 weeks of development. After this, the embryo is known as a fetus and the next 30 weeks are devoted mainly to growth and to the maturation of the tissues and organs. A description of the stages in human development can be found on the website listed in the 'General further reading' section. The photographs show (from top): a fertilized human egg just before the first cleavage; a human embryo; a baby.

Top photograph courtesy Alpesh Dashi. Middle photograph reproduced courtesy of the MRC/Wellcome-funded Human Developmental Biology Resource. Bottom photograph reproduced via the Creative Commons attribution 20.0 generic license, © Tarotastic.

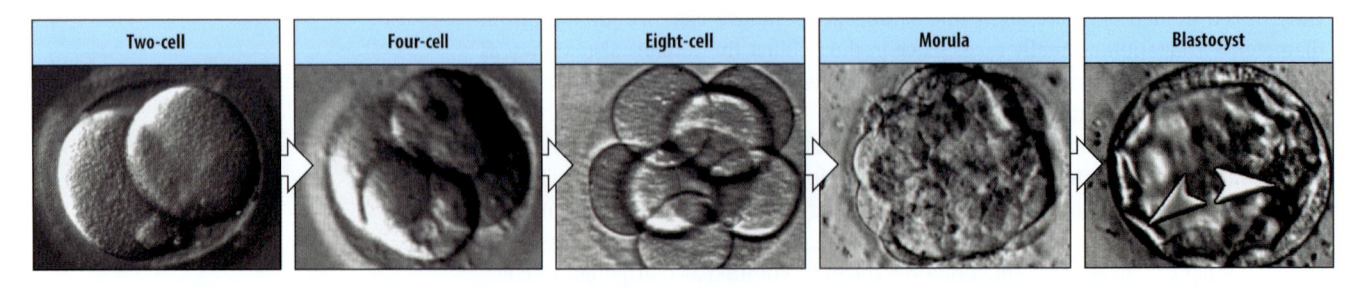

Fig. 3.61 Cleavage in a human embryo developing *in vitro*. The photographs show cleavage of a human egg fertilized *in vitro* from the two-cell stage through to the formation of a blastocyst. The stages are very similar to those in the early mouse embryo (see Fig. 3.48). In the last panel, the gray arrow points to the trophectoderm, the white arrow to the inner cell mass.

*Reproduced from Niakan, K.K., et al.: **Human pre-implantation embryo development.** Development 2012, **139**: 829-841.*

Fig. 3.62 The human embryo at the beginning of gastrulation. Upper panel: the two-layered embryo (the bilaminar germ disc) composed of epiblast (blue) and hypoblast (yellow), and some of its surrounding membranes after implantation in the uterine wall (gray) and before gastrulation has begun. The cavity above the epiblast is the amniotic cavity and the cavity below the hypoblast is the yolk sac, and the whole structure is surrounded by a layer of extra-embryonic mesoderm. This structure is suspended in the chorionic cavity, which is only partly shown here (see Fig. 3.63), and is connected to the wall of the cavity by a stalk of extra-embryonic mesoderm. Lower panel: a cutaway view of the surface of the embryo. The extra-embryonic mesoderm is not shown for simplicity. At gastrulation, the primitive streak begins to form as a furrow in the epiblast, which lies on top of the hypoblast. As in mouse and chick embryos, the node marks the furthest extent of the primitive streak. The oval depression shown in the upper surface of the epiblast marks the position at which the future mouth will form.

mammals, for example rabbits, cows, and pigs, also have flat embryos—rodents are the exception. The embryo proper comes from the epiblast, which after implantation arranges itself into a flat disc of epithelial cells. The underlying epithelial layer is referred to as the hypoblast, which is the equivalent of the mouse visceral endoderm and lines the yolk sac cavity. Another important difference from the mouse embryo is the emergence of the amnion and the amniotic cavity before gastrulation; in the mouse, the amnion forms during gastrulation.

The primitive streak becomes apparent in the upper surface of the epiblast at around 14 days, the beginning of the third week of development. It starts on one side of the epiblast, which now becomes the posterior end of the embryo (see Fig. 3.62, lower panel). A subset of epiblast cells ingress through the primitive streak in the same way as the epiblast cells in chick and mouse embryos, and give rise to the mesoderm and the endoderm. The epiblast cells that do not ingress form the ectoderm. In humans and other mammals, gastrulation is concomitant with rapid cell proliferation that generates the mass of cells that are distributed between the different germ layers.

During gastrulation, the main axes of the body become apparent and the neural plate is induced and folds up to form the neural tube. As the embryo continues to grow and the primordia of the different organs start to form, it undergoes folding, as in the chick and the mouse, followed by ventral closure, to generate the three-dimensional form of the body. The expansion of the embryo also forces it into a convex shape, with both ends of the main body axis curling round; the amniotic cavity enlarges substantially and comes to surround the embryo, whereas the yolk sac cavity becomes much reduced (Fig. 3.63).

Fluid from the amniotic cavity can be collected by a procedure known as amniocentesis between 14 and 16 weeks of gestation. Amniocentesis is one technique that can be used for **prenatal diagnosis** if there are thought to be problems with the pregnancy. The levels of metabolic products in the amniotic fluid that have been produced by the fetus can be measured, and cells that have been sloughed off the fetus into the fluid can be grown in culture and analyzed—for example, for chromosomal abnormalities.

The early stages in human development take longer than the same stages in the mouse. Blastocyst formation, for example, does not occur in humans until 5–6 days after fertilization, whereas in the mouse it occurs between 3 and 4 days; gastrulation takes place at 14 days in the human embryo but at 6 days in the mouse. The laying down of the body plan and the formation of organs is even slower, and it takes about 8 weeks until the structure of the body and organs has been laid down in miniature. After 8 weeks, the human embryo is known as a **fetus**, and during the rest of gestation, known as the fetal period, there is substantial growth accompanied by further maturation of the organs (see Fig. 3.60).

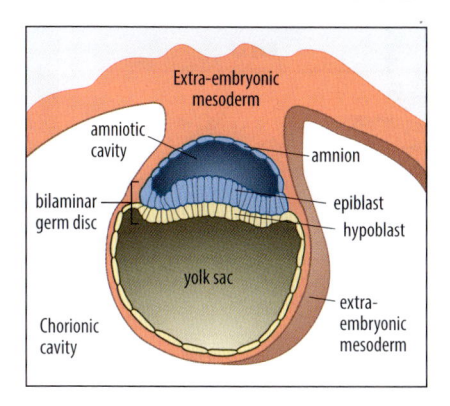

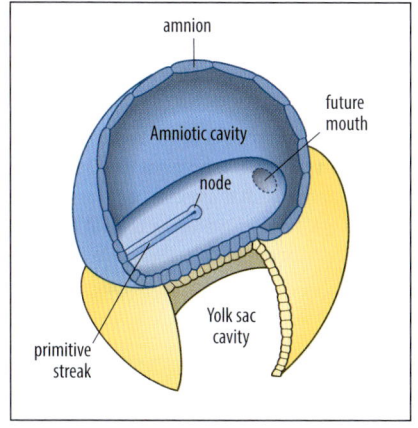

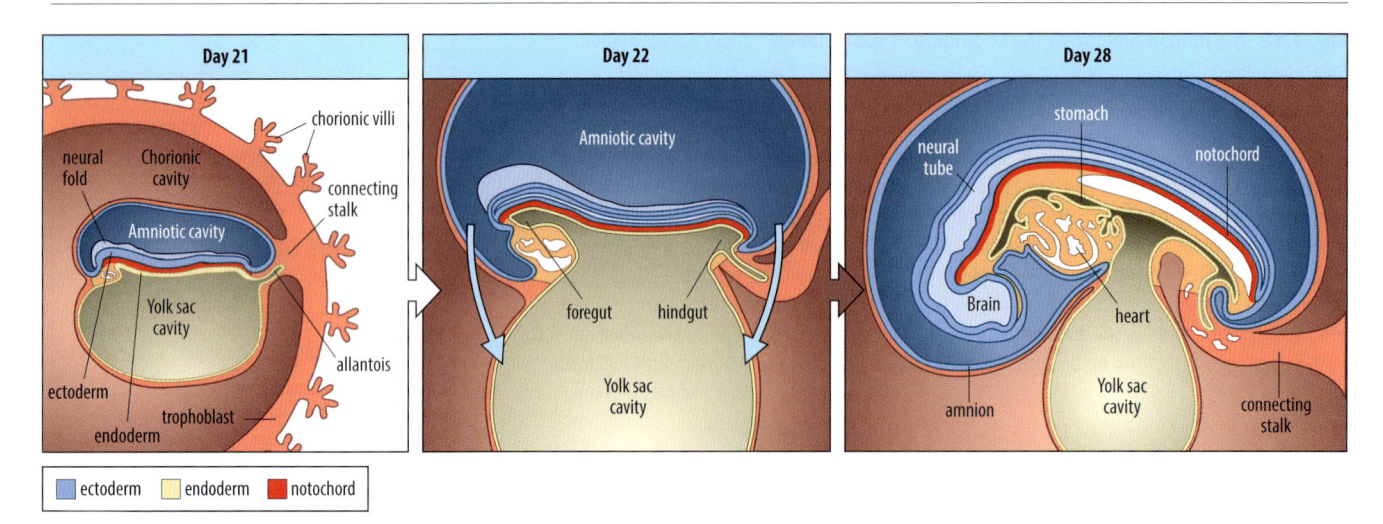

Fig. 3.63 The process of folding in the human embryo and the expansion of the amniotic cavity. Between day 21 and day 28 of development, the flat embryo becomes converted into the three-dimensional form of the body. Lateral folding brings the edges of the flat embryo together ventrally, while expansion of the embryo also causes both ends of the main body axis to curl round (blue arrows in middle panel). This flexure of the main body axis is accompanied by expansion of the amniotic cavity, which comes to surround the embryo, while the yolk sac becomes very reduced (right panel).

MEDICAL BOX 31 Pre-implantation genetic diagnosis

We now take *in vitro* fertilization (IVF) of human eggs as a treatment for infertility almost for granted. The first IVF baby, Louise Brown, was born in 1978. The procedure was developed by UK scientists Robert Edwards and Jean Purdy and the clinician Patrick Steptoe. The development of the technique followed on from the pioneering work of the developmental biologist Anne McLaren, who was the first researcher to accomplish IVF in mice, and involved finding out how to culture and fertilize human oocytes and grow human blastocysts *in vitro*. The development of IVF ushered in the beginning of human developmental biology, as opposed to the descriptive embryology and histology of preserved human clinical material. Edwards received a Nobel prize in 2010, but Purdy and Steptoe had died some years earlier and so could not share the prize.

The successful development of IVF has led to further assisted reproductive technologies that provide clinical treatments. For example, it has become possible to determine the genotype of embryos produced by IVF before implantation without harming the embryo. This procedure is called pre-implantation genetic diagnosis and was developed in the late 1980s for use with fertile couples at risk of passing on a serious genetic disease to their children. The aim was to provide an alternative to pre-natal diagnosis and the potential termination of an affected pregnancy.

For pre-implantation genetic diagnosis, one blastomere is removed from an IVF-produced embryo during its early cleavage *in vitro* at day 3 (Fig. 1). This can be done without affecting

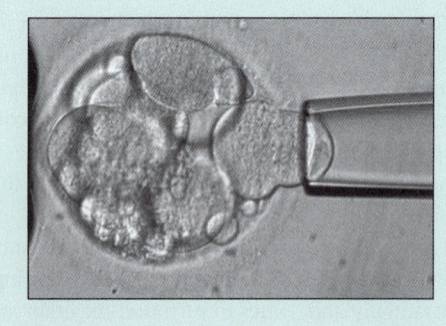

Fig. 1
Photograph courtesy of Dr. Malpani, MD, Malpani Infertility Clinic, Mumbai, India. www.drmalpani.com.

subsequent development because human embryos, like early mouse embryos, are highly regulative. The DNA from this blastomere can be amplified *in vitro* and tested for the presence or absence of mutations known to cause disease. Where there is a known high risk that the parents will pass on a particular genetic disease—for example, when both parents are carriers for the cystic fibrosis gene—pre-implantation genetic diagnosis can be used to ensure that an embryo that would develop the disease is not implanted into the mother (Fig. 2). Biopsies of trophectoderm cells of day 5 human embryos grown in culture are increasingly being used for pre-implantation genetic diagnosis because there are better clinical outcomes.

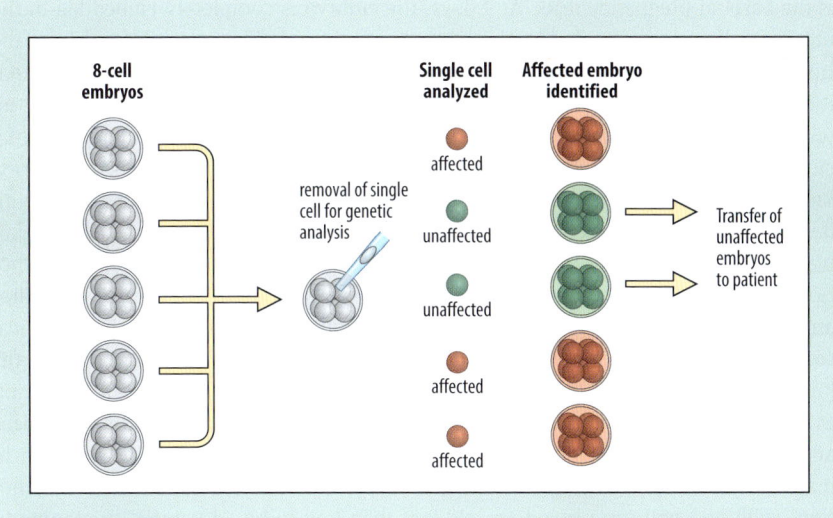

Fig. 2

The demand for pre-implantation genetic diagnosis is likely to increase, because it can be used not only to identify embryos with mutations that will inevitably cause a potentially fatal disease in infancy or childhood, but also to identify embryos with mutations that predispose an individual to disease in later life. An example is the gene *BRCA1*, in which certain mutations predispose women to develop breast and ovarian cancer. *BRCA1* mutations account for 80% of these tumors in women with a genetically inherited predisposition (5–10% of all breast and ovarian cancers). In men, mutations in *BRCA1* are linked to an increased susceptibility to prostate cancer. By determining whether an embryo from a high-risk family carries a *BRCA1* disease allele, the genetic susceptibility to these cancers could, in principle, be eliminated from a family. In the UK, pre-implantation genetic diagnosis has been licensed by the Human Fertilisation and Embryology Authority (HFEA) for more than 60 genetic diseases. Some high-risk couples are now choosing IVF and selection by pre-implantation genetic diagnosis to ensure that they do not pass on a *BRCA1* disease allele to their children.

There are practical and ethical questions in relation to pre-implantation genetic diagnosis, such as which genetic conditions it should be applied to. In the UK, the HFEA made a general ruling allowing decisions to be made on a case-by-case basis for conditions not on their list. An example of the kind of ethical question that has arisen is illustrated by parents who wish to select an IVF embryo with the best human leukocyte antigen (HLA) match to a sibling suffering from a rare blood disease, so that the child conceived through IVF could eventually donate stem cells to their sibling. Cases like this have been approved for pre-implantation genetic diagnosis in the UK.

Another set of assisted reproductive technologies—IVF with mitochondrial replacement therapy—was licensed in the UK in 2017. This therapy is designed to prevent transmission of maternally transmitted genetic diseases caused by mutations in mitochondrial DNA. It involves transfer of the nuclear DNA from an affected egg or zygote to a corresponding enucleated egg or zygote that has normal mitochondria.

3.29 The timing of formation and the anatomy of the human placenta differ from that in the mouse

The formation of the placenta in mammalian embryos is crucial because it attaches the embryo to the mother and provides a fetal–maternal interface that ensures an adequate supply of nutrients and the removal of toxic waste. As in the mouse, the trophectoderm gives rise to the trophoblast cells that make the human placenta, and direct contact is established between the trophoblast cells and the maternal blood. But as both the timing of placenta formation and its anatomy are different in humans, the mouse is not a good model for this aspect of human development.

After the human blastocyst has adhered to the uterine epithelium, the trophectoderm cells that make up the blastocyst wall proliferate and differentiate into trophoblast cells. The attachment of the human blastocyst to the uterine epithelium and invasion of maternal tissue leading to implantation is via the polar trophectoderm, whereas in the mouse these processes occur via the mural trophectoderm. The polar trophoblast cells produce cells that fuse together to form giant **syncytiotrophoblast cells** that invade the uterine wall. The syncytiotrophoblast cells secrete the hormone human chorionic gonadotrophin, which is required to maintain the pregnancy and is the basis of pregnancy tests. At 9 days, the embryo is completely embedded in the uterine wall and surrounded by the single layer of trophoblast cells derived from the trophectoderm—the **cytotrophoblast**—and the thick layer of syncytiotrophoblast that it has produced. Within the syncytiotrophoblast, vacuoles have formed that join up with maternal capillaries, and these blood-filled spaces establish a utero-placental circulation. By the beginning of gastrulation, the embryo is suspended in a cavity, known as the chorionic cavity, that has developed between the embryo together with its attached amniotic and yolk sac cavities and the chorion, the layer of cytotrophoblast cells that originally formed the wall of the blastocyst (see Fig. 3.62). The chorionic cavity is lined by a layer of extra-embryonic mesoderm which also continues around the embryo and forms the connecting stalk that gives rise to the umbilical cord. By 21 days, local proliferation of the cytotrophoblast has produced outgrowths known as chorionic villi (see Fig. 3.63). The chorionic villi fulfill the same function as the placental labyrinth in the mouse. The villi extend into the blood-filled spaces and undergo branching, increasing the surface area of fetal tissue that is bathed in maternal blood. By this stage, the extra-embryonic mesoderm has invaded the chorionic villi and will form blood vessels that then link to blood vessels developing in the embryo.

Figure 3.64 shows the detailed structure of the human placenta at around 12 weeks when the maternal arterial circulation has been established. By this time the amniotic cavity has enlarged, and the amnion has fused with the chorion. The chorionic villi are covered by an outer layer of syncytiotrophoblast cells which are generated from the underlying single layer of cytotrophoblast cells (also known as villous trophoblast). The syncytiotrophoblast is the fetal tissue that directly contacts the maternal blood in the intervillous spaces and is where maternal–fetal transport takes place. At the tip of anchoring villi that extend across the placenta to the uterine wall, the cytotrophoblast generates cells of another type—extravillous trophoblast cells. The extravillous trophoblast cells migrate into the uterine wall, helping to anchor the placenta and remodeling uterine arteries, which increases the maternal blood flow into

Fig. 3.64 Human placenta and umbilical cord. Blood vessels in the chorionic villi connect with blood vessels in the extraembryonic mesoderm of the chorion (the chorionic plate) and umbilical cord and hence to blood vessels in the fetus. The chorionic villi are covered by an outer layer of syncytiotrophoblast cells that have differentiated from the inner layer of villous cytotrophoblast cells. The syncytiotrophoblast cells are in direct contact with maternal blood in the intervillous spaces and transport nutrients and waste. At the tip of anchoring villi, the cytotrophoblast gives rise to extravillous trophoblast cells that migrate into the uterine wall (blue arrows) and remodel uterine arteries so as to increase blood flow into the intervillous spaces. Uterine veins return blood from the placenta to the mother.

Adapted from Novakovic, B., Saffrey R.: **DNA methylation profiling highlights the unique nature of the human placental epigenome.** Epigenomics 2010, **2**: 627-638.

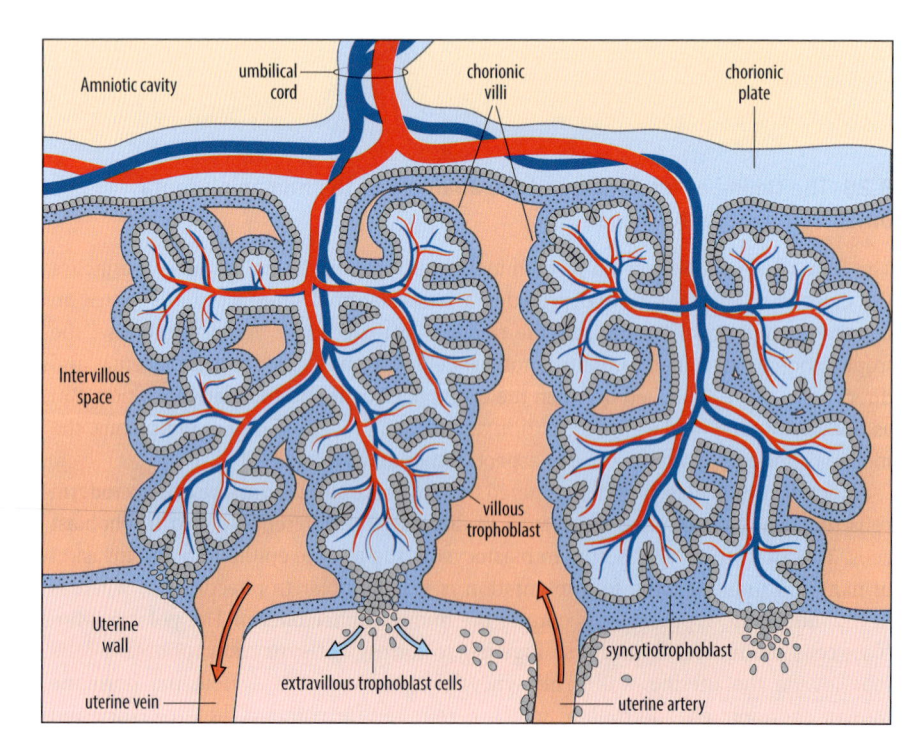

the intravillous spaces. These functions of the extravillous trophoblast cells are crucial for the establishment of a fully functioning placenta. Defective placental development can lead to serious clinical conditions including pre-eclampsia, intrauterine growth retardation, miscarriage, and stillbirth.

In vitro models are now revolutionizing the study of the human placenta. Culture methods pioneered with gut cells have been used to make three-dimensional cultures called organoids that can mimic some properties of human tissues and organs (discussed in Chapter 7). Organoids containing human cytotrophoblast stem cells can be used to study how the stem cells are maintained, and how they differentiate into either syncytiotrophoblast or extravillous trophoblast cells. Such organoids can also be used to study responses to drugs and hormones.

Identical twins are produced when an early embryo splits in half and the extent to which monozygotic twins share the placenta indicates the stage at which twinning occurred (Box 3J). If each fetus has its own placenta and occupies its own chorionic cavity, this indicates that twinning took place at the two-cell stage, whereas if the fetuses share a placenta and a chorionic cavity, twinning must have taken place after blastocyst formation. In some rare cases, fetuses also share a common amnion, indicating that twinning occurred even later by splitting of a single inner cell mass before gastrulation—at the latest, around 14 days after fertilization.

3.30 Some studies of human development are possible but are subject to strict laws

We depend very much on studies of model organisms to help us understand how humans develop, but as we have pointed out, there are some differences between human and other organisms, even other mammals, that require investigation, particularly as they might lie at the root of disease. In some instances, non-human primates could be used as experimental models for humans, but even in this case differences exist, such as the time at which the embryo implants. In the twentieth century, much of our knowledge of human development relied on the analysis of collections of preserved human embryos, most famously the Carnegie collection (see 'General further reading'). These collections have had a great impact on medical embryology and enabled the basic comparison of human embryos and model organism embryos at equivalent developmental stages. There are now many new techniques that can provide insights into the cell and molecular biology of human embryos. For these studies, it is possible to collect some information directly from human embryos with ethical and legal oversight. In most countries such work is subject to strict laws and ethical guidelines that govern assisted reproductive technologies, human embryo research, and the use of human tissues.

As outlined earlier, the ability to carry out IVF with human eggs provides a means of observing and studying cleavage stages and blastocyst formation, and culture techniques for blastocysts have more recently extended the period of development that can be studied. The expression of developmentally important genes has been mapped during these stages. As in mice, genome-wide expression analyses have been carried out on human eggs and two-cell embryos through to the blastocyst stage and revealed both similarities and differences in gene-expression profiles between human and mouse. The more recent advances in culturing later post-implantation stages have enabled study of the way in which the epiblast cells in human embryos arrange themselves to form a flat cell sheet.

Until recently, it was not possible to culture living human embryos beyond about 9 days. And even had it been possible, the experiment would have had to be terminated at 13 days. This is because of the 14-day rule, the limit beyond which human embryos cannot be cultured. This rule is encoded in laws governing research on living human embryos in many countries, including the USA and the UK. The justification

BOX 3J Identical twins

The existence of identical, or monozygotic, twins shows that early human embryos, like early-stage mouse embryos, can regulate. Identical twins are produced when an early embryo splits in half—unlike non-identical dizygotic twins, which are the result of the fertilization of two separate eggs.

Monozygotic twinning can occur at various stages in early development and the stage at which it has occurred can be deduced from the arrangement of the extra-embryonic membranes and the placenta. Identical twins with separate amniotic and chorionic cavities and separate placentas are produced by the splitting of the morula during cleavage and before implantation, as shown here for twinning at the two-cell stage (Fig. 1, left). More commonly, twins have their own amniotic cavities but share a chorionic cavity and a placenta, which represents splitting of the very early inner cell mass inside the blastocyst (Fig. 1, center). A very small percentage of identical twins (about 4%) share both amniotic and chorionic cavities and a placenta, which means that twinning must have occurred quite late, with splitting of the inner cell mass at epiblast stages, between 9 and 15 days after fertilization (Fig. 1, right).

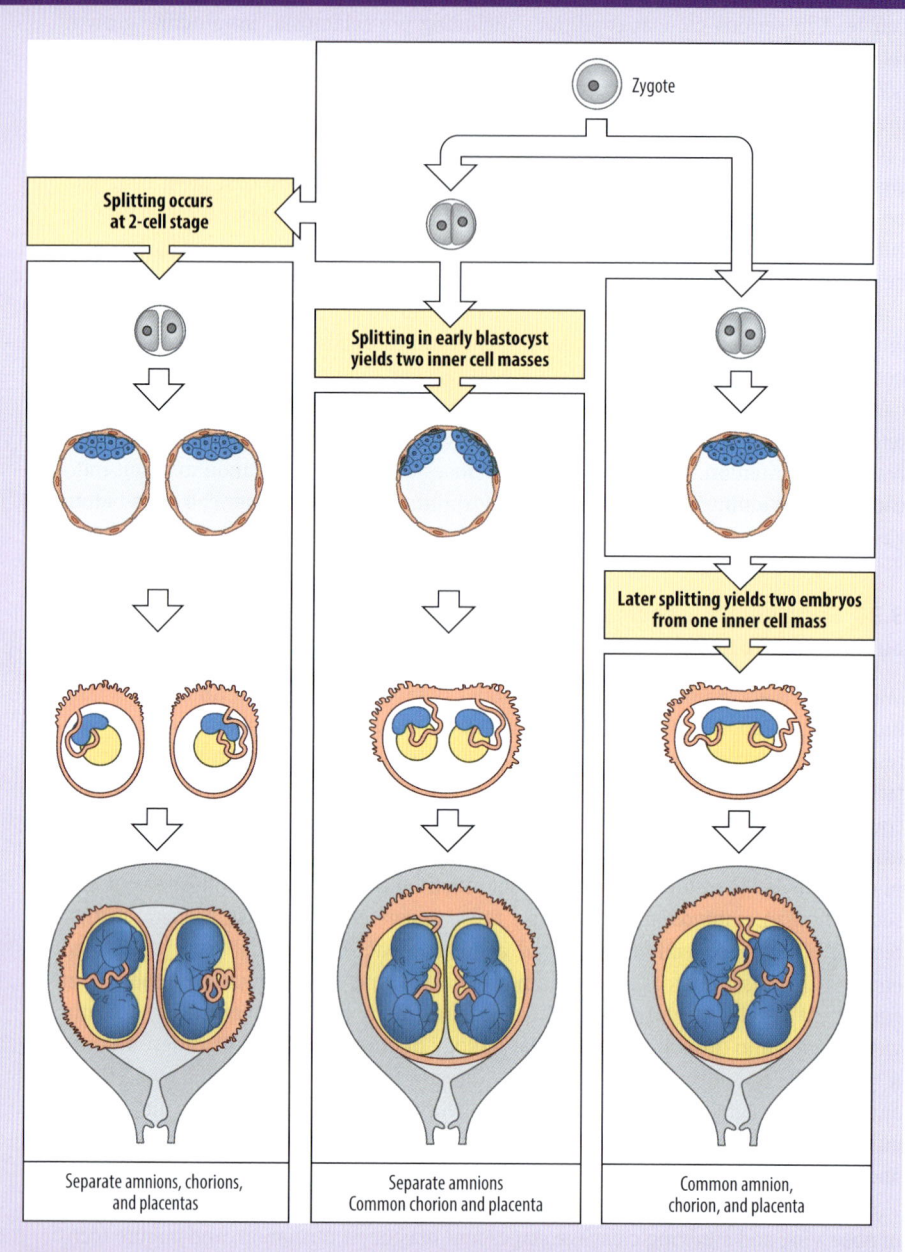

Fig. 1

Adapted from Larsen, W.J.: Human Embryology. 2nd edn. Edinburgh: Churchill Livingstone, 1997.

for this time limit is that it is the latest stage at which twinning can occur, and therefore it can be argued that this is the stage at which the individual is determined. Nevertheless, the recent success in culturing human embryos for 12–13 days has ignited debate about whether the 14-day rule should be revised.

The collection and use of later-stage embryonic and fetal human tissue from clinical interventions and terminated pregnancies are governed by other laws. Studies using modern techniques can be carried out on these donated embryos. For example, a single-cell transcriptomic analysis has been carried out on a gastrulating human embryo (embryos at this stage are rarely obtained for study) and showed that the general outline of events during gastrulation is similar to that in the mouse, although differences in the signal proteins used were found. Gene-expression analyses in embryos and fetal tissues at later stages using, for example, *in situ* hybridization, again have revealed detailed differences between mouse and human. These differences mean that we should not assume that findings on mouse embryos will always apply to human embryos and so we need to study human embryos directly. A large international program aims to build a human developmental cell atlas using imaging, single-cell transcriptomics, and computation, with the ambition of integrating spatial information over time at all stages from fertilized egg to birth. Human tissue can also be used for CHIP-seq analysis to identify gene targets of transcription factors.

As we described for our model vertebrates, several genetics-based techniques can be used to study development, and some of these approaches can be applied to study human development. New or inherited mutations in humans that lead to congenital conditions have uncovered developmentally important genes—an example of traditional forward genetics. Because the human population is very large, exceptionally rare events can be detected, and because differences in human patients are scrutinized in detail, mutations that produce rather subtle phenotypes can be identified. In this way, clinical genetics studies have been able to match specific gene mutations with very specific phenotypes in humans. In some cases, mutations that result in the substitution of a single amino acid in a protein by another amino acid have dramatically different effects, depending on which amino acid is changed and which amino acid replaces it. Examples of this are mutations that result in different amino acids being included in the receptor protein for a signal protein that governs limb development. Precise changes in the amino acid sequence are associated with specific disease phenotypes in which the fingers are fused together to different extents in affected people.

The CRISPR-Cas9 genome-editing technique, which provides a simple and efficient way to carry out reverse genetics by inducing mutation in a given gene (see Box 1C), has been applied to fertilized human eggs *in vitro*. In a landmark study, Oct4, one of the pluripotency-determining transcription factors with a central role in early mouse development was targeted in a fertilized human egg. Development was compromised from the eight-cell stage onwards, whereas when the gene encoding Oct4 was similarly targeted in a fertilized mouse egg, development was not comprised until blastocyst formation. These results suggest that Oct4 functions earlier in human development than in mouse development.

The CRISPR-Cas9 system also has the potential to repair existing mutations in human eggs. However, given the possibility of off-target effects and the risks involved in modifying the human genome, such procedures are not considered either ethical or legal at present. In 2018, a Chinese scientist explained to a shocked audience that he had used CRISPR-Cas9 in IVF embryos to try to modify the genomes of the two resulting girl babies to make them more resistant to HIV. This generated a storm of protest from researchers and the public; the scientist lost his job, and was given a large fine and a prison sentence of 3 years by a Chinese court for illegal medical practice. Most laws governing assisted reproductive technologies ban any genetic modification of human embryos.

Given the ethical concerns about carrying out research on living human embryos, increasing efforts are being made to develop *in vitro* systems such as gastruloids, neuruloids, and organoids derived from human stem cells to study human development (discussed in Chapter 7).

178 Chapter 3 Vertebrate development I: establishing the body plan

> **SUMMARY**
>
> For most of the twentieth century, human developmental biology could only be studied in a limited way, through collections of clinical material. The advent of IVF enabled the very earliest stages of human development, from the fertilized egg to the blastocyst stage, to be observed. The stages of early human development that lay down the body plan are similar to those in the mouse, but the rate of development is slower. One major difference from the mouse is that, like the chick and most mammals other than rodents, the human epiblast is flat, not cup-shaped. The timing of formation and the anatomy of the placenta are also different from that in the mouse. Although the general set of signal proteins and transcription factors involved in laying down the body plan is similar to that in the mouse, there are differences. There is an increasing emphasis on the need to study human development directly. Such studies raise ethical issues and *in vitro* systems are being devised to study human development without using embryos.

Summary to Chapter 3

- All vertebrates have a similar basic body plan. The defining vertebrate structures are the vertebral column, which surrounds the spinal cord, and the bony or cartilaginous skull that forms the head and encloses the brain.
- The vertebrate body has three main axes of symmetry—the antero-posterior axis running from head to tail, the dorso-ventral axis running from back to belly, and bilateral symmetry around the dorsal midline, so that outwardly the right and left sides are mirror images of each other. Some internal organs, such as lungs, kidneys, and gonads, are also present as symmetrically paired structures, although their position and morphology may differ, but single organs, such as the heart and liver, are arranged asymmetrically with respect to the dorsal midline.
- Early vertebrate embryos pass through a set of developmental stages—cleavage, gastrulation, and neurulation—to form embryos that broadly resemble each other in a stage known as the phylotypic stage.
- There are considerable differences in the development of the different model organisms before the phylotypic stage; the differences relate to how and when the axes are set up, and how the germ layers are established. These differences are mainly due to the different modes of reproduction of the animals, which affect the amount of yolk in the egg and hence the consequent form of the earliest embryo.
- Structures typical of the different vertebrate groups, such as fins, beaks, wings, and tails, develop after the phylotypic stage.
- The body pattern and the initial organ primordia are laid down when the embryo is still very small. Growth in size takes place at later stages.
- The mechanisms underlying development can only be studied by disturbing the normal developmental process in specific ways and observing what happens.
- Techniques for interfering with development can be broadly divided into two types: experimental embryological techniques that manipulate the embryo by removing or adding cells to cleavage-stage embryos or transplanting blocks of cells from one embryo to another; and genetics-based techniques that disturb the expression of developmentally important genes by mutation, gene silencing, overexpression, or misexpression.
- In *Xenopus* and zebrafish, early development involves localized maternal determinants and cell-cell interactions; in chick, maternal determinants may be present at the very early stages but any role is unknown. Maternal determinants have not been identified in the mouse embryo and development is governed by very early zygotic gene activation and cell-cell interactions.
- It is possible to construct a fate map in the early embryo for the three germ layers—mesoderm, endoderm, and ectoderm. The fate maps of our four model vertebrates have strong

similarities, the major differences being due to the amount of yolk in the egg, which affects the cleavage divisions, and the way in which gastrulation occurs.

- In *Xenopus*, ectoderm and endoderm are specified by maternal determinants, but the mesoderm is induced from the animal region in the blastula by signals from the vegetal region before gastrulation begins. The dorsal-most mesoderm is established as the embryonic organizer (the Spemann organizer). The equivalent is the shield region in zebrafish, Hensen's node in the chick, and the node in the mouse. Transplants of these regions to another early embryo can induce a complete (or partial in the case of the mouse) secondary embryo.
- In chick and mouse, specification of mesoderm and endoderm occurs later than in *Xenopus* and zebrafish, at gastrulation.
- The process of gastrulation involves extensive shape changes and cell movements and converts the very early embryo (blastula, blastoderm, or epiblast) into the three-dimensional vertebrate body plan, with the mesoderm and endoderm internalized and ectoderm on the outside.
- In *Xenopus*, gastrulation occurs by the involution of coherent sheets of mesoderm and endoderm cells at the blastopore. In zebrafish, gastrulation occurs by the involution of the rim of the dome-shaped blastoderm. In chick, mouse, and human, gastrulation occurs by the ingression of epiblast cells along the future dorsal midline of the embryo, forming a furrow called the primitive streak. The internalized cells become mesoderm or endoderm and spread under the epiblast to form the germ layers and their derivatives.
- Overlapping with gastrulation is the specification of the central nervous system (brain and spinal cord)–from part of the ectoderm. The remaining ectoderm becomes the epidermis.
- The establishment of the antero-posterior axis and patterning of all three germ layers along the dorso-ventral axis depends on signaling by the Spemann organizer (the embryonic organizer) in *Xenopus*, and by the shield region in zebrafish. In chick and mouse the antero-posterior axis is determined by different mechanisms.
- When grafted to the ventral side of an early *Xenopus* gastrula, the Spemann organizer induces a second body axis with a head and a complete dorso-ventral axis, and a twinned embryo develops. The shield can likewise induce a twinned fish embryo with a second body axis, and Hensen's node is also capable of inducing a complete body axis, including a second neural plate when transplanted to another chick embryo. Transplantation of the mouse node, however, produces a twinned embryo with a trunk but no head, as the head is specified by a different mechanism.
- Early stages in human development are similar to mouse development but the mouse is not a good model for development of the placenta.
- There is an increasing emphasis on the need to study human development directly, but such studies raise ethical issues.

End of chapter questions

Check your understanding of the content of this chapter by attempting the following long-answer concept questions.

1. All vertebrates share certain features; what are those features? Referring to Figure 3.2, identify those features in the row of illustrations that shows the phylotypic stage.

2. At the neurula stage in vertebrate embryos, the precursors to several adult structures first form. What structures will form from: neural plate; somites; intermediate mesoderm; splanchnopleural mesoderm; notochord?

3. How does the way the embryo is nourished affect early development? Compare all four model vertebrates. Include discussion of the relative amounts of yolk in the eggs and the importance of extra-embryonic tissues.

4. What are some of the features that would make an animal species a good 'model organism' for the study of development? For example, what features make a chicken an attractive model for development compared with a frog? By contrast, what advantages has the mouse offered over the chick?

5. Briefly summarize two ways of producing transgenic mice. What additional technique is now widely used to make knock-out mice and what are its advantages?

6. You have discovered that a particular gene encoding a transcription factor is expressed in the developing liver at a certain stage in both mouse and chick embryos. How could you set about finding its function? Describe the experiments you would carry out in both model organisms.

7. After fertilization, all embryos undergo cleavage. Explain how, in the *Xenopus* embryo, the planes of cleavage mean that future dorsal and ventral regions are already specified by the four-cell stage.

8. The organization of the *Xenopus* embryo begins with the establishment of the dorso-ventral axis. A crucial component of this event is the accumulation of β-catenin in cells of the dorsal marginal zone. Describe the series of events that lead to this accumulation, starting with the maternal deposition of proteins and RNAs in the oocyte.

9. The Nieuwkoop center was defined by various experiments in *Xenopus* embryos. Describe the results of the following experiments:

(a) Cells from the animal region of a late blastula are cultured together with cells from the dorsal vegetal region.

(b) Cells from the dorsal vegetal region of a 32-cell embryo are grafted into the ventral side of another 32-cell stage embryo.

10. What is the result of the following experiment? The dorsal lip of the blastopore from an early *Xenopus* gastrula is grafted into the ventral marginal zone of another early gastrula (see also Section 1.4 for a classic experiment carried out with newt embryos). What is the explanation of this result?

11. Describe the developmental roles of the VegT protein found in fertilized eggs and early embryos in *Xenopus* embryonic development. Include in your answer what kind of protein it is; whether it is maternally supplied or is transcribed and translated from the embryo's own genes; where it is located in the fertilized egg; and what germ layers it is involved in specifying, whether alone or in combination with other proteins.

12. Early amphibian development is driven largely by maternally supplied factors until the mid-blastula transition. What is the mid-blastula transition? What is the model that explains how it occurs? What experimental evidence supports this model?

13. Describe how the process of gastrulation in *Xenopus* leads to the establishment of the three germ layers in their proper positions in the final body plan and leads to the development of the main body axes.

14. What is the posterior marginal zone of the chick embryo? What is its significance to chick development? Describe the external forces that lead to its formation and the signal molecules associated with it.

15. Distinguish between the following regions in the chick embryo: (1) epiblast and hypoblast; (2) posterior marginal zone, Koller's sickle, and Hensen's node; (3) notochord and somites; (4) head process mesoderm and notochord.

16. Compare the activities of the Spemann organizer of the amphibian embryo, the shield region of the zebrafish, and the chick Hensen's node. How were these activities discovered?

17. What are the key differences in the way the mesoderm forms in the chick and the mouse, in comparison to *Xenopus*?

18. Distinguish between the inner cell mass and the trophectoderm in the mouse blastocyst. Outline the events that will lead to formation of the epiblast and of the primitive streak.

19. Describe the events that lead to the establishment of the antero-posterior axis in the mouse.

20. What are the main similarities and differences between an early mouse embryo and an early human embryo?

21. Describe the development and function of human placental cytotrophoblast cells and their derivatives. Why is it important to study organoids that contain human cytotrophoblast cells?

22. What is the legal limit (in days) for the culture of a living human embryo *in vitro* in many countries? What are the scientific reasons behind this limit, and why are some scientists now pressing for the law to be changed to allow a longer limit? What alternatives now exist for studying aspects of early human development *in vitro* without using human embryos?

■ General further reading

Bard, J.B.L.: *Embryos: Color Atlas of Development*. London: Wolfe, 1994.

Carlson, B.M.: *Patten's Foundations of Embryology*. New York: McGraw-Hill, 1996.

Descriptions of developmental stages of frog and zebrafish embryos, and other resources

Xenopus: developmental stages: Xenbase—Nieuwkoop and Faber stage series [http://www.xenbase.org/anatomy/alldev.do] (date accessed 3 September 2024); movies of cleavage and gastrulation [http://www.xenbase.org] (look under Anatomy & Development: date accessed 3 September 2024).

Zebrafish: developmental stages: Karlstrom Lab ZFIN Embryonic Developmental Stages [https://zfin.org/zf_info/zfbook/stages] (date accessed 3 September 2024); developmental movie: Karlstrom, R.O., Kane, D.A.: **A flipbook of zebrafish embryogenesis**. *Development* 1996, **123**: 1–461; Karlstrom Lab [http://www.bio.umass.edu/biology/karlstrom/] (date accessed 3 September 2024).

Chick: list and description of Hamilton and Hamburger developmental stages: UNSW Embryology—Chicken Developmental Stages [https://embryology.med.unsw.edu.au/embryology/index.php/Chicken_stages] (date accessed 3 September 2024).

Mouse: Edinburgh Mouse Atlas Project [http://www.emouseatlas.org] (date accessed 3 September 2024).

Human: The Multidimensional Human Embryo [embryo.soad.umich.edu/] (date accessed 3 September 2024); UNSW Embryology: Carnegie staging of human embryos [http://embryology.med.unsw.edu.au] (date accessed 3 September 2024).

deBakker, B.S., deJong, K.H., Hagoort, J., deBree, K., Besselink, C.T., deKanter, F.E., Veldhuis, T., Bais, B., Schildmeijer, R., Ruijter, J.M., Oostra, R.J., Christoffels, V.M., Moorman, A.F.: **An interactive three-dimensional digital atlas and quantitative database of human development**. *Science* 2016, **354**: aag0053. [https://www.3dembryoatlas.com] (date accessed 3 September 2024).

Haniffa, M., *et al.*; Human Cell Atlas Developmental Biological Network. **A roadmap for the Human Developmental Cell Atlas**. *Nature* 2021, **597**: 196–205.

Schoenwolf, G.C., Bleyl, S.B., Brauer, P.R., Francis-West, P.H.: *Larsen's Human Embryology*. 6th edn. Philadelphia, PA: Elsevier; 2021.

International Mouse Phenotyping Consortium [https://www.mousephenotype.org] (date accessed 3 September 2024).

■ Section further reading

3.1 All vertebrate embryos pass through a broadly similar set of developmental stages

Bard, J.B.L.: *Embryos: Color Atlas of Development*. London: Wolfe, 1994.

Carlson, B.M.: *Patten's Foundations of Embryology*. New York: McGraw-Hill, 1996.

3.2 Normal development can be visualized by various types of imaging, by detecting patterns of gene and protein expression, and by making fate maps

Caviglia, S., Unterweger, I.A., Gasiūnaitė, A., Vanoosthuyse, A.E., Cutrale, F., Trinh, L.A., Fraser, S.E., Neuhauss, S.C.F., Ober, E.A.: **FRaeppli: a multispectral imaging toolbox for cell tracing and dense tissue analysis in zebrafish**. *Development* 2022, **149**: dev199615.

Chatterjee, K., Pratiwi, F.W., Wu, F.C.M., Chen, P., Chen, B.-C.: **Recent progress in light sheet microscopy for biological applications**. *Appl. Spectrosc.* 2018, **72**: 1137–1169.

Choi, H.M.T., *et al.*: **Mapping a multiplexed zoo of mRNA expression**. *Development* 2016, **143**: 3632–3637.

Crosetto, N., Bienko, M., vanOudenaarden, A.: **Spatially resolved transcriptomics and beyond**. *Nat. Rev. Genet.* 2015, **16**: 57–66.

Huisken, J., Swoger, J., DelBene, F., Wittbrodt, J., Stelzer, E.H.: **Optical sectioning deep inside live embryos by selective plane illumination microscopy**. *Science* 2004, **305**: 1007–1009.

Lohoff, T., Ghazanfar, S., Missarova, A., Koulena, N., Pierson, N., Griffiths, J.A., Bardot, E.S., Eng, C.-H.L., Tyser, R.C.V., Argelaguet, R., Guibentif, C., Srinivas, S., Briscoe, J., Simons, B.D., Hadjantonakis, A.-K., Göttgens, B., Reik, W., Nichols, J., Cai, J., Marioni, J.C.: **Integration of spatial and single-cell transcriptomic data elucidates mouse organogenesis**. *Nat. Biotechnol.* 2022, **40**: 74–85.

Marx, V.: **Method of the Year: spatially resolved transcriptomics**. *Nat. Methods* 2021, **18**: 9–14.

Metscher, B.D.: **MicroCT for developmental biology: a versatile tool for high-contrast 3D imaging at histological resolutions**. *Dev. Dyn.* 2009, **238**: 632–640.

Pillai, S., Chellappan, S.P.: **ChIP on chip and ChIP-Seq assays: genome-wide analysis of transcription factor binding and histone modifications**. *Methods Mol. Biol.* 2015, **1288**: 447–472.

Sharpe, J.: **Optical projection tomography as a new tool for studying embryo anatomy**. *J. Anat.* 2003, **202**: 175–181.

Southall, T.D., Brand, A.H.: **Chromatin profiling in model organisms**. *Brief. Funct. Genomics Proteomics* 2007, **6**: 133–140.

Tang, F., Barbacioru, C., Wang, Y., Nordman, E., Lee, C., Xu, N., Wang, X., Bodeau, J., Tuch, B.B., Siddiqui, A., Lao, K., Surani, M.A.: **mRNA-seq whole-transcriptome analysis of a single cell**. *Nat. Methods* 2009, **6**: 377–382.

Truong, T.V., Supatto, W., Koos, D.S., Choi, J.M., Fraser, S.E.: **Deep and fast live imaging with two-photon scanned light-sheet microscopy**. *Nat. Methods* 2011, **8**: 757–760.

Turnbull, D.H., Mori, S.: **MRI in mouse developmental biology**. *NMR Biomed.* 2007, **20**: 265–274.

Wang, Z., Gerstein, M., Snyder, M.: **RNA-Seq: a revolutionary tool for transcriptomics**. *Nat. Rev. Genet.* 2009, **10**: 57–63.

Box 3A Fate mapping and lineage tracing reveal which cells in which parts of the early *Xenopus* embryo give rise to particular structures in the tadpole

Dale, L., Slack, J.M.: **Fate map for the 32-cell stage of *Xenopus laevis***. *Development* 1987, **99**: 527–551.

Zernicka-Goetz, M., Pines, J., Ryan, K., Siemering, K.R., Haseloff, J., Evans, M.J., Gurdon, J.B.: **An indelible lineage marker for *Xenopus* using a mutated green fluorescent protein**. *Development* 1996, **122**: 3719–3724.

3.3 Disrupting the developmental process in specific ways helps to discover underlying mechanisms

Eisen, J.S., Smith, J.C.: **Controlling morpholino experiments: don't stop making antisense**. *Development* 2008, **135**: 1735–1743.

Wang, H., Yang, H., Shivalila, C.S., Dawlaty, M.M., Cheng, A.W., Zhang, F., Jaenisch, R.: **One-step generation of mice carrying mutations in multiple genes by CRISPR/Cas-mediated genome engineering**. *Cell* 2013, **153**: 910–918.

3.4 Not all techniques are equally applicable to all vertebrates

Bedzhov, I., Leung, C.Y., Bialecka, M., Zernicka-Goetz, M.: *In vitro* **culture of mouse blastocysts beyond the implantation stages**. *Nat. Protoc.* 2014, **9**: 2732–2739.

Behringer, R., Gertsenstein, M., Vintersten Nagy, K., Nagy, A.: *Manipulating the Mouse Embryo: A Laboratory Manual*. 4th edn. New York: Cold Spring Harbor Laboratory Press, 2014.

Sharpe, P., Mason, I.: *Molecular Embryology Methods and Protocols*. 2nd edn. New York: Humana Press, 2008.

Stern, C.D.: **The chick: a great model system just became even greater**. *Dev. Cell* 2004, **8**: 9–17.

Wu, R.S., Lam, I.I., Clay, H., Duong, D.N., Deo, R.C., Coughlin, S.R.: **A rapid method for directed gene knockout for screening in G0 zebrafish**. *Dev. Cell* 2018, **46**: 112–125.

Box 3B Ways of making transgenic mice

Lanigan, T.M., Kopera, H.C., Saunders, T.L.: **Principles of genetic engineering**. *Genes* 2020, **11**: 291.

Box 3C Large-scale mutagenesis screens for recessive mutations in zebrafish

Stemple, D.: **Tilling—a high throughput harvest for functional genomics**. *Nat. Rev. Genet.* 2004, **5**: 1–6.

3.5 The frog *Xenopus laevis* is the model amphibian for studying development of the body plan

Hausen, P., Riebesell, H.: *The Early Development of* Xenopus laevis. Berlin: Springer-Verlag, 1991.

Nieuwkoop, P.D., Faber, J.: *Normal Tables of* Xenopus laevis. Amsterdam: North Holland, 1967.

3.6 The earliest stages in *Xenopus* development are controlled by maternal factors

King, M.L., Messitt, T.J., Mowry, K.L.: **Putting RNAs in the right place at the right time: RNA localization in the frog oocyte**. *Biol. Cell* 2005, **97**: 19–33.

Sheets, M.D., Fox, C.A., Dowdle, M.E., Blaser, S.I., Chung, A., Park, S.: **Controlling the messenger: regulated translation of maternal mRNAs in *Xenopus laevis* development**. *Adv. Exp. Med. Biol.* 2017, **953**: 49–82.

Tokmakov, A.A., Stefanov, V.E., Sato, K.I.: **Dissection of the ovulatory process using *ex vivo* approaches**. *Front. Cell Dev. Biol.* 2020, **8**: 605379.

White, J.A., Heasman, J.: **Maternal control of pattern formation in *Xenopus laevis***. *J. Exp. Zool. B Mol. Dev. Evol.* 2008, **310**: 73–84.

3.7 Cortical rotation specifies the future dorsal side of the *Xenopus* embryo

Cha, S.W., Tadjuidje, E., Tao, Q., Wylie, C., Heasman, J.: **Wnt5a and Wnt11 interact in a maternal Dkk1-regulated fashion to activate both canonical and non-canonical signaling in *Xenopus* axis formation**. *Development* 2008, **135**: 3719–3729.

Gerhart, J., Danilchik, M., Doniach, T., Roberts, S., Browning, B., Stewart, R.: **Cortical rotation of the *Xenopus* egg: consequences for the antero-posterior pattern of embryonic dorsal development**. *Development (Suppl.)* 1989, **107**: 37–51.

Houston, D.W.: **Cortical rotation and messenger RNA localization in *Xenopus* axis formation**. *WIREs Dev. Biol.* 2012, **1**: 371–388.

Houston, D.W., Elliott, K.L., Coppenrath, K., Wlizla, M., Horb, M.E.: **Maternal Wnt11b regulates cortical rotation during *Xenopus* axis formation: analysis of maternal-effect *wnt11b* mutants**. *Development* 2022, **149**: dev200552.

Rowning, B.A., Wells, J., Wu, M., Gerhart, J.C., Moon, R.T., Larabell, C.A.: **Microtubule-mediated transport of organelles and localization of beta-catenin to the future dorsal side of *Xenopus* eggs**. *Proc. Natl Acad. Sci. USA* 1997, **94**: 1224–1229.

Tao, Q., Yokota, C., Puck, H., Kofron, M., Birsoy, B., Yan, D., Asashima, M., Wylie, C.C., Lin, X., Heasman, J.: **Maternal *wnt11* activates the canonical wnt signaling pathway required for axis formation in *Xenopus* embryos**. *Cell* 2005, **120**: 857–871.

Weaver, C., Kimelman, D.: **Move it or lose it: axis specification in *Xenopus***. *Development* 2004, **131**: 3491–3499.

Box 3E The Wnt/β-catenin signaling pathway

Nusse, R., Varmus, H.: **Three decades of Wnts: a personal perspective on how a scientific field develops**. *EMBO J.* 2012, **31**: 2670–2684.

Wiese, K.E., Nusse, R., vanAmerongen, R.: **Wnt signalling: conquering complexity**. *Development* 2018, **145**: dev165902.

3.8 The Nieuwkoop center, a signaling center, develops on the dorsal side of the blastula

Ding, Y., Ploper, D., Sosa, E.A., Colozza, G., Moriyama, Y., Benitez, M.D.J., Zhang, K., Merkurjev, D., DeRobertis, E.M.: **Spemann organizer transcriptome induction by early beta-catenin, Wnt, Nodal, and Siamois signals in *Xenopus laevis***. *Proc. Natl Acad. Sci. USA* 2017, **114**: E3081–E3090.

Glimich, R.L., Gerhart, J.C.: **Early cell interactions promote embryonic axis formation in *Xenopus laevis***. *Dev. Biol.* 1984, **104**: 117–130.

Nieuwkoop, P.D.: **The formation of the mesoderm in urodelean amphibians. I. Induction by the endoderm**. *Wilhelm Roux Arc. Entw Mech. Org.* 1969, **162**: 341–373.

Vonica, A., Gumbiner, B.M.: **The *Xenopus* Nieuwkoop center and Spemann–Mangold organizer share molecular components and a requirement for maternal Wnt activity**. *Dev. Biol.* 2007, **312**: 90–102.

3.9 Control of development is transferred from the maternal to the zygotic genome at the mid-blastula transition

Amodeo, A.A., Jukam, D., Straight, A.F., Skotheim, J.M.: **Histone titration against the genome sets the DNA-to-cytoplasm threshold for the *Xenopus* midblastula transition**. *Proc. Natl Acad. Sci. USA* 2015, **112**: E1086–E1095.

Jukam, D., Kapoor, R.R., Straight, A.F., Skotheim, J.M.: **The DNA-to-cytoplasm ratio broadly activates zygotic gene expression in *Xenopus***. *Curr. Biol.* 2021, **31**: 4269–4281.e8.

Tadros, W., Lipshitz, H.D.: **The maternal-to-zygotic transition: a play in two acts**. *Development* 2009, **136**: 3033–3042.

3.10 The axes of the blastula are transformed by gastrulation into the body axes of the tailbud embryo

Martinez Arias, A., Steventon, B.: **On the nature and function of organizers**. *Development* 2018, **145**: dev159525.

GerhartJ.: **Changing the axis changes the perspective**. *Dev. Dyn.* 2002, **225**: 380–383.

3.11 Mesoderm is induced and its patterning is initiated by signals from vegetal and dorsal regions of the blastula

Agius, E., Oelgeschläger, M., Wessely, O., Kemp, C., DeRobertis, E.M.: **Endodermal Nodal-related signals and mesoderm induction in *Xenopus***. *Development* 2000, **127**: 1173–1183.

Charney, R.M., Paraiso, K.D., Blitz, I.L., Cho, K.W.Y.: **A gene regulatory program controlling early *Xenopus* mesendoderm formation: network conservation and motifs**. *Semin. Cell Dev. Biol.* 2017, **66**: 12–24.

De Robertis, E.M., Larrain, J., Oelgeschläger, M., Wessely, O.: **The establishment of Spemann's organizer and patterning of the vertebrate embryo**. *Nat. Rev. Genet.* 2000, **1**: 171–181.

Dupont, S., Zacchigna, L., Cordenonsi, M., Soligo, S., Adorno, M., Rugge, M., Piccolo, S.: **Germ-layer specification and control of cell growth by Ectodermin, a Smad4 ubiquitin ligase**. *Cell* 2005, **121**: 87–99.

Heasman, J.: **Patterning the early *Xenopus* embryo**. *Development* 2006, **133**: 1205–1217.

Kimelman, D.: **Mesoderm induction: from caps to chips**. *Nat. Rev. Genet.* 2006, **7**: 360–372.

Kofron, M., Demel, T., Xanthos, J., Lohr, J., Sun, B., Sive, H., Osada, S.-I., Wright, C., Wylie, C., Heasman, J.: **Mesoderm induction in *Xenopus* is a zygotic event regulated by maternal VegT via TGFβ growth factors**. *Development* 1999, **126**: 5759–5770.

Mir, A., Kofron, M., Zorn, A.M., Bajzer, M., Haque, M., Heasman, J., Wylie, C.C.: **FoxI1e activates ectoderm formation and controls cell position in the *Xenopus* blastula**. *Development* 2007, **134**: 779–788.

Xanthos, J.B., Kofron, M., Wylie, C., Heasman, J.: **Maternal VegT is the initiator of a molecular network specifying endoderm in *Xenopus laevis*.** *Development* 2001, **128**: 167–180.

Box 3F Signaling by members of the TGF-β family of growth factors

Massagué, J.: **How cells read TGF-β signals.** *Nat Rev. Mol. Cell Biol.* 2000, **1**: 169–178.

Schier, A.F.: **Nodal signaling in vertebrate development.** *Annu. Rev. Cell. Dev. Biol.* 2003, **19**: 589–621.

Schier, A.F.: **Nodal signals.** *Cold Spring Harb. Perspect. Biol.* 2009, **1**: a003459.

Zinski, J., Tajer, B., Mullins, M.C.: **TGF-β family signaling in early vertebrate development.** *Cold Spring Harb. Perspect. Biol.* 2017, **10**: a033274.

Box 3G Timing and cell–cell interactions in mesoderm specification

Gurdon, J.B., Lemaire, P., Kato, K.: **Community effects and related phenomena in development.** *Cell* 1993, **75**: 831–834.

3.12 Members of the TGF-β family have been identified as mesoderm inducers and induce formation of the organizer

Jones, C.M., Kuehn, M.R., Hogan, B.L., Smith, J.C., Wright, C.V.: **Nodal-related signals induce axial mesoderm and dorsalize mesoderm during gastrulation.** *Development* 1995, **121**: 3651–3662.

Osada, S.I., Wright, C.V.: ***Xenopus* nodal-related signaling is essential for mesendodermal patterning during early embryogenesis.** *Development* 1999, **126**: 3229–3240.

White R.J., Sun, B.I., Sive, H.L., Smith, J.C.: **Direct and indirect regulation of derrière, a *Xenopus* mesoderm-inducing factor, by VegT.** *Development* 2002, **129**: 4867–4876.

Box 3H Investigating receptor function using dominant-negative proteins

Dyson, S., Gurdon, J.B.: **Activin signalling has a necessary function in *Xenopus* early development.** *Curr. Biol.* 1997, **7**: 81–84.

3.13 Threshold responses to gradients of signal proteins are likely to pattern the mesoderm

Fletcher, R.B., Harland, R.M.: **The role of FGF signaling in the establishment and maintenance of mesodermal gene expression in *Xenopus*.** *Dev. Dyn.* 2008, **237**: 1243–1254.

Green, J.B.A., New, H.V., Smith, J.C.: **Responses of embryonic *Xenopus* cells to activin and FGF are separated by multiple dose thresholds and correspond to distinct axes of the mesoderm.** *Cell* 1992, **71**: 731–739.

Gurdon, J.B., Standley, H., Dyson, S., Butler, K., Langon, T., Ryan, K., Stennard, F., Shimizu, K., Zorn, A.: **Single cells can sense their position in a morphogen gradient.** *Development* 1999, **126**: 5309–5317.

Piepenburg, O., Grimmer, D., Williams, P.H., Smith, J.C.: **Activin redux: specification of mesodermal pattern in *Xenopus* by graded concentrations of endogenous activin B.** *Development* 2004, **131**: 4977–4986.

Saka, Y., Smith, J.C.: **A mechanism for the sharp transition of morphogen gradient interpretation in *Xenopus*.** *BMC Dev. Biol.* 2007, **7**: 47.

3.14 Regulation is still possible at the blastula stage

DeRobertis, E.M.: **Spemann's organizer and self-regulation in amphibian embryos.** *Nature Rev. Mol. Cell Biol.* 2006, **7**: 296–302.

Meistermann, D., *et al*.: **Integrated pseudotime analysis of human pre-implantation embryo single-cell transcriptomes reveals the dynamics of lineage specification.** *Cell Stem Cell* 2021, **28**: 1625–1640.

Snape, A., Wylie, C.C., Smith, J.C., Heasman, J.: **Changes in states of commitment of single animal pole blastomeres of *Xenopus laevis*.** *Dev. Biol.* 1987, **119**: 503–510.

Wylie, C.C., Snape, A., Heasman, J., Smith, J.C.: **Vegetal pole cells and commitment to form endoderm in *Xenopus laevis*.** *Dev. Biol.* 1987, **119**: 496–502.

3.15 The zebrafish embryo develops around a large yolk cell

Kimmel, C.B., Ballard, W.W., Kimmel, S.R., Ullmann, B., Schilling, T.F.: **Stages of embryonic development of the zebrafish.** *Dev. Dyn.* 1995, **203**: 253–310.

Warga, R.M., Nusslein-Volhard, C.: **Origin and development of the zebrafish endoderm.** *Development* 1999, **26**: 827–838.

Westerfield, M. (Ed.): *The Zebrafish Book: A Guide for the Laboratory Use of Zebrafish* (Brachydanio rerio). Eugene, OR: University of Oregon Press, 1989.

3.16 Genetic studies show that the axes of the zebrafish blastoderm are established by maternal determinants

Dosch, R., Wagner, D.S., Mintzer, K.A., Runke, G., Wiemelt, A.P., Mullins, M.C.: **Maternal control of vertebrate development before the midblastula transition: mutants from the zebrafish I.** *Dev. Cell* 2004, **6**: 771–780.

Fuentes, R., Tajer, B., Kobayashi, M., Pelliccia, J.L., Langdon, Y., Abrams, E.W., Mullins, M.C.: **The maternal coordinate system: molecular-genetics of embryonic axis formation and patterning in the zebrafish.** *Curr. Top. Dev. Biol.* 2020, **140**: 341–389.

Lu, F.I., Thisse, C., Thisse, B.: **Identification and mechanism of regulation of the zebrafish dorsal determinant.** *Proc. Natl Acad. Sci. USA* 2011, **108**: 15876–15880.

Pelegri, F.: **Maternal factors in zebrafish development.** *Dev. Dyn.* 2003, **228**: 535–554.

Schier, A.F., Talbot, W.S.: **Molecular genetics of axis formation in zebrafish.** *Annu. Rev. Genet.* 2005, **39**: 561–613.

Tran, L.D., Hino, H., Quach, H., Lim, S., Shindo, A., Mimori-Kiyosue, Y., Mione, M., Ueno, N., Winkler, C., Hibi, M., Sampath, K.: **Dynamic microtubules at the vegetal cortex predict the embryonic axis in zebrafish.** *Development* 2012, **139**: 3644–3652.

3.17 The germ layers are specified in the zebrafish blastoderm by signals similar to those in *Xenopus*

Agathon, A., Thisse, C., Thisse, B.: **The molecular nature of the zebrafish tail organizer.** *Nature* 2003, **424**: 448–452.

Fauny, J.D., Thisse, B., Thisse, C.: **The entire zebrafish blastula-gastrula margin acts as an organizer dependent on the ratio of Nodal to BMP activity.** *Development* 2009, **136**: 3811–3819.

Harvey, S.A., Smith, J.C.: **Visualisation and quantification of morphogen gradient formation in the zebrafish.** *PLoS Biol.* 2009, **7**: e101.

Kimmel, C.B., Warga, R.M., Schilling, T.F.: **Origin and organization of the zebrafish fate map.** *Development* 1990, **108**: 581–594.

Kudoh, T., Concha, M.L., Houart, C., Dawid, I.B., Wilson, S.W.: **Combinatorial Fgf and Bmp signalling patterns the gastrula ectoderm into prospective neural and epidermal domains.** *Development* 2004, **131**: 3581–3592.

Montague, T.G., Schier, A.F.: **Vg1-Nodal heterodimers are the endogenous inducers of mesendoderm.** *eLife* 2017, **6**: e28183.

Saúde, L., Woolley, K., Martin, P., Driever, W., Stemple, D.L.: **Axis-inducing activities and cell fates of the zebrafish organizer.** *Development* 2000, **127**: 3407–3417.

Tajer, B., Mullins, M.C.: **Heterodimers reign in the embryo.** *eLife* 2017, **6**: e33682.

3.18 The chick is a classic embryological model, especially from gastrulation to organogenesis stages

Bellairs, R., Osmond, M.: *An Atlas of Chick Development.* 2nd edn. London: Academic Press, 2005.

Chuai, M., Weijer, C.J.: **The mechanisms underlying primitive streak formation in the chick embryo.** *Curr. Top. Dev. Biol.* 2008, **81**: 135–156.

Chuai, M., Zeng, W., Yang, X., Boychenko, V., Glazier, J.A., Weijer, C.J.: **Cell movement during chick primitive streak formation.** *Dev. Biol.* 2006, **296**: 137–149.

Hamburger, V., Hamilton, H.L.: **A series of normal stages in the development of a chick.** *J. Morph.* 1951, **88**: 49–92.

Lillie, F.R.: *Development of the Chick: An Introduction to Embryology.* New York: Holt, 1952.

Patten, B.M.: *The Early Embryology of the Chick.* 5th edn. New York: McGraw-Hill, 1971.

Stern, C.D.: **Cleavage and gastrulation in avian embryos (version 3.0).** *Encyclopedia of Life Sciences* 2009 [https://doi.org/10.1002/9780470015902.a0001075.pub3].

3.19 In the chick, the main reference axis for development of the body plan is the antero-posterior axis

Kochav, S., Eyal-Giladi, H.: **Bilateral symmetry in chick embryo determination by gravity.** *Science* 1971, **171**: 1027–1029.

Nagai, H., Sezaki, M., Kakiguchi, K., Nakaya, Y., Lee, H.C., Ladher, R., Sasanami, T., Han, J.Y., Yonemura, S., Sheng, G.: **Cellular analysis of cleavage-stage chick embryos reveals hidden conservation in vertebrate early development.** *Development* 2015, **142**: 1279–1286.

Seleiro, E.A.P., Connolly, D.J., Cooke, J.: **Early developmental expression and experimental axis determination by the chicken Vg-1 gene.** *Curr. Biol.* 1996, **11**: 1476–1486.

3.20 Mesodermal transcription factors are upregulated during primitive streak formation

Chapman, S.C., Matsumoto, K., Cai, Q., Schoenwolf, G.C.: **Specification of germ layer identity in the chick gastrula.** *BMC Dev. Biol.* 2007, **7**: 91–107.

Bertocchini, F., Skromne, I., Wolpert, L., Stern, C.D.: **Determination of embryonic polarity in a regulative system: evidence for endogenous inhibitors acting sequentially during primitive streak formation in the chick embryo.** *Development* 2004, **131**: 3381–3390.

Khaner, O., Eyal-Giladi, H.: **The chick's marginal zone and primitive streak formation. I. Coordinative effect of induction and inhibition.** *Dev. Biol.* 1989, **134**: 206–214.

3.21 A fate map for the chick can be made after primitive streak formation

Garcia-Martinez, V., Alvarez, I.S., Schoenwolf, G.C.: **Locations of the ectodermal and nonectodermal subdivisions of the epiblast at stages 3 and 4 of avian gastrulation and neurulation.** *J. Exp. Zool.* 1993, **267**: 431–446.

Lopez-Sanchez, C., Garcia-Martinez, V., Schoenwolf, G.C.: **Localization of cells of the prospective neural plate, heart and somites within the primitive streak and epiblast of avian embryos at intermediate primitive-streak stages.** *Cells Tissues Organs* 2001, **169**: 334–346.

3.22 The mouse is the major model for mammalian development

Cross, J.C., Werb, Z., Fisher, S.J.: **Implantation and the placenta: key pieces of the developmental puzzle.** *Science* 1994, **266**: 1508–1518.

Hemberger, M., Hanna, C.W., Dean, W.: **Mechanisms of early placental development in mouse and humans.** *Nat. Rev. Genet.* 2020, **21**: 27–43.

Hu, D., Cross J.C.: **Development and function of trophoblast giant cells in the rodent placenta.** *Int. J. Dev. Biol.* 2010, **54**: 341–354.

Kaufman, M.H.: *The Atlas of Mouse Development.* 2nd edn. London: Academic Press, 1992.

Kaufman, M.H., Bard, J.B.L.: *The Anatomical Basis of Mouse Development.* London: Academic Press, 1999.

Nahaboo, W., Migeotte, I.: **Cleavage and gastrulation in the mouse embryo.** *eLS* (John Wiley, 2018). [https://doi.org/10.1002/9780470015902.a0001068.pub3] (date accessed 31 December 2022).

Thowfeequ, S., Srinivas, S.: **Mammalian embryo: establishment of the embryonic axes.** *eLS* (John Wiley, 2021). [https://doi.org/10.1002/9780470015902.a0029305] (date accessed 31 December 2022).

Wu, D., Dean, J.: **Maternal factors regulating preimplantation development in mice.** *Curr. Top. Dev. Biol.* 2020, **140**: 317–340.

3.23 Blastomeres in the morula are specified as trophectoderm or inner cell mass depending on their position and 3.24 The inner cell mass becomes differentiated into primitive endoderm and cells that will form the embryo proper

Arnold, S.J., Robertson, E.J.: **Making a commitment: cell lineage allocation and axis patterning in the early mouse embryo.** *Nature Mol. Cell Biol. Rev.* 2009, **10**: 91–103.

Chazaud, C., Yamanaka, Y.: **Lineage specification in the mouse preimplantation embryo.** *Development* 2016, **143**: 1063–1074.

Deb, K., Sivaguru, M., Yul Yong, H., Roberts, M.: **Cdx2 gene expression and trophectoderm lineage specification in mouse embryos.** *Science* 2006, **311**: 992–996.

Hillman, N., Sherman, M.I., Graham, C.: **The effect of spatial arrangement on cell determination during mouse development.** *J. Embryol. Exp. Morph.* 1972, **28**: 263–278.

Manzanares, M., Rodriguez, T.A.: **Hippo signaling turns the embryo inside out.** *Curr. Biol.* 2013, **23**: R559–R561.

Posfai, E., Petropoulos, S., deBarros, F.R.O., Schell, J.P., Jurisica, I., Sandberg, R., Lanner, F., Rossant, J.: **Position- and Hippo signaling-dependent plasticity during lineage segregation in the early mouse embryo.** *eLife* 2017, **6**: e22906.

Rossant, J., Tam, P.P.L.: **Blastocyst lineage formation, early embryonic asymmetries and axis patterning in the mouse.** *Development* 2009, **136**: 701–713.

Stephenson, R.O., Rossant, J., Tam, P.P.: **Intercellular interactions, position and polarity in establishing blastocyst cell lineages and embryonic axes.** *Cold Spring Harb. Perspect. Biol.* 2012, **4**: a008235.

3.25 The first asymmetry in mouse development appears in the blastocyst

Chen, Q., Shi, J., Tao, Y., Zernicka-Goetz, M.: **Tracing the origin of heterogeneity and symmetry breaking in the early mammalian embryo.** *Nat. Commun.* 2018, **9**: 1819.

Lim, H.Y.G., Plachta, N.: **Cytoskeletal control of early mammalian development.** *Nat. Rev. Mol. Cell Biol.* 2021, **22**: 548–562.

Wu, D., Dean, J.: **Maternal factors regulating preimplantation development in mice.** *Curr. Top. Dev. Biol.* 2020, **140**: 317–340.

3.26 The antero-posterior body axis of the mouse embryo is predicted by the movement of the anterior visceral endoderm

Bardot, E.S., Hadjantonakis, A.K.: **Mouse gastrulation: coordination of tissue patterning, specification and diversification of cell fate.** *Mech. Dev.* 2020, **163**: 103617.

Bischoff, M., Parfitt, D.E., Zernicka-Goetz, M.: **Formation of the embryonic–abembryonic axis of the mouse blastocyst: relationships between orientation of early cleavage divisions and pattern of symmetric/asymmetric divisions.** *Development* 2008, **135**: 953–962.

Molè, M.A., Weberling, A., Zernicka-Goetz, M.: **Comparative analysis of human and mouse development: from zygote to pre-gastrulation.** *Curr. Top. Dev. Biol.* 2020, **136**: 113–138.

Rodriguez, T.A., Srinivas, S., Clements, M.P., Smith, J.C., Beddington, R.S.: **Induction and migration of the anterior visceral endoderm is regulated by the extra-embryonic ectoderm.** *Development* 2005, **132**: 2513–2520.

Srinivas, S., Rodriguez, T., Clements, M., Smith, J.C., Beddington, R.S.P.: **Active cell migration drives the unilateral movements of the anterior visceral endoderm.** *Development* 2004, **131**: 1157–1164.

Takaoka, K., Yamamoto, M., Hamada, H.: **Origin and role of distal visceral endoderm.** *Nature Cell Biol.* 2011, **13**: 743–752.

3.27 Apart from its topology the fate map of the mouse is similar to that of the chick

Beddington, R.S., Robertson, E.J.: **Anterior patterning in mouse.** *Trends Genet.* 1998, **14**: 277–284.

Kinder, S.J., Tsang, T.E., Wakamiya, M., Sasaki, H., Behringer, R.R., Nagy, A., Tam, P.P.: **The organizer of the mouse gastrula is composed of a dynamic population of progenitor cells for the axial mesoderm.** *Development* 2001, **128**: 3623–3634.

3.28 The early development of a human embryo is generally similar to that of the mouse but the topology is more like that of the chick

Belle, M., Godefroy, D., Couly, G., Malone, S.A., Collier, F., Giacobini, P., Chédotal, A.: **Tri-dimensional visualization and analysis of early human development.** *Cell* 2017, **169**: 161–173.

deBakker, B.S., deJong, K.H., Hagoort, J., deBree, K., Besselink, C.T., deKanter, F.E., Veldhuis, T., Bais, B., Schildmeijer, R., Ruijter, J.M., Oostra, R.J., Christoffels, V.M., Moorman, A.F.: **An interactive three-dimensional digital atlas and quantitative database of human development.** *Science* 2016, **354**: aag0053.

Drews, U.: *Color Atlas of Embryology*. Stuttgart and New York: Thieme, 1995.

Niakan, K.K., Han, J., Pedersen, R.A., Simon, C., Pera, R.A.R.: **Human pre-implantation embryo development.** *Development* 2012, **139**: 829–841.

Schoenwolf, G.C., Belyl, S.B., Brauer, P.R., Francis-West, P.H.: *Larsen's Human Embryology*. 4th edn. London and New York: Churchill Livingstone, 2008.

Shahbazi, M.N., Jedrusik, A., Vuoristo, S., Recher, G., Hupalowska, A., Bolton, V., Fogarty, N.N.M., Campbell, A., Devito, L., Ilic, D., Khalaf,Y., Niakan, K.K., Fishel, S., Zernicka-Goetz, M.: **Self-organization of the human embryo in the absence of maternal tissues.** *Nat. Cell Biol.* 2016, **18**: 700–708.

Tyser, R.C.V., Mahammadov, E., Nakanoh, S., Vallier, L., Scialdone, A., Srinivas, S.: **Single-cell transcriptomic characterization of a gastrulating human embryo.** *Nature* 2021, **600**: 285–289.

Zhu, Z., Huangfu, D.: **Human pluripotent stem cells: an emerging model in developmental biology.** *Development* 2013, **140**: 705–717.

Box 3I Pre-implantation genetic diagnosis

Brezina, P.R., Kutteh, W.H.: **Clinical applications of preimplantation genetic testing.** *Br. Med. J.* 2015, **350**: g7611.

Greenfield, A., Braude, P., Flinter, F., Lovell-Badge, R., Ogilvie, C., Perry, A.C.F. **Assisted reproductive technologies to prevent human mitochondrial disease transmission.** *Nat. Biotechnol.* 2017, **35**: 1059–1068.

Reigstad, M.M., Storeng, R.: **Development of *in vitro* fertilization, a very important part of human reproductive medicine, in the last 40 years.** *Int. J. Womens Health Wellness* 2019, **5**: 089.

Yan, L., Yang, M., Guo, H., Yang, L., Wu, J., Li, R., Liu, P., Lian, Y., Zheng, X., Yan, J., Huang, J., Li, M., Wu, X., Wen, L., Lao, K., Li, R., Qiao, J., Tang, F.: **Single-cell RNA-Seq profiling of human preimplantation embryos and embryonic stem cells.** *Nat. Struct. Mol. Biol.* 2013, **20**: 1131–1139.

3.29 The timing of formation and the anatomy of the human placenta differs from that in the mouse

Hemberger, M., Hanna, C.W., Dean, W.: **Mechanisms of early placental development in mouse and humans.** *Nat. Rev. Genet.* 2020, **21**: 27–43.

Maltepe, E., Fisher, S.J.: **Placenta: the forgotten organ.** *Annu. Rev. Cell Dev. Biol.* 2015, **31**: 523–552.

Schoenwolf, G.C., Belyl, S.B., Brauer, P.R., Francis-West, P.H.: *Larsen's Human Embryology*. 4th edn. London and New York: Churchill Livingstone, 2008.

Turco, M.Y., Moffett, A.: **Development of the human placenta.** *Development* 2019, **146**: dev163428.

3.30 Some studies of human development are possible but are subject to strict laws

Focus on Humans 2.0. *Nat. Biotechnol.* 2017 [https://www.nature.com/collections/czfsycmfws] (date accessed 3 September 2024).

Fogarty, N.M.E., McCarthy, A., Snijders, K. E., Powell, B. E., Kubikova, N., Blakeley, P., Lea, R., Elder, K., Wamaitha, S. E., Kim, D., Maciulyte, V., Kleinjung, J., Kim, J. S., Wells, D., Vallier, L., Bertero, A., Turner, J. M. A., Niakan, K. K.: **Genome editing** reveals a role for OCT4 in human embryogenesis. *Nature* 2017, **550**: 67–73.

Gerrelli, D., Lisgo, S., Copp, A.J., Lindsay, S.: **Enabling research with human embryonic and fetal tissue resources.** *Development* 2015, **142**: 3073–3076.

Hyun, I., Wilkerson, A., Johnston, J.: **Embryology policy: revisit the 14-day rule.** *Nature* 2016, **533**: 169–171.

Rossant, J.: **Implantation barrier overcome.** *Nature* 2016, **533**: 182–183.

Stamatiadis, P., Boel, A., Cosemans, G., Popovic, M., Bekaert, B., Guggilla, R., Tang, M., DeSutter, P., VanNieuwerburgh, F., Menten, B., Stoop, D., Chuva de Sousa Lopes, S.M., Coucke, P., Heindryckx, B.: **Comparative analysis of mouse and human preimplantation development following POU5F1 CRISPR/Cas9 targeting reveals interspecies differences.** *Hum. Reprod.* 2021, **36**: 1242–1252.

Vassena, R., Boué, S., González-Roca, E., Aran, B., Auer, H., Veiga, A., Izpisúa Belmonte, J.C.: **Waves of early transcriptional activation and pluripotency program initiation during human preimplantation development.** *Development* 2011, **138**: 3699–3709.

Vertebrate development II: completing the body plan

- The *Xenopus* Spemann organizer patterns the dorso-ventral and antero-posterior body axes
- The *Xenopus* Spemann organizer induces neural tissue and initiates neural patterning
- Organizer activity in the zebrafish, chick, and mouse
- Elongation of the body and the origin of posterior tissues
- Somite formation and the Hox code
- in antero-posterior patterning
- The origin, behavior, and patterning of neural crest
- Determination of left-right asymmetry

In this chapter, we look at how the vertebrate body plan is established during gastrulation under the control of the embryonic organizer and is then completed. We first consider gastrulation in Xenopus and how the Spemann organizer coordinates patterning along the antero-posterior and dorso-ventral body axes and induces the future nervous system. We then consider how the body plan is established during gastrulation in our other model organisms, and the extent to which their embryonic organizers are functionally equivalent to the Spemann organizer. We go on to look at how some features of the vertebrate body plan are completed, focusing mainly on the chick and the mouse. We describe the formation of the somites, the role of the Hox genes in the regionalization of the antero-posterior body axis, and the development of the neural crest—a population of cells unique to vertebrates that gives rise to a wide variety of different tissues, including the peripheral nervous system. Finally, we look at the intriguing question of how the vertebrate body's left-right asymmetry—as revealed by the positioning of the heart and some other organs—is established. By the end of this chapter, it will have become clear how the recognizable features of the vertebrate body form.

Introduction

The laying down of the vertebrate body plan is a continuous developmental process but can be broken down into several stages, each building on the previous one. In Chapter 3, we considered how the cells generated by divisions of the fertilized egg become different, how the embryo is initially patterned, and how the embryonic organizer is specified. In this chapter, we look at how patterning controlled by the organizer continues in the ensuing stage of gastrulation. Gastrulation establishes the germ layers in their proper positions for further development, and lays down the pattern of structures along the main body axes. A key event accompanying gastrulation is the induction of the central nervous system from the dorsal ectoderm under the influence of signals from the organizer. We explain how the embryonic organizer harmonizes different processes across space and time so that the vertebrate organism is patterned, and the body plan is correctly laid down.

The organizer was first discovered in amphibian embryos through its ability to induce a second body axis and is known in amphibians as the Spemann organizer. Organizers with similar properties and with a crucial role in organizing the body plan have been identified in our other model vertebrate embryos, and their formation is introduced in Chapter 3. The mechanics of gastrulation and the way in which the organizer controls the cell rearrangements that establish the germ layers in their proper positions and lead to elongation of the body along the antero-posterior axis are discussed in Chapter 5, as is the process of neurulation, in which the neural tissue forms a tube that will develop into the brain and spinal cord. In this chapter we concentrate on the function of the organizer in patterning the body. This is accomplished by the organizer producing secreted proteins that establish morphogen gradients that provide positional information. As a result of the activity of the organizer, the embryo is concomitantly patterned along its dorso-ventral and antero-posterior body axes and, at the same time, the nervous system is induced and roughly patterned. The precise distribution of morphogen signals and provision of positional information across each axis is not established instantaneously, and entire body axes cannot be patterned all at once. Changes in organizer activity over time enable the body axes to be patterned gradually, and underlie the ability of the organizer to induce an almost complete second body axis.

We know most about the workings of the organizer in *Xenopus* and so the first two parts of the chapter will focus on the role of the Spemann organizer in *Xenopus* embryos in the patterning of the main body axes and the induction of neural tissue from the ectoderm. We then look at organizer function in our other model organisms. Although its general function in coordinating the laying down of the body plan is the same, there are differences from *Xenopus* in how this is achieved, especially in the chick and the mouse.

In the remainder of the chapter, we examine how the body plan is completed after gastrulation and neurulation, with the focus on the chick and the mouse. Events that occur around the phylotypic stage generate key dorsal tissues of the trunk, notably the spinal cord and the somites (blocks of tissue formed from mesoderm on either side of the notochord). These will give rise to the vertebral column, the skeletal muscles of the trunk, and some of the dermis of the skin. We describe how somites are generated in a periodic manner in response to a 'molecular clock', and how their position within the body plan reflects the time at which they are generated.

Both the spinal cord and the somites have distinct antero-posterior identities that are controlled by the Hox genes, which are related to the Hox genes responsible for antero-posterior patterning in *Drosophila* (see Chapter 2). Hox genes and their role in antero-posterior patterning are common to all vertebrates, but their function has been particularly well studied in the mouse (see Box 3B), in which techniques for knocking out specific genes are particularly well developed, and we will see how the Hox genes are involved in the antero-posterior patterning of the developing somites.

The neural crest is a population of cells that is unique to vertebrates, and is derived from cells at the border between the neural plate and non-neural ectoderm. Neural crest cells start to migrate from this border region as the neural tube forms during neurulation. The cells migrate to different locations in the body to give rise to a wide range of tissues, including the autonomic and sensory nervous systems, and in the head, the connective tissues of the face. We examine the generation and behavior of neural crest cells, and how Hox gene expression specifies the regional identity of neural crest cells that populate the head. The migration of the neural crest is discussed in

Chapter 5, and later patterning and the development of the central nervous system, including the brain, are considered in Chapter 10.

Finally, we briefly consider how the left–right asymmetry of the vertebrate body plan is established. Although the vertebrate body is outwardly symmetric about the midline, there is left–right internal asymmetry, which is indicated, for example, by the positions of the heart and liver. As we shall see, this asymmetry originates early in embryonic life.

The *Xenopus* Spemann organizer patterns the dorso-ventral and antero-posterior body axes

The Spemann organizer is a group of cells in the dorsal part of the equatorial region (the mesodermal marginal zone) of the early gastrula that have a global organizing function. Spemann organizer cells from an early gastrula cause the formation of a second body axis with distinct head and trunk and complete dorso-ventral axis if transplanted to another embryo at an appropriate stage, resulting in a twinned embryo (see Section 3.7 and Fig. 1.9). Axial mesoderm and anterior endoderm in the second body axis develop from the graft, but the rest of the second body axis is induced and patterned in the host tissue by the organizer cells. The Spemann organizer therefore directs global patterning along both the dorso-ventral and antero-posterior axes.

The way in which the Spemann organizer acts can be illustrated through its effects on dorso-ventral patterning of the mesoderm. Recall that mesoderm in *Xenopus* is induced from cells of the animal region by signals from the vegetal region of the blastula (see Section 3.11). In the early blastula, dorsal cells of the marginal zone are instructed to become the Spemann organizer, while the other equatorial cells acquire a non-organizer mesodermal identity. Fate maps show that the mesoderm cells of the Spemann organizer give rise to the axial mesoderm—the prechordal mesoderm and the notochord—which define the dorsal midline of the body, with the prechordal mesoderm ending up in the most anterior position in the body plan as a result of the cell rearrangements during gastrulation. The fate maps also show what non-organizer mesoderm gives rise to. The mesoderm immediately adjacent to the Spemann organizer is called **paraxial mesoderm**, and gives rise to somites (embryonic tissue that produces the muscle and the skeleton of the trunk); then comes lateral plate mesoderm (which gives rise to the heart) and intermediate mesoderm (from which the kidneys develop), and then ventral-most mesoderm which gives rise to blood (see Fig. 3.19 and Box 3A, Fig. 2).

The Spemann organizer is responsible for specifying the pattern of non-organizer mesoderm along the dorso-ventral body axis. This function was first shown by experiments in which a fragment of dorsal marginal zone from a late blastula, which will form the organizer, was combined with a fragment of ventral mesoderm. The ventral fragment now forms substantial amounts of muscle, whereas ventral mesoderm isolated and cultured on its own produces mainly blood-forming tissue. In this part of the chapter, we shall examine the properties of the organizer that enable it to induce a second body axis and to coordinate patterning along both the dorso-ventral and antero-posterior axes.

4.1 Secreted proteins from the organizer pattern the dorso-ventral axis by antagonizing the effects of ventral signals

The Spemann organizer patterns the mesoderm along the dorso-ventral axis by secreting proteins that counteract so-called ventralizing signals secreted throughout the late blastula (Fig. 4.1). This results in a ventral (high)-to-dorsal (low) gradient of ventralizing signals (Fig. 4.1). This gradient provides positional information, effectively

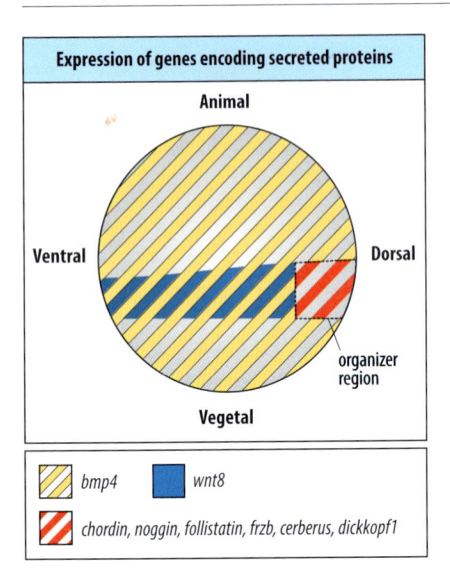

Fig. 4.1 Distribution of mRNAs encoding secreted proteins in the *Xenopus* late blastula. Bone morphogenetic (BMP)-4 and Wnt-8 signal proteins are made throughout the late blastula/early gastrula and in the marginal zone respectively. The secreted proteins made in the organizer include Chordin, Noggin, and Follistatin, which block the action of BMPs, Dickkopf1 and Frizzled-related protein (Frzb), which block the action of Wnts, and Cerberus, which blocks the action of both BMPs and Wnts.

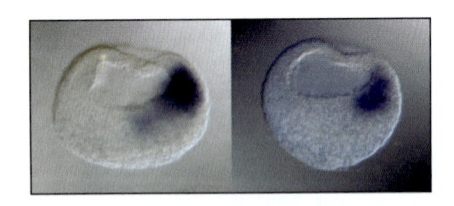

Fig. 4.2 Expression of *noggin* and *chordin* in the *Xenopus* late blastula. Blue staining reveals mRNA for *chordin* (left panel) and *noggin* (right panel); transcripts for each are specifically located in the region defined as the Spemann organizer.

From Khokha, M.K., et al.: ***Depletion of three BMP antagonists from Spemann's organizer leads to a catastrophic loss of dorsal structures***. Dev. Cell. 2005, 8: 401–411.

switching on genes that specify different cell fates and behaviors. (See Section 3.13 for an example of how a diffusible protein could pattern a tissue by turning on different genes at different threshold concentrations.)

The most prominent ventralizing signals are members of the bone morphogenetic protein (BMP) family. Before the Spemann organizer forms, BMP-4 and other BMPs are produced in cells throughout the embryo (see Fig. 4.1). However, once the Spemann organizer is induced in dorsal mesoderm, it begins to make and secrete proteins that inhibit, or antagonize, BMP activity. In this way, a morphogen-like gradient of BMP is established, with the highest point of BMP at the ventral side of the embryo. The BMP gradient can be visualized by immunofluorescent labeling for nuclear phosphorylated Smad1/5, a direct intracellular readout of BMP signaling (see Box 3F). BMP is acting as a morphogen whose spatial concentration varies and to which cells respond differently at different threshold concentrations to pattern a tissue. If BMP-4 is overexpressed on the dorsal side of the embryo, then dorsal mesoderm, such as notochord and muscle, fails to form. Conversely, when the action of BMPs is blocked throughout the embryo by introducing a dominant-negative receptor (see Box 3H), the embryo is dorsalized, with ventral cells now differentiating as notochord and muscle.

The genes encoding the antagonists of BMPs are expressed specifically in the Spemann organizer, under the influence of Spemann organizer-specific transcription factors, such as Siamois and Goosecoid (Gsc) (see Fig. 3.27). The first antagonists to be discovered were the proteins Noggin and Chordin, both of which inhibit BMP activity. Noggin, for example, was discovered in a screen for factors that could rescue ultraviolet-irradiated ventralized *Xenopus* embryos, in which a dorsal region had not been specified by the usual activity of the β-catenin pathway (see Section 3.7), and the *noggin* gene was found to be transcribed specifically in cells of the Spemann organizer (Fig. 4.2). Noggin protein cannot induce mesoderm in animal-cap explants, but it can dorsalize explants of ventral marginal zone tissue, making it a good candidate for a signal that patterns the mesoderm along the dorso-ventral axis.

The secretion of antagonists that restrict signaling to a particular region, or modulate the level of signaling, is a common strategy in embryonic development. We have already come across an example in Section 3.7, where a Wnt antagonist helps to localize activation of Wnt/β-catenin signaling to the dorsal side of the very early *Xenopus* embryo, and we shall come across other examples of signal antagonists in Section 4.2 and in Chapter 10.

Noggin, Chordin, and Follistatin proteins interact with BMP proteins and prevent them from binding to their receptors. This sets up functional BMP gradients along the dorso-ventral axis of the *Xenopus* gastrula, with their highest point ventrally and little or no activity dorsally. The BMP gradients are not simply due to BMP antagonists diffusing from the organizer along the dorso-ventral axis, but to a shuttling mechanism. Chordin produced dorsally binds to BMPs in this region, inhibiting their action and forming a complex. The Chordin–BMP complexes diffuse away from the dorsal region of the embryo. In the ventral region, an extracellular metalloproteinase, ADMP, is produced that degrades the Chordin portion of the complex, thereby depositing BMP (Fig. 4.3). The metalloproteinase also acts as a clearing agent for Chordin, reducing the extent of its long-range diffusion and helping to maintain Chordin high, dorsally.

The action of Chordin in creating a gradient of BMPs, including BMP-4, mirrors that of the *Drosophila* Chordin homolog Sog on the BMP-4 homolog Decapentaplegic (Dpp) in the patterning of the dorso-ventral axis in the fly (see Section 2.17). In the fly, however, the dorso-ventral axis is inverted compared with that of vertebrates, and so Dpp specifies a dorsal fate, whereas BMP-4 specifies a ventral fate (the inversion of the dorso-ventral axis during evolution is discussed in Section 13.10). Other components of the mechanism that forms the gradient of Dpp activity in *Drosophila* are also conserved in vertebrates, with the *Xenopus* metalloproteinase being related to the *Drosophila* metalloproteinase Tolloid.

4.2 Secreted proteins from the organizer pattern the antero-posterior axis by antagonizing the effects of posterior signals

At the same time that distinct mesoderm cells are being specified along the dorso-ventral axis, distinct mesoderm cells are also being specified along the antero-posterior axis. The Spemann organizer acts as a reference point for global patterning of the antero-posterior axis of the body because, as gastrulation proceeds, organizer cells that were at the dorsal blastoderm lip in the early gastrula come to occupy the most anterior position in the body plan, at the head end of the embryo. The mechanism that provides positional information for the antero-posterior axis parallels the mechanism that provides positional information along the dorso-ventral axis discussed in the previous section. The Spemann organizer secretes proteins that counteract posteriorizing signals, creating an antero-posterior morphogen gradient of posteriorizing signals, with the high point posteriorly.

The main posteriorizing signals are members of the Wnt family. In Chapter 3 we saw that in the mid-blastula embryo, Wnt signaling is localized dorsally and induces the Spemann organizer, and in the late blastula, zygotic Wnt-8 is expressed throughout the marginal zone which will give rise to the mesoderm (see Fig. 4.1). However, during gastrula stages, cells of the dorsal lip/dorsal organizer start to make and secrete a cocktail of Wnt signal antagonists, such as Frzb, Dickkopf1, and Cerberus (see Fig. 4.1). Wnt signaling is excluded from the dorsal organizer cells by the activity of the Wnt antagonists and is now localized to more ventral mesoderm that will involute only later during gastrulation. This change in localization coincides with a dramatic change in Wnt function, and it now promotes posterior tissue development. Embryos deficient in Wnt signaling lose trunk and tail, but have an enlarged head, whereas embryos in which Wnt signaling is increased lose head structures. Wnt antagonists are produced only by cells at the dorsal lip of the early gastrula. This population comprises the cells that were at the dorsal margin of the blastula and have just involuted through the blastopore (Fig. 4.4, first panel, orange and brown regions). Fine-grained fate maps show that these cells are the future anterior endoderm and prechordal mesoderm. Mesoderm cells that were positioned more ventrally in the blastula and will give rise to the anterior-most part of the notochord (red in Fig. 4.4) are yet to involute, and do not express Wnt antagonists.

As gastrulation proceeds, expression of the Wnt antagonists Cerberus and Dickkopf persists in the prospective anterior endoderm and prechordal mesoderm as they move further inside to take up their position at the head end of the lengthening embryo (see Fig. 4.4, second panel). In this way, by the end of gastrulation, a gradient of Wnt signaling is established that spans the entire antero-posterior body axis (Fig. 4.5). Head

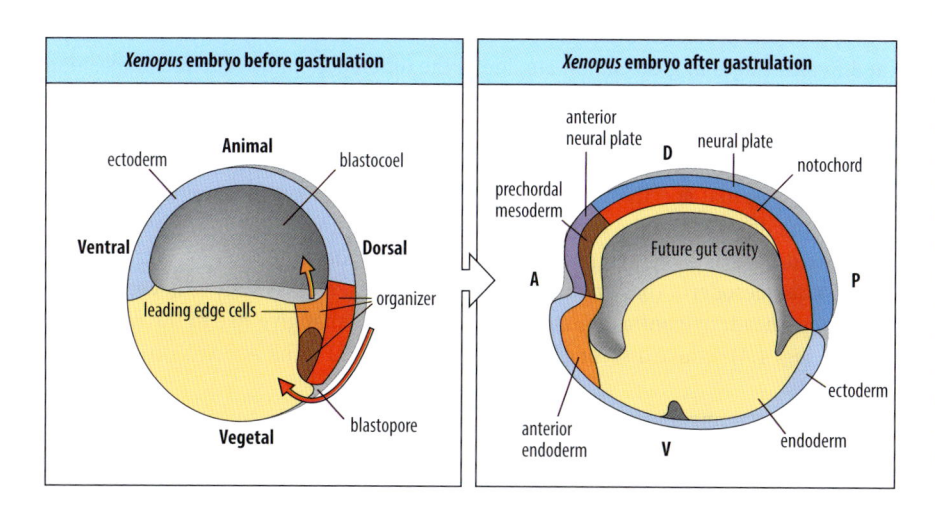

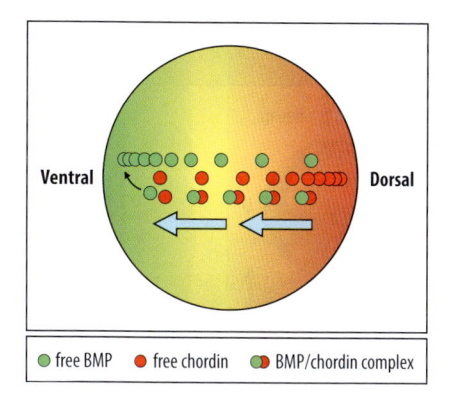

Fig. 4.3 A shuttling mechanism generates a ventral-dorsal gradient of bone morphogenetic protein (BMP) signaling in the early gastrula. Chordin (red) secreted from the organizer binds to BMP (green) in the dorsal region of the embryo. Some BMPs and BMP-like signal molecules are secreted by cells throughout the early gastrula; some are secreted by the organizer itself. The Chordin–BMP complexes diffuse ventrally. In the ventral region, Chordin is cleaved by a metalloproteinase, ADMP, which is made in this region, and this releases BMP from the complex and deposits it ventrally. This shuttling mechanism generates a gradient of BMP signaling running from high ventrally (green background) to low dorsally (red background) across the embryo.

*After Lewis, J.: **From signals to patterns: space, time, and mathematics in developmental biology**. Science 2008, **322**: 399-403.*

Fig. 4.4 Different regions of the dorsal lip of the blastopore in the early gastrula give rise to different tissues and differentially express Wnt antagonists. The left panel shows the early gastrula just after gastrulation begins. Cells at the dorsal lip of the gastrula are colored. The cells at the leading edge (orange) of the tissue that has already moved inside give rise to anterior endoderm at the neurula stage (right panel). The mesoderm cells that were at the dorsal margin of the blastula and had internalized next (brown) give rise to the prechordal mesoderm, which underlies the anterior neural plate, the future brain. The remainder of the organizer gives rise to notochord (red). Wnt antagonists are expressed in prospective and definitive anterior endoderm and prechordal mesoderm. A, anterior; D, dorsal; P, posterior; V, ventral.

*Adapted from Kiecker, C., Niehrs, C.: **A morphogen gradient of Wnt/β-catenin signalling regulates anteroposterior neural patterning in Xenopus**. Development 2001, **128**: 4189-4201.*

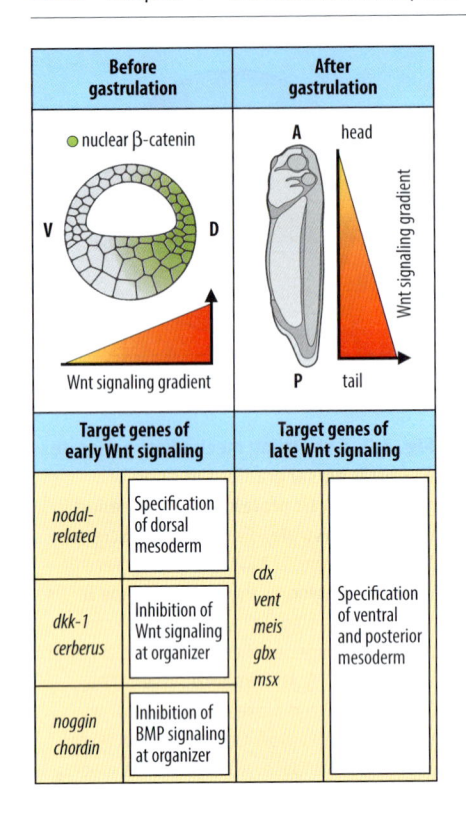

Fig. 4.5 Wnt signaling specifies the dorso-ventral axis in the early *Xenopus* embryo before gastrulation, whereas a second round of Wnt signaling after gastrulation patterns the antero-posterior axis. Before gastrulation, accumulation of nuclear β-catenin following cortical rotation specifies the future dorsal region of the embryo (shown on the left). At these early stages, Wnt signaling is graded with its high point at the dorsal side of the embryo, where it participates in specifying the Nieuwkoop center and the organizer by activating genes that specify dorsal mesoderm and genes expressed in the organizer, including Wnt antagonists. During gastrulation, Wnt antagonists secreted by the subset of organizer cells which give rise to the anterior mesoderm and prechordal mesoderm begin to locally inhibit Wnt signaling. After gastrulation, Wnt signaling is now graded along the antero-posterior axis of the gastrula with its high point at the posterior of the embryo. Target genes activated include those for transcription factors involved in specification of posterior mesoderm that later gives rise to the tail of the tadpole (shown on the right) and also transcription factors involved in specifying ventral mesoderm. D, dorsal; V, ventral; A, anterior; P, posterior; BMP, bone morphogenetic protein.

After Hikasa, H., Sokol, S.Y.: **Wnt signaling in vertebrate axis specification.** *Cold Spring Harb. Perspect. 2013,* **5**: *a007955.*

formation at the anterior is specified by low levels of Wnt signaling due to high levels of antagonists such as Cerberus and Dickkopf, whereas tail formation at the posterior, where there are low levels of these antagonists, is specified by high levels of Wnt signaling (see Fig. 4.5).

Expression of Wnt antagonists is controlled by a sophisticated network of transcription factors, including Goosecoid, Otx2, and Lim1. Gene knock-out studies show that these transcription factors are required for the development of head structures in the normal course of development. Other transcription factors, such as Not, are expressed exclusively in the later-involuting regions of the blastopore. Not is required for specification of the notochord, and hence for the particular set of secreted proteins made by the notochord.

Classic experiments in which the dorsal lip of the blastopore of embryos at later stages of gastrulation was used as an organizer graft showed that its inductive properties change over time. When the blastopore dorsal lip from a very early gastrula was grafted to the ventral side of the marginal zone of another early gastrula, it induced a second embryo with a well-defined head, a central nervous system, a trunk region, and tail (Fig. 4.6, left panels, and see Fig. 3.15) and was said to function as a 'head' organizer. (Note that grafting the Nieuwkoop center to the ventral side of a blastula produced a similar result (see Fig. 3.17) because the Nieuwkoop center induces a new Spemann organizer region with head organizer function.) By contrast, a grafted dorsal lip from a mid-gastrula induces a trunk and tail but no head and was said to function as a 'trunk' organizer, and a grafted dorsal lip from a late gastrula induced only a tail (Fig. 4.6, right panels).

We now know that the differences in the inductive abilities of grafts of the dorsal lip at different stages in gastrulation are due to the differences in its cellular composition and the consequent level of Wnt antagonists. At mid-gastrula, the dorsal lip of the blastopore is composed of dorsal mesoderm cells that will form the notochord and do not produce Wnt antagonists, and therefore the grafts do not result in the low levels of Wnt signaling that specify head formation. The results of grafting subpopulations of cells of the dorsal lip of the early gastrula are consistent with this explanation. Grafts composed of cells that give rise to the anterior endoderm (orange in Fig. 4.4) and prechordal mesoderm (brown in Fig. 4.4) can induce a second body axis with a well-developed head; by contrast, grafts composed of cells that give rise to the notochord (red in Fig. 4.4) induce a second body axis consisting of trunk and tail structures but not heads.

The observations and experimental results outlined in this and the previous section have led to a model of early *Xenopus* development that incorporates orthogonal

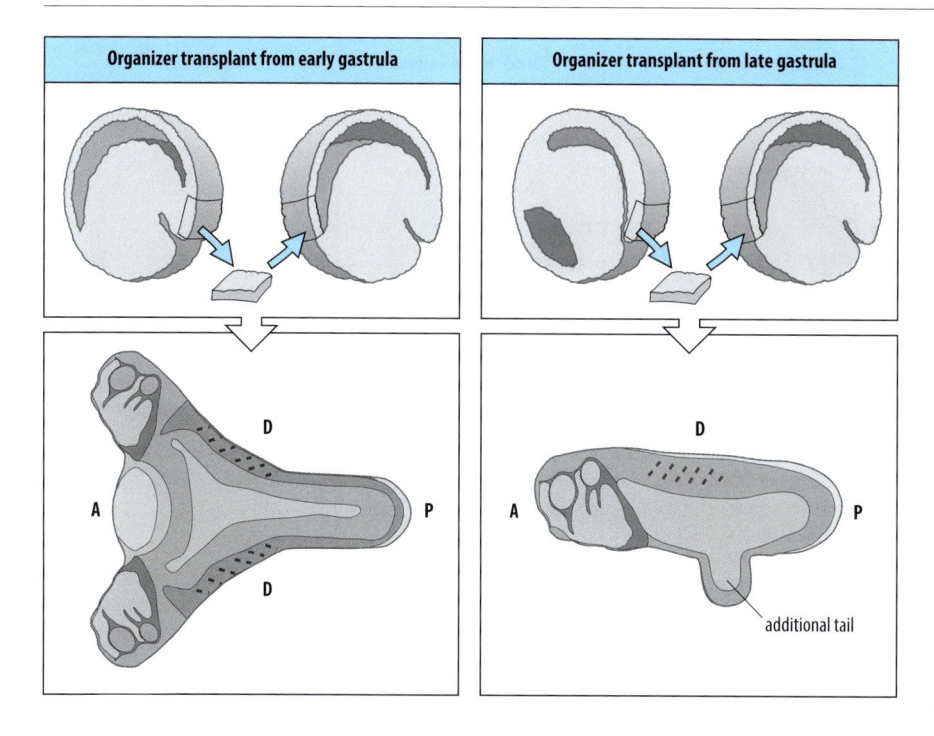

Fig. 4.6 The inductive properties of the organizer change during gastrulation. A graft of the organizer region—the dorsal lip of the blastopore of an early *Xenopus* gastrula—to the ventral side of another early gastrula, results in the development of a twinned embryo with the second embryo induced by the graft having a head (left panels). A graft from the dorsal-lip region of a late gastrula to an early gastrula only induces tail structures (right panels). A, anterior; P, posterior; D, dorsal.

gradients of the morphogens BMP and Wnt into a three-dimensional coordinate system of positional information that is established and maintained over time. The model is an oversimplification, because the patterning processes along the two axes must be integrated and other signals such as Nodal are involved, but it provides the framework for understanding how the body axes are patterned in the gastrula (Fig. 4.7). Morphogen gradients at right angles to each other also operate to pattern the *Drosophila* wing (discussed in Chapter 9) and represent an important way of patterning an embryonic tissue.

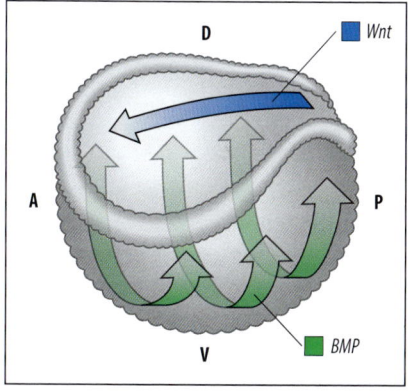

Fig. 4.7 Double-gradient model of embryonic axis formation. The model shows how gradients of Wnts (blue) and BMPs (green) perpendicular to each other regulate antero-posterior and dorso-ventral patterning in *Xenopus*. The color scales of the arrows indicate the signaling gradients; arrows indicate the direction of spreading of the signals. For example, the formation of the tail requires a higher level of Wnt signaling than does the head. Patterning begins at gastrula stages, but for clarity is depicted here in an early neurula. A, anterior; D, dorsal; P, posterior; V, ventral.

Adapted from Niehrs, C.: **Regionally specific induction by the Spemann-Mangold organiser.** Nat. Rev. Genet. 2004, **5:** 425-434.

SUMMARY

Patterning of the early *Xenopus* embryo, along both the antero-posterior and dorso-ventral axes, depends on the action of the Spemann organizer and its morphogenesis during gastrulation. The organizer secretes growth-factor antagonists, which inhibit the actions of BMPs and Wnts. This sets up gradients of BMP signaling from ventral (high) to dorsal (low) that pattern the dorso-ventral axis, and a gradient of Wnt signaling from posterior (high) to anterior (low) that patterns the embryo along the antero-posterior axis.

The *Xenopus* Spemann organizer induces neural tissue and initiates neural patterning

During gastrulation, the ectoderm in the dorsal region of the embryo becomes specified as neural ectoderm (**neuroectoderm**) and forms the **neural plate**, which will give rise to the nervous system. This flat plate of cells then rolls up into the **neural tube** in the process of neurulation (Fig. 4.8). A key challenge in the early embryo is to ensure that development of the nervous system is coordinated with the development of the rest of the body. This is achieved through the action of the Spemann organizer: some

Fig. 4.8 Neurulation in amphibians. Top row: the neural plate develops neural folds just as the notochord begins to form in the midline (see center row). Center and bottom rows: the neural folds come together in the midline to form the neural tube, from which the brain and spinal cord will develop. During neurulation, the embryo elongates along the antero-posterior axis. The left panel shows sections through the embryo in the planes indicated by the blue dashed lines in the center panel. The center panel shows dorsal surface views of the embryo. The right panel shows sections through the embryo in the planes indicated by the green dashed lines in the center panel. This diagram shows neurulation in a urodele amphibian embryo rather than *Xenopus*, as the neural folds in urodele embryos are more clearly defined. A, anterior; P, posterior.

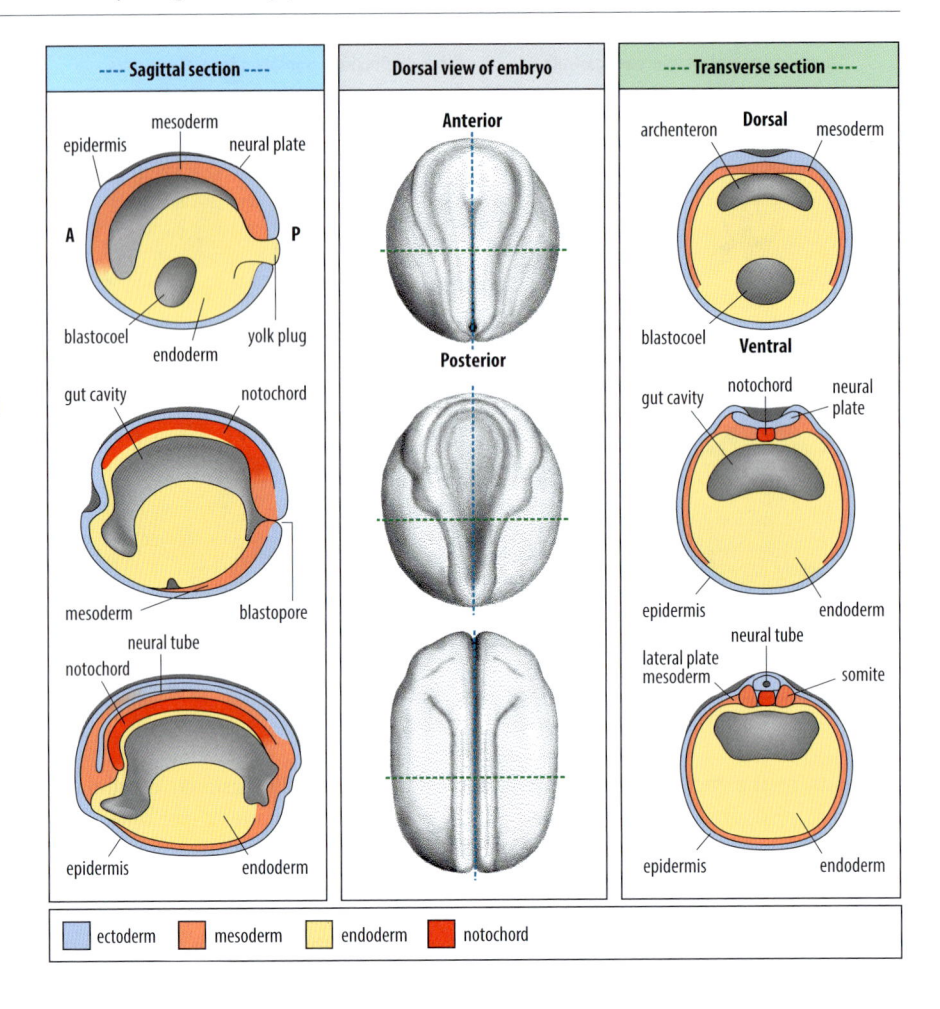

of the same organizer-derived factors that confer general pattern on the mesoderm of the dorso-ventral and antero-posterior axes also induce the neural plate and begin to pattern it along its axes.

4.3 The neural plate is induced from the ectoderm through the action of BMP antagonists

In the Spemann organizer transplant experiment (illustrated in Figs 1.9 and 3.15), a nervous system develops within the second embryonic axis. This suggested that factors emanating from the organizer can direct the formation of neural tissue in adjacent ectoderm that would otherwise give rise to epidermis. This hypothesis was confirmed by experiments in which the ectoderm that normally forms neural tissue (neuroectoderm) was replaced with ectoderm that normally forms epidermis; the transplanted ectoderm epidermis now develops into neural tissue (Fig. 4.9).

An enormous amount of effort was devoted in the 1930s and 1940s to trying to identify the signals involved in neural induction in amphibians. However, the key discovery was made only in the 1980s, when it was found that the BMP antagonist Noggin could induce neural differentiation in ectoderm explants from *Xenopus* blastulas. Noggin was one of the first secreted proteins to be isolated from the organizer and is instrumental in mesodermal dorso-ventral patterning (see Section 4.1 and Fig. 4.2). In the late *Xenopus* blastula, BMPs are expressed throughout the ectoderm (see Fig. 4.1), but their expression is subsequently lost in the ectoderm adjacent to the Spemann organizer—that is, ectoderm that will form the neural plate. These

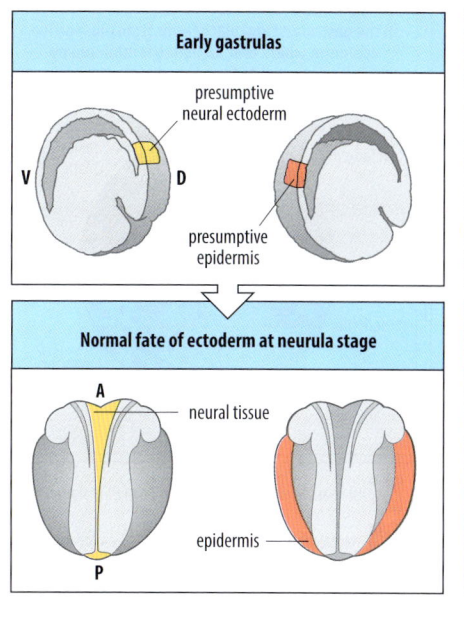

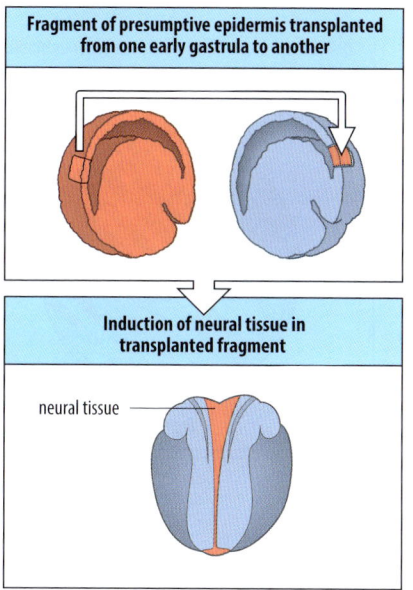

Fig. 4.9 The nervous system of *Xenopus* is induced during gastrulation. The left panels show the normal developmental fate of ectoderm at two different positions in the early gastrula. The right panels show the transplantation of a piece of ventral (V) ectoderm, which normally forms epidermis, from the ventral side of an early gastrula to the dorsal (D) side of another, where it replaces a piece of dorsal ectoderm which normally forms neural tissue. In its new location, the transplanted ectoderm develops not as epidermis, but as neural tissue, and forms part of a normal nervous system. This shows that the ventral tissue has not yet been determined at the time of transplantation, and that neural tissue is induced during gastrulation. A, anterior; P, posterior.

results suggested that neural plate can only develop if BMP signaling is absent. This conclusion is supported by the dramatic effect of using antisense morpholino oligonucleotides to simultaneously knock down expression of BMP-4 and the three other BMPs expressed in the late blastula. In the absence of BMP signaling, the embryo becomes covered in neural tissue (Fig. 4.10). As discussed in Section 4.1, a number of BMP antagonists were then found to be produced by, and secreted from, the organizer, including (in addition to Noggin), Chordin, Follistatin, and Cerberus, and these became attractive candidates for inhibiting BMP signaling in the prospective neural plate and allowing neural induction to proceed. As BMP signaling itself maintains expression of the BMP genes, antagonizing BMP signaling will also shut down BMP gene expression.

These observations led to the so-called default model for neural induction in *Xenopus*. This proposed that the default state of the dorsal ectoderm is to develop as neural tissue, but that this pathway is blocked by the presence of BMPs, which promote the epidermal fate (Fig. 4.11). The role of the organizer is to lift this block by producing proteins that inhibit BMP activity. The region of ectoderm that comes under the influence of the organizer-derived antagonists will then develop as neural ectoderm. In this model, BMP antagonists secreted from the organizer act on ectoderm that lies adjacent to the organizer at the beginning of gastrulation. As gastrulation proceeds, internalized cells derived from the organizer—which are developing into anterior endoderm, prechordal mesoderm, and anterior-most notochord—continue to secrete these antagonists, maintaining the ectoderm that now overlies them as a 'BMP-free' zone. Although, traditionally, the Spemann organizer is said to 'induce' neural tissue, the default model suggests, rather, that the organizer unveils a latent neural program. Nonetheless, for the sake of simplicity, throughout this book we use the more traditional term.

Elimination of individual organizer antagonists in amphibians has rather modest effects on neural induction. But when combinations of the BMP antagonists, such as a combination of Chordin, Noggin, and Follistatin, or of Cerberus, Chordin, and Noggin, are simultaneously depleted in the organizer by antisense morpholino oligonucleotides, there is a dramatic failure of neural and other dorsal development, and an expansion of ventral and posterior fates. These experiments show that BMP inhibitors emanating from the Spemann organizer are required for neural induction.

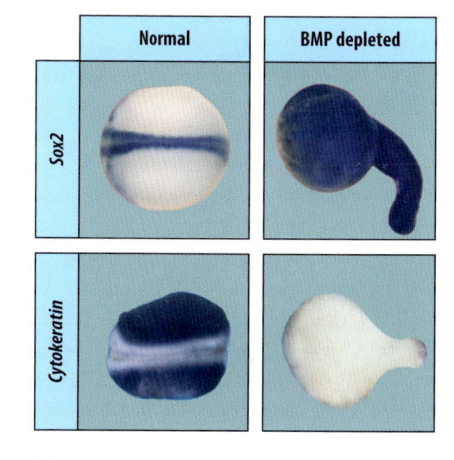

Fig. 4.10 In the absence of bone morphogenetic protein (BMP) signaling, the embryo becomes covered in neural tissue. Left panels: normal neurula-stage *Xenopus* embryos stained blue for expression of *Sox2* mRNA, a neural marker (upper panel) or *cytokeratin* mRNA, an epidermal marker. Right panels: neurula-stage *Xenopus* embryos that have been simultaneously depleted of all four BMP mRNAs by injection of antisense morpholinos at the 2-4-cell stage and are stained similarly to the embryos on the left. The BMP-depleted embryos are completely covered in neural tissue.

From De Robertis, E.M.: **Spemann's organizer and the self-regulation of embryonic fields.** *Mech. Dev. 2009, 126: 925–941.*

Fig. 4.11 Inhibition of bone morphogenetic protein (BMP) signaling is required for induction of the neural plate. Left panel: BMP (green) is expressed throughout the blastula including the ectoderm, and the organizer is just beginning to produce BMP antagonists. Right panel: in the early gastrula, BMP antagonists secreted from the Spemann organizer inhibit BMP signaling in adjacent regions in all three germ layers. This contributes to the induction of the neural plate—the formation of neural tissue from the ectoderm.

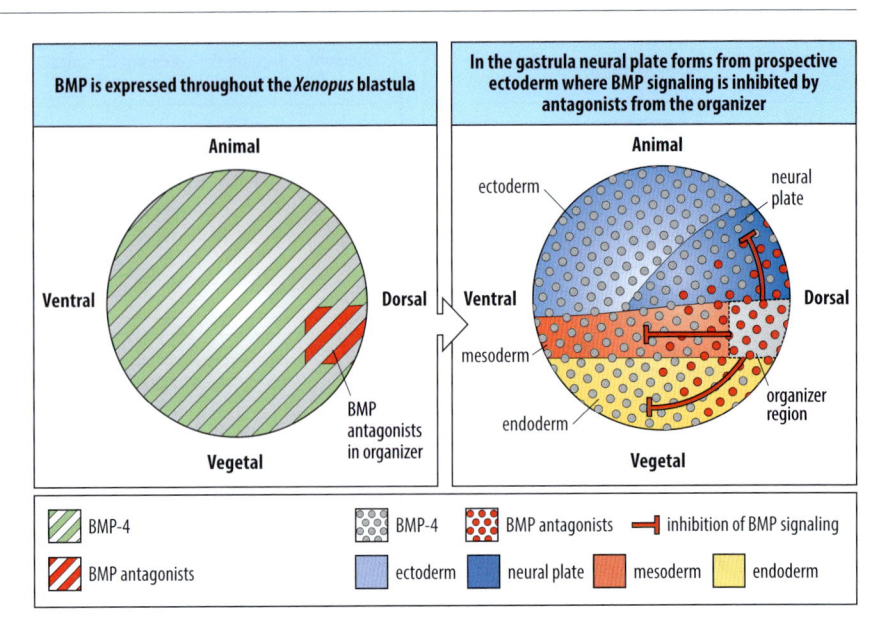

The classic default model for neural induction, mediated by the antagonism of BMPs by BMP antagonists, is, however, an over-simplification, as the growth factor FGF also contributes to neural development in *Xenopus*. FGF signals are produced by cells in the blastula and are thought to act as early neural inducers. Classic studies had shown that explanted newt ectoderm can spontaneously differentiate into neural tissue, and this is thought to be caused by activation of the Ras–mitogen-activated protein kinase (Ras–MAPK) pathway. Activation of this pathway occurs as a result of FGF signaling (Box 4A) and treatment of explanted newt ectoderm with small-molecule inhibitors of the pathway prevents neural differentiation. The Ras–MAPK signaling pathway is also activated when *Xenopus* ectodermal cells are dissociated, and in fact, simply dissociating *Xenopus* ectoderm cells for a few hours can result in their differentiation as neural cells after reaggregation. Induction of the neural plate therefore involves FGF signaling, not just antagonism of BMP signaling. There may also be cross-talk between the two signaling pathways. MAPK, activated as a result of receipt of an FGF signal, can interfere with BMP signaling by causing an inhibitory phosphorylation of Smad1, which is a component of the intracellular signaling pathway stimulated by BMPs (see Box 3F). Therefore, FGF signaling could also contribute to the inhibition of BMP signaling and could help to allow neural induction in this indirect way.

Although neural induction is a complex process, the role of the organizer in this process illustrates how neural induction and patterning of the ectoderm along the dorso-ventral axis of the body are coordinated. BMP antagonists secreted from the Spemann organizer establish a morphogen-like gradient of BMP activity along the dorso-ventral axis in the ectoderm, just as they do in the mesoderm. This gradient provides positional information. Ectoderm cells exposed to the highest levels of BMP antagonists are specified to form neural tissue; ectoderm cells exposed to the highest levels of BMPs are specified to form epidermis; and as we will describe later, ectoderm cells that are exposed to medium levels of BMPs are specified to form a specialized population of cells known as the **neural plate border**, which immediately adjoins the neural tissue and which generates the neural crest (discussed in Section 4.15). How endoderm is patterned is much less well understood, but potentially, the organizer might also pattern this germ layer by antagonizing BMP signaling (see Fig. 4.11).

> **CELL BIOLOGY BOX 4A** The fibroblast growth factor signaling pathway
>
> The developmentally important growth factors, fibroblast growth factor (FGF) and epidermal growth factor (EGF), along with many other growth and differentiation factors, signal through transmembrane receptors with intracellular tyrosine kinase domains, known as receptor tryrosine kinases. Several different signaling pathways operate downstream of these receptors, to control many aspects of cell behavior throughout an animal's life. The signaling pathway shown here is a simplified version of a pathway triggered by FGF that leads to changes in gene expression that promote cell survival, cell growth, cell division, or differentiation, depending on the cells involved and the developmental context. There are multiple different forms of FGF and several different FGF receptors, which are also used in different contexts. For example, in *Xenopus*, FGF-4 is involved in maintaining *TbxT* expression in the mesoderm, and in the chick, FGF-8 is involved in signaling from the organizing centers in the developing brain (Chapter 10), whereas FGF-10 is crucial for limb development (Chapter 9).
>
> The main intracellular signaling pathway leading from the FGF receptors is the Ras-MAPK pathway, so-called because of the key involvement of the small G protein Ras and the activation of a cascade of serine/threonine kinases that ends in the activation of a mitogen-activated protein kinase (MAPK). The name comes from the fact that FGF and other growth factors can act as mitogens, agents that stimulate cell proliferation. In one variation or another, the central Ras-MAPK module occurs in the signaling pathways emanating from many different tyrosine kinase receptors (Fig. 1).
>
> Binding of extracellular FGF to its receptor results in dimerization of two ligand-bound receptor molecules, activating the intracellular tyrosine kinase domains, which then phosphorylate each other. The phosphorylated receptor tails recruit adaptor proteins (Grb and Sos), which, in turn, recruit and activate Ras at the plasma membrane. This results in the binding and activation of
>
> **Fig. 1**
>
> the first serine/threonine kinase in the cascade, called Raf. This phosphorylates and activates the next protein kinase, MAPK kinase (MAPKK), which phosphorylates and activates MAPK. MAPK may then phosphorylate other kinases and can also enter the nucleus and phosphorylate transcription factors, thus activating gene expression. There are several different MAPKs in vertebrate cells that are used in different pathways and target different transcription factors.

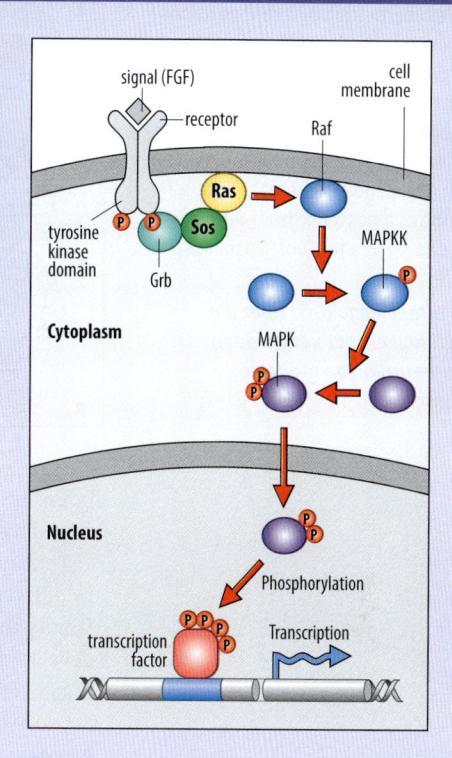

4.4 Factors that pattern the antero-posterior body axis also pattern the antero-posterior axis of the nervous system

The Spemann organizer and its derivatives coordinate the patterning of the antero-posterior axis of the central nervous system with that of the antero-posterior axis of the body, so that the brain will develop in the head and the spinal cord in the trunk. These ideas were first demonstrated through classic grafting experiments in the newt, which showed that pieces of anterior (prechordal) mesoderm and anterior endoderm taken from a neurula-stage embryo and placed in the blastocoel of an early newt embryo induce a brain (within a head), whereas pieces of posterior mesoderm (including notochord) induce a spinal cord (within a trunk) (Fig. 4.12). These grafting experiments were interpreted as showing that the mesoderm underlying the neural plate in different positions along the antero-posterior axis produces different factors, with those produced by anterior mesoderm and endoderm promoting

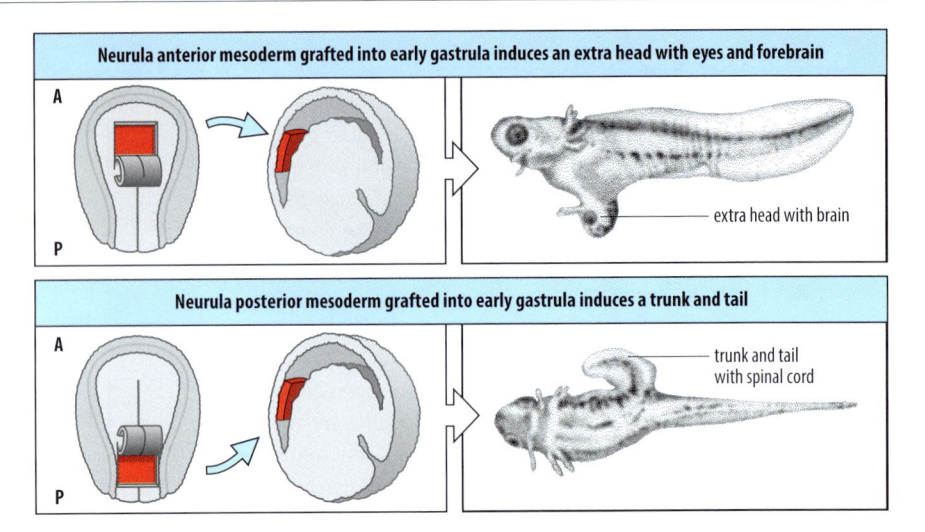

Fig. 4.12 Induction of the nervous system by the mesoderm is region specific. Mesoderm from different positions along the dorsal antero-posterior axis of early newt neurulas induces neural structures appropriate to that region when transplanted to ventral regions of early gastrulas. Anterior mesoderm/anterior endoderm forms a head with a brain (top panels), whereas posterior mesoderm forms a posterior trunk with a spinal cord ending in a tail (bottom panels). A, anterior; P, posterior.

Illustration after Mangold, O.: Über die induktionsfahigkeit der verschiedenen bezirke der neurula von urodelen. Naturwissenschaften *1933,* **21***: 761-766.*

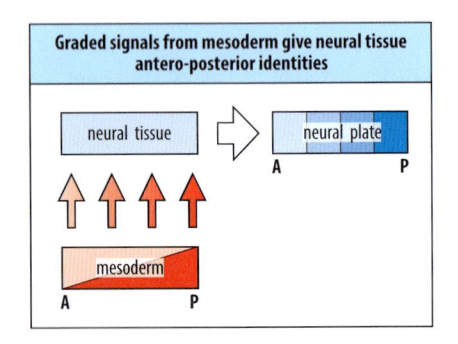

Fig. 4.13 Graded signals pattern the neural plate. Quantitative differences in a signal along the antero-posterior axis help to pattern the neural plate. For example, the levels of Wnt/β-catenin signaling increase from anterior to posterior, and the high levels at the posterior end confer a more posterior positional value on the neural plate.

Illustration after Kelly, O.G., Melton, D.A.: **Induction and patterning of the vertebrate nervous system.** Trends Genet. *1995,* **11***: 273-278.*

brain development, and those produced by posterior mesoderm promoting spinal cord development.

This interpretation was found to be correct when the nature of the factors produced in different regions of mesoderm was uncovered. Importantly, it was found that the factors that direct 'head' and 'trunk' organization also direct brain and spinal cord development. Recall that BMP antagonists are expressed throughout the Spemann organizer (see Section 4.1), and that they induce neural tissue in the gastrula (see Section 4.3). And that Wnt antagonists such as Cerberus, Dickkopf1, and Frzb are expressed in a subpopulation of organizer cells and their derivatives, and direct head organization by antagonizing posteriorizing Wnt signals (see Section 4.2). At late gastrula and neurula stages, expression of these antagonists persists in mesoderm/endoderm cells that derive from the organizer. Wnt and BMP antagonists are therefore secreted from head endoderm and prechordal mesoderm underlying the neural plate at the anterior (head) end of the embryo, and BMP antagonists are secreted from notochord underlying the neural plate in more posterior parts of the embryo. The combined action of BMP and Wnt antagonists promotes the development of the brain. When the Wnt antagonist Dickkopf1 is overexpressed, it is able to induce extra brains when in combination with BMP antagonists.

The gradient of Wnt activity along the antero-posterior axis of the neurula (see Fig. 4.7) embeds a fine-grained pattern onto the neural plate, with higher levels of Wnt signaling conferring more posterior positional values (Fig. 4.13). In animal cap explants taken from embryos injected with mRNA for the BMP antagonist Noggin, and used to represent anterior neuralized tissue, increasing levels of Wnt signaling induce the expression of progressively more posterior neural marker genes.

Additional signals including FGFs and retinoic acid also act as posteriorizing morphogens, differentially activating posterior neural genes in a concentration-dependent manner. Thus, both quantitative and qualitative differences in signaling can account for antero-posterior neural patterning, and several different signals are involved.

In addition to establishing a crude pattern along the antero-posterior axis of the neural plate, factors that are first produced by cells when they were in the organizer begin to establish a pattern along the dorso-ventral axis of the neural tube that is formed by the folding of the neural plate. This occurs because axial mesoderm cells that derive from the Spemann organizer come to lie under the neural tube, adjacent to its ventral midline. Their continued expression of BMP antagonists, such as Chordin and Noggin, together with the initiation of expression of another secreted morphogen, Sonic hedgehog (Shh), will be essential to establishing ventral pattern in the neural tube. This is discussed in Chapter 10.

4.5 The final *Xenopus* body plan emerges by the end of gastrulation and neurulation

Gastrulation transforms the spherical *Xenopus* blastula into an embryo with the three germ layers in their final positions in the body (Fig. 4.14, left panel). The cells that involuted first through the dorsal lip of the blastopore form anterior endoderm and prechordal mesoderm. The cells that involuted next form the notochord that runs along the dorsal midline of the body. The notochord is flanked immediately on each side by more ventral mesoderm, the paraxial mesoderm, which by the end of gastrulation is beginning to segment to form blocks of somites, starting at the anterior end. The neural plate is induced at late blastula and gastrula stages, and then undergoes neurulation. The edges of the neural plate roll upwards to form the neural folds, which meet over the midline to form a tubular structure, the neural tube, which is then covered by ectoderm (see Fig. 4.8). The internal structure of the *Xenopus* embryo just after the end of neurulation is illustrated in Figure 4.14 (center and right panels). The main structures that can be recognized are the neural tube, the notochord, the somites, the lateral plate mesoderm (which lies ventral to the somites on either side), the endoderm lining the gut, and the ectoderm covering the whole embryo.

Figure 4.15 summarizes which signaling pathways are active and which are antagonized in the different germ layers at the late blastula stage. The four main signaling pathways are the transforming growth factor (TGF)-β pathways (Nodals and BMPs), the FGFs, and the Wnts. For all three germ layers, we can see that their specification and patterning involves not only activating particular signaling pathways, but also ensuring that other pathways are inactive. Although this diagram is a useful summary, bear in mind that early development of the body plan in *Xenopus* does not simply depend on the presence or absence of particular signals in different regions of the embryo, but also on graded signaling and the duration of signaling.

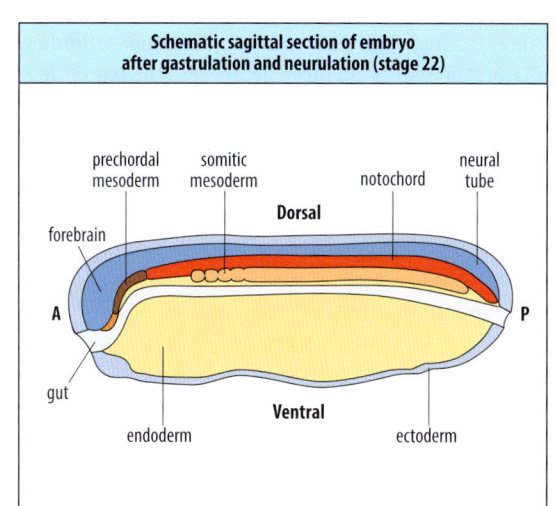

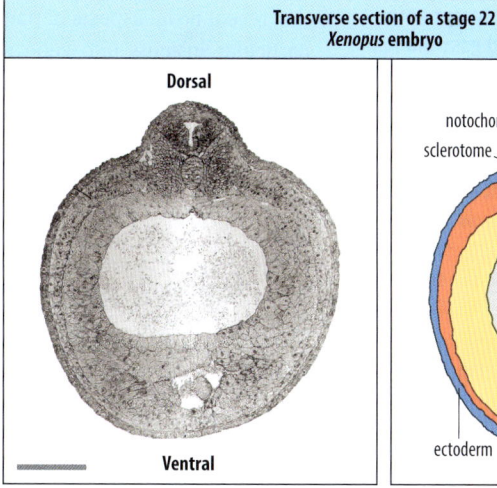

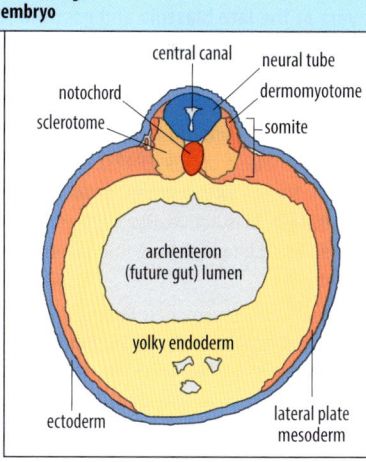

Fig. 4.14 The final body plan of the embryo emerges during gastrulation and neurulation. Left panel: schematic diagram of a sagittal section through a tailbud-stage (stage 22) *Xenopus* embryo after gastrulation and neurulation. The germ layers are now all in place for future development and organogenesis. The mesoderm gives rise to the prechordal mesoderm (brown), notochord (red), somites (pale orange), and lateral plate mesoderm (not shown in this view). The endoderm (yellow) has moved inside to line the gut. The neural tube (dark blue) forms from dorsal ectoderm, and the ectoderm that will form the epidermis (light blue) covers the whole embryo. The antero-posterior axis has emerged, with the head at the anterior (A) end. P, posterior. Center and right panels, a cross-section through a stage-22 *Xenopus* embryo. The most dorsal parts of the somites have already begun to differentiate into the dermomyotome, which will give rise to the trunk and limb muscles and the dermis, while the ventral parts of the somites, the sclerotome, will give rise to the vertebral column. Scale bar: center and right panels = 0.2 mm. Left panel not to same scale.

Photograph reproduced with permission from Hausen, P., Riebesell, M.: The Early Development of Xenopus laevis. Berlin: Springer-Verlag, 1991.

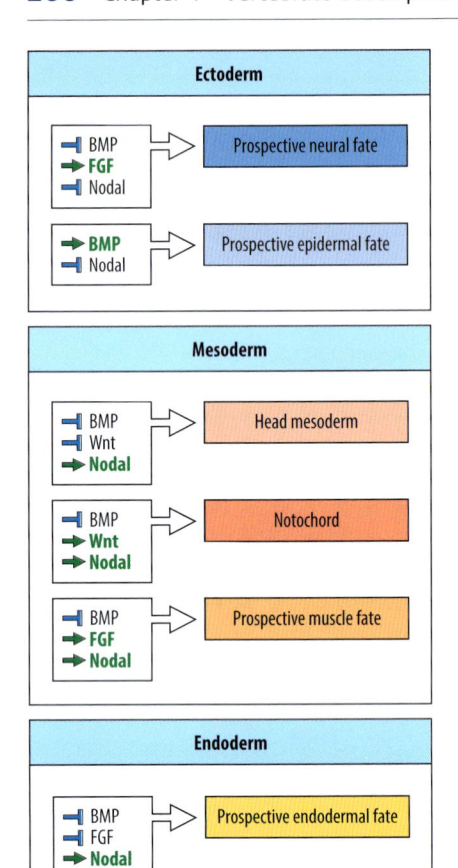

Fig. 4.15 Summary of active and antagonized signals in the different germ layers at the late blastula and gastrula stages. Four main families of secreted signal proteins are involved in germ-layer specification and patterning in *Xenopus*: Wnts, fibroblast growth factors (FGFs), bone morphogenetic proteins (BMPs), and Nodals. Green lettering with a green arrow indicates that the signal is active. Black lettering with a blue barred line indicates that the signal is inhibited.

Adapted from Heasman, J.: **Patterning the early *Xenopus* embryo**. Development 2006, **133**: 1205-1217.

SUMMARY

Development of the vertebrate nervous system, which forms from the neural plate, depends both on early signals from the organizer and on signals from the mesoderm that comes to lie beneath the neural plate ectoderm during gastrulation. Inhibition of BMP signaling by BMP antagonists produced by the organizer, such as Noggin, is required for neural induction. Patterning of the neural plate involves both quantitative and qualitative differences in signaling by the underlying mesoderm, with high levels of Wnt, FGF, and retinoic acid signaling specifying more posterior positional values.

Organizer activity in the zebrafish, chick, and mouse

We now look at patterning of the dorso-ventral and antero-posterior body axes and neural induction in our other model organisms. As a reminder, around the early stages of gastrulation, there are considerable topological differences among the four embryos. However, comparison of their fate maps shows that the relative location of cells that will give rise to equivalent tissues in vertebrate embryos is highly conserved, despite the very different shapes of the embryos (Fig. 4.16; the individual fate maps are discussed in more detail in Chapter 3). And indeed, many of the proteins that direct development of the dorso-ventral and antero-posterior body axes in *Xenopus* are conserved in other vertebrates. So although differences in geometry result in 'organizer' activity being somewhat dispersed in space and time in embryos of different species, the zebrafish shield and the chick Hensen's node are functionally equivalent to the Spemann organizer. Each can induce a second body axis if transplanted to another embryo. In this part of the chapter we touch on events and mechanisms that are similar in other model organisms to those in *Xenopus*, but focus mainly on studies that reveal novel aspects of patterning. For instance, studies in zebrafish have afforded particular insights into gene regulatory networks involved in mesoderm patterning and have revealed that in this organism, patterning of the orthogonal antero-posterior and dorso-ventral axes are temporally linked. We also focus on features in which the other model organisms differ from *Xenopus*. These include the role of proteins other than BMP antagonists that operate in neural induction, and the limited inductive capacity of the mouse node.

4.6 The shield in zebrafish has organizing activity

The zebrafish shield can induce an almost-complete second body axis if transplanted to another embryo at an appropriate stage (see Fig. 3.33), which demonstrates that it can confer pattern along the dorso-ventral and anterio-posterior axes of the host and can direct the formation of neural tissue. Dorso-ventral patterning of the mesoderm and induction of neural tissue (at least that of the head and trunk) are very similar in zebrafish and *Xenopus*. Under the action of BMP antagonists secreted from the shield, initially widespread BMP proteins become restricted to the ventral half of the embryo, and act as morphogens to specify the development of different structures at discrete dorso-ventral positions. High BMP signaling activity specifies ventral tissues such as epidermis and blood, intermediate BMP signaling levels specify more dorsal tissue structures such as neural crest and kidney, whereas the absence of BMP signaling specifies dorsal tissues such as neural tissue and somites. As in *Xenopus*, experimentally inhibiting BMP signaling results in loss of ventral tissues such as epidermis, pronephros (the kidney precursor), and blood, and the concurrent expansion of dorsal tissues such as neural tissue and somites.

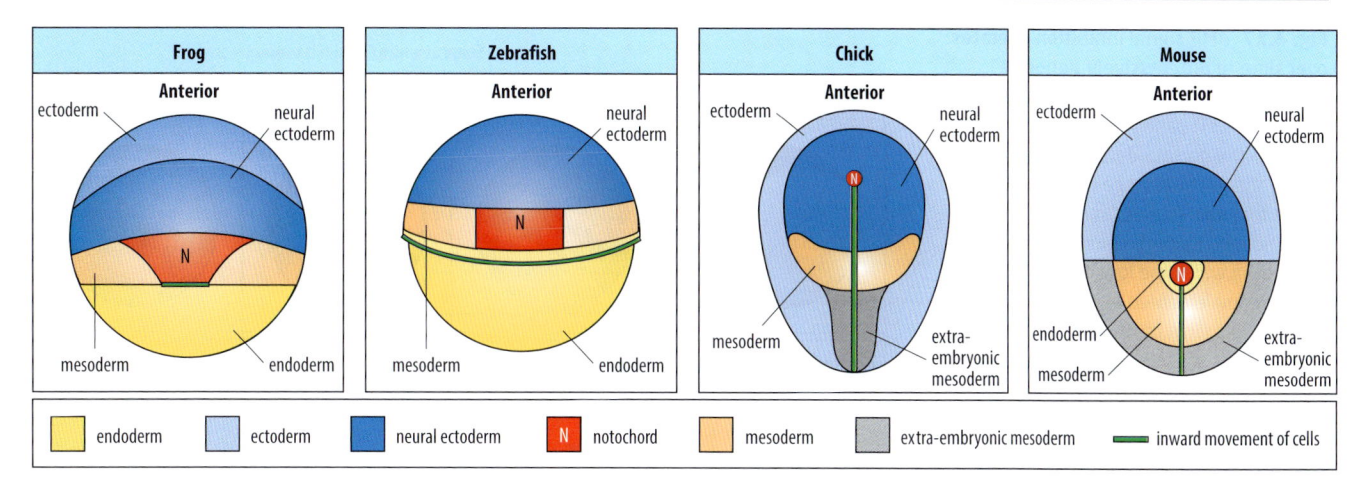

Fig. 4.16 The fate maps of vertebrate embryos at comparable developmental stages. In spite of all the differences in early development, the fate maps of vertebrate embryos at stages equivalent to a late blastula or early gastrula show strong similarities when we look at the relationship between the germ layers and the site of inward movement of cells at gastrulation. The relative location of cells that will give rise to equivalent tissues in vertebrate embryos is also conserved, despite the very different shapes of the embryos. All maps are shown with the future axial mesoderm (notochord and prechordal mesoderm) in the center. The prospective neural ectoderm lies adjacent to the axial mesoderm, with the rest of the ectoderm anterior to it. The mouse fate map depicts the late gastrula stage. The epidermal ectoderm of the zebrafish is on its ventral side. The endoderm has already ingressed in the chick. Individual fate maps are discussed in more detail in Box 3A (*Xenopus*), Section 3.17 (zebrafish), Section 3.21 (chick), and Section 3.27 (mouse).

However, there are differences. In *Xenopus*, organizer activity for heads, trunks, and tails is all spatially contained within the Spemann organizer at the dorsal lip where all the mesoderm moves inside (see Fig. 4.16), whereas in zebrafish, in which the mesoderm involutes in a ring around the edge of the blastoderm (see Fig. 4.16), such activity is spatially distributed in the mesoderm: transplantation of the shield induces heads and trunks; transplantation of the intermediate mesoderm induces just trunks; and transplantation of the ventral mesoderm induces just tails.

In the mesoderm, many transcription factors are conserved between *Xenopus* and zebrafish, underlining their essential roles. Their study in zebrafish has advanced our understanding of the gene regulatory networks that operate to refine mesoderm and to confer particular cell behaviors on differentiating mesoderm populations. The zebrafish *no tail* gene, which is homologous to the *Xenopus* T-box gene *TbxT* (*brachyury*), is one of the earliest genes expressed in the mesoderm. Using the technique of chromatin immunoprecipitation followed by DNA sequencing (ChIP-seq, see Section 3.2), the transcription factor No tail has been shown to activate a network of genes encoding transcription factors and signal molecules involved in mesoderm patterning, mesoderm differentiation, and mesoderm cell behaviors (Box 4B).

Cell specification along the antero-posterior axis of zebrafish is directed through the same mechanisms that operate in *Xenopus*. Gradients of Wnt and BMP signaling are established, with their high points in posterior and ventral tissues. The gradients are shaped by antagonists of BMP and Wnt, which are secreted from the dorsal shield, and by Nodal, which induced the shield (discussed in Chapter 3). The head is induced in regions where BMP and Wnt signaling is low or absent, and the tail is induced where BMP and Wnt signaling is high. The ratio of Nodal to BMP signaling is also a major determining factor for specifying different cells along the axis. High Nodal-to-BMP ratios induce posterior head structures, medium ratios induce the trunk, and low Nodal-to-BMP ratios organize the tail. The requirement for inhibition of Wnt signaling for head formation is confirmed by a mutation in the zebrafish version of a gene for the transcription factor T-cell-specific factor 3 (TCF3). TCF3 represses Wnt target genes when the pathway is inactive (see Box 3E). When *Tcf3* is mutated, this

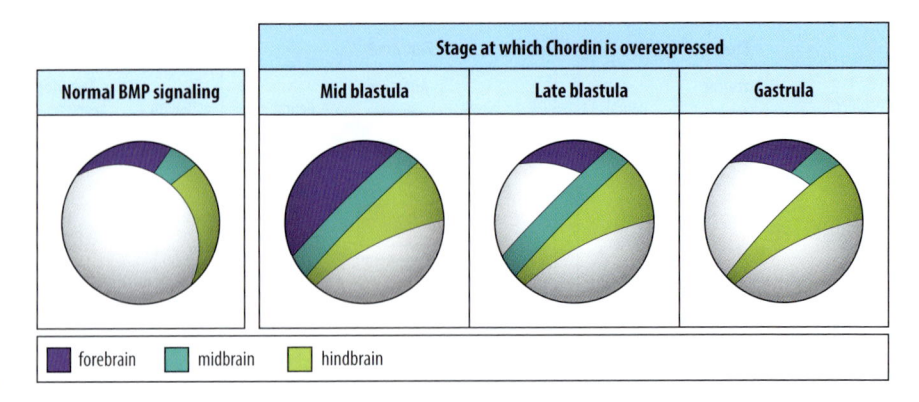

Fig. 4.17 BMP signal inhibition operates over time to progressively generate anterior to posterior neural tissue in the zebrafish. Left panel: the neural plate forms on the dorsal side of the embryo and by the gastrula stage consists of future forebrain (purple), midbrain (blue), and hindbrain (green) regions. Right panel: embryos in which the loss of BMP signaling has been implemented at different stages by overexpression of the BMP antagonist Chordin. Loss of BMP signaling at the mid-blastula stage causes a radial expansion of all brain regions. Loss at the late-blastula stage leads to a radial expansion of midbrain and hindbrain, but the forebrain is restricted as in the wild-type embryo. Loss of BMP signaling in the early gastrula leads to a radial expansion of hindbrain, but the forebrain and midbrain are restricted as in the wild type.

Adapted from Tuazon, F.B., Mullins, M.C.: *Temporally coordinated signals progressively pattern the anteroposterior and dorsoventral body axes.* Semin. Cell Dev. Biol. 2015, **42**: 118-133.

enhances Wnt activity and the mutant lacks a head (hence, the mutation is named *headless*). The importance of Nodal and BMPs was shown by transplanting ectopic sources of Nodal and BMP, which caused the induction of an almost complete secondary embryonic axis.

Neural induction in the zebrafish also closely parallels that in *Xenopus*. BMP antagonists are secreted from the zebrafish shield, and induce neural tissue in adjacent ectoderm. In zebrafish, BMP signaling can be inhibited over very discrete time windows by using a transgene expressing the BMP antagonist Chordin under control of a heat-shock promoter. Studies using this construct demonstrate that the extent of time over which BMPs are inhibited in ectoderm dictates the type of neural tissue that is induced. If BMP signaling is inhibited on the ventral side of the embryo, there is an expansion of neural tissue. When BMP is inhibited in the mid-blastula, markers of forebrain, midbrain, and hindbrain are all expanded. However, when BMP is inhibited a short time later, in the late blastula, only markers of midbrain and hindbrain are expanded, showing that forebrain-like tissue has already been induced. Similarly, when BMP is inhibited in the gastrula, only hindbrain markers are expanded (Fig. 4.17). Similar experiments using mesodermal markers for position along the antero-posterior axis show that the mesoderm is also patterned in a progressive, temporal manner from anterior to posterior. Thus, in the zebrafish, BMPs have a role in integrating positional information along two orthogonal axes, to progressively pattern the embryo over time.

Although the inhibition of BMP signaling is involved in induction of neuroectoderm that develops into the brain, the neuroectoderm that develops into the spinal cord is some distance away from the shield on the ventral-vegetal side of the zebrafish embryo, where organizer signals do not reach (Fig. 4.18). Here the initiator of neural development is FGF, and BMP signals, which are high in the ventral region, seem to push the neural ectoderm towards a posterior fate.

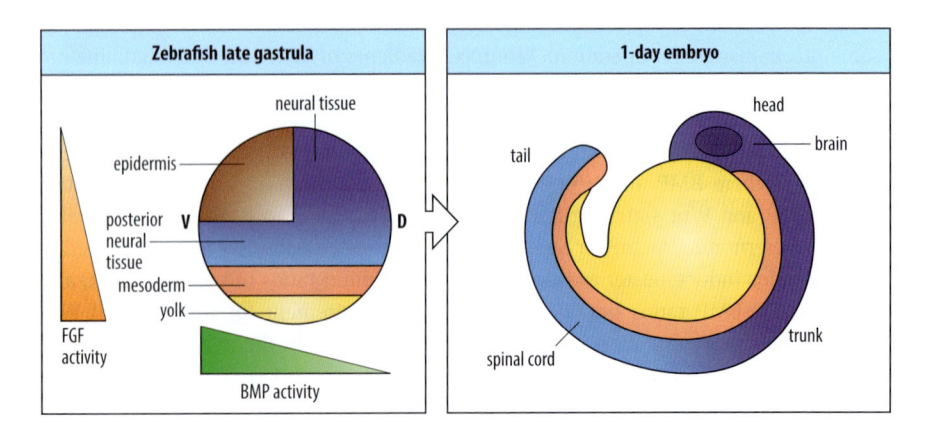

Fig. 4.18 Prospective spinal cord in the zebrafish embryo is distant from the organizer. In the late gastrula, the ectoderm that will form the spinal cord is situated on the other side of the gastrula from the organizer, in the ventral ectoderm, and is too far away to be influenced by signals from the organizer. Fibroblast growth factor (FGF) signals in the ventral (V) region induce this ectoderm as prospective neural tissue and bone morphogenetic proteins (BMPs) promote its development to form neural tissue (spinal cord). The epidermis is not shown in the 1-day embryo for simplicity. By this stage it covers the whole embryo. D, dorsal.

Illustration from Kudoh, T., et al.: *Combinatorial Fgf and Bmp signalling patterns the gastrula ectoderm into prospective neural and epidermal domains.* Development 2004, **131**: 3581-3592.

EXPERIMENTAL BOX 4B A zebrafish gene regulatory network

The zebrafish transcription factor No tail (Ntl) is a homolog of the *Xenopus* and mouse mesodermal transcription factor TbxT (Brachyury), and the technique of chromatin immunoprecipitation followed by DNA sequencing (ChIP-seq) has been used to uncover the targets of Ntl in mesoderm specification, patterning, and differentiation. Ntl activates a network of genes encoding transcription factors (e.g. Snail and other T-box (Tbx) proteins), intercellular signals (e.g. FGFs, Notch, and Wnt-11 in its later role as a signal directing cell rearrangements that lengthen the embryo during gastrulation), and genes involved in the differentiation of specific cell types (e.g. genes encoding the transcription factors Pax3 and Myf5 involved in muscle differentiation, discussed in Section 6.6). The network is shown in Figure 1. Genes are grouped to represent different aspects of Ntl activity in the mesoderm: control of morphogenetic movements; notochord specification; muscle specification; specification of posterior tissue; and left-right patterning. Expression of all the genes shown here is activated by Ntl. Lines ending in arrows indicate activation; those ending in barred lines inhibition. Double arrowheads leading from a gene indicate that the gene product is an intercellular signal molecule. Solid arrowed lines joining two genes indicate that the interaction has been verified both genetically and by direct binding of the encoded transcription factor to the target gene control region. Dashed lines indicate that the interaction has been verified either genetically (in the case of signal molecules) or by transcription factor binding to its target. Shaded boxes indicate that additional assays have also shown direct regulation of the target gene by Ntl. Genes whose encoded proteins regulate other genes have been colored for clarity. Note the number of gene products that feed back to regulate expression of the *ntl* gene itself.

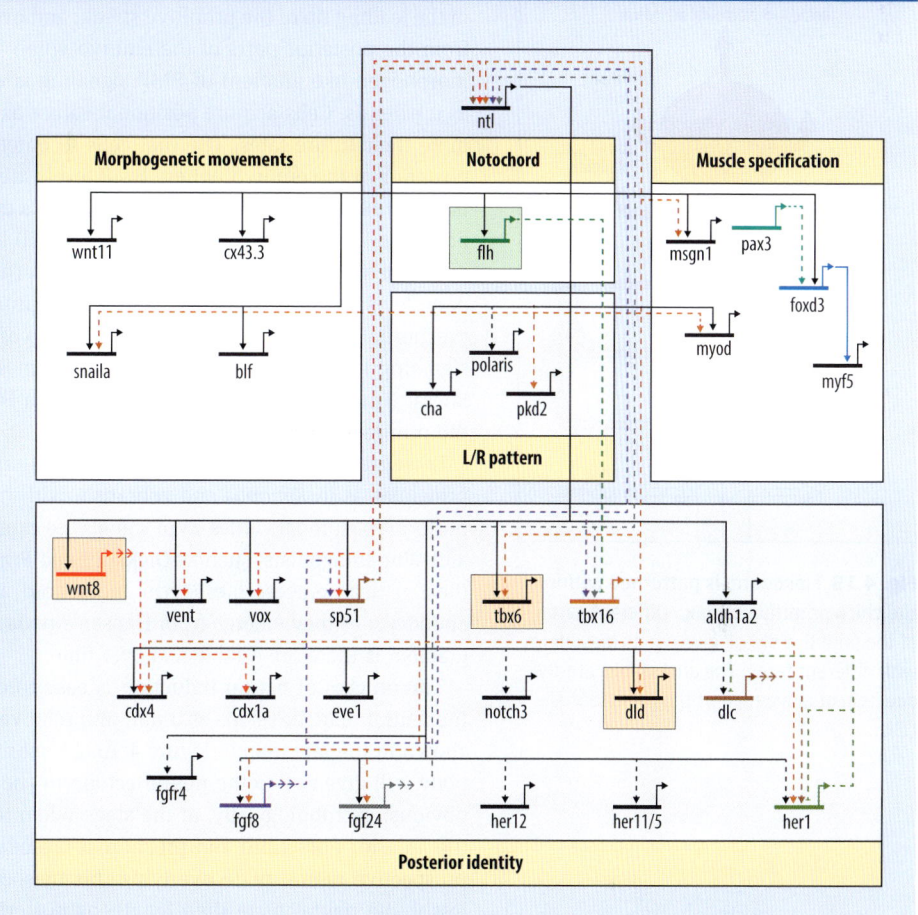

Fig. 1
From Morley, R.H., et al. *A gene regulatory network directed by zebrafish No tail accounts for its roles in mesoderm formation.* Proc. Natl Acad. Sci. USA 2009, **106**: 3829–3834.

4.7 Hensen's node in chick is the embryonic organizer

The avian equivalent of the Spemann organizer is Hensen's node, the region that forms at the anterior end of the fully extended primitive streak (see Fig. 3.39, lowest panel). Hensen's node has organizing activity and can induce a second axis when transplanted to another embryo at the appropriate stage (see Fig. 3.40). The second axis is patterned along both dorso-ventral and antero-posterior axes and has its own nervous system. The advancing tip of the primitive streak also has organizing activity, and this means that the patterning of the axes and the directed development of neural tissue has begun even before the fully extended streak stage.

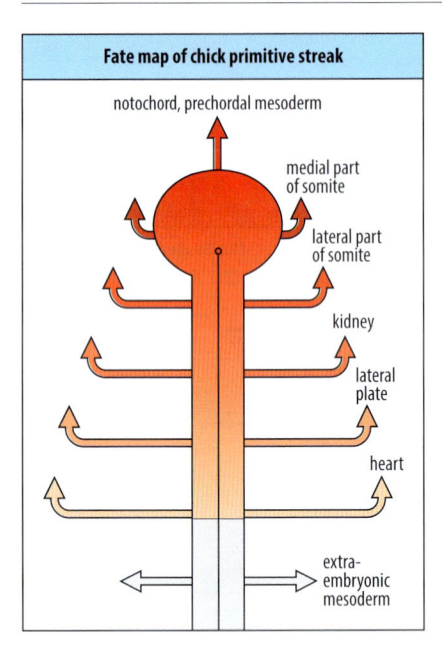

204 Chapter 4 Vertebrate development II: completing the body plan

Fig. 4.19 Mesoderm is patterned within the chick primitive streak. Different parts of the primitive streak give rise to mesoderm with different fates. The arrows indicate the direction of movement of the mesodermal cells.

Fig. 4.20 Neural tissue formation in the chick embryo. Chick embryos at successive stages of development (Hamburger Hamilton (HH) 4–6), overlaid with schematics of neural development. First panel: at HH4, the primitive streak has extended maximally and Hensen's node forms at its anterior limit. Prospective neural tissue is induced around Hensen's node. Second panel: at HH5, the node has started to regress (arrow indicates direction of movement). Head-process mesoderm, composed of anterior notochord and prechordal mesoderm (green), extends forwards from Hensen's node, beneath the induced neural plate (blue), now obvious as a thickened region. Third panel: at HH6, the primitive streak and Hensen's node are regressing, depositing posterior notochord in their wake. As the primitive streak regresses, the neural plate lengthens. Neural plate above the head-process mesoderm gives rise to the brain; neural plate above the posterior notochord gives rise to anterior spinal cord.

Patterning of the dorso-ventral axis in the chick follows the same principles as in *Xenopus* and zebrafish. BMP antagonists, such as Noggin and Chordin, are expressed in the leading tip of the primitive streak, and over time become increasingly distanced from the posterior parts of the embryo where BMPs are expressed. Consequently, a morphogen-like gradient of BMP signaling is established along the primitive streak as it extends. Cells acquire positional values as they enter the streak and these determine their future fates. The first cells to enter the streak as it starts to form at the posterior of the embryo, where BMP levels are highest, give rise to extra-embryonic mesoderm. As the streak progresses and cells enter from the epiblast at more anterior positions (albeit where BMP signaling is still high), kidneys and lateral plate mesoderm are specified next; paraxial mesoderm (future somites) is specified by intermediate levels of BMP signaling; and axial (future dorsal) mesoderm—notochord and prechordal mesoderm—is specified at the tip of the fully extended streak, where BMP signaling is low or absent (Fig. 4.19). So, mesoderm is patterned along the medio-lateral (future dorso-ventral) axis through a BMP gradient that is established along the primitive streak.

Likewise, patterning along the antero-posterior axis in the chick follows the same principles as in *Xenopus* and zebrafish, with high levels of Wnt signaling specifying posterior positional values. Wnt signals are expressed in posterior parts of the embryo, and Wnt antagonists, such as Dickkopf and Frzb, are expressed in Hensen's node and then in head-process mesoderm (notochord and prechordal mesoderm), and head endoderm as they emerge from Hensen's node and extend forwards. A Wnt signaling gradient is therefore established over time.

The process of neural induction is easy to observe in the chick embryo, because the neural plate develops as a flat, and relatively large, sheet of neuroepithelium. As shown in the fate map in Figure 4.16, a region of the epiblast lying around Hensen's node will give rise to the neural ectoderm (neural plate). The neural plate becomes obvious, morphologically, at the stage when head-process mesoderm that will form the anterior notochord and prechordal mesoderm emerges from Hensen's node: the prospective neural plate overlying this mesoderm begins to visibly thicken. Seamlessly, the neural plate then lengthens posteriorly as the primitive streak regresses (Fig. 4.20). Neural plate tissue generated anterior to Hensen's node will become forebrain, midbrain, and hindbrain, whereas neural plate generated as the primitive streak regresses will become spinal cord. Unlike in *Xenopus* therefore, the spinal

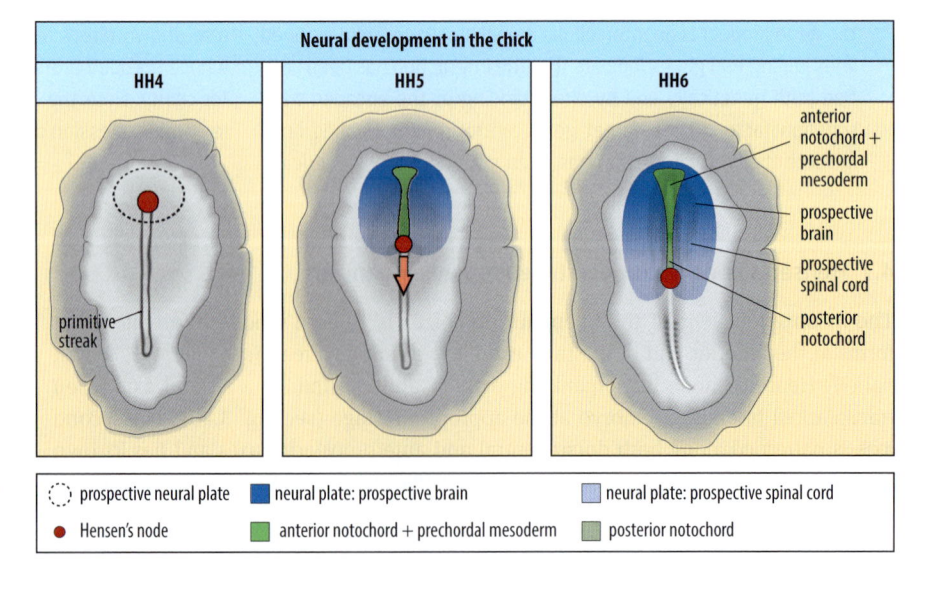

Fig. 4.21 Cross-talk between the BMP and FGF signaling pathways. A simplified view of the interaction between the BMP and FGF signaling pathways, by which FGF induces a neural fate. (1) When BMPs bind BMP receptors, Smad1/5 is activated by phosphorylation at its C-terminal end and neural fate is inhibited. (2) Binding of receptor tyrosine kinases by FGFs can lead to activation of the Ras-MAPK pathway, which in turn can cause phosphorylation of the Smad1 linker domain. This results in inhibition of Smad1 and so alleviates the inhibition of neural fate. (3) FGFs can also stimulate Churchill (ChCh), a transcriptional activator, which upregulates the gene for Smad-interacting protein (*Sip1*). Sip1 protein suppresses BMP signaling, and so also relieves the inhibition of neural fate.

cord is progressively generated from cells located in the epiblast next to the primitive streak.

The detailed neural-directing properties of the avian node have been investigated by transplanting quail nodes to chick embryos. Quail cells can be distinguished from chick cells by their heavily staining nuclei. When a node from a gastrula-stage quail embryo is grafted to a chick embryo at the same stage of development, the transplanted node can induce a complete additional axis with a head (see Fig. 3.40). However, when the node is taken from an older embryo, after the head-process mesoderm has emerged, it only induces trunk neural tissue (see Fig. 3.40). Thus, the properties of the node change over time during gastrulation in the same way as those of the *Xenopus* Spemann organizer (see Section 4.2).

The BMP antagonists Chordin and Noggin, which have such an important role in *Xenopus* and zebrafish neural induction, are produced by Hensen's node, but in fact, the crucial neural-inducing factor secreted by Hensen's node is FGF. Nonetheless, FGF operates to induce neural tissue through cross-talk with the BMP signaling pathway. FGF signaling can repress BMP genes themselves, or can repress components of the BMP signaling pathway. One way it does this is via the Ras–MAPK pathway, which converges on BMP signaling by phosphorylating the linker domain of Smad1, and preventing translocation of Smad1 to the nucleus (Fig. 4.21). In this way, it effectively inhibits BMP signaling, and so promotes a neural fate.

One of the key genes activated by FGF signaling in neural induction in chick embryos is *Churchill*, which encodes a zinc finger transcription factor. When *Churchill* expression is downregulated using an antisense morpholino oligonucleotide, the neural plate does not form. In the normal embryo, *Churchill*, once expressed, activates the Smad-interacting protein (Sip1), which results in the further suppression of BMP signaling and activation of the gene for the transcription factor Sox2, one of the earliest definitive markers of the neural plate.

The double cross-talk between FGF and BMP signaling pathways means that an initial differentiation trajectory (i.e. towards neural fate) is reinforced. Cross-talk between different signaling pathways, and the reinforcement of an initial bias, are each common strategies that are deployed by the embryo in many tissues and at many times.

As in *Xenopus*, antero-posterior patterning of the chick neural plate also depends on interactions with the mesoderm that comes to underlie it. Anterior endoderm and prechordal mesoderm pattern the forebrain and midbrain, whereas anterior notochord patterns the hindbrain (see Chapter 10).

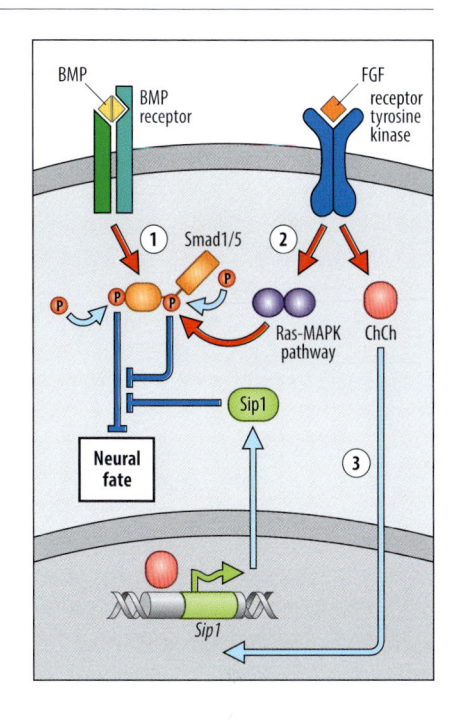

The role of FGF in neural induction in the chick embryo provides another good illustration of how the same signaling factor can be deployed by the embryo for different tasks at different times. We have already seen how Wnt signaling specifies the dorso-ventral axis of the *Xenopus* embryo before gastrulation, and then patterns the antero-posterior body axis later (see Section 4.2). In the chick, FGF is initially produced before gastrulation by Koller's sickle and at that time induces mesoderm (see Fig. 3.45). Epiblast cells are prevented from taking on a neural fate at that time because the chromosomal regions encoding neural-inducing genes are in a 'silenced' state in which genes cannot be transcribed. The lifting of this inhibition during gastrulation is associated with the removal of repressor proteins from the regulatory regions of such genes and the remodeling of the chromatin into an active state by protein complexes called **chromatin-remodeling complexes**. Epiblast cells that receive FGF signals but have not formed mesoderm by this stage can now form neural plate.

Overall, then, there is an essential similarity in the mechanism of neural induction between *Xenopus* and chick. Indeed, Hensen's node can induce neural gene expression in *Xenopus* ectoderm. This suggests that there has been an evolutionary conservation of inducing signals and confirms the essential similarity of Hensen's node and the Spemann organizer. Moreover, early nodes induce gene expression characteristic of anterior amphibian neural structures, whereas older nodes induce expression typical of posterior structures. These results are in line with the hypothesis that the changing tissue composition of Hensen's node over time, described earlier in this section, means that it specifies different antero-posterior positional values in the neural plate at different times.

4.8 Mesoderm patterning in mouse is similar to that in other model organisms, but antero-posterior organizer functions are spatially and temporally separated

The induction of mesoderm and patterning along the dorso-ventral axis in the mouse follows similar principles to those in chick. Mesoderm induction occurs in the primitive streak, which is initiated in the posterior part of the epiblast. Recall that in mouse, an antero-posterior axis is apparent at this early stage of development, as a result of the determination of the anterior end of the embryo by the distal visceral endoderm (DVE) and anterior visceral endoderm (AVE) (see Section 3.26). Posterior epiblast cells, which are maintained as such by virtue of their proximity to high BMP signals, begin to express mesodermal transcriptional factors, such as TbxT, and converge and internalize at the point of convergence to initiate the primitive streak (see Fig. 3.51). As in other vertebrates, Wnt and Nodal signaling are crucial for mesoderm induction. In the mouse, a key Wnt signal is Wnt-3, which is initially expressed in the posterior visceral endoderm and then in the posterior region of the epiblast (see Fig. 3.58). Wnt-3 has two important roles. First, it activates expression of *TbxT*. Second, it maintains *Nodal* expression. This means that, whereas *Nodal* was expressed generally in the epiblast in the pre-gastrula embryo, at gastrulation *Nodal* expression now becomes confined—albeit transiently—to the posterior region (see Fig. 3.58). *Nodal* is only expressed in the cells that will ingress through the primitive streak and is required for mesoderm formation. In mutants lacking *Nodal* function, mesoderm does not form. Nodal, then, is yet another signal protein that acts at different times in development to direct different events, as we have already seen for Wnts and for FGFs in chick. Such powerful and multipurpose signal proteins must be tightly regulated and finely tuned to ensure that the right amounts are present in the right place at the right time (Box 4C).

In the mouse, a common experimental technique to analyze the function of a protein is to knock out the gene that encodes it. However, when the same signals are deployed again and again to effect different developmental events at different times, a straightforward knock-out is not always informative for analyses of later functions. This barrier has been overcome through elegant genetic strategies such as the Cre/*loxP* system, which have been developed to target gene knock-outs to a particular tissue and/or to a particular time in development (Box 4D).

As in the chick, dorso-ventral mesodermal patterning occurs in the mouse streak during gastrulation, with cells as they move through different regions of the streak being specified to form different mesodermal tissues along the dorso-ventral axis (Fig. 4.22). Cells at the very posterior end of the streak are the first to ingress and give rise to extra-embryonic mesoderm. The cells that ingress next in the posterior region form lateral plate mesoderm followed by cells in a slightly more anterior position that

CELL BIOLOGY BOX 4C Fine-tuning Nodal signaling

Mouse Nodal and the corresponding Nodal-related proteins in other vertebrates are secreted signal proteins that are essential for the induction and patterning of the mesoderm and the establishment of left-right asymmetry in early embryos. Signal molecules that have such powerful effects on development must be tightly regulated to ensure that the right amounts are present in the right place at the right time, and the strength of Nodal signaling can be modulated and regulated at several different levels as indicated in Fig. 1.

A first level of control is the production of a functional protein (1). Like many other growth factors, Nodal is synthesized and secreted as an inactive precursor protein that has to be converted to the active form by proteolytic cleavage; this is carried out by enzymes called proprotein convertases. If these are not present, then even if the *Nodal* gene is expressed, no functional protein will be made.

A second level of control is due to secreted extracellular inhibitors that bind to Nodal and/or its receptors (2) and prevent the signal from reaching the target cell, as discussed in this chapter. Like other transforming growth factor-β family members, Nodal binds to heterodimeric receptors known as type I/II receptors, and initiates an intracellular signaling cascade in which proteins called Smads are phosphorylated and enter the nucleus to regulate the expression of target genes (see Box 3F). Extracellular antagonists of Nodal signaling include the proteins Cerberus-1 (Cer1), Cerberus-like 2 (Cerl2), Lefty-1, and Lefty-2.

A third level of control of Nodal signaling is through regulation of transcription of the *Nodal* gene, which is itself a target of Nodal signaling (3). The expression of *Nodal*, as well as the expression of *Lefty1/2* and *Cer1/Cerl2*, can all be increased in response to Nodal signaling, and this can lead to either positive or negative feedback on the amount of Nodal present. A negative-feedback loop, in which Nodal increases the expression of *Lefty*, fine tunes the level of Nodal signaling in the zebrafish blastoderm and in the proximo-distal patterning of the mouse epiblast. In the mouse epiblast, Nodal induces the anterior visceral endoderm (AVE), which then secretes Cer1 and Lefty-1, which, in turn, downregulate *Nodal* expression in an adjacent area of the epiblast, thus enabling this region to become the anterior end of the antero-posterior axis. A positive-feedback loop, in which Nodal signaling upregulates the expression of *Nodal*, operates in the establishment of left-right asymmetry and leads to high levels of Nodal on the left side of the node in mouse embryos (see Section 4.17). A reaction-diffusion type of mechanism may also operate (see Box 9E). Yet another level of fine tuning of Nodal signaling discovered more recently involves the specific inhibition of translation of *Nodal* and *Lefty* mRNAs by microRNAs (4 on Fig. 1).

Surprisingly many of the molecules that have key roles in specifying the vertebrate body plan act as extracellular antagonists, or inhibitors, of intercellular signaling molecules, and block or modulate signaling. Other widely used developmental signals such as Wnts and BMPs have antagonists that counteract their expression or activity at some level, and also make positive- and negative-feedback loops with these antagonists in a similar way to Nodal. In *Xenopus*, for example, the dorsalizing signals produced by the Spemann organizer include the proteins Noggin and Chordin, which block the action of the ventralizing signal BMP-4 (see Section 4.1). In antero-posterior patterning of the epiblast in mouse embryos, the Wnt posteriorizing signal is blocked by the anteriorly produced extracellular antagonist Dickkopf-1 (see Section 3.26).

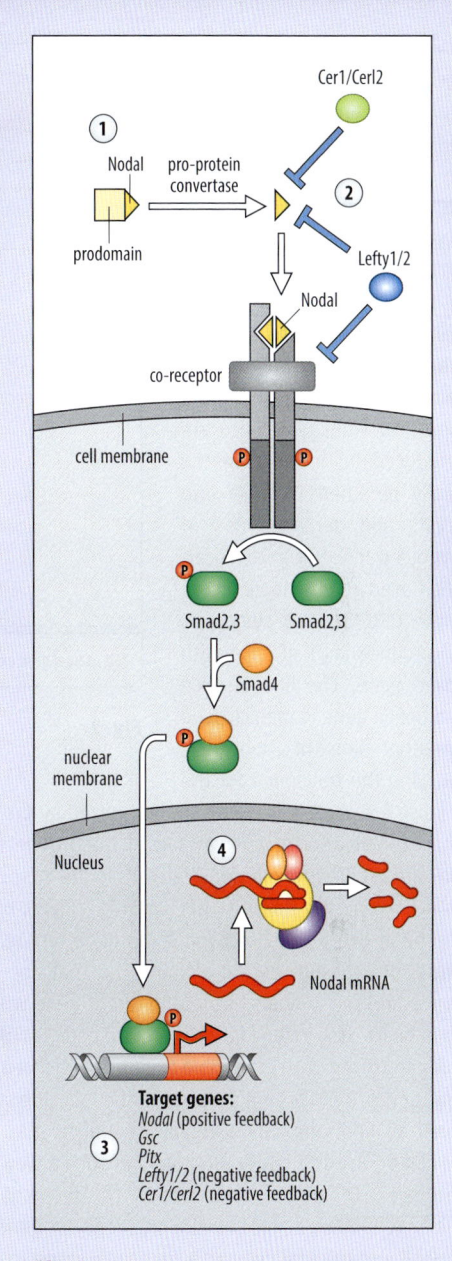

Fig. 1

EXPERIMENTAL BOX 4D The Cre/loxP system: a strategy for making gene knock-outs in mice

One powerful and widely used way of targeting a gene knock-out to a specific tissue and/or a particular time in development is provided by the **Cre/*loxP* system** (Fig. 1). The target gene is first 'floxed' by inserting a *loxP* sequence of 34 base pairs on either side of the gene (hence 'floxed', for 'flanked by *lox*'). This targeting has, until recently, been carried out using homologous recombination in ES cells in culture (see Box 3B). Transgenic mice are then produced by introducing the genetically modified ES cells into blastocysts and deriving lines of homozygous transgenic mice all carrying the floxed gene. In more recent times, the CRISPR-Cas9 system is used to produce lines of mice with the floxed gene. The line of mice with the floxed gene is then crossed with another line of transgenic mice carrying the transgene for the recombinase Cre. *loxP* sequences are recognized by Cre, and the resulting recombination reaction excises all the DNA between the two *loxP* sites. If Cre is expressed in all cells in the offspring, then all cells will excise the floxed target, causing a knock-out of the target gene in every cell. However, if the gene for Cre is under the control of a tissue-specific promoter, so that, for example, it is only expressed in heart tissue, the target gene will only be excised in heart tissue (see Fig. 1). If the *Cre* gene is linked to an inducible control region, it is possible to induce excision of the target gene, at will, by exposing the mice to the inducing stimulus.

As well as knocking out a coding sequence, the Cre/*loxP* method can be used to target promoters in such a way that the associated gene is permanently turned on. This technique is used in cell-lineage studies in mice to turn on a reporter gene such as *GFP* or *lacZ* in particular cells at a particular stage in development. The progeny of the labeled cells can then be distinguished in later-stage embryos or after birth. Transgenic mouse embryos expressing a reporter gene linked to a suitable promoter, or in a particular set of cells, under the control of the Cre/*loxP* system provide a convenient way of tracing cell lineage. Techniques such as these are continually being refined to give better resolution and are particularly useful for tracing cell lineage during later events in embryogenesis, such as organ development, the development of the nervous system, and cell behavior in epithelia in which cells are continually being replaced, such as skin and gut. Clonal cell-lineage analysis is now possible using a transgenic mouse in which a drug-inducible Cre recombinase (linked to the reporter gene *lacZ* (see Box 1E) or to *GFP*) is expressed in every cell; the pregnant mouse can then be treated with an appropriate dose of the drug such that a low frequency of recombination occurs and clones of marked cells are generated in the embryos. This strategy is conceptually similar to clonal analysis in *Drosophila* and led to the discovery of dorso-ventral compartments in the developing mouse limb.

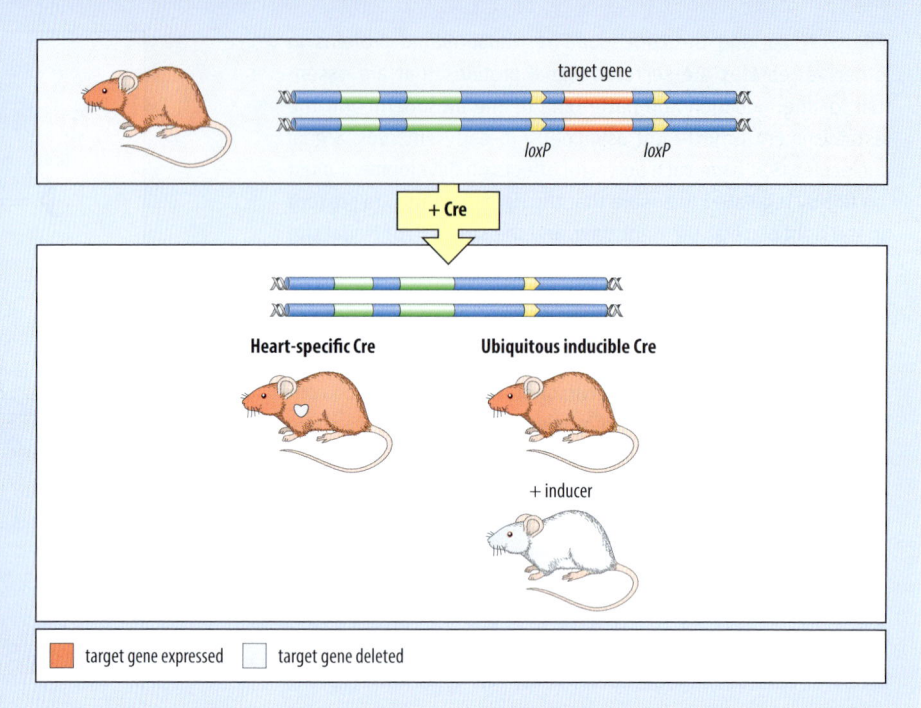

Fig. 1

form intermediate mesoderm. Cells that ingress in more anterior regions of the streak form paraxial mesoderm, while cells that ingress at the anterior-most end of the streak form axial mesoderm and anterior endoderm. These cells ingress together and are therefore frequently known as **axial mesendoderm**. But whether there are bipotential cells that can give rise to either mesoderm or endoderm is still not resolved.

In mouse, the main signal proteins responsible for dorso-ventral mesoderm patterning are BMPs as in other model vertebrates, but Nodal also has a key role. At

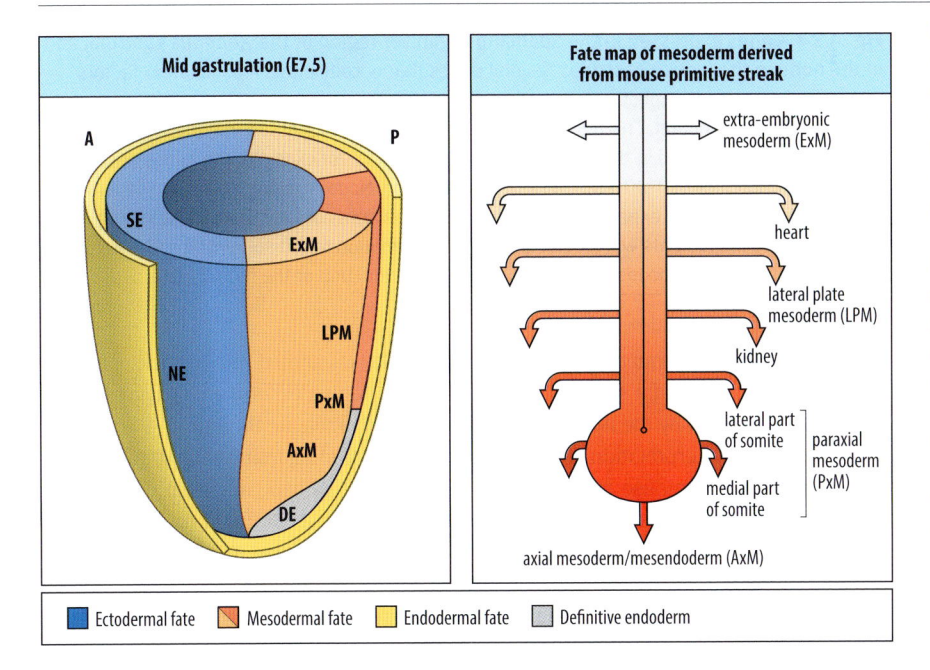

Fig. 4.22 Prospective fates of different regions of the mouse embryo at mid-gastrulation stage. The left panel shows the mouse cup-shaped embryo as shown in Fig. 3.19, with the beginning of the primitive streak at the top, and omitting the more proximal extra-embryonic tissues. Prospective fates: AxM, axial mesoderm/mesendoderm; DE, definitive endoderm; ExM, extra-embryonic mesoderm; LPM, lateral plate mesoderm; NE, neuroectoderm, PxM, paraxial mesoderm (prospective somites), SE, surface ectoderm (prospective epidermis). The right panel shows the patterning of the mesoderm within the mouse primitive streak. Different parts of the primitive streak give rise to mesoderm with different fates. The arrows indicate the direction of movement of the mesodermal cells.

Left panel adapted from Bardot, E.S., Hadjantonakis, A.K.: **Mouse gastrulation: coordination of tissue patterning, specification and diversification of cell fate.** Mech. Dev. *2020,* **163***: 103617.*

around E6.5, the primitive streak forms in a region where there are high levels of BMP, Wnt, FGF, and Nodal signaling (see Fig. 3.58). As the primitive streak elongates and extends distally, *Nodal* expression becomes confined to the anterior region, eventually becoming restricted to the node, while BMP signaling remains high at the posterior of the embryo (Fig. 4.23). This results in opposing gradients of Nodal and BMP signaling along the extending streak. High levels of Nodal in the anterior streak specify axial mesendoderm, while high levels of BMP in the posterior streak specify lateral plate mesoderm. These pathways act in a cross-inhibitory fashion. The activity of Nodal, for example, is attenuated by BMPs, and therefore high levels of BMPs in posterior parts of the embryo help to keep Nodal activity low here and reinforce the Nodal gradient. Levels of BMP signaling in the anterior streak are also kept low by the production of BMP antagonists by the node that include Chordin and Noggin. Overall, mesoderm dorso-ventral patterning is not so fundamentally different in mouse to that in the chick, but the main counteracting factors are Nodal and BMPs, rather than BMP antagonists and BMPs.

To what extent is antero-posterior patterning in mouse also like that in other model organisms? For many years it was thought that the mouse node is functionally identical to the *Xenopus* Spemann organizer and chick Hensen's node in its ability to induce and pattern the antero-posterior axis. Each expresses several homologous genes (Fig. 4.24), the node of both mouse and chick develop at the anterior tip of the primitive streak (see Fig. 3.51), and when a mouse node is grafted into a chick embryo, it induces a complete second axis. In fact, a similar experiment carried out in the 1930s by the British embryologist Conrad Waddington had first suggested that organizers from mammalian

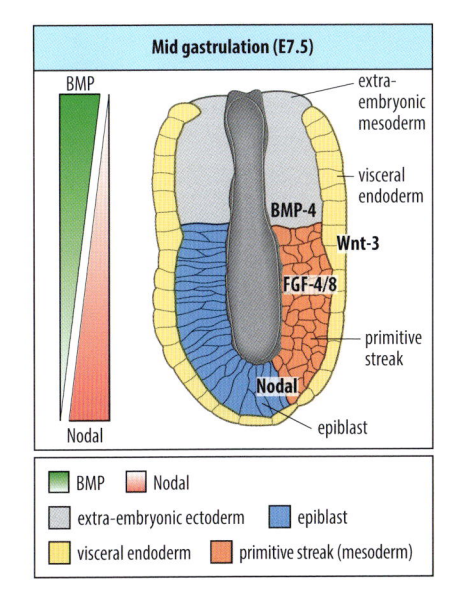

Fig. 4.23 Signals that specify mesoderm patterning within the mouse primitive streak. Mesodermal fates are determined in gastrulating mice by the signaling environment that cells are exposed to as they ingress through and migrate through the primitive streak. The primitive streak develops in regions of the embryo where there are high levels of Wnt-3 and FGF-4/8. Nodal is maintained at high levels in the anterior part of the streak, whereas BMP-4 is maintained at high levels in posterior parts of the embryo and the extra-embryonic ectoderm. The counteracting gradients of BMP and Nodal signaling pattern the mesoderm.

Adapted from Bardot, E.S., Hadjantonakis, A.K.: **Mouse gastrulation: coordination of tissue patterning, specification and diversification of cell fate.** Mech. Dev. *2020,* **163***: 103617.*

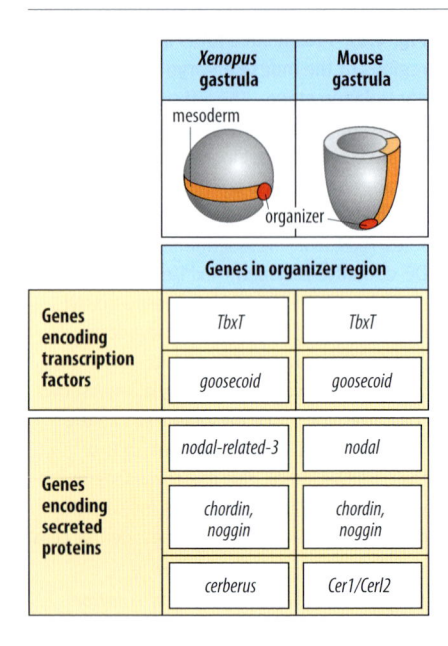

Fig. 4.24 Genes expressed in the Spemann organizer region of the *Xenopus* gastrula, and in the node in the mouse gastrula. Several genes that encode either transcription factors or secreted proteins are specifically expressed in vertebrate organizers and are known to be required for their function. The figure lists some of the genes expressed in both the Spemann organizer of *Xenopus* and in the mouse node. These include, for example, the genes encoding the transcription factors TbxT and Goosecoid. In addition, Nodal signal proteins and several secreted antagonists are expressed that will produce gradients of BMP and Wnt signaling in the embryo. Such gradients will provide positional information for dorso-ventral and antero-posterior patterning of the mesoderm and also for the induction of the neural plate.

	Xenopus gastrula	Mouse gastrula
	mesoderm / organizer	organizer
	Genes in organizer region	
Genes encoding transcription factors	TbxT	TbxT
	goosecoid	goosecoid
Genes encoding secreted proteins	nodal-related-3	nodal
	chordin, noggin	chordin, noggin
	cerberus	Cer1/Cerl2

and bird embryos are equivalent: when a rabbit node was grafted into a chick embryo it caused the formation of a second chick axis. These grafting experiments suggested that the signals produced by the node in chick, mouse, and rabbit are the same.

However, this interpretation was questioned by the finding that, when grafted into a mouse embryo, the mouse node can induce only a partial axis, made up of trunk tissue that includes neural tissue and somites but lacks a head. Grafts of node precursors from earlier streak stages also have inducing ability, but likewise never induce a head. Instead, the AVE, which was established in distal parts of the pre-gastrula stage embryo (see Section 3.26), induces heads. It does so because, like the 'head organizer' region of the Spemann organizer, the AVE produces factors, including antagonists of Wnt and BMP signaling such as Dickkopf-1 and Cerberus-1 that initiate anterior (head) pattern and prevent posterior (trunk and tail) pattern. In other words, in mouse, the AVE acts as a 'head organizer' and the node as a 'trunk organizer'. The same molecular principles operate: a gradient of Wnt signaling with its high point at the posterior of the embryo provides positional information along the antero-posterior axis. Mouse embryos mutant for the Wnt antagonist Dickkopf-1 are headless, indicating that Wnt inhibition is essential for head formation. But in *Xenopus*, zebrafish, and chick, organizing molecules for head and trunk are, at least transiently, all located in the Spemann organizer, the shield, and Hensen's node, respectively, whereas in the mouse, the cell populations that secrete 'head organizing' versus 'trunk organizing' factors are always spatially separate.

The very different shapes of the embryos of each model organism, and slight variations in the relative locations of cells that will give rise to equivalent tissues provide a possible explanation for whether 'head' and 'trunk' organizers are spatially segregated, or not. An imagined 'flattened' mouse egg cylinder produces the equivalent of the chick blastoderm. Likewise, if one imagines the chick blastoderm rolled up with the epiblast on the outside, one can also see that the vegetal cells in the blastocoel floor of a *Xenopus* blastula and the dorsal yolk syncytial layer in the zebrafish could be considered equivalent to the anterior hypoblast of the chick (Fig. 4.25). This kind of representation also makes it easier to see, at early gastrula stages, the similar spatial relationship of different tissues, such as the node (or Spemann organizer in *Xenopus*) and the prospective germ layers and neural tissue in the different embryos. One possibility is that the relative positions of cells are slightly shifted in different organisms: in the chick, for instance, prechordal mesoderm underlies the forming forebrain. Another possibility is that Hensen's node, the Spemann organizer, and the zebrafish shield retain a remnant of, respectively, chick hypoblast cells, vegetal blastocoel floor cells, and dorsal yolk syncytial layer cells. If these fulfill the same role as the AVE in the mouse, they might contribute to the ability of organizer transplants from *Xenopus*, zebrafish, and chick to induce full duplications.

In the mouse, cell proliferation continues during gastrulation, and the embryo anterior to the node grows rapidly in size. The epiblast that lies anterior to the node will give rise to the neural plate, which will go on to form the forebrain, midbrain, hindbrain, and anterior spinal cord. Neural induction in the mouse is initiated in the anterior epiblast around E6.5 to E7.0 under the influence of anteriorizing and neural-inducing factors that, at this stage, are secreted from the AVE. As in the chick, the main neural-inducing agent in the mouse is FGF, and the main anteriorizing factors are BMP and Wnt antagonists. As neural-plate induction proceeds, anterior definitive endoderm and anterior mesendoderm (prechordal mesoderm that, in the mouse, is closely associated with endoderm) migrate anteriorly from the node and displace, or intermingle with, the AVE. Consequently, from E7.5 to E8.5, these tissues lie beneath the anterior neural plate and, behind these, anterior notochord that has extended from the node lies beneath the developing midbrain, hindbrain, and anterior spinal cord. Development from E6.0 to E8.5 is shown in Figure 4.26. Through their expression of Wnt inhibitors, the anterior definitive endoderm and prechordal mesendoderm maintain forebrain specification that was initially promoted by the AVE. This type of event—where one tissue takes on the signaling properties of a near neighbor—is seen repeatedly in the developing embryo and we shall see further examples in Chapter 10. It can bring robustness to patterning, against the background of growth.

SUMMARY

The zebrafish shield and the chick Hensen's node serve functions similar to those of the Spemann organizer. Each can promote dorso-ventral patterning and each can specify a new antero-posterior axis. In the mouse, the node promotes dorso-ventral patterning and can specify a new axis apart from a head, for which previous exposure of the epiblast to the anterior visceral endoderm (AVE) is required. The zebrafish shield, the chick Hensen's node, and the mouse node plus the AVE produce and secrete antagonists to signal molecules that result in gradients of signaling that provide cells in the embryo with positional information. The crucial gradients are the same as those in *Xenopus*. High levels of BMP signaling specify ventral, low levels dorsal. High levels of Wnt signaling specify posterior, low levels anterior. The head forms where levels of BMP, Wnt, and Nodal signaling are low. The brain and anterior spinal cord, which develop from the neural plate, are induced by FGF signals in chick and mouse, through the ability of FGF to inhibit BMP signaling. Patterning of the neural plate involves both quantitative and qualitative differences in signaling by the mesoderm, with high levels of Wnt and FGF signals specifying the more posterior structures.

Elongation of the body and the origin of posterior tissues

In *Xenopus* and zebrafish, the antero-posterior body axis is patterned along much of its length during gastrulation; there is little growth, and convergent–extension movements of cells (discussed in Chapter 5) lead to elongation of the embryo in the head and trunk regions; growth only plays a part in the elongation of the tail. By contrast, in chick and mouse embryos, growth occurs during the early patterning of the anterior of the embryo during

Fig. 4.25 Comparison of the topology of the different regions of pre-gastrula and early gastrula stage embryos of mouse, chick, *Xenopus*, and zebrafish. It is possible to see that simple topological transformations reveal a similarity in the organization of the different embryos. Flattening of the mouse egg cylinder produces the equivalent of the chick blastoderm. Rolling up the chick blastoderm with the epiblast on the outside and closing off the blastocoel produces the *Xenopus* embryo. An increase in hypoblast mass (in this case the yolk syncytial layer) produces an approximation of the zebrafish embryo. The key refers to mouse tissues. A, anterior; P, posterior; D, dorsal; V, ventral; CNS, central nervous system.

From Beddington, R.S., Robertson, E.J.: **Anterior patterning in mouse.** Trends Genet. 1998, **14**: 277–284.

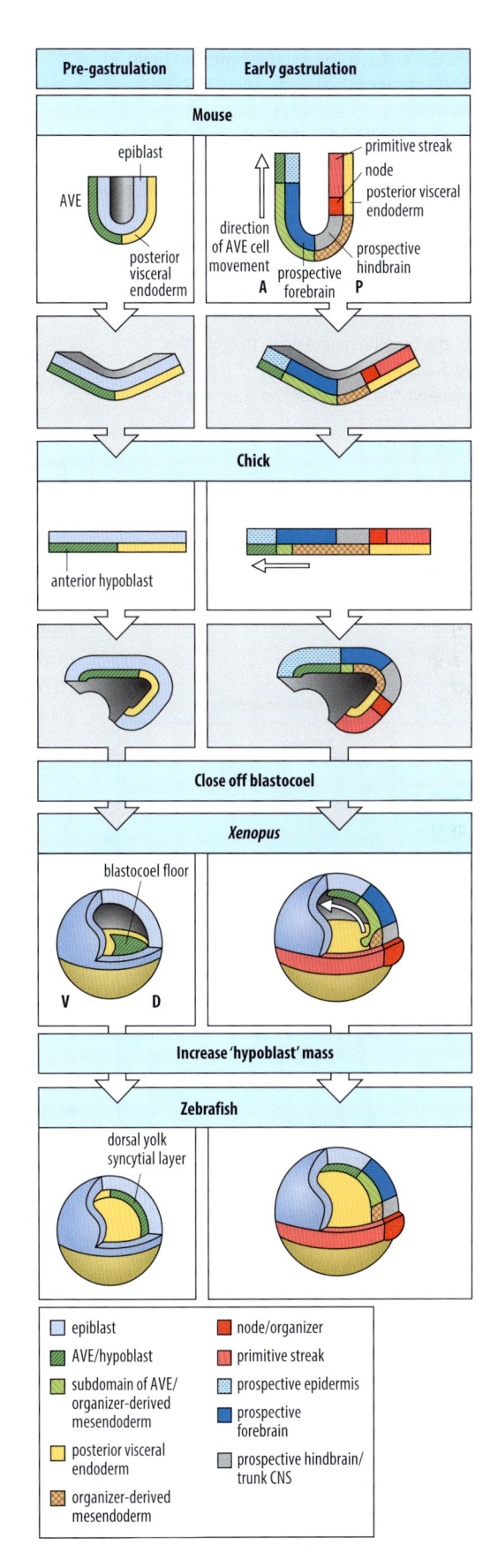

Fig. 4.26 Induction and specification of the neural plate in the post-implantation mouse embryo and establishment of brain and anterior spinal cord. At E6.0–E7.0 the AVE begins to specify neural tissue in the overlying epiblast. By E7.5, the node has formed at the anterior limit of the primitive streak, and the anterior movements of the axial mesoderm/mesendoderm (anterior mesendoderm (AME) and notochord) mean that AME displaces and intermingles with the AVE. The forebrain domain is specified and patterned within the anterior portion of the forming neural plate that overlies AVE and AME tissues. From E7.5 to E8.5, more posterior parts of the brain and the anterior spinal cord are underlain by anterior notochord.

Adapted from Levine, A.J., Brivanlou, A.H.: **Proposal of a model of mammalian neural induction.** Dev. Biol. 2007, **308**: 247–256.

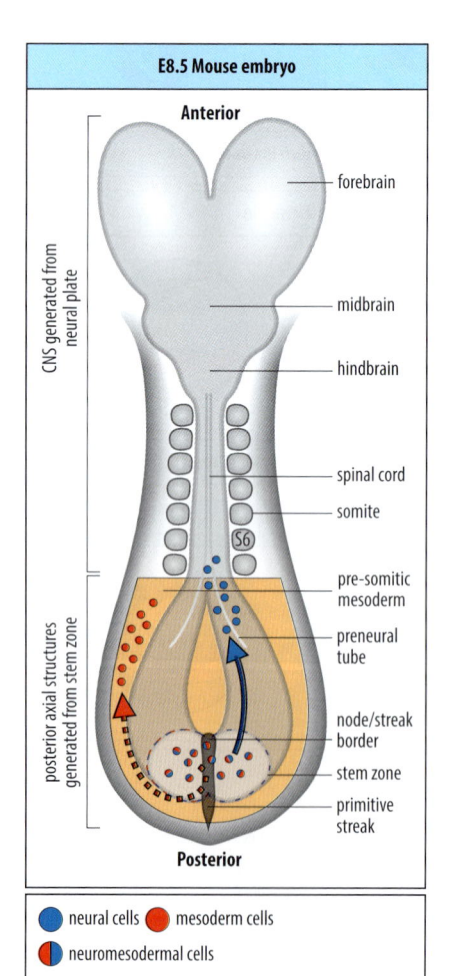

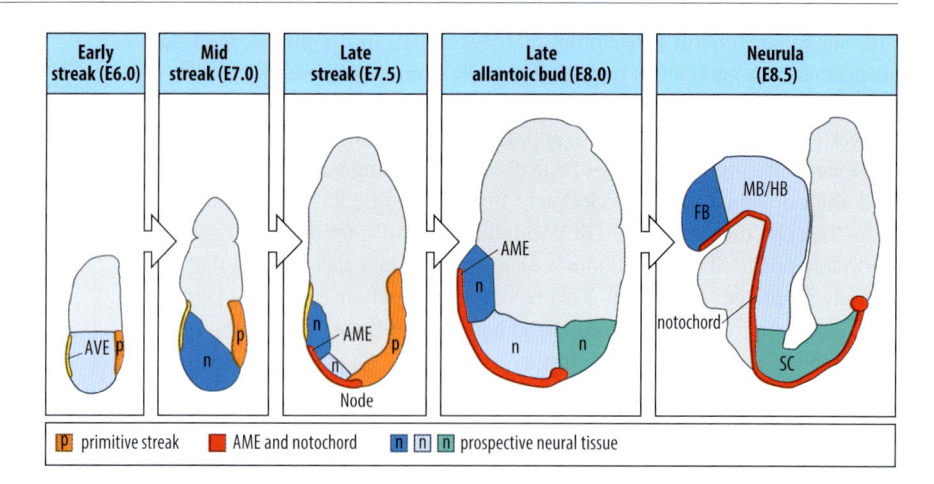

gastrulation, and most of the body axis is then generated over a period of time as the node and primitive streak gradually regress and the embryo continues to grow and elongate. So, in chick and mouse, the early patterning of the neural plate establishes the brain and anterior spinal cord. But the posterior neural tube, which will form the rest of the spinal cord, and other axial tissues of the trunk are laid down in an anterior-to-posterior sequence by self-renewing multipotent stem cells that originate from the node and from an arc of epiblast on either side of the primitive streak immediately posterior to the node called the **stem zone** (Fig. 4.27). These tissues include the notochord of trunk and tail regions, the paraxial mesoderm that will contribute to the somites (termed **pre-somitic mesoderm**) and some lateral plate mesoderm. Descendants of the primitive streak and neighboring epiblast later form the tail bud, from which the most posterior region of the body develops. This last phase in axis formation is known as **secondary body formation** and differs in some respects from the earlier processes that form the head and trunk, which are known as **primary body formation**. We will focus here on the generation of axial tissues during primary body formation.

4.9 Axial structures in chick and mouse are generated from self-renewing cell populations

The cells in the node and the stem zone that lay down axial tissues during primary body formation constitute self-renewing populations and have stem-cell-like properties (see Chapter 1). Long-term fate-mapping studies show that descendants of a cell

Fig. 4.27 Location of cell populations that generate axial tissues during body elongation in mouse embryos. The schematic diagram shows the dorsal view of a E8.5 mouse embryo at an early stage in axis formation in which seven somites have already been laid down; eventually 65 somites are formed, the last 35 will be formed from the tail bud during secondary body formation. At E8.5, the three regions of the brain and the anterior spinal cord have already been specified in the neural plate, and the neural tube that will give rise to the rest of the spinal cord in the body is being generated. Posteriorly, the neural tube is still open (preneural tube). The emerging trunk notochord is generated from a self-renewing population of cells in the node, while almost the entire length of the neural tube and the **medial** parts (those nearest the midline) of the somites in the rest of the body are generated from self-renewing populations of cells in an arc of epiblast on either side of the streak immediately posterior to the node. This region of epiblast is known as the stem zone (also called the caudal lateral epiblast) and consists of multipotent neuromesodermal cells, which can give rise to cells of the neural tube (blue arrow) or ingress through the streak to give rise to mesodermal cells that contribute to formation of the somites (broken red arrow). CNS, central nervous system.

From Henrique, D., et al.: **Neuromesodermal progenitors and the making of the spinal cord.** Development 2015, **142**: 2864–2875.

in the stem zone can either remain *in situ* or can contribute to tissues along almost the entire length of the antero-posterior axis. Furthermore, detailed cell-lineage analyses show that the cells in the stem zone are multipotent neuromesodermal progenitor cells and express genes that are characteristic of both early neural tissue (*Sox2*) and early mesoderm (*TbxT*). Some descendants of these multipotent cells remain in the upper layer of cells and give rise to neural tissue, whereas others ingress through the streak and give rise to pre-somitic mesoderm that will generate the medial part of the somites—the part nearest to the dorsal midline.

The other tissues of the extending body axis, which include the intermediate mesoderm and parts of the lateral plate mesoderm that will give rise to the kidneys, the gonads, and the limb buds, for example, are not generated from the stem zone. The mesodermal cells that form the lateral parts of the somites and the lateral plate mesoderm come from cells that ingressed in more-posterior regions of the streak as the streak extended forwards. The gut endoderm is largely generated by the expansion of endoderm formed by cells that ingressed through the early streak.

Although the way in which trunk axial tissues are generated differs in chick and mouse compared with *Xenopus* and zebrafish, the same signals—Wnt and FGF— are involved. Mutations in mice in genes encoding components of these signaling pathways, for example Wnt-3a and the FGF receptor FGFR1, cause truncation of the axis. As we have seen, a key target for Wnt and FGF signaling is the gene for the transcription factor TbxT, the earliest marker for mesoderm formation; Wnt and FGF signaling also regulate expression of the gene for the transcription factor Sox2, an early marker for neural tissue. Spontaneous and induced mutations in the *TbxT* gene have long been known to cause axis truncation in mice (see Chapter 1).

The combination of Wnt and FGF signaling by cells in the posterior region of chick and mouse embryos maintains a pool of undifferentiated proliferating cells in the stem zone. An anterior (low)–posterior (high) gradient of FGF and Wnt signaling is set up in the region between the last-formed somite and the regressing node with the high point at the node (Fig. 4.28). The FGF signaling gradient is not formed entirely by diffusion. Instead, cells in the node and primitive streak transcribe the *Fgf8* gene and the *Fgf8* mRNA is then gradually degraded in the cells that have ingressed to form the mesoderm. The result is that mesoderm cells at increasing distances from the node produce progressively lower amounts of FGF-8, and so an FGF-8 activity gradient is generated. The high levels of FGF/Wnt signaling posteriorly maintain the pool of self-renewing neuromesodermal cells in the stem zone.

Fig. 4.28 Opposing gradients of fibroblast growth factor (FGF)/Wnt and retinoic acid (RA) control the generation of axial tissues from the stem zone in chick and mouse embryos. Upper panel: opposing gradients of RA and FGF/Wnt are established and maintained in the posterior region of the embryo between the last-formed somite and the regressing node. The high levels of FGF/Wnt signaling posteriorly maintain the pool of self-renewing neuromesodermal cells in the stem zone; the high levels of RA anteriorly promote the differentiation of cells leaving the stem zone into neural tissue in the spinal cord and into mesoderm that contributes cells to the somites. Lower panel: the production of RA from newly formed somites and the production of FGF and Wnt signals from the stem zone is regulated by mutual antagonism, so that the opposing signaling gradients are robust. In both chick and mouse embryos, RA represses expression of *Fgf8* in the stem zone (*Fgf8* and *Wnt3a* in the stem zone in the mouse). In the chick, FGF-8 represses expression of *Raldh2*—the gene encoding the enzyme that generates RA in the somites. In the mouse, FGF-8 promotes expression of *Cyp26a1*—the gene encoding an enzyme that metabolizes RA—in the posterior region of the embryo (metabolism indicated by dotted line), ensuring that RA signaling is kept low posteriorly. Only a somite on the left side is shown for simplicity.

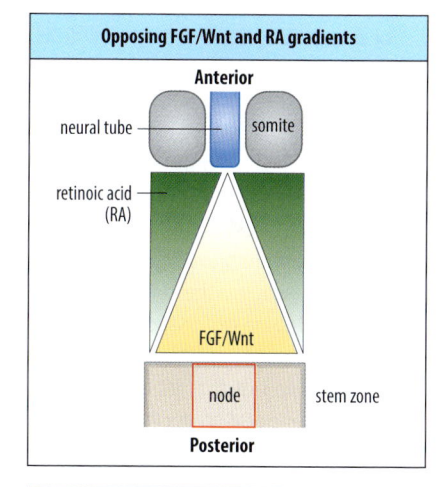

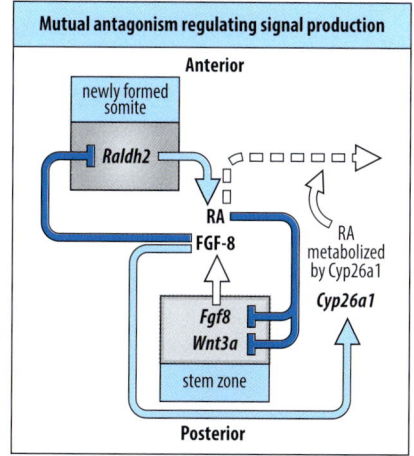

CELL BIOLOGY BOX 4E Retinoic acid: a small-molecule intercellular signal

Retinoic acid is a small diffusible molecule derived from vitamin A (retinol). Animals obtain vitamin A through their diet, and both too little or too much can lead to developmental defects. Sadly, babies with craniofacial malformations have been born to women, unaware that they were pregnant, who were treated for severe skin diseases with synthetic vitamin A derivatives.

Retinol is present at high levels in chicken eggs, and in mammalian embryos is supplied by the maternal circulation. Retinol binds to retinol-binding proteins in the extracellular space (Fig. 1) and is transferred into cells by a retinol transport receptor. Once inside the cell, retinol is bound by an intracellular retinol-binding protein (CRBP). Retinoic acid is produced in a two-step reaction, in which retinol is first metabolized to retinaldehyde by alcohol dehydrogenase (ADH) or retinol dehydrogenases (mainly RDH10); retinaldehyde is then metabolized to retinoic acid by retinaldehyde dehydrogenases (RALDH). There are three vertebrate retinaldehyde dehydrogenases, encoded by the genes *Raldh1*, *Raldh2*, and *Raldh3*, and expression of these genes is tightly regulated both temporally and spatially in vertebrate embryos to ensure that retinoic acid is produced at very precise locations and specific times during development.

Retinoic acid is broken down by enzymes of the P450 cytochrome system, mainly by the subfamily of CYP26 enzymes, which, like the RALDHs, are expressed in very precise patterns in the embryo. Intricate homeostatic buffering mechanisms involving regulation of expression of the genes encoding retinoic-acid-metabolizing enzymes and end-product feedback serve to maintain appropriate levels of retinoic acid. If retinoic acid levels increase, *Raldh* gene expression is reduced and expression of *Cyp26* genes is increased; when retinoic acid levels fall, the reverse happens.

Retinoic acid acting as an intercellular signal is shown in Figure 1, although it probably also acts as a signal within the cells that produce it. Retinoic acid is lipid-soluble and so can pass through cell membranes. Inside a cell it is bound by proteins that transport it to the nucleus, where it binds to receptors that act directly as transcriptional regulators. Mammals and birds have three retinoic acid receptor (RAR) proteins—α, β, and γ—which act as dimers in combination with the retinoid X receptor proteins (RXR; also α, β, and γ). In the absence of retinoic acid, the heterodimeric RAR:RXR receptors are bound to regulatory DNA sequences known as retinoic acid response elements (RAREs) and repress the associated genes. Binding of retinoic acid to its nuclear receptor leads to activation of the gene associated with the response element. Gene activation by retinoic acid can involve not only the direct initiation of transcription, but also the recruitment of co-activator proteins that induce chromatin remodeling, which enables the gene to be expressed.

Among the genes with a RARE is the gene for RARβ (*Rarb*), and so one direct response to retinoic acid is a switching on of *Rarb*. Expression of *Rarb* can therefore be used as a reporter of retinoic acid signaling, and transgenic mice have been made in which the reporter gene *lacZ* (encoding β-galactosidase; see Box 1E) is hooked up to the *Rarb* RARE. β-galactosidase can then be revealed in embryos and tissues by a histochemical reaction that gives a blue color, showing which cells are responding to retinoic acid signaling. Figure 2 shows an E9.5 mouse embryo stained in this way.

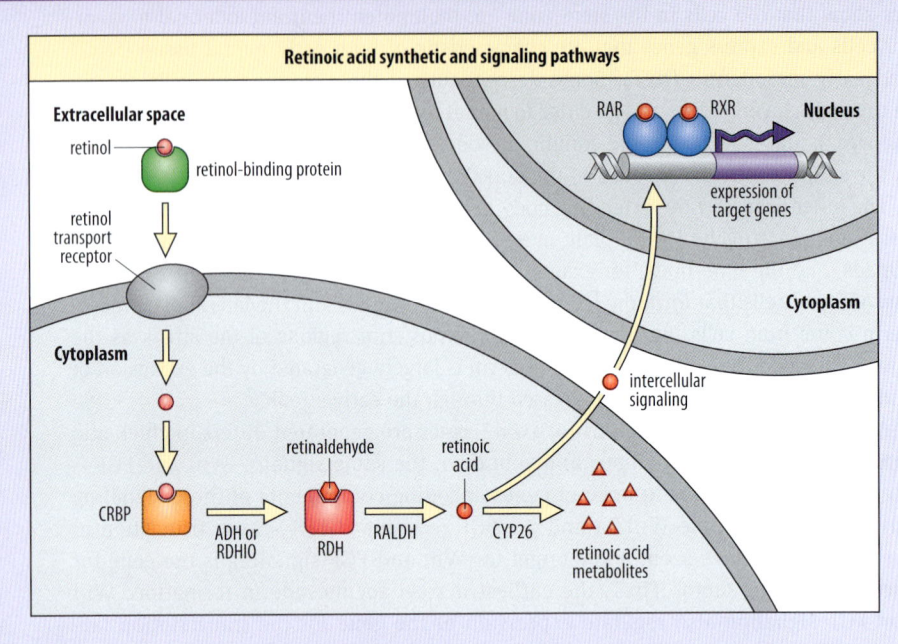

Fig. 1

Fig. 2 Photograph courtesy of Pascal Dolle from Ribes, V., et al. **Rescue of cytochrome P450 oxidoreductase (Por) mouse mutants reveals functions in vasculogenesis, brain and limb patterning linked to retinoic acid homeostasis.** Dev. Biol. 2007, **303**: 66–81.

Other important developmental signals that act through intracellular receptors of the same superfamily as the RAR and RXR proteins are the steroid hormones (e.g. estrogen and testosterone, discussed in Chapters 8 and 11) and the thyroid hormones (thyroxine and tri-iodothyronine; discussed in Chapter 11). Like retinoic acid, these molecules are lipid-soluble; they diffuse through the plasma membrane unaided and bind to nuclear receptor proteins. The ligand–receptor complex is then able to act as a transcriptional regulator, binding directly to specific response elements in the DNA to activate (or in some cases repress) transcription of several genes simultaneously.

The differentiation of the neuromesodermal cells as either neural tissue or mesoderm is controlled by a gradient of the secreted signal molecule **retinoic acid** (Box 4E) that runs in the opposite direction to the FGF gradient. Retinoic acid is synthesized by mesoderm cells in the newly formed somites and diffuses posteriorly. When the cells leave the stem zone, they will be exposed to progressively lower levels of FGF signaling and progressively higher levels of retinoic acid signaling, and this promotes their differentiation into either the neural tissue of the spinal cord or into mesoderm cells that contribute to the somites.

Mutual antagonism regulates the production of the posterior FGF/Wnt signals and the production of the anterior retinoic acid signal, helping to ensure that the opposing gradients are robust (see Fig. 4.28). This is a general mechanism: we have previously seen that opposing gradients of BMPs and BMP antagonists, or BMPs and Nodal, pattern the dorso-ventral axis in chick and mouse, respectively, and we will encounter a similar principle in Chapter 10, when we consider dorso-ventral patterning of the spinal cord. The details of how the mutual antagonism between FGF/Wnt and retinoic acid in the posterior axis is established are slightly different in chick and mouse, but the outcome is the same. Retinoic acid represses transcription of *Fgf8* in the node and stem zone (also of *Wnt3a* in the mouse), whereas FGF in the chick represses the expression of the gene for the retinoic-acid-metabolizing enzyme Raldh2 (a retinaldehyde dehydrogenase), which generates retinoic acid in the somites. In the mouse, FGF signaling in the posterior region of the embryo regulates the expression of the gene for the retinoic-acid-metabolizing enzyme Cyp26a1, and this ensures that retinoic acid signaling is kept low in cells in the stem zone. Mouse embryos completely lacking *Cyp26a1* gene function have axis truncations, as do mouse embryos treated with excess retinoic acid.

Finally, in addition to the patterning and proliferation mechanisms described above, mechanical forces contribute to the elongation of the body axis. These are outlined in Section 5.13.

After primary body formation, the body axis in mouse and chick is completed by the generation of the most-posterior structures from the tail bud, which develops once 30 somites and 22 somites, respectively, have formed. At this stage, the final plan of the body has emerged and the chick and mouse embryos look similar to those of frogs and fish (see Fig. 3.2). Establishment of the body plan has also fixed the positions in which various organs will develop, even though there are, as yet, few overt signs of organ formation. Nevertheless, numerous grafting experiments have shown that the potential to form a given organ is now confined to specific regions of the embryo. Each of these regions has, however, considerable capacity for regulation, so that if part of the region is removed, a normal structure can still form.

> **SUMMARY**
>
> In *Xenopus* and zebrafish, the prechordal mesoderm (or mesendoderm in zebrafish) and much of the notochord derive from the organizer, or its equivalent (the shield). Signals from the organizer/shield, and their derivatives, direct formation of almost the entire body axis, and induce neural tissue during gastrulation. By contrast, in the chick and the mouse, only the cells of the prechordal mesoderm and anterior notochord are derived from the node that is formed at the tip of the primitive streak when it reaches its most anterior position. Signals from these cells induce a neural plate in the overlying ectoderm that gives rise to the brain and anterior spinal cord. The rest of the spinal cord, the notochord, the somites of the trunk and the tailbud are generated from self-renewing cell populations in the regressing node and an arc of epiblast surrounding it, called the stem zone, which are maintained by Wnt and FGF signaling.

Somite formation and the Hox code in antero-posterior patterning

Vertebrates are defined by the presence of the vertebral column, whose characteristic feature is that it is segmented. The vertebral column derives from the two strips of paraxial mesodermal tissue that lie one on each side of the notochord. Specifically, it derives from the posterior paraxial mesoderm, otherwise known as pre-somitic mesoderm—the term we will use here. One of the first morphogenetic events that will lead to development of the vertebral column is the division of the pre-somitic mesoderm into regularly sized chunks of tissue called somites. This process is called **somitogenesis**. The process of somitogenesis imposes an important segmental organization on other features of the vertebrate body such as the spinal nerves and ganglia of the trunk nervous system (see Section 5.22) and the vasculature. Somite segmentation does not occur all at once; instead, single somite pairs successively and rhythmically detach from the anterior border of the pre-somitic mesoderm and are laid down in the anterior to posterior direction.

In addition to generating the vertebral column, somites give rise to the skeletal muscles, and to the dermis of the skin. In the limb-forming regions, they supply the myogenic cells for the muscles of the limbs (see Chapter 9). Each somite block has a medial (future dorsal) and a lateral (future ventral) portion that are initially specified in adjacent regions of the pre-gastrula embryo (see Fig. 4.19), and then brought together during gastrulation. In chick and mouse, the cells of the medial part of the somite come from the stem zone and are specified by low-level BMP signaling. The cells in the lateral part come from a slightly more-posterior region of the primitive streak and are specified by slightly higher levels of BMP signaling.

The pre-somitic mesoderm is supplied with positional information by the expression of the Hox genes, and this determines the future regional identity of the vertebral column along the antero-posterior axis of the body. This regional identity results in the formation of the appropriate vertebrae in the proper places. For example, the vertebrae that form at different positions along the spine have characteristic shapes. In this part of the chapter, we will first examine the development of the somites, how their periodic formation is controlled, and what determines their number. We will then consider how their different regional identities along the antero-posterior axis are specified by Hox gene expression. Finally, we discuss how different regions within an individual somite are specified and which structures they give rise to.

4.10 Somites are formed in a well-defined order along the antero-posterior axis

Somitogenesis begins in pre-somitic (posterior paraxial) mesoderm that lies adjacent to the posterior hindbrain; the more anterior paraxial mesoderm will not segment, but instead will form facial muscles. The pre-somitic mesoderm is dynamic: somites are sequentially generated, beginning at the anterior end; as this happens, the posterior end is replenished with new cells entering from the stem zone (see Section 4.9), together with cells that ingressed in the posterior streak. In this way, as somites are generated, the length of the pre-somitic mesoderm remains more or less constant but the body lengthens, and the distance between the hindbrain and the unsegmented pre-somitic mesoderm increases.

Somitogenesis, then, is very tightly linked to the proliferation of the self-renewing population of multipotent cells of the stem zone, and many of the same signals that operate to govern the proliferation–maturation gradient in the stem zone (see Section 4.9) simultaneously direct somite formation. Somites are formed in pairs, a somite on either side of the notochord, with the two somites in a pair forming simultaneously. The first-formed somite is called somite 1, the second, somite 2, and so on (Fig. 4.29). In the chick, somites 1–5 contribute to the posterior part of the skull and to some of the head and neck muscles. Somites 6–19 give rise to the cervical vertebrae, somites 20–26 to the thoracic vertebrae, and somites 27–41 to the lumbosacral and caudal vertebrae.

Pre-somitic mesoderm is composed of relatively immature mesenchymal cells, whereas somites are more epithelial spheres within which only a small cavity is filled with undifferentiated mesenchymal cells. The formation of the somite unit from the pre-somitic mesoderm therefore requires a **mesenchymal-to-epithelial transition (MET)**, in which the relatively loosely packed mesenchymal cells transition to a state where they adhere tightly to each other to form a more rigid epithelial structure. This involves coordinated changes in the cell's cytoskeleton, in the adherens junctions that bind cells together (discussed in Chapter 5, Box 5A), and in the extracellular matrix, that together result in changes in cell movements, and the formation of anterior and posterior somite borders.

Somitogenesis is a tightly time-regulated event: each pair of somites forms at a precise time interval that is species-specific: every 90 minutes in the chick, every 120 minutes in the mouse, every 45 minutes in *Xenopus*, every 30 minutes in zebrafish, and every 6 hours in humans. This begs the question of what is the mechanism underlying this periodicity?

The prevailing view is based on a theoretical model that was first proposed in the 1970s, which is called the **clock-and-wavefront model**. The model proposes two interacting components that together direct the formation of periodic structures: (1) a clock in the pre-somitic mesoderm composed of a gene whose expression oscillates (on–off); and (2) a wavefront of cell maturation that moves from anterior to posterior (head to tail). Cells behind the wavefront (i.e. anterior) are mature, and cells in front of the wavefront (i.e. posterior) are immature. Only cells in the mature state are capable of forming a somite. Like any wave, the wavefront moves in pulses, but importantly, the pulses are induced by the pre-somitic mesoderm clock. The progression of the wavefront is stopped, and somite formation can occur, when the clock is on; the progression of the wavefront restarts when the clock is off. Regularly spaced and sized somites will therefore form periodically as the clock oscillates at a constant period and the wavefront progresses toward the tail (Fig. 4.30).

Support for the clock-and-wavefront model for somitogenesis came first through studies in the chick. *In situ* hybridization studies showed that components of the Notch signaling pathway (Box 4F) show oscillatory expression (on–off) in the pre-somitic mesoderm and that the periodicity of the oscillations equals the time it takes

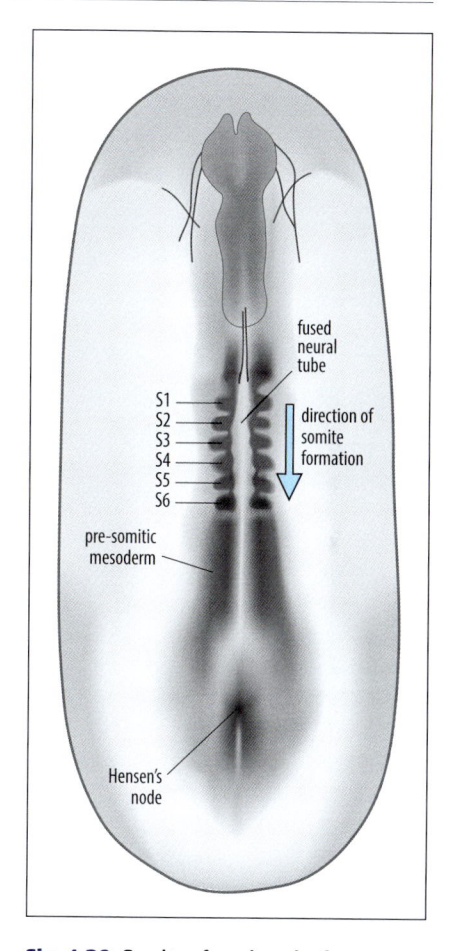

Fig. 4.29 Somites form in pairs from the paraxial mesoderm. Somites are transient blocks of tissue that form from the posterior paraxial mesoderm (also termed the pre-somitic mesoderm), which lies on either side of the notochord. They are made in an anterior to posterior direction from unsegmented pre-somitic mesoderm. The medial region of each somite is produced from a pool of stem cells around the node and the lateral region from cells from the more posterior regions of the streak. The first five somites give rise to the posterior part of the skull; the remaining somites produce the muscles and bones of the trunk and tail and the muscles of the limbs.

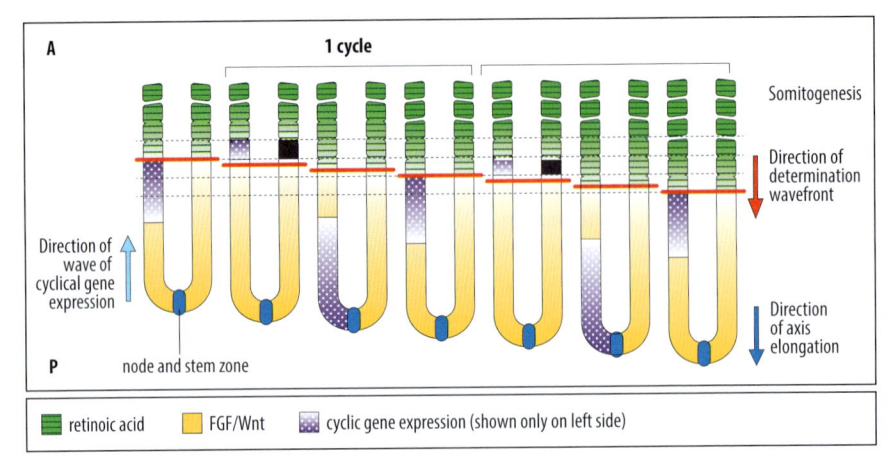

Fig. 4.30 The clock-and-wavefront model of somite formation. Two cycles of somite specification are illustrated. Retinoic acid (green) produced by the somites forms a gradient extending from anterior (A) to posterior (P) in the mesoderm and meets a gradient of fibroblast growth factor (FGF)/Wnt signaling (peach) running from posterior to anterior. A wavefront of somite determination (red), specified by a threshold level of FGF/Wnt signaling, advances through the mesoderm from anterior to posterior as the mesoderm elongates. At the same time, cycles of gene expression (purple, the sequential expression of one gene is illustrated) sweep through the mesoderm from posterior to anterior. Cycling gene expression is shown only on the left-hand side of the mesoderm for clarity. During each cycle a pair of new somites is specified (black box, shown on the right-hand side only for clarity) in mesoderm that has experienced both the determination signal and expression of the cycling gene. Gene expression characteristic of a somite 'prepattern' is switched on, and somite segmentation and differentiation (somitogenesis) eventually ensues. Somites specified in previous cycles, but which have not yet differentiated, are indicated in the anterior mesoderm (shown in more detail in Fig. 4.31).

From Gomez, C., Pourquié, O.: **Developmental control of segment numbers in vertebrates.** J. Exp. Zool. B Mol. Dev. Evol. *2009*, **312**: *533-544.*

for a pair of somites to form (90 minutes in the chick). Expression of the transcription factor gene *Hairy-1*—a target of Notch signaling—oscillates, as does the presence of the cleaved Notch intracellular domain—the hallmark of Notch signaling. The oscillations can be visualized experimentally as a wave that sweeps through the pre-somitic mesoderm from posterior to anterior, where the somites are being formed (see Fig. 4.30). It is important to appreciate that what travels is the wave of gene expression or of Notch activity, and not the cells expressing the genes or the cells in which Notch signaling is taking place. In other words, cells in the pre-somitic mesoderm express a given gene for a given period of time, undergo Notch signaling and then stop, and this process then repeats. Subsequently, it was found that mutations in the genes *Lunatic fringe* and *Delta-like 1* (whose protein products are both involved in Notch signaling; Box 4F) in both mouse and zebrafish resulted in segmentation defects. A prevailing view, therefore, is that Notch signaling oscillations constitute—or at least form important components of—the pre-somitic molecular clock.

The wavefront is formed by FGF and Wnt signaling, which show graded expression in the pre-somitic mesoderm, with high levels posteriorly and low levels anteriorly (see Figs 4.28 and 4.30). High levels of FGF/Wnt keep pre-somitic mesoderm cells in an undifferentiated mesenchymal state, and somite formation occurs where FGF/Wnt signaling drops below a threshold level that enables cells to switch to a mature, somite-competent state (Fig. 4.31). The source of FGF/Wnt is located in the node and stem zone, and consequently the wavefront moves posteriorly as the embryo grows.

CELL BIOLOGY BOX 4F The Notch signaling pathway

The family of Notch transmembrane proteins act as receptors and initiate the Notch intracellular signaling pathway. This pathway is involved in numerous developmental decisions that determine cell fate, ranging from determination of neuronal precursors in *Drosophila* and vertebrates to the differentiation of epidermal cells in mammalian skin. The ligands that bind to Notch are also membrane-bound proteins, and so Notch signaling requires direct cell-cell contact. This means that the Notch signaling pathway can be activated in a single cell out of a population of several cells, and that this cell can then have a different fate from surrounding cells. We shall see an example of this when we look at the development of neuroblast specification in Chapter 10.

The ligands that bind to Notch are members of the Delta and Serrate/Jagged1 families. Binding of ligand on one cell to Notch on an adjacent cell initiates intracellular signaling from the receptor (Fig. 1). Notch has a large extracellular domain with several glycosylation sites that are specifically modified by glycosyltransferases, including those of the Fringe family, which includes Lunatic Fringe. Glycosylation modulates Notch signaling by affecting the ability of the extracellular domain to bind to different ligands, facilitating its binding to Delta ligands while reducing its ability to bind to Serrate ligands.

Activation of Notch by its ligand causes the enzymatic cleavage of the cytoplasmic tail of the receptor (the Notch intracellular domain) and its translocation to the nucleus, where it binds and activates a transcription factor of the CSL family (CSL is named after the Notch-activated transcription factors CBF1/RBPkJ in mammals, Suppressor of Hairless (Su(H)) in *Drosophila*, and LAG-1 in *Caenorhabditis elegans*. In the absence of Notch signaling, CSL is held in a complex with repressor proteins. Activation of the CSL complex by binding of the Notch intracellular domain involves release of the repressors and recruitment of the co-activator protein Mastermind and other co-activators. The outcome of the Notch signaling pathway is the transcription of specific genes. Notch signaling is very versatile. In different organisms and in different developmental circumstances, activation of the Notch pathway is used to switch a wide range of different genes on or off.

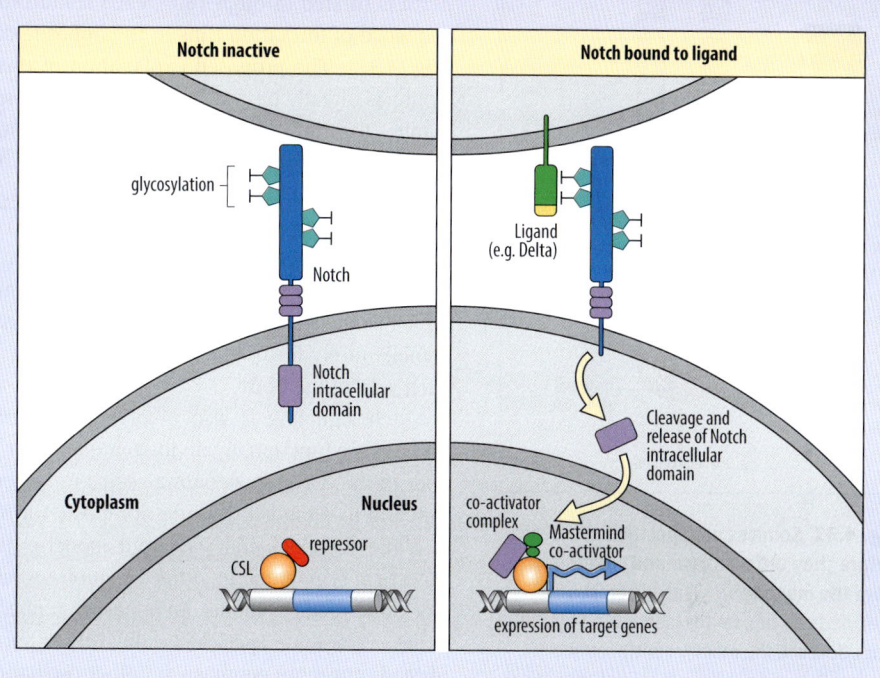

Fig. 1

As the wavefront extends posteriorly, it successively defines regions in the pre-somitic mesoderm that become competent to form somites—that is, they will go on to segment and form somites. In mice, such competence occurs through upregulation of the gene for the transcription factor Mesp2. The anterior border of *Mesp2* expression marks the location of a new somite boundary. The posterior border of *Mesp2* expression is defined by FGF, and therefore the wavefront will specify the location of somite boundaries. The timing of *Mesp2* expression is determined by Notch oscillation, and so the clock sets the pace of somite formation. The periodicity of the clock and the gradual regression of the wavefront together ensure the formation of regularly sized and spaced somites. The gradient of retinoic acid that runs in the opposite direction to that of FGF and Wnts (see Fig. 4.28) has an antagonistic effect and represses transcription of the genes encoding FGF and Wnt and so prevents the pre-somitic region from continually getting longer.

220 Chapter 4 Vertebrate development II: completing the body plan

The clock-and-wavefront model provides an explanation for experimental observations in the chick. If a relatively anterior piece of the unsegmented pre-somitic mesoderm is rotated through 180°, each somite still forms at the normal time, so that the sequence of formation runs in the opposite direction to normal in the inverted tissue (Fig. 4.32). The proposed explanation of this result is that the determination wavefront has already passed over before the pre-somitic mesoderm was rotated and so the somites form in their normal sequence within the rotated mesoderm.

A fascinating question is how the oscillations in gene expression are synchronized between neighboring cells, so that they behave as a unit at the tissue level. Genetic studies in mouse and zebrafish point to an important role for Delta–Notch signaling in this process. In mutants of the Notch signaling pathway, loss of synchrony is associated with the disruptions of the segmentation process. In mouse embryos in which Delta–Notch signaling is either made constitutively active or abolished entirely, somites rarely form, but if they do, they vary in size and shape and are different on each side of the body.

Notch signaling is also involved in determining a boundary within the somite. Somites are functionally divided into anterior and posterior halves with different fates, and, in mice, Notch signaling seems to be required downstream of its possible clock function to establish this anterior–posterior boundary.

While the clock-and-wavefront model is the most widely accepted model for the process of segmentation, other variations of the model have been proposed, including a clock-and-gradient model, an opposing gradient model, and a reaction–diffusion model. Some of these are species-specific; others put less emphasis on the role of Notch, and instead, put more emphasis on other oscillating factors, including Wnts and FGFs; yet others put less emphasis on the wavefront. In spite of these variations, all the models are based on the existence of a gene regulatory network that establishes a pattern of gene expression that predicts somitogenesis. This type of early pattern that is laid down at one stage but is only revealed through a later event is termed a **prepattern**.

An important developmental question is what determines the final number of somites. Somite number is highly variable between different vertebrates: birds and humans have around 50, whereas snakes have up to several hundred. One answer

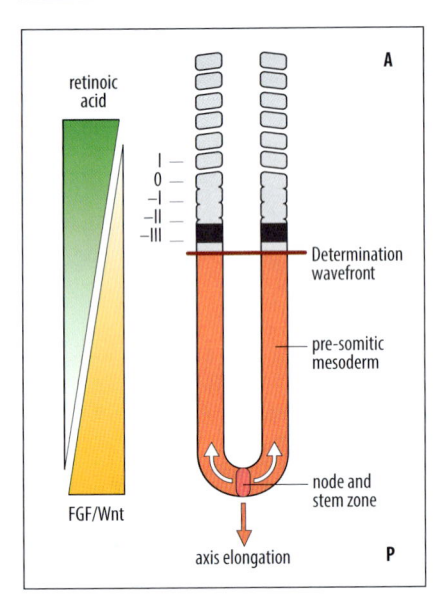

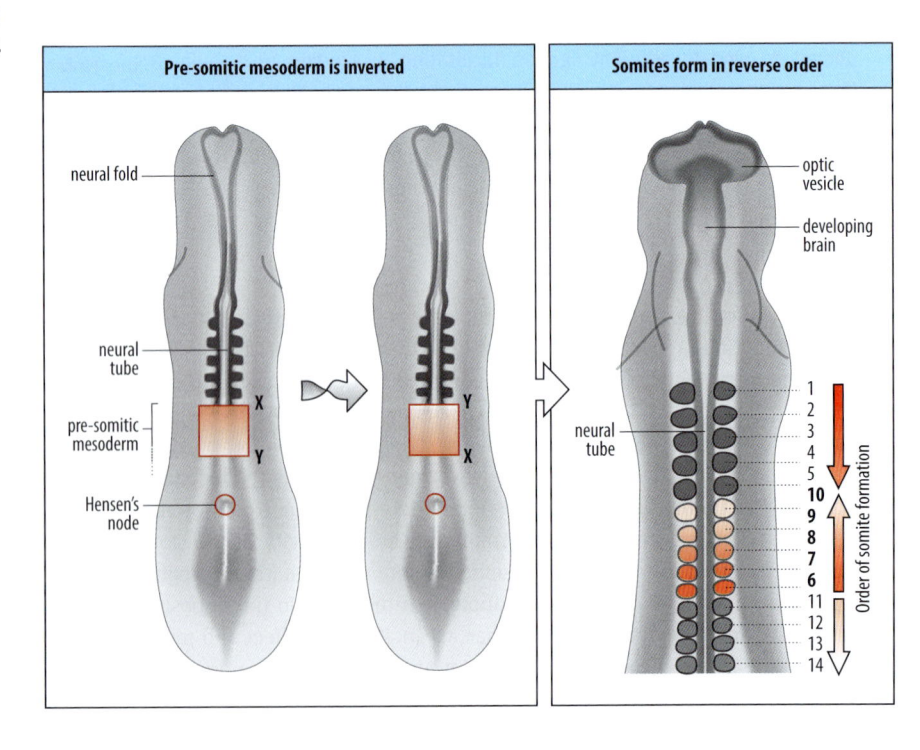

Fig. 4.31 Somites are specified some time before they differentiate and segment from the mesoderm. The most recently specified pair of somites is indicated by the black boxes. I, the most recently formed somite; 0, somite in the process of formation; -I, -II, and -III, blocks of cells that have been specified and will form somites during successive cycles. The position of somite formation is specified as a result of threshold levels of FGF/Wnt signaling (pale brown) and retinoic acid signaling (green), which form opposing gradients in the mesoderm.
A, anterior; P, posterior.

From Gomez, C., Pourquié, O.: **Developmental control of segment numbers in vertebrates.** J. Exp. Zool. B Mol. Dev. Evol. 2009, *312*: 533-544.

Fig. 4.32 The temporal order of somite formation is specified before somites are morphologically obvious. Somite formation in the chick proceeds in an antero-posterior direction. Somites form sequentially in the pre-somitic region between the last-formed somite and Hensen's node, which moves posteriorly. If a relatively anterior piece of pre-somitic mesoderm is rotated through 180°, as shown by the arrow, the temporal order of somite formation is not altered—somite 6 still develops before somite 10.

could be that the total number of somites is determined by the length of pre-somitic mesoderm available to form somites. The length of the pre-somitic mesoderm will depend on the extent of axis elongation, and this varies between embryos.

A larger number of somites could also be generated from the same length of pre-somitic mesoderm if each somite were smaller. This could be accomplished by making the clock go faster so that more, smaller, segments form in a given time period. Alternatively, the pace of the clock could be kept the same and the advance of the determination wave slowed down, so that more oscillations of the clock take place in a length of pre-somitic mesoderm before the cells become committed to form a segment.

Experiments in chick embryos have shown that implanting beads soaked in FGF in the pre-somitic mesoderm, which would be expected to decelerate the speed of advance of the determination wavefront, does indeed lead to the formation of smaller somites. Corn snakes have 260 somites, which are each a third of the size of chick or mouse somites, and this seems to be due to acceleration of the pace of the clock and thus a shortening of the time allowed for one somite to form.

4.11 Identity of somites along the antero-posterior axis is specified by Hox gene expression

Somites differentiate into particular axial structures depending on their position along the antero-posterior axis. This can be seen in the vertebral column, where vertebrae have well-defined anatomical characteristics depending on their antero-posterior position. Anterior-most somites form anterior-most vertebrae that are specialized for attachment and articulation of the skull; then come somites that will form the cervical vertebrae of the neck, followed by more posterior somites that develop as the rib-bearing thoracic vertebrae. And the most-posterior somites develop into caudal vertebrae of the lumbar, sacral, and caudal regions.

Patterning of the vertebral column (among many other structures) is based on the paraxial mesodermal cells acquiring a positional value that reflects their position along the axis and so determines their subsequent development. Mesodermal cells that will form thoracic vertebrae, for example, have different positional values from those that will form cervical vertebrae. The concept of positional value has important implications for developmental strategy; it implies that a cell or a group of cells in the embryo acquires a unique state related to its position at a given time. This state is then interpreted to specify regional identity and this then determines later development (see Section 1.15). Positional values that lead to acquisition of regional identity are specified before somite formation itself begins. This was revealed when unsegmented somitic mesoderm from the prospective thoracic region of the chick embryo was grafted in place of the prospective mesoderm of the neck region, and still went on to form thoracic vertebrae with ribs (Fig. 4.33). How then is the pre-somitic mesoderm patterned so that somites acquire their identity and form particular vertebrae?

Patterning along the antero-posterior axis in all animals involves the expression of **Hox genes** that specify regional identity along the axis. The Hox genes are members of the large family of homeobox genes that are involved in many aspects of development (Box 4G) and they are homologous to the homeotic genes that specify segment identity along the antero-posterior axis in *Drosophila* (see Chapter 2). Hox genes and other homeobox genes pattern different types of tissue along the vertebrate antero-posterior axis, including mesodermal structures (discussed in this chapter) and the nervous system (discussed in Chapter 10).

The Hox genes are the most striking example of a widespread conservation of developmental genes—that is, genes that play an important part in development—in animals. Although it was widely believed that common mechanisms underlie the development of all animals, it seemed at first unlikely that genes involved in specifying the identity of the segments of *Drosophila* would specify identity of the somites

Fig. 4.33 The pre-somitic mesoderm has acquired a positional value before somite formation. Pre-somitic mesoderm in a chick embryo that will give rise to thoracic vertebrae is grafted to an anterior region of a younger embryo that would normally develop into cervical vertebrae. The grafted mesoderm develops according to its original position and forms ribs in the cervical region.

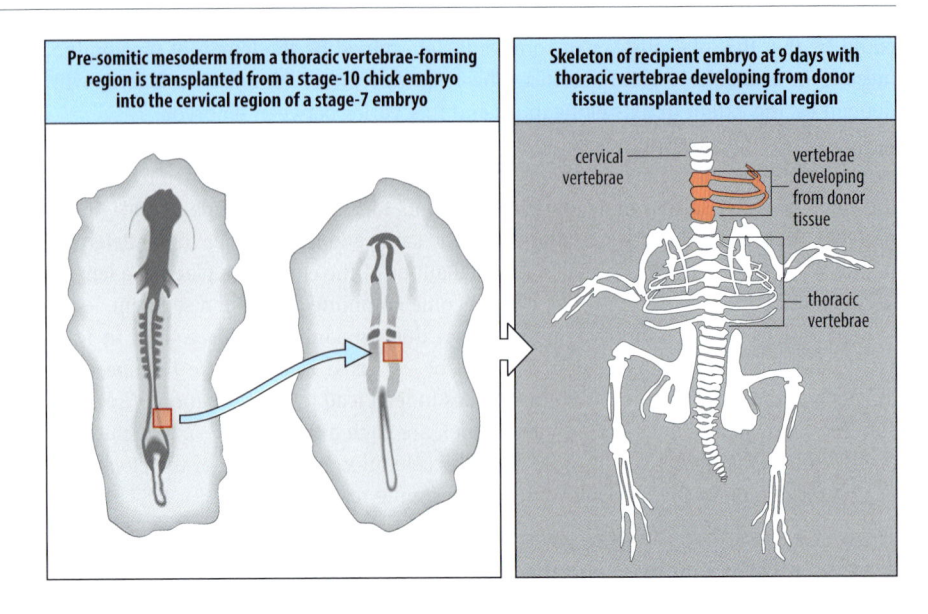

in vertebrates. However, it is now widely appreciated that numerous developmental genes that were first identified in *Drosophila* have homologs in vertebrates.

Most vertebrates have four separate clusters of Hox genes—the **Hox clusters** Hoxa, Hoxb, Hoxc, and Hoxd—that are thought to have arisen by duplications of the gene clusters themselves during vertebrate evolution. Some of the Hox genes within each cluster might also have arisen by duplication (see Box 4G). The zebrafish is unusual in having seven Hox gene clusters, as a result of a further duplication. A particular feature of Hox gene expression in both insects and vertebrates is that the genes in the clusters are expressed in a spatial order in developing tissues that reflects their order on the chromosome. The Hox clusters are the only known case where a spatial arrangement of genes on a chromosome corresponds to a spatial pattern of expression in the embryo. Hox genes in vertebrates are also activated in a temporal sequence that reflects their order on the chromosome. Both the spatial order and the temporal order of Hox gene expression are intrinsic to the structure of the regions of the genome that encode these genes. They therefore provide an example of a mechanism through which a genome-intrinsic timing mechanism can lead to a predictable and ordered spatial pattern of gene expression—a conceptually important point.

A simple idealized model illustrates the key features by which a Hox gene cluster could record spatial regional identity by a combinatorial mechanism. Consider four genes—I, II, III, and IV—arranged along a chromosome in that order (Fig. 4.34). The genes are expressed in a corresponding order along the antero-posterior axis of the body, linked to the time at which each gene was first transcribed. Thus, gene I is expressed throughout the body with its anterior boundary at the head end. Gene II has its anterior boundary in a more posterior position and expression continues posteriorly. The same principles apply to the two other genes. This pattern of expression defines four distinct regions, each expressing a different combination of genes. If the amount of gene product is varied within each expression domain, for example by more sophisticated gene regulatory networks, many more regions can be specified.

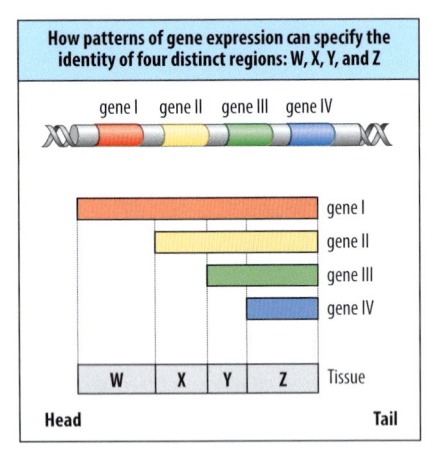

Fig. 4.34 Gene activity can specify regional identity. The model shows how the pattern of gene expression along the body can specify the distinct regions W, X, Y, and Z. For example, only gene I is expressed in region W, but all four genes are expressed in region Z.

The function of the Hox genes in vertebrate axial patterning has been best studied in the mouse, because it is possible to either knock out particular Hox genes or alter their expression in transgenic mice (see Box 3B). As in all vertebrates, the Hox genes start to be expressed in mesoderm cells of early mouse embryos during gastrulation when these cells begin to move internally away from the primitive streak and towards the anterior. Hox genes are also expressed in regions of the developing neural tube that will give rise to hindbrain and spinal cord (see Chapter 10). In both

BOX 4G The Hox genes

The vertebrate Hox genes encode proteins belonging to a large group of gene-regulatory proteins that all contain a similar DNA-binding region of around 60 amino acids. The DNA-binding region, known as the **homeodomain**, contains a helix-turn-helix DNA-binding motif that is characteristic of many DNA-binding proteins. The homeodomain is encoded by a DNA motif of around 180 base pairs termed the **homeobox**, a name that came originally from the fact that this gene family was discovered through mutations in *Drosophila* that produce a homeotic transformation—a mutation in which one structure replaces another. For example, in one homeotic mutant in *Drosophila*, a segment in the fly's body that does not normally bear wings is transformed to resemble the adjacent segment that does bear wings, resulting in a fly with four wings instead of two (see Fig. 2.46).

The Hox genes in *Drosophila* are organized into two distinct gene clusters or gene complexes—the Antennapedia complex and the bithorax complex (known collectively as the HOM-C complex) (Fig. 1). In vertebrates, including humans, the Hox genes are organized into four clusters, known as the **Hox clusters** or the **Hox complexes**. (Zebrafish have seven Hox clusters rather than four as a result of a partial duplication of the whole genome in their evolutionary history.) The homeoboxes of the genes in each Hox cluster are related in DNA sequence to the homeoboxes of genes of the HOM-C complex in *Drosophila*. In each Hox cluster, the order of the genes from 3' to 5' in the DNA is the order and position in which they are expressed along the antero-posterior axis and specify regional identity (see Fig. 1). In the mouse, the four Hox clusters, designated Hoxa, Hoxb, Hoxc, and Hoxd (originally called Hox1, Hox2, Hox3, and Hox4), are located on chromosomes 6, 11, 15, and 2, respectively (see Fig. 1). All Hox genes resemble each other to some extent; the homology is most marked within the homeobox and less marked in sequences outside it. The four Hox gene clusters are thought to have arisen by duplication of an ancestral gene cluster. The corresponding genes in the different clusters (e.g. Hoxa4, Hoxb4, Hoxc4, Hoxd4) are known as **paralogous genes**, and collectively as a paralogous gene subgroup. In the mouse and human there are 13 paralogous gene subgroups.

The Hox genes and their role in development are of ancient origin. The Hox genes in each model organism are similar to each other and to those of *Drosophila*, both in their coding sequences and in their order on the chromosome. In both *Drosophila* and vertebrates, these homeotic genes are involved in specifying regional identity along the antero-posterior axis. Many other genes contain a homeobox but they do not, however, belong to a homeotic complex, nor are they involved in homeotic transformations. Other subfamilies of homeobox genes in vertebrates include the **Pax genes**, which contain a homeobox typical of the *Drosophila* gene *paired*. All these genes encode transcription factors with various functions in development and cell differentiation.

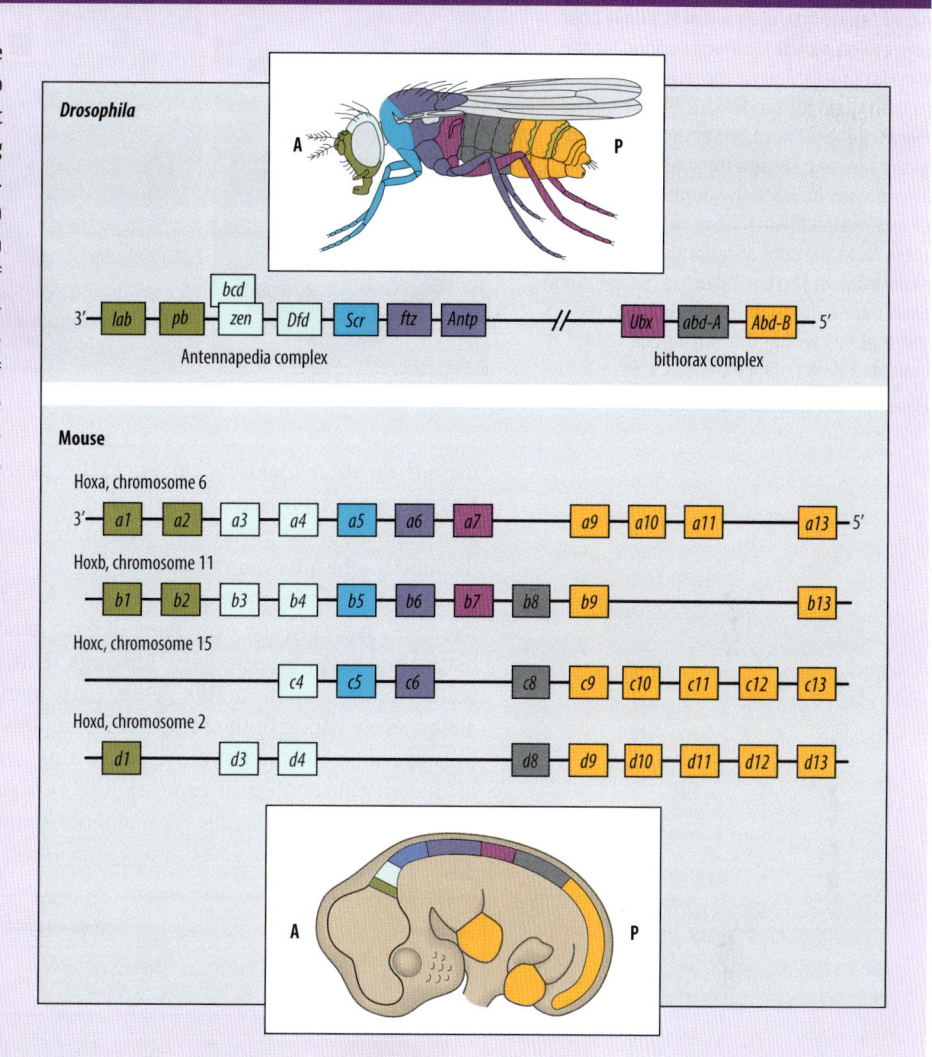

Fig. 1

224　Chapter 4　Vertebrate development II: completing the body plan

Fig. 4.35 Hox gene expression in the mouse embryo after neurulation. The three panels show lateral views of 9.5 days post-conception mouse embryos stained for the product of a *lacZ* gene reporter (blue) linked to promoters for the *Hoxb1*, *Hoxb4*, and *Hoxb9* genes. These genes are expressed in the neural tube and the mesoderm. The arrowheads indicate the anterior boundary of expression of each gene within the neural tube. Note that the anterior boundary of expression in the mesoderm is not the same as in the neural tube. The position of the three genes within the Hoxb gene cluster is indicated (inset). Scale bar = 0.5 mm.
Photographs courtesy of A. Gould.

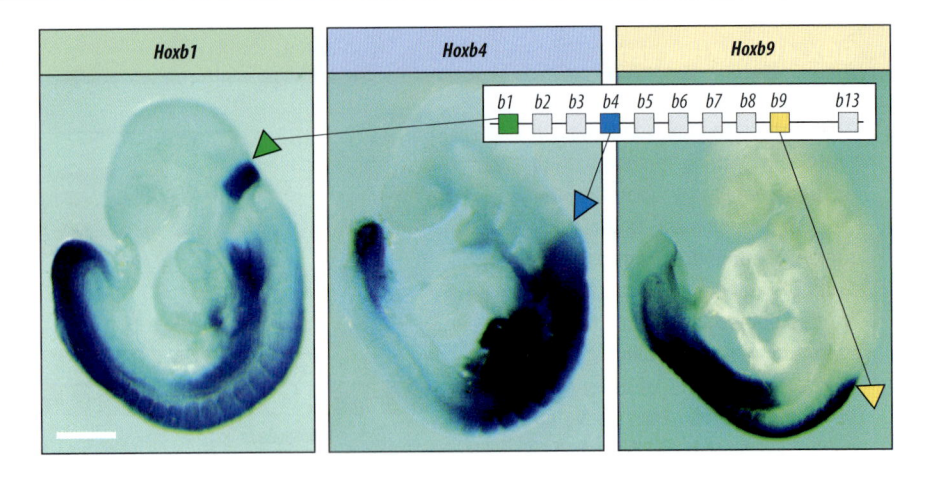

mesoderm and neural tube, 'anterior' Hox genes are expressed first. More 'posterior' Hox genes are expressed as gastrulation proceeds and clearly defined patterns of Hox gene expression are most easily seen in the mesoderm after somite formation and in the neural tube after neurulation.

Figure 4.35 shows the expression patterns of three Hox genes in mouse embryos at day E9.5. Typically, the expression of each gene is characterized by a sharp anterior border and extends posteriorly. Although there is considerable overlap in the spatial pattern of expression of Hox genes, particularly of neighboring genes in a cluster, almost every region in the mesoderm along the antero-posterior axis is characterized by a specific combination of expressed Hox genes, as shown by a summary picture of the anterior borders of expression of Hox genes in mouse embryonic mesoderm (Fig. 4.36). For example, the most anterior somites are characterized by expression of

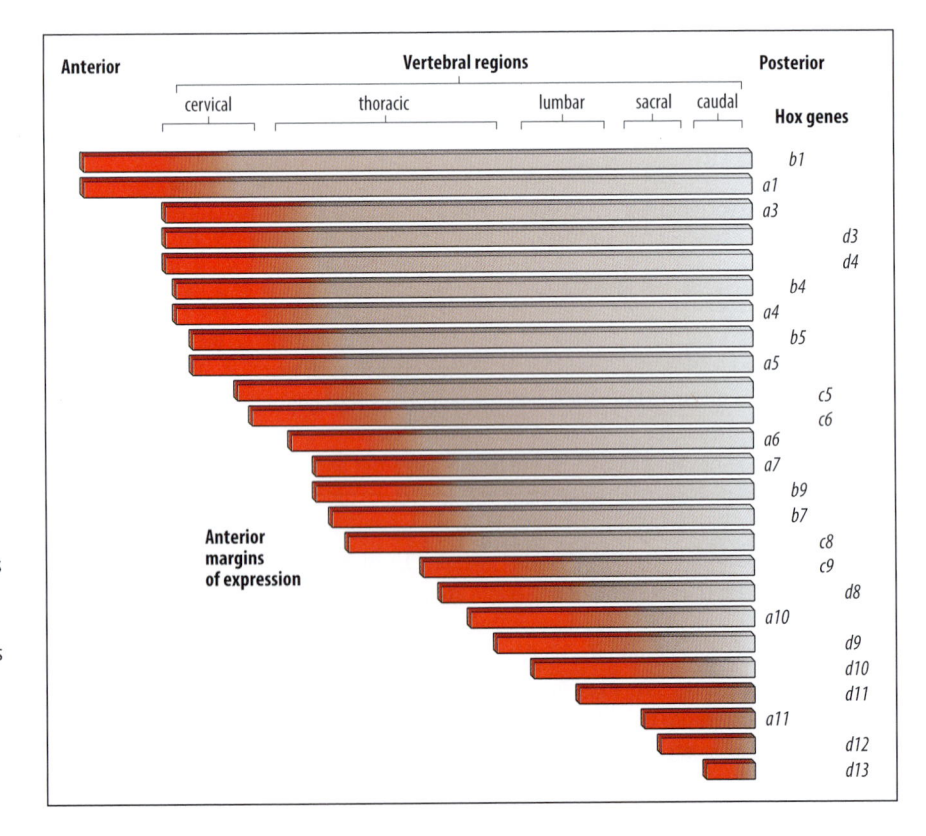

Fig. 4.36 A summary of Hox gene expression along the antero-posterior axis of the mouse mesoderm. The anterior border of expression of each gene is shown by the red blocks. Expression usually extends backwards some distance but the posterior margin of expression may be poorly defined. The pattern of Hox gene expression specifies the regional identity of the tissues at different positions. This figure represents an overall picture of Hox gene expression and is not a snapshot of gene expression at any particular time.

Hoxa1 and *Hoxb1*, and no other Hox genes are expressed in this region. By contrast, all Hox genes are expressed in the more posterior regions. The Hox genes can therefore provide a code for regional identity. The spatial and temporal order of expression is similar in the mesoderm and the ectodermally derived nervous system, but the positions of the boundaries between regions of gene expression in the two germ layers do not always correspond (see Fig. 4.35). The most anterior region of the body in which Hox genes are expressed is the hindbrain; other homeobox genes such as *Pax*, *Emx*, and *Otx*, but not Hox genes, are expressed in more anterior regions of the vertebrate body—midbrain, forebrain, and the anterior part of the head (see Chapter 10).

If we focus on just the Hoxa cluster of genes, we see that the most anterior border of expression in the mesoderm is that of *Hoxa1* in the posterior head mesoderm, while *Hoxa11*, a very posterior gene in the Hoxa cluster, has its anterior border of expression in the sacral region (see Fig. 4.36). This exceptional correspondence, or **co-linearity**, between the order of the genes on the chromosome and their order of spatial and temporal expression (**spatial** and **temporal co-linearity**) along the antero-posterior axis is typical of all the Hox clusters. The genes of each Hox cluster are expressed in an orderly sequence, with the gene lying most 3' in the cluster being expressed the earliest and in the most anterior position. The correct expression of the Hox genes is dependent on their position within the cluster, and 'anterior' genes (3') are expressed before more 'posterior' (5') genes. This typical sequential activation of Hox genes in the order they occur along the chromosome is likely to be due, at least in part, to changes in chromatin status (see Box 6B).

Support for the idea that the Hox genes are involved in specifying regional identity in vertebrates comes from comparing their patterns of expression in mouse and chick with the well-defined anatomical regions—cervical, thoracic, sacral, and lumbar (Fig. 4.37). Hox gene expression corresponds well with the different regions. For

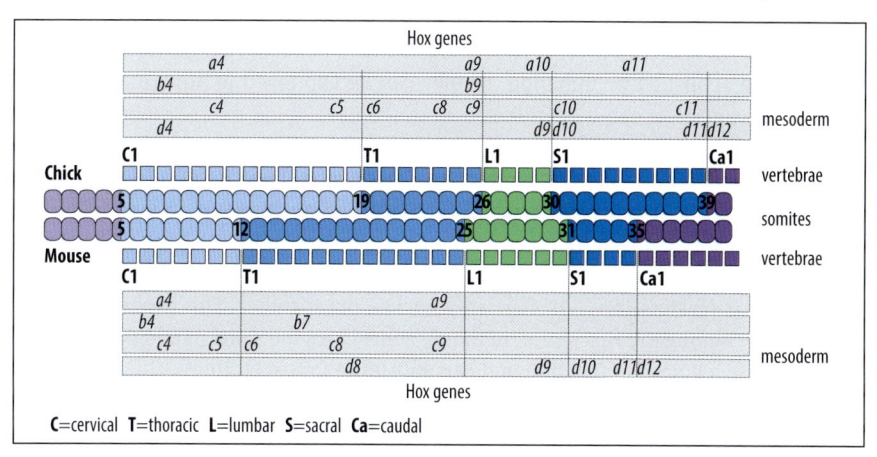

Fig. 4.37 Patterns of Hox gene expression in the mesoderm of chick and mouse embryos, and their relation to regionalization. The pattern of Hox gene expression varies along the antero-posterior axis. The vertebrae (squares) of the spinal column are derived from the embryonic somites (circles), 40 of which are shown here. The vertebrae have characteristic shapes in each of the five regions along the antero-posterior axis: cervical (C), thoracic (T), lumbar (L), sacral (S), and caudal (Ca). Which somites form which vertebrae differs in chick and mouse. For example, thoracic vertebrae start at somite 19 in the chick but at somite 12 in the mouse. Despite this difference, the transition from one region to another corresponds to a similar pattern of Hox gene expression in chick and mouse, such that the anterior boundaries of expression of *Hoxc5* and *Hoxc6* lie on either side of the transition from cervical to thoracic vertebrae (light blue to mid-blue) in both species. Similarly, the anterior boundaries of *Hoxd9* and *Hoxd10* are at the transition between the lumbar and sacral regions (green to dark blue) in both chick and mouse.

Illustration after Burke, A.C., et al.: **Hox genes and the evolution of vertebrate axial morphology**. *Development 1995,* **121***: 333-346.*

example, even though the number of cervical vertebrae in birds (14) is twice that of mammals, the anterior boundaries of *Hoxc5* and *Hoxc6* gene expression in the somites in both chick and mouse lie on either side of the cervical/thoracic boundary. A correspondence between Hox gene expression and regional identity is also similarly conserved among vertebrates at other anatomical boundaries.

It must be emphasized that the summary picture of Hox gene expression given in Figure 4.36 does not represent a 'snapshot' of expression at a particular time but rather an integrated picture of the overall pattern of expression. Some genes are switched on early and expression is then reduced, whereas others are switched on considerably later; the most 'posterior' Hox genes, such as *Hoxd12* and *Hoxd13*, are expressed in the tail. Moreover, this summary picture reflects only the general expression of the genes in any particular region; not all Hox genes expressed in a region are expressed in all the cells. Nevertheless, the overall pattern suggests that the combination of Hox genes expressed specifies regional identity. In the cervical region, for example, each somite, and thus each vertebra, could be specified by a unique pattern of Hox gene expression or **Hox code**.

As we saw in Figure 4.33, grafting experiments show that the character of the somites is already determined in the pre-somitic mesoderm, and that pre-somitic tissue transplanted to other levels along the axis retains its original identity. As would be expected if the Hox code specifies vertebral identity, the transplanted pre-somitic mesoderm also retains its original pattern of Hox gene expression.

4.12 Deletion or overexpression of Hox genes causes changes in axial patterning

If the Hox genes specify regional identity that determines a region's subsequent development, then one would expect morphological changes if their pattern of expression is altered. This is, indeed, the case. To investigate the function of Hox genes, their expression can be abolished by mutation, or they can be expressed in abnormal positions. Hox gene expression can be eliminated from the developing mouse embryo by gene knock-out. Experiments along these lines indicate that the absence of given Hox genes affects patterning in a way that agrees with the idea that they are normally providing cells with regional identity. In some cases, however, knocking out an individual Hox gene produces little effect, or affects only one tissue, and more substantial changes in patterning are only seen when two or more genes, particularly **paralogous genes** from different Hox clusters (see Box 4G), are knocked out together. This probably reflects redundancy among the Hox genes, in that in some cases the same paralogs in different clusters have a common function and can substitute for each other.

Hoxd3 has an anterior limit in cervical regions (see Fig. 4.36). Mice in which *Hoxd3* is deleted show structural defects in the first and second cervical vertebrae, and in the sternum (breastbone). More posterior structures, where the inactivated gene is also normally expressed together with other Hox genes, show no evident defects, however. This observation illustrates a general principle of Hox gene expression, which is that more posteriorly expressed Hox genes tend to override the function of more anterior Hox genes whenever they are co-expressed. This phenomenon is known as **posterior dominance** or **posterior prevalence**. It has been proposed that the action of microRNAs (miRNAs) encoded by genes within Hox clusters might explain posterior prevalence. The encoded miRNAs in the Hox clusters are likely to be involved in post-transcriptional regulation of Hox protein expression. More of the predicted target genes of a given miRNA tend to lie on the 3′ side of the miRNA gene rather than the 5′ side. Given this arrangement, the Hox miRNAs might help to repress the function of more anterior Hox genes and thus explain the general principle of posterior prevalence.

Another general point illustrated by the *Hoxd3* knock-out is that the effects of a Hox gene knock-out can be tissue specific, in that only certain tissues in which the

Hox gene is normally expressed are affected, whereas other tissues at the same position along the antero-posterior axis appear normal. For example, although *Hoxd3* is expressed in neural tube, **branchial arches**, and paraxial mesoderm at the same antero-posterior level in the body, only the vertebrae are defective in a *Hoxd3* knock-out. The apparent absence of an effect in such cases may be due to redundancy, with paralogous genes from another cluster being able to compensate. *Hoxa3* is expressed strongly in the same tissues as *Hoxd3* in the cervical region, but when *Hoxa3* alone is knocked out there are no defects in cervical vertebrae; instead, other defects are observed, including reductions in cartilage elements derived from the second branchial arch. When both *Hoxd3* and *Hoxa3* are knocked out together, much more severe defects are seen in both vertebrae and branchial arch derivatives, suggesting that the two genes normally function together in these tissues.

Loss of Hox gene function often results in **homeotic transformation**—the conversion of one body part into another, as most spectacularly and clearly seen in *Drosophila* (see Fig. 2.46). This is the case in the *Hoxd3* knock-out mouse we have just considered. Detailed analysis of the cervical region in these mice revealed that the atlas vertebra (the first cervical vertebra) appears to be transformed into the basal occipital bone at the base of the skull, and that the axis vertebra (the second cervical vertebra) has acquired morphological features resembling the atlas. Thus, in the absence of *Hoxd3* expression, cells acquire a more anterior regional identity and develop into more anterior structures. The double knock-out of *Hoxa3* and *Hoxd3* results in a deletion of the atlas. The complete absence of this bone in these double knock-outs suggests that one target of Hox gene action is the cell proliferation required to build such a structure from the somite cells. Unfortunately, rather few direct gene targets transcriptionally regulated by Hox proteins have yet been identified. Indeed, how different Hox proteins, with similar DNA-binding properties, regulate different developmental programs along the antero-posterior axis is poorly understood. ChIP-seq approaches to characterize the genome-wide binding properties of Hoxb1 protein suggest that it functions, at least in part, through a repressive role in gene expression.

Another example of homeotic transformation is seen in *Hoxc8* knock-out mice. In normal embryos, *Hoxc8* is expressed in the thoracic and more posterior regions of the embryo from late gastrulation onward (see Fig. 4.36). Mice homozygous for deletion of *Hoxc8* die within a few days of birth and have abnormalities in patterning between the seventh thoracic vertebra and the first lumbar vertebra. The most obvious homeotic transformations are the attachment of an eighth pair of ribs to the sternum and the development of a fourteenth pair of ribs on the first lumbar vertebra (Fig. 4.38). Thus, as with the knock-outs discussed above, the absence of *Hoxc8* has led to some of the cells that would normally express it acquiring a more anterior regional identity, and developing accordingly.

Homeotic transformations can affect entire axial domains and occur when all the paralogous genes in a subgroup are knocked out together. When all the Hox10 paralogous genes are absent, there are no lumbar vertebrae in mice and there are ribs on all posterior vertebrae; in the absence of all the Hox11 paralogous genes several sacral vertebrae become lumbar. These homeotic transformations do not occur if only some members of the paralogous subgroup are mutated, again suggesting redundancy of Hox gene function.

In contrast to the effects of knocking out Hox genes, the ectopic expression of Hox genes in anterior regions that normally do not express them can result in transformations of anterior structures into structures that normally develop in a more posterior position. For example, when *Hoxa7* in the mouse, whose normal anterior border of expression is in the thoracic region, is expressed throughout the whole antero-posterior axis, the basal occipital bone of the skull is transformed into a pro-atlas structure, normally the next more posterior skeletal structure.

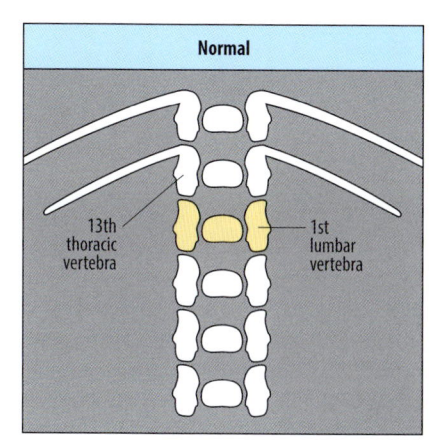

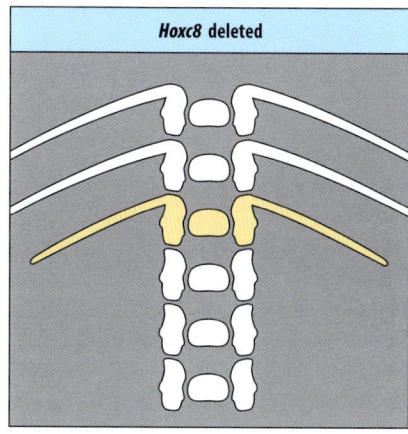

Fig. 4.38 Homeotic transformation of vertebrae due to deletion of *Hoxc8* in the mouse. In loss-of-function homozygous mutants of *Hoxc8*, the first lumbar vertebra (yellow) is transformed into a rib-bearing 'thoracic' vertebra. The mutation has resulted in the transformation of the lumbar vertebra into a more anterior structure.

4.13 Hox gene expression is activated in an anterior to posterior pattern

In all vertebrates, the Hox genes begin to be expressed at an early stage of gastrulation. The genes expressed most anteriorly, which correspond to those at the 3′ end of a cluster, are expressed first. If an 'early' *Hoxd* gene is relocated to the 5′ end of the *Hoxd* cluster, for example, its spatial and temporal expression pattern then resembles that of the neighboring *Hoxd13* (see Box 4G). This shows that the structure of the Hox cluster is crucial in determining the pattern of Hox gene expression.

Unlike the situation in *Drosophila*, where the activation of Hox gene transcription clearly depends on factors unequally distributed along the antero-posterior axis, the activation of Hox gene transcription in vertebrates is more complex and occurs in a strict temporal order. Transcription of the genes of all four clusters is activated in the primitive streak in mouse embryos. The first Hox gene transcripts can be detected in the most posterior region of the fully extended primitive streak at around E7.0–E7.5. The most-3′ Hox genes are transcribed first, starting with Hox1 paralogous genes and their expression domains then spread anteriorly to encompass the node. Subsequently, the more-5′ genes are transcribed one after another and their expression domains also spread anteriorly so that the cells in the node and stem zone from which the trunk axial structures will be formed express more-posterior genes over time. Sequential gene activation is terminated by transcription of the Hox13 paralogous genes, the most-5′ genes in the clusters, in the tail bud. The expression patterns of Hox genes are maintained when the cells leave the stem zone to form the spinal cord and pre-somitic mesoderm, and in this way the temporal pattern of Hox gene activation can be converted into regional identity; the cells expressing different Hox genes are generated progressively along the antero-posterior axis as structures are laid down in an antero-posterior sequence.

A model for the way in which the Hox genes in a cluster are transcribed in a precise temporal order in the primitive streak of a mouse embryo is shown in Figure 4.39. The model is based on detailed studies of the Hoxa cluster, including studies on stem cells derived from the epiblast. Transcription of the Hoxa genes is controlled by *cis*-regulatory regions that lie 3′ to the Hoxa cluster, within the cluster itself, and just 5′ to it. Just before gastrulation, the 3′ *cis*-regulatory region and *Hoxa1* are primed to become active. Then, after gastrulation has started, Wnt signaling in the posterior streak activates *Hoxa1–Hoxa3* transcription in the primitive streak via enhancers in the 3′ regulatory region. This is followed by activation of transcription of *Hoxa4–Hoxa9* as a result of the homeodomain protein Cdx2 acting on *cis*-regulatory elements in the middle part of the cluster. Lastly, transcription of the 5′-most genes, *Hoxa11–Hoxa13*, is activated via *cis*-regulatory enhancers 5′ to the cluster. The protein product of the most posterior gene in the cluster, Hoxa13, antagonizes the inducing action of Cdx2 and thus completes the process. The Hoxd cluster has a similar structure to the Hoxa cluster and the temporal order of expression of Hoxd genes is likely to be controlled in the same way; different mechanisms are likely to operate in the other two clusters. The Hoxa and Hoxd clusters are also expressed in precise temporal and spatial patterns in the developing limb (discussed in Chapter 9), and here again we shall see how the structure of the clusters is crucial in determining the pattern of Hox expression.

Retinoic acid has been shown experimentally in mice to alter the expression of Hox genes, but it is not clear whether it is involved in the initial activation of Hox gene transcription in the early embryo. The gene for the retinoic-acid-metabolizing enzyme, Cyp26A1, is expressed in the anterior half of the mouse embryo during gastrulation stages, whereas the gene for the retinoic-acid-generating enzyme Ralhd2 is expressed in the posterior half. This leads to graded retinoic acid signaling from low levels anteriorly to high levels posteriorly. Later in development, as already discussed in Section 4.9, the gene for Cyp26A1 is expressed in the posterior region of the

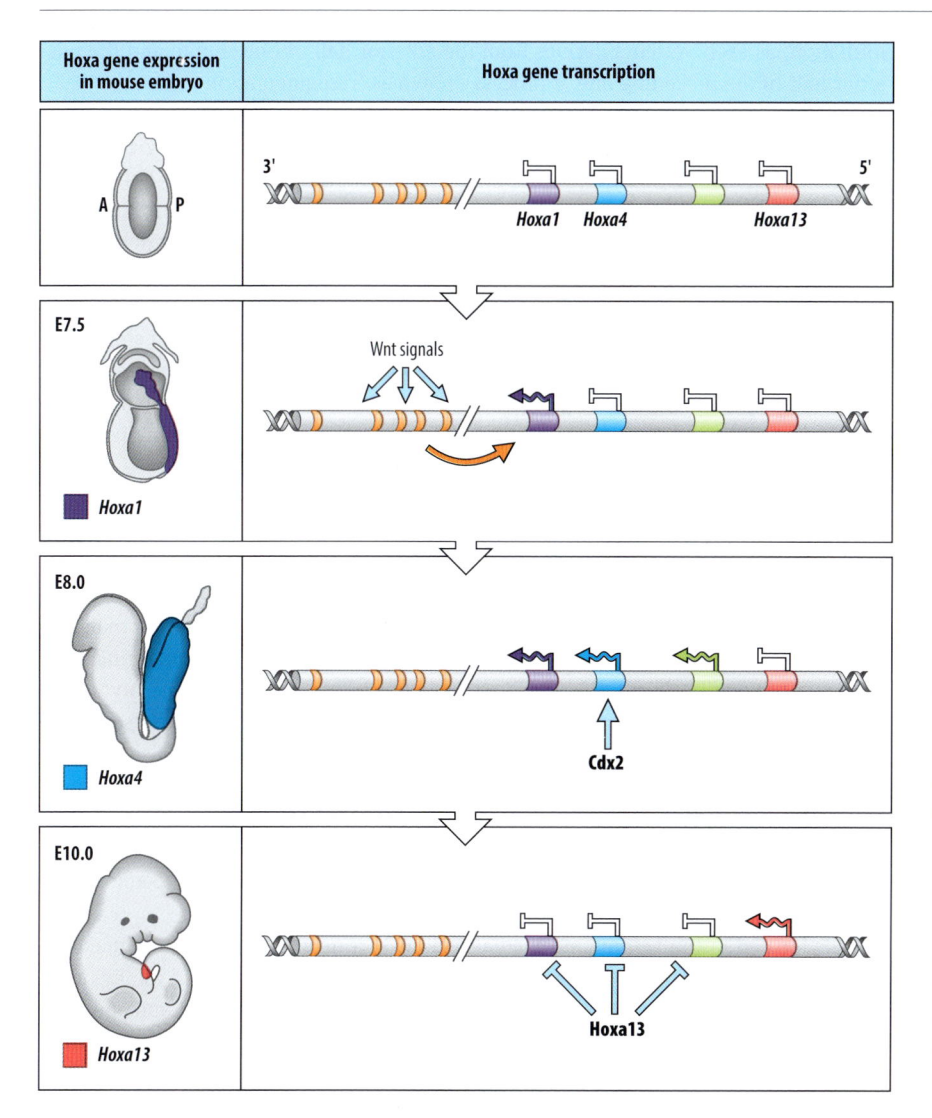

Fig. 4.39 A model for the way in which the genes in the Hoxa cluster are transcribed in a temporal order in developing mouse embryos, leading to an overlapping pattern of expression along the main body axis. The left panels show the spatial pattern of individual Hoxa genes at different stages in development. The first gene in the cluster, *Hoxa1*, is expressed at E7.0–E7.5 in the primitive streak and allantois; by E8.0, when the axial structures are being generated from the stem zone, *Hoxa4–Hoxa9* are sequentially expressed in the extending posterior region of the embryo; *Hoxa13* is expressed last of all at E10.0 in the tail bud. The right panels show how transcription of the genes along the cluster is regulated. Transcription of *Hoxa1* at the 3′ end of the cluster is activated (arrowed line) by enhancers (orange) in a *cis*-regulatory region 3′ to the cluster under the control of Wnt signaling (blue arrows). At this stage the other genes in the cluster are not transcribed (barred lines). Expression of *Hoxa4–Hoxa9* genes is regulated by the activity of Cdx2 protein, allowing transcription from enhancers located in this region of the cluster; *Hoxa13* transcription is under the control of enhancers located in the region 5′ to the cluster (not shown here). Hoxa13 protein antagonizes transcription of more-3′ genes in the cluster (barred blue lines) thus ensuring that these genes are not expressed out of order. Not all the genes in the cluster are shown for clarity.

From Neijts, R., et al.: ***Cdx is crucial for the timing mechanism driving colinear Hox activation and defines a trunk segment in the Hox cluster topology.*** Dev. Biol. 2017, 422: 146-154.

elongating body axis as the somites are being generated, and the retinoic-acid gradient then runs from anterior to posterior. Mouse embryos lacking the gene *Cyp26a1* not only have a truncated axis, but also have posterior homeotic transformations, with each vertebra taking on the identity of its neighboring more-posterior vertebra—for example, the last cervical vertebra is transformed into a vertebra carrying ribs. These transformations occur in regions where retinoic acid signaling is elevated and where *Hoxb4* expression has expanded anteriorly. The influence of retinoic acid on Hox gene expression and development of the axial skeleton can also be seen in zebrafish. The zebrafish mutation *giraffe*, which affects the *Cyp26* gene, has anteriorly extended Hox gene expression and vertebral transformations.

4.14 The fate of somite cells is determined by signals from the adjacent tissues

We now return to the individual somite and consider how it is patterned along its antero-posterior, dorso-ventral, and medio-lateral axes. The differences between the anterior and posterior halves of the somite are later used by neural crest cells and motor nerves as guidance cues to generate the periodic arrangement of spinal nerves and ganglia (discussed in Chapter 5). The subdivision is also important for vertebra formation, as each vertebra derives from the anterior half of one somite and the posterior half of the preceding one, a process known as 'resegmentation'.

The antero-posterior patterning process within an individual somite is independent of the antero-posterior patterning of the whole pre-somitic mesoderm by Hox gene expression. Individual somites are subdivided into anterior and posterior halves with different properties. In the mouse, this subdivision is linked to the transcription factor Mesp2, which is initially expressed in the pre-somitic mesoderm that will give rise to the next somite, but which then becomes confined to just the anterior half of the somite (Fig. 4.40). The restricted expression of *Mesp2* is thought to be a consequence of positive- and negative-feedback loops between Mesp2 and Notch–Delta signaling.

Earlier in this chapter, we commented that cells that contribute to different medio-lateral regions of the somites originate from slightly different areas: medial parts of the somite come from the stem zone, and lateral parts come from a slightly more posterior region and are specified by slightly higher levels of BMP signaling. However, the different history of somite cells does not specify their later fate; instead, once somitogenesis has occurred, other signals pattern somites along the dorso-ventral axis and medio-lateral axis and begin to confer positional information that will result in cells in different parts of the somite giving rise to different tissues.

Cells located in the dorsal and lateral regions of a newly formed somite make up the **dermomyotome**, which expresses the gene *Pax3*, a homeobox-containing gene of the *paired* family (Fig. 4.41). The dermomyotome gives rise first to the **myotome**, from which muscle cells originate, and then forms the epithelial sheet, dorsal to the myotome, from which the dermis (skin) originates. The dermomyotome also contains cells that contribute to the vasculature. Cells from the medial region of the myotome form mainly axial and back (epaxial) muscles, whereas cells from the lateral region of the myotome cells migrate to form abdominal and limb (hypaxial) muscles. All committed muscle precursors express muscle-specific transcription factors of the MyoD family.

Cells located in the ventral and medial regions of a newly formed somite make up the **sclerotome**, which expresses *Pax1*, a second member of the *paired* family. Most sclerotome cells will migrate and eventually surround the notochord and much of the neural tube. They will form cartilage, and ultimately develop into the vertebral column and ribs. Cells at the boundary of the sclerotome and myotome specifically express the transcription factor Scleraxis and will give rise to tendons.

Which cells will form cartilage, muscle, or dermis is not yet determined at the time of somite formation. This is clearly shown by experiments in which the dorso-ventral

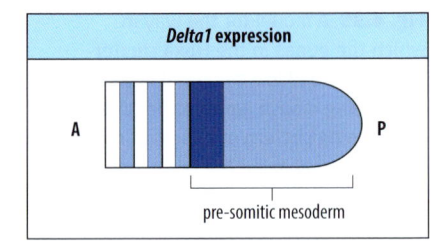

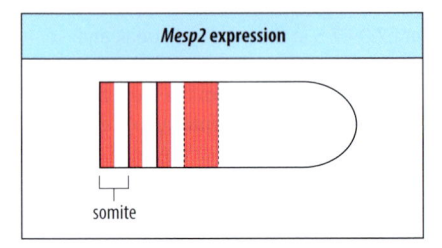

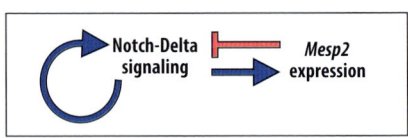

Fig. 4.40 Division of somites into anterior and posterior halves. Top and center panels: dorsal views of the tail regions of E9.5 mouse embryos. The Notch ligand Delta-1 and the transcription factor Mesp2 are expressed in a somite-wide stripe in pre-somitic mesoderm that is about to generate a somite. As somites form, Mesp2 expression becomes confined to the anterior (A) half, and Delta expression to the posterior (P) half. Bottom panel: positive- and negative-feedback loops between Notch signaling and Mesp2 are essential for dividing the somite into an anterior half and a posterior half.

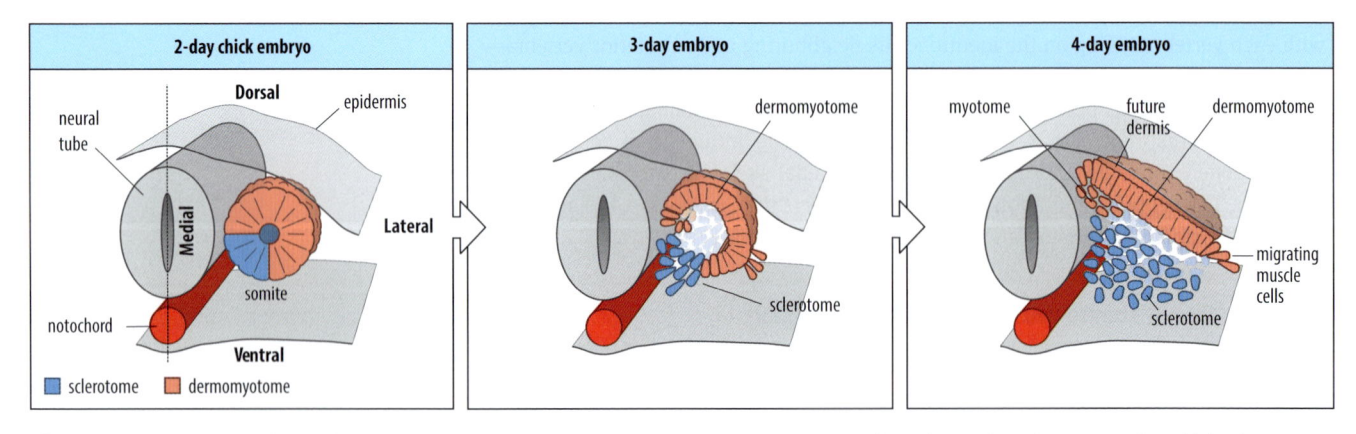

Fig. 4.41 The fate map of a somite in the chick embryo. The ventral medial quadrant (blue) gives rise to the sclerotome cells, which migrate to form the cartilage of the vertebrae. The rest of the somite—the dermomyotome—forms the myotome, which gives rise to all the trunk muscles and the epithelial dermomyotome region that gives rise to the dermis of the skin. The dermomyotome also gives rise to muscle cells that migrate into the limb buds.

orientation of newly formed somites is inverted but normal development nevertheless ensues. Instead, specification of the fate of somite tissue requires signals from adjacent tissues. The neural tube, the notochord, and the lateral plate mesoderm all produce signals that pattern the somite and are required for its future development. The notochord and floor plate region of the ventral neural tube specify the sclerotome; the dorsal neural tube and overlying ectoderm specify the medial part of the dermomyotome; the lateral plate mesoderm specifies the lateral dermomyotome (Fig. 4.42). Their patterning effects can be demonstrated through grafting experiments. For instance, when an extra notochord is implanted in an unusually dorsal position, almost the entire somite now forms sclerotome and cartilage precursors (Fig. 4.43).

Many of the signals that derive from each tissue to specify different regions of the somite have been identified. In the chick, both the notochord and the floor plate of the neural tube express Sonic hedgehog (Shh), a secreted protein that is a key molecule for positional signaling in many developmental situations. We encountered Shh in Chapter 1 as a signal involved in craniofacial development (see Box 1H) and will meet it later in this chapter with respect to left–right asymmetry. We shall meet it again in Chapter 9 (see Box 9C, for the Shh signaling pathway) in connection with limb development, and in Chapter 10 in the dorso-ventral patterning of the neural tube. In somite patterning, high levels of Shh signaling specify the ventral region of the somite and are required for sclerotome development. Wnts provided by the dorsal neural tube and ectoderm specify the dermomyotome, acting together with additional BMP-4 provided by the lateral plate mesoderm (see Fig. 4.42).

Shh, BMPs, and Wnts play an ongoing role over time to specify fates that are increasingly restricted. For example, *Pax3* is initially expressed throughout the dermomyotome, in response to BMP-4 and Wnt. Over time, its expression is further modulated by BMP-4 and Wnt proteins so that it becomes confined to muscle precursors. It is then further downregulated in cells that differentiate as the muscles of the back but remains switched on in the migrating presumptive muscle cells that populate the limbs. Mice that lack a functional *Pax3* gene—*Splotch* mutants—lack limb muscles. This suggests that other genes act with Pax3 to specify dermomyotome and myotome, but that Pax3 is absolutely required for the final stage of this differentiation trajectory, which is limb muscle.

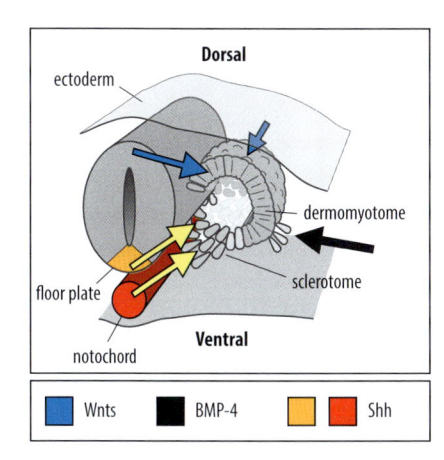

Fig. 4.42 The somite is patterned by signals secreted by adjacent tissues. The sclerotome is specified by a diffusible signal, Sonic hedgehog protein (Shh; yellow arrows), produced by the notochord (red) and the floor plate (orange) of the neural tube. Signals from the dorsal neural tube and ectoderm such as Wnts (blue arrows) specify the dermomyotome, together with lateral signals such as BMP-4 (black arrow) from the lateral plate mesoderm.

Illustration after Johnson, R.L., et al.: ***Ectopic expression of Sonic hedgehog alters dorsal-ventral patterning of somites***. *Cell 1994, **79**: 1165-1173.*

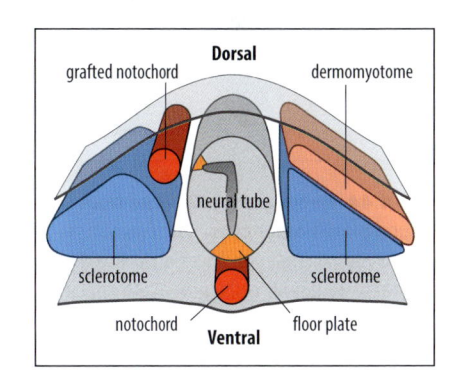

Fig. 4.43 A signal from the notochord induces sclerotome formation. A graft of an additional notochord next to the dorsal region of a somite in a 10-somite chick embryo suppresses the formation of the dermomyotome from the dorsal portion of the somite, and instead induces the formation of sclerotome, which develops into cartilage. Note the additional notochord also alters dorso-ventral pattern of the neural tube, inducing an additional floor plate (see Chapter 10).

SUMMARY

Somites are blocks of mesodermal tissue that are formed after gastrulation. They form sequentially in pairs, a somite on either side of the notochord, starting adjacent to the posterior region of the hindbrain. The somites give rise to the vertebrae, to the muscles of the trunk and limbs, and to most of the dermis of the skin. The periodic formation of the somites can be viewed in terms of the clock-and-wavefront model, in which segmentation is driven by an oscillatory molecular clock running through the pre-somitic mesoderm, with the size of a somite being determined by the distance that a wave of determination travels during one oscillation of the clock. The pre-somitic mesoderm is patterned with respect to the antero-posterior axis during gastrulation stages. The manifestation of this pattern is the expression of the Hox genes in the pre-somitic mesoderm. The regional identity of the somites is specified by the combinatorial expression of genes of the Hox clusters along the antero-posterior axis, from next to the hindbrain to the posterior end of the embryo, with the order of expression of these genes along the axis corresponding to their order along the chromosome. Mutation or overexpression of a Hox gene results, in general, in localized defects in the anterior parts of the regions in which the gene is expressed and can cause homeotic

> transformations. We can think of Hox genes as encoding positional information that specifies the identity of a repeated structure and hence its later development. They act on downstream gene targets and, mechanistically, can function as repressors of gene activity.
>
> Individual somites are patterned by signals from the notochord, neural tube, and ectoderm, which induce regions of each somite to give rise to muscle, cartilage, or dermis.

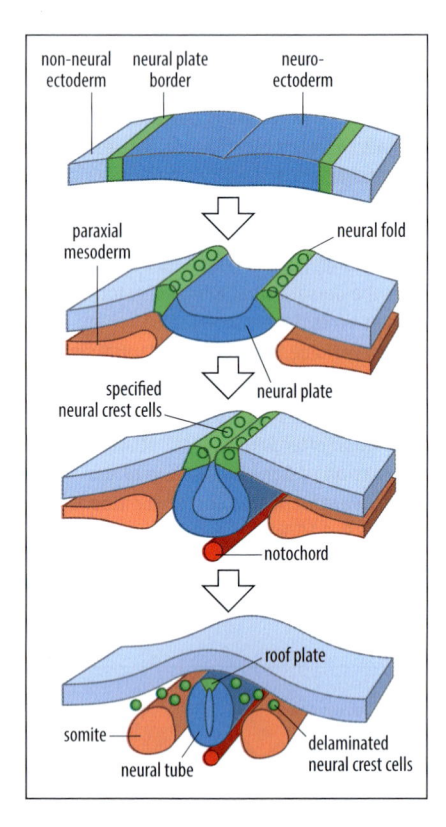

Fig. 4.44 Neural crest cells arise from neural plate border cells. The diagram shows neural crest specification in the chick. Neural plate border cells (green) are specified at the edges of the neural plate (dark blue). During neurulation, neural plate border cells are carried to the dorsal crests of the neural tube, and begin to be specified as neural crest cells. Around the time of neural tube closure many neural crest cells have been specified. These undergo an epithelial-to-mesenchymal transition, delaminate, and migrate away from the neural tube. This diagram shows the delamination of neural crest cells in the trunk; cranial neural crest starts to delaminate before fusion of the neural folds. A subset of neural plate border cells do not form neural crest, but are retained as roof plate cells in the dorsal-most aspect of the neural tube.

The origin, behavior, and patterning of neural crest

In this part of the chapter we consider the origin, properties, and an example of patterning of the neural crest, which is a population of cells that is specific to vertebrates. Neural crest cells arise in the neural plate border region. This region rises up at neurulation to form the neural folds that ultimately fuse to give the neural tube. A network of transcription factors directs neural crest cell specification and induces expression of genes that regulate the cell behaviors that characterize the neural crest. These include genes that direct cell proliferation and genes that govern an epithelial-to-mesenchymal cell transition. Once specified, neural crest cells delaminate from the neural tube and migrate away from it. The migration of the cranial neural crest from the hindbrain to the branchial arches is touched on briefly in this chapter in relation to the antero-posterior patterning of the neural crest. The migration of neural crest cells is described in more detail in Chapter 5.

4.15 Neural crest cells arise from neural plate border cells and migrate to give rise to a wide range of different cell types

Neural plate border cells form at the boundary between the neural plate and non-neural ectoderm, where levels of BMP are just high enough to prevent neural plate specification (Fig. 4.44 and see also Section 4.3). As the neural plate starts to form the neural tube during neurulation, the neural folds carry the neural plate border cells to the dorsal crests of the forming neural tube. Analyses from single-cell RNA-seq studies in the chick show that the neural crest lineage is gradually specified and segregated from the neural plate border during neurulation, from around HH7 onwards. Neural crest cell specification from neural plate border cells involves many changes in cell behavior. Developing neural crest cells exhibit the properties of stem cells, in that they self-renew. In addition, they proliferate extensively, and undergo an epithelial-to-mesenchymal transition, enabling them to delaminate from the neural tube and migrate away. Neural crest cells give rise to an enormous range of different tissues and cell types, and are sometimes referred to as the fourth germ layer. Neural plate border cells are heterogeneous, and while most give rise to neural crest cells, in the cranial region some border cells give rise to ectodermal placodes that develop into structures such as the ears, nose, and the lens of the eye. Another subset of neural plate border cells are retained as **roof plate** cells in the dorsal-most aspect of the neural tube (see Fig. 4.44). Roof plate cells will go on to pattern the dorsal aspect of the neural tube, as discussed in Chapter 10.

Single-cell RNA transcriptional studies in chick embryos show that a network of transcription factors is established over time, and progressively allocates neural plate border cells to defined neural crest cell lineages. These transcription factors ultimately control different aspects of neural crest cell behaviors, including proliferation, multipotency, survival, **delamination**, and migration (Fig. 4.45).

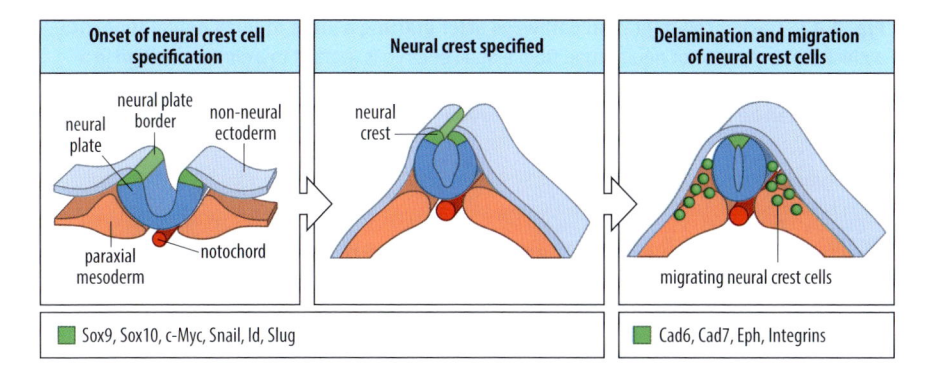

Fig. 4.45 Transcription factors that control cell proliferation and epithelial-to-mesenchymal transition are upregulated during neural crest cell specification. As neural crest cells are specified during neurulation, they immediately upregulate genes for transcription factors such as Sox9/10, c-Myc, Snail, Slug, and Id that promote proliferation, prevent apoptosis, and begin to promote epithelial-to-mesenchymal transitions (first and second panels). These transcription factors regulate the expression of genes for various cell-adhesion proteins, some of which are noted on the right. Changes in cell adhesion enable neural crest cells to delaminate and migrate away from the dorsal neural tube (third panel).

Specification of neural crest is marked by expression of the genes for the transcription factors Sox9, Sox10, c-Myc, Id, Slug, and Snail. These proteins support functions that characterize neural crest cell behaviors. The c-Myc and Id proteins are involved in the acquisition of multipotency and proliferation. The Sox9, Sox10, and Snail proteins prevent apoptosis, and Snail also promotes epithelial-to-mesenchymal transition. At the same time, these genes control expression of many other genes, including the cell-adhesion proteins cadherins, ephrins, and integrins, that enable the morphological transition of the epithelial neural plate border cell into the migratory mesenchymal neural crest cell. Snail is the vertebrate version of the *Drosophila* protein *snail*, which directs a similar epithelial-to-mesenchymal transition during *Drosophila* gastrulation (see Section 5.10); such epithelial-to-mesenchymal transitions occur in many situations during animal development, for example in the ingression of epiblast cells into the primitive streak in bird and mammalian embryos. *Sox10* expression provides a marker for neural crest and the gene continues to be expressed in migrating neural crest cells.

Once neural crest cells have detached from the neural tube they migrate to many sites, where they differentiate into a remarkable number of different cell types (Fig. 4.46). Neural crest cells are, for example, the origin of the neurons and glia of the autonomic nervous system, the glial Schwann cells of the peripheral nervous system, the pigment cells (melanocytes) of the skin, and the adrenaline (epinephrine)-producing cells of the adrenal glands. Surprisingly, much of the bone, cartilage, and connective tissue of the face—tissue types that are more commonly of mesodermal origin—are of neural crest origin, and the head mesenchyme derived from neural crest is called **ectomesenchyme**. Cells of the neural crest also contribute to the outflow tract of the developing heart (see Chapter 9).

The tremendous differentiation potential of the neural crest was originally discovered by removing the neural folds from amphibian embryos and noting which

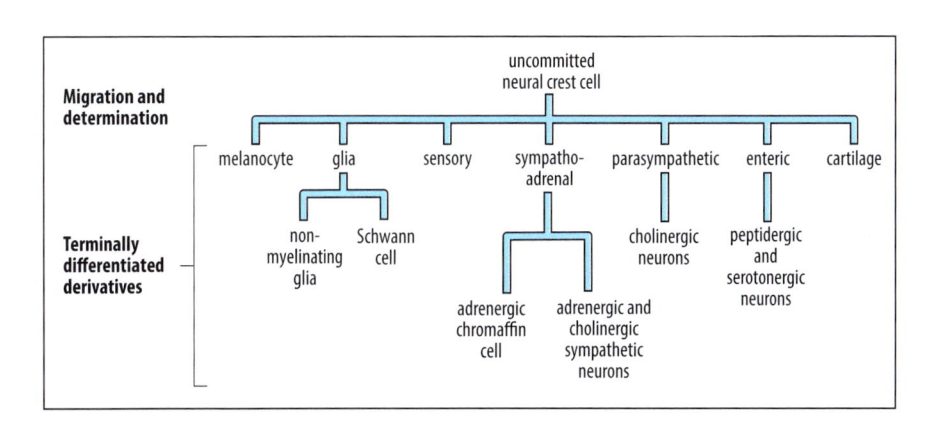

Fig. 4.46 Derivatives of the neural crest. Neural crest cells give rise to a wide range of cell types that include melanocytes, cartilage, glia, and neurons of the sensory, sympathetic, parasympathetic, and enteric parts of the nervous system. Each type of neuron is distinguished by their functional specializations and the neurotransmitters they produce. Cholinergic neurons use acetylcholine as their neurotransmitter, adrenergic neurons principally use noradrenaline (norepinephrine), and peptidergic and serotonergic neurons produce peptide neurotransmitters and serotonin, respectively.

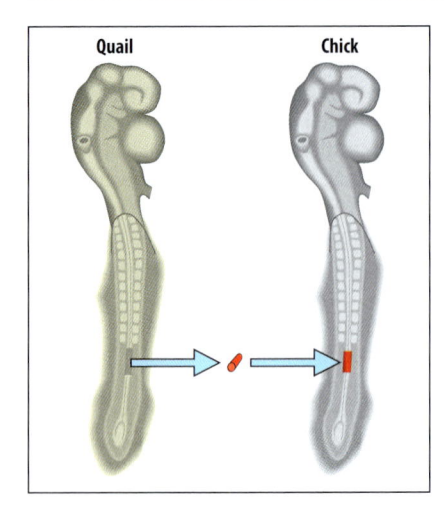

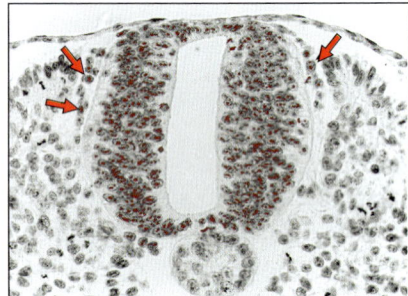

Fig. 4.47 Tracing cell-migration pathways after grafting a piece of quail neural tube into a chick host. A piece of neural tube from a quail embryo is grafted to a similar position in a chick host. The photograph shows migration of the quail neural crest cells (red arrows). Their migration can be traced because quail cells have a nuclear marker that distinguishes them from chick cells.

Photograph courtesy of N. Le Douarin.

structures failed to develop. This was later followed by an extensive series of experiments using quail neural crest transplanted into chick embryos (Fig. 4.47). As we have already discussed, it is possible to trace the transplanted quail cells in the chick embryos and see which tissues they contribute to and what cell types they differentiate into. To construct a fate map of the neural crest, small regions of quail dorsal neural tube were grafted to equivalent positions in chick embryos of the same age, at a stage before neural crest migration. The fate map shows a general correspondence between the position of crest cells along the antero-posterior axis and the axial position of the cells and tissues they will give rise to (Fig. 4.48). For example, the neural crest cells that give rise to tissues in the facial and branchial regions (the branchial region is also called the **pharyngeal region** in mammals) are derived from the cranial crest, whereas ganglion cells of the sympathetic system come from crest between somites 7 and 28.

To what extent is the fate of neural crest cells fixed before migration? Some crest cells are unquestionably multipotent. Single neural crest cells injected with a tracer shortly after they have left the neural tube can be seen to give rise to different cell types, both neuronal and non-neuronal. Also, by changing the position of neural crest before cells start to migrate, it has been shown that the developmental potential of these cells is greater than their normal fate would suggest. Multipotent stem cells have been isolated from neural crest that can give rise to neurons, glia, and smooth muscle. Similar multipotent cells have even been isolated from the peripheral nerves of mammalian embryos days after neural crest migration.

Fig. 4.48 Fate map of the neural crest in a chick embryo. There is general correspondence between the position of neural crest cells along the antero-posterior axis and the position of the structure to which these cells give rise. For example, craniofacial cartilage and bone comes from the cranial crest. The cardiac crest is marked in red, the two regions of the crest that give rise to the parasympathetic ganglia of the gut are in green, and the trunk crest in orange. A summary of the derivatives of the neural crest from each region is shown. However, their developmental potential is greater than their presumptive fate. FB, forebrain; HB, hindbrain; MB, midbrain.

Reproduced from Gandhi, S., Bronner, M.E.: **Insights into neural crest development from studies of avian embryos.** *Int. J. Dev. Biol. 2018, 62: 183-194 with the permission of UPV/EHU Press.*

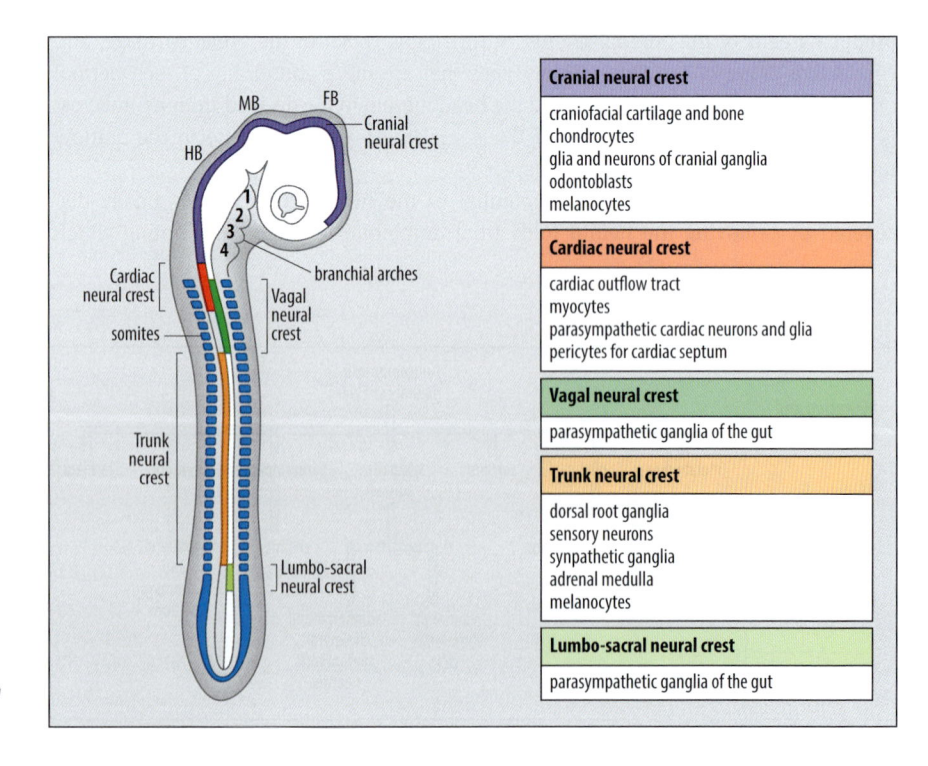

Cranial neural crest
craniofacial cartilage and bone chondrocytes
glia and neurons of cranial ganglia
odontoblasts
melanocytes

Cardiac neural crest
cardiac outflow tract myocytes
parasympathetic cardiac neurons and glia
pericytes for cardiac septum

Vagal neural crest
parasympathetic ganglia of the gut

Trunk neural crest
dorsal root ganglia
sensory neurons
synpathetic ganglia
adrenal medulla
melanocytes

Lumbo-sacral neural crest
parasympathetic ganglia of the gut

Almost all the cell types to which the neural crest can give rise will differentiate in tissue culture, and the developmental potential of single neural crest cells has been studied in this way. Most of the clones derived from single neural crest cells taken at the time of migration contain more than one cell type, showing that the cell was multipotent at the time it was isolated. As neural crest cells migrate, however, their potential progressively decreases: the size of clones derived from these cells gets smaller and fewer cell types are produced. Thus, shortly after the beginning of migration, the neural crest is a mixed population of multipotent cells and cells whose potential is already restricted.

4.16 Neural crest cells migrate from the hindbrain to populate the branchial arches

Neurons and neural crest cells obtain their antero-posterior regional identity from the Hox genes they express as part of the neural tube and, in the case of the neural crest cells, they retain this Hox gene expression when they migrate. This can be illustrated by the cranial neural crest of the hindbrain, which migrates to populate the branchial arches, where it will contribute to mesenchymal tissues (bones, cartilage, and connective tissues) of the face and skull.

The cranial neural crest that migrates from the dorsal region of the hindbrain has a segmental arrangement, correlating closely with the segmentation of this region of the brain into units called rhombomeres. We will discuss the way in which this organization develops in the brain in Chapter 10. The migration pathways have been followed by marking chick neural crest cells *in vivo*. Crest cells from rhombomeres 2, 4, and 6 populate the first, second, and third branchial arches, respectively (Fig. 4.49).

The crest cells have already acquired a regional identity before they begin to migrate. When crest cells of rhombomere 4 are replaced by cells from rhombomere 2 taken from another embryo, these cells enter the second branchial arch but develop into structures characteristic of the first branchial arch, to which they would normally have migrated. This can result in the development of an additional lower jaw in the chick embryo. However, neural crest cells have some developmental plasticity and their ultimate differentiation depends on signals from the tissues into which they migrate.

The involvement of Hox genes in patterning the neural crest of the hindbrain region has also been demonstrated by gene knock-outs in mice. The results are not always easy to interpret, as knock-out of a particular Hox gene can affect multiple

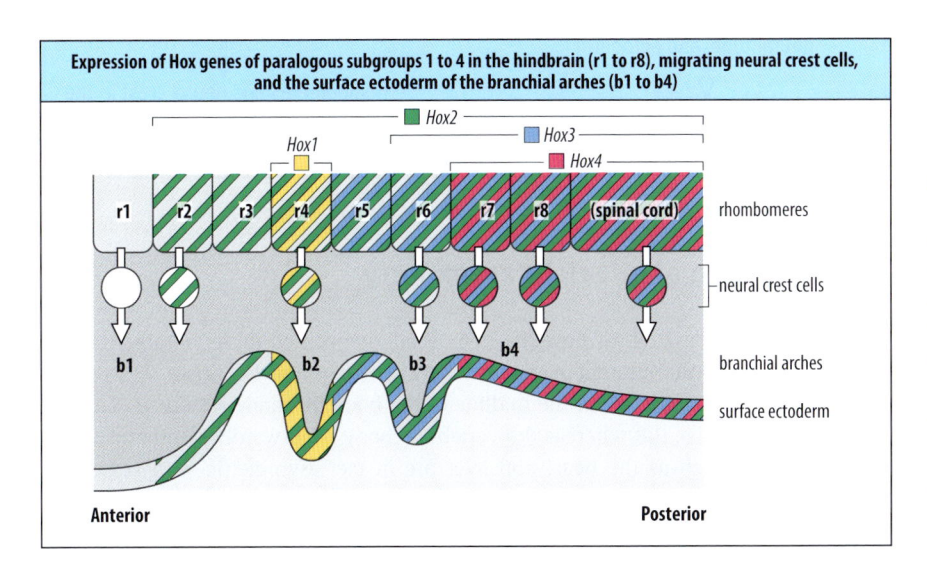

Fig. 4.49 Expression of Hox genes in the branchial region of the head. The expression of Hox genes in the hindbrain (rhombomeres r1–r8), neural crest, branchial arches (b1–b4), and surface ectoderm is shown. The arrows indicate the migration of neural crest cells into the branchial arches. There is little neural crest migration from r3 and r5.

Illustration after Krumlauf, R.: ***Hox genes and pattern formation in the branchial region of the vertebrate head.*** *Trends Genet.* 1993, *9: 106–112.*

populations of neural crest cells, for instance cranial neural crest that will form ectomesenchyme as well as cranial neural crest that will form the cranial ganglia. Knockout of the *Hoxa2* gene, for example, results in skeletal defects in the region of the head corresponding to the normal domain of expression of the gene, which extends from rhombomere 2 backwards. Skeletal elements in the second branchial arch, all of which come from neural crest cells derived from rhombomere 4, are abnormal and some of the skeletal elements normally formed by the first branchial arch develop in the second arch. Thus, suppression of *Hoxa2* causes a partial homeotic transformation of one arch into another. This shows that Hox genes encode positional information that specifies the regional identity of the segmental cranial neural crest cells in the same way as the Hox genes do in the somites.

SUMMARY

Neural crest cells arise from neural plate border cells and migrate away from the forming neural tube to differentiate into a wide range of cell types throughout the body, including the sympathetic, parasympathetic, and enteric nervous systems, the melanocytes of the skin, and some of the bones of the face. Before they leave the crest, some crest cells still have a broad developmental potential, whereas others are already restricted in their fate. Neural crest cells that migrate from the hindbrain populate the branchial arches in a position-dependent fashion and positional information is provided by a Hox gene code.

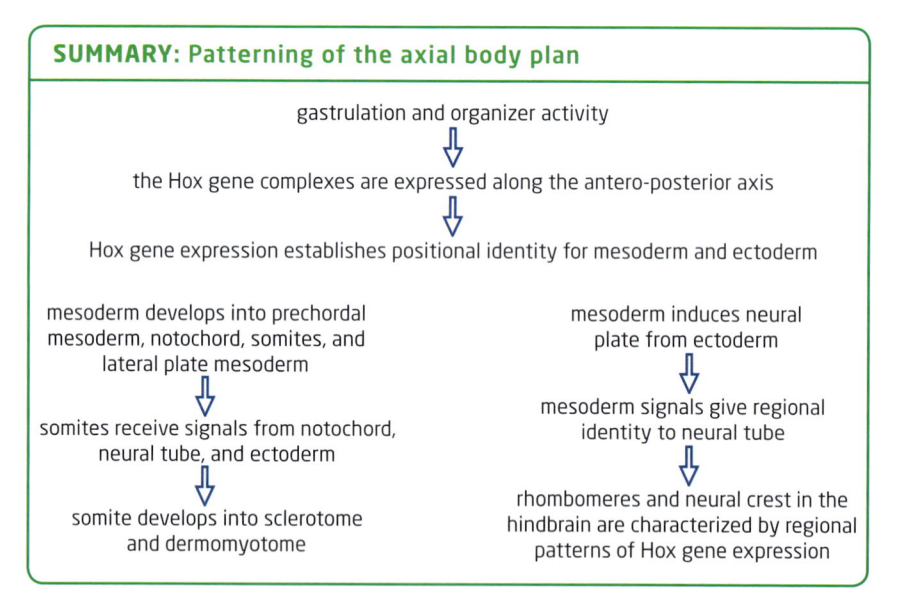

SUMMARY: Patterning of the axial body plan

gastrulation and organizer activity
⇩
the Hox gene complexes are expressed along the antero-posterior axis
⇩
Hox gene expression establishes positional identity for mesoderm and ectoderm

mesoderm develops into prechordal mesoderm, notochord, somites, and lateral plate mesoderm	mesoderm induces neural plate from ectoderm
⇩	⇩
somites receive signals from notochord, neural tube, and ectoderm	mesoderm signals give regional identity to neural tube
⇩	⇩
somite develops into sclerotome and dermomyotome	rhombomeres and neural crest in the hindbrain are characterized by regional patterns of Hox gene expression

Determination of left–right asymmetry

We end our consideration of how the body plan is specified in vertebrate embryos by looking at how the internal asymmetry between right and left arises. Vertebrates are bilaterally symmetric about the midline of the body for many structures, such as eyes, ears, and limbs. But whereas the vertebrate body is outwardly symmetric, most internal organs such as the heart and liver are in fact asymmetric with respect to left and right, as are the brain hemispheres and lobes of the lung. This is known as **left–right asymmetry**.

4.17 The bilateral symmetry of the early embryo is broken to produce left-right asymmetry of internal organs

Specification of left and right is fundamentally different from specification of the other axes of the embryo, as left and right have meaning only after the antero-posterior and dorso-ventral axes have been established. If one of these axes were reversed, then so too would be the left–right axis (this is the reason that handedness is reversed when you look in a mirror: your dorso-ventral axis is reversed, and so left becomes right and vice versa). One suggestion to explain the development of left–right asymmetry is that an initial asymmetry at the molecular level is converted to an asymmetry at the cellular and multicellular level. If that were so, the asymmetric molecules or molecular structure would need to be oriented with respect to both the antero-posterior and dorso-ventral axes.

The breaking of left–right asymmetry occurs early in development. The initial symmetry-breaking mechanism leads to a cascade of events that lead to organ asymmetry. Left–right differences in gene expression are established in the lateral plate mesoderm in the early embryo around the end of gastrulation at the time the first somites are forming. These differences in gene expression are then translated into asymmetric positioning and shape of organs. In mice and humans, for example, the heart is on the left side, the right lung has more lobes than the left (see Chapter 9), the stomach and spleen lie towards the left, and the bulk of the liver towards the right. This handedness of organs is remarkably consistent, but there are rare individuals, about one in 10,000 in humans, who have the condition known as **situs inversus**, a complete mirror-image reversal of handedness. Such people are generally asymptomatic, even though all their organs are reversed. A similar condition is produced in mice that carry the *iv* mutation, in which organ asymmetry is randomized (Fig. 4.50). In this context, 'randomized' means that some *iv* mutant mice have the normal organ asymmetry and some the reverse. In another mouse mutant, *inversion of turning* (*inv*), laterality is consistently reversed.

The *iv* mutation in mice affects the gene that encodes left–right dynein, a protein associated with microtubules that acts as a directional microtubular motor that is required for ciliary motion and intracellular transport. Patients with Kartagener's

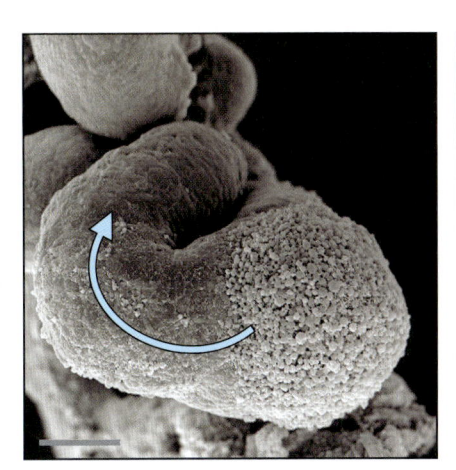

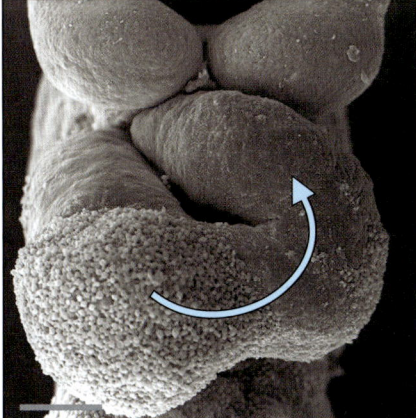

Fig. 4.50 Left-right asymmetry of the mouse heart is under genetic control. Each photograph shows a mouse heart viewed frontally after the loops have formed. The normal asymmetry of the heart results in it looping to one side, as indicated by the arrow (left panel). Fifty percent of mice that are homozygous for the mutation in the *iv* gene have hearts that loop to the other side (right panel). Scale bar = 0.1 mm.
Photographs courtesy of N. Brown.

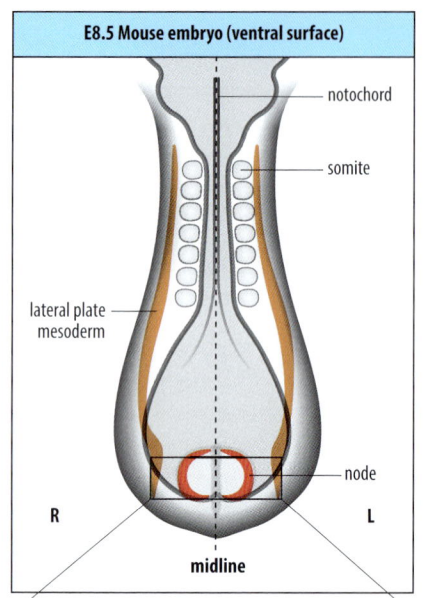

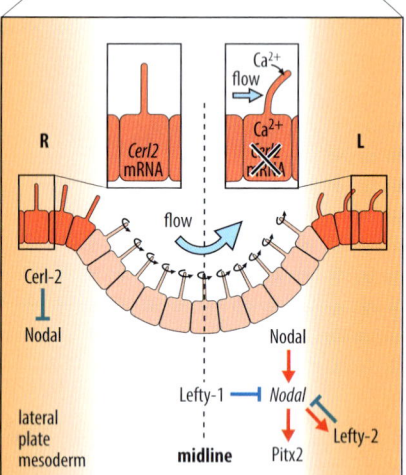

Fig. 4.51 In the mouse embryo, cilia-directed leftward flow of extracellular fluid generates left–right asymmetry across the midline and leads to specification of left and right. Top panel: ventral view of an E8.5 mouse embryo showing location of the node that acts as the left–right organizer (note that left (L) and right (R) are designated as if looking from the ventral side of the body). Bottom panel: section through node showing the two types of cell that line the pit below it: crown cells (dark red) with immotile cilia and pit cells (pale red) with motile cilia. The cilia on the pit cells generate a leftward flow of fluid. The diagram shows the mechanosensory model in which the cilia on the crown cells on the left side of the node respond to the flow and extracellular Ca^{2+} moves into the cells. This results in the decay of the mRNA for the gene *Cerberus-like 2* (*Cerl2*) in these cells, whereas crown cells on the right side of the node express *Cerl2*. The Cerl2 protein is an antagonist of Nodal. Crown cells on both sides of the node express Nodal, but Nodal signaling on the right side is blocked by Cerl2. Nodal diffuses and/or is transported to the lateral plate mesoderm on the left where it activates *Nodal* gene expression and expression of the genes encoding the Nodal antagonist Lefty-2 and the transcription factor Pitx2. Lefty-2 limits the range of Nodal signaling and together with Lefty-1 produced by midline structures confines Nodal signaling to the lateral plate mesoderm on the left. This is a simplified version of a more extensive gene regulatory network specifying 'left' which contains other components including Shh.

Illustration adapted from Thowfeequ, S., Srinivas, S.: **Mammalian embryo: establishment of the embryonic axes.** eLS *2021*, **2**: 1–15; and Hamada, H.: **Molecular and cellular basis of left-right asymmetry in vertebrates.** Proc. Jpn Acad. Ser. B Phys. Biol. Sci. *2020*, **96**: 273–296.

syndrome characterized by situs inversus have abnormal cilia, and many other human syndromes in which ciliary development is abnormal—known as **ciliopathies**—are also characterized by left–right asymmetry defects. This suggested that in mice and humans the normal functioning of cilia is required for left–right patterning.

It is now well established in mouse, zebrafish, and *Xenopus* that ciliary activity is critical in breaking left–right symmetry and inducing the asymmetric expression of genes that specify left and right. In these embryos at early somite stages, a 'leftward' flow of extracellular fluid is stimulated across the embryonic midline by transient populations of ciliated cells at the posterior end of the notochord (Fig. 4.51). In the mouse, ciliated cells are found on the ventral surface of the node, in a pit where there is no underlying endoderm, and create fluid flow from right to left across the midline. Reversal of the leftward flow in the ventral node by an imposed fluid movement reverses left–right asymmetry. In the zebrafish, the ciliated epithelium lines a fluid-filled vesicle known as Kupffer's vesicle, while in *Xenopus* a triangular region of ciliated mesoderm, the amphibian gastrocoel roof plate, straddles the midline in the roof of the archenteron. Disrupting the action of the cilia in cells in either Kupffer's vesicle or the gastrocoel roof plate by, for example, using antisense morpholinos to knock down expression of ciliary dynein genes, leads to a breakdown in left–right asymmetry.

In chick embryos, motile cilia do not appear to be present on the ventral surface of the node. Instead, molecular asymmetries at Hensen's node are created by leftward cell movements that occur transiently.

There are two types of ciliated cell on the ventral surface of the mouse node. Central 'pit cells' with motile cilia generate the flow, and peripheral 'crown cells' with immotile cilia respond to it. Coordinated rotational beating of the central cilia generates a flow of extracellular fluid directed leftwards—a result of the alignment of the basal bodies of the cilia. One hypothesis is that the immotile cilia act as mechanosensors for the direction of flow and their stimulation results in the influx of Ca^{2+} in the crown cells on the left side of the node, initiating the cascade of left–right asymmetric gene expression (see Fig. 4.51). Other hypotheses propose that the leftward flow leads to the accumulation of morphogen molecules on the left-hand side of the node, and that this morphogen directly or indirectly causes the asymmetric gene expression. Irrespective of which of these hypotheses is correct, or whether movement of cells breaks the symmetry at the midline, the result is that Nodal signaling is enhanced on the left and not on the right.

Nodal signaling specifies 'left' in the embryos of all vertebrates and some invertebrates. Nodal is expressed symmetrically on both sides of the node in mouse embryos, and adjacent to Kupffer's vesicle in zebrafish, and in the gastrocoel roof plate in *Xenopus*. By contrast, a gene encoding a secreted antagonist of Nodal signaling, *Cerberus-like 2* (*Cerl2*; see Box 4C), comes to be asymmetrically expressed, with lower expression on the left side leading to higher Nodal signaling to the left. In mice, the decay of *Cerl2* mRNA might be linked to the influx of Ca^{2+} in the crown cells on the left of the node in response to the leftward flow of fluid (see Fig. 4.51). In the chick, *Nodal* comes to be expressed in the cells on the left side of the node because of the transient leftward movement of cells around Hensen's node. These cells express the secreted signal protein Shh. As a consequence of the transient leftward movements, the initial bilateral expression of *Shh* becomes restricted to the left side. Shh signaling promotes Nodal. If a pellet of cells secreting Shh is placed on the right side, then organ asymmetry is randomized. *Shh* mutant mice also have left–right patterning defects, but in the mouse Shh signaling is not required for *Nodal* expression in the node but instead later in the lateral plate mesoderm.

Asymmetric Nodal signaling on the left side of the midline in all the vertebrate embryos leads to widespread and stable activation of *Nodal* expression in the left lateral plate mesoderm. *Nodal* expression becomes amplified as a result of a positive-feedback loop, because one effect of Nodal signaling is to activate the *Nodal* gene, and so produce more Nodal protein (see Box 4C). Nodal signaling also induces expression of the long-range secreted antagonists Lefty-1 and Lefty-2. As we saw in mesoderm induction in zebrafish (see Section 3.17), Lefty proteins limit the range of Nodal signaling, and this ensures that Nodal signaling occurs in the lateral plate mesoderm only on the left. Restriction of Nodal signaling is also helped by cells of the notochord and floor plate of the neural tube in the midline, which secrete Lefty proteins.

Nodal signaling in left lateral plate mesoderm activates expression of the gene for the transcription factor Pitx2, a key determinant of leftness. This left-sided pattern of *Pitx2* gene expression in the lateral plate mesoderm that gives rise to internal organs such as the heart is highly conserved, being found in the mouse, zebrafish, *Xenopus*, and chick. It has been shown in chick embryos that if the expression of either *Nodal* or *Pitx2* in the lateral plate mesoderm is made symmetric, then organ asymmetry is randomized.

The 'left' gene regulatory pathway outlined above is a simplified version of a more extensive network that contains other components including Shh. Very little is known about the gene regulatory pathway that specifies 'right'. The transcription factor Snail, which can act as a repressor, is expressed in the lateral plate mesoderm on the right and represses *Nodal* expression on the right.

Left–right symmetry breaking might be initiated within cells of the early embryo. An early indication of left–right asymmetry in *Xenopus* is the asymmetric activity of a proton–potassium pump (H^+/K^+-ATPase). This ATPase can be detected as early as the third cleavage division—the eight-cell stage—in *Xenopus*, and treatment of early embryos with drugs that inhibit the pump cause randomization of asymmetry. How the asymmetry of the H^+/K^+-ATPase itself originates is not known, but one possibility is that the mRNA encoding the ion transporter and/or the ion transporter protein itself might be moved by motor proteins along oriented cytoskeletal elements in the fertilized egg. In *Xenopus*, disruption of cortical actin at the first cleavage division can affect left–right asymmetry. This asymmetry established during early cleavage stages could lead to the subsequent asymmetric alignment of the basal bodies of the cilia on the cells in the gastrocoel roof plate, and so to a cascade of further asymmetries.

> **SUMMARY**
>
> Consistent left-right organ asymmetry found in vertebrates is specified by the expression of the secreted protein Nodal and the transcription factor Pitx2 on the left side only. This asymmetric gene expression is generated by leftward flow of fluid generated by the beating of cilia in mouse, humans, zebrafish, and *Xenopus*, and by leftward movement of cells in chick. Sonic hedgehog signaling is required for correct left-right patterning in the embryos of all four model vertebrates, but its precise function varies.

Summary to Chapter 4

- During gastrulation, the germ layers specified during the earlier stages of embryonic development become further patterned along the antero-posterior and dorso-ventral axes.
- Patterning of all three germ layers along the dorso-ventral axis, together with the establishment of the antero-posterior axis and neural induction during gastrulation, depends on signaling by the Spemann organizer in *Xenopus*, by the shield region in zebrafish, by Hensen's node in chick, and by the combined action of the AVE and the node in mouse.
- When grafted to the ventral side of an early *Xenopus* gastrula, the Spemann organizer induces a second body axis with a head and a complete dorso-ventral axis, and a twinned embryo develops. The shield and Hensen's node can likewise induce a twinned fish, or chick embryo with a second body axis.
- In the mouse embryo, axis-inducing activity is located in cells that are spatially separate, so the AVE can induce a head, and the node can induce trunk.
- In all vertebrates, dorso-ventral patterning is achieved through the graded activity of BMP signaling. High levels of BMP signaling specify ventral cells, and low levels specify dorsal cells.
- In *Xenopus*, zebrafish, and chick, the main factors that derive from dorsal regions to shape a BMP gradient are Chordin and Noggin, but in the mouse the main factor that counteracts BMPs is Nodal.
- In all vertebrates, antero-posterior patterning is achieved through Wnt signaling. High levels of Wnt signaling specify posterior identities and low levels specify anterior identities. In all vertebrates, low Wnt signaling is maintained in anterior regions by Wnt antagonists. Wnt antagonists are first made by cells in the hypoblast/yolk syncytium/AVE, and are then maintained in tissues that underlie the developing forebrain, including the anterior endoderm and prechordal mesoderm.
- In *Xenopus*, neural tissue that will give rise to much of the brain and spinal cord is induced by BMP antagonists that derive from the organizer.
- In zebrafish, mouse, and chick, neural tissue that will give rise to the brain is induced by factors that derive from the shield, or node. In zebrafish, the main neural-inducing factors are BMP antagonists. In chick and mouse, the main neural-inducing factors are FGFs, which act in part by antagonizing BMPs.
- There are many similarities in the way in which the germ layers are specified and patterned in all vertebrates, despite the different topologies of the embryos.

- At the end of gastrulation, the basic body plan has been laid down and the prospective neural tissue has been induced from the dorsal ectoderm.
- In *Xenopus* and the zebrafish, axial mesendoderm that extends from the head to the trunk derives from the organizer or shield, but in chick and mouse, only axial mesendoderm that underlies the head derives from Hensen's node or the node.
- In the chick and the mouse, the trunk notochord, spinal cord, and part of the somites are generated from self-renewing populations of cells around the node as the body axis elongates. Opposing gradients of FGF and retinoic acid control the balance of proliferation and differentiation.
- Somites are formed as a series of blocks of mesodermal tissue on either side of the notochord with the periodicity of segmentation being driven by a molecular clock.
- Specific regions of each somite give rise to cartilage (the forerunner of the skeleton), muscle, and dermis, and these regions are specified by signals from the surrounding tissues, notochord, neural tube, lateral plate mesoderm, and ectoderm.
- Regional identity of cells along the antero-posterior axis of vertebrates is encoded by the expression of combinations of genes of the four Hox gene clusters.
- There is both spatial and temporal co-linearity between the order of Hox genes on the chromosomes and the order in which they are expressed along the axis from the level of the hindbrain to the posterior end of the embryo.
- Inactivation or overexpression of Hox genes can lead both to localized abnormalities and to homeotic transformations of one 'segment' of the axis into another, indicating that these genes are crucial in specifying regional identity.
- The neural crest arises from neural plate border cells that are found at the lateral borders of the neural plate and the dorsal crests of the neural tube. Neural crest cells undergo an epithelial-to-mesenchymal cell transition and migrate to give rise to a wide range of different cell types, including cells of the peripheral nervous system, pigment cells, and head connective tissue cells.
- Left-right asymmetry involves an initial symmetry-breaking event leading to asymmetric gene expression on the left and right sides of the embryo, which is translated into asymmetric positioning and morphogenesis of organs.

End of chapter questions

Check your understanding of the content of this chapter by attempting the following long-answer concept questions.

1. What is the developmental function of the TGF-β family member BMP-4 in the early *Xenopus* gastrula?

2. The *Xenopus* organizer coordinates patterning along both the dorso-ventral and antero-posterior axes. Describe how secreted factors that derive from the organizer act over time to achieve this.

3. Describe how the pattern of Wnt signaling in the *Xenopus* embryo changes between blastula and gastrula stages, and explain how this affects (a) dorso-ventral patterning and (b) antero-posterior patterning.

4. The TGF-β signaling pathway is an example of the remarkable conservation of signaling pathways during evolution. Compare the activities of BMP-4 in *Xenopus* and Decapentaplegic in *Drosophila*. Describe how a gradient of BMP signaling is established in the *Xenopus* gastrula. To what extent is the mechanism conserved in *Drosophila*?

5. Review what we have learned about the specification of the anterior-most mesoderm in the *Xenopus* gastrula including *goosecoid* activation in response to TGF-β family signaling. What is the role of this mesoderm in subsequent regionalization of neural tissue? (Refer to Section 4.4 and Chapter 10.)

6. Zebrafish embryos are good models for genetic analysis of development. Which zebrafish mutants have given insights into the laying down of the vertebrate body plan? Explain why.

7. The 'default model' for neural induction in *Xenopus* proposes that the development of dorsal ectoderm as neural tissue is the default pathway. What is the evidence for this model and what is the proposed molecular basis? How has the default model been modified in the light of more recent work? Include FGF signaling in your answer.

8. What is the evidence that Wnt signaling and Wnt antagonism govern regionalization along the antero-posterior axis?

9. Why are the mouse AVE plus the mouse node together considered functionally equivalent to the *Xenopus* Spemann organizer?

10. Describe the morphological events that establish the axial mesoderm in the chick embryo. Explain the impact of these events on development of the antero-posterior axis.

11. In development, signaling pathways frequently interact with one another. Describe the manner in which the FGF and BMP signaling pathways can interact, and discuss the importance of this interaction in neural induction.

12. In the chick and the mouse, several tissues in posterior parts of the body are generated from self-renewing population of cells in and around the regressing node. Describe experiments that (a) have revealed this, and (b) have revealed the key signals involved.

13. Distinguish between the notochord, the spinal cord, and the vertebral column, and describe the relationships between the following structures: neural plate, neural folds, neural tube, and neural crest.

14. Somite formation in chick embryos proceeds from anterior to posterior in response to a gradient of FGF-8. Describe this gradient: where does the FGF-8 originate? Why does it form a gradient? How does the gradient trigger somite formation? What other gradients are present in the embryo during this period and what are their functions?

15. Somites are formed from the pre-somitic mesoderm in a periodic fashion going from anterior to posterior. What mechanism has been proposed to explain this regular periodicity? Explain in principle how it works.

16. The Hox gene clusters in mice (Box 4G) illustrate several mechanisms that operate in genome evolution: genes can become duplicated to give rise to paralogs, the entire cluster can become duplicated to give rise to paralogous groups of genes, and within any given cluster, genes can be lost. Give an example of each of these three mechanisms.

17. The Hox genes are expressed in the mesoderm of vertebrate embryos both temporally and spatially in order, from one end of the Hox gene cluster to the other. Refer to Figure 4.36 and compare Hox gene expression in the mouse thoracic region to that in the lumbar/sacral region. How is the temporal activation of Hox genes along the clusters translated into the spatial patterns of Hox genes along the main body axis? How do the gene expression patterns illustrate how different combinations of gene expression can produce spatial differences in regional identity along an axis?

18. Compare the consequences of *Hoxc8* gene deletion in the mouse (Fig. 4.38) with deletion of the *Ubx* gene in *Drosophila* (Fig. 2.47). Are the transformations seen toward anterior or posterior identities? What do these transformations say about the roles of Hox genes in identity specification? Do these transformations in mice qualify as homeotic transformations?

19. Outline how axial and paraxial tissues are generated during primary body formation in mouse and chick embryos. What are the key signals involved and the mechanisms that ensure that the system is robust?

20. Describe the effects of the experiment shown in Figure 4.43, in which a notochord from a donor embryo is grafted adjacent to the dorsal region of the neural tube in a host embryo. What signaling molecule is proposed to be responsible for the observed induction? What kind of proteins are the Pax proteins? What is the role of Pax3 in the somite?

21. What are neural plate border cells, neural crest cells, and roof plate cells? How does the transcriptional profile of neural crest cells predict their cellular behavior?

22. How have chick–quail chimeras been used in the study of the neural crest? How have experiments on neural crest cells in culture added to our knowledge of the development of neural crest derivatives *in vivo*?

23. The establishment of left–right asymmetry is associated with higher levels of the TGF-β family signaling molecule Nodal on the left side of the embryo, relative to the right. What are the influences that led to this asymmetry of Nodal concentration? Include positive feedback, Notch signaling, Sonic hedgehog signaling, and the Lefty protein in your answer.

■ Section further reading

4.1 Secreted proteins from the organizer pattern the dorso-ventral axis by antagonizing the effects of ventral signals

De Robertis, E.M.: **Spemann's organizer and self-regulation in amphibian embryos**. *Nature Mol. Cell Biol. Rev.* 2006, **7**: 296–302.

De Robertis, E.M., Moriyama, Y., Colozza, G.: **Generation of animal form by the Chordin/Tolloid/BMP gradient: 100 years after D'Arcy Thompson**. *Dev. Growth Differ.* 2017, **59**: 580–592.

Piccolo, S., Sasai, Y., Lu, B., De Robertis, E.M.: **Dorsoventral patterning in *Xenopus*: inhibition of ventral signals by direct binding of chordin to BMP-4**. *Cell* 1996, **86**: 589–598.

Zimmerman, L.B., De Jesús-Escobar, J.M., Harland, R.M.: **The Spemann organizer signal noggin binds and inactivates bone morphogenetic protein 4**. *Cell* 1996, **86**: 599–606.

4.2 Secreted proteins from the organizer pattern the antero-posterior axis by antagonizing the effects of posterior signals

De Robertis, E.M., Moriyama, Y.: **The Chordin morphogenetic pathway**. *Curr. Top. Dev. Biol.* 2016, **116**: 231–245.

Glinka, A., Wu, W., Delius, H., Monaghan, P., Blumenstock, C., Niehrs, C.: **Dickkopf-1 is a member of a new family of secreted proteins and functions in head induction**. *Nature* 1998, **391**: 357–362.

Glinka, A., Wu, W., Onichtchouk, D., Blumenstock, C., Niehrs, C.: **Head induction by simultaneous repression of Bmp- and Wnt signalling in *Xenopus***. *Nature* 1997, **389**: 517–519.

Niehrs, C.: **Regionally specific induction by the Spemann-Mangold organizer**. *Nat. Rev. Genet.* 2004, **5**: 425–434.

Niehrs, C.: **The role of *Xenopus* developmental biology in unraveling Wnt signaling and antero-posterior axis formation**. *Dev. Biol.* 2022, **482**: 1–6.

Piccolo, S., Agius, E., Leyns, L., Bhattacharya, S., Grunz, H., Bouwmeester, T., De Robertis, E.M.: **The head inducer Cerberus is a multifunctional antagonist of Nodal, BMP and Wnt signals**. *Nature* 1999, **397**: 707–710.

4.3 The neural plate is induced from the ectoderm through the action of BMP antagonists

De Robertis, E.M.: **Spemann's organizer and self-regulation in amphibian embryos**. *Nat. Rev. Mol. Cell Biol.* 2006, **4**: 296–302.

De Robertis, E.M., Kuroda, H.: **Dorsal-ventral patterning and neural induction in *Xenopus* embryos**. *Annu. Rev. Dev. Biol.* 2004, **20**: 285–308.

Reversade, B., De Robertis, E.M.: **Regulation of ADMP and BMP2/4/7 at opposite embryonic poles generates a self-regulating embryonic field**. *Cell* 2005, **123**: 1147–1160.

Box 4A The fibroblast growth factor signaling pathway

Dorey, K., Amaya, E.: **FGF signalling: diverse roles during early vertebrate embryogenesis**. *Development* 2010, **137**: 3731–3742.

4.4 Factors that pattern the antero-posterior body axis also pattern the antero-posterior axis of the nervous system

Kiecker, C., Niehrs, C.: **A morphogen gradient of Wnt/beta-catenin signaling regulates anteroposterior neural patterning in *Xenopus***. *Development* 2001, **128**: 4189–4201.

Pera, E.M., Ikeda, A., Eivers, E., De Robertis, E.M.: **Integration of IGF, FGF, and anti-BMP signals via Smad1 phosphorylation in neural induction**. *Genes Dev.* 2003, **17**: 3023–3028.

Polevoy, H., Gutkovich, Y.E., Michaelov, A., Volovik, Y., Elkouby, Y.M., Frank, D.: **New roles for Wnt and BMP signaling in neural anteroposterior patterning**. *EMBO Rep.* 2019, 20: e45842.

4.5 The final *Xenopus* body plan emerges by the end of gastrulation and neurulation

Bouwmeester, T.: **The Spemann-Mangold organizer: the control of fate specification and morphogenetic rearrangements during gastrulation in *Xenopus***. *Int. J. Dev. Biol.* 2001, **45**: 251–258.

4.6 The shield in zebrafish has organizing activity

Fauny, J.D., Thisse, B., Thisse, C.: **The entire zebrafish blastula-gastrula margin acts as an organizer dependent on the ratio of Nodal to BMP activity**. *Development* 2009, **136**: 3811–3819.

Gonzalez, E.M., Fekany-Lee, K., Carmany-Rampey, A., Erter, C., Topczewski, J., Wright, C.V., Solnica-Krezel, L.: **Head and trunk in zebrafish arise via coinhibition of BMP signaling by bozozok and chordino**. *Genes Dev.* 2000, **14**: 3087–3092.

Kudoh, T., Concha, M.L., Houart, C., Dawid, I.B., Wilson, S.W.: **Combinatorial Fgf and Bmp signalling patterns the gastrula ectoderm into prospective neural and epidermal domains**. *Development* 2004, **131**: 3581–3592.

Tuazon, F.B., Mullins, M.C.: **Temporally coordinated signals progressively pattern the anteroposterior and dorsoventral body axes**. *Semin. Cell Dev. Biol.* 2015, **42**: 118–133.

Xu, P.F., Houssin, N., Ferri-Lagneau, K.F., Thisse, B., Thisse, C.: **Construction of a vertebrate embryo from two opposing morphogen gradients**. *Science* 2014, **344**: 87–89.

Box 4B A zebrafish gene regulatory network

Morley, R.H., Lachani, K., Keefe, D., Gilchrist, M.J., Flicek, P., Smith, J.C., Wardle, F.C.: **A gene regulatory network directed by zebrafish No tail accounts for its roles in mesoderm formation**. *Proc. Natl Acad. Sci. USA* 2009, **106**: 3829–3834.

4.7 Hensen's node in chick is the embryonic organizer

Joubin, K., Stern, C.D.: **Molecular interactions continuously define the organizer during the cell movements of gastrulation**. *Cell* 1999, **98**: 559–571.

Stern, C.: **Neural induction: old problems, new findings, yet more questions**. *Development* 2005, **132**: 2007–2021.

Storey, K., Crossley, J.M., De Robertis, E.M., Norris, W.E., Stern, C.D.: **Neural induction and regionalization in the chick embryo**. *Development* 1992, **114**: 729–741.

Storey, K.G., Selleck, M.A., Stern, C.D.: **Neural induction and regionalisation by different subpopulations of cells in Hensen's node**. *Development* 1995, **121**: 417–428.

Streit, A., Berliner, A.J., Papanayotou, C., Sirulnik, A., Stern, C.D.: **Initiation of neural induction by FGF signalling before gastrulation**. *Nature* 2000, **406**: 74–78.

4.8 Mesoderm patterning in mouse is similar to that in other model organisms, but antero-posterior organizer functions are spatially and temporally separated

Andoniadou, C.L., Martinez-Barbera, J.P.: **Developmental mechanisms directing early anterior forebrain specification in vertebrates**. *Cell. Mol. Life Sci.* 2013, **70**: 3739–3752.

Bardot, E.S., Hadjantonakis, A.K.: **Mouse gastrulation: coordination of tissue patterning, specification and diversification of cell fate**. *Mech. Dev.* 2020, **163**: 103617.

Beddington, R.S.: **Induction of a second neural axis by the mouse node**. *Development* 1994, **120**: 613–620.

Kinder, S.J., Tsang, T.E., Wakamiya, M., Sasaki, H., Behringer, R.R., Nagy, A., Tam, P.P.: **The organizer of the mouse gastrula is composed of a dynamic population of progenitor cells for the axial mesoderm**. *Development* 2001, **128**: 3623–3634.

Martinez Arias, A., Steventon, B.: **On the nature and function of organizers**. *Development* 2018, **145**: dev159525.

Mukhopadhyay, M., Shtrom, S., Rodriguez-Esteban, C., Chen, L., Tsuku, T., Gomer, L., Dorward, D.W., Glinka, A., Grinberg, A., Huang, S.-P., Niehrs, C., Izpisúa Belmonte, J.-C., Westphal, H.: **Dickkopf1 is required for embryonic head induction and limb morphogenesis in the mouse**. *Dev. Cell* 2001, **1**: 423–434.

Ozair, M.Z., Kintner, C., Brivanlou, A.H.: **Neural induction and early patterning in vertebrates**. *WIREs Dev. Biol.* 2013, **2**: 479–498.

Robertson, E.J.: **Dose-dependent Nodal/Smad signals pattern the early mouse embryo**. *Semin. Cell Dev. Biol.* 2014, **32**: 73–79.

Box 4C Fine-tuning Nodal signaling

Hill, C.S.: **Spatial and temporal control of NODAL signaling**. *Curr Opin Cell Biol.* 2018, **51**: 50–57.

Box 4D The Cre/*loxP* system: a strategy for making gene knock-outs in mice

Lewandoski, M.: **Conditional control of gene expression in the mouse**. *Nat. Rev. Genet.* 2001, **2**: 743–755.

4.9 Axial structures in chick and mouse are generated from self-renewing cell populations

Delfino-Machin, M., Lunn, J.S., Breitkreuz, D.N., Akai, J., Storey, K.G.: **Specification and maintenance of the spinal cord stem zone**. *Development* 2005, **132**: 4273–4283.

Diez del Corral, R., Olivera-Martinez, I., Goriely, A., Gale, E., Maden, M., Storey, K.: **Opposing FGF and retinoid pathways control ventral neural pattern, neuronal differentiation, and segmentation during body axis extension**. *Neuron* 2003, **40**: 65–79.

Dubrulle, J., Pourquié, O.: ***fgf8* mRNA decay establishes a gradient that couples axial elongation to patterning in the vertebrate embryo**. *Nature* 2004, **427**: 419–422.

Henrique, D., Abranches, E., Verrier, L., Storey, K.G.: **Neuromesodermal progenitors and the making of the spinal cord**. *Development* 2015, **142**: 2864–2875.

Sakai, Y., Meno, C., Fujii, H., Nishino, J., Shiratori, H., Saijoh, Y., Rossant, J., Hamada, H.: **The retinoic acid-inactivating enzyme CYP26 is essential for establishing an uneven distribution of retinoic acid along the anteroposterior axis within the mouse embryo**. *Genes Dev.* 2006, **15**: 213–225.

Solovieva, T., Wilson, V., Stern, C.D.: **A niche for axial stem cells—a cellular perspective in amniotes**. *Dev. Biol.* 2022, **490**: 13–21.

Box 4E Retinoic acid: a small-molecule intercellular signal

Niederreither, K., Dolle, P.: **Retinoic acid in development: towards an integrated view**. *Nat. Rev. Genet.* 2008, **9**: 541–553.

Rossant, J., Zirngibl, R., Cado, D., Shago, M., Giguère, V.: **Expression of a retinoic acid response element-hsplacZ transgene defines specific domains of transcriptional activity during mouse embryogenesis**. *Genes Dev.* 1991, **5**: 1333–1344.

4.10 Somites are formed in a well-defined order along the antero-posterior axis

Carrieri, F.A., Murray, P.J., Ditsova, D., Ferris, M.A., Davies, P., Dale, J.K.: **CDK1 and CDK2 regulate NICD1 turnover and the periodicity of the segmentation clock**. *EMBO Rep.* 2019, **20**: e46436.

Dequeant, M.L., Pourquié, O.: **Segmental patterning of the vertebrate embryonic axis**. *Nat. Rev. Genet.* 2008, **9**: 370–382.

Maroto, M., Bone, R.A., Dale, J.K.: **Somitogenesis**. *Development* 2012, **139**: 2453–2456.

McGrew, M.J., Dale, J.K., Fraboulet, S., Pourquié, O.: **The lunatic fringe gene is a target of the molecular clock linked to somite segmentation in avian embryos**. *Curr. Biol.* 1998, **8**: 979–982.

Oates, A.C., Morelli, L.G., Ares, S.: **Patterning embryos with oscillations: structure, function and dynamics of the vertebrate segmentation clock**. *Development* 2012, **139**: 625–639.

Pourquié, O.: **Somite formation in the chicken embryo**. *Int. J. Dev. Biol.* 2018, **62**: 57–62.

Pourquié, O.: **A brief history of the segmentation clock**. *Dev. Biol.* 2022, **485**: 24–36.

Uriu, K., Liao, B.K., Oates, A.C., Morelli, L.G.: **From local resynchronization to global pattern recovery in the zebrafish segmentation clock**. *eLife* 2021, **10**: e61358.

Box 4F The Notch signaling pathway

Boareto, M.: **Patterning via local cell-cell interactions in developing systems**. *Dev. Biol.* 2020, **460**: 77–85.

Bray, S.J.: **Notch signalling in context**. *Nat. Rev. Mol. Cell Biol.* 2016, **17**: 722–735.

Hori, Y., Sen, A., Artavanis-Tsakonas, S.: **Notch signaling at a glance**. *J. Cell Sci.* 2013, **126**: 2135–2140.

4.11 Identity of somites along the antero-posterior axis is specified by Hox gene expression

Burke, A.C., Nelson, C.E., Morgan, B.A., Tabin, C.: **Hox genes and the evolution of vertebrate axial morphology**. *Development* 1995, **121**: 333–346.

Song, J.Y., Pineault, K.M., Dones, J.M., Raines, R.T., Wellik, D.M.: ***Hox* genes maintain critical roles in the adult skeleton**. *Proc. Natl Acad. Sci. USA* 2020, **117**: 7296–7304.

Weldon, S.A., Münsterberg, A.E.: **Somite development and regionalisation of the vertebral axial skeleton**. *Semin. Cell Dev. Biol.* 2022, **127**: 10–16.

Wellik, D.M.: **Hox patterning of the vertebrate axial skeleton**. *Dev. Dyn.* 2007, **236**: 2454–2463.

Wellik, D.M.: **Hox genes and vertebrate axial pattern**. *Curr. Top. Dev. Biol.* 2009, **88**: 257–278.

Box 4G The Hox genes

Hrycaj, S.M., Wellik, D.M.: **Hox genes and evolution**. *F1000Res.* 2016, **5**: F1000 Faculty Rev-859.

Montavon, T.: **Hox genes: embryonic development**. *eLS.* Chichester: John Wiley & Sons, 2015. [https://doi.org/10.1002/9780470015902.a0005046.pub2] (date accessed 21 December 2022).

Pearson, J.C., Lemons, D., McGinnis, W.: **Modulating Hox gene functions during animal body patterning**. *Nat. Rev. Genet.* 2005, **6**: 893–904.

4.12 Deletion or overexpression of Hox genes causes changes in axial patterning

Condie, B.G., Capecchi, M.R.: **Mice with targeted disruptions in the paralogous genes *Hoxa3* and *Hoxd3* reveal synergistic interactions**. *Nature* 1994, **370**: 304–307.

Favier, B., Le Meur, M., Chambon, P., Dollé, P.: **Axial skeleton homeosis and forelimb malformations in Hoxd11 mutant mice**. *Proc. Natl Acad. Sci. USA* 1995, **92**: 310–314.

Kessel, M., Gruss, P.: **Homeotic transformations of murine vertebrae and concomitant alteration of the codes induced by retinoic acid**. *Cell* 1991, **67**: 89–104.

Mallo, M., Wellik, D.M., Deschamps, J.: **Hox genes and regional patterning of the vertebrate body plan**. *Dev. Biol.* 2010, **344**: 7–15.

Singh, N.P., De Kumar, B., Paulson, A., Parrish, M.E., Scott, C., Zhang, Y., Florens, L., Krumlauf, R.: **Genome-wide binding analyses of HOXB1 revealed a novel DNA binding motif associated with gene repression**. *J. Dev. Biol.* 2021, **9**: 6.

van den Akker, E., Fromental-Ramain, C., de Graaff, W., Le Mouellic, H., Brûlet, P., Chambon, P., Deschamps, J.: **Axial skeletal patterning in mice lacking all paralogous group 8 Hox genes**. *Development* 2001, **128**: 1911–1921.

Wellik, D.M., Capecchi, M.R.: **Hox10 and Hox11 genes are required to globally pattern the mammalian skeleton**. *Science* 2003, **301**: 363–367.

Yekta, S., Tabin, C., Bartel, D.P.: **MicroRNAs in the Hox network: an apparent link to posterior prevalence**. *Nat. Rev. Genet.* 2008, **9**: 789–796.

4.13 Hox gene expression is activated in an anterior to posterior pattern

Deschamps, J., Duboule, D.: **Embryonic timing, axial stem cells, chromatin dynamics, and the Hox clock**. *Genes Dev.* 2017, **31**: 1406–1416.

Emoto, Y., Wada, H., Okamoto, H., Kudo, A., Imai, Y.: **Retinoic acid-metabolizing enzyme Cyp26a1 is essential for determining territories of hindbrain and spinal cord in zebrafish**. *Dev. Biol.* 2005, **278**: 415–427.

Neijts, R., Amin, S., van Rooijen, C., Deschamps, J.: **Cdx is crucial for the timing mechanism driving colinear Hox activation and defines a trunk segment in the Hox cluster topology**. *Dev. Biol.* 2017, **422**: 146–154.

Neijts, R., Deschamps, J.: **At the base of colinear Hox gene expression: *cis*-features and *trans*-factors orchestrating the initial phase of Hox cluster activation**. *Dev. Biol.* 2017, **428**: 293–299.

Nolte, C., De Kumar, B., Krumlauf, R.: **Hox genes: downstream 'effectors' of retinoic acid signaling in vertebrate embryogenesis**. *Genesis* 2019, **57**: e23306.

4.14 The fate of somite cells is determined by signals from the adjacent tissues

Brand-Saberi, B., Christ, B.: **Evolution and development of distinct cell lineages derived from somites**. *Curr. Topics Dev. Biol.* 2000, **48**: 1–42.

Chal, J., Pourquié, O.: **Making muscle: skeletal myogenesis *in vivo* and *in vitro***. *Development* 2017, **144**: 2104–2122.

Olivera-Martinez, I., Coltey, M., Dhouailly, D., Pourquié, O.: **Mediolateral somitic origin of ribs and dermis determined by quail-chick chimeras**. *Development* 2000, **127**: 4611–4617.

Pourquié, O., Fan, C.-M., Coltey, M., Hirsinger, E., Watanabe, Y., Bréant, C., Francis-West, P., Brickell, P., Tessier-Lavigne, M., Le Douarin, N.M.: **Lateral and axial signals involved in avian somite patterning: a role for BMP-4**. *Cell* 1996, **84**: 461–471.

4.15 Neural crest cells arise from neural plate border cells and migrate to give rise to a wide range of different cell types

Dupin, E., Calloni, G.W., Coelho-Aguiar, J.M., Le Douarin, N.M.: **The issue of the multipotency of the neural crest cells**. *Dev. Biol.* 2018, **444**: S47–S59.

Gandhi, S., Bronner, M.E.: **Insights into neural crest development from studies of avian embryos**. *Int. J. Dev. Biol.* 2018, **62**: 183–194.

Huang, X., Saint-Jeannet, J-P.: **Induction of the neural crest and the opportunities of life on the edge**. *Dev. Biol.* 2004, **275**: 1–11.

Kelsh, R.N., Erickson, C.A.: **Neural crest: origin, migration and differentiation**. *eLS*. Chichester: John Wiley & Sons, 2013 [https://doi.org/10.1002/9780470015902.a0000786.pub2] (date accessed 21 December 2022).

Le Douarin, N., Kalcheim, C.: *The Neural Crest*. 2nd edn. Cambridge: Cambridge University Press, 1999.

Williams, R.M., Lukoseviciute, M., Sauka-Spengler, T., Bronner, M.E.: **Single-cell atlas of early chick development reveals gradual segregation of neural crest lineage from the neural plate border during neurulation**. *eLife* 2022, **11**: e74464.

4.16 Neural crest cells migrate from the hindbrain to populate the branchial arches

Amin, S., Donaldson, I.J., Zannino, D.A., Hensman, J., Rattray, M., Losa, M., Spitz, F., Ladam, F., Sagerström, C., Bobola, N.: **Hoxa2 selectively enhances Meis binding to change a branchial arch ground state**. *Dev. Cell* 2015, **32**: 265–277.

Grammatopoulos, G.A., Bell, E., Toole, L., Lumsden, A., Tucker, A.S.: **Homeotic transformation of branchial arch identity after Hoxa2 overexpression**. *Development* 2000, **127**: 5355–5365.

Krumlauf, R.: **Hox genes and pattern formation in the branchial region of the vertebrate head**. *Trends Genet.* 1993, **9**: 106–112.

Parker, H.J., Pushel, I., Krumlauf, R.: **Coupling the roles of Hox genes to regulatory networks patterning cranial neural crest**. *Dev. Biol.* 2018, **444**: 547–559.

Rijli, F.M., Mark, M., Lakkaraju, S., Dierich, A., Dolle, P., Chambon, P.: **A homeotic transformation is generated in the rostral branchial region of the head by disruption of *Hoxa2*, which acts as a selector gene**. *Cell* 1993, **75**: 1333–1349.

4.17 The bilateral symmetry of the early embryo is broken to produce left-right asymmetry of internal organs

Blum, M., Beyer, T., Weber, T., Vivk, P., Andre, P., Bitzer, E., Schweickert, A.: ***Xenopus*, an ideal model system to study vertebrate left-right asymmetry**. *Dev. Dyn.* 2009, **238**: 1215–1225.

Brennan, J., Norris, D.P., Robertson, E.J.: **Nodal activity in the node governs left-right asymmetry**. *Genes Dev.* 2002, **16**: 2339–2344.

Brown, N.A., Wolpert, L.: **The development of handedness in left/right asymmetry**. *Development* 1990, **109**: 1–9.

Grimes, D.T., Burdine, R.D.: **Left-right patterning: breaking symmetry to asymmetric morphogenesis**. *Trends Genet.* 2017, **33**: 616–628.

Gros, J., Feistel, K., Viebahn, C., Blum, M., Tabin, C.J.: **Cell movements at Hensen's node establish left/right asymmetric gene expression in the chick**. *Science* 2009, **324**: 941–944.

Hamada, H.: **Molecular and cellular basis of left–right asymmetry in vertebrates**. *Proc. Jpn Acad. Ser. B Phys. Biol. Sci.* 2020, **96**: 273–296.

Little, R.B., Norris, D.P.: **Right, left and cilia: how asymmetry is established**. *Semin. Cell Dev. Biol.* 2021, **110**: 11–18.

McGrath, J., Somlo, S., Makova, S., Tian, X., Brueckner, M.: **Two populations of node monocilia initiate left-right asymmetry in the mouse**. *Cell* 2003, **114**: 61–73.

Monsoro-Burq, A.H., Levin, M.: **Avian models and the study of invariant asymmetry: how the chicken and the egg taught us to tell right from left**. *Int. J. Dev. Biol.* 2018, **62**: 63–77.

Rana, A.A., Barbera, J.P., Rodriguez, T.A., Lynch, D., Hirst, E., Smith, J.C., Beddington, R.S.P.: **Targeted deletion of the novel cytoplasmic dynein mD2LIC disrupts the embryonic organiser, formation of body axes and specification of ventral cell fates**. *Development* 2004, **131**: 4999–5007.

Raya, A., Izpisua Belmonte, J.C.: **Insights into the establishment of left-right asymmetries in vertebrates**. *Birth Defects Res. C Embryo Today* 2008, **84**: 81–94.

Schlueter, J., Brand, T.: **Left-right axis development: examples of similar and divergent strategies to generate asymmetric morphogenesis in chick and mouse embryos**. *Cytogenet. Genome Res.* 2007, **117**: 256–267.

Shen, M.M.: **Nodal signaling: developmental roles and regulation**. *Development* 2007, **134**: 1023–1034.

Thowfeequ, S., Srinivas, S.: **Mammalian embryo: establishment of the embryonic axes**. *eLS* 2021, **2**: 1–15 [https://doi.org/10.1002/9780470015902.a0029305] (date accessed 18 October 2022).

Vandenberg, L.N., Levin, M.: **A unified model for left-right asymmetry? Comparison and synthesis of molecular models of embryonic laterality**. *Dev Biol.* 2013, **379**: 1–15.

Yoshiba, S., Hamada, H.: **Roles of cilia, fluid flow, and Ca^{2+} signaling in breaking of left-right symmetry**. *Trends Genet.* 2014, **30**: 10–17.

Morphogenesis: change in form in the early embryo

- Cell adhesion
- Cleavage and formation of the blastula
- Gastrulation movements
- Neural tube formation
- Formation of tubes and branching morphogenesis
- Cell migration

Changes in form in animal embryos are brought about by cellular forces that are generated in various ways. These include cell division, changes in cell shape and size, rearrangement of cells within tissues, the development of hydrostatic pressure within structures, and the migration of individual cells and groups of cells from one part of the embryo to another. Cellular forces are often generated within epithelial sheets by a contraction of the cells' internal cytoskeleton; resisting these forces and enabling the epithelial sheet to bend or fold are the cells' own structure and their adhesion to other cells and to the extracellular matrix, which hold them together in tissues. The major morphogenetic event in early animal embryos is gastrulation, when a wholesale reorganization of cells in the embryo occurs. Other examples of morphogenesis in vertebrates discussed in this chapter are the formation of tubular structures such as the neural tube and the branched tubular network of the blood system, and the migration of neural crest cells and fish lateral line cells to specific locations.

Introduction

So far, we have discussed early development mainly from the viewpoint of the assignment of cell fate. In this chapter, we look at embryonic development from a different perspective—the generation of form, or **morphogenesis**. All animal embryos undergo a dramatic change in shape during their early development and the most important of these changes occurs during gastrulation—the process that transforms a uniform two-dimensional sheet of cells into a three-dimensional arrangement of layers of cells with different identities. Gastrulation involves extensive rearrangements and directed movement of cells from one location to another, and in animals with three germ layers yields a universal organization of cells into ectoderm, mesoderm, and endoderm, with an antero-posterior patterns (see Chapters 3 and 4).

If pattern formation can be likened to painting, morphogenesis is more akin to modeling a formless lump of clay into a recognizable shape. An important general principle to remember is that individual cells on their own cannot drive morphogenesis. It is their organization into coherent tissues that enables cells to coordinate their behavior and generate forces that model the embryo into a different shape. Change in form is largely a problem of how the behavior of individual cells affects the mechanical properties of the tissues of which they are a part, and how different tissues use these properties to position themselves relative to each other. This means that understanding morphogenesis requires an appreciation of the forces generated by cells both individually and collectively, as well as of the causes and consequences of cell and tissue shape changes associated with these forces.

Three key properties of cells are involved in the generation of animal embryonic form. The first is **cell adhesiveness**, the tendency of cells to stick to one another and to their environment. In tissues, animal cells adhere to one another, and to the extracellular matrix—a protein-based complex material secreted by the cells—through interactions involving cell-surface proteins known collectively as **cell-adhesion molecules**. Changes in the adhesion molecules at the cell surface can alter the strength of cell–cell adhesion and its specificity, with far-reaching effects on the tissue of which the cells are a part. Cell–cell adhesion is an essential feature of epithelia, the building blocks of many tissues and organs in animals.

The second property is **cell shape**: cells can actively change shape by means of internal contractions and constrictions and by interactions with other cells. Changes in cell shape, which are caused by cytoskeletal rearrangements, are crucial in the activities of tissues in many developmental processes. Folding or rolling of an epithelial cell sheet—a very common feature in embryonic development—is caused by coordinated changes in the shape of selected cells within the sheet. In this chapter we will see how the embryonic neural tube is formed in this way.

The third is **cell migration**, the ability of many types of cells to travel as individuals or groups from one location to another. Many developmental processes, such as the formation of the vasculature, the homing of the germ cells to their location in the gonads (discussed in Chapter 8), and the development of structures derived from neural crest, involve the guided coordinated migration of individual cells, or groups of cells, from their site of origin to their final location.

The coordination of all these cell activities is central to morphogenesis: we are just beginning to understand that this coordination depends on interactions between chemical and mechanical signals, and between intracellular dynamics and interactions with other cells and the extracellular matrix. The organization of cells into tissues amplifies the potential of these key cell properties, enabling the forces that they generate to operate over larger dimensions. A cell shrinking or expanding within a tissue will not only affect its neighbors, but will also have an effect several cell diameters away. Changes in cell adhesiveness are crucial to both cell-shape changes and cell migration. Not surprisingly, there are many points of interaction between the molecular machinery of cell adhesion and the machinery controlling the cytoskeleton. Some of the cell-adhesion molecules we will discuss in this chapter act as signaling receptors, transmitting signals from the extracellular environment into the cell to effect changes in the cytoskeleton. Cell adhesion, cell-shape changes, and cell migration can therefore be controlled in a coordinated fashion by the developmental history of the cell (which determines, for example, which cell-adhesion molecules it expresses) acting together with extracellular signals.

An additional force that operates during morphogenesis, more particularly in plants, but also in some aspects of animal embryogenesis, is hydrostatic pressure, which is generated by osmosis and fluid accumulation. In plants there is no cell movement during growth, and changes in form are generated by oriented cell division and cell expansion.

Changes in embryonic form are therefore the final consequence of the precise spatio-temporal expression of proteins that control cell adhesion, cell motility, oriented cell division, and the generation of hydrostatic pressure. Gastrulation, for example, can be thought of as the animation of a detailed map of instructions that has been progressively laid down on the canvas that is the early embryo. The earlier patterning process determines which cells will express those proteins that are required to generate and harness the appropriate forces.

In this chapter we shall discuss some examples of the morphological changes that occur during the development of the animal body plan and their underlying molecular causes. First, we will look at how cleavage of the zygote gives rise to the simple shape of the early embryo, of which the spherical blastocyst of the mouse and the blastulas of sea urchin and amphibian are good examples. We then consider the movements that occur during gastrulation and during neurulation—the formation of the neural tube in vertebrates—which involve rearrangement of cell layers and folding of cell sheets. The formation and branching of tubular epithelia—known as branching morphogenesis—occurs early in animal development and during the formation of many different organs. In this chapter we discuss the development of the tracheal system in insects and the network of blood vessels of the vascular system in vertebrates. In vertebrates, migration of cells from the neural crest after neurulation generates various structures in the trunk and head, and we consider how these cells are guided to their correct sites. A mass movement of cells also occurs in the formation of the lateral line sense organ in fishes. In **directed dilation**, hydrostatic pressure is the force that drives changes in shape. Other morphogenetic mechanisms, such as cell growth, cell proliferation, and cell death, are considered in relation to the development of particular organs and the growth of an organism as a whole in Chapters 9, 10, and 11. Some aspects of growth in plants are considered in Chapter 12.

We begin by considering how cells adhere to each other, and how differences in the strength and the specificity of adhesion are involved in maintaining boundaries between tissues.

Cell adhesion

Adhesive interactions between cells, and between cells and the extracellular matrix, are an important element of cell behavior in morphogenesis. They enable cells to assemble into groups that will behave coordinately, and differences in cell adhesiveness help to set up and maintain boundaries between different tissues and structures.

All cells have a surface tension at their membrane, where they encounter the extracellular environment. This surface tension results in isolated cells adopting a spherical shape in aqueous medium, just as water forms drops in air, because this minimizes the energy of interaction with their surroundings. Surface tension also keeps the surface of a film of water intact, as interactions between water molecules are energetically favored compared with the interactions between them and air. These same principles drive cells when they touch to adhere and spread on each other spontaneously, as a way of minimizing their overall surface tension by reducing the amount of surface exposed. The strength and specificity of cell adhesion are controlled by cell-adhesion molecules, which are proteins on the cell surface that can bind strongly to proteins on other cell surfaces or in the extracellular matrix (Box 5A).

The particular cell-adhesion molecules expressed by a cell will determine which cells it can stick to, a decision that is central to many developmental phenomena; for example, keeping different epithelia separate might be important to build a particular tissue, and therefore each epithelium might use specific cell-adhesion molecules to keep 'like with like'. Adjacent cells in epithelia, for example, are joined together by specialized structures called **adherens junctions** that incorporate cell-adhesion

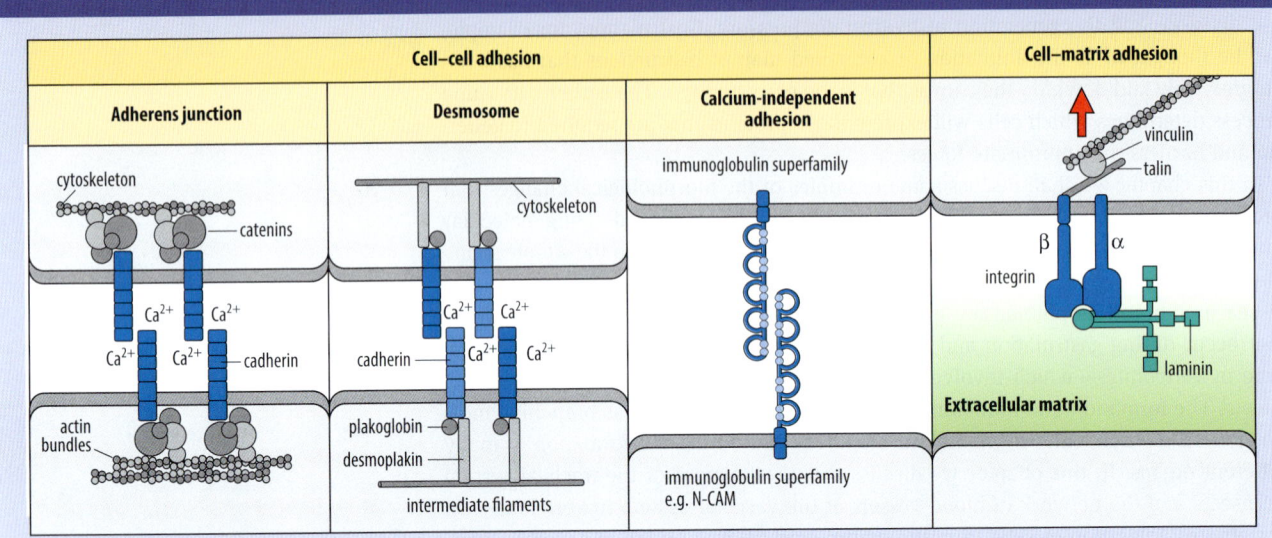

CELL BIOLOGY BOX 5A Cell-adhesion molecules and cell junctions

Fig. 1

Three classes of **adhesion molecules** are particularly important in development (Fig. 1). The **cadherins** are transmembrane proteins that, in the presence of calcium ions (Ca^{2+}), adhere to cadherins on the surface of another cell. Calcium-independent cell–cell adhesion involves a different structural class of transmembrane proteins, which are members of the large **immunoglobulin superfamily**. The neural cell-adhesion molecule (N-CAM), which was first isolated from neural tissue, is a typical member of this family; another member is L-CAM. Some immunoglobulin superfamily members, such as N-CAM, bind to similar molecules on other cells; others bind to a third class of adhesion protein, the transmembrane **integrins**. Integrins also act as receptors for molecules of the extracellular matrix, mediating adhesion between a cell and its substratum.

About 30 different types of cadherins have been identified in vertebrates. Cadherins bind to each other through one or more binding sites located within the extracellular amino-terminal 100 amino acids. In general, a cadherin binds only to another cadherin of the same type, but they can also bind to some other molecules. Cadherins are the adhesive components in both **adherens junctions** (adhesive cell–cell junctions that are present in many tissues) and **desmosomes** (cell–cell junctions present mainly in epithelia). As cells approach and touch one another, the cadherins cluster at the site of contact. They interact with the intracellular cytoskeleton through the connection of their cytoplasmic tails with **catenins** (α-, β-, and γ-) and other proteins, and thus can be involved in transmitting signals to the cytoskeleton (see Fig. 1). (This role of the protein β-catenin is quite separate from its role as a key element in the Wnt intracellular signaling pathway; see Box 3E). The interaction of cadherins with the cytoskeleton is required for strong cell–cell adhesion, and enables the coordination of its activity between cells. In adherens junctions the connection is to cytoskeletal actin filaments, whereas in desmosomes it is to intermediate filaments such as keratin. Nectin, a member of the immunoglobulin superfamily of adhesion molecules, also clusters to adherens junctions in mammalian tissues and connects to the actin cytoskeleton.

An integrin molecule is composed of two different subunits, an α subunit and a β subunit. Twenty-four different integrins are known so far in vertebrates, made up from combinations of nine different β subunits and 24 different α subunits. Through their extracellular face, integrins bind to extracellular matrix proteins such as collagen, fibronectin, laminin, and tenascin, and to proteoglycans. Through their intracellular face they associate with the cell's cytoskeleton through complexes of proteins (see Fig. 1). Integrins transmit information from the matrix that affects cell shape, motility, metabolism, and cell differentiation. As well as this 'outside-in' signaling, integrins can mediate 'inside-out' signaling in response to intracellular signals, changing the conformation of their extracellular portions which affects their binding to the extracellular matrix.

Hemidesmosomes are integrin-based junctions that rivet epithelial cells firmly and stably to the extracellular matrix of the basement membrane through extracellular binding to a matrix molecule and intracellular association with keratin intermediate filaments. Integrins can also make **focal adhesions**—contacts between the cell and its substratum in which the intracellular face of the integrin interacts with the actin cytoskeleton. Focal adhesions can be made and broken more easily than hemidesmosomes and are involved in cell motility (see Box 5B). Integrins also mediate cell–cell adhesion, binding either to adhesion molecules of the immunoglobulin superfamily or via a shared ligand to integrins on another cell surface.

A type of cell junction present in some, but not all, epithelia is the **tight junction**, which forms a seal that prevents water and other molecules passing between the epithelial cells. Multiple lines of transmembrane proteins called claudins and occludins in adjoining cell membranes form a continuous seal around the apical end of the cell (see Fig. 5.15).

molecules. Adhesive interactions affect the surface tension at the cell membrane, reflected by a flattening of the cell at the point of contact with another cell or with the substratum.

Different surface tensions are what keep two immiscible liquids such as water and oil separate, and embryologists had long noted the 'liquid-like' behavior of the tissues in gastrulating frog embryos, observing how the separate germ layers formed and remained distinct from each other, slid over each other, spread out, and underwent internal rearrangement to shape the embryo. These observations led to the **differential adhesion hypothesis**, which explains these liquid-like tissue dynamics as the consequences of tissue surface tensions that arise from differences in intercellular adhesiveness.

5.1 Sorting out of dissociated cells demonstrates differences in cell adhesiveness in different tissues

Differences in cell adhesiveness can be illustrated by experiments in which different tissues are confronted with one another in an artificial setting. Two pieces of early endoderm from an amphibian blastula will fuse to form a smooth sphere if placed in contact. In contrast, when a piece of early endoderm and a piece of early ectoderm are combined, they initially fuse, but in time, the pieces of endoderm and ectoderm do not remain together but separate, until only a narrow bridge connects the two types of tissue (Fig. 5.1).

A different result is obtained when a piece of early ectoderm is placed in contact with a piece of early mesoderm. Ectoderm and mesoderm normally adhere to each other within the blastula, and the tissues do not separate from each other like ectoderm and endoderm. Instead, the pieces of tissue remain in contact and the mesoderm spreads over the ectoderm and eventually envelops it. When the different tissues from blastulas or gastrulas are disaggregated into single cells, and the cells then are mixed and allowed to reaggregate, cells from the different tissues sort out, so that within the aggregate, like cells associate preferentially with like cells—ectoderm with ectoderm, and mesoderm with mesoderm. Furthermore, ectoderm cells are on the inside of the aggregate and mesoderm on the outside (Fig. 5.2), exactly the same arrangement as adopted by the fused pieces of tissue. Similar experiments with disaggregated cells from the tissues of zebrafish embryos have shown that ectoderm also ends up on the inside of the aggregate and mesendoderm on the outside. When cells of amphibian prospective epidermis and prospective neural plate are dissociated, mixed, and left to reaggregate, they sort out to re-form the two different tissues. In this case, the epidermal cells are eventually found on the outer face of the aggregate, surrounding a mass of neural cells.

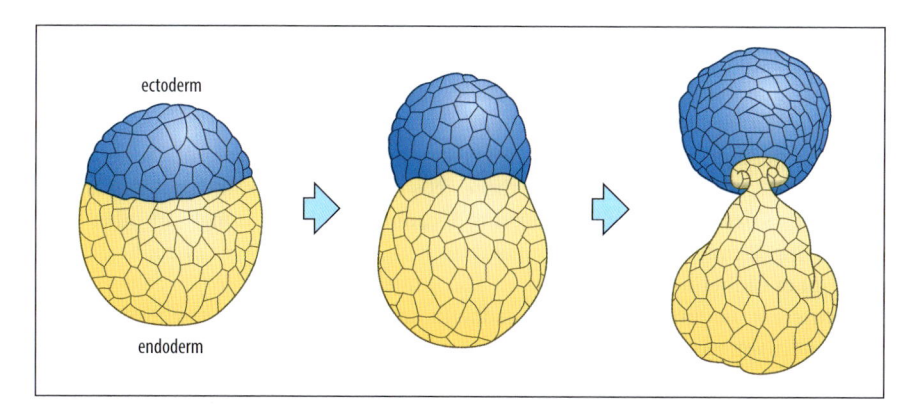

Fig. 5.1 Separation of embryonic tissues with different adhesive properties. When pieces of early ectoderm (blue) and early endoderm (yellow) from an amphibian blastula are placed together, they initially fuse but then separate until only a narrow strip of tissue joins the two tissues.

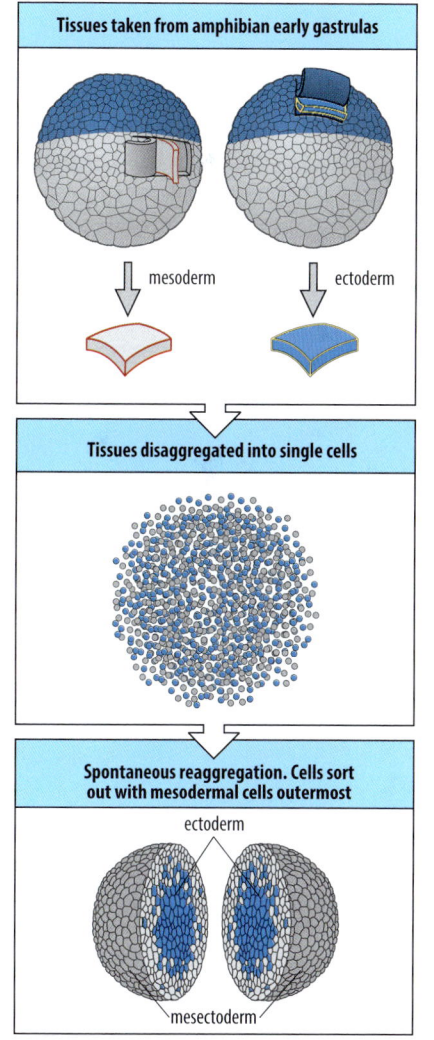

Fig. 5.2 Sorting out of different cell types. Ectoderm (gray) and mesoderm (blue) from early amphibian gastrulas are disaggregated into single cells by treatment with an alkaline solution. The cells, when mixed together, sort out with the mesoderm cells on the outside.

The sorting out behavior is the combined result of differences in surface tension between cells of the different types due to differences in intercellular adhesiveness, so that the overall energy of interactions between cell surfaces within the cell mass is eventually minimized. Cells of the same type tend to adhere preferentially to each other. Initially, cells in the mixed aggregate start to exchange weaker for stronger adhesions. Over time, the adhesive interactions between cells produce different degrees of surface tension that are sufficient to generate the sorting-out behavior, just as two immiscible liquids, such as oil and water, separate out when mixed. As the constituent cells form stable contacts with each other, the tissues become reorganized and the strength of intercellular binding in the system is maximized. In general, if the adhesion between 'unlike' cells is weaker than the average of the adhesions between 'like' cells, the cells will segregate according to type, with the more adhesive cell type on the inside of the aggregate. In terms of the surface tension, those cells with the higher surface tension become surrounded by those with lower surface tension.

It is important to note that experiments on isolated cells and tissues *in vitro*, like those described above, simply indicate the different adhesiveness and different surface tensions of the tissues in isolation; they do not define which tissue will be on the inside or the outside in the embryo itself. This is because in a real embryo, tissues are under pre-existing developmental constraints that define their location and the specialization of their adhesive properties and surface tension, and they may also adhere to other tissues. Nevertheless, these *in vitro* experiments show how differential cell adhesion and different surface tensions can, in principle, create and stabilize boundaries between tissues. Furthermore, there are a few examples where sorting out occurs during development, as in the early mouse embryo, where sorting out of prospective embryonic and extra-embryonic precursors contributes an alternative strategy to positional information for generating a pattern (see Section 3.24).

5.2 Cadherins can provide adhesive specificity

Differential adhesiveness between cells is the result of differences in the types and numbers of adhesion molecules they have on their surfaces. The main types of adhesion molecule we will be concerned with in this chapter are outlined in Box 5A. One class comprises the **cadherins**, which depend on calcium for their function and are components of a type of adhesive junction called the **adherens junction**, which is the junction that holds cells together in all epithelial tissues. There are many kinds of cadherins and, in many instances, they are cell-type specific. Evidence that the cadherins provide adhesive specificity comes from studies in which cells with different cadherins on their surface are mixed together. Fibroblasts of the mouse L cell line do not express cadherins on their surface and do not adhere strongly to each other. But if the gene encoding E-cadherin is transfected into L cells and expressed, the cells produce that cadherin on their surface and stick together, forming a structure resembling a compact epithelium. The adherence is both calcium-dependent, which indicates that it is due to the cadherin, and specific, as the transfected cells do not adhere to untreated L cells lacking surface cadherins.

When populations of L cells are transfected with different types of cadherin and mixed together in suspension, only those cells expressing the same cadherin adhere strongly to each other: cells expressing E-cadherin adhere strongly to other cells expressing E-cadherin, but only weakly to cells expressing P- or N-cadherin. The amount of cadherin on the cell surface also influences adhesion. When cells expressing different amounts of the same cadherin are mixed, those cells with more cadherin on their surface end up on the inside of the reaggregate, surrounded by the cells with less surface cadherin (Fig. 5.3). This is a direct demonstration that quantitative differences in cell-adhesion molecules lead to differential cell adhesion and sorting out.

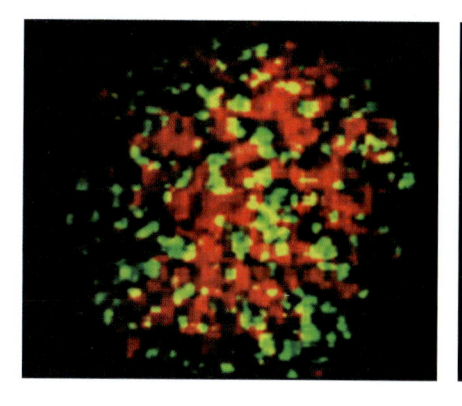

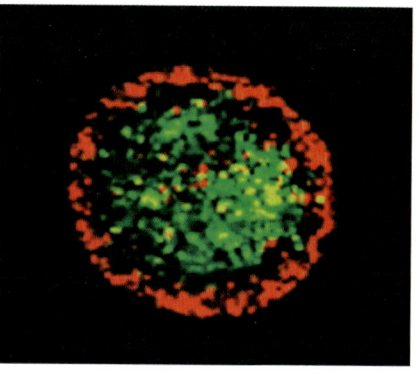

Fig. 5.3 Sorting out of cells carrying different amounts of cell-adhesion molecules on their surfaces. When two cell lines with different amounts of N-cadherin on their surface are mixed, they sort out, with the cells containing the most N-cadherin (green-stained cells) ending up on the inside.

Photographs courtesy of M. Steinberg, from Foty, R.A., Steinberg, M.S.: **The differential adhesion hypothesis: a direct evaluation.** Dev. Biol. 2005, **278**: 255-263.

Cadherin molecules bind to each other through their extracellular domains, whose structure is strictly dependent on calcium binding, but adhesion is not exclusively controlled by these domains. The cytoplasmic domain of a cadherin associates with the actin filaments of the cytoskeleton by means of a protein complex containing α- and β-catenins (see Box 5A), and failure to make this association results in weak adhesion. The initial binding of the extracellular cadherin domains transmits a signal to the cytoskeleton, which then stabilizes the interaction. This means that, just as there is information flowing from outside the cell to the inside via the cadherin–catenin complex, there is also information flowing from the inside of the cell outwards.

In the developing spinal cord of the zebrafish, domains of progenitor cells that will give rise to different types of neurons are precisely organized along the dorso-ventral axis. This organization is manifest in the expression of domain-specific transcription factors (see Fig. 10.10), yet the precision of the pattern suggests some mechanism that allows cells of the same type to stay together and separate from the other cell types to avoid intermingling. In the zebrafish, this appears to be mediated by a combinatorial code of cadherins, which is controlled by domain-specific transcription factors that result from the interpretation of a gradient of the signal protein Sonic hedgehog (Shh). From dorsal to ventral, one can identify three domains of neuronal progenitors—p0, pMN, and p3—each of which can be characterized by a specific transcription factor combination. Cells from each of these domains exhibit differential adhesion towards cells from the other domains, and genetic analysis revealed three cadherins—Cdh1, Cdh2, and Pcdh19—as mediating these differences. The cadherins are expressed in different amounts in the different domains, and sorting experiments reveal a cadherin code for each region. The p0 domain is maintained by a combination of Cdh2–Cdh2 and Cdh2–Pcdh19 interactions, pMN by Cdh1–Cdh2 interactions, and p3 by interactions between Pcdh19 and Cdh2 (Fig. 5.4). Alterations in the gradient of Shh, or in the

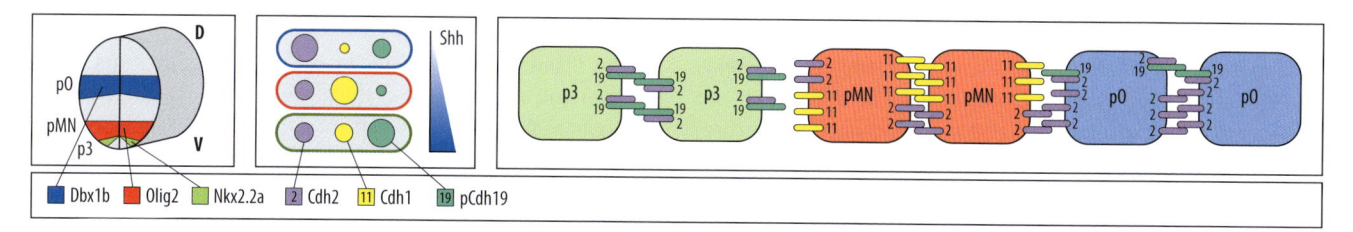

Fig. 5.4 Combinations of cadherin molecules expressed by neuronal progenitors help to produce the dorso-ventral organization of the developing spinal cord. Left panel: domains of motor neuron organization in the zebrafish spinal cord. Center panel: distribution of the different cadherins across the different domains. Each domain is outlined by its color in the first panel and the sizes of the colored spots represent the level of expression of each cadherin. Right panel: interactions between the principal cadherins in each domain and how differential adhesion between the neuronal progenitors within and between the domains leads to their dorso-ventral organization in the spinal cord.

Adapted from Tsai, T.Y.-C., et al.: **An adhesion code ensures robust pattern formation during tissue morphogenesis.** Science 2020, **370**: 113-116.

responding domain-specific transcription factors, result in changes in the adhesion landscape and the disruption of the neuronal pattern.

5.3 The activity of the cytoskeleton regulates the mechanical properties of cells and their interactions with each other

Differential adhesion alone cannot explain how cells actively change shape or deform under pressure from neighboring cells in a coherent manner to drive the large-scale directional movements that occur, for example, during gastrulation. The shapes of cells in tissues are determined by counteracting forces: adhesion between cells tends to increase the area of surface contact, whereas the contractile cytoskeleton limits the expansion of contacts. In addition, external interactions between cells and their substrate, such as integrin-mediated interactions with the extracellular matrix, allow cells to spread over the substrate, and in this way affect the outcome of morphogenetic processes. The dynamics of morphogenetic movements, such as cell contraction and expansion, relies on the activity of the cytoskeleton, in particular the motor proteins associated with it, and on the mechanical properties of the cellular components and the extracellular matrix, such as stiffness and viscosity.

It is important to remember that a cell is not a passive bag of chemicals, but is a collection of active molecular agents that are constantly producing forces and are able to change the cell's mechanical behavior in reaction to forces exerted on the cell. In particular, the cytoskeleton provides the engine to create, control, and modulate cell-shape changes (Box 5B). The cytoskeleton is a key sensing device for the cell that, through its linkage to proteins such as cadherins, simultaneously senses and reacts to the magnitude and direction of tensile forces and adhesions that drive cell-shape changes, cellular rearrangements, and cell movements during morphogenesis. Furthermore, because the cytoskeleton of one cell can be effectively linked to the cytoskeleton of an adjacent cell via cadherin–catenin complexes, this enables the coordination of these changes across tissues, as we shall see in the examples of gastrulation and neurulation.

CELL BIOLOGY BOX 5B The cytoskeleton, cell-shape change, and cell movement

Cells actively undergo major changes in shape during development. The two main changes are associated with cell migration and with the folding of epithelial sheets. Changes in shape are generated by the cytoskeleton, an intracellular protein framework that also controls changes in cell movement. There are three principal types of protein polymers in the cytoskeleton—**microtubules**, **actin filaments (microfilaments)**, and **intermediate filaments**—as well as many other proteins that interact with them. Microtubules and actin filaments are dynamic structures, polymerizing and depolymerizing according to the cell's requirements. Intermediate filaments are more stable. They form rope-like structures that transmit mechanical forces, spread mechanical stress, and provide mechanical stability to the cell.

Microtubules are polymers of the globular protein tubulin that are organized into long tubules (about 25 nm in diameter and up to 50 μm and longer in the axons of nerve cells) with a structural role. Microtubules are often dynamic structures and can rapidly assemble and disassemble. They are the main component of cilia, play an important part in maintaining cell shape, cell asymmetry, and polarity, and provide tracks along which motor proteins deliver molecules, and even organelles, to particular locations in the cell. They also form the spindles that segregate chromosomes at mitosis and meiosis.

Actin filaments are fine threads of protein about 7 nm in diameter and are polymers of the globular protein actin. They are organized into bundles and three-dimensional networks, which, in most cells, lie mostly just beneath the plasma membrane, forming a gel-like cell cortex. Numerous actin-binding proteins are associated with actin filaments and are involved in bundling the filaments together, forming networks, and aiding the polymerization and depolymerization of the actin subunits. Actin filaments can form rapidly by polymerization of actin subunits and can be equally rapidly depolymerized.

This provides the cell with a versatile system for assembling actin filaments in various ways and in different locations as required. Polymerization and depolymerization of actin filaments at the leading edges of cells are involved in changes in cell shape and cell movements, and actin filaments also associate with the motor protein myosin to form contractile bundles of **actomyosin**, which are primarily responsible for force-generating contractions within cells and which act as 'miniature muscles' within cells. For example, contraction of the **contractile ring**, bundles of actomyosin arranged in a ring in the cell cortex, pinches animal cells in two at cell division and during cleavage (see Fig. 1, left panel). The contraction of a similar ring of actomyosin around the apical end of a cell leads to apical constriction and, often, to a consequent elongation of the cell (see Fig. 1, center panel).

Many embryonic cells can migrate over a solid substratum, such as the extracellular matrix. They move by extending a thin sheet-like layer of cytoplasm known as a **lamellipodium** (see Fig. 1, right panel), or long, fine cytoplasmic processes called **filopodia**. Both these temporary structures are pushed outwards from the cell by the assembly of actin filaments, which are cross-linked to form bundles. Contraction of the actomyosin network at the rear of the cell and disassembly of integrin-based focal adhesions at the rear then moves the cell body forward. To do this, the contractile system must be able to exert a force on the substratum and this occurs at the focal adhesions, points at which the advancing filopodia or lamellipodium are anchored to the surface over which the cell is moving. At focal adhesions, integrins adhere to extracellular matrix molecules through their extracellular domains, while their cytoplasmic domains provide an anchoring point for actin filaments. Integrins transmit signals from the extracellular matrix across the plasma membrane at focal adhesions, enabling the cell to sense the environment over which it is traveling and to adjust its movement accordingly.

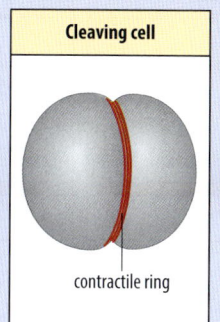

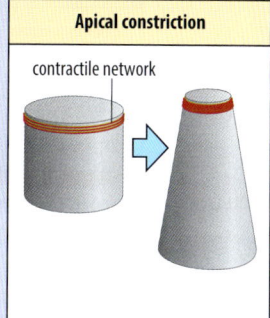

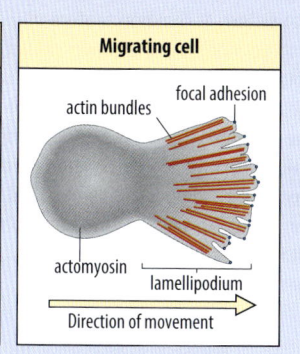

Fig. 1

5.4 Transitions of tissues from an epithelial to a mesenchymal state, and vice versa, involve changes in adhesive junctions

Morphogenesis is driven by interactions within and between two kinds of supracellular architectures—epithelial and mesenchymal. In an epithelium, adherens junctions hold the cells tightly together to form a continuous sheet. Mesenchyme has a much looser organization. Associations between individual mesenchymal cells are weaker, and in some cases, the cells lack mature adhesive junctions. Individual cells are often partly surrounded by extracellular matrix, tend to contact each other through very thin processes, and become connected by **gap junctions**. These are protein pores in the apposed cell membranes of two adjacent cells through which ions and some small molecules can pass directly from the cytoplasm of one cell to the other. Gap junctions are also found between epithelial cells.

As well as serving distinct functions in differentiated tissues, such as uptake and transport of nutrients, epithelia are present within early embryos as dynamic and transient structures, as we have seen in previous chapters. The first morphogenetic event in an embryo is the formation of an epithelium. Epithelia serve as the starting tissues from which different cell types emerge, and they can be sculpted into different forms by the coordinated actions of the cells within them.

A frequent event in early development is the conversion of an epithelium into a more loosely connected mass of mesenchyme, or even into individual mesenchymal cells that can migrate. This change, which involves the

Fig. 5.5 Epithelial-to-mesenchymal transition. A common feature of embryogenesis is the conversion of an epithelium into more loosely organized mesenchyme. This occurs by dissolution of the adherens junctions between adjoining epithelial cells, the detachment of cells from the basement membrane, and their conversion through intermediate phenotypes into individual or loosely adhering mesenchymal cells.

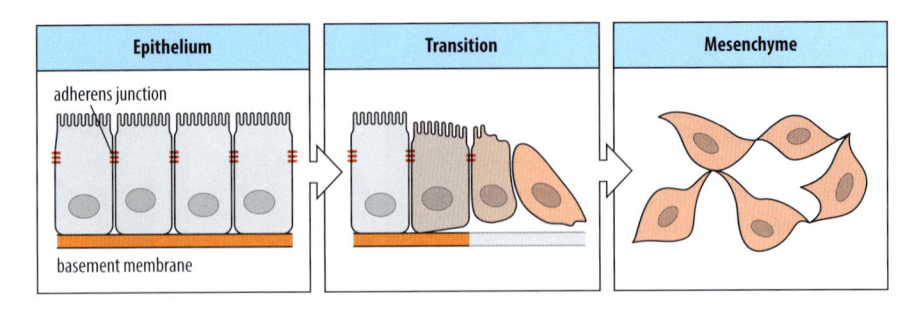

dissolution of the adherens junctions between the epithelial cells, is known as an **epithelial-to-mesenchymal transition (EMT)** (Fig. 5.5). We have already seen such transitions at work in gastrulation. In *Drosophila* and the sea urchin, mesoderm cells detach from the ectoderm once inside the embryo. In chick and mouse embryos, cells of the epiblast detach from the epithelium and move inside the embryo through the primitive streak. In this chapter we shall look more closely at the mechanisms underlying the EMT in these situations.

The reverse process, a **mesenchymal-to-epithelial (MET)** transition, is also found in embryogenesis. For example, mesenchymal mesoderm condenses into blocks of epithelia to form the somites (see Chapter 4), and into tubules of epithelium to form blood vessels (see Sections 5.19 and 5.20) and kidney tubules. This type of transition involves the formation of adherens junctions between the cells. Later in their development, cells from the somites will undergo EMTs as they give rise to prospective muscle cells.

SUMMARY

The adhesion of cells to each other and to the extracellular matrix maintains the integrity of tissues and the boundaries between them. The associations of cells with each other are determined by the cell-adhesion molecules that they express on their surface: cells bearing different adhesion molecules, or different quantities of the same molecule, sort out into separate tissues. This is due to intercellular adhesiveness on the cortical cytoskeleton and thus on cell-surface tension and interfacial interactions between different tissues.

Cell-cell adhesion is due mainly to two classes of surface proteins: the cadherins, which bind in a calcium-dependent manner to identical cadherins on another cell surface; and members of the immunoglobulin superfamily, some of which bind to similar molecules on other cells. The binding of this second class of surface protein is calcium independent. Adhesion to the extracellular matrix is mediated by a third class of adhesion molecules—the integrins—which can also bind to members of the immunoglobulin superfamily. Stable cell–cell adhesion through cadherins involves both physical contact between the external domains of cadherins on adjacent cells, and connection of the cytoplasmic domain of the cadherin molecule to the cell cytoskeleton via a protein complex containing catenins. Because adhesion molecules are connected to the cytoskeleton, they can transmit signals from the extracellular environment to the cytoskeleton to cause, for example, changes in cell shape, and are also sensitive to the internal dynamics of the individual cells. In tissues, cell-cell adhesions allow the coordination of tissue-level cell rearrangements and movements. An important feature of morphogenesis is the transition of cells from an epithelial organization to form the more loosely organized mesenchyme, a change known as an epithelial-to-mesenchymal transition. This involves the dissolution of adhesive junctions. The opposite transition is the condensation of mesenchymal cells into epithelial structures, called a mesenchymal-to-epithelial transition.

Cleavage and formation of the blastula

The first step in animal embryonic development is the division of the fertilized egg by cleavage into a number of smaller cells (blastomeres), followed, in many animals, by the formation of a hollow sphere of cells—the blastula (see Chapter 3). In early embryos, cleavage involves short mitotic cell cycles, in which cell division and mitosis succeed each other repeatedly without intervening periods of cell growth. During cleavage, cell size gets smaller and, therefore, the mass of the embryo does not increase.

Early cleavage patterns vary widely between different groups of animals (Fig. 5.6). In **radial cleavage**, the divisions occur at right angles to the egg surface and the first few cleavages produce tiers of blastomeres that sit directly over each other. This type of cleavage is characteristic of the deuterostomes, such as sea urchins and vertebrates. The eggs of mollusks and annelids, which are protostomes, illustrate another cleavage pattern, called **spiral cleavage**, in which successive divisions are at planes at slight angles to each other, producing a spiral arrangement of cells. In both radial and spiral cleavage, some divisions may be unequal. The first three cleavages in the sea urchin, for example, give rise to equal-sized blastomeres, whereas the later cleavage that forms the micromeres is unequal, with one daughter cell being smaller than the other. In nematodes, the very first cleavage of the egg is unequal, producing two cells of different sizes (see Fig. 5.6). In the early *Drosophila* embryo, the nuclei undergo repeated divisions without cell division, forming a syncytium, and separation into individual cells only occurs later, with the growth of cell membranes between the nuclei (see Section 2.1).

The amount of yolk in the egg can influence the pattern of cleavage. In yolky eggs undergoing otherwise equal cleavages, a cleavage furrow develops in the least yolky region and gradually spreads across the egg, but its progress is slowed or even halted by the presence of the yolk. In these embryos, cleavage may thus be incomplete for some time. This effect is most pronounced in the heavily yolked eggs of birds and zebrafish, where complete divisions are restricted to a region at one end of the egg, and the embryo is formed as a cap of cells sitting on top of the yolk (see Fig. 3.39). Even in moderately yolky eggs, such as those of amphibians, the presence of yolk can

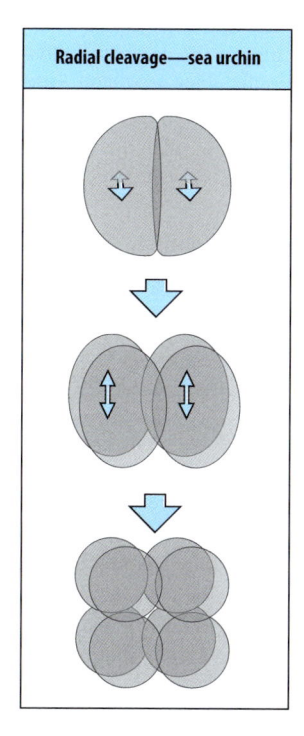

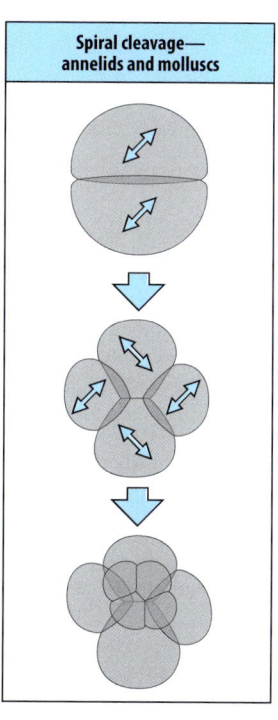

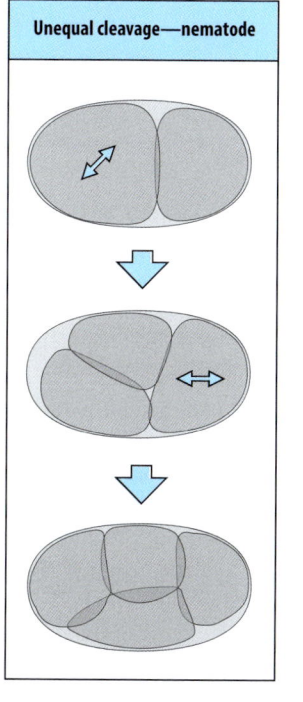

Fig. 5.6 Different patterns of early cleavage are found in different animal groups. Left panel: radial cleavage (e.g. sea urchin) gives tiers of blastomeres sitting directly above each other. Center panel: in spiral cleavage (e.g. mollusks and annelid worms), the mitotic apparatus is oriented at a slight angle to the long axis of the cell, resulting in a spiral arrangement of blastomeres. Right panel: unequal cleavage (e.g. nematode) results in one daughter cell being larger than the other. The blue arrows indicate the orientation of the mitotic spindle. Cleavage occurs at right angles to this axis.

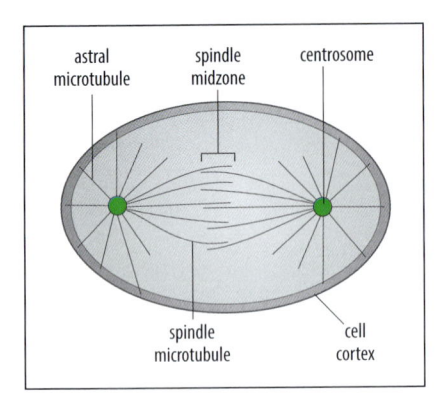

258 Chapter 5 Morphogenesis: change in form in the early embryo

Fig. 5.7 The mitotic spindle. The mitotic spindle is the structure constructed by the cell that will segregate the duplicated chromosomes into two identical sets. It is composed of microtubules that emanate from two centrosomes that become positioned opposite to each other in the cell and which form the two poles of the spindle. One set of microtubules, the spindle microtubules, grows out from each centrosome to meet and overlap with the spindle microtubules from the opposite pole at the spindle midzone. The spindle midzone marks the point at which the cell will divide. A second set of microtubules, the astral microtubules, anchors each centrosome to the cell cortex. They are so called because they form a star-shaped structure, called an aster, at each pole. Chromosomes are not shown for simplicity.

influence cleavage patterns. In frog eggs, for example, the later cleavages are unequal and asynchronous, resulting in an animal region that is composed of a mass of small cells and a yolky vegetal region composed of fewer and larger cells.

Two key questions arise in relation to the mechanics of early cleavages. How are the positions of the cleavage planes determined? And how does cleavage lead to a hollow blastula (or its equivalent), which has a clear inside–outside polarity (i.e. the outer surface of the blastula epithelium is markedly different from the inner surface)?

5.5 The orientation of the mitotic spindle determines the plane of cleavage at cell division

Mitosis and cell division depend on the formation of a **mitotic spindle**, a structure formed by microtubules that grows out from two **centrosomes** that are positioned opposite each other in the cell and form the two poles of the spindle. The microtubules from each pole overlap where they meet to form the **spindle midzone**, which is, in general, equidistant from each spindle pole. Other microtubules radiate away from the centrosome with their ends anchored to motor proteins in the actomyosin-rich cortex of the cell, forming an arrangement called an **aster** at each pole (Fig. 5.7). The duplicated chromosomes become attached to spindle microtubules at the midzone, and the chromatids of each pair are subsequently pulled back to opposite ends of the cell by shortening of these microtubules. Two new nuclei are formed, and the cell divides to form two new cells by cleaving at right angles to the spindle axis at the position of the spindle midzone. In animal cells, cleavage is achieved by the constriction of an actomyosin-based **contractile ring**, and the position of cleavage becomes visible on the cell surface as the cleavage furrow (see Box 5B).

The orientation and position of the plane of cleavage at cell division is of great importance in embryonic development. They not only determine the spatial arrangement of daughter cells relative to each other, and whether the cells will be equal or unequal in size, but the position of the cleavage plane can also result in the apportionment of different cytoplasmic determinants to different daughter cells, thus giving them different fates, as, for example, in the first cleavage division in nematodes.

The orientation of cleavage in general follows an old rule, the Hertwig rule, which says that a cell orients the mitotic spindle along the longer axis of the cell before division. Cleavage then occurs perpendicular to this axis. All cells, even epithelial cells, round up during division, which would erase any cue of asymmetry and orientation. However, it seems that the actin cytoskeleton maintains a memory of the original shape of the cell by leaving fine, actin-based retraction fibers attached to the cell's original points of attachment to the extracellular matrix. The guide for the spindle to orient itself, therefore, is a footprint of the shape of the cell before division, and this positions the spindle accordingly. Spindle orientation is particularly important in the cells of an epithelium, because to maintain a continuous single layer, the plane of cleavage must be at right angles to the epithelial surface. The rule of the long axis can be overridden by signals delivered through the actin cytoskeleton that target the spindle and orient it in another direction.

A common form of early animal embryo is the hollow spherical blastula, composed of an epithelial sheet enclosing a fluid-filled interior. The development of such a structure from a fertilized egg depends both on particular patterns of cleavage and on changes in the way cells pack together, as shown in schematic form in Figure 5.8. If the plane of cell division is always at right angles to the surface, the cells will remain in a single layer. As cleavage proceeds, individual cells get smaller, but the surface area of the cell sheet increases and a space—the blastocoel—forms in the interior, which increases in volume at each round of cell division. This is essentially how the sea urchin blastula is formed.

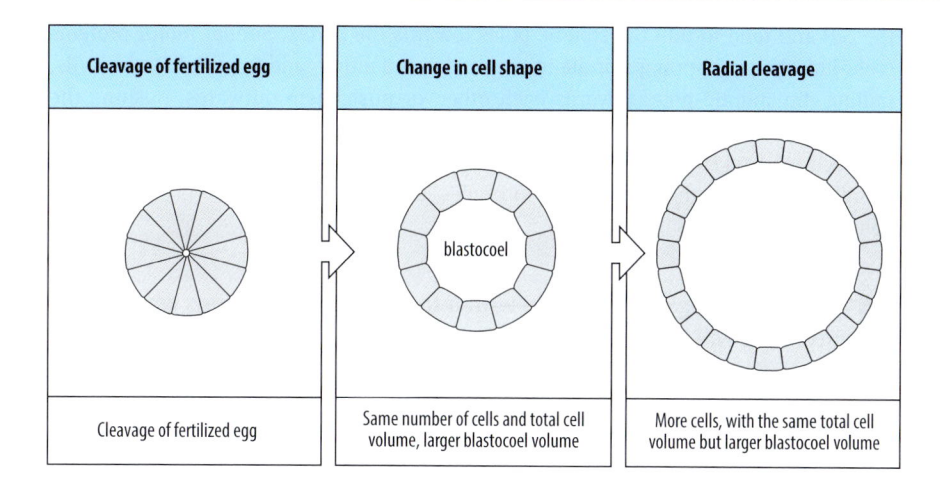

Fig. 5.8 Specifically oriented cell divisions convert a small ball of cells into a hollow spherical blastula. Left panel: when adjacent cells in a prospective epithelium, such as the early sea urchin blastula, contact each other over large areas of cell surface, the overall volume of the cell aggregate is relatively small as there is little space at its center. Middle panel: with a decrease in intercellular contact, the size of the internal space (the blastocoel), and thus the overall volume of the blastula, is greatly increased without any corresponding increase in cell number or total cell volume. Right panel: if the number of cells is increased by radial cleavage, and the packing remains the same, the blastocoel volume will increase further, again without any increase in total cell volume.

Cleavage of fertilized egg	Change in cell shape	Radial cleavage
Cleavage of fertilized egg	Same number of cells and total cell volume, larger blastocoel volume	More cells, with the same total cell volume but larger blastocoel volume

The orientation of the spindle and the resulting plane of cleavage are determined by the behavior of the centrosomes. Before mitosis, the single centrosome in the cell becomes duplicated and the daughter centrosomes move to opposite sides of the nucleus and form the asters, whose interactions with cortical microtubules will anchor and help to orient the mitotic spindle (see Fig. 5.7). The pattern of centrosome duplication and movement usually results in successive planes of cell division being at right angles to each other. An example of this is the division of the nematode AB cell, one of the pair of cells formed by the first cleavage of the zygote, where the mitotic spindle in the dividing AB cell is oriented roughly at right angles to that of the zygote. However, in the other product of the first cleavage, the P_1 cell, the centrosomes initially migrate to the same positions as in the AB cell, but the nucleus and its associated centrosomes then rotate 90° towards the anterior of the cell, so that the eventual plane of cleavage of P_1 is the same as that of the zygote (Fig. 5.9). The different planes of cleavage in the AB and P_1 cells are likely to be under the ultimate control of cytoplasmic factors, such as the PAR proteins, that are differentially distributed at the first cleavage and have effects on spindle positioning.

The poles of the mitotic spindle are positioned by a balance between forces that tend to pull the centrosomes apart and those that tend to pull them towards each other. Microtubule-associated motor proteins anchor the astral microtubules in the cortex of

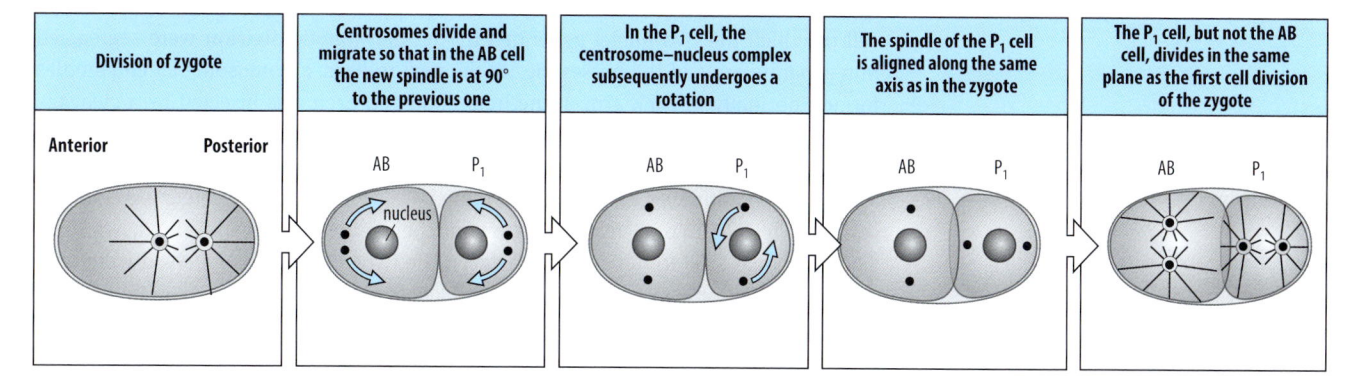

Division of zygote	Centrosomes divide and migrate so that in the AB cell the new spindle is at 90° to the previous one	In the P_1 cell, the centrosome–nucleus complex subsequently undergoes a rotation	The spindle of the P_1 cell is aligned along the same axis as in the zygote	The P_1 cell, but not the AB cell, divides in the same plane as the first cell division of the zygote
Anterior Posterior	AB P_1	AB P_1	AB P_1	AB P_1

Fig. 5.9 Different planes of cleavage in different cells are determined by the behavior of the centrosomes. The first cleavage of the nematode zygote divides it unequally into a larger anterior AB cell and a smaller posterior P_1 cell. Before the next division, the duplicated centrosomes of the AB cell move apart in such a way that the plane of cleavage will be at right angles to the first division. The duplicated centrosomes in the P_1 cell initially move into a similar alignment, but before mitosis, the nucleus and its associated centrosomes rotate 90° so that cleavage is in the same plane as the first cleavage.

*Illustration after Strome, S.: **Determination of cleavage planes.** Cell 1993, **72**: 3-6. Copyright © 1993.*

the cell and tend to pull the spindle poles towards the cortex. Similar motor proteins cross-link the overlapping spindle microtubules and either tend to pull them inwards, pulling the spindle poles towards each other, or push them outwards, pushing the spindle poles away from each other. The actomyosin-rich cell cortex, by undergoing contraction or relaxation, can also exert force on the centrosomes and the subsequent spindle poles through the astral microtubules, and thus influence the position of the spindle in the cell.

5.6 The positioning of the spindle within the cell also determines whether daughter cells will be the same or different sizes

The position of the cleavage site in relation to the axis of the cell determines whether the daughter cells are equal or unequal in size. When the mitotic spindle is positioned centrally within the cell, and each arm of the spindle is of equal length, the cleavage plane at the spindle midzone will be equidistant from the two ends of the cell, and two cells of equal size will be produced at division. If the two cells have the same contents and the same developmental potential, such a division is called a **symmetric division**. Most cell proliferation is of this type. However, depending on the location of a cytoplasmic determinant in the cell, geometrically equal divisions can sometimes lead to cells with different content, and thus with different fates. In the developmental sense, these are considered to be **asymmetric divisions**, because the result is two cells with different content and often with different fates.

If the spindle moves away from the center of the cell, and/or one of its arms becomes extended so that the structure is no longer symmetric, the spindle midzone will no longer correspond to the center of the cell's long axis and cleavage will produce daughter cells of unequal sizes. Such divisions are generally, but not necessarily, asymmetric, and they lead to cells with different developmental fates as a result of the differential distribution of cytoplasmic determinants. The position of the cleavage plane depends on the forces acting on the poles of the spindle. In an equal division, the forces are equal on both poles, and the spindle is symmetrical, whereas in an unequal division, signals from the cortex at one pole lead to a change in the position of the pole, resulting in an asymmetrically positioned spindle.

Early experiments investigating the role of the spindle in cell division used ingenious ways of manipulating dividing echinoderm eggs so that spindles were moved out of their normal position or placed in unusual configurations. In one such experiment, a glass bead on the end of a needle was inserted into a fertilized sand dollar egg so that it interrupted the first cleavage furrow. This created a horseshoe-shaped cell in which the two nuclei generated by the first mitotic division were segregated into separate arms of the horseshoe. At the next mitosis, chromosome-laden spindles formed normally in each arm of the horseshoe and each was bisected by a cleavage furrow, but a third furrow also formed between the two adjacent asters at the top of the horseshoe (Fig. 5.10).

This observation was originally interpreted as showing that the asters were able to direct the formation of a cleavage furrow independently of the spindle by delivering a signal of some kind to the cortex. However, the two centrosomes involved were later shown to have some microtubules that, unlike astral microtubules, extend in the cytoplasm towards the plane of the cleavage furrow, where they meet rather like spindle microtubules, and these could be exerting an influence.

Subsequent experiments indicated that the position of cleavage might be determined by signals from the spindle midzone—directing the formation of the contractile ring in the cortex nearest to the midzone. In one set of experiments, a physical barrier placed between the spindle midzone and the cortex was found to block cleavage-furrow formation in both sand dollar eggs and cultured mammalian cells. However, the asters might also influence contractile ring positioning indirectly. In preparation

Cleavage and formation of the blastula | **261**

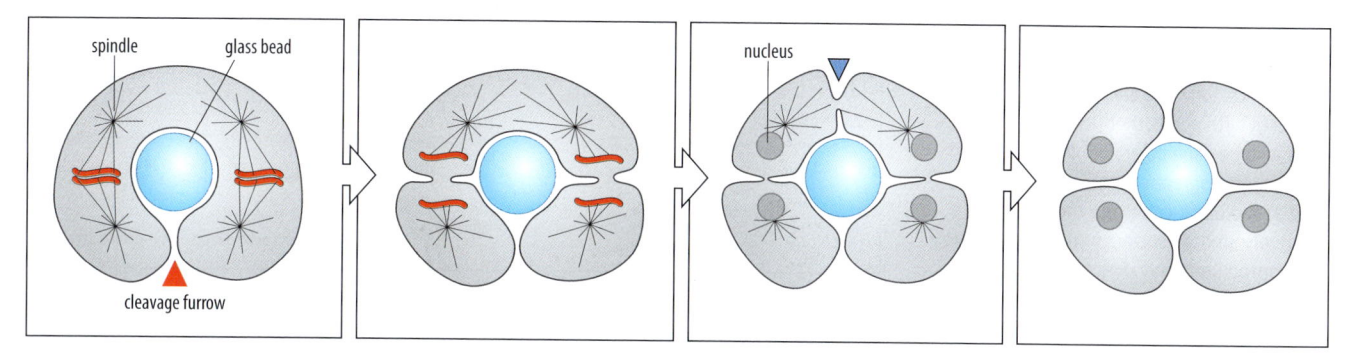

Fig. 5.10 The role of asters and spindles in determining the position of the cleavage furrow. If the mitotic apparatus of a fertilized sand dollar egg is displaced at the first cleavage by a glass bead, the cleavage furrow forms only on the side of the egg to which the mitotic apparatus has been moved. At the next cleavage, furrows bisect each mitotic spindle as expected (only a single chromosome is shown on the spindle for simplicity), but an additional furrow, indicated by the blue triangle, forms in the absence of chromosomes between the two adjacent asters of different mitotic spindles. This was originally interpreted as the asters being able to direct the formation of a cleavage furrow independently of the presence of a spindle. However, a few microtubules from each of these centrosomes do grow out into the cytoplasm like spindle microtubules (and unlike astral microtubules) and meet at the cleavage position, and could therefore have a role in defining the position of the cleavage furrow.

for mitosis, tension increases throughout the cell cortex because of cytoskeletal contraction, which causes the cell to round up. It has been suggested that the interaction of astral microtubules with the cortex at each spindle pole relaxes the cortex equally in the area around each pole, leaving a region of greater contractility midway between the two poles—in which the contractile ring will form. It is likely, therefore, that the position of the contractile ring is determined by a combination of signals from various sources acting on the cortex.

5.7 Cells become polarized in the sea urchin blastula and the mouse morula

The emergence of polarity in the cells of early embryos is an important event in development, as in some cases it is instrumental in assigning differences in cell fate by the polarized localization of cytoplasmic determinants and their consequent unequal distribution at cell division. One type of polarity found in epithelia is **apico-basal polarity**, which describes the differences between the two faces of an epithelium.

The sea urchin egg cleaves in such a way that the surface of the egg, which is covered in non-motile actin-based protrusions called microvilli, becomes the outer surface of the blastula. A layer of extracellular matrix called the **hyaline layer**, which contains the proteins hyalin and echinonectin, is secreted onto the external surface and the cells—the blastomeres—adhere to this sheet. Adhesive cell–cell junctions develop between adjacent blastomeres so that they form an epithelium, and the cell contents become polarized in an apical–basal direction, with the Golgi apparatus oriented towards the apical surface, which in this case is the outer surface of the blastula. In the sea urchin blastula the cells are held together laterally by cadherin-based adherens junctions and by **septate junctions**, a type of cell junction found in invertebrates but not in chordates. The septate junction is thought to provide the same type of permeability barrier as the vertebrate **tight junction**—sealing the epithelium and so preventing water, small molecules, and ions from permeating freely between the cells. The epithelial cells can then control, through their cell membranes, what enters and exits the blastula. On the inner (basal) surface of the epithelium, a basement membrane (an organized layer of extracellular matrix) is laid down. As the blastula develops, cilia are produced on the outer surface (Fig. 5.11).

In the mouse embryo, the earliest sign of structural differentiation is at the eight-cell stage, when the morula undergoes a process known as **compaction** (Fig. 5.12). Until

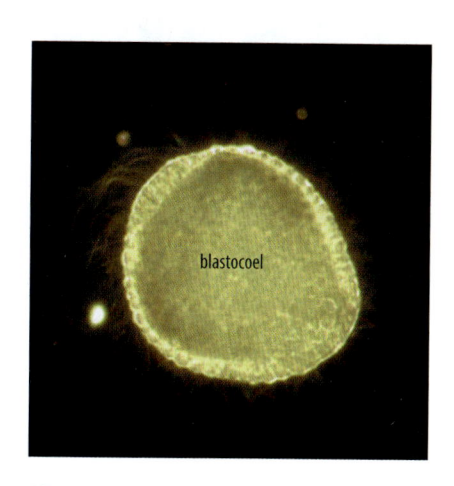

Fig. 5.11 Blastula of a sea urchin embryo, *Lytechinus pictus*. A single-layered epithelium surrounds the hollow blastocoel. The epithelial cells bear cilia on their outer surface. Scale bar = 10 μm.

From Raff, E.C., et al.: **Experimental taphonomy shows the feasibility of fossil embryos.** *Proc. Natl. Acad. Sci. USA 2006, 103: 5846–5851. Copyright 2006 National Academy of Sciences, USA.*

Fig. 5.12 Compaction of the mouse embryo. At the eight-cell stage, compaction takes place, with the cells forming extensive contacts with each other (top panel). The cells also become polarized. The scanning electron micrographs (bottom panels) show that at the eight-cell stage the cells have relatively smooth surfaces and microvilli are distributed uniformly over the surface. At compaction, microvilli are confined to the outer surface and cells have increased their area of contact with one another. Scale bar in bottom panel = 10 μm.

*Photograph courtesy of T. Bloom reproduced with permission from Bloom T.L.: **The effects of phorbol ester on mouse blastomeres: a role for protein kinase C in compaction?** Development 1989, **106**: 159-171.*

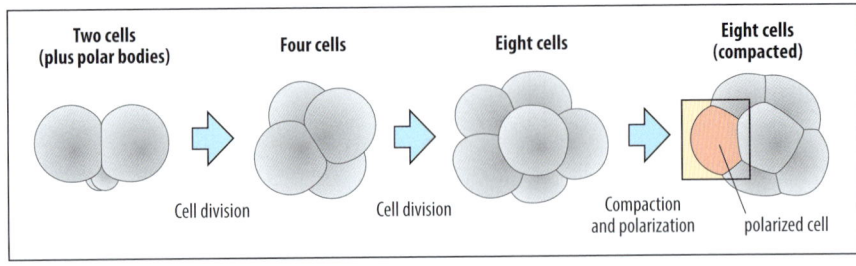

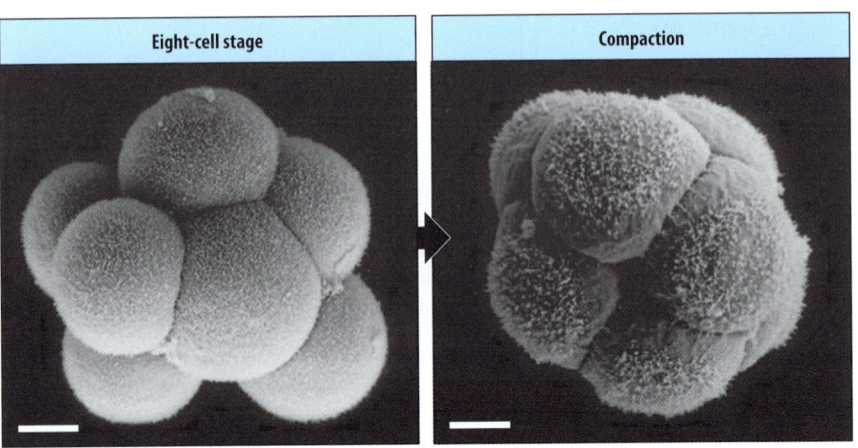

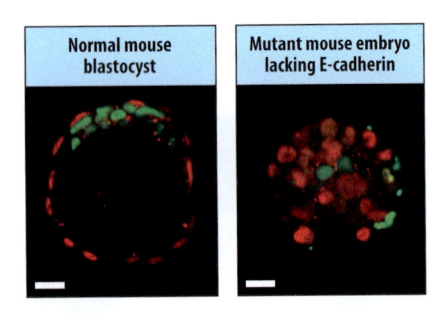

Fig. 5.13 Mouse embryos lacking E-cadherin do not develop into a blastocyst. The first panel shows a normal mouse blastocyst. The inner cell mass (green, cells expressing the pluripotency factor Oct4) lies at one end of a fluid-filled space surrounded by the trophectoderm (red, cells expressing Cdx2). The second panel shows the disorganized mass of cells that develops from cleavage of a mutant zygote lacking E-cadherin. Most of the cells express the trophectodermal marker Cdx2, with few cells expressing Oct4. Scale bars, 20 μm.

*From Stephenson, R.O., et al.: **Disorganized epithelial polarity and excess trophectoderm cell fate in preimplantation embryos lacking E-cadherin**. Development 2010, **137**: 3383-3391.*

then, cells divide equally without a preferred orientation and the blastomeres form a loosely packed ball, with each cell surface uniformly covered by microvilli. At compaction, blastomeres flatten against each other, maximizing cell–cell contact, and cells develop an apico-basal polarity, which is defined by the presence of E-cadherin-based adherens junctions on the lateral faces and by microvilli confined to the outer apical surface (see Fig. 5.12). Compaction is associated with a radical remodeling of the cell cortex. Microvilli disappear from the regions of the surface associated with intercellular contact while actin, the motor protein myosin, and another cytoskeletal protein, spectrin, are cleared from the cortex in these regions to become concentrated in a band in the cortex around the apical face of the polarized cell. Conversely, E-cadherin becomes localized within the adhesive contacts.

The formation of cadherin-based adherens junctions is a major feature of compaction. At the two-cell and four-cell stages, E-cadherin is uniformly distributed over blastomere surfaces and contact between cells is not extensive. At the eight-cell stage, however, E-cadherin becomes restricted to regions of intercellular contact, where it now acts for the first time as an adhesion molecule and forms adherens junctions (see Box 5A).

The transition from 8 to 16 cells is most important in mammalian development. This round of division generates two cell populations with different developmental fates. Outer polarized cells with epithelial characteristics express *Cdx2* and form the trophectoderm that will give rise to the placenta. By contrast, a group of inner non-polarized cells expresses the pluripotency markers Nanog and Oct4 and forms the inner cell mass, which will give rise to the embryo proper and the yolk sac (see Section 3.23). The formation of the outer epithelial layer of trophectoderm cells linked by E-cadherin is essential for further development; in embryos that lack E-cadherin it does not form, resulting in a disorganized mass of cells with abnormal ratios of cells expressing the pluripotency factor Oct4 or the Cdx2 protein (Fig. 5.13).

The way these two populations emerge highlights the importance of mechanical forces in morphogenesis. At the eight-cell stage, each cell divides along one of two possible planes: radial cleavages (perpendicular to the embryo's surface) produce two polarized cells that will stay on the outside (symmetric divisions), but some cleavages now occur tangentially (parallel to the embryo's surface), each producing one outer polarized cell and one inner non-polarized cell (asymmetric divisions). For some time it was thought that the orientation of cell divisions is the way the trophectoderm and inner cell populations emerged; however, it has become clear that radial divisions can give rise to cells of the inner cell mass. It is now thought that the decision to be inner or outer is initially determined by differences in surface tension. Cells inheriting the apical domain of a compacted blastomere (most often both the cells generated by radial cleavages and the outer cells generated by tangential cleavages) have a lower surface tension than cells that do not inherit the apical domain (usually the inner cells generated by tangential cleavages). The differences in surface tension result from the activity of actomyosin networks within the cell cortex (see Box 5A), with the cortex in the apical domain of a polarized blastomere at the eight-cell stage being less contractile than the cortex in the non-apical domain (Fig. 5.14). The cells with high surface tension move inside the embryo to interact with cells in a similar state (see Section 5.1). Cells with lower surface tension remain on the outside, where the adherens junctions stabilize the outer epithelium. A proof of this notion is that cells lacking myosin activity, and therefore having low cortical contractility, tend to stay on the outside and become trophectoderm.

One important consequence of differences in surface tension is their effect on the **Hippo pathway**. This signaling pathway (see Box 11A) acts as a mechanosensing system that transmits the effects of forces to the cell's transcriptional machinery. High surface tension activates the intracellular signaling protein Hippo and results in the inactivation of the transcriptional effector Yap/Taz. The most important consequence of this is that *Cdx2* expression is not maintained in the cells of the inner cell mass

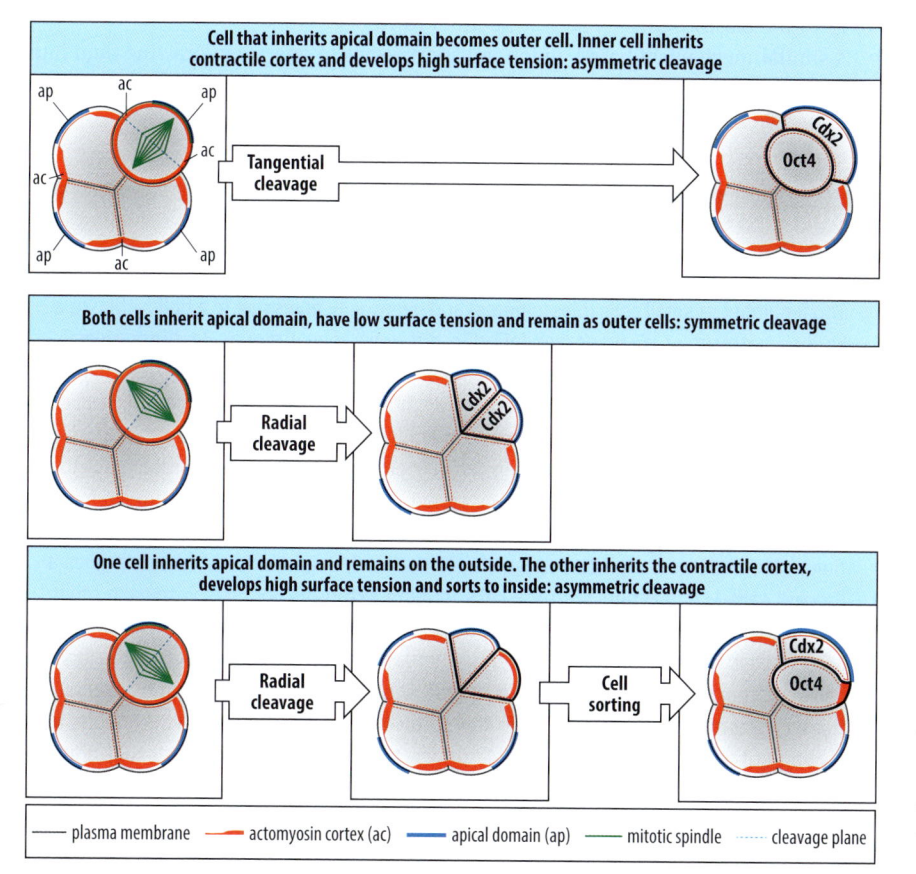

Fig. 5.14 Surface tension determines the location of cells within the 16-cell mouse embryo. Cell divisions from the 8- to the 16-cell stage can be tangential (top panel) or radial (center and bottom panels). The divisions can also be asymmetric (top panel) or symmetric (center and bottom panels) regarding the inheritance of the apical domain. Because the apical domain is less contractile than the non-apical cortex, cells inheriting it will have a lower surface tension and will stay on the surface. Top panel: a tangential cleavage results in the cell that does not inherit the apical domain moving into the interior of the embryo. Center panel: a symmetric radial cleavage will result in two cells that share an apical domain and stay at the surface of the embryo. Bottom panel: an asymmetric radial cleavage results in just one of the cells inheriting the apical domain and staying at the surface. The higher surface tension of the other cell results in its moving into the interior of the embryo.

Adapted from Maître, J.L.: **Mechanics of blastocyst morphogenesis**. *Biol. Cell 2017, 109: 323-338.*

and expression of the pluripotency genes is maintained instead (see Section 3.23). Because the Hippo pathway is not active in the outer cells, which have low surface tension, expression of *Cdx2*, which specifies trophectoderm, is maintained.

5.8 Fluid accumulation as a result of tight-junction formation, ion transport, and hydraulic fracturing forms the blastocoel of the mammalian blastocyst

Accumulation of fluid in the interior space, or **blastocoel**, of a blastocyst or blastula exerts an outward pressure on the blastula wall, and this hydrostatic pressure is one of the forces that forms and maintains the spherical shape. The blastocoel is formed in the preimplantation mammalian embryo by a cavitation process in which fluid accumulates in the interior of the developing blastocyst, separating most of the epithelial trophectoderm from the remaining mass of cells, which forms the inner cell mass. The inflow of water is linked to the active transport of sodium ions and other electrolytes into the extracellular spaces.

At the eight-cell stage in the mouse, tight junctions (Fig. 5.15) start to be formed between the polarized cells of the outer layer of the morula. By the 32-cell stage this barrier is fully formed. At the same time, the Na+/K+-ATPase sodium pump and other membrane transport proteins become active in the basolateral membranes of the outer cells, transporting sodium and other electrolytes into the extracellular spaces, from which the blastocoel will form. As the ion concentration in the blastocoel fluid increases, water is drawn into the blastocoel by osmosis through aquaporin water channels in the cell membranes (Fig. 5.16).

More recent work has shown that the formation of the blastocoel starts with the formation of hundreds of microdroplets within the basolateral extracellular spaces between the cells of the morula and, as the fluid keeps accumulating, the pressure generated breaks cell–cell contacts and eventually drives the fluid into a single lumen to form the blastocoel (see Fig. 5.16, bottom row).

A similar mechanism based on ion transport seems to operate in drawing fluid into the blastocoel in the *Xenopus* blastula. In the case of *Xenopus*, a small blastocoel can be observed by electron microscopy as early as the two-cell stage.

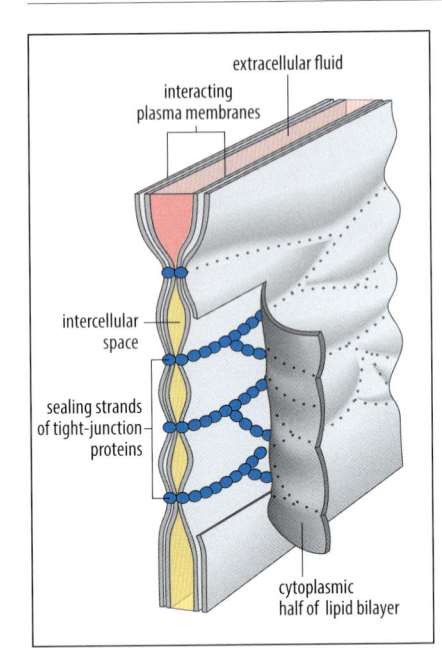

Fig. 5.15 The structure of the tight junction. The apposing membranes of adjacent epithelial cells are held closely together by the tight junction, which is located towards the apex of the cell and forms a tight seal between the extracellular and intracellular environment, preventing water, small molecules, and ions from permeating freely between the cells. The membranes are held together by horizontal rows (sealing strands) of membrane proteins (occludins and claudins), which bind tightly to each other. The pink shading represents the space outside the epithelium; the yellow shading represents the spaces between cells inside the epithelium.

'Figure 19-26a', from Molecular Biology of the Cell, *Fifth Edition by Bruce Alberts, et al. Copyright © 2008, 2002 by Bruce Alberts, Alexander Johnson, Julian Lewis, Martin Raff, Keith Roberts, and Peter Walter. Used by permission of W. W. Norton & Company, Inc.*

SUMMARY

In many animals, the fertilized egg undergoes a cleavage stage that divides the egg into a number of small cells (blastomeres) and eventually gives rise to a hollow blastula. Various patterns of cleavage are found in different animal groups. In some animals, such as nematodes and mollusks, the plane of cleavage is of importance in determining the position of particular blastomeres in the embryo, and the distribution of cytoplasmic determinants. The plane of cleavage in animal cells is determined by the orientation of the mitotic spindle, which is, in turn, determined by the final positioning of the centrosomes. At the end of the cleavage period, the spherical blastulas of, for example, sea urchins and *Xenopus*, essentially consist of a polarized epithelium surrounding a fluid-filled blastocoel. During cleavage in mammals, the eight-cell embryo undergoes a process called compaction, with the cells forming extensive contacts with each other, and the outer layer of cells becoming polarized in the apicobasal direction. The outer layer of cells will become trophectoderm, whereas the inner cells will become the inner cell mass that gives rise to the embryo proper. The allocation of cells to the two populations is determined by differences in the surface tension of the cells, which leads to sorting out, with cells having a higher surface tension ending up on the inside. Fluid accumulation in the center of the mammalian blastocyst to form the blastocoel is due to active ion transport into the blastocoel, drawing water in after it by osmosis.

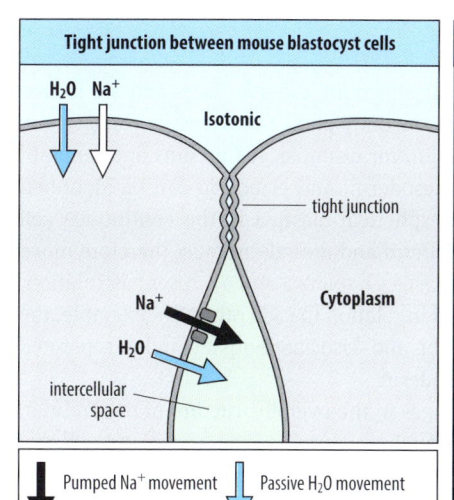

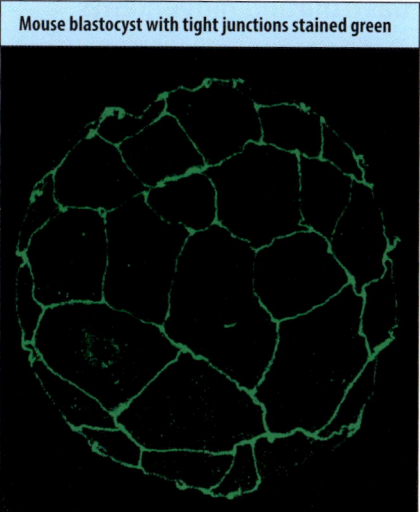

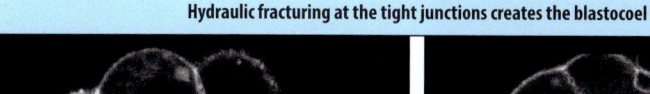

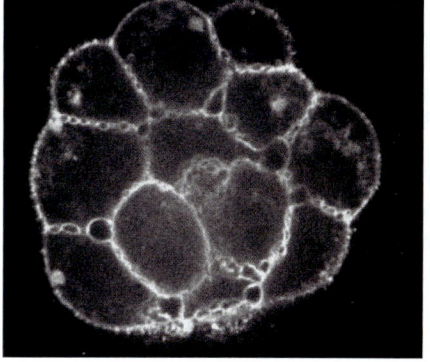

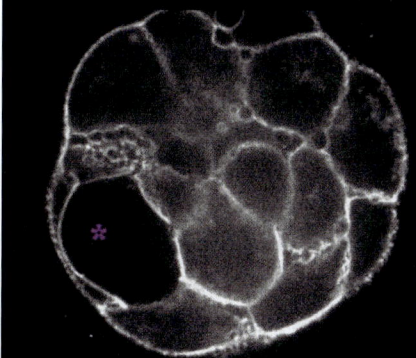

Cleavage and formation of the blastula 265

Fig. 5.16 The blastocoel of the mammalian blastocyst is formed by the inflow of water, creating an internal fluid-filled cavity. Top left panel: active transport of sodium ions (Na+) across the basolateral membranes of trophectoderm cells into the extracellular space, which will be continuous with the blastocoel cavity of the mouse blastocyst (yellow arrow), results in an increase in ion concentration in the extracellular space, which causes water to flow in from the cytoplasm by osmosis. The sodium in the trophectoderm cells is replaced by sodium from the isotonic fluid surrounding the blastocyst (white arrow) and water also enters the cell by osmosis. The blastocoel is sealed off from the outside by tight junctions between the trophectoderm cells at their outer edge, and so the inflow of water creates hydrostatic pressure that enlarges the blastocoel. Top right panel: a surface view of a mouse blastocyst shows the tight junctions between epithelial cells. Bottom row: the photograph on the left shows the formation of microdroplets a few micrometers across between the basolateral faces of the cells in the mouse morula, formed under pressure as water is drawn into the morula. The photograph on the right shows the formation of a lumen (starred) as the microdroplets are driven to coalesce. This will enlarge to become the blastocoel.

Photograph in top right panel from Eckert, J.J., Fleming, T.P.: **Tight junction biogenesis during early development.** *Biochim. Biophys. Acta 2008,* **1778**: *717-728. Photographs in bottom row from a movie in Dumortier, J.G., et al.:* **Hydraulic fracturing and active coarsening position the lumen of the mouse blastocyst.** *Science 2019,* **365**: *465-468.*

SUMMARY: Blastula formation

cleavage divides fertilized egg into a ball of smaller cells

Mouse → tangential and radial cleavages: inner and outer cells determined by differences in surface tension → sodium ions pumped into blastocoel, inflation of blastocyst by osmosis → blastocyst comprises a single layer of trophectoderm cells enclosing inner cell mass and fluid-filled blastocoel

Sea urchin → radial cleavages only → inflation of internal volume by radial cell division → blastula comprises a single-layer epithelium enclosing the fluid-filled blastocoel

Gastrulation movements

During gastrulation, the cells from the epiblast acquire different fates and are placed into their appropriate position in relation to the body plan. The necessity for gastrulation is clear from fate maps of the blastula in, for example, sea urchins and amphibians, where the presumptive endoderm, mesoderm, and ectoderm can be identified as adjacent regions on the outside of the spherical blastula in the continuous cell sheet that encloses the blastocoel; the endoderm and mesoderm must therefore move from the surface to inside the early embryo (see Chapters 3 and 4). After gastrulation, these tissues become completely rearranged in relation to each other. For example, the endoderm develops into an internal gut tube and associated organs and is separated from the outer ectoderm by a layer of mesoderm.

Gastrulation thus involves dramatic changes in the overall structure of the embryo, converting it into a complex three-dimensional structure. During gastrulation, a program of cell activity involving changes in cell shape and adhesiveness remodels the embryo, so that the endoderm and mesoderm move inside and only ectoderm remains on the outside. The primary force for gastrulation is provided by changes in cell shape, as well as the ability of cells to migrate (see Box 5B). In some embryos, all this remodeling occurs with little or no accompanying increase in cell number or total cell mass.

In this part of the chapter, we first consider the relatively simple process of gastrulation in sea urchins and insects. *Xenopus* serves as our main model for the more complex gastrulation process in vertebrates. Because of the shape of the very early embryo, gastrulation in chick and mouse has some notable differences from that in *Xenopus*, even though the eventual body plan is very similar (discussed in Chapters 3 and 4). We will finish this part of the chapter by briefly comparing and contrasting the types of gastrulation movements in chick and mouse with those in *Xenopus*.

5.9 Gastrulation in the sea urchin involves an epithelial-to-mesenchymal transition, cell migration, and invagination of the blastula wall

Just before gastrulation begins, the sea urchin blastula consists of a single-layered ciliated epithelium surrounding a fluid-filled blastocoel. The future mesoderm occupies the most vegetal region, with the future endoderm adjacent to it (Fig. 5.17). The rest of the embryo gives rise to ectoderm. The epithelium is polarized in an apico-basal direction: on its apical surface is the hyaline layer, whereas the basal surface facing the blastocoel is lined with a basement membrane, or basal lamina. The epithelial cells are attached laterally to each other by septate junctions and adherens junctions. Gastrulation involves various cell movements and cell behaviors that make use of the adhesive and motile properties of embryonic cells discussed in previous sections.

Gastrulation begins with an epithelial-to-mesenchymal transition (EMT), in which the most vegetal mesodermal cells leave the blastula epithelium and become motile and mesenchymal in form (Fig. 5.18). They become detached from each other and from the hyaline layer, and migrate into the blastocoel as single cells that have lost both their epithelial apico-basal polarity and their cuboidal shape. They are now known as **primary mesenchyme**. The transition to primary mesenchyme cells and entry into the blastocoel is foreshadowed by intense pulsatory activity on the inner face of these cells, while still part of the epithelium, and on occasion a small transitory infolding or invagination is seen in the surface of the blastula before migration begins properly. Cell internalization requires loss of cell–cell adhesion and is associated with repression of cadherin expression, the removal of cadherin from the cell surface by endocytosis, and the disappearance of the α- and β-catenins that link cadherins to the cytoskeleton (see Box 5A).

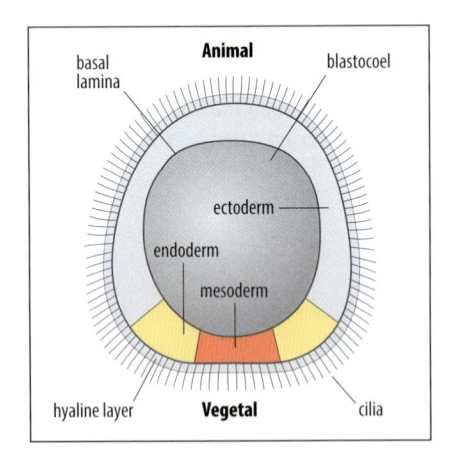

Fig. 5.17 The sea urchin blastula before gastrulation. The prospective endoderm (yellow) and mesoderm (orange) are at the vegetal pole. The rest of the blastula is prospective ectoderm (gray). There is an extracellular hyaline layer and a basal lamina lines the blastocoel. The blastula surface is covered with cilia.

Gastrulation movements   **267**

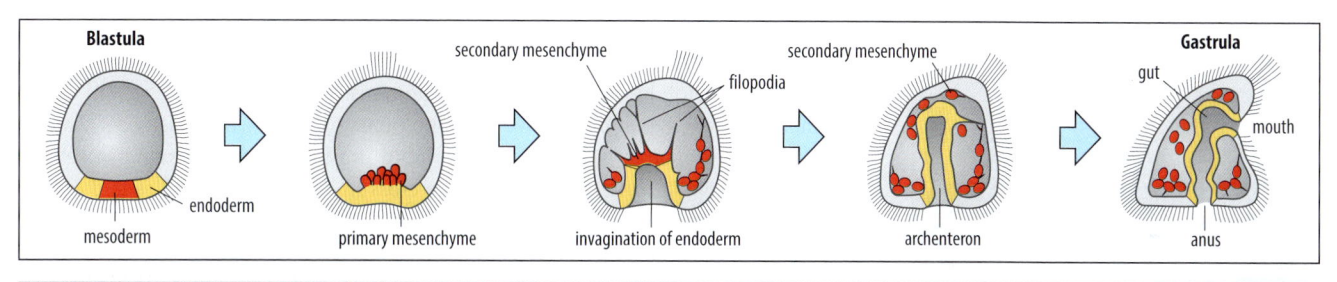

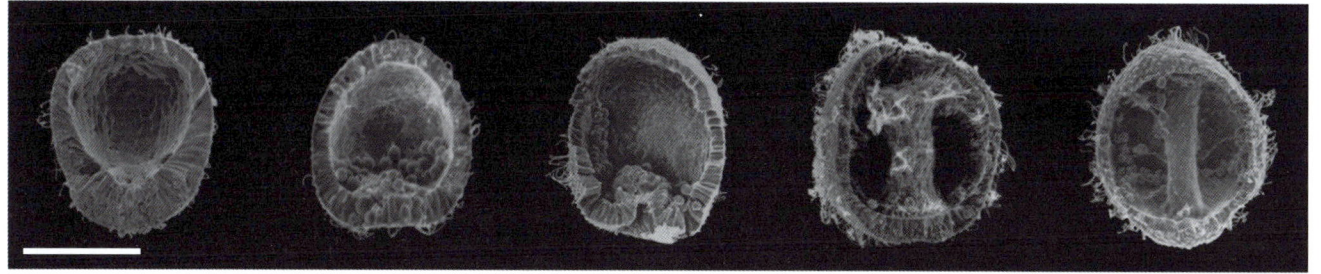

Fig. 5.18 Sea urchin gastrulation. Cells of the vegetal mesoderm undergo the transition to primary mesenchyme cells and enter the blastocoel at the vegetal pole. This is followed by an invagination of the endoderm, which extends inside the blastocoel towards the animal pole, forming the embryonic gut, or archenteron. Filopodia extending from the secondary mesenchyme cells at the tip of the invaginating endoderm contact the blastocoel wall and draw the invagination to the site of the future mouth, with which it fuses, forming the gut. Scale bar = 50 μm.

Photographs courtesy of J. Morrill.

This EMT is regulated, in part, by the sea urchin version of the gene *snail*, which encodes a transcription factor and which, like many other developmental genes, was first identified in *Drosophila*. Experiments using antisense morpholinos to knock down *snail* expression in early sea urchin embryos show that it is required for the repression of cadherin gene expression and for cadherin endocytosis in the EMT. This results in the dissolution of the epithelium, thus enabling the EMT. The role of the Snail family of transcription factors in this type of cellular change is highly conserved across the animal kingdom, as we shall see in this chapter. The actions of Snail and a related transcription factor, Slug, are also of medical interest. A similar EMT takes place in cancer cells, enabling them to undergo **metastasis** and migrate from the original site of the tumor to set up secondary cancers elsewhere in the body. In these metastatic cells, the expression of *snail* and *slug*, which is low in non-metastatic cells, is increased.

In the sea urchin, the primary mesenchyme cells migrate within the blastocoel to form a characteristic pattern on the inner surface of the blastocoel wall. They first become arranged in a ring around the gut in the vegetal region at the ectoderm–endoderm border. Some then migrate to form two extensions towards the animal pole on the ventral (oral) side (Fig. 5.19). The migration path of individual cells varies

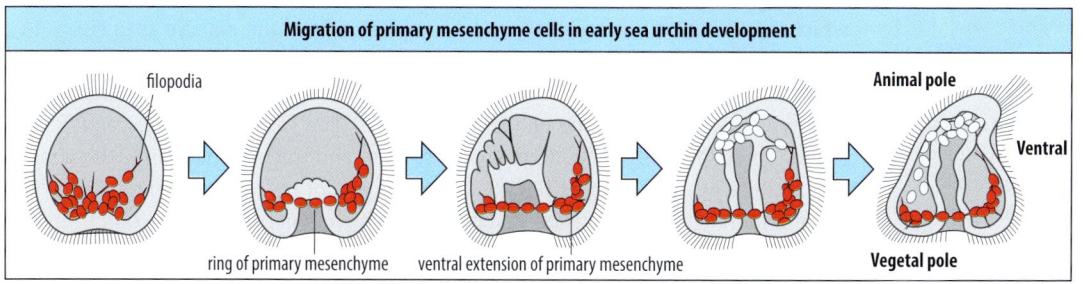

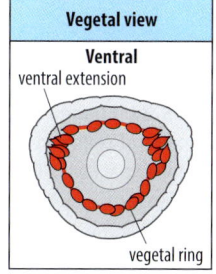

Fig. 5.19 Migration of primary mesenchyme in early sea urchin development. The primary mesenchyme cells (red) enter the blastocoel at the vegetal pole and migrate over the blastocoel wall by filopodial extension and contraction. Within a few hours they take up a well-defined ring-like pattern in the vegetal region with extensions along the ventral side.

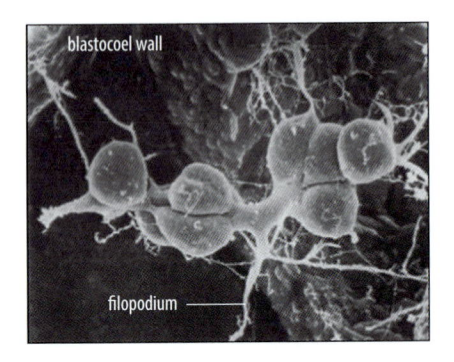

Fig. 5.20 Filopodia of sea urchin mesenchyme cells. The scanning electron micrograph shows a group of primary mesenchyme cells moving over the blastocoel wall by means of their numerous filopodia, which can extend and contract. As illustrated here, the filopodia of adjacent cells fuse together.

Photograph courtesy of J. Morrill, from Morrill, J.B., Santos, L.L.: ***A scanning electron micrographical overview of cellular and extracellular patterns during blastulation and gastrulation in the sea urchin, Lytechinus variegatus.*** *In* The Cellular and Molecular Biology of Invertebrate Development. Sawyer, R.H., Showman, R.M. (Eds). University of South Carolina Press, 1985: 3-33.

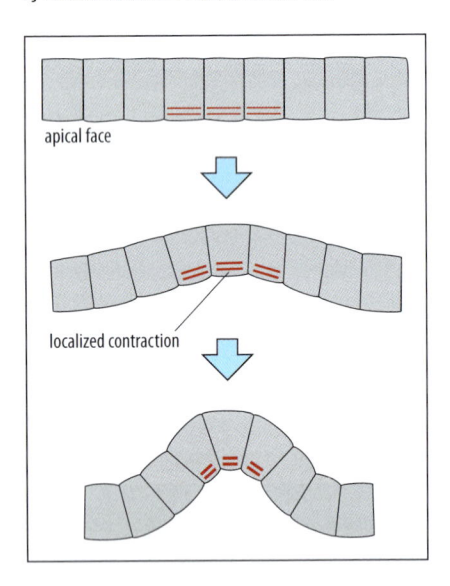

Fig. 5.21 Change in cell shape in a small number of cells can cause invagination of endodermal cells. Bundles of contractile filaments composed of actin and myosin contract at the outer edge of a small number of adjacent cells, making them wedge-shaped. As long as the cells remain mechanically linked to each other and to adjacent cells in the sheet, this local change in cell shape draws the sheet of cells inward at that point.

considerably from embryo to embryo, but their final pattern of distribution is fairly constant. The primary mesenchyme cells later lay down the skeletal rods of the sea urchin endoskeleton by secretion of matrix proteins.

Primary mesenchymal cells move over the inner surface of the blastocoel wall by means of fine filopodia, which can be up to 40 μm long and can extend in several directions. Filopodia contain actin filaments cross-linked to form bundles and extend forward by the rapid assembly of actin filaments that push out the leading edge of the filopodium (see Box 5B). At any one time, each cell has, on average, six filopodia, most of which are branched. When filopodia contact, and adhere to, the basal lamina lining the blastocoel wall, they retract, drawing the cell body towards the point of contact. Each cell extends several filopodia, some or all of which may retract on contact with the wall, and there seems to be competition between the filopodia, the cell being drawn towards that region of the wall where the filopodia make the most stable contact. The movement of the primary mesenchymal cells therefore resembles a random search for the most stable attachment. As the cells migrate, their filopodia fuse with each other, forming cable-like cellular extensions (Fig. 5.20).

If primary mesenchyme cells are artificially introduced by injection at the animal pole, they move in a directed manner to their normal positions in the vegetal region. This suggests that guidance cues, possibly graded, are distributed in the basal lamina. Even cells that have already migrated will migrate again to form a similar pattern when introduced into a younger embryo. The stability of contacts between filopodia and the blastocoel wall is one factor determining the pattern of cell migration. The most stable contacts are made in the regions where the cells finally accumulate, namely the vegetal ring and the two ventro-lateral clusters (see Fig. 5.19).

The signal proteins fibroblast growth factor A (FGF-A) and vascular epithelial growth factor (VEGF) have been implicated as guidance cues. In the blastula, FGF-A is expressed in ectodermal regions to which the primary mesenchyme will migrate and the FGF receptor is expressed on the primary mesenchyme cells. If signaling by FGF-A is inhibited, then so is cell migration and skeleton formation. VEGF is expressed by the prospective ectoderm cells at the most ventral sites to which the mesenchyme cells migrate, and the VEGF receptor is again expressed exclusively in the prospective primary mesenchyme. As with FGF, if VEGF or its receptor are lacking, the cells do not reach their correct positions and the skeleton does not form.

The entry of the primary mesenchyme is followed by the invagination and extension of the endoderm to form the embryonic gut (the archenteron). The endoderm invaginates as a continuous sheet of cells (see Fig. 5.18). Formation of the gut occurs in two phases. During the initial phase, the endoderm invaginates to form a short, squat cylinder extending up to halfway across the blastocoel. There is then a short pause, after which extension continues. In this second phase, the cells at the tip of the invaginating gut, which will later detach as the secondary mesenchyme, form long filopodia, which contact the blastocoel wall. Filopodial extension and contraction pull the elongating gut across the blastocoel until it meets and fuses with the mouth region, which forms a small invagination on the ventral side of the embryo (see Fig. 5.18). During this process the number of invaginating cells doubles, and this is, in part, due to cells around the initial site of invagination contributing to the hindgut.

How is invagination of the endoderm initiated? The simplest explanation is that a change in shape of the endodermal cells causes, and initially maintains, the change in curvature of the cell sheet (Fig. 5.21). At the site of invagination, the cuboid cells adopt a more elongated, wedge-shaped form, which is narrower at the outer (apical) face. This change in cell shape is the result of constriction of the cell at its apex by the contraction of actomyosin in the cortex (see Box 5B). Such a shape change is sufficient to pull the outer surface of the cell sheet inward and to maintain invagination, as shown by a computer simulation in which apical constriction, modeled to spread over the vegetal pole, results in an invagination (Fig. 5.22).

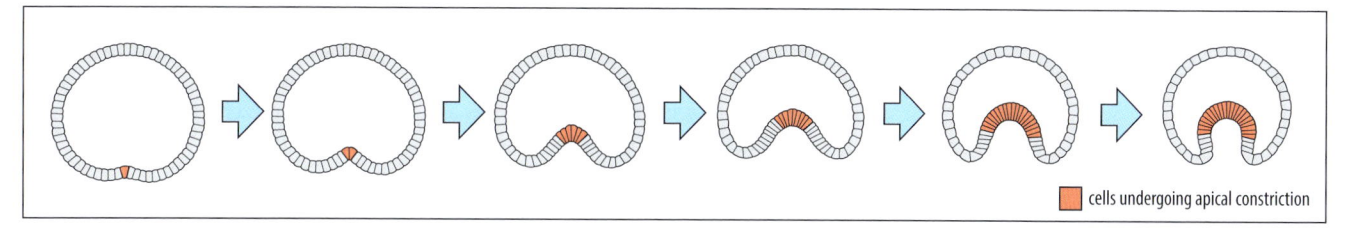

Fig. 5.22 Computer simulation of the role of apical constriction in invagination. Computer simulation of the spreading of apical constriction over a region of a cell sheet shows how this can lead to an invagination.

Illustration after Odell, G.M., et al.: **The mechanical basis of morphogenesis. I. Epithelial folding and invagination.** *Dev. Biol. 1981,* **85**: *446-462.*

The primary invagination only takes the gut about a third of the way to its final destination. The second phase of gastrulation involves two different mechanisms: contraction of filopodia in contact with the future mouth region, as described above, and a filopodia-independent process. Treatments that interfere with filopodial attachment to the blastocoel wall result in failure of the gut to elongate completely, but it still reaches about two-thirds of its complete length. Filopodia-independent extension is due mainly to directed cell movement and active rearrangement of cells within the endodermal sheet, with cells intercalating parallel to the direction of elongation. This intercalation of endodermal cells has been visualized by transplanting a macromere from a sea urchin embryo expressing fluorescent markers into a normal embryo so that a sector of cells in the vegetal endoderm is labeled. This sector of cells is seen to become a long, narrow strip as the gut extends (Fig. 5.23).

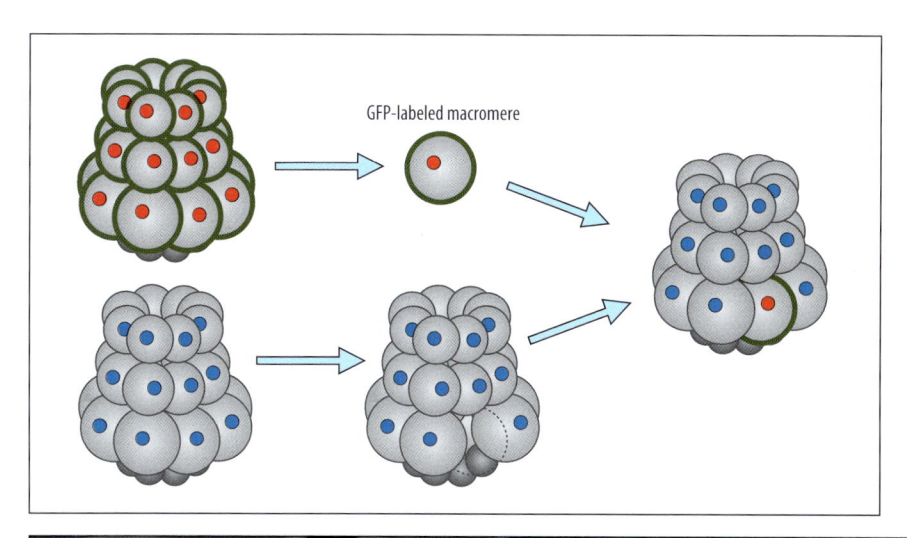

Fig. 5.23 Gut extension during sea urchin gastrulation. A single green fluorescent protein (GFP)-labeled Veg macromere from a sea urchin embryo is transplanted to an equivalent location in an unlabeled embryo. The dotted outline in the unlabeled embryo indicates where the macromere has been removed and will be replaced with the GFP-labeled macromere. The micrographs show the transplanted cell first dividing, and the progeny cells then sliding by one another parallel to the axis of extension, intercalating between each other to form a ribbon of cells that will form the gut.

From Martik, M.L., McClay, D.R.: **New insights from a high-resolution look at gastrulation in the sea urchin,** *Lytechinus variegatus.* Mech. Dev. *2017,* **148**: *3-10.*

The sheet of cells has elongated in one direction while becoming narrower in width, a process of cell rearrangement known as **convergent extension**. We will look at other examples of convergent extension in more detail later in this chapter, in relation to *Drosophila* germ-band extension and amphibian and zebrafish elongation of the body axis, although in these cases cell movement is mainly perpendicular to the direction in which the tissue will extend rather than being parallel to it.

What guides the tip of the gut to the future mouth region? As the long filopodia at the tip of the gut initially explore the blastocoel wall, they make more stable contacts at the animal pole and then in the region where the future mouth will form. Filopodia making contact there remain attached for 20–50 times longer than when attached to other sites on the blastocoel wall. Gastrulation in the sea urchin clearly shows how changes in cell shape, changes in cell adhesiveness, and cell migration all work together to cause a major change in embryonic form.

5.10 Mesoderm invagination in *Drosophila* is due to changes in cell shape controlled by genes that pattern the dorso-ventral axis

At the beginning of gastrulation, the *Drosophila* embryo consists of a blastoderm of about 6000 cells, which forms a superficial single-celled layer (see Section 2.1). Gastrulation begins with the invagination of a longitudinal strip of future mesodermal cells (9–10 cells wide) on the ventral side of the embryo, to form a ventral furrow and then a transient tube inside the body. The tube breaks up into individual cells, which spread out to form a single layer of mesoderm on the interior face of the ectoderm (Fig. 5.24). The gut develops slightly later, by invaginations of prospective endoderm near the anterior and posterior ends of the embryo, which we will not consider here.

Invagination of the mesoderm occurs in two phases. First, the central strip of cells develops flattened and smaller apical surfaces, and their nuclei move away from the apical surface. These changes in cell shape are coordinated along the antero-posterior axis and result in formation of the furrow, which then folds into the interior of the embryo to form a tube, the central cells forming the tube proper while the peripheral cells form a 'stem' (see Fig. 5.24, third panel). This first phase takes about 30 minutes. During the second phase, which takes about an hour, the tube dissociates into individual cells, which proliferate and spread out laterally. The entry of cells into cell division is delayed during invagination and this delay is essential for the cell-shape changes to occur.

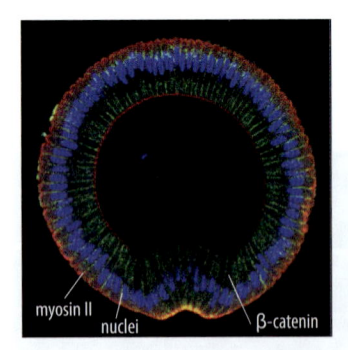

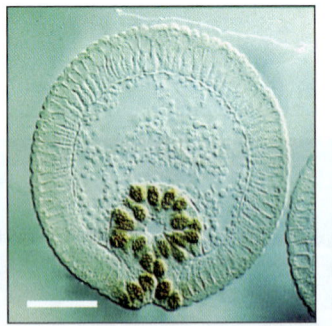

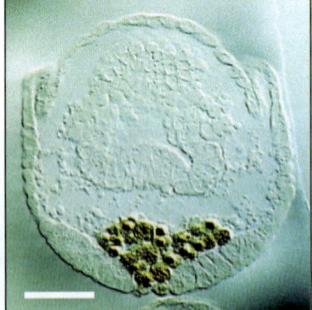

Fig. 5.24 *Drosophila* **gastrulation.** The first panel shows a transverse section through a *Drosophila* embryo in the early stages of gastrulation, showing the apical constriction of the mesodermal cells highlighted by the high levels of myosin II (red) and β-catenin (green). Nuclei are stained blue. The other three panels show the internalization of the mesoderm. The mesodermal cells (stained brown) are present in a longitudinal strip in the epithelium on the ventral side of the embryo. A change in their shape from cuboidal to wedge-shaped causes an invagination, which develops into a ventral furrow (left panel). Further apical constriction of the cells results in the mesoderm forming a tube on the inside of the embryo (center panel). The mesodermal cells then start to migrate individually to different sites (right panel). Scale bars = 50 μm.

Based on Leptin, M., et al.: **Mechanisms of early Drosophila mesoderm formation.** Development Suppl. *1992*, 23–31.

Mutant *Drosophila* embryos that have been dorsalized or ventralized show that during gastrulation each mesodermal cell behaves as a unit and would undergo apical constriction to try to form a furrow if isolated—that is, the behavior is cell-autonomous and is not affected by adjacent tissues. In dorsalized embryos, which have no mesoderm, no cells undergo nuclear migration or apical constriction. In ventralized embryos, in which most of the cells are mesoderm, these changes occur throughout the whole of the dorso-ventral axis. Apical constriction of prospective mesoderm cells alone, however, cannot account for the subsequent folding of the furrow into the interior of the embryo, and changes in the shape of other cells in the embryo produce global mechanical forces that contribute to this process and to its coordination.

Drosophila gastrulation provided the first direct link between the action of patterning genes and morphogenesis, in this case through the involvement of such genes in causing changes in cell shape and cell adhesiveness. The transcription factor genes *twist* and *snail* are expressed in the prospective mesoderm before gastrulation as part of the patterning of the dorso-ventral axis and mutations in these patterning genes abolish mesoderm invagination. *Twist* and *snail* are also associated with EMTs.

When the outlines of ventral cells in normal *Drosophila* embryos are visualized by labeling cell membranes with green fluorescent protein, apical constriction of the prospective mesoderm cells can be seen to occur. Labeling of myosin and actin showed that the constriction is due to a succession of pulsed contractions of an actomyosin network that forms under the apical surface of the cell. No contractile network forms at all in loss-of-function *snail* mutants, whereas in *twist* mutants, pulsed contractions still occur but they do not produce apical constriction. Further investigations confirmed that the Snail transcription factor is required to initiate the whole process, whereas Twist is required to stabilize the state of contraction between pulses. In the absence of Twist function, the cell surface relaxes back to its original area after each contraction.

Twist and Snail activities also control the second stage of mesoderm internalization. When the central part of the mesoderm is fully internalized, its cells make contact with the ectoderm. Adherens junctions that had been established in the blastoderm are now lost—one of the results of Snail activity—and the mesoderm undergoes an EMT, dissociating into individual cells that spread out to cover the inner side of the ectoderm as a single-cell layer. As in the sea urchin, FGF signals from the ectoderm acting on an FGF receptor on the mesodermal cells are needed for the invaginating cells to spread. Failure to activate the FGF signaling pathway results in the mesodermal cells remaining near the site of invagination. Expression of the FGF receptor in the mesoderm is induced by Twist and Snail, while Snail also represses expression of FGF in the mesoderm, thus restricting FGF to the ectoderm.

Mesodermal cell spreading involves the cells switching from expression of ectodermal E-cadherin to N-cadherin, so that the mesoderm cells no longer adhere to the ectodermal cells. The change from E- to N-cadherin is also under the control of Snail and Twist. The actions of Snail suppress the production of E-cadherin in the mesoderm, whereas expression of Twist induces that of N-cadherin.

5.11 Germ-band extension in *Drosophila* involves myosin-dependent remodeling of cell junctions and cell intercalation

Another dramatic change in shape that occurs in the *Drosophila* embryo at the time of gastrulation is the extension of the germ band (see Fig. 2.4), which leads to the thorax and abdomen of the embryo almost doubling in length.

The germ band comprises the mesoderm, ventral ectoderm (which will give rise to the nervous system), and dorsal epidermis, but not the amnioserosa. The main driver of the extension is not cell division or the elongation of individual cells, but a process of cell rearrangement in the epidermis called **convergent extension**. Convergent extension in the germ band is a highly stereotyped process in which adjacent cells pass between each other, or intercalate, in a lateral direction towards the midline of the tissue (the medio-lateral direction). As a consequence, the tissue narrows, causing it to extend in the antero-posterior direction. In the *Drosophila* embryo this remodeling occurs over the whole of the epidermis, but is most clearly seen in the ventro-lateral region. Studies at the level of single cells have shown that the remodeling is due to regulated changes in the cadherin-containing adherens junctions that normally hold the cells tightly together. These changes lead to a highly stereotyped series of alterations in cell packing within small groups of cells that spreads throughout the epidermis (Fig. 5.25).

At the beginning of germ-band extension, the epidermal cells are packed in a regular hexagonal pattern, with their boundaries either parallel to the dorso-ventral axis or at 60° to it, as shown in Figure 5.25. When extension starts, the adherens junctions on the faces parallel to the dorso-ventral axis shrink and disappear, and the cells become diamond-shaped. New junctions parallel to the antero-posterior axis appear, making the boundaries hexagonal again and intercalating the cells along the dorso-ventral axis—which makes the tissue extend along the antero-posterior axis. As this remodeling spreads throughout the epidermis, the embryo lengthens along the antero-posterior axis, concomitant with a narrowing in the dorso-ventral direction.

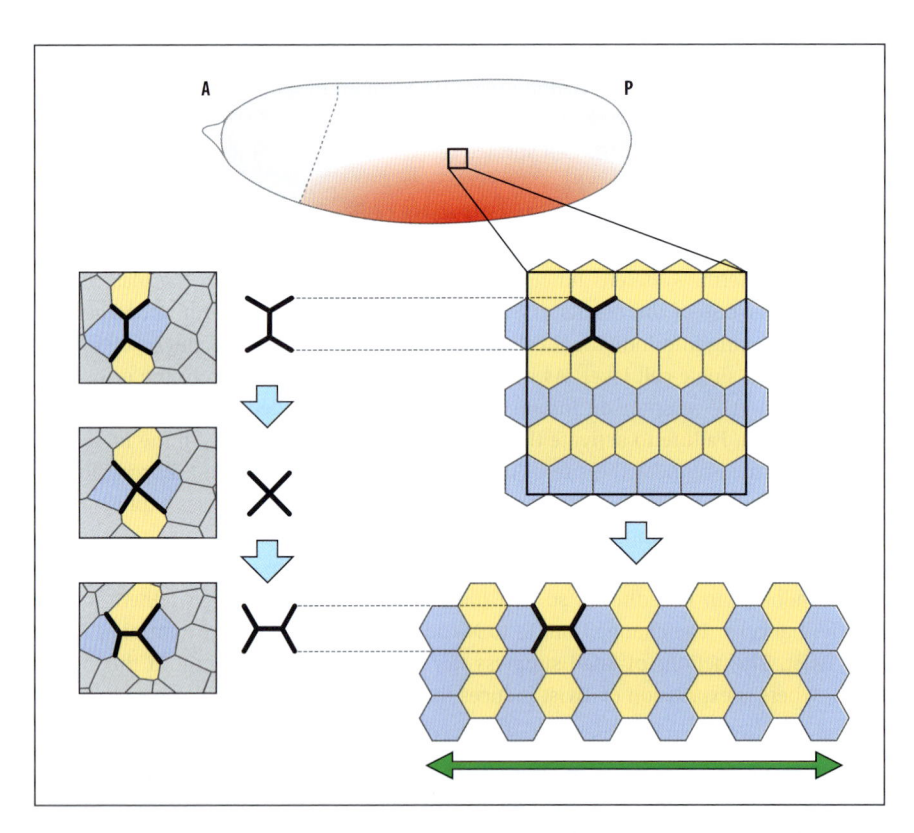

Fig. 5.25 Junctional remodeling enables cells to undergo intercalation in *Drosophila* germ-band extension. The disassembly and reassembly of intercellular contacts in the germ-band epithelium drives the tissue to elongate along the antero-posterior axis and to narrow along the dorso-ventral axis. The left-hand diagrams show how four cells in contact with each other change their intercellular contacts during junction remodeling. The right-hand diagram indicates schematically how this remodeling elongates the tissue. A, anterior; P, posterior.

Adapted from Bertet, C., et al.: **Myosin-dependent junction remodelling controls planar cell intercalation and axis elongation.** *Nature 2004, **429**: 667-671.*

The mechanism of intercalation in the germ band involves the regulated localization and activity of the motor protein myosin, which co-localizes with β-catenin–E-cadherin-actin complexes at the adherens junctions and is enriched in shrinking junctions. The proposed mechanism is that regulated actomyosin contraction at the junctions prevents E-cadherin from holding the cells together, enabling new contacts to be made that result in cell intercalation. The spread of this activity throughout the epithelium is effective because it has a global directionality—that is, the intercalations occur in the same direction in all groups of cells (see Fig. 5.25).

The basis for this directionality, and the way it is linked to the activity of the cytoskeleton is not yet clear. One obvious link is likely to be the proteins involved in planar cell polarity (discussed in Section 5.12), some of which are organized in longitudinal stripes along the extending field of cells. But how the underlying polarization of the germ-band epithelial cells along the antero-posterior axis influences the cells' behavior is not yet understood.

Genetic screens for mutations that affect directed intercalation have revealed a role for three members of the *Drosophila* Toll family of receptors (see Box 2B)—Toll-2, Toll-6, and Toll-8—in helping to set directionality. These receptors are expressed in overlapping transverse stripes, which thus create an anterior–posterior code with a clear directionality. In triple mutants for the three Toll receptors, polarized actomyosin contractility and cell intercalation are impaired, and the germ band does not extend properly.

5.12 Planar cell polarity confers directionality on a tissue

To build an embryo, cells and tissues need to have directional as well as positional information—information that determines, for example, in what direction a tissue will extend and in what direction cells will migrate when they need to. The coordinated cell rearrangements that occur in germ-band extension in the *Drosophila* embryo reflect the underlying antero-posterior pattern of the cells in the epidermal sheet which gives each cell an individual polarity. This is the phenomenon of **planar cell polarity (PCP)**, first described in epithelia, in which cells acquire a 'directionality' or orientation in the plane of the epithelium, perpendicular to their apico-basal organization. PCP represents a molecular system that provides information to the cell on its alignment relative to the body axes, and helps to ensure a coordination across fields of cells so that convergent extension and other cell movements occur in the correct direction. It is a near-universal characteristic of embryonic epithelia and can also be seen in vertebrate mesenchyme.

PCP in epithelia is most easily visible in the orientation of surface structures such as the bristles on the legs of adult *Drosophila*, the hairs on their wings and abdomen, and the hair follicles in your own skin, which consistently point in the same direction within a given area (Fig. 5.26). This reflects an underlying polarized organization of the cytoskeleton and other organelles. In most cases, all the cells in a particular epithelium, or region of an epithelium, will have the same polarity, thus giving a directionality to the epithelial sheet that can influence its further development. PCP is not restricted to epithelia, and is important in processes that require coordinated movements of large groups of cells in defined directions, as when building tissues and organs.

PCP is an essential feature of all cells, not only of visibly polarized cells. In a polarized cell, one end of the cell differs structurally, or sometimes just molecularly, from the other—for example, in the formation of filopodia or lamellipodia, the asymmetric distribution of the cytoskeleton, or the location of surface proteins. PCP was first investigated in *Drosophila* in relation to the orientation of the cuticular hairs on the cells of the wing and the abdomen, but is found in all animals and in different manifestations. For example, it is involved in the outgrowth of vertebrate limb buds; in the correct

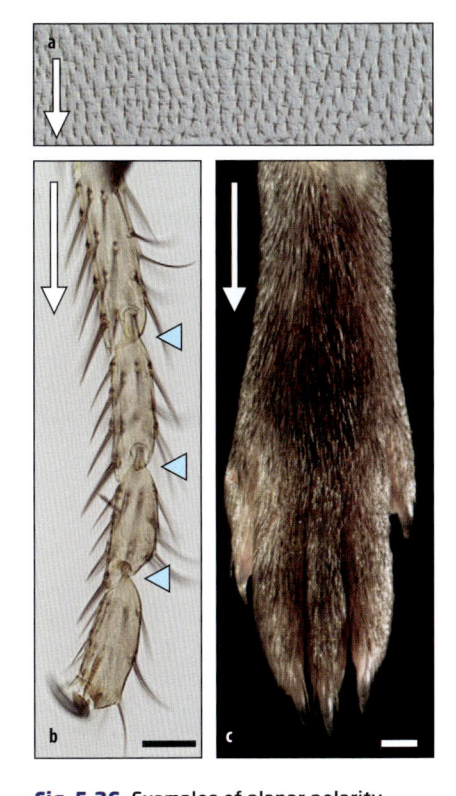

Fig. 5.26 Examples of planar polarity.
(a) Hairs on adult *Drosophila* abdominal epidermis. (b) Bristles and joints (blue triangles) on a *Drosophila* leg. (c) Fur on a mouse paw.

(a) Courtesy Peter Lawrence. (b, c) From Goodrich, L.V., Strutt, D.: **Principles of planar polarity in animal development.** *Development 2011,* **138***: 1877-1892.*

274 Chapter 5 Morphogenesis: change in form in the early embryo

PCP proteins in *Drosophila* and vertebrates		
Drosophila	**Mammals**	**Type of protein**
Core PCP module		
Frizzled 1 (Fz)	Frizzled 3, 6 and 7	Transmembrane (TM) receptor protein
Starry night (Stan)/ Flamingo (Fmi)	Cadherin EGF Lag Seven-pass G-type Receptor (Celsr) 1, 2, 3	TM protein, atypical cadherin
Van Gogh (Vang)/ Strabismus (Stbm)	Van Gogh-like (Vangl) 1,2	TM protein
Prickle (Pk)	Prickle-like (Pkl) 1, 2	Intracellular protein
Dishevelled (Dsh)	Dishevelled-like (Dvl) 1, 2 and 3	Intracellular protein
Diego (Dgo)	Inversin	Intracellular protein
FtDsFj module		
Dachsous (Ds)	Dachsous-like (Dsl)	Atypical cadherin
Fat (Ft)	Fat-j	Atypical cadherin
Four-jointed (Fj)	Fjx	TM protein

Fig. 5.27 *Drosophila* **and vertebrate PCP proteins.**

placement of uterine crypts in the mouse uterus, which is required for fertility; in the ciliated epithelia of hair cells involved in hearing and balance in the inner ear, where PCP aligns the cilia to point in a given orientation; and in ciliated epithelia generally. In this chapter we shall also see PCP at work in the process of convergent extension in vertebrates that underlies body-axis extension during gastrulation, the formation and extension of the notochord, and the narrowing and elongation of the neural plate that is a prerequisite for neural tube closure. The establishment of PCP is also required for the collective migrations of cells that form the lateral line in fish. How the coordination of cell polarity in PCP is achieved in tissues made up of hundreds or thousands of cells is a fascinating developmental problem that is still far from being resolved.

The establishment of PCP in any tissue depends on the prior establishment of the directional axes of the body and the initial patterning along these axes. In the case of the *Drosophila* embryo, for example, planar polarity in the epidermis, and germ-band extension, are impaired in embryos mutant for genes such as the gap gene *Krüppel* or the pair-rule gene *eve*, which generate antero-posterior pattern. This suggests that these transcription factors play a role in setting up PCP for germ band extension.

Genetic analysis in *Drosophila* has identified a set of genes responsible for the PCP underlying abdominal and wing hair patterns, and has enabled vertebrate counterparts of these genes and proteins to be identified (Fig. 5.27). These proteins fall into two functional groups. One is a 'core PCP module' associated with the polarity of individual cells and local signaling that propagates an identical polarity from cell to cell across a tissue. The other is the 'FtDsFj module' (for the atypical cadherins Fat and Dachsous and the membrane protein kinase Four-jointed), which is thought to provide a parallel local polarity system independent of the core PCP system in at least some cases of PCP, and which can also provide information that determines what the direction of polarity across the tissue will be.

The core PCP module in *Drosophila* includes three transmembrane proteins: the atypical cadherin Starry night (Stan; also known as Flamingo); the receptor protein Frizzled (Fz); and the protein Van Gogh (Vang). Cytoplasmic proteins associated with the core PCP module are Prickle (Pk), Dishevelled (Dsh), and Diego. Experiments using genetic mosaics to study PCP in *Drosophila* adult abdominal epidermis and wing established some basic principles about how the proteins of the PCP modules interact with each other to establish cell polarity. Clones of cells lacking or overexpressing the core PCP transmembrane proteins Stan, Vang, and Frizzled in various combinations can be made, and the alterations in polarity they cause in the surrounding tissue can be easily observed by changes in the direction in which the hairs point. These experiments established that Frizzled forms a complex with Stan (Fz:Stan) in the membrane of one cell, which preferentially binds, via Stan, to Stan on its own (or in a complex with Vang) in the facing membrane of a neighboring cell, forming asymmetric bridges between the cells (Fig. 5.28). Vang is not absolutely required for the establishment of PCP in this system, but has an important role in stabilizing the intercellular bridges once they are made.

On their own, these interactions would not confer polarity on the cell or tissue. An additional cue, such as a gradient of morphogen or the graded expression or activity of one of the PCP components, is required to set the direction of polarity across the tissue. On the basis of the genetic mosaic experiments, a simple model based on a graded concentration of Frizzled in response to a morphogen gradient has been proposed for the establishment of PCP in the *Drosophila* abdominal epidermis (see Fig. 5.28). The initial cue is the direction of slope of an extracellular gradient of a secreted morphogen, which is thought to be translated into an intercellular gradient of Frizzled activity across an abdominal segment, with the high point at the anterior of the segment. The polarity of the individual cells is set by the interactions between one cell and its immediate anterior and posterior neighbors. As Frizzled concentration is graded from anterior to posterior, any given cell will have more Frizzled than

Fig. 5.28 The core PCP proteins make asymmetric bridges between neighboring cells. Frizzled (Fz) forms a complex with Starry night (Fz:Stan) in the membrane of one cell, which preferentially binds, via Stan, to Stan on its own (or in a complex with Van Gogh (Vang)) in the facing membrane of a neighboring cell, forming asymmetric bridges between the cells. One model of how this arrangement can be converted into information for planar polarity in a particular direction proposes that Fz activity is graded in that direction, in response to an extracellular gradient of a morphogen, for example. The difference in Fz activity between neighboring cells leads to a graded formation of intercellular bridges between neighboring cells, which gives signals for the direction of polarity (as shown by the orientation of the hairs on the cells).

Lawrence, P.A., Casal, J.: **Planar cell polarity: two genetic systems use one mechanism to read gradients.** Development 2018, **145**: dev168229.

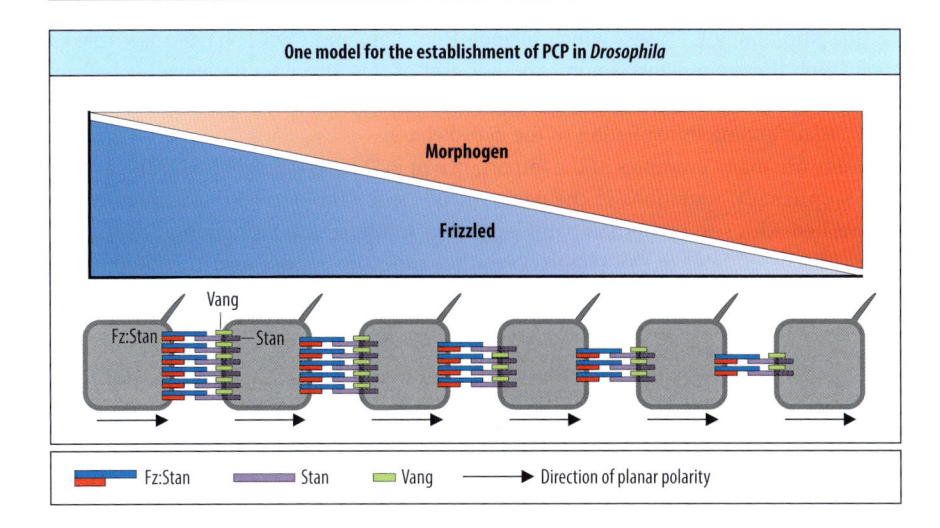

its immediate posterior neighbor, and so will make most of its Fz:Stan–Stan bridges with its posterior neighbor. In turn, that cell will make most Fz:Stan–Stan bridges with its posterior neighbor, and so the direction of polarity is set across the tissue. For example, if the gene for Fz is turned off in a small clone of epidermal cells in a wild-type background, the unaffected cells immediately posterior to the clone reverse polarity, so that the hairs point towards the clone, showing that polarity is influenced by intercellular communication. It is proposed that each cell reads out its own polarity by comparing the amounts and orientations of the Fz:Stan–Stan intercellular bridges on its anterior and posterior faces; how it does so is still unknown, however.

The DsFtFj system is proposed to work in essentially the same way: Daschous in the membrane of one cell binds to Fat in the opposing membrane of the adjacent cell, and Four-jointed acts to stabilize the interaction. In the *Drosophila* tissues studied, the direction of the polarity depends on the graded expression of Daschous.

A similar disposition and orientation of PCP molecules and intercellular bridges is also seen in polarized vertebrate tissues, indicating that the PCP system is highly conserved. Whether the mechanism for the establishment of PCP in vertebrates has similarities with the model outlined above is, however, a matter of debate, and there are likely to be variations in different systems. Other models have also been proposed for *Drosophila*. How the overall directionality of a tissue or a field of cells is set is not clear in most cases. Gradients of extracellular signals or graded gene expression, as in the model above, are one idea. Another is that a short-range 'domino effect' spreads the polarity sequentially across the field of cells.

The cellular outcome of PCP is cytoskeletal rearrangements that lead, for example, to changes in cell shape, directional cell motility, directed growth, or the orientation of cellular structures, depending on the tissue. The function of the PCP protein complexes is, principally, to regulate the activity of the cytoskeleton and, because the PCP complexes make physical interactions with the complexes on adjacent cells, the cytoskeletons of adjacent cells are essentially linked to each other functionally, enabling highly coordinated multicellular activity.

The link between the disposition of the transmembrane proteins of the core PCP module and the cytoskeleton is thought to be mediated by intracellular proteins associated with the transmembrane PCP proteins, in particular Dsh, which associates with Frizzled. In PCP, Dsh signals using a pathway that is quite different from the familiar Wnt signaling pathway involving β-catenin. In this alternative pathway, often called the **non-canonical Wnt pathway** or the **PCP pathway**, signaling by Frizzled via Dsh activates the Rho-family GTPase RhoA and the Rho kinase Rok2, leading to changes in the cell's actin cytoskeleton (see Box 5C). The small GTP-binding proteins known

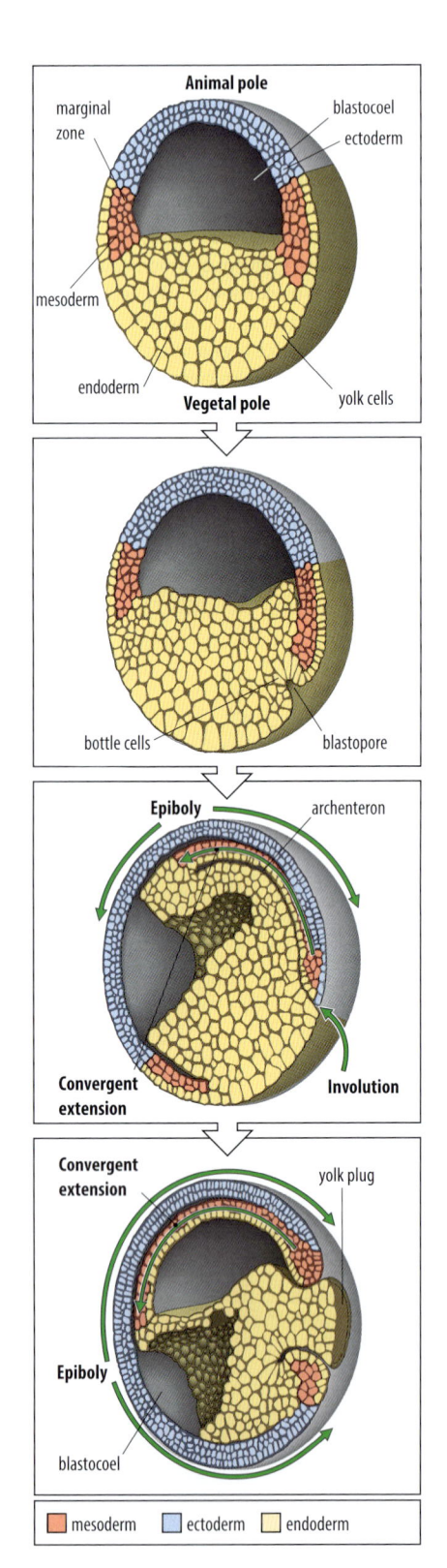

276 Chapter 5 Morphogenesis: change in form in the early embryo

as the Rho-family GTPases, which include Rho, Rac, and Cdc42, have a key role in regulating the actin cytoskeleton. Another arm of this signaling pathway leads from Dsh via mitogen-activated protein kinase (MAP kinase) and the N-terminal Jun kinase JNK, again ending in regulation of the cytoskeleton.

There are many examples of the importance of PCP in morphogenesis and, perhaps not surprisingly, mutations in the genes encoding the proteins of these modules are associated with several human pathologies. For example, the formation of the neural tube (discussed later in this chapter) requires exquisite control over the direction of cell movements and convergent extension within the neural plate, and mutations in several genes encoding PCP proteins have been associated with human neural tube defects (see Box 5E). Another example of a human syndrome associated with a failure of PCP is the very rare Robinow syndrome, a severe skeletal dysplasia characterized by short limbs and craniofacial differences, some forms of which are associated with mutations in the genes for Wnt-5A or the human versions of Dsh.

The establishment of PCP is essential for the process of convergent extension, which is involved in gastrulation and axis elongation in vertebrates, and which we consider next.

5.13 Gastrulation in amphibians and fish involves involution, epiboly, and convergent extension

Gastrulation in vertebrates involves a much more dramatic and complex rearrangement of tissues than in sea urchins, because of the numbers of cells involved and the need to produce a more complex body plan. In amphibians, fish, and birds, there is also the added complication of the presence of large amounts of yolk. But the outcome is the same: the transformation of a two-dimensional sheet of cells into a three-dimensional embryo, with ectoderm, mesoderm, and endoderm in the correct positions relative to each other for further development of body structure. We shall first consider gastrulation in amphibians and fish. The main movements of gastrulation are **involution**, which is the rolling-in of the coherent sheet of endoderm and mesoderm at the blastopore, as occurs in *Xenopus* (Fig. 5.29); **epiboly**, which is the spreading and thinning of the ectodermal sheet as the endoderm and mesoderm move inside and which can be observed in *Xenopus* and zebrafish; and convergent extension, which elongates the body axis not only in *Xenopus* and zebrafish, but also in chick and mouse embryos (Box 5C). We have already seen a simple example of convergent extension in the cell intercalation that elongates the germ band in *Drosophila* (see Section 5.11).

In the late *Xenopus* blastula, the presumptive endoderm extends from the most vegetal region to cover the presumptive mesoderm. During gastrulation, the presumptive endoderm moves inside through the blastopore to line the gut, while all the cells in

Fig. 5.29 Tissue movements during gastrulation of *Xenopus*. First panel: the late blastula basically consists of a continuous sheet of cells surrounding the blastocoel cavity. The future mesoderm (red) is in the marginal zone, overlaid by the future endoderm (yellow). The future ectoderm is shown in blue. Second panel: gastrulation is initiated by the formation of bottle cells in the blastopore region. Third panel: this is followed by the involution of marginal zone endoderm and of mesoderm that moves inside over the dorsal lip of the blastopore. The marginal zone endoderm, which was on the surface of the blastula, now lies ventral to the mesoderm and forms the roof of the archenteron (the future gut). At the same time, the ectoderm of the animal cap spreads downward. The mesoderm converges and extends along the antero-posterior axis. Fourth panel: the region of involution spreads ventrally to include more endoderm, and forms a circle around a plug of yolky vegetal cells. The ectoderm spreads by epiboly.

Illustration after Balinsky, B.I.: An Introduction to Embryology. 4th edn. Philadelphia, PA: W.B. Saunders, 1975.

Fig. 5.30 Formation of the blastopore and gastrulation in *Xenopus*. This figure shows details of tissue movements around the blastopore in a gastrulating *Xenopus* embryo as illustrated in the first three panels of Fig. 5.29. First panel: before gastrulation, in the marginal zone, prospective endoderm (yellow) overlies the prospective mesoderm (red). Second panel: bottle cells at the site of the blastopore undergo apical constriction and elongate, causing the involution of surrounding cells and the formation of a groove that defines the dorsal lip of the blastopore. Third panel: as gastrulation proceeds, the prospective endoderm and mesoderm involute and move anteriorly under the ectoderm (blue). Fourth panel: the archenteron—the future gut lined by endoderm—starts to form and the mesoderm converges and extends along the antero-posterior axis. The ectoderm continues to move down by epiboly to cover the whole embryo.

*After Hardin, J.D., Keller, R.: **The behaviour and function of bottle cells during gastrulation of** Xenopus laevis. Development 1988, **103**: 211-230.*

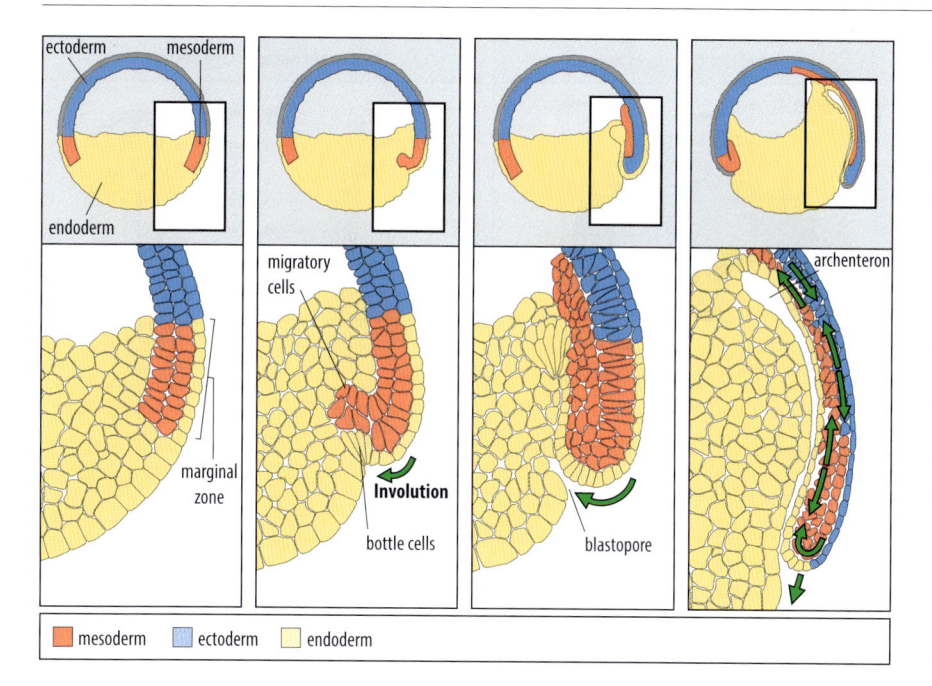

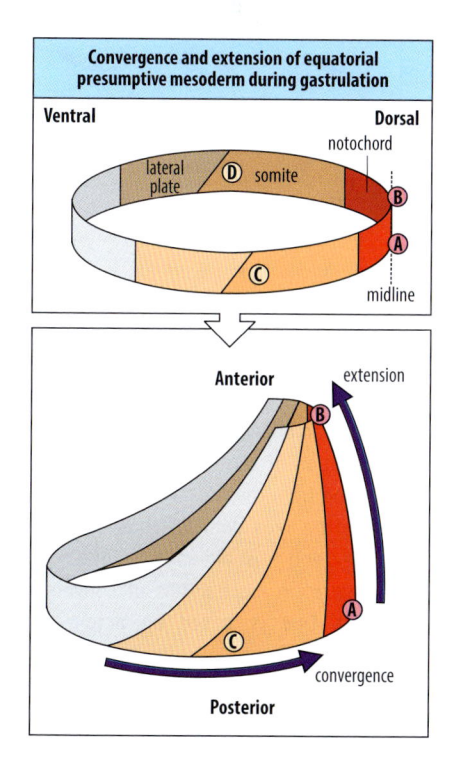

the equatorial band of mesoderm move inside to form a layer of mesoderm underlying the ectoderm, extending antero-posteriorly along the dorsal midline of the embryo (Fig. 5.30).

Gastrulation in *Xenopus* starts at a site on the dorsal side of the vegetal half of the blastula. The first visible sign is the conversion of some of the presumptive endodermal cells into bottle-shaped cells as a result of apical constriction (see Box 5B). As in the sea urchin and *Drosophila*, this change in cell shape forms a groove in the blastula surface. In *Xenopus*, this starts as a small groove—the blastopore—whose dorsal lip is the site of the Spemann organizer (see Section 3.5). The coherent sheet of mesoderm and endoderm starts to involute around the blastopore, rolling into the interior of the blastula against its own undersurface (see Fig. 5.30). As gastrulation proceeds, the region of involution spreads laterally and ventrally to involve the vegetal endoderm, and the blastopore eventually forms a circle around a plug of yolky cells. In time, the blastopore contracts, forcing the yolky cells into the interior where they form the floor of the gut.

The first mesodermal cells to move inside migrate as individual cells along the surface of the blastocoel roof to give rise to the anterior-most mesodermal structures in the head. Behind them is mesoderm, which enters with the overlying endoderm as a single multilayered sheet. For this cell sheet, going through the narrow blastopore is rather like going through a funnel, and the cells become rearranged by convergent extension, which is a major feature of gastrulation.

The mesoderm in *Xenopus* is initially in the form of an equatorial ring, but during gastrulation it converges and extends along the antero-posterior axis (Fig. 5.31). Thus, cells initially on opposite sides of the embryo come to lie very close to each other (Fig. 5.32), whereas others become separated from each other along the long axis of the embryo. The migrating mesoderm cells are polarized in the direction of the animal pole, and their migration depends on interaction with fibrils of the protein fibronectin in the

Fig. 5.31 Convergent extension of the mesoderm in *Xenopus*. The mesoderm (red and shades of orange) is initially located in an equatorial ring, but during gastrulation it converges and extends along the antero-posterior axis. A-D are reference points used to illustrate the extent of movement during convergent extension. In the bottom panel, D is hidden behind C.

Fig. 5.32 The rearrangement of mesoderm and endoderm during gastrulation in *Xenopus*. First panel: the prospective mesoderm (red and orange) is present in a ring around the blastula underlying the endoderm (yellow), which is shown peeled back. Second panel: during gastrulation, both these tissues move inside the embryo through the blastopore and completely change their shape, converging and extending along the antero-posterior axis, so that points A and B move apart. The endoderm that forms the gut is shown in the lateral view; it is hidden under the notochord and somites in dorsal view. Thus, the two points C and D, originally on opposite sides of the blastula, come to lie very close to each other. In the neural ectoderm (blue) there is also convergent extension, as shown by the movement of points E and F. Note that little cell division or cell growth has occurred in these stages, and all these changes have occurred by rearrangement of cells within the tissues.

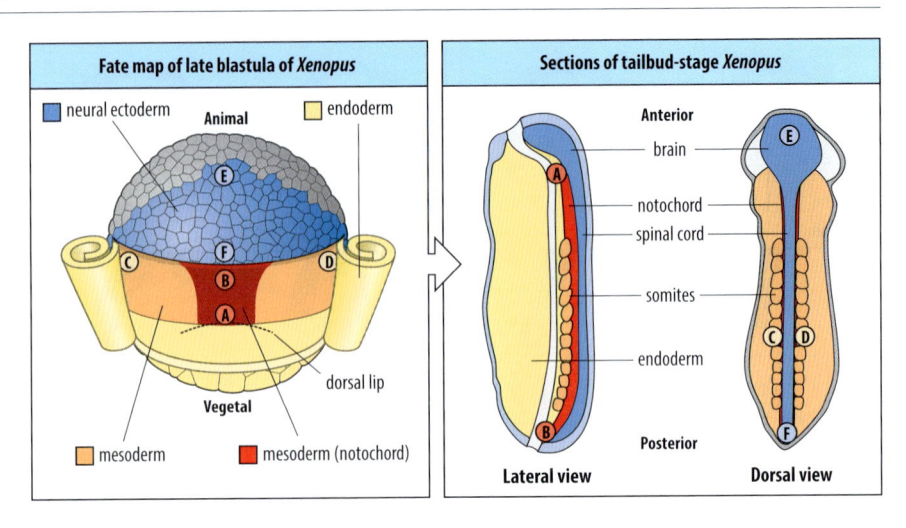

extracellular matrix lining the blastocoel roof. The assembly of the fibrils from fibronectin dimers in the matrix is associated with an increase in surface tension of the blastocoel roof epithelium. This is caused by an increase in cell–cell adhesion as a result of the formation of adherens junctions (see Box 5A) between the epithelial cells, which starts at the beginning of gastrulation. The surface tension is transmitted to the extracellular matrix via integrins in the epithelial cell membranes that interact with the fibronectin dimers and promote fibril formation.

BOX 5C Convergent extension

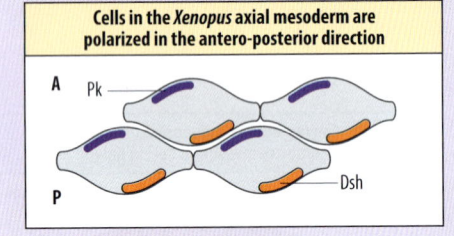

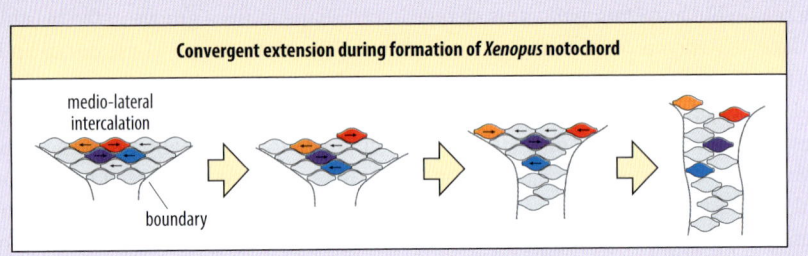

Fig. 1

Convergent extension has a key role in gastrulation and other morphogenetic processes. It is a mechanism for elongating a sheet of cells in one direction while narrowing its width, and occurs by rearrangement of cells within the sheet, rather than by long-range cell migration or cell division. In vertebrates it can be observed, for example, in the elongation of the antero-posterior axis in amphibian, zebrafish, mouse, and chick embryos, in the elongation of the notochord, and in the formation of the neural tube from the neural plate. For convergent extension to take place, the axes along which the cells will intercalate and the tissue extend must already have been defined (see Section 5.12).

Two main mechanisms of convergent extension have been found. One is based on actomyosin-driven remodeling of adherens junctions between cells, as described in Section 5.11 for the extension of the *Drosophila* germ band, an epithelial tissue. Cells remain attached to each other and intercalate between each other in the required direction by the shrinkage and reassembly of selected adherens junctions. The other mechanism of convergent extension was in fact the first to be discovered—in the developing *Xenopus* notochord, a mesenchymal tissue. It involves the cells becoming more loosely attached to each other, becoming motile, and shuffling past each other, eventually on the midline. Both types of convergent extension depend on an underlying planar cell polarity (PCP) having first been established within the cells by core PCP proteins. It was thought that convergent extension by junctional remodeling was typical of epithelia, whereas the shuffling mechanism was typical of mesenchyme, but it is becoming evident that both types can occur in both types of tissue, in both *Drosophila* and vertebrates—and even in the same tissue, as in convergent extension in the mouse neural plate (see Section 5.16).

In the developing *Xenopus* notochord (Fig. 1), PCP in the individual mesodermal cells is established in the antero-posterior direction, as seen by the localization of core PCP proteins (represented here by Prickle (Pk) and Dishevelled (Dsh), see Section 5.12). The cells become elongated in a direction at right angles to the direction in which tissue extension will occur, and become aligned parallel to each other. When extension will be along the antero-posterior axis, as in the notochord, cells become elongated in the medio-lateral direction. Cells then pass between each other medially and laterally, causing the tissue to elongate in the antero-posterior direction and narrow medio-laterally.

Active movement is largely confined to each end of these elongated bipolar cells, which form filopodia (Fig. 2), enabling them to exert traction on neighboring cells on each side and on the underlying substratum and to shuffle in between each other—to intercalate—always along the medio-lateral axis, with some cells moving medially (towards the midline) and some laterally (away from the midline). At the tissue boundary, one end of the cell is captured and movement ceases, but cells can still exert traction on each other to draw the boundaries closer together.

The boundary of a tissue undergoing convergent extension is maintained by the behavior of cells at the boundary. When an active cell tip is at the boundary of the tissue, its movement ceases and the cell becomes monopolar (see Fig. 2), with only the tip facing into the tissue still active. It is as if the cell tip that has reached the boundary has become fixed there. Precisely what defines a boundary is not clear, but boundaries between mesoderm, ectoderm, and endoderm in *Xenopus* seem to be specified before gastrulation commences. Once the boundary is fixed, the cells undergo two different processes in a coordinated manner: cell crawling and cell contraction. Cell crawling is mediated by actin-based protrusions located basolaterally, and cell contraction relies on adherens junction remodeling mediated by actomyosin contraction (see Section 5.11).

This process of **medio-lateral intercalation** extends and narrows a tissue. A mechanically similar process, but which causes thinning of a multilayered cell sheet, is called **radial intercalation**.

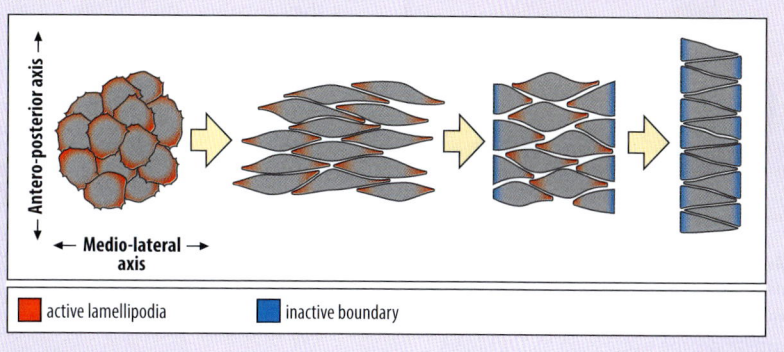

Fig. 2

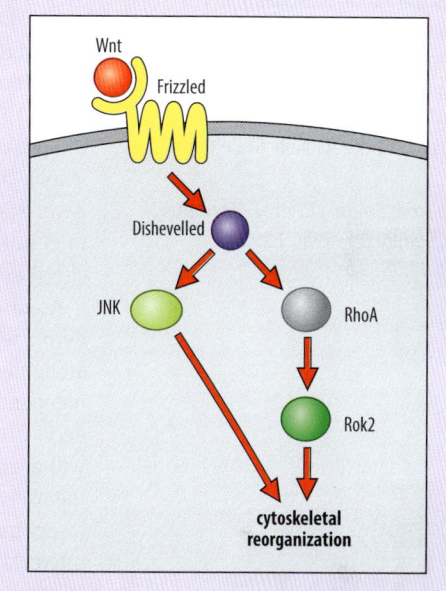

Fig. 3

As the sheet becomes thinner it extends around the edges. In the multilayered ectoderm of the animal cap of *Xenopus* and zebrafish embryos, the cells intercalate outwards in a direction perpendicular to the surface, moving from one layer into the one immediately above. This leads to a thinning of the cell sheet and an increase in its surface area, and is, in part, the cause of the epiboly of the ectoderm seen during frog and zebrafish gastrulation (see Fig. 5.34).

Convergent extension in many cases involves Wnt signaling targeting the cytoskeleton through Frizzled via the non-canonical Wnt pathway (Fig. 3), to act on the cell's cytoskeleton. Signaling via Dsh activates RhoA and the Rho kinase Rok2, leading to changes in the cell's actin cytoskeleton. Another arm of the pathway leads to the Jun N-terminal kinase, which again, in this context, is thought to mediate effects on the cytoskeleton.

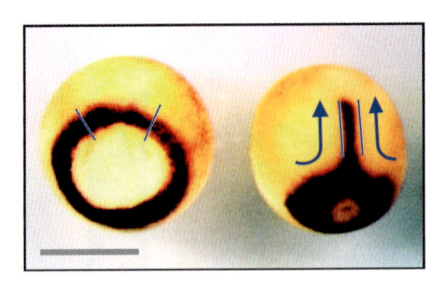

Fig. 5.33 Expression of *TbxT* (*brachyury*) during *Xenopus* gastrulation illustrates convergent extension. Left: before gastrulation, the expression of *TbxT* (dark stain) marks the future mesoderm, which is present as an equatorial ring when viewed from the vegetal pole. Right: as gastrulation proceeds, the mesoderm that will form the notochord (delimited by blue lines) converges and extends along the midline. Scale bar = 1 mm.

*Photograph from Smith, J.C., et al.: Xenopus Brachyury. Semin. Dev. Biol. 1995, **6**: 405–410. © 1995.*

In *Xenopus*, convergent extension occurs in both the mesoderm and endoderm as they involute, and in the future neural ectoderm that gives rise to the brain and the spinal cord. Together with the elongation of the notochord, all these processes elongate the embryo in an antero-posterior direction. Cells in the tissues undergoing convergent extension along the antero-posterior axis become elongated in the medio-lateral direction and develop filopodia at both tips (see Box 5C). The contractile forces generated at the tips of the elongated cells both move the cells towards the midline and extend the tissue. One can visualize the process of convergent extension in terms of lines of cells pulling on one another in a direction at right angles to the direction of extension. Because the cells at the boundary are trapped at one end, this tensile force causes the tissue to narrow and hence extend anteriorly.

The dramatic nature of convergent extension can be appreciated by viewing the early gastrula from the blastopore. The future mesoderm can be identified by the expression of the transcription factor gene *TbxT*, and can be seen as a narrow ring around the blastopore (Fig. 5.33). As the *TbxT*-expressing ring of tissue enters the gastrula, it converges into a narrow band along the dorsal midline of the embryo, extending in the antero-posterior direction, and *TbxT* expression is eventually restricted to the prospective notochordal mesoderm along the midline itself. Genetic analysis has shown that *TbxT* activates a network of genes that includes genes involved in mesoderm specification and genes encoding proteins involved in convergent extension, thus linking patterning and the behavior of the cells during gastrulation and notochord extension.

As gastrulation proceeds, the mesoderm comes to lie immediately beneath the ectoderm, while the endoderm immediately below the mesoderm lines the roof of the archenteron—the future gut (see Fig. 5.29). The mesoderm and the ectoderm move independently of each other. As the mesoderm and endoderm move inside, the ectoderm of the animal cap region spreads down over the vegetal region by epiboly and will eventually cover the whole embryo. Epiboly involves both stretching of the cells themselves and cell rearrangements, in which cells lower down in the ectodermal layers intercalate between the cells immediately above them—a process called radial intercalation (see Box 5C). Together, these processes lead to thinning of the ectoderm layer and an increase in its surface area.

Gastrulation in the zebrafish has both similarities and differences to that of *Xenopus* (see Fig. 3.32 for an outline of zebrafish gastrulation). It begins by the blastoderm cells at the animal pole starting to spread out over the yolk cell and down towards the vegetal pole. As in *Xenopus*, this epibolic spreading is due to cells of the deep layer undergoing radial intercalation, leading to the layer becoming thinner and thus increasing its surface area (Fig. 5.34, left panel). By the time the blastoderm margin reaches about half-way down the yolk cell, the deep-layer cells have accumulated around the blastoderm edge to form a thickening known as the embryonic germ ring, and have formed two cell layers—prospective ectoderm overlying mesendoderm—under the superficial enveloping layer. The mesendodermal cells now move internally at the edge of the germ ring and then move upwards under the ectoderm towards the future anterior end of the embryo (Fig. 5.34, center panel). The ectoderm continues to undergo epiboly and move downwards. The next step is convergent extension of the mesendodermal and ectodermal layers towards the dorsal midline and along the future antero-posterior axis, which elongates the embryo (Fig. 5.34, right panel).

In fish, notochord elongation initially involves directed migration of individual mesodermal cells from lateral regions towards the midline rather than medio-lateral intercalation of cells in a coherent cell sheet as in *Xenopus*. This directed migration of mesodermal cells in fish embryos is known as **dorsal convergence**, and is followed by medio-lateral intercalation as mesodermal cells become incorporated into the future notochord at the midline.

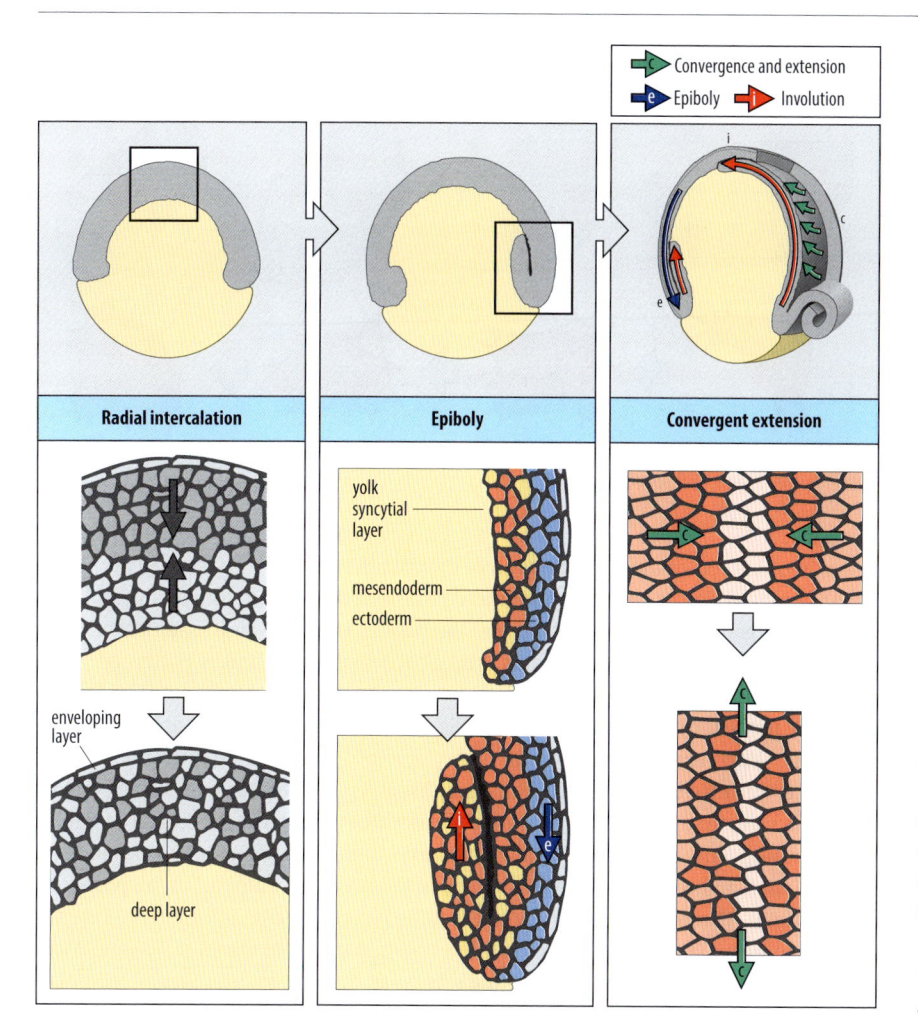

Fig. 5.34 Zebrafish gastrulation movements. Left panel: the first sign of gastrulation in the zebrafish embryo is spreading of the blastoderm over the surface of the yolk cell. This epiboly is caused by a thinning of the deep layer by a radial intercalation process. Center panel: when the blastoderm has reached about half the way down the yolk cell, separate layers of prospective mesendoderm (brown and yellow cells) and ectoderm (blue) have formed and the mesendoderm starts to internalize by involution (red arrow). Ectoderm does not internalize and will continue to move down and spread out over the yolk cell by epiboly (blue arrow). It will eventually cover the whole embryo. Right panel: convergent extension (green arrows) of the mesendoderm towards the dorsal midline results in elongation of the embryo.

*Illustration after Montero, J.A., Heisenberg, C.P.: **Gastrulation dynamics: cells move into focus.** Trends Cell Biol. 2004, **14**: 620-627.*

5.14 *Xenopus* notochord development illustrates the dependence of medio-lateral cell elongation and cell intercalation on a pre-existing antero-posterior polarity

In *Xenopus*, the prospective notochord mesoderm is among the first of the tissues to internalize during gastrulation, and the notochord is the first of the axial structures to differentiate. It will eventually form a solid rod of cells running along the dorsal midline of the embryo, where it has a structural supporting function and also provides signals for somite patterning (see Section 4.14). During gastrulation, the notochord mesoderm elongates along with the body axis by the process of medio-lateral intercalation, leading to convergent extension along the antero-posterior axis (see Box 5C). This intercalation depends on the pre-existing antero-posterior polarity set up by previous patterning processes in the notochord mesoderm.

The notochord can initially be distinguished from the adjacent somitic mesoderm by the closer packing of its cells, and by the slight gap that forms the boundary between the two tissues, possibly reflecting differences in cell adhesiveness. After an initial elongation by medio-lateral intercalation and convergent extension, the *Xenopus* notochord undergoes a further dramatic narrowing in width accompanied by an increase in height (Fig. 5.35). The cells become elongated at right angles to the antero-posterior axis of the notochord and intercalate between their neighbors again, causing further convergent extension and an increase in height of the notochord. The notochord eventually forms a rod, in which the cells have a 'pizza-slice' arrangement

Fig. 5.35 Changes in cell arrangement and cell shape during notochord development in *Xenopus*. After the notochord's initial phase of convergent extension (not shown here, see Box 5C), it undergoes further changes in shape. The cells become elongated in the medio-lateral direction and intercalate between each other in a second phase of convergent extension. This narrows the notochord further and increases its height. Later cell rearrangements result in a rod-like notochord with 'pizza-slice'-shaped cells. The top row of the figure shows the arrangement of cells in the notochord at the various stages, and the bottom row the changes in shape of an individual cell.

*After Keller, R., et al.: **Cell intercalation during notochord development in** Xenopus laevis. J. Exp. Zool. 1989, **251**: 134-154.*

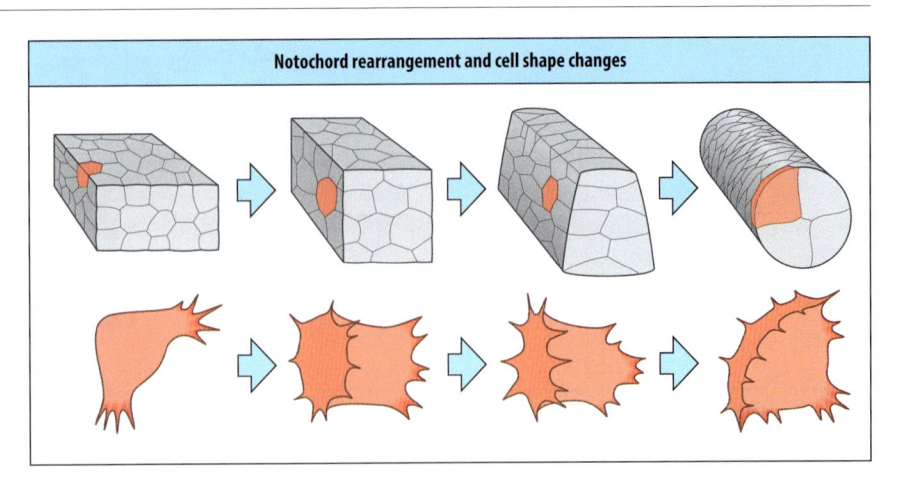

(see Fig. 5.35). A later stage in notochord development, which stiffens and further elongates this rod of mesoderm, involves directed dilation.

Antero-posterior patterning of the mesoderm is a prerequisite for cell intercalation and convergent extension. In *Xenopus*, the prospective notochord mesoderm originates from the organizer at the dorsal lip of the blastopore and is patterned along the antero-posterior axis, with the cells nearest the dorsal lip (which will form the anterior notochord) involuting before cells that will form posterior notochord (see Fig. 4.4). When explants of 'anterior' and 'posterior' prospective notochord from the organizer of a *Xenopus* early gastrula are disaggregated and the cells then mixed, the cells recognize each other by their different antero-posterior positional identities and sort out. Only combined explants of anterior and posterior tissue could undergo intercalation and convergent extension; combinations of 'all anterior' or 'all posterior' cells did not. These experiments suggest that the patterning along the antero-posterior axis is an essential prerequisite for cell intercalation and convergent extension. More recent work suggests that this patterning sets up an antero-posterior planar cell polarity (PCP) in individual notochord cells that, by inducing cytoskeletal rearrangements, drives their medio-lateral elongation (see Box 5C). When the PCP pathway is rendered non-functional, *Xenopus* mesodermal cells fail to elongate or to develop polarized lamellipodia. How cells integrate the cues they receive from antero-posterior patterning and the PCP system has still to be worked out.

5.15 Gastrulation in chick and mouse embryos involves the separation of individual cells from the epiblast and their ingression through the primitive streak

The events associated with gastrulation in frogs and fishes repeat themselves in amniotes (birds and mammals) and the endpoint is remarkably similar in all cases (see Fig. 3.2). But the different topologies and organization of chick and mouse embryos compared with those of frogs and fishes impose mechanical constraints that have led to adaptation of the movements.

Gastrulation in mammals and birds involves the formation of the structure known as the **primitive streak**, through which epiblast cells migrate into the interior of the embryo, becoming specified as mesoderm and endoderm as they are internalized (described in Chapters 3 and 4). The primitive streak can be likened functionally to the blastopore of amphibian embryos, but in the essentially flat embryos of chick and mouse, the internalization movements differ from those in the spherical, hollow *Xenopus* blastula. The cells, instead of internalizing through an orifice—the blastopore—carve a furrow, the streak. In chick, mouse, and human embryos, instead of the involution movements of mesodermal and endodermal cells as continuous cell sheets, as in fish and frog embryos, epiblast cells undergo a complete separation from the epiblast epithelium. They move through the streak and into the interior as individual cells, a process called **ingression** (Fig. 5.36).

In the chick embryo, the primitive streak is initiated from a point in the posterior region of the blastodisc as a dynamic population of cells, and streak formation progresses forward, while epiblast cells ingress behind the moving end to give rise to the endoderm and the different mesodermal derivatives in an orderly manner (see Fig. 3.39). The epiblast cells remaining on the surface become the ectoderm. Once internalized, cells migrate internally to form the mesodermal and endodermal layers, and the embryo elongates by a combination of cell proliferation and convergent extension.

Primitive-streak formation involves many of the cell behaviors and morphogenetic movements we have looked at already. Separation of cells from the epiblast and movement of epiblast cells in through the streak is achieved by a combination of processes—specifically, coordinated epithelial-to-mesenchymal transitions (EMTs) (a combination of apical constriction, loss of adherens junctions, and breakdown of the basement membrane) that start at the posterior end of the blastoderm and progress forward, creating a moving furrow. The *TbxT* gene is expressed in the streak and, as in *Xenopus* (see Fig. 5.33), identifies the ingressing cells.

From a mechanical point of view, cell flow and tissue deformation observed in the chick embryo during streak formation and elongation are related to convergent extension at the posterior of the embryo. High-resolution movies of chick epiblast show that as well as apical constriction and separation of epithelial cells at the primitive streak, convergent-extension movements in the epiblast have a role in the morphogenesis of the primitive streak, by drawing epithelial cells towards the midline.

Such a combination of cell behavior in the posterior region of the chick is likely to account for the formation and elongation of the primitive streak and for the concomitant large-scale movements observed across the entire chick epiblast. As the streak forms and elongates, bilateral and symmetrical counter-rotational movements occur throughout the epiblast (Fig. 5.37). These movements are known as 'polonaise' movements, as they resemble a formal eighteenth-century dance of that name. As with axis elongation in amphibians and zebrafish, this example of convergent extension in the chick involves the PCP signaling pathway (see Section 5.12), and inhibition of this pathway blocks streak formation.

Under the primitive streak, cells form a loose mass of mesenchyme that spreads out under the ectoderm of the epiblast to populate the whole embryo. Cells that enter the streak give rise to endoderm and mesoderm. Precursors for the various tissues come out of the streak underneath the epiblast in a defined order. Endoderm is produced early, with different mesodermal derivatives in order: cardiac, lateral, intermediate,

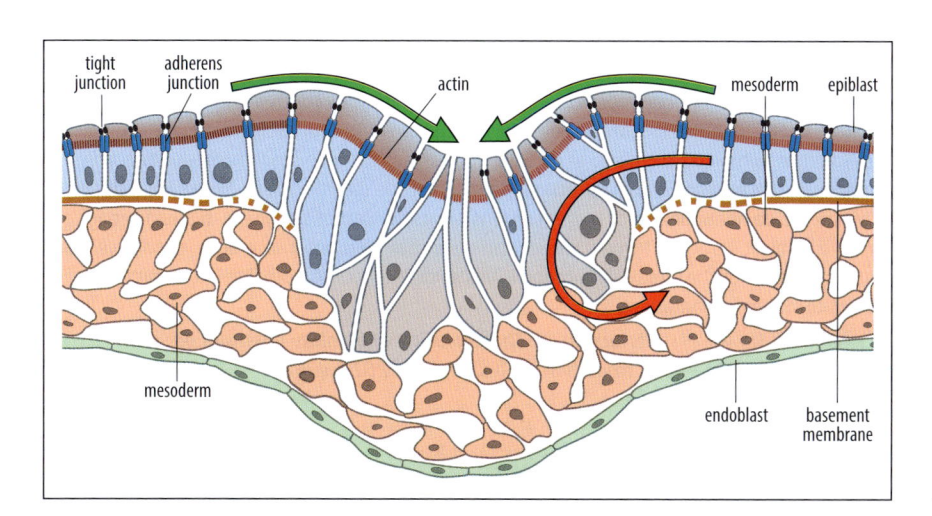

Fig. 5.36 Separation of cells from the epiblast and movement through the primitive streak involves apical constriction and an epithelial-to-mesenchymal transition. The primitive streak is a furrow in the epithelial epiblast created by apical constriction of the cells in this region. The cells become wedge-shaped as a result of the contraction of actin microfilaments that run along just under the apical surfaces of the epiblast cells. Cells in the streak undergo an epithelial-to-mesenchymal transition that involves loss of cell junctions, including adherens junctions, and the basement membrane in this region also breaks down. This allows cells to separate from the epiblast and then migrate as individuals beneath it (red arrow). The cells that ingress in this way will form the mesoderm and the endoderm. The cells that remain in the epiblast will form the ectoderm. As cells ingress through the streak, the epiblast on either side is drawn towards the streak (green arrows).
From Nayaka, Y., Sheng, G.: **An amicable separation: chick's way of doing EMT.** Cell Adh. Migr. *2009, 3: 160-163.*

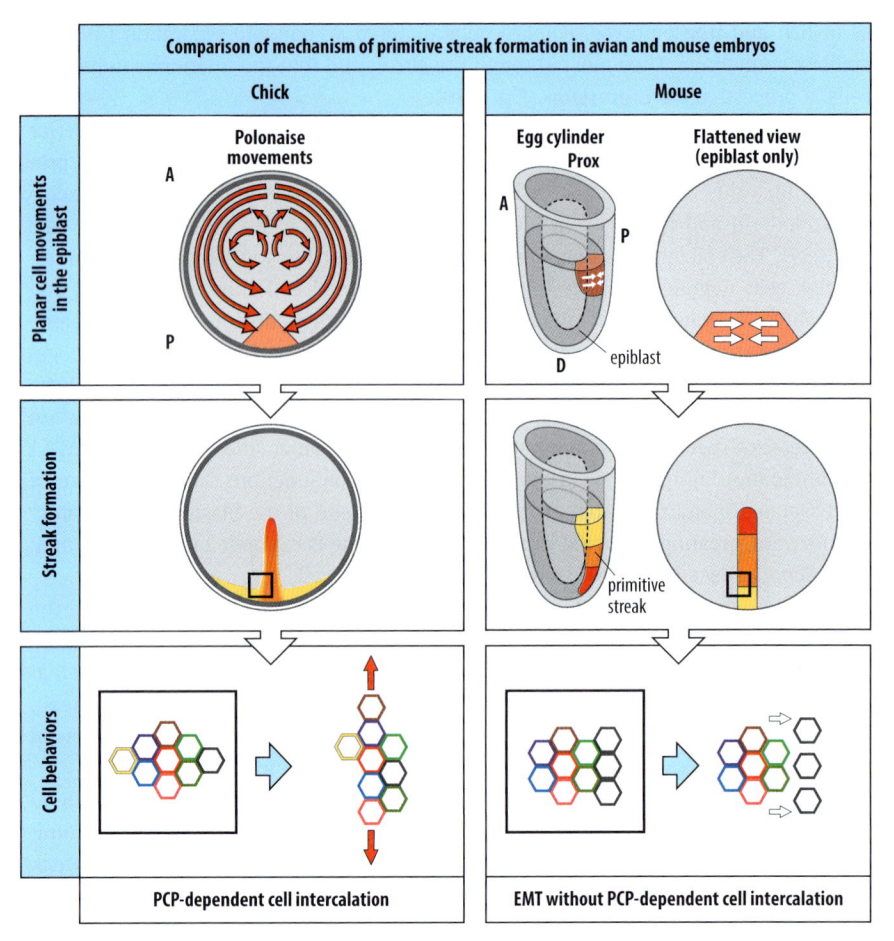

284 Chapter 5 Morphogenesis: change in form in the early embryo

Fig. 5.37 Comparison of mechanisms of streak formation in the chick and the mouse. Top left: chick embryo at the start of gastrulation. Around 6–7 hours after laying, cells in the chick epiblast are undergoing characteristic extensive 'polonaise' movements (red arrows) outwards from the midline and back towards the position at which the primitive streak will start (brown triangle). Top right: mouse embryo at the start of gastrulation (stage E6.5). The left-hand diagram shows the cup-shaped epiblast; the right-hand diagram shows the epiblast flattened out to compare with the flat chick epiblast. The start of the primitive streak on the posterior side of the embryo, at the edge of the epiblast below the extra-embryonic ectoderm, is indicated in brown. Epiblast cells do not make extensive cell movements towards the start of the streak in the mouse. Middle left: chick embryo around 12–13 hours after laying when the streak has extended about half-way. Cells in the streak (red, anterior; orange middle and posterior regions) and in the region around the site of formation (yellow) express the gene *TbxT*. Bottom left: boxed section of the diagram above. Epiblast cells undergo convergent extension as a result of planar cell polarity (PCP)-dependent cell intercalation as they approach the forming end of the primitive streak. To enter the streak, cells will undergo an epithelial-to-mesenchymal transition (EMT). Middle right: mouse embryo at E7.0. The streak has extended in the anterior direction almost to its full extent. Bottom right: boxed section of the diagram above. Unlike in chick, epiblast cells do not undergo convergent extension at the start of the streak, but just separate from the epiblast via an EMT to enter and form the streak. Red, anterior; orange, middle; yellow, posterior regions of the streak. A, anterior; P, posterior. Mouse: Prox, proximal; D, distal (with respect to site of implantation).

Adapted from Stower, M.J., Bertocchini, F.: **The evolution of amniote gastrulation: the blastopore-primitive streak transition.** *WIREs Dev. Biol. 2017, e262.*

paraxial, and, finally, axial (the notochord) (see Fig. 4.19). As in sea urchin and *Drosophila* embryos, the migration of the mesoderm cells is directed by FGFs. These signals control the motility of the ingressing cells, and in their absence the cells do not form a streak. This requirement for cell behavior obscures any effect that the FGFs might have on patterning. Remarkably, FGF signaling as a mechanism for controlling mesodermal cell migration appears to have been highly conserved during evolution.

The process of gastrulation in the mouse is superficially similar to that in chick in that there is a streak that is the source of endodermal and mesodermal cells—but there are significant differences. For example, the fact that the mouse epiblast is a hollow cup as opposed to a disc (see Fig. 5.37), results in a different topology, which has consequences for the progress of gastrulation, in particular there will be different mechanical constraints acting on individual cells. Most significantly, at the onset of gastrulation the mouse embryo has only about 600 cells, in contrast to the more than 30,000 cells characteristic of the chick embryo. This means that during gastrulation the mouse embryo needs to generate increased cell numbers while they are being allocated to specific fates and positions.

The mouse primitive streak starts at a point on the upper rim of the cup, near the trophoblast, which identifies the posterior region of the embryo, and cells enter it from the inside of the cup (see Fig. 3.51). This means that the endoderm ends up as the outer layer of the early mouse embryo, rather than the inner. Much as in the chick, the streak advances anteriorly; in its wake, ingressing cells form endoderm and the various mesodermal derivatives in order as they migrate out under the epiblast, which will form the ectoderm (see Fig. 5.37).

Like the chick, gastrulation in mice is associated with the expression of *TbxT* in the streak, EMTs of epiblast cells, and their separation from the rest of the epiblast (see Fig. 5.37). Mutational screens in the mouse have confirmed the involvement of EMTs, directed cell movement, and cell–matrix interactions in gastrulation, because mutations in genes affecting the dynamics of the cytoskeleton, apico-basal polarity, and the composition of the extracellular matrix abort the process. Such screens have also revealed an important role for Wnt signaling in this process, in both maintaining *TbxT* expression via canonical Wnt signaling and contributing to PCP of the directed cell movement via the non-canonical Wnt pathway.

Advances in microscopy have allowed the filming of mouse gastrulation at the level of single cells. Together with image analysis, these films have allowed a detailed analysis of the movement and interaction of single cells during the process, as well as the establishment of the origin of the primordia of the different germ layer derivatives.

The mouse epiblast is wrapped by the visceral endoderm (see Fig. 3.58), which will blend with the endoderm coming off the streak to together contribute to the definitive endoderm. This unusual 'endoderm outside' configuration is resolved later in development when the mouse embryo turns inside out (see Fig. 3.53). This turning process is exclusive to rodents. In humans and most mammals other than mice and rats, where a flat blastoderm is topologically more like that of the chick, gastrulation superficially resembles that of the chick, with the primitive streak forming in the upper surface of the epiblast. However, as in the mouse, the pre-gastrulation human embryo has a small number of cells and must undergo a large expansion in cell numbers during gastrulation, a feature that sets mammalian gastrulation apart from gastrulation in the chick.

Measurements of cell motility, tissue rigidity, and the compliance of the extracellular space (its ability to undergo elastic deformation) all suggest that mechanical forces, acting at both the cellular and the tissue level, contribute to elongation of the body axis in mouse and chick and suggest similarities associated with gastrulation in the overall mechanism driving extension, once it has started. Recall that in chick and mouse, early patterning events direct head formation, whereas the trunk and tail form through elongation by addition of cells coming from the stem zone (see Fig. 4.27); in zebrafish (and likely in *Xenopus*), early patterning and convergent extension lead to

Fig. 5.38 Mechanisms contributing to body elongation in different vertebrates. Left panel: initial patterning and convergent extension make a greater contribution to body axis elongation in zebrafish and *Xenopus* than in the mouse and the chick, whereas extension due to growth by cell proliferation forms most of the trunk and tail of mouse and chick, but is restricted to the tail in zebrafish and *Xenopus*. Right panel: opposing gradients of tissue rigidity/cell density and extracellular space/cell motility in the pre-somitic mesoderm help propel elongation of the body axis in a posterior direction in the chick embryo. For clarity, each gradient is shown on one side of the embryo only, whereas in reality they occur together on both sides of the embryo. Similar gradients occur in other elongating tissues such as the notochord.

Adapted from Mongera, A., et al.: **Mechanics of anteroposterior axis formation in vertebrates.** Annu. Rev. Cell Dev. Biol. *2019, 35: 259-283.*

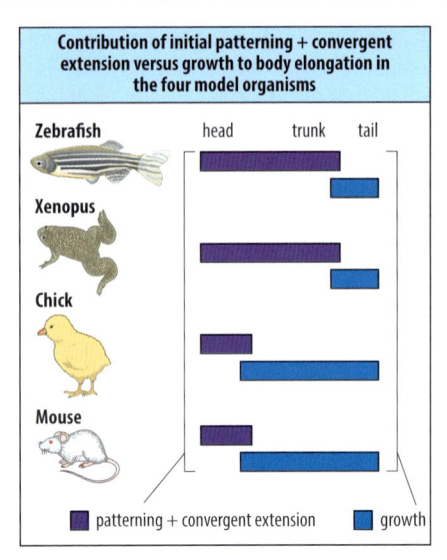

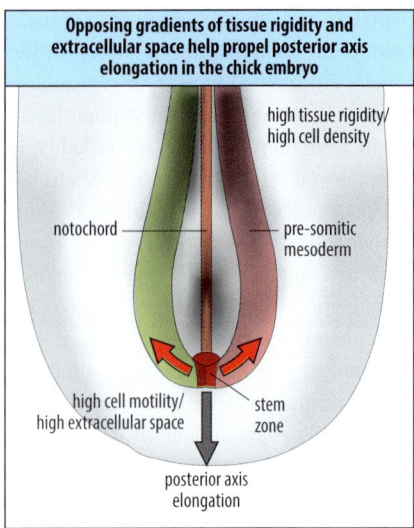

head and trunk formation, and only the tail is formed through elongation by growth (see Fig. 5.38, left panel).

Axial elongation is associated with coupled antero-posterior gradients of extracellular space and cell motility, with highest amounts of extracellular space and highest cell motility in posterior regions. Mathematical modeling studies suggest that the high cell motility exerts an active pressure that propels the axis posteriorly. Targeted ablation studies in the zebrafish show that at the same time, tissue rigidification, which is highest anteriorly, has a role in maintaining the axis extended, and in propelling it posteriorly. In chick and mouse, the process is sustained by the provision of a constant stream of new cells from the stem zone (see Fig. 5.38, right panel).

5.16 Differences in gastrulation between vertebrate species result from the global control of conserved mechanical features of cell populations

As we have seen, there are different modes of gastrulation in vertebrates: in anamniote embryos (amphibians or fish), the process is characterized by a blastopore and the involution of cells, whereas in amniotes (birds and mammals) it is driven by cell migration along the primitive streak. An important and surprising feature of the process is that, regardless of the starting cellular organization of the embryo, gastrulation leads to a similar body plan (see Fig. 3.2). The question then arises of how such disparate starting conditions and apparently different mechanisms can lead to the same body plan.

Despite the differences, there are common features in all cases—in particular, the emergence of a coordinate system with antero-posterior and dorso-ventral axes as well as a midline. Also, and most significantly, the mechanisms that mediate cell behavior during gastrulation are conserved: actomyosin contraction and cell migration are coordinated at the level of cell populations by planar cell polarity and mechanisms of collective migration. Gastrulation is, essentially, a process mediated and controlled by cellular forces.

One significant difference between anamniote and amniote embryos is the prominent presence of extra-embryonic tissues in the latter that serves as a reference for the behavior of cell populations. The primitive streak can be seen as the manifestation of a pre-existing midline, probably outlined through mechanical interactions between embryonic and extra-embryonic tissues that define a line where forces resulting from their interactions are balanced—the midline. This is what is revealed by cell movements that, probably, respond to the mechanics of that domain. In this process it is likely that

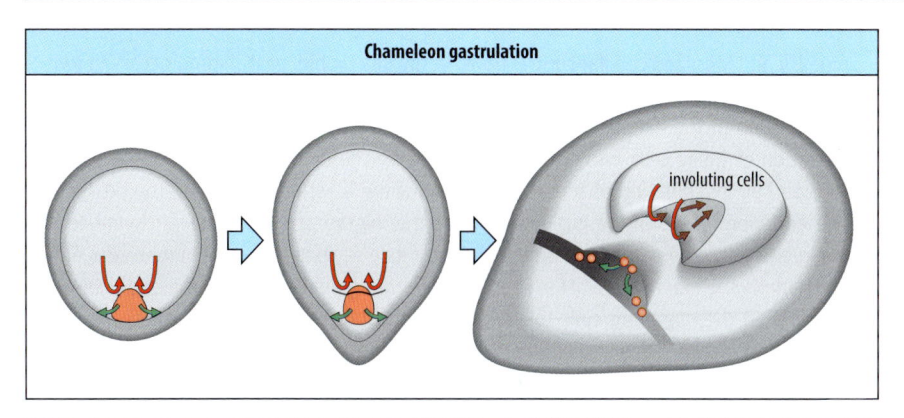

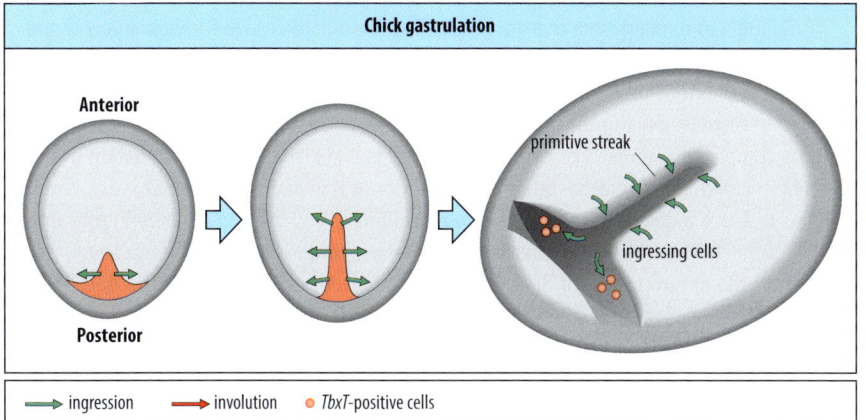

Fig. 5.39 Gastrulation in a reptile compared with that in the chick. Gastrulation in the chameleon is representative of the process in reptiles and displays a combination of a blastopore, characteristic of anamniotes (e.g. amphibia), anteriorly and a hint of an amniote primitive streak, posteriorly. In the chick, the dominant structure is the primitive streak. The transitional nature of the reptilian arrangement can be seen in the expression of *TbxT* only at the rudimentary primitive streak, which is unlike the situation in the amphibian blastopore, where mesodermal cells express *TbxT* before and after involution, and more like the situation in the amniote primitive streak.

From Stower, M.J., et al.: *Bi-modal strategy of gastrulation in reptiles.* Dev. Dyn. 2015, 244: 1144–1157.

the presence of extra-embryonic tissues influences the behavior of the cells and limits their movements. In reptiles, in which extra-embryonic membranes first appear in evolution, gastrulation displays features of both a blastopore and a primitive streak—both involution and cell migration (Fig. 5.39), reflecting the beginning of an evolutionary transition. Gastrulation starts at the posterior end with a structure that resembles the beginning of the primitive streak in amniotes but, shortly afterwards, a blastopore develops anterior to it. It is not difficult to see how in birds, which are closely related to reptiles, these two structures fused and gave rise to the primitive streak.

What this suggests is that the vertebrate body plan is, in all cases, latent, and that the differences in gastrulation we observe reflect the mechanical influence of the environment of the embryo on the organization and behavior of its cells. This has been shown in a series of experiments in which various interferences with the behavior of the cells in the chick epiblast transform the primitive streak into structures and gastrulation modes that resemble reptile and frog gastrulation.

SUMMARY

At the end of cleavage, the animal embryo is essentially a continuous sheet of cells, which is often in the form of a sphere enclosing a fluid-filled interior. Gastrulation converts this sheet into a solid three-dimensional embryonic animal body. During gastrulation, cells move from the exterior into the interior of the embryo, and the endodermal and mesodermal germ layers take up their appropriate positions inside the embryo. Gastrulation results from a well-defined spatio-temporal pattern of changes in cell shape, cell movements, and changes in cell adhesiveness. The main forces underlying the morphological changes associated with gastrulation are generated by changes in cell motility and localized cellular contractions. In the sea urchin, gastrulation occurs in two phases. In the first, changes in shape and adhesion of prospective mesoderm cells in a small region of the blastula wall result in an epithelial-to-mesenchymal transition and migration of these cells into the interior and along the inner face of the blastula wall, followed by invagination of the adjacent part of the blastula epithelium, the prospective endoderm, to form the gut. Filopodia on the migrating primary mesenchyme cells explore the environment, drawing the cells to those regions of the wall where the filopodia make the most stable attachment; in this way, guidance cues in the extracellular matrix lead the migrating cells to their correct locations. In the second phase, extension of the gut to reach the prospective oral (mouth) region on the opposite side of the embryo occurs first by cell rearrangement within the endoderm, which causes the tissue to narrow and lengthen (convergent extension), and subsequently by traction of filopodia extending from the gut tip against the blastocoel wall. Invagination of the mesoderm in *Drosophila* occurs by a similar mechanism to that in sea urchin, whereas germ-band extension is due to myosin-driven remodeling of cell junctions and intercalation of cells, leading to convergent extension. Gastrulation in vertebrates involves more complex movements of cells and cell sheets and also results in elongation of the embryo along the antero-posterior axis. In *Xenopus* it involves three processes: involution, in which a double-layered sheet composed of endoderm and mesoderm rolls into the interior over the lip of the blastopore; convergent extension of endoderm and mesoderm in the antero-posterior direction to form the roof of the gut, and the notochord and somitic mesoderm, respectively; and epiboly—the spreading of the ectoderm from the animal cap region to cover the whole outer surface of the embryo. Both convergent extension and epiboly are due to cell intercalation, the cells becoming rearranged with respect to their neighbors. Convergent extension of the antero-posterior axis at gastrulation involves the vertebrate version of the PCP signaling pathway and requires pre-existing antero-posterior patterning of the mesoderm. Gastrulation in the chick and the mouse involves the separation of individual cells from the epiblast by epithelial-to-mesenchymal transitions and their ingression through the primitive streak into the interior of the embryo, where they form mesoderm and endoderm. Elongation of the body axis in chick and mouse after gastrulation involves both convergent extension and growth.

Neural tube formation

Neurulation in vertebrates results in the formation of the **neural tube**—a tube of epithelium derived from the dorsal ectoderm—which develops into the brain and spinal cord. Following induction by the mesoderm during gastrulation (see Chapter 4), the ectoderm that will give rise to the neural tube initially appears as a wide thickened plate of epithelium—the **neural plate**—in which the cells have become more columnar than the cells of the adjacent ectoderm. The neural plate elongates, narrows, and then rolls up into a tube from which the nervous system will emerge. The cells at the boundary of the neural plate with the adjacent ectoderm give rise to the **neural crest**. This population of cells leaves the dorsal surface of the neural tube on both sides just before and after closure of the neural tube, and migrates to form a wide range of different tissues. Neural crest cells give rise to the peripheral nervous system, the melanocytes of the skin, and even to some of the tissues of the face; neural crest and its migration are discussed later in this chapter. First, we will look at the formation of the neural tube itself. The further development of the nervous system from the cells of the neural tube is considered in Chapter 10.

> **SUMMARY: Gastrulation in sea urchin, *Xenopus*, chick, and mouse**
>
> **Sea urchin**
>
> formation of blastopore: mesodermal cells migrate individually into the interior; endoderm moves inside by invagination of the epithelial sheet
>
> ⇩
>
> convergent extension of endoderm and traction via filopodia completes gut extension
>
> ***Xenopus* and zebrafish**
>
> formation of blastopore: mesoderm and endoderm move inside by involution or internalization
>
> ⇩
>
> convergent extension of mesoderm along antero-posterior axis
>
> ⇩
>
> extension of ectoderm over whole surface by epiboly
>
> **Chick and mouse**
>
> formation of primitive streak: epiblast cells migrate individually into the interior (ingression) through the streak and become specified as mesoderm and endoderm; remaining epiblast becomes ectoderm
>
> ⇩
>
> mesoderm cells migrate laterally and anteriorly under the epiblast
>
> ⇩
>
> gut is formed by fusion of ventral edges of embryo

The vertebrate neural tube is formed by two different mechanisms in different regions of the body. The anterior neural tube forms the brain and the central nervous system in the trunk, and is essentially formed by the folding of the neural plate into a tube, a process called **primary neurulation**. In amphibians, all the head and trunk central nervous system derives from the neural plate. In birds and mammals, a stem-cell-like population of cells makes an important contribution to the nervous system of the trunk (see Section 4.9). By contrast, the posterior neural tube beyond the lumbosacral region, and the tail in *Xenopus* tadpoles, develops from stem-like cells in the tailbud, which form a solid rod of mesenchymal cells that then undergoes a mesenchymal-to-epithelial transition and develops an interior cavity or lumen that connects with that of the anterior neural tube. This process is known as **secondary neurulation**. There are, however, variations in different vertebrates: in zebrafish and other fishes, for example, the whole of the neural plate first forms a solid rod, which later becomes hollow in a similar manner. Here we shall concentrate on primary neurulation.

5.17 Neural tube formation is driven by changes in cell shape and convergent extension

The basic morphological events that occur in neural tube formation in a mammalian embryo are summarized in Figure 5.40. Neurulation starts when the neural plate bends along the midline to form the **neural furrow**. The edges of the neural plate become raised above the plane of the epithelium, forming two parallel **neural folds** with a

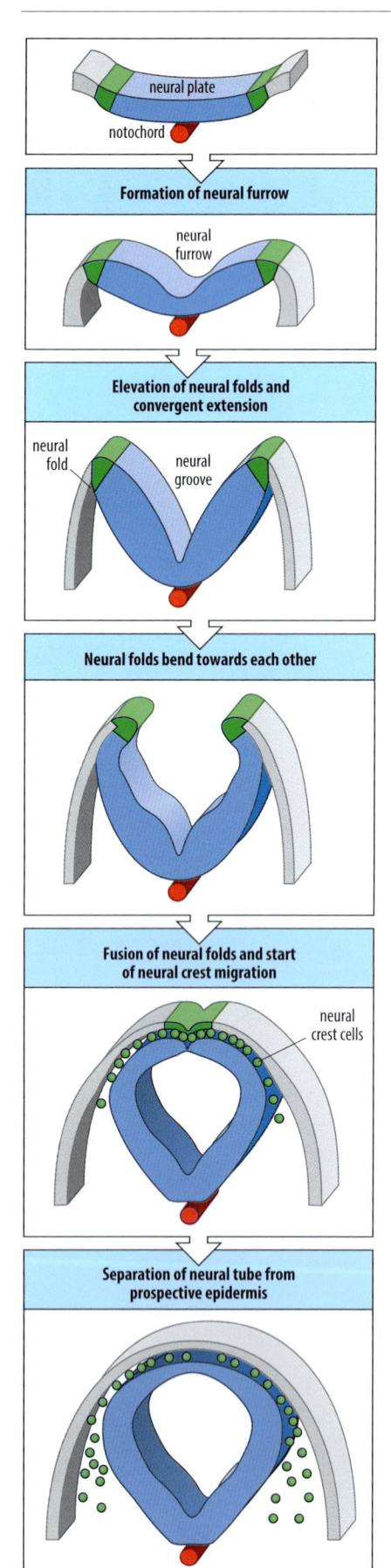

Fig. 5.40 Neural tube formation results from the bending of the neural plate and fusion of the neural folds. Bending of the neural plate inwards at both sides creates the neural groove, with the neural folds at the edges. The neural plate undergoes convergent extension (not shown), which narrows and elongates it. The neural folds come together and fuse to form a tube of epithelium—the neural tube. This detaches from the rest of the ectoderm, which becomes the epidermis. Neural crest cells detach from the dorsal neural tube and migrate away from it. In the mouse, the cranial neural crest cells start to delaminate before fusion of the neural folds.

Adapted from Yamaguchi, Y., Miura, M.: **How to form and close the brain: insight into the mechanism of cranial neural tube closure in mammals.** *Cell. Mol. Life Sci. 2013,* **70***: 3071-3186.*

depression—the **neural groove**—between them. The neural folds eventually come together along the dorsal midline of the embryo and fuse at their edges to form the neural tube, which then separates from the adjacent ectoderm. Neural crest cells detach from the dorsal neural tube and begin to migrate. The scanning electron micrographs in Figure 5.41 show successive stages in neurulation in a chick embryo in the cranial region. In amphibians, the neural tube closes along its whole length almost simultaneously; in the chick, closure starts at the midbrain region and proceeds both anteriorly and posteriorly; and in mammals, closure has been suggested to occur in different regions at different times.

A key process in the initiation of neurulation is convergent extension of the neural plate—similar to the process that takes place during gastrulation. This reduces the width of the neural plate, so that the neural folds are sufficiently close to each other that they can subsequently fuse. Convergent extension also elongates the forming neural tube along the antero-posterior axis as the rest of the body elongates, and is a prerequisite for neural plate bending. Elements of the non-canonical Wnt signaling pathway (see Box 5C) drive convergent extension in response to signals from midline structures, especially the underlying notochord. The cellular mechanisms mediating convergent extension in the neural plate are different in different organisms. In *Xenopus* it is driven by medio-lateral intercalation (see Box 5C), whereas in the mouse and chick it is associated with the remodeling of adherens junctions and the formation and rearrangement of 'rosettes' of cells, much as in the *Drosophila* embryonic epidermis during germ-band extension (see Section 5.11).

Following the convergent extension movements, the bending of the neural plate is driven in two stages by apical constrictions of defined cell populations. First, cells along the midline of the neural plate become wedge-shaped as a result of apical constriction, resulting in folding of the neural plate at the so-called **median hinge point** to form the neural furrow (see Fig. 5.41). The forces generated by the cell-shape changes in this constrained situation lead to the elevation of the edges of the neural plate to form the neural folds. This is a very important step in neural tube formation, and failure to elevate the folds as a result of a failure of planar cell polarity (PCP)—and, therefore, of convergent extension—leads to a complete failure of neurulation.

Cells on the dorso-lateral faces of the folds also become wedge-shaped, in part through apical constriction, forming two more hinge points, known as the **dorso-lateral hinge points** (see Fig. 5.41). These cause the neural folds to buckle inwards, so that the tips are now pointing towards each other. As in many other situations, apical constriction in the cells at the midline and the dorso-lateral hinge points is thought to occur by the purse-string mechanism of actomyosin contraction illustrated in Box 5B.

The last stage of neural tube formation is **neural tube closure**. The cells at the tips of the folds develop lamellipodia and filopodia that interdigitate when the folds meet. The edges of the neural folds then fuse with each other and the tube separates from

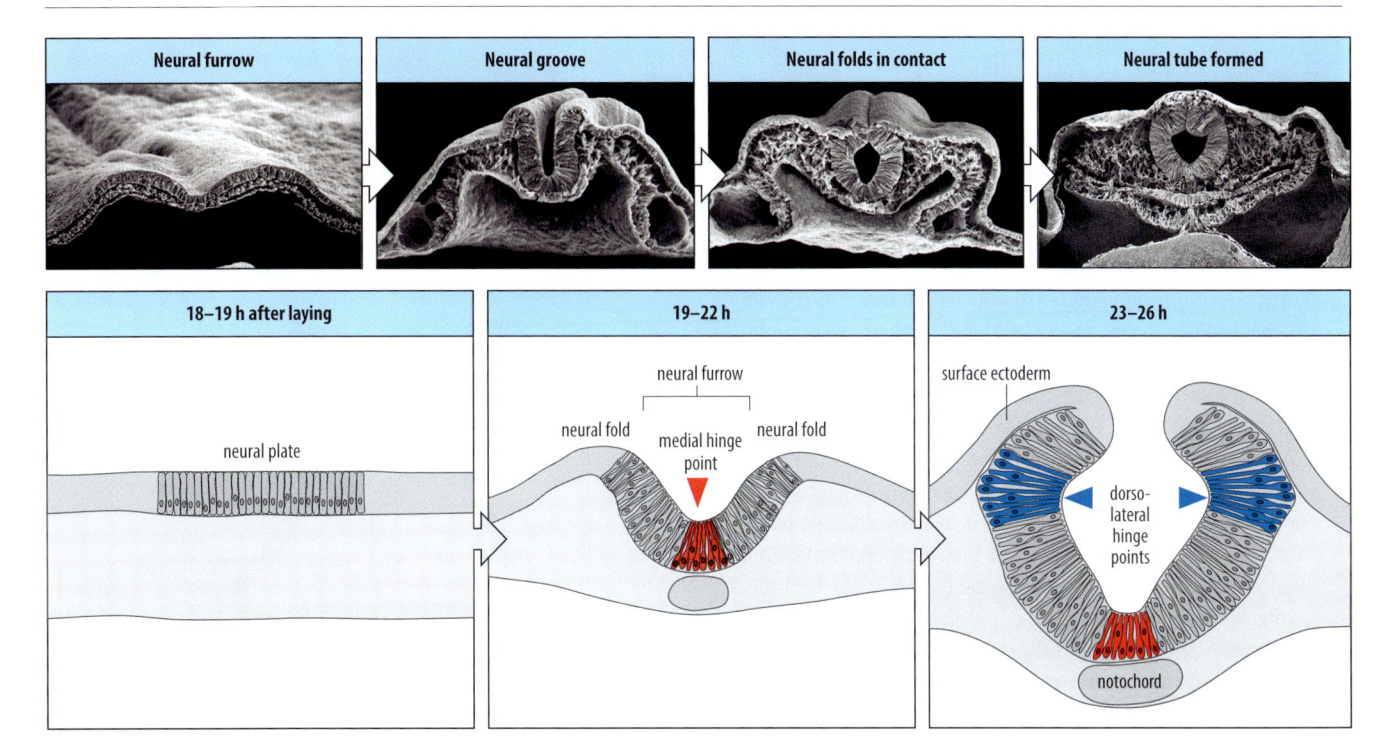

Fig. 5.41 Chick neurulation and changes in cell shape in the neural plate. The electron micrographs show successive stages in neurulation in the chick embryo in sections taken at the approximate level of the future midbrain. The diagrams show that cells in the center of the neural plate become wedge-shaped as a result of apical constriction, defining a median 'hinge point' at which the neural plate bends. Additional dorso-lateral hinge points with wedge-shaped cells also form on the sides of the furrow and help to bring the neural folds together.

Photographs courtesy of G. Schoenwolf from Colas, J.-F., Schoenwolf, G.C.: **Towards a cellular and molecular understanding of neurulation**. *Dev. Dyn. 2001,* **221***: 117-145. Illustration after Schoenwolf, G.C., Smith, J.L.:* **Mechanisms of neurulation: traditional viewpoint and recent advances**. *Development 1990,* **109***: 243-270.*

the rest of the ectoderm, which closes above it (see Fig. 5.40). The folds zip together so closely that the lumen almost disappears; it only opens again after the folds fuse.

The neural tube, which is initially part of the ectoderm, separates from the presumptive epidermis as a result of changes in cell adhesiveness. In the chick, neural plate cells, like the rest of the ectoderm, initially have the adhesion molecule L-CAM on their surface. However, as the neural folds develop, the neural plate ectoderm begins to express both N-cadherin and N-CAM, whereas the adjacent ectoderm expresses E-cadherin. These differences may help the neural tube to separate from the surrounding ectoderm and allow it to sink beneath the surface, the rest of the ectoderm re-forming over it in a continuous layer.

The closing of the neural tube is a complex process that requires the integration of the activity of many different kinds of proteins—adhesion proteins such as the cadherins, and signaling proteins such as the ephrins and their receptors (Box 5D), cytoskeletal components, and regulators of cytoskeletal activity—in addition to changes in gene expression driven by transcription factors.

Because so many cell biological processes have to function perfectly for neural-fold elevation and closure to succeed, it is perhaps not surprising that the process goes wrong relatively frequently, resulting in a failure of neural tube closure. Neurulation defects, or **neural tube defects** are among the most common human birth defects (around 1 in 1000 births), and in their most severe forms are lethal, resulting in stillbirth or death within a few days of birth. The type of neural tube defect depends on the region of the neural tube involved (Box 5E).

CELL BIOLOGY BOX 5D Eph receptors and their ephrin ligands

Ephrins and their receptors (Ephs) are transmembrane signaling proteins that interact with each other on adjacent cells and can both signal to their respective cells. They direct a variety of cell–cell interactions and behaviors, including repulsive and adhesive interactions between cells, mediated by the ability to increase cortical tension and to decrease cell–cell adhesion. Such changes in mechanical forces at cell surfaces lead to transcriptional changes in the cells through **mechanotransduction**—the ability of a cell to sense, integrate, and convert mechanical stimuli into intracellular changes. An adhesive interaction between an ephrin and its receptor is required for neural tube closure in mice, but they are best known for producing repulsion. Because ephrins and Ephs are both attached to a cell surface, they can only act through direct cell–cell contact, like Notch and Delta. Ephrins also act as contact-dependent guidance molecules for growing axons in the nervous system. For example, they guide the axons of retinal neurons to their correct locations in the brain's visual-processing centers; their role in axon guidance is discussed in Chapter 10. And according to work in mice, Ephs and ephrins are required to correctly set up the local neuronal circuits in the spinal cord that enable you to walk with alternate legs, rather than having to hop like a rabbit. They also act as guidance molecules for migrating neural crest cells.

The 14 different Eph receptors in mammals fall into two structural classes: EphAs (EphA1–EphA8 and EphA10) and EphBs (EphB1–EphB4 and EphB6). Ephrins similarly can be classified into ephrin As (ephrins A1–A5) and ephrin Bs (ephrins B1–B3). The A-type receptors typically bind to ephrin As, and the B-type receptors to ephrin Bs, but EphA4 is an exception, being able to bind both ephrin A and B ligands. Ephrins binding to their Eph receptors can generate a bidirectional signal, in which both the Eph and the ephrin send a signal to their own cell (Fig. 1). The main target of Eph–ephrin signaling is the actin cytoskeleton, whose rearrangement causes the changes in cell morphology and behavior that underlie repulsion.

Ephs are receptor tyrosine kinases, with an intracellular kinase domain, whereas ephrins signal by associating with and activating a cytoplasmic kinase. Ephrin–Eph contact is best known for producing repulsion between the interacting cells, as occurs at the interface between different rhombomeres in the hindbrain, but Eph–ephrin interactions can also result in attraction and adhesion. Whether a particular Eph–ephrin pairing repels or attracts is not straightforward, and seems to depend on a range of different factors, including the cell types involved, the numbers of receptors present, and their degree of clustering.

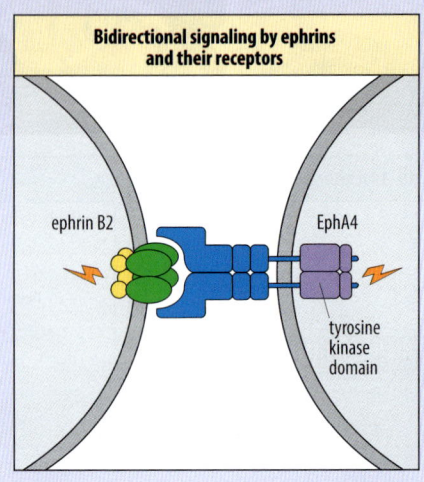

The repulsion commonly mediated by Eph–ephrin pairs is something of a paradox, given their inherent affinity for each other. One solution to the paradox was the discovery that repulsion by ephrin As involves cleavage of the extracellular domain, which enables cell disengagement. Internalization of Eph–ephrin complexes by endocytosis is another possible solution. Repulsive interactions seem generally to require a relatively high level of signaling from the Eph tyrosine kinase. If this is lacking or at a low level, then the binding of Eph and ephrin will tend to promote cell adhesion.

An example of this has been found in the closure of the neural tube in mice, in which a form of EphA7 lacking its kinase domain is produced by the cells of the neural folds as a result of alternative RNA splicing. The neural fold cells also express ephrin A5, and the adhesive interaction between EphA7 and ephrin A5 on opposing folds is essential for neural tube closure. Mouse mutants that lack ephrin A5 have neural tube defects resembling the human condition of anencephaly, in which the cranial neural tube fails to close, the forebrain does not develop, and the fetus is usually stillborn. In tissues other than the neural tube, in which EphA7 is produced with an intact tyrosine kinase domain, its interaction with ephrin A5 causes repulsion between the interacting cells.

Fig. 1 Bidirectional signaling by ephrins and their receptors (ephrin B2, EphA4, tyrosine kinase domain).

MEDICAL BOX 5E Neural tube defects

Neural tube defects are one of the most frequent congenital conditions in newborn babies. Neurulation defects have been studied extensively in the mouse, which has spontaneous mutations that affect neural tube development, and in which large-scale mutagenesis screens can be carried out. More than 60 different genes have been found that, when mutated, cause neural tube defects similar to those found in humans. On the basis of the clustering of mouse and human neural tube defects at particular positions, it has been proposed that neural tube closure occurs at multiple sites. According to this model, in mice, closure starts at E8.5 at a site (1 on Fig. 1) at the hindbrain-cervical boundary, from which it proceeds in both directions. Closure progresses posteriorly along the length of the spinal cord, with completion at the posterior neuropore, which marks the end of the trunk neural tube (the neural tube in the tail region is formed by secondary neurulation). Closure is also initiated a little later at a second site, at the forebrain-midbrain boundary (site 2). A third wave of neural tube closure starts at the anterior-most end of the neural tube (site 3) and proceeds posteriorly to meet closure from the second site at the anterior neuropore. However, recent work proposes that in humans at least, there is just one site of closure, which is located just below the hindbrain-cervical boundary in the region of the neural tube opposite the third and fourth somite pairs. This is still under debate.

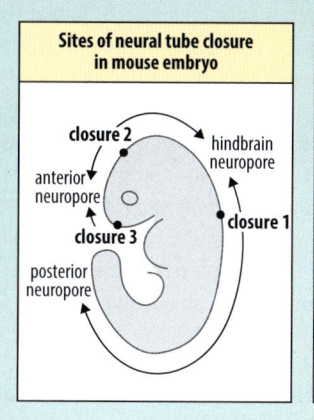

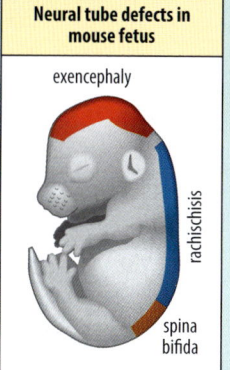

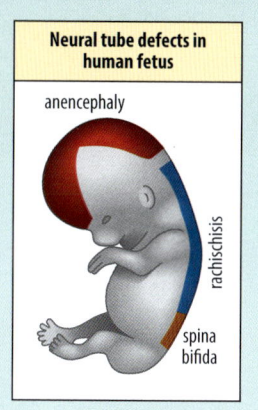

Failure in neural fold elevation or closure causes a range of different defects, including anencephaly, rachischisis, craniorachischisis, and various forms of spina bifida, depending on the regions involved, as shown in the second and third panels of Figure 1. Anencephaly (called exencephaly in the mouse) is a lethal defect that results from a failure of proper neural tube formation in the fore- and midbrain regions. Rachischisis represents the failure of the spinal cord to close, and craniorachischisis results from the neural tube remaining open from the mid-brain to the lower spine. Spina bifida results from a failure to complete neural tube closure at the posterior neuropore, and the consequent failure of the spinal column to form correctly at that point. In humans, the clinically most severe form of this condition is known as myelomeningocele, in which the spinal cord and meninges protrude through the fault in the spine. This form of spina bifida is, in general, seriously disabling even after corrective treatment, as some spinal cord function has been lost. In humans, neural tube defects originate at 4 weeks' gestation, the time at which the neural tube is being formed.

Mutations in genes encoding elements of the core PCP system are associated with neural tube defects in both mice and humans. For example, mouse embryos lacking the vertebrate PCP pathway components Dvl, Celsr, or Vangl2 (see Fig. 5.27) display craniorachischisis. The failure of neural tube closure in these mouse mutants is due to a preceding failure of convergent extension in the neural plate as a result of the absence of PCP. In the absence of convergent extension, the neural plate does not narrow, and so the neural folds are not brought close enough together to fuse to form the closed neural tube. The association of defects in the core PCP pathway with defects in neural tube formation indicates that elements of the PCP pathway that impinge on the cytoskeleton, as well as regulators of actomyosin assembly and function involved in the convergent extension and bending of the neural plate and closure of the neural tube, are essential for correct neural tube development.

Mutations in PCP genes are risk factors for neural tube defects in humans. Defects in the gene *VANGL1* were the first to be associated with various forms of spina bifida and, since then, mutations in most of the core PCP protein genes have been found associated with the condition.

The other main identified risk factors for neural tube defects are defects in genes that control folate metabolism. In the 1970s, an association between a deficiency of the vitamin folic acid (folate) in mothers with affected pregnancies was noticed, and efforts were started to improve the folate status of mothers-to-be. Some countries fortify food with folic acid and this seems to have reduced the incidence of neural tube defects. However, most mothers with an affected pregnancy have folate levels within the normal range, and it is now thought likely that folate deficiency is a risk factor for neural tube defects if there are also genetic predisposing factors. How folate promotes normal neural tube closure is not yet understood. It has a direct effect on neurulation, as mouse embryos genetically predisposed to neural tube defects as a result of defective folate metabolism can be 'rescued' by folic acid treatment *in vitro*.

Fig. 1
First panel based on Copp AJ. **Neurulation in the cranial region—normal and abnormal.** *J Anat. 2005, **207**: 623-635. Second and third panels from Juriloff, D.M., Harris, M.J.:* **Mouse models for neural tube closure defects.** *Hum. Mol. Genet. 2000, **9**: 993-1000.*

294 Chapter 5 Morphogenesis: change in form in the early embryo

> **SUMMARY**
>
> In vertebrates, induction of neural tissue by the mesoderm during gastrulation is followed by neurulation—the formation of the neural tube, which will eventually form the brain and at least some of the spinal cord. Neurulation in mammals, birds, and amphibians involves both the folding of the neural plate into a tube (in the head and trunk), and the formation of a central cavity in a solid rod of cells (in the tail). The formation of the neural folds and their coming together in the midline to form the neural tube is driven by convergent extension and changes in cell shape in the neural plate. Separation of the neural tube from the overlying ectoderm involves changes in cell adhesiveness. The complexity of the process of neural tube closure means that defects are not uncommon, giving rise to conditions such as spina bifida.

Formation of tubes and branching morphogenesis

A feature of development that we have not yet considered is the formation and branching of tubular epithelia. Epithelia, in which the cells adhere tightly together to form a cell sheet, are the most common type of tissue organization in animals and can form tubular structures. Many organs, such as the lungs, kidneys, and mammary glands in mammals, contain functionally specialized tubular epithelia. A common feature of these organs is that the epithelium forms branching tubes. This phenomenon is known as **branching morphogenesis** (Fig. 5.42).

Extensive branching occurs during the development of the network of tubules in the tracheal system of insects and the complex network of interconnecting blood vessels in the vascular system of vertebrates, and we will discuss these examples in this

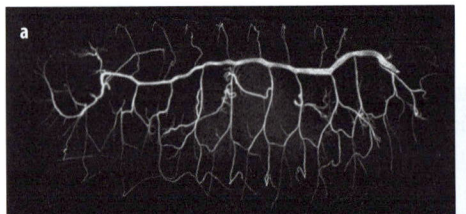

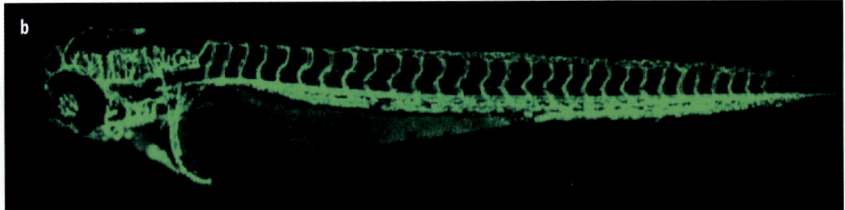

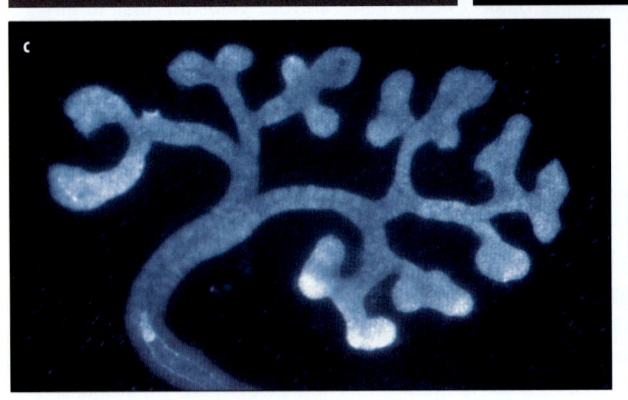

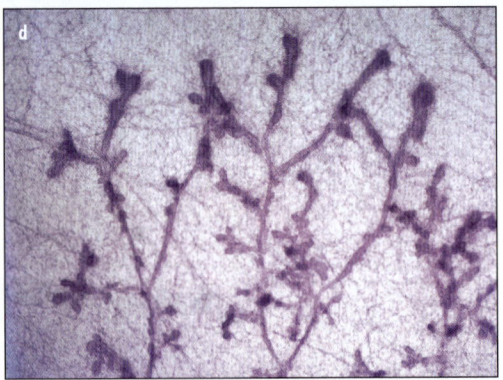

Fig. 5.42 Examples of branching morphogenesis. (a) *Drosophila* embryonic tracheal system. (b) Zebrafish embryonic vascular system. (c) Mouse embryonic kidney. (d) Rat virgin mammary gland. The branching epithelia are visualized by antibody staining (panels a, c) or a histological stain (d) or by a fluorescent reporter gene (b).

(a) Reproduced from Luschnig, S., et al.: **Serpentine and vermiform encode matrix proteins with chitin binding and deacetylation domains that limit tracheal tube length in Drosophila.** Curr. Biol. *2006,* **16***: 186-194. (b) Courtesy Ochoa-Espinosa, A., Affolter, M.:* **Branching morphogenesis: from cells to organs and back.** Cold Spring Harb. Perspect. Biol. *2012,* **4***: a008243. (c) Image courtesy of J. Davies. (d) From Schedin, P., et al.:* **Microenvironment of the involuting mammary gland mediates mammary cancer progression.** J. Mamm. Gland Biol. Neoplasia *2007,* **12***: 71-82.*

chapter. As we will see, branching morphogenesis in both these cases is driven by **collective cell migration**. This is when a group of cells undergoes migration together as a collective entity, remaining linked to each other by adhesive junctions. The same mechanism is also involved in branching morphogenesis of the developing mammary gland. Branching morphogenesis in the vertebrate lung employs a different strategy. Branching morphogenesis in both the mammary gland and the vertebrate lung are discussed in Chapter 9. Branching morphogenesis in the kidney is similar to that in the lung.

5.18 The *Drosophila* tracheal system is a prime example of branching morphogenesis

The development of the *Drosophila* **tracheal system**, which delivers oxygen to the tissues, provides an excellent model for branching morphogenesis. The formation of this network of fine tubules initially involves invagination of epithelial cells brought about by changes in cell shape, similar to those involved in the internalization of *Drosophila* mesoderm cells (see Section 5.10)—followed by collective cell migration and cell rearrangements.

Air enters the tracheal system of the *Drosophila* larva through openings in the body wall called spiracles, and oxygen is delivered to the tissues by some 10,000 or so fine tubules, which develop during embryogenesis from 20 **placodes** of thickened ectodermal epithelium (10 on each side). Each pair of placodes produces the tracheal system for one complete segment in the larva. At the start of germ-band retraction, towards the end of embryonic development, the placode ectoderm invaginates to form a hollow sac of around 80 cells, which gives rise through successive branching to hundreds of fine terminal branches. Remarkably, the extension of the sacs to form branched tubes does not involve any further cell proliferation, but is all achieved by collective cell migration, cell rearrangement by intercalation, and changes in cell shape. As development proceeds, branches from different placodes fuse to form a body-wide network of interconnected tubes (see Fig. 5.42a).

The initial invagination of a tracheal placode involves constriction of apical cell surfaces and changes in cell shape, and the first tubules are formed when some cells in the wall of the sac develop filopodia. These cells become migratory and move towards a source of chemoattractant, drawing an elongating tube of cells behind them as a collective (Fig. 5.43). These tubes then branch by a combination of cell intercalation and remodeling of intercellular junctions to produce secondary tubules composed of

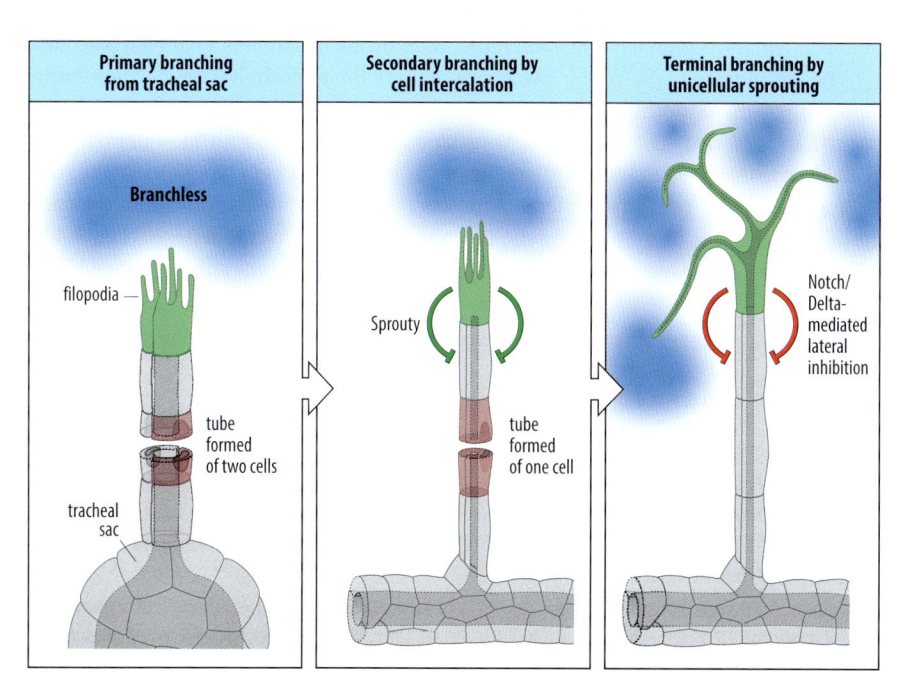

Fig. 5.43 Branching in the *Drosophila* tracheal system is the result of collective cell migration and cell rearrangements. In the first stage of tracheal development, localized secretion of Branchless protein (blue) causes cells of the tracheal sac epithelium to form filopodia and move towards the source of Branchless. Collective migration creates a primary branch with walls formed of two cells edge to edge (left panel). A secondary branch is formed by some cells on the sides of the primary branches putting out filopodia and becoming migratory, generating pulling forces that result in cells in the stalk intercalating between each other and wrapping round themselves to form a tube of single cells placed end to end, with a lumen running through the center (center panel). At the tip of the secondary branch, expression of the gene *sprouty* is induced, and the Sprouty protein (green) inhibits branching further away from the source of Branchless. Tip cells that will undergo terminal branching are then specified by Notch-Delta signaling. A tip cell puts out fine cytoplasmic extensions that develop a lumen and branch extensively (right panel). Cells behind the tip are inhibited by the Notch signaling (red barred lines) from adopting a tip-cell fate.

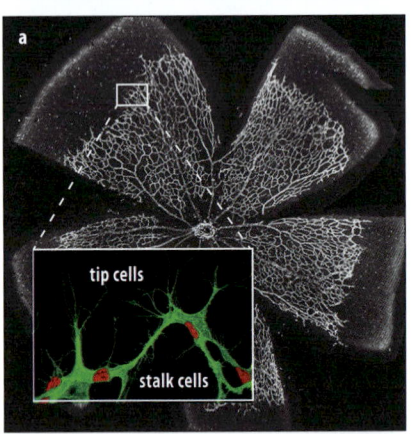

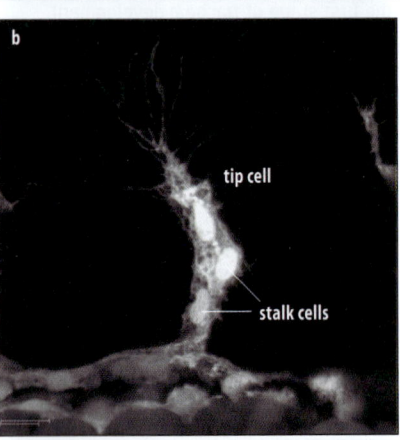

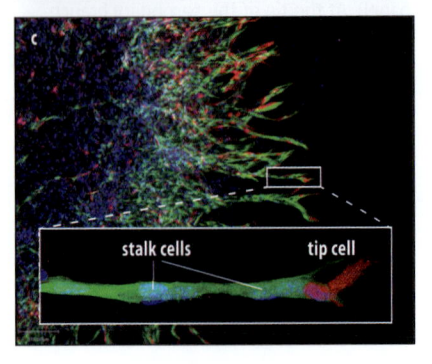

single cells joined end to end, with each cell wrapped around itself to form the tubule lumen. Unlike cell intercalation in the insect germ band (see Section 5.11), cell intercalation in the tracheal system is not myosin-dependent but seems to be driven by global pulling forces generated by collective cell migration.

The chemoattractant that guides tube extension by collective migration is the protein Branchless, which is the *Drosophila* version of mammalian FGF. It is expressed in clusters of overlying epidermal cells in a pattern determined by the previous anteroposterior and dorso-ventral patterning of the embryo (see Fig. 5.43). Branchless acts via the *Drosophila* version of an FGF receptor called Breathless, which is expressed in the tracheal cells and which is also involved in initiating branching. Genes such as *breathless* and *branchless* were identified by mutations that disrupted tracheal morphology—hence their names. Another such gene is *sprouty*, which is expressed by the tracheal cells. Sprouty protein prevents excess branching by antagonizing Breathless protein signaling; this prevents the more proximal stalk cells away from the tip of the tubules from forming branches. In the absence of Sprouty, many more secondary branches form than normal. It has been estimated from screens for mutations that affect the growth and branching of tracheal tubes that more than 200 patterning and morphogenesis genes are required to build the *Drosophila* tracheal system.

As the tubules extend into the larval body, low oxygen levels lead to local expression of *branchless* in the epidermis. Breathless signaling in response induces cells at the tips of the secondary tubules to sprout to form the fine terminal branches of the tracheal system. Each tip cell forms a much-branched unicellular sprout, with the lumen of the tubule formed within the cell itself (see Fig. 5.43). Excessive terminal branching is prevented by Notch–Delta signaling, which assigns tip-cell fate and prevents the cells away from the tip becoming tip cells and being able to form terminal branches.

5.19 The vertebrate vascular system develops by vasculogenesis followed by sprouting angiogenesis

The vascular system, including blood vessels and blood cells, is among the first organ systems to develop in vertebrate embryos. Its early development is not surprising, because oxygen and nutrients need to be delivered to the rapidly developing tissues, and subsequently to developing organs. The development of the vascular system involves extensive branching morphogenesis, which employs a similar mechanism to that in the developing insect tracheal system. The initial formation of tubes, however, occurs by a mesenchymal-to-epithelial transition rather than by invagination of epithelial placodes.

The defining cell type of the vascular system is the **endothelial cell**, which forms the lining—the **endothelium**—of the entire circulatory system, including the heart, veins, and arteries. The development of blood vessels starts with mesodermal cells called **angioblasts**: these are the precursors of endothelial cells. Angioblasts migrate to form chains of cells that undergo a mesenchymal-to-epithelial transition to form a primitive network of tubular vessels, in a process called **vasculogenesis** (Fig. 5.44). The primary blood vessels such as the dorsal aorta arise from cords of angioblasts

Fig. 5.44 Vasculogenesis and angiogenesis. Top panel: angioblasts migrating in chains, the first step in vasculogenesis that then results in the formation of simple tubes. Individual angioblasts at time point 1 are indicated in red and at time point 2 in green to show movement, with arrows indicating direction. Ringed cells are immotile. Scale bar = 200 μm. The other photographs show sprouting during angiogenesis in model systems used to study the process. Collective migration is involved in which tip cells extend filopodia-like processes and direct the migration of stalk cells in the following processes: (a) vascularization of mouse retina; (b) formation of a zebrafish intersomitic vessel; (c) formation of blood vessels in culture from an aggregate of mouse embryonic stem cells.

Top panel from Sato, Y., et al.: **Dynamic analysis of vascular morphogenesis using transgenic quail embryos.** PLoS One *2010,* **5***: e12674. © 2010 Sato et al. (a–c) From Guedens, I., Gerhardt, H.:* **Coordinating cell behavior during blood vessel formation.** Development *2011,* **138***: 4569–4583.*

located in the lateral plate mesoderm that then remodel themselves to form a tube. In blood vessel development, the initial vessels formed by vasculogenesis are then elaborated into a vascular system that ramifies throughout the body. This is achieved by the process of **angiogenesis**, in which new vessels grow from existing blood vessels via collective migration and branch to form extensive networks of fine capillaries and of the thicker venules and arterioles that can go on to form veins and arteries. The vessels also fuse with each other and undergo anastomosis (the bringing together and joining of two previously separated branches) to create complex interconnected networks of blood vessels. The endothelial cells that make up the first, delicate, thin-walled tubular structures of the vasculature are covered by connective-tissue cells called pericytes. Venules, arterioles, veins, and arteries are enveloped by additional layers of smooth muscle and outer fibrous connective tissue.

Arteries and veins are defined by the direction of blood flow, as well as by structural and functional differences. Evidence from lineage tracing shows that angioblasts are specified as arterial or venous before they form blood vessels; their identities are, however, still labile. Arterial and venous capillaries join each other in capillary beds, which form the interchange sites between the arterial and venous blood systems. The cell-sorting and guidance molecule ephrin B2 (see Box 5D) is expressed in arterial blood vessels, whereas its receptor EphB4 is expressed in venous vessels. Interaction between them in primitive capillary networks may be required for endothelial cells to sort out into distinct arterial and venous vessels.

Lymphatic vessels are also developmentally part of the vascular system and originate by budding from veins. The expression of the homeobox gene *Prox1* marks an early stage of commitment to the lymphatic lineage. Angioblast differentiation into endothelial cells requires the growth factor VEGF and its receptors. VEGF is also a potent mitogen for endothelial cells, stimulating their proliferation. VEGF is secreted by axial structures, such as somites, and its expression is driven by Shh signaling in the notochord. Expression of the *Vegf* gene is induced by lack of oxygen or hypoxia, and thus an active organ using up oxygen promotes its own vascularization.

5.20 New blood vessels are formed from pre-existing vessels in angiogenesis

Angiogenesis is very similar to tracheal branching, although here the chemoattractant is VEGF rather than Branchless, and cell proliferation is involved in growth and branching of the capillaries, whereas in the tracheal system growth is due solely to cell rearrangements (Fig. 5.45). Cells at the tip of a capillary sprout extend filopodia-like processes and undergo collective migration that guides and extends the sprout. The response of the tip cells to VEGF is to express the Notch ligand Delta-like 4, which then activates Notch in adjacent proximal cells in the stalk. Notch signaling blocks expression of the VEGF receptor. This is one of the mechanisms that confines outgrowth to the tip of the tube and is reminiscent of the Delta–Notch feedback mechanism deployed in *Drosophila* tracheal development (see Section 5.18).

During their development, blood vessels navigate along specific paths towards their targets, with the filopodia at the leading edge of the capillary sprouts responding to both attractant and repellent cues on other cells and in the extracellular matrix. Cues in the extracellular environment are provided by proteins of the netrin and semaphorin families, which can act on the tips of growing blood vessels to block filopodial activity. Netrins and semaphorins are also involved in guidance of axons (discussed in Chapter 10), and there is a striking similarity in the signals and mechanisms that guide developing blood vessels and neurons.

Blood vessel morphogenesis also requires modulation of adhesive interactions between endothelial cells and the extracellular matrix and between the endothelial cells themselves. Changes in integrin-mediated cell adhesion to the extracellular matrix through focal contacts are particularly important in this context. VEGF

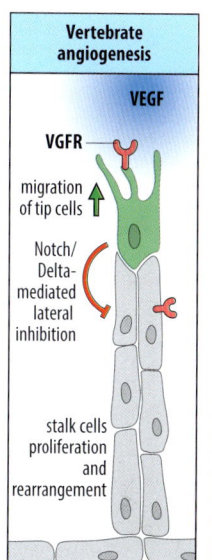

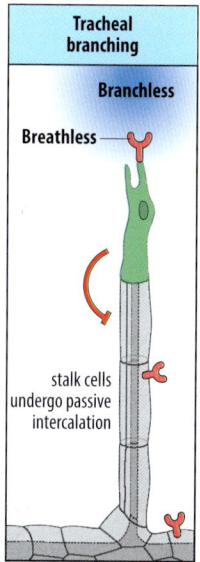

Fig. 5.45 Outgrowth and branching of new vessels in angiogenesis. Left panel: in the first stage of angiogenesis, localized secretion of vascular epithelial growth factor (VEGF; blue) causes nearby endothelial cells of pre-existing blood vessels to form filopodia and move towards the source of VEGF. VEGF also stimulates proliferation and rearrangement of the endothelial cells, so forming a new branch. Delta production is enhanced in the tip cells (green) of the sprout, leading to activation of Notch in neighboring cells, which prevents them from becoming tip cells. These cells become stalk cells and a lumen forms. Right panel: the mechanism of branching of the *Drosophila* tracheal system is very similar, apart from the fact that in this case cell proliferation is not involved. VGFR, VEGF receptor.

From Ochoa-Spinosa, A., Affolter, M.: ***Branching morphogenesis: from cells to organs and back.*** *Cold Spring Harb. Perspect. Biol. 2012,* **4***: a008243.*

stimulates endothelial cells to degrade the surrounding basement membrane matrix and to migrate and proliferate. In the skin of mice, embryonic nerves form a template that directs the growth of arteries. The nerves secrete VEGF, which can both attract blood vessels and specify them as arteries. Endothelial cells also respond to the mechanical forces exerted by blood flow, which influence cell migration and shape changes during development and remodeling of the vascular system.

Angiogenesis is not just a property of the embryo. It can occur throughout life and, properly regulated, is the means of repairing damaged blood vessels. Excessive or abnormal angiogenesis is, however, the hallmark of many human diseases, including cancer, obesity, psoriasis, atherosclerosis, and arthritis. Human disorders relating to the nervous system that involve abnormalities in the vasculature include vascular dementia, where blockage of the vasculature is a major cause of disease symptoms, and motor neuron disease. Many solid tumors produce VEGF and other growth factors that stimulate angiogenesis, and preventing new vessel formation would be a means of reducing tumor growth. More than 10 drugs that target the VEGF pathway have been approved for clinical use in various solid tumors. Other drugs that target molecules involved in blood vessel stability are also being developed.

SUMMARY

The insect tracheal system and the vertebrate vascular system both consist of a network of branched tubular epithelia. Tubular structures can be formed *de novo* by two basic mechanisms: invagination of a pre-existing epithelium, as in the *Drosophila* tracheal system, or by mesenchymal cells undergoing a mesenchymal-to-epithelial transition, as in the formation of blood vessels during vasculogenesis in vertebrates. The branching morphogenesis of both the insect tracheal system and the vertebrate vascular system during angiogenesis involves collective cell migration. The direction of migration depends on local cues from the surrounding tissues, as well as the response of the cells to air flow or blood flow, respectively. Negative feedback from the tips of the epithelial sprouts prevents excessive branching.

Cell migration

Cell migration is a major feature of animal morphogenesis, with cells sometimes moving over relatively long distances from one site to another to integrate into new territories. Here we shall describe three examples of cell migration—**neural crest cells** in vertebrates, **lateral line cells** in fishes, and an example of motile cells drawing an epithelial sheet together in the phenomenon of **dorsal closure** in *Drosophila* and of ventral closure of the body wall in *Caenorhabditis* and in avian and mammalian embryos.

Other examples of cell migration can be found elsewhere in the book. The migration of germ cells from their site of formation to their final location in the gonads is described for zebrafish and mouse in Chapter 8, the migration of cells from the somites into the developing chick limb to form muscle cells is discussed in Chapter 9, and the migration of immature neurons within the mammalian central nervous system in Chapter 10.

5.21 Embryonic neural crest gives rise to a wide range of different cell types

Neural crest cells are a hallmark of vertebrates and produce an enormously diverse range of cell types (see Section 4.15). They have their origin at the edges of the neural plate and first become recognizable during neurulation. They arise from the dorsal neural tube just before and after fusion of the neural folds, and are recognizable as individual cells on their emergence from the neural epithelium. Once neural crest cells have detached from the neural tube they migrate to many sites, where they differentiate into a remarkable number of different cell types.

Neural crest cells give rise to the neurons and glia of the peripheral nervous system (which includes the sensory and autonomic nervous systems), endocrine cells such as the chromaffin cells of the adrenal medulla, some structures in the heart, and melanocytes, which are pigmented cells found in the skin (see Fig. 4.46). In mammals and birds, the cranial neural crest also gives rise to bone, cartilage, and dermis in the head—tissues that are formed elsewhere in the body by mesoderm.

5.22 Neural crest migration is controlled by environmental cues

Neural crest cells of vertebrates first become recognizable during neurulation through the expression of a suite of transcription factors that includes Snail and Slug (see Fig. 4.45). The expression of Snail and Slug triggers an epithelial-to-mesenchymal transition (see Fig. 5.5) in the neural crest cells, which detach as individual cells from the neural tube (see Fig. 4.45). The neural crest cells then migrate away from the neural tube on both sides under the guidance of chemical cues, and recent work suggests that, in some regions at least, they may migrate as a collective. Figure 5.46 shows neural crest cells migrating in culture.

Figure 5.47(a, b) shows streams of migrating neural crest cells in zebrafish embryos visualized by *in situ* hybridization for mRNAs for neural crest cell markers. Migrating neural crest cells can be tracked in a living embryo by labeling them with a fluorescent dye such as DiI, or by expressing a photoactivatable green fluorescent protein (pa-GFP) in the neural tube. pa-GFP can then be made to fluoresce when required by a precisely focused beam of light, and the migration patterns of a small group of neural crest cells, or even single cells, can be followed. Figure 5.47(c) shows migrating neural crest cells in a live zebrafish embryo, carrying a reporter transgene in which a GFP coding sequence is linked to a promoter for a gene expressed in neural crest cells (*Sox10*).

There are two main migratory pathways for neural crest cells in the trunk of the chick embryo (Fig. 5.48). One goes dorso-laterally under the ectoderm and over the somites: cells that migrate this way mainly give rise to pigment cells, which populate the skin and feathers. The other pathway is more ventral, primarily giving rise to sympathetic and sensory ganglion cells. Some crest cells move to the somites to form dorsal root ganglia; others migrate through the somites to form sympathetic ganglia

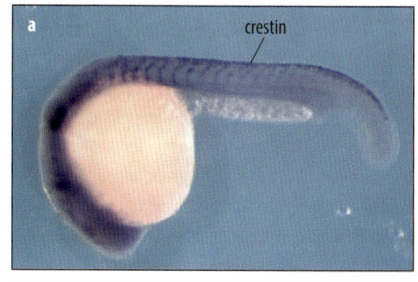

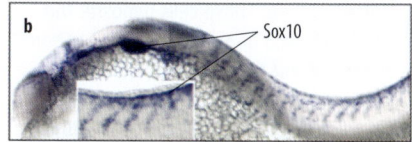

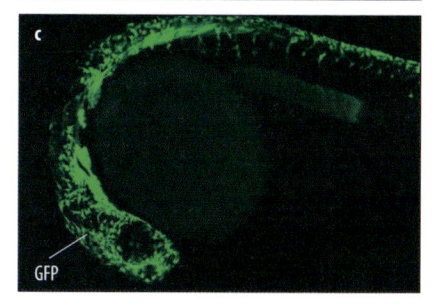

Fig. 5.47 Migrating neural crest cells in zebrafish embryos. (a) Zebrafish embryo (25-somite stage, around 19 hours post-fertilization) in which the expression of RNA for the neural crest marker *crestin* (blue) in migrating neural crest cells is detected by *in situ* nucleic acid hybridization in a fixed embryo. (b) The expression of the neural crest gene *Sox10* mRNA (blue) in neural crest cells detected by *in situ* hybridization in a fixed embryo (24 hours post-fertilization). (c) A live transgenic zebrafish embryo at the same stage of development that carries a reporter transgene in which a green fluorescent protein (GFP) coding sequence is attached to the promoter region of *Sox10*. GFP is therefore expressed in the neural crest cells that express *Sox10*. Note the contribution of the neural crest to the head.

Photographs (a) courtesy of Roberto Mayor; (b, c) Robert Kelsh. (b) Reproduced with permission from Kelsh, R., et al.: **Sorting out Sox10 functions in neural crest development.** *BioEssays 2006, **28**: 788–789.*

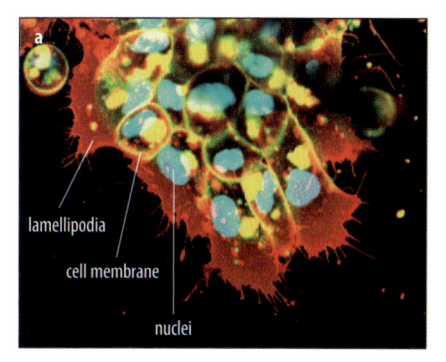

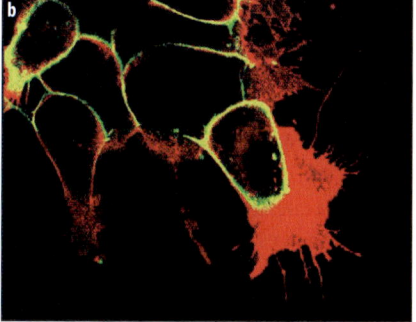

Fig. 5.46 *Xenopus* **neural crest cells migrating in culture.** Lamellipodia at the leading edges of the cells are stained red, cell membranes yellow, and nuclei blue. The photograph on the right shows a higher-magnification view.

Images courtesy of Roberto Mayor.

Fig. 5.48 Neural crest migration in the trunk of the chick embryo. One group of cells (1) migrates under the ectoderm to give rise to pigment cells (shown in dotted outline). The other group of cells (2) migrates over the neural tube and then through the anterior half of the somite; the cells do not migrate through the posterior half of the somite. Those that migrate along this pathway give rise to dorsal root ganglia, sympathetic ganglia, and cells of the adrenal cortex, and their future sites are also shown in dotted outline. Those cells opposite the posterior regions of a somite migrate in both directions along the neural tube until they come to the anterior region of a somite. This results in a segmental pattern of migration and is responsible for the segmental arrangement of ganglia.

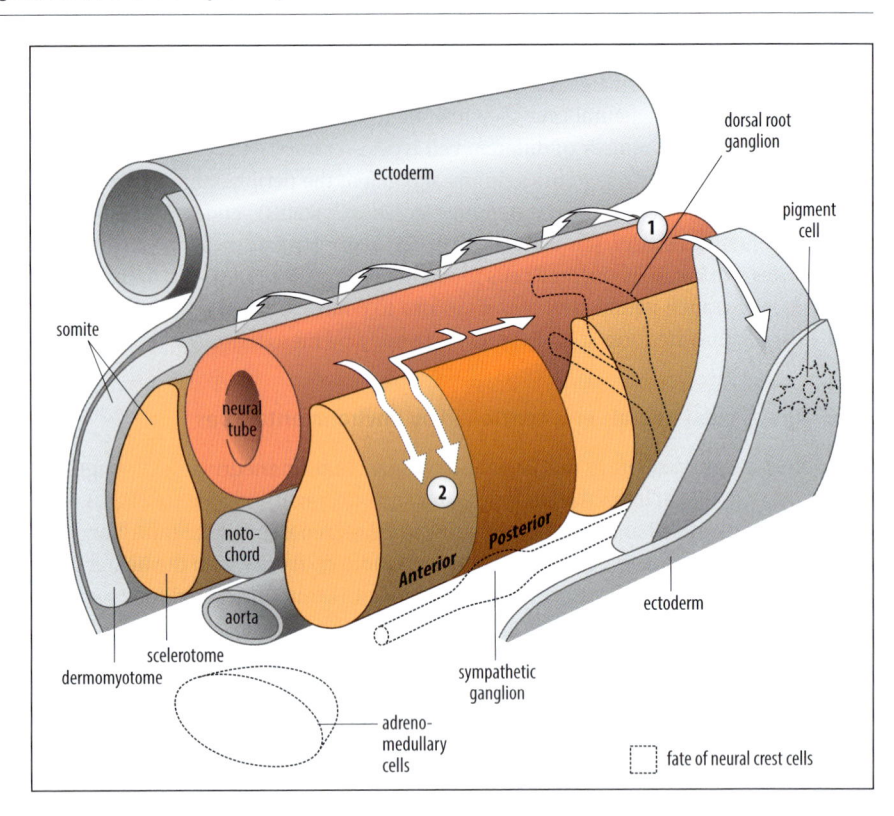

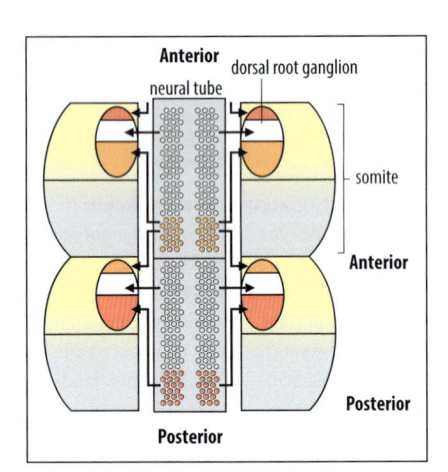

Fig. 5.49 Segmental arrangement of dorsal root ganglia is due to the migration of crest cells through the anterior half of the somite only. Neural crest cells cannot migrate through the posterior (gray) half of a somite but can migrate in either direction along the neural tube and through the anterior half of somites (yellow). The dorsal root ganglion in a given segment is thus made up of crest cells from the posterior region of the segment anterior to it (dark orange), neural crest cells immediately adjacent to it (white), and crest cells from the posterior of its own segment (light orange).

and adrenal medulla, but seem to avoid the region around the notochord. Trunk neural crest selectively migrates through the anterior half of the somite and not through the posterior half. Within each somite, neural crest cells are found only in the anterior half, even when they originate in neural crest adjacent to the posterior half of the somite. This behavior is unlike that of the neural crest cells taking the dorsal pathway, which migrate over the whole dorso-lateral surface of the somite. Neural crest cells that will form the enteric ganglia (the gut nervous system) arise from an anterior site in the neural tube (vagal neural crest) and a posterior site (sacral neural crest) (see Fig. 4.48) and populate the entire length of the gut.

Neural crest migration through the anterior of each somite results in the distinct segmental arrangement of spinal ganglia in vertebrates, with one pair of ganglia for each pair of somites in the embryo (Fig. 5.49). The segmental pattern of migration is due to different guidance cues on the two halves of each somite. If the somites are rotated through 180° so that their antero-posterior axis is reversed, the crest cells still migrate through the original anterior halves only.

Interactions between ephrins and their Eph receptors are likely to contribute to the expulsion of neural crest cells from the posterior halves of the somites. In the chick, ephrin B1 is localized to cells in the posterior halves of somites and acts as a repulsive guidance cue, interacting with EphB3 on the neural crest cells, whereas the cells in the anterior halves of the somites, through which the neural crest cells do travel, express EphB3. Membrane proteins called **semaphorins** are also involved in repelling neural crest cells from the posterior half of the somite. Interactions such as these provide a molecular basis for the segmental arrangement of the spinal ganglia.

Both the neural tube and notochord also influence neural crest migration. If the early neural tube is inverted through 180°, before neural crest migration starts, so that the dorsal surface is now the ventral surface, one might think that the cells that normally migrate ventrally, now being nearer to their destination, would move ventrally. But this is not the case, and many of the crest cells move upward through the sclerotome in a ventral-to-dorsal direction, staying confined to the anterior half of each somite. This suggests that the neural tube somehow influences the direction of migration of neural crest cells. The notochord also exerts an influence, inhibiting neural crest cell migration over about 50 μm, and thus preventing the cells from approaching it.

The migration of streams of neural crest cells in particular directions mainly results from the cells becoming polarized in the direction of travel, probably as the result of the action of the PCP signaling pathway (see Section 5.12). Inhibition of components of this pathway, such as non-canonical Wnts and their receptor Frizzled, inhibit neural crest migration. The direction of neural crest migration is, however, also thought to involve chemotaxis towards specific chemical cues, including stromal cell-derived factor 1 (SDF-1) and FGF.

Neural crest cells are guided by interactions with the extracellular matrix over which they are moving, as well as by cell–cell interactions. Neural crest cells can interact with extracellular matrix molecules by means of their cell-surface integrins. Avian and *Xenopus* neural crest cells cultured *in vitro* adhere to, and migrate efficiently on, fibronectin, laminin, and various collagens. Preventing adhesion of neural crest cells to fibronectin or laminin by blocking the integrin $β_1$ subunit *in vivo* causes severe defects in the head region but not in the trunk, suggesting that the crest cells in these two regions adhere by different mechanisms, probably involving other integrins. It is striking how neural crest cells in culture will preferentially migrate along a track of fibronectin, although the role of this extracellular matrix molecule in guiding the cells in the embryo is still unclear.

These observations have suggested that gradients of stiffness in the extracellular matrix might work together with chemotactic signals to guide neural crest cell migration. This synergy has been observed during migration of the cranial neural crest in *Xenopus*, which is guided by placodes adjacent to the migrating neural crest. The neural crest cells will contribute to the craniofacial skeleton, while the placodes give rise to a variety of structures including the inner ear, the lens of the eye, the olfactory epithelium, and some cranial nerves. The interactions between placodes and neural crest are essential for correct craniofacial development.

A placode is a source of the chemoattractant SDF-1 (also called CXCL12), which initially attracts the neural crest cells to crawl towards the placode over a bed of fibronectin extracellular matrix that separates the placode from the migrating neural crest. SDF-1 provides a direction to the movement by inhibiting actomyosin contractions at the front end of the cells, thus promoting a rear-to-front propulsion. As the neural crest cells reach the placodal cells they interact with them, creating a cadherin-mediated repulsion that causes the placodal cells to move out of the way. Interactions between the migrating neural crest and the placode cells result in a shallow gradient of stiffness being generated in the placodal tissue and the overlying extracellular matrix, with the stiffness highest nearest to the neural crest (Fig. 5.50). Neural crest cells can sense the gradient and this enhances their migratory properties. The ability to sense the stiffness of a substrate is called **durotaxis** and is likely to have a role not only in the migration of the neural crest but in that of other cells, such as the migration of limb bud mesoderm into limb primordia.

In the case of the neural crest, chemotaxis and durotaxis synergize. *In vitro*, stiffness or chemical gradients on their own were found to promote inefficient neural crest cell migration, but together they made migration more effective.

5.23 The formation of the lateral-line primordium in fishes is an example of collective cell migration

A good example of collective migration is the **lateral-line primordium** in fishes, which gives rise to a series of sensory organs along the length of the body—the lateral line—that will form a major sensory apparatus. The primordium is a group of more

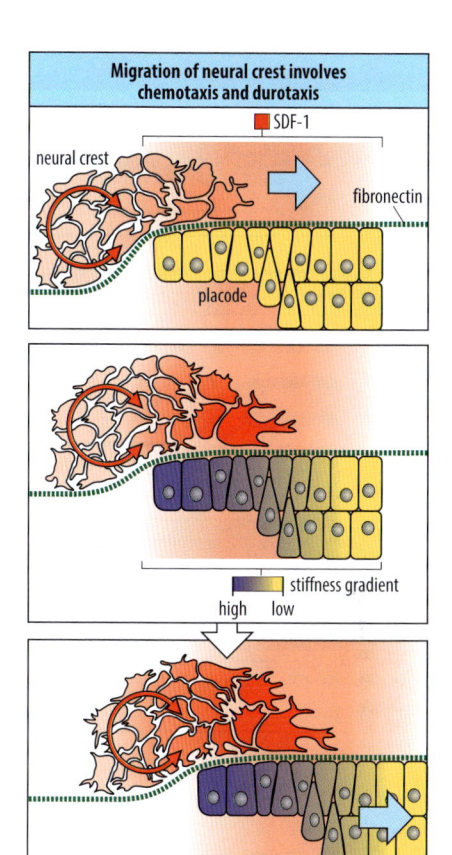

Fig. 5.50 Migration of neural crest involves chemotaxis and durotaxis. Top panel: the initial movement of the neural crest cells towards the placode is driven by chemotaxis mediated by the chemoattractant stromal cell-derived factor 1 (SDF-1) (pale red) secreted by the placode. The red arrow in the neural crest represents actomyosin-mediated forces generated in response to SDF-1 that propel the neural crest cells forward. The dark green dotted line is fibronectin-rich extracellular matrix. Center panel: when the neural crest cells interact with the placodal cells, a stiffness gradient is formed in the placode. Bottom panel: the neural crest cells sense and respond to the stiffness gradient by enhancing actomyosin-driven forward movement. Movement in response to differences in the stiffness of a tissue is called durotaxis. The placodal cells move forwards with the neural crest cells so that the gradient is maintained.

From Shellard, A., Mayor, R. **Collective durotaxis along a self-generated stiffness gradient** in vivo. Nature 2021, **600**: 690-694.

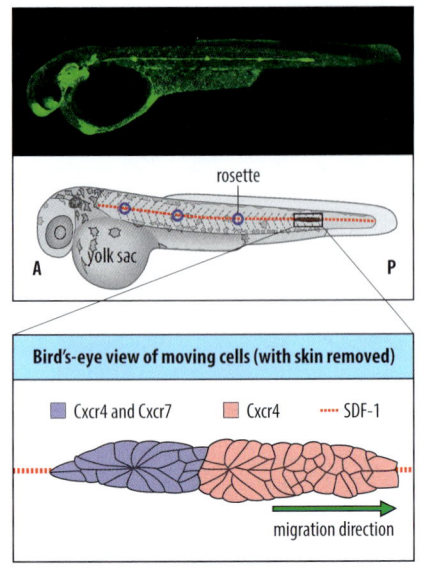

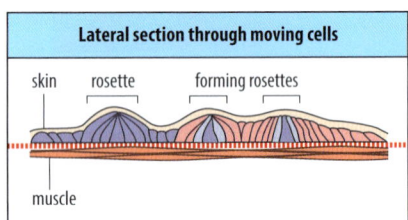

Fig. 5.51 Collective cell migration in zebrafish lateral-line formation. Top panel: photograph and explanatory drawing of zebrafish embryo showing the location of the developing lateral line. A, anterior; P, posterior. The transgenic embryo carries a glycosyl phosphatidylinositol-green fluorescent protein that labels membranes and enables visualization of the developing lateral line in the transparent embryo. As the lateral-line primordium moves along the route from the head, guided by a pre-patterned line of chemoattractant stromal cell-derived factor 1 (SDF-1), rosettes of cells that will form the lateral-line sensory organs are left behind at regular intervals. Center panel: a bird's-eye view of the migrating cells (the overlying skin has been removed). Cells at the tip of the migrating primordium express CXCR4, the receptor for SDF-1 (dotted pink line). Trailing cells also express the SDF-1 receptor CXCR7. Bottom panel: a section through the lateral-line primordium showing the beginning of rosette formation.

From Friedl, P., Gilmour, D.: *Collective cell migration in morphogenesis, regeneration and cancer.* Nat. Rev. Mol. Cell Biol. 2009, **10**: 445-457.

than a hundred cells, looking rather like a garden slug (Fig. 5.51), which moves as a collective along the flank of the embryo under the skin. It derives from the cranial ectoderm and moves from anterior to posterior along a track of the chemoattractant SDF-1 laid down along the myoseptum, the membrane connecting the underlying muscle cells. SDF-1 is sensed by its receptor CXCR4, which is expressed in all cells of the primordium but is only active in those at the leading edge, which drive the movement. In the primordium, cells are attached to each other by junctional complexes that are looser and more dynamic at the leading edge, where the cells have undergone a partial epithelial-to-mesenchymal transition. The structure is dynamic because as it moves along, the cells within it are dividing, and at regular intervals a small cluster of cells is left behind to form a sensory organ.

As we have seen, similar collective cell migrations occur during the development of the highly branched system of tracheal tubules in *Drosophila*, which conveys oxygen into the tissues (see Section 5.18). Leading cells at the tip of each branch follow a trail of signals, dragging the extending tracheal system behind them. Collective migration also occurs during branching morphogenesis of the vascular system in vertebrates (see Section 5.19) and in neural crest migration.

5.24 Body wall closure occurs in *Drosophila*, *Caenorhabditis*, mammals, and chick

Some morphogenetic events involve the movement of an intact epithelial sheet. The process of dorsal closure in the *Drosophila* embryo provides a good example of these movements. Half-way through *Drosophila* embryogenesis, when cell proliferation ceases and the germ band retracts, there is a gap in the dorsal epidermis. This gap is covered by a squamous epithelium, the amnioserosa, which is continuous with the epidermis. The gap is closed by the process of dorsal closure, which involves the coordinated movement of the two halves of the epidermis towards each other, accompanied by internalization of the amnioserosa in the embryo, where it undergoes apoptosis (Fig. 5.52).

The process relies on the specific structure and function of the cytoskeleton in the cells of the epidermis and the amnioserosa. In the epidermis, the cells in contact with the amnioserosa remodel their junctions in the plane of the epithelium and form a large supracellular actinomyosin cable whose contractions drive the coordinated expansion of the cells towards the midline, and hence the movement of the two sides of the epidermis towards each other. The process is aided by contraction of the cells of the amnioserosa, which provide a complementary force by pulling on the epidermal sheets. As the amnioserosa contracts, its cells fall inside the embryo and eventually undergo apoptosis and disappear.

Fig. 5.52 Dorsal closure in *Drosophila*. (a) Embryo at the beginning of dorsal closure. Expression of the gene *puckered* (brown), which encodes a phosphatase that regulates Jun N-terminal kinase signaling, defines cells at the leading edge of the epidermis. Once these cells have been defined they trigger an elongation of the epidermis (blue arrows) and a contraction of the cells of the amnioserosa (red arrows) which drives the organized closure of the epidermal gap. (b, c) During the final phases, the epidermal cells zip up the gap from both ends by means of filopodia developed on the leading-edge cells. (d) Visualization of expression of the gene *engrailed* (green) and *patched* (red) shows the precise alignment of cells at the join, with cells expressing *engrailed* fusing only with each other and *patched*-expressing cells fusing only with each other, thus preserving segment patterning across the join.

(a–c) Courtesy of Alfonso Martinez Arias; (d) from Millard, T.H., Martin, P.: **Dynamic analysis of filopodial interactions during the zippering phase of** *Drosophila* **dorsal closure**. Development 2008, **135**: 621–626.

The cells at the leading edge of the epidermis develop filopodia and lamellipodia, extending up to 10 μm from the edge, that have cadherin molecules on their surface. These structures do not provide force for the process but play a central part in the zippering of the two sides when they meet each other at the midline. We have also seen this general mechanism operating in neural tube closure.

The small GTPase Rac has a key role in organizing actin filaments at the leading edges of the cells, and is also involved through the Jun N-terminal kinase (JNK) intracellular signaling pathway in controlling changes in cell shape and in cell adhesion. JNK activates the expression of the transforming growth factor (TGF)-β family member Decapentaplegic (Dpp) in the dorsal-most epidermal cells, and Dpp acts as a secreted signal to control the elongation of the lateral epidermis. Dorsal closure also requires the segment patterning on the two sides to be in precise register with each other, and this is mediated by the actin-based extensions. For example, the gene *engrailed* is expressed in a series of transverse stripes along the length of the embryo, corresponding to the posterior compartment of each segment (see Section 2.23). Cells expressing *engrailed* on one edge interact only with cells expressing *engrailed* on the other edge, and cells expressing the gene *patched*, which form a different set of stripes, fuse only with *patched*-expressing cells on the other edge, thus neatly matching up the stripes across the seam (see Fig. 5.52).

The *Caenorhabditis elegans* embryo undergoes a process similar to dorsal closure, but on the ventral surface. At the end of gastrulation, the epidermis only covers the dorsal region and the ventral region is bare. The epidermis then spreads around the embryo until its edges finally meet along the ventral midline. Time-lapse studies show that the initial ventral migration is led by four cells, two on each side, which extend filopodia towards the ventral midline. Blocking filopodial activity by laser ablation or by cytochalasin D, which disrupts actin filaments and inhibits actin polymerization, prevents ventral closure, showing that the filopodia provide the driving force. Closure probably occurs by a zipper mechanism similar to that in the fly.

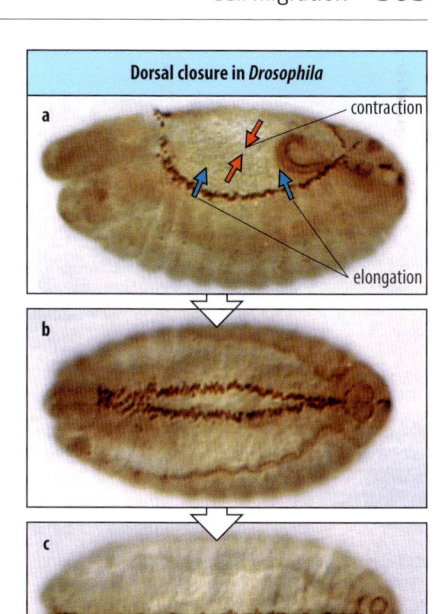

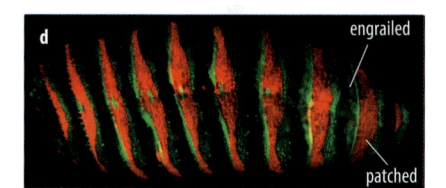

Ventral body wall closure also takes place in avian and mammalian embryos and is mediated by mechanisms related to those underlying dorsal closure in *Drosophila* and ventral closure in *C. elegans*. Closure is needed to encase the endodermal derivatives, most notably the gut, and involves the coordinated activity of several tissue elements, most importantly skin and muscle cells. It also requires the spatial coordination of cellular behavior along the antero-posterior and dorso-ventral axes. In mammals, the gut is at first covered by an epithelium and a thin layer of mesoderm known as the primary body wall. This is reinforced first by a wave of migrating myofibroblasts and then by cells such as muscle cells that migrate towards the midline. As in *Drosophila*, members of the TGF-β family of signaling molecules are involved in promoting the polarized migration of the cells—mesodermal derivatives in this case. Detailed measurements of human embryos suggest that preferential growth of the medio-lateral body wall is also a factor in ventral body closure. Failure of ventral closure in human embryos results in a large number of syndromes, one of which is gastroschisis, in which the gut does not return inside the body. A secondary consequence of the process of ventral closure is the closure of the umbilical cavity, and failures in the process lead to umbilical hernias.

SUMMARY

Cell migration is an important force in morphogenesis and can be observed in individual cells as well as in groups and even in epithelia. It allows the redistribution of cells or the reorganization of tissues. In the case of the neural crest, streams of cells migrate from both sides of the dorsal neural tube to form a wide range of different tissues in different parts of the body. Their migration is guided by signals from other cells and by interactions with the extracellular matrix. In neural crest cells migrating in the trunk, repulsive interactions between ephrins in the posterior halves of the somites and Eph receptors on the neural crest cells prevent their migration through the posterior halves. Thus, neural crest cells that will give rise to the dorsal root ganglia of the peripheral nervous system accumulate adjacent to the anterior half of each somite, resulting in a segmental arrangement of paired ganglia along the antero-posterior axis. The development of the lateral line in the fish is an example of collective migration of a group of cells that moves as a slug and sows sensory organs in the epidermis of the fish embryos. In both cases, the cells originate from an epithelium and undergo epithelial-to-mesenchymal transitions, complete in the case of the neural crest, partial in the case of the lateral line. In some instances, of which dorsal closure in *Drosophila* is one, what moves is a whole epithelium.

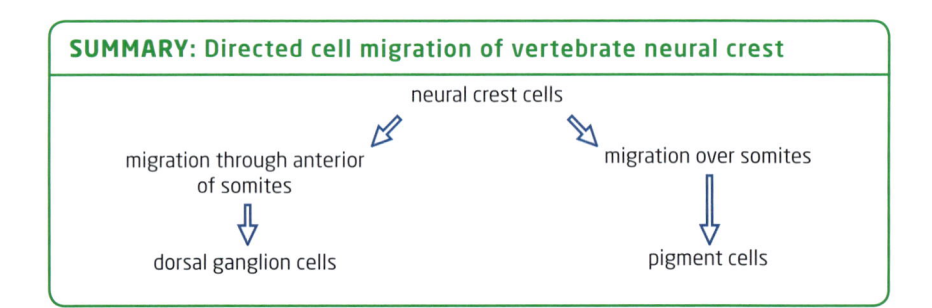

SUMMARY: Directed cell migration of vertebrate neural crest

neural crest cells

migration through anterior of somites → dorsal ganglion cells

migration over somites → pigment cells

Summary to Chapter 5

- Changes in the shape of animal embryos as they develop are mainly due to forces generated by differences in cell adhesiveness, changes in cell shape, and cell migration.
- Epithelia and mesenchyme are the two main types of cellular organization in embryos, and epithelial-to-mesenchymal transitions are a common feature of embryonic development. Mesenchymal-to-epithelial transitions also occur.
- Formation of a hollow, spherical blastula or blastocyst is the result of particular patterns of cell division, cell packing, and cell polarization.
- The spherical shape of a blastula or blastocyst is maintained by flow of fluid into the interior and the hydrostatic pressure generated.
- Gastrulation involves major movements of cells, such that the future endoderm and mesoderm move inside the embryo to their appropriate positions in relation to the main body plan.

- The process of convergent extension results in the narrowing and elongation of a tissue. It has a key role in many developmental contexts, such as elongation of the antero-posterior body axis during vertebrate gastrulation, elongation of the notochord, lengthening and closure of the neural tube, and germ-band extension in *Drosophila*.
- A prerequisite for convergent extension is planar cell polarity (PCP), a property of cells that gives them a polarity in the plane of the tissue and confers directionality on a tissue. PCP is based on a dedicated set of proteins, which regulate the spatial organization and activity of the cytoskeleton at the level of the single cell and which also coordinate the activity of individual cells at the tissue level.
- Formation of the vertebrate neural tube from the neural plate involves convergent extension that narrows the neural plate, and local cell-shape changes that cause bending of the plate and the edges to curve upwards to form the neural folds that then fuse.
- The development of the insect tracheal system and the vertebrate vascular system involves the formation of tubules via invagination of a pre-existing epithelium and a mesenchymal–epithelial transition, respectively; in both cases the tubules undergo branching morphogenesis via collective migration.
- Neural crest cells leave the dorsal neural tube and migrate to give rise to a wide range of different tissues throughout the body.
- The formation of the lateral line in fishes involves collective cell migration.
- Dorsal closure in *Drosophila* embryos and ventral closure in *Caenorhabditis* embryos is driven by changes in the cytoskeleton of the epidermal cells that brings two edges of epidermis together. Ventral closure in mammals and chick involves the migration of myofibroblasts and muscle cells.

■ End of chapter questions

Check your understanding of the content of this chapter by attempting the following long-answer concept questions.

1. Name the four types of adhesive interactions between cells and between cells and the extracellular matrix. What are the transmembrane proteins that characterize each of these adhesive interactions? What are the particular features of an adherens junction: what cell adhesion molecule is used, what cytoskeletal component is involved, and what links the adhesion molecule to the cytoskeleton?

2. Describe the results of an experiment in which epidermis and neural plate from a neurula stage amphibian embryo are first disaggregated into single cells, then mixed together and reaggregated. How is this experiment interpreted in terms of the relative strength of 'self' adhesiveness in these two tissues? How does this result reflect the natural organization of these tissues in the embryo?

3. Define 'radial cleavage'. In the radial cleavage of an embryo such as the sea urchin, how are the centrosomes positioned at the first cleavage, with respect to the animal–vegetal axis? The second cleavage? The third cleavage?

4. What is compaction of the mouse embryo? How is compaction related to two important subsequent events in mouse embryogenesis: the formation of the blastocoel and the formation of the inner cell mass?

5. Contrast the properties of an epithelial tissue with those of a mesenchymal tissue. How are cadherins related to the transition of epithelium to mesenchyme; how is this transition regulated, or triggered (think: '*snail*')?

6. Refer to the fate maps of sea urchins and *Xenopus* (Box 3A, Fig. 2): in sea urchins, mesoderm is most vegetal and endoderm lies between mesoderm and ectoderm; in *Xenopus*, endoderm is most vegetal and mesoderm lies between the endoderm and the ectoderm. Can you briefly describe the differences in gastrulation in these two organisms that explain these differing arrangements?

7. Contrast epiboly and convergent extension. How are these two processes used during *Xenopus* gastrulation?

8. If cells in a flat sheet constrict one surface (the apical surface), the sheet will develop a depression in that region. Give examples of how this mechanism is used to initiate gastrulation in sea urchins, *Drosophila*, and *Xenopus*.

9. How do convergent extension and changes in cell shape collaborate in the formation of the neural tube?

10. Describe the migration route of the neural crest cells that will form the dorsal root ganglia. What part do Ephs and ephrins play in this migratory pathway?

11. Summarize the general processes involved in blood vessel morphogenesis in vertebrates. What is the role of the VEGF pathway?

12. In your own words, describe the cell movements that occur during gastrulation, from the first appearance of the blastopore, to the first appearance of the neural folds. Do this first for *Xenopus*, then describe the major differences seen in chicks and zebrafish.

General further reading

Alberts, B., Heald, R., Johnson, A., Morgan, D., Raff, M., Roberts, K., Walter, P., Wilson, J., Hunt, T.: *Molecular Biology of the Cell.* 7th edn. New York: W.W. Norton & Company, 2022.

Aman, A., Piotrowski, T.: **Cell migration during morphogenesis**. *Dev. Biol.* 2010, **341**: 20–33.

Bray, D. *Cell Movements: From Molecules to Motility.* 2nd edn. Oxford: Routledge, 2000.

Engler, A.J., Humbert, P.O., Wehrle-Haller, B., Weaver, V.M.: **Multiscale modeling of form and function**. *Science* 2009, **324**: 209–212.

Lecuit, T., Lenne, P.F.: **Cell surface mechanics and the control of cell shape, tissue patterns and morphogenesis**. *Nat. Rev. Mol. Cell Biol.* 2007, **8**: 633–644.

Mattila, P.K., Lappalainen, P.: **Filopodia: molecular architecture and cellular functions**. *Nat. Rev. Mol. Cell Biol.* 2009, **9**: 446–454.

Stern, C.D. (Ed.): *Gastrulation: From Cells to Embryos.* New York: Cold Spring Harbor Laboratory Press, 2004.

Section further reading

5.1 Sorting out of dissociated cells demonstrates differences in cell adhesiveness in different tissues

Davis, G.S., Phillips, H.M., Steinberg, M.S.: **Germ-layer surface tensions and 'tissue affinities' in *Rana pipiens* gastrulae: quantitative measurements**. *Dev. Biol.* 1997, **192**: 630–644.

Krieg, M., Arboleda-Estudillo, Y., Puech, P.-H., Käfer, J., Graner, F., Müller, D.J., Heisenberg, C.-P.: **Tensile forces govern germ-layer organization in zebrafish**. *Nat. Cell Biol.* 2008, **10**: 429–436.

Lecuit, T., Lenne, P.-F.: **Cell surface mechanics and the control of cell shape, tissue patterns and morphogenesis**. *Nat. Rev. Mol. Cell Biol.* 2007, **8**: 633–644.

Steinberg, M.S.: **Differential adhesion in morphogenesis: a modern view**. *Curr. Opin. Genet. Dev.* 2007, **17**: 291–296.

Box 5A Cell-adhesion molecules and cell junctions and 5.2 Cadherins can provide adhesive specificity

Brizzi, M.F., Tarone, G., Defilippi, P.: **Extracellular matrix, integrins, and growth factors as tailors of the stem cell niche**. *Curr. Opin. Cell Biol.* 2012, **24**: 645–651.

Halbleib, J.M., Nelson, W.J.: **Cadherins in development: cell adhesion, sorting, and tissue morphogenesis**. *Genes Dev.* 2006, **20**: 3199–3214.

Hynes, R.O.: **Integrins: bidirectional allosteric signaling mechanisms**. *Cell* 2002, **110**: 673–699.

Hynes, R.O.: **The extracellular matrix: not just pretty fibrils**. *Science* 2009, **326**: 1216–1219.

Tepass, U., Truong, K., Godt, D., Ikura, M., Peifer, M.: **Cadherins in embryonic and neural morphogenesis**. *Nat Rev. Mol. Cell Biol.* 2000, **1**: 91–100.

Thiery, J.P.: **Cell adhesion in development: a complex signaling network**. *Curr. Opin. Genet. Dev.* 2003, **13**: 365–371.

Tsai, T.Y.-C., Sikora, M., Xia, P., Colak-Champollion, T., Knaut, H., Heisenberg, C.-P., Megason, S.G.: **An adhesion code ensures robust pattern formation during tissue morphogenesis**. *Science* 2020, **370**: 113–116.

Box 5B The cytoskeleton, cell-shape change, and cell movement and 5.3 The activity of the cytoskeleton regulates the mechanical properties of cells and their interactions with each other

Duguay, D., Foty, R.A., Steinberg, M.S.: **Cadherin-mediated cell adhesion and tissue segregation: qualitative and quantitative determinants**. *Dev. Biol.* 2003, **253**: 309–323.

Fagotto, F.: **Regulation of cell adhesion and cell sorting at embryonic boundaries**. *Curr. Top. Dev. Biol.* 2015, **112**: 19–64.

Levine, E., Lee, C.H., Kintner, C., Gumbiner, B.M.: **Selective disruption of E-cadherin function in early *Xenopus* embryos by a dominant negative mutant**. *Development* 1994, **120**: 901–909.

Munjal, A., Lecuit, T.: **Actomyosin networks and tissue morphogenesis**. *Development* 2014, **141**: 1789–1793.

Takeichi, M., Nakagawa, S., Aono, S., Usui, T., Uemura, T.: **Patterning of cell assemblies regulated by adhesion receptors of the cadherin superfamily**. *Proc. R. Soc. Lond. B* 2000, **355**: 995–996.

5.4 Transitions of tissues from an epithelial to a mesenchymal state, and vice versa, involve changes in adhesive junctions

Campbell, K., Casanova, J.: **A common framework for EMT and collective cell migration**. *Development* 2016, **143**: 4291–4300.

Ferrer-Vaquer, A., Viotti, M., Hadjantonakis, A.K.: **Transitions between epithelial and mesenchymal states and the morphogenesis of the early mouse embryo**. *Cell Adh. Migr.* 2010, **4**: 447–457.

Lim, J., Thiery, J.P.: **Epithelial–mesenchymal transitions: insights from development**. *Development* 2012, **139**: 3471–3486.

Nakaya, Y., Sheng, G.: **Epithelial to mesenchymal transition during gastrulation: an embryological view**. *Dev. Growth Differ.* 2008, **50**: 755–766.

Nieto, M.A.: **The ins and outs of the epithelial to mesenchymal transition in health and disease**. *Annu. Rev. Cell Dev. Biol.* 2011, **27**: 347–376.

Yang, J., Weinberg, R.A.: **Epithelial–mesenchymal transition: at the crossroads of development and tumor metastasis**. *Dev. Cell* 2008, **14**: 818–829.

5.5 The orientation of the mitotic spindle determines the plane of cleavage at cell division

Glotzer, M.: **Cleavage furrow positioning**. *J. Cell Biol.* 2004, **164**: 347–351.

Théry, M., Racine, V., Pépin, A., Piel, M., Chen, Y., Sibarita, J.-B., Bornens, M.: **The extracellular matrix guides the orientation of the cell division axis**. *Nat. Cell Biol.* 2005, **7**: 947–953.

5.6 The positioning of the spindle within the cell also determines whether daughter cells will be the same or different sizes

Galli, M., van den Heuvel, S.: **Determination of the cleavage plane in early *C. elegans* embryos**. *Annu. Rev. Genet.* 2009, **42**: 399–411.

Gönczy, P., Rose, L.S.: **Asymmetric cell division and axis formation in the embryo**. In *WormBook*. The *C. elegans* Research Community (Ed.). 15 October 2005. [http://www.wormbook.org/chapters/www_asymcelldiv/asymcelldiv.html] (date accessed 3 September 2024).

5.7 Cells become polarized in the sea urchin blastula and the mouse morula

Anani, S., Bhat, S., Honma-Yamanaka, N., Krawchuk, D., Yamanaka, Y.: **Initiation of Hippo signaling is linked to polarity rather than to cell position in the pre-implantation mouse embryo**. *Development* 2014, **141**: 2813–2824.

Korotkevich, E., Niwayama, R., Courtois, A., Friese, S., Berger, N., Buchholz, F., Hiiragi, T.: **The apical domain is required and sufficient for the first lineage segregation in the mouse embryo**. *Dev. Cell* 2017, **40**: 235–247.

Maître, J.L., Turlier, H., Illukkumbura, R., Eismann, B., Niwayama, R., Nédélec, F., Hiiragi, T.: **Asymmetric division of contractile domains couples cell positioning and fate specification**. *Nature* 2016, **536**: 344–348.

Maître, J.L.: **Mechanics of blastocyst morphogenesis**. *Biol. Cell* 2017, **109**: 323–338.

Nishioka, N., et al.: **The Hippo signaling pathway components Lats and Yap pattern Tead4 activity to distinguish mouse trophectoderm from inner cell mass**. *Dev. Cell* 2009, **16**: 398–410.

Pauken, C.M., Capco, D.G.: **Regulation of cell adhesion during embryonic compaction of mammalian embryos: roles for PKC and beta-catenin**. *Mol. Reprod. Dev.* 1999, **54**: 135–144.

Samarage, C.R., White, M.D., Álvarez, Y.D., Fierro-González, J.C., Henon, Y., Jesudason, E.C., Bissiere, S., Fouras, A., Plachta, N.: **Cortical tension allocates the first inner cells of the mammalian embryo**. *Dev. Cell* 2015, **34**: 435–447.

White, M.D., Plachta, N.: **How adhesion forms the early mammalian embryo**. *Curr. Top. Dev. Biol.* 2015, **112**: 1–17.

5.8 Fluid accumulation as a result of tight-junction formation, ion transport, and hydraulic fracturing forms the blastocoel of the mammalian blastocyst

Barcroft, L.C., Offenberg, H., Thomsen, P., Watson, A.J.: **Aquaporin proteins in murine trophectoderm mediate transepithelial water movements during cavitation**. *Dev. Biol.* 2003, **256**: 342–354.

Dumortier, J.G., Le Verge-Serandour, M., Tortorelli, A.F., Mielke, A., de Plater, L., Turlier, H., Maître, J.-L.: **Hydraulic fracturing and active coarsening position the lumen of the mouse blastocyst**. *Science* 2019, **365**: 465–468.

Eckert, J.J., Fleming, T.P.: **Tight junction biogenesis during early development**. *Biochim. Biophys. Acta* 2008, **1778**: 717–728.

Fleming, T.P., Sheth, B., Fesenko, I.: **Cell adhesion in the preimplantation mammalian embryo and its role in trophectoderm differentiation and blastocyst morphogenesis**. *Front. Biosci.* 2001, **6**: 1000–1007.

Watson, A.J., Natale, D.R., Barcroft, L.C.: **Molecular regulation of blastocyst formation**. *Anim. Reprod. Sci.* 2004, **82–83**: 583–592.

5.9 Gastrulation in the sea urchin involves an epithelial-to-mesenchymal transition, cell migration, and invagination of the blastula wall

Davidson, L.A., Oster, G.F., Keller, R.E., Koehl, M.A.R.: **Measurements of mechanical properties of the blastula wall reveal which hypothesized mechanisms of primary invagination are physically plausible in the sea urchin *Strongylocentrotus purpuratus***. *Dev. Biol.* 1999, **204**: 235–250.

Duloquin, L., Lhomond, G., Gache, C.: **Localized VEGF signaling from ectoderm to mesenchyme cells controls morphogenesis of the sea urchin embryo skeleton**. *Development* 2007, **134**: 2293–2302.

Ettensohn, C.A.: **Cell movements in the sea urchin embryo**. *Curr. Opin. Genet. Dev.* 1999, **9**: 461–465.

Gustafson, T., Wolpert, L.: **Studies on the cellular basis of morphogenesis in the sea urchin embryo: directed movements of primary mesenchyme cells in normal and vegetalized larvae**. *Exp. Cell Res.* 1999, **253**: 299–295.

Mattila, P.K., Lappalainen, P.: **Filopodia: molecular architecture and cellular functions**. *Nat. Rev. Mol. Cell Biol.* 2009, **9**: 446–454.

Miller, J.R., McClay, D.R.: **Changes in the pattern of adherens junction-associated β-catenin accompany morphogenesis in the sea urchin embryo**. *Dev. Biol.* 1997, **192**: 310–322.

Odell, G.M., Oster, G., Alberch, P., Burnside, B.: **The mechanical basis of morphogenesis. I. Epithelial folding and invagination**. *Dev. Biol.* 1991, **95**: 446–462.

Raftopoulou, M., Hall, A.: **Cell migration: Rho GTPases lead the way**. *Dev. Biol.* 2004, **265**: 23–32.

Röttinger, E., Saudemont, A., Duboc, V., Besnardeau, L., McClay, D., Lepage, T.: **FGF signals guide migration of mesenchymal cells, control skeletal morphogenesis and regulate gastrulation during sea urchin development**. *Development* 2009, **135**: 353–365.

5.10 Mesoderm invagination in *Drosophila* is due to changes in cell shape controlled by genes that pattern the dorso-ventral axis

Martin, A.C.: **The physical mechanisms of *Drosophila* gastrulation: mesoderm and endoderm invagination**. *Genetics* 2020, **214**: 543–560.

Martin, A.C., Kaschube, M., Wieschaus, E.F.: **Pulsed contractions of an actin–myosin network drive apical contraction**. *Nature* 2009, **457**: 495–499.

Rauzi, M., Krzic, U., Saunders, T.E., Krajnc, M., Ziherl, P., Hufnagel, L., Leptin, M.: **Embryo-scale tissue mechanics during *Drosophila* gastrulation movements**. *Nat. Commun.* 2015, **6**: 8677.

5.11 Germ-band extension in *Drosophila* involves myosin-dependent remodeling of cell junctions and cell intercalation

Bertet, C., Sulak, L., Lecuit, T.: **Myosin-dependent junction remodelling controls planar cell intercalation and axis elongation**. *Nature* 2004, **429**: 667–671.

Paré, A.C., Vichas, A., Fincher, C.T., Mirman, Z., Farrell, D.L., Mainieri, A., Zallen, J.A.: **A positional Toll receptor code directs convergent extension in *Drosophila***. *Nature* 2014, **515**: 523–527.

5.12 Planar cell polarity confers directionality on a tissue

Butler, M.T., Wallingford, J.B.: **Planar cell polarity in development and disease**. *Nat. Rev. Mol. Cell Biol.* 2017, **18**: 375–388.

Davey, C., Moens, C.: **Planar cell polarity in moving cells: think globally act locally**. *Development* 2017, **144**: 187–200.

Devenport, D.: **The cell biology of planar cell polarity**. *J. Cell Biol.* 2018, **207**: 171–179.

Ewen-Campen, B., Comyn, T., Vogt, E., Perrimon, N.: **No evidence that Wnt ligands are required for planar cell polarity in *Drosophila***. *Cell Rep.* 2020, **32**: 108121.

Goodrich, L.V., Strutt, D.: **Principles of planar polarity in animal development**. *Development* 2011, **138**: 1877–1892.

Jülicher, F., Eaton, S.: **Emergence of tissue shape changes from collective cell behaviours**. *Semin. Cell Dev. Biol.* 2017, **67**: 103–112.

Lawrence, P.A, Casal, J.: **The mechanisms of planar cell polarity, growth and the Hippo pathway: some known unknowns**. *Dev. Biol.* 2013, **377**: 1–8.

Lawrence, P.A., Casal, J.: **Planar cell polarity: two genetic systems use one mechanism to read gradients**. *Development* 2018, **145**: dev168229.

Lawrence, P.A., Casal, J., Struhl, G.: **Cell interactions and planar polarity in the abdominal epidermis of *Drosophila***. *Development* 2004, **131**: 4651–4664.

Lawrence, P.A., Struhl, G., Casal, J.: **Planar polarity: one or two pathways**. *Nat. Rev. Genet.* 2008, **8**: 555–563.

Ma, D., Yang, C., McNeill, H., Simon, M.A., Axelrod, J.D.: **Fidelity in planar cell polarity signalling**. *Nature* 2003, **421**: 543–547.

Simons, M., Mlodzik, M.: **Planar cell polarity signaling: from fly development to human disease**. *Annu. Rev. Genet.* 2008, **42**: 517–540.

Strutt, D.: **The planar polarity pathway**. *Curr. Biol.* 2008, **16**: R898–R902.

Strutt, D., Strutt, H.: **Differential activities of the core planar polarity proteins during *Drosophila* wing patterning**. *Dev. Biol.* 2007, **302**: 181–194.

Wu, J., Mlodzik, M.: **The Frizzled extracellular domain is a ligand for Van Gogh/Stm during nonautonomous planar cell polarity signaling**. *Dev. Cell* 2008, **15**: 462–469.

Wu, J., Roman, A.C., Carvajal-Gonzalez, J.M., Mlodzik, M.: **Wg and Wnt4 provide long-range directional input to planar cell polarity orientation in *Drosophila***. *Nat. Cell Biol.* 2013, **15**: 1045–1055.

Yang, Y., Mlodzik, M.: **Wnt-Frizzled/planar cell polarity signaling: cellular orientation by facing the wind (Wnt)**. *Annu. Rev. Cell Dev. Biol.* 2015, **31**: 623–646.

Yu, J.J.S., Maugarny-Calès, A., Pelletier, S., Alexandre, C., Bellaiche, Y., Vincent, J.-P., McGough, I.J.: **Frizzled-dependent planar cell polarity without secreted Wnt ligands**. *Dev. Cell* 2020, 54: 583–592.

5.13 Gastrulation in amphibians and fish involves involution, epiboly, and convergent extension

Dzamba, B.J., Jakab, K.R., Marsden, M., Schwartz, M.A., DeSimone, D.W.: **Cadherin adhesion, tissue tension, and noncanonical Wnt signaling regulate fibronectin matrix organization**. *Dev. Cell.* 2009, **16**: 421–432.

Goto, T., Davidson, L., Asashima, M., Keller, R.: **Planar cell polarity genes regulate polarized extracellular matrix deposition during frog gastrulation**. *Curr. Biol.* 2005, **15**: 787–793.

Heisenberg, C.P., Solnica-Krezel, L.: **Back and forth between cell fate specification and movement during vertebrate gastrulation**. *Curr. Opin. Genet. Dev.* 2008, **18**: 311–316.

Keller, R.: **Cell migration during gastrulation**. *Curr. Opin. Cell Biol.* 2005, **17**: 533–541.

Keller, R., Davidson, L.A., Shook, D.R.: **How we are shaped: the biomechanics of gastrulation**. *Differentiation* 2003, **71**: 171–205.

Montero, J.A., Carvalho, L., Wilsch-Brauninger, M., Kilian, B., Mustafa, C., Heisenberg, C.P.: **Shield formation at the onset of zebrafish gastrulation**. *Development* 2005, **132**: 1197–1199.

Montero, J.A., Heisenberg, C.P.: **Gastrulation dynamics: cells move into focus**. *Trends Cell Biol.* 2004, **14**: 620–627.

Ninomiya, H., Elinson, R.P., Winklbauer, R.: **Antero-posterior tissue polarity links mesoderm convergent extension to axial patterning**. *Nature* 2004, **430**: 364–367.

Ninomiya, H., Winklbauer, R.: **Epithelial coating controls mesenchymal shape change through tissue-positioning effects and reduction of surface-minimizing tension**. *Nat. Cell Biol.* 2009, **10**: 61–69.

Shih, J., Keller, R.: **Gastrulation in *Xenopus laevis*: involution—a current view**. *Semin. Dev. Biol.* 1994, **5**: 85–90.

Wacker, S., Grimm, K., Joos, T., Winklbauer, R.: **Development and control of tissue separation at gastrulation in *Xenopus***. *Dev. Biol.* 2000, **224**: 429–439.

Box 5C Convergent extension

Keller, R.: **Shaping the vertebrate body plan by polarized embryonic cell movements**. *Science* 2002, 298: 1950–1954.

Keller, R., Davidson, L., Edlund, A., Elul, T., Ezin, M., Shook, D., Skoglund, P.: **Mechanisms of convergence and extension by cell intercalation**. *Proc R. Soc. Lond. B* 2000, **355**: 897–922.

Torban, E., Kor, C., Gros, P.: **Van Gogh-like2 (Strabismus) and its role in planar cell polarity and convergent extension in vertebrates**. *Trends Genet.* 2004, **20**: 570–577.

Wallingford, J.B., Fraser, S.E., Harland, R.M.: **Convergent extension: the molecular control of polarized cell movement during embryonic development**. *Dev. Cell* 2002, **2**: 695–706.

Zallen, J.A.: **Planar polarity and tissue morphogenesis**. *Cell* 2007, **129**: 1051–1063.

5.14 *Xenopus* notochord development illustrates the dependence of medio-lateral cell elongation and cell intercalation on a pre-existing antero-posterior polarity

Keller, R., Cooper, M.S., D'Anilchik, M., Tibbetts, P., Wilson, P.A.: **Cell intercalation during notochord development in *Xenopus laevis***. *J. Exp. Zool.* 1989, **251**: 134–154.

5.15 Gastrulation in chick and mouse embryos involves the separation of individual cells from the epiblast and their ingression through the primitive streak

Arkell, R.M., Fossat, N., Tam, P.P.: **Wnt signalling in mouse gastrulation and anterior development: new players in the pathway and signal output**. *Curr. Opin. Genet. Dev.* 2013, **23**: 454–460.

Chuai, M., Hughes, D., Weijer, C.J.: **Collective epithelial and mesenchymal cell migration during gastrulation**. *Curr. Genomics* 2012, **13**: 267–277.

McDole, K., Guignard, L., Amat, F., Berger, A., Malandain, G., Royer, L.A., Turaga, S.C., Branson, K., Keller, P.J.: **In toto** imaging and reconstruction of post-implantation mouse development at the single-cell level. *Cell* 2018, **175**: 859–876.

Mongera, A., Michaut, A., Guillot, C., Xiong, F., Pourquié, O.: **Mechanics of anteroposterior axis formation in vertebrates.** *Annu. Rev. Cell Dev. Biol.* 2019, **35**: 259–283.

Saykali, B., Mathiah, N., Nahaboo, W., Racu, M.-L., Hammou, L., Defrance, M., Migeotte, I.: **Distinct mesoderm migration phenotypes in extra-embryonic and embryonic regions of the early mouse embryo.** *eLife* 2019, **8**: e42434.

Stower, M.J., Bertocchini, F.: **The evolution of amniote gastrulation: the blastopore-primitive streak transition.** *WIREs Dev. Biol.* 2017, e262.

Sun, X., Meyers, E.N., Lewandoski, M., Martin, G.R.: **Targeted disruption of Fgf8 causes failure of cell migration in the gastrulating mouse embryo.** *Genes Dev.* 1999, **13**: 1834–1846.

Voiculescu, O., Bertocchini, F., Wolpert, L., Keller, R.E., Stern, C.D.: **The amniote primitive streak is defined by epithelial cell intercalation before gastrulation.** *Nature* 2007, **449**: 1049–1052.

Williams, M., Burdsal, C., Periasamy, A., Lewandoski, M., Sutherland, A.: **Mouse primitive streak forms *in situ* by initiation of epithelial to mesenchymal transition without migration of a cell population.** *Dev. Dyn.* 2012, **241**: 270–283.

5.16 Differences in gastrulation between vertebrate species result from the global control of conserved mechanical features of cell populations

Sheng, G., Martinez Arias, A., Sutherland, A.: **The primitive streak and cellular principles of building an amniote body through gastrulation.** *Science* 2021, **374**: eabg1727.

Stower, M.J., Diaz, R.E., Carrera Fernandez, L., Crother, M.W., Crother, B., Marco, A., Trainor, P.A., Srinivas, S., Bertocchini, F.: **Bi-modal strategy of gastrulation in reptiles.** *Dev. Dyn.* 2015, **244**: 1144–1157.

5.17 Neural tube formation is driven by changes in cell shape and convergent extension

Colas, J.F., Schoenwolf, G.C.: **Towards a cellular and molecular understanding of neurulation.** *Dev. Dyn.* 2001, **221**: 117–145.

Copp, A.J., Greene, N.D.E., Murdoch, J.N.: **The genetic basis of mammalian neurulation.** *Nat. Rev. Genet.* 2003, **4**: 784–793.

Copp, A.J., Stanier, P., Greene, N.D.: **Neural tube defects: recent advances, unsolved questions, and controversies.** *Lancet Neurol.* 2013, **12**: 799–810.

Davidson, L.A., Keller, R.E.: **Neural tube closure in *Xenopus laevis* involves medial migration, directed protrusive activity, cell intercalation and convergent extension.** *Development* 1999, **126**: 4547–4556.

Haigo, S.L., Hildebrand, J.D., Harland, R.M., Wallingford, J.B.: **Shroom induces apical constriction and is required for hingepoint formation during neural tube closure.** *Curr. Biol.* 2003, **13**: 2125–2137.

Massarwa, R., Ray, H.J., Niswander, L.: **Morphogenetic movements in the neural plate and neural tube: mouse.** *Wiley Interdiscip. Rev. Dev. Biol.* 2014, **3**: 59–68.

Nikolopoulou, E., Galea, G.L., Rolo, A., Greene, N.D.E., Copp, A.J.: **Neural tube closure: cellular, molecular and biomechanical mechanisms.** *Development* 2017, **144**: 552–566.

Pai, Y.-J., Abdullah, N.L., Mohd-Zin, S.W., Mohammed, R.S., Rolo, A., Greene, N.D.E., Abdul-Aziz, N.M., Copp, A.J.: **Epithelial fusion during neural tube morphogenesis.** *Birth Defects Res. A Clin. Mol. Teratol.* 2012, **94**: 817–823.

Ray, H.J., Niswander, L.: **Mechanisms of tissue fusion during development.** *Development* 2012, **139**: 1701–1711.

Rolo, A., Skoglund, P., Keller, R.: **Morphogenetic movements during neural tube closure in *Xenopus* require myosin IIB.** *Dev. Biol.* 2009, **327**: 327–339.

Wallingford, J.B., Harland, R.M.: **Neural tube closure requires Dishevelled-dependent convergent extension of the midline.** *Development* 2002, **129**: 5915–5925.

Box 5D Eph receptors and their ephrin ligands

Holder, N., Klein, R.: **Eph receptors and ephrins: effectors of morphogenesis.** *Development* 1999, **126**: 2033–2044.

Poliakoff, A., Cotrina, M., Wilkinson, D.G.: **Diverse roles of Eph receptors and ephrins in the regulation of cell migration and tissue assembly.** *Dev. Cell* 2004, **7**: 465–490.

Xu, Q., Mellitzer, G., Wilkinson, D.G.: **Roles of Eph receptors and ephrins in segmental patterning.** *Proc. R. Soc. Lond. B* 2000, **353**: 993–1002.

Box 5E Neural tube defects

Butler, M.T., Wallingford, J.B.: **Planar cell polarity in development and disease.** *Nat. Rev. Mol. Cell Biol.* 2017, **18**: 375–388.

De Marco, P., Merello, E., Piatelli, G., Cama, A., Kibar, Z., Capra, V.: **Planar cell polarity gene mutations contribute to the etiology of human neural tube defects in our population.** *Birth Defects Res. A Clin. Mol. Teratol.* 2014, **100**: 633–641.

Greene, N.D.E., Stanier, P., Copp, A.J.: **Genetics of human neural tube defects.** *Hum. Mol. Genet.* 2009, **19**: R113–R129.

Juriloff, D.M., Harris, M.J.: **Mouse models for neural tube closure defects.** *Hum. Mol. Genet.* 2000, **9**: 993–1000.

Juriloff, D.M., Harris, M.J.: **A consideration of the evidence that genetic defects in planar cell polarity contribute to the etiology of human neural tube defects.** *Birth Defects Res. A Clin. Mol. Teratol.* 2012, **94**: 824–840.

Kibar, Z., Capra, V., Gros, P.: **Toward understanding the genetic basis of neural tube defects.** *Clin. Genet.* 2007, **71**: 295–310.

Torban, E., Patenaude, A.M., Leclerc, S., Rakowiecki, S., Gauthier, S., Andelfinger, G., Epstein, D.J., Gros, P.: **Genetic interaction between members of the Vang1 family causes neural tube defects in mice.** *Proc. Natl. Acad. Sci. USA* 2009, **105**: 3449–3454.

Wallingford, J.B., Niswander, L.A., Shaw, G.M., Finnell, R.H.: **The continuing challenge of understanding, preventing, and treating neural tube defects.** *Science* 2013, **339**: 1222002.

5.18 The *Drosophila* tracheal system is a prime example of branching morphogenesis

Affolter, M., Caussinus, E.: **Branching morphogenesis in *Drosophila*: new insights into cell behaviour and organ architecture.** *Development* 2008, **135**: 2055–2064.

Ochoa-Espinosa, A., Harmansa, S., Caussinus, E., Affolter, M.: **Myosin II is not required for *Drosophila* tracheal branch elongation and cell intercalation**. *Development* 2017, **144**: 2961–2968.

Ribeiro, C., Neumann, M., Affolter, M.: **Genetic control of cell intercalation during tracheal morphogenesis in *Drosophila***. *Cell* 2004, **14**: 2197–2207.

Spurlin, J.W., Nelson, C.M.: **Building branched structures: from single cell guidance to coordinated construction**. *Phil. Trans. R. Soc. B* 2017, **372**: 20150527.

5.19 The vertebrate vascular system develops by vasculogenesis followed by sprouting angiogenesis

Coultas, L., Chawengsaksophak, K., Rossant, J.: **Endothelial cells and VEGF in vascular development**. *Nature* 2005, **438**: 937–945.

Geudens, I., Gerhardt, H.: **Coordinating cell behavior during blood vessel formation**. *Development* 2011, **138**: 4569–4583.

Harvey, N.L., Oliver, G.: **Choose your fate: artery, vein or lymphatic vessel?** *Curr. Opin. Genet. Dev.* 2004, **14**: 499–505.

Larrivée, B., Freitas, C., Suchting, S., Brunet, I., Eichmann, A.: **Guidance of vascular development: lessons from the nervous system**. *Circ. Res.* 2009, **104**: 428–441.

Rossant, J., Hirashima, M.: **Vascular development and patterning: making the right choices**. *Curr. Opin. Genet. Dev.* 2003, **13**: 408–412.

Sato, Y., Poynter, G., Huss, D., Filla, M.B., Czirok, A., Rongish, B.J., Little, C.D., Fraser, S.E., Lansford, R.: **Dynamic analysis of vascular morphogenesis using transgenic quail embryos**. *PLoS One* 2010, **5**: e12674.

Spurlin, J.W., Nelson, C.M.: **Building branched structures: from single cell guidance to coordinated construction**. *Phil. Trans. R. Soc. B* 2017, **372**: 20150527.

Wang, H.U., Chen, Z.-F., Anderson, D.J.: **Molecular distinction and angiogenic interaction between embryonic arteries and veins revealed by ephrin-B2 and its receptor Eph-B4**. *Cell* 1998, **93**: 741–753.

5.20 New blood vessels are formed from pre-existing vessels in angiogenesis

Betz, C., Lenard, A., Belting, H.G., Affolter, M.: **Cell behaviors and dynamics during angiogenesis**. *Development* 2016, **143**: 2249–2260.

Jain, R.K., Carmeliet, P.: **Snapshot: tumor angiogenesis**. *Cell* 2012, **149**: 1408–1408.

5.21 Embryonic neural crest gives rise to a wide range of different cell types and 5.22 Neural crest migration is controlled by environmental cues

Bronner-Fraser, M.: **Mechanisms of neural crest migration**. *BioEssays* 1993, **15**: 221–230.

Kuan, C.Y., Tannahill, D., Cook, G.M., Keynes, R.J.: **Somite polarity and segmental patterning of the peripheral nervous system**. *Mech. Dev.* 2004, **121**: 1055–1069.

Kuriyama, S., Mayor, R.: **Molecular analysis of neural crest migration**. *Phil. Trans. R. Soc. Lond. B Biol. Sci.* 2009, **363**: 1349–1362.

Nagawa, S., Takeichi, M.: **Neural crest emigration from the neural tube depends on regulated cadherin expression**. *Development* 1999, **125**: 2963–2971.

Stark, D.A., Kulesa, P.M.: **An *in vivo* comparison of photoactivatable fluorescent proteins in an avian embryo model**. *Dev. Dyn.* 2007, **236**: 1583–1594.

Thevenean, E., Marchant, L., Kuriyama, S., Gull, M., Moeppo, B., Parsons, M., Mayor, R.: **Collective chemotaxis requires contact-dependent cell polarity**. *Dev. Cell* 2010, **19**: 39–53.

Tucker, R.P.: **Neural crest cells: a model for invasive behavior**. *Int. J. Biochem. Cell Biol.* 2004, **36**: 173–177.

5.23 The formation of the lateral-line primordium in fishes is an example of collective cell migration

Friedl, P., Gilmour, D.: **Collective cell migration in morphogenesis, regeneration and cancer**. *Nat. Rev. Mol. Cell Biol.* 2009, **10**: 445–457.

Haas, P., Gilmour, D.: **Chemokine signaling mediates self-organizing tissue migration in the zebrafish lateral line**. *Dev. Cell* 2006, **10**: 673–680.

5.24 Body wall closure occurs in *Drosophila*, *Caenorhabditis*, mammals, and chick

Aldeiri, B., Roostalu, U., Albertini, A., Wong, J., Morabito, A., Cossu, G.: **Transgelin-expressing myofibroblasts orchestrate ventral midline closure through TGFβ signalling**. *Development* 2017, **144**: 3336–3348.

Chin-Sang, I.D., Chisholm, A.D.: **Form of the worm: genetics of epidermal morphogenesis in *C. elegans***. *Trends Genet.* 2000, **16**: 544–551.

Jacinto, A., Woolner, S., Martin, P.: **Dynamic analysis of dorsal closure in *Drosophila*: from genetics to cell biology**. *Dev. Cell* 2002, **3**: 9–19.

Millard, T.H., Martin, P.: **Dynamic analysis of filopodial interactions during the zippering phase of *Drosophila* dorsal closure**. *Development* 2008, **135**: 621–626.

Peralta, X.G., Toyama, Y., Hutson, M.S., Montague, R., Venakides, S., Kiehart, D.P., Edwards, G.S.: **Upregulation of forces and morphogenic asymmetries in dorsal closure during *Drosophila* development**. *Biophys. J.* 2007, **92**: 2583–2596.

Woolner, S., Jacinto, A., Martin, P.: **The small GTPase Rac plays multiple roles in epithelial sheet fusion—dynamic studies of *Drosophila* dorsal closure**. *Dev. Biol.* 2005, **292**: 163–173.

6

Cell differentiation and stem cells

- **The control of gene expression**
- **Cell differentiation and stem cells**
- **The plasticity of the differentiated state**

The differentiation of unspecialized cells into many different cell types occurs first in the developing embryo and continues after birth and throughout adulthood. The character of specialized cells, such as blood, muscle, or nerve cells, is the result of a particular pattern of gene activity, which determines which proteins are synthesized. Thus, how this particular pattern of gene expression develops and how it is maintained are central questions in cell differentiation. These questions are discussed in this chapter in relation to early mouse development and well-studied examples of mammalian cell differentiation. Gene expression is under a complex set of controls that include the actions of transcription factors and of epigenetic mechanisms such as chemical modification of DNA and post-translational modifications to chromatin proteins. Extracellular signals have a key role in differentiation by triggering intracellular signaling pathways that affect gene expression. We shall see how signaling proteins and transcription factors are organized into gene regulatory networks that exert tight control over how cells make decisions about cell fate and differentiation and integrate these processes over time. Techniques have now been developed that enable us to survey which genes are expressed by a single cell during development. These technical advances provide an unprecedented vision of the process of cell-type assignment over time. We shall also see how, after birth, many tissues are renewed throughout life from their own adult stem cells, which are capable of both self-renewing and differentiating into the range of cell types needed to renew that tissue. Another intriguing question in development is the degree of plasticity of the differentiated state: can a differentiated cell change into another cell type? Such changes are possible, although rare under normal circumstances. And it is now well established that a nucleus from a differentiated cell can be reprogrammed when transferred into an enucleated egg and become able to direct embryonic development—the process that underlies animal cloning.

Chapter 6 Cell differentiation and stem cells

Introduction

Embryonic cells start out morphologically similar to each other but eventually become different, acquiring distinct identities and specialized functions. This is a progressive process and, as we have seen in previous chapters, it involves transient changes in cell shape, patterns of gene activity, and proteins synthesized. **Cell differentiation** involves the gradual emergence of specialized cell types that have a stable clear-cut identity and specialized function, such as nerve cells, red blood cells, and fat cells (Fig. 6.1). Initially, embryonic cells fated to become different cell types can be distinguished from each other only by small differences in patterns of gene expression, and thus in the proteins produced. The process of differentiation results in a gradual restriction of potential fates over successive cell generations as cells gradually acquire structural and functional features associated with their specialized functions (Box 6A).

As with earlier processes in development, the central feature of cell differentiation is a progressive change in the set of genes being expressed by the cell. All cells continue to produce the housekeeping proteins concerned with general metabolism, but as cells differentiate, they also begin to express proteins tailored to the specialized function of the differentiated cell. Red blood cells produce the oxygen-carrying protein hemoglobin; skin epidermal cells produce the fibrous protein keratin, which helps to give skin its resilience; neurons produce voltage-gated ion channels and the various neurotransmitters that enable them to produce, receive, and interpret electrical signals; and muscle cells produce muscle-specific actins and myosins that enable contraction. It is important to remember that out of the several thousand genes active

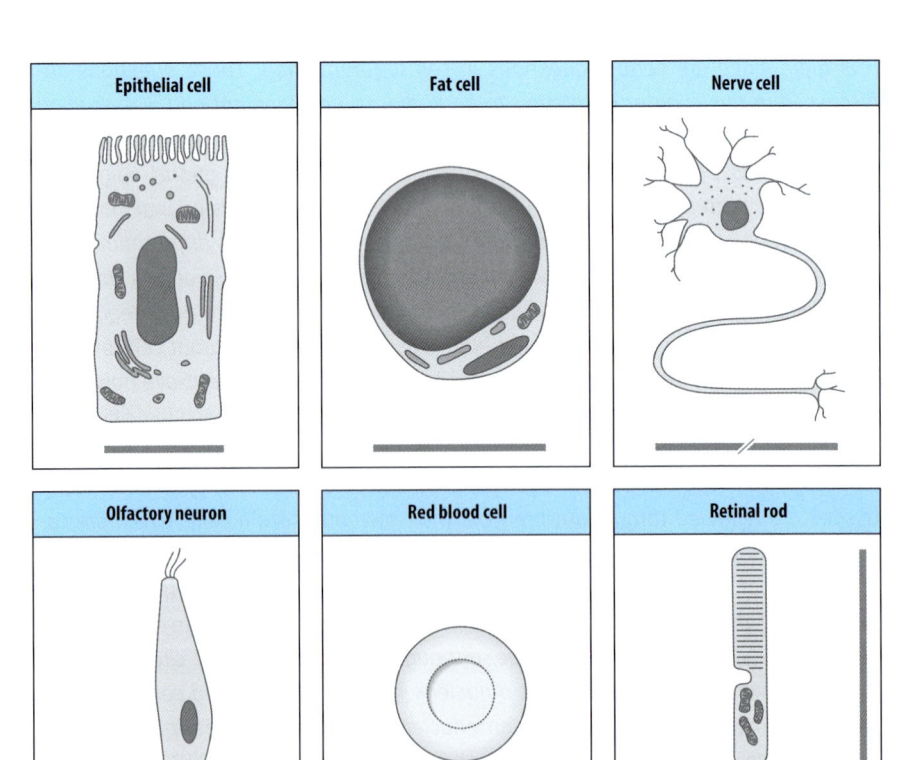

Fig. 6.1 Differentiated cell types.
Mammalian cell types come in various shapes and sizes. Scale bars: epithelial cell = 15 μm; fat cell = 100 μm; nerve cell = 100 μm-1 m; olfactory neuron = 8 μm; red blood cell = 8 μm; retinal rod = 20 μm.

BOX 6A Conrad Waddington's 'epigenetic landscape' provides a framework for thinking about how cells differentiate

It is sometimes difficult to visualize the process of cell differentiation. In the 1940s, the developmental biologist Conrad Waddington suggested how cell-fate choices might occur in the developing embryo and imagined differentiating cells (the blue marbles in Fig. 1) as entities rolling down a hill in which bifurcating furrows separate and channel them from less- to more-specialized states; he described these scenarios as the 'epigenetic landscape'. At the bottom of the hill, deep basins represent the differentiated states; cells cannot progress beyond them. The descent represents the differentiation process, and as the cells progress along it, they encounter bifurcations at which they make choices by falling into one of alternative furrows that will channel them to specific basins.

Figure 1 shows two different pathways that could be followed by the progeny of the original uncommitted, undifferentiated embryonic cell.

Waddington coined the term 'epigenetics' ('above genetics') in the context of development, but he was not referring to modifications of DNA and histone proteins, which were not known at the time, but to the observation that the choices and transitions that cells undergo during development do not result from the action of individual genes, but to a higher-order cooperative activity of their products. The 'epigenetic landscape' was a metaphor that synthesized many observations and thoughts but has become a very helpful representation of the process of differentiation. For example, the hierarchical and bifurcating process that unfolds during blood cell differentiation (discussed in this chapter) can be mapped onto it. The landscape has two implicit principles: one is that fate choices occur at special bifurcation points, which are binary; the other is that during development, cells follow paths—the furrows—that might be predetermined. The first is, as you will see in this chapter, mostly true. The second probably reflects the existence of gene-expression programs of differentiation. Molecular biological analysis of gene expression in differentiating cells has, in general, supported Waddington's prescient ideas.

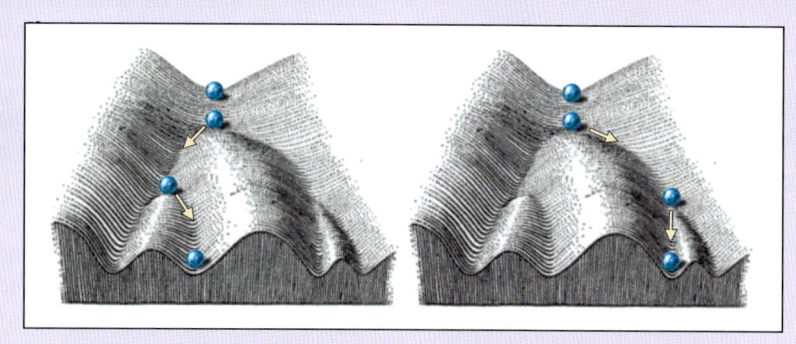

Fig. 1

From Noble, D.: *Conrad Waddington and the origin of epigenetics.* J. Exp. Biol. 2015, **218**: 816-818; Waddington, C.H.: The Strategy of the Genes. London: George, Allen and Unwin, 1957. Reprinted 2014.

---

in any cell in the embryo at any given time, only a small number might be involved in specifying cell fate or differentiation, most of which encode transcription factors; the rest are involved in the construction and maintenance of the cell. The genes that are being expressed in a particular tissue or at a particular stage of development can be detected by the sequencing and identification of the mRNAs produced—a technique called RNA-seq. When finer discrimination is needed, it is now possible to analyze mRNA expression at the level of single cells.

In contrast to the transitory differences in gene activity characteristic of earlier stages in development, fully differentiated cells achieve a stable state in which they undergo no further changes and do not change into another cell type. In this state they are called **terminally differentiated**. A fully differentiated cell has often undergone considerable structural change. For example, mature mammalian red blood cells lose their nucleus and become biconcave discs stuffed with hemoglobin, whereas neutrophils, a type of white blood cell, develop a multi-lobed nucleus and a cytoplasm filled with secretory granules containing proteins involved in protecting the body against infection. Yet both cells are derived from the same early lineage of blood-forming (hematopoietic) cells.

The first step in differentiation is for a cell to become committed, or determined, to becoming a particular cell type (see Section 1.12). This process begins at gastrulation when the progenitors of all tissues are laid out. At the earliest stages, progenitors are still multipotent and each can usually give rise to several different cell types in the tissue. As development progresses, cells become determined for a particular cell type and become precursor cells that will only differentiate into that type. At this time, the earliest precursors have no obvious structural differences from each other. The precursors of cartilage and muscle cells, for example, look the same and might be described as undifferentiated, but will differentiate as cartilage and muscle, respectively. There are, however, subtle differences in the sets of genes expressed by these precursor cells, and thus in the proteins produced, which control their future development. **Determination** is a key stage preceding differentiation. Cells that have become determined with respect to their eventual fate will form only the appropriate cell types, even if grafted to a different site in the embryo: that is, they retain their identity (see Fig. 1.23). Once determined for a particular fate, cells pass on that determined state to all their progeny.

In some cases, expression of a single gene can determine how a cell will commit to differentiation. The gene *Myod*, for example, encodes a transcription factor, MyoD, that is essential for skeletal muscle differentiation, and if it is introduced experimentally into fibroblasts—a cell type that does not normally express *Myod*—the fibroblasts will express muscle-specific genes and develop into muscle cells. *Myod* is therefore often called a **master regulatory gene** for muscle development and was the first such gene to be identified. Master genes usually encode transcription factors, and their expression is both necessary and sufficient to trigger activation of many other genes in a specific program leading to the development of a particular tissue or organ.

The initiation and progress of cell differentiation is under strict control by extracellular signals produced by other cells. These range from cell-surface proteins to secreted signal proteins, such as growth factors, and molecules of the extracellular matrix. These signals result in the activation of a core set of transcription factors, which leads to the expression of the specific genes that implement the differentiation process. It is important to remember that, while the extracellular signals that stimulate differentiation are often referred to as being 'instructive', they are, in general, 'selective', in the sense that, in practice, the number of developmental options open to a cell at any given time is limited. These options are set by the cell's internal state—the transcription factors it expresses—which, in turn, reflects its developmental history. Extracellular signals do not, for example, convert an endodermal cell into a muscle or nerve cell. However, extracellular signals acting at a crucial time, when the cell is choosing between alternative fates, can tip the cell into one fate or another. They do so by triggering a sequence of intracellular molecular events that reach the nucleus and modulates the activity of the transcription factors it expresses. These restrictions mean that the signals we have seen acting in early vertebrate development, such as Wnts, fibroblast growth factors (FGFs), Notch, and members of the transforming growth factor (TGF)-β family (see Box 3F), can continue to be used in determination and differentiation with different effects, as these will depend on the transcriptional state of the cell.

One feature of fully differentiated cells is that they divide far less frequently than undifferentiated embryonic cells or cease to divide altogether. Embryonic cells continue to proliferate after they have been determined, but cessation of cell division is necessary for full cell differentiation to occur. Some cells, such as skeletal muscle and many nerve cells, do not divide at all after they have become fully differentiated. In cells that can divide after differentiation, such as fibroblasts and liver cells, the differentiated state, like the determined state, is passed on through subsequent cell divisions. As the pattern of gene activity is the key feature in cell differentiation, this raises the question of how a particular pattern of gene activity is first established, and then how it is passed on to daughter cells.

We therefore start this chapter by considering the general mechanisms by which a pattern of gene activity can be established and maintained in a specific cell lineage.

We then look at recent advances in the analysis of cell-fate decisions at a high level of resolution and show one application of this to the earliest points of cell commitment in mammalian embryos during gastrulation. Further on, we look at the molecular basis of terminal cell differentiation in specific instances, with muscle cells and blood cells providing our main examples.

This chapter concentrates on the mechanisms of specification and differentiation in animal cells. As we will see in Chapter 12, differentiated plant cells do not have a permanently determined state, and a small piece of tissue, or even a single somatic cell, can readily give rise to a whole plant.

The control of gene expression

Every nucleus in the body of a multicellular organism is derived from the single zygotic nucleus in the fertilized egg and thus carries the same genome, which in humans comprises around 20,000 distinct protein-coding genes. But the subsets of these genes that are active in differentiated cells vary enormously from one cell type to another. The egg itself has a pattern of gene activity that is different from that of the embryonic cells to which it gives rise. So how do distinct and unique combinations of genes come to be expressed in each differentiated cell type? Although there are only a few hundred morphologically recognizable basic mammalian cell types, there are, in fact, millions of functionally distinct cells in the mammalian body, each with its particular pattern of gene activity. Think, for example, of the number of functionally different neurons required to enable the nervous system to work. There are fewer genes in the genome than types of differentiated cells that need to be specified; for example, in the human brain alone, there are about 100 billion neurons, most of them unique, which have to be specified by a genome of around 20,000 genes. Each distinct cell type must therefore be defined by the expression of a combination of genes rather than by a single gene (Fig. 6.2). This raises the question of how the particular pattern of gene activity in a differentiated cell is specified and how it is inherited.

To understand the molecular basis of cell differentiation, we first need to know how a gene can be expressed in a cell-specific manner—why a given gene gets switched on in one cell and not in another. A second question is how the particular combination of genes expressed by a differentiated cell comes to be selected, and how their expression is coordinated. A final, and difficult, question is how the genes expressed at any given time in a particular cell are related to its phenotype. We focus here on the regulation of **transcription**, which is the first (and generally the most significant) step in the expression of a gene and the subsequent synthesis of a protein (see Section 1.10). **RNA splicing** and **translation** of the mRNA to produce a protein,

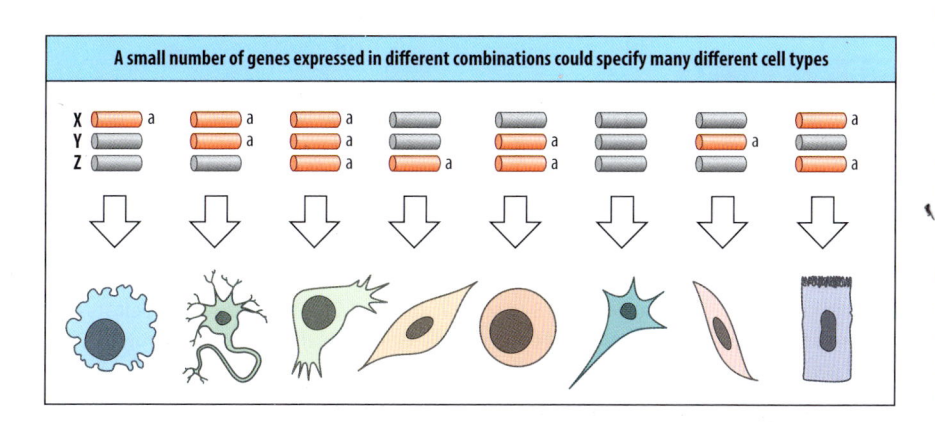

Fig. 6.2 Different cell types are the result of the expression of different combinations of genes. To make the most efficient use of the relatively modest numbers of genes in the genome, different functional cell types are specified by the expression of different combinations of genes. This enables many more cell types to be specified than if each gene was only able to uniquely specify one cell type. This hypothetical example shows how a set of three genes (X, Y, and Z) could specify eight different cell types by the unique combinations of those genes that are active (indicated by the red bars) and inactive (gray bars) in each cell type.

however, are also subject to several layers of control. We have seen an example of translational regulation in early *Drosophila* development, where the Nanos protein controls the production of Hunchback protein by preventing translation of maternal *hunchback* RNA (see Section 2.9). **Alternative splicing**, the splicing of an initial RNA transcript in different ways to give mRNAs encoding different proteins, is another important means of adding to the diversity of proteins that can be produced from a relatively limited number of genes. A dramatic example of alternative splicing in *Drosophila* is in the Down syndrome cell-adhesion molecule gene (*Dscam*), which is involved in the specification of neuronal connections. This gene has 95 exons, from which alternative splicing could potentially generate 38,016 different proteins. But despite these additional layers of control, it is, for the most part, transcriptional regulation of gene expression that underlies the process of differentiation.

6.1 Control of transcription involves both general and tissue-specific transcriptional regulators

Most of the key genes in development are initially in an inactive state and require activating transcription factors, or **activators**, to turn them on. The inactivity often results either because of active repression—which is very common—or because the gene is only activated when many different activator proteins come together on the same piece of DNA in the same cell at the same time. These activators bind to sites in the *cis*-regulatory regions in the DNA surrounding the gene (the '*cis*' simply refers to the fact that the regulatory region is on the same DNA molecule as the gene it controls). The control site to which an individual transcription factor binds is a short stretch of nucleotides with a sequence specifically recognized by the factor. Sites to which transcription factors bind are often clustered within **control regions** called **enhancers**. The importance of control regions in tissue-specific gene expression was clearly demonstrated by the observation that their deletion abolishes the expression of the gene they are associated with, and also by experiments in which the control region of one tissue-specific gene was replaced by the control region of another (Fig. 6.3). We have already seen how such replacements are routinely used in experimental developmental biology to force the expression or misexpression of a gene in a particular tissue, and to mark cells expressing a particular gene in order to investigate its developmental function (see Box 3B and Box 2D).

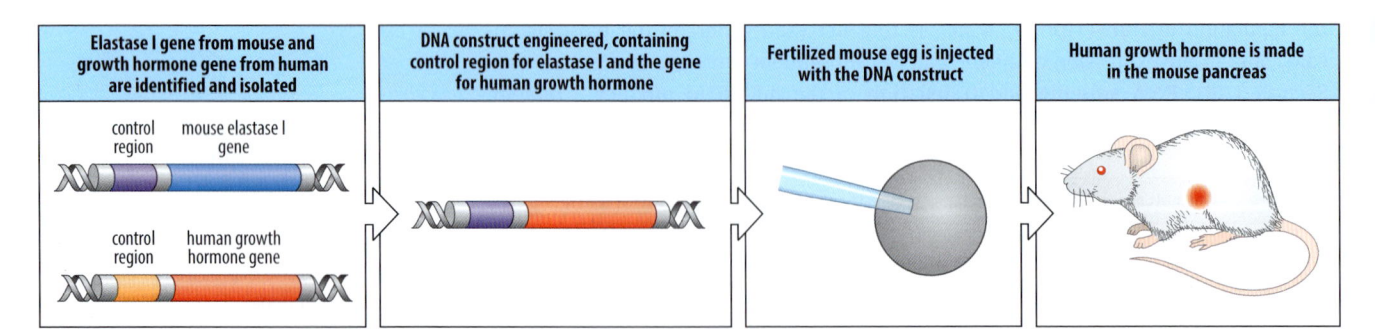

Fig. 6.3 Tissue-specific gene expression is controlled by regulatory regions. Growth hormone is normally made in the pituitary and the enzyme elastase I only in the pancreas. To demonstrate that the regulatory regions of a gene determine its tissue-specific expression, the control region of the mouse elastase I gene was joined to a DNA sequence coding for human growth hormone. This DNA construct was injected into the nucleus of a fertilized mouse egg, where it becomes integrated into the genome. When the mouse develops, human growth hormone is made in the pancreas. A 213-base-pair stretch of DNA that includes the elastase I promoter and other control regions is sufficient to drive growth hormone expression in the pancreas.

In eukaryotic cells, most regulated protein-coding genes are transcribed by RNA polymerase II. The polymerase binds to a control region in the gene called the **promoter**, where it is positioned so that it can start transcription in the correct place. The binding of RNA polymerase II to the promoter requires the cooperation of a set of **general transcription factors**, so called because they are required by all genes transcribed by RNA polymerase II. They form a **transcription initiation complex** with the polymerase at the promoter. For highly regulated genes, such as those involved in development, this complex cannot be formed and activated without additional help. This is provided by the binding of regulatory proteins to specific enhancers, defined by nucleotide sequences in the control regions of the gene, which help to attract the general transcription factors and polymerase, and position them correctly on the promoter (Fig. 6.4). Some control sites, including those within some enhancers, are alternatively bound by **repressors**, gene-regulatory proteins that prevent a gene being expressed by blocking access to the transcription initiation complex. The activation process requires the eviction of these repressors or the transformation of their activity.

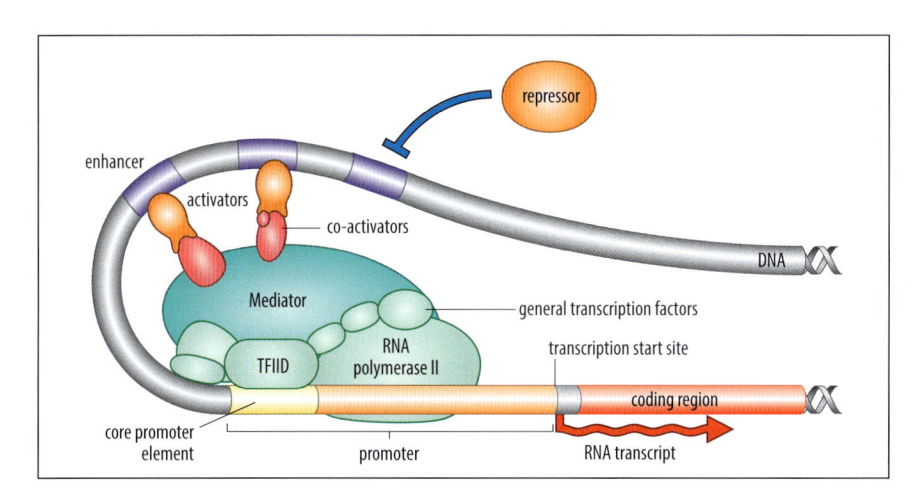

Fig. 6.4 Gene expression is regulated by the coordinated action of gene-regulatory proteins that bind to regulatory regions in DNA. Most protein-coding genes in eukaryotic cells, including all highly regulated developmental genes, are transcribed by RNA polymerase II. The transcriptional machinery itself, composed of the polymerase and its associated general transcription factors (the core transcriptional machinery, e.g. TFIID and the large multiprotein complex called Mediator), recognizes the promoter region by binding to specific sequences within it that are common to all genes transcribed by RNA polymerase II. These core promoter sequences are platforms for the assembly of the core transcriptional complex and determine the nucleotide at which transcription will start (the transcription start site). Other control elements are gene specific and can be occupied by specific activator or repressor proteins together with their associated co-factors. These proteins interact with the polymerase and associated proteins to either initiate or prevent transcription. Thus, the particular combination of activators and repressors occupying the binding sites within a control element such as an enhancer determines whether a gene is expressed or not in a particular cell. Control elements can be located adjacent to the promoter, or further upstream, or sometimes downstream, of the coding region of the gene. For some genes, control elements can be many kilobases away from the start point of transcription. The proteins bound to these distant sites are brought into contact with the transcription machinery by looping of the DNA. The regulation of gene expression requires the connection of activators and repressors with the RNA polymerase II complex, and this is effected by the Mediator complex. The fully operational initiation complex activates the RNA polymerase and releases it to start transcription.

*Illustration after Tjian, R.: **Molecular machines that control genes.** Sci. Am. 1995, **272**: 54-61.*

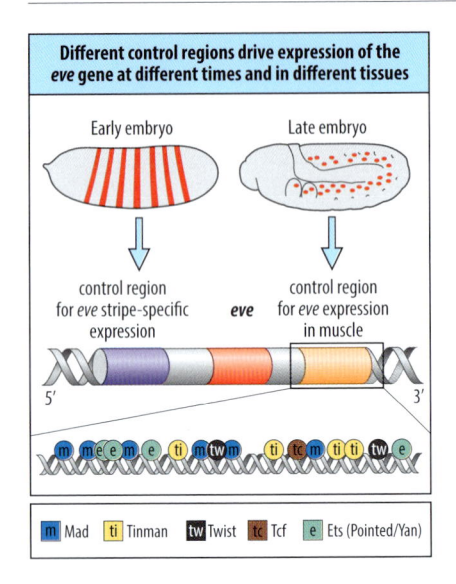

Fig. 6.5 Changes in the expression of the *Drosophila even-skipped* gene throughout development are due to the use of different control regions. The *even-skipped* (*eve*) gene in *Drosophila* is expressed in a pattern of transverse stripes in the early embryo, and this expression is controlled by regulatory modules (one for each stripe) located on the 5' side of the gene (described in Fig. 2.37, not shown in detail here). In the late embryo, *eve* is expressed in muscle precursors, under the control of a different regulatory region on the 3' side of the gene. Illustrated here is the combination of transcription factors that binds to that 3' region to regulate gene expression in a particular muscle group in each segment. Five different transcription factors bind at 17 different sites in this region to switch on *eve* expression. Some of the transcription factors are effectors of signaling pathways: Mad (BMP, TGF-β), Tcf (Wnt), and Ets (FGF). Different combinations of factors control *eve* expression in the other muscle groups.

Fig. 6.6 A single transcription factor can regulate the expression of different genes in a positive or negative fashion. Transcription factors can activate or repress the activity of genes by binding to their control regions. In the illustration, the production of transcription factor X in the cell eventually leads to the production of four new proteins (A, B, C, and E), and the repression of protein D production.

Illustration after Alberts B., et al.: Molecular Biology of the Cell. 2nd edn. New York: Garland Publishing, 1989.

For any given gene, when and where it is expressed at any given stage of development is determined by particular combinations of gene-regulatory proteins binding to individual sites in the gene's control regions (Fig. 6.5). These sites are typically 7–10 nucleotides long. At least 1000 different transcription factors are encoded in the genomes of *Drosophila* and *Caenorhabditis elegans*, and as many as 3000 in the human genome. On average, around five different transcription factors act together at the control region of a gene, and in some cases considerably more. The efficient use of the transcription factor repertoire is maximized by the fact that many transcription factors act on the control regions of more than one gene. This means that a single transcription factor can help to coordinate the activity of these genes, activating some and repressing others (Fig. 6.6).

The combinatorial action of gene-regulatory proteins is a key principle in the control of gene expression and is fundamental to the exquisite control and complexity of the gene expression that drives development. Complex control regions that regulate the temporal and spatial expression of a gene during embryonic development are shown in Figure 2.37 and Figure 6.5 for the *Drosophila eve* gene. The gene is expressed at two different times in development and in two different sets of cells, and each of these patterns is under specific regulation. The spatially delimited expression of *eve* in the second of seven transverse stripes in the embryo involves four different gene-regulatory proteins, which include activators and repressors, acting at 11 distinct sites. The later expression in muscle precursors integrates the responses to various extracellular signals. These are integrated by different combinations of transcription factors acting at multiple control sites in *eve* in different muscle groups (see Fig. 6.5).

Sites within the promoter region determine the start site of transcription and bind the general transcription factors and the RNA polymerase (see Fig. 6.4). Tissue-specific or developmental-stage expression is controlled by sites outside the immediate promoter region, such as those at which tissue-specific gene-regulatory proteins act. These sites vary greatly in type and position from gene to gene and may be thousands of base pairs away from the start point of transcription. These distant sites are able to control gene activity because DNA can form loops, thus bringing the sites and their bound proteins into close proximity to the promoter. Proteins bound at distant sites can thus make contact with proteins at the promoter, forming a fully active transcription initiation complex (see Fig. 6.4). A famous example of this is the ZRS (zone of polarizing activity regulatory sequence) enhancer of the vertebrate *Shh* gene involved in patterning the digits; this is located one megabase (1 Mb = 1000 bases) from the gene itself, in the intron of another gene. Mutations in this enhancer result in activation of the

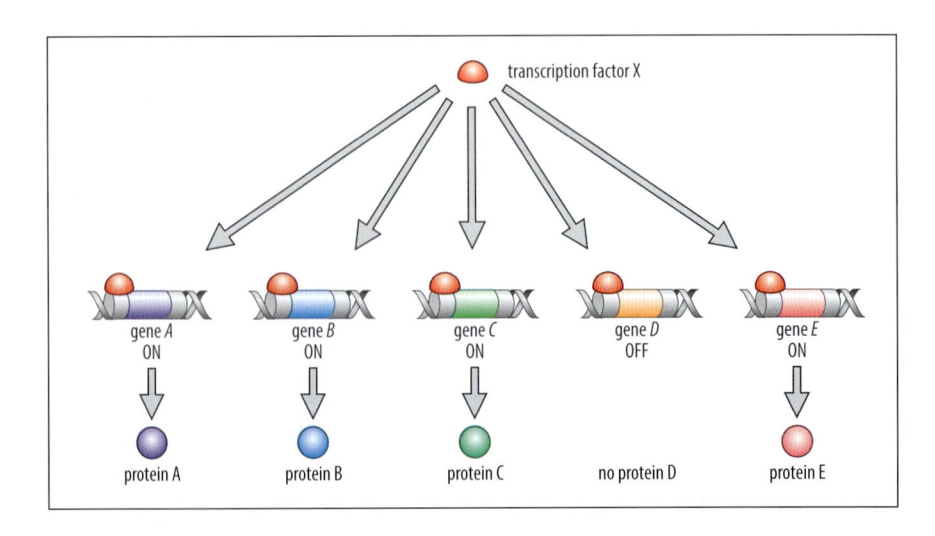

Shh gene in the wrong place in the developing hand or paw and the appearance of extra digits (discussed in Box 9B).

Another important class of gene-regulatory proteins comprises the **co-activators** and **co-repressors**, which do not bind DNA themselves but link the transcriptional machinery to the DNA-binding activator and repressor proteins (see Fig. 6.4). One co-activator we have already encountered is β-catenin (whose activity is the end product of the Wnt signaling pathway), which binds transcription factors of the T-cell-specific factor (TCF)/lymphoid enhancer-binding factor (LEF) family (see Box 3E). In the absence of Wnt signaling, TCFs bound to **control elements** in genes that are targets of Wnt signaling are bound by a co-repressor, called Groucho in *Drosophila* and transducin-like enhancer (TLE) in humans, and transcription is repressed. β-catenin accumulating in the nucleus as the result of Wnt signaling displaces Groucho/TLE and binds to TCF, turning it into a transcriptional activator. This basal repression of the target genes of Wnt signaling ensures a tight control of their expression, abolishing spurious transcription so that only defined levels of β-catenin will activate gene expression. The basal repression strategy is very common and lies at the core of the Hedgehog (see Box 2E) and Notch (see Box 4F) signaling pathways.

Transcriptional regulators fall into two main groups: those that are required for the transcription of a wide range of genes, and which are found in many different cell types; and those that are required for a particular gene or set of genes whose expression is tissue restricted, or tissue specific, and which are only found in one or a very few cell types. We shall encounter both types of transcription factor in the following sections, when we look at the regulation of muscle-specific genes in muscle precursor cells and the differentiation of red blood cell precursors. In general, it can be assumed that activation of each gene involves a unique combination of transcription factors.

As noted earlier, developmental genes generally have complex control regions, containing binding sites for a range of activating or inhibitory transcription factors, so whether or not a gene is activated depends on the precise combinations, and levels, of these transcription factors, and their associated co-activators and co-repressors, in the cell. The particular combination of gene-regulatory proteins present in a given embryonic cell at a given time is the result of its developmental history, and determines the next step in its development.

6.2 Gene expression is also controlled by epigenetic chemical modifications to DNA and histone proteins that alter chromatin structure

A central feature of development in general, and cell differentiation in particular, is that in any given cell some genes are maintained in an active state, whereas others are repressed and inactive. As development and differentiation proceed, genes that were previously inactive become active in particular cells, while previously active genes become permanently shut down. Transcriptional control by transcription factors and post-transcriptional controls such as RNA processing and translational regulation are not the only means available to the cell for controlling gene expression. In eukaryotes, transcription depends on large-scale mechanisms by which genes can be either made accessible for transcription or kept permanently or semi-permanently inactive. These mechanisms rely on the organization and regulation of the structure of **chromatin**— the complex of DNA, histones, and other proteins of which the chromosomes are made. In chromatin, the DNA is wrapped around complexes of histone proteins to form structures called nucleosomes, and the composition of the histone core (i.e. the combinations of different histones), post-translational modifications of the histones, and the positioning of the nucleosomes relative to the DNA sequence have an impact on whether a gene is transcribed or not. For some genes, the chromatin is in an open state in which the DNA is accessible to the transcriptional machinery and to control by gene-regulatory proteins such as those discussed in Section 6.1. Genes that are not required for a particular cell type are packed away, or **silenced**, in those cells by changes in the structure of the chromatin that render the DNA inaccessible to regulatory proteins. In this way, large numbers of genes can be shut down over the long term by a general mechanism, without requiring the continued production of a unique set of dedicated regulatory proteins for each gene. Chromatin that is packaged into a relatively compact structure that cannot be transcribed is called **heterochromatin**. There are two kinds of heterochromatin: **constitutive**, encompassing DNA that is unlikely to ever be transcribed and usually involves repetitive DNA and transposable elements; and **facultative**, DNA that although hidden from the transcriptional machinery at any given time, retains the ability to be transcribed in a regulated manner.

Whether a stretch of DNA in a chromosome can be transcribed or not therefore depends, in the first instance, on its chromatin structure, which determines the accessibility of the DNA to transcription factors and RNA polymerase. When the DNA is wrapped around the histones, the regulatory sequences cannot be easily reached by transcriptional regulators. Special types of transcription factors called '**pioneer factors**' are able to recognize some of the sequences, loosen them from the histones, and make them available to other proteins that will drive the assembly of the transcription initiation complexes.

Over the past two decades much progress has been made in understanding the spatial organization of chromatin at the local level, and at the higher level along a whole chromosome, by using **chromosome conformation capture** techniques—such as the Hi-C technique, which can capture all pairwise interactions within a chromosomal region. These studies have revealed two levels of control over the accessibility of DNA.

At the local level, the structure of the chromatin is controlled by covalent chemical modifications to the DNA and its associated histone proteins (Box 6B). Unlike mutations, these modifications do not change the DNA sequence and for this reason are known as **epigenetic** ('above genetic') modifications. They are carried out by dedicated sets of enzymes—some that add the chemical tags, and some that remove them. Although epigenetic modifications are normally permanent for the lifetimes of a somatic cell and its progeny, they can be reversed given the right environment, as shown by the reprogramming of a somatic cell nucleus when transplanted into an enucleated oocyte (see Fig. 1.20). These covalent modifications determine whether a gene can be expressed or not.

The other level of chromatin organization is very large scale and involves the physical organization of the chromatin within the chromosomes. This organization involves discrete regions of chromatin called **topologically associating domains (TADs)**. Within each of these domains the strand of chromatin is folded and looped up on itself, and the genomic sequences interact with themselves more frequently than with sequences elsewhere in the genome. The technique of chromosome conformation capture is used to delineate TADs. The lengths of individual TADs are in the range of megabases. TADs are separated by regions of chromatin in which very few interactions occur. The boundaries of the TADs are controlled by large protein complexes based on the DNA-binding protein CTCF, which binds to sites at the TAD boundaries. It is thought that a TAD serves to define a suite of genes under the control of enhancers located within the domain, and also protects genes from spurious control by enhancers located in different TADs.

TADs are dynamic structures, and their structure changes when the cell receives signals that the genes within the TADs should be expressed—during cell differentiation, for example. When genes are not being expressed, the chromatin in a TAD is tightly folded and typically carries repressive histone marks such as H3K27me3 that prevent transcription (see Box 6B). When the cell receives a differentiation signal, however, the appropriate chromatin unfolds, the histone marks change to those that permit transcription, and transcription factors are now able to bind to sites within the control regions and switch on the associated genes (Fig. 6.7). TADs can also provide

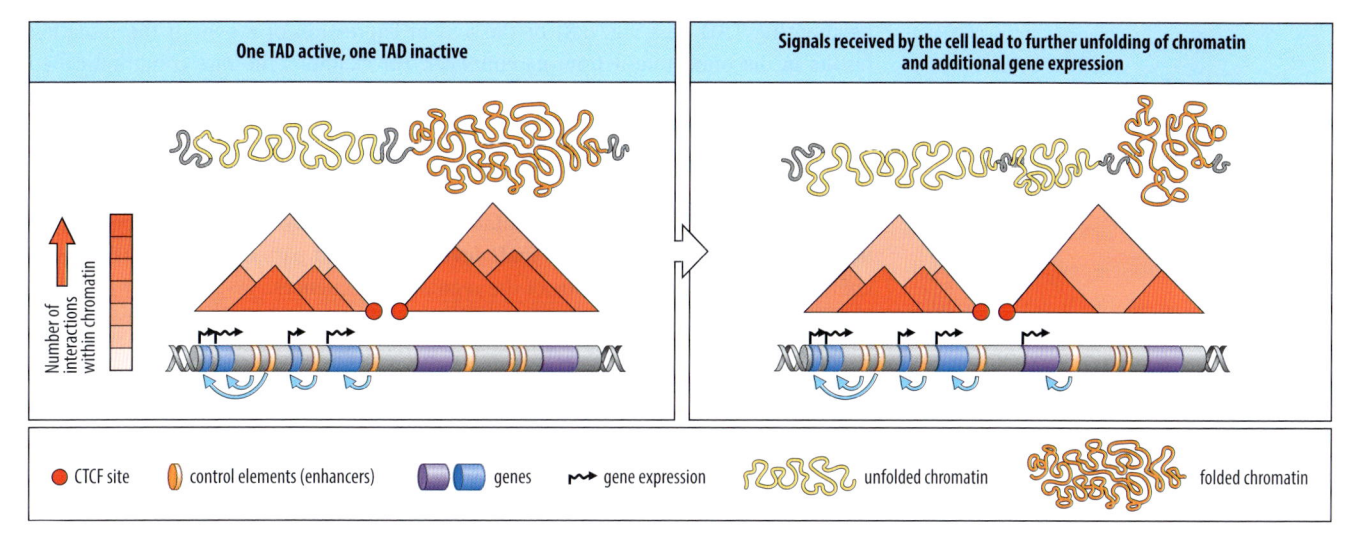

Fig. 6.7 Topologically associating domains are dynamic structures that change their conformation when the genes within them are required to be expressed. Left panel: throughout a chromosome, there are discrete regions where the strand of chromatin (shown at the top of the panel) is folded and looped up on itself. These regions of the genome are called topologically associating domains (TADs) and are demarcated by binding sites for the DNA-binding protein CTCF (red circles). The density of interactions that the genomic sequences within a TAD make with each other is conventionally denoted by shades of red, with the darkest red indicating the greatest number of interactions. Two hypothetical TADs are shown here. The DNA in a TAD typically contains several control elements (yellow rings) for one or more genes (blue or purple boxes). Genes in the left-hand TAD are being expressed. When genes are not being expressed, as in the right-hand TAD in this panel, the chromatin is tightly folded and carries repressive histone marks such as H3K27me3 (see Box 6B). Right panel: when signals are received for genes to be expressed—during cell differentiation, for example—these cause changes in TAD structure that enable additional genes to be expressed. The chromatin unfolds and modifications to the chromatin such as histone acetylation and the replacement of H3K27me3 with H3K4me3 enable gene expression. Note that part of the second TAD remains inactive and the gene within this region is not expressed.
*From Aranda, S., et al.: **Regulation of gene transcription by Polycomb proteins.** Sci. Adv. 2015, **1**: e1500737.*

a basis for the long-term shutdown of large cohorts of genes, which can be placed in regions of the nucleus where they are not accessible to the transcriptional machinery.

CTCF seems to be the 'gate keeper' of those domains. This role is revealed if CTCF-binding sites between an active and an inactive TAD are deleted. The active TAD expands, leading to the misactivation of a previously inactive gene that now comes under the control of a nearby enhancer (Fig. 6.8).

Disruption of the structure of TADs is associated with some human pathologies, including some developmental disorders and cancers. For example, the deletion or inversion of a chromosomal region that includes a TAD boundary can lead to misexpression of a gene as a result of it coming under the control of an enhancer in a

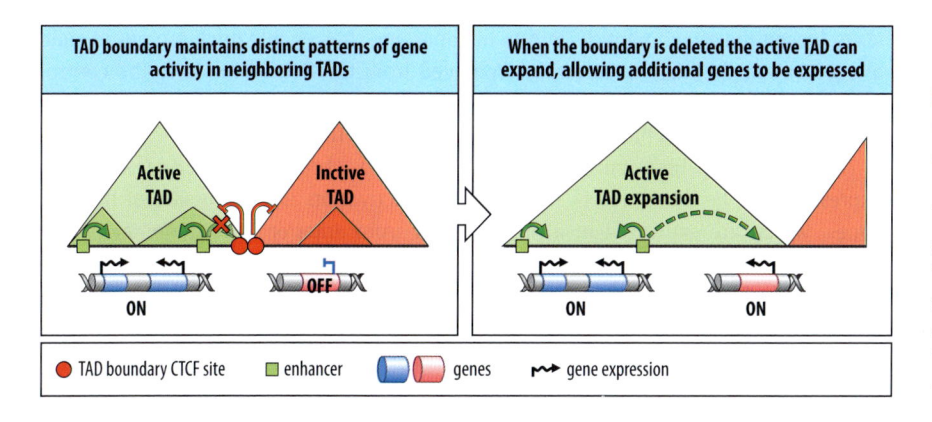

Fig. 6.8 CTCF is the gatekeeper at TAD boundaries. Left panel: when the CTCF sites at the boundary between two TADs are intact, signals from an enhancer in an active TAD (green) are prevented from turning on gene expression in a neighboring inactive TAD (pink). Right panel: when the CTCF sites are lost, the active TAD can expand to encompass a gene formerly in the inactive TAD, and this gene can now be turned on by the enhancer.

neighboring TAD (see Fig. 6.8) or the loss of expression of a gene if the mutation results in disconnecting it from its enhancer. For example, the rare condition called acheiropodia, in which people are born with truncations of the limbs, is due to a mutation that affects the regulation of the expression of the *SHH* gene. People with this condition have a deletion of CTCF sites around the ZRS enhancer that disconnects it from the *SHH* gene, leading to a loss of *SHH* expression in the developing limb (see Box 9B). An increasing number of mutations associated with this genomic region have been discovered, associated with limb development and patterning, emphasizing the complex nature and interactions of the regulatory sequences.

An important local epigenetic modification that keeps genes silenced in many organisms, especially in mammals and in plants, is **DNA methylation**. Cytosines occurring in cytosine–guanine (CpG) sequences at certain sites in the DNA are methylated by the enzyme **DNA methylase**. DNA methylation preferentially occurs in promoters and enhancers and is correlated with the absence of transcription in the neighboring genes. DNA methylation is one way in which certain genes are rendered inactive in the maternal or paternal genome in the phenomenon of **genomic imprinting** (discussed in Section 8.8). The inactive X chromosome in the cells of female mammals also has a pattern of DNA methylation that differs from that of the active X, and this is likely to play a part in keeping it inactive (discussed in Section 8.17). By contrast, embryonic stem cells have very low levels of methylation in their DNA, concurrent with their ability to adopt any fate. The pattern of methylation can be faithfully inherited when the DNA replicates, thus providing a mechanism for passing on a pattern of gene activity to daughter cells.

Chemical modifications of the tails of the histone proteins that make up the nucleosome cores are another important component of transcriptional regulation. The histone tails carry reproducible patterns of covalent modifications such as acetylation and methylation along their length that predict whether a gene is poised to be expressed, is being expressed, or is repressed. These modifications are carried out by specific enzymes—for example, histone acetyltransferases and histone methyltransferases—and can be reversed by other enzymes—histone deacetylases and histone demethylases—which remove the chemical tag. Like DNA methylation, the pattern of histone modification can be passed on to daughter cells, thus creating a lasting memory of the transcriptional status of a region of DNA, but the patterns of histone modification tend to be more subject to change than those of DNA methylation, and changes in these patterns are important in the process of differentiation. Enhancers, promoters, and protein-coding regions tend to have histones with specific modifications, and it is the specific histone-modification fingerprint at a particular gene locus that determines the gene's activity (see Box 6B).

Another type of epigenetic chromatin modification is carried out by multiprotein complexes called **chromatin-remodeling complexes**. Such complexes contain ATPase activity and can be recruited to control regions in the DNA, usually by pre-existing chemical modifications to chromatin such as DNA methylation and histone methylation and acetylation (see Box 6B). They are thought to use the energy of ATP hydrolysis to loosen the attachment of DNA to histones and thus enable nucleosomes to slide along DNA or the histone core to be removed from the DNA. This frees the control region DNA to bind other gene-regulatory proteins, RNA polymerase, and the rest of the transcriptional machinery, thus enabling the gene to be transcribed.

The relationship between chromatin modification, chromatin structure, and gene expression can be illustrated in the case of the activation and expression of the Hox genes, which are expressed in a precise order from anterior to posterior along the body axis of most animal phyla (see Box 4G and Fig. 4.36). The expression of genes of the Hoxa cluster in mouse embryos illustrates this by showing how the progressive activation of the members of the cluster is associated with changes in the large-scale structure of the chromatin and in the histone modifications (Fig. 6.9). The onset of Hox gene expression in the Hoxa cluster during early development in the mouse provides a

CELL BIOLOGY BOX 6B Epigenetic control of gene expression by chromatin modification

In chromatin, DNA is typically packaged together with four different histone proteins into structures called nucleosomes, in which the DNA is wrapped around the outside of a core composed of eight histone molecules. There are four core histone units (H2A, H2B, H3, and H4), but other histones are incorporated into nucleosomes associated with particular transcriptional states, and which loosen or tighten the interactions of the histone core with DNA. The histones in chromatin can undergo post-translational covalent modifications such as methylation, acetylation, or phosphorylation, on specific amino acid residues in their amino-terminal tails, particularly on lysine (K) (Fig. 1). These modifications are closely linked to the regulation of gene activity. Methylation and acetylation have been the most studied, and their patterns within the histones associated with a gene creates a code that determines whether the gene is ON or OFF: it is now possible to gauge the pattern of activity of a gene in a cell during development from the pattern of modifications of its associated histones. Histone acetylation, for example, tends to be associated with regions of chromatin that are being transcribed, whereas the effect of methylation depends on the lysine residue that is methylated and the number of methyl groups covalently attached to it. Up to three methyl groups can be added to a lysine. On some lysine residues, for example K4 in histone H3, methylation is associated with gene expression, with the precise effect depending on the number of methyl groups added. But on the K9 and K27 residues of the same histone, monomethylation is associated with expression, whereas trimethylation promotes the formation of transcriptionally silent heterochromatin and is associated with repression. Figure 2 shows the distributions of a few key epigenetic modifications in active and inactive genes. The first three rows show an active gene and the bottom two rows an inactive gene. The top row shows the concentration of RNA polymerase II (Pol II, red) along the gene, which represents the progress of transcription. It also shows the relation between DNA methylation (green; see text) and transcription. At the promoter, very low levels of DNA methylation are correlated with high levels of occupancy of the promoter by Pol II and the initiation of transcription. The level of DNA methylation is often higher in the body of a gene, and correlates with a lower occupancy of Pol II, but does not prevent ongoing transcription and the elongation of the transcript. The second and third rows represent the concentrations of various histone marks along an active gene. Acetylated histones (pale green) and the three forms of methylated H3K4 (H3K4me3, pink; H3K4me2 and H3K4me1, orange) are concentrated at the promoter, whereas monomethylated H3K9 and H3K27 (H3K9me1 and H3K27me1, blue), and other acetylation marks (e.g. H3K12ac and H4K16ac (yellow)), are less discriminatory and are spread throughout the whole region being transcribed. In an inactive gene (bottom two rows) the promoter DNA is methylated along with the rest of the gene, and the repressive di- and trimethylated versions of H3K9 and H3K27 (turquoise) are present along the length of the gene.

The histone modification proteins can be divided into three classes: 'writers', which modify the histone residues and include acetyltransferases, methylases, and also ubiquitinases and kinases; 'erasers', which eliminate these modifications and are the enzymatic reversers, such as deacetylases; and 'readers', which recognize the modifications, bind to their locations, and are able to recruit activator and repressor proteins to exert changes in transcriptional activity. Thus, modifications to histones exert their effects on chromatin structure and gene expression by the recruitment of proteins that recognize the altered sites and maintain the transcriptional status of the chromatin. Histone methylation can recruit proteins that either initiate heterochromatin formation or promote transcription, depending on the residue methylated, whereas histone acetylation recruits proteins that keep the DNA available for transcription. These proteins, in turn, can recruit other protein complexes that will, for example, specifically affect gene expression.

We have seen an example of this in *Drosophila*, where proteins of the Polycomb group maintain the repression of Hox genes in cells in which they have been turned off, whereas proteins of the Trithorax group maintain the expression of Hox genes in cells in which they have been turned on (see Section 2.27). Polycomb group proteins form a large multi-protein complex that promotes the trimethylation of histone H3 on lysine 27 (H3K27me3). This creates a repressive mark on the histones and thus suppresses gene expression. The Trithorax group is a multi-protein complex with opposing activity to that of Polycomb. It places activating marks on histones, for example the trimethylation of histone H3 on lysine 4 (H3K4me3), and can also remodel the chromatin by changing the placement of the histones. Some of the Trithorax group proteins interact with acetyltransferases which, by promoting histone acetylation, promote gene expression. The function of the Trithorax group is to keep genes available for transcription.

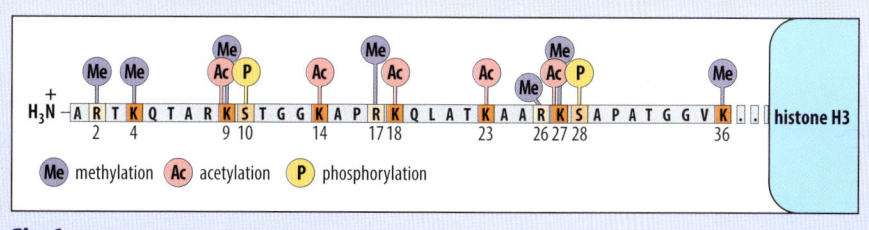

Fig. 1

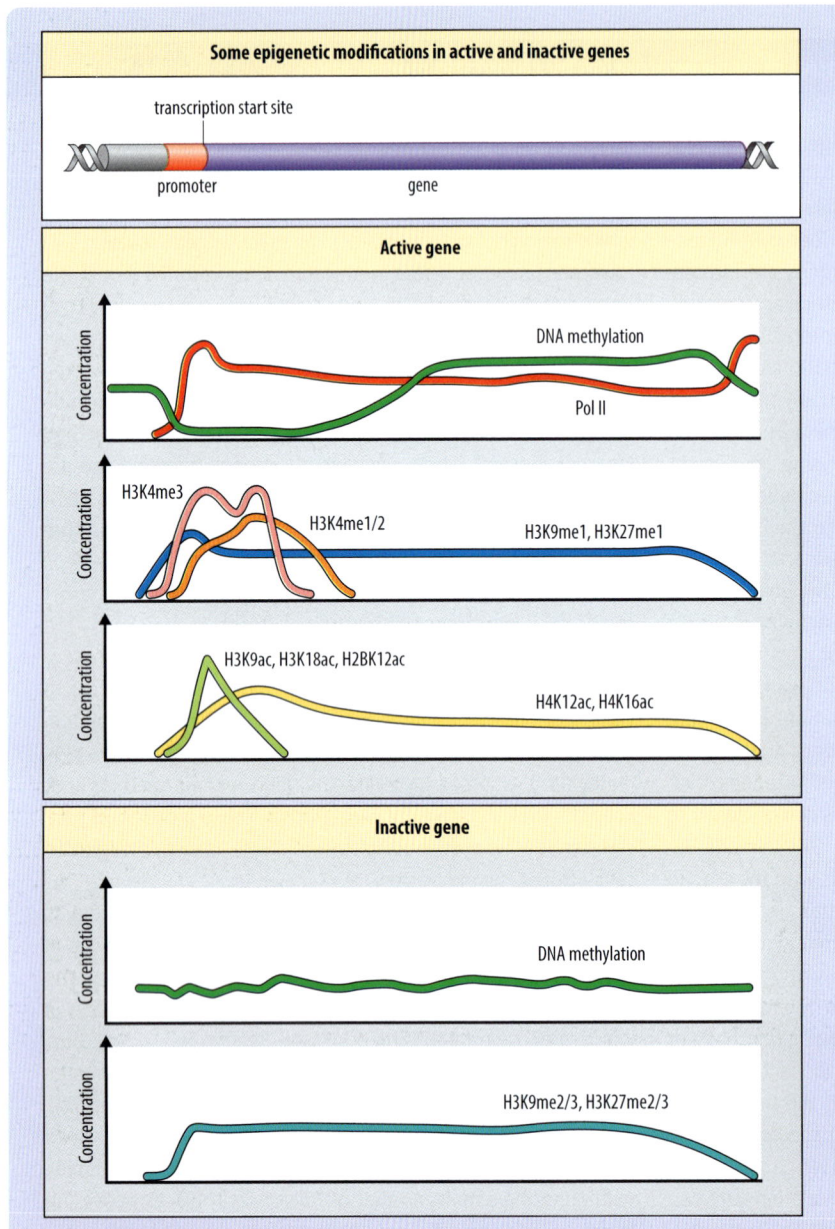

Some epigenetic modifications in active and inactive genes

Active gene

Inactive gene

Fig. 2
Adapted from Botchkarev, V.A., et al.: **Epigenetic regulation of gene expression in keratinocytes.** J. Invest. Dermatol. 2012, **132**: 2505-2521.

The regulation of chromatin structure in terms of DNA methylation, nucleosome composition, and histone modifications is a shared feature of plants and animals, and the modifications and the enzymes are conserved. Changes in chromatin status are often self-propagating because histone modifications, such as acetylation/deacetylation or methylation/demethylation, recruit the corresponding chromatin-modifying enzymes to the chromatin, causing a 'domino effect' that automatically propagates the change in chromatin structure along the chromosome.

good example of how the expression of Hox genes is temporally regulated and linked to the development of the body plan. The cluster lies between two TADs that contain control elements (enhancers) that drive its expression. Just before the onset of gastrulation at around E6.0, the chromatin is condensed and marked with the repressive histone modification H3K27me3, and the Hoxa genes are not expressed. As the posterior region of the embryo emerges during gastrulation, the cluster opens up under the influence of Wnt signaling acting on a Wnt-responsive enhancers located in the 3′ TAD.

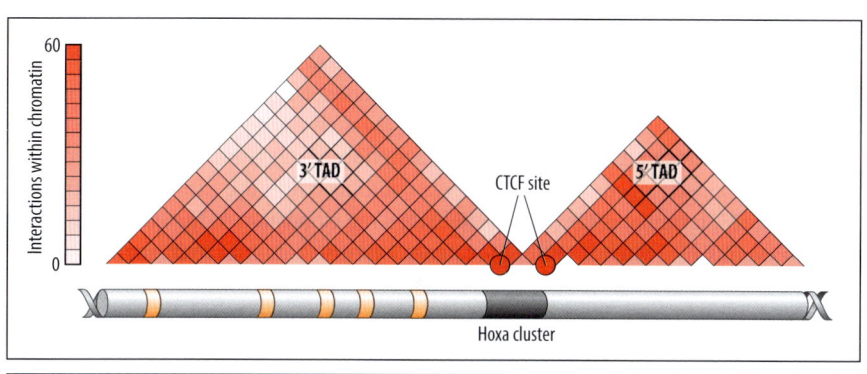

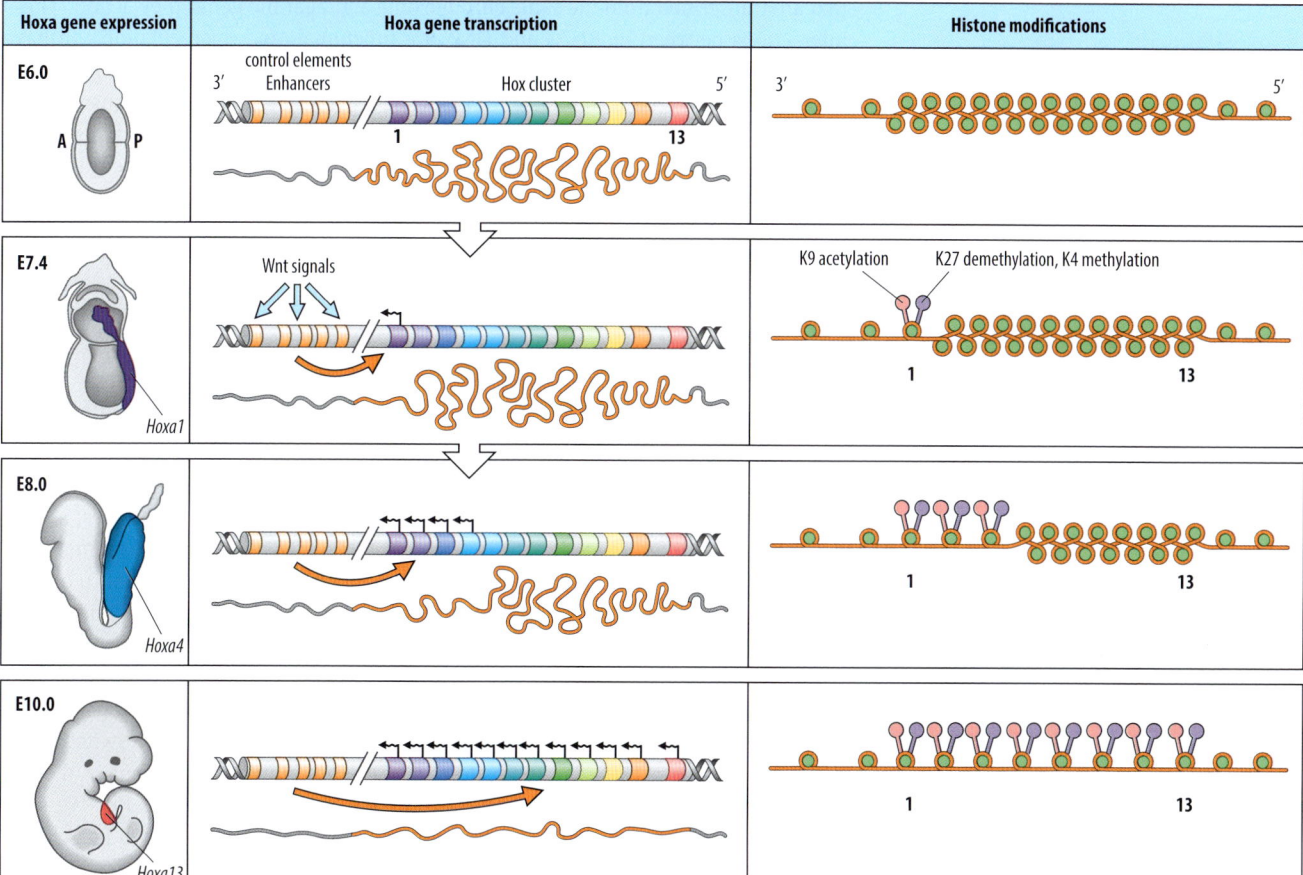

Fig. 6.9 The sequential expression of genes in a Hox cluster depends on epigenetic changes in the state of the chromatin. Top panel: the Hoxa cluster lies between two topologically associating domains (TADs) that contain the control elements (enhancers) that drive its expression. The schematic diagrams below show Hox gene expression in the mouse embryo at different developmental stages and the associated changes in histone modifications. The left panels show the domains of Hoxa gene expression in the mouse embryo at different stages. The center panels show the relation between chromatin state (folded up or unfolded) and the ability of the Hox genes to be expressed. The right panels show the associated changes in histone modification. At E6.0 just before the onset of gastrulation, the chromatin is condensed, the repressive chromatin mark H3K27me3 (not shown) is abundant, and the Hoxa genes are not expressed. Then as gastrulation takes place, Wnt signaling acting on the Wnt-responsive enhancers located in the 3′ TAD starts to open up the cluster, enabling the expression of *Hoxa1*. H3K27 methylation is progressively removed and replaced by H3K4 methylation, which is characteristic of active genes. Histones are also acetylated on H3K9, another mark of an active gene. These epigenetic changes continue along the cluster until the whole cluster is opened up and all the Hoxa genes are available for transcription. A, anterior; P, posterior.

From Dixon, J.R., et al.: **Topological domains in mammalian genomes identified by analysis of chromatin interactions.** Nature 2012, **485**: 376-380.

The chromatin progressively decondenses, and the Hox genes begin to be expressed, starting with *Hoxa1*. Consistent with this, H3K27 methylation is progressively removed and replaced by H3K4 methylation, which is characteristic of active genes. Histones are also acetylated on H3K9, another mark of an active gene. Locally acting enhancers within the Hoxa cluster switch on transcription of the Hoxa genes in the correct order as the cluster opens up. This process continues until the expression of the last gene, *Hoxa13*, terminates the process and stops the elongation of the embryo. Hoxa13 protein eventually switches off the Hox genes expressed anterior to it (see Fig. 4.39).

Gene expression is also regulated by non-coding RNAs, which act by interfering with both transcription and translation. A particularly surprising class of regulatory RNAs in both animals and plants comprises the **long non-coding RNAs (lncRNAs)** that are transcribed from the same chromosome and near the genes they act on. These interactions between lncRNAs and DNA are functional, and misregulation can have dramatic consequences. For example, large deletions of a region of chromosome 1 next to the *EN1* gene in humans are associated with a syndrome that causes differences in the patterning of the limb, including the appearance of ventral nails on the digits. This deletion was shown to remove a lncRNA that is required for the expression of *EN1*, which itself is involved in the dorso-ventral patterning of the limb. The correspondence between the lncRNA, called Maenli, *EN1*, and the limb difference was proved by engineering the mutation into mice and showing that it caused a similar syndrome.

6.3 Patterns of gene activity can be inherited by persistence of gene-regulatory proteins or by maintenance of chromatin modifications

Once committed to a particular differentiation pathway, embryonic cells continue to proliferate, usually only ceasing when they enter the final stages of differentiation. And some cell types also continue dividing after differentiation. In both cases, a cell's determined or differentiated state is transmitted unchanged to their daughter cells over many cell generations. The question then is: how can a particular pattern of gene activity be inherited intact after DNA replication and the disruption of cell division?

One way of maintaining gene activity is to ensure the continued production of the required gene-regulatory proteins in the differentiated cell and its progeny. If the gene product itself acts as the positive regulatory protein, all that is required for gene activity to be maintained is the initial event that first activates the gene; once switched on, the gene remains active. The requisite regulatory protein will be present in the cell after division and so can restart gene expression immediately (Fig. 6.10). We shall see this type of positive-feedback control of gene expression occurring in muscle cell differentiation, where the MyoD protein acts as an activator of its own gene, *Myod*. In *Drosophila*, the selector gene *engrailed* remains active in the posterior compartment of segments throughout embryonic and larval development and into the adult (see Chapter 2). Its expression is at least partly maintained by a similar positive feedback of the Engrailed transcription factor on the transcription of its own gene, long after the pair-rule transcription factors that initiated expression of *engrailed* in the embryo have disappeared.

Another way of maintaining a particular pattern of active and inactive genes and transmitting it to daughter cells is through the maintenance of chromatin structure. Evidence that changes in the packing state of chromatin can be inherited through many cell divisions and can keep genes inactive over a long period was first provided by the phenomenon of X-chromosome inactivation in female mammals. One of the two X chromosomes in each somatic cell in females becomes highly condensed and inactive early in embryogenesis and is maintained in this state throughout the individual's lifetime. In this case the silencing mechanism primarily involves the production of a non-coding RNA that coats the chromosome (described in Section 8.17). The inactive and active X chromosomes also have different DNA methylation patterns. Inactivation

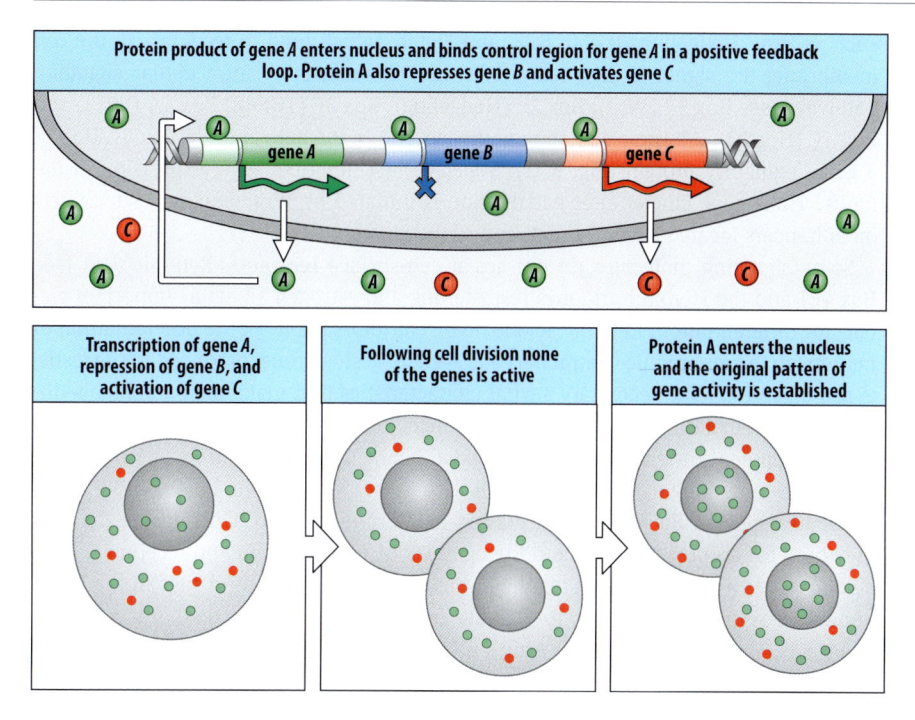

Fig. 6.10 Continued expression of gene-regulatory proteins can maintain a stable pattern of gene activity in a differentiated cell and its progeny. Top panel: transcription factor A is produced by gene A, and acts as a positive gene-regulatory protein for its own control region. Once activated, therefore, gene A remains switched on and the cell always contains A. Transcription factor A also acts on the control regions of genes B and C to repress and activate them, respectively, setting up a cell-specific pattern of gene expression. Bottom panels: after cell division, the cytoplasm of both daughter cells contains sufficient amounts of protein A to reactivate gene A, and thus maintain the pattern of expression of genes B and C.

is not irreversible, however, as the inactive X is reactivated in the germline during the formation of egg cells, which each end up containing a single X chromosome.

Both the pattern of DNA methylation and the pattern of histone modifications on chromatin can be transmitted to daughter cells. In the case of DNA methylation, DNA replication produces a double-stranded DNA with the original methylation pattern on the parental strand. DNA methylase then recognizes the methylated cytosines in CpG sequences and methylates the corresponding cytosine on the newly synthesized strand to restore the pattern on that strand.

Patterns of histone modification are thought to be inherited according to a similar principle. When the DNA is replicated, nucleosomes are displaced, and new nucleosomes then reassemble on the two new DNA molecules. Some of the 'parental' marked histones are retained at the DNA during replication and are distributed at random to the two daughter DNA molecules. They reassemble with newly synthesized unmarked histones to form new nucleosomes. As noted in Section 6.2, chromatin states can be self-propagating because modified histones recruit the corresponding chromatin-modifying enzymes to the chromatin. This is thought to be the way that the original state of a chromatin region is reconstructed after chromosome replication.

6.4 Changes in patterns of gene activity during differentiation can be triggered by extracellular signals

As we have seen in previous chapters, the effect of developmentally important cell–cell signaling interactions is almost always to induce a change in gene expression in the responding cell. We have already encountered extracellular signal proteins such as Wnt, Hedgehog, TGF-β, FGF, and Notch, which bind to receptors in the cell membrane; the signal is then relayed to the cell nucleus by intracellular signaling pathways (see Box 3E (Wnt), Box 2E (Hedgehog), Box 3F (TGF-β), Box 4A (FGF), and Box 4F (Notch)). All these signaling pathways end in the activation or inactivation of a gene-regulatory protein (Fig. 6.11). We have just described how expression of the mouse Hoxa gene cluster after gastrulation is initiated by Wnt signaling, which acts on enhancers located beyond the 3' end of the cluster.

Some signaling molecules do not act at cell-surface receptors. Retinoic acid (see Box 4E) and the steroid hormones testosterone and estrogen all act as important signals for differentiation. Retinoic acid is involved, for example, in the differentiation of neurons in the embryonic central nervous system. Testosterone, produced in the testis, is responsible for the secondary sexual characteristics that make male mammals different from females, whereas estrogen, produced mainly by the ovaries, is involved in the development of female secondary sexual characteristics and in female reproductive function, among many other physiological roles. Non-protein, lipid-soluble signal molecules such as these cross the plasma membrane unaided. Once inside the cell, they bind to receptor proteins, and the complex of signal protein and receptor is then able to act as a transcriptional regulator, binding directly to control sites in the DNA, known as response elements, to activate (or in some cases repress) transcription. In insects, the steroid hormone ecdysone is responsible for metamorphosis; it acts through similar intracellular receptors and induces the differentiation of a wide range of cell types. In all these cases, the hormone coordinates the expression of a whole set of genes, turning some on and some off by acting, via its intracellular receptors, at control sites that are present in all these genes.

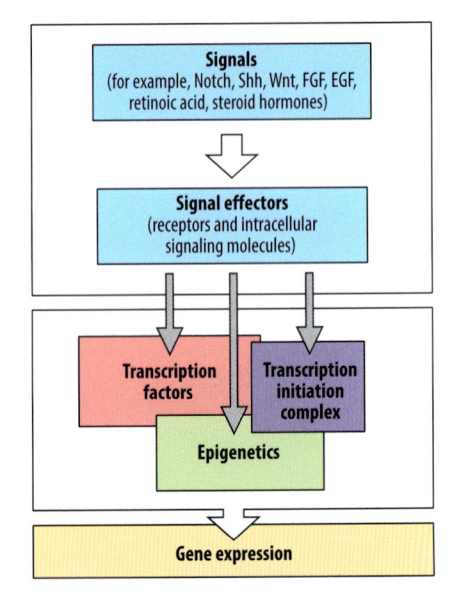

Fig. 6.11 Extracellular signals trigger changes in gene expression. Extracellular signals act through various signal effectors—the receptors and intracellular signaling proteins that transduce the signal within the cell to eventually affect gene expression. These effectors can influence various inputs to gene expression, including the expression or activation of transcription factors, effects on the transcription initiation complex, and effects on the epigenetic factors controlling gene expression, such as DNA methylases and the enzymes that add or remove chemical markers to and from histone proteins (see Box 6B).

> ## SUMMARY
>
> Transcriptional control is a key feature of cell differentiation. Tissue-specific expression of a eukaryotic gene depends on control elements located in the *cis*-regulatory regions flanking the gene. These elements comprise both the promoter sequences adjacent to the start site of transcription, which bind RNA polymerase, and more distant elements such as enhancers that control tissue-specific or developmental-stage-specific gene expression. The combination of different regulatory proteins bound to sites in these control elements determines whether a gene is active or not, and so even small differences between two genes in the number and type of control sites can produce quite different patterns of expression. Gene-regulatory proteins interact not only with DNA, but also with each other, forming multi-component transcription complexes that are responsible for initiating transcription. The chromatin is organized in very large domains called topologically associating domains (TADs) that compartmentalize regulatory regions and protein–protein interactions to achieve fine regulation. Once established, the pattern of gene activity in a differentiated cell can be maintained over long periods of time and is transmitted to the cell's progeny. Mechanisms that maintain the pattern include the continued action and production of gene-regulatory proteins, and long-term localized epigenetic changes in the structure and properties of chromatin that are due to chemical modifications to DNA and to histone proteins, and to the higher-order organization of chromatin into TADs.
>
> Some extracellular signals, such as protein growth factors, influence gene expression by acting at cell-surface receptors to stimulate intracellular signaling pathways. Other, lipid-soluble, signal molecules, such as retinoic acid or steroid hormones, diffuse through cell membranes to form a complex with an intracellular receptor protein. This complex then acts as a transcription factor to influence gene expression.

> **SUMMARY: Control of specific gene expression by extracellular signals**
>
> peptide and protein signals acting on cell-surface receptors
> ⇩
> signal transduction
> ⇘
> modification of a combination of transcription factors
>
> steroid hormones and retinoic acid
> ⇙
> activation of intracellular receptor which acts as a transcription factor
>
> transcription factors bind to *cis*-control regions in DNA
> ⇩
> gene activity switched on or off
> ⇩
> pattern of gene activity is passed on to daughter cells by epigenetic and other mechanisms
> ⇙   ⇩   ⇘
> positive-feedback control of gene expression and interaction with other genes   DNA methylation   changes in chromatin state (e.g. in inactivated X chromosome)

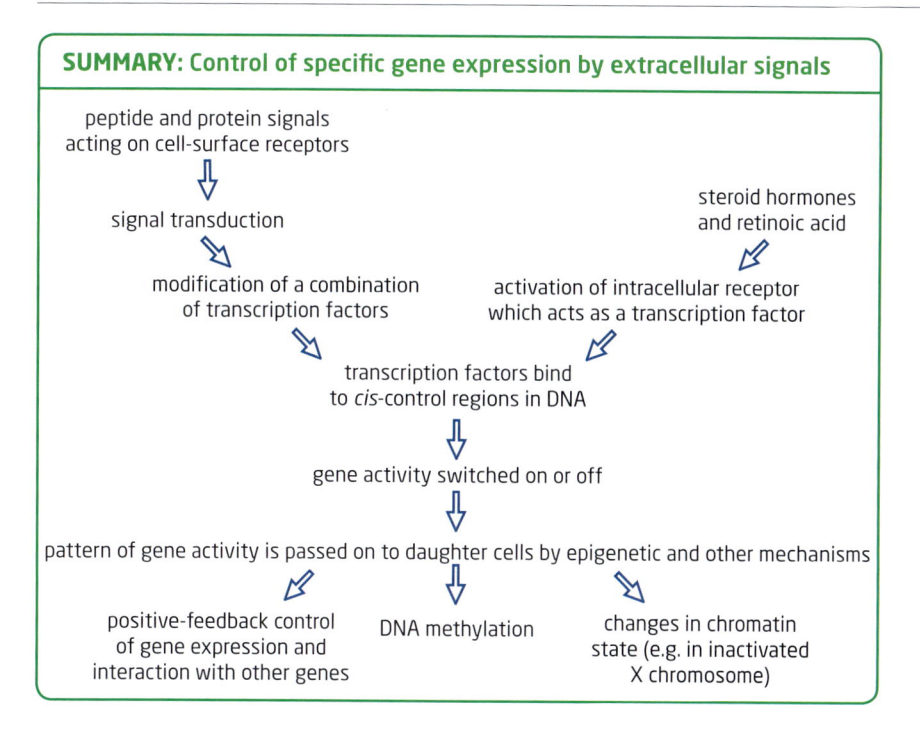

Cell differentiation and stem cells

In this part of the chapter we look first at the signs of cell differentiation detected during gastrulation. We then consider some particularly well-studied examples of terminal cell differentiation, which illustrate many of the principles underlying the process. The differentiation of mammalian striated skeletal muscle from undifferentiated embryonic mesodermal cells in the somites (see Chapter 4) is an example of the differentiation of a single cell type. But cellular differentiation is not confined to the embryo. Most differentiated tissues do not last forever, and most can be renewed from undifferentiated stem cells, specific to each tissue. Among the best-studied examples of cell differentiation and the properties of 'adult' stem cells are the blood, skin, and the gut epithelial lining, which are subject to continual wear and tear and cell loss and are renewed throughout life. We then return to skeletal muscle to see how it is repaired, when necessary, in adulthood from a muscle-specific stem-cell population set aside during embryonic development. We also see that even the central nervous system has a limited capacity for self-renewal via stem cells.

6.5 The early stages of cell differentiation during gastrulation can be detected by single-cell transcriptional analysis

Cells start to become different from each other at gastrulation, as cells begin to be allocated to different germ layers and then to cell types, with reference to a system of coordinates that emerges during gastrulation and provides positional information in relation to the antero-posterior, dorso-ventral, and medio-lateral axes of the embryo (see Chapters 3 and 4). In the mouse, gastrulation takes 2 days and transforms a group of about 600 cells with a shared signature of gene expression and lacking spatial organization into a collection of more than 60,000 cells, within which you can identify the primordia of most tissues and organs by their expression of cell-type-associated transcriptional regulators and their position in the embryo. At the end of gastrulation not only is there a collection of cells with different patterns of gene expression, but there are also morphological differences between different regions of the embryo (see Chapters 3 and 4).

EXPERIMENTAL BOX 6C Single-cell analysis of cell-fate decisions

During development, differentiating cells progress from a multipotent state towards terminally differentiated states, in which they express a specific set of genes that are associated with the specialized structure and function of the fully differentiated cells. This progression is generally envisaged as a sequence of binary choices that progressively restrict cell fate. It is a cell-autonomous process—that is, the decisions at the choice points are taken by individual cells, with extracellular signals coordinating these choices across cell populations. The key factor in the fate restrictions is a change in the genes that a cell expresses, particularly those that encode transcription factors, as it is the genes that are regulated by these transcription factors that will determine the state—the phenotype—of the differentiated cell.

An exciting technical advance in gene-expression analysis is the ability to determine all the genes being expressed in a single cell. The bulk transcriptional profile of a population of cells averages out gene expression and hides differences in individual cells that might reflect important developmental stages. To study what is happening in differentiation, individual cells are isolated from a population of differentiating cells at different stages of the process. Each cell's total mRNA (its transcriptome) is isolated, and the RNA amplified and sequenced. The transcriptomes of thousands of cells can be characterized simultaneously in this manner, thus giving an unprecedentedly detailed view of the changes in gene expression that are occurring during differentiation.

Single-cell transcriptome analysis has been extensively applied to many different developmental systems and is allowing us to track the differentiation of many cell types from common progenitors in great detail. This work has revealed that progenitor cell populations that seemed to be homogeneous, as judged by their phenotype, are heterogeneous in their patterns of gene expression (Fig. 1). Populations of cells at different stages of differentiation are conventionally defined by phenotypes that consist of particular combinations of cell-surface proteins. But when individual cells from one of these populations are isolated and their gene expression patterns determined, many different combinations of gene expression patterns are identified. This is making researchers rethink the idea of a phenotype reflecting just one specific state of gene expression.

Statistical and bioinformatics tools have been developed that cluster the single cells into different groups on the basis of the ensemble of genes they express and the levels of expression of each gene. This analysis gives rise to so-called principal components analysis (PCA) or uniform manifold approximation and projection (UMAP) plots (Fig. 2) that represent in two dimensions the relationship between different cells to each other on the basis of the expression of thousands of genes. What these techniques achieve is to look at the thousands of genes that each cell expresses and use them as coordinates, in which their levels of expression are recorded. Then all cells are compared with each other in terms of the genes they express and cells with similar patterns are put

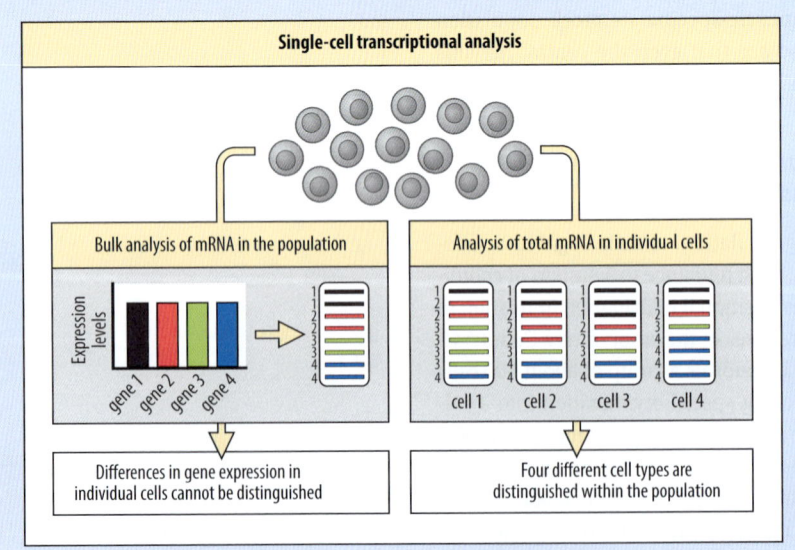

Fig. 1

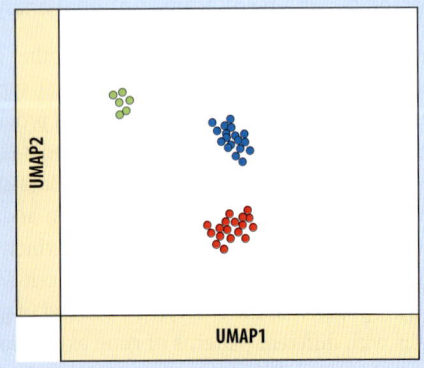

Fig. 2

together into clusters in a high-dimensional space. However, a space with thousands of coordinates is abstract and difficult to visualize; what techniques such as PCA and UMAP do is transform this high-dimensional space into one with two or three coordinates that preserve the clusters and their relationship to facilitate their visualization. The resulting maps identify distinct cell populations in terms of similarity in gene expression using thousands of genes in the comparison. The UMAP plots in Fig. 6.12 are examples of this approach used to analyze the emergence of clusters of new cell types with different patterns of gene expression as the mouse embryo goes through gastrulation.

This 'big data' approach is also beginning to establish definitive progenitor, neuronal, and glial cell-type taxonomies in the brain using a three-stage analysis. First, the transcriptomes of thousands of individual brain cells are grouped through techniques such as UMAP or PCA. Second, the resulting cell groups are compared with each other to identify differentially expressed marker genes. Finally, transcriptomic classifications are compared with known cell types to build a sophisticated picture of brain cell types.

Single-cell RNA analysis was first applied to look at cell differentiation in tissues such as the blood (see Section 6.8) and provides a much fuller picture of all the changes in gene expression that occur in the progress of a particular cell population towards full differentiation than previous techniques could do. The 'relatedness' of the different groups to each other in terms of gene expression is determined, and the groups can then be arranged in a series from less- to more-differentiated that reveals the changes in gene expression associated with this progression. For example, cells going from state A, defined by expression of the set of genes $a1, a2, a3, \ldots$, to a more differentiated state B, defined by expression of genes $b1, b2, b3, \ldots$, do not suddenly switch off all the a genes and turn on all the b genes. Instead, the cells in the population are distributed along the path from A to B because they are gradually and continuously changing their pattern of gene expression from the A combination to the B one.

Single-cell analyses reveal that the change of a cell from one state to a more differentiated state over time is a progressive one, and results from a progressive and ordered accumulation of genes expressed. Progress has to be inferred from time-series data on the cells collected at discrete points in real time, which is then converted into a sequence of cells ordered by their progress through differentiation: this sequence is therefore often called a 'pseudotime' series and reflects biological progress (Fig. 3). The reason for calling it pseudotime is that a given real-world time point might contain cells at different stages of differentiation and therefore might not tell much about the path or progress

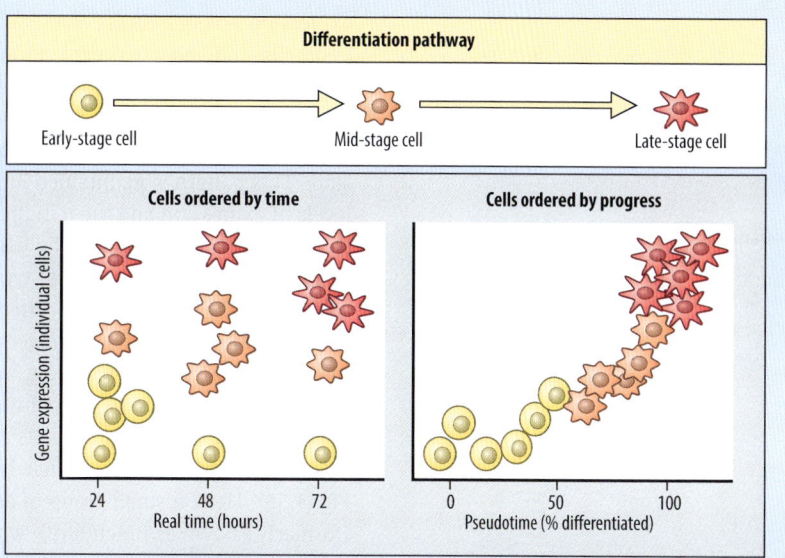

Fig. 3

Adapted from Trapnell, C.: **Defining cell types and states with single-cell genomics.** Genome Res. *2015*, **25**: 1491-1498.

of differentiation. The pseudotime arrangement focuses on the changes in gene expression that a cell undergoes.

Another technique derived from single-cell analysis provides an idea of the direction of differentiation of individual cells by tracking the history of the transcriptional process in the cell population. It is called RNA velocity and it looks at the ratio between unspliced (U) and spliced (S) RNA as a measure of whether a gene is being, or has been, transcribed. If U is greater than S, the cell is transcribing the gene or genes; the reverse says that transcription is over, with the ratio being a measure of how active the transcriptional process is. Thus the ratio provides a measure of the 'velocity' of the cell in the UMAP—as U decreases, the speed of transcription decreases—and, using this as a proxy for the state of the cell, it is possible to use it to measure the direction of differentiation: the less transcription, the more differentiated the cell is.

One can envisage these ordered sequences of cells in different states of gene expression (different sub-stages of differentiation) filling the paths in the Waddington landscape and reaching the bifurcations at different times. The bifurcations represent points in differentiation where cells become more limited in their fate (see Fig. 6.18). The trajectories of the cells and the bifurcations can be deduced from the single-cell analysis data. First, computational methods arrange the cells in series in which one cell is placed in between two cells with expression profiles that are slightly different, the ordering being determined by the degree of similarity. Thus, a cell X is placed between two cells A and B with different profiles and B being more differentiated than A. If the profile of X is closer to that of A, this creates the series A ⇒ X ⇒ B. Now if B is a bifurcation point out of which there are two very different cells, Y and Z, which cannot be arranged linearly, new cells C, D, E, ... etc. will be arranged with respect to Y and Z as X was between A and B, thus filling in a space and creating the continuum of the bifurcation.

In mouse embryos, differences between cells accompanied by changes in gene expression emerge earlier than gastrulation—during implantation, when the blastocyst cells undergo a reorganization to form the epiblast (see Fig. 3.50). In the blastocyst, the cells that will form the embryo are characterized by the activity of a gene regulatory network dominated by the expression and interactions of three transcription factors—Oct4, Nanog, and Sox2—that keeps the cells locked in a pluripotent state. This pattern is maintained in the epiblast after implantation, but with decreased levels of expression and the initiation of the expression of another transcription factor, Otx2. This condition makes the cells receptive to instruction by extracellular signals and able to start expressing cell-type specific genes.

Towards the end of day 5 after fertilization (E5.75), BMP-4 signaling from the extra-embryonic ectoderm triggers changes that lead to the onset of gastrulation through the activation of Nodal and Wnt signaling in the epiblast. Antagonists of these three pathways secreted from the anterior visceral endoderm restrict the signaling events to a group of cells located next to the extra-embryonic ectoderm on one side of the epiblast, which becomes the posterior region of the embryo (see Fig. 3.58). Here, a small group of cells starts expressing the transcription factor TbxT (formerly known as Brachyury), which controls genes involved in cell movement and adhesion and promotes an epithelial-to-mesenchymal transition. A gene regulatory network involving TbxT, Nodal, and the Wnt signaling ligand Wnt-3 propels this transition in an anterior direction to generate the primitive streak, with cells ingressing from the epiblast as the streak progresses (see Fig. 3.51). In parallel, during the next 2 days, there is an increase in the total number of cells of the embryo that is concomitant with an increase in the number of different cell types, as represented by the gene-expression patterns. One can see this in single-cell transcriptomic studies, which are discussed below.

Single-cell studies survey the whole genome in every cell and enable statistical analyses that place cells into clusters with similar global patterns of gene expression, creating maps of the processes under study (see Box 6C). The most strongly expressed genes within each cluster can then be used as a reference to define the cell-type identity of the group. One finding from these studies is that cells within a cluster exhibit very heterogeneous levels of gene expression, a clear demonstration that a cell type is defined by a combination of genes, and not simply by one or a few. In most cases, the defining genes code for transcription factors, and computational analysis can reveal networks involved in cell-fate decisions. Single-cell studies can also reveal activity at regulatory sequences that define the time and place of gene expression, and thus help us to understand the regulatory logic underlying the process of cell-type specification.

Throughout mouse gastrulation, expression of BMP-4 remains restricted to the extra-embryonic ectoderm, whereas the expression of Nodal and Wnt-3 progresses with the primitive streak. Over time, this organization creates gradients of signals that control the emergence of cell fates. The first cells that enter the primitive streak are under the influence of high levels of BMP and give rise to extra-embryonic mesoderm. This will produce the allantois (which contributes to the umbilical cord), the amnion, and the yolk-sac mesoderm, and remains associated with the posterior of the streak (see Fig. 3.50). Importantly, the primordial germ cells are also specified in these early stages (see Section 8.2).

As the primitive streak moves anteriorly, and cells entering the streak are farther away from the extra-embryonic ectoderm, they receive a weaker BMP signal, but stronger Nodal and Wnt signals. This positional information leads to the emergence of different cell fates in sequence (see Fig. 4.19 for this illustrated for chick) under the control of transcriptional master regulators.

This progression emerges in the single-cell analysis of the gastrulating mouse embryo. Figure 6.12 shows UMAP plots (see Box 6C) of single-cell RNA analysis of mouse embryos at different stages.

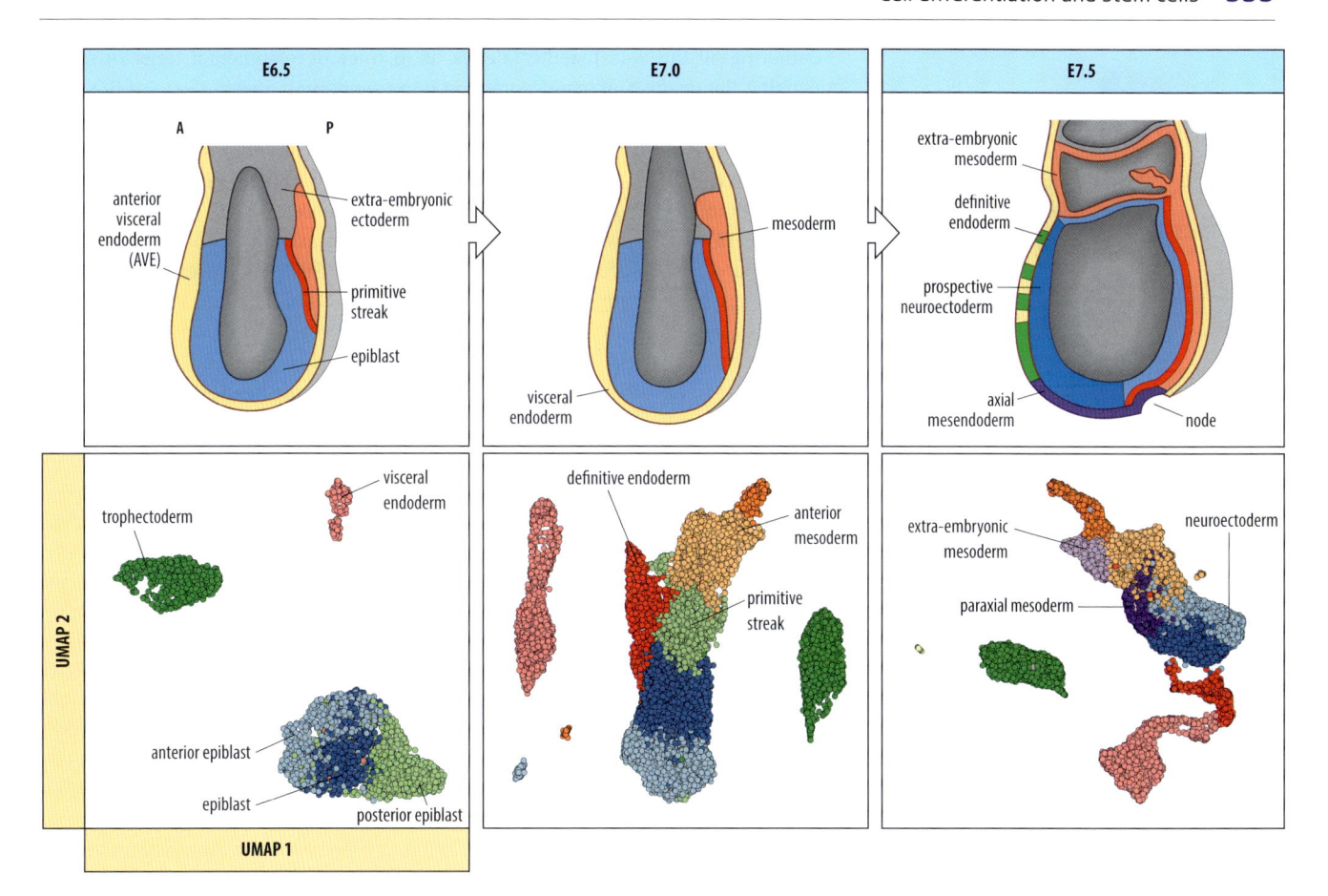

Fig. 6.12 Single-cell transcriptomic analysis of mouse embryos at different stages of gastrulation shows cell proliferation and the emergence of different prospective tissues. UMAPs representing clusters of single cells sharing gene-expression profiles, representing different stages of gastrulation in the mouse. Embryos at three different stages were dissected, single-cell suspensions were made, and, after genetically barcoding (labeling) each individual cell, they were processed for sequencing and identification of the genes they express. Analysis of the sequences and statistical analysis allows cells to be clustered by similarities in their patterns of gene expression, as indicated, and related to specific germ layers and tissues on the basis of known genes associated with specific cell types. The distance between the clusters is a loose measure of the genetic relatedness between their component cells. The color key at the bottom of the panels identifies different cell types. For simplicity, not all the clusters have been identified. As one can see, the complexity in terms of cell types increases as gastrulation proceeds. Note that the colors in the schematic diagrams do not match the tissues identified in the UMAP plots.

UMAP plots generated by Gabriel Torregrosa from data from Pijuan-Sala, B., et al.: ***A single-cell molecular map of mouse gastrulation and early organogenesis.*** Nature 2019, **566**: 490-495.

As the embryo grows, there are more cells in every cluster, and new prospective cell types emerge in the clusters representing the epiblast. At E6.5, we can distinguish epiblast (with different patterns of gene expression emerging for different regions), extra-embryonic ectoderm, and visceral endoderm (see Fig. 6.12, left panel). Between E6.5 and E7, the gene *TbxT*, which marks the primitive streak and mesoderm, starts being expressed in a population of cells. As time progresses, more and more cells express mesodermal markers (see Fig. 6.12, center panel). Later, a cluster of cells expressing *Tbx6*, a marker for paraxial mesoderm can be distinguished (see Fig. 6.12, right panel). By E7.5, many different cell clusters can be detected, expressing transcription factors that are associated with the progenitors of different tissues, including the anterior neural tissues.

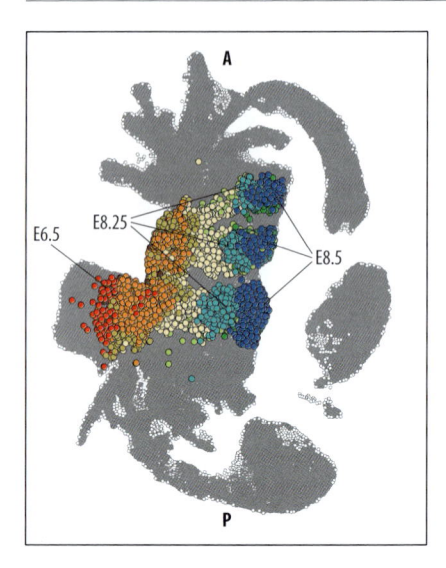

334 Chapter 6 Cell differentiation and stem cells

Fig. 6.13 Gene-expression trajectories associated with the anterior paraxial mesoderm in the mouse. This plot represents a sum of all stages associated with mouse gastrulation as determined by single-cell transcriptional analysis. The cell clusters in color are the prospective paraxial mesoderm (defined by the expression of various genes, such as *Tbx6* and *Tbx2*) at different stages, from pre- to post-gastrulation. The earliest cluster (red) represents E6.5, when gastrulation is just beginning. By E8.25 (green), three distinct regions of anterior paraxial mesoderm can be distinguished, which are consolidated by E8.5 (dark blue). Going from anterior (A) to posterior (P), these will form the muscles of the head, the neck, and the thorax, respectively.

From Guibentif, C., et al.: **Diverse routes toward early somites in the mouse embryo.** Dev. Cell *2021*, **56**: 141-153.

Gathering all the data together allows us to trace developmental trajectories, by organizing the changes of gene expression into pseudotime trajectories (see Box 6C). Thus, it is possible to see, for example, how all the different types of mesoderm are rooted in *TbxT*-expressing cells and how the paraxial mesoderm becomes transformed into somitic cells. The analysis complements more conventional genetic studies and allows us to follow cell-fate decisions at the single-cell level.

Figure 6.13 shows how such plots can be used to trace gene-expression trajectories associated with specific developmental events. This plot represents a sum of all stages associated with mouse gastrulation. The cells highlighted in color are the paraxial mesoderm at different stages, from pre- to post-gastrulation. It shows that three regions of paraxial mesoderm can be distinguished post-gastrulation. Going from anterior to posterior, they represent the muscles of the head, the neck, and the thorax, respectively.

Single-cell transcriptomic studies of mouse gastrulation are a useful platform for understanding similar processes in human embryos, because donated gastrulating human embryos are rarely available. A study of a gastrulating human embryo has been made, however, and shows that the general outline of events during gastrulation is similar to that in the mouse, although differences in the signaling proteins used were found.

However, one must remember that these representations, which identify and follow genes involved in early cell-fate decisions, do not represent developmental lineages, because we cannot tell from this technique whether cells expressing similar combinations of genes, and thus similar phenotypes, are derived from the same founder cell. Also, some cells at these very early stages can still be expressing some genes characteristic of different tissues. However, techniques are being developed that combine traditional genetic lineage tracing (see Box 3A) with single-cell transcriptomics to provide more detailed representations of cell and tissue ancestry. And when the data from single-cell transcriptomics are linked with spatial studies of gene expression in the technique of spatial transcriptomics, it is possible to see cell-fate decisions happening not only over time but also in space.

6.6 Muscle differentiation is determined by the MyoD family of transcription factors

An important mesodermal cell lineage arising during gastrulation is the paraxial mesoderm, which will give rise to the somites (see Section 4.10). As we saw in Section 6.5, its specification involves different transcription factors from those that specify the lateral plate, the heart, or the intermediate mesoderm. Some cells in the somites will give rise to striated skeletal muscle. As this differentiation process can be studied in both the organism and in cell culture, it provides a valuable model system for the study of cell differentiation.

A subset of somite cells expresses the transcription factor Pax3, which has a key role in initiating the differentiation of skeletal muscle cells—**myogenesis**—from these cells. These Pax3-expressing muscle stem cells represent the first wave of myogenesis and differentiate into primary muscle fibers that will act as a scaffold for fetal and adult myofibers. Muscle stem cells in the somites are initially uncommitted and are also called **myogenic progenitors**. As they begin to differentiate they lose their stem-cell-like character and become dedicated skeletal muscle **precursor cells**. After this first wave, a large subset of Pax3-expressing stem cells in the somites expresses Pax7, which also has an overlapping and significant role in myogenesis. Some of the Pax3/Pax7 cells downregulate Pax3 and become precursors of fetal secondary myofibers that differentiate alongside the primary fibers. In the late fetal period, some Pax7-expressing cells form specialized muscle stem cells called **satellite cells** that will participate in muscle growth and repair in the adult. The growth of muscle continues after birth by enlargement of the muscle fibers and addition of new myofibrils.

In response to extracellular signals, myogenic progenitors express members of the MyoD (myoblast-determination) family of proteins and become committed, but as yet undifferentiated, muscle precursor cells called **myoblasts**. These cells can be isolated from chick or mouse embryos and grown in culture, and mouse myoblasts can be cloned to produce cultures of cells derived from a single cell by repeated cell division. Myoblasts will proliferate until they reach high density or until growth factors are removed from the culture medium, whereupon proliferation ceases and overt differentiation into muscle begins. This behavior illustrates the key importance of withdrawal from the cell cycle in promoting the final stages of differentiation. The myoblasts begin to synthesize muscle-specific proteins such as actin, myosin II, and tropomyosin, which are part of the contractile apparatus, and muscle-specific enzymes such as creatine phosphate kinase.

Myoblasts also undergo morphological changes during differentiation. They first become bipolar in shape as a result of reorganization of the cytoskeleton, and then fuse to form multinucleate **myotubes** (Fig. 6.14). Cell–cell fusion *in vivo* requires the presence of a number of factors, including the muscle-specific fusion proteins known as Myomaker and Myomixer (which is also known as Myomerger or Minion). In culture, about a week after withdrawal of growth factors, typical striated **muscle fibers** can be seen.

MyoD is a member of a family of basic helix–loop–helix (bHLH) transcription factors that are expressed in myogenic progenitor and muscle precursor cells. The proteins in the MyoD family of myogenic regulatory factors (MRFs) are crucial transcription factors that are required for muscle cell commitment and differentiation, as they switch on muscle-specific genes and lead to lineage progression to the differentiated state. As well as the *MyoD* gene, the family includes two other determination genes, *Mrf4* and *Myf5*, and one differentiation gene, *myogenin* (*Myog*). If these genes are transfected into cultured fibroblasts or other non-muscle cells that do not normally express these genes nor make muscle-specific structural proteins, they can induce differentiation into muscle cells. This is an example of a phenomenon known as **transdifferentiation**, in which one differentiated cell type is transformed into another cell type.

Myf5, *Mrf4*, and *MyoD* are the first muscle-specific genes to be switched on in myogenic progenitors in mammals, in response to extracellular signals acting with Pax3 and Pax7, and they confer a skeletal muscle fate on uncommitted progenitors (Fig. 6.15). In some cases these genes can be activated directly by Pax3, whose expression becomes restricted to muscle-cell precursors in the early stages of development. Once turned on, some of these factors maintain their own expression by means of positive-feedback loops of the type illustrated in Figure 6.15. To different degrees, the MyoD and Mrf4 proteins can activate *Myog*, which is involved in

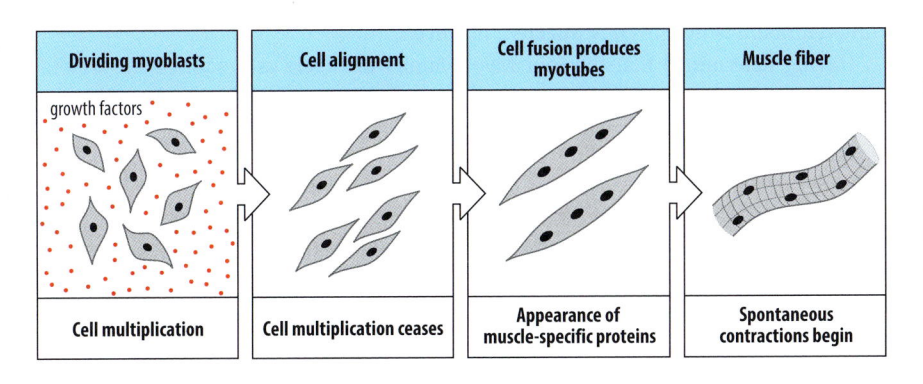

Fig. 6.14 Differentiation of striated muscle in culture. Myoblasts are cells that are committed to becoming muscle but have yet to show signs of differentiation. In the presence of growth factors they continue to multiply but do not differentiate. When growth factors are removed, the myoblasts stop dividing; they align and fuse into multinucleate myotubes, which develop into muscle fibers that contract spontaneously.

Fig. 6.15 Key features of the differentiation of skeletal muscle in mammals. Skeletal muscle cells are derived from stem-cell-like myogenic progenitor cells in the somites. The first muscle cells formed in the embryo are derived from a subset of somitic cells expressing the transcription factor Pax3. These produce primary muscle fibers that act as a scaffold for the subsequent development of secondary muscle fibers in the fetus. The secondary fibers are derived from myogenic progenitors expressing both Pax3 and Pax7. In the development of primary and secondary muscle, extracellular signals, together with expression of the Pax transcription factors, lead to activation of the genes *Myf5*, *Mrf4*, and *MyoD* in the muscle progenitor cells, which become committed skeletal muscle precursor cells (myoblasts), and initiate differentiation of the myoblasts into muscle fibers. One of these genes tends to be expressed preferentially (depending on the animal species) and their activity is generally self-sustaining. The transcription factor products of *Mrf4* and *MyoD* activate muscle-differentiation genes such as *Myog*, which, in turn, activate expression of muscle-specific proteins, such as the proteins that make up the contractile apparatus.

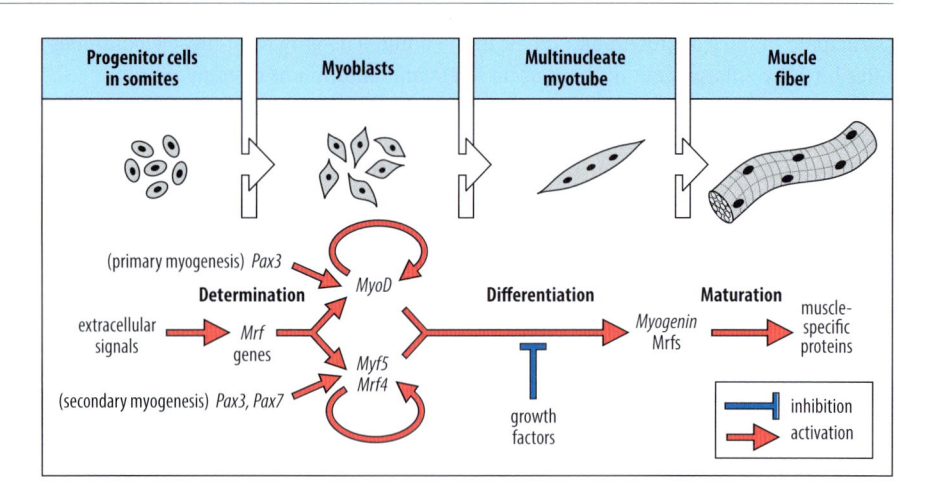

muscle structural differentiation and functional maturation. *Mrf4*, *MyoD*, and *Myf5* are expressed in proliferating, undifferentiated myogenic cells, whereas *Myog* is only expressed in non-dividing, differentiating cells; *Mrf4* and *MyoD* also have later roles in muscle fiber differentiation.

Despite the demonstrated muscle-determining effects of MyoD, gene knock-out experiments in transgenic mice show that mice lacking *MyoD* make skeletal muscle if *Myf5* is active, suggesting some redundancy: that is, one protein can functionally cover for the absence of another. Redundancy is not uncommon in development and has likely evolved to serve as a back-up mechanism to ensure the continuation of development in the face of fluctuations in gene expression (see Section 1.19). Mice lacking *Myf5* alone also make skeletal muscle and are viable, whereas mice lacking both *Myf5* and *Mrf4* have other defects, most notably shortening of the ribs, which suggests non-cell-autonomous effects. Mice in which both *Myf5* and *MyoD* are lacking, however, die before birth, although in such embryos that still express *Mrf4*, some embryonic muscle is still made.

Other types of experiments have confirmed the muscle-determining activity of these transcription factors. Individual prospective muscle cells can be identified experimentally by making mice carrying mutant versions of *Myf5* and *MyoD* that become activated as normal (so that the mRNA can be detected) but which do not make functional Myf5 or MyoD protein. These cells do not become myoblasts: they are found, instead, away from their normal locations, and can be integrated into other differentiation pathways, such as those that produce cartilage and dermis.

In contrast to what happens when the *MyoD* gene is knocked out, mice in which *Myog* has been knocked out form myoblasts, but most skeletal muscle is absent. Myogenin is thus considered to be required for muscle differentiation, rather than being essential for muscle determination. When *Myog* is inactivated in the adult, muscle differentiation is not severely affected, however.

Having determined that the four MyoD family members are required for skeletal muscle specification and differentiation, what we want to know next is how they act together to orchestrate changes in gene expression. Organization of genes into TADs, along with epigenetic changes, is one way of coordinately regulating gene expression. TADs containing muscle-specific genes have been identified—for example, one domain contains the *Myog* gene and several other myogenic-specific genes—and during myogenesis, these domains come closer together towards the center of the nucleus. There are also extensive epigenetic changes associated with the differentiation of myoblasts into myotubes, with widespread loss of histone acetylation, accompanied by loss of the repressive methylation H3K27me3 marks on myogenic genes but retention of these marks on non-myogenic genes.

6.7 The differentiation of muscle cells involves withdrawal from the cell cycle, but is reversible

The proliferation and differentiation of myoblasts are mutually exclusive. Skeletal myoblasts that are multiplying in culture do not differentiate. Only when proliferation ceases does differentiation begin. In the presence of protein growth factors, myoblasts that are expressing both MyoD and Myf5 proteins continue to proliferate and do not differentiate. This means that the presence of MyoD and Myf5 is not in itself sufficient for muscle differentiation, and that an additional signal or signals are required. In culture, this stimulus to differentiate can be supplied by the removal of growth factors from the medium. The myoblasts then withdraw from the cell cycle, cell fusion occurs, and differentiation takes place.

There is an intimate relationship between the proteins that control progression through the cell cycle and the muscle determination and differentiation factors. MyoD and Myf5 are phosphorylated by cyclin-dependent protein kinases, which become activated at key points in the cell cycle (see Fig. 11.4). Phosphorylation affects the activity of MyoD and Myf5 by making them more likely to be degraded. Thus, in actively proliferating cells, the levels of the myogenic factors will be kept low.

The MyoD family of transcription factors activates transcription of muscle-specific genes by binding to a nucleotide sequence called the E-box, which is present in the regulatory regions of these genes. MyoD forms a heterodimer with products of the *E2.2* gene such as E12, and this binds to the E-box. A number of proteins interact with myogenic factors or with their active partners such as E12, and inhibit their activity. One interacting protein is the transcription factor Id, which is present at high concentrations in proliferating cells, and which probably helps to prevent premature activation of muscle-specific genes.

The myogenic factors themselves actively interfere with the cell cycle to cause cell-cycle arrest. Myogenin and MyoD activate transcription of members of the p21 family of proteins, which block cell-cycle progression, causing the cell to withdraw from the cell cycle and to differentiate. Another key protein in preventing progression through the cell cycle is the retinoblastoma protein RB. RB is phosphorylated in proliferating cells by the cyclin-dependent kinase 4 (Cdk4), and dephosphorylation of RB is essential both for withdrawal from the cell cycle and for muscle differentiation. MyoD can interact strongly and specifically with Cdk4 to inhibit RB phosphorylation, which would also help to cause cell-cycle arrest.

A more general mechanism for shutting down genes involved in cell proliferation has also been found to operate in muscle cells. Rather unexpectedly, this involves changing components of the core transcriptional machinery. One of the general transcription factors in the basic promoter-recognition complex described in Figure 6.4 is TFIID, which is a multi-subunit protein that recognizes the core promoter element. TFIID is used in the promoter-recognition complex that controls the expression of genes involved in progression through the cell cycle and cell division. When the myoblasts undergo terminal differentiation, however, TFIID is dismantled and is replaced by a different protein complex, which controls the expression of differentiation genes such as *Myog* (Fig. 6.16). This switch selectively turns on one transcriptional program in the cell while effectively shutting down another, and could be a general mechanism for shutting down cell proliferation and initiating differentiation.

The opposition between cell proliferation and terminal differentiation makes developmental sense. In those tissues where terminally differentiated cells do not divide further, it is essential that sufficient cells are produced to make a functional structure, such as a muscle, before differentiation begins. The reliance on extracellular signals to induce differentiation, or to permit a cell already poised for differentiation to start differentiating, is a way of ensuring that differentiation only occurs in the appropriate conditions.

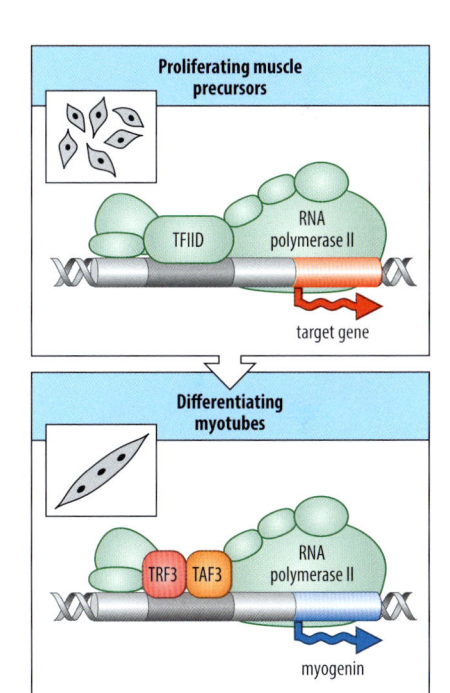

Fig. 6.16 The cessation of myoblast proliferation and the onset of muscle cell differentiation is accompanied by a change in the core transcription complex. In myoblasts, as in most other cells studied, the core transcription complex contains the general transcription factor TFIID, which recognizes the core promoter element in gene promoters. In differentiating myotubes, TFIID is dismantled and is thought to be replaced by a complex containing the proteins TRF3 and TAF3. Along with specific activators, this complex recognizes the *Myog* gene promoter and initiates transcription.

Despite the apparently terminal nature of muscle cell differentiation, muscle cells provide an example of a differentiated cell that can undergo **dedifferentiation**—the loss of differentiated characteristics—and re-enter the cell cycle. Muscle cell dedifferentiation occurs in the formation of the blastema in regenerating newt limbs. The tracing of single cells has shown that the dedifferentiated muscle cells then redifferentiate to form the muscle of the new limb (discussed in Chapter 11).

Once formed, skeletal muscle cells can enlarge by cell growth but do not divide. They can, however, be replaced, as we shall see later in this chapter, from the specialized satellite cells that persist into adulthood.

6.8 All blood cells are derived from multipotent stem cells

Hematopoiesis, or blood formation, is a particularly well-studied example of differentiation, partly because hematopoietic cells at all stages of differentiation are relatively accessible in adult animals and can be grown in culture, and partly because of its medical importance. The process is continuous: in order to keep roughly 25 trillion (25×10^{12}) blood cells circulating in our bodies, most of which are red blood cells, we make 2 million new blood cells per second. In hematopoiesis, we are looking at the differentiation of a **multipotent** stem cell—a stem cell that can give rise to a large but limited range of differentiated cell types (see Section 1.17). We first look at the general process of hematopoiesis and at the extracellular signals that influence it, and then consider the control of gene expression in one cell type—the developing red blood cell.

All the blood cells in an adult mammal originate from **hematopoietic stem cells** located in the bone marrow. Hematopoietic cells have different origins at different stages of development. The first source is extra-embryonic mesoderm in the yolk-sac blood islands, which provides transient progenitors, with limited self-renewal potential, that oxygenate and feed the early embryo. Later, half way through embryogenesis, self-renewing hematopoietic stem cells are found in the embryo, in a region of the aorta and then in a number of definitive blood-forming sites, including the fetal liver and bone marrow. Bone marrow remains the site of origin of blood cells after birth.

Hematopoietic stem cells are both self-renewing and give rise to cells called progenitor cells that, like the early ones in the yolk sac, are more limited in the lineages they can give rise to, and proliferate to act as small amplifying compartments for particular lineages. Hematopoiesis generates a large number of structurally and functionally diverse cells that are essential for the survival of the organism and, for this reason, there is a continual cell turnover, and hematopoiesis must continue throughout life.

The cells circulating in the mammalian bloodstream comprise numerous fully differentiated cell types, together with immature cells at various stages of differentiation (Fig. 6.17). Blood cell types are classified into three main lineages—**erythroid**, **lymphoid**, and **myeloid**. The erythroid lineage yields the red blood cells, or erythrocytes, and the megakaryocytes, which give rise to blood platelets.

The lymphoid and myeloid lineages produce the white blood cells, or leukocytes. The lymphoid lineage produces the lymphocytes, which include the two antigen-specific cell types of the immune system, the B lymphocytes (B cells) and T lymphocytes (T cells). In mammals, B cells develop in the bone marrow, whereas T cells develop in the thymus, from precursors that originate from the multipotent stem cells in the bone marrow and migrate from the bloodstream into the thymus. Both B cells and T cells undergo a further terminal differentiation after they encounter antigen. Terminally differentiated B cells become antibody-secreting plasma cells, whereas T cells differentiate into at least six functionally distinct effector T cells. The B and T cells of the immune system are the only known examples in vertebrates in which the DNA in the differentiated cell becomes irreversibly altered, in a somatic recombination process that potentially enables billions of different antigen receptors to be generated.

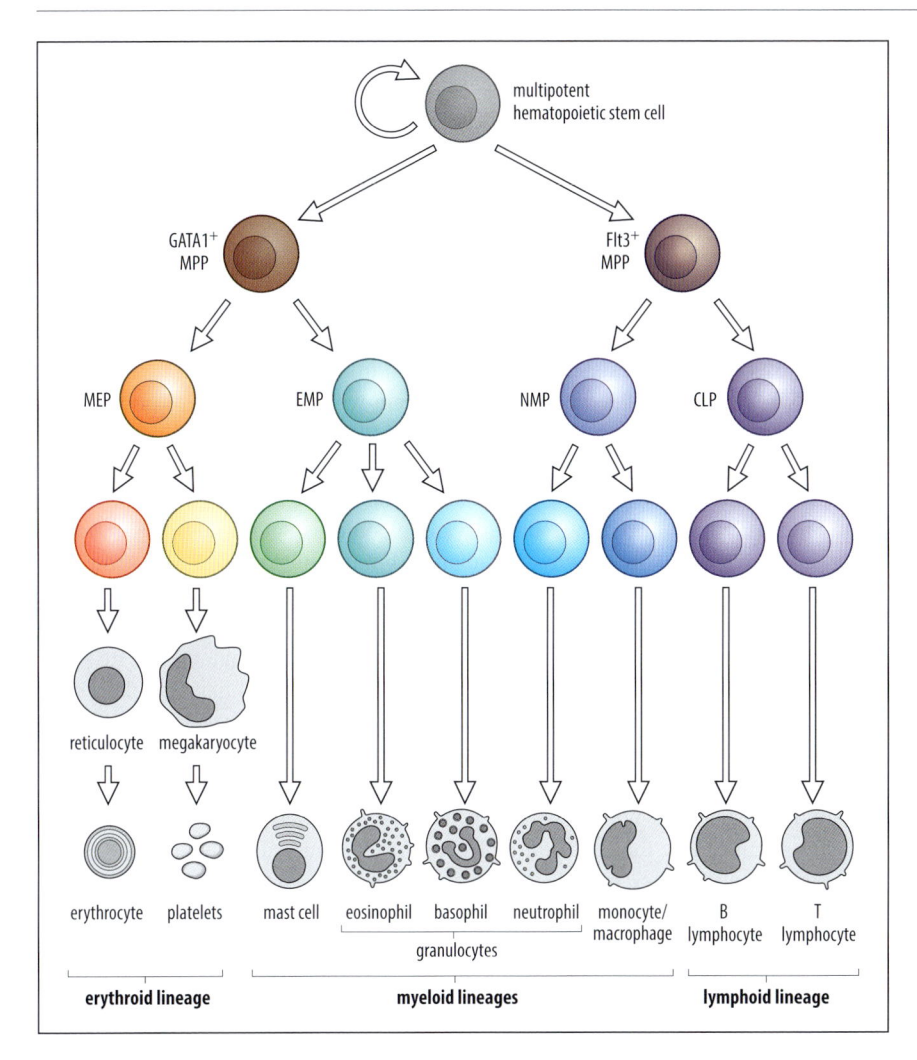

Fig. 6.17 Blood cells derive from multipotent hematopoietic stem cells. Multipotent stem cells (the hematopoietic stem cells) in the bone marrow give rise to all cells that circulate in the blood, as well as to some cells that reside in tissues (such as the dendritic cells of the immune system and the osteoclasts that remodel bone (not shown here)). As well as renewing themselves, the hematopoietic stem cells give rise to progenitor cells with more restricted potential, which, in turn, give rise to committed precursor cells for each blood cell type. Only a few of the progenitor and precursor cells that have been identified or are deduced from experimental data are shown on this highly simplified diagram. Relationships within the blood cell hierarchy are still under revision as a result of data from single-cell transcriptional analysis (discussed in Box 6C). The hematopoietic stem cell is now thought to give rise to two distinct multipotent progenitor cells (MMPs) that express the transcription factor GATA1 and the receptor tyrosine kinase Flt3, respectively. The GATA1⁺ MPP (plum) gives rise to the erythroid lineage, which produces erythrocytes and megakaryocytes. It also gives rise to a lineage of myeloid cells comprising tissue mast cells and the circulating eosinophils and basophils. The Flt3⁺ MPP (brown) gives rise to the lymphoid lineage, which produces the T cells, B cells, and natural killer cells (not shown here), and also to a lineage of myeloid cells comprising the monocytes and the tissue macrophages that derive from them, and the neutrophils. CLP, common lymphoid progenitor; EMP, eosinophil–mast cell progenitor; MEP, megakaryocyte–erythrocyte progenitor; NMP, neutrophil–monocyte progenitor.

Based on data from Drissen, R., et al.: **Distinct myeloid progenitor-differentiation pathways identified through single-cell RNA sequencing.** Nat. Immunol. 2016, **17**: 666-676.

The myeloid lineage includes the eosinophils, neutrophils, and basophils (collectively known as granulocytes or polymorphonuclear leukocytes), and the monocytes and macrophages. Macrophages are resident in the tissues: some are produced in the embryo, but after birth, macrophages are replaced as needed by monocytes that leave the blood and enter tissues. Other myeloid cells that mainly reside in tissues include mast cells, dendritic cells, and osteoclasts (bone-remodeling cells). Recent work indicates that myeloid cells differentiate as two distinct lineages, one of which gives rise to eosinophils and mast cells, whereas the other produces the neutrophils and monocytes (see Fig. 6.17).

The hematopoietic system can thus be considered as a hierarchical system with the multipotent hematopoietic stem cell at the top. This multipotent cell has not been directly observed, but its existence can be inferred from the ability of bone marrow to reconstitute a complete blood and immune system when transplanted into individuals whose own bone marrow has been destroyed. The experiment that showed this capacity, some 60 years ago, was the transfusion of a suspension of bone marrow cells into a mouse that had received a potentially lethal dose of X-irradiation. The mouse would normally have died as a result of lack of blood cells, which, like other proliferating cells, are particularly sensitive to radiation, but transfusion of bone marrow cells reconstituted the hematopoietic system and the mouse recovered. Later experiments showed that this regenerative ability resided in single cells. Hematopoietic stem cells are exploited therapeutically in the use of bone marrow transplantation to treat diseases of the blood and the immune system. Additional sources of hematopoietic stem cells such as umbilical cord blood are now also routinely used in transplantation.

Within the bone marrow, the different types of blood cells and their precursors are intimately mixed up with connective-tissue cells—the **bone marrow stromal cells**. The ability of stem cells both to self-renew and to differentiate into different cell types is tightly regulated by the cells and molecules in their immediate environment—the so-called niche that they inhabit and on which they depend for their survival and function. In the case of hematopoiesis, this **stem-cell niche** is provided by the bone marrow stromal cells. Secreted signal proteins of the Wnt and bone morphogenetic protein (BMP) families, cell-surface Notch ligands (which stimulate the Notch pathway in responding cells on cell–cell contact), retinoic acid, and prostaglandin E_2 have been implicated in maintaining proliferation and self-renewal of hematopoietic stem cells in the stem-cell niche of the bone marrow.

A central question in considering the behavior of stem cells is how a single stem cell can divide to produce two daughter cells, one of which remains a stem cell while the other gives rise to a lineage of differentiating cells. One possibility is that there is an intrinsic difference between the two daughter cells because the stem-cell division is an asymmetric division that results in the two cells acquiring a different complement of proteins. A clear example of this type of division is neuroblast development in *Drosophila*, which is considered in Chapter 10. The second possibility is that extracellular signals make the daughter cells different: a daughter cell that remains in the stem-cell niche continues to renew itself, whereas one that ends up outside the niche differentiates. In the case of hematopoiesis, both mechanisms may operate. On the one hand, four proteins have been found to segregate asymmetrically at cell division when human hematopoietic cells are cultured without the bone marrow stroma that provides the stem-cell niche *in vivo*. On the other hand, factors produced by bone marrow stromal cells that control stem-cell renewal have also been identified.

The multipotent stem cell generates a range of progenitor cells that have different potentials for further development, and they or their progeny become irreversibly committed to particular lineages, after which they undergo further rounds of proliferation and further commitment. Progenitor cells' potential for development is tested by isolating single cells at the various stages and seeing what they can differentiate into, given the appropriate growth factors. There seems to be an early commitment to either the erythroid or the lymphoid lineage, whereas both these progenitor classes

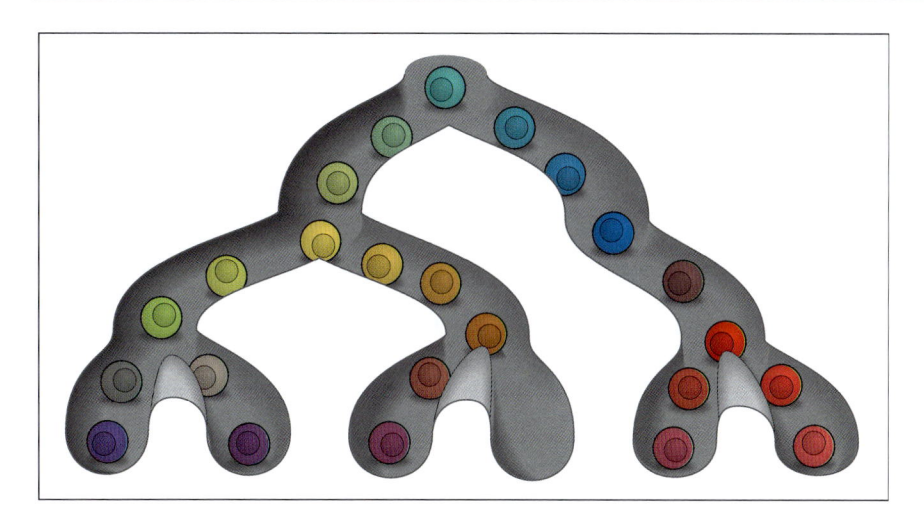

Fig. 6.18 Differentiating cells undergo gradual transitions from one stage to the next. Cells differentiating along a particular pathway are at different sub-stages of the differentiation program, indicated here by the gradual color transitions.

are able to produce myeloid progenitors. This early split is followed by commitment to particular lineages within these broad classes and overt differentiation into recognizable, functionally distinct cell types (see Fig. 6.17).

All stages of the differentiation process are regulated by extracellular signals in the form of hematopoietic growth factors. For those blood cells differentiating in the bone marrow, most of the signal molecules are produced by the bone marrow stromal cells. In this way, the numbers of the different types of blood cells can be regulated according to the individual's physiological need. Loss of blood results in an increase in red cell production, for example, whereas infection leads to an increase in the production of lymphocytes and other white blood cells. Several types of leukemia are caused by white blood cells continuing to proliferate instead of differentiating. These cells become stuck at a particular immature stage in their normal development, a stage that can be identified by the molecules expressed on their cell surfaces.

Thanks to advances in molecular biology techniques, especially the ability to amplify very small amounts of RNA and rapidly sequence it, it is now possible to determine all the genes being expressed in a single cell (see Box 6C). This has been especially useful in investigating blood cell differentiation, and the hematopoietic hierarchy of differentiation shown in Figure 6.17 is based on single-cell transcriptional analysis in the mouse. This type of analysis has also produced some surprising results: in many cases, groups of cells at differentiation stages that seem phenotypically similar have been found to be highly heterogeneous in their gene expression, and, in fact, contain cells at different sub-stages along the differentiation pathway. Furthermore, it seems that changes in transcriptional state are not caused by abrupt large-scale changes in the pattern of gene expression, but that the pattern instead changes gradually from one stage to the next. This can be visualized by borrowing Waddington's 'epigenetic landscape' diagram (see Box 6A). Figure 6.18 shows a hypothetical example in which the marbles rolling down the hill between bifurcations represent cells in which these gradual changes are taking place. The bifurcations represent more pronounced changes in gene expression that occur, often reflecting a restriction in cell fate.

6.9 Intrinsic and extrinsic changes control differentiation of the hematopoietic lineages

Hematopoiesis can be characterized by the activities of a hierarchy of transcription factors, reflecting the hierarchical organization of the different cell types and whose overlapping patterns of expression specify the various cell lineages. Some are only expressed in immature cells and are not lineage specific. Somewhat surprisingly,

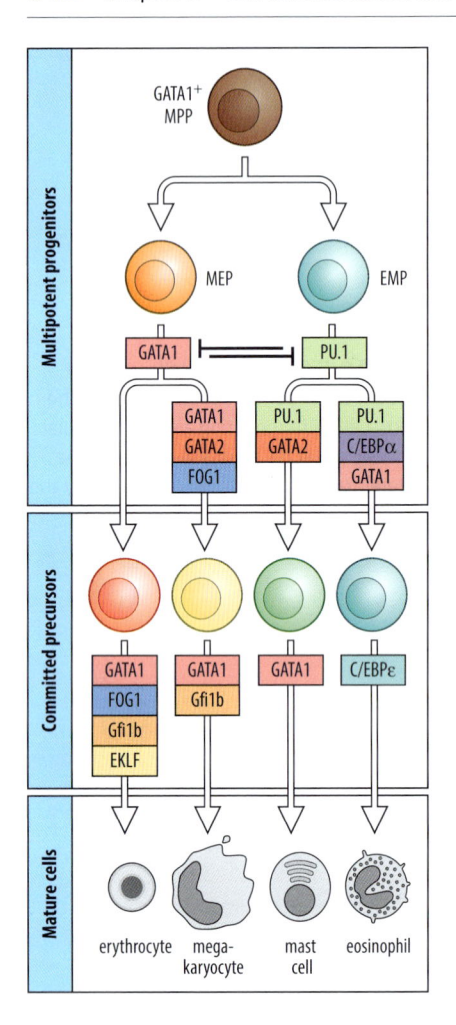

342 Chapter 6 Cell differentiation and stem cells

Fig. 6.19 Hierarchical requirements for lineage-specific differentiation of cell types within the hematopoietic pathway. The pathways leading from the GATA1-positive multipotent progenitor (GATA1⁺ MPP) to the erythroid lineage and one of the myeloid lineages (see Fig. 6.17) are shown here. At the top of the hierarchy, GATA1⁺ MPP cells choose between the megakaryocyte-erythrocyte erythroid lineage and the mast cell-eosinophil myeloid lineage, and they do this by the levels of the transcription factors GATA1 and PU.1 that arise in individual cells. Progenitor cells express low levels of these transcription factors, which repress each other's expression at the same time as they reinforce their own, and this leads to the stabilization of expression of just one of them in different cells. GATA1 and PU.1 then drive sequential programs of differentiation as indicated, through combinatorial transcriptional codes. Note that the same transcription factor can contribute to different cell types through being a member of different transcription factor combinations.

Adapted from Orkin, S.H., Zon L.I.: *Hematopoiesis: an evolving paradigm for stem cell biology.* Cell 2008, **132**: 631-644.

hematopoietic precursors express low levels of some of the genes that will be expressed later only in the myeloid or erythroid lineages. Thus, gene repression, as well as activation, must occur when a cell becomes committed to a particular lineage. Some of the transcription factor combinations that are required for differentiation of particular cell types in the erythroid and myeloid lineages are illustrated in Figure 6.19.

In blood cell precursors, the transcription factor GATA1, for example, is necessary for the differentiation of the erythroid lineage, whereas another key transcription factor, PU.1, drives differentiation of the myeloid lineages. The relative concentration of the two proteins determines cell fate, because GATA1 and PU.1 inhibit each other's ability to regulate the transcription of lineage-specific genes. When the concentration of GATA1 is low, PU.1 will activate the myeloid program and repress the erythroid program and the reverse will happen when the concentration of PU.1 is low, and that of GATA1 is high, resulting in the activation of the erythroid program. As with all cell types, it is the combination of transcription factors, and not any particular one, that is responsible for the specific pattern of gene expression that leads to differentiation into a given cell type.

How then is the activity of all these factors controlled? Signals delivered by extracellular protein growth factors and differentiation factors have a key role. Studies on blood cell differentiation in culture have identified at least 20 extracellular proteins, known generally as **colony-stimulating factors** or **hematopoietic growth factors**, which, together with some of the more general signaling molecules, can affect cell proliferation and cell differentiation at various points in hematopoiesis (Fig. 6.20). Both stimulators and inhibitors of differentiation have been identified. Not all the factors controlling blood cell production are made by blood cells or stromal cells. The protein erythropoietin, for example, which induces the differentiation of committed precursor red blood cells, is mainly produced by the kidney in response to physiological signals that indicate red cell depletion.

Despite the well-established role of these factors in blood cell proliferation and differentiation, it is possible that the commitment of a cell to one or other pathway in

Fig. 6.20 Hematopoietic growth factors and their target cells.

Some hematopoietic growth factors and their target cells	
Type of growth factor	Responding hematopoietic cells
Erythropoietin (EPO)	Erythroid progenitors
Granulocyte colony-stimulating factor (G-CSF)	Granulocytes, neutrophils
Granulocyte-macrophage colony-stimulating factor (GM-CSF)	Granulocytes, monocytes/macrophages
Interleukin-3	Multipotent precursor cells
Stem cell factor (SCF)	Stem cells
Macrophage colony-stimulating factor (M-CSF)	Macrophages, granulocytes

the myeloid lineage might be a chance event, with the growth factors simply promoting the survival of particular cell lineages. Evidence for this comes from experiments where the two daughters of a single early progenitor cell (which can give rise to a range of different blood cell types) are cultured separately but under the same conditions. Both daughter cells usually give rise to the same combination of cell types, but in about 20% of cases they give rise to dissimilar combinations.

Among the plethora of growth factors, it is difficult to distinguish those that might be exerting specific effects on differentiation from those that may be required for the survival and proliferation of a lineage. However, the contrasting functions of three growth factors—**granulocyte–macrophage colony-stimulating factor (GM-CSF)**, **macrophage colony-stimulating factor (M-CSF)**, and **granulocyte colony-stimulating factor (G-CSF)**—are well established. GM-CSF is required for the development of most myeloid cells from the earliest progenitors that can be identified. In combination with G-CSF, however, it tends to stimulate only granulocyte (principally neutrophil) formation from the common granulocyte–macrophage progenitor. This is in contrast to M-CSF, which, in combination with GM-CSF, tends to stimulate differentiation of monocytes (which will ultimately become macrophages) from the same progenitors (Fig. 6.21).

These growth and differentiation factors have no strict specificity of activity, in the sense of each factor exerting a specific effect on one type of target cell only; rather, they act in different combinations on target cells to produce different results, the outcome also depending on the developmental history of their target cell, which determines which proteins are already bound to DNA and which chromatin modifications are present.

6.10 Developmentally regulated globin gene expression is controlled by regulatory sequences far distant from the coding regions

We now focus on one type of blood cell, the red blood cell or erythrocyte, to consider in more detail how the production of one cell-specific protein, the oxygen-carrying hemoglobin, is regulated within a differentiating lineage. The main feature of red blood cell differentiation is the synthesis of large amounts of hemoglobin. All the hemoglobin contained in a fully differentiated red blood cell is produced before its terminal differentiation. Terminally differentiated mammalian red blood cells lose their nuclei, whereas those of birds retain the nucleus but it becomes transcriptionally inactive.

Vertebrate hemoglobin is a tetramer of two identical α-type and two identical β-type globin chains, and so its synthesis involves the coordinated control of two different sets of globin genes. The genes encoding α-globin and β-globin belong to different multigene families. Each family consists of a cluster of genes, and the two families are located on different chromosomes. In mammals, different members of each family are expressed at various stages of development so that distinct hemoglobins are produced during embryonic, fetal, and adult life. Hemoglobin switching is a mammalian adaptation to differing requirements for oxygen transport at different stages of life; human fetal hemoglobin, for example, has a higher affinity for oxygen than adult hemoglobin, and so is able to efficiently take up the oxygen that has been released by maternal hemoglobin in the placenta. In turn, the oxygen affinity of adult hemoglobin is sufficient for it to pick up oxygen in the relatively oxygen-rich environment of the lungs. Here, we look at the control of expression of the β-globin genes, as examples of genes that are not only uniquely cell specific, but are also developmentally regulated.

The human β-globin gene cluster contains five genes—ε, γ_G, γ_A, δ, and β—and is located within a TAD (see Section 6.2) in the short arm of chromosome 11. These genes are expressed at different times during development: ε is expressed in the early embryo in the embryonic yolk sac; the two γ genes, which encode proteins differing by only one amino acid, are expressed in the fetal liver; and the δ and β genes are expressed in erythroid precursors in adult bone marrow. The protein products of all

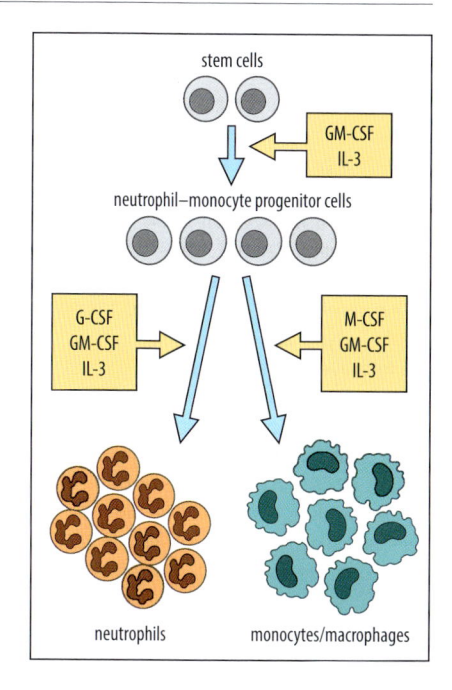

Fig. 6.21 Colony-stimulating factors can direct the differentiation of neutrophils and macrophages. Neutrophils and macrophages are derived from a common neutrophil-monocyte progenitor cell. The choice of differentiation pathway can be determined by, for example, the growth factors granulocyte colony-stimulating factor (G-CSF) and macrophage colony-stimulating factor (M-CSF). The growth factors granulocyte-macrophage colony stimulating factor (GM-CSF) and interleukin (IL)-3 are also required, in combination, to promote the survival and proliferation of cells of the myeloid lineages generally.

Illustration after Metcalf, D.: **Control of granulocytes and macrophages: molecular, cellular, and clinical aspects.** Science 1991, 254: 529-533.

Fig. 6.22 The human globin genes. Human hemoglobin is made up of two identical α-type globin subunits and two identical β-type globin subunits, which are encoded by α-family and β-family genes on chromosomes 16 and 11, respectively. The composition of human hemoglobin changes during development. In the embryo it is made up of ζ (α-type) and ε (β-type) subunits; in the fetal liver, of α and γ subunits; and in the adult most of the hemoglobin is made up of α and β subunits, but a small proportion is made up of α and δ subunits. The regulatory regions α-LCR and β-LCR are locus control regions involved in switching on the hemoglobin genes at different stages. LCR, locus control region.

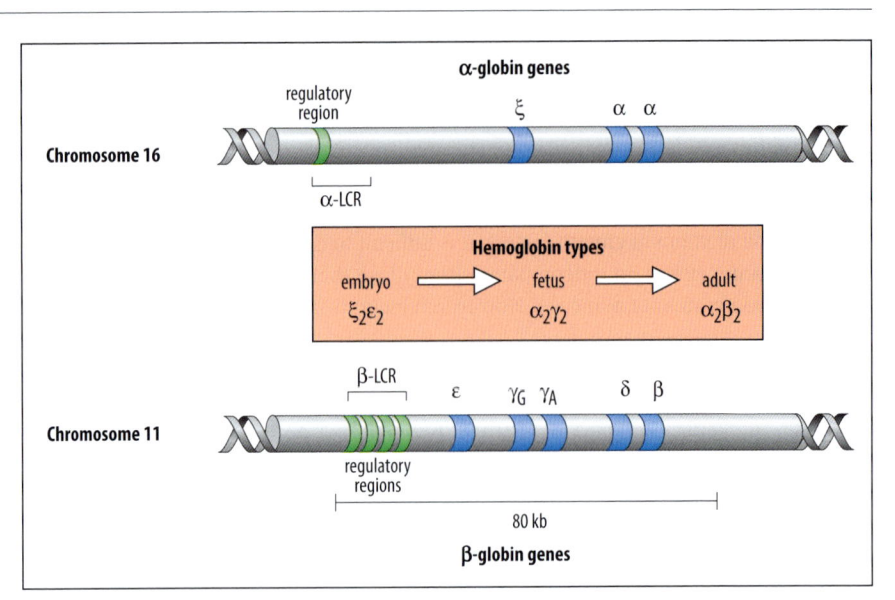

these genes combine with globins encoded by the α-globin complex to form physiologically different hemoglobins at each of these three stages of development (Fig. 6.22).

Mutations in the globin genes are the cause of several relatively common inherited blood diseases. One of these, sickle-cell disease, is caused by a point mutation in the gene coding for β-globin. In people homozygous for the mutation, the abnormal hemoglobin molecules aggregate into fibers, forcing the cells into the characteristic sickle shape. These cells cannot pass easily through fine blood vessels and tend to block them, causing many of the symptoms of the disease. They also have a much shorter lifetime than normal red cells. Together, these effects cause anemia—a lack of functional red blood cells. Sickle-cell disease is one of the few genetic diseases where the link between a mutation and the subsequent developmental effects on health is fully understood.

The control regions that regulate gene expression are often very complex and extensive, and we highlight this here with the example of the β-globin gene cluster. Each gene in the cluster has a promoter and control sites immediately upstream (to the 5′ side) of the transcription start point, and there is also an enhancer downstream (to the 3′ side) of the β-globin gene, which is the last gene in the cluster (Fig. 6.23). β-globin expression is primarily driven by the transcription factor GATA1, which has numerous binding sites in the control regions of the gene. But these local control sequences, which contain binding sites for transcription factors specific for erythroid cells, as well as for other more widespread transcriptional activators, are not sufficient to provide properly regulated expression of the β-globin genes. As in other instances, there is a need for a combination of inputs.

The successive switching on and off of different globin genes during development is an intriguing feature of globin gene expression. This developmentally regulated expression of the β-globin gene cluster depends on a region a considerable distance upstream from the ε gene. This is the **locus control region** (LCR; see Fig. 6.23), which stretches over some 10,000 base pairs at around 5000 base pairs from the 5′ end of the ε gene. The LCR confers high levels of expression on any β-family gene linked to it, and is also involved in directing the developmentally correct sequence of expression of the whole β-globin gene cluster, even though the β-globin gene itself, for example, is around 50,000 base pairs away from the extreme 5′ end of the LCR. A similar control region is located upstream of the α-globin gene cluster (see Fig. 6.22).

An attractive model for the control of globin gene switching envisages an interaction of LCR-bound proteins with proteins bound to the promoters of successive globin

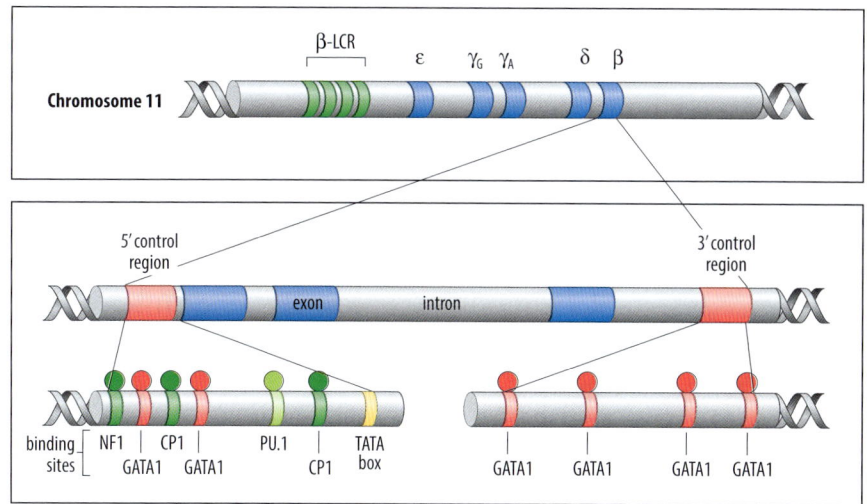

Fig. 6.23 The control regions of the β-globin gene. The β-globin gene is part of a complex of other β-type globin genes and is only expressed in the adult. Binding sites for the relatively erythroid-specific transcription factor GATA1, as well as for other non-tissue-specific transcription factors, such as NF1 and CP1, are located in the control regions. Upstream of the whole β-globin cluster, in the locus control region (LCR), are additional control regions required for high-level expression and full developmental regulation of the β-globin genes.

genes (Fig. 6.24). The chromatin between the LCR region and the globin genes is thought to loop in such a way that proteins binding to the LCR can physically interact with proteins bound to the globin gene promoters. Thus, in erythrocyte precursors in the embryonic yolk sac, the LCR would interact with the ε promoter; in fetal liver it would interact with the two γ promoters; and in the adult bone marrow with the β-gene promoter. One proposed mechanism is that the β-globin LCR enhances transcription by delivering pre-assembled core transcription initiation complexes to the appropriate gene, which are then activated at the gene and start elongation of the RNA transcript. Chromatin looping between the LCR and the β-gene promoter was a key observation that led to the appreciation of the importance of three-dimensional chromatin interactions in regulating gene expression, and to the subsequent discovery of topologically associating domains involved in the regulation of many other genes (see Section 6.2). In addition, the control of globin gene switching involves epigenetic changes involving histone and DNA modifications that are developmentally regulated. Sickle-cell disease only becomes apparent after the fetal–adult switch, and potential therapies being explored are directed at preventing or reversing the silencing of the fetal γ-globin gene. Box 6D illustrates the principle underlying these therapies.

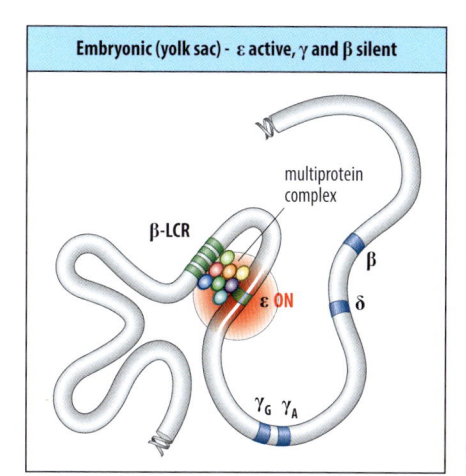

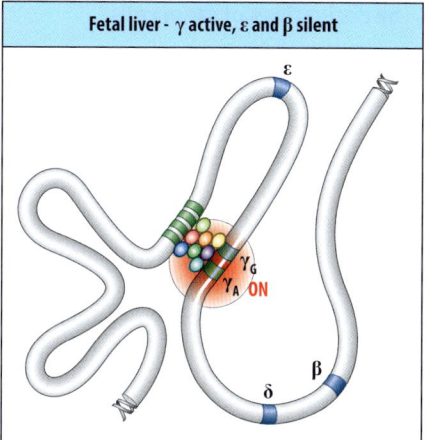

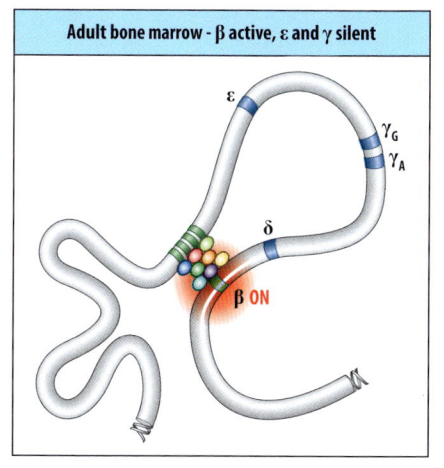

Fig. 6.24 The successive activation of β-family globin genes by the locus control region (LCR) during development. The LCR makes contact with the promoter of each gene in succession at different stages of development, by looping out of the intervening DNA, thus controlling their temporal expression.

Based on Lee, W.S., et al.: *Epigenetic interplay at the β-globin locus.* Biochim. Biophys. Acta 2017, **1860**: 393-404.

MEDICAL BOX 6D Towards a therapy for β-globin gene defects

Two pioneering gene therapies to target the root causes of sickle-cell disease and β-thalassaemia build on previous work that has used genetic strategies to cure blood-related diseases. The aim in the case of the β-globin-related diseases is to boost the expression of fetal γ-globin (the fetal equivalent of β-globin), which is normally switched off postnatally, to compensate for the defective β-globin. In the case of sickle-cell disease, hematopoietic stem cells from individuals with the condition were taken from the bone marrow, genetically modified in culture, and then reinfused back into the individuals, who had been treated with drugs to ablate the remaining faulty blood stem cells.

One modification strategy used the CRISPR-Cas9 technique (see Box 1C) to alter the regulatory sequences of the gene *BCL11A*, whose expression is required to switch off production of fetal γ-globin. CRISPR-Cas9 made a mutation in the enhancer that controls *BCL11A*, so that the gene was not expressed. Figure 1 shows how a specially designed guide RNA directs CRISPR-Cas9 to the GATA1 binding site in the erythroid-specific enhancer region of *BCL11A* where it makes a mutation in the binding site (blue) that prevents the enhancer binding GATA1 and the gene being activated. The gold boxes depict the five *BCL11A* exons. Following reinfusion of the modified stem-cell population, red blood cells were produced that expressed fetal hemoglobin (HbF, composed of α-globin and γ-globin).

A different approach modified hematopoietic stem cells from people with sickle-cell disease or with β-thalassemia using a lentiviral vector encoding a short hairpin RNA targeting *BCL11A* mRNA,

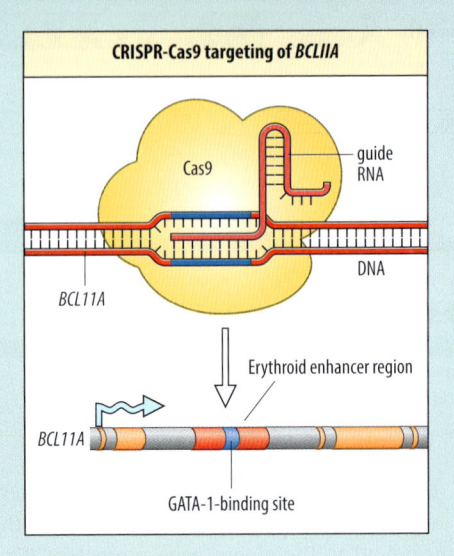

Fig. 1
Adapted from Frangoul, H., et al.: *CRISPR-Cas9 gene editing for sickle cell disease and β-thalassemia*. N. Engl. J. Med. *2021*, *384*: 252-260.

thereby allowing erythroid-lineage-specific post-transcriptional silencing of *BCL11A*.

Although experimental in nature, and successful so far with just a small number of patients, these approaches provide exciting possibilities for combined genetic modification and stem-cell based therapies.

6.11 The epidermis of adult mammalian skin is continually being replaced by derivatives of stem cells

The epidermis of the skin is in constant contact with the harsh external environment and is subject to the continual loss of large numbers of cells from its surface. Like the blood, it needs to be constantly replaced. This task is carried out by tissue-specific stem cells, which are embedded at the base of the epidermis and provide a constant supply of new cells.

Mammalian skin is composed of two cellular layers—the **dermis**, which mainly contains fibroblasts, and the protective outer **epidermis**, which contains mainly keratin-filled cells called **keratinocytes**. We are concerned here with the epidermis, which is renewed from stem cells. Damage to the dermis is repaired by proliferation of fibroblasts and regrowth of damaged blood vessels. A **basal lamina** or **basement membrane** composed of extracellular matrix separates the epithelial epidermis from the dermis. The epidermis comprises a multilayered epithelium of keratinocytes that provides a tough, watertight, and resilient barrier, protecting internal tissues from the outside environment. Associated with this epithelium are evenly spaced hair follicles and sebaceous and sweat glands. The epidermis between the hairs is known as the **interfollicular epidermis**. Cells are being continuously lost from the outer surface of the epidermis and must be replaced—every

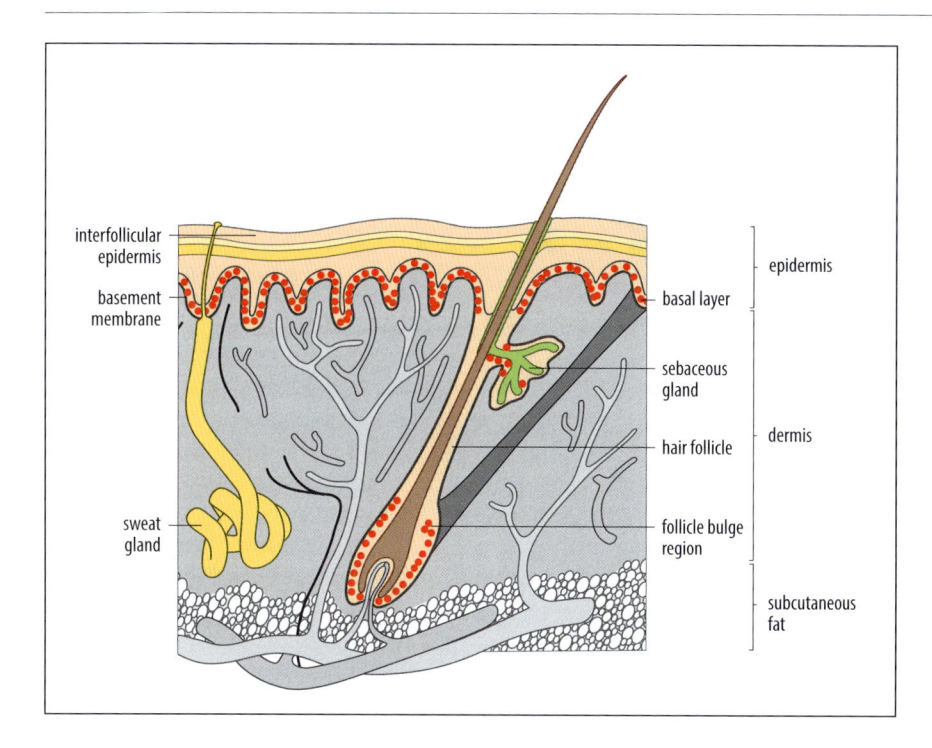

Fig. 6.25 Section through human skin showing epidermal structures. The skin is composed of an outer epidermis (pale orange) separated from the underlying dermis by a basement membrane. Structures that are epidermal in origin are shown in color here; dermal structures are in gray. The locations of stem cells are shown in red. In the case of injury, the stem cells at the base of the hair follicles and in the sebaceous gland can migrate to contribute to a new basal layer.

4 weeks we have a brand new epidermis. New cells of the interfollicular epidermis are produced throughout life by the differentiation of stem cells that reside in the **basal layer** of the epidermis in contact with the basement membrane. Each hair has its own system of support in the form of stem cells at the base of each follicle (Fig. 6.25).

Basal layer cells depend on stimuli received from fibroblasts in the dermis and the extracellular molecules of the basement membrane to remain proliferative. The stem-cell niche for epidermal stem cells therefore comprises the basement membrane and dermis. Once a cell leaves this stem-cell niche, it becomes committed to differentiate. After becoming committed, the epidermal cells differentiate further after leaving the basal layer. One terminal differentiation pathway culminates in the production of keratinocytes, which form the outermost, cornified epidermal layers that provide the skin's protective covering (Fig. 6.26). In the interfollicular epidermis, differentiating keratinocytes are displaced towards the outer layers of the epidermis as they mature. When they reach the outermost layer they are completely filled with the fibrous protein keratin and their membranes are strengthened with the protein involucrin. The dead cells eventually flake off the surface. The terminally differentiated cells of the sebaceous glands, by contrast, are lipid-filled sebocytes that eventually burst, releasing their contents onto the surface of the skin. Differentiation of a hair follicle is even more complex, comprising eight different cell lineages.

In living epidermal cells, keratins are an essential cytoskeletal component, forming the intermediate filaments that connect to desmosomes, the adhesive junctions through which epidermal cells are attached to each other (see Box 5A), and giving skin its resilience and elasticity. Basal layer cells express keratin types 5 and 14, and mutations in the genes encoding these keratins are found in patients with one form of epidermolysis bullosa, a disease in which the epidermis is not anchored properly to the basement membrane and so is very fragile and easily blisters. Spinous cells expressing keratins 1 and 10 are in the suprabasal layer of the epidermis and the switch from expression of keratin types 5 and 14 to expression of keratins 1 and 10 is a reliable indicator that cells have become committed to terminal differentiation (see Fig. 6.26). Several signals are now known to regulate this switch. Activation of

Fig. 6.26 Differentiation of keratinocytes in human epidermis. Descendants of basal layer cells, which will become the keratinocytes of the epidermis, detach from the basement membrane, divide several times, and leave the basal layer before beginning to differentiate. Basal layer cells express the intermediate filament proteins keratin 14 and keratin 5, and also the integrin $\alpha_6\beta_4$, required for hemidesmosome formation and adhesion to the basement membrane. Differentiation into keratinocytes involves a loss of integrin expression, a change in the keratin genes expressed, and the production of large amounts of keratin 1 and keratin 10. In the intermediate layers, the cells are still large and metabolically active, whereas in the outer layers of the epidermis, the cells lose their nuclei, become filled with keratin filaments, and their membranes become insoluble owing to deposition of the protein involucrin. The dead cells are eventually shed from the skin surface. EGFR, epidermal growth factor receptor.

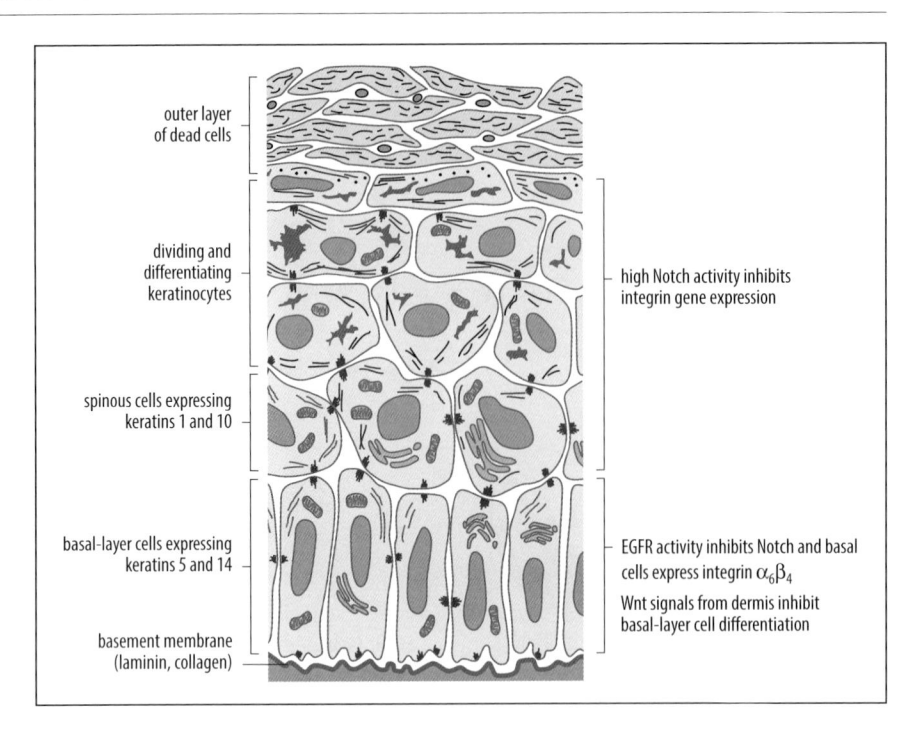

Notch signaling in basal layer cells promotes differentiation, and is inhibited by various mechanisms, including signaling through the epidermal growth factor receptor (EGFR), which, as a consequence, contributes to maintaining basal cell fate. Wnt produced by the dermis and acting on the basal layer cells also contributes to the inhibition of differentiation. Cell-adhesion molecules also have a key role. The basal cell layer is attached to the basement membrane by specialized cell junctions that include integrin-containing hemidesmosomes and focal adhesions (see Box 5A). Basal cells express the integrin $\alpha_6\beta_4$, which is required for hemidesmosome formation and interacts with collagen and laminin in the extracellular matrix. These cells also express high levels of integrin $\alpha_3\beta_1$, which is a receptor for laminin. Detachment of a cell from the basement membrane is accompanied by a decrease in the amount of integrins on the cell surface, which is a result of activation of Notch.

Mutations in the collagen associated with epidermal hemidesmosomes and in laminin have been found in a subset of patients with a type of epidermolysis bullosa known as junctional epidermolysis bullosa. A young boy with junctional epidermolysis bullosa had almost the whole of his skin successfully replaced with healthy skin grown *in vitro* from his epidermal stem cells, which had been genetically corrected to replace the faulty gene (Box 6E).

The interfollicular epidermis can also be generated from the hair follicle, where stem cells have been located in a region known as the bulge (see Fig. 6.25). Lineage-tracing experiments in transgenic mice in which, for example, green fluorescent protein is expressed under the control of the promoter for the keratin 5 gene, showed that the cells that give rise to new hair follicles originate in the bulge. Under certain conditions, usually as a result of injury, bulge cells can also contribute to the sebaceous gland and epidermis. The descendants of single-follicle bulge cells from mouse epidermis cultured in the laboratory can give rise to hair follicles, sebaceous glands, and epidermis when combined with dermal fibroblasts and used as skin grafts. These experiments show that within mouse epidermis, there are different populations of stem cells that can be brought into play when required. They have also highlighted several modes of stem-cell division, which we discuss next.

MEDICAL BOX 6E Treatment of junctional epidermolysis bullosa with skin grown from genetically corrected stem cells

The power of skin stem cells to renew the epidermis was illustrated by the successful treatment of a young boy with a particularly severe case of a rare genetic skin disease—junctional epidermolysis bullosa. This is an example of the potential of tissue engineering to build replacement tissues and body parts from stem cells.

The skin has two components that need to interact in order to protect internal tissues and organs and to defend against infections: the epidermis, which is a barrier to the environment, and underneath it the dermis, which anchors and provides mechanical support for the epidermis (see Fig. 6.25). In junctional epidermolysis bullosa, the interaction between the dermis and epidermis is impaired and minor injuries cause severe damage to the epidermis, which ruptures, leading to blisters and ulcers. The condition is often associated with mutations in the components of laminin and in a specific form of collagen—proteins involved in the interaction between epidermal cells and the extracellular matrix.

In June 2015, a 7-year-old boy was admitted to a hospital in Bochum in Germany with a very severe manifestation of junctional epidermolysis bullosa that had led to a loss of more than 60% of his skin. He carried a mutation in the gene for laminin β3 (*LAMB3*), one of the subunits of laminin-332, which is often affected in this condition. A team of physicians together with stem-cell biologists undertook the challenge of regenerating the boy's skin from his own epidermal stem cells (the dermis in junctional epidermolysis bullosa remains relatively intact). Transplants of epidermis derived from stem cells *in vitro* have been a staple of burn treatment for a number of years, but this work went further. The team took a 4-cm² piece of epidermis from the remnants of the boy's intact skin and, after growing the cells in culture, used a retroviral vector to introduce a normal copy of the *LAMB3* gene into the cells to correct laminin function. They then proceeded to grow cell sheets totaling 0.85 m² of repaired epidermis, which was used to cover the body of the child in several separate transplant operations (Fig. 1). Nearly 6 years later, the boy is fully recovered and playing normally, his skin is healthy, and the usual childhood injuries repair themselves.

The cure is remarkable in many ways. Retroviral-based repair of a mutated gene carries a risk of triggering unwanted genomic activity, but that has not happened in this case. Having overcome this hurdle, there was no assurance that the engineered cells would cover such a large surface area of the body. The fact that they did is witness to the power of skin stem cells, which renew your whole epidermis once a month.

The treatment also addressed an interesting biological question about the long-term renewal potential of the various types of clonogenic cells in the epidermis. As well as the self-renewing stem cells, the epidermis contains cells that will proliferate for a while and then become terminally differentiated. During the genetic repair, the retrovirus inserts a normal copy of a gene in different genomic locations in different cells, thus creating a specific label for each cell. For the transplants here, the surgeons used clones derived from differently labeled cells, and thus they could determine whether, after a time, the new skin was a patchwork of many different clones arising from all the clonogenic cells that had been initially transplanted, or whether just a relatively few long-lived stem cells would continue to renew the epidermis. By 8 months after the transplants, there were far fewer different clones in the new epidermis than there had been originally. One explanation is that most of the clonogenic cells that were initially transplanted soon became terminally differentiated, leaving only a relatively few long-lived stem cells to continue to renew the epidermis. An alternative explanation is that initially there are many clones with different proliferative abilities and differentiation balances that compete with each other and, in the end, the progeny of one will take over (see Section 6.12).

It is early days, but this is an example of the possibilities that lie ahead when responsible basic research is combined with good medicine. In many ways, this kind of repair has been done for years in the case of blood transfusions and, in the future, it may be possible with intestinal organoids (see Box 7D).

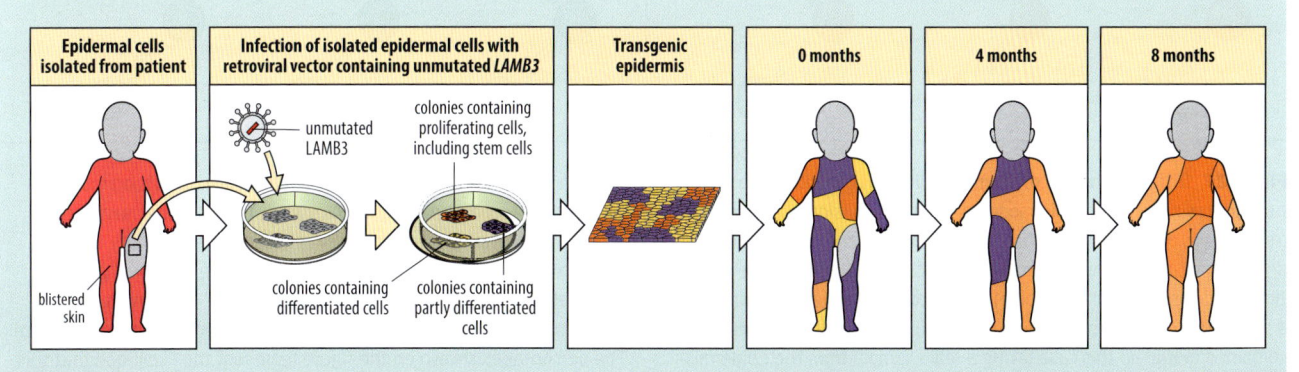

Fig. 1

Adapted from Aragona, M., Blanpain, C.: **Transgenic stem cells replace skin**. Nature 2017, **551**: 306-307.

6.12 Stem cells use different modes of division to maintain tissues

The precise identification of the stem cells in the basal layer of the epidermis, and whether all the cells within the basal layer can, under certain circumstances such as injury, act as stem cells, is still not resolved. Basal epidermal cells are heterogeneous, with some cells proliferating rapidly and others dividing far less frequently. It has been suggested that the slowly cycling cells are the stem cells and that the surrounding proliferating basal cells, known as **transit-amplifying cells** or sometimes as progenitor cells, are committed to terminal differentiation. These observations fit the classic model of stem-cell division, which is one of invariant asymmetric cell division. Each stem-cell division results in one stem cell and one cell committed to differentiate, which then undergoes several rounds of amplifying divisions before differentiating (Fig. 6.27).

However, other investigations tracking the fate of individually marked basal cells observed continuous proliferation of these cells in certain parts of the epidermis, which suggested that in these regions all the basal cells could be equivalent in their developmental potential—that is, that they were all acting as progenitor cells rather than classic stem cells. These observations led to additional models of basal cell division in which the cells can divide symmetrically or asymmetrically, and the proportions of the different types of division keeps the balance between stem/progenitor and

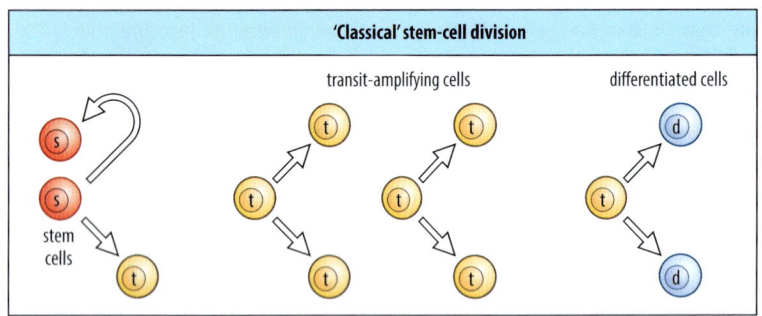

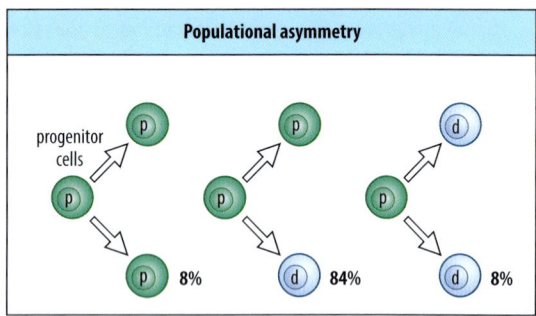

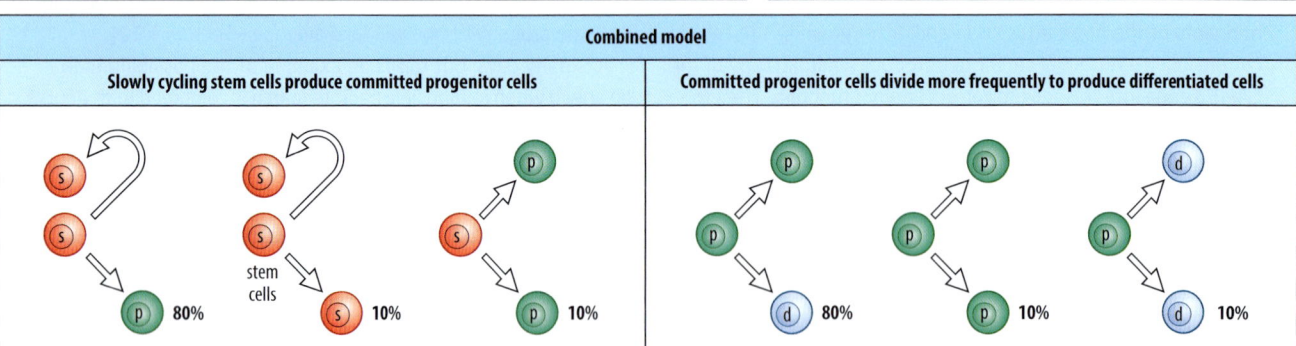

Fig. 6.27 Different modes of stem-cell division contribute to tissue homeostasis. In the 'classic' model of stem-cell division, the stem cell (red) always divides asymmetrically to give rise to another stem cell and a transit-amplifying cell (orange) that is committed to divide several times and then differentiate (blue). In the populational asymmetry model, a progenitor cell (green) can divide in the three modes as indicated. In the combined model—in this case taking a real example from the epidermis—very slowly dividing stem cells (undergoing only a few divisions per year) produce progenitor cells that then divide much more rapidly, following the rules of populational asymmetry, to renew the tissue. To achieve tissue homeostasis, the proportions of each type of division must be correctly balanced; in particular, the symmetric divisions must occur in equal proportions. Thus, if a stem cell (S) can divide in any of three ways to give S + S, D + D, or S + D, where D is a cell that will go on to differentiate, then as long as the number of S + S divisions is equal to the number of D + D divisions, the system will remain at a constant size. The percentages shown on the figure represent the likelihood that this particular division will occur within the starting population of cells.

*From Mascré, G., et al.: **Distinct contribution of stem and progenitor cells to epidermal maintenance.** Nature 2012, **489**: 257-262.*

differentiated cells and maintains tissue homeostasis (see Fig. 6.27). According to the **populational asymmetry** model, basal layer cells are progenitor cells that can divide in any of three ways: symmetrically to give rise to two cells that remain progenitor cells and do not differentiate; symmetrically to give rise to two cells that go on to differentiate; or asymmetrically to give one of each type (as in the classic model; see Fig. 6.27). In this mode of tissue renewal there is no need for an amplifying phase, as the progenitor cells are dividing continuously. And so long as there is a correct balance between the three kinds of divisions, the tissue is maintained. The populational asymmetry model, however, does not account for the presence of the slowly cycling cells in the epidermal basal layer. Analysis of the abundant experimental and observational data on epidermal renewal suggests that a combined model (see Fig. 6.27)—in which classic stem cells (the slowly cycling cells) produce progenitor cells that then proliferate and differentiate according to the rules of populational asymmetry—best fits the observed dynamics of epidermal renewal over time.

6.13 The lining of the gut is another epithelial tissue that requires continuous renewal

Like the outer layer of the epidermis, the epithelium lining the gut is exposed to a relatively harsh environment—in this instance the lumen of the gut and its contents—and undergoes continuous cell loss. It is also replaced from stem cells, and there are considerable similarities to epidermal renewal, especially in the signaling molecules used to regulate stem-cell differentiation. The stem cells in the gut have also been observed to divide by all the modes of division outlined in Section 6.12.

In the small intestine, the endoderm-derived epithelial cells form a single-layered epithelium that is folded to form villi that project into the gut lumen and crypts that penetrate the underlying mesoderm-derived connective tissue. Cells are continuously shed from the tips of the villi, while the stem cells that produce their replacements lie in the base of the crypts. The new cells generated by the stem cells move upwards towards the tips of the villi. En route, they multiply in the lower half of the crypt and constitute a transit-amplifying population (Fig. 6.28). The turnover of cells in the mammalian small intestine is about 4 days.

A single crypt in the small intestine of the mouse contains about 250 cells comprising four main differentiated cell types. Three are secretory cells: Paneth cells secrete antimicrobial peptides, goblet cells secrete mucus, and enteroendocrine cells secrete

Fig. 6.28 Epithelial cells lining the mammalian gut are continually replaced. The upper panel shows the villi, which are epithelial extensions covering the wall of the small intestine. The lower panel is a detail of a crypt and villus. The stem cells, from which the intestinal epithelium is renewed, are found in the base of the crypts. One type of stem cell (yellow) is characterized by a slow rate of division and is found at some distance from the bottom of the crypt, while a second type of stem cell (red) is interspersed between the Paneth cells in the base of the crypt. These stem cells divide more frequently than the first type and are the ones mostly involved in routinely renewing the epithelium. The stem cells give rise to cells that proliferate and move upwards. By the time they have left the crypt, they have differentiated into epithelial cell types. They continue to move upwards on the projecting villus and are finally shed from the tip. The total time of transit is about 4 days. The photograph shows a section through mouse small intestine. Individual stem cells at the base of each crypt were initially labeled in different colors. As the crypts were renewed, each was, at first, a mixture of clones of all the stem cells (not shown). But, over time, competition between the different clones results in each crypt gradually becoming populated by the descendants of a single stem cell, as shown here.

*From Snippert, H.J., et al.: **Intestinal crypt homeostasis results from neutral competition between symmetrically dividing Lgr5 stem cells.** Cell 2010, **143**: 134–144.*

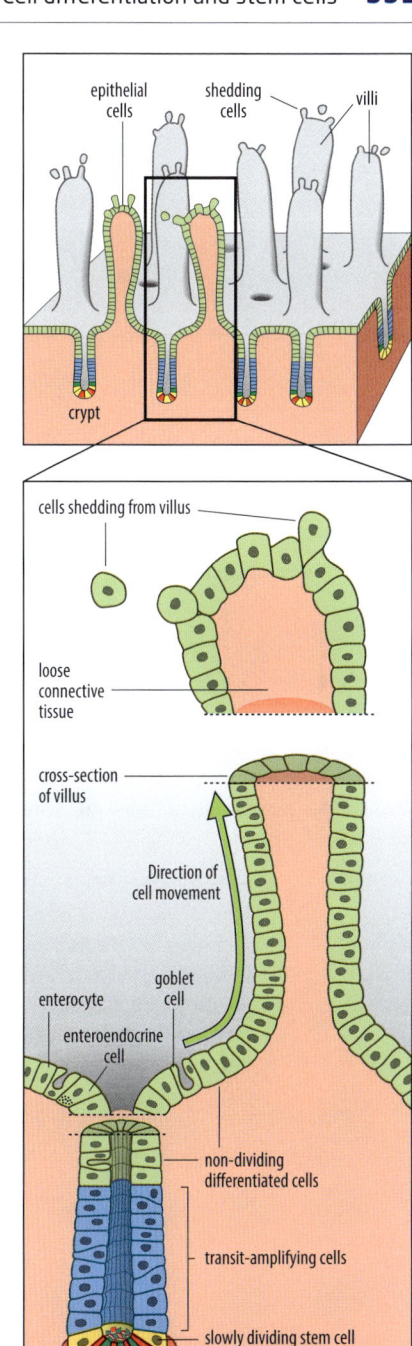

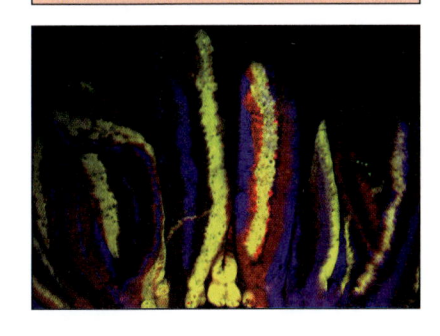

various peptide hormones. The fourth cell type, the enterocyte, is the predominant cell in the epithelium of the small intestine and absorbs nutrients from the gut contents. About 150 of the crypt cells are proliferative, dividing about twice a day so that the crypt produces 300 new cells each day; about 12 cells move from the crypt into the villus each hour.

The base of each crypt has a very stereotyped structure and contains two types of stem cell. The stem-cell niche is composed of the Paneth cells in the crypt and the mesenchymal cells surrounding the base of the crypt. One type of stem cell, originally identified by its slow rate of division, is located a short distance above the bottom of the crypt, at the edge of the stem-cell niche. The other stem cells, which are considerably more proliferative, are slender columnar cells—between four to six cells per crypt, that are interspersed between Paneth cells at the bottom of the crypt and divide every 24 hours (see Fig. 6.28). Both sets of stem cells have distinctive patterns of gene expression. The slowly dividing cells express high levels of Bmi1, a component of the Polycomb complex of chromatin-binding proteins. Long-term lineage tracing of cells expressing *Bmi1* in transgenic mice using LacZ as a marker resulted in the labeling of entire crypts and villi after 12 months, showing that the *Bmi1*-expressing cells are multipotent stem cells. The other population of stem cells expresses a Wnt target, the gene encoding Lgr5, a G-protein-coupled receptor whose ligand is R-spondin-1, a secreted protein that acts as a Wnt signal enhancer. Experiments in which *Lgr5*-expressing cells and their progeny were marked irreversibly by expression of LacZ showed that a single one of these columnar cells could give rise to an entire crypt and to a ribbon of cells on adjacent villi after 2 months (Fig. 6.29). Thus, like the slowly cycling cells, the *Lgr5*-expressing columnar cells are multipotent stem cells, and are the cells that generate the new cells for the epithelium of the villi during normal tissue homeostasis. More recently, another type of very slowly cycling cell has been detected in the bottom of the crypt that expresses both *Bmi1* and *Lgr5*. Slowly dividing stem cells—at least those in the base of the crypts—provide a reserve of stem cells for regeneration.

The multipotency of the *Lgr5*-expressing stem cells was confirmed by isolating them and growing them in culture. When cultured under the right conditions in a three-dimensional system, a single *Lgr5*-expressing cell can give rise to organized intestinal villus–crypt tissue known as an **organoid**, which contains all four differentiated cell types. In addition, the organoid also contains stem cells, and these are maintained in organoids that have been passaged in culture for more than 18 months. Thus, this protocol provides a way of culturing adult stem cells, and similar protocols have been devised to culture adult stem cells from other tissues. We discuss organoids and their potential clinical applications in more detail in Chapter 7.

The maintenance of stem cells in both the epidermis and the small intestine is under the control of the same signals. In both instances, Wnt signaling through

Fig. 6.29 Cell lineage tracing of *Lgr5*-expressing cells in the mouse small intestine. Transgenic mice that expressed LacZ from the *Lgr5* promoter were used to trace the origin and movement of *Lgr*-expressing cells. The photos show histological analysis of LacZ activity (blue-staining cells) in small intestine: (a) 1 day after induction; (b) 5 days after induction; and (c) 60 days after induction.

From Barker, N., et al.: ***Identification of stem cells in small intestine and colon by marker gene* Lgr5.** Nature *2007,* **449***: 1003–1007.*

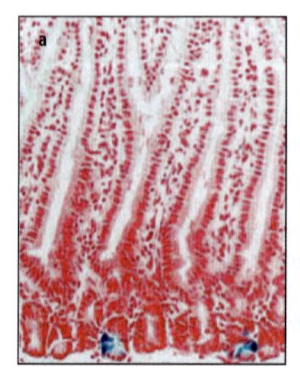

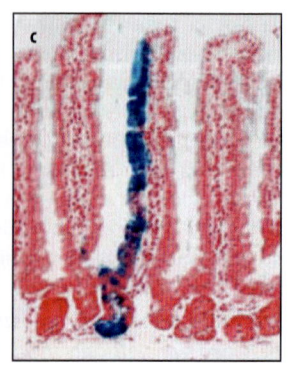

β-catenin promotes stem-cell renewal, whereas BMP inhibits it. Populations of stem cells at particular sites in the tissue can be kept in a quiescent state by the action of BMP and inhibition of Wnt signaling, while other populations are stimulated to divide by Wnt signaling in the presence of BMP antagonists (Fig. 6.30). Notch signaling is essential to maintain the stem-cell population in both tissues. It also has a later role in the small intestine in providing the switch that determines whether a cell leaving the crypt becomes an enterocyte or a goblet cell. This interplay between Wnt, BMP, and Notch signaling pathways in the maintenance and differentiation of stem-cell pools is common to other types of adult stem cell as well. The importance of these pathways in the maintenance and differentiation of stem-cell populations is highlighted by the incidence of mutations in components of the pathways in several types of cancer.

6.14 Skeletal muscle and neural cells can be renewed from stem cells in adults

Muscles are subject to injury and are often in need of repair, but these structurally highly differentiated cells do not divide. Instead, mammals have a reserve of muscle stem cells, called **satellite cells**, that are associated with mature skeletal muscle fibers, and that become activated in response to muscle injury. When satellite cells are activated, they proliferate and develop into myoblasts, which then differentiate into new muscle fibers (see Section 6.6).

Satellite cells emerge during the differentiation of skeletal muscle in the embryo and generate a potential reservoir of muscle tissue. Set aside from the myoblast pool before the final stages of differentiation, they are small mononuclear cells that lie between the basement membrane and the plasma membrane of mature muscle fibers, and can be recognized by their expression of characteristic proteins, such as the cell-surface protein CD34 and the transcription factor Pax7 (Fig. 6.31). Under favorable conditions, the satellite cells associated with one muscle fiber (about seven satellite cells), when grafted into an irradiated muscle in a mouse, can generate more than 100 new fibers and will also undergo self-renewal. The irradiation kills the mouse's own satellite cells, thus making it impossible that the muscle could have regenerated on its own. Even a single muscle stem cell transplanted into irradiated mouse muscle can produce both muscle fibers and more satellite cells.

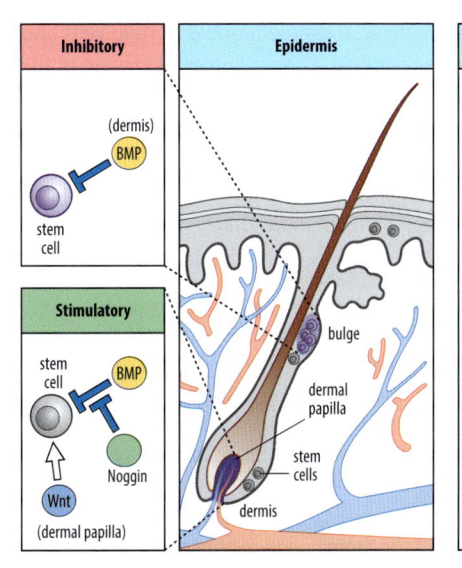

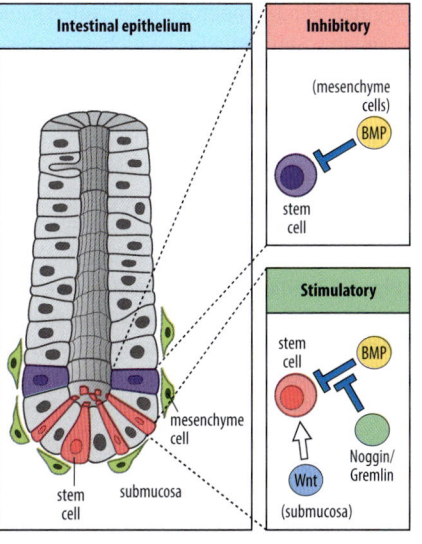

Fig. 6.30 Signaling in the stem-cell niche. Some signaling pathways perform conserved functions in different stem-cell populations. For example, Wnt promotes stem-cell division and BMP inhibits it in both the skin and the intestine. Stem-cell populations in different locations within the tissue are kept quiescent by the action of BMP and the inhibition of Wnt signaling, or are activated by the action of Wnt and the inhibition of BMP signaling by the BMP antagonists Noggin or Gremlin.
From Li, L., Clevers, H.: **Coexistence of quiescent and active adult stem cells in mammals.** *Science 2010,* **327***: 542-545.*

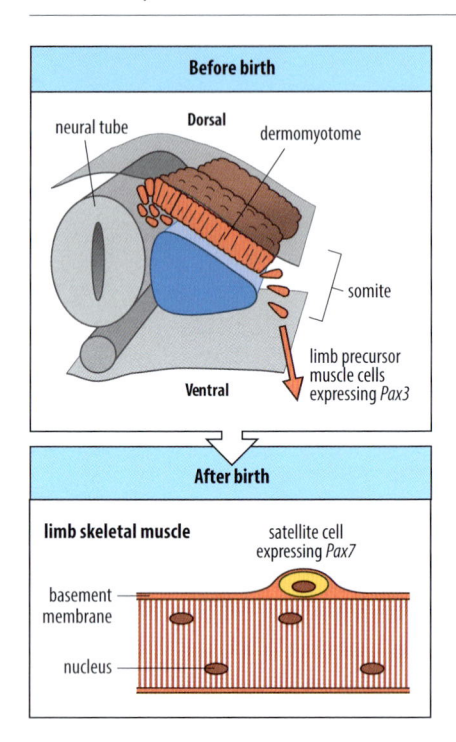

Fig. 6.31 Muscle satellite cells. During fetal development, some *Pax7*-expressing muscle cell progenitors remain associated with the multinucleate muscle fibers as muscle stem cells or satellite cells. These are the source of new myoblasts for muscle growth and repair in the adult.

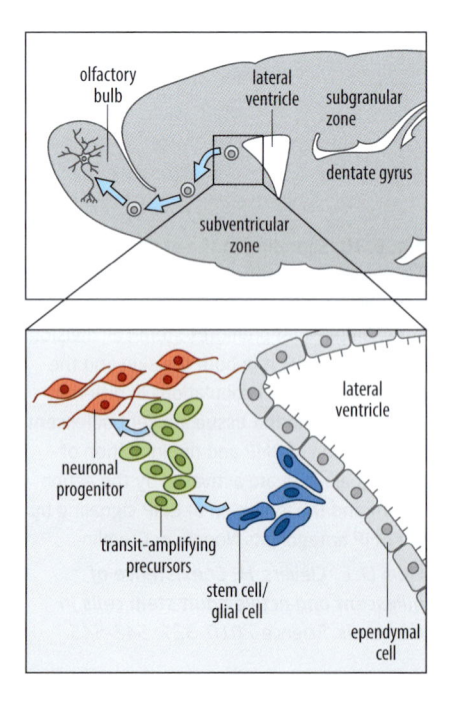

The function of Pax7 seems to overlap with that of the closely related Pax3. Just as Pax3 is essential for early determination of muscle cell precursors in the embryo, as discussed in Section 6.6, Pax7 is essential for the maintenance of the satellite cell population during regeneration of adult muscle. In muscle satellite cells, the transcription of myogenic genes such as *Myf5* is repressed. Activation of their transcription and differentiation of the cells into muscle depends on epigenetic modifications brought about, for example, by Pax7 recruiting chromatin modifiers such as the Trithorax complex, which methylates the histone H3K4 (see Box 6B).

It is becoming clear that the existence of stem cells to support the normal life of an adult tissue, as in the skin and intestine, or as a reservoir in case of injury, as in muscle, is very widespread, and that just about every tissue and organ in the body has a population of cells which, upon need, can be activated to repair damage. This is also true, although to a more limited extent, of the nervous system. We will see in Chapter 10 how neurons and other nervous system cells develop from embryonic neural stem cells. Once fully differentiated, however, neurons in the mammalian brain do not divide, and for many years it was thought that no new brain neurons at all were produced in the adult mammal, although adult neurogenesis occurs in the brains of songbirds, reptiles, amphibians, and fish. However, the production of new neurons, or **neurogenesis**, has now been demonstrated as a normal occurrence in the adult mammalian brain, and neural stem cells have been identified in adult mammals that can generate neurons, astrocytes, and oligodendrocytes in culture.

These adult neural stem cells are a population of slowly dividing radial glial cells, like those in the embryonic cortex, that produce neurons—at least in the adult rodent. Neurogenesis in adult mammals occurs in just two regions: the subventricular zone of the lateral ventricle, and the subgranular zone of the dentate gyrus in the hippocampus. New neurons generated in the subventricular zone migrate to give rise to **interneurons** in the olfactory bulb, whereas the stem cells in the hippocampus give rise to new neurons of the dentate granule layer of the hippocampus. The radial glial cells in the dentate gyrus express some astrocyte-specific genes and stem-cell markers. They can give rise to both neurons and astrocytes, and there is increasing evidence that they are a heterogeneous cell population, although the details of the cell lineages have yet to be worked out. In the subventricular zone of the lateral ventricle in adult mice, these stem cells give rise to proliferating transit-amplifying cells, which then mature into neural progenitors (Fig. 6.32) and also give rise to astrocyte and oligodendrocyte glial cells. Other glial cells adjacent to these regions of the adult brain that contain stem cells provide a stem-cell niche, producing signals that control neural stem-cell behavior. Wnt signaling through the β-catenin pathway in neural stem cells has an important role in regulating adult neurogenesis. Notch signaling determines the number of stem cells and regulates Shh signaling, which promotes their survival. Other sources of signals in the adult brain include the ependymal cells lining the ventricle, the blood vessels, and blood components. As in the adult stem cells in skeletal muscle, chromatin modifications are crucial in regulating the differentiation of adult neural stem cells.

Evidence as to whether new neurons are born also in the adult human brain has remained sparse and contentious. However, confirmation that new neurons are,

Fig. 6.32 Sites of neurogenesis in the adult mammalian brain. Neurogenesis occurs mainly in the subgranular layer of the dentate gyrus in the hippocampus and the anterior part of the subventricular zone, which is shown in more detail in this figure. Radial glial cells act as neural stem cells, giving rise to rapidly dividing transit-amplifying cells, which differentiate into neurons. New neuronal cells generated in the dentate gyrus migrate to the granular layer, whereas new neuronal cells generated in the subventricular zone migrate to the olfactory bulb and differentiate there into interneurons. Endothelial cells of nearby blood vessels also provide signals for stem-cell differentiation (not shown).

Illustration after Taupin, P.: **Adult neurogenesis in the mammalian central nervous system: functionality and potential clinical interest.** *Med. Sci. Monit. 2005,* **11***: 247–252.*

indeed, produced in the human hippocampus (in large numbers—possibly as many as 1400 a day—and throughout life) has been provided by an ingenious study that exploited the fact that the testing of atomic bombs during the 1950s and early 1960s caused a spike in atmospheric ^{14}C. Starting with its uptake and fixation by plants, the carbon radioisotope was incorporated into newly born neurons through the food chain, providing a 'natural' form of pulse labeling that has provided a reasonably accurate marker of neuronal birth date. The hippocampus is a part of the cortex that is crucial to the acquisition and storage of new memory, and it is tempting to ascribe a related functional significance to the persistence of neurogenesis in this brain region.

SUMMARY

Cell differentiation is controlled by complex combinations of transcription factors, whose expression and activity are influenced and temporally guided by external signals. Epigenetic mechanisms involving chemical modifications to DNA and histones and the three-dimensional structure of the chromatin also have key roles. Some transcription factors are common to many cell types, but others have a very restricted pattern of expression. Combinations of transcription factors define individual differentiated cell types. Muscle differentiation, for example, can be brought about by the expression of transcription factors specific to the muscle lineage, such as MyoD, which initiate the differentiation program and bind to the control regions of muscle-specific genes.

Several adult tissues, such as the blood, the epidermis, and the lining of the gut, are continuously being replaced by the differentiation of new cells from stem cells. Hematopoiesis illustrates how a single progenitor cell, the multipotent stem cell, can be both self-renewing and generate a range of different cell lineages. The derivation of blood cells from the multipotent stem cell involves a successive restriction of developmental potential as cells become committed to the different lineages. The differentiation of the various lineages of blood cells seems to depend on extracellular signals, although there is some evidence for an intrinsic tendency to diversification. Extracellular factors are, however, essential for the survival and proliferation of the particular cell types.

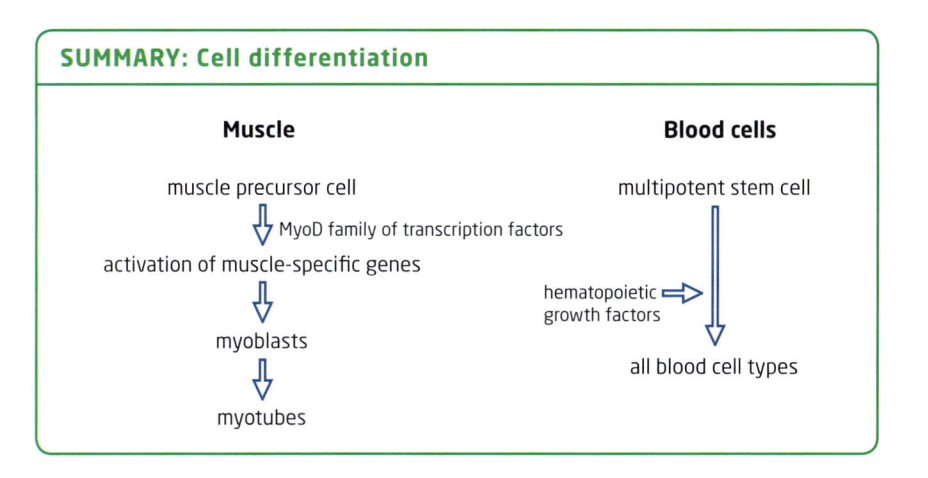

SUMMARY: Cell differentiation

Muscle

muscle precursor cell
⬇ MyoD family of transcription factors
activation of muscle-specific genes
⬇
myoblasts
⬇
myotubes

Blood cells

multipotent stem cell
⬇
hematopoietic ⇒
growth factors
⬇
all blood cell types

The plasticity of the differentiated state

In earlier parts of this chapter we have looked at examples of genes being turned on and off during development and have considered some of the mechanisms involved in cell differentiation. These chiefly involve the binding of different combinations of gene-regulatory proteins to the control regions of genes, and the epigenetic modifications to chromatin and its three-dimensional organization. But how reversible is cell differentiation? We have seen several examples of multipotent adult stem cells that both self-renew and give rise to a range of different cell types. If such stem cells could be produced reliably and in sufficient numbers, it might be possible to use them to replace cells that have been damaged or lost by disease or injury. This is one of the main aims of the field of **regenerative medicine**. The therapeutic use of stem cells will depend on understanding precisely how gene activity can be controlled in stem cells to give the desired cell type, and just how plastic stem cells are. Could, for example, blood stem cells give rise to neurons under appropriate conditions? How easy is it to alter the pattern of gene activity? Is it possible to make pluripotent stem cells from differentiated cells?

We start the discussion of cell plasticity by showing that the pattern of gene activity in differentiated cells can be changed experimentally. One way of finding out is to place the nucleus of a differentiated cell in a different cytoplasmic environment, one that contains a different set of gene-regulatory proteins. We then look at a few natural instances where differentiated cells change to another cell type. Finally, we look at the process of somatic cell nuclear transfer, where the pattern of gene activity in the nucleus of a differentiated cell reverts to the pattern present in a fertilized egg and can support development.

6.15 Patterns of gene activity in differentiated cells can be changed by cell fusion

It is possible to place the nucleus of one cell type in the cytoplasm of another by fusing the two cells. In this process, which can be brought about by treatment with certain chemicals or viruses, the plasma membranes fuse and nuclei from the different cells come to share a common cytoplasm. A cell-division inhibitor is then used to ensure that the two nuclei remain separate.

A striking example of the reversibility of gene activity after cell fusion is provided by the fusion of chick red blood cells with human cancer cells in culture. Unlike mammalian red blood cells, mature chick red blood cells have a nucleus. Gene activity in the nucleus is, however, completely turned off and it produces no mRNA. When a chick red blood cell is fused with a cell from a human cancer cell line, gene expression in the red blood cell nucleus is reactivated, and chick-specific proteins are expressed. This chick-specific expression shows not only that the chick genes can be reactivated, but that human cells contain cytoplasmic factors capable of doing this, yet another striking example of the conservation of the crucial molecules within the vertebrates that regulate cell differentiation.

Fusion of differentiated cells with striated muscle cells from a different species provides further evidence for reversibility of gene expression in differentiated cells (Fig. 6.33). Multinucleate striated muscle cells are good partners for cell fusion studies as they are large, and muscle-specific proteins can easily be identified. Differentiated human cells representative of each of the three germ layers have been fused with mouse multinucleate muscle cells, so that the human cell nuclei are exposed to the mouse muscle cytoplasm. This results in muscle-specific gene expression being switched on in the human nuclei. For example, human liver-cell nuclei in mouse muscle cytoplasm no longer express liver-specific genes; instead, their muscle-specific genes are activated and human muscle proteins are made. Furthermore, expression of genes such as human *MYOD1*, which can initiate muscle differentiation, is activated, showing that reprogramming of the liver-cell nuclei in mouse muscle cytoplasm seems to proceed through the same steps as those taken when a muscle cell differentiates.

These results clearly show that patterns of gene expression in differentiated cells can be changed, and, furthermore, that reprogramming factors are present in the

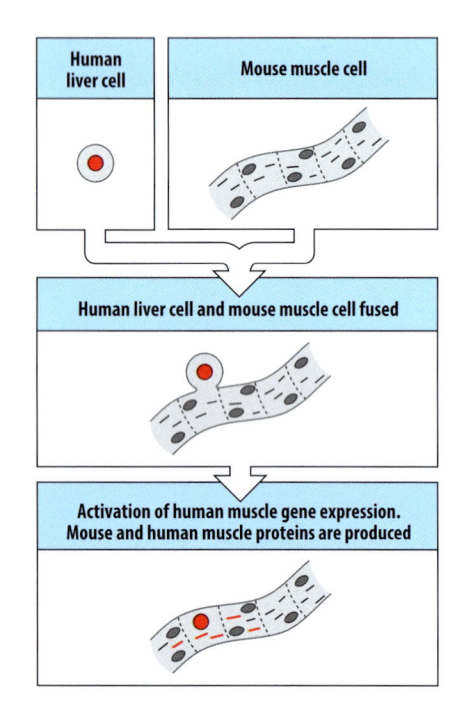

Fig. 6.33 Cell fusion shows the reversibility of gene inactivation during differentiation. A human liver cell is fused with a mouse muscle cell. The exposure of the human nucleus (red) to the mouse muscle cytoplasm results in the activation of muscle-specific genes and repression of liver-specific genes in the human nucleus. Human muscle proteins are made together with mouse muscle proteins. This shows that the inactivation of muscle-specific genes in human liver cells is not irreversible.

cytoplasm of highly differentiated cells such as striated muscle. At least some of these factors are transcription factors, and this leads to the conclusion that the differentiated state may be at least partly maintained by the continuous action of transcription factors on their target genes (see Section 6.3). We have already seen in our discussion on muscle cell differentiation, how manipulating the expression of transcription factors can convert cultured fibroblasts into myogenic cells (see Section 6.6). Other direct cell conversions of one cell type into another have also been achieved by a similar strategy. But before we discuss this, we will consider some instances in which reprogramming of differentiated cells occurs naturally.

6.16 The differentiated state of a cell can change by transdifferentiation

A fully differentiated cell is generally stable, which is essential if it is to serve a particular function in the mature animal. If it has retained the ability to divide, the cell passes its differentiated state on to all its descendants. In some long-lived cells, such as neurons, which do not divide once they are differentiated, the differentiated state must remain stable for many years. Plant cells maintain their differentiated state while in the plant, but do not maintain it when placed in cell culture (discussed in Chapter 12). But under certain conditions, differentiated animal cells are not stable and this provides yet another demonstration of the potential reversibility of patterns of gene activity. The change of one differentiated cell type into another is known as **transdifferentiation**. The allied phenomenon in which there is a switch of committed, but not yet differentiated, progenitor cells into a different lineage is known as **transdetermination**. The best-studied examples of transdetermination occur in regenerating *Drosophila* imaginal discs, where rare transdetermination events result in the homeotic transformation of one type of adult structure into another.

Transdifferentiation occurs very rarely as part of normal development in vertebrates and is more commonly seen in regeneration (discussed in Chapter 11) and certain pathological conditions. There are some examples of transdifferentiation in normal invertebrate development, however. In *Caenorhabditis elegans*, in which the fate of every cell can be followed, a few cells seem to undergo transdifferentiation. An epithelial cell (known as Y), which forms part of the rectum, is generated in the embryo and has the morphological and molecular hallmarks of a differentiated epithelial cell. In the second larval stage, this cell leaves the rectum epithelium and migrates to form a motor neuron (known as PDA), which now possesses an axonal process and makes synaptic connections. Thus, a fully differentiated epithelial cell with no detectable neuronal features undergoes transdifferentiation into a neuron with no residual epithelial characteristics. It appears that the Y cell first dedifferentiates and then redifferentiates as the PDA cell, rather than dedifferentiation and redifferentiation occurring at the same time.

A role for the extracellular matrix in determining the state of cell differentiation is suggested by a particularly interesting example of transdifferentiation in jellyfish striated muscle. This cell type can be made to transdifferentiate into two different types in succession. When a small patch of striated muscle and its associated extracellular matrix (mesogloea) is cultured, the striated muscle state is maintained. However, if the cultured tissue is treated with enzymes that degrade the extracellular matrix, the cells form an aggregate, and within 1–2 days some have transdifferentiated into smooth muscle cells, which have a different cellular morphology. This is followed by the appearance of a second cell type—nerve cells. This example indicates a role for the extracellular matrix in maintaining the differentiated state of the striated muscle.

The classic example of transdifferentiation in vertebrates is lens regeneration in the adult newt eye (Fig. 6.34). When the lens is completely removed by surgery (lentectomy), a new lens vesicle starts to form around 10 days post-lentectomy from the dorsal region of the pigmented epithelium of the iris. The iris cells first dedifferentiate—they

Fig. 6.34 Lens regeneration in adult newt.
Top panels: removal of the lens from the eye of a newt results in regeneration of a new lens from the dorsal pigmented epithelium of the iris. Bottom panels: sections through an intact and a regenerating newt eye viewed by scanning electron microscopy. (a) A section through an intact eye. Note that the aqueous anterior chamber (aq) is clear, whereas the vitreous chamber (v) is dominated by compacted layers of extracellular matrix (pink). Lens is gray, cornea is shown as light blue, the retina as purple, and the ligaments and matrix associated with the attachment to the lens are colored brown. (b) A section through an eye 5 days post-lentectomy. The aqueous chamber has been invaded by extracellular matrix and the iris has become thickened. (c) A section through an eye 25 days post-lentectomy. The aqueous chamber is gradually clearing, the lens, which developed from the dorsal rim of the iris, is now attached to the ventral iris (arrow), and extracellular matrix in the vitreous chamber is becoming reorganized. Magnification ×50.

*Micrographs from Tsonis, P.A., et al.: **A newt's eye view of lens regeneration.** Int. J. Dev. Biol. 2004, **48**: 975–980.* © UPV/EHU Press.

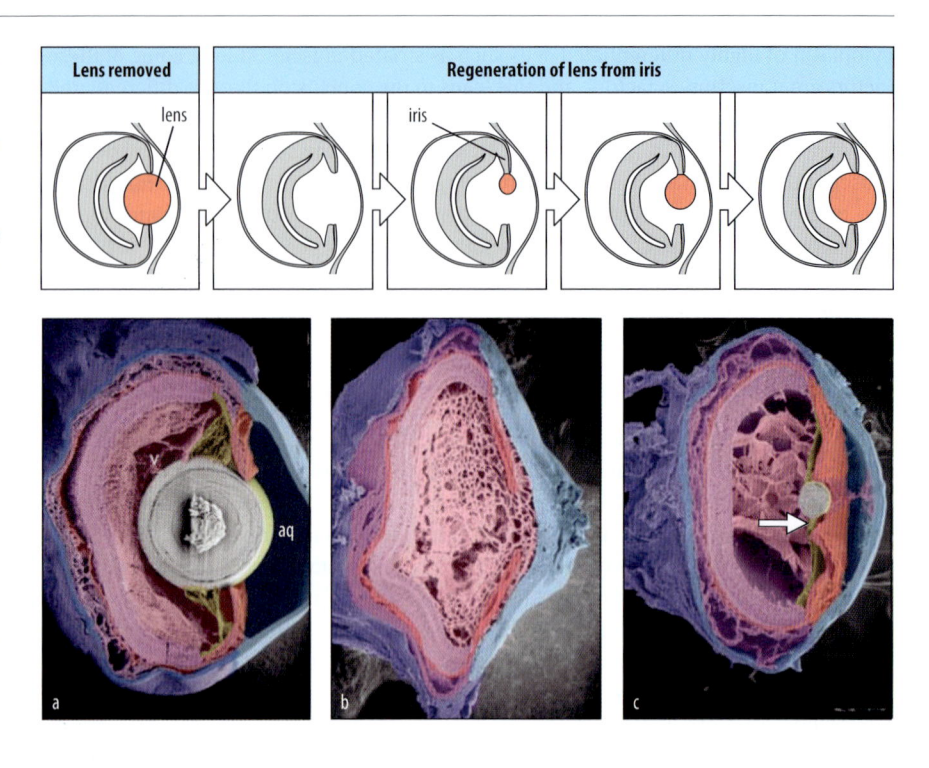

lose their pigmentation and alter their shape from a flat to a columnar epithelium—and start to proliferate. They then redifferentiate to form a new lens.

The lens of the adult mammalian or avian eye cannot regenerate *in vivo*, but the pigmented epithelial cells of the embryonic chick retina can be induced to transdifferentiate in culture. A single pigmented cell from the embryonic retina can be grown in culture to produce a monolayer of pigmented cells. On further culture in the presence of hyaluronidase, serum, and phenylthiourea, the cells lose their pigment and retinal cell characteristics. If cultured at a high density with ascorbic acid, the cells then start to take on the structural characteristics of lens cells, and to produce the lens-specific protein crystallin. It should be noted that, in both these examples in the newt and the chick, transdifferentiation occurs to a developmentally related cell type. Retinal and iris pigment cells and lens cells are all derived from the ectoderm (vertebrate eye development is discussed in Chapter 9).

Transdifferentiation could also underlie well-documented examples in humans, usually associated with tissue damage and repair, in which rare differentiated cells of one type appear in an inappropriate location in another tissue. Such examples include the appearance of hepatic cells in the pancreas and vice versa, and gastric-type epithelium in the esophagus in place of stratified squamous epithelium. In the latter example, known as Barrett's metaplasia, the change predisposes the person to developing esophageal cancer, and illustrates the clinical importance of transdifferentiation.

The earliest demonstration of experimentally induced transdifferentiation was the reprogramming of fibroblasts into muscle by the introduction of the gene for the muscle-specific transcription factor MyoD (see Section 6.6). Since then, other differentiated cell types have been converted directly into another differentiated cell type using the same strategy of introducing transcription factor genes. As examples, cell types have been switched within the hematopoietic system and cultured mouse and human fibroblasts have been directly converted into neurons. Remarkably, the conversion of one cell type to another has also been accomplished *in vivo*, raising a possibility that transdifferentiation could be harnessed in a clinical setting.

There have been reports over the years from experiments in mice that adult hematopoietic stem cells will differentiate into cell types quite different from their normal range when transplanted into another type of tissue. However, it was subsequently shown that the hematopoietic stem cells had not been able to give rise to cells outside their normal repertoire but instead had fused with existing cells.

6.17 Nuclei of differentiated cells can support development

The most dramatic experiments addressing the reversibility of differentiation investigated the ability of diploid nuclei from cells at different stages of development to replace the nucleus of an egg and to support normal development. If they can do this, it would indicate that no irreversible changes have occurred to the genome during differentiation. It would also show that a particular pattern of nuclear gene activity is determined by whatever transcription factors and other regulatory proteins are being synthesized in the cytoplasm of the cell. Such experiments were first carried out using the eggs of amphibians, which are large and particularly robust, and therefore very suitable for experimental manipulation.

In the unfertilized eggs of *Xenopus*, the nucleus lies directly below the surface at the animal pole. A dose of ultraviolet radiation directed at the animal pole destroys the DNA within the nucleus, and thus effectively removes all nuclear function. These enucleated eggs can then be injected with a diploid nucleus taken from a cell at a later stage of development to see whether it can function in place of the inactivated nucleus. The results are striking: nuclei from early embryos and from some types of differentiated cells of larvae and adults, such as gut and skin epithelial cells, can support the development of an embryo up to the tadpole stage (Fig. 6.35), and in a small number of cases even into an adult. The organisms that result are a clone of the animal from which the somatic cell nucleus was taken, as they have the identical genetic constitution. The process by which they are produced is called **cloning**. In 2012, the British biologist John Gurdon was awarded a share in the Nobel Prize in Physiology or Medicine for his work on nuclear transplantation.

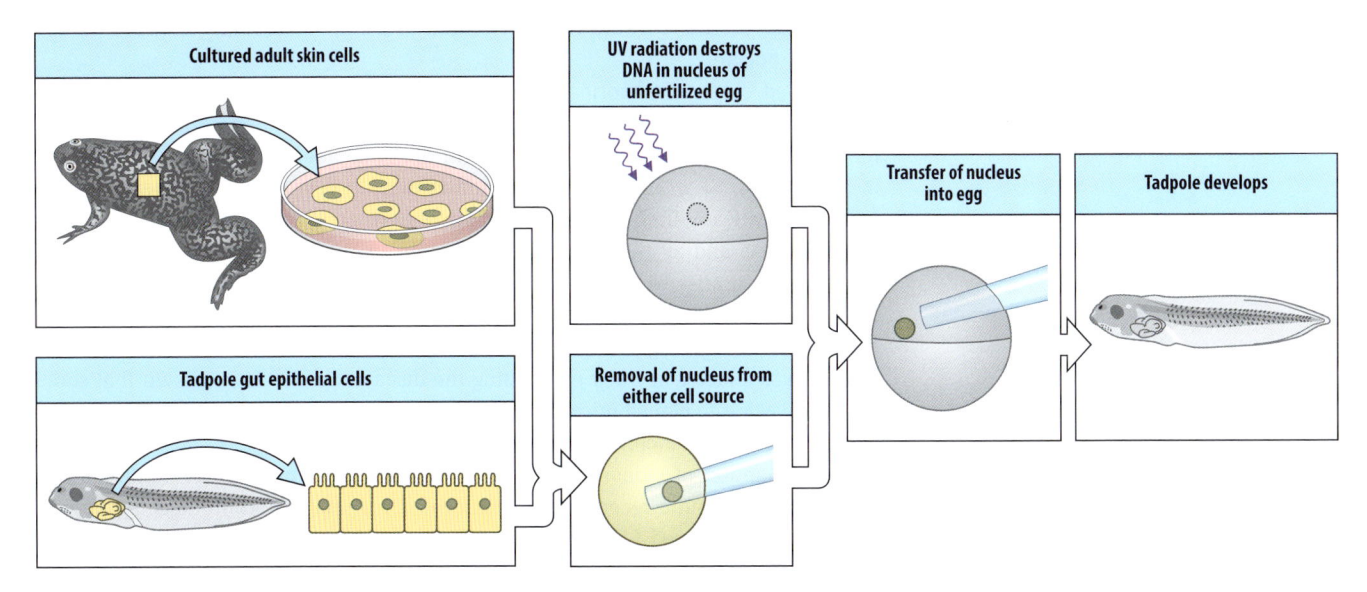

Fig. 6.35 Nuclear transplantation. The haploid nucleus of an unfertilized *Xenopus* egg is rendered non-functional by treatment with ultraviolet (UV) radiation. A diploid nucleus taken from an epithelial cell lining the tadpole gut or from a cultured adult skin cell is transferred into the enucleated egg, and can support the development of an embryo until at least the tadpole stage.

Nuclei from adult skin, kidney, heart, and lung cells, as well as from intestinal cells and cells in the myotome of *Xenopus* tadpoles, can support development when transplanted into enucleated eggs. This technique is known as **somatic cell nuclear transfer**. However, the success rate with nuclei from somatic cells of adults is very low, with only a small percentage of nuclear transplants developing past the cleavage stage. In general, the later the developmental stage the nuclei come from, the less likely they are to be able to support development. Transplantation of nuclei taken from blastula cells is much more successful; by transplanting nuclei from the same blastula into several enucleated eggs, clones of genetically identical frogs can be obtained (Fig. 6.36). An even greater success rate is achieved if the process is repeated with nuclei from the blastula stage of a cloned embryo. These results show that, at least in those differentiated cell types tested, the genes required for development are not irreversibly altered. More importantly, when exposed to the egg's cytoplasmic factors, they behave as the genes in the zygotic nucleus of a fertilized egg would. So in this sense at least, many of the nuclei of the embryo and the adult are equivalent, and their behavior is determined entirely by factors present in the cell.

What about organisms other than *Xenopus*? The genetic equivalence of nuclei at different developmental stages has also been demonstrated in ascidians. In plants, single somatic cells can give rise to fertile adult plants, demonstrating complete reversibility of the differentiated state at the cellular level. The first mammal to be cloned by somatic cell nuclear transfer was a sheep, the famous Dolly. In this case, the nucleus was taken from a cell line derived from the udder. Mice have been similarly cloned using, for example, nuclei from the somatic cumulus cells that surround the recently ovulated egg, and several generations of these mice have been produced by repeating the procedure.

In general, the success rate of cloning by somatic cell nuclear transfer in mammals is extremely low, and the reasons for this are not yet well understood. Most cloned mammals derived from nuclear transplantation die before birth, and those that survive are usually abnormal in some way. The most likely cause of failure is incomplete reprogramming of the donor nucleus to remove all the epigenetic modifications, such as DNA methylation and histone modification, that are involved in determining and maintaining the differentiated cell state (see Section 6.2). In such a case the DNA would not be restored to a state resembling that of a newly fertilized oocyte.

A related cause of abnormality may be that the reprogrammed genes have not gone through the normal **imprinting** process that occurs during germ cell development, where different genes are silenced in the male and female parents (discussed in Section 8.8). The abnormalities in adults that do develop from cloned embryos include early death, limb deformities, and hypertension in cattle, and immune impairment in mice, and all these defects are thought to be due to abnormalities of gene expression that arise from the cloning process. Dolly the sheep developed arthritis at 5 years old and was humanely put down at the age of 6 years (relatively young for a sheep) after developing a progressive lung disease, although there is no evidence that cloning was a factor in Dolly contracting the disease. Studies have shown that some 5% of the genes in cloned mice are not correctly expressed and that almost half of the imprinted genes are incorrectly expressed.

The generation of cloned blastocysts following the transfer of an adult somatic cell nucleus has more recently been achieved for primates and humans by modifying the protocols used for nuclear transfer in other animals. In both cases, nuclei from adult skin cells were transferred into enucleated oocytes, many of which then underwent cleavage, with a small number eventually forming blastocysts. Primates (macaque monkeys) have been cloned by nuclear transfer from fetal fibroblasts, but

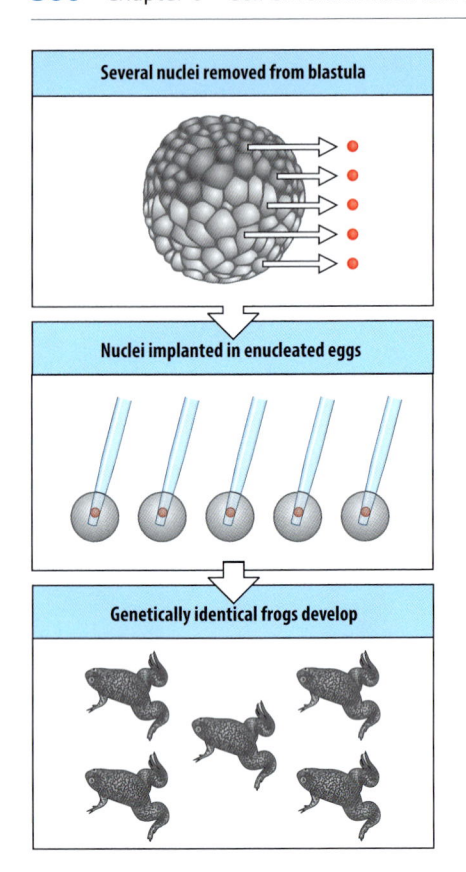

Fig. 6.36 Cloning by nuclear transfer. Nuclei from cells of the same blastula are transferred into enucleated, unfertilized *Xenopus* eggs. The frogs that develop are a clone, as they all have the same genetic constitution.

the success rate was extremely low. Cloning a human being by these means, even if it could be achieved, is banned in most countries on both practical and ethical grounds. Despite reports in the media that humans have been cloned, none of these reports has been verified.

One of the reasons for producing cloned blastocysts of mice and other mammals is to generate embryonic stem cells of known genetic constitution from the cells of the inner cell mass to study disease. Embryonic stem cells have been isolated from cloned monkey blastocysts and from human blastocysts produced by somatic nuclear transfer.

In Chapter 7 we will see how embryonic stem cells are routinely isolated from early embryos grown in culture, and how they are being used to provide models of very early embryonic development and are being explored for their potential uses in regenerative medicine.

SUMMARY

Cell fusion studies and the transplantation of nuclei from differentiated animal cells into fertilized eggs show that the pattern of gene expression in the nucleus of a differentiated cell can be reversed, implying that it is determined by factors supplied by the cytoplasm and that no genetic material has been lost. Although the differentiated state of an animal cell *in vivo* is usually extremely stable, some cases of differentiation are reversible. Transdifferentiation of one differentiated cell type into another has been shown to occur during regeneration in some cases and in cultured cells. Amphibians and some mammals can be cloned by nuclear transplantation from an adult somatic cell into an enucleated egg.

Summary to Chapter 6

- Cell differentiation leads to distinguishable cell types, such as blood cells, nerve cells, and muscle cells, whose specialized functions and properties are determined by their pattern of gene activity and thus by the specialized proteins they produce.
- The first step in differentiation is the commitment, or determination, of the cell to that particular fate. Subsequent differentiation occurs gradually, usually over several cell generations.
- Differentiation reflects the selection of a particular subset of genes to be expressed and the silencing of other genes.
- The pattern of gene activity in a differentiated cell can be maintained over long periods of time and can be transmitted to the cell's progeny. Maintenance is likely to involve a combination of several mechanisms, including the continued action of gene-regulatory proteins and epigenetic chemical modifications to chromatin, such as DNA methylation or the acetylation or methylation of histone proteins.
- It is now possible to analyze the complete transcriptomes of a population of thousands of single cells simultaneously. This single-cell RNA sequence analysis has made for more accurate charting of the pathways taken by differentiating cells. What was taken to be a homogeneous cell population by previous techniques can now be separated into various subpopulations of cells with different differentiation trajectories.
- The single-cell RNA analysis technique has also been applied to mouse and human embryos undergoing gastrulation and has provided information about the order in which different cell fates arise in the embryo.

- Many adult tissues, such as blood, epidermis, and gut lining, are continually being renewed from undifferentiated, self-renewing cells called stem cells. In other tissues, such as skeletal muscle, stem cells are present and can be mobilized to produce new tissue after injury.
- The differentiated state of an animal cell *in vivo* is usually extremely stable, but in some cases transdifferentiation of one differentiated cell type into another can occur.
- The cloning of amphibians and some mammals by somatic cell nuclear transfer into enucleated eggs shows that the nuclei of differentiated cells can be reprogrammed and restored to a state in which they are able to direct embryonic development.

End of chapter questions

Check your understanding of the content of this chapter by attempting the following long-answer concept questions.

1. Use the Glossary to review the terms 'fate', 'specified', and 'determination'. How do these cellular states differ from the topic of this chapter, differentiation?

2. Contrast the concepts of 'housekeeping' genes (and proteins) with the cell-specific genes (and proteins) that characterize, and define, a differentiated cell type. Propose an RNA-seq experiment that would distinguish the tissue-specific gene expression from housekeeping gene expression in muscle cells and neurons.

3. Gene function is often studied by misexpression of the protein in an abnormal site in the embryo or abnormal tissue in the adult body to see what effects the protein has. What are the essential features of the transcriptional apparatus that make misexpression experimentally possible (include the role of both *cis*-regulatory DNA sequences and transcription factors)?

4. Create two lists: one of peptide signaling molecules, and one of lipid-soluble signaling molecules. How do their mechanisms of action differ? What generalization(s) can you make regarding the way they activate a transcription factor?

5. Cell division entails the separation of the two strands of the DNA during DNA replication, which disrupts patterns of gene expression that have been previously established in the cell. Nonetheless, the two daughter cells must somehow 'remember' the differentiated state of the parental cell. Discuss how the differentiated state of a cell is remembered, using transcription factors, chromatin structure, and DNA methylation.

6. Hematopoiesis involves the development of lymphocytes (B cells and T cells), red blood cells, macrophages, and neutrophils (among other blood cell types) from cells in the bone marrow. Outline the general stages that lead to each of the cell types named, giving the name of the cell from which they all derive.

7. What is meant by 'stem-cell niche' and what is its role? For one of the stem-cell niches described in this chapter give examples of the signal molecules that enable it to carry out this role.

8. Differentiation of the erythroid lineage requires signaling by erythropoietin (Epo), a peptide signal molecule made by the kidney in response to red blood cell deficiency. One end product of the Epo signaling pathway is the GATA1 transcription factor, and the Epo gene is itself a target of GATA1. The β-globin gene is also a target of GATA1. Using these facts, propose a pathway from a physiological need for more red blood cells, through the multipotent stem cell, to the globin-expressing red blood cell.

9. Describe the choice between differentiation as a neutrophil and differentiation as a macrophage at the level of the transcription factors involved. Describe the same choice at the level of the signaling molecules involved.

10. Why are different globins required at different stages of development? How is the switching between globin types thought to be controlled?

11. Describe the process by which epithelial cells of the intestine are replaced from a stem-cell population. What is the role of Wnt signals in this process?

12. Despite the ability of MyoD to act as a master regulator of muscle differentiation, mouse knock-outs of *MyoD* are viable. Refer to Figure 6.15, and propose a reason that *MyoD* is not essential. By contrast, knock-out of both *MyoD* and *Myf5* is lethal; knock-out of *Mrf4* is also lethal; how would you interpret this result?

13. The differentiation of muscle cells requires withdrawal from the cell cycle. Elaborate on the following points regarding growth versus differentiation of myoblasts: (1) removal of growth factors causes muscle precursor cells to differentiate in culture; (2) MyoD activates expression of p21; (3) loss of RB will allow myoblasts to re-enter the cell cycle.

14. In the light of modern advances in single-cell analysis of gene expression, how would you define a cell type?

15. Describe the process referred to as cloning by somatic cell nuclear transfer. How does the success of this process illustrate the importance of the complement of transcription factors present in any given cell? By contrast, how does the difficulty of the process illustrate the importance of epigenetic mechanisms in development?

Section further reading

Box 6A Conrad Waddington's 'epigenetic landscape' provides a framework for thinking about how cells differentiate

Moris, N., Pina, C., Martinez Arias, A.: **Transition states and cell fate decisions in epigenetic landscapes**. *Nat. Rev. Genet.* 2016, **17**: 693–703.

Slack, J.: **Hal Waddington: the last Renaissance biologist?** *Nat. Rev. Genet.* 2002, **3**: 889–895.

Waddington, C.H.: *Organisers and Genes*. Cambridge: Cambridge University Press, 1940.

6.1 Control of transcription involves both general and tissue-specific transcriptional regulators

Andersson, R.: **Promoter or enhancer, what's the difference? Deconstruction of established distinctions and presentation of a unifying model**. *BioEssays* 2015, **37**: 314–323.

Calo, E., Wysocka, J.: **Modification of enhancer chromatin: what, how, and why?** *Mol. Cell* 2013, **49**: 825–837.

Furlong, E.E.M., Levine, M.: **Developmental enhancers and chromosome topology**. *Science* 2018, **361**: 1341–1345.

Long, H.K., Prescott, S.L., Wysocka, J.: **Ever-changing landscapes: transcriptional enhancers in development and evolution**. *Cell* 2016, **167**: 1170–1187.

Spitz, F., Furlong, E.E.: **Transcription factors: from enhancer binding to developmental control**. *Nat. Rev. Genet.* 2012, **13**: 613–626.

Ushiki, A., Zhang, Y., Xiong, C., Zhao, J., Georgakopoulos-Soares, I., Kane, L., Jamieson, K., Bamshad, M.J., Nickerson, D.A., University of Washington Center for Mendelian Genomics, Shen, Y., Lettice, L.A., Silveira-Lucas, E.L., Petit, F., Ahituv, N.: **Deletion of CTCF sites in the *SHH* locus alters enhancer–promoter interactions and leads to acheiropodia**. *Nat. Commun.* 2021, **12**: 2282.

6.2 Gene expression is also controlled by epigenetic chemical modifications to DNA and histone proteins that alter chromatin structure and 6.3 Patterns of gene activity can be inherited by persistence of gene-regulatory proteins or by maintenance of chromatin modifications

Allou, L., *et al*.: **Non-coding deletions identify Maenli lncRNA as a limb-specific En1 regulator**. *Nature* 2021, **592**: 93–98.

Andrey, G., Mundlos, S.: **The three-dimensional genome: regulating gene expression during pluripotency and development**. *Development* 2017, **144**: 3646–3658.

de Laat, W., Duboule, D.: **Topology of mammalian developmental enhancers and their regulatory landscapes**. *Nature* 2013, **502**: 499–506.

Deschamps, J., Duboule, D.: **Embryonic timing, axial stem cells, chromatin dynamics**. *Genes Dev.* 2017, **31**: 1406–1416.

Ho, L., Crabtree, G.R.: **Chromatin remodelling during development**. *Nature* 2010, **463**: 474–484.

Krijger, P.H., de Laat, W.: **Regulation of disease-associated gene expression in the 3D genome**. *Nat. Rev. Mol. Cell. Biol.* 2016, **17**: 771–782.

Robson, M.I., Ringel, A.R., Mundlos, S.: **Regulatory landscaping: how enhancer-promoter communication is sculpted in 3D**. *Mol. Cell* 2019, **74**: 1110–1122.

Soshnikova, N., Duboule, D.: **Epigenetic temporal control of mouse Hox genes *in vivo***. *Science* 2009, **324**: 1320–1323.

Spielmann, M., Lupiáñez, D.G., Mundlos, S.: **Structural variation in the 3D genome**. *Nat. Rev. Genet.* 2018, **19**: 453–467.

Box 6B Epigenetic control of gene expression by chromatin modification

Barth, T.K., Imhof, A.: **Fast signals and slow marks: the dynamics of histone modifications**. *Trends Biochem. Sci.* 2010, **35**: 618–626.

Calo, E., Wysocka, J.: **Modification of enhancer chromatin: what, how and why?** *Mol. Cell* 2013, **49**: 825–837.

6.4 Changes in patterns of gene activity during differentiation can be triggered by extracellular signals

Barolo, S., Posakony, J.: **Three habits of highly effective signaling pathways: principle of transcriptional control by developmental cell signalling**. *Genes Dev.* 2002, **16**: 1167–1181.

6.5 The early stages of cell differentiation during gastrulation can be detected by single-cell transcriptional analysis

Crosetto, N., Bienko, M., van Oudenaarden, A.: **Spatially resolved transcriptomics and beyond**. *Nat. Rev. Genet.* 2015, **16**: 57–66.

Guibentif, C., Griffiths, J.A., Imaz-Rosshandler, I., Ghazanfar, S., Nichols, J., Wilson, V., Göttgens, B., Marioni, J.C.: **Diverse routes toward early somites in the mouse embryo**. *Dev. Cell* 2021, **56**: 141–153.

Kester, L., van Oudenaarden, A.: **Single-cell transcriptomics meets lineage tracing**. *Cell Stem Cell* 2018, **23**: 166–179.

Marx, V.: **Method of the Year: spatially resolved transcriptomics**. *Nat. Methods* 2021, **18**: 9–14.

Nowotschin, S., Setty, M., Kuo, Y.Y., Liu, V., Garg, V., Sharma, R., Simon, C.S., Saiz, N., Gardner, R., Boutet, S.C., Church, D.M., Hoodless, P.A., Hadjantonakis, A.K., Pe'er, D.: **The emergent landscape of the mouse gut endoderm at single-cell resolution**. *Nature* 2019, **569**: 361–367.

Pijuan-Sala, B., Griffiths, J.A., Guibentif, C., Hiscock, T.W., Jawaid, W., Calero-Nieto, F.J., Mulas, C., Ibarra-Soria, X., Tyser, R.C.V., Ho, D.L.L., Reik, W., Srinivas, S., Simons, B.D., Nichols, J., Marioni, J.C., Göttgens, B.: **A single-cell molecular map of mouse gastrulation and early organogenesis**. *Nature* 2019, **566**: 490–495.

Tyser, R.C.V., Mahammadov, E., Nakanoh, S., Vallier, L., Scialdone, A., Srinivas, S.: **Single-cell transcriptomic characterization of a gastrulating human embryo**. *Nature* 2021, **600**: 285–289.

Box 6C Single-cell analysis of cell-fate decisions

Huang, S.: **Non-genetic heterogeneity of cells in development: more than just noise**. *Development* 2009, **136**: 3853–3862.

Martinez Arias, A., Hadjantonakis, A.K.: **Single-cell approaches: Pandora's box of developmental mechanisms**. *Dev. Cell* 2016, **38**: 574–578.

Pijuan-Sala, B., Guibentif, C., Göttgens, B.: **Single-cell transcriptional profiling: a window into embryonic cell-type specification**. *Nat. Rev. Mol. Biol.* 2018, **19**: 399–412.

Tanay, A., Regev, A.: **Scaling single-cell genomics from phenomenology to mechanism**. *Nature* 2017, **541**: 331–338.

Trapnell, C.: **Defining cell types and states with single cell genomics**. *Genome Res.* 2015, **25**: 1491–1498.

6.6 Muscle differentiation is determined by the MyoD family of transcription factors

Asp, P., Blum, R., Vethantham, V., Parisi, F., Micsinai, M., Cheng, J., Bowman, C., Kluger, Y., Dynlacht, B.D.: **Genome-wide remodeling of the epigenetic landscape during myogenic differentiation**. *Proc. Natl Acad. Sci. USA* 2011, **108**: E149–E158.

Comai, G., Tajbakhsh, S.: **Molecular and cellular regulation of skeletal myogenesis**. *Curr. Top. Dev. Biol.* 2014, **110**: 1–73.

Graf, T.: **Historical origins of transdifferentiation and reprogramming**. *Cell Stem Cell* 2011, **9**: 504–516.

Kassar-Duchossoy, L., Gayraud-Morel, B., Gomes, D., Rocancourt, D., Buckingham, M., Shinin, V., Tajbakhsh, S.: **Mrf4 determines skeletal muscle identity in Myf5: MyoD double-mutant mice**. *Nature* 2004, **431**: 466–471.

Neems, D.S., Garza-Gongora, A.G., Smith, E.D., Kosaka, S.T.: **Topologically associated domains enriched for lineage-specific genes reveal expression-dependent nuclear topologies during myogenesis**. *Proc. Natl Acad. Sci. USA* 2016, **113**: E1691–E1700.

Tapscott, S.J.: **The circuitry of a master switch: MyoD and the regulation of skeletal muscle transcription**. *Development* 2005, **132**: 2685–2695.

6.7 The differentiation of muscle cells involves withdrawal from the cell cycle, but is reversible

Deato, M.D., Tjian, R.: **An unexpected role of TAFs and TRFs in skeletal muscle differentiation; switching of core promoter complexes**. *Cold Spring Harb. Symp.* 2008, **73**: 217–225.

Novitch, B.G., Mulligan, G.J., Jacks, T., Lassar, A.B.: **Skeletal muscle cells lacking the retinoblastoma protein display defects in muscle gene expression and accumulate in S and G_2 phases of the cell cycle**. *J. Cell Biol.* 1996, **135**: 441–456.

6.8 All blood cells are derived from multipotent stem cells

Drissen, R., Buza-Vidas, N., Woll, P., Thongjuea, S., Gambardella, A., Giustacchini, A., Mancini, E., Zriwil, A., Lutteropp, M., Grover, A., Mead, A., Sitnicka, E., Eirik, S., Jacobsen, W., Nerlov, C.: **Distinct myeloid progenitor–differentiation pathways identified through single-cell RNA sequencing**. *Nat. Immunol.* 2016, **17**: 666–676.

Ema, H., Nakauchi, H.: **Self-renewal and lineage restriction of hematopoietic stem cells**. *Curr. Opin. Genet. Dev.* 2003, **13**: 508–512.

Phillips, R.L., Ernst, R.E., Brunk, B., Ivanova, N., Mahan, M.A., Deanehan, J.K., Moore, K.A., Overton, G.C., Lemischka, I.R.: **The genetic program of hematopoietic stem cells**. *Science* 2000, **288**: 1635–1640.

Zon, L.I.: **Intrinsic and extrinsic control of haematopoietic stem-cell renewal**. *Nature* 2008, **453**: 306–313.

6.9 Intrinsic and extrinsic changes control differentiation of the hematopoietic lineages

Anguita, E., Hughes, J., Heyworth, C., Blobel, G.A., Wood, W.G., Higgs, D.R.: **Globin gene activation during haemopoiesis is driven by protein complexes nucleated by GATA-1 and GATA-2**. *EMBO J.* 2004, **23**: 2841–2852.

Dahlin, J.S., Hamey, F.K., Pijuan-Sala, B., Shepherd, M., Lau, W.W.Y, Nestorowa, S., Weinreb, C., Wolock, S., Hannah, R., Diamanti, E., Kent, D.G., Göttgens, B., Wilson, N.K.: **A single-cell hematopoietic landscape resolves 8 lineage trajectories and defects in Kit mutant mice**. *Blood* 2018, **131**: e1–e11.

Kluger, Y., Lian, Z., Zhang, X., Newburger, P.E., Weissman, S.M.: **A panorama of lineage-specific transcription in hematopoiesis**. *BioEssays* 2004, **26**: 1276–1287.

6.10 Developmentally regulated globin gene expression is controlled by regulatory sequences far distant from the coding regions

Engel, J.D., Tanimoto, K.: **Looping, linking, and chromatin activity: new insights into β-globin locus regulation**. *Cell* 2000, **100**: 499–502.

Kiefer, C.M., Hou, C., Little, J.A., Dean, A.: **Epigenetics of beta-globin gene regulation**. *Mutat. Res.* 2008, **647**: 68–76.

Lee, W.S., McColl, B., Maksimovic, J., Vadolas, J.: **Epigenetic interplay at the β-globin locus**. *Biochim. Biophys. Acta* 2017, **1860**: 393–404.

Tolhuis, B., Palstra, R.J., Splinter, E., Grosveld, F., de Laat, W.: **Looping and interaction between hypersensitive sites in the active beta-globin locus**. *Mol. Cell* 2002, **10**: 1453–1465.

Box 6D Towards a therapy for β-globin gene defects

Esrick, E.B.: **Post-transcriptional genetic silencing of *BCL11A* to treat sickle cell disease**. *N. Engl. J. Med.* 2021, **384**: 205–215.

Frangoul, H., et al.: **CRISPR-Cas9 gene editing for sickle cell disease and β-thalassemia**. *N. Engl. J. Med.* 2021, **384**: 252–260.

Hacein-Bey-Abina, S., et al.: **A modified γ-retrovirus vector for X-linked severe combined immunodeficiency**. *N. Engl. J. Med.* 2014, **371**: 1407–1417.

6.11 The epidermis of adult mammalian skin is continually being replaced by derivatives of stem cells

Clayton, E., Doupé, D.P., Klein, A.M., Winton, D.J., Simons, B.D., Jones, P.H.: **A single type of progenitor cell maintains normal epidermis**. *Nature* 2007, **446**: 185–189.

Coulombe, P.A., Kerns, M.L., Fuchs, E.: **Epidermolysis bullosa simplex: a paradigm for disorders of tissue fragility**. *J. Clin. Invest.* 2009, **119**: 1784–1793.

Fuchs, E.: **Finding one's niche in the skin**. *Cell Stem Cell* 2009, **4**: 499–502.

Fuchs, E.: **The tortoise and the hair: slow-cycling cells in the stem cell race**. *Cell* 2009, **137**: 811–819.

Fuchs, E., Nowak, J.A.: **Building epithelial tissues from skin stem cells**. *Cold Spring Harb. Symp. Quant. Biol.* 2008, **73**: 333–350.

Jones, P.H., Simons, B.D., Watt, F.M.: **Sic transit gloria: farewell to the epidermal transit amplifying cell?** *Cell Stem Cell* 2007, **1**: 371–381.

Mascré, G., Dekoninck, S., Drogat, B., Youssef, K.K., Broheé, S., Sotiropoulou, P.A., Simons, B.D., Blanpain, C.: **Distinct contribution of stem and progenitor cells to epidermal maintenance**. *Nature* 2012, **489**: 257–262.

Schepeler, T., Page, M.E., Jensen, K.B.: **Heterogeneity and plasticity of epidermal stem cells**. *Development* 2014, **141**: 2559–2567.

Box 6E Treatment of junctional epidermolysis bullosa with skin grown from genetically corrected stem cells

Hirsch, T., et al.: **Regeneration of the entire human epidermis using transgenic stem cells**. *Nature* 2017, **551**: 327–332.

6.12 Stem cells use different modes of division to maintain tissues

Li, X., Upadhyay, A.K., Bullock, A.J., Dicolandrea, T., Xu, J., Binder, R.L., Robinson, M.K., Finlay, D.R., Mills, K.J., Bascom, C.C., Kelling, C.K., Isfort, R.J., Haycock, J.W., MacNeil, S., Smallwood, R.H.: **Skin stem cell hypotheses and long term clone survival explored using agent-based modeling**. *Sci. Rep.* 2013, **3**: 1904.

Mascré, G., Dekoninck, S., Benjamin Drogat, B., Youssef, K.K., Broheé, S., Sotiropoulou, P.A., Simons, B.D., Blanpain, C.: **Distinct contribution of stem and progenitor cells to epidermal maintenance**. *Nature* 2012, **489**: 257–262.

Rue, P., Martinez Arias, A.: **Cell dynamics and gene expression control in tissue homeostasis and development**. *Mol. Sys. Biol.* 2015, **1**: 792.

Simons, B.D., Clevers, H.: **Strategies for homeostatic stem cell self-renewal in adult tissues**. *Cell* 2011, **145**: 851–862.

6.13 The lining of the gut is another epithelial tissue that requires continuous renewal

Barker, N., van Es, J.H., Kuipers, J., Kujala, P., van den Born, M., Cozijnsen, M., Haegebarth, A., Korving, J., Begthel, H., Peters, P.J., Clevers, H.: **Identification of stem cells in small intestine and colon by marker gene *Lgr5***. *Nature* 2007, **449**: 1003–1008.

Sato, T., Vries, R.G., Snippert, H.J., van de Wetering, M., Barker, N., Stange, D.E., van Es, J.H., Abo, A., Kujala, P., Peters, P.J., Clevers, H.: **Single *Lgr5* stem cells build crypt villus structures *in vitro* without a mesenchymal cell niche**. *Nature* 2009, **459**: 262–265.

Van der Flier, L.G., Clevers, H.: **Stem cells, self-renewal and differentiation in the intestinal epithelium**. *Annu. Rev. Physiol.* 2009, **71**: 241–260.

6.14 Skeletal muscle and neural cells can be renewed from stem cells in adults

Berg, D.A., Bond, A.M., Ming, G.L., Song, H.: **Radial glial cells in the adult dentate gyrus: what are they and where do they come from?** *F1000Res.* 2018, **7**: 277.

Collins, C.A., Partridge T.A.: **Self-renewal of the adult skeletal muscle satellite cell**. *Cell Cycle* 2005, **4**: 1338–1341.

Dhawan, J., Rando, T.A.: **Stem cells in postnatal myogenesis: molecular mechanisms of satellite cell quiescence, activation and replenishment**. *Trends Cell Biol.* 2005, **15**: 663–673.

Gonzales-Roybal, G., Lim, D.A.: **Chromatin-based epigenetics of adult subventricular zone neural stem cells**. *Front. Genet.* 2013, **4**: 194.

Kemperman, G.: **New neurons for 'survival of the fittest'**. *Nat. Rev. Neurosci.* 2012, **13**: 727–736.

Li, G., Pleasure, S.J.: **Ongoing interplay between the neural network and neurogenesis in the adult hippocampus**. *Curr. Opin. Neurobiol.* 2010, **20**: 126–133.

Lie, D.-C., Colamarino, S.A., Song, H.-J., Desire, L., Mira, H., Consiglio, A., Lein, E.S., Jessberger, S., Lansford, H., Dearier, A.R., Gage, F.H.: **Wnt signalling regulates adult hippocampal neurogenesis**. *Nature* 2005, **437**: 1370–1375.

Ninkovic, J., Gotz, M.: **Signaling in adult neurogenesis: from stem cell niche to neuronal networks**. *Curr. Opin. Neurobiol.* 2007, **17**: 338–344.

Relaix, F., Rocancourt, D., Mansouri, A., Buckingham, M.A.: ***pax3/pax7* dependent population of skeletal muscle progenitor cells**. *Nature* 2005, **435**: 948–953.

Sacco, A., Doyonnas, R., Kraft, P., Vitorovic, S., Blau, H.M.: **Self-renewal and expansion of single transplanted muscle stem cells**. *Nature* 2008, **456**: 502–506.

Spalding, K.L., Bergmann, O., Alkass, K., Bernard, S., Salehpour, M., Huttner, H.B., Boström, E., Westerlund, I., Vial, C., Buchholz, B.A., Possnert, G., Mash, D.C., Druid, H., Frisén J.: **Dynamics of hippocampal neurogenesis in adult humans**. *Cell* 2013, **153**: 1219–1227.

6.15 Patterns of gene activity in differentiated cells can be changed by cell fusion

Blau, H.M.: **How fixed is the differentiated state? Lessons from heterokaryons**. *Trends Genet.* 1989, **5**: 268–272.

Blau, H.M., Baltimore, D.: **Differentiation requires continuous regulation**. *J. Cell Biol.* 1991, **112**: 781–783.

Blau, H.M., Blakely, B.T.: **Plasticity of cell fates: insights from heterokaryons**. *Cell Dev. Biol.* 1999, **10**: 267–272.

Pomerantz, J.H., Mukherjee, S., Palermo, A.T., Blau, H.M.: **Reprogramming to a muscle fate by fusion recapitulates differentiation**. *J. Cell Sci.* 2009, **122**: 1045–1053.

6.16 The differentiated state of a cell can change by transdifferentiation

Eguchi, G., Eguchi, Y., Nakamura, K., Yadav, M.C., Millán, J.L., Tsonis, P.A.: **Regenerative capacity in newts is not altered by repeated regeneration and ageing**. *Nat. Commun.* 2011, **2**: 384.

Graf, T.: **Historical origins of transdifferentiation and reprogramming**. *Cell Stem Cell* 2011, **9**: 504–516.

Hajduskova, M., Ahier, A., Daniele, T., Jarriault, S.: **Cell plasticity in *Caenorhabditis elegans*: from induced to natural cell reprogramming**. *Genesis* 2012, 50: 1–17.

Horb, M.E., Shen, C.N., Tosh, D., Slack, J.M.: **Experimental conversion of liver to pancreas**. *Curr. Biol.* 2003, **13**: 105–115.

Jarriault, S., Shwab, Y., Greenwald, I.A.: ***Caenorhabditis elegans* model for epithelial-neuronal transdifferentiation**. *Proc. Natl Acad. Sci. USA* 2008, **105**: 3790–3795.

Slack, J.M.W.: **Metaplasia and transdifferentiation from pure biology to the clinic**. *Nat. Rev. Mol. Cell Biol.* 2007, **8**: 369–378.

Tsonis, P.A., Madhavan, M., Tancous, E.E., Del Rio-Tsonis, K.: **A newt's eye view of lens regeneration**. *Int. J. Dev. Biol.* 2004, **48**: 975–980.

6.17 Nuclei of differentiated cells can support development

Eggan, K., Baldwin, K., Tackett, M., Osborne, J., Gogos, J., Chess, A., Axel, R., Jaenisch, R.: **Mice cloned from olfactory sensory neurons**. *Nature* 2004, **428**: 44–49.

Gurdon, J.B.: **Nuclear transplantation in eggs and oocytes**. *J. Cell Sci. Suppl.* 1986, **4**: 287–318.

Gurdon, J.B., Melton, D.A.: **Nuclear reprogramming in cells**. *Science* 2008, **322**: 1811–1815.

Humpherys, D., Eggan, K., Akutsu, H., Friedman, A., Hochedlinger, K., Yanagimachi, R., Lander, E.S., Golub, T.R., Jaenisch, R.: **Abnormal gene expression in cloned mice derived from embryonic stem cell and cumulus cell nuclei**. *Proc. Natl Acad. Sci. USA* 2002, **99**: 12889–12894.

Liu, Z., Cai, Y., Wang, Y., Nie, Y., Zhang, C., Xu, Y., Zhang, X., Lu, Y., Wang, Z., Poo, M., Sun, Q.: **Cloning of macaque monkeys by somatic cell nuclear transfer**. *Cell* 2018, **172**: 881–887.

Rhind, S.M., Taylor, J.E., De Sousa, P.A., King, T.J., McGarry, M., Wilmut, I.: **Human cloning: can it be made safe?** *Nat. Rev. Genet.* 2003, **4**: 855–864.

Wilmut, I., Taylor, J.: **Primates join the club**. *Nature* 2007, **450**: 485–486.

7

Embryonic stem cells, embryo models, and regenerative medicine

- Embryonic stem cells and induced pluripotent stem cells
- Embryo models
- Medical applications of stem cells

We saw in Chapter 6 that the multipotent stem-cell populations responsible for tissue renewal in adults are restricted in the range of cell types they give rise to and require specific niches to maintain their stem-cell state, such as the bone marrow stroma that supports hematopoietic stem cells. However, there is a different type of stem cell whose developmental and differentiation potentials are not restricted: this is the embryonic stem cell. These cells are derived from the mammalian blastocyst and can be grown in culture to give rise to clonal populations of pluripotent cells—cells that can generate all cell types in the organism. Originally obtained from mouse embryos, embryonic stem cells have now been derived from a range of species including rats, pigs, and humans. Over the past 20 years, it has become possible to differentiate embryonic stem cells in vitro under defined chemical conditions to give a variety of specialized cell types, and to show that the sequence of molecular and cellular events in this process mirrors the differentiation paths that they would follow in vivo. Stem cells called induced pluripotent stem cells can also be derived from differentiated cells by forced expression of a set of transcription factors that return the cell to a pluripotent state. Embryonic and adult stem cells are the basis for the development of organoids—three-dimensional stem-cell-based aggregates that mimic many features of embryonic and adult organs. Under certain experimental conditions, embryonic stem cells can organize into structures that mimic gastrulation called embryoids and gastruloids. These structures have ushered in opportunities to model disease and also to understand normal development, particularly in the case of humans, in whom it is not possible to carry out such experiments. Altogether, cultured stem cells have become a new model system, to complement work in vivo and to reveal significant aspects of cell-fate choices and morphogenesis.

Introduction

Early mammalian embryos are very different from other vertebrate embryos, as we saw in Chapters 3 and 4. Mammalian eggs are small (about 0.1 mm in diameter compared with the 40 mm of an average hen's egg), and although they still have nutrients that are associated with the yolk in other eggs, they have reduced and compressed them. The reduction in nutrients means that the mammalian embryo relies, exclusively, on nutrients derived from interaction with its mother. This is reflected in the fact that, in contrast to fish and frog embryos, mammalian embryos have very little input from maternal factors. In mammals, zygotic gene expression starts very early, at the two-cell stage in mouse and the four-cell stage in humans. During cleavage of the zygote, two cell-fate decisions segregate the precursors of cell lineages that will not form part of the embryo itself. One is the trophectoderm, which contributes to the placenta and forms the outer surface of the blastocyst. The other is the primitive endoderm, which gives rise to the yolk sac, a structure that provides nutrients and gas exchange for the embryo until the chorion, and later the placenta, take over these functions. The remaining cells of the blastocyst are pluripotent and will go on to form the embryo, or epiblast (see Section 3.24).

In the mouse embryo, epiblast cells emerge in the blastocyst at around day E4.0 and remain uncommitted until E6.0, when cell assignments start during the process of gastrulation. Prospective primitive endoderm cells and epiblast cells are initially mixed in the inner cell mass. When blastocysts or inner cell masses are placed in culture under defined conditions, individual prospective epiblast cells generate clonal populations that will divide indefinitely and maintain the pluripotent state. These are the **embryonic stem cells (ES cells)** and have enormous biomedical potential.

In this chapter we discuss the origin of different kinds of ES cells in mice and humans and how they can be harnessed to study the early stages of mammalian development. We shall also look at how ES cells are used to generate three-dimensional models of organ development *ex vivo* and the medical potential of these structures.

Embryonic stem cells and induced pluripotent stem cells

In this part of the chapter, we see how pluripotent ES cells are isolated from the blastocyst and maintained in culture, and we look at the properties of the various types of mouse and human ES cells that have been obtained. We also see how another type of pluripotent stem cell, called the **induced pluripotent stem cell (iPS cell)**, can be produced by the introduction of so-called pluripotency factors into adult differentiated somatic cells, and we will see how all these pluripotent stem cells can be induced to differentiate into various cell types. We start by looking at the properties that characterize ES cells.

7.1 Embryonic stem cells are pluripotent cells derived from mammalian blastocysts and can contribute to normal development *in vivo*

In this section we describe some general properties of ES cells. Most of this information has been derived from mouse ES cells. Human ES cells can also be produced, but have some important differences, which will be discussed in Section 7.2.

When mouse blastocysts between stages E4.0 and E5.0 (i.e. a time before they would implant in the uterus) are dissociated into cells and plated on a layer of fibroblasts, individual surviving cells proliferate and generate colonies. These colonies of pluripotent ES cells can either be maintained in culture, or by changing the

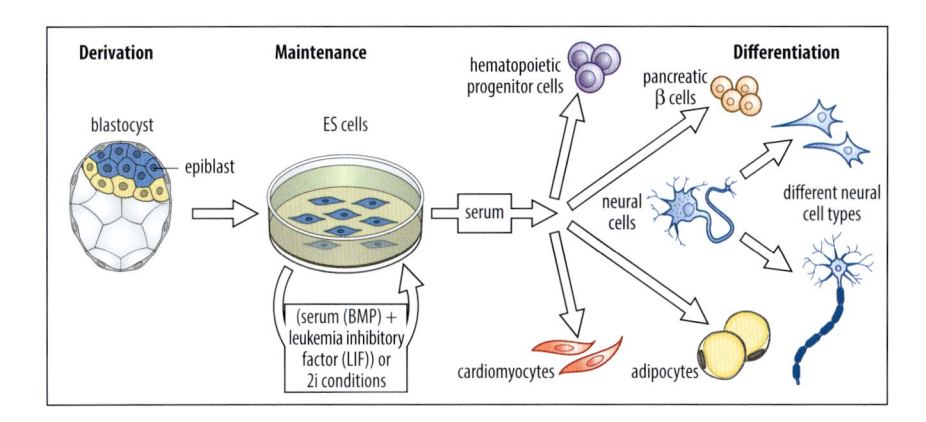

Fig. 7.1 Isolation and differentiation of mouse embryonic stem cells. Mouse embryonic stem cells (ES cells) can be obtained by culturing whole mouse blastocysts or the isolated inner cell mass in a dish and selecting for cells that can grow in the presence of serum and the signal protein leukemia inhibitory factor (LIF). After a few days, colonies emerge, derived from single cells, which can be propagated without any sign of differentiation. The factor in serum that helps maintain pluripotency has been found to be the signal protein BMP-4. Alternatively, serum can be replaced by inhibitors of the protein kinases MAPK and GSK3 (known as the 2i condition), because signaling by MAPK and GSK3 promotes cell differentiation. On removal of LIF and serum (or removal of the 2i condition), ES cells can be differentiated into many different cell types and tissues in culture by the application of various cocktails of growth factors and culture times tailored to produce specific cell fates. These are described in more detail in Section 7.4.

culture conditions can be induced to differentiate into a wide variety of cell types (Fig. 7.1).

When ES cells are injected into a mouse morula or a blastocyst which is then returned to the uterus of a surrogate mother (see Box 3B, Fig. 1), the cells integrate with those of the embryo and form a chimeric blastocyst, in which the cultured cells are shown to contribute to the normal development of the embryo and to all the tissues, including the germ line (although not to the extra-embryonic lineages). For this reason, ES cells are said to be pluripotent—they can give rise to all cell types of the organism. Such experiments have shown that just two or three ES cells can contribute to a whole embryo. Another assay for pluripotency is the tetraploid chimeric assay, in which the host blastocyst is tetraploid (formed from the fusion of the blastomeres of a two-cell embryo in culture). The tetraploid cells will form the placenta and the embryo will derive from the ES cells only.

As we saw for stem cells in Chapter 6, ES cells can proliferate to produce more ES cells, a feature called self-renewal. This means that pluripotent ES cells can be maintained and expanded in culture for many cell generations. The fibroblasts on which mouse ES cells were initially isolated can be substituted by serum (a poorly defined source of nutrients and signal molecules) and the signal protein leukemia inhibitory factor (LIF), which together inhibit their differentiation (see Fig. 7.1). A search for the specific component in serum that complements LIF in the maintenance of pluripotency yielded the signal protein BMP-4. An alternative derivation procedure replaces the serum with inhibitors of the protein kinases MAPK and GSK3 because signaling by MAPK and GSK3 promotes cell differentiation. This is known as the 2i condition. These culture conditions can be used to derive ES cells from rat, as well as mouse, blastocysts, and the 2i condition has turned out to be a good general approach to deriving pluripotent ES cells from other mammals.

All these signals act on cell-intrinsic mediators of pluripotency—a small set of transcription factors that are essential for maintaining cultured mouse ES cells in an undifferentiated state. These factors are Nanog, Oct4, Sox2, Esrrb, and Klf4, and are all expressed in the embryonic cells of the blastocyst before implantation, which shows that ES cells are a good representation of the state of the pluripotent cells in the embryo. However, in culture the expression of the genes for these factors can be maintained over time, whereas in the embryo it is transient, a feature that is reflected in the rapid downregulation of these genes as the embryo approaches the start of gastrulation. This is particularly clear in the case of *Nanog*, whose expression is completely restricted to the epiblast, and later the germline, and which is quickly downregulated as the embryo begins to gastrulate (Fig. 7.2).

The main function of the transcription factors that mediate pluripotency is to prevent differentiation, and to ensure this, they form a gene regulatory network—the

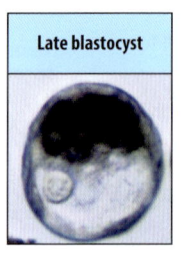

Fig. 7.2 Expression of the mRNA for the pluripotency factor Nanog in the mouse embryo. *Nanog* mRNA transcripts are at their maximum between the late morula and the mid-blastocyst stages and are downregulated just before implantation. During the formation of the blastocyst, the expression of *Nanog* is first restricted to the inner cell mass and later, as this becomes subdivided into the primitive endoderm and the epiblast, it is restricted to the latter. *Nanog* expression is lost around the time of implantation. At all times the expression of *Nanog* correlates with the potential of cells to give rise to ES cells.

From Chambers, I., et al.: ***Functional expression cloning of Nanog, a pluripotency sustaining factor in embryonic stem cells.*** Cell 2003, **113**: 643-655.

pluripotency network—where they maintain each other's expression and suppress the expression of differentiation genes. For this reason, loss of the function of any of these does not result in the death of the cells, but in their differentiation into specific cell types (Fig. 7.3).

ES cells also make their own signals, and two of them, Wnt and FGF, are crucial for the stability of the pluripotency transcriptional network and the maintenance of pluripotency in mouse ES cells. Active β-catenin produced by the Wnt intracellular signaling pathway promotes pluripotency, whereas FGF signaling via the intracellular protein kinase ERK (see Box 4A) promotes differentiation. Wnt acts through inhibition of the protein kinase GSK3, which forms part of the destruction complex that prevents β-catenin activation (see Box 3E). It is likely that the fluctuations in gene expression observed in cultures of ES cells are due to fluctuations in the synthesis and secretion of these signals. When Wnt signaling is promoted by the small-molecule agonist CHIR99021 and FGF signaling is inhibited by the small molecule PD03, differentiating cells disappear from the culture, the heterogeneities in gene expression are eliminated, and the culture becomes a homogeneous collection of pluripotent cells (Fig. 7.4); this is the basis for the 2i conditions.

These experiments show that mouse ES cells grown in LIF and BMP-4 (or in 2i conditions) are poised for differentiation, and this is highlighted by the existence of both repressive and activating chromatin marks (see Box 6B) associated with the regulatory regions of many of the genes that are expressed in ES cells. These histone marks become resolved into cell-type-specific sets of marks as differentiation signals lead to the activation or inactivation of these genes. The genes expressed in ES cells are also organized into topologically associating domains (TADs, see Section 6.2), which undergo rearrangements on differentiation.

Self-renewing cells have also been derived from the extra-embryonic lineages of the pre-implantation mouse embryo—the primitive endoderm and the trophectoderm (see Fig. 3.49). Those derived from the trophectoderm are called **trophoblast stem cells (TS cells)**, and those from the primitive endoderm become **extra-embryonic endoderm cells (XEN cells)**. When injected into morula-stage embryos, they incorporate into and contribute to their lineages of origin only.

7.2 Human ES cells differ from mouse ES cells in several of their properties

The first human ES cells were produced in 1998, and since then other types of human ES cells with slightly different properties have been isolated. The original type of

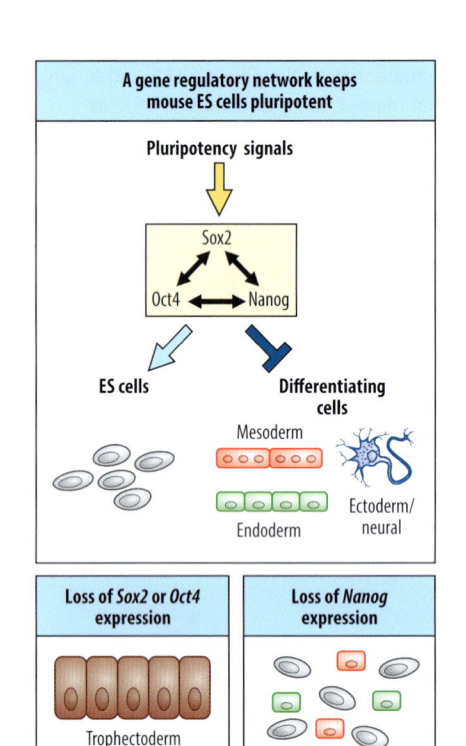

Fig. 7.3 A gene regulatory network keeps mouse ES cells pluripotent. Upper panel: signals from the culture medium and from the ES cells themselves act on an intracellular pluripotency gene regulatory network comprising the transcription factors Oct4, Sox2, and Nanog and their genes. These transcription factors activate each other as shown, to maintain their presence in the cells and thus the pluripotent state. Other transcription factors are also involved, e.g. Esrrb and Klf4, but experiments have shown that these three are crucial. Lower panels: if ES cells in culture are mutated to lose *Oct4* or *Sox2* expression they differentiate into trophectoderm, whereas in cells mutant for *Nanog* the pluripotent state is not stable and ES cells differentiate easily.

Fig. 7.4 A balance between Wnt and FGF signals maintains the pluripotency network.
ES cells produce both Wnt and FGF. Wnt promotes the maintenance of pluripotency by inhibiting the protein kinase GSK3, which is part of a 'destruction complex' that prevents the activation of β-catenin (see Box 3E). β-catenin promotes pluripotency. FGF tends to promote differentiation via activation of the protein kinase ERK, which is part of the FGF signaling pathway (see Box 4A). These two signals act in a balance with each other to maintain pluripotency. When the Wnt signaling pathway is kept activated by the Wnt agonist CHIR, and the FGF signaling pathway is inhibited by the inhibitor PD03, signs of fluctuating gene expression disappear from the culture and it becomes a homogeneous culture of pluripotent stem cells.

human ES cells were isolated from 5-day blastocysts and express the same transcription factors as do mouse ES cells, but the signal molecules that support their self-renewal in culture are different: FGF and the TGF-family members Activin or Nodal support self-renewal and pluripotency (Fig. 7.5). LIF has no effect on these cells and BMP promotes differentiation, which is a different situation from what happens in mouse ES cells. Over the past few years, human cells with many characteristics of mouse ES cells, in particular their ability to grow in the absence of MAPK signaling, have been derived from human embryos or human ES cells. These cells, and their counterpart mouse ES cells, are known as **naive ES cells**. Remarkably, when the developmental potential of these naive human ES cells was explored, they were found to be **totipotent**. That is, in addition to their ability to differentiate into embryonic cell types, they can differentiate into trophectoderm and primitive endoderm. This highlights differences between the embryonic cells of mouse and human early embryos—the mouse naive ES cells are not totipotent—and suggests that the process of pre-implantation lineage allocation might be different in these two species. In the mouse it is progressive. A totipotent cell in the morula first spins off trophectoderm. The remaining inner cell mass cells choose between primitive endoderm and epiblast. However, despite the differences in the ES cells of the two species, single-cell RNA sequencing has shown that this order of cell-fate decisions is similar in humans.

The differences between naive mouse ES cells and the original human ES cells have been resolved to some extent by the discovery that, in mice, pluripotent stem cells can also be derived from epiblasts between E5.0 and E6.0, a more advanced developmental stage at which the blastocyst would have implanted in the uterus. These cells are called **epiblast stem cells (EpiSCs)** and have many similarities to human ES cells, such as the fact that FGF and Activin or Nodal promote pluripotency and self-renewal (see Fig. 7.5).

EpiSCs also have some of the properties of naive mouse ES cells in the transcription factors they express and their ability to differentiate into a wide range of cell types. Yet, they cannot contribute to blastocyst chimeras (see Fig. 7.5). If transplanted to post-implantation epiblasts, however, they will engage in development. It seems that EpiSCs do not have enough cadherin to attach to the pre-implantation epiblast and, indeed, if cadherin is overexpressed in EpiSCs, they will make blastocyst chimeras.

The pluripotency of EpiSCs is tested by injecting them subcutaneously in adult immunodeficient mice, where, because of their high rate of proliferation, they give rise to tumors known as **teratomas** (Fig. 7.6). These unusual tumors can be cancerous (in which case they are called teratocarcinomas) or non-cancerous, and contain a mixture of differentiated cells with representatives of the three germ layers. Because the chimera test is not possible with human ES cells, they are also tested for pluripotency by their ability to make teratomas in mice.

Despite their similarities, a significant difference between human ES cells and mouse EpiSCs is the ability of the human cells to make primordial germ cells, which has been lost in EpiSCs. Thus, human ES cells seem to be at an intermediate stage relative to mouse naive ES cells and EpiSCs (see Fig. 7.5).

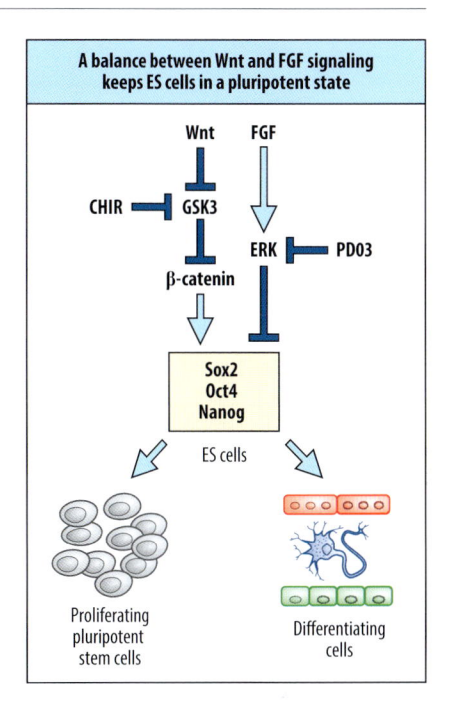

372 Chapter 7 Embryonic stem cells, embryo models, and regenerative medicine

Mouse		Human	Human
Blastocyst (4.5 days)	**Post-implantation embryo (5 days)**	**Blastocyst (5 days)**	**Blastocyst (5 days)**
trophectoderm, epiblast, primitive endoderm (inner cell mass)	ectoplacental cone, extra-embryonic ectoderm, visceral endoderm, epiblast, parietal endoderm		
serum (BMP) + LIF or 2i + LIF	FGF + Activin/Nodal	FGF + Activin/Nodal	LIF + 2i + aPKCi
Naive mouse ES cells	**Mouse EpiSCs**	**Human ES cells**	**Naive human ES cells**
Express the transcription factors Nanog, Oct4, Sox2, Klf4	Express the transcription factors Nanog, Oct4, Sox2	Express the transcription factors Nanog, Oct4, Sox2	Express the transcription factors Nanog, Oct4, Sox2
Can contribute to chimeras	Cannot contribute to chimeras	Can form teratomas	They are totipotent
Can form teratomas	Can form teratomas	Can differentiate into a wide range of cell types *in vitro*	Can differentiate into: trophectoderm, primitive endoderm and embryonic tissues
Can differentiate into all cell types *in vitro*	Can differentiate into a wide range of cell types *in vitro*	Can make PGCs	Can form teratomas
Can make PGCs	Cannot make PGCs		

(Mouse ES cells can contribute to chimeras as shown by arrow back to blastocyst.)

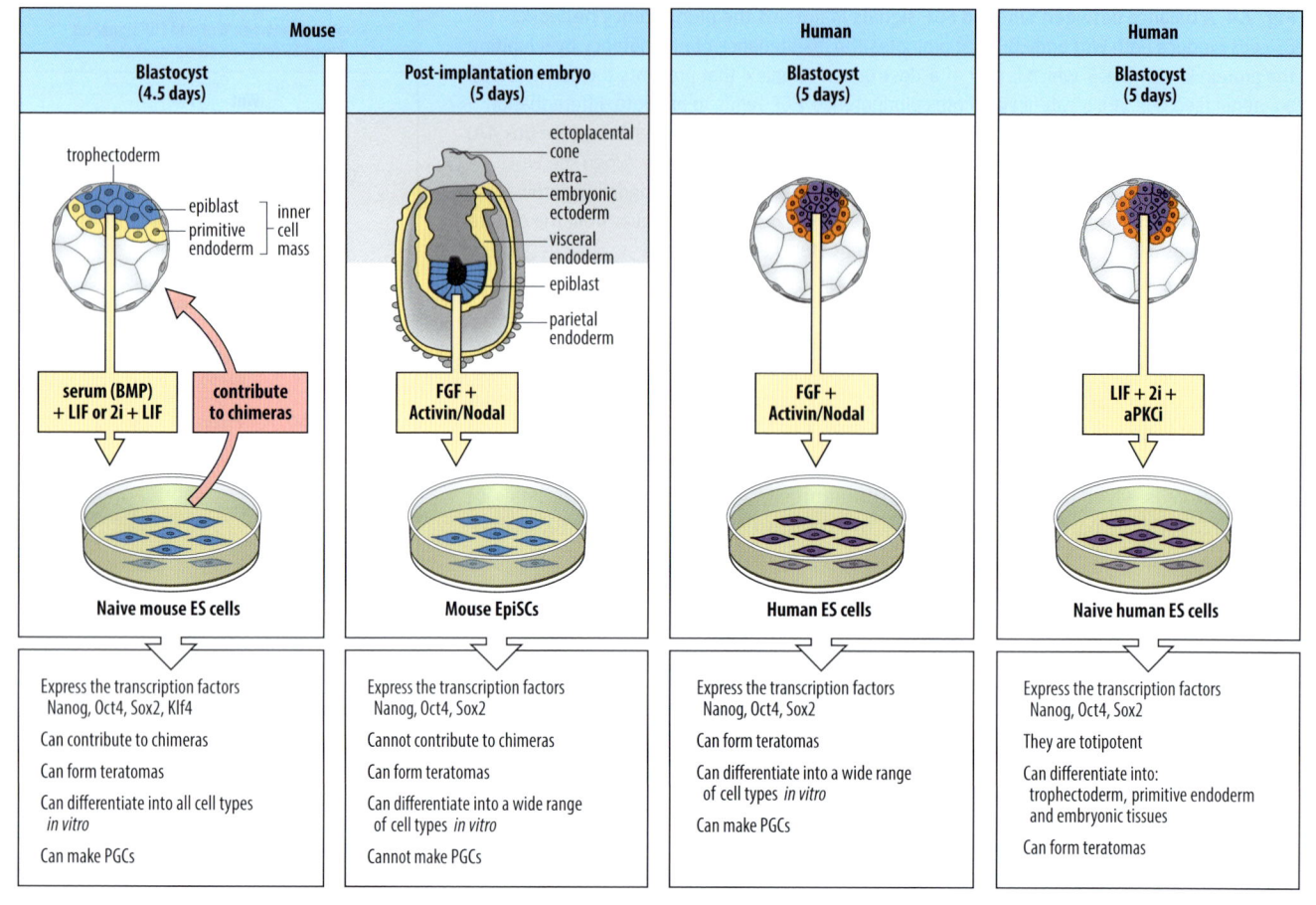

Fig. 7.5 A comparison of different types of mouse and human embryonic stem cells and their properties. Mouse embryonic stem cells (ES cells) and mouse epiblast stem cells (EpiSCs) are isolated from pre-gastrulation embryos and have different growth requirements. The mouse ES cells derived from the blastocyst at E4.5 are often called naive mouse ES cells to distinguish them from EpiSCs. EpiSCs are derived from post-implantation epiblasts and, in contrast to naive cells, require FGF signaling. Also, unlike naive cells, they cannot give rise to primordial germ cells (PGCs). These cells are often referred to as 'primed' ES cells. Human ES cells can be derived from human blastocysts and their growth conditions are similar to those of mouse EpiSC. However, unlike these, human ES cells can give rise to PGCs. More recently, a second type of human ES cell has been isolated from the blastocyst using different culture conditions. These cells are at an earlier stage of development, are called naive human ES cells, and are totipotent, that is, they can give rise to extra-embryonic and embryonic tissues. aPKCi, inhibitor of atypical protein kinase C; LIF, leukemia inhibitory factor.

7.3 Adult differentiated cells can be reprogrammed to form pluripotent stem cells

In *Xenopus*, the nucleus of a differentiated cell transplanted into an enucleated egg can support the normal development of the egg (see Section 6.17). This showed that the differentiated state can be reverted and genes required for development reactivated, in this case by cytoplasmic factors in the egg. Furthermore, when adult differentiated somatic cells are fused with ES cells, they are transformed into pluripotent cells, showing that factors present in the ES cells are able to confer ES-cell-like properties on differentiated cells. It has also been shown that fibroblasts, for example, can be converted into muscle cells by introducing the gene for the muscle master transcription factor MyoD (see Section 6.6). All these observations led to a landmark discovery: that adult differentiated mouse cells can be reprogrammed to an ES-cell-like state by introducing into them the genes for four transcription factors associated with pluripotency in ES cells: *Sox2*, *Oct4*, *Klf4*, and *c-Myc* (Box 7A).

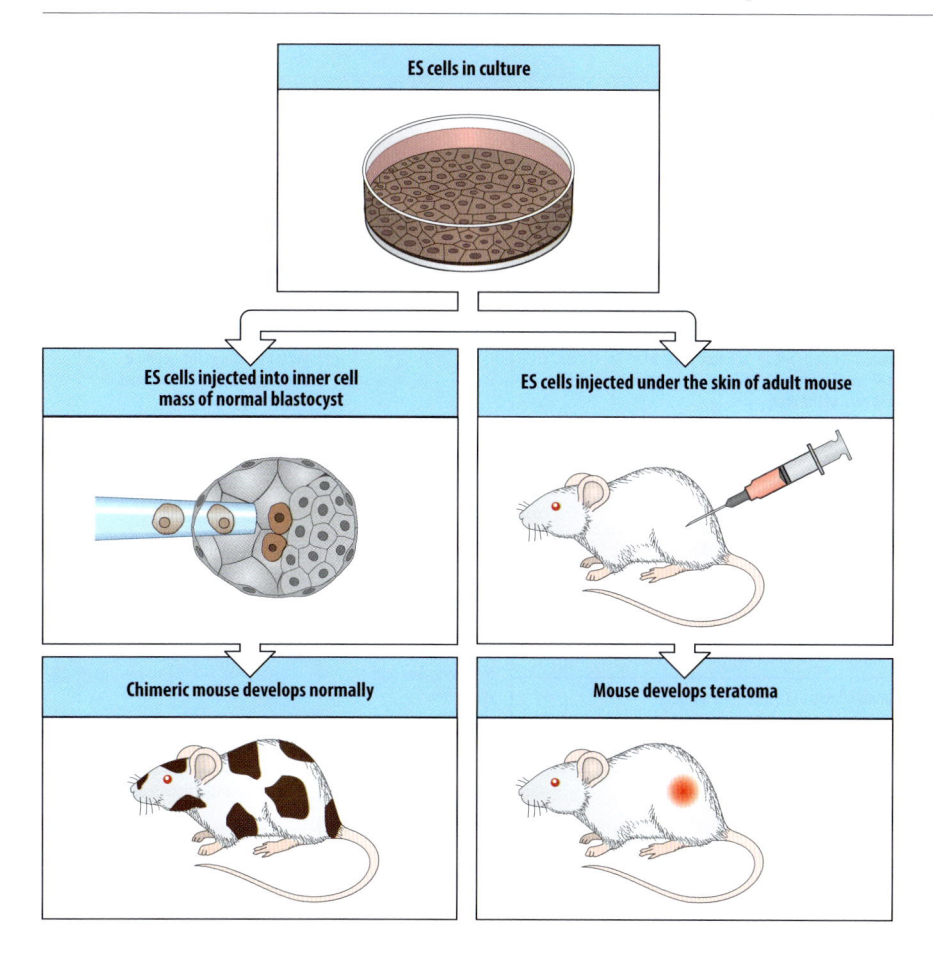

Fig. 7.6 Mouse embryonic stem cells can develop normally or into a tumor, depending on the environmental signals they receive. Cultured embryonic stem cells (ES cells) originally obtained from an inner cell mass of a mouse contribute to a healthy chimeric mouse when injected into an early embryo (left panels), but if injected under the skin of an immunodeficient adult mouse, the same cells will develop into a teratoma (right panels). Human ES cells are also tested for pluripotency in this way.

EXPERIMENTAL BOX 7A Induced pluripotent stem cells

The first induced pluripotent stem cells (iPS cells) were produced from skin fibroblasts isolated from transgenic mice that carried a drug-selectable marker linked to expression of a gene associated with pluripotency, either *Oct4* or *Nanog*. The genes encoding four transcription factors associated with pluripotency in mouse ES cells—*Oct4, Sox2, Klf4,* and *c-Myc*—were introduced into the adult fibroblasts using retroviral vectors (Fig. 1). The transfected cells were then treated with the appropriate drugs to select for the rare iPS cells. Most transfected cells did not express either *Oct4* or *Nanog* and therefore had no drug resistance and died. The surviving colonies of cells resembled ES cells and could be isolated and expanded in culture. The inserted genes in the iPS cells shut themselves off once the reprogramming process was completed, and the cells reactivated their own pluripotency genes to maintain the pluripotent state. Even though *Nanog* is not part of the iPS cocktail, its expression in the final stages of the reprogramming is essential for obtaining true iPS cells. Only cells that expressed *Nanog* are truly pluripotent by the criteria of self-renewal, being able to differentiate in culture and, most significantly, to contribute to fertile chimeras (see Section 7.1). Although the same set of transcription factor genes can reprogram adult human fibroblasts, the conditions needed to expand human iPS cell populations are different, with mouse cells requiring LIF and human cells FGF. Human iPS cells have also been derived by different research groups using a slightly different mixture of transcription factors.

Advances have been made over the past few years in inducing pluripotency without the need to introduce a persisting viral vector. This will be a prerequisite for any medical use for iPS cells in cell-replacement therapy. Transposon-based vectors have been produced that snip themselves and the genes they carry cleanly out of the genome after the transformation to pluripotency is complete. Other approaches involve the discovery of small-molecule drugs that will act as co-activators of a cell's own pluripotency genes.

An equal effort is going into reducing the number of genes that need to be introduced. Adult neural stem cells have been reprogrammed to pluripotency by introducing and expressing the gene

Chapter 7 Embryonic stem cells, embryo models, and regenerative medicine

Some methods for deriving induced pluripotent stem cells (iPS cells)			
Method of delivery or activation of pluripotency genes	**Integration into genome**	**Advantages**	**Disadvantages**
Viral vectors			
Retrovirus-derived vector. Original method of Takahashi and Yamanaka (2006). Multiple viruses each carrying one transcription factor gene	Yes. Transgenes usually become silenced in the iPS cells after reprogramming and integration	Efficient and stable reprogramming	Vector and transgenes remain in genome and might be reactivated in differentiated cells
Lentivirus-derived vector; vectors often encode multiple transcription factors	Yes. Transgenes may remain active after reprogramming and integration, but can be made excisable using a Cre/lox system (see Box 4E) or silenced using a doxycyclin-inducible promoter and withdrawal of the drug		The long terminal repeats of the vector remain in the genome and might activate oncogenes if they integrate near by
Retroviral and lentiviral vectors are widely used to generate iPS cells for research purposes, but because of the persistence of the vector and transgenes in the genome they cannot be used to generate iPS cells for potential clinical use			
DNA plasmids and transposons			
Modified *piggybac* transposon + plasmid carrying PB transposase	Vector integrates but is excised by the transposase	Efficient reprogramming. Differentiated cells transgene-free and vector free	Some cells might still carry small pieces of integrated DNA
Linear DNA plasmid with *loxP* sites	Vector integrates but is excised in the presence of Cre recombinase	Differentiated cells transgene-free and vector free	Less efficient than viral vectors
Non-integrative plasmids	No. Plasmids diluted out as iPS cells divide		
Non-DNA methods			
Transfection with synthetic RNAs (RNA-induced iPS cells (RiPS cells)	Differentiated cells free of any exogenous DNA		Repeated transfections required
Transcription factor proteins			Slow and less efficient than viral vectors
Activation of endogenous pluripotency genes by treatment with small-molecule drugs (chemically induced iPS cells (CiPS cells))			Relatively new method: complete reprogramming using only chemical inducers first reported in 2013. Slow

Fig. 1

for just one of the transcription factors, Oct4. Adult neural stem cells already express high levels of Sox2 and c-Myc, however, and this may explain why introduction of Oct4 alone is sufficient to induce pluripotency.

In these experiments, after a few days, these cells had stopped expressing the genes that conferred the differentiated state and had assembled the pluripotency gene regulatory network by activating the endogenous genes for these transcription factors. It was surprising that so few transcription factor genes needed to be introduced into adult somatic cells to induce pluripotency.

Cells produced in this way are known as **induced pluripotent stem cells (iPS cells)** and were first derived from mouse fibroblasts in 2006 by Shinya Yamanaka, who later shared the Nobel prize with John Gurdon 'for the discovery that mature cells can be reprogrammed to become pluripotent'. The same set of transcription factors was very soon found to be able to reprogram adult human fibroblasts and many other adult cell types, opening up exciting opportunities for generating particular cell types from patients' cells for clinical applications.

The pluripotency of mouse iPS cells was tested by injecting them into normal mouse blastocysts to generate chimeras, which were then replaced in a surrogate mother. The iPS cells contributed to all cell types in the resulting embryos, including the germline.

When these iPS chimeric mice were crossed with normal mice, iPS-derived gametes successfully participated in fertilization and live mice were born. More recently, fertile adult mice generated entirely from iPS cells have been derived from tetraploid blastocysts into which fibroblast-derived iPS cells were injected. Also, iPS cells, like ES cells, can differentiate into a wide range of cell types *in vitro*. Thus, iPS cells have all the same properties as ES cells. In the case of human iPS cells, it is not possible to generate chimeras for obvious ethical reasons. In this instance, pluripotency is demonstrated, as in the case of the ES cells, through the formation of teratomas in mice (see Fig. 7.6).

The reprogramming experiments start with millions of cells, but only a small fraction—often less than 1% of the initial population—become pluripotent stem cells. Because these cells have a proliferative advantage, it is always possible to recover large numbers of iPS cells. Nevertheless, it is important to highlight that the reprogramming does not depend exclusively on the expression of the Yamanaka cocktail, but on some property of the cell state that, for the moment, remains unknown. What has been learnt from following the process over time is that it involves profound alterations to the epigenome and, in the final stage, the activation of the endogenous pluripotency gene regulatory network.

iPS cells have opened up the possibility of deriving pluripotent cells from an individual's own somatic cells. This could help the development of regenerative medicine tailored to the individual.

7.4 Pluripotent stem cells are used as models of cell differentiation

Mouse and human ES cells can be made to differentiate into all cell types in adherent culture—that is, when cells form a single layer adhering to the surface of the culture dish (see Fig. 7.1). When plated in two dimensions like this, and steered by the controlled addition of signals to the medium, ES cells can be coaxed to develop into representatives of the three germ layers and their derivatives, following the paths they would do in an embryo. For example, as they exit the pluripotent state, exposure of ES cells to Wnt and Nodal signals pushes them towards mesoderm and endoderm, whereas inhibitors of these pathways promote neuroectodermal differentiation, as happens in the embryo (see Chapter 4).

Figure 7.7 shows one set of pathways mouse EpiSCs can take as they differentiate in culture. When mouse EpiSCs are exposed to FGF and Wnt signals, they express the gene *TbxT* and first enter a state similar to that of cells in the primitive streak that give rise to endodermal and mesodermal derivatives. They progress to resemble cells like those in the stem zone (the caudal lateral epiblast) (labeled EpiCE in Fig. 7.7), which can become different kinds of mesoderm—lateral plate, intermediate, paraxial, and axial—and can also give rise to nerve cells (see Fig. 4.27). Continuation in this condition leads to downregulation of Oct4 and to a multipotent state that mimics that of the tailbud in mouse/human and chick. Exposure of these cells to BMP-4 gives rise to lateral plate mesoderm, whereas exposure to Wnt signals and suppression of BMP signaling promotes a neuromesodermal progenitor state—that is, a mixture of cells primed to become paraxial mesoderm and spinal cord progenitors. Depending on the culture conditions, these cells can, in turn, give rise to either nerve cells, such as motor neurons and interneurons, or derivatives of the paraxial mesoderm such as muscle (see Fig. 7.7).

Although an initial choice between neuroectoderm and mesoderm is readily achieved, differentiation into specific cell types requires further precise changes in the composition of the growth medium over time. The different conditions needed to derive various types of mesodermal cells from the neuromesodermal progenitor illustrate this. Elevation of Wnt signaling in the neuromesodermal progenitor population leads to formation of paraxial mesoderm cell types. Remarkably, the oscillations in gene expression associated in the embryo with the somite segmentation clock

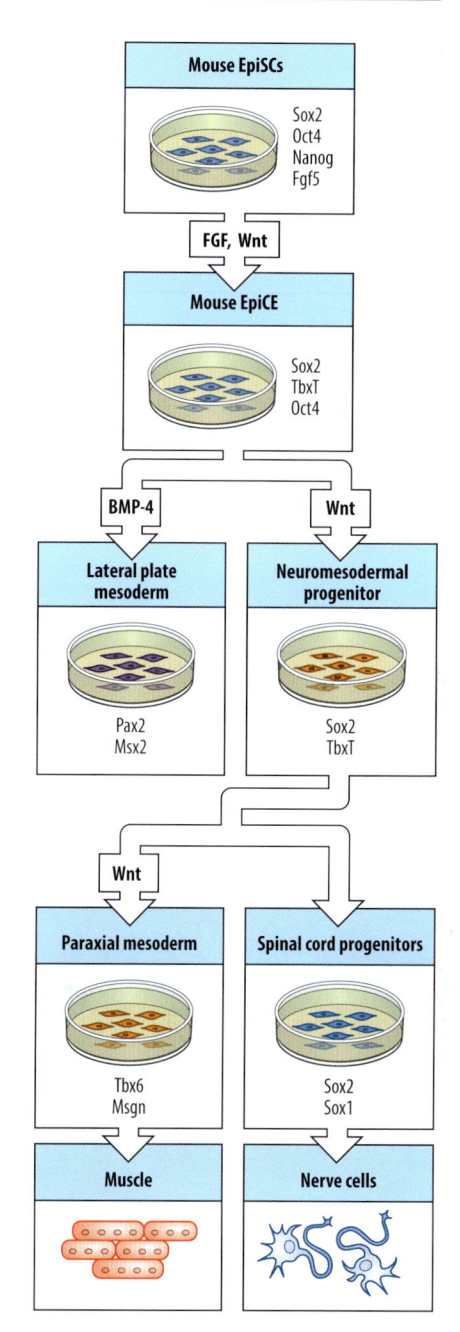

Fig. 7.7 Differentiation of mouse epiblast stem cells in culture into derivatives of the stem zone. When exposed to FGF and Wnt signaling, epiblast stem cells (EpiSCs) differentiate into a population (EpiCE) that resembles that of the stem zone (the caudal lateral epiblast) of the mouse embryo. Exposure of this population to BMP-4 induces the differentiation into lateral plate mesoderm and its derivatives, whereas exposure to Wnt signaling creates a population of bipotent stem cells that, depending on further signals, will generate spinal cord or paraxial mesoderm progenitors.

(see Section 4.10) are recapitulated in individual paraxial mesoderm cells in culture and become synchronized over time. Further changes of medium, in particular specific regimes of exposure to Shh, BMP, and TGF-β agonists, lead to the generation of the sclerotome (which gives rise to the vertebral column) and dermomyotome (from which the dermis of the skin and some muscles will derive).

These experiments reveal the existence of cell-autonomous programs of gene expression that are followed by the cultured cells, including complex behaviors such as the oscillations in gene expression. Similar experiments are being done with human cells.

Despite the similarity in the sequence of cell states, there are two significant differences between the cell populations in culture and those in embryos. The first one is that in the dish, cells are exposed to non-patterned and high levels of signals, which contrasts with the existence of defined sources of signals in the embryo; nevertheless, the cells respond by developing specific fates. In addition, in many cases, the same signals seem to elicit different cell-fate responses at different times in culture. For example, Wnt signaling promotes expression of *TbxT* early on and paraxial mesoderm markers, such as *Tbx6* later. This suggests that the effect of signals depends on the context on which they act, as we have seen elsewhere, and that they have an important role in cell-fate decisions independently of their *in vivo* role in patterning, as patterning is absent in the cultured cells. The timing effect also suggests that signals affect cell-fate decisions by modulating cell-intrinsic programs of gene expression, most strikingly shown here by the oscillation in gene expression in paraxial mesoderm cells in adherent culture. The situation in culture therefore reflects the representation of development in the Waddington landscape, with signals acting at the bifurcations (see Fig. 6.18), as what is seen is that differentiation progresses through a sequence of binary decisions.

The second difference between cells in culture and in embryos is the obvious lack of spatial organization and relative proportions of the different cell fates that are so important in the make-up of tissues and organs. These two features missing in adherent culture are hallmarks of embryos and suggest that patterning during development is not a simple consequence of cell-fate decisions and perhaps requires cell densities and spatial organizations that are not achieved in adherent culture.

To mimic the situation in the embryo more closely, cells are seeded on small surfaces of defined sizes and patterns coated with cell-adherent substances that restrict the cells' organization and growth. When ES cells are plated in this manner on circular patterns and exposed to signals, they differentiate as they would in classic adherent culture but now exhibit clear organization into patterns of different cell types. These experiments were first done with human ES cells; mouse ES cells give similar results.

When grown on a circularly patterned matrix, ES cells in culture can produce a structured version of the cell-fate decisions that occur in gastrulation. As shown in Figure 7.8, when human ES cells grown on such a pattern are exposed to BMP, they reproducibly differentiate as concentric rings, each representing a germ layer as formed during gastrulation. In this experiment, the inner circle expresses the transcription factor *SOX2* and corresponds to epiblast or neuroectoderm. Surrounding and overlapping with this is a ring of cells expressing *TBXT*, mimicking the primitive streak (mesoderm). On the outermost edge of the circle is a ring of *CDX2*-expressing cells corresponding to trophectoderm. In other experiments a ring of endoderm is also formed between the trophectoderm and mesoderm rings. Studies of the emergence of these patterns have shown that the effect of BMP depends on Wnt and Nodal signaling, which interact to organize the patterns—a situation that mimics what happens in mouse embryos (see Fig. 3.58), and that strengthens the known similarities in gastrulation between human and mouse.

These experiments also show the importance of cell organization and the density of the cell population for the patterning of cell fates. For example, changing the diameter of the patterned circles in the matrix alters the organization of the germ layers. Experimental and theoretical studies of the patterns have shown that they are self-organized

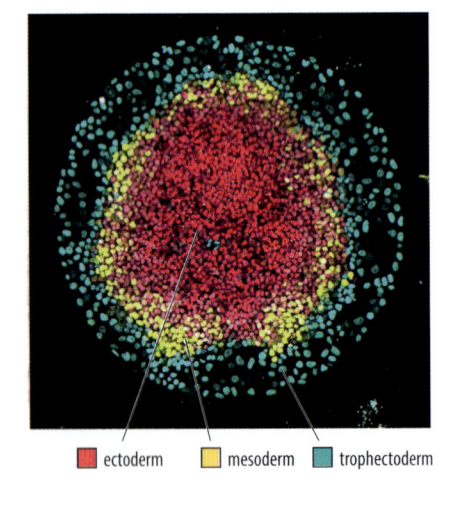

Fig. 7.8 Human ES cells grown on a circularly micropatterned substrate differentiate into spatially patterned germ layers. Human ES cells are plated in circular adhesive micopatterns that restrict their growth. After 42 hours of exposure to BMP-4, the colony becomes patterned into components of the early embryo shown by immunostaining for Sox2 (ectoderm, red), TbxT (mesoderm, yellow), and Isl1 (trophectoderm, blue).

Photograph courtesy of Aryeh Warmflash.

(in that they are responding to an underlying chemical pattern that arises in a Turing-like manner; see Box 9E), require a high density of cells, and depend on cell–cell interactions.

> **SUMMARY**
>
> Embryonic stem cells (ES cells) are pluripotent stem cells derived from the inner cell mass of mammalian blastocysts. They can be grown indefinitely in culture and can differentiate into many different cell types depending on the signals that they are exposed to. Their continued pluripotency depends on the expression of a small set of transcription factors. Different types of these pluripotent stem cells have been distinguished in mice and humans depending on the stage at which they are isolated and the culture conditions required to isolate and maintain them. Epiblast stem cells (EpiSCs), for example, are pluripotent stem cells isolated from the mouse blastocyst at a later time than the so-called naive mouse ES cells and have some different properties. Notably, mouse ES cells can contribute to forming a chimeric mouse and can make primordial germ cells, whereas EpiSCs cannot. There are also differences between human ES cells and mouse ES cells in the conditions required to isolate them and in some of their properties. Adult differentiated cells of both mice and humans can be reprogrammed to the ES state by the induced expression of the so-called pluripotency genes. These cells are called induced pluripotent cells (iPS cells) and because they can be produced from a patient's own cells have potential for various clinical applications such as testing drug therapies and regenerative medicine.

Embryo models

The discovery that ES cells in culture could be differentiated into all the cell types of an embryo, but not to form organs and tissues, raised the question of what was missing that would enable them to assemble into a structure resembling an embryo or, at least, an early embryo. As we have seen, the differentiation processes *in vitro* lack coordination across cell populations and do not result in the relative proportions of different cell types required to organize the tissues and organs that configure an embryo. The missing factor was most likely to be something present in the embryo, because when ES cells are placed in blastocysts they contribute to normal development. The obvious differences between the ES cell cultures and an embryo are the dimensionality (cultures are two-dimensional whereas embryos are three-dimensional) and the number of cells in the culture that tend to be many more than in an embryo.

Taking these ideas on board, the past few years have seen the development of three-dimensional culture systems in which the number and arrangement of the starting cells, as well as their chemical environment, are carefully balanced. The result is a collection of ES-cell-based models of early mammalian development that are providing new insights into the mechanisms by which cells build embryos.

7.5 ES cells can be used to provide models of early development

The existence of cultured stem-cell populations representing the three lineages of the mammalian blastocyst (trophectoderm, embryo proper, and primitive endoderm) has led to attempts to reconstruct a blastocyst from these component stem-cell populations. This approach allows researchers to study the role of cell interactions in the assembly of the early embryo experimentally and without being restricted by the number of embryos available, which is always limiting when studying mammalian development. Developmental mechanisms can also be studied from an engineering perspective using this approach: rather than looking for what is necessary for a

378 Chapter 7 Embryonic stem cells, embryo models, and regenerative medicine

particular process, which is what genetics tells us, the engineering perspective can ask what is sufficient—what is the minimum number of elements required for a process to occur or a pattern to emerge.

When large numbers of mouse ES cells in suspension are aggregated and grown further in the presence of serum, **embryoid bodies** are formed and consist of a disorganized mass of multiple cell types. If a similar experiment is done with trophoblast stem cells (TS cells), these form hollow epithelial balls called **trophospheres**, whose formation is driven by the epithelial nature of the TS cells. When ES cells and TS cells are cultured together, cell sorting ensues, and the TS cells create an epithelial coating around the embryoid body (Fig. 7.9, first three rows). If the experiment is done more precisely, starting with defined numbers of cells of each type (e.g. 10 ES and 20 TS cells), and adding them sequentially, the cells can self-assemble into a geometry that triggers them to develop together. In a carefully controlled culture medium, this leads, after 2.5 days, to a structure that resembles a blastocyst in its organization and size: it contains a cavity surrounded by an epithelial layer of TS cells within which the cluster of ES cells has been displaced to one side, in the same way as the inner cell mass in a natural blastocyst (see Fig. 7.9, bottom row). These structures are called **blastoids**. These also form an underdeveloped primitive endoderm, that occasionally arises in the same position as in the blastocyst (see Fig. 3.49). When these blastoids are placed into a mouse uterus, they attach and instruct the uterus to wrap around them. This is recapitulating the beginning of implantation, although blastoids do not progress towards a post-implantation epiblast. Compared with blastoids, trophospheres have a diminished potential to instruct the uterus, which suggests that inducers secreted by the ES cells make TS cells competent for instructing the uterus. The formation of a blastoid is a clear example of self-assembly (Box 7B), and, as long as the numbers

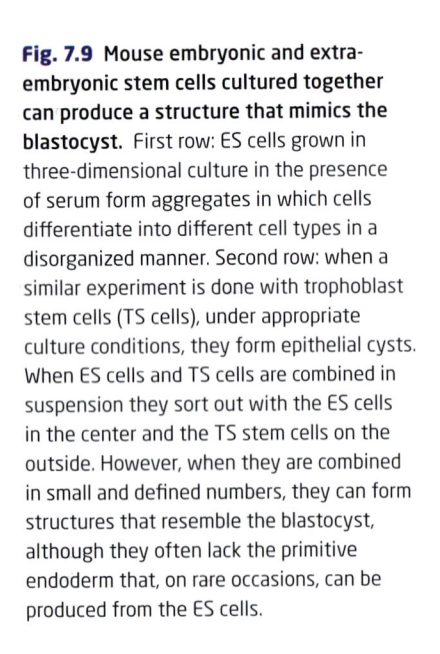

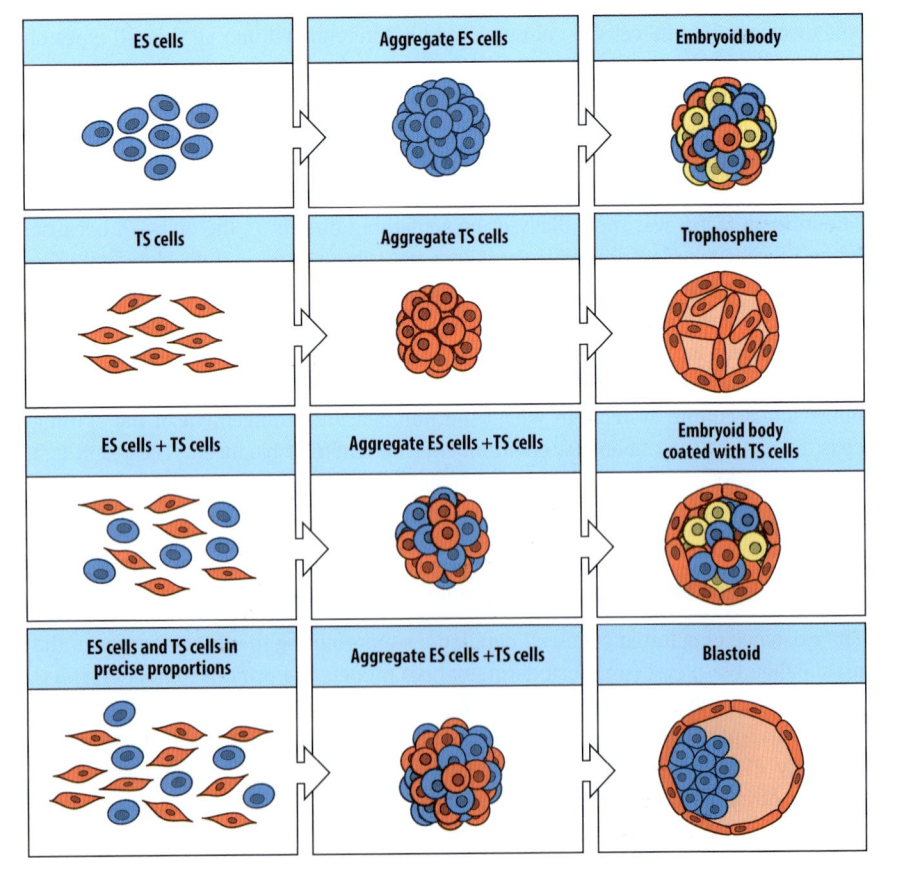

Fig. 7.9 Mouse embryonic and extra-embryonic stem cells cultured together can produce a structure that mimics the blastocyst. First row: ES cells grown in three-dimensional culture in the presence of serum form aggregates in which cells differentiate into different cell types in a disorganized manner. Second row: when a similar experiment is done with trophoblast stem cells (TS cells), under appropriate culture conditions, they form epithelial cysts. When ES cells and TS cells are combined in suspension they sort out with the ES cells in the center and the TS stem cells on the outside. However, when they are combined in small and defined numbers, they can form structures that resemble the blastocyst, although they often lack the primitive endoderm that, on rare occasions, can be produced from the ES cells.

BOX 7B Self-organization and self-assembly

The ability of stem cells to generate structures resembling adult organs and early embryos in the laboratory has brought the notion of **self-organization** in developmental biology into greater prominence. Self-organization refers to the process whereby a collection of equivalent constituent elements of a system, such as cells, organize themselves into well-defined structures in the absence of external templates or guidance (Fig. 1, top row). This ability is intrinsic to the elements and requires constant expenditure of energy: the formation of clouds, sand dunes, and snowflakes are examples.

Self-organization contrasts with **self-assembly**, a process whereby the emergence of organization is implicit in the structure of the component elements (Fig. 1, bottom row). An example of this is the assembly of a piece of furniture from a kit in which there is only one way that the components can fit together into the final structure. Self-assembly processes tend to thermodynamic equilibrium.

Organoids are often said to self-organize. But although the early stages of their development might well fit the strict definition, it is probably the case that early on an asymmetry appears in the multicellular system that guides the subsequent development of the organoid and, when this occurs, the rest of the process is said to be 'guided'. Turing patterns (see Box 9E) are examples of pure self-organization, whereas the morphogen gradients that are formed in an embryo represent guided self-organization.

Human and mouse gastruloids (described in Section 7.6) offer examples of both self-organization and guided self-organization. In the original version of the technique for making gastruloids, ES cells are aggregated and, after 2 days' growth in a basic culture medium, they are treated with a Wnt agonist for 24 hours to mimic gastrulation. This leads to a spontaneous breaking of symmetry and the emergence of an axis, marked by the polarized expression of the gene *TbxT*, that mimics the anteroposterior axis of the embryo and guides the subsequent patterning events in the aggregate that produce the gastruloid (Fig. 2, top row). This is an example of self-organization. Note that if we could orient all the initial aggregates the same way, the axis would arise at different positions in the individual aggregates.

In another version of the gastruloid system, a group of ES cells is treated with BMP to mimic the early events of gastrulation—namely, when BMP from the trophectoderm induces the expression of Nodal and Wnt in the embryo that acts as a trigger for gastrulation (see Section 4.8). When these BMP-induced cells are placed next to an aggregate of pluripotent cells that have not been exposed to any signals, they initiate a polarized primitive-streak-like structure that guides the further patterning of the aggregate (Fig. 2, bottom row). This is an example of guided self-organization. In this case, if we could orient all

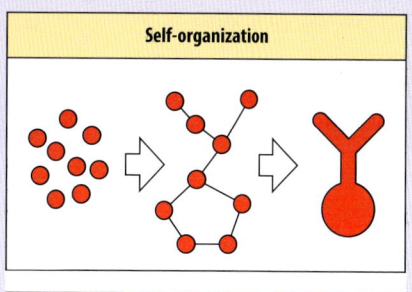

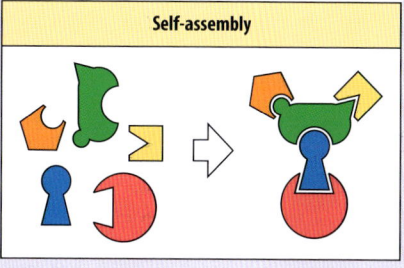

Fig. 1

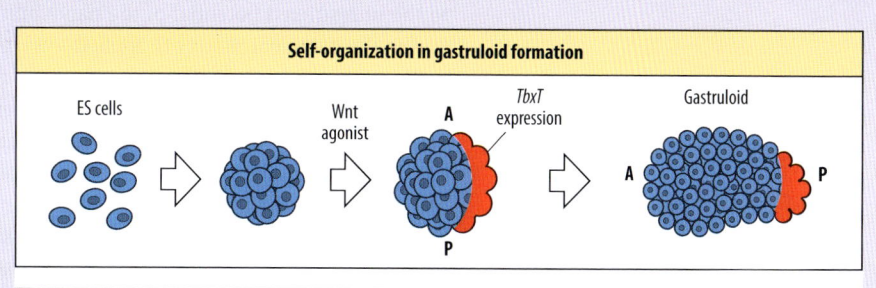

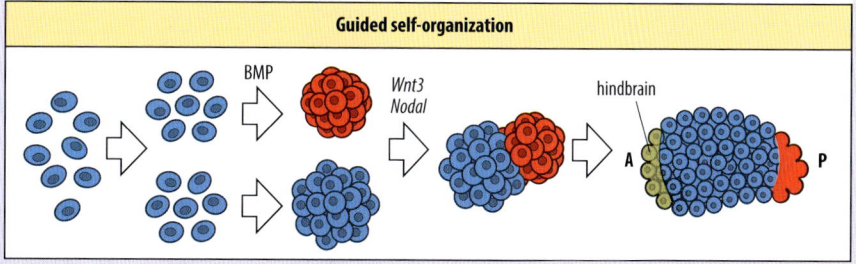

Fig. 2

the initial aggregates the same way, the axis would always emerge in the same position in individual aggregates, with the BMP-induced cells always positioned at the posterior end of the resulting gastruloid.

The outcome is very similar in both cases but the origin of the initial asymmetry is different. In the self-organized gastruloid, the axis will be different in different aggregates, whereas in the guided gastruloids, the axis will always emerge in the same direction.

What these experimental systems tell us is that cells have an ability to self-organize, which in the embryo is controlled—guided—by spatially organized interactions between different tissues.

and ratios of cells of each type are carefully measured, it can be achieved more than 50% of the time.

Studies of blastoids have revealed details of the signaling interactions that occur during the segregation of early cell lineages and that would be very difficult to discover through standard genetic analysis in the embryo. For example, comparison of trophospheres and blastoids has revealed that Wnt, BMP-4, and Nodal secreted by the ES cells help to induce the epithelization and self-renewal of the trophectoderm. This complements previously observed interactions between the embryo and the trophectoderm in which FGF induces cell proliferation. These roles had not been noticed in genetic studies, possibly because they were masked by the robustness and redundancies of molecular pathways. Another useful observation is that TS cells cultured in two dimensions and as trophospheres show differences in gene expression when compared with the blastocyst trophectoderm, but in blastoids, where they are exposed to the inducers secreted by the ES cells, their gene-expression profile becomes much more similar to that of trophectoderm in the blastocyst.

Blastoids can be made from human ES cells and TS cells in the same way as those from mouse. However, if naive human ES cells are grown in defined culture conditions, they can, on their own, form blastoids with the three lineages (trophectoderm, epiblast, and primitive endoderm) that closely resemble human blastocysts in their gene expression and organization. Thus, in contrast to mouse ES cells, which cannot easily form TS cells, naive human ES cells can. It is likely that this difference is due to the unrestricted potential of the naive human ES cells, as described in Section 7.2, and to the chemical protocols applied.

Blastoids are an important model system for studying the early stages of development. The fact that, unlike blastocysts, they are formed from established lines of stem cells, means that they do not necessitate the repeated use of embryos for some studies of pre-implantation development. This reduces the ethical concerns normally associated with the use of human embryos in research (see Section 3.30) and enables the formation of large numbers of blastoids for experimental studies. Furthermore, blastoids can be made from naive iPS cells, which removes ethical concerns about the use of embryos as a source of the stem cells.

Combination of mouse ES cells and TS cells has a very different outcome if it occurs in the presence of the extracellular matrix material Matrigel rather than in suspension. In this situation, both types of cells form epithelia that, in a small number of cases, create a lumen and develop into structures that resemble the post-implantation epiblast. These structures are called **ET embryoids**. They form at a greater frequency if the assembly of the two cell types is precisely controlled (Fig. 7.10, top row). Surprisingly, although ET embryoids lack primitive endoderm, they exhibit polarized expression of *TbxT* in cells undergoing an epithelial-to-mesenchymal transition in a manner resembling the initiation of the primitive streak. However, this structure never undergoes a polarized movement, as it does in the embryo, and the embryoids develop into something that has the anterior and posterior poles of

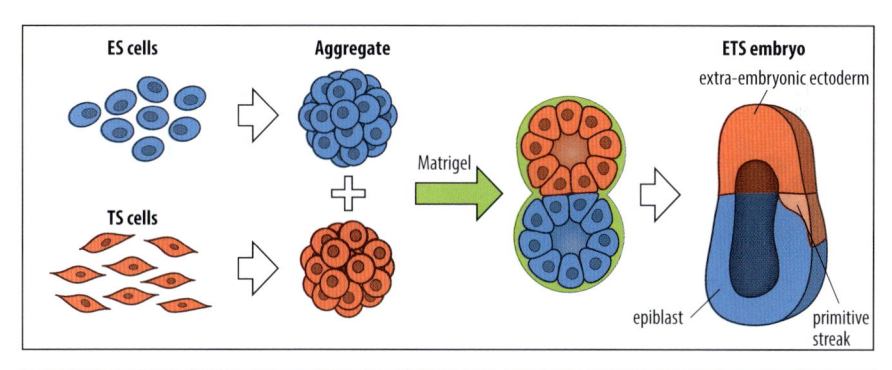

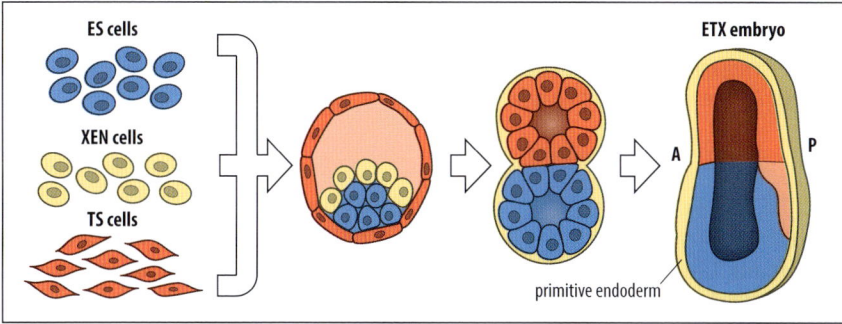

Fig. 7.10 Mouse ES cells and TS cells grown in the presence of extracellular matrix form structures that resemble the post-implantation epiblast. Top row: mouse ES cells and TS cells are grown separately in suspension and small pellets of each type of cell are prepared. These are combined in medium with the extracellular matrix material Matrigel added. If the manner of assembly is controlled, both types of cells form epithelial structures resembling the epiblast and the extra-embryonic ectoderm of an E5.75 embryo, and develop further to form the beginning of a primitive streak. These structures are called ET embryoids. Bottom row: if ES cells and TS cells are grown with XEN cells (endoderm stem cells) they form a more organized structure that develops primitive endoderm and a more robust primitive streak. This structure is called an ETX embryoid and does not require Matrigel in the culture medium as the extracellular matrix components are thought to be supplied by the primitive endoderm.

the embryo, but lacks much in between. These experiments also revealed that the expression of *TbxT* and the formation of the rudimentary primitive streak require BMP-4 supplied by the TS cells, as this can be provided by beads coated with BMP-4 instead of TS cells.

In a more complete model of the mouse post-implantation epiblast, primitive endoderm cells are added to the ET embryos, either by supplying XEN cells to the ES- and TS-cell combination (see Section 7.1), or by inducing primitive endoderm in the ES cell population through ectopic expression of transcription factors involved in the specification of this cell type. Under these conditions, an epiblast-like structure appears that has a closer resemblance to the normal epiblast and a more robust primitive-streak-like structure. These structures do not require Matrigel, very likely because the primitive endoderm cells provide the necessary 'organizing' components, mostly the extracellular matrix associated with a basal membrane. These structures are called **ETX embryoids**, and display a well-organized epiblast with an AVE-like structure at one side of the primitive endoderm (see Fig. 3.58) and the initiation of a primitive streak at the opposite side (see Fig. 7.10, bottom row). Nevertheless, they do not develop much further, at least for now.

ETX embryoids have not yet been generated from human ES cells.

7.6 Pluripotent stem cells can model the process of gastrulation

Despite the similarities between embryos and ES cell models of pre-gastrulation development, and the observation that some of the embryo models discussed in the previous section initiate gastrulation, they do not complete it. In the past few years, however, structures that show some post-gastrulation features have been produced at high frequency from both human and mouse stem cells *in vitro*.

If small and defined numbers of mouse ES cells are aggregated in suspension in controlled chemical conditions, and are exposed to a 24-hour pulse of Wnt signaling on

Fig. 7.11 Mouse ES cells cultured under particular conditions can form gastruloids, structures that resemble the embryo after gastrulation. When defined numbers of mouse ES cells are aggregated in suspension and, after leaving the pluripotent state, are exposed to a pulse of the Wnt agonist CHIR99021, they organize themselves into elongated structures that exhibit an organization similar to that of the E8.5 embryo with the three orthogonal axes. h, hours; A, anterior; D, dorsal; V, ventral; P, posterior.

the third day of culture, they will form structures with many features of an embryo after gastrulation after a further 3 or 4 days in culture (Fig. 7.11). These structures are called **gastruloids**, because they look as if they have gone through the process of gastrulation. The initial number of cells is crucial in these experiments, as these structures only form if the initial aggregate contains around 300 ES cells. After 5 days in culture, gastruloids lack extra-embryonic tissues, but gene-expression analysis shows an organization of derivatives of the three germ layers with regard to the three main body axes, which suggests that the epiblast has an intrinsic ability to break symmetry and organize itself. The antero-posterior axis is outlined by the localization of *TbxT* expression and a node-like structure towards one pole, with a nested and polarized expression of Hox genes (Fig. 7.12). The dorso-ventral axis is revealed by the organization of spinal cord at the midline with mesodermal cells on either side, as we will see later in this chapter. Gastruloids also exhibit left–right asymmetries in gene expression, suggesting that they have a functional node. They also have a bilateral organization of the main mesodermal derivatives: lateral, intermediate, and paraxial (Box 7C, Fig. 1). Altogether, these features suggest that the aggregates undergo a gastrulation-like process and that the pulse of Wnt signaling mimics the role of Wnt signaling in gastrulation (see Chapter 4).

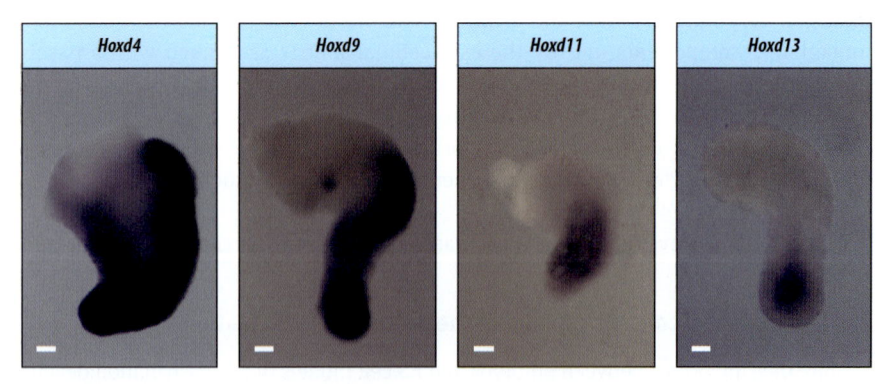

Fig. 7.12 Hox gene expression in gastruloids. Expression of four members of the Hoxd cluster in 120-hour-old mouse gastruloids. The pattern is nested, with the position of the genes along the antero-posterior body axis matching their position in the chromosome (see Box 4G).

Photographs courtesy of Leonardo Beccari and Denis Duboule. From Beccari, L., et al.: **Multi-axial self-organization properties of mouse embryonic stem cells into gastruloids.** Nature *2018, **562**: 272-276.*

BOX 7C Somitogenesis in Matrigelled gastruloids

When mouse gastruloids are grown in suspension, after 5 days in culture they exhibit the hallmarks of a mammalian body plan in terms of organization of gene expression with regard to three orthogonal axes (antero-posterior, dorso-ventral, and left-right) in a manner that resembles the embryo at E8.5, although they lack a brain. This organization can be clearly seen in the extending tail, where there is a progression of gene expression in the mesoderm that mimics the organization of somitogenesis (Fig. 1, left panel). *TbxT* at the posterior indicates undifferentiated mesoderm, followed by more-anterior domains of *Tbx6* and *Mesogenin* expression, which indicate paraxial mesoderm, and anterior to these, a domain in which genes associated with somites, such as *Tcf15*, *Uncx4*, and *Pax3*, are expressed. As in somitogenesis in the embryo, the boundary between the pre-somitic and somitic mesoderm is identified by a narrow domain of expression of *Mesp1* and *Ripply2*.

Furthermore, it is possible to observe the oscillations in gene expression characteristic of somitogenesis (see Fig. 1, left panel). The graph shows oscillation of *Lfng* expression. However, although these domains of gene expression match in organization and extent those observed in the embryo, these gastruloids lack somites, and the different territories cannot be distinguished from each other in cell shape and organization. This is surprising, as one might think that gene expression means cell identity, and that this is translated directly into morphogenesis, but this is not the case.

However, if gastruloids are embedded in Matrigel after just 3 days in culture, in the next few days they exhibit the periodic appearance of cyst-like structures that resemble somites in the anterior-most region, overlapping the domain of somitic gene expression (see Fig. 1, right panel). Furthermore, these structures become periodically segmented, as somites do, and later go on to differentiate into dermomyotome and sclerotome. In some instances, a well-organized spinal cord structure also appears and this improves the quality of the emerging somites.

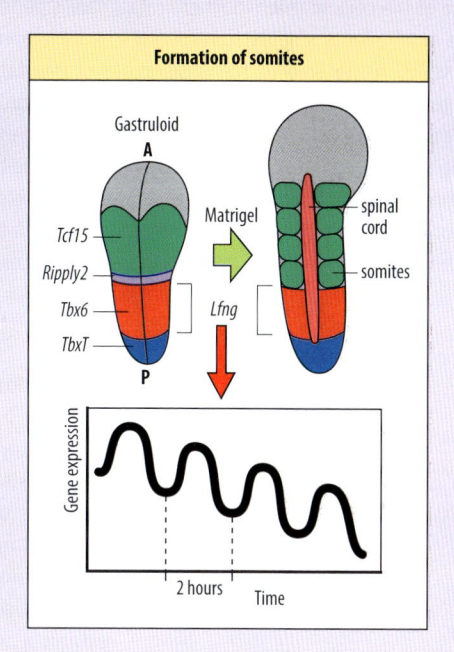

Fig. 1

Figure 2a shows a mouse gastruloid cultured without Matrigel. Pre-somitic mesoderm is labeled in red and neural tissue in blue. Figure 2b shows a mouse gastruloid cultured in Matrigel. Somites are now visible in the mesoderm (red) and an organized spinal cord (blue) is visible. Figure 2c shows a human gastruloid cultured in Matrigel. Well-formed somites (red) are evident along its length.

The differences in gene expression between gastruloids in suspension and those in Matrigel reveal the emergence of gene-expression patterns in the latter associated with extracellular matrix deposition and sensing; for example, the genes for integrin, laminin, and related receptors display elevated expression after exposure to Matrigel. A similar observation has been made

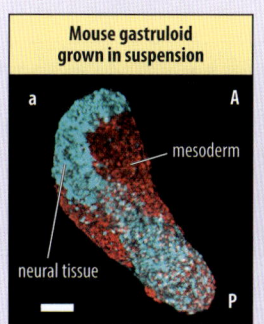

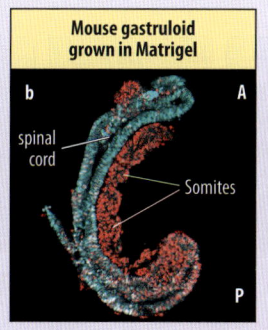

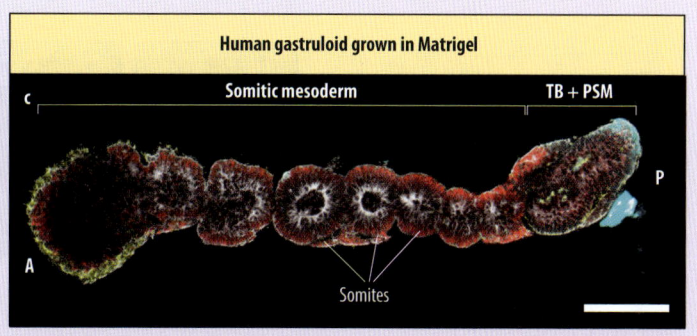

Fig. 2 A, anterior; P, posterior; TB, tail bud; PSM, pre-somitic mesoderm.
(a, b) From Veenvliet, J.V., et al.: **Mouse embryonic stem cells self-organize into trunk-like structures with neural tube and somites**. Science 2020, **370**: eaba4937. (c) From Yamanaka, Y., et al.: **Reconstituting human somitogenesis in vitro**. Nature 2023, **614**: 509-520.

with human gastruloids, suggesting that this disconnect and the response of the cells to mechanochemical signals is general.

These experiments suggest that during the development of some tissues in the embryo the assignment of cell fates and morphogenesis are two separate processes that are coordinated, most probably by interactions between cells and the extracellular matrix that constrains them. This conclusion would be difficult to come to through a purely genetic approach where somites do not form because one gene, and all the ones it regulates, fail. This is what happens in *Tcf15* mutants in which somites do not form. In the case of gastruloids, all genes are present and the response to Matrigel can, therefore, be studied as a response of the whole genome. This opens up new experimental avenues to explore not only which components in Matrigel trigger somite formation but how cell fate and morphogenesis are coordinated. It also introduces the possibility of exploring the nature of the signals that mediate somite formation. For example, gastruloids could be embedded in extracellular matrices of defined composition to study the contribution of particular components to the formation of somites.

Gastruloids have two features that distinguish them from embryos: they lack a brain, which is likely to be due to the early exposure of the aggregates to Wnt signaling, which suppresses brain development (see Chapter 4), and they lack much of the morphogenesis characteristic of an embryo. As the spatial organization of gene expression coincides with that of an embryo, this observation *suggests* a disconnect between cell fates and morphogenesis in the absence of the particular conditions within an embryo; this had not been revealed by the genetic analysis of development.

This disconnect is clearly seen in the case of somitogenesis, where gastruloids exhibit a well-organized and proportioned gene-expression pattern characteristic of paraxial mesoderm but do not form somites. However, when gastruloids are embedded in Matrigel after 3 days in culture, somites do emerge in the region of somitic gene expression, indicating the importance of the extracellular matrix and mechanical constraints in morphogenesis in the embryo (Box 7C). In some cases a single cardiac-like structure emerges at the anterior end.

The remarkable organization of gastruloids in the absence of extra-embryonic tissues, in particular the appearance of the positional coordinate system, raises the question of the role of those tissues in early embryogenesis. It is very likely that what they do is to bias the intrinsic self-organization of the embryonic cells, ensuring that gastrulation starts always in the same place near the trophectoderm.

This is a rapidly developing field fueled by the surprising self-organizing properties of ES cells. Over the past 2 years, structures have been derived from mouse ES cells that resemble post-gastrulation embryos. These structures have been obtained by making some alterations to the protocols described above and have been called **stem-cell embryo models (SCEMs)**, though the term 'embryo' should be interpreted

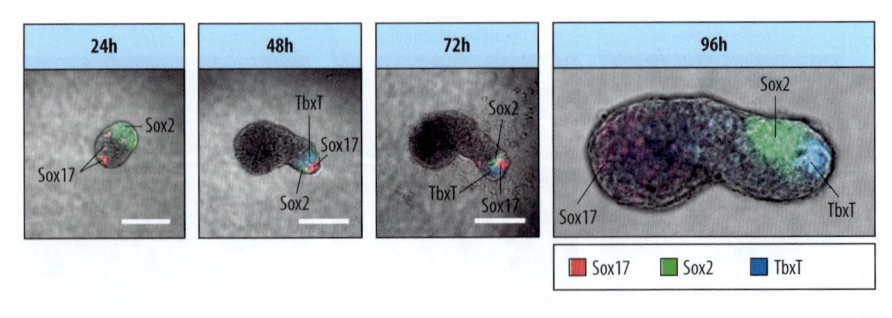

Fig. 7.13 Stages in the formation of a human gastruloid. Formation of a human gastruloid from human ES cells showing the organization of gene expression corresponding to three germ layers (*Sox2*, ectoderm; *Sox17*, endoderm; and *TbxT*, mesoderm) concomitant with the antero-posterior elongation of the aggregate. h, hours.

Photographs courtesy of Naomi Moris.

with care as the structures created in this manner exhibit significant differences with 'natural' embryos. The way these structures have been obtained is by bringing together wild-type mouse ES cells with two mouse ES cell lines carrying inducible transgenes. One cell line carries a transgene for the transcription factor GATA4, to trigger the development of primitive endoderm. The other carries a transgene for the transcription factor Cdx2, which leads to the emergence of trophectoderm (see Section 3.23). The three cell lines are cultured together under carefully defined conditions and the expression of the transgenes induced. On induction of the transgenes the three cell lines assemble into an epiblast that undergoes gastrulation and, when transferred into a device engineered to sustain the development of early mouse embryos outside the uterus, a small number of these continue to develop, much as a mouse embryo does, until it forms a structure resembling an E8.5 embryo (Fig. 7.14). The frequency of success is very low, less than 1% of the starting cultures, but it does make the point that mammalian embryos display both self-assembly and self-organization (see Box 7B).

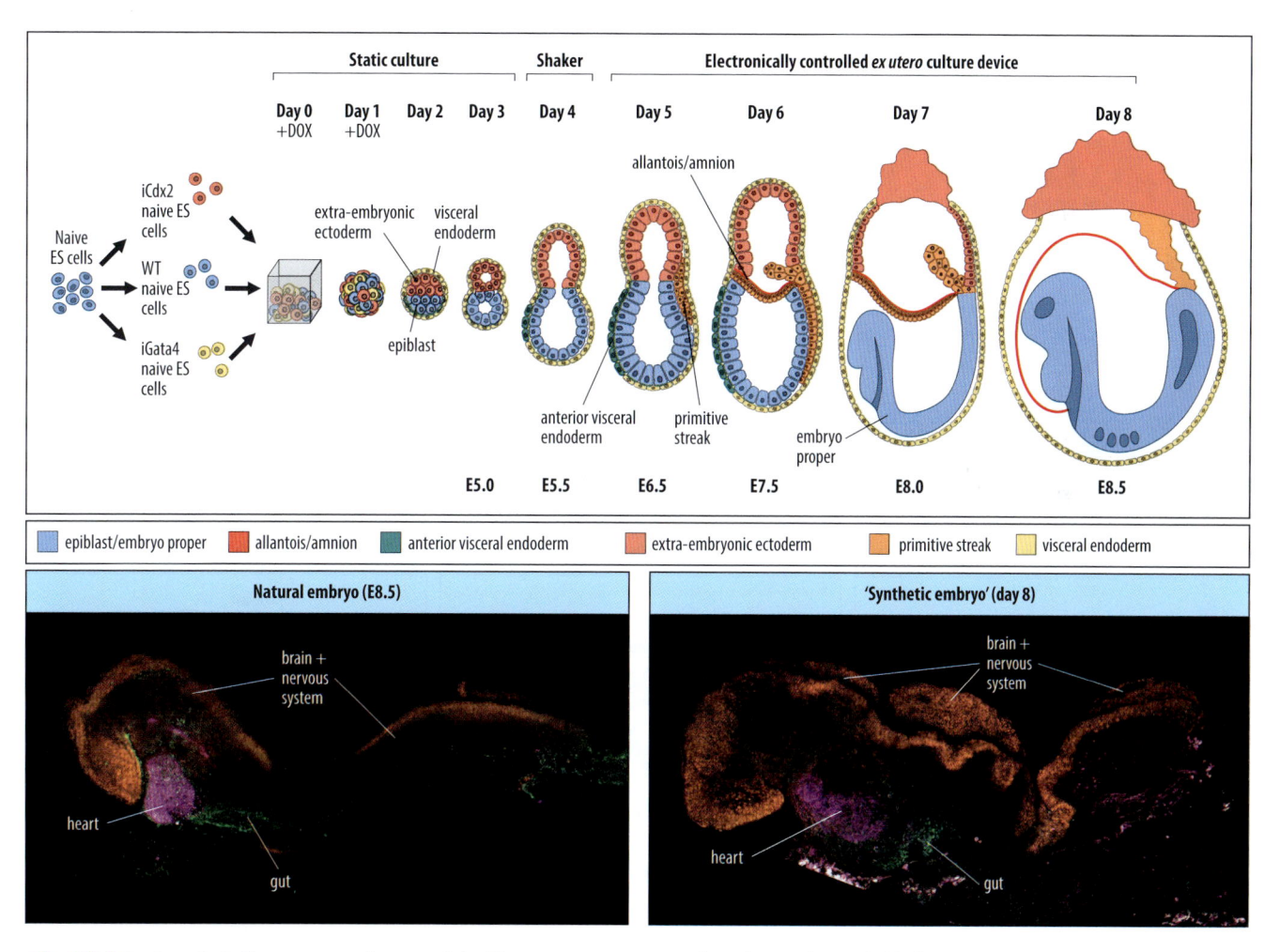

Fig. 7.14 Post-gastrulation mouse embryo models. Three populations of naive mouse ES cells are prepared: one in which the *Cdx2* gene is expressed to promote the formation of trophectoderm, a second in which the *Gata4* gene is expressed to promote primitive endoderm, and a third of standard wild-type (WT) ES cells. After a brief period of induction of gene expression, varied proportions of the different ES cells are sown at random in special wells in culture media that favors differentiation. Grown under these conditions, in a small number of cases (around 1% of starting cultures) they form structures resembling a full outline of the E8.5 mouse embryo body plan. The photographs show a comparison of a natural and an ES-cell-derived embryo, the latter also called a 'stem-cell embryo'. Orange labels the brain and nervous system, purple the heart, and green the gut.

Photographs courtesy of Jacob Hanna. From Tarazi, S., et al.: **Post-gastrulation synthetic embryos generated ex utero from mouse naive ESCs.** *Cell 2022, 185: 3290-3306.*

386 Chapter 7 Embryonic stem cells, embryo models, and regenerative medicine

In the case of human ES cells, when cultured embedded in certain synthetic matrices that mimic the extracellular matrix, they form a structure that resembles the amniotic sac and even show signs of the initiation of gastrulation and the emergence of primordial germ cells. These experiments have shown that even a single human ES cell can produce this structure, but that it forms more readily if it is started from small numbers of cells confined within a microfluidic device with BMP (an initiator of differentiation in human ES cells) being pumped in. In the embryo—and in both mouse and human ES cells when they start differentiating—BMP promotes posterior fates in the primitive streak with a very special role in the specification of primordial germ cells, as we shall see in Chapter 8.

In a remarkable advance, in 2023 researchers obtained a structure resembling the ensemble of extra-embryonic and embryonic tissues in humans at day 14, the onset of gastrulation. The structure is self-assembled from naive human ES cells, which in contrast to those of mice are totipotent, without any genetic manipulation (Fig. 7.15). The structures reproduced the development and organization of the ensemble of embryonic epiblast with the trophoblast and the yolk sac (from the

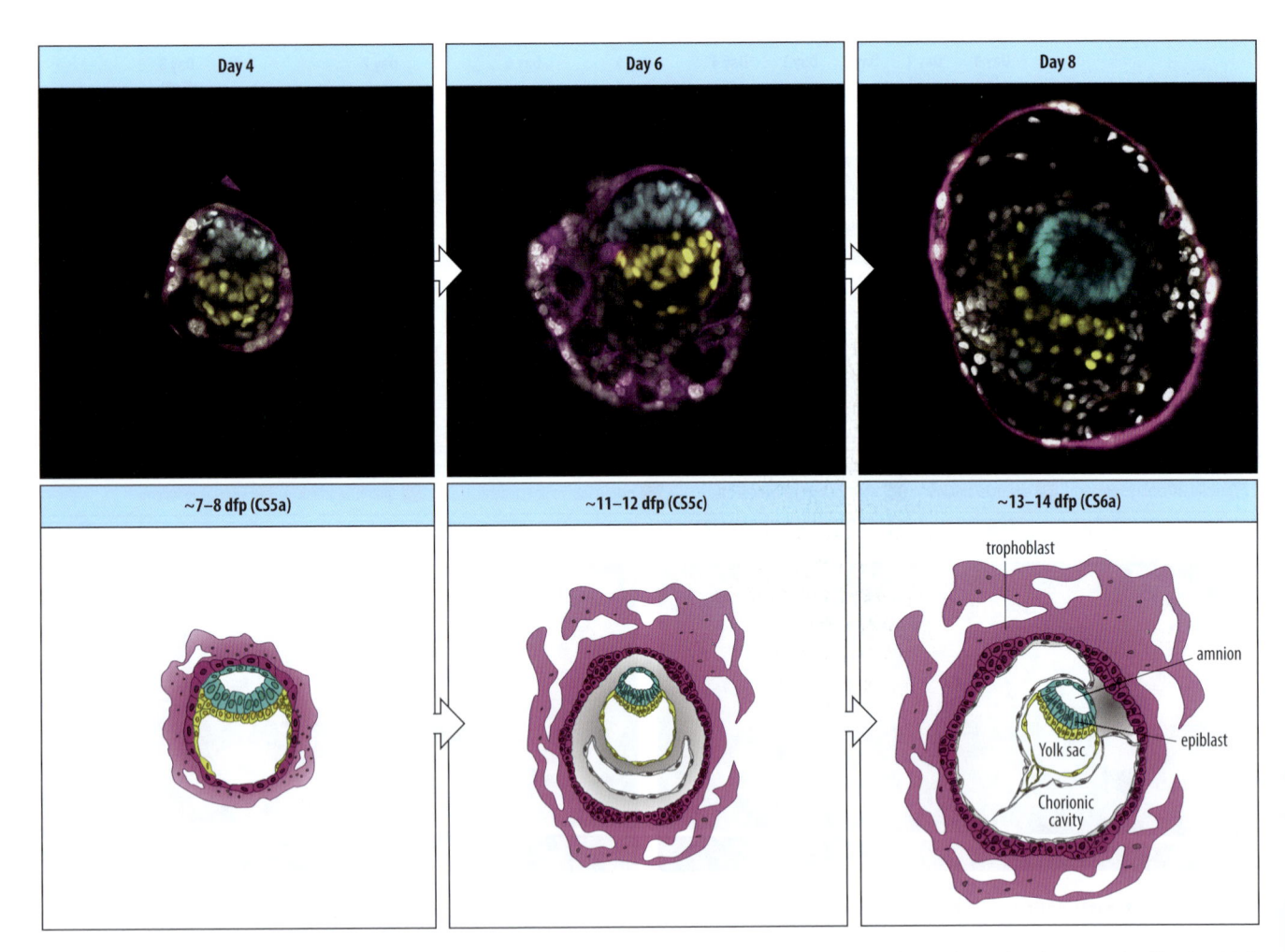

Fig. 7.15 Human stem-cell embryo model (SCEM) corresponding to a day-14 human embryo. Top row: from left to right, immunofluorescence images of SCEMs from days 4, 6, and 8 of culture in the *in vitro* protocol, showing OCT4 (cyan), SOX17 (yellow), CK7 (magenta), and nuclei (DAPI, white) labeling different embryonic structures (marked with arrows). Bottom row: labeled drawings of the structures above. The headings refer to the days post-fertilization (dpf) and Carnegie stages (CS) of a natural embryo corresponding to these structures. Scale bars, 50 μm.

Photograph courtesy of Jacob Hanna. From Oldak, B., et al.: **Complete human day 14 post-implantation embryo models from naïve ES cells.** Nature *2023,* **622***: 562-573.*

trophectoderm and the primitive endoderm, respectively), thus providing a window into the stages of implantation, something impossible to study *in vivo*. Like the mouse ES-cell-derived structures, these human SCEMs also occur at a low frequency, but the fact that they occur at all will let us explore the mechanisms of cell interactions around the gastrulation stages in humans. However, as these SCEMs improve, it is important that we consider the ethical issues associated with them and their relationship with embryos, so that we can take advantage of the benefits that they bring to our understanding of early human development. This is an area being developed and we should proceed with care for now; although the structures display remarkable similarities to early embryos, they have yet to be shown to develop their potential. For the moment, the 14-day rule (see Section 3.30) would apply to SCEMs as it applies to natural cultured embryos.

> **SUMMARY**
>
> The realization that *in vitro* ES cells are capable of differentiating along the same lines as they would in the embryo has opened up their use as a new model system for the study of cell-fate choices and morphogenesis. It has revealed that cells can follow precise pathways of differentiation without assembling into organs and tissues. When, however, ES cells are aggregated in three dimensions under controlled culture conditions, they assemble structures that mimic many aspects of early mammalian development: blastocyst and epiblast formation in blastoids and ETX embryoids as well as the emergence of the body plan through gastrulation in gastruloids. These models of early development have revealed surprising self-organizing abilities of ES cells and opened up new experimental possibilities. In contrast to genetic analysis, which reveals what is necessary for the emergence and development of an early embryo, these models mimic engineering approaches and reveal the minimal number and arrangement of elements necessary to organize an embryo. Much more clearly than differentiation of ES cells in adherent cultures, these three-dimensional models have revealed the disconnect between cell fates and morphogenesis. This is most obvious in the process of somitogenesis. In suspension culture, gastruloids exhibit the temporal pattern of gene expression associated with somitogenesis, with a domain of somite gene expression—but without somites. However, contact with Matrigel—a complex matrix that provides mechanical and chemical signals—leads to the appearance of somites.
>
> These studies suggest that during development there are two parallel, and somewhat independent, processes going on. One comprises the programs of gene expression, which are largely driven by gene regulatory networks. The other comprises tissue-specific morphogenetic programs that require mechanical and chemical interactions between cells and between cells and their environment. Correct development requires an integration of both.

Medical applications of stem cells

Regenerative medicine is the branch of medicine that aims to repair or replace damaged cells and tissues to restore normal function. One possible strategy employs **cell-replacement therapies**, the restoration of tissue function by the introduction of new healthy cells. Because stem cells can proliferate and differentiate into a wide range of cell types, they have tremendous potential for providing the appropriate cells for making repairs. This could involve transplanting the replacement cells, either alone or seeded on a scaffold (a three-dimensional support made out of an inert material). Adult stem cells could also be used for repairs in specific tissues. An alternative

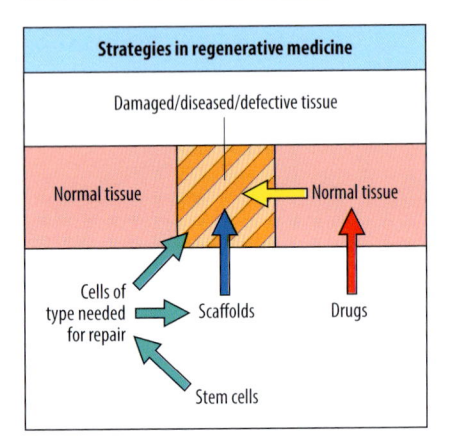

Fig. 7.16 Strategies in regenerative medicine. Damaged, diseased, or defective tissue could be repaired by cell-replacement therapies in which stem cells are the source for producing cells for transplantation, either alone or on scaffolds (turquoise and blue arrows). We can already regenerate a complete blood or immune system by transplanting hematopoietic stem cells (see Chapter 6). Embryonic stem cells, adult stem cells, or iPS cells could be used for this purpose. An alternative strategy is to develop drugs that could be applied to nearby undamaged tissue (red arrow) to stimulate endogenous adult stem cells and/or induce transdifferentiation in order to generate cells to replace the damaged tissue (yellow arrow).

strategy is to apply drugs to stimulate endogenous adult stem cells and/or to induce transdifferentiation of nearby undamaged tissue to generate replacement cells (Fig. 7.16). These cell-replacement therapies might eventually offer an alternative to conventional organ transplantation from a donor, with its attendant problems of rejection and shortage of organs, and might also be able to restore function to tissues such as the brain.

7.7 Stem cells could be a key to regenerative medicine

ES cells and iPS cells have the advantage over adult stem cells as a source of cells for cell-replacement therapies in that they can differentiate into a wide range of different cell types and could, in theory, be used to repair any tissue. But because of the proliferative capacity and pluripotency of these stem cells, the risk of embryonal tumors—tumors derived from embryonic cells—is high. For example, cultured mouse ES cells introduced into another mouse embryo will develop normally, but when mouse or human ES cells (or mouse EpiSCs) are introduced under the skin of an adult mouse they can give rise to a teratocarcinoma (see Fig. 7.6). Stringent selection procedures will be needed to ensure that no undifferentiated cells are present in populations of differentiated cells derived from pluripotent stem cells before transplantation into a patient.

ES cells, adult stem cells, and iPS cells all have been studied with cell-replacement therapy in mind. Some cell-therapy treatments are already being tested in people for their safety and efficacy. Clinical trials of a stem-cell-based therapy for certain types of age-related macular degeneration have been carried out.

Age-related macular degeneration is a degenerative disease of the eye in which there is irreversible cell loss, initially of retinal pigment epithelial cells, that eventually leads to blindness. The main purpose of the first trials was to assess the safety of the procedure. Retinal pigment epithelial cells derived from human ES cells or iPS cells were injected into the eyes of people with various types of macular degeneration. Transplantation of the cells did not cause any ill effects and a slight improvement in vision was detected in some participants. More recently, small patches of retinal pigment epithelium grown *in vitro* from retinal pigment cells derived from human ES cells were inserted microsurgically into the eyes of two people with severe 'wet' age-related macular degeneration. The patches were well tolerated and visual acuity improved. These early results are encouraging and the research team is now planning larger-scale trials, which are needed to confirm the effectiveness and safety of the treatment before it can be used routinely.

Similar approaches are being used to target other tissues but much progress will have to be made. A major challenge is the reliable generation of adult differentiated cell phenotypes from iPS and ES cells. Indeed, fetal or perinatal phenotypes are often the best that can be obtained from these cells, underscoring the problems involved in regulating the chronological clock in culture. This issue also affects the modeling of age-related diseases in culture, where it is desirable to accelerate the clock to obtain phenotypes that only appear after several decades in humans.

It is rightly claimed that there are ethical issues associated with the use of ES cells. To make human ES cells, the blastocyst must be destroyed, and there are those who believe that this is a destruction of a human life. There is good evidence that the blastocyst does not necessarily represent an individual at this very early stage, because it is still capable of giving rise to twins at a later stage (see Box 3J). And in practice, many early embryos are lost during assisted reproduction involving *in vitro* fertilization (IVF), a widely accepted medical intervention. Acceptance of IVF and rejection of the use of ES cells could be seen as a contradiction. Alternative approaches, such as transdifferentiation, and especially the use of iPS cells, could produce cells of a person's own tissue type but avoid the ethical issues associated with the use of human ES cells.

Adult tissues are another source of stem cells for use in cell-replacement therapy, but in many tissues these cells are rare and relatively inaccessible. For many years, however, adult hematopoietic stem-cell transplants in the form of bone marrow transplants from donors have been successfully used as therapy for certain immunological diseases and cancers, despite associated problems such as graft-versus-host reactions. Epidermis grown *in vitro* from patients' own epidermal stem cells has been used as transplants to repair burns. Epidermal stem cells from a child with severe junctional epidermolysis bullosa and genetically corrected *in vitro*, have even been cultured to provide enough epidermis to successfully replace the child's damaged skin (see Box 6E).

Over the past few years there have great advances in growing adult stem cells from other organs such as the small intestine. The first advances in growing such adult stem cells came from experiments in which single stem cells from the crypts of the small intestine from adult mice (see Section 6.13) were cultured to test whether they were multipotent. The techniques pioneered there led to the development of potential new routes for generating cells for regenerative medicine in which both adult stem cells and ES cells are grown under conditions in which they reorganize themselves into tissues and structures called **organoids**, as discussed later in the chapter.

At present, the most immediate clinical uses for stem cells are for investigating disease mechanisms and for screening the effectiveness of drug treatments. Although ES cells derived from mouse models of human diseases can be differentiated into the cell types affected in the disease and the cells studied and experimented on, it is stem cells derived from people with the disease that are most useful, because animal models do not always mimic the human disease completely. Both iPS cells and organoids derived from patients are being used for this purpose. For example, iPS cells have been derived from skin fibroblasts from a child with genetically determined spinal muscular atrophy, and differentiated into motor neurons in culture. Neuron production in the child's iPS cell cultures differed from neuron production in cultures of iPS cells made from the child's healthy mother, with the number and size of motor neurons produced decreasing over time in cultures derived from the child. iPS cells have also been made successfully from people with the neurodegenerative disease Parkinson's disease. Such cell cultures could be used to test the effects of drugs, which could lead to new treatments. More generally, human iPS cells could be useful for screening compounds for toxicity and for teratogenicity in drug development.

7.8 Various approaches could be used to generate differentiated cells for cell-replacement therapies

The challenge facing cell-replacement therapy is to devise differentiation protocols that will produce the particular cell types needed—whether from ES cells, adult stem cells, iPS cells, or by transdifferentiation of pre-existing cell types—in sufficient numbers to effect repairs and that are safe to use. The generation of insulin-producing pancreatic β cells to replace those destroyed in type 1 diabetes is a prime medical target, and we shall use it to illustrate the experimental approaches that are being developed for cell-replacement therapy generally.

In type 1 diabetes, the insulin-producing β cells of the endocrine pancreas (the pancreatic islets) are destroyed by an autoimmune reaction, resulting in non-production of insulin and the potentially lethal build-up of sugar in the blood. Unlike blood or muscle, the adult pancreas does not contain dedicated stem cells from which it can regenerate, and people with type 1 diabetes face a lifetime of dependence on insulin injections. Pancreas or islet cell transplants from cadaver donors have had some success, but tissue for transplantation is scarce, and treatment with powerful immunosuppressant drugs is necessary to prevent rejection. Transplantation of insulin-producing pancreatic cells derived from the patient's own cells just might eventually provide a long-term solution.

To achieve the goal of generating insulin-producing pancreatic cells for transplantation, protocols that direct the differentiation of pluripotent human stem cells towards making pancreatic β cells had to be devised. This proved to be very difficult and it took many years of painstaking work to devise a protocol that produced large numbers of insulin-producing glucose-responsive cells.

The pancreas develops from the endoderm, and using knowledge of the signals that induce endoderm and pancreas development in mouse embryos, the first successful protocol—a 9-day laboratory protocol involving stepwise treatments of human ES cells with different signal molecules—was drawn up. Human ES cells put through this protocol differentiate into endoderm, expressing the endoderm-specific transcription factor *FOXA2*, but only a few of the cells in these cultures (around 5%) express *PDX1*, a transcription factor expressed in pancreatic progenitor cells and essential for pancreas development. The next step, therefore, was to find a way of boosting the differentiation of pancreatic progenitor cells. A library of 5000 organic compounds was screened and one compound, indolactam V, was found to stimulate the differentiation of *PDX1*-positive cells. The percentage of *PDX1*-expressing cells increased to 45% when endoderm cultures derived from the human ES cells were treated with indolactam V together with FGF-10, one of the signals involved in embryonic pancreatic development. When the *PDX1*-expressing cells were transplanted under the kidney capsule in a mouse, a site that allows continued differentiation, some of them differentiated into insulin-secreting cells (Fig. 7.17). This pioneering work provided the foundation for the development of more complex culture protocols—including steps involving three-dimensional cultures—for the large-scale production of differentiated insulin-secreting, glucose-responsive cells that very closely resemble human β islet cells. Both ES and iPS cell lines have been used successfully to generate these cells, which when transplanted into experimental mouse models of diabetes can suppress the build-up of sugar in the blood. Human pancreatic progenitor cells derived from pluripotent stem cells are now in clinical trials for cell-replacement therapy.

The success in generating insulin-producing cells from human iPS cell lines opens up the possibility of using the fibroblasts from an individual patient to make iPS cells, which could then be differentiated into insulin-producing cells and used to treat patients. But it is not clear whether this could ever become a routine procedure, because the culture conditions would need to be optimized for the cells of each person. Another route that could, in principle, be used to generate patient-specific cells for cell-replacement therapy is via somatic nuclear transfer and the isolation of ES cells from the resultant blastocyst (see Fig. 7.17). But this strategy is even less likely to be used clinically, not only because of the technical challenges, but also because of the ethical issues it raises.

Induced transdifferentiation may be yet another way of generating insulin-producing cells for cell-replacement therapy (see Fig. 7.17). The advantage of this approach is again that the cells generated would be from the patient themself. The feasibility of this approach for generating insulin-producing endocrine cells *in vivo* from liver cells and from pancreatic exocrine cells has been demonstrated experimentally.

The liver and the pancreas both derive from endoderm, and arise from adjacent regions of endoderm in the embryo, but the essential pancreatic transcription factor Pdx1 is not expressed in the liver. However, when a gene encoding an activated form of Pdx1 is expressed transiently in the liver in transgenic *Xenopus* tadpoles under the control of a liver-specific promoter, part or all of the liver is converted to both exocrine and endocrine pancreatic tissue, while proteins characteristic of differentiated liver cells disappear from the cells in the converted regions. Transdifferentiation of liver to pancreatic islet cells has also been achieved in adult mice by infecting the liver with viral vectors expressing the transcription factor Neurogenin, which is another pancreatic transcription factor.

Pancreatic exocrine cells have also been induced to transdifferentiate into endocrine β cells in adult mice. The exocrine cells synthesize and secrete digestive enzymes and,

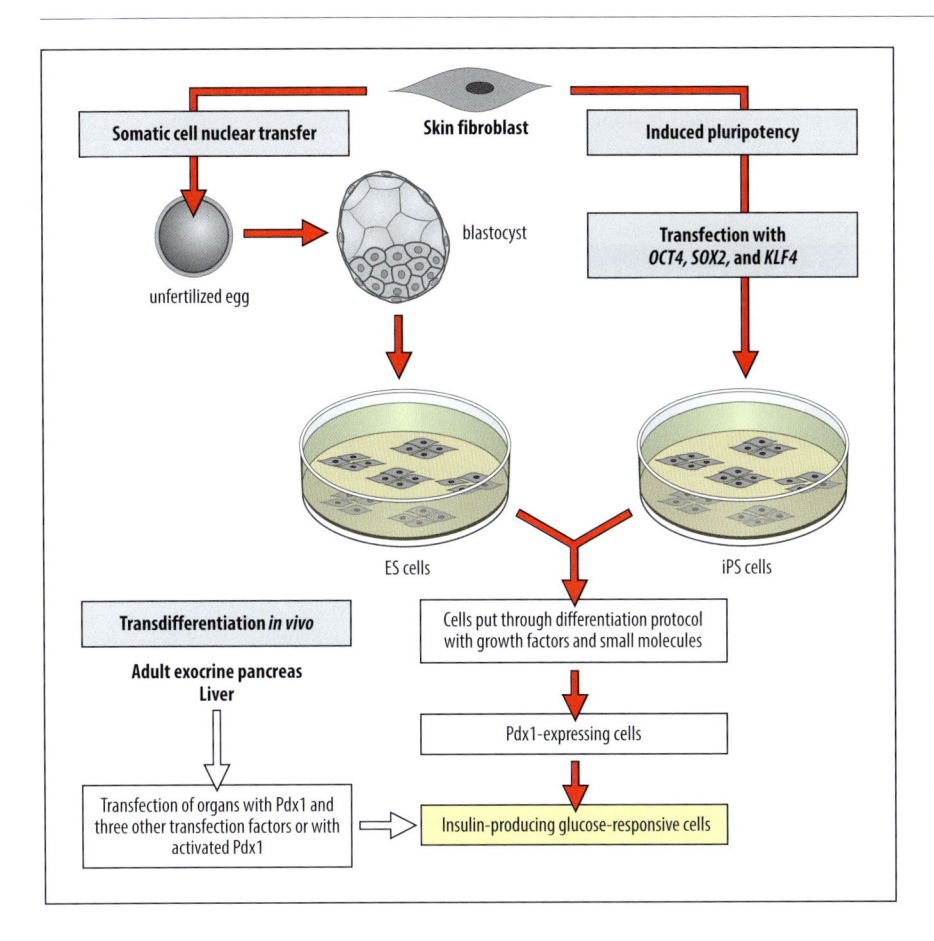

Fig. 7.17 Three experimental routes for making insulin-producing cells by nuclear reprogramming. Two routes start with a skin fibroblast. In somatic cell nuclear transfer, the nucleus of the fibroblast is introduced into an unfertilized egg and embryonic stem cells (ES cells) are isolated from the resulting blastocyst. Alternatively, cultured fibroblasts can be transfected with the genes for the pluripotency transcription factors Oct4, Sox2, and Klf4 to produce induced pluripotent stem (iPS) cells. The ES cells and iPS cells can then be differentiated into insulin-producing cells *in vitro*. A third possible route is the transdifferentiation of adult exocrine pancreatic cells or liver cells into insulin-producing cells *in vivo* by the targeted transfection of these organs with genes (including that for the transcription factor Pdx1) required for differentiation of endocrine pancreatic cells. Steps in red have been accomplished for human cells, although the success rate for deriving ES cells from human blastocysts made by somatic cell nuclear transfer is extremely low. This means that this route for making human insulin-producing cells is not feasible at present. Insulin-producing glucose-responsive cells have been made starting with human ES cells derived from a normal blastocyst and differentiated *in vitro*.

Adapted from Gurdon, J.B., Melton, D.A.: ***Nuclear reprogramming in cells.*** *Science 2008, **322**: 1811-1815.*

unlike the endocrine pancreas, can readily regenerate. A combination of three transcription factors required for endocrine pancreatic cell differentiation, including Pdx1, were delivered into the pancreas using genetically modified adenoviruses. Endocrine cells were induced and, although they were not found in islets, these induced cells closely resembled their normal counterparts and had the ability to secrete insulin. Transdifferentiation was efficient (around 20%) and relatively fast, with the first insulin-positive cells appearing on day 3 after treatment.

There are, however, many obstacles to overcome before any potential therapeutic benefits of transdifferentiation can be realized. So far attempts to transdifferentiate human exocrine pancreatic cells to endocrine pancreatic cells *in vitro* have not been very successful. In addition, there are safety issues about the use of viral vectors to introduce the genes designed to mediate transdifferentiation. Viral vectors can integrate randomly into the genome and might activate genes that cause cancer, and so alternative ways of inducing expression of the necessary transcription factors will need to be found before medical applications are possible.

7.9 ES cells can be used to generate organoids and tissues

When multipotent stem cells from particular organs are cultured in suspension, in the context of natural or synthetic extracellular matrix material, they can develop into three-dimensional structures called **organoids**, which are composed of a mixture of cell types and resemble smaller replicas of tissues or organs, although they are not vascularized or innervated. Organoids have now been derived from adult stem cells or progenitor cells from various mouse and human tissues, including intestine, liver, and lung. These include multiple cell types, and mimic, but not perfectly and on a smaller scale, organs or tissues in their cellular organization, although they are not vascularized or innervated. Over the past decade there have been great advances in this area and some important successes in terms of resemblance to the proper organ and reproducibility of the organoid-making technique. Intestinal organoids are the most successful and advanced organoids so far, and can recapitulate the organization of stem cells and their differentiation in the intestinal crypt (Box 7D). Human intestinal organoids can

EXPERIMENTAL BOX 7D Intestinal organoids can be produced from adult stem cells or ES cells

The formation of organoids from adult stem cells was first observed when individual intestinal stem cells from mice were embedded within the extracellular matrix material Matrigel and grown in defined culture conditions. After a few days, individual cells had generated cyst-like structures that resembled the intestine, with different cell types spatially organized as they are in a crypt. Figure 1 shows an intestinal organoid (scale bar = 50 µm) and Figure 2 shows the procedure for generating it from a stem cell.

Cultures derived from the combination of a stem cell and a Paneth cell, which in the intestinal environment cooperate to maintain homeostasis in the tissue, were even more efficient than single stem cells at making organoids. The intestinal organoids reproduce many of the structural features of the intestine and are studied as models of its behavior *in vivo*. They can also be propagated to produce large numbers of organoids, and these mouse intestinal organoids have been fragmented and successfully transplanted back into the intestines to produce functional intestinal tissue. Mouse intestinal organoids can also be produced from ES cells grown in three-dimensional culture in the presence of specific combinations of growth factors. These do not mature *in vitro* but when transplanted into a mouse they develop crypts.

Human intestinal organoids can be made from biopsies of intestinal tissue. This opens up the possibility of growing human organoids for regenerative medicine—if large-scale methods of production can be developed—and using them to repair damaged intestines. This is for the future, but intestinal organoids grown from the cells of a person with a particular disease can already be used to study that disease. Intestinal organoids have been made from people with cancer, and these responded to different treatments in the same way that the individual people did. Having established the principle, this strategy could improve cancer therapy in general, as treatments could be explored rapidly via the patient's organoids *in vitro* before applying the treatment to the patient.

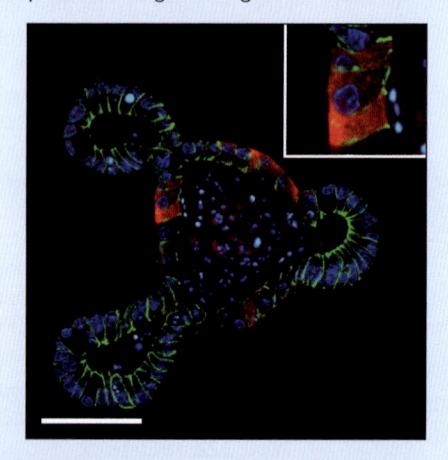

Fig. 1

Photograph from Gjorevski, N., et al.: **Designer matrices for intestinal stem cell and organoid culture.** Nature 2016, **539**: 560-564.

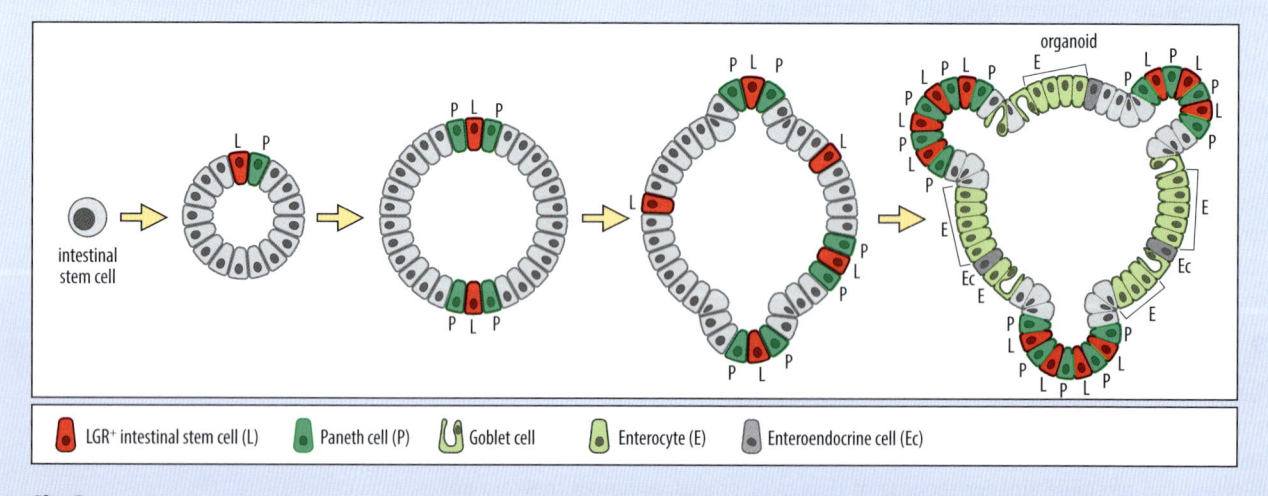

Fig. 2

Adapted from Sato, T., Clevers H.: **Growing self-organizing mini-guts from a single intestinal stem cell: mechanism and applications.** Science 2013, **340**: 1190-1194.

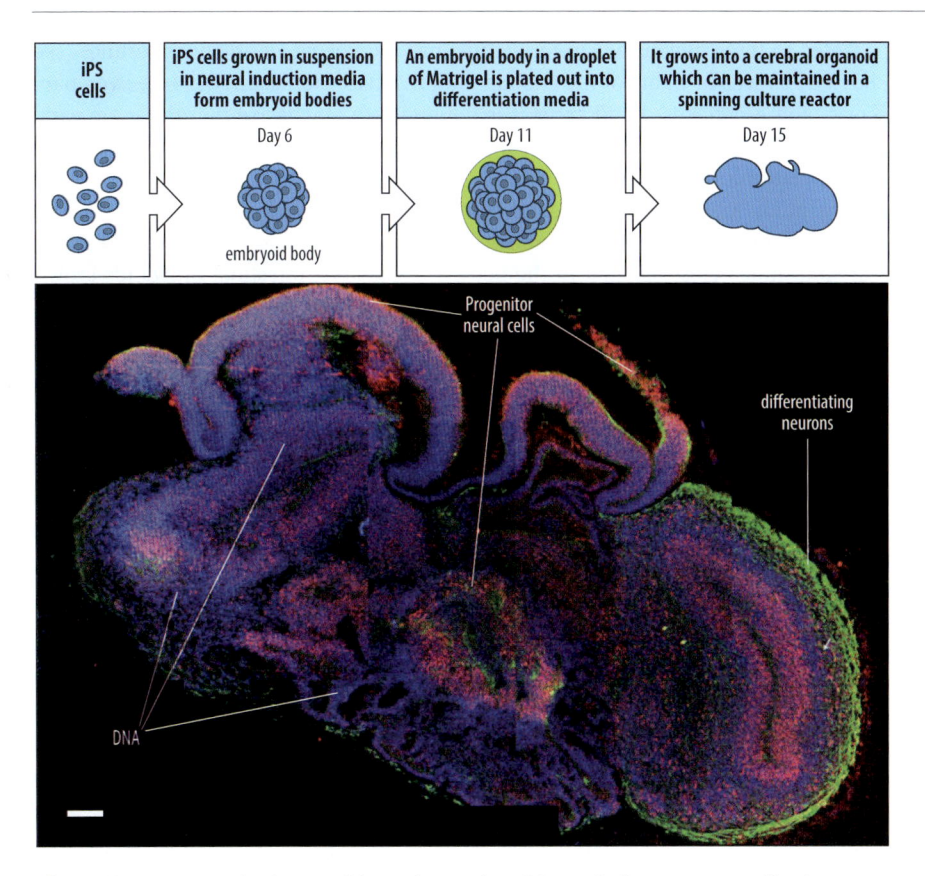

Fig. 7.18 Human cerebral organoids can be produced from pluripotent stem cells. Large numbers of iPS cells are aggregated under conditions that promote neural induction. After 10 days, they are placed in Matrigel, which promotes epithelialization and triggers the initiation of cortical development. Many of the organoids develop structures very similar to the cerebral cortex of the human brain. The photograph shows a cross-section through a cerebral organoid, which reveals morphological similarity between *in vitro* grown organoids and the human fetal brain. Progenitor neural cells are in red (labeled by anti-Sox2), differentiating neurons are in green (labeled by anti-TuJ1), DNA is in blue. The image is a composite of multiple individual confocal recordings. Cortical tissue is visible on the right side of the image, choroid plexus in the lower parts of the image. Scale bar = 100 μm.

From Lancaster, M.A., et al.: *Cerebral organoids model human brain development and microcephaly.* Nature *2013,* **501***: 373-379.*

also be derived from biopsy material, opening up the possibility of generating cells from a patient for their own cell-replacement therapy.

ES cells are an important starting point for generating a variety of organoids that cannot be derived from adult stem cells or progenitor cells, such as human cerebral tissue. For example, if the ES cells in culture are not exposed to any special signals, they develop into neural structures that are called 'minibrains' because of their resemblance to the cortex of the forebrain (Fig. 7.18). Cerebral organoids often have eye cups, the precursor structures to the eyes, revealing the developmental connection between the forebrain and the eye primordium (see Section 9.33) and recently, in a small variation of the protocol, rudimentary eyes and brain tissue have emerged in the same organoid. The eyes seem to come in pairs and to be light sensitive.

A more complicated differentiation process involving combinations of growth factors can be used to generate organoids of other tissues. For example, a culture protocol that will steer mouse ES cells into differentiating into intestinal cells can be adapted to three dimensions to form structures mimicking embryonic intestines (see Box 7D) as well as many of the endoderm derivatives—for example, lung, liver, pancreas, and esophagus. In all cases it is important to emphasize that these are early days—that these organoids are far from being functional organs but that they represent a platform for the future.

Organoids are particularly useful to model diseases and screen drugs because the behavior of cells in the organoids is different from that in two-dimensional culture and much closer to that in the body. For example, intestinal organoids grown from the cells of a person with a particular disease can be used to study that disease and also screen drug treatments. Screening drugs on organoids from individual patients will allow treatments to be tailored to that individual—a goal of personalized medicine.

The ability of stem cells to differentiate into organized structures resembling tissues and organs is often referred to as self-organization (see Box 7B). It clearly reflects important influences of the spatial environment on differentiation that we do not yet understand. How cells sense whether they are in two or three dimensions is an important question for the future.

The pluripotent potential of ES cells and our increased understanding of the process of differentiation in culture suggest the possibility of eventually using organoids to obtain replacements for all tissues, although there are many challenges, including developing methods for large-scale production. Particularly in the context of iPS cells, which would allow immunological matching with a patient, many possibilities open up. For example, organoids could be tissue-engineered to produce replacement body parts. A recent study in mice describes how artificial bile ducts were made by growing the epithelial cells from healthy ducts as organoids on tubular scaffolds, which were then transplanted into mice to replace damaged bile ducts.

7.10 Stem cells might be used to grow human organs in other animals for transplantation

Because of the shortage of human organs for transplantation, there is interest in the possibility of using ES cells to grow human organs in other animals. Although any clinical applications are very far off and have attendant ethical issues, the approach has been illustrated by growing a rat pancreas successfully in a mouse that lacked a vital gene for pancreatic development. The rat organ was produced by injecting rat iPS cells into mutant blastocysts that lacked *Pdx1*, so that no mouse cells could give rise to the pancreas in the chimeric embryo. Homozygous *Pdx1*-mutant mice would normally die soon after birth because of the lack of a pancreas. The chimeric blastocysts were replaced into surrogate female mice and the pancreas in the resulting chimeric mice developed entirely from rat cells. In the few chimeric mice that survived to adulthood, the pancreas was apparently performing normally and the response to glucose was normal (Fig. 7.19).

Similar strategies could be used with animals closer to humans as hosts—for example, the pig. The pig is being studied as a possible source of organs for transplantation to humans, because of its similar size, physiology, and ability to be genetically

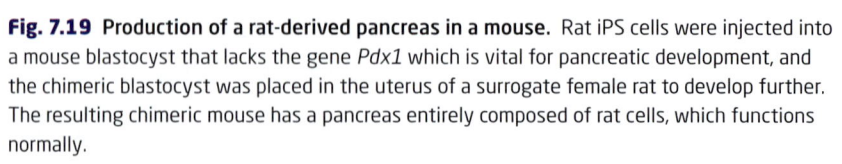

Fig. 7.19 Production of a rat-derived pancreas in a mouse. Rat iPS cells were injected into a mouse blastocyst that lacks the gene *Pdx1* which is vital for pancreatic development, and the chimeric blastocyst was placed in the uterus of a surrogate female rat to develop further. The resulting chimeric mouse has a pancreas entirely composed of rat cells, which functions normally.

Based on Kobayashi, T., et al.: **Generation of rat pancreas in mouse by interspecific blastocyst injection of pluripotent stem cells.** *Cell 2010,* **142***: 787-799.*

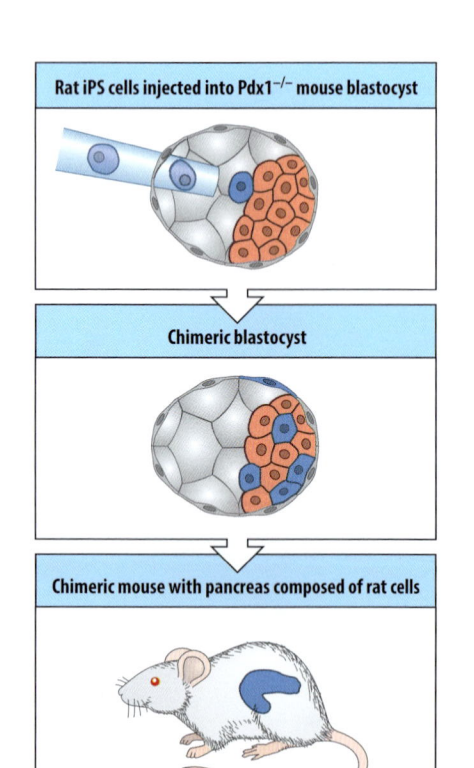

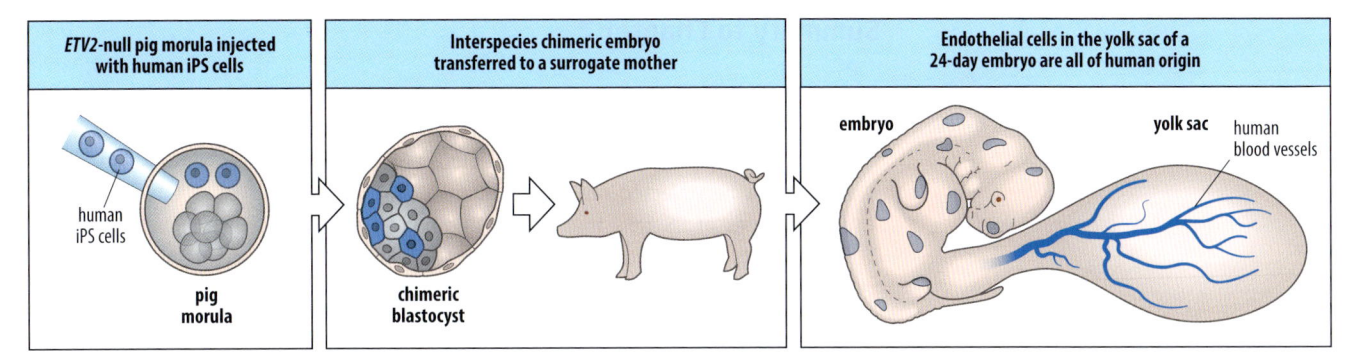

Fig. 7.20 Human iPS cells can complement a pig embryo deficient in the endothelial cell lineage. Pig oocytes mutant for the gene *ETV2*, which is essential for blood vessel development, were initially generated by somatic cell nuclear transfer from ETV-mutant fibroblasts (not shown). Parthenogenetic morulas grown from these oocytes *in vitro* were injected with small numbers of GFP-labeled human iPS cells and the resulting embryos transferred to surrogate mothers to develop further. Analysis of embryos collected at day 18 showed that endothelial markers (von Willebrand factor and the protein TIE) co-localized with the human cells, showing that the human iPS cells had integrated into the embryo and had differentiated into endothelial cells.

modified to avoid some of the problems of incompatibility, and so might also be a good host for growing human tissues. One can imagine a similar experiment as that with mice and rats performed with pigs as hosts and human ES cells as donors, which would give rise to a human pancreas in a pig. That this might be possible has been demonstrated in a proof-of-concept experiment in which an early pig embryo mutant for the gene *Etv2*, which would die due to a lack of the cell lineages that form blood vessel endothelium, was complemented with human iPS cells. *Etv2*-null pig morulas were made by CRISPR mutation of pig fibroblasts followed by somatic cell nuclear transfer into oocytes (see Section 6.17). The morulas were injected with human iPS cells and the chimeric embryos placed in a female pig uterus to develop further. Embryos were collected at day 18, by which time the human iPS cells had formed all the endothelial cells of the chimeric embryo. The experiments were performed starting with parthenogenetically activated pig oocytes, which produce embryos that cannot complete development. In this manner ethical risks associated with the experiment are eliminated (Fig. 7.20).

SUMMARY

Both multipotent and pluripotent stem cells are being investigated with a view to being used in cell-replacement therapy and other clinical uses, especially human iPS cells which avoid many of the ethical concerns around the use of human ES cells and allow a personalized approach to drug testing, diagnosis, and study of disease. Transdifferentiation of pre-existing cell types into the desired type of cell is also being studied as an alternative approach to stem-cell-based therapies. Both adult stem cells and pluripotent human iPS cells and ES cells have been used to produce organoids, miniature *in vitro* mimics of human tissues and organs, which are used to study aspects of development not easily accessible in human embryos, and which also have clinical potential as a source of cells for repair and to model diseases and screen drugs. A more speculative use of stem cells could be to grow human organs in other animals for transplantation.

Summary to Chapter 7

- Pluripotent embryonic stem cells (ES cells) can be isolated from the inner cell mass of the mammalian blastocyst and grown in culture. They can be induced to differentiate into a great range of cell types in culture, and mouse ES cells can produce all the cell types in the animal when injected into a mouse morula.
- Various types of ES cells from mouse and human have been isolated, with different requirements for isolation and culture and with different developmental potentials. So-called naive human ES cells are totipotent—that is, they can produce cells of the extra-embryonic lineage as well as all the different cell types in the embryo.
- Adult differentiated somatic cells can also be induced to develop into pluripotent cells, called induced pluripotent stem cells (iPS cells), by the expression of four transcription factors associated with pluripotency in ES cells.
- Pluripotent stem cells grown in two-dimensional culture can differentiate into many different cell types but cannot develop structures typical of embryos. When grown on circularly micropatterned plates, however, they will develop concentric rings of cells with gene expression typical of the three germ layers.
- Three-dimensional culture systems have been developed that enable models of early embryonic development, such as blastoids (which mimic blastocysts), to be grown from mixtures of trophoblast (trophectoderm) stem cells and ES cells. These allow researchers to study the role of cell interactions in the assembly of the early embryo.
- The addition of an extracellular material such as Matrigel to suspensions of cultured pluripotent stem cells enables embryo models to develop further than the blastoid stage. Endoderm stem cells (XEN cells) can substitute for Matrigel because they develop into cells that secrete extracellular matrix. These embryo models show the need for additional chemical and mechanical signals for the three-dimensional development of the embryo.
- If small and defined numbers of mouse ES cells are aggregated in suspension in controlled chemical conditions and are exposed to Wnt signals at the appropriate time, they will form structures called gastruloids, which have many features of an embryo after gastrulation.
- Stem cells are being used to try to produce cells of various types for medical applications, such as retinal cells to treat macular degeneration and glucose-responsive insulin-producing pancreatic cells to treat type 1 diabetes.
- Pluripotent stem cells or organ-specific stem cells cultured in suspension can develop into miniature three-dimensional structures called organoids, which are composed of a mixture of cell types and have features of the tissues or organs that the stem cells would normally have developed into. Organoids made from a patient's own cells can be used to study features of disease and test drug treatments.
- Pluripotent stem cells can be used to grow human tissues in other animals.

■ End of chapter questions

Check your understanding of the content of this chapter by attempting the following long-answer concept questions.

1. How many classes of mouse embryonic stem cells (ES cells) are there? What are their characteristics?

2. What are the differences between embryonic and adult stem cells?

3. Summarize similarities and differences between mouse and human ES cells.

4. Totipotency, pluripotency, and multipotency are terms associated with different classes of stem cells. What do they mean? Give some examples of each kind.

5. What are induced pluripotent stem cells (iPS cells)? How are they obtained? What can they be used for?

6. ES cells maintain their potency in culture, in principle, indefinitely. How are they obtained? Do they exist in the embryo?

7. ES cells can be differentiated into all cell types of the organism *in vitro*. How is this done? How does this process resemble or differ from what happens in the embryo?

8. How many models of mouse pre-implantation development do you know? Describe their composition and what each tells us about embryos.

9. What is a gastruloid? How do you make one?

10. What are the differences between gastruloids and embryos?

11. Under certain culture conditions, a combination of extra-embryonic stem cells and ES cells can be coaxed to make structures that are called 'synthetic embryos'. Describe these structures and discuss whether the name 'synthetic embryos' is appropriate for them.

12. Large mammals are being considered as hosts to make human organs for transplantation. Using the example of the production of a mouse with a rat pancreas, describe in principle how you might make a chimeric organism of this sort.

13. Pigs are being considered as hosts in which to generate human organs for transplantation, but a human organ with a blood supply from the pig would be immediately rejected. Describe a proof-of-concept experiment that shows how this might be prevented. In this preliminary experiment, how was the birth of a piglet with human characteristics avoided?

14. What is an organoid? Describe how organoids are used in research and what their potential applications are. What are the advantages of using them for these purposes rather than differentiated cells?

Section further reading

European Union. *Regulation of Stem Cell Research in Europe* [https://www.eurostemcell.org/regulation-stem-cell-research-europe] (date accessed 20 October 2022).

Greenfield, A.: **Embryos, genes & regulation: the impact of scientific progress on future reproductive medicine and the HFEA**. UK Human Fertilisation & Embryology Authority [https://www.hfea.gov.uk/about-us/30th-anniversary-expert-series/embryos-genes-regulation-the-impact-of-scientific-progress-on-future-reproductive-medicine-and-the-hfea] (date accessed 20 October 2020).

International Society for Stem Cell Research. *Guidelines for Stem Cell Research and Clinical Translation* [https://www.isscr.org/guidelines] (date accessed 20 October 2022).

National Institutes of Health. *Introduction to Stem Cells* [https://stemcells.nih.gov/info/basics] (date accessed 20 October 2022).

National Institutes of Health. *NIH Guidelines for Human Stem Cell Research* [https://stemcells.nih.gov/research-policy-and-faqs] (date accessed 20 October 2022).

7.1 Embryonic stem cells are pluripotent cells derived from mammalian blastocysts and can contribute to normal development *in vivo*

Chambers, I., Silva, J., Colby, D., Nichols, J., Nijmeijer, B., Robertson, M., Vrana, J., Jones, K., Grotewold, L., Smith, A.: **Nanog safeguards pluripotency and mediates germline development**. *Nature* 2007, **450**: 1230–1234.

Evans, M.J., Kaufman, M.H.: **Establishment in culture of pluripotential cells from mouse embryos**. *Nature* 1981, **292**: 154–156.

Kojima, Y., Kaufman-Francis, K., Studdert, J.B., Steiner, K.A., Power, M.D., Loebel, D.A., Jones, V., Hor, A., de Alencastro, G., Logan, G.J., Teber, E.T., Tam, O.H., Stutz, M.D., Alexander, I.E., Pickett, H.A., Tam, P.P.: **The transcriptional and functional properties of mouse epiblast stem cells resemble the anterior primitive streak**. *Cell Stem Cell* 2013, **14**: 107–120.

Martin, G.R.: **Isolation of a pluripotent cell line from early mouse embryos cultured in medium conditioned by teratocarcinoma stem cells**. *Proc. Natl Acad. Sci. USA* 1981, **78**: 7634–7638.

Morgani, S., Nichols, J., Hadjantonakis, A.K.: **The many faces of pluripotency:** *in vitro* **adaptations of a continuum of** *in vivo* **states**. *BMC Dev. Biol.* 2017, **17**: 7.

Nichols, J., Smith, A.: **The origin and identity of embryonic stem cells**. *Development* 2011, **138**: 3–8.

Weinberger, L., Ayyash, M., Novershtern, N., Hanna, J.H.: **Dynamic stem cell states: naive to primed pluripotency in rodents and humans**. *Nat. Rev. Mol. Cell Biol.* 2016, **17**: 155–169.

Wray, J., Kalkan, T., Gomez-Lopez, S., Eckardt, D., Cook, A., Kemler, R., Smith, A.: **Inhibition of glycogen synthase kinase-3 alleviates Tcf3 repression of the pluripotency network and increases embryonic stem cell resistance to differentiation**. *Nat. Cell Biol.* 2011, **13**: 838–845.

Ying, Q.L., Wray, J., Nichols, J., Batlle-Morera, L., Doble, B., Woodgett, J., Cohen, P., Smith, A.: **The ground state of embryonic stem cell self-renewal**. *Nature* 2008, **453**: 519–523.

7.2 Human ES cells differ from mouse ES cells in several of their properties

Brons, I.G., Smithers, L.E., Trotter, M.W., Rugg-Gunn, P., Sun, B., Chuva de Sousa Lopes, S.M., Howlett, S.K., Clarkson, A., Ahrlund-Richter, L., Pedersen, R.A., Vallier, L.: **Derivation of pluripotent epiblast stem cells from mammalian embryos**. *Nature* 2007, **448**: 191–195.

Guo, G., Stirparo, G.G., Strawbridge, S.E., Spindlow, D., Yang, J., Clarke, J., Dattani, A., Yanagida, A., Li, M.A., Myers, S., Özel, B.N., Nichols, J., Smith, A.: **Human naive epiblast cells possess unrestricted lineage potential**. *Cell Stem Cell* 2021, **28**: 1040–1056.

Kinoshita, M., Barber, M., Mansfield, W., Cui, Y., Spindlow, D., Stirparo, G.G., Dietmann, S., Nichols, J., Smith, A.: **Capture of mouse and human stem cells with features of formative pluripotency**. *Cell Stem Cell* 2021, **28**: 453–471.

Meistermann D., et al.: **Integrated pseudotime analysis of human pre-implantation embryo single-cell transcriptomes reveals the dynamics of lineage specification**. *Cell Stem Cell* 2021, **28**: 1625–1640.

Radley, A., Corujo-Simon, E., Nichols, J., Smith, A., Dunn, S.J.: **Entropy sorting of single-cell RNA sequencing data reveals the inner cell mass in the human pre-implantation embryo**. *Stem Cell Rep.* 2022, **18**: 47–63.

Tesar, P.J., Chenoweth, J.G., Brook, F.A., Davies, T.J., Evans, E.P., Mack, D.L., Gardner, R.L., McKay, R.D.: **New cell lines from mouse epiblast share defining features with human embryonic stem cells**. *Nature* 2007, **448**: 196–199.

Thomson, J.A., Itskovitz-Eldor, J., Shapiro, S.S., Waknitz, M.A., Swiergiel, J. J., Marshall, V.S., Jones, J.M.: **Embryonic stem cell lines derived from human blastocysts**. *Science* 1998, **282**: 1145–1147.

7.3 Adult differentiated cells can be reprogrammed to form pluripotent stem cells and Box 7A Induced pluripotent stem cells

Okita, I., Ichisaka, T., Yamanaka, S.: **Generation of germline-competent induced pluripotent stem cells**. *Nature* 2007, **448**: 313–317.

Takahashi, K., Yamanaka, S.: **Induction of pluripotent stem cells from mouse embryonic and adult fibroblast cultures by defined factors**. *Cell* 2006, **126**: 663–676.

Woltjen, K., Michael, I.P., Mohseni, P., Desai, R., Mileikovsky, M., Hämäläinen, R., Cowling, R., Wang, W., Liu, P., Gertsenstein, M., Kaji, K., Sung, H.K., Nagy, A.: **piggyBac transposition reprograms fibroblasts to induced pluripotent stem cells**. *Nature* 2009, **458**: 766–770.

Yusa, K., Rad, R., Takeda, J., Bradley, A.: **Generation of transgene-free induced pluripotent mouse stem cells by the piggyBac transposon**. *Nat. Methods* 2009, **6**: 363–369.

7.4 Pluripotent stem cells are used as models of cell differentiation

Camacho-Aguilar, E., Warmflash, A.: **Insights into mammalian morphogen dynamics from embryonic stem cell systems**. *Curr. Top. Dev. Biol.* 2020, **137**: 279–305.

Deglincerti, A., Etoc, F., Guerra, M.C., Martyn, I., Metzger, J., Ruzo, A., Simunovic, M., Yoney, A., Brivanlou, A.H., Siggia, E., Warmflash, A.: **Self-organization of human embryonic stem cells on micropatterns**. *Nat. Protocol* 2016, **11**: 2223–2232.

Edri, S., Hayward, P., Jawaid, W., Martinez Arias, A.: **Neuromesodermal progenitors (NMPs): a comparative study between pluripotent stem cells and embryo-derived populations**. *Development* 2019, **146**: dev180190.

Faial, T., Bernardo, A.S., Mendjan, S., Diamanti, E., Ortmann, D., Gentsch, G.E., Mascetti, V.L., Trotter, M.W., Smith, J.C., Pedersen, R.A.: **Brachyury and SMAD signalling collaboratively orchestrate distinct mesoderm and endoderm gene regulatory networks in differentiating human embryonic stem cells**. *Development* 2015, **142**: 2121–2135.

Gadue, P., Huber, T.L., Paddison, P.J., Keller, G.M.: **Wnt and TGF-beta signaling are required for the induction of an in vitro model of primitive streak formation using embryonic stem cells**. *Proc. Natl Acad. Sci. USA* 2006, **103**: 16806–16811.

Gouti, M., Tsakiridis, A., Wymeersch, F.J., Huang, Y., Kleinjung, J., Wilson, V., Briscoe, J.: **In vitro generation of neuromesodermal progenitors reveals distinct roles for wnt signalling in the specification of spinal cord and paraxial mesoderm identity**. *PLoS Biol.* 2014, **12**: e1001937.

Siggia, E.D., Warmflash, A.: **Modeling mammalian gastrulation with embryonic stem cells**. *Curr. Top. Dev. Biol.* 2018, **129**: 1–23.

Spagnoli, F.M., Hemmati-Brivanlou, A.: **Guiding embryonic stem cells towards differentiation: lessons from molecular embryology**. *Curr. Opin. Genet. Dev.* 2006, **16**: 469–475.

Tsakiridis, A., Huang, Y., Blin, G., Skylaki, S., Wymeersch, F., Osorno, R., Economou, C., Karagianni, E., Zhao, S., Lowell, S., Wilson, V.: **Distinct Wnt-driven primitive streak-like populations reflect in vivo lineage precursors**. *Development* 2014, **141**: 1209–1221.

7.5 ES cells can be used to provide models of early development

Amadei, G., Handford, C.E., Qiu, C. De Jonghe, J., Greenfeld, H., Tran, M., Martin, B.K., Chen, D.Y., Aguilera-Castrejon, A., Hanna, J.H., Elowitz, M.B., Hollfelder, F., Shendure, J., Glover, D.M., Zernicka-Goetz, M.: **Embryo model completes gastrulation to neurulation and organogenesis**. *Nature* 2022, **610**: 143–153.

Amadei, G., Lau, K.Y.C., De Jonghe, J., Gantner, C.W., Sozen, B., Chan, C., Zhu, M., Kyprianou, C., Hollfelder, F., Zernicka-Goetz, M.: **Inducible stem-cell-derived embryos capture mouse morphogenetic events in vitro**. *Dev. Cell* 2021, **56**: 366–382.

Harrison, S.E., Sozen, B., Christodoulou, N., Kyprianou, C., Zernicka-Goetz, M.: **Assembly of embryonic and extraembryonic stem cells to mimic embryogenesis in vitro**. *Science* 2017, **356**: eaal1810.

Kagawa, H., Javali, A., Khoei, H.H., Maria Sommer, T.M., Sestini, G., Novatchkova, M., Scholte op Reimer, Y., Castel, G., Bruneau, A., Maenhoudt, N., Lammers, J., Loubersac, S., Freour, T., Vankelecom, H., David, L., Rivron, N.: **Human blastoids model blastocyst development and implantation**. *Nature* 2022, **601**, 600–605.

Posfai, E., Lanner, F., Mulas, C., Leitch, H.G.: **All models are wrong, but some are useful: establishing standards for stem cell-based embryo models**. *Stem Cell Rep.* 2021, **16**: 1117–1141.

Rivron, N.C., Frias-Aldeguer, J., Vrij, E.J., Boisset, J.-C., Korving, J., Vivié, J., Truckenmüller, R.K., van Oudenaarden, A., van Blitterswijk, C.A., Geijsen, N.: **Blastocyst-like structures generated solely from stem cells**. *Nature* 2018, **557**: 106–111.

Shahbazi, M.N., Siggia, E.D., Zernicka-Goetz, M.: **Self-organization of stem cells into embryos: a window on early mammalian development**. *Science* 2019, **364**: 948–951.

Sozen, B., Amadei, G., Cox, A., Wang, R., Na, E., Czukiewska, S., Chappell, L., Voet, T., Michel, G., Jing, N., Glover, D.M., Zernicka-Goetz, M.: **Self-assembly of embryonic and two extra-embryonic stem cell types into gastrulating embryo-like structures**. *Nature Cell Biol.* 2018, **20**: 979–989.

Tarazi, S., *et al.*: **Post-gastrulation synthetic embryos generated ex utero from mouse naive ESCs**. *Cell* 2022, **185**: 3290–3306.

Box 7B Self-organization and self-assembly

Halley, J.D., Winkler, D.A.: **Consistent concepts of self-organization and self-assembly**. *Complexity* 2008, **14**: 10–17.

Wedlich-Söldner, R., Betz, T.: **Self-organization: the fundament of cell biology**. *Phil. Trans. R. Soc. B* 2018, **373**: 20170103.

7.6 Pluripotent stem cells can model the process of gastrulation

Beccari, L., Moris, N., Girgin, M., Turner, D.A., Baillie-Johnson, P., Cossy, A.C., Lutolf, M.P., Duboule, D., Martinez Arias, A.: **Multiaxial self-organization properties of mouse embryonic stem cells into gastruloids**. *Nature* 2018, **562**: 272–276.

Martinez Arias, A., Marikawa, Y., Moris, N.: **Gastruloids: pluripotent stem cell models of mammalian gastrulation and embryo engineering**. *Dev. Biol.* 2022, **488**: 35–46.

Moris, N., Anlas, K., van den Brink, S.C., Alemany, A., Schröder, J., Ghimire, S., Balayo, T., van Oudenaarden, A., Martinez Arias, A.: **An *in vitro* model of early anteroposterior organization during human development**. *Nature* 2020, **582**: 410–415.

Oldak, B., *et al.*: **Complete human day 14 post-implantation embryo models from naïve ES cells**. *Nature* 2023, **622**: 562–573.

Rivron, N., Martinez Arias, A., Pera, M., Moris, N., M'Hamdi, H.I.: **An ethical framework for human embryology with embryo models**. *Cell* 2023, **186**: 3548–3557.

van den Brink, S.C., Alemany, A., van Batenburg, V., Moris, N., Blotenburg, M., Vivié, J., Baillie-Johnson, P., Nichols, J., Sonnen, K.F., Martinez Arias, A., van Oudenaarden, A.: **Single-cell and spatial transcriptomics reveal somitogenesis in gastruloids**. *Nature* 2020, **582**: 405–409.

Veenvliet, J.V., Bolondi, A., Kretzmer, H., Haut, L., Scholze-Wittler, M., Schifferl, D., Koch, F., Guignard, L., Kumar, A.S., Pustet, M., Heimann, S., Buschow, R., Wittler, L., Timmermann, B., Meissner, A., Herrmann, B.G.: **Mouse embryonic stem cells self-organize into trunk-like structures with neural tube and somites**. *Science* 2020, **370**: eaba4937.

Xu, P.F., Borges, R.M., Fillatre, J., de Oliveira-Melo, M., Cheng, T., Thisse, B., Thisse, C.: **Construction of a mammalian embryo model from stem cells organized by a morphogen signalling centre**. *Nat. Commun.* 2021, **12**: 3277.

Zheng, Y., Xue, X., Shao, Y., Wang, S., Esfahani, S.N., Li, Z., Muncie, J.M., Lakins, J.N., Weaver, V.M., Gumucio, D.L., Fu, J.: **Controlled modelling of human epiblast and amnion development using stem cells**. *Nature* 2019, **573**: 421–425.

Box 7C Somitogenesis in Matrigelled gastruloids

Veenvliet, J.V., Bolondi, A., Kretzmer, H., Haut, L., Scholze-Wittler, M., Schifferl, D., Koch, F., Guignard, L., Kumar, A.S., Pustet, M., Heimann, S., Buschow, R., Wittler, L., Timmermann, B., Meissner, A., Herrmann, B.G.: **Mouse embryonic stem cells self-organize into trunk-like structures with neural tube and somites**. *Science* 2020, **370**: eaba4937.

Yamanaka, Y., Yoshioka-Kobayashi, K., Hamidi, S., Munira, S., Sunadome, K., Zhang, Y., Kurokawa, Y., Mieda, A., Thompson, J.L., Kerwin, J., Lisgo, S., Yamamoto, T., Moris, N., Martinez-Arias, A., Tsujimura, T., Alev, C.: **Reconstituting human somitogenesis in vitro**. *Nature* 2003, **614**: 509–520.

7.7 Stem cells could be a key to regenerative medicine

da Cruz, L., *et al.*: **Phase 1 clinical study of an embryonic stem cell-derived retinal pigment epithelium patch in age-related macular degeneration**. *Nat. Biotechnol.* 2018, **36**: 328–337.

Ma, H., *et al.*: **Abnormalities in human pluripotent cells due to reprogramming mechanisms**. *Nature* 2014, **511**: 177–183.

Maeda, T., Sugita, S., Kurimoto, Y., Takahashi, M.: **Trends of stem cell therapies in age-related macular degeneration**. *J. Clin. Med.* 2021, **10**: 1785.

Pera, M.F., Trounson, A.O.: **Human embryonic stem cells: prospects for development**. *Development* 2004, **131**: 5515–5525.

Protze, S.I., Lee, J.H., Keller, G.M.: **Human pluripotent stem cell-derived cardiovascular cells: from developmental biology to therapeutic applications**. *Cell Stem Cell* 2019, **25**: 311–327.

Sampaziotis, F., *et al.*: **Reconstruction of the mouse extrahepatic biliary tree using primary human extrahepatic cholangiocyte organoids**. *Nat. Med.* 2017, **23**: 954–963.

Schwartz, S.D., *et al.*: **Human embryonic stem cell-derived retinal pigment epithelium in patients with age-related macular degeneration and Stargardt's macular dystrophy: follow-up of two open-label phase 1/2 studies**. *Lancet* 2015, **385**: 509–516.

7.8 Various approaches can be used to generate differentiated cells for cell-replacement therapies

Chen, S., Borowiak, M., Fox, J.L., Maehr, R., Osafune, K., Davidow, L., Lam, K., Peng, L.F., Schreiber, S.L., Rubin, L.L., Melton, D.: **A small molecule that directs differentiation of human ESCs into the pancreatic lineage**. *Nat. Chem. Biol.* 2009, **5**: 258–265.

Kroon, E., Martinson, L.A., Kadoya, K., Bang, A.G., Kelly, O.G., Eliazer, S., Young, H., Richardson, M., Smart, N.G., Cunningham, J., Agulnick, A.D., D'Amour, K.A., Carpenter, M.K., Baetge, E.E.: **Pancreatic endoderm derived from human embryonic stem cells generates glucose-responsive insulin-secreting cells in vivo**. *Nat. Biotechnol.* 2008, **26**: 443–452.

Maehr, R., Chen, S., Snitow, M., Ludwig, T., Yagasaki, L., Goland, R., Leibel, R.L., Melton, D.A.: **Generation of pluripotent stem cells from patients with type 1 diabetes**. *Proc. Natl Acad. Sci. USA* 2009, **106**: 15768–15773.

Pagliuca, F.W., Millman, J.R., Gürtler, M., Segel, M., Van Dervort, A., Ryu, J.H., Peterson, Q.P., Greiner, D., Melton, D.A.: **Generation of functional human pancreatic β cells in vitro**. *Cell* 2014, **159**: 428–439.

Siehler, J., Blöchinger, A.K., Meier, M., Lickert, H.: **Engineering islets from stem cells for advanced therapies of diabetes**. *Nat. Rev. Drug Discov.* 2021, **20**: 920–940.

Yamaguchi, T., Sato, H., Kato-Itoh, M., Goto, T., Hara, H., Sanbo, M., Mizuno, N., Kobayashi, T., Yanagida, A., Umino, A., Ota, Y., Hamanaka, S., Masaki, H., Rashid, S.T., Hirabayashi, M., Nakauchi, H.: **Interspecies organogenesis generates autologous functional islets**. *Nature* 2017, **542**: 191–196.

Zhou, Q., Brown, J., Kanarek, A., Rajagopal, J., Melton, D.A.: **In vivo reprogramming of adult pancreatic exocrine cells to beta-cells**. *Nature* 2008, **455**: 627–632.

Zhou, Q., Melton., D.A.: **Pancreas regeneration**. *Nature* 2018, **557**: 351–358.

7.9 ES cells can be used to generate organoids and tissues and Box 7D Intestinal organoids can be produced from adult stem cells or ES cells

Bartfeld, S., Clevers, H.: **Stem cell-derived organoids and their application for medical research and patient treatment**. *J. Mol. Med.* 2017, **95**: 729–738.

Drost, J., Clevers, H.: **Translational applications of adult stem cell-derived organoids**. *Development* 2017, **144**: 968–975.

Gabriel, E., *et al.*: **Human brain organoids assemble functionally integrated bilateral optic vesicles**. *Cell Stem Cell* 2021, **28**: 1740–1757.

Lancaster, M.A., Renner, M., Martin, C.A., Wenzel, D., Bicknell, L.S., Hurles, M.E., Homfray, T., Penninger, J.M., Jackson, A.P., Knoblich, J.A.: **Cerebral organoids model human brain development and microcephaly**. *Nature* 2013, **501**: 373–379.

Sasai, Y., Eiraku, M., Suga, H.: **In vitro organogenesis in three dimensions: self-organising stem cells**. *Development* 2012, **139**: 4111–4121.

Sato, T., Clevers H. **Growing self-organizing mini-guts from a single intestinal stem cell: mechanism and applications**. *Science* 2013, **340**: 1190–1194.

Turner, D., Baillie-Johnson, P., Martinez Arias, A.: **Organoids and the genetically-encoded self-assembly of embryonic stem cells**. *BioEssays* 2016, **38**: 181–191.

7.10 Stem cells might be used to grow human organs in other animals for transplantation

Das, S., Koyano-Nakagawa, N., Gafni, O., Maeng, G., Singh, B.N., Rasmussen, T., Pan, X., Choi, K.-D., Mickelson, D., Gong, W., Pota, P., Weaver, C.V., Kren, S., Hanna, J. H., Yannopoulos, D., Garry, M.G., Garry, D.J.: **Generation of human endothelium in pig embryos deficient in *ETV2***. *Nat. Biotechnol.* 2020, **38**: 297–302.

Garry, D.J., Garry, M.G.: **Interspecies chimeras as a platform for exogenic organ production and transplantation**. *Exp. Biol. Med.* 2021, **246**: 1838–1844.

Kobayashi, T., Yamaguchi, T., Hamanaka, S., Kato-Itoh, M., Yamazaki, Y., Ibata, M., Sato, H., Lee, Y.-S., Usui, J.-I., Knisely, A.S., Hirabayashi, M., Nakauchi, H.: **Generation of rat pancreas in mouse by interspecific blastocyst injection of pluripotent stem cells**. *Cell* 2010, **142**: 787–799.

8

Germ cells, fertilization, and sex determination

- **The development of germ cells**
- **Fertilization**
- **Determination of the sexual phenotype**

Animal embryos develop from a single cell, the fertilized egg or zygote, which is the product of the fusion of an egg and a sperm. In previous chapters we have considered how the basic body plan of animal embryos is generated by the development of the somatic cells and their assignment to the three germ layers. The zygote, in addition, gives rise to the germline cells—or germ cells—that develop into eggs and sperm, and through which an organism's genes are passed on to the next generation. As we shall discuss in this chapter, in animals, germline cells are specified and set apart in the early embryo, although functional mature eggs and sperm are only produced by the adult. An important property of germ cells is that unlike somatic cells they retain characteristics that enable them to combine to produce a totipotent cell able to give rise to all the different types of cells in the body and any extra-embryonic membranes. Eggs and sperm in mammals have certain genes differentially silenced during germ cell differentiation by a process known as genomic imprinting, and we will look at the implications of this for development. Embryonic development is initiated by the process of fertilization, and we shall discuss mechanisms in sea urchins and mammals that ensure that only one sperm enters the egg. In the animals we cover in this chapter, sex is determined by the chromosomal constitution, and we shall see that mechanisms of sex determination and of compensating for the unequal complement of sex chromosomes between males and females differ considerably in different species.

Introduction

In sexually reproducing organisms, there is a fundamental distinction between the **germline cells (germ cells)** and the somatic or body cells (see Section 1.2): the former give rise to the **gametes**—eggs and sperm in animals—whereas somatic cells make no genetic contribution to the next generation. Germ cells have three key functions: the preservation of the genetic integrity of the germline, such as the prevention of aging; the generation of genetic diversity; and the transmission of genetic information to the next generation.

So far in this book, we have looked at the somatic development of our model organisms. Not surprisingly, a great deal of the biology of animals and plants is devoted to reproduction and sex, and in this chapter we look at germ cell formation, fertilization of the egg by the sperm, and sex determination, mainly in the mouse, *Drosophila*, and *Caenorhabditis elegans*. As germ cells are the cells that will give rise to the next generation, their development is crucial, and in animals, germ cells are usually specified and set aside early in embryonic development. Plants, although reproducing sexually, differ from most animals in that their germ cells are not specified early in embryonic development, but during the development of the flowers (discussed in Chapter 12). It is worth noting that some simple animals, such as the coelenterate *Hydra*, can reproduce asexually, by budding, and that even in some vertebrates, such as turtles, the eggs can develop without being fertilized, a process known as **parthenogenesis**.

We will begin this chapter by considering how the germ cells are specified in the early animal embryo and how they differentiate into eggs and sperm. Then, we will look at the **fertilization** and activation of the egg by the sperm—the vital step that initiates development. Finally, we turn to **sex determination**—why males and females are different from each other. In all the animals we look at here, sex is determined genetically, by the number and type of specialized chromosomes known as the **sex chromosomes**. Male and female embryos initially look the same, with sexual differences emerging only as a result of the activity of sex-determining genes located on the sex chromosomes.

The development of germ cells

In all but the simplest animals, the cells of the germline are the only cells that can combine to give rise to a new organism. So, unlike somatic cells, which eventually all die, germ cells in a sense outlive the bodies that produced them in that they produce the next generation. The outcome of germ cell development is either a male gamete (the sperm in animals) or a female gamete (the egg). The fertilized egg is a particularly remarkable cell, as it ultimately gives rise to all the cells in an organism. In species whose embryos receive no nutrition from the mother after fertilization, the egg must also provide everything necessary for development, as the sperm contributes virtually nothing to the organism other than its chromosomes.

Animal germ cells typically differ from somatic cells by dividing less often during early embryogenesis. Later, they are the only cells to undergo the type of nuclear division called **meiosis**, by which gametes with half the number of chromosomes of the germ cell precursor are produced. The halving of chromosome number at meiosis means that when two gametes come together at fertilization to form the zygote, the diploid number of chromosomes is restored. If there were no reduction division in germ cell formation, the number of chromosomes in the somatic cells would double in each generation.

The precursors of germ cells arise early in development. From the time they are first specified as germline cells until the time that they enter the **gonads**—the specialized reproductive organs—they are known as **primordial germ cells**. In the gonads, the germ cells enter meiosis and differentiate into eggs in the **ovary** of female embryos and differentiate into sperm in the **testis** in male embryos. In many animals, including *Drosophila* and vertebrates, the primordial germ cells are first specified in a region quite distant from where the gonads will eventually form, and then migrate from their site of origin into the forming gonads. In addition, primordial germ cells are often formed in locations that seem to protect them from the inductive signals that determine the fate of somatic cells, and instead shut down the somatic developmental program. In *Caenorhabditis*, for example, there is evidence of a general repression of transcription in the primordial germ cells. In some germlines, other mechanisms that help suppress mutation and maintain genetic integrity are present. In *Drosophila*, for example, the movement of transposons in the genome, which can cause insertion mutations, is prevented by an RNA silencing pathway involving the Piwi-interacting small RNAs (piRNAs).

Germ cells are specified very early in some animals, although not in mammals, by cytoplasmic determinants present in the egg. We therefore start our discussion of germ cell development by looking at the specification of primordial germ cells by special cytoplasm—the **germplasm**—which is already present in these eggs.

8.1 Germ cell fate is specified in some embryos by a distinct germplasm in the egg

In flies, nematodes, fish, and frogs, molecules localized in specialized cytoplasm in the egg are involved in specifying the germ cells and this may also be true in the chick. The clearest example is in *Drosophila*, where primordial germ cells known as **pole cells** become distinct at the posterior pole of the egg about 90 minutes after fertilization, more than an hour before the cellularization of the rest of the embryo (see Fig. 2.2). The cytoplasm at the posterior pole is called the **pole plasm** and is distinguished by large organelles, the polar granules, which contain both proteins and RNAs. Two key experiments demonstrate that there is something special about the posterior cytoplasm. First, if the posterior end of the egg is irradiated with ultraviolet light, which destroys the pole plasm activity, no germ cells develop, although the somatic cells in that region do develop. Second, if pole plasm from an egg is transferred to the anterior pole of another embryo, the nuclei that become surrounded by the pole plasm are specified as germ cell nuclei. If these anterior cells containing pole plasm are transplanted into the posterior pole region of a third embryo, they develop as functional germ cells (Fig. 8.1).

In Chapter 2, we saw how the main axes of the *Drosophila* egg are specified by the follicle cells in the ovary, and how the mRNAs for proteins such as Bicoid and Nanos become localized in the egg (see Section 2.14). The pole plasm also becomes localized at the posterior end of the egg under the influence of the follicle cells. Several maternal genes are involved in pole plasm formation in *Drosophila*. Mutations in any of at least eight genes result in the affected homozygous individual being 'grandchildless'. Its offspring lack a proper pole plasm, and although they may develop normally in other ways, they lack germ cells and are therefore sterile. One of these eight genes is *oskar*, which plays a central role in the organization and assembly of the pole plasm; of the genes involved in pole plasm formation, *oskar* is the only one to have its mRNA localized at the posterior pole. The signal for localization is contained in the 3′ untranslated region of the mRNA. Staufen protein is required for the localization of *oskar* mRNA and may act by linking the mRNA to the microtubule system that is polarized along the antero-posterior axis of the egg (see Section 2.14). If the region of the *oskar* gene that codes for the 3′ localization signal is replaced by the *bicoid* 3′ localization signal, transgenic flies made with this DNA construct have *oskar* mRNA localized at the anterior end of the egg in addition to its

Fig. 8.1 Transplanted pole plasm can induce germ cell formation in *Drosophila*. Pole plasm from a fertilized egg of genotype P (pink) is transferred to the anterior end of an early cleavage-stage embryo of genotype Y (yellow). After cellularization, cells containing pole plasm induced at the anterior end of embryo Y are transferred to the posterior end (a site from which germ cells can migrate into the gonad) of another embryo, of genotype G (green). The adult fly that develops from embryo G contains germ cells of genotype Y as well as those of G.

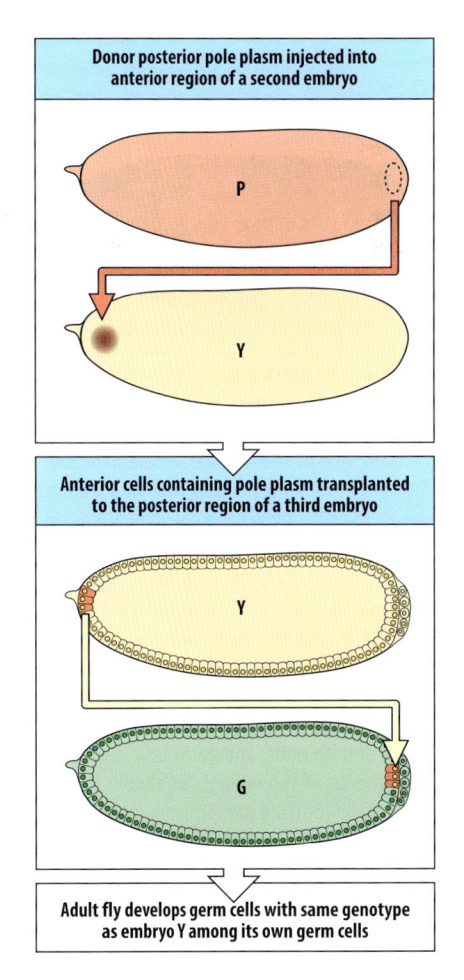

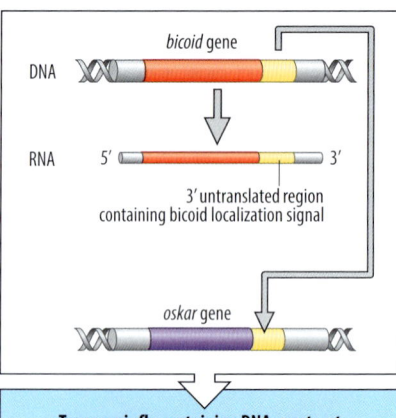

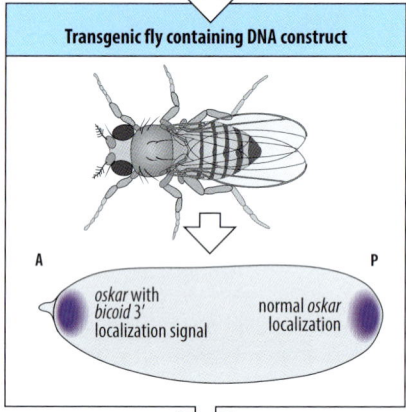

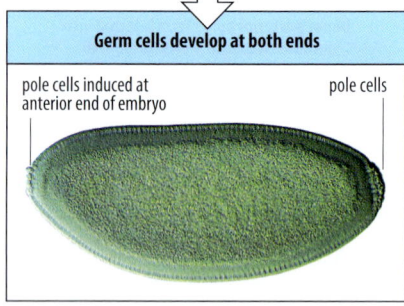

Fig. 8.2 The gene *oskar* is involved in specifying the germplasm in *Drosophila*. In normal eggs, *oskar* mRNA is localized at the posterior end of the embryo, whereas *bicoid* mRNA is at the anterior end. The localization signals for both *bicoid* and *oskar* mRNA are in their 3′ untranslated regions. By manipulating the *Drosophila* DNA, the localization signal of *oskar* can be replaced by that of *bicoid* (top panel). A transgenic fly is made containing a construct of the modified DNA. In its egg, *oskar* becomes localized at the anterior end (middle panel). The egg therefore has *oskar* mRNA at both ends, and germ cells develop at both ends of the embryo, as shown in the photograph (bottom panel). Thus, *oskar* alone is sufficient to initiate the specification of germ cells. A, anterior; P, posterior.

Photograph courtesy of R. Lehmann.

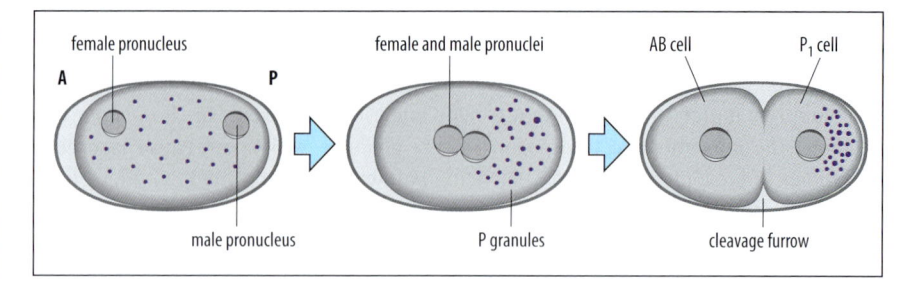

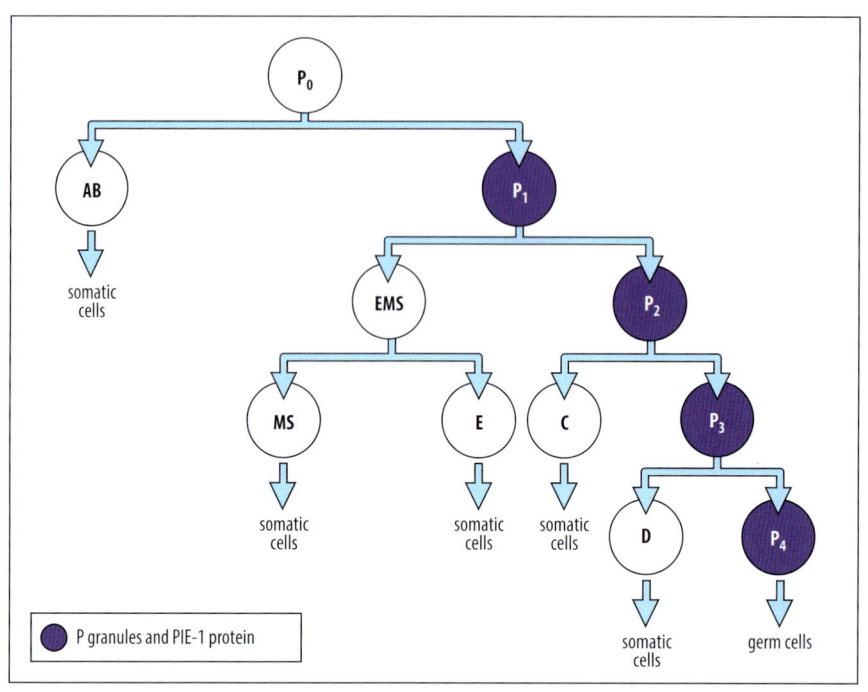

Fig. 8.3 P granules and PIE-1 protein become asymmetrically distributed to germline cells during cleavage of the nematode egg. Before fertilization, P granules (purple) are distributed throughout the egg. After fertilization, P granules become localized at the posterior (P) end of the egg. At the first cleavage, they are only included in the P_1 cell (top panel), and thus become confined to the P-cell lineage. The PIE-1 protein is present only in P cells. All germ cells are derived from P_4, which is formed at the fourth cleavage (bottom panel). A, anterior.

normal posterior localization. As a consequence, germ cells develop at both ends of the embryo (Fig. 8.2).

In the nematode, a germ cell lineage is set up at the end of the fourth cleavage division, with all the germ cells being derived from the P_4 blastomere. The P_4 cell is derived from three stem-cell-like divisions of the P_1 cell. At each of these divisions, one daughter produces somatic cells, whereas the other divides again to produce a somatic cell progenitor and a P cell. The egg contains P granules in its cytoplasm that become asymmetrically distributed before the first cleavage division and are subsequently confined to the P-cell lineage (Fig. 8.3). The association of germ cell formation with the P granules suggested that they might have a role in germ cell specification, and at least one P-granule component, the product of the *pgl-1* gene, has been shown to be necessary for germ cell development. PGL-1 protein may specify germ cells by regulating some aspect of mRNA metabolism.

In both the fly and the nematode, repression of transcription is necessary to specify a germ cell fate. In the nematode, the gene *pie-1* is involved in maintaining the stem-cell property of the P blastomeres. It encodes a nuclear protein that is expressed maternally and is not a component of P granules; the PIE-1 protein is present only in the germline blastomeres and represses new transcription of zygotic genes in these blastomeres until it disappears at around the 100-cell stage. This general repression may protect the germ cells from the actions of transcription factors that promote development into somatic cells.

Xenopus eggs also have distinct germplasm. Distinct yolk-free patches of cytoplasm aggregate at the yolky vegetal pole after fertilization. When the blastomeres at the vegetal pole cleave, this cytoplasm is distributed asymmetrically, so that it is retained only in the most vegetal daughter cells, from which the germ cells are derived. Ultraviolet irradiation of the vegetal cytoplasm abolishes the formation of germ cells, and transplantation of fresh vegetal cytoplasm into an irradiated egg restores germ cell formation. At gastrulation, the germplasm is located in cells in the floor of the blastocoel cavity, among the cells that give rise to the endoderm. Cells containing the germplasm are, however, not yet determined as germ cells and can contribute to all three germ layers if transplanted to other sites. At the end of gastrulation, the primordial germ cells are determined, and migrate out of the presumptive endoderm and into the **genital ridge**, which develops from mesoderm lining the abdominal cavity and will form the gonads. There are, however, quite big differences in germ cell origin in different groups of amphibians. In urodeles—tailed amphibians that include newts and the axolotl—there is no evidence for cytoplasmic localization of germplasm, and the germ cells arise from lateral plate mesoderm. The evolutionary significance of this difference is not known.

In the zebrafish, germplasm containing several maternal mRNAs is present in several locations in the fertilized egg and becomes localized to the distal ends of the cleavage furrows in the first few cleavages. Its fate can be followed during early development by tracking one of its components, such as maternal *vasa* mRNA (Fig. 8.4). By the 32-cell stage, germplasm has been segregated into four cells. These cells then continue to divide asymmetrically, with the germplasm segregating to one daughter cell at each division, so that even by the 512-cell stage there are still only four prospective germline cells in the embryo. At the sphere stage of the embryo (about 3.8 hours post-fertilization, see Section 3.15), the germplasm expands to fill the cells, and equal divisions begin, eventually producing a total of around 30 primordial germ cells, which will migrate to the prospective ovaries or testes.

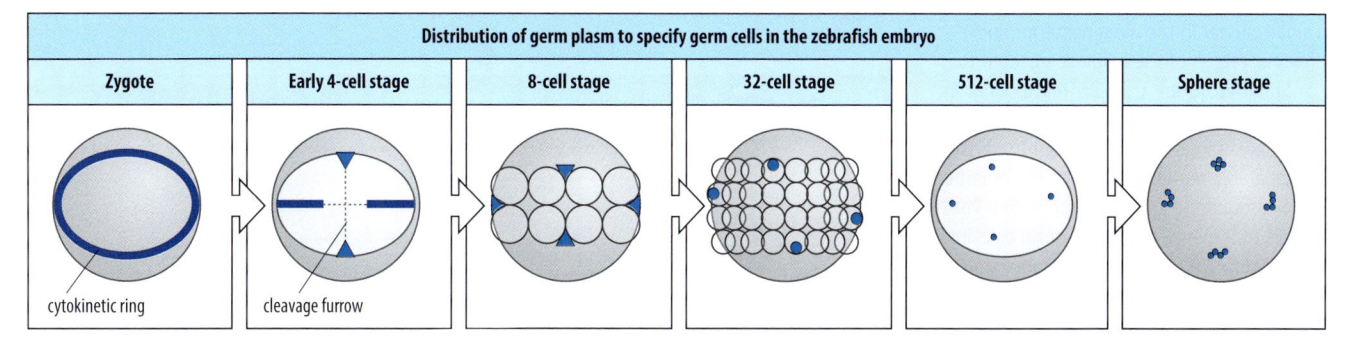

Fig. 8.4 Distribution of germplasm in the zebrafish embryo during cleavage. In this bird's-eye view of the animal pole of the egg, maternal *vasa* mRNA (blue) is a marker for the germplasm. The *vasa* mRNA becomes localized to the cleavage furrows at first and second divisions. As a furrow completes division the *vasa* mRNA forms tight clumps at the distal ends, at or just outside the cell boundary but under the yolk cell membrane. At the 32-cell stage, the *vasa* aggregates ingress into four cells and segregate asymmetrically at subsequent divisions, ending up in four widely separated cells at the 512-cell stage. At the sphere stage (cleavage cycle 13), the *vasa* mRNA becomes evenly distributed throughout the four cells, and subsequent cell divisions produce a total of around 30 primordial germ cells in the four locations.

Adapted from Pelegri, F.: **Maternal factors in zebrafish development.** Dev. Dyn. 2003, **228**: 535–554.

8.2 In mammals germ cells are induced by cell-cell signaling during development

There is no evidence for germplasm in the mouse or other mammals including humans. Because germ cell specification in the mouse involves cell–cell interactions (i.e. cell–cell signaling), cultured embryonic stem cells (ES cells) can, when injected into the inner cell mass of the blastocyst, give rise to both germ cells and somatic cells (see Fig. 7.5). Cell–cell interactions are also involved in specifying germ cells in many other animals including axolotls (tailed amphibians), turtles, and some insects such as crickets.

Mammalian germ cells were previously identified morphologically and by their high concentration of the enzyme alkaline phosphatase, which provides a convenient means of detecting them histochemically. Now that genes involved in germline specification have been identified, it is possible to recognize primordial germ cells at a much earlier stage by detecting the expression of such genes.

In the mouse, the earliest detectable primordial germ cells can be identified in the proximal region of the epiblast just before the beginning of gastrulation, when the primitive streak is initiated. They are specified from a subset of *TbxT*-expressing cells by a combination of BMP signaling from neighboring extra-embryonic ectoderm and Wnt signaling from the streak and express the gene encoding the transcriptional repressor protein Blimp1 (also known as Prmd1). The *Blimp1*-expressing cells also express the pluripotency genes *Nanog*, *Oct4*, and *Sox2*. By E6.25, they form a cluster of 6–8 cells in the most posterior and proximal region of the mouse epiblast adjacent to the primitive streak (Fig. 8.5). During gastrulation, the Blimp1-positive cells proliferate. By E7.25 there are around 40 primordial germ cells at the base of the allantois in the neighborhood of the primitive streak. These represent the full complement of primordial germ cells that will eventually migrate to the mouse gonads. Blimp1 and two other transcription factors, TFAPC2 and Stella, that come to be expressed by the primordial germ cells act in combination to repress expression of somatic genes and to activate expression of genes characteristic of germ cells.

Unlike in the somatic cells at this stage, the Hox genes and other typical somatic developmental genes are repressed in the primordial germ cells, which may explain their escape from a somatic cell fate. Instead, the primordial germ cells express pluripotency genes as noted earlier and undergo epigenetic reprogramming so that they maintain pluripotency. They also express many genes involved in cell adhesion and cell migration.

In humans, the specification of the primordial germ cells occurs at a comparable stage of development, during weeks 2–3 of development in the posterior region at the

Fig. 8.5 Cell–cell signaling for mammalian germ cell induction. Left panel: in the mouse, BMP and Wnt signals induce a small number of primordial germ cells (PGCs) (pale orange) expressing *Blimp1*. These cells form a small cluster in the most posterior and proximal region of the epiblast adjacent to the primitive streak at E6.25. Right panel: in humans, the precursors of primordial germ cells (pale orange) are first thought to arise in a comparable region adjacent to the primitive streak around the start of gastrulation. The signals that induce primordial germ cells in human embryos *in vivo* are not known but it is postulated that BMP and WNT signals are produced by tissues at the edges of the bilaminar disc. AVE, anterior visceral endoderm.

Adapted from Marlow, F.: **Primordial germ cell specification and migration.** F1000Res. 2015, **4**: 1462; Tang, W.W., et al.: **Specification and epigenetic programming of the human germ line.** Nat. Rev. Genet. 2016, **17**: 585–600.

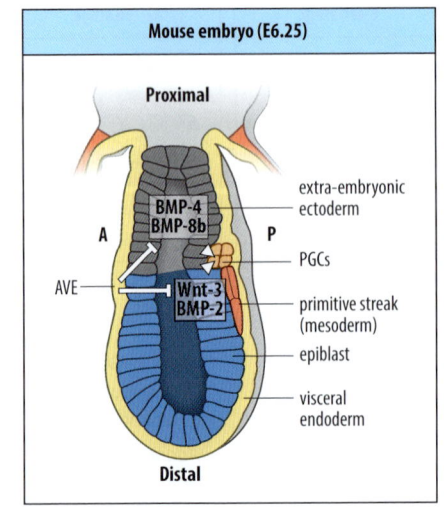

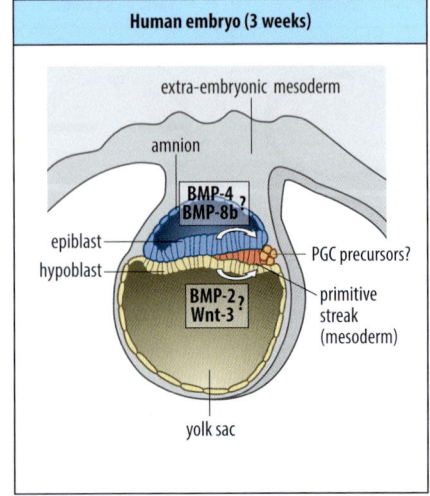

interface between the amnion and the epiblast (see Fig. 8.5). BMP and WNT signaling are also postulated to be involved. But there is a significant difference: human primordial germ cells express and require SOX17, which is not the case for mouse primordial germ cells. This was discovered in studies that differentiated human ES cells into primordial germ cells in adherent cultures or in embryo models. It has also been corroborated in studies with pigs and non-human primates and suggests a uniqueness of mouse embryos among mammals. It also shows the importance of looking directly at human development.

8.3 Germ cells migrate from their site of origin to the gonad

In many animals, the primordial germ cells develop at some distance from the gonads, and only later migrate to them, where they become determined as germ cells and differentiate into eggs or sperm. The reason for this separation of site of origin from final destination is not known, but it may be a mechanism for excluding germ cells from the general developmental upheaval involved in laying down the body plan, or alternatively a mechanism for selecting the healthiest germ cells, namely those that survive migration. The migration path of primordial germ cells is controlled by their environment; in *Xenopus*, for example, primordial germ cells transplanted to the wrong place in the blastula do not end up in the gonad.

The vertebrate gonad develops from mesoderm lining the abdominal cavity, which is known as the genital ridge. In *Xenopus*, primordial germ cells that have translocated with the endoderm to the floor of the blastocoel migrate to the future gonad along a cell sheet that joins the gut to the genital ridge. Only a small number of cells start this journey, dividing about three times before arrival, so that about 30 germ cells colonize the gonad—about the same number as in the zebrafish (see Fig. 8.4). By contrast, in the mouse, the number of primordial germ cells that arrive at the genital ridge is about 8000, starting from a population of around 40 at the base of the allantois at the posterior end of the primitive streak. Mouse primordial germ cells enter the hindgut endoderm, and migrate to the dorsal mesentery, the cell sheet that suspends the gut in the body cavity, and then across it to reach the genital ridges (Fig. 8.6).

In chick embryos, the pattern of migration is different again: the germ cells originate in the epiblast and then migrate to the head end of the embryo. Most arrive at their destination in the gonads by circulating in the blood, leaving the bloodstream at the hindgut and then migrating along an epithelial sheet to the gonad.

In zebrafish, small clusters of primordial germ cells are specified in four separate positions relative to the embryonic axis (see Fig. 8.4). These cells migrate to form bilateral lines of cells in the trunk at the level of somites 1–3 at around 6 hours post-fertilization, and then reach their final position at the level of somites 8–10 at around 12 hours, where they and the surrounding somatic cells form the gonads (Fig. 8.7). Zebrafish primordial germ cells face a challenging journey, as they originate in four

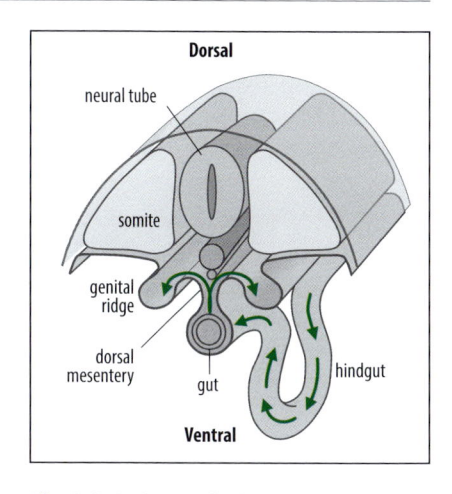

Fig. 8.6 Pathway of primordial germ cell migration in the mouse embryo. In the final stage of migration, in E10.5 embryos, the cells move from the gut tube into the genital ridge, via the dorsal mesentery.

Illustration after Alberts, B., et al.: Molecular Biology of the Cell. 2nd edn. New York: Garland Publishing, 1989.

Fig. 8.7 Primordial germ cells in zebrafish migrate to the gonads in two stages. The diagrams show a dorsal view of the embryo. The chemoattractant stromal-cell-derived factor 1 (SDF-1; yellow) is expressed initially in a broad domain that includes the four sites at which the primordial germ cells (PGCs; blue) are formed (see Fig. 8.4). PGCs migrate towards the region of highest SDF-1, which is initially at the level of the first somite. Alterations in the pattern of SDF-1 expression then cause the PGCs to migrate posteriorly to the level of the eighth and tenth somites, where the gonads develop.

Adapted from Weidinger, G., et al.: Regulation of zebrafish primordial germ cell migration by attraction towards an intermediate target. Development 2002, 129: 25-36.

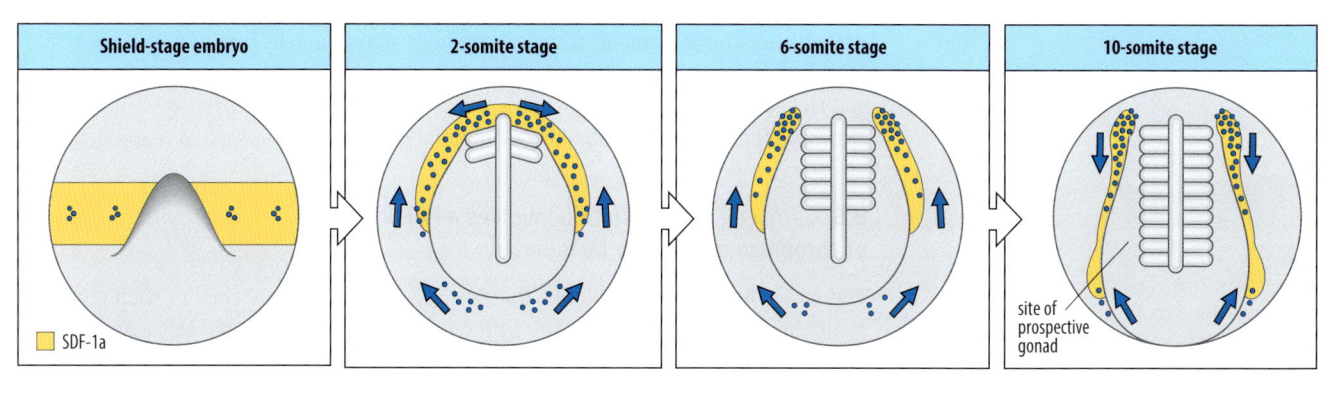

separate locations. The journey is broken up into two distinct stages. This may explain why germ cells taken from the margin of a mid-blastula zebrafish embryo find their way to the gonad even after transplantation to the head region (the animal pole) of a similar-stage embryo.

8.4 Germ cells are guided to their destination by chemical signals

All migrating germ cells are continuously receiving signals for guidance, survival, and proliferation from the tissues through which they migrate. Some of these signals have been identified; if they are lacking, germ cells are absent from the gonads or their numbers are greatly reduced. The main guidance cue both in zebrafish and in mice seems to be the chemoattractant protein stromal-cell-derived factor 1 (SDF-1). We have come across this chemoattractant in relation to the collective migration of the cells that give rise to the lateral line in zebrafish (see Section 5.23). SDF-1 is secreted into the extracellular matrix by tissue cells in the regions through which the germ cells migrate in zebrafish and is thought to be the long-range guidance cue for the clusters of primordial germ cells to reach their final positions from their different starting points (see Fig. 8.7). Experiments that change the spatial expression of the gene for SDF-1 change the germ cell migration patterns correspondingly. Also, knockdown by antisense RNA of the expression of either SDF-1 or its receptor, CXCR4, which is found on the surface of primordial germ cells, disrupts migration.

In the mouse, the primordial germ cells in the primitive streak migrate anteriorly and become incorporated into the hindgut endoderm, which they eventually leave to migrate into the genital ridge (see Fig. 8.6). The mouse version of SDF-1 seems to act as a guidance cue for at least this final part of the journey, as in animals lacking either SDF-1 or CXCR4, most germ cells get no further than the hindgut. SDF-1 and CXCR4 are not only of interest to embryos; they are also expressed in some invasive human tumors, where they contribute to the ability of the tumor cells to spread to other sites in the body.

In *Drosophila*, the primordial germ cells are carried into the embryo by invagination of the midgut primordium. They then migrate across the dorsal midgut epithelium to the adjacent mesoderm, moving along the dorsal surface of the midgut until they split up into two groups to join with somatic cells on either side of the embryo to form the gonads. Screening for mutations that affect the direction of germ cell migration has identified genes which encode proteins that keep the cells on track. Two related proteins, Wunen (WUN) and WUN2, which are enzymes that modify lipids, are involved in repelling germ cells from the rest of the gut, so preventing them from dispersing before they reach the gonadal mesoderm. Extracellular lipid signaling also repels migrating primordial germ cells in zebrafish, suggesting that this mechanism may be widely conserved. In *Drosophila*, the expression of the enzyme HMG-CoA reductase in the prospective gonad is needed for attracting the germ cells. This enzyme—which has a key role in the metabolic pathway that produces cholesterol and related lipids—is involved in regulating the activity of a chemoattractant. A candidate for this attractant is Hedgehog, more familiar as an extracellular signal protein that provides positional information (see Chapter 2): HMG-CoA potentiates Hedgehog signaling in *Drosophila*, and Hedgehog is produced by the prospective gonad. Whether Hedgehog is, indeed, the attractant that directs the migrating primordial germ cells in *Drosophila* is still debated.

8.5 Germ cell differentiation involves a halving of chromosome number by meiosis

The sperm and egg are **haploid**—that is, they contain just one copy of each chromosome, in contrast to the two copies present in the **diploid** somatic cells. At fertilization, therefore, the diploid chromosome number is reinstated. The primordial germ

Fig. 8.8 Meiosis produces haploid cells. Meiosis reduces the number of chromosomes from the diploid to the haploid number. Only one pair of homologous chromosomes is shown here for simplicity. Before the first meiotic division (meiosis I) the DNA replicates, so that each chromosome entering meiosis is composed of two identical chromatids. The paired homologous chromosomes (known as a bivalent) undergo crossing over and recombination, and align on the meiotic spindle at the metaphase of the first meiotic division. The homologous chromosomes separate, and each is segregated into a different daughter cell at the first cell division. There is no DNA replication before the second meiotic division (meiosis II). The daughter chromatids of each chromosome separate and segregate at the second cell division. The chromosome number of the resulting daughter cells is thus halved from the starting diploid number and is haploid.

cells described in the previous sections are diploid, and reduction from the diploid to the haploid state during germ cell development occurs after the germ cells have reached the gonads. The number of chromosomes is reduced to the haploid number during a specialized nuclear division known as meiosis. Meiosis comprises two cell divisions. The chromosomes are replicated before the first division (meiosis I), but not before the second (meiosis II), and this results in the number of chromosomes being reduced by half (Fig. 8.8). The cell divisions during meiosis to produce the egg can be unequal and such an unequal division gives rise to a small structure called a **polar body** (Box 8A).

During prophase of the first meiotic division, replicated homologous chromosomes pair up and undergo **recombination**, in which corresponding DNA sequences are exchanged between the homologs (see Fig. 8.8). Homologous chromosomes usually each carry different versions, or alleles, of many of the genes, and so meiotic recombination generates chromosomes with new combinations of alleles. Meiosis therefore results in gametes whose chromosomes carry different combinations of alleles than the parent. This means that when two gametes come together at fertilization, the resulting animal will differ in genetic constitution from either of its parents. So although we might resemble our parents, we never look identical to them. Meiotic recombination is the main source of the genetic diversity present within populations of sexually reproducing organisms, including humans and other vertebrates.

Development of eggs and sperm follow different courses, even though they both involve meiosis, and the course is determined by the sex of the embryo. In *Drosophila*, there is a continuous production of both eggs and sperm from a population of stem cells in females and males, respectively. As the development of the egg begins with the division of a stem cell (see Fig. 2.16), there is no intrinsic limitation to the number of eggs that a female fly can produce. This contrasts with the way in which eggs develop in mammals, which limits the number of eggs a female mammal can produce.

The development of the egg is known as **oogenesis**, and the main stages of mammalian oogenesis are shown in Figure 8.9 (left panel). In mammals, germ cells undergo proliferative mitotic cell divisions as they migrate to the gonad. In the case of developing eggs, the diploid germline cells, now known as **oogonia**, continue to divide mitotically for a short time within the ovary. After entry into meiosis, they are called **primary oocytes**. Within the ovary, individual oocytes are each enclosed in a sheath of somatic ovarian cells called the **follicle**. Mammalian primary oocytes enter meiosis in the embryo but become arrested in the prophase of the first meiotic division, the stage at which homologous chromosomes pair up and recombination occurs (see Fig. 8.8). Arrest depends on high levels of cyclic AMP within the oocyte, which are generated as a result of signaling from G-protein-coupled receptors. The first meiotic division will not be completed until just before ovulation in the adult, and the second meiotic division not until after fertilization.

Oocytes never proliferate again after entry into meiosis; thus, the number of oocytes at this embryonic stage is generally considered to be the maximum number of eggs a female mammal can ever have. In humans, many oocytes degenerate before puberty,

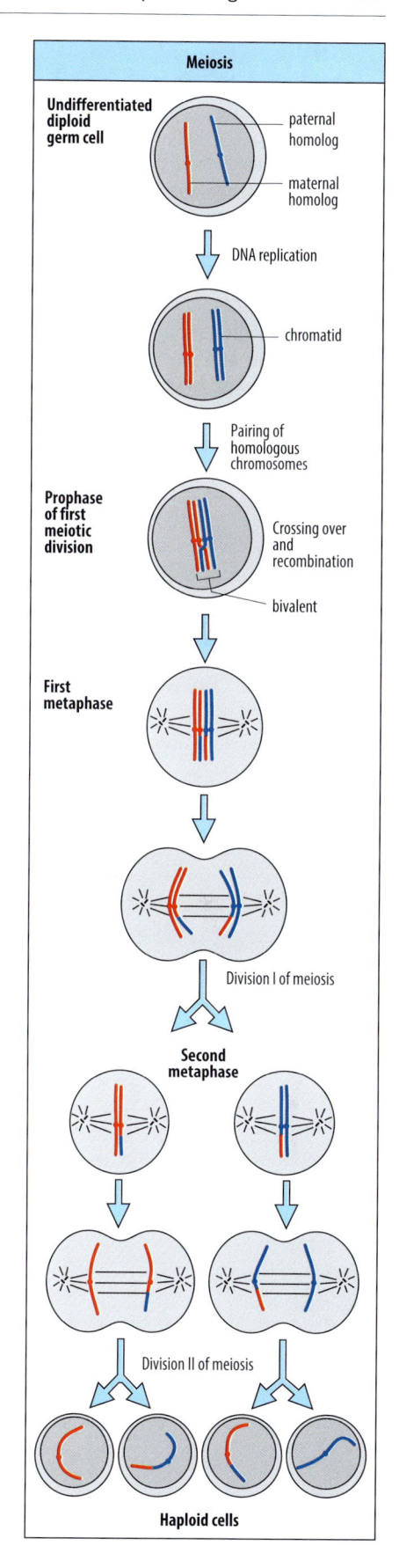

BOX 8A Polar bodies

Polar bodies are small cells formed by meiosis during the development of an **oocyte** into an egg. In this highly schematic illustration (Fig. 1), the segregation of only one pair of chromosomes is shown for simplicity. There are two cell divisions associated with meiosis, and one daughter from each division is almost always very small compared with the other, which becomes the egg.

The timing of fertilization in relation to the development of the oocyte varies in different animals, and in some species meiosis is completed and the second polar body formed only after fertilization (Fig. 1). In general, polar body formation is of little importance for later development, but in some animals the site of formation is a useful marker for the embryonic axes.

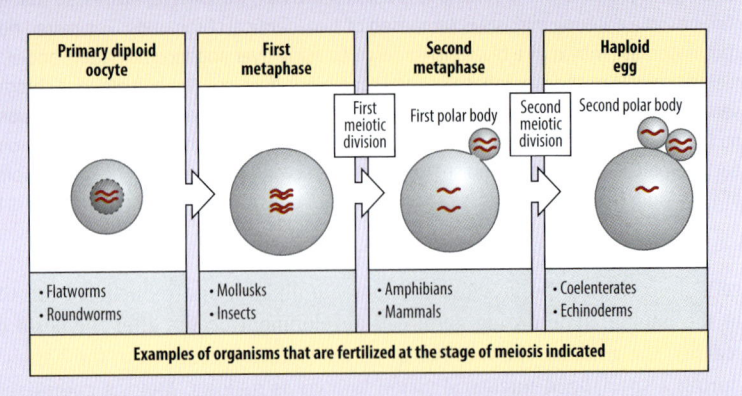

Primary diploid oocyte	First metaphase	Second metaphase	Haploid egg
• Flatworms • Roundworms	• Mollusks • Insects	• Amphibians • Mammals	• Coelenterates • Echinoderms

Examples of organisms that are fertilized at the stage of meiosis indicated

Fig. 1

leaving about 400,000 out of the 700,000 or so present at birth. This number declines with age, with the decline becoming steeper after the mid-thirties until menopause, typically in the fifties (Fig. 8.10). In mammals and many other vertebrates, oocyte development is held in suspension at the first meiosis prophase for months (mice) or for years (humans) after birth. When the female becomes sexually mature, the oocytes start to undergo further development—maturation—as the result of hormonal stimuli. During maturation they grow up to 10-fold in size in mammals, and very much larger in some animals, such as frogs. In some vertebrates, such as frogs and fish, large numbers of oocytes become mature and are released, or ovulated, simultaneously at the end of each reproductive cycle. In mammals, only a relatively small number of oocytes mature during each cycle and are released under the influence of luteinizing hormone. In mammals, meiosis resumes just before ovulation to produce **secondary oocytes** in which the first meiotic division, with the production of one polar body, has been completed and meiosis has proceeded as far as the metaphase of the second meiotic division. Here meiosis is arrested again. The secondary oocytes are ovulated into the female reproductive tract and the second meiotic division is only completed after fertilization, with the production of the second polar body. The time of fertilization in relation to the stage of oocyte development differs in different animal groups (see Box 8A).

A major error in human oocyte meiosis results in the oocyte having an extra chromosome 21, known as a **trisomy** of chromosome 21, as the result of failure of the pair of homologous chromosomes 21 to separate at the first meiotic division. This trisomy is the cause of Down syndrome and is one of the most common genetic causes of congenital conditions and learning difficulties. Trisomies and other errors in chromosome segregation are relatively common in human eggs, far more so than in human sperm, and their incidence rises exponentially as oocytes age. Most human embryos with too many or too few chromosomes are not viable, because of the imbalance in the expression of large numbers of genes. They undergo developmental arrest at the pre-implantation stage, or fail to implant, or are spontaneously aborted during gestation. Chromosomal abnormalities such as those in Down syndrome can be detected prenatally by techniques such as amniocentesis (see Section 3.28) or chorionic villus sampling, but because of the risk of inducing a miscarriage, these invasive techniques are only used if the chance of a fetus with Down syndrome is high. Although the

The development of germ cells **411**

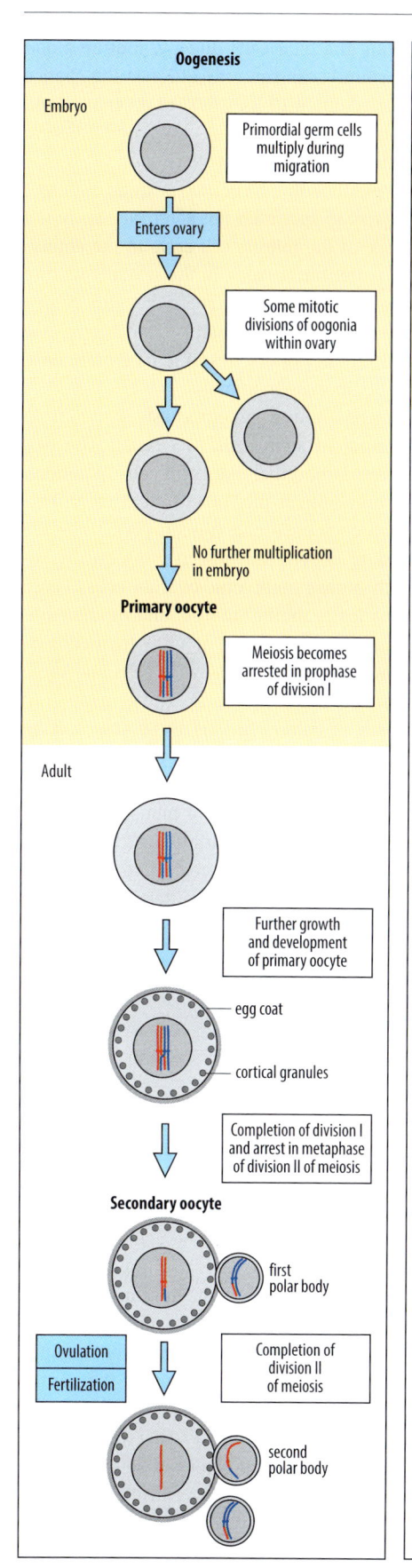

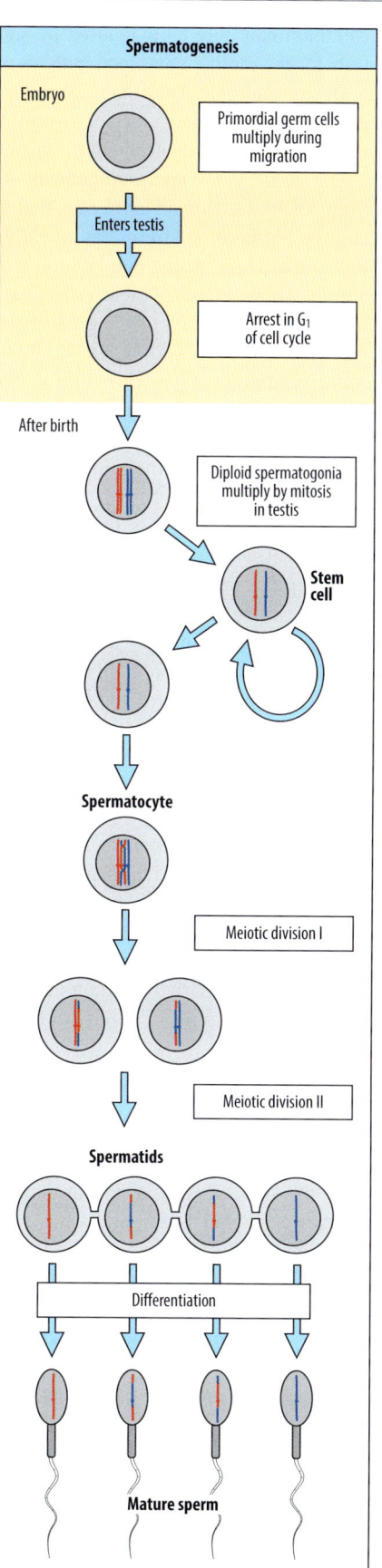

Fig. 8.9 Oogenesis and spermatogenesis in mammals. Left panel: after germ cells enter the embryonic ovary, where they will form oocytes, they divide mitotically a few times and then enter the prophase of the first meiotic division. No further cell multiplication occurs. Further development occurs in the sexually mature adult female. This includes a 10-fold increase in mass (not shown here), the formation of external cell coats, and the development of a layer of cortical granules located under the oocyte plasma membrane. In each reproductive cycle, a group of follicles starts to grow, and oocyte growth and maturation follow; a few eggs are ovulated but most degenerate. Eggs continue to mature in the ovary under hormonal influences, but become blocked in the second metaphase of meiosis, which is only completed after fertilization. Polar bodies are formed at meiosis (see Box 8A). Right panel: germ cells enter the embryonic testis, where they will become sperm, and become arrested at the G_1 stage of the cell cycle. After birth, they begin to divide mitotically again, forming a population of stem cells (spermatogonia). These give off cells, spermatocytes, that then undergo meiosis and differentiate into sperm. Sperm can therefore be produced indefinitely. The final stages of maturation in which sperm become motile take place in the epididymis.

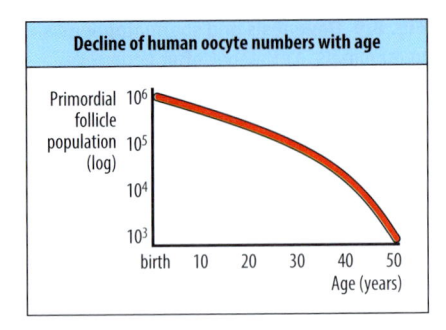

Fig. 8.10 Decline in human oocyte numbers with age. The graph shows the changes in numbers of oocytes in the human ovary with age, with approximately 700,000 present at birth and then decreasing by apoptotic cell death. At puberty there remain about 400,000 oocytes and only a very small number of these, between 400 and 500, are released throughout a woman's lifetime. At menopause, only about 1000 oocytes are left.

From Wallace, W.H., Kelsey, T.W.: **Ovarian reserve and reproductive age may be determined by measurement of ovarian volume by transvaginal sonography.** Hum. Reprod. *2004,* **19***: 1612-1617.*

condition can only be diagnosed by sampling the fetal cells directly, routine blood tests for the mother combined with an ultrasound scan can now estimate the chance of Down syndrome. The combined test is carried out at 10–14 weeks of pregnancy and can also estimate the chance of trisomy 18 and trisomy 13; these other conditions are extremely rare and babies with these conditions usually do not survive.

The strategy for **spermatogenesis**—the production of sperm—is quite different from oogenesis. Diploid germ cells that give rise to sperm do not enter meiosis in the embryo but become arrested at an early stage of the mitotic cell cycle in the embryonic testis. They resume mitotic proliferation after birth. Later, in the sexually mature animal, spermatogonial stem cells give rise to differentiating spermatocytes, which undergo meiosis, each forming four haploid spermatids that mature into sperm (see Fig. 8.9, right panel). Thus, unlike the fixed number of oocytes in female mammals, sperm continue to be produced throughout life in male mammals.

Considerable progress has been made in reconstituting gametogenesis *in vitro* from both pluripotent ES cells and induced pluripotent stem cells (iPS cells). The first step is to differentiate the pluripotent cells into primordial-like germ cells in culture. This has been achieved with both mouse and human pluripotent stem cells. This work revealed differences in the way in which mouse and human germ cells are specified (see Section 8.3). In mice, secondary oocytes have been produced by co-culturing the primordial-like germ cells with mouse embryonic ovarian cells. These oocytes can be fertilized, re-implanted into a foster mother mouse, and produce offspring. Spermatogonial stem cells in mice have been produced at a low frequency by co-culturing the primordial-like germ cells with embryonic testicular cells and these cells can complete spermatogenesis when transplanted into testes. The generation of mature sperm from spermatogonial stem cells in culture has yet to be achieved. In humans, the procedures for producing differentiated germ cells are not as advanced as in mice although early oocytes and spermatogonial stem cells have been produced by co-culturing human primordial-like germ cells with cells from mouse embryonic ovaries and testes, respectively. Therefore, there are many hurdles to overcome before the complete process of human gametogenesis can be reconstituted *in vitro*. The realization of this long-term aim would impact medical issues such as infertility and open up the possibility of using human gametes generated *in vitro* for reproduction. As in the application of other stem-cell technologies, there are crucial issues around safety, which are particularly important as germline cells are being generated, and there are also ethical issues about the use of human gametes produced *in vitro*.

8.6 Oocyte development can involve gene amplification and contributions from other cells

Eggs vary enormously in size among different animals, but they are always larger than the somatic cells. A typical mammalian egg cell is about 0.1 mm in diameter, a frog's egg about 1 mm, and a hen's egg about 3 cm (i.e. the yolk; the white of the egg is extracellular material) (see Fig. 3.2). To achieve growth to such a large size, various mechanisms have evolved, some organisms using several of them together. One strategy is to increase the overall number of gene copies in the developing oocyte, as this proportionately increases the amount of mRNA that can be transcribed, and thus the amount of protein that can be synthesized. Vertebrate oocytes, for example, are arrested in prophase of the first meiotic division and so have double the normal diploid number of genes. Transcription and oocyte growth continue while meiosis is halted.

In addition, insects and amphibians produce many extra copies of selected genes whose products are needed in very large quantities in the egg. During amphibian oocyte development, the rRNA genes are amplified from hundreds of copies to millions; their subsequent translation produces enough ribosomal proteins to ensure sufficient ribosomes for protein synthesis during oocyte growth and early embryonic development. In insects, the genes that encode the proteins of the egg membrane, the chorion, become amplified in the surrounding follicle cells, which produce the egg membrane.

Yet another strategy is for the oocyte to rely on the synthetic activities of other cells. In insects, the nurse cells adjacent to the oocyte make many mRNAs and proteins and deliver them to the oocyte (see Fig. 2.16). Yolk proteins in birds and amphibians are made by liver cells, and are carried by the blood to the ovary, where the proteins enter the oocyte by endocytosis and become packaged into yolk platelets. In *C. elegans*, yolk proteins are made in intestinal cells, and in *Drosophila* in the fat body, from where they are transported to the oocytes. In amphibian eggs, the oocyte is polarized from an early stage onwards, and the yolk platelets accumulate at the vegetal pole. In animals that depend on maternal determinants for their early development, these mRNAs are produced by the mother and delivered into the egg, where they become localized to the appropriate position by microtubule-based transport (see, e.g., Section 3.6).

8.7 Factors in the cytoplasm maintain the totipotency of the egg

Like other cells in the body, mature gametes are specialized cells that have undergone a program of differentiation. As we considered in detail in Chapter 6, except in rare instances cell differentiation does not involve any alteration in the sequence or amount of DNA, but instead involves epigenetic changes. These are chemical modifications to DNA and to the chromosomal proteins associated with it that change the structure of the chromosome locally, so that some genes are selectively silenced while others can still be transcribed. Some somatic cell types, such as muscle cells and nerve cells, do not divide further once differentiation is complete. In cell types that retain the ability to divide, such as liver cells (hepatocytes) or connective tissue fibroblasts, the epigenetic changes associated with that cell type are passed on to the daughter cells, so that they are also hepatocytes or fibroblasts, respectively. Once differentiation is under way, none of these somatic cell types can in normal circumstances give rise to a completely different type of differentiated cell, let alone to the whole range of cell types needed to build a new organism.

Unlike differentiated somatic cells, mature gametes retain characteristics that enable them to combine in fertilization and produce the zygote, a totipotent cell that can develop into a new individual. Factors present in the cytoplasm of the egg are crucial for establishing a totipotent state. This is demonstrated most dramatically by the ability of the cytoplasm of an unfertilized egg to 'reprogram' the nucleus from a differentiated somatic cell so that a totipotent cell is produced that can develop into a new organism. This forms the basis of animal cloning by somatic cell nuclear transfer, which has been achieved in amphibians and some species of mammals, including mice, and is described in Section 6.17.

8.8 In mammals some genes controlling embryonic growth are 'imprinted'

Cloning mammals turned out to be much more difficult to achieve than with some other animals, and this is most probably due to the phenomenon of **genomic imprinting**—by which certain genes are switched off in either the egg or the sperm during their development and remain silenced in the genome of the early embryo. Among the vertebrates, imprinting occurs only in mammals. Evidence for imprinting came initially from the demonstration that the maternal and paternal genomes make different contributions to embryonic development.

Mouse eggs can be manipulated by nuclear transplantation to have either two paternal genomes or two maternal genomes, and can be re-implanted into a female mouse for further development. The embryos that result are known as **androgenetic**

Fig. 8.11 Paternal and maternal genomes are both required for normal mouse development. A normal biparental embryo has both paternal and maternal genomes (contained in pronuclei) in the zygote after fertilization (left panel). Using nuclear transplantation, an egg can be constructed with two paternal or two maternal genomes from an inbred strain. Embryos that develop from an egg with two maternal genomes—gynogenetic embryos (center panel)—have underdeveloped extra-embryonic structures. This results in development being blocked, although the embryo itself is relatively normal and well developed. Embryos that develop from eggs with two paternal genomes—androgenetic embryos (right panel)—have normal extra-embryonic structures, but the embryo itself only develops to a stage where a few somites have formed.

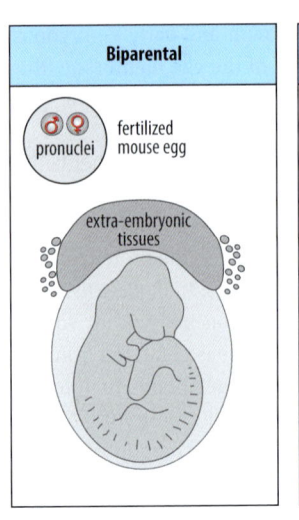

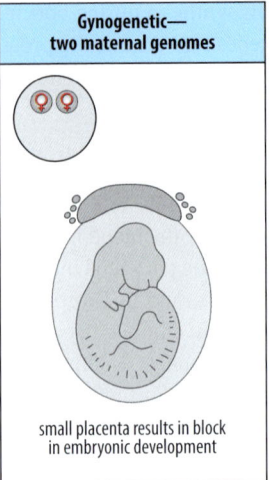

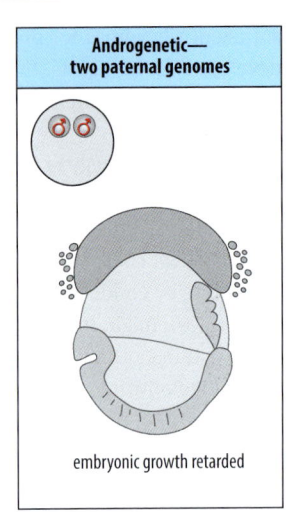

and **gynogenetic** embryos, respectively. Although both kinds of embryo have a diploid number of chromosomes, their development is abnormal. The embryos with two paternal genomes have well-developed extra-embryonic tissues, but the embryo itself is abnormal, and does not proceed beyond a stage at which several somites are present. By contrast, the embryos with diploid maternal genomes have relatively well-developed embryos, but the extra-embryonic tissues—placenta and yolk sac—are poorly developed (Fig. 8.11). These results clearly show that both the maternal and the paternal genomes are necessary for normal mammalian development: the two parental genomes make different contributions, and both are required for the normal development of the embryo and the placenta. This could explain why adult mammals cannot be produced parthenogenetically by activation of an unfertilized egg—because they lack a paternal genome.

Such observations suggested that the paternal and maternal genomes must be epigenetically modified during germ cell differentiation. The paternal and maternal genomes contain the same set of genes, but the imprinting process turns on or off certain genes in either the sperm or the egg, so that they are, or are not, expressed during development. Imprinting implies that the affected genes carry a 'memory' of being in a sperm (paternally derived) or an egg (maternally derived).

Imprinting is a reversible process. The reversibility is important, because in the next generation any of the chromosomes (apart from the Y in mammals) could end up in male or female germ cells. Inherited imprinting is probably erased during early germ cell development, and imprinting is later established afresh during germ cell differentiation (Fig. 8.12). When mammals are cloned using donor nuclei from differentiated somatic cells, these nuclei will not have gone through the normal reprogramming and imprinting process that occurs during germ cell formation, and this could account for the large number of failures and abnormalities in the cloned embryos. Being able to replicate this process of erasing, and then resetting, imprinting accurately in pluripotent stem cells is one of the hurdles that has to be overcome to achieve the goal of reconstituting the entire process of spermatogenesis in culture (see Section 8.5).

The imprinted genes affect not only early development, but also the later growth of the embryo. Further evidence that imprinted genes are involved in embryonic growth comes from studies on chimeras made between normal embryos and androgenetic or gynogenetic embryos. When inner cell mass cells from gynogenetic embryos are injected into normal embryos, growth is retarded by as much as 50%. But when androgenetic inner cell mass cells are introduced into normal embryos, the chimera's

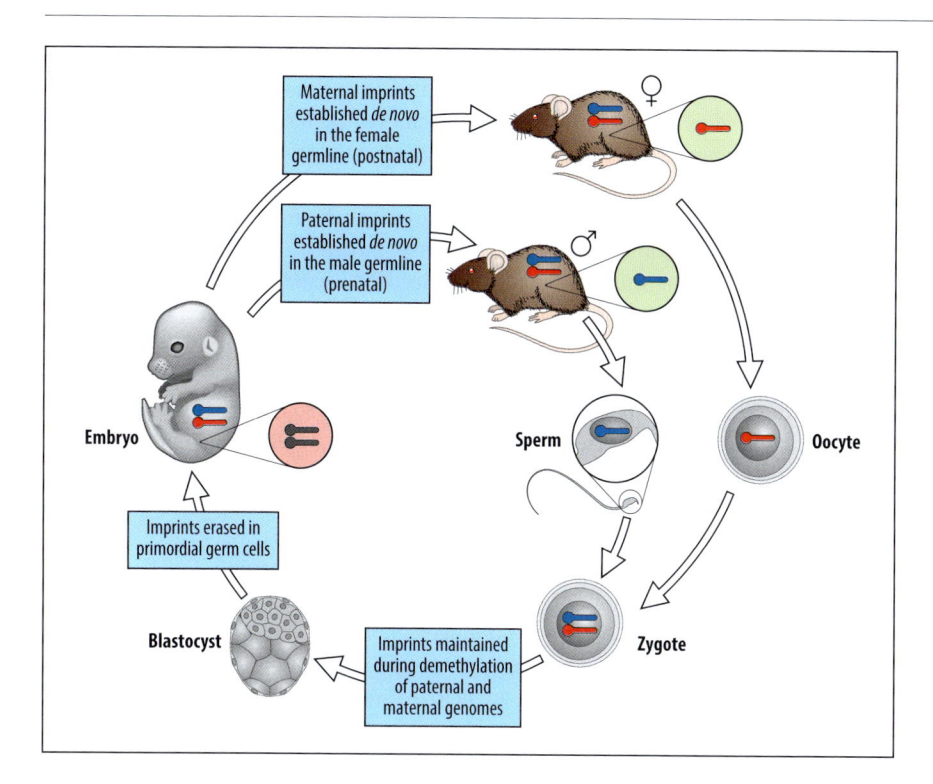

Fig. 8.12 Establishment, maintenance, and erasure of genomic imprints during mouse life cycle. Imprints are acquired in a sex-specific manner in the mature germline (pale green), with male imprints (illustrated by a single blue chromosome for clarity) being established prenatally and female imprints (red chromosome) established postnatally. The imprints are maintained in the embryo despite global changes in demethylation of male and female genomes that occur after fertilization and are also maintained in somatic cells. Imprints are erased in primordial germ cells (gray chromosomes) and then re-set for the next generation.

Adapted from Plasschaert, R.N., Bartolomei, M.S.: **Genomic imprinting in development, growth, behavior and stem cells.** *Development* 2014, **141**: 1805–1813.

growth is increased by up to 50%. The imprinted genes on the male genome thus significantly increase the growth of the embryo.

About 150 imprinted genes have been identified in mammals, some of which encode **non-coding RNAs** (ncRNAs), RNAs with gene-regulatory roles such as short microRNAs, and long ncRNAs. Some imprinted genes are involved in growth control. The insulin-like growth factor IGF-2 is required for embryonic growth; its gene, *Igf2*, is imprinted in the maternal genome—that is, it is turned off, so that only the paternal gene is active (Fig. 8.13). Direct evidence for the *Igf2* gene being imprinted comes from the observation that when sperm carrying a mutated, and thus defective, *Igf2* gene fertilize a normal egg, small offspring result. This is because only very low levels of IGF-2 are produced from the imprinted maternal gene and this is not enough to make up for the loss of IGF-2 expression from the paternal genome. By contrast, if the non-functional mutant gene is carried by the egg's genome, development is normal, with the required IGF-2 activity being provided from the normal paternal gene.

Lying close to the *Igf2* gene on mouse chromosome 7 is the gene *H19*, which encodes an ncRNA that generates two microRNAs. *H19* is imprinted in the opposite direction to the *Igf2* gene and is expressed from the maternal chromosome but not from the paternal one (see Fig. 8.13). By manipulating the chromosomes of an egg, it was possible to make a diploid egg with two maternal genomes, but with one genome having the normal female imprinting of *Igf2* and *H19*, while in the other *Igf2* was not imprinted, as in the male germ cell (*H19* was absent in this genome). This egg could develop normally, unlike an embryo with two normally imprinted maternal genomes. Eighty percent of identified imprinted genes are found in clusters and the arrangement in the imprinted gene cluster containing the *Igf2* gene just outlined illustrates a general trend found in clusters; the cluster consists of several protein-coding genes expressed from the same parental chromosome and an ncRNA expressed from the other parental chromosome.

A possible evolutionary explanation for the reciprocal imprinting of genes that control growth invokes parental conflict theory: the theory that the reproductive strategies of the father and mother are different. Paternal imprinting promotes embryonic

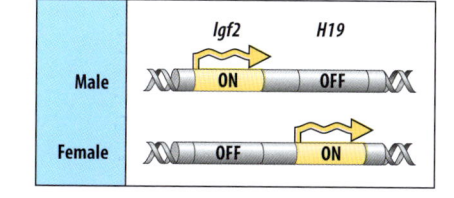

Fig. 8.13 Imprinting of genes controlling embryonic growth. In mouse embryos, the paternal gene for insulin-like growth factor 2 (*Igf2*) is on but the gene on the maternal chromosome is off. By contrast, the closely linked gene *H19* is on in the maternal genome but is off in the paternal genome.

growth, whereas maternal imprinting reduces it, for example. The father wants to have maximal growth for his own offspring, so that his genes have a good chance of surviving and being carried on. This can be achieved by having a large placenta, as a result of producing growth hormone, whose production is stimulated by IGF-2. The mother, who may mate with different males, benefits more by spreading her resources over all her offspring, and so wishes to prevent too much growth in any one embryo. Thus, a gene that promotes embryonic growth is turned off in the mother. Paternal genes expressed in the offspring could be selected to extract more resources from mothers, because an offspring's paternal genes are less likely to be present in the mother's other children. There are, however, many effects of imprinted genes other than on growth.

Imprinting occurs during germ cell differentiation, and so a mechanism is required both for maintaining the imprinted condition throughout development and for erasing it during the next cycle of germ cell development. One mechanism for maintaining imprinting is DNA methylation, an epigenetic modification in which methyl groups are attached to cytosines in DNA. DNA methylation is usually associated with gene silencing. Evidence that methylation of cytosine in cytosine-guanine (CpG) sequences is required for imprinting comes from mice in which the methylation process is aberrant. In these mice, the *Igf2* gene is no longer imprinted and is expressed from the maternal chromosome, as well as the paternal one. In addition to DNA methylation, other factors implicated in imprinting are ncRNAs, repressor proteins of the Polycomb group, and the chemical modification of histone proteins, the proteins that help package DNA in the chromosomes (see Box 6B).

Some ncRNAs are involved in imprinting other genes in the same cluster. The long ncRNA *Airn* (*antisense Igfr2 RNA*), for example, is encoded in the *Igf2r* cluster of imprinted genes. *Igfr2* (which encodes an IGF receptor) and the other two genes in the cluster are expressed from the maternal chromosome. *Airn* itself, however, is expressed from the paternal chromosome and represses expression of the three surrounding genes; it is therefore responsible for the paternal imprinting of the cluster.

Several developmental conditions in humans are associated with imprinted genes. One is Prader–Willi syndrome, which is linked to a loss of expression of the paternal copy of a gene on chromosome 15, usually because of a deletion of a small region of the chromosome containing that gene. Infants fail to thrive and later can become extremely obese; they also show intellectual differences and mental disturbances such as obsessional-compulsive behavior. Angelman syndrome results from the loss of the same region of the maternal chromosome 15. The effect in this case is severe motor and intellectual differences. Beckwith–Wiedemann syndrome is due to a generalized disruption of imprinting on a region of chromosome 11, with one of the genes affected being *IGF2*. There is excessive fetal overgrowth and an increased predisposition to cancer.

SUMMARY

In most of our model organisms, the germline cells are specified by localized cytoplasmic determinants in the egg, whose localization is controlled in the mother by cells surrounding the oocyte. By contrast, germ cells in mammals are not specified by maternal determinants, but by intercellular interactions in the embryo. Once the germ cells are determined, they migrate from their site of origin to the gonads, where further development and differentiation takes place. In mammals, the diploid germ cells in the gonads in female embryos enter the first meiotic prophase, eventually producing the haploid germline cells—the eggs; diploid germ cells in the gonads in male embryos become arrested at an early stage in mitosis and do not undergo meiosis to produce the haploid germline cells—the sperm—until after birth in the mature animal. The number of oocytes in a female mammal is fixed before birth, whereas sperm production in male mammals is continuous throughout adult life. Eggs are always larger than somatic cells, and in some animal groups are very large indeed. To achieve this increased size, specialized somatic cells surrounding the developing oocyte can provide some of the constituents, such as yolk; in addition, some genes producing materials required in large amounts can be amplified in the oocyte.

Both maternal and paternal genomes are necessary for normal mammalian development. Embryos with either diploid maternal or diploid paternal genomes develop abnormally. Certain genes in eggs and sperm are imprinted, so that the activity of the same gene is different depending on whether it is of maternal or paternal origin. Several imprinted genes are involved in growth control of the embryo. Inappropriate imprinting can lead to developmental conditions in humans.

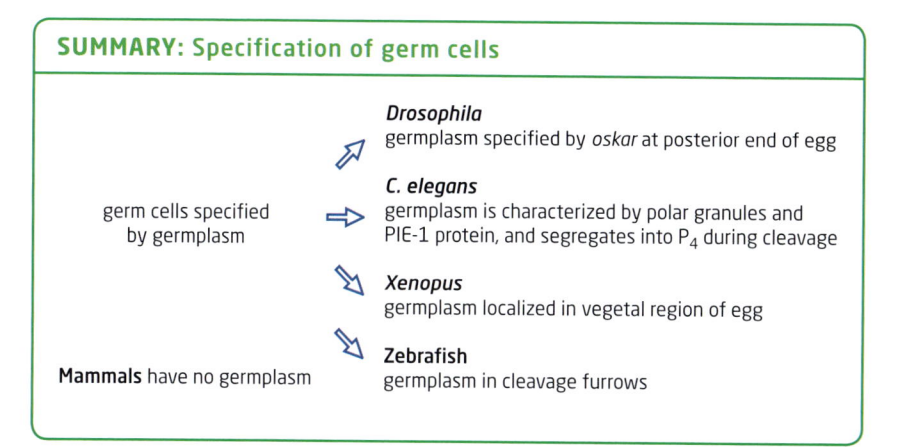

SUMMARY: Specification of germ cells

germ cells specified by germplasm

Drosophila
germplasm specified by *oskar* at posterior end of egg

C. elegans
germplasm is characterized by polar granules and PIE-1 protein, and segregates into P_4 during cleavage

Xenopus
germplasm localized in vegetal region of egg

Zebrafish
germplasm in cleavage furrows

Mammals have no germplasm

Fertilization

Fertilization—involving the fusion of egg and sperm—initiates development. In many marine organisms, such as sea urchins, sperm released into the water by the males are attracted to eggs by chemotaxis—the sperm swim up a gradient of chemical released by the egg. The membranes of the egg and sperm fuse, and the contents of the sperm enter the egg cytoplasm, with the sperm nucleus becoming the male **pronucleus**. In mammals and many other animals, fertilization triggers the completion of meiosis in the egg. One set of maternal chromosomes is retained in the egg and becomes the female pronucleus, while the other set is expelled in the second polar body (see Box 8A). The male and female pronuclei in mammals (see Fig. 1.1) converge and their nuclear membranes break down, allowing the parental chromosomes to mingle, and the egg starts dividing and embarks on its developmental program. Fertilization can be either external, as in frogs and sea urchins, or internal, as in *Drosophila*, mammals, and birds. Of all the sperm released by the male, only one fertilizes each egg. In many animals, including mammals, sperm penetration activates a blocking mechanism in the egg that prevents any further sperm entering. This is necessary because if more than one sperm nucleus enters the egg there will be additional sets of chromosomes, resulting in abnormal development. In humans, embryos with such abnormalities fail to develop. As we shall see in this chapter, there are multiple overlapping mechanisms for ensuring that only one sperm nucleus contributes to the zygote. In all mammals, except rodents, the sperm centrosome enters the egg and becomes the active zygotic centrosome. Centrosomal defects are one cause of sperm-based infertility in humans. In *Xenopus*, the sperm centrosome enters the egg and the formation of its aster influences the cortical rotation that specifies the dorso-ventral axis of the blastula (see Section 3.7).

Eggs and sperm are structurally specialized for fertilization. The specializations of the egg are directed to preventing fertilization by more than one sperm, and then to starting development, whereas those of the sperm are directed to penetrating the egg. The unfertilized egg is usually surrounded by several protective layers outside the plasma membrane, and the eggs of many organisms have a layer of **cortical granules** just beneath the plasma membrane, whose contents are released on fertilization and help to block the entry of more sperm. These barriers together serve to make the egg impenetrable to more than one sperm.

8.9 Fertilization involves cell-surface interactions between egg and sperm

Sperm are motile cells, typically designed for activating the egg and at the same time delivering their nucleus into the egg cytoplasm. Mammalian sperm essentially consist of a nucleus, a few hundred mitochondria to provide an energy source, a centrosome, and a flagellum for movement, and have virtually no cytoplasm (Fig. 8.14). The anterior end is highly specialized to aid penetration. The sperm of *Caenorhabditis elegans* and some other invertebrates are unusual as they more closely resemble tissue cells and move by ameboid movement.

After sperm have been deposited in the mammalian female reproductive tract, they undergo a process known as **capacitation**, which facilitates fertilization and involves membrane remodeling and removal of certain inhibitory factors. There are very few mature eggs—usually one or two in humans and about 10 in mice—waiting to be fertilized, and fewer than a hundred of the millions of sperms deposited reach these eggs.

The sperm must penetrate several physical barriers to enter the egg (Fig. 8.15). In mammalian eggs, the first is a sticky layer of hyaluronic acid and embedded somatic **follicle cells**, which surrounds the egg when it is released. These cells are called **cumulus cells**, as in low-power microscopic images of a released egg they appear to form a 'cloud' around it. The first cloned mouse was called Cumulina as she was produced using the nucleus of a cumulus cell. Hyaluronidase activity on the surface of the sperm head helps it to penetrate this cumulus layer. The sperm next encounters the **zona pellucida**, a layer of fibrous glycoproteins secreted by the oocyte. This also acts as a physical barrier, but sperm are helped to penetrate it by the **acrosomal reaction**—the release of enzymes contained in the **acrosome**, a Golgi-derived vesicle located in the sperm head (see Fig. 8.14).

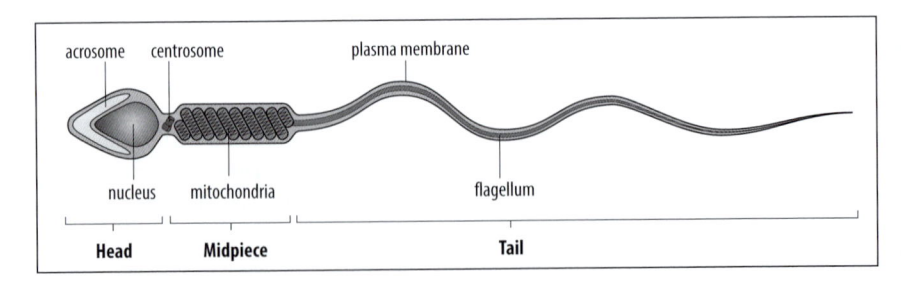

Fig. 8.14 A human sperm. The acrosome at the anterior end of the sperm contains enzymes that are used to digest the protective coats around the egg. The plasma membrane on the head of the sperm contains various specialized proteins that bind to the egg coats and facilitate entry. The sperm moves by its single flagellum, which is powered by mitochondria. The overall length from head to tail is about 60 μm.

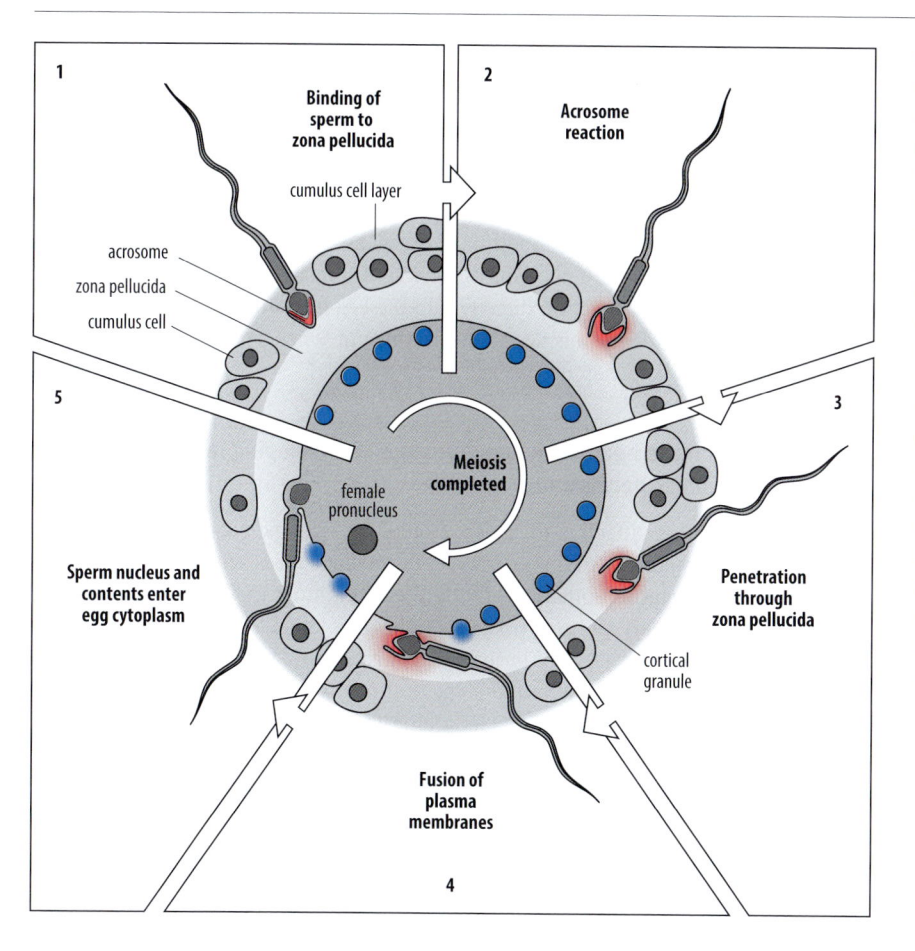

Fig. 8.15 Fertilization of a mammalian egg. After penetrating the follicle-derived cumulus layer, the sperm binds to the zona pellucida (1). This triggers the acrosome reaction (2), in which enzymes are released from the acrosome and break down the zona pellucida. This enables the sperm to penetrate the zona pellucida (3) and bind to the egg plasma membrane. The plasma membrane of the sperm head fuses with the egg plasma membrane and within minutes cortical granules begin to be released from the egg to produce a block to polyspermy (4). (This block is completely established within about 30 minutes.) The contents of the sperm including the nucleus then enter the egg (5). By this time meiosis has been completed in the egg and a female pronucleus has been formed.

Illustration after Alberts, B., et al.: Molecular Biology of the Cell. 2nd edn. New York: Garland Publishing, 1989.

Mammalian sperm carry various cell-surface-associated proteins that are involved in binding to, and penetrating, the zona pellucida. The zona pellucida of human eggs is made up of four glycoproteins (the mouse has three) and failure to assemble a proper zona pellucida from these proteins results in female infertility. ZP2 is a good candidate for the sperm receptor although it is not clear which cell-surface protein on the sperm acts as the ligand. At some point during the penetration of the layers surrounding the egg, a signaling pathway is triggered in the sperm that results in release of the contents of the acrosome by exocytosis. The exact point at which the acrosome reaction is triggered is still debated. The acrosome enzymes break down the oligosaccharide side chains on the zona pellucida glycoproteins, making a hole in the zona pellucida that helps the sperm to approach the egg plasma membrane. In the sperm of many invertebrates, such as the sea urchin, the acrosome reaction also results in the extension of a rod-like acrosome process. This forms by polymerization of actin in the sperm cytoplasm and facilitates contact with the egg plasma membrane.

The acrosome reaction exposes proteins on the sperm surface that can bind to the egg plasma membrane and are involved in the sperm and the egg recognizing each other and allowing fusion of sperm and egg plasma membranes. One of these proteins, called Izumo1, after a Japanese marriage shrine, is a cell-surface protein expressed on the sperm that is essential for sperm–egg binding and successful fertilization in mice. More recently, four new sperm proteins have also been shown to be essential for fertilization in mice, although their precise functions are not yet known. The ligand for Izumo1 on the egg plasma membrane is a folate receptor, renamed Juno, after the Roman goddess of fertility and marriage. Female transgenic mice whose eggs lack Juno and male transgenic mice whose sperm lack Izumo1 are both infertile. The same pair of proteins is involved in binding of sperm to the egg in humans. In mice, two other proteins that span the egg plasma membrane and might organize fusion-competent membrane domains have been identified as being required for fertilization. The molecule that mediates fusion of the sperm and egg plasma membranes is still unknown.

Human eggs can be fertilized in culture and the very early pre-blastocyst embryo transferred to the mother's womb, where it implants and develops normally (see Box 3I). This procedure of *in vitro* fertilization has been of great help to couples who have, for various reasons, difficulty in conceiving. A human egg can even be fertilized by injecting a single intact sperm directly into the egg in culture, a technique known as intracytoplasmic sperm injection, which is useful when infertility is due to the sperm being unable to penetrate the egg.

8.10 Changes in the egg plasma membrane and enveloping layers at fertilization block polyspermy

Although many sperm attach to the layers around the egg, and even reach the plasma membrane, it is important that only one sperm nucleus fuses with the egg nucleus, as the unbalanced chromosomal constitution resulting from the fusion of more than one sperm nucleus (**polyspermy**) would be ultimately lethal to the embryo. Different organisms have different ways of ensuring fertilization by only one sperm. In birds, for example, many sperm penetrate the egg, but only one sperm nucleus fuses with the egg nucleus; the other sperm nuclei are destroyed in the cytoplasm. It is likely in this case that the sperm that enters immediately into the germinal vesicle—the region containing the egg chromosomes—acts as the fertilizing sperm. The DNA of sperm entering the cytoplasm outside this region is probably degraded by cytoplasmic DNases, whose presence has been detected in quail eggs. In sea urchins, amphibians, and mammals, however, the entry of more than one sperm is prevented. This general strategy is called the **block to polyspermy**.

The eggs of sea urchins and *Xenopus* have a two-stage block to polyspermy: an initial rapid reaction followed by a slower one. Both these eggs are externally fertilized and so are likely to be bombarded simultaneously by large numbers of sperm. An immediate means of signaling that one sperm has fused with the egg plasma membrane is therefore desirable. The **rapid block to polyspermy** in sea urchins is triggered within seconds by a transient depolarization of the egg plasma membrane that occurs on sperm–egg fusion. The electrical membrane potential across the plasma membrane goes from −70 mV to +20 mV within a few seconds of sperm entry (Fig. 8.16). The membrane potential then slowly returns to its original level. If depolarization is prevented, polyspermy occurs, but how depolarization blocks polyspermy is not yet clear. A similar rapid electrical block to polyspermy occurs in *Xenopus*.

As the plasma membrane of the sea urchin egg repolarizes, an impenetrable membrane called the **fertilization membrane** is formed around the egg. This is the result

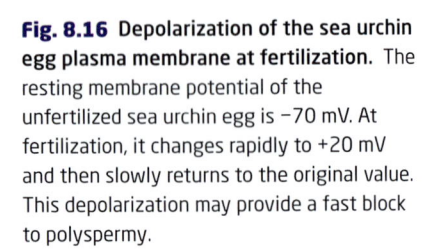

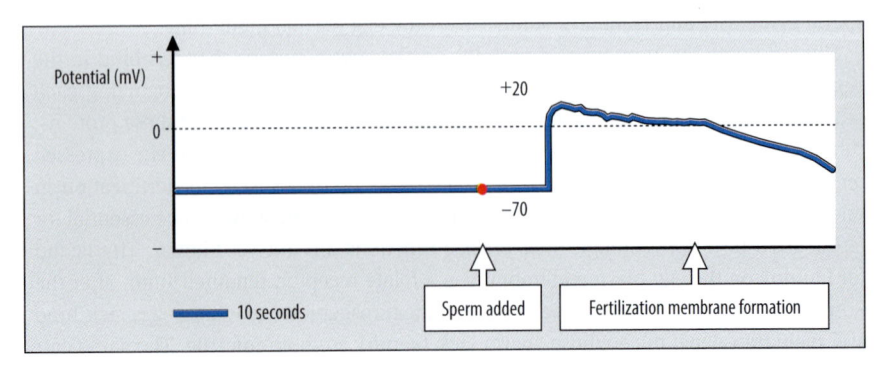

Fig. 8.16 Depolarization of the sea urchin egg plasma membrane at fertilization. The resting membrane potential of the unfertilized sea urchin egg is −70 mV. At fertilization, it changes rapidly to +20 mV and then slowly returns to the original value. This depolarization may provide a fast block to polyspermy.

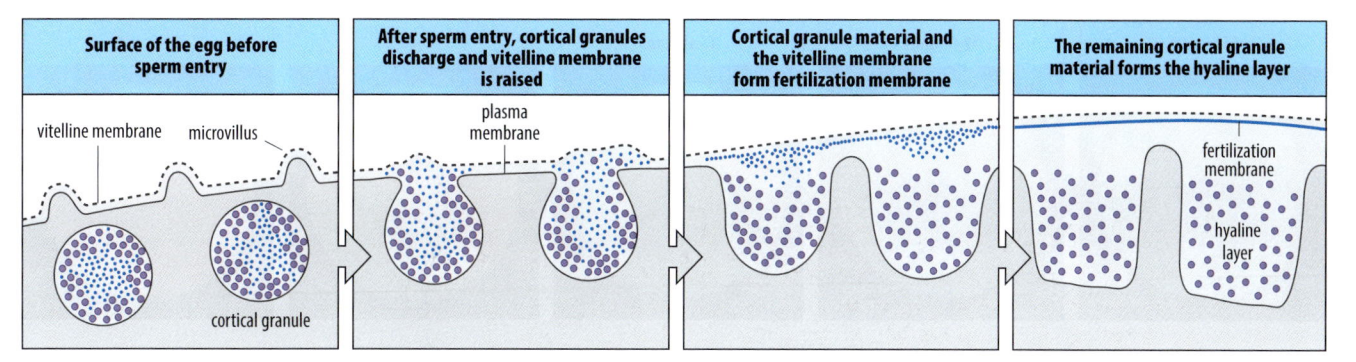

of the **slow block to polyspermy**; it is triggered by a wave of calcium release in the egg due to sperm entry and leads to the cortical granules releasing their contents to the outside of the plasma membrane by exocytosis. The wave of calcium also activates development, as we shall see in the next section. The unfertilized sea urchin egg is surrounded by a **vitelline membrane**, which corresponds to the zona pellucida of mammalian oocytes. Release of the granule contents into the space between the egg plasma membrane and the vitelline membrane causes the vitelline membrane to lift off the plasma membrane. The cortical granule contents cross-link molecules in the vitelline membrane to produce a 'hardened' fertilization membrane, and also provide an additional jelly-like hyaline layer between it and the egg plasma membrane (Fig. 8.17). Some of the released enzymes also cleave sperm-binding proteins. Together, these changes prevent additional sperm from gaining access to the egg. The sea urchin fertilization membrane dissolves at the blastula stage, and the blastula 'hatches' out.

In mammals, including humans, the passage of sperm through the reproductive tract is regulated and relatively few sperm—a few thousand out of the millions of sperm deposited during coitus—reach the oviducts (also called the **Fallopian tubes** in mammals) where fertilization takes place. Polyspermy is prevented in mammals by changes that occur to both the zona pellucida and the egg plasma membrane once the first sperm has entered the egg. There is no rapid electrical block to polyspermy in mammalian eggs, but there is a slow block very similar to that in the sea urchin and *Xenopus*, which develops within 30 minutes to an hour after fertilization. After the first sperm fuses with the plasma membrane, the granules in the egg cortex are released by exocytosis and their contents form a layer immediately outside the egg plasma membrane. One component of the cortical granules is the protease ovastacin, and this cuts ZP2, one of the proteins in the zona pellucida, so that sperm can longer bind to it. However, mechanisms other than this cortical reaction must be involved in the zona pellucida block to polyspermy, as transgenic mice that express ZP2 that cannot be cut have only reduced fertility.

Mammalian eggs also have a membrane block to polyspermy that develops over a similar time scale to the zona pellucida block. The membrane block has been demonstrated in mouse eggs whose zona pellucida has been removed before fertilization *in vitro*. Challenge with a second dose of sperm after the initial fertilization results in very few, if any, of these sperm entering the egg. Juno, the receptor that we noted earlier as essential for sperm–egg recognition, disappears from the egg cell plasma membrane about 40 minutes after fertilization and this is a plausible mechanism for the membrane block to polyspermy in mammalian eggs. When eggs are fertilized *in vitro* by intracytoplasmic sperm injection, the zona pellucida can be removed and the eggs re-fertilized. This is because Juno remains on the egg plasma membrane under these conditions. It is not known how the disappearance of Juno from the egg plasma membrane that normally follows fertilization is triggered.

Fig. 8.17 The cortical reaction at fertilization in the sea urchin. The egg is surrounded by a vitelline membrane, which lies outside the plasma membrane. Membrane-bound cortical granules lie just beneath the egg plasma membrane. At fertilization, the cortical granules fuse with the plasma membrane, and some of the contents are extruded by exocytosis. These join with the vitelline membrane to form a tough fertilization membrane, which then lifts off the egg surface and prevents further sperm entry. Other cortical granule constituents give rise to a hyaline layer, which surrounds the egg under the fertilization membrane.

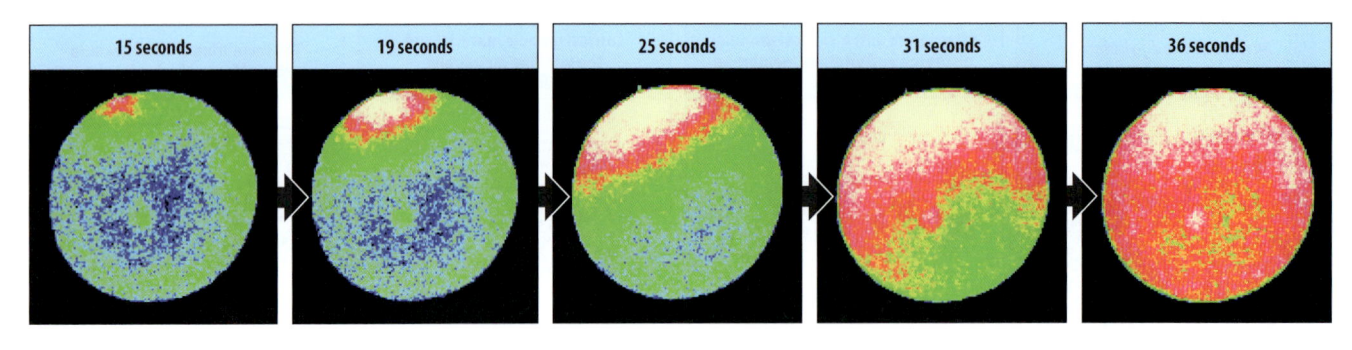

Fig. 8.18 Calcium wave at fertilization.
A series of images showing an intracellular calcium wave at fertilization in a sea urchin egg. The fertilizing sperm has fused just to the left of the top of the egg and triggered the wave. Calcium ion concentration is monitored with a calcium-sensitive fluorescent dye, using confocal fluorescence microscopy. Calcium concentration is shown in false color: red is the highest concentration, then yellow, green, and blue. Times shown are seconds after sperm entry.
Photographs courtesy of M. Whitaker.

8.11 Sperm-egg fusion causes a calcium wave that results in egg activation

At fertilization the egg becomes activated, and a series of events is initiated that result in the start of development. In the sea urchin egg, for example, there is a severalfold increase in protein synthesis, and there are often changes in egg structure, such as the cortical rotation that occurs in amphibian eggs (see Section 3.7). Amphibian and mammalian eggs, which are arrested in second meiotic metaphase before fertilization (see Box 8A), now complete meiosis, after which the egg and sperm pronuclei come together to form the diploid zygotic genome and the fertilized egg enters mitosis. The pronuclear membranes disappear before the pronuclei come together in mice and humans. In mammals, the sperm mitochondria are destroyed, and so all mitochondria in the fertilized egg are of maternal origin. The mitochondrial DNA genome is, therefore, inherited unchanged down the female line apart from the occasional change in DNA sequence as a result of mutation. The rare DNA sequence changes accumulated in the mitochondrial genomes of present-day humans have been used to trace the movements of our earliest human ancestors out of Africa and their spread to different parts of the world.

Fertilization and egg activation are associated with an explosive release of free calcium ions (Ca^{2+}) from stores within the egg, producing a wave of Ca^{2+} that travels across it (Fig. 8.18). The calcium release is triggered by sperm entry. In sea urchin eggs, the wave starts at the point of sperm entry and crosses the egg at a speed of 5–10 μm per second. In mammals, oscillations in calcium concentration occur and these continue for several hours after fertilization until pronucleus formation. Ca^{2+} release at fertilization is triggered by the activity of a sperm-specific enzyme, phospholipase C zeta. This initiates a signaling pathway that leads to production of the second messenger inositol 1,4,5-trisphosphate, which acts on receptors in intracellular membranes to release calcium stored in the endoplasmic reticulum.

The sharp increase in free Ca^{2+} is the natural trigger for egg activation, although eggs produced by transgenic female mice lacking phospholipase C zeta that do not exhibit Ca^{2+} release are still activated. The eggs of many animals can be activated if the Ca^{2+} concentration in the egg cytosol is artificially increased—for example, by direct injection of Ca^{2+}. Conversely, preventing calcium increase by injecting agents that bind it, such as the calcium chelator EGTA, blocks activation. It has been known for many years that *Xenopus* eggs can be activated simply by prodding them with a glass needle; this is due to a local influx in calcium at the site of prodding that triggers a calcium wave.

Calcium initiates the completion of meiosis in the fertilized egg by acting on proteins that control the cell cycle. The unfertilized *Xenopus* egg is maintained in the metaphase of the second meiotic division by the presence of high levels of a protein complex called maturation-promoting factor (MPF), which is a complex of a cyclin-dependent kinase (Cdk), Cdk1, and its partner cyclin B. The effects of MPF are due to phosphorylation of various protein targets by the kinase. For the egg to complete meiosis, the level of MPF activity must be reduced (Fig. 8.19). Similar Cdk–cyclin B

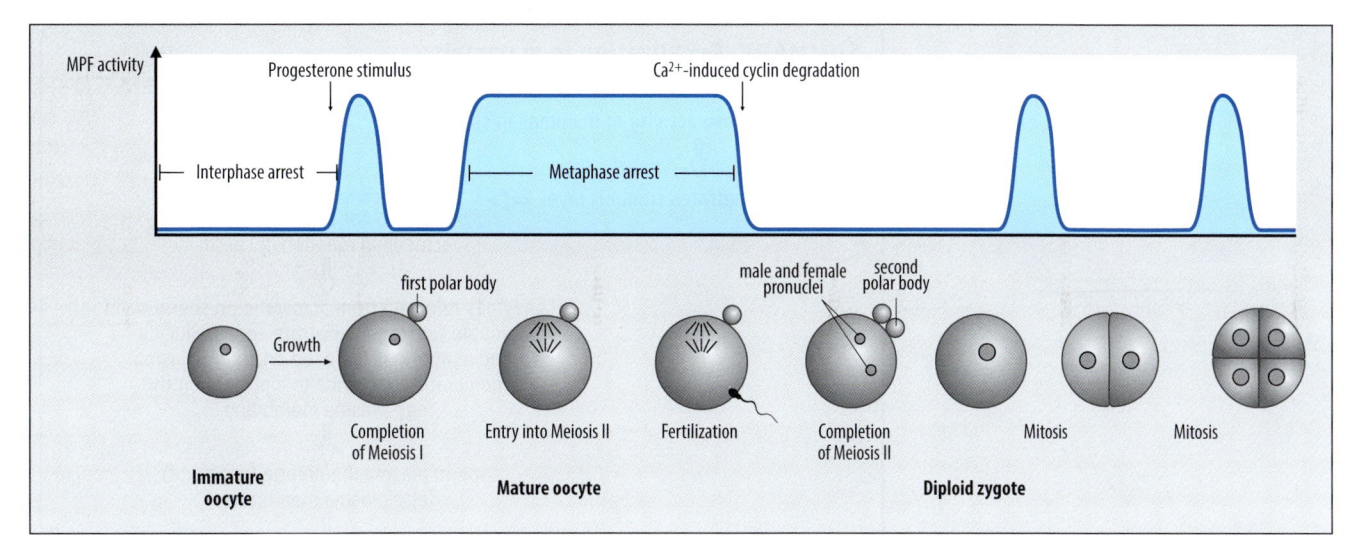

Fig. 8.19 Profile of maturation-promoting factor (MPF) activity in early *Xenopus* development. In the immature *Xenopus* oocyte the cell cycle is arrested. On receipt of a hormonal progesterone stimulus it enters, and completes, the first meiotic division, with the formation of the first polar body. It enters the second meiotic division but becomes arrested again in metaphase. The egg is laid at this point. At fertilization, a monotonic calcium wave leads to completion of meiosis, and the second polar body is formed. The zygote starts to cleave rapidly by mitotic divisions. MPF activity rises sharply just before each division of the meiotic and mitotic cell cycles, remains high during mitosis, and then decreases abruptly and remains low between successive mitoses.

complexes control the mitotic cell cycle (see Section 11.2). The calcium wave results in the activation of the enzyme calmodulin-dependent protein kinase II. The activity of this kinase indirectly results in the degradation of the cyclin component of MPF, which deactivates its Cdk1 kinase and allows the egg to complete meiosis. Following this, sperm and egg genomes undergo global changes in demethylation and male and female pronuclei are formed. The pronuclei then come together and the zygote moves on to the next stage of its development, which is entry into the mitotic cell cycles of cleavage.

SUMMARY

The fusion of sperm and egg at fertilization stimulates the egg to start dividing and developing. Both sperm and egg have specialized structures relating to fertilization. The initial binding of sperm to egg is mediated by molecules in the zona pellucida (in mammals) or the corresponding vitelline membrane (in sea urchins). Sperm binding leads to the release of the contents of the sperm acrosome, which facilitates the penetration of the sperm through the layers surrounding the egg and allows it to reach the egg plasma membrane. A protein on the sperm plasma membrane binds to a protein on the egg plasma membrane and this is followed by fusion of the two cells. A block to polyspermy allows only one sperm to fuse with the egg and deliver its nucleus into the egg cytoplasm. In mammals, one block results from the release of egg cortical granule contents to the exterior that modify the zona pellucida, and the second involves changes in the egg plasma membrane. A key role in egg activation after fertilization is played by the release of available calcium ions into the cytosol, which spread in a wave from the site of sperm fusion. In mammals, and most other vertebrates, fertilization triggers the completion of the second meiotic division; the male and female haploid pronuclei give rise to the zygote nucleus, and the fertilized egg divides.

SUMMARY: Fertilization in mammals

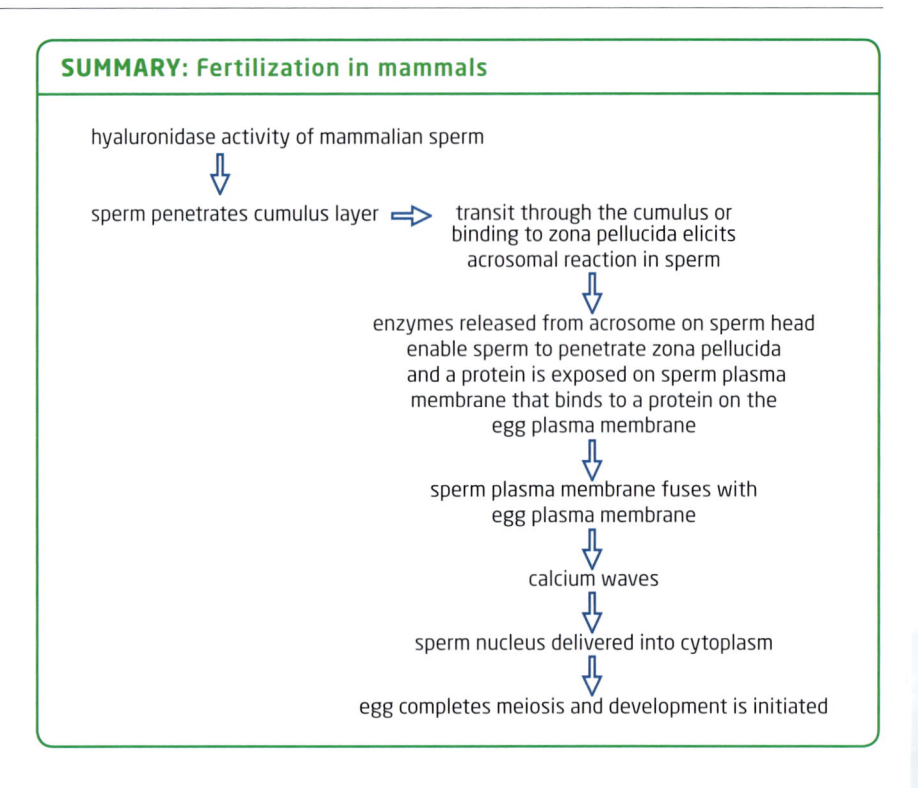

hyaluronidase activity of mammalian sperm
⇩
sperm penetrates cumulus layer ⇒ transit through the cumulus or binding to zona pellucida elicits acrosomal reaction in sperm
⇩
enzymes released from acrosome on sperm head enable sperm to penetrate zona pellucida and a protein is exposed on sperm plasma membrane that binds to a protein on the egg plasma membrane
⇩
sperm plasma membrane fuses with egg plasma membrane
⇩
calcium waves
⇩
sperm nucleus delivered into cytoplasm
⇩
egg completes meiosis and development is initiated

Determination of the sexual phenotype

In organisms that produce two phenotypically different sexes, sexual development involves the modification of a basic developmental program. Early development is similar in both male and female embryos, with sexual differences only appearing at later stages. In the organisms considered here, somatic sexual phenotype—that is, the development of the individual as either male or female—is genetically fixed at fertilization by the chromosomal content of the gametes that fuse to form the fertilized egg. In mammals, for example, sex is determined by the Y chromosome: males are XY, females are XX.

Even among vertebrates, however, sex is not always determined by which chromosomes are present; in alligators, it is determined by the environmental temperature during incubation of the embryo, and some fish can switch sex as adults in response to environmental conditions. In insects, there is a wide range of different sex-determining mechanisms. Intriguing though these are, we focus here on those organisms in which the genetic and molecular basis of sex determination is best understood—mammals, *Drosophila*, and *Caenorhabditis elegans*—in all of which sex is determined by chromosomal content, although by quite different mechanisms.

We first consider the determination of the somatic sexual phenotype. We then deal with the determination of the sex of the germ cells—whether they become eggs or sperm—and finally consider how the embryo compensates for the difference in chromosomal composition between males and females.

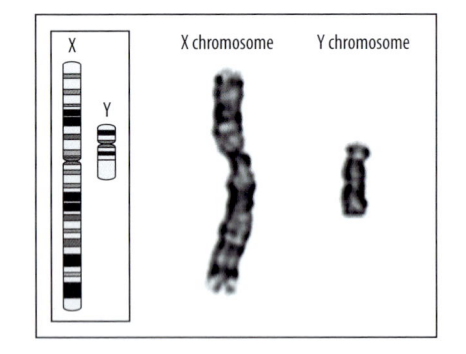

Fig. 8.20 The sex chromosomes in humans. If two X chromosomes (XX) are present, a female develops, whereas the presence of a Y chromosome (XY) leads to development of a male. The inset shows a diagrammatic representation of the banding in the chromosomes, which represent regions of increased chromatin condensation.

8.12 The primary sex-determining gene in mammals is on the Y chromosome

The genetic sex of a mammal is established at the moment of conception, when the sperm introduces either an X or a Y chromosome into the egg (Fig. 8.20). Eggs contain one X chromosome; if the sperm introduces another X the embryo will be female, if a Y it will be male. The presence of a Y chromosome causes testes to develop, and the hormones they produce switch the development of all somatic tissues to a characteristic male pathway and suppress female development. In the absence of a Y chromosome, the development of somatic tissues is along the female pathway. Specification of a gonad as a testis is controlled by a single gene on the Y chromosome, the **sex-determining region of the Y chromosome** (*SRY* in humans and *Sry* in mice), whose product was formerly known as testis-determining factor.

Evidence that a region on the Y chromosome actively determines maleness first came from two unusual human syndromes: Klinefelter syndrome, in which individuals have two X chromosomes and one Y (XXY), but are still males; and Turner syndrome, in which individuals have just one X chromosome (XO) and are female. Both these types of individuals have some differences; those with Klinefelter syndrome are infertile males with small testes, whereas females with Turner syndrome do not produce eggs. There are also rare cases of XY individuals who are phenotypically female, and XX individuals who are phenotypically male. This is due to part of the Y chromosome being lost (in XY females) or to part of the Y chromosome being transferred to the X chromosome (in XX males). This can happen during meiosis in the male germ cells as the X and Y chromosomes are able to pair up and crossing over can occur between them. On rare occasions, this crossing over transfers the *SRY* gene from the Y chromosome onto the X (Fig. 8.21), thus leading to sex reversal.

The sex-determining region alone is sufficient to specify maleness, as shown by an experiment in which the mouse equivalent of the *SRY* gene (*Sry*) was introduced into the eggs of XX mice. These transgenic embryos developed as males, even though they lacked all the other genes on the Y chromosome. The presence of the *Sry* gene, which encodes a transcription factor, resulted in these XX embryos developing testes instead of ovaries. In these embryos, as in normal males, *Sry* was expressed in the somatic cells of the developing gonad just before it started to differentiate and drove testis development by triggering the differentiation of these bipotential somatic cells into testis-specific Sertoli cells rather than ovary-specific granulosa cells. However, these XX + *Sry* males were not completely normal but were infertile, as other genes on the Y chromosome are necessary for the development of the sperm.

Sertoli cells, whose differentiation is triggered by *Sry* expression, are essential for testis formation and for spermatogenesis, as they instruct and 'nurse' the germ cells that migrate into the gonad. They also secrete **Müllerian-inhibiting substance**, a member of the transforming growth factor-β family of signaling molecules, which suppresses a female sexual phenotype by causing regression of an embryonic precursor of the female reproductive organs, the Mullerian duct (see Section 8.13). Production of the transcription factor Sox9 is increased in Sertoli cells. *Sox9* is a direct gene target of SRY, and recently the enhancer (see Chapter 6) that regulates *Sox9* expression and is essential to initiate testis development has been identified. When the enhancer is deleted in transgenic mice, this results in levels of *Sox9* expression in the gonads of XY females comparable to that seen in XX gonads. Fibroblast growth factor (FGF)-9 is also required for Sertoli cell differentiation and male mice lacking the *Fgf9* gene develop as females. This is because FGF represses expression of *Wnt4*, which promotes ovary development. Therefore, development of a testis needs active repression of the female developmental pathway (Fig. 8.22).

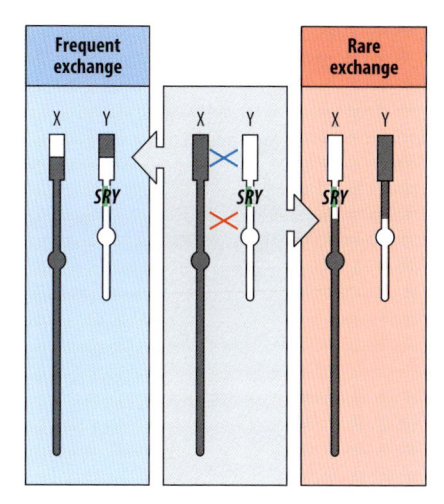

Fig. 8.21 **Sex reversal in humans due to chromosomal exchange.** At meiosis in male germ cells, the X and Y chromosomes pair up (center panel). Crossing over of the distal 'pseudo-autosomal' region (upper blue cross) does not affect sexual development (left panel). On rare occasions, crossing over can involve a larger segment that includes the *SRY* gene (lower red cross), so that the X chromosome now carries this male-determining gene (right panel).

Illustration after Goodfellow, P.N., Lovell-Badge, R.: **SRY and sex determination in mammals.** *Ann. Rev. Genet. 1993,* **27***: 71–92.*

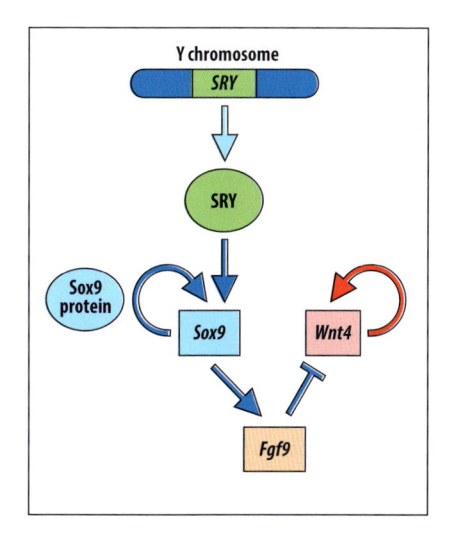

Fig. 8.22 Genetic interactions that determine the sex of the gonads in male mammals. In males, SRY protein upregulates *Sox9* expression in cells in the developing testis, and *Sox9* expression becomes self-regulating. Sox9 protein upregulates expression of *Fgf9*. FGF-9 signaling in males represses *Wnt4* expression, thus preventing Wnt-4 signaling which would lead to ovary development.

Adapted from Jameson, S.A., et al.: **Testis development requires the repression of Wnt4 by Fgf signaling.** *Dev. Biol. 2012,* **370***: 24–32.*

8.13 Mammalian sexual phenotype is regulated by gonadal hormones

Although the sex of the gonads is genetically determined, all the other cells in the mammalian body are neutral, irrespective of their chromosomal sex. It does not matter if they are XX or XY, as any future sex-specific development they undergo is controlled by hormones. The primary role of the testis in directing male development was originally demonstrated by removing the prospective gonadal tissue from early rabbit embryos. All the embryos developed as females, irrespective of their chromosomal constitution. To develop as a male, therefore, a testis must be present. The testis exerts its effect on sexual differentiation of somatic tissues primarily by secreting the hormone testosterone, which is produced by the interstitial Leydig cells. The differentiation of these cells in the testis is promoted by signals from the Sertoli cells that include Desert Hedgehog, one of the three vertebrate Hedgehog signal proteins.

The gonads in mammals develop in close association with the **mesonephros**; this is an embryonic kidney that contributes to both male and female reproductive organs. Associated with the mesonephros on each side of the body are the **Wolffian ducts**, which run down the body to the cloaca, an undifferentiated opening. Another pair of ducts, the **Müllerian ducts**, run parallel to the Wolffian ducts and also open into the cloaca. In early mammalian development, before gonadal differentiation, both sets of ducts are present (Fig. 8.23). In females, in the absence of the testes, the Müllerian ducts develop into the **oviducts**, also called the Fallopian tubes in mammals, which transport eggs from the ovaries to the uterus, while the Wolffian ducts degenerate. In males, the Müllerian ducts degenerate as a result of the production of Müllerian-inhibiting substance, while—under the influence of testosterone—the Wolffian ducts develop into a vas deferens on each side, which are the ducts that carry sperm to the penis. Recent work suggests that the degeneration of the Wolffian ducts in female

Fig. 8.23 Development of the gonads and related structures in mammals. Left panel: early in development, there is no difference between males and females in the structures that give rise to the gonads and related organs. The future gonads lie adjacent to the mesonephros, which are embryonic kidneys that are not functional in adult mammals. (The true kidney develops from the metanephros, from which the ureter carries urine to the bladder.) Two sets of ducts are present: the Wolffian ducts, which are associated with the mesonephros, and the Müllerian ducts. Both ducts enter the cloaca. Top right panel: after testes develop in the male, their secretion of Müllerian-inhibiting substance results in degeneration of the Müllerian duct by programmed cell death, whereas the Wolffian duct becomes the vas deferens, carrying sperm from the testis. Bottom right panel: in females, the Wolffian duct disappears, also by programmed cell death, and the Müllerian duct becomes the oviduct. The uterus forms at the end of the Müllerian ducts.

Illustration after Higgins, S.J., et al.: **Induction of functional cytodifferentiation in the epithelium of tissue recombinants II. Instructive induction of Wolffian duct epithelia by neonatal seminal vesicle mesenchyme.** *Development* 1989, **106**: 235-250.

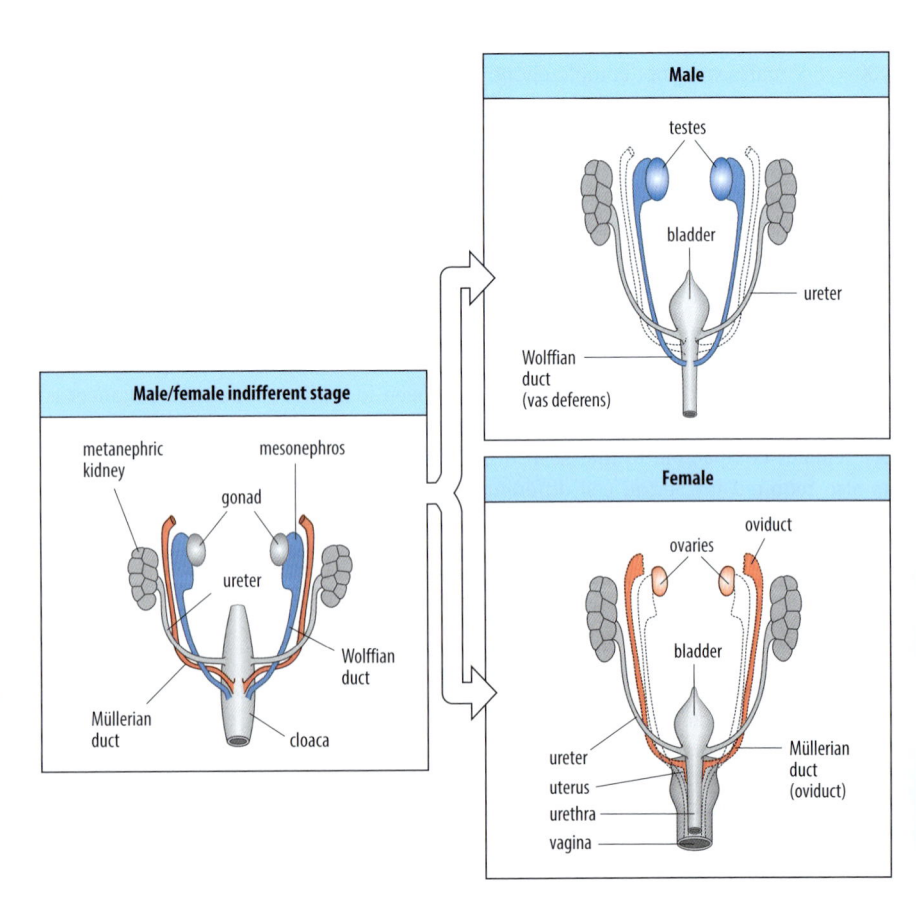

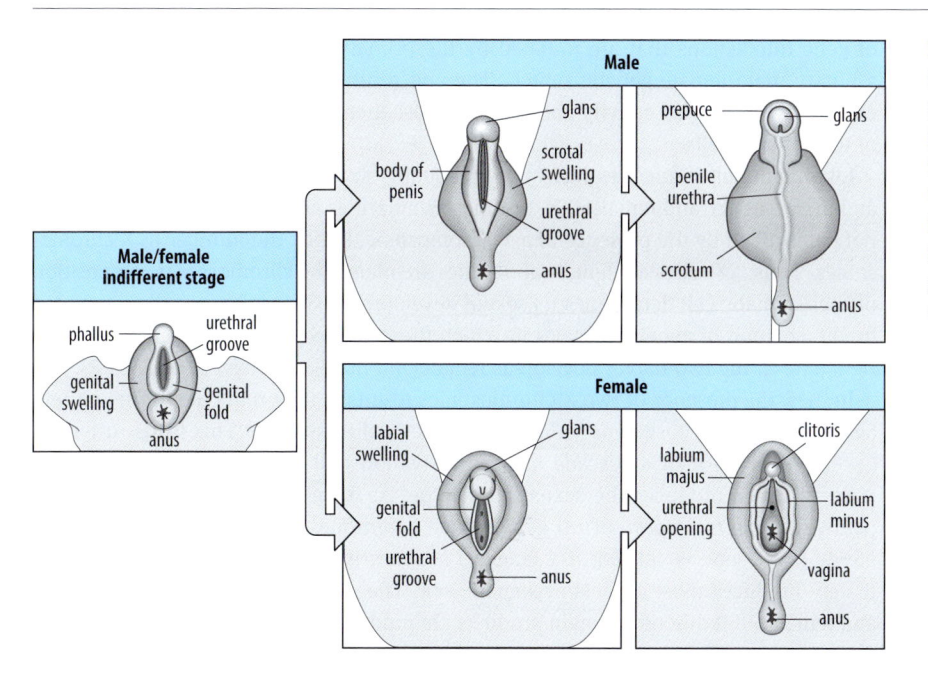

Fig. 8.24 Development of the genitalia in humans. At an early embryonic stage, the genitalia are the same in males and females (left panel). After testis formation in males, the phallus and the genital fold give rise to the penis (top right panels), whereas in females they give rise to the clitoris and the labium minus (bottom right panels). The genital swelling forms the scrotum in males and the labium majus in females.

embryos is not simply due to the lack of male-specific hormones but is actively promoted, arguing against the long-held view that female is the default sexual phenotype.

The main secondary sexual characters that distinguish males and females are the reduced size of mammary glands in males, and the development of a penis and a scrotum in males instead of the clitoris and labia of females (Fig. 8.24). At early stages of embryonic development, the genital regions of males and females are indistinguishable. Differences only arise after gonad development, because of the action of testosterone in males. For example, in humans, the phallus gives rise to the clitoris in females and the end of the penis in males. The full development of the secondary sexual characteristics takes place at puberty under the control of testosterone produced by the testes in males and estrogen produced by the ovaries in females, and the testes start to produce sperm and the ovaries start to release eggs.

The role of hormones in sexual development is illustrated by rare cases of unusual differences in sexual development. Certain XY males develop as phenotypic females in external appearance, even though they have testes and secrete testosterone. They have a mutation that renders them insensitive to testosterone because they lack the testosterone receptor, which is normally present throughout the body. Conversely, genetic females with a completely normal XX constitution can develop as phenotypic males in external appearance if they are exposed to male hormones during their embryonic development.

Sex-specific behavior is also affected by the hormonal environment as a result of the effects of hormones on the brain. For example, male rats castrated after birth develop the sexual behavioral characteristics of genetic females.

8.14 The primary sex-determining factor in *Drosophila* is the number of X chromosomes and is cell autonomous

The external sexual differences between *Drosophila* males and females are mainly in the genital structures, although there are also some differences in bristle patterns and pigmentation, and male flies have a sex comb on the first pair of legs. In flies, sex determination of the somatic cells is cell autonomous—that is, it is specified on a cell-by-cell basis—and there is no process resembling the control of somatic sexual differentiation by hormones. Somatic sexual development is the result of a series

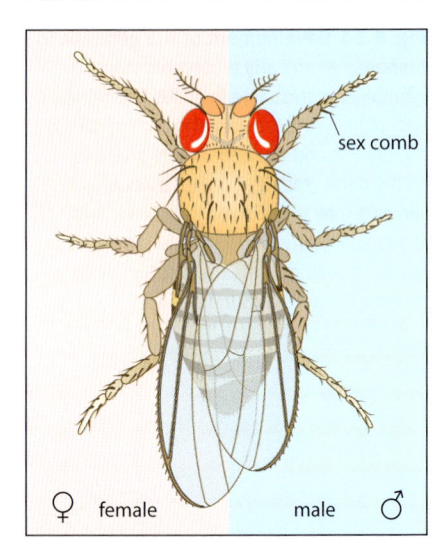

428 Chapter 8 Germ cells, fertilization, and sex determination

Fig. 8.25 A *Drosophila* female/male genetic mosaic. The left side of the fly is composed of XX cells and develops as a female, whereas the right side is composed of X cells and develops as a male. The male fly has smaller wings, a special structure—the sex comb—on its first pair of legs, and different genitalia at the end of the abdomen (not shown).

of gene interactions that are initiated by the primary sex signal, and which act on a binary male/female genetic switch. The end result is the expression of just a few effector genes, whose activity controls the subsequent male or female differentiation of the somatic cells.

Like mammals, fruit flies have two unequally sized sex chromosomes, X and Y, and males are XY and females XX. But these similarities are misleading. In flies, sex is not determined by the presence of a Y chromosome, but by the number of X chromosomes. Thus, XXY flies are female and X flies are male. The chromosomal composition of each somatic cell determines its sexual development. This is beautifully illustrated by the creation of genetic mosaics in which the left side of the animal is XX and the right side X: the two halves develop as female and male, respectively (Fig. 8.25).

In flies, the presence of two X chromosomes results in the production of the protein Sex-lethal (Sxl), whose gene is located on the X chromosome. This leads to female development through a cascade of gene activation that first determines the sexual state and then produces the sexual phenotype. At the end of the sex-determination pathway is the *transformer* (*tra*) gene, which determines how the mRNA of the *doublesex* (*dsx*) gene is spliced; *dsx* encodes a transcription factor whose activity ultimately produces most aspects of somatic sex. The *dsx* gene is active in both males and females, but different protein products are produced in the two sexes as a result of the sex-specific RNA splicing (Fig. 8.26). Males and females thus express similar but distinct Doublesex (Dsx) proteins, which act in somatic cells to induce expression of sex-specific genes, as well as to repress characteristics of the opposite sex. Production of the male form of the protein in the absence of Sxl results in development as a male, but in the presence of Sxl, *tra* mRNA undergoes RNA splicing, and this, together with the actions of the Transformer-2 protein, leads to the female form of the Dsx protein being made, resulting in development as a female.

Once the *Sxl* gene is activated in females, it remains activated through an autoregulatory mechanism, which results in the Sxl protein being synthesized throughout female development. Early expression of *Sxl* in females occurs through activation of a promoter, P_e, at about the time of syncytial blastoderm formation. Sxl protein is synthesized and accumulates in the blastoderm of female embryos. At the cellular blastoderm stage, another promoter for *Sxl*, P_m, becomes active in both males and females

Sex determination in *Drosophila*			
Primary signal	Stable binary genetic switch	Transducers of the sexual state	Effectors of sexual phenotype
X chromosomes	*Sxl*	e.g. *tra*	splicing *dsx* → dsx^f Female / dsx^m Male
XX	ON	ON	dsx^f ♀
X	OFF	OFF	dsx^m ♂

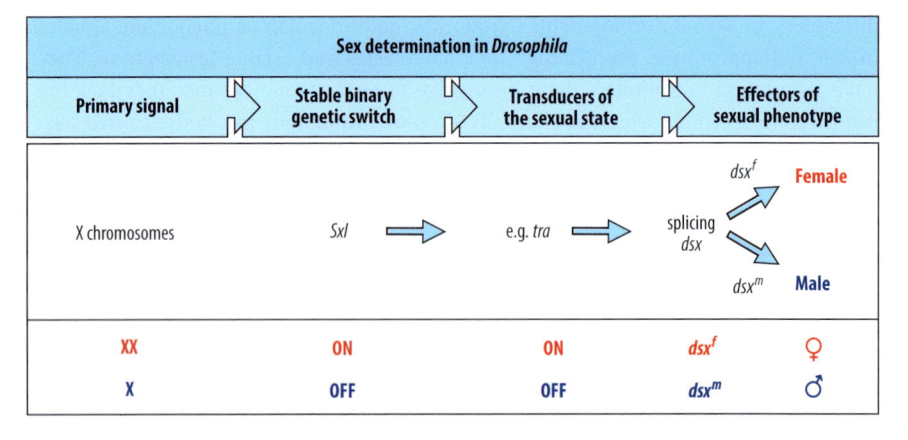

Fig. 8.26 Outline of the sex-determination pathway in *Drosophila*. The number of X chromosomes is the primary sex-determining factor, and in females the presence of two X chromosomes activates the gene *Sex-lethal* (*Sxl*). This produces Sex-lethal protein, whereas no Sex-lethal protein is made in males, which have only one X chromosome. The activity of *Sex-lethal* is transduced via the *transformer* gene (*tra*) and causes sex-specific splicing of *doublesex* RNA (dsx^f), such that the cells follow a female developmental pathway. In the absence of Sex-lethal protein, the splicing of *doublesex* RNA to give dsx^m RNA leads to male development.

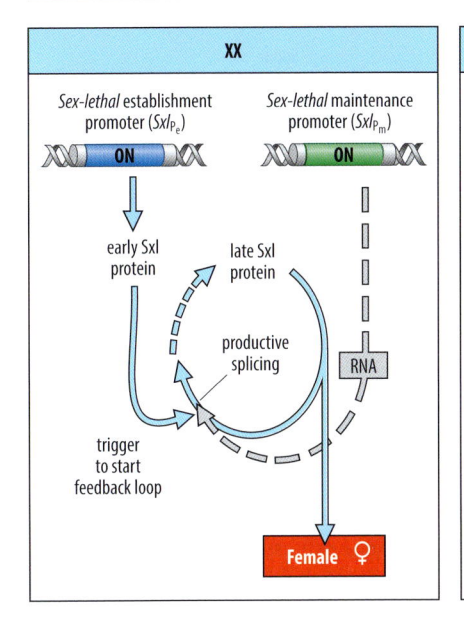

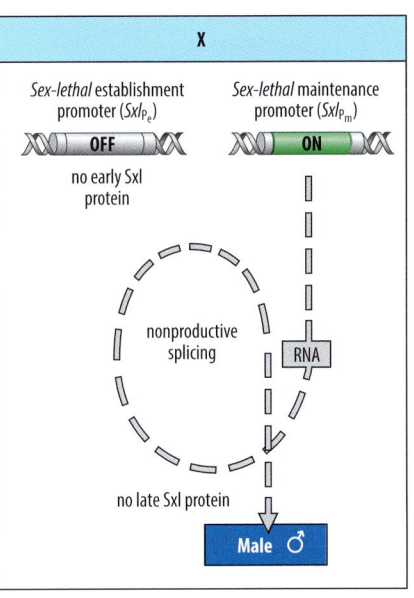

Fig. 8.27 Production of Sex-lethal protein in *Drosophila* sex determination. When two X chromosomes are present, the early establishment promoter (P$_e$) of the *Sex-lethal* (*Sxl*) gene is activated at the syncytial blastoderm stage in future females, but not in males. This results in the production of Sxl protein. Later, at the blastoderm stage, the maintenance promoter (P$_m$) of *Sxl* becomes active in both females and males, and P$_e$ is turned off. The *Sxl* RNA is only correctly spliced if Sxl protein is already present, which is only in females. A positive-feedback loop for Sxl protein production is thus established in females. The continued presence of Sxl protein initiates a cascade of gene activity leading to female development. If no Sxl protein is present, male development ensues.
Illustration after Cline, T.W.: The Drosophila *sex determination signal: how do flies count to two?* Trends Genet. *1993, 9: 385-390.*

and P$_e$ is shut off, but the sex is already determined. Splicing of the RNA transcribed from P$_m$ into functional *Sxl* mRNA needs some Sxl protein to be present already, and so can only happen in females (Fig. 8.27).

How does the number of X chromosomes control these key sex-determining genes? In *Drosophila*, the mechanism involves interactions between the products of so-called numerator genes on the X chromosome and of genes on the autosomes, as well as maternally specified factors. Essentially, in females, the double dose of numerator proteins activates *Sxl* by binding to sites in the P$_e$ promoter (see Fig. 8.27), overcoming the repression of Sxl by autosomally encoded proteins that would occur in males.

The pathway outlined in Figure 8.26 is an oversimplification, as Dsx does not control all aspects of somatic sexual differentiation in *Drosophila*. There is an additional branch of the sex-differentiation pathway downstream of *tra*, which controls sexually dimorphic aspects of the nervous system and sexual behavior. This branch includes the gene *fruitless*, whose activity has been shown to be necessary for male sexual behavior.

In other dipteran insects, the same general strategy for sex determination is used, but there are marked differences at the molecular level. *dsx* has only been found in dipterans distantly related to *Drosophila*; *Sxl* has been found in other dipterans but is not involved in sex determination in them.

8.15 Somatic sexual development in *Caenorhabditis* is determined by the number of X chromosomes

In the nematode *C. elegans*, the two sexes are self-fertilizing hermaphrodite (essentially a modified female) and male (Fig. 8.28), although in other nematodes they are male and female. Hermaphrodites produce a limited amount of sperm early in development, with the remainder of the germ cells developing into oocytes. Sex in *C. elegans* (and other nematodes) is determined by the number of X chromosomes: the hermaphrodite (XX) has two X chromosomes, whereas the presence of just one X chromosome leads to development as a male (XO). One of the primary sex signals for hermaphrodite development is the SEX-1 protein, which is encoded on the X chromosome and is a key element in 'counting' the number of X chromosomes present. SEX-1 is a nuclear hormone receptor that represses the sex-determining gene *XO lethal* (*xol-1*), also present on the X chromosome. In the presence of a double dose of SEX-1 produced from the two X chromosomes, *xol-1* expression is inhibited, resulting in the

Fig. 8.28 Hermaphrodite and male *Caenorhabditis elegans*. The hermaphrodite has a 'two-armed' gonad and initially makes sperm, which is then stored in the spermatheca. It then switches to making eggs. The eggs are fertilized internally. The male makes sperm only.

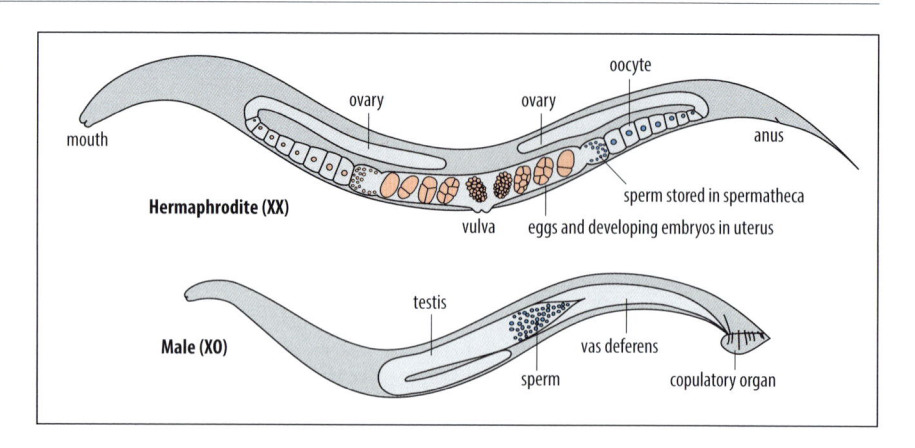

development of a hermaphrodite. When only one X chromosome is present, *xol-1* is expressed at a high level and the embryo develops as a male.

A cascade of gene activity converts the level of *xol-1* expression into the somatic sexual phenotype (Fig. 8.29). Unlike *Drosophila*, sex determination in *C. elegans* requires intercellular interactions, as at least one of the genes involved encodes a secreted protein. At the end of the cascade is the gene *transformer-1* (*tra-1*), which encodes a transcription factor. Expression of TRA-1 protein is both necessary and sufficient to direct all aspects of hermaphrodite (XX) somatic cell development, as a gain-of-function mutation in *tra-1* leads to hermaphrodite development in an XO animal, irrespective of the state of any of the regulatory genes that normally control its activity. Mutations that inactivate *tra-1* lead to complete masculinization of XX hermaphrodites.

8.16 Determination of germ cell sex depends on both genetic constitution and intercellular signals

Determination of the sex of animal germ cells—that is, whether they will develop into eggs or sperm—is strongly influenced by the signals they receive when they become part of a gonad. In the mouse, for example, their future development is determined

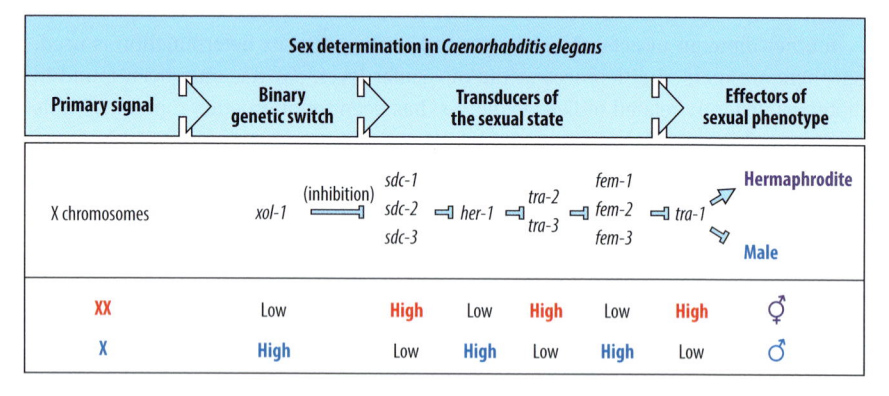

Fig. 8.29 Outline of the somatic sex-determination pathway in *Caenorhabditis elegans*. The primary factor in sex determination is the number of X chromosomes. When two X chromosomes are present, the expression of the gene *XO lethal* (*xol-1*) is low, leading to hermaphrodite development, whereas *xol-1* is expressed at a high level in males. There is a cascade of gene expression starting from *xol-1* that leads to the gene *transformer-1* (*tra-1*), which codes for a transcription factor. If *tra-1* is active, development as a hermaphrodite occurs, but if it is expressed at low levels, males develop. The product of the *hermaphrodite-1* (*her-1*) gene is a secreted protein, which probably binds to a receptor encoded by *transformer-2* (*tra-2*), inhibiting its function.

Fig. 8.30 The timing of meiosis in the germ cells differs considerably between males and females in mammals. Top panel: migrating germ cells that localize in the ovary enter meiotic prophase and start developing as oocytes. Migrating germ cells that localize in the testis are prevented from entering meiotic prophase and mitotic division is blocked. Bottom panel: in male mice after birth, immature diploid spermatogonia in the testis enter and complete meiosis in the testis to produce haploid germ cells, which eventually mature into sperm (left). In female mice after birth, oocytes complete their first meiotic division and enter the second meiotic division in the ovary just before ovulation. The egg only completes meiosis after fertilization.

largely by the sex of the gonad in which they reside, and not by their own chromosomal constitution. In reality, chromosomal constitution and gonadal signals will almost always coincide, but it has been shown that germ cells from male mouse embryos can develop into oocytes rather than sperm if grafted into female embryonic gonads and vice versa.

There is a distinct difference in the timing of meiosis in male and female mammals. In male mouse embryos, the diploid germ cells stop dividing when they enter the testis, becoming arrested in the G_1 phase of the mitotic cell cycle. They start dividing mitotically again after birth and enter meiosis some 7–8 days after birth. In female mouse embryos, the primordial diploid germ cells undergo a few rounds of mitotic division after arriving in the ovary and then enter prophase of the first meiotic division; they then arrest at this primary oocyte stage until the mouse becomes a sexually mature female, about 9 weeks after birth, when at each reproductive cycle, selected oocytes complete the first meiotic division and start the second meiotic division (Fig. 8.30). Meiosis is only completed after fertilization.

Germ cells, whether XX or XY, that fail to arrive in the gonads and instead end up in adjacent tissues such as the embryonic adrenal gland or mesonephros, enter meiosis and begin developing as oocytes in both male and female embryos. This observation was originally interpreted as showing that the default germ cell sex is female, but we now know that this interpretation was incorrect. Retinoic acid is the intercellular signal that induces the germ cells in the developing ovary in female mouse embryos to enter meiotic prophase and become primary oocytes. The main source of this retinoic acid is the mesonephros, which is closely associated with the gonads. The high levels of retinoic acid in most tissues of the embryo explain why germ cells that end up in the wrong places begin developing into oocytes. In male mouse embryos, meiotic entry of the germ cells in the developing testis is prevented by production of the enzyme Cyp26b1, which degrades retinoic acid (see Box 4E). Retinoic acid is in fact also the trigger for entering meiosis in males, but this occurs much later—in spermatocytes at puberty. Once mouse germ cells enter meiosis they are committed to an intrinsic pathway of differentiation. All mouse germ cells that enter meiosis before birth develop as eggs, whereas those not entering meiosis until after birth develop as sperm.

Despite the crucial influences of the environment in determining germ cell sex in mice, the genetic constitution of the germ cells does play some part, because when chromosomal constitution and gonadal signals do not coincide, functional eggs or sperm are not produced. XX/XY chimeric mouse embryos can be made by combining male and female four-cell embryos and, because of the presence of one Y chromosome, can develop testes. In these XX/XY embryos, XX germ cells that are surrounded by testis cells start to develop along the spermatogenesis pathway, but the later development of the gametes is abnormal. In a particular strain of sex-reversed XY mice that develops ovaries rather than testes, XY germ cells appear to develop quite normally as oocytes, but fail to develop further after fertilization as the spindle does not assemble properly for the second meiotic division. Experiments in which XY oocyte nuclei were transferred into normal XX oocytes produced healthy offspring, indicating that the fault lies in the XY cytoplasm.

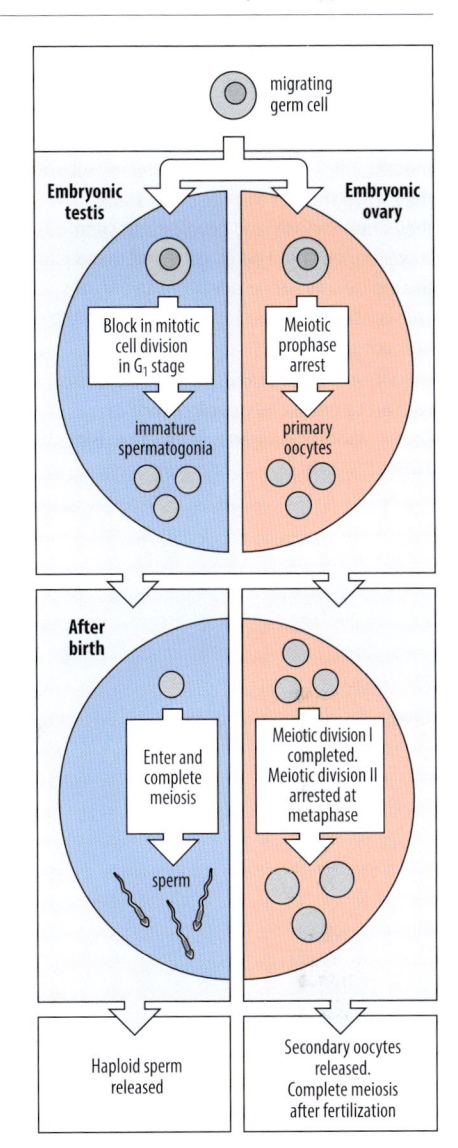

Fig. 8.31 Determination of germ cell sex in the hermaphrodite *Caenorhabditis elegans* gonad. Upper panel: during the larval stage, germ cells in a zone close to the distal tips of the gonad proliferate; when they leave this zone in the larval stage they enter meiosis and develop into sperm. Lower panel: in the adult, cells that leave the proliferative zone develop into oocytes. The eggs are fertilized as they pass into the uterus.

*Illustration after Clifford, R., et al.: **Somatic control of germ cell development in Caenorhabditis elegans**. Semin. Dev. Biol. 1994, **5**: 21-30.*

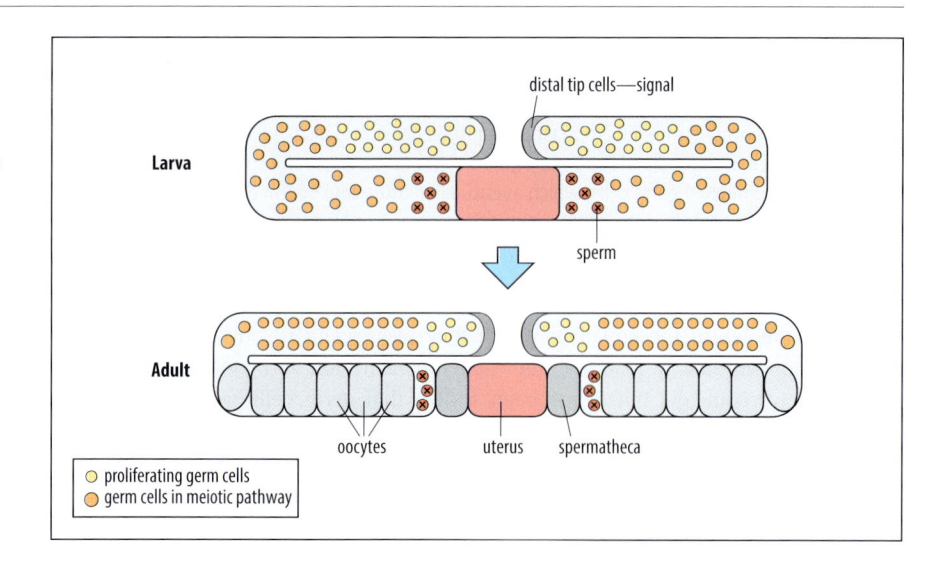

In *Drosophila*, the difference in the behavior of XY and XX germ cells depends initially on the number of X chromosomes, as in somatic cells; the *Sxl* gene again has an important role, although most other elements in the sex-determination pathway can differ from those in somatic cells. Both chromosomal constitution and cell interactions are involved in the development of germ cell sexual phenotype. Transplantation of genetically marked pole cells (see Section 8.1) into a *Drosophila* embryo of the opposite sex shows that male XY germ cells in a female XX embryo become integrated into the ovary and begin to develop as sperm; that is, their behavior is autonomous with respect to their genetic constitution. By contrast, XX germ cells in a testis attempt to develop as sperm, showing a role for environmental signals. In neither case, however, are functional sperm produced.

The hermaphrodite of *C. elegans* provides a particularly interesting example of germ cell differentiation, as both sperm and eggs develop within the same gonad (Fig. 8.31). Unlike the somatic cells of the adult nematode, which have a fixed lineage and number, the number of germ cells is indeterminate, with about 1000 germ cells in each 'arm' of the gonad. At hatching of the first-stage larva, there are just two founder germ cells, which proliferate to produce the germ cells. The germ cells are flanked on each side by cells called distal tip cells, and their proliferation is controlled by a signal from the distal tip cells. This signal is the protein LAG-2, which is homologous to the Notch ligand Delta. The receptor for LAG-2 on the germ cells is GLP-1, which is similar both to nematode LIN-12, which is involved in vulva formation, and to Notch.

In *C. elegans*, entry of germ cells into meiosis from the third larval stage onward is controlled by the distal tip signal. In the presence of this signal, the cells proliferate, but as they move away from it, they enter meiosis and develop as sperm (see Fig. 8.31, top). In the hermaphrodite gonad, all the cells that are initially outside the range of the distal tip signal develop as sperm, but cells that later leave the proliferative zone and enter meiosis develop as oocytes (see Fig. 8.31, bottom). The eggs are fertilized by stored sperm as they pass into the uterus. The male gonad has similar proliferative and meiotic regions, but all the germ cells develop as sperm.

Sex determination of the nematode germ cells is somewhat similar to that of the somatic cells, in that the chromosomal complement is the primary sex-determining factor and many of the same genes are involved in the subsequent cascades of gene expression. The terminal regulator genes required for spermatogenesis are called *fem* and *fog*. In hermaphrodites, there must be a mechanism for activating the *fem* genes in some of the XX germ cells, so allowing them to develop as sperm.

Determination of the sexual phenotype 433

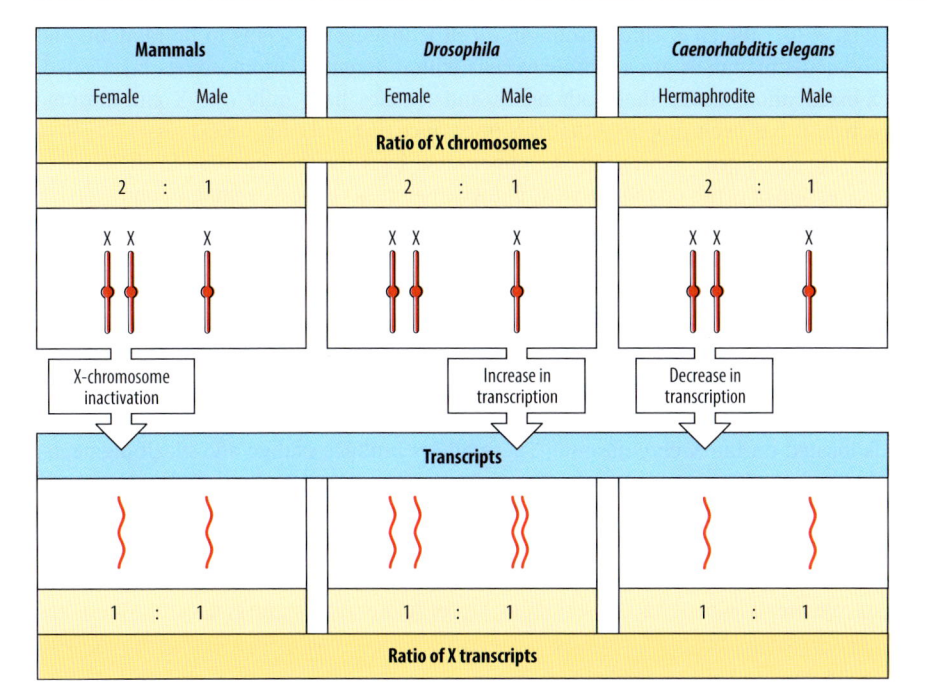

Fig. 8.32 Mechanisms of dosage compensation. In mammals, *Drosophila*, and *Caenorhabditis elegans* there are two X chromosomes in one sex and only one in the other. Mammals inactivate one of the X chromosomes in females; in *Drosophila* males there is an increase in transcription from the single X chromosome; and in *C. elegans* there is a decrease in transcription from the X chromosomes in hermaphrodites. The result of these different dosage-compensation mechanisms is that the level of X chromosome transcripts is approximately the same in males and females.

8.17 Various strategies are used for dosage compensation of X-linked genes

In all the animals we have considered in this chapter there is an imbalance of X-linked genes between the sexes. One sex has two X chromosomes, whereas the other has one. This imbalance has to be corrected to ensure that the level of expression of genes carried on the X chromosome is the same in both sexes. The mechanism by which the imbalance in X-linked genes is dealt with is known as **dosage compensation**. Failure to correct the imbalance leads to abnormalities and arrested development. Different animals deal with the problem of dosage compensation in different ways (Fig. 8.32).

In mammals, such as mice and humans, the female has two copies of the large X chromosome that has many genes while the male has only one X chromosome and a small Y chromosome that has very few genes (see Fig. 8.20). To achieve dosage compensation, one of the X chromosomes in each cell in female embryos is inactivated after the blastocyst has implanted in the uterine wall. Once an X chromosome has been inactivated in an embryonic cell, this chromosome is maintained in the inactive state in all the resulting somatic cells, and inactivation persists throughout the life of the organism (Fig. 8.33). The inactive X chromosome is replicated at each cell division but remains transcriptionally inactive and in a different physical state from the other chromosomes during the rest of the cell cycle. At mitosis all chromosomes become highly condensed. During the interphase of the cell cycle—the period between successive mitoses—all the other chromosomes decondense into extended threads of chromatin—the complex of DNA and proteins of which chromosomes are made—and are no longer visible under the light microscope. The inactive X chromosome, however, remains condensed and is visible in human cells as the Barr body

Fig. 8.33 Inheritance of an inactivated X chromosome. In early mammalian female embryos one of the two X chromosomes, either the paternal X (X_p) or the maternal X (X_m), is randomly inactivated. In the figure, X_p is inactivated and this inactivation is maintained through many cell divisions. The inactivated chromosome becomes highly condensed.
Illustration after Alberts, B., et al.: Molecular Biology of the Cell. 2nd edn. New York: Garland Publishing, 1989.

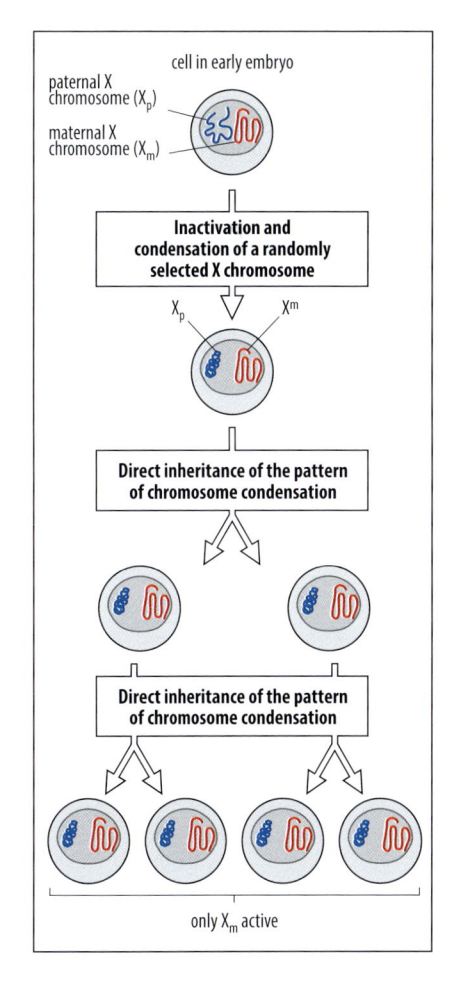

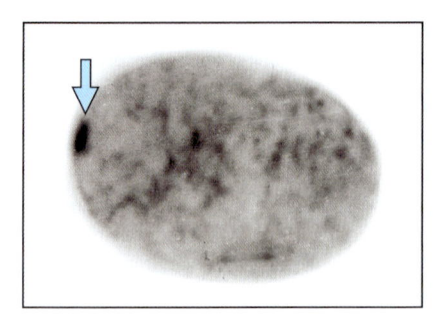

Fig. 8.34 Inactivated X chromosome (the Barr body). The photograph shows the Barr body (arrow) in the interphase nucleus of a female human buccal cell.

Photograph courtesy of J. Delhanty.

Fig. 8.35 The coloring of tortoiseshell cats is due to mosaicism of X-linked alleles due to X inactivation.

Foxfire, photograph courtesy of Bruce Goatly.

(Fig. 8.34). Which X chromosome in a cell is inactivated seems to be random, and so female mammals are a mosaic of cells with different X chromosomes inactivated. X-inactivation means that both males and females have only one X chromosome active. It has been proposed that this imbalance of X-linked gene expression relative to autosomal gene expression is compensated by a twofold increase in the level of expression of genes on the active X, but whether this is the case remains controversial.

The mosaic effect of X inactivation is sometimes visible in the coats of female mammals. Female mice heterozygous for a coat pigment gene carried on the X chromosome have patches of color on their coat, produced by clones of epidermal cells that express the X chromosome carrying a functional pigment gene. The rest of the epidermis is composed of cells in which that X chromosome has been inactivated. The coat pattern of tortoiseshell cats is also due to X-linked mosaicism. They carry two co-dominant alleles, *orange* (X^O) and *black* (X^B), of the main coat color gene, which is located on the X chromosome. These alleles produce orange and black pigments, respectively. Pigment genes are expressed in melanocytes that derive from neural crest cells that migrate to the skin. Cells in which the chromosome carrying X^O is inactivated express the X^B allele and vice versa. In bi-colored tortoiseshell cats (Fig. 8.35), the two cell types are intermingled, producing the characteristic brindled appearance. In calico cats, which have distinct patches of white, orange, and black fur, there has been less intermingling and larger clones of cells of one color develop, along with white patches with no pigment.

How cells count and choose chromosomes for inactivation is not yet fully understood. But the finding that tetraploid cells have two active X chromosomes has led to the following proposal, which depends on the levels of a signal produced by the X chromosomes themselves and invokes a negative-feedback mechanism between autosomes and the X chromosomes. Before inactivation, the X chromosomes are producing a signal that binds to the other chromosomes in the cell. When all the binding sites are occupied, the autosomal chromosomes, in turn, produce a signal that can bind to and inactivate X chromosomes. The first X chromosome to acquire a critical mass of bound signal initiates its own inactivation. The output of the X chromosome signal is therefore reduced and the autosomes stop producing the X-inactivating signal, with the result that just one X chromosome is inactivated per diploid genome. This type of mechanism can also accommodate the long-standing observation that in diploid female cells with an abnormal sex chromosome constitution, such as XXY or XXXY, there is only one active X chromosome per somatic cell, with all the other X chromosomes inactivated.

X inactivation in mammals depends on a small region of the X chromosome, the X-inactivation center. The center contains many genes, including the gene for the long non-coding RNA *Xist*, which is the major regulator of X-inactivation. Another long non-coding RNA, *Tsix*, which negatively regulates *Xist*, is found in the center in mice. Before inactivation, both RNAs are transcribed at low levels from both X chromosomes. *Xist* expression then dramatically increases on one or other of the X chromosomes and ceases on the other. What causes this to happen is still unknown, but a candidate might be the still-hypothetical inactivating signal from the autosomes noted above. A ubiquitin ligase encoded by a gene on the X chromosome initiates an increase in *Xist* expression. This so-called competence factor could be responsible for sensing the number of X chromosomes present in the cell and ensuring that only one X chromosome is inactivated in a female cell. The *Xist* transcripts spread from their site of synthesis to coat the whole X chromosome on which they are made, blocking transcription, including that of *Tsix*, and inactivating the chromosome. On the chromosome that is not inactivated, *Tsix* remains active for a little time, but soon after X inactivation is complete both *Tsix* and *Xist* are switched off on the active chromosome. *Xist* RNA continues to keep the inactivated chromosome inactive through subsequent cell divisions by recruiting proteins such as those that modify histones. The inactive X chromosome also has a pattern of DNA methylation that differs from that of the active X, and this methylation is likely to help keep it inactive. DNA methylation is one of the mechanisms used in mammals for long-term silencing of genes, and we have already seen how it can contribute to the differential imprinting of maternal and paternal genes in the germline (see Section 8.8). DNA methylation and other epigenetic mechanisms such as histone modifications for controlling gene expression are discussed in more detail in Box 6B.

Dosage compensation in *Drosophila* works in a different way from that in mice and humans (see Fig. 8.32). Instead of repression of the 'extra' X activity in females, transcription of the X chromosome in males is increased nearly twofold. A set of male-specific genes, the MSL complex, controls most dosage compensation, and these genes are repressed in females by Sxl protein, thereby preventing excessive transcription of the X chromosome. The increased activity in males is regulated by the primary sex-determining signal, which results in the dosage-compensation mechanism operating when *Sxl* is 'off'. In females, where *Sxl* is 'on', it turns off the dosage-compensation mechanism. As in the mouse, this regulatory mechanism involves non-coding RNAs.

In *C. elegans*, dosage compensation is achieved by reducing the level of X chromosome expression in XX animals to that of the single X chromosome in XO males (see Fig. 8.32). The number of X chromosomes is communicated by a set of X-linked genes that can repress the master gene *xol-1*. A key event in initiating nematode dosage compensation is the expression of the protein SDC-2, which occurs only in hermaphrodites. It forms a complex specifically with the X chromosome and triggers assembly of a protein complex called the dosage-compensation complex, which binds to a specific region on the X chromosome and reduces transcription.

SUMMARY

The development of early embryos of both sexes is very similar. A primary sex-determining factor then sets off development towards one or the other sex, and in mammals, *Drosophila*, and *C. elegans*, this factor is the chromosomal constitution of the fertilized egg. In mammals, the presence of the Y chromosome is important: the *Sry* gene on the Y chromosome is responsible for the embryonic gonad developing into a testis and producing hormones that determine male sexual characteristics. In *C. elegans* and *Drosophila*, the primary factor is the number of X chromosomes. In *Drosophila*, the presence of two X chromosomes activates the gene *Sex-lethal*, which is located on the X chromosome, such that this gene is turned on in females but not in males; the Sex-lethal protein then regulates the sex-specific splicing of RNA encoded by genes that determine the sexual phenotype. In *C. elegans*, activity of the gene *XO lethal* is low in hermaphrodites (which have two X chromosomes) and high in males (which have just one), eventually leading to sex-specific expression of the gene *transformer-1*, which determines the sexual phenotype.

Somatic sexual differentiation in *Drosophila* is cell autonomous and is controlled by the number of X chromosomes; in *C. elegans*, cell-cell interactions are also involved. In mammals, the signals the germ cells receive while in the gonad determine whether they ultimately develop into oocytes or sperm. Male germ cells in *Drosophila* develop along the sperm pathway even in an ovary, but female germ cells develop along the sperm pathway when placed in a testis. Most *C. elegans* adults are hermaphrodites, and produce both sperm and eggs from the same gonad.

Various strategies of dosage compensation are used to correct the imbalance of X chromosomes between males and females. In female mammals, one of the X chromosomes is inactivated; in *Drosophila* males the activity of the single X chromosome is upregulated; and in *C. elegans* the activity of the X chromosomes in XX hermaphrodites is downregulated to match that from the single X chromosome in males.

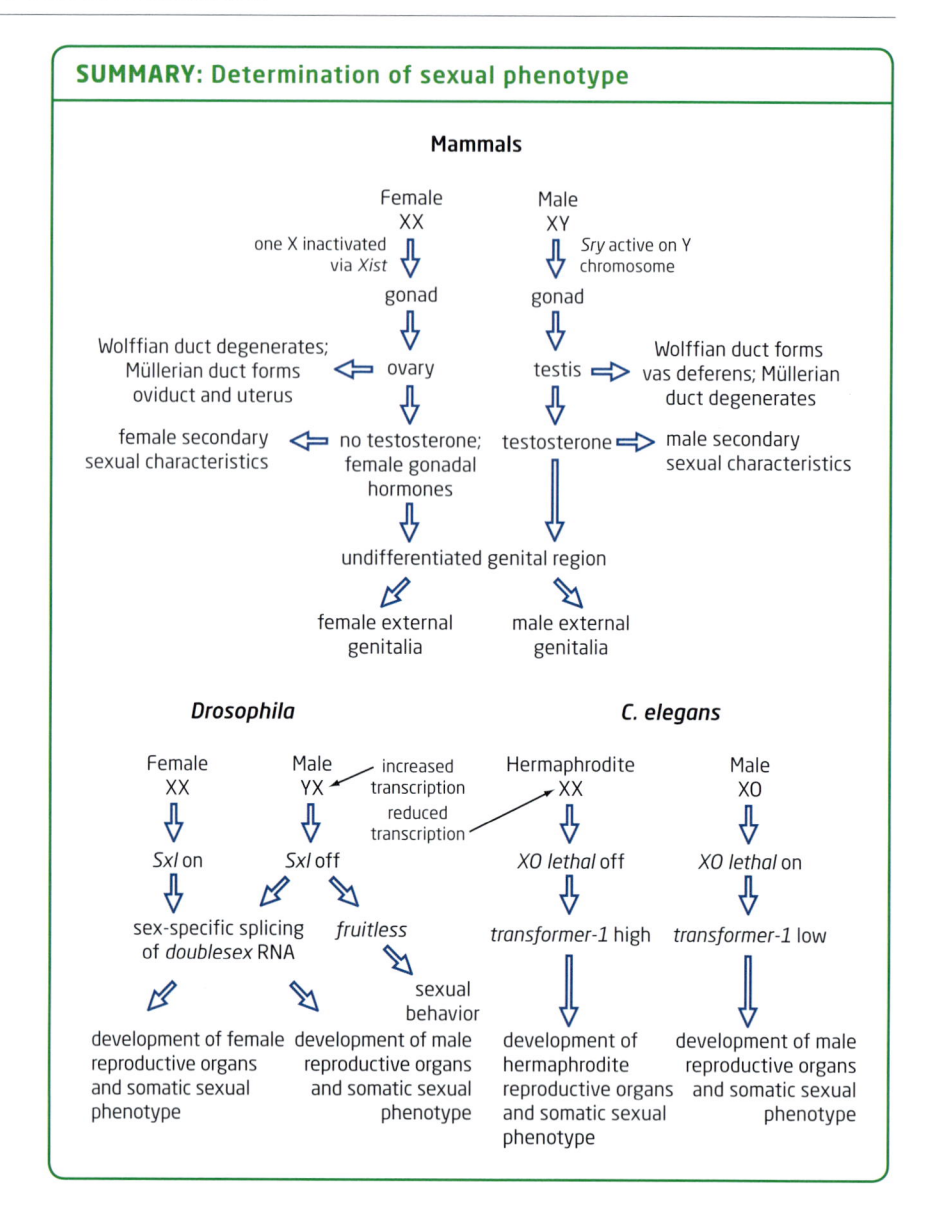

436 Chapter 8 Germ cells, fertilization, and sex determination

---

SUMMARY: Determination of sexual phenotype

Mammals

Female XX — one X inactivated via *Xist* → gonad → ovary → no testosterone; female gonadal hormones → female secondary sexual characteristics
Wolffian duct degenerates; Müllerian duct forms oviduct and uterus

Male XY — *Sry* active on Y chromosome → gonad → testis → testosterone → male secondary sexual characteristics
Wolffian duct forms vas deferens; Müllerian duct degenerates

undifferentiated genital region → female external genitalia / male external genitalia

Drosophila

Female XX → *Sxl* on → sex-specific splicing of *doublesex* RNA → development of female reproductive organs and somatic sexual phenotype

Male YX (increased transcription / reduced transcription) → *Sxl* off → *fruitless* → sexual behavior → development of male reproductive organs and somatic sexual phenotype

C. elegans

Hermaphrodite XX → *XO lethal* off → *transformer-1* high → development of hermaphrodite reproductive organs and somatic sexual phenotype

Male XO → *XO lethal* on → *transformer-1* low → development of male reproductive organs and somatic sexual phenotype

---

Summary to Chapter 8

- In most of our model organisms, the future germ cells are specified by localized cytoplasmic determinants in the egg. Germ cells in mice and other mammals including humans are specified entirely by cell–cell interactions.
- In animals, the development of germ cells into sperm or egg depends both on chromosomal constitution and on interactions with the cells of the gonad.
- At fertilization, fusion of sperm and egg initiates development, and there are mechanisms to ensure that only one sperm enters an egg.
- Both maternal and paternal genomes are required for normal mammalian development, as some genes are imprinted; for such genes, whether they are expressed or not during development depends on whether they are derived paternally, from the sperm, or maternally, from the egg.
- In many animals, the chromosomal constitution of the embryo determines which sex will develop. In mammals, the Y chromosome is male-determining; it specifies the

development of a testis, and the hormones produced by the testis cause the development of male sexual characteristics. In the absence of a Y chromosome, the embryo develops as a female.

- In *Drosophila* and *C. elegans*, sexual development is initially determined by the number of X chromosomes, which sets in train a cascade of gene activity. In flies, somatic sexual differentiation is cell autonomous; in nematodes, the somatic sexual phenotype is determined by cell–cell interactions.
- Animals use various strategies of dosage compensation to correct the imbalance in the number of X chromosomes in males and females.

End of chapter questions

Check your understanding of the content of this chapter by attempting the following long-answer concept questions.

1. Distinguish between the germline, gametes, and somatic cells. What key functions are germ cells responsible for?

2. Summarize the evidence for the existence of a special germplasm in *Drosophila* and *Xenopus*. Is the presence of germplasm a universal feature of animal development?

3. Draw a schematic of two generations of *Drosophila*: start with a male and a female, each of which is heterozygous for a loss-of-function recessive *oskar* mutation. Mate them together to get homozygous *oskar* males and females. Mate these *oskar* mutants to normal (wild-type) flies. How do these matings illustrate why such mutations are referred to as 'grandchildless'? Does it matter whether the homozygous *oskar* mutant is a male or a female? Explain.

4. What is the role of SDF-1 and CXCR4 in germ cell migration? To what extent is this role conserved among vertebrates?

5. Discuss the following aspects of meiosis: (1) Why is a 'reduction division' (meiosis II) important? (2) What is the significance of recombination in sexual reproduction? (3) Why do four sperm cells result from each cell undergoing meiosis during spermatogenesis, yet only one egg cell results during oogenesis—where do the other 'eggs' go?

6. When in human development do primary oocytes form? What happens to a primary oocyte between this time and its maturation into an egg ready to be fertilized? At what point in meiosis is the oocyte when fertilization occurs?

7. What is meant by 'genomic imprinting'? What are the mechanisms that maintain imprinting?

8. Describe the process by which the nucleus and contents of the sperm enter the mammalian egg. Include acrosome enzymes, Juno, and Izumo1 in your answer.

9. Fertilization of a sea urchin egg leads to a wave of calcium that sweeps across the egg. Is this calcium intracellular or extracellular in origin? What are two consequences of this increase in calcium?

10. What is MPF? Summarize the role of MPF-like complexes in the control of the various stages of the cell cycle.

11. Explain how chromosomal abnormalities can be used as evidence for the following statements. (1) Determination of the male sex in humans is controlled by the presence of the Y chromosome, not two X chromosomes. (2) Determination of female sex in *Drosophila* is controlled by the presence of two X chromosomes and not the presence of a Y chromosome. (3) One X chromosome in humans is sufficient to determine the female sex in humans. (4) One X chromosome is sufficient to determine the male sex in *Drosophila*. (5) *SRY* is the gene responsible for male sex determination in humans.

12. How does *SRY* trigger male development in mammals? Include Sox9, Sertoli cells, Leydig cells, Müllerian-inhibiting substance, and testosterone in your answer.

13. Summarize the development of the gonads and related structures in mammals, focusing on the Wolffian and Müllerian ducts.

14. Summarize the information presented in Figures 8.26 and 8.27 on how cell-autonomous sexual differentiation occurs in *Drosophila*. Do not just copy the two figures; instead, incorporate the information into one comprehensive schematic.

15. What is different about the sexes in *C. elegans* compared with the sexes in our other model organisms? How does this affect the worm's method of reproduction?

16. Briefly compare and contrast the sex-determination pathways in *Drosophila* and *C. elegans*. Are there similarities in the strategies used? Are there differences in the mechanisms involved?

17. Say how X chromosome inactivation will achieve dosage compensation in humans with these chromosomal compositions: XY, XX, XXY, or XXX.

18. How does *Xist* control X chromosome inactivation in mammals? Speculate why the other gene involved in this process, *Tsix*, might have been given this name.

General further reading

Chadwick, D., Goode, J.: *The Genetics and Biology of Sex Determination 2002* (Novartis Foundation Symposium, Vol. 244). New York: John Wiley, 2002.

Lehmann, R.: *The Immortal Germline* (Current Topics in Developmental Biology, Vol. 135). Amsterdam: Elsevier, 2019.

Zarkower, D.: **Establishing sexual dimorphism: conservation amidst diversity?** *Nat. Rev. Genet.* 2001, **2**: 175–185.

Section further reading

8.1 Germ cell fate is specified in some embryos by a distinct germplasm in the egg

Extavour, C.G., Akam, M.: **Mechanisms of germ cell specification across the metazoans: epigenesis and preformation.** *Development* 2003, **130**: 5869–5884.

Matova, N., Cooley, L.: **Comparative aspects of animal oogenesis.** *Dev. Biol.* 2001, **231**: 291–320.

Mello, C.C., Schubert, C., Draper, B., Zhang, W., Lobel, R., Priess, J.R.: **The PIE-1 protein and germline specification in *C. elegans* embryos.** *Nature* 1996, **382**: 710–712.

Micklem, D.R., Adams, J., Grunert, S., St. Johnston, D.: **Distinct roles of two conserved Staufen domains in *oskar* mRNA localisation and translation.** *EMBO J.* 2000, **19**: 1366–1377.

Ray, E.: **Primordial germ-cell development: the zebrafish perspective.** *Nat. Rev. Genet.* 2003, **4**: 690–700.

Williamson, A., Lehmann, R.: **Germ cell development in *Drosophila*.** *Annu. Rev. Cell Dev. Biol.* 1996, **12**: 365–391.

8.2 In mammals germ cells are induced by cell-cell signaling during development

Hansen, C.L., Pelegri, F.: **Primordial germ cell specification in vertebrate embryos: phylogenetic distribution and conserved molecular features of preformation and induction.** *Front. Cell Dev. Biol.* 2021, **9**: 730332.

Magnusdottir, E., Surani, M.A.: **How to make a primordial germ cell.** *Development* 2014, **141**: 245–252.

McLaren, A.: **Primordial germ cells in the mouse.** *Dev. Biol.* 2003, **262**: 1–15.

Ohinata, Y., Payer, B., O'Carroll, D., Ancelin, K., Ono, Y., Sano, M., Barton, S.C., Obukhanych, T., Nussenzweig, M., Tarakhovsky, A., Saitou, M., Surani, M.A.: **Blimp1 is a critical determinant of the germ cell lineage in mice.** *Nature* 2005, **436**: 207–213.

8.3 Germ cells migrate from their site of origin to the gonad and 8.4 Germ cells are guided to their destination by chemical signals

Barton, L.J., LeBlanc, M.G., Lehmann, R.: **Finding their way: themes in germ cell migration.** *Curr. Opin. Cell Biol.* 2016, **42**: 128–137.

Deshpande, G., Zhou, K., Wan, J.Y., Friedrich, J., Jourjine, N., Smith, D., Schedl, P.: **The *hedgehog* pathway gene *shifted* functions together with the *hmgcr*-dependent isoprenoid biosynthetic pathway to orchestrate germ cell migration.** *PLoS Genet.* 2013, **9**: e1003720.

Doitsidou, M., Reichman-Fried, M., Stebler, J., Koprunner, M., Dorries, J., Meyer, D., Esguerra, C.V., Leung, T., Raz, E.: **Guidance of primordial germ cell migration by the chemokine SDF-1.** *Cell* 2002, **111**: 647–659.

Kim, J.H., Hanlon, C.D., Vohra, S., Devreotes, P.N., Andrew, D.J.: **Hedgehog signaling and Tre1 regulate actin dynamics through PI(4,5)P$_2$ to direct migration of Drosophila embryonic germ cells.** *Cell Rep.* 2021, **34**: 108799.

Knaut, H., Werz, C., Geisler, R., Nusslein-Volhard, C., Tubingen 2000 Screen Consortium. **A zebrafish homologue of the chemokine receptor Cxcr4 is a germ-cell guidance receptor.** *Nature* 2003, **421**: 279–282.

Richardson, B.E., Lehmann, R.: **Mechanisms guiding primordial germ cell migration: strategies from different organisms.** *Nat. Rev. Mol. Cell Biol.* 2010, **11**: 37–49.

Weidinger, G., Wolke, U., Köprunner, M., Thisse, C., Thisse, B., Raz, E.: **Regulation of zebrafish primordial germ cell migration by attraction towards an intermediate target.** *Development* 2002, **129**: 25–36.

8.5 Germ cell differentiation involves a halving of chromosome number by meiosis

Hikabe, O., Hamazaki, N., Nagamatsu, G., Obata, Y., Hirao, Y., Hamada, N., Shimamoto, S., Imamura, T., Nakashima, K., Saitou, M., Hayashi, K.: **Reconstitution *in vitro* of the entire cycle of the mouse female germ line.** *Nature* 2016, **539**: 299–303.

Hultén, M.A., Patel, S.D., Tankimanova, M., Westgren, M., Papadogiannakis, N., Jonsson, A.M., Iwarsson, E.: **The origins of trisomy 21 Down syndrome.** *Mol. Cytogenet.* 2008, **1**: 21–31.

Mehlmann, L.M.: **Stops and starts in mammalian oocytes: recent advances in understanding the regulation of meiotic arrest and oocyte maturation.** *Reproduction* 2005, **130**: 791–799.

Pacchierottia, F., Adler, I.-D., Eichenlaub-Ritter, U., Mailhes, J.B.: **Gender effects on the incidence of aneuploidy in mammalian germ cells.** *Environ. Res.* 2007, **104**: 46–69.

Phillips, B.T., Gassei, K., Orwig, K.E.: **Spermatogonial stem cell regulation and spermatogenesis.** *Philos. Trans. R. Soc. Lond. B Biol. Sci.* 2010, **365**: 1663–1678.

Saitou, M., Hayashi, K.: **Mammalian *in vitro* gametogenesis.** *Science* 2021, **374**: eaaz6830.

Vogta, E., Kirsch-Volders, M., Parry, J., Eichenlaub-Rittera, U.: **Spindle formation, chromosome segregation and the spindle checkpoint in mammalian oocytes and susceptibility to meiotic error.** *Mutat. Res.* 2008, **651**: 14–29.

8.6 Oocyte development can involve gene amplification and contributions from other cells

Browder, L.W.: **Oogenesis.** New York: Plenum Press, 1985.

Choo, S., Heinrich, B., Betley, J.N., Chen, A., Deshler, J.O.: **Evidence for common machinery utilized by the early and late RNA localization pathways in *Xenopus* oocytes.** *Dev. Biol.* 2004, **278**: 103–117.

8.7 Factors in the cytoplasm maintain the totipotency of the egg

Gurdon, J.B.: **Nuclear transplantation in eggs and oocytes.** *J. Cell Sci. Suppl.* 1986, **4**: 287–318.

8.8 In mammals some genes controlling embryonic growth are 'imprinted'

Barlow, D.P., Bartolomei, M.S.: **Genomic imprinting in mammals.** *Cold Spring Harb. Perspect. Biol.* 2014, **6**: a018382.

Bartolomei, M.S: **Genomic imprinting: employing and avoiding epigenetic processes.** *Genes Dev.* 2009, **23**: 2124–2133.

Plasschaert, R.N., Bartolomei, M.S.: **Genomic imprinting in development, growth, behavior and stem cells.** *Development* 2014, **141**: 1805–1813.

Reik, W., Walter, J.: **Genomic imprinting: parental influence on the genome.** *Nat. Rev. Genet.* 2001, **2**: 21–32.

Wood, A.J., Oakey, R.J.: **Genomic imprinting in mammals: emerging themes and established theories.** *PLoS Genet.* 2006, **2**: e147.

8.9 Fertilization involves cell-surface interactions between egg and sperm

Avella, M.A., Xiong, B., Dean, J.: **The molecular basis of gamete recognition in mice and humans.** *Mol Hum Reprod.* 2013; **19**: 279-289.

Bianchi, E., Doe, B., Goulding, D., Wright, G.J.: **Juno is the egg Izumo receptor and is essential for mammalian fertilization.** *Nature* 2014, **508**: 483-487.

Bianchi, E., Wright, G.J.: **Sperm meets egg: the genetics of mammalian fertilization.** *Annu. Rev. Genet.* 2016, **50**: 93-111.

Bianchi, E., Wright, G.J.: **Find and fuse: unsolved mysteries in sperm-egg recognition.** *PLoS Biol.* 2020, **18**: e3000953.

8.10 Changes in the egg plasma membrane and enveloping layers at fertilization block polyspermy

Bianchi, E., Wright, G.J.: **Sperm meets egg: the genetics of mammalian fertilization.** *Annu. Rev. Genet.* 2016, **50**: 93-111.

Burkart A.D., Xiong, B., Baibakov, B., Jiménez-Movilla, M., Dean, J.: **Ovastacin, a cortical granule protease, cleaves ZP2 in the zona pellucida to prevent polyspermy.** *J. Cell Biol.* 2012, **197**: 37-44.

Wong, J.L., Wessel, G.M.: **Defending the zygote: search for the ancestral animal block to polyspermy.** *Curr. Top. Dev. Biol.* 2006, **72**: 1-151.

8.11 Sperm-egg fusion causes a calcium wave that results in egg activation

Kashir, J., Nomikos, M., Lai, F.A., Swann, K.: **Sperm-induced Ca^{2+} release during egg activation in mammals.** *Biochem. Biophys. Res. Commun.* 2014, **450**: 1204-1211.

Suzuki, T., Yoshida, N., Suzuki, E., Okuda, E., Perry, A.C.: **Full-term mouse development by abolishing Zn^{2+}-dependent metaphase II arrest without Ca^{2+} release.** *Development* 2010, **137**: 2659-2669.

Whitaker, M.: **Calcium signaling in early embryos.** *Philos. Trans. R. Soc. Lond. B Biol. Sci.* 2008, **363**: 1401-1418.

8.12 The primary sex-determining gene in mammals is on the Y chromosome and 8.13 Mammalian sexual phenotype is regulated by gonadal hormones

Capel, B.: **The battle of the sexes.** *Mech. Dev.* 2000, **92**: 89-103.

Jameson, S.A., Lin, Y.T., Capel, B.: **Testis development requires the repression of *Wnt4* by Fgf9 signaling.** *Dev. Biol.* 2012, **370**: 24-32.

Koopman, P.: **The genetics and biology of vertebrate sex determination.** *Cell* 2001, **105**: 843-847.

Koopman, P., Sinclair, A., Lovell-Badge, R.: **Of sex and determination: marking 25 years of Randy, the sex-reversed mouse.** *Development* 2016, **143**: 1633-1637.

Lin, Y.T., Capel, B.: **Cell fate commitment during mammalian sex determination.** *Curr. Opin. Genet. Dev.* 2015, **32**: 144-152.

Schafer, A.J., Goodfellow, P.N.: **Sex determination in humans.** *BioEssays* 1996, **18**: 955-963.

Svingen, T., Koopman, P.: **Building the mammalian testis: origins, differentiation and assembly of the component cell populations.** *Genes Dev.* 2013, **27**: 2409-2426.

Swain, A., Lovell-Badge, R.: **Mammalian sex determination: a molecular drama.** *Genes Dev.* 1999, **13**: 755-767.

Vainio, S., Heikkila, M., Kispert, A., Chin, N., McMahon, A.P.: **Female development in mammals is regulated by Wnt-4 signalling.** *Nature* 1999, **397**: 405-409.

Zhao, F., Franco, H.L., Rodriguez, K.F., Brown, P.R., Tsai, M.J., Tsai, S.Y., Yao, H.H.: **Elimination of the male reproductive tract in the female embryo is promoted by COUP-TFII in mice.** *Science* 2017, **357**: 717-720.

8.14 The primary sex-determining factor in *Drosophila* is the number of X chromosomes and is cell autonomous

Brennan, J., Capel, B.: **One tissue, two fates: molecular genetic events that underlie testis versus ovary development.** *Nat. Rev. Genet.* 2004, **5**: 509-520.

Hodgkin, J.: **Sex determination compared in *Drosophila* and *Caenorhabditis*.** *Nature* 1990, **344**: 721-728.

MacLaughlin, D.T., Donahoe, M.D.: **Sex determination and differentiation.** *N. Engl. J. Med.* 2004, **350**: 367-378.

8.15 Somatic sexual development in *Caenorhabditis* is determined by the number of X chromosomes

Carmi, I., Meyer, B.J.: **The primary sex determination signal of *Caenorhabditis elegans*.** *Genetics* 1999, **152**: 999-1015.

Meyer, B. J.: **X-chromosome dosage compensation.** In *WormBook*. The C. elegans Research Community (Ed.). 25 June 2005. [http://www.wormbook.org/chapters/www_dosagecomp/dosagecomp.html] (date accessed 3 September 2024).

Raymond, C.S., Shamu, C.E., Shen, M.M., Seifert, K.J., Hirsch, B., Hodgkin, J., Zarkower, D.: **Evidence for evolutionary conservation of sex-determining genes.** *Nature* 1998, **391**: 691-695.

8.16 Determination of germ cell sex depends on both genetic constitution and intercellular signals

Childs, A.J., Saunders, P.T.K., Anderson, R.A.: **Modelling germ cell development *in vitro*.** *Mol. Hum. Reprod.* 2008, **14**: 501-511.

McLaren, A.: **Signaling for germ cells.** *Genes Dev.* 1999, **13**: 373-376.

Obata, Y., Villemure, M., Kono, T., Taketo, T.: **Transmission of Y chromosomes from XY female mice was made possible by the replacement of cytoplasm during oocyte maturation.** *Proc. Natl Acad. Sci. USA* 2008, **105**: 13918-13923.

Seydoux, G., Strome, S.: **Launching the germline in *Caenorhabditis elegans*: regulation of gene expression in early germ cells.** *Development* 1999, **126**: 3275-3283.

Spiller, C., Koopman, P., Bowles, J.: **Sex determination in the mammalian germline.** *Annu. Rev. Genet.* 2017, **51**: 265-285.

8.17 Various strategies are used for dosage compensation of X-linked genes

Avner, P., Heard, E.: **X-chromosomes inactivation: counting, choice and initiation.** *Nat. Rev. Genet.* 2001, **2**: 59-67.

Brockdorff, N., Turner, B.M.: **Dosage compensation in mammals.** *Cold Spring Harb. Perspect. Biol.* 2015, **7**: a019406.

Csankovszki, G., McDonel, P., Meyer, B.J.: **Recruitment and spreading of the *C. elegans* dosage compensation complex along X chromosomes**. *Science* 2004, **303**: 1182–1185.

Navarro, P., Avner, P.: **An embryonic story: analysis of the gene regulative network controlling *Xist* expression in mouse embryonic stem cells**. *BioEssays* 2010, **32**: 581–588.

Panning, B.: **X-chromosome inactivation: the molecular basis of silencing**. *J. Biol.* 2008, **7**: 30.

Pollex, T., Heard, E.: **Recent advances in X-chromosome inactivation research**. *Curr. Opin. Cell Biol.* 2012, **24**: 825–832.

Sahakyan, A., Yang, Y., Plath, K.: **The role of *Xist* in X-chromosome dosage compensation**. *Trends Cell Biol.* 2018, **28**: 999–1013.

Starmer, J., Magnuson, T.: **New model for random X chromosome inactivation**. *Development* 2009, **136**: 1–10.

9

Organogenesis

- The insect wing and leg
- The vertebrate limb
- Teeth
- Branching organs: the vertebrate lung and the mammalian mammary gland
- The vertebrate heart
- The vertebrate eye

Once the basic animal body plan has been laid down, the development of organs as varied as insect wings and vertebrate eyes begins. The positions in which these organs will develop have already been specified as part of the process of laying down the body plan, but the subsequent development of the organ primordia is essentially autonomous. Organogenesis depends on the same basic mechanisms as those used in early embryogenesis but is more complex. An organ is composed of different types of tissues and their development must be integrated. Nevertheless, many of the mechanisms used, such as positional information, and cell activities, such as migration and differential adhesion, are the same as those used in earlier development, and certain signals are used again and again.

Introduction

So far, we have concentrated almost entirely on the laying down of the basic body plan in various organisms, and on early morphogenesis and cell differentiation. We now turn to the development of specific organs and structures—**organogenesis**—which is a crucial phase of development that will eventually lead to the embryo becoming a fully functioning organism, capable of independent survival.

The developmental processes involved in organogenesis are essentially similar to those encountered in earlier stages: they include positional information and cellular differentiation, as well as inductive events and changes in cell shape and multicellular organization. Many of the genes and signal molecules involved will be familiar from earlier chapters. The development of certain organs has been studied in great detail and they provide excellent models for looking at these fundamental developmental processes. In this chapter, we first consider some classic model systems—the *Drosophila* wing and leg, and the vertebrate limb. We touch on the induction of teeth and patterning of the dentition, which are based on similar principles to the induction and patterning of vertebrate limbs. We then consider some internal organs, focusing on the vertebrate lung and heart, and the mammalian mammary gland, all of which are based on tubular structures. The development of other major vertebrate organs, such as the gut, kidneys, liver, and pancreas, involves no new principles and so we do not consider them here. The liver, however, is interesting as an example of a vertebrate organ that can regenerate, and some aspects of liver growth and regeneration are discussed in Chapter 11. Finally, we consider the vertebrate eye as an example of how a complex sense organ develops. Some general developmental principles apply to all organs. Each organ begins as a small, undifferentiated rudimentary structure called an organ **primordium** (plural **primordia**), which arises at a particular position in the embryo, and has been assigned a specific fate. The initial specification of an organ primordium—for example, a vertebrate **limb bud** or a *Drosophila* wing primordium—is crucial, because once it has formed it can develop largely autonomously, without reference to the rest of the embryo. The identity of an organ primordium is encoded by 'master' genes and/or combinations of genes that encode transcription factors and other proteins that activate the gene regulatory network required for the development of that particular organ. The primordium will eventually develop into a functional organ. Cells differentiate into the specialized tissues of which the organ is composed, and pattern formation directs how these differentiated cells and tissues are arranged. We will look at some of these processes, and also consider how organs are shaped as the primordia grow. The control of organ growth is considered in Chapter 11.

Pattern formation in organs involves signaling centers that provide positional information, as in the laying down of the body plan. The signal proteins produced can also act as morphogens in organ development, and their signaling pathways control the expression of genes for organ-specific transcription factors which, in turn, regulate the genes that govern organ growth and development. An important principle in the patterning of an organ is that although the same strategies are used to provide positional information in, for example, the forelimb and the hindlimb of a mouse, or in a mouse limb and a mouse tooth, the interpretation of that information depends on the developmental history of the responding cells. Another key feature of organogenesis is cell–cell interactions, which frequently involve interactions between an epithelium and a mesenchyme as, for example, in the vertebrate limb and in teeth. On the evolutionary scale, we shall see that the genes and the molecular networks that specify the identity of functionally similar structures, such as eyes, are often unexpectedly conserved across a wide range of species, even though the structure of the organs produced is totally different.

The insect wing and leg

The adult organs and appendages of *Drosophila*, such as the wings, legs, eyes, and antennae, develop from imaginal discs (see Fig. 2.6), which are excellent experimental systems for analyzing pattern formation. The discs are simple pouches of epithelium that invaginate from the ectoderm early in embryonic development and remain as such, growing by cell division until metamorphosis. All imaginal discs superficially look rather similar, but they develop differently, according to the segment in which they are located.

We will first look at how insect wings, legs, and other appendages develop on particular segments, and then consider the development of wings and legs in more detail. Although insect wings and legs look so different, they are partially homologous structures, and the strategy of their patterning is similar.

9.1 Imaginal discs arise from the ectoderm in the early *Drosophila* embryo

In *Drosophila*, wings develop on thoracic segment 2 (T2), halteres—appendages that balance the fly during flight—on segment 3 (T3), and legs on each of the three thoracic segments (T1–T3). The imaginal discs in which the primordia that give rise to the wings, halteres, and legs of the adult fly emerge are formed early in embryogenesis (see Fig. 2.6). In each thoracic segment, the region of epidermis that will give rise to the imaginal discs, is first specified as a single unit. This small patch of embryonic epidermis expresses *Distal-less* (*Dll*) at the intersection of the influence of the signal molecules Wingless (Wg) and Decapentaplegic (Dpp) (shown for T2 in Fig. 9.1). In T2 and T3, each unit then splits into two patches, one located ventrally and one dorsally. The ventral patches continue to express *Dll* and become leg imaginal discs; the dorsal ones express the transcription factor Escargot (Esg) and become discs for the segment-specific appendages: the wing disc in T2 and the haltere disc in T3.

The wing and leg discs are specified in the embryo as clusters of 20–40 cells. During larval development, cell division increases the number of cells about 1000-fold. Imaginal discs are highly regulative. If some of their cells are removed or destroyed, the disc will regrow to the correct size and develop normally.

Why do *Drosophila* wings develop in only one particular thoracic segment, legs in all three, and antennae on the head? The answer is that the types of appendages

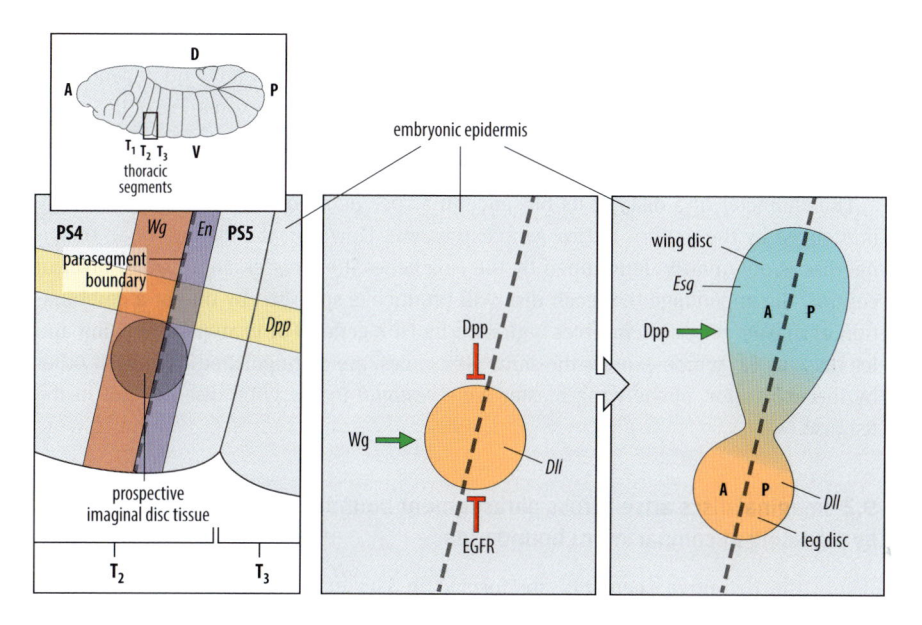

Fig. 9.1 Specification of imaginal discs in the thoracic segments in *Drosophila*. The thoracic imaginal discs are specified in the epidermis of the ventral flank of the embryo (inset). A late-stage embryo, in which the discs have formed, is depicted here. The specification of the discs in the second thoracic segment (T2) is shown in more detail in the next three panels. First panel: at around 8 hours of embryonic development, an antero-posterior stripe of cells secreting the signal protein Decapentaplegic (Dpp) intersects the dorso-ventral stripes of cells secreting the signal protein Wingless (Wg), and the transcription factor Engrailed (En), at the parasegment (PS) boundary (dashed line) (see Section 2.24). A small patch of cells receiving Wg signals is specified as imaginal disc. The disc is formed across the parasegment boundary and this boundary remains as a compartment boundary—a boundary of cell-lineage restriction (see Section 2.23)—as the disc grows. Second panel: the extent of the patch is limited dorsally by Dpp signals and ventrally by epidermal growth factor receptor (EGFR)-like signals, which inhibit imaginal disc fate. The prospective imaginal disc cells express the leg imaginal disc marker Distal-less (Dll). Third panel: as the cells in the disc proliferate and the patch extends dorsally, the dorsal-most cells come under the influence of strong Dpp signals. They then express the transcription factor Escargot (Esg), a marker for the wing and haltere imaginal discs. The patch of cells splits in two to give two individual discs—the leg disc ventrally and the wing disc dorsally. Each patch will sink under the epidermis to form an epithelial pouch connected by a thin stem to the epidermis.

444 Chapter 9 Organogenesis

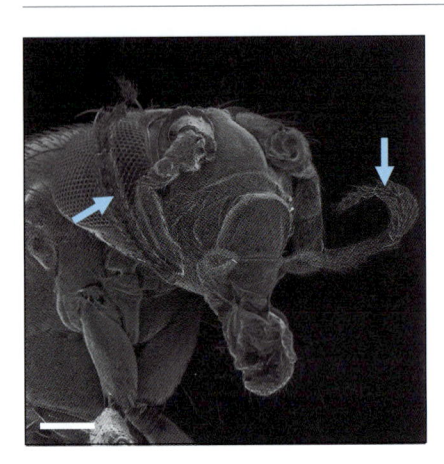

Fig. 9.2 Scanning electron micrograph of *Drosophila* carrying the *Antennapedia* mutation. Flies with this mutation have the antennae converted into legs (arrows). Scale bar = 0.1 mm.

Photograph by D. Scharfe, from Science Photo Library.

produced by each segment reflect the identity conferred on that segment by the pattern of Hox gene expression along the antero-posterior axis in the embryo (see Section 2.26). The role of the Hox genes is strikingly illustrated in regard to leg and antenna. The Hox gene *Antennapedia* is normally expressed in segments T2 and T3, allowing legs to develop on these segments, and in T2 *Antennapedia* confers identity on the imaginal discs for the second pair of legs. (Development of legs on T1 is due to expression of a different Hox gene, *Sex combs reduced*.) However, if *Antennapedia* is ectopically expressed in the head region (where it is not normally expressed), this converts the whole antennal imaginal disc to a leg disc and leads to the development of an ectopic leg, which is a T2 leg (Fig. 9.2). This observation reminds us that the Hox genes exert their homeotic effects by transforming pre-existing structures rather than creating new ones.

A second example of the way in which Hox genes work is given by their effects on the development of the wing and haltere imaginal discs in adjacent thoracic segments (T2 and T3, respectively). These segments are distinguished by the expression of the Hox gene *Ultrabithorax* (*Ubx*) in T3, which produces the haltere, but not in T2, which produces the wing. Mutations that destroy *Ubx* function lead to a wing on T3, instead of a haltere, indicating that the way *Ubx* works normally is to transform a potential wing disc into a haltere disc. This principle of suppressing or transforming a pre-existing structure is also at work in the abdominal segments, which have no appendages, not because positional information is missing, but because the expression of the genes required for formation of leg and wing discs is suppressed there by other Hox genes, such as *Ubx, abdominal-A* (*abd-A*) and *Abdominal-B* (*Abd-B*) (see Section 2.27).

The character of a disc, or its presence in a specific segment, is therefore strongly influenced by the pattern of Hox gene expression. However, the identity of an imaginal disc is only partly determined by the Hox genes it expresses. In a further layer of control, the appendage that each disc will produce is specified by one or a combination of master genes, themselves regulated by Hox genes. For example, the wing and leg discs in T2, which express the same Hox genes, are distinguished from each other by the expression of the wing master gene *vestigial* in the wing disc but not in the leg disc.

9.2 Imaginal discs arise across parasegment boundaries and are patterned by signaling at compartment boundaries

Before the definitive segments are formed in late embryogenesis, the *Drosophila* embryo is divided along the antero-posterior axis into developmental units called parasegments (see Section 2.23). The parasegment boundaries delineate areas of cell-lineage restriction called compartments, and when the segments form, the parasegment boundaries become compartment boundaries that divide each segment into an anterior and a posterior part. The imaginal discs are formed across parasegment boundaries, which means that when each disc is formed, it is divided by a compartment boundary (see Fig. 9.1). This boundary is maintained as the disc cells proliferate during larval growth. Compartments define units of growth, with the compartment boundaries defining the limits of these units.

During the larval growth phase, the compartment boundary acts as a crucial signaling center for patterning the disc, in the same way that the parasegment boundary operates in patterning the embryonic segment (see Section 2.25). Compartment boundaries persist into the adult fly, where they can be shown experimentally to demarcate these units of growth (Fig. 9.3). It is interesting that the compartment boundary in the wing is remarkably sharp and straight, and does not correspond to any of the wing's structural features such as the veins. The compartment boundary is under genetic control; we know this because it does not form in the absence of the *engrailed* gene (see Fig. 9.3).

Compartments were, in fact, first discovered in the derivatives of the *Drosophila* wing disc. They were identified in experiments intended to understand cellular growth during the emergence of appendages. Individual cells were marked genetically in the larval disc so that their descendants could be distinguished. It was found that the clones arising from the marked cells never crossed a 'boundary' located about half-way across the disc (and half-way across the wing that derives from it)—this is the compartment boundary. The phenomenon is even more obvious when the marked cells are given a growth advantage by making them mutant for the gene *Minute*. The clones produced by *Minute* cells cover a greater area of the wing (see Fig. 9.3), but they still respect the boundary and do not invade the adjacent patch of cells.

The experiment illustrated in Figure 9.3 also shows that the pattern of the wing does not depend on cell lineage. A single marked embryonic cell can give rise to about a twentieth of the cells of the adult wing or, using the *Minute* technique to increase clone size, about half the wing. The lineage of the wing cells in each case is quite different, yet the wing's pattern is normal. This indicates that groups of cells are being patterned; that is, patterning does not occur on an individual cell-by-cell basis.

9.3 The adult wing emerges at metamorphosis after folding and evagination of the wing imaginal disc

The adult insect wing is a largely epidermal structure in which two epithelial layers—the dorsal and ventral surfaces—are close together. Mesodermal cells are also associated with the disc, and are located in the non-wing part of the disc that contains the primordium for the dorsal part of the thoracic wall, also called the notum, and gives rise to adult muscles. The wing is joined to the notum by a 'hinge' region, which is also specified within the imaginal disc. At metamorphosis, by which time the imaginal discs are almost completely patterned, the wing disc undergoes a series of profound morphological changes. Essentially, the epithelial pouch is turned inside out, as its cells differentiate and change shape. The pouch extends and folds so that one half comes to lie beneath the other to form the double-layered wing structure (Fig. 9.4).

The wing imaginal disc initially contains cells that will form the notum, but not the hinge or the wing. The wing primordium—the group of cells that will give rise to the wing—emerges within the imaginal disc in the first larval instar. Genes specifying the body wall are repressed in the wing primordia, thus permitting the appendages to develop. The wing (and leg) discs are each divided into an anterior and a posterior compartment. In the wing-disc epithelium, a second compartment boundary, between the dorsal and ventral regions, develops during the second and third larval instars.

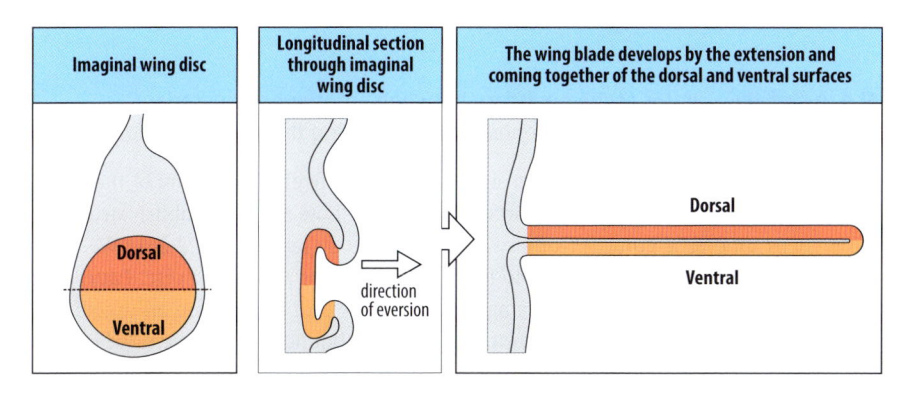

Fig. 9.4 Schematic representation of the emergence of the wing blade from the imaginal disc. Initially, the future dorsal and ventral surfaces of the wing are in the same plane within the imaginal disc (first panel). At metamorphosis, the pouch turns inside out and extends outwards (second and third panels); the dorsal and ventral surfaces of the wing come together.

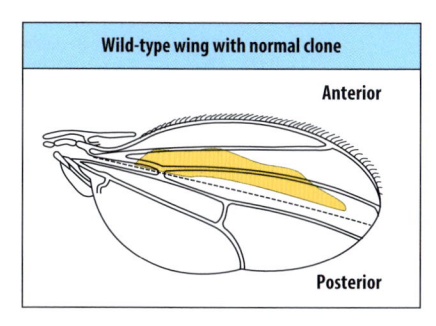

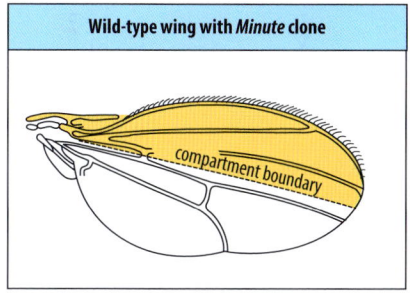

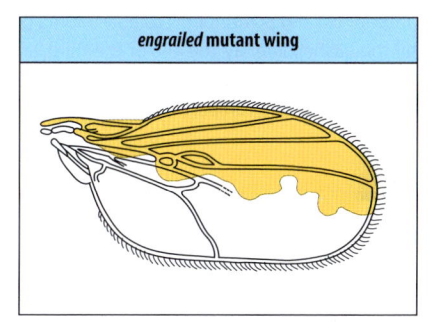

Fig. 9.3 The boundary between anterior and posterior compartments in the *Drosophila* wing can be demonstrated by marked cell clones. Top panel: clones marked experimentally by mitotic recombination give the marked cells a phenotype different from other cells of the wing. In a wild-type wing, the clones produced are too small to fill the compartment and demonstrate the compartment boundary unequivocally. Middle panel: cells heterozygous for the Minute mutation (M/+) grow more slowly than wild-type cells. In this fly, mitotic recombination leads to M/M cells that die and +/+ cells (yellow) that have a growth advantage over the surrounding M/+ cells and gives clones that are large enough to demonstrate that cells from one compartment do not cross the boundary into an adjoining compartment. Bottom panel: in the wing of an *engrailed* mutant in which the Engrailed protein is not produced, there is no posterior compartment or boundary. Clones in the anterior part of the wing cross over into the posterior region, and the posterior region is transformed into a more anterior-like structure, bearing anterior-type hairs on its margin.

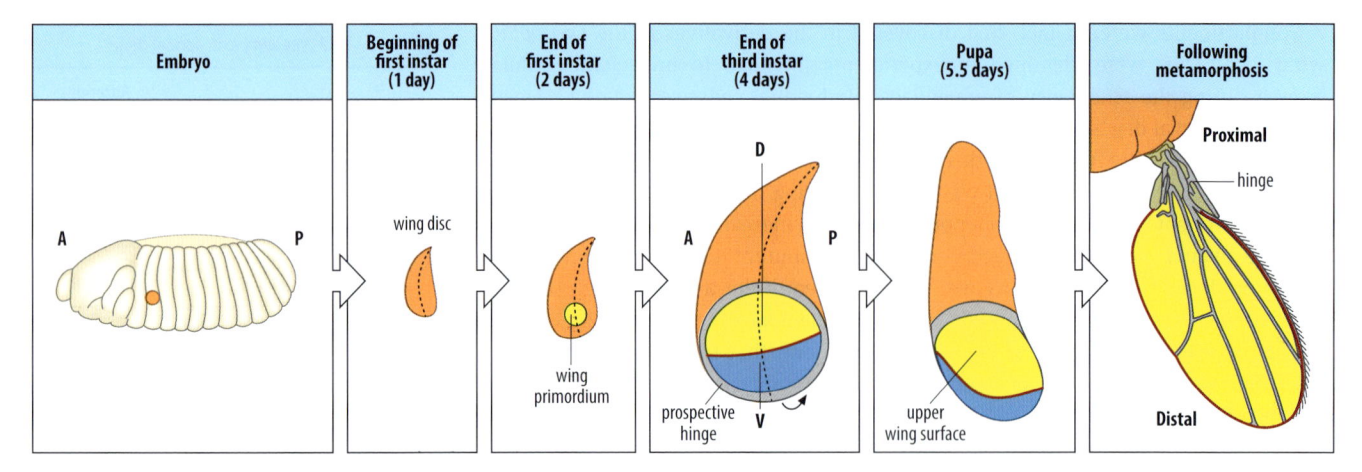

Fig. 9.5 Development and fate map of the wing imaginal disc of *Drosophila*. The wing primordium emerges in the early stages of larval growth as a small cluster of 20–30 cells in one half of the wing imaginal disc (orange); these cells then proliferate rapidly to form the ovoid wing primordium. The wing disc is bisected by the interface between the anterior (A) and posterior (P) compartments (see Section 9.1), shown by the dashed line. By the end of the third larval instar, the future wing is an ovoid epithelial sheet taking up around half of the imaginal disc, and a second compartment boundary (red), perpendicular to the anterior–posterior compartment boundary, has developed between the future dorsal (D, yellow) and ventral (V, blue) surfaces of the wing. At metamorphosis, the ventral surface folds under the dorsal surface (arrow) in the manner outlined in Figure 9.4, giving rise to the adult wing. The wing disc also contains the precursors to the hinge region (gray), and part of the body wall—the notum (orange)—which develop from the proximal part of the imaginal disc.

From Martinez Arias, A., Stewart, A.: Molecular Principles of Animal Development. *Oxford: Oxford University Press, 2002.*

The proximal parts of the wing (and leg) imaginal discs give rise to parts of the thoracic body wall (Fig. 9.5).

The development and growth of the *Drosophila* wing is completely dependent on the function of the gene *vestigial*, which encodes a transcriptional co-activator whose expression is first restricted to the wing disc by an interaction between the segment-specific Hox gene activity in the disc and the positional information provided by Wingless and Dpp signaling (see Fig. 9.1). *vestigial* is expressed at a low level throughout the early wing disc, and the induction of the wing primordium is correlated with an increase in *vestigial* expression in a small group of about 30 cells in the middle of the wing imaginal disc in response to Notch and Wingless signaling. These cells will become the wing margin (see Fig. 9.5) and will seed the outgrowth of the wing. Wingless is expressed in the wing disc from an early stage. Reduction of Wingless function results in a failure to specify the wing primordium, and therefore the fly has no wings; the gene is named after this adult phenotype.

9.4 A signaling center at the boundary between anterior and posterior compartments patterns the *Drosophila* wing disc along the antero-posterior axis

The compartment boundaries act as signaling centers for the patterning of the discs, and their activity has provided important clues about how signals establish structural differences among a group of otherwise similar cells (see Section 2.25). We shall look first at this process in relation to the antero-posterior axis of the disc. Patterning along this axis occurs in the early stages of larval life and arises from an organizing center located at the anterior–posterior compartment boundary. The *engrailed* gene is expressed in the posterior compartment of the imaginal disc, a pattern of expression inherited from the embryonic parasegment from which the disc derives (Fig. 9.6). The cells that express *engrailed* also express the gene *hedgehog* (see Section 2.25).

The Hedgehog signal protein is expressed throughout the posterior compartment and acts through its receptor, Patched, to activate the transcription factor Ci

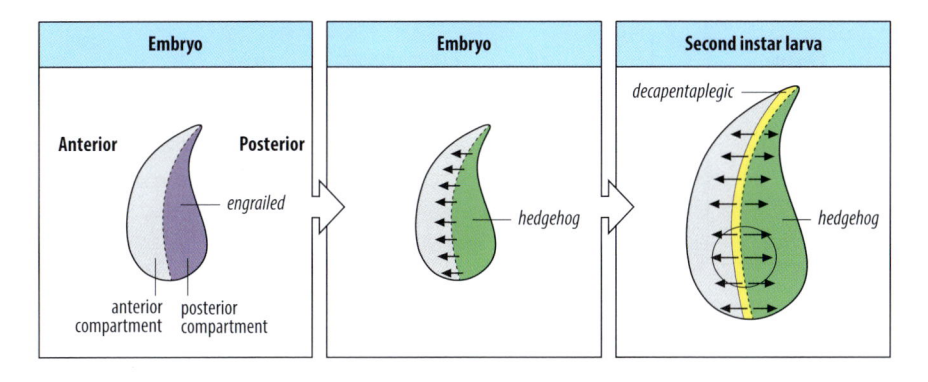

Fig. 9.6 Establishment of a signaling center in the wing imaginal disc at the anterior-posterior compartment boundary. The gene *engrailed* is expressed in the posterior compartment, where the cells also express the gene *hedgehog* and secrete the Hedgehog protein. Where high levels of Hedgehog protein signal to anterior compartment cells, a stripe of *dpp* gene expression is activated, and Dpp protein is secreted into both compartments, as indicated by the black arrows. The black circle indicates the wing primordium.

(see Box 2E). The Hedgehog protein secreted from the cells in the posterior compartment creates a gradient that reaches about 10 cell diameters into the anterior compartment, about half the width of the disc at the early stage. The Hedgehog protein is heavily modified by lipids, and this will restrict the distance over which it can diffuse. Its transcriptional effector, Ci, is only expressed in the anterior compartment, and this ensures that only anterior cells respond to Hedgehog. Signaling generates graded activation of Ci in cells at different distances from the compartment boundary, and this leads to different responses in adjacent cells along the gradient. Hedgehog is therefore acting as a morphogen, patterning the future wing tissue along the antero-posterior axis.

A very important target of Ci is the *decapentaplegic* (*dpp*) gene, which encodes a member of the bone morphogenetic protein (BMP) family of signal proteins (see Box 3F). In response to high levels of Hedgehog signaling, *dpp* is expressed in a narrow stripe along the anterior side of the anterior–posterior compartment boundary (see Fig. 9.6), leading to the secretion of Dpp protein, which forms symmetric concentration gradients in the anterior and posterior compartments. Dpp is the positional signal for patterning both compartments along the antero-posterior axis.

The gradient of Dpp across the wing disc provides a long-range signal that controls growth by cell proliferation. Within the area expressing *vestigial*, Dpp controls the localized expression of the genes for the transcription factors Spalt (Sal), Optomotor-blind (Omb), and Brinker in a concentration-dependent manner (Fig. 9.7). These transcription factors implement the patterning functions of Dpp, and refine the initial coarse patterning of the wing. Thus, Dpp acts as a morphogen (Fig. 9.8): low levels

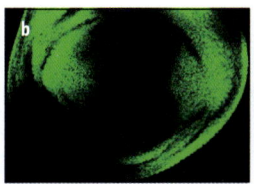

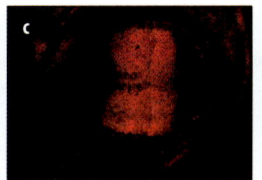

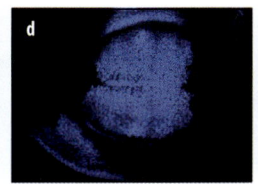

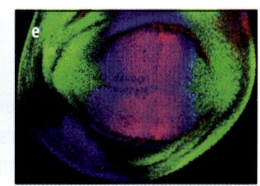

Fig. 9.7 Domains of gene expression initiated by Dpp signaling pattern the *Drosophila* wing disc. Panel (a) shows the expression of *decapentaplegic* (blue) at the anterior-posterior compartment boundary of the wing blade. Dpp signaling represses expression of the gene *brinker* (green, panel (b)), which is therefore expressed only in two regions on either side of the zone of *dpp* expression. Brinker protein represses the expression of *spalt* (red, panel (c)) and *omb* (blue, panel (d)), which are therefore expressed in a wide stripe over the anterior-posterior compartment boundary, where *brinker* is not expressed. Panel (e) shows a composite overlay of (b), (c), and (d).

(a) Photograph reproduced with permission from Nellen, D., et al.: **Direct and long-range action of a dpp morphogen gradient.** *Cell 1996, **85**: 357-368. © 1996, Cell Press. (b-e) Photographs reproduced with permission from Moser, M., Campbell, G.:* **Generating and interpreting the Brinker gradient in the Drosophila wing.** *Dev. Biol. 2005, **286**: 647-658.*

Fig. 9.8 A model for patterning the antero-posterior axis of the wing disc. The Decapentaplegic protein (Dpp) is produced along the anterior (A) side of the anterior-posterior compartment boundary. The schematic cross-section of the wing disc (X–Y) shows how the gradient in Dpp in both anterior and posterior (P) compartments activates the genes *spalt* and *omb* at their threshold concentrations. A lower Dpp threshold (red dashed line) allows cell proliferation and growth to continue throughout the wing.

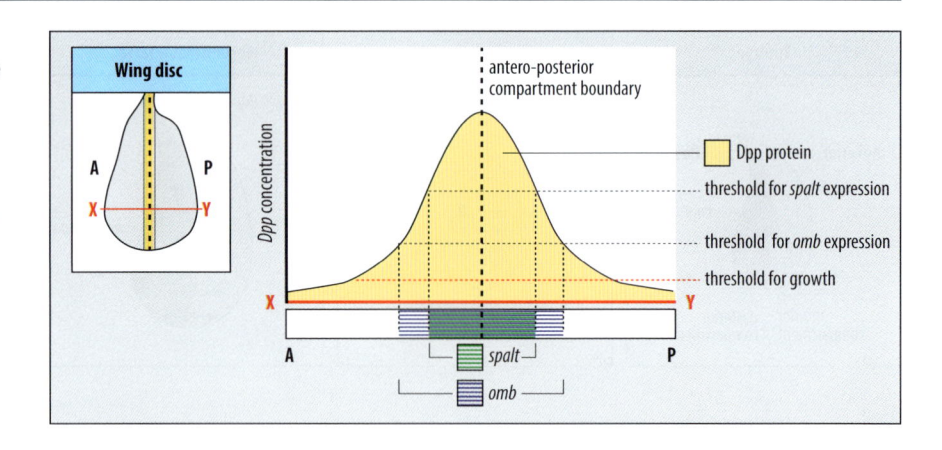

of Dpp induce *omb*, whereas higher levels are required to induce *spalt*. *omb* and *spalt* are related to the *Tbx* and *Sall* genes, respectively, which are involved in vertebrate limb patterning. *omb* and *spalt* expression in the wing disc determine the position of various veins in the adult wing. Brinker is a repressor of *spalt* and *omb* that is antagonized by the Dpp signal, and thus provides fine-tuned control of their expression.

Very low levels of the Dpp protein can be found at all times across the developing wing, and it has been found that at these low levels, Dpp promotes cell proliferation. The dependence of growth and patterning on different levels of Dpp allows these processes to be coupled and coordinated.

Several types of experiments demonstrate that Dpp is a morphogen in the wing disc. First, clones of cells unable to respond to a Dpp signal do not express *spalt* or *omb*. Second, ectopic expression of the *dpp* gene in a region that does not normally express *spalt* and *omb* leads to the localized activation of these genes around the *dpp*-expressing cells. Formation of the long-range Dpp gradient in the wing disc is a complex process and is not yet fully understood. The glypican Dally, present on the cell surface, is involved in both shaping the Dpp gradient and influencing Dpp signaling via the Dpp receptor Thick veins. Various ways of producing a graded signal are outlined in Box 9A.

The pattern of veins on the adult wing blade is one of the ultimate outcomes of the antero-posterior patterning activity of Hedgehog and Dpp in the wing disc (Fig. 9.9). This was shown by ectopically expressing the *hedgehog* gene in genetically marked cell clones generated at random in wing discs. When such clones form in the posterior compartment they have little effect (as *hedgehog* is normally expressed throughout this compartment in the disc) and development of the wing is more or less normal. With *hedgehog*-expressing clones in the anterior

Fig. 9.9 Ectopic expression of *hedgehog* and *decapentaplegic* alters wing patterning. Left panel: in the normal wing, Decapentaplegic protein (Dpp) is made at the boundary of the compartment in which *hedgehog* is expressed. Right panel: when *hedgehog* is expressed in a clone of cells in the anterior compartment, a new source of Dpp is set up, and a new wing pattern develops in relation to this new source. The numbers indicate the main wing veins.

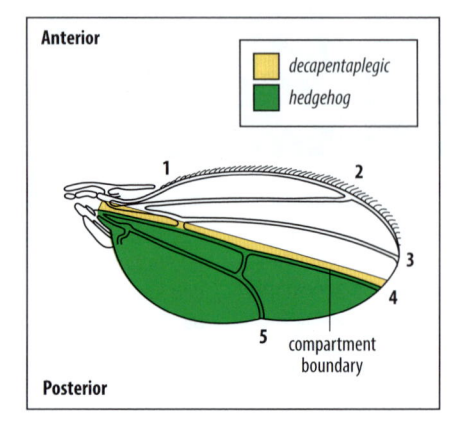

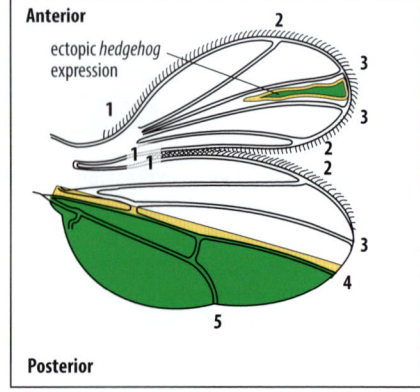

The insect wing and leg 449

BOX 9A Positional information and morphogen gradients

Pattern formation in development can be specified by positional information, which, in turn, can be defined by a gradient of some molecular property. The basic idea, related to the French flag problem (see Section 1.15), is that a specialized group of cells at a boundary of an area to be patterned—for example, a field of responding cells—secretes a molecule whose concentration then decreases with distance from this source, thus forming a gradient. A cell at any point along the gradient then 'reads' the local concentration and interprets it to respond in a manner appropriate to its position. Molecules that can specify cell fate in this way are known as morphogens and are said to form 'gradients of positional information'. These gradients have some important properties: in particular, they can provide a measure for the size of a tissue in terms of the slope and the decay length of the gradient (a measure of how far the gradient extends), and also harbor the possibility of scaling—a most important property of biological systems, according to which the proportions of the pattern elements within a tissue are independent of its size. Morphogens are used for various tasks during development: regionalization of the antero-posterior and dorso-ventral axes and patterning of segments and imaginal discs in insects (see Chapter 2 and this chapter); mesoderm patterning in vertebrates (see Chapters 3 and 4); vertebrate limb patterning (this chapter); and patterning along the dorso-ventral axis of the vertebrate neural tube (see Chapter 10), among others.

Gradient formation depends on localized production and secretion of the morphogen, its transport across the field of responding cells, and its degradation. In the model related to the French flag problem, the morphogen is removed by a sink at the opposite end of the field to the source and the resulting gradient is linear (see Section 1.15). If, however, the morphogen is removed throughout the field by, for example, receptor-mediated endocytosis, which seems to occur in most cases, the gradient has an exponential form (Fig. 1).

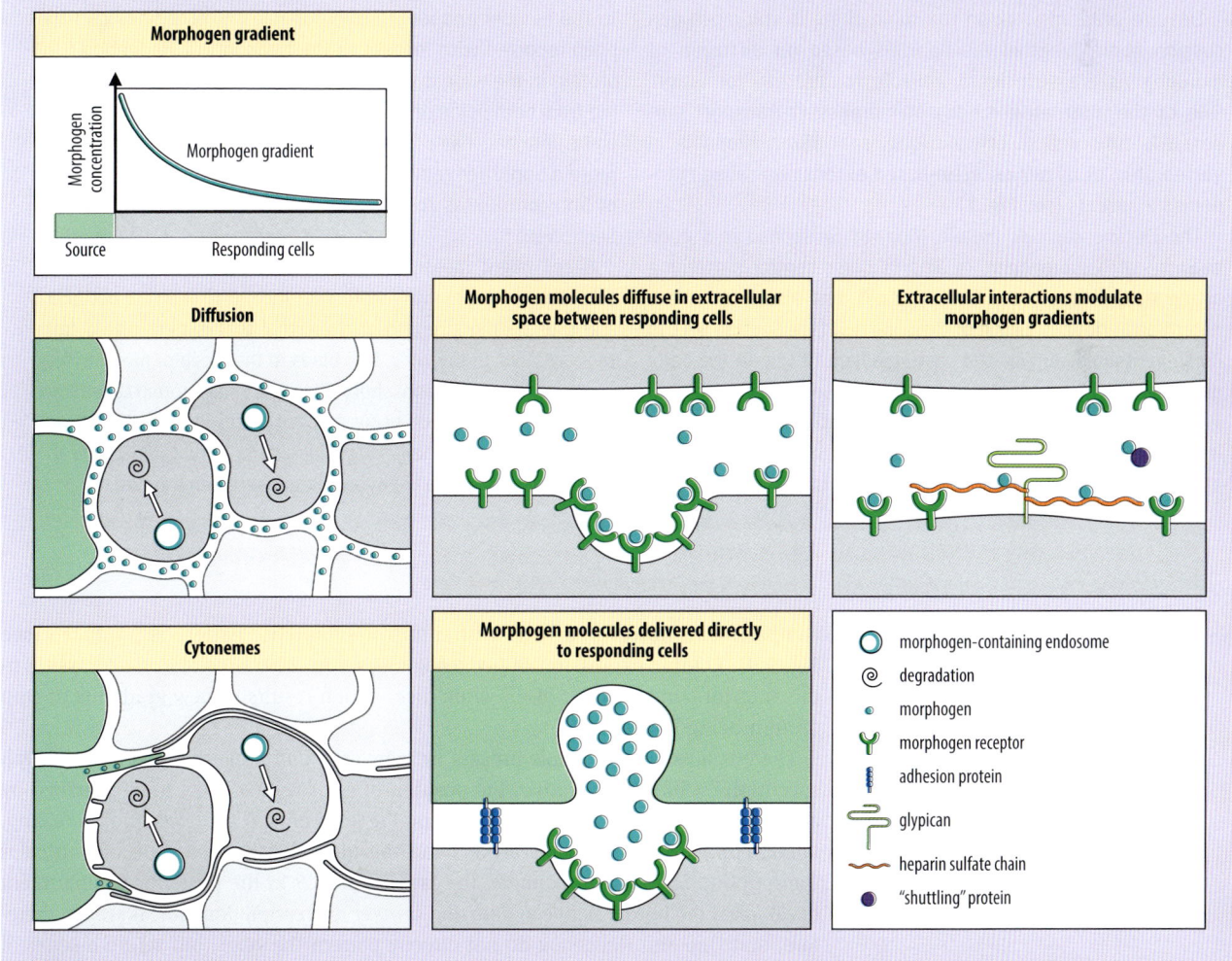

Fig. 1
Adapted from Stapornwongkul, K.S., Vincent, J.P.: **Generation of extracellular morphogen gradients: the case for diffusion**. Nat. Rev. Genet. 2021, **22**: 393-411.

Two main modes of morphogen transport have been envisaged—extracellular diffusion of the morphogen or its movement along cellular extensions. Morphogens could travel by simple diffusion through extracellular spaces and be degraded by the responding cells throughout the tissue so that a gradient forms. However, a diffusing morphogen is likely to encounter other extracellular proteins present in the tissue and interact with them. These interactions modulate the shape of the morphogen gradient. They could, for example, hinder diffusion and reduce the range of the gradient or they could lead to facilitated diffusion or 'shuttling' in which an extracellular protein binds to the morphogen and actively transports it to form a gradient (see, e.g., the formation of the ventral-dorsal BMP gradient in the *Xenopus* embryo, Fig. 4.3). The architecture of the tissue will also affect the ability of a morphogen to form a reliable informational gradient. For example, a morphogen diffusing freely within the plane of a single-layered epithelium, such as in a *Drosophila* imaginal disc (this chapter), would be prone to leak out of the cell layer.

Despite years of work on the topic, there is still much discussion about whether mechanisms based on diffusion of morphogens are sufficiently reliable for pattern formation. One of the alternative modes of morphogen transport proposed is from cell to cell via long, specialized filopodia in vertebrates, or by similar actin-based extensions called cytonemes in insects (see Fig. 1).

The Decapentaplegic protein (Dpp) in *Drosophila* is a good example of a morphogen. It acts in dorso-ventral patterning in the early embryo (see Section 2.17), and also later in the antero-posterior patterning of the wing disc (see Section 9.4). In dorso-ventral patterning, the sharp gradient of Dpp in the early blastoderm is formed by the interaction of the diffusing morphogen with other proteins, both secreted extracellular proteins and its own cell-surface receptors. In the wing disc, the formation of an extracellular Dpp gradient has been visualized directly by using a fusion of green fluorescent protein (GFP) to the Dpp protein and expressing it in its normal pattern.

Recent experiments using an inert protein (GFP) have shown that diffusion can produce a functional gradient of Dpp receptor activation and almost normal antero-posterior pattern formation in the *Drosophila* wing imaginal disc *in vivo*. Using clever genetic techniques, researchers engineered the disc to produce GFP instead of Dpp and modified both the cell-surface receptors for Dpp and membrane-anchored extracellular proteins called glypicans in the wing imaginal disc so that they bound GFP instead of Dpp. An almost normally patterned wing was produced. Binding to glypicans (see Fig. 1), which are membrane-bound non-signaling receptors and which reduces leakage, was crucial in forming the near-normal pattern.

Cytonemes have also been seen extending from responding cells in the wing disc to the Dpp source. In a later round of Dpp signaling in wing development, Dpp has been shown to move along cytonemes towards recipient cells in the wing disc, and there are indications that the range of Dpp signals is related to the number and length of cytonemes. Similar findings have been reported with respect to the transport of Hedgehog (Hh) protein in the wing disc. Filopodia have also been seen extending from both polarizing-region cells and responding cells in the chick limb and have been shown to transport the limb signal protein Sonic hedgehog (Shh). It is not clear how this transport of Shh contributes to Shh gradient formation in the limb (see Section 9.16).

A cell might measure a particular morphogen concentration according to how many receptors for the morphogen are occupied, and thus by the strength of the signal that is transmitted from the receptors. In the wing, Dpp binds to the receptor protein Thick veins, leading to the phosphorylation of a Smad protein called Mad, which can be detected using a specific anti-P antibody (P-Mad). Superimposition of the intracellular distribution of P-Mad and the Dpp gradient shows a linear correlation between the two.

compartment, however, a wing with a mirror-symmetric pattern is produced (see Fig. 9.9). Ectopic expression of *hedgehog* has set up new sites of *dpp* expression in the anterior compartment of the wing disc, which results in new gradients of Dpp protein being formed.

There is no single or simple process by which the morphogens Hedgehog and Dpp specify the vein pattern: rather, the position of each vein is likely to be specified by a unique combination of factors, including the level of Dpp and Hedgehog signaling, which compartment the cells are in, and the induction or repression of particular transcription factors. For example, the lateral vein L5 in the posterior compartment develops at the border between *omb* and *brinker* expression domains. It seems, therefore, that Dpp and Hedgehog do not directly pattern the veins, but set up a series of cell–cell interactions that specify positional information across the wing disc, perhaps even to the level of single cells. It is striking how precisely the pattern of veins matches on both wings of a fly.

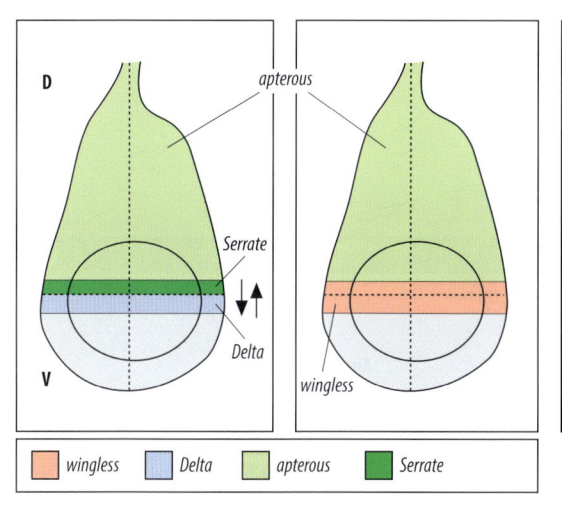

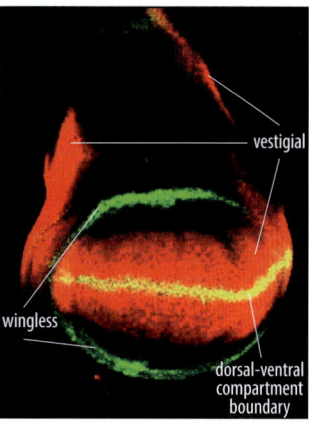

Fig. 9.10 Wingless is the signal protein at the boundary between dorsal and ventral compartments in the wing disc. The dorsal (D)–ventral (V) compartment boundary is established late in the second-instar larva, when a dorsal compartment is delineated by expression of the gene for the transcription factor Apterous. Apterous induces Notch signaling by first inducing Serrate expression in its own domain, which leads to Delta expression on the adjacent ventral cells (Delta and Serrate proteins are ligands for Notch). Serrate and Delta enhance each other's expression and lead to the expression of *wingless* in a stripe at the boundary. The combination of these signals increases expression of the gene *vestigial*, which encodes a key regulator of wing development, in a broad stripe at the dorsal-ventral boundary. The photograph shows *wingless* (green) and *vestigial* (red) expression in a wing disc from a third-instar larva. Both *wingless* and *vestigial* are expressed at the dorsal-ventral compartment boundary, as indicated by the yellow stripe.
Photograph reproduced with permission from Zecca, M., et al.: Direct and long-range action of a wingless morphogen gradient. Cell 1996, 87: 833–844. © 1996 Cell Press.

9.5 A signaling center at the boundary between dorsal and ventral compartments patterns the *Drosophila* wing along the dorso-ventral axis

During the second larval instar, the wing disc epithelium also becomes subdivided into dorsal and ventral compartments, which correspond to the future dorsal and ventral surfaces of the adult wing (Fig. 9.10). The dorsal and ventral compartments were originally identified by cell-lineage studies, but they are also defined by expression of the selector gene *apterous*, which is confined to the dorsal compartment and defines the dorsal state.

Like the anterior–posterior compartment boundary, the boundary between dorsal and ventral compartments acts as an organizing center, in this case with Wingless as the signal protein (see Fig. 9.10). Apterous begins to be expressed in the dorsal compartment shortly after the wing primordium is established. Later, at the dorsal–ventral compartment boundary, Apterous induces the expression of the gene encoding Serrate, a Notch ligand, in dorsal cells, and restricts the expression of the gene for the other Notch ligand, Delta, to ventral cells. Notch signaling (see Box 4F) at the compartment boundary induces expression of *wingless* along the boundary. Wingless protein, in turn, induces expression of Delta and Serrate at the boundary, and progressively refines its own expression as well as that of Notch signaling. As the patterns of Wingless and Notch signaling change, they focus expression of *vestigial* in a broad stripe centered on the dorsal–ventral boundary (see Fig. 9.10). Genetic analysis of Wingless function suggests that it may not be acting as a long-range morphogen, but may rather be collaborating with other genes that act locally, or cell-autonomously—particularly genes encoding gene-regulatory proteins, most significantly *vestigial*.

9.6 Vestigial is a key regulator of wing development that acts to specify wing identity and control wing growth

The signaling pathways we have seen in action above are used in the development and patterning of all imaginal discs. In the wing disc, their activity is channeled by Vestigial, which is a transcriptional co-factor that confers wing-specific activity on the transcription factor to which it binds (Fig. 9.11). The gene *vestigial* was initially identified by genetic analysis as the key gene controlling the establishment, growth, and patterning of the wing. Loss- and gain-of-function experiments showed that in the complete absence of *vestigial* expression the wing does not develop (see Fig. 1.12), whereas *vestigial* mutants with partial function lead to wing defects. However, the ectopic expression of *vestigial* in other imaginal discs (eye, leg, or genital) turns these tissues into wings.

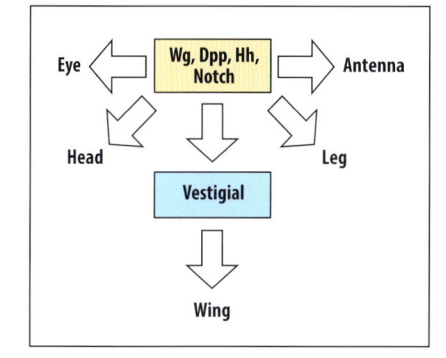

Fig. 9.11 The wing-specific transcriptional co-factor Vestigial channels signals common to all imaginal discs into making a wing. All imaginal discs use the same signaling pathways in their development and patterning. In the wing disc, the interaction of these signaling pathways with the *vestigial* gene directs their effects to activating genes that specify wing-specific development.

Fig. 9.12 vestigial integrates different signals via different enhancers in its control regions. The *vestigial* gene contains two different enhancers, the 'boundary enhancer' (BE) and the 'quadrant enhancer' (QE), which respond to different inputs and result in an increase in *vestigial* expression at different times and in different areas of the wing disc. The enhancers are located in introns within the gene. BE responds to Notch signaling and Wingless (Wg) at the dorsal-ventral (D-V) boundary to increase *vestigial* expression at this boundary. QE is activated later, at the beginning of the third instar, in response to Wg signaling at the dorsal-ventral boundary and Dpp signaling at the anterior-posterior (A-P) boundary, and drives *vestigial* expression in the four quadrants of the wing disc (but not at the dorsal-ventral or anterior-posterior boundaries). This latter expression then controls the growth and further development of the wing blade.

*Data from Kim, J., et al.: **Integration of positional signals and regulation of wing formation and identity by Drosophila vestigial gene.** Nature 1996, 382: 133-138.*

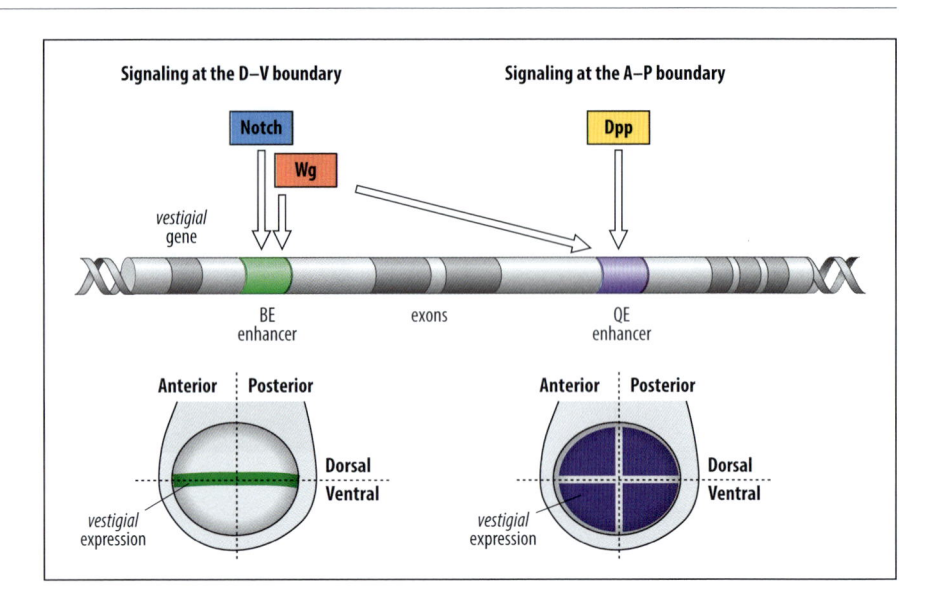

Vestigial protein is located in the nucleus. It does not have a DNA-binding domain and its ability to regulate gene expression is mediated by its interactions with Scalloped, a widely expressed transcription factor that is also essential for wing development. Although few of the individual targets of the Vestigial–Scalloped complex are known, the effects of mutations in either gene show that the complex must regulate a large battery of genes that ensure that the growing cells make wing structures.

The levels and pattern of expression of *vestigial* in the imaginal disc vary throughout wing development, and this reflects the response of the prospective wing tissue to the various molecular inputs that govern its patterning and growth. *vestigial* integrates these signals through position-specific enhancer sites in its control regions that respond to different inputs.

The initiation of wing development depends on the expression of *vestigial* in a small group of about 30 cells in the middle of the wing disc in the first-instar larva (see Section 9.3). This expression is under the control of an enhancer ('boundary enhancer' (BE)) that responds to Notch and Wingless signaling in the disc at this stage. As Wingless expression within the wing disc develops into a particular spatial pattern, *vestigial* expression becomes concentrated at the dorsal–ventral compartment boundary of the wing primordium at the beginning of the second-instar larva. These cells will become the wing margin (see Fig. 9.5). Later, at the beginning of the third instar, a combination of Wingless signaling from the dorsal–ventral boundary and Dpp signaling from the anterior–posterior compartment boundary switches on *vestigial* expression throughout the remainder of the wing primordium via another enhancer ('quadrant enhancer' (QE)). If a *lacZ* reporter gene is put under the control of the QE enhancer, it shows very accurately the growth of the wing during the larval instars. The function of Vestigial protein in promoting wing-cell proliferation and wing outgrowth during the later larval instars is due to the combination of the QE and the BE patterns of expression (Fig. 9.12).

The expression of *vestigial* under the control of the QE enhancer is entirely dependent on its previous expression within the disc, and on the presence of Vestigial protein; it is therefore an example of keeping a gene active by autoregulation and ensuring that it remains active in daughter cells (see Section 6.3). The QE enhancer also integrates antero-posterior and dorso-ventral positional information by being able to respond to inputs from both Wingless and Dpp signaling. It is likely that other discs in *Drosophila*, and organs in other organisms, develop under the influence of proteins functionally similar to Vestigial—that is, disc- or organ-specific regulatory proteins that channel the activity of widely expressed signal molecules and transcription factors towards the genes that produce the final structure and pattern particular to that organ. In some instances, a combination of proteins is more likely than a single protein to perform this function. In the *Drosophila* leg, for example, there is no single 'leg-specific' gene corresponding to *vestigial*.

The patterning processes described in the previous sections divide the wing disc into overlapping domains of gene expression. These domains specify the position of pattern elements, such as veins, and they also specify the size of the various parts of the wing, although how they do this is still not completely understood. The development and positioning of these domains of gene expression can be related to the actions of the antero-posterior and dorso-ventral signaling centers, but the patterns of growth within the wing, although dependent to a certain degree on Dpp and Wingless, are not obviously related to the gradients of these signals or to the patterns of expression of their target genes. The wing disc grows in size by cell division, which occurs throughout the disc and then ceases uniformly before metamorphosis when the correct size is reached. Organs, in general, also grow until they reach a given size, and imaginal discs have been used as a simple model system to investigate how a structure monitors its growth to reach the correct size (also discussed in Chapter 11). The adult wing is produced at metamorphosis by changes in cell shape without further cell division.

It is becoming clear that cells not only receive biochemical inputs that stimulate, or inhibit, their proliferation, but are also subject to physical forces and mechanical signals associated with the number of cells in a specific area and the geometric constraints of the tissue. The boundary of the circular wing disc, for example, is demarcated by deep furrows that create physical forces (strains and stresses) in particular spatial patterns. These stresses also affect growth rates and directions of tissue outgrowth, and can thus shape the tissue.

Genetic analysis has identified the Hippo signaling pathway (see Box 11A) as an important element in sensing changes in cell shape, adhesion, and density, and relating them to cell growth. Hippo signaling suppresses cell proliferation and promotes apoptosis. Both Wingless and Dpp can interact with this pathway, and this might be one way of channeling their inputs into directing growth in the wing. A possible clue to the control of wing growth is that Scalloped, the transcription factor that interacts with Vestigial, is a member of the TEAD transcription factor family, whose activity is controlled by the Hippo pathway. Scalloped is essential for wing growth and when the Hippo pathway is inactive it can act in a complex with the co-activator Yki (a component of the pathway) to turn on genes involved in cell proliferation and the suppression of apoptosis (discussed in Section 11.4).

9.7 The *Drosophila* wing disc is also patterned along the proximo-distal axis

Along the proximo-distal axis, the wing imaginal disc gives rise to various structures in addition to the wing itself (see Fig. 9.5). The proximo-distal axis of the wing itself is less easy to visualize in the flat imaginal disc than the antero-posterior and dorso-ventral axes. The prospective wing-blade tissue can be considered the most **distal** element of the wing imaginal disc. It is surrounded by cells that will give rise to the hinge—a structure by which the wing is attached to the notum, which comes from the most proximal part of the imaginal disc. During the development and growth of the disc, these structures must be differentiated from the wing primordium and patterned appropriately. There is evidence that Wingless activity specifies these structures sequentially from proximal (notum) to distal (wing blade), and that both short-range and long-range Wingless signaling are involved in patterning the wing blade itself along the proximo-distal axis. As well as the general asymmetry of wing

Fig. 9.13 *Drosophila* **leg disc extension at metamorphosis.** The disc epithelium, which is an extension of the epithelium of the body wall, is initially folded internally. At metamorphosis it extends outward, as if pulled out from the center. The red arrow in the first panel is the viewpoint that produces the concentric rings shown in Figure 9.14.

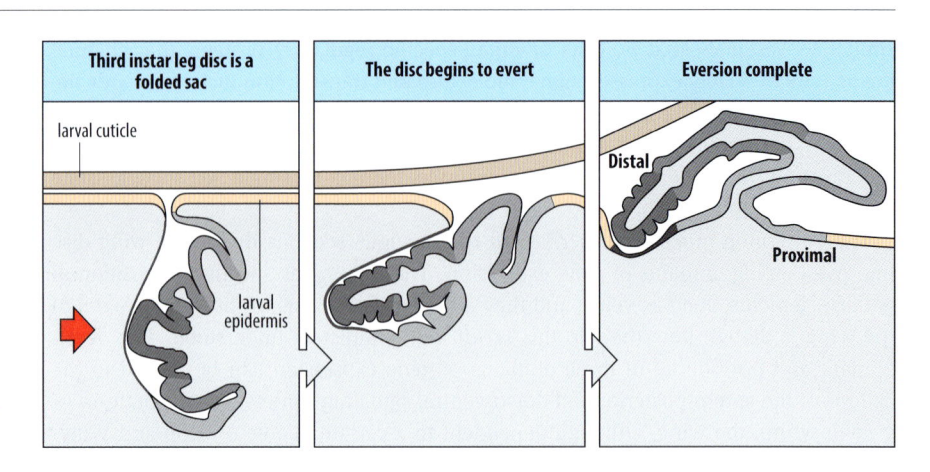

structure, which is due to patterning along the various axes, the individual cells in the adult *Drosophila* wing epidermis become polarized in a proximo-distal direction; this is reflected by the pattern of hairs (trichomes) on the wing surface. Each of the approximately 30,000 cells in the wing epidermis produces a single hair, which points distally (see Fig. 9.5, which shows only those hairs that would be visible on the edge of the wing). This is an example of planar cell polarity (see Section 5.12).

9.8 The leg disc is patterned in a similar manner to the wing disc, except for the proximo-distal axis

Insect legs are essentially jointed tubes of epidermis and thus have a quite different structure from those of vertebrates. The epidermal cells secrete the hard outer cuticle that forms the exoskeleton. Internally, there are muscles, nerves, and connective tissues. A change in the shape of the epithelial cells of the leg imaginal disc is responsible for the outward extension of the leg at metamorphosis. The process is rather like pulling the disc out from the center, with the result that the center of the disc ends up as the distal end, or tip, of the leg (Fig. 9.13).

The easiest way of relating the leg imaginal disc to the adult leg is to think of it as a collapsed cone. Looking down on the disc, one can imagine it as a series of concentric rings, each of which will form a proximo-distal segment of leg (Fig. 9.14). The outermost ring gives rise to the base of the leg, which is attached to the body, and the rings nearer the center give rise to the more distal structures.

The leg disc contains some 30 cells at its initial formation in the embryo but grows to contain more than 10,000 cells by the third instar. Unlike the wing disc, it contains some cells specified to become leg cells at its formation. The first steps in the patterning of the leg disc along its antero-posterior axis are the same as in the wing. The *engrailed* gene is expressed in the posterior compartment and induces expression of

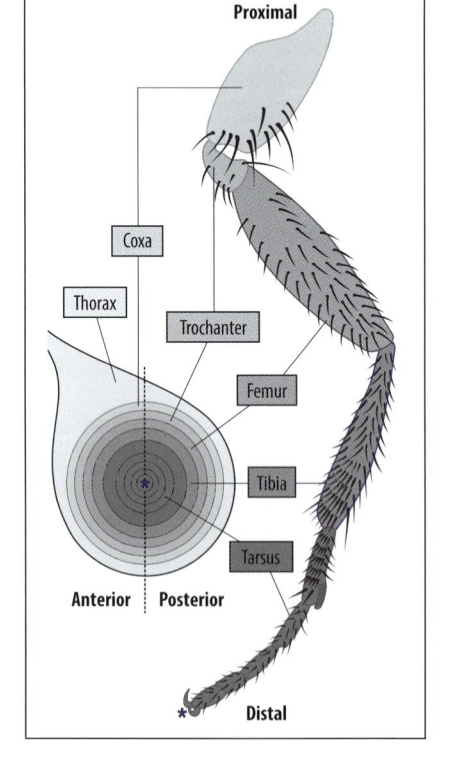

Fig. 9.14 Fate map of the leg imaginal disc of *Drosophila*. The disc is a roughly circular epithelial sheet, which becomes transformed into a tubular leg at metamorphosis. The center of the disc becomes the distal tip of the leg and the circumference gives rise to the base of the leg—this defines the proximo-distal axis. The tarsus is divided into five tarsal segments. The prospective regions of the adult leg, such as the tibia, are thus arranged as a series of circles with the future tip at the center. A compartment boundary divides the disc into anterior and posterior regions. There is no division of the leg disc into dorsal and ventral compartments as there is in the wing.

Illustration after Bryant, P.J.: **The polar coordinate model goes molecular.** Science 1993, **259**: 471-472.

Fig. 9.15 Establishment of signaling centers at the anterior-posterior compartment boundary in the leg disc, and the specification of the distal tip. The gene *engrailed* is expressed in the posterior compartment and induces expression of *hedgehog*. Where Hedgehog protein meets and signals to anterior compartment cells, the gene *decapentaplegic* (*dpp*) is expressed in dorsal regions and the gene *wingless* is expressed in ventral regions. Both of these genes encode secreted signal proteins. Expression of the gene *Distal-less*, which specifies the proximo-distal axis, is activated where the Wingless and Dpp proteins meet.

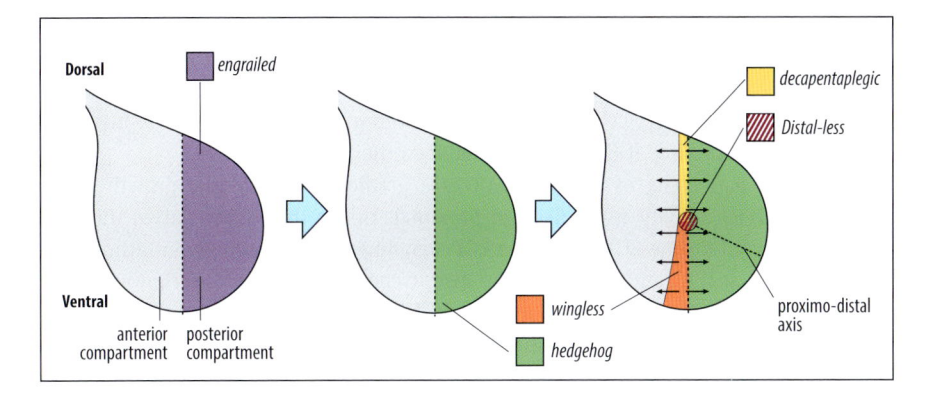

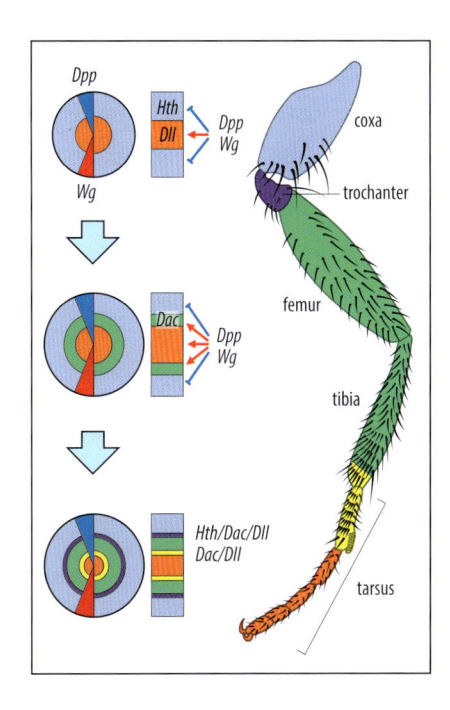

hedgehog. The Hedgehog protein induces a signaling region at the anterior–posterior compartment border. In the dorsal region of the leg disc, the expression of *dpp* is induced in the anterior compartment, as it is in the wing. In the ventral region, however, Hedgehog induces expression of *wingless* instead of *dpp* in the anterior compartment along the compartment border, and Wingless protein acts, as in the wing, to coordinate the sequence of short-range signaling events associated with the growth of the leg (Fig. 9.15).

The patterning of the proximo-distal axis is much better understood in the leg disc than in the wing and involves interactions between Wingless and Dpp signaling. The distal end of the proximo-distal axis is the point at which Wingless and Dpp expression meet and is marked by expression of the homeodomain transcription factor *Distal-less*. Although *Distal-less* is also expressed in *Drosophila* wing development, it appears to have no function there, because if the gene is removed, there is no effect on wing development. It is, however, crucial for correct leg development.

Patterning along the proximo-distal axis takes place sequentially, and produces a two-dimensional pattern in the leg disc that will be converted at metamorphosis into the three-dimensional tubular leg. Early expression of *Distal-less* is governed by a particular module in the gene's control region and lasts for just a few hours. Later expression of *Distal-less* in the future distal region is directed by other *cis*-regulatory modules. The gene for another homeodomain transcription factor, *homothorax*, is expressed in the peripheral region surrounding *Distal-less* expression, and marks the future proximal region (Fig. 9.16). The actions of Distal-less and homothorax proteins lead to the expression of the gene *dachshund*, which also encodes a transcription factor, in a ring between *Distal-less* and *homothorax*, which eventually leads to some overlap of the expression of these three genes. Each of these genes is required

Fig. 9.16 Regional subdivision of the *Drosophila* leg along the proximo-distal axis. The pattern of gene expression in the leg disc over time is shown on the right. The stages of gene expression are shown in the two columns on the left, and are viewed as if looking down on the disc. Because of the way in which the leg extends from the disc, the center corresponds to the future tip of the leg, and the successive rings around the center correspond to more proximal regions of the leg. *decapentaplegic* (*dpp*) and *wingless* (*wg*) are initially expressed in a graded manner along the anterior-posterior compartment boundary and together induce *Distal-less* (*Dll*) and repress *homothorax* (*hth*) in the center; *hth* is expressed in the outer region. *dpp* and *wg* then induce *dachshund* (*dac*) in a ring between *Dll* and *hth*. Further signaling leads to these domains overlapping. While *hth* expression corresponds to proximal regions and *Dll* to distal regions, there is no simple relation between the other genes and the leg segments.

After Milan, M., Cohen, S.M.: **Subdividing cell populations in the developing limbs of Drosophila: do wing veins and leg segments define units of growth control?** Dev. Biol. 2000, **217**: 1–9.

for the formation of particular leg regions, but the expression domains do not correspond precisely with leg segments. Expression of *dachshund*, for example, corresponds to femur, tibia, and proximal tarsus. There is no evidence of a proximo-distal compartment boundary; cells expressing *homothorax* and *Distal-less* are, however, prevented from mixing at the interface between their two territories.

Additional patterning of the leg involves a gradient of epidermal growth factor receptor (EGFR) activity from distal to proximal. In the third-instar larva, the Vein protein (a ligand for the EGFR) and the EGF signaling pathway component Rhomboid are expressed at the central point of the disc. Genes coding for transcription factors Bric-a-brac and Bar are expressed at different levels in the tarsal segments and may determine their identity. The level of activity of these genes might be determined by the gradient of EGFR activity. Leg-joint formation requires Notch signaling. Delta and Serrate, ligands for Notch, are expressed as a ring in each leg segment and their activation of Notch results in specification of the joint-forming cells.

9.9 Different imaginal discs can have the same positional values

The patterning of legs and wings involves similar signals, yet the actual pattern that develops is very different. This implies that the wing and leg discs interpret positional signals, such as Wingless, Dpp, and Hedgehog, in different ways. Using the technique of **mitotic recombination**, a small clone of *Antennapedia*-expressing cells can be generated within a normal antennal disc. These cells develop as leg cells, and exactly which type of leg cell depends on the cells' position along the proximo-distal axis in the disc. If, for example, the cells are at the tip, they form a claw. This suggests that the positional values in the antenna and leg discs are similar, and that the difference between the two structures lies in the interpretation of these values, which is governed by the expression or non-expression of the *Antennapedia* gene (Fig. 9.17). This is yet another example of cells developing according to both their position and their developmental history, which determines which genes they are expressing at any given time.

The question of how expression of a single transcription factor such as Antennapedia can transform an antenna into a leg is still to be resolved, and will need identification of the Antennapedia protein's target genes. Some evidence suggests that Antennapedia acts as a repressor of antennal identity in the leg by, for example, preventing the co-expression of the genes *homothorax* and *Distal-less* in the 'femur' region. These two genes are expressed together in the corresponding region in the antennal disc, but in the leg disc they are expressed in adjacent and non-overlapping domains (see Fig. 9.16). In evolutionary terms, this indicates that the antennal state may be the ground state for limb development.

Identical positional information can also be demonstrated in the wing and haltere discs. If a small clone of cells containing a mutation affecting *Ubx* expression (such as *bx*) is generated in the haltere imaginal disc, the cells in the clone make wing structures, which correspond exactly to those that would form in a similar position in a wing. This indicates that the positional values in haltere and wing discs are identical, and all that has been altered in the mutant is how this positional information is interpreted.

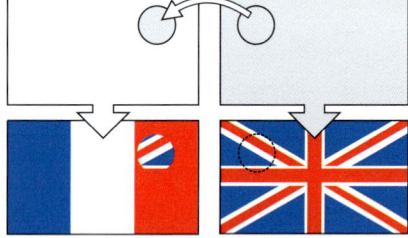

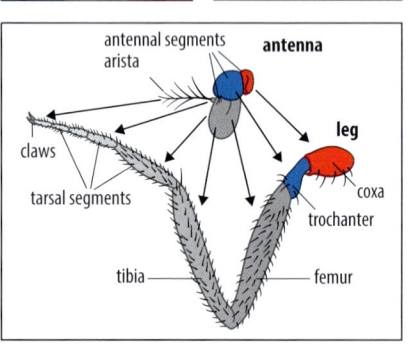

Fig. 9.17 Cells interpret their position according to their developmental history and genetic make-up. If two flags used the same positional information to produce different patterns, then a graft from one to the other would result in the graft developing according to its new position, but with its original pattern (upper panels). Imaginal discs similarly use the same positional information to produce differently patterned appendages (lower panel).

SUMMARY

The legs and wings of the adult *Drosophila* develop from epithelial sheets—imaginal discs—that are set aside in the embryo. Hox genes acting in the parasegments specify which sort of appendage will be made. The leg and wing imaginal discs are divided at an early stage into anterior and posterior compartments. The boundary between the compartments is an organizing center and a source of signal proteins that pattern the disc. In the wing disc, expression of *decapentaplegic* is activated by the Hedgehog protein at the anterior-posterior compartment boundary and the Decapentaplegic and Hedgehog proteins act as antero-posterior patterning morphogens. The wing disc also has dorsal and ventral compartments, with Wingless produced at the dorsal-ventral boundary acting as a patterning signal. The essential wing-specific regulatory gene *vestigial* integrates this positional information through its control regions and channels it towards genes that build wing-specific structures and that control wing growth. In the leg disc, the anterior-posterior compartment boundary is established in a very similar way, except that Hedgehog activates *wingless* instead of *decapentaplegic* in the ventral region, and the Wingless protein acts as the patterning signal in this region. There is no evidence of dorsal and ventral compartments in the leg disc. The proximo-distal axis of the leg is specified by the interaction between Decapentaplegic and Wingless signaling, which activates genes such as *dachshund* and *homothorax* in concentric domains corresponding to regions along the leg.

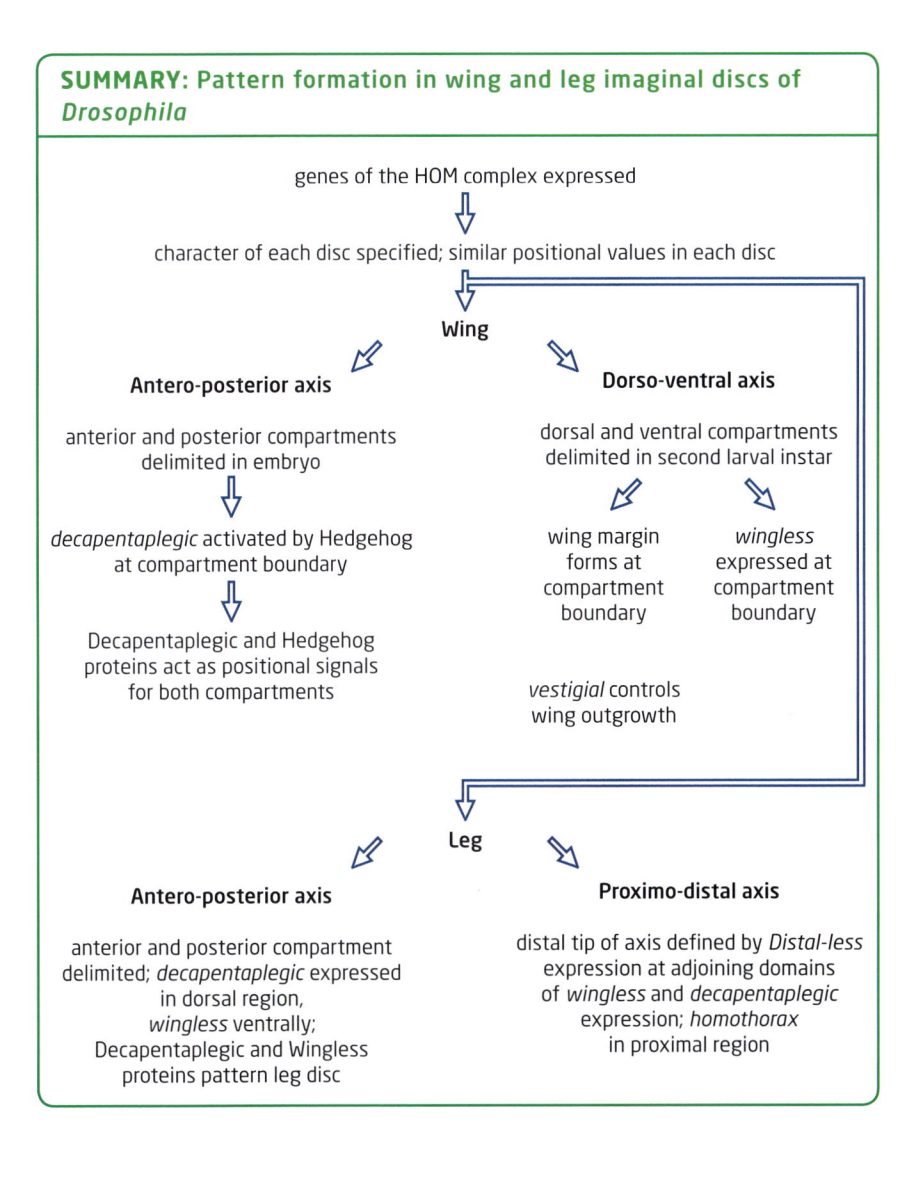

SUMMARY: Pattern formation in wing and leg imaginal discs of *Drosophila*

genes of the HOM complex expressed
⇩
character of each disc specified; similar positional values in each disc
⇩
Wing

Antero-posterior axis

anterior and posterior compartments delimited in embryo
⇩
decapentaplegic activated by Hedgehog at compartment boundary
⇩
Decapentaplegic and Hedgehog proteins act as positional signals for both compartments

Dorso-ventral axis

dorsal and ventral compartments delimited in second larval instar

wing margin forms at compartment boundary

wingless expressed at compartment boundary

vestigial controls wing outgrowth

⇩
Leg

Antero-posterior axis

anterior and posterior compartment delimited; *decapentaplegic* expressed in dorsal region, *wingless* ventrally; Decapentaplegic and Wingless proteins pattern leg disc

Proximo-distal axis

distal tip of axis defined by *Distal-less* expression at adjoining domains of *wingless* and *decapentaplegic* expression; *homothorax* in proximal region

458 Chapter 9 Organogenesis

The vertebrate limb

The embryonic vertebrate limb is a particularly good system in which to study general developmental principles. Development of the limb involves some simple changes in shape, and the basic pattern of the limb skeleton can be easily recognized. The limb is also a good model for studying cellular interactions within a structure containing many cells, and for elucidating the role of cell–cell signaling in development. Mice are used to study aspects of limb development, mainly through spontaneous mutants and artificial gene knock-outs. But the basic principles of limb morphogenesis and pattern formation have been most extensively investigated in chick embryos, because here the developing limbs are easily accessible for microsurgical manipulation. A window can be made in the eggshell and the limb buds manipulated while the embryo remains in the egg. After surgery, the window can be re-sealed with adhesive tape and the embryo allowed to continue development, so that the effects of the manipulations on limb development can be assessed. The early development of zebrafish pectoral fin buds is very similar to that of limb buds, and zebrafish are increasingly being used to study these early stages.

In chick embryos, the first signs of limb development can be seen around the third day after the egg is laid, when the structures of the main body axis are already well established. Two pairs of small protrusions—the limb buds—arise from the body wall of the embryo; the upper pair give rise to the wings, the lower pair to the legs (Fig. 9.18). By 10 days, the main features of the limbs are well developed. Figure 9.19 shows the pattern of the skeletal elements in the wing; they first form as cartilage and are later replaced by bone (how bones grow is described in Section 11.9). Feather buds have formed and the limb at this stage also has muscles and tendons. The limb has three developmental axes. The proximo-distal axis runs from the base of the limb to the tip. The antero-posterior axis of a limb runs parallel with the body axis—in the chick wing it goes from digit 2 to digit 4, in the human hand from the thumb (anterior) to the little finger (posterior). The dorso-ventral axis is the third axis—in the human hand it runs from the back of the hand to the palm.

9.10 The chick wing is a model for limb development

The early chick wing bud has two main components—a core of a loose meshwork of mesenchymal cells and an outer layer of epithelial cells (Fig. 9.20). The mesenchyme comprises two separate lineages—cells derived from the lateral plate mesoderm,

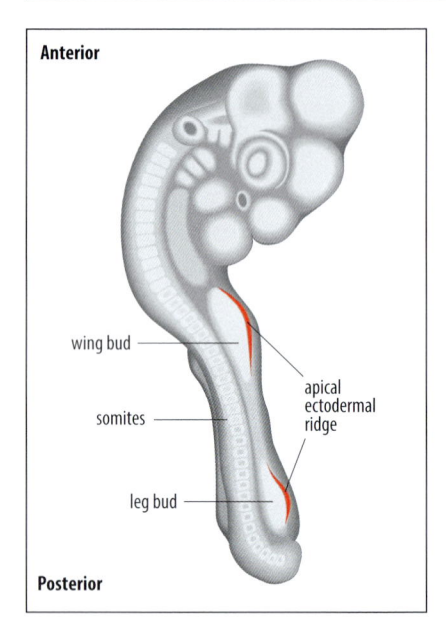

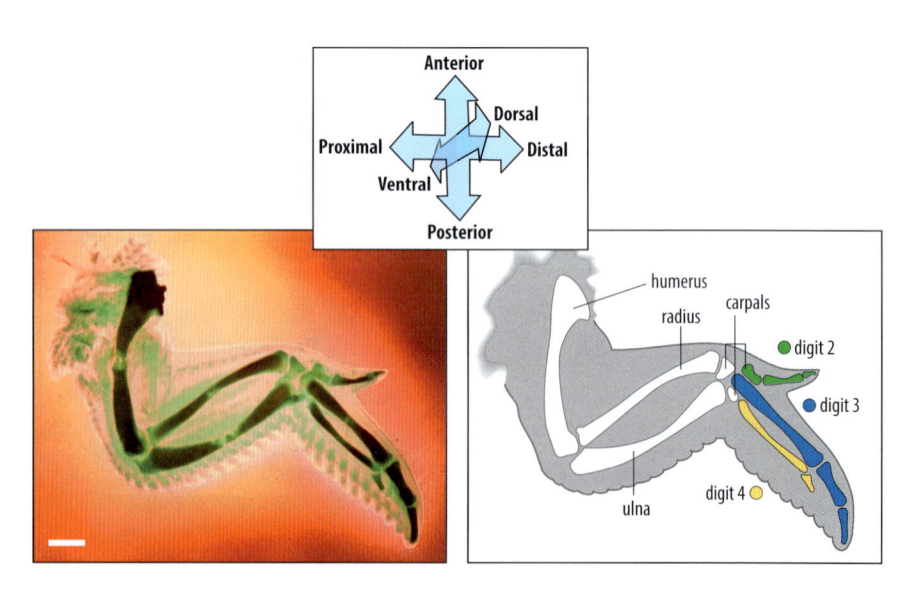

Fig. 9.18 The limb buds of the chick embryo. Limb buds appear at 3 days of incubation after the egg has been laid (only the limb buds on the right side are shown here). They are composed of mesoderm, with an outer covering of ectoderm. Along the tip of each runs a thickened ridge of ectoderm, the apical ectodermal ridge.

Fig. 9.19 The embryonic chick wing. The photograph shows a whole mount of the wing of a chick embryo at 10 days of incubation after the egg has been laid. The wing has been stained to show the pattern of differentiated cartilage tissue. By this time, the main skeletal elements (e.g. humerus, radius, and ulna) have been laid down in cartilage. They later become ossified to form bone. The muscles and tendons are also well developed at this stage but cannot be seen in this type of preparation. Feather buds can be seen, particularly along the posterior margin of the limb. The three developmental axes of the limb are proximo-distal, antero-posterior, and dorso-ventral, as shown in the top panel. Note that the chick wing has only three digits, which have traditionally been called 2, 3, and 4 (in relation to the five-digit tetrapod limb) although evolutionary studies now identify them as 1, 2, and 3. Scale bar = 1 mm.

which are multipotent and give rise to the skeletal elements and other connective tissues of the limb, and cells derived from the somites, which migrate into the limb bud and give rise to the myogenic cells that will form muscle (see Section 6.6). The limb bud is vascularized and the vasculature is also formed by cells derived from the somites. The epithelial cells are derived from the ectoderm and give rise to the epidermis of the skin (see Section 6.11).

At the very tip of the chick wing bud is a thickening in the ectoderm—the **apical ectodermal ridge** or **apical ridge**, which runs along the boundary between dorsal and ventral ectoderm (Fig. 9.21). The apical ectodermal ridge is asymmetric, being thicker over the posterior part of the bud—that is, the part nearest to the tail. Directly beneath the apical ectodermal ridge lies a region composed of undifferentiated mesenchyme, whose cells only begin to differentiate when they leave this zone. As the bud grows out, the cells left behind start to differentiate, and the cartilaginous elements of the skeleton begin to appear in the mesenchyme. The proximal part of the limb bud—that is, the part nearest to the body—is the first to differentiate, and differentiation proceeds distally as the limb bud grows out, with the digits differentiating last. Of the structures found in the developing limb, the pattern of the cartilage elements has been the best studied, as cartilage can be stained and easily seen in whole mounts of the embryonic limb (see Fig. 9.19). The disposition of muscles and tendons is more intricate, and although this can now be studied in whole mounts by staining with antibodies for tissue-specific proteins; earlier studies involved histological examination of serial sections through the limb.

The first sign of cartilage differentiation to form a skeletal element is the increased local packing together of the mesenchyme cells, a process known as condensation. The cartilage elements are laid down in a proximo-distal sequence. In the developing chick wing bud, the humerus is laid down first, followed by the radius and the ulna, then the wrist elements (carpals), and finally three easily distinguishable digits, 2, 3, and 4 (see Fig. 9.19). This traditional numbering of the wing digits will have been used in papers up to about 2011, and we will use this numbering system when we discuss the classic work on chick wing development to avoid confusion. More recently they have been designated 1, 2, and 3 on the basis of evolutionary studies.

Figure 9.22 shows how the shape of a chick wing bud changes over these stages of development. An indentation appears in the anterior margin of the elongated bud marking the 'elbow', and later the distal region of the limb bud broadens and forms the flattened **digital plate** (**hand plate**) in which the digits arise. The figure also compares this sequence in the development of a chick wing with the similar sequence in the development of a mouse forelimb.

The chick wing bud at 3 days is about 1 mm wide by 1 mm long, but by 10 days it has grown around 10-fold, mostly in length. The basic pattern has been laid down well before then, but even at 10 days the wing is still very small compared with the size of the wing when the chick hatches. The apical ridge disappears as soon as all the basic elements of the wing are in place, and growth occupies most of the subsequent development of the wing, both before and after hatching. During the growth phase, the cartilaginous elements are largely replaced by bone. Nerves only enter the wing after the cartilage has been laid down, at around 4.5 days of incubation, and we shall discuss this in Chapter 10. The fundamental change in shape that occurs during early wing development is elongation of the bud, and the general mechanisms of how this elongation is accomplished are thought to be similar in other regions of the embryo that undergo outgrowth—for example, the posterior part of the main body axis (see Chapter 4).

The generation of the complex arrangement of differentiated cells and tissues in the developing chick wing is an excellent example of pattern formation. Positional information has been fundamental to understanding pattern formation in the developing limb, although limbs also have some capacity for self-organization. Central to the idea of positional information is the distinction between positional specification and

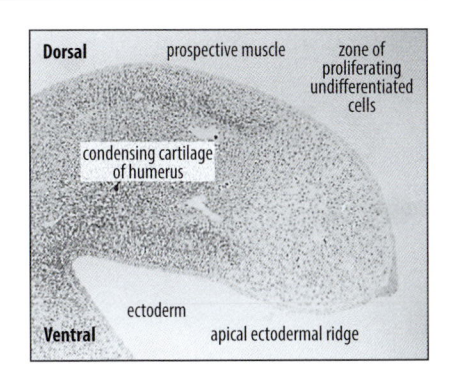

Fig. 9.20 Cross-section through an embryonic chick wing bud. The thickened apical ectodermal ridge is at the tip. Beneath the apical ridge is a region of undifferentiated cells. Proximal to this region, mesenchyme cells condense and differentiate into cartilage. Prospective muscle cells migrate into the limb from the adjacent somites and form dorsal and ventral muscle masses. Scale bar = 0.1 mm.

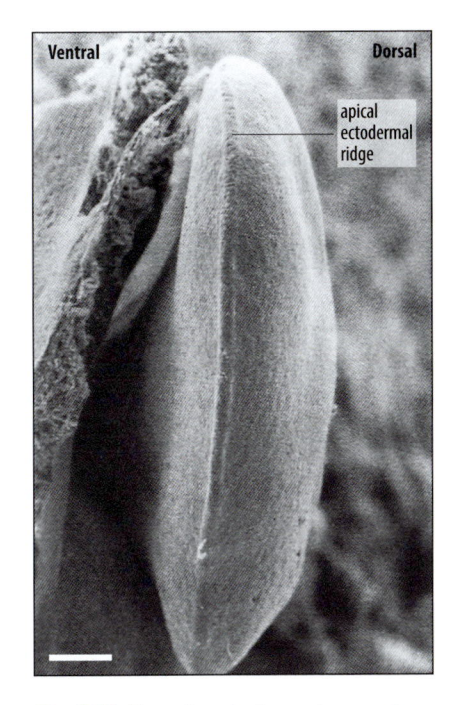

Fig. 9.21 Scanning electron micrograph of a limb bud of a chick embryo at 4.5 days incubation after laying, showing the apical ectodermal ridge. A thickened apical ectodermal ridge is also present at the tip of mouse limb buds. Scale bar = 0.1 mm.

460 Chapter 9 Organogenesis

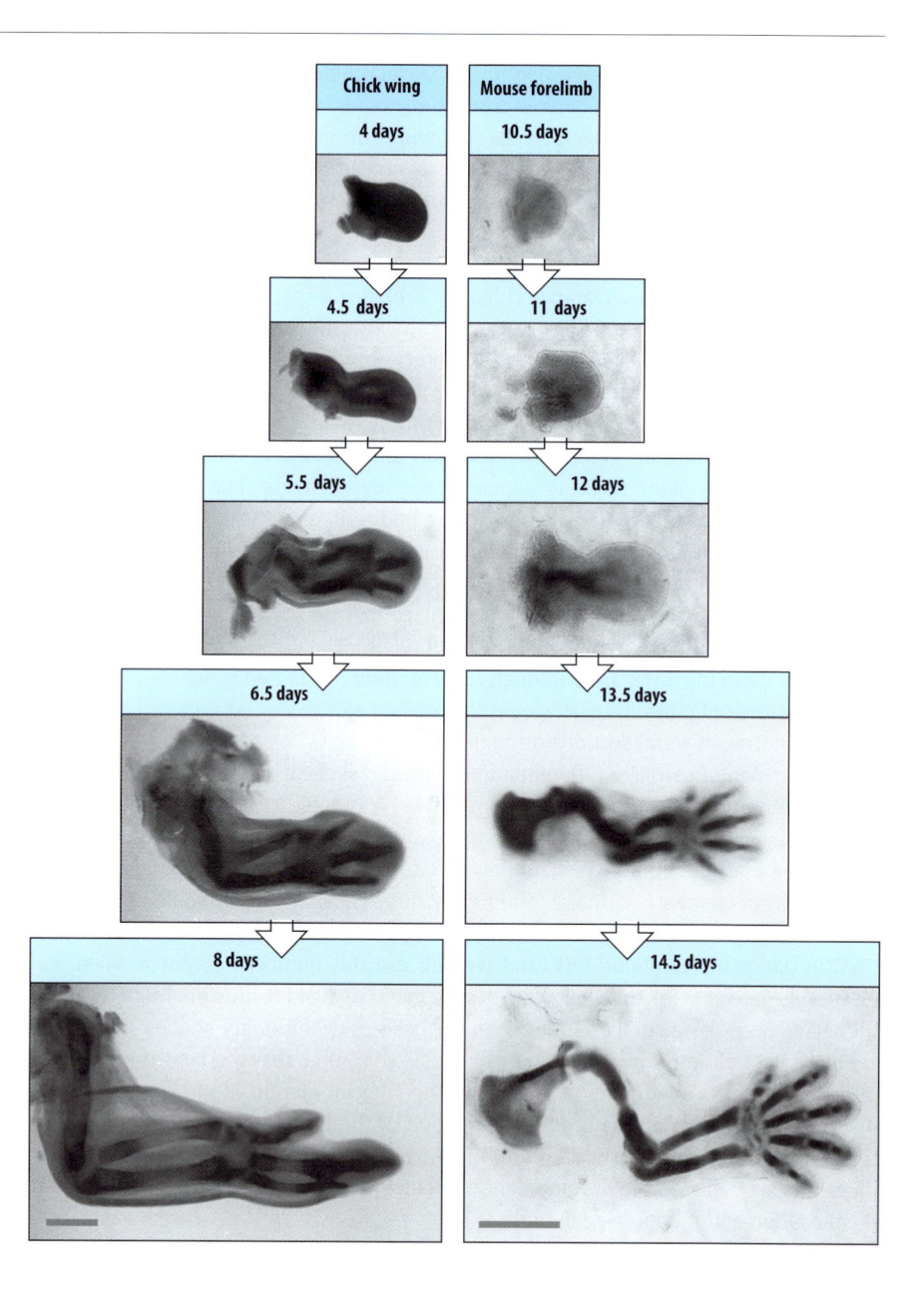

Fig. 9.22 The development of the chick wing and mouse forelimb is similar. The skeletal elements are laid down, as cartilage, in a proximo-distal sequence as the limb bud grows outward. The cartilage of the humerus is laid down first, followed by the radius and ulna, then wrist elements, and finally digits. Scale bar = 1 mm.

interpretation. Cells acquire positional information first, and then interpret these positional values according to their developmental history. Positional information is signaled in wing and leg buds in the same way, and it is the differences in developmental history that make chick embryo wings and legs different. A good demonstration of the interchangeability of positional signals comes from grafting limb-bud tissue to different positions along the proximo-distal axis of the bud. If tissue that would normally give rise to the thigh is grafted from the proximal part of an early chick leg bud to the tip of an early wing bud, it develops into toes with claws (Fig. 9.23). The tissue has acquired a more distal positional value after transplantation but interprets this value according to its own developmental program, which is to make leg structures. This is the same principle we saw with respect to insect imaginal discs when a clone of cells with genetic mutations in a Hox gene that specifies the character of one disc is created in a different disc (see Section 9.9).

The first consideration, however, is how limb buds develop at the appropriate places on the body and how their polarity and identity are specified.

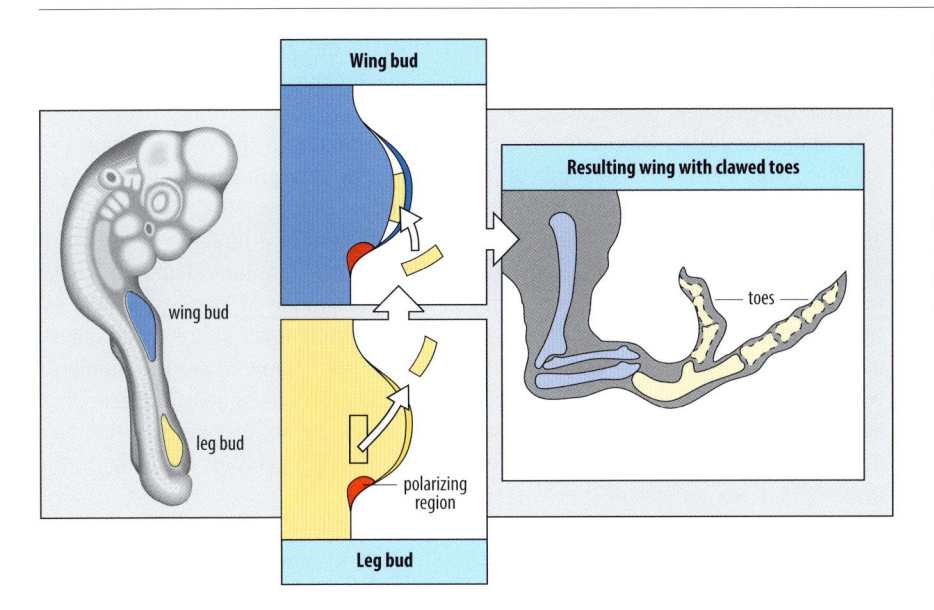

Fig. 9.23 Chick proximal leg bud cells grafted to a distal position in a wing bud acquire distal positional values. Proximal tissue from a leg bud that would normally develop into a thigh is grafted to the tip of a wing bud underneath the apical ectodermal ridge. It acquires more distal positional values and interprets these in terms of leg structures, forming clawed toes at the tip of the wing.

9.11 Genes expressed in the lateral plate mesoderm induce limb development and specify limb polarity and identity

The forelimbs and hindlimbs of vertebrates arise at precise positions along the antero-posterior axis of the body. Transplantation experiments in chick embryos have shown that the lateral plate mesoderm becomes determined to form limbs in these positions long before limb buds are visible, and that the antero-posterior polarity of the limb bud is determined around the same time.

As in the pre-somitic mesoderm, Hox gene expression becomes regionalized in the lateral plate mesoderm of the trunk along the antero-posterior axis, with 3' genes being expressed more anteriorly than 5' Hox genes. For example, Hox genes in paralogous subgroups 4 and 5 (see Box 4G) are expressed in the region of the lateral plate mesoderm that will form the wings/forelimbs, and Hox genes in subgroups 8, 9, and 10 are expressed more posteriorly. A combinatorial Hox code determines the position in which limb development is induced by controlling the expression of the gene for the T-box transcription factor Tbx5 in the prospective wing/forelimb region (Fig. 9.24) and the gene for the related transcription factor Tbx4 in the prospective

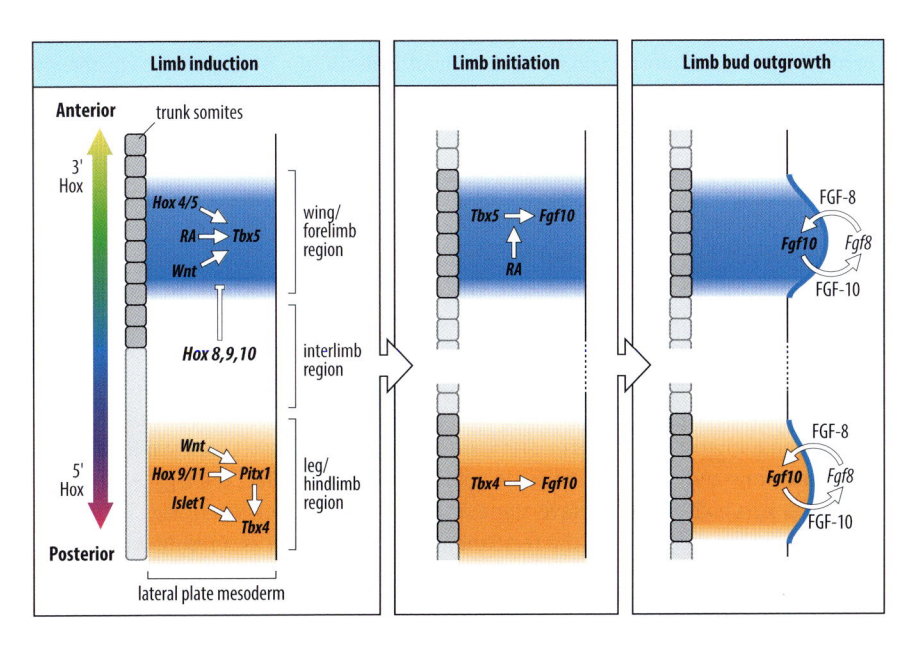

Fig. 9.24 Schematic diagram of limb initiation in chick and mouse embryos. A combinatorial Hox code in the lateral plate mesoderm induces wing/forelimb development by activating expression of *Tbx5* in the lateral plate mesoderm in the appropriate position. Wnt and retinoic acid (RA) signaling are also involved. In the region where the leg/hindlimb will develop, a Hox code regulates the expression of the gene encoding the transcription factor, Pitx1. Pitx1 together with Islet1, another transcription factor expressed in this region, are responsible for the induction of *Tbx4* expression. Wnt signaling, but not RA signaling, is also involved. Tbx4 and Tbx5 proteins induce *Fgf10* gene expression in the limb-forming regions and this initiates limb bud formation. Retinoic acid is also required for activation of *Fgf10* expression in wing/forelimb. Fibroblast growth factor FGF-10 protein signaling in turn induces expression of *Fgf8* in the overlying ectoderm that will form the apical ectodermal ridge. FGF-8 protein signals from the apical ridge then maintain *Fgf10* expression in the mesoderm. This establishes a positive-feedback loop of FGF signaling, which is common to both wings/forelimbs and legs/hindlimbs and drives limb bud outgrowth.

Based on Nishimoto, S., Logan, M.P.: ***Subdivision of the lateral plate mesoderm and specification of the forelimb and hindlimb forming domains.*** *Semin. Cell. Dev. Biol. 2016,* **49***: 102–108; Feneck, E., Logan, M.P.:* ***The role of retinoic acid in establishing the early limb bud.*** *Biomolecules 2020,* **10***: 312.*

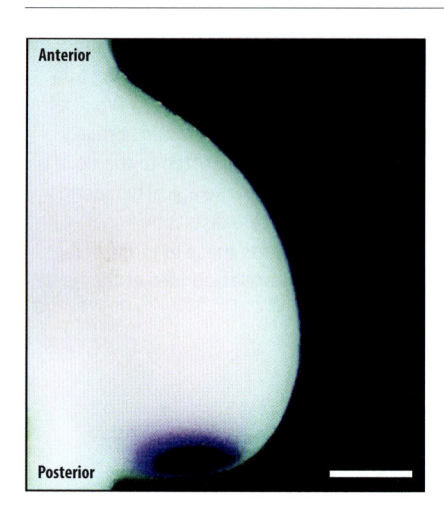

Fig. 9.25 The polarizing region or zone of polarizing activity (ZPA) of a chick limb bud is located at the posterior margin of the bud. The polarizing region expresses *Sonic hedgehog* (*Shh*) transcripts (blue stain). Scale bar = 0.1 mm.

Photograph courtesy of C. Tabin.

leg/hindlimb region. Tbx4 and Tbx5 are related to the mesodermal transcription factor TbxT (formerly known as Brachyury), and initiate leg/hindlimb and wing/forelimb development respectively.

In mouse embryos, *Tbx5* expression becomes restricted to the region where the forelimb will develop as the result of its activation by Hox proteins of subgroups 4 and 5, combined with repression by more 'posterior' Hox proteins (Hoxc8, Hoxc9, and Hoxc10). *Tbx5* expression and subsequent formation of a forelimb also require retinoic acid and Wnt/β-catenin signaling. The way in which retinoic acid acts is controversial. One scenario is that it contributes to the activation of *Tbx5* expression by Hox proteins as shown in Figure 9.24. Alternatively, it might act indirectly by relieving repression of *Tbx5* expression. In the mouse, in the region where the hindlimb will develop, Hox proteins of subgroups 9–11 regulate the expression of the gene that encodes the transcription factor Pitx1, which together with another transcription factor, Islet1, results in restriction of *Tbx4* expression to this region. Wnt signaling is also involved, but not retinoic acid.

Hox genes also determine the antero-posterior polarity of the buds that give rise to the forelimbs. Sophisticated experiments in mice, in which all the Hox genes in the same paralogous subgroup were simultaneously deleted, showed that proteins encoded by subgroup-5 Hox genes are required for specifying the anterior part of the bud, whereas proteins encoded by subgroup-9 Hox genes are required for specifying the posterior part of the bud. Antero-posterior polarity is essential for the subsequent establishment of one of the main organizing regions of the limb bud in the mesoderm at its posterior margin. This region is the **polarizing region** or **zone of polarizing activity (ZPA)**. It expresses the gene encoding the signal protein Sonic hedgehog (Shh) (Fig. 9.25), which is crucial for specifying the pattern of structures that develops across the antero-posterior axis of the limb. In the absence of function of all subgroup-5 Hox genes, *Shh* is expressed at the anterior margin of the forelimb bud in addition to the posterior, whereas in the absence of function of all the paralogous Hox9 genes, *Shh* is not expressed at all. *Shh* is also expressed at the posterior margin of hindlimbs and *Islet1*, which encodes a transcription factor, is required for specifying the posterior part of the hindlimb bud.

Many components of the gene regulatory networks downstream of the subgroup-5 and -9 Hox genes have been uncovered. These include other Hox genes in the Hoxa and Hoxd clusters, which are involved in specifying the posterior part of the forelimb bud. Genetic deletion of all Hoxa and Hoxd genes in the early mouse forelimb bud results in the absence of *Shh* expression and severe limb truncations. Conversely, genetic manipulations that lead to inversions of the Hoxd cluster or delete large parts of the 3′ regulatory region—so that Hoxd genes in the 5′ part of the cluster are expressed in the patterns of 3′ Hoxd genes—result in *Shh* expression both anteriorly and posteriorly, and the expression of *Hoxd13*, for example, throughout the early forelimb bud instead of being restricted to the posterior. The transcription factor Sall4 acts in the gene regulatory networks that specify the anterior part of the bud in both forelimbs and hindlimbs. Like the 5′ Hoxa and Hoxd genes, *Sall4* is expressed later in limb development and is involved in patterning the limb.

The transcription factors Tbx4 and Tbx5 were originally thought to specify limb identity, but when the *Tbx5* gene is inactivated in mice and *Tbx4* is expressed in the forelimb bud in its place, a limb with forelimb characteristics still develops. Pitx1 is now known to have a key role in determining the difference between hindlimb and forelimb. Mutations in both the *TBX5* and *PITX1* genes are associated with limb differences in humans. Mutations in *TBX5* are responsible for Holt–Oram syndrome, and affect the arms and the heart, whereas mutations in *PITX1* affect the legs. A very rare condition in which the arms have some anatomical characteristics of legs is thought to be caused by alterations in the regulation of *PITX1*, which, as a result, would have been expressed in the arm buds during embryonic development. A partial alteration in limb identity involving changes in the regulation of the expression of *Pitx1* and *Tbx5* can account for feathered feet in fancy breeds of pigeons and chickens.

Tbx4 and Tbx5 activate *Fgf10* expression in the lateral plate mesoderm in the prospective limb-forming regions (see Fig. 9.24). In chick embryos, this activation of *Fgf10* expression in the wing-forming region has been shown to require retinoic acid. Rather remarkably, local application of FGF to the interlimb region between the prospective wing and prospective leg buds in chick embryos can induce an additional limb. Wings develop when FGF is applied to the anterior interlimb region, whereas legs develop from application to the posterior interlimb region (Fig. 9.26). The entire interlimb region is competent to form limbs. Knock-out of either *Fgf10* or the gene encoding its receptor in mice results in embryos lacking limb buds. Wnt signal proteins produced and secreted by the mesoderm have a key role in determining where FGFs are expressed and maintained. For example, cells producing Wnt-2b or Wnt-8c can induce *Fgf10* expression when applied to the interlimb regions of chick embryos and lead to the formation of additional limbs.

FGF-10 signaling by the mesoderm is crucial for establishing the apical ectodermal ridge, which is, in turn, essential for limb-bud outgrowth and for pattern formation along the proximo-distal axis of the limb. The apical ridge arises at a boundary between dorsal and ventral compartments of cell-lineage restriction (see Section 2.23 and Section 9.2 for an explanation of compartments), which are present in the trunk ectoderm of early embryos before any buds are visible. Dorso-ventral patterning of the trunk mesoderm is specified by graded BMP signaling, which is then imposed on the overlying ectoderm. The cells that will form the apical ridge are initially scattered throughout both the dorsal and ventral ectoderm, but then, as a result of coordinated cell rearrangements, move to the compartment boundary, where they form a ridge of closely packed columnar epithelial cells. The apical ridge begins to express the *Fgf8* gene in response to FGF-10 protein signals from the mesenchyme and a positive-feedback loop is established in which FGF-8 from the apical ridge maintains *Fgf10* gene expression in the mesenchyme (see Fig. 9.24). FGF signals from the apical ridge also maintain *Shh* expression in the polarizing region. In turn, Shh protein signals from the polarizing region maintain the apical ridge over the posterior part of the limb bud. Once a limb bud has formed, equipped with its own signaling regions, it can develop autonomously—that is, on its own without reference to neighboring parts of the embryo.

9.12 The apical ectodermal ridge is required for limb bud outgrowth and the formation of structures along the proximo-distal axis of the limb

As we have just seen, FGF signaling by limb mesoderm establishes the apical ectodermal ridge. The apical ectodermal ridge is essential for outgrowth of the limb bud and the progressive formation of skeletal elements along the proximo-distal axis of the limb. When the apical ridge is surgically removed from a chick wing bud, outgrowth is significantly reduced, and the wing that develops is anatomically truncated, with distal parts missing. The proximo-distal level at which the wing is truncated depends on the time at which the apical ridge is removed (Fig. 9.27). The earlier the apical ridge is removed, the greater the effect; removal at a late stage results only in loss of the digits, whereas when the apical ridge is removed at an early stage, most of the wing is missing. Following apical ridge removal, proliferation of cells at the tip of the wing bud is greatly reduced, and some cells in the bud die. The importance of signaling by the apical ridge in promoting outgrowth has also been shown by grafting an apical ridge dissected from an early wing bud to the dorsal surface of another early chick wing bud. The result is an outgrowth from the dorsal surface, which develops into a new wing bud tip containing cartilaginous elements, including digits. The apical ridge from a mouse forelimb can support the normal development of wing-type

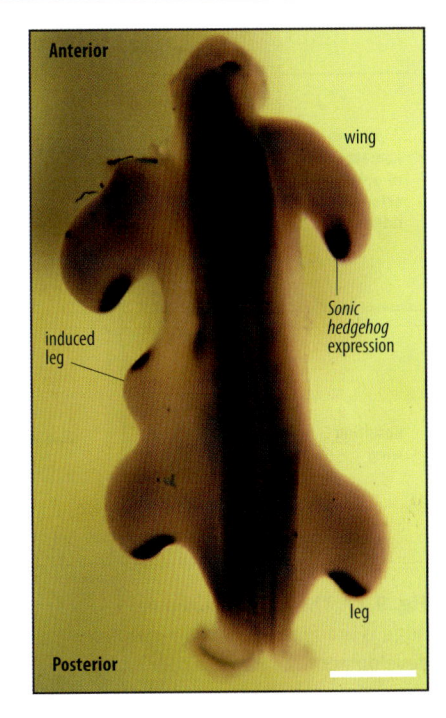

Fig. 9.26 Result of local application of fibroblast growth factor (FGF) to the interlimb region of a chick embryo. A bead soaked in FGF is implanted in the flank close to the prospective leg bud region, where it induces outgrowth of an additional leg bud. The dark regions indicate *Sonic hedgehog* (*Shh*) transcripts. *Shh* is expressed at the posterior margin of the normal wing and leg buds, but at the anterior of the additional leg bud, and the leg that subsequently develops has an opposite polarity to that of the normal limbs. It is likely that *Shh* is expressed in this way in the additional leg bud because, due to its position along the antero-posterior axis of the body, the subgroup-9 Hox paralogs will be expressed at its anterior margin. Scale bar = 1 mm.

Photograph reproduced with permission from Cohn, M.J., et al.: **Fibroblast growth factors induce additional limb development from the flank of chick embryos.** Cell 1995, **80:** 739-746. © 1995 Cell Press.

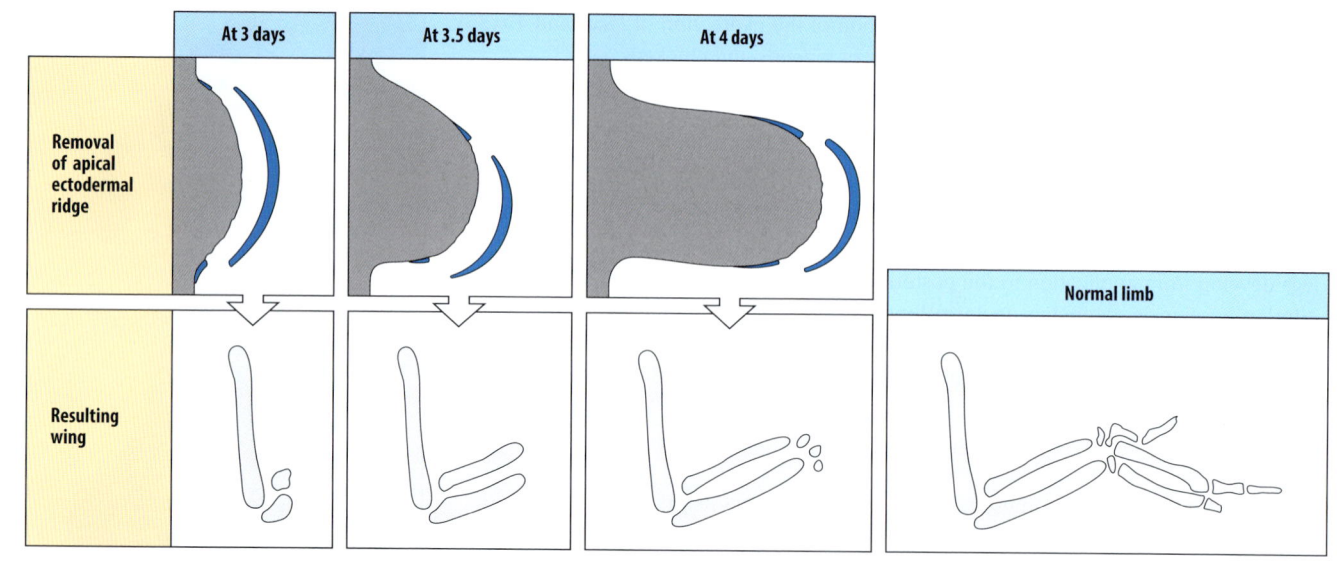

Fig. 9.27 The apical ectodermal ridge is required for proximo-distal development. Limbs develop in a proximo-distal sequence. Removal of the apical ridge from a developing wing bud leads to truncation of the wing; the later the apical ridge is removed, the more complete the resulting wing.

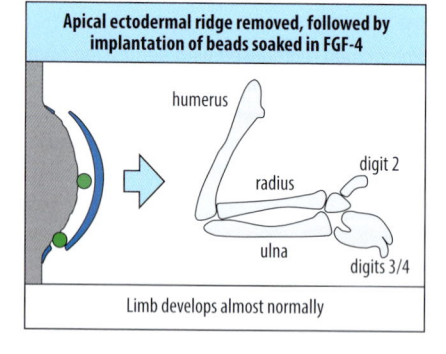

Fig. 9.28 Fibroblast growth factor (FGF)-4 can substitute for the apical ectodermal ridge. After the apical ridge is removed from a chick wing bud, placing heparin beads soaked in FGF-4 at the tip of the bud results in almost normal development.

structures when recombined with the mesodermal core of a chick wing bud, showing that the signals from the apical ridge are the same in both birds and mammals.

A key signal from the apical ectodermal ridge is provided by FGFs, although Wnts are also produced and help to maintain the undifferentiated state of the underlying proliferating mesenchyme cells. The gene for FGF-8 is expressed throughout the apical ridge, and genes for FGF-4 and two other FGFs are expressed in the posterior region of the apical ridge. Experiments in chick wing buds show that FGF-8 (or FGF-4) can act as a functional substitute for an apical ridge. When the apical ridge is removed and beads soaked in FGF are placed at the tip of the wing bud in its place, more-or-less normal outgrowth of the wing bud continues. If enough FGF is provided to the outgrowing cells, a fairly normal wing develops although the digits tend to be bunched (Fig. 9.28). Experiments in transgenic mice in which FGF genes were knocked out specifically in the apical ridge, also show that FGF signaling is required for limb-bud outgrowth and normal development. The targeting of FGF gene knock-outs specifically to the apical ridge (see Box 4D for a technique) is needed, because FGFs have key roles earlier in development. Mice in which *Fgf4*, for example, is knocked out, do not survive to limb-bud stages. Of the four different FGF genes that are expressed in the apical ridge, *Fgf8* alone is enough for normal limb development. Complete limb truncations are produced when *Fgf8* and two of the other FGF genes are simultaneously knocked out. The simplest explanation is that the individual FGF genes expressed in the apical ridge are, to a large extent, functionally equivalent, but that having several FGFs involved ensures the robustness of apical ridge signaling.

9.13 Formation and outgrowth of the limb bud involves oriented cell behavior

We have just seen that the apical ectodermal ridge controls outgrowth of the limb, but how does the bud form in the first place and then acquire its elongated shape? In the first 24–30 hours of chick wing-bud development, when a well-defined apical ridge is present, the length of the bud along the proximo-distal axis increases about threefold, whereas the dimensions of its antero-posterior and dorso-ventral axes are almost unchanged.

In the initial formation of a chick limb bud, mesodermal cells from the somatopleural coelomic epithelium lining the lateral plate are recruited to the lateral plate mesoderm in the limb-forming regions through an epithelial-to-mesenchymal transition

(EMT; see Section 5.4). Subsequently, mesenchymal cell proliferation is maintained in the limb-bud region but decreases in the neighboring inter-limb region. As the bud elongates, cell proliferation occurs throughout the bud, regulated by synergistic Wnt and FGF signals produced by the apical ridge. Although most studies have shown that cell proliferation is higher distally and near the ectoderm, computer modeling of mouse limb buds reveals that these local differences, which are seen in both chick and mouse limb buds, are not enough to explain bud outgrowth. Instead, cell polarization and oriented cell movements in the direction of the apical ridge, and cell divisions oriented along the proximo-distal axis, make an important contribution to the formation of the bud and its elongation.

Live imaging of mouse embryos has revealed dramatic changes in mesenchyme cell polarization as limb buds form. Just before limb-bud formation, the cells in the body wall are elongated parallel to the main head-to-tail body axis as a result of elongation of the body, with the orientation of cell division in the same direction. Individual cells point towards the tail of the embryo, as indicated by the position of the Golgi apparatus, which is always in front of the nucleus in moving cells, with movement oriented in this direction (Fig. 9.29, first panel). As the bud begins to form, the cells just proximal to the bud become reoriented perpendicular to the body axis and become aligned with the proximo-distal axis of the emerging limb bud. Cell divisions also become reoriented along the proximo-distal axis, as does cell movement, with cells moving towards the apical ridge. At the same time, cells move from positions in the body wall anterior to the limb bud into the distal region of the bud and across to the posterior side (see Fig. 9.29, second panel), with the speed of movement increasing as they enter the distal region. Successive cell divisions and the movement of mesenchyme cells into the proximal region push the limb bud outwards.

In buds that have begun to elongate, the polarization of the mesenchyme cells becomes more complex. In both mouse and chick limb buds, the central cells are still aligned along the proximo-distal axis. Cells at the periphery of the bud become polarized almost perpendicularly to the ectoderm, with divisions oriented along this axis and then the daughter cells reorienting along the proximo-distal axis. Cells in the central and distal regions generally show oriented cell divisions and movement towards the tip of the limb bud, with distal cells moving faster than proximal cells, and so continuing to push the limb bud outwards (see Fig. 9.29, third panel).

The oriented behavior of mesenchyme cells in early limb buds is regulated by a combination of Wnt signaling setting up planar cell polarity (PCP) in the cells (see Section 5.12), and FGF signaling. Wnt-5a, one of the Wnts that signals via the

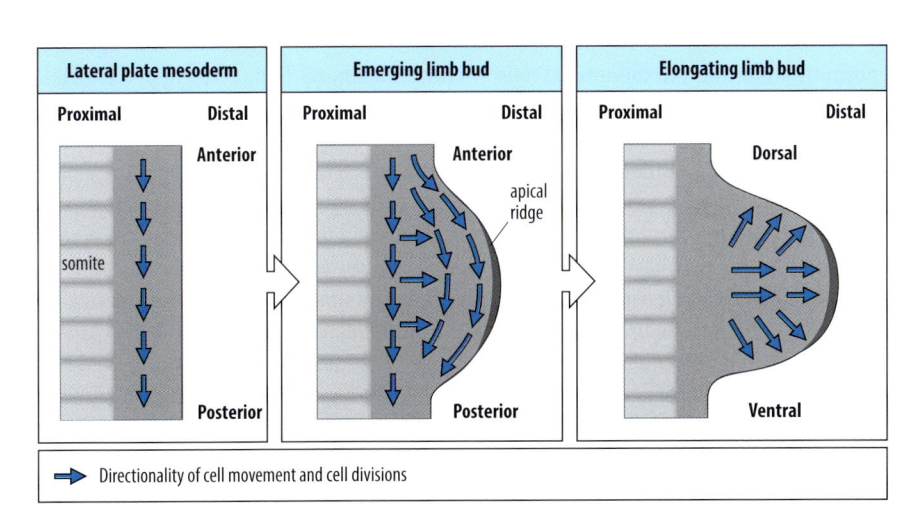

Fig. 9.29 Mesenchymal cell orientation during formation and outgrowth of the limb bud. First panel: before limb-bud formation, the mesenchymal cells in the lateral plate mesoderm are polarized parallel to the antero-posterior axis of the body. They point posteriorly (as indicated by the direction of the arrows) and move in this direction. Cell divisions are also oriented along the antero-posterior axis. Second panel: the early limb bud forms as a result of anterior cells of the lateral plate mesoderm moving posteriorly, and proximal cells becoming polarized perpendicular to the antero-posterior main body axis and moving in the direction of budding. Third panel: in the elongating bud, mesenchymal cells in the central region and at the tip are polarized parallel to the proximo-distal axis and move distally towards the apical ridge, with cell divisions oriented along the proximo-distal axis. Cells in dorsal and ventral regions are polarized towards the ectoderm almost perpendicular to the direction of outgrowth. The two panels on the left are based on observations in developing mouse embryos, the right panel on the developing chick limb.

First and second panels: based on Wyngaarden, L.A., et al.: ***Oriented cell motility and division underlie early limb bud morphogenesis.*** *Development 2010,* ***137****: 2551-2558; third panel: Gros, J., et al.:* ***Wnt5a/Jnk and FGF/Mapk pathways regulate the cellular events shaping the vertebrate limb.*** *Curr. Biol. 2010,* ***20****: 1993-2002.*

non-canonical pathway used in PCP, is expressed in a graded fashion in limb-bud mesenchyme, with the highest level of transcripts at the tip. Mutant mice lacking *Wnt5a* have short limbs as a result of reduction in the length of the proximo-distal axis, and the distal parts of the digits are missing, whereas the width of the bud along the dorso-ventral axis is increased. Live imaging of limb buds from this mutant revealed that the mesenchymal cells are not polarized, and that cell movement and cell division are only weakly oriented. This suggests that in the normal limb bud, the perpendicularly polarized cells adjacent to the ectoderm in the dorso-ventral regions undergo convergent extension under the influence of Wnt-5a signals, thus contributing to bud elongation and to maintaining the correct dorso-ventral width of the bud. Convergent extension is explained in Box 5C.

Other experiments show that Wnt-5a might act as a chemoattractant in early limb-bud formation and orient the migration of mesenchyme cells towards the forming bud. *Wnt5a* is expressed at the tips of all regions of the embryo that undergo outgrowth—not only the limb buds, but also the buds that give rise to the jaws, the bud that gives rise to the genital tubercle, and the posterior region of the main body axis. So the orientation of mesenchyme cells by Wnt-5a is likely to be a general mechanism for elongation.

FGF-4, which is produced by the apical ectodermal ridge, has also been shown to act as a chemoattractant for limb mesenchyme cells. Another FGF produced by the ridge, FGF-8, controls the velocity of cell movement, explaining why distal mesenchyme cells move faster than proximal ones in both mouse and chick limb buds. FGF signaling is also required for elongation of the main body axis (see Section 4.9) and, in chick embryos, controls a gradient of cell motility in the extending pre-somitic mesoderm, with the highest motility posteriorly.

9.14 Positional values along the proximo-distal axis of the limb are specified by a combination of graded signaling and a timing mechanism

The structures along the proximo-distal axis of the limb are laid down in a proximal-to-distal sequence as the limb bud grows out, with proximal structures (humerus or femur) beginning to differentiate first. Wnt signals, together with FGFs produced by the apical ridge, maintain the region of proliferating undifferentiated mesenchyme at the tip of the limb bud. The way in which proximo-distal values are specified has been a matter of debate, and opposing models, one involving timing and the other graded signals, have been proposed. It now seems likely that distal positional values are specified by timing, whereas proximal positional values are specified in the early limb bud by graded signals. The fact that both mechanisms are involved might explain why it has been so difficult to distinguish between the opposing models.

We first consider the two opposing models. It was proposed some time ago that position along the proximo-distal axis is based on timing (Fig. 9.30, upper panels). This long-standing model derives from work on chick limb buds and proposes that proximo-distal positional values are specified by the length of time that cells spend in the zone of undifferentiated cells at the tip of the bud. This region was therefore called the '**progress zone**'. As the limb bud grows out, cells will leave the progress zone, and once they have left the zone, their positional value will be fixed. This means that cells that leave early will have proximal positional values, whereas cells that leave later will have more-distal positional values. The experimental observation that removal of the apical ridge results in a distally truncated limb (see Fig. 9.27) would be explained by the fact that the progress zone is no longer maintained, and so the more-distal structures do not form.

By contrast, the second model proposed that proximo-distal positional values are specified in the early wing bud by graded signals. As the limb bud grows out, the prespecified regions of the limb would expand successively in turn, to differentiate

The vertebrate limb **467**

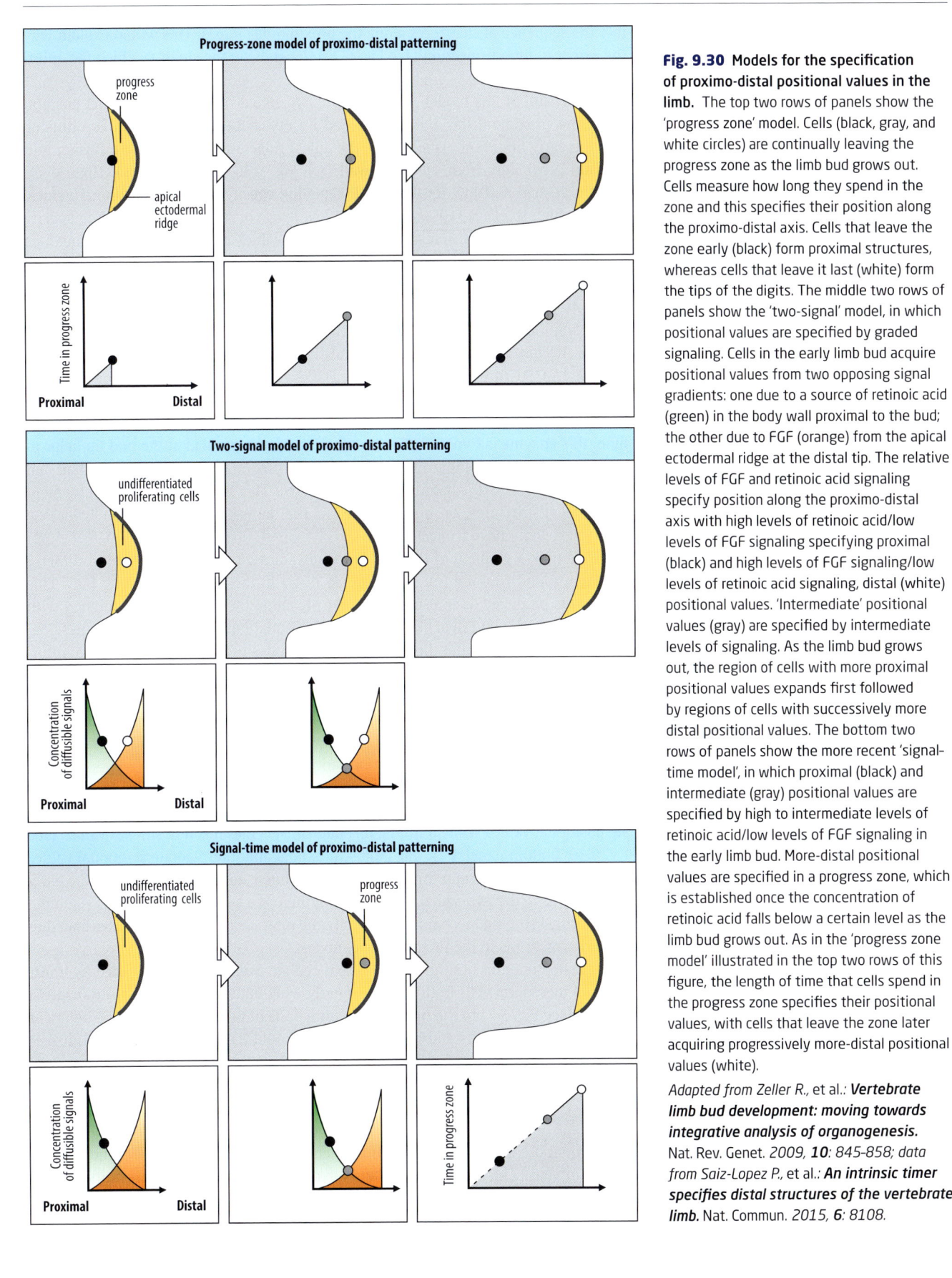

Fig. 9.30 Models for the specification of proximo-distal positional values in the limb. The top two rows of panels show the 'progress zone' model. Cells (black, gray, and white circles) are continually leaving the progress zone as the limb bud grows out. Cells measure how long they spend in the zone and this specifies their position along the proximo-distal axis. Cells that leave the zone early (black) form proximal structures, whereas cells that leave it last (white) form the tips of the digits. The middle two rows of panels show the 'two-signal' model, in which positional values are specified by graded signaling. Cells in the early limb bud acquire positional values from two opposing signal gradients: one due to a source of retinoic acid (green) in the body wall proximal to the bud; the other due to FGF (orange) from the apical ectodermal ridge at the distal tip. The relative levels of FGF and retinoic acid signaling specify position along the proximo-distal axis with high levels of retinoic acid/low levels of FGF signaling specifying proximal (black) and high levels of FGF signaling/low levels of retinoic acid signaling, distal (white) positional values. 'Intermediate' positional values (gray) are specified by intermediate levels of signaling. As the limb bud grows out, the region of cells with more proximal positional values expands first followed by regions of cells with successively more distal positional values. The bottom two rows of panels show the more recent 'signal-time model', in which proximal (black) and intermediate (gray) positional values are specified by high to intermediate levels of retinoic acid/low levels of FGF signaling in the early limb bud. More-distal positional values are specified in a progress zone, which is established once the concentration of retinoic acid falls below a certain level as the limb bud grows out. As in the 'progress zone model' illustrated in the top two rows of this figure, the length of time that cells spend in the progress zone specifies their positional values, with cells that leave the zone later acquiring progressively more-distal positional values (white).

Adapted from Zeller R., et al.: **Vertebrate limb bud development: moving towards integrative analysis of organogenesis.** Nat. Rev. Genet. 2009, **10**: 845-858; *data from Saiz-Lopez P., et al.:* **An intrinsic timer specifies distal structures of the vertebrate limb.** Nat. Commun. 2015, **6**: 8108.

into the sequence of structures along the proximo-distal axis. This would explain why the extent of limb truncation (see Fig. 9.27) depends on the stage at which the apical ridge is removed.

One version of this model, known as the two-signal model, proposes that position along the proximo-distal axis is specified by opposing gradients of FGFs diffusing into the early limb bud distally from the apical ridge and retinoic acid diffusing into the bud proximally from the body wall (see Fig. 9.30, middle panels). High levels of retinoic acid and low levels of FGF signaling specify proximal positional values, whereas high levels of FGF signaling and low levels of retinoic acid signaling specify distal positional values. Intermediate positional values are specified by intermediate levels of signaling.

An interplay between FGF and retinoic acid signaling is not unique to the developing limb. Opposing gradients of retinoic acid and FGF have been proposed to operate during the growth and antero-posterior patterning of the main body axis, and to control the formation of the somites and the differentiation of the spinal cord (see Section 4.9). As in the primitive streak, FGF signaling antagonizes retinoic acid signaling in the limb bud. FGF signals from the apical ridge control the expression of a gene encoding a retinoic-acid-metabolizing enzyme in the adjacent mesoderm. In the limb, this enzyme is Cyp26b1. The metabolism of retinoic acid at the bud tip helps to create a gradient of retinoic acid signaling in the proximal region of the limb bud and ensures that the distal region is retinoic-acid-free. This is important for development of the distal structures, as shown by the fact that mouse embryos completely lacking Cyp26b1 activity have truncated limbs.

In addition, a gradient of the homeodomain transcription factors Meis1 and Meis2 has been found running from proximal to distal in the early mouse limb bud. This gradient is produced as a result of positive signaling by retinoic acid combined with inhibitory activity by FGFs, likely due to their regulation of *Cyp26b1* gene expression. It has been suggested that the Meis gradient encodes proximo-distal positional information. Consistent with this view, elimination of *Meis* gene function in mouse limb buds leads to loss of proximal structures. But misexpression of *Meis* genes in chick wing buds leads to distal abnormalities or truncations, rather than transforming distal structures into proximal structures, which would be expected if the Meis gradient encodes proximo-distal positional information.

The most recent model for proximo-distal pattern formation, based on further work in chick wing buds, proposes that graded signaling specifies proximal and intermediate positional values while a timing mechanism specifies distal positional values (see Fig. 9.30, bottom panels). In this model, proximal and intermediate positional values are specified above certain threshold levels of retinoic acid/FGF signaling in the early limb bud (as in the two-signal model). Then, there is a switch from extrinsic signaling to an intrinsic timing mechanism to specify the more-distal positional values linked to bud outgrowth (as in the progress zone model).

Intrinsic properties of limb bud cells that change over time were detected by transplanting mesoderm cells from the tips of early wing buds of chick embryos transgenic for GFP (see Fig. 3.7) to the tips of older wing buds of normal embryos. The behavior of the transplanted cells and their descendants could then be followed. The intrinsic timed properties include cell-cycle dynamics and activation of Hox gene expression. As we shall see, 5′ Hoxa and Hoxd genes control regional identity along the proximo-distal axis of the limb, and the timing of activation of expression of the Hox genes that control the development of distal structures was maintained in the transplanted cells. In addition, oscillatory expression of the gene *hairy-2* during the development of the distal limb has been detected. Oscillatory expression of the related gene *hairy-1* in the pre-somitic mesoderm was the first molecular evidence for the operation of a 'clock' in somitogenesis (see Section 4.10), and so finding the same behavior in the limb is consistent with a timing mechanism acting here as well.

In the signal–time model for proximo-distal patterning in the limb, the proposal is that the switch from extrinsic graded signaling to an intrinsic timing mechanism for specifying positional values is triggered by falling levels of retinoic acid in the distal limb bud. This fall is due not only to the activity of retinoic-acid-metabolizing enzymes, but also to continued bud outgrowth away from the body wall where the retinoic acid is produced.

9.15 The polarizing region specifies position along the limb's antero-posterior axis

The specification of positional information along the antero-posterior axis of the limb bud is most clearly seen with respect to the digits, as it gives them their identities. Here we see how the polarizing region at the posterior margin of the limb bud specifies the pattern of the three digits in the chick wing. The polarizing region has a pivotal function in the developing limb because it is also responsible for maintaining the apical ectodermal ridge over the posterior region of the limb bud where the digits arise.

The polarizing region of a vertebrate limb bud has organizing properties almost as striking as those of the Spemann organizer in amphibians. When the polarizing region from an early chick wing bud is grafted to the anterior margin of another early chick wing bud, a wing with a mirror-image pattern develops: instead of the normal pattern of digits—2 3 4—the pattern 4 3 2 2 3 4 develops (Fig. 9.31, compare the first and second rows). A polarizing-region graft can also specify another ulna anteriorly, showing that the polarizing region influences antero-posterior positional values throughout the region of the wing distal to the elbow.

The additional digits come from the host wing bud and not from the graft, showing that the grafted polarizing region has altered the developmental fate of the host cells in the anterior region of the wing bud. The wing bud widens in response to the polarizing graft, which enables the additional digits to be accommodated. Widening of the wing bud is associated with an increase in cell proliferation and the maintenance of the apical ectodermal ridge over the anterior region of the bud.

One way that the polarizing region could specify position along the antero-posterior axis is by producing a morphogen that forms a posterior-to-anterior gradient. The local concentration of morphogen would provide a positional value with respect to the polarizing region located at the posterior margin of the limb. Cells could then interpret their positional values by developing specific structures at particular threshold concentrations of morphogen. Digit 4, for example, would develop at a high concentration, digit 3 at a lower one, and digit 2 at an even lower one (see Fig. 9.31, top row). According to this model, a graft of an additional polarizing region to the anterior margin would set up a mirror-image gradient of morphogen, which would result in the 4 3 2 2 3 4 pattern of digits that is observed (see Fig. 9.31, center row).

If the action of the polarizing region in specifying the identity of a digit is due to the concentration of the signal, then when the signal is weakened, the pattern of digits should be altered in a predictable manner. Indeed, grafting small numbers of polarizing region cells to a wing bud results in the development of only an additional digit 2 (see Fig. 9.31, bottom row). The same result can be obtained by leaving the polarizing-region graft in place for a short time and then removing it. When the graft is left in place for 15 hours, an additional digit 2 develops, whereas the graft must be in place for 17–24 hours for additional digits 3 and 4 to develop.

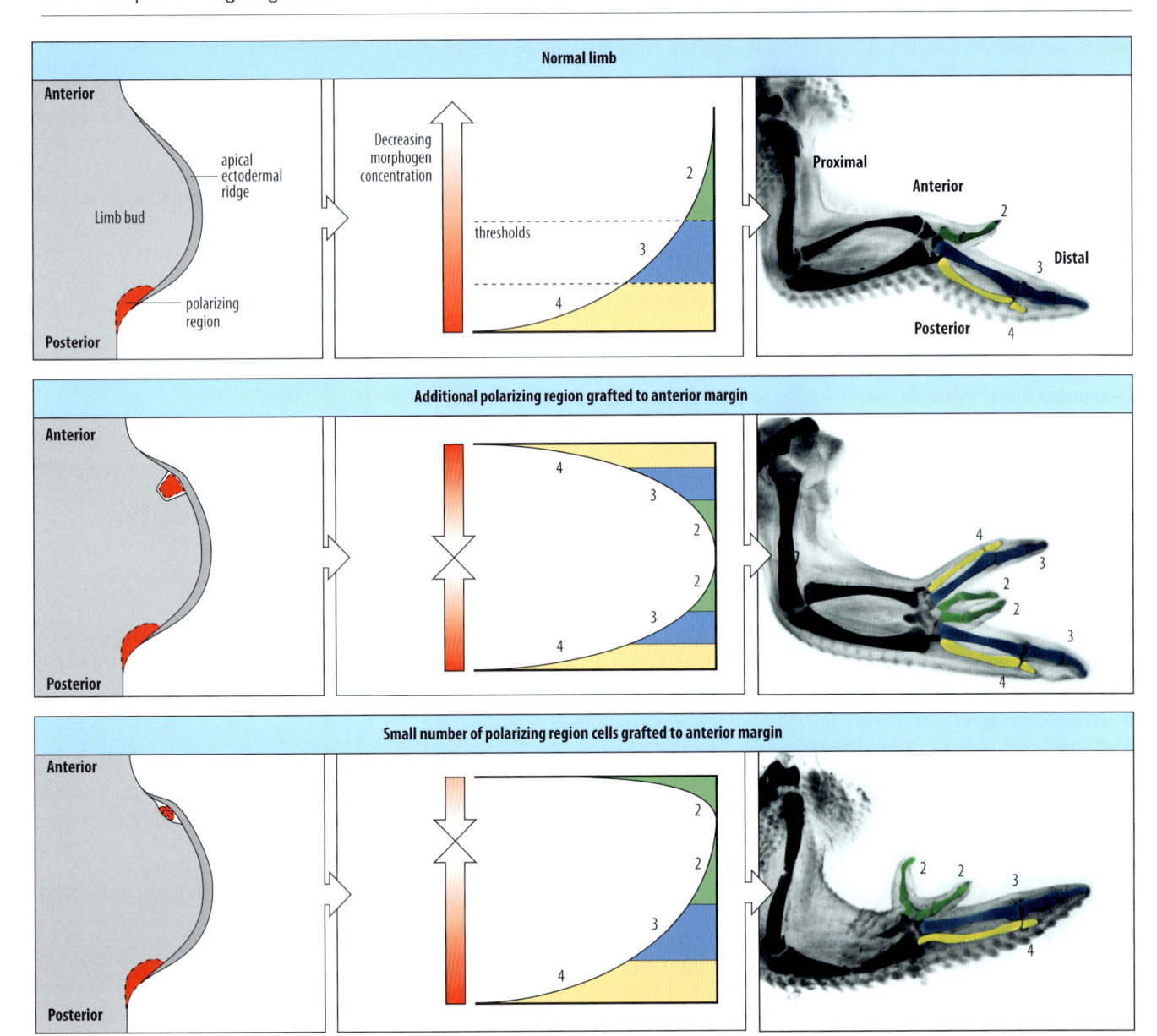

Fig. 9.31 The polarizing region specifies positional information along the antero-posterior axis. If the polarizing region is the source of a graded morphogen, the different digits could be specified at different threshold concentrations of signal, as shown for a normal chick wing bud in the top panels, with digit 4 developing where morphogen concentration is high and digit 2 where it is low. Threshold concentrations of morphogen that give rise to the different digits are indicated in the center panel of each row and the digits in the right-hand panels are color-coded to match. Grafting an additional polarizing region to the anterior margin of a wing bud (center row of panels) would result in a mirror-image gradient of signal, and thus the observed mirror-image duplication of digits. A small number of polarizing region cells grafted to the anterior margin of the wing bud produces only a weak signal, and so only an extra digit 2 is produced (bottom row of panels).

The limb buds of other vertebrates, including mice, pigs, ferrets, turtles, and even humans, have been shown to have a polarizing region. When the posterior margin of a limb bud from an embryo of these species is cut out and grafted to the anterior margin of a chick wing bud, additional digits are produced. The additional digits induced are chick wing digits, showing that, although the polarizing-region signal is conserved among vertebrates, its interpretation depends on the responding cells. This is the same principle we saw with respect to inter-species grafts of the node between different vertebrate embryos (see Section 4.8).

9.16 Sonic hedgehog is a good candidate for the polarizing region morphogen

The secreted protein Sonic hedgehog (Shh) is a good candidate for the morphogen that patterns the antero-posterior axis of vertebrate limbs. The Sonic hedgehog gene (*Shh*) is expressed in the polarizing region of chick limb buds (see Fig. 9.25) and the Shh protein is present in a graded distribution across the posterior region of the limb bud (the region of the limb bud where the digits will form) at a distance from the cells in the polarizing region that produce it. *Shh* is also expressed at the posterior margin of the limb buds of mouse embryos and other vertebrate embryos that have been studied, including human limb buds and fish fin buds. Shh protein is involved in numerous patterning processes elsewhere in vertebrate development: for example, in somite patterning (see Section 4.14); in the establishment of internal left–right asymmetry in chick embryos (see Section 4.17); and in the patterning of the neural tube (discussed in Chapter 10). As we saw earlier in this chapter, the related Hedgehog protein is a key signal molecule in patterning *Drosophila* wings and legs.

Shh has all the properties expected of the polarizing-region morphogen. Cultured chick fibroblasts transfected with a retrovirus containing the *Shh* gene acquire the properties of a polarizing region; when grafted to the anterior margin of a chick wing bud, they cause the development of a mirror-image wing. A bead soaked in Shh protein will give the same result. The bead must be left in place for 16–24 hours for extra digits to be specified, and the pattern of digits depends on Shh concentration. Shh also has a direct effect on mesenchymal cell proliferation in chick wing buds by controlling the expression of genes encoding cell-cycle regulators, thus revealing a mechanism that contributes to the widening of the wing bud following a polarizing graft.

Retinoic acid was the first defined chemical signal that was found to mimic the signaling of the polarizing region of the chick wing, and it can induce mirror-image duplications of the digits when applied to the anterior margin of a chick wing bud. But it was later shown that retinoic acid induces *Shh* expression in anterior tissue, and this explains why retinoic acid induces the digit duplications. We have already discussed the possibility that retinoic acid signaling patterns the proximal part of the limb, but how this would be related to its ability to induce *Shh* expression is not clear.

Evidence that Shh has a crucial role in specifying the pattern of digits in mammalian limbs comes from studies of mutations in mice that cause **pre-axial polydactyly**. '**Polydactyly**' denotes limbs with extra digits and 'pre-axial' means that the extra digits form anteriorly. These mutations affect antero-posterior patterning, and *Shh* transcripts are present at both anterior and posterior margins of the mutant mouse limb buds. Similar mutations have been found in individuals with pre-axial polydactyly (Box 9B). By contrast, in Shh-deficient mouse embryos, the limbs are truncated and no digits form in the forelimb and only a single digit in the hindlimb.

The concentration gradient of Shh protein provides a good explanation for the way in which the positional values for the three digits of the chick wing are specified, although it is still not clear exactly how this Shh gradient across the limb bud is established (see Box 9A). Specification of the positional values for the five digits of the mouse forepaw is, however, more complicated, as the two posterior digits (and part of the middle digit) have been shown to be derived from the polarizing region itself. It has been proposed that although Shh concentration might specify anterior mouse digits, posterior mouse digits are specified by a timing mechanism based on the duration of intrinsic Shh signaling: this is known as the 'temporal-expansion model'. As the limb bud expands across the antero-posterior axis, cells progressively leave the polarizing region and stop expressing *Shh*. The time that cells leave the polarizing region would determine their positional value, with cells that leave last forming the most posterior digit.

A key question is how limb bud cells measure the Shh concentration or duration of Shh signaling. The intracellular signaling pathway that is activated in response to

MEDICAL BOX 9B   Too many fingers: mutations that affect antero-posterior patterning can cause polydactyly

Mutations have been discovered in mice that affect antero-posterior limb patterning and cause pre-axial polydactyly. 'Polydactyly' denotes limbs with extra digits and 'pre-axial' the fact that the extra digits form anteriorly. These mouse mutants provide more evidence for the crucial role of Shh in specifying the pattern of digits, as the formation of extra anterior digits is due to additional anterior expression of the *Shh* gene in the limb buds. The polydactylous Sasquatch mouse mutant has been particularly informative, because the mutation maps to a long-range *cis*-regulatory region in the mouse genome, which is now known as the ZRS (zone of polarizing activity regulatory sequence). The ZRS governs limb-specific Shh expression, even though it is located one megabase away from the *Shh* gene. In normal development, the ZRS integrates both positive and negative inputs, so that *Shh* expression is localized at the posterior margin of the limb bud. The mutation in the Sasquatch ZRS results in an additional anterior domain of *Shh* expression in the limb buds.

The discovery that an individual in Japan with pre-axial polydactyly had a chromosomal translocation breakpoint that mapped to the ZRS showed the importance of Shh signaling in human limb development and provided an explanation for this condition. Affected people in families with pre-axial polydactyly have mutations in the ZRS located upstream of the human *SHH* gene, and several other polydactylous conditions in humans are associated with micro-duplications within this control sequence. Mutations in the ZRS have also been found in polydactylous cats with extra anterior digits, including the celebrated tribe of cats that inhabit the novelist Ernest Hemingway's old home in Key West, Florida (Fig. 1, and look out for one of these cats in the film *Licence to Kill*). The tribe is descended from a polydactylous cat given to Hemingway by a sea captain, and all the polydactylous individuals have the same single-nucleotide substitution in the ZRS. ZRS mutations have now been found in several other polydactylous animals, including the ancient Silkie breed of chickens, which have an extra toe.

In contrast to the extra digits that develop from limb buds with additional anterior *Shh* gene expression, limbs of mouse embryos in which the *Shh* gene has been functionally deleted lack distal structures, although proximal structures are still present. At best, one very small digit develops in the hindlimb.

Shh-deficient mouse embryos also have other defects, including craniofacial defects such as holoprosencephaly (see Box 1H) and patterning defects in the neural tube. Complete deletion of the ZRS in mouse embryos also leads to loss of posterior *Shh* expression in the limb buds and results in limb truncations resembling those in Shh-deficient embryos. In embryos with the ZRS deletion, however, Shh function is deleted just in the limb buds, and no other parts of the embryo are affected.

Loss of distal limb structures is also seen in the very rare inherited human condition acheiropodia, in which structures below the elbow or knee fail to develop but no other parts of the body are affected. Acheiropodia has been shown to be due to a chromosomal deletion near the ZRS that deletes CTFC-binding sites (see Section 6.2) and disconnects the enhancer from the *SHH* gene so that it is not expressed in the polarizing region.

Fig. 1
Photograph courtesy of Lettice, L.A., et al.: **Point mutations in a distant Shh cis-regulator generate a variable regulatory output responsible for preaxial polydactyly.** Hum. Mol. Genet. 2008, *17*: 978–985.

The long-range control of *Shh* expression by the ZRS is similar to the developmental control of β-globin expression in red blood cells by the locus control region (see Section 6.10). The robustness of the interactions between the ZRS and the *Shh* gene promoter is due to the three-dimensional architecture of the chromosome, which not only promotes these interactions but also prevents non-specific interactions. Chromosomal rearrangements can disrupt the three-dimensional architecture, so that genes come under the influence of a 'foreign' *cis*-regulatory region. An example of such a disruption that altered the location of the *SHH* gene was found in a person with severe limb and face differences. The effects of this relocation were modeled in transgenic mice. A 'foreign' *cis*-regulatory region near the human *SHH* gene in the affected person was shown to drive an abnormal pattern of *Shh* expression in the mouse limb bud, resulting in limb defects. This provides an explanation for the way in which the limb differences were caused in the affected individual.

Shh involves the Gli transcription factors Gli1, Gli2, and Gli3, which are analogous to Ci in the Hedgehog signaling pathway in *Drosophila* (see Box 2E). The vertebrate Shh signaling pathway is shown in Box 9C. In the presence of Shh, the Gli proteins function as transcriptional activators, whereas in its absence, Gli2 and Gli3 are processed to short forms and function as transcriptional repressors. Therefore, the ratio of Gli3 activator to Gli3 repressor in a cell reflects the level of Shh signaling.

CELL BIOLOGY BOX 9C Sonic hedgehog signaling and the primary cilium

The secreted signal protein Sonic hedgehog (Shh) is a vertebrate homolog of the Hedgehog protein that has a major role in the patterning of the *Drosophila* embryo (see Chapter 2) and the wing imaginal disc (see Section 9.4). Mice and humans have three genes for hedgehog proteins—SONIC HEDGEHOG (SHH), INDIAN HEDGEHOG (IHH), and DESERT HEDGEHOG (DSS); zebrafish have a fourth *hedgehog* gene, *Tiggywinkle*, as a result of genome duplication. The signaling pathway stimulated by Shh and the other vertebrate hedgehog proteins (Fig. 1) has many similarities to the Hedgehog signaling pathway in *Drosophila* (see Box 2E).

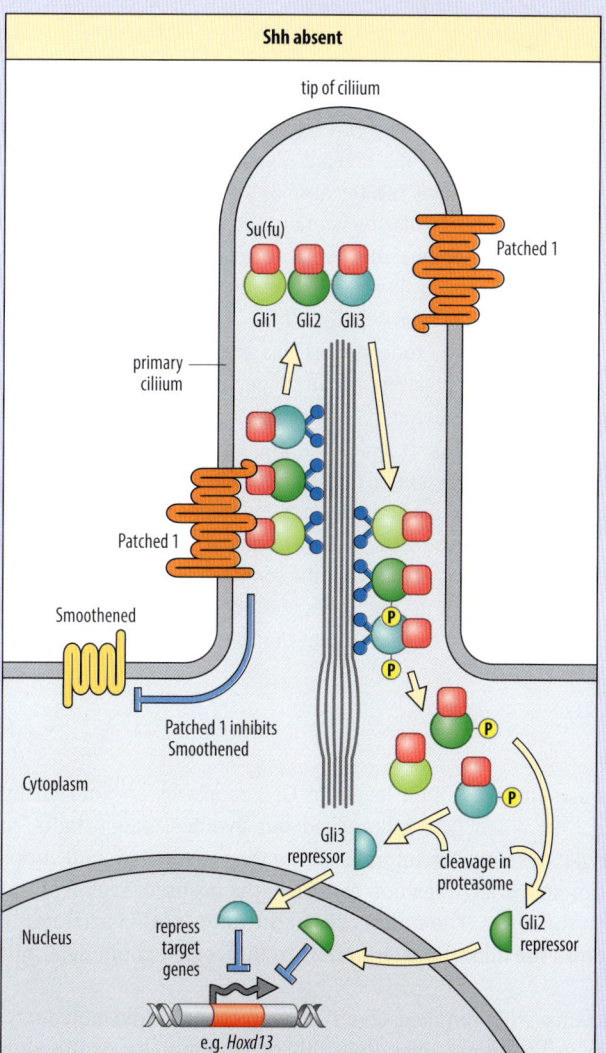

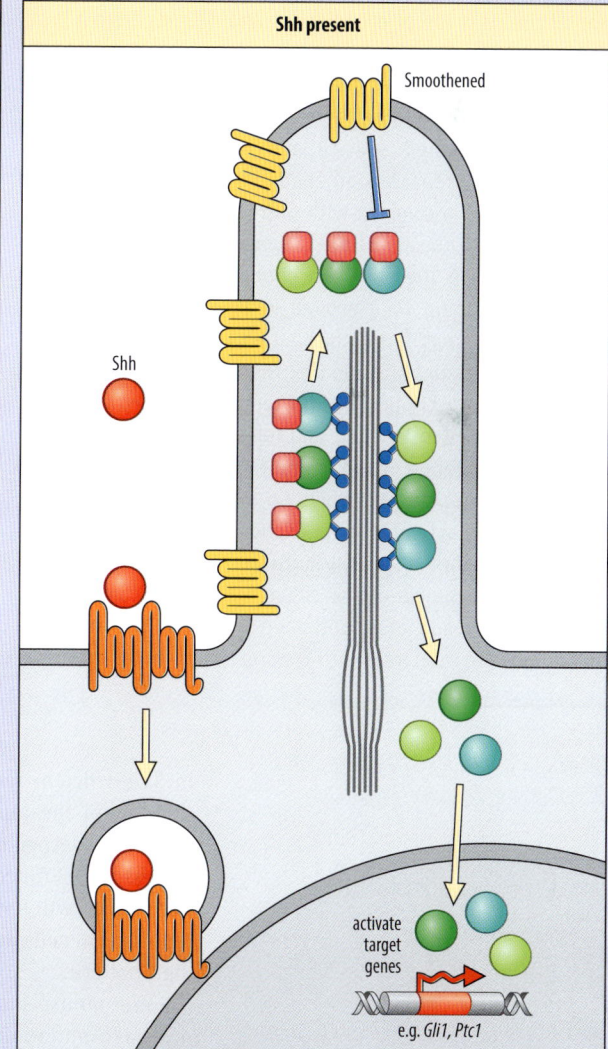

Fig. 1

There are vertebrate homologs of the membrane proteins Patched (Patched1 is the version expressed in limb mesenchyme) and Smoothened, which sense the presence or absence of a hedgehog signal. The vertebrate counterparts of the *Drosophila* transcription factor Cubitus interruptus (Ci) are the Gli transcription factors Gli1, Gli2, and Gli3. (The name Gli comes from the initial identification of one of these genes as being amplified and highly expressed in a human glioma.) Like Ci, the Gli proteins are processed so that they act as transcriptional activators in the presence of a hedgehog signal and Gli2 and Gli3 are processed to transcriptional repressors in its absence. Proteins homologous with Cos2 and Su(fu) in *Drosophila* (see Box 2E) form complexes with the Gli proteins and keep them in an inactive state.

A major difference between a vertebrate Hedgehog signaling pathway and Hedgehog signaling in *Drosophila* is that, in vertebrates, Gli processing takes place in a cell structure called the **primary cilium**. Most vertebrate cells have a primary cilium, which, like all cilia, is made of microtubules and extends from a basal body derived from a centrosome. Unlike most other types of cilia, the primary cilium is non-motile.

In the absence of a Hedgehog ligand such as Shh, Patched inhibits Smoothened activity and moves into the cilium membrane, whereas Smoothened does not. The Gli3 and Gli2 proteins in a complex with Su(fu) are transported along microtubules to the tip of the cilium and then back again to the base, where they are phosphorylated by PKA and CKI, which target them for cleavage in the proteasome. Cleavage produces short forms that act as transcriptional repressors. The repressors Gli3 and Gli2 then enter the nucleus and prevent the transcription of Shh target genes. When Shh is present, it binds to Patched and releases Smoothened from inhibition. Hedgehog ligand bound to Patched is internalized in membrane vesicles and plays no further part in signaling. Smoothened is activated by phosphorylation and moves into the cilium membrane. The Gli/Su(fu) complexes accumulate at high levels at the tip of the cilium, where the complex dissociates as a result of Smoothened signaling. The activated Gli proteins are then transported along microtubules down the cilium to the cell body, where they become fully activated by phosphorylation. They then enter the nucleus and activate transcription of another set of Shh target genes.

The first clues to the importance of the primary cilium in Shh signaling came from a genetic screen in mice for embryonic patterning mutants following mutagenesis with ethylnitrosourea. Two mutants were identified with defects in Shh signaling in the neural tube, one of which survived long enough for limb polydactyly to be observed. Most unexpectedly, a search against the databases revealed that the genes mutated in these embryos encoded proteins similar to those that transport material within a flagellum (a longer version of a cilium), and which were known to be essential for the growth and maintenance of flagella in the unicellular alga *Chlamydomonas*. Subsequently, mutations in genes encoding subunits of intraflagellar dynein and kinesin motor proteins were also found to result in defective Shh signaling. Further analysis of fish and chicken mutants showed that the involvement of primary cilia in hedgehog signaling is universal among vertebrates.

All known mutations that prevent formation of the primary cilium cause patterning defects in the limbs and the neural tube. In the absence of the primary cilium, both Gli repressor and Gli activator functions fail. Because of the importance of Gli repressor function in antero-posterior limb patterning this leads to polydactyly, whereas the neural tube is dorsalized as a result of the importance of Gli activator function in its dorso-ventral patterning (discussed in Chapter 10).

Defects in the formation of cilia and basal bodies have been linked to several dozen human congenital conditions, collectively known as ciliopathies. For example, people with Bardet-Biedl syndrome (BBS), in which the primary cilia are affected, have a range of differences that can include polydactyly, polycystic kidney disease, and hearing loss. Mutations in any of 21 known genes can cause BBS, but this number will increase because these 21 genes account for the condition in only 75% of people with BBS. Some of the genes identified so far encode proteins that are part of the basal body, thus providing an explanation for defects in ciliogenesis.

In *Shh*-deficient mouse embryos, high levels of Gli3 repressor are present in cells throughout the bud, but when *Gli3* is knocked out in mice, many digits, all the same unpatterned type, develop. Therefore Shh signaling in the limb bud suppresses repressor Gli3 function in a graded fashion in cells in the posterior region of the limb bud, which will form the digits. The differences in the ratios of Gli3 activator to Gli3 repressor in cells across the antero-posterior axis in this region could determine the digit pattern.

Some mouse mutants with reduced levels of repressor Gli and polydactyly do not have mutations in the genes encoding Shh or components of the signaling pathway, but instead have mutations in genes essential for the formation of a cell structure called the primary cilium. This immotile microtubule-based structure is essential for Shh signaling and the processing of Gli proteins in vertebrate cells, as described in Box 9C.

Among the gene targets of the Gli transcription factors in the early limb bud are 5′ Hoxd genes that specify regional identity in the limb, and a signaling cascade involving BMP-2 and genes encoding the transcription factors Sall1, Tbx2, and Tbx3. BMP-2 is related to *Drosophila* Dpp, and the *Sall* and *Tbx* genes to the *Drosophila spalt* and *omb* genes, respectively. This means that, rather surprisingly, this cascade in antero-posterior patterning of the vertebrate limb has the same components as the signaling cascade downstream of Hedgehog signaling in the antero-posterior patterning of the *Drosophila* wing (see Section 9.4). In *Drosophila*, Dpp acts as a morphogen and it is possible that BMP-2 acts as a secondary signal in specifying positional values in the vertebrate limb, because various manipulations that alter the level of BMP signaling in early chick wing buds can transform digit identity.

The digits develop from condensations of mesenchymal cells expressing the transcription factor Sox9, which arise at specific positions in the digital plate. Sox9 controls cartilage cell differentiation. The tissue between each condensation is known as the interdigital region. BMP-2 is expressed in the posterior of the early limb bud, but also with other BMPs in the interdigital regions. Each forming digit is characterized by a specific level of activity of Smad transcription factors (see Box 3F), which provides a read-out of the response to the strength of BMP signals from the adjacent interdigital region. Removing interdigital tissue, or implanting beads soaked in the BMP antagonist Noggin into the interdigital regions in the chick leg, anteriorizes the adjacent digit. BMP signaling in the forming digit therefore seems to activate genes involved in late stages of morphogenesis as well as potentially being involved in specifying digit identity at earlier stages.

In the chick leg, *Tbx2* and *Tbx3* are expressed in an overlapping pattern in the region of the digital plate that contains the condensations for the two posterior digits. It has been reported that overexpression of either *Tbx2* or *Tbx3* can change one type of digit into the other in the chick leg. It seems likely that the combined activity of these transcription factors and other transcription factors such as the Sall proteins and the Hox proteins specifies regional identity across the antero-posterior axis of the limb. Mutations in the human versions of these transcription factor genes are associated with human syndromes that include limb differences. Mutations in *TBX3* are associated with ulnar-mammary syndrome, and mutations in *SALL1* and *SALL4* with Townes–Brocks syndrome and Okihiro/Duane radial ray syndrome, respectively. Okihiro/Duane radial ray syndrome, together with Holt–Oram syndrome (associated with mutations in *TBX5*) (see Section 9.11), can be confused with the effects of the harmful drug thalidomide (Box 9D).

9.17 The dorso-ventral pattern of the limb is controlled by the ectoderm

The chick wing has a well-defined pattern along the dorso-ventral axis: large feathers are only present on the dorsal surface, and muscles and tendons have a complex dorso-ventral organization. Flexor muscles develop on the ventral side, extensors on the dorsal.

The development of pattern along the dorso-ventral axis in the mesoderm has been studied by recombining ectoderm taken from left limb buds with mesoderm from right limb buds in such a way that the dorso-ventral axis of the ectoderm is reversed with respect to the underlying mesoderm, but the antero-posterior axis is unchanged. The right and left wing buds are removed from a chick embryo and, after treatment with cold trypsin, the ectoderm can be pulled off the mesodermal core just like taking off a glove. Ectoderm from a left bud is then recombined with mesoderm from a right bud so that the dorso-ventral axis of the ectoderm runs

MEDICAL BOX 9D External agents and the consequences of damage to the developing embryo

External agents, such as a chemical or an infection, can interfere with embryonic development, leading to damaging effects that are seen when the baby is born. Damage to an embryo can also be due to heritable gene mutations, and it is sometimes difficult to distinguish between these two causes.

One of the best-documented chemicals that interferes with development is the drug thalidomide. This was widely prescribed for morning sickness during early pregnancy in the late 1950s and early 1960s, first in Germany and then throughout Europe and other countries. It is estimated that at least 10,000 severely affected babies were born worldwide as the result of thalidomide (of which around 40% did not survive beyond their first birthday). The most obvious effects are seen in the limbs, although other organs may have been affected, including the heart, eyes, and ears. Typically, the limbs, most commonly the arms, are either completely absent or proximal structures are reduced or missing but distal structures are relatively unaffected. Figure 1 shows an X-ray of the arms of someone whose mother took thalidomide. The lower arm is absent, but the fingers are present. The period during which a harmful agent can interfere with development of specific organs is known as the 'critical period'. The critical period for thalidomide damage in human embryos is during days 20–36 of development, when structures such as the arms and heart are being formed.

So how did a drug that causes so much harm to developing embryos get approved for clinical use? All drugs must go through safety tests before they are approved. Safety tests have become much more stringent since the 1950s, in part at least, because of the catastrophic effects of thalidomide. But mice and rats, which are commonly used to screen drugs for effects on embryos, are not very sensitive to thalidomide. Once it was realized that thalidomide was damaging unborn children, it was withdrawn from sale to treat pregnant women, and research to classify thalidomide injuries and to find out how they are caused began immediately. It is still important to know how thalidomide damages the developing embryo, as the drug has been reintroduced to treat leprosy. Although thalidomide is not now recommended for use by the World Health Organization (WHO) in patients with leprosy because safer drugs are available, a few babies affected by thalidomide continue to be born every year in Brazil.

Even today, there is no consensus on how thalidomide interferes with limb development. One theory is that thalidomide kills cells. X-irradiation of early chick wing buds causes extensive

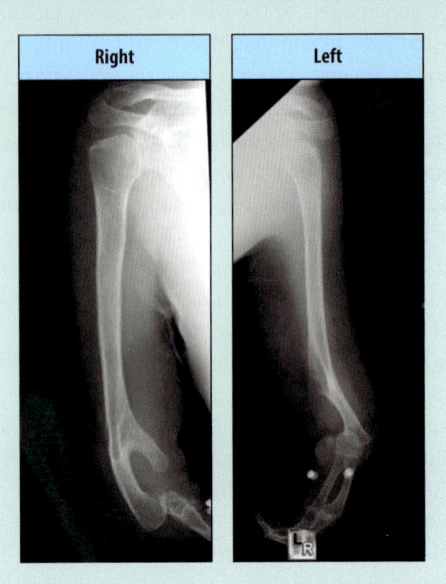

Fig. 1
Published with permission from
LearningRadiology.com.

death, and this also results in the loss of proximal structures. It has been suggested that distal structures are relatively unaffected because cells spend longer at the tip of the wing bud in order to repopulate the progress zone, and thus acquire more-distal positional values. A similar explanation has been advanced for thalidomide-induced loss of proximal limb structures in humans. An alternative explanation is that proximal cells are selectively eliminated.

Another theory is based on the anti-angiogenic effects of thalidomide that block the growth of blood vessels. This property has led to the reintroduction of thalidomide and its analogs as anticancer drugs, for example, for multiple myeloma, with appropriate precautions. An analog of thalidomide with anti-angiogenic properties has been shown to lead to the loss of wings in treated chick embryos. The analog specifically interferes with the development of the delicate blood capillaries in the early wing bud, and this interruption in the blood supply leads to cell death in the bud. It has been suggested that vascularization of the limbs might be affected in human embryos in response to thalidomide, and that this could explain why limbs are absent in some children whose mothers took the drug.

the opposite way to that of the mesoderm—the equivalent of putting a left-hand glove on the right hand (Fig. 9.32). The 'recombinant' limb buds are grafted to the flank of a host embryo and allowed to develop. The proximal regions of the limbs that develop have dorso-ventral polarity corresponding to that of the mesoderm, whereas the distal regions, in particular the digital regions, have a reversed dorso-ventral axis, with the pattern of muscles and tendons reversed and corresponding to

Yet another theory is that thalidomide leads to the production of reactive oxygen species which could damage macromolecules or interfere with signaling pathways required for normal limb development.

A recent discovery is that thalidomide binds to the protein cereblon, and this might mediate its effects on the developing limb. Cereblon is part of a protein complex that promotes the ubiquitination of proteins, making them susceptible to degradation. Bound thalidomide changes the activity of the complex so that new proteins are recruited and targeted for degradation. One of the new proteins recruited to cereblon in human embryonic stem cells treated with thalidomide is the transcription factor SALL4. This is a very significant finding because Sall4 is known to be involved in limb development in mouse and chick embryos. Furthermore, mutations in *SALL4* have been found in individuals with Okihiro syndrome and some other syndromes with overlapping phenotypes, which are characterized by limb differences similar to those caused by thalidomide (see Section 9.16). Indeed, there have been cases in which people, diagnosed as being affected by thalidomide, in fact have a mutation in *SALL4*. It was important to uncover this misdiagnosis because there had been reports that babies born to parents who had themselves been affected by thalidomide also had damaged limbs. This had suggested that thalidomide might be a mutagen, causing a permanent and heritable change in a gene. This possibility can now be discounted. Another important aspect of the discovery of the potential involvement of SALL4 is that it explains the species specificity of thalidomide. In mouse cells, the cereblon complex is unable to tag Sall4 for degradation because of differences in the amino acid sequences in both mouse cereblon and Sall4 proteins.

A difference between the phenotypes of individuals with *SALL4* mutations and of individuals whose mothers took thalidomide is that only the arms are affected in the genetic condition, whereas both arms and legs can be affected by thalidomide. This suggests that there are other targets of thalidomide, including other proteins targeted for destruction when thalidomide binds to the cereblon complex, that are also involved in producing limb differences. One of these other proteins is the transcription factor p63, which maintains the apical ridge of the limb bud and is required for normal limb development in mice. The limb phenotypes of p63 knock-outs in mouse embryos and a human condition caused by mutations in p63 have some features in common with the effects of thalidomide.

Limb differences occur in around 5 per 10,000 live births (precise figures vary), but only a small percentage can be confidently ascribed to environmental factors, such as exposure to harmful chemicals before birth. More than 50% are estimated to be due to genetic causes. In more than 25% of cases, the cause is unknown and is likely to be multifactorial—the result of a combination of genetic and environmental factors. Just as the genotype of a woman determines her reaction to drugs used to treat disease—the basis of personalized medicine—her genotype similarly could determine sensitivity to environmental factors that could harm the development of her unborn child. This could explain why some pregnant women exposed to environmental factors have affected babies, whereas others exposed to the same factors do not.

Thalidomide is probably the best-known example of a chemical that has deleterious effects on development. The list of chemicals proved to affect development is fairly short and includes anticonvulsants (such as trimethadione and valproic acid), folic-acid antagonists, retinoids (derivatives of vitamin A), ethanol, and dioxins. One important issue regarding all such chemicals is to understand how they are metabolized, as the products can be more, or less, harmful. Either a deficiency or an excess of vitamin A is harmful to animal embryos and, despite the thalidomide catastrophe, affected babies were born to women in the United States who were being treated with retinoid-based drugs for severe skin conditions. Ethanol can also damage developing embryos when consumed in large amounts by the mother (see Box 10A), and this may be due to its being metabolized by the same pathway that produces retinoic acid.

Infectious agents can also harm embryos. A well-known example in humans is the virus that causes rubella, commonly known as German measles. If a woman catches German measles in the first 16 weeks of pregnancy—the critical period for this agent—damage to the embryo can affect ears, eyes, the heart, and the brain, as these organs are developing at this time. How the virus causes this spectrum of effects, known as congenital rubella syndrome, is not understood. Vaccination against rubella has largely eliminated the syndrome in developed countries, but elsewhere it is still a problem, with the WHO estimating that about 110,000 babies worldwide are born every year with the syndrome. Another virus that affects development is the Zika virus, which is transmitted by mosquitos. An outbreak of Zika occurred in Brazil in 2014 and rapidly spread throughout Latin America and the Caribbean. The effect is microcephaly, a condition in which the baby's head is small and the brain underdeveloped. Finally, exposure to physical agents such as high doses of ionizing radiation can also interfere with embryonic development.

the dorso-ventral polarity of the ectoderm. The ectoderm covering the sides of the limb bud can therefore specify dorso-ventral pattern.

Genes controlling the pattern of structures along the dorso-ventral axis in vertebrate limbs have been identified from mutations in mice. The dorso-ventral pattern in mouse limbs can be distinguished because the ventral surface of the paw normally has no fur, whereas the dorsal surface does. The dorsal surfaces of the digits also

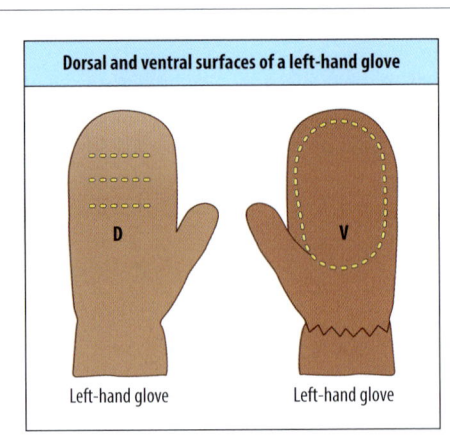

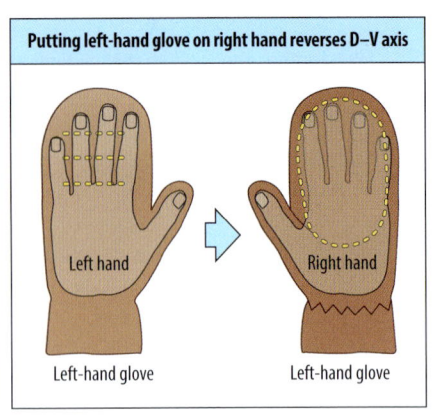

Fig. 9.32 Reversing the dorso-ventral axis of limb-bud ectoderm. The only way a left-hand glove (representing the ectoderm from a left-hand limb bud) can fit on a right hand (the mesoderm of a right hand) is by turning it over so that the former ventral side (patterned) is now uppermost. The dorso-ventral polarity of the glove is now reversed in relation to the dorso-ventral axis of the hand. The antero-posterior relationship remains the same. D, dorsal; V, ventral.

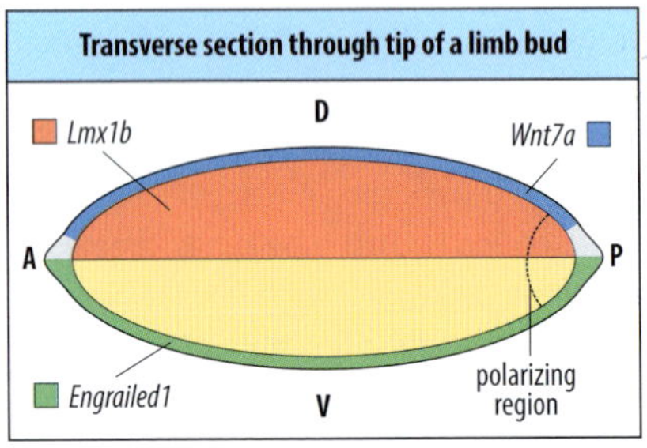

Fig. 9.33 The ectoderm controls dorso-ventral pattern in the developing limb bud. The gene for the secreted signal Wnt-7a is expressed in the dorsal ectoderm and BMP signaling ventrally induces expression of the gene *Engrailed1* in the ventral ectoderm and ventral half of the apical ridge (shown here in cross-section at both anterior and posterior edges of the limb bud). Expression of the gene for the transcription factor Lmx1b is induced in the dorsal mesoderm compartment by Wnt-7a signaling, and Lmx1b is involved in specifying dorsal structures. A, anterior; D, dorsal; P, posterior; V, ventral.

have claws (nails in humans). *Wnt7a* is expressed in the dorsal ectoderm (Fig. 9.33), and mutations that inactivate the *Wnt7a* gene result in limbs in which many of the dorsal tissues adopt ventral fates to give a double ventral limb, the two halves being mirror images. The finding that *Wnt7a* mutants have a double ventral limb suggests that the ventral pattern may be the ground state and is modified dorsally by the dorsal ectoderm, with the secreted Wnt-7a protein diffusing into the dorsal mesoderm and playing a key part in patterning it. The ventral ectoderm is characterized by high levels of BMP signaling, and this induces expression of the gene *Engrailed1* (*En1*) in the ventral ectoderm and ventral half of the apical ectodermal ridge. *En1* encodes a homeodomain transcription factor first described in *Drosophila* (see Section 2.23 and Section 9.1). Mutations that destroy *En1* function in mice result in *Wnt7a* being expressed in ventral, as well as dorsal, ectoderm, giving a double dorsal limb.

Wnt-7a protein induces expression of the gene encoding the LIM homeobox transcription factor Lmx1b in the mesoderm underlying the dorsal ectoderm (see Fig. 9.33). Clonal analysis of marked cells in mouse limb buds has revealed dorso-ventral compartments of lineage restriction in the mesoderm, with the progeny of dorsal cells being restricted to the dorsal half of the limb bud and the progeny of ventral cells to the ventral half. This is the only instance in which cell-lineage restrictions have been demonstrated in mesenchymal tissue; all other compartments discovered so far in both vertebrates and invertebrates are in epithelia—for example, the *Drosophila* epidermis (see Section 2.23 and Section 9.2), the trunk ectoderm of vertebrate embryos (see Section 9.11), and the rhombomeres of the vertebrate hindbrain (see Section 10.4). The compartmentalization of the mesenchyme might serve to control gene expression; *Lmx1b*, for example, is expressed exactly in the dorsal compartment. In chick embryos, *Lmx1b* has been shown to specify a dorsal pattern in the mesoderm. Ectopic expression of *Lmx1b* in the ventral mesoderm of the chick limb results in the cells adopting a dorsal fate, leading to a mirror-image dorsal limb. This compartmentalization is also essential for establishing the pattern of muscles and innervation of the developing limb.

Lmx1b protein is related to the Apterous transcription factor in *Drosophila*, which defines the dorsal compartment in the wing (see Section 9.5). Loss-of-function mutations in human *LMX1B* cause an inherited condition called nail–patella syndrome, in which these dorsal structures (nails and kneecaps) are not well formed or absent.

9.18 Development of the limb is integrated by interactions between signaling centers

The polarizing region has a pivotal role in the chick wing: it not only signals antero-posterior positional information (see Section 9.15) but also maintains the apical ectodermal ridge over the posterior part of the bud. It is crucial that development along all three limb axes is integrated, so that the correct anatomy of the limb is generated. This integration is accomplished by interactions between signaling generated by Wnt-7a from dorsal ectoderm, FGFs from the apical ectodermal ridge, and Shh from the polarizing region (Fig. 9.34). In mice mutant for loss of *Wnt7a* function, posterior digits are frequently missing—showing that *Wnt7a* is also required for normal antero-posterior patterning—and expression of *Shh* in the polarizing region is reduced. *Shh* expression is similarly reduced when the dorsal ectoderm of the wing bud is removed in chick embryos.

Development along the proximo-distal and antero-posterior axes is coordinated during limb bud outgrowth by a positive-feedback loop between Shh signaling by the polarizing region and FGF signaling by the apical ectodermal ridge. Shh signaling maintains the level of *Fgf* gene expression in the apical ridge over the posterior part of the bud, while in turn, FGF signaling by the apical ridge maintains *Shh* gene expression in the polarizing region.

The way in which Shh maintains *Fgf* expression in the apical ridge is by positively regulating expression of the gene that encodes the protein Gremlin, which is an antagonist of BMP signaling. BMP-4 is first involved in limb development in the formation of the apical ridge (see Section 9.11) and is then widely expressed in the limb bud and will repress *Fgf* expression in the apical ridge. Therefore, to maintain *Fgf* expression in the apical ridge over the posterior part of the limb bud, BMP signaling must be repressed in this part of the bud, and this is achieved by the action of Gremlin. *gremlin* gene expression and subsequent production of Gremlin is regulated by two interlinked feedback loops (see Fig. 9.34). One just mentioned is the positive FGF–Shh feedback loop in which Shh maintains *gremlin* expression in the mesoderm in the posterior region of the limb bud. This feedback loop takes about 12 hours to be established. The second is a rapidly established negative-feedback loop in which *gremlin* gene expression is induced in the limb bud mesoderm by BMP-4. The Gremlin protein produced then antagonizes BMP activity, keeping it low. This self-regulating system of interlinked feedback loops is an excellent example of a gene regulatory network that makes development so reliable, containing feedback control and employing more than one way of achieving the required outcome (see Section 1.19).

9.19 Hox genes control regional identity in the limb

As discussed in Section 9.11, a combinatorial code of Hox genes in the lateral plate mesoderm induces limb development in the appropriate positions along the antero-posterior axis of the vertebrate body. Hox genes also determine the antero-posterior polarity of at least the forelimb buds and are components of the gene regulatory network that ensures that *Shh* is expressed at the posterior margin of the limb bud (see Section 9.11). Hox genes are then expressed in the developing limb buds, in response to signals in the bud that include retinoic acid, FGFs, Wnts, Shh, and BMPs. It has been suggested that this response is mediated by a gradient of Meis transcription factors interacting directly with the chromatin of Hox clusters and regulating expression of Hox genes within the cluster (see Section 9.14).

At least 23 different Hox genes are expressed during mouse and chick limb development. Attention has largely focused on the genes of the *Hoxa* and *Hoxd* gene clusters, because genes in the 5′ regions of these clusters are expressed at high levels in both forelimbs/wings and hindlimbs/legs and have been shown to control regional identity.

There are two phases of expression of the 5′ *Hoxa* and *Hoxd* genes during limb development. As we have seen in other developmental examples, the spatial and temporal patterns of expression of the Hox genes in the limb correspond to the order of the genes on the chromosome (see Sections 2.29 and 4.11). In the first phase of Hox gene expression in the limb, the genes of the *Hoxa* and *Hoxd* clusters are

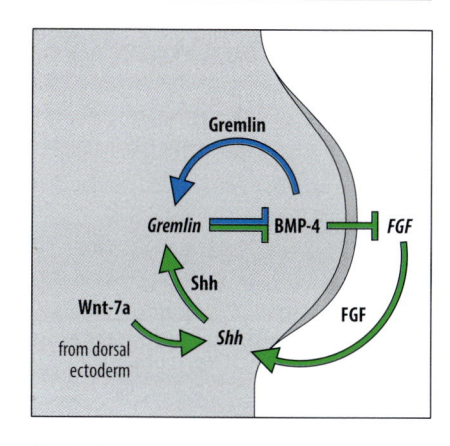

Fig. 9.34 Integration of limb development signals by a gene regulatory network containing interlinked feedback loops. FGF protein signals from the apical ectodermal ridge maintain *Shh* expression in the polarizing region. In turn, the Shh protein signal from the polarizing region maintains *Fgf* expression in the apical ectodermal ridge. The maintenance of *Fgf* expression in the apical ridge is accomplished by the Shh signal maintaining expression of the gene encoding the BMP antagonist Gremlin in the limb-bud mesoderm. Gremlin keeps BMP activity low, which lifts its repression of *Fgf* expression and allows a positive-feedback loop to be generated between Shh and FGF production. Another way in which Gremlin is produced is via a negative-feedback loop in which the signal protein BMP-4 itself induces *gremlin* expression. The signal protein Wnt-7a produced by the dorsal ectoderm also helps to maintain *Shh* expression in the polarizing region. Arrows indicate positive control, in which a signal protein promotes gene expression, while barred lines indicate either negative control of signaling by an antagonist or repression of gene expression. Blue indicates the negative-feedback loop that produced Gremlin and keeps BMP activity low. Green indicates the FGF-Shh positive-feedback loop in which the Shh signal protein maintains *gremlin* expression.

Adapted from Bénazet, J., Zeller, R.: **Vertebrate limb development: moving from classical morphogen gradients to an integrated 4-dimensional patterning system.** *Cold Spring Harb. Perspect. Biol. 2009,* **1***: a001339.*

Fig. 9.35 Patterns of Hox gene expression in developing mouse limb buds. The left panels show the first phase of 5′ Hoxd and Hoxa gene expression in early mouse limb buds. The 5′ Hoxd genes (left upper panel) come to be expressed in a nested overlapping pattern along the proximo-distal axis, centered on the posterior tip of the early limb bud. *Hoxd9* is expressed throughout the limb bud but is not shown for clarity. *Hoxd13* is expressed most posteriorly and distally. The 5′ Hoxa genes (left lower panel) are expressed in a similar nested overlapping pattern along the proximo-distal axis of the bud. *Hoxa13* is expressed most distally. The right panels show the patterns of expression of 5′ Hoxd (right upper panel) and Hoxa (right lower panel) genes in later mouse limb buds when a second phase of Hox gene expression has been initiated in the digital plate, with *Hoxd13* and *Hoxa13* being expressed throughout this region and *Hoxd12–10* being expressed in overlapping domains centered on the posterior margin. A, anterior; P, posterior.

Adapted from Tanaka, M.: **Evolution of vertebrate limb development.** eLS 2017, 1-10.

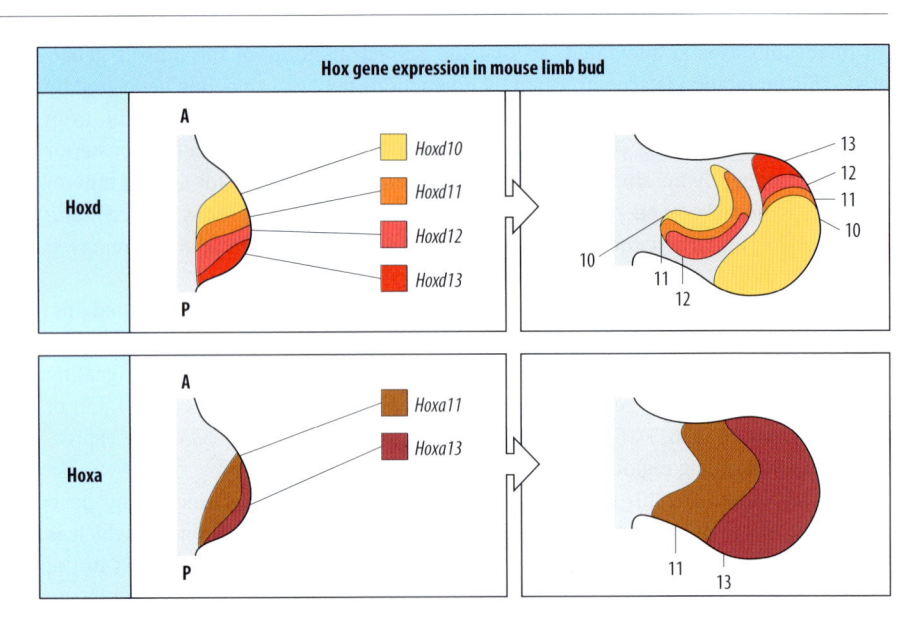

expressed in their order along the chromosomes in the 3′ to 5′ direction. *Hoxd9* and *Hoxd10*, the more 3′ genes, are expressed first in the lateral plate mesoderm as it begins to thicken to become a limb bud. Then expression of *Hoxd11*, *Hoxd12*, and *Hoxd13*, the more 5′ genes, is initiated in rapid succession in progressively more-restricted domains centered on the posterior distal region of both forelimbs and hindlimbs. These domains then extend along the limb bud as it continues to grow out, as shown for the mouse forelimb in Figure 9.35 (left, upper panel). Expression of *Hoxa9–13* is initiated slightly later in the early limb bud, and then gives rise to similar overlapping domains of expression along the proximo-distal axis, but without a marked posterior bias, and these domains extend along the limb bud as it grows out (see Fig. 9.35, left, lower panel). *Hoxa9* is expressed throughout the limb bud and successive genes are expressed in more and more distal regions, with *Hoxa11* expression becoming stronger in the intermediate region of the limb, where *Hoxd11* is also strongly expressed; *Hoxd13* and *Hoxa13* are expressed in the distal region of the limb bud.

As the digital plate begins to form, a second phase of Hoxd gene expression is initiated in the distal part of the limb (see Fig. 9.35, right panels) and in this phase, the genes are expressed in the opposite order along the chromosome in a 5′ to 3′ direction. *Hoxd13* expression is initiated first throughout the digital plate, with progressively smaller domains of *Hoxd12–10* expression then being initiated successively within this region, except in the anterior-most part. These Hoxd genes, together with *Hoxa13*, expressed throughout the digital plate, are involved in digit formation. The gap in Hoxd gene expression between the domain of strong expression of genes such as *Hoxd10* in the intermediate region of the limb and the domain of strong Hoxd expression in the digital plate corresponds to the region that will form the wrist or ankle.

Genetic experiments in mice, together with analyses of post-translational histone modifications associated with gene activity and chromosomal conformation (see Box 6B), have identified remote *cis*-regulatory regions in the two topologically associating domains (TADs, see Section 6.2) extending in gene-poor genomic intervals (gene deserts) of about one megabase on either side of the Hoxd and Hoxa gene clusters. These regulatory regions act as enhancers and drive the two phases of Hox gene expression in the limb bud. The same regulatory logic governs transcription of both Hoxa and Hoxd gene clusters in the developing limb, so we will consider

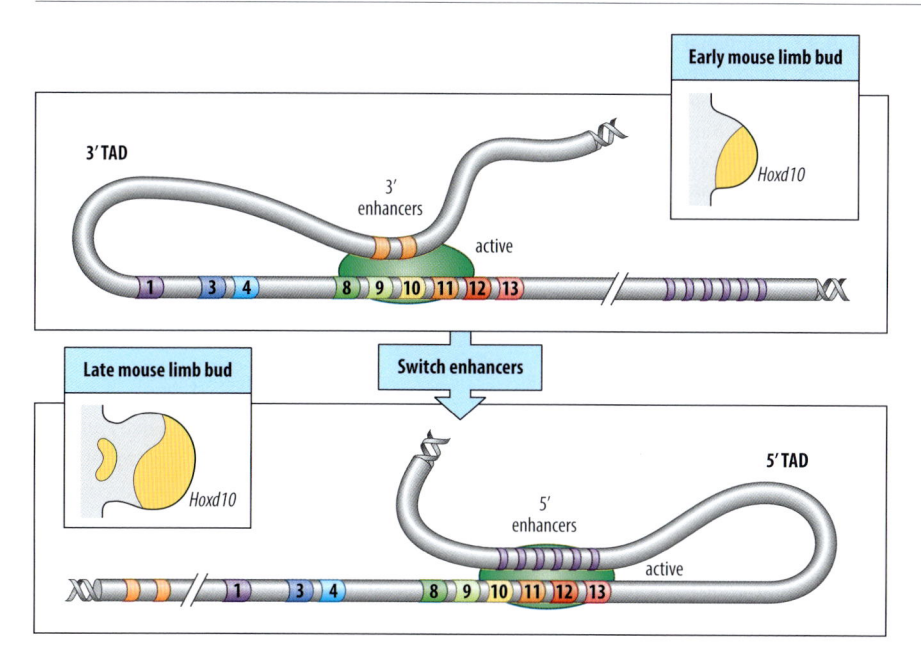

Fig. 9.36 A switch between 3′ and 5′ enhancers controls Hoxd gene expression in the developing mouse limb. Upper panel: in cells in the early limb bud, enhancers (orange) in the TAD on the 3′ side of the Hoxd cluster are brought into contact with genes in the cluster by chromosomal looping, activate them (green) and control the first phase of Hox gene expression, as illustrated for *Hoxd10* (top right). Lower panel: in cells in the digital plate, enhancers (purple) in the TAD on the 5′ side of the cluster are brought into contact with genes towards the 5′ end of the cluster and control the second phase of Hox gene expression, as illustrated for *Hoxd10* (bottom left). Expression of *Hoxd10*, which is not on the periphery of the cluster, is controlled by both sets of enhancers, but the patterns of expression are different.

Adapted from Andrey, G., et al.: **A switch between topological domains underlies HoxD genes collinearity in mouse limbs.** Science 2013, **340**: 1234167.

just the Hoxd cluster in more detail. Two enhancers in the TAD on the 3′ side of the Hoxd cluster drive the nested pattern of Hoxd gene expression in early limb buds. Later, multiple enhancers lying in the 5′ TAD drive Hoxd gene expression in the digital plate (Fig. 9.36).

Thus, during limb development, there is a switch from control of gene expression via the 3′ enhancers to control via the 5′ enhancers. Genes such as *Hoxd10* that are not on the periphery of the cluster are therefore expressed in two distinct domains, in the intermediate part of the limb bud early in development and in the distal part of the limb bud later (see Fig. 9.35). It has been suggested that this switch results in a zone of low Hox gene expression in the developing limb, and that this zone marks the position in which the wrist or ankle will develop. Experiments deleting the 3′ and 5′ control regions in turn show that only deletion of the entire control region completely abolishes Hoxd gene expression in the digital plate, which highlights the complexity of gene regulation. In addition, as we saw for the expression of Hoxa cluster genes along the main axis (see Section 4.13), Hoxa13 (and Hoxd13) proteins antagonize transcription of the more-3′ genes in the cluster, reinforcing the switch in expression from the 3′ to the 5′ enhancers.

As a result of the two phases of Hox gene expression, the proximo-distal pattern of expression of Hoxa and Hoxd genes eventually comes to correspond broadly with the three main proximo-distal regions of the limb (see Fig. 9.35). The patterns of expression of these Hox genes are generally similar in the forelimb and hindlimb, although there are a few differences. In general, *Hoxa9* and *Hoxd9* are expressed in the proximal part of the limb, *Hoxa11* and *Hoxd11* are expressed in the intermediate part, and *Hoxa13* and *Hoxd13* are expressed in the digital plate. The phenotypes of mice mutant for genes in the same paralogous subgroup show that the different Hox genes influence the formation of the structures that develop from the regions of the limb in which they are expressed (Fig. 9.37).

For example, the *Hoxa10* and *Hoxd10* genes control the development of proximal elements—humerus or femur—whereas *Hoxa11* and *Hoxd11* genes control formation of the radius and ulna in the forelimb (tibia and fibula in the hindlimb). Inactivation of Hox13 paralogs disrupts digit development, as one might expect from the expression of *Hoxd13* and *Hoxa13* in the distal region of the limb.

482 Chapter 9 Organogenesis

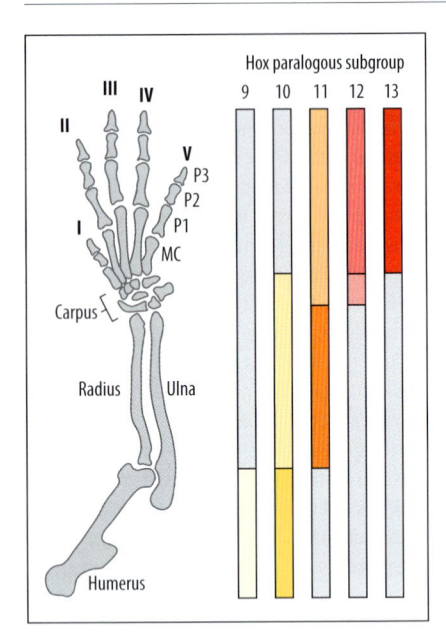

Fig. 9.37 Functional domains of Hox gene expression in the patterning of the mouse forelimb. Colored regions denote regions of Hox gene influence on the proximo-distal limb pattern, as determined from gene knock-out experiments. Hox9 (pale yellow) and Hox10 (yellow) paralogs function together for the upper forelimb (humerus); Hox10 paralogs also affect the middle forelimb (radius and ulna), as represented by the lighter yellow shading. Hox11 paralogs (orange) are mostly involved in formation of this middle region, with some influence on the wrist and 'hand' region (lighter orange shading). According to this analysis, Hox12 paralogs (pink) predominantly pattern the wrist, although Hoxd gene activity is low here, and Hox13 paralogs (red) predominantly pattern the hand. MC, metacarpal; P, phalanx.
Adapted from Wellik, D.M., Capecchi, M.R.: **Hox10 and Hox11 genes are required to globally pattern the mammalian skeleton.** *Science 2003, **301**: 363-367.*

Different combinations of Hox genes are expressed across the antero-posterior axis in the digital plate, and the dosage of Hoxd and Hoxa proteins is lower anteriorly. Only *Hoxd13* is expressed in the region that will give rise to the thumb and controls its development.

How these patterns of Hox genes are interpreted to give the specific anatomies of the different regions of the limb is still not understood. Among the known gene targets, for example, are genes that encode proteins involved in cell adhesion, but exactly how they might be involved is not yet known.

Mutations in the human HOX genes are known to underlie conditions in which there are limb differences. Mutations in *HOXD13* lead to fusion of digits and some types of polydactyly, whereas mutations in *HOXA13* lead to hand–foot–genital syndrome, in which the thumb and big toe are short. In both conditions, it is the digits of the limbs that are affected.

9.20 Self-organization might be involved in the development of the limb

Although positional information is a fundamental mechanism in limb pattern formation, the developing limb also has some capacity for self-organization (self-organization is defined in Box 7B). This was first revealed by experimental manipulations of chick limb buds. Cells from a disaggregated chick limb bud can be reaggregated and will develop digits, even in the absence of a polarizing region. In this experiment, mesodermal cells disaggregated from early chick limb buds are thoroughly mixed to disperse the polarizing region, or the mesodermal cells are taken from just the anterior half of a limb bud, which does not contain the polarizing region. The disaggregated cells are placed in an ectodermal jacket (prepared as described in Section 9.17), and the 'recombinant' limb bud is grafted to a site such as the dorsal surface of an older limb, where it can acquire a blood supply but which does not provide polarizing signals. Limb-like structures develop from these recombinant limb buds, even though there is no polarizing region. Figure 9.38 shows a limb that developed from a recombinant limb bud of

Fig. 9.38 Reaggregated limb bud cells form digits in the absence of a localized polarizing region. Mesodermal cells from a chick leg bud are disaggregated, mixed to disperse all the cells, including the polarizing region, and then reaggregated, placed in an ectodermal jacket, and grafted to a neutral site. The photograph shows the results of the graft: well-formed toes develop distally.

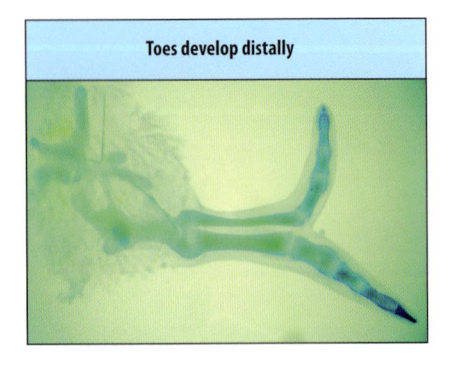

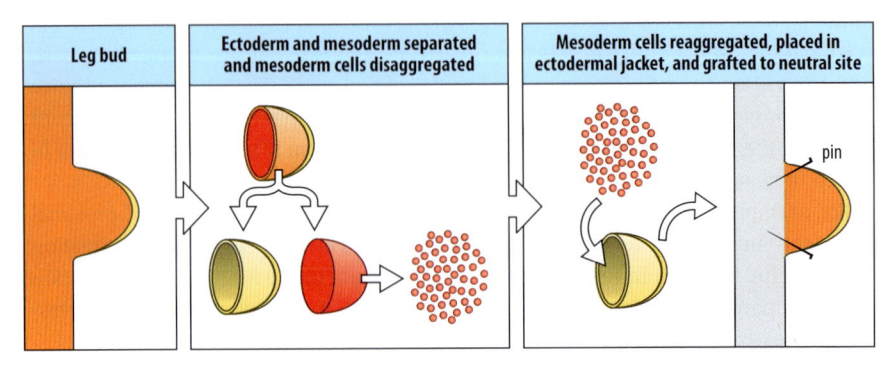

reaggregated cells from chick leg buds. Several long cartilaginous elements have formed in the more proximal regions of the limb, although none of these elements can be easily identified with normal structures. More distally, however, the reaggregated hindlimb-bud cells have formed identifiable toes. The fact that well-formed cartilaginous elements can develop at all in the absence of a discrete polarizing region shows that the bud has some capacity for self-organization. Even recombinant limb buds made by reaggregating limb-bud mesodermal cells that have been grown in culture for several days and then stuffed into an ectodermal jacket can produce some limb-like structures.

These results suggest that the limb bud might be able to generate a basic prepattern consisting of cartilaginous elements that are all equivalent to each other in that they have no distinct positional identity. A **prepattern** is a basic organization that is generated autonomously in the embryo and can be identified in advance of the later development of a similar pattern of structures. The pattern that eventually develops does not have to follow the prepattern exactly but can be a modified form of it. For example, the cartilaginous elements of the limb prepattern could be given their positional identities and further refined in response to positional information that involves signals such as Shh and the subsequent activation of Hox genes and other downstream gene targets (see Sections 9.16 and 9.19).

The mechanism for generating the prepattern of a series of repeated structures, such as the digits, could be based on reaction–diffusion. For example, if a reaction–diffusion mechanism generated a periodic pattern of cartilage condensations across the digital plate, merely widening the limb bud by some small developmental accident without altering the periodicity would enable a further condensation to form (Box 9E). This is indeed what is seen in mouse *Gli3* mutants, which lack antero-posterior positional information and have broadened limb buds with many unpatterned digits (see Section 9.16). When the Hox genes that are expressed in the digital plate (*Hoxa13* and *Hoxd11–Hoxd13*) are progressively deleted in the *Gli3* mutants, the number of digits progressively increases, although the size of the digital plate remains the same. In *Gli3*-mutant mice in which *Hoxd11–13* are completely deleted and only one copy of *Hoxa13* remains, the limb has around 12–14 thin digits crammed into the digital plate. This finding was initially unexpected, but the idea now is that in normal development Hox genes regulate the reaction–diffusion mechanism, thereby ensuring that the correct number of digits are formed. The components of the reaction–diffusion system that generates the periodic pattern of equivalent cartilaginous elements have been suggested to be BMPs, Wnts, and the transcription factor Sox9.

9.21 Limb muscle is patterned by the connective tissue

So far, we have considered pattern formation in the limb in terms of the skeleton, but the limb also has a complex pattern of muscles. There are, for example, 13 named muscles in the forearm of the chick wing, some of which are visible in the cross-section in Figure 9.39. When a polarizing region is grafted to the anterior margin of a chick wing bud (or beads soaked in retinoic acid or Shh are implanted in the anterior margin) not only are the cartilage elements distal to the elbow duplicated but also the muscles and tendons. This suggests that the antero-posterior pattern of the muscles and tendons is specified in the same way as the antero-posterior pattern of the cartilage elements; that is, by the cells that will form the individual muscles and tendons acquiring positional values. But it turns out to be more complicated in the case of the muscles. Each individual muscle is composed of bundles of differentiated muscle fibers wrapped up in connective tissues. The two tissues have separate developmental origins. The cells that give rise to the myogenic cells of the limb muscles and differentiate into the muscle fibers migrate into the limb bud from the somites, and the cells of the muscle connective tissues arise from the lateral plate mesoderm.

The origin of the limb myogenic cells in the somites was shown by making chick–quail chimeras (see Section 3.4). If the somites opposite the site where the wing bud

BOX 9E Reaction-diffusion mechanisms

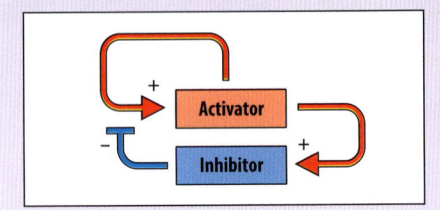

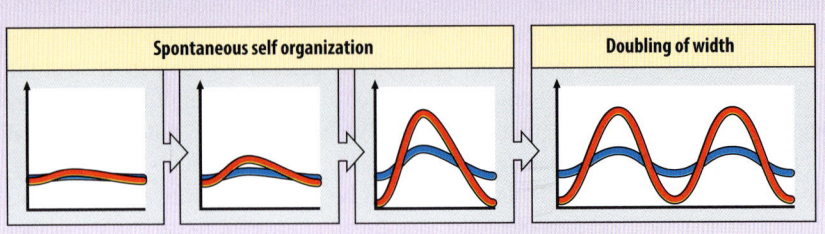

Fig. 1

Fig. 2 Spontaneous self organization — Doubling of width

Fig. 3
From Meinhardt, H.: **Turing's theory of morphogenesis of 1952 and the subsequent discovery of the crucial role of local self-enhancement and long-range inhibition.** Interface Focus 2012, **2**: 407–416.

Fig. 4
From Sheth, R., et al.: **Hox genes regulate digit patterning by controlling the wavelength of a Turing-type mechanism.** Science 2012, **338**: 1476–1480.

Sox9 expression in the developing mouse hand-plate: Normal | $Hoxd11$–$13^{-/-}$, $Gli3^{-/-}$

Some chemical systems, such as the famous Belousov–Zhabotinsky reaction, are self-organizing and spontaneously generate spatial patterns of concentration of their components. In two dimensions, the initial distribution of the molecules is uniform, but over time the system forms expanding wave-like patterns. The essential features of such self-organizing systems can be mathematically modeled by chemical reactions between two or more molecules with different rates of diffusion. Such systems are called **reaction-diffusion** systems or **Turing mechanisms** (after the mathematician Alan Turing, who was the first to prove that simple repetitive patterns could arise in this way).

For example, consider a closed two-component system of diffusible activator and inhibitor molecules in which the activator molecule stimulates both its own synthesis and that of the inhibitor, which, in turn, inhibits synthesis of the activator (Fig. 1), and the inhibitor diffuses faster than the activator. A type of lateral inhibition will occur such that synthesis of activator becomes confined to one region, forming a peak of activator concentration with a given wavelength. If the wavelength stays the same when the size of the system is increased, two peaks will eventually develop, then three, and so on, as the system continues to grow (Fig. 2). In three dimensions, a pattern of peaks of high concentration of activator can develop like that in Figure 3. If the system stays the same size, a shortening of the wavelength is needed to generate more peaks.

Reaction-diffusion interests biologists because, given the appropriate conditions and components, it could, in theory, generate periodic patterns such as patterns of repeated structures like the sepals and petals of flowers and the spots and stripes on animals.

The limb has considerable capacity for self-organization, and it has been suggested that reaction-diffusion mechanisms could generate the periodic pattern of skeletal elements, such as the series of digits. In the digital plate of mouse embryos (Fig. 4, left panel), five *Sox9*-expressing stripes across the antero-posterior axis fan out from the proximal wrist region towards the wider distal tip of the digital plate and prefigure the five digits. In a reaction-diffusion system, the stripes would represent high concentrations of activator, and the fan-like pattern would be accomplished by changing the wavelength of stripe spacing across the antero-posterior axis so that it is longer distally than proximally, thus preventing the stripes splitting into two at the wider tip of the digital plate. Another, more recent, model has suggested that the reaction-diffusion mechanism generates an initial periodic pattern of spots of *Sox9* expression which mark where digits will form, and that other mechanisms are involved in producing the subsequent stripes.

In $Gli3^{-/-}$ mouse limb buds that lack antero-posterior positional information (see Section 9.16), the digital plate is broader and, as predicted in a reaction-diffusion system, an increased number of *Sox9*-expressing stripes is formed, which also fan out from the wrist as in the normal limb. When the dose of Hox genes expressed in the digital plate of $Gli3^{-/-}$ mouse embryos is reduced, many more, thinner, *Sox9*-expressing stripes are produced, which are all crammed into the same-sized digital plate, and these may split into two distally (Fig. 4, right panel). This pattern of stripes matches the activator pattern predicted computationally for a reaction-diffusion system in which Hox genes modulate inhibitor production and thus shorten the wavelength of the peaks of activator. The interacting protein components of the Turing gene regulatory network in the digital plate have been suggested to be BMPs, Sox9, and Wnts.

will develop in the chick embryo are replaced with those from a quail embryo, the myogenic cells of the wing that forms are of quail origin, but the connective tissue cells associated with the muscles are of chick origin. Further transplantation experiments showed that if somites are grafted from the future neck region of the early chick embryo to replace the wing-level somites, a normal pattern of limb muscles still develops. This indicates that the myoblasts that will give rise to the myogenic cells of the muscles are equivalent and do not have a distinct positional identity. The muscle pattern is determined by the connective tissue into which the myoblasts migrate, rather than by the myoblasts themselves acquiring positional values.

The initial migration of myoblasts into the limb is restricted so that they enter either the dorsal or the ventral region; once in the limb, myoblasts do not cross the dorsal–ventral compartment boundary, which is defined by the dorsal expression of the transcription factor Lmx1b (see Fig. 9.33). The prospective muscle-associated connective tissue might have surface or adhesive properties that the myoblasts recognize. If the pattern of adhesiveness in the connective tissue changed over time, this could result in associated changes in the migration of the myoblasts. The myoblasts then multiply and initially form two blocks of presumptive muscle—the dorsal muscle mass and the ventral muscle mass (Fig. 9.39). The cells in these blocks then undergo a series of reorganizations and differentiate to give rise to bundles of muscle fibers that are the templates for the named muscles of the limb. Muscle bundle formation involves orientation of differentiating muscle fibers, followed by clustering of the fibers and finally compaction. Most individual muscle bundles are formed during the clustering and compaction of initially large domains of oriented fibers. A few muscle bundles result from the splitting of compacted bundles. These changes in muscle cell organization are mediated by modifications of the extracellular matrix by the associated connective tissue cells.

Hoxa11 is expressed in myoblasts entering the limb, whereas cells in the posterior region of the dorsal and ventral muscle masses express *Hoxa13*. These patterns of Hox gene expression differ from those in the surrounding connective tissue and could signify that the myoblasts have by now acquired their own positional values. In the ventral muscle mass, Shh directly controls myogenic differentiation, and in the absence of Shh signaling in the developing limbs of mouse embryos, ventral paw muscles do not develop.

The myoblasts proliferating in the limb express *Pax3*. When *Pax3* is downregulated, cell proliferation stops and the cells differentiate. Myogenesis is initiated in the muscle masses by the MyoD family of transcription factors (see Section 6.6). Signals from the overlying ectoderm, including BMP-4, prevent premature differentiation. The development of tendons is marked by the expression of the transcription factor Scleraxis.

9.22 The initial development of cartilage, muscles, and tendons is autonomous

As we saw in Section 9.21, the patterns of cartilaginous elements, tendons, and muscles in the limb are all specified by the same set of signals that provide positional information to the cells of the limb bud derived from the lateral plate mesoderm. But there appears to be little or no communication between these different elements—that is, the specification of each element is autonomous. If, for example, the tip of an early chick wing bud is removed and grafted to the flank of a host embryo, it initially develops into normal distal structures with a wrist and three digits. The long tendon that normally runs along the ventral surface of digit 3 differentiates, even though both its proximal end and the muscle to which it attaches are absent. The tendon later degenerates, however, because it does not make the necessary connection to a muscle, and so is not put under tension. The tendons in the proximal part of the limb behave differently. Although they are also specified autonomously and start to develop in the absence of muscles, they depend on attachment to muscle to differentiate.

The mechanism whereby the correct connections between tendons, muscles, and cartilage are established has still to be determined. It is clear, however, that there is

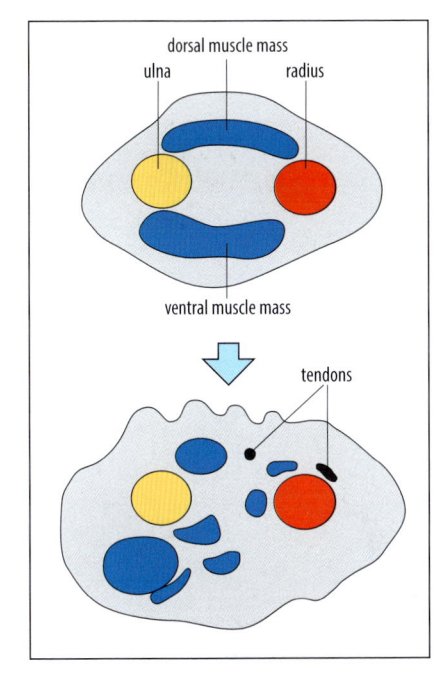

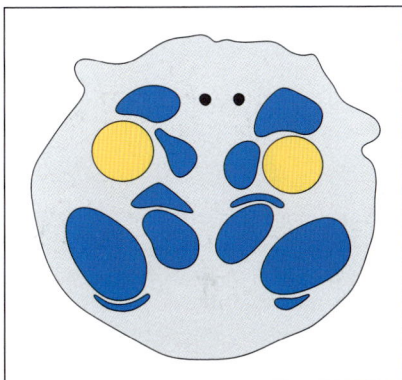

Fig. 9.39 Development of muscle in the chick wing. A cross-section through the chick wing in the region of the radius and ulna shortly after cartilage formation shows presumptive muscle cells present as two blocks—the dorsal muscle mass and the ventral muscle mass (upper panel). The cells in these blocks then undergo a series of changes in organization and differentiate to give rise to individual muscles. A polarizing-region graft to the anterior margin of a chick wing bud causes the formation of a mirror-image pattern of muscles (lower panel). Tendons are indicated in black.

From Shellswell, G.B., Wolpert, L.: **The pattern of muscle and tendon development in the chick wing.** *In Limb and Somite Morphogenesis. Ede, D.A., et al. (Eds). Cambridge: Cambridge University Press, 1977: 71-86.*

little or no specificity involved in making such connections; if the tip of a developing limb is inverted dorso-ventrally, dorsal and ventral tendons can join up with inappropriate muscles and tendons. They are promiscuous and simply make connections with those muscles and tendons nearest to their free ends.

9.23 Joint formation involves secreted signals and mechanical stimuli

The elbow/knee joints are specified by positional information in the same way as the cartilage elements. The periodic pattern of the joints in the digits, however, might involve a Turing-like reaction–diffusion mechanism (see Box 9E). An early event in joint formation is the formation of an 'interzone' at the site of the prospective joint, in which the cartilage-producing cells become fibroblast-like and cease producing the cartilage matrix, thus separating one cartilage element from another (Fig. 9.40). At this time, genes for Wnt-9A (previously known as Wnt-14) and the BMP-related protein GDF-5 are expressed in the prospective joint region.

Ectopic expression of Wnt-9A can induce the early steps in joint formation, and mice lacking GDF-5 have some of their joints missing. Indian hedgehog signaling is also required for joint formation. A little later, the musculature begins to function and muscle contraction occurs. At this point, high levels of hyaluronan, a component of the proteoglycans that lubricate joint surfaces, are made by the cells of the interzone; hyaluronan is secreted in vesicles that coalesce to form the joint cavity, which separates the articulating surfaces of the joint from each other.

In chick embryos in which muscle activity has been inhibited by drugs, the joints fuse, indicating that joint formation depends on the mechanical stimulation that results from muscle contractions. Muscle contraction promotes the synthesis of hyaluronan. Cells isolated from the interzone region and grown in culture respond to mechanical stimulation by increased secretion of hyaluronan when the substratum to which they are attached is stretched. In mouse mutants with muscles that are unable to contract or in mutants that lack limb muscles, some, but not all, of the limb joints are fused. In the affected joints, Wnt signaling is reduced during the earliest stages in their formation, and the joint cells differentiate into cartilage instead of undergoing their normal program of differentiation to form joint tissues. Restricted fetal movements in humans, for example, due to neuromuscular disorders or low volume of amniotic fluid can result in joint fusions and other changes in skeletal development. The very rare human condition arthrogryposis, in which joints are bent or angled differently, is associated with a reduction in, or absence of, fetal movements.

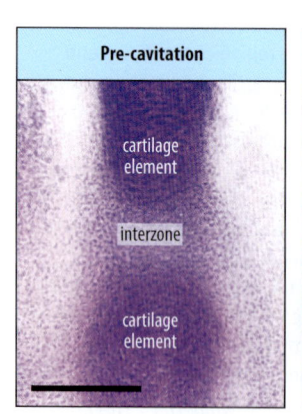

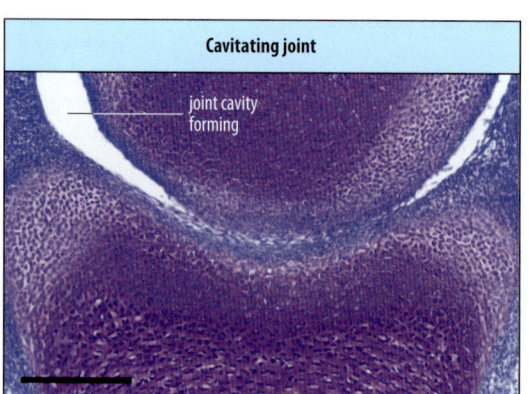

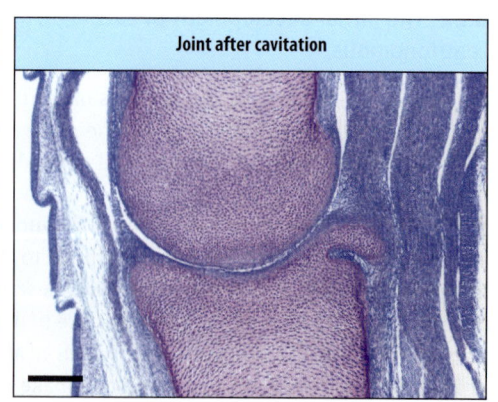

Fig. 9.40 Formation of a joint. Sections through a developing proximal interphalangeal joint in the limb of a chick embryo stained with toluidine blue, showing different stages in development. The appearance of the interzone region of fibroblast-like cells between two cartilaginous skeletal elements is an early morphological sign of joint formation (left panel) and cells in this region express Wnt-9A and GDF-5. The cavity of the joint (middle panel) begins to form as interzone cells secrete hyaluronan. The completely cavitated joint (right panel) has a continuous cavity between the articulating surfaces of the two cartilage elements. Scale bar = 50 μm.
Photographs courtesy of A.S. Pollard and A.A. Pitsillides.

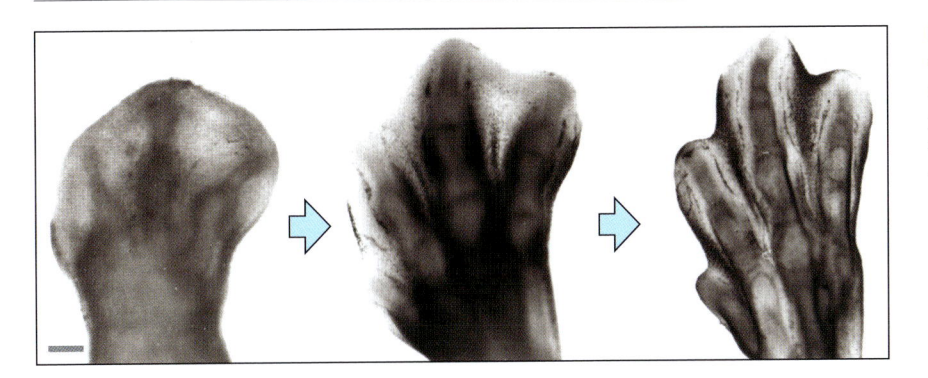

Fig. 9.41 Cell death during leg development in the chick. Programmed cell death in the interdigital region results in separation of the toes. The dying cells are the tiny dark speckles clearly visible between the developing digits in the middle and right images. Scale bar = 1 mm.

Photograph reproduced with permission from Garcia-Martinez, V., et al.: **Internucleosomal DNA fragmentation and programmed cell death (apoptosis) in the interdigital tissue of embryonic chick leg bud.** J. Cell Sci. *1993, 106: 201-208.*

9.24 Separation of the digits is the result of programmed cell death

Programmed cell death by apoptosis has a key role in molding the form of chick and mammalian limbs, especially the digits. Separation of the digits depends on the death of the cells in the interdigital regions between the developing cartilaginous elements (Fig. 9.41). BMPs are involved in signaling this cell death. If the functioning of BMP receptors is blocked in the developing chick leg, cell death does not occur and the digits are webbed. Perhaps not surprisingly, other signals are also involved. For example, FGF-8 from the overlying ectoderm functions as a survival signal for interdigital cells, and it is the balance between cell death and survival signals that ensures the appropriate extent and location of cell death.

Programmed cell death is a normal part of development. The fact that ducks and other waterfowl have webbed feet, whereas chickens do not, is simply the result of less cell death between the digits of waterfowl. When chick-leg mesoderm is recombined with duck-leg ectoderm, cell death between the digits occurs as in the chick (Fig. 9.42). It is, therefore, the mesoderm that determines the patterns of cell death. In amphibians, digit separation is not due to cell death, but results from the digits growing more than the interdigital regions.

Programmed cell death also occurs in other regions of the developing limb, such as the anterior margin of the limb bud, and between the radius and ulna. In the chick wing bud, there is substantial cell death in the polarizing region. Indeed, it was the investigation of the control of cell death in this region of a chick wing bud, by transplanting it to the anterior margin of another chick wing bud, that led to the discovery that it acts as a signaling center. One suggestion for the role of cell death in the polarizing region is that it controls the number of cells expressing *Shh*.

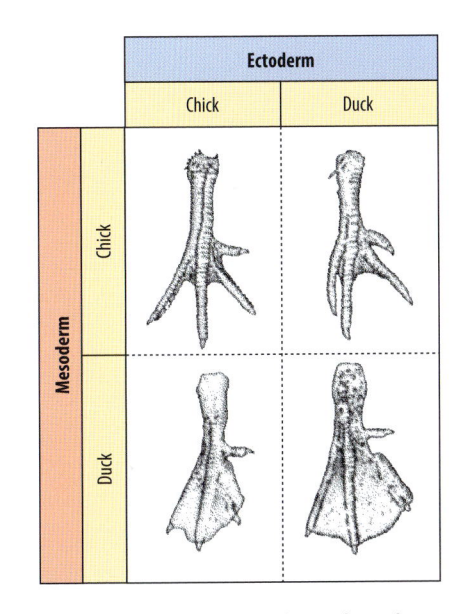

Fig. 9.42 The mesoderm determines the pattern of cell death. In the developing webbed feet of ducks and other water birds, there is less cell death between the digits than in birds without webbed feet. When the mesoderm and ectoderm of embryonic chick and duck limb buds are exchanged, webbing develops only when duck mesoderm is present.

Based on Kieny, M., Patou, M.-P.: **Rôle du mêsodernie dans l'évolution morphogénétique de la membrane interdigitale du Poulet et du Canard.** C. R. Acad. Sci. Paris Ser. D *1967, 246: 3030-3033.*

SUMMARY

Patterning of the vertebrate limb is largely carried out by cell–cell signaling that provides the cells with positional information. The induction of expression of *Tbx4* and *Tbx5* genes that initiate development of the hindlimbs and forelimbs, respectively, is determined by a combinatorial Hox code in the lateral plate mesoderm. The antero-posterior polarity of the forelimbs is also determined by Hox genes. Another transcription factor, Pitx1, determines hindlimb identity. Two key signaling regions are established within the limb bud. One is the apical ectodermal ridge, which is required for limb-bud outgrowth; the second is the polarizing region at the posterior margin of the limb bud, which controls development along the antero-posterior axis. The signals from the apical ridge are FGFs and are essential for limb-bud outgrowth, which involves oriented cell movements and cell divisions. Positional values along the proximo-distal axis are specified by opposing gradients of retinoic acid and FGFs together with an intrinsic timing mechanism. Shh protein is produced by the polarizing region and is graded across the antero-posterior axis of the limb bud. This gradient, together with a timing mechanism based on the duration of Shh signaling, is thought to determine the identities of the digits. An underlying prepattern based on reaction-diffusion could generate the periodic pattern of the condensations of cells that form the digits. Pattern along the dorso-ventral axis is specified by the ectoderm with Wnt-7a produced by dorsal ectoderm specifying dorsal structures.

Signaling along each of the three limb axes is integrated by a gene regulatory network that involves interlinked feedback loops. Hox genes are expressed in complex spatio-temporal patterns within the limb bud and control regional identity. Hox genes later regulate digit spacing in the digital plate. The myogenic cells of the limb muscles are not generated within the limb but migrate in from the somites and are patterned by the limb connective tissue. The development of joints involves local production of signal molecules and requires mechanical stimuli. In birds and mammals, separation of the digits is achieved by programmed cell death.

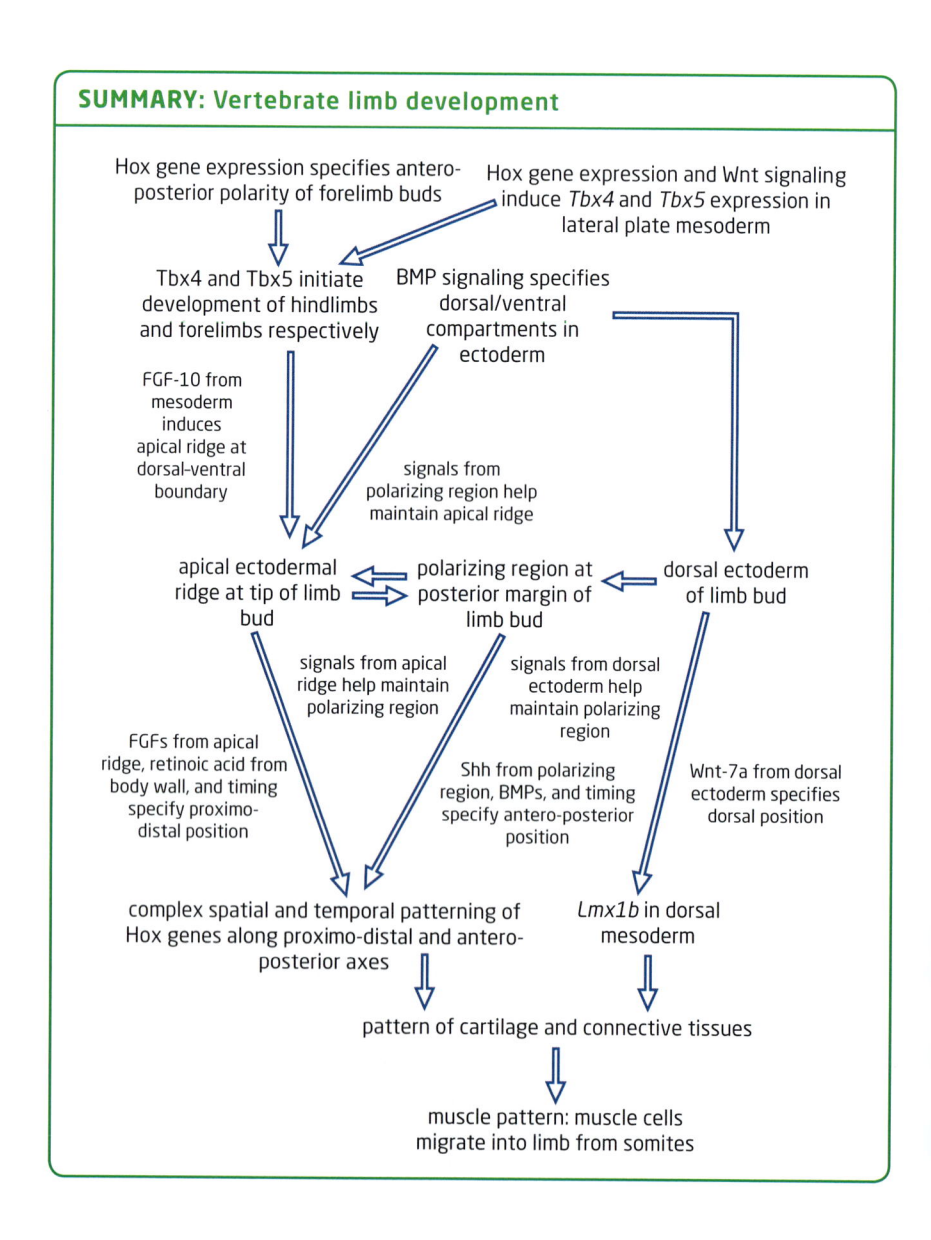

SUMMARY: Vertebrate limb development

Hox gene expression specifies antero-posterior polarity of forelimb buds
↓
Hox gene expression and Wnt signaling induce *Tbx4* and *Tbx5* expression in lateral plate mesoderm
↓
Tbx4 and Tbx5 initiate development of hindlimbs and forelimbs respectively

BMP signaling specifies dorsal/ventral compartments in ectoderm

FGF-10 from mesoderm induces apical ridge at dorsal-ventral boundary

signals from polarizing region help maintain apical ridge

apical ectodermal ridge at tip of limb bud ⇌ polarizing region at posterior margin of limb bud ← dorsal ectoderm of limb bud

signals from apical ridge help maintain polarizing region

signals from dorsal ectoderm help maintain polarizing region

FGFs from apical ridge, retinoic acid from body wall, and timing specify proximo-distal position

Shh from polarizing region, BMPs, and timing specify antero-posterior position

Wnt-7a from dorsal ectoderm specifies dorsal position

complex spatial and temporal patterning of Hox genes along proximo-distal and antero-posterior axes

Lmx1b in dorsal mesoderm
↓
pattern of cartilage and connective tissues
↓
muscle pattern: muscle cells migrate into limb from somites

Teeth

We next consider the development of mammalian teeth, which has significant similarities to the development of the limbs. Both teeth and limbs represent a series of repeated organs. As we have seen for the limbs, the key issues are how position, identity, and number are determined. Within a tooth or a limb itself, a series of repeated structures arise—cusps and digits, respectively—and the same issues apply to them. The development of a tooth, like a limb, involves epithelial–mesenchymal interactions and many of the same signal molecules are involved. In a tooth, a well-defined series of reciprocal interactions takes place between the oral epithelium, which, in mammals, is of ectodermal origin, and the mesenchyme of the facial primordia, which is of neural crest origin.

The first sign of tooth development is a horseshoe-shaped band of thickened oral epithelium called the **dental lamina**. The development of individual teeth in the dental lamina is then initiated at specific positions along it by the formation of **placodes**, discrete regions of thickened epithelium. A transient signaling center known as the initiation knot develops within the placodes of incisors and molars and this regulates proliferation of the neighboring epithelial placodal cells, which then invaginate to form a bud. The epithelial bud is known as the enamel organ. The surrounding mesenchyme—the dental mesenchyme—condenses around the epithelial bud to form a tooth primordium (Fig. 9.43). In each enamel organ, a specialized group of cells—the primary enamel knot—arises at the tip. The primary enamel knot acts as a signaling center and regulates the growth and folding of the inner enamel epithelium so that it forms cap and bell shapes. In multi-cusped teeth, such as molars, the primary enamel knot determines the positions of secondary enamel knots, which mark the tips of the future cusps. It has been suggested that the cusp patterns of different teeth are generated by reaction–diffusion mechanisms (see Box 9E). The mesenchymal component of the tooth primordium, known as the **dental papilla**, determines tooth identity. During cell differentiation, the inner epithelial cells of the enamel organ differentiate into **ameloblasts** and secrete the enamel, whereas the underlying mesenchyme cells differentiate into **odontoblasts** and secrete dentin. The dental papilla also gives rise to the dental pulp, which contains nerves and blood vessels. The mouse is used as the model for human tooth development, although it has a simpler dentition with just two types of teeth—incisors and molars—and has only a single set of teeth in its lifetime.

The pattern of the permanent human dentition in each half of the jaw (from anterior to posterior) consists of two incisors, one canine, two premolars, and three molars (Fig. 9.43, upper left photograph). The pattern of the mouse dentition consists of one incisor and three molars separated by a region in which no teeth form (upper right photograph). The lower panels of Figure 9.43 show stages in initiation of tooth development, formation of the dental placode and its growth and morphogenesis, differentiation of epithelial and mesenchymal cells to produce enamel and dentin, respectively, and the final stages of eruption.

9.25 Tooth development involves epithelial-mesenchymal interactions and a homeobox gene code specifies tooth identity

The basic pattern of the dentition is set up before any external signs of tooth formation. The oral epithelium produces signals that establish a spatial pattern of homeobox gene expression in the facial mesenchyme that provides a code for regional identity. We saw in Section 9.19 that, in a similar way, the spatial pattern of Hox gene expression controls regional identity in the limb.

We will consider the development of the teeth in the lower jaw of a mouse embryo, as our example. The lower jaw arises from the fused mandibular primordia. The lateral region of each mandibular primordium will give rise to the posterior region

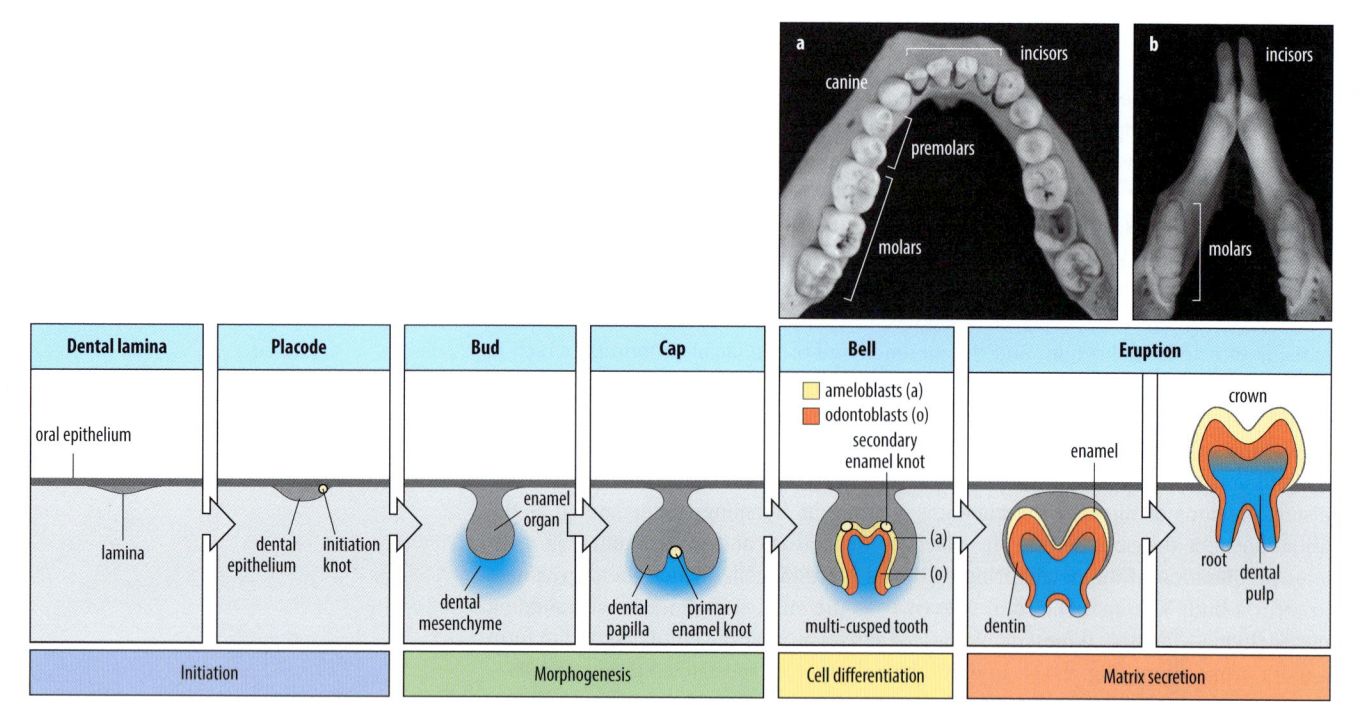

Fig. 9.43 The human and mouse dentition and the stages of tooth development. The photographs show (a) the upper dentition of an adult human and (b) the upper dentition of the mouse. Premolars and canines are absent in the mouse. The diagram outlines the stages of tooth development. The first sign of tooth formation is a thickening of the oral epithelium in the tooth-forming region to form the dental lamina. The initiation of development of an individual tooth is marked by formation of a dental placode within the dental lamina (at E11.5 in mouse). An initiation knot arises in the placode and drives the invagination of the dental epithelium to form a bud—the enamel organ. The dental mesenchyme condenses around the bud. The epithelium then extends further into the mesenchymal tissue and starts to enclose the condensing mesenchyme—the dental papilla—to form the cap; the primary enamel knot is formed. During the bell stage (E18.5 in mouse), cusp patterns emerge, and the crown of the tooth is formed. In multi-cusped mammalian teeth, secondary enamel knots form at the positions of future cusps. Final growth and matrix secretion follows (postnatally in mouse), during which time the inner enamel epithelium differentiates into ameloblasts, which produce enamel, and the underlying mesenchymal cells differentiate into odontoblasts that secrete dentin. The tooth erupts through the surface of the jaw. The pulp in the center of the tooth is derived from the dental papilla and contains nerves and blood vessels. Roots continue to develop during eruption.

Reproduced with permission from Jussila, M., Thesleff, I.: **Signaling networks regulating tooth organogenesis and regeneration, and the specification of dental mesenchymal and epithelial cell lineages.** *Cold Spring Harb. Perspect. Biol. 2012, **4**: a0084245. Modified according to Thesleff, I.:* **Current understanding of the process of tooth formation: transfer from the laboratory to the clinic.** *Aust. Dent. J. 2014, **59** Suppl 1: 48-54.*

of the future jaw, where the molars are located; the central regions, where the primordia fuse, will give rise to the anterior region of the future jaw, where the incisors are located. *Fgf8* is expressed in the oral epithelium in the lateral regions and *Bmp4* is expressed in the oral epithelium in the central regions. FGF signaling positively regulates the expression of *Barx1* and *Dlx2* which encode homeodomain transcription factors in the underlying lateral mesenchyme. BMP-4 signaling positively regulates the expression of *Msx1* and *Msx2* which encode two related homeodomain transcription factors in the underlying central mesenchyme, while at the same time negatively regulating *Barx1* expression. *Islet1* which encodes another homeodomain transcription is expressed exclusively in the oral epithelium in the central region and has a positive reciprocal interaction with *Bmp4*. We have already come across *Islet1* through its involvement in hindlimb initiation (see Section 9.11) and we will meet it again in neural development. *Lhx6* and *Lhx7* which encode Lim homeodomain transcription factors are also expressed in the mesenchyme throughout the oral half of the mandibular primordia, and their expression marks the region where teeth will form (Fig. 9.44).

Fig. 9.44 The expression domains of homeobox genes in the lower jaw before initiation of tooth primordia. Schematic diagram of a cut-away section of the fused mandibular primordia of a mouse embryo, viewed from the back of the oral cavity. The lateral regions of the mandibular primordia will give rise to the posterior regions of the lower jaws, while the central region will give rise to the anterior region. *Barx1* and *Dlx2* expression are induced in the lateral mesenchyme by FGF signaling from the overlying epithelium and expression is confined to this region by inhibitory BMP signaling from central oral ectoderm. *Msx1* and *Msx2* expression in the central mesenchyme is induced by BMP signaling from overlying epithelium. *Lhx6* and *Lhx7* are expressed throughout the mesenchyme of the oral half of the mandibular primordia. *Islet1* is expressed in the central oral epithelium. Expression of the *Islet1* and *Bmp4* genes is maintained by mutual positive interactions. Dotted circles indicate positions in which incisors and molars will develop. Homeobox gene code for incisor: *Msx1, Msx2, Lhx6, Lhx7*. Code for molar: *Barx1, Dlx2, Lhx6, Lhx7*. Arrows indicate positive signaling; barred lines indicate inhibition.

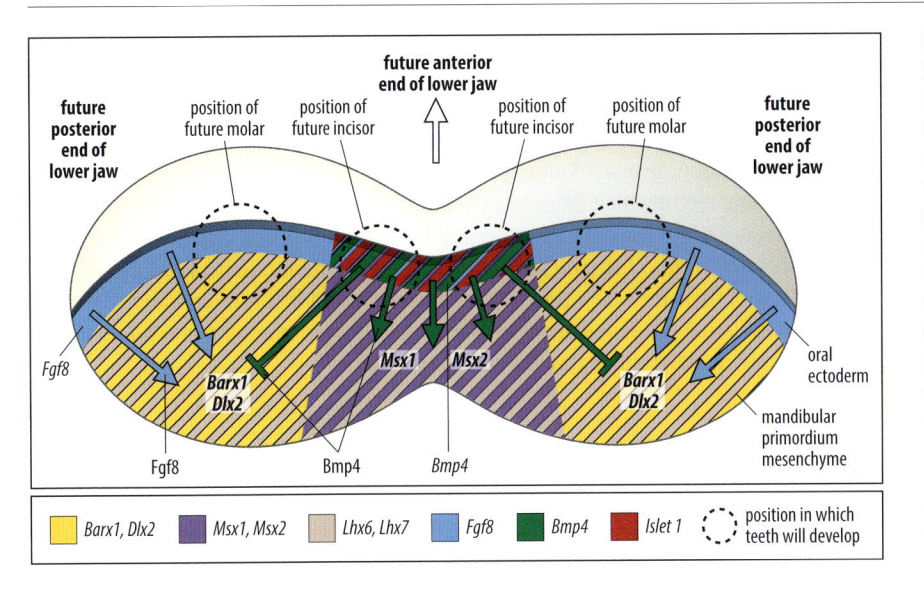

Dlx2, Barx1, Lhx6, and *Lhx7* are expressed in mesenchyme cells that will form molars, whereas *Msx1, Msx2, Lhx6*, and *Lhx7* are expressed in mesenchyme cells that will form incisors. The importance of these transcription factors in determining tooth identity was shown by experiments in which beads soaked in the BMP antagonist Noggin were placed in cultured explants of the presumptive incisor region of a mouse mandibular primordium. This resulted in loss of *Msx1* expression and promotion of *Barx1* expression; on further culture a molar tooth developed rather than an incisor. In mouse mutants lacking both *Dlx2* and *Dlx1* (a similar gene that can substitute for *Dlx2* in development and is also expressed in the facial mesenchyme), molars do not develop in the upper jaw, whereas the incisors are unaffected.

Early tooth development is controlled by reciprocal interactions between the oral epithelium and the underlying mesenchyme. The same families of signal molecules—Wnt, FGF, Hedgehog, and the transforming growth factor (TGF)-β family that includes BMPs—are involved as in the limb. *Shh* is expressed in the dental lamina, and later becomes restricted to the initiation knots in the placodes, where the Shh protein acts as a mitogen to induce local proliferation of the epithelium. Many Wnt ligands are also expressed in the oral epithelium and Wnt signaling is essential for placode formation. Dental placodes fail to form in mutant mice overexpressing the gene for the Wnt antagonist Dkk1, whereas activation of Wnt signaling can induce additional placodes in the dental lamina, which then develop into supernumerary teeth. BMP-4 and FGF-8 signaling by the oral epithelium not only induces position-dependent patterns of expression of transcription factors in the facial mesenchyme, but also induces expression of the genes for signal molecules that include BMP-4, FGFs, and Wnts. At this stage, the direction of signal changes, so that the signals secreted by the mesenchyme direct the development of the epithelium and lead to the formation of the enamel knot that controls tooth morphogenesis. The enamel knot produces signals including Shh, FGFs, Wnts, and BMPs. FGFs stimulate the neighboring epithelium to proliferate, whereas the enamel knot cells themselves do not proliferate, and this shapes the developing tooth.

Developing mouse teeth have the capacity for self-organization. Functional teeth have been generated by recombining aggregates of epithelial cells and mesenchymal cells disaggregated from embryonic tooth primordia, culturing the reconstituted tooth primordia in organ culture to allow epithelial–mesenchymal interactions to take place, and then transplanting them to adult mice to continue development. Such bioengineered teeth can form functional teeth when transplanted into the oral cavity.

Functional teeth have even been produced using cells from adult tissues in place of one of the populations of cells from embryonic tooth primordia. This is a step towards making bioengineered cells for use in dentistry. But it has not yet been possible to generate teeth entirely from adult cells. If this challenge could be overcome and suitable sources of cells from people identified, this would fulfill the ambition of being able to replace missing teeth, which dates back centuries. The anatomist John Hunter, a leading exponent of tooth transplantation in the eighteenth century, obtained living teeth for transplantation into his patients by advertising for donors who were prepared to sell him one of their teeth at a price. Other approaches to modern reparative dentistry include stimulating resident stem cells in adult teeth to differentiate into odontoblasts which secrete new dentin so that the tooth repairs itself.

> **SUMMARY**
>
> Teeth develop from buds of oral epithelium and the underlying neural crest-derived mesenchyme. A homeobox gene code in the mesenchyme specifies tooth identity, and the further development of a tooth is controlled by reciprocal signaling between epithelium and mesenchyme.

Branching organs: the vertebrate lung and the mammalian mammary gland

Like teeth, many internal organs in vertebrates—such as the lungs and mammary glands—develop from epithelial buds. But in these organs, unlike teeth, the buds grow out and undergo branching morphogenesis to give a tree-like system of epithelial tubules. Some general principles of branching morphogenesis are illustrated in Chapter 5 in relation to the development of vertebrate blood vessels and the insect tracheal system.

The position in which the buds develop and initiate organ and gland formation, is determined by the mesoderm as part of the laying down of the vertebrate body plan. Thus, as in the developing limb, the mesoderm carries the positional information that determines where organs develop.

In this part of the chapter we will look at the lungs and mammary glands as examples of organs that develop by branching morphogenesis. Reciprocal interactions between the epithelial bud and its associated mesoderm are required for outgrowth of the buds in these organs and determine their branching morphogenesis. In the lungs, the epithelial bud is derived from the endoderm, whereas in the mammary gland it is derived from the ectoderm. We first consider how lung development is initiated and branching morphogenesis occurs. We then examine development of the mammary gland and note that understanding of the development of the embryonic gland sheds light on how breast cancers form (Box 9F).

9.26 The vertebrate lung develops from buds of endoderm

The paired lungs of vertebrates arise from the foregut endoderm. The trachea is formed by subdivision of the early foregut along the middle of the dorso-ventral axis, and the first sign of lung development is the formation of two primary lung buds, which form, independently of the trachea, by budding of the ventral foregut (Fig. 9.45). The buds then grow out to form the primary bronchi. The lung is one of the internal organs that has left–right asymmetry and this becomes evident very early in its development. In the mouse, the primary lung bud on the left gives rise to one primary

MEDICAL BOX 9F What developmental biology can tell us about cancer

It has long been recognized that fundamental developmental processes, such as growth, are instrumental in the formation of tumors. As we shall see in Chapter 11, most cancers originate from a single cell that has mutations in genes that control cell proliferation. Tumors also produce factors that stimulate angiogenesis, and unless a tumor becomes vascularized it is unable to grow larger than about 1–2 mm^3.

A hallmark of malignant tumors is their invasiveness. Cells from the tumor invade the surrounding tissues either as cords of cells or single cells. They can gain access to blood vessels and lymph vessels and undergo metastasis, resulting in the formation of secondary tumors at sites far from the site of the original tumor. Invasiveness is also a characteristic of the branching morphogenesis seen in many developing organs, such as the lung and the mammary gland.

Most tumors are carcinomas—tumors that originate in epithelia. Reciprocal interactions between an epithelium and the surrounding mesenchyme are striking features of organ development, and these interactions are mediated by signal proteins. So it is not surprising that some key signal proteins with roles in cancer were first discovered as developmentally important genes, for example *hedgehog* in *Drosophila*. Conversely, other signal proteins important in development were first discovered in tumors, for example FGF-4 in human stomach tumors. The Wnt family of signal proteins illustrates how these two routes of discovery converged (see Box 3E). The *wingless* gene in *Drosophila* is homologous to the vertebrate gene *Int-1*, which is located at a site in the mouse genome where integration of a virus causes cancer. Developmentally important transcription factors that determine cell behavior are also relevant to cancer. Examples are the transcription factors Slug and Snail, which regulate epithelial-to-mesenchymal transitions, which are involved in both embryonic development and in cancer metastasis.

Figure 1 compares mechanisms involved in the development of the mammary gland and in the formation of breast cancers. Breast cancer is the most common cancer in the United Kingdom and the United States. Most breast cancers are of epithelial origin and can invade and metastasize. One special feature common to both the developing gland and breast cancer is hormonal regulation.

The data on the developing mammary gland in Figure 1 comes from studies in mouse embryos. For the future, we need to obtain similar information for the human mammary gland. This knowledge gap is being addressed by studying mammary gland organoids grown from human cells, and this will enable direct comparisons between the developing gland and breast cancer.

Comparison between features of the developing mouse mammary gland and breast cancers		
Features in common	**Developing mouse mammary gland**	**Breast cancers**
Invasiveness	Invasive branching of solid cords or buds of epithelial cells occurs during normal embryonic and postnatal development. Invasion is driven by proliferation and directed by mechanical constraints.	Invasiveness can involve cords of cancer cells and collective cell migration, but can switch to invasion by single cells generated by an epithelial-to-mesenchymal transition regulated by Snail and Slug transcription factors.
	Gland lumen formed in invasive branches in embryo by fusion of small crevices (microlumina) between epithelial cells.	Similar arrangement of microlumina (cribriform pattern) seen in some breast cancers.
Epithelial–mesenchymal cell interactions	Reciprocal signaling between epithelium and mesenchyme at all stages in gland development.	Mammary mesenchyme from embryonic glands can "normalize" mouse mammary carcinoma cells.
		Breast cancer cells can stimulate mesenchymal cells in stroma (connective tissue associated with tumor) to produce cytokines that enhance the invasive behavior of cancer cells
Signal molecules and transcription factors	FGFs, Wnts, Neuroregulin-3 involved in gland initiation and at later stages in embryonic gland together with BMPs.	FGFs and Wnts overexpressed in some breast cancers.
	FGF-10 involved in branching in postnatal gland.	Neuroregulin-3 signals through ErbB4, which heterodimerizes with ErbB2. ERBB2 amplified in some breast cancers and is a therapeutic target.
	Tbx3 essential for mammary gland development; Hox genes have multiple inputs.	TBX3 overexpressed in some breast cancers Changes in Hox gene expression detected in tumors. Breast tumors have different patterns of Hox gene expression from those shared by tumors of endodermal origin (e.g. colon, prostate, and lung).
Hormonal regulation	Parathyroid hormone-related peptide involved in embryonic gland; testosterone in males.	Growth is hormone-dependent in three-quarters of breast cancers.
	Estrogen and growth hormone involved in elaboration of postnatal gland in puberty; progesterone in pregnancy.	
Epithelial cell differentiation	Cells in the mammary placode are already determined to form the luminal cells of the adult gland. Epithelial cell lineages have been traced in embryonic and adult gland. Stem cells and cell types have been characterized, most recently by single-cell transcriptomics.	Characterization of different cell types and epithelial cell lineages provide molecular markers that can reveal the cellular origin of the tumor and extent of differentiation. This is important information for prognosis and treatment.

Fig. 1

494 Chapter 9 Organogenesis

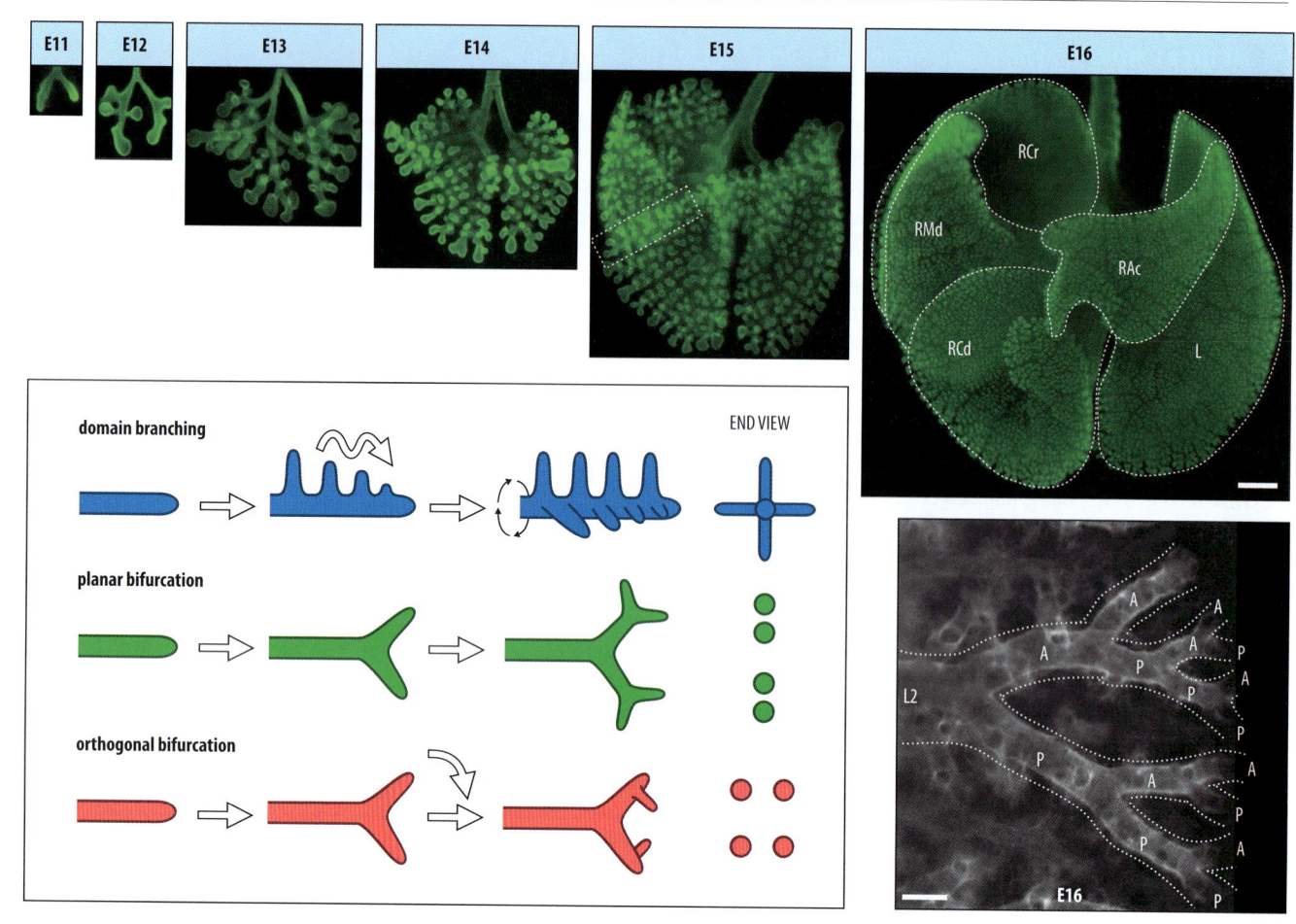

Fig. 9.45 Morphogenesis of the mouse lung involves outgrowth and branching of the primary lung buds. The lung airway epithelium develops by successive branching of the primary lung buds in response to signals from the surrounding mesenchyme. The upper panels show whole mounts of lungs (ventral view) at different embryonic stages immunostained for E-cadherin to show airway epithelium. L, left lobe; RAc, accessory lobe; RCr, right cranial lobe; RCd, right caudal lobe; RMd, right middle lobe. Scale bar = 500 μm. Three basic modes of branching (left lower panel) are used repeatedly and in combination. The simplest of these is planar bifurcation, shown here in the lung of an E16 mouse embryo (right lower panel). L2, lineage from the initial left branch. Sequential branching occurs in the same plane. A, anterior branches; P, posterior branches. Scale bar = 100 μm.

Images kindly provided by R. Metzger from Metzger, R.J., et al.: **The branching programme of lung development.** *Nature* 2008, **453**: 745-750.

bronchus, whereas the bud on the right branches to give rise to four primary bronchi (see Fig. 9.45). The number of primary bronchi determines the number of lobes. The primary bronchi then undergo extensive budding and branching to form an arborized system of conducting airways known in mice as bronchioles. These become finer and finer and eventually give rise to millions of thin-walled sacs known as alveoli. The alveoli are intimately associated with a highly branched system of blood capillaries which develops at the same time.

The lung epithelium is patterned along the proximo-distal axis from bronchi to alveoli. The epithelial cells lining the conducting airways differentiate into multi-ciliated and mucus secretory cells. The epithelial cells lining the alveoli differentiate either into cells that secrete surfactant or into cells that are flattened and specialized for gas exchange across their surfaces: oxygen is taken up by the blood and carbon dioxide is released back into the lung. During all stages of development, the lung mesoderm interacts with the lung epithelium to promote morphogenesis and patterning. The mesoderm also gives rise to many cell types associated with the airways, including the supporting cartilage and smooth muscle and to specialized fibroblasts in the alveoli.

Fig. 9.46 Interactions between mesoderm and endoderm specify the region of ventral foregut endoderm that will form the lung. Cut-away cross-section through endoderm and the splanchnopleural mesoderm in the region that will form the lung. Retinoic acid signaling in the mesoderm specifies positional information along the antero-posterior axis and induces expression of vertebrate *Hedgehog* genes in the foregut endoderm, which produce Hedgehog signal proteins that, in turn, signal back to the mesoderm. In response to Hedgehog signaling via the Gli2/3 transcriptional effectors, the ventral mesoderm expresses Wnt and BMP that then pattern the foregut endoderm along the dorso-ventral axis. Wnt signaling promotes Nkx2.1 expression in ventral foregut endoderm (arrow), while BMP signaling represses the inhibitory action of the transcription factor Sox2 which specifies dorsal endoderm (not shown). Retinoic acid signaling is also required for foregut endoderm to respond to these signals (broken arrow). Solid arrows indicate main sets of interactions. In *Xenopus*, the vertebrate *Hedgehog* genes expressed in the foregut endoderm are *Shh* and *Dhh*, while in mice *Shh* and *Ihh* are expressed (see Box 9C).

Based on Rankin, S.A., et al.: *A retinoic acid-hedgehog cascade coordinates mesoderm-inducing signals and endoderm competence during lung specification.* Cell Rep. 2016, **16**: 66–78.

The blood vessels associated with the alveoli arise from the trunk vasculature by angiogenesis (see Section 5.19).

The region of the ventral foregut that will give rise to the lungs is specified by Wnt and BMP signals from the adjacent splanchnopleural lateral plate mesoderm (Fig. 9.46). The local production of these signals is the result of a cascade of interactions between lateral plate mesoderm and endoderm, with retinoic acid produced by the mesoderm first patterning the endoderm along the antero-posterior axis into discrete regions and specifying the foregut region in which the lung will develop. Retinoic acid then induces expression of vertebrate *Hedgehog* genes including *Shh* in the foregut endoderm and these produce Hedgehog signal proteins, which, in turn, signal back to the mesoderm. In response to Hedgehog signaling, the ventral mesoderm produces Wnt and BMP signals. Wnt signaling promotes *Nkx2.1* expression in the ventral foregut endoderm, whereas BMP signaling represses the inhibitory action of the transcription factor Sox2, which specifies dorsal endoderm. Retinoic acid signaling is also required for the ventral foregut endoderm to respond to the Wnt and BMP signals. This gene regulatory network involved in specification of foregut endoderm to form the lungs was worked out in *Xenopus* embryos. Essentially the same network is involved in mammals. Nkx2.1 is a marker for respiratory endoderm, and *Nkx2.1*-null mice have a rudimentary bronchial tree, as well as defects in the thyroid and forebrain, where *Nkx2.1* is also expressed. As initiation of lung development still takes place in the absence of Nkx2.1, this suggests that other transcription factors are involved in specifying lung identity.

The dissection of this complex sequence and interplay of signals involved in specifying the ventral foregut endoderm has been crucial for devising differentiation protocols for producing lung tissues from induced pluripotent stem cells derived from human patients (see Chapter 7). Single-cell transcriptomic analysis is providing more detailed information about the signaling networks and cell types involved in lung specification. This information will improve these protocols for producing models for human lung development and monitoring the cell types they contain.

9.27 Morphogenesis of the lung involves three modes of branching

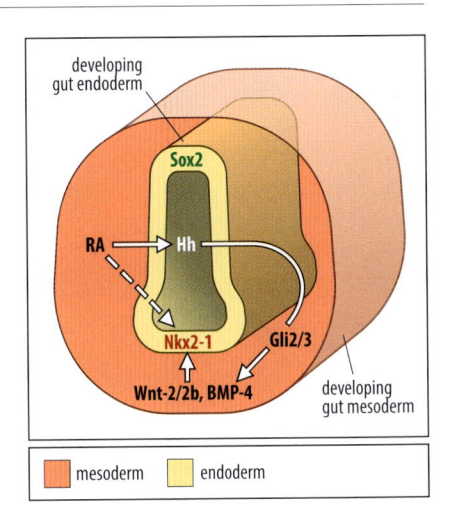

The pattern of tubule branching in the developing mouse lung has been analyzed in some detail and is highly stereotyped—every lung branches in the same pattern with very little variation between individuals or mouse strains. Just three modes of branching, used in different combinations and at different times, can account for the complex tree-like system of tubules that develops (see Fig. 9.45). The initial sprouting and outgrowth of the primary lung buds from the foregut depends on signals from the surrounding mesenchyme. The main driver of this sprouting is secretion of FGF-10 which is expressed specifically in the surrounding mesoderm cells. FGF-10-deficient mice fail to form bronchi although the trachea still develops. FGF-10 interacts with the receptor FGFR2b expressed by the lung epithelial cells and has been implicated in further branching of the bronchi and lung tubules. Bifurcation involves buckling of the lung epithelium which is constrained by the associated layer of smooth muscle.

Activation of FGFR2b induces expression of the *Sprouty* gene in the lung epithelial cells. *Sprouty* was first identified by a mutation in *Drosophila* that causes many more branches than normal to form in the tracheal system (see Section 5.18). Sprouty protein is controlled by FGF signaling and, in turn, antagonizes FGF activity. In the lung, it is thought that this negative-feedback loop could help to prevent the cells in the main tube away from the tip of the tubule from forming branches.

Another signal protein essential for lung branching morphogenesis is Shh, which is expressed in the endodermal cells at the tips of the extending tubules in response to FGF-10 signaling. Shh, in turn, inhibits expression of the gene *Fgf10* in the mesoderm. A reaction–diffusion model (see Box 9E) based on these FGF–Shh interactions can mimic the bifurcation mode of lung branching. BMPs, Wnts, Notch, and retinoic acid are also components of the complex network of interacting signal molecules at the tip of each extending tubule that organizes outgrowth and branching.

Unlike the insect tracheal system and the vertebrate vasculature, outgrowth of the bronchial tubules in the vertebrate lung is due to cell proliferation, which is greatest towards the tips of the advancing tubules. Wnt-5a, which signals via the planar polarity pathway (see Box 5C), is expressed at the tips of the lung tubules. Mice lacking Wnt-5a have a shortened trachea and although the two primary buds generate the correct number of lobes, there is excess branching of distal tubules. We saw that Wnt-5a controls outgrowth of the limb by controlling oriented cell behavior (see Section 9.13) and it may control lung-tubule outgrowth in a similar way, although this has yet to be shown. In mouse lungs, the differentiated airway smooth muscle controls airway diameter.

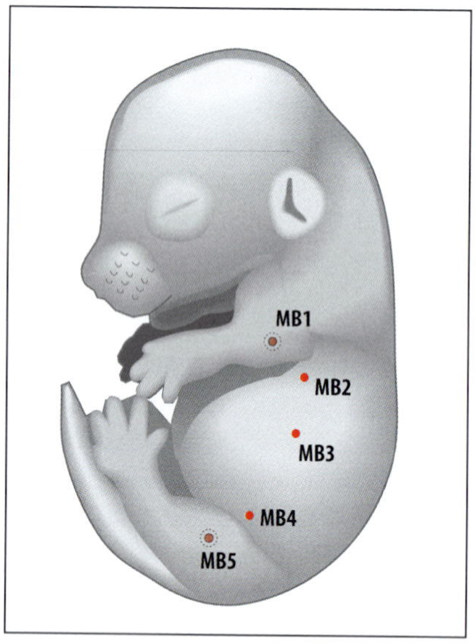

9.28 The mammary glands develop from buds of ectoderm

Studies of mice have furnished most of our current knowledge about how breasts develop. In the embryo, the mammary glands form from distinct ectodermal placodes at precise positions along mammary 'lines', which run antero-posteriorly along either side of the ventral region of the body. Five placodes form on either side—three in the armpit and anterior region, and two in the groin (Fig. 9.47). (In humans, typically only one gland develops on each side, although extra nipples sometimes develop elsewhere along the mammary line.) The epithelial placodes sink into the underlying lateral plate mesenchyme to form a mammary bud (MB). As the bud sinks deeper to form the mammary bulb, it remains connected to the ectoderm by a stalk and becomes surrounded by circumferentially oriented mesenchyme, known as the mammary mesenchyme. In female mice, the mammary bulb then elongates to form a sprout that invades the fat pad where, under its influence, it undergoes branching morphogenesis (Fig. 9.48). At the same time, the lumen of the gland begins to develop through the fusion of many microlumina—small crevices that appear within the branches because non-adhesive surfaces form on the epithelial cells. Cell death also occurs, so that the ducts of the gland are lined by two layers of cells: an inner luminal cell layer and a basal myoepithelial layer. At birth, the gland consists of a tree-like structure of branching ducts terminating in solid end buds consisting of luminal cells surrounded by a layer of myoepithelial cells. Unlike the lung, branching of the mammary gland is stochastic and there is a random pattern of branches. During puberty, the terminal end

Fig. 9.47 The location of the mammary buds in a mouse embryo. Five mammary buds (MBs) develop on each side of the embryo. MB1 and MB5 (ringed with dashed circles) are, in fact, located behind the limbs, in the 'armpit' and 'groin', respectively. Mouse embryo (E12.5) viewed from left side.

Modified from Veltmaat, J.M., Mailleux, A.A., Thiery, J.P., Bellusci, S.: **Mouse embryonic mammogenesis as a model for the molecular regulation of pattern formation.** Differentiation *2003,* **71***: 1-17.*

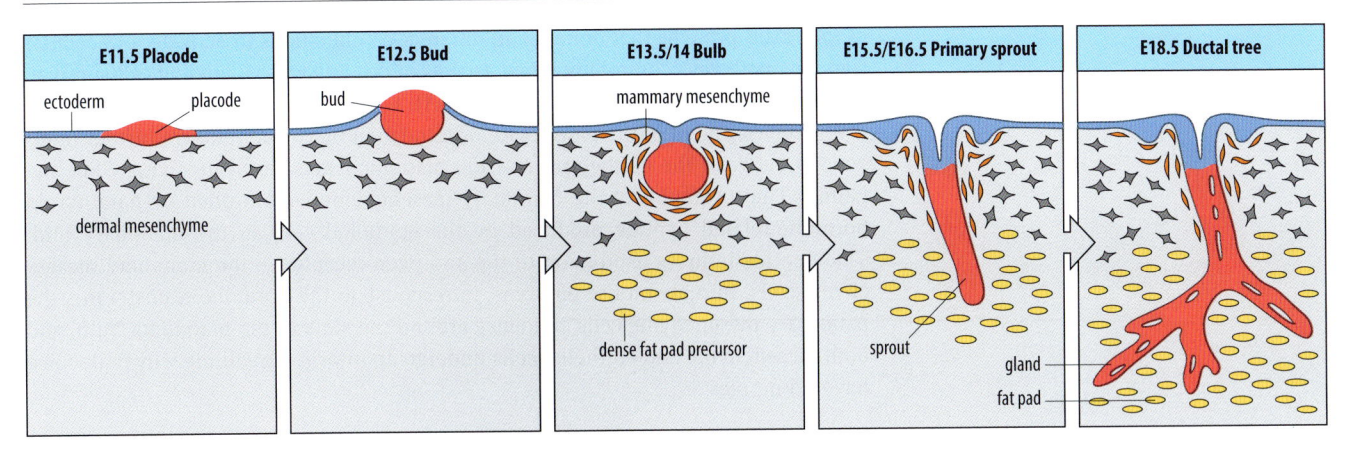

| E11.5 Placode | E12.5 Bud | E13.5/14 Bulb | E15.5/E16.5 Primary sprout | E18.5 Ductal tree |

Fig. 9.48 Schematic of mammary gland development in a mouse embryo. The diagram outlines the stages in embryonic development of a mouse mammary gland. At E11.5 the ectodermal placode forms. By E12.5 the placode has formed a bud which sinks into the underlying mesenchyme. By E13.5/14 the bulb has sunk deeper into the mesenchyme but remains attached to the ectoderm by a stalk and the mammary mesenchyme associated with the gland can be distinguished. In later stages of development in female mice, the bulb first elongates to form a primary sprout (E15.5/E16.5) which then begins to branch and develop a lumen (E18.5).

*Modified from Veltmaat, J.M., Mailleux, A.A., Thiery, J.P., Bellusci, S.: **Mouse embryonic mammogenesis as a model for the molecular regulation of pattern formation.** Differentiation 2003, **71**: 1-17.*

buds undergo a further burst of branching, and in pregnancy, the luminal epithelium expands to form alveoli which produce milk.

The antero-posterior extent of the mammary line in the ectoderm is determined by the domain of expression of Hox genes of paralogous subgroup 8 in the underlying mesoderm. Later, *Hox8* and *Hox9* genes are expressed in the mammary mesenchyme. Differences can be detected between the pairs of mammary glands in the mouse that arise in different positions along the mammary line, suggesting that the glands have positional identities. In addition, subtle differences in development between glands in the same pair are probably a reflection of the fundamental left–right asymmetry of the body plan which is specified very early in development and is most clearly seen in internal organs. There are also consistent reports of a slightly higher incidence of cancer in the left breast, although the basis for this is unknown. Box 9F compares mechanisms involved in the development of the mouse mammary gland and in the formation of breast cancers.

Key signals that initiate placode formation are members of the Wnt and FGF families. They induce expression of the gene encoding the transcription factor, Tbx3, which is essential for mammary gland development. *Tbx3* is expressed in the mesoderm underlying the mammary line and later in the epithelial cells of the mammary gland throughout embryonic development. *Tbx3* is also expressed in developing limbs. Mutations in human *TBX3* cause mammary–ulnar syndrome, in which both the arms and the mammary glands are affected. Hox genes and *Tbx3* are also expressed in adult mouse mammary glands. Hox9 genes, for instance, are required for the further development of the gland during pregnancy.

Another important signaling pathway in mammary gland initiation was discovered from studies of a strain of mice which lack the third mammary bud but sometimes have additional mammary glands in abnormal positions. Researchers dubbed the gene responsible *Scaramanga*, after the villain with an extra nipple in the James Bond film *The Man with the Golden Gun*. *Scaramanga* was found to encode Neuregulin-3, a growth factor that is secreted by the mesoderm along with the other signal molecules involved in mammary gland initiation. In the *Scaramanga* strain of mice, *Neuregulin3* expression is reduced in the lateral plate mesoderm in the region where mammary glands usually form. Neuregulin-3 binds to a receptor tyrosine kinase, ErbB4—one of four ErbB proteins that dimerize in different combinations to activate intracellular signaling pathways controlling many aspects of cell behavior, including cell survival, proliferation, and differentiation (see Box 4A).

The development of the placode to form the mammary bud requires further interactions between the epithelium and the mammary mesoderm involving BMP signaling. The formation of the mammary mesenchyme is controlled by parathyroid hormone-related peptide produced by the epithelial cells in the mammary bud. This peptide induces expression of the androgen receptor in the mammary mesenchyme cells. In male mouse embryos, in response to the hormone testosterone, the mammary mesenchyme cells condense around the stalk of the mammary bulb, and both mesenchyme and epithelial cells undergo apoptosis, explaining why male mice have no nipples.

9.29 The branching pattern of the mammary gland epithelium is determined by the mesenchyme

As with other glands and organs, the process of branching morphogenesis is controlled by signaling back and forth between mesenchyme and epithelium. The crucial influence of the mesenchyme on branching in the embryonic mammary gland was highlighted by a classic experiment in which the epithelial sprout of a mammary gland was combined with the mesenchyme associated with a developing salivary gland. The branching patterns of the epithelium in the mammary gland and the salivary gland are quite different, and mammary gland epithelium combined with salivary mesenchyme branched like a salivary gland.

Invasion—penetration into surrounding tissue—is a normal feature of mammary gland branching. The embryonic mammary bud proliferates to form an elongated sprout that then invades the surrounding tissue. Invasion might involve the process of directed dilation, which is driven by proliferation and directed by the mechanical forces exerted on expanding tissue. The initial branches in the embryonic gland invade as solid cords of epithelial cells. Later, in postnatal development, invasive branching is driven by the growth of the luminal epithelial cells in the terminal end buds at the tips of the ducts and is shaped by constraints imposed by the basal layer of myoepithelial cells. The hormones estrogen and growth hormone regulate branching of the adolescent gland. Progesterone regulates branching of the adult gland. In postnatal development, FGF-10 is produced by the mesenchyme associated with end buds and is involved in their branching, as in the developing lung.

Tissue-recombination experiments show that not only is the branching pattern dictated by the mesenchyme but also that the cells of the mouse mammary placode are already programmed to differentiate into luminal cells that can produce milk. Under suitable conditions, cells transplanted from the embryonic mammary gland into the fat pad can self-organize and reconstitute a complete gland. Lineage-tracing experiments in mouse embryos have identified bipotential stem cells in the mammary bud that can give rise to both luminal and myoepithelial cells, and a complete mouse mammary gland has been reconstituted from just one cell of an adult gland. However, this could be due to the plasticity of cell differentiation (see Chapter 6) rather than to the existence of multipotent stem cells in the adult. Instead, lineage-tracing experiments in the adult gland suggest that it contains unipotent stem cells that give rise to just luminal or just myoepithelial cells. Genes that are expressed by cells of the different lineages have been identified, and as in the developing lung, single-cell transcriptomics is providing a wealth of new information about the different cell types in both developing and adult mammary glands. Such analyses are defining cell types not only in the mammary gland epithelium but also in the associated connective tissues.

> **SUMMARY**
>
> The development of both the vertebrate lung and the mammalian mammary gland is initiated by mesodermal signals. These signals lead to a series of interactions between mesoderm and endoderm in lung development, and between mesoderm and ectoderm in mammary gland development. Both organs are composed of branched epithelial tubes and the branching pattern develops in response to local cues from the surrounding mesodermal cells. The molecules involved in regulating branching morphogenesis of the embryonic mammalian lung and the postnatal mammary gland are the same as those involved in the early branching of the *Drosophila* larval respiratory system. In the lung, branching is accomplished by buckling of the epithelial lining, whereas in the embryonic and postnatal mammary glands, branching is accomplished by growth of solid cords of epithelial cells.

The vertebrate heart

The vascular system—the heart, blood vessels, and blood cells—is the first organ system to develop in vertebrate embryos. This early development is not surprising, because oxygen and nutrients need to be delivered to the rapidly developing tissues and subsequently to developing organs. The development of the vascular system involves not only extensive branching morphogenesis, but also the formation of tubes by a mesenchymal-to-epithelial transition. In Chapter 5 we discussed the parallels between the development of the vertebrate vascular system and the tracheal system in insects, which have the same function of conveying oxygen to the tissues but very different developmental origins. Here we consider how the vertebrate heart arises from a tube established ventral to the foregut.

9.30 The development of the vertebrate heart involves morphogenesis and patterning of a mesodermal tube

The vertebrate heart is mainly of mesodermal origin and is first established by ventral midline fusion of the two arms of the cardiac crescent, also known as the cardiogenic crescent. The cardiac crescent is composed of cardiogenic (heart) cells located in the splanchnopleural lateral plate mesoderm underlying the head folds on either side of the body. The cells in the two arms of the crescent move towards the midline and form a tube (Fig. 9.49). Mutations have been found in zebrafish that disrupt this process and result in two laterally positioned hearts—a condition known as cardia bifida. One of the mutated genes is called *miles apart*, and codes for a receptor that binds lyso-sphingolipids; sphingosine 1-phosphate is the likely ligand in this case. *miles apart* is not expressed in the migrating heart cells themselves, but in cells on either side of the midline, and thus may direct the migration of the heart cells.

The linear heart tube resulting from fusion of the cardiac crescent consists of two layers of cells—the inner endocardium, which is a sheet of endothelium, and the outer myocardium, which will become the contractile cardiac muscle cells. These two cell layers are separated by a layer of extracellular matrix, known as cardiac jelly, produced by the myocardium. A third layer of cells, called the epicardium, develops later, as an epithelial sheet covering the myocardium. Once formed, the primitive cardiac tube soon becomes connected to blood vessels and, in human embryos, starts to pump blood by day 20. The dynamics of blood flow is an important factor in regulating subsequent stages in heart morphogenesis. The heart tube then elongates and different regions of the tube which will give rise to the chambers of the heart are mapped

500 Chapter 9 Organogenesis

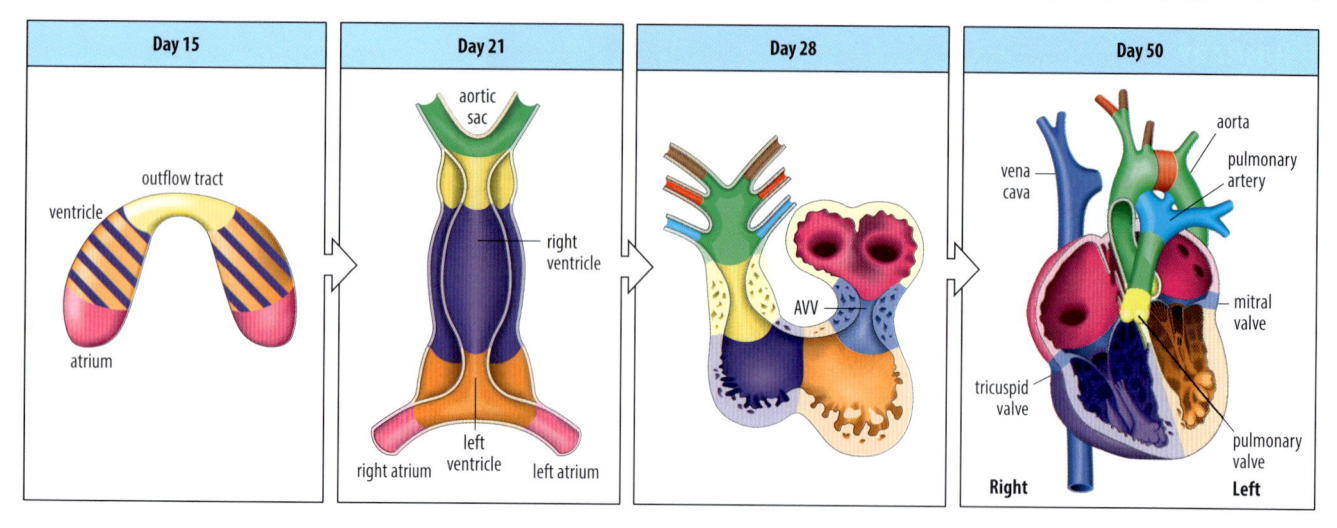

Fig. 9.49 Schematic of human heart development. By day 15 of human embryonic development, cardiogenic cells have formed a crescent, as shown in the first panel. The two arms of the crescent fuse along the midline to give a linear heart tube, which elongates and becomes patterned along the antero-posterior axis so that the regions and chambers of the mature heart are mapped out (second panel). After looping (third panel), these regions are disposed approximately in their eventual positions. Later development results in further patterning (fourth panel), and the formation of septa and valves between, for example, the atria and ventricles. AVV, atrioventricular valve region.
After Srivastava, D., Olson, E.N.: **A genetic blueprint for cardiac development.** *Nature 2000, **407**: 221–226.*

out (see Fig. 9.49, second panel). Next, the tube changes shape and undergoes a complex morphogenetic process known as looping, which brings the regions of the heart tube that will form the future chambers of the heart into their proper positions (see Fig. 9.49, third panel). Later development involves separation of the chambers by formation of septa and valves between, for example, the atria and ventricles (see Fig. 9.49, fourth panel). The valves control the direction of blood flow and stop the blood flowing backwards.

Heart looping normally occurs to the right and is related to the left–right asymmetry of the internal organs of the embryo (see Fig. 4.50). The basis for this asymmetric looping is not well understood. The extracellular matrix molecule flectin is expressed earlier on the left side compared with the right, and experiments in chick embryos suggest that this might be the result of the expression of the transcription factor Pitx2 on the left.

A two-chambered heart is the basic adult form in fish, but in birds and mammals, further partitioning of atrium and ventricle gives rise to a four-chambered heart. Nevertheless, zebrafish have proved to be very useful in mutant screens to uncover the genetics of heart development, because their embryos and larvae do not require a beating heart to survive. In humans, about 8 in 1000 live-born infants have a congenital heart abnormality; *in utero*, heart abnormalities lead to death of the embryo in between 5% and 10% of conceptions (the different numbers reflect different studies, and it could even be as high as 30%).

9.31 Several different cell populations contribute to making a heart

Heart mesoderm in *Xenopus* is initially specified by signals from the organizer during the process of mesoderm induction and dorso-ventral patterning. In chick and mouse embryos, cells that will eventually become heart cells ingress through the primitive streak and become part of the lateral plate mesoderm. A few hours after ingression, the cells have become committed to being heart cells and will differentiate into cardiac muscle cells if isolated.

Fig. 9.50 The populations of cells that make up a heart. In the earliest stage of heart development (E7.5), illustrated here in the mouse embryo, different populations of cardiogenic cells form the cardiac crescent (red) and the second heart field (green), which extend across the ventral midline on either side. The two arms of the crescent fuse in the midline below the forming foregut to make a simple tube that then begins to loop (E8). Cells of the second heart field lie all along the length of the heart tube and extend the tube by contributing to both arterial (dark green) and venous (purple) poles (E8.5). In addition, cardiac neural crest cells (yellow) migrate from the pharyngeal arches to the arterial pole. The pro-epicardial organ (blue) forms near the venous pole (E9.5) and migrates to form the epicardium that covers the entire heart. The contributions of the various cell populations to the chambers of the looped heart tube are indicated (E10.5). By E14.5, the heart chambers have formed and the heart is mature. AA, aortic arch; Ao, aorta; EC, endocardial cushion; IVC, inferior caval vein; IVS, interventricular septum; LA, left atrium; LV, left ventricle; OFT, outflow tract; PT, pulmonary trunk; PV, pulmonary vein; RA, right atrium; RV, right ventricle; SVC, superior caval vein.

From Vincent, S.D., Buckingham, M.: *How to make a heart: the origin and regulation of cardiac progenitor cells.* Curr. Top. Dev. Biol. 2010, **90**: 1–41.

Lineage tracing of cells in the early heart region in mouse embryos revealed the existence of two distinct lineages of heart cells that are derived from a common precursor very early in development at, or before, gastrulation. These cell lineages both differentiate into heart muscle. The first lineage to differentiate gives rise to the myocardial cells of the linear heart tube and later to the primitive left ventricle, whereas the other lineage is the source of the myocardial cells of the outflow tract and the right ventricle. The atria are colonized by both lineages. The first differentiating myocardial cells are found in the cardiac crescent, which is therefore known as the '**first heart field**', whereas the second lineage of myocardial cells lies medially and posteriorly to the crescent and is known as the '**second heart field**' (Fig. 9.50).

Clonal analysis in mice shows that cells in the second heart field give rise not only to myocardial cells of the heart, but also to the myogenic cells of several head muscles (Fig. 9.51). Known as branchiomeric muscles, these include muscles involved in chewing and facial expression. Myogenic cells of other head muscles have a separate origin: the myogenic cells of most tongue muscles, for example, migrate into the head from trunk somites. At first sight, a common cellular origin for heart and facial myogenic cells seems surprising, but the linear heart tube forms at the same level along the antero-posterior axis as the future face and later becomes displaced more posteriorly.

The second heart field makes a major contribution to elongation of the early heart tube. As a result of morphogenetic movements, it comes to lie behind the primitive heart tube all along its length (see Fig. 9.50). Heart cells move from a proliferative zone within the second heart field to contribute to both the anterior end of the tube (arterial/outflow end) and the posterior end (venous/inflow end), called the arterial and venous poles, respectively. The cells for the different regions of the heart seem to acquire specific patterns of gene expression according to their position along the antero-posterior axis, and evidence suggests that this is accomplished by graded retinoic acid signaling, which establishes domains of Hox gene expression. Retinoic acid signaling also defines the posterior limit of the second heart field.

The addition of cells from the second heart field to the heart tube accompanies looping of the tube and the recruitment of cardiac neural crest cells to the outflow region of the heart. The cells of neural crest origin are essential for the formation of the pulmonary artery and the aorta from the single embryonic outflow tract (see Fig. 9.50). There may also be important interactions between the neural crest cells and the second heart field cells in this region of the developing heart, because when the cardiac neural crest is ablated there is over-proliferation of the cells of the second heart field. Defective development of the embryonic outflow tract accounts for 30%

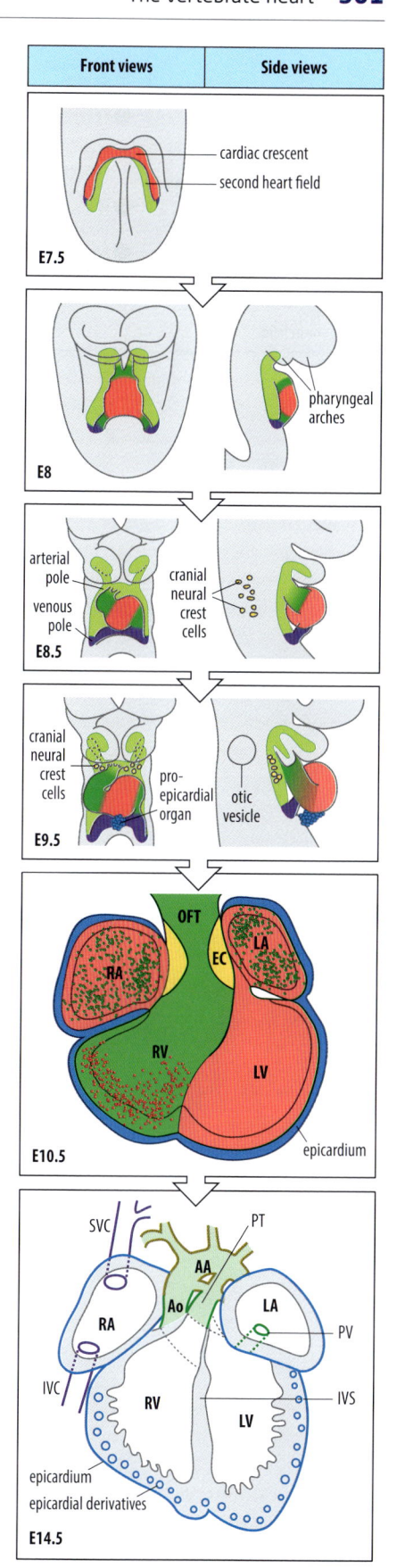

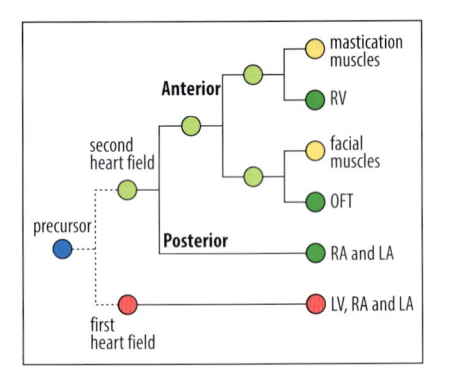

Fig. 9.51 Lineage relationships of the cells that give rise to the mouse heart. The cells of both the first heart field (red) and the second heart field (green) are derived from a common precursor (blue); dotted lines indicate that this precursor has yet to be identified. The first heart field generates cells that make up the left ventricle (LV) and contribute to both left and right atria (LA and RA). The posterior region of the second heart field generates cells that contribute to both left and right atria, while the anterior region of the second heart field not only generates cells that make up the right ventricle (RV) and the outflow tract (OFT), but also gives rise to the myogenic cells of mastication and facial muscles.

After Diogo, R., et al.: ***A new heart for a new head in vertebrate cardiopharyngeal evolution.*** Nature 2015, **520**: 466–473.

of congenital heart abnormalities in humans, some of which are the result of developmental perturbations in neural crest cell specification and migration.

Just after heart looping, the heart tube acquires its outer epithelial layer, the epicardium, which is essential for the normal development and growth of the heart. The epicardium develops from a separate population of mesodermal cells outside the heart tube near its posterior end, known as the pro-epicardial organ (see Fig. 9.50), which grows over the myocardium. Some epicardial cells later undergo an epithelial-to-mesenchymal transition and invade the myocardium to give rise to the cardiac connective tissues—fibroblasts, interstitial cells, and cells that support the coronary arteries. The endothelial cells of the coronary vessels migrate from the endocardium of the inflow region into the ventricular region after the formation of the epicardium. Some cells stay just beneath the epicardium on the surface of the myocardium and form the coronary veins, while others invade the myocardium and form the coronary arteries. Reciprocal interactions between the epicardium and the myocardium involving protein signals that include FGFs, Shh, Wnt, and Notch, control maturation and growth of the heart. The cardiac fibroblasts are of great medical importance, because they give rise to the massive scarring that occurs as the result of a heart attack.

The heart tube then forms the separate chambers—four in the case of the mammalian heart. Continuous structures, known as septa, are established between them. The septa that separate the atrial and ventricular spaces form from localized bulges in precise locations on opposite sides of the inner wall of the heart known as endocardial cushions. In the cushions, endocardial cells undergo an epithelial-to-mesenchymal transition and migrate into the cardiac jelly, so that when two opposing cushions come into contact and break down, the mesenchyme cells form a bridge that stabilizes the fusion of the two cushions. By contrast, the atrial and ventricular septa that separate the right and left chambers of the heart develop from growing spurs of myocardium with mesenchyme cells at their leading edges that eventually fuse with mesenchyme cells of the endocardial cushions. We have already come across tissue fusion as a morphogenetic mechanism in neural tube closure (see Section 5.17).

The endocardial cushions also give rise to the heart valves and are therefore crucial to normal heart function. Their development is particularly dependent on blood flow. For example, in the zebrafish *silent heart* mutant, which lacks a heartbeat due to mutation of a gene encoding a protein essential for myocardial activity, the endocardial cushions fail to develop.

9.32 Key transcription factors specify development of the heart and cardiac differentiation

An increasing number of transcription factors involved in specifying development of the heart and the differentiation of cardiac muscle cells have been identified. One of the first markers of early heart cells is the homeodomain transcription factor Nkx2.5; it is expressed in both the cardiac crescent and the second heart field. Mutations in the *Nkx2.5* gene in mice and humans result in heart abnormalities. Transcription factors of the T-box family also have key functions in heart development. Tbx1 is required for the development of the derivatives of the second heart field, and deficiency of Tbx1 is a likely cause of DiGeorge syndrome, one of the most common human congenital syndromes, which is characterized by heart and craniofacial differences. Another T-box transcription factor, Tbx5, is also expressed in the second heart field. Its involvement in myocardial differentiation is shown by the fact that when Tbx5 and the transcription factor Gata4 are ectopically expressed in mouse embryos in the presence of a chromatin-modeling component, they can induce beating myocardial tissue. Chromatin remodeling gives Gata4 access to regions of DNA that encode genes involved in cardiac differentiation. We came across Tbx5 in relation to its essential function for forelimb development in mice, and the effects of mutations in *TBX5* in humans, which cause Holt–Oram syndrome (see Section 9.11). In people with Holt–Oram syndrome the heart is affected in addition to the arms. As well as its role in myocardial differentiation, Tbx5 has a key function in controlling heart septation: in 40% of people with Holt–Oram syndrome atrial septation is affected.

There are remarkable similarities—although we should no longer be surprised—between the genes involved in heart development in *Drosophila* and in vertebrates. The homeobox gene *tinman* is required for heart formation in *Drosophila* and is a homolog of vertebrate *Nkx2.5*. When vertebrate *tinman*-like genes are expressed in *Drosophila* they can substitute for *Drosophila tinman* and rescue some of the abnormalities caused by lack of normal *tinman* functions. In *Drosophila*, signaling by Dpp maintains *tinman* expression in the dorsal mesoderm, and in vertebrates, BMPs, which are homologs of Dpp, have been shown to induce *Nkx2.5*. Homologs of other key transcription factors in vertebrate heart development are also involved in *Drosophila* heart formation—Dorsocross (a Tbx5/6 homolog) and Pannier (a GATA-4/6 homolog).

SUMMARY

The heart develops from a tube which is patterned along its antero-posterior axis into atrial and ventricular regions. The simple tube undergoes looping to bring the future chambers into their proper positions. The heart loops to the right, reflecting its left-right asymmetry. Several different cell populations give rise to the heart, including neural crest cells, which are essential for partitioning the outflow tract. The mammalian heart is divided into four chambers by the formation of septa. The transcription factors required for vertebrate heart development are the same as those required for *Drosophila* heart development.

The vertebrate eye

In this part of the chapter, we will consider the vertebrate eye as an example of the development of a sense organ. The development of the eye conforms to the general principles already discussed with respect to other organs. For example, its development requires extensive cell–cell interactions between different components and employs the same families of signal molecules as other organs. The morphogenesis of the eye involves complex folding of epithelia, but nevertheless the eye has a surprising capacity for self-organization. A 'master' gene for eye identity is widely conserved across species and the signal molecules involved in pattern formation and cell differentiation are the same as those we have come across in other developing organs.

The structure of the vertebrate eye is shown in Figure 9.52. Light enters through the pupil at the front of the eye and passes through a convex transparent lens, which

Fig. 9.52 Vertebrate eyes. Vertebrate eyes are 'camera' eyes with a single lens that focuses light on photoreceptor cells in the neural retina lining the back of the eyeball. The space between the lens and the neural retina is filled with vitreous humor. The photoreceptor cells connect to retinal neurons, whose axons form the optic nerve, which connects the eye to the brain. The different cell layers of the neural retina are indicated. A, anterior; P, posterior.

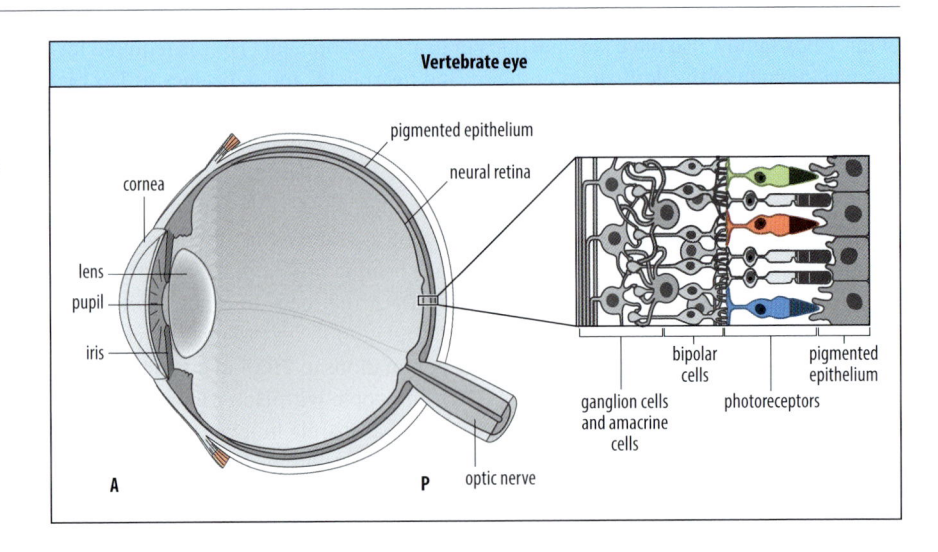

focuses the light on the photosensitive retina lining the back of the eyeball. Photoreceptor cells in the retina—the rods and cones—register the incoming photons and pass signals on to nerve cells, which transmit them via the optic nerve to the brain, where they are decoded. We will consider in Chapter 10 the way in which the neurons from the retina make ordered connections with the visual centers in the brain.

A long-standing puzzle is how the photoreceptors function, because they occupy the innermost layer of the retina rather than the surface adjacent to the vitreous humor. If they were adjacent to the vitreous humor they would get more light and produce a sharper image, as light would not be dispersed and diffracted by the overlying layers of cells (see Fig. 9.52). A potential solution to the puzzle is the demonstration that radial glia called Müller cells, which span the whole thickness of the retina, act as fiber-optic light guides, efficiently transmitting the image pixel by pixel from the surface to the photoreceptors.

9.33 Development of the vertebrate eye involves interactions between an extension of the forebrain and the ectoderm of the head

In developmental terms, the vertebrate eye is essentially an extension of the forebrain, together with a contribution from the overlying ectoderm and surrounding mesenchyme, mainly migrating neural crest cells. The development of an eye starts at E8.5 in the mouse and around 22 days in human embryos with the formation of a bulge, or **evagination**, in the epithelial wall of the posterior forebrain, in the region called the diencephalon. This evagination is called the **optic vesicle**. An optic vesicle is formed in the lateral wall on either side of the forebrain and extends to meet the surface ectoderm (Fig. 9.53). The optic vesicle interacts with the ectoderm to induce formation of the lens placode, which is a thickened region of ectoderm from which the lens will develop. The lens placode is part of a larger region of head ectoderm that gives rise to the epithelial placodes of some other sensory organs, including the placode that gives rise to the semi-circular canals, cochlea, and endolymphatic duct of the ear, and the placode that gives rise to the olfactory epithelium in the nose.

After induction of the lens placode, the tip of the optic vesicle, adjacent to the placode, invaginates to form a two-layered cup, the optic cup. The inner epithelial layer of the optic cup will form the neural retina, while the outer layer will form the retinal pigment epithelium. The invagination of the lens placode is coordinated with the invagination of the optic vesicle. The placode then detaches from the surface ectoderm to form a small hollow sphere of epithelium, the lens vesicle, that will develop

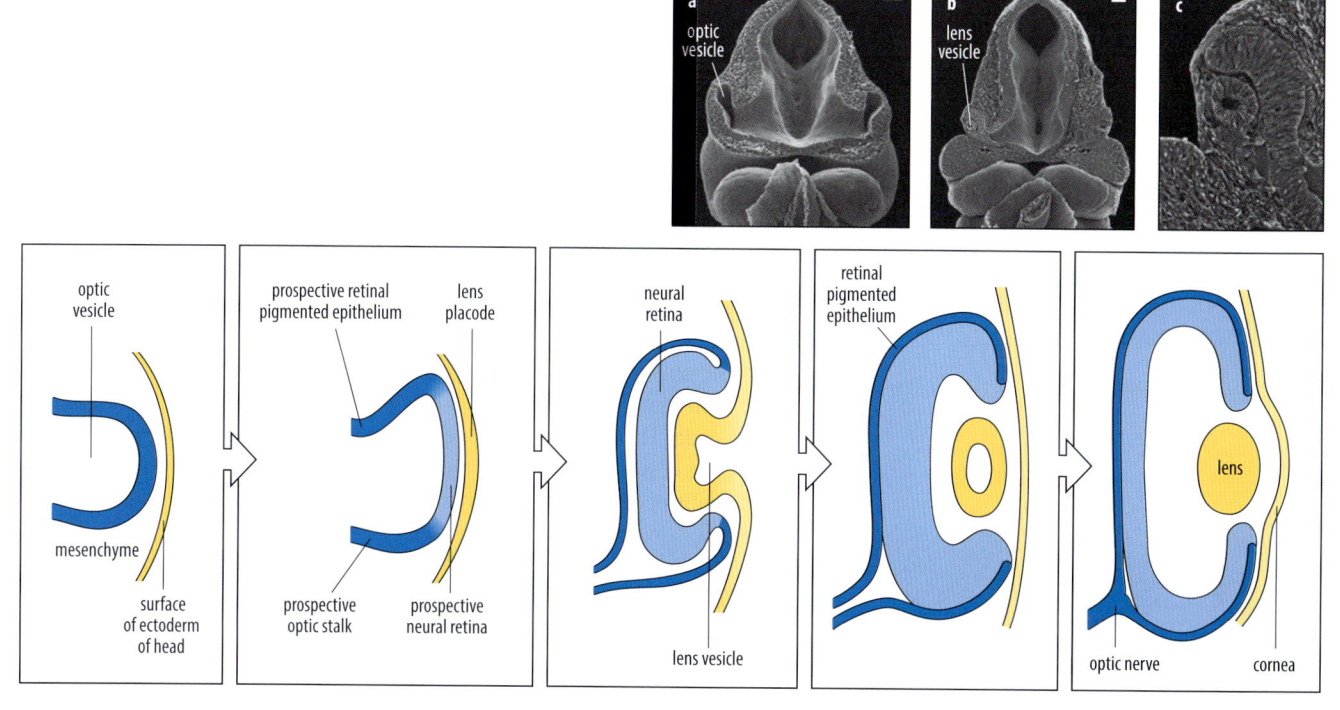

Fig. 9.53 The main stages in the development of the vertebrate eye. The schematic diagrams of the development of the optic cup and lens show that the optic vesicle (blue) develops as an outgrowth of the forebrain and induces the lens placode in the surface ectoderm (yellow) of the head. The optic vesicle invaginates to form a two-layered cup, the optic cup, around the developing lens (the lens vesicle). The inner layer of the optic cup forms the neural retina (pale blue), and the outer layer forms the pigmented epithelium underlying the retina (dark blue). The lens vesicle detaches from the surface ectoderm to form the lens and the remaining overlying ectoderm forms the outer layer of the cornea. The inner layer of the cornea (not shown) is formed by neural crest cells that migrate into the eye. The iris (not shown) develops from the rim of the optic cup. The fluid-filled space between the iris and the cornea is the anterior chamber of the eye. The scanning electron micrographs show frontal sections through the head of a mouse embryo, showing (a) formation of the optic vesicle (E9-9.5); (b) formation of the lens vesicle and optic cup (E10.5); and (c) high-magnification view of the optic cup and lens vesicle in (b).

Diagram adapted from Adler, R., Valeria Canto-Soler, M.: **Molecular mechanisms of optic vesicle development: complexities, ambiguities, and controversies.** Dev. Biol. 2007, **305**: 1-13. Electron micrographs from Heavner, W., Pevny, L.: **Eye development and retinogenesis.** Cold Spring Harb. Perspect. Biol. 2012, **4**: a008391.

into the lens, while the remaining surface ectoderm fuses to give rise to the outer layer of the cornea.

The lens is formed by proliferation of cells of the epithelium on the anterior side of the lens vesicle—that is, the side nearer the cornea—with the new cells moving into the center of the lens, where they start to manufacture crystallin proteins. The cells eventually lose their nuclei, mitochondria, and internal membranes to become completely transparent lens fibers filled with crystallin. Renewal of lens fibers continues after birth but proceeds much more slowly than in the embryo. In the adult chicken, the transformation of an epithelial cell into a crystallin-filled fiber takes 2 years.

The cornea is a transparent epithelium that seals the front of the eye. It is composed of inner and outer layers with different developmental origins. The inner layer is formed by mesenchymal neural crest cells that migrate into the anterior eye chamber to form a thin layer initially overlying the lens. Neural crest cells also contribute to other structures in the anterior part of the eye. The outer layer of the cornea is derived from the surface ectoderm adjacent to the eye. Most of the iris develops from the rim of the optic cup—also known as the ciliary margin—with neural crest contributing to the anterior iris. In the adult mammalian eye, potential stem cells have been identified in the ciliary margin, but whether they are able to self-renew and to give rise to all the cell types in the eye is as yet unclear. Formation of the lens is a crucial step in the development of anterior eye structures such as the cornea and the iris.

The neural retina in the vertebrate eye develops three distinct layers of cells, with the photoreceptor cells forming the innermost layer, underneath a layer of ganglion and amacrine cells and a layer of bipolar cells (see Fig. 9.52). This contrasts with the camera eyes of octopus and squid, in which the photoreceptor cells are on the surface of the neural retina. Visual signals are transmitted from the eye to the brain via the optic nerve, which is formed by the axons of the neural retina ganglion cells. In Chapter 10 we will see how the neurons of the optic nerve connect with precise positions in the visual-processing centers in the brain in order to produce a 'map' of the retina.

The morphogenesis of the optic cup involves outgrowth and coordinated folding of epithelial cell sheets. Although the optic vesicle and the lens placode are tightly apposed when invagination begins, the optic vesicle can still invaginate in the absence of the lens placode. This conclusion came from classic work carried out in the 1930s on eye development in amphibian embryos. When an optic vesicle is transplanted to the trunk, the optic vesicle can develop into an optic cup, even though trunk ectoderm is unable to form a lens.

Self-organization of the optic vesicle takes place in three-dimensional aggregates of mouse embryonic stem (ES) cells that have been induced to differentiate into retinal epithelial cells and cultured with basement membrane components (Fig. 9.54). After about a week in culture, the ES cells aggregate to form hollow epithelial spheres, and hemispherical bulges of epithelium begin to evaginate from the main body of the spheres, thus mimicking the evagination of the optic vesicle from the diencephalon at the start of eye development in the embryo. Even more remarkably, over the next 2 days of culture the hemispherical bulges undergo further dynamic changes in shape to give rise to two-walled cup-like structures mimicking the invagination of the optic vesicle to give rise to the optic cup in the embryo. Cells in the inner wall express genes characteristic of neural retina and cells in the outer wall those characteristic of pigmented retina. The cells in the inner wall subsequently differentiate into all the main neuronal cell types in the retina, including photoreceptors, arranged in appropriate layers, whereas the cells in the outer wall become pigmented.

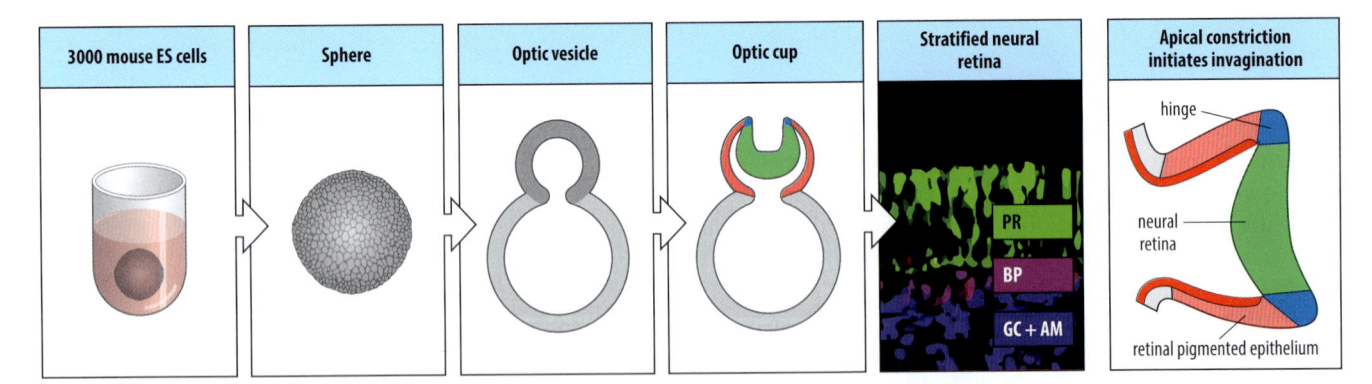

Fig. 9.54 The generation of an optic cup in culture. From left: floating aggregates of mouse embryonic stem cells (ES cells) that have been induced to differentiate into retinal epithelial cells are grown in serum-free medium containing extracellular matrix components including laminin. They form spheres. A retinal epithelial hemisphere subsequently evaginates from the sphere to give an optic-vesicle-like structure. The distal region of the vesicle then undergoes invagination to form an optic cup. The photomicrograph shows a cross-section of the stratified neural retina of the self-organized optic cup with various recognizable cell types. PR, photoreceptors; BP, bipolar cells; GC + CM, ganglion cells + amacrine cells. Right panel: the hinge region (blue) at the junction between the region of epithelium that will give rise to the pigmented retina (pink and red) and the neural retina (green). Apical constriction of cells at the hinge initiates invagination.

*Photomicrograph from Eiraku, M., et al.: **Self-organizing optic-cup morphogenesis in three-dimensional culture.** Nature 2011, **472**: 51-56.*

The entire process of generation of the optic cup can be followed in some detail by imaging the living three-dimensional cell aggregates in which cells in the 'eye field' (the region across the center of the anterior neural plate from which the eyes will form) and later cells in the neural retina express a GFP reporter gene. The movies revealed, for example, that curvature of the epithelium is associated with changes in the shape of the epithelial cells. In places where the epithelium becomes folded at an acute angle, hinge points with wedge-shaped cells like those seen during neurulation (see Section 5.17) are observed.

Self-formation of a two-walled cup and development of a stratified neural retina can similarly be produced from aggregates of human ES cells. Brain organoids of human induced pluripotent stem cells cultured under certain conditions can also develop eye-like structures (see Section 7.9). These structures are connected to the rest of the organoid that has developed into forebrain-resembling visual-processing centers. Strikingly, two bilaterally positioned 'eyes' frequently develop in each organoid and consist not only of neural retina and pigmented retina but also lens and cornea-like epithelium. Remarkably, the connections between these structures and the forebrain in the organoids are functional. The organoids respond to light levels within the range to which the eye normally responds, and electrical recordings from the organoids indicate that this response is processed. These organoids could provide a model for human eye development as the stages in their development recapitulate the stages in normal eye development and take a similar length of time. In addition, they might provide useful material for studying degenerative diseases of the human eye.

9.34 Eye-specifying transcription factors are conserved

Although evagination of the optic vesicles does not occur until neural tube closure is almost completed, the specification of cells as eye cells occurs much earlier, in the neural plate. Genes encoding key conserved eye-specifying transcription factors, such as Pax6, Six3, and Otx2, are expressed in anterior neural plate at late gastrula stages as part of the initial antero-posterior patterning of the neural plate. They continue to be expressed in the optic vesicle epithelium and the lens placode, as well as in other precursors of sensory organs such as the olfactory placode. Retinal precursor cells are first specified as a single eye field across the center of the anterior neural plate. The eye field eventually becomes separated into two lateral regions of cells that will give rise to the optic vesicles after neural tube formation. Separation is achieved by downregulating expression of Pax6 and the other eye-specifying transcription factors along the midline. Failure of this separation could be one of the causes of cyclopia in human embryos, in which a single abnormal central eye develops. The normal downregulation of the expression of Pax6 and other transcription factors in the forebrain midline is likely to be mediated by Shh signaling, which is known to be required for the correct specification of ventral midline structures in the brain. A failure of Shh signaling can result in the condition of holoprosencephaly, which, in its most severe form, results in failure of the forebrain to divide into right and left hemispheres and loss of midline facial structures, resulting in cyclopia (see Box 1H).

Later, dorso-ventral patterning of the optic cup involves Shh signaling ventrally and BMP-4 signaling dorsally, as it does throughout the entire central nervous system (see Chapter 10). Other signals, including Notch, various BMPs, Wnts, FGFs, TGF-β, and retinoic acid, some of them produced by the neural crest surrounding the optic cup, are important in inducing and maintaining differentiation of the cells in different regions of the optic cup. Thus, for example, *Fgf9* is expressed in the region of the optic cup destined to be neural retina. Expression of ectopic *Fgf9* leads to a duplicate neural retina; however, when *Fgf9* is lacking, the pigmented retinal epithelium extends into the region that would normally form neural retina.

The same set of transcription factors is essential for eye formation throughout the animal kingdom. *Pax6* is the classic example of a 'master' gene with a conserved basic function and is required for the development of light-sensing structures in all bilaterian animals (animals with bilateral symmetry)—from the simple light-sensing organs of planarians to the compound eyes of insects and the camera eyes of vertebrates and cephalopods (discussed in Section 13.6).

Pax6 was initially identified from the genetic analysis of mutations causing abnormal eye development in mice and humans and is homologous with the *Drosophila* gene *eyeless*. Heterozygous mouse *Pax6* mutant embryos have smaller eyes than normal and homozygous *Pax6* mutants have no eyes at all. People heterozygous for mutations in *PAX6* have various eye malformations collectively known as **aniridia**, because of the partial or complete absence of the iris; these individuals also have cognitive differences, as *PAX6* has several roles in brain development other than eye formation. In the very rare cases of homozygous mutations in *PAX6*, eyes are absent, and this may be associated with other severe abnormalities that can be incompatible with life.

In *Xenopus*, injection of *Pax6* mRNA into an animal pole blastomere at the 16-cell stage leads to formation of ectopic eye-like structures, with well-formed lenses and epithelial optic cups, in the tadpole head. Even more amazingly, the ectopic expression of *Pax6* from mouse, *Xenopus*, ascidians, or squid in *Drosophila* imaginal discs causes the development of *Drosophila*-type compound eye ommatidia (the individual visual units of a compound eye), on adult structures such as antennae (Fig. 9.55).

Another transcription factor, Six3, is also involved in eye development in both vertebrates and *Drosophila* (the *Drosophila* transcription factor is encoded by the *sine oculis* gene) and can induce eye structures when overexpressed in fish embryos. Thus, there seems to be a conserved network of a small number of 'eye' transcription factors that govern development of this sense organ.

SUMMARY

The vertebrate eye develops as an extension of the forebrain. Evagination of the forebrain lateral wall produces the optic vesicle, which, in turn, induces formation of the lens from the surface ectoderm. Invagination of the tip of the optic vesicle produces a two-layered optic cup, which surrounds the lens and develops into the eyeball, with the inner epithelial layer forming the neural retina, and the outer epithelium developing into the pigmented epithelium at the back of the retina. The optic vesicle has a remarkable capacity for self-organization but the lens is needed for the eye to develop further. The transcription factor Pax6 is essential for eye development in animals as diverse as vertebrates, cephalopods, and *Drosophila*.

Fig. 9.55 *Pax6* **is a master gene for eye development.** (a) The ectopic expression of mouse *Pax6* in a *Drosophila* antennal disc results in compound eye structures developing on the antenna. (b) A high-power view of the eye-like structure.

Photographs reproduced with permission from Gehring, W.J.: **New perspectives on eye development and the evolution of eyes and photoreceptors.** *J. Hered. 2005,* **96***: 171-184.*

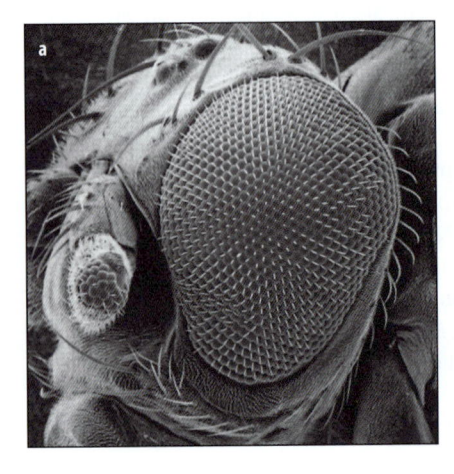

Summary to Chapter 9

- In *Drosophila*, adult appendages develop from imaginal discs formed in the embryo, and the segment in which an imaginal disc arises helps to determine appendage identity. Segment identity is defined by a combinatorial Hox code, and within a segment, discs corresponding to different structures (e.g. leg and wing discs in thoracic segment 2) are distinguished by the expression of different sets of transcription factors.
- In vertebrate embryos, the position at which organs develop is specified as part of the process of laying down the body plan. For example, a combinatorial Hox code in the lateral plate mesoderm induces development of the limbs; a combinatorial homeobox gene code in the neural crest cells of the facial primordia specifies tooth position and identity.
- The identity of organs is encoded by 'master' or 'selector' genes and/or specific combinations of genes for transcription factors and other proteins that activate the gene regulatory network required for subsequent development of that particular organ. *vestigial*, for example, which encodes a gene-regulatory co-activator, is the master gene for *Drosophila* wing development; *Pax6*, which encodes a transcription factor, is the master gene for vertebrate eye development.
- There is a remarkable degree of conservation in the transcription factors involved in the development of particular organs in vertebrates and *Drosophila*. The transcription factors encoded by the gene *Nkx2.5* in vertebrates and its homolog *tinman* in *Drosophila* are required for heart formation; the homolog of *Pax6* in *Drosophila* is required for insect eye development; the transcription factor encoded by *apterous* specifies dorsal identity in the *Drosophila* wing, and its vertebrate homolog, *Lmx1b*, has the same function in the vertebrate limb.
- Cell-cell interactions between different components, mainly mesenchymal and epithelial components, have a key role in organ development. In vertebrates, the limb bud ectoderm imposes dorso-ventral pattern on the mesoderm, and lung mesenchyme influences the branching of lung epithelium. In the eye, interactions between different epithelia induce the formation of the lens.
- Pattern formation in organs involves the interpretation of positional information. Organ primordia become equipped with signaling regions, which produce morphogens that specify positional values. Examples of these signaling regions are the compartment boundaries in wing and leg imaginal discs in *Drosophila* and the polarizing region in a vertebrate limb bud.
- The same families of signal molecules act as morphogens in different organs, the differences in interpretation being due to the different developmental histories of the responding cells. For example, Shh signaling is involved in the development of all the vertebrate organs we have considered here—limbs, teeth, lungs, heart, and eyes—and Hedgehog signaling also acts in wing development in *Drosophila*.
- Cell differentiation generates specialized cells for specific organ functions. The same differentiated cell types can be generated from cells with different developmental histories (e.g. myogenic cells of the limb muscles are derived from the somites, whereas the myogenic cells of some of the facial muscles originate from the lateral plate mesoderm) or a unique cell type can differentiate in a particular organ (photoreceptors in the eye).
- Organ morphogenesis involves many of the same mechanisms used during early embryogenic development—for example, oriented cell movements and cell divisions (in the limb) and folding of cell sheets (in the eye). A key process in the development of many organs is branching morphogenesis, illustrated in this chapter by the lung and the mammary gland.
- Organs have a remarkable capacity for self-organization (e.g. limb digits, teeth, optic vesicle), and reaction-diffusion mechanisms have been implicated in generating periodic prepatterns (digits in the limb, cusps in teeth, epithelial branches in the lung).
- Mutations in developmentally important genes involved in organogenesis can lead to human congenital differences. Development can also be adversely affected by exposure to harmful environmental agents (e.g. the drug thalidomide) at critical periods during organogenesis.

End of chapter questions

Check your understanding of the content of this chapter by attempting the following long-answer concept questions.

1. How is the identity of the different *Drosophila* imaginal discs established in the embryo—for example, how is it determined whether a disc will form a wing, a leg, or some other structure? How does misexpression of *Antennapedia* in the head region illustrate this?

2. The gene *engrailed* is expressed in the posterior compartment of the wing disc in the *Drosophila* larva and in the posterior compartment of segments in the embryo. Compare and contrast the signals downstream of posterior compartment *engrailed* expression in the wing disc and in segmentation (refer back to Chapter 2 for information about segmentation).

3. The *Drosophila* genes *wingless*, *apterous* (pteron is Greek for wing), and *vestigial* can all reduce or eliminate the formation of the wing when mutated. On the basis of the roles of these genes in the wing imaginal disc, propose mechanisms by which these mutations cause the wing defects.

4. Describe an experiment that showed that fibroblast growth factors (FGFs) have a critical role in the initiation of limb-bud formation in vertebrates. What causes the expression of FGFs? Propose a linear gene expression pathway from Hox gene expression to limb bud formation based on these facts.

5. Describe the location and extent of the apical ectodermal ridge with respect to the three main axes of the limb. How does this structure arise?

6. What is the consequence of removing the apical ectodermal ridge at different times during limb-bud development? How do you interpret these results? Cite evidence for your interpretation.

7. Describe the models for specifying regional identity along the proximo-distal axis in the developing limb. What is the proposed function of FGF signaling in each of these models?

8. Where in the limb bud is the polarizing region (also known as the zone of polarizing activity or ZPA)? What properties of this region make it an organizing region (see Fig. 9.31)? Through which signal molecule does it exert its effects? What is the evidence for the importance of this molecule?

9. What is Gli3? What is its potential role in specifying digit identity, and through which downstream genes does it appear to exert this role?

10. What is the consequence of grafting the ectoderm from a left limb bud onto a right mesodermal core, such that the dorso-ventral axis is reversed without altering the antero-posterior axis? What conclusion is drawn from this experiment? How are Wnt and Engrailed involved in this patterning, and how do we know?

11. How is the activity of the signaling centers in the limb bud—dorsal ectoderm, apical ectodermal ridge, and polarizing region—integrated? What is the role of Gremlin in this process?

12. What is the origin of the muscle cells in a limb? What is the evidence that the limb mesoderm controls the pattern of muscles that forms, so that the musculature is appropriate to the limb?

13. Teeth form a repeated pattern. What ensures that the appropriate type of tooth forms in its correct position within the jaw? Take as your example the formation of incisors and molars in the mouse.

14. Outline the interactions between endoderm and mesoderm that specify the region of the ventral foregut that will form the lung. Suggest the essential features of a protocol based on these interactions which could be used to generate lung tissue from induced pluripotent stem cells.

15. What is known about the molecules involved in lung bud formation and outgrowth and how is the branching pattern established?

16. How are the three cell layers of the heart formed? Include endocardium, myocardium, and epicardium. What is the importance of cell–cell interactions between the epicardium and the myocardium?

17. Discuss how the discovery of developmentally important genes in heart development has given insights into human congenital heart conditions.

18. What type of protein does the gene *Pax6* encode? Discuss its role in eye development. What human conditions result from alterations in the normal function or expression of *Pax6*?

Section further reading

9.1 Imaginal discs arise from the ectoderm in the early *Drosophila* embryo; 9.2 Imaginal discs arise across parasegment boundaries and are patterned by signaling at compartment boundaries; and 9.3 The adult wing emerges at metamorphosis after folding and evagination of the wing imaginal disc

Morata, G.: **How *Drosophila* appendages develop**. *Nat. Rev. Mol. Cell Biol.* 2001, **2**: 89–97.

9.4 A signaling center at the boundary between anterior and posterior compartments patterns the *Drosophila* wing disc along the antero-posterior axis

Crozatier, M., Glise, B., Vincent, A.: **Patterns in evolution: veins of the *Drosophila* wing**. *Trends Genet.* 2004, **20**: 498–505.

Entchev, E.V., Schwabedissen, A., González-Gaitán, M.: **Gradient formation of the TGF-β homolog Dpp**. *Cell* 2000, **103**: 981–991.

Han, C., Belenkaya, T.Y., Wang, B., Lin, X.: ***Drosophila* glypicans control the cell-to-cell movement of Hedgehog by a dynamin-independent process**. *Development* 2004, **131**: 601–611.

Kruse, K., Pantazis, P., Bollenbach, T., Julicher, F., González-Gáitan, M.: **Dpp gradient formation by dynamin-dependent endocytosis: receptor trafficking and the diffusion model**. *Development* 2004, **131**: 4843–4856.

Moser, M., Campbell, G.: **Generating and interpreting the Brinker gradient in the *Drosophila* wing**. *Dev. Biol.* 2005, **286**: 647–658.

Muller, B., Hartmann, B., Pyrowolakis, G., Affolter, M., Basler, K.: **Conversion of an extracellular Dpp/BMP morphogen gradient into an inverse transcriptional gradient**. *Cell* 2003, **113**: 221–233.

Tabata, T.: **Genetics of morphogen gradients**. *Nat. Rev. Genet.* 2001, **2**: 620-630.

Wartlick, O., Mumcu, P., Kicheva, A., Bittig, T., Seum, C., Jülicher, F., González-Gaitán, M.: **Dynamics of Dpp signaling and proliferation control**. *Science* 2011, **331**: 1154-1159.

Box 9A Positional information and morphogen gradients

Ashe, H.L., Briscoe, J.: **The interpretation of morphogen gradients**. *Development* 2006, **133**: 385-394.

Kerszberg, M., Wolpert, L.: **Specifying positional information in the embryo: looking beyond morphogens**. *Cell* 2007, **130**: 205-209.

Kicheva, A., González-Gaitán, M.: **The decapentaplegic morphogen gradient: a precise definition**. *Curr. Opin. Cell Biol.* 2008, **20**: 137-143.

Kornberg, T.B.: **Distributing signaling proteins in space and time: the province of cytonemes**. *Curr. Opin. Genet. Dev.* 2017, **45**: 22-27.

Müller, P., Rogers, K.W., Yu, S.R., Brand, M., Schier, A.F.: **Morphogen transport**. *Development* 2013, **140**: 1621-1638.

Roy, S., Huang, H., Liu, S., Kornberg, T.B.: **Cytoneme mediated contact-dependent transport of the *Drosophila* decapentaplegic signaling protein**. *Science* 2014, **343**: 1244624.

Sanders, T.A., Llagostera, E., Barna, M.: **Specialized filopodia direct long-range transport of SHH during vertebrate tissue patterning**. *Nature* 2013, **497**: 628-632.

Simsek, M.F., Özbudak, E.M.: **Patterning principles of morphogen gradients**. *Open Biol.* 2022, **12**: 220224.

Stapornwongkul, K.S., de Gennes, M., Cocconi, L., Salbreux, G., Vincent, J.P.: **Patterning and growth control *in vivo* by an engineered GFP gradient**. *Science* 2020, **370**: 321-327.

Stapornwongkul, K.S., Vincent, J.P.: **Generation of extracellular morphogen gradients: the case for diffusion**. *Nat. Rev. Genet.* 2021, **22**: 393-411.

9.5 A signaling center at the boundary between dorsal and ventral compartments patterns the *Drosophila* wing along the dorso-ventral axis

Baeg, G.H., Selva, E.M., Goodman, R.M., Dasgupta, R., Perrimon, N.: **The Wingless morphogen gradient is established by the cooperative action of Frizzled and heparan sulfate proteoglycan receptors**. *Dev. Biol.* 2004, **276**: 89-100.

Bollenbach, T., Pantazis, P., Kicheva, A., Bökel, C., González-Gaitán, M., Jülicher, F.: **Precision of the Dpp gradient**. *Development* 2008, **135**: 1137-1146.

Fujise, M., Takeo, S., Kamimura, K., Matsuo, T., Aigaki, T., Izumi, S., Nakato, H.: **Dally regulates Dpp morphogen gradient formation in the *Drosophila* wing**. *Development* 2003, **130**: 1515-1522.

Hayward, P., Kalmar, T., Martinez Arias, A.: **Wnt/Notch signalling and information processing during development**. *Development* 2008, **135**: 411-424.

Kicheva, A., González-Gaitán, M.: **The Decapentaplegic morphogen gradient: a precise definition**. *Curr. Opin. Cell Biol.* 2008, **20**: 137-143.

Lawrence, P.: **Morphogens: how big is the picture?** *Nat. Cell Biol.* 2001, **3**: E151-E154.

Martinez Arias, A.: **Wnts as morphogens? The view from the wing of *Drosophila***. *Nat. Rev. Mol. Cell. Biol.* 2003, **4**: 321-325.

Milán, M., Cohen, S.M.: **A re-evaluation of the contributions of Apterous and Notch to the dorsoventral lineage restriction boundary in the *Drosophila* wing**. *Development* 2003, **130**: 553-562.

Piddini, E., Vincent, J.P.: **Interpretation of the Wingless gradient requires signaling-induced self-inhibition**. *Cell* 2009, **136**: 296-307.

Rauskolb, C., Correia, T., Irvine, K.D.: **Fringe-dependent separation of dorsal and ventral cells in the *Drosophila* wing**. *Nature* 1999, **401**: 476-480.

Zecca, M., Basler, K., Struhl, G.: **Sequential organizing activities of engrailed, hedgehog and decapentaplegic in the *Drosophila* wing**. *Development* 1995, **121**: 2265-2278.

9.6 Vestigial is a key regulator of wing development that acts to specify wing identity and control wing growth

Kim, J., Sebring, A., Esch, J.J., Kraus, M.E., Vorwerk, K., Magee, J., Carroll, S.B.: **Integration of positional signals and regulation of wing formation and identity by *Drosophila* *vestigial* gene**. *Nature* 1996, **382**: 133-138.

Klein, T., Martinez Arias, A.: **The *vestigial* gene product provides a molecular context for the interpretation of signals during the development of the wing in *Drosophila***. *Development* 1999, **126**: 913-925.

9.7 The *Drosophila* wing disc is also patterned along the proximo-distal axis

Klein, T., Martinez Arias, A.: **Different spatial and temporal interactions between Notch, wingless and vestigial specify proximal and distal pattern elements of the wing in *Drosophila***. *Dev. Biol.* 1998, **194**: 196-212.

9.8 The leg disc is patterned in a similar manner to the wing disc, except for the proximo-distal axis

Emerald, B.S., Cohen, S.M.: **Spatial and temporal regulation of the homeotic selector gene Antennapedia is required for the establishment of leg identity in *Drosophila***. *Dev. Biol.* 2004, **267**: 462-472.

Estella, C., Voutev, R., Mann, R.S.: **A dynamic network of morphogens and transcription factors patterns the fly leg**. *Curr. Top. Dev. Biol.* 2012, **98**: 173-198.

Galindo, M.I., Bishop, S., Greig, S., Couso, J.P.: **Leg patterning driven by proximal-distal interactions and EGFR signaling**. *Science* 2002, **297**: 258-259.

Kojima, T.: **The mechanism of *Drosophila* leg development along the proximodistal axis**. *Dev. Growth Differ.* 2004, **46**: 115-129.

9.9 Different imaginal discs can have the same positional values

Carroll, S.B.: **Homeotic genes and the evolution of arthropods and chordates**. *Nature* 1995, **376**: 479-485.

Morata, G.: **How *Drosophila* appendages develop**. *Nat. Rev. Mol. Cell Biol.* 2001, **2**: 89-97.

Si Dong, P.D., Chu, J., Panganiban, G.: **Coexpression of the homeobox genes Distal-less and homothorax determines *Drosophila* antennal identity**. *Development* 2000, **127**: 209-216.

9.10 The chick wing is a model for limb development

Davey, M.G., Towers, M., Vargesson, N., Tickle, C.: **The chick limb: embryology, genetics and teratology.** *Int. J. Dev. Biol.* 2018, **62**: 85–95.

McQueen, C., Towers, M.: **Establishing the pattern of the vertebrate limb.** *Development* 2020, **147**: dev177956.

Saunders, J.W., Gasseling, M.T., Cairns, J.M.: **The differentiation of prospective thigh mesoderm grafted beneath the apical ectodermal ridge of the wing bud in the chick embryo.** *Dev. Biol.* 1959, **1**: 281–301.

Tickle, C.: **An historical perspective on the pioneering experiments of John Saunders.** *Dev. Biol.* 2017, **429**: 374–381.

9.11 Genes expressed in the lateral plate mesoderm induce limb development and specify limb polarity and identity

Altabef, M., Clarke, J.D.W., Tickle, C.: **Dorso-ventral ectodermal compartments and origin of apical ectodermal ridge in developing chick limb.** *Development* 1997, **124**: 4547–4556.

Cohn, M.J., Izpisúa-Belmonte, J.C., Abud, H., Heath, J.K., Tickle, C.: **Fibroblast growth factors induce additional limb development from the flank of chick embryos.** *Cell* 1995, **80**: 739–746.

DeLaurier, A., Schweitzer, R., Logan, M.: **Pitx1 determines the morphology of muscle, tendon, and bones of the hindlimb.** *Dev. Biol.* 2006, **299**: 22–34.

Domyan, E.T., Kronenberg, Z., Infante, C.R., Vickrey, A.I., Stringham, S.A., Bruders, R., Guernsey, M.W., Park, S., Payne, J., Beckstead, R.B., Kardon, G., Menke, D.B., Yandell, M., Shapiro, M.D.: **Molecular shifts in limb identity underlie development of feathered feet in two domestic avian species.** *eLife* 2016, **15**: e12115.

Feneck, E., Logan, M.: **The role of retinoic acid in establishing the early limb bud.** *Biomolecules* 2020, **10**: 312.

Minguillon, C., Buono, J.D., Logan, M.P.: **Tbx5 and Tbx4 are not sufficient to determine limb-specific morphologies but have common roles in initiating limb outgrowth.** *Dev. Cell* 2005, **8**: 75–84.

Nishimoto, S., Logan, M.P.: **Subdivision of the lateral plate mesoderm and specification of the forelimb and hindlimb forming domains.** *Semin. Cell Dev. Biol.* 2016, **49**: 102–108.

Nishimoto, S., Minguillon, C., Wood, S., Logan, M.P.: **A combination of activation and repression by a colinear Hox code controls forelimb-restricted expression of Tbx5 and reveals Hox protein specificity.** *PLoS Genet.* 2014, **10**: e1004245.

Tickle, C.: **How the embryo makes a limb: determination, polarity and identity.** *J. Anat.* 2015, **227**: 418–430.

Xu, B., Hrycaj, S.M., McIntyre, D.C., Baker, N.C., Takeuchi, J.K., Jeannotte, L., Gaber, Z.B., Novitch, B.G., Wellik, D.M.: **Hox5 interacts with Plzf to restrict Shh expression in the developing forelimb.** *Proc. Natl Acad. Sci. USA* 2013, **110**: 19438–19443.

Xu, B., Wellik, D.M.: **Axial Hox9 activity establishes the posterior field in the developing forelimb.** *Proc. Natl Acad. Sci. USA* 2011, **108**: 4888–4891.

9.12 The apical ectodermal ridge is required for limb bud outgrowth and the formation of structures along the proximo-distal axis of the limb

Fernandez-Teran, M., Ros, M.A.: **The apical ectodermal ridge: morphological aspects and signaling pathways.** *Int. J. Dev. Biol.* 2008, **52**: 857–871.

Niswander, L., Tickle, C., Vogel, A., Booth, I., Martin, G.R.: **FGF-4 replaces the apical ectodermal ridge and directs outgrowth and patterning of the limb.** *Cell* 1993, **75**: 579–587.

Verheyden, J.M., Sun, X.: **Embryology meets molecular biology: deciphering the apical ectodermal ridge.** *Dev. Biol.* 2017, **429**: 387–390.

9.13 Formation and outgrowth of the limb bud involves oriented cell behavior

Boehm, B., Westerberg, H., Lesnicar-Pucko, G., Raja, S., Rautschka, M., Cotterell, J., Swoger, J., Sharpe, J.: **The role of spatially controlled cell proliferation in limb bud morphogenesis.** *PLoS Biol.* 2010, **8**: e1000420.

Gros, J., Hu, J.K., Vinegoni, C., Feruglio, P.F., Weissleder, R., Tabin, C.J.: **Wnt5a/Jnk and FGF/Mapk pathways regulate the cellular events shaping the vertebrate limb.** *Curr. Biol.* 2010, **20**: 1993–2002.

Gros, J., Tabin, C.J.: **Vertebrate limb bud formation is initiated by localized epithelial-to-mesenchymal transition.** *Science* 2014, **343**: 1253–1256.

Wyngaarden, L.A., Vogeli, K.M., Ciruna, B.G., Wells, M., Hadjantonakis, A.K., Hopyan, S.: **Oriented cell motility and division underlie early limb bud morphogenesis.** *Development* 2010, **137**: 2551–2558.

9.14 Positional values along the proximo-distal axis of the limb are specified by a combination of graded signaling and a timing mechanism

Cooper, K.L., Hu, J.K., ten Berge, D., Fernandez-Teran, M., Ros, M.A., Tabin, C.J.: **Initiation of proximo-distal patterning in the vertebrate limb by signals and growth.** *Science* 2011, **332**: 1083–1086.

Delgado, I., López-Delgado, A.C., Roselló-Díez, A., Giovinazzo, G., Cadenas, V., Fernández-de-Manuel, L., Sánchez-Cabo, F., Anderson, M.J., Lewandoski, M., Torres, M.: **Proximo-distal positional information encoded by an Fgf-regulated gradient of homeodomain transcription factors in the vertebrate limb.** *Sci Adv.* 2020, **6**: eaaz0742.

Rosello-Diez, A., Ros, M.A., Torres, M.: **Diffusible signals not autonomous mechanisms determine the main proximodistal limb subdivision.** *Science* 2011, **332**: 1086–1088.

Saiz-Lopez, P., Chinnaiya, K., Campa, V.M., Delgado, I., Ros, M.A., Towers, M.: **An intrinsic timer specifies distal structures of the vertebrate limb.** *Nat. Commun.* 2015, **6**: 8108.

Summerbell, D., Lewis, J.H., Wolpert, L.: **Positional information in chick limb morphogenesis.** *Nature* 1973, **244**: 492–496.

9.15 The polarizing region specifies position along the limb's antero-posterior axis

Tickle, C.: **Making digit patterns in the vertebrate limb.** *Nat. Rev. Mol. Cell Biol.* 2006, **7**: 1–9.

Tickle, C., Summerbell, D., Wolpert, L.: **Positional signaling and specification of digits in chick limb morphogenesis**. *Nature* 1975, **254**: 199–202.

9.16 Sonic hedgehog is a good candidate for the polarizing region morphogen

Dahn, R.D., Fallon, J.F.: **Interdigital regulation of digit identity and homeotic transformation by modulated BMP signaling**. *Science* 2000, **289**: 438–441.

Harfe, B.D., Scherz, P.J., Nissim, S., Tian, H., McMahon, A.P., Tabin, C.J.: **Evidence for an expansion-based temporal Shh gradient in specifying vertebrate digit identities**. *Cell* 2004, **118**: 517–528.

Litingtung, Y., Dahn, R.D., Li, Y., Fallon, J.F., Chiang, C.: **Shh and Gli3 are dispensable for limb skeleton formation but regulate digit number and identity**. *Nature* 2002, **418**: 979–983.

Riddle, R.D., Johnson, R.L., Laufer, E., Tabin, C.: **Sonic hedgehog mediates polarizing activity of the ZPA**. *Cell* 1993, **75**: 1401–1416.

Suzuki, T., Hasso, S.M., Fallon, J.F.: **Unique SMAD1/5/8 activity at the phalanx-forming region determines digit identity**. *Proc. Natl Acad. Sci. USA* 2008, **105**: 4185–4190.

Suzuki, T., Takeuchi, J., Koshiba-Takeuchi, K., Ogura, T.: **Tbx genes specify posterior digit identity through Shh and BMP signaling**. *Dev. Cell* 2004, **6**: 43–53.

te Welscher, P., Zuniga, A., Kuijper, S., Drenth, T., Goedemans, H.J., Meijlink, F., Zeller, R.: **Progression of vertebrate limb development through SHH-mediated counteraction of GLI3**. *Science* 2002, **298**: 827–830.

Tickle, C., Towers, M.: **Sonic hedgehog signaling in limb development**. *Front. Cell Dev. Biol.* 2017, **5**: 14.

Towers, M., Mahood, R., Yin, Y., Tickle, C.: **Integration of growth and specification in chick wing digit-patterning**. *Nature* 2008, **452**: 882–886.

Vokes, S.A., Ji, H., Wong, W.H., McMahon, A.P.: **A genome-scale analysis of the *cis*-regulatory circuitry underlying Shh-mediated patterning of the mammalian limb**. *Genes Dev.* 2008, **22**: 2651–2663.

Yang, Y., Drossopoulou, G., Chuang, P.T., Duprez, D., Marti, E., Bumcrot, D., Vargesson, N., Clarke, J., Niswander, L., McMahon, A., Tickle, C.: **Relationship between dose, distance and time in Sonic Hedgehog-mediated regulation of anteroposterior polarity in the chick limb**. *Development* 1997, **124**: 4393–4404.

Box 9B Too many fingers: mutations that affect antero-posterior patterning can cause polydactyly

Hill, R.E., Lettice, L.A.: **Alterations to the remote control of Shh gene expression cause congenital abnormalities**. *Trans. R. Soc. B* 2013, **368**: 20120357.

Lettice, L.A., Daniels, S., Sweeney, E., Venkataraman, S., Devenney, P.S., Gautier, P., Morrison, H., Fantes, J., Hill, R.E., FitzPatrick, D.R.: **Enhancer-adoption as a mechanism of human developmental disease**. *Hum. Mutat.* 2011, **32**: 1492–1499.

Lettice, L.A., Hill, A.E., Devenney, P.S., Hill, R.E.: **Point mutations in a distant Shh *cis*-regulator generate a variable regulatory output responsible for preaxial polydactyly**. *Hum. Mol. Genet.* 2008, **17**: 978–985.

Box 9C Sonic hedgehog signaling and the primary cilium

Bangs, F., Anderson, K.V.: **Primary cilia and mammalian hedgehog signaling**. *Cold Spring Harb. Perspect. Biol.* 2017, **9**: a028175.

Huangfu, D., Liu, A., Rakeman, A.S., Murcia, N.S., Niswander, L., Anderson, K.V.: **Hedgehog signalling in the mouse requires intraflagellar transport proteins**. *Nature* 2003, **426**: 83–87.

Lee, R.T., Zhao, Z., Ingham, P.W.: **Hedgehog signalling**. *Development* 2016, **143**: 367–372.

Box 9D External agents and the consequences of damage to the developing embryo

Donovan, K.A., An, J., Nowak, R.P., Yuan, J.C., Fink, E.C., Berry, B.C., Ebert, B.L., Fischer, E.S.: **Thalidomide promotes degradation of SALL4, a transcription factor implicated in Duane radial ray syndrome**. *eLife* 2018, **7**: e38430.

Ito, T., Ando, H., Suzuki, T., Ogura, T., Hotta, K., Imamura, Y., Yamaguchi, Y., Handa, H.: **Identification of a primary target of thalidomide teratogenicity**. *Science* 2010, **327**: 1345–1350.

Ito, T., Handa, H.: **Molecular mechanisms of thalidomide and its derivatives**. *Proc. Jpn Acad. Ser. B Phys. Biol. Sci.* 2020, **96**: 189–203.

Tabin, C.J.: **A developmental model for thalidomide defects**. *Nature* 1998, **396**: 322–323.

Therapontos, C., Erskine, L., Gardner, E.R., Figg, W.D., Vargesson, N.: **Thalidomide induces limb defects by preventing angiogenic outgrowth during early limb formation**. *Proc. Natl Acad. Sci. USA* 2009, **106**: 8573–8578.

Vargesson, N.: **The teratogenic effects of thalidomide on limbs**. *J. Hand Surg. Eur.* 2019, **44**: 88–95.

Wilkie, A.O.: **Why study human limb malformations?** *J. Anat.* 2003, **202**: 27–35.

9.17 The dorso-ventral pattern of the limb is controlled by the ectoderm

Arques, C.G., Doohan, R., Sharpe, J., Torres, M.: **Cell tracing reveals a dorsoventral lineage restriction plane in the mouse limb bud mesenchyme**. *Development* 2007, **134**: 3713–3722.

Geduspan, J.S., MacCabe, J.A.: **The ectodermal control of mesodermal patterns of differentiation in the developing chick wing**. *Dev. Biol.* 1987, **124**: 398–408.

Riddle, R.D., Ensini, M., Nelson, C., Tsuchida, T., Jessell, T.M., Tabin, C.: **Induction of the LIM homeobox gene *Lmx1* by Wnt-7a establishes dorsoventral pattern in the vertebrate limb**. *Cell* 1995, **83**: 631–640.

9.18 Development of the limb is integrated by interactions between signaling centers

Bénazet, J.D., Bischofberger, M., Tiecke, E., Gonçalves, A., Martin, J.F., Zuniga, A., Naef, F., Zeller, R.: **A self-regulatory system of interlinked signaling feedback loops controls mouse limb patterning**. *Science* 2009, **323**: 1050–1053.

Niswander, L., Jeffrey, S., Martin, G.R., Tickle, C.: **A positive feedback loop coordinates growth and patterning in the vertebrate limb**. *Nature* 1994, **371**: 609–612.

Pizette, S., Niswander, L.: **BMPs negatively regulate structure and function of the limb apical ectodermal ridge**. *Development* 1999, **126**: 883–894.

Zeller, R., Lopez-Rios, J., Zuniga, A.: **Vertebrate limb development: moving towards integrative analysis of organogenesis**. *Nat. Rev. Genet.* 2009, **10**: 845–855.

Zuniga, A., Zeller, R.: **Dynamic and self-regulatory interactions among gene regulatory networks control vertebrate limb bud morphogenesis**. *Curr. Top. Dev. Biol.* 2020, **139**: 61–88.

9.19 Hox genes control regional identity in the limb

Andrey, G., Montavon, T., Mascrez, B., Gonzalez, F., Noordermeer, D., Leleu, M., Trono, D., Spitz, F., Duboule, D.: **A switch between topological domains underlies HoxD genes collinearity in mouse limbs**. *Science* 2013, **340**: 1234167.

Beccari, L., Yakushiji-Kaminatsui, N., Woltering, J.M., Necsulea, A., Lonfat, N., Rodríguez-Carballo, E., Mascrez, B., Yamamoto, S., Kuroiwa, A., Duboule, D.: **A role for HOX13 proteins in the regulatory switch between TADs at the HoxD locus**. *Genes Dev.* 2016, **30**: 1172–1186.

Bolt, C.C., Duboule, D.: **The regulatory landscapes of developmental genes**. *Development* 2020, **147**: dev171736.

Goodman, F.R.: **Limb malformations and the human HOX genes**. *Am. J. Med. Genet.* 2002, **112**: 256–265.

Montavon, T., Duboule, D.: **Chromatin organization and global regulation of Hox gene clusters**. *Phil. Trans R. Soc.* 2013, **368**: 20120367.

Nelson, C.E., Morgan, B.A., Burke, A.C., Laufer, E., DiMambro, E., Muytaugh, L.C., Gonzales, E., Tessarollo, L., Parada, L.F., Tabin, C.: **Analysis of Hox gene expression in the chick limb bud**. *Development* 1996, **122**: 1449–1466.

Wellik, D.M., Capecchi, M.R.: **Hox10 and Hox11 genes are required to globally pattern the mammalian skeleton**. *Science* 2003, **301**: 363–367.

9.20 Self-organization might be involved in the development of the limb

Hardy, A., Richardson, M.K., Francis-West, P.N., Rodriguez, C., Izpisúa-Belmonte, J.C., Duprez, D., Wolpert, L.: **Gene expression, polarising activity and skeletal patterning in reaggregated hind limb mesenchyme**. *Development* 1995, **121**: 4329–4337.

Raspopovic, J., Marcon, L., Russo, L., Sharpe, J.: **Modeling digits. Digit patterning is controlled by a Bmp-Sox9-Wnt Turing network modulated by morphogen gradients**. *Science* 2014, **345**: 566–570.

Box 9E Reaction–diffusion mechanisms

Hiscock, T.W., Tschopp, P., Tabin, C.J.: **On the formation of digits and joints during limb development**. *Dev. Cell* 2017, **41**: 459–465.

Kondo, S., Miura, T.: **Reaction-diffusion model as a framework for understanding biological pattern formation**. *Science* 2010, **329**: 1616–1620.

Meinhardt, H.: **Turing's theory of morphogenesis of 1952 and the subsequent discovery of the crucial role of local self-enhancement and long-range inhibition**. *Interface Focus* 2012, **2**: 407–416.

Meinhardt, H., Gierer, A.: **Pattern formation by local self-activation and lateral inhibition**. *BioEssays* 2000, **22**: 753–760.

Raspopovic, J., Marcon, L., Russo, L., Sharpe, J.: **Modeling digits: digit patterning is controlled by a Bmp-Sox9-Wnt Turing network modulated by morphogen gradients**. *Science* 2014, **345**: 566–570.

Sheth, R., Marcon, L., Bastida, F.M., Junco, M., Quintana, L., Dahn, R., Kmita, M., Sharpe, J., Ros, M.A.: **Hox genes regulate digit patterning by controlling the wavelength of a Turing-type mechanism**. *Science* 2012, **338**: 1476–1480.

9.21 Limb muscle is patterned by the connective tissue

Anderson, C., Williams, V.C., Moyon, B., Daubas, P., Tajbakhsh, S., Buckingham, M.E., Shiroishi, T., Hughes, S.M., Borycki, A.G.: **Sonic hedgehog acts cell-autonomously on muscle precursor cells to generate limb muscle diversity**. *Genes Dev.* 2012, **26**: 2103–2117.

Besse, L., Sheeba, C.J., Holt, M., Labuhn, M., Wilde, S., Feneck, E., Bell, D., Kucharska, A., Logan, M.P.O.: **Individual limb muscle bundles are formed through progressive steps orchestrated by adjacent connective tissue cells during primary myogenesis**. *Cell Rep.* 2020, **30**: 3552–3565.

Hashimoto, K., Yokouchi, Y., Yamamoto, M., Kuroiwa, A.: **Distinct signaling molecules control Hoxa-11 and Hoxa-13 expression in the muscle precursor and mesenchyme of the chick limb bud**. *Development* 1999, **126**: 2771–2783.

Hu, J.K., McGlinn, E., Harfe, B.D., Kardon, G., Tabin, C.J.: **Autonomous and nonautonomous roles of Hedgehog signaling in regulating limb muscle formation**. *Genes Dev.* 2012, **26**: 2088–2102.

Robson, L.G., Kara, T., Crawley, A., Tickle, C.: **Tissue and cellular patterning of the musculature in chick wings**. *Development* 1994, **120**: 1265–1276.

Schweiger, H., Johnson, R.L., Brand-Sabin, B.: **Characterization of migration behaviour of myogenic precursor cells in the limb bud with respect to Lmx1b expression**. *Anat. Embryol.* 2004, **208**: 7–18.

Shellswell, G.B., Wolpert, L.: **The pattern of muscle and tendon development in the chick wing**. In *Limb and Somite Morphogenesis*. Ede, D.A., Hinchliffe, J.R., Balls, M. (Eds). Cambridge: Cambridge University Press, 1977: 71–86.

9.22 The initial development of cartilage, muscles, and tendons is autonomous

Huang, A.H.: **Coordinated development of the limb musculoskeletal system: tendon and muscle patterning and integration with the skeleton**. *Dev. Biol.* 2017, **429**: 420–428.

Kardon, G.: **Muscle and tendon morphogenesis in the avian hind limb**. *Development* 1998, **125**: 4019–4032.

Ros, M.A., Rivero, F.B., Hinchliffe, J.R., Hurle, J.M.: **Immunohistological and ultrastructural study of the developing tendons of the avian foot**. *Anat. Embryol.* 1995, **192**: 483–496.

Shellswell, G.B., Wolpert, L.: **The pattern of muscle and tendon development in the chick wing**. In *Limb and Somite Morphogenesis*. Ede, D.A., Hinchliffe, J.R., Balls, M. (Eds). Cambridge: Cambridge University Press, 1977: 71–86.

9.23 Joint formation involves secreted signals and mechanical stimuli

Hartmann, C., Tabin, C.J.: **Wnt-14 plays a pivotal role in inducing synovial joint formation in the developing appendicular skeleton.** *Cell* 2001, **104**: 341–351.

Kahn, J., Shwartz, Y., Blitz, E., Krief, S., Sharir, A., Breitel, D.A., Rattenbach, R., Relaix, F., Maire, P., Rountree, R.B., Kingsley, D.M., Zelzer, E.: **Muscle contraction is necessary to maintain joint progenitor cell fate.** *Dev. Cell* 2009, **16**: 734–743.

Khan, I.M., Redman, S.N., Williams, R., Dowthwaite, G.P., Oldfield, S.F., Archer, C.W.: **The development of synovial joints.** *Curr. Top. Dev. Biol.* 2007, **79**: 1–36.

Koyama, E., Ochiai, T., Rountree, R.B., Kingsley, D.M., Enomoto-Iwamoto, M., Iwamoto, M., Pacifici, M.: **Synovial joint formation during mouse limb skeletogenesis: roles of Indian hedgehog signaling.** *Ann. NY Acad. Sci.* 2007, **1116**: 100–112.

Scoones, J.C., Hiscock, T.W.: **A dot-stripe Turing model of joint patterning in the tetrapod limb.** *Development* 2020, **147**: dev183699.

9.24 Separation of the digits is the result of programmed cell death

Garcia-Martinez, V., Macias, D., Gañan, Y., Garcia-Lobo, J.M., Francia, M.V., Fernandez-Teran, M.A., Hurle, J.M.: **Internucleosomal DNA fragmentation and programmed cell death (apoptosis) in the interdigital tissue of the embryonic chick leg bud.** *J. Cell. Sci.* 1993, **106**: 201–208.

Hernandez-Martinez, R., Covarrubias, L.: **Interdigital cell death function and regulation: new insights on an old programmed cell death model.** *Dev. Growth Diff.* 2011, **53**: 245–258.

Lorda-Diez, C.I., Montero, J.A., Garcia-Porrero, J.A., Hurle, J.M.: **Interdigital tissue regression in the developing limb of vertebrates.** *Int. J. Dev. Biol.* 2015, **59**: 55–62.

9.25 Tooth development involves epithelial–mesenchymal interactions and a homeobox gene code specifies tooth identity

Cobourne, M.T., Sharpe, P.T.: **Making up the numbers: the molecular control of mammalian dental formula.** *Semin. Cell Dev. Biol.* 2010, **21**: 314–324.

Jernvall, J., Thesleff, I.: **Tooth shape formation and tooth renewal: evolving with the same signals.** *Development* 2012, **139**: 3487–3497.

Jussila, M., Thesleff, I.: **Signaling networks regulating tooth organogenesis and regeneration, and the specification of dental mesenchymal and epithelial cell lineages.** *Cold Spring Harb. Perspect. Biol.* 2012, **4**: a008425.

Mogollón, I., Moustakas-Verho, J.E., Niittykoski, M., Ahtiainen, L.: **The initiation knot is a signaling center required for molar tooth development.** *Development* 2021, **148**: dev194597.

Oshima, M., Tsuji, T.: **Whole tooth regeneration using a bioengineered tooth.** In *New Trends in Tissue Engineering and Regenerative Medicine—Official Book of the Japanese Society for Regenerative Medicine.* Hibi, H. (Ed.). London: InTech, 2014.

Sharpe, P.: **Regenerative dentistry.** *Front. Dent. Med.* 2020, **1**: 3.

Tucker, A., Sharpe, P.: **The cutting-edge of mammalian development: how the embryo makes teeth.** *Nat. Rev. Genet.* 2004, **5**: 499–508.

9.26 The vertebrate lung develops from buds of endoderm

Han, L., Chaturvedi, P., Kishimoto, K., Koike, H., Nasr, T., Iwasawa, K., Giesbrecht, K., Witcher, P.C., Eicher, A., Haines, L., Lee, Y., Shannon, J.M., Morimoto, M., Wells, J.M., Takebe, T., Zorn, A.M.: **Single cell transcriptomics identifies a signaling network coordinating endoderm and mesoderm diversification during foregut organogenesis.** *Nat. Commun.* 2020, **11**: 4158.

Herriges, M., Morrisey, E.E.: **Lung development: orchestrating the generation and regeneration of a complex organ.** *Development* 2014, **141**: 502–513.

Rankin, S. A., Han, L., McCrackern, K.W., Kenny, A.P., Anglin, C.T., Grigg, E.A., Crawford, C.M., Wells, J.M., Shannon, J.M., Zorn, A.M.: **A retinoic acid-hedgehog cascade coordinates mesoderm-inducing signals and endoderm competence during lung specification.** *Cell Rep.* 2016, **16**: 66–78.

9.27 Morphogenesis of the lung involves three modes of branching

Barkauskas, C.E., Chung, M.I., Fioret, B., Gao, X., Katsura, H., Hogan, B.L.: **Lung organoids: current uses and future promise.** *Development* 2017, **144**: 986–997.

Goodwin, K., Nelson, C.M.: **Branching morphogenesis.** *Development* 2020, **147**: dev184499.

Metzger, R.J., Klein, O.D., Martin, G.R., Krasnow, M.A.: **The branching program of mouse lung development.** *Nature* 2008, **453**: 745–750.

Ochoa-Espinosa, A., Affolter, M.: **Branching morphogenesis: from organs to cells and back.** *Cold Spring Harb. Perspect. Biol.* 2012, **4**: a008243.

Pepicelli, C.V., Lewis, P.M., McMahon, A.P.: **Sonic hedgehog regulates branching morphogenesis in the mammalian lung.** *Curr. Biol.* 1998, **8**: 1083–1086.

Spurlin, J.W., Nelson, C.M.: **Building branched tissue structures: from single cell guidance to coordinated construction.** *Phil. Trans. R. Soc. Lond. B Biol. Sci.* 2017, **372**: 20150527.

Young, R.E., Jones, M.K., Hines, E.A., Li, R., Luo, Y., Shi, W., Verheyden, J.M., Sun, X.: **Smooth muscle differentiation is essential for airway size, tracheal cartilage segmentation, but dispensable for epithelial branching.** *Dev. Cell* 2020, **53**: 73–85.

9.28 The mammary glands develop from buds of ectoderm

Howard, B., Panchal, H., McCarthy, A., Ashworth, A.: **Identification of the *scaramanga* gene implicates Neuregulin3 in mammary gland specification.** *Genes Dev.* 2005, **19**: 2078–2090.

Tickle, C., Jung, H.-S.: **Embryonic mammary gland development.** In *eLS*. Chichester: John Wiley & Sons, 2016. [https://doi.org/10.1002/9780470015902.a0026057] (date accessed 25 November 2022).

Veltmaat, J.M., Mailleux, A.A., Thiery, J.P., Bellusci, S.: **Mouse embryonic mammogenesis as a model for the molecular regulation of pattern formation.** *Differentiation* 2003, **71**: 1–17.

Veltmaat, J.M., Ramsdell, A.F., Sterneck, E.: **Positional variations in mammary gland development and cancer.** *J. Mamm. Gland Biol. Neoplasia* 2013, **18**: 179–188.

Box 9F What developmental biology can tell us about cancer

Howard, B., Ashworth, A.: **Signalling pathways implicated in early mammary gland morphogenesis and breast cancer**. *PLoS Genet.* 2006, **2**: e112.

Ilina, O., Gritsenko, P.G., Syga, S., Lippoldt, J., La Porta, C.A.M., Chepizhko, O., Grosser, S., Vullings, M., Bakker, G.J., Starruß, J., Bult, P., Zapperi, S., Käs, J.A., Deutsch, A., Friedl, P. **Cell-cell adhesion and 3D matrix confinement determine jamming transitions in breast cancer invasion**. *Nat. Cell Biol.* 2020, **22**: 1103–1115.

Karnoub, A.E., Dash, A.B., Vo, A.P., Sullivan, A., Brooks, M.W., Bell, G.W., Richardson, A.L., Polyak, K., Tubo, R., Weinberg, R.A.: **Mesenchymal stem cells within tumour stroma promote breast cancer metastasis**. *Nature* 2007, **449**: 557–563.

Mohan, S.C., Lee, T.Y., Giuliano, A.E., Cui, X.: **Current status of breast organoid models**. *Front. Bioeng. Biotechnol.* 2021, **9**: 745943.

9.29 The branching pattern of the mammary gland epithelium is determined by the mesenchyme

Goodwin, K., Nelson, C.M.: **Branching morphogenesis**. *Development* 2020, **147**: dev184499.

Sakakura, T., Nishizuka, Y., Dawe, C.J.: **Mesenchyme-dependent morphogenesis and epithelium-specific cytodifferentiation in mouse mammary gland**. *Science* 1976, **194**: 1439–1441.

Shackleton, M., Vaillant, F., Simpson, K.J., Stingl, J., Smyth, G.K., Asselin-Labat, M.L., Wu, L., Lindeman, G.J., Visvader, J.E.: **Generation of a functional mammary gland from a single stem cell**. *Nature* 2006, **439**: 84–88.

Sternlicht, M.D.: **Key stages in mammary gland development: the cues that regulate ductal branching morphogenesis**. *Breast Cancer Res.* 2006, **8**: 201.

Twigger, A.J., Khaled, W.T.: **Mammary gland development from a single cell 'omics view**. *Semin. Cell Dev. Biol.* 2021, **114**: 171–185.

9.30 The development of the vertebrate heart involves morphogenesis and patterning of a mesodermal tube

Courchaine, K., Rykiel, G., Rugonyi, S.: **Influence of blood flow on cardiac development**. *Prog. Biophys. Mol. Biol.* 2018, **137**: 95–110.

Linask, K.K., Yu, X., Chen, Y., Han, M.D.: **Directionality of heart looping: effects of Pitx2c misexpression on flectin asymmetry and midline structures**. *Dev. Biol.* 2002, **246**: 407–417.

Srivastava, D., Olson, E.N.: **A genetic blueprint for cardiac development**. *Nature* 2000, **407**: 221–226.

9.31 Several different cell populations contribute to making a heart

Diogo, R., Kelly, R.G., Christiaen, L., Levine, M., Ziermann, J.M., Molnar, J.L., Noden, D.M., Tzahor, E.: **A new heart for a new head in vertebrate cardiopharyngeal evolution**. *Nature* 2015, **520**: 466–473.

Kelly, R.G., Buckingham, M.E., Moorman, A.F.: **Heart fields and cardiac morphogenesis**. *Cold Spring Harb. Perspect. Med.* 2014, **4**: a015750.

Meilhac, S.M., Esner, M., Kelly, R.G., Nicolas, J.F., Buckingham, M.E.: **The clonal origin of myocardial cells in different regions of the embryonic mouse heart**. *Dev. Cell* 2004, **6**: 685–698.

Perez-Pomares, J.M., de la Pompa, J.L.: **Signalling during epicardium and coronary vessel development**. *Circ. Res.* 2011, **109**: 1429–1442.

Ray, H.J., Niswander, L.: **Mechanisms of tissue fusion during development**. *Development* 2012, **139**: 1701–1711.

Takeuchi, J.K., Bruneau, B.G.: **Directed transdifferentiation of mouse mesoderm to heart tissue by defined factors**. *Nature* 2009, **459**: 708–711.

Vincent, S.D., Buckingham, M.E.: **How to make a heart: the origin and regulation of cardiac progenitor cells**. *Curr. Top. Dev. Biol.* 2010, **90**: 1–41.

9.32 Key transcription factors specify development of the heart and cardiac differentiation

Bruneau, B.G.: **The developmental genetics of congenital heart disease**. *Nature* 2008, **451**: 943–948.

Bruneau, B.G.: **The developing heart: from *The Wizard of Oz* to congenital heart disease**. *Development* 2020, **147**: dev194233.

9.33 Development of the vertebrate eye involves interactions between an extension of the forebrain and the ectoderm of the head

Adler, R., Canto-Soler V.: **Molecular mechanisms of optic vesicle development: complexities, ambiguities and controversies**. *Dev. Biol.* 2007, **305**: 1–13.

Eiraku, M., Takata, N., Ishibashi, H., Kawada, M., Sakakura, E., Okuda, S., Sekiguchi, K., Adachi, T., Sasai, Y.: **Self-organizing optic-cup morphogenesis in three-dimensional culture**. *Nature* 2011, **427**: 51–56.

Gabriel, E., et al.: **Human brain organoids assemble functionally integrated bilateral optic vesicles**. *Cell Stem Cell* 2021, **28**: 1740–1757.

Gehring, W.J.: **New perspectives on eye development and the evolution of eyes and photoreceptors**. *J. Hered.* 2005, **96**: 171–184.

Heavner, W., Pevney, L.: **Eye development and retinogenesis**. *Cold Spring Harb. Perspect. Biol.* 2012, **4**: a008391.

Nakano, T., Ando, S., Takata, N., Kawada, M., Muguruma, K., Sekiguchi, K., Saito, K., Yonemura, S., Eiraku, M., Sasai, Y.: **Self-formation of optic cups and storable stratified neural retina from human ESCs**. *Cell Stem Cell* 2012, **10**: 771–785.

Sasai, Y., Eiraku, M., Suga, H.: **In vitro organogenesis in three-dimensional self-organizing stem cells**. *Development* 2012, **139**: 4111–4121.

Streit, A.: **The preplacodal region: an ectodermal domain with multipotential progenitors that contribute to sense organs and cranial sensory ganglia**. *Int. J. Dev. Biol.* 2007, **51**: 447–461.

9.34 Eye-specifying transcription factors are conserved

Takahashi, S., Asashima, M., Kurata, S., Gehring, W.J.: **Conservation of Pax6 function and upstream activation by Notch signaling in eye development of frogs and flies**. *Proc. Natl Acad. Sci. USA* 2002, **99**: 2020–2025.

10

Development of the nervous system

- Specification of cell identity in the nervous system
- The formation and migration of neurons
- Axon navigation and mapping
- Synapse formation and map refinement

The nervous system is the most complex of all the organ systems in the animal embryo. In mammals, for example, billions of nerve cells, or neurons, develop into highly organized networks that characterize all parts of the functioning nervous system, including the brain. Many hundreds, even thousands, of different types of neuron are generated throughout development, differing in their location and identity. Once specified, neurons extend axons that are guided to their targets, forming connections that become refined over time. Nervous system development can be understood in terms of fundamental processes, including cell specification, proliferation, differentiation, migration, and survival. As we decipher these processes, we gain insight into human congenital and degenerative conditions as well as regenerative approaches and human behaviors.

Introduction

The nervous system is the most complex of all the organ systems in the animal embryo and its formation in embryogenesis represents a truly remarkable feat of self-organization. The human brain, for example, has nearly 100 billion **nerve cells**, or **neurons**, each forming thousands of connections. Remarkable progress over the past decades means that—while our knowledge is still patchy—we are coming to understand how, in development, these billions of neurons form into organized subsets and come together to generate the circuits of the nervous system that underlie function and behavior.

In all nervous systems—vertebrate and invertebrate—electrically excitable neurons that differ in their location, shape, size, and patterns of connectivity form highly organized networks that provide a living communication system (Fig. 10.1). A neuron

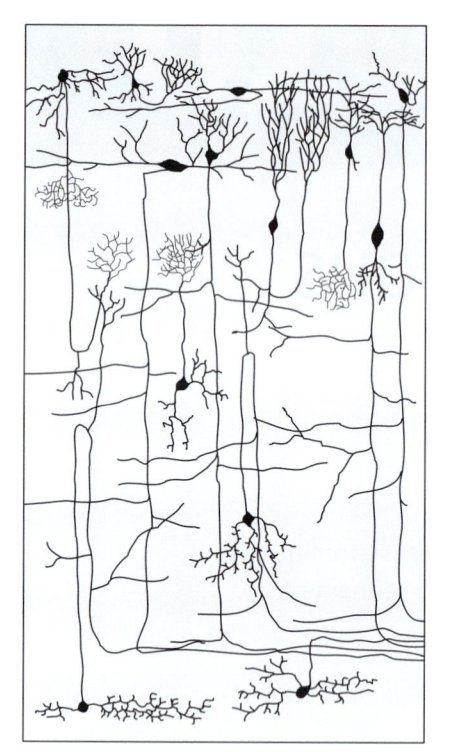

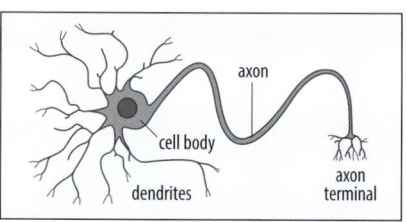

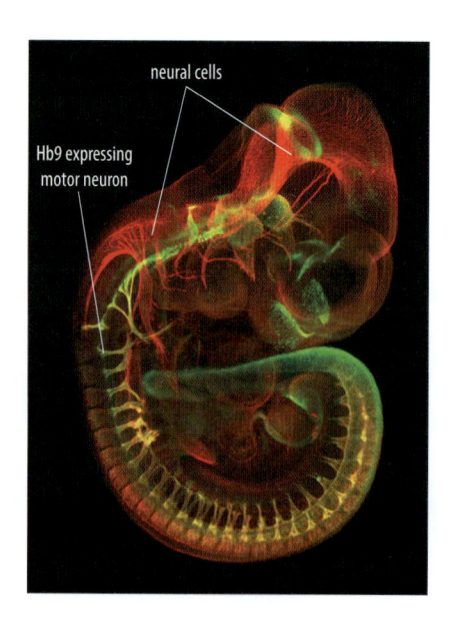

Fig. 10.1 Neurons come in many shapes and sizes. Upper panel: a montage of the various shapes and axon trajectories of neurons found in one part of the chick brain. Only some of the many neuronal connections are shown here. Lower panel: a simple type of neuron. This single neuron consists of a small, rounded cell body that contains the nucleus, a single axon, and a 'tree' of much-branched dendrites. Signals from other neurons are received at the dendrites, are processed and integrated, and an outgoing signal—a nerve impulse—is generated and sent down the axon to the axon terminal.

receives multiple inputs from its environment, including other neurons, through its highly branched **dendrites**. If strong enough, the inputs generate an electrical signal—a **nerve impulse**, or **action potential**—at the **cell body**, which is conducted along the **axon** to the **axon terminal**. The extensively branched axon terminals form junctions—**synapses**—with the dendrites or cell bodies of other neurons, or other cells. Axon terminals and dendrites of individual neurons are extensively branched, and a single neuron in the central nervous system can form hundreds, and receive thousands, of synaptic connections. At a synapse, the electrical signal is converted into a chemical signal, which alters the properties—chemical, electrical, or both—of the target cell. Different types of neuron relay information using various short peptide-based chemical signals, mainly **neurotransmitters**, but in some neurons, **neuropeptides** and **neurohormones**.

In addition to harboring extensive neuronal networks, the nervous system contains billions of non-neuronal cells, called **glial cells**. The types and functions of glial cells (collectively called **glia**) are rich and varied, but frequently, they provide support and protection to neurons. Neurons and glia are collectively termed neural cells. It is important to remember that the term 'neural' relates generally to nervous system tissue, including both neurons and glia and/or their progenitors, whereas the term 'neuronal' refers only to neurons.

The functioning nervous system—responsible for behaviors, emotions, and memories—emerges through the integrated activity of its different types of cell. The nervous system can only function properly if each neuronal type is present, correctly connected, adequately supported and protected, and making the appropriate chemical signal. Central questions in nervous system development are: how many unique or similar neuron subsets there are; how particular types of neuron form in the right place in the nervous system; how connections develop with the appropriate specificity; how many glial subsets exist and how do they support neuronal development. As we will see later in this chapter, increasing evidence indicates that altered developmental processes during the formation of the nervous system in humans can result in complex conditions, including neuropsychiatric conditions in later life.

Anatomically, the nervous system is divided into the **central nervous system (CNS)**, comprising the brain and spinal cord, and the **peripheral nervous system (PNS)**, the nerve cells and ganglia that lie outside the brain and spinal cord and connect the rest of the body to the CNS (Fig. 10.2). Neurons in the spatially delimited CNS receive sensory inputs, process them, integrate them centrally, and produce outputs that guide all behaviors, as wide-ranging as movement and sleep. Such an organization broadly defines the CNS of animals as far apart evolutionarily as annelid worms and mammals, and is in contrast to the diffuse nervous system, or nerve net, of cnidarians

Fig. 10.2 The nervous system of a mouse embryo at 11 days of gestation. Different parts of the nervous system of a mouse embryo have been revealed by immunocytochemical labeling with an antibody that detects all neural cells (red) and an antibody that detects the motor neuron protein Hb9 (green). The brain, the cranial sensory ganglia, and the segmental array of spinal nerves are particularly prominent.
Photograph courtesy of I. Lieberam and T. Jessell.

(*Hydra* and sea anemones), which receives sensory input and produces output without central integration.

In this chapter, we shall examine nervous system development using examples from the CNS of both vertebrate and invertebrate model organisms. For the sake of brevity, we focus largely on neuronal development, and mention only a few glial cell types. In all animals, the nervous system develops in early embryogenesis, a period when the mechanisms used for building it show extensive conservation between organisms. In humans, the main regions of the CNS are apparent by 12 weeks post-fertilization. Much of our conceptual understanding of core mechanisms has been gained through observations and experiments in the simpler nervous systems of invertebrates. This raises the question: do the CNS of vertebrates and invertebrates trace back to a common ancestral type? Several lines of evidence suggest that they might. First, in both vertebrates and invertebrates, development of the nervous system begins when a subset of ectoderm cells are set aside as neuroectoderm cells (see Sections 2.5 and 4.3). Second, in all the Bilateria—animals with bilateral symmetry that include worms, flies, and vertebrates—the brain and associated sensory organs develop in a common position, at the anterior (head) end of the embryo. Third, in the annelid worm *Platynereis*, the neuroectoderm is divided longitudinally into domains of neuronal progenitor cells (cells that are specified to give rise to neurons) that correspond to similar domains in the vertebrate neural tube. Finally, invertebrates and vertebrates use some of the same signals to pattern the developing CNS, and to direct neuron formation. All this evidence suggests that a CNS architecture was present in the last common ancestor of the bilaterians, and supports a common origin for the nervous system in Bilateria.

We can recognize eight stages of nervous system development: the induction of neuroectoderm and neurulation; the formation of brain and spinal cord; the formation of the CNS and the PNS; the specification of individual neural cell identity, and the switch from a proliferating progenitor or precursor cell to a **post-mitotic neuron** that will not divide again; the further differentiation of neurons and their migration; the outgrowth of axons to their targets; the formation of synapses with target cells; and the refinement of synaptic connections through the elimination of axon branches and neuronal death. Despite its complex development, the nervous system is the product of the same kinds of process as those involved in the development of other organs: cell specification, cell movements in morphogenesis, cell differentiation, cell migration, and cell survival. These processes—described collectively as cell behaviors—operate throughout these eight stages.

We have already considered some aspects of the development of the vertebrate nervous system, including neural induction (see Section 4.3), the formation of the neural tube (neurulation) (see Section 5.17), formation of the brain versus the spinal cord (see Chapter 4), and the formation of the PNS from neural crest cells (see Section 4.15). We will revisit some of these events in this chapter. A single chapter, such as this one, cannot cover all aspects of nervous system development, and the general focus here is on the mechanisms that regionalize the CNS, control the identity of neurons, and pattern their connections. We selectively illustrate how developmental approaches are providing insight into congenital and degenerative conditions, regenerative approaches, and human behavior.

Specification of cell identity in the nervous system

Neural tissue is formed when the body plan is being established, as discussed in Chapters 3 and 4. At this stage it is composed of proliferating progenitor cells. We now consider the next stage in the process of generating a nervous system: how the major regions of the CNS develop, each with its characteristic size and shape, and repertoire

Fig. 10.3 Local signaling centers regionalize the developing vertebrate brain. Lateral view of the brain of a 2-day chick embryo (Hamilton and Hamburger (HH) stage 13; left panel) and of a 4.5-day chick embryo (stage HH24; right panel). Local signaling centers—the midbrain-hindbrain boundary (MHB), the zona limitans intrathalamica (ZLI), the anterior neural ridge (ANR), and the prechordal mesendoderm—are indicated. Bidirectional signaling from the MHB and ZLI, and unidirectional signaling from the ANR and prechordal mesendoderm (indicated by black arrows) results in the patterning of adjacent brain regions, and involves secretion of fibroblast growth factor (FGF)-8 and Wnt-1 (red) from the MHB, FGF-8 (blue) from the ANR, Sonic hedgehog (Shh, green) from the ZLI, and Shh and Nodal from the prechordal mesendoderm. The boundary between expression of the transcription factor genes *Otx2* in the midbrain (violet) and *Gbx2* in the hindbrain (yellow) is linked to the formation of the MHB.

Adapted from Kiecker, C., Lumsden, A.: Compartments and their boundaries in vertebrate brain development. Nat. Rev. Neurosci. 2005, 6: 553-564.

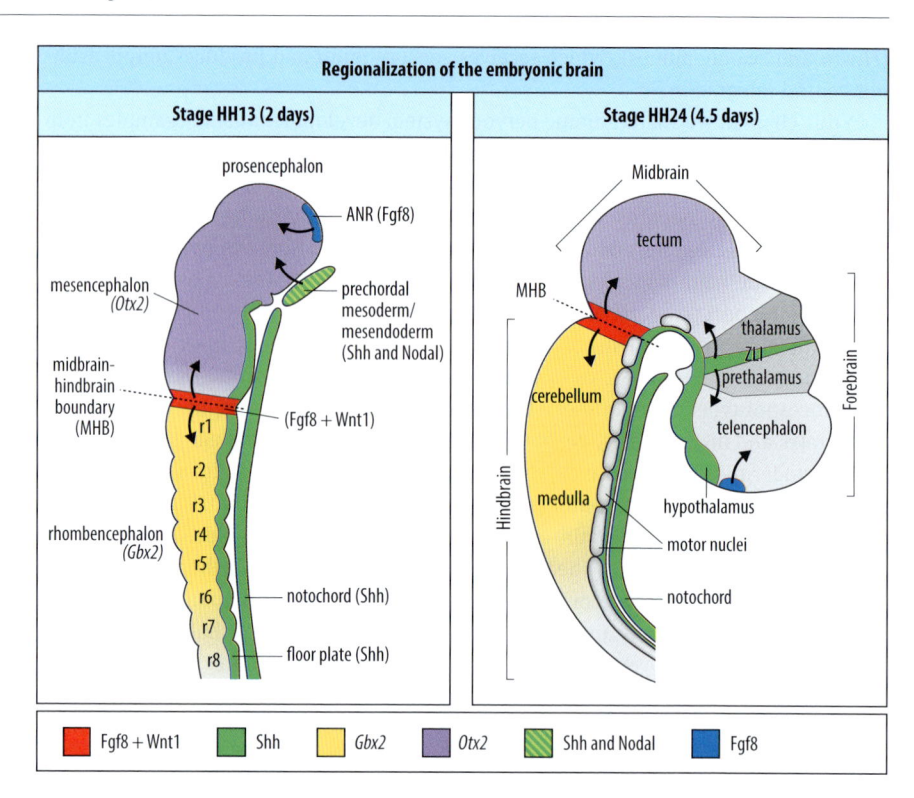

of neuronal cell types. How, for example, does the early neural tissue give rise to forebrain, midbrain, and hindbrain? And then, how do the many distinct post-mitotic neuronal types present in the CNS acquire their correct individual identities, each in its correct position with respect to the whole?

10.1 The vertebrate brain is organized into three main regions

In Chapters 3 and 4 we defined two major types of neural progenitor cell—neuroectoderm cells of the neural plate (formed through neural induction) and neuromesodermal progenitors. In many vertebrates, including mammals, progenitor cells in the neural plate give rise to the brain, whereas neuromesodermal progenitors generate most of the spinal cord (see Fig. 4.27). We start this chapter by looking at how the brain begins to develop from the neural plate and divides into three main regions—the **forebrain** (prosencephalon), **midbrain** (mesencephalon), and **hindbrain** (rhombencephalon) (Fig. 10.3). A considerable amount of cell proliferation, migration, and morphogenesis occurs over this period, changing the shape of the developing brain: for instance, lateral expansions develop from the embryonic forebrain—the rudiments of the optic vesicles—from which the eyes will form (see Section 9.33).

The embryonic forebrain gives rise to the **pallium** and basal ganglia (the **telencephalon**), the **thalamus**, and the **hypothalamus**. The pallium of mammalian embryos develops into the outer layers of the bilateral cerebral hemispheres, the **cerebral cortex**, where the highest-level processing centers for sensory information, motor control, learning, and memory are located. The thalamus is a relay station that distributes incoming sensory information to the appropriate region of the **cerebral cortex**, while the hypothalamus controls a broad range of behaviors and homeostatic physiological processes, including the sleep–wake cycle, hunger, thirst, response to stress, and a range of social, sexual, and emotional behaviors. The embryonic midbrain gives rise to the dorsally located **tectum**, a relay station, and to ventrally located nuclei such as the substantia nigra, that control movement. The embryonic hindbrain gives rise to the **cerebellum** (rhombomere 1), the pons (rhombomeres 2 and 3), and the **medulla oblongata**, which, with the hypothalamus, regulate basic body activities that are independent of conscious control.

At present, many aspects of our understanding of brain development are in their infancy. We still do not know, for instance, how many functionally distinct neuronal types there are in each brain region, nor how many distinct neural progenitor cell types can be defined. However, new technologies such as single-cell RNA sequencing (see Box 6C) and multiplex *in situ* hybridization (see Fig. 3.5) are rapidly filling this knowledge gap by enabling a more accurate categorization of the different neuronal and glial subtypes, and their progenitor cells, in the developing nervous system. These new studies will build on the strong foundations laid by studies performed over the past few decades, which have begun to show how distinct post-mitotic neuronal types present in distinct brain regions acquire their correct individual identities, each in its correct position. This occurs through a process of continuous and progressive steps that transform progenitor cells into coarsely regionalized tissue, then into finer subdomains and even finer progenitor subsets and, ultimately, defined post-mitotic neuronal types.

10.2 The vertebrate brain is regionalized by local signaling centers

Local signaling centers (sometimes called **local organizers** or **secondary organizers**), some located next to, and some within the developing brain, direct the specification of the brain's major sub-regions (see Fig. 10.3). The **prechordal mesendoderm** (composed of prechordal mesoderm and associated head endoderm), lying ventral to the developing forebrain, induces the hypothalamus. The **anterior neural ridge (ANR)**, which lies where the anterior-most neural tissue meets non-neural ectoderm, is responsible for telencephalic development. The **midbrain–hindbrain boundary (MHB)**, a conspicuous constriction that lies between these developing brain regions, patterns the posterior midbrain (tectum) and the anterior hindbrain (cerebellum). The **zona limitans intrathalamica (ZLI)** in the posterior forebrain regulates patterning of the thalamus and prethalamus. The MHB, prechordal mesendoderm, and the ANR are already major signaling centers in the 2-day chick brain and are still active in the brain of 4.5-day chick embryos, whereas the ZLI is formed later and acts as a signaling center in the 4.5-day chick brain.

Ablation and grafting experiments in chick embryos provided the first evidence that each of these centers provides signals that confer broad regional identity on adjacent neural tissue. As an example, if prechordal mesendoderm is removed, hypothalamic cell types do not form; conversely, if prechordal mesendoderm is grafted next to the anterior-dorsal forebrain, it causes ectopic hypothalamic cells to form there. The signals deriving from each local organizer were subsequently identified through a combination of *in situ* hybridization, or immunohistochemical techniques (to determine the genes that are transcribed and translated in each center), and pharmacological and/or genetic approaches that test the function of each gene. For instance, *Shh* and *Nodal* are expressed in prechordal mesendoderm, and transgenic mouse lines genetically altered to lack components of either the Shh or Nodal signaling pathways fail to develop a hypothalamus. Similarly, pharmacological blockade of either Shh or Nodal signaling pathways prevents hypothalamic development, whereas overexpression of Shh or Nodal signaling pathways leads to ectopic hypothalamic tissue. Therefore Shh and Nodal work together to control the development of the hypothalamus. The pathways that cause the hypothalamus to form are evolutionarily conserved, and these animal studies have provided great insight into human **congenital conditions** (Box 10A).

In the case of the ZLI and the MHB, a single signal diffuses in two directions into neuroectoderm that is differently pre-patterned on each side. In this way a single signal can direct the development of two different brain regions. Shh, diffusing from the ZLI, induces the prethalamus anteriorly and the thalamus posteriorly. Similarly, FGF-8, diffusing from the MHB, directs the tectum anteriorly and the cerebellum posteriorly

MEDICAL BOX 10A Alcohol exposure and brain development

Many common human congenital conditions arise from a poorly understood combination of genetic and environmental factors. Environmental agents are thought to affect the penetrance of a predisposing mutation (i.e. the likelihood that a phenotype will be observed), but this has been difficult to demonstrate. A relatively common congenital condition known as holoprosencephaly arises when the hypothalamus fails to develop properly and affects development of the brain and face (see Box 1H). Mutations in *SHH*, and in the genes in Shh and Nodal signaling pathway components, increase the risk of holoprosencephaly. Environmental factors, such as exposure to alcohol in the womb, can also contribute to this condition, and there is a 'critical' period in gestation when alcohol increases the risk of holoprosencephaly (see also critical period in Box 9D). A study in mice has shown that the combination of alcohol and a mutation in a gene called *Cdon*, encoding a co-receptor which can influence Shh signaling, changes Nodal signaling, and this combination of gene and environment, occurring in a critical time window, can lead to holoprosencephaly (Fig. 1). This understanding could lead to new strategies to decrease the risk of holoprosencephaly in the future.

Figure 1 shows the experiment that identified that alcohol increases the risk of holoprosencephaly in mice in which the Shh pathway is already compromised. Developing mouse embryos that lack *Cdon* (*Cdon-/-*) were exposed to a saline control or to alcohol (ethanol) at embryonic day E7.25 or E7.5. Embryos were allowed to develop for a week, until E14.0, when features of the brain and face are becoming obvious. Frontal views of E14.0 embryos show that treatment of *Cdon-/-*embryos with alcohol at E7.25, but not at E7.5, results in holoprosencephaly. The *Cdon-/-* embryo treated with alcohol at E7.25 has a fused upper lip and a single nostril (white arrow), whereas the *Cdon-/-*embryo treated with alcohol at E7.5 does not, and looks the same as a saline control (red arrow). The critical time period is when Nodal and Shh induce the hypothalamus.

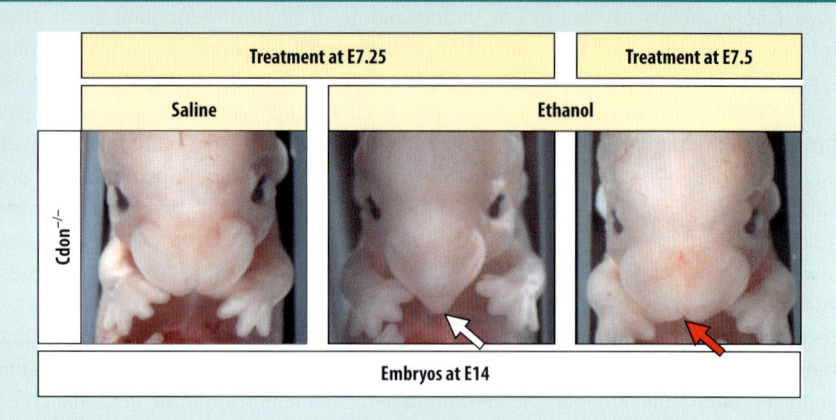

Fig. 1
Adapted with permission from Hong, M., et al.: ***Cdon* mutation and fetal alcohol converge on Nodal signaling in a mouse model of holoprosencephaly.** eLife 2020, 9: e60351.

(see Fig. 10.3). If an FGF-8-soaked bead is grafted into the anterior midbrain, the presumptive thalamus is converted into tectum, whereas if grafted to the posterior hindbrain, presumptive medulla is converted into cerebellum. As part of its action, FGF-8 controls the activity of a second signal, Wnt-1, activating it in cells just anterior to the boundary, and repressing it in cells posterior to the boundary. This shows how a simple signaling center can rapidly develop into a more sophisticated one.

Many of the diffusible signals released from each local signaling center act as morphogens that induce the expression of different transcription factors in neighboring cells depending on the concentration of morphogen the cell is exposed to (see Section 1.15 and Box 9A). These transcription factors are readouts of a cell's positional value and are involved in translating this positional value into the cell's final identity.

The signals released from each local signaling center can interact with the transcription factors they induce, to govern tissue size as well as tissue identity. For instance, FGF-8 deriving from the anterior neural border induces the expression of the gene *Foxg1* (encoding the transcription factor Foxg1) that is essential for development of the telencephalon. Foxg1 in turn maintains expression of *Fgf8*, and the relative strength of the FGF-8 signal affects the size and nature of telencephalic subdomains, as it governs cell proliferation as well as cell patterning. This positively reinforcing circuit, consisting of a transcription factor and a signal molecule, is thought to be key to the evolution of the telencephalon, the part of the brain that will give rise to the cortex in mammals.

10.3 Parts of the nervous system develop through a process of self-organization

In many parts of the developing nervous system, cell behaviors and organization at one point in time predict and prefigure cell behaviors and cell organization at a successive time point. The process through which a tissue—or part of a tissue—develops without the need for additional inputs from an external source is called **self-organization** (see also Box 7B).

As an example, antero-posterior patterning of the neural plate (also commonly termed the neuroectoderm or neuroepithelium) is achieved through graded cell signaling, mediated by retinoic acid and members of the FGF and Wnt families, which regulates the spatial expression of transcription factors that specify broad regional identity. As a result, the transcription factor Otx2 is expressed in the anterior neural plate, while the transcription factor Gbx2 is expressed in the posterior neural plate (Fig. 10.4). The interface of Otx2 and Gbx2 prefigures and predicts the MHB because genes whose expression characterizes the MHB, such as *Fgf8*, are upregulated at their interface. As described in Section 10.2, FGF-8 then diffuses out of the MHB to act on cells that express Otx2 (on the anterior side) and Gbx2 (on the posterior side), altering their transcriptomes and so initiating the 'next step'. Gbx2-positive cells posterior to the border now begin to show a transcriptome that defines the developing cerebellum, whereas cells anterior to the border begin to show transcriptomes that define the tectum and the substantia nigra (see Fig. 10.4). This two-way conversation, where neighboring cell groups communicate back and forth to take each other to the next step of development, explains how a self-organizing system develops accurately and predictably, and explains how early events prefigure later events that assemble the complex nervous system. Among the genes induced in the developing substantia nigra by FGF-8 is that for semaphorin, a long-range guidance cue. Semaphorin repels the axons of post-mitotic midbrain dopaminergic neurons—that were also generated in response to FGF-8—towards their correct targets (see Fig. 10.4).

Fig. 10.4 Promotion of self-organization at the midbrain-hindbrain boundary of the developing chick brain. At neural plate stages of development, a transcription factor boundary forms at the interface of two crudely regionalized progenitor cell types, expressing either *Otx2* or *Gbx2* (first panel). The boundary will become the midbrain-hindbrain boundary (MHB), characterized by upregulation of a signal (FGF-8) (second panel). FGF-8 signals back to the *Otx2*- and *Gbx2*-expressing progenitor cells to initiate new transcriptional responses, refining and increasing progenitor types (third panel). In the forming midbrain, FGF-8 acts with the ventral morphogen, Shh, to induce post-mitotic midbrain dopaminergic neurons of the substantia nigra. FGF-8 also upregulates semaphorin, a diffusible protein that repels dopaminergic axons, guiding them towards their targets in the telencephalon (fourth panel). HH, Hamilton and Hamburger.

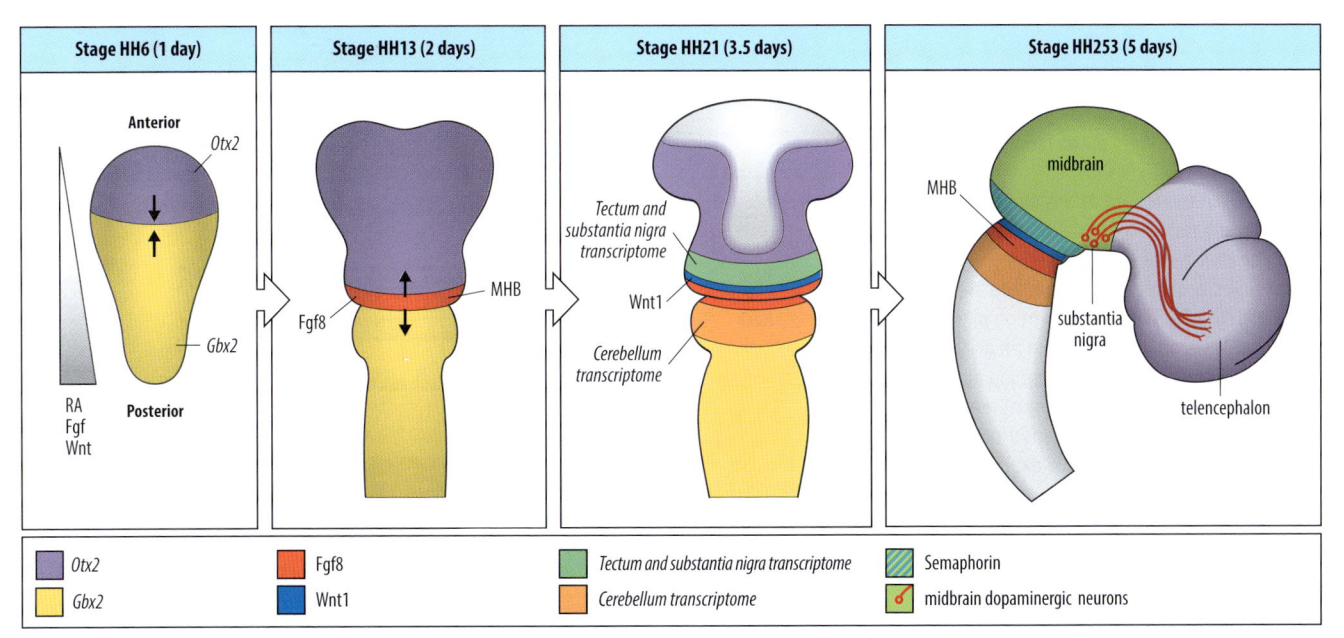

10.4 Specialized boundaries formed in the developing brain delimit areas of cell-lineage restriction

Specialized boundary cells have many roles in the morphogenesis of the developing brain. In addition to providing positional information (see Section 10.3), the MHB provides a lineage-restriction boundary that prevents the intermingling of cells with different developmental fates. Lineage-tracing studies show that at neural plate stages, Otx2-positive and Gbx2-positive progenitor cells can intermingle, but once the MHB has formed, cells are confined to either midbrain or hindbrain, respectively. This ensures that midbrain and hindbrain fates remain segregated while cells within each unit continue to proliferate and reorganize. In this respect, the midbrain and hindbrain are behaving like *Drosophila* compartments (see Section 2.23). Later in development, MHB cells can act as guides for neuronal migration and axon pathfinding (see Fig. 10.4).

Boundary cells are important, not just at the MHB, but throughout the length of the developing hindbrain. The neuroepithelium of the hindbrain shows a series of transient swellings along the antero-posterior axis, called **rhombomeres** (see Fig. 10.3; r1–r8). Each constitutes a compartment of cell-lineage-restricted neural progenitor cells, separated by boundaries. Progenitor cells within each rhombomere acquire positional values and will give rise to the segmentally arranged post-mitotic cranial motor nerves that innervate the face and neck (Fig. 10.5), as well as to cranial neural crest cells (see Fig. 4.49).

We have a good understanding of the mechanisms underlying rhombomere boundary formation. The graded signals that pattern the neural plate—such as FGF and retinoic acid—lead to the spatial expression of a network of transcription factors, including Egr2 (Krox20) and Hox family members. These specify the antero-posterior identity of each hindbrain rhombomere segment: for example, r3 and r5 are specified by Egr2, and r4 is specified by Hoxb1 (Fig. 10.6). In the early embryo, the borders between different rhombomere subdivisions are ragged, and there is some overlap in expression of transcription factors that (ultimately) confer different identities. This imprecision is likely to be caused because the graded signals that control expression of *Egr2* and the Hox genes are noisy, and evoke a fuzzy response. In addition, cells in the developing neural tube proliferate and intercalate, likely becoming jumbled. The ragged borders become progressively razor sharp through two mechanisms: **cell segregation** and **cell-identity switching**.

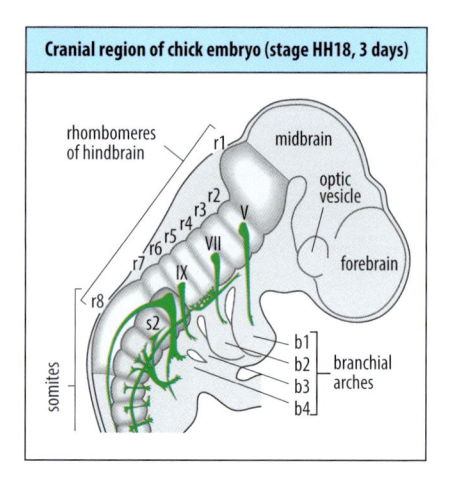

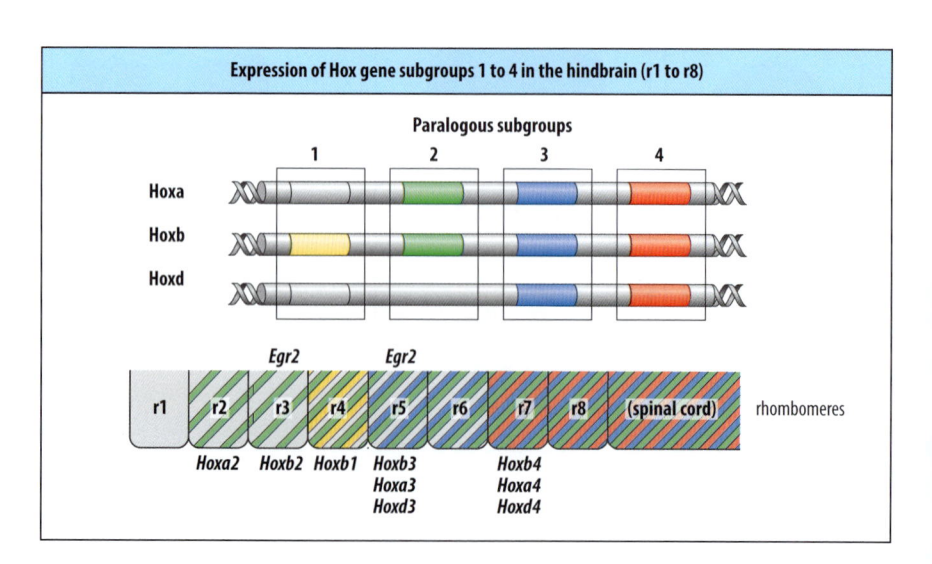

Fig. 10.5 Rhombomeres in a 3-day chick embryo (Hamilton and Hamburger (HH) stage 18). At this stage the hindbrain is divided into the rhombomeres (r). The positions of three (V, VII, IX) of the nine cranial nerves that arise from the hindbrain are shown in green; b1-b4 are the four branchial arches; s, somite.

Illustration adapted with permission from Lumsden, A.: **Cell lineage restrictions in the chick embryo hindbrain.** *Phil. Trans. R. Soc. Lond. B 1991, **331**: 281-286.*

Fig. 10.6 Expression of Hox genes in the hindbrain. The expression of genes of three paralogous Hox clusters in the hindbrain (rhombomeres r1-r8) is shown. *Hoxa1* and *Hoxd1* are not expressed at this stage. Note that there are no Hox genes expressed in r1. The Hox genes whose anterior expression starts in each rhombomere are listed underneath. With the exception of *Hoxb1*, which is only expressed in r4, all the Hox genes listed are also expressed in all rhombomeres posterior to their most anterior expression. The gene for the transcription factor Egr2 is also shown and is expressed in rhombomeres 3 and 5.

Illustration after Krumlauf, R.: **Hox genes and pattern formation in the branchial region of the vertebrate head.** *Trends Genet. 1993, **9**: 106-112.*

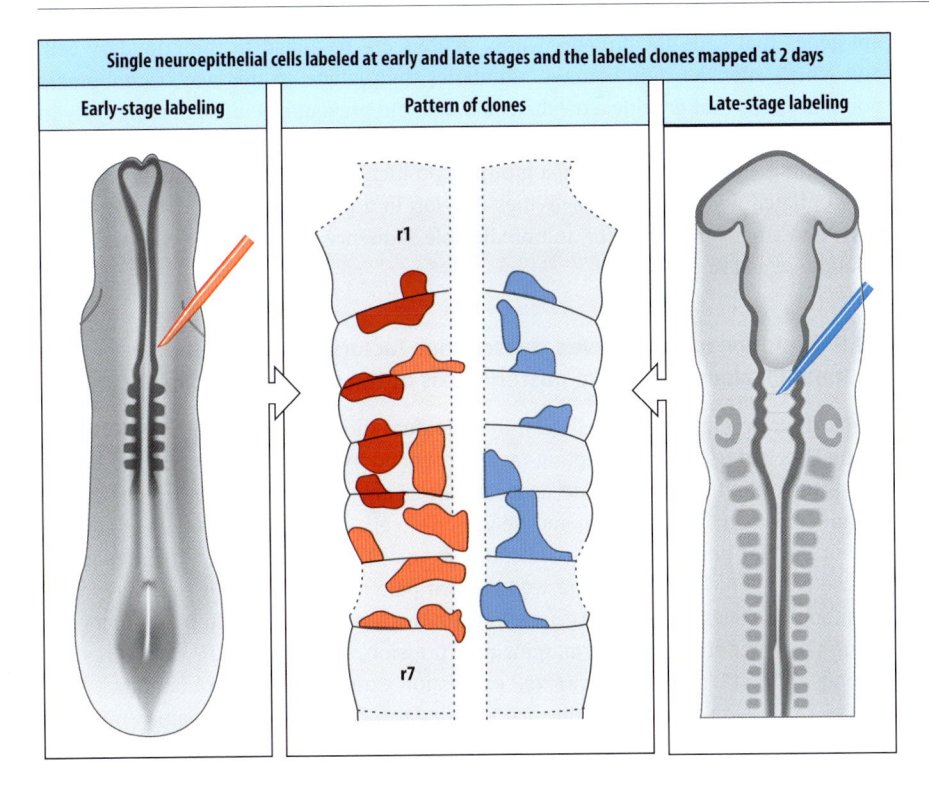

Fig. 10.7 Lineage restriction in rhombomeres of the embryonic chick hindbrain. Single neural precursor cells are injected with a label (rhodamine-labeled dextran) before (left panel) or after (right panel) rhombomeres are seen, and their descendants are traced 2 days later (center panel). Cells injected before rhombomere boundaries form can give rise to progeny that span two segments (dark red) or cross into the midline (orange). Cells injected after boundaries form do not show intermingling across segment borders (blue).

Illustration adapted with permission from Lumsden, A.: **Cell lineage restrictions in the chick embryo hindbrain.** Phil. Trans. R. Soc. Lond. B *1991,* **331***: 281–286.*

Clonal analyses in the chick show that before boundary formation, cells can intermingle into adjacent segments, but as boundary cells form, their distinct cellular properties cause rhombomere cells to segregate (Fig. 10.7). Cell segregation occurs through intercellular signaling at the interface of even and odd rhombomeres. This is mediated by members of the **ephrin** family and their **Eph receptors** (see Box 5D), which are expressed separately in alternate rhombomeres, and are themselves regulated by genes such as *Egr2*. The changing transcriptomes in turn trigger new cell behaviors, such as increased cell repulsion. In this way, the EphA4 receptor, which is strongly expressed in rhombomeres r3 and r5, and its ligand ephrin B2, which is expressed in r2, r4, and r6, trigger repulsive interactions at each rhombomere boundary and so increase border sharpness and prevent cell mixing.

A second mechanism that operates to sharpen boundaries is cell-identity switching. Clonal analyses in the chick hindbrain first implied that cells might be able to switch their antero-posterior identity. Direct evidence then came from transplantation experiments in mouse and zebrafish embryos. When single cells are transplanted between hindbrain segments, they downregulate markers of their site of origin and switch to the identity of their new location. This happens through local retinoic acid signaling. If cells encounter a discontinuity in local signaling, they alter their expression of transcription factors that regulate segmental identity.

In contrast to single cells, groups of cells transplanted between segments maintain their original identity. Such 'community effects' occur because positive feedback within a small group of cells results in levels of the local signal (in this case, retinoic acid) being kept above the discontinuity threshold. Community effects (see also Box 3G) might be generally required during development, to maintain homogeneous regional identity within streams of migratory cells, or when one tissue arises from different lineages at different times.

The rhombomere borders undergo further transcriptional changes as development proceeds. Genes that code for proteins that act outside the boundary cell, including signals that regulate further patterning of the rhombomere cells, are upregulated, as are genes that code for proteins whose actions are restricted to the boundary cell, and which alter the rhombomere boundaries themselves. In this way, rhombomere boundary cells that provide a mechanical barrier to prevent cell intermingling early in development then become a signaling hub. Later still, they provide a scaffold for axon pathfinding and, finally, become a group of proliferating progenitors that give rise to differentiated neurons. These activities develop in a partly overlapping manner over time, and so ensure a smooth and predictable sequence of building events—another example of self-organization.

10.5 Evolutionarily conserved transcription factors provide positional information along the antero-posterior axis of the developing brain

The further development of cells within individual rhombomeres is under the control of the Hox genes. The transcription factors encoded by these genes, which are highly conserved in evolution, provide at least part of the molecular basis for the future identities of both rhombomere and neural crest cells at different positions along the hindbrain. Hox genes from four paralogous groups (1, 2, 3, and 4) are expressed in the mouse embryonic hindbrain in a well-defined pattern, which closely correlates with rhombomeric segmental pattern (see Fig. 10.6). In general, the different paralogous groups have different anterior margins of expression. For example, *Hoxa2* expression extends to the r1/r2 boundary, *Hoxb2* expression extends to the r2/r3 boundary, and *Hoxb3* expression extends to the r4/r5 boundary. Complex *cis*-regulatory modules control Hox gene expression in the individual rhombomeres. The regulatory region of the *Hoxb2* gene, for example, carries two separate *cis*-regulatory modules, one of which regulates its expression in r3 and r5 (activated in part by Erg2), and the second of which regulates expression in r4.

Evidence that Hox genes determine the fate and future development of cells in the rhombomeres comes from misexpression experiments. Each pair of rhombomeres harbors post-mitotic motor neurons whose axons project to a single branchial arch: axons from r2/r3 project to the first branchial arch, whereas those from r4/r5 project to the second arch (see Fig. 10.5). *Hoxb1* is normally expressed in r4 but not in r2. If *Hoxb1* is misexpressed in r2, however, these axons now extend to the second arch. This is yet another example of a homeotic transformation produced by misexpression of Hox genes (see Sections 2.27 and 4.12 for others).

Hox genes are not expressed in anterior-most neural tissue—the forebrain, midbrain, and r1 of the hindbrain. Instead, homeodomain transcription factors such as Otx are expressed here, and provide positional information to cells in these regions (see Fig. 10.3). Like the Hox genes, *Otx* genes have been highly conserved in evolution. The *Drosophila orthodenticle* gene, homologous to the vertebrate *Otx* genes, is expressed in the future *Drosophila* brain, and mutations in the gene result in a greatly reduced brain. In mice, mutation in *Otx1* leads to brain abnormalities and epilepsy. Mice with a defective *Otx* gene can be partly rescued by replacing it with *orthodenticle*, even though the sequence similarity of the two proteins is confined to the homeodomain region. Human *OTX* can even rescue *orthodenticle* mutants in *Drosophila*.

10.6 Opposing morphogen gradients pattern the spinal cord along the dorso-ventral axis

We have seen how a coarse pattern of regional specification occurs in the developing brain as a result of signals from local organizers that activate transcription factors that, in turn, act as regional or sub-regional determinants of fate. The next challenge facing the embryo is to ensure, at a finer level of patterning, that individual post-mitotic neurons of particular types and functions differentiate in the correct positions in the neural tube. We best understand these events through studies in the

Fig. 10.8 Post-mitotic neurons in the spinal cord are found in predictable locations along the dorso-ventral axis. Twelve post-mitotic neuronal subtypes develop in predictable locations along the dorso-ventral axis of the developing spinal cord relative to the dorsal midline roof plate and the ventral midline floor plate. Most of these are interneurons, which relay impulses between sensory neurons and motor neurons. Only a half of the spinal cord is labeled for simplicity. Gray ovals: migrating neural crest cells. Green circles: dorsal interneurons DI1, DI2, DI3, DI4–6. Purple circles: ventral interneurons V0, V1, V2a, V2b. Blue circle: motor neuron (MN). Orange circle: ventral interneuron V3.

developing spinal cord. This experimentally accessible system, in which relatively few (12) neuronal types differentiate in characteristic dorso-ventral locations (Fig. 10.8), has helped us work out the principles through which morphogen-mediated patterning leads to neuronal fate specification.

Each of the 12 neuronal subtypes within the spinal cord is located at a characteristic position relative to two specialized glial cell types—the ventral midline **floor plate**, located directly above the notochord, and the dorsal midline **roof plate**, located beneath dorsal surface ectoderm (future epidermis). Roof plate cells arise from neural plate border cells (see Section 4.15); floor plate cells in the anterior and posterior parts of the neural tube arise from medial-most neural plate cells, and cells in the stem zone, respectively (see Section 4.9). Tissue-grafting and tissue-removal studies in the chick embryo first showed that the position at which post-mitotic neurons differentiate along the dorso-ventral axis is determined by signals provided by the ventrally located notochord and floor plate, and the dorsally located dorsal surface ectoderm and roof plate (Fig. 10.9). When a piece of notochord is grafted lateral to the neural tube, a second floor plate is induced in the adjacent neural tube and additional **motor neurons** on each side (the experimental set-up is described in Fig. 4.43); at the same time, markers of dorsal cell types are suppressed. Conversely, if the notochord is removed, the embryo develops a neural tube composed of only dorsal cell types. Similar types of experiment have been performed to show the patterning activities of floor plate, roof plate, and surface ectoderm. Notochord, floor plate, roof plate, and surface ectoderm are therefore secondary organizing centers.

Subsequent studies showed that dorsal and ventral organizing centers are the source of opposing morphogen signals (see Fig. 10.9) that act on proliferating neural progenitor cells, conferring distinct dorsal–ventral positional information. Dorso-ventral identity is maintained as progenitor cells undergo their final division to form post-mitotic neurons. Shh, emanating first from the mesodermal notochord and then from ventral midline floor plate cells, induces ventral progenitor identities, whereas members of the BMP and Wnt families, secreted from surface ectoderm and dorsal midline roof plate cells, promote dorsal progenitor identities (Fig. 10.10). Shh acts as a morphogen: *in situ* hybridization studies show that *Shh* mRNA is detected in the notochord and floor plate, and immunohistochemical studies show that Shh protein diffuses away from these to establish a ventral-to-dorsal gradient in the neural tube. It is still uncertain how BMPs and Wnts form a dorsal-to-ventral gradient: one possibility is that they become locally induced in other dorsal cells; a second is that cytonemes (see Box 9A)—long filamentous cellular processes that can extend for hundreds of micrometers—deliver packages of BMPs and Wnts to neural progenitors at a distance from the roof plate. Whatever the mechanism, Shh and BMPs/Wnts antagonize one

Fig. 10.9 Secondary organizers pattern the spinal cord along its dorso-ventral axis. The floor plate and roof plate are specialized glial cells that occupy the ventral and dorsal midline of the neural tube; many of their characteristic properties are induced by signals from the notochord and surface ectoderm, respectively. BMPs and Wnts from dorsal organizers (surface ectoderm, then roof plate; blue) oppose Shh from ventral organizers (notochord, then floor plate; green) to pattern the dorso-ventral axis.

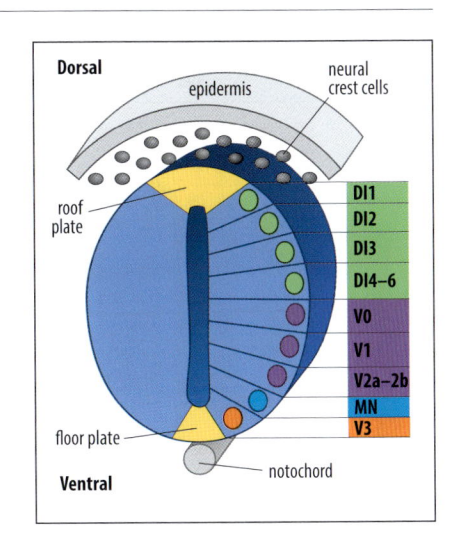

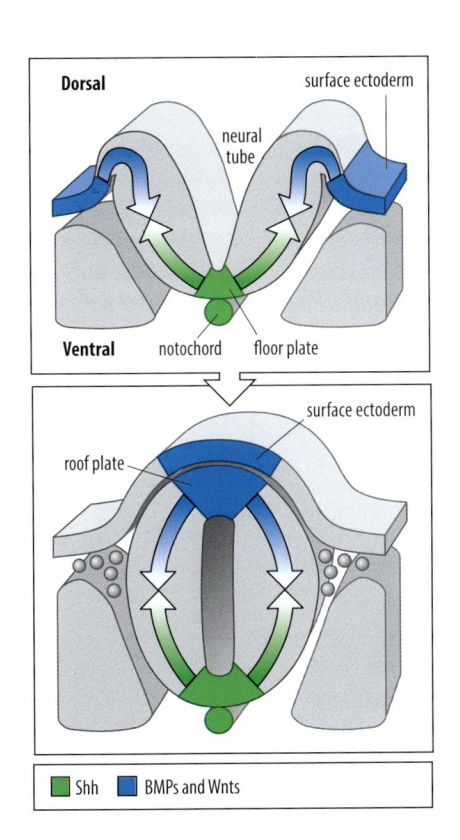

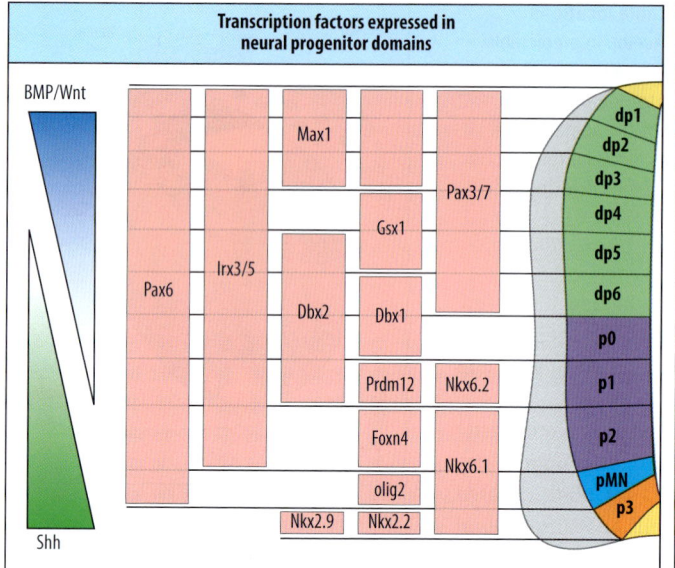

528 Chapter 10 Development of the nervous system

Fig. 10.10 Patterning the spinal cord along the dorso-ventral axis. Left panel: opposing gradients of Shh and BMPs/Wnts lead to differential activation of transcription factors along the dorso-ventral axis (pink boxes) that together define 11 neural progenitor domains (dp1–dp6, p0–p2, motor neuron progenitors (pMN), p3). These are located between the dorsal roof plate and ventral floor plate (yellow triangles), whose midline positions enable them to pattern the developing neural tube with bilateral symmetry around the central lumen. Right panel: the neural progenitor cells give rise to 12 distinct post-mitotic neuronal types (dorsal interneurons DI1–DI6; ventral interneurons V0, V1, V2a, V2b, V3; and motor neurons (MN)), each characterized by the expression of specific transcription factors (blue boxes).

Based on Sagner, A., Briscoe, J.: **Establishing neuronal diversity in the spinal cord: a time and a place.** *Development 2019,* **146***: dev182154.*

another, and the induction of distinct progenitor cell types along the dorso-ventral axis of the neural tube is a consequence of their opposing actions.

In the early neural tube, Shh, BMPs, and Wnts begin to induce distinct progenitor cell types because the incremental changes in their concentration levels evoke incremental changes in the levels of their signaling-pathway effectors in the responding neural progenitor cells. Different concentrations of Shh are translated into a ventral-to-dorsal gradient of the activator Gli proteins (see Box 9C), whereas different concentrations of BMPs evoke a dorsal-to-ventral gradient of phosphorylated SMADs (the effectors of BMP signaling, see Box 3F). Shh, BMP, and Wnt signaling effectors—activated to a different degree in cells along the dorso-ventral axis—then direct differential transcription in a cell-autonomous manner, orchestrating the expression of homeodomain (HD) and basic helix–loop–helix (bHLH) transcription factors.

In this way, the opposing morphogen gradients are translated into spatially delimited and partially overlapping domains of transcription factor expression that define six dorsal (dp1–dp6) and five ventral (p0–p2, pMN, p3) neural progenitor domains (see Fig. 10.10). Gain- and loss-of-function studies show that the transcriptional code—the combination of transcription factors expressed by a progenitor cell—determines the specific type of post-mitotic neuron that it generates: DI1–DI6 and V0–V3 interneurons and motor neurons (see Fig. 10.10).

10.7 Stable progenitor domains in the ventral spinal cord are established over time

Morphogens operate over time, and in growing tissues. Studies in the ventral spinal cord show that the relationship between the extracellular concentration of Shh and the level of transcriptional activation produced by Gli proteins is not straightforward.

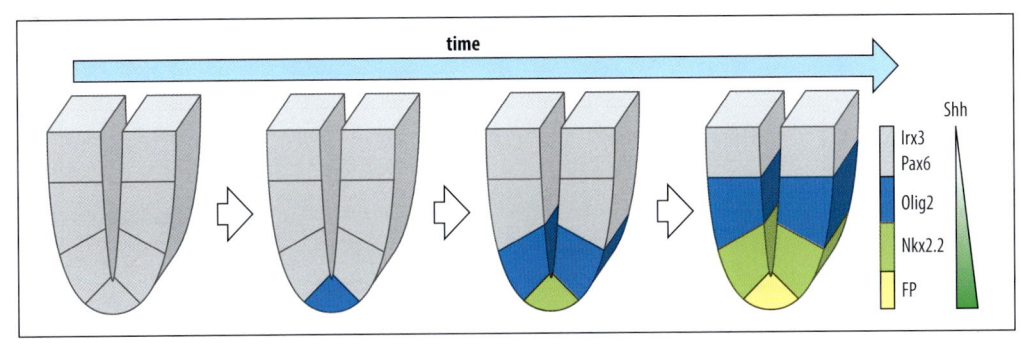

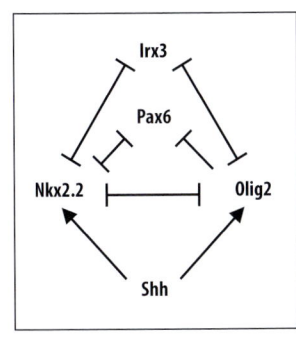

Fig. 10.11 Sequential induction of progenitor markers in the ventral spinal cord and stabilization of progenitor domains. Left panel: at early stages of development, neural progenitors throughout the dorso-ventral axis express the transcription factors Pax6 and Irx3 (gray). Over time, increasing levels and duration of Shh signaling result in the sequential induction of the transcription factors Olig2 (blue) and Nkx2.2 (green), giving rise to progenitor domains p3 (defined by Nkx2.2 expression), motor neuron progenitors (pMN) (defined by expression of Olig2 and low levels of Pax6), and p2 (defined by expression of Irx3 and high levels of Pax6). Right panel: these domains are stabilized by cross-repression between these transcription factors: Nkx2.2 and Olig2 repress transcription of all the others; Irx3 represses transcription of Olig2 and Nkx2.2; and Pax6 represses transcription of Nkx2.2. The duration of Shh signaling is partly responsible for the distinction between Olig2 and Nkx2.2 induction. At an early time point, Shh signaling leads to activation of the Gli transcriptional activators (see Box 9C), and expression of Olig2 is upregulated. If the levels of Shh signaling are sustained, Nkx2.2 is induced and Olig2 is repressed. But if the levels of Gli in a cell decline before this time point, Olig2 expression is consolidated. The response of cells is therefore a function of both Gli activity and the duration of Shh exposure. In this view, the induction of each progenitor state requires exposure to a concentration of Shh above a defined threshold for a distinct period of time. The dynamics of this gene regulatory network means that lower levels and duration of Shh favor expression of Olig2, whereas high levels of Shh lead to expression of Nkx2.2. FP, floor plate.

*Based on Sagner, A., Briscoe, J.: **Establishing neuronal diversity in the spinal cord: a time and a place.** Development 2019, **146**: dev182154.*

With time, cells adapt and become desensitized to the ongoing Shh exposure, resulting in a decline in Gli activity. At the same time, a complex gene regulatory network, in which Shh-responsive transcription factors that are expressed in adjacent progenitor cells mutually repress one another's transcription, provides a mechanism to produce discrete spatial switches in gene expression, and discrete progenitor domains (Fig. 10.11). In addition, the gene regulatory network defines the dynamics of the response to Gli activity and the temporal aspect of neural tube patterning, promoting cells through sequential ventral progenitor identities (see Fig. 10.11). This challenges the traditional view of Shh as a morphogen, as it demonstrates that positional values are not simply conferred through interpretation of the Shh concentration. By responding to both the levels of Shh and the duration of its action, each gene regulatory network ensures that the right progenitor domains form and are in the correct places relative to each other (see Fig. 10.11).

The ongoing activity of the gene regulatory networks also means that stable progenitor domains are maintained as the neural tube grows in size along its dorso-ventral axis. At early developmental stages, when the neural tube is small, a cell's position on morphogen gradients can be decoded with high precision, but the gene regulatory networks become important when the growth of the tissue means that many cells are simply too far away from the source of the morphogen to sense it accurately. The concept of progenitor-domain stabilization during the extensive growth of the embryo is of general importance to all developing vertebrate organs.

The transcription factor codes that specify and maintain distinct progenitor identities then direct the specification of discrete post-mitotic neuronal identities. They do so in two ways. One is by inducing domain-specific combinations of additional transcription factors (see Fig. 10.10) that activate programs of gene expression that specify particular neuronal subtypes. The other is by repressing downstream genes associated with alternative progenitor and neuronal subtypes. This double mechanism prevents the

MEDICAL BOX 10B Directed differentiation of induced pluripotent stem cells towards functional neuronal types

Here we shall look at two examples of the directed differentiation *in vitro* of induced pluripotent stem cells (iPS cells) into different types of neuron. Motor neurons control muscle functions, including movement, posture, and breathing. In motor neuron disease, motor neurons degenerate, leading to a loss of muscle innervation which results in muscle weakness, difficulty in swallowing and breathing, and eventually death. Animal studies have provided outstanding insight into how motor neurons are generated in the embryo. These studies have led to the development of protocols for the directed differentiation of human iPS cells to human motor neurons. Specific concentrations of factors that switch on the Shh signaling pathway are used as part of the differentiation protocol, and readouts of motor neuron transcription factors ascertain the accuracy of the *in vitro* differentiation.

Figure 1 shows the steps involved in the directed differentiation of iPS cells to motor neurons. White arrows show key signaling factors used to direct each step in the differentiation program. Transcription factors expressed, characteristic of each cell type, are shown underneath the cells.

In the hypothalamus of the brain, specialized neurons produce neuropeptides that regulate behavior. The neuropeptide hypocretin (also called orexin) is produced by neurons of the lateral hypothalamus. It has an important role in sleep: loss of hypocretin neurons causes the sleep condition, narcolepsy, in which people experience overwhelming drowsiness and fall asleep suddenly. Animal studies show that the transcription factor Lhx9 acts in a cell-autonomous way to upregulate hypocretin. This knowledge has been used to develop protocols for the production of hypocretin neurons *in vitro* from iPS cells, providing a future platform to better understand and treat narcolepsy.

Figure 2 shows schematics of coronal sections through the human brain indicating the number of hypocretin-producing neurons in the hypothalamus (black dots). The number of hypocretin-producing neurons is much reduced in people with narcolepsy.

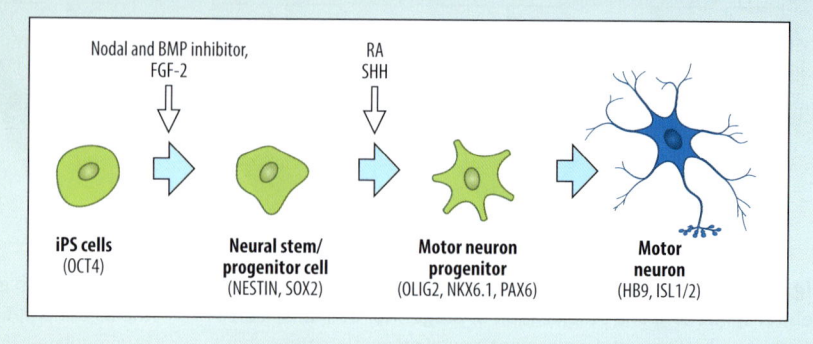

Fig. 1

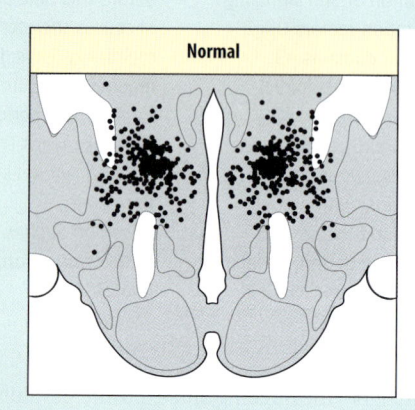

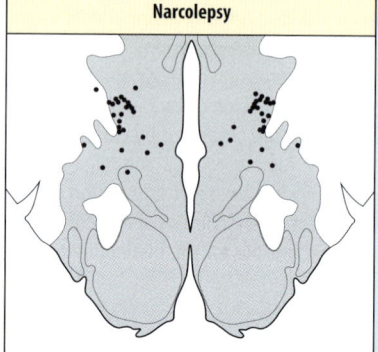

Fig. 2

specification of mixed neuronal identities and results in the appropriate differentiation of discrete post-mitotic neuronal types along the dorso-ventral axis of the spinal cord.

The details of how different neuronal cell types are specified have important and widespread medical relevance. Knowledge of how the embryo builds particular neurons is harnessed and applied to direct the differentiation of human induced pluripotent stem cells (iPS cells) to defined neuronal identities (Box 10B, and see Chapter 7).

10.8 Hox genes specify motor neuron subtypes in the spinal cord

The mechanisms that direct the differentiation of neural progenitors into post-mitotic neurons explain how broadly different types of spinal cord neuron are formed (see Sections 10.6 and 10.7), but these post-mitotic neurons subsequently diversify into many molecularly and functionally distinct subtypes. We now look at how this can occur, using the generation of different motor neuron subtypes along the antero-posterior axis of the forming spinal cord as an example.

Antero-posterior specification of motor neuron function in the spinal cord was illustrated some 40 years ago by experiments in which a section of the spinal cord that would normally innervate wing muscles was transplanted from one chick embryo into the lumbar region (which normally innervates the legs) of another embryo. Chicks developing from the grafted embryos spontaneously activated both legs together as though they were trying to flap wings, rather than activating each leg alternately as if walking. These studies showed that motor neurons generated at a given antero-posterior level in spinal cord have intrinsic properties characteristic of that position.

We now know how developing motor neurons in the spinal cord acquire distinct identities along the antero-posterior axis. These identities are conferred by Hox genes, which, as we learned in Chapter 4, specify regional identity along the antero-posterior axis. In the same way as occurs in the mesoderm (see Chapter 4) and the hindbrain (see Section 10.5), the ordered expression of Hox genes along the antero-posterior axis of the spinal cord subdivides progenitors and neurons into three broad domains along the antero-posterior axis: Hox4–7 genes define cervical cell identities, Hox9 genes thoracic, and Hox10–11 genes lumbar (Fig. 10.12). As elsewhere in the embryo, the Hox code is established in response to signals present in the early embryo. In the spinal cord these include retinoic acid, FGFs, and growth differentiation factor 11 (GDF-11). As the spinal cord lengthens, the patterns of Hox gene expression are maintained by cross-repression similar to that described in dorso-ventral patterning (see Section 10.7). The patterns of Hox gene expression define distinct motor columns that innervate specific muscles. These include the phrenic motor column (PMC), which innervates the diaphragm, the **lateral motor columns** (LMC), which innervate limb muscles, and the preganglionic and hypaxial motor columns (PGC and HMC) that innervate sympathetic ganglia and hypaxial muscles—the muscles of the body wall (see Fig. 10.12).

These columns are further subdivided into specific pools of cells that are built over time. For example, the LMC is divided into two divisions—lateral nearest the outer edge of the spinal cord, and medial nearest the midline. Early-born LMC neurons secrete retinoic acid, which recruits later-born motor neurons into the LMC, but in these neurons, retinoic acid downregulates the transcription factor Isl1 and upregulates the transcription factor Lhx1. Neurons expressing Lhx1 settle in the lateral division, whereas those expressing Isl1 settle in the medial division. Experiments tracing axon outgrowth show that lateral LMC neurons send axons into the dorsal limb-bud muscles, whereas medial LMC neurons send their axons into the ventral limb-bud muscles.

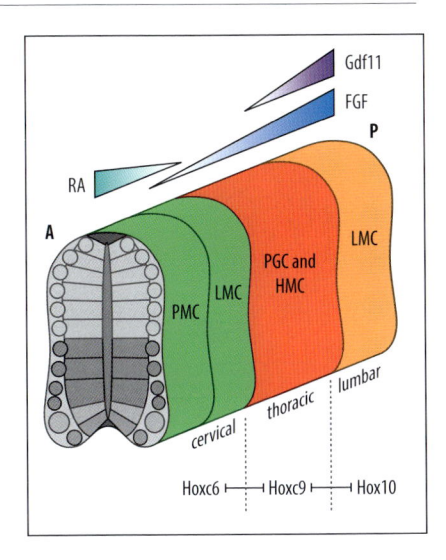

Fig. 10.12 Hox genes pattern the spinal cord along the antero-posterior axis. Opposing morphogen gradients induce distinct patterns of Hox gene expression along the anterior-posterior (A-P) axis of the spinal cord. Only a subset of the Hox genes expressed at each level are shown. Green, pink, and orange boxes indicate Hox genes expressed at cervical, thoracic, and lumbar regions of the spinal cord, respectively. Cross-repression between Hox genes gives rise to specific domains, and defines distinct neuronal subtypes along the A-P axis. HMC, hypaxial motor column; LMC, lateral motor column; PGC, preganglionic motor column; PMC, phrenic motor column.

Based on Sagner, A., Briscoe, J.: **Establishing neuronal diversity in the spinal cord: a time and a place.** Development 2019, **146**: dev182154.

10.9 Temporal mechanisms contribute to the generation of neural cell diversity

So far we have considered how the expression of signal molecules on orthogonal axes results in the expression of transcription factors that determine regional and cellular fate at antero-posterior and dorso-ventral positions. Cells acquire their unique morphological and functional identities according to their grid reference in this Cartesian coordinate system of positional information. However, the nervous system is built up over time, and evidence now indicates that a temporal transcription factor code works alongside the spatial transcription factor code to further diversify neuronal types. In both the spinal cord and the retina, for instance, early-born neurons express Onecut family transcription factors, whereas later-born neurons express Nfia and Nfib. And—so far shown only in the retina—early-born neurons depend on Onecut whereas later-born neurons depend on Nfia/b for their generation. This raises the question of how such a temporal transcription factor code could be controlled. One possibility is that different extracellular cues acting at distinct time points induce a different transcriptional signature in uncommitted neural progenitor cells. Alternatively, neural progenitor cells might progressively change their transcriptional profile through cell-autonomous changes, resulting in changed competence and the generation of different neuronal subtypes over time. Such a mechanism would resemble that observed in neuroblast lineages in the *Drosophila* embryo (discussed in Section 10.11). We know that temporal mechanisms provide a means for the ordered establishment of neuronal circuitry; the birth order of neurons underlies the specificity in neuronal connectivity and circuit formation, as we shall see later in the chapter.

Another example of the influence of time is that throughout the CNS, a temporal switch leads neural progenitor cells first to give rise to neurons and later to give rise to glia. This temporal switch depends on the sequential induction of members of the SoxE and NFI families of transcription factors and is regulated by multiple signaling pathways, most notably those of Notch, FGF, and transforming growth factor-β (TGF-β).

Together, the spatially and temporally restricted expression of transcriptional factors in neural progenitors results in the spatially segregated generation of distinct neuronal subtypes that form over time. This is the first step in the assembly of functional neuronal circuits that underlie behavior. We will return to this later in the chapter, where we will look at how transcription factors that control neuronal identity also control characteristics that enable their axons to grow towards their targets.

SUMMARY

The developing brain is divided into three regions—forebrain, midbrain, and hindbrain. Secondary organizer regions produce signals that induce and pattern each brain region, governing the expression of evolutionarily conserved transcription factors that provide positional values within each emerging region. Specialized boundaries between the midbrain and hindbrain and within the hindbrain establish discrete developmental units. Patterning of neuronal cell types along the dorso-ventral axis of the spinal cord similarly involves signals secreted from ventral and dorsal secondary organizers that establish 11 discrete progenitor subsets along the dorso-ventral axis. The transcription factor code in the ventral half of the neural tube acts as a gene regulatory network, enabling cells to respond to both the levels of Shh and the duration of its action. Changes in transcriptional programs within progenitor subtypes then specify different neuronal subtypes along the dorso-ventral axis of the spinal cord. Neuronal identity along the antero-posterior axis of the cord is specified by the combinatorial expression of Hox genes. Timing mechanisms support the building of increasingly diverse cell types.

The formation and migration of neurons

We will now consider further how progenitor cells that have acquired their positional identity give rise to neurons. The process of forming a neuron is called **neurogenesis**. We will first look at *Drosophila*, which has revealed some key developmental processes in neurogenesis. We then look at how lessons learned through *Drosophila* enable an understanding of the more complex process of neurogenesis in the vertebrate spinal cord and brain.

10.10 Neurons in *Drosophila* arise from proneural clusters

In *Drosophila*, the cells that will give rise to the CNS of the larva are specified at an early stage of embryonic development, as part of the patterning process that divides the embryo into different regions along the dorso-ventral and antero-posterior axes.

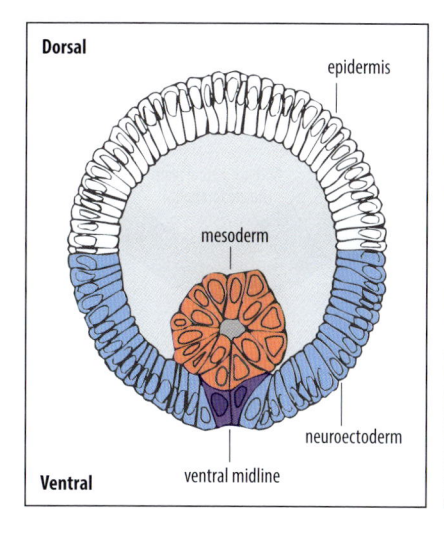

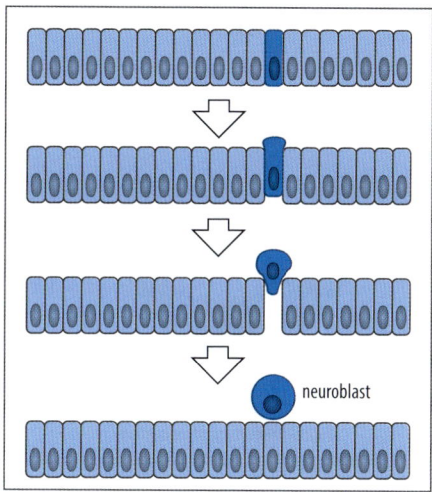

Fig. 10.13 Cross-section of a *Drosophila* early gastrula. Left panel: at around 3 hours after fertilization, the neuroectoderm (pale blue) is located on either side of the ventral midline (purple). The mesoderm (orange), which originally lies along the ventral midline, has already been internalized. Right panel: soon after, individual neuroblasts (dark blue) begin to be specified and delaminate from the neuroectoderm into the interior of the embryo.

In insects, the main nerve cord runs ventrally, rather than dorsally as in vertebrates, and the future CNS is specified as two longitudinal regions of neuroectoderm in the ventral half of the embryo, either side of the ventral midline. After gastrulation and internalization of the mesoderm, the neuroectoderm remains on the outside of the embryo on either side of the ventral midline (Fig. 10.13, left panel). Cells of the neuroectoderm, or neurogenic zone, are fated to form either neural cells (neurons or glia) or epidermis. As individual cells become specified as neural cells, called **neuroblasts**, they move (delaminate) from the surface into the interior of the embryo (see Fig. 10.13, right panel), where they divide further and differentiate into neurons and glia, forming two tracts of axons running longitudinally on either side of the ventral midline, connected at intervals by **commissures**. These are defined as regions where axons cross the midline and thus help link the two sides of the body together.

Despite very great differences between the insect CNS and that of vertebrates, there are intriguing parallels between the genes that pattern both systems at a very early stage. *Drosophila* neuroblasts form three longitudinal columns of cells on each side of the ventral midline, and the identity of these columns is specified while the cells are still in the neuroectoderm (Fig. 10.14). The homeodomain transcription factor genes *msh*, *ind*, and *vnd* are expressed in dorsal to ventral order in the *Drosophila* neuroectoderm, in response to the actions of earlier dorso-ventral patterning genes, such as *rhomboid* (see Section 2.16), and the activation of the *Drosophila* epidermal growth factor signaling pathway. The developing vertebrate neural plate (the future brain: see Section 10.2) also has three longitudinal domains of prospective neural cells (see Fig. 10.14), which express genes homologous to those in the fly—*Msx*, *Gsh*, and *Nkx2.1*, respectively. This is a striking example of the evolutionary conservation of a mechanism for regional specification.

How do neuroblasts form? In response to previous patterning events, cells of the *Drosophila* neuroectoderm form an orthogonal pattern of **proneural clusters** along the antero-posterior and dorso-ventral axes, each composed of small numbers of cells, which can be distinguished by the expression of genes known as **proneural genes** (Fig. 10.15). Initially, all the cells within a proneural cluster are capable of becoming a neuroblast (a neural precursor cell), but one cell, apparently randomly, produces a signal that promotes its own development as a neuroblast and prevents adjacent cells from following that fate. This cell will become the single neuroblast in the cluster, and can be distinguished, for example, by a high level of expression of a proneural gene such as *achaete*, whereas the surrounding cells no longer express *achaete* and will develop as epidermis.

534 Chapter 10 Development of the nervous system

Fig. 10.14 The neuroepithelium of both *Drosophila* and vertebrates is organized into three columns of neural precursors on either side of the midline. The developing *Drosophila* embryonic central nervous system (left) and the vertebrate embryonic neural plate (future forebrain, right) are both organized into three longitudinal domains of gene expression; in *Drosophila* these are on either side of the ventral midline (purple); in vertebrates these are on either side of the ventral midline of the anterior neural tube. Similar genes are expressed in the same medial-lateral (ventro-dorsal) order in *Drosophila* and in vertebrates, with *vnd/Nkx2.1* expressed nearest the ventral midline, *ind/Gsh* in an intermediate position, and *msh/Msx* expressed furthest from the ventral midline.

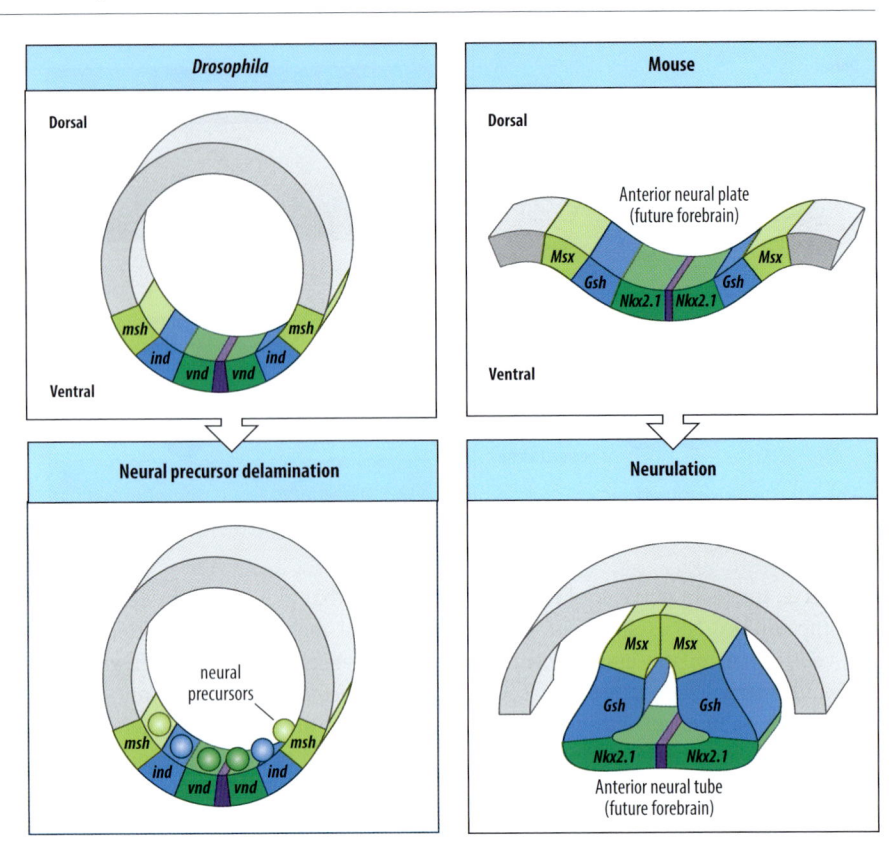

The singling out of the prospective neuroblast occurs through **lateral inhibition** (Fig. 10.16) and is the result of signaling between the prospective neuroblast and its neighbors via the transmembrane protein Notch and its transmembrane ligand Delta (see Box 4F for the Notch–Delta signaling pathway). All the cells in a proneural cluster initially express equal amounts of both Notch and Delta. Delta expression is controlled in part by proneural genes, such as those of the *achaete–scute* gene complex. One cell will, however, start to express Delta sooner or more strongly than the others (it is unclear whether this is a random event or occurs in response to a local signal). Delta interacts with Notch on the less advanced neighbor cells, upregulating the Notch signaling pathway. Notch signaling represses the proneural genes in these cells and so both inhibits their further development as neural cells and also suppresses their expression of Delta (see Fig. 10.16, bottom row). As a consequence, Notch in the first cell will not receive a Delta signal, meaning that genes that negatively regulate the proneural genes will be turned off, the proneural genes will become hyperactivated, and so the first cell will now become a neuroblast. Through this process of

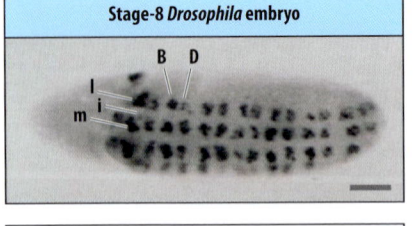

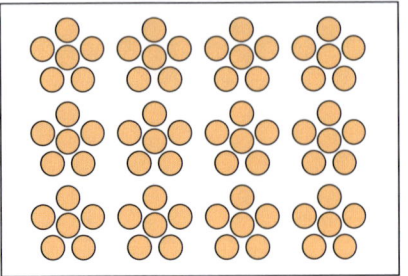

Fig. 10.15 Proneural clusters in the neuroectoderm of the *Drosophila* embryo. Upper panel: a ventro-lateral view of a stage-8 *Drosophila* embryo. Proneural clusters, labeled here for expression of the proneural Achaete protein, are arrayed orthogonally along the antero-posterior and dorso-ventral axes. Proneural gene expression initially divides each hemisegment into a grid consisting of four rows (only rows B and D are marked) and three columns, medial (m), intermediate (i), and lateral (l). Not all proneural clusters are revealed by this labeling but the overall pattern is clear. Scale bar = 50 μm. Lower panel: schematic view of proneural clusters.

Adapted from Skeath, J.B., et al.: **Gene regulation in two dimensions.** Genes Dev. 1992, **6**; 2606-2619.

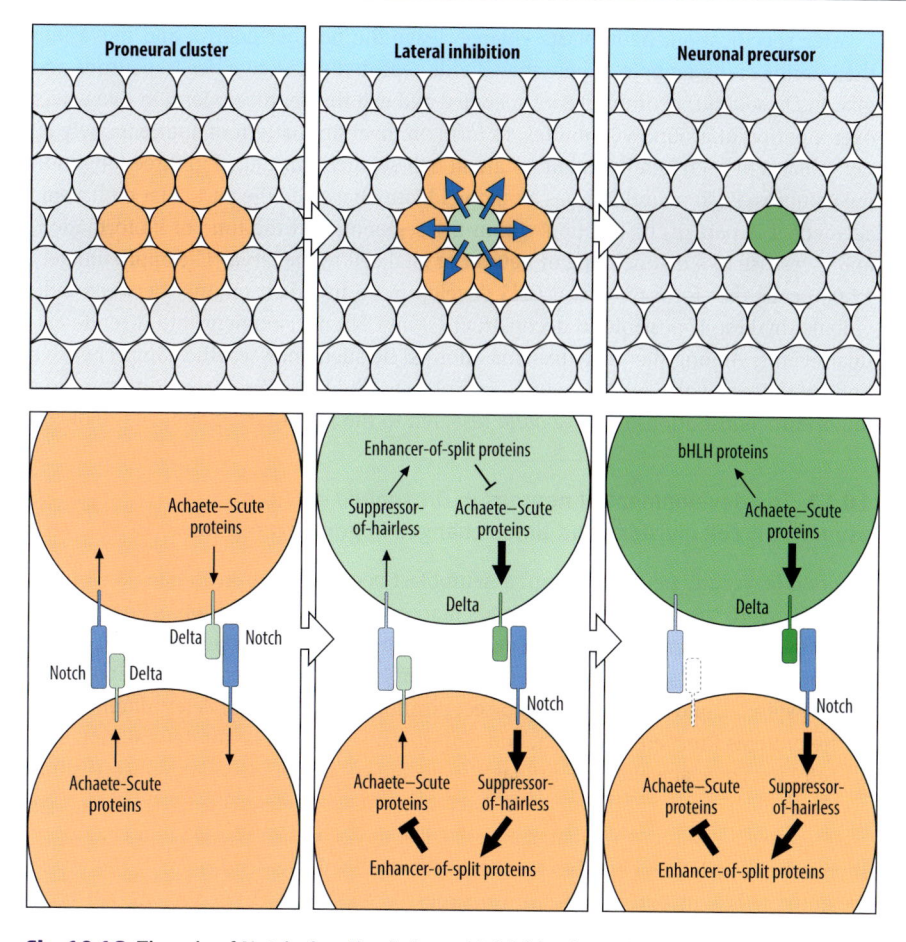

Fig. 10.16 The role of Notch signaling in lateral inhibition in embryonic neural development in *Drosophila*. As shown in the upper row, the proneural cluster (pale orange) gives rise to a single neuroblast (green) by means of lateral inhibition. The rest of the cells in the cluster become epidermal cells. As shown in the lower panels, the presence of Achaete–Scute proteins drives Notch signaling between cells of the proneural cluster. The level of signaling is initially similar and keeps all cells in the proneural state (first panel). An imbalance develops when one cell (pale green) begins to express higher levels of Delta, the ligand for Notch, and thus activates Notch signaling to a higher level in neighboring cells. This initial imbalance in the level of Notch signaling is rapidly amplified by a feedback pathway involving two sets of transcription factors, Suppressor of hairless and the Enhancer of split proteins (second panel). Their activities repress the production of Achaete–Scute proteins and Delta protein in the affected cells, which prevents these cells from proceeding along the pathway of neuronal development and from delivering inhibitory Delta signals to the first cell. Proneural genes in that cell are hyperactivated and the cell becomes a prospective neuroblast (dark green) (third panel). Basic helix-loop-helix (bHLH) transcription factors upregulated in the prospective neuroblast turn on genes required for neuronal development.

After Kandel, E.R., et al.: Principles of Neural Science. 2nd edn. New York: McGraw-Hill, 2000.

feedback inhibition, a small bias is quickly stabilized into a 'one versus another' fate. In this way, initially equivalent cells of the proneural cluster are resolved by lateral inhibition into one neuroblast per cluster. The other cells of the proneural cluster go on to become epidermis. In the *Drosophila* ectoderm, if either Notch or Delta function is inactivated, the neuroectoderm makes many more neuroblasts and fewer epidermal cells than normal, and conversely, if a proneural gene such as *achaete or scute* is inactivated the neuroectoderm makes no neuroblasts and more epidermal cells than normal.

After specification, neuroblasts enlarge, leave the neuroectoderm, and move into the interior of the embryo, where they divide repeatedly and give rise to neurons and glia. In *Drosophila*, neuroblasts are specified and exit the neuroectoderm in five waves over a period of around 90 minutes, to form an invariant pattern of approximately 30 neuroblasts in each hemisegment (a lateral half-segment) arranged in three longitudinal columns. Each neuroblast has a unique identity based on the particular embryonic segment it is part of, its position within the segment, and the time of its formation, and gives rise to a distinct type of neuron or glial cell in the larva. The approximately 15,000 cells that form the larval CNS develop in an invariant manner between individuals, making it possible to deconstruct larval CNS development into discrete cellular events. Among the many insights afforded by such analyses, the comprehensive understanding of the way in which neuroblasts divide represents one of the most important contributions of *Drosophila* research to the field of developmental biology.

10.11 The development of neurons in *Drosophila* involves asymmetric cell divisions and timed changes in gene expression

Before it leaves the neuroectoderm, a neuroblast becomes polarized in an apico-basal direction, with the apical and basal ends marked by the accumulation of specific sets of cytoplasmic determinants. Determinants at the apical end, such as the Par complex proteins, help set the plane of cell division by influencing the orientation of the mitotic spindle, and also direct the correct positioning of the basal determinants, such as the proteins Prospero, Miranda, and Numb, which are crucial for determining neural cell fate. Once specified, neuroblasts move out of the neuroectodermal epithelium and then behave as neural stem cells. Each neuroblast divides asymmetrically to give a larger apical cell, which remains a neuroblast that acts as a neural stem cell, and a smaller basal cell—the **ganglion mother cell**, which will divide once more and whose daughter cells will differentiate into neurons or glia (Fig. 10.17). Among the basal cytoplasmic determinants distributed to the ganglion mother cell by this asymmetric division are proteins of the Miranda–Prospero complex. Miranda segregates with both the transcription factor Prospero, and the Numb complex. Prospero regulates hundreds of target genes involved in inhibiting neural stem-cell behavior and promoting differentiation. The Numb complex inhibits Notch signaling.

The neuroblasts at different locations along the body continue to divide according to the fly's developmental program, giving rise to a ganglion mother cell and a neural stem cell at each division. As more ganglion mother cells are generated, the older cells are pushed further into the embryo, giving the final nerve cord a layered structure.

Fig. 10.17 Formation of neuronal cells from neuroblasts in *Drosophila* by asymmetric cell division. Generation of ganglion mother cells of the central nervous system by the asymmetric division of neuroblasts in the *Drosophila* embryo. Neuroblasts acquire apico-basal polarity under the influence of the Par protein complex (first panel). Par proteins (red) are located apically and Miranda and Prospero proteins (green) are located basally (second panel). Prospero and Miranda are required for the proper localization of Numb proteins. The daughter cell that receives Par proteins is retained as a neuroblast stem cell. The daughter that receives Prospero, Miranda and Numb becomes the ganglion mother cell that will give rise to neurons and glia (third and fourth panels).

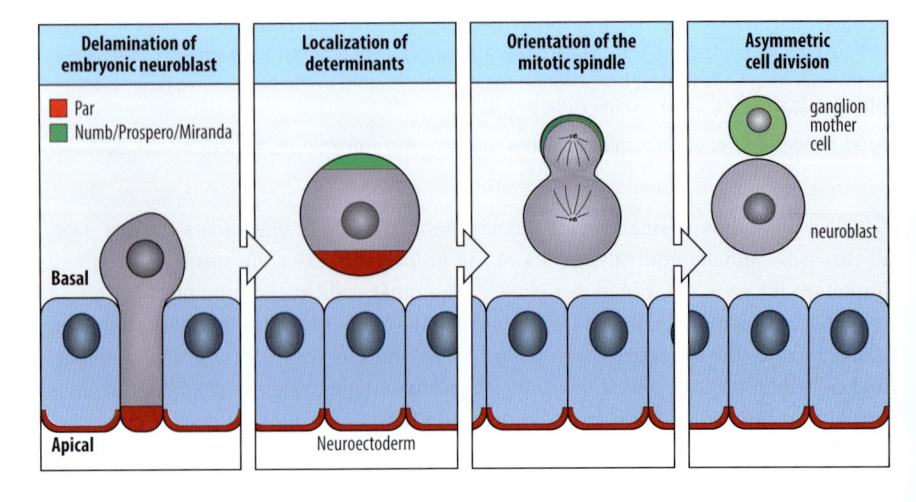

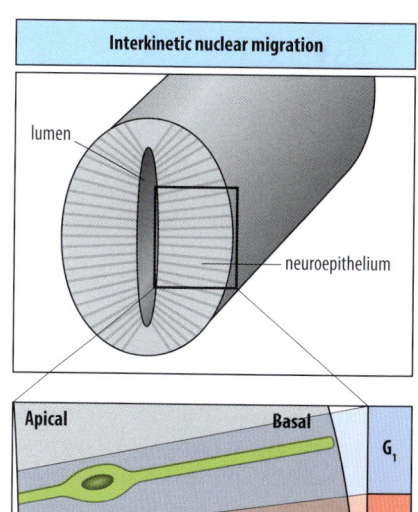

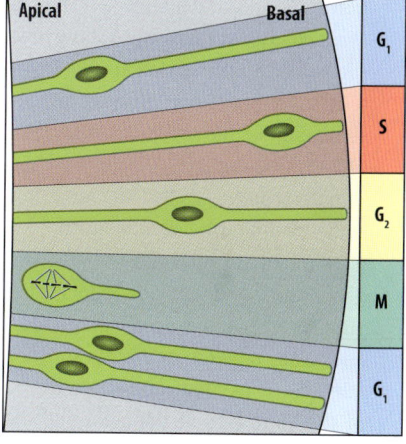

Fig. 10.18 Interkinetic nuclear migration during neuroepithelial cell proliferation in the chick embryo neural tube. Neuroepithelial cells span the width of the neural tube. The blow-up shows cells at successive stages of the cell cycle (from the top down G_1, S, G_2, M, then an additional G_1). A cell retracts its apical and basal processes when it undergoes a mitotic division at the lumen, and then re-extends these processes at the onset of the next cell cycle. Only one cell is shown in the schematic, but interkinetic nuclear migration occurs in all the neuroepithelial cells. Adjacent cells are often in different stages of their cell cycle, and so their nuclei are in different locations relative to the lumen, giving the neuroepithelium a pseudostratified appearance.

Based on Spear P.C., Erickson C.A.: *Apical movement during interkinetic nuclear migration is a two-step process.* Dev. Biol. 2012, **370**: 33–41.

As neurogenesis proceeds, the neuroblast undergoes successive changes in the expression of a particular set of transcription factors, which gives the neurons produced at different times their specific identities. This means that neurons with different functions develop in the correct positions within the ventral nerve cord. Thus the precise timing of the transition from expression of one transcription factor to the next in the neuroblasts is crucial for the correct formation of the nerve cord. The timing of the first transition seems to be linked to neuroblast division, whereas later transitions are linked to an intrinsic timing mechanism in the neuroblast that is independent of cell division.

The embryonic phase of neuroblast division and neurogenesis provides a functional CNS for the first-instar *Drosophila* larva, but these neurons comprise only about 10% of the neurons that are eventually required by the adult fly. A further extended period of neurogenesis occurs before metamorphosis and produces the remaining 90%.

10.12 In the vertebrate neural tube, neurons are formed in the proliferative zone and migrate outwards

The vertebrate neural tube is initially a single-layered neuroepithelium, arranged with bilateral symmetry around a central lumen. The lumen will form the brain ventricles and the central canal of the spinal cord. The basal surfaces of the neuroepithelial cells form the lateral edges of the neural tube and their apical surfaces line the lumen. As neuroepithelial cells progress through the cell cycle, their nuclei move in a stereotypic manner called interkinetic nuclear movement, in which mitosis occurs apically and S phase occurs basally. Symmetric mitotic divisions predominate in the early stages of CNS development, and mean that each neuroepithelial cell gives rise to two neuroepithelial progeny (Fig. 10.18). Figure 10.19 shows consecutive frames from a movie of this process in the developing spinal cord. As more cells get packed in side by side, the neural tube expands. At this stage of development, the neuroepithelial cells are patterned progenitor cells—for instance, the type discussed in Section 10.6 in the developing spinal cord.

In most regions of the CNS, the single-layered neuroepithelial cells change shape soon after their early patterning, becoming spindle-shaped cells known as

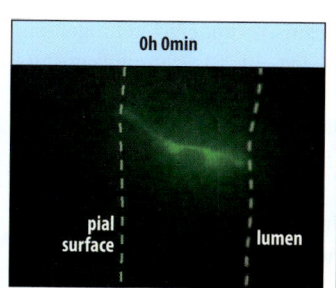

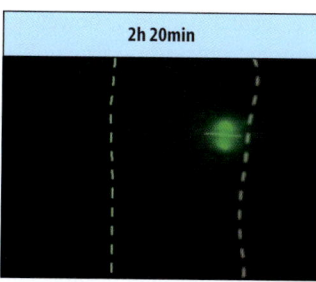

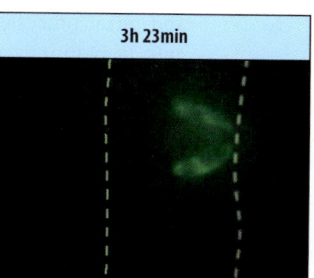

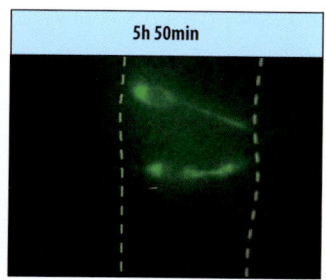

Fig. 10.19 Symmetric divisions expand the spinal cord neuroepithelial cell pool. Images show consecutive frames from a movie of neuroepithelial cells dividing in the prospective chick spinal cord (see https://www.jove.com/v/3920/high-resolution-live-imaging-cell-behavior-developing). A single neuroepithelial cell, labeled experimentally using a green fluorescent protein construct (frame 1), retracts its basal process (frame 2), and divides symmetrically (frame 3). Each daughter cell re-extends apical and basal processes and re-establishes contact with the membrane called the pia on the outer (basal) surface of the neural tube (frame 4). Dashed lines show inner (lumen, L) and outer edges of the neural tube. h, hours; min, minutes.

Reproduced with permission from K. Storey.

CELL BIOLOGY BOX 10C Glial cells in the nervous system

Although this chapter largely focuses on the development of neurons, we should not forget the other major cell type of the nervous system—the non-neuronal **glial cell**, found in both vertebrate and invertebrate nervous systems. The types and functions of glial cells (collectively called **glia**) are rich and varied. In the early nervous system of both invertebrates and vertebrates, midline glial cells have an important role in early patterning events. In the vertebrate spinal cord, floor plate and roof plate glia pattern the dorso-ventral axis of the neural tube (see Section 10.6) and later are the source of axon-guidance cues. In the *Drosophila* larva, midline glia show functional similarities to the vertebrate floor plate and have a central role in axon pathfinding and separation of the major commissures of the CNS axon scaffold.

In vertebrates, as development proceeds, some neuroepithelial progenitor cells are maintained as **radial glial stem cells**. Radial glial stem cells provide pools of neural stem-like cells and neural progenitor cells, and also provide scaffolds for neuronal migration. Ependymal cells line the surface of the ventricles of the brain and the central canal of the spinal cord, and have a role in cerebrospinal fluid homeostasis. Other types of glia—the Schwann cells of the PNS and the oligodendrocytes of the CNS—provide survival factors to nearby neurons and form the myelin sheaths that envelop and insulate neurons, allowing the rapid propagation of nerve impulses. The astrocytes of the brain (so called because they are star-shaped) and satellite cells of the PNS are also glial cells. They provide nourishment (trophic support) to neurons, and, by stabilizing microtubules within neuronal dendrites, can determine critical periods for neural-circuit formation. The birth of glial cells, called **gliogenesis**, mostly occurs after neurogenesis. Similar principles govern the development of glial cells and neurons. For instance, oligodendrocytes in the CNS develop from Olig2-expressing progenitor cells after the major period of neurogenesis. Microglia are the major immune cells of the CNS, responding to infection and inflammation. Figure 1 shows some of the major glial cell types of the nervous system.

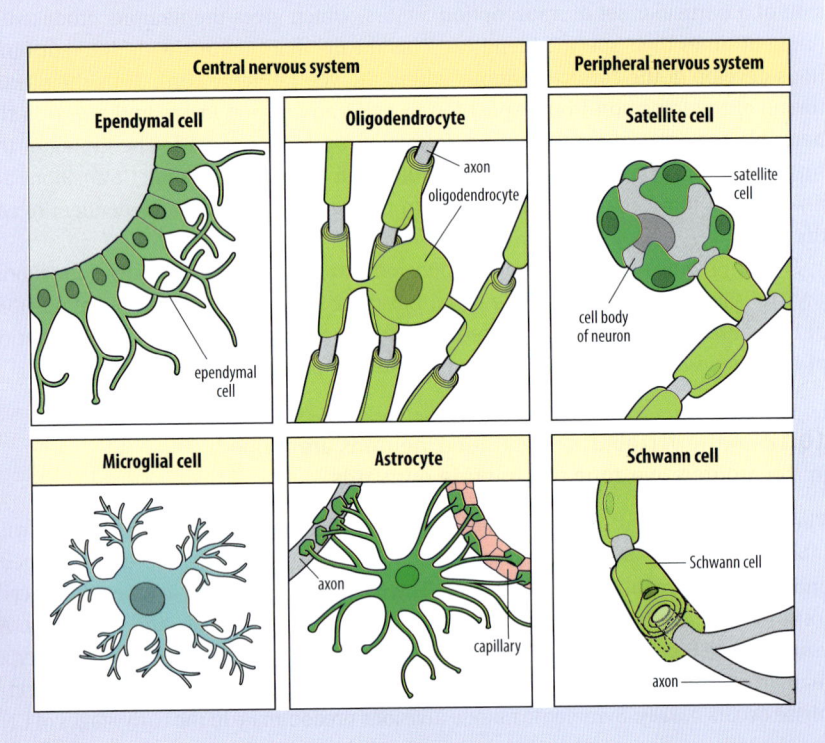

Fig. 1

---

radial glial cells (Box 10C). Like neuroepithelial cells, these span the entire width of the neural tube. Both neuroepithelial cells and radial glial cells act as neural stem cells. They can either divide symmetrically and give rise to two neuroepithelial or **radial glial stem cells**, or they can divide asymmetrically. Following an asymmetric division, one daughter remains a radial glial stem cell and re-establishes its apical and basal process. The second daughter, however, begins to undergo neurogenesis and develop into a neuronal precursor—a neuroblast (Fig. 10.20). Neuronal precursors retract their apical process as they differentiate, and migrate laterally to the pial surface along a scaffold provided by the process of their sister radial glial cell. During this migration, a neuronal precursor can undergo a final symmetric division, giving rise to post-mitotic neurons.

Symmetric divisions gradually give way to increasing numbers of asymmetric divisions as neurogenesis gathers pace, but throughout the period of spinal cord development, a pool of multipotent neural stem or progenitor cells are retained at the lumen.

Fig. 10.20 Formation of spinal cord neurons from vertebrate radial glial cells by asymmetric cell division. Radial glial cells span the width of the neural tube. The blow-up shows cells at successive stages of the cell cycle (from the top down G_1, S, G_2, and M), showing how they retract their apical and basal processes when they undergo a mitotic division at the lumen. After an asymmetric cell division, one daughter is a radial glial cell, which then re-extends its apical and basal processes, and one daughter (purple) is a neuronal precursor. This migrates along the scaffold provided by its sister, undergoes terminal division and differentiates into a post-mitotic neuron that settles just beneath the pial (basal) surface.
Based on Spear P.C., Erickson C.A.: *Apical movement during interkinetic nuclear migration is a two-step process.* Dev. Biol. 2012, 370: 33–41.

This zone is called the **ventricular zone**, and the zone in which post-mitotic neurons are found is called the mantle zone. Beyond this, the marginal zone is made up of axons that extend from neuronal cell bodies (Fig. 10.21). The marginal zone also harbors various types of glia, which are generated from radial glial cells after the major period of neurogenesis. The generation of neurons and glia occurs over an extensive period of time, and results in a great increase in the width of the spinal cord. In the adult, a remnant of neural stem or progenitor cells—now called ependymal cells—are still retained around the central canal. Around these, and composing much of the adult spinal cord, is the gray matter, harboring the cell bodies of the many neurons that have now differentiated. An outermost layer, the white matter, is composed of myelinated axons.

10.13 Specialized features in the cortex support high levels of neurogenesis

Maintenance of a pool of multipotent neural stem or progenitor cells during embryonic development is essential for growth of the CNS, especially in the brain, where billions of neurons are generated. In the cerebral cortex, or **neocortex**, a region of the brain that develops from the embryonic dorsal telencephalon (the pallium) and is unique to mammals, specialized features support high levels of neurogenesis. A core arrangement of developing progenitors and newborn neurons in the developing cortex is similar to that in the spinal cord. Radial glial cells lie at the ventricular zone and extend a process to the pia, and during neurogenesis, neurons migrate along the radial glial cell scaffold (Fig. 10.22). However, two unique features enable the generation of increased numbers of neurons in the cortex, compared with the spinal cord. First, the cortex harbors an additional layer of proliferating progenitor cells that lie in a region called the **subventricular zone** (see Fig. 10.22). This zone is particularly thick in humans. Second, in the cerebral cortex, the relationship between gray and white matter is reversed compared with the spinal cord: the cortex has most of its neuronal cell bodies located at the outer surface, whereas the axons extend inward to form an intermediate zone of white matter adjacent to the subventricular zone. By keeping the cell bodies outside the more rigid myelinated axons, the mammalian cortex has acquired great potential for expansion during evolution.

In the cortex, wave after wave of neurogenesis occurs over a protracted period of embryogenesis, building up six layers (I–VI) of differentiated neurons in the cortical plate (see Fig. 10.22). Each layer contains neurons with distinctive shapes and connections. For example, large pyramidal cells that project long distances to sub-cortical targets are concentrated in layer V, and smaller neurons that process visual information predominate in layer IV.

The cortical layer to which newborn neurons migrate relates to the time when the neuron is born, and the time of birth specifies the identity of a cortical neuron. The experiments that originally established the relation between the birth-time of a neuron and its final layer destination are described in Box 10D. The newly formed neuron is still immature; later it extends an axon and dendrites (see Fig. 10.2), and assumes the morphology of a mature neuron.

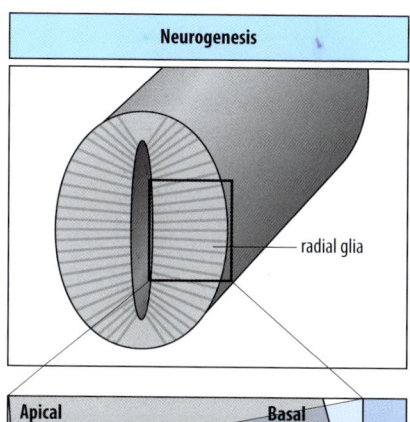

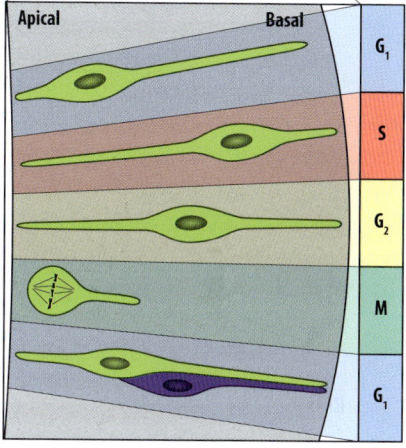

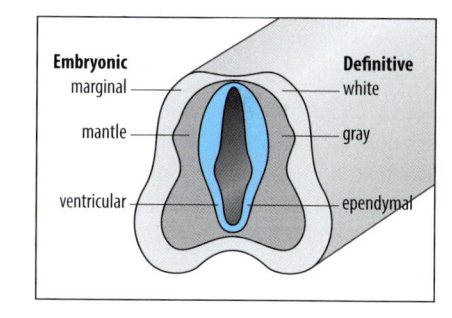

Fig. 10.21 Zones in the embryonic and adult spinal cord. The schematic shows the different zones in the spinal cord, and the terms used to define them in the embryo (left) and in the adult (right).

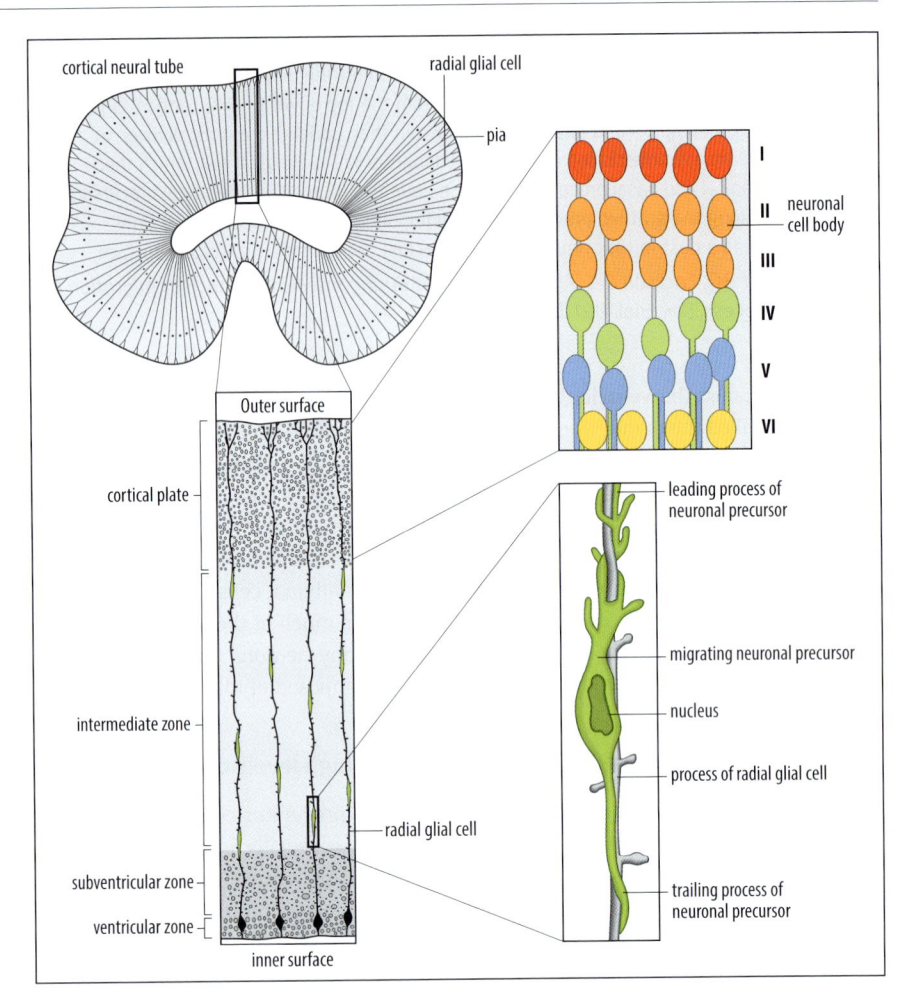

Fig. 10.22 Cortical neurons migrate along radial glial cells. Neurons generated in the ventricular or subventricular zone migrate to their final locations in the cortex along radial glial cells that span the entire extent of the developing cortex. Neurons are generated in waves that build up six layers (I–VI) in the cortical plate.
Illustration after Rakic, P.: **Mode of cell migration to the superficial layers of fetal monkey neocortex.** *J. Comp. Neurol. 1972,* **145***: 61–83.*

Neurons born at early stages of cortical development migrate to layers closest to their site of birth, whereas those born later end up further away, in more superficial layers (see Fig. 10.23). Later-born 'younger' neurons therefore migrate through early-born 'older' neuronal layers. The exception is the first-formed layer, which remains as the outer layer. Mouse mutations that disrupt cortical neuronal migration have provided considerable insight into the process. Mice with a mutation in the *reeler* gene that causes a loss of Reeler protein show disrupted radial neuronal migration. Successive waves of neurons are unable to pass through the layer formed by their predecessors and so the cortical layers are arranged in reverse order. As the name implies, *reeler* mutant mice have very poor motor coordination.

10.14 Inhibitory interneurons undergo tangential migration into the cortex

All of the cortical excitatory neurons, which use glutamate as their transmitter, have their origin in the pallial (dorsal telencephalic) ventricular zone. By contrast, the numerous, small GABAergic inhibitory interneurons of the cortex are generated in the

EXPERIMENTAL BOX 10D Timing the birth of cortical neurons

The time at which a neuron is born determines its future cortical layer. A neuron's time of birth is defined by the last mitotic division of its progenitor cell. If progenitor cells are given a pulse of a labeled compound such as tritiated thymidine, BrDU, or EdU that can be incorporated into DNA, those cells that undergo their last round of replication immediately after taking up the compound will have the most heavily labeled DNA, as it will not be diluted out through further rounds of mitosis. Cells that stop dividing immediately after taking up the label can therefore be distinguished and their final destinations detected.

Figure 1 shows the results of an experiment of this type that determined the order in which neurons reached their final layers in the monkey **visual cortex**. Tritiated thymidine ([3H]thymidine), which becomes incorporated into replicating DNA when a cell is in S-phase, was injected into the developing neural tube at different times during development. The red bars represent the distribution of heavily labeled neurons found after each injection. Neurons born earliest remain closest to their site of birth, the proliferative ventricular zone. Neurons born later become part of successively higher cortical zones.

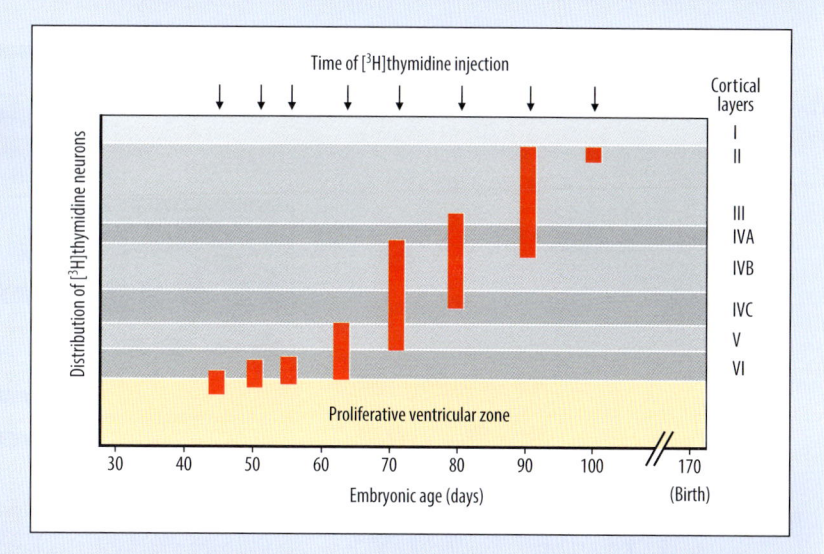

Fig. 1

---

subpallium (also known as the medial ganglionic eminence)—a ventral telencephalic structure—then undergo **tangential migration** dorsally into the pallium (Fig. 10.24) where they eventually intersperse among the glutamatergic neurons of the forming cortical plate (see Fig. 10.22). Migration is directed by both short- and long-range chemoattractive cues, the **neuregulins**, which diffuse from the lateral ganglionic eminence and cortical plate, respectively. These proteins act on ErbB4 receptors carried by the interneurons. As well as being attracted by pathway and target cells, the migrating interneurons are repulsed by another group of proteins, the **semaphorins**, which are secreted by the medial ganglionic eminence and act on the interneurons through neuropilin receptors.

A simple explanation for this remarkable process of tangential migration is that while both glutamatergic (excitatory) and GABAergic (inhibitory) neurons are needed close to one another for correct network function, the mechanism of cell specification in the forebrain can only generate each of these distinct cell types at different overall positions along the dorso-ventral axis of the developing telencephalon. In other words, in parallel with the acquisition of specific behaviors, developing neurons become functionally distinct (excitatory or inhibitory) as a feature of earlier positional values. Therefore, migration is needed to bring these two types of functionally distinct neuron together.

10.15 Delta–Notch signaling and lateral inhibition govern vertebrate neurogenesis

In *Drosophila*, lateral inhibition mediated by Delta–Notch signaling is used to distinguish neuronal progenitors from a sheet of epidermal cells (see Fig. 10.16). In vertebrates, lateral inhibition mediated by Delta–Notch signaling is also crucial for

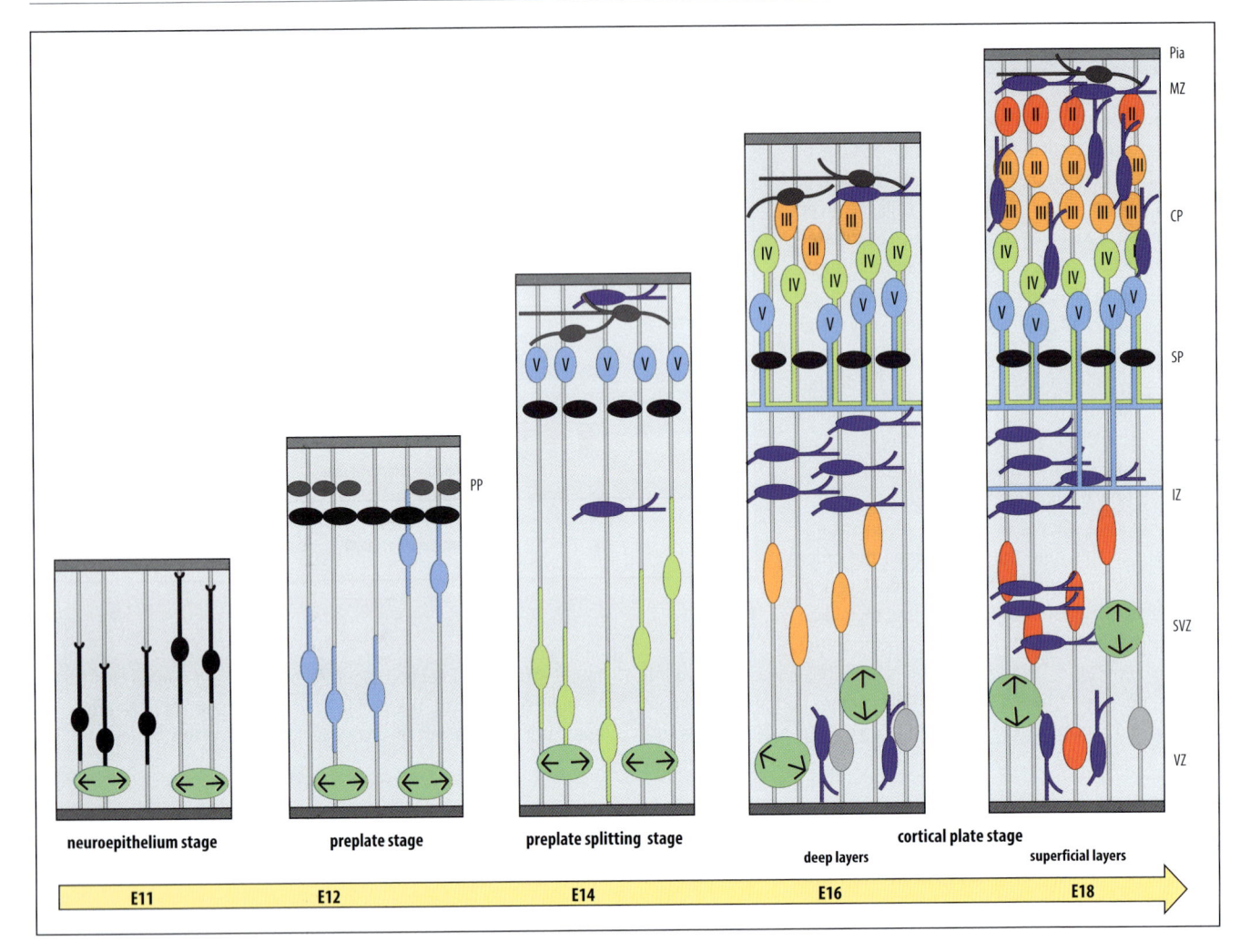

Fig. 10.23 The generation of neuronal layers in the mammalian cortex. Neurogenesis begins when a first set of cells (black) migrate from the ventricular zone (VZ) to the pial surface and form the preplate (PP). Next, a second wave of neurogenesis occurs (blue cells). These newborn neurons migrate to the preplate, which they penetrate, splitting the preplate into two layers, an upper marginal zone (MZ) and a lower subplate (SP). The latter will form the deepest cortical layer, layer VI. In each successive wave of neurogenesis (green, orange, red), cortical neurons migrate through the subplate and the previous neuronal layers, and stop just beneath the marginal zone. The end result is the stratification of cortical neurons by birth day. The oldest neurons, those of the preplate, are found in layer I (derived from the marginal zone) and deep portions of layer VI (cells derived from the subplate) of the developing cortex. The remaining cortical layers are 'inside-out' with respect to birth dates; layer II contains the youngest cells, while layer VI contains the oldest. Once all six layers of the cortical plate are in place, the axons of the major projecting layers (V and VI) grow inwards towards the ventricular zone and turn into the intermediate zone (IZ) of future white matter. The intermediate zone contains inhibitory interneurons (purple) migrating in from the subpallium. CP, cortical plate; SVZ, subventricular zone. The time line is for mouse development, days *in utero*.

Adapted from Honda, T., et al.: **Cellular and molecular mechanisms of neuronal migration in neocortical development.** Semin. Cell Dev. Biol. *2003*, **14**: 169-174.

neurogenesis, but operates in tissue that is already specified as neural tissue, playing a major role in the decision of a cell to be retained as a progenitor cell (a neuroepithelial or radial glial cell) or to differentiate into a neuron. Genes that direct lateral inhibition and neuroblast formation in *Drosophila* (see Fig. 10.16) have vertebrate homologs that function in an almost identical manner to their invertebrate counterparts. For instance, the vertebrate Notch target genes, *Hairy Enhancer of Split* (*Hes*) and related *Hey* genes, are homologs of the *Drosophila Enhancer of split*. And, as in *Drosophila*, the Hes and Hey transcription factors inhibit neuronal cell specification (Fig. 10.25,

Fig. 10.24 Origin and tangential migration of cortical interneurons. Schematic transverse section through one side of the forebrain of a mouse embryo. Colored regions show major sites of neurogenesis: glutamatergic neurons are born in the pallium (cortex (Cx) and hippocampus (H), purple); GABAergic neurons are born in the subpallium (lateral and medial ganglionic eminences (LGE and MGE, respectively, green)); cholinergic neurons form in the pre-optic area of the hypothalamus (POA). Green arrows show the migration paths taken by GABAergic neurons from the subpallium into the pallium, reaching as far as the prospective hippocampus.

Adapted from Marin, O., Rubenstein, J.L.R.: **A long, remarkable journey: tangential migration in the telencephalon.** *Nat. Rev. Neurosci. 2001,* **2***: 780–790.*

first two panels). Similarly, Achaete–Scute homologs (encoded by the ASCL genes) and other proneural proteins, such as the transcription factor **neurogenin**, are homologs of *Drosophila* proneural transcription factors. Cells that express neurogenin go on to express genes characteristic of differentiating neurons, such as *NeuroD*.

The fate of a cell to become a neuron or a progenitor is tightly linked to asymmetric cell division. Like *Drosophila* neural progenitors, vertebrate neural progenitors (and radial glial stem cells) are polarized along the apico-basal axis, and apico-basal determinants govern the inheritance of Delta–Notch pathway components. Single-cell RNA-seq studies show that Delta is expressed in newborn neurons, and transgenic overexpression of Delta in the neural plate or expression of a mutant Notch that signals continuously both result in a reduction in the number of neurons formed. Conversely, inhibition of Delta, silencing of Notch, or overexpression of proneural genes, leads to an increase in the number of neurons. This suggests that, in a similar manner to that described in *Drosophila*, the higher levels of Delta present in the newborn neuron activate the Notch signaling pathway in surrounding cells, and so maintain these neighbors as proliferative progenitors—in effect, the lateral inhibition of the neuronal fate. While the details of this are complicated, the overall idea is that asymmetric divisions promote neurons that have higher levels of Delta, and these influence sister-cell fate through activation of the Notch signaling pathway (Fig. 10.25, third panel).

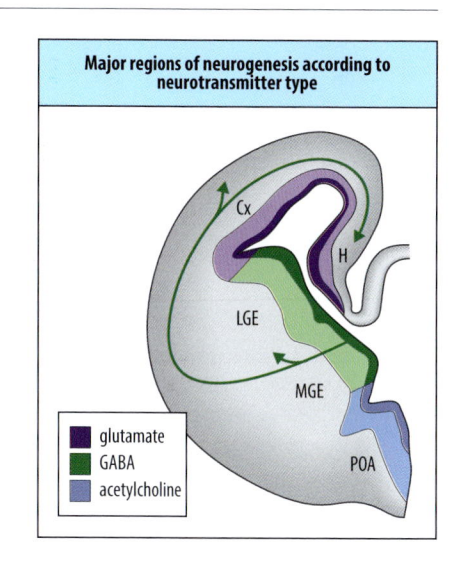

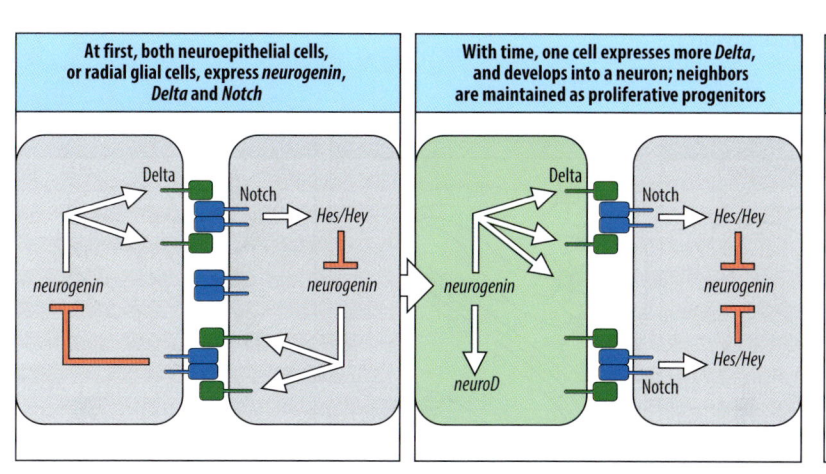

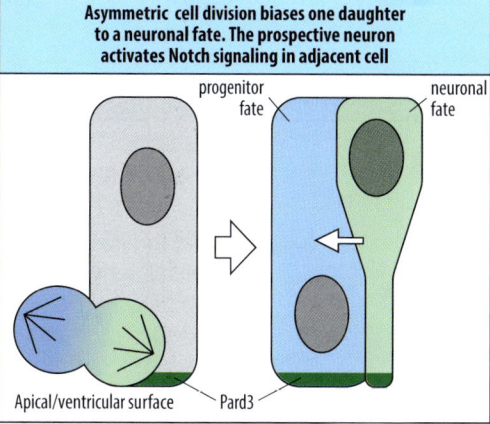

Fig. 10.25 Notch signaling and lateral inhibition govern vertebrate neurogenesis. First two panels: Notch–Delta signaling works in a similar way to that described in *Drosophila* (see Fig. 10.16) to direct neuronal versus proliferating progenitor fates. Hes and Hey genes are homologs of the *Drosophila* gene *Enhancer of split*. The gene *neurogenin* encodes a proneural transcription factor. Third panel: the apical polarity protein, Pard3 (dark green), is localized to the apical surface of the neuroepithelium. During mitosis, a daughter cell that inherits the apical attachment also inherits the Delta modulator Mindbomb, which biases the cell towards a neuronal fate (green). The daughter that transiently loses the apical attachment but retains the basal attachment is biased towards a progenitor fate (blue). At the same time, this bias is strengthened because Delta in the neuronal daughter activates Notch signaling in its sister and neighboring cells (white arrow), further promoting the progenitor fate.

SUMMARY

In *Drosophila*, prospective neural tissue is specified early in development as a longitudinal band of neuroectoderm in the ventral part of the dorso-ventral axis. Within the neuroectoderm, expression of proneural genes distinguishes clusters of neuroectoderm cells with the potential to form neural cells. As a result of lateral inhibition involving Notch and Delta, only one cell of the cluster finally gives rise to a neuroblast. Neuroblasts that give rise to the CNS behave as stem cells, undergoing repeated asymmetric division, each giving rise to a neuroblast and a daughter cell that can give rise to neurons. Delta-Notch signaling and lateral inhibition likewise play an important part in neurogenesis in vertebrates, both in the maintenance of neural progenitor pools and in neuronal diversification. Neurons are born in the proliferative zone on the inner surface of the neural tube and then migrate to different locations within the developing brain and spinal cord.

Axon navigation and mapping

We shall now look at a feature of development that is unique to the nervous system—the guidance of axons to their final targets. This stage of neuronal differentiation occurs after immature post-mitotic neurons have migrated to their eventual locations. Each neuronal cell body extends an axon and dendrites, through which it will, respectively, send and receive signals. Even in the most primitive nervous systems, all neurons participate in synaptically interconnected networks, or circuits, which are built up as developing axons follow very precise paths to their targets—including other neurons—where they make numerous and precise connections (Fig. 10.26). In the rest of this chapter we will look at how these connections are set up. We will first consider axon outgrowth and the mechanisms that guide the growing axon to its target, illustrating key points with reference to specific parts of the nervous system. Then we will look at how axons are mapped to their targets so that the CNS makes sense of the external world.

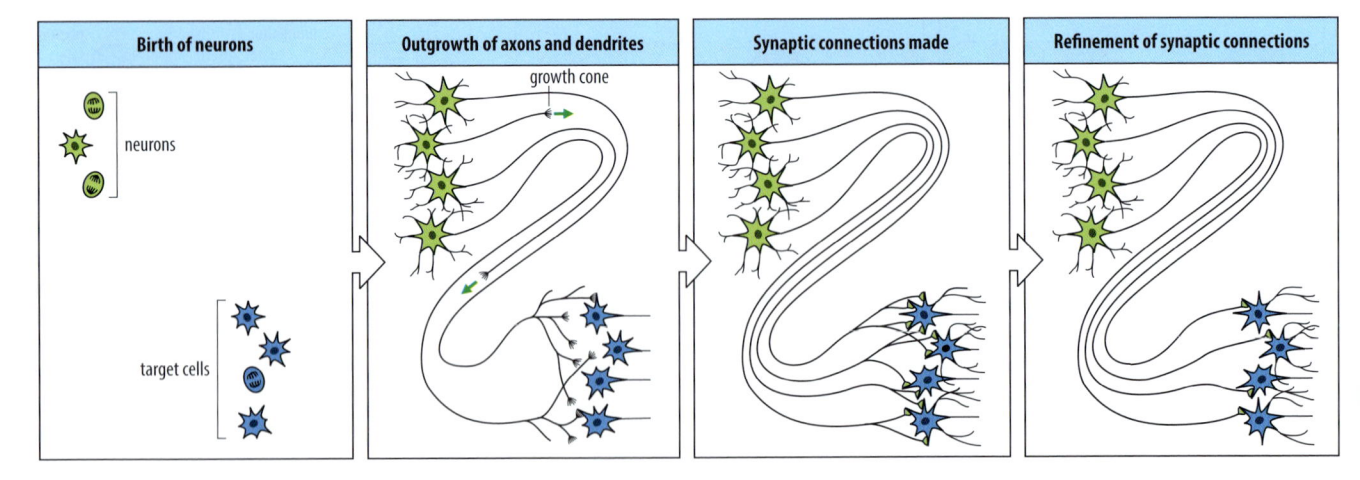

Fig. 10.26 Neurons make precise connections with their targets. Neurons (green) and their target cells (blue) usually develop in different locations. Connections between them are established by axonal outgrowth, guided by the movement of the axon tip (growth cone). As described in the last part of this chapter, the initial set of relatively non-specific synaptic connections is then refined to produce a more precise pattern of connectivity.

Illustration after Alberts, B., et al.: Molecular Biology of the Cell. 2nd edn. New York: Garland Publishing, 1989.

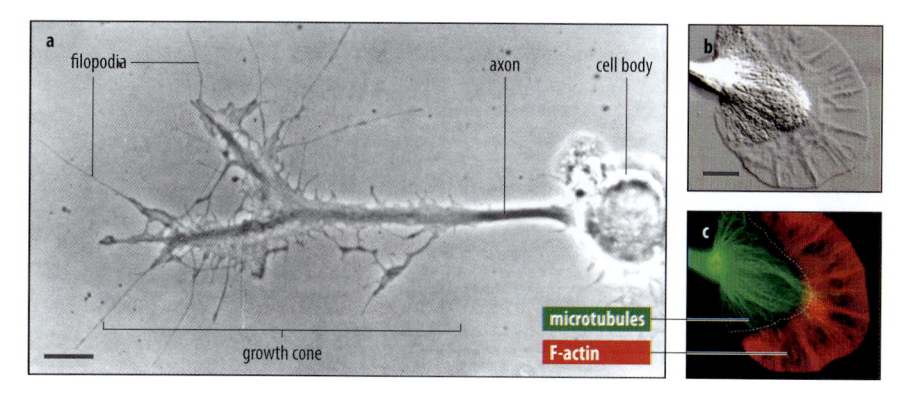

Fig. 10.27 A developing axon and growth cone. (a) The axon growing out from a dorsal root ganglion neuronal cell body ends in the growth cone. (b) High-resolution differential interference contrast micrograph of a growth cone from the sea-slug *Aplysia*. In the outer area of cytoplasm, filopodia and lamellipodia continually extend and retract to explore the surrounding environment. c, the same growth cone fixed and labeled to show F-actin (red) and microtubules (green). Scale bars = 10 μm.

Photographs courtesy of (a) P. Gordon-Weeks and (b, c) Suter, D.M., Forscher, P.: **Substrate-cytoskeletal coupling as a mechanism for the regulation of growth cone motility and guidance.** J. Neurobiol. *2000,* **44***: 97-113.*

10.16 The growth cone's response to guidance cues controls the path taken by a growing axon

A newborn neuron exhibits polarity and completes its differentiation by extending an axon from one side and dendrites from the other. Internally, axons and dendrites are composed of microtubules that are highly polarized along the length of the axon but less ordered in dendrites, a feature of the different types of microtubule-associated protein expressed in each. The tip of the growing axon is called the **growth cone**. This specialized motile part of the cell (Fig. 10.27) is rich in receptors that sense guidance cues in its environment. As in other cells capable of migration, such as neural crest cells, the growth cone continually extends and retracts filopodia at its leading edge, which make and break connections with the underlying substratum. Between the filopodia, the edge of the growth cone forms thin ruffles—lamellipodia—similar to those on a moving fibroblast (see Box 5B). Indeed, in its ultrastructure and mechanism of movement, the growth cone closely resembles the leading edge of a fibroblast crawling over a surface. Unlike a moving fibroblast, however, the extending axon also grows in length and diameter, with an accompanying increase in the total surface area of the neuron's plasma membrane. The additional membrane is provided by intracellular vesicles, which fuse with the plasma membrane.

Growth cones navigate the route from the neuronal cell body to the target in response to guidance cues—environmental signals that direct growing axons. Guidance cues are divided into different functional categories, depending on whether they operate as long-range or short-range cues, and whether they attract or repel axons (Fig. 10.28). Long-range attraction and long-range repulsion involve diffusible **chemoattractant** or **chemorepellent** molecules that the growth cone can detect at a distance from their source. Short-range guidance depends on contact between the growth cone and the substratum; the latter might be the extracellular matrix or another axon.

Genetic and biochemical studies have led to the discovery of several different classes of axon guidance molecules and their receptors. The **semaphorins** are a prominent and evolutionarily conserved family of axon-guidance proteins. There are seven subfamilies of semaphorins: three secreted, and four associated with the cell membrane.

Fig. 10.28 Axon-guidance mechanisms. Four general types of mechanism contribute to controlling the direction of a growth cone: long-range attraction, long-range repulsion, short-range attraction, and short-range repulsion. Long-range cues are provided by secreted molecules (e.g. netrins and semaphorins) that form gradients in the extracellular matrix. Short-range cues, such as the cadherins and the ephrins, require contact between the growth cone and the substratum.

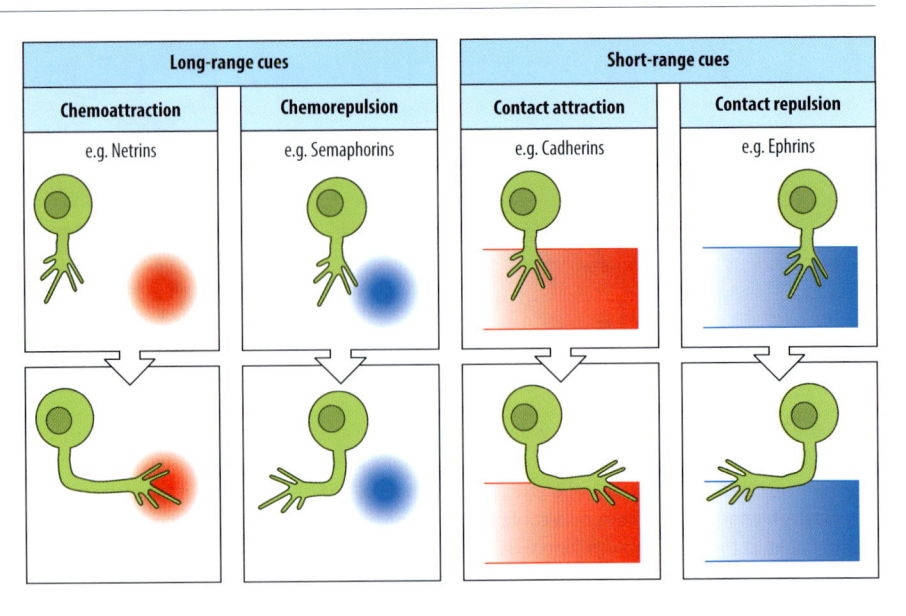

Semaphorins are primarily axon repellents that activate complexes of neuronal receptors called the **plexins** and the **neuropilins**. Semaphorins can, in some instances, attract growth cones, depending on the nature of the neuron and on the type of semaphorin receptor present in the neuron membrane. Different regions of the same neuron can respond in opposite ways to the same semaphorin—the apical dendrites of cortical pyramidal neurons grow towards a source of semaphorin 3A, whereas their axons are repelled. This enables the receiving end of the neuron and the transmitting end of the neuron to be oriented in opposite directions by the same signal. This difference in response seems to be due to the presence of guanylate cyclase, a component of the semaphorin signaling pathway in the dendrites, but not in the axon. The role of semaphorins in directing the formation of a simple neural circuit is discussed in Box 10E.

The **netrin** family of guidance proteins are also bifunctional, in that they attract some axons and repel others. These opposing actions are mediated by two classes of receptors: the DCC (deleted in colorectal cancer) receptors mediate attraction, and the UNC5 (uncoordinated 5) receptors mediate repulsion. Netrins are secreted and mainly act as long-range guidance cues, but may also operate as short-range contact-dependent cues.

The **Slit proteins** are secreted glycoproteins that normally repel growth cones by acting on **Robo** (Roundabout) receptors, but can also stimulate elongation and branching in sensory axons.

The **ephrins** are cell-surface proteins that activate **Eph receptors** on the surface of adjacent cells (see Box 5D) and so act as short-range contact cues. They mainly direct repulsive interactions in the nervous system, but an ephrin–Eph interaction can, in some cases, lead to an attractive response. Ephrins can also act as receptors by transducing a signal into the expressing cell; in this case, Ephs act as the ligands.

The gradients of morphogens that originate in secondary organizers and pattern progenitor cells in the early neural tube, including BMPs, Wnts, Shh, and FGFs, are reused later in development to act as axon-guidance cues.

Three classes of **cell-adhesion molecules (CAMs)** are important in axon navigation: the cadherins; the immunoglobulin superfamily CAMs; and the integrins. Their mechanisms of action are described in Box 5A. CAMs can modulate the action of other guidance cues. They can also serve to couple the growth cone to the substratum. Integrins, for example, support adhesion of a cell to the extracellular matrix, one end

BOX 10E Putting it together: the development of a simple neuronal circuit

Neuronal circuits underlie behavior. Neuronal circuits form through a self-organization process that unfolds over time. This involves the mechanisms considered in this chapter: progenitor and neuronal specification and migration; axon navigation; and neuron-neuron connectivity. We look here at the formation of a simple circuit that mediates a simple behavior, the knee-jerk reflex.

The knee-jerk reflex is an example of a general muscle-stretch reflex that detects and responds to changes in muscle length (proprioception). This reflex is responsible for maintaining limb and body position (although probably most familiar because it can be elicited by a tap to the patellar tendon just below the knee that causes the quadriceps muscle to lengthen and the leg to jerk upwards). In the knee-jerk circuit (Fig. 1, left panel), stretch receptors, called muscle spindles (orange) in the quadriceps, detect an increase in muscle length. They are innervated by proprioceptive **sensory neurons**, called 1a afferents (blue), whose cell bodies are in a dorsal root ganglion (DRG) next to the spinal cord. The sensory neuron axons synapse directly with the dendrites of motor neurons (green) in the ventral horn of the spinal cord whose axons innervate the quadriceps muscle and cause its contraction. This monosynaptic circuit operates simultaneously with a disynaptic circuit, which includes inhibitory interneurons (red) that reciprocally prevent contraction of the antagonistic muscle, in this case, the hamstrings. How is this simple circuit built?

The component neurons are born in early development. Sensory neurons originate from neural crest cells that have migrated and coalesced to form the DRG (see Section 5.22). Motor and inhibitory neurons derive from progenitor cells that are patterned in the spinal cord through the antagonistic actions of Shh and BMPs. The Hox code determines the position of each of these neurons along the antero-posterior axis. After the birth of DRG sensory neurons, their axons begin to extend bilaterally. There are functionally and molecularly distinct classes of DRG sensory neurons, but all have a central projection that enters the dorsal horn of the spinal cord (Fig. 1, right panel). Those that relay pain and temperature (purple) terminate close to their entry point; those that relay proprioception, the 1a afferents (blue), extend further ventrally to synapse with the dendrites of the appropriate motor neurons (green). Semaphorin 3, expressed in the ventral spinal cord, repels all ingrowing sensory axons except the 1a afferents.

Proprioceptive sensory afferents face the challenging task of forming selective connections with only a few of dozens of motor pools innervating limb muscles, for instance, ignoring those motor neurons that innervate antagonistic muscles. This is achieved through a strategy that relies on a progressive exclusion of inappropriate targets, thereby constraining choice, as shown in Figure 2.

Figure 2 shows how a proprioceptive sensory neuron (green) makes the correct connection with a target motor neuron. This is a blow-up of one side of the ventral region of the spinal cord as shown in Fig. 10.21. First, the laminar (tier-like) organization of the motor pools—a function of earlier dorsoventral patterning—reduces the number of targets (Fig. 2, first panel). Then, sensory axons and motor dendrites grow towards a common meeting point, defined by a coordinate system of positional information and by dendritic geometry (Fig. 2, second panel). The ability to detect and respond appropriately to the positional information, and to show intrinsic patterns of

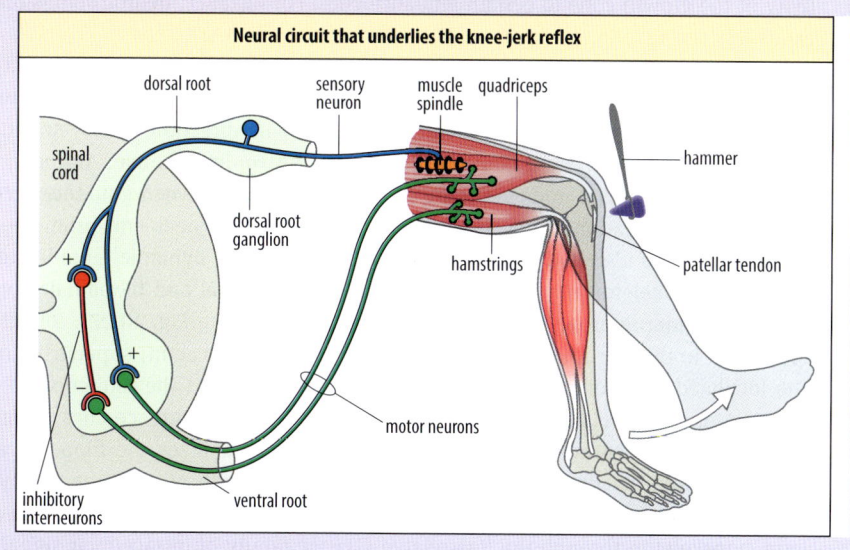

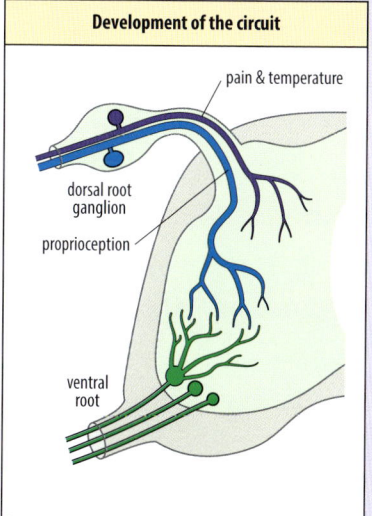

Fig. 1

548 Chapter 10 Development of the nervous system

Strategies used by sensory neurons to make the correct connections with their target motor neurons

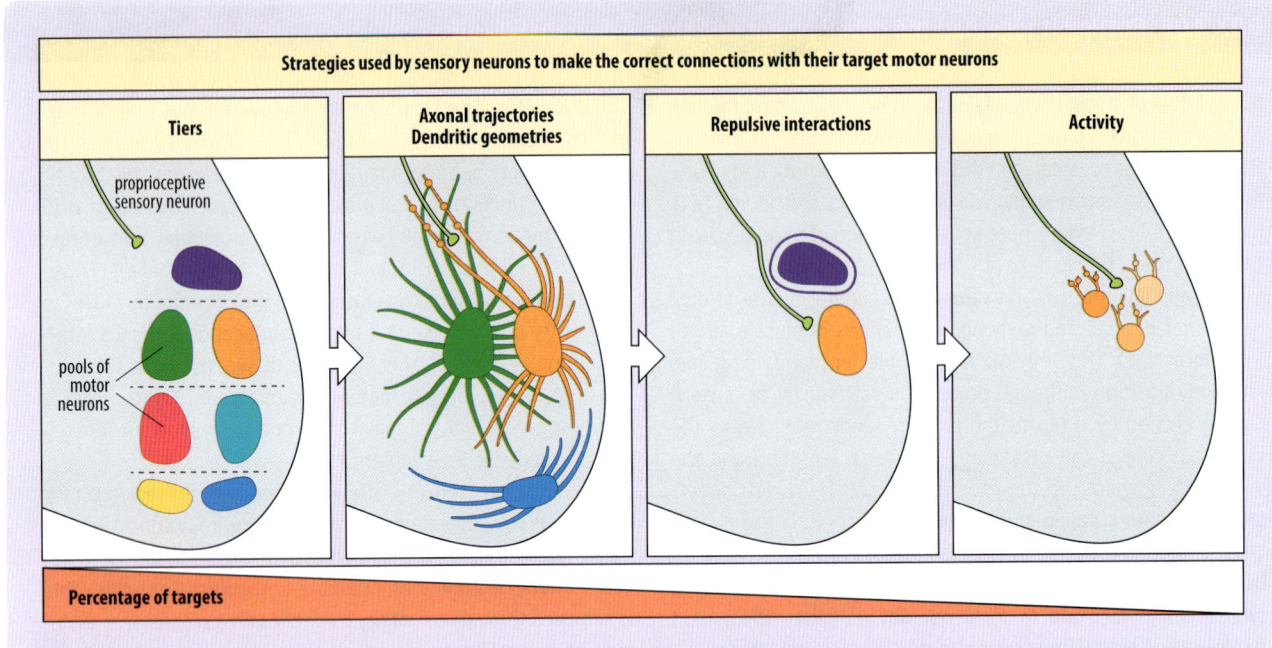

| Tiers | Axonal trajectories / Dendritic geometries | Repulsive interactions | Activity |

Percentage of targets

Fig. 2

dendritic growth, is part of each neuron's specification. At this meeting point, additional recognition mechanisms come into play; for instance, repulsive signals exclude non-appropriate partners (Fig. 2, third panel). Once sensory-motor connections have formed, neural activity fine tunes their distribution (Fig. 2, fourth panel).

of the molecule binding to fibronectin, laminin, and tenascin in the matrix and the other end associating with the cytoskeleton of the growth cone.

Other classes of extracellular molecules act closely with guidance cues to support and modulate directed axon growth. The **neurotrophins** provide trophic support, and stimulate survival and the speed of axon growth. They can also influence the sensitivity of the growth cone to certain guidance cues. For instance, neurotrophins make axons less sensitive to repulsion mediated by semaphorin 3A.

The general concept is that when a receptor on a growth cone is stimulated by a guidance cue, it activates pathways that result in reorganization of the growth-cone cytoskeleton, causing the growth cone to alter its direction of movement. In this way, the guidance cue is converted into a **chemotropic response**. Although we only have a superficial understanding of the association between guidance receptors and the regulation of the cytoskeletal dynamics that steer and promote the growth cone, the idea is that receptor activation leads to a dynamic remodeling of the actin cytoskeleton that directly affects the lamellipodial and filopodial protrusions that underlie growth-cone motility and guidance (see Fig. 10.27). Repellent cues result in the disruption and loss of F-actin (filamentous actin) superstructures. The localized disassembly of the F-actin cytoskeleton causes the retraction of the filopodia and the 'collapse' of the growth cone. Attractive guidance cues promote the asymmetric incorporation of F-actin, which both stabilizes the filopodia and drags microtubules into their back ends, promoting growth of the axon. Many of the extracellular signals that affect axon guidance are known to have effects on the actin cytoskeleton through the Ras-related GTPase family of intracellular signaling proteins.

10.17 Axon guidance is a dynamic process

A growth cone encounters many guidance cues as it grows towards its final target cell. At any point in its trajectory, the growth cone sums and integrates all the information from the various incoming cues to make a decision on the direction of growth. The response to a given set of cues is not always the same, however, because the properties of the growth cone undergo intrinsic changes over time. For instance, a neuron might begin to place a particular receptor in its growth cone only some time after the axon has begun to extend, causing a change in the response of the growth cone to a particular cue.

An active area of axon-guidance research is to examine how guidance receptors appear in the growth-cone membrane at particular points in the growing axon's route. In theory, this could occur due to changes at the transcriptional or the translational level. It used to be thought that receptors were always transcribed and translated in the cell body, then transported down the growing axon to the growth cone, focusing attention on motor proteins that would rapidly transport receptors from the cell body to the growth cone. But the fact that the growth cone can be a very long distance from the cell body prompted questions as to whether this mechanism was always feasible. In experiments with *Xenopus* retinal ganglion cells, when the growth cone was separated from the cell body by severing the extending axon, the growth cone continued to navigate correctly. Subsequent studies showed that local protein translation and degradation can occur in growth cones independently of gene transcription in the cell nucleus, and, importantly, can be triggered by guidance cues (Fig. 10.29). The growth cone stores mRNAs that code for receptors and intracellular signaling proteins involved in cytoskeletal remodeling. Guidance cues can then activate translation of particular subsets of these mRNAs. Such local protein translation can explain the rapid and changing responses of the growth cone to the cues that it encounters en route.

Why might a growth cone need to change its response? In both invertebrates and vertebrates, early axon pathways are formed from **pioneer axons** that navigate a route when the embryo is relatively tiny. Even so, the axon needs to cover a long distance, and for many pioneering axons, the route from the cell body to the final target is not direct. The long distance that needs to be covered, and the indirectness of the route, mean that many axons use **intermediate targets** (also called **choice points** or **guidepost cells**) that signpost a route. Axons are attracted to intermediate targets, but rather than staying there, they switch trajectory, and grow to the next guidepost, until they reach their target destination. The behavior of a growth cone at the intermediate target requires a switch from attraction to repulsion. By changing the expression of its surface receptors, the growth cone can respond rapidly and appropriately to an intermediate target.

Pioneer axons establish a scaffold, which later-born **follower axons** join, ensuring they can navigate to targets in the ever-growing embryo. This process is called **fasciculation** and is commonly mediated by homophilic interactions involving the CAMs. Over time, axon pathways are built into **nerve tracts**, or **nerve fascicles**. The changing nature of receptors in the growth cone also dictates whether an axon can join a fascicle, or leave the fascicle, once it reaches its target.

10.18 Axons that cross the ventral midline provide a model system for axon guidance

Much of the early work in identifying the location and nature of axon-guidance cues was done in insects, whose nervous systems are relatively simple. For instance, in the grasshopper, almost every neuron in the embryonic nerve cord has been identified, allowing a map of axon tracts and guidepost cells to be made. Individual guidepost

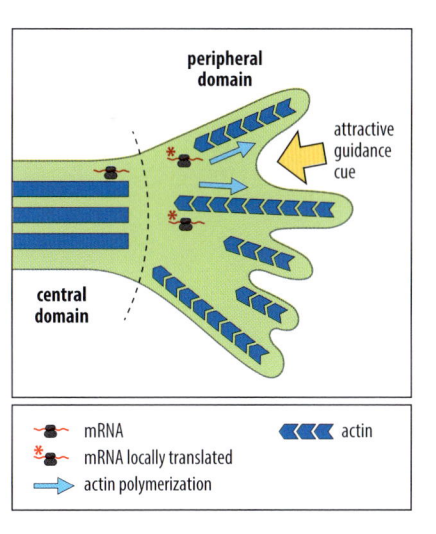

Fig. 10.29 Sites of local mRNA translation in a growth cone. In response to attractive guidance factors, mRNAs are localized to the stimulated side of the growth cone, followed by local translation, which extends the axon in the desired direction.

Adapted from Sasaki, Y.: *Local translation in growth cones and presynapses, two axonal compartments for local neuronal functions.* Biomolecules 2020, **10**: 668.

Fig. 10.30 Pathway of commissural axons in the spinal cord. In the vertebrate spinal cord, the axons of spinal cord commissural neurons extend ventrally, and then towards the floor plate cells. They cross the midline at the floor plate and then grow anteriorly towards the brain. The sharp anterior turn after crossing the midline is thought to be due to a gradient of Wnt proteins in the floor plate with the high point anterior.

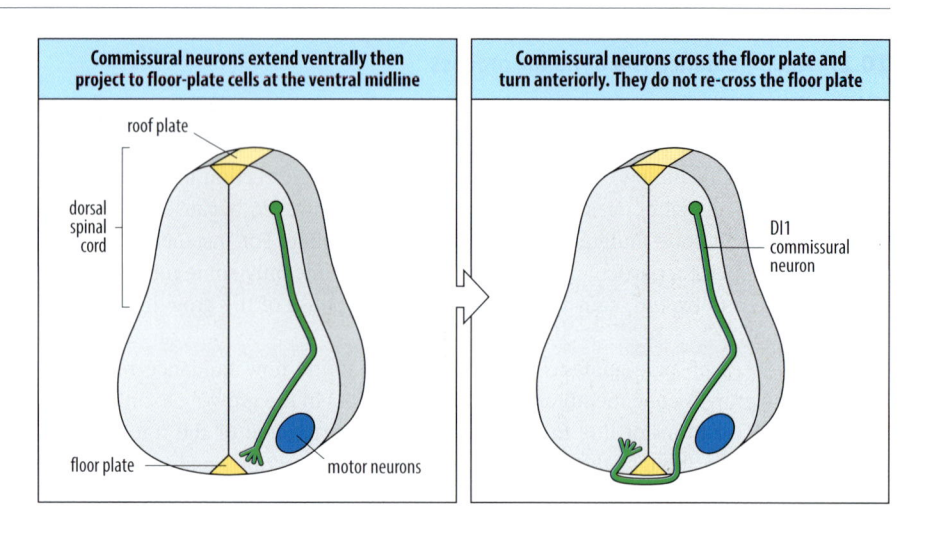

cells, carrying potential cues, were then laser-ablated, and the consequences for axon navigation examined. And since the 1980s, model organisms that are amenable to genetic manipulation have been instrumental in determining how mutations in particular guidance molecules, or their receptors, cause axons to make navigational errors. At the same time, gain- and loss-of-function experiments in model organisms whose genetics are not well known but which are easily accessible to experimental manipulation at particular times (i.e. when axons are growing) have enabled the deconstruction of many axon pathways. Many axon pathways are now understood in detail, and in this section and the next, we focus on two whose extensive study has revealed many of the key principles discussed in Sections 10.15, 10.16, and 10.17.

In both invertebrates and vertebrates, development of the nervous system is largely symmetrical about the midline of the body, and proceeds independently on either side. But one half of the body needs to know what the other half is doing. This is achieved by **commissural neurons**, whose axons cross the ventral midline of the developing CNS to connect the two halves.

Axon growth of dorsal commissural interneurons (DI1) of the vertebrate spinal cord has been studied extensively, especially with respect to their behavior at the ventral midline floor plate, which serves as an intermediate target. The cell bodies of DI1 commissural interneurons are located next to the roof plate (see Section 10.6), but they send their axons ventrally along the lateral margin of the spinal cord. Signals from the roof plate, including BMPs, repel the axons. Halfway down the dorso-ventral axis of the spinal cord, the axons turn and project to the ventral midline as they start to encounter a floor-plate-derived chemoattractant (Fig. 10.30). They cross the floor plate, then make a sharp 90° turn and project anteriorly to the brain; they do not re-cross the midline because they are no longer attracted to it.

Commissural axons are initially attracted to the ventral midline by the chemoattractant Netrin-1, which is secreted at high levels from ventral midline floor plate cells. Knock-out of the mouse *netrin-1* gene, or the gene for its receptor DCC, results in abnormal commissural axon pathways (Fig. 10.31). Other floor plate-derived molecules, including Shh, also play some part in attracting commissural axons to the midline, explaining why some axons still manage to navigate there in the *netrin*-null mouse.

A third guidance cue, Slit, is also present in midline floor plate cells. When Slit binds to the Robo receptors Robo1 and Robo2 it repels axons from the midline. However, as vertebrate commissural axons approach the midline, they express high levels of a splice variant of a third Robo receptor, Robo3 sv1 (sometimes called Rig1). Robo3 sv1 does not bind Slit, and, through a *cis*-interaction, interferes with the effect of Slit signal repulsion via Robo1/2 (Fig. 10.32). Therefore, the chemorepellent effect of Slit

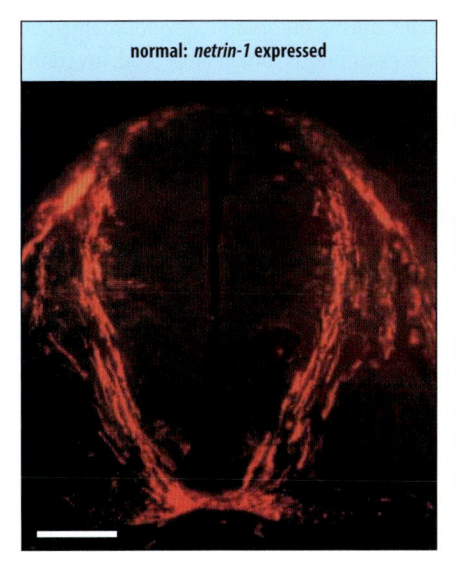

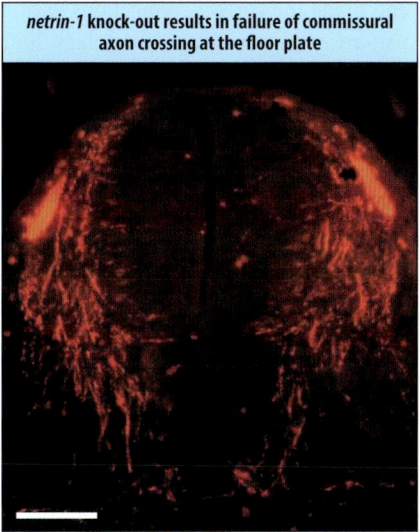

Fig. 10.31 Effect of *netrin-1* gene knock-out in mice. In mice lacking Netrin-1, commissural axons do not grow properly to the floor plate. Scale bar = 0.1 mm.

Photographs courtesy of M. Tessier-Lavigne, reproduced with permission from Serafini, T., et al.: **Netrin-1 is required for commissural axon guidance in the developing vertebrate nervous system.** *Cell* 1996, **87**: 1001–1014. © 1996 Cell Press.

is overcome. Once the commissural axons cross the midline, Robo3 sv1 ceases to be expressed, and instead a non-interfering splice variant (Robo3 sv2) is now inserted into the growth cone, alongside high levels of Robo1 and Robo2. The growth cone can now be repelled from the midline by Slit proteins, which prevents it crossing back. The change in the number and type of Robo receptors inserted into the cell membrane of the growth cone therefore elicits the switch from attraction to repulsion. Floor plate-derived contact-mediated cues, including Shh, stimulate changes in the growth cone that lead to the Robo changes. Experimental evidence for this comes from observations of cultured commissural neurons, which respond to Slit by growth-cone collapse only if they have contacted the floor plate.

Netrin, Robo, and Slit have been highly conserved in evolution, and mediate the same activities—to cross or not to cross the ventral midline—in invertebrates. In fact, the Slits and their Robo receptors were first identified in *Drosophila*.

10.19 Motor neuron innervation of the chick limb is controlled by patterns of gene expression

The muscular system enables an animal to move—a crucial behavior. Appropriate and coordinated movement depends on motor neurons having taken the correct path

Fig. 10.32 Competing chemoattractant and chemorepellent signals enable commissural axons to cross the midline and not to cross back. Commissural axons are attracted to the midline by the chemoattractant protein Netrin (green circle) acting through its receptor DCC (green). As they grow toward the midline, the chemorepellent activity of Slit (yellow diamond) is overcome, because a splice variant of Robo3 (Robo3 sv1, orange), which does not bind Slit, interferes with Robo1 and Robo2 (pale orange). Once commissural axons reach and cross the midline, alternative splicing produces Robo3 sv2 (purple), which does not interfere with Robo1/2. Many Robo1/2 receptors are now inserted into the growth cone, and so repulsion by Slit can now prevent the axons from re-crossing the midline. Non-commissural axons do not express Robo3 and so are continuously repelled from the midline by Slit.

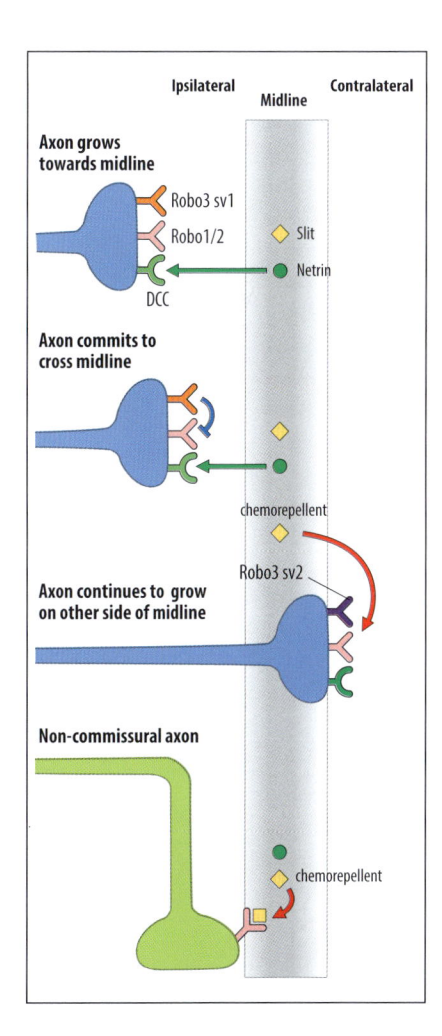

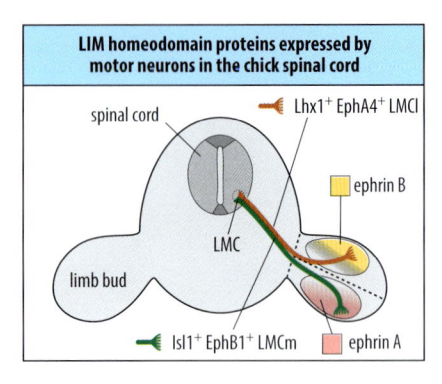

Fig. 10.33 Different motor neurons in the chick spinal cord express different LIM homeodomain proteins and project to different limb muscle targets. Lateral motor column neurons separate into lateral pools (LMCl) and medial pools (LMCm), distinguished by expression of the LIM homeodomain proteins Lhx1 and Isl1, respectively. These control expression of the ephrin receptors EphA4 and EphB1 on the growth cones, which determine the pathway of each axon through repulsive interactions mediated through ephrin A and ephrin B, respectively, in the limb.

and made the correct connections with muscles during embryonic development. The pattern of innervation of the chick limb, for example, is predictable and invariant (Fig. 10.33). As we saw in Section 10.8, different groups of motor neurons in the lateral motor column (LMC), distinguished by the expression of particular combinations of LIM and Hox genes, innervate the limb muscles. Lateral and medial LMC axons both project in a bundle to the base of the limb, but once there, lateral LMC neurons send axons into dorsal muscles in the limb bud, whereas medial LMC neurons send their axons into ventral muscles in the limb bud.

Local guidance cues in the limb are responsible for this sorting out. In experiments in chick embryos, a small section of the spinal cord whose motor neurons will innervate the hindlimb was inverted (rotated by 180° along the antero-posterior axis) before the axons grew out. After inversion, motor axons started to grow incorrectly, but then reoriented to innervate the correct muscles. So, even when the axon bundles entered in reverse antero-posterior order, the correct relationship between motor neurons and muscles was achieved. The short-range contact-dependent Eph–ephrin guidance cues are part of the guidance system that enables this matching of axons and muscle targets. EphA4 signaling in the lateral LMC axons mediates a repulsive interaction with its ligand, ephrin A, which is restricted to the ventral limb. At the same time, EphB1 in the medial LMC axons mediates a repulsive interaction with ephrin B ligands, which are restricted to the dorsal limb.

In the early part of this chapter we introduced the concept of self-organization, the idea that the organization of the developing nervous system at any time point predicts and prefigures its organization at a successive point in time. The projection of lateral LMC motor neuron axons to the chick wing nicely illustrates this concept. Lateral LMC motor neuron identity is controlled by the transcription factor Lhx1, and medial LMC motor neuron identity is controlled by the transcription factor Isl1 (see Section 10.8). These same transcription factors control expression of the Eph genes. Lhx1 upregulates the *EphA4* gene, whereas Isl1 both downregulates the *EphA4* gene and upregulates *EphB1*. Thus, transcription factors that control neuronal identity also control cues that direct the next step—appropriate axon guidance.

10.20 The retino-tectal system is an example of a topographic map

The CNS is confined within the skull and vertebral column, isolated from both the outside world and the internal body. Vitally, it must connect to each, and be able to make sense both of the external world and of the body. It achieves this by representing sensory information in the form of maps within the CNS, called **topographic maps**. These are systems in which groups of neurons in one tissue project axons to a target tissue in an organized manner that maintains spatial relationships. In other words, neurons next to one another in the first tissue innervate adjacent regions of their target tissue.

The ordered connectivity of a topographic map is one of the most formidable tasks facing the developing nervous system. The fundamental problem is how to establish precise connectivity within a target structure in a way that creates a map— a spatially ordered representation of an object. Because precise connectivity is intimately linked to the directed growth of axons, clues to the mechanism of topographic map formation came from examining the order and manner in which developing axons innervate their target.

In the vertebrate visual system, the highly organized projection of retinal neurons to neurons in the **optic tectum** is one of the best models to understand the development of topographic maps. There are millions of individual photoreceptor cells in a vertebrate retina. Each is continuously recording a minute part of the eye's visual field and the signals must be sent to the brain in an orderly manner so that they can be assembled into a coherent picture. Small sets of adjacent photoreceptor cells activate

Fig. 10.34 The larval amphibian retino-tectal topographic map. Retinal neurons from the right eye connect to the left-side optic tectum, and retinal neurons from the left eye connect to the right-side optic tectum, the axons crossing over each other at the optic chiasm. The relative spatial relationships of neuronal cell bodies and final target projections are maintained (shown for each eye in dark blue, medium blue, light blue and red, pink, white). T, temporal; N, nasal; A, anterior; P, posterior; L, lateral; M, medial.

*Illustration after Goodman, C.S., Shatz, C.J.: **Developmental mechanisms that generate precise patterns of neuronal connectivity.** Cell Suppl. 1993, **72**: 77-98.*

individual retinal ganglion neurons whose axons exit the eye in the optic nerve (see Section 9.33 for a brief overview of the development of the eye itself). In birds and amphibians, the optic nerve connects the retina to the optic tectum—a part of the midbrain—where visual signals are processed. Neurons from the left retina and left optic nerve connect to the right optic tectum, and those from the right retina and right optic nerve connect to the left tectum, crossing over at the **optic chiasm** (Fig. 10.34).

Retinal neurons map in a highly ordered manner onto their target neurons, with a point-to-point correspondence between the position of a neuron on the retina and its target in the tectum. Figure 10.34 shows how neurons in different positions along the **naso-temporal axis** (roughly the antero-posterior axis) of the tadpole retina map in a reverse direction to neurons along the antero-posterior axis of the optic tectum. A similar ordering, with similarly reversed polarization, is seen in the mapping of neurons along the lateral–medial axis (roughly the same as the dorso-ventral axis) of the retina: dorsal retinal neurons project to the lateral (ventral) region of the tectum and neurons from the ventral region of the retina project to the medial (dorsal) region of the tectum (Fig. 10.35). Together, this means that the outside visual world is represented as a two-dimensional topographic map in the tectum.

Experiments in frogs provided the first clue as to how the tectum topographic map is set up. In frogs, axons can regenerate after the optic nerve is severed. Experiments in which the optic nerve is severed, and the eye is then rotated dorso-ventrally through 180°, resulted in regenerated axons making connections with their original sites of contact in the tectum, and consequently, the animals behaved as if they were seeing the world upside down (Fig. 10.36). Such experiments, performed in the 1940s, led Roger Sperry (who later won a Nobel prize) to propose that each retinal neuron and each tectal neuron carries a molecular tag—a postcode that enables them to connect reliably. This is known as the **chemoaffinity hypothesis** of connectivity.

Clearly, the genetic cost of a one-tag–one-axon-target code would be impossibly high. There are approximately 20,000 genes in a human genome, and even though cell-surface molecules implicated in axon guidance such as the CAMs and protocadherins have multiple splice variants (for instance, the *Drosophila* CAM Dscam is theoretically estimated to be able to produce 38,000 distinct forms through alternative splicing), they could still not account for the billions of discretely mapped connections seen in the nervous system. Indeed, as far as we know, there is only one direct mapping system—the olfactory map—in which approximately 1000 genes encode different olfactory receptor proteins, each of which determines its target. Instead, in general, it is thought that mapping is achieved through opposing gradients of signaling molecules and their receptors that provide complementary positional information to axons and target cells.

Ephrins and their Eph receptors are likely to operate in this way to contribute to retino-tectal mapping. Across the antero-posterior (naso-temporal) axis of the retina, there is a low-to-high gradient of Eph receptor expression. At the same time, there is a reversed gradient of ephrin ligand in the optic tectum, highest posteriorly and lowest anteriorly. The ephrin–Eph receptor interaction is a repulsive cue, as first demonstrated in experiments carried out in chick embryos (Fig. 10.37), meaning that as axons advance towards the posterior tectum, they become increasingly repelled.

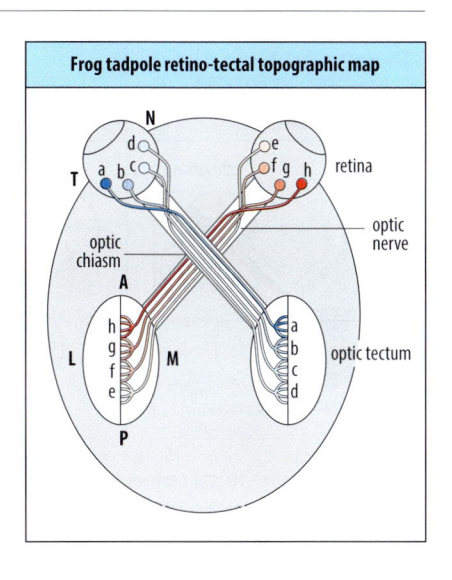

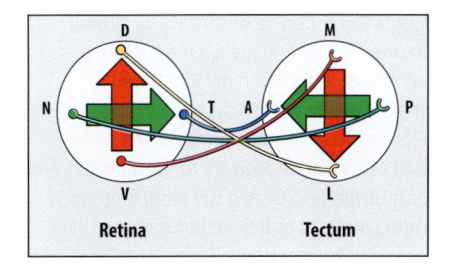

Fig. 10.35 The retina maps onto the tectum. In the amphibian tectum, dorsal (D) retinal neurons connect to the lateral (L) side of the tectum and ventral (V) retinal neurons connect to the medial (M) tectum. Similarly, temporal (T) (or posterior) retinal neurons connect to anterior (A) tectum and nasal (N) (or anterior) retinal neurons connect to the posterior (P) tectum.

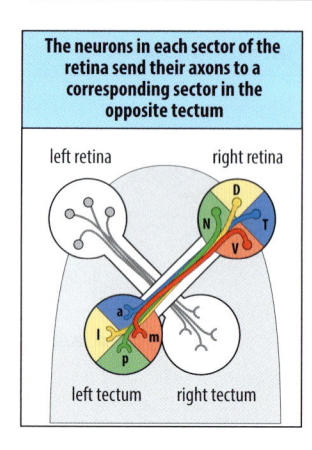

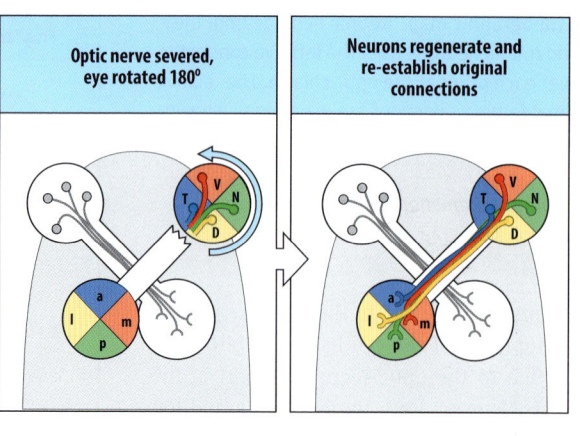

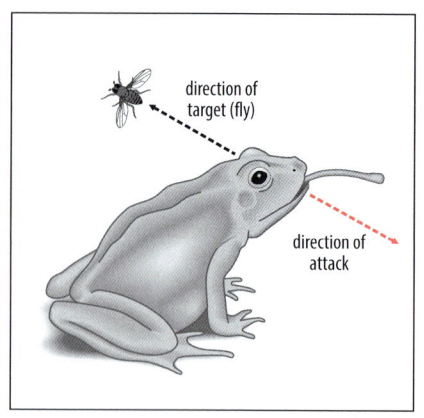

Fig. 10.36 Retino-tectal connections in amphibians are re-established in the original arrangement after severance of the optic nerve and rotation of the eye. First panel: neurons in the optic nerve from the left eye connect to right optic tectum and those from the right eye to the left optic tectum. There is a point-to-point correspondence between neurons from different regions in the retina (nasal (N), temporal (T), dorsal (D), and ventral (V)) and their connections in the tectum (posterior (p), anterior (a), lateral (l), and medial (m), respectively). Second and third panels: if one optic nerve of a frog is cut and the eye rotated dorso-ventrally through 180°, the severed ends of the axons degenerate. When the neurons regenerate, they make connections with their original sites of contact in the tectum. However, because the eye has been rotated, the image falling on the tectum is upside down compared with normal. When the frog sees a fly above its head, it thinks the fly is below it, and moves its head downward to try to catch it (fourth panel).

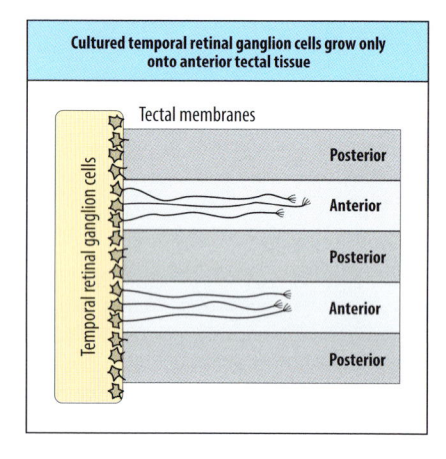

Fig. 10.37 Choice of targets by retinal axons. An axon-guidance mechanism based on repulsion was first demonstrated in the developing visual system of the chick embryo. Normally, the temporal (posterior) half of the retina projects to the anterior part of the tectum and the nasal (anterior) retina projects to the posterior tectum. If pieces of temporal retina are placed next to a 'carpet' of alternating stripes (90 μm wide) of anterior and posterior tectal-cell membranes, temporal axons show a preference for anterior tectal-cell membranes. This choice is mediated by repulsion of the axons—as shown by the collapse of their growth cones—by a factor located on the surface of posterior tectal cells.

Growth cones from the anterior (nasal) part of the retina express lowest levels of Eph receptor, giving them a low sensitivity to Ephrin-mediated repulsion and allowing them to project to the posterior tectum. Growth cones from the posterior (temporal) part of the retina express high levels of Eph receptor and so are repelled by the posterior tectum and stop in the anterior tectum.

Although ephrin–Eph receptor interactions contribute to retino-tectal mapping, however, this is not a hard-wired molecular postcode. For example, if part of the retina is surgically removed, the remaining retinal axons expand their termination zones to fill the tectum while maintaining the retino-tectal topographic map, pointing to a plasticity in map formation.

Indeed, in birds and mammals, such plasticity is likely to have an important role in the formation of many topographic maps. In some topographic systems (such as in fish), axons grow to their targets in a very precise manner, but in most mammalian systems axons branch exuberantly once they reach their target before being pruned back. This is seen in the projection of retinal axons to the **superior colliculus**—the mammalian equivalent of the tectum (Fig. 10.38). DiI labeling experiments in mice show that the axons from both the temporal and nasal parts of the retina initially branch extensively, axons from each retina initially invading the entire superior colliculus in an antero-posterior direction. Only later do they become pruned, and restricted to anterior and posterior regions of the superior colliculus. In this case, EphB and ephrin B mediate attractive, rather than repulsive, interactions, and are thought to operate together with additional cues that provide repulsive interactions: a gradient of the signaling protein Wnt3 in the tectum, and a receptor, Ryk, on the retinal axons. Similarly to axon guidance, then, the integration of multiple cues determines the fine mapping of axons. As we shall see, coordinated electrical activity also contributes to topographic mapping.

So far, we have looked at how axons from a single region map onto their target. But in many instances, two or more structures innervate the same target. Their orderly assembly at the target is crucial for the information they deliver to be interpreted coherently. For example, in mammals, the projection of retinal axons to the superior colliculus is only a minor pathway. Most retinal axons project to the

lateral geniculate nuclei (LGN) in the thalamus, from which other neurons then convey the signals to layer IV of the visual cortex, in which most visual processing occurs (Fig. 10.39). In mammals with full binocular and stereoscopic vision, neurons from the temporal half of the retina project to the LGN on the same side as the eye, and those from the nasal half cross over to the opposite LGN at the optic chiasm (see Fig. 10.39). In other words, each LGN will receive inputs from both eyes. However, as in the cortex, the neurons are arranged in layers and each layer receives input from retinal axons from either the right or the left eye, but not both. Thus, inputs from left and right eyes must be sorted into eye-specific layers (see Fig. 10.39). The way this works is similar to the sorting described in the retino-collicular system. Initially, retino-geniculate axons branch throughout the developing LGN and then retract into eye-specific zones or layers. Eph–ephrin interactions and neuronal activity both have a crucial role in this mapping.

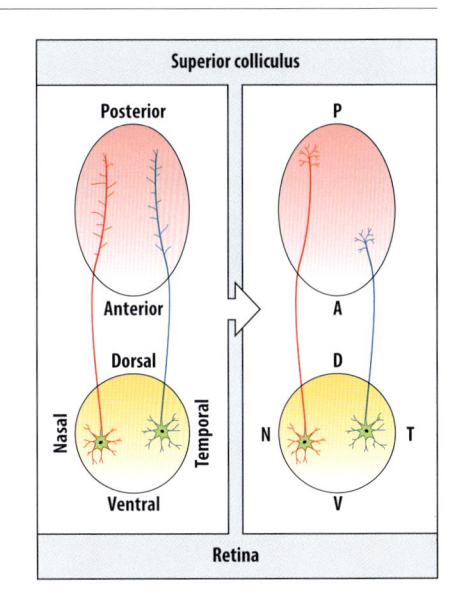

Fig. 10.38 Refinement of the retino-collicular map. Axons from the retina initially branch throughout the colliculus but then become refined, so that axons from the nasal side of the retina terminate in the posterior colliculus and axons from the temporal side of the retina terminate in the anterior colliculus.

SUMMARY

After neurons are born, they extend an axon that grows with precision to its target cell. Growth cones at the tip of the extending axon are rich in receptors for the environmental guidance cues that dictate its path. Guidance is achieved through the integrated activity of many different classes of guidance cues and their receptors. Guidance cues can operate at long range or short range, and can attract or repel the growth cone. Guidance cues can act on the growth cone to regulate the local translation of mRNAs involved in directional growth. Attraction and repulsion control axons crossing the midline in both vertebrates and *Drosophila*. In the development of motor neurons that innervate vertebrate limb muscles, combinations of transcription factors give each motor neuron an identity and determine its axon pathway. Topographic maps ensure

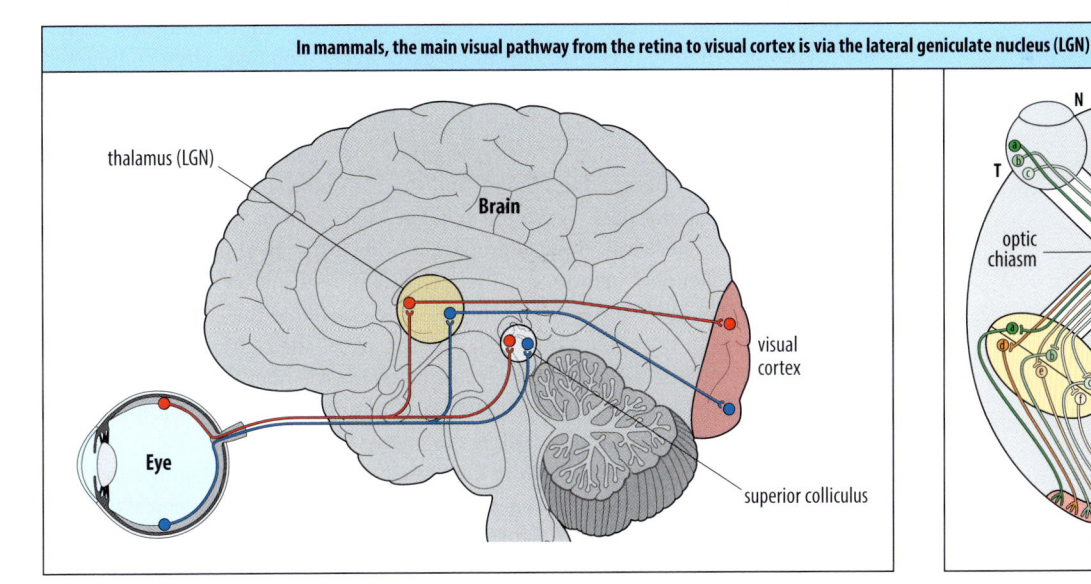

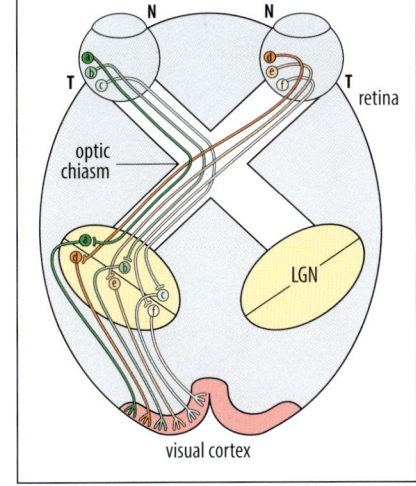

Fig. 10.39 Topographic mapping in the mammalian visual system. Left panel: in mammals, the main visual pathway is to the lateral geniculate nucleus (LGN), from which neurons project to layer IV of the visual cortex. Right panel: neurons from the temporal half of the retina project to the LGN on the same side as the eye, and those from the nasal half cross over to the opposite LGN at the optic chiasm. For the sake of clarity, the neurons from only one half of the retina are shown. The relative spatial relationships of neuronal cell bodies and final target projections are maintained for each eye, but axons from each eye project to a slightly different layer. Axon pathways from left eye shown as dark blue, medium blue, and light blue; axon pathways from right eye shown as red, pink, and white. N, nasal; T, temporal; L, lateral; M, medial; A, anterior; P, posterior.

> that groups of neurons in one tissue project axons to a target tissue in an organized manner that maintains spatial relationships. Topographic mapping in the visual retino-tectal system involves opposing gradients in cell-surface molecules on both the tectal neurons and the retinal axons.

Synapse formation and map refinement

Once at their targets, axons form specialized connections—**synapses**—which are essential for signaling between neurons and their target cells. The connections may be made with other nerve cells, with muscles, and also with certain glandular tissues. The formation of precise, stable synaptic connections is a critical step in the formation of the functional circuits that underlie behavior, and understanding synapse formation and refinement is central to our understanding of nervous system function. In this part of the chapter we look at the development and stabilization of synapses at **neuromuscular junctions**—the junctions between motor neurons and muscle cells in vertebrates, and discuss how these have guided our understanding of synapse formation between neurons. We then introduce the importance of apoptosis in the nervous system—only neurons that make appropriate connections survive. Survival depends on the neuron receiving **neurotrophic factors** or **neurotrophins**, which are produced by the target tissue and for which neurons compete. Fine-tuning of synaptic connections also depends on neuronal activity, a refinement that also involves competition for neurotrophins. We will first look at synapse formation and the role of neurotrophins in making or breaking early connections.

10.21 Synapse formation involves reciprocal interactions

The neuromuscular junction is one of the most intensively studied and best-understood types of synapse. It is made up of an extensively modified nerve ending and pre-patterned parts of the muscle-cell membrane that are rich in acetylcholine receptors (AChRs) and other receptors (including those for the neurotransmitters GABA and glycine) (Fig. 10.40). The synapse comprises the axon plasma membrane, the opposing muscle-cell plasma membrane, and the cleft between—the **synaptic cleft**. Electrical signals cannot pass across the synaptic cleft, and for the neuron to signal to the muscle, the electrical impulse propagated down the axon is converted at the terminal into a chemical neurotransmitter. This neurotransmitter—acetylcholine in motor neurons that innervate skeletal muscle—is released into the synaptic cleft from **synaptic vesicles** in the axon terminal. It diffuses across the cleft and interacts with receptors on the muscle-cell membrane, causing the muscle fiber to contract. Because the signal travels from nerve to muscle, the axon terminal is called the **pre-synaptic** part of the junction and the muscle cell is the **post-synaptic** partner.

The pre-synaptic terminals form from the growth cone of the developing axon. The transition from growth cone to pre-synaptic terminals involves morphological changes such as filopodial retraction and tight junction formation, and differentiation such as the upregulation of new membrane synthesis and signal proteins. Development of a neuromuscular junction is progressive, and occurs through reciprocal signaling interactions between pre-synaptic terminals and post-synaptic cells (Fig. 10.41). The signals are complex, and include components of the basal lamina, such as the glycoprotein laminin β_2, growth factors, such as neuregulin-1, and CAMs.

The neuromuscular junction is a highly specialized type of synapse, but synapse formation within the vast network of neurons that form the CNS follows similar principles. Synapses are made between the axon terminal of one neuron and a dendrite,

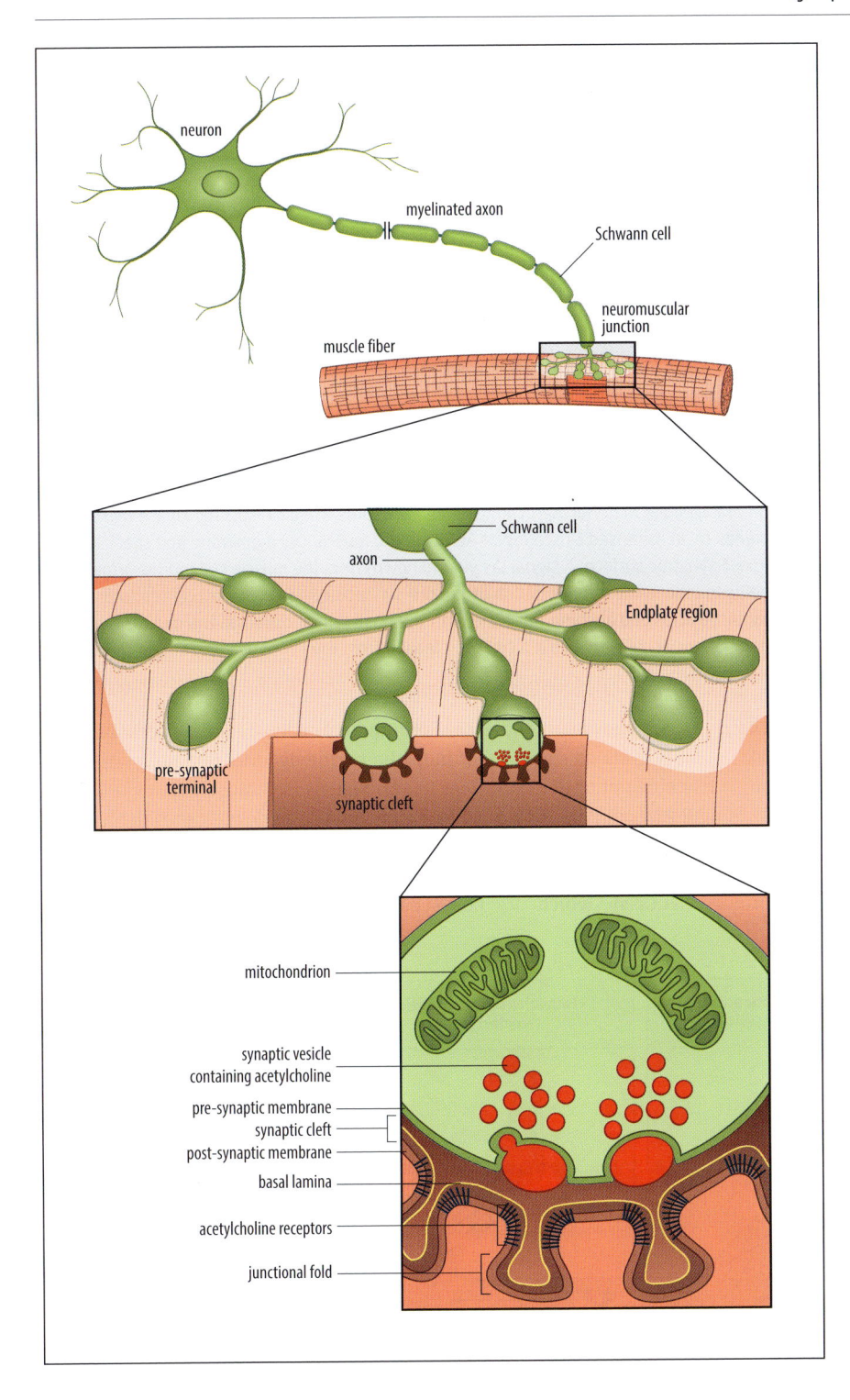

Fig. 10.40 Structure of the vertebrate neuromuscular junction. The motor neuron axon, which is covered in a myelin sheath produced by the glial Schwann cell, branches at its terminal. Each branch ends in a swelling called the pre-synaptic terminal, which is in contact with a special endplate region on the muscle fiber. The plasma membranes of the axon and muscle are separated by the synaptic cleft, filled with extracellular material secreted by both the neuron and the muscle cell. Communication between nerve and muscle is by release of the neurotransmitter acetylcholine into the synaptic cleft. Acetylcholine diffuses across the cleft and binds to acetylcholine receptors on the muscle cell membrane.

Illustration after Kandell, E.R., et al.: Principles of Neural Science. *3rd edn. New York: Elsevier Science Publishing Co., Inc., 1991.*

cell body, or, more rarely, the axon of another neuron. Developing axons can branch extensively, giving rise to numerous sprouts—each with a growth cone at the end—each of which can initiate a synapse. Although contacts made by the growth cone can, and do, initiate synapse formation, other parts of both the pre- and post-synaptic cell can also do this. Ultimately, an individual neuron will receive input from many other neurons on its dendrites and cell body. As with neuromuscular junctions, interneuronal synapse formation is likely to involve reciprocal signaling between axon and

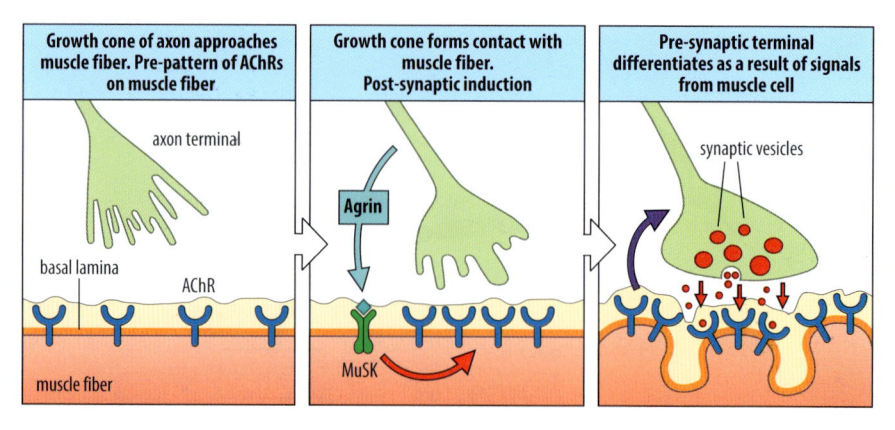

Fig. 10.41 Development of the neuromuscular junction. Before contact by an axon, acetylcholine receptors (AChRs) are pre-clustered at low density in the muscle endplate. When a motor axon growth cone contacts a muscle fiber it releases the proteoglycan Agrin (blue arrow) into the extracellular material. Agrin binds to and activates a muscle-specific kinase (MuSK), which leads to an increased clustering of AChRs in the muscle cell membrane and specialization of the post-synaptic surface. A signal (as yet unknown) from the muscle cell (purple arrow) then helps to induce the differentiation of the pre-synaptic terminal, promoting the formation of synaptic vesicles and aligning this zone with the post-synaptic area on the muscle. Synaptic signaling by acetylcholine (small red circles) released from the synaptic vesicles can then begin.

dendrite. Synapses between neurons can form quite rapidly, on a time scale of about an hour, and observations suggest that filopodia on the dendrite initiate synapse formation by reaching out to the axon (Fig. 10.42). Signals from the axon, including the release of neurotransmitters such as GABA, may initially guide the dendrite filopodia.

Many other factors have been shown to promote synapse formation, including CAMs and Wnt signaling. CAMs implicated in synapse formation include the **neuroligins (NLGNs)**, which are post-synaptic membrane proteins that bind to

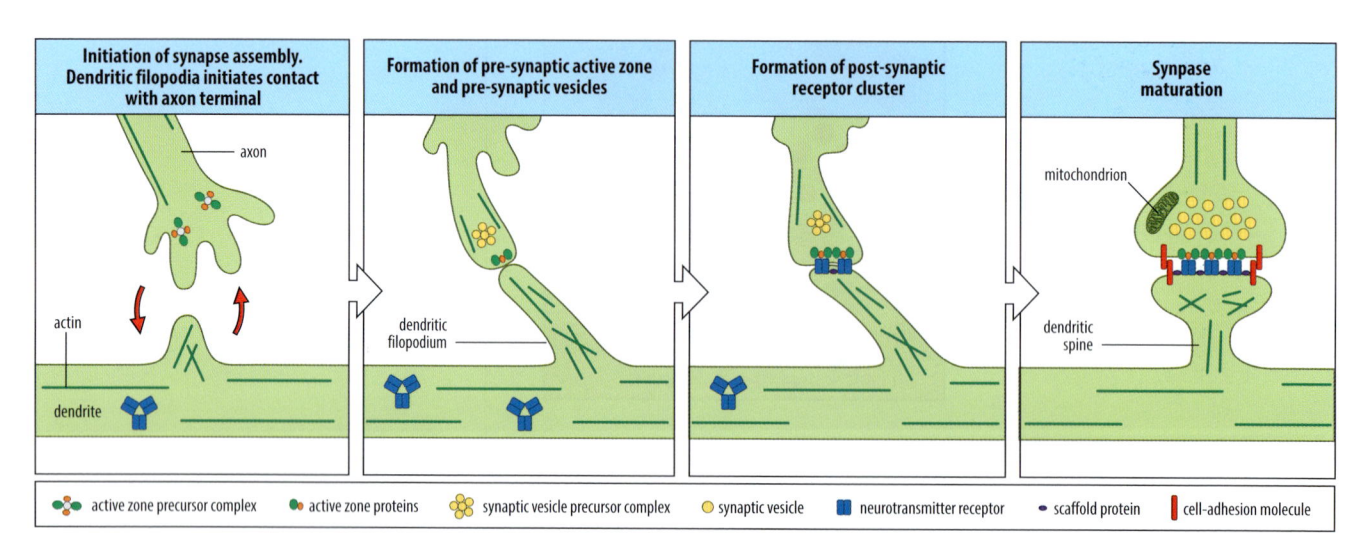

Fig. 10.42 Interneuronal synapse formation. A dendrite sends out filopodia that contact the axon terminal. Reciprocal signaling (red arrows) initiates synapse formation aided by cell-adhesion molecules. In the axon, proteins that comprise the pre-synaptic active zone gather at the site of contact along with synaptic vesicle precursors. In the dendrite, neurotransmitter receptors are synthesized and inserted into the post-synaptic membrane along with post-synaptic scaffold proteins. The post-synaptic connection develops into a dendritic spine.

Illustration after Li, Z., Sheng, M.: **Some assembly required: the development of neuronal synapses.** *Nat. Rev. Mol. Cell Biol. 2003, 4: 833–841.*

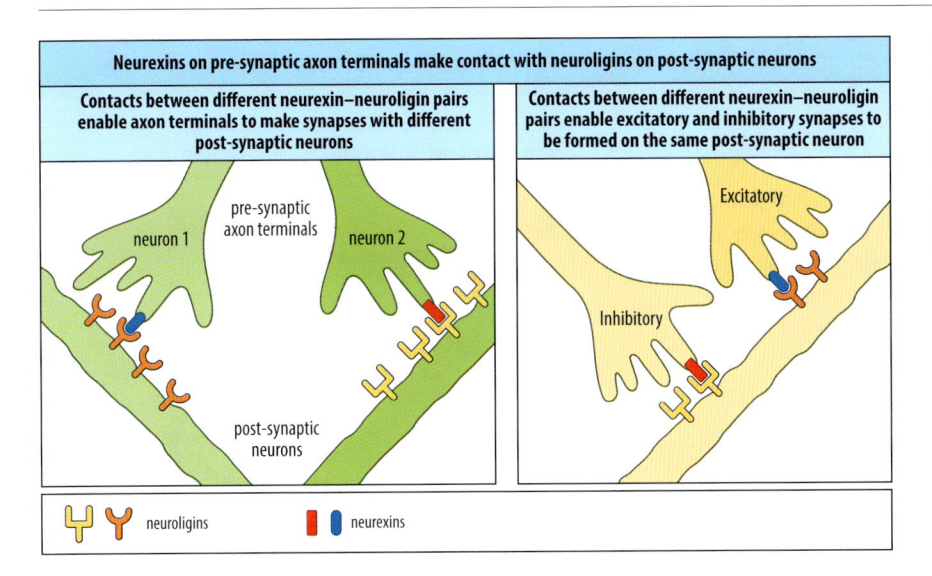

Fig. 10.43 Specificity of neuron–neuron connections. The differential expression of neuroligins and neurexins allows different pre-synaptic partners to select between different post-synaptic partners (left panel), and allows the separate innervation of a single neuron by both excitatory and inhibitory pre-synaptic neurons (right panel).

proteins called β-neurexins on the pre-synaptic axon surface, causing the neurexins to cluster, which induces differentiation in the pre-synaptic terminal. Depending on their specific subtype, the neuroligins and neurexins can promote either inhibitory or excitatory synapses. For example, NLGN1 mainly promotes the formation of excitatory synapses, whereas NLGN2 preferentially induces the formation of inhibitory synapses. Their differential expression during development is therefore important in allowing different pre-synaptic partners to select different post-synaptic partners, and allows the separate innervation of a single neuron by both excitatory and inhibitory pre-synaptic neurons (Fig. 10.43). Defects in interneuronal synapse formation and/or plasticity, including those due to altered function of neuroligins and neurexins, are thought to be a major contributor to autism and autism spectrum conditions (Box 10F).

10.22 Many neurons die through apoptosis during normal development

Setting up the organization of a complex nervous system in vertebrates involves refining an initially rather imprecise organization. This is a problem of morphogenesis, and key among the morphogenetic activities that sculpt the final number of neurons and the pattern of their connections is **apoptosis**—the process of programmed cell death. In spite of the high metabolic cost of producing a cell, the nervous system keeps only a fraction of the neurons that it generates during development, and vast numbers of newborn neurons are eliminated. For example, in the developing chick, about half of the motor neurons that initially innervate the leg undergo apoptosis.

This large amount of neuronal cell death is important for many different aspects of nervous system development. It leads to sexual differences in the nervous system, for instance, by controlling, in a sexually dimorphic manner (different in males and females), the numbers of hypothalamic neurons that regulate reproductive behaviors. It removes structures that are required only transiently in embryogenesis—for instance the subplate of the mammalian cortex (see Fig. 10.23). But more generally, apoptosis appropriately matches the size of a neuronal population to the size of its target. Excess neurons are generated and only those that make the required connections are selected for survival.

Most commonly in the developing nervous system, the process of apoptosis occurs as, or soon after, axon terminals reach their targets (Fig. 10.44). Experiments that examine motor axon innervation of the chick leg bud support the idea that neuronal survival depends on the target. If the leg bud—containing muscle cell targets—is removed, the number of motor neurons that survive decreases sharply compared with normal,

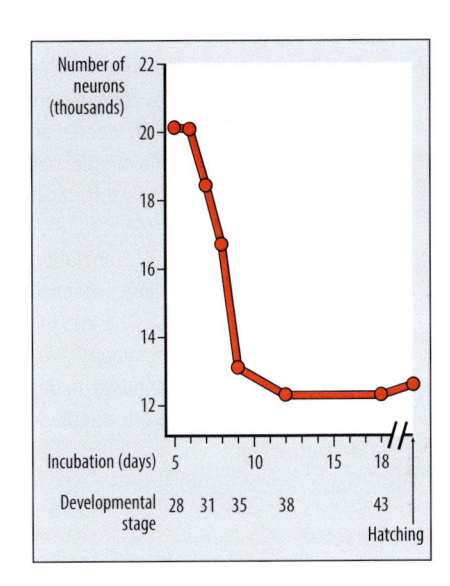

Fig. 10.44 Motor neuron death is a normal part of development in the chick spinal cord. The number of spinal cord motor neurons innervating a chick limb decreases by about half before hatching, as a result of programmed cell death during development. Most of the neurons die over a period of 4 days.

MEDICAL BOX 10F Autism: a developmental condition that involves synapse dysfunction

The brain works through the release of neurotransmitter molecules at synapses, which interconnect neurons in highly complex networks. Correct synapse formation and refinement during development is therefore of crucial importance to cognitive function. Autism is a common (1 in around 120 live births) and pervasive condition of developmental origin that is characterized by challenges with communication and social interaction together with repetitive behaviors and focused interests.

Human **genome-wide association studies (GWAS)** have identified hundreds of genes that when mutated or present in an abnormal number of copies are associated with autism. A large proportion of these genes act in developmental pathways that converge on synapse function, with the mutant genes interfering with synapse formation and/or synaptic homeostasis—the cell-wide management of synaptic activity to keep a suitable balance between excitation and inhibition. Among this group of genes, those encoding proteins located at the synapse itself predominate; some of these proteins are known to be involved in neurotransmission, others in cell-cell adhesion. For example, neurexins and their partner neuroligins (see Section 10.21) are frequently found mutated in autism. The scaffolding protein SHANK3 is found at the post-synaptic membrane in excitatory neurons, where it is associated with neuroligins and other proteins (the glutamic acid receptor *N*-methyl-D-aspartate receptor (NMDAR); the metabotropic glutamate receptor (mGLUR); and the adaptor proteins guanylate kinase-like protein (GKAP), postsynaptic density protein 95 (PSD-95), and homer). Mutations in the *SHANK3* gene are also frequent in autism.

The genetics of autism is complex, but there are several monogenic human conditions with Mendelian inheritance that include an autistic phenotype. For example, fragile X mental retardation, a common form of intellectual difference, is caused by mutations in the *FMR1* gene. This encodes fragile X mental retardation protein (FMRP), a suppressor of translation that targets the mRNAs for multiple genes encoding synaptic proteins that ultimately regulate synaptic plasticity. The loss of these proteins especially affects mGLUR-dependent synaptic depression, a weakening of synaptic strength that is thought to be essential to learning and memory, acting alongside the complementary process of synaptic strengthening.

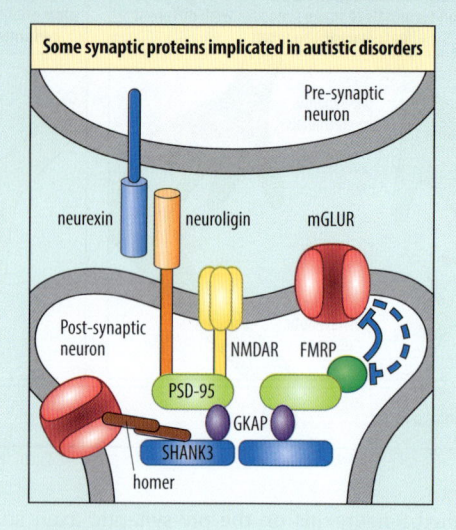

Fig. 1

Rett syndrome is a severe X-linked neurodevelopmental condition that is often associated with autistic behaviors, and results from mutations in the *MeCP2* gene, which encodes a methylcytosine-binding protein. *MeCP2* knock-out mice exhibit a typical Rett's phenotype—including abnormal motor function, tremors, and seizures. Remarkably, subsequent reactivation of *MeCP2* in adult animals can lead to resolution of the abnormalities. Thus, while conventional wisdom suggests that alterations in brain development will be unresponsive to treatments initiated in adulthood, this suggests otherwise, and instead points to considerable synaptic plasticity (see Section 10.22). Interestingly, treatment of the *MeCP2* knock-out mice with insulin-like growth factor-1 (IGF-1) also leads to reversal of many aspects of the phenotype, including synaptic abnormalities. Induced pluripotent stem cells (iPS cells; see Box 7A) derived from patients with Rett syndrome, in which *MeCP2* is silenced, have been induced to differentiate into neurons in culture, providing an *in vitro* means of studying the syndrome. The neurons display synaptic abnormalities, particularly disruptions to the balance between excitation and inhibition. As with the mutant mice, application of IGF-1 to these iPS-cell-derived neurons rescues the reduction in the number of excitatory synapses, bringing closer the possibility of an effective molecular therapy.

---

and many more dying cells are detected. Conversely, when an additional limb bud is grafted at the same level as the leg, providing additional muscle targets for the axons, fewer apoptotic cells are seen and the number of motor neurons that survive increases.

But how might this promote a matching in number of motor neurons and target cells? Innervating neurons compete for a limited supply of a neurotrophic factor that promotes survival. This is called the **neurotrophic hypothesis**. The actions of neurotrophic factors are widely responsible for determining the extent of neuronal survival throughout the nervous system.

But additional mechanisms contribute to dictating the final number of surviving neurons. Once a contact is established, the motor neuron activates the muscle, and this activation now leads to the death of a proportion of the other motor neurons approaching the muscle cell. This can be demonstrated using the drug curare, which prevents neuromuscular signal transmission, and so prevents muscle activation. Exposure of embryonic limb buds to curare results in a large increase in the number of motor neurons that survive. So both target-derived trophic support and target activity contribute to neuron–target matching.

10.23 Neuronal cell death and survival involve both intrinsic and extrinsic factors

Apoptosis can be triggered or prevented either by an intrinsic developmental program or by external factors. For example, in the developing nematode, particular neurons are intrinsically programmed to die, and the specificity of cell death is under genetic control. In the vertebrate nervous system, it remains unclear how programmed cell death is triggered. On the one hand, neurons of a particular lineage may be programmed to die as part of their normal differentiation program. In this context, a large group of 'death receptors'—known as **dependence receptors**—has been identified. These are defined through their ability to fire a death signal in the absence of ligand: the death signal can only be suppressed if a neurotrophic ligand binds to the death receptor. On the other hand, the apoptotic pathway might be triggered by a signal that originates from a target cell.

In either event, the genes and proteins that control the apoptotic pathway in vertebrates show extensive homology to those in *Caenorhabditis elegans* (see Box 14A). Members of the Bcl-2 protein family control the apoptotic pathway in vertebrates. In the nervous system, the original family member, Bcl2, prevents apoptosis, but other family members, such as Bax (Bcl-2-associated X protein), promote neuronal apoptosis; mice mutant for *Bax* have increased numbers of motor neurons in the face. A protein called Survivin inhibits caspase activation (which promotes programmed cell death), and so inhibits apoptosis during early neurogenesis. Mutant mice in which Survivin is deleted are born with smaller brains than normal, with numerous foci of apoptosis, and die soon after birth.

What are the neurotrophic factors that repress or inhibit the apoptotic pathway? The first to be identified was **nerve growth factor (NGF)**, whose discovery by Rita Levi-Montalcini was due to the serendipitous observation that a mouse tumor implanted into a chick embryo evoked extensive growth of nerve fibers towards the tumor. This suggested that the tumor was producing a secreted diffusible factor that promoted axonal outgrowth. The factor was eventually identified as a protein, NGF, using axon outgrowth in culture as an assay. For her work on NGF, Levi-Montalcini shared the 1986 Nobel Prize in Physiology or Medicine with Stanley Cohen, who discovered epidermal growth factor (EGF). NGF is necessary for the survival of a number of types of neuron, particularly those of the sensory and sympathetic nervous systems. It can also guide growth cones, meaning that it is both a trophic and a tropic factor.

NGF is a member of a family of proteins known as the neurotrophins, secreted factors that can act at a short distance from their source. Although originally identified as a neurotrophic factor, NGF and the other neurotrophins can promote either survival or cell death, according to context. Each neurotrophin has a high-affinity Trk receptor. Binding of a neurotrophin to a Trk receptor results in activation of the Akt pathway, which inhibits apoptosis. However, all the neurotrophins also bind to a low-affinity receptor, p75-NTR, a **dependence receptor** that triggers cell death. Different types of neuron require different neurotrophins for their survival, and the requirement for certain neurotrophins also changes during development. In each case, the outcome (survival or death) is context dependent, cell-type dependent, and dependent on the balance between the different ligands and their receptors.

10.24 Neural activity refines synapses and topographic maps

So far, we have considered how guidance cues direct growth cones to appropriate targets, and how neurotrophic factors and target activity begin to eliminate unnecessary neurons, to closely match innervating neurons and their target cells. Even so, there is still an abundance of synaptic connections at early stages of development. At the neuromuscular junction, single muscle fibers are innervated by axon terminals from several different motor neurons. With time, most of these connections are eliminated, until each muscle fiber is innervated by the axon terminals from just one motor neuron (Fig. 10.45). In other words, synapses as well as cells are lost during development. This is due to competition between the synapses, an activity-dependent process in which the most powerful input to the target cell destabilizes the less powerful inputs to the same target. Coordinated electrical activity between the pre- and post-synaptic cells increases synaptic survival, whereas uncoordinated firing increases synaptic loss.

Electrical activity can be converted into survival through the ability of neurotrophins to act as 'punishment and reward' signals. The neurotrophin brain-derived neurotrophic factor (BDNF), its high-affinity receptor TrkA, and its low-affinity receptor p75-NTR, are all expressed in the neuromuscular junction. BDNF is present in an unprocessed form (pro-BDNF) and a mature form; processing of pro-BDNF to BDNF is controlled by matrix metalloproteinases (MMPs). Coordinated electrical activity between a pre- and post-synaptic cell leads to high MMP expression, which processes more BDNF. Mature BDNF binds preferentially to TrkB, promotes survival, and triggers synaptic strengthening ('potentiation') and maturation. However, a less coordinated activity between a pre- and post-synaptic cell leads to low MMP expression and more pro-BDNF. This immature form of BDNF signals via p75-NTR to suppress synaptic transmission and cause axonal retraction. In this way, a weak bias can be rapidly amplified to strengthen those synapses where pre- and post-synaptic cell activity is coordinated, and to weaken and eliminate pre-synaptic connections whose activity is not coordinated with the target.

This type of activity-dependent coordination of neurons and their targets is not limited to synapse elimination but operates more widely within the context of cell elimination in neuron–neuron matching. In the rodent cerebral cortex a large wave of programmed neuronal cell death occurs during early postnatal development and ultimately determines the final number of cortical neurons. Programmed cell death in the developing cortex is particularly dependent on neuronal activity. Such matching is crucial for function. For instance, pyramidal cells and interneurons of the cortex adjust their numbers in a synchronized manner, an event which is essential for the establishment of balanced networks of excitatory and inhibitory neurons.

The development and function of the nervous system depends, then, not only on synapse formation, but also on the regulated disassembly of synaptic connections. We have already considered how, in mammals, axons from the retina make connections with the superior colliculus. Initially, axon terminals from neighboring cells in the retina branch profusely over a large area of the superior colliculus before becoming more finely tuned. Fine-tuning results in part through the action of repulsive cues (see Section 10.18). But, as in muscle, fine-tuning also results from the withdrawal of axon terminals from most of the initial contacts, in a manner that requires neuronal activity.

Neuronal activity is important in refining connections throughout the nervous system. One well-studied region is the retinal–LGN–cortical pathway, the main visual pathway in mammals with full binocular and stereoscopic vision (see Section 10.20). We previously discussed how layers in the LGN receive inputs from either the left or the right eye, a sorting that is partly dependent on repulsive cues. Neuronal activity also has a role in the eye-specific layer segregation. This was shown in animal experiments where tetrodotoxin (TTX), which blocks the conduction of action potentials, was injected into the brain. TTX injections prevented the segregation of the inputs

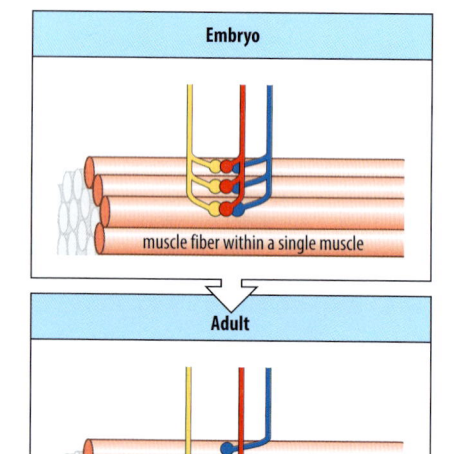

Fig. 10.45 Refinement of muscle innervation by neural activity. Initially, several motor neurons innervate the same muscle fiber. Elimination of synapses means that each fiber is eventually innervated by only one neuron.

Illustration after Goodman, C.S., Shatz, C.J.: **Developmental mechanisms that generate precise patterns of neuronal connectivity.** Cell Suppl. 1993, **72**: 77-98.

Fig. 10.46 Development of ocular dominance columns. Initially, neurons from the lateral geniculate nucleus, representing projections from both eyes and stimulated by the same visual stimulus, project to the same region of the visual cortex. (Only projection to layer IV is shown here.) Over time the neuronal connections separate out into columns, each representing innervation from only one eye. If neuronal activity is blocked, ocular dominance columns do not form.

*Illustration after Goodman, C.S., Shatz, C.J.: **Developmental mechanisms that generate precise patterns of neuronal connectivity.** Cell Suppl. 1993, 72: 77-98.*

from the two eyes. This segregation occurs before birth and it is thought to be driven through spontaneous electrical activity in the retina. In each eye, bursts of activity that spread to neighboring cells lead to the simultaneous activation of groups of cells, and waves of activity. Retinal waves would cause cells in each eye to fire together, ultimately leading to segregation of eye inputs in the LGN through a mechanism that we discuss below.

In the next part of the retinal–LGN–cortical pathway, inputs from left and right eyes that were segregated in the LGN now each make connections with layer IV of the visual cortex. Inputs from corresponding positions in both eyes—that is, from the same part of the visual field—arrive at approximately the same location in the cortex (see Fig. 10.39). Indeed, at birth, the nerve endings from the two eyes overlap and are mixed at their common final location, but with time the inputs from left and right eyes become separated into blocks of cortical cells which are known as **ocular dominance columns** (Fig. 10.46). Adjacent columns respond to the same stimulus in the visual field; one column responding to signals from the left eye, and the next to signals from the right eye. This arrangement enables good binocular and stereoscopic vision. The columns can be detected and mapped through electrophysiological recordings. They can also be directly observed by injecting a tracer, such as radioactive proline, into one eye. The tracer is taken up by retinal neurons, transported by the optic nerve to the LGN and from there to the visual cortex, where its pattern can be detected by autoradiography. This reveals a striking array of stripes representing the input from one eye (Fig. 10.47). As with the retinal waves, patterns of spontaneous activity are observed in the LGN, and could mediate the segregation of ocular dominance columns. If neural activity is blocked during development by the injection of TTX, ocular dominance columns do not develop, and the inputs from the two eyes into the visual cortex remain mixed.

The favored explanation for activity-based segregation is based on competition between incoming axons. The cortex can initially receive input from both eyes. Within a particular region, there will, therefore, be overlap of stimuli originating from the two eyes, and this overlap has to be resolved. As explained above, neighboring cells carrying input from the same eye fire simultaneously; if they both innervate the same target cell, they can thus cooperate to excite it. As in muscle, stimulation of electrical activity in the target cell tends to strengthen the active synapses and suppress those that are not active at the time—cells that fire together, wire together. As there is competition between neurons for targets, this could generate discrete regions of cortical cells that

Fig. 10.47 Visualization of ocular dominance columns in the monkey visual cortex. A radioactive tracer is injected into one eye, from where it is transported to the visual cortex through the neurons. Tracer injected at birth is broadly distributed in the cortex. Tracer injected at later times becomes confined to alternating columns of cortical cells (brighter stripes), representing the ocular dominance columns for that eye, as seen in the photograph. Scale = 1 mm.

Illustration after Kandel, E.R., et al.: Essentials of Neural Science and Behavior. Norwalk, CT: Appleton & Lange, 1991.

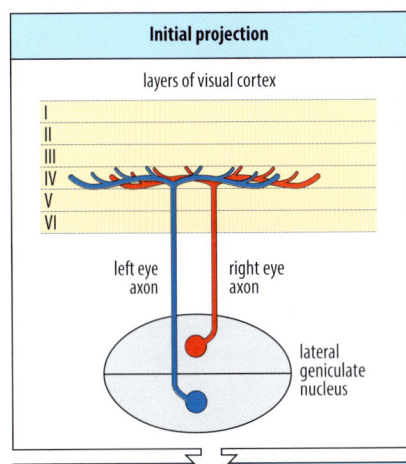

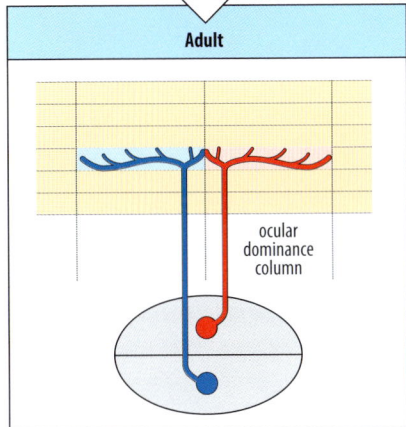

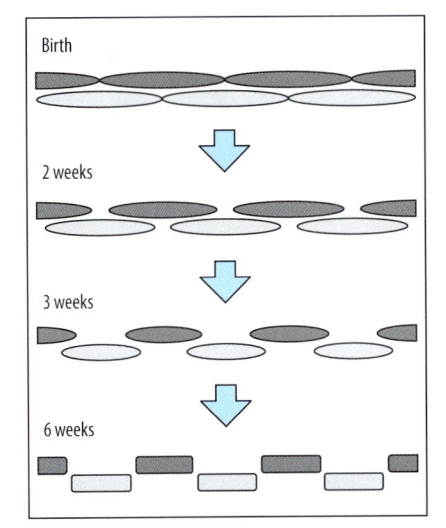

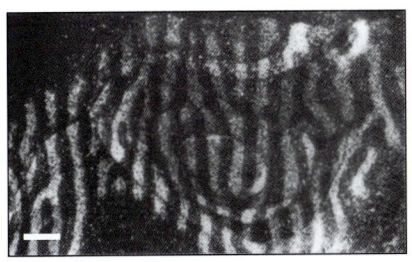

respond only to one eye or the other, and so form the ocular dominance columns. One possible mechanism for refining connections in response to neuronal activity involves local release of neurotrophins. A certain level of activation, or activation by two axons simultaneously, could induce the release of neurotrophins from the target cells, and only those axons that have been recently active might be able to respond to them.

Finally, although the formation of the ocular dominance columns occurs before birth, and so is not dependent on visual experience, the relative size of the columns can be dramatically altered by visual experience. If, after birth, the input from one eye is blocked, the territory in the visual cortex occupied by the other eye's input expands at the expense of the blocked eye. This is referred to as **experience-dependent development**, or **plasticity**. This type of development occurs throughout the nervous system, and conventionally is thought to be confined to distinct time windows, known as sensitive or critical periods. It is widely thought to rely on **synaptic plasticity**—the ability of a neuron to change synaptic weight or efficacy, that is, the amplitude of depolarization or hyperpolarization associated with a certain level of activation of a given synapse. **Long-term potentiation (LTP)** is defined as a long-term increase in synaptic efficacy, resulting from coordinated activity between pre- and post-synaptic cells. It is important to note that synaptic potentiation occurs widely, and can occur postnatally as well as during embryonic life.

10.25 The mammalian brain has a high degree of plasticity, even in adults

For many years it was thought that no new neurons could be generated in the adult CNS of mammals, although adult neurogenesis was known to occur in the brains of songbirds, reptiles, amphibians, and fish. Neural stem cells are now known to exist in limited regions in the adult mammalian brain, and can give rise to new neurons and glial cells, as described in Section 6.14. Establishing the existence and extent of neurogenesis in the adult brain would transform our understanding of how the brain works, and how to tackle brain damage and disease. As with so much of developmental biology, key insights are likely to come from studies in a wide range of model organisms, including *Drosophila*.

Understanding not just the extent to which new cells can be added to the nervous system through life, but also the extent to which new projections grow, form, and strengthen, will have an enormous impact on our understanding of the brain. Experiences shape the brain: as neurons experience new tasks, they grow, and form and strengthen new connections, so that learning becomes hard-wired. Major research efforts are currently under way to understand how experience shapes structural plasticity, and how, conversely, stress, depression, aging, and disease can have the opposite effect, triggering neurons to break down and die. Importantly, processes that build and strengthen neurons and synaptic connections in the embryo can be redeployed to maintain a dynamic nervous system throughout life.

SUMMARY

Neurons communicate with each other and with target cells such as muscles by means of specialized junctions called synapses. The formation of a neuromuscular junction depends on reciprocal signaling between muscle and nerve cell. Reciprocal interactions also occur in the formation of synapses between axons and dendrites. Cell-adhesion molecules have a crucial role in synapse formation, and defects in synapse formation are a major contributor to autism. In development, an excessive number of neurons are born, and many die through apoptosis. Those that survive do so because they make functional connections with their target cells. In other words, the establishment of a connection between a neuron and its target is essential not only for the functioning of the nervous system but also for the very survival of neurons and the refinement of neuronal maps. Many neurons depend on neurotrophins, such as nerve growth factor, for their survival, with different classes of neuron requiring different neurotrophins. In early development there is also an abundance of synaptic connections, many of which are then disassembled. Brain function relies on both the assembly and the regulated disassembly of synaptic connections. Neural activity refines synaptic connections. Initially, most mammalian muscle fibers are innervated by two or more motor axons, but nervous activity results in competition between synapses, so that a single fiber is eventually innervated by only one motor neuron. Coordinated activity strengthens neurotrophic signaling that promotes synaptic strengthening, whereas uncoordinated activity leads to axonal retraction. Neuronal activity has a major role in refining the connections between the eye and the brain, and in mammals is required for the sorting of inputs to eye-specific layers of the lateral geniculate nucleus and the development of ocular dominance columns in the visual cortex. The ability of the nervous system to generate new cells and to form new synaptic connections is important in experience-based learning through life.

Summary to Chapter 10

- The processes involved in the development of the nervous system are superficially similar to those found in other developmental systems but involve the acquisition of individual cell identities such that a vast number of highly specific cell–cell connections can be generated, often over long distances.
- The early neuroectoderm is first patterned into broad regions with distinct and diverse fates by the action of local signaling centers, before region-specific neuronal cell types appear.
- Signal molecules confer positional information on a responding field by forming a diffusion gradient that induces the expression of transcription factors that act as region or cell-type determinants.
- Prospective neural tissue is specified early in development—during gastrulation in vertebrates and when the dorso-ventral axis is patterned in *Drosophila*.
- Within the neuroectodermal tissue, the specification of cells that give rise to neural cells involves lateral inhibition.
- The further development of neurons from neuronal precursors involves both asymmetric cell divisions and cell–cell signaling.
- The patterning of different types of neuron within the vertebrate spinal cord is due to both ventral and dorsal signals.
- As they develop, neurons extend axons and dendrites. The axons are guided to their destination by growth cones at their tips.
- Guidance is due to the growth cone's response to attractive and repulsive signals, which may be diffusible or bound to the substratum.
- Gradients in such molecules can guide the axons to their destination, as in the retino-tectal system of vertebrates.
- The functioning of the nervous system depends on the establishment of specific synapses between axons and their targets.
- Specificity appears to be achieved by an initial overproduction of neurons that compete for targets, with many neurons dying during development. Refinement of synaptic connections involves further competition.
- Neural activity has a major role in refining connections, such as those between the eye and the brain.

End of chapter questions

Check your understanding of the content of this chapter by attempting the following long-answer concept questions.

1. How is positional information and positional value assigned during regionalization of the vertebrate neuroectoderm?

2. The response of the posterior midbrain to FGF-8 signaling from the midbrain–hindbrain boundary differs from that of the anterior hindbrain. How might this asymmetry be controlled?

3. What are the distinguishing features of local signaling centers or organizers? How do they differ from the primary (Spemann) organizer?

4. What are rhombomeres? What are the similarities between rhombomeric segmentation and segmentation in the *Drosophila* larva?

5. Describe an experiment that demonstrates a homeotic transformation of rhombomeres. Outline the strategy used and compare this with other strategies that result in homeotic transformations, drawing your examples from both vertebrates and invertebrates.

6. The formation of motor neurons in the developing spinal cord depends on a gradient of Sonic hedgehog (Shh) protein. What is the source of this Shh gradient? What homeodomain protein genes are active in response to the precise level of this gradient required for motor neuron progenitors (pMN) to become specified as motor neurons (MN)? What role do the LIM homeodomain proteins have in motor neuron identity?

7. In *Drosophila*, the central nervous system (CNS) is ventral, whereas in vertebrates, the CNS is dorsal. Despite these differences, what similarities exist that would indicate a common evolutionary origin?

8. Outline the mechanism of lateral inhibition in the proneural cluster of *Drosophila*, in which the expression of Delta in one cell leads to loss of Delta in a neighboring cell. Compare the *Drosophila* pathway with that in vertebrates.

9. What are radial glial cells? Include in your answer: the embryonic structure in which they are found, their location in that structure, and their roles in development of the nervous system (name two roles).

10. How do apico-basal and Notch–Delta proteins collaborate to cause one daughter cell to become a neuron after: (a) division of the neuroblast in the *Drosophila* embryo and (b) division of a neuroepithelial or radial glial cell in the vertebrate neural tube?

11. What is the evolutionary advantage of the distinct pattern of neurogenesis and neuronal migrations that distinguish the mammalian cortex from, for example, the spinal cord?

12. Although the adult mammalian brain cannot in general produce new neurons, the dentate gyrus of the hippocampus is an exception. Relate this observation to the embryological origin of neurons in the neural tube.

13. Compare and contrast netrins, semaphorins, cadherins, and ephrins: in what way are netrins and cadherins similar in effect, and in what way are they different? In what way are semaphorins and ephrins similar, and in what way do they differ?

14. What is meant by the phrase 'the retina maps onto the tectum' (see Fig. 10.35)? How does the chemoaffinity hypothesis attempt to explain this phenomenon? How do the ephrins and their receptors fit into this hypothesis?

15. Describe the results of the experiment in which the optic nerve to a frog's eye is cut and the eye rotated by 180°. How is the subsequent response of the frog to a visual stimulus explained?

16. What is the role of neurotrophic factors, such as nerve growth factor, in refining the connections between neurons and their targets? Give a specific example of such action.

17. Draw a schematic diagram of a synapse, labeling the pre-synaptic cell and the post-synaptic cell and indicating the direction of signal transmission. Label the synaptic cleft and say why it prevents neurons from passing electrical signals directly to each other. How do neurons communicate across the synaptic cleft?

18. Describe the steps involved in the formation of the neuromuscular junction, starting with the Agrin signal from the neuronal growth cone.

19. Cell death (apoptosis) in the developing nervous system is under the control of both intrinsic and extrinsic influences. Give examples of intrinsic (i.e. intracellular) and extrinsic (i.e. cell–cell signaling) proteins that control apoptosis in the nervous system during development.

20. Contrast the retinal inputs to the lateral geniculate nucleus in the mammalian brain with those to the tectum in the amphibian brain. What are ocular dominance columns, and why are they important? What is meant by 'cells that fire together, wire together'?

General further reading

Kandel, E.R., Koester, J.D., Mack, S.H., Siegelbaum, S.A.: *Principles of Neural Science*. 6th edn. New York: McGraw-Hill, 2021.

Sanes, D., Reh, T., Harris, W., Landgraf, M.: *Development of the Nervous System*. 4th edn. Cambridge, MA: Academic Press, 2019.

Section further reading

10.1 The vertebrate brain is organized into three main regions

Kiecker, C., Lumsden, A.: **The role of organizers in patterning the nervous system**. *Annu. Rev. Neurosci.* 2012, **35**: 347–367.

10.2 The vertebrate brain is regionalized by local signaling centers

Epstein, D.J.: **Regulation of thalamic development by sonic hedgehog**. *Front. Neurosci.* 2012, **6**: 57.

Houart, C., Westerfield, M., Wilson, S.W.: **A small population of anterior cells patterns the forebrain during zebrafish gastrulation**. *Nature* 1998, **391**: 788–792.

Pera, E.M., Kessel, M.: **Patterning of the chick forebrain anlage by the prechordal plate**. *Development* 1997, **124**: 4153–4162.

Rhinn, M., Brand, M.: **The midbrain-hindbrain boundary organizer**. *Curr. Opin. Neurobiol.* 2001, **11**: 34–42.

Rubenstein, J.L.: **Development of the cerebral cortex: implications for neurodevelopmental disorders**. *J. Child Psychol. Psychiatr.* 2011, **52**: 339–355.

Sagai, T., Amano, T., Maeno, A., Ajima, R., Shiroishi, T.: **SHH signaling mediated by a prechordal and brain enhancer controls forebrain organization**. *Proc. Natl Acad. Sci. USA* 2019, **116**: 23636–23642.

Shimamura, K., Rubenstein, J.L.R.: **Inductive interactions direct early regionalization of the mouse forebrain**. *Development*, 1997, **124**: 2709–2718.

Box 10A Alcohol exposure and brain development

Hong, M., Christ, A., Christa, A., Willnow, T.E., Krauss, R.S.: *Cdon* **mutation and fetal alcohol converge on Nodal signaling in a mouse model of holoprosencephaly**. *eLife* 2020, **9**: e60351.

10.3 Parts of the nervous system develop through a process of self-organization

Broccoli, V., Boncinelli, E., Wurst, W.: **The caudal limit of *Otx2* expression positions the isthmic organizer**. *Nature* 1999, **401**: 164–168.

Crossley, P.H., Martinez, S., Martin, G.R.: **Midbrain development induced by FGF8 in the chick embryo**. *Nature* 1996, **380**: 66–68.

Millet, S.: **A role for *Gbx2* in repression of *Otx2* and positioning the mid/hindbrain organizer**. *Nature* 1999, **401**: 161–164.

Yamauchi, K., Mizushima, S., Tamada, A., Yamamoto, N., Takashima, S., Murakami, F.: **FGF8 signaling regulates growth of midbrain dopaminergic axons by inducing semaphorin 3F**. *J. Neurosci.* 2009, **29**: 4044–4055.

10.4 Specialized boundaries formed in the developing brain delimit areas of cell-lineage restriction

Addison, M., Xu, Q., Cayuso, J., Wilkinson, D.G.: **Cell identity switching regulated by retinoic acid signaling maintains homogeneous segments in the hindbrain**. *Dev. Cell* 2018, **45**: 606–620.

Fraser, S., Keynes, R.J., Lumsden, A.: **Segmentation in the chick embryo hindbrain is defined by cell lineage restrictions**. *Nature* 1990, **344**: 636–638.

Kesavan, G., Machate, A., Hans, S., Brand, M.: **Cell-fate plasticity, adhesion and cell sorting complementarily establish a sharp midbrain-hindbrain boundary**. *Development* 2020, **147**: dev186882.

Krumlauf, R., Wilkinson, D.G.: **Segmentation and patterning of the vertebrate hindbrain**. *Development* 2021, **148**: dev186460.

Xu, Q., Mellitzer, G., Robinson, V., Wilkinson, D.W.: **In vivo cell sorting in complementary segmental domains mediated by Eph receptors and ephrin ligands**. *Nature* 1999, **399**: 267–271.

10.5 Evolutionarily conserved transcription factors provide positional information along the antero-posterior axis of the developing brain

Bell, E., Wingate, R., Lumsden, A.: **Homeotic transformation of rhombomere identity after localized *Hoxb1* misexpression**. *Science* 1999, **284**: 2168–2171.

Chambers, D., Wilson, L.J., Alfonsi, F., Hunter, E., Saxena, U., Blanc, E., Lumsden, A.: **Rhombomere-specific analysis reveals the repertoire of genetic cues expressed across the developing hindbrain**. *Neural Dev.* 2009, **4**: 6.

Simeone, A., Avantaggiato, V., Moroni, M.C., Mavilio, F., Arra, C., Cotelli, F., Nigro, V., Acampora, D.: **Retinoic acid induces stage-specific antero-posterior transformation of rostral central nervous system**. *Mech. Dev.* 1995, **51**: 83–98.

Wassef, M.A., Chomette, D., Pouilhe, M., Stedman, A., Havis, E., Trin-Dihn-Desmarquet, C., Schneider-Maunoury, S., Gilardi-Hebenstreit, P., Charnay, P., Ghislain, J.: **Rostral hindbrain patterning involves the direct activation of a *Krox20* transcriptional enhancer by Hox/Pbx and Meis activators**. *Development* 2008, **135**: 3369–3378.

10.6 Opposing morphogen gradients pattern the spinal cord along the dorso-ventral axis

Briscoe, J., Chen, Y., Jessell, T.M., Struhl, G.: **A hedgehog-insensitive form of patched provides evidence for direct long-range morphogen activity of sonic hedgehog in the neural tube**. *Mol. Cell* 2001, **7**: 1279–1291.

Chamberlain, C.E., Jeong, J., Guo, C., Allen, B.L., McMahon, A.P.: **Notochord-derived Shh concentrates in close association with the apically positioned basal body in neural target cells and forms a dynamic gradient during neural patterning**. *Development* 2008, **135**: 1097–1106.

Dessaud, E., McMahon, A.P., Briscoe, J.: **Pattern formation in the vertebrate neural tube: a sonic hedgehog morphogen-regulated transcriptional network**. *Development* 2008, **135**: 2489–2503.

Liem, K.F. Jr, Jessell, T.M., Briscoe, J.: **Regulation of the neural patterning activity of sonic hedgehog by secreted BMP inhibitors expressed by notochord and somites**. *Development* 2000, **127**: 4855–4866.

Liem, K.F. Jr, Tremml, G., Jessell, T.M.: **A role for the roof plate and its resident TGFbeta-related proteins in neuronal patterning in the dorsal spinal cord**. *Cell* 1997, **91**: 127–138.

Placzek, M., Briscoe, J.: **Sonic hedgehog in vertebrate neural tube development**. *Int. J. Dev. Biol.* 2018, **62**: 225–234.

Stamataki, D., Ulloa, F., Tsoni, S.V., Mynett, A., Briscoe, J.: **A gradient of Gli activity mediates graded Sonic Hedgehog signaling in the neural tube**. *Genes Dev.* 2005, **19**: 626–641.

Yamada, T., Pfaff, S.L., Edlund, T., Jessell, T.M.: **Control of cell pattern in the neural tube: motor neuron induction by diffusible factors from notochord and floor plate**. *Cell* 1993, **73**: 673–686.

10.7 Stable progenitor domains in the ventral spinal cord are established over time

Briscoe, J., Pierani, A., Jessell, T.M., Ericson, J.: **A homeodomain protein code specifies progenitor cell identity and neuronal fate in the ventral neural tube**. *Cell* 2000, **101**: 435–445.

Delile, J., Rayon, T., Melchionda, M., Edwards, A., Briscoe, J., Sagner, A.: **Single cell transcriptomics reveals spatial and temporal dynamics of gene expression in the developing mouse spinal cord**. *Development* 2019, **146**: dev173807.

Dessaud, E., Yang, L.L., Hill, K., Cox, B., Ulloa, F., Ribeiro, A., Mynett, A., Novitch, B.G., Briscoe, J.: **Interpretation of the sonic hedgehog morphogen gradient by a temporal adaptation mechanism**. *Nature* 2007, **450**: 717–720.

Jessell, T.M.: **Neuronal specification in the spinal cord: inductive signals and transcriptional controls**. *Nat. Rev. Genet.* 2000, **1**: 20–29.

Xiong, F., Tentner, A.R., Huang, P., Gelas, A., Mosaliganti, K.R., Souhait, L., Rannou, N., Swinburne, I.A., Obholzer, N.D.,

Cowgill, P.D., Schier, A.F., Megason, S.G.: **Specified neural progenitors sort to form sharp domains after noisy Shh signaling**. *Cell* 2013, **153**: 550–561.

Box 10B Directed differentiation of induced pluripotent stem cells towards functional neuronal types

Merkle, F.T., Maroof, A., Wataya, T., Sasai, Y., Studer, L., Eggan, K., Schier, A.F.: **Generation of neuropeptidergic hypothalamic neurons from human pluripotent stem cells**. *Development* 2015, **142**: 633–643.

Wichterle, H., Lieberam, I., Porter, J.A., Jessell, T.M.: **Directed differentiation of embryonic stem cells into motor neurons**. *Cell* 2002, **110**: 385–397.

10.8 Hox genes specify motor neuron subtypes in the spinal cord

Catela, C., Shin, M.M., Lee, D.H., Liu, J.P., Dasen, J.S.: **Hox proteins coordinate motor neuron differentiation and connectivity programs through *Ret/Gfrα* genes**. *Cell Rep.* 2016, **14**: 1901–1915.

Dasen, J.S., Jessell, T.M.: **Hox networks and the origins of motor neuron diversity**. *Curr. Top. Dev. Biol.* 2009, **88**: 169–200.

Philippidou, P., Dasen, J.S.: **Hox genes: choreographers in neural development, architects of circuit organization**. *Neuron* 2013, **80**: 12–34.

10.9 Temporal mechanisms contribute to the generation of neural cell diversity

Sagner, A., Briscoe, J.: **Establishing neuronal diversity in the spinal cord: a time and a place**. *Development* 2019, **146**: dev182154.

Sagner, A., Zhang, I., Watson, T., Lazaro, J., Melchionda, M., Briscoe, J.: **A shared transcriptional code orchestrates temporal patterning of the central nervous system**. *PLoS Biol.* 2021, **19**: e3001450.

10.10 Neurons in *Drosophila* arise from proneural clusters

Arefin, B., Parvin, F., Bahrampour, S., Stadler, C.B., Thor, S.: ***Drosophila* neuroblast selection is gated by Notch, Snail, SoxB, and EMT gene interplay**. *Cell Rep.* 2019, **29**: 3636–3651.

Bossing, T., Udolph, G., Doe, C.Q., Technau, G.M.: **The embryonic central nervous system lineages of *Drosophila melanogaster*. I. Neuroblast lineages derived from the ventral half of the neuroectoderm**. *Dev. Biol.* 1996, **179**: 41–64.

Bourouis, M., Heitzler, P., el Messal, M., Simpson, P.: **Mutant *Drosophila* embryos in which all cells adopt a neural fate**. *Nature* 1989, **341**: 442–444.

Cornell, R.A., Ohlen, T.V.: ***Vnd/nkx, ind/gsh,* and *msh/msx*: conserved regulators of dorsoventral neural patterning?** *Curr. Opin. Neurobiol.* 2000, **10**: 63–71.

Skeath, J.B.: **At the nexus between pattern formation and cell-type specification: the generation of individual neuroblast fates in the *Drosophila* embryonic central nervous system**. *BioEssays* 1999, **21**: 922–931.

10.11 The development of neurons in *Drosophila* involves asymmetric cell divisions and timed changes in gene expression

Doe, C.Q.: **Temporal patterning in the *Drosophila* CNS**. *Annu. Rev. Cell Dev. Biol.* 2017, **33**: 219–240.

Grosskortenhaus, R., Pearson, B.J., Marusich, A., Doe, C.Q.: **Regulation of temporal identity transitions in *Drosophila* neuroblasts**. *Dev. Cell* 2005, **8**: 193–202.

Harding, K., White, K.: ***Drosophila* as a model for developmental biology: stem cell-fate decisions in the developing nervous system**. *J. Dev. Biol.* 2018, **6**: 25.

Loyer, N., Januschke, J.: **Where does asymmetry come from? Illustrating principles of polarity and asymmetry establishment in *Drosophila* neuroblasts**. *Curr. Opin. Cell Biol.* 2020, **62**: 70–77.

Pollington, H.Q., Seroka, A.Q., Doe, C.Q.: **From temporal patterning to neuronal connectivity in *Drosophila* type I neuroblast lineages**. *Semin. Cell Dev. Biol.* 2023, **142**: 4–12.

10.12 In the vertebrate neural tube, neurons are formed in the proliferative zone and migrate outwards

Anthony, T.E., Klein, C., Fishell, G., Heintz, N.: **Radial glia serve as neuronal progenitors in all regions of the central nervous system**. *Neuron* 2004, **41**: 881–890.

Casingal, C.R., Descant, K.D., Anton, E.S.: **Coordinating cerebral cortical construction and connectivity: unifying influence of radial progenitors**. *Neuron* 2022, **110**: 1100–1115.

Hansen, D.V., Lui, J.H., Parker, P.R.L., Kriegstein, A.R.: **Neurogenic radial glia in the outer subventricular zone of human neocortex**. *Nature* 2010, **464**: 554–561.

Renner, M., Lancaster, M.A., Bian, S., Choi, H., Ku, T., Peer, A., Chung, K., Knoblich, J.A.: **Self-organized developmental patterning and differentiation in cerebral organoids**. *EMBO J.* 2017, **36**: 1316–1329.

Box 10C Glial cells in the nervous system

Jäkel, S., Dimou, L.: **Glial cells and their function in the adult brain: a journey through the history of their ablation**. *Front. Cell Neurosci.* 2017, **11**: 24.

10.13 Specialized features in the cortex support high levels of neurogenesis

Cárdenas, A., Villalba, A., de Juan Romero, C., Picó, E., Kyrousi, C., Tzika, A.C., Tessier-Lavigne, M., Ma, L., Drukker, M., Cappello, S., Borrell, V.: **Evolution of cortical neurogenesis in amniotes controlled by Robo signaling levels**. *Cell* 2018, **174**: 590–606.

Li, Y., Muffat, J., Omer, A., Bosch, I., Lancaster, M.A., Sur, M., Gehrke, L., Knoblich, J.A., Jaenisch, R.: **Induction of expansion and folding in human cerebral organoids**. *Cell Stem Cell* 2017, **20**: 385–396.

Pilz, G.A., Shitamukai, A., Reillo, I., Pacary, E., Schwausch, J., Stahl, R., Ninkovic, J., Snippert, H.J., Clevers, H., Godinho, L., Guillemot, F., Borrell, V., Matsuzaki, F., Götz, M.: **Amplification of progenitors in the mammalian telencephalon includes a new radial glial cell type**. *Nat. Commun.* 2013, **4**: 2125.

Box 10D Timing the birth of cortical neurons

Rakic, P.: **Neurons in rhesus monkey visual cortex: systematic relation between time of origin and eventual disposition**. *Science* 1974, **183**: 425–427.

10.14 Inhibitory interneurons undergo tangential migration into the cortex

Anderson, S.A., Eisenstat, D.D., Shi, L., Rubenstein, J.L.R.: **Interneuron migration from basal forebrain to neocortex: dependence on *Dlx* genes**. *Science* 1997, **278**: 474–476.

Lepiemme, F., Stoufflet, J., Javier-Torrent, M., Mazzucchelli, G., Silva, C.G., Nguyen, L.: **Oligodendrocyte precursors guide interneuron migration by unidirectional contact repulsion**. *Science* 2022, **376**: eabn6204.

Marín, O.: **Cellular and molecular mechanisms controlling the migration of neocortical interneurons**. *Eur. J. Neurosci.* 2013, **38**: 2019–2029.

Marin, O., Rubenstein, J.L.R.: **A long, remarkable journey: tangential migration in the telencephalon**. *Nat. Rev. Neurosci.* 2001, **2**: 780–790.

Shi, Y., Wang, M., Mi, D., Lu, T., Wang, B., Dong, H., Zhong, S., Chen, Y., Sun, L., Zhou, X., Ma, Q., Liu, Z., Wang, W., Zhang, J., Wu, Q., Marín, O., Wang, X.: **Mouse and human share conserved transcriptional programs for interneuron development**. *Science* 2021, **374**: eabj6641.

Southwell, D.G., Nicholas, C.R., Basbaum, A.I., Stryker, M.P., Kriegstein, A.R., Rubenstein, J.L., Alvarez-Buylla, A.: **Interneurons from embryonic development to cell-based therapy**. *Science* 2014, **344**: 1240622.

10.15 Delta-Notch signaling and lateral inhibition govern vertebrate neurogenesis

Chitins, A., Henrique, D., Lewis, J., Ish-Horowitcz, D., Kintner, C.: **Primary neurogenesis in *Xenopus* embryos regulated by a homologue of the *Drosophila* neurogenic gene *Delta***. *Nature* 1995, **375**: 761–766.

Engler, A., Zhang, R., Taylor, V.: **Notch and neurogenesis**. *Adv. Exp. Med. Biol.* 2018, **1066**: 223–234.

Garcez, P.P., Diaz-Alonso, J., Crespo-Enriquez, I., Castro, D., Bell, D., Guillemot, F.: **Cenpj/CPAP regulates progenitor divisions and neuronal migration in the cerebral cortex downstream of Ascl1**. *Nat. Commun.* 2015, **6**: 6474.

Ma, Q., Kintner, C., Anderson, D.J.: **Identification of *neurogenin*, a vertebrate neuronal determination gene**. *Cell* 1996, **87**: 43–52.

Manning, C.S., Biga, V., Boyd, J., Kursawe, J., Ymisson, B., Spiller, D.G., Sanderson, C.M., Galla, T., Rattray, M., Papalopulu, N.: **Quantitative single-cell live imaging links HES5 dynamics with cell-state and fate in murine neurogenesis**. *Nat. Commun.* 2019, **10**: 2835.

Moore, R., Alexandre, P.: **Delta–Notch signaling: the long and the short of a neuron's influence on progenitor fates**. *J. Dev. Biol.* 2020, **8**: 8.

10.16 The growth cone's response to guidance cues controls the path taken by a growing axon

Atkins, M., Nicol, X., Fassier, C.: **Microtubule remodelling as a driving force of axon guidance and pruning**. *Semin. Cell Dev. Biol.* 2023, **140**: 35–53.

Dudanova, I., Klein, R.: **Integration of guidance cues: parallel signaling and crosstalk**. *Trends Neurosci.* 2013, **36**: 295–304.

Lowery, L.A., Van Vactor, D.: **The trip of the tip: understanding the growth cone machinery**. *Nat. Rev. Mol. Cell. Biol.* 2009, **10**: 332–343.

Roig-Puiggros, S., *et al.*: **Construction and reconstruction of brain circuits: normal and pathological axon guidance**. *J. Neurochem.* 2020, **153**: 10–32.

Tessier-Lavigne, M.: **Wiring the brain: the logic and molecular mechanisms of axon guidance and regeneration**. *Harvey Lect.* 2002–2003, **98**: 103–143.

Yam, P.T., Charron, F.: **Extracellular phosphorylation in axon pathfinding**. *Nat. Chem. Biol.* 2019, **15**: 1030–1031.

Zang, Y., Chaudhari, K., Bashaw, G.J.: **New insights into the molecular mechanisms of axon guidance receptor regulation and signaling**. *Curr. Top. Dev. Biol.* 2021, **142**: 147–196.

Box 10E Putting it together: the development of a simple neural circuit

Arber, S., Costa, R.M.: **Connecting neuronal circuits for movement**. *Science* 2018, **360**: 1403–1404.

Balaskas, N., Ng, D., Zampieri, N.: **The positional logic of sensory-motor reflex circuit assembly**. *Neuroscience* 2020, **450**: 142–150.

Li, W.-C., Cooke, T., Sautois, B., Soffe, S.R., Borisyuk, R., Roberts, A.: **Axon and dendrite geography predict the specificity of synaptic connections in a functioning spinal cord network**. *Neural Dev.* 2007, **2**: 17.

Surmeli, G., Akay, T., Ippolito, G.C., Tucker, P.W., Jessell, T.M.: **Patterns of spinal sensory-motor connectivity prescribed by a dorsoventral positional template**. *Cell* 2011, **147**: 653–665.

10.17 Axon guidance is a dynamic process

Campbell, D.S., Holt, C.E.: **Chemotropic responses of retinal growth cones mediated by rapid local protein synthesis and degradation**. *Neuron* 2001, **32**: 1013–1026.

Deglincerti, A., Liu, Y., Colak, D., Hengst, U., Xu, G., Jaffrey, S.R.: **Coupled local translation and degradation regulate growth cone collapse**. *Nat. Commun.* 2015, **6**: 6888.

Koppers, M., Cagnetta, R., Shigeoka, T., Wunderlich, L.C., Vallejo-Ramirez, P., Qiaojin Lin, J., Zhao, S., Jakobs, M.A., Dwivedy, A., Minett, M.S., Bellon, A., Kaminski, C.F., Harris, W.A., Flanagan, J.G., Holt, C.E.: **Receptor-specific interactome as a hub for rapid cue-induced selective translation in axons**. *eLife* 2019, **8**: e48718.

10.18 Axons that cross the ventral midline provide a model system for axon guidance

Chédotal, A.: **Further tales of the midline**. *Curr. Opin. Neurobiol.* 2011, **21**: 68–75.

Gorla, M., Bashaw, G.J.: **Molecular mechanisms regulating axon responsiveness at the midline**. *Dev. Biol.* 2020, **466**: 12–21.

Kennedy, T.E., Serafini, T., de la Torre, J.R., Tessier-Lavigne, M.: **Netrins are diffusible chemotropic factors for commissural axons in the embryonic spinal cord**. *Cell* 1994, **78**: 425–435.

Simpson, J.H., Bland, K.S., Fetter, R.D., Goodman, C.S.: **Short-range and long-range guidance by Slit and its Robo receptors: a combinatorial code of Robo receptors controls lateral position**. *Cell* 2000, **103**: 1019–1032.

Wu, Z., Makihara, S., Yam, P.T., Teo, S., Renier, N., Balekoglu, N., Moreno-Bravo, J.A., Olsen, O., Chédotal, A., Charron, F., Tessier-Lavigne, M.: **Long-range guidance of spinal commissural axons by Netrin1 and Sonic Hedgehog from midline floor plate cells**. *Neuron* 2019, **101**: 635-647.

10.19 Motor neuron innervation of the chick limb is controlled by patterns of gene expression

Tosney, K.W., Hotary, K.B., Lance-Jones, C.: **Specifying the target identity of motoneurons**. *BioEssays* 1995, **17**: 379–382.

Xu, N.-J., Henkemeyer, M.: **Ephrin reverse signaling in axon guidance and synaptogenesis**. *Semin. Cell Dev. Biol.* 2012, **23**: 58–64.

10.20 The retino-tectal system is an example of a topographic map

Klein, R.: **Eph/ephrin signaling in morphogenesis, neural development and plasticity**. *Curr. Opin. Cell Biol.* 2004, **16**: 580–589.

Löschinger, J., Weth, F., Bonhoeffer, F.: **Reading of concentration gradients by axonal growth cones**. *Phil. Trans. R. Soc. Lond. B* 2000, **355**: 971–982.

McLaughlin, T., Lim, Y.S., Santiago, A., O'Leary, D.D.: **Multiple EphB receptors mediate dorsal-ventral retinotopic mapping via similar bi-functional responses to ephrin-B1**. *Mol. Cell Neurosci.* 2014, **63**: 24–30.

Petros, T.J., Shrestha, B.R., Mason, C.: **Specificity and sufficiency of EphB1 in driving the ipsilateral retinal projection**. *J. Neurosci.* 2009, **29**: 3463–3474.

Suetterlin, P., Marler, K.M., Drescher, U.: **Axonal ephrinA/EphA interactions, and the emergence of order in topographic projections**. *Semin. Cell Dev. Biol.* 2012, **23**: 1–6.

10.21 Synapse formation involves reciprocal interactions

Bailey, C.H., Kandel, E.R., Harris, K.M.: **Structural components of synaptic plasticity and memory consolidation**. *Cold Spring Harb. Perspect. Biol.* 2015, **7**: 1–29.

Blockus, H., Polleux, F.: **Developmental mechanisms underlying circuit wiring: novel insights and challenges ahead**. *Curr. Opin. Neurobiol.* 2021, **66**: 205–211.

Buffelli, M., Burgess, R.W., Feng, G., Lobe, C.G., Lichtman, J.W., Sanes, J.R.: **Genetic evidence that relative synaptic efficacy biases the outcome of synaptic competition**. *Nature* 2003, **424**: 430–434.

Clarke, L.E., Barres, B.A.: **Emerging roles of astrocytes in neural circuit development**. *Nat. Rev. Neurosci.* 2013, **14**: 311–321.

Colón-Ramos, D.A.: **Synapse formation in developing neural circuits**. *Curr. Top. Dev. Biol.* 2009, **87**: 53–79.

Davis, G.W., Schuster, C.M., Goodman, C.S.: **Genetic analysis of the mechanisms controlling target selection: target-derived Fasciclin II regulates the pattern of synapse formation**. *Neuron* 1997, **19**: 561–573.

Jessell, T.M., Kandel, E.R.: **Synaptic transmission: a bidirectional and self-modifiable form of cell-cell communication**. *Cell* 1993, **72**: 1–30.

Schmucker, D., Clemens, J.C., Shu, H., Worby, C.A., Xiao, J., Muda, M., Dixon, J.E., Zipursky, S.L.: ***Drosophila* Dscam is an axon guidance receptor exhibiting extraordinary molecular diversity**. *Cell* 2000, **101**: 671–684.

Südhof, T.C.: **Towards an understanding of synapse formation**. *Neuron* 2018, **100**: 276–293.

Box 10F Autism: a developmental disorder that involves synapse dysfunction

De Rubeis, S., *et al.*: **Synaptic, transcriptional and chromatin genes disrupted in autism**. *Nature* 2014, **515**: 209–215.

Ebert, D.H., Greenberg, M.E.: **Activity-dependent neuronal signalling and autism spectrum disorder**. *Nature* 2013, **493**: 327–337.

Geschwind, D.H.: **Autism: many genes, common pathways?** *Cell* 2008, **135**: 391–395.

Guy, J.G., Gan, J., Selfridge, J., Cobb, S., Bird, A.: **Reversal of neurological defects in a mouse model of Rett syndrome**. *Science* 2007, **315**: 1143–1147.

Satterstrom, F.K., *et al.*: **Large-scale exome sequencing study implicates both developmental and functional changes in the neurobiology of autism**. *Cell* 2020, **180**: 568–584.

Sudhof, T.: **Neuroligins and neurexins link synaptic function to cognitive disease**. *Nature* 2008, **455**: 903–911.

Walsh, C.A., Morrow, E.M., Rubenstein, J.L.R.: **Autism and brain development**. *Cell* 2008, **135**: 396–400.

10.22 Many neurons die through apoptosis during normal development

Oppenheim, R.W.: **Cell death during development of the nervous system**. *Annu. Rev. Neurosci.* 1991, **14**: 453–501.

Yamaguchi, Y., Miura, M.: **Programmed cell death in neurodevelopment**. *Dev. Cell.* 2015, **32**: 478–490.

10.23 Neuronal cell death and survival involve both intrinsic and extrinsic factors

Burden, S.J.: **Wnts as retrograde signals for axon and growth cone differentiation**. *Cell* 2000, **100**: 495–497.

Harrington, A.W., Ginty, D.D.: **Long-distance retrograde neurotrophic factor signalling in neurons**. *Nat. Rev. Neurosci.* 2013, **14**: 177–187.

Li, G., Hidalgo, A.: **The Toll route to structural brain plasticity**. *Front. Physiol.* 2021, **12**: 679766.

Park, H., Poo, M.M.: **Neurotrophin regulation of neural circuit development and function**. *Nat. Rev. Neurosci.* 2013, **14**: 7–23.

Reichardt, L.F.: **Neurotrophin-regulated signalling pathways**. *Philos. Trans. R. Soc. Lond. B Biol. Sci.* 2006, **361**: 1545–1564.

10.24 Neural activity refines synapses and topographic maps

Huberman, A.D.: **Mechanisms of eye-specific visual circuit development**. *Curr. Opin. Neurobiol.* 2007, **7**: 73–80.

Katz, L.C., Crowley, J.C.: **Development of cortical circuits: lessons from ocular dominance columns**. *Nat. Rev. Neurosci.* 2002, **3**: 34–42.

Katz, L.C., Shatz, C.J.: **Synaptic activity and the construction of cortical circuits**. *Science* 1996, **274**: 1133–1138.

Marin, I.A., Gutman-Wei, A.Y., Chew, K.S., Raissi, A.J., Djurisic, M., Shatz, C.J.: **The nonclassical MHC class I Qa-1 expressed in layer 6 neurons regulates activity-dependent plasticity via microglial CD94/NKG2 in the cortex**. *Proc. Natl Acad. Sci. USA* 2022, **119**: e2203965119.

10.25 The mammalian brain has a high degree of plasticity, even in adults

Gage, F.H.: **Adult neurogenesis in mammals**. *Science* 2019, **364**: 827–828.

Homem, C.C., Knoblich, J.A.: ***Drosophila* neuroblasts: a model for stem cell biology**. *Development* 2012, **139**: 4297–4310.

11

Growth, post-embryonic development, and regeneration

- Growth
- Molting and metamorphosis
- Regeneration
- Aging and senescence

Patterning of the embryo occurs on a small scale and is followed by growth. The control of growth, and therefore of size, is a key problem in development, involving at its core the control of cell proliferation. Such control is crucial in adult life in preventing cancer—the result of uncontrolled cell proliferation. The genetic programs that determine why different species grow to different characteristic sizes are, however, still a complete mystery. In this chapter we look at the balance between intrinsic programs of cell proliferation during embryonic development, and growth, usually at later stages, which is stimulated by extrinsic signals such as circulating hormones. One important aspect of post-embryonic development in many animals is metamorphosis, in which the form of the animal is completely changed between the larval stage and the adult. In this chapter we shall look at some aspects of metamorphosis in insects and in the vertebrate amphibians. Another intriguing feature of some animals is their capacity as adults to regrow various body parts—such as tails, limbs, and even the heart—when injured. Using examples from amphibians, insects, and the zebrafish, we shall examine the adult regeneration process focusing on the source of cells for regeneration and how patterning and growth of the regenerated structures is controlled. The final stage of post-embryonic development in most organisms is aging, which results from the accumulation of cellular damage that outstrips the ability of the body to repair itself. This age-related decline in repair—senescence—is genetically programmed.

Introduction

Development does not stop once the embryonic phase is complete. Most, but by no means all, of the growth in animals and plants occurs in the post-embryonic period, when the basic form and pattern of the organism has already been established. Growth is a central aspect of all developing systems, determining the final size and

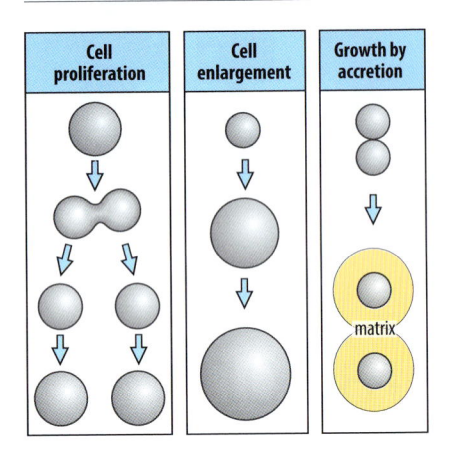

Fig. 11.1 The three main mechanisms for growth in vertebrates. The most common mechanism is cell proliferation—cell growth followed by division. A second mechanism is cell enlargement, in which cells increase their size without dividing. The third mechanism is to increase size by accretionary growth, such as secretion of a matrix.

shape of the organism and its parts. In many animals, the embryonic stage is immediately succeeded by a free-living larval or immature adult stage, in which growth occurs. In others, such as mammals, considerable growth occurs during a late embryonic or fetal period (see life cycles in Chapter 3), while the fetus is still dependent on maternal resources. Growth then continues after birth. In animals with a larval stage, the larva not only grows in size but eventually undergoes **metamorphosis**, in which it is transformed into the adult form. Metamorphosis often involves a radical change in form and the development of new organs. Unlike the situation in animals, where the embryo is essentially a miniature version of the free-living larva or adult, plant embryos bear little resemblance to the mature plant. Most of the adult plant structures are generated after germination by the shoot or root meristems, which have a capacity for continual growth. Some aspects of plant growth are discussed in Chapter 12.

A related phenomenon to growth is that of **regeneration**. This is the ability of the fully developed juvenile or adult organism to replace tissues, organs, and even most of the body, by the regrowth or repatterning of somatic tissue. Plants have remarkable regenerative ability (discussed in Section 12.4) and all animals can repair damaged tissues to some extent. But in mammals, the ability to regrow lost limbs, for example, or regenerate injured organs such as the heart, is very restricted. In this chapter we shall look at a few classic examples of regeneration in animals—the regrowth of amphibian limbs, insect appendages, and the zebrafish heart. Like all vertebrates, mammals renew some tissues throughout their lives from stem cells (see Chapter 6) and can regenerate liver tissue if not too much is removed, as we shall see in this chapter. But they cannot regrow lost limbs or hearts. Why animals such as amphibians and fishes, with much the same basic anatomy as mammals, can do so is an intriguing question. Finally, we also look at the last stage of post-embryonic development—the phenomenon of aging.

We start this chapter by considering first what we mean by 'growth', and then look at the roles of intrinsic growth programs and of extrinsic factors, such as growth factors and hormones, in controlling both embryonic and post-embryonic growth. In this part of the chapter, we shall also touch on the disturbances of growth control that lead to cancer.

Growth

Growth is defined as an increase in the mass or overall size of a tissue or organism; this increase may result from cell proliferation, cell enlargement without division, or by accretion of extracellular material, such as bone matrix or even water (Fig. 11.1). In animals, the basic body pattern is laid down when the embryo is still very small, of the order of millimeters in size. The intrinsic growth program—that is, how large an organism or an individual organ can grow, and how it responds to growth stimulants such as hormones—may also be specified at an early stage in development, as we shall see later in relation to finger length. Overall growth of the organism mainly occurs in the post-embryonic period after the basic form and pattern of the body has been established; there are, however, many examples where earlier organogenesis involves localized growth, as in the vertebrate limb bud and the lung (see Chapter 9), in which cell proliferation mediates outgrowth and morphogenesis. Localized growth also underlies some of the size differences in the brain between species. In the embryonic mouse brain, for example, the neuronal progenitor cells that give rise to the neocortex divide only around 11 times before differentiating into neurons, whereas in primates they undergo a greater number of cell divisions before differentiating, expanding this cell population and so enabling it to give rise to the much more extensive primate neocortex. In this way, genetically programmed differences in the rate of growth in different parts of the body, or at different times, during early development profoundly affect the shape of organs and the organism.

Growth will determine the size of the different organs and the overall size of the body of the organism. The size of the organs must also fit the size of the body, and the same principles are involved in determining both organ size and overall body size. One important question is: how do individual organs or an organism know when the correct size has been reached and therefore know when to stop growing? The enormous differences in body size in different mammals result mainly from differences in cell number, not in cell size. A 70-kilogram person is composed of some 10^{13} cells and has a body mass about 3000 times greater than that of a mouse, which is composed of about 3×10^9 cells. In principle, therefore, overall body size could be controlled by counting cell divisions and cell number. But there is also evidence that body size can be determined by monitoring overall dimensions, rather than the absolute number of cells or cell divisions. In species in which the cells are either larger or smaller than normal, as a result of differences in the number of chromosomes, the overall body size is not affected. Thus, for example, some naturally tetraploid salamanders with cells twice as large as diploid cells grow to the same size as their diploid relatives but have only half the number of cells.

Growth depends on both intrinsic and extrinsic factors. Although considerable progress has been made in identifying extrinsic factors—for example, the growth factors and hormones that control growth—the genetic basis for differences in the sizes of animals of different species is completely unknown. A classic illustration of an intrinsic growth program, which is genetically determined, comes from the limb and the results of grafting limb buds between large and small species of salamanders of the genus *Ambystoma*. A limb bud from the larger species grafted to the smaller species initially grows slowly, but eventually ends up at its normal size, which is much larger than any of the limbs of the host animal (Fig. 11.2). This indicates that whatever the circulating factors, such as hormones, that influence growth, the intrinsic response of a tissue to them is crucial.

Fig. 11.2 The size of limbs is genetically programmed in salamanders. An embryonic limb bud from a large species of salamander, *Ambystoma tigrinum*, grafted to the embryo of a smaller species, *Ambystoma punctatum*, grows much larger than the host limbs—to the size it would have grown in *Ambystoma tigrinum*.

From Harrison, R.G.: Organization and Development of the Embryo. New Haven, CT: Yale University Press, 1969.

11.1 Tissues can grow by cell proliferation, cell enlargement, or accretion

Although growth often occurs through **cell proliferation**, which increases the number of cells by mitotic cell division, this is only one of three main ways by which growth in size of an organ or body structure can occur (see Fig. 11.1). There is also a significant amount of programmed cell death in many growing tissues, and the overall growth rate is therefore determined by the balance between the rates of cell death and cell proliferation.

A second way in which growth occurs is through **cell enlargement**—that is, by individual cells increasing their mass and getting bigger. This is the case in *Drosophila* larval growth—except for the imaginal discs, in which cells proliferate during the larval stages. And differences in size between closely related species of *Drosophila* are partly the result of the larger species having larger cells. In mammals, skeletal and heart muscle cells and neurons divide very rarely, if at all, once differentiated, although they do increase in size. Neurons grow by the extension and growth of axons and dendrites, whereas muscle growth in neonatal and juvenile animals involves an increase in mass by fusion of satellite cells to pre-existing muscle fibers. Much growth involves a combination of cell proliferation and cell enlargement. For example, the cells in the lens of the eye are produced by cell division from a proliferative zone for an extended period, whereas their differentiation involves considerable enlargement.

The third growth mechanism, **accretionary growth**, involves an increase in the volume of the extracellular space, which is achieved by the secretion of large quantities of extracellular matrix by cells. Accretionary growth occurs in both cartilage and bone, where most of the tissue mass is extracellular.

Intracellular signaling pathways controlling growth in cell size and cell proliferation have been discovered (Fig. 11.3). The target of rapamycin (TOR) pathway, whose

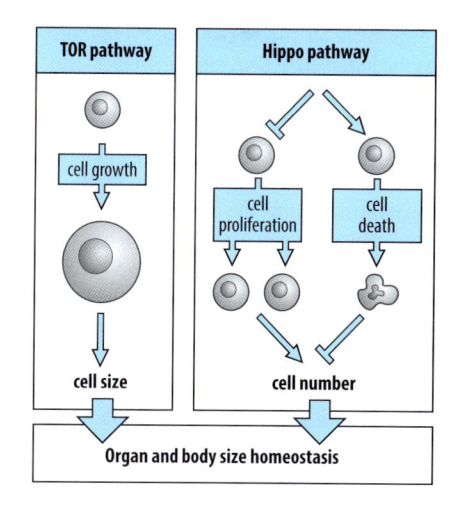

Fig. 11.3 Intracellular pathways controlling cell size and cell number together can determine tissue growth. Target of rapamycin (TOR) is a serine-threonine protein kinase that regulates organ size by stimulating cell growth, thereby increasing cell size. Its general function is as a central control hub inside cells, integrating information on the availability of energy and nutrients and responding to growth factor signaling. Hippo is a serine-threonine protein kinase that acts in a signaling pathway (outlined in Box 11A) that controls organ size by restricting cell number through the suppression of proliferation and promotion of apoptosis.

From Tumaneng, K., et al.: *Organ size control by Hippo and TOR pathways.* Curr. Biol. 2012, **22**: R368-R370.

activity is regulated by nutrient supply, positively controls cell size, whereas the Hippo signaling pathway controls cell number by suppressing proliferation and promoting cell death. We will come across both pathways later in this chapter and examine the Hippo signaling pathway in more detail.

11.2 Cell proliferation is controlled by regulating entry into the cell cycle

When a eukaryotic cell duplicates itself by mitosis it goes through a fixed sequence of phases called the **cell cycle**, which were described in Box 1B. The cell grows in size (phases G_1 and G_2), the DNA is replicated (S phase), and the replicated chromosomes undergo mitosis (M phase) and become segregated into two daughter nuclei. The cell then divides (cytokinesis) to give two daughter cells. The cell cycle shown in Box 1B is the standard cell cycle of a dividing somatic animal cell. At different stages of development, or in specialized cell types, different phases of the cell cycle are absent, or the length of time taken to complete them varies. During cleavage of a fertilized amphibian egg, for example, which produces smaller and smaller cells, growth phases G_1 and G_2 are virtually absent.

The timing of events in the cell cycle is controlled by a set of 'central' timing mechanisms (Fig. 11.4). Proteins known as **cyclins** control the passage through key transition points in the cycle. Cyclin concentrations oscillate during the cell cycle, and these oscillations correlate with transitions from one phase of the cycle to the next. Cyclins act by forming complexes with, and helping to activate, protein kinases known as **cyclin-dependent kinases (Cdks)**. These kinases phosphorylate proteins that trigger the events of each phase, such as DNA replication in S phase or mitosis in M phase.

Once a cell has entered the cycle and progressed beyond a checkpoint commonly known as 'Start', which occurs in mid-to-late G_1, it will continue through and complete the cycle without needing any further external signal. Transitions into successive phases are marked by **cell-cycle checkpoints** at which the cell monitors progress to ensure, for example, that an appropriate size has been reached, that DNA replication is complete, and that any DNA damage has been repaired. If such criteria are not met, progress into the next stage is delayed until all the necessary processes have been completed. If the cell has suffered some damage that cannot be repaired, the cell cycle will be arrested, and the cell will usually undergo apoptosis. These mechanisms are

Fig. 11.4 Progression through the cell cycle is regulated by the levels of cyclins. Different cyclins (G_1/S, S, and M) regulate progression through different phases of the cell cycle. The expression of each cyclin rises and falls throughout the cell cycle. As cyclin protein levels increase, the cyclins bind to their corresponding inactive cyclin-dependent kinases (Cdks; not shown), forming active enzymes. These active kinases are ultimately responsible for driving cell-cycle events such as the initiation of DNA replication (S phase) and the entry into mitosis (M phase). There are two checkpoints in the cell cycle (red bars), in addition to a 'Start' checkpoint, that ensure that the cell does not progress to the next stage until the previous stage has been completed successfully. At the end of mitosis, existing cyclins are targeted to the proteasome and destroyed, so that the new cells formed at cell division are reset to early G_1 phase.

Adapted from Morgan, D.O.: The Cell Cycle. London: New Science Press Ltd, 2007.

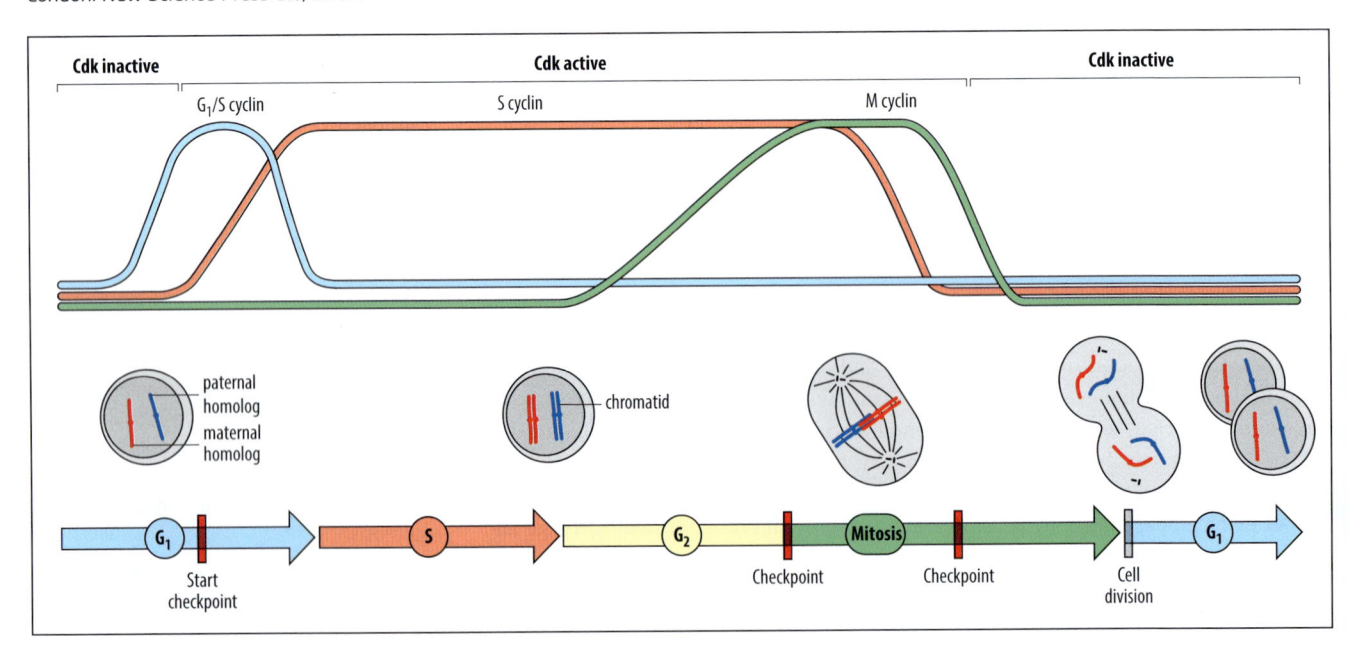

intrinsic to all normal eukaryotic cells and, for example, prevent them from continuing to divide with damaged DNA.

Extracellular signals, such as those provided by growth factors, control cell proliferation by stimulating entry into the cell cycle. Studies of animal cells in culture show that growth factors are essential for cells to multiply, with the growth factor or factors required depending on the cell type. When somatic cells are not stimulated to proliferate, they are usually in a state known as G_0, into which they withdraw after mitosis (see Box 1B). Growth factors enable the cell to proceed out of G_0 and progress through the cell cycle. Most cells in adult organisms are not actively proliferating and so need an external signal before they can re-enter the cell cycle. Many of the extracellular signal molecules we met in earlier chapters in their roles as patterning agents, such as fibroblast growth factors (FGF) and members of the transforming growth factor (TGF)-β family, were initially discovered through their ability to stimulate cell proliferation. They can also act in this way during embryonic and fetal growth, and in the control of cell proliferation in the adult. Other growth factors have specialized roles in directing the proliferation of particular tissues. One example is erythropoietin, which promotes the proliferation of red blood cell precursors.

Cells must receive signals, such as growth factors, not only for them to divide, but also simply to survive. In the absence of all such stimulation, cells commit suicide by apoptosis, as a result of activation of an internal cell death program.

11.3 Cell division in early development can be controlled by an intrinsic developmental program

Embryonic cells proliferate much more freely compared to the cells of adult organisms, and in many organisms, the pattern of cell division in the earliest embryonic stages is controlled by a cell-autonomous developmental program that does not depend on stimuli such as growth factors. One well-understood example comes from *Drosophila*, in which the early cell cycles are controlled by the proteins that pattern the embryo; these proteins exert their effects by interacting with components of the cell-cycle control system.

The first cell cycles in the *Drosophila* embryo are represented by rapid and synchronous nuclear divisions, without any accompanying cell divisions, which create the syncytial blastoderm (see Fig. 2.2). There are virtually no G phases, just alternations of DNA synthesis (S phase) and mitosis. But at cycle 14, there is a major transition to a different type of cell cycle, a transition like the mid-blastula transition in frogs (see Section 3.9). A well-defined G_2 phase is seen in cycle 14, and the blastoderm becomes cellularized. A G_1 phase is seen at cycle 16. After cycle 17 or 18, cells in the epidermis and mesoderm stop dividing and differentiate. The cessation of cell proliferation is caused by the exhaustion of maternal cyclin E reserves originally laid down in the egg.

At the cell cycle 14, distinct spatial domains with different cell-cycle times can be seen in the *Drosophila* blastoderm (Fig. 11.5). This patterning of cell cycles is produced by a change in the synthesis and distribution of a protein phosphatase called String, which exerts control on the cell cycle by dephosphorylating and activating a cyclin-dependent kinase and controlling the transition from G_2 to M. In the fertilized egg, the String protein is of maternal origin and is uniformly distributed. It therefore produces a synchronized pattern of nuclear division throughout the embryo. After cycle 13, the maternal String protein disappears and the zygotic expression of String protein becomes the controlling factor for entry into M phase.

Zygotic *string* gene transcription occurs in a complex spatial and temporal pattern. Only cells in which the *string* gene is expressed enter mitosis. This results in variation in the rate of cell division in different parts of the blastoderm, which ensures that the correct number of cells is generated in different tissues. The pattern of zygotic *string* gene expression is controlled by transcription factors encoded by

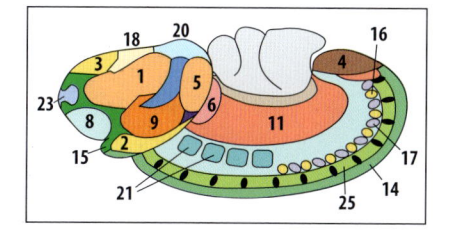

Fig. 11.5 Domains of mitosis in the *Drosophila* blastoderm. Areas composed of cells that divide at the same time are indicated by the various colors, and the numbers indicate the order in which these domains undergo mitosis at the fourteenth cell cycle, when zygotic String protein is first expressed. The schematic illustrates a lateral view of the embryo corresponding to a stage after mesoderm has been internalized and segmentation has begun (segments are indicated by the black marks along the lower surface). Anterior is to the left and dorsal is up. The gray region on the dorsal surface is the amnioserosa. The internalized mesoderm (domain 10) and some other domains are not visible in this view.

Illustration after Foe, V.E.: **Mitotic domains reveal early commitment of cells in Drosophila embryos.** Development *1989,* **107***: 1-22.*

the early patterning genes, such as the gap and pair-rule genes, and those patterning the dorso-ventral axis.

One exception to the rule that expression of *string* leads to cell proliferation is the presumptive mesoderm, which is the first domain in which *string* is expressed but the tenth to start cell division. The delay in cell division is due to the expression of the gene *tribbles* in this region, as Tribbles protein degrades String. The delay is necessary to allow ventral furrow formation and mesoderm invagination (see Section 5.10), because cell division inhibits ventral furrow formation. The prospective mesodermal cells start to proliferate after they have been internalized. Another protein phosphatase related to String—called Twine—is also expressed maternally, and Twine is now believed to be the key determinant of the mid-blastula transition in *Drosophila*. Although the early cell cycles in *Drosophila* development are quite unusual as they occur in a syncytium, they provide a good example of how a pattern of cell division can be under genetic control and the proteins involved have vertebrate homologs.

11.4 Extrinsic signals coordinate cell division, cell growth, and cell death in the developing *Drosophila* wing

The *Drosophila* wing has proved an interesting model system for studying many aspects of growth and especially how the size of an organ might be determined. In contrast to the intrinsic embryonic program in the blastoderm, the growth in the wing imaginal disc and other discs is modulated by extracellular signals produced by the previous patterning process, and involves the coordination of cell growth, cell proliferation, and cell death.

At its formation, the wing disc is initially composed of about 40 cells, and normally grows in the larva to about 50,000 cells. Cell division occurs throughout the disc, and then ceases uniformly when the correct size is reached. By the end of the pupal stage, the disc has developed the structure of the adult wing, which is produced at metamorphosis by changes in cell shape with only minimal further cell division (see Fig. 9.4). Early experiments investigating wing growth showed that the final size of the wing does not depend on the imaginal disc undergoing a fixed number of cell divisions, or attaining a particular number of cells. Instead, final size seems to be controlled by some mechanism that monitors the overall size of the developing wing disc and adjusts cell division and cell size accordingly. Experiments show that there is no restriction on how much of the wing the progeny of a given cell can make; the clonal descendants of a single cell can contribute from a tenth to as much as a half of the wing.

Wing growth can also be experimentally decoupled from cell proliferation, as demonstrated by the fact that if cell division is blocked in either the anterior or posterior compartment in the larval wing disc, the final size of the wing is normal, but the individual cells are larger. In the same way, the size of a region of the wing need not be determined by the rate of cell division. This can be shown by making mosaics of wild-type cells and slower-growing *Minute* cells. Within a wing compartment, the faster-dividing wild-type cells contribute more cells than the slow-growing *Minute* cells, but the compartment remains the same size as if all the cells were wild type.

In normal growth, the final size of the wing (and that of any other organ) is likely to be achieved by a balance between cell proliferation and apoptosis. An important coordinator of these cell activities is the Hippo signaling pathway, which was first discovered in *Drosophila* but has since been found to operate in mammals and other vertebrates (see Box 11A). The *Drosophila* Hippo pathway involves a cascade of serine–threonine protein kinases, of which Hippo is one, and whose activity results in the inactivation of a transcriptional co-activator called Yorkie (Yki). When the pathway is inactive, Yki translocates to the nucleus and forms a complex with cell-type-specific transcription factors to induce the expression of genes encoding proteins that promote cell proliferation and suppress apoptosis. By contrast, when Hippo signaling is activated, it phosphorylates Warts (Wts), another protein kinase in the cascade, and

CELL BIOLOGY BOX 11A The core Hippo signaling pathways in *Drosophila* and mammals

The outcome of activation of the Hippo pathway is the inactivation of the transcriptional co-activator Yorkie (Yki) in *Drosophila* (see Fig. 1, left panels) and the homologous transcriptional co-activators Yap (Yes-associated protein) or Taz in mammals (see Fig. 1, right panels). When the pathway is inactive, Yki or Yap/Taz translocate to the nucleus and help to activate expression of genes that promote cell proliferation and suppress apoptosis. When the pathway is active, upstream regulators, including Merlin (Mer), Expanded (Ex), and Kibra in *Drosophila* (neurofibromatosis 2 (NF2), FRMD, and Kibra, respectively, in mammals), activate the protein kinase Hippo (Hpo) (Mst1 or Mst2 in mammals), which then phosphorylates the protein kinase Warts (Lats 1 or Lats 2 in mammals), which, in turn, phosphorylates Yki (Yap or Taz in mammals). Phosphorylated Yki (or Yap or Taz) is retained in the cytoplasm and inactivated. Expression of cell-proliferation genes is suppressed, and expression of apoptosis genes is promoted. Salvador (Sav) and Mats in *Drosophila* (Mob1 in mammals) are co-factors for the core protein kinases in the Hippo pathway.

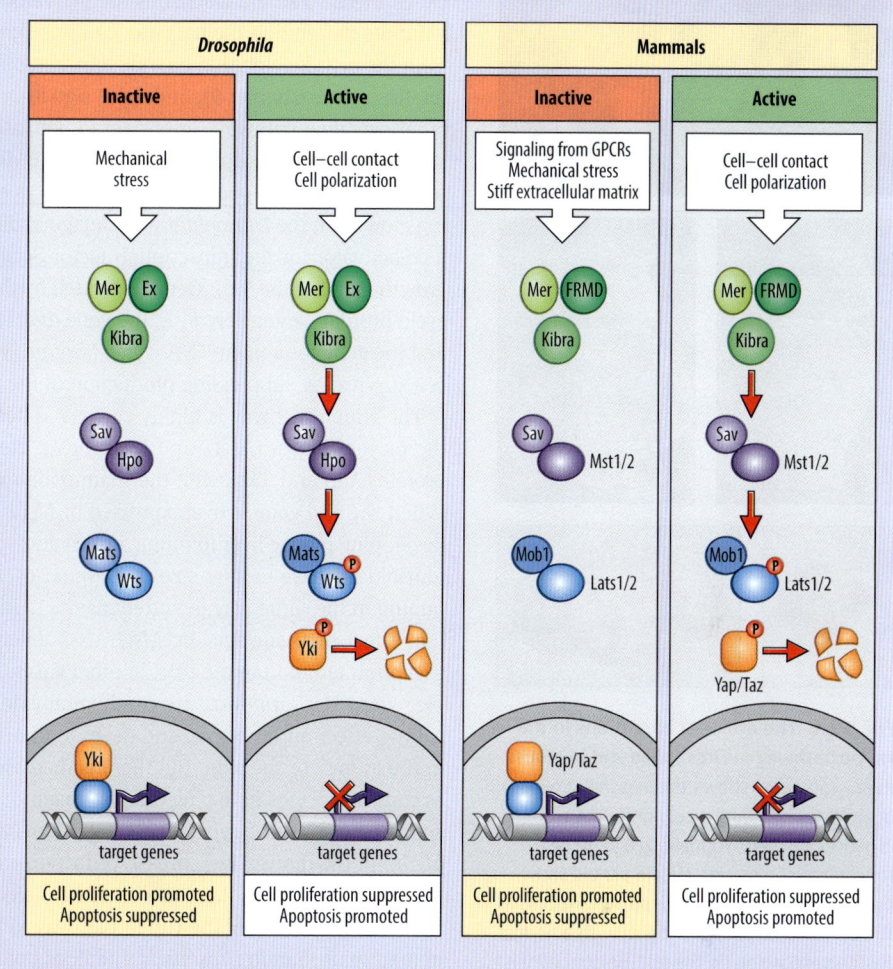

Fig. 1

this, in turn, phosphorylates Yki. Phosphorylated Yki is retained in the cytoplasm and therefore cannot translocate to the nucleus to act as a transcriptional activator. Loss-of-function mutations in either Hippo or Warts or the overexpression of Yki in *Drosophila* wing discs lead to a dramatic increase in cell proliferation and a decrease in apoptosis, resulting in wing discs growing to almost eight times the normal size (Fig. 11.6). The inactivation of Yki by the Hippo signaling network is therefore a mechanism for suppressing growth, and the Hippo pathway could be one means of terminating growth when an organ reaches its required size.

It is becoming clear that the Hippo pathway acts as a sensor of the overall state of a tissue and thus could ensure that it does not grow beyond a certain size. How exactly it achieves this is not known, but it is not surprising that there are multiple inputs into the Hippo pathway that are integrated to regulate the activity of the pathway both positively and negatively. Many of these inputs comprise signal molecules associated with adhesion and the cytoskeleton, which can transmit not only chemical information, but also mechanical information resulting from cell–cell contacts.

In *Drosophila*, inputs that activate the pathway comprise growth-inhibitory signals, including signals from molecules in septate junctions (these signals act on the kinase cascade), and signals from molecules localized in apical domains of epithelial cells (e.g. Crumbs) that provide information about tissue organization (these signals act on the upstream regulators of the kinase cascade). There is also evidence that the atypical cadherins Fat and Dachsous are activators of the Hippo pathway.

Inputs that inhibit the pathway comprise growth-promoting signals, including signals from the cytoskeleton that provide information about the mechanical forces generated in the tissue (see Chapter 5), and these signals act on Yki. The pathway also interacts with other signaling pathways important in regulating growth and development, such as the Wnt, TGF-β, bone morphogenetic protein (BMP), Hedgehog, and Notch pathways. At the output end of the Hippo pathway, Yki acts as a co-activator of genes involved in regulating growth. In the *Drosophila* wing imaginal disc, for example, Yki primarily functions as a co-activator for the transcription factor Scalloped, which is essential for normal wing growth (see Section 9.6). Genes activated by the Yki–Scalloped complex include the cell-cycle-promoting gene *cyclin E*, the gene *diap1*, which encodes an inhibitor of apoptosis, and the gene for a microRNA called *bantam*, which promotes both cell proliferation and cell survival by suppressing production of the pro-apoptotic protein Hid.

The Hippo pathway is highly conserved between invertebrates and vertebrates, with the protein kinases Mst1 and Mst2 being the mammalian versions of Hippo, and the proteins Yap and Taz being the mammalian versions of Yki (see Box 11A for details). When the *Yap* gene is overexpressed or *Mst1* and *Mst2* are conditionally deleted in the developing mouse liver to mimic inactivation of the pathway, there is a striking three- to fourfold increase in liver size (see Fig. 11.6), consistent with the Hippo pathway functioning in the same way in vertebrates as in insects and regulating organ size. However, this effect of tissue-specific *Mst1/Mst2* deletion on the developing liver is more the exception rather than the rule. In other cases, there is no effect on organ size. For example, when *Mst1* and *Mst2* are conditionally deleted in the developing mouse limb, there is little effect on limb size, and deletion of *Mst1* and *Mst2* has no apparent effect at all in the kidney or the skin, but there is increased proliferation of specific cell types such as cardiomyocytes in the heart and progenitor cells in the intestine. Furthermore, tissue-specific deletions of genes encoding other pathway components, such as the deletion of *Yap* in the mouse liver, breast, or intestine does not lead to the predicted decrease in organ size, although there are bile duct defects in the liver. All these findings suggest that the functioning of the Hippo pathway in regulating organ size is more complicated in mammals than in *Drosophila* and does not even seem to be involved in some organs.

The regulation of the Hippo pathway is also highly conserved in mammals and *Drosophila*, with the same activating and inhibiting inputs (see Box 11A). Studies on vertebrate cells in culture have shown, for example, that the pathway is activated by cell–cell contact and is therefore responsible for density-dependent inhibition of cell proliferation, a well-known property of normal cells in culture. Cell polarization also activates Hippo signaling. Both these inputs act directly on the transcriptional co-activators, Yap and Taz. The pathway is inhibited by signaling of G-protein-coupled receptors to which extracellular growth-promoting molecules may bind, in addition to mechanical cues transmitted by the cytoskeleton. These cues include those that depend on the stiffness of the extracellular matrix, and thus the Hippo pathway is responsible for anchorage-dependent cell proliferation, another well-known property of normal cells in culture. A picture is emerging of a multitude of signals conveying information about a cell's environment converging on the Hippo pathway to determine whether that cell will divide or not, and whether it will live or die. It remains to be determined whether any of these signals control organ size in mammals.

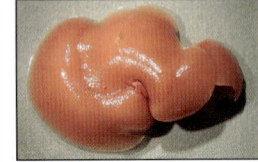

Fig. 11.6 The effect of alterations in the Hippo pathway in *Drosophila* and in mouse liver. Top panel shows the massive growth of a *Drosophila* wing disc (right) when *Yki* is overexpressed. Normal-sized wing disc for comparison on left. Middle panels show overgrowth of clones of *Hippo* mutant cells in the cuticle of *Drosophila* (right) with a normal fly for comparison on the left. Bottom panels show the large increase in the size of the liver of a mouse in which both *Mst1* and *Mst2*, the two mammalian *Hippo* homologs, were conditionally inactivated during embryonic development (right). Liver of a normal mouse for comparison on left. It should be noted, however, that other mouse organs in which *Mst1/2* were deleted appear normal.

Top panel from Huang, J., et al.: **The Hippo signaling pathway coordinately regulates cell proliferation and apoptosis by inactivating Yorkie, the Drosophila homolog of YAP.** Cell 2005, **122**: 421-434. Middle and bottom panels from Halder, G., Johnson, R.L.: **Hippo signaling; growth control and beyond.** Development 2011, **138**: 9-22.

11.5 Cancer can result from mutations in genes that control cell proliferation

Most cancers in adults—more than 85%—occur in epithelia. This is not surprising as many epithelia (such as the epidermis of the skin and the lining of the gut) are renewed from stem cells that have the potential to divide indefinitely (see Chapter 6). In normal epithelia, stem-cell division is under strict regulation and the cells generated by stem cells soon stop dividing. By contrast, cancerous epithelial cells continue to divide, although not necessarily more rapidly, and usually fail to differentiate. Another feature of cancer cells is that they are genetically unstable and often gain or lose chromosomes when they divide.

Most cancers derive from a single abnormal cell that has acquired mutations that increase its rate of proliferation. The progression of such a cell to becoming a malignant tumor cell, known as **tumor progression**, is an evolutionary process, involving both further mutation and selection of those cells best able to proliferate. A greater frequency of division increases the cell's chance of acquiring more mutations. In almost all cancers, the cells are found to have a mutation in at least one, and usually many, genes. On average, 63 mutations are found in a pancreatic cancer cell. Genes that when mutated can lead to cancer fall into two classes—**proto-oncogenes** and **tumor-suppressor genes**.

Proto-oncogenes are involved in driving cell proliferation in normal cells. If a proto-oncogene becomes permanently switched on or abnormally expressed because of mutation or a chromosomal rearrangement, this can lead to the overstimulation of cell division and thus contribute to cancer formation. Mutation of one copy of a proto-oncogene in a cell can be sufficient to promote uncontrolled proliferation. At least 70 proto-oncogenes are known in mammals; the first to be identified in humans was Ras, a small GTPase involved in intracellular signaling pathways that promote cell proliferation.

Tumor-suppressor genes, in contrast, are involved in suppressing cell proliferation in normal cells. Inactivation or deletion of both copies of the gene in a cell removes these restraints to proliferation and is required for a cell to become cancerous. The classic example of a human tumor-suppressor gene is the *retinoblastoma* (*RB*) gene that encodes a protein, RB, that is involved in inhibiting cell-cycle progression.

Retinoblastoma is a childhood tumor of retinal cells, and although it is normally very rare, certain families have an inherited predisposition to it. This led to the discovery of the *RB* gene. In some of these families, the predisposing genetic defect was identified as a deletion of a particular region on one copy of chromosome 13. A deletion on one copy of the chromosome does not, on its own, cause retinal cells to become cancerous. However, if a retinal cell also acquires a deletion of the same region on the other copy of chromosome 13, a retinal tumor develops. The gene in this region that is responsible for susceptibility to retinoblastoma is *RB*. Both copies of *RB* must be lost or inactivated for a cell to become cancerous (Fig. 11.7).

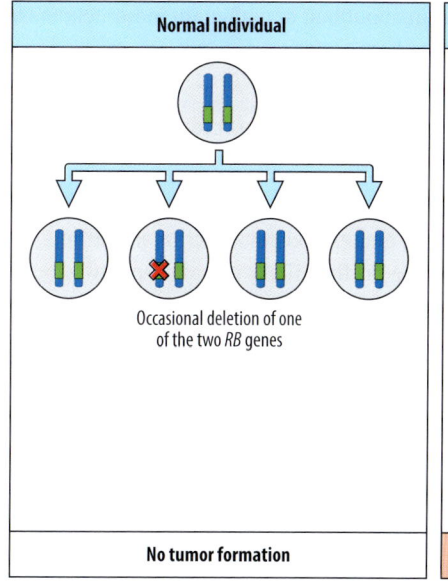

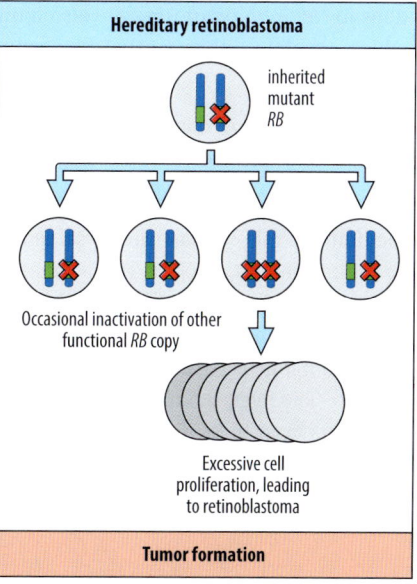

Fig. 11.7 The retinoblastoma gene *RB* is a human tumor-suppressor gene. If only one copy of *RB* is lost or inactivated by mutation, no tumor develops (left panel). In individuals already carrying an inherited mutant *RB*, if the other copy of the gene is also lost or inactivated in a retinal cell, that cell will generate a retinal tumor (right panel). Individuals with an inherited *RB* mutation are thus at a much greater risk of developing retinoblastoma, and usually do so at a young age.

The tumor-suppressor gene *p53* has a key role in many cancers; about half of all human tumors contain a mutated form of *p53*. This gene is not required for development of tumors per se, but when cells are exposed to agents that damage DNA, then *p53* is activated and arrests the cell cycle, giving the cell time to repair the DNA. The p53 protein thus prevents the cell from replicating damaged DNA and giving rise to mutant cells. Instead, p53 will cause the cell to die by apoptosis if the damage is too severe to be repaired. The mutant forms of *p53*, found in many cancers, do not promote apoptosis, and so the affected cells are more likely to accumulate mutations.

Genes encoding components of the Wnt and Hedgehog signaling pathways are frequently found mutated in human cancers. Indeed, the first mammalian member of the Wnt family, mouse *Int-1*, which is involved in brain development, was discovered through its ability to act as a proto-oncogene. The Wnt pathway component adenomatous polyposis coli (APC), however, is a tumor suppressor. Wnt signaling in the stem-cell niche is necessary for the proliferation of intestinal cells, along with those of many other stem cell pools (see Fig. 6.30), and APC normally helps to keep the Wnt pathway inhibited in the absence of Wnt signaling (see Box 3E). Mutations that inactivate *APC* result in the constitutive activation of the pathway, leading to unregulated cell proliferation; such mutations are the cause of the inherited syndrome familial adenomatous polyposis coli (hence the gene name), in which pre-cancerous polyps form in the colon. Inactivation of *APC* on its own is not sufficient to produce a malignant tumor, but greatly increases the risk of further mutations leading to polyps developing into colorectal cancer.

Mutations in Hedgehog signaling pathway genes can also lead to tumors. The link between Hedgehog signaling and cancer was first recognized by the finding that mutations in the gene for the Hedgehog receptor Patched (see Boxes 2E and 9C) are the cause of the rare inherited Gorlin syndrome in humans, one feature of which is multiple cancers of epidermal basal cells. Patched acts as a tumor suppressor, keeping the Hedgehog pathway switched off in the absence of Hedgehog. Abnormal activation of the Hedgehog pathway is also present in nearly all pancreatic cancers, where it is thought to maintain cancer stem cells and to be involved in cancer progression.

The Hippo pathway has been found to be dysregulated in many solid tumors, but in most cases, this is not due to mutations in the genes encoding Hippo pathway components. Instead, dysregulation is most probably due to mutations in genes involved in the Wnt and Hedgehog signaling pathways or to aberrant signaling by G-protein-coupled receptors. Mutations in the gene encoding Neurofibromatosis-2 (the homolog of Merlin; see Box 11A) are, however, associated with growths in the nervous system arising particularly from Schwann cells. These growths are non-cancerous, but mutations in *Neurofibromatosis-2* are also found in about half of malignant mesotheliomas, which are almost always caused by exposure to asbestos.

11.6 The relative contributions of intrinsic and extrinsic factors in controlling size differ in different mammalian organs

As we have seen, intrinsic genetic programs, such as those that determine the size of the limbs in different species of salamanders, and extracellular factors that stimulate or inhibit growth, such as those that impinge on the Hippo signaling pathway, can both contribute to organ size. But the relative importance of these two mechanisms in different mammalian organs varies a good deal. This is nicely illustrated by comparing what happens to the liver and the pancreas after parts of them are destroyed in the embryo. The liver, which is the main detoxifying organ in mammals, can regrow and restore its mass to its normal size in both the embryo and the adult after damage. The size of the pancreas, however, is intrinsically determined, and if part of it is destroyed in the embryo it ends up smaller than normal.

Experiments that show these different properties of the liver and the pancreas make use of the fact that organ precursor cells can be specifically destroyed in embryonic mice using transgenic techniques. A strain of mice is generated that carries an introduced gene for diphtheria toxin under the control of a tetracycline-repressible promoter. Part of this promoter comes from a gene that is only expressed in the precursor cells of the particular organ to be targeted, and so only directs expression of the toxin in those cells. If a pregnant mouse of this strain is continuously fed tetracycline, the toxin is not expressed at all in the embryos she carries. If tetracycline is withdrawn, the toxin is produced in the targeted precursor cells, killing them but leaving other tissues in the embryo untouched.

To investigate the regenerative powers of the embryonic liver, tetracycline was withdrawn for a short period at the appropriate time during gestation, with the result that a proportion of the liver precursor cells in the embryos are destroyed. The longer the period without tetracycline, the greater the number of precursor cells lost. After such treatment, the embryonic liver grows back to a normal size, indicating that it does not arise from a fixed number of cells (Fig. 11.8).

By contrast, if some of the cells of the pancreas in a mouse embryo are destroyed (by the technique described above) after the pancreatic 'bud' has formed, a smaller than normal pancreas develops (see Fig. 11.8). This finding argues against a mechanism in the pancreas that senses organ size, and instead suggests that the size of the embryonic pancreas is largely under intrinsic control. One intrinsic mechanism could involve counting the number of cell divisions, and each pancreatic precursor cell might only be capable of giving rise to a fixed number of cells. Another organ with intrinsic growth control is the thymus gland. If multiple fetal thymus glands are transplanted into a developing mouse embryo, each one grows to full size. If the same experiment is done with spleens, however, each spleen grows much smaller than normal, so that the final total mass of the spleens is equivalent to one normal spleen. It seems, therefore, that the spleen, like the liver, is regulating its final size by some kind of negative-feedback mechanism, responding to factors in the tissue environment.

Because of the importance of the liver in human health and the relatively high incidence of liver damage, there have been many studies of regeneration of the mammalian liver. Liver cells can be induced to divide by injury to the liver or other stimuli, such as removal of part of the organ. In the regenerating human liver, hepatocytes start to divide a day after surgery (e.g. to remove a liver tumor). In adult mammals, the liver can regenerate after as much as two-thirds of it has been surgically removed. In rats and mice, for example, if two of the five liver lobes are removed, these lobes do not regrow, but instead the three remaining lobes increase in size to restore the normal mass of liver tissue.

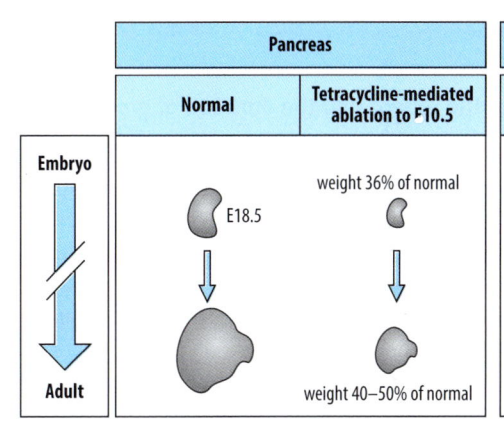

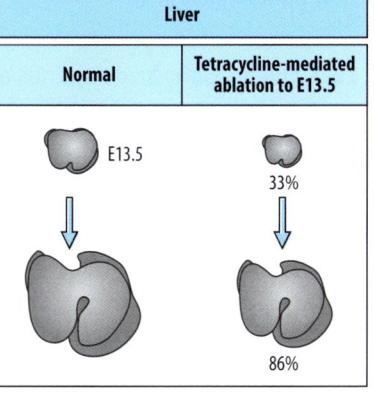

Fig. 11.8 Tetracycline-controlled cell killing in embryonic pancreas and liver in mice reveals their different potentials for regeneration. The pancreas does not regenerate when its cells are killed in the embryo, and remains small, whereas the liver can regenerate, and rapidly regains almost its normal size when its cells are killed in the embryo.

Adapted from Liu, J.C., Baron, J.: **Mechanisms limiting body growth in mammals.** Endocr. Rev. 2011, **32**: 422–440.

So what switches on growth of the liver when part of it is removed, and how does it know when it has grown enough? Although various growth-promoting factors are known to be required for liver regeneration, the size-regulating mechanism is still unclear. One negative-feedback mechanism for regulating liver size could be by sensing when the liver is big enough to carry out its function of removing bile acids from the circulation. Bile acids are made in the liver and released into the small intestine to aid digestion. They are potentially toxic and so are subsequently resorbed into the blood vessels draining the small intestine and returned directly to the liver, where they are taken up by hepatocytes and reused. An increase in the concentration of bile acids in the blood produced, for example, by feeding mice a diet containing bile acid, signals that the liver is not large enough to maintain the proper levels of bile in the blood, and thus stimulates regeneration, whereas draining the bile duct before removing liver lobes delays regeneration, because the levels of bile acids in the circulation are low and so signal that the liver is large enough.

A similar feedback mechanism may also contribute to regulating the size of the kidney. Removal of a kidney leads to an increase in size of the remaining kidney, perhaps in response to a temporary rise in the concentration of creatinine in the serum, which signals the need to increase kidney function. In the kidney, however, the increase in size is mainly the result of cell enlargement rather than cell proliferation.

One hypothesis for the regulation of liver size is that organs such as the liver produce molecules that negatively regulate their growth. Such factors were first postulated some 40 years ago and called 'chalones'. Because the concentration of these factors would depend on the number of cells in an organ, this could provide a mechanism to stop the organ growing once the correct number of cells had been produced. But these factors remained unidentified and the concept of chalones fell out of favor. However, secreted proteins with the properties of the classic chalones are now beginning to be identified, although no chalone-like molecule has so far been identified in the liver.

Skeletal muscle is the best example so far of a tissue in which size is regulated by negative feedback mediated directly by a 'chalone'-like protein. This protein is myostatin, a TGF-β-family protein that is produced and secreted by myoblasts (immature muscle cells) and which inhibits muscle growth. The importance of myostatin in regulating muscle size was shown by functionally inactivating the *myostatin* gene in mice. There is a significant increase in muscle mass in the mutant mice; the number of muscle fibers and their size are both increased. Similarly, 'double-muscled' domesticated animals, for example Belgian blue cattle (Fig. 11.9, top panel) and certain breeds of sheep, have been shown to lack myostatin function. So-called 'bully' whippets with greatly enlarged muscles are also homozygous for a loss-of-function *myostatin* mutation. The mutation is maintained in the heterozygous state in the whippet breed, as one copy of the mutant gene makes the dog a faster racer (Fig. 11.9, bottom panel).

Fig. 11.9 'Double-muscled' breeds of cattle and dogs have inherited inactivating mutations in the *myostatin* gene. *Top panel: this file is licensed under the Creative Commons Attribution-Share Alike 3.0 Unported license. Author: Stoolhog. Bottom panel: ©2011 Stuart Isett. All rights reserved.*

11.7 Overall body size depends on the extent and the duration of growth

Overall body size depends not only on the rate of growth, but on how long growth continues. Human growth during the embryonic, fetal, and postnatal periods is typical of the different phases of growth in mammals. The human embryo increases in length from 150 μm at implantation to about 50 cm over the 9 months of gestation. During the first 8 weeks after conception, the embryonic body does not increase greatly in size, but the basic human form is laid down in miniature. After 8 weeks the human embryo is technically known as a fetus.

In humans, the greatest rate of growth occurs at about 4 months of intrauterine development, when the fetus grows as much as 10 cm per month. Growth after birth follows a well-defined pattern (Fig. 11.10, left panel). During the first year after birth,

Growth   **583**

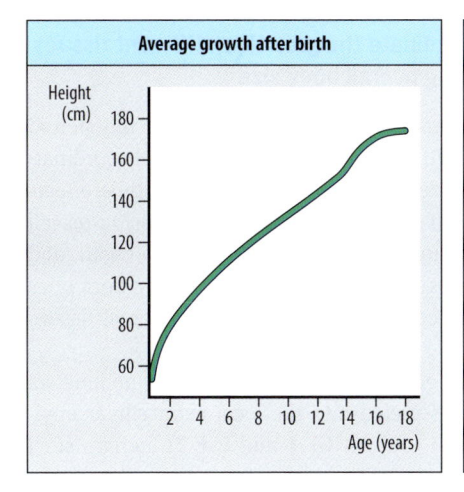

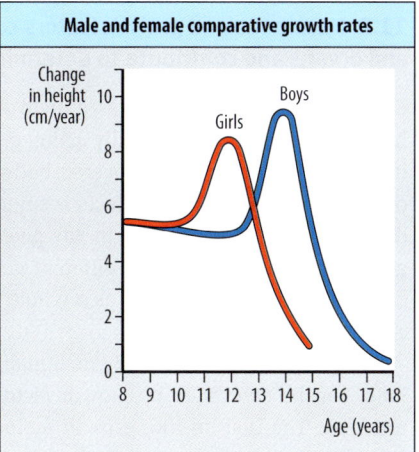

Fig. 11.10 Normal human growth. Left panel: an average growth curve for a human male after birth. Right panel: comparative growth rates of boys and girls. There is a growth spurt at puberty in both sexes, which occurs earlier in girls.

growth occurs at a rate of about 2 cm per month. The growth rate then declines steadily until the start of a characteristic adolescent growth spurt at puberty at about 11 years in girls and 13 years in boys (Fig. 11.10, right panel) followed by a sharp decline. In Pygmies, sexual maturation at puberty is not accompanied by this adolescent growth spurt, hence their characteristic short stature. There are similar declines in growth rate in other mammals after birth, but in rodents, for example, the decline takes place over a few weeks rather than years.

Different tissues and organs grow at different rates during development, and this affects the way that the proportions of the different parts of the body change in relation to each other during gestation and after birth (Fig. 11.11). At 9 weeks of development, the head of a human embryo, for example, is more than one-third of the length of the whole embryo, whereas at birth it is only one-quarter. After birth, the rest of the body grows much more than the head, which is only about one-eighth of the body length in an adult.

Fig. 11.11 Different parts of the human body grow at different rates. At 9 weeks of development the head is relatively large but, with time, other parts of the body grow much more than the head.

Illustration after Gray, H.: Gray's Anatomy. Edinburgh: Churchill-Livingstone, 1995.

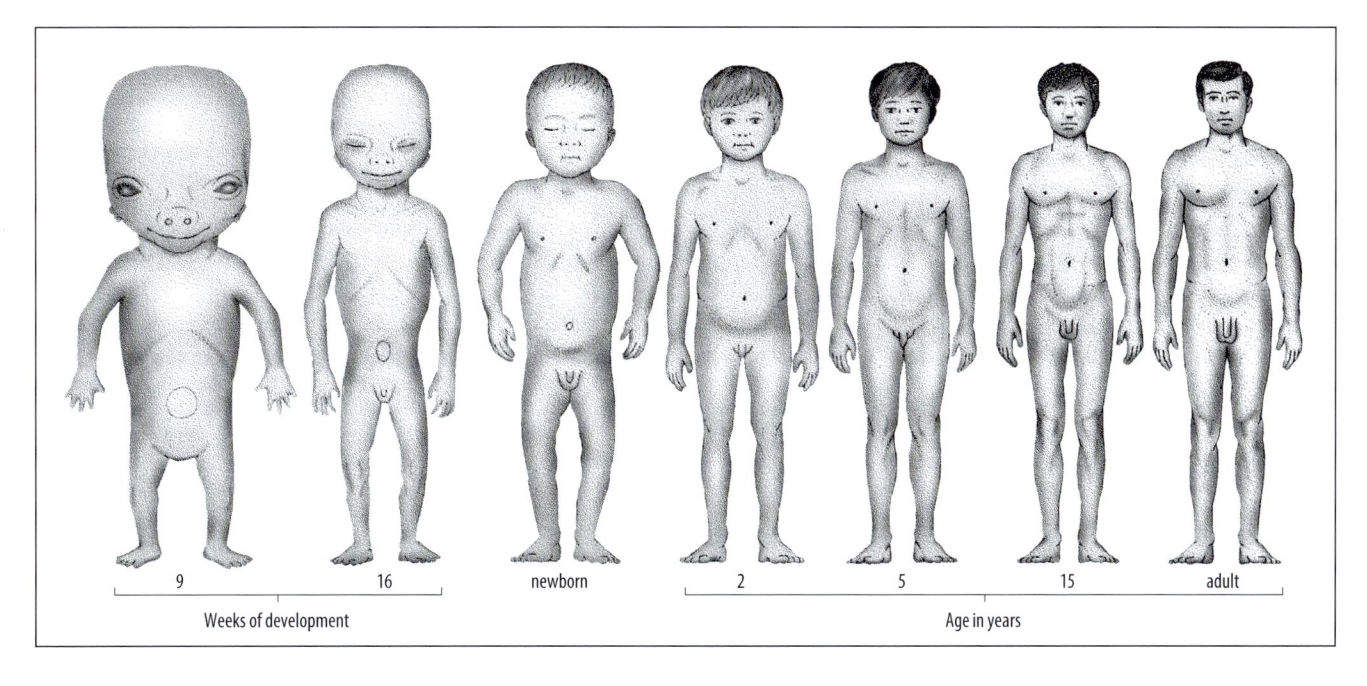

11.8 Hormones and growth factors coordinate the growth of different tissues and organs and contribute to determining overall body size

Circulating hormones, such as growth hormone, thyroid hormone, and the steroid hormones, together with locally acting growth factors, have a major role in coordinating growth and controlling overall body size. Most hormones have multiple effects on cells, but one of these effects is often to stimulate or inhibit cell division or cell differentiation. The response to any given hormone differs from organ to organ, and generally is also different at different stages of development. In this way hormones can act throughout the body as a whole to coordinate the growth of different tissues and organs.

Cell proliferation in the early mammalian embryo, when it is only a few millimeters long, can be controlled by growth factors secreted into the local extracellular environment. The **insulin-like growth factors 1 and 2 (IGF-1 and IGF-2)** are two such growth factors with key roles in promoting cell proliferation and growth during embryonic development. IGF-1 and IGF-2 are single-chain protein growth factors that closely resemble the hormone insulin and each other in their amino acid sequence and produce their effects on cells through a shared insulin-signaling pathway. Newborn mice lacking a functional *Igf1* gene develop relatively normally but weigh only 60% of the normal newborn body weight. Mice in which the *Igf2* gene has been inactivated are also growth retarded. Both the factors and their receptors can be detected as early as the eight-cell stage of mouse development. *Igf2* is one of the genes that are imprinted in mammals (see Section 8.8); it is inactivated in maternal germ cells and is only expressed in the embryo from the paternal genome. Like many growth factors, the IGFs also have important roles in postnatal growth and in controlling cell proliferation in adult mammals. In humans, IGF-1 is produced after birth mainly by the liver under the stimulation of **growth hormone**, a circulating hormone produced by the pituitary (Fig. 11.12). It is also produced in growth hormone-responsive tissues, where it can act locally. *IGF1* and genes encoding other components of the growth hormone–IGF-1 axis have been identified as major genetic determinants of body size in dogs (see Box 13F).

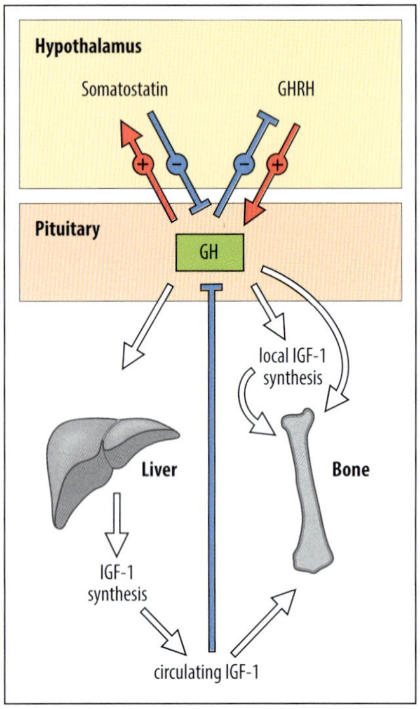

Growth hormone is crucial to the growth of humans and other mammals after birth. Within the first year of birth, the pituitary gland begins to secrete growth hormone. A child with insufficient growth hormone grows less than normal, but if the synthetic form of the hormone is given regularly, normal growth is restored. In this case, there is a catch-up phenomenon, with a rapid initial response that tends to restore the growth curve to its original trajectory. Catch-up growth should be contrasted with the compensatory growth we have discussed, for example, in the liver, which restores organ mass.

Production of growth hormone in the pituitary is under the control of two hormones produced in the hypothalamus: **growth hormone-releasing hormone**, which promotes growth hormone synthesis and secretion, and **somatostatin**, which inhibits its production and release. As we have just noted, growth hormone produces many of its effects by inducing synthesis of IGF-1, but it can also act directly on some tissues such as the growth plates of long bones (see Fig. 11.12). Postnatal growth, in addition to embryonic growth, is therefore largely due to the actions of the insulin-like growth factors, and to the complex hormonal regulatory circuits that control their production.

Puberty is also initiated by activity in the hypothalamus, which governs the intermittent release of **gonadotropin-releasing hormone (GnRH)** by hypothalamic neurons. The mechanism determining this timing is not known. One of the results of a pulse of GnRH is a sharp increase in the secretion of gonadotropins (luteinizing hormone and follicle-stimulating hormone) by the pituitary; these cause increased production of the steroid sex hormones—the estrogens and androgens. These, in turn, stimulate the production of pulses of growth hormone, which are responsible for the growth spurts at puberty in girls and boys.

Fig. 11.12 Growth hormone production is under the control of the hypothalamic hormones. Growth hormone is made in the pituitary gland and is secreted. Growth hormone-releasing hormone (GHRH) from the hypothalamus promotes the synthesis of growth hormone (GH), whereas somatostatin inhibits it. Growth hormone controls its own release by feedback signals to the hypothalamus. Growth hormone stimulates the synthesis of the insulin-like growth factor (IGF)-1, and this, in turn, has a negative effect on the production of growth hormone in the pituitary. IGF-1 is mainly made in the liver but can also be made locally. Long bone growth is controlled directly by the action of growth hormone and by both circulating and locally produced IGF-1.

Fig. 11.13 Comparative growth of the cartilaginous elements in the embryonic chick wing. When first laid down, the cartilaginous elements of the humerus, ulna, and wrist are the same size, but the humerus and ulna then grow much more than the wrist element.

11.9 Elongation of the long bones illustrates how growth can be determined by a combination of an intrinsic growth program and extracellular factors

Patterning of the embryo occurs while the organs are still very small. For example, human limbs have their basic pattern established when they are less than 1 cm long. Over the years, the limb grows to be at least 100 times longer and this growth is driven by elongation of the long bones. The long bones in the arm are the humerus, radius, and ulna, and in the leg, the femur, fibula, and tibia. So how is this growth controlled?

Each of the skeletal elements in the chick wing has its own intrinsic growth program, which is specified during embryonic development. The cartilaginous elements representing the long bones—the humerus and the ulna, for example—are initially similar in size to the elements in the wrist (Fig. 11.13). Yet, with growth, the humerus and ulna increase many times in length compared with the wrist bones. These growth programs are specified when the elements are initially patterned and involve both cell proliferation and matrix secretion (accretion). Each skeletal element follows its own growth program even when grafted to a developmentally neutral site, provided that a good blood supply is established. It is likely that this differential growth is due at least in part to the different Hox genes that come to be expressed in the different limb 'segments'. In addition, the radius/ulna region of the chick wing (and the tibia/fibula region of the chick leg) expresses the *Shox* (*Short stature homeobox*) gene. *SHOX* deficiencies in humans are associated with Turner syndrome and various short stature conditions, which are characterized by the forearms and the lower legs being disproportionately short.

One intriguing effect of the sex hormones testosterone and estrogen is on differences in relative finger length between the two sexes. This growth program is apparently established in a narrow developmental window in the embryo as a result of the differing levels of these hormones in male and female embryos (Box 11B).

The growth of the long bones occurs in the **growth plates**. Long bones are initially laid down as cartilaginous elements (see Section 9.10). In both fetal and postnatal growth, the cartilage is replaced by bone in a process known as **endochondral ossification**, in which ossification starts in the centers of the long bones and spreads outward (Fig. 11.14). Secondary ossification centers then develop at each end of the bone, known as the epiphyses. The adult long bones thus have a bony shaft (diaphysis) with cartilage confined to the articulating surfaces at each end, and to two regions near each end—the growth plates—in which growth occurs. In the growth plates, the cartilage cells, or **chondrocytes**, are usually arranged in columns, and various zones can be identified. Just next to the bony epiphysis is a narrow germinal zone, which contains stem cells. Next is a proliferative zone of cell division, followed by a zone of maturation, and a hypertrophic zone, in which the cartilage cells increase in size. Finally, there is a zone in which the cartilage cells die and are replaced by bone laid down by cells called **osteoblasts**. Some osteoblasts are derived from hypertrophic cartilage—an example of transdifferentiation (see Section 6.16)—whereas others differentiate from mesenchyme cells that form the perichondrium surrounding the cartilage. This stage involves Wnt signaling. There is a marked similarity to the development of the epidermis of the skin, where basal stem cells give rise to dividing cells, which differentiate into keratinocytes and finally die (see Section 6.11).

The proliferation of chondrocytes at the ends of the long bones, and later in the growth plate, is controlled so that at a given distance from the end of the bone they stop dividing and enlarge. In mice, the proliferation of chondrocytes is controlled

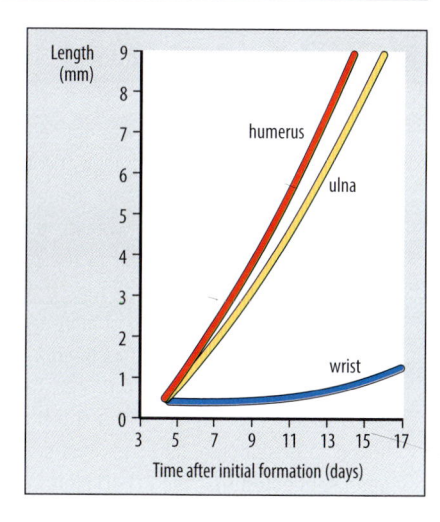

BOX 11B Digit length ratio is determined in the embryo

More than 100 years ago, it was noticed that the relative lengths of different fingers vary between the sexes. In men, the fourth finger (ring finger) is generally longer than the second finger (index finger), whereas in women, it is shorter or the same length. The effect is stronger in the right hand. These proportions are usually expressed as the **2D:4D digit ratio** (the ratio is less than 1 in men; the same or more than 1 in women) (Fig. 1).

In the past decade there has been a resurgence of interest in this ratio and a low 2D:4D ratio has been correlated with various attributes, including greater sporting prowess, and has been linked to developmental conditions such as autism. It has also been suggested that the 2D:4D digit ratio could be a bio-marker for some diseases such as osteoarthritis, where the risk is greater in one sex than the other.

Sexual dimorphism in finger length is determined before birth. It has been detected in human fetuses and was proposed to reflect the degree of prenatal exposure to the 'male' hormone testosterone. This proposal has been tested experimentally in mice. Mice also show the same sex differences in the 2D:4D digit ratio as humans and, like humans, the effect is stronger on right limbs. These differences can first be detected in the embryo when the chondrocytes (cartilage cells) in the developing digits have just begun to hypertrophy, leading to the mean 2D:4D digit ratio in the right paws of male embryos being significantly smaller than that in female embryos. Digits 2 and 4 in mammalian limbs have similar anatomy, but the differences in growth show that they have different identities (see Section 9.15 for how digit identities are specified during limb development).

To test whether this ratio is controlled by testosterone, and whether the 'female' sex hormone estrogen was also involved, mice were created in which either the androgen receptor (the receptor for testosterone) or the estrogen receptor were deleted genetically in the developing mouse limbs. When the androgen-receptor gene is deleted, the digit ratio in mutant males is significantly higher, consistent with testosterone being required for a masculine (low) digit ratio. In contrast, when the estrogen-receptor gene is deleted, the digit ratio in mutant males is significantly lower, consistent with estrogen being required for a feminine (high) digit ratio. Thus, estrogen, as well as testosterone, influences digit growth and their effects are opposite.

There is a narrow window in digit development when the balance of hormones determines whether there is a masculine or feminine digit ratio. Surprisingly, the differences in digit ratios in the hormone-treated offspring are due entirely to effects on the length of digit 4. The length of this digit increases in female offspring of mothers treated with testosterone, whereas it decreases in male offspring of mothers treated with estrogen. The reason that digit 4 is more responsive to hormone levels than digit 2 is that cells in digit 4 have many more receptors for both testosterone and estrogen (Fig. 2).

These experiments explain how the 2D:4D ratio arises during normal embryonic development. In female embryos, which have higher levels of estrogen than of testosterone, there will be higher levels of activated estrogen receptor in digit 4 and its growth will decrease. In male embryos, which have higher levels of testosterone than estrogen, there will be higher levels of activated androgen receptor in digit 4 and its growth will increase.

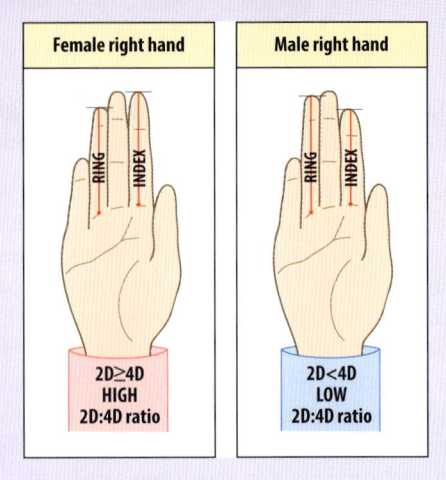

Fig. 1

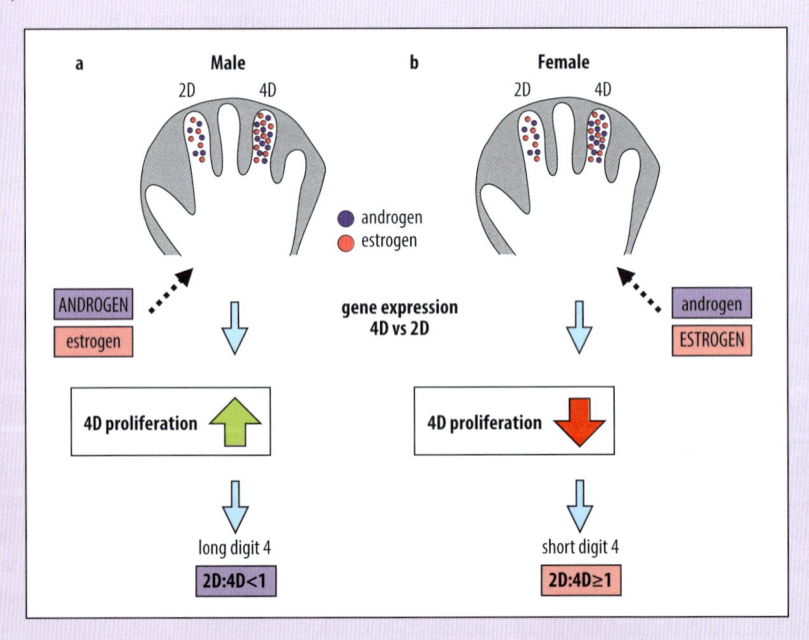

Fig. 2

Zheng, Z., Cohn, M.J.: **Developmental basis of sexually dimorphic digit ratios.** Proc. Natl Acad. Sci. USA 2011, **108**: 16289-16294.

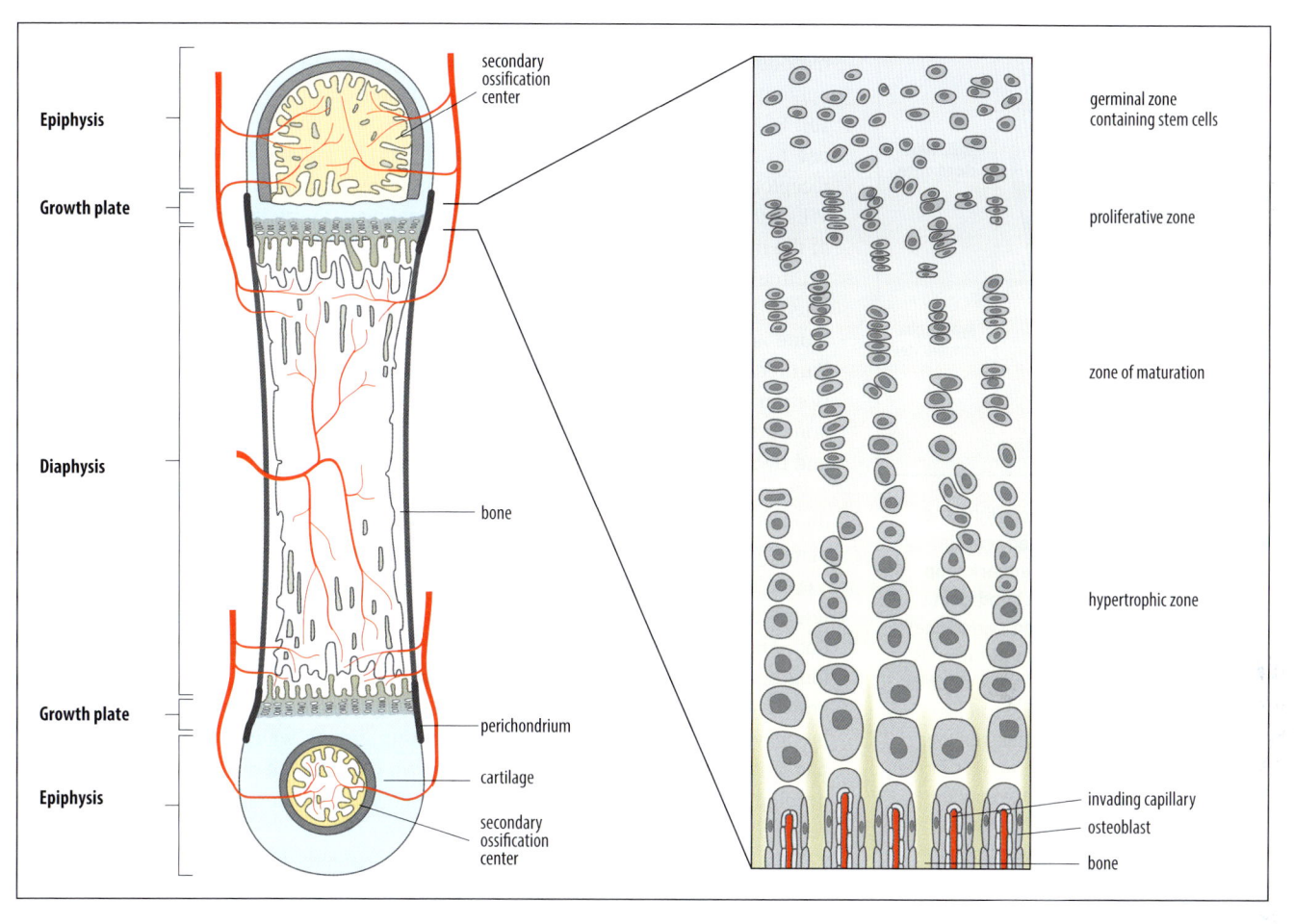

Fig. 11.14 Growth plates and endochondral ossification in the long bone of a vertebrate. The long bones of vertebrate limbs increase in length by growth from cartilaginous growth plates that lie between the epiphysis and the shaft of the bone. In the figure, bone has already replaced cartilage in the diaphysis, and more bone is being added at the growth plates. Within the growth plates, cartilage cells multiply in the proliferative zone, then mature and undergo hypertrophy. They are then replaced by bone, which is laid down by osteoblasts derived from both perichondrial cells and hypertrophic chondrocytes. **Osteoclasts** invade the bone along with blood vessels and are involved in resorbing the cartilage and remodeling bone. Secondary sites of ossification are located within the epiphyses.

Illustration after Walls, G.A. **Here today, bone tomorrow.** *Curr. Biol. 1993,* **3**: *687–689.*

by the secreted signaling proteins parathyroid hormone-related protein (PHRP) and Indian hedgehog (Ihh). Ihh belongs to the family of Hedgehog signal proteins (see Box 9C). PHRP is secreted by the chondrocytes and cells of the perichondrium at the ends of the prospective bones and stimulates chondrocytes to proliferate, which prevents them from expressing *Ihh*. Once chondrocytes move out of the zone of influence of PHRP they stop proliferating, start to express *Ihh*, and become hypertrophic (Fig. 11.15). Ihh protein diffuses back into the pool of proliferating chondrocytes, where it increases their rate of proliferation. Ihh also, by some mechanism as yet unknown, stimulates production of PHRP by the cells at the ends of the bone and this ensures that the proliferating zone of cells is maintained. Ihh also acts on the adjacent perichondrial cells to form bone-producing osteoblasts. In the absence of the *Ihh* gene in mice there is an accelerated rate of hypertrophy, resulting in short, stubby limbs, and cell proliferation is also suppressed.

Other key signal molecules that promote chondrocyte proliferation and bone growth together with Ihh are members of the BMP family. BMPs were originally identified by their ability to induce ectopic bone but are also important signal molecules used over

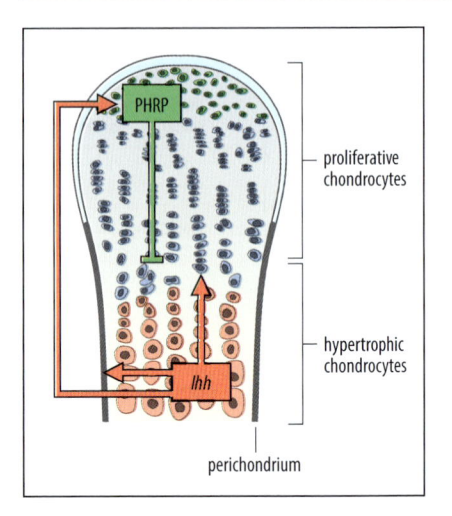

Fig. 11.15 Indian hedgehog (Ihh) and parathyroid hormone-related protein (PHRP) form a negative-feedback loop that controls the onset of chondrocyte differentiation. PHRP (green, barred line) secreted by the chondrocytes at the ends of developing long bones acts on proliferating chondrocytes (blue) to maintain proliferation and delay the expression of the gene for Indian hedgehog (*Ihh*). Chondrocytes farther away from the end of the bone (orange) escape the influence of PHRP, express *Ihh*, and differentiate into hypertrophic chondrocytes. Ihh protein acts on adjacent chondrocytes still in the proliferative zone to increase the rate of proliferation and on perichondrial cells to form osteoblasts. In some way that is not yet understood, production of Ihh also stimulates PHRP synthesis by the chondrocytes at the end of the bone.

Illustration after Kronenberg, H.M.: **Developmental regulation of the growth plate.** *Nature (Insight) 2003,* **423***: 332-336.*

and over again during development in, for example, dorso-ventral patterning of the mesoderm (see Chapter 4), patterning of the somites (see Chapter 4), limb bud development (see Chapter 9), and dorso-ventral patterning of the neural tube (see Chapter 10). FGF signaling has opposite effects to BMP signaling. The genetic defect that gives rise to achondroplasia (short-limbed dwarfism) is a dominant mutation in the FGF receptor-3, whose normal function is to limit, rather than promote, bone formation. When this receptor is mutated, the FGF signaling pathway is activated in the absence of the ligand and therefore the inhibition of growth is greatly exaggerated.

Circulating growth hormone affects bone growth by acting on the growth plates. The cells in the germinal zone have receptors for growth hormone, and growth hormone is probably directly responsible for stimulating these stem cells to proliferate. Further growth, however, is mediated by IGF-1, whose production in the growth plate is stimulated by growth hormone. Thyroid hormones are also necessary for optimal bone growth; they act both by increasing the secretion of growth hormone and IGF-1, and by stimulating hypertrophy of the cartilage cells.

The rate of increase in the length of a long bone is equal to the rate of new cell production per column multiplied by the mean height of an enlarged cell in the growth plate. The rate of new cell production depends both on the time cells take to complete a cycle in the proliferative zone, and the size of this zone. Different bones grow at different rates, and this can reflect the size of the proliferative zone, the rate of proliferation, and the degree of cell enlargement in the growth plate. The rate of growth of the long bones after birth also follows a similar pattern of decline as seen for the rate of growth of the whole body (see Section 11.7). If bone growth is prevented, for example by heavy loading, then there is some catch-up growth when the load is removed. It has been suggested that bone catch-up growth is due to the resumption of an intrinsic developmental program, which determines how the proliferation rate of growth-plate cells changes in a cell-autonomous fashion over time.

This hypothesis about catch-up growth has recently been tested by blocking the proliferation of cells in the growth plates of mouse embryonic limbs by cartilage-specific expression of an inducible cell-cycle suppressor, p21, driven by an enhancer that normally controls *Pitx2* expression. As we saw in Section 4.17, *Pitx2* is a key determinant of leftness, so cell proliferation was blocked preferentially in the growth plate of left limbs and the right limbs served as controls. This inducible system leads to a mosaic pattern of *p21* expression: around 60% of the chondrocytes in the left tibia expressed *p21*, but only 23% at most in the right tibia expressed the protein. The long bones of the left limbs in which cartilage-cell proliferation was blocked were short, but not as short as would be expected given the proportion of cells with a cell-cycle block. Therefore, catch-up growth must have occurred in a non-cell-autonomous way by compensatory proliferation of the cells not expressing *p21*. More surprisingly, there was also a general reduction in growth, with the growth of the right limbs being reduced so that they ended up the same length as the left limbs. These results show that extrinsic factors rather than intrinsic factors ensure that body proportions are correctly maintained during catch-up growth. The cross-talk between left and right limbs might help to explain how human bones in limbs on opposite sides of the body can grow for some 15 years apparently independently of each other, and yet eventually match to an accuracy of about 0.2%.

When growth of a bone ceases the growth plate ossifies, and this occurs at different times for different bones. Ossification of growth plates occurs in a strict order in different bones and can therefore be used to provide a measure of physiological age. The timing of growth cessation in the growth plate appears to be intrinsic to the plate itself rather than to hormonal influence. The cessation of growth is due to the cessation of cell proliferation, and this timing may be programmed in the cells. Cell senescence, and thus growth cessation, may be due to the chondrocyte stem cells having only a finite potential for division.

The length of the long bones dictates the length of the limbs and the growth of other tissues in the limbs including the muscles. The number of striated (skeletal) muscle fibers in a muscle is determined during embryonic development. Once differentiated, striated muscle cells lose the ability to divide. Post-embryonic growth of muscle tissue results from an increase in individual fiber size, both in length and girth, during which the number of myofibrils within the enlarged muscle fiber can increase more than tenfold. Additional nuclei to support the functioning of the much-enlarged cell are provided by the fusion of satellite cells with the fiber. Satellite cells, which are undifferentiated cells lying adjacent to the differentiated muscle, also act as a reserve population of stem cells that can replace damaged muscle (see Section 6.14).

The increase in the length of a muscle fiber is associated with an increase in the number of sarcomeres—the functional contractile units—it contains. For example, in the soleus muscle of the mouse leg, as the muscle increases in length, the number of sarcomeres increases from 700 to 2300 at 3 weeks after birth. This increase in number seems to depend on the growth of the long bones putting tension on the muscle through its tendons. If the soleus muscle is immobilized by placing the leg in a plaster cast at birth, sarcomere number increases slowly over the next 8 weeks, but then increases rapidly when the cast is removed. One can, therefore, see how bone and muscle growth are mechanically coordinated.

11.10 The amount of nourishment an embryo receives can have profound effects in later life

Whatever the type of growth program, no animal will reach its full potential size if inadequately nourished as an embryo and during the postnatal growth period. In mammals, inadequate nutrition or poor nutrition of the embryo not only has direct effects on embryonic and fetal growth, but can have serious, and in some ways unexpected, effects in adult life. Population studies in developed countries, such as the United Kingdom, have associated smaller size or relative thinness at birth (due to either maternal undernutrition or premature birth) and during early infancy with an increased risk of developing coronary heart disease, stroke, or type 2 diabetes in adult life.

When undernutrition during early development is followed by improved nutrition later, whether later in gestation or in the early postnatal period, the period of catch-up growth can incur a cost, as the body's resources are diverted from 'repair and maintenance' to growth. Premature birth itself, independent of size for gestational age, has been associated with insulin intolerance and glucose intolerance in pre-pubertal children that can continue into young adulthood and can be accompanied by high blood pressure. Catch-up growth seems to predispose to overweight and even obesity, and this may partly explain the effects on health in later life. One proposal is that in response to undernutrition *in utero* the fetus lays down more fat cells as a precaution. That is a good strategy if conditions after birth are indeed hard and food is short, but in conditions of plentiful food, this leads to undesirable consequences.

Experiments in animals support the effects of undernutrition and an unbalanced maternal diet observed in humans. The embryos of pregnant rats fed a low-protein, but otherwise calorie-sufficient, diet during the preimplantation period (0–4.5 days) showed altered development in multiple organ systems. If the pregnancy was allowed to go to full term, the offspring had low birth weight, increased postnatal growth, or adult-onset high blood pressure.

Obesity is associated with numerous diseases in later life, including type 2 diabetes and heart disease. Although much obesity in children and adults is due to overeating and lack of exercise, early developmental nutritional experience and genetic background can contribute. Human fatty tissue comprises some 40 billion adipose cells, with most of the bulk stashed under the skin, and obesity represents both greater numbers of adipose cells compared with lean people and excessive deposition of fat in these cells, which increases their size. Humans are born with a certain number of adipose cells, with females generally having more than males. The number of adipose cells increases throughout late childhood and early puberty, and after this remains fairly constant. However, the number of adipose cells increases more rapidly in genuinely obese children than in lean children, meaning that they end up with more adipose cells. Once fat cells develop in the body, they remain there for life and they seldom die, although there is some turnover: each month about 1% of adipose cells in the human body die and are replaced. Thus, adult obesity is often linked to childhood obesity.

> **SUMMARY**
>
> Growth in animals mainly occurs after the basic body plan has been laid down and the organs are still very small. Final organ size can be controlled both by external signals and by intrinsic growth programs. In some cases, organ size can be determined by monitoring the dimensions of the growing organ rather than being set by cell number or cell size; in others, organ size is regulated via negative feedback of circulating factors. In animals, growth can occur by cell multiplication, cell enlargement, and secretion of large amounts of extracellular matrix. The TOR and Hippo signaling pathways integrate multiple inputs from the tissue environment to modulate proliferation and apoptosis of excess unwanted cells. Cancer is the result of the loss of growth control and differentiation. In mammals, insulin-like growth factors are required for normal embryonic growth, and they also mediate the effects of growth hormone after birth. Human postnatal growth is largely controlled by growth hormone, which is made in the pituitary gland. Growth in the long bones is both intrinsically programmed and controlled by local factors, and, in addition, occurs in response to growth hormone stimulation of the cartilaginous growth plates at either end of the bone. Catch-up growth is controlled by extrinsic factors rather than a cell-autonomous program.

Molting and metamorphosis

Many animals do not develop directly from an embryo into an 'adult' form, but into a larva from which the adult eventually develops by metamorphosis. The changes that occur at metamorphosis can be rapid and dramatic, the classic examples being the metamorphosis of a caterpillar into a butterfly, a maggot into a fly, and a tadpole into a frog. Another striking example of metamorphosis is the transformation of the pluteus larva of the sea urchin into the adult. In some cases, it is hard to see any resemblance between the animal before and after metamorphosis. The adult fly does not resemble the larva at all, because adult structures develop from the imaginal discs and so are completely absent from the larval stages (see Chapters 2 and 9). In frogs, the most obvious external changes at metamorphosis are the regression of the tadpole's tail and the development of limbs, although many other structural changes occur. In arthropods and nematodes, increase in size in larval and pre-adult stages requires shedding of the external cuticle, which is rigid, a process known as **molting**.

Several features distinguish early embryogenesis from molting, metamorphosis, and other aspects of post-embryonic development. Whereas the signal molecules in early development act over a short range and are typically protein growth factors, many signals in post-embryonic development are produced by specialized endocrine cells and include both protein and non-protein hormones. The synthesis of these hormones is orchestrated by the central nervous system in response to environmental cues, and there is complex feedback between the endocrine glands and their secretions.

11.11 Arthropods must molt in order to grow

Arthropods have a rigid outer skeleton, the cuticle, which is secreted by the epidermis. This makes it impossible for the animals to increase in size gradually. Instead, increase in body size takes place in steps, associated with the loss of the old outer skeleton and the deposition of a new larger one. This process is known as **ecdysis** or molting. The stages between molts are known as instars. *Drosophila* larvae have three instars and molts. The increase in overall size between molts can be striking, as illustrated in Figure 11.16 for the tobacco hornworm, the caterpillar of the tobacco hawkmoth (*Manduca*).

At the start of a molt, the epidermis separates from the cuticle in a process known as apolysis, and a fluid (molting fluid) is secreted into the space between the two (Fig. 11.17). The epidermis then increases in area by cell proliferation or cell enlargement and becomes folded. It begins to secrete a new cuticle, and the old cuticle is partly digested away, eventually splits, and is shed.

Molting is under hormonal control. Stretch receptors that monitor body size are activated as the animal grows, and this results in the brain secreting **prothoracicotropic hormone**. This activates the prothoracic gland to release the steroid maturation hormone **ecdysone**, which is the hormone that causes molting. A similar hormonal circuit controls metamorphosis.

11.12 Insect body size is determined by the rate and duration of larval growth

When an insect larva has reached a particular stage, it does not grow and molt any further but undergoes a more radical metamorphosis into the adult form. Larval tissues, such as gut, salivary glands, and certain muscles, undergo programmed cell death. The imaginal discs now develop into rudimentary adult appendages, such as wings, legs, and antennae. The nervous system is also remodeled.

Growth occurs only in the larval stages and therefore the size of the adult is determined by the size the larva reaches before it stops feeding and becomes a **pupa**. In normal circumstances the larva pupates when it reaches a size typical for the species—this is genetically controlled, but if the larval stage is prolonged or shortened experimentally, adults that are, respectively, larger or smaller than usual can be produced. The same hormonal circuit, involving prothoracicotropic hormone, that controls molting also controls pupation and metamorphosis, with a pulse of ecdysone being the universal signal that terminates larval development (Fig. 11.18).

How body size is monitored by insect larvae and how the timing of the ecdysone signal is determined are still not entirely understood, and the mechanisms appear different in different groups of insects. Various factors, including the levels of nutrition, temperature, and light, can regulate the synthesis of ecdysone by acting on the neurosecretory cells in the brain to control the release of prothoracicotropic

Fig. 11.16 Growth and molting of the caterpillar of the tobacco hawkmoth (*Manduca sexta*). The caterpillar, known as the tobacco hornworm, goes through a series of molts. The tiny hatchling (1)—indicated by the arrow—molts to become a caterpillar (2), and then undergoes three further molts (3, 4, and 5). The increase in size between molts is about twofold. The caterpillars are sitting on a lump of caterpillar food. Scale bar = 1 cm.
Photograph courtesy of S.E. Reynolds.

Fig. 11.17 Molting and growth of the epidermis in arthropods. The cuticle is secreted by the epidermis. At the start of molting, the cuticle separates from the epidermis—apolysis—and a fluid is secreted between them. The epidermis grows, becomes folded, and begins to secrete a new cuticle. Enzymes weaken the old cuticle, which is shed.

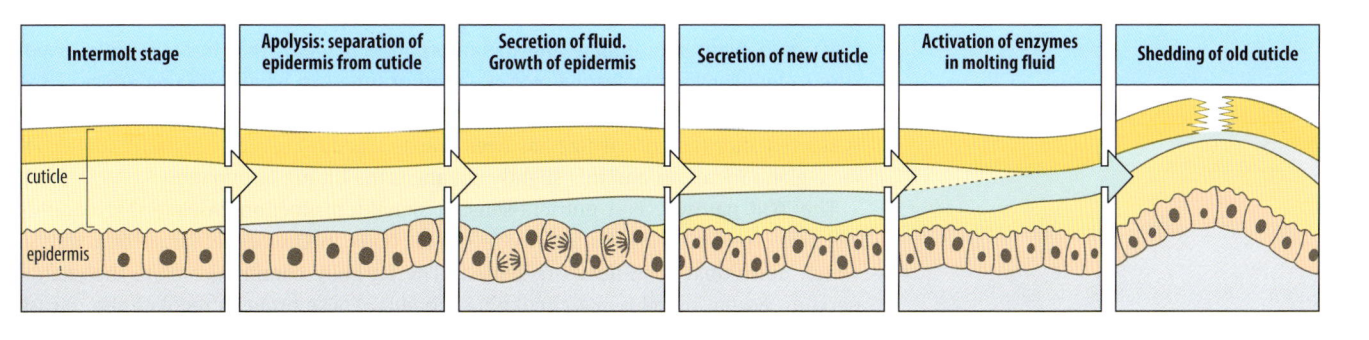

| Intermolt stage | Apolysis: separation of epidermis from cuticle | Secretion of fluid. Growth of epidermis | Secretion of new cuticle | Activation of enzymes in molting fluid | Shedding of old cuticle |

Fig. 11.18 Insect metamorphosis.
The corpora allata of a butterfly larva secrete juvenile hormone, which inhibits metamorphosis. In response to environmental changes, such as an increase in light and temperature, the brain of the final instar larva begins to produce prothoracicotropic hormone (PTTH), which is released from the corpora allata. This acts on the prothoracic gland to stimulate the secretion of ecdysone, the hormone that overcomes the inhibition by juvenile hormone and causes metamorphosis.

*Illustration after Tata, J.R.: **Gene expression during metamorphosis: an ideal model for post-embryonic development.** BioEssays 1993, **15**: 239-248.*

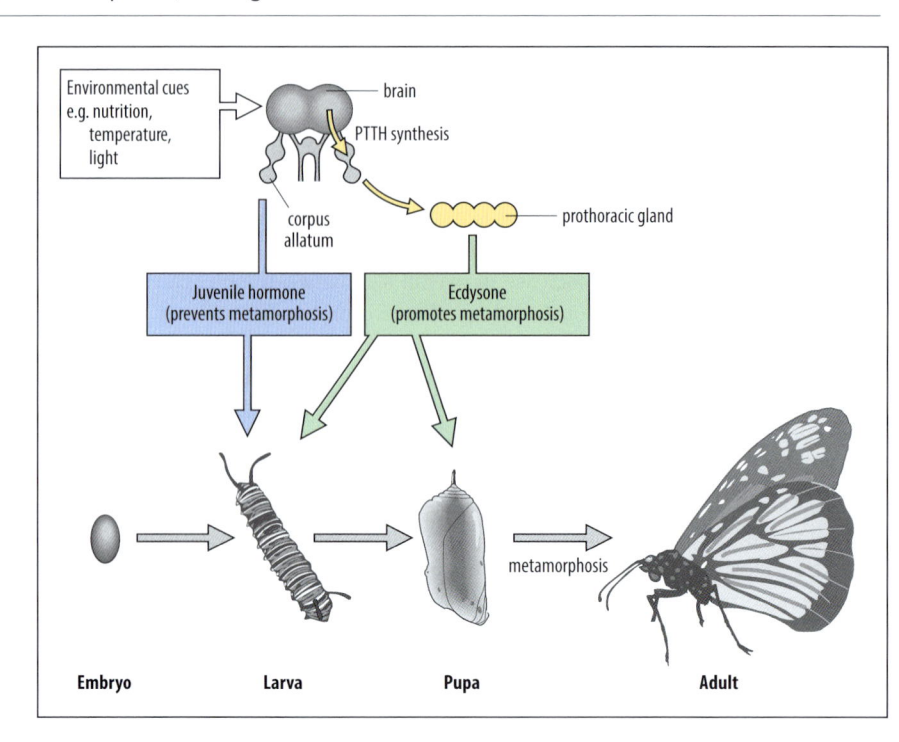

hormone. In *Drosophila*, for example, nutrition must be adequate for the larva to grow to a minimum critical size, as otherwise it will not survive metamorphosis. In normal circumstances this minimum size is attained about half-way through the last instar, coinciding with a pulse of ecdysone, and the larva continues to feed and grow throughout the terminal growth period until pupation (Fig. 11.19). If the larva is starved after it reaches the critical size, it cannot grow during the terminal growth period. Under these circumstances, larvae still undergo metamorphosis but produce smaller adults than normal. However, if the larvae are kept on a low-nutrient diet throughout the third instar, the terminal growth period is extended and pupation is delayed, enabling the larvae to grow to nearly normal size (see Fig. 11.19).

The larva of the *Manduca* moth also attains the minimum critical size about half-way through the final instar. In this case, however, as in other butterflies and moths, the pulse of ecdysone at the critical size is triggered by a drop in the level of another hormone—**juvenile hormone**—produced by the corpora allata, a pair of endocrine glands located just behind the brain. As its name implies, juvenile hormone maintains the larval state and inhibits metamorphosis. There is also another difference between *Drosophila* and *Manduca*, in that the length of terminal growth phase in *Manduca* is constant irrespective of nutrient levels. Therefore, if nutrients are scarce during the final instar in a *Manduca* larva, a smaller adult moth is produced (see Fig. 11.19).

The delay in pupation in *Drosophila* when nutrients are low throughout the final instar is accompanied by downregulation of the TOR pathway in the prothoracic gland, and this has the effect of keeping ecdysone production at a low level. As the larva does eventually pupate and metamorphose, it seems that in *Drosophila*, exposure to low levels of ecdysone over a period of time can have the same effect as the distinct pulse of ecdysone that is the usual signal for pupation. By contrast, if the larva is starved after the critical size has been reached, ecdysone production in the prothoracic gland increases and this slightly advances the timing of pupation.

The TOR pathway is a nutrient-sensing signal-transduction cascade that controls cell size (see Fig. 11.3). Its activity in the **fat body** in *Drosophila* larvae is a way in which nutritional conditions are sensed. The fat body communicates to the central nervous system by releasing circulating proteins. These proteins regulate the release of peptides from neurosecretory cells in the brain that resemble insulin/insulin-like

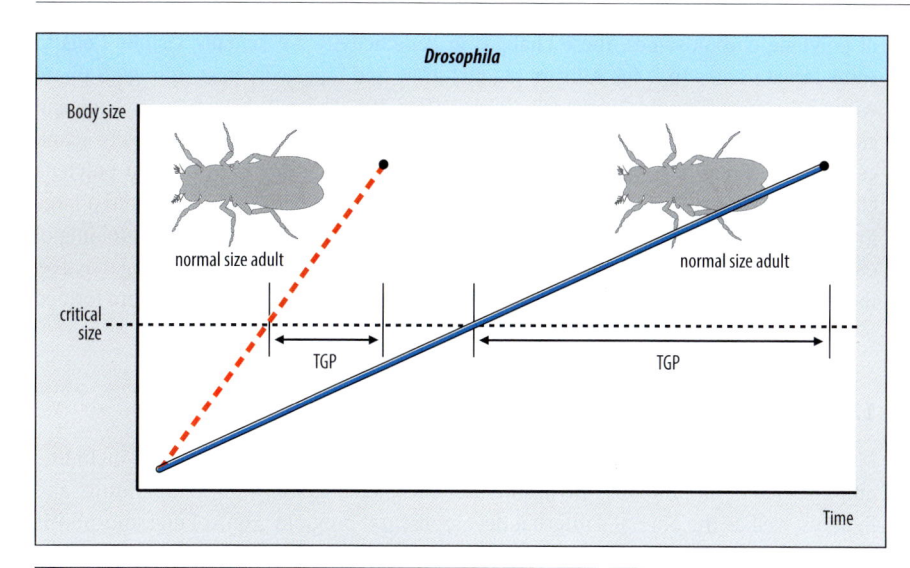

Fig. 11.19 Nutrition-dependent regulation of insect body size. The level of nutrition available regulates adult body size in different ways in different insects. Upper panel: in *Drosophila*, once the larva has reached a critical size, if nutrients are scarce the subsequent time the larva spends feeding (the terminal growth period; TGP) is simply prolonged until the larva reaches a normal size. It then undergoes metamorphosis, producing a normal-sized adult. Lower panel: in the tobacco hornworm *Manduca* (in which the adult is a moth), the larval terminal growth period (after reaching a critical size) remains the same whether nutrients are plentiful or scarce. This means that when nutrients are restricted a smaller adult is produced.

From Nijhout, H.F.: *Size matters (but so does time), and it's OK to be different.* Dev. Cell 2008, **15**: 491-492.

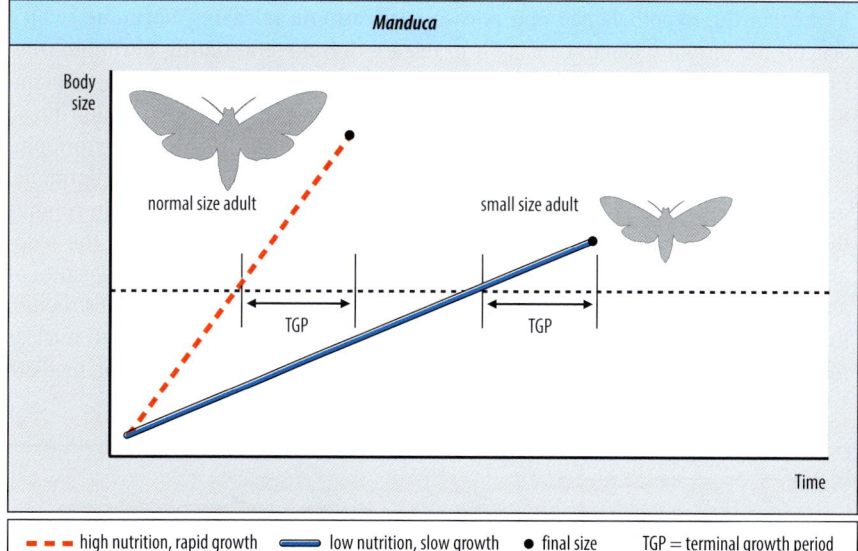

growth factors and act as circulating hormones. These peptides not only activate the insulin/TOR signaling pathways in the prothoracic gland that regulate the synthesis of ecdysone, they are also potent activators of growth, and their signaling pathway strongly resembles the vertebrate insulin/IGF signaling pathway. Thus, an equivalent major hormonal pathway controls growth in both *Drosophila* and vertebrates.

Another pathway acts in parallel with the TOR pathway in the fat body to control larval growth in *Drosophila*. This pathway is activated by the increase in ecdysone when the larva is starved after the critical size is reached. Ecdysone signaling in the fat body leads to increased production of an insulin-like growth-factor-binding protein, which binds to and inactivates the insulin-like peptides. This has the effect of reducing growth and explains why the adults are smaller under these nutrient conditions.

Ecdysone crosses the plasma membrane, where it interacts with intracellular ecdysone receptors that belong to the steroid hormone receptor superfamily. These receptors are transcription factors that belong to the class of nuclear receptors, which are activated upon binding their hormone ligand, similarly to the receptors for retinoic acid (see Box 4E). The hormone–receptor complex binds to the regulatory regions of a number of target genes, inducing a new pattern of gene activity characteristic of metamorphosis.

There are alterations in the expression of many genes—several hundred at least—during *Drosophila* metamorphosis. In *Drosophila*, because of the special characteristics

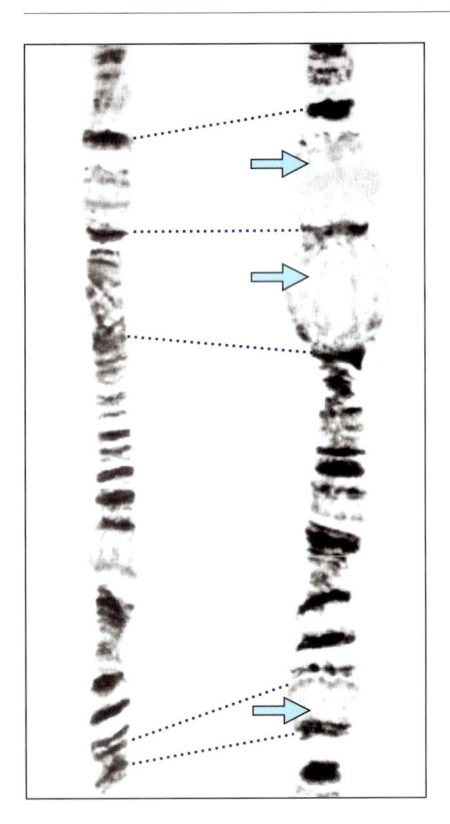

Fig. 11.20 Gene activity seen as puffs on the polytene chromosomes of *Drosophila*. A region of a chromosome is shown from a young third instar (left), and from an older larva (right), after ecdysone has induced puffs at three loci (arrowed).

Photograph courtesy of M. Ashburner.

of polytene chromosomes, these changes in gene activity are actually visible. Cells in some larval tissues (e.g. the salivary glands) grow and repeatedly pass through S phase without undergoing mitosis and cell division (Box 1B). The cells become very large and can have several thousand times the normal complement of DNA. In salivary gland cells, many copies of each chromosome are packed side by side to form giant polytene chromosomes. When a gene is active, the chromosome at that site expands into a large localized 'puff', which is easily visible (Fig. 11.20). The puff represents the unfolding of chromatin and the associated transcriptional activity. When the gene is no longer active, the puff disappears. During the last days of larval life, a large number of puffs are formed in a precise sequence, a pattern that is under the direct influence of ecdysone.

11.13 Metamorphosis in amphibians is under hormonal control

Metamorphosis occurs in many animal groups other than arthropods, including amphibians. In amphibians, environmental cues, such as nutrition, temperature, and light, as well as the animal's internal developmental program, control metamorphosis through their effects on neurosecretory cells in the brain. The neurosecretory cells are located in the hypothalamus and release **corticotropin-releasing hormone**, which acts on the pituitary gland, causing it to release **thyroid-stimulating hormone** (thyrotropin). (This action of corticotropin-releasing hormone is peculiar to non-mammalian vertebrates, and in *Xenopus* only occurs at the tadpole stage. In adult frogs, as in mammals, the release of thyroid-stimulating hormone is due to thyrotropin-releasing hormone produced by the hypothalamus.) Thyroid-stimulating hormone, in turn, acts on the thyroid gland to stimulate the secretion of the **thyroid hormones** that bring about metamorphosis (Fig. 11.21). The thyroid hormones are the iodo-amino acids thyroxine (T_4) and tri-iodothyronine (T_3), which are signal molecules of ancient origin, occurring even in plants. Although very different in chemical structure to ecdysone, they, too, pass through the plasma membrane and interact with nuclear receptors that belong to the same superfamily of nuclear receptors that mediate

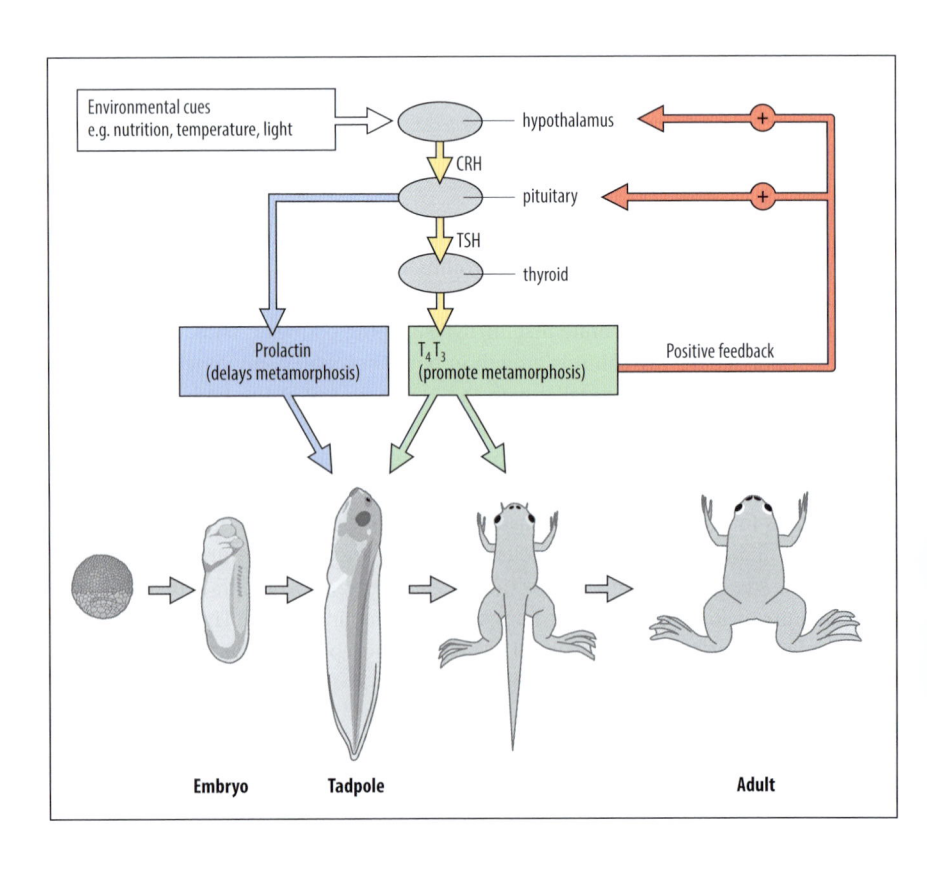

Fig. 11.21 Amphibian metamorphosis. Changes in the environment, such as an increase in nutritional levels, cause the secretion of corticotropin-releasing hormone (CRH) from the larval hypothalamus, which acts on the pituitary to release thyroid-stimulating hormone (TSH). This, in turn, acts on the thyroid glands to stimulate secretion of the thyroid hormones thyroxine (T_4) and tri-iodothyronine (T_3), which cause metamorphosis. The thyroid hormones also act on the hypothalamus and pituitary to maintain synthesis of CRH and TSH. Prolactin is also produced by the pituitary and was traditionally thought to delay metamorphosis.

*Illustration after Tata, J.R.: **Gene expression during metamorphosis: an ideal model for post-embryonic development.** BioEssays 1993, **15**: 239–248.*

ecdysone signaling. The pituitary also produces the protein hormone prolactin, which was originally thought to be an inhibitor of metamorphosis; however, overexpression of prolactin does not prolong tadpole life but reduces tail resorption.

A striking feature of the hormones that stimulate metamorphosis is that, in addition to affecting a wide range of tissues, they affect different tissues in different ways, their effects varying from the subtle to the gross. In the tadpole limb, for example, thyroid hormones promote development and growth, whereas they cause cell death and degeneration in the tail. A specific class of skeletal muscles, the fast muscles are the first to go, and later the notochord collapses. Yet in all these cases, the hormone produces these very different effects by binding to the same nuclear receptor. The difference in outcomes is due to the hormone–receptor complex switching on or off the expression of different sets of target genes in different tissues, as a result of their different developmental histories. Each tissue has its own response to the hormones that cause metamorphosis, and some of these effects can be reproduced in culture. When excised *Xenopus* tadpole tails are cultured, exposure to thyroid hormones causes cell death and complete tissue regression. Metamorphosis also leads to changes in the responsiveness of cells to other signals; for example, in *Xenopus*, estrogen can only induce the synthesis of vitellogenin, a protein required for the egg yolk, after metamorphosis and the attainment of sexual maturity.

SUMMARY

Arthropod larvae grow by undergoing a series of molts in which the rigid cuticle is shed. Metamorphosis during the post-embryonic period can result in a dramatic change in the form of an organism. In insects, it is hard to see any resemblance between the animal before and after metamorphosis, whereas in amphibians, the change is somewhat less dramatic. Environmental and hormonal factors control metamorphosis. Thyroid hormones cause metamorphosis in amphibians, and ecdysone does the same in insects. In insects, the size of the adult body is determined by the size of the larva when it begins metamorphosis. In *Drosophila*, gene activity during metamorphosis can be monitored by localized puffing on the giant polytene chromosomes.

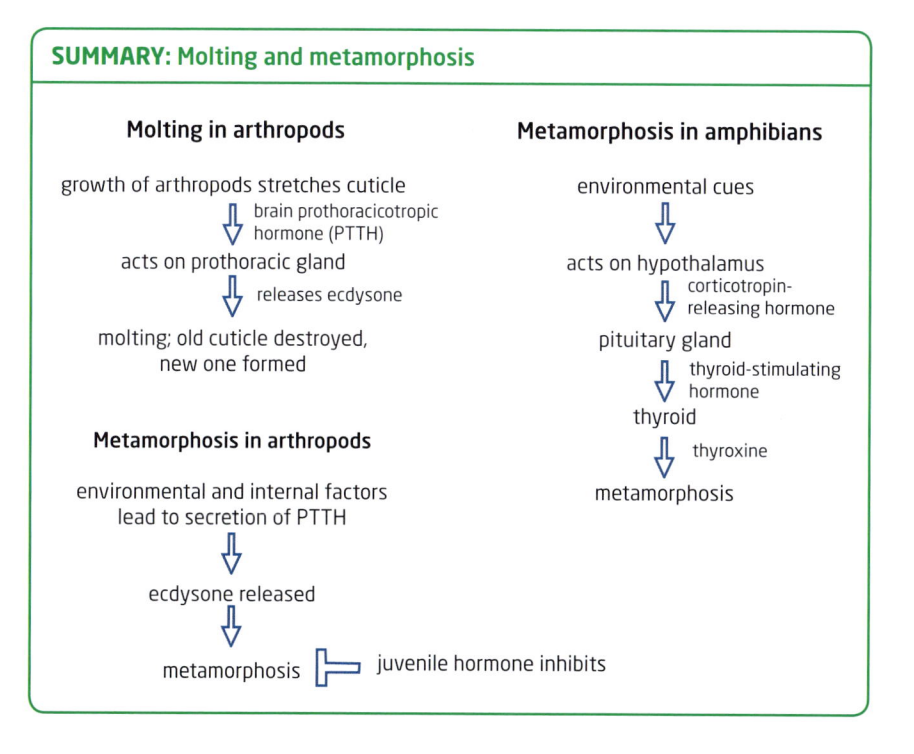

SUMMARY: Molting and metamorphosis

Molting in arthropods

growth of arthropods stretches cuticle
⇩ brain prothoracicotropic hormone (PTTH)
acts on prothoracic gland
⇩ releases ecdysone
molting; old cuticle destroyed, new one formed

Metamorphosis in arthropods

environmental and internal factors lead to secretion of PTTH
⇩
ecdysone released
⇩
metamorphosis ⊣ juvenile hormone inhibits

Metamorphosis in amphibians

environmental cues
⇩
acts on hypothalamus
⇩ corticotropin-releasing hormone
pituitary gland
⇩ thyroid-stimulating hormone
thyroid
⇩ thyroxine
metamorphosis

Fig. 11.22 The capacity for regeneration in urodele amphibians. The emperor newt can regenerate its dorsal crest (1), limbs (2), retina and lens (3 and 4), jaw (5), and tail (not shown).

Regeneration

Regeneration is the repair and replacement of missing or damaged parts. As we have seen, many animal embryos can regulate and develop normally after parts are removed (see Chapters 3 and 4), but the ability to regenerate and rebuild missing tissues and organs as an adult is much rarer. By contrast, plants can regenerate very well—this is the reason that cuttings, for example, can produce an entire new plant (see Chapter 12).

In land-living animals, growth both in the embryo and during the postnatal period determines the final size of the adult body and the shape and size of its constituent parts. Once the adult size has been attained, growth stops. But, as we have seen, some organs, such as the liver in mammals, can regenerate even in the adult. But in liver regeneration, the missing parts are not replaced, the remaining parts just get bigger to restore the mass of tissue. In contrast, some adult insects and other arthropods, and even some vertebrates, can regenerate new, fully functional organs. Insects can regenerate lost appendages, such as legs, as can urodele (tailed) amphibians (salamanders) such as newts and the axolotl salamander (the Mexican salamander *Ambystoma mexicanum*). These urodele amphibians have a remarkable capacity for regeneration even as adults, being able to regenerate not only limbs, but also new tails, jaws, the retina and lens of the eye, and some internal tissues (Fig. 11.22). Regeneration of the lens in the newt eye from the pigmented epithelium of the iris is an example of transdifferentiation (see Section 6.16). In contrast to the extensive regenerative ability of adult salamanders, frogs (anuran amphibians such as *Xenopus*) can regenerate at larval stages but generally lose the ability at metamorphosis. Another adult vertebrate that can regenerate new organs is the zebrafish, which can regenerate fins, and even the heart after removal of part of the ventricle. By contrast, mammals cannot regenerate lost limbs, although they have the capacity to replace the very ends of the digits, and are unable to regenerate the heart.

As a rule, the efficiency of wound healing decreases with age, and the same was thought to be true of regeneration. But unexpectedly, it has been shown that regeneration of the newt lens can occur repeatedly, and that its efficiency does not diminish with age. In this long-term experiment, the lens was removed 18 times in the same animal, with successful regeneration still occurring when the animal was 30 years old.

Some invertebrate animals also show great ability to regenerate: small fragments of animals such as starfish, planarians (flatworms), and *Hydra* can give rise to a whole animal (Fig. 11.23). This ability to regenerate a whole animal may be related to the ability of these animals to reproduce asexually; that is, to produce a complete new individual by budding or fission.

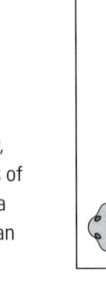

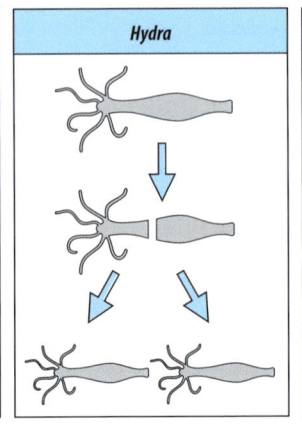

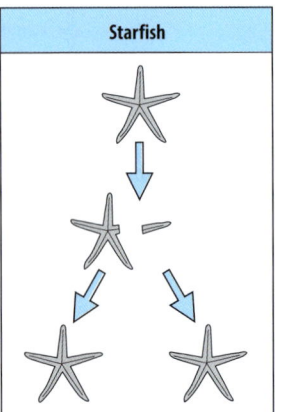

Fig. 11.23 Regeneration in some invertebrate animals. A planarian, *Hydra*, and a starfish all show remarkable powers of regeneration. When parts are removed or a small fragment isolated, a whole animal can be regenerated.

The issue of regeneration raises several major questions. What is the origin of the cells that give rise to the regenerated structures? Do differentiated cells dedifferentiate and start dividing? Are stem cells involved in regeneration? What mechanisms pattern the regenerated tissue and how are these related to the patterning processes that occur in embryonic development?

We will first look at regeneration in *Hydra* and planarians and then focus on regeneration of limbs in urodele amphibians and insects. We will also look briefly at regeneration of heart muscle in zebrafish. Understanding regeneration in these systems could help in the development of medical ways of repairing tissues such as those of the mammalian heart.

11.14 Regeneration involves repatterning of existing tissues and/or growth of new tissues

Regeneration requires, at the minimum, the production of a population of cells whose growth, differentiation, and patterning are regulated. A distinction has been drawn between two types of regeneration. In one—**morphallaxis**—there is little new cell division and growth, and regeneration of structure occurs mainly by the repatterning of existing tissue and the re-establishment of boundaries. Regeneration of the whole body in the freshwater cnidarian *Hydra* is a classic example of morphallaxis, even though the body column of *Hydra* is rich in adult stem cells ready to divide and differentiate after injury. By contrast, regeneration of a newt limb depends on the growth of completely new, correctly patterned structures, and this is known as **epimorphosis**. Both types of regeneration can be illustrated with reference to the French flag pattern (Fig. 11.24). In morphallaxis, new boundary regions are first established and then new positional values are specified in relation to them; in epimorphosis, new positional values are linked to growth from the cut surface. In planarians, a population of adult stem cells that exists throughout the body produces the regenerated tissue (Box 11C), and regeneration in this organism involves both morphallaxis and epimorphosis.

11.15 Regeneration in *Hydra* involves repatterning of existing tissues

The remarkable ability of the small freshwater cnidarian *Hydra* to regenerate the missing region of the body when cut transversely was discovered more than 250 years ago. Growth and repair in *Hydra* relies on adult stem cells, paused in G_2, ready to differentiate or undergo mitosis rapidly after injury. As growth follows patterning, it is a good example of morphallaxis. In morphallaxis, the remaining tissue is repatterned so that the missing region is replaced. Regeneration in *Hydra* has also been an important model for testing the principle of positional information.

Hydra consists of a hollow tubular body about 0.5 cm long. It has distinct apical–basal polarity, with a head at one end and a basal region at the other, by which it sticks to a surface (Fig. 11.25). It grows continuously, in that there is dynamic turnover of cells. Both endoderm and ectoderm cells in the gastric region of the body act as unipotent stem cells and divide every 3–4 days, with the daughter cells getting displaced towards the head or the basal disc where they terminally differentiate and are progressively lost. The multipotent interstitial cells in the gastric region also self-renew but at a faster rate. These three stem-cell lineages are distinct but all the stem cells escape the G_1 phase of the cell cycle and pause in G_2. In the adult animal, the cells are continually changing their relative positions and are forming new structures as they move up or down the body column. Cells are repatterned during this dynamic process, an ability that gives *Hydra* its remarkable capacity for regeneration.

If the body column of a *Hydra* is cut transversely at mid-gastric level, the lower piece regenerates a head and the upper piece a foot. Two small *Hydra* are produced, each half the size of the original. Even more dramatically, a mixture of *Hydra* cells

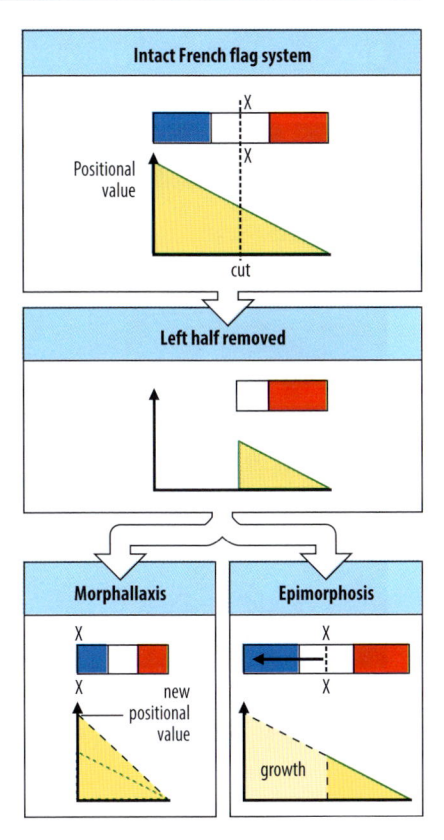

Fig. 11.24 Morphallaxis and epimorphosis. A pattern such as the French flag can be specified by a gradient in positional value. If the system is cut in half, it can regenerate in one of two ways. In regeneration by morphallaxis, a new boundary is established at the cut and the positional values are changed throughout. In regeneration by epimorphosis, new positional values are linked to growth from the cut surface.

BOX 11C Planarian regeneration

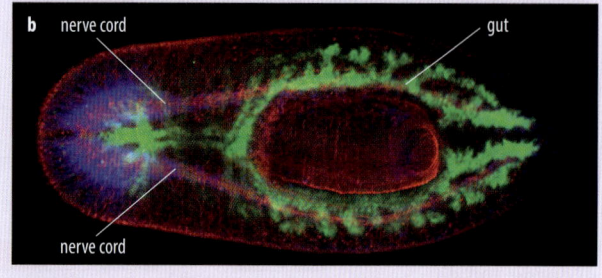

Fig. 1

From (a) Adell, T., et al.: **Gradients in planarian regeneration and homeostasis.** *CSH Perspect. 2012, **2**: a000505. (b) Rink, J.C., Gurley, K.A., Elliott, S.A., Sánchez Alvarado, A.:* **Planarian Hh signaling regulates regeneration polarity and links Hh pathway evolution to cilia.** *Science 2009, **326**: 1406–1410.*

Planarians are free-living flatworms with three body layers—ectoderm, mesoderm, and endoderm—but no coelom. They have a complex anatomy with various organ systems, including a nervous system, kidneys, reproductive organs, and gut. Some species of planarians are unique among bilaterian animals in having a large pool of adult stem cells that enables them to regenerate any part of the body, including the brain. Planarian regeneration is unparalleled among animals. When one of these animals is cut into many small pieces, each piece will regenerate and give rise to a complete new animal. The minimum size of fragment that can regenerate is 1/279th (0.3%) of an adult animal—about 1000 cells.

The species most used for studying regeneration is the freshwater planarian *Schmidtea mediterranea*, which is about 0.1–2 cm long (Fig. 1). The body is bilaterally symmetrical and has distinct antero-posterior polarity, with a well-developed head at one end and a tail at the other, and also distinct dorso-ventral polarity, with, for example, two eye spots on the dorsal side of the head (see Fig. 1a). The mouth is in the middle of the ventral surface, through which a muscular pharynx protrudes during feeding. Figure 1b shows a ventral view with the gut (green), the pharynx in the center, and the nerve cords (red) coming from the head end.

Adult planarians constantly replace their tissues. In planarians, the source of new cells is a population of small, undifferentiated cells known as **neoblasts**, which are the only cells in the adult animal that are mitotically active. Neoblasts constitute about 25–30% of the total number of cells and are scattered throughout the body. They not only self-renew, but can also differentiate into different cell types—epidermis, muscle, neurons, and germ cells.

Both morphallaxis and epimorphosis (see Section 11.14) are involved in planarian regeneration. Morphallaxis is clearly seen by the remodeling of fragments of the body that regenerate to form a complete small animal. But heavily irradiated planarians are unable to regenerate because regeneration of the head, for example, following amputation, involves epimorphosis, with the early proliferation and migration of neoblasts to form a blastema. The blastema resembles that of a urodele limb blastema in that it consists of a mass of undifferentiated cells covered by a specialized wound epithelium and, as in urodele limbs, the wound epithelium is required for blastema formation. The head blastema provides at least some of the replacement parts needed for the new head, including the brain, which forms over the next week or so (Fig. 2). Neoblasts are particularly sensitive to radiation and this explains why irradiated planarians are unable to regenerate.

A key issue in planarian regeneration is whether the neoblasts are pluripotent and able to give rise to all the cell types in the regenerated tissue, or whether the population of neoblasts is heterogeneous, with individual cells being lineage restricted. This question can be addressed by transplanting a single neoblast cell into an adult planarian that has been irradiated to destroy its own population of neoblasts and following the fate of the transplanted cell's progeny. Some transplanted neoblasts were indeed found to be pluripotent and produced clones of neoblasts

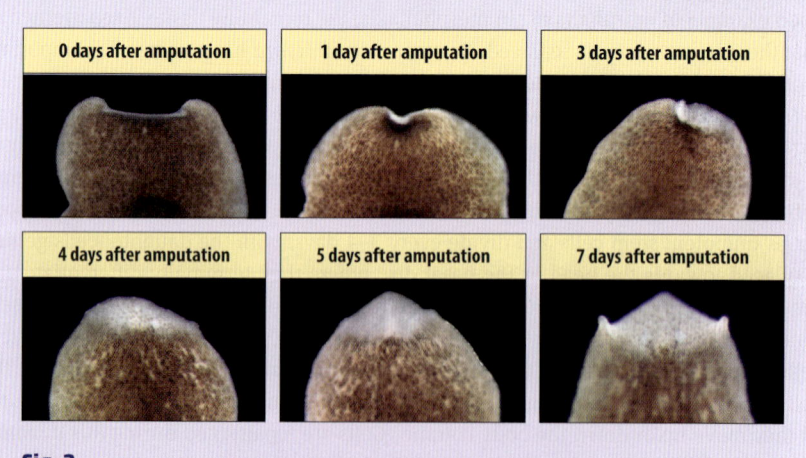

Fig. 2

From Newmark, P.A., Sánchez-Alvarado, A.: **Regeneration in planaria.** *Encyclopedia of Life Sciences. Wiley, 2001.*

that differentiated to form neurons, intestinal cells, and other cell types. Furthermore, a single neoblast cell can rescue a lethally irradiated animal. Not all neoblasts have this ability, however, and some neoblasts appear to be lineage restricted. For example, during eye regeneration, specialized neoblasts give rise to pigment cells and other specialized neoblasts give rise to neurons (Fig. 3).

Single-cell RNA sequencing has recently characterized the neoblast cell population in planarians. The population is heterogeneous with a pluripotent subpopulation. Single cells from this subpopulation transplanted into irradiated hosts are much more efficient in rescuing lethally irradiated planarian hosts than single cells taken from the whole neoblast population. The identification of pluripotent neoblasts will enable future studies to be focused on these cells and their behavior during regeneration.

The planarian body has intrinsic polarity. Antero-posterior polarity is maintained even in small fragments during regeneration, so that a head regenerates at an anterior-facing cut surface whereas a tail regenerates at the posterior cut surface. Around the beginning of the twentieth century, it was suggested that this polarity could be based on opposing gradients of 'head stuffs' and 'tail stuffs'. More than 100 years later, a gradient of Wnt signaling that is critical for maintaining antero-posterior polarity was discovered.

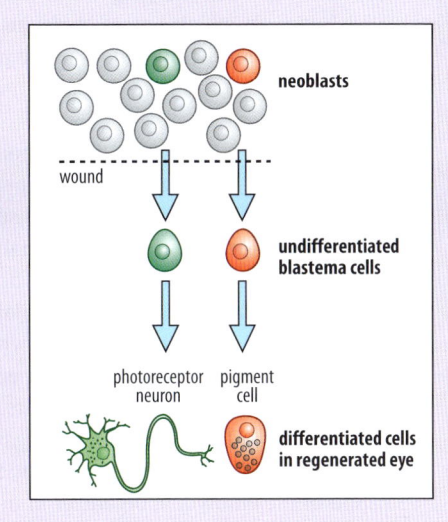

Fig. 3
From Reddien, P.W.: *Specialized progenitors and regeneration.* Development 2013, **140**: 951-957.

A planarian Wnt signal protein (Wnt-P1) is produced in a small number of cells in the tail in the adult animal and diffuses anteriorly to form a gradient. Wnt signaling suppresses head formation along this gradient except at the anterior end where it is weakest. When Wnt signaling is reduced in an uninjured planarian by knocking down β-catenin expression for an extended period of time, ectopic heads emerge all over the body. In a regenerating planarian, knockdown of β-catenin expression results in a head instead of a tail developing from a posterior-facing cut surface, giving a two-headed animal. When Wnt signaling was reduced in non-regenerative species of planarians, they were able to regenerate a fully functional head.

Planarians can also regenerate the missing lateral half of the body if cut in half longitudinally. BMP-4, is normally expressed in a stripe along the dorsal midline and controls dorso-ventral pattern of the body. Bmp signaling is not only required for medial-lateral regeneration but also more generally for blastema formation. All in all, the remarkable ability of adult planarians to regenerate seems to depend on using the signals that normally maintain tissue homeostasis.

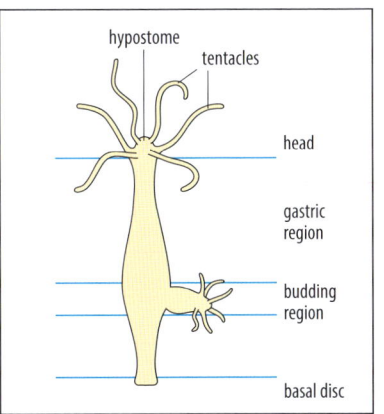

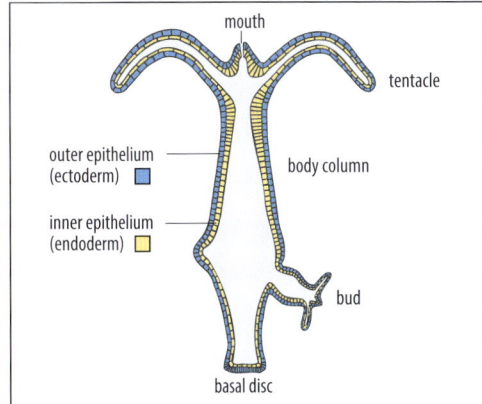

Fig. 11.25 The anatomy of *Hydra*. *Hydra* has distinct apical-basal polarity, with a head at one end and a basal region at the other. The head consists of a conical hypostome with an opening from the gut surrounded by a number of tentacles. Unlike most of the animals discussed in this book, which have three germ layers, *Hydra* has only two. The body wall is composed of an outer epithelium, which corresponds to the ectoderm, and an inner epithelium lining the gut cavity, which corresponds to the endoderm. These two layers are separated by a gel-like basement membrane or mesoglea. There is also a population of adult multipotent stem cells, called interstitial cells, which give rise to a dozen different specialized cell types, including neurons, secretory cells, and the stinging cells, or nematocytes, that are used to capture prey. *Hydra* has no central nervous system—although an apical nerve ring is clearly visible in some species—but has a network of neurons distributed throughout the body with a much higher density at the apical and basal poles.

Photograph reproduced with permission from W.A. Müller.

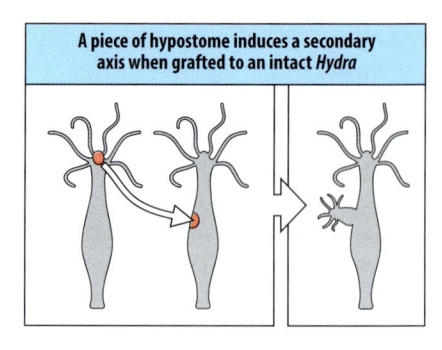

Fig. 11.26 A piece of hypostome induces a secondary axis when grafted to an intact *Hydra*. When a small fragment of the hypostome region is grafted into the gastric region of another *Hydra*, a new body axis is induced complete with head and tentacles. This experiment showed the existence of an organizing region at the head end.

obtained by tissue disaggregation will reaggregate to reform an animal. Heavily irradiated *Hydra*, in which no cell divisions occur, can regenerate more or less normally, showing that regeneration does not require growth. The ectoderm and endoderm cells are the main contributors to regeneration, as *Hydra* lacking interstitial cells can still regenerate.

Hydra has a well-defined overall polarity and this determines that the appropriate structure is regenerated. A head regenerates at an apical-facing cut surface, whereas a foot regenerates at a basal-facing cut surface. The basis for this polarity was revealed by grafting experiments. When a small fragment of the hypostome region is grafted into the gastric region of another *Hydra*, a new body axis is induced complete with head and tentacles (Fig. 11.26). Similarly, transplantation of a fragment of the basal region induces a new body column with a basal disc at its end. *Hydra* therefore has two organizing regions, one at each end, which give the animal its overall polarity. The first of these grafting experiments, performed more than 100 years ago by Ethel Browne, led to the concept of an organizer. There is evidence that her experiments influenced the famous experiments by Spemann and Mangold that led to the identification of the embryonic organizer in newt embryos (see Section 1.4).

Additional grafting experiments revealed that the hypostome is also the source of a long-range diffusible inhibitor of head formation, which prevents the inappropriate formation of extra heads by lateral inhibition (see Section 1.16). The basal disc is similarly thought to produce an inhibitor of foot formation. The organizing regions at the ends of the body therefore each produce a pair of signals, which form opposing gradients along the body column: one signal from each organizer specifies a gradient in positional values and the other signal gradient inhibits formation of either a head or a foot.

A simple model for head regeneration in *Hydra* is based on the pair of signal gradients produced by the head organizer (Fig. 11.27). Both gradients are proposed to be linear, their values decreasing at a constant rate with distance from the head. The model proposes that, in the intact animal, provided the level of inhibitor (indicated by the gradient labeled I) at every point along the body is greater than the threshold set by the positional value at those points (indicated by the gradient labeled P), head formation is inhibited. When the head is removed, however, the concentration of inhibitor falls at the cut surface, and when the inhibitor falls below the threshold set

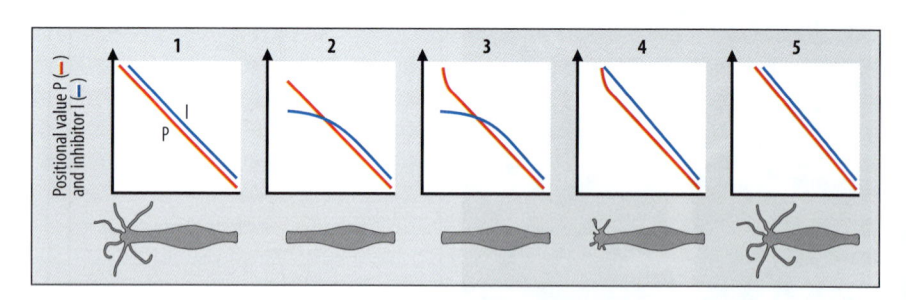

Fig. 11.27 A model for head regeneration in *Hydra*. The model is based on a pair of signal gradients produced by the head organizer whose values decrease in a linear fashion with distance from the head. One (red line) sets the positional values (P), while the other (blue line) indicates levels of an inhibitor of head formation (I). In the intact animal (1), provided the level of inhibitor at every point along the body is greater than the threshold set by the positional value at those points, head formation is inhibited. When the head is removed, however, the concentration of inhibitor falls at the cut surface (2), and when the inhibitor falls below the threshold set by the positional value at that point, the positional value of cells at the cut surface increases to that of a normal head region (3), and a new head region is specified. The new head then starts to make inhibitor and the inhibitory gradient is re-established (4 and 5).

by the positional value at that point, the positional value of cells at the cut surface increases to that of a normal head region. Thus, the first key step is the specification of a new head region at the cut surface. The new head then starts to make inhibitor and the inhibitory gradient is re-established. The gradient in positional value also returns to normal, but this can take more than 24 hours. In an alternative model that can also account for head regeneration, reaction–diffusion (see Box 9E) involving chemical interactions between a short-range head activator and the long-range head inhibitor generates logarithmic signal gradients with a high point at the head region.

Genes homologous with those associated with vertebrate embryonic organizers are expressed in the adult *Hydra* head organizer, suggesting an ancient evolutionary origin for such organizers. The *Wnt3* homolog together with other *Wnt* homologs is expressed at the tip of the hypostome in an adult *Hydra* (Fig. 11.28). Within a couple of hours after amputation of the head region, *Wnt3* is expressed at high levels throughout the cut surface. This local Wnt signaling is involved in setting up the head organizer and is required for regeneration, because in its absence the head does not regenerate.

Wnt3 expression is at low levels in the mid-gastric region of the body column. Bisection of a *Hydra* in this region leads to upregulation of *Wnt3* expression within a few hours in endodermal cells at the head-regenerating surface. This upregulation is in response to Wnt3 signals released by interstitial cells, which undergo apoptosis at the amputation site. When apoptosis is inhibited, endodermal *Wnt3* expression is not induced and regeneration is blocked. Therefore injury-induced apoptosis is needed to induce the head organizer in the body column. This explains why, in transgenic *Hydra* in which Wnt signaling is activated in cells throughout the body, only multiple tentacles—not complete heads—are initially generated all along the body column. The involvement of apoptosis in regeneration may be a general phenomenon. Senescent cells that can subsequently undergo apoptosis are induced in regenerating amphibian limbs although their function is as yet unknown.

Fig. 11.28 Wnt signaling is involved in setting up the head organizer. *Wnt3* (dark blue stain) is expressed at the tip of the hypostome in the adult *Hydra* (left panel). When the head region is amputated, *Wnt3* becomes expressed at the cut surface (right panel). A head will eventually regenerate from the cut surface.

Photographs courtesy of B. Hobmayer and T.W. Holstein, from Hobmayer, B., et al.: **WNT signalling molecules act in axis formation in the dipoblastic metazoan** *Hydra.* Nature 2000, **407**: 186-189.

11.16 Amphibian limb regeneration involves cell dedifferentiation and new growth

Amputation of an adult urodele limb is followed by a rapid migration of epidermal cells from the edges of the wound over the wound surface. The wound epidermis then becomes specialized to form the **apical epidermal cap** or **apical epithelial cap**, which is essential for subsequent regeneration; if formation of the wound epidermis is prevented by suturing the edges of the stump together, there is no regeneration. The specialization of the wound epithelium to form the apical epithelial cap is dependent on the presence of nerves. It has been suggested that the apical epithelial cap may have a similar function to that of the apical ectodermal ridge in embryonic limb bud development (see Section 9.12).

An apical epithelial cap forms after amputation of limbs of *Xenopus* frog larvae at stages when they can regenerate. Single cell transcriptomics show that the transcriptional profile of these apical epithelial cap cells is very similar to the profile of cells of the apical ectodermal ridge of developing *Xenopus* limbs. For example, the gene *Fgf8*, which is expressed in the apical ectodermal ridge and required for embryonic limb bud outgrowth (see Chapter 9), is also expressed in the apical epithelial cap. By contrast, following amputation of *Xenopus* limbs at later regeneration-incompetent stages, only an unspecialized wound epidermis forms. Experiments in which regeneration-competent limbs were co-cultured with regeneration-incompetent limbs showed that under these conditions, the regeneration-competent limbs did not form an apical epithelial cap. This suggests that secreted factors in regeneration-incompetent limbs inhibit formation of an apical epithelial cap.

The apical epithelial cap controls the formation of a mass of undifferentiated cells called the **blastema** underneath it, and this mass of cells gives rise to the regenerated

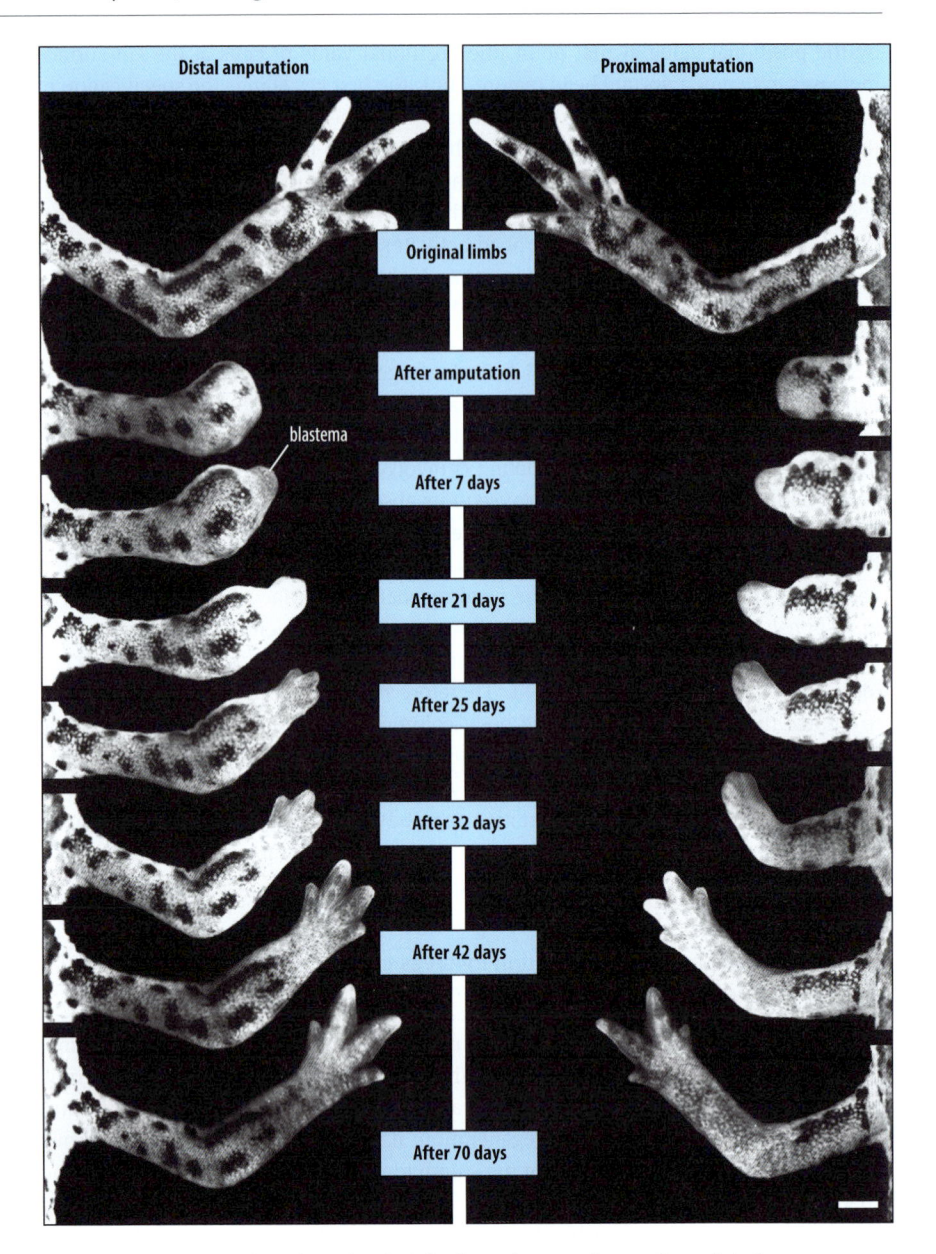

Fig. 11.29 Regeneration of the forelimb in the red-spotted newt *Notophthalmus viridescens*. The left panel shows the regeneration of a forelimb after amputation at a distal (mid-radius/ulna) site. The right panel shows regeneration after amputation at a proximal (mid-humerus) site. At the top, the limbs are shown before amputation. Successive photographs were taken at the times shown after amputation. Note that the blastema gives rise to structures distal to the cut. Scale bar = 1 mm.

limb (Fig. 11.29). Wnt signaling is activated during formation of the blastema. Then the blastema cells start to divide, eventually forming an elongated cone. Over a period of weeks, as the limb regenerates, the blastema cells differentiate into cartilage, muscle, and connective tissue. The undifferentiated cells of the early blastema are derived locally by breakdown of the extracellular matrix of the stump tissues close to the site of amputation. Cues that promote this dedifferentiation may be provided by the novel composition of the extracellular matrix of the blastema. The blastema cells come from the connective tissues, dermis in particular, but also from muscle.

The ability of muscle fibers to give rise to proliferating cells in the blastema is particularly intriguing, as vertebrate skeletal muscle cells do not normally divide after becoming fully differentiated. A general feature of vertebrate muscle differentiation is the withdrawal from the cell cycle prior to fusion of myocytes to produce myotubes (see Section 6.6). This withdrawal involves the dephosphorylation of the cell-cycle control protein retinoblastoma (RB) protein (see Section 11.5). In myotubes of the regenerating newt limb, RB protein is inactivated by phosphorylation. The myotube then fragments to form mononucleate cells that divide. Re-entry into the cell cycle is also associated with the activation of thrombin in the blastema. Indeed, thrombin could be a key factor in regeneration in several systems, as the regeneration of the newt lens from the dorsal margin of the iris also correlates with thrombin activity in that region. Thrombin is a proteolytic enzyme that is more familiar as part of the blood-clotting cascade, but also seems to be involved in providing an environment for dedifferentiation. Multinucleate post-mitotic newt muscle cells can be returned to the cell cycle in culture in the presence of thrombin in the culture medium.

Recently, it has been shown that thrombin and another blood-clotting protease, plasmin, promote newt muscle dedifferentiation and re-entry into the cell cycle by increasing the potency of BMPs, with BMP-4/BMP-7 heterodimers being most potent. Other factors that are extracellularly released upon wounding may also be involved. Dedifferentiation of the newt muscle cells also involves expression of the homeodomain transcription factor Msx1. This multifunctional transcriptional factor is expressed in the region of undifferentiated mesenchyme cells at the tip of developing vertebrate limb buds and also contributes to the homeobox code for tooth identity (see Section 9.25). Rather surprisingly, the dedifferentiation of skeletal muscle to contribute proliferating cells to a limb blastema does not occur in adult axolotls nor in newt larvae before metamorphosis. Instead, cells that express Pax7, functionally similar to the satellite stem cells that regenerate skeletal muscle fibers in adult mammals, contribute to the blastema. It is not clear why the source of muscle cells in the blastema changes in newts after metamorphosis or why newts and axolotls differ, particularly because satellite stem cells are present in adult newt muscle.

One of the key questions in limb regeneration is whether the limb blastema cells are multipotent or whether they are lineage restricted. Can blastema cells derived from muscle, for example, give rise to all the differentiated cell types in the regenerated limb? In other words, can they transdifferentiate? Or do they just give rise to muscle? Although classic work suggested that cells in the amphibian blastema are multipotent, it has now been shown that blastema cells retain a restricted developmental potential related to their origin (Fig. 11.30). Genetic lineage analysis revealed that blastema

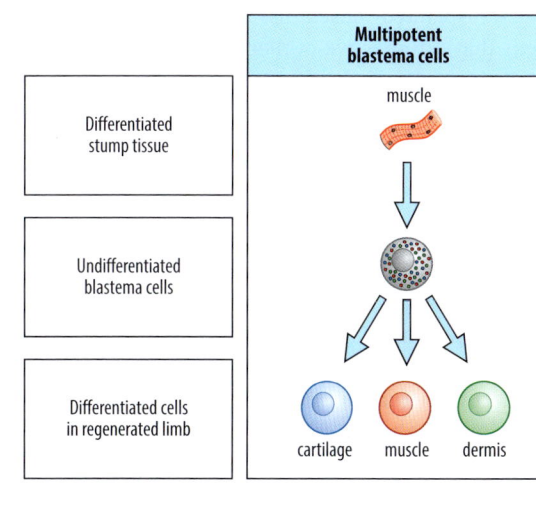

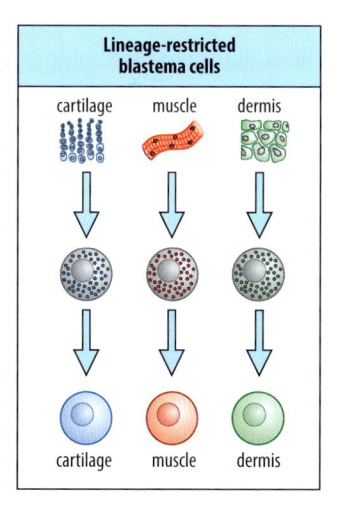

Fig. 11.30 Are blastema cells multipotent or are they lineage restricted? Multipotent blastema cells derived from differentiated tissues would be able to give rise to all the tissues in the limb—for example, cartilage, muscle, and dermis. Lineage-restricted blastema cells would only be able to give rise to their own tissue of origin.

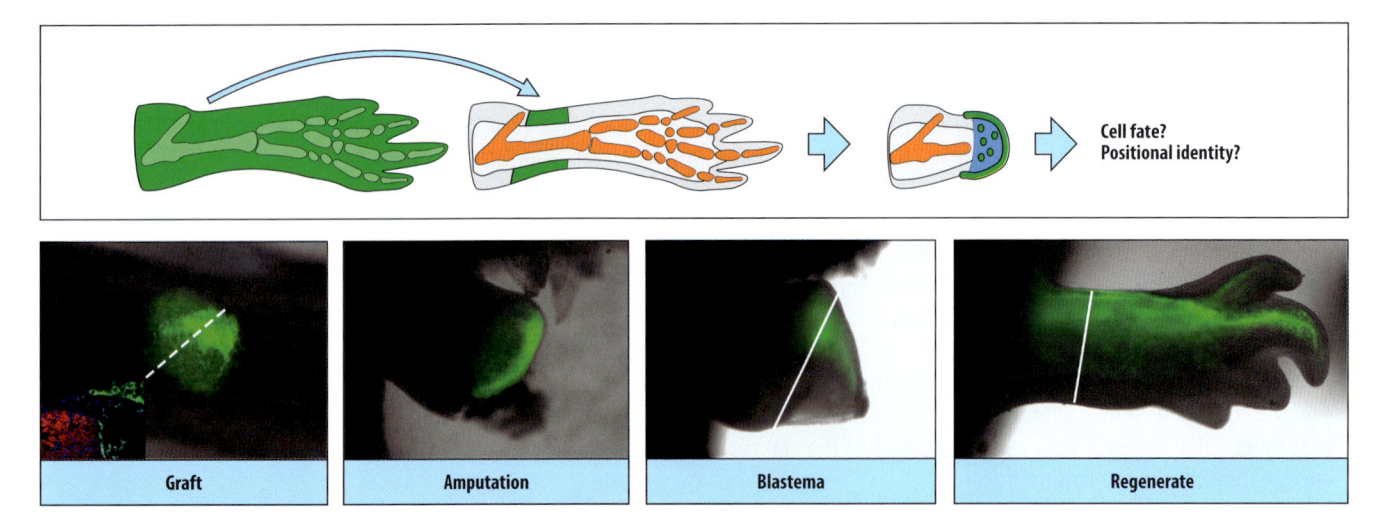

Fig. 11.31 Cells that regenerate the axolotl limb have restricted developmental potential. The experimental procedure is shown in the top row and photographs of the limb at various stages are shown in the bottom row. Cells of a particular tissue type (e.g. connective tissue represented by dermis as shown here) and position (e.g. proximal as shown here) in the limb of an axolotl transgenic for green fluorescent protein were transplanted into the limb of a non-transgenic animal and the limb amputated across the graft. In the regenerating limb that grew from the blastema, the transplanted cells and their progeny can be detected by their green fluorescence. These cells could then be tested for their tissue type. In the case illustrated, the transplanted dermis cells give rise to new connective tissues, dermis, and cartilage but not to muscle. Figure 11.32 shows the lineage restrictions discovered. Cells taken from a proximal position in the limb could give rise to distal structures (as seen here by the labeled cells in the digits).

From Kragl, M., et al.: **Cells keep a memory of their tissue origin during axolotl limb regeneration.** Nature *2009,* **460***: 60-65.*

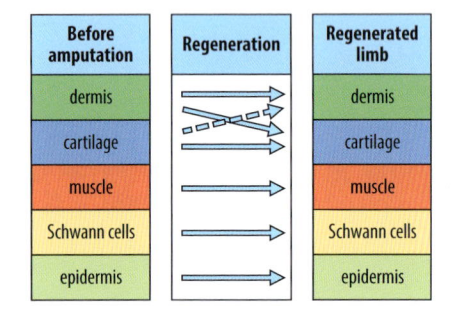

Fig. 11.32 Lineage restriction of cells in the axolotl limb blastema. Blastema cells derived from particular tissues are lineage restricted and only contribute to the same tissue in the regenerated limb. Blastema cells derived from the connective tissues—for example, the dermis—contribute to both dermis and cartilage and there is a small contribution by blastema cells derived from cartilage to the dermis. The same lineage restrictions are also seen in the adult newt limb blastema, although in the newt no contribution of cartilage/bone to the dermis was seen.

cells are lineage restricted. The experiments involved transplanting tissue from lines of transgenic animals expressing a fluorescent protein in all their cells and then following the fate of the transplanted tissue in the limb that regenerated after amputation. A transgenic line of axolotls expressing green fluorescent protein (GFP) or a transgenic line of newts expressing mCherry fluorescent protein were used. A patch of tissue of a particular type, such as muscle, epidermis, or connective tissue—dermis or cartilage/bone, for example—was transplanted into the forelimbs of non-transgenic animals of the same species. The forelimbs of these animals were then amputated across the site of the transplant, and the fate of the glowing green or red transplanted cells could then be traced over time as the limb regenerated (Fig. 11.31). The tissue type made by the fluorescent cells after redifferentiation was then identified by testing for tissue-specific markers.

These experiments showed that transplanted muscle tissue in both axolotl and newt limbs produced blastema cells that, irrespective of their origin, only give rise to muscle, and transplanted connective tissues—dermis and cartilage—only give rise to connective tissue (Fig. 11.32). Epidermal cells are also restricted in their further development, as are the glial Schwann cells, which provide the myelin coating of peripheral nerve fibers. Although connective tissue is restricted in its developmental potential, one type of connective tissue can give rise to several different types of connective tissue. Cells of the dermis could contribute to both new dermis and new cartilage in the skeleton in both axolotl and newt limbs, and cartilage could also contribute to dermis. In normal regeneration, cells derived from the connective tissues make up the mesenchymal component of the early blastema; cells derived from the muscle and Schwann cells enter the blastema at later stages.

The detailed behavior of the connective cell lineage during regeneration has been investigated in a transgenic line of axolotls in which mCherry fluorescent protein is specifically expressed in all the cells of the connective tissues in the limb. Single-cell

RNA sequencing was used to characterize the cells of the connective tissue lineage in amputated limbs at different stages during regeneration. Shortly after amputation, the cells of the various mature connective tissues—dermis, tendon, and ligament—all dedifferentiate to form a relatively homogeneous population of cells (Fig. 11.33). This population of cells is characterized by the expression of genes that encode proteins involved in the inflammatory response to wounding and involved in the breakdown of extracellular matrix. As the blastema forms and begins to grow, the transcriptional profile of the cells comes to resemble the transcriptional profile of early embryonic limb bud cells. Early embryonic limb buds consist of cells derived from the lateral plate mesoderm which give rise to the connective tissues of the limb (see Section 9.10). The genes expressed include those for the 5′ Hoxd transcription factors and the secreted signal protein Sonic hedgehog (Shh) which are involved in growth of the developing limb bud and specifying positional values. Analysis of the differentiation trajectories of individual cells in later blastemas revealed that they are multipotent and can contribute to multiple different types of connective tissue in the regenerated limb. This was confirmed by clonal analysis of cells of limb blastemas at the same stage and of cells of blastemas of regenerating digit tips using transgenic axolotls in which individual cells express different combinations of fluorescent proteins: a single blastemal cell can give rise to cells in skeletal (cartilage/bone) and periskeletal tissues, dermis, and tendon. The close similarity between the transcriptional profiles of the cells of connective tissue cell lineage in the blastema and the limb bud and the multipotency of the connective tissue cell lineage in both regenerating adult limbs and developing limb buds suggests that regeneration of the limb recapitulates embryonic development of the limb.

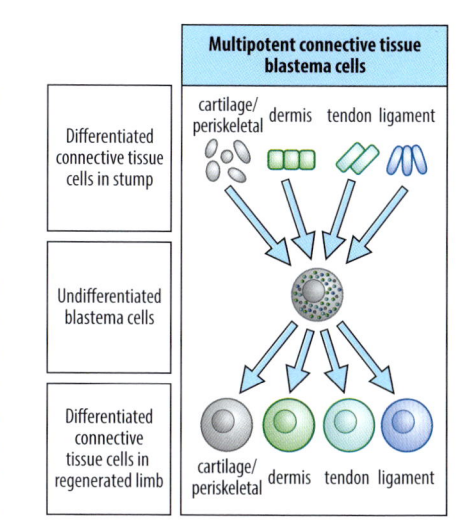

Fig. 11.33 Connective tissue cells give rise to multipotent connective tissue cells in the blastema. Blastema cells that dedifferentiate from the various connective tissues are multipotent and can give rise to all the different connective tissue cell types in the regenerated limb. The multipotent blastema cells resemble the cells of the embryonic limb bud that arise from the lateral plate mesoderm and give rise to all the connective tissues of the limb during development.

11.17 Limb regeneration in amphibians depends on the presence of nerves

Growth and development of the amphibian limb blastema depends on the overlying apical epithelial cap, which develops from the wound epidermis. The development of the apical epithelial cap depends on the presence of nerves in the stump of the amputated limb (Fig. 11.34). This is different from embryonic limb development, in which nerves are not required. In amphibian limbs in which the nerves have been cut before amputation, a blastema forms but fails to grow. The nerves have no influence on the character or pattern of the regenerated structure; it is the amount of innervation—not the type of nerve—that matters. In regenerating limbs of adult newts, the nerves provide a growth factor called newt anterior gradient protein (nAG) which contributes to the regenerative capacity of the blastema. nAG is secreted initially by the glial Schwann cells of the incoming nerves and later by the glandular cells of the wound epidermis. Overexpression of nAG in the blastema of a denervated limb suffices to rescue complete regeneration.

An interesting phenomenon is that if embryonic newt limbs are denervated very early in their development (so-called aneurogenic limbs), and so do not become exposed to the influence of nerves, they can regenerate in the complete absence of any nerve supply (see Fig. 11.34, fourth panel). If an aneurogenic limb is subsequently innervated, however, it rapidly becomes dependent on nerves for regeneration. This suggests that dependence on innervation is imposed on the limb only after the nerves grow into it. It is now known that these phenomena can be attributed to the levels of nAG. During normal limb development, nAG is highly expressed in the epidermis, but its expression is reduced when the nerves grow in and remains low in the adult. In denervated limbs, this reduction in epithelial expression of nAG does not occur, and high levels of nAG persist in the adult, which explains why these limbs can regenerate even though they are not innervated. When an aneurogenic limb is subsequently innervated, nAG is downregulated in the epidermis and thus the limb once more becomes dependent on nerves for regeneration.

Fig. 11.34 Innervation and limb regeneration. Normal limbs require a nerve supply to regenerate (first panels). Limbs denervated before amputation will not regenerate (second panels). However, if the limb is amputated and the cut surface is electroporated with plasmid DNA encoding newt anterior gradient protein (nAG), regeneration is rescued (third panels). Limbs that have never been innervated, because the nerve was removed during development, can regenerate normally in the absence of innervation (fourth panels).

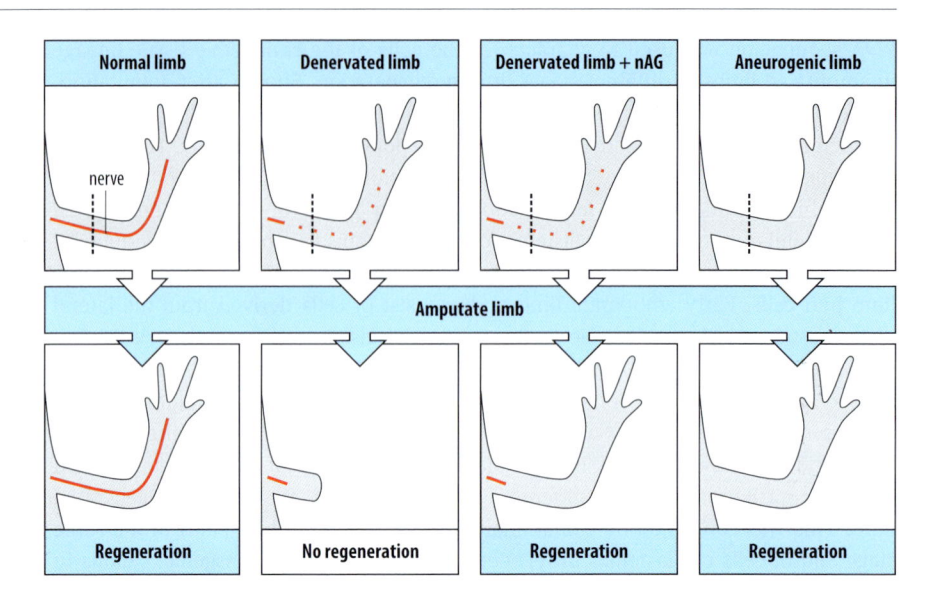

A striking instance of the influence of nerves on regeneration of amphibian limbs is that if a major peripheral nerve, such as the sciatic nerve, is cut and the branch inserted into a wound on a limb or on the surface of the adjacent flank, a supernumerary limb develops at that site. This experimental system provides an opportunity to study limb regeneration in the absence of the considerable cell damage caused by amputation.

11.18 The limb blastema gives rise to structures with positional values distal to the site of amputation

Regeneration always proceeds in a direction distal to the cut surface, enabling replacement of the lost part. If the hand is amputated at the wrist, only the carpals and digits are regenerated, whereas if the limb is amputated through the middle of the humerus, everything distal to the cut (including the distal humerus) is regenerated. Positional value along the proximo-distal axis is therefore of great importance and is at least partly retained in the blastema. The experiments in the axolotl illustrated in Figure 11.31 show that positional values are registered in the connective tissues. Cells from cartilage or dermis transplanted from a proximal position in the original limb give rise to more distal structures in the regenerating limb, whereas cells from these tissues transplanted from distal regions do not give rise to more proximal structures. By contrast, muscle cells do not seem to possess positional values, suggesting that, as

Fig. 11.35 Limb regeneration is always in the distal direction. The distal end of a limb is amputated, and the limb inserted into the belly. Once vascular connections are established, a cut is made through the humerus. Both cut surfaces regenerate the same distal structures even though, in the case of one of the regenerating limbs, distal structures are already present.

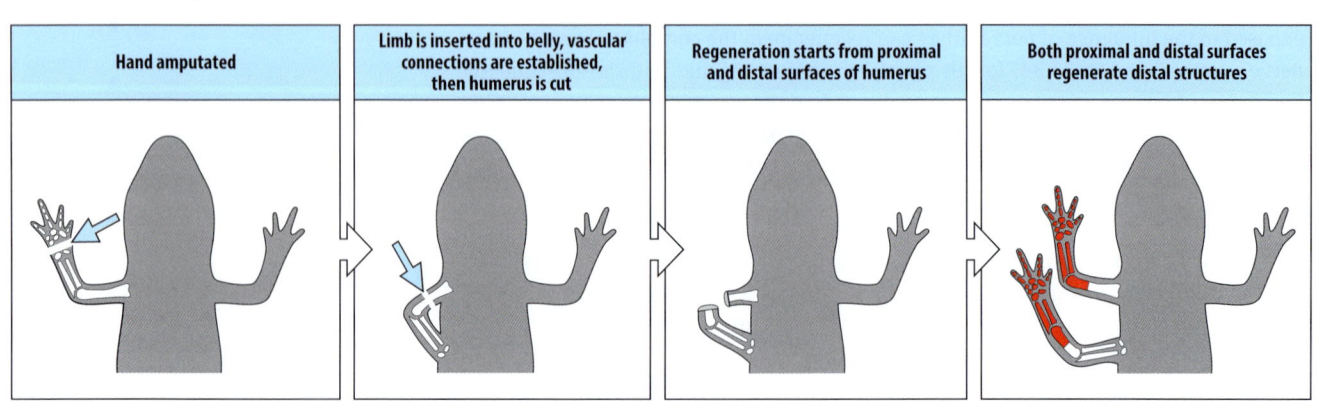

in limb development, the connective tissues guide muscle patterning in regenerating limbs. The blastema has considerable morphogenetic autonomy. If it is transplanted to a neutral location that permits growth, such as the dorsal crest of a newt larva or even the anterior chamber of the eye, it gives rise to a regenerated structure appropriate to the position from which it was taken.

The growth of the blastema and the nature of the structures it gives rise to depend on the site of the amputation and not on the nature of the more proximal tissues. The limb is not, however, simply 'trying' to replace missing parts. This was shown in a classic experiment in which the distal end of a newt limb that had been amputated at the wrist was inserted into the belly of the same animal, so as to establish a blood supply to it. The limb was then cut mid-humerus. Both surfaces regenerated distally, even though the part attached to the belly already had a radius and ulna (Fig. 11.35).

Regeneration can best be understood in terms of an adult limb having a set of positional values along its proximo-distal axis, which have been set up during embryonic development. The regenerating limb in some way reads the positional value at the site of the amputation and then regenerates all positional values distal to it. The ability of cells to recognize a discontinuity in positional values is illustrated by grafting a distal blastema to a proximal stump. In this experiment, the forelimb's stump and blastema have different positional values, corresponding to shoulder and wrist, respectively. The result is a normal limb in which structures between the shoulder and wrist have been generated by **intercalary growth**, predominantly from the proximal stump, whereas the cells from the wrist blastema mostly give rise to the hand (Fig. 11.36). A fundamental question in any discussion of pattern formation is how the proposed positional information is encoded molecularly. A major advance on this front has been the identification of Prod1, a cell-surface protein that is expressed in a graded manner along the proximo-distal axis of the newt limb, with higher levels of expression proximally. Proximo-distal values in the blastema can be altered by treatment with retinoic acid, as discussed in more detail in the next section, and retinoic acid treatment was used to compare gene expression in blastemas whose proximo-distal values had been altered. Treatment with retinoic acid increases the level of expression of the *Prod1* gene and expression is higher in 'proximal' blastemas compared with 'distal' blastemas. Prod1 was therefore a candidate for encoding positional values. The growth factor nAG, which contributes to the regenerative capacity (see Section 11.17), is a ligand for Prod1. It is thought that nAG has no input into generating proximo-distal pattern, but acts through Prod1 to promote cell proliferation.

The gradient in Prod1 fits very well with the observation that cell-surface properties are involved in regeneration. When mesenchyme from two blastemas from different proximo-distal sites are confronted in culture, the more proximal mesenchyme engulfs the distal, whereas mesenchyme from two blastemas from similar sites maintains a stable boundary (Fig. 11.37, left panels). This behavior suggests that there is a graded difference in cell adhesiveness along the axis, with adhesiveness being highest distally. We have seen such behavior in mixtures of explants of ectoderm, mesoderm, and endoderm from amphibian blastulas, in which tissues normally adjacent and adherent to each other, but with different degrees of cohesion, will envelop each other (see Section 5.1). In the blastema explants, the cells in the distal explant remain more tightly bound to each other than do the cells from the proximal blastema, which therefore spreads to a greater extent. Prod1 is involved in this process because when a blocking antibody to Prod1 is added to the culture, the proximal explant fails to spread over the distal explant (see Fig. 11.37, center panels).

A difference in the adhesiveness of tissues along the proximo-distal axis is also suggested by the behavior of a distal blastema when grafted to the dorsal surface of a proximal blastema so that their mesenchymal cells are in contact. Under these conditions, the distal blastema moves during limb regeneration to end up at the site from which it originated (Fig. 11.37, right panels). This suggests that its cells adhere more

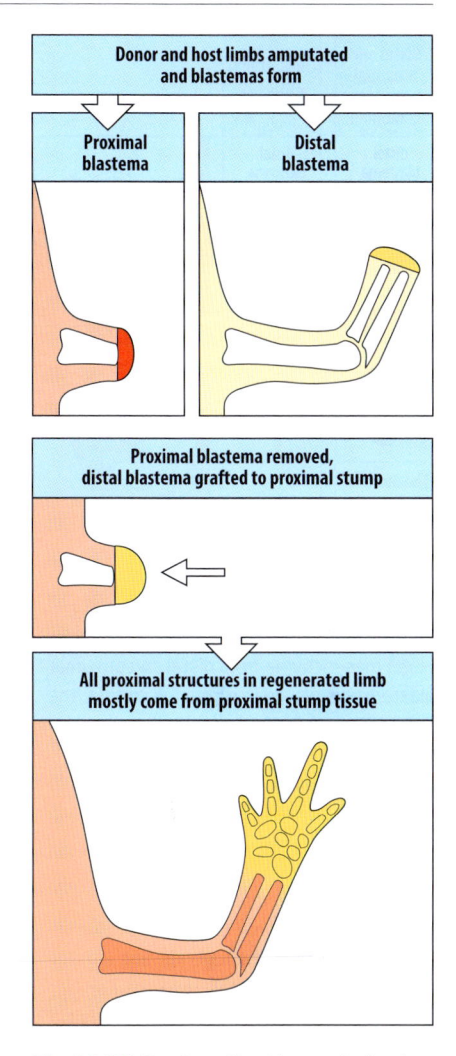

Fig. 11.36 Proximo-distal intercalation in limb regeneration. A distal blastema grafted to a proximal stump results in intercalation of all the structures proximal to the distal blastema. Almost all of the intercalated tissue comes from the proximal stump.

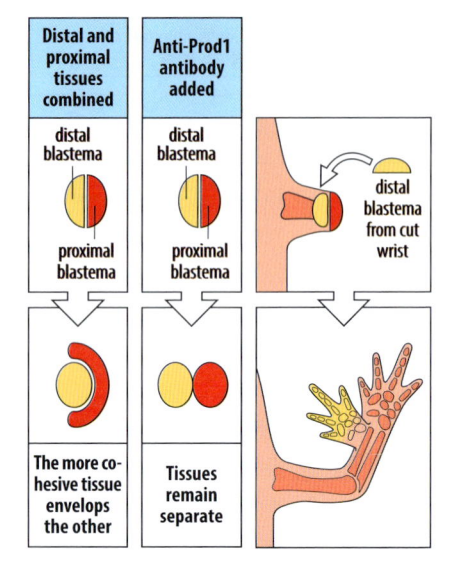

Fig. 11.37 Cell-surface properties vary along the proximo-distal axis. Left panels: when mesenchyme from distal and proximal blastemas is placed in contact in culture, the proximal mesenchyme engulfs the distal mesenchyme, which has greater adhesion between its cells. Center panels: if an anti-Prod1 antibody is added to the cultured blastemas, the proximal tissue does not envelop the distal tissue. Right panels: if a distal blastema (in this case from a cut wrist) is grafted to the dorsal surface of a more proximal blastema, the regenerating wrist blastema will move distally to a position on the host limb that corresponds to its original level and regenerates a hand.

strongly to the regenerated wrist region than they do to the proximal region to which it was grafted. When a shoulder-level blastema is grafted to a shoulder stump, it does not move but produces a normal distal outgrowth from the shoulder level.

All these experiments suggest that proximo-distal positional values in limb regeneration in urodele amphibians are encoded as a graded property, probably in part at the cell surface, and that cell behavior—growth, movement, and adhesion—relevant to axial specification is a function of the expression of this property, relative to neighboring cells.

Maintaining the continuity of positional values by intercalation is a fundamental property of regenerating epimorphic systems, and later in the chapter we will consider it in relation to the cockroach leg. It has been suggested that new positional values might be intercalated during normal regeneration of the amphibian limb between the cells at the level of amputation and those with the most distal positional values at the tip of the blastema, which are proposed to be specified by the wound epidermis. But as we discussed in Section 11.16, the current view is that positional values are specified in the regenerating limb in the same way as in the developing limb—in a proximo-distal direction and employing the same patterning molecules (see Chapter 9). As in developing limbs, Shh signaling, for example, is involved, but as the blastema is much larger than the embryonic limb bud—it can be 10 times larger in terms of cell numbers—this and other secreted signals are most unlikely to be able to diffuse right across it. Instead, these signal molecules will act more locally. Another similarity to the developing limb is that Hox genes are re-expressed in the regenerating limbs in patterns that suggest they encode regional identity.

11.19 Retinoic acid can change proximo-distal positional values in regenerating limbs

In Chapter 9 we saw how experimental treatment with retinoic acid can alter positional values in the developing chick limb. It also has striking effects on regenerating amphibian limbs.

When a regenerating limb is exposed to retinoic acid, the blastema becomes proximalized; that is, the limb regenerates as if it had originally been amputated at a more proximal site. For example, if a limb is amputated through the wrist, treatment with retinoic acid can result not only in the regeneration of the elements distal to the cut, but also in the production of an extra complete radius and ulna and humerus (Fig. 11.38).

Fig. 11.38 Retinoic acid can proximalize positional values. A forelimb amputated at the level of the hand, as indicated by the dotted line, and then treated with retinoic acid, regenerates structures that would normally arise from amputating at the level of the proximal end of the humerus. Scale bar = 1 mm.

Photograph courtesy of M. Maden.

The effect of retinoic acid is dose-dependent, and with a high dose it is possible to regenerate a whole extra limb, including part of the shoulder girdle, on a limb from which only the hand has been amputated. Retinoic acid therefore alters the proximo-distal positional value of the blastema, making it more proximal. It probably does this by affecting various pathways for specifying positional information, in particular increasing the expression of *Prod1* (see Section 11.18). Retinoic acid also appears to shift positional values in the blastema in a proximal direction through the activation of *Meis* homeobox genes, which are involved in specifying proximal positional values in the developing limb and are normally repressed in distal regions. Retinoic acid can also, under some experimental conditions, shift positional values along the antero-posterior axis in a posterior direction and digit duplications can be produced.

Retinoic acid can act through several different nuclear receptors, but only one (RAR δ2) is involved in changes in proximo-distal positional value in the regenerating newt limb. By constructing a chimeric receptor from a RAR δ2 DNA-binding portion and a thyroxine receptor hormone-binding portion, the retinoic acid receptor can be selectively activated by thyroxine, so that the effects of its activation can be studied experimentally (Fig. 11.39). Cells in the distal blastema are first transfected with the chimeric receptor gene and then grafted to a proximal stump, which is then treated with thyroxine. The transfected cells behave as if they have been treated with retinoic acid. The result is a movement of the transfected blastemal cells to more proximal regions in the intercalating regenerating blastema, which shows that activation of the retinoic acid pathway can proximalize the positional values of the cells, which then respond by movement to more proximal sites. *Prod1*, which was first identified as a retinoic-acid-responsive gene in newt blastemas, is a very strong candidate for mediating this change, as when the concentration of Prod1 protein is increased, it too can cause cells to move proximally.

Another remarkable effect of retinoic acid is its ability to bring about a homeotic transformation of tails into limbs in tadpoles of the frog *Rana temporaria*. If the tail of a tadpole is removed, it will regenerate. But treatment of regenerating tails with retinoic acid at the same time as the hindlimbs are developing results in the appearance of additional hindlimbs in place of a regenerated tail (Fig. 11.40). There is, as yet, no satisfactory explanation for this result, but it has been speculated that the retinoic acid alters the antero-posterior positional value of the regenerating tail blastema to that of the site along the antero-posterior axis where hindlimbs would normally develop.

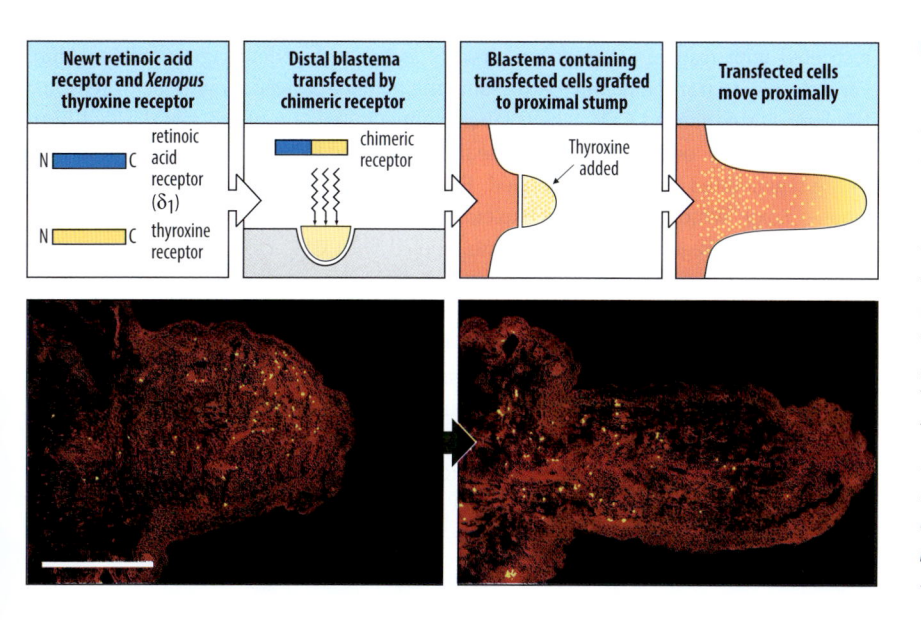

Fig. 11.39 Retinoic acid proximalizes the positional value of individual cells. Some of the cells of a newt distal blastema are transfected by a chimeric receptor, through which retinoic acid receptor function can be activated by thyroxine. This distal blastema is grafted to a proximal stump and treated with thyroxine. During intercalary growth, the transfected cells, which have been labeled, move proximally because their positional values have been proximalized by the activation of retinoic acid receptor function. The photographs illustrate proximalization of the transfected cells. Scale bar = 0.5 mm.

*Photographs reproduced with permission from Pecorino, L.T., et al.: **Activation of a single retinoic acid receptor isoform mediates proximodistal respecification.** Curr. Biol. 1996, **6**: 563–569. Copyright Elsevier.*

Fig. 11.40 Retinoic acid can induce additional limbs in the regenerating tail of a frog tadpole. After amputation, the regenerating tail stump of a tadpole of *Rana temporaria* is treated with retinoic acid at the time when hindlimbs are developing. This results in the appearance of additional hindlimbs in place of a new tail. Scale bar = 5 mm.

Photograph courtesy of M. Maden.

11.20 Mammals can regenerate the tips of the digits

Although mammals cannot regenerate whole limbs, many mammals, including mice and humans, even as adults can regenerate the tips of their digits, provided the nail-generating tissue—the nail organ—is still present. In mice and humans, the level from which digits can regenerate is limited to the base of the claw or nail, respectively. This probably reflects the presence of connective tissue cells under the nail organ that express the transcription factor Msx1, which is associated with undifferentiated mesenchymal cells (see Section 11.16). In mammalian limb development, *Msx1* is expressed at the tip of the embryonic limb bud, and in mice it continues to be expressed in the tips of the digits even after birth. The region in which digit regeneration in mice can occur corresponds with the domain of *Msx1* expression; more proximal regions of the digits cannot regenerate. BMP signaling is required for digit-tip regeneration in mice and the application of BMPs to mouse digits amputated more proximally can elicit a regenerative response. The way in which the tissues are generated in response to this BMP treatment is, however, different from that involved in normal digit-tip regeneration.

The cells that proliferate to regenerate the digit tip in the mouse seem to be lineage restricted with respect to connective tissues as in the regenerating amphibian limb. For example, in transgenic mice expressing the platelet-derived growth factor receptor labeled with enhanced GFP, which labels connective-tissue fibroblasts derived from neural crest that form nerve bundle sheaths protecting nerves, both bone and dermis in the regenerated digit tip are labeled. However, in transgenic mice expressing GFP-labeled Sox9, which would label skeletal cell precursors, although bone cells in the skeleton of the regenerated digit tip are labeled with GFP, cells in other connective tissues such as tendons and dermis are unlabeled.

The significance of the regenerative capacity of the tips of mammalian digits is not clear. Single-cell RNA sequencing shows that the transcriptional profile of the cells in the regenerating digit tip does not resemble that of the cells in developing mouse digits. This suggests that, unlike regenerating amphibian limbs, in which regeneration recapitulates embryonic development (see Section 11.16), a distinct mechanism is involved in mammalian digit tip regeneration. It is important to consider how the ability to regenerate appendages may have evolved, because this will help us to answer the question of why we cannot regenerate our limbs (Box 11D).

BOX 11D Why can't we regenerate our limbs?

Examples of adult animals that are able to regenerate complex organs are found throughout the animal kingdom. Regenerative capacity is not confined to animals with simple body plans, and not all simple animals can regenerate. The nematodes, for example, have no regenerative capacity. This is probably in part because their bodies have high internal hydrostatic pressure, which only makes healing possible after limited cell puncture but not after extensive cell loss, and also because adult nematodes do not have any proliferating somatic cells.

Among vertebrates, adult salamanders such as newts and the axolotl can regrow multiple body parts, including limbs, the lens of the eye, and the heart, whereas mammals can only regrow the very tips of the digits and repair the heart only in the first week of postnatal life. So why do newts and other salamanders show these remarkable regenerative capacities whereas mammals, including humans, have such a limited capacity? The view that prevailed until recently was that the ability of some adult vertebrates to regenerate is a fundamental vertebrate property that we mammals and other vertebrates have lost during our evolutionary history. Indeed, studies in invertebrates have shown that the variability in regenerative ability in a group of closely related annelids is most likely to be due to the evolutionary loss of regenerative ability in some species.

In adult salamanders, however, there is growing evidence that regenerative ability evolved locally in this group of animals. The first indication of this was the finding that Prod1, the key protein that encodes proximo-distal positional values in regenerating amphibian limbs (see Section 11.18), is a salamander-specific protein and has no mammalian homolog. Further analysis of the transcriptome of newts has identified several more genes encoding novel proteins that have no apparent homologs in mammals. The levels of transcripts of some of these genes have already been shown to change during regeneration (Fig. 1).

This emerging focus on novel newt proteins involved in regeneration contrasts with much of the previous research on regeneration, which concentrated on molecules and pathways already known to be conserved during embryonic development. So how do novel proteins interact with these pathways during limb regeneration? Some insights into this question have come from studies on Prod1, as it has been shown that expression of the newt *Prod1* gene is regulated by the homeoprotein Meis. Meis was already implicated in specifying proximal positional values in chick and mouse developing limbs (see Section 9.14). Like *Prod1*, it is expressed at higher levels in proximal blastemas compared with distal blastemas in regenerating amphibian limbs, although unlike *Prod1* it is not expressed in a graded fashion in the intact limb. The interaction between *Prod1* and Meis that occurs following limb amputation could be involved in specifying proximal positional values in regenerating limb blastemas.

This perspective on the question of why newts have such great regenerative abilities has implications for thinking about ways of tackling our inability to regenerate our limbs. Other important issues are to do with the size of human limbs and the length of time they would take to 're-grow'.

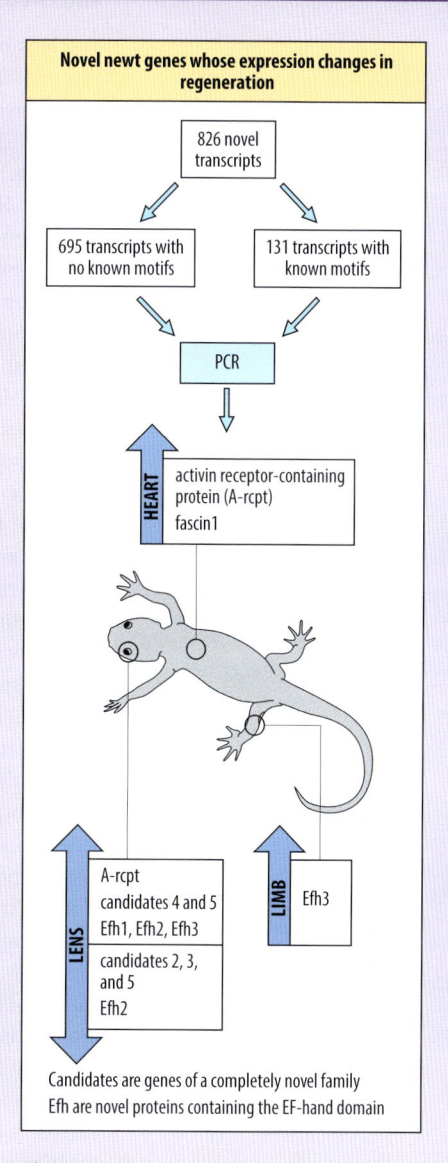

Fig. 1

Adapted from Mihaylova, Y., Aboobaker, A.A.: *What is it about eye of newt?* Genome Biol. 2013, **14**: 106.

11.21 Insect limbs intercalate positional values by both proximo-distal and circumferential growth

The legs of some insects, such as the cockroach and cricket, can regenerate. Unlike *Drosophila*, these insects do not undergo a complete metamorphosis, and the juvenile stages (nymphs) are more like miniature adults. When the tibia of a third instar nymph of the cricket *Gryllus bimaculatus* is amputated, it takes about 40 days to restore the adult leg. Wingless and Decapentaplegic signaling activate epidermal growth factor receptor signaling in a distal-to-proximal gradient in the cricket blastema, which patterns the regenerating distal leg.

In *Drosophila*, signaling through the cadherins Fat and Dachsous and the associated Hippo pathway coordinates cell proliferation and apoptosis in imaginal discs, determining their final size and thus the size of adult wings and legs (see Section 11.4). Experiments using RNA interference to investigate the role of Fat, Dachsous, and the Hippo pathway in cricket limb regeneration suggest that they also determine the final size of legs in insects whose legs develop directly. Knockdown of either Fat or Dachsous results in a shorter than normal regenerated limb, whereas knockdown of another set of genes in the Hippo pathway produces a longer leg than normal. It is proposed that a Dachsous/Fat gradient exists along the limb and that the steepness of the gradient could control cell proliferation.

Regenerating insect legs follow an epimorphic process of blastema formation and outgrowth that intercalates missing positional values. This contrasts with the current view of the way in which the proximo-distal axis of the regenerating amphibian limb is specified, which is in a proximo-distal sequence (see Section 11.18). However, when cells with disparate positional values are placed next to one another in amphibian limbs, intercalary growth occurs in order to regenerate the missing positional values. Such intercalation of positional values seems to be a general property of epimorphically regenerating systems and is particularly clearly illustrated by limb regeneration in the cockroach. It also occurs in *Drosophila* leg and wing imaginal discs which regenerate after they have been cut into fragments.

A cockroach leg is made up of several distinct segments, arranged along the proximo-distal axis in the order: coxa, femur, tibia, tarsus. Each segment seems to contain a similar set of proximo-distal and circumferential positional values and will intercalate missing positional values during regrowth. When a distally amputated tibia is grafted onto a host tibia that has been cut at a more proximal site, localized growth occurs at the junction between graft and host, and the missing central regions of the tibia are intercalated (Fig. 11.41, left panels). In contrast to amphibian regeneration, there is a predominant contribution from the distal piece. As in amphibians, however, regeneration is a local phenomenon and the cells are indifferent to the overall pattern of the tibia. Thus, when a proximally cut tibia is grafted onto a more distal site, making an abnormally long tibia, regenerative intercalation again restores the missing positional values, making the tibia even longer (Fig. 11.41, right panels). The regenerated portion is in the reverse orientation to the rest of the limb, as indicated by the direction in which the bristles point, suggesting that the gradient in positional values also specifies cell polarity, as in insect body segments. These results show that when cells with non-adjacent positional values are placed next to each other, the missing values are intercalated by growth to re-establish a set of continuous positional values.

A similar set of positional values is present in each segment of the limb. Thus, a mid-tibia amputation, when grafted to the mid-femur of a host, will heal without intercalation. By contrast, a distally amputated femur grafted onto a proximally amputated host tibia results in intercalation, largely femur in type. There must be other factors making each segment different, rather like the body segments of the insect larva.

Intercalary regeneration also occurs in a circumferential direction. When a longitudinal strip of epidermis is removed from the leg of a cockroach, normally

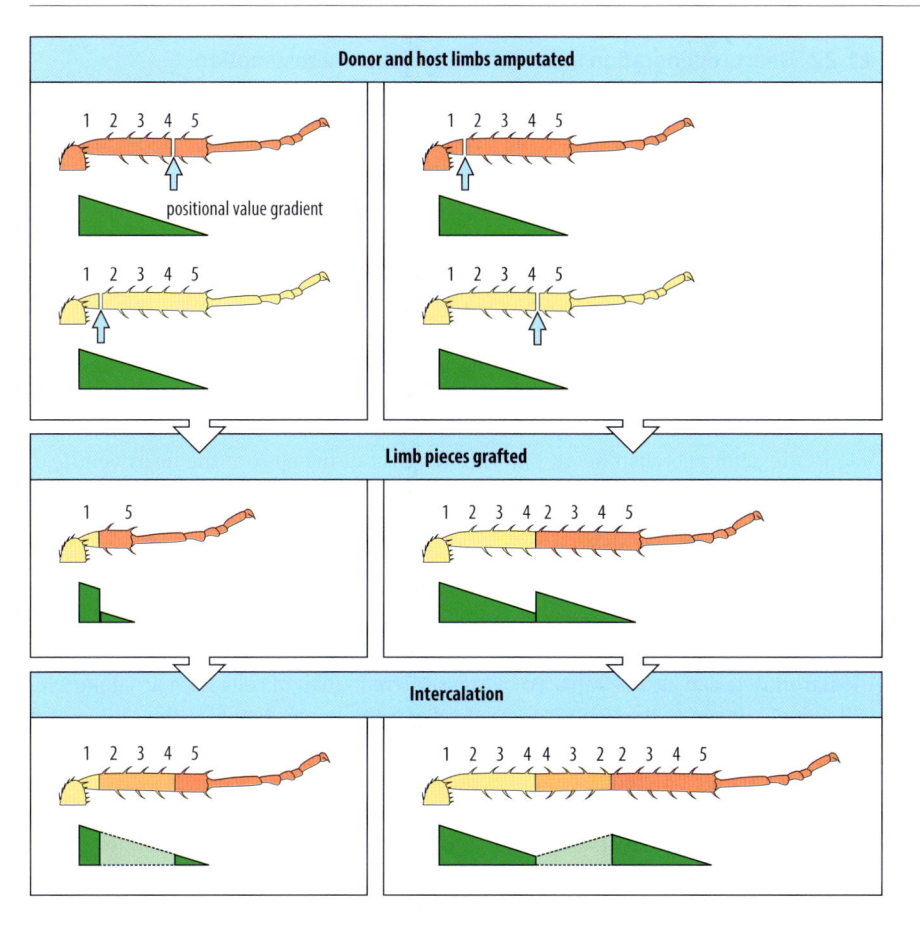

Fig. 11.41 Intercalation of positional values by growth in the regenerating cockroach leg. Left panels: when a distally amputated tibia (5) is grafted to a proximally amputated host tibia (1), intercalation of the positional values 2-4 occurs, irrespective of the proximo-distal orientation of the grafts, and a normal tibia is regenerated. Right panels: when a proximally amputated tibia (1) is grafted to a distally amputated host tibia (4), however, the regenerated tibia is longer than normal and the regenerated portion is in the reverse orientation to normal, as judged by the orientation of surface bristles. The reversed orientation of regeneration is due to the reversal in positional value gradient. The proposed gradient in positional value is shown under each figure.

*Illustration after French, V., et al.: **Pattern regulation in epimorphic fields.** Science 1976, **193**: 969-981.*

non-adjacent cells come into contact with one another, and intercalation in a circumferential direction occurs after molting (Fig. 11.42). Cell division occurs preferentially at sites of mismatch around the circumference. One can treat positional values in the circumferential direction as a clock face, with values going continuously 12, 1, 2, 3, ... 6, ... 9, ... 11. As in the proximo-distal axis, there is intercalation of the missing positional values.

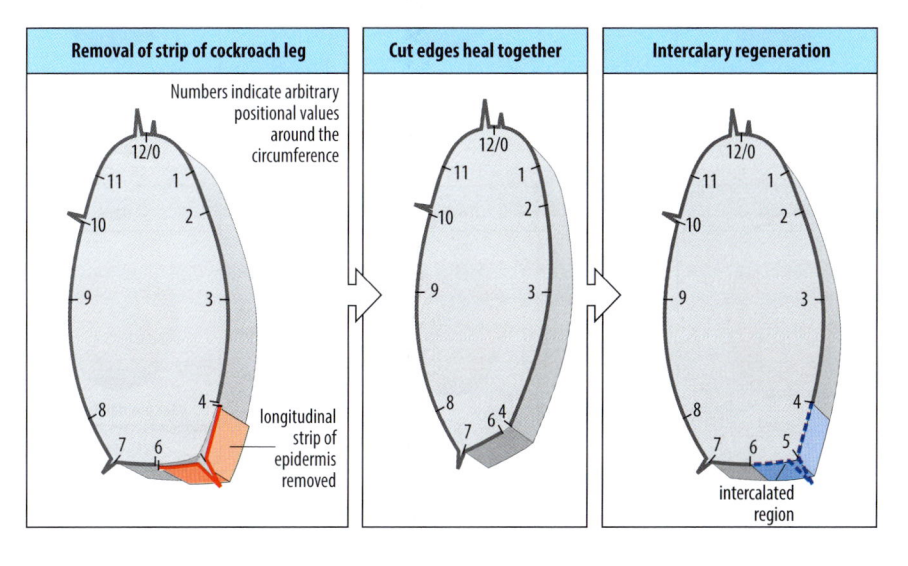

Fig. 11.42 Circumferential intercalation in the cockroach leg. The leg is seen in transverse section. When a piece of cockroach ventral epidermis is removed (left panel), the cut edges heal together (center panel). When the insect molts and the cuticle regrows, circumferential positional values are intercalated (right panel). The positional values are arranged around the circumference of the leg, rather like the hours on a clock face.

*Illustration after French, V., et al.: **Pattern regulation in epimorphic fields.** Science 1976, **193**: 969-981.*

11.22 Heart regeneration in zebrafish involves the resumption of cell division by cardiomyocytes

Some vertebrates can regenerate the heart. For example, if 20% of the ventricle of an adult zebrafish heart is removed, it will regrow, and the regenerated tissue will become functionally integrated with the existing heart tissue (Fig. 11.43). This phenomenon is also seen in newts, but is of particular interest in the zebrafish, as in this model organism regeneration can be studied genetically.

One of the key questions in zebrafish heart regeneration, as in the regenerating limb, is the origin of the cells that make up the regenerated tissue. Genetic fate-mapping experiments show that existing mature muscle cells of the adult myocardium, the **cardiomyocytes**, are the main source of proliferating cells for regeneration, rather than undifferentiated progenitor cells. Cardiomyocytes proliferate at a very low rate in the adult zebrafish heart, but after removal of the apex of the heart ventricle, cardiomyocytes in and around the wounded area in the ventricle dedifferentiate and start to divide. The new cardiomyocytes then migrate into the wound area and at the same time the regenerated tissue becomes vascularized. The new myocardial tissue becomes electrically coupled to the cardiomyocytes in the undamaged region of the heart between 2 and 4 weeks after injury.

In an embryonic model of zebrafish heart regeneration, however, damage to ventricular muscle can also be repaired by transdifferentiation of cells from atrial muscle. Following damage to the ventricle, atrial cells at the border between the atrium and the ventricle dedifferentiate and start to divide. They then migrate to the damaged region where they differentiate into ventricular muscle.

Regeneration of the myocardium in the adult zebrafish heart involves interactions with cells in the other layers of the heart, the endocardium and epicardium (see Section 9.31). As early as 1 hour after injury, the entire endocardium, the endothelial lining of the heart, begins to express *raldh2*, which encodes a retinoic-acid-synthesizing enzyme (see Fig. 11.43). Within the next day or so, *raldh2* is also expressed throughout

Fig. 11.43 Time course of regeneration after the amputation of the apex of the ventricle of the adult zebrafish heart. Endocardial activation occurs throughout the heart within a few hours of amputation, epicardial activation a few days later marked by expression of the gene for the retinoic acid synthesizing enzyme Raldh2, and genes involved in other signaling pathways. By 7 days after amputation, activation of the epicardium is localized to the wound area. Cardiomyocytes are stimulated to activate regulatory sequences controlling expression of Gata4, a transcription factor involved in heart development, and to proliferate. The regenerating muscle is vascularized over the next 7 days. By about 30 days, the myocardium has regenerated and is electrically coupled with the rest of the heart.
Adapted from Gemberling, M., et al.: *The zebrafish as a model for complex tissue regeneration.* Trends Genet. 2013, **29**: 611–620.

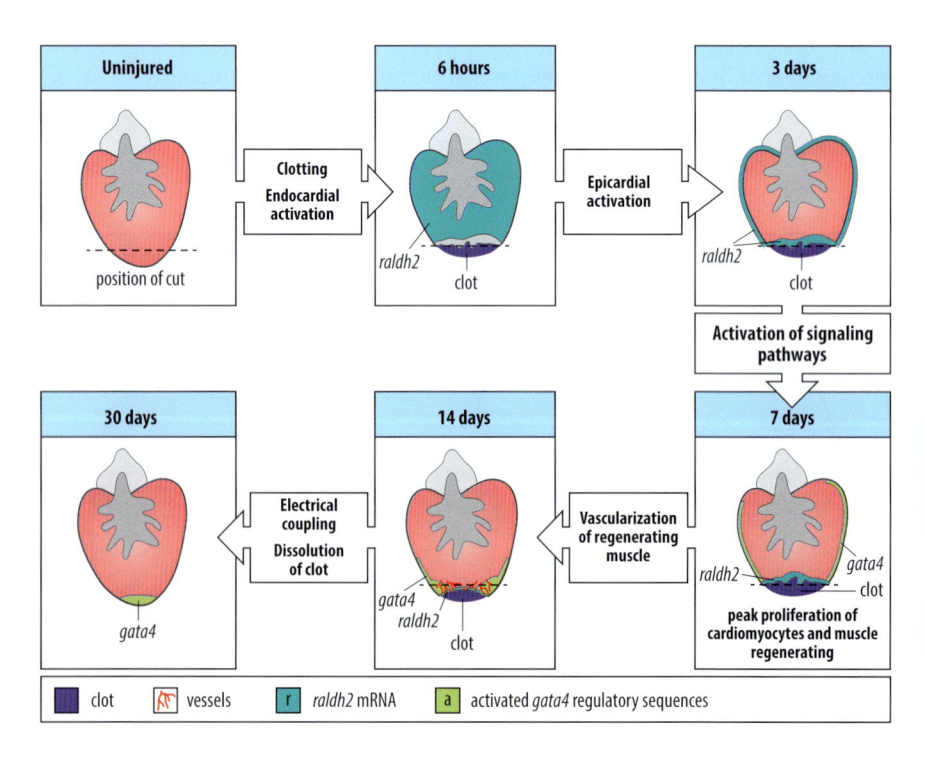

the entire epicardium, the thin outer layer covering the myocardium, and then later becomes confined to the wound area. The epicardial cells proliferate and surround the regenerating muscle, releasing retinoic acid and other signals, including Shh and IGF-2, that aid cardiomyocyte proliferation. However, the signal that initiates adult cardiomyocyte proliferation has not yet been identified, although a few molecules that act extracellularly have been shown to stimulate cardiomyocyte proliferation in the absence of injury: Neuregulin 1, vitamin D, and VEGFA (vascular endothelial growth factor A). The epicardial cells give rise to vascular supporting cells in the regenerated tissue, just as they do during normal heart development. The peak of cell proliferation occurs 7 days after injury and at this time, regulatory sequences controlling expression of *Gata4*, which encodes a transcription factor involved in heart development, are activated throughout the outer layer of the myocardium. The new cardiac muscle is made up largely of the progeny of cells in which these *Gata4*-associated sequences were activated. Other genes are expressed that are also involved in the normal development of cardiac muscle, such as *Nkx2.5*, an early marker of prospective heart cells (see Section 9.32).

In adult mammals, growth of the heart after birth is due to cell enlargement, and the rates of cell proliferation are extremely low. Unlike the adult zebrafish heart, the mammalian heart cannot regenerate and produce fully functional heart tissue after injury. If adult mammalian heart muscle is damaged, it forms a fibrous scar. However, the cardiomyocytes in an adult mouse heart can resume cell division when specific signaling pathways are activated. Even then, because very few cells are involved, this is not sufficient to restore adult function. There is, however, in the mouse, a brief period just after birth when a fully functioning heart can regenerate after surgical removal of the apex of the left ventricle (15% of the ventricular myocardium is removed). Genetic lineage analysis has shown that just as in the adult zebrafish, the new cardiomyocytes are derived from pre-existing ones. Another similarity to adult heart regeneration in zebrafish is that epicardial activation occurs. Thus, although the zebrafish heart differs from the mammalian heart in having only one ventricle, understanding the mechanisms involved in regeneration of the zebrafish heart will be relevant to mammalian heart regeneration.

Cardiomyocyte proliferation is stimulated in mice when the Hippo pathway is inactivated (see Section 11.4). A large extracellular matrix proteoglycan, Agrin, which is present at high levels in the neonatal mouse heart but not in the adult heart, regulates the Hippo pathway, and this may explain why heart regeneration is only possible in the period just after birth. Agrin is best known for its role in the formation of the neuromuscular junction in embryos. In the neonatal heart, Agrin binds to the dystrophin–glycoprotein complex on the surface of cardiomyocytes. The dystrophin–glycoprotein complex is found on the surface of both skeletal and cardiac muscle cells and mediates interactions between the cell surface, the extracellular matrix, and the cytoskeleton, in addition to providing mechanical strength. Mutations in genes encoding proteins in the complex lead to muscular dystrophy and cardiomyopathy. The binding of cardiomyocytes to Agrin results in inhibition of the Hippo pathway, and as a result Yap translocates to the nucleus, switching on genes that promote the cell proliferation required for regeneration. In adult hearts where levels of Agrin are low, Yap is bound to the intracellular face of the dystrophin–glycoprotein complex and is unable to translocate to the nucleus. Under these conditions, proliferation is suppressed and the heart cannot regenerate. The identification of Agrin as a key molecule that promotes cardiomyocyte proliferation might eventually lead to clinical applications, as administration of Agrin can stimulate the regeneration and repair of damaged hearts in adult mice.

The ability of the mouse heart to regenerate is lost when the mouse is 7 days old. One of the molecules that blocks cardiomyocytes from entering the cell cycle is Meis1, a homeodomain transcription factor required for embryonic heart development. When

Meis1 function is deleted specifically in adult mouse cardiomyocytes, they enter the cell cycle and proliferate without undergoing hypertrophy. Conversely, postnatal overexpression of *Meis1* reduces cardiomyocyte proliferation and prevents the regeneration normally seen in the first week after birth. There is evidence that Meis1 acts by directly controlling expression of genes encoding cell-cycle inhibitors.

> **SUMMARY**
>
> Adult urodele amphibians can regenerate amputated limbs and tails. Regeneration depends on the specialized apical epithelial cap which develops from the epidermis covering the wound and on the presence of nerves. The nerves provide a growth factor that contributes to the regenerative response, but limbs that have never been innervated can regenerate. Stump tissues at the site of amputation dedifferentiate to form a blastema. The dedifferentiated cells of the blastema give rise to the various cell types within the regenerated limbs and are lineage-restricted, with cells giving rise to their own tissue of origin. Blastemal cells derived from connective tissues are multipotent and give rise to the various different connective tissues in the regenerated limb. In this they resemble embryonic limb bud cells. Regeneration always gives rise to structures with positional values more distal than those at the site of amputation. When a blastema is grafted to a stump with different positional values, proximo-distal intercalation of the missing positional values occurs. Positional values may be related to a proximo-distal gradient in a cell-surface protein. Retinoic acid changes the positional values of the cells of the blastema, giving them more proximal values. Limb regeneration in mammals is limited to the tips of the digits. Insect limbs can also regenerate, with intercalation of missing positional values occurring in both proximo-distal and circumferential directions. The adult heart can regenerate in amphibians and fish. In the first week of birth, the mouse heart can also regenerate as a result of interactions with the extracellular matrix that inhibit Hippo signaling. In both adult zebrafish and neonatal mice, the new myocardial heart tissue is produced by proliferation of pre-existing cardiomyocytes that dedifferentiate, and this is facilitated by interactions with epicardial cells in the outer covering of the heart.

> **SUMMARY: Regeneration of an amphibian limb**
>
> amputation of a newt limb
> ⇩
> formation of apical epithelial cap
> ⇩
> local dedifferentiation of stump tissue to form a blastema
> ⇩
> the blastema grows to form distal structures
> provided nerves are present
> ⇩
> graft of distal blastema to proximal stump
> ⇩
> intercalation of missing proximo-distal positional values from stump tissue growth

Aging and senescence

Organisms are not immortal, even if they escape disease or accidents. With aging—which occurs over the passage of time—comes an increasing impairment of physiological functions, which reduces the body's ability to deal with various stresses, and an increased susceptibility to disease. This age-related decline in function is known as **senescence**. Aging is observed in most multicellular animals, but there are notable exceptions, such as sea anemones, *Hydra*, and planarians. All sexually reproducing animals age, whereas most asexual animals do not. Several *Hydra* species, whose capacity for regeneration is discussed in Section 11.15, do not age even when they undergo sexual differentiation, while in at least one species, *Hydra oligactis*, gametogenesis seems to trigger aging. The phenomenon of senescence raises many questions as to the underlying mechanisms, and these are still largely unanswered, but we can at least consider some general questions, such as whether senescence is part of an organism's post-embryonic developmental program or whether it is simply the result of wear and tear. Germ cells do not age; if they did, the species would die out.

Although individuals can vary in the time at which particular aspects of aging appear, the overall effect is summed up as an increased probability of dying in most animals, including humans, with increased age. This life pattern, which is illustrated in relation to *Drosophila* (Fig. 11.44), is typical of many animals. There are, however, exceptions such as the naked mole rat, which lives more than 30 years in captivity (mice live up to 4 years) and shows no increased probability of dying with age. In the wild, the probability of dying depends on environmental factors—for example, 90% of wild mice die during their first year. Another interesting example is the Pacific salmon, in which death does not come after a process of gradual aging but is linked to a certain stage in the life cycle, in this case to spawning.

One view of aging is that it is due to an accumulation of damage that eventually outstrips the ability of the body to repair itself, and so leads to the loss of essential functions. For example, some old elephants die of starvation because their teeth have worn out. The nematode *Caenorhabditis elegans* lives for an average of around 20 days and the major cellular change that occurs with age is the progressive deterioration of muscle. This deterioration has a random effect on the life span, which varies from 10 to 30 days. Nevertheless, there is clear evidence that senescence is under genetic control, as different species age at vastly different rates, as shown by their different life spans (Fig. 11.45). An elephant, for example, is born after 21 months' embryonic development, and at that point shows few, if any, signs of aging, whereas a 21-month-old mouse is already well into middle age and beginning to show signs of senescence.

The genetic control of aging can be understood in terms of the 'disposable soma' theory, which puts it into the context of evolution. The disposable soma theory proposes that natural selection tunes the **life history** of the organism so that sufficient resources are invested in maintaining the repair mechanisms that prevent aging, at least until the organism has reproduced and cared for its young. Thus mice, which start reproducing when just a few months old, need to maintain their repair mechanisms for much less time than do elephants, which only start to reproduce when around 13 years old. In most species of animals in the wild, few individuals live long enough to show obvious signs of senescence and senescence need only be delayed until reproduction is complete. Cells have numerous mechanisms to delay aging, which are quite like the mechanisms used to prevent malignant transformation. These cellular mechanisms protect the cell from internal damage by reactive chemicals and routinely repair damage to DNA, which is occurring continually in living cells even when they are not actively dividing. They are particularly active in germ cells.

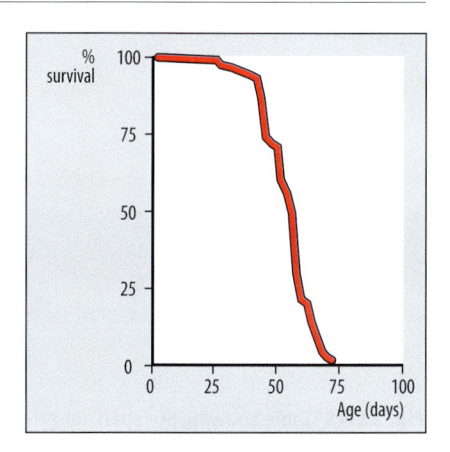

Fig. 11.44 Aging in *Drosophila*. The probability of dying increases rapidly at older ages.

618 Chapter 11 Growth, post-embryonic development, and regeneration

Longevity and time to attain sexual maturity for various mammals			
	Maximum lifespan (months)	Length of gestation (months)	Age at sexual maturity (months)
Human	1440	9	153
Fin whale	1368	12	87
Indian elephant	780	21	156
Horse	712	8	112
Chimpanzee	684	11	30
Brown bear	480	3.3 (not including period of delayed implantation)	43
Rhesus monkey	480	5.5	39
Squirrel monkey	362	5	32
Cat	360	2	9
Pig	324	4	10
Dog	288	2	17
Gray squirrel	283	1.5	11
Sheep	274	5	18
Cattle	240	9	15
Guinea-pig	144	2	2
European rabbit	108	1	23.5 (in the wild)
Mouse	48	0.6	1.5
Golden hamster	47	0.5	1.5
Brown rat	46	0.7	3

Fig. 11.45 Table showing life span, length of gestation, and age at sexual maturity for various mammals. The numbers for the life spans are the maximum life span reliably recorded for the species; average life spans are usually much shorter.

Data from AnAge: The Animal Aging and Longevity Database (http://genomics.senescence.info/species).

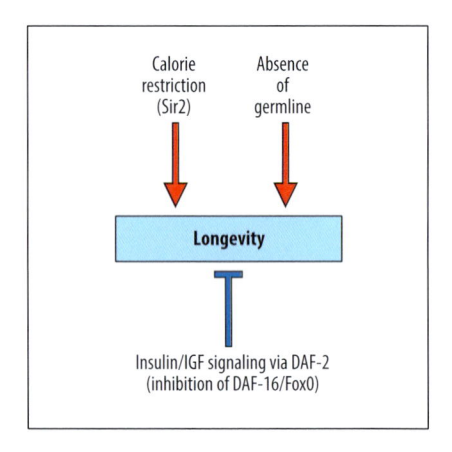

Fig. 11.46 Potential inputs to aging in *Caenorhabditis elegans* and *Drosophila*. Calorie restriction and the absence of a germline promote an extended life span in both animals. In *Drosophila*, the calorie-restriction effect has been shown to require the histone deacetylase Sir2, which is likely to act on gene expression through its ability to modify chromatin (see Box 6B). In *C. elegans*, insulin/insulin-like growth factor (IGF) signaling has been shown to regulate life span via its inhibitory effect on expression of the gene for the transcription factor DAF-16 (a member of the FoxO family of transcription factors).

11.23 Genes can alter the timing of senescence

The maximum recorded life spans of animals show dramatic differences (see Fig. 11.45). Humans can live as long as 120 years, some owls 68 years, cats 28 years, *Xenopus* 15 years, mice 4 years, *Drosophila melanogaster* 3 months, and the nematode about 25 days. Mutations in genes that affect life span have been identified in *C. elegans*, *Drosophila*, mice, and humans, and may give clues to the mechanisms involved: the ability to resist damage to DNA and the effects of oxygen radicals are important.

A *C. elegans* worm that hatches as a first-instar larva in an uncrowded environment with ample food grows to adulthood and can survive for 25 days. In crowded conditions and when food is short, however, the animal enters a quiescent third-instar larval state known as the **dauer** state, where it neither eats nor grows until food becomes available again. When conditions become favorable, the dauer larva molts and becomes a fourth-instar larva. The dauer state can last for 60 days and has no effect on the post-dauer life span: it is therefore considered to be a state in which the larva does not age. An insulin/IGF-1 signaling system plays an important part both in controlling entry into the dauer state in response to stress and in the overall control of fertility, life span, and metabolism in *C. elegans* and in *Drosophila*. Reduction in insulin/IGF-1 signaling increases longevity (Fig. 11.46). This is the same hormonal system that is responsible for controlling metabolism and promoting growth in *Drosophila* and vertebrates, as we saw earlier (see Section 11.8). The intracellular pathway that transduces insulin/IGF-1 signaling is linked to the TOR pathway (Fig. 11.47) that controls cell size and whose activity is regulated by nutrient supply (see Section 11.1).

In *C. elegans*, mutations that cause a strong reduction in the expression of *daf-2*, which encodes a receptor in the insulin-signaling pathway, arrest development in the dauer state. The normal role of DAF-2 protein is to antagonize the activity of DAF-16, a transcription factor whose activity lengthens life span and increases resistance to some types of stress. A partial loss of DAF-2 function results in longer adult life span after the dauer state but with reduced fertility and viability of the progeny, whereas a further reduction in DAF-2 at the larval stage by RNA interference increases life span even more, without shortening the dauer state. The presence of a germline seems to have a negative effect on life span extension. Removal of germline precursor cells in *daf-2* mutant embryos resulted in their having a mean life span of 125 days and remaining quite healthy—in humans, this would equate to a life span of 500 years. But when *daf-2* mutants were cultured alongside wild-type worms, the mutants became extinct in just a few generations, partly as a result of their reduced fertility. Microarray analysis of the effects of DAF-16 shows that it activates stress-response and antimicrobial genes. In laboratory conditions it seems that aging animals in which DAF-16 is inhibited are killed by the bacteria on which they feed.

A similar system regulates aging in *Drosophila*. Mutations disabling the insulin/IGF-1 pathway almost double the life span of this animal, and calorie restriction also extends life span. The mutant flies, rather like the nematode dauer larvae, enter a state of reproductive diapause or quiescence. These effects on life span may be due to resistance to oxidative stress. Ablation of germline cells in *Drosophila* also extends life span, probably through the effects of the germline on insulin signaling. Aging in mice can be retarded by reduction in pituitary activity, and there is also evidence that female mice with a mutation in the gene for the IGF-1 receptor (which is the receptor activated by IGF-1 and IGF-2) can live 33% longer.

There is evidence that oxidative damage accelerates aging and that reactive oxygen radicals are key players in causing cell damage. Reduction in food intake increases life span in other animals; rats on a minimal diet live about 40% longer than rats allowed to eat as much as they like. This is thought to be partly due to a reduced exposure to free radicals, which are formed during the oxidative breakdown of food. Free radicals are highly reactive and can damage both DNA and proteins. A long-lived rodent species generates less reactive oxygen than the laboratory mouse. Oxidative stress also exerts its effects by damaging mitochondria.

Humans who are homozygous for a recessive gene defect leading to Werner syndrome age prematurely. Growth is retarded at puberty, and by their early twenties people with this condition have gray hair and have various illnesses, such as heart disease, that are typical of old age. Most people affected die before the age of 50. The gene affected in Werner syndrome has been identified, and encodes a protein involved in unwinding DNA. Such unwinding is required for DNA replication, DNA repair, and gene expression. The inability to carry out DNA repair properly in people with Werner syndrome could subject the genetic material to a much higher level of damage than normal. The link between Werner syndrome and DNA thus fits with the possibility that aging is linked to the accumulation of damage in DNA. Aging may also be related to cell senescence. Hutchinson–Gilford progeria syndrome (progeria means premature aging) is even more severe in its effects. The gene affected in this syndrome encodes the intranuclear protein lamin A, and the primary cellular defect is an instability in the structure of the nuclear envelope, which means that cells are more likely to die prematurely.

11.24 Cell senescence blocks cell proliferation

One might think that when cells are isolated from an animal, plated out in a culture dish, and provided with adequate medium and growth factors, they would continue to proliferate almost indefinitely. But this is not the case for mammalian fibroblasts. These cells from connective tissues will only go through a limited number of cell doublings in culture; the cells then stop dividing, however long they are cultured (Fig. 11.48). For normal fibroblasts, the number of cell doublings depends both on the species and the age of the animal from which they are taken. Fibroblasts taken from a human fetus go through about 60 doublings, those from an 80-year-old about 30, and

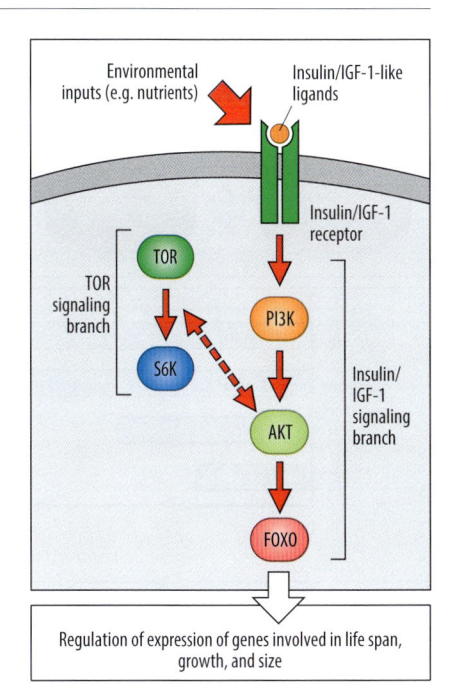

Fig. 11.47 A highly simplified version of the insulin/IGF-1 and TOR nutrient-sensing network. The network is activated in response to environmental inputs, for example, nutrients. Insulin/IGF-like ligands bind to a receptor on the plasma membrane triggering phosphorylation of the protein kinase PI3K. Phosphorylation of PI3K indirectly leads to phosphorylation of another protein kinase AKT. Activity of AKT is also regulated by interactions with the TOR pathway, which includes the protein kinase S6K. AKT phosphorylates members of the FOXO transcription factor family, preventing them from entering the nucleus. Reduction in the activity of the pathway by, for example, restricting nutrients, allows FOXO transcription factors to enter the nucleus and regulate their gene targets, which include those involved in determining life span, growth, and size. The network is highly conserved, but the components have different names in different species. In *C. elegans*, the insulin/IGF receptor is known as DAF-2, PI3K as AGE-1 and FOXO as DAF-16. Note that vertebrates, including humans, have two receptors, an insulin receptor and an insulin-like receptor.

Adapted from Flatt, T., Partridge, L.: **Horizons in the evolution of aging.** *BMC Biol. 2018, 16: 93.*

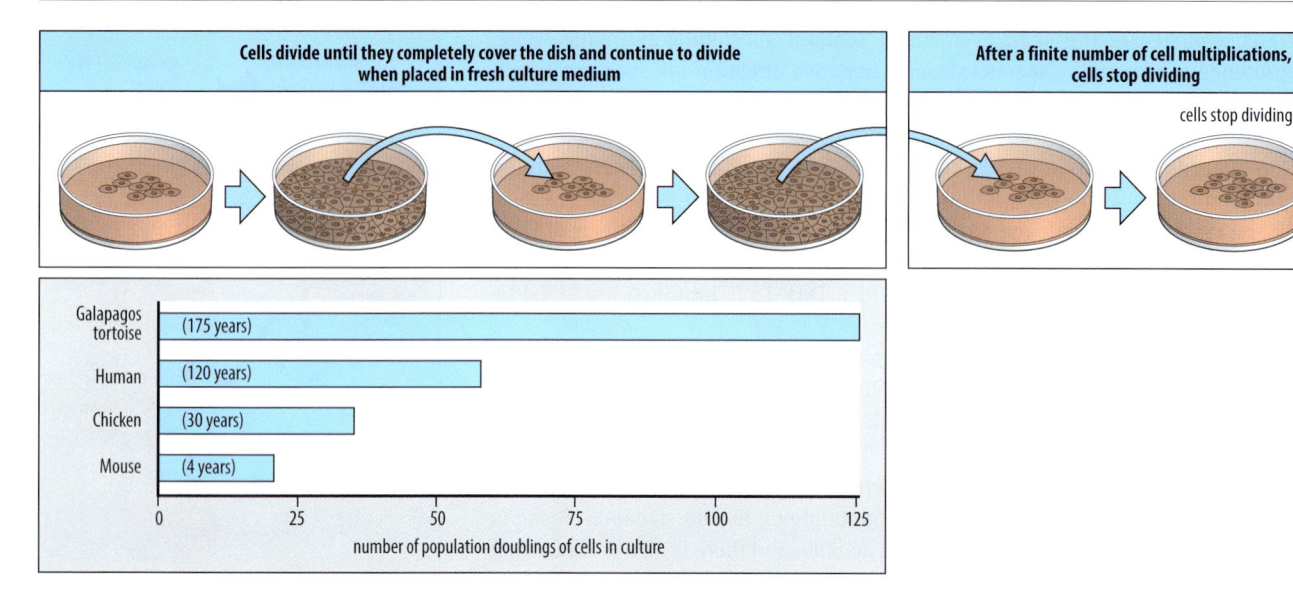

Fig. 11.48 Vertebrate fibroblasts can only go through a limited number of divisions in culture. Fibroblasts placed in culture are sub-cultured until they stop growing (top panels). The number of cell doublings in culture before they stop dividing is related to the maximal life span of the animal from which the cells were taken, as indicated by the figures in brackets on the graph (bottom panel).

those from an adult mouse about 12–15. When the cells stop dividing, they seem to be healthy but are stuck at some point in the cell cycle, often G_0; this phenomenon is known as **cell senescence**. Cells taken from patients with Werner syndrome, who show an acceleration of many features of normal aging (see Section 11.23), make significantly fewer divisions in culture than normal cells. However, it is far from clear how this behavior of cells in culture is significant for the aging of the organism and to what extent it reflects the way the cells are cultured.

A feature shared by senescent cells in culture and *in vivo* is shortening of the **telomeres**. Telomeres are the repetitive DNA sequences at the ends of the chromosomes that preserve chromosome integrity and ensure that chromosomes replicate themselves completely without loss of information-encoding DNA at the ends. The length of the telomeres is reduced in older human cells. Telomere length is found to decrease at each DNA replication, suggesting that they are not completely replicated at each cell division and this may be related to senescence. If the enzyme telomerase, which maintains telomere length and is normally absent from cells in culture, is expressed in those cells, senescence in culture does not occur. Embryonic stem cells also express telomerase and proliferate indefinitely. However, it is not yet clear whether telomere shortening is a major cause of aging in somatic cells, as certain rodent cells such as Schwann cells can, under appropriate conditions, proliferate indefinitely and their telomeres do not control replication.

Senescent cells that have stopped dividing have changes in the expression of proteins that are involved in cell-cycle control. The proteins p21 and p16 are often expressed; they are part of the tumor-suppressor pathways governed by p53 and RB in response to DNA damage (see Section 11.5). They act as inhibitors of cyclin-dependent kinases and prevent a cell with damaged DNA from entering the cell cycle. This helps to prevent senescent cells, with their DNA damage, from becoming cancerous. Activation of p53 also eliminates cells with DNA damage by apoptosis. Senescence and apoptosis seem to be two different cell fates and can both be triggered by various forms of damage and stress. The difference between the two fates is that senescent cells secrete and produce a range of cytokines and matrix remodeling proteins—known as the senescence-associated secretory phenotype—which can affect neighboring cells, whereas apoptotic cells tend to behave more autonomously, because the DNA fragments and other contents of cell breakdown are retained within the cell membrane.

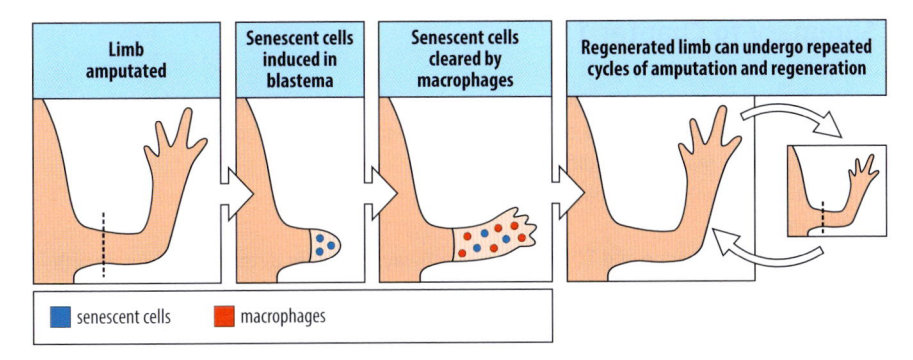

Fig. 11.49 The induction and clearance of senescent cells during salamander limb occurs during repeated cycles of regeneration. First panel: newt limb is amputated. Second panel: senescent cells that secrete cytokines and matrix remodeling proteases are induced in the blastema that arises from the amputated limb. Third panel: at later stages in regeneration the senescent cells are cleared by macrophages so that they do not accumulate in the regenerated limb. Fourth panel: the limb can undergo repeated cycles of regeneration throughout the lifetime of the animal.

11.25 The ability of adult salamanders to regenerate tissues does not diminish with age

In most species, including mammals, senescent cells accumulate in adult tissues over time and this is associated with age-related deterioration in organ function and repair processes such as wound healing. The ability of adult salamanders to regenerate tissues such as the lens does not, however, diminish with age. Surprisingly, senescent cells are induced in limb blastemas in both newts and axolotls and then later eliminated as the limb regenerates (Fig. 11.49). The function of the senescent cells in the blastema is not known, but it could be related to the cytokines and extracellular matrix proteases they secrete. The subsequent rapid and efficient elimination of senescent cells at later stages in regeneration is accomplished by macrophages. Macrophages are routinely recruited to wounds in mammals, and this leads to scarring. Their recruitment to the wound epidermis of a limb blastema is therefore surprising but seems to be essential for regeneration. Depletion of macrophages in an axolotl during the few days before limb amputation blocks regeneration completely, but the stumps can regenerate when re-amputated once macrophages have been replenished. It is not clear whether this requirement for macrophages is because they are needed to clear the senescent cells or whether they have some other function. The ability of adult salamanders to clear senescent cells prevents their accumulation in tissues and this might be related to the salamanders' ability to maintain their regenerative capacity as they age. It has been suggested that targeting senescent cells in mammals might prevent the deterioration of tissues over time and increase life span.

SUMMARY

Aging is largely caused by damage to cells, particularly by reactive oxygen with consequences at several levels (DNA damage, protein instability, mitochondrial destruction). Senescence is the age-related decline in the ability to carry out repairs and is under genetic control. Many normal cells in culture can only undergo a limited number of cell divisions, which correlate with their age at isolation and the normal life span of the animal from which they came. Senescent cells accumulate in tissues in mammals but not in salamanders, which can regenerate new fully functional organs repeatedly into old age. Genes that can increase life span have been identified in *C. elegans* and *Drosophila* and may act by increasing the animals' resistance to oxidative stress.

Summary to Chapter 11

- The form of many animals, such as mammals and birds, is laid down in miniature during embryonic development, and they then grow, keeping the basic body form, although different regions may grow at different rates.
- Growth may involve cell proliferation, cell enlargement, and the laying down of extracellular material.
- The intracellular Hippo and TOR signaling pathways together determine tissue growth.
- Cancer can be viewed as an aberration of growth, as it usually results from mutations that lead to excessive cell proliferation and failure of cells to differentiate.
- In vertebrates, structures have an intrinsic growth program, which is under hormonal control.
- The larvae of arthropods and some other groups, such as frogs among the vertebrates, do not resemble the adult and undergo metamorphosis to reach the adult form. Arthropod larvae are covered by a relatively inextensible chitin cuticle and so must undergo successive molts as they grow.
- Regeneration is the ability of an organism to replace a lost part of its body and can involve growth. The capacity to regenerate varies greatly among different groups of organisms and even between different species within a group. Adult mammals have very limited powers of regeneration, whereas adult newts can regenerate limbs, jaws, and the lens of the eye.
- Regeneration of adult urodele amphibian limbs involves dedifferentiation of cells to form a blastema, which grows and forms the regenerate. Dedifferentiated cells derived from connective tissues make up the early blastema and are multipotent, giving rise to the various connective tissues, and resemble embryonic limb bud cells. Dedifferentiated cells derived from other tissues such as muscle enter the blastema later and are lineage restricted. Regeneration requires the presence of nerves.
- A proximo-distal gradient in positional values is likely to be present in regenerating amphibian and insect limbs because intercalation of positional values occurs when normally non-adjacent values are placed next to each other.
- The adult zebrafish heart can regenerate as a result of dedifferentiation of cardiomyocytes and their proliferation to form new tissue. The neonatal mouse heart in the first week after birth can also regenerate by the same mechanism, but this ability is subsequently lost because of changes in the extracellular matrix.
- The symptoms of aging (senescence) seem mainly to be caused by damage to cells that accumulates over time. Aging is also under genetic control, as shown by aging syndromes in humans and by the effect of mutations in the nematode.

■ End of chapter questions

Check your understanding of the content of this chapter by attempting the following long-answer concept questions.

1. Define 'growth'. What are the main mechanisms of growth? How does cell death fit into a discussion of growth?

2. Why is it inaccurate to say that the cell cycle starts when cytokinesis produces two cells after M phase? According to molecular analysis of the cell cycle, each round of the cycle 'starts' at what point in the cycle? What are the molecular events that define this start point? (The cell-cycle diagram in Box 1B might also be helpful.)

3. In what way is the standard cell cycle modified during the first 13 pre-cellular divisions in *Drosophila* embryogenesis, to allow very rapid cell divisions? What are the maternally supplied proteins that drive these early divisions; and what are these proteins' functions?

4. The final size of some organs, such as liver and spleen, is regulated by extrinsic factors, whereas the size of other organs, such as pancreas and thymus, is under the control of a cell-intrinsic developmental program. Describe the experiments that reveal these two different mechanisms of growth control.

5. In *Drosophila*, the Hippo protein and the *bantam* microRNA have opposing functions, as their names imply. What are the functions of these two regulators of growth? Outline the pathway that initiates signaling to Hippo, and signaling to the *bantam* gene.

6. Address the following questions about human growth hormone: what is growth hormone? Where is it produced? How is its synthesis and secretion controlled? What other growth factor is a major mediator of growth hormone's effect on cellular proliferation?

7. Review the various signal molecules and signaling pathways discussed in the chapter, and indicate which could be classified as proto-oncogenes, and which as tumor suppressors.

8. In one experiment, a frog tadpole is exposed to thyroxine in the water in which it is being reared. In another experiment, the thyroid gland is removed from a tadpole. What results would you predict in each case? Justify your predictions.

9. Outline experiments designed to test the potency of the blastema cells in regenerating axolotl limbs. Are the blastema cells multipotent or lineage restricted? What has single-cell transcriptomics revealed about the behavior of the blastema cells over time?

10. What is the effect of denervation on regeneration of a salamander limb? What is the role of the nerve supply? What protein can substitute for the nerves in enabling regeneration? Why can a limb that has never been innervated regenerate?

11. Refer to Figure 11.35. Describe in your own words what the experiment illustrated was intended to investigate, what was observed, and what interpretation can be drawn from results of the experiment.

12. The dedifferentiated cells of the salamander blastema will regenerate only structures distal to the site of the amputation, which requires some kind of memory of proximo-distal position in the intact limb. Summarize the information that Prod1 may be involved in registering proximo-distal position; be sure to include the results of retinoic acid treatment on *Prod1* expression.

13. Describe the effect of retinoic acid on regeneration; for example, if a limb is cut at the wrist level and treated with retinoic acid, how will its regeneration be altered, compared to regeneration without such treatment?

14. In the regeneration of the cockroach leg, proximo-distal values will be restored through intercalation if two pieces of leg are apposed by grafting (see Fig. 11.41). How does this process of intercalation explain the results of the experiments shown in the figure?

15. Discuss the importance to biomedical science of the ability of portions of the adult zebrafish heart to regenerate (you may wish to review some of the advantages of the zebrafish as a model organism from Chapter 3).

16. How is muscle size regulated? Discuss whether the same principles apply to regulation of the size of other organs such as the liver.

17. What are the prospects for devising ways to stimulate repair and regeneration of human limbs?

18. Discuss the evidence for and against the statement: 'The changes that occur with aging are an inevitable part of an animal's genetically determined postembryonic developmental program.'

■ General further reading

Birnbaum, K.D., Sánchez Alvarado, A.: **Slicing across kingdoms: regeneration in plants and animals**. *Cell* 2008, **132**: 697–710.

Brockes, J.P., Kumar, A.: **Principles of appendage regeneration in adult vertebrates and their implications for regenerative medicine**. *Science* 2005, **310**: 1919–1923.

Brockes, J.P., Kumar, A.: **Comparative aspects of animal regeneration**. *Annu. Rev. Cell Dev. Biol.* 2008, **24**: 525–549.

Galliot, B., Tanaka, E., Simon, A.: **Regeneration and tissue repair**. *Cell. Mol. Life Sci* 2008, **65**: 3–7.

Goss, R.J.: *The Physiology of Growth*. New York: Academic Press, 1978.

Hanahan, D., Weinberg, R.A.: **Hallmarks of cancer: the next generation**. *Cell* 2011, **144**: 646–674.

Kirkwood, T.B.L.: **Understanding the odd science of aging**. *Cell* 2005, **120**: 437–447.

Laudet, V.: **The origins and evolution of vertebrate metamorphosis**. *Curr. Biol* 2011, **21**: R726–R737.

Lopez-Otin, C., Blasco, M.A., Partridge, L., Serrano, M., Kroemer, G.: **The hallmarks of aging**. *Cell* 2013, **153**: 1194–1217.

Melzer, D., Pilling, L.C., Ferrucci, L.L.: **The genetics of human ageing**. *Nat. Rev. Genet.* 2020, **21**: 88–101.

Nature Insight: **Cell division and cancer**. *Nature* 2004, **432**: 293–341.

Partridge, L.: **The new biology of ageing**. *Phil. Trans R. Soc. B* 2010, **365**: 147–154.

Yin, H., Price, F., Rudnicki, M.A.: **Satellite cells and the muscle stem cell niche**. *Physiol. Rev.* 2013, **93**: 23–67.

■ Section further reading

11.1 Tissues can grow by cell proliferation, cell enlargement, or accretion

Tumaneng, K., Russell, R.C., Guan, K.-L.: **Organ size control by TOR and Hippo pathways**. *Curr. Biol.* 2012, **22**: R368–R379.

11.2 Cell proliferation is controlled by regulating entry into the cell cycle

Morgan, D.O.: *The Cell Cycle: Principles of Control*. London: New Science Press Ltd, 2007.

11.3 Cell division in early development can be controlled by an intrinsic developmental program

Edgar, B., Lehner, C.F.: **Developmental control of cell cycle regulators: a fly's perspective**. *Science* 1996, **274**: 1646–1652.

Follette, P.J., O'Farrell, P.H.: **Connecting cell behavior to patterning: lessons from the cell cycle**. *Cell* 1997, **88**: 309–314.

Lasko, P.: **Development: new wrinkles on genetic control of the MBT**. *Curr. Biol.* 2013, **23**: R65–R67.

11.4 Extrinsic signals coordinate cell division, cell growth, and cell death in the developing *Drosophila* wing

de la Cova, C., Abril, M., Bellosta, P., Gallant, P., Johnston, L.A.: ***Drosophila* myc regulates organ size by inducing cell competition**. *Cell* 2004, **117**: 107–116.

Halder, G., Johnson, R.L.: **Hippo signaling: growth control and beyond**. *Development* 2011, **138**: 9–22.

Hariharan, I.K.: **Organ size control: lessons from *Drosophila***. *Dev. Cell* 2015, **34**: 255–265.

Huang, J., Wu, S., Barrera, J., Matthews, K., Pan, D.: **The Hippo signaling pathway coordinately regulates cell proliferation and apoptosis by inactivating Yorkie, the *Drosophila* homolog of YAP**. *Cell* 2005, **122**: 421–434.

Martin, F.A., Herrera, S.C., Morata, G.: **Cell competition, growth and size control in the *Drosophila* wing imaginal disc.** *Development* 2009, **136**: 3747–3756.

Willecke, M., Hamaratoglu, F., Sansores-Garcia, L., Tao, C., Halder, G.: **Boundaries of Dachsous cadherin activity modulate the Hippo signaling pathway to induce cell proliferation.** *Proc. Natl Acad. Sci USA* 2008, **105**: 14897–14902.

Yu, F.X., Zhao, B., Guan, K.L.: **Hippo pathway in organ size control, tissue homeostasis, and cancer.** *Cell* 2015, **163**: 811–828.

Box 11A The core Hippo signaling pathways in *Drosophila* and mammals

Tumaneng, K., Russell, R.C., Guan, K.-L.: **Organ size control by TOR and Hippo pathways.** *Curr. Biol.* 2012, **22**: R368–R379.

Zheng, Y., Pan, D.: **The Hippo signaling pathway in development and disease.** *Dev. Cell* 2019, **50**: 264–282.

11.5 Cancer can result from mutations in genes that control cell proliferation

Beachy, P.A., Karhadkar, S.S., Berman, D.M.: **Tissue repair and stem cell renewal in carcinogenesis.** *Nature* 2004, **432**: 324–331.

Harvey, K.F., Zhang, X., Thomas, D.M.: **The Hippo pathway and human cancer.** *Nat. Rev. Cancer* 2013, **13**: 246–257.

Hunter, T.: **Oncoprotein networks.** *Cell* 1997, **88**: 333–346.

Jones, S., et al.: **Core signaling pathways in human pancreatic cancers revealed by global genomic analyses.** *Science* 2008, **321**: 1801–1806.

Shilo, B.Z.: **Dispatches from *patched*.** *Nature* 1996, **382**: 115–116.

Taipale, J., Beachy, P.A.: **The hedgehog and Wnt signalling pathways in cancer.** *Nature* 2001, **411**: 349–353.

Van Dyke, T.: **p53 and tumor suppression.** *N. Engl. J. Med.* 2007, **356**: 79–92.

Yu, F.X., Zhao, B., Guan, K.L.: **Hippo pathway in organ size control, tissue homeostasis, and cancer.** *Cell* 2015, **163**: 811–828.

11.6 The relative contributions of intrinsic and extrinsic factors in controlling size differ in different mammalian organs

Amthor, H., Huang, R., McKinnell, I., Christ, B., Kambadur, R., Sharma, M., Patel, K.: **The regulation and action of myostatin as a negative regulator of muscle development during avian embryogenesis.** *Dev. Biol.* 2002, **251**: 241–257.

Lui, J.C., Baron, J.: **Mechanisms limiting body growth in mammals.** *Endocr. Rev.* 2011, **32**: 422–440.

Michalopoulos, G.K., Bhushan, B.: **Liver regeneration: biological and pathological mechanisms and implications.** *Nat. Rev. Gastroenterol. Hepatol.* 2021, **18**: 40–55.

Penzo-Méndez, A.I., Stanger, B.Z.: **Organ-size regulation in mammals.** *Cold Spring Harb. Perspect. Biol.* 2015, **7**: a019240.

Rodriguez, J., Vernus, B., Chelh, I., Cassar-Malek, I., Gabillard. J., Hadj Sassi, A., Seiliez, I., Picard, B., Bonnieu, A.: **Myostatin and the skeletal muscle atrophy and hypertrophy signaling pathways.** *Cell. Mol. Life Sci.* 2014, **71**: 4361–4371.

Shingleton, A.W.: **Body-size regulation: combining genetics and physiology.** *Curr. Biol.* 2005, **15**: R825–R827.

Stanger, B.Z.: **The biology of organ size determination.** *Diabetes Obes. Metab.* 2008, **10**: 16–26.

Stanger, B.Z., Tanaka, A.J., Melton, D.A.: **Organ size is limited by the number of embryonic progenitor cells in the pancreas but not in the liver.** *Nature* 2007, **445**: 886–891.

Taub, R.: **Liver regeneration: from myth to mechanism.** *Nat. Rev. Mol. Cell Biol.* 2004, **5**: 836–847.

11.8 Hormones and growth factors coordinate the growth of different tissues and organs and contribute to determining overall body size

Efstratiadis, A.: **Genetics of mouse growth.** *Int. J. Dev. Biol.* 1998, **42**: 955–976.

Lui, J.C., Baron, J.: **Mechanisms limiting body growth in mammals.** *Endocr. Rev.* 2011, **32**: 422–440.

Lupu, F., Terwilliger, J.D., Lee, K., Segre, G.V., Efstratiadis, A.: **Roles of growth hormone and insulin-like growth factor 1 in mouse postnatal growth.** *Dev. Biol.* 2001, **229**: 141–162.

Sanders, E.J., Harvey S.: **Growth hormone as an early embryonic growth and differentiation factor.** *Anat. Embryol.* 2004, **209**: 1–9.

11.9 Elongation of the long bones illustrates how growth can be determined by a combination of an intrinsic growth program and extracellular factors

Aghajanian, P., Mohan, S.: **The art of building bone: emerging role of chondrocyte-to-osteoblast transdifferentiation in endochondral ossification.** *Bone Res.* 2018, **6**: 19.

DiGirolamo, D.J., Kiel, D.P., Esser, K.A.: **Bone and skeletal muscle: neighbors with close ties.** *J. Bone Miner. Res.* 2013, **28**: 1509–1518.

Galea, G.L., Zein, M.R., Allen, S., Francis-West, P.: **Making and shaping endochondral and intramembranous bones.** *Dev. Dyn.* 2021, **250**: 414–449.

Kember, N.F.: **Cell kinetics and the control of bone growth.** *Acta Paediatr. Suppl.* 1993, **391**: 61–65.

Kronenberg, H.M.: **Developmental regulation of the growth plate.** *Nature* 2003, **423**: 332–336.

Nilsson, O., Baron, J.: **Fundamental limits on longitudinal bone growth: growth plate senescence and epiphyseal function.** *Trends Endocrinol. Metab.* 2004, **8**: 370–374.

Roselló-Díez, A., Joyner, A.L.: **Regulation of long bone growth in vertebrates; it is time to catch up.** *Endocr. Rev.* 2015, **36**: 646–680.

Roselló-Díez, A., Madisen, L., Bastide, S., Zeng, H., Joyner, A.L. **Cell-nonautonomous local and systemic responses to cell arrest enable long-bone catch-up growth in developing mice.** *PLoS Biol.* 2018, **16**: e2005086.

Roush, W.: **Putting the brakes on bone growth.** *Science* 1996, **273**: 579.

Schultz, E.: **Satellite cell proliferative compartments in growing skeletal muscles.** *Dev. Biol.* 1996, **175**: 84–94.

Williams, P.E., Goldspink, G.: **Changes in sarcomere length and physiological properties in immobilized muscle.** *J. Anat.* 1978, **127**: 450–468.

Box 11B Digit length ratio is determined in the embryo

Manning, J.T.: *The Finger Book*. London: Faber and Faber, 2008.

Manning, J.T.: **Resolving the role of prenatal sex steroids in the development of digit ratio.** *Proc. Natl Acad. Sci. USA* 2011, **108**: 16143–16144.

Zheng, Z., Cohn, M.J.: **Developmental basis of sexually dimorphic digit ratios**. *Proc. Natl Acad. Sci. USA* 2011, **108**: 16289–16294.

11.10 The amount of nourishment an embryo receives can have profound effects in later life

Barker, D.J.: **The Wellcome Foundation Lecture, 1994. The fetal origins of adult disease**. *Proc. R. Soc. Lond.* 1995, **262**: 37–43.

Gluckman, P.D., Hanson, M.A., Cooper, C., Thornburg, K.L.: **Effect of *in utero* and early-life conditions on adult health and disease**. *N. Engl. J. Med.* 2008, **359**: 61–73.

11.12 Insect body size is determined by the rate and duration of larval growth

Andersen, D.S., Colombani, J., Leopold, P.: **Coordination of organ growth: principles and outstanding questions from the world of insects**. *Trends Cell Biol.* 2013, **23**: 336–344.

De Loof, A.: **Ecdysteroids, juvenile hormone and insect neuropeptides: recent successes and remaining major challenges**. *Gen. Comp. Endocrinol.* 2008, **155**: 3–13.

Lee, G.J., Han, G., Yun, H.M., Lim, J.J., Noh, S., Lee, J., Hyun, S.: **Steroid signaling mediates nutritional regulation of juvenile body growth via IGF-binding protein in *Drosophila***. *Proc. Natl Acad. Sci. USA* 2018, **115**: 5992–5997.

Mirth, C.K., Riddiford, L.M.: **Size assessment and growth control: how adult size is determined in insects**. *BioEssays* 2007, **29**: 344–355.

Mirth C.K., Shingelton, A.W. **Integrating body and organ size in *Drosophila*: recent advances and outstanding problems**. *Front. Endocr.* 2012, **3**: 1–13.

Nijhout, H.F.: **Size matters (but so does time), and it's OK to be different**. *Dev. Cell* 2008, **15**: 491–492.

Stern, D.: **Body-size control: how an insect knows it has grown enough**. *Curr. Biol.* 2003, **13**: R267–R269.

Thummel, C.S.: **Flies on steroids—*Drosophila* metamorphosis and the mechanisms of steroid hormone action**. *Trends Genet.* 1996, **12**: 306–310.

11.13 Metamorphosis in amphibians is under hormonal control

Brown, D.D., Cai, L.: **Amphibian metamorphosis**. *Dev. Biol.* 2007, **306**: 20–33.

Huang, H., Brown, D.D.: **Prolactin is not juvenile hormone in *Xenopus laevis* metamorphosis**. *Proc. Natl Acad. Sci. USA* 2000, **97**: 195–199.

Tata, J.R.: *Hormonal Signaling and Postembryonic Development*. Heidelberg: Springer, 1998.

11.14 Regeneration involves repatterning of existing tissues and/or growth of new tissues

Agata, K., Saito, Y., Nakajima, E.: **Unifying principles of regeneration I: epimorphosis versus morphallaxis**. *Dev. Growth Differ.* 2007, **49**: 73–78.

Allen, J.M., Ross, K.G., Zayas, R.M.: **Regeneration in invertebrates: model systems**. In *eLS* (John Wiley & Sons, 2016) [https://doi.org/10.1002/9780470015902.a0001095.pub2] (date accessed 16 October 2022).

Ferretti, P.: **Regeneration of vertebrate appendages**. In *eLS* (John Wiley & Sons, 2013) [https://doi.org/10.1002/9780470015902.a0001099.pub3] (date accessed 16 October 2022).

Khan, S.J., Schuster, K.J., Smith-Bolton, R.K.: **Regeneration in crustaceans and insects**. In *eLS* (John Wiley & Sons, 2016) [https://doi.org/10.1002/9780470015902.a0001098.pub2] (date accessed 16 October 2022).

King, R.S., Newmark, P.A.: **The cell biology of regeneration**. *J. Cell Biol.* 2012, **196**: 553–562.

Sánchez Alvarado, A., Tsonis, P.A. **Bridging the regeneration gap: genetic insights from diverse animal models**. *Nat. Rev. Genet.* 2006, **7**: 873–884.

11.15 Regeneration in *Hydra* involves repatterning of existing tissues

Bode, H.R.: **The head organizer in *Hydra***. *Int. J. Dev. Biol.* 2012, **56**: 473–478.

Broun, M., Gee, L., Reinhardt, B., Bode, H.R.: **Formation of the head organizer in *Hydra* involves the canonical Wnt pathway**. *Development* 2005, **132**: 2907–2916.

Browne, E.N.: **The production of new hydranths in *Hydra* by the insertion of small grafts**. *J. Exp. Zool.* 1909, **7**: 1–37.

Chera, S., Ghila, L., Dobretz, K., Wenger, Y., Bauer, C., Buzgariu, W., Martinou, J.C., Galliot, B.: **Apoptotic cells provide an unexpected source of Wnt3 signaling to drive hydra head regeneration**. *Dev. Cell* 2009, **17**: 279–289.

Galliot, B.: **Regeneration in *Hydra***. In *eLS* (John Wiley & Sons, 2013) [https://doi.org/10.1002/9780470015902.a0001096.pub3] (date accessed 16 October 2022).

Gierer, A., Meinhardt, H.: **A theory of biological pattern formation**. *Kybernetik* 1972, **12**: 30–39.

Hicklin, J., Wolpert, L.: **Positional information and pattern regulation in *Hydra*: the effect of gamma-radiation**. *J. Embryol. Exp. Morph.* 1973, **30**: 741–752.

Hobmayer, B., Rentzsch, F., Kuhn, K., Happel, C.M., von Laue, C.C., Snyder, P., Rothbächer, U., Holstein, T.W.: **WNT signalling molecules act in axis formation in the diploblastic metazoan *Hydra***. *Nature* 2000, **407**: 186–189.

Müller, W.A.: **Pattern formation in the immortal *Hydra***. *Trends Genet.* 1996, **12**: 91–96.

Takahashi, T., Fujisawa, T.: **Important roles for epithelial cell peptides in hydra development**. *BioEssays* 2009, **31**: 610–619.

Vogg, M.C., Galliot, B., Tsiairis, C.D.: **Model systems for regeneration: *Hydra***. *Development* 2019, **146**: dev177212.

Wolpert, L., Hornbruch, A., Clarke, M.R.B.: **Positional information and positional signaling in *Hydra***. *Am. Zool.* 1974, **14**: 647–663.

Box 11C Planarian regeneration

Aboukhatwa, E., Aboobaker, A.: **An introduction to planarians and their stem cells**. In *eLS* (John Wiley & Sons, 2015) [https://doi.org/10.1002/9780470015902.a0001097.pub2] (date accessed 16 October 2022).

Adell, T., Cebrià, F., Saló, E.: **Gradients in planarian regeneration and homeostasis**. *Cold Spring Harb. Perspect. Biol.* 2010, **2**: a000505.

Adell, T., Saló, E., Boutros, M., Bartscherer, K.: **Smed-Evi/Wntless is required for beta-catenin-dependent and -independent processes during planarian regeneration**. *Development* 2009, **136**: 905–910.

De Robertis, E.M.: **Wnt signaling in axial patterning and regeneration: lessons from planaria**. *Sci. Signal.* 2010, **3**: pe21.

Reddien, P.W.: **Specialized progenitors and regeneration**. *Development* 2013, **140**: 951-957.

Reddien, P.W., Sánchez Alvarado, A.: **Fundamentals of planarian regeneration**. *Annu. Rev. Cell Dev. Biol.* 2004, **20**: 725-757.

Scimone, M.L., Cloutier, J.K., Maybrun, C.L., Reddien, P.W.: **The planarian wound epidermis gene equinox is required for blastema formation in regeneration**. *Nat. Commun.* 2022, **13**: 2726.

Simon, A.: **On with their heads**. *Nature*, 2013, **500**: 32-33.

Wagner, D.E., Wang, I.E., Reddien, P.W.: **Clonogenic neoblasts are pluripotent adult stem cells that underlie planarian regeneration**. *Science*, 2011, **332**: 811-816.

Zeng, A., HuaLi, H., Guo, L., Gao, X., McKinney, S., Wang, Y. Yu, Z., Park, J., Semerad, C., Ross, E., Cheng, L.C., Davies, E., Lei, K., Wang, W., Perera, A., Hall., K., Peak, A., Box, A., Sánchez Alvarado, A.: **Prospectively isolated tetraspanin + neoblasts are adult pluripotent stem cells underlying planaria regeneration**. *Cell* 2018, **173**: 1593-1608.

11.16 Amphibian limb regeneration involves cell dedifferentiation and new growth

Alvarado, A.S.: **A cellular view of regeneration**. *Nature* 2009, **460**: 39-40.

Aztekin, C.: **Tissues and cell types of appendage regeneration: A detailed look at the wound epidermis and its specialized forms**. *Front. Physiol.* 2021, **12**: 771040.

Aztekin, C., Hiscock, T.W., Gurdon, J., Jullien, J., Marioni, J., Simons, B.D.: **Secreted inhibitors drive the loss of regeneration competence in *Xenopus* limbs**. *Development* 2021, **148**: dev199158.

Bassat, E., Tanaka, E.M.: **The cellular and signaling dynamics of salamander limb regeneration**. *Curr. Opin. Cell Biol.* 2021, **73**: 117-123.

Brockes, J.P., Kumar, A.: **Plasticity and reprogramming of differentiated cells in amphibian regeneration**. *Nat. Rev. Mol. Cell Biol.* 2002, **3**: 566-574.

Currie, J.D., Kawaguchi, A., Traspas, R.M., Schuez, M., Chara, O., Tanaka, E.M.: **Live imaging of axolotl digit regeneration reveals spatiotemporal choreography of diverse connective tissue progenitor pools**. *Dev. Cell* 2016, **39**: 411-423.

Frasch, M.: **Dedifferentiation, redifferentiation, and transdifferentiation of striated muscles during regeneration and development**. *Curr. Top. Dev. Biol.* 2016, **116**: 331-355.

Gerber, T., Murawala, P., Knapp, D., Masselink, W., Schuez, M., Hermann, S., Gac-Santel, M., Nowoshilow, S., Kageyama, J., Khattak, S., Currie, J.D., Camp, J.G., Tanaka, E.M., Treutlein, B.: **Single-cell analysis uncovers convergence of cell identities during axolotl limb regeneration**. *Science* 2018, **362**: eaaq0681.

Godwin, J.W., Pinto, A.R., Rosenthal, N.A.: **Macrophages are required for adult salamander limb regeneration**. *Proc. Natl Acad. Sci. USA* 2013, **110**: 9415-9420.

Imokawa, Y., Simon, A., Brockes, J.P.: **A critical role for thrombin in vertebrate lens regeneration**. *Phil. Trans. R. Soc. Lond. B Biol. Sci.* 2004, **359**: 765-776.

Kragl, M., Knapp, D., Nacu, E., Khattak, S., Maden, M., Epperlein, H.H., Tanaka, E.M.: **Cells keep a memory of their tissue origin during axolotl limb regeneration**. *Nature* 2009, **460**: 60-65.

Kumar, A., Velloso, C.P., Imokawa, Y., Brockes, J.P.: **The regenerative plasticity of isolated urodele myofibers and its dependence on MSX1**. *PLoS Biol.* 2004, **2**: E218.

Nacu, E., Glausch, M., Le, H. Q., Damanik, R.F.F., Schuez, M., Knapp, D., Khattak, S., Richter, T., Tanaka, E.M.: **Connective tissue cells, but not muscle cells, are involved in establishing the proximo-distal outcome of limb regeneration in the axolotl**. *Development* 2013, **140**: 513-518.

Sandoval-Guzmán, T., Wang, H., Khattak, S., Schuez, M., Roensch, K., Nacu, E., Tazaki, A., Joven, A., Tanaka, E.M., Simon, A.: **Fundamental differences in dedifferentiation and stem cell recruitment during skeletal muscle regeneration in two salamander species**. *Cell Stem Cell* 2013, **14**: 174-187.

Satoh, A., Graham, G.M.C., Bryant, S.V., Gardiner, D.M.: **Neurotrophic regulation of epidermal dedifferentiation during wound healing and limb regeneration in the axolotl (*Ambystoma mexicanum*)**. *Dev. Biol.* 2008, **319**: 321-355.

Stocum, D.L.: **Mechanisms of urodele limb regeneration**. *Regeneration* 2017, **4**: 159-200.

Tanaka, E.M., Drechel, D.N., Brockes, J.P.: **Thrombin regulates S-phase re-entry by cultured newt myotubes**. *Curr. Biol.* 1999, **9**: 792-799.

Tanaka, H.V., Ng, N.C., Yang Yu, Z., Casco-Robles, M.M., Maruo, F., Tsonis, P.A., Chiba, C.: **A developmentally regulated switch from stem cells to dedifferentiation for limb muscle regeneration in newts**. *Nat. Commun.* 2016, **7**: 11069.

Wagner, I., Wang, H., Weissert, P.M., Straube, W.L., Shevchenko, A., Gentzel, M., Brito, G., Tazaki, A., Oliveira, C., Sugiura, T., Shevchenko, A., Simon, A., Drechsel, D.N., Tanaka, E.M.: **Serum proteases potentiate BMP-induced cell cycle re-entry of dedifferentiating muscle cells during newt limb regeneration**. *Dev. Cell* 2017, **40**: 608-617.

11.17 Limb regeneration in amphibians depends on the presence of nerves

Kumar, A., Delgado, J.P., Gates, P.B., Neville, G., Forge, A., Brockes, J.P.: **The aneurogenic limb identifies developmental cell interactions underlying vertebrate limb regeneration**. *Proc. Natl Acad. Sci. USA* 2011, **108**: 13588-13593.

Kumar, A., Godwin, J.W., Gates, P.B., Garza-Garcia, A.A., Brockes, J.P.: **Molecular basis for the nerve dependence of limb regeneration in an adult vertebrate**. *Science* 2007, **318**: 772-777.

11.18 The limb blastema gives rise to structures with positional values distal to the site of amputation

Da Silva, S., Gates, P.B., Brockes, J.P.: **New ortholog of CD59 is implicated in proximodistal identity during amphibian limb regeneration**. *Dev. Cell* 2002, **3**: 547-551.

Echeverri, K., Tanaka, E.M.: **Proximodistal patterning during limb regeneration**. *Dev. Biol.* 2005, **279**: 391-401.

Roensch, K., Tazaki, A., Chara, O., Tanaka, E.M.: **Progressive specification rather than intercalation of segments during limb regeneration**. *Science* 2013, **342**: 1375-1379.

11.19 Retinoic acid can change proximo-distal positional values in regenerating limbs

Maden, M.: **The homeotic transformation of tails into limbs in *Rana temporaria* by retinoids**. *Dev. Biol.* 1993, **159**: 379-391.

Mercader, N., Tanaka, E.M., Torres, M.: **Proximodistal identity during limb regeneration is regulated by Meis homeodomain proteins**. *Development* 2005, **132**: 4131-4142.

Pecorino, L.T., Entwistle, A., Brockes, J.P.: **Activation of a single retinoic acid receptor isoform mediates proximo-distal respecification.** *Curr. Biol.* 1996, **6**: 563–569.

Scadding, S.R., Maden, M.: **Retinoic acid gradients during limb regeneration.** *Dev. Biol.* 1994, **162**: 608–617.

11.20 Mammals can regenerate the tips of the digits

Han, M., Yang, X., Farrington, J.E., Muneoka, K.: **Digit regeneration is regulated by *Msx1* and BMP4 in fetal mice.** *Development* 2003, **130**: 5123–5132.

Han, M., Yang, X., Lee, J., Allan, C.H., Muneoka, K.: **Development and regeneration of the neonatal digit tip in mice.** *Dev. Biol.* 2008, **315**: 125–135.

Rinkevich, Y., Lindau, P., Ueno, H., Longaker, M.T., Weissman, I.L.: **Germ-layer and lineage-restricted stem/progenitors regenerate the mouse digit tip.** *Nature* 2011, **476**: 409–413.

Seifert, A.W., Muneoka, K.: **The blastema and epimorphic regeneration in mammals.** *Dev. Biol.* 2018, **433**: 190–199.

Storer, M.A., Mahmud, N., Karamboulas, K., Borrett, M.J., Yuzwa, S.A., Gont, A., Androschuk, A., Sefton, M.V., Kaplan, D.R., Miller, F.D.: **Acquisition of a unique mesenchymal precursor-like blastema state underlies successful adult mammalian digit tip regeneration.** *Dev. Cell* 2020, **52**: 509–524.

Takeo, M., Chou, W.C., Sun, Q., Lee, W., Rabbani, P., Loomis, C., Taketo, M.M., Ito, M.: **Wnt activation in nail epithelium couples nail growth to digit regeneration.** *Nature* 2013, **499**: 228–232.

Box 11D Why can't we regenerate our limbs?

Alibardi, L.: **Limb regeneration in humans: dream or reality?** *Ann. Anat.* 2018, **217**: 1–6.

Bely, A.E.: **Evolutionary loss of animal regeneration: pattern and process.** *Integr. Comp. Biol.* 2010, **50**: 515–527.

Garza-Garcia, A.A., Driscoll P.C., Brockes, J.P.: **Evidence for the local evolution of mechanisms underlying limb regeneration in salamanders.** *Integr. Comp. Biol.* 2010, **50**: 528–535.

Mihaylova, Y., Aboobaker, A.A.: **What is it about eye of newt?** *Genome Biol.* 2013, **14**: 106.

Shaikh, N., Gates, P.B., Brockes, J.P.: **The Meis homeoprotein regulates the axolotl *Prod 1* promoter during limb regeneration.** *Gene* 2011, **484**: 69–74.

Slack, J.M.W.: **Animal regeneration: ancestral character or evolutionary novelty?** *EMBO Rep.* 2017, **18**: 1497–1508.

11.21 Insect limbs intercalate positional values by both proximo-distal and circumferential growth

Bando, T., Mito, T., Maeda, Y., Nakamura, T., Ito, F., Watanabe, T., Ohuchi, H., Noji, S.: **Regulation of leg size and shape by the Dachsous/Fat signalling pathway during regeneration.** *Development* 2009, **136**: 2235–2245.

French, V.: **Pattern regulation and regeneration.** *Phil. Trans. R. Soc. Lond. B Biol. Sci.* 1981, **295**: 601–617.

Nakamura, T., Mito, T., Miyawaki, K., Ohuchi, H., Noji, S.: **EGFR signaling is required for re-establishing the proximodistal axis during distal leg regeneration in the cricket *Gryllus bimaculatus* nymph.** *Dev. Biol.* 2008, **319**: 46–55.

11.22 Heart regeneration in zebrafish involves the resumption of cell division by cardiomyocytes

Bassat, E., Mutlak, Y.E., Genzelinakh, A., Shadrin, I.Y., Baruch Umansky, K., Yifa, O., Kain, D., Rajchman, D., Leach, J., Riabov Bassat, D., Udi, Y., Sarig, R., Sagi, I., Martin, J.F., Bursac, N., Cohen, S., Tzahor, E.: **The extracellular matrix protein agrin promotes heart regeneration in mice.** *Nature* 2017, **547**: 179–184.

Eroglu, E., Chien, K.R.: **Heart regeneration 4.0: matrix medicine.** *Dev. Cell* 2017, **42**: 7–8.

Gemberling, M., Bailey, T.J., Hyde, D.R., Poss, K.D.: **The zebrafish as a model for complex tissue regeneration.** *Trends Genet.* 2013, **29**: 611–620.

Itou, J., Oishi, I., Kawakami, H., Glass, T.J., Richter, J., Johnson, A., Lund, T.C., Kawakami, Y.: **Migration of cardiomyocytes is essential for heart regeneration in zebrafish.** *Development* 2012, **139**: 4133–4142.

Lepilina, A., Coon, A.N., Kikuchi, K., Holdway, J.E., Roberts, R.W., Burns, C.G., Poss, K.D.: **A dynamic epicardial injury response supports progenitor cell activity during zebrafish heart regeneration.** *Cell* 2006, **127**: 607–619.

Mahmoud, A.I., Kocabas, F., Shalini A., Muralidhar, S.A., Kimura, W., Koura, A.S., Thet, S., Porrello, E.R., Sadek, H.A.: **Meis1 regulates postnatal cardiomyocyte cell-cycle arrest.** *Nature* 2013, **497**: 249–253.

Oyama, K., El-Nachef, D., MacLellan, WR.: **Regeneration potential of adult cardiac myocytes.** *Cell Res.* 2013, **23**: 978–979.

Porrello, E.R., Mahmoud, A.I., Simpson, E., Hill, J.A., Richardson, J.A., Olson, E.N., Sadek, H.A.: **Transient regenerative potential of the neonatal mouse heart.** *Science* 2011, **331**: 1078–1080.

Poss, K.D., Wilson, L.G., Keating, M.T.: **Heart regeneration in zebrafish.** *Science* 2002, **298**: 2188–2190.

Zhang, R., Han, P., Yang, H., Ouyang, K., Lee, D., Lin, Y.-F., Ocorr, K., Kang, G., Chen, J., Stainier, D.Y.R., Yelon, D., Chi, N.C.: **In vivo cardiac reprogramming contributes to zebrafish heart regeneration.** *Nature* 2013, **498**: 497–501.

11.23 Genes can alter the timing of senescence

Arantes-Oliviera, N., Berman, J.R., Kenyon, C.: **Healthy animals with extreme longevity.** *Science* 2003, **302**: 611.

Campisi, J., d'Adda di Fagagna, F.: **Cellular senescence: when bad things happen to good cells.** *Nat. Rev. Mol. Cell Biol.* 2007, **8**: 729–740.

Finkel, T., Serrano, M., Blasco, M.A.: **The common biology of cancer and ageing.** *Nature* 2007, **448**: 767–774.

Flatt, T., Partridge, L.: **Horizons in the evolution of aging.** *BMC Biol.* 2018, **16**: 93.

Harper, M.E., Bevilacqua, L., Hagopian, K., Weindruch, R., Ramsey, J.J.: **Ageing, oxidative stress, and mitochondrial uncoupling.** *Acta Physiol. Scand.* 2004, **182**: 321–331.

Kenyon, C.: **The plasticity of aging: insights from long-lived mutants.** *Cell* 2005, **120**: 449–460.

Kipling, D., Davis, T., Ostler, E.L., Faragher, R.G.: **What can progeroid syndromes tell us about human aging?** *Science* 2004, **305**: 1426–1431.

Kudlow, B.A., Kennedy, B.K., Monnat, R.J.: **Werner and Hutchinson–Gilford progeria syndromes: mechanistic basis of progerial diseases**. *Nat. Rev. Mol. Cell Biol.* 2007, **8**: 394–404.

Murphy, C.T., Partridge, L., Gems, D.: **Mechanisms of ageing: public or private**. *Nat Rev. Genet.* 2002, **3**: 165–175.

Partridge, L.: **Some highlights of research on aging with invertebrates**. *Aging Cell* 2008, **7**: 605–608.

Ram, J.L., Conn, P.M. (Eds). *Conn's Handbook of Models for Human Aging*. 2nd edn. Cambridge, MA: Academic Press, 2018.

Ruby, J.G., Smith, M., Buffenstein, R.: **Naked mole-rat mortality rates defy gompertzian laws by not increasing with age**. *eLife* 2018, **7**: e31157.

Tatar, M., Bartke, A., Antebi, A.: **The endocrine regulation of aging by insulin-like signals**. *Science* 2003, **299**: 1346–1351.

Tomczyk, S., Fischer, K., Austad, S., Galliot, B.: **Hydra, a powerful model for aging studies**. *Invertebr. Reprod. Dev.* 2015, **59**: 11–16.

Weindruch, R.: **Caloric restriction and aging**. *Science* 1996, **274**: 46–52.

11.24 Cell senescence blocks cell proliferation

Childs, B.G., Baker, D.J., Kirkland, J.L., Campisi, J., van Deursen, J.M.: **Senescence and apoptosis: dueling or complementary cell fates?** *EMBO Rep.* 2014, **15**: 1139–1153.

Shay, J.W., Wright, W.E.: **When do telomeres matter?** *Science* 2001, **291**: 839–840.

Sherr, C.J., DePinho, R.A.: **Cellular senescence: mitotic clock or culture shock**. *Cell* 2000, **102**: 407–410.

11.25 The ability of adult salamanders to regenerate tissues does not diminish with age

Yun, M.H., Davaapil, H., Brockes, J.P.: **Recurrent turnover of senescent cells during regeneration of a complex structure**. *eLife* 2015, **4**: e05505.

12

Plant development

- Introduction to *Arabidopsis thaliana*
- Embryonic development
- Meristems
- Flower development and control of flowering

In this chapter we will give an overview of the mechanisms of plant development. Plants offer an insight into the design principles of development; although plants evolved independently of animals, many of the developmental mechanisms are similar. There are unique features to plant development, including the lack of cell movement, rigid cell walls, and the fact that plant development continues post-embryonically and is highly responsive to environmental cues. Despite these differences, plants, just as with animals, determine cell fate in development through a combination of positional signals and intercellular communication.

Introduction

Plants are vital to all life on Earth, providing essential ecosystem services, as well as being the major source of food for people and other animals. The plant kingdom is very large, ranging from the algae, many of which are unicellular, to the multicellular land plants, which exist in a prodigious range of forms. Plants and animals are thought to have evolved the process of multicellular development independently, their last common ancestor being a unicellular eukaryote that lived some 1.6 billion years ago. The study of plant development is, therefore, of interest not only for its agricultural importance, but also because it sheds light on the way that developmental mechanisms in two groups of multicellular organisms have evolved independently and under different sets of developmental constraints.

Do plants and animals use the same developmental mechanisms? As we shall see in this chapter, the logic behind the spatial layouts of gene expression that pattern a developing flower is similar to that of Hox gene action in patterning the body axis in animals, but the genes involved are completely different. Such similarities between plant and animal development are due to the fact that the basic means of regulating gene expression are the same in both, and thus similar general mechanisms for patterning gene expression in a multicellular tissue are bound to arise. As we shall see, many of the general control mechanisms we have encountered in animal development, such as **asymmetric cell division**, the response to positional signals, **lateral inhibition** and changes in gene expression in response to extracellular signals, are all present in plants. Differences between plants and animals in the way development is controlled arise from some different ways in which plant cells can communicate with each other compared with animal cells, from the existence of rigid cell walls and the lack of large-scale cell movements, and to the fact that the environment has a much greater impact on plant development than on that of animals.

One general and crucial difference between plant and animal development is that most of the development occurs not in the embryo, but in the growing plant. The mature plant embryo inside a seed is not simply a smaller version of the organism it will become. All the 'adult' structures of the plant—shoots, roots, stalks, leaves, and flowers—are produced after germination from localized groups of undifferentiated cells known as **meristems**. Two meristems are established in the embryo: one at the tip of the root and the other at the tip of the shoot. These persist in the adult plant, and almost all the other meristems, such as those in flower shoots, are derived from them. Cells within meristems can divide repeatedly and can potentially give rise to all plant tissues and organs. This means that developmental patterning within meristems to produce organs such as leaves and flowers continues throughout a plant's life.

Plant and animal cells share many common internal features and much basic biochemistry, but there are some key differences that have a bearing on plant development. One of the most important is that plant cells are surrounded by a framework of relatively rigid cell walls. There is, therefore, virtually no cell migration in plants, and major changes in the shape of the developing plant cannot be achieved by the movement of sheets of cells. In plants, form is largely generated by differences in rates of cell division and by division in different planes, followed by directed enlargement of the cells. When a plant cell divides, it is not pinched in half by a contractile actomyosin ring as in an animal cell. Instead, a new cell membrane and cell wall is formed *in situ* midway between the two poles of the mitotic spindle. In plants, the plane of division seems to be determined before mitosis begins, as a circumferential band of microtubules and actin filaments transiently forms in the position of the future cell division.

As in animal development, one of the main questions in plant development is how cell fate is determined. Many structures in plants normally develop by stereotyped patterns of cell division, but despite this observation, which implies the importance of lineage, cell fate is, in many cases, also known to be determined by factors such as position in the meristem and cell–cell signaling. The cell wall would seem to impose a barrier to the passage of large signal molecules, such as proteins, although it is very thin in some regions, such as the meristems; but most of the known plant extracellular signal molecules—such as ethylene, auxins, gibberellins, cytokinins, steroids, and peptides—are small molecules that penetrate cell walls. Steroids and peptides have transmembrane receptors of the protein kinase type, which are numerous in plants. The receptors for other types of plant signals are mainly intracellular. Plant cells also communicate with each other through fine cytoplasmic channels known as **plasmodesmata**, which link neighboring plant cells through the cell wall, and they may be the channels by which some developmentally important gene-regulatory proteins, and even mRNAs, move directly between cells. The size of the channels varies, with those in shoot cells having the largest diameter.

Another important difference between plants and most animals is that a complete, fertile plant can develop from a single differentiated somatic cell and not just from a fertilized egg. This suggests that, unlike the differentiated cells of adult animals, some differentiated cells of the adult plant may remain **totipotent**. Perhaps they do not become fully determined in the sense that adult animal cells do, or perhaps they are able to escape from the determined state, although how this could be achieved is as yet unknown. In any case, this difference between plants and animals illustrates the dangers of the wholesale application to plant development of concepts derived from animal development. Nevertheless, the genetic analysis of development in plants is turning up instances of genetic strategies for developmental patterning rather similar to those of animals.

The small cress-like weed *Arabidopsis thaliana* has become the model plant for genetic and developmental studies, and will provide many of the examples in this chapter. We will begin by describing the main features of its morphology, life cycle, and reproduction.

Introduction to *Arabidopsis thaliana*

12.1 The model plant *Arabidopsis thaliana* has a short life cycle and a small diploid genome

The equivalent of the model animals *Drosophila* or the mouse for the study of plant development is the small crucifer *Arabidopsis thaliana*. Although its common name is thale cress, most researchers refer to it simply as *Arabidopsis*. *Arabidopsis* is well suited to genetic and developmental studies. It is a diploid (unlike many plants, which are polyploid) and has a relatively small, compact genome that has been sequenced and which contains about 27,000 protein-coding genes. It is an annual, flowering in the first year of growth, and develops as a small ground-hugging rosette of leaves, from which a branched flowering stem is produced with a flowerhead, or **inflorescence**, at the end of each branch. It develops rapidly, with a total life cycle in laboratory conditions of some 6–8 weeks, and like all flowering plants, mutant strains and lines can easily be stored in large quantities in the form of seeds. The sequences and information about each gene can be easily viewed on online portals such as Arabidopsis.org. The life cycle of *Arabidopsis* is shown in Figure 12.1.

Fig. 12.1 Life cycle of *Arabidopsis*. In flowering plants, egg cells are contained separately in ovules inside the carpels. Fertilization of an egg cell by a male nucleus from a pollen grain takes place inside the ovule. The egg then develops into an embryo contained within the ovule coat, forming a seed. *Arabidopsis* is a dicotyledon, and the mature embryo has two wing-like cotyledons (storage organs) at the apical (shoot) end of the main axis—the hypocotyl—which contains a shoot meristem at one end and a root meristem at the other. Following germination, the seedling develops into a plant with roots, a stem, leaves, and flowers. The photograph shows five mature *Arabidopsis* plants.

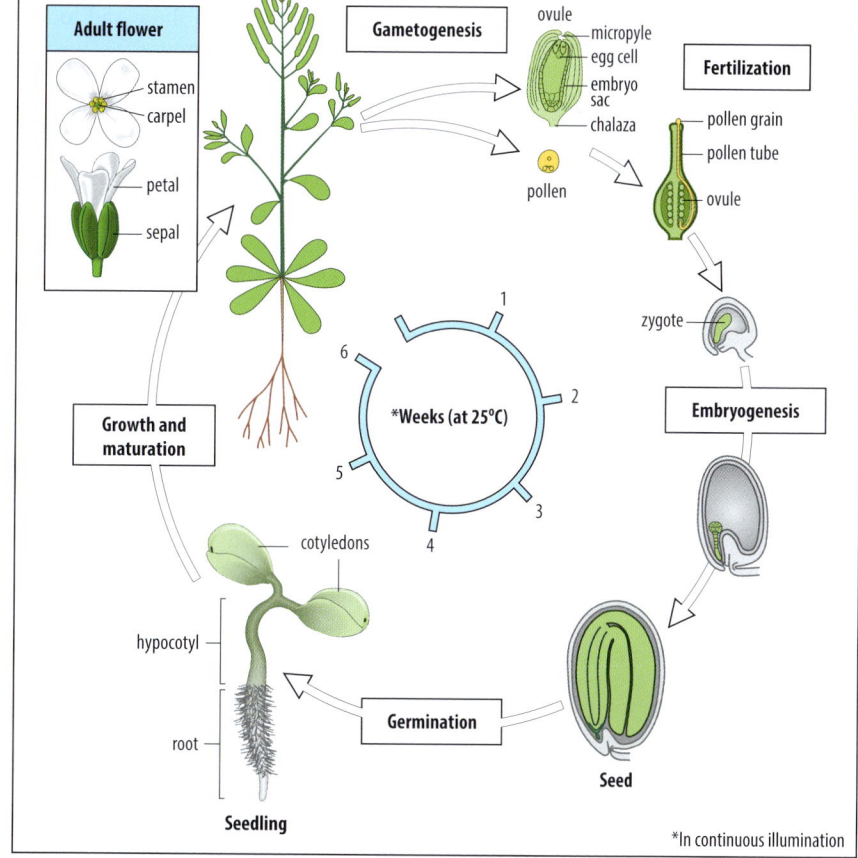

Fig. 12.2 An individual flower of *Arabidopsis*. Scale bar = 1 mm.
From Meyerowitz, E. et al.: *A genetic and molecular model for flower development in Arabidopsis thaliana.* Development 1991, **113**: 157-167.

Each *Arabidopsis* flower (Fig. 12.2) consists of four sepals surrounding four white petals; inside the petals are six stamens, the male sex organs, which produce pollen containing the male gametes. Petals, sepals, and the other floral organs derive evolutionarily from modified leaves. At the center of the flower are the female sex organs, which consist of an **ovary** of two fused carpels, which contain the **ovules**. Each ovule contains an egg cell. Fertilization of an egg cell occurs when a pollen grain deposited on elongated cells at the tip of the carpel, known as papillae, grows a tube that penetrates the carpel and delivers two haploid pollen nuclei to an ovule. One nucleus fertilizes the egg cell, while the other fuses with two other nuclei in the ovule. This forms a triploid cell that will develop into a specialized nutritive tissue—the **endosperm**—that surrounds the fertilized egg cell and provides the food source for embryonic development.

Following fertilization, the embryo develops inside the ovule, taking about 2 weeks to form a mature seed, which is shed from the plant. The seed will remain dormant until suitable external conditions trigger germination. The early stages of germination and seedling growth rely on food supplies stored in the **cotyledons** (seed leaves), which are storage organs developed by the embryo. *Arabidopsis* embryos have two cotyledons, and thus belong to the large group of plants known as dicotyledons (dicots for short). The other large group of flowering plants is the monocotyledons (monocots), which have embryos with one cotyledon; monocots typically have long, narrow leaves and include many important staple crops, such as wheat, rice, and maize. Agriculturally important dicots include soy, potato, tomato, sugar beet, and most other vegetables.

At germination, the shoot and root elongate and emerge from the seed. Once the shoot emerges above ground it starts to photosynthesize and forms the first true leaves at the shoot apex. About 4 days after germination the seedling is a self-supporting plant. Flower buds are usually visible on the young plant 3–4 weeks after germination and will open within a week. Under ideal conditions, the complete *Arabidopsis* life cycle thus takes about 6–8 weeks (see Fig. 12.1).

Embryonic development

Formation of the embryo, or embryogenesis, occurs inside the ovule, and the end result is a mature, dormant embryo enclosed in a seed awaiting germination. During embryogenesis, the shoot–root polarity of the plant body, which is known as the **apical–basal axis**, is established, and the shoot and root meristems are formed. Plants also possess **radial symmetry**, as seen in the concentric arrangement of the different tissues in a plant stem, and this **radial axis** is also set up in the embryo. Development of the *Arabidopsis* embryo involves a rather invariant pattern of cell division (which is not the case in all plant embryos) and this enables structures in the *Arabidopsis* seedling to be traced back to groups of cells in the early embryo to provide a fate map.

12.2 Plant embryos develop through several distinct stages

Arabidopsis belongs to the angiosperms, or flowering plants, one of the two major groups of seed-bearing plants; the other is the gymnosperms, which includes conifers. The typical course of embryonic development in angiosperms is outlined in Box 12A. Like an animal zygote, the fertilized plant egg cell undergoes repeated cell divisions, cell growth, and differentiation to form a multicellular embryo. The first division of the zygote is at right angles to the long axis, dividing it into an apical cell and a basal cell,

Embryonic development **633**

BOX 12A Angiosperm embryogenesis

In flowering plants (angiosperms), the egg cell is contained within an ovule inside the ovary in the flower (right inset in Fig. 1). At fertilization, a pollen grain deposited on the surface of the stigma puts out a pollen tube, down which two male gametes migrate into the ovule. One male gamete fertilizes the egg cell while the other combines with another cell inside the ovule to form a specialized nutritive tissue, the endosperm, which surrounds, and provides the food source for, the developing embryo.

The small annual weed *Capsella bursa-pastoris* (shepherd's purse) is a typical dicotyledon that shows very similar embryogenesis to *Arabidopsis*. The first, asymmetric, cleavage divides the zygote transversely into an apical and a basal cell (Fig. 1). The basal cell then divides several times to form a single row of cells—the suspensor—which, in many angiosperm embryos, takes no further part in embryonic development, but may have an absorptive function; in *Capsella*, however, it contributes to the root meristem. Most of the embryo is derived from the apical cell. This undergoes a series of stereotyped divisions, in which a precise pattern of cleavages in different planes gives rise to the heart-shaped embryonic stage typical of dicotyledons. This develops into a mature embryo that consists of a cylindrical body with a meristem at either end, and two cotyledons.

The early embryo becomes differentiated along the radial axis into three main tissues—the outer epidermis, the prospective vascular tissue, which runs through the center of the main axis and cotyledons, and the ground tissue (prospective cortex) that surrounds it.

The ovule containing the embryo matures into a seed (left inset in Fig. 1), which remains dormant until suitable external conditions trigger germination and growth of the seedling. A typical dicotyledon seedling comprises the shoot apical meristem, two cotyledons, the trunk of the seedling (the hypocotyl), and the root apical meristem (Fig. 2). The seedling may be thought

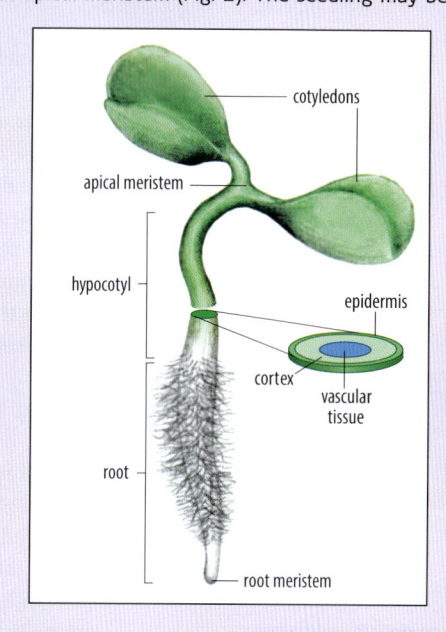

Fig. 2

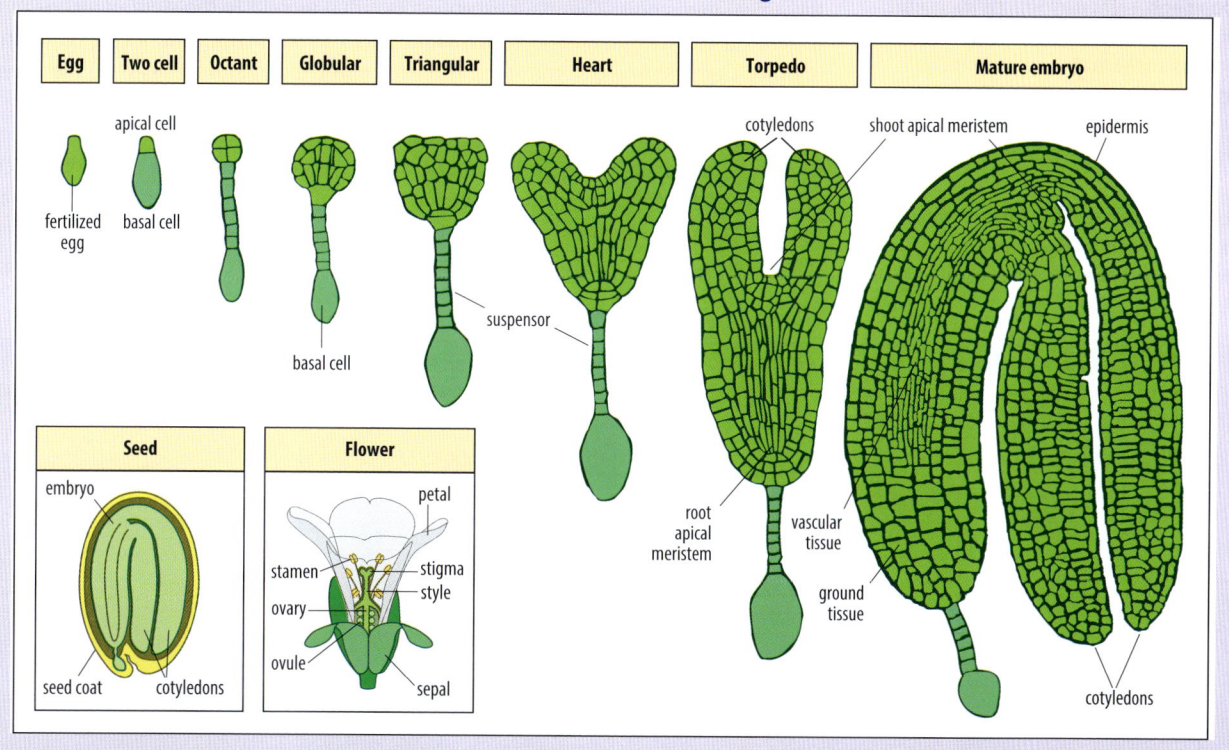

Fig. 1

634 Chapter 12 Plant development

of as the phylotypic stage of flowering plants (see Fig. 3.2 for phylotypic stage of vertebrate embryos). The seedling body plan is simple. One axis—the apical-basal axis—defines the main polarity of the plant. The shoot forms at the apical pole and the root at the basal pole. The meristems at the tips of shoots and roots in seedlings and adult plants are called apical meristems. A plant stem also has a radial axis, evident in the radial symmetry in the hypocotyl and continued in the root and shoot. In the center is the vascular tissue, surrounded by cortex, and an outer covering of epidermis. At later stages, structures such as leaves and other organs have a dorso-ventral axis running from the upper surface to the lower surface.

and establishing an initial polarity that is carried over into the apical–basal polarity of the embryo and into the apical–basal axis of the plant. In many species, the first zygotic division is unequal, with the basal cell larger than the apical cell. The basal cell divides to give rise to the **suspensor**, which may be several cells long. This attaches the embryo to maternal tissue and is a source of nutrients. The apical cell divides vertically to form a two-celled **proembryo**, which will give rise to the rest of the embryo. In some species, the basal cell contributes little to the further development of the embryo, but in others, such as *Arabidopsis*, the topmost suspensor cell is recruited into the embryo, where it is known as the **hypophysis**, and contributes to the embryonic root meristem and root cap.

The next two divisions produce an eight-cell **octant-stage** embryo, which develops into a **globular-stage** embryo of around 32 cells (Fig. 12.3). The embryo elongates and the cotyledons start to develop as wing-like structures at one end, while an embryonic root forms at the other. This stage is known as the **heart stage**. Two groups of undifferentiated cells capable of continued division are located at each end of this axis; these are known as the **apical meristems**. The meristem lying between the cotyledons gives rise to the shoot, while the one at the opposite end of the axis, towards the end of the embryonic root, will drive root growth at germination. The region in between the embryonic root and the future shoot will become the seedling stem or **hypocotyl**. Almost all above-ground adult plant structures are derived from the apical meristems. The main exception is radial growth in the stem, which is most evident in woody plants, and which is produced by the **cambium**, a ring of secondary meristematic tissue in the stem. After the embryo is mature, the apical meristems remain quiescent until germination.

Seedling structures can be traced back to groups of cells in the early embryo to provide a fate map (Fig. 12.4). In *Arabidopsis*, patterns of cleavage up to the 16-cell stage are highly reproducible, and even at the octant stage it is possible to make a fate map for the major regions of the seedling along the apical–basal axis. The upper tier of cells gives rise to the cotyledons and the shoot meristem, the next tier is the origin

Fig. 12.3 *Arabidopsis* **embryonic development.** Light micrographs (Nomarski optics) of cleared wild-type seeds of *Arabidopsis thaliana*. The cotyledons can already be seen at the heart stage. The embryo proper is attached to the seed coat through a filamentous suspensor. Scale bar = 20 μm.

Photographs courtesy of D. Meinke, from Meinke, D.W.: **Seed development in Arabidopsis thaliana.** *In* Arabidopsis *(CSHLP, 1994).*

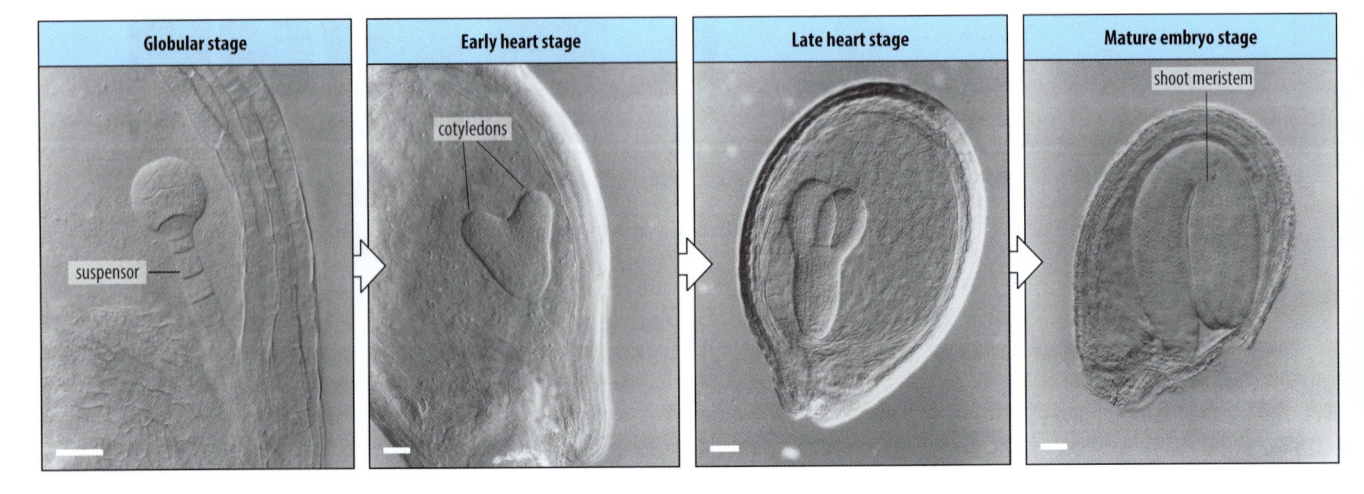

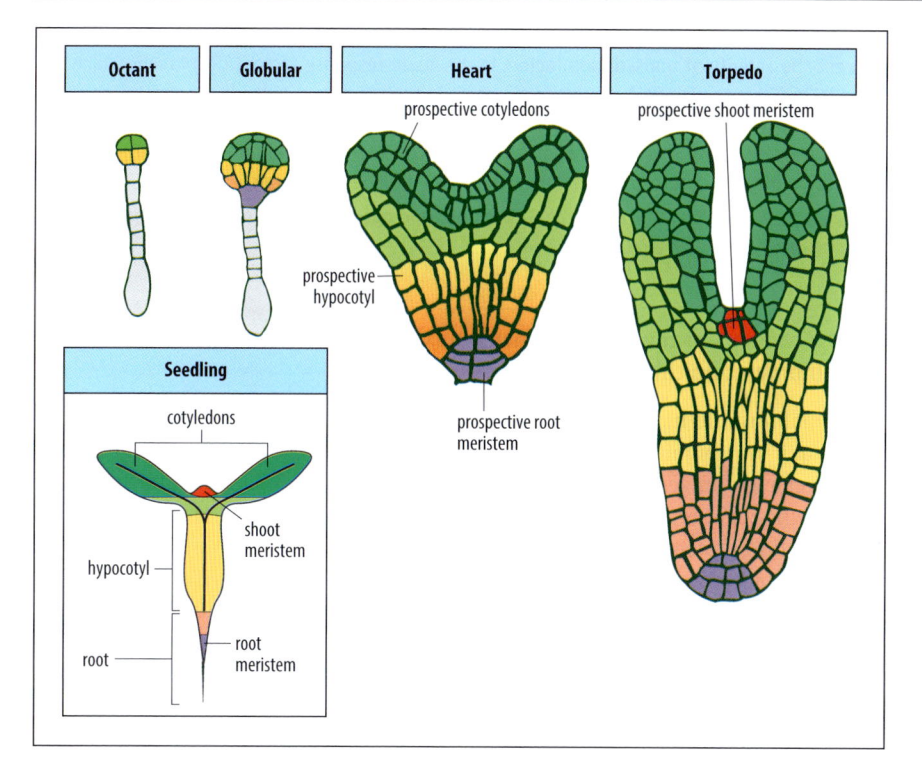

Fig. 12.4 Fate map of the *Arabidopsis* embryo. The stereotyped pattern of cell division in dicotyledon embryos means that at the globular stage it is already possible to map the three regions that will give rise to the cotyledons (dark green) the hypocotyl (yellow), and the root meristem (purple). Shoot apical meristem in the torpedo stage (red).

*Illustration after Scheres, B., et al.: **Embryonic origin of the** Arabidopsis **primary root and root meristem initials**. Development 1994, 120: 2475-2487.*

of the hypocotyl, and the bottom tier together with the region of the suspensor where it joins the embryo will give rise to the root. At the heart stage, the fate map is clear.

The radial pattern in the embryo comprises three concentric rings of tissue: the outer **epidermis**, the ground tissue (**cortex** and **endodermis**), and the vascular tissue at the center. This radial axis appears first at the octant stage, when adaxial (central) and abaxial (peripheral) regions become established. Subsequently, **periclinal** divisions, in which the plane of division is parallel to the outer surface, give rise to the different rings of tissue, and **anticlinal** divisions, in which the plane of cell division is at right angles to the outer surface, increase the number of cells in each ring of tissue (Fig. 12.5). At the 16-cell stage the epidermal layer, or dermatogen, is established.

It is not known to what extent the fate of cells is determined at this stage or whether their fate is dependent on the early pattern of asymmetric cell divisions. It seems that cell lineage is not crucial, as mutations have been discovered that uncouple cell division from pattern formation. The *fass* mutation in *Arabidopsis* disrupts the regular pattern of cell division by causing cells to divide in random orientation, and produces a much fatter and shorter seedling than normal. But although the *fass* seedling is misshapen, it has roots, shoots, and even eventually flowers in the correct places, and the correct pattern of tissues along the radial axis is maintained.

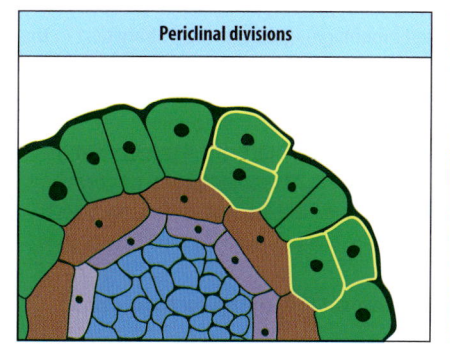

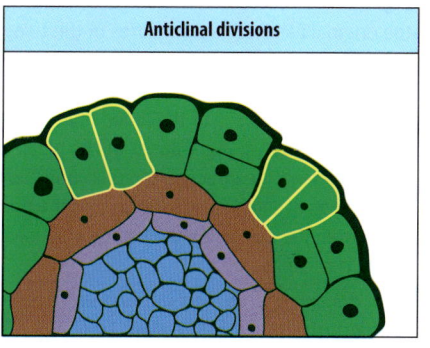

Fig. 12.5 Periclinal and anticlinal divisions. Periclinal cell divisions are parallel to the organ surface, whereas anticlinal divisions are at right angles to the surface.

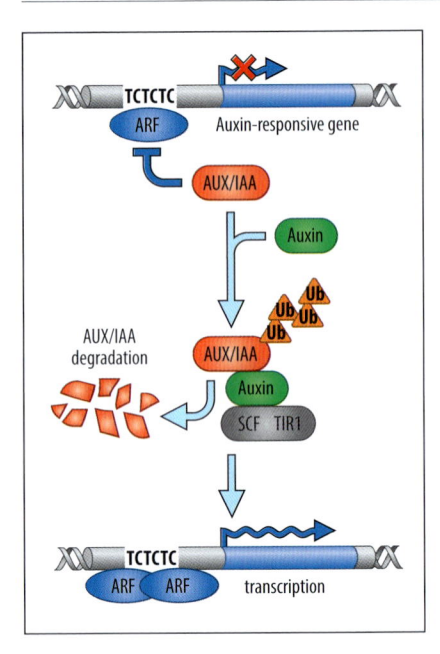

Fig. 12.6 Auxin signaling pathway. In the absence of auxin, AUX/IAA proteins bind to and repress the activity of transcription factors in the auxin-response factor (ARF) family, which bind TGTCTC-containing DNA sequence elements in the promoters of auxin-responsive genes. Auxin targets the ubiquitin ligase complex SCF/TIR1 to the AUX/IAA protein, causing it to be ubiquitinated (Ub). This modification targets AUX/IAA for degradation and thus lifts the repression of the auxin-responsive gene, which can then be transcribed.

Adapted from Chapman, E.J., Estelle, M.: *Cytokinin and auxin intersection in root meristems.* Genome Biol. 2009, **10**: 210.

12.3 Gradients of the signal molecule auxin establish the embryonic apical–basal axis

The small organic molecule **auxin** (indole-3-acetic acid or IAA) is one of the most important and ubiquitous chemical signals in plant development and plant growth. It causes changes in gene expression by promoting the ubiquitination and degradation of transcriptional repressors known as **AUX/IAA proteins**. In the absence of auxin, these bind to proteins called **auxin-response factors (ARFs)** to block the transcription of so-called **auxin-responsive genes**. Auxin-stimulated degradation of AUX/IAA proteins frees the ARFs, which can then activate, or in some cases repress, these genes (Fig. 12.6). Although additional mechanisms have been proposed for auxin signaling, most auxin-mediated phenotypes can be accounted for by auxin-stimulated ubiquitination of AUX/IAAs. Auxin-responsive genes include genes involved in the regulation of cell division and cell expansion, as well as genes involved in specifying cell fate. In some cases auxin appears to be acting as a classic **morphogen**, forming a concentration gradient and specifying different fates according to a cell's position along the gradient.

The earliest known function of auxin in *Arabidopsis* is in the very first stage of embryogenesis, where it establishes the apical–basal axis. Immediately after the first division of the zygote, auxin is actively transported from the basal cell into the apical cell, where it accumulates. It is transported out of the basal cell by the auxin-efflux protein PIN7, which is localized in the apical plasma membrane of the basal cell. The auxin is required to specify the apical cell, which gives rise to all the apical embryonic structures such as shoot meristem and cotyledons. Through the subsequent cell divisions, transport of auxin continues up through the suspensor cells and into the cell at the base of the developing embryo until the globular-stage embryo of about 32 cells. The apical cells of the embryo then start to produce auxin, and the direction of auxin transport is suddenly reversed. PIN7 proteins in the suspensor cells move to the basal faces of the cells. Other *PIN* genes are activated, and the concerted actions of the PIN proteins cause auxin from the apical region to be transported into the basal region of the globular embryo, from which will develop the hypocotyl, root apical meristem, and embryonic root (Fig. 12.7).

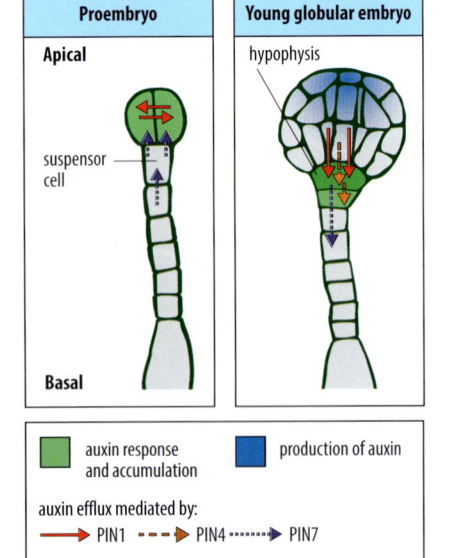

Fig. 12.7 The role of auxin in patterning the early embryo. Left panel: auxin produced in the original basal cell accumulates in the two-celled proembryo (green) through transport in the basal to apical direction mediated by the protein PIN7, which is located in the apical membranes of the basal and suspensor cells (purple arrows). Another PIN protein, PIN1, transports auxin between the two cells of the proembryo (red arrows). Cell division distributes auxin into the developing embryo. Right panel: at the globular stage, free auxin starts to be produced at the apical pole and the direction of auxin transport is reversed. PIN7 becomes localized in basal membranes of the suspensor cells and the proteins PIN1 and PIN4 (orange arrows) transport auxin from the apical region into the most basal cell of the embryo, known as the hypophysis, where it accumulates.

From Friml, J., et al.: *Efflux-dependent auxin gradients establish the apical-basal axis of Arabidopsis.* Nature 2003, **426**: 147–153.

Expression of the homeobox genes *WOX2* and *WOX8*, which encode transcription factors, in the asymmetrically dividing zygote seems to be essential for the formation of the embryonic auxin gradient. Both genes are expressed in the zygote, but after division, *WOX2* expression is restricted to the apical cell and its descendants, whereas *WOX8* is expressed only in the basal cell lineage. Expression of *WOX8* appears to be required for the continued expression of *WOX2* and for the establishment of an auxin gradient, most probably by effects of the WOX proteins on expression of the PIN proteins that direct auxin transport.

The importance of auxin in specifying the basal region of the embryo is illustrated by the effects of mutations in the cellular machinery for auxin transport or auxin response. The gene *MONOPTEROS* (*MP*), for example, encodes one of the ARFs, ARF5. Embryos mutant for *MP* lack the hypocotyl and the root meristem, and abnormal cell divisions are observed in the regions of the octant-stage embryo that would normally give rise to these structures.

An early step in axis formation in the embryo is the specification of cells as potential shoot or root. The mutation *topless-1* causes a dramatic switching of cell fate, transforming the whole of the apical region into a second root region and abolishing development of cotyledons and shoot meristem. The normal function of the TOPLESS protein is to repress expression of root-promoting genes in the top half of the early embryo. It does this by acting as a transcriptional co-repressor, forming a complex together with Aux/IAA proteins and the ARFs they regulate.

The *topless-1* mutation is temperature-sensitive, and so can be made to act at different times during development by simply changing the temperature at which the embryos are grown. Such manipulations show that the apical region can be re-specified as root between the globular and heart stage, even after it has begun to show molecular signs of developing cotyledons and shoot meristem. This suggests that even though apical cell fate is normally specified by this stage of embryonic development, the decision is not irreversible.

The shoot apical meristem is characterized by the expression of a number of genes, among them *SHOOT MERISTEMLESS* (*STM*), which encodes a transcription factor that is also required to maintain cells in the pluripotent state in the adult shoot meristem. Mutations in *STM* completely block the formation of the shoot apical meristem but have no effect on the root meristem or other parts of the embryo. The pattern of *STM* expression develops gradually, which is also typical of several other genes that characterize the shoot apical meristem. Expression is first detected in the globular stage in one or two cells and only later in the central region between the two cotyledons (Fig. 12.8).

Fig. 12.8 Section through a late heart-stage *Arabidopsis* embryo showing expression of *SHOOT MERISTEMLESS* (*STM*). At this stage, the *STM* RNA (stained red) is expressed in cells located between the cotyledons. Scale bar = 25 μm.

Photograph courtesy of K. Barton. From Long, J.A., Barton, K.B.: **The development of apical embryonic pattern in** Arabidopsis. *Development* 1999, **125**: 3027-3035.

12.4 Plant somatic cells can give rise to embryos and seedlings

As gardeners know well, plants have amazing powers of regeneration. A complete new plant can develop from a small piece of stem or root, or even from the cut edge of a leaf. This reflects an important difference between the developmental potential of plant and animal cells. In animals, with few exceptions, cell determination and differentiation are irreversible. By contrast, many somatic plant cells remain totipotent. Cells from roots, leaves, stems, and even, for some species, a single, isolated protoplast—a cell from which the cell wall has been removed by enzymatic treatment—can be grown in culture and induced by treatment with the appropriate growth hormones to give rise to a new plant (Fig. 12.9). Plant cells proliferating in culture can give rise to cell clusters that pass through a stage resembling normal embryonic development, although the pattern of cell divisions is not the same. These 'embryoids' can then develop into seedlings. Plant cells can also form callus, an apparently disorganized mass of cells, and the callus can form new shoot or root apical meristems, and therefore new shoots and roots. The ability of single somatic cells to give rise to

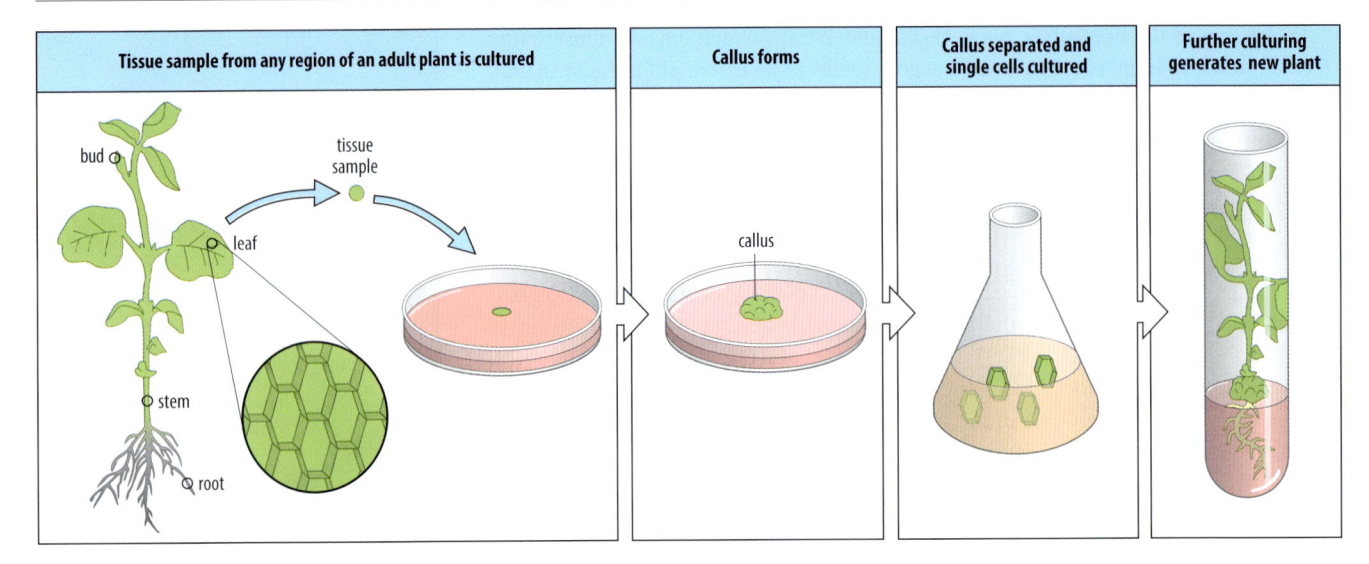

Fig. 12.9 Cultured somatic cells from a mature plant can form an embryo and regenerate a new plant. The illustration shows the generation of a plant from single cells. If a small piece of tissue from a plant stem or leaf is placed on a solid agar medium containing the appropriate nutrients and growth hormones, the cells start to divide to form a disorganized mass of cells known as a callus. The callus cells are then separated and grown as single cells in liquid culture, again containing the appropriate growth hormones. In suspension culture, some callus cells divide to form small cell clusters. These cell clusters can resemble the globular stage of a dicotyledon embryo, and with further culture on solid medium, develop through heart-shape and later stages to regenerate a complete new plant.

whole plants has two important implications for plant development. The first is that maternal determinants may be of little or no importance in plant embryogenesis, as it is unlikely that many somatic cells would still be carrying such determinants. Second, it suggests that many cells in the adult plant body are not fully determined with respect to their fate, but remain totipotent. Of course, this totipotency seems only to be expressed under special conditions, but it is, nevertheless, quite unlike the behavior of animal cells. It is as if such plant cells have no long-term developmental memory, or that such memory is easily reset.

The ability of plants to regenerate from somatic cells means that transgenic plants can be easily generated from small pieces of leaf tissue into which foreign DNA has been introduced. **Plant transformation** is the general name given to techniques that introduce DNA into plant cells so that they express new or modified genes. There are two main ways of delivering DNA across the barrier of the cell wall. One is through infection of plant tissue in culture with the bacterium *Agrobacterium tumefaciens*, the causal agent of crown gall tumors. *Agrobacterium* is a natural genetic engineer. It carries a plasmid—the Ti plasmid—that contains the genes required for the proliferation of infected cells to form a callus. During infection, a portion of this plasmid—the T-DNA—is transferred into the genome of the plant cell, where it can become stably integrated (Fig. 12.10). Foreign genes inserted into the T-DNA will therefore also be transferred into the plant cell chromosomes. Ti plasmids, modified so that they do not cause uncontrolled cell proliferation but still retain the ability to transfer T-DNA, are widely used as vectors for gene transfer. The genetically modified cultured cells can then be grown into a complete new transgenic plant that carries the introduced gene (the transgene) in all its cells and can transmit it to the next generation. Genetic transformation of plant cells using *Agrobacterium* infection is called agroinfection. Monocotyledons, which include the agriculturally important grasses such as maize, rice, and wheat, are not naturally infected by *Agrobacterium*, but refinements of the basic technique, such as

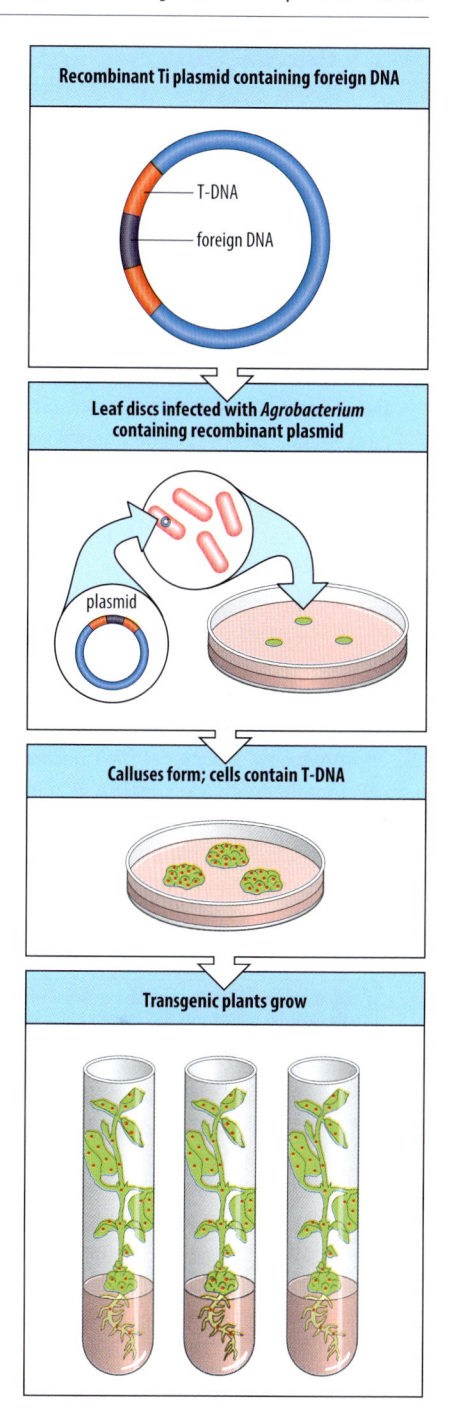

Fig. 12.10 Generation of transgenic plants using the Ti plasmid of *Agrobacterium tumefaciens.* Leaf discs are infected with agrobacteria that contain a Ti plasmid into which the desired gene has been inserted. The gene is inserted into the T-DNA region of the plasmid (shown in red) and is transferred into the plant cells along with the T-DNA. The transformed leaf discs form callus (as in Fig. 12.9), from which transgenic plants carrying the introduced gene in all their cells can be grown.

injecting or infiltrating the plant tissue with a suspension of *Agrobacterium* (agro-inoculation and agroinfiltration, respectively), have made it possible to efficiently transform at least some of these species by this route.

Another way of getting DNA into a plant cell is by particle bombardment, also known as biolistics. This uses the physical force of a 'gene gun' to fire fine particles of a heavy metal (tungsten or gold) coated with nucleic acid into plant tissue in culture. This technique can, in principle, be applied to any species and was used, for example, to produce commercial lines of genetically modified maize resistant to the corn borer, a major insect pest.

Protoplasts, which have no cell wall, can be directly transfected by DNA in the same way as animal cells, but relatively few plants of agricultural interest can currently be regenerated from protoplasts.

A drawback of all these plant transformation techniques is that they insert the foreign DNA randomly into the plant genome. Over the past decade, a new generation of genome-editing technologies has been developed that enable changes to the genome to be targeted more precisely, and these are now being applied to plants (Box 12B).

12.5 Cell enlargement is a major process in plant growth and morphogenesis

Plants grow in size by a combination of cell division and cell expansion. Cell expansion can provide up to a 50-fold increase in the volume of a growing plant tissue, and much of the growth in size of plants and their organs is due to cell enlargement. The driving force for expansion is the hydrostatic pressure—turgor pressure—exerted on the cell wall as the protoplast swells as a result of the entry of water into cell vacuoles by osmosis (Fig. 12.11). Plant cell expansion involves synthesis and deposition of new cell wall material, and is an example of a process called **directed dilation**, in which an increase in internal hydrostatic pressure causes a distinct change in shape in a particular direction. Directed dilation is also seen in animal development—in the development of the vertebrate notochord and elongation of the nematode body.

In plants, the direction of cell growth is determined by the orientation of the cellulose fibrils in the cell wall. Enlargement occurs primarily in a direction at right angles to the fibrils, where the wall is weakest. The orientation of cellulose fibrils in the cell wall is determined by the microtubules of the cell's cytoskeleton, which are responsible for positioning the enzyme assemblies that synthesize cellulose at the cell wall. Plant growth hormones such as ethylene and gibberellins alter the orientation in which the fibrils are laid down and so can alter the direction of expansion. Auxin aids expansion by loosening the structure of the cell wall.

The development of a leaf in a particular shape involves a complex pattern of cell division and cell elongation along the proximo-distal and medio-lateral axes of the outgrowing leaf, which is thought to be organized by a region at the base of the leaf. Cell elongation plays a central part in the expansion of the leaf blade in the later stages of leaf outgrowth. Two mutations that affect the shape of the blade by affecting the direction of cell elongation have been identified. Leaves of the *Arabidopsis* mutant

EXPERIMENTAL BOX 12B Plant transformation and genome editing

The conventional plant transformation techniques such as agro-infection, biolistics, or protoplast transfection described in Section 12.4 result in the random insertion of the introduced DNA into the plant genome. Until recently it has proved difficult to target the incoming DNA to specific genes in plants by modifying the DNA constructs themselves. The strategy of targeting a gene by homologous recombination, which has been used to produce knock-out mice, for example (see Box 3B), is less effective in most plants. Plants typically preferentially use the alternative repair pathway of non-homologous end-joining (NHEJ) to integrate DNA into the genome.

More recent methods of generating targeted mutations, such as zinc-finger nucleases, TALENS, and CRISPR-Cas9 technologies, are rapidly changing the picture (see Section 3.3). The newest of these genome-editing techniques—CRISPR-Cas9 and its offshoots—is increasingly being used to produce targeted mutations in plants, as well as in animals, because of its ease of use, low cost, and effectiveness. The technique and its general applications are described in Box 1C, but it essentially relies on introducing a bacterial DNA nuclease—Cas9—into the plant cell, together with a 'guide RNA' that makes a complex with the nuclease and targets it to a genomic sequence complementary to the guide RNA. The enzyme makes a double-strand break at a precise place in the genomic DNA, which is then repaired by the cell. Repair by NHEJ usually introduces small insertions and deletions into the original sequence, which inactivate the gene. In plants, the nuclease and its guide RNA are generally delivered, via agroinfection or particle bombardment, as DNA plasmids that encode the enzyme and the RNA. The introduced genes are then expressed by the plant cell and the targeted gene is mutated. Once the mutation is made, the Cas9 and guide RNA are no longer needed, and so for many applications transient expression of these genes is more desirable than their stable integration into the genome, where their continued expression could lead to higher rates of 'off-target' mutations. The nuclease and guide RNA have also been introduced into plant cells directly in the form of ribonucleoprotein (RNP) particles, without a DNA vector, which has the advantage that no foreign DNA enters the cell (Fig. 1). This technique, using particle bombardment to deliver the RNPs into immature embryos, has been successful in maize and in wheat, which is hard to transform by conventional means.

The next challenge for plant transformation is to make these new techniques more precise and to tailor them better to plants. CRISPR-Cas9-based techniques have been developed, and are being used in animal cells, that can make single base changes in a gene, that can introduce several mutations at the same time, or that use template-directed homology-directed repair to repair existing mutations and replace one gene, or part of a gene, with another sequence.

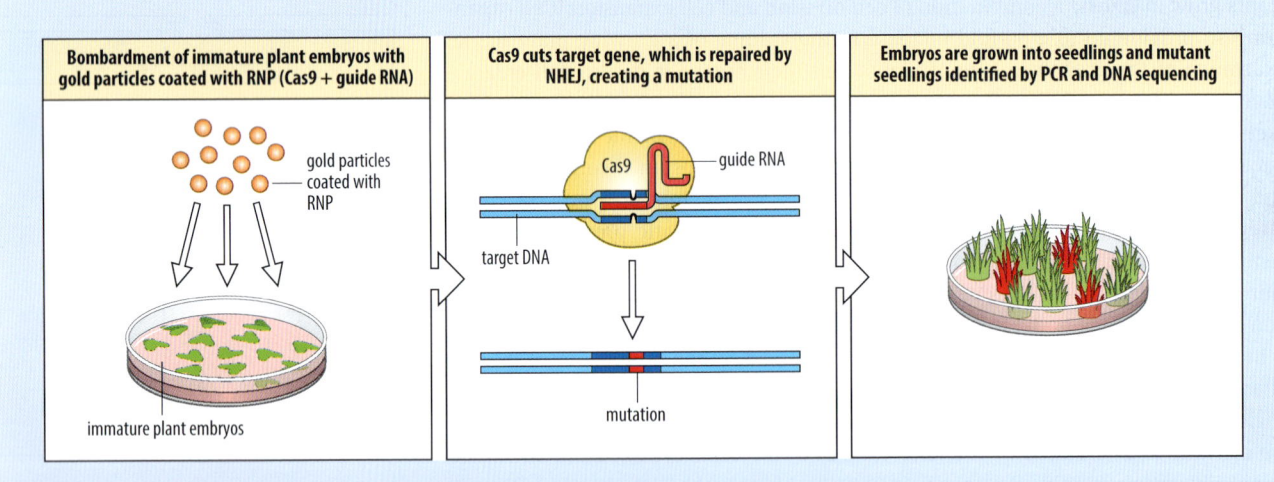

Fig. 1
Adapted from Yin, K., et al.: *Progress and prospects in plant genome editing.* Nat. Plants 2017, **3**: 17107.

angustifolia are similar in length to the wild type but are much narrower (Fig. 12.12). By contrast, *rotundifolia* mutations reduce the length of the leaf relative to its width. Neither of these mutations affects the number of cells in the leaf. Examination of the cells in the developing leaf shows that these mutations are affecting the direction of elongation of the enlarging cells.

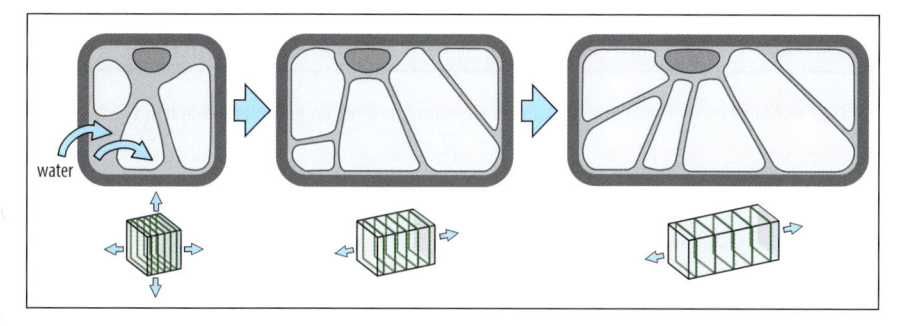

Fig. 12.11 Enlargement of a plant cell. Plant cells expand as water enters the cell vacuoles and thus causes an increase in intracellular hydrostatic pressure. The cell elongates in a direction perpendicular to the orientation of the cellulose fibrils in its cell wall.

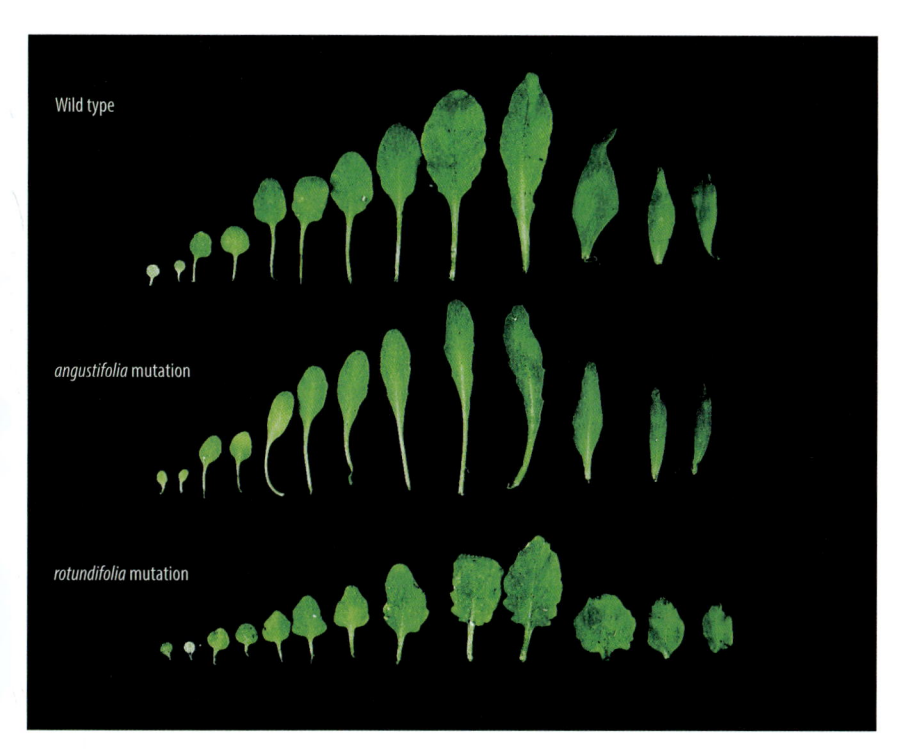

Fig. 12.12 The shape of the leaves of *Arabidopsis* is affected by mutations affecting cell elongation. The *angustifolia* mutation results in cells with narrow leaves, whereas *rotundifolia* mutations cause short, fat leaves to develop.

Photograph courtesy of H. Tsukaya, from Tsuge, T., et al.: **Two independent and polarized processes of cell elongation regulate blade expansion in** Arabidopsis thaliana *(L.) Heynh.* Development *1996,* ***122****: 1589–1600.*

SUMMARY

Early embryonic development in many plants is characterized by asymmetric cell division of the zygote, which specifies apical and basal regions. Embryonic development in flowering plants establishes the shoot and root meristems from which the adult plant develops. Auxin gradients are involved in specifying the apical–basal axis of the *Arabidopsis* embryo. One major difference between plants and animals is that, in culture, a single somatic plant cell can develop through an embryo-like stage and regenerate a complete new plant, indicating that some differentiated plant cells retain totipotency.

SUMMARY: Early development in flowering plants

first asymmetric cell division and auxin signaling in embryo establishes apical-basal axis
⇩
embryonic cell fate is determined by position
⇩
shoot and root meristems of seedling give rise to all adult plant structures

Meristems

In plants, most of the adult structures are derived from just two regions of the embryo, the embryonic shoot and root meristems, which are maintained after germination. The embryonic shoot meristem, for example, becomes the shoot apical meristem of the growing plant, giving rise to all the stems, leaves, and flowers. As the shoot grows, lateral outgrowths from the meristem give rise to leaves and to side shoots. In flowering shoots, the vegetative meristem becomes converted into one capable of producing **floral meristems** that make flowers, not leaves. In *Arabidopsis*, for example, the shoot apical meristem changes from a vegetative meristem that makes leaves around it in a spiral pattern to an **inflorescence meristem** that then produces floral meristems, and thus flowers, around it in a spiral pattern. The first stages of future organs are known as primordia (singular primordium). Each primordium consists of a small number of **founder cells** that produce the new structure by cell division and cell enlargement, accompanied by differentiation.

There is usually a time delay between the initiation of two successive leaves in a shoot apical meristem, and this results in a plant shoot being composed of repeated modules. Each module consists of an **internode** (the cells produced by the meristem between successive leaf initiations), a **node** and its associated leaf, and an axillary bud (Fig. 12.13). The axillary bud itself contains a meristem, known as the **lateral shoot meristem**, which forms at the base of the leaf. The growth of lateral buds immediately below the apical bud is suppressed, the phenomenon known as **apical dominance**. Apical dominance is caused by the hormone auxin, which is produced by the shoot apex and moves back to inhibit the activity of lateral meristems (see Section 12.13).

Root growth is not so obviously modular, but similar considerations apply, as new lateral meristems initiated behind the root apical meristem give rise to lateral roots.

Shoot apical meristems and root apical meristems operate on the same principles, but there are some significant differences between them. We will use the shoot apical meristem to illustrate the basic principles of meristem structure and properties, and then discuss roots.

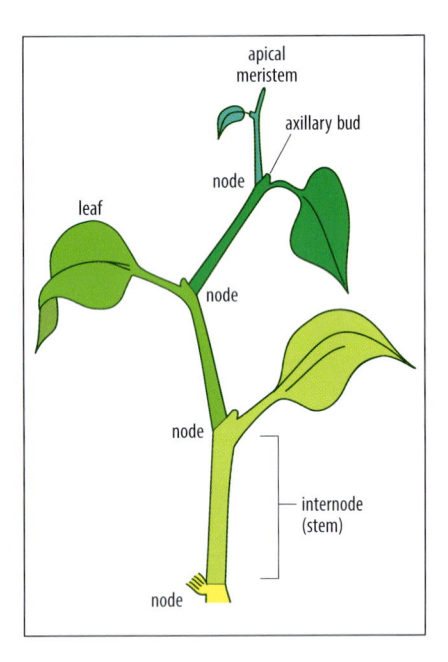

Fig. 12.13 Plant shoots grow in a modular fashion. The shoot apical meristem produces a repeated basic structural module. The vegetative shoot module typically consists of internode, node, leaf, and axillary bud (from which a side branch may develop). Successive modules are shown here in different shades of green. As the plant grows, the internodes behind the meristem lengthen and the leaves expand.

12.6 A meristem contains a small central zone of self-renewing stem cells

Shoot apical meristems are rarely more than 250 µm in diameter in angiosperms and contain a few hundred relatively small, undifferentiated cells that are capable of cell division. Most of the cell divisions in normal plant development occur within meristems, or soon after a cell is displaced from the meristem, and much of a plant's growth in size is due to cell elongation and enlargement. Cells leave the periphery of the meristem to form organs such as leaves or flowers, and are replaced from a small central zone of slowly dividing, self-renewing stem cells or **initials** at the tip of the meristem (Fig. 12.14). In *Arabidopsis* this zone comprises around 12–20 cells. Initials behave in the same way as animal stem cells (see Chapter 6). They can divide to give one daughter that remains a stem cell and one that loses the stem-cell property. This daughter cell continues to divide and its descendants are displaced towards the peripheral zone of the meristem, where they become founder cells for a new organ or internode, leave the meristem, and differentiate. A very small number of long-term stem cells at the meristem center may persist for the whole life of the plant.

Meristem stem cells are maintained in the self-renewing state by cells underlying the central zone that form the **organizing center**. As we shall see, it is the microenvironment maintained by the organizing center that gives stem cells their identity.

The undetermined state of meristem stem cells is confirmed by the fact that meristems are capable of regulation. If, for example, a seedling shoot meristem is divided into two or four parts by vertical incision, each part becomes reorganized into a complete meristem, which gives rise to a normal shoot. Provided that some subpopulation of organizing center cells plus overlying stem cells is present, a normal meristem will regenerate. If a shoot apical meristem is completely removed, no new apical meristem forms, but the incipient lateral shoot meristem at the base of the leaf is now able to develop and form a new side shoot. In the presence of the original meristem, this prospective meristem remains inactive, as a result of the inhibitory effect of auxin provided by the shoot apex (see Section 12.13). This regulative behavior is in line with cell–cell interactions being a major determinant of cell fate in the meristem.

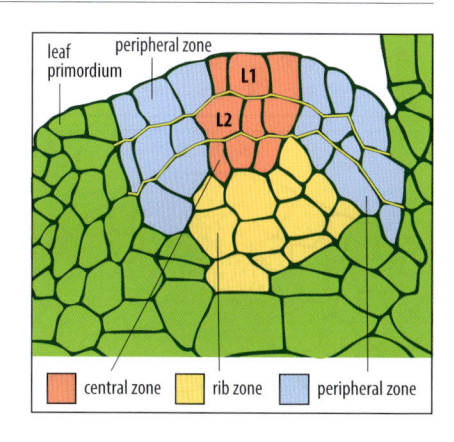

Fig. 12.14 Organization of the *Arabidopsis* shoot meristem. A longitudinal section is shown. The meristem has three main layers, L1, L2, and an inner layer, as indicated by the yellow lines, and is divided functionally into a central zone (orange), a rib zone (yellow), and a peripheral zone (blue). The stem cells or initials lie in the central zone, while the peripheral zone contains proliferating cells that will give rise to leaves and side shoots. The rib zone gives rise to the central tissues of the plant stem.

12.7 The size of the stem-cell area in the meristem is kept constant by a feedback loop to the organizing center

Numerous genes control the behavior of the cells in the meristem. The gene *STM*, which is involved in specifying the shoot meristem in the development of the *Arabidopsis* embryo (see Section 12.3) is, for example, expressed throughout adult shoot meristems but is suppressed as soon as cells become part of an organ primordium. Its role seems to be to maintain meristematic cells in an undifferentiated state, as loss of *STM* function results in all the meristem cells being incorporated into organ primordia, which causes the meristem to terminate.

In *Arabidopsis*, cells of the organizing center express a homeobox transcription factor, WUSCHEL (WUS), which is required to produce a signal that gives the overlying cells their stem-cell identity. Although *WUS* transcription and translation is restricted to the organizing center, the protein can move through plasmodesmata to the central zone above where it promotes stem-cell identity. Mutations in *WUS* result in termination of the shoot meristem and cessation of growth as a result of the loss of stem cells, while its overexpression increases stem-cell numbers. A key target of WUS is *CLAVATA3* (*CLV3*), which is expressed in the central zone. Opposite to the effect of mutations in *WUS*, *CLV3* mutants have increased meristem size as a result of an increase in the number of stem cells.

CLV3 functions as part of a negative feedback loop to control *WUS* activity in the organizing center. *CLV3* encodes a small peptide that diffuses to the organizing center and indirectly represses *WUS* expression. This feedback loop controls *WUS* activity in the organizing center and suppresses *WUS* activation in neighboring cells, thus limiting the extent of *WUS* expression (Fig. 12.15). In turn, this regulates the extent of the stem-cell zone above the organizing center. If stem-cell numbers temporarily fall, for example, less CLV3 is produced and *WUS* activity increases, with a consequent increase in stem-cell numbers. More CLV3 is then produced and limits the extent of *WUS* activity. Other CLAVATA proteins are involved in the feedback loop. Normally, this feedback loop keeps the number of stem cells in a meristem roughly the same throughout a plant's life, despite the continual displacement of cells from the stem-cell pool. However, recent findings show that environmental

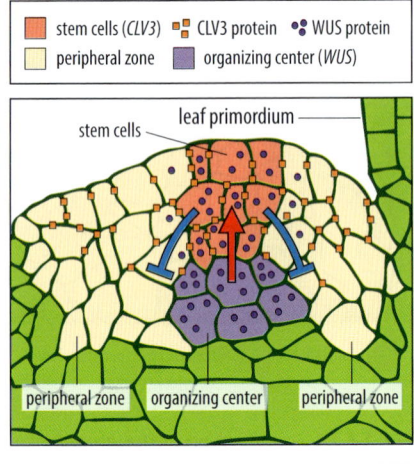

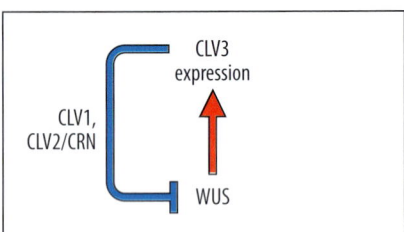

Fig. 12.15 Regulation of the stem-cell population in a shoot meristem. Intercellular signals control the position and size of the stem-cell population in the *Arabidopsis* shoot meristem. Top panel: the organizing center (mauve) expresses the transcription factor WUS. The WUS protein (purple dots) can move between cells via plasmodesmata and acts as an intercellular signal (red arrow) that helps specify and maintain the overlying cells as stem cells (orange). The stem cells express and secrete the signal protein CLAVATA3 (CLV3; orange dots), which moves laterally and downwards, and indirectly represses transcription of the *WUS* gene in the surrounding cells, acting through its cell-surface receptor proteins CLV1 and a complex of CLV2 and the protein CORYNE (CRN). CLV3 thus limits the extent of the area specified as stem cells (blue barred lines). The descendants of the stem cells are continuously displaced into the peripheral zone of the meristem (pale yellow), where they are recruited into leaf primordia. Bottom panel: the negative-feedback loop whereby WUS expression is restricted by CLV3. WUS proteins produced in the organizing center cells move into the overlying cells, where they induce the expression of CLV3. CLV3 in turn signals through CLV1/2 to suppress the expression of *WUS*.

Adapted from Brand U., et al.: **Dependence of stem cell fate in Arabidopsis on a feedback loop regulated by CLV3 activity.** Science 2000, **289**: 635-644.

factors, such as light levels and nutrients in the soil, can modulate meristem size via the hormone cytokinin.

Meristems can be induced elsewhere in the plant by the misexpression of genes involved in specifying stem-cell identity, yet another indication that stem-cell identity is conferred by cell–cell interactions and not by an embryonically specified cell lineage.

12.8 The fate of cells from different meristem layers can be changed by changing their position

Other evidence that the fate of a meristematic cell is determined by its position in the meristem, and thus the intercellular signals it is exposed to, comes from observing the fates of cells in the different meristem layers. As well as being organized into central and peripheral zones, the apical meristem of a dicotyledon, such as *Arabidopsis*, is composed of three distinct layers of cells (Fig. 12.16). The outermost layer, L1, is just one cell thick. Layer L2, just beneath L1, is also one cell thick. In both L1 and L2, cell divisions are anticlinal—that is, the new wall is in a plane perpendicular to the layer—thus maintaining the two-layer organization. The innermost layer is L3, in which the cells can divide in any plane. L1 and L2 are often known as the tunica, and L3 as the corpus.

To find which tissues each layer can give rise to, the fates of cells in the different layers can be followed by marking one layer with a distinguishable mutation, such as a change in pigmentation or a polyploid nucleus. When a complete layer is genetically different from the others in this way, the organism is known as a **periclinal chimera** (Fig. 12.17). As we saw in relation to animal embryos, chimeras are organisms composed of cells of two different genotypes, and cells of either genotype can be traced. Plant chimeras can be made by inducing mutation in the apical meristem of a seed or shoot tip by X-irradiation, or by treatment with chemicals such as colchicine that induce polyploidy.

Layer L1 gives rise to the epidermis that covers all structures produced by the shoot, while L2 and L3 both contribute to cortex and vascular structures. Leaves and flowers are produced mostly from L2; L3 contributes mainly to the stem. Although the three layers maintain their identity in the central region of the meristem over long periods of growth, cells in either L1 or L2 occasionally divide periclinally; new cell walls are formed parallel to the surface of the meristem, and thus one of the new cells invades an adjacent layer. This migrant cell now develops according to its new position, showing that cell fate is not necessarily determined by the meristem layer in which the cell originated, and that intercellular signaling is involved in specifying or changing its fate in its new position. The anticlinal pattern of cell

Meristems **645**

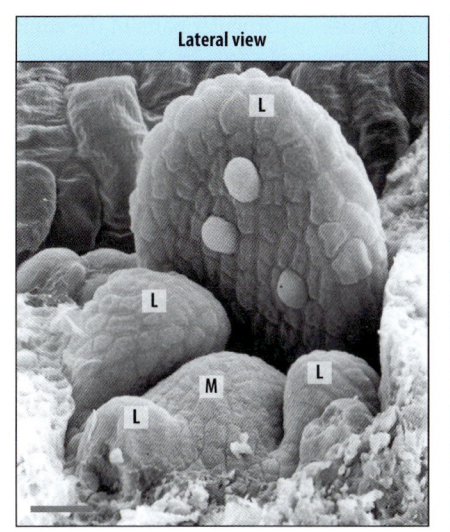

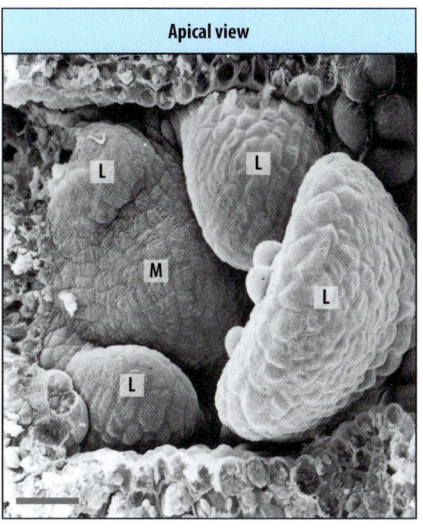

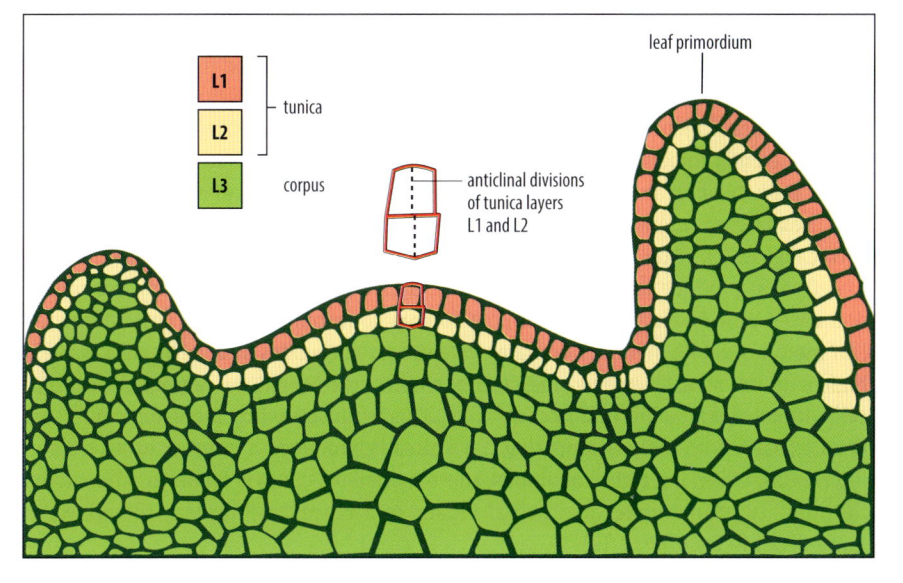

Fig. 12.16 Apical meristem of *Arabidopsis*. Top panels: scanning electron micrographs showing the organization of the meristem at the young vegetative apex of *Arabidopsis*. The plant is a *clavata1* mutant, which has a broadened apex that allows for a clearer visualization of the leaf primordia (L) and the meristem (M). Scale bar = 10 μm. Bottom panel: diagram of a vertical section through the apex of a shoot. The three-layered structure of the meristem is apparent in the most apical region. In layer 1 (L1) and layer 2 (L2), the plane of cell division is anticlinal; that is, at right angles to the surface of the shoot. Cells in layer 3 (L3, the corpus) can divide in any plane. A leaf primordium is shown forming at one side of the meristem.

Reprinted with permission from Bowman J. (Ed.): Arabidopsis: An Atlas of Morphology and Development. *Springer-Verlag, 1994.*

division in L2 becomes disrupted when leaf primordia start to form, when the cells divide periclinally, as well as anticlinally.

The transcription factor Knotted-1 in maize is homologous to *Arabidopsis* STM and, like STM, it is expressed throughout the shoot meristem to keep cells in an undifferentiated state. It is one example of a transcription factor that moves directly from cell to cell. The *KNOTTED-1* gene is expressed in all layers except L1, but the protein is also found in L1, suggesting that it can move between cells, perhaps via plasmodesmata. In *Knotted-1* gain-of-function mutants, in which the gene is misexpressed in leaves, Knotted-1 protein fused to green fluorescent protein has also been observed to move from the inner layers of the leaf to the epidermis, but not in the opposite direction.

12.9 A fate map for the embryonic shoot meristem can be deduced using clonal analysis

Much of our knowledge about the general properties of meristems comes from studies some decades ago that determined how the embryonic shoot apical meristem is related to the development of a plant over its whole lifetime. What was the fate of individual 'embryonic initials', as the stem cells were known at the time? Did

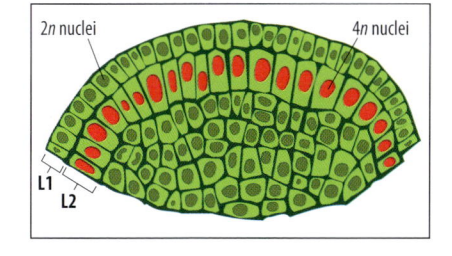

Fig. 12.17 A periclinal meristem chimera composed of cells of two different genotypes. In L1 the cells are diploid, whereas the cells of L2 are tetraploid—that is, they have double the normal chromosome number—and are larger and easily recognized.

Illustration after Steeves, T.A., Sussex, I.M.: Patterning in Plant Development. *Cambridge University Press, 1989.*

Fig. 12.18 Tobacco plant mericlinal chimera. This plant has grown from an embryonic shoot meristem in which an albino mutation has occurred in a cell of the L2 layer. The affected area occupies about a third of the total circumference of the shoot, suggesting that there are three apical initial cells in the embryonic shoot meristem.

Photograph courtesy of S. Poethig.

particular regions in the embryonic meristem give rise to particular parts of the adult plant? Individual initials can be marked by mutagenesis of seeds by X-irradiation or transposon activation, to give cells with a different color from the rest of the plant, for example. If the marked meristem cell and its immediate progeny populate only part of one layer of the meristem (in contrast to a periclinal chimera), then this area will give rise to visible sectors of marked cells in stems and organs as the plant grows; this type of chimera is known as a **mericlinal chimera** (Fig. 12.18). The fate of individually marked initials in mericlinal chimeras can be determined using clonal analysis in a manner similar to its use in *Drosophila*.

In maize, the marked sectors usually start at the base of an internode and extend apically, terminating within a leaf. Some sectors include just a single internode and leaf, representing the progeny of an embryonic initial that is lost from the meristem before the generation of the next leaf primordium. Others, however, extend through numerous internodes, showing that some embryonic initials remain in the meristem for a long time, contributing to a succession of nodes and internodes. In sunflowers, marked clones have been observed to extend through several internodes up into the flower, showing that a single initial can contribute to both leaves and flowers.

From the analysis of hundreds of mericlinal chimeras, fate maps of the embryonic shoot meristems of several species were constructed, which shed light on the properties of the shoot meristem and how it behaves during normal development. These fate maps are probabilistic because it was not possible to know the location of the marked cell in the embryonic meristem, which is inaccessible inside the seed.

The probabilistic fate map for the maize embryonic shoot apical meristem indicates that the three most apical cells in L1 give rise to the male inflorescence (the tassel and spike; Fig. 12.19). The remainder of the maize meristem can be divided into five tiers of cells that produce internodes and leaves, and which form overlapping concentric domains on the fate map. The outermost domain contributes to the earliest internode–leaf modules, while the inner domains give rise to internodes and leaves successively higher up the stem. However, the fates of meristem cells can be reprogrammed and cells do not rigidly adhere to this fate map. This is illustrated by meristems that have been excised from maize plants that have already produced a number of leaves and internodes. Given the right conditions, these meristems can still go on to produce the full number of internodes and lateral organs of a normal adult plant (described in more detail in Section 12.10).

A fate map has been similarly constructed for the embryonic shoot apical meristem of *Arabidopsis* (Fig. 12.20). Most of the *Arabidopsis* embryonic meristem gives rise to the first six leaves, whereas the remainder of the shoot, including all the flowerheads, is derived from a very small number of embryonic cells at the center of the meristem. Unlike maize, the number of leaves in *Arabidopsis* is not fixed—its growth is said to be **indeterminate**. There is no relation between particular cell lineages and particular structures, which indicates that position in the meristem is crucial in determining cell fate. One exception is that germ cells always arise from L2. The L2 layer is a clone, and so there is, in the case of germ cells, a relation between fate and a particular cell lineage.

The conclusions from the clonal analysis experiments are that the initials that contribute to a particular structure are simply those that happen to be in the appropriate region of the meristem at the time; they have not been prespecified in the embryo as, say, flower or leaf.

12.10 Meristem development is dependent on signals from other parts of the plant

To what extent does the behavior of a meristem depend on other parts of the plant? It seems to have some autonomy, because if a meristem is isolated from adjacent tissues by excision it will continue to develop, although often at a much slower rate. Excised

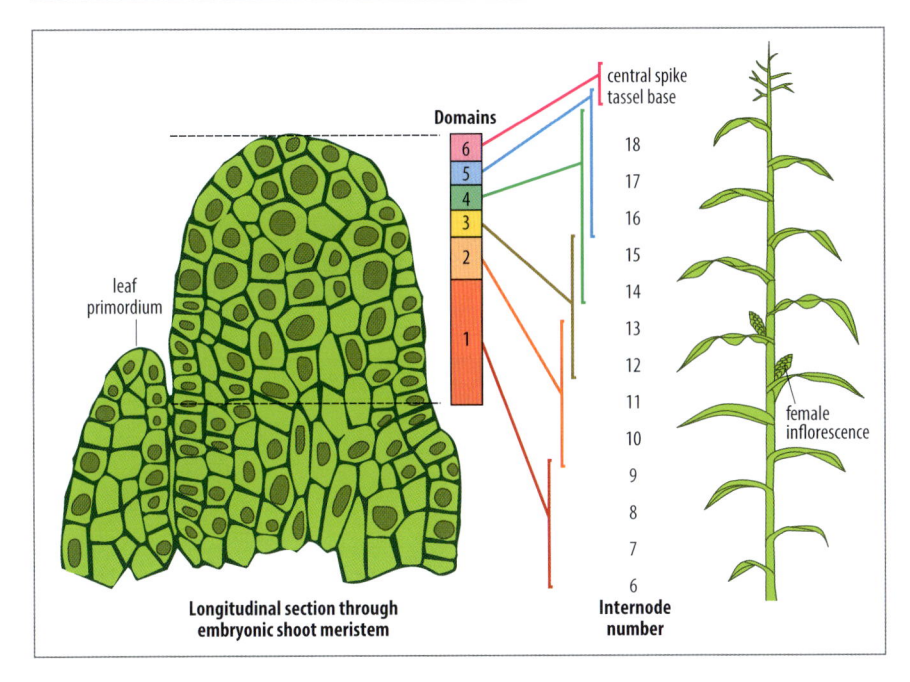

Fig. 12.19 Probabilistic fate map of the shoot apical meristem in the mature maize embryo. In maize, the primordia of the first six leaves are already present in the mature embryo and are excluded from the analysis. At the time the cells were marked, the meristem contained about 335 cells, which will give rise to 12 more leaves, the female inflorescences, and the terminal male inflorescence (the tassel and spike). A longitudinal section through the embryonic apical dome—the shoot meristem—is shown on the left. Clonal analysis shows that it can be divided into six vertically stacked domains, each comprising a set of initials that can give rise to a particular part of the plant. The number of initials in layers L1 and L2 of each domain at the embryonic stage can be estimated from the final extent of the corresponding marked sectors in the mature plant (see text). The fate of each domain, as estimated from the collective results of clonal analysis of many different plants, is shown on the right. Domain 6, comprising the three most apical L1 cells of the meristem, will give rise to the terminal male inflorescence. The fate of cells in the other domains is less circumscribed. Domain 5, for example, which consists of around eight L1 cells surrounding domain 6 and the underlying four or so L2 cells, can contribute to nodes 16–18; domain 4 to nodes 14–18; and domain 3 to nodes 12–15. The female inflorescences, which develop in the leaf axils, are derived from the corresponding domains.

Illustration after McDaniel, C.N., Poethig, R.S.: **Cell lineage patterns in the shoot apical meristem of the germinating maize embryo.** Planta 1988, *175*: 13–22.

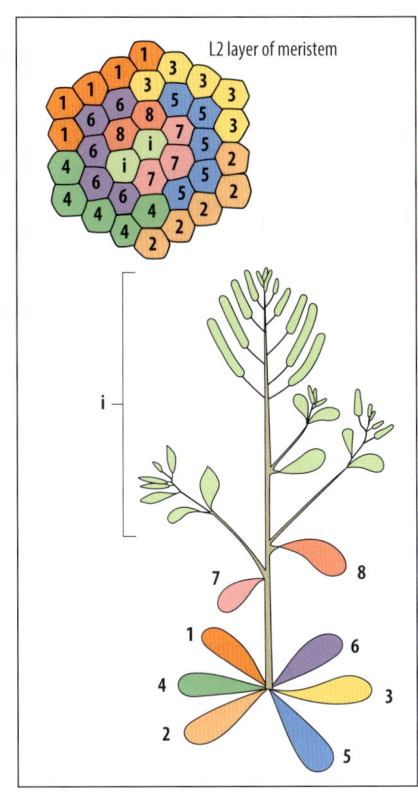

Fig. 12.20 Probabilistic fate map of the embryonic shoot meristem of *Arabidopsis*. The L2 layer of the meristem is depicted as if flattened out and viewed from above. The numbers indicate the leaf, as shown on the plant below, to which each group of meristem cells contributes, and indicate the sequence in which the leaves are formed. The inflorescence shoot (i) is derived from a small number of cells in the center of the layer.

After Irish, V.E.: **Cell lineage in plant development.** Curr. Opin. Genet. Dev. 1991, *1*: 169–173.

shoot apical meristems from a variety of plants can be grown in culture, where they will develop into shoots complete with leaves if the growth hormones auxin and cytokinin are added. The behavior of the meristem *in situ*, however, is influenced in more subtle ways by interactions with the rest of the plant.

As we saw earlier, the shoot apical meristem of maize gives rise to a succession of nodes, and terminates in the male flower. The number of nodes before flowering is usually between 16 and 22. This number is not controlled by the meristem alone, however. Evidence comes from culturing shoot tips consisting of the apical meristem and one or two leaf primordia. Meristems taken from plants that have already formed up to 10 nodes still develop into normal maize plants with the full number of nodes. The isolated maize meristem has no memory of the number of nodes it has already formed, and repeats the process from the beginning. The meristem itself is therefore not determined in the embryo with respect to the number of nodes it will form. In the plant, control over the number of nodes that are formed must therefore involve signals from the rest of the plant to the meristem.

648 Chapter 12 Plant development

Fig. 12.21 Leaf phyllotaxis. In shoots where single leaves are arranged spirally up the stem, the leaf primordia arise sequentially in a mathematically regular pattern in the meristem. Leaf primordia arise around the sides of the apical dome, just outside the meristematic region. A new leaf primordium is formed slightly above and at a fixed radial angle from the previous leaf, often generating a helical arrangement of primordia visible at the apex. Top panel: lateral views of the shoot apex. Bottom panel: view looking down on cross-sections through the apex near the tip, at successive stages from the top panel.

Top panel, illustration after Poethig, R.S., Sussex, I.M.: **The cellular parameters of leaf development in tobacco: a clonal analysis.** Planta *1985,* **165***: 170-184. Bottom panel, illustration after Sachs, T.:* Pattern Formation in Plant Tissues. *Cambridge: Cambridge University Press, 1994.*

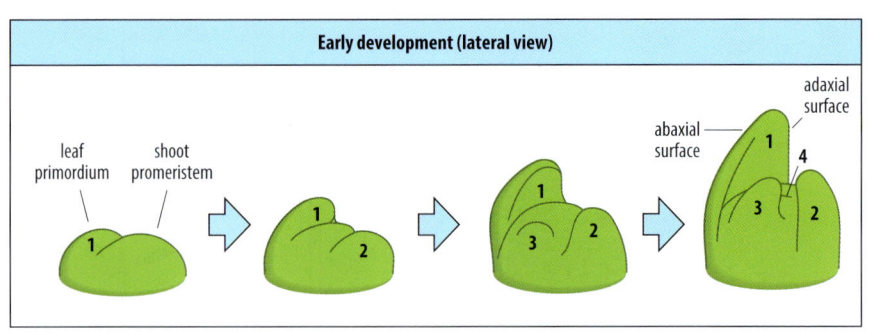

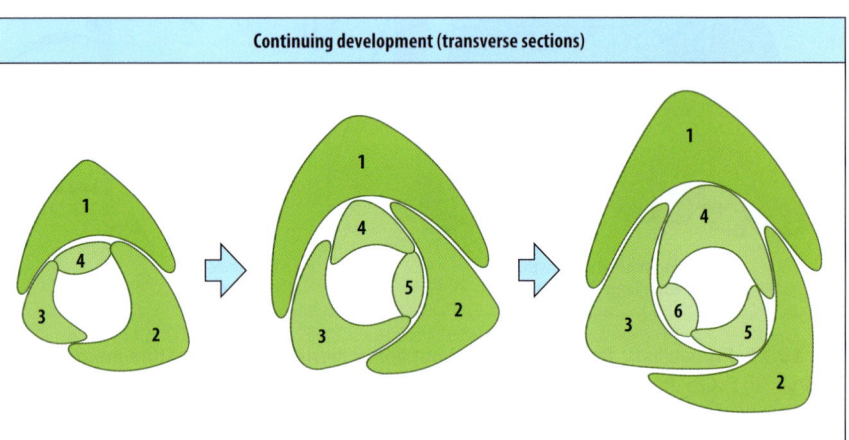

12.11 Gene activity patterns the proximo-distal and adaxial-abaxial axes of leaves developing from the shoot meristem

Leaves develop from groups of founder cells within the peripheral zone of the shoot apical meristem. The first indication of leaf initiation in the meristem is usually a swelling of a region to the side of the apex, which forms the **leaf primordium** (Fig. 12.21). This small protrusion is the result of increased localized cell multiplication and altered patterns of cell division. It also reflects changes in polarized cell expansion.

Three new axes that relate to the future leaf are established in a leaf primordium. These are the **proximo-distal axis** (leaf base to leaf tip), **medio-lateral axis** (midvein to margin), and the **adaxial–abaxial axis** (upper surface to lower surface, sometimes called dorsal to ventral). The latter is termed adaxial–abaxial as it is related to the radial axis of the shoot. The upper surface of the leaf derives from cells near the center of this axis (adaxial), whereas the lower surface derives from more peripheral cells (abaxial). The two leaf surfaces carry out different functions and have different structures, with the top surface being specialized for light capture and photosynthesis. In *Arabidopsis*, flattening of the leaf along the adaxial–abaxial axis occurs after leaf primordia begin developing, but in monocots such as maize the leaf is flattened as it emerges. The establishment of adaxial–abaxial polarity makes use of positional information along the radial axis of the meristem.

Arabidopsis leaf primordia emerge from the shoot meristem with distinct programs of development in the adaxial and abaxial halves. This asymmetry can be seen from the beginning, as the leaf primordium has a crescent shape in cross-section, with a convex outer (abaxial) side and a concave inner (adaxial) surface (see Fig. 12.21). Different genes are expressed in the future adaxial and abaxial sides, and these genes control both cell type and organ shape. For example, members of the *KANADI* (*KAN*) family of transcription factor genes are normally expressed in the abaxial side of the leaf primordium, and specify an abaxial cell fate. Ectopic expression of *KANADI1* throughout the leaf primordium can cause all cells to adopt an abaxial cell fate, and the leaf develops as a cylindrical structure. The specification of adaxial cell fate in *Arabidopsis* involves genes *PHABULOSA* (*PHB*), *PHAVOLUTA* (*PHV*), and *REVOLUTA* (*REV*). These encode HD-ZIP III transcription factors and are initially expressed in the shoot meristem and in the adaxial side of the primordium. The balance between the adaxial and abaxial cell identities is kept by mutually antagonistic regulation between adaxially and abaxially expressed genes. This antagonism between the two regions of the leaf primordium involves small RNAs. **MicroRNAs** are involved in the restriction of *PHB*, *PHV*, and *REV* expression to the adaxial side. miRNA165 and 166 are expressed in the abaxial epidermis and form a gradient across this side of the leaf (Fig. 12.22). These miRNAs target and destroy *PHB*, *PHV*, and *REV* mRNAs on the abaxial side, thus limiting their activity to the adaxial side. In turn, PHB, PHV, and REV inhibit the expression of miRNA165 and 166, creating a mutual antagonism. Gain-of-function mutations in *PHB* or *PHV* that prevent their regulation by miRNA165 and 166 result in their expression throughout the leaf. These leaves develop a cylindrical structure with only adaxial cell types characteristic of the upper side of the leaf.

Another type of small RNA that can move intracellularly, called *trans*-acting **short interfering RNAs** (ta-siRNAs), is expressed on the adaxial side and forms a gradient along the adaxial–abaxial axis. ta-siRNAs repress the expression of *AUXIN RESPONSE FACTOR 2, 3,* and *4* (*ARF2, ARF3,* and *ARF4*) which function with *KAN* genes to specify abaxial identity. This causes the elimination of *ARF2/3/4* expression from the adaxial domain, where ta-siRNA levels are highest (see Fig. 12.22).

It has been suggested that, in normal plants, the interaction between adaxial and abaxial initial cells at the boundary between them initiates lateral growth, resulting in the formation of the leaf blade and the flattening of the leaf. Indeed, the formation of the adaxial–abaxial boundary leads to the expression of *WOX1* and *PRS* (two WOX-gene family members) within the middle domain of the *Arabidopsis* leaf. The expression of *YABBY* (*YAB*) genes in the abaxial side of the leaf and within the middle domain is also dependent on adaxial–abaxial boundary formation (see Fig. 12.22). Both classes of gene are needed for normal lateral growth of the leaf. The importance of boundaries in controlling pattern and form has been seen in the parasegments of *Drosophila* (see Section 2.23) and further examples can be found in Chapter 9.

Development along the leaf proximo-distal axis also seems to be under genetic control. Like other grasses, a maize leaf primordium is composed of prospective leaf-sheath tissue proximal to the stem and prospective leaf-blade tissue distally. Mutations in certain genes result in distal cells taking on more proximal identities; for

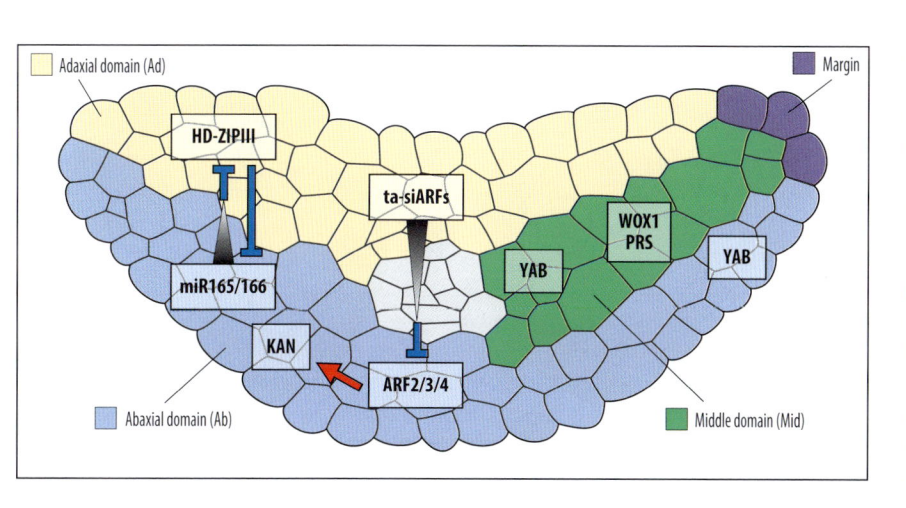

Fig. 12.22 Molecular mechanisms of leaf patterning along the adaxial–abaxial axis. The left side of the figure shows a simplified schematic of interactions between transcription factors (HD-ZIPIII, KANADI (KAN), ARF2/3/4) and small RNAs (miR165/166, ta-siARFs) that set up the adaxial (yellow) and abaxial (blue) domains. The right side illustrates the middle domain (green) which is positioned at the adaxial and abaxial boundary. *PRESSED FLOWER* (*PRS*) and *WUSCHEL-RELATED HOMEOBOX 1* (*WOX1*) are expressed in the middle domain. *YABBY* genes (*YAB*) are expressed in the abaxial domain and middle domain. *PRS*, *WOX1*, and *YAB* are involved in growth of the leaf along the medio-lateral axis. The leaf margin is specified where the middle domain overlaps with the epidermis of the leaf (purple).

Adapted from Du, F., Guan, C., Jiao, Y.: **Molecular mechanisms of leaf morphogenesis.** Mol. Plant 2018, **11**: 1117–1134.

example, making sheath in place of blade. Similar proximo-distal shifts in pattern occur in *Arabidopsis* as a result of mutation. Regional identity along the proximo-distal axis may reflect the developmental age of the cells, with distal cells adopting a different fate from proximal cells because they mature later.

12.12 The regular arrangement of leaves on a stem is generated by regulated auxin transport

As the shoot grows, leaves are generated within the meristem at regular intervals and with a particular spacing. Leaves are arranged along a shoot in various ways in different plants, and the particular arrangement, known as **phyllotaxy** or **phyllotaxis**, is reflected in the arrangement of leaf primordia in the meristem. Leaves can occur singly at each node, in pairs, or in whorls of three or more. A common arrangement is the positioning of single leaves spirally up the stem, which can sometimes form a striking helical pattern in the shoot apex.

In plants in which leaves are borne spirally, a new leaf primordium forms at the center of the first available space outside the central region of the meristem and above the previous primordium (see Fig. 12.21). This pattern suggests a mechanism for leaf arrangement based on **lateral inhibition** (see Section 1.16), in which each leaf primordium inhibits the formation of a new leaf within a given distance. In this model, inhibitory signals emanating from recently initiated primordia prevent leaves from forming close to each other. There is some experimental evidence for this. In ferns, leaf primordia are widely spaced, allowing experimental microsurgical interference. Destruction of the site of the next primordium to be formed results in a shift toward that site by the future primordium whose position is closest to it (Fig. 12.23). However, competition, as well as inhibition, is thought to be responsible for the spacing.

As we saw in the determination of apical–basal polarity in the embryo, auxin is transported out of cells with the help of proteins such as PIN1 (see Section 12.3). In the shoot, auxin produced in the shoot tip below the meristem is transported upwards into the meristem, through the epidermis and the outermost meristem layer. The direction of auxin flow in the shoot apex is controlled by PIN1 and follows a simple rule: the side of the cell on which PIN1 is found is the side nearest the neighbor cell with the highest auxin concentration. Thus auxin transport is always towards a region of higher concentration. A high concentration of auxin is a primordium activator, and initially, auxin is pumped towards a new primordium that is developing at a site of high auxin concentration (Fig. 12.24). This, however, depletes a zone of cells around the primordium of

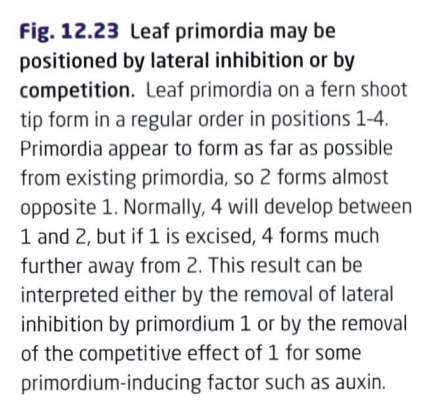

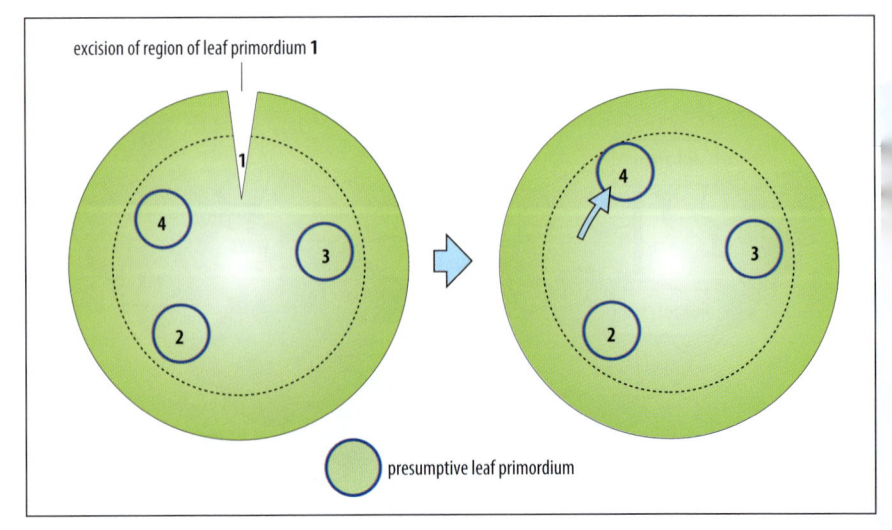

Fig. 12.23 Leaf primordia may be positioned by lateral inhibition or by competition. Leaf primordia on a fern shoot tip form in a regular order in positions 1-4. Primordia appear to form as far as possible from existing primordia, so 2 forms almost opposite 1. Normally, 4 will develop between 1 and 2, but if 1 is excised, 4 forms much further away from 2. This result can be interpreted either by the removal of lateral inhibition by primordium 1 or by the removal of the competitive effect of 1 for some primordium-inducing factor such as auxin.

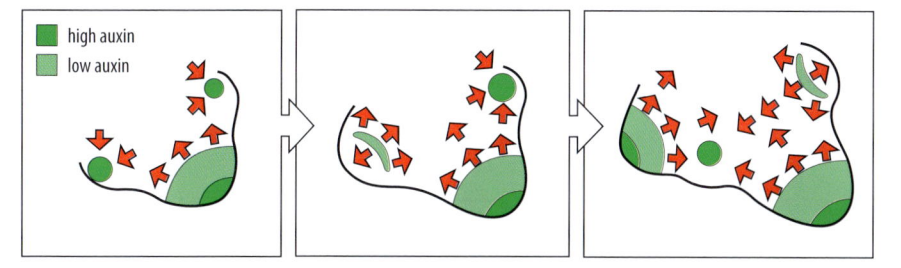

Fig. 12.24 Auxin-dependent mechanism of phyllotaxis in *Arabidopsis*. Auxin is transported through PIN proteins towards areas of high auxin concentration (dark green), at which an organ primordium will form. Cells around the developing primordium become depleted of auxin (light green), which causes the polarity of the PIN proteins to reverse so that auxin then flows away from the primordium. The red arrows indicate the direction of PIN polarity. It is proposed that this pattern of auxin circulation could set up the regular pattern of leaf and flower formation from meristems.

Adapted from Heisler, M.G., et al.: **Patterns of auxin transport and gene expression during primordium development revealed by live imaging of the** *Arabidopsis inflorescence meristem.* Curr. Biol. 2005, **15**: 1899–1911.

auxin, such that cells nearer the center of the meristem now have more auxin than the cells adaxial to the new primordium. This activates a feedback mechanism that causes PIN1 to move to the other side of these cells, and auxin now flows out of the new primordium towards the meristem, creating a new spot of high auxin concentration in the meristem farthest away from any new primordium. This leads to auxin peaks occurring sequentially, at the regular positions later occupied by new leaves.

12.13 The outgrowth of secondary shoots is under hormonal control

Auxin also has a key role in the repression of lateral buds, which develop at the same time as leaf primordia but remain dormant until a later stage of development. Auxin produced at the shoot apical meristem acts over long distances to inhibit lateral bud growth. The existence of a diffusible inhibitor produced by the shoot apex was demonstrated by putting an excised apex in contact with an agar block, which could then, on its own, inhibit lateral growth. Auxin can reproduce the effect of the excised apex in this experiment, as application of auxin to the cut tip replaces the suppressive effect of the shoot apex (Fig. 12.25). The suppressive effect of auxin on bud outgrowth is indirect, as experiments using radiolabeled auxin showed that auxin applied apically

Fig. 12.25 Apical dominance in plants. Left panels: the growth of lateral buds is inhibited by the apical meristem above them. If the upper part of the stem is removed, lateral buds start growing. Right panels: an experiment to show that apical dominance is due to inhibition of lateral bud outgrowth by a substance secreted by the apical region. This substance is the plant hormone auxin.

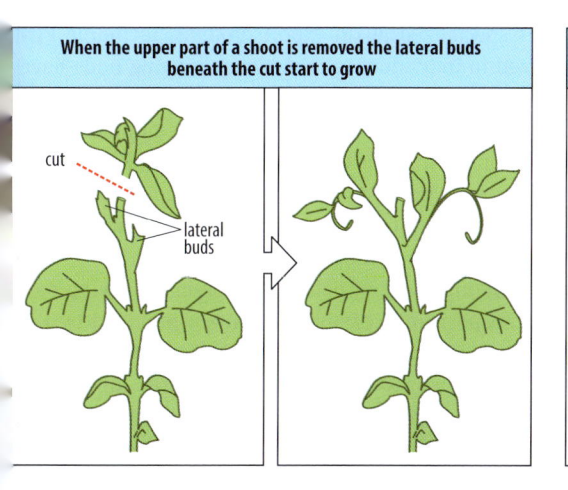

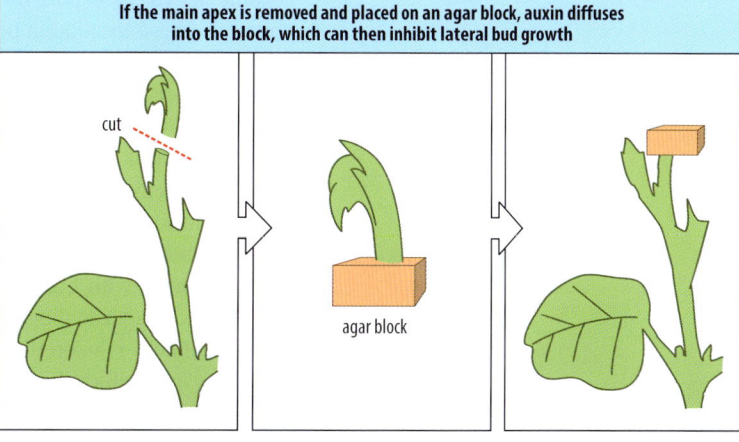

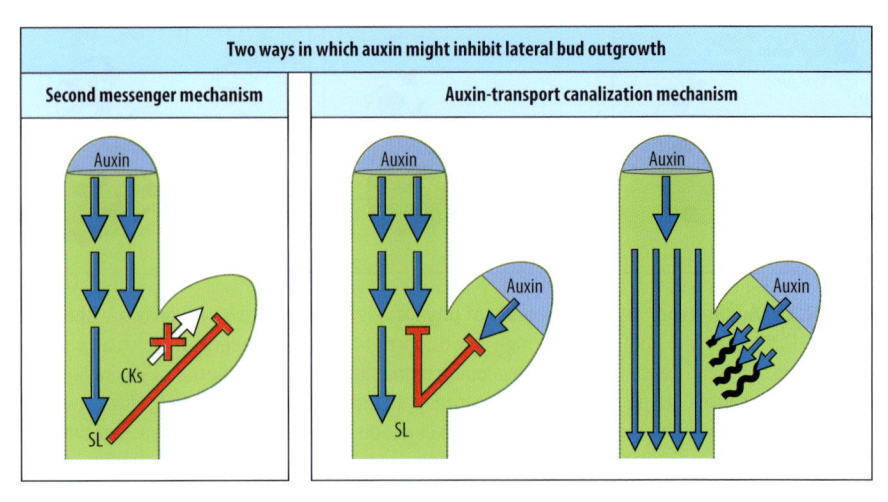

Fig. 12.26 Two ways in which auxin might control lateral bud outgrowth and shoot branching. Left panel: in the proposed second messenger mechanism, auxin produced by the shoot apex is transported down the stem (blue arrows) and promotes the production of the plant hormone strigolactone (SL), which inhibits bud outgrowth (red barred line). Cytokinins (CKs) produced in the main stem promote bud outgrowth (white arrow), but high levels of auxin inhibit their production, thus inhibiting bud outgrowth near the shoot apex. Right panel: in the proposed auxin-transport canalization mechanism, auxin produced in a lateral bud must be transported out of the bud before it can start to grow. Strigolactone inhibits this process (left), partly by promoting the removal of PIN proteins required for auxin transport. This means that, in the presence of a growing shoot apex that actively produces auxin, the canalization of auxin transport out of the bud fails (right). Auxin flow out of the bud does not connect with that down the main stem and the bud does not grow out.

Adapted from Kebrom, T.H.: **A growing stem inhibits bud outgrowth—the overlooked theory of apical dominance.** *Front. Plant Sci. 2017, **8**: 1874.*

does not enter buds. Auxin is likely to mediate this indirect effect on lateral bud outgrowth in two ways, both of which involve interactions with the hormones cytokinin and strigolactone. In the first mechanism, auxin inhibits the synthesis of cytokinin (which promotes lateral bud outgrowth) and promotes the synthesis of strigolactone (which inhibits bud outgrowth). Both of these hormones can move into lateral buds and, because their levels reflect the level of auxin, they are likely to function as 'second messengers', conveying information to the lateral buds about the level of auxin moving down the stem from the shoot apex (Fig. 12.26, left panel).

The second mechanism focuses on the regulation of auxin transport out of lateral buds. Buds need to be able to export auxin in order for the lateral meristems to become active. Its export involves specialized auxin export proteins, including PIN proteins, whose location is polarized in individual cells. Actively growing buds have narrow channels of cells with highly polarized PIN proteins. These auxin-transport pathways carry auxin away from the bud and connect with pathways of polar auxin transport in the main stem. It is thought that these pathways are established by a positive-feedback loop, in a process referred to as 'canalization' (Fig. 12.26, right panel). Auxin moves along an initial concentration gradient from the bud, which acts as an auxin source, to the stem, which acts as an auxin sink. Auxin reinforces its own transport in that direction by upregulating levels of polar PIN proteins, allowing narrower channels with high auxin transport to be established. Auxin provided by the growing shoot apex is thought to block canalization and inhibit bud growth by filling up the channels of auxin transport in the main stem. Strigolactones partly inhibit bud growth by promoting the removal of PIN proteins from the plasma membrane and weakening the auxin-mediated positive feedback that is needed for buds to establish their auxin export pathways. Recent evidence suggests that, similarly, cytokinin promotes bud outgrowth partly by increasing the levels of PIN proteins on the plasma membrane, thus making it easier for buds to export auxin.

It has also been proposed that sugars play an important part in the activation of bud outgrowth. This idea hypothesizes that bud outgrowth is promoted by sugar availability and that the growing shoot apex acts as a strong sink for sugars, thus preventing sugar accumulation in buds. A recent study in pea showed that, upon decapitation, rapid changes in sugar distribution occur, allowing sugars to accumulate in buds. It is proposed that it is this sugar accumulation, rather than auxin depletion, that causes bud outgrowth upon decapitation.

In addition to the shoot and root apical meristems and the lateral meristems in axillary buds, plant meristematic tissue also exists in the form of a cylinder of tissue, referred to as cambium, running within each stem. The cambium is responsible for growth in stem thickness (most noticeable in trees). The same three hormones involved in bud outgrowth also regulate cambium activity. Cytokinin promotes meristematic activity in cambium, as it does in buds. Auxin and strigolactone, however, have opposite effects in the two contexts, as although both inhibit axillary buds, they promote cambium growth. These opposing activities may provide a mechanism to ensure coordinated activity of both cambial and apical meristems. When the main apex dominates and buds are repressed owing to high strigolactone and auxin levels in the main stem, the stem will increase in thickness, promoting the formation of a single strong shoot. The levels of strigolactone and cytokinin are modulated in response to nutrient availability, providing a mechanism for the observation that shoots can adopt different forms in different environments.

12.14 Root tissues are produced from *Arabidopsis* root apical meristems by a highly stereotyped pattern of cell divisions

The organization of tissues in the *Arabidopsis* root tip is shown in Figure 12.27. The radial pattern comprises single layers of epidermal, cortical, endodermal, and pericycle cells, with vascular tissue in the center (protophloem and protoxylem). Root apical meristems resemble shoot apical meristems in many ways and give rise to the root in a similar manner to shoot generation. But there are some important differences between the root and shoot apical meristems. The shoot apical meristem is at the extreme tip of the shoot, whereas the root apical meristem is covered by a root cap (which is itself derived from one of the layers of the meristem). Also, unlike the shoot apical meristem, which produces new lateral organs from the undifferentiated cells at its flanks, the root apical meristem does not produce lateral organs directly from its flanks. Instead, subsets of cells within the pericycle layer become competent to initiate lateral root primordia close to the root tip, just after being displaced from the meristematic region, but only go on to form a new lateral root primordium once they are displaced to a greater distance from the tip.

The root is set up early (see Section 12.2) and a well-organized embryonic root can be identified in the late heart-stage embryo (Fig. 12.28). An antagonistic interaction between auxin and cytokinin controls the establishment of the root stem-cell niche. Clonal analysis has shown that the seedling root meristem can be traced back to a set of embryonic initials that arise from a single tier of cells in the heart-stage embryo.

As in the shoot apical meristem, a root apical meristem is composed of an organizing center, called the **quiescent center** in roots, in which the cells divide only very rarely, and which is surrounded by stem-cell-like initials that give rise to the root tissue (see Fig. 12.28). The quiescent center is essential for meristem function. When parts of the meristem are removed by microsurgery, it can regenerate, but regeneration is always preceded by the formation of a new quiescent center. Laser destruction of individual quiescent-center cells shows that, as in the shoot meristem, a key function of the

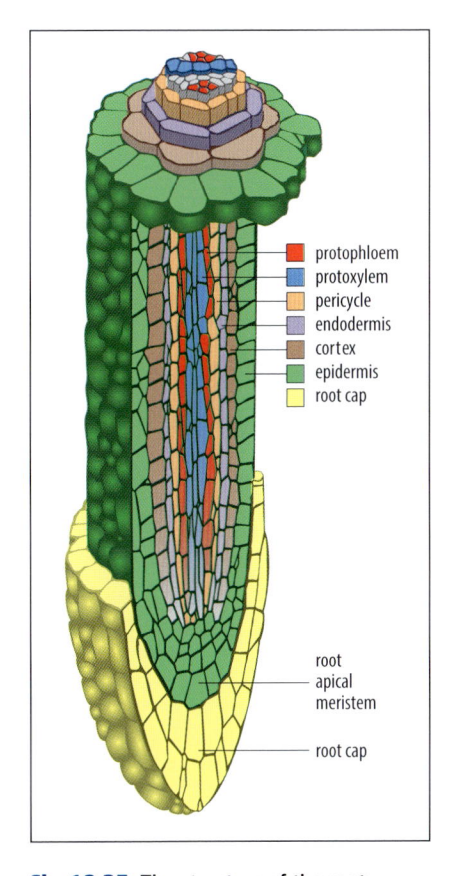

Fig. 12.27 The structure of the root tip in *Arabidopsis*. Roots have a radial organization. In the center of the growing root tip is the future vascular tissue (protoxylem and protophloem). This is surrounded by further tissue layers.

Fig. 12.28 Fate map of root regions in the heart-stage *Arabidopsis* embryo. The root grows by the division of a set of initial cells. The root meristem comes from a small number of cells in the heart-shaped embryo. Each tissue in the root is derived from the division of a particular initial cell. At the center of the root meristem is a quiescent center, in which cells rarely divide.

Illustration after Scheres, B., et al.: **Embryonic origin of the** *Arabidopsis* **primary root and root meristem initials.** *Development 1994, 120: 2475-2487.*

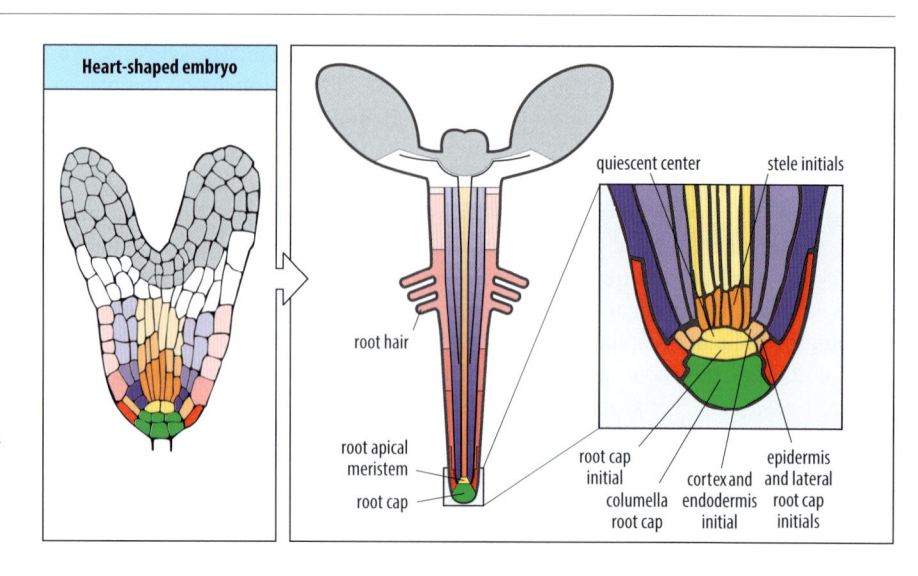

quiescent center is to maintain the immediately adjacent initials in the stem-cell state and prevent them from differentiating.

Each initial undergoes a stereotyped pattern of cell divisions to give rise to a number of columns, or **files**, of cells in the growing root (see Fig. 12.27); each file of cells in the root thus has its origin in a single initial. Some initials give rise to both endodermis and cortex, whereas others give rise to both epidermis and the root cap. Before it is displaced from the meristem, therefore, the undifferentiated progeny of an endodermis/cortex initial, for example, will divide asymmetrically to give one daughter that produces cortex and one that produces endodermis. The genes *SCARECROW* (*SCR*) and *SHORTROOT* (*SHR*) are necessary to confer this asymmetry on the dividing cell, and mutations in these genes result in roots with just a single layer of cells in the endodermis/cortex lineage, instead of having both endodermis and cortex. In *scr* mutants this single-cell layer has characteristics of both endodermis and cortex, but in the *shr* mutant it lacks endodermal characteristics. The mixed identity of the *scr* mutant endodermis/cortex cell layer has been confirmed by single-cell RNA sequencing (scRNAseq), which showed that these cells have a cortex-like identity in their early developmental stages and then acquire an endodermal-like identity through time. Thus, scRNAseq is revealing unprecedented detail of plant cell fates, just as it is doing in animals. The technique of scRNAseq is described in Box 6C.

The normal pattern of cell divisions is not obligatory, however. As discussed earlier, *fass* mutants, which have disrupted cell divisions, still have relatively normal patterning in the root. In addition, laser ablation of individual meristem cells does not lead to an abnormal root. The remaining initials undergo new patterns of cell division that replace the progeny of the cells that have been destroyed. Such observations show that, as in the shoot meristem, the fate of cells in the developing root meristem depends on their recognition of positional signals and not on their lineage.

As we saw in Section 12.3, auxin gradients have a major role in patterning the embryo and specifying the root region, and mutations that affect auxin localization lead to root defects. At the globular stage of embryonic development, the auxin-transport protein PIN1 is localized in cells in the future root region and the highest level of auxin is found adjacent to where the quiescent center will develop.

The role of auxin in root development continues into the adult plant. Auxin in the root is transported out of cells via the PIN proteins, and enters adjacent cells. Cells with raised auxin levels transport auxin better, probably due to an increase in the number of PIN proteins in the membrane because auxin prevents their endocytosis and recycling.

Auxin has a key role in patterning the growing root. Modeling of auxin movement, using the known distribution of PIN proteins in the membranes of different cell types, has shown that PIN-directed flow can explain the formation and maintenance of a stable auxin maximum at the quiescent center. Auxin is transported in a reverse-fountain pattern, downwards in the central vascular tissue of the root tip, outwards and upwards through the outer layers of the root tip, and inwards again through intermediate cell files (endodermis and pericycle; Fig. 12.29).

Modeling suggests that this pattern of transport forms an auxin concentration maximum at the quiescent center in the root tip, with a decreasing concentration along the apical–basal axis. The auxin gradient forms because as auxin is transported up the root in the outer layers, it is also simultaneously transported back to the central downwards stream via the intermediate cell files, causing the auxin concentration to drop higher up the root tip. High levels of auxin biosynthesis in the quiescent center region at the root tip also contribute to the gradient formation.

An idea of how the auxin gradients might be translated into effects on cell fate and cell behavior is provided by the graded distribution of the four PLETHORA-family transcription factors in the root. These transcription factors are required for proper root development and are expressed in a graded manner along the apical–basal axis, with expression maxima for all at the quiescent center—the region of maximum auxin concentration. Here they are essential for stem-cell maintenance and function. Lower concentrations of PLETHORA proteins correspond to the meristem region, where cells are proliferating, and even lower concentrations appear to be necessary for exit from the meristem and cell differentiation in the elongation zone. Recent evidence shows that the gradient of PLETHORA proteins is produced by transcription of *PLETHORA* genes in the quiescent center region, in response to prolonged high auxin concentrations. PLETHORA proteins then form a gradient through intercellular movement and by being diluted through the growth of descendants of the stem-cell niche.

Other influences on root patterning are the interplay between auxin and the hormone cytokinin. Cytokinin helps regulate root meristem size and set the position of the transition zone—where cells stop dividing and start elongating and differentiating—by repressing both auxin transport and cellular responses to auxin in this region. This is mediated by the Aux/IAA *SHORT HYPOCOTYL 2* (*SHY2*), which is expressed at the transition zone and in the differentiation zone above as it is promoted by cytokinin and inhibited by auxin. *SHY2* inhibits the expression of PIN proteins in the outer cell layers of the root, reducing the upwards transport of auxin from the root tip (see Fig. 12.29).

Auxin is also involved in the ability of plants to regenerate from a small piece of stem. In general, roots form from the end of the stem that was originally closest to the root, whereas shoots tend to develop from dormant buds at the end that was nearest to the shoot. This polarized regeneration is related to vascular differentiation and to the polarized transport of auxin. Transport of auxin from its source in the shoot tip towards the root leads to an accumulation of auxin at the 'root' end of the stem cutting, where it induces the formation of roots. One hypothesis suggests that polarity is both induced and expressed by the oriented flow of auxin.

Fig. 12.29 Zonation of the root along the apical-basal axis. An auxin gradient with a maximum at the quiescent center (QC) is formed as a result of the localization of auxin biosynthesis around this region combined with the pattern of polar auxin transport driven by PIN proteins in the root tip (black arrows). Auxin concentration is low in the transition zone where cytokinin inhibits PIN protein expression via SHY2. The gradient in auxin concentration (purple) causes the formation of a gradient of PLETHORA proteins (green). The concentration of PLETHORA proteins specifies the zone of the root.

After Del Bianco, M., Giustini, L., Sabatini, S.: **Spatiotemporal changes in the role of cytokinin during root development.** New Phytol. 2013, **199**: 324-338.

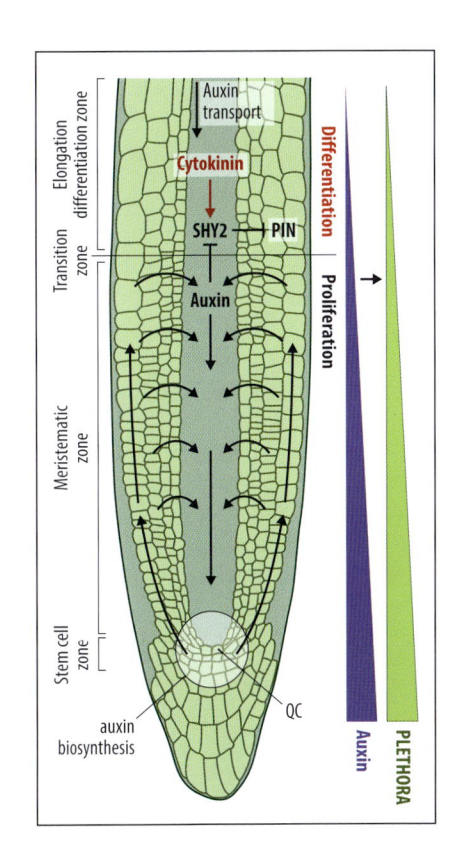

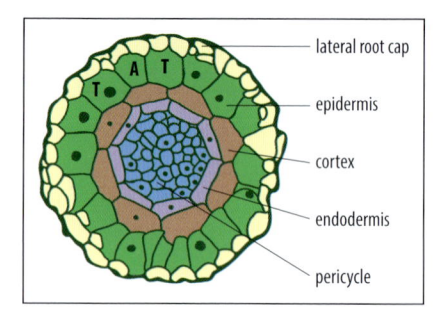

Fig. 12.30 Organization of cell types in the root epidermis. The epidermis is composed of two types of cells: trichoblasts (T), which will form root hairs, and atrichoblasts (A), which will not. Trichoblasts overlie the junction between two cortical cells and atrichoblasts are located over the outer tangential wall of cortical cells.

After Dolon, L., Scheres, B.: **Root pattern: shooting in the dark?** *Semin. Cell Dev. Biol. 1998, 9: 201-206.*

One of the best examples of developmentally important transcription factor movement from one cell to another is found in roots. As noted earlier, expression of the gene *SHR* is required for root cells to adopt an endodermal fate. SHR protein is, however, not synthesized in the prospective endodermal cells, but in the adjacent cells on the inner side. SHR is transported from these cells outwards into the prospective endodermis, and this movement seems to be regulated, and not simply due to diffusion. The expression of the gene *SCR* is upregulated in the endodermis in response to SHR protein. The SCR and SHR proteins bind to each other and this confines SHR to the nuclei of endodermal cells, preventing it from moving to the next cell layer outwards in the root. This ensures that only a single file of endodermal cells is specified by SHR.

12.15 Root hairs are specified by a combination of positional information and lateral inhibition

Root hairs are formed from epidermal cells at regular intervals around the root, and this regularity is thought to be achieved by a combination of responses to positional information and lateral inhibition by the movement of transcription factors between cells. Files of cells that will make root hairs alternate with files of non-hair-producing cells on the surface of the developing root. The importance of position is shown by the fact that if an epidermal cell overlies a junction between two cortical cells it forms a root hair, whereas if it contacts just one cortical cell it does not (Fig. 12.30). And if a cell changes its position in relation to the cortex, its fate will also change, from a potential hair-forming cell to a non-hair-forming cell and vice versa. Most cell divisions in the future epidermis are horizontal, increasing the number of cells per file, but occasionally a vertical anticlinal division occurs, pushing one of the daughter cells into an adjacent file. The daughter cell then assumes a fate corresponding to its new position in relation to the adjacent cortical cells.

The positional cues, as yet unknown, are thought to be detected by the epidermal cells through the SCRAMBLED protein, which is a receptor-like protein kinase. SCRAMBLED activity influences the activity of a network of transcription factors that control cell fate. Two groups of transcription factors have been identified by mutation experiments, one group promoting a root-hair fate and one suppressing it. A key transcription factor that seems to be regulated by SCRAMBLED activity is WEREWOLF, whose expression is suppressed in presumptive hair-forming cells, presumably in response to the positional signal. As well as promoting an atrichoblast fate, however, WEREWOLF is also required for the expression of transcription factors (CAPRICE, TRYPTYCHON, and ENHANCER OF TRYPTYCHON AND CAPRICE1) that are needed to specify root-hair cells. After positional signaling, these proteins will only be produced in the presumptive atrichoblasts, but CAPRICE proteins move laterally into the adjacent epidermal cells and promote these cells' differentiation as trichoblasts by inhibiting genes that would otherwise give an atrichoblast fate.

SUMMARY

Meristems are the growing points of a plant. The apical meristems, found at the tips of shoots and roots, give rise to all the plant organs—roots, stem, leaves, and flowers. They consist of small groups of a few hundred undifferentiated cells that are capable of repeated division. The center of the meristem is occupied by self-renewing stem cells, which replace the cells that are lost from the meristem when organs are formed. The fate of a cell in the shoot meristem depends upon its position in the meristem and interactions with its neighbors, as when a cell is displaced from one layer to another it adopts the fate of its new layer. Meristems can also self-regulate when parts are removed, in line with cell-cell interactions determining cell fate. Fate maps of embryonic shoot meristems show that they can be divided into domains, each of which normally contributes to the tissues of a particular region of the plant, but the fate of the embryonic initials is not fixed. The shoot meristem gives rise to leaves in species-specific patterns—phyllotaxy—which seem best accounted for by regulated transport of auxin. Lateral inhibition is involved in the regular spacing of hairs on root and leaf surfaces. In the root meristem, the cells are organized rather differently from those in the shoot meristem, and there is a much more stereotyped pattern of cell division. A set of initial cells maintains root structure by dividing along different planes.

SUMMARY: Meristems give rise to all adult tissues

Shoot meristem

cell fate is determined by position
⇩
shoot meristem gives rise to stem internodes, leaves, and flowers
⇩
arrangement of leaves on stem may be due to an auxin-based mechanism at shoot apex

Root meristem

stereotyped division of initials along different planes
⇩
root meristem gives rise to all root structures

Flower development and control of flowering

Flowers contain the reproductive cells of higher plants and develop from the shoot meristem. In most plants, the transition from a vegetative shoot meristem to a floral meristem that produces a flower is largely, or absolutely, under environmental control, with day length and temperature being important determining factors. In a plant such as *Arabidopsis*, in which each flowering shoot produces multiple flowers, the vegetative shoot meristem first becomes converted into an inflorescence meristem, which then forms floral meristems, each of which develops completely into a single flower (Fig. 12.31). Floral meristems are thus determinate, unlike the indeterminate shoot apical meristem. Flowers, with their arrangement of floral organs (sepals, petals, stamens, and carpels), are rather complex structures, and it is a major challenge to understand how they arise from the floral meristem.

The conversion of a vegetative shoot meristem into one that makes flowers involves the induction of so-called **meristem identity genes**. A key regulator of floral induction in *Arabidopsis* is the meristem identity gene *LEAFY* (*LFY*); a related gene in

Fig. 12.31 Scanning electron micrograph of an *Arabidopsis* inflorescence meristem. The central inflorescence meristem (shoot apical meristem; SAM) is surrounded by a series of floral meristems (FM) of varying developmental ages. The inflorescence meristem grows indeterminately, with cell divisions providing new cells for the stem below, and new floral meristems on its flanks. The floral meristems (or floral primordia) arise one at a time in a spiral pattern. The most mature of the developing flowers is on the right (FM1), showing the initiation of sepal primordia surrounding a still-undifferentiated floral meristem. Eventually such a floral meristem will also form petal, stamen, and carpel primordia.

*Photograph reproduced with permission from Meyerowitz, E.M., et al.: **A genetic and molecular model for flower development in Arabidopsis thaliana**. Development Suppl. 1991, 157–167.*

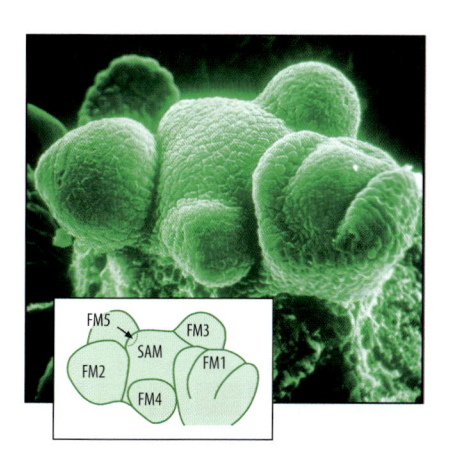

Antirrhinum is *FLORICAULA* (*FLO*). How environmental signals, such as day length, influence floral induction is discussed later. We will first consider the mechanisms that pattern the flower, in particular those that specify the identity of the floral organs.

12.16 Homeotic genes control organ identity in the flower

The individual parts of a flower each develop from a **floral organ primordium** produced by the floral meristem. Unlike leaf primordia, which are all identical, the floral organ primordia must each be given a correct identity and be patterned according to it. An *Arabidopsis* flower has four concentric whorls of structures, which reflect the arrangement of the floral organ primordia in the meristem. The sepals (whorl 1) arise from the outermost ring of meristem tissue, and the petals (whorl 2) from a ring of tissue lying immediately inside it. An inner ring of tissue gives rise to the male reproductive organs—the stamens (whorl 3). The female reproductive organs—the carpels (whorl 4)—develop from the center of the meristem. In a floral meristem of *Arabidopsis*, there are 16 separate primordia, giving rise to a flower with four sepals, four petals, six stamens and a pistil made up of two carpels (Fig. 12.32).

The primordia arise at specific positions within the meristem, where they develop into their characteristic structures. After the emergence of the primordia in *Antirrhinum*, cell lineages become restricted to particular whorls, rather like the lineage restriction to compartments in *Drosophila* (see Section 2.23). Lineage restriction occurs at the time when the pentagonal symmetry of the flower becomes visible and genes that give the different floral organs their identity are expressed. The lineage compartments within the floral meristem appear to be delineated by narrow bands of non-dividing cells.

Like the homeotic selector genes that specify segment identity in *Drosophila*, mutations in floral identity genes cause homeotic mutations in which one type of flower part is replaced by another. In the *Arabidopsis* mutant *apetala2*, for example, the sepals are replaced by carpels and the petals by stamens; in the *pistillata* mutant, petals are replaced by sepals and stamens by carpels. These mutations identified the floral organ identity genes, and have enabled their mode of action to be determined.

Homeotic floral mutations in *Arabidopsis* fall into three classes, each of which affects the organs of two adjacent whorls (Fig. 12.33). The first class of mutations, of which *apetala2* is an example, affect whorls 1 and 2, giving carpels instead of sepals in whorl 1, and stamens instead of petals in whorl 2. The phenotype of the flower, going from the outside to the center, is therefore carpel, stamen, stamen, carpel. The second class of homeotic floral mutations affects whorls 2 and 3. In this class, *apetala3* and *pistillata* give sepals instead of petals in whorl 2 and carpels instead of stamens in whorl 3, with a phenotype sepal, sepal, carpel, carpel. The third class of mutations affects whorls 3 and 4, and gives petals instead of stamens in whorl 3 and sepals or variable structures in whorl 4. The mutant *agamous*, which belongs to this class, has an extra set of sepals and petals in the center instead of the reproductive organs.

These mutant phenotypes can be accounted for by an elegant model in which overlapping patterns of gene activity specify floral organ identity in a manner highly reminiscent of the way in which *Drosophila* homeotic genes specify segment identity

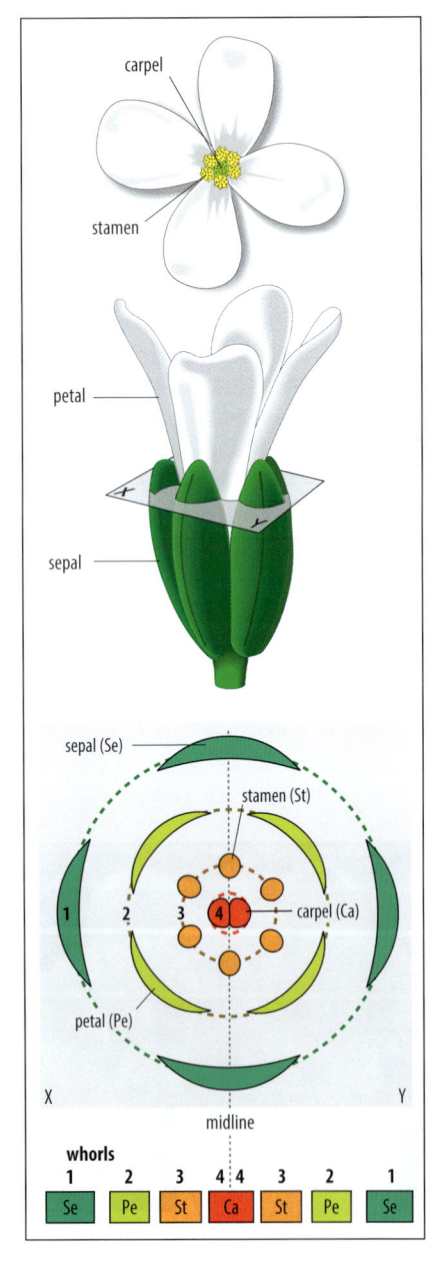

Fig. 12.32 Structure of an *Arabidopsis* flower. *Arabidopsis* flowers are radially symmetrical and have an outer ring of four identical green sepals enclosing four identical white petals, within which is a ring of six stamens, with two carpels in the center. Bottom: floral diagram of the *Arabidopsis* flower representing a cross-section taken in the plane indicated in the top diagram. This is a conventional representation of the arrangement of the parts of the flower, showing the number of flower parts in each whorl and their arrangement relative to each other.

Illustration after Coen, E.S., Meyerowitz, E.M.: **The war of the whorls: genetic interactions controlling flower development.** *Nature* 1991, **353**: 31–37.

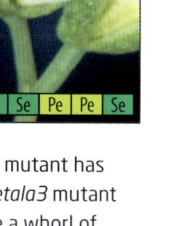

Fig. 12.33 Homeotic floral mutations in *Arabidopsis*. Left panel: an *apetala2* mutant has whorls of carpels and stamens in place of sepals and petals. Center panel: an *apetala3* mutant has two whorls of sepals and two of carpels. Right panel: *agamous* mutants have a whorl of petals and sepals in place of stamens and carpels. Transformations of whorls are shown inset, and can be compared to the wild-type arrangement, as shown in Figure 12.32.

*Left and right panels based on Meyerowitz, E.M., et al.: **A genetic and molecular model for flower development in Arabidopsis thaliana**. Development Suppl. 1991, 157-167. Center panel from Bowman, J.L., et al.: **Genes directing flower development in Arabidopsis**. Plant Cell 1989, **1**: 37-52. Published by permission of The American Society of Plant Physiologists.*

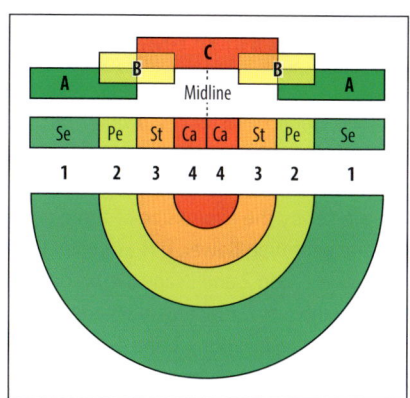

Fig. 12.34 The three overlapping regions of the *Arabidopsis* floral meristem that have been identified by the homeotic floral identity mutations. Region A corresponds to whorls 1 and 2, B to whorls 2 and 3, and C to whorls 3 and 4.

along the insect's body (Fig. 12.34). In detail, however, there are many differences, and quite different genes are involved. In this instance, plants and animals have, perhaps not surprisingly, independently evolved a similar approach to patterning a multicellular structure, but have recruited different proteins to carry it out.

In essence, the floral meristem is divided by the expression patterns of the homeotic genes into three concentric overlapping regions—A, B, and C—which partition the meristem into four non-overlapping regions corresponding to the four whorls. Each of the A, B, and C regions corresponds to the zone of action of one class of homeotic genes and the particular combinations of *A*, *B*, and *C* functions give each whorl a unique identity and so specify organ identity. Of the genes mentioned in Figure 12.33, *APETALA1* (*AP1*) and *APETALA2* (*AP2*) are *A*-function genes, *APETALA3* (*AP3*) and *PISTILLATA* (*P1*) are *B*-function genes, and *AGAMOUS* (*AG*) is a *C*-function gene. The expression of *AP3* and *AG* in the developing flower is shown in Figure 12.35. All the homeotic genes, also known as **floral organ identity genes**, encode transcription factors, and the *B*- and *C*-function proteins such as AP3 and AG contain a conserved DNA-binding sequence known as the MADS box. MADS box genes are present in animals and yeast, but a role in development is known mainly in plants—although a MADS box transcription factor, MEF2, is involved in muscle differentiation in animals. The original simple model for specifying floral organ identity is presented in more detail in Box 12C. Since the model was first proposed, more has become known

Fig. 12.35 Expression of *APETALA3* and *AGAMOUS* during flower development. *In situ* hybridization shows that *AGAMOUS* is expressed in the central whorls (left panel), whereas *APETALA3* is expressed in the outer whorls that give rise to petals and stamens (right panel).

*Reprinted with permission from Meyerowitz, E.M.: **The genetics of flower development**. Sci. Am. 1994, **271**: 40-47.*

BOX 12C The basic model for the patterning of the *Arabidopsis* flower

The floral meristem is divided into three overlapping regions (see Fig. 12.34), with each region corresponding to a class of homeotic mutations, as shown in Figure 12.33. These regions are set by the expression of three gene functions: A, B, and C. In the wild-type flower, it is assumed that A is expressed in whorls 1 and 2, B in 2 and 3, and C in whorls 3 and 4. In addition, A function inhibits C function in whorls 1 and 2 and C function inhibits A function in whorls 3 and 4—that is, A and C functions are mutually exclusive. A alone specifies sepals, A and B together specify petals, B and C stamens, and C alone carpels (Fig. 1, left panel).

The homeotic mutations eliminate the functions of A, B, or C, and alter the regions within the meristem where the various functions are expressed. Mutations in A, such as *apetala2*, result in an absence of function A, and C spreads throughout the meristem, resulting in the half-flower pattern of carpel, stamen, stamen, carpel (see Fig. 1, center panel). Mutations in B, such as *apetala3* (see Fig. 12.33), result in only A functioning in whorls 1 and 2, and C in whorls 3 and 4, giving sepal, sepal, carpel, carpel. Mutations in C genes (such as *agamous*) result in A activity in all whorls, giving the phenotype sepal, petal, petal, sepal (see Fig. 1, right panel).

All the floral homeotic mutants discovered so far in *Arabidopsis* can be quite satisfactorily accounted for by this model (although there are variations in gene numbers and expression patterns in other species that allow mutant phenotypes not seen in *Arabidopsis*), and particular genes can be assigned to each controlling function. Function A corresponds to the activity of genes such as *APETALA2*, B to *APETALA3* and *PISTILLATA*, and C to *AGAMOUS*. The model also accounts for the phenotype of double mutants, such as *apetala2* with *apetala3*, and *apetala3* with *pistillata*, as shown in Figure 2.

This system emphasizes the similarity in function between the homeotic genes in animals and those controlling organ

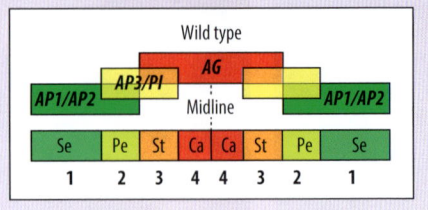

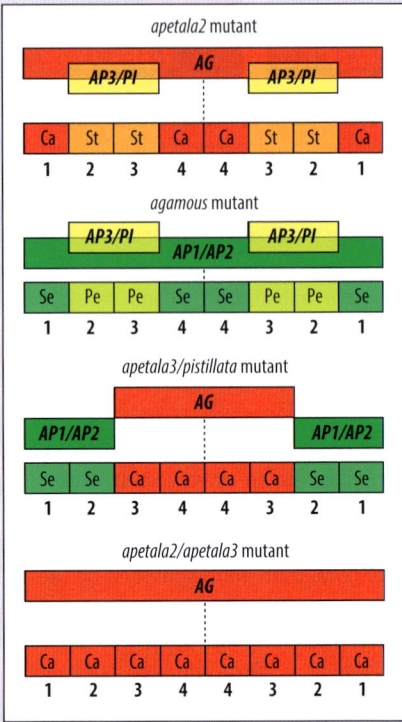

Fig. 2

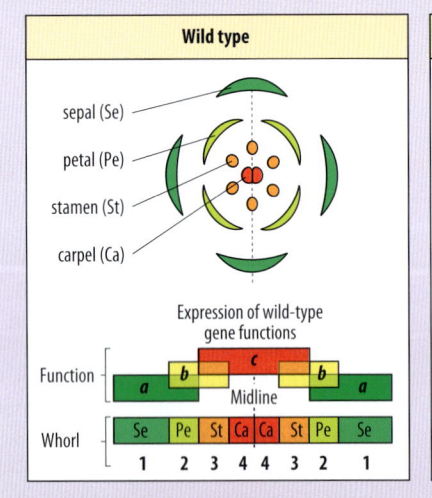

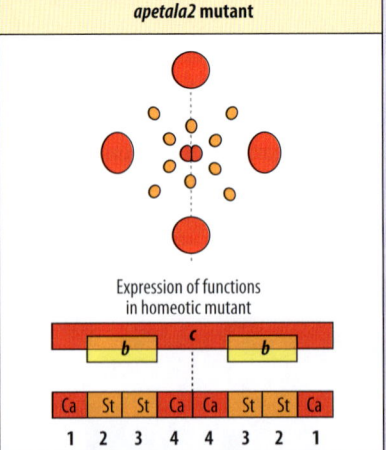

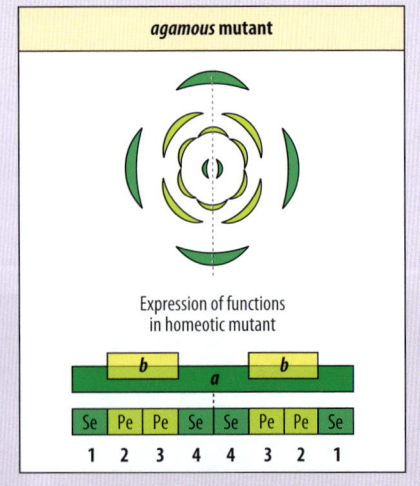

Fig. 1

identity in flowers, although the genes themselves are completely different. The functional similarity with the HOM-C complex of *Drosophila* is further illustrated by the role of the *CURLY LEAF* gene of *Arabidopsis*, which is necessary for the stable repression of a floral homeotic gene. *CURLY LEAF* is related to the Polycomb family of animal genes, which are similarly required for stable repression of homeotic genes in the fly and in mammals.

about the activities and functions of the genes identified by the homeotic mutations, more genes controlling flower development have been discovered, and more 'functions' added.

Although very useful conceptually, the original ABC model is incomplete. For example, the *ABC* genes only showed their homeotic properties in the floral meristem and do not convert leaves into floral organs when artificially overexpressed in vegetative meristems, as might be expected for homeotic genes of this type. The discovery of the *SEPALLATA* (*SEP*) genes, which also encode MADS box proteins, addressed this issue.

When the *SEP* genes are overexpressed alongside *ABC* genes in vegetative meristems the emerging leaves are converted into floral organs. The *SEP* genes are required for the *A*, *B*, and *C* functions and are only active in floral meristems. SEP proteins form tetrameric complexes with *A*, *B*, and *C* gene products, and it is these tetrameric complexes which regulate gene expression and specify floral organ identity. Thus, the *SEP* genes provide the information that the tissue context is a floral meristem, allowing *A*, *B*, and *C* gene products to function. This is further illustrated by the fact that in the absence of *SEP*s, although the *ABC* genes remain expressed in the floral meristems, the floral meristems form leaves instead of floral organs. An updated view of the mechanism for specifying floral organ identity is shown in Figure 12.36.

There is a better understanding of the functions and patterning of the floral homeotic genes than when the ABC model was first proposed. In particular, the positioning of the A and C regions appears to be more complex than just mutual repression between them. The MADS box homeotic A-class gene *AP1* has been found to have a dual role: it acts early with other genes to specify general floral meristem identity and only later contributes to *A* function. It is induced by the meristem identity gene *LFY*, which is expressed throughout the meristem, and *AP1* is actively inhibited in the central regions of floral meristems by *AGAMOUS*. The expression of the *A*-function gene *APETALA2* (*AP2*), however, is not directly repressed by *AGAMOUS*, but is translationally repressed by a microRNA, which keeps the AP2 protein at a low level. *APETALA3* and *PISTILLATA1* are thought to be activated as a result of a meristem identity gene called *UNUSUAL FLORAL ORGANS* (*UFO*), which is expressed in the meristem in a pattern similar to that of *B*-function genes (see Fig. 12.36). *UFO* encodes a component of ubiquitin ligase, and is thought to exert its effects on flower development by targeting specific proteins for degradation. As we saw in animals, in relation to the control of β-catenin degradation (see Chapter 3), regulated degradation of proteins can be a powerful developmental mechanism. In the center of the floral meristem, the expression of *AGAMOUS* is partly controlled by *WUS*, which as we saw earlier, is expressed in the organizing center of the vegetative shoot meristem and continues to be expressed in floral meristems.

Fig. 12.36 Updated status of the ABC model of floral organ identity. The regulatory genes *LEAFY*, *WUSCHEL* (*WUS*), and *UNUSUAL FLORAL ORGANS* (*UFO*) are expressed in specific domains in the floral meristem, which, together with repression of *APETALA1* by *AGAMOUS*, results in the pattern of *ABC* functions. ABC proteins and the co-factor SEP proteins assemble into complexes that specify the different organ identities.

Adapted from Lohmann, J.U., Weigel, D.: **Building beauty: the genetic control of floral patterning.** *Dev. Cell 2002,* **2***: 135–142.*

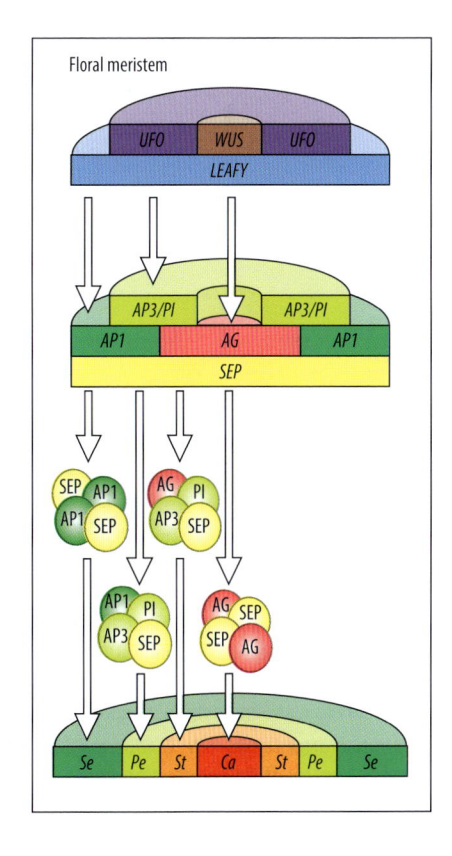

662 Chapter 12 Plant development

Fig. 12.37 Mutations in *CYCLOIDEA* make the *Antirrhinum* flower symmetrical. In the wild-type flower (left) the petal pattern is different along the dorso-ventral axis. In the mutant (right) the flower is symmetrical. All the petals are like the most ventral one in the wild type and are folded back.

Photograph reproduced with permission from Coen, E.S., Meyerowitz, E.M.: **The war of the whorls: genetic interactions controlling flower development.** *Nature* 1991, **353**: 31-37. © 1991 Macmillan Magazines Ltd.

Another group of genes that help pattern the floral organ primordia are genes that control cell division. The gene *SUPERMAN* is one example, controlling cell proliferation in stamen and carpel primordia, and in ovules. Plants with a mutation in this gene have stamens instead of carpels in the fourth whorl. *SUPERMAN* is expressed in the third whorl, and maintains the boundary between the third and fourth whorls.

12.17 The *Antirrhinum* flower is patterned dorso-ventrally as well as radially

Like *Arabidopsis* flowers, those of *Antirrhinum* consist of four whorls, but unlike *Arabidopsis*, they have five sepals, five petals, four stamens, and two united carpels (Fig. 12.37, left). Floral homeotic mutations similar to those in *Arabidopsis* occur in *Antirrhinum*, and floral organ identity is specified in the same way. Several of the *Antirrhinum* homeotic genes have extensive homology with those of *Arabidopsis*, the MADS box, in particular, being well conserved.

An extra element of patterning is required in the *Antirrhinum* flower, which has a bilateral symmetry imposed on the basic radial pattern common to all flowers. In whorl 2, the upper two petal lobes have a shape quite distinct from the lower three, giving the flower its characteristic snapdragon appearance. In whorl 3, the uppermost stamen is absent, as its development is aborted early on. The *Antirrhinum* flower therefore has a distinct dorso-ventral axis. Another group of homeotic genes, different from those that govern floral organ identity, appear to act in this dorso-ventral patterning. For example, mutations in the gene *CYCLOIDEA*, which is expressed in the dorsal region, abolish dorso-ventral polarity and produce flowers that are more radially symmetrical (Fig. 12.37, right).

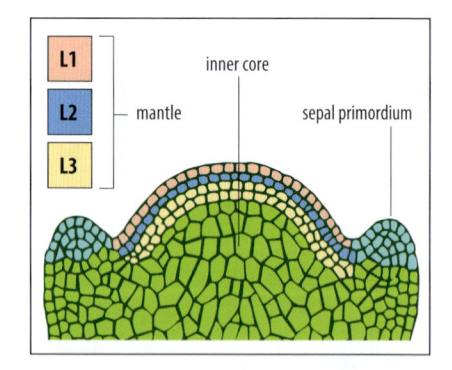

Fig. 12.38 Floral meristem. The meristem is composed of layers L1, L2, and L3. The inner core cells are derived from L3. The sepal primordia are just beginning to develop.

Illustration after Drews, G.N., Goldberg, R.B.: **Genetic control of flower development.** *Trends Genet.* 1989, **5**: 256-261.

12.18 The internal meristem layer can specify floral meristem patterning

All three layers of a floral meristem are involved in organogenesis (Fig. 12.38). However, the contribution of cells from each layer to a particular structure may be variable. Cells from one layer can become part of another layer without disrupting normal morphology, suggesting that a cell's position in the meristem is the main determinant of its future behavior. Some insight into positional signaling and patterning in the floral meristem can be obtained by making periclinal chimeras (see Section 12.8) from cells that have different genotypes and that give rise to different types of flower. From such chimeras, one can find out whether the cells develop autonomously according to their own genotype, or whether their behavior is controlled by signals from other cells.

As well as being produced by mutation, chimeras can also be generated by grafting between two plants of different genotypes. A new shoot meristem can form at the junction of the graft, and sometimes contains cells from both genotypes. Such chimeras can be made between wild-type tomato plants and tomato plants carrying the mutation *fasciated*, in which the flower has an increased number of floral organs per whorl. This phenotype is also found in chimeras in which only layer L3 contains *fasciated* cells (Fig. 12.39). The increased number of floral organs is associated with an overall increase in the size of the floral meristem, and in the chimeric plants this cannot be achieved unless the *fasciated* cells of layer L3 induce the wild-type L1 cells to divide more frequently than normal. The mechanism of intercellular signaling between L3 and L1 is not yet known. In *Antirrhinum*, the abnormal expression of *FLORICAULA* in only one meristem layer can result in flower development. These results illustrate the importance of signaling between layers in flower development.

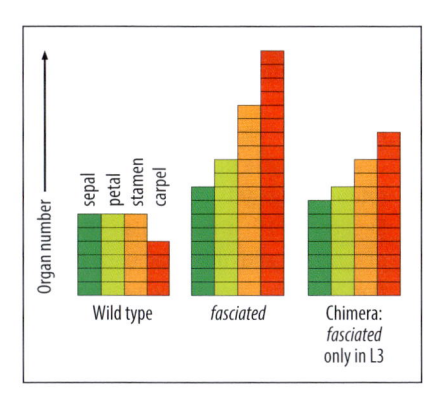

Fig. 12.39 Floral organ number in chimeras of wild-type and *fasciated* tomato plants. In the *fasciated* mutant there are more organs in the flower than in wild-type plants. In chimeras in which only layer L3 of the floral meristem contains *fasciated* mutant cells, the number of organs per flower is still increased, showing that L3 can control cell behavior in the outer layers of the meristem.

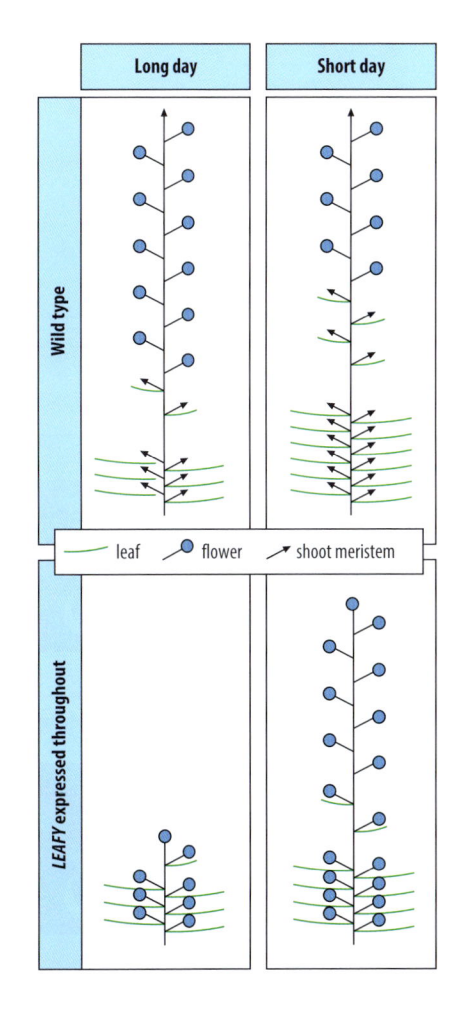

12.19 The transition of a shoot meristem to a floral meristem is under environmental and genetic control

Flowering plants first grow vegetatively, during which time the shoot apical meristem generates leaves. Then, triggered by environmental signals such as increasing day length, the plant switches to a reproductive phase and from then on the apical meristem gives rise only to flowers. There are two types of transition from vegetative growth to flowering. In the determinate type, the inflorescence meristem becomes a terminal flower, whereas in the indeterminate type the inflorescence meristem gives rise to a number of floral meristems. *Arabidopsis* is of the indeterminate type. A primary response to floral inductive signals in *Arabidopsis* is the expression of floral meristem identity genes such as *LEAFY* and the dual-function *AP1*, which are necessary and sufficient for this transition (see Section 12.16). *LEAFY* potentially activates *AP1* throughout the meristem while also activating *AGAMOUS* in the center of the flower. *AGAMOUS* then represses the expression of *AP1* in the center, helping to restrict its floral organ identity function to region A (see Fig. 12.35). Mutations in floral meristem identity genes partly transform flowers into shoots. In a *leafy* mutant, which lacks LEAFY function, the flowers are transformed into spirally arranged sepal-like organs along the stem, whereas expression of *LEAFY* throughout a plant is sufficient to confer a floral fate on lateral shoot meristems and they develop as flowers (Fig. 12.40, lower panels).

In *Arabidopsis*, flowering is promoted by increasing day length, which predicts the end of winter and the onset of spring and summer (Fig. 12.40). This behavior is called **photoperiodism**. Grafting experiments have shown that day length is sensed not by the shoot meristem itself, but by the leaves. When the period of continuous light reaches a certain length, a diffusible flower-inducing signal is produced that is

Fig. 12.40 Flowering can be controlled by day length and LEAFY expression. Upper panels: when wild-type *Arabidopsis* is grown under long-day conditions (left), few lateral shoots are formed before the apical shoot meristem begins to form floral meristems. When grown under short-day conditions, flowering is delayed and there are in consequence more lateral shoots. Lower panels: the gene *LEAFY* is normally expressed only in inflorescence and floral meristems, but if it is expressed throughout the plant, all shoot meristems produced are converted to floral meristems in both day lengths.

BOX 12D The circadian clock coordinates plant growth and development

Many organisms have evolved a circadian clock to predict the day-night cycle of light and temperature that is due to Earth's rotation. In plants, as well as the transition to flowering (see Section 12.19), the circadian clock regulates multiple aspects of growth and development. These include seed germination, seedling growth, and stress responses. In fact, the clock has been shown to drive 24-hour rhythms of transcription in around 30% of *Arabidopsis* genes under constant light and temperature conditions. As in other eukaryotes, these 24-hour rhythms are generated by multiple interlocking feedback loops of proteins (Fig. 1). These feedback loops receive input from environmental cues, which allows the clock rhythm to be synchronized, or entrained, to the day-night cycle. Output pathways from this core oscillator allow the peak expression of processes to be timed precisely to a particular point of the day or night.

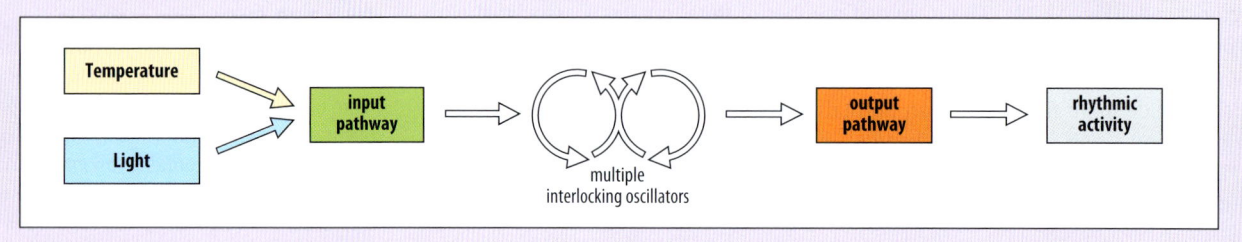

Fig. 1
From Gardner, M.J., Hubbard, K.E., Hotta, C.A., Dodd, A.N., Webb, A.A.R.: **How plants tell the time.** Biochem. J. 2006, **397**: 15–24.

transmitted through the phloem to the shoot meristem. The pathway that triggers flowering involves the plant's **circadian clock** (Box 12D), the internal 24-hour timer that causes many metabolic and physiological processes to vary throughout the day and night. One of the genes regulated by the circadian clock is *CONSTANS* (*CO*), which is a key gene in controlling the onset of flowering and provides the link between the plant's day length-sensing mechanism and production of the flowering signal. The expression of *CO* oscillates on a 24-hour cycle under the control of the circadian clock, and its timing is such that the peak *CO* expression occurs towards the end of the afternoon. This means that in longer days, peak expression occurs in the light, whereas in short days, it will already be dark at this time. In the dark, the CO protein is degraded and so the circadian control ensures that CO only accumulates to high enough levels to trigger the flowering pathway when days are sufficiently long (Fig. 12.41).

CO is a transcription factor that activates a gene known as *FLOWERING LOCUS T* (*FT*), producing the FT protein, which appears to act as the flowering signal. The FT protein is thought to travel from the leaf through the phloem to the shoot apical meristem, where it acts in a complex with the transcription factor FLOWERING LOCUS D (FD), which is expressed in the meristem, to turn on the expression of genes such as *AP1* that promote flowering (Fig. 12.42). If *FT* is activated in a single leaf, this is sufficient to induce flowering.

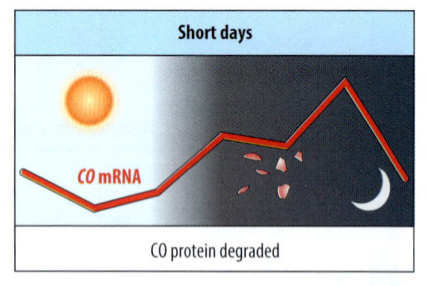

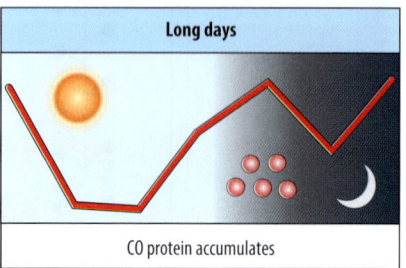

Fig. 12.41 The initiation of flowering is under the dual control of day length and the circadian clock. The transcription factor CONSTANS (CO) is required for production of the flowering signal and is expressed in leaves under the control of the circadian clock. In short days, expression of the *CO* gene (the red line represents the *CO* mRNA) peaks in the dark and the CO protein is rapidly degraded. In long days, peak expression occurs in the light, and the CO protein accumulates.

Adapted from Hayama, R., Coupland, G.: **Shedding light on the circadian clock and the photoperiodic control of flowering.** Curr. Opin. Plant Biol. 2003, **6**: 13–19.

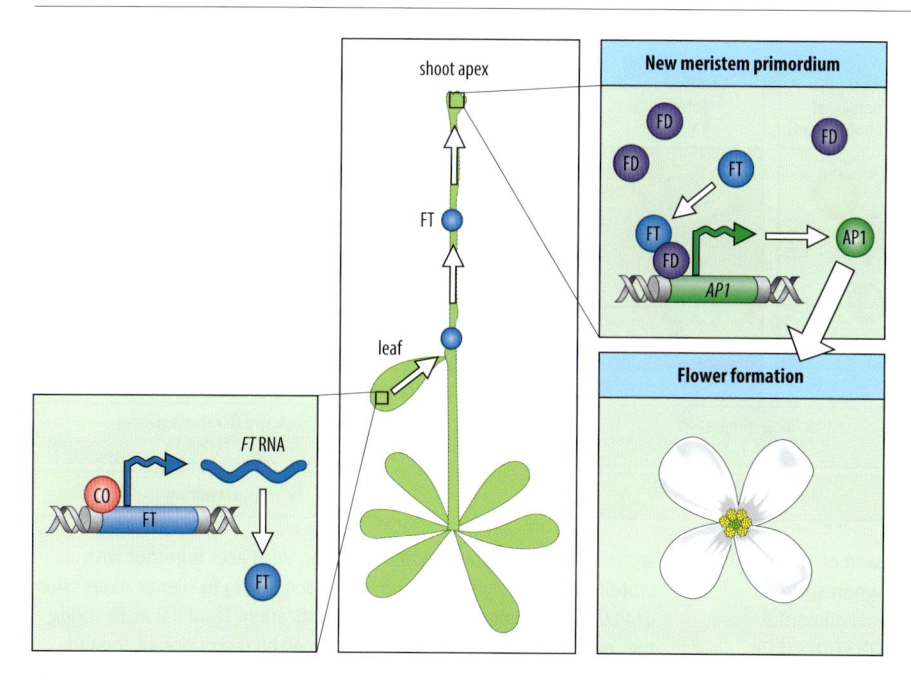

Fig. 12.42 Signals that initiate flowering in *Arabidopsis*. When the day length becomes longer after winter, the transcription factor CO accumulates in the leaf, which, in turn, activates the transcription of the *FT* gene in leaf phloem cells. The FT protein travels through the phloem to the shoot apex. It interacts there with the transcription factor FD to form a complex, which acts together with the transcription factor LEAFY (whose own expression is upregulated in the shoot apex by FT) to activate key floral meristem identity genes such as *AP1*, which convert the vegetative meristem into one that will produce floral meristems.

*Illustration from Blazquez, M.A.: **The right time and place for making flowers.** Science 2005, 309: 1024-1025.*

12.20 Vernalization reflects the epigenetic memory of winter

In some strains of *Arabidopsis*, and in many other plants, a period of winter cold is required for flowering to be triggered as the days lengthen in spring. This phenomenon is known as **vernalization**. It prevents seedlings that germinate and start growing in late summer and fall (autumn) from flowering until the next spring, but still allows flowering to start when conditions improve. As well as the appropriate day length, the induction of flowering requires the downregulation of a set of floral repressor genes, which includes *FLOWERING LOCUS C* (*FLC*). These genes suppress the transition from a vegetative to a flowering state until the positive signals to flower are received. *FLC* encodes a protein that binds to the gene encoding the flowering signal *FT* and suppresses its expression (see Section 12.19); the level of FLC protein therefore determines whether a plant is capable of flowering in response to increasing day lengths.

The level of FLC is set early in the plant's development, and is controlled by two opposing factors. One is the so-called Autonomous pathway, which is independent of environmental inputs. The pathway consists of a series of interconnecting genes that repress FLC expression and so render the plant capable of flowering if day length is right. The other factor is the protein FRIGIDA (FRI)—a component of a transcriptional activator complex that promotes the expression of *FLC*, and thus prevents flowering even if day length is appropriate (Fig. 12.43). The level of FLC in seedlings determines whether or not they need to overwinter before flowering (i.e. whether they require vernalization). Whether an *Arabidopsis* strain requires vernalization to flower depends on which version of the *FRI* gene it carries. Strains requiring vernalization have *FRI* alleles that result in high levels of *FLC* expression,

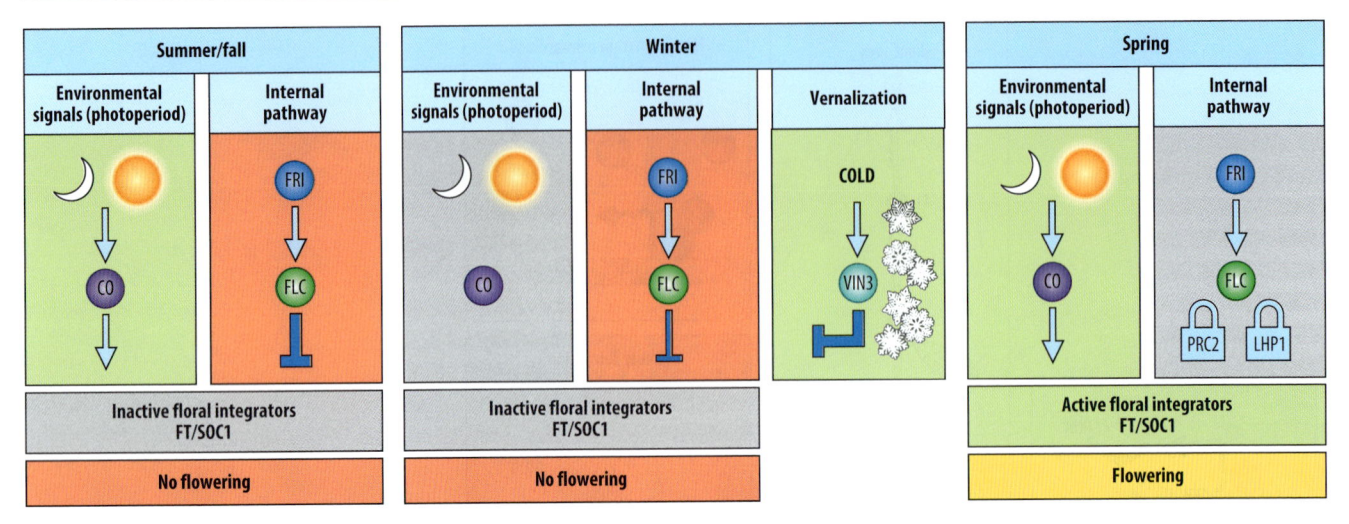

Fig. 12.43 Vernalization downregulates expression of the floral repressor gene *FLC* so that the plant can flower when spring arrives. First panel: in summer and fall (autumn), environmental signals (photoperiod) induce the accumulation of CONSTANS (CO) protein and could potentially induce flowering (see Section 12.19). Plants that require vernalization are prevented from flowering by an intrinsic pathway, independent of environmental signals, in which the protein FRIGIDA (FRI) promotes the expression of high levels of the floral repressor protein FLC. Second panel: the winter cold induces vernalization. The homeodomain protein VIN3 acts together with proteins of the Polycomb group (PCR2 and LHP1) to silence expression of *FLC* through epigenetic histone modification. Third panel: in spring, the resulting low levels of FLC, maintained by repression of the FRI-FLC pathway by PCR2 and LHP1, allow the plant to flower in response to environmental signals.

After Amasino, R.M., Michaels, S.D.: **The timing of flowering.** *Plant Physiol. 2010,* **154***: 516–520.*

whereas others that flower rapidly without vernalization carry mutations at the *FRI* locus that result in lower *FLC* expression.

When plants are chilled for a long period, *FLC* expression is stably downregulated by epigenetic mechanisms. This process involves multiple phases in which different mechanisms operate to silence *FLC* expression. First, exposure to an initial period of cold leads to a decrease in *FLC* transcription and an increase in expression of **long non-coding RNAs** (lncRNAs; see Section 6.2) transcribed from a promoter called *COOLAIR*. The *COOLAIR* promoter is located at the 3' end of the *FLC* locus, oriented towards the beginning of the *FLC* gene, and it drives the production of a number of lncRNAs, which are antisense transcripts of FLC. At an individual *FLC* locus, there is mutually exclusive transcription of *COOLAIR* and *FLC*; thus, upregulation of *COOLAIR* in the cold is correlated with reduced *FLC* transcription. Without functional *COOLAIR*, it takes longer for plants to switch off *FLC* transcription in response to cold. Also, *COOLAIR* transcripts are needed for the removal of the activating histone mark H3K36me3 from the *FLC* gene, which is an important step in *FLC* switching to an epigenetically silenced state.

After the initial transcriptional response, as the duration of cold that plants experience increases, stable epigenetic changes at the *FLC* locus are brought about by the Polycomb protein PRC2, which functions in a complex with plant homeodomain proteins, including VIN3, and causes high levels of the repressive histone mark H3K27me3 to be added at the *FLC* gene. The level of stable downregulation is set by the duration of the cold period, enabling plants to ignore transient fluctuations in temperature. During the cold, each cell independently switches from an FLC 'on' to an FLC 'off' epigenetic state and the 'off' state is maintained upon return to warm conditions. The fraction of 'off' FLC cells increases as the cold period increases, rather than all cells gradually diminishing their FLC expression level at the same time. Mathematical models suggest that it is this digital memory that stops plants from initiating flowering after a short cold snap in fall. High levels of H3K27me3 at the *FLC* locus ensure that expression of *FLC* remains low when higher temperatures return in spring. This stable memory of winter at the *FLC* locus therefore enables plants to rapidly induce FT expression in response to increased photoperiod in spring. The integration of both photoperiod and vernalization responses is important for cereal crops such as winter wheat, which can be sown in fall (autumn), uses seasonal rainfall to grow and increase in biomass, but delays flowering until spring when the risk of frost damage has passed.

12.21 Most flowering plants are hermaphrodites, but some produce unisexual flowers

Unlike animals, plants do not set aside germ cells in the embryo, and germ cells are only specified when a flower develops. Any meristem cell can, in principle, give rise to a germ cell of either sex, and there are no sex chromosomes. The great majority of flowering plants (angiosperms) give rise to flowers that contain both male and female sexual organs, in which meiosis occurs. The male sexual organs are the stamens; they produce pollen, which contains the male gamete nuclei corresponding to the sperm of animals. The female reproductive structures are the carpels, which are either free or are fused to form a compound ovary. Carpels are the site of ovule formation, and each ovule produces an egg cell.

Angiosperms, such as *Arabidopsis*, undergo so-called double fertilization. Each pollen grain contains two sperm nuclei, which are delivered into the ovule by the growth of a vegetative pollen tube. One nucleus fuses with the haploid egg cell, forming the zygote that develops into the embryo. The second nucleus fuses with the diploid 'central cell' in the ovule, forming a triploid cell that proliferates and develops into a storage tissue—the endosperm. In some plants, such as *Arabidopsis*, the endosperm nourishes the growth of the embryo, while in others it is broken down at germination to produce nutrients for the seedling.

Most flowering plants are hermaphrodite, bearing flowers with functional male and female sexual organs, as in *Arabidopsis*. In about 10% of flowering plant species, flowers of just one sex are produced. Flowers of different sexes may occur on the same plant, or be confined to different plants. The development of male or female flowers usually involves the selective degeneration of either the stamens or pistil after they have been specified and have started to grow. In maize, for example, male and female flowers develop at particular sites on the shoot. The tassel at the tip of the main stem only bears flowers with stamens; the 'ears' at the ends of the lateral branches bear female flowers containing pistils. Sex determination becomes visible when the flower is still small, with the stamen primordia being larger in males and the pistil longer in females. The smaller organs eventually degenerate. The plant hormone gibberellic acid may be involved in sex determination, as differences in gibberellin concentration are associated with the different sexual organs. In the maize tassel, gibberellin concentration is 100-fold lower than in the developing ears. If the concentration of gibberellic acid is increased in the tassel, pistils can develop.

Genomic imprinting (see Section 8.8) occurs in flowering plants, with the bulk of imprinting occurring in the endosperm, rather than in the embryo or somatic cells of the plant. In the endosperm, imprinting events take place in the central-cell genome and the sperm genome. In plants, as in mammals, imprinting involves gene silencing due to DNA methylation, histone modifications, Polycomb proteins, and non-coding RNAs. Unlike in mammals, however, the imprinting occurring in the endosperm does not have to be removed to produce a new generation, as the endosperm is a temporary tissue that does not contribute cells to the embryo. As well as the silencing of paternal genes in sperm by DNA methylation and histone modification, a feature of angiosperm imprinting is the DNA demethylation of previously silenced maternal genes in the central cell, which then enables their expression. The activated genes include members of the Polycomb group, which then maintain the repression of incoming silenced paternal alleles and silence the maternal alleles of certain other genes. Imprinting in angiosperms could have evolved together with double fertilization as a means of preventing endosperm proliferation in the absence of fertilization.

SUMMARY

Before flowering, which is triggered by environmental conditions such as day length, the vegetative shoot apical meristem becomes converted into an inflorescence meristem, which either then becomes a flower or produces a series of floral meristems, each of which develops into a single flower. Genes involved in the initiation of flowering and patterning of the flower have been identified in both *Arabidopsis* and *Antirrhinum*. Flowering is induced by day length acting together with the plant's natural circadian rhythms of gene expression to turn on a gene in the leaves that produces a flowering signal that is transported to the shoot meristem. This signal turns on the expression of meristem identity genes that are required for the transformation of the vegetative shoot meristem to an inflorescence meristem and the formation of floral meristems from the inflorescence meristem. Some plants undergo vernalization, a process by which a period of winter cold is required for the initiation of flowering as soon as the appropriate signals are received in spring. Homeotic floral organ identity genes, which specify the organ types found in the flowers, have been identified from mutations that transform one flower part into another. On the basis of these mutations, a model has been proposed in which the floral meristem is divided into three concentric overlapping regions, in each of which certain floral identity genes act in a combinatorial manner to specify the organ type appropriate to each whorl. Studies with chimeric plants have shown that different meristem layers communicate with each other during flower development and that transcription factors can move between cells. Unlike animals, plants do not set aside germ cells in the embryo, and germ cells are only specified when a flower develops. Genomic imprinting occurs in flowering plants, but mainly in the endosperm.

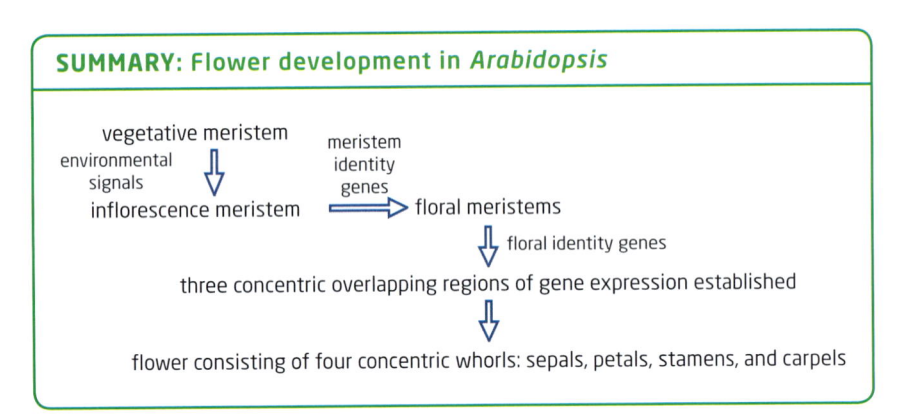

SUMMARY: Flower development in *Arabidopsis*

environmental signals ⇩ vegetative meristem → (meristem identity genes) → inflorescence meristem ⇒ floral meristems ⇩ floral identity genes

three concentric overlapping regions of gene expression established

⇩

flower consisting of four concentric whorls: sepals, petals, stamens, and carpels

---

Summary to Chapter 12

- Distinctive features of plant development are the presence of relatively rigid cell walls and the absence of any cell migration.
- Another distinctive feature compared with animals is that a single, isolated somatic cell from a plant can regenerate into a complete new plant.
- Early embryonic development is characterized by asymmetric cell division of the fertilized egg, which specifies the future apical and basal regions.
- During early development of flowering plants, both asymmetric cell division and cell–cell interactions are involved in patterning the body plan. During this process, the shoot and root apical meristems are specified and these meristems give rise to all the organs of the plant—stems, leaves, flowers, and roots.
- The shoot apical meristem gives rise to leaves in well-defined positions, a process involving regulated transport of a morphogen, auxin.

- The shoot apical meristem eventually becomes converted to an inflorescence meristem, which either becomes a floral meristem (in determinate inflorescences) or gives rise to a series of floral meristems, retaining its shoot meristem identity indefinitely (in indeterminate inflorescences).
- In floral meristems, each of which develops into a flower, homeotic floral organ identity genes act in combination to specify the floral organ types.
- Increasing day length induces the synthesis of a flowering signal in the leaves that is transported to the shoot meristem where it induces flower formation.
- In some plants, a period of winter cold is required for flowering to be triggered as the days lengthen in spring. This phenomenon is known as vernalization and involves downregulation of the *FLC* gene, which encodes a repressor of flowering, over the winter so that flowering can occur in spring.
- Unlike animals, plants do not set aside germ cells in the embryo, and germ cells are only specified when a flower develops. Genomic imprinting occurs in flowering plants, but mostly occurs in the endosperm and not in the embryo.

End of chapter questions

Check your understanding of the content of this chapter by attempting the following long-answer concept questions.

1. What features led to the adoption of *Arabidopsis thaliana* as the predominant model for plant development?

2. Distinguish between the following parts of a plant: shoot, root, node, leaf, meristem, sepal, petal, stamen, carpel.

3. What is the role of auxin before the 32-cell stage in *Arabidopsis* embryogenesis? What is the mechanism by which a gradient in auxin concentrations is generated? What is the mechanism by which auxins influence gene expression?

4. Describe the biolistics and the *Agrobacterium*-based methods for plant transformation.

5. What is a meristem? Describe the structure of the shoot meristem of *Arabidopsis*.

6. Contrast the roles of the homeobox genes *WUSCHEL* (*WUS*) and *SHOOT MERISTEMLESS* (*STM*) in formation and maintenance of shoot apical meristems.

7. What has the study of mericlinal chimeras revealed about cell specification during plant embryogenesis? Are cells specified in the embryo to form leaves and flowers, in a way analogous, for example, to the specification of cells as dorsal mesoderm in animal embryos?

8. What are the terms applied to the upper and lower surfaces of a leaf, in reference to the radial axis of the shoot? How are the transcription factors PHB, PHV, and REV restricted to the top surface and what happens in mutants where they are not?

9. How does auxin control the positioning of leaf primordia? Include the role of the PIN transporter in your answer.

10. What is SHORT-ROOT? How does it come to be present in endodermal cells? Might plasmodesmata be involved?

11. What is the nature of the homeotic mutations that can occur in *Arabidopsis*? What genes are involved?

12. What is the ABC model for flower development? How does it illustrate a combinatorial gene code for cellular identity?

13. The MADS-box is named for the prototypic proteins in which it was found: MCM1 (*Saccharomyces*), AGAMOUS (*Arabidopsis*), DEFICIENS (*Antirrhinum*), and SRF (*Homo*). What is the function of the MADS-box in a protein? (Note that the MADS-box proteins are not related to Smads; see Box 3F.)

14. What modification of the ABC model derived from studies of *Arabidopsis* is required to explain flower development in *Antirrhinum*?

15. Through what mechanism is the photoperiod interpreted to trigger flower development in *Arabidopsis*?

General further reading

Angel, A., Song, J., Dean, C., Howard, M.: **A Polycomb-based switch underlying quantitative epigenetic memory**. *Nature* 2011, **476**: 105–108.

Koornneef, M., Meinke, D.: **The development of *Arabidopsis* as a model plant**. *Plant J.* 2010, **6**: 909–921.

Meyerowitz, E.M.: **Plants compared to animals: the broader comparative view of development**. *Science* 2002, **295**: 1482–1485.

Sena, G., Birnbaum K.D.: **Built to rebuild: in search of organizing principles in plant regeneration**. *Curr. Opin. Genet. Dev.* 2010, **20**: 460–465.

Section further reading

12.1 The model plant *Arabidopsis thaliana* has a short life cycle and a small diploid genome and 12.2 Plant embryos develop through several distinct stages

Lloyd, C.: **Plant morphogenesis: life on a different plane**. *Curr. Biol.* 1995, **5**: 1085–1087.

Mayer, U., Jürgens, G.: **Pattern formation in plant embryogenesis: a reassessment**. *Semin. Cell Dev. Biol.* 1998, **9**: 187–193.

Meyerowitz, E.M.: **Genetic control of cell division patterns in developing plants**. *Cell* 1997, **88**: 299–308.

Torres-Ruiz, R.A., Jürgens, G.: **Mutations in the *FASS* gene uncouple pattern formation and morphogenesis in *Arabidopsis* development**. *Development* 1994, **120**: 2967–2978.

12.3 Gradients of the signal molecule auxin establish the embryonic apical–basal axis

Breuninger, H., Rikirsch, E., Hermann, M., Ueda, M., Laux, T.: **Differential expression of *WOX* genes mediates apical–basal axis formation in the *Arabidopsis* embryo**. *Dev. Cell* 2008, **14**: 867–876.

Friml, J., Vieten, A., Sauer, M., Weijers, D., Schwarz, H., Hamann, T., Offringa, R., Jürgens, G.: **Efflux-dependent auxin gradients establish the apical–basal axis of *Arabidopsis***. *Nature* 2003, **426**: 147–153.

Jenik, P.D., Barton, M.K.: **Surge and destroy: the role of auxin in plant embryogenesis**. *Development* 2005, **132**: 3577–3585.

Jürgens, G.: **Axis formation in plant embryogenesis: cues and clues**. *Cell* 1995, **81**: 467–470.

Long, J.A., Moan, E.I., Medford, J.I., Barton, M.K.: **A member of the knotted class of homeodomain proteins encoded by the *STM* gene of *Arabidopsis***. *Nature* 1995, **379**: 66–69.

Szemenyei, H., Hannon, M., Long, J.A.: **TOPLESS mediates auxin-dependent transcriptional repression during *Arabidopsis* embryogenesis**. *Science* 2008, **319**: 1384–1386.

12.4 Plant somatic cells can give rise to embryos and seedlings

Zimmerman, J.L.: **Somatic embryogenesis: a model for early development in higher plants**. *Plant Cell* 1993, **5**: 1411–1423.

Box 12B Plant transformation and genome editing

Shah, T., Andleeb, T., Lateef, S., Noor, M.A.: **Genome editing in plants: advancing crop transformation and overview of tools**. *Plant Physiol. Biochem* 2018, **131**: 12–21.

Yin, K., Gao, C., Qiu, J.-L.: **Progress and prospects in plant genome editing**. *Nat. Plants* 2017, **3**: 17107.

12.5 Cell enlargement is a major process in plant growth and morphogenesis

Backues, S.K., Konopka, C.A., McMichael, C.M., Bednarek, S.Y.: **Bridging the divide between cytokinesis and cell expansion**. *Curr. Opin. Plant Biol.* 2007, **10**: 607–615.

Kuchen, E.K., Fox, S., Barbier de Reuille, P., Kennaway, R., Bensmihen, S., Avondo, J., Calder, G.M., Southam, P., Robinson, S., Bangham, A., Coen, E.: **Generation of leaf shape through early patterns of growth and tissue polarity**. *Science* 2012, **335**: 1092–1096.

Tsuge, T., Tsukaya, H., Uchimiya, H.: **Two independent and polarized processes of cell elongation regulate leaf blade expansion in *Arabidopsis thaliana* (L.) Heynh**. *Development* 1996, **122**: 1589–1600.

12.6 A meristem contains a small central zone of self-renewing stem cells

Barton, M.K.: **Twenty years on: the inner workings of the shoot apical meristem, a developmental dynamo**. *Dev. Biol.* 2010, **341**: 95–113.

Byrne, M.E., Kidner, C.A., Martienssen, R.A.: **Plant stem cells: divergent pathways and common themes in shoots and roots**. *Curr. Opin. Genet. Dev.* 2003, **13**: 551–557.

Großhardt, R., Laux, T.: **Stem cell regulation in the shoot meristem**. *J. Cell Sci.* 2003, **116**: 1659–1666.

12.7 The size of the stem-cell area in the meristem is kept constant by a feedback loop to the organizing center

Brand, U., Fletcher, J.C., Hobe, M., Meyerowitz, E.M., Simon, R.: **Dependence of stem cell fate in *Arabidopsis* on a feedback loop regulated by *CLV3* activity**. *Science* 2000, **289**: 635–644.

Carles, C.C., Fletcher, J.C.: **Shoot apical meristem maintenance: the art of dynamical balance**. *Trends Plant Sci.* 2003, **8**: 394–401.

Clark, S.E.: **Cell signalling at the shoot meristem**. *Nat Rev. Mol. Cell Biol.* 2001, **2**: 277–284.

Landrein, B., Formosa-Jordan, P., Malivert, A., Schuster, C., Melnyk, C.W., Yang, W., Turnbull, C., Meyerowitz, E.M., Locke, J.C.W., Jönsson, H.: **Nitrate modulates stem cell dynamics in *Arabidopsis* shoot meristems through cytokinins**. *Proc. Natl Acad. Sci. USA* 2018, **115**: 1382–1387.

Lenhard, M., Laux, T.: **Stem cell homeostasis in the *Arabidopsis* shoot meristem is regulated by intercellular movement of CLAVATA3 and its sequestration by CLAVATA1**. *Development* 2003, **130**: 3163–3173.

Reddy, G.V., Meyerowitz, E.M.: **Stem-cell homeostasis and growth dynamics can be uncoupled in the *Arabidopsis* shoot apex**. *Science* 2005, **310**: 663–667.

Schoof, H., Lenhard, M., Haecker, A., Mayer, K.F.X., Jürgens, G., Laux, T.: **The stem cell population of *Arabidopsis* shoot meristems is maintained by a regulatory loop between the *CLAVATA* and *WUSCHEL* genes**. *Cell* 2000, **100**: 635–644.

Vernoux, T., Benfey, P.N.: **Signals that regulate stem cell activity during plant development**. *Curr. Opin. Genet. Dev.* 2005, **15**: 388–394.

Yadav, R.K., Perales, M., Gruel, J., Girke, T., Jönsson, H., Reddy, G.V.: **WUSCHEL protein movement mediates stem cell homeostasis in the *Arabidopsis* shoot apex**. *Genes Dev.* 2011, **25**: 2025–2030.

12.8 The fate of cells from different meristem layers can be changed by changing their position

Castellano, M.M., Sablowski, R.: **Intercellular signalling in the transition from stem cells to organogenesis in meristems**. *Curr. Opin. Plant Biol.* 2005, **8**: 26–31.

Gallagher, K.L., Benfey, P.N.: **Not just another hole in the wall: understanding intercellular protein trafficking**. *Genes Dev.* 2005, **19**: 189–195.

Laux, T., Mayer, K.F.X.: **Cell fate regulation in the shoot meristem**. *Semin. Cell Dev. Biol.* 1998, **9**: 195–200.

Sinha, N.R., Williams, R.E., Hake, S.: **Overexpression of the maize homeobox gene *knotted-1*, causes a switch from determinate to indeterminate cell fates**. *Genes Dev.* 1993, **7**: 787–795.

Turner, I.J., Pumfrey, J.E.: **Cell fate in the shoot apical meristem of *Arabidopsis thaliana***. *Development* 1992, **115**: 755–764.

Waites, R., Simon, R.: **Signaling cell fate in plant meristems: three clubs on one tousle**. *Cell* 2000, **103**: 835–838.

12.9 A fate map for the embryonic shoot meristem can be deduced using clonal analysis

Irish, V.F.: **Cell lineage in plant development**. *Curr. Opin. Genet. Dev.* 1991, **1**: 169–173.

12.10 Meristem development is dependent on signals from other parts of the plant

Doerner, P.: **Shoot meristems: intercellular signals keep the balance**. *Curr. Biol.* 1999, **9**: R377–R380.

Irish, E.E., Nelson, T.M.: **Development of maize plants from cultured shoot apices**. *Planta* 1988, **175**: 9–12.

Sachs, T.: *Pattern Formation in Plant Tissues*. Cambridge: Cambridge University Press, 1994.

12.11 Gene activity patterns the proximo-distal and adaxial-abaxial axes of leaves developing from the shoot meristem

Bowman, J.L.: **Axial patterning in leaves and other lateral organs**. *Curr. Opin. Genet. Dev.* 2000, **10**: 399–404.

Chitwood, D.H., Nogueira, F.T., Howell, M.D., Montgomery, T.A., Carrington, J.C., Timmermans, M.C.: **Pattern formation via small RNA mobility**. *Genes Dev.* 2009, **23**: 549–554.

Du, F., Guan, C., Jiao, Y.: **Molecular mechanisms of leaf morphogenesis**. *Mol. Plant* 2018, **11**: 1117–1134.

Emery, J.F., Floyd, S.K., Alvarez, J., Eshed, Y., Hawker, N.P., Izhaki, A., Baum, S.F., Bowman, J.L.: **Radial patterning of *Arabidopsis* shoots by class III HD-ZIP and KANADI genes**. *Curr. Biol.* 2003, **13**: 1768–1774.

Kidner, C.A., Martienssen, R.A.: **Spatially restricted microRNA directs leaf polarity through *ARGONAUTE1***. *Nature* 2004, **428**: 81–84.

Yoshikawa, M.: **Biogenesis of trans-acting siRNAs, endogenous secondary siRNAs in plants**. *Genes Genet. Syst.* 2013, **88**: 77–84.

12.12 The regular arrangement of leaves on a stem is generated by regulated auxin transport

Berleth, T., Scarpella, E., Prusinkiewicz, P.: **Towards the systems biology of auxin-transport-mediated patterning**. *Trends Plant Sci.* 2007, **12**: 151–159.

Heisler, M.G., Ohno, C., Das, P., Sieber, P., Reddy, G.V., Long, J.A., Meyerowitz, E.M.: **Patterns of auxin transport and gene expression during primordium development revealed by live imaging of the *Arabidopsis* inflorescence meristem**. *Curr. Biol.* 2005, **15**: 1899–1911.

Jönsson, H., Heisler, M.G., Shapiro, B.E., Meyerowitz, E.M., Mjolsness, E.: **An auxin-driven polarized transport model for phyllotaxis**. *Proc. Natl Acad. Sci. USA* 2006, **103**: 1633–1638.

Mitchison, G.J.: **Phyllotaxis and the Fibonacci series**. *Science* 1977, **196**: 270–275.

Reinhardt, D.: **Regulation of phyllotaxis**. *Int. J. Dev. Biol.* 2005, **49**: 539–546.

Scheres, B.: **Non-linear signaling for pattern formation**. *Curr. Opin. Plant Biol.* 2000, **3**: 412–417.

Schiefelbein, J.: **Cell-fate specification in the epidermis: a common patterning mechanism in the root and shoot**. *Curr. Opin. Plant Biol.* 2003, **6**: 74–78.

Smith, L.G., Hake, S.: **The initiation and determination of leaves**. *Plant Cell* 1992, **4**: 1017–1027.

Stoma, S., Lucas, M., Chopard, J., Schaedel, M., Traas, J., Godin, C.: ***Flux-based* transport enhancement as a plausible unifying mechanism for auxin transport in meristem development**. *PLoS Comput. Biol.* 2008, **4**: 1000207.

12.13 The outgrowth of secondary shoots is under hormonal control

Agusti, J., Herold, S., Schwarz, M., Sanchez, P., Ljung, K., Dun, E.A., Brewer, P.B., Beveridge, C.A., Sieberer, T., Sehr, E.M., Greb, T.: **Strigolactone signaling is required for auxin-dependent stimulation of secondary growth in plants**. *Proc. Natl Acad. Sci. USA* 2011, **108**: 20242–20247.

Domagalska, M.A., Leyser, O.: **Signal integration in the control of shoot branching**. *Nat. Rev. Mol. Cell Biol.* 2011, **12**: 211–221.

Leyser, O.: **Auxin, self-organisation, and the colonial nature of plants**. *Curr. Biol.* 2011, **9**: R331–R337.

Mason, M.G., Ross, J.J., Babst, B.A., Wienclaw, B.N., Beveridge, C.A.: **Sugar demand, not auxin, is the initial regulator of apical dominance**. *Proc. Natl Acad. Sci USA* 2014, **111**: 6092–6097.

Ursache, R., Nieminen, K., Helariutta, Y.: **Genetic and hormonal regulation of cambial development**. *Physiol. Plant.* 2013, **147**: 36–45.

Waldie, T., Leyser, O.: **Cytokinin targets auxin transport to promote shoot branching**. *Plant Physiol.* 2018, **177**: 803–818.

12.14 Root tissues are produced from *Arabidopsis* root apical meristems by a highly stereotyped pattern of cell divisions

Costa, S., Dolan, L.: **Development of the root pole and patterning in *Arabidopsis* roots**. *Curr. Opin. Genet. Dev.* 2000, **10**: 405–409.

Del Bianco, M., Giustini, L., Sabatini, S.: **Spatiotemporal changes in the role of cytokinin during root development**. *New Phytol.* 2013, **199**: 324–338.

Dello Ioio, R., Nakamura, K., Moubayidin, L., Perilli, S., Taniguchi, M., Morita, M.T., Aoyama, T., Costantino, P., Sabatini, S.: **A genetic framework for the control of cell division and differentiation in the root meristem**. *Science* 2008, **322**: 1380–1384.

Dolan, L.: **Positional information and mobile transcriptional regulators determine cell pattern in *Arabidopsis* root epidermis**. *J. Exp. Bot.* 2006, **57**: 51–54.

Grieneisen, V.A., Xu, J., Marée, A.F.M., Hogeweg P., Scheres, B.: **Auxin transport is sufficient to generate a maximum and gradient guiding root growth**. *Nature* 2007, **449**: 1008–1013.

Moreno-Risueno, M.A., Van Norman, J.M., Moreno, A., Zhang, J., Ahnert, S.E., Benfey, P.N.: **Oscillating gene expression determines competence for periodic *Arabidopsis* root branching**. *Science* 2010, **329**: 1306–1311.

Sabatini, S., Beis, D., Wolkenfeldt, H., Murfett, J., Guilfoyle, T., Malamy, J., Benfey, P., Leyser, O., Bechtold, N., Weisbeek, P., Scheres, B.: **An auxin-dependent distal organizer of pattern and polarity in the *Arabidopsis* root**. *Cell* 1999, **99**: 463–472.

Scheres, B., McKhann, H.I., van den Berg, C.: **Roots redefined: anatomical and genetic analysis of root development**. *Plant Physiol.* 1996, **111**: 959–964.

van den Berg, C., Willemsen, V., Hendriks, G., Weisbeek, P., Scheres, B.: **Short-range control of cell differentiation in the *Arabidopsis* root meristem**. *Nature* 1997, **390**: 287–289.

Veit, B.: **Plumbing the pattern of roots**. *Nature* 2007, **449**: 991–992.

12.15 Root hairs are specified by a combination of positional information and lateral inhibition

Kwak, S.-H., Schiefelbein, J.: **The role of the SCRAMBLED receptor-like kinase in patterning the *Arabidopsis* root epidermis.** *Dev. Biol.* 2007, **302**: 118–131.

Schiefelbein, J., Kwak, S.-H., Wieckowski, Y., Barron, C., Bruex, A.: **The gene regulatory network for root epidermal cell-type pattern formation in *Arabidopsis*.** *J. Exp. Bot.* 2009, **60**: 1515–1521.

12.16 Homeotic genes control organ identity in the flower

Bowman, J.L., Sakai, H., Jack, T., Weigel, D., Mayer, U., Meyerowitz, E.M.: **SUPERMAN, a regulator of floral homeotic genes in *Arabidopsis*.** *Development* 1992, **114**: 599–615.

Breuil-Broyer, S., Morel, P., de Almeida-Engler, J., Coustham, V., Negrutiu, I., Trehin, C.: **High-resolution boundary analysis during *Arabidopsis thaliana* flower development.** *Plant J.* 2004, **38**: 182–192.

Coen, E.S., Meyerowitz, E.M.: **The war of the whorls: genetic interactions controlling flower development.** *Nature* 1991, **353**: 31–37.

Irish, V.F.: **Patterning the flower.** *Dev. Biol.* 1999, **209**: 211–220.

Krizek, B.A., Fletcher, J.C.: **Molecular mechanisms of flower development: an armchair guide.** *Nat Rev. Genet.* 2005, **6**: 688–698.

Krizek, B.A., Meyerowitz, E.M.: **The *Arabidopsis* homeotic genes *APETALA3* and *PISTILLATA* are sufficient to provide the B class organ identity function.** *Development* 1996, **122**: 11–22.

Lohmann, J.U., Weigel, D.: **Building beauty: the genetic control of floral patterning.** *Dev. Cell* 2002, **2**: 135–142.

Ma, H., dePamphilis, C.: **The ABCs of floral evolution.** *Cell* 2000, **101**: 5–8.

Meyerowitz, E.M.: **The genetics of flower development.** *Sci. Am.* 1994, **271**: 40–47.

Meyerowitz, E.M., Bowman, J.L., Brockman, L.L., Drews, G.M., Jack, T., Sieburth, L.E., Weigel, D.: **A genetic and molecular model for flower development in *Arabidopsis thaliana*.** *Development Suppl.* 1991, **1**: 157–167.

Sakai, H., Medrano, L.J., Meyerowitz, E.M.: **Role of *SUPERMAN* in maintaining *Arabidopsis* floral whorl boundaries.** *Nature* 1994, **378**: 199–203.

Vincent, C.A., Carpenter, R., Coen, E.S.: **Cell lineage patterns and homeotic gene activity during *Antirrhinum* flower development.** *Curr. Biol.* 1995, **5**: 1449–1458.

Wagner, D., Sablowski, R.W.M., Meyerowitz, E.M.: **Transcriptional activation of *APETALA 1*.** *Science* 1999, **285**: 582–584.

12.17 The *Antirrhinum* flower is patterned dorso-ventrally as well as radially

Coen, E.S.: **Floral symmetry.** *EMBO J.* 1996, **15**: 6777–6788.

Luo, D., Carpenter, R., Vincent, C., Copsey, L., Coen, E.: **Origin of floral asymmetry in *Antirrhinum*.** *Nature* 1996, **383**: 794–799.

12.18 The internal meristem layer can specify floral meristem patterning

Szymkowiak, E.J., Sussex, I.M.: **The internal meristem layer (L3) determines floral meristem size and carpel number in tomato periclinal chimeras.** *Plant Cell* 1992, **4**: 1089–1100.

12.19 The transition of a shoot meristem to a floral meristem is under environmental and genetic control

An, H., Roussot, C., Suarez-Lopez, P., Corbesier, L., Vincent, C., Pineiro, M., Hepworth, S., Mouradov, A., Justin, S., Turnbull, C., Coupland, G.: **CONSTANS acts in the phloem to regulate a systemic signal that induces photoperiodic flowering of *Arabidopsis*.** *Development* 2004, **131**: 3615–3626.

Becroft, P.W.: **Intercellular induction of homeotic gene expression in flower development.** *Trends Genet.* 1995, **11**: 253–255.

Blázquez, M.A.: **The right time and place for making flowers.** *Science* 2005, **309**: 1024–1025.

Corbesier, L., Vincent, C., Jang, S., Fornara, F., Fan, Q., Searle, I., Giakountis, A., Farrona, S., Gissot, L., Turnbull, C., Coupland, G.: **FT protein movement contributes to long-distance signaling in floral induction of *Arabidopsis*.** *Science* 2007, **316**: 1030–1033.

Hake, S.: **Transcription factors on the move.** *Trends Genet.* 2001, **17**: 2–3.

Jaeger, K.E., Wigge, P.A.: **FT protein acts as a long-range signal in *Arabidopsis*.** *Curr. Biol.* 2007, **17**: 1050–1054.

Kobayashi, Y., Weigel, D.: **Move on up, it's time for change—mobile signals controlling photoperiod-dependent flowering.** *Genes Dev.* 2007, **21**: 2371–2384.

Lohmann, J.U., Hong, R.L., Hobe, M., Busch, M.A., Parcy, F., Simon, R., Weigel, D.: **A molecular link between stem cell regulation and floral patterning in *Arabidopsis*.** *Cell* 2001, **105**: 793–803.

Putterill, J., Laurie, R., Macknight, R.: **It's time to flower: the genetic control of flowering time.** *BioEssays* 2004, **26**: 363–373.

Valverde, F., Mouradov, A., Soppe, W., Ravenscroft, D., Samach, A., Coupland, G.: **Photoreceptor regulation of CONSTANS protein in photoperiodic flowering.** *Science* 2004, **303**: 1003–1006.

12.20 Vernalization reflects the epigenetic memory of winter

Angel, A., Song, J., Yang, H., Questa, J.I., Dean, C., Howard, M.: **Vernalizing cold is registered digitally at FLC.** *Proc. Natl Acad. Sci. USA* 2015, **112**: 4146–4151.

Rosa, S., Duncan, S., Dean, C.: **Mutually exclusive sense–antisense transcription at FLC facilitates environmentally induced gene repression.** *Nat. Commun.* 2016, **7**: 13031.

Sheldon, C.C., Hills, M.J., Lister, C., Dean, C., Dennis, E.S., Peacock, W.J.: **Resetting of *FLOWERING LOCUS C* expression after epigenetic repression by vernalization.** *Proc. Natl Acad. Sci. USA* 2008, **105**: 2214–2219.

Song, J., Angel, A., Howard, M., Dean, C.: **Vernalization—a cold-induced epigenetic switch.** *J. Cell Sci.* 2012, **125**: 3723–3731.

12.21 Most flowering plants are hermaphrodites, but some produce unisexual flowers

Huh, J.H., Bauer, M.J., Hsieh, T.-F., Fischer, R.L.: **Cellular programming of plant gene imprinting.** *Cell* 2008, **132**: 735–744.

Irisa, E.N.: **Regulation of sex determination in maize.** *BioEssays* 1996, **18**: 363–369.

13

Evolution and development

- The evolution of development
- The diversification of body plans
- Evolutionary modification of specialized characters
- Changes in the timing of developmental processes

The evolution of multicellular organisms is fundamentally linked to embryonic development, for it is through development that genetic changes cause changes in body form that can be passed on to future generations. This raises important questions as to how evolutionary and developmental processes are linked. How did development itself evolve? How has embryonic development been modified during animal evolution? How important are changes in the timing of developmental processes? In this chapter, we discuss these general questions mostly using particular examples from animal evolution.

Introduction

All life on Earth today is thought to have evolved from a common ancestor. The history of evolution is a long one, taking place over thousands of millions of years, and it can only be accessed indirectly—through the study of fossils and by comparisons between living organisms. Figure 13.1 shows a simplified timeline for evolution. In regard to the multicellular animals and plants that are the subject of this book, it is thought that around 1.5 billion years ago, three separate eukaryotic lineages—the ancestors of multicellular animals, plants, and fungi—evolved from a common unicellular ancestor.

As Charles Darwin was the first to realize, evolution is the result of heritable changes in life forms and the selection of those that reproduce best (Box 13A). The evolution of multicellular forms of life is the result of changes in embryonic development, and these changes are, in turn, due to genetic changes that control cell fates and cell behavior in the embryo. It is also true, however, as the evolutionary biologist Theodosius Dobzhansky once said, that nothing in biology makes sense unless viewed in the light of evolution. Certainly, it would be very difficult to make sense of many aspects of development without an evolutionary perspective. For example, in our consideration of vertebrate development we have seen how, despite different modes of very early development, all vertebrate embryos develop through a rather

674 Chapter 13 Evolution and development

Some major events in metazoan evolution			
Period	Million years ago (Mya)	Paleontology (fossil evidence)	Genetic, cellular and molecular evolution
Quaternary	0.2 2.6	Earliest fossils of modern humans (*Homo sapiens*)	
Neogene (Tertiary)	4–5 ~8 23	Earliest hominin fossils (e.g. *Ardipithecus* (debated); *Australopithecus*)	Estimated date of chimpanzee–human last common ancestor from DNA analysis
Paleogene (Tertiary)	66	Radiations of mammals, birds and insects	
Cretaceous	145	Cretaceous–Tertiary mass extinction; extinction of the dinosaurs Earliest angiosperms (flowering plants) Earliest birds	
Jurassic	200	Dinosaurs dominant	
Triassic	252	End-Triassic mass extinction Earliest mammals Earliest dinosaurs	
Permian	298	Permian–Triassic mass extinction Radiation of reptiles	
Carboniferous	359	Earliest reptile (first amniote) Earliest tetrapods on land	Evolution of keratin (the material of scales, feathers, and hair)
Devonian	419	Late Devonian mass extinction Earliest amphibians Diversification of bony fish Earliest tetrapods Earliest insects Earliest bony fish Invasion of land by arthropods	Evolution of limbs from fins Evolution of bone
Silurian	445	Jawed fish	
Ordovician	485	Ordovician–Silurian mass extinction	
Cambrian	541	Earliest land plants Agnathan (jawless fish) fossils Earliest vertebrate fossil known—a jawless craniate (an animal with a backbone and distinct skull) The Cambrian explosion: fossils from China (524 Mya), the Burgess Shale (508 Mya) and elsewhere include early arthropods, echinoderms, chordates, and possible jawless vertebrates Earliest definitive arthropod fossils, echinoderms, shelled molluscs Widespread biomineralization	Hemoglobin duplication into α and β chains Genome duplication in the lineage leading to vertebrates Evolution of neural crest Body plans of all extant phyla established during the Cambrian Phylotypic organization of metazoans Segmentation Hox gene clusters
Precambrian	~600	Fossil reef-building animals with mineralized skeletons Earliest fossil of a bilaterian organism Earliest cnidarian fossils Abundant 'Ediacaran' fossils, the earliest evidence for complex multicellular life. Experts still disagree about what many of these enigmatic remains represent Earliest sponge fossils	Hox genes Evolution of collagen (characteristic of multicellular animals)
	1200 1500 2000 3500 ~4600	First evidence for sexual reproduction (in a red alga) Major radiation of single-celled, colonial and simple multicellular eukaryotes: ancestors of plants, animals and fungi Earliest eukaryotes Earliest reasonable evidence for life (microbial mats) Earth formed	

Fig. 13.1 Some selected events in evolution. The lengths of the geological periods are not to scale. Paleontological events within each period are listed in the order of their occurrence in the fossil record.

Modified from Gerhardt, J., Kirschner, M.: Cells, Embryos, and Evolution: Towards a Cellular and Developmental Understanding of Phenotypic Variation and Evolutionary Adaptability. Oxford: Blackwell, 1997.

similar stage, after which their development diverges again (see Fig. 3.2). This shared stage, the phylotypic or tailbud stage, which is the embryonic stage after neurulation and the formation of the somites, is a stage through which some distant ancestor of the vertebrates passed. It has persisted ever since, to become a fundamental characteristic of the development of all vertebrates, whereas the stages before and after this stage have evolved differently in different vertebrates.

Over the course of evolution, genetically based, heritable changes in development that generated more successful modes of reproduction, or adult forms better adapted to their environment, have arisen and have been selected for. In vertebrate development, for example, changes occurring before neurulation are more often associated with changes in reproduction; those occurring after neurulation are more associated with the evolution of animal form. Genetic variability resulting from mutations and other genomic alterations, sexual reproduction, and genetic recombination is present in the populations of all the organisms we have looked at in this book and provides new phenotypes upon which selection can act.

Changes in both the sequences of the control regions of genes and in the protein-coding regions that generate novel protein functions have played a part in evolution. There is ample evidence that changes in protein structure have had a direct impact on animal biochemistry and have an important role in producing key physiological differences, as we will see in dogs. And in the case of genes encoding transcription factors, alterations in their coding sequences, and thus in the structure and binding properties of the protein they encode, can lead to the factor being able to regulate a different set of downstream target genes. By contrast, changes in the sequences of the control regions of genes alter the level, the timing, or the tissue-specificity of their expression, and these types of changes are particularly important in evolution. Many developmental genes have acquired many separate modules in their control regions during their evolutionary history, each of which drives expression of the gene in a different spatial pattern or in a different tissue at various times in development—for example, the *eve* gene in *Drosophila* (see Fig. 6.5). Therefore, it is easy to see how mutations in control regions, by changing where and when a gene is expressed, could be a major force for evolutionary change.

Gene duplication followed by changes in the sequences of gene control regions and/or protein-coding regions is one of the main ways in which genes can diverge and proteins with new functions can be produced. Another mechanism for generating new proteins involves the fusion of genes that originally encoded separate protein domains. Tandem duplication of a gene can occur by various mechanisms during DNA replication and provides the embryo with an additional copy of the gene. This copy can diverge in the nucleotide sequences of both its protein-coding region and its control regions, thus changing the gene's function and its pattern of expression without depriving the organism of the function of the original gene (Fig. 13.2). Gene duplication has been fundamental in the evolution of new proteins and new patterns of gene expression; it is clear, for example, that the different hemoglobins in humans (see Fig. 6.22) have arisen as a result of gene duplication. Whole-genome duplication has taken place several times during evolution in both animals and plants. Although extensive gene loss occurs following such events, whole-genome duplications can increase the number of developmentally important genes, which can then diverge. Now that genomes of individual organisms can be readily sequenced, it has become easier to look for the differences between members of different species that can give information on their evolutionary history.

Throughout this book we have emphasized the conservation of some developmental mechanisms at the cellular and molecular level among distantly related organisms. The widespread use of the Hox gene complexes and of the same few families of protein signal molecules provides excellent examples of this and has become a hallmark of the animal kingdom. It is these basic similarities in molecular mechanisms that have made developmental biology so exciting in recent years; it has meant that discoveries of genes in one animal have had important implications for understanding development in others. It seems that when a useful developmental mechanism evolved, it was retained and redeployed in very different species, and at different times and places in the same organism.

In the rest of this book, we have looked at the early development of a range of different organisms and found some similarities, as well as several differences. In this chapter we mainly confine our attention to two phyla—the arthropods (which include the insects and the crustaceans, e.g. crabs, lobsters, and woodlice) and the

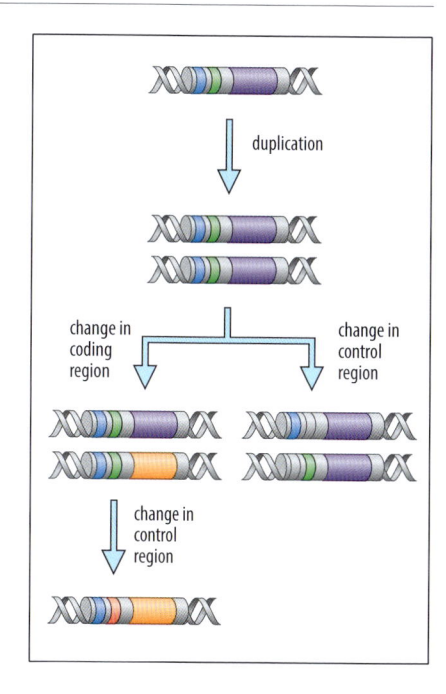

Fig. 13.2 Gene duplication and diversification. Once a gene has become duplicated, the second copy can evolve new functions and/or new expression patterns. Illustrated is a hypothetical example of a gene (purple) with two control regions (blue and green) that each confer expression in a different tissue. Left pathway: after duplication of the complete gene and its control regions, mutation in the coding region of one of the copies can generate a gene with a new and useful function that is selected for. Both copies will be retained in the genome as they both fulfill useful functions. Subsequent mutation in a control region could lead to the new gene acquiring a new expression pattern. Right pathway: alternatively, a mutation in one control region in one copy and the other control region in the other copy can generate two copies of the gene that retain the original function but are now only expressed in a single tissue each. Both will be retained in the genome as they are both now needed to fulfill the function of the original gene. Examples of all these types of duplication and divergence are found in animal genomes.

BOX 13A Darwin's finches

'Darwin's finches' are a good example of the evolutionary role of development and of changes in gene expression. Charles Darwin visited the Galapagos Islands in 1835 and collected a group of finches, distinguishing at the time 13 closely related species. What he found particularly striking was the variation in their beaks. He wrote in *The Voyage of the Beagle*: 'It is very remarkable that a nearly perfect gradation of structure in this one group can be traced in the form of the beak, from one exceeding in dimensions that of the largest gros-beak, to another differing but little from that of a warbler.' The shapes of the beaks reflected differences in the birds' diets and how they got their food. And Darwin later commented: 'Seeing this gradation and diversity of structure in one small, intimately related group of birds, one might really fancy that from an original paucity of birds in this archipelago, one species had been taken and modified for different ends.' Figure 1 shows the great variation in the shape and size of the beaks of closely related Galapagos finches as a result of adaptation to different sources of food. The drawings are by the English ornithologist John Gould from specimens collected by Darwin on the Galapagos Islands.

Modern investigations have uncovered clues to the developmental basis of these differences in beak form. The species with broader, deeper beaks relative to length express higher levels of the gene encoding the signal protein bone morphogenetic protein (BMP)-4 in the beak growth zone compared with species with long, pointed beaks. Experiments in which *Bmp4* was overexpressed in the developing beak of a chick embryo resulted in the beak growing broader and deeper. A subsequent DNA microarray analysis of differences in the levels of all the mRNAs expressed in the different finch beaks revealed another key player that shapes beak form. The gene encoding the calcium-binding protein calmodulin, a component of a calcium-dependent signal transduction pathway, is expressed at higher levels in the long, pointed beaks than in the short, broad ones, and was also able to cause the upper beak of a chick embryo to elongate when overexpressed. This is an example of **heterometry**, a general mechanism that is being recognized for producing evolutionary variations by changing the levels of gene transcription.

But what are the heritable genetic changes that have led to these variations in levels of gene transcription in the different finch species? This question has been addressed by whole-genome sequencing, which has pinpointed regions of the genome associated with differences in beak shape and beak size. In one study looking at beak shape, the genomes of more than 120 individual birds from the different finch species and two other closely related bird species were sequenced. Detailed comparisons between the genomes of finches with blunt beaks and those with pointed beaks identified regions of the genome associated with differences in beak shape. The candidate genes in these regions included *calmodulin* but, even more strikingly, *Alx homeobox1* (*Alx1*), which encodes a transcription factor important in craniofacial development. Mutations in *ALX1* have been found in human patients with severe craniofacial defects. Differences were identified not only in the coding sequence of *Alx1* in blunt-beaked versus pointed-beaked finches, but also in the sequences of nearby conserved non-coding regions, suggesting that changes in gene regulation might also be involved.

In another study looking at beak size, genome sequence data were collected from a further 60 individual finches. They included medium-sized ground finches in which birds with smaller beaks had been selected for during a recent drought because this trait reduced competition for food with larger finches. Genomic analysis revealed that the most likely candidate for the gene affecting beak size in these populations of finches is one called *HMGA2*, with the small-beak allele of the gene being more common in finches that survived the drought. *HMGA2* encodes a chromatin-associated protein that can regulate gene expression indirectly by altering the state of the chromatin and thus affecting the activity of transcription factors (see Section 6.2). It has been implicated in regulating growth and craniofacial development in humans, dogs, mice, and rabbits. The next challenge will be to unravel how the gene-regulatory proteins encoded by *Alx1* and *HMGA2* impinge on the activity of developmental signaling pathways. This should give us a more complete picture of how the differences in beak form that fascinated Darwin evolved.

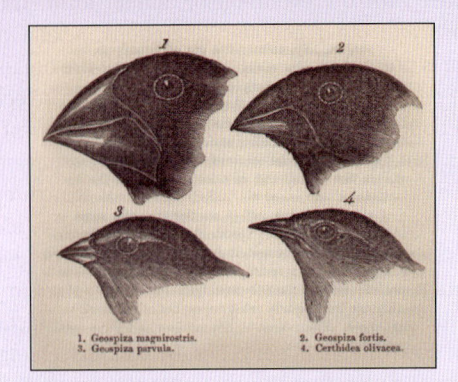

Fig. 1

1. *Geospiza magnirostris*. 2. *Geospiza fortis*.
3. *Geospiza parvula*. 4. *Certhidea olivacea*.

---

chordates (which include the vertebrates). We focus on those differences that distinguish the members of a large group of related animals, such as the insects or the vertebrates, from each other. We will look at the relationship between the development of the individual organism (its **ontogeny**) and the evolutionary history of the species or group (the **phylogeny**): why, for example, do mammalian embryos pass through an apparently fish-like stage that has structures resembling gill slits? We discuss the many variations that occur on the theme of a basic, segmented, body plan: what determines the different numbers and positions of paired appendages, such as legs and wings, in different groups of segmented organisms? We consider the timing of developmental events, and how variations in growth can have major effects on the shape and form of an organism. In all cases, we ultimately want to understand the changes in the developmental processes and their genetic basis that have resulted in the extraordinary range of different multicellular animals. This is an exciting area of study in which many problems remain to be solved. But first we will briefly consider how development itself might have evolved. What is the origin of the egg, and how might processes such as pattern formation and gastrulation have evolved?

The evolution of development

To look for the origins of development we must consider how multicellularity evolved, what we know about the closest living unicellular relatives of animals and about the ancestors of multicellular animals. Multicellularity has evolved independently numerous times in the lineages leading to animals, fungi, and plants. Its origins are difficult to investigate, as fossils of early multicellular organisms are rare and their identification sometimes contentious. The earliest unequivocal fossils of multicellular eukaryotes are dated to about 1.2 billion years ago and have been identified as red algae. Recognizable multicellular animals—the sponges—enter the fossil record at around 600 million years ago (see Fig. 13.1). The earliest multicellular animals would have been small and soft-bodied, and therefore unlikely to have been preserved as fossils. Genes associated with multicellularity, such as those for cell-adhesion molecules, intracellular signaling proteins, and some types of receptor, have been identified in the unicellular relatives of animals, and now that individual genomes can be sequenced relatively easily and rapidly, the transition to multicellularity can be investigated in more detail by this means. In this part of the chapter, we also look at the origins of gastrulation, some general principles underlying the evolution of the body plan, how new anatomical features can arise from modification of existing structures, and the phenomenon of parallel evolution.

13.1 Multicellular organisms evolved from single-celled ancestors

For many years, the question of how animals evolved from a unicellular ancestor was addressed by studying the modern choanoflagellates—a group of unicellular and facultatively colonial organisms. Modern choanoflagellates possess a single flagellum surrounded by a 'collar' of villi, and resemble choanocytes, the collar cells that make up the digestive system of sponges. The choanoflagellates have now been confirmed by molecular phylogenetic evidence as a 'sister group' to animals and their closest living relatives. However, recent phylogenetic studies have identified two additional unicellular groups that have 'animal cell' features and can claim to be close cousins of animals. These are the Filasterea, which are unicellular ameboid protists with long slender filopodia, and the Ichthyosporea, unicellular protist parasites of fish. These three groups—the choanoflagellates, the filastereans, and the ichthyosporeans—together with the metazoans, comprise the Holozoa, the group of organisms that includes animals and their close relatives (Fig. 13.3). This is the group that will shed light on the origins of animals and of animal multicellularity (Box 13B).

What had to be invented for the transition to multicellularity? And how did embryonic development from an egg evolve? The key requirements for embryonic development, as we have seen, are a program of gene activity, signal production and transduction (so that cells can communicate with each other), cell motility and cohesion (so that overall form can change), and cell differentiation into functionally distinct cell types.

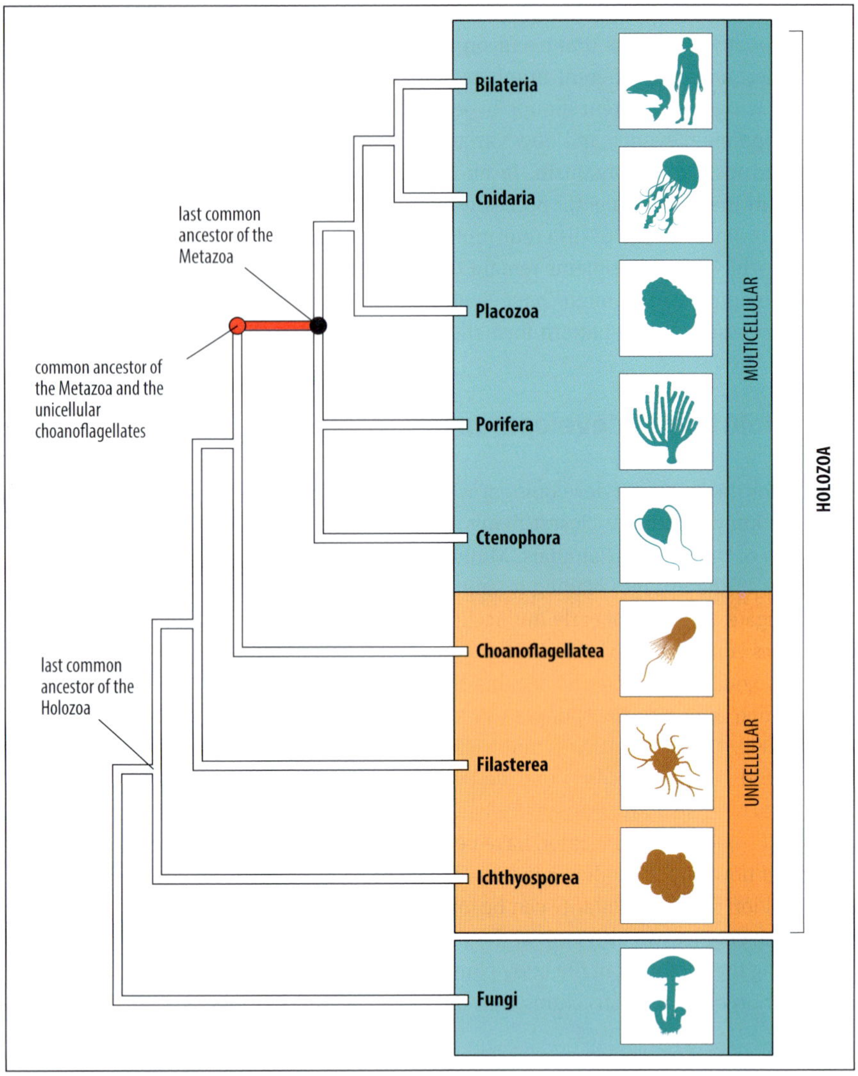

Fig. 13.3 A family tree of the Holozoa. The Holozoa includes the multicellular animals (the Metazoa—Bilateria, Cnidaria, Placozoa, Porifera, and Ctenophora; see Box 13B) and their unicellular relatives (Choanoflagellatea, Filasterea, and Ichthyosporea). The nearest unicellular relatives to animals are the choanoflagellates. Multicellularity in the animal lineage evolved somewhere in the evolutionary time period represented by the red line joining the points marking the common ancestor of the Metazoa and the choanoflagellates, which would have been unicellular, and the last common ancestor of the Metazoa, which would have been multicellular by definition. The Fungi are included here as a separate group of organisms that contains the nearest living relatives of the Holozoa.

Adapted from Sebé-Pedrós, A., et al.: *The origin of Metazoa: a unicellular perspective.* Nat. Rev. Genet. 2017, **18**: 498-512.

Unicellular eukaryotes can move, adhere, and respond to signals. Judging by modern unicellular eukaryotes such as the choanoflagellates and filastereans, the single-celled organisms ancestral to animals would have possessed all these features in primitive form, and little new would have had to be invented, although much had to evolve further and acquire different functions for multicellularity to evolve. The genomes of the modern choanoflagellate *Monosiga brevicollis* and of the filasterean *Capsaspora owczarzaki* contain sequences encoding some cell-adhesion and signaling proteins that are otherwise present only in metazoans. Both have integrins and extracellular-matrix-like proteins, as well as cadherin-like proteins, but not the cadherin–β-catenin adhesion module characteristic of metazoans. Of particular note, these organisms have a broad repertoire of genes for 'metazoan' transcription factors, including T-box factors, p53, and members of the STAT family. They also have a broad repertoire of receptor tyrosine kinase genes, but the receptors lack clear extracellular components. Some genes, such as that for the T-box-containing transcription factor TbxT, are absent from *Monosiga* but present in *Capsaspora*. At a minimum, the ancestor of multicellular organisms would have already possessed the characteristic eukaryotic cell cycle, which entails a complex program of gene activity, the ability to undergo cellular differentiation, signaling by receptor tyrosine kinases, and cadherin-mediated cell adhesion.

What then was the origin of multicellularity and embryos? One possibility is that mutations resulted in the progeny of a single-celled organism not separating after cell division, leading to a loose colony of identical cells that occasionally fragmented to give new 'individuals'. This behavior can be seen today in facultatively colonial choanoflagellates, which can live both as single cells and as a colony. It is thought that the colonial mode of life might enhance the ability to gain food from the surrounding water. Another advantage of a colony might originally have been that when food was in short supply, the cells could feed off each other, and so the colony survived. This could have been the origin of both multicellularity and the requirement for cell death in multicellular organisms.

The egg cell might subsequently have evolved as a cell fed by other cells; in sponges, for example, the egg phagocytoses neighboring cells, and in *Drosophila* the surrounding nurse cells provide large quantities of proteins and RNAs to the oocyte (see Fig. 2.16). The evolutionary advantage of an organism developing from a single cell—the egg—is that all the cells of the organism will have the same genes. This is a prerequisite for the development of complex patterns and body plans, as patterning requires communication between cells, and that can only be reliable if the cells obey the same rules because they have the same genetic instructions. There are 'multicellular' organisms that develop from more than one cell, such as slime molds, but they have never evolved numerous complex forms.

Once multicellularity evolved, it opened up all sorts of new possibilities, such as cell specialization for different functions. Some cells could specialize in providing motility, for example, and others in feeding, as we see in the nematocytes, the stinging cells of *Hydra* that are used to capture prey. The origin of pattern formation—that is, cell differentiation in a spatially organized arrangement—is not known but might have depended on asymmetries set up by external influences such as inside–outside differences. Molecular analysis of the genomes of choanoflagellates and filastereans reveals that the emergence of the Metazoa is accompanied by the appearance of Wnt, Notch, and bone morphogenetic protein (BMP)/Smad signals, which are not present in the single-celled relatives but are present in sponges, placozoans, and ctenophores. There is also an increase in the repertoire of cell-surface receptors and molecules mediating cell–cell interactions in the Metazoa. The signaling pathways uniquely associated with metazoans are therefore likely to be associated not so much with the emergence of different cell types but with the coordination of their activities.

13.2 Genomic evidence is throwing light on the evolution of animals

How animals evolved is a tough problem. From genomic evidence and studies of the simplest animals, the last common ancestors of the bilaterians—the Urbilateria (see Box 13B)—must already have been complex creatures, possessing most of the developmental gene pathways used by present-day bilaterians. Fossils of adult bilaterians make their appearance suddenly, early in the Cambrian period around 540 million years ago, a time called the 'Cambrian explosion' because of the sudden appearance of an enormous range of animal life. Fossils from this time reveal the existence of segmented organisms with bilateral symmetry and a well-differentiated head. Interestingly, it is estimated that land plants appeared during the same period, about 500 million years ago. What developmental mechanisms urbilaterians might have used and their genomic complexity are key questions in the evolution–development (**evo-devo**) field. Another central challenge is to explain how conserved gene networks already present in the archetypal ancestor were modified to generate the wonderful diversity of animal life on Earth today.

The genome sequences of species from all the non-bilaterian groups of the Metazoa have been determined: the cnidarians *Hydra magnipapillata* and the starlet sea anemone *Nematostella vectensis*; the sponge *Amphimedon queenslandica*; *Trichoplax*

BOX 13B The metazoan family tree

The animal kingdom is conventionally divided into three main groups in terms of basic body structure—the bilaterally symmetrical Bilateria, the radially symmetrical Cnidaria and Ctenophora, and the Parazoa (the sponges (Porifera) and the Placozoa), which are the simplest in terms of their morphology and cellular differentiation and lack body symmetry. The Cnidaria and the Bilateria are likely 'sister groups'—that is, both groups are considered to descend from a common ancestor that gave rise only to these groups. The order of origin of the other groups—the Placozoa, Ctenophora, and Porifera—is still a matter of debate (note the slight differences in the placement of these groups in the tree in Fig. 13.3). Genomic evidence tends to suggest that the Ctenophora is the sister group of all other metazoans, whereas other evidence places the Porifera in that position. Together, all these groups constitute the Metazoa—the multicellular animals—and are considered to derive from a common ancestor that only gave rise to the Metazoa (Fig. 1). The tree shown here is based on ribosomal DNA and Hox gene sequences, and comparisons of whole-genome sequences.

The largest group of animals is the Bilateria, which have bilateral symmetry across the main body axis in at least some stage of development and have a characteristic pattern of Hox gene expression along the antero-posterior axis. The position of the Urbilateria, the last common ancestors of all bilaterians, is represented here by a red circle. Note that Urbilateria is also the last common ancestor of the chordates and the arthropods, whose development is often compared in evolutionary studies. The Bilateria includes vertebrates and other chordates, echinoderms (despite the apparent radial symmetry of the adults), arthropods, annelids, mollusks, and nematodes. These animals are **triploblasts**, as they have the three germ layers—endoderm, mesoderm, and ectoderm. The Cnidaria (corals, jellyfish, hydras, and their relatives) are **diploblasts**, as are the Ctenophora (the comb jellies). They have only two germ layers (ectoderm and endoderm), and generally have radial symmetry. Some cnidarians, such as *Nematostella*, do, however, show evidence of bilateral symmetry, and studies of Hox gene expression in these animals have suggested to some researchers that the common ancestor of the Bilateria and the Cnidaria might have had bilateral symmetry.

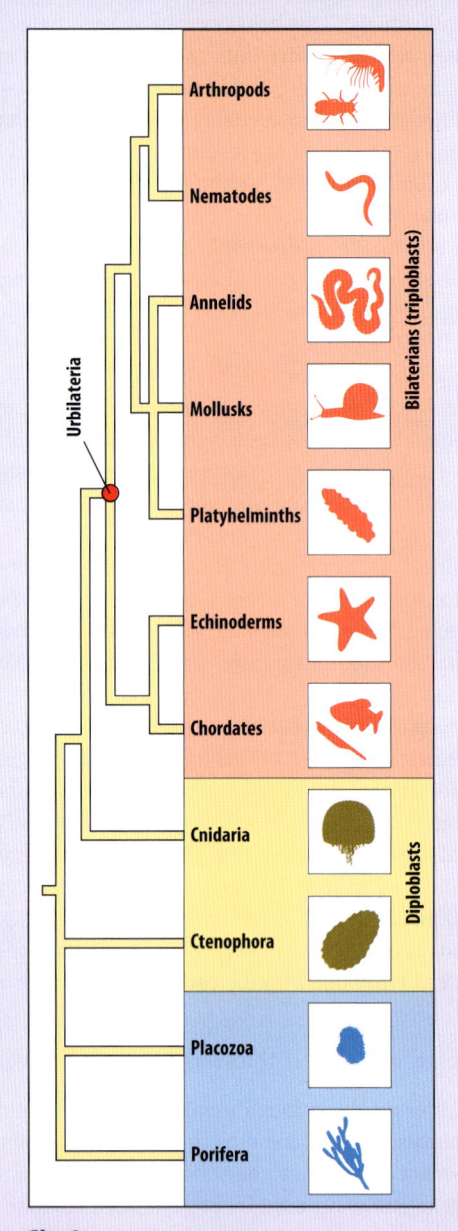

Fig. 1

adhaerens from the Placozoa; and the comb jelly *Pleurobrachia bachei* representing the ctenophores. These genome sequences are filling gaps in our knowledge of the evolution of key developmental genes, such as the Hox genes and the genes for signal proteins such as Wnt and Hedgehog, and reveal that the full complement of proteins characteristic of animal cell types emerged very early in the evolution of multicellular animals. Some typical developmental signal molecules present in these phyla and the Bilateria are shown in Figure 13.4. Genome analysis also reveals that some of the genes that encode these signal molecules, for example, the *hedgehog* gene, evolved by gene fusion. DNA sequences encoding proteins that represent one of the two domains of the Hedgehog protein are found in sponges, and even in choanoflagellates.

The *Nematostella* genome reveals an extensive gene repertoire more similar to that of vertebrates than to flies or nematodes, implying that the genome of the ancestor of all animals was similarly complex. Nearly one-fifth of the inferred genes in the *Nematostella* genome are novelties—not known from other animal groups—and the genome is rich in genes for animal functions such as cell–cell signaling, cell adhesion, and synaptic transmission.

Taking all the evidence from bilaterians and cnidarians together, we can speculate that the ancestor of the Eumetazoa (all present-day animals other than the sponges), lived perhaps 700 million years ago and would have had flagellate sperm, development through a process of gastrulation, multiple germ layers, true epithelia lying on a basement membrane, a lined gut, neuromuscular and sensory systems, and fixed body axes. *Nematostella*'s genome contains about 18,000 protein-coding genes, a similar number to that of *Caenorhabditis*. Genes representing 56 families of homeodomain-containing gene-regulatory proteins are present in *Nematostella*, including three putative Hox genes and representatives of four different classes of Pax genes—genes that in bilaterians are involved in many different aspects of development such as eye development, segmentation, and neural patterning. In regard to signaling, cnidarians have a large complement of Wnt genes, more similar to that of vertebrates than of invertebrates. For example, *Nematostella* has versions of 11 of the 12 Wnt genes present in vertebrates, and both *Hydra* and *Nematostella* possess the Wnt antagonist Dickkopf (Dkk), which, other than in cnidarians, is only present in chordates. More remarkably for an animal with only two germ layers, at least seven genes associated with the formation of mesoderm in bilaterians are expressed in the developing endoderm of *Nematostella* and might contribute to a form of ancestral mesoderm, or to cells performing this function. Thus, the genetic machinery for germ-layer specification was already present in the ancestors of both bilaterians and cnidarians.

The sponges are very different from the Cnidaria and the Bilateria in their morphology and level of structural differentiation. They have a few different cell types, some of them sensory and others digestive, but they share many features with other animal groups at the molecular level, such as the presence of signal proteins (Notch, Wnt, and BMP); some homeobox genes, but not Hox genes; and neural-like cell types. All this suggests that the ancestor of all metazoans already had, at the very least, some of the cell types and proteins familiar from modern animals.

There is one very simple free-living marine animal called *Trichoplax*, which is so unusual it has been placed in a phylum of its own, the Placozoa. Its relationship to other animals is not yet clear, but genomic comparisons place it as a very simple member of the Metazoa. *Trichoplax* is composed of two layers of epithelium, which form a flat disc with no gut, and it has just four different cell types. It mainly reproduces by fission. Nevertheless, in line with the genomes of other animals, the *Trichoplax* genome contains an estimated 11,500 protein-coding genes, which encode a rich array of transcription factors, including homeodomain-containing proteins, T-box proteins similar to TbxT, and components of the Wnt/β-catenin signaling pathway.

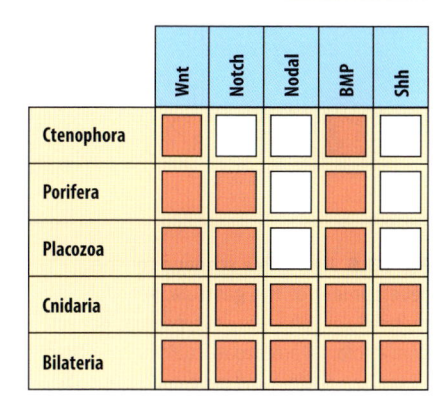

Fig. 13.4 Signaling pathways present in different animal groups. The signaling pathways that are present in each phylum, as judged by their genome sequences.

13.3 How gastrulation evolved is not known

Gastrulation is a key event in the early embryonic development of almost all animals, in which cells and tissues undergo extensive rearrangement to establish the germ layers in their correct positions and to form the gut. Cnidarians and ctenophores, as well as bilaterians, undergo gastrulation. It is not implausible to consider a scenario for the evolution of gastrulation in which a hollow sphere of cells, the common ancestor of

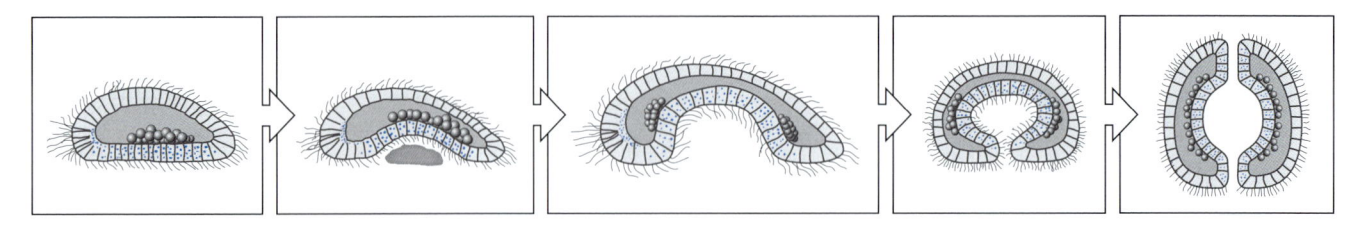

Fig. 13.5 A possible scenario for the development of the gastrula. A hollow multicellular sphere of cells, perhaps derived from a colonial protozoan, could have settled on the sea bottom and developed a gut-like invagination to aid feeding. Eventually, the invagination could have reached the other side of the sphere and formed a through gut.

Based on Jägersten, G.: **The early phylogeny of the metazoa. The bilatero-gastrea theory.** Zool. Bidrag. (Uppsala) *1956*, **30**: 321–354; Martindale, M.Q.: **The evolution of metazoan axial properties.** Nat. Rev. Genet. 2005, **6**: 917–927.

all multicellular animals, changed its form to assist feeding (Fig. 13.5). This ancestor may, for example, have sat on the ocean floor, ingesting food particles by phagocytosis. A small invagination developing in the body wall could have promoted feeding by forming a primitive gut. Movement of cilia could have swept food particles more efficiently into this region, where they would be phagocytosed. Once the invagination formed, it is not too difficult to imagine how it could eventually extend right across the sphere to fuse with the other side and form a continuous gut, which would be the endoderm. At a later stage in evolution, cells migrating inside, between the gut and the outer epithelium, would give rise to the mesoderm.

Gastrulation provides a good example of developmental change during evolution. Although there is considerable similarity in the process of gastrulation in many different animals, there are also significant differences, as discussed throughout this book in relation to our model animals. Bilaterians are traditionally subdivided into the **protostomes** and the **deuterostomes**, according to how the mouth originates in relation to gastrulation. The protostomes develop the mouth during gastrulation, close to the invaginating endoderm and mesoderm. They have a ventral nerve cord which emanates from two dorsal ganglia in the head that serve as a 'brain'. The two branches from the ganglia extend ventrally on either side of the foregut, and in most protostomes then fuse to form a single ventral cord. Of the animals covered in this book, the arthropods, such as *Drosophila* (see Chapter 2), and the nematodes, such as *Caenorhabditis*, are protostomes.

In deuterostomes, the mouth develops independently of the gastrulation process, which defines the anus, and the nervous system develops dorsally. Deuterostomes include the chordates, such as the vertebrates (see Chapters 3 and 4), the ascidians, and the echinoderms, such as the sea urchin. But how these differences between protostomes and deuterostomes evolved and what could have been the adaptive nature of the intermediate forms is a difficult question.

13.4 More general characteristics of the body plan develop earlier than specializations

Comparisons of vertebrate embryos have suggested two important generalizations about development. First, that the more general characteristics of the body plan of a group of animals (i.e. those shared by all members of the group) usually appear earlier in their embryos than the more specialized ones, and probably arose earlier in evolution; and, second, that as the embryo becomes more specialized, it diverges more and more from other embryos. These principles, laid down by Ernst von Baer in the early nineteenth century, summarize the most important features of developing embryos. Germ layers and axial organization are established early in development and all vertebrate embryos have, for example, branchial arches, which quickly become specialized in a species-specific manner and precede the emergence of the much more specialized characteristics such as scales and fur, which appear later in development. A good example of a general characteristic of the vertebrate body plan is the notochord, which is common to all vertebrates and is also found in other

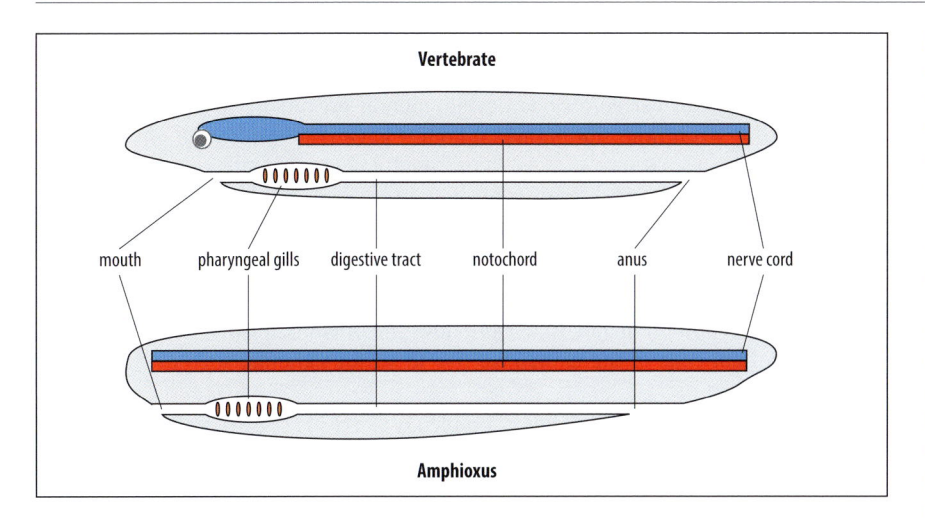

Fig. 13.6 The cephalochordate amphioxus compared with a hypothetical ancestral vertebrate similar to a present-day jawless vertebrate, the lamprey. The overall construction of the two organisms is very similar, with a dorsal nerve cord (blue), the axial rod-like notochord (red) below it, and a ventral digestive tract. Both animals have gills in the pharyngeal region, structures that are designed to capture food and also take in oxygen from the water. The notochord extends right to the anterior end of the amphioxus, but the vertebrate has a prominent head at the anterior end, extending beyond the notochord. In the jawed vertebrates (fishes, amphibians, reptiles, birds, and mammals), the notochord is present only in the embryo, being replaced by the vertebral column.

Modified from Finnerty, J.R.: **Evolutionary developmental biology: head start.** *Nature* 2000, **408**: 6814.

chordate embryos. The phylum Chordata comprises three subphyla—the vertebrates; the cephalochordates, such as the amphioxus (*Branchiostoma*) (Fig. 13.6); and the urochordates, such as the ascidians. They all have, at some embryonic stage, a notochord flanked by muscle, a dorsal neural tube, and branchial arches. The amphioxus, in particular, has many features believed to be ancestral to vertebrates, even when adult: a dorsal hollow nerve cord, in this case supported by the persistent notochord rather than a bony spine; and segmental muscles that derive from somites.

All vertebrate embryos pass through a phylotypic stage at which they all more or less resemble each other and show the specific embryonic features of the phylum to which they belong (see Fig. 3.2). It is a stage at which structures common to all vertebrates—such as somites, notochord, and neural tube—have developed. After the phylotypic stage, vertebrate embryos diversify, giving rise to the diverse forms of the different vertebrate classes and acquiring the more-specialized characteristics of that class. The phylotypic stage itself is somewhat variable, however, as can be seen in Figure 13.7, which shows that the number of somites can be very different, and that some organs, such as limb buds, are at different stages of development at this time. It should be noted that the stages in development of the different vertebrate classes before the phylotypic stage are also highly divergent, because of their very different modes of reproduction. For example, very early mammalian embryos form trophectoderm and an inner cell mass. This is an example of a special feature of development that arose late in vertebrate evolution and is related to the nutrition of the embryo through a placenta rather than a yolky egg.

13.5 Embryonic structures have acquired new functions during evolution

If two groups of animals that differ greatly in their adult structure and habits (e.g. fishes and mammals) pass through a very similar embryonic stage, this could indicate that they are descended from a common ancestor. One example is a stage present in all vertebrate embryos, including humans, in which branchial arches and clefts are located just behind the head on either side. These structures are

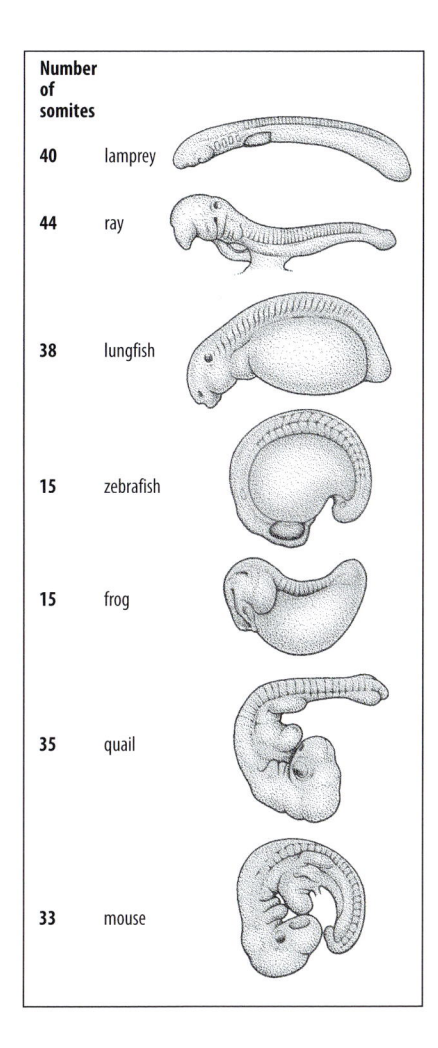

Fig. 13.7 Vertebrate embryos at the phylotypic (tailbud) stage show considerable variation in somite number. Not to scale.

Modified from Richardson, M.K., et al.: **There is no highly conserved embryonic stage in the vertebrates: implications for current theories of evolution and development.** *Anat. Embryol.* 1997, **196**: 91-106.

Number of somites	
40	lamprey
44	ray
38	lungfish
15	zebrafish
15	frog
35	quail
33	mouse

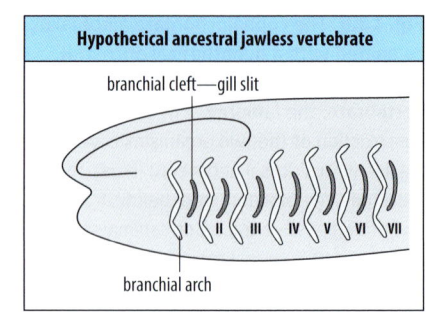

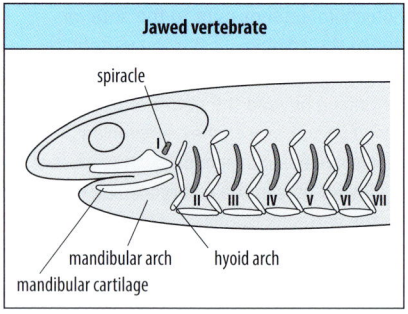

Fig. 13.8 Modification of the branchial arches during the evolution of jaws in vertebrates. The ancestral jawless fish had a series of at least seven gill slits, also called branchial clefts, supported by cartilaginous or bony arches, the branchial arches. Jaws developed from a modification of the first arch to give the mandibular arch, with the mandibular cartilage of the lower jaw, and the hyoid arch behind it. The spiracle is a respiratory opening behind the eye in present-day cartilaginous fish.

not, however, the relics of the gill arches and gill slits of an adult fish-like ancestor but represent structures that would have been present in the embryo of the fish-like ancestor of vertebrates as developmental precursors to gill slits and gill arches. During evolution, the branchial arches gave rise both to the gill arches of the primitive jawless fishes and, in a later modification, to gills and jaw elements in later-evolved fishes (Fig. 13.8). With time, the arches became further modified, and in mammals they now give rise to various structures in the face and neck (Fig. 13.9), many of which are derived from the neural crest cells that migrate into the branchial arches early in development (see Section 4.16). The cleft between the first and second branchial arches provides the opening for the eustachian tube, and endodermal cells in the clefts give rise to various glands, such as the thyroid and thymus.

Evolution rarely, if ever, generates a completely novel structure out of the blue. New anatomical features usually arise from modification of an existing structure. One can therefore think of much of evolution as a 'tinkering' with existing structures, which gradually fashions something different. This is possible because many structures are modular; that is, animals have anatomically distinct parts that can evolve independently of each other. Vertebrae are modules, for example, and can evolve independently of each other; so too are limbs; and so are digits.

A nice example of a modification of an existing structure is provided by the evolution of the mammalian middle ear. This is made up of three bones—malleus, incus, and stapes—that transmit sound from the eardrum (the tympanic membrane) to the inner ear. In the reptilian ancestors of mammals, the joint between the skull and the lower jaw was between the quadrate bone of the skull and the articular bone of the lower jaw, which were also involved in transmitting sound (Fig. 13.10). The vertebrate lower jaw was originally composed of several bones, but during mammalian evolution one of these bones, the dentary, increased in size and came to comprise the whole lower jaw, and the other bones were lost. The articular bone was no longer attached to the lower jaw, and by changes in their development, the articular and the quadrate in mammals were modified into two bones, the malleus and incus, respectively, whose function was now to help transmit sound from the tympanic membrane. The quadrate is evolutionarily and developmentally homologous with the dorsal cartilage of the first branchial arch in the ancestral vertebrate, and the stapes with the dorsal cartilage of the second.

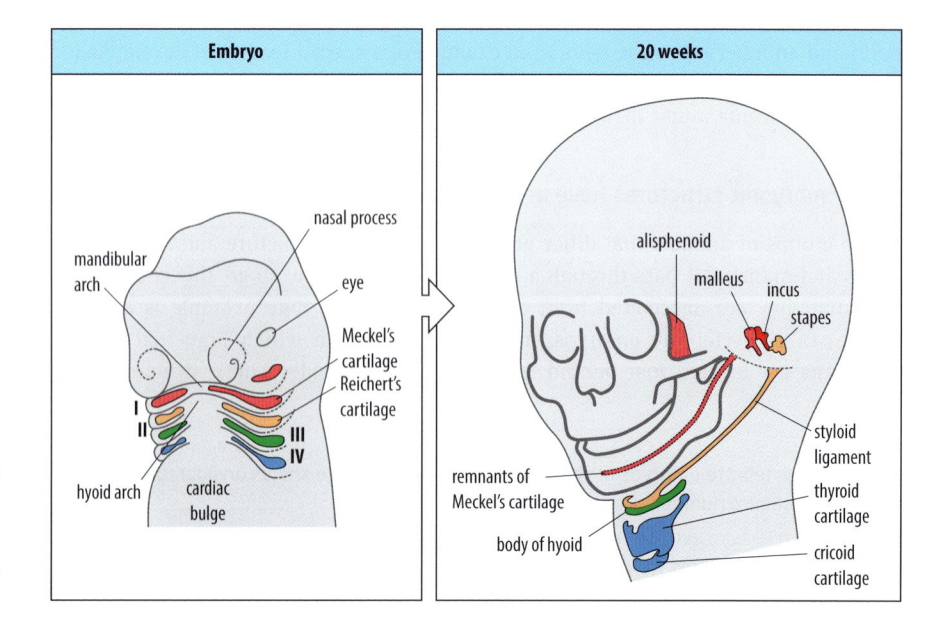

Fig. 13.9 Fate of branchial arch cartilage in humans. In the embryo, cartilage develops in the branchial arches, which gives rise to elements of the three auditory ossicles, the hyoid, and the pharyngeal skeleton. The fate of the various elements is shown by the color coding.

Illustration after Larsen, W.J.: Human Embryology. New York: Churchill Livingstone, 1993.

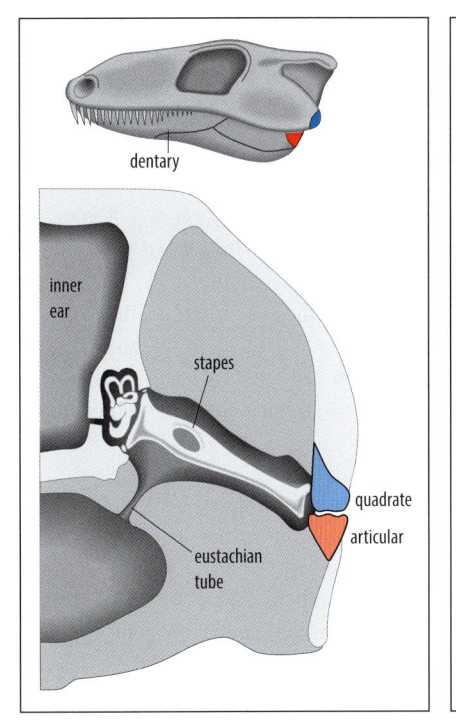

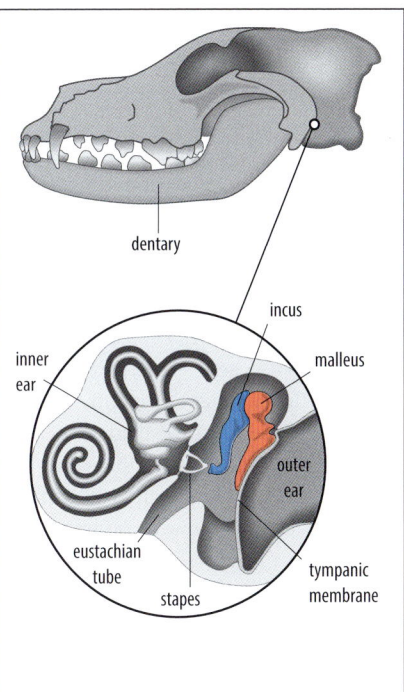

Fig. 13.10 Evolution of the bones of the mammalian middle ear. Left panel: the articular and quadrate bones of ancestral reptiles were part of the lower jaw articulation. Sound was transmitted to the inner ear via these bones and their connection to the stapes. Right panel: when the lower jaw of mammals became a single bone (the dentary), the articular bone became the malleus, and the quadrate bone the incus of the middle ear, acquiring a new function in transmitting sound to the inner ear from the tympanic membrane. The eustachian tube forms between branchial arches I and II.

Illustration after Romer, A.S.: The Vertebrate Body. Philadelphia, PA: W.B. Saunders, 1949.

These examples provide evidence of a key relationship between evolution and development, the gradual change of a structure into a different form. In many cases, however, we do not understand how intermediate forms were adaptive and gave a selective advantage to the animal. Consider, for example, the intermediate forms in the transition of the first branchial arch to jaws; what was the adaptive advantage? We do not know, and because of the passage of time and our current ignorance of the ecology of ancient organisms, we may never know.

13.6 Evolution of different types of eyes in different animal groups is an example of parallel evolution

Charles Darwin considered the evolution of the vertebrate eye as one of the most difficult problems in evolution because of the eye's complex structure. The very different eyes found in various animal groups are now thought to be examples of parallel evolution, which has adapted certain cell specializations that were present in the ancestors of all bilaterian animals or even earlier. The evolution of multicellularity opened up the possibility for cells to become specialized into differentiated cell types. This would have eventually led to simple types of photoreceptor cells, such as those in the larvae of present-day jellyfish (which belong to the Cnidaria), which have individual photosensitive cells in the ectoderm. Jellyfish photoreceptor cells not only use opsins as the photoreceptor proteins, as do all animals and some modern unicellular organisms, but also contain the pigment melanin, which is the pigment in the retinal pigmented epithelium of the vertebrate eye. The function of the pigment is to absorb any light that passes through the photoreceptor layer and so prevent it reflecting back onto the photoreceptors and producing an imprecise signal.

The next step in the evolution of the eye would have been the differentiation of the single light-sensing cell into two cell types: a sensory photoreceptor cell and a pigment cell (Fig. 13.11). The idea that the origin of the eye involved two such cells goes back to Darwin. Simple two-celled 'pigment cup' light-sensitive organs are present in the larvae of many bilaterian animals, whereas the more complex frontal eye of the larval amphioxus (a cephalochordate) consists of a pigment cup and two rows of photoreceptor cells connected by neurons to the central nervous system, foreshadowing the eyes of vertebrates.

Modern molecular biology has thrown some light on the problem of the evolution of the eye, and in the process has uncovered a remarkable example of conservation of gene function, suggesting that adaptation of this circuitry was involved in parallel evolution. As we have seen, the conservation of *Pax6/eyeless* function is demonstrated by the fact that *Pax6* from the mouse can substitute for *eyeless* and induce ectopic eye structures in *Drosophila* if expressed in the imaginal disc for the antenna (see Fig. 9.55). Whatever the source of the *Pax6* gene, the eye structures that form in *Drosophila* are always *Drosophila* ommatidia, not eyes like those of the animal from which the *Pax6* gene came. *Pax6* is acting as a master gene that turns on *Drosophila*'s own eye-development program. When and how *Pax6* became such a crucial gene in eye development is unknown.

The crystallin proteins that give the eye lenses of cephalopods (octopuses and squids) and vertebrates their transparency also have an intriguing evolutionary history. The crystallins were originally thought to be unique to the lens and to have evolved for this special function, but they are now thought to originate in metabolic enzymes and stress-related proteins that have become recruited for this function by changes in their regulatory regions, sometimes accompanied by gene duplication. Alpha-crystallins, the major crystallins in humans, are thought to have derived through gene duplication from a small heat-shock protein which is very stable, and it is this stability, and the protein's biophysical properties, that allowed it to evolve its role in the lens. This situation in which a protein involved in one function acquires a different one because its properties make it suitable for a different role is called **co-option**. Another example of co-option, although at a different level, is feathers, which are thought to have evolved initially as insulation but later could provide an adaptation to flight.

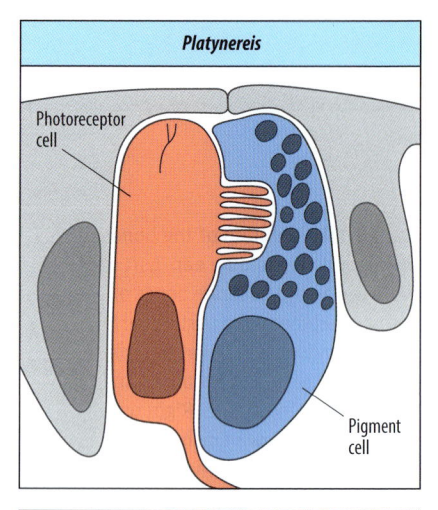

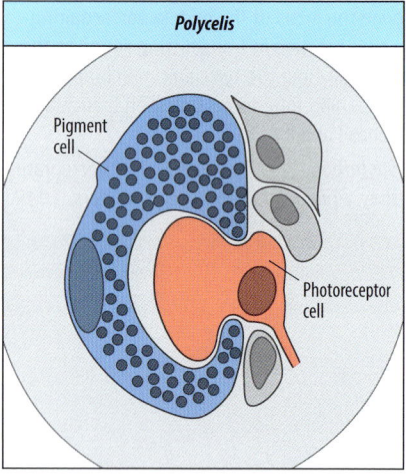

Fig. 13.11 Simple eyes like those of some invertebrates are thought to resemble the prototypical bilaterian eye. Illustrated here are the larval eye of the polychaete worm *Platynereis* (top panel), and one of the numerous simple eyes of the planarian *Polycelis auricularia* (bottom panel). They both consist of a photoreceptor cell shielded on one side by a pigment cell.

Top panel from Arendt, D.: **Evolution of eyes and photoreceptor cell types.** Int. J. Dev. Biol. 2003, **47**: 563-571. Bottom panel from Gehring, W.J.: **Chance and necessity in eye evolution.** Genome Biol. Evol. 2011, **3**: 1053-1056.

SUMMARY

The development of an embryo provides insights into the evolutionary origin of the animal, as the comparison of embryos of different species reveals the twists and turns that change one morphology into another. The embryo arose during the evolution of multicellular organisms from their single-celled ancestors, which have most of the cellular properties required by embryonic development and much of the molecular machinery to implement these activities. Multicellularity, having arisen, might have persisted originally because a colony of cells could provide a source of food for some of the members in times of scarcity. The large, nutrient-rich egg cell might have evolved as the cell that is fed by other cells and provides the resources for the development of complex structures. All but the very simplest multicellular animals undergo gastrulation, an extensive rearrangement of cells and tissues by which the germ layers are established in their correct positions inside the embryo, ready for further development, and a gut is formed. Comparison of vertebrate embryos shows that they all pass through a stage, the phylotypic stage, at which they more or less resemble each other. Specialized features that distinguish the different groups (fur, scales, etc.) develop later. Evolution rarely generates a completely novel structure out of the blue. During evolution, the development of structures can be altered so that they acquire new functions, as has happened in the evolution of the mammalian middle ear from a reptilian jaw bone, which itself evolved from a skeletal element in an ancestral branchial arch. In the evolution of disparate structures that have the same function, for example, eyes, there is a remarkable conservation of the genetic circuitry involved in their development.

The diversification of body plans

The largest group of animals is the Bilateria, which evolved a characteristic body plan with bilateral symmetry across the main body axis in at least some stage of development and three germ layers (see Section 13.2). We and other vertebrates are bilaterians, as are other chordates, echinoderms, arthropods, annelids, mollusks, and nematodes. The sister group to the Bilateria, the Cnidaria—the corals, jellyfish, hydra, and their relatives—have a radially symmetrical body axis and two body layers (see Box 13B). Hox genes are a hallmark of bilaterians, although they appear for the first time in cnidarians, and are tightly linked to bilaterian antero-posterior organization. They are key genes in the control of development and are expressed regionally along the antero-posterior axis of the embryo. The nature of the Hox genes and their pattern of expression is generally conserved in all bilaterian embryos.

The highly conserved organization of Hox gene expression along the body axis suggests that changes in their pattern of expression would have important consequences for the morphology of the animal. The Hox genes code for transcription factors that exert their influence by controlling the expression of target genes. These target genes are the 'effector' genes that interpret the positional information encoded by the Hox genes, and any individual Hox gene will control many downstream target genes. Mutations in the targets of the Hox genes are equally likely to be a major source of morphological change in evolution, especially mutations that affect how positional information is interpreted. A number of genes targeted by the Hox transcription factors are now known, but many still remain to be identified.

In this part of the chapter we will consider how Hox genes evolved in the genome and how changes in their expression and their target genes can help to explain the rich diversity of the body plans of the animals that exist today. We have discussed the function of Hox genes in the development of the antero-posterior axis of *Drosophila* and of the embryos of our model vertebrates, particularly the mouse (see Chapters 2 and 4 respectively). We will build on this knowledge and the advent of the CRISPR-Cas9 system, which can rapidly produce targeted mutations (see Box 1C) that enable the roles of the Hox genes to be tested in a wide range of non-model animals. We use, as our examples, first the evolutionary modifications of the arthropod body plan, and, in particular, the differences in the type and distribution of paired appendages along the body; and, second, the evolutionary modifications of the vertebrate body plan with reference to the differences in the pattern of the vertebrae along the spine. In contrast to the similarities in the way the antero-posterior axis of the body is specified in arthropods and vertebrates, the body plan along the dorso-ventral axis of the body plan is inverted in vertebrates compared with that of arthropods, and we will consider how this might have evolved.

13.7 Hox gene complexes have evolved through gene duplication

The broad similarities in the pattern of Hox gene expression in vertebrates and arthropods, whose evolution diverged hundreds of millions of years ago, are taken as good supporting evidence for the idea that all multicellular animals descend from a common ancestor. By comparing the organization and structure of the Hox genes in insects and vertebrates, we can determine how one class of important developmental genes has changed during evolution. The Hox gene complexes provide one of the clearest examples of the importance of gene duplication in the evolution of developmental systems. The Hox genes are members of the homeobox gene superfamily, which is characterized by a short 180-base-pair motif, the homeobox, encoding a DNA-binding helix–turn–helix domain called the homeodomain that is involved in transcriptional regulation. Two features characterize all Hox genes: the individual genes are organized into one or more gene clusters (gene complexes), and they display spatial co-linearity:

that is, the spatial order of expression of individual genes along the antero-posterior axis of the body is the same as their sequential order in the gene cluster (see Box 4G).

The simplest Hox gene complexes are found in invertebrates, and comprise a relatively small number of sequence-related genes carried on one chromosome. By contrast, vertebrates typically have four clusters of Hox genes carried on four different chromosomes, suggesting two rounds of wholesale duplication of an ancestral Hox gene complex, in line with the generally accepted idea that large-scale duplications of the genome have occurred during vertebrate evolution. There have also been further duplications and losses of Hox genes within each cluster. In an ancestor of teleost fish there was one further round of genome duplication; for example, the zebrafish and the Japanese puffer fish *Takifugu rubripes* have seven Hox clusters each, the result of an extra round of duplication and the subsequent loss of one cluster—one of the duplicated Hoxd clusters. Genes have also been preferentially lost from the remaining pairs of Hox clusters.

Comparisons of Hox genes of various arthropod and chordate species indicate that the common ancestor of arthropods and vertebrates most probably had a simple Hox cluster of seven genes. As we get nearer to the vertebrates, the cephalochordate amphioxus has just one Hox cluster, containing 15 genes, and one can think of this as most closely resembling the ancestor of the four vertebrate Hox gene clusters: Hoxa, Hoxb, Hoxc, and Hoxd. Possible ways in which both the vertebrate and *Drosophila* Hox clusters could have evolved from a simpler ancestral gene cluster by gene duplication have been reconstructed (Fig. 13.12). In *Drosophila*, for example, successive tandem duplications of an ancestral *Antp*-like gene could have given rise to the genes *abdominal-A* (*abd-A*), *Ultrabithorax* (*Ubx*), and *Antennapedia* (*Antp*).

Sequence comparisons between *Drosophila* and mouse suggest that the multiple mouse Hox genes most similar in sequence to the *Drosophila* Abdominal-B (*Abd-B*) gene do not have direct homologs within the *Drosophila* HOM-C complex (see Fig. 13.12). This means they probably arose by tandem duplication from an ancestral gene after the split of insect and vertebrate lineages, but before duplication of the whole cluster in vertebrates. The likely advantage of duplication was that the embryo had more Hox genes to control downstream targets and in different tissues, and so could make a more complicated body.

Fig. 13.12 Gene duplication and Hox gene evolution. A suggested evolutionary relationship between the Hox genes of a hypothetical common ancestor and *Drosophila* (an arthropod), amphioxus (a cephalochordate), and the mouse (a vertebrate). Duplications of genes of the ancestral cluster (red) could have given rise to the additional genes in *Drosophila* and amphioxus. Two duplications of the whole cluster in a chordate ancestor of the vertebrates could have given rise to the four separate Hox gene clusters in vertebrates. There has also been a loss of some of the duplicated genes in vertebrates. In *Drosophila*, the two Hox gene clusters comprising the HOM-C complex are called the Antennapedia complex and the bithorax complex, and their co-linearity has been broken as a result of a chromosomal rearrangement. The paralogous subgroup Hox14 has been found in the coelacanth and the horn shark, as well as in amphioxus, but Hox15 has so far only been found in amphioxus species, and has no relationship to the other paralogous subgroups.

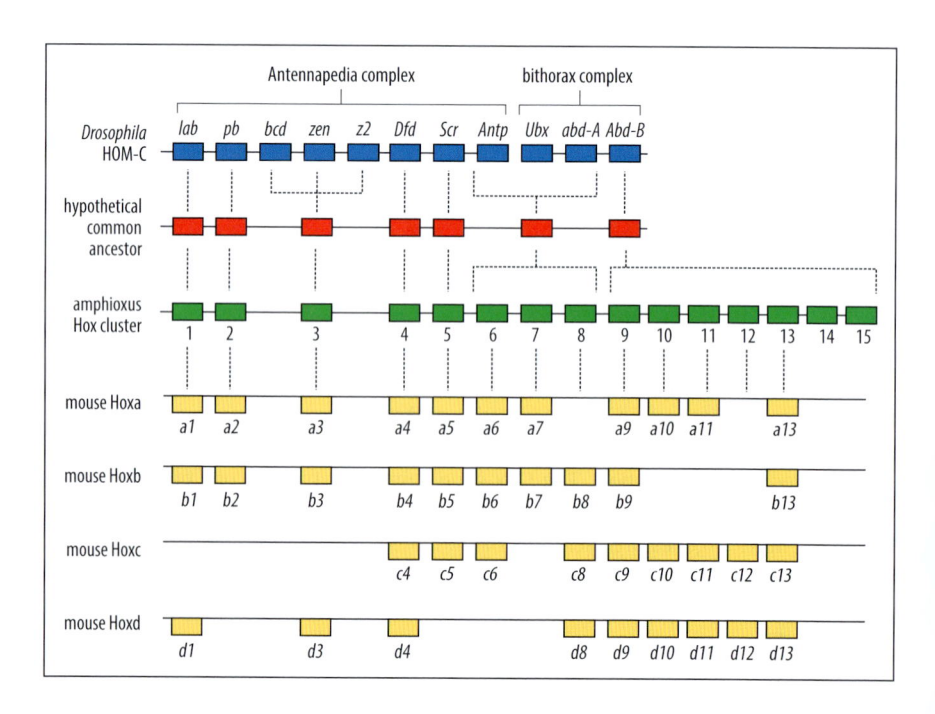

13.8 Differences in Hox gene expression determine the variation in position and type of paired appendages in arthropods

The function of Hox genes in the development and patterning of the body plan in *Drosophila* has already been discussed (see Chapters 2 and 9), but changes in patterns of Hox gene expression can also help to explain the evolution of the great diversity of arthropod body plans, examples of which are shown in Figure 13.13. A segmented body is distinguishable in all arthropod embryos, even though in some classes, such as the spiders, the segmented structure is not easily discernible in the adult. Efforts to

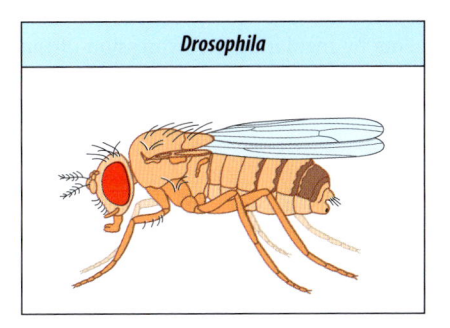

Drosophila

Insecta

Head						Thorax			Abdomen									
Oc	An	In	Mn	Mx	La	T1	T2	T3	A1	A2	A3	A4	A5	A6	A7	A8	A9	A10
						leg	leg	leg										
							wing	wing										

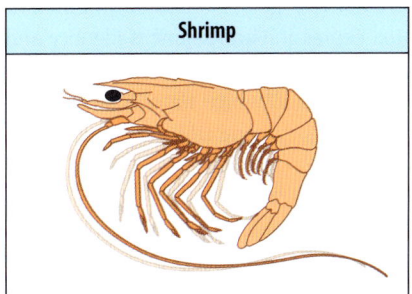

Shrimp

Crustacea (Malacostraca: Decapoda)

Head						Thorax (Pereon)								Abdomen (Pleon)						
Oc	An1	An2	Mn	Mx1	Mx2	T1	T2	T3	T4	T5	T6	T7	T8	A1	A2	A3	A4	A5	A6	A7
						mxp	mxp	mxp	leg	leg	leg	leg	leg	sw	sw	sw	sw	sw		

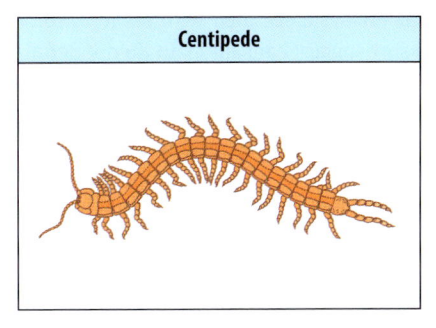

Centipede

Myriapoda

Head							Trunk									
Oc	An	In	Mn	Mx1	Mx2	Mxp	T1	T2	T3	T4	T5	T6	T7	T8	T9	
						mxp	leg	leg	leg	leg	leg	leg	leg	leg	leg	

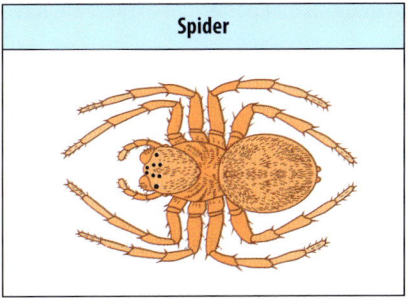

more posterior segments added with growth (all with legs)

Spider

Chelicerata

Prosoma							Opisthosoma								
Oc	Ch	Pp	L1	L2	L3	L4	O1	O2	O3	O4	O5	O6	O7	O8	O9
			leg	leg	leg	leg									

Fig. 13.13 The variety of arthropod body plans. Shown here are examples from the four main arthropod classes and their segment pattern. Segments that bear maxillipeds (mxp), legs, swimmerets (sw), and wings are noted. The head segments bear a variety of specialized appendages such as antennae and mouthparts. An, antenna; Ch, chelicera; In, intercalary segment; La, labium; Mn, mandible; Mx, maxilla; Oc, ocular; Pp, pedipalp.

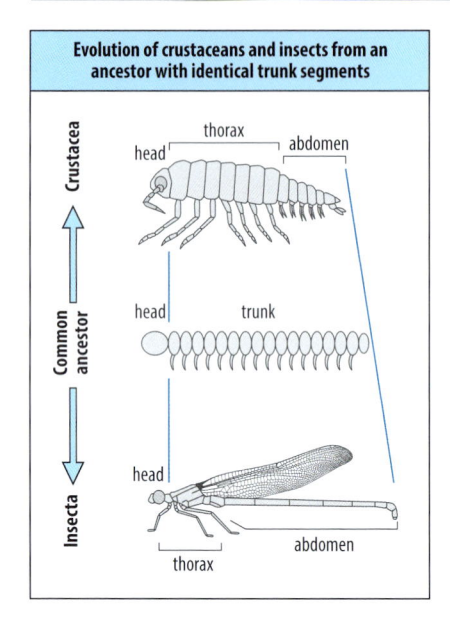

Fig. 13.14 Insects and crustaceans evolved from an ancestor that probably had an adult body composed of more-or-less uniform segments.

From Abzhanov, A., Kaufman, T.C.: **Crustacean (malacostracan) Hox genes and the evolution of the arthropod trunk.** Development *2000,* **127***: 2239-2249.*

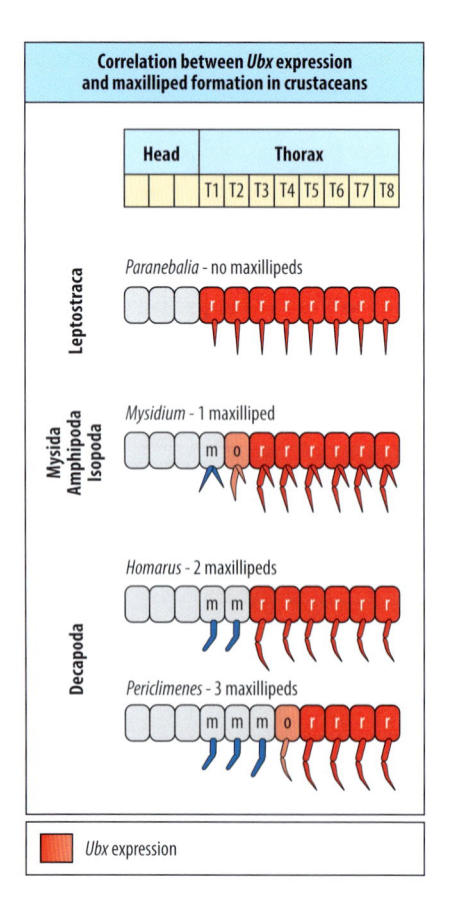

determine the homology of different types of segments between the different arthropod groups, and thus to identify the nature of the changes that have occurred during their evolution, have proved difficult on anatomical grounds alone.

We shall focus here on insects and crustaceans, in which most is known about patterns of Hox gene expression. Insects and crustaceans evolved from an ancestor that probably had an adult body composed of more-or-less uniform segments, all bearing similar paired leg-like appendages (Fig. 13.14). Evolution has restricted the legs to particular segments. Differences in the type and distribution of the paired appendages along the body in crustaceans have provided a good test of the role of Hox genes in controlling this process and thus their probable role in the evolution of arthropod body plans. If the product of the Hox gene is actively 'instructing' the identity of a particular segment or contiguous group of segments by regulating the expression of the target genes that specify the structural features of that segment, then the same change in segmental morphology in different groups should be accompanied by a consistent change in the expression pattern of that Hox gene.

Crustaceans are particularly well suited to such a study. Their thoracic segments vary in number between different groups, but each segment bears one of two kinds of appendages: legs (which can be of several types) or specialized 'feeding' legs (maxillipeds), which resemble mouthparts in structure and are used to manipulate food. Different orders of crustaceans have a different complement of legs and maxillipeds on the thorax, and a comparison of the patterns of Hox gene expression in different species reveals a clear relationship between Hox expression and segment identity and function, as judged by the appendages they carry. Figure 13.15 shows the four different patterns found among the malacostracan crustaceans, which have eight thoracic segments, and the correlation with expression of the Hox gene *Ubx*. The other main groups of crustaceans either carry legs on all the thoracic segments (the branchiopods, such as the brine shrimp *Artemia*) or have one maxilliped (the maxillipods, such as the American tadpole shrimp *Triops*), and show the same correlation between the loss of *Ubx* expression in a thoracic segment and the presence of a maxilliped instead of a leg.

The presence of maxillipeds on a segment is always associated with an absence of *Ubx* expression. However, correlations of this sort suggest causality, but do not definitively prove it. As we saw in *Drosophila*, causality can be tested by deliberately altering the spatial pattern of Hox gene expression in an individual species and analyzing the effects (see Fig. 2.47). RNA interference (RNAi) and, more recently, the CRISPR-Cas9 system have been used to manipulate *Ubx* expression in the amphipod *Parhyale* (a malacostracan crustacean). Lowering *Ubx* expression allowed the emergence of maxilliped-like structures instead of legs, whereas expressing *Ubx* in a thoracic segment in which it is normally not expressed led to the appearance of legs instead of maxillipeds (Fig. 13.16). In this case, therefore, we can definitely say that changes in the regulation of *Ubx* that resulted in a change in its spatial pattern of expression have

Fig. 13.15 The presence and position of maxillipeds (feeding legs) in different species of crustaceans depends on evolutionary changes in the pattern of *Ubx* and *abd-A* expression. Four examples from the class Malacostraca are shown here, as this class shows the greatest diversity. The number and position of maxillipeds (blue) on thoracic segments varies between different orders (e.g. the Decapoda and the Amphipoda), and their presence clearly correlates with the anterior boundary of expression of the Hox genes *Ubx* and *abd-A*. A monoclonal antibody that recognizes both the Ubx and abd-A proteins was used to stain the larval stages of different species. Red indicates higher levels of the two proteins, which correlates with the presence of walking legs, and orange indicates lower-level expression, which correlates with grasping legs. Maxillipeds develop where *Ubx* and *abd-A* are not expressed.

Adapted from Averof, M., Patel, N.H.: **Crustacean appendage evolution associated with changes in Hox gene expression.** Nature *1997,* **388***: 682-685.*

Fig. 13.16 Manipulation of *Ubx* expression in the amphipod *Parhyale hawaiensis*. First panel: wild-type *P. hawaiensis* bears a single pair of maxillipeds on the first thoracic segment, a type of leg called a "grasping leg" on the next two segments, and walking legs on the remaining five thoracic segments. Second panel: when *Ubx* expression was lowered in the embryo by RNA interference (RNAi) through injection of short-interfering RNAs, maxillipeds appeared on the first three thoracic segments. Third panel: a stronger phenotype was observed by the application of CRISPR-Cas9 to inactivate *Ubx*, which gave an additional partial transformation of T4 and T5 towards T1 (maxilliped), but with some distal characteristics of T2/T3, perhaps due to changes in the expression of other Hox genes, *Scr* and *abd-A* in particular. These experiments show that the normal low level of expression of *Ubx* in T2 and T3 (see first panel) makes them the most sensitive segments to its knockdown, and that a phenotypic change in T4 and T5 requires a more complete loss of function, as afforded by CRISPR-Cas9. Fourth panel: when transgenic *Ubx* was expressed ectopically in all thoracic segments under the control of a heat-shock gene promoter, legs of the type normally found on T2 appeared on T1 instead of a maxilliped, reflecting the transformation of the segment to a more posterior fate. In this last experiment, because uniform high-level *Ubx* transgene expression throughout the body in addition to the underlying wild-type expression was lethal, the transgene was expressed for only short periods each day. The outcome illustrated here was observed in around 10% of hatchlings. In a smaller proportion of hatchlings (3%) legs of the T4 type developed on all thoracic segments. As in Figure 13.15, red indicates higher levels of *Ubx* expression, and shades of orange lower levels.

Data from Liubicich, D.M., et al.: **Knockdown of Parhyale Ultrabithorax recapitulates evolutionary changes in crustacean appendage morphology.** Proc. Natl Acad. Sci. USA 2009, **106**: 13892-13896; Pavlopoulos, A., et al.: **Probing the evolution of appendage specialization by Hox gene misexpression in an emerging model crustacean.** Proc. Natl Acad. Sci. USA 2009, **106**: 13897-13902; Martin, A., et al.: **CRISPR/Cas9 mutagenesis reveals versatile roles of Hox genes in crustacean limb specification and evolution.** Curr. Biol. 2016, **26**: 14-26.

driven evolutionary change in crustaceans. Notably, complete transformations only affected the first three thoracic segments. When *Ubx* expression was lowered throughout the thorax, maxillipeds appeared on the first three segments; a leg developed as usual on the fourth and following thoracic segments, whereas when *Ubx* function is knocked-out completely by CRISPR-Cas9 the legs on the fourth and fifth thoracic segments were partially transformed into maxilliped-like appendages.

Despite the enormous variations in appearance among extant insects, their basic body plans are less diverse than those of crustaceans. All adult insects have just three thoracic segments, bearing a pair of legs on each. Wings, when present, appear on the second and third segments, or just the second in the case of flies, where the wing on the third thoracic segment has been modified into a haltere, a small organ that allows balancing during flight. Insect fossils, however, display a much wider range of patterns in the position and number of their legs and wings. Some insect fossils have appendages on every segment, whereas others only have legs in a distinct thoracic region. The number of abdominal segments bearing legs varies, as does the size and shape of the legs. Wings arose later than legs in insect evolution. Wing-like appendages are present on all the thoracic and abdominal segments of some insect fossils but are restricted to the thorax in others. To see, in principle, how these different patterns of appendages could have arisen during insect evolution we will compare two orders of modern insects, the Lepidoptera (butterflies and moths) and the Diptera (flies, including *Drosophila*).

The basic pattern of Hox gene expression along the antero-posterior axis is similar in all present-day insect species that have been studied. Yet caterpillars (larvae) of Lepidoptera have small walking appendages (pro-legs) on the abdomen and the thorax, and the adults have two pairs of wings, whereas the more recently evolved

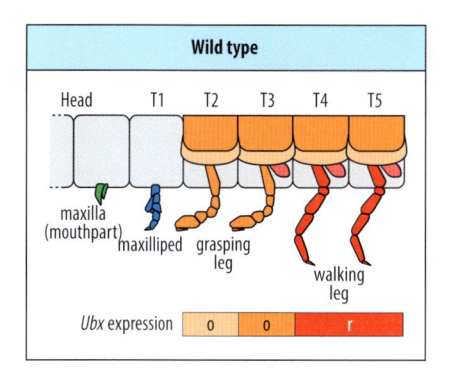

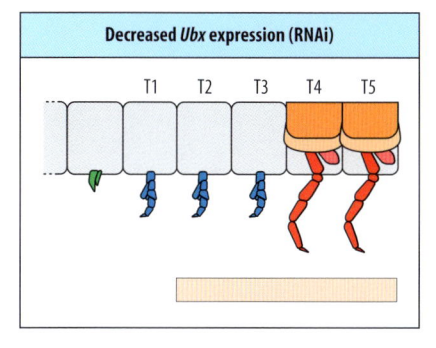

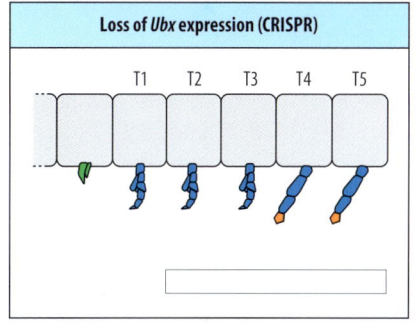

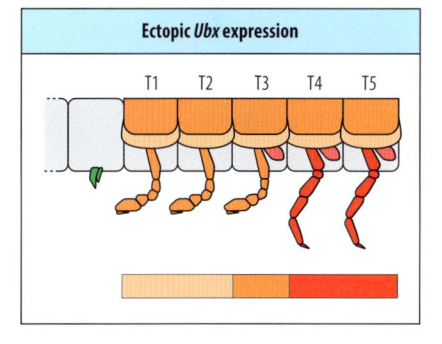

Diptera have no legs on the abdomen in the larva, and only one pair of wings in the adult, with the second pair of wings having been modified into small balancing organs, the halteres. How are these differences related to differences in Hox gene activity in the two groups of insects?

In *Drosophila*, the *Distal-less* gene, which is required for leg formation (see Section 9.8), is a target of the Hox genes of the bithorax complex, which are expressed in the abdomen. The Hox proteins repress *Distal-less* expression and thus suppress leg formation in the abdomen. If one removes the bithorax complex completely, every abdominal segment in the embryo bears a leg disc and a wing disc (this mutation is lethal so larvae do not develop). This suggests that the potential for appendage development is still present in every segment in insects, even in the Diptera, and is under active repression in the fly abdomen. This evidence further supports the idea that the ancestral arthropod from which insects evolved had appendages on all its segments (see Fig. 13.14). The gene responsible for this repression is *Ubx*.

During the embryonic development of lepidopterans, in contrast, the genes *Ubx* and *abd-A* (a member of the bithorax complex) are turned off in the ventral parts of the embryonic abdominal segments; this results in *Distal-less* being expressed and prolegs developing on the abdomen in the caterpillar. The presence or absence of legs on the abdomen is thus determined by whether or not particular Hox genes are expressed there, which is yet more evidence that changes in the pattern of Hox gene expression have had a key role in evolution.

Nevertheless, differences in the downstream targets of the Hox genes are likely to be even more important in evolution of novel body plans. Although the sequence encoding the DNA-binding homeodomain is highly conserved between Hox genes in different species, the coding sequences outside the homeodomain, which affect which gene-control regions and gene-regulatory proteins the Hox proteins can interact with, have undergone considerable change during evolution. Members of the bithorax complex repress appendage development in the insect abdomen, but in crustaceans these genes are expressed in the thorax (see Fig. 13.15) and in the abdomen, yet appendages (legs on the thorax and swimmerets on the abdomen) still develop. Comparison of the structure of the protein encoded by the *Ubx* gene between crustaceans and insects reveals a transcriptional repression domain at the carboxy-terminus of the protein, present in insects but not in crustaceans, which is responsible for the suppression of appendages in the abdomen. This is another example of **heterotypy**, changes in protein function associated with alterations in protein structure, which is being recognized as an element driving evolutionary novelty (see Box 13A).

Evolutionary changes in the targets of Hox genes also help to explain why flies have two wings, while butterflies have four. In *Drosophila*, expression of *Ubx* in the third thoracic segment (T3) specifies the appendage on that segment as a haltere, rather than the wing that develops on the neighboring T2, in which *Ubx* is not expressed at the time that the wing is specified. Genes known to be involved in patterning the wing are repressed by *Ubx* and are therefore not expressed in segment T3, explaining the differences between the appendages on neighboring segments. But in butterflies, because of changes in the control regions of these patterning genes during evolution, only some of them are repressed by *Ubx*, and this results in the development of a hindwing on segment T3 rather than a haltere.

Genes other than Hox genes are involved in the basic specification and patterning of the arthropod head. The head regions of different types of arthropods are very different, and indeed spiders do not seem to have a head at all (see Fig. 13.13). Their body is divided into a front and a back region (the prosoma and the opisthosoma), the front end having segments that bear the spider's fangs (the chelicerae after which the class is named), pedipalps, and four pairs of walking legs; the front region thus acts as a head and thorax combined. Despite these differences, typical 'head genes' are expressed in the anterior-most region in both spiders and insects, including the homeobox gene *orthodenticle*, which is involved in specifying the larval head and anterior brain in *Drosophila* (see Section 2.30). This comparison shows how much morphology can change as a result of changes in the downstream targets of genes. It is interesting to note that in vertebrates, the *Otx* genes of the orthodenticle gene family are required for specifying anterior structures, including the brain (see Chapter 10).

13.9 Changes in Hox gene expression and their target genes contributed to the evolution of the vertebrate axial skeleton

In vertebrates, an easily visible evolutionary modification of the body plan that involves Hox genes is the pattern of the axial skeleton—the vertebral column and ribs. This pattern is characterized by the number and type of vertebrae in the main anatomical regions—cervical (neck), thoracic (rib-bearing), lumbar, sacral, and caudal (Fig. 13.17). The number of vertebrae of a particular type varies considerably among the different vertebrate classes—mammals, with rare exceptions, have seven cervical vertebrae, whereas birds can have between 13 and 15, and certain reptiles, such as snakes, can have no cervical vertebrae but more than 200 rib-bearing vertebrae.

How do these differences arise? A comparison between the mouse and the chick shows that the patterns of Hox gene expression in the somites along the body differ. The anterior limit of expression of, for example, *Hoxc6*, one of the Hox genes that has been shown experimentally to specify thoracic regional identity, is at somite 12 in the mouse, which has seven cervical vertebrae, but at somite 19 in the chick, which has 13 cervical vertebrae (see Fig. 4.37; the first five somites do not give rise to vertebrae). Moreover, the anterior limit of *Hoxc6* expression is at somite 22/23 in geese, which have three more cervical vertebrae than chickens. These observations fit with the idea that diversification of axial patterning in vertebrates evolved through changes in Hox gene expression.

Analysis of Hox gene expression and regulation in different snake species has been carried out to explore how changes in their body plan could have evolved. Most of the snake body consists of an elongated trunk, and the trunk axial skeleton consists of a very large number of morphologically similar vertebrae, which all have ribs, as can be seen in the skeleton of the python embryo in Figure 13.18. The arrow in Figure 13.18 shows the position of the hindlimb rudiments, which have been removed in this specimen. Rib-bearing vertebrae lie anterior to the arrow and extend all the way up to the two vertebrae at the base of the skull. In pythons, Hoxc6 and Hoxc8 proteins are detected in the somites throughout the trunk, representing an expansion of Hoxc6 and Hoxc8 domains compared with the chick, which reflects the expansion of thoracic identity in the python. This apparently homogeneous Hox protein distribution corresponds with the apparent lack of morphological regionalization of the axial skeleton, suggesting that changes in Hox gene expression could be involved in the evolution of the snake body plan. But in the North American corn snakes, which belong to a family of completely limbless snakes, a more comprehensive analysis of Hox gene transcripts in the somites revealed that the distribution is not as homogeneous as in the python. There is some evidence of regionalization of Hox gene expression in the somites that will give rise to the rib-bearing vertebrae, although the anterior limit of expression is more diffuse. More recent morphometric analysis of snake skeletons has revealed subtle morphological differences in the vertebrae along the spine that could be specified by some sort of Hox code based on the patterns of Hox gene expression seen in the corn snake embryos.

A change in a Hox target gene has been found in corn snakes that could explain why all the trunk vertebrae bear ribs. Some Hox10 paralogs, *Hoxa10* for example, that specify lumbar vertebral identity in mice, are unexpectedly expressed in the somites in the posterior region of the corn snake trunk. Experiments have shown that the reason that rib-bearing thoracic vertebrae develop in this region in these snakes is not because the functioning of the Hoxa10 protein is changed. When tested in pre-somitic mesoderm in transgenic mice, corn snake *Hoxa10* can repress rib formation

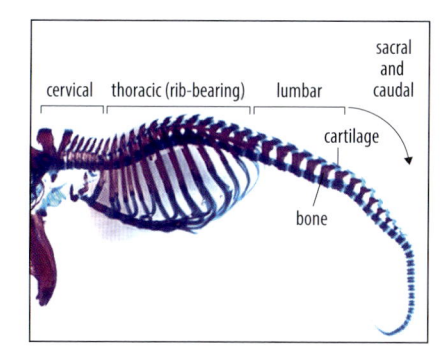

Fig. 13.17 Vertebrae come in many different types. Lateral view of an E18.5 mouse skeleton stained with Alcian blue and Alizarin red (for cartilage and bone, respectively), showing the pattern of vertebrae in the main anatomical regions: cervical (neck) seven; thoracic (rib-bearing) 13; lumbar (lower back) six; sacral (hip region) four; caudal (tail), approximately 30 (although not all of these are easy to make out in this specimen). Note that the pectoral and pelvic girdles have been removed.

Photograph courtesy D.M. Wellik, from Wellik, D.M.: **Hox genes and vertebrate axial pattern.** *Curr. Topics Dev. Biol. 2009,* **88***: 257–278.*

694 Chapter 13 Evolution and development

Fig. 13.18 Comparison of Hox protein and gene expression in python and chick embryos. The photograph shows a skeleton of a python embryo at 24 days' incubation stained with Alcian blue and Alizarin red. The arrow marks the position of hindlimb primordia, which have been removed in this preparation. Note how similar the vertebrae are in the trunk, which is anterior to the arrow. The right panel shows a schematic comparison of domains of *Hoxb5* (green), *Hoxc8* (blue), and *Hoxc6* (red) in chick and python embryos (in pythons it was the presence of Hox proteins that was detected). The expansion of the Hoxc8 and Hoxc6 domains in the python correlates with the expansion of thoracic identity in the axial skeleton. The domains of Hox proteins are also expanded in the python lateral plate mesoderm, correlating with flank identity.

Adapted from Cohn, M.J., Tickle, C.: **Developmental basis of limblessness and axial patterning in snakes.** Nature 1999, **399**: 474-479.

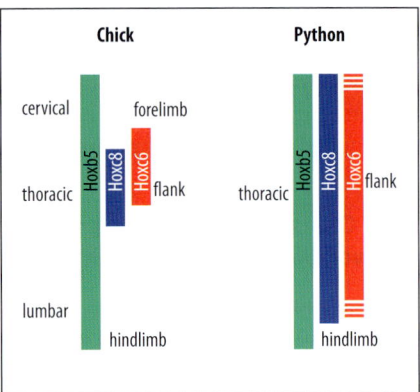

on thoracic vertebrae just as mouse *Hoxa10* does in the same test. Instead, what has happened in the corn snake is a change in the interpretation of the Hox code. A single nucleotide change in the enhancer of one of the target genes of Hoxa10 protein has been identified. The gene is required for rib formation, and in mice its expression is inhibited in the lumbar region by Hoxa10 protein. As a result of the mutation in corn snakes, Hoxa10 protein is unable to bind to the enhancer and repress expression of the rib-forming gene, and so ribs develop even when Hoxa10 protein is present. Remarkably, the identical genetic change is also found in animals, such as African elephants, that have elongated rib cages. These findings point up how important changes in Hox target genes can be in animal evolution.

13.10 The basic body plan of arthropods and vertebrates is similar but the dorso-ventral axis is inverted

The comparison between the body plans of arthropods and vertebrates reveals that in spite of many similarities—both have an anterior head, a nerve cord, and a gut running anterior to posterior—the dorso-ventral axis of vertebrates (and of other chordates) is inverted when compared with that of arthropods. The most obvious manifestation of this is that the main nerve cord runs ventrally in arthropods and dorsally in vertebrates (Fig. 13.19, left panels).

One explanation for this, first proposed in the nineteenth century, is that during the evolution of the vertebrates from their last common ancestor with the arthropods, the dorso-ventral axis was turned upside down, so that the ventral nerve cord of the ancestor became dorsal. This inversion may have been dictated by the position of the mouth. The mouth defines the ventral side, and a change in the position of the mouth away from the side of the nerve cord would have resulted in the reversal of the dorso-ventral axis in relation to the mouth. The position of the mouth is specified during gastrulation in protostomes (to which the arthropods belong) but not in deuterostomes (to which the chordates belong), where it develops as a secondary independent opening, and thus it is not difficult to imagine how if the position of the mouth moved, other changes in body structure and the nervous system also occurred.

The startling idea that the dorso-ventral axis was turned upside down during the evolution of the vertebrates has recently found some support from molecular evidence showing that the same genes are expressed along the dorso-ventral axis in both insects and vertebrates, but in opposite directions. In *Xenopus*, the protein Chordin is one of the signals that specifies the dorsal region, whereas the growth factor BMP-4 specifies a ventral fate. In *Drosophila*, the pattern of gene expression is the opposite: the protein Decapentaplegic, which is closely related to BMP-4, is the dorsal signal, and the protein short gastrulation (Sog), which is related to Chordin, is the ventral signal (Fig. 13.19, right panels, and see Section 2.17). These signal molecules are experimentally

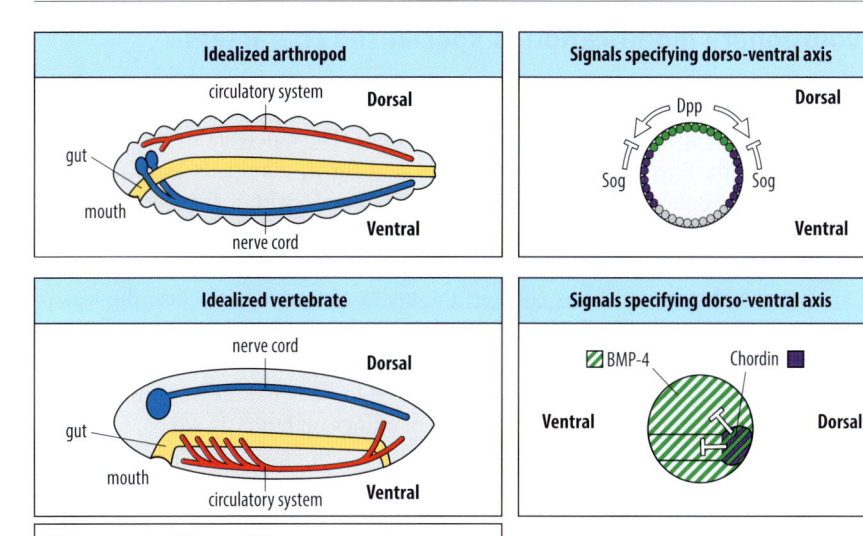

Fig. 13.19 The vertebrate and *Drosophila* dorso-ventral axes are related but inverted. Left panels: in arthropods, the nerve cord is ventral, whereas in vertebrates it is dorsal—with ventral being defined by the position of the mouth. Right panels: in arthropods, illustrated here by *Drosophila* (transverse section of blastoderm), and in vertebrates (*Xenopus*, surface view of blastula), the signals specifying the dorso-ventral axis are similar but are expressed in inverted positions. The protein Chordin, a dorsal specifier in vertebrates, is related to Sog, which is a ventral specifier in *Drosophila*, and the vertebrate ventral specifier BMP-4 is related to *Drosophila* Decapentaplegic (Dpp), which specifies dorsal.

*Illustration after Ferguson, E.L.: **Conservation of dorsal-ventral patterning in arthropods and chordates.** Curr. Opin. Genet. Dev. 1996, 6: 424–431.*

interchangeable between insects and frogs. Chordin can promote ventral development in *Drosophila*, and Decapentaplegic promotes ventral development in *Xenopus*. The molecules and mechanisms that set up the dorso-ventral axes in the two groups of animals are thus homologous, strongly suggesting that the divergence in the body plans of the arthropods and vertebrates involved an inversion of this axis. Thus, our image of the common ancestor of chordates and arthropods is an animal in which the signals establishing the two major body axes, antero-posterior and dorso-ventral, were similar to those in present-day animals.

SUMMARY

The basic body plan of all bilaterian animals is associated with the patterns of Hox gene expression that encode positional information. Both the patterns of Hox gene expression and their target genes that interpret positional information have changed in evolution. This is particularly evident in the variation in position and type of appendages in arthropods, which can be investigated by techniques for altering gene expression. The Hox genes themselves have undergone considerable evolution by gene duplication and divergence and are a good example of Darwin's description of evolution as 'descent with modification'. In vertebrates, differences in Hox gene expression in the various groups can be seen in the differences in the number and type of vertebrae in their axial skeleton. Arthropods and vertebrates are both bilaterians and so have some similarities in their body plans. But comparison of patterns of dorso-ventral gene expression in the specification of the body plan suggests that during the evolution of the vertebrates, the dorso-ventral axis of an invertebrate ancestor was inverted.

SUMMARY: The diversification of body plans

Hox gene duplication, diversification, and changes in interpretation

⇩

different body plans

Signals for *Xenopus* and *Drosophila* dorso-ventral axes are homologous but axes are inverted: BMP-4–Decapentaplegic; Chordin–Short gastrulation

Evolutionary modification of specialized characters

It has long been recognized that more general characteristics of the body plan develop earlier than more specialized ones (see Section 13.4). The nineteenth-century embryologist Ernst von Baer gave scales in reptiles and fur in mammals as examples of **specialized characters** of particular groups of animals within the vertebrates. A specialized character is defined as an adaptation of an organ or part of an organ for a particular function or environment. We have already come across examples of specialized characters in this chapter: the beaks of Darwin's finches, the legs of crustaceans, and the wings of flies. Here we discuss in detail some examples of specialization. We look first at the paired appendages of vertebrates—how limbs originated in tetrapod vertebrates and then evolved to carry out different functions in different environments. Next we consider the markings on butterfly wings, which are involved in sexual selection and adaptation to the environment. We finally consider adaptive evolution within a single species, taking three-spine stickleback fish and cavefish as our examples.

13.11 Limbs evolved from fins

The limbs of tetrapod vertebrates are specialized characters that develop after the phylotypic stage. Amphibians, reptiles, birds, and mammals have limbs, whereas fish have fins. The limbs of the first land vertebrates evolved from the pelvic and pectoral fins of fish-like ancestors, which had lobe fins; ray-finned fish, including present-day teleosts, evolved from a separate lineage of bony fish (Fig. 13.20). The basic limb pattern is highly conserved in both the forelimbs and hindlimbs of all tetrapods, although there are differences both between forelimbs and hindlimbs, and between different

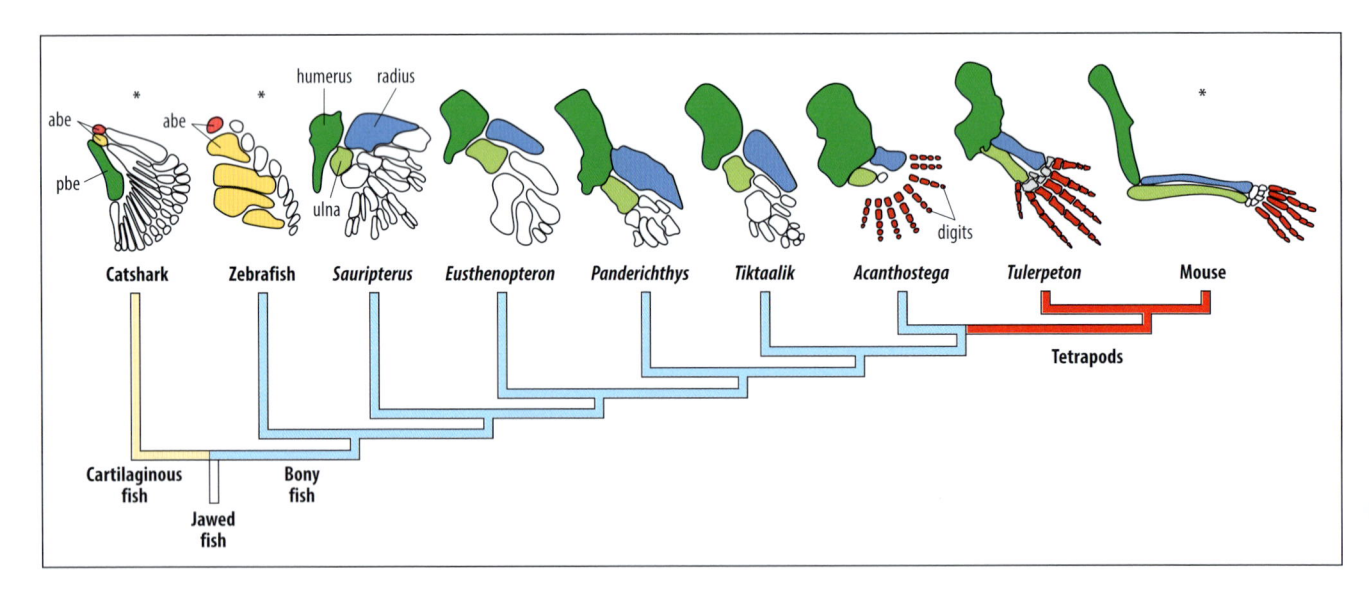

Fig. 13.20 A family tree showing skeletal patterns of fins or limbs of jawed vertebrates, including both present-day and fossil animals. The family tree shows the evolutionary origins of present-day cartilaginous fish, bony fish, and tetrapods, represented by catshark, zebrafish, and mouse, respectively (asterisks). Note that the tetrapods evolved from lobe-finned ancestors such as *Panderichthys* and the teleosts from a separate lineage of bony fish. The schematics are of the skeletal patterns of anterior appendages. Colors indicate the posterior basal element/humerus (dark green), the more anterior basal elements (yellow and pink), ulna (light green), radius (blue), and digits (red). The original sources for the skeletal patterns are cited in the papers listed below.

Based on Tanaka, M.: **Fins into limbs: autopod acquisition and anterior elements reduction by modifying gene networks involving 5′ Hox, Gli3, and Shh.** Dev. Biol. 2016, **413**: 1–7. See also Lebedev O.A., Coates M.I.: **The postcranial skeleton of the Devonian tetrapod Tulerpeton curtum Lebedev.** Zool. J. Linn. Soc. 1995, **114**: 307–348.

tetrapods. Fins evolved from paired appendages, but how paired appendages evolved in the first place is far less clear. Several scenarios have been proposed, including the evolution of paired appendages by the splitting of a lateral fin fold. The evolution of paired appendages made use of signal molecules, such as Sonic hedgehog (Shh) and fibroblast growth factor (FGF), and of transcription factors, such as the Hox proteins, which were already being used to pattern the body.

Other key transcription factors co-opted for the development of paired appendages are those encoded by the Tbx genes *Tbx4* and *Tbx5*, whose ancestral role is likely to have been in heart development (see Section 9.32). *Tbx5* is expressed in the anterior pair of appendages and *Tbx4* in the posterior pair of appendages in both fish and tetrapods. The amphioxus, which lacks paired appendages, has only a single *Tbx4/5* gene, but this can function in the same way as the vertebrate genes in limb formation. A dramatic demonstration is that expression of the amphioxus *Tbx4/5* gene in transgenic mice in which *Tbx5* has been knocked out rescues forelimb development.

Cartilaginous fish such as present-day sharks and rays evolved from a lineage that emerged before that of bony fish. Shark fins have three main proximal bones (basal bones), whereas the fins of lobe-finned fish (such as *Panderichthys*)—from which tetrapods evolved and which are represented today by coelacanths and lungfish—have a single large proximal bone attached to a pectoral/pelvic girdle (see Fig. 13.20). *Shh* is expressed at the posterior margin of catshark fin buds, as it is in mouse and other vertebrate limb buds, although at later stages. In addition, the expression of genes normally restricted to anterior domains in mouse limb buds is extended more posteriorly in shark fin buds and is associated with changes in *Gli3* expression. When Shh signaling is increased experimentally in shark fin buds, the fins that develop lack anterior skeletal elements and have only a single proximal element, illustrating how changes in Shh signaling could have been important in the evolution of antero-posterior pattern in fins and limbs. As the shark fin bud develops, a periodic pattern of *Sox9*-positive spots later forms across the antero-posterior axis of the bud and prefigures a series of distal skeletal elements in the fin. This is reminiscent of the periodic pattern of *Sox9*-positive stripes that develops in the hand plate of mouse embryos, which prefigures the formation of a series of digits. As in the mouse limb, the periodic pattern in the catshark fin can be modeled by reaction–diffusion (see Box 9E), although the details differ. These observations suggest that the basic mechanisms used to pattern the antero-posterior axis of tetrapod limbs may have been present very early in the evolution of paired appendages.

The fossil record suggests that the transition from lobe fins to limbs occurred in the Devonian period, between 400 and 360 million years ago. The transition probably occurred in the fish ancestors of tetrapod vertebrates that lived in shallow waters. The fins of Devonian lobe-finned fishes, such as *Panderichthys*, are probably ancestral to tetrapod limbs, an early example of which is the limb of the Devonian tetrapod *Tulerpeton* (see Fig. 13.20). The proximal skeletal elements in the pectoral fin of *Panderichthys* are thought to correspond to the humerus, radius, and ulna in the upper limb of *Tulerpeton*. Digits have long been considered to be a tetrapod novelty but analysis by computed tomography scanning has revealed that the distal region of the *Panderichthys* pectoral fin contains separate skeletal elements. So fingers may have evolved from skeletal elements present in lobe fins.

13.12 Studies of zebrafish fin development give insights into the origins of tetrapod limbs

To gain insights into the basis of the transition from fin to limb, researchers have turned to the zebrafish, a ray-finned teleost, which although it arose in a separate lineage from tetrapods (see Fig. 13.20) is a model organism in which pectoral fin

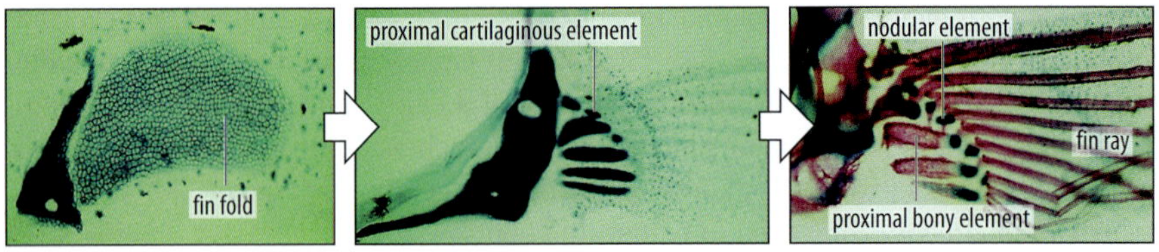

Fig. 13.21 The development of the pectoral fin of the zebrafish *Danio*. Left panel: the pectoral girdle and fin fold. Middle panel: four proximal cartilaginous elements and several more distal nodular cartilaginous elements have developed, and fin rays have formed in the fin fold at the tip (the rays are very faint in this image). Right panel: four proximal bony elements and 6–8 smaller distal nodular elements support the fin ray region in the adult fish.

Photographs courtesy of D. Duboule, from Sordino P., et al.: **Hox gene expression in teleost fins and the origin of vertebrate digits.** *Nature 1995, **375**: 678-681.*

development can be followed in detail and techniques for genetic manipulation are well established. The fin buds of the zebrafish embryo are initially morphologically like tetrapod limb buds, but important differences soon arise during development. The proximal part of the fin bud gives rise to four main skeletal elements, the proximal radials, by the subdivision of a cartilaginous sheet, with 6–8 smaller nodular skeletal elements called the distal radials developing more distally (Fig. 13.21). All these skeletal elements undergo **endochondral ossification** (see Section 11.9). The most striking difference between zebrafish fin and tetrapod limb development is in the distal region of the buds. In the zebrafish fin bud, an ectodermal fin fold forms at the distal tip of the bud, and, within this fold, fin rays develop made up of dermal bone (bone formed directly by connective tissue cells rather than by endochondral ossification). The extensive flattened region supported by fin rays that subsequently develops forms the major part of the fin. By contrast, the distal tip of the tetrapod limb bud is rimmed by the apical ectodermal ridge and digits made of endochondral bone develop from underlying bud mesenchyme cells.

Evolutionary relationships between the structures in the zebrafish fin and tetrapod limb can be inferred from Figure 13.20. The proximal skeletal elements of the zebrafish fin are thought to have evolved from the anterior basal elements of an ancestral fish fin whereas the proximal skeletal elements of tetrapod limbs (e.g. humerus/femur) have evolved from the posterior basal element of the ancestral fish fin. More distal structures in tetrapod limbs such as the digits are proposed to have evolved in the lobe-finned fish lineage and therefore would not correspond to any structures in the zebrafish fin.

Hox genes control regional identity in developing mouse limbs and changes in Hox gene expression and function may have been involved in the fin/limb transition. In the mouse limb, *Hoxd13* together with *Hoxa13* control distal regional identity and the formation of digits (see Section 9.19). They are expressed throughout the distal region of the limb bud in a second phase of Hox gene expression as the digital plate forms. Zebrafish fin buds also express genes in the 5′ regions of the Hoxd and Hoxa gene clusters. Initially the second phase of Hoxd gene expression was not detected in zebrafish fin buds. But more comprehensive studies subsequently showed that, as in mouse limb buds, there is a later phase of Hox gene expression, sometimes called the third phase in zebrafish, in the distal region of the buds. Furthermore, the same regulatory mechanisms that control Hox gene expression in the mouse limb are present in zebrafish. In zebrafish, as in mice, *cis*-regulatory regions on the 5′ side of the Hoxa and Hoxd clusters interact with genes towards the 5′ end of the clusters to drive the late phase of Hoxa and Hoxd gene expression in the fin bud; *cis*-regulatory regions on the 3′ side of the Hox gene cluster interact with genes towards the 3′ end of the cluster and drive an earlier phase of Hox gene expression. The same genomic organization has been found in the recently sequenced genome of the spotted gar (*Lepisosteus oculatus*). The spotted gar is a ray-finned fish that evolved in a separate lineage of bony fish from the teleosts before the whole-genome duplication in that lineage, and its genome provides a bridge between those of teleosts and tetrapods.

To explore whether individual enhancers within 5′ cis-regulatory regions are conserved between ray-finned fish and tetrapods, enhancers from zebrafish and gar were tested in reporter assays in transgenic mice (Box 13C). Whereas zebrafish 5′ enhancers only drive expression in the proximal region of the mouse limb bud, an enhancer in the 5′ control region of the Hoxa gene cluster from the spotted gar can drive reporter gene expression in the digit-forming region of mouse limb buds as effectively as the related mouse enhancer. The ability of the spotted gar enhancer to function in the mouse limb suggests that the regulation of expression of *Hoxa13* and *Hoxd13* genes

EXPERIMENTAL BOX 13C Transgenic tests of the function of Hox digit enhancers

Enhancers located 5′ to the Hoxa and Hoxd clusters that drive the late phase of expression of *Hoxa13* and *Hoxd13* genes in the distal region of the limb buds which will form the digits were first discovered in mice. Some of these enhancer sequences have also been found in the genomes of zebrafish and the spotted gar. The enhancers known as Island1 and CsB are located 5′ to the Hoxd cluster, whereas the enhancer known as e16 is located 5′ to the Hoxa cluster. The function of these enhancer sequences can be tested in transgenic mice and zebrafish by linking them to a reporter gene and then examining the expression of the reporter in limb buds and fin buds, respectively. Figure 1 shows the results of such transgenic tests of enhancer function. These tests were carried out not only in the same animal (same-species transgenic, Fig. 1, left panels), but also across species (between-species transgenic, Fig. 1, right panels). The expression pattern of the reporter driven by mouse enhancers is shown in blue and the expression pattern driven by the fish sequences in red. The orientation of the buds is indicated by P (proximal) and D (distal). The fish enhancer sequences (from both zebrafish and gar) drive reporter expression at the distal tip of the zebrafish fin buds. But while the zebrafish enhancers only drive reporter expression in the proximal regions of mouse limb buds, if at all, the gar enhancers drive reporter expression in the distal region of mouse limb buds just as effectively as the mouse enhancers. The e16 enhancer located 5′ to the Hoxa cluster, for example, from either the mouse or the gar drives reporter gene expression throughout the entire digital plate of the mouse limb bud.

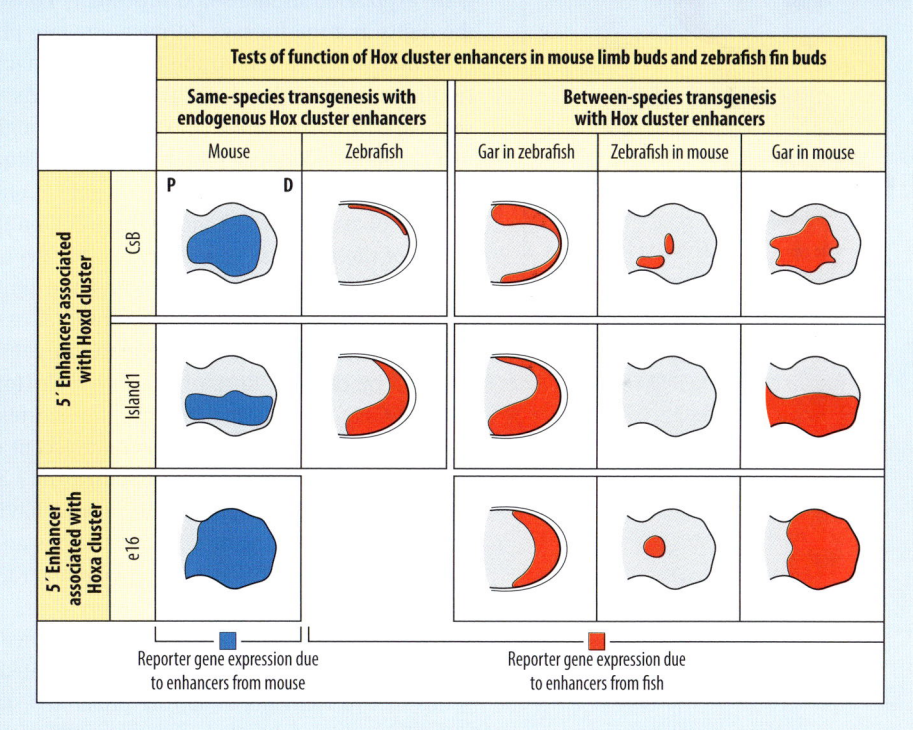

Fig. 1

Adapted from Gehrke, A.R., Shubin, N.H.: **Cis-regulatory programs in the development and evolution of vertebrate paired appendages.** Semin. Cell Dev. Biol. 2016, **57**: 31–39.

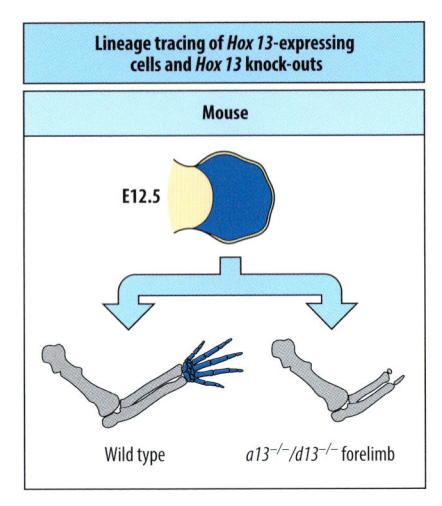

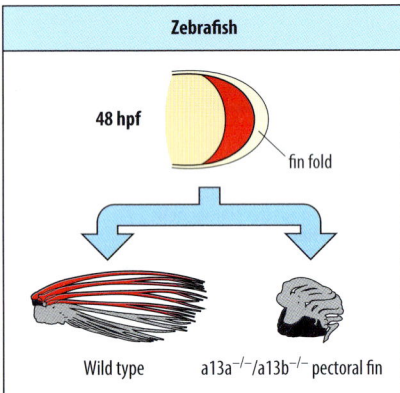

Fig. 13.22 Lineage tracing of *Hoxa13*-expressing cells and effects of knocking out Hox13 paralogs in mouse limb buds and zebrafish fin buds. Cells that were expressing *Hoxa13* in the digital plate of E12.5 mouse forelimb buds end up in the wrist and digits, and the forelimbs of mice in which both *Hoxa13* and *Hoxd13* are inactivated lack digits (upper panel). Cells that were expressing *Hoxa13* in the distal tip of zebrafish fin buds 48 hpf (hours post-fertilization) end up in the fin rays, and in zebrafish in which both *Hoxa13a* and *Hoxa13b* are deleted the fin rays are extremely reduced. Endochondral bones in the mouse limb and the zebrafish fin are shown in gray.

Adapted from Nakamura, T., et al.: **Digits and fin rays share common developmental histories.** *Nature 2016, **537**: 225–228.*

is highly conserved even down to the level of individual enhancers in ray-finned fish and tetrapods. Only the 3′ Hox control regions are present in the amphioxus, suggesting that the control of Hox clusters by enhancers on both sides was acquired during the evolution of the vertebrates.

Recent experiments suggest that not only are the mechanisms that control Hox gene expression in limbs and fins highly conserved, but also the cellular origins of the digits and the fin rays. In mice, the late phase of *Hoxa13* expression marks the cells that will give rise to the digits, and inactivation of both *Hoxa13* and *Hoxd13* results in loss of the digits (Fig. 13.22). As shown in Box 13C, a gar enhancer sequence can drive reporter expression throughout the distal regions of both zebrafish fin buds and mouse limb buds. Genetic lineage analysis using this gar enhancer linked to a reporter in transgenic zebrafish showed that cells that were expressing the late phase of *Hoxa13* in the distal region of the fin bud give rise to the dermal bones of the fin rays. This indicates that even though the distal skeleton develops in a completely different way in fins and limbs, both fin rays (formed by dermal bone) and digits (formed by endochondral ossification) arise from a population of *Hoxa13*-expressing cells. Furthermore, in zebrafish lacking *Hoxa13a* and *Hoxa13b* (created by crossing lines of fish in which Hox13 paralogs had been deleted using CRISPR-Cas9), the dermal fin-rays are shortened (see Fig. 13.22). This disruption of the development of the rays is reminiscent of the disruption of the development of the digits when both *Hoxa13* and *Hoxd13* are knocked out in mouse limbs (see Section 9.19). This points to Hox genes having similar roles in specifying distal positional values in the development of fins and limbs, but with differences in interpretation, due to differences in downstream targets, leading to formation of either fin rays or digits.

Two zebrafish mutants have been identified in chemical mutagenesis screens in which the proximal fin radials are segmented, producing an endoskeleton with a series of two elements along the proximo-distal axis. The pattern is reminiscent of that of the proximal region of the tetrapod limb in which the humerus/femur articulates with radius/tibia and ulna/fibula. Furthermore, the formation of these additional elements requires the input of Hox genes. In mouse limbs, *Hoxa11* and *Hoxd11* are expressed in the intermediate region of the limb bud and control formation of radius and ulna in the forelimb (tibia and fibula in the hind limb). In the zebrafish mutants, *Hoxa11* is expressed at higher levels compared with normal fin buds and deletion of Hox11 paralogs restores normal fin morphology. These observations are consistent with there being a latent ability in zebrafish to interpret the expression of Hox11 paralogs to produce intermediate limb-like structures in the fins. Thus, the system in which Hox genes specify position along the proximo-distal axis of paired appendages was probably present in the ancestors of bony fish.

13.13 Limbs have evolved to fulfill different specialized functions

The great range of anatomical specializations in the limbs of mammals that adapt them for different functions (Fig. 13.23) is due to changes both in limb patterning and in the differential growth of parts of the limbs during embryonic development, but the basic underlying pattern of skeletal elements is the same. This is an excellent example of the modularity of the skeletal elements. If one compares the forelimb of a bat and a horse, one can see that, although both retain the basic pattern of limb bones, the limb has been modified to provide a specialized function in each animal. In the bat, the limb is adapted for flying and the digits are greatly lengthened to support a membranous wing. In the horse, the limb is adapted for running and the lateral digits are reduced, the central metacarpal (a hand bone in humans) is lengthened, and the radius and ulna are fused for greater strength. The whale flipper is modified for balance and steering, with the elongated flippers having digits with many phalanges.

A striking feature of vertebrate limb evolution is that, while reduction in digit number is common (there are only three in the chick wing and reduction is common in lizards), there are no modern species with more than five digits. The giant panda's 'thumb', which appears to be a sixth digit, is, in fact, a modified wrist bone. Species with more than five digits appear in the fossil record of the earliest tetrapods, such as *Acanthostega* (see Fig. 13.20), but it is not clear whether there were more than five morphologically distinct types of digits. In present-day limbs with polydactyly as a result of mutations, at least two of the digits are the same (see Box 9B). It was originally suggested that the five-digit pattern might be due to there being five Hoxd genes expressed in the early limb bud, providing only five discrete genetic programs for giving a digit an identity. While this is an attractive idea, it is likely to be an over-simplification.

Reduction in the number of digits has occurred repeatedly in tetrapod evolution. In limbs of different species of the Australian lizard *Hemiergis* that have different numbers of digits, a reduced number of digits correlates with a decrease in the extent and duration of Shh signaling by the polarizing region. Digit reduction has also occurred in hoofed mammals, the ungulates, which are classified according to whether they have an even or an odd number of toes. In cows and pigs, which are even-toed, Shh signaling is also altered, but in camels, which are closely related to pigs, digit loss seems to be due to extensive cell death at later stages in limb development, as it is also in horses, odd-toed ungulates.

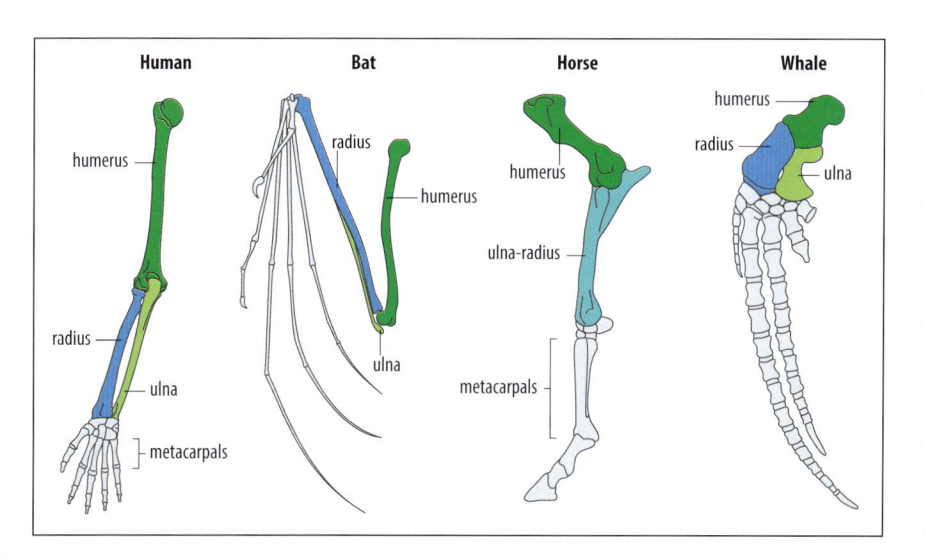

Fig. 13.23 Diversification of vertebrate limbs. The basic pattern of bones in the forelimb is conserved throughout the vertebrates, but there are changes in the proportions of the different bones, as well as fusion and loss of bones. In the horse limb, the radius and ulna have become fused into a single bone, and the central metacarpal (a hand bone in humans) is greatly elongated. In addition, there has been loss and reduction of the digits in the horse. In the bat wing, by contrast, the digits have become greatly elongated to support the membranous wing. In the whale flipper, there has been digit reduction, and elongation of two of the digits is accompanied by an increase in the number of phalanges.

In the early limb buds of cattle—cows have only two toes (digits III and IV)—there is reduced expression of the bovine version of the *Patched-1* (*Ptch1*) gene compared with mouse limb buds. *Ptch1* encodes the receptor for Shh and a change in a *cis*-regulatory region of *Ptch1* has been identified and suggested to underlie the decrease in its expression and the evolution of digit reduction. Binding of Shh to Patched-1 not only activates Shh signaling, but also affects how far Shh can spread across the limb bud. In the absence of high levels of Patched-1 in bovine limb buds, Shh spreads further across the limb bud, and this would be expected to affect the number and type of digits that develop. Alterations in *Ptch1* expression have also been identified in the developing limb buds of pigs—pigs have four toes in a symmetric pattern, two central weight-bearing toes (digits III and IV) and a lateral toe on either side (digits II and V). Comparative open chromatin profiling in developing pig and mouse limbs has identified widespread regulatory changes associated with genes involved in limb development, including those functioning in Shh and other signaling pathways.

Reduction in the number of phalanges in the digits is also common. Whales and dolphins are exceptional in having elongated digits with additional phalanges and it has been suggested that this is due to persistence of the apical ectodermal ridge, which prolongs outgrowth of the bud. By contrast, in bats in which the elongated digits III, IV, and V support the membranous wing, the number of phalanges has not increased but each metacarpal and phalanx is longer (see Fig. 13.23). This is due to rapid fetal and postnatal elongation of the bones. Comparative studies of the growth of the third metacarpal in mice and bats has shown that elongation of this metacarpal in the bat is due to an increase in the number and activity of the cells in the **growth plate** (see Section 11.9).

13.14 The evolution of limblessness in snakes is associated with changes in axial gene expression and mutations in a limb-specific enhancer

The loss of limbs altogether has occurred independently during evolution in different tetrapods, including cetacean mammals, birds, and amphibians, but is characteristic of snakes. Snakes evolved from a four-legged lizard-like ancestor. A fossil snake from the Cretaceous period (around 145 million years ago) lacked forelimbs but had relatively normal hindlimbs, suggesting that limb reduction occurred progressively during evolution, with forelimbs being lost before hindlimbs. In many present-day snakes, including cobras and rattlesnakes, there are no signs of limb-bud formation and both forelimbs and hindlimbs are completely absent. In others, such as pythons and boas, there are small buds that develop into rudimentary hindlimbs consisting of a pelvic girdle and a femur capped by a claw. The fact that forelimb buds have never been observed in embryos of present-day snakes suggests that the loss of forelimbs and hindlimbs occurred by different mechanisms during evolution.

How might the loss of forelimbs have evolved? As we saw in Chapter 9, a combinatorial Hox code in the lateral plate mesoderm in mouse and chick embryos has been suggested to determine the position in which forelimbs develop. More posteriorly expressed Hox genes (*Hoxc8*, *Hoxc9*, and *Hoxc10*) repress activity of a control region of the T-box gene *Tbx5*, and so that gene's expression is restricted to the prospective forelimb region, where it is essential for forelimb development. In python embryos, only the distribution of Hoxc8 protein has been investigated so far; it is expressed along the entire length of the trunk (see Fig. 13.18), presumably restricting *Tbx5* expression completely. Thus, repression of *Tbx5* by Hoxc8 (and possibly by other posterior Hox proteins) could provide a mechanism whereby changes in Hox gene expression might have led to loss of forelimbs in snakes.

By contrast, the progressive loss of hindlimbs in snake evolution is associated with changes in the control region that regulates expression of the *Shh* gene in the polarizing region of limb buds. *Shh* is expressed only transiently in the hindlimb buds of python embryos, and this leads to subsequent degeneration of the apical ectodermal ridge and hence to the development of the rudimentary limb. Studies in mice have identified a control region that regulates *Shh* expression specifically in the limb bud, called the zone of polarizing activity regulatory sequence (ZRS; see Box 9B). Complete deletion of the ZRS in mouse embryos leads to loss of *Shh* expression in the polarizing region and results in severe limb truncations, which the rudimentary python hindlimbs resemble. Analysis of the genomes of different snakes has identified mutations in the ZRS, and the sequence changes are more extensive in completely limbless snakes such as cobras than in snakes with rudimentary hindlimbs such as pythons. Tests in transgenic mice have shown that the snake ZRS is non-functional—it does not drive *Shh* expression in the mouse limb bud and the mouse limbs that subsequently develop are severely truncated, like the limbs of mouse embryos in which the ZRS is deleted. Furthermore, the mutations in the snake ZRS affect sites in the DNA that, in mouse limb buds, bind transcription factors involved in regulating *Shh* expression. These findings suggest that progressive loss of the ZRS underlies the evolution of hindlimb loss in snakes (Box 13D). It is unlikely that degeneration of the ZRS is a consequence of hindlimb loss because other genes involved in limb development are still expressed in early python hindlimbs and their expression is only reduced after the end of the transitory pulse of *Shh* expression.

Why have hindlimbs not been lost as readily as forelimbs during evolution? One possible reason is that the development of the hindlimb is closely linked with the development of the external genitalia. Not only do these different structures arise from adjacent cell populations in chick, mouse, and lizard embryos, but their development also involves many of the same genes, whose expression is often regulated by shared control regions. This sharing of control regions means that any mutations in these regions could affect development of both hindlimbs and external genitalia and will therefore be less likely to be selected for.

13.15 Butterfly wing markings have evolved by redeployment of genes previously used for other functions

Butterfly wing markings are a striking example of a specialized characteristic that has adaptive functions. The variety and pattern of color markings on the wings is a beautiful and distinctive feature of the 20,000 or so known species of butterflies. Many of these patterns are variations on a basic 'ground plan' of color bands and concentric eyespots (Fig. 13.24). The basis for these patterns is the arrangement of overlapping cuticular scales that cover the wing and that are colored by pigment synthesized and deposited by the underlying epidermal cells. Eyespots are an evolutionary novelty within the Lepidoptera, the order of insects to which butterflies belong, and are thought to have evolved in response to sexual selection, and also as a means of avoiding predators. By startling or deceiving the predator, the butterfly can escape. Phylogenetic studies of *Bicyclus* species suggest that eyespots on the dorsal forewing surface are involved in sexual signaling, whereas eyespots on the ventral wing surface, which is exposed when the butterfly is at rest, are those mainly involved in predator avoidance.

In the cases where the mechanism of eyespot development is best understood, eyespot specification and patterning is an example of the iterative use of a 'tool kit' of genes selected from those used for previous patterning events in the larva. Comparison between different species shows that the spatial control of the output of these genes to produce different patterns is a fertile substrate for selection.

Fig. 13.24 Butterfly wing pattern. The African butterfly *Bicyclus anynana* (female shown here), showing ventral wing color pattern and prominent eyespots.

Photograph courtesy of William Piel and Antónia Monteiro.

EXPERIMENTAL BOX 13D Using CRISPR-Cas9 genome-editing techniques to test the functioning of the snake ZRS

The zone of polarizing activity regulatory sequence (ZRS) is an upstream control region, first discovered in mice, that governs the transcription of the *Shh* gene specifically in the limb buds. To investigate the functioning of the snake ZRS, CRISPR-Cas9 was used to replace the mouse ZRS with that from a python (pZRS) or a cobra (cZRS) by homology-driven repair (see Box 1C). For comparison, the mouse ZRS was also replaced by either a human (hZRS) or a coelacanth ZRS (fZRS), and all the results were compared to the effect of simply deleting the mouse ZRS using CRISPR-Cas9 (ΔZRS) (Fig. 1). The ZRS from both humans and coelacanths could drive *Shh* expression at the posterior margin of mouse limb buds, even though the mammalian and coelacanth lineages are estimated to have diverged more than 400 million years ago. By contrast, the ZRS from the snakes was unable to drive *Shh* expression in mouse limb buds. Hindlimb skeletons were normal in mouse embryos in which the ZRS had been replaced by a human or coelacanth ZRS, but were greatly reduced if replaced by a cobra or python ZRS and resembled the hindlimbs of mouse embryos in which the ZRS had been completely deleted (see Fig. 1; scale bars = 1 mm for Shh expression, and 2 mm for hindlimb skeleton).

To narrow down the region of the snake ZRS that renders it inactive, ZRS sequences from eight different snake species were compared. A 17-kb sequence was identified which is largely conserved in the genomes of a range of different vertebrates, including human, chick, coelacanth, and elephant shark, but which is absent in all the snake genomes. Insertion of this short sequence into the python ZRS was sufficient to rescue hindlimb development when tested in transgenic mice. This region of the ZRS contains binding sites for transcription factors that are known to regulate the activity of the ZRS in mice, and this explains why, when this region is deleted, the ZRS is unable to drive *Shh* expression in the limb bud.

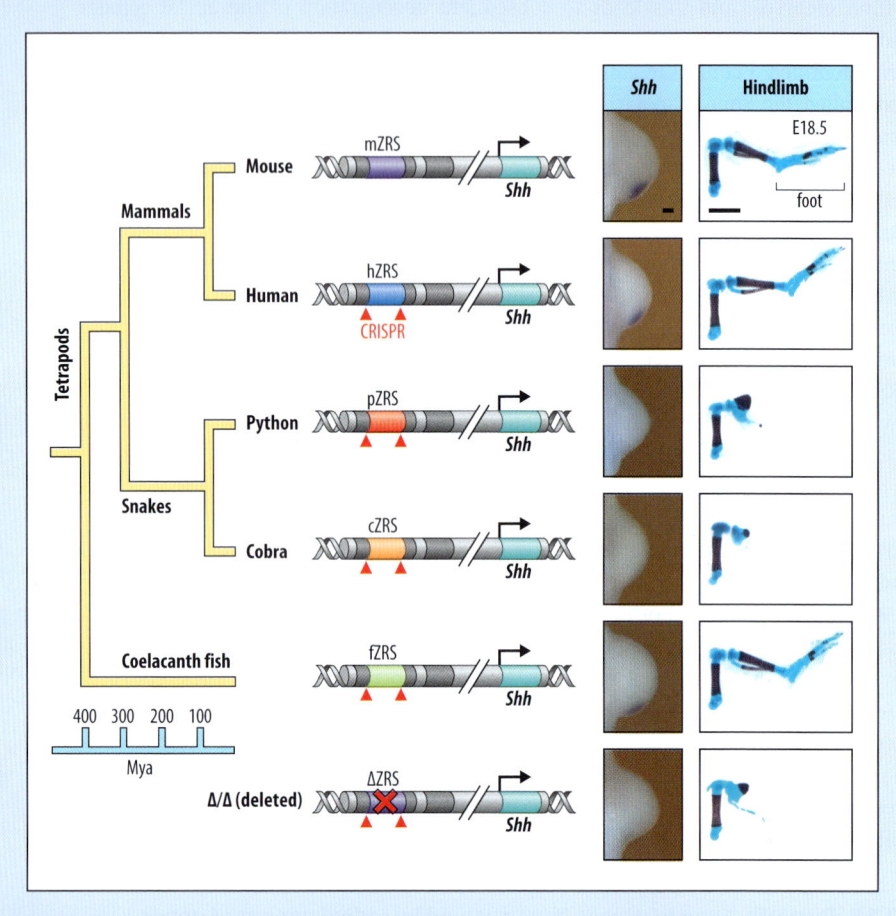

Fig. 1
From Kvon, E.Z.: *Progressive loss of function in a limb enhancer during snake evolution.* Cell 2016, **167**: 633–642.

Butterfly wings develop from imaginal discs in the caterpillar in a similar way to *Drosophila* wings, and the global patterning of the discs is similar to that of *Drosophila*. A number of the genes that control wing development in *Drosophila*, such as *apterous* and *wingless*, are also expressed in the butterfly in a spatial and temporal pattern similar to that in the fly. Thus, as in *Drosophila*, the shape and structure of the butterfly wing is patterned by a field of positional information. The patterning of the pigmentation, however, involves the establishment of additional smaller fields of positional information and a redeployment of genes for specific purposes. The eyespots are the clearest manifestation of these fields. They appear late in development, during the development of the pupa, and spread radially from an initial central focus. Surgical manipulation has shown that this focus acts as an organizing center for the spot: if it is destroyed, the eyespot does not develop, and if it is transplanted to a location that lacks an eyespot, a new eyespot will emerge.

Analysis of gene expression during the emergence of eyespots reveals a hierarchy of events in the evolution of the patterns. In the Satyrinae subfamily of butterflies, which includes *Bicyclus anynana*, the specification of the focus is typically associated with the expression of the genes for the transcription factors Distal-less (Dll), Antennapedia (Antp), and Spalt, and for the signaling receptor Notch (Fig. 13.25), all of which are involved in earlier patterning events in the butterfly wing; in the case of *Antp*, a Hox gene that specified segment identity in the early embryo is now being deployed as a

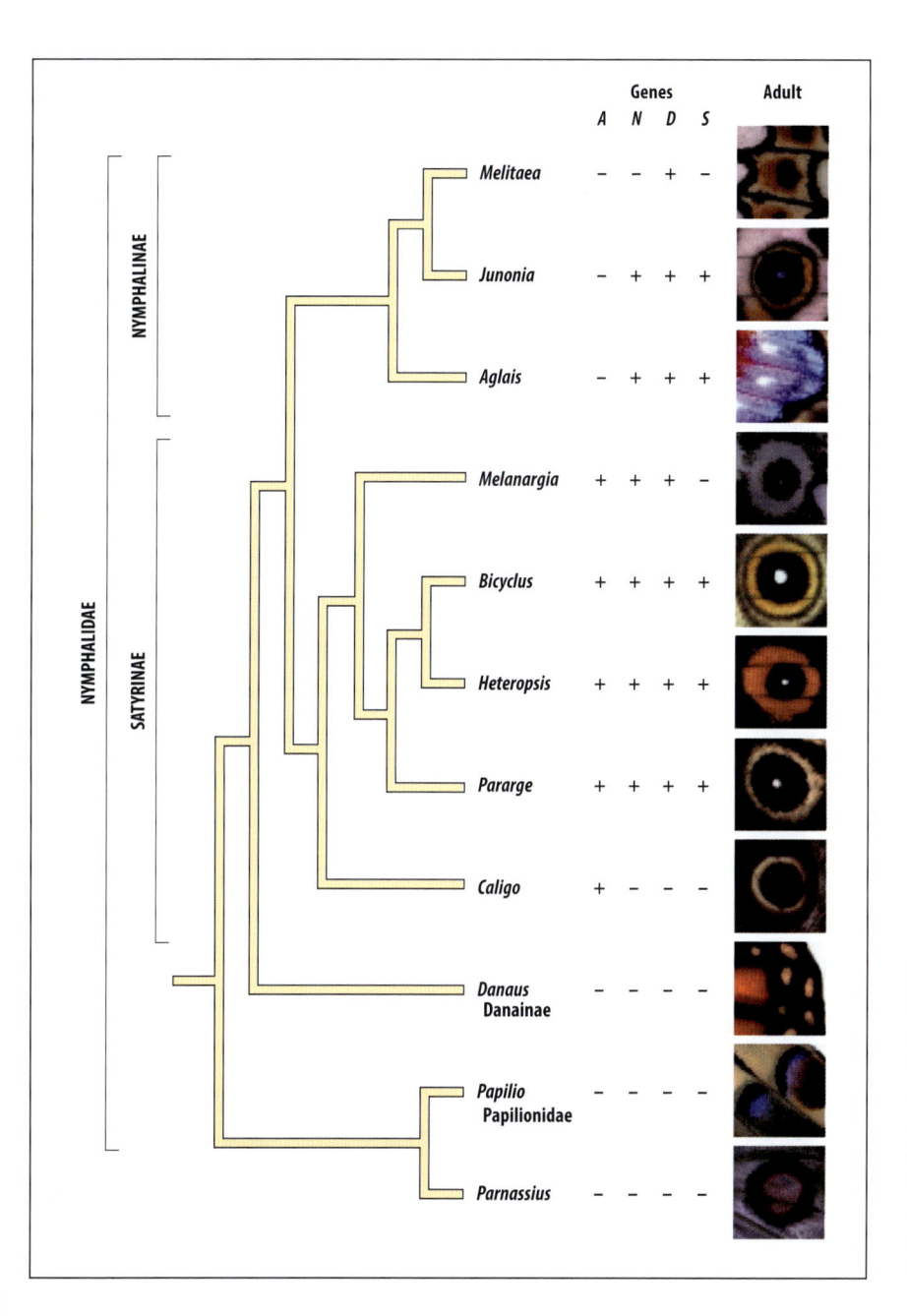

Fig. 13.25 Phylogenetic tree of butterfly eyespot patterns and usage of the Antennapedia (A), Notch (N), Distal-less (D), and Spalt (S) module.

*From Saenko, S.V.: **Chapter 6. Butterfly wing pattern evolution: insights from comparative analysis of the eyespot developmental network**. 2010. [https://scholarlypublications.universiteitleiden.nl/handle/1887/16039] (date accessed 4 December 2022).*

patterning gene on a very local level. *Notch*, *Dll*, and *Spalt* are also widely deployed in eyespot development in other groups of butterflies, such as the Nymphalinae. Use of RNA interference and, more recently, CRISPR-Cas9 genome editing to reduce gene expression has revealed the function of these genes in the patterning of the eyespot. In the painted lady butterfly *Vanessa cardui*, a member of the Nymphalinae, *Spalt* promotes and *Dll* represses eyespot formation, probably by contributing to the positioning of the focus, as there is a strict correlation between their pattern of expression and the location of the eyespot in different species. What regulates these locations is not known, although it is likely to be related to the global positional information system of the wing.

Once the focus of the eyespot is specified, it acts as the source of an as-yet-unknown diffusible factor that will pattern radially, as can be seen in the expression of genes such as those for the transcription factors Engrailed (En) and Spalt, the signaling protein Hedgehog (Hh), Patched (Ptc, the receptor for Hedgehog), and Ci (a transcriptional effector used in the Hh pathway) in the focus and in concentric patterns around it (Fig. 13.26). This pattern of gene expression most probably regulates the expression of the genes that control pigment production. Comparison of gene expression in the developing eyespots of *B. anynana* and *Junonia coenia*, a member of the Nymphalinae, shows that different lineages of butterflies may have evolved to use different selections from what is thought to be the original tool kit for eyespot production. In both *B. anynana* and *J. coenia*, Hedgehog and its signaling pathway components play their usual part in global wing growth, but in *J. coenia*, Hedgehog is also expressed in developing eyespots, together with En, Ci, and Ptc in the eyespot centers (see Fig. 13.26). In *B. anynana* in contrast, only En and Ci are expressed in the eyespots at a comparable larval stage. Knockdown of Hh function by an anti-Hh antibody in the larvae resulted in smaller wings and smaller eyespots in both species, as expected from the role of Hh in global wing growth. But whereas the eyespot is

Fig. 13.26 Differences in usage of Hedgehog signaling pathway components in eyespot development in different butterfly species. The eyespot patterns on the hindwing surface of the butterflies *Junonia coenia* and *Bicyclus anynana*, which belong to different subfamilies, are shown on the left. The larval hindwing imaginal discs are on the right. mRNA and protein expression data show that, in both species, *hedgehog* (*hh*) and *engrailed* (*en*) are expressed in the posterior compartment of the hindwing disc in the same conserved pattern as in the non-lepidopteran *Drosophila* (see Fig. 9.6). *Patched* (*ptc*), which encodes a receptor for Hh, is expressed along the compartment boundary. In *J. coenia*, *hh* is also expressed in small domains flanking the developing eyespot centers, and En, Ci (a transcription factor that is a component of the Hh signaling pathway, see Box 2E), and *ptc* are expressed in the eyespot centers themselves. In *B. anynana*, however, *hh* is not involved in eyespot development.

Adapted from Tong, X., et al.: **Differential involvement of Hedgehog signaling in butterfly wing and eyespot development.** PLoS One 2012, **7**: e51087. © 2012 Tong et al.

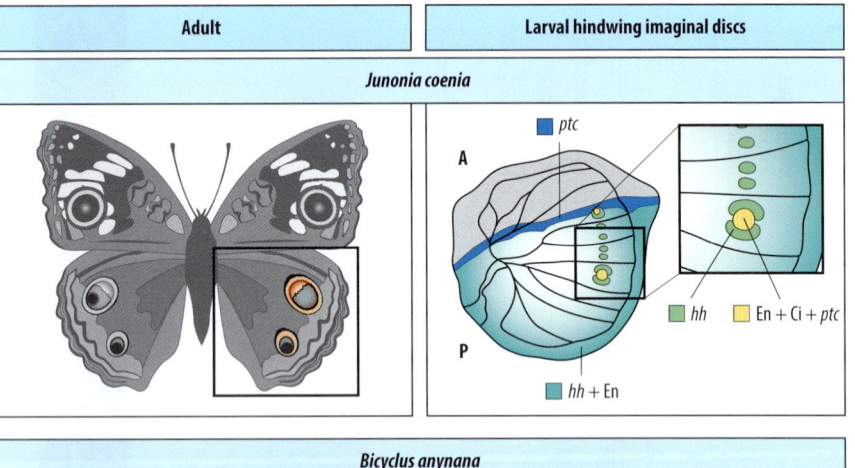

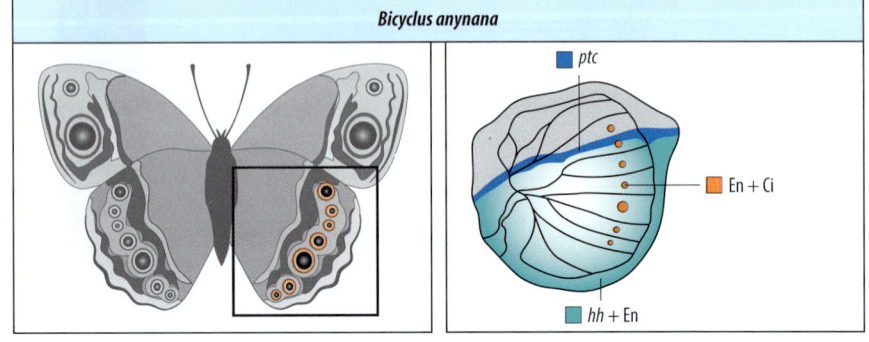

proportionately reduced in relation to wing size in *B. anynana*, it is disproportionately smaller in *J. coenia*, indicating that Hh does have an extra role in eyespot development in this species.

Two important principles emerge from these observations. The first is that proteins involved in different patterning events previously (e.g. signaling receptors such as Notch, and patterning transcription factors such as Spalt and Dll) can become redeployed to form a gene regulatory network that creates additional patterns—the eyespots. The second principle is that once assembled, a tool kit of genes can be deployed during evolution, most likely in a selection-dependent manner, for the creation of multiple patterns that are variations on the basic theme. Variations in the pattern can be created by adding or removing elements of the network, as in the Nymphalinae, whereas in other cases, such as the Satyrinae, variations in the pattern are created mainly by controlling the spatial deployment of the genes involved. The mechanism underlying this spatial variation is unknown, but is likely to involve yet another combination of patterning elements that act in other developmental events.

13.16 Adaptive evolution within the same species provides a way of studying the developmental basis for evolutionary change

The three-spine stickleback fish *Gasterosteus aculeatus* exists in very different morphological forms within the same species as a result of **adaptive evolution**. It is, therefore, a good model in which to study the developmental mechanisms involved in evolutionary change. Stickleback populations that became isolated when they colonized new lakes and streams formed as the glaciers melted after the ice ages about 20,000 years ago have undergone reduction or loss of the pelvic spine (a modified pelvic fin) together with other morphological changes—loss of body armor and reduction in the number of dorsal spines (Fig. 13.27).

Paired pelvic spines are present in all marine populations of three-spine sticklebacks, but the pelvic spine has been lost repeatedly in several freshwater populations, as a result of failure of pelvic fin-bud development. By making genetic crosses between marine and freshwater sticklebacks, the cause of pelvic reduction was traced to mutations in a control region of the gene for the homeodomain transcription factor Pitx1, which regulates tissue-specific expression of *Pitx1* in the pelvic spine (Box 13E).

Pelvic-spine loss in sticklebacks is an interesting example of **convergent evolution**, because different populations have different mutations in the *Pitx1* control region, which have been selected independently in each population. Pitx1 has a key role

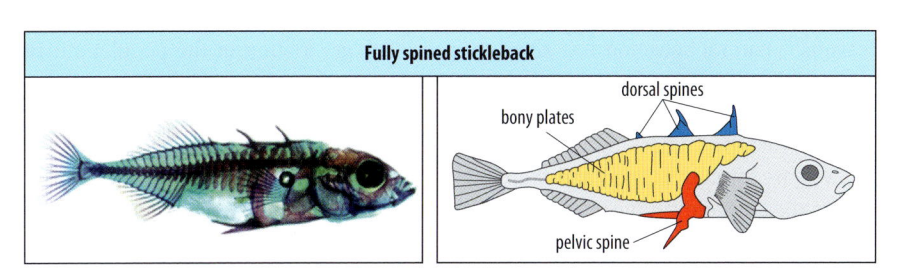

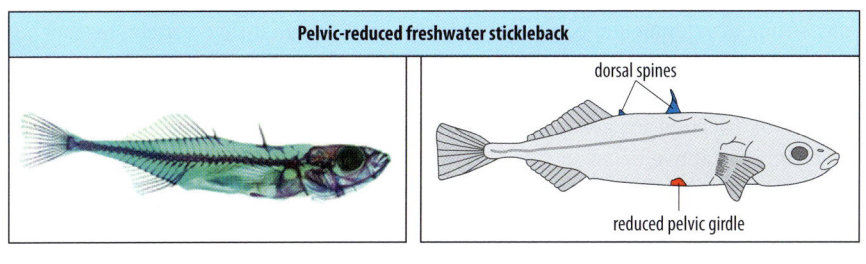

Fig. 13.27 Loss of the pelvic spines in three-spine sticklebacks (*Gasterosteus aculeatus*). Upper row: marine populations and many freshwater populations of three-spine sticklebacks have a pair of pelvic spines (red in the drawing). Fish in marine populations also have bony plates along their sides (yellow in the drawing) and three dorsal spines (blue in the drawing). Lower row: in fish in some freshwater populations, the pelvic spines have been lost and the pelvic girdle very reduced; in addition, the bony plates along their sides are absent or very reduced and the number of dorsal spines is reduced.

*From Tickle, C., Cole, N.J.: **Morphological diversity: taking the spine out of three-spine stickleback.** Curr. Biol. 2004, **14**: R422–R424.*

in vertebrate hindlimb development. When the *Pitx1* gene is knocked out in mice, hindlimb development is reduced, with the right hindlimb being more affected than the left. This asymmetry in limb reduction is likely to be due to the protein encoded by the closely related gene, *Pitx2*, which is expressed on the left-hand side of the body, partially compensating for lack of *Pitx1*. Interestingly, pelvic-spine reduction in sticklebacks is also asymmetric, being more marked on the right than on the left.

The main genes associated with the loss of body armor and dorsal spines in freshwater sticklebacks have also been identified. Loss of armor is due to mutations in a *cis*-regulatory region that controls tissue-specific expression of the gene *EDA* in the flank region of the fish. *EDA* encodes the protein ectodysplasin A, which is known to be essential for normal development of ectodermal organs in mammals. Mutations in the gene are associated with hypohidrotic ectodermal dysplasia in humans, a condition in which hair follicles, teeth, and other ectodermal structures fail to develop properly. Reduction in development of the dorsal spines of the stickleback is caused by a mutation that leads to a splicing change in the *MSX2A* gene transcript that results in quantitative differences in the number of full-length and truncated transcripts produced in marine and freshwater fish. *MSX2A* encodes a transcription factor involved in the normal development of ectodermal organs and organ regeneration in zebrafish. *EDA* and *MSX2A* are located on the same chromosome and this explains why loss of body armor and dorsal spines tend to go together.

But why have these bony structures been lost in freshwater sticklebacks? Two environmental factors that may allow spine-reduced sticklebacks to be more successful in freshwater environments are the limited availability of calcium required for bone formation and the reduced need for protection against predation.

Another example of adaptive evolution within the same species is found in the teleost fish *Astyanax mexicanus*. This exists in two radically different forms: a sighted, pigmented, surface-dwelling form and a blind, unpigmented, cave-dwelling form known as a cavefish (Fig. 13.28). At least 30 different populations of *Astyanax* cavefish are present in limestone caverns in Mexico, generally thought to have been isolated from their surface relations for the past few million years. Although functional eyes are lacking in these cavefish adults, the embryos begin to develop small eye primordia consisting of an optic cup and lens, which subsequently degenerate. The major cause of eye degeneration is apoptosis of the developing lens cells, which then prevents the growth of other optic tissues, including the cornea. The non-development of the eye is, in turn, the cause of other craniofacial differences between cavefish and surface fish, although some other sensory changes may be independent of this. Lens apoptosis is induced by increased activity of the Shh signaling system along the cavefish embryonic midline compared with the embryos of surface-dwelling *Astyanax*. However, the genetic basis for this increased activity is unknown. Recent work suggests that, in cavefish, natural selection has acted on pre-existing variation in the population. In surface-dwelling forms of cavefish, variations in eye size seem to be buffered by the heat-shock protein HSP60. The proposal is that the environment in caves disturbs this system, thereby unmasking variations in eye size on which natural selection can act.

Eyes are initially formed and then degraded during larval or adult development in all sightless cave-dwelling vertebrates. It is surprising that eyes begin to develop at all. This may be because early steps in eye development are required for other essential steps in development, and the elimination of these steps would be fatal. Because of this strong developmental constraint, it is thus unlikely that cave vertebrates lacking embryonic eyes will be discovered. But why have eyes been lost in cavefish? The answer is not known. A recent suggestion is that loss of eyes is an indirect consequence of selection for genes required for the development of the sensory receptors involved in adaptive behavior that makes foraging in the dark easier. Evolution of eye loss may also have occurred to conserve energy, because maintaining the retina and the visual parts of the brain is very energy expensive.

Fig. 13.28 The two different forms of *Astyanax mexicanus*. Surface-living fish have eyes and are pigmented (top photo). In cave-dwelling fish (cavefish; bottom photo) the eyes do not develop and pigmentation has been lost.

Photographs courtesy of A. Strickle, Y. Yamamoto, and W. Jeffery.

EXPERIMENTAL BOX 13E Pelvic reduction in sticklebacks is based on mutations in a gene control region

Marine three-spine sticklebacks have a substantial pair of serrated pelvic spines on the ventral side—their scientific name, *Gasterosteus aculeatus*, means 'bony stomach with spines'. But populations of three-spine stickleback which spread into freshwater environments in different parts of the world, including Alaska, Canada, and Scotland, after the last Ice Age, have evolved into a form in which the pelvic structures—the pelvic girdle and associated spines—are reduced or completely absent. Genetic crosses between fish from these different natural populations identified a region in the genome containing the gene *Pitx1* that is linked with the variance in pelvic spine size, but no mutations in the *Pitx1* protein-coding region were found. However, no *Pitx1* expression could be detected in the developing pelvic region of pelvic-reduced fish, and this suggested that mutations in a control region of the *Pitx1* gene that activates its expression in the pelvic region could explain the missing spines. This control region was suspected to lie in an intergenic region upstream of the *Pitx1* gene that showed significant differences in pelvic-reduced fish but which is highly conserved in zebrafish and other teleosts.

To test the function of this region, reporter transgenes composed of the coding sequence for green fluorescent protein (GFP) linked to a 2.5-kilobase DNA sequence from the intergenic region upstream of the *Pitx1* coding sequence from both marine and pelvic-reduced fish were made. The reporter DNA was injected into fertilized eggs of marine sticklebacks, which were then allowed to develop. When intergenic DNA from marine fish was used in the reporter transgene, GFP was strongly expressed in the developing pelvic region, whereas with the intergenic DNA from spine-reduced fish, no GFP was expressed (Fig. 1; region outlined with a white dotted line). The square outlined with a white dotted line in the left panels indicates the region shown at higher magnification in the panels on the right. These experiments suggested that the DNA upstream of the *Pitx1* gene contains an enhancer that normally activates expression of the gene in the pelvic region but that this enhancer was not functional in pelvic-reduced fish. To test this hypothesis, the *Pitx1* control region from marine sticklebacks was linked to the *Pitx1* coding region and this construct was injected into fish with pelvic reduction. This resulted in the dramatic rescue of pelvic development and provided direct evidence that pelvic reduction is due to mutations in the control region of the *Pitx1* gene (Fig. 2).

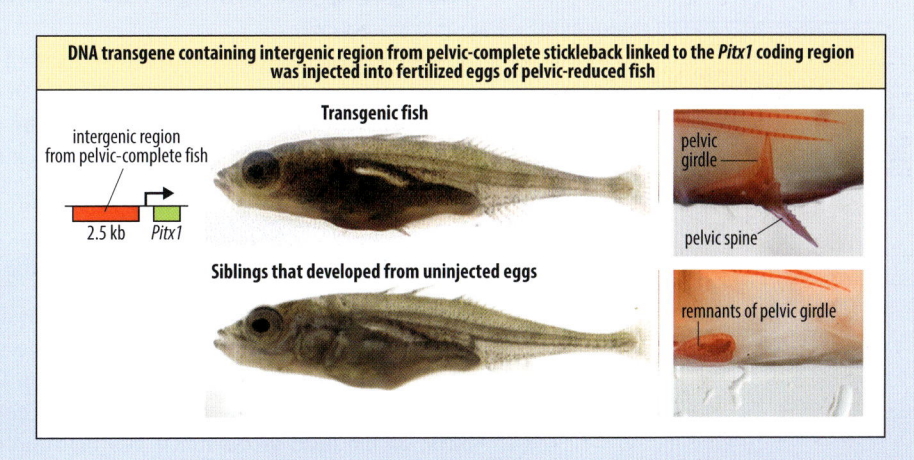

Fig. 1

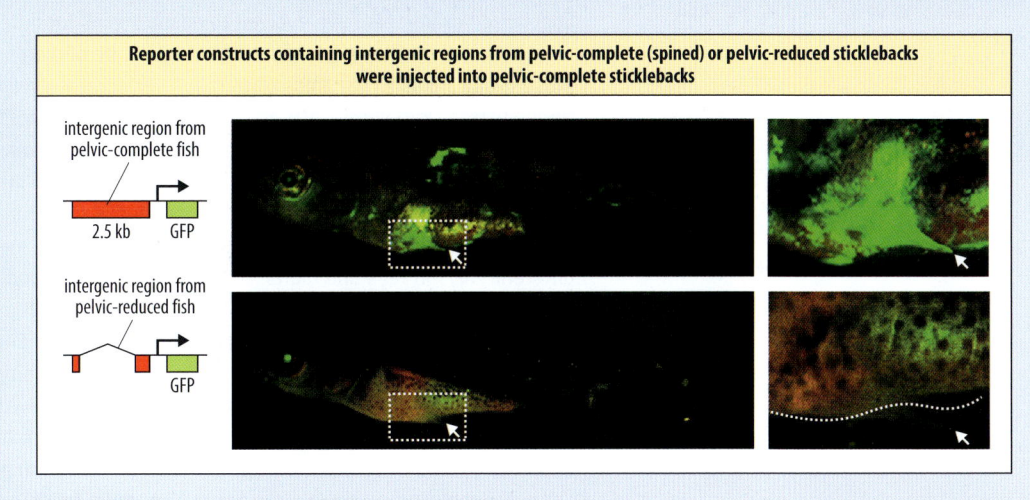

From Chan, Y.F., et al.: ***Adaptive evolution of pelvic reduction in sticklebacks by recurrent deletion of a Pitx1 enhancer.*** Science 2010, **327**: 302-305.

Fig. 2

From Chan, Y.F., et al.: ***Adaptive evolution of pelvic reduction in sticklebacks by recurrent deletion of a Pitx1 enhancer.*** Science 2010, **327**: 302-305.

SUMMARY

Specialized characters have evolved to adapt organs for different functions and environments, and this has given rise to the huge range of morphological diversity seen in present-day animals. The limbs of tetrapod vertebrates evolved from the fins of lobe-finned fish and initially arose as an adaptation for life on land. The digits of the tetrapod limb were thought to be an evolutionary novelty, but the complex regulatory organization of the Hox cluster that controls Hox gene expression in the digits now seems to be evolutionarily ancient. The role of Hox genes is similar in the development of the fin rays in ray-finned fish and the digits in tetrapod limbs, even though bone forms in different ways in rays and digits. Diversification of tetrapod limbs generally involves reductions. Loss of hindlimbs in snakes is associated with mutations in a *cis*-regulatory region that controls limb-bud-specific expression of *Shh*. The eyespots present on the wings of many butterflies are an evolutionary novelty within the Lepidoptera. Their specification and development involve genes that have been used previously for global patterning of the insect body and wing, and that are now deployed in different combinations and different spatial patterns of expression to generate the many different eyespot patterns. Sticklebacks are a good model for studying evolutionary mechanisms, and mutations in *cis*-regulatory control regions that regulate tissue-expression of key developmental genes underlie some changes in the body plan, whereas mutations that affect RNA splicing underlie another. Cavefish are another model in which adaptive evolution occurs in the same species, leading to loss of eyes.

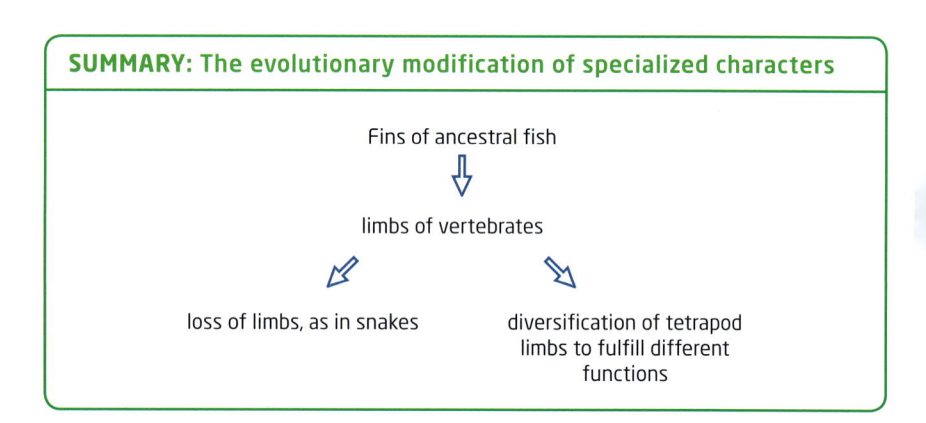

SUMMARY: The evolutionary modification of specialized characters

Fins of ancestral fish
⇩
limbs of vertebrates
↙ ↘
loss of limbs, as in snakes   diversification of tetrapod limbs to fulfill different functions

Changes in the timing of developmental processes

In the previous part of the chapter we focused on changes in spatial patterning that have occurred during evolution. But changes in the timing of developmental processes can also have major effects. In this part of the chapter, we look at some examples of how changes in the timing of growth and sexual maturation can affect animal form and behavior.

13.17 Changes in growth rates can modify the basic body plan

Differential growth rates during development are a major way in which parts of the body can be modified during evolution for specialized functions. Many of the changes that occur during evolution reflect changes in the relative dimensions of parts of the body. We have seen how growth can alter the proportions of the human baby after birth, as the head grows much less than the rest of the body (see Fig. 1.17). The different breeds of dogs come in a great variety of shapes and sizes and provide a good example of the effects of differential growth after birth. The various face shapes in different breeds are also due to differential growth. All dogs are born with rounded faces; some, such as bulldogs and pugs, keep this shape and have disproportionately short muzzles, but in others such as greyhounds and pointers, the nasal regions and jaws elongate during growth.

Dogs provide a unique resource for investigating the genetic determinants of body size and proportions, face shape, and other morphological features such as coat texture, and how these evolved. All dogs are descended from a common ancestor, the gray wolf, about 40,000 years ago, and were then put through intensive breeding programs for the last 200–300 years, which selected for specific characteristics that were present in early domesticated dogs. This selection has resulted in a small amount of genetic variation between dogs of the same breed but a large amount of variation between dogs of the different breeds. This means that there is a relatively small amount of genetic variation shared between dogs of different breeds with the same phenotypic characteristic—for example, body size. The advent of genome sequencing and genome-wide comparative methods has enabled the discovery of where the genetic determinants of particular characteristics are located in the dog genome and hence the identification of the key genes that determine them (Box 13F).

Because individual structures, such as bones, can grow at different rates, the overall shape of an organism can be changed substantially during evolution by heritable changes in the duration of growth, which also leads to an increase in overall size of the organism. In the horse, for example, the central digit of the ancestral horse grew faster than the digits on either side, so that it ended up longer than the lateral digits (Fig. 13.29). As horses continued to increase in overall size during evolution, this discrepancy in growth rates resulted in the relatively smaller lateral digits no longer touching the ground because of the much greater length of the central digit. At a later stage in evolution, the now-redundant lateral digits became reduced even further in

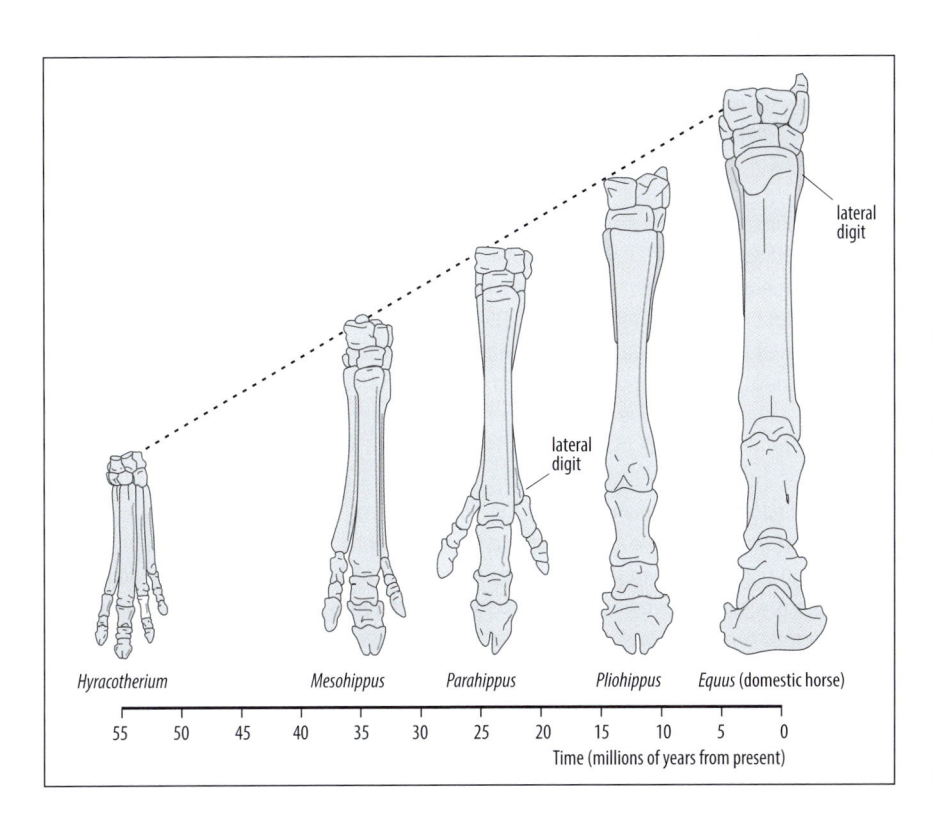

Fig. 13.29 Evolution of the forelimb in horses. *Hyracotherium*, the first true equid, was about the size of a large dog. Its forefeet had four digits, of which one (the third digit in anatomical terms) was slightly longer, as a result of a faster growth rate. All digits were in contact with the ground. As equids increased in size, the lateral digits lost contact with the ground, as the result of the relatively greater increase in length of metacarpal 3 (the proximal bone of the third digit). At a later stage, the lateral digits became even shorter, because of a separate genetic change.

Illustration after Gregory, W.K.: Evolution Emerging. *New York: Macmillan, 1957.*

BOX 13F Origins of morphological diversity in dogs

Domestic dogs are said to show the largest range of sizes of any living land animal. The dog (*Canis canis*) is estimated from studies of ancient DNA from dog fossils to have become a species distinct from its gray wolf (*Canis lupus*) ancestor around 40,000 years ago and was domesticated sometime between then and 20,000 years ago. Archeological evidence suggests that there were already size differences in the earliest domestic dogs. Some present-day breeds, such as Greyhounds and the Pharaoh Hound, look very similar to dogs depicted in 8000-year-old temple paintings, although it is not clear whether our dogs are their direct descendants. Intensive selective breeding over the past 200–300 years has led to the 400 or so distinct breeds that we know today, with enormous differences in size and shape: compare Toby, a Lurcher (a Greyhound cross) with Roly, the Norfolk Terrier (Fig. 1). A Mastiff, one of the large breeds of dog, is 30 times larger overall than a small breed, the Yorkshire Terrier, and is 50 times heavier than a Chihuahua. This size variation is matched by a large amount of genetic variation between breeds, whereas dogs of the same breed have a small amount of genetic variation because they are, effectively, closed breeding populations. This makes dogs ideal subjects for searching for genes that control complex features such as body size and proportions and face shape.

So what is the genetic basis for these enormous differences in body size? All domestic dogs possess the same genome of 2.8 billion base pairs, and the full genome sequence of a female Boxer is now used as the reference dog genome. To identify regions of the genome associated with size, DNA sequences were compared in detail not only between dogs of different sizes within a single breed (Portuguese Water Dogs), but also between dogs from small and giant breeds. The first success was the identification of a primary region of the genome responsible for differences in size, which was then refined to a single gene, *IGF1*, which encodes insulin-like growth factor 1 (IGF-1). IGF-1 is known to be an important growth factor and hormone in controlling growth during embryonic and fetal development and postnatally (discussed in Section 11.8). All small Portuguese Water Dogs were homozygous for the same *IGF1* allele (a particular version of the gene), whereas larger Portuguese Water Dogs had a different *IGF1* allele. In addition, small Portuguese Water Dogs have lower levels of IGF-1 protein in their serum than large dogs of this breed. The same characteristic mutation in *IGF1* found in small Portuguese Water Dogs was found in virtually all the breeds of small dogs tested, even in breeds that are not closely related, and was hardly ever found in giant dog breeds, such as the Saint Bernard or the Irish Wolfhound. This indicates that this allele is likely to have evolved early in the history of dog domestication and was subsequently selected for by the breeding programs.

Extensive genome-wide comparisons between dogs from different breeds and different sizes have now identified six key genes in locations in the dog genome associated with size. One of these identified in this way was again *IGF1*, which together with the other five genes account for 50% of the variation of body size. These genes include the IGF receptor and the growth hormone receptor, consistent with the growth hormone–IGF axis (see Sections 11.8 and 11.9) being a major determinant of body size. IGF-1 is one of the main effectors of growth hormone action, because growth hormone stimulates its synthesis. Another of these six body-size genes encodes the chromatin-associated protein HMGA2, which is also a candidate for the gene affecting beak size in Darwin's finches (see Box 13A).

More recently, genes associated with other characteristics in dogs have been identified using similar methods. Disproportionate muzzle shortening, as in Pekingese and Bulldog, is associated with variation in a single gene, which encodes a bone morphogenetic protein, while short legs, as in Basset Hound and Dachshund, are due to a single genomic event—a DNA insertion. Like those associated with size, these genetic changes seem to have evolved early in the domestication of dogs and were then selected for.

The small number of locations in the genome associated with size in dogs should be contrasted with the large number of locations in the human genome implicated in size determination. In the human genome, 180 locations—potentially 180 different genes—have been found to be significantly associated with height, but even then only account for 10% of the variation in height among people. The reason for this difference between dogs and humans is due to the intense artificial selection of dogs for particular characteristics from a very small number of founders. For example, all Golden Retrievers are descended from the puppies that resulted from a single mating of a male Yellow Retriever with a female Water Spaniel in the 1860s.

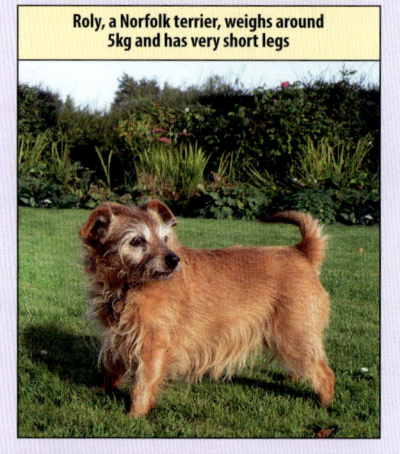

Fig. 1

Photograph of Toby courtesy Monique Welten. Photograph of Roly courtesy R. Mumford.

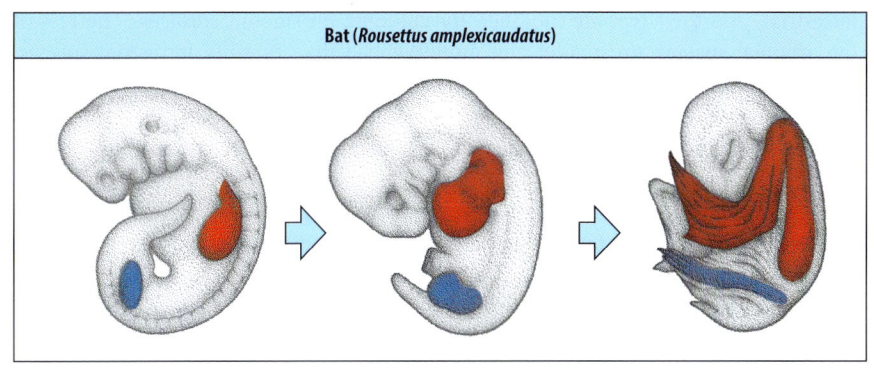

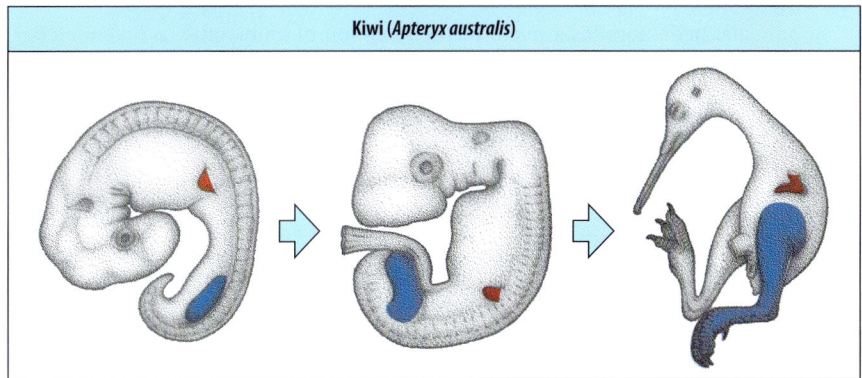

Fig. 13.30 Developmental expression of adult limb size. Top row: the bat *Rousettus amplexicaudatus* has large forelimbs and smaller hindlimbs when adult. The forelimb bud is large at all stages, relative to the hindlimb. Bottom row: the flightless kiwi (*Apteryx australis*) has relatively small forelimbs and large hindlimbs at all stages of development, even in the early embryo.

Adapted from Richardson, M.K.: **Vertebrate evolution: the developmental origins of adult variation.** *Bioessays* 1999, **21**: 604-613.

size because of a separate genetic change that led to extensive cell death on both sides of the digital plate (see Section 13.13).

We have already seen that the bones in the digits in the bat wing are greatly lengthened (see Fig. 13.23). By contrast, the relative lengths of the different bones of the bat hindlimb are similar to those in the mouse hindlimb. In bats, there are differences in size of fore- and hindlimbs from a very early embryonic stage, although the large increase in the length of the digits in the bat wing occurs relatively late, once the main pattern of the skeleton is laid down. In the flightless kiwi, the wing buds are relatively small compared with the leg buds, even at early embryonic stages (Fig. 13.30).

The homeobox transcription factor Prx1 is expressed at high levels in the distal region of the developing bat wing and this may contribute to digit elongation. When the enhancer from the mouse *Prx1* gene is replaced by the bat *Prx1* enhancer in a transgenic mouse, there is a small but significant increase in the length of the mouse forelimb. Elongation of the bat wing digits may also be promoted by a second wave of Shh signaling; *Shh* expression has been detected at later stages in developing bat wings after the primordia of the digits have been laid down. FGF-8 is also produced and could act to prolong outgrowth of the digits. This postulated mechanism is similar to that invoked to explain elongation of the digits in the whale flipper (see Section 13.13), although in the case of the flipper, prolonging outgrowth of the bud has led to the formation of additional phalanges rather than to their elongation.

13.18 Evolution can be due to changes in the timing of developmental events

Differences among species in respect of the time at which developmental processes occur relative to one another, and relative to their timing in an ancestor, can have dramatic effects on both morphology and behavior. Differences in the feet of members of a genus of tropical salamanders illustrate the effect of a change in developmental timing on both the morphology and ecology of different species. Many species of the salamander genus *Bolitoglossa* are arboreal (tree living), rather than typically

714 Chapter 13 Evolution and development

terrestrial, and their feet are modified for climbing on smooth surfaces. The feet of the arboreal species are smaller and more webbed than those of terrestrial species, and their digits are shorter (Fig. 13.31). These differences seem to be mainly the result of the development and growth of the foot ceasing at an earlier stage in the arboreal species than in the terrestrial species. The term used to describe such differences in timing is **heterochrony**.

There are different kinds of heterochronies depending on how developmental time is altered in comparison to an ancestral or related species. They fall into two broad classes: **pedomorphosis** and **peramorphosis**. Figure 13.32 illustrates the general features of these classes using hypothetical examples of the changes the various types of pedomorphosis and peramorphosis would make to plant embryogenesis—the period between fertilization and germination of the seed.

In animals, pedomorphosis refers to the retention of embryonic or juvenile characters during part of the life of the organism. Amphibians provide many examples of pedomorphosis; some species can become sexually mature while still retaining tadpole features such as gill slits. Pedomorphosis can be subdivided into two types: **neoteny** and **progenesis**. One example of neoteny is the retention of juvenile

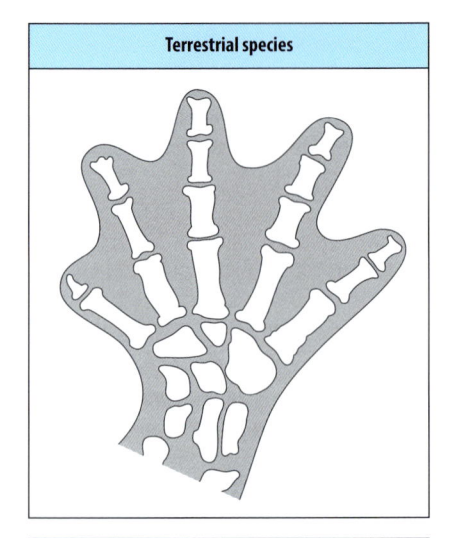

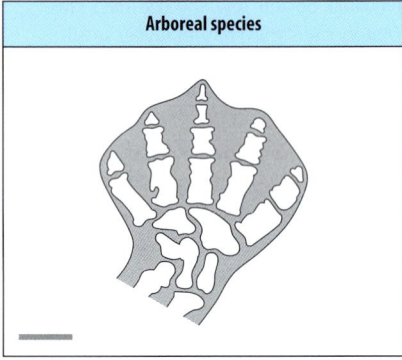

Fig. 13.31 Heterochrony in salamanders. In terrestrial species of the salamander *Bolitoglossa* (top panel), the foot is larger, has longer digits, and is less markedly webbed than in those that live in trees (bottom panel). This difference can be accounted for by foot growth ceasing at an earlier stage in the arboreal species. Scale bar = 1 mm.

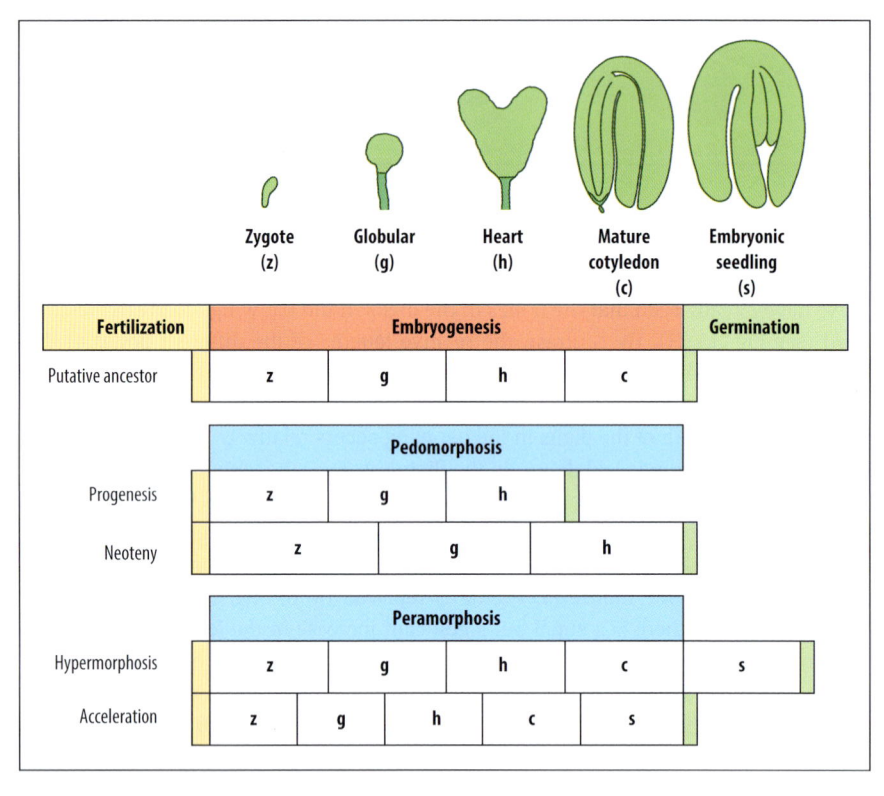

Fig. 13.32 Pedomorphosis and peramorphosis. The principal differences between various types of pedomorphosis and peramorphosis are illustrated here by hypothetical examples in the development of a plant embryo from fertilization to the point of germination. Embryogenesis similar to that of *Arabidopsis* has been taken as the 'ancestral state'. Pedomorphosis is illustrated by progenesis, where germination would be reached more rapidly and at an earlier stage than in the ancestor, and by neoteny, where germination also occurs at an earlier stage, but the rate of development is slower. Peramorphosis is illustrated by examples of hypermorphosis, where the rate of development is very similar to that of the ancestor but an additional stage has been added before germination, and by acceleration, in which an extra stage has also been added, but development is speeded up so that each stage takes less time than in the ancestor.

From Buendía-Monreal, M., Stewart Gillmor, C.: **The times they are a-changin': heterochrony in plant development and evolution.** Front. Plant Sci. *2018,* **9**: *1349.*

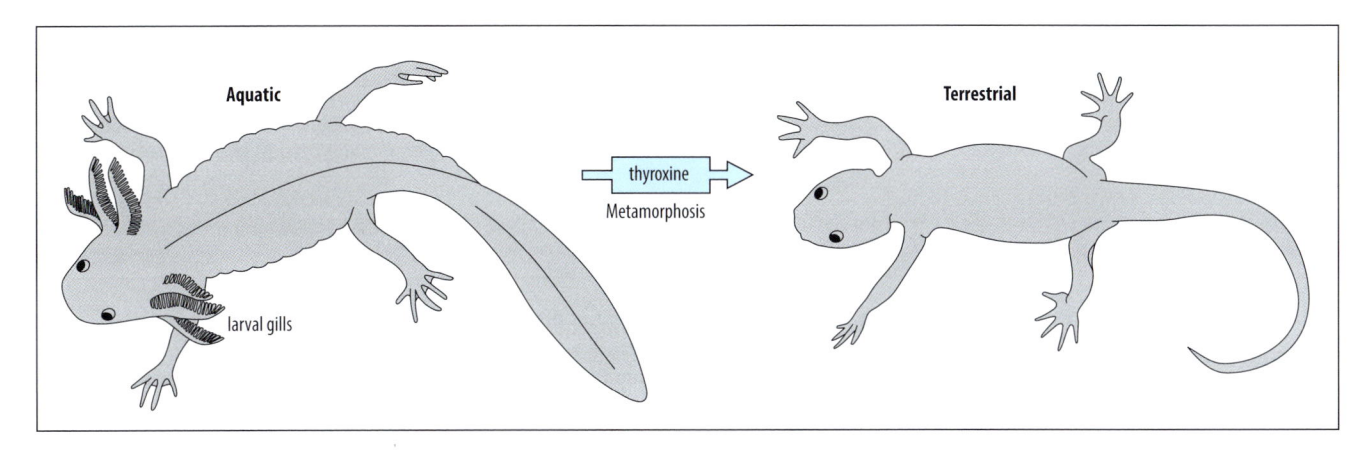

Fig. 13.33 Neoteny in salamanders. The sexually mature axolotl (*Ambystoma mexicanum*) retains larval features, such as gills, and remains aquatic. This neotenic form metamorphoses into a typical terrestrial adult salamander if treated with thyroxine, which it does not produce, and which is the hormone that causes metamorphosis in other amphibians (see Section 11.13).

characters in an adult, which means that reproduction happens in a youthful context; axolotls are examples of this. The axolotl, *Ambystoma mexicanum*, (see Chapter 11) grows in size as a larva and matures sexually at around the normal age, but does not undergo metamorphosis. In this instance, the somatic development of the animal, although not its growth, is retarded in relation to the maturation of the reproductive organs. The sexually mature form remains aquatic and looks like an overgrown larva. However, the axolotl can be induced to undergo metamorphosis by treatment with the hormone thyroxine (Fig. 13.33). The retention of some juvenile traits in humans (such as our small jaws and teeth, flat face, and rounded cranium) into adulthood compared with our close relatives the great apes is another example of neoteny. The other type of pedomorphosis, progenesis, refers to the attainment of sexual maturity more rapidly than the ancestor or related species. The outcome is similar to neoteny—the animal attains sexual maturity while still in the juvenile period—but the cause of the change is different.

Peramorphosis refers to the case where a species shows greater development compared with its ancestor or related species. There are two main classes of peramorphosis. **Hypermorphosis**, or addition, describes the situation in which the rate of development is conserved between species but there is an additional or extended stage in some species. For example, hypermorphosis allows the growth of elaborate antlers in some species of deer. Another type of peramorphosis is **acceleration**, which occurs when growth and development occur at a faster rate in some features.

Direct development is another example of heterochrony. Direct development can be seen in frogs of the genus *Eleutherodactylus*, which unlike the more typical amphibians *Rana* and *Xenopus* develop directly into adults and have no aquatic tadpole stage, the eggs being laid on land. Typical tadpole features such as gills and cement glands do not develop, and prominent limb buds appear shortly after the formation of the neural tube (Fig. 13.34). In the embryo, the tail is modified into a respiratory organ. Such direct development requires a large supply of yolk to the egg in order to support development through to an adult without a tadpole feeding stage. This increase in yolk may itself be an example of heterochrony, involving a longer or more rapid period of yolk synthesis in the development of the egg.

Another example in which direct development has evolved is in some species of sea urchin. Most sea urchins have a larval stage that takes a month or more to become adult. But there are some species that no longer go through a functional larval stage. Such species have large eggs and, as a result of their very rapid development, become juvenile urchins within 4 days. This has involved changes in early development such

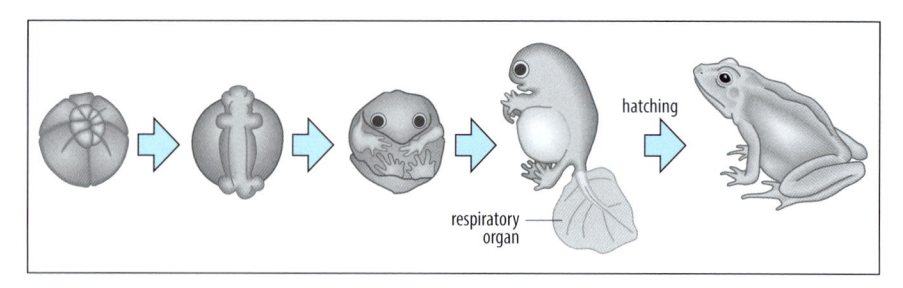

Fig. 13.34 Direct development of the frog *Eleutherodactylus*. Typical frogs, such as *Xenopus* or *Rana*, lay their eggs in water and develop through an aquatic tadpole stage, which undergoes metamorphosis into the adult. Frogs of the genus *Eleutherodactylus* lay their eggs on land and the frog hatches from the egg as a miniature adult, without going through an aquatic free-living larval stage. The embryonic tail is modified as a respiratory organ.

that the directly developing embryo gives rise to a 'larval' stage that lacks a gut and cannot feed, and which metamorphoses rapidly into the adult form.

13.19 The evolution of life histories has implications for development

Animals and plants have very diverse life histories: small birds breed in the spring following their birth, and continue to do so each year until their death; Pacific salmon breed in a suicidal burst at 3 years of age; and oak trees require 30 years of growth before producing acorns, which they eventually produce by the thousand. In order to understand the evolution of such life histories, evolutionary ecologists consider them in terms of probabilities of survival, rates of reproduction, and optimization of reproductive effort. These factors have important implications for the evolution of developmental strategies, particularly in relation to the rate of development. For example, a characteristic feature of many animal life histories is the presence of a free-living larval stage that is distinct from, and usually simpler in form than, the sexually mature adult, and which feeds in a different way.

Consideration of environmental pressures and life histories helps us to understand the evolution of the complex developmental mechanisms that lay down the body plan in long-germ insects, such as *Drosophila*, which are of more recent origin than short-germ insects, such as grasshoppers. In **long-germ development**, the whole insect body plan is present in the blastoderm (Fig. 13.35); in **short-germ development**, only part of the insect body plan is present in the blastoderm and the more posterior segments are added sequentially as the embryo grows (see Box 13G). The flour beetle *Tribolium* represents a short-germ insect in which anterior and some thoracic tissue is specified in the blastoderm, while the grasshopper *Schistocerca* is an example of an extreme short-germ insect, with only a very small amount of anterior tissue specified in the blastoderm (see Fig. 13.35). Grasshoppers do not have a larval stage and hatch directly into small, immature adult-like forms called nymphs. Embryonic development in the grasshopper takes 5–6 days before the egg hatches and the immature grasshopper begins to feed. By contrast, *Drosophila* develops very rapidly into a feeding larva, taking only 24 hours to start feeding. It is easy to imagine conditions in which there would have been a selective advantage to insects whose larvae begin to feed as quickly as possible. It is also likely that embryos are more vulnerable than adults, and so there would be selection for making the embryonic stage shorter. Thus, the complex developmental mechanisms of long-germ insects—the system for setting up the complete antero-posterior axis in the egg—probably evolved as a result of selection pressure for rapid development. In the case of fruit flies such as *Drosophila*, this was perhaps related to the ability to feed on fast-disappearing fruit.

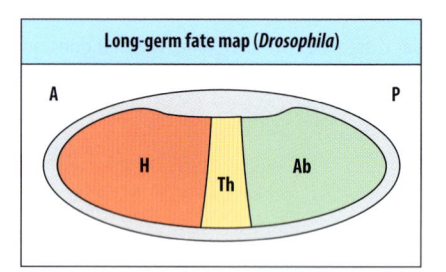

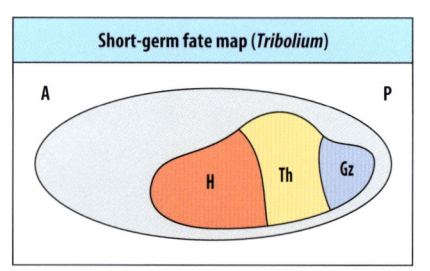

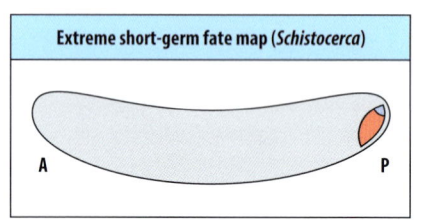

Fig. 13.35 Differences in the development of long-germ and short-germ insects. Top panel: the general fate map of the blastoderm of long-germ insects such as *Drosophila* shows that the whole of the body plan—head (H), thorax (Th), and abdomen (Ab)—is present at the time of initial germ-band formation. Middle panel: in a short-germ insect such as the flour beetle *Tribolium*, only anterior (A) regions (H and Th) are present at this embryonic stage. The abdominal segments develop later, after gastrulation, from a posterior (P) growth zone (Gz). Bottom panel: in an extreme short-germ insect such as the grasshopper *Schistocerca*, only the anterior-most region is present. The rest of the body segments will be added by growth from the growth zone after gastrulation.

Egg size can also be best understood within the context of life histories. If we assume that the parent has limited energy resources to put into reproduction, the question is how should these resources best be invested in making gametes, particularly eggs; is it more advantageous to make lots of small eggs or a few large ones? In general, it seems that the larger the egg, and thus the larger the offspring at birth, the better are the chances of the offspring surviving. This would seem to suggest that in most circumstances an embryo needs to give rise to a hatchling as large as possible. Why then, do some species lay many small eggs capable of rapid development? One possible answer is that a parental investment in large eggs might reduce the parents' own chance of surviving and laying more eggs. This strategy of laying many small eggs might be especially successful in variable environmental conditions, where populations can suddenly crash. The rapid development of a larval stage enables early feeding and dispersal to new sites in such circumstances.

BOX 13G Long- and short-germ development in insects

Drosophila belongs to an evolutionarily advanced group of insects, in which all the segments are specified more or less at the same time in development, as shown both by the striped patterns of pair-rule gene expression in the syncytial blastoderm and the appearance of all the segments shortly after gastrulation (see Chapter 2). This type of development is known as long-germ development, because the fate map of the blastoderm corresponds to the whole of the future embryo, and leads to a larval form bearing imaginal discs that will give rise to the adult. Many other insects, such as the red flour beetle *Tribolium castaneum*, have a short-germ development; that is, the fate map of the blastoderm corresponds to only a region of the embryo, usually only the anterior segments. In an extreme case of short-germ development, the grasshopper *Schistocerca*, the blastoderm gives rise only to the most anterior structures (see Fig. 13.35). In short-germ insects, the posterior segments are generated sequentially from the growth zone, a distinct region of growth and patterning at the posterior end of the embryo, after completion of the blastoderm stage and gastrulation. The short-germ type of development is thought to be the ancestral mode of development in insects, with long-germ development arising independently within some insect groups. In spite of the early differences between them, the mature germ-band stages of both long- and short-germ insect embryos look similar—a common stage through which all insect embryos develop.

So what new had to evolve so that long-germ insects could specify all segments at once? This is still a subject of active research. It is likely that the long-germ mode of development developed independently in different orders of insects (it occurs in the Coleoptera (beetles) and the Hymenoptera (bees and wasps), for example, as well as in the Diptera). In the Diptera, long-germ development in the group of flies that includes *Drosophila* is associated with the appearance of the *bicoid* gene, which is not present in any other insects. Bicoid protein seems to have taken over the primary patterning role of the much more ancient protein Orthodenticle, which in *Tribolium*, for example, forms a maternally derived anterior-to-posterior gradient in the early blastoderm. Some long-germ insects have retained and adapted this mechanism. In the long-germ embryos of the parasitoid wasp *Nasonia*, for example, maternally derived gradients of Orthodenticle pattern both the anterior and posterior regions of the blastoderm, in a manner similar to the action of Bicoid.

Drosophila is at the extreme of the developmental strategy in which all of the segments are specified at once, and this perhaps, explains the presence of Bicoid, which represents a mechanism for measuring the length of the embryo. The Bicoid gradient is scaled to match the initial length of the blastoderm (which will vary slightly between individuals) in order to produce the correct developmental outcome.

From the point of view of the molecular mechanisms mediating segmentation, a clear difference between long- and short-germ insects is that the patterning of the body plan in *Drosophila* takes place before cell boundaries form, whereas much of the body plan in short-germ insects is laid down progressively at later stages, because the posterior segments are generated as the embryo grows. At that stage the embryo is cellular rather than syncytial. So are the same genes involved?

Over the past few years, increasing numbers of genes involved in segmentation and patterning have been studied in detail in insects other than *Drosophila*, with a particular emphasis on *Tribolium*, in which genetic analysis is possible. Figure 1 compares the earliest embryonic stage (top row; regions that are fated at this stage to give rise to the embryo are shaded in gray); the expression patterns of a gap gene (second row; green); a pair-rule gene (third row; blue); and a segment polarity gene (bottom row; purple), at comparable embryonic stages in *Drosophila*, *Tribolium*, and *Schistocerca*, which shows that many of the same genes and developmental processes are, indeed, involved. For

718 Chapter 13 Evolution and development

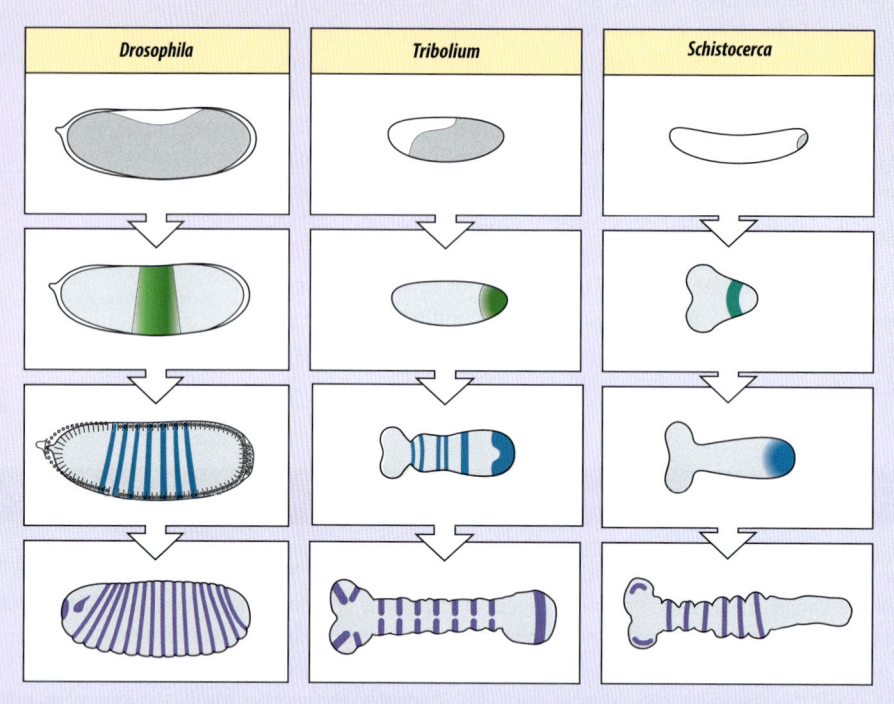

Fig. 1
Adapted from Nagy, L.M.: **Insect segmentation: a glance posterior.** *Curr. Biol. 1994,* **4***: 811-814; Dearden, P.K., Akam, M.:* **Early embryo patterning in the grasshopper,** Schistocerca gregaria: wingless, decapentaplegic *and* caudal *expression.* Development 2001, **128**, 3435-3444.

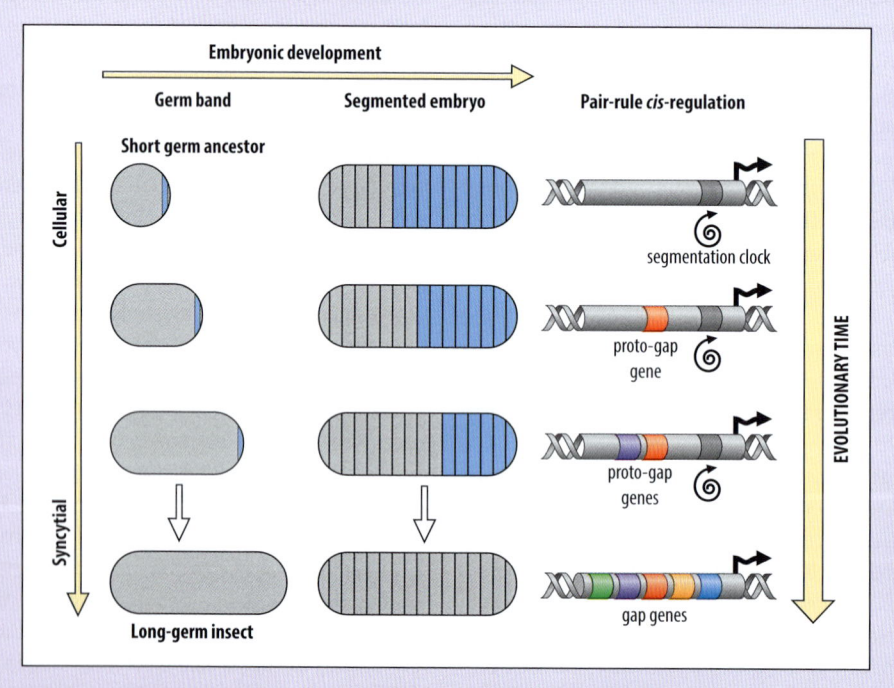

Fig. 2
Adapted from Peel, A., Akam, M.: **Evolution of segmentation: rolling back the clock.** *Curr. Biol. 2003,* **13***: R708-R710.*

example, the gap gene *Krüppel* is initially expressed at the posterior end of the *Tribolium* embryo at the blastoderm stage and only later moves into the middle region, as in *Drosophila*. (The expression of the gap gene *hunchback* is shown in the *Schistocerca* heart-stage blastoderm.)

Loss of Krüppel activity has a very different outcome in *Drosophila* and *Tribolium*: in the fly it leads to the loss of all central segments, whereas in the beetle it only affects the more posterior ones. Krüppel therefore seems to be specifying the corresponding parts of the blastoderm in the two insects, but because of the different growth strategies, the outcome of loss of function is different. Similarly, only two repeats of the *eve* pair-rule stripes are present at the blastoderm stage in *Tribolium* (the original stripes split into two soon after they form), together with a posterior cap of pair-rule gene expression, in contrast to the seven repeats in *Drosophila* (see Fig. 1). In *Schistocerca*, only a posterior cap of *eve* expression, in the growth zone, is present at a comparable stage, and it seems to act as a gap gene rather than a pair-rule gene.

The segmentation genes *wingless* and *engrailed* are expressed in *Tribolium* and in *Schistocerca* in a similar relation to that in *Drosophila*, which reflects the conservation of the Wingless-Engrailed parasegment boundary as a patterning reference (see Section 2.24).

The sequential emergence of segments in short-germ insects is shared by other arthropods, for example spiders and crustaceans, and is reminiscent of the process of somitogenesis in vertebrate embryos (see Chapter 4). This feature has led to the suggestion that a segmentation mechanism similar to that of the vertebrate somite-formation clock also exists in invertebrates (Fig. 2), and that differences in the mechanism of segmentation in different insects are associated with the progressive replacement of this mechanism in some lineages by the one represented by Bicoid and the gap genes in *Drosophila*.

SUMMARY

Changes in the timing of developmental processes that have occurred during evolution can alter the form of the body, for if different regions grow faster than others or continue to grow for longer, their size is proportionately increased if the animal gets larger. The increase in size of the central digit of horses is partly due to this type of developmental change. Different breeds of dog provide examples of how differential growth can affect the size of the body and the proportions of its different parts. The advent of genomic resources for dogs has revealed that genetic variations arose early in the evolution of dogs and then were selected for by intensive breeding. Because the breeds are essentially closed breeding populations, dogs have provided a unique resource for identifying the key genes responsible for body size and proportions. Speed of development and egg size also have important evolutionary implications. Changing the time at which an animal becomes sexually mature can result in adults with larval characteristics. Some animals, such as frogs and sea urchins, which usually have a larval form, have evolved species that develop directly into the adult without a larval stage. The life histories of animals also have implications for evolution and development. A different strategy for setting up the antero-posterior axis has evolved in long-germ insects to allow rapid development. Although long-germ insects and short-germ insects undergo different modes of development, they use many of the same core developmental genes.

SUMMARY: Changes in timing of developmental processes

increase in size
⇩
faster-growing regions get proportionately larger
e.g. third digit in the horse

retention of juvenile traits in adults
⇩
neoteny

elimination of larval stage
⇩
rapid development to adult

Summary to Chapter 13

- Many developmental processes have been conserved during evolution. Although many questions relating to evolution and development remain unanswered, it is clear that development reflects the evolutionary history of ancestral embryos.
- The origins of multicellular animals (the Metazoa) and their embryos from a single-celled ancestor are still highly speculative. Analysis of complete genomes of the earliest evolved metazoans (e.g. sponges, cnidarians, and ctenophores) show that the earliest ancestor of multicellular animals must have already had many of the molecular signaling pathways and other cell biological features present in modern animals.
- All vertebrate embryos pass through a conserved phylotypic developmental stage, where they share many morphological features, although there can be considerable divergence both earlier and later in development.
- The pattern of Hox gene expression along the body axis of the different types of arthropods is conserved, and variations in expression of Hox genes between different arthropod classes correlate closely with differences in body plan and the numbers and positions of appendages.
- Differences in the number and type of vertebrae in the spinal column in the different classes of vertebrates correlate with the different patterns of Hox gene expression along the antero-posterior axis, indicating how changes in Hox gene expression could have led to modifications of the basic body plan. This is seen most dramatically in snakes, where the number of rib-bearing vertebrae is greatly increased. Changes in both Hox gene regulation and the interpretation of a Hox code could have led to major morphological changes during evolution.
- The signals involved in patterning the dorso-ventral axis of the body in vertebrates and arthropods show remarkable similarity and conservation, and show that the dorso-ventral axis became inverted in the evolution of vertebrates from their common ancestor with arthropods.
- The limbs of tetrapods evolved from the fins of their ancient lobe-finned fish ancestors. The basic mechanisms to pattern the antero-posterior axis of tetrapod limbs may have been present very early in the evolution of paired appendages. The organization of the Hox gene complexes that regulate the expression of the Hox genes in the digits is also evolutionarily ancient, being functionally conserved in the fins of bony fish and the limbs of tetrapods.
- The loss of forelimbs in snakes is associated with changes in expression of the Hox genes. The loss of hindlimbs has been traced to changes in a key control region for *Shh*, which render it non-functional.
- The variety of colorful eyespots on the wings of different butterfly species has evolved by deploying in new spatial patterns a relatively small set of genes involved in previous global patterning of the embryo.
- Genetic analysis of recent evolutionary changes in natural populations of organisms such as sticklebacks and cavefish throw light on the mechanisms of evolution. In sticklebacks, the evolution of major morphological changes is based on changes in the tissue-specific control regions of key developmental genes; an example of a morphological change which is based on changes in RNA splicing has also been discovered.
- Alterations in the timing of developmental events have played an important role in evolution.
- Such changes can alter the overall form of an organism as a result of differences in growth rates of different structures, and can also result in sexual maturation at larval stages.
- Dogs provide a model for understanding how differences in body size and proportions evolve.
- Short-germ and long-germ insects undergo different modes of development with respect to timing but use many of the same core developmental genes.

End of chapter questions

Check your understanding of the content of this chapter by attempting the following long-answer concept questions.

1. Trace the possible steps in the evolution of the Metazoa from a single-celled ancestor. What were the key innovations that led to the evolution of the Metazoa, and especially of bilaterally symmetric animals?

2. Why has the phylotypic stage been so strongly conserved during evolution? What events and processes are occurring at this point in development that make this stage so central to animal development?

3. Give an example of a change in protein structure and function that occurred after gene duplication. Give an example of a change in gene expression that occurred after gene duplication.

4. Tetrapod limbs evolved from fins. What conclusions about how this might have occurred emerge from (1) studying the fossil record and (2) studying fin development in zebrafish. If you could get hold of embryos of an early tetrapod such as *Panderichthys*, what experiments would you carry out to investigate the molecular basis of the fin-to-limb transition?

5. *Drosophila* larvae have no legs, yet lepidopteran larvae have legs on each thoracic and abdominal segment. Explain how Hox genes and the *Distal-less* gene mediate this difference.

6. Snakes have evolved an elongated body with many morphologically similar vertebrae. How are these changes related to changes in Hox gene function? Consider both changes in Hox gene expression and changes in how the information encoded by Hox genes is interpreted.

7. 'Darwin's finches' are a classic example of the evolutionary role of development. What is currently known about the developmental basis for the adaptive evolution of beak shape? What are the remaining outstanding questions and what approaches could you use to find answers?

8. Synthesize the information in Figures 13.8, 13.9, and 13.10 and provide a narrative for the evolution of the incus and malleus bones in the inner ear of mammals.

9. Describe what is meant by the term heterochrony and give an example of an evolutionary change that can be described as being due to heterochrony.

10. *Drosophila* has played a central part in our understanding of insect development, but is not necessarily representative of all insects. How does the long-germ development of *Drosophila* differ from the short-germ type of development seen in beetles and grasshoppers?

11. Give examples from this chapter of how CRISPR-Cas9 genome editing has been applied to investigating the development of non-model organisms and what new insights have emerged. What limits the usefulness of this technology?

General further reading

Carroll, S.B.: **Evo-devo and an expanding evolutionary synthesis: a genetic theory of morphological evolution**. *Cell* 2008, **134**: 25–36.

Carroll, S.B., Grenier, J.K., Weatherbee, S.D.: *From DNA to Diversity*. 2nd edn. Malden, MA: Blackwell Science, 2005.

Finnerty, J.R., Pang, K., Burton, P., Paulson, D., Martindale, M.Q.: **Origins of bilateral symmetry: Hox and *dpp* expression in a sea anemone**. *Science* 2004, **304**: 1335–1337.

Hoekstra, H.E., Coyne, J.A.: **The locus of evolution: evo devo and the genetics of adaptation**. *Evolution* 2007, **61**: 995–1016.

Holland, P.W., Marlétaz, F., Maeso, I., Dunwell, T.L., Paps, J.: **New genes from old: asymmetric divergence of gene duplicates and the evolution of development**. *Philos. Trans. R Soc. Lond. B Biol. Sci.* 2017, **372**: 20150480.

Kirschner, M., Gerhart, J.: **Evolvability**. *Proc. Natl Acad. Sci. USA* 1998, **95**: 8420–8427.

Kirschner, M., Gerhart, J.: *The Plausibility of Life*. New Haven, CT: Yale University Press, 2005.

Raff, R.A.: *The Shape of Life*. Chicago, IL: University of Chicago Press, 1996.

Richardson, M.K.: **Vertebrate evolution: the developmental origins of adult variation**. *BioEssays* 1999, **21**: 604–613.

Wray, G.A.: **The evolutionary significance of *cis*-regulatory mutations**. *Nat. Rev. Genet.* 2007, **8**: 206–216.

Box 13A Darwin's finches

Abzhanov, A., Kuo, W.P., Hartmann, C., Grant, B.R., Grant, P.R., Tabin, C.J.: **The calmodulin pathway and evolution of elongated beak morphology in Darwin's finches**. *Nature* 2006, **442**: 563–567.

Almén, M.S., Lamichhaney, S., Berglund, J., Grant, B.R., Grant, P.R., Webster, M.T., Andersson, L.: **Adaptive radiation of Darwin's finches revisited using whole genome sequencing**. *BioEssays* 2015, **38**: 14–20.

Lamichhaney, S., Berglund, J., Almén, M.S., Maqbool, K., Grabherr, M., Martinez-Barrio, A., Promerová, M., Rubin, C.J., Wang, C., Zamani, N., Grant, B.R., Grant, P.R., Webster, M.T., Andersson, L.: **Evolution of Darwin's finches and their beaks revealed by genome sequencing**. *Nature* 2015, **518**: 371–375.

Lamichhaney, S., Han, F., Berglund, J., Wang, C., Almén, M.S., Webster, M.T., Grant, B.R., Grant, P.R., Andersson, L.: **A beak size locus in Darwin's finches facilitated character displacement during a drought**. *Science* 2016, **352**: 470–474.

Section further reading

13.1 Multicellular organisms evolved from single-celled ancestors

Brooke, N.M., Holland P.W.H.: **The evolution of multicellularity and early animal genomes**. *Curr. Opin. Genet. Dev.* 2003, **6**: 599–603.

King, N.: **The unicellular ancestry of animal development**. *Dev. Biol.* 2004, **7**: 313–325.

King, N., *et al.*: **The genome of the choanoflagellate *Monosiga brevicollis* and the origin of metazoans**. *Nature* 2008, **451**: 783–788.

Miller, D.J., Ball, E.E.: **Animal evolution: the enigmatic phylum Placozoa revisited**. *Curr. Biol.* 2005, **15**: R26–R28.

Rudel, D., Sommer, R.J.: **The evolution of developmental mechanisms**. *Dev. Biol.* 2003, **264**: 15–37.

Sebé-Pedrós, A., Degnan, B.M., Ruiz-Trillo, I.: **The origin of Metazoa: a unicellular perspective**. *Nat. Rev. Genet.* 2017, **18**: 498–512.

Sebé-Pedrós, A., de Mendoza, A., Lang, B.F., Degnan, B.M., Ruiz-Trillo, I.: **Unexpected repertoire of metazoan transcription factors in the unicellular holozoan *Capsaspora owczarzaki*.** *Mol. Biol. Evol.* 2011, **28**: 1241–1254.

Szathmary, E., Wolpert, L.: **The transition from single cells to multicellularity**. In *Genetic and Cultural Evolution of Cooperation*. Hammersteen, P. (Ed.). Cambridge, MA: MIT Press, 2004: 271–289.

Wolpert, L., Szathmary, E.: **Evolution and the egg**. *Nature* 2002, **420**: 745.

13.2 Genomic evidence is throwing light on the origin of animals and Box 13B The metazoan family tree

Babonis, L.S., Martindale, M.Q.: **Phylogenetic evidence for the modular evolution of metazoan signalling pathways**. *Philos. Trans. R. Soc. Lond. B Biol. Sci.* 2017, **372**: 20150477.

Guder, C., Philipp, I., Lengfeld, T., Watanabe, H., Hobmayer, B., Holstein, T.W. **The Wnt code: cnidarians signal the way**. *Oncogene* 2006, **25**: 7450–7460.

Paps, J., Holland, P.W.H.: **Reconstruction of the ancestral metazoan genome reveals an increase in genomic novelty**. *Nat. Commun.* 2018, **9**: 1730.

Petersen, C.P., Reddien, P.W.: **Wnt signaling and the polarity of the primary body axis**. *Cell* 2009, **139**: 1056–1068.

Putnam, N.H., Srivastava, M., Hellsten, U., Dirks, B., Chapman, J., Salamov, A., Terry, A., Shapiro, H., Lindquist, E., Kapitonov, V.V., Jurka, J., Genikhovich, G., Grigoriev, I.V., Lucas, S.M., Steele, R.E., Finnerty, J.R., Technau, U., Martindale, M.Q., Rokhsar, D.S.: **Sea anemone genome reveals ancestral eumetazoan gene repertoire and genomic organization**. *Science* 2007, **317**: 86–94.

Richards, G.S., Degnan, B.M.: **The dawn of developmental signaling in the metazoa**. *Cold Spring Harb. Symp. Quant. Biol.* 2009, **74**: 81–90.

Ryan, J.F., Pang, K., Schnitzler, C.E., Nguyen, A.D., Moreland, R.T., Simmons, D.K., Koch, B.J., Francis, W.R., Havlak, P., NISC Comparative Sequencing Program, Smith, S.A., Putnam, N.H., Haddock, S.H., Dunn, C.W., Wolfsberg, T.G., Mullikin, J.C., Martindale, M.Q., Baxevanis, A.D.: **The genome of the ctenophore *Mnemiopsis leidyi* and its implications for cell type evolution**. *Science* 2013, **342**: 1242592.

Srivastava, M., *et al.*: **The *Trichoplax* genome and the nature of placozoans**. *Nature* 2008, **454**: 955–960.

Technau, U., Steele, R.E.: **Evolutionary crossroads in developmental biology: Cnidaria**. *Development* 2011, **138**: 1447–1458.

13.3 How gastrulation evolved is not known

Jägersten, G.: **The early phylogeny of the metazoa. The bilaterogastrea theory**. *Zool. Bidrag. (Uppsala)* 1956, **30**: 321–354.

Martindale, M.Q.: **The evolution of metazoan axial properties**. *Nat. Rev. Genet.* 2005, **6**: 917–927.

Wolpert, L.: **Gastrulation and the evolution of development**. *Development (Suppl.)* 1992, 7–13.

13.4 More general characteristics of the body plan develop earlier than specializations

Abzhanov, A.: **von Baer's law for the ages: lost and found principles of developmental evolution**. *Trends Genet.* 2013, **29**: 712–722.

Hazkani-Covo, E., Wool, D., Graur, D.: **In search of the vertebrate phylotypic stage: a molecular examination of the developmental hourglass model and von Baer's third law**. *J. Exp. Zool. B Mol. Dev. Evol.* 2005, **304**: 150–158.

13.5 Embryonic structures have acquired new functions during evolution

Anthwal, N., Tucker, A.S.: **Q&A: morphological insights into evolution**. *BMC Biol.* 2017, **15**: 83.

Cerny, R., Lwigale, P., Ericsson, R., Meulemans, D., Epperlein, H.H., Bronner-Fraser, M.: **Developmental origins and evolution of jaws: new interpretation of 'maxillary' and 'mandibular'**. *Dev. Biol.* 2004, **276**: 225–236.

Cohn, M.J.: **Lamprey Hox genes and the origin of jaws**. *Nature* 2002, **416**: 386–387.

De Robertis, E.M.: **Evo-devo: variations on ancestral themes**. *Cell* 2008, **132**: 185–195.

Erwin, D.H.: **Early origin of the bilaterian developmental toolkit**. *Philos. Trans. R. Soc. Lond. B Biol. Sci.* 2009, **364**: 2253–2261.

Romer, A.S.: *The Vertebrate Body*. Philadelphia, PA: W.B. Saunders, 1949.

13.6 Evolution of different types of eyes in different animal groups is an example of parallel evolution

Arendt, D.: **Evolution of eyes and photoreceptor cell types**. *Int. J. Dev. Biol.* 2003, **47**: 563–571.

Gehring, W.J.: **Chance and necessity in eye development**. *Genome Biol. Evol.* 2011, **3**: 1053–1056.

Halder, G., Callaerts, P., Gehring, W.J.: **Induction of ectopic eyes by targeted expression of the eyeless gene in *Drosophila***. *Science* 1995, **267**: 1788–1792.

Shubin, N., Tabin, C., Carroll, S.: **Deep homology and the origins of evolutionary novelty**. *Nature* 2009, **457**: 818–823.

Wawersik S., Maas, R.L.: **Vertebrate eye development as modeled in *Drosophila***. *Hum. Mol. Genet.* 2000, **9**: 917–925.

Wistow, G.: **The human crystallin gene families**. *Hum. Genom.* 2012, **6**: 26.

13.7 Hox gene complexes have evolved through gene duplication

Brooke, N.M., Gacia-Fernandez, J., Holland, P.W.H.: **The ParaHox gene cluster is an evolutionary sister of the Hox gene cluster**. *Nature* 1998, **392**: 920–922.

Duboule, D.: **The rise and fall of Hox gene clusters**. *Development* 2007, **134**: 2549–2560.

Ferrier, D.E.K., Holland, P.W.H.: **Ancient origin of the Hox gene cluster**. *Nat. Rev. Genet.* 2001, **2**: 33–34.

Holland, L.Z., *et al.*: **The amphioxus genome illuminates vertebrate origins and cephalochordate biology**. *Genome Res.* 2008, **18**: 1100–1111.

Pascual-Anaya, J., D'Aniello, S., Kuratani, S., Garcia-Fernandez, J.: **Evolution of Hox gene clusters in deuterostomes**. *BMC Dev. Biol.* 2013, **13**: 26.

Prince, V.E., Pickett, F.B.: **Splitting pairs: the diverging fates of duplicated genes**. *Nat. Rev. Genet.* 2002, **3**: 827–837.

Valentine, J.W., Erwin, D.H., Jablonski, D.: **Developmental evolution of metazoan bodyplans: the fossil evidence**. *Dev. Biol.* 1996, **173**: 373–381.

13.8 Differences in Hox gene expression determine the variation in position and type of paired appendages in arthropods

Angelini, D.R., Kaufman, T.C.: **Comparative developmental genetics and the evolution of arthropod body plans**. *Annu. Rev. Genet.* 2005, **39**: 95–119.

Averof, M.: **Origin of the spider's head**. *Nature* 1998, **395**: 436–437.

Carroll, S.B., Weatherbee, S.D., Langeland, J.A.: **Homeotic genes and the regulation and evolution of insect wing number**. *Nature* 1995, **375**: 58–61.

Galant, R., Carroll, S.B.: **Evolution of a transcriptional repression domain in an insect Hox protein**. *Nature* 2002, **415**: 910–913.

Levine, M.: **How insects lose their wings**. *Nature* 2002, **415**: 848–849.

Liubicich, D.M., Serano, J.M., Pavlopoulos, A., Kontarakis, Z., Protas, M.E., Kwan, E., Chatterjee, S., Tran, K.D., Averof, M., Patel, N.H.: **Knockdown of *Parhyale* Ultrabithorax recapitulates evolutionary changes in crustacean appendage morphology**. *Proc. Natl Acad. Sci. USA* 2009, **106**: 13892–13896.

Martin, A., Serano, J.M., Jarvis, E., Bruce, H.S, Wang, J., Ray, S., Barker, C.A., O'Connell, L.C., Patel, N.H.: **CRISPR/Cas9 mutagenesis reveals versatile roles of Hox genes in crustacean limb specification and evolution**. *Curr. Biol.* 2016, **26**: 14–26.

Pavlopoulos, A., Akam, M.: **Hox gene Ultrabithorax regulates distinct sets of target genes at successive stages of *Drosophila* haltere morphogenesis**. *Proc. Natl Acad. Sci. USA* 2011, **108**: 2855–2860.

Pavlopoulos, A., Kontarakis, Z., Liubicich, D.M., Serano, J.M., Akam, M., Patel, N.P., Averof, M.: **Probing the evolution of appendage specialization by Hox gene misexpression in an emerging model crustacean**. *Proc. Natl Acad. Sci. USA* 2009, **106**: 13897–13902.

Weatherbee, S.D., Carroll, S.: **Selector genes and limb identity in arthropods and vertebrates**. *Cell* 1999, **97**: 283–286.

Weatherbee, S.D., Nijhout, H.F., Grunnert, L.W., Halder, G., Galant, R., Selegue, J., Carroll, S.: **Ultrabithorax function in butterfly wings and the evolution of insect wing patterns**. *Curr. Biol.* 1999, **9**: 109–115.

13.9 Changes in Hox gene expression and their target genes contributed to the evolution of the vertebrate axial skeleton

Cohn, M.J., Tickle, C.: **Developmental basis of limblessness and axial patterning in snakes**. *Nature* 1999, **399**: 474–479.

Guerreiro, I., Nunes, A., Woltering, J.M., Casaca, A., Nóvoa, A., Vinagre, T., Hunter M. E., Duboule, D., Mallo, M.: **Role of a polymorphism in a Hox/Pax-responsive enhancer in the evolution of the vertebrate spine**. *Proc. Natl Acad. Sci. USA* 2013, **110**: 10682–10686.

Hrycaj, S.M., Wellik, D.M.: **Hox genes and evolution**. *F1000Res.* 2016, **5**: 859.

Mansfield, J.H.: ***cis*-regulatory change associated with snake body plan evolution**. *Proc. Natl Acad. Sci. USA* 2013, **110**: 10473–10474.

Woltering, J.M., Vonk, F.J., Müller, H., Bardine, N., Tuduce, I.L., de Bakker, M.A., Knöchel, W., Sirbu, I.O., Durston, A.J., Richardson, M.K.: **Axial patterning in snakes and caecilians: evidence for an alternative interpretation of the Hox code**. *Dev. Biol.* 2009, **332**: 82–89.

13.10 The basic body plan of arthropods and vertebrates is similar but the dorso-ventral axis is inverted

Arendt, D., Nübler-Jung, K.: **Dorsal or ventral: similarities in fate maps and gastrulation patterns in annelids, arthropods and chordates**. *Mech. Dev.* 1997, **61**: 7–21.

Davis, G.K., Patel, N.H.: **The origin and evolution of segmentation**. *Trends Biochem. Sci.* 1999, **24**: M68–M72.

Gerhart, J., Lowe, C., Kirschner, M.: **Hemichordates and the origins of chordates**. *Curr. Opin. Genet. Dev.* 2005, **15**: 461–467.

Holley, S.A., Jackson, P.D., Sasai, Y., Lu, B., De Robertis, E., Hoffman, F.M., Ferguson, E.L.: **A conserved system for dorso-ventral patterning in insects and vertebrates involving *sog* and *chordin***. *Nature* 1995, **376**: 249–253.

Mizutani, C.M., Bier, E.: **EvoD/Vo: the origins of BMP signalling in the neuroectoderm**. *Nat. Rev. Genet.* 2008, **9**: 663–677.

13.11 Limbs evolved from fins

Boisvert, C.A., Mark-Kurik, E., Ahlberg, P.E.: **The pectoral fin of *Panderichthys* and the origin of digits**. *Nature* 2008, **456**: 636–638.

Minguillon, C., Gibson-Brown, J.J., Logan, M.P.: **Tbx4/5 gene duplication and the origin of vertebrate paired appendages**. *Proc. Natl Acad. Sci. USA* 2009, **106**: 21726–21730.

Onimaru, K., Kuraku, S., Takagi, W., Hyodo, S., Sharpe, J., Tanaka, M.: **A shift in anterior-posterior positional information underlies the fin-to-limb evolution**. *eLife* 2015, **4**: e07048.

Onimaru, K., Marcon, L., Musy, M., Tanaka, M., Sharpe, J.: **The fin-to-limb transition as the re-organization of a Turing pattern**. *Nat. Commun.* 2016, **7**: 11582.

Tanaka, M.: **Fins into limbs: autopod acquisition and anterior elements reduction by modifying gene networks involving 5′ *Hox*, *Gli3*, and *Shh***. *Dev. Biol.* 2016, **413**: 1–7.

13.12 Studies of zebrafish fin development give insights into the origins of tetrapod limbs

Ahn, D., Ho, R.K.: **Tri-phasic expression of posterior Hox genes during development of pectoral fins in zebrafish: implications for the evolution of vertebrate paired appendages**. *Dev. Biol.* 2008, **322**: 220–233.

Gehrke, A.R., Shubin, N.H.: **Cis-regulatory programs in the development and evolution of vertebrate paired appendages**. *Semin. Cell Dev. Biol.* 2016, **57**: 31–39.

Hawkins, M.B., Henke, K., Harris, M.P.: **Latent developmental potential to form limb-like skeletal structures in zebrafish**. *Cell* 2021, **184**: 899–911.

Hoff, M.: **A footnote to the evolution of digits**. *PLoS Biol.* 2014, **12**: e1001774.

Nakamura, T., Gehrke, A.R., Lemberg, J., Szymaszek, J., Shubin, N.H.: **Digits and fin rays share common developmental histories**. *Nature* 2016, **537**: 225–228.

Sordino, P., van der Hoeven, F., Duboule, D.: **Hox gene expression in teleost fins and the origin of vertebrate digits**. *Nature* 1995, **375**: 678–681.

Tulenko, F.J., Currie, P.D.: **Bones of contention: skeletal patterning across the fin-to-limb transition**. *Cell* 2021, **184**: 854–856.

Woltering, J.M., Noordermeer, D., Leleu, M., Duboule, D.: **Conservation and divergence of regulatory strategies at *Hox* loci and the origin of tetrapod digits**. *PLoS Biol.* 2014, **12**: e1001773.

Box 13C Transgenic tests of the function of Hox digit enhancers

Gehrke, A.R., Shubin, N.H.: **Cis-regulatory programs in the development and evolution of vertebrate paired appendages**. *Semin. Cell Dev. Biol.* 2016, **57**: 31–39.

13.13 Limbs have evolved to fulfill different specialized functions

Cooper, K.L., Sears, K.E., Uygur, A., Maier, J., Baczkowski, K.S., Brosnahan, M., Antczak, D., Skidmore, J.A., Tabin, C.J.: **Patterning and post-patterning modes of evolutionary digit loss in mammals**. *Nature* 2014, **511**: 41–45.

Lopez-Rios, J., *et al.*: **Attenuated sensing of SHH by Ptch1 underlies evolution of bovine limbs**. *Nature* 2014, **511**: 46–51.

Richardson, M.K., Oelschläger, H.H.: **Time, pattern, and heterochrony: a study of hyperphalangy in the dolphin embryo flipper**. *Evol. Dev.* 2002, **4**: 435–444.

Saxena, A., Towers, M., Cooper, K.L.: **The origins, scaling and loss of tetrapod digits**. *Philos. Trans. R Soc. Lond. B Biol. Sci.* 2017, **372**: 20150482.

Shapiro, M.D., Hanken, J., Rosenthal, N.: **Developmental basis of evolutionary digit loss in the Australian lizard *Hemiergis***. *J. Exp. Zool. B Mol. Dev. Evol.* 2003, **297**: 48–56.

13.14 The evolution of limblessness in snakes is associated with changes in axial gene expression and mutations in a limb-specific enhancer

Cohn, M.J., Tickle C.: **Developmental basis of limblessness and axial patterning in snakes**. *Nature* 1999, **399**: 474–479.

Infante, C.R., Mihala, A.G., Park, S., Wang, J.S., Johnson, K.K., Lauderdale, J.D., Menke, D.B.: **Shared enhancer activity in the limbs and phallus and functional divergence of a limb-genital cis-regulatory element in snakes**. *Dev. Cell* 2015, **35**: 107–119.

Kvon, E.Z., Kamneva, O.K., Melo, U.S., Barozzi, I., Osterwalder, M., Mannion, B.J., Tissières, V., Pickle, C.S., Plajzer-Frick, I., Lee, E.A., Kato, M., Garvin, T.H., Akiyama, J.A., Afzal, V., Lopez-Rios, J., Rubin, E.M., Dickel, D.E., Pennacchio, L.A., Visel, A.: **Progressive loss of function in a limb enhancer during snake evolution**. *Cell* 2016, **167**: 633–642.

Leal, F., Cohn, M.J.: **Loss and re-emergence of legs in snakes by modular evolution of Sonic hedgehog and HOXD enhancers**. *Curr. Biol.* 2016, **26**: 2966–2973.

Leal, F., Cohn, M.J.: **Developmental, genetic, and genomic insights into the evolutionary loss of limbs in snakes**. *Genesis* 2018, **56**: e23077.

Box 13D Using CRISPR-Cas9 genome-editing techniques to test the functioning of the ZRS from snakes

Kvon, E.Z., Kamneva, O.K., Melo, U.S., Barozzi, I., Osterwalder, M., Mannion, B.J., Tissières, V., Pickle, C.S., Plajzer-Frick, I., Lee, E.A., Kato, M., Garvin, T.H., Akiyama, J.A., Afzal, V., Lopez-Rios, J., Rubin, E.M., Dickel, D.E., Pennacchio, L.A., Visel, A.: **Progressive loss of function in a limb enhancer during snake evolution**. *Cell* 2016, **167**: 633–642.

13.15 Butterfly wing markings have evolved by redeployment of genes previously used for other functions

Brakefield, P.M., French, V.: **Butterfly wings: the evolution of development of colour patterns**. *BioEssays* 1999, **21**: 391–401.

French, V., Brakefield, P.M.: **Pattern formation: a focus on Notch in butterfly spots**. *Curr. Biol.* 2004, **14**: R663–R665.

Saenko, S.V.: **Chapter 6. Butterfly wing pattern evolution: insights from comparative analysis of the eyespot developmental network**. 2010. [https://scholarlypublications.universiteitleiden.nl/handle/1887/16039] (date accessed 3 September 2024).

Tong, X., Lindemann, A., Monteiro, A.: **Differential involvement of Hedgehog signaling in butterfly wing and eyespot development**. *PLoS One* 2012, **7**: e51087.

13.16 Adaptive evolution within the same species provides a way of studying the developmental basis for evolutionary change

Chan, Y.F., Marks, M.E., Jones, F.C., Villarreal, G., Jr, Shapiro, M.D., Brady, S.D., Southwick, A.M., Absher, D.M., Grimwood, J., Schmutz, J., Myers, R.M., Petrov, D., Jónsson, B., Schluter, D., Bell, M.A., Kingsley, D.M.: **Adaptive evolution of pelvic reduction in sticklebacks by recurrent deletion of a *Pitx1* enhancer**. *Science* 2010, **327**: 302–305.

Colosimo, P.F., Hosemann, K.E., Balabhadra, S., Villarreal, G., Jr, Dickson, M., Grimwood, J., Schmutz, J., Myers, R.M., Schluter, D., Kingsley, D.M.: **Widespread parallel evolution in sticklebacks by repeated fixation of Ectodysplasin alleles**. *Science* 2005, **307**: 1928–1933.

Howes, T.R., Summers, B.R., Kingsley, D.M.: **Dorsal spine evolution in threespine sticklebacks via a splicing change in *MSX2A***. *BMC Biol.* 2017, **15**:115.

Jeffery, W.R.: **Evolution and development in the cavefish *Astyanax***. *Curr. Top. Dev. Biol.* 2009, **86**: 191–221.

Krishnan, J., Rohner, N.: **Cavefish and the basis for eye loss**. *Philos. Trans. R. Soc. Lond. B Biol. Sci.* 2017, **372**: 20150487.

Peichel, C.L., Marques, D.A.: **The genetic and molecular architecture of phenotypic diversity in sticklebacks**. *Philos. Trans. R. Soc. Lond. B Biol. Sci.* 2017, **372**: 20150486.

Rohner, N., Jarosz, D.F., Kowalko, J.E., Yoshizawa, M., Jeffery, W.R., Borowsky, R.L., Lindquist, S., Tabin, C.J.: **Cryptic variation in morphological evolution: HSP90 as a capacitor for loss of eyes in cavefish**. *Science* 2013, **342**: 1372–1375.

Shapiro, M.D., Marks, M.E., Peichel, C.L., Blackman, B.K., Nereng, K.S., Jonsson, B., Schluter, D., Kingsley, D.M.: **Genetic and developmental basis of evolutionary pelvic reduction in threespine sticklebacks**. *Nature* 2004, **428**: 717–723.

Yamamoto, Y., Stock, D.W., Jeffery, W.R.: **Hedgehog signaling controls eye degeneration in blind cavefish**. *Nature* 2004, **431**: 844–847.

Yoshizawa, M., Yamamoto, Y., O'Quin, K.E., Jeffery, W.R.: **Evolution of an adaptive behavior and its sensory receptors promotes eye regression in blind cavefish**. *BMC Biol.* 2012, **10**: 108.

Box 13E Pelvic reduction in sticklebacks is based on mutations in a gene control region

Chan, Y.F., Marks, M.E., Jones, F.C., Villarreal, G., Jr, Shapiro, M.D., Brady, S.D., Southwick, A.M., Absher, D.M., Grimwood, J., Schmutz, J., Myers, R.M., Petrov, D., Jónsson, B., Schluter, D., Bell, M.A., Kingsley, D.M.: **Adaptive evolution of pelvic reduction in sticklebacks by recurrent deletion of a *Pitx1* enhancer**. *Science* 2010, **327**: 302–305.

13.17 Changes in growth rates can modify the basic body plan

Behringer, R.R., Rasweiler, J.J., Chen, C.H., Cretekos, C.J.: **Genetic regulation of mammalian diversity**. *Cold Spring Harb. Symp. Quant. Biol.* 2009, **74**: 297–302.

Hockman, D., Cretekos, C.J., Mason, M.K., Behringer, R.R., Jacobs, D.S., Illing, N.: **A second wave of *Sonic hedgehog* expression during the development of the bat limb**. *Proc. Natl Acad. Sci. USA* 2008, **105**: 16982–16987.

Richardson, M.K.: **Vertebrate evolution: the developmental origins of adult variation**. *BioEssays* 1999, **21**: 604–613.

Box 13F Origins of morphological diversity in dogs

Botigué, L.R., Song, S., Scheu, A., Gopalan, S., Pendleton, A.L., Oetjens, M., Taravella, A.M., Seregély, T., Zeeb-Lanz, A., Arbogast, R.M., Bobo, D., Daly, K., Unterländer, M., Burger, J., Kidd, J.M., Veeramah, K.R.: **Ancient European dog genomes reveal continuity since the Early Neolithic**. *Nat. Commun.* 2017, **8**: 16082.

Boyko, A.R., et al.: **A simple genetic architecture underlies morphological variation in dogs**. *PLoS Biol.* 2010, **8**: e1000451.

Ostrander, E.A.: **Genetics and the shape of dogs**. *Am. Sci.* 2007, **95**: 406–413.

Ostrander, E.A., Wayne, R.K., Freedman, A.H., Davis, B.W.: **Demographic history, selection and functional diversity of the canine genome**. *Nat. Rev. Genet.* 2017, **18**: 705–720.

Rimbault, M., Ostrander, E.A.: **So many doggone traits: mapping genetics of multiple phenotypes in the domestic dog**. *Hum. Mol. Genet.* 2012, **21**: R52–R57.

Rimbault, M., Beale, H.C., Schoenebeck, J.J., Hoopes, B.C., Allen, J.J., Kilroy-Glynn, P., Wayne, R.K., Sutter, N.B., Ostrander, E.A.: **Derived variants at six genes explain nearly half of size reduction in dog breeds**. *Genome Res.* 2013, **23**: 1985–1995.

Schoenebeck, J.J., Ostrander, E.A.: **The genetics of canine skull shape variation**. *Genetics* 2013, **193**: 317–325.

Shearin, A.L., Ostrander, E.A.: **Canine morphology: hunting for genes and tracking mutations**. *PLoS Biol.* 2010, **8**: e1000310.

Sutter, N.B., et al.: **A single *IGF-1* allele is a major determinant of small size in dogs**. *Science* 2007, **316**: 112–115.

13.18 Evolution can be due to changes in the timing of developmental events

Alberch, P., Alberch, J.: **Heterochronic mechanisms of morphological diversification and evolutionary change in the neotropical salamander *Bolitoglossa occidentales* (Amphibia: Plethodontidae)**. *J. Morphol.* 1981, **167**: 249–264.

Lande, R.: **Evolutionary mechanisms of limb loss in tetrapods**. *Evolution* 1978, **32**: 73–92.

McNamara, K.J.: **Heterochrony: the evolution of development**. *Evo. Edu. Outreach* 2012, **5**: 203–218.

Raynaud, A.: **Developmental mechanism involved in the embryonic reduction of limbs in reptiles**. *Int. J. Dev. Biol.* 1990, **34**: 233–243.

Wray, G.A., Raff, R.A.: **The evolution of developmental strategy in marine invertebrates**. *Trends Evol. Ecol.* 1991, **6**: 45–56.

13.19 The evolution of life histories has implications for development

Partridge, L., Harvey, P.: **The ecological context of life history evolution**. *Science* 1988, **241**: 1449–1455.

Stearns, S.C.: **Trade-offs in life-history evolution**. *Funct. Ecol.* 1989, **3**: 259–268.

Box 13G Long- and short-germ development in insects

Akam, M., Dawes, R.: **More than one way to slice an egg**. *Curr. Biol.* 1992, **8**: 395–398.

Angelini, D.R., Liu, P.Z., Hughes, C.L., Kaufman, T.C.: **Hox gene function and interaction in the milkweed bug *Oncopeltus fasciatus* (Hemiptera)**. *Dev. Biol.* 2005, **287**: 440–455.

Brown, S.J., Parrish, J.K., Beeman, R.W., Denell, R.E.: **Molecular characterization and embryonic expression of the *even-skipped* ortholog of *Tribolium castaneum***. *Mech. Dev.* 1997, **61**: 165–173.

French, V.: **Segmentation (and *eve*) in very odd insect embryos**. *BioEssays* 1996, **18**: 435–438.

French, V.: **Insect segmentation: genes, stripes and segments in 'Hoppers'**. *Curr. Biol.* 2001, **11**: R910–R913.

Lynch, J., Desplan, C.: **Evolution of development: beyond Bicoid**. *Curr. Biol.* 2003, **13**: R557–R559.

Lynch, J.A., Brent, A.E., Leaf, D.S., Pultz, M.A., Desplan, C.: **Localized maternal *orthodenticle* patterns anterior and posterior in the long germ wasp *Nasonia***. *Nature* 2006, **439**: 728–732.

Sander, K.: **Pattern formation in the insect embryo**. In *Cell Patterning* (Ciba Foundation Symposium, Vol. 29). Porter, R., Rivers, J. (Eds.). London: Ciba Foundation, 1975: 241–263.

Tautz, D., Sommer, R.J.: **Evolution of segmentation genes in insects**. *Trends Genet.* 1995, **11**: 23–27.

Glossary

2D:4D digit ratio the ratio of the lengths of the second and fourth fingers in humans, which differs between the sexes, being less than 1 in men; the same or more than 1 in women.

abdomen in insects, the most posterior of the three distinct parts of the body (head, thorax, and abdomen).

abembryonic pole see **embryonic-abembryonic axis**.

aboral refers to the side of a sea urchin embryo opposite to the mouth.

acceleration a type of heterochrony in which growth and development occur at a faster rate in some features.

accretionary growth increase in mass due to secretion of large quantities of extracellular matrix by cells, which enlarges the volume of extracellular space. It occurs, for example, in cartilage and bone, where most of the tissue mass is extracellular.

acrosomal reaction the release of enzymes and other proteins from the **acrosome** of the sperm head that occurs once a sperm has bound to the zona pellucida surrounding the mammalian egg. It helps the sperm to penetrate the zona pellucida enabling it to approach the egg plasma membrane.

acrosome vesicular structure on the sperm head that contains enzymes and other proteins. See also **acrosomal reaction**.

actin filament one of the three principal types of protein filaments in the cytoskeleton. Actin filaments are involved in cell movement and changes in cell shape, and are also part of the contractile apparatus of muscle cells. Also called microfilaments.

action potential the electrical signal generated in a neuron when it receives sufficient stimulatory inputs. It propagates along the length of the axon and stimulates the release of neurotransmitter from the axon terminal. Also called a nerve impulse.

activator gene-regulatory protein that helps to turn a gene on when it binds to a particular site in a gene's control region.

actomyosin an assembly of actin filaments and the motor protein myosin that can undergo contraction.

adaptive evolution the result of selection of characters that fit a particular environment.

adaxial–abaxial axis in plants, the axis that runs from the center of the plant stem to the circumference. In plant leaves it runs from the upper surface to the lower surface (dorsal to ventral).

adherens junction type of adhesive cell junction in which the adhesion molecules linking the two cells together are transmembrane cadherins that are linked intracellularly to the actin cytoskeleton.

adhesion molecule see **cell-adhesion molecule**.

affinity the strength of binding of a molecule at a single site on another molecule.

allantois a set of extra-embryonic membranes that develops in many vertebrate embryos. In bird and reptile embryos the allantois collects liquid waste and acts as a respiratory surface, whereas in mammals its blood vessels carry blood to and from the placenta.

allele a particular version of a gene. In diploid organisms, two alleles of each gene are present (one on each chromosome of a homologous pair), which may or may not be identical.

alternative splicing RNA splicing that generates an alternative form of an mRNA by using different splicing.

ameloblast cell that secretes enamel during tooth development.

amnion extra-embryonic membrane in birds, reptiles, and mammals that forms a fluid-filled sac, which encloses and protects the embryo. It is derived from extra-embryonic ectoderm and mesoderm.

amnioserosa extra-embryonic membrane on the dorsal side of the *Drosophila* embryo.

amniotes vertebrates whose embryos have an amnion. They comprise the mammals, birds, and reptiles.

anamniotes vertebrates whose embryos do not have an amnion. They comprise fish and amphibians.

androgenetic describes an embryo in which the two sets of homologous chromosomes are both paternal in origin.

angioblast mesodermal precursor cells that will give rise to the endothelial cells of blood vessels.

angiogenesis the process by which small blood vessels sprout from the larger vessels.

animal cap the tissue around the animal pole in *Xenopus* eggs.

animal pole see **animal region**.

animal region the 'upper' hemisphere of amphibian eggs and blastulas away from the yolk wherein the egg, the nucleus, resides. The region in the blastula from which the ectoderm will develop. The most terminal part of this region is the animal pole, which is directly opposite the vegetal pole at the terminal part of the vegetal region.

animal–vegetal axis axis that runs from the animal pole to the vegetal pole in an egg or early embryo.

aniridia the absence of an iris in the eye.

Antennapedia complex a cluster of genes that comprises one part of the Hox gene clusters (the HOM-C complex) in *Drosophila*.

anterior at, or in the direction of, the head end of an embryo or animal.

anterior neural ridge (ANR) a signaling center in the developing brain of the early chick embryo that lies immediately adjacent to and within the anterior-most forebrain, and is responsible for patterning of the telencephalon (the part of the forebrain that gives rise to the pallium and the basal ganglia).

anterior visceral endoderm (AVE) an extra-embryonic tissue in the early mouse embryo that is involved in inducing anterior regions of the embryo.

antero-posterior axis the axis running from the head end to the tail end of an animal. The head is anterior and the tail posterior. In the vertebrate limb, this axis runs from the thumb to little finger.

anticlinal describes cell divisions in planes at right angles to the outer surface of a tissue.

antisense RNA an RNA complementary in nucleotide sequence to an mRNA, which blocks expression of a protein by binding to its mRNA and blocking translation.

apical–basal axis the axis running from shoot tip to root tip in a plant. The shoot end is apical and the root end is basal.

apical dominance the phenomenon in plants that buds in the axils of nodes behind the tip of a plant shoot will not form side stems when the tip is intact. It is due to the production of the hormone auxin by the shoot tip which suppresses outgrowth of axillary buds.

apical ectodermal ridge, apical ridge a thickening of the ectoderm at the distal end of the developing chick and mammalian limb bud, which is essential for limb bud outgrowth and for correct patterning along the proximo-distal axis of the limb.

apical epidermal cap the epidermis that forms over the cut surface of an amputated amphibian limb. Its formation is essential for subsequent regeneration. Also called wound epidermis.

apical epithelial cap see **apical epidermal cap**.

apical meristem the region of dividing cells at the tip of a growing plant shoot or root, from which all new tissue is produced.

apico-basal polarity cell polarity seen in an epithelium, in which the apical and basal faces of the cells have different properties.

apoptosis type of cell death that occurs widely during development. Also called programmed cell death. The cell is induced by an external or internal signal to commit 'cell suicide', which involves fragmentation of the DNA and shrinkage of the cell, without release of the cell contents. Apoptotic cells are removed by the body's scavenger cells and, unlike necrosis, their death does not cause damage to surrounding cells.

archenteron the cavity formed inside the embryo when the endoderm and mesoderm invaginate during gastrulation. It forms the gut.

area opaca the outer mottled area of the chick blastoderm.

area pellucida the central translucent area of the chick blastoderm.

aster structure made of microtubules radiating away from the centrosome and anchored in the cortex, found at each pole of the mitotic or meiotic spindle.

asymmetric cell division, asymmetric division a cell division in which the daughter cells are different from each other because some cytoplasmic determinant(s) has been distributed unequally between them.

autonomous describes any developmental process that can continue without a requirement for extracellular signals to be continuously present. See also **cell autonomous**.

AUX/IAA protein any of a family of proteins that, by binding to auxin-response factors, block the expression of auxin-responsive genes in the absence of the plant hormone auxin.

auxin small organic molecule that is an important plant hormone in almost all aspects of plant development. Auxin acts by regulating the expression of auxin-responsive genes by stimulating the degradation of AUX/IAA proteins and thus enabling gene expression.

auxin-response factor (ARF) DNA-binding transcriptional activator (or in a few cases a repressor) that binds to auxin-responsive elements in the control region of auxin-responsive genes.

auxin-responsive gene any plant gene whose expression is controlled by the plant hormone auxin.

AVE See **anterior visceral endoderm**.

axial mesendoderm prechordal mesoderm, notochord, and underlying endoderm.

axial mesoderm mesodermal structures that form along the dorsal midline of a vertebrate embryo—notochord and prechordal mesoderm.

axon long cell process of a neuron that conducts nerve impulses away from the cell body. The end of an axon—the axon terminal—forms contacts (synapses) with other neurons, muscle cells, or glandular cells.

axon terminal see **axon**.

β-catenin protein that functions both as a transcriptional co-activator and as a protein that links adhesive cell junctions to the cytoskeleton. β-catenin acts as a transcriptional co-activator in early development of many embryos as the end result of a Wnt signaling pathway.

β-neurexin transmembrane protein in the pre-synaptic membrane of some synapses, which interacts with the protein neuroligin in post-synaptic membranes to promote the functional differentiation of the synapse.

basal lamina see **basement membrane**.

basal layer the deepest layer of the epidermis of the skin, which contains epidermal stem cells from which the epidermis is renewed.

base editing a technique by which single-base changes in targeted genes can be made in an organism's genome, using specialized nucleases.

basement membrane sheet of extracellular matrix that separates an epithelial layer from the underlying tissues. For example, the epidermis of the skin is separated from the dermis by a basement membrane. Also called the basal lamina.

bilaminar germ disc a two-layered sheet of epithelial cells in the human embryo, the upper epiblast which gives rise to the embryo and the lower hypoblast which gives rise to extraembryonic structures.

bilateral symmetry type of symmetry possessed by animals with a main axis of symmetry running from head to tail and the two sides of the body being mirror images of each other.

bithorax complex a cluster of genes that comprises one part of the Hox gene complex (HOM-C complex) in *Drosophila*.

blastema the mass of cells that develops at the site of amputation of the amphibian limb. It is formed from the dedifferentiation and proliferation of cells beneath the wound epidermis, and gives rise to the regenerated limb.

blastocoel fluid-filled cavity that develops in the interior of a blastula.

blastocyst the stage of a mammalian embryo that corresponds in form to the blastula stage of other animal embryos, and is the stage at which the embryo implants in the uterine wall.

blastoderm a post-cleavage embryo composed of a solid layer of cells rather than a spherical blastula, as found in early chick, zebrafish, and *Drosophila* embryos.

blastodisc alternative name for the blastoderm in the chick embryo.

blastoid structure produced *in vitro* from a combination of embryonic stem cells and trophoblast stem cells which mimics the mammalian blastocyst.

blastomere any of the cells formed by the cleavage divisions of the fertilized egg.

blastopore slit-like or circular invagination on the surface of amphibian and sea urchin embryos, at which the mesoderm and endoderm move inside the embryo at gastrulation.

blastula early stage in the development of some embryos (e.g. amphibians, sea urchin), which is the outcome of cleavage. It is a hollow ball of cells, composed of an epithelial layer of cells enclosing a fluid-filled cavity—the blastocoel.

block to polyspermy the mechanism in some animals that prevents fertilization of an egg by more than one sperm. In sea urchins and *Xenopus* it is composed of two stages—an immediate rapid electrical block to polyspermy that involves depolarization of the egg plasma membrane, and a subsequent slow block to polyspermy, which involves the formation of an impenetrable fertilization membrane around the egg.

body plan the overall organization of an organism; for example, the relative positions of the head and tail, and the plane of bilateral symmetry, where it exists. The body plan of most animals is organized around two main axes: the antero-posterior axis and the dorso-ventral axis.

bone marrow stromal cells cells in bone marrow that form the stem-cell niche that supports and maintains the hematopoietic stem cells.

branchial arches structures that develop on each side of the embryonic head and give rise to the gill arches in fishes, and to the jaws and other facial structures in other vertebrates. Often called the pharyngeal arches in mammals.

branching morphogenesis describes the development and growth of structures such as blood vessels or the bronchi and bronchioles of the lung, which develop by the epithelium forming a succession of branches.

cadherin any of a family of cell-adhesion molecules with important roles in development. Cadherins are the adhesion molecules in adherens junctions.

cambium a ring of meristem in the stems of plants that gives rise to new stem tissue, which increases the diameter of the stem.

canonical Wnt/β-catenin pathway an intracellular signaling pathway stimulated by members of the Wnt family of signaling proteins that leads to the stabilization of β-catenin and its entry into nuclei, where it acts as a transcriptional co-activator.

capacitation the functional maturation of sperm after they have been deposited in the female reproductive tract.

cardiomyocyte a mature fully differentiated heart muscle cell.

caspase any of a family of intracellular proteases, some of which are involved in apoptosis.

catenin any of a family of proteins that are part of adherens junctions, involved in linking cadherins to the cytoskeleton. One member of the family, β-catenin, also acts as a transcriptional activator in the canonical Wnt/β-catenin signaling pathway.

cell-adhesion molecule (CAM) any of several different families of proteins that bind cells to each other and to the extracellular matrix. The main classes of adhesion molecules important in development are the cadherins, the immunoglobulin superfamily, and the integrins.

cell adhesiveness the property of cells that causes them to stick together. It is mediated by adhesive cell junctions composed of cell-adhesion molecules that bind to similar or different types of cell-adhesion molecules on the neighboring cell.

cell autonomous describes the effects of a gene that only affects the cell the gene is expressed in.

cell body the part of a neuron that contains the nucleus, and from which the axon and dendrites extend.

cell–cell interaction, cell–cell signaling general terms that describe various types of intercellular communication by which one cell influences the behavior of another cell. Cells can communicate with each other via cell contact and interaction of membrane-bound signal molecules with receptors on the contacting cell, or by the secretion of signal molecules that act on specific receptors on (or in) other cells nearby or at a distance to influence their behavior.

cell cycle the sequence of events by which a cell duplicates its DNA and divides in two.

cell-cycle checkpoint any one of several points that occur during the cell cycle at which the cell monitors progress through the cycle and ensures that a previous stage has been completed before embarking on the next.

cell differentiation the process by which cells become functionally and structurally different from one another and become distinct cell types, such as muscle or blood cells.

cell enlargement an increase in cell size without division, which is one way in which a structure or embryo can grow in size. Growth by cell enlargement is especially common in plants.

cell-fate determinants proteins and RNAs present in a cell that determine its fate.

cell-identity switching a mechanism that neatens the boundaries between rhombomeres in the developing brain, in which some cells at the boundaries switch fate.

cell-lineage restriction the situation where all the descendants of a particular group of cells remain within a 'boundary' and never mix with an adjacent group of cells of a different lineage. The area within such a boundary is known as a compartment.

cell migration the active movement of cells from one location to another in the embryo by 'crawling' along the extracellular matrix, often under the guidance of molecules previously laid down in the matrix.

cell proliferation the growth and division of cells to form new cells, which is one way in which a structure or embryo can grow in size.

cell-replacement therapy the correction of disease by replacing damaged cells with new healthy cells, including stem cells. So far, mainstream cell-replacement therapy using stem cells is limited to hematopoietic stem cell transplants (such as bone marrow transplants), which use naturally occurring sources of donor stem cells, but therapies using other types of stem cells are being researched.

cell segregation a mechanism that neatens the boundaries between rhombomeres in the developing brain in which cells at the boundary segregate to an adjacent rhombomere as a result of cell signaling through ephrins and Eph receptors.

cell senescence see **senescence**.

cell shape one of three key cell properties that are crucial in embryonic development. Cells can change shape by means of rearrangements of the internal cytoskeleton.

central nervous system (CNS) the brain and spinal cord.

centrosome the organizing center for microtubule growth within a cell. It duplicates before mitosis or meiosis, and each centrosome forms one end of the microtubule spindle. It consists of a pair of centrioles, one of which gives rise to the basal body of cilia, including primary cilia which are found on most vertebrate cells.

cerebellum part of the hindbrain concerned with cognition, emotion, and motor control, including muscle tone and balance.

cerebral cortex the outer layers of the cerebral hemispheres, which develop from the pallium of the embryonic forebrain and in which the highest-level processing centers for sensory information, motor control, learning, and memory are located.

chemoaffinity hypothesis the idea that each retinal neuron carries a chemical label that enables it to connect reliably with an appropriately labeled cell in the optic tectum.

chemoattractant a molecule that attracts cells or axons to move towards it.

chemorepellent a molecule that repels cells, causing them to move away from it.

chemotropic response the response of the growth cone of an axon to chemotactic cues.

chimera an organism or tissue that is composed of cells from two or more sources with different genetic constitutions.

ChIP-chip (or ChIP-on-chip) and ChIP-seq techniques for determining which sites in chromosomal DNA are bound by particular proteins *in vivo*. They involve immunoprecipitation of chromatin fragments by antibodies against the protein of interest followed by analysis of the bound DNA by microarrays (ChIP-chip) or DNA sequencing (ChIP-seq).

choice point see **intermediate target**.

chondrocyte a differentiated cartilage cell.

chorion the outermost of the extra-embryonic membranes in birds, reptiles, and mammals. It is involved in respiratory gas exchange. In birds and reptiles it lies just beneath the shell. In mammals it is part of the placenta and is also involved in nutrition and waste removal. The chorion of insect eggs has a different structure.

chromatin the material of which chromosomes are made. It is composed of DNA packaged into nucleosomes by histone proteins, and is also associated with many other proteins that influence its structure. Chemical modifications to DNA and histone proteins affect the structure of chromatin and the proteins that are associated with it and determine whether a given gene is available for transcription or has been packed away so that it is inaccessible. Such chemical modifications are known as epigenetic modifications as, unlike mutation, they affect gene expression without changing the nucleotide sequence of the DNA.

chromatin-remodeling complex enzyme complex that acts on chromatin to modify it and alter the ability of the DNA to be transcribed.

chromosome conformation capture various techniques that map the physical interactions made between different regions of DNA in the genome. These techniques have identified chromosomal structural features such as topologically associating domains— regions in which the DNA makes many interactions with itself and which are separated by regions in which there are many fewer interactions.

ciliopathy any of a number of human syndromes, which may affect many different organs, in which the function of cilia is defective. Some ciliopathies, such as Kartagener's syndrome, show abnormal right–left asymmetry of organs such as heart and liver (situs inversus) in a proportion of cases.

circadian clock an internal 24-hour timer present in living organisms that causes many metabolic and physiological processes, including the expression of some genes, to vary in a regular manner throughout the day.

cis-**regulatory module** see *cis*-**regulatory region**.

cis-**regulatory region** the non-coding sequences flanking a gene and containing sites at which the expression of that gene can be controlled by the binding of transcription factors (activators and repressors). Also called control regions. Many regulatory regions in developmental genes contain a series of different *cis*-**regulatory modules**, which are short regions containing multiple binding sites for different transcription factors; the combination of factors bound determines whether the gene is switched on or off. Different regulatory modules control gene expression in different locations or times during development.

cleavage a series of rapid cell divisions without cell growth that occurs after fertilization and divides the embryo up into a number of small cells.

clock-and-wavefront model a model that has been proposed to explain the periodicity of somite formation. It proposes that a 'clock' represented by regularly cycling gene expression in the pre-somitic mesoderm interacts with a wavefront of somite determination moving anterior to posterior to specify the formation of each somite.

clone (1) a collection of genetically identical cells derived from a single cell by repeated cell division. (2) The genetically identical offspring of a single individual produced by asexual reproduction or artificial cloning techniques.

cloning procedure by which an individual genetically identical to a 'parent' is produced by transplantation of a parental somatic cell nucleus into an unfertilized oocyte.

co-activator gene-regulatory protein that promotes gene expression but does not bind to DNA itself. It binds specifically to other DNA-binding transcription factors, enabling them to act as activators.

coding region that part of a gene that encodes a polypeptide or functional RNA.

coelom fluid-filled, mesoderm-lined body cavity that surrounds the internal organs in many groups of metazoans. In vertebrate embryos it initially develops within the lateral plate mesoderm.

co-linearity the correspondence between the order of Hox genes on a chromosome and their temporal and spatial order of expression in the embryo.

collective cell migration a type of cell movement in which a group of cells undergoes migration together as a collective entity, remaining linked to each other by adhesive junctions.

colony-stimulating factor any protein that drives the differentiation of blood cells.

commissural neuron neuron in the spinal cord of vertebrates, or the ventral nerve cord of insects, whose axon crosses the midline, forming connections with neurons in the contralateral side of the cord.

commissure tract of axons that cross the midline, forming connections with neurons on the contralateral side.

community effect the phenomenon that induction of cell differentiation in some tissues depends on there being a sufficient number of responding cells present.

compaction in the mouse embryo, the stage in early cleavage at which the blastomeres flatten against each other and microvilli become confined to the outer surface of the ball of cells.

compartment a discrete area of an embryo that contains all the descendants of a small group of founder cells and shows cell-lineage restriction. Cells in compartments respect the compartment boundary and do not cross over into an adjacent compartment. Compartments tend to act as discrete developmental units.

competence the ability of a tissue to respond to an inducing signal. Embryonic tissues only remain competent to respond to a particular signal for a limited period of time.

competent able to respond to an inducing signal.

conditional mutation a mutation whose effects only become manifest in a particular condition—for example, at a higher temperature than normal.

congenital condition a medical condition present at birth, such as spina bifida. Often, but not always, due to a genetic mutation.

constitutive in relation to heterochromatin, refers to DNA that is unlikely ever to be transcribed.

contractile ring contractile structure composed of actomyosin that develops around the circumference of an animal cell at the plane of cleavage. Contraction of the ring forms the cleavage furrow that divides the cell in two.

control element *see* **control region**.

control region a non-coding region of a gene to which regulatory proteins bind and determine whether or not the gene is transcribed. Also called *cis*-regulatory region, control element.

convergent evolution evolution in two different lineages or populations that independently produces a similar outcome in both.

convergent extension the process by which a sheet of cells changes shape by extending in one direction and narrowing—converging—in a direction at right angles to the extension, caused by the cells intercalating between each other. It contributes, for example, to elongation of the notochord, elongation of the body in amphibian and zebrafish embryos, and extension of the germ band in *Drosophila* embryos.

cooperativity the phenomenon in which the binding of a ligand to one site on another molecule (e.g. the binding of a transcription factor to a site in DNA) makes the binding of subsequent ligands to other sites on the same molecule more likely to occur.

co-option the situation in evolution in which a protein or a structure with one function acquires a different one because its properties make it suitable for a different role.

co-repressor gene-regulatory protein that acts to repress gene expression but does not bind to DNA itself. It acts in a complex with DNA-binding transcription factors.

cortex (1) the outer layer of the cytoplasm of an animal cell, just underneath the plasma membrane, which is more gel-like in consistency than the rest of the cytoplasm and is rich in actin and myosin. (2) The tissue between the epidermis and central vascular tissue in plant stems and roots. (3) *See* **cerebral cortex**.

cortical granules granules present in the cortex of some eggs, which release their contents on fertilization to form the fertilization membrane.

cortical rotation a movement of the cortex of the fertilized amphibian egg that occurs immediately after fertilization. The cortex rotates with respect to underlying cytoplasm away from the point of sperm entry.

corticotropin-releasing hormone peptide hormone released by neurosecretory cells in the hypothalamus. In non-mammalian vertebrates it acts on the pituitary gland to cause the release of thyroid-stimulating hormone and consequent metamorphosis.

cotyledon the embryonic leaf in seed-bearing plants that acts as a food storage organ.

Cre/*loxP* system a transgenic modification of mice that enables a gene to be expressed in a particular tissue or at a particular time in development.

CRISPR-Cas9 system a gene-editing system derived from bacteria that can be introduced into animal cells to make precise targeted mutations.

cumulus cells somatic cells that surround the developing mammalian oocyte and are shed with it at ovulation.

cyclin any of a family of proteins that periodically rise and fall in concentration during the cell cycle and are involved in controlling progression through the cycle. Cyclins act by binding to and activating cyclin-dependent kinases (Cdks).

cyclin-dependent kinases (Cdks) protein kinases that have specific roles in particular phases of the cell cycle and are activated by the binding of cyclins. Because cyclin concentration varies periodically in the cell cycle, so do the activities of the cyclin-dependent kinases.

cytoplasmic determinant a cytoplasmic protein or mRNA that has a specific influence on development. The unequal distribution of cytoplasmic determinants to daughter cells by an asymmetric cell division is one way of producing cells with different developmental fates.

cytoplasmic localization the non-uniform distribution of some protein factor or determinant in a cell's cytoplasm, so that when the cell divides, the protein is unequally distributed to the daughter cells.

cytoskeleton the network of protein filaments—microfilaments, microtubules, and intermediate filaments—that gives cells their shape, enables them to move, and provides tracks for transport of materials in the cell.

cytotrophoblast the single layer of trophoblast cells derived from the trophectoderm that surrounds the mammalian embryo once it is implanted in the uterine wall. It will form the chorion.

dauer a larval state in *Caenorhabditis elegans* that arises in response to starvation conditions and in which the larvae neither eat nor grow until food is again available.

dedifferentiation loss of the structural characteristics of a differentiated cell, which may result in the cell then differentiating into a new cell type.

deep layer the layer several cells deep under the outer enveloping layer of the zebrafish blastoderm. It gives rise to the embryo.

definitive endoderm the endoderm that forms the lining of the gut.

delamination the process by which epithelial cells leave an epithelium as individual cells. It occurs, for example, in the primitive streak and in the movement of neural crest cells out of the neural tube.

dendrite an extension from the body of a nerve cell that receives stimuli from other nerve cells.

dental lamina band of thickened oral epithelium in the jaw within which individual teeth will develop.

dental papilla condensation of ectomesenchyme cells that gives rise to dentin and the soft connective tissue in the center of the tooth (the dental pulp).

denticle small tooth-like outgrowth of the cuticle on insect larvae.

dependence receptor any of a large group of 'death receptors' in neurons that can fire a signal promoting apoptosis in the absence of ligand. They act in the programmed cell death of neurons.

dermis the connective tissue layer of the skin lying beneath the epidermis, from which it is separated by a basement membrane.

dermomyotome the region of the somite that will give rise to both muscle and dermis.

desmosome type of cell junction in which the cadherin cell-adhesion molecules are linked intracellularly to keratin filaments of the cytoskeleton.

determinant factor that can be asymmetrically distributed at cell division and so influence how the daughter cells develop. See also **cytoplasmic determinant**.

determination a stable change in the internal state of a cell such that its fate is now fixed, or **determined**. A determined cell will follow that fate if grafted into other regions of the embryo.

deuterostomes animals, such as chordates and echinoderms, that have radial cleavage of the egg, and in which the primary invagination of the gut at gastrulation forms the anus, with the mouth developing independently.

developmental biology the study of the development of a multicellular organism from its origin in a single cell—the fertilized egg—until it is fully adult.

developmental gene any gene that specifically controls a developmental process.

differentiation see **cell differentiation**.

differential adhesion hypothesis a hypothesis to explain the movement of cells in developing embryos (e.g. at gastrulation) in terms of the differences in the strength of adhesion of different types of cells for each other.

differential gene expression the turning on and off of different genes in different cells in a multicellular organism, thus generating cells with different developmental and functional properties.

digital plate the region at the end of the developing vertebrate limb from which the digits will develop. Also called the hand plate.

diploblasts animals with two germ layers (endoderm and ectoderm) only; they include cnidarians such as *Hydra* and jellyfish.

diploid describes cells that contain two sets of homologous chromosomes, one from each parent, and thus two copies of each gene.

direct development type of development in some species of amphibians, where they develop directly into adults without a larval (tadpole) stage.

directed dilation the expansion of a structure (in animals) or an individual cell (in plants) in a particular direction due to an increase in internal hydrostatic pressure inside a constraining sheath or membrane.

distal describes the end of a structure, such as a limb or wing, that is furthest away from the point of attachment to the body.

distal visceral endoderm (DVE) the endoderm located at the distal end of the cup-shaped mouse epiblast, which moves and extends away from the posterior side of the embryo, upon which it becomes the anterior visceral endoderm (AVE).

DNA methylase enzyme that adds methyl groups to DNA.

DNA methylation covalent attachment of methyl groups to DNA, which alters the ability of the DNA to be transcribed.

DNA microarray analysis technique for detecting and measuring the expression of large numbers of genes simultaneously, by hybridization of cellular RNA or cDNA to arrays of oligonucleotides representing gene sequences.

dominant describes an allele that determines the phenotype even when present in only a single copy.

dorsal describes the back of an embryo or animal, the side opposite to the belly or underside. In plants, in relation to leaves, it describes the upper surface of the leaf.

dorsal closure in *Drosophila*, the bringing together of the two edges of the dorsal epidermis over the amnioserosa, to fuse along the dorsal midline of the embryo.

dorsal convergence in zebrafish, directed migration of individual mesodermal cells from lateral regions towards the midline to form the notochord.

dorsal ectoderm one of the four main regions into which the dorso-ventral axis of the early *Drosophila* embryo is divided before gastrulation. This region will give rise to epidermis.

dorsalized describes embryos that have developed greatly increased dorsal regions at the expense of ventral regions, as a result of experimental manipulation.

dorsalizing factor in the early vertebrate embryo, a protein or RNA that promotes the formation of dorsal structures, such as maternal components of the Wnt signaling pathway in *Xenopus*.

dorsal lip the dorsal edge of the forming blastopore in *Xenopus* embryos.

dorso-lateral hinge point site at each side of the developing neural tube at which it bends to bring the tips of the neural folds towards each other, to eventually fuse and close the neural tube.

dorso-ventral axis the body axis running from the upper surface or back (dorsal) to the under surface (ventral) of an organism or structure. The mouth is always on the ventral side.

dosage compensation the mechanism that ensures that, although the number of X chromosomes in males and females is different, the level of expression of X-chromosome genes is the same in both sexes. Mammals, insects, and nematodes all have different dosage-compensation mechanisms.

durotaxis the ability of cells to sense the stiffness of a substrate and to migrate along a stiffness gradient.

dynamic fate map fate maps that trace over time what each cell gives rise to.

ecdysis molting in arthropods, in which the external cuticle is shed to allow for growth.

ecdysone a steroid hormone in insects that is responsible for initiating molting and also the transition to pupation and metamorphosis.

ectoderm the embryonic germ layer that gives rise to the epidermis and the nervous system.

ectomesenchyme cells of ectodermal origin (neural crest) that give rise to bone, cartilage, and connective tissue of the face.

ectoplacental cone extra-embryonic structure formed from the polar trophectoderm in mammalian embryos.

egg chamber the structure in a female *Drosophila* within which an oocyte develops surrounded by its nurse cells and follicle cells.

egg cylinder the cylindrical structure comprising the epiblast covered by visceral endoderm in early post-implantation mouse embryogenesis.

embryogenesis the process of development of the embryo from the fertilized egg.

embryoid body structures that can be formed from embryonic stem cells *in vitro*, consisting of a disorganized mass of multiple cell types.

embryology the study of the development of an embryo.

embryonic germ ring in zebrafish embryos, a thickening around the blastoderm edge formed by deep-layer cells.

embryonic organizer alternative name for the Spemann organizer in amphibians and similar organizing regions in other vertebrates (such as the node in chick) that can direct the development of a complete embryo.

embryonic pole the site on the blastocyst wall where the inner cell mass is attached to the trophectoderm.

embryonic stem cells (ES cells) pluripotent cells derived from the inner cell mass of a mammalian embryo, most commonly mouse embryos, that can be indefinitely maintained in culture. When injected into another blastocyst, they combine with the inner cell mass and can potentially contribute to all the tissues of the embryo.

embryonic–abembryonic axis in the mammalian blastocyst the axis running from the site of attachment of the inner cell mass (the embryonic pole), to the opposite pole (the abembryonic pole).

endoblast in the chick embryo, the layer of cells that grows out from the posterior marginal zone prior to primitive streak formation, and replaces the hypoblast underlying the epiblast.

endochondral ossification the replacement of cartilage with bone in the growth plates of vertebrate embryonic skeletal elements, such as the long bones of the limbs.

endoderm the embryonic germ layer that gives rise to the gut and associated organs, such as the lungs and liver in vertebrates.

endodermis a tissue layer in plant roots interior to the cortex and outside the vascular tissue.

endomesoderm *see* **mesendoderm**.

endosperm nutritive tissue in higher plant seeds that serves as a source of food for the embryo.

endothelial cell cell type that makes up the endothelium lining blood vessels.

endothelium the epithelium lining blood vessels.

enhancer a gene control region to which transcription factors bind to switch on the gene, especially in respect of highly regulated tissue-specific genes. Enhancers often contain multiple sites for binding both activating and repressive transcription factors.

enhancer-trap technique used in *Drosophila* to turn on the expression of a specific gene in a particular tissue or stage in development.

Eph receptor cell-surface protein that is the receptor for an **ephrin** protein. Interactions of ephrins and their receptors can cause repulsion of cells or attraction and adhesion, and are involved, for example, in delimiting compartments in rhombomeres and in axonal guidance.

ephrin *see* **Eph receptor**.

epiblast in mouse and chick embryos, a group of cells within the blastocyst or blastoderm, respectively, that gives rise to the embryo proper. In the mouse, it develops from cells of the inner cell mass.

epiblast stem cells (EpiSCs) stem cells derived from a later-stage mouse epiblast than the inner cell mass. They can be maintained indefinitely in culture and can give rise to a wide range of cell types, but unlike embryonic stem cells they cannot give rise to chimeras when injected into another blastocyst.

epiboly the spreading and movement of cell sheets during gastrulation in amphibians and fish embryos. In amphibians, the ectoderm extends to cover the whole of the embryo.

epidermis in vertebrates, insects, and plants the outer layer of cells that forms the interface between the organism and its environment. Its structure is quite different in the different organisms.

epigenesis an old term for the formation of an embryo by the progressive development of structures, as opposed to the now discredited idea of preformation, in which all the structures are there in miniature from the beginning.

epigenetic describes mechanisms of gene regulation that involve the modification of chromatin by, for example, DNA methylation, histone methylation, and histone acetylation without changes to the DNA sequence itself, and which can be passed on to daughter cells.

epimorphosis a type of regeneration in which the regenerated structures are formed by new growth.

EpiSCs *see* **epiblast stem cells**.

epithelial-to-mesenchymal transition (EMT) the process by which epithelial cells lose adhesiveness and detach from the epithelium as single cells to form mesenchyme.

epithelium (plural **epithelia**) a sheet of cells tightly bound to each other by adhesive cell junctions.

equatorial region the belt-like region at the equator of the late blastula of an amphibian embryo which will form mesoderm. *See also* **marginal zone**.

erythroid describes the lineage of blood cells that gives rise to red blood cells (erythrocytes) and megakaryocytes.

ES cells *see* **embryonic stem cells**.

ET embryoid structure resembling the post-implantation epiblast formed *in vitro* from a combination of mouse ES cells and trophoblast stem cells grown in the presence of the extracellular matrix material Matrigel.

ETX embryoid epiblast-like structure with initiation of a primitive streak formed *in vitro* from a combination of mouse primitive endoderm cells and ES cells.

evagination the turning of a pouch or tubular structure inside out and its protrusion from the surface of a structure or the body.

evo-devo an informal term for the study of the evolution of development.

experience-dependent development aspects of neural development, such as ocular dominance columns, that can be modified by sensory experience.

explant small fragment of tissue dissected out from an embryo and cultured *in vitro*.

extra-embryonic describes tissues external to the embryo proper that are involved in its protection and nutrition. In mammals they include the amnion, chorion, and placental tissues.

extra-embryonic ectoderm in mammals, embryonic tissue that contributes to the formation of the placenta.

extra-embryonic endoderm cells (XEN cells) self-renewing cells that can be derived from primitive endoderm *in vitro*.

facultative in reference to heterochromatin, DNA that although hidden from the transcriptional machinery at any given time, retains the ability to be transcribed in a regulated manner.

Fallopian tubes the name given to the oviducts in mammals.

fasciculation the formation of axons (nerve fibers) into bundles (nerve tracts) that follow a particular pathway.

fat body an organ in *Drosophila* akin to our liver and adipose tissue.

fate describes what a cell will normally develop into. Having a particular fate does not, however, imply that a cell could not develop differently if placed in a different environment.

fate map a schematic showing what a particular region of an embryo or cell population will give rise to at a later time.

fertilization the fusion of sperm and egg to form the zygote.

fertilization membrane a membrane formed around the eggs of some species after fertilization to prevent entry of further sperm.

fetus the name given to a human embryo after 8 weeks of gestation, and to other mammalian embryos after a similar stage.

file in a plant root, a vertical column of cells that originates from a single initial in the root meristem.

filopodium (plural **filopodia**) long cytoplasmic extension put out by some cells, which enables them to pull themselves along a substratum.

first heart field the cardiac crescent, a region of prospective heart cells located in the splanchnic mesoderm on either side of the body underlying the head folds. These regions come together to form the heart tube. Cells from the cardiac crescent are the first to differentiate into heart muscle cells.

floor plate a small region of the developing neural tube at the ventral midline that is composed of non-neural cells. It is involved in patterning the ventral part of the neural tube.

floral meristem a region of dividing cells at the tip of a shoot that gives rise to a flower.

floral organ identity gene any of a number of genes in plants that give the different parts of a flower (petal, stamen, etc.) their individual identities.

floral organ primordium parts of a floral meristem that have become specified to give rise to the individual parts of a flower (petal, stamen, etc.).

focal adhesions contacts between a cell and its substratum, formed by integrins at the cell surface interacting with the extracellular matrix.

follicle the structure in the vertebrate ovary that contains an egg cell and its supporting somatic cells (follicle cells). In *Drosophila*, follicle cells are the somatic cells that surround the oocyte and nurse cells in the egg chamber during egg development.

follicle cells *see* **follicle**.

follower axons axons that follow a scaffold established by pioneer neurons to a particular target in the brain, forming a nerve tract.

forebrain the anterior part of the vertebrate embryonic brain that will give rise to the cerebral hemispheres, the thalamus, and the hypothalamus. Also called the prosencephalon.

forward genetics type of genetic analysis in which a mutant organism is first identified by its unusual phenotype and then genetic experiments are done to discover which gene is responsible for the mutant phenotype.

founder cells the small number of cells deriving from the apical meristem that give rise to a plant organ such as a flower or leaf.

gain of function describes a mutation that confers a gain of function on the organism.

gamete cell that carries the genes into the next generation—in animals it is the egg or sperm.

ganglion mother cell a cell formed by division of a neuroblast in *Drosophila* and which gives rise to neurons.

gap gene any of a number of zygotic genes coding for transcription factors expressed in early *Drosophila* development that subdivide the embryo into regions along the antero-posterior axis.

gap junction type of cell junction in which there is direct communication between the cytoplasms of adjoining cells, and through which ions and small molecules can pass from cell to cell.

gastrula the embryo at the stage in animal development at which prospective endodermal and mesodermal cells of the blastula or blastoderm move inside the embryo.

gastrulation the process in animal embryos in which prospective endodermal and mesodermal cells move from the outer surface of the embryo to the inside, where they give rise to internal organs.

gastruloid an embryo model grown from embryonic stem cells *in vitro*.

gene cluster, gene complex a group of related genes closely linked on a chromosome that usually constitute a functional unit. Examples are the Hox gene clusters in animals, in which sequential expression of genes of each cluster specify positional values along the antero-posterior axis of the embryo.

gene expression the process of gene activation, transcription, and translation that produces a functional protein or RNA from a gene. Transcription of the gene can be detected by *in situ* nucleic acid hybridization while production of the protein is detected by *in situ* immunostaining.

gene knockdown *see* **gene silencing**.

gene knock-in the introduction of a new functional gene into the genome using techniques involving homologous recombination and transgenesis.

gene knock-out the complete and permanent inactivation or deletion of a particular gene in an organism by means of genetic manipulation.

general transcription factor gene-regulatory protein involved in the expression of many different genes. These transcription factors form part of the transcription initiation complex; they bind to RNA polymerase and position it in the correct place on the DNA to start transcription.

generative program the instructions for development contained in the genome that determine when and where the proteins that control cell behavior, and thus development, are produced.

gene regulatory network (GRN) a group of genes, especially genes encoding transcription factors and signaling molecules, whose activities interact with each other over space and time to produce developmental outcomes such as the assignment of cell or tissue fate.

gene-regulatory protein general term for any protein involved in switching genes on and off.

gene silencing the switching off of a gene (or block of genes) by epigenetic modifications to chromatin, by microRNAs, or by RNA interference. Unlike gene knock-out, it does not affect the structure of the gene itself but prevents its transcription or the translation of the mRNA.

genetic equivalence the fact that all the somatic cells in the body of a multicellular organism contain the same set of genes.

genetic lineage analysis a type of fate mapping in which a small number of cells are marked with a genetic marker such as the gene encoding green fluorescent protein, so that they can be followed during development.

genital ridge the region of mesoderm lining the abdominal cavity from which the gonads develop in vertebrates.

genome the hereditary information of a particular organism including its complete set of genes and non-coding DNA sequences.

genome editing a technique by which precise targeted mutations can be made in an organism's genome using targeted DNA nucleases such as the CRISPR-Cas9 system.

genomic imprinting the process in which certain genes are switched off in either the egg or the sperm during their development, and remain silenced in the genome of the early embryo. This imprinting occurs during gamete formation, and is probably due to differential DNA methylation of the gene during egg or sperm production.

genome-wide association studies (GWAS) genetic studies that scan genetic markers across the genomes of many people to find genetic variations associated with a particular disease.

genotype the exact genetic constitution of a cell or organism in terms of the alleles it possesses for any given gene.

germ band the central body of the early *Drosophila* embryo, from which most of the trunk region of the embryo will develop.

germ-band extension the extension of the germ band of the *Drosophila* embryo towards the posterior and onto the dorsal side, which occurs during gastrulation. The germ band later retracts as development is completed.

germ cells cells in an animal that give rise to eggs or sperm.

germ layers the regions of the early animal embryo that will give rise to distinct types of tissue. Most animals have three germ layers—ectoderm, mesoderm, and endoderm.

germ ring in zebrafish embryos, the thickened edge of the blastoderm when it has spread out to cover the upper half of the yolk.

germarium reproductive structure in the adult female *Drosophila* containing stem cells that give rise to a succession of egg chambers, each containing an oocyte.

germline cells the cells that give rise to the gametes—eggs and sperm.

germline cyst the 16-cell structure in adult female *Drosophila* that contains an oocyte precursor and nurse cells, before meiosis occurs.

germplasm the special cytoplasm in some animal eggs, such as those of *Drosophila*, that is involved in the specification of germ cells.

glia, glial cells the non-neuronal cells of the nervous system, such as the Schwann cells of the peripheral nervous system and the astrocytes of the brain.

gliogenesis the formation of glial cells.

globular stage the stage at which a plant embryo is a ball of around 32 cells.

glycosylation the addition of carbohydrate side chains to a protein, a common modification in membrane and extracellular proteins that occurs after their translation.

gonad reproductive organ in animals. The gonads comprise the ovary (female) and the testis (male).

gonadotropin-releasing hormone (GnRH) protein hormone secreted by the hypothalamus that acts on the pituitary to stimulate the release of gonadotropins, which, in turn, increase the production of the steroid sex hormones at puberty.

granulocyte colony-stimulating factor (G-CSF) hematopoietic growth factor that helps promote the differentiation of granulocytes from the granulocyte–macrophage progenitor cell.

granulocyte–macrophage colony-stimulating factor (GM-CSF) hematopoietic growth factor that is required for the development of most myeloid cells from the earliest progenitors that can be identified.

growth an increase in size, which can occur by cell multiplication, increase in cell size, and deposition of extracellular material.

growth cone the region at the end of the extending axon of a developing neuron. The growth cone crawls forward on the substratum and senses its environment by means of filopodia.

growth hormone protein hormone produced by the pituitary gland that is essential for the post-embryonic growth of humans and other mammals.

growth hormone-releasing hormone protein hormone produced by the hypothalamus that stimulates the production of growth hormone by the pituitary.

growth plate cartilaginous region in the long bones of vertebrates at which growth of the bone occurs. The cartilage grows in size and is eventually replaced by bone by the process of endochondral ossification.

guidepost cell *see* **intermediate target**.

gynogenetic an embryo in which the two sets of homologous chromosomes are both maternal in origin.

hand plate *see* **digital plate**.

haploid in diploid organisms, describes cells that contain a single set of chromosomes and that are derived from diploid cells by meiosis. In most animals the only haploid cells are the gametes—the sperm or egg.

head the structure located at the anterior end of a bilaterally symmetrical animal, such as an arthropod or a vertebrate, that typically houses the brain (or equivalent), various sense organs, and the mouth.

head fold an infolding of the three germ layers in the head region of the gastrula in chick and mammalian embryos that indicates the start of the development of the pharynx and the foregut.

head process the prechordal mesoderm and anterior end of the notochord (head notochord) that projects into the head of mammalian and avian embryos.

heart stage a stage in embryogenesis in dicotyledonous plants in which the cotyledons and embryonic root are starting to form, giving a heart-shaped embryo.

Hedgehog a secreted signaling protein, first identified in *Drosophila*, that is a member of an important family of developmental signaling proteins that includes Sonic hedgehog in vertebrates.

hematopoiesis the formation of blood cells from a multipotent stem cell. In adult vertebrates this occurs mainly in the bone marrow.

hematopoietic growth factor any of various secreted signaling proteins that induce differentiation of blood cell progenitors into the various blood cell types.

hematopoietic stem cell a multipotent stem cell in the bone marrow of vertebrates that gives rise to all the blood cells.

hemidesmosome type of adhesive cell junction through which cells adhere to the extracellular matrix. Integrins in the cell membrane bind to proteins such as laminin in the extracellular matrix.

Hensen's node a condensation of cells at the anterior end of the primitive streak in chick embryos. It corresponds to the Spemann organizer in amphibians. The equivalent region in mammals is just called the node. Cells of the node give rise to the prechordal mesoderm and the notochord in chick and mouse embryos.

hermaphrodite an organism that possesses both male and female gonads and produces both male and female gametes.

heterochromatin the state of chromatin in which transcription of the DNA is not possible.

heterochronic *see* **heterochrony**.

heterochrony an evolutionary change in the timing of developmental events. A mutation that changes the timing of a developmental event is called a **heterochronic** mutation.

heterometry an evolutionary change in the amount of something, usually the level of transcription of a gene and thus a change in the level of its product.

heterotypy an evolutionary change in the function of a protein or body part as a result of changes in its structure.

heterozygous describes the state in which a diploid organism carries two different alleles of a given gene, one inherited from the father and one from the mother.

hindbrain the most posterior part of the embryonic brain, which gives rise to the cerebellum, the pons, and the medulla oblongata. It is also called the rhombencephalon.

Hippo pathway an intracellular signaling pathway that controls cell proliferation and apoptosis. One of its components is the protein kinase Hippo first discovered in *Drosophila*. Mutations in Hippo give rise to a hippopotamus-like phenotype. The pathway has subsequently been identified in vertebrates including mammals.

histoblasts undifferentiated cells set aside in the *Drosophila* embryo, from which some adult structures, such as the abdominal epidermis and internal organs, develop during metamorphosis. *See also* **imaginal discs**.

HOM-C complex the name of the Hox gene clusters in *Drosophila*.

homeobox a short region of DNA in certain genes that encodes a DNA-binding domain called the homeodomain. Genes containing this motif are known generally as homeobox genes. The homeodomain is present in a large number of transcription factors that are important in development, such as the products of the Hox genes and the Pax genes.

homeobox gene *see* **homeobox**.

homeodomain *see* **homeobox**.

homeosis the phenomenon in which one structure is transformed into another, homologous, structure as a result of a mutation. An example of such a homeotic transformation is the development of legs in place of antennae in *Drosophila* as a result of mutation.

homeotic gene a gene which when mutated can result in a homeotic transformation. Examples are the Hox genes.

homeotic selector genes in *Drosophila*, genes that specify the identity and developmental pathway of a group of cells. They encode homeodomain transcription factors and act by controlling the expression of other genes. Their expression is required throughout development. The *Drosophila* gene *engrailed* is an example of a homeotic selector gene.

homeotic transformation *see* **homeosis**.

homologous genes genes in different species that share significant similarity in their nucleotide sequence and are derived from a common ancestral gene.

homologous recombination the recombination of two DNA molecules at a specific site of sequence similarity.

homology morphological or structural similarity due to common ancestry.

homozygous describes the state in which a diploid organism carries two identical alleles of a given gene, one from the mother and one from the father.

Hox clusters, Hox complex *see* **Hox genes**.

Hox code the combinations of Hox genes that specify different regions along the antero-posterior axis.

Hox genes a family of homeobox-containing genes that are present in all animals except sponges (as far as is known) and are involved in patterning the antero-posterior axis of the body. In many animals they are clustered on the chromosomes in one or more gene clusters (gene complexes). Combinatorial expression of different Hox genes characterizes different regions or structures along the main antero-posterior axis and also along the proximo-distal axis of vertebrate limbs.

hyaline layer a layer of extracellular matrix secreted onto the surface of the sea urchin blastula.

hypermorphosis a type of heterochrony in which an extra stage is added before the end of development, compared to the ancestral species.

hypoblast a sheet of cells in the early chick embryo that covers the yolk under the blastoderm and gives rise to extra-embryonic structures such as the stalk of the yolk sac. In early human embryos it is the sheet of epithelial cells that lie below the epiblast in the bilaminar disc and gives rise to extraembryonic structures.

hypocotyl the seedling stem that develops from the region between the embryonic root and the future shoot.

hypodermis the name given to the epidermis in nematodes.

hypophysis a cell in some plant embryos that is recruited from the suspensor and contributes to the embryonic root meristem and root cap.

hypothalamus region of the brain containing many neurons including those that produce and secrete neuropeptides and neurohormones that act on cells in the pituitary to stimulate the release of hormones.

IGF *see* **insulin-like growth factors**.

imaginal discs small sacs of epithelium present in the larva of *Drosophila* and other insects, which at metamorphosis give rise to adult structures such as wings, legs, antennae, eyes, and genitalia.

immunoglobulin superfamily large family of molecules that contain immunoglobulin-like domains, and which includes some cell-adhesion molecules such as N-CAM.

immunohistochemistry technique that uses specific fluorescently labeled antibodies to visualize proteins in cells.

imprinting *see* **genomic imprinting**.

indeterminate describes growth in plants that do not make a fixed number of leaves or flowers.

induced pluripotent stem cells (iPS cells) pluripotent stem cells that are produced by the conversion of differentiated somatic cells by the introduction and expression of a few specific transcription factor genes that induce the cell to revert to a state of pluripotency.

induction is the process whereby one group of cells signals to another group of cells in the embryo and so affects how they will develop.

inflorescence a flowerhead or flowering shoot in plants.

inflorescence meristem meristems that develop from vegetative shoot meristems and produce flowering shoots.

ingression the movement of individual cells from the outside of the embryo into the interior during gastrulation. In mammalian and avian embryos cells detach from the epiblast surface and ingress through the primitive streak.

initial any cell in the meristem of a plant that is able to divide continuously, giving rise both to dividing cells that stay within the meristem and to cells that leave the meristem and go on to differentiate.

initiation of transcription a crucial step in the expression of a gene. In most developmental genes it is under tight control so that genes are not expressed at the wrong time and place.

inner cell mass a discrete mass of cells in the blastocyst of the early mammalian embryo which is derived from the inner cells of the morula, and which will give rise to the embryo proper and some extra-embryonic membranes.

***in situ* DNA hybridization chain reaction** type of DNA hybridization technique that enables multiple target mRNAs to be detected.

***in situ* nucleic acid hybridization** technique used to detect where in the embryo particular genes are being transcribed. The mRNA is detected by its hybridization to a labeled single-stranded complementary DNA probe.

instar the larval phase between each molt in animals in which the larva goes through successive phases of growth and molting before developing into an adult.

instructive describes induction in which cells respond differently to different concentrations of the inducing signal.

insulin-like growth factors 1 and 2 (IGF-1 and 2) polypeptide growth factors that stimulate cell division in the early embryo and which after birth mediate many of the effects of growth hormone and are essential for post-natal growth in mammals.

integrin any of a family of cell-adhesion molecules by which cells attach to the extracellular matrix.

intercalary growth growth that occurs in animals capable of epimorphic regeneration when two pieces of tissue with different positional values are placed next to each other. The intercalary growth replaces the intermediate positional values.

interfollicular epidermis the epidermis of the skin between hair follicles.

intermediate filament one of the three principal types of protein filaments of the cytoskeleton. Intermediate filaments are involved in strengthening tissues such as epithelia.

intermediate mesoderm the region of mesoderm between the paraxial mesoderm and the lateral plate mesoderm that develops into the kidneys.

intermediate target in axon navigation, a decision point between the beginning of an axon's outgrowth and its target where the axon can switch its direction of movement. Also called a guidepost.

interneuron a type of neuron in the central nervous system that relays signals from one neuron to another.

internode that portion of a plant stem between two nodes (sites at which a leaf or leaves form).

intracellular signaling the process by which an extracellular signal received by a receptor at the cell's surface is relayed onward to its final destination inside the cell by a series of intracellular signaling pathway proteins that interact with each other.

***in vitro* fertilization** the technique of fertilizing eggs completely *in vitro*. The fertilized eggs can be replaced in a surrogate mother to complete their development.

invagination the local inward deformation of a sheet of embryonic epithelial cells to form a bulge-like structure, as in early gastrulation in the sea urchin embryo.

involution a type of cell movement that occurs at the beginning of amphibian gastrulation, when a sheet of cells enters the interior of the embryo at the blastopore by rolling in under itself.

iPS cells *see* **induced pluripotent stem cells**.

juvenile hormone hormone in insects that maintains the larval state and prevents premature metamorphosis. Metamorphosis follows a drop in the level of juvenile hormone.

keratinocyte differentiated epidermal skin cell that produces keratin, eventually dies, and is shed from the skin surface.

Koller's sickle a crescent-shaped region of small cells lying at the front of the posterior marginal zone in the chick blastoderm.

lamellipodium broad cytoplasmic extension put out by a cell which enables it to move over a substrate.

lateral geniculate nuclei in mammals, paired regions (one on each side) in the brain at which most of the axons from the retina terminate.

lateral inhibition the mechanism by which cells inhibit neighboring cells from developing in a similar way to themselves.

lateral line cells sensory organ precursor cells that migrate from the cranial ectoderm to form the lateral-line organs in fishes.

lateral-line primordium group of cells arising from the cranial ectoderm in fishes that migrates collectively along the route of the future lateral line, depositing sensory-organ precursor cells, that will form the lateral-line organs, as it moves.

lateral motor column longitudinal column of motor neurons, present in both sides of the spinal cord in the brachial and lumbar regions. Its neurons innervate the fore- and hindlimbs.

lateral plate mesoderm mesoderm in vertebrate embryos that lies lateral and ventral to the somites and gives rise to the tissues of the heart, gonads, blood, and the limb connective tissues.

lateral shoot meristem meristem that arises from the apical shoot meristem and gives rise to lateral shoots.

leaf primordium small set of cells at the edge of the shoot apical meristem from which a leaf develops.

left–right asymmetry the bilateral asymmetry of arrangement and structure of most internal organs in vertebrates. In mice and humans, for example, the heart is on the left side, the right lung has more lobes than the left, and the stomach and spleen lie to the left.

life history the life cycle of an organism or species viewed in terms of its reproductive strategy and its unique ecology or interaction with the environment.

ligand general term for any molecule that binds to another molecule.

ligand-mediated apoptosis programmed cell death induced by an external signal molecule interacting with a cell-surface receptor.

light-sheet fluorescence microscopy (LSFM) a non-invasive imaging technique that uses a plane of light to optically section and view tissues with subcellular resolution.

limb bud small structures that grow out from the flank of the vertebrate embryo and develop into limbs.

lineage (1) the ancestry of a given cell, that is, the progenitor cells and the sequence of cell divisions that gave rise to the cell in question. (2) The genealogy of an organism or species.

local organizer general term for any signaling center with localized effects on neighboring tissue (compared with the embryonic organizer), for example, the signaling centers at the midbrain–hindbrain boundary, the anterior neural ridge, and the zona limitans intrathalamica in the developing vertebrate brain.

locus control region *cis*-regulatory region that controls the sequential expression of the genes in a multigene locus, such as a globin gene cluster or a Hox cluster, and which is located quite far from the genes that it controls.

long-germ development type of insect development in which the embryonic blastoderm gives rise to the whole of the future embryo, as in *Drosophila*.

long non-coding RNAs (lncRNAs) a class of RNA molecules that are longer than 200 nucleotides and, sometimes, do not encode proteins.

long-term potentiation (LTP) a long-term increase in synaptic efficacy, resulting from coordinated activity between pre- and post-synaptic cells.

loss of function describes a mutation that causes a loss of a function in the organism that carries it.

lymphoid describes the lymphocytes of the immune system, which derive from the multipotent hematopoietic stem cell via the lymphoid progenitor cell.

macromere the larger of the cells that result from an unequal cleavage division in certain embryos, such as those of sea urchins.

macrophage colony-stimulating factor (M-CSF) hematopoietic differentiation factor that promotes the differentiation of macrophages from the granulocyte–macrophage progenitor cell.

magnetic resonance imaging (MRI) a non-invasive imaging technique that uses radio waves and strong magnetic fields to create detailed images of the inside of an organism.

marginal zone the belt-like region at the equator of the late blastula of an amphibian embryo which will form the mesoderm.

master regulatory gene gene, usually encoding a transcription factor, whose expression is necessary and sufficient to trigger activation of many other genes in a specific program leading to the development of a particular cell type, tissue, or organ.

maternal-effect gene *see* **maternal-effect mutation**.

maternal-effect mutation a mutation in the mother that has no effect on the phenotype of the mother but affects the development of the egg and later the embryo. Genes affected by such mutations are called maternal-effect genes.

maternal factor protein or RNA that is deposited in the egg by the mother during oogenesis. The production of these maternal proteins and RNAs is under the control of so-called **maternal genes**.

maternal gene in general terms, any gene inherited from the mother. *See also* **maternal-effect gene**.

maternal zygotic transition (MZT) the period in which the embryo shifts from relying on maternal genetic material to its own.

mechanotransduction the ability of a cell to sense, integrate, and convert mechanical stimuli into intracellular changes.

medial at or towards the midline.

median hinge point the midline of the neural plate at which the plate bends to start to form the neural tube.

medio-lateral axis in vertebrates the axis that runs from the midline of the body to the periphery, and in the leaf of a plant from midvein to periphery.

medio-lateral intercalation cell movements that occur during convergent extension in amphibian gastrulation. Cells push in sideways between their neighbors, resulting in the extension and narrowing of the cell sheet.

medulla oblongata brain region that develops from part of the hindbrain and which is involved in regulating unconscious activities such as heartbeat, breathing, and blood pressure.

meiosis a special type of cell division that occurs during formation of sperm and eggs, and in which the number of chromosomes is halved from diploid to haploid.

mericlinal chimera plant in which a genetically marked cell gives rise to a sector of an organ or of the whole plant.

meristem a group of undifferentiated, dividing cells that persist at the growing tips of plants. They give rise to all the adult structures—shoots, leaves, flowers, and roots.

meristem identity genes in plants, genes that specify whether a meristem is a vegetative or an inflorescence meristem.

mesectoderm tissue that can give rise to both ectoderm and mesoderm.

mesenchymal-to-epithelial transition (MET) the aggregation of mesenchyme cells to form an epithelium. Occurs, for example, in the formation of somites and blood capillaries.

mesenchyme loose connective tissue, usually of mesodermal origin, whose cells are capable of migration; some epithelia of ectodermal origin, such as the neural crest, undergo an epithelial-to-mesenchymal transition.

mesendoderm tissue that gives rise to both endoderm and mesoderm.

mesoderm germ layer that gives rise to the skeleto-muscular system, connective tissues, the blood, and internal organs such as the kidney and heart.

mesonephros an embryonic kidney in mammals that contributes to the male and female reproductive organs.

messenger RNA (mRNA) the RNA molecule that specifies the sequence of amino acids in a protein. It is produced by transcription from DNA.

metamorphosis the process by which a larva is transformed into an adult. It often involves a radical change in form and the development of new organs, such as wings in butterflies and limbs in frogs.

metastasis the movement of cancer cells from their site of origin to invade underlying tissues and to spread to other parts of the body.

microfilament *see* **actin filament**.

micromere the smaller cell that results from an unequal cleavage division during early animal development.

microRNAs (miRNAs) small noncoding RNAs that suppress the expression of specific genes.

microtubule one of the three principal protein filaments of the cytoskeleton. They are involved in the transport of proteins and RNAs within cells.

mid-blastula transition in amphibian embryos, the stage in development when the embryo's own genes begin to be transcribed, cleavages become asynchronous, and various other changes occur.

midbrain the middle section of the embryonic vertebrate brain that gives rise to the tectum (in amphibians and birds) and similar structures in mammals, which are the sites of integration and relay centers for signals coming to and from the hindbrain, and also for inputs from the sensory organs. Also called the mesencephalon.

midbrain–hindbrain boundary (MHB) the site of a signaling center in the embryonic brain that helps to pattern the brain along the antero-posterior axis.

miRNAs *see* **microRNAs**.

misexpression screening large-scale screens that look systematically for genes whose overexpression or misexpression in a particular tissue causes a mutant phenotype.

mitosis the nuclear division that occurs during the proliferation of somatic diploid cells, and results in both daughter cells having the same diploid complement of chromosomes as the parent cell.

mitotic recombination an exchange of regions of chromatids between homologous chromosomes during mitosis.

mitotic spindle the microtubule-based apparatus in a dividing cell that partitions the chromosomes equally to daughter nuclei at mitosis.

model organisms the small number of species that are commonly studied in developmental biology and about whose development most is known.

molting the shedding of an external cuticle when arthropods and nematodes grow, and its replacement with a new one.

morphallaxis a type of regeneration that involves repatterning of existing tissues without growth.

morphogen any substance essential for pattern formation whose spatial concentration varies and to which cells respond differently at different threshold concentrations.

morphogenesis the processes involved in bringing about changes in form in the developing embryo.

morpholino antisense RNA a type of antisense RNA composed of stable morpholino nucleotide analogs.

morula the very early stage in a mammalian embryo when cleavage has resulted in a solid ball of cells.

mosaic a term used historically to describe the development of organisms that appeared to develop mainly by distribution of localized cytoplasmic determinants.

motor neuron neuron that innervates muscle and controls muscle contraction and thus movement.

Müllerian ducts tubules that run adjacent to the Wolffian ducts in the mammalian embryo and become the oviducts in females.

Müllerian-inhibiting substance protein secreted by the developing testis that induces regression of the Müllerian ducts in males.

multipotent describes a cell that can give rise to many different types, but not all types, of differentiated cell.

mural trophectoderm the trophectoderm in the blastocyst that is not in contact with the cells of the inner cell mass.

muscle fiber fully differentiated muscle cell, a multinucleate cell that produces and organizes muscle-specific proteins such as muscle actin and myosin that enable the muscle cell to contract.

mutagenesis screening screening for developmental genes that aims to treat large enough populations with the mutagen so that, overall, a mutation is induced in every gene in the genome.

myeloid describes a lineage of blood cells such as granulocytes, monocytes/macrophages, and mast cells which derive from the multipotent hematopoietic stem cell via the erythroid–myeloid progenitor cell.

myoblast a committed but still undifferentiated muscle cell.

myogenesis the development of muscle.

myogenic progenitor cell in the somites that will develop into muscle cells but is not yet committed.

myotome that part of the somite that gives rise to muscle.

myotube multinucleate cell that is an intermediate stage in muscle cell differentiation. It develops from the fusion of myoblasts and will develop into a muscle fiber.

naive ES cells human embryonic stem cells (ES cells) with many of the characteristics of mouse ES cells.

naso-temporal axis in the retina, the axis corresponding roughly to the antero-posterior axis.

necrosis a type of cell death due to pathological damage in which cells break up, releasing their contents.

negative feedback, negative-feedback loop a type of regulation in which the end-product of a pathway or process inhibits an earlier stage.

neoblasts a population of small, undifferentiated cells in planarians, which are the only cells in the adult that are mitotically active. They not only self-renew but also include a subpopulation of cells that can differentiate into all the different cell types—epidermis, muscle, neurons, and germ cells.

neocortex *see* **cerebral cortex**.

neoteny the phenomenon in which an animal acquires sexual maturity while still in larval form.

nerve cell *see* **neuron**.

nerve fascicle *see* **nerve tract**.

nerve growth factor (NGF) a signal protein that promotes the survival and growth of neurons.

nerve impulse *see* **action potential**.

nerve tract a bundle of axons (nerve fibers) that follow a particular pathway in the nervous system.

netrin a secreted protein that act as a guidance molecule for neurons in the nervous system; netrins can be either attractant or repellent.

neural crest, neural crest cells in vertebrates, a population of cells derived from the edge of the neural plate. They migrate from the dorsal neural tube to different regions of the body and give rise to a wide variety of tissues, including the autonomic nervous system, the sensory nervous system, the pigment cells of the skin, and some cartilage in the head.

neural folds the two folds that rise up at each edge of the neural plate at the beginning of neurulation and will eventually fuse to form the neural tube, which gives rise to the nervous system.

neural furrow a longitudinal infolding along the midline of the neural plate epithelium, caused by localized changes in cell shape, which is one of the first signs of neurulation.

neural groove in neural tube formation, the longitudinal depression in the neural plate between the two neural folds.

neural plate an area of thickened dorsal ectodermal epithelium at the anterior of a vertebrate embryo that gives rise to the nervous system through the process of neurulation.

neural plate border cells cells at the border of the neural plate that will give rise to the neural crest.

neural tube the ectodermally derived tubular structure that forms along the dorsal midline of a vertebrate embryo and gives rise to the nervous system.

neural tube closure the coming together and fusion of the dorsal tips of the neural folds to form the neural tube that occurs during neurulation.

neural tube defect any of a number of congenital conditions, such as spinal bifida, that are caused by localized failures of the neural tube to close properly.

neuregulin any of a family of chemoattractant proteins that are involved in guiding the tangential migration of neurons in the developing brain.

neuroblast an embryonic cell that will give rise to neural tissue (neurons and glia).

neuroectoderm embryonic ectoderm with the potential to form neural cells.

neurogenesis the formation of neurons from their precursor cells.

neurogenin a transcription factor specifically expressed in prospective neurons.

neurohormone chemical signal used to transmit information from neurons to other cells including neurons.

neuroligin (NLGN) protein present in the post-synaptic membrane of some synapses, which interacts with the transmembrane protein β-neurexin in pre-synaptic membranes to promote the functional differentiation of the synapse.

neuromuscular junction the specialized synapse made between a motor neuron and a muscle fiber, at which the neuron stimulates muscle activity.

neuron the electrically excitable cell type in the nervous system, also called a nerve cell, which conveys information in the form of electrical signals to other neurons, to muscles and to some glandular cells. Neurons communicate with other cells through specialized structures called synapses, at which the electrical signal stimulates production of a chemical signal (a neurotransmitter) by the pre-synaptic cell. The neurotransmitter diffuses across a narrow space between the two cell membranes and binds to receptors on the post-synaptic cell.

neuropeptide chemical signal used to transmit information from neurons to other cells including neurons.

neuropilin a receptor for semaphorins.

neurotransmitter short-lived chemical signal used to transmit information between neurons at chemical synapses.

neurotrophic hypothesis the idea that competition for a limiting supply of a neurotrophic factor that promotes survival promotes a matching in number of motor neurons and target cells.

neurotrophin, neurotrophic factor general name for a protein that is necessary for neuronal survival, such as nerve growth factor.

neurula the stage of vertebrate embryonic development at the end of gastrulation when the neural tube is forming.

neurulation the process in vertebrates in which the future brain and spinal cord are formed from the ectodermal neural plate. As a result largely of localized changes in cell shape, the neural plate develops a central groove (the neural groove) with folds rising up on either side (neural folds). The folds eventually meet and fuse along the midline to form a tubular structure (the neural tube) that develops into the brain and spinal cord. In birds and mammals, the neural plate gives rise to the brain and the spinal cord is formed from the stem zone.

Nieuwkoop center a signaling center on the dorsal side of the early *Xenopus* embryo. It forms in the dorsal vegetal region of the blastula as a result of cortical rotation, and helps to specify the position of the Spemann organizer, which is formed just above it.

Nodal, Nodal-related proteins a subfamily of the TGF-β family of signaling proteins of vertebrates. They are involved in all stages of development, but particularly in early mesoderm induction and patterning.

node (1) in avian and mammalian embryos, the embryonic organizing center analogous to the Spemann organizer of amphibians. It is also known as Hensen's node in birds. (2) In plants, that part of the stem at which leaves and lateral buds form.

non-autonomous, non-cell-autonomous describes gene expression in a cell that affects cells other than the cell in which the gene is expressed. Genes for secreted signal proteins, for example, have non-autonomous effects.

non-canonical Wnt pathway an intracellular signaling pathway in which signals from Frizzled (the receptor for Wnts) are relayed via the intracellular protein Dishevelled to a pathway that ends in regulating the activity of the cytoskeleton. It is particularly associated with planar cell polarity.

non-coding RNA any of a large number of RNAs encoded in the genome that have no protein-coding function. Many act as gene-regulatory RNAs and includes short (small) noncoding RNAs, microRNAs, long non-coding RNAs.

notochord a transient stiff, rod-like cellular structure in vertebrate embryos that runs from head to tail and lies centrally beneath the neural tube. It is derived from mesoderm and its cells eventually become incorporated into the vertebral column.

nurse cells germline cells surrounding the developing oocyte in *Drosophila* which synthesize proteins and RNAs that are deposited in the oocyte.

octant stage the eight-cell stage of a plant embryo.

ocular dominance columns columns of neurons in the visual cortex that respond to the same visual stimulus from either the left or the right eye.

odontoblast cell that secretes dentin in a developing tooth.

ontogeny the development of an individual organism.

oocyte an immature egg.

oogenesis the process of egg formation in the female.

oogonia diploid germ cells that divide by mitosis within the ovary before entering meiosis to produce the oocytes.

optic chiasm the site at which the retinal nerve fibers from the right eye cross over fibers from the left eye on their way to opposite sides of the optic tectum (chicks and amphibians). In mammals, the optic chiasm is the site at which the nerve fibers from an eye divide, some taking a contralateral course to the opposite side of the lateral geniculate nucleus (LGN), whereas some connect with the ipsilateral side of the LGN.

optic tectum the region of the brain in amphibians and birds where the axons from the retina terminate.

optic vesicle the precursor of the retina of the vertebrate eye and forms as an outgrowth from the wall of the forebrain.

optical projection tomography (OPT) a non-invasive, three-dimensional imaging technique that uses visible light to create optical sections of large biological samples.

oral referring to the mouth, or to the side of the body on which the mouth is located.

oral–aboral axis in sea urchins and other radially symmetrical organisms, an axis that runs from the centrally situated mouth to the opposite side of the body.

organizer, organizing region, organizing center a signaling center that directs the development of the whole embryo or of part of the embryo, such as a limb. In amphibians, the organizer usually refers to the Spemann organizer. The organizing center in plants refers to the cells underlying the central zone of the meristem, which maintains the stem cells of the central zone.

organogenesis the development of specific organs such as limbs, eyes, and heart.

organoid three-dimensional structure grown from stem cells that includes multiple differentiated cell types and mimics an organ or tissue in its cellular organization.

osteoblast cell that makes and secretes bone matrix, leading to bone formation and repair.

osteoclast cell that is involved with cartilage resorption and bone remodeling.

ovariole the string of egg chambers produced in the *Drosophila* ovary.

ovary the internal reproductive structure in female animals and in plants, it contains ovules that each contain an egg cell.

oviduct the tube in female birds and mammals that transports the eggs from the ovaries to the uterus.

ovule the structure in plants that contains an egg cell.

P granules granules that become localized to the posterior end of the fertilized egg of *Caenorhabditis elegans*.

pair-rule gene any of a number of genes in *Drosophila* that are involved in delimiting parasegments. They are expressed in transverse stripes in the blastoderm, each pair-rule gene being expressed in alternate parasegments.

pallium part of the embryonic forebrain that develops into the cerebral cortex, the outer layers of the cerebral hemispheres, where the highest-level processing centers for sensory information, motor control, learning, and memory are located.

PAR (partitioning) proteins a group of proteins, initially discovered in *Caenorhabditis elegans*, that are required for the correct positioning and orientation of mitotic spindles in the early cleavages to ensure division at the required place in the cell and in the required plane.

paralogous genes genes within a species that have arisen by duplication and divergence. Examples are the Hox genes in vertebrates, which comprise several paralogous subgroups made up of paralogous genes.

parasegments developmental units arranged along the body of the developing *Drosophila* embryo and which give rise to the segments of the larva and adult.

paraxial mesoderm mesoderm lying on either side of the dorsal midline and which gives rise to the somites.

parietal endoderm layer on the inner surface of the mural trophectoderm formed from cells from the primitive endoderm that migrate to cover the surface.

parthenogenesis the ability of the eggs of some species to develop into an embryo without fertilization.

pattern formation the process by which cells in a developing embryo acquire identities that lead to a well-ordered spatial pattern of cell activities.

Pax genes family of genes encoding transcriptional regulatory proteins that contain both a homeodomain and another protein motif, the paired motif.

PCP pathway an intracellular signaling pathway associated with planar cell polarity (PCP) in which signals from Frizzled (the receptor for Wnts) are relayed via the intracellular protein Dishevelled to a pathway that ends in regulating the activity of the cytoskeleton.

pedomorphosis retention of embryonic or juvenile characters during part of the life of the organism.

peramorphosis the case where a species shows greater development compared with its ancestor or related species.

periclinal describes cell divisions in a plane parallel to the surface of the tissue.

periclinal chimera in plants, chimera in which one of the three meristem layers has a genetic marker that distinguishes it from the other two.

peripheral nervous system (PNS) all of the nervous system apart from the brain and spinal cord.

perivitelline space the space between the vitelline membrane lining the egg case and the egg plasma membrane in the fertilized eggs and early embryos of insects and other animals.

permissive describes induction in which a cell makes only one kind of response to an inducing signal, and makes it when a given level of signal is reached.

pharyngeal region the name given to the branchial region in mammals.

phenotype the observable or measurable characters and features of a cell or an organism.

photoperiodism the response of an organism to relative day length; in plants it is responsible for promoting flowering as days become longer.

phyllotaxis, phyllotaxy the arrangement of leaves along a shoot.

phylogeny the evolutionary history of a species or group.

phylotypic stage the stage at which the embryos of the different vertebrate groups closely resemble each other. This is the stage at which the embryo possesses a distinct head, a neural tube, and somites.

pioneer axon axons that are the first to navigate a particular pathway to a target.

pioneer factors specialized transcription factors that bind to sequences in chromatin and loosen the DNA from histones and make it available for transcription.

placenta a structure that forms in the uterine wall at which the blood systems of mother and embryo form an interface with each other. Mammalian embryos (with the exception of the monotremes such as the egg-laying duck-billed platypus and echidna) are nourished by the mother by the passage of nutrients through the placenta.

placental labyrinth in the mouse, the layer of the placenta that is the transport interface for nutrients and toxic waste, derived from the inner cells of the ectoplacental cone.

placode a region of thickened epithelium, usually on the surface of the embryo, that gives rise to a particular structure. An example is the lens placode, which gives rise to the lens of the eye.

planar cell polarity the polarization of cells in the plane of the tissue.

plant transformation general name given to techniques that introduce DNA into plant cells so that they express new or modified genes.

plasmodesmata (singular **plasmodesma**) threads of cytoplasm that run through the cell wall and connect adjacent plant cells.

plasticity in development, the capacity of embryonic tissues or cells to change from their normal fate if placed in a different environment where they are exposed to different signals.

plexin a receptor for semaphorins.

pluripotent describes a stem cell, such as an embryonic stem cell, that can give rise to all the types of cells in the body.

pluripotency network a gene regulatory network formed from the transcription factor genes responsible for pluripotency and which maintains pluripotency.

pluteus the larval stage of the sea urchin.

polar body a small cell that is a product of meiosis during egg development. Polar bodies take no part in embryonic development.

polar follicle cells cells that are specified at the anterior and posterior ends of a *Drosophila* egg follicle. They induce stalk formation between egg chambers and help position the oocyte at the posterior end of the follicle.

polar trophectoderm the trophectoderm in contact with the inner cell mass in the mammalian blastocyst.

polarity the property of a cell, structure, or organism in which one end is different from the other.

polarizing region in developing chick and mouse limb buds, an area at the posterior margin of the bud that produces a signal specifying position along the antero-posterior axis. Also called the zone of polarizing activity (ZPA).

pole cells cells formed at the posterior end of the *Drosophila* blastoderm that are precursors of the germ cells.

pole plasm the cytoplasm at the posterior end of the *Drosophila* egg that is involved in specifying germ cells.

polydactyly the occurrence of extra digits on hands or feet.

polyspermy the entry of more than one sperm into the egg.

polytene chromosome giant chromosome that is formed by repeated DNA replication in the absence of cell division.

populational asymmetry a pattern of stem-cell division that has been proposed to explain the dynamics of the renewal of the epidermis.

positional information molecular information, in the form, for example, of a gradient of an extracellular signal molecule, that can be interpreted by cells to provide the basis for pattern formation in the embryo.

positional value the property acquired by a cell by virtue of its position with respect to the boundaries of a field of positional information. The cell then interprets this positional value according to its genetic constitution and developmental history, and develops accordingly.

positive feedback, positive-feedback loop a type of regulation in which the end-product of a pathway or process can activate an earlier stage.

posterior at, or in the direction of, the tail end of an embryo or animal.

posterior dominance, posterior prevalence the fact that the more posteriorly expressed Hox genes can inhibit the action of more anteriorly expressed Hox genes when they are expressed in the same region.

posterior marginal zone a dense region of cells at the edge of the blastoderm of the chick embryo that will give rise to the primitive streak.

post-implantation refers to a mammalian embryo after it has implanted in the uterine wall.

post-mitotic neuron a neuron after its formation from a precursor cell, so-called because most neurons do not divide further once they are formed.

post-synaptic describes the side of a synapse that receives the signal.

post-translational modification any modification that occurs to a protein after it has been synthesized. The protein can, for example, be enzymatically cleaved, glycosylated, or acetylated.

preaxial polydactyly the occurrence of extra digits on the anterior side of the hands or feet, that is on the thumb or big toe side.

prechordal mesendoderm tissue composed of prechordal mesoderm and underlying endoderm in vertebrates.

prechordal mesoderm the anterior-most mesoderm in the vertebrate embryo, located anterior to, and continuous with, the notochord. It gives rise to various ventral tissues of the head.

prechordal plate *see* **prechordal mesendoderm**.

precursor cells cells that have become committed to become a cell type, such as muscle, but are not yet differentiated.

pre-implantation refers to a mammalian embryo before it has implanted in the uterine wall.

pre-implantation genetic diagnosis a means of determining the genotype of embryos produced by *in vitro* fertilization before implantation without harming the embryo, which can be used if the embryo is known to be at risk of carrying a mutation for a genetic disease. One blastomere is removed from the embryo during its early cleavage *in vitro*, which does not affect its subsequent development. The DNA from this blastomere can be amplified and tested for the presence or absence of mutation.

prenatal diagnosis detection of potentially disease-causing mutations while the fetus is still within the mother.

prepattern a basic organization generated autonomously in a structure that can be identified in advance of the later development of a similar set of structures. It may subsequently be modified during development.

pre-somitic mesoderm the unsegmented mesoderm between the node (in chick and mouse) and the already formed somites. It will form somites from its anterior end.

pre-synaptic describes that side of a synapse that generates the signal.

primary body formation in avian and mammalian embryos, the formation of the head and trunk.

primary cilium an immotile microtubule-based structure found on most vertebrate cells; it is the site of Sonic hedgehog signaling.

primary mesenchyme in sea urchin embryos, the first mesodermal cells to enter the blastula at gastrulation, and which migrate along the interior wall and eventually lay down the skeletal rods of the sea urchin larval endoskeleton by secretion of matrix proteins.

primary neurulation the folding of the neural plate to form the neural tube.

primary oocyte female germ cell that has entered meiosis.

primitive endoderm in mammalian embryos that part of the inner cell mass that contributes to extra-embryonic membranes.

primitive streak the site of gastrulation in avian and mammalian embryos and the forerunner of the antero-posterior axis. It is a strip of ingressing cells that extends into the epiblast from the posterior margin. Epiblast cells move through the streak into the interior of the embryo to form mesoderm and endoderm.

primordial germ cells precursor cells in the early embryo that represent the germline and will produce the germ cells.

primordium (plural **primordia**) small group of undifferentiated cells that will give rise to a structure such as an insect wing, a tooth, leaf, flower, or floral organ.

proembryo the two-celled stage in a plant embryo.

progenesis a type of heterochrony in which the animal attains sexual maturity more rapidly than the ancestor or related species.

programmed cell death *see* **apoptosis**.

progress zone in the timing model of vertebrate limb development, an area at the tip of the limb bud where cells acquire positional values.

promoter a region of DNA immediately preceding the coding sequence to which RNA polymerase binds to begin transcription of a gene.

proneural cluster small cluster of cells within the neuroectoderm in which one cell will eventually become a neuroblast.

proneural gene a gene that promotes a neural fate in neuroectoderm cells.

pronucleus the haploid nucleus of sperm or egg after fertilization but before nuclear fusion and the first mitotic division.

prothoracicotropic hormone protein hormone secreted by the insect brain that causes the secretion of the steroid hormone ecdysone, and the initiation of molting or pupation and metamorphosis.

proto-oncogene a gene that is involved in the regulation of cell proliferation and that can cause cancer when mutated or expressed under abnormal control.

protostomes those animals, such as insects, in which cleavage of the zygote is not radial and in which gastrulation primarily forms the mouth.

proximo-distal axis the axis of a limb or other appendage (e.g. a leaf in a plant) that runs from the point of attachment to the body or stem (proximal) to the tip of the appendage (distal).

pupa in *Drosophila* and other insects that undergo metamorphosis, a stage following the larval stages in which the organism can remain dormant for long periods and in which metamorphosis occurs.

quiescent center a central group of cells in a plant root-tip meristem that divide rarely but are essential for meristem function.

radial axis the axis running from the center of a structure to the circumference.

radial cleavage a type of cleavage that occurs at right angles to the egg surface and produces blastomeres sitting directly over each other.

radial glial cell elongated glial cell that spans the whole width of the neural tube wall and provides scaffold tracks for migrating neurons.

radial glial stem cell multipotent stem-cell-like radial glia.

radial intercalation process that occurs in the multilayered ectoderm of an amphibian gastrula in which cells intercalate in a direction perpendicular to the surface, thus thinning and extending the cell sheet.

radial symmetry the symmetry around the central axis in cylindrical structures such as plant stems and roots.

rapid block to polyspermy *see* **block to polyspermy**.

reaction–diffusion mechanism that can produce self-organizing patterns of chemical concentrations, which has been proposed to underlie some types of patterning in development.

recessive describes a mutation that only changes the phenotype when both copies of a gene carry the mutation.

reciprocal negative feedback control mechanism in which two transcription factors control the expression of each other's gene.

recombination exchange of DNA between homologous chromosomes during meiosis, which shuffles parental genes into new combinations in the haploid gametes.

redundancy the existence of different genes or pathways that can substitute for each other during development.

regeneration the ability of a fully developed organism to replace lost parts.

regenerative medicine an approach that aims to use stem cells and their derivatives to replace diseased tissue with healthy tissue.

regulation the ability of the embryo to develop normally even when parts are removed or rearranged. Embryos that can undergo regulation are described as **regulative**.

regulative *see* **regulation**.

repressor gene-regulatory protein that helps to suppress gene activity when it binds to a particular site in a gene's control region.

retinoic acid small non-protein secreted signaling molecule with many roles in development.

reverse genetics an approach to genetic analysis that starts with the nucleotide sequence of a gene or amino acid sequence of a protein, and then uses that information to determine the gene's function.

rhombomere any of a sequence of compartments of cell-lineage restriction along the hindbrain of chick and mice embryos.

RNA interference (RNAi) a means of suppressing gene expression by promoting the destruction of a given mRNA by targeting a nuclease to it by means of a short complementary RNA called a short interfering RNA (siRNA).

RNA polymerase an enzyme that catalyzes the synthesis of RNA from a DNA template.

RNA processing the process in eukaryotic cells in which newly transcribed RNAs are modified in various ways to make a functional messenger RNA or structural RNA. It includes RNA splicing, which removes introns from the transcript to leave a continuous coding messenger RNA.

RNA seq a technique used to determine which genes are expressed in a particular tissue or at a particular time in development, in which the total RNA in the cells is converted into cDNA and then sequenced.

RNA splicing the process in eukaryotic cells that removes introns from a newly transcribed messenger RNA to leave a continuous coding sequence that can be translated into a protein.

Robo family of receptors on neurons that bind the chemorepellent Slit proteins.

robust the characteristic of development that it can withstand large variations in conditions and still continue normally.

roof plate a small strip of non-neural cells in the dorsal-most part of the neural tube around the midline, which is involved in patterning the dorsal part of the tube.

satellite cell undifferentiated stem cell present in postnatal and adult muscle that can be reactivated to produce more skeletal muscle cells if the muscle is damaged.

sclerotome that part of a somite that will give rise to the cartilage of the vertebrae.

second heart field a second lineage of cells that gives rise to the heart, and exclusively to the outflow tract. These cells lie behind the cardiac crescent (the first heart field) in the pharyngeal mesoderm.

secondary body formation formation of the most posterior region of the body from the tailbud in mouse and chick embryos.

secondary neurulation in vertebrates, the formation of the posterior neural tube, beyond the lumbo-sacral region, as a solid rod of cells, derived from stem-like cells in the tailbud, which then develops an interior cavity or lumen that connects with that of the anterior neural tube.

secondary oocyte a stage in oocyte development in which the first meiotic division has been completed and meiosis has proceeded as far as the metaphase of the second meiotic division.

secondary organizer see **local organizer**.

segmentation the division of the body of an organism, or a particular structure, into a succession of morphological units—the segments—along the antero-posterior axis.

segmentation gene any of a number of genes in *Drosophila* involved in patterning the parasegments and segments.

selector gene a gene whose expression determines the behavior or properties of a group of cells, and whose continued expression is required to maintain that behavior.

self-assembly a process in which the emergence of organization is implicit in the structure of the component elements.

self-organization the process whereby a collection of equivalent constituent elements of a system, such as cells, organize themselves into well-defined structures in the absence of external templates or guidance.

semaphorin any of a family of secreted proteins that act as guidance cues for neurons and axons.

semi-dominant describes a mutation that affects the phenotype when just one allele carries the mutation, but where the effect on the phenotype is much greater when both alleles carry the mutation.

senescence the impairment of function associated with aging. Cells that are senescent have stopped dividing and produce a range of cytokines and matrix remodeling proteins which can affect neighboring cells.

sensory neuron a type of neuron that is triggered by signals from within the body (e.g. muscle stretch) or from the environment (e.g. touch, heat), and carries those signals to the central nervous system.

septate junction a type of cell junction in invertebrate cells with a similar function to the vertebrate tight junction.

sex chromosome any chromosome that specifically determines sexual phenotype (e.g. the X and Y chromosomes in mammals).

sex determination the genetic and developmental process by which an organism's sex is specified.

sex-determining region of the Y chromosome (SRY) a genetic locus on the Y chromosome that determines maleness by specifying the gonad as a testis.

shield region the name given to the shield-shaped embryonic organizer tissue in zebrafish embryos.

shield stage the stage in in zebrafish embryos in which the embryonic organizer (the shield) has been formed on one side of the blastoderm.

short interfering RNA (siRNA) in the phenomenon of RNA interference, a short single-stranded RNA that is exactly complementary to the mRNA of a given gene, and which becomes incorporated into a nuclease that can then target the mRNA for destruction.

short-germ development type of insect embryonic development in which the blastoderm only gives rise to the anterior segments of the embryo and the remaining segments are added sequentially during growth.

signal transduction the process by which a cell converts an extracellular signal received in one form, such as the binding of a molecule to a cell-surface receptor, into an intracellular signal of a different form, such as the phosphorylation of a cytoplasmic protein.

signaling center a localized region of the embryo that exerts a special influence on surrounding cells, usually by means of secreted signal proteins, and thus determines how those cells develop.

silenced see **gene silencing**.

single-cell RNA seq (scRNA-seq) a method that analyzes the RNA of individual cells in a given population by simultaneously measuring the RNA concentration of hundreds to thousands of genes to gain insight into the complexity of different cell types within a tissue.

single-cell transcriptional analysis detection of the total mRNAs being produced by a single cell.

situs inversus in humans a rare condition in which there is complete mirror-image reversal of the positions of the internal organs.

Slit proteins family of secreted proteins that act as guidance cues to repel growing axons in the developing nervous system.

slow block to polyspermy reaction that follows fertilization in some animals, in which an impenetrable membrane (the fertilization membrane) is formed around the egg, preventing the entry of sperm.

somatic cell any cell in the body other than germ cells. In most animals, the somatic cells are diploid.

somatic cell nuclear transfer cloning of an animal by the transfer of a somatic cell nucleus into an enucleated egg, and the subsequent development of the egg into a new individual. The embryo that develops will have an identical nuclear genetic constitution to the somatic cell donor.

somatic transgenesis the introduction of an mRNA or a DNA construct directly into somatic cells to test the effect of overexpression of that gene. Also called transient transgenesis.

somatopleural layer the outer layer of the lateral plate mesoderm in vertebrate embryos, which contributes to the connective tissues of the body walls and limbs.

somatostatin a protein hormone produced by the hypothalamus that inhibits the production and release of growth hormone by the pituitary.

somites blocks of mesoderm that segment from the paraxial mesoderm on either side of the notochord. They give rise to trunk and limb muscles, the vertebral column and ribs, and the dermis.

somitogenesis formation of somites.

spatial co-linearity *see* **co-linearity**.

spatial transcriptomics techniques that can integrate the single-cell transcriptomic data with spatial information on the location of the cells within the developing tissues.

specialized character an adaptation of an organ or part of an organ for a particular function or environment.

specified describes the developmental status of a group of cells that when isolated and cultured in a minimal medium will develop according to their normal fate.

Spemann organizer, Spemann–Mangold organizer a signaling center on the dorsal side of the amphibian early embryo that acts as the main embryonic organizer. Signals from this center can organize new antero-posterior and dorso-ventral axes.

sperm the haploid male gamete in animals.

sperm aster microtubule structure produced from the sperm centrosome that enters the egg at fertilization. In *Xenopus*, the sperm aster microtubules direct cortical rotation.

spermatogenesis the production of sperm.

sphere stage in zebrafish embryos the developmental stage at which the embryo consists of a hemispherical blastoderm of around 1000 cells lying over a spherical yolk cell.

spindle midzone the structure formed by the overlap of the microtubules of the two halves of a mitotic spindle; its position along the long axis of the cell marks the location of cell division.

spiral cleavage type of cleavage typical of mollusks and annelid worms, in which the plane of cell division is at a slight angle to the egg surface and blastomeres end up in a spiral arrangement.

splanchnopleural layer the inner layer of the lateral plate mesoderm in vertebrate embryos, which gives rise to the heart and blood vessels, among other tissues such as the connective tissue layer surrounding the gut.

stem cells undifferentiated cells that are both self-renewing and also give rise to differentiated cell types. Stem cells are found in most adult tissues and may contribute to their repair. *See also* **embryonic stem cells (ES cells)**.

stem-cell embryo model (SCEM) embryo model generated *in vitro* from combinations of stem cells and that resemble the post-gastrulation epiblast.

stem-cell niche the cells adjacent to stem cells in tissues, and which provide a specialized environment that supports stem-cell maintenance and self-renewal.

stem zone in avian and mammalian embryos, an arc of self-renewing epiblast cells on either side of the primitive streak immediately posterior to the regressing node that give rise to the trunk neural tube and the medial parts of the somites.

subgerminal space the cavity that develops under the area pellucida in the early chick blastoderm.

subpallium the medial ganglionic eminence (part of the ventral telencephalon) in which the GABAergic inhibitory neurons of the cortex are generated.

subventricular zone a layer of proliferating cells under the ventricular layer in the forebrain neural tube which also gives rise to neurons and glial cells.

superior colliculus region of the brain in mammals to which some retinal neurons project. It corresponds to the optic tectum of amphibians and birds.

suspensor structure that attaches the plant embryo to maternal tissue and is a source of nutrients.

symmetric division a cell division in which the daughter cells are identical to each other and have the same fates.

synapse specialized cell–cell junction at which the axon terminal of a neuron transmits a signal to another neuron, muscle cell, or glandular cell by means of a chemical neurotransmitter that diffuses across a small space (synaptic cleft) between the apposed cell membranes where it binds to receptors on the receiving cell. The transmitting cell is called the pre-synaptic neuron and the receiving cells are called the post-synaptic cells.

synaptic cleft the small space between the cell membrane of the transmitting cell and that of the receiving cell at a synapse.

synaptic plasticity the ability of a neuron to change synaptic efficacy.

synaptic vesicle small neurotransmitter-filled membrane-bounded vesicle in the axon terminals of neurons, which releases its contents into the synaptic cleft when an action potential (nerve impulse) generated at the cell body arrives at the terminal.

syncytial blastoderm the very early embryo in *Drosophila*, in which nuclear division after fertilization is not accompanied by cytoplasmic division, resulting in a blastoderm containing many nuclei in a common cytoplasm. The nuclei are arranged around the periphery of the embryo.

syncytiotrophoblast cell giant cells derived from the trophectoderm of the blastocyst in mammals that invade the uterine wall and in humans secrete chorionic gonadotropin.

syncytium a cell with many nuclei in a common cytoplasm.

TALENS transcription activator-like effector nucleases. Nucleases that can be engineered to make targeted mutations in DNA.

tangential migration the migration of brain neurons parallel to the brain surface.

targeted gene expression techniques that turn on the expression of a gene in a particular place and time during development.

tectum brain structure, deriving from the embryonic midbrain, that is the site of integration and relay center for signals coming to and from the hindbrain, and also for inputs from the sensory organs, such as the eye.

telencephalon that part of the embryonic forebrain that gives rise to the pallium and the basal ganglia.

telomere structure composed of repetitive non-coding DNA at each end of a chromosome that prevents chromosomes sticking to each other and prevents gene loss at DNA replication. Telomeres typically become shorter at each cell division.

temperature-sensitive mutation a mutation that only causes a change in phenotype at a different temperature than normal, most commonly a higher temperature.

temporal co-linearity *see* co-linearity.

teratoma a capsulated solid tumor containing a mixture of tissues, often with differentiated cells, that are characteristic of more than one germ layer. It may arise from germ cells or pluripotent embryonic cells. Undifferentiated human embryonic stem cells injected under the skin of an immunocompromised mouse form teratomas, and this property has been used as a test of pluripotency.

terminally differentiated describes a cell that has become fully structurally and functionally differentiated and will not undergo any further differentiation. Terminally differentiated cells often do not undergo further cell division.

testis the internal male reproductive organ in animals; it produces the male germ cells, the sperm.

tetraploid describes a cell that contains four sets of chromosomes, or a species whose somatic cells contain four sets of chromosomes.

thalamus brain region, derived from the embryonic forebrain, which is a major relay station that distributes incoming sensory information to the appropriate region of the cortex.

thorax in insects, the middle part of the distinctly three-part body (head, thorax, and abdomen). It follows the head, and in adults carries the legs and the wings, where the latter are present. In vertebrates, the thorax is the chest region.

threshold concentration that concentration of a chemical signal that can elicit a particular response from a cell.

thyroid hormones thyroxine (T_4), and tri-iodothyronine (T_3), produced by the thyroid gland, which are required generally for growth, and for metamorphosis in *Xenopus*.

thyroid-stimulating hormone hormone released by the pituitary gland that acts on the thyroid gland to stimulate production of thyroid hormones.

tight junction a type of adhesive cell junction that binds epithelial cells very tightly together to form an epithelium and that seals off the environment on one side of the epithelium from the other.

TILLING targeting-induced local lesions in genomes: a technique for detecting mutations by hybridizing the mutated DNA against unmutated DNA and detecting mismatched bases, which indicate the site of a mutation.

tissue-specific describes the expression of a gene or protein only in a particular tissue(s) or cell type(s).

topographic map the orderly projection of neurons from one region of the nervous system to another so that nearest-neighbor relations are maintained. An example is the highly organized projection of neurons from the retina via the optic nerve to the thalamus (in mammals) or tectum (in birds and amphibians).

topologically associating domain (TAD) type of large-scale physical organization of chromatin within a chromosome, which consists of a region in which the DNA makes many interactions with itself. Such domains are separated by regions in which there are many fewer interactions.

totipotent the capacity of a cell to develop into a new organism. The only truly totipotent cell in most animals is the fertilized egg, which gives rise to the embryo and any extra-embryonic membranes, including the placenta in mammals. In plants, somatic cells remain potentially totipotent, being able to form a complete new plant if cultured under the appropriate conditions.

tracheal system in insects, a system of fine tubules that deliver air (and thus oxygen) to the tissues.

transcription the copying of the DNA sequence of a gene into a complementary RNA sequence. In some genes, the RNA is the end-product, but in protein-coding genes, the RNA is then translated to produce a protein.

transcription factor a protein required to initiate or regulate the transcription of a gene into RNA. Transcription factors act within the nucleus of a cell by binding to specific sites in the regulatory regions of the gene.

transcription initiation complex a protein complex consisting of a set of 'general' transcription factors and the RNA polymerase that assembles on the promoter and ensures that the polymerase is positioned to start transcription at the appropriate place.

transdetermination the process by which a committed, but not yet differentiated, cell can become redetermined as a different cell type. The best-known examples occur in regenerating *Drosophila* imaginal discs, where rare transdetermination events occur, resulting in the homeotic transformation of one type of adult structure into another.

transdifferentiation the process in which a differentiated cell dedifferentiates and redifferentiates into a different cell type.

transfection technique by which mammalian and other animal cells are induced to take up foreign DNA molecules. The introduced DNA sometimes becomes inserted permanently into the host cell's DNA.

transgene an additional or altered gene that has been inserted into an organism's genome.

transgenesis the technique by which an additional or altered gene is introduced into an organism's genome with the aim of altering its phenotype. The gene transferred may be from a different species.

transgenic describes an organism whose genetic make-up is the result of the deliberate introduction of new DNA—for example, new genes, genes with specific mutations, or DNAs that cause the inactivation of specific genes. The term is also used to describe the various genetic engineering techniques that can be used to accomplish this.

transient transgenesis see **somatic transgenesis**.

transit-amplifying cells in continually renewing tissues such as epidermis and the gut lining, rapidly dividing cells that are produced from stem cells and after several divisions will differentiate into the specialized cell types of the tissue.

translation the process by which messenger RNA directs the order of amino acids in a protein during protein synthesis on ribosomes.

triploblast an animal with three germ layers—endoderm, mesoderm, and ectoderm.

trisomy the presence of three copies of a chromosome, rather than the normal two, which is characteristic of some conditions, such as Down syndrome.

trophectoderm the outer layer of cells of the early mammalian embryo that contributes to the placenta and to the protective membranes around the embryo.

trophoblast giant cells cells derived from the mural trophectoderm of the mammalian blastocyst after its implantation in the uterine wall, which invade the uterus wall and provide an interface with the maternal tissue.

trophoblast stem cells (TS cells) self-renewing cells derived from the trophectoderm.

trophosphere hollow ball of epithelium formed when trophoblast stem cells are grown in suspension.

tumor progression the transition of a tumor through a series of stages that transform cells from normal to cancerous.

tumor-suppressor gene gene that can cause a cell to become cancerous when both copies of the gene have been inactivated. In non-cancerous cells, tumor-suppressor genes are involved in suppressing cell proliferation.

Turing mechanisms see **reaction–diffusion**.

unequal describes cell division that produces two unequally sized daughter cells.

vasculogenesis the initial stages in the formation of blood vessels, comprising the condensation of angioblasts to form a tubular vessel.

vegetal pole is the central point on the surface of the vegetal region, directly opposite the animal pole. See **vegetal region**.

vegetal region the yolky 'lower' hemisphere of amphibian eggs and blastulas, and the region from which the endoderm will develop.

ventral describes the underside or belly of an embryo or animal, the side opposite to the dorsal side (back or upper side). In plants, in relation to leaves, it describes the under surface of the leaf.

ventral closure the coming together and closing of the sides of the chick or mammalian embryo on the ventral side of the body to internalize the gut and form the ventral body wall.

ventral ectoderm one of the four main regions into which the dorso-ventral axis of the early *Drosophila* embryo is divided before gastrulation. This region will give rise to the nervous system and some epidermis.

ventral lip the edge of the blastopore in *Xenopus* on the opposite side of the blastopore to the dorsal lip.

ventralized describes embryos that are deficient in dorsal regions and have much increased ventral regions.

ventricular zone a layer of proliferating cells lining the lumen of the vertebrate neural tube, from which neurons and glia are formed.

vernalization the phenomenon by which flowering in spring is accelerated after the plant has been exposed to a long period of cold temperature.

vertebral column the backbone or spine of vertebrates, composed of a succession of vertebrae.

visceral endoderm tissue derived from the primitive endoderm that develops on the surface of the egg cylinder in the mammalian blastocyst.

visual cortex that part of the mammalian cerebral cortex to which visual signals are sent after the lateral geniculate nuclei and are processed to produce a visual perception.

vitelline envelope, vitelline membrane extracellular layer surrounding the eggs of animals. In the sea urchin it gives rise to the fertilization membrane.

Wingless secreted signaling protein in *Drosophila* with a role in segment patterning and wing development. It is a member of the Wnt family of signaling proteins.

Wnt family family of secreted signaling proteins present in all Metazoa, including the sponges, members of which act in many aspects of development. It includes the Wingless protein in *Drosophila* and the Wnt proteins of vertebrates. Wnt proteins can signal through various pathways including the canonical Wnt/β-catenin pathway and the Wnt planar polarity pathway.

Wnt/β-catenin pathway see **canonical Wnt/β-catenin pathway**.

Wolffian ducts paired ducts associated with the mesonephros in mammalian embryos. They become the vas deferens in males.

X-chromosome inactivation the random inactivation of one copy of the X chromosome that occurs in the somatic cells of female mammals.

yolk sac an extra-embryonic structure with an internal cavity in birds and mammals. In the chick embryo the cavity contains yolk.

yolk syncytial layer a continuous layer of multinucleate non-yolky cytoplasm underlying the blastoderm in zebrafish embryos.

zinc-finger nucleases (ZFNs) nucleases that are used in a technique to specifically cleave and disrupt genes *in vivo* in a targeted fashion.

zona limitans intrathalamica (ZLI) a signaling center located in the forebrain of the vertebrate embryo, which helps pattern the brain along the antero-posterior axis.

zona pellucida a layer of glycoprotein surrounding the mammalian egg that serves to prevent polyspermy.

zone of polarizing activity (ZPA) *see* **polarizing region**.

zygote the fertilized egg. It is diploid and contains chromosomes from both the male and female parents.

zygotic gene any gene present in the fertilized egg and which is expressed in the embryo itself.

Index

Note: Figures and boxes are indicated by an italic *f* and *b* following the page number.

2D:4D digit ratio 586*b*, 586*f*

A

ABC model of floral organ identity 659–61, 661*f*
abdominal-A (*abd-A*) gene and product 91–2, 92*f*, 688, 688*f*, 690*f*, 692
abdominal-B (*abd-B*) gene and product 91–2, 92*f*, 688*f*
abembryonic pole 165
Acanthostega 701
acceleration 715
accretionary growth 572*f*, 572, 573
acetylcholine 556, 557*f*, 558*f*
 receptor 556, 557*f*, 558*f*
achaete gene 533, 534*f*, 535*f*, 535, 543
acheiropodia 472*b*
achondroplasia 588
acrosome reaction 418–19, 418*f*
actin filaments 254*b*
activators 316
Activin 133*b*, 137, 137*f*, 139–40, 140*f*
 receptor 138, 138*b*
actomyosin 254*b*
adaptive evolution 707, 708
adaxial–abaxial axis, leaves 648–9, 649*f*
adenomatous polyposis coli (APC) 580
adherens junctions 249–51, 250*f*, 250*b*, 252
adhesive cell junctions 249–51, 255–6
AGAMOUS (*AG*) gene and product 658, 659*f*, 659, 660*b*, 660*f*, 661, 661*f*, 663
age-related macular degeneration 388
aging 572, 617–21, 617*f*, 618*f*
 cell senescence 619–20
 genetic influences 618–19
agrin 615
Agrobacterium tumefaciens 638–9, 639*f*
agroinfection 638, 639*f*, 640*b*
agroinfiltration 639
agroinoculation 639
Airn gene 416
alcohol exposure and brain development 522*b*, 522*f*
allantois 153, 154*f*, 332
alleles 12, 13, 172*b*, 409, 434*f*, 434, 667, 712*b*
alternative RNA splicing 22, 316

Ambystoma mexicanum 596, 603, 604*f*, 604, 715*f*, 715
Ambystoma punctatum 573, 573*f*
Ambystoma tigrinum 573, 573*f*
ameloblasts 489, 490*f*
amniocentesis 171, 410
amnion 153, 154*f*, 171*f*, 171, 332
amnioserosa, *Drosophila* 52, 52*f*
amniotes 105
amphibians
 fate map 131*f*, 132, 201*f*
 gastrulation 121*f*
 limb regeneration 601–10, 602*f*, 610*f*
  nerve influence 605–6, 606*f*
 metamorphosis 590, 594–5, 594*f*
 neurulation 194*f*
 senescent cell elimination 621*f*, 621
 visual system 553*f*, 553, 554*f*
  eye determination 27–8, 29*f*
 see also Xenopus
Amphimedon queenslandica 679
amphioxus 683*f*, 683, 688, 688*f*, 697, 700
anamniotes 105
androgenetic embryos 413–15, 414*f*
anencephaly 293*b*, 293*f*
Angelman syndrome 416
angioblasts 296–7, 296*f*
angiogenesis 296–8, 296*f*, 297*f*, 493*b*
angiosperms 632, 633*b*, 633*f*
angustifolia mutation 640, 641*f*
animal pole 3*f*, 3*b*, 119, 123
animal–vegetal axis
 Xenopus 119, 122–7
 zebrafish 145
aniridia 508
Antennapedia (*Antp*) gene 92, 444*f*, 444, 456, 688, 688*f*, 705–6, 705*f*
Antennapedia homeotic gene complex 90–1, 91*f*, 93, 223*b*, 444
anterior gradient protein (AG) 605, 606*f*, 607
anterior neural ridge (ANR) 520*f*, 521
anterior visceral endoderm (AVE) 166–7, 167*f*, 207*b*
antero-posterior axis 18, 18*f*, 286
 brain 526
 Drosophila 53*f*, 54–7, 62–5, 63*f*, 73, 73*f*
 egg 62–5, 63*f*

spinal cord 531, 531*f*
vertebrates 102, 191–3, 213
 chick 154–5, 155*f*, 156*f*
 limb axes 458*f*, 458, 465*f*, 469–75, 470*f*, 472*b*, 472
 mouse 165–7
 somite formation 216–21, 217*f*, 218*f*, 220*f*
 somite identity 221–6, 222*f*, 224*f*, 225*f*
 Xenopus 197–8, 198*f*
 see also body axes, establishment of; pattern formation
anticlinal divisions 635, 635*f*
Antirrhinum 658, 662*f*, 662, 663
antisense RNA 23
APETALA genes and products 658, 659*f*, 659, 660*b*, 660*f*, 661, 661*f*
apical–basal axis 18
 Arabidopsis 634, 636–7
 see also pattern formation
apical ectodermal ridge 459, 459*f*
 role in vertebrate limb development 463–4, 464*f*
apico-basal polarity 19, 261, 266, 273
apolysis 591
apoptosis
 digit separation role 487*f*, 487
 Drosophila 576–8, 578*f*
 eye loss in *Astyanax* 708
 neuronal cell death 559–61
apterous gene and product 451*f*, 451, 704
Arabidopsis thaliana 2, 10, 11*f*, 631, 632*f*
 embryonic development 632–42
 fate map 634–5, 635*f*, 654*f*
 fertilization 631*f*, 632
 flower development 657, 658*f*, 658, 659*f*, 660*b*, 660*f*, 665*f*, 665, 667, 668
 flowering control 663–4, 663*f*, 664*b*, 665*f*, 665
 homeotic gene role 658–62, 659*f*
 germination 632
 leaf axes 648–9, 650
 leaf directed dilation 639–40, 641*f*
 life cycle 631–2, 631*f*
 meristems 643, 643*f*
  cell fate 644, 645*f*, 646, 647*f*
  stem cells 643–4, 644*f*

Arabidopsis thaliana (Continued)
  organizing center 643-4, 644*f*
  phyllotaxis 651*f*
  root tissues 653, 653*f*
  see also plant development
archenteron 121*f*, 121, 199*f*, 238, 267*f*, 268, 276*f*, 280
area opaca 149, 151*f*
area pellucida 149, 150*f*, 151*f*
Aristotle 4
Artemia 690
arteries 297
arthrogryposis 486
arthropods
  evolution 689-93, 689*f*, 690*f*
  molting 590, 591, 591*f*
astrocytes 538*b*, 538*f*
Astyanax mexicanus 708*f*, 708
asymmetric cell division 36, 36*f*
  early plant development 637
  morphogenesis 260
  neurogenesis 536-7, 536*f*
  stem cells 36-7, 37*f*, 350-1, 350*f*
autism 560*b*
Aux/IAA proteins *see* auxin
auxin
  gradients 636-7, 636*f*
  leaf arrangement on stem and 650-1, 651*f*
  root development and 653, 654-5, 655*f*
  secondary shoots and 651-3, 651*f*, 652*f*
  signaling pathway 636*f*
auxin-response factors (ARFs) 636*f*, 636, 637
axes *see* body axes, establishment of
Axin protein 87*b*, 128*b*
axolotl 596, 603, 604*f*, 604, 715*f*, 715
axons 518*f*, 518
  commissural 550*f*, 550, 551*f*, 551
  follower 549
  growth cone 544*f*, 545*f*, 545-9, 546*f*, 549*f*
  intermediate targets 549
  navigation and mapping 544-56
    dynamics 549
    guidance 545-8, 546*f*
    midline crossing 549-51, 550*f*, 551*f*
  pioneer 549

B

B lymphocytes 338
Baer, Ernst von 682, 696
bantam microRNA 578
Bar transcription factor 456
Bardet-Biedl syndrome (BBS) 473*b*
Barr body 433, 434*f*
Barrett's metaplasia 358
Barx1 gene 490, 491*f*, 491
basal ganglia 520
basal lamina 346
basal layer of epidermis 347, 348

base editing 15*b*
basement membrane 346
bat limb 701, 701*f*, 702, 713*f*, 713
Bax gene 561
Bcl-2 gene and product 561
Beckwith-Wiedemann syndrome 416
Belousov-Zhabotinsky reaction 484*b*
β-catenin 125*f*, 127, 128*b*, 128*f*, 137, 319
  planarians 598*b*
β-globin gene expression 343-5, 344*f*, 345*f*
  defects 346*b*, 346*f*
β-neurexins 559*f*, 559
β-thassalaemia 346*b*
bicoid gene and product 54-6, 54*f*, 55*f*, 61, 65-6, 66*f*, 73, 73*f*, 74*f*, 74, 75*f*
Bicoid protein 53*f*, 55-6, 55*f*, 56*f*, 57, 81*f*
Bicyclus 703, 703*f*, 705, 706-7, 706*f*
bilaminar germ disc 105
bilateral symmetry 679, 680*b*, 687
Bilateria 678*f*, 679, 680*b*, 681, 681*f*, 687
bile ducts 394
biolistics 639, 640*b*
birds 148
  see also chick
bithorax homeotic gene complex 90-1, 91*f*, 92*f*, 94, 223*b*
bithorax mutation 91*f*, 91
blastema 598*b*, 601-5, 602*f*, 603*f*, 604*f*, 606-9, 607*f*, 608*f*, 609*f*
blastocoel 3*f*, 3*b*, 105, 120, 121*f*
  fluid accumulation 264
blastocyst 105, 105*f*, 158*f*, 159, 159*f*, 163*f*, 163-6, 164*f*, 164, 165*f*, 166*f*, 167, 169, 171, 332
blastoderm 142-3, 143*f*, 149, 150*f*, 151*f*, 211*f*
  antero-posterior axes 154-5, 155*f*, 156*f*
blastoids 378-80, 378*f*
blastomere 103, 122
  microsurgical removal 113
  specification 162-4, 163*f*, 164*f*
blastopore 8, 9*f*, 120, 121*f*, 122, 276*f*, 276, 277*f*, 280
blastula 3*f*, 3*b*, 105, 107, 120, 121*f*, 122, 123, 200*f*, 258, 259*f*
  fate map 131*f*, 132
  formation of 261-5
Blimp1 protein 406, 406*f*
block to polyspermy 420-1, 420*f*, 421*f*
blood cells 338-46
blood vessels 296-8, 296*f*, 297*f*
Bmi1 expression 352
body axes, establishment of 18, 18*f*
  double-gradient model 193*f*
  Drosophila 51-61, 52*f*, 62-5, 63*f*
    wing 446-51, 453-4
  leg disc 454-6
  vertebrates
    chick 154-5, 155*f*, 203

    limb axes 458*f*, 458, 464-5, 465*f*, 466-70, 467*f*, 470*f*, 475-8, 478*f*
    mouse 165-7
    Xenopus 122-31
    zebrafish 145
  see also pattern formation
body plan
  arthropods 689*f*, 694-5, 695*f*
  evolutionary changes 687-95, 710-13
  vertebrates 102*f*, 102-3, 118-31, 199, 199*f*, 682-3, 694-5, 695*f*
Bolitoglossa 713-14, 714*f*
bone growth 585-9, 587*f*, 588*f*
bone marrow 338, 340, 341
  transplants 389
bone morphogenetic proteins (BMPs) 133*b*, 200*f*, 200, 201, 202*f*, 386, 587-8
  antagonism 189-90, 196, 198, 231
  BMP-4 71, 73, 375, 490, 694
    cell differentiation 332
    chordin interaction 73, 190, 191*f*
    mesoderm patterning 190*f*, 190
    somite patterning 231
    trophectoderm 380
  cell differentiation 332
  gastruloids 379*b*
  limb development and 475, 478*f*, 478, 479, 479*f*, 487
  lung development 495*f*, 495, 498
  mammary gland development 498
  nervous system development 527-8
  neural plate induction 194-6, 195*f*, 196*f*, 198
bottle cells 277*f*, 277
brachyury gene *see TbxT* gene
brain-derived neurotrophic factor (BDNF) 562
brain development 520
  alcohol exposure 522*b*, 522*f*
  boundaries 524*f*, 524-6, 525*f*
  cortical neurons 539-40, 540*f*, 541*b*, 541*f*, 542*f*, 562
  Hox gene role 524
  main regions 520-1
  plasticity 564
  retino-tectal system 552-5, 553*f*, 554*f*, 555*f*
  self-organization 523, 523*f*
  signaling centers 520*f*, 521-3
  visual centers 552-5, 553*f*, 554*f*, 555*f*, 562-4, 563*f*
branchial arches 524*f*, 526
  evolutionary changes 683-4, 684*f*, 685
  Hox gene expression 235-6, 235*f*
  neural crest cell migration to 235-6, 235*f*
branching morphogenesis 294-8, 294*f*, 295*f*
  lung development 494*f*, 495-6
branchiomeric muscles 501
branchless gene and product 295*f*, 296

Index **753**

BRCA1 gene 172*b*
breast cancer 493*b*, 493*f*, 497
breathless gene and product 296
Bric-a-brac transcription factor 456
brinker gene and product 447–8, 447*f*, 450
Browne, Ethel 8
butterfly
  metamorphosis 590, 592*f*, 592
  wing development 691–2
    markings 703–7, 703*f*, 705*f*, 706*f*
buttonhead gene and product 94

C

Cactus protein 59, 60*b*
cadherins 63–4, 250*b*, 252–4, 253*f*, 546*f*, 546
  E-cadherin 252, 262*f*, 262, 273
  N-cadherin 252, 253*f*, 271
  P-cadherin 252
  Starry night (Stan/Flamingo) 274*f*, 274, 275*f*, 275
Caenorhabditis 2, 10, 11*f*, 681, 682
  aging 618*f*, 618–19
  apoptosis 561
  dauer state 618–19
  developmental genes 22–3
  dosage compensation 433*f*, 435
  germ cell development 403
  mosaic development 28
  sex determination 429–30, 430*f*, 432*f*, 432
  transdifferentiation 357
  ventral closure 303
calcium wave, egg activation 422*f*, 422–3, 423*f*
cambium 634, 653
Cambrian explosion 679
camel 701
canalization 652*f*, 652
cancer 493*b*, 578–80, 579*f*
  breast 493*b*, 493*f*, 497
cancer cells 2, 578–80
  metastasis 267
canonical Wnt/β-catenin pathway 85, 123, 128*b*, 128*f*, 145, 198*f*, 198
  Hydra development role 601
  see also β-catenin; Wnt protein signaling
capillaries 297
CAPRICE protein 656
Capsaspora owczarzaki 678
Capsella 633*b*, 633*f*
cardia bifida 499
cardiomyocytes 614, 614*f*, 615, 616
cartilage 485–6, 585*f*, 585, 586*b*, 587*f*, 588
cat, X, inactivation 434*f*, 434
catenins 250*b*
catshark 697
caudal gene and product 57
Cdon gene 522*b*

Cdx2 transcription factor 163–4, 164*f*, 263–4, 385
cell adhesion 248, 249–56
  cell-adhesion molecules 21, 248, 249–51, 250*b*, 546–8
    neurulation and 291
    synapse formation and 556, 558–9, 558*f*
  differential adhesion hypothesis 251
  epithelial–mesenchymal transitions 255–6, 256*f*
  regeneration and 607
  tissue variation 251–2, 251*f*
cell-autonomous action 29
cell-autonomous gene effects 29
cell–cell interactions 8, 17
  cell–cell signaling 21
  germ cell development 406–7, 406*f*
cell complexity 38–9
cell cycle 6*b*, 6*f*, 574*f*, 574–5
  checkpoints 574*f*, 574
  regulation 574*f*, 574–5
cell death *see* apoptosis
cell determination 27–8, 28*f*
cell differentiation 20, 311–15, 312*f*, 313*b*, 313*f*
  cell fusion effects 356*f*, 356–7
  evolution 685
  models 338–55
    blood cells 338–46
    epithelia 346–53
    muscle 334–8, 335*f*, 336*f*, 337*f*
    pluripotent stem cells as models of 375–7, 375*f*, 376*f*
  reversibility 356, 359–60
    dedifferentiation 338, 597, 602, 603
  senescence effect 619–20
  spinal cord 526–8, 527*f*, 529*f*
  transdifferentiation 335, 357–9, 358*f*, 389, 390, 391, 603
    replacement cell generation 390–1, 391*f*
  vertebrate limb 459
cell division
  anticlinal 635, 635*f*
  asymmetric 36, 36*f*
    neurogenesis 536–7, 536*f*
  control, *Drosophila* wing 576–8, 578*f*
  coordination 576–8, 578*f*
  differentiation and 314
  periclinal 635, 635*f*
  see also meiosis; mitosis
cell enlargement 572*f*, 572, 573
cell fate 26*b*, 27–9, 28*f*
  germ cells 403–5, 403*f*, 404*f*, 405*f*
  plant meristems 644–6, 645*f*, 646*f*, 647*f*
  single-cell analysis 329–34, 330*b*, 333*f*
  somite cell fate determination 229–31
  see also fate maps
cell fusion 356*f*, 356–7
cell-identity switching 524, 525

cell lineage 27
  restriction 82–3, 83*f*
    brain development 524*f*, 524–6, 525*f*
  tracing 109, 110*b*, 110*f*
cell migration 248, 295, 298
  dorsal closure in *Drosophila* 302–3, 303*f*
  germ cells 407–8, 407*f*
    chemical signals 408
  lateral-line primordium 301–2, 302*f*
  neural crest cells 232*f*, 232–6, 233*f*, 234*f*, 298–301, 299*f*, 300*f*, 304
  sea urchin gastrulation 266–70, 267*f*
  ventral closure in *Caenorhabditis* 303–4
cell motility 249
cell orientation during limb bud outgrowth 464–6, 465*f*
cell packing 258
cell proliferation 572*f*, 572, 573
  control of 574*f*, 574–5
  cancer 578–80, 579*f*
cell-replacement therapy 373, 387–90, 388*f*
  cell generation 390, 393
cell segregation 524–5
cell senescence 619–20
cell specification 27, 28*f*
cell theory 5
centipede, body plan 689*f*
centrosome 258
cephalochordates 688
cephalopods 686
Cerberus-like 2 (*Cerl2*) gene 239
Cerberus protein 190*f*, 191, 195, 198, 207*b*
cerebellum 520, 521–2
cereblon 476*b*
cerebral cortex 520
cerebral organoids 393*f*, 393
chalones 582
chameleon 287*f*
chemoaffinity hypothesis of connectivity 553
chemotropic response 548
chick 148–57
  antero-posterior axis 213
  axial structures 212–15, 213*f*
  blastoderm 105
  body axes 154–5, 155*f*, 156*f*
  brain development 520*f*
    boundaries 524*f*, 525*f*, 525
    local signaling centers 520*f*, 521
    self-organization 523*f*
  cartilage growth 585*f*, 585
  cleavage 103, 149, 149*f*, 151*f*
  dorso-ventral axis 215
  egg 103, 149, 149*f*
  embryo 4*f*, 4, 10–11, 104*f*
    development 107
  extra-embryonic structures 153, 154*f*
  eye 505
  fate map 113, 168, 201*f*

Index

chick *(Continued)*
   gastrulation 149, 150f, 152, 282–6, 283f, 284f, 286f, 289
   germ cell development 403, 407
   heart 500
   Hensen's node 156f, 156, 203–6, 204f
   Hox genes 221, 225–6, 225f, 693, 693, 694f
   in situ hybridization 108f
   left–right asymmetry 238, 239
   life cycle 150f
   limb development *see* vertebrate limb development
   lineage tracing 113
   mesoderm induction and patterning 204f, 204
   microsurgical manipulation 112, 112f, 113
   motor neuron innervation 551–2, 552f
   neural crest migration 299–300, 300f
   neural induction 204–6, 204f
   neuronal cell death 559–60, 559f
   neurulation 152, 291f
   organizer region *see* Hensen's node
   somite fate map 230f
      patterning within somite 231
   somite formation 217–21, 217f, 218f, 220f
   spinal cord 527, 531
   teratogens 476b
   transgenic technique 113f, 113, 114f, 118
   transient transgenesis studies 118
   visual system 553–4, 554f
   wing as model for limb development 458f, 458–61, 460f, 461f
   wing development 701
   see also vertebrate limb development
chimera 29, 29f, 113, 115b, 115f, 162–3
   mericlinal 646f, 646
   periclinal 644, 645f
ChIP-chip technique 109
ChIP-seq technique 109, 177
cis-regulatory control region 710
   see also gene control region, zone of polarizing activity regulatory sequence (ZRS), locus control region (LCR)
choanocytes 677
choanoflagellates 677, 678f, 678, 679, 681
chondrocytes 585–7, 586b, 587f, 588f, 588
Chordata 683
Chordin protein 190f, 190, 191f, 195, 198, 202f, 202, 204, 205, 694–5, 695f
   BMP-4 interaction 73, 190, 191f
chorion 153, 154f, 174
chorionic cavity 174, 175
chorionic villi 174, 174f
chromatin 25, 206, 323b
   gene expression regulation 319–26
   immunoprecipitation techniques 109
   modification 230, 323b, 326–7

chromatin-remodeling complexes 206, 323b
chromosome conformation capture 320
Churchill gene and product 205
ciliopathies 238, 473b
circadian clock 664b, 664f, 664
CLAVATA (*CLV*) genes 643, 644f
cleavage 3f, 3b, 6, 6f, 17, 257–60, 259f
   Drosophila 45f
   furrow 258, 260, 261f
   plane of 258–60, 259f
   radial 257, 257f
   spiral 257, 257f
   vertebrates 103
     chick 103, 149, 149f, 151f
     human 105, 169, 170f
     mouse 105, 158–9, 159f, 164
     Xenopus 17, 18f, 103, 119–20, 120f
     zebrafish 103, 142, 143f
clock and wavefront model of somite formation 217–21, 218f
cloning 359–60, 360f
   mammals 360–1
   somatic cell nuclear transfer 359–61, 359f, 360f
Cnidaria 678f, 680b, 681, 681f, 685, 687
co-activators 319
cobra 703, 704b, 704f
cockroach limb regeneration 612, 613f
coelom 106f, 107, 151
Cohen, Stanley 561
collective cell migration 301, 408
colony-stimulating factors 342
colorectal cancer 580
commissural axons 550f, 550, 551f, 551
community effect 135b, 135f
compaction 261–2, 262f
compartments 83–5, 84f, 85f, 444–5, 445f, 457, 524
complexity 38–9
condensation 459
congenital abnormalities 2, 521, 522b
CONSTANS (*CO*) gene and product 664f, 664, 665f, 666f
constriction 268f, 268, 270f, 271
contraction 254b, 268, 268
convergent evolution 707
convergent extension
   gastrulation 270, 269f, 271–2, 273, 276–80, 277f, 278f, 278b, 279f, 285, 286f
   neurulation 290
COOLAIR promoter 666
co-repressors 319
corn snake 693–4
cornea 505f, 505–6
cortical granules 418
corticotropin-releasing hormone 594, 594f
cotyledons 632
cow 701, 702

craniorachischisis 293b
Cre/loxP system 206, 208b, 208f
cricket limb regeneration 612
CRISPR-Cas9 system 14, 15b, 15f, 16f, 74, 111, 690, 691f
   β-globin gene defects 346b, 346f
   butterfly wing markings 706
   humans 177
   mouse 15b, 16f, 115b, 118
   plants 640b, 640f
   snake, zone of polarizing activity regulatory sequence 704b, 704f
   zebrafish 15b, 118
critical period 476b, 522b
crustaceans, paired appendages 689f, 690f, 690–3
crystallins 686
CTCF protein 320, 321f, 322
Ctenophora 678f, 680b, 681f
cumulus cells 418
CURLY LEAF gene 661b
CXCR4 receptor 408
cyclin-dependent kinases (Cdks) 574f, 574
   Cdk4 337
cyclins 574f, 574
CYCLOIDEA gene 662f, 662
cyclopia 32b, 507
Cyclops protein 148
Cyp26a1 gene 228–9
cystic fibrosis 172b
cytokinins 652f, 652, 653, 655, 655f
cytoplasmic determinants 36–7, 36f
cytoskeleton 30, 254b
   activity 254
cytotrophoblast 174

D

dachshund gene 455–6, 455f
daf-2 gene and product 618–19
Dally glypican 448
Darwin, Charles 673, 676b, 685
Darwin's finches 676b, 676f
Daschsous (Ds) cadherin 275, 578, 612
dauer state 618–19
DCC receptor 550
decapentaplegic (*dpp*) gene and product 69f, 70f, 70, 71–3, 72f, 447–50, 447f, 448f, 449b, 452, 453, 455f, 455, 695f, 695
dedifferentiation 338
   limb regeneration and 597, 603
deep layer 105
Deformed gene 93
Delta–Notch signaling pathway 63f, 63, 220
   see also Delta protein; Notch protein
Delta protein 124f, 124b, 451f, 451
   leg development and 456
   neurogenesis role 534–5, 535f, 541–3, 543f
dendrites 518f, 518

dental lamina 489, 490f
dental papilla 489, 490f
denticle pattern, *Drosophila* 48, 48f, 49, 86f, 86-8, 87b, 88f, 89f
dependence receptors 561
dermis 346
dermomyotome 230, 230f, 231
Derrière protein 137, 137f
desmosomes 250b
determinants 6, 6f
  cytoplasmic localization 36-7, 36f
determination 27-8, 28f, 140-1, 313-14
  amphibian eye region 27-8, 29f
deuterostomes 682
developmental genes 22
  Caenorhabditis 22-3
  control of expression 23-6, 25f
    cis-regulatory modules 21-2, 22f, 25
  Drosophila 22, 23
    genetic screening 49-51, 50b, 50f
developmental stages 104f
Dharma transcription factor 145, 146f
diabetes 389-90
Dickkopf1 127, 190f, 191, 198, 204
Diego protein 274f, 274
differentiation *see* cell differentiation
DiGeorge syndrome 503
digits
  identity specification 471-5, 472b
  polydactyly 471, 472b, 472f, 473b, 474, 482
  programmed cell death role in separation 487f, 487
dioxins 476b
diploblasts 680b
diploidy 5, 12, 408-9, 409f
direct development 715-16
directed dilation 249, 282, 639
  notochord 282
  plant growth 639-40, 641f
Dishevelled (Dsh) protein 125f, 127, 128b, 274f, 274, 275, 276
disposable soma theory 617
distal-less gene and product 455f, 455-6, 692
  butterfly wing markings 705, 705f, 706, 707
distal visceral endoderm (DVE) 166, 167f
Dlx1 gene 491
Dlx2 gene 490, 491f, 491
DNA
  control regions 23-5
  methylation 322-6, 323b, 416
  microarray analysis 109
  mid-blastula transition 130f, 130
  mitochondrial 422
  repair 619
Dobzhansky, Theodosius 673
dog 582f, 584, 710, 712b

dolphin 702
dorsal closure, *Drosophila* 302-3, 303f
dorsal convergence 280
dorsal gene and product 59, 60f, 60b, 68-71, 69f, 70f
dorsal root ganglia 299, 300f
dorsalized embryos 59, 70
dorsalizing factors 125f, 126, 127, 190
dorso-lateral hinge points 290
dorso-ventral axis 18, 18f, 286
  Drosophila 58-9, 60f, 68-71, 69f
  egg 67, 67f
  wing 446-51, 453-4
  spinal cord 526-8, 527f, 528f, 529f
  vertebrates 102, 215
    chick 154
    limb axes 458f, 458, 475-8, 478f
    versus arthropods 694-5, 695f
    Xenopus 125f, 125-9, 189-90, 190f, 215, 694, 695f, 695
    zebrafish 145
  see also body axes, establishment of; pattern formation
dosage compensation 433f, 433-5
doublesex (*dsx*) gene and product 428, 428f
Down syndrome 410-12
Driesch, Hans 7-8, 7f
Drosophila 2, 9, 10, 11f, 11, 43-4, 679, 682
  aging 617, 617f, 618f, 618, 619, 619f
  body axes 51-61, 52f
    antero-posterior axis 53f, 54-7, 62-5, 63f, 73, 73f
    dorso-ventral axis 58-9, 60f, 67, 67f, 68-71, 69f
  body plan 689f
  cell cycle regulation 575-6, 575f
  chromatin modification 323b
  cleavage 45f
  compound eye 508, 508f
  cytoplasmic determinants 36, 37
  denticles 48, 48f, 49
    patterns 86f, 86-8, 87b, 88f, 89f
  developmental genes 22, 23
    genetic screening 49-51, 50b, 50f
  dorsal closure 302-3, 303f
  dorso-ventral axis 694-5, 695f
  dorso-ventral patterning 451f, 451
  dosage compensation 433f, 435
  eye development 686
  gastrulation 46-7, 47f, 48f, 270-2, 270f
  gene expression regulation 316, 318f, 318, 319, 323b, 326
  gene regulatory networks and cell-fate decisions 26b, 26f
  germ-band extension 48, 48f, 271-3, 272f
  germ cell development 402, 403-5, 403f, 404f, 408
  growth 573, 576-8, 577b, 577f, 578f
  head region 94, 693

heart 503
histoblasts 48, 49f
Hox genes 90-5, 221-2, 223b, 223f, 323b, 444, 688, 688f, 689f
imaginal discs 48, 49f, 84, 442-54, 456
  leg 454f, 454-6
  fate map 454f
  origins 442-4
  parasegments 444-5
  patterning 444-5
  positional values 456f, 456
  size determination 576, 578
  specification 443f
  wing 442-54, 446f
instar 48, 48f
intercellular signals 65f
larva 48-9, 48f
leg development 442-4, 454-6, 457, 692
life cycle 44-51, 45f, 716f, 716, 717b, 718f
maternal genes 51, 53-61
  mutation effects 54-5, 54f
mesoderm 46-7, 47f, 52, 52f, 69
  invagination 270-2, 270f
metamorphosis 48-9, 49f, 445-6, 445f, 446f, 454, 592-4, 593f, 594f
misexpression screening 76f, 76b
neurogenesis 532-7, 533f, 534f, 535f, 541-3
oogenesis 61-9, 62f
  antero-posterior polarity 62-5, 63f, 67
  dorso-ventral polarity 67
  oocyte cyotskeleton reorganization 65-7
pattern formation 68-77
  denticle patterns 86f, 86-8, 87b, 88f, 89f
planar cell polarity 273-6, 273f
pole cells 46
pupa 48
segmentation 48, 48f
  compartments 83-5, 84f, 85f, 444-5, 445f
  parasegment boundary stabilization 84-9, 444-5
  parasegment establishment 78-82, 79f, 444-5
  segment identity specification 90-5
  segmentation gene expression 82-90, 85f
sex determination 427-9, 428f, 429f, 432
syncytial blastoderm 45-6, 52, 53f, 56
targeted gene expression 76b
terminal region specification 58
tracheal system 294f, 295-6, 295f
transgenic 76b
wing development 442-54, 576-8, 578f, 658-9, 660b
  antero-posterior axis 446-50, 447f, 448f, 449f
  axis establishment 446-51, 453-4

756 Index

Drosophila (Continued)
    compartment boundaries 444–50
    metamorphosis 445–6, 445*f*, 446*f*
    pattern formation 446–51, 453–4
    proximo-distal axis patterning 453–4
    vestigial gene 444, 451–3, 451*f*, 452*f*
DsFtFj system 275

E

E-box 337
ecdysone 328, 591–2, 592*f*, 593, 594*f*, 594
echinoderms *see* sea urchins
ectoderm 3*f*, 3*b*, 19, 19*b*, 19*f*
    Drosophila 46, 47, 47*f*, 442–4
    limb development and 475–8, 478*f*
    mammary gland development 496–8, 497*f*
    mouse 159
    vertebrates 102, 144, 148, 149
        Xenopus 132–4, 133*f*, 134*f*, 137*f*, 196
        zebrafish 146, 147, 147*f*
Ectodermin 132–3
ectodysplasin A protein 708
ectomesenchyme 233
EDA gene 708
Edwards, Robert 172*b*
EGF *see* epidermal growth factor (EGF)
egg 402, 408–13, 417–20
    changes at fertilization 420–3
        activation 422*f*, 422–3, 423*f*
        polyspermy prevention 420–1, 420*f*, 421*f*
    chick 103, 149, 149*f*
    cleavage and 257–8
    Drosophila 61–2, 63*f*
    evolution of 679
    germplasm 403–5, 403*f*, 404*f*, 405*f*
    human 169, 170*f*
    mouse 158
    size 410, 412, 715, 717
    totipotent potential 37, 413
    Xenopus 103, 119*f*, 122, 123
    zebrafish 103, 142
    see also fertilization; oogenesis
egg cylinder 105
Egr2 gene and product 524, 524*f*, 526
electroporation 113
Eleutherodactylus 715, 716*f*
embryo models 377
    early development, embryonic stem cells 377–81, 378*f*, 381*f*
    gastrulation, pluripotent stem cells 381–7, 382*f*, 383*b*, 383*f*, 384*f*, 385*f*, 386*f*
embryogenesis 2
    evolution 683–5
    generative program 37–8
    human embryo shape changes 20*f*
    nutrition effects in later life 589–90
    phylotypic stage 683, 683*f*
embryoid bodies 378, 378*f*

embryological manipulation *see* microsurgical manipulation
embryology 2
embryonic organizer 105
embryonic pole 165
embryonic stem cells (ES)
    human 370–1, 372*f*, 375, 376*f*, 376, 379*b*, 380, 383*b*, 384*f*, 386–7, 386*f*, 390, 392*b*
    models of early development 377–81, 378*f*, 381*f*
    naive 371
    organoids 391–4, 392*b*
empty spiracles gene and product 94
endoblast 155
endocardial cushions 502
endocardium 499
endochondral ossification 585, 587*f*, 698
endoderm 3*f*, 3*b*, 19, 19*b*, 19*f*
    Drosophila 46, 47, 47*f*
    lung development 492–5, 494*f*
    vertebrates 102, 142–3, 148, 375
        chick 149
        mouse 159, 160–2, 160*f*, 161*f*, 164–7, 165*f*, 167*f*, 285
        Xenopus 129, 129*f*, 132, 132*f*, 133*f*, 134*f*, 134, 196, 280
        zebrafish 146, 147, 147*f*, 148
endosperm 632
endothelial cells 296, 297–8
endothelium 296
engrailed gene and product 79*f*, 82–5, 83*f*, 85*f*, 86*f*, 88*f*, 303*f*, 303, 454, 455*f*
    butterfly wing markings 706, 706*f*
    compartment boundaries and 83–4, 84*f*, 446, 447*f*
    expression regulation 326
Engrailed1 (*En1*) gene 478*f*, 478
ENHANCER OF TRYPTYCHON AND CAPRICE1 protein 656
enhancer-trap technique 76*b*
enhancers 316
ependymal cells 538*b*, 538*f*, 539
Eph receptors 292*b*, 292*f*, 525, 546*f*, 546, 552*f*, 552, 553–4
ephrins 128*f*, 292*b*, 292*f*, 525, 546*f*, 546, 552*f*, 552, 553–4
epiblast 148, 149, 151*f*, 152, 153*f*, 157, 158, 159, 160, 160*f*, 161, 161*f*, 164–5, 165*f*, 171*f*, 171, 204, 282–6, 284*f*, 285
    polonaise movements 283, 284*f*
    stem cells (EpiSCs) 371, 372*f*, 375, 375*f*
epiboly 276*f*, 276, 280
    Xenopus 122
    zebrafish 144*f*, 144, 280
epicardium 499, 502
epidermal growth factor (EGF) 124*f*, 561
epidermal growth factor receptor (EGFR) 348*f*, 348, 456, 612

epidermis 346–51, 347*f*
epidermolysis bullosa 347, 389
    junctional 348, 349*b*, 349*f*
epigenesis 4, 5, 413
epigenetic modifications 320, 323*b*
epigenetics 313*b*
epimorphic regeneration 612
epimorphosis 597, 597*f*
    planarians 598*b*
epithelia 294, 578–9
    mammary gland development 496–8
epithelial-to-mesenchymal transition (EMT) 233, 255–6, 256*f*, 266–8
ErbB proteins 497–8
erythrocytes 338, 343–5, 344*f*
estrogen 328, 498
ET embryoids 380–1, 381*f*
ethanol 476*b*
ETX embryoids 381*f*, 381
Eumetazoa 681
even-skipped (*eve*) gene and product 76*b*, 79*f*, 80–1, 80*f*, 81*f*, 81, 83, 83*f*
    regulation of expression 318*f*, 318
evolution 39, 673–7, 674*f*
    adaptive 708
    body plans, diversification of 687–95
        arthropods and vertebrates compared 694–5
        Hox gene changes 687–94
    convergent 707
    of development 677–86
    gastrulation 681–2, 682*f*
    multicellular organisms 677–9
    parallel 685–6
    specialized characters 696–710
        butterfly wing markings 703–7, 703*f*, 705*f*, 706*f*
        CNS 534*f*
        developmental basis 707–8, 707*f*, 708*f*
        limblessness 702–3
        limbs 696–702, 696*f*, 699*b*, 699*f*, 700*f*
    timing of developmental processes 710–19
exencephaly 293*b*, 293*f*
experience-dependent development 564
extra-embryonic endoderm cells (XEN cells) 370
extra-embryonic membranes 105
extra-embryonic structures 154*f*, 157, 159, 286–7
eye
    age-related macular degeneration 388
    determination 27–8, 29*f*
    evolution 685–6, 686*f*
    lens 503–4, 503*f*, 686
        formation 504–6, 505*f*
        placode 504, 506, 507
        regeneration, newt 357–8, 358*f*, 596
    non-development in *Astyanax* 708*f*, 708
    vertebrates 503–8

development 504–7, 505*f*, 506*f*
structure 503–4, 503*f*
transcription factors 507–8, 508*f*
see also retina
eyeless gene 508, 686

F

familial adenomatous polyposis coli 580
fasciated mutation 663, 663*f*
fasciculation 549
fass mutation 635, 654
Fat cadherin 578, 612
fate *see* cell fate; fate maps
fate maps 109
    Arabidopsis 634–5, 635*f*, 654*f*
    Drosophila imaginal discs 446*f*, 454*f*
    neural crest cells 234*f*, 234
    plants 645–6, 647*f*, 654*f*
    somites 230*f*
    vertebrates
        amphibians 110*b*, 110*f*, 113, 131*f*, 132, 201*f*
        chick 113, 168, 201*f*
        mouse 113–14, 168*f*, 168, 201*f*
        zebrafish 113, 146*f*, 146–7, 201*f*
    see also cell fate
fem gene 432
fertilization 3*f*, 3*b*, 5, 417–24
    Arabidopsis 631*f*, 632
    cell-surface interactions 418–20, 419*f*
    egg activation 422*f*, 422–3, 423*f*
    human 169
    polyspermy prevention 420–1, 420*f*, 421*f*
    sea urchin 420–1, 420*f*, 421*f*, 422
    Xenopus 103, 123, 125, 125*f*, 126, 420, 422–3, 423*f*
fertilization membrane 420–1, 420*f*, 421*f*
Fgf8 gene 490
Fgf9 gene 507
fibroblast growth factor (FGF) 124*b*, 124*f*, 139, 370, 375, 588, 697
    brain development 524
    FGF-8 520*f*, 521–2, 523, 523*f*
    FGF-9 425
    FGF-10 496
    FGF-A 268
    mammary gland development 497
    neural development and 196, 198, 531, 532
    receptors (FGFR) 296
    signaling pathway 197*b*, 197*f*
    somite formation and 218
    trophectoderm 380
    vertebrate limb development 463, 463*f*, 464*f*, 464, 465, 466, 467*f*, 479, 479*f*
fibroblasts 314
    senescence effect 619–20, 620*f*
Filasterea 677, 678*f*, 678, 679
filopodia 254*b*

Caenorhabditis 303
Drosophila 303*f*, 303
sea urchin 268*f*, 268, 269, 270
synapse formation and 558, 558*f*
finger lengths 586*b*, 586*f*
fins 696–700, 696*f*, 698*f*, 699*b*, 699*f*, 700*f*
    zebrafish 697–700, 698*f*, 699*b*, 699*f*, 700*f*
floor plate 527*f*, 527, 528*f*, 529*f*
floral organ identity genes 659
FLORICAULA (*FLO*) gene 658, 663
flour beetle 716*f*, 716, 717*b*, 718*f*
FLOWERING LOCUS C (*FLC*) gene and product 665–6, 666*f*
FLOWERING LOCUS D (*FD*) gene and product 664
FLOWERING LOCUS T (*FT*) gene and product 664, 665*f*
flowers 657–8, 658*f*
    development 658–63, 665*f*, 665–7, 666*f*
    flowering control 663–4, 663*f*, 664*f*, 665*f*
FMR1 gene and product 560*b*
fog gene 432
folic-acid antagonists 476*b*
follicle cells 62, 62*f*, 409
Follistatin protein 190*f*, 190, 195
follower axons 549
forebrain 520*f*, 520
    see also brain development
forward genetics 14, 111, 112, 114
Foxg1 gene 522
Foxl transcription factors 133*f*, 133–4, 390
fragile X mental retardation 560*b*
FRIGIDA (FRI) protein 665–6, 666*f*
Fringe protein 219*b*
Frizzled (Fz) receptor 128*b*, 274*f*, 274–5, 275*f*
Frizzled 7 (Fz7) receptor 127
Frizzled-related protein (Frzb) 190*f*, 191, 198, 204
fruit fly *see Drosophila*
fruitless gene 429
fushi tarazu gene and product 79*f*, 80, 81, 83, 83*f*

G

gain-of-function experiments 109–10
Gal4 transcription factor 76*b*
gametes 401, 402
ganglion mother cell 536–7
gap genes
    expression 73–7, 73*f*, 74*f*, 75*f*, 77*f*, 81
    head gap genes 94
gap junctions 30255
Gasterosteus aculeatus 707, 707*f*, 708, 709*b*, 709*f*
gastroschisis 304
gastrula 7, 9*f*, 105, 198*f*, 211*f*, 251*f*, 280, 507, 682*f*
    amphibian eye determination 28, 29*f*

chick 149
Drosophila 533*f*
mouse 161*f*
Xenopus 127*f*, 127, 135, 135*b*, 137*f*, 139, 139*f*, 141, 146*f*, 189, 190*f*, 191–3, 191*f*, 200*f*, 282
zebrafish 146*f*, 146, 201*f*, 202*f*, 202
gastrulation 3*f*, 3*b*, 19–20, 105, 266–88
    cell differentiation 329–34, 330*b*, 333*f*
    Drosophila 46–7, 47*f*, 48*f*, 270–2, 270*f*
    evolution 681–2, 682*f*
    human 171*f*, 171
    movements 266–88
    pluripotent stem cells model 381–7, 382*f*, 384*f*, 385*f*, 386*f*
    sea urchin 20*f*, 266–70, 267*f*, 269*f*, 289
    vertebrates 102, 276–87, 287*f*
        chick 149, 150*f*, 152, 282–6, 283*f*, 284*f*, 286*f*, 289
        human 334
        mouse 158*f*, 161–2, 161*f*, 282–3, 284*f*, 285, 286*f*, 289, 329, 332–4, 333*f*, 334*f*
        Xenopus 120–2, 121*f*, 140, 189, 191–3, 193*f*, 199, 199*f*, 276*f*, 276–80, 277*f*, 278*b*, 278*f*, 279*f*, 281–2, 282*f*, 285–6, 286*f*, 289
        zebrafish 144*f*, 144, 280, 281*f*, 285–6, 286*f*, 289
gastruloids 37, 379*b*, 379*f*, 382*f*, 382, 383*b*, 383*f*, 384, 384*f*
GATA1 transcription factor 342*f*, 342, 344, 346*b*
GATA4 transcription factor 385*f*, 385*f*, 503
Gata6 transcription factor 165*f*, 165
Gbx2 transcription factor 520*f*, 523, 523*f*
GDF-1 signal protein 123, 123*f*, 137, 147, 155, 156*f*
GDF-5 protein 486
gene control region 709*b*
gene duplication 687–8, 688*f*
gene expression 21
    control of 20–6, 315–29
        chromatin changes 319–26
        external signals 327–8, 328*f*, 329
        feedback loops 26, 27*f*
        transcriptional regulators 316–19
    differential 23
    DNA microarray and RNA seq analysis 109
    evolutionary changes 675, 675*f*
    hindbrain 524, 524*f*
    misexpression screening 76*f*, 76*b*
    motor neuron innervation of chick limb 551–2, 552*f*
    organizer region 206
    plasticity 355–61
    protein synthesis 21–3, 22*f*
    targeted 76*b*
    visualizing in embryos 24*b*
gene knockdown 14, 111–12

gene knock-ins 111
gene knock-outs 14, 111, 115b, 116f, 118
　　Cre/loxP system 206, 208b, 208f
　　Hox genes 226-7, 235-6
gene regulatory networks 26b, 26f, 26, 203b, 479, 529f
gene silencing 14, 111-12, 320, 416
genes 10, 14
　　control of development 17, 21-3
　　homologous 12
　　identification 11-12
　　　　spontaneous mutations 12-14, 13f, 14f
　　maternal 51
　　selector 84
　　see also developmental genes
genetic equivalence 23
genetic screening
　　developmental genes, Drosophila 49-51, 50b, 50f
　　misexpression screening 76f, 76b
　　mutagenesis screen, zebrafish 114, 116, 116b, 117f, 147
　　preimplantation 172b, 172f, 173f
genetics 8-10
　　forward 14, 111, 112, 114
　　reverse 14, 111, 112, 114, 116, 118, 177
genital ridge 405, 407
genitalia 427f, 427
genome 12, 679-81
　　editing 111
genomic imprinting 360, 413-16, 415f
　　plants 667
genotype 10f, 10
germ-band extension, Drosophila 48, 48f, 271-3, 272f
germ cells 5, 5f, 66, 401-2
　　development of 402-17
　　　　cell-cell interactions 406-7, 406f
　　　　cell fate specification 403-5, 403f, 404f, 405f, 417
　　　　meiosis 408-12, 409f
　　　　migration to gonad 407-8, 407f
　　　　chemical signals 408
　　　　primordial 372, 402
　　see also sex determination
germ layers 3b, 19, 19b, 19f
　　chick and mouse 168
　　Xenopus 132-42
　　zebrafish 146-8, 146f, 147f
　　see also ectoderm; endoderm; mesoderm
germline cyst 62, 62f
germplasm 403-5, 403f, 404f, 405f
giant gene and product 73, 73f, 75, 80, 80f, 81f
giant panda 701
gibberellic acid 667
giraffe mutation 229
Gli genes and products 473-5, 473b, 483, 484b, 528-9, 529f

glial cells 518
　　radial 538b, 538, 539f, 539
gliogenesis 538b
globin gene expression 343-5, 344f, 345f
　　defects 346b, 346f
　　locus control region (LCR) 344f, 344-5, 345f
glycosylation 22, 22f
gonadotropin-releasing hormone (GnRH) 584
gonads 402, 425, 425f, 426, 426f
goosecoid gene and product 139, 139f, 140f, 140, 190, 192
Gorlin syndrome 580
granulocyte colony-stimulating factor (G-CSF) 343, 343f
granulocyte-macrophage colony-stimulating factor (GM-CSF) 343, 343f
grasshopper 716f, 716, 717b, 718f
gray matter 539
green fluorescent protein (GFP) 24b, 56f, 56, 109, 110f, 110b, 113f, 113, 164, 348, 604f, 604, 709b
Gremlin gene and product 479, 479f
Groucho 319
growth 20f, 20, 571-90
　　coordination 576-8, 578f
　　Hydra 597
　　intercalary 607, 609f, 612-13
　　long bones 585-9, 587f, 588f
　　molting 590-1, 591f
　　muscle 582f, 582, 589
　　nutrition effects 589-90
　　　　embryo nutrition 589-90
　　organ size control 580-2, 581f, 582f, 589
　　tissue growth 573-4, 573f
growth cone 545
growth differentiation factor (GDF) 137, 531
growth factors 342-3, 342f, 575
　　hematopoietic 342, 342f
growth hormone 498, 584f, 584, 588
growth hormone-releasing hormone 584f, 584
growth plates 585, 587f, 588
Gryllus bimaculatus 612
GSK3 369f, 369
Gurdon, John 23, 359, 374
gurken gene and product 63f, 64-5, 67f, 67
gynogenetic embryos 414f, 414

H

H19 gene 415
H3K27me3 histone mark 320, 321f, 323b, 325f, 326, 336, 666
Hairy Enhancer of Split (Hes) gene 542, 543f
hairy gene and product 81
hairy-1 gene 218, 468-9
hairy-2 gene 468
Haltere mimic mutation 91f

hand-foot-genital syndrome 482
handedness see left-right asymmetry
haploidy 5, 408-9, 409f
head
　　arthropods 692-3
　　Drosophila 94, 693
　　Hydra 597, 600-1, 600f
head gap genes 94
head organizer 192
heart
　　regeneration
　　　　mammalian heart 615-16
　　　　zebrafish 596, 614-15, 614f
　　vertebrates 499-503, 500f, 501f, 502f
hedgehog gene and product 84-5, 85f, 86f, 86, 88f, 89, 89f, 124f, 446-7, 447f, 448-50, 448f, 455f, 455
Hedgehog signaling pathway 31b, 87b, 87f, 681
　　butterfly wing markings 706, 706f
　　cancer and 580
　　lung development 495
hematopoiesis 338-43
hemidesmosomes 250b, 348f, 348
Hemigeris 701
hemoglobin 343-4, 344f
Hensen's node 149-50, 150f, 151f, 152f, 156f, 156
　　regression 150, 153f
　　transplantation studies 205
hermaphrodites 429-30, 430f, 432f, 432
heterochromatin 323b, 320
heterochrony 714f, 714-16
heterotypy 676, 692
Hey gene 542, 543f
Hi-C technique 320
hindbrain 520f, 520
　　gene expression 524, 524f
　　Hox gene role 524, 524f, 526
　　neural crest cell migration 235-6, 235f
　　rhombomeres 235-6, 524f, 524-6, 525f
　　see also brain development
Hippo signaling pathway 163, 263-4, 453, 576-8, 577b, 577f, 578f, 612, 615
histoblasts, Drosophila 48, 49f
histones 320, 323b, 324, 325f, 326
holoprosencephaly 32b, 507, 522b
Holozoa 677, 678f
Holt-Oram syndrome 462, 475, 503
homeobox genes 223b
　　nervous system development
　　　　plants 637
　　tooth identity specification 489-92, 491f
　　see also homeotic genes; Hox genes
homeodomain 94, 223b
homeosis 91
homeotic genes 90-5, 91f
　　flower development role 658-62, 659f
　　see also homeobox genes; Hox genes

homeotic transformation 227, 227f
homologous genes 12
homologous recombination 115b, 116f
homothorax gene 455-6, 455f
homunculus 4, 5f
hormones 328, 584
　see also growth factors
horse limbs 701, 701f, 711-13, 711f
housekeeping activities 21
Hox code 226
Hox genes 90-5, 223b, 322-6, 323b, 675, 681
　activation pattern 228-9
　arthropods 689-93, 689f, 690f
　axial patterning 221-6
　branchial arches 235-6, 235f
　breast cancer 497
　chick 221, 225-6, 225f
　deletion/overexpression studies 226-7
　Drosophila 90-5, 221-2, 223b, 223f, 323b, 444, 688, 688f, 689f
　evolutionary changes 687-94
　　crustacean paired appendages 689f, 690f, 690-3
　　gene duplication 687-8, 688f
　　insect paired appendages 689f, 690f, 690-3
　　vertebrate axial skeleton 693-4, 694f
　fin development 698-700, 699b, 699f, 700f
　gastruloids 382, 382f
　limb development roles 461-2, 461f, 479-82, 480f, 481f, 482f, 483, 485, 609, 698-700, 699b, 699f, 700f
　　limb regeneration 609
　limblessness 702
　mammary gland development 497
　mouse 222-6, 223b, 223f, 224f, 225f, 227, 227f, 322-6, 325f, 688, 688f, 693, 694, 698, 699b, 699f, 700f, 700
　nervous system development roles 524, 530-1, 531f
　　hindbrain 524, 524f, 526
　order of expression 93, 221-6, 224f
　somite identity and 221-6, 222f, 224f, 225f
　targets 693-4
　zebrafish 222, 223b, 688, 698, 699b, 699f, 700f, 700
　see also homeobox genes; homeotic genes
human
　amniotic cavity expansion 171, 172f, 174
　bilaminar germ disc 105
　cell differentiation 334
　cleavage 105, 169, 170f
　digit tip regeneration 610
　egg 169, 170f
　embryo development 2f, 2, 104f, 107, 169-78
　　shape changes 20f
　embryonic folding 171, 172f
　embryonic stem (ES) cells 370-1, 372f, 375, 376f, 376, 379b, 380, 383b, 384f, 386-7, 386f, 390, 392b
　fertilization 169
　gastrulation 171f, 171
　germ cell development 406-7, 406f
　growth 582-3, 583f, 584, 585, 588
　　fingers 586b, 586f
　　nutrition effects 589-90
　left-right asymmetry 237-8
　legal restrictions on developmental studies 175-7
　life cycle 169, 170f
　limbs 701f
　sex determination 424f, 425, 425f, 427f, 427
　somite formation 217
hunchback gene and product 57, 57f, 73, 73f, 74f, 74-7, 75f, 77f, 80-1, 80f, 81f
Hunter, John 492
Hutchinson-Gilford progeria syndrome 619
hyaline layer 261
hyaluronan 486, 486f
Hydra 519, 679, 681
　aging 617
　growth 597
　polarity 599f, 600
　regeneration 596, 596f, 597-601, 599f, 600f, 601f
　　genetic control 601
　　hypostome region 600f, 600, 601, 601f
hydrostatic pressure 249, 264, 265f
hypermorphosis 715
hypoblast 149, 151f, 171f, 171
hypocotyl 634
hypocretin 530b
hypophysis 634
hypothalamus 520, 521
Hyracotherium 711f

I

Ichthyosporea 677, 678f
identical twins 10f, 10, 175, 176b, 176f
IGF1 gene 712b
imaginal discs, *Drosophila* 48, 49f, 84, 442-54, 456
　leg 454-6
　　fate map 454f
　origins 442-4
　parasegments 444-5
　patterning 444-5
　positional values 456f, 456
　size determination 576, 578
　specification 443f
　wing 442-54
　　fate map 446f
immunoglobulin superfamily 250b
immunohistochemistry 108
imprinting 360, 413-16, 415f
　plants 667
in situ hybridization 24b, 108f, 108, 177, 521
in vitro fertilization (IVF) 169, 172b, 175, 177, 388, 420, 421
indeterminate growth 646
Indian hedgehog protein 587, 588f
indoleacetic acid (IAA) *see* auxin
induced pluripotent stem cells (iPS cells) 37, 373b, 374f, 374-5, 388, 389, 394, 394f, 395f, 395, 507
　directed differentiation towards functional neuronal types 530b, 530f
　naïve 380
induction and inducing signals 8, 29-33, 30f
　competence 30
　mesoderm
　　chick 204f, 204
　　mouse 204
　　Xenopus 132-40, 134f, 136f, 137f
　　zebrafish 146-8
　neural
　　chick 204-6, 204f
　　Xenopus 193-6
　　zebrafish 202
　permissive/instructive 30
　response to 33
ingression 149, 283
Inhibin 133b, 137
inner cell mass, mouse 162-5, 163f, 164f
insects
　larval growth 591-4, 592f, 593f, 594f
　limb regeneration 612-13, 613f
　metamorphosis 590, 591-2, 592f
　paired appendages 689f, 690f, 690-3
　see also Drosophila
instar, *Drosophila* 48, 48f
insulin 618, 619, 619f
insulin-like growth factor-1 (IGF-1) 584f, 584, 588, 618, 619, 619f
insulin-like growth factor-2 (IGF-2) 415-16, 415f, 584, 619
insulin-producing cell generation 389-91, 391f
integrins 250b, 546-8
intercalary growth 607, 609f, 612-13
intercalation 272-3, 272f
　circumferential 612-13, 613f
　medio-lateral 278b, 281
　proximo-distal 607f, 608, 612
　radial 278b, 280
intercellular signals
　Drosophila 65f
　vertebrates 124b, 124f
interfollicular epidermis 346
interkinetic nuclear migration 537f
intermediate filaments 254b

760 Index

intermediate mesoderm 105, 122
intestinal epithelia 351-3, 351f, 352f
intestinal organoids 389, 392-3, 392b, 392f, 393-4
intracellular signaling 30, 31b, 31f
intracytoplasmic sperm injection (ICSI) 420
invagination
    Drosophila 270-2, 270f
    gastrulation evolution and 682f, 682
    sea urchin 268f, 268
involution 120, 276f, 276, 280
ionizing radiation 476b
iris 505-6
Irx3 transcription factor 529f
Isl1 transcription factor 531
Islet1 transcription factor 461f, 462, 490, 491f
Izumo1 protein 419-20

J

JAK-STAT pathway 63, 64b, 64f, 65
jellyfish, transdifferentiation 357
Johannsen, Wilhelm 10
joint formation 486, 486f
Jun N-terminal kinase (JNK) 276, 303
junctional epidermolysis bullosa 348, 349b, 349f
Juno protein 419-20, 421
Junonia coenia 706, 706f, 707
juvenile hormone 592f, 592

K

KANADI genes 648-9, 649f
keratinocytes 346, 347, 348f
keratins 347, 348f, 348
kidney 582
kiwi 713f, 713
Klinefelter syndrome 425
knee-jerk reflex 547b, 547f
knirps gene and product 73, 73f, 75
knock-outs *see* gene knock-outs
KNOTTED-1 gene and product 645
Koller's sickle 149, 151f, 155, 156f, 156
Krox20 gene and product 524, 524f
Krüppel gene and product 73, 73f, 75-7, 75f, 77f, 80, 80f, 81f, 717b

L

lacZ gene 452
LAG-2 protein 432
lamellipodium 254b
laminin 556
lamprey 683f
larva 715, 716
    Drosophila 48, 48f
    rate and duration of growth 591-4, 592f, 593f, 594f
lateral geniculate nuclei (LGN) 555, 555f, 562-3

lateral inhibition 35-6, 36f, 541-3, 543f
    leaves 650
    neural development 534-5, 535f
    root hairs 656f, 656
lateral motor column (LMC) 531, 552f, 552
lateral plate mesoderm 105-6, 107, 122, 150, 153f, 458, 499, 500
LEAFY (*LFY*) gene and product 657, 661f, 663, 663f, 665f
leaves
    arrangement on stem 650-1, 650f
    axes 648-50
    directed dilation 639-40
    phyllotaxis 648f
    primordia 647, 648f, 648-51, 650f
left-right asymmetry 236-40
    chick 238, 239
    human 237-8
    mouse 237, 237f, 238f, 238, 239
    Xenopus 238, 239-40
    zebrafish 238, 239
Lefty protein 147, 166-7, 167f, 207b, 239
lens 503-4, 503f, 686
    formation 504-6, 505f
    placode 504, 506, 507
    regeneration, newt 357-8, 358f, 596
Lepidoptera *see* butterfly
Lepisosteus oculatus 699, 699b, 699f
leukemia inhibitory factor (LIF) 369f, 369, 370, 371, 373b
leukocytes 338
Levi-Montalcini, Rita 561
Lewis, Edward 49, 90
Lfng expression 383b, 383f
Lgr5 gene and product 352, 352f
Lhx1 transcription factor 531, 552
Lhx6 gene 490, 491f, 491
Lhx7 gene 490, 491f, 491
Lhx9 transcription factor 530b
life cycle
    Arabidopsis 631-2, 631f
    Drosophila 44-51, 45f, 716f, 716, 717b, 718f
    evolution of 716-17
    sea urchin 715-16
    vertebrates 103-7
        chick 150f
        human 169, 170f
        mouse 158f
        Xenopus 119f
        zebrafish 143f
    see also metamorphosis
ligands 31b
light-sheet fluorescence microscopy (LSFM) 108, 112-13
limb buds 458f, 458-61, 460f, 461f
    outgrowth 464-6, 465f
    patterning 466
    self-organization 482-3, 482f
    see also vertebrate limb development

limblessness 702-3
limbs
    evolution 39, 691-2, 696-702, 696f, 699b, 699f, 701f
    insect leg development 442-4, 454-6, 457, 691-2
    regeneration 34, 596, 601-13, 611b, 616
        amphibians 601-10, 602f, 610f
        insects 612-13, 613f
    see also vertebrate limb development
lineage *see* cell lineage
liver 390, 580-2
    regeneration 580-1, 581f, 596
Lmx1 genes and products 478f, 478
Lmx1b transcription factor 479f, 479, 485
locus control region (LCR) 344f, 344-5, 345f
long-germ development 716f, 716, 717b, 718f
long non-coding RNAs (lncRNAs) 326, 415-16, 666
long-term potentiation (LTP) 564
longevity 618f, 618-19
loss-of-function experiments 109-10
Lunatic fringe enzyme 218, 219b
lung development 295
    vertebrates 492-6
    endoderm 492-5, 494f
    morphogenesis 495-6
lymphatic vessels 297
lymphocytes 338

M

macrophage colony-stimulating factor (M-CSF) 343, 343f
macrophages 621
macular degeneration 388
Mad protein 449b
MADS box 659, 660
magnetic resonance imaging (MRI) 108
maize 645, 647f, 647, 667
Malpighi, Marcello 4f, 4
mammary glands 492, 496-8
    breast cancer 493b, 493f
    development 496-8, 497f
    epithelium branching pattern determination 498
mammary-ulnar syndrome 497
Manduca sexta 591, 591f, 592, 593f
Mangold, Hilde 8, 9f
mantle zone 539
marginal zone 133, 539
master regulatory genes 314, 508, 508f
maternal factors 53
    Drosophila 53-77
        body axis establishment 53-61
        maternal determinant localization during oogenesis 61-9
    Xenopus 122-3, 125
    zebrafish 145
maternal genes 51, 53-5, 53f, 54f

Matrigel 380, 381, 383–4
matrix metalloproteinases (MMPs) 562
maturation-promoting factor (MPF) 422–3, 423*f*
MeCP2 gene and product 560*b*
median hinge point 20
medio-lateral axis, leaves 648
medulla oblongata 521
MEF2 transcription factor 659
meiosis 5, 402, 408–12, 409*f*
    timing 431*f*, 431
Meis gene family 468, 611*b*, 615–16
Mendel, Gregor 5, 8, 9
meristems 630, 642–57
    cell fate 644–6, 645*f*, 646*f*, 647*f*
    floral meristem 642, 657, 657*f*, 659*f*, 662*f*, 662–4, 663*f*
    identity genes 657–8
    inflorescence meristem 631, 642, 657, 657*f*
    lateral shoot meristem 642
    root apical meristem 642, 653–6, 653*f*, 654*f*
    shoot apical meristem 642*f*, 642–3, 643*f*, 645–6, 647*f*, 647, 653
        transition to floral meristem 663–4, 663*f*, 664*f*, 665*f*
    stem cells 643–4, 644*f*
mesectoderm, *Drosophila* 69
mesenchymal-to-epithelial transition (MTE) 217
mesenchyme 149
    mammary gland epithelium branching pattern 498
    primary 266–8, 267*f*
    transition to epithelium 233, 255–6, 256*f*, 266–8
mesendoderm 142–3, 144, 146, 146*f*, 147
    prechordal 521
mesoderm 3*f*, 3*b*, 19, 19*b*, 19*f*
    Drosophila 46–7, 47*f*, 52, 52*f*, 69
        invagination 270–2, 270*f*
    induction *see* mesoderm induction
    lung development 494–5, 495*f*
    mammary gland development 497–8
    nervous system patterning and 197–8, 198*f*
    patterning of 139–40, 189–90
        threshold responses 139–40, 139*f*, 140*f*
    vertebrates 102, 143, 148, 375
        axial 105
        chick 149, 150
        heart 499–500, 502, 503
    vertebrates
        intermediate mesoderm 105, 122
        lateral plate mesoderm 105–6, 107, 122, 150, 153*f*, 499, 500
        mouse 204, 333, 334*f*
        paraxial mesoderm 199, 204, 375–6
        prechordal mesoderm 105, 150

pre-somitic mesoderm 216, 217–21, 217*f*, 220*f*
Xenopus 129, 129*f*, 132–41, 132*f*, 133*f*, 134*f*, 135*f*, 136*f*, 137*f*, 280
zebrafish 146, 147, 147*f*, 148
mesoderm induction
    chick 204
    mouse 204
    Xenopus 132–40, 134*f*, 136*f*, 137*f*
        timing 135*b*
    zebrafish 147
mesonephros 426*f*, 426, 431
Mesp1 expression 383*b*
Mesp2 gene and product 230*f*, 230
message transport organizer region (METRO) 123
messenger RNA (mRNA) 21, 22*f*
metamorphosis 572, 590, 595, 715*f*, 715, 716*f*, 716
    control of 591–5
    Drosophila 48–9, 49*f*, 445–6, 445*f*, 446*f*, 454, 592–4, 593*f*, 594*f*
metastasis 267
Metazoa 677, 678*f*, 678
    family tree 680*b*, 680*f*
    origin of 679–81
microcephaly 476*b*
microfilaments 254*b*
microglia 538*b*, 538*f*
microRNAs (miRNA) 22, 22*f*, 415, 649
microscopy 108, 112–13
microsurgical manipulation
    chick 112, 112*f*, 113
    mouse 112*f*
    suitability of different vertebrate embryos 112*f*
    Xenopus 112, 112*f*
    zebrafish 112*f*, 113
microtubules 254*b*
    reorganization in *Drosophila* oocyte 65–6, 66*f*
mid-blastula transition 130*f*, 130
midbrain 520*f*, 520
    see also brain development
midbrain–hindbrain boundary (MHB) 520*f*, 521, 523, 523*f*
middle ear 684, 685*f*
miles apart gene 499
'minibrains' 393*f*, 393
Minute technique 445, 445*f*, 576
Miranda protein 536*f*, 537
misexpression screening 76*f*, 76*b*
mitochondrial DNA 422
mitogen-activated protein kinase (MAPK) 196, 276, 369*f*, 369, 371
mitosis 5, 575, 575*f*
    see also cell cycle
mitotic spindle 258*f*, 258–61
model organisms 10–12

phylogenetic tree 11*f*
molting 590–1, 591*f*, 595
MONOPTEROS (*MP*) gene 637
Monosiga brevicollis 678
Morgan, Thomas Hunt 8, 9–10
morphallaxis 597, 597*f*
    Hydra 597
    planarians 598*b*
morphogenesis 19–20, 247–9
    branching 294–8, 294*f*, 295*f*
    lung development 494*f*, 495–6
morphogens 34*f*, 34, 55
    Drosophila body plan 55, 71
    Drosophila wing 446
    gradients 449*b*, 449*f*
    nervous system development 522, 523*f*, 526–9, 531*f*
    threshold concentrations 34*f*, 34, 56
    vertebrate limb 471
morpholino antisense RNA 112
morpholinos 23
morula 159*f*, 159, 162–5, 261–2
mosaic development 7*f*, 7, 28
    X inactivation effect 434*f*, 434
motor neurons 527, 530*b*, 530–1, 531*b*
    chick 551–2, 552*f*
    neuronal cell death 559–61, 559*f*
mouse 10, 11*f*, 11, 104*f*, 157–69
    antero-posterior axis 213
    axial structures 212–15, 212*f*, 213*f*
    axonal navigation 551*f*
    blastocyst 105, 105*f*
    body axes 165–7
        patterning 222–5
    body plan 102*f*
    brain development
        cell-identity switching 525
        hindbrain 526
        holoprosencephaly 522*b*, 522*f*
    breast cancer 493*b*, 493*f*, 497
    cell specification in early embryo 162–4
    chimeric 29, 29*f*, 115*b*, 115*f*, 162–3
    cleavage 105, 158–9, 159*f*, 164
    compaction 261–2, 262*f*
    CRISPR-Cas9 system 15*b*, 16*f*, 115*b*, 118
    digit tip regeneration 611
    dorso-ventral axis 215
    egg 158
    embryo development 107
    embryonic kidney 294*f*
    embryonic stem (ES) cells 368–77, 369*f*, 370*f*, 371*f*, 372*f*, 373*f*, 373*b*, 375*f*, 378*f*, 379*b*, 380, 381*f*, 381–2, 382*f*, 383*b*, 383*f*, 384, 385, 385*f*, 386, 388, 389, 390–1, 392*b*, 393–4, 394*f*, 506, 506*f*
    epidermis 348
    extra-embryonic structures 165*f*, 165, 166
    eye 504, 505*f*, 506*f*, 508, 508*f*
    fate map 113–14, 168*f*, 168, 201*f*

Index

mouse (Continued)
  gastrulation 158f, 161–2, 161f, 282–3, 284f, 285, 286f, 289, 329, 332–4, 333f, 334f
  gene expression regulation 316f
  germ cell development 406, 406f, 407, 407f
  heart 501f, 501, 502f, 502–3
    regeneration 615–16
  Hox genes 222–6, 223b, 223f, 224f, 225f, 227, 227f, 322–6, 325f, 688, 688f, 693, 694, 698, 699b, 699f, 700f, 700
  intestinal epithelia 351–2, 351f
  left–right asymmetry 237, 237f, 238f, 238, 239
  life cycle 158f
  light-sheet microscopy 113
  limb development 458, 459, 460f, 461f, 462, 463–4, 465–6, 470, 471, 472b, 473b, 474–5, 477–8, 479–82, 480f, 481f, 482f, 483, 486, 698, 699b, 699f, 700f, 700, 702
  limb evolution 713
  lineage tracing 113–14
  lung development 496
  mammary glands 496f, 496–8, 497f
  mesoderm induction and patterning 204f, 204
  microsurgical manipulation 112f
  muscle growth 589
  mutagenesis screen 114
  nervous system 518f
  neurogenesis 540, 542f, 543f
  neurulation 162
  polarity 261–4, 262f
  post-implantation development 159–67, 160f, 162f
  sex determination 425, 430–1, 431f
  somite formation 217
  somite patterning 230
  teeth 489–92, 490f, 491f
  teratogens 476b
  transgenic techniques 112, 114f, 114, 115b, 115f
  turning of the embryo 162, 162f
  vertebrae 693, 693f
  visual system 554
  zone of polarizing activity regulatory sequence 704b, 704f
Mrf4 gene and product 335–6, 336f
MSL genes 435
Msx1 gene and product 490, 491f, 491, 603, 610
Msx2 gene 490, 491f, 491
MSX2A gene 708
Müller cells 504
Müllerian ducts 426f, 426
Müllerian-inhibiting substance 425, 426f, 426

muscle 334–8, 335f, 336f, 337f, 483–5, 485f
  gene expression, Xenopus 135b, 135f
  growth 582f, 582, 589
  limb muscle patterning 483–5, 485f
muscle-specific kinase (MuSK) 558f
mutagenesis
  Drosophila 50b, 50f
  mouse 114
  Xenopus 114
mutagenesis screens 111, 114
  mouse 114
  Xenopus 114
  zebrafish 114, 116, 116b, 117f, 147
mutagens 476b
mutations 12–14, 13f, 14f
  conditional 14
  dominant 13
  maternal-effect 50–51
  recessive 13f, 13
  semi-dominant 13f, 13, 14f
  spontaneous 111
  temperature-sensitive 14
myelomeningocele 293b
Myf5 gene and product 335–6, 336f, 337
myoblasts 335, 335f, 336f, 336, 337, 337f, 485
myocardium 499, 502, 614–15, 614f
MyoD gene and product 115b, 314, 326, 334–6, 336f, 356, 372
myogenesis 334
myogenic progenitors 334
myogenin (Myog) gene and product 335–6, 336f, 337, 337f
myostatin gene and product 582f, 582
myotome formation 230
myotubes 335, 335f, 337f

N

nail–patella syndrome 478
naïve embryonic stem cells 371
Nanog transcription factor 147, 163, 165f, 165, 332, 373b
nanos gene and product 54–5, 54f, 56–7, 57f, 61–2, 67
narcolepsy 530b, 530f
Nasonia 717b
naso-temporal axis 553
negative feedback 26, 27f, 38, 644
nematocytes 679
nematodes
  germ cell development 404f, 404–5
  see also Caenorhabditis
Nematostella 679, 680b, 681
neoblasts 598b, 599f
neocortex 539
neoteny 714–15, 714f, 715f
nerve fascicles 549
nerve growth factor (NGF) 561
nerve tracts 549

nervous system 517–19, 518f
  axon navigation and mapping 544–56
  cell identity specification 519–32
    Drosophila 532–7, 533f, 534f, 535f
    vertebrates 537–44
  glial cells 538b, 538f
  neural cell diversity 531–2
  self-organization 523, 523f
  see also brain development; neurons; neurulation; spinal cord
netrins 297, 546f, 546, 550, 551f
neural crest 288
  cells 106
  derivatives 233, 233f
neural folds 106, 289–91, 291f
neural furrow 289, 291f
neural groove 290, 291f
neural induction
  chick 204–6, 204f
  Xenopus 193–6
  zebrafish 202
neural plate 106, 151, 288, 290, 527, 533, 534f, 543
  cell shape changes 291f
  induction 194–6
  patterning 197–8, 198f
neural tube 3b, 106f, 106, 151, 153f, 288–94, 290f, 537–9, 537f, 538b, 539f
  defects 291, 293b, 293f
  formation see neurulation
  polar cell polarity 276
neuregulin-1 556
neuregulin-3 497–8
neuregulins 541
neuroblasts 536–7, 536f, 538
NeuroD gene 543
neuroectoderm 519, 520, 523
  Drosophila 52, 52f, 69
  vertebrates 375
neuroepithelium 534f
Neurofibromatosis-2 mutations 580
neurogenesis 354, 354f, 532
  cortical features 539–40, 540f
  Drosophila 532–7, 533f, 534f, 535f
  vertebrates 537–44
neurogenin 543, 543f
neuroligins 558–9, 559f
neuromuscular junctions 556, 557f, 557
  development 556–8, 558f
neurons 517–18, 518f
  migration 537–9, 539f, 540–1, 542f
  motor neurons 527, 530b, 530–1, 531b
  neuronal cell death 559–61, 559f
  post-mitotic 538
  retinal 504, 505f, 506f
  retino-tectal system 552–5, 553f, 554f, 555f
  see also axons; neurogenesis
neuropilins 546

Index

neurotrophic hypothesis 560
neurotrophins 548, 556, 561, 564
neurula 106–7, 122
neurulation 3f, 3b, 106, 289–90
   amphibians 194f
   chick 152, 291f
   mouse 162
   primary 289
   secondary 289
   Xenopus 106f, 106, 122, 199, 199f
   zebrafish 145
newt
   lens regeneration 357–8, 358f, 596
   limb regeneration 602f, 603
newt anterior gradient protein (nAG) 605, 606f, 607
NFI transcription factor 532
Nfia transcription factor 531
Nfib transcription factor 531
Nieuwkoop center 129f, 129–30, 136f, 136
Nkx2.1 transcription factor 495
Nkx2.2 transcription factor 529f
Nkx2.5 gene and product 502
No tail (*Ntl*) gene and product 201, 203b, 203f
Nodal protein 239
   brain development 521, 522b
   gastruloids 379b, 379f
   signaling 147–8, 147f, 202, 207b, 207f, 239
      cell differentiation 332
   trophectoderm 380
Nodal-related proteins (Ndr) 132–3, 137, 137f, 147
   see also Xenopus Nodal-related (*Xnr*) genes and products
node *see* Hensen's node
Noggin gene and product 190f, 190, 194, 195, 198, 204, 205
non-canonical Wnt pathway 275–6
non-cell-autonomous action 29
non-cell-autonomous gene effects 29
non-coding RNAs (ncRNAs) 415, 416
   long (lncRNAs) 326
   microRNAs 22, 22f, 415, 649
Notch–Delta signaling pathway *see* Delta–Notch signaling pathway
Notch protein 219b
   leg development and 456
   neural development and 532
   neurogenesis role 534–5, 535f, 541–3, 543f
   signaling pathway 31b, 218–19, 219b, 219f, 220, 297, 348, 451f, 451, 452, 496
      butterfly wing markings 705, 705f, 706, 707
notochord 105, 106f, 107, 122, 682–3, 683f
   chick 150, 153f, 204, 204f, 205, 212

directed dilation 280
   neural crest migration 300, 301
   somitic cell fate and 231
   Xenopus 281–2, 282f
Notophthalmus viridensis 602f
nucleosomes 322, 323b
Numb protein 536f, 537
nurse cells 62, 62f, 65–6
Nüsslein-Volhard, Christiane 49

O

obesity 589–90
Oct4 transcription factor 163–4, 164f, 332, 373b
ocular dominance columns 563f, 563–4
odontoblasts 489, 490f
Okihiro/Duane radial ray syndrome 475
Okihiro syndrome 476b
Olig2 transcription factor 529f
oligodendrocytes 538b, 538f
omb gene and product 447–8, 447f, 448f, 450
ommatidia 508
oncogenes 579, 580
Onecut transcription factor 531
ontogeny 676
oocyte 409–13, 410b
   decline in numbers with age 410, 412f
   primary 409
   secondary 410
   see also egg
oogenesis 409–12, 411f
   Drosophila 61–9, 62f
      antero-posterior polarity 62–5, 63f, 67
      dorso-ventral polarity 67
      oocyte cytoskeleton reorganization 65–7
oogonia 409
optic chiasm 553f, 553
optic cup 504, 505f, 506–7, 506f
optic nerve 504f, 504, 505f, 506, 553, 554f
optic tectum 552–5, 553f, 554f, 555f
optic vesicle 504, 505f, 506, 506f, 507
optical projection tomography (OPT) 108
orexin 530b
organ primordium 442
organizer region
   butterfly wing markings 705
   genes expressed 206
   Hydra hypostome region 600
   plants 643–4, 644f, 653–4, 654f, 655f
   proteins expressed 191–2
   see also Hensen's node; Spemann organizer
organogenesis 3f, 3b, 107, 441–2
   organ size control 580–2, 581f, 582f, 589
   Xenopus 122
organoids 37, 175, 352, 379b, 389, 391–4, 392b
orthodenticle gene and product 94, 526, 693

oskar gene 61–2, 65, 66f, 66–7, 403–4, 404f
osteoblasts 585, 587f, 588f
Otx2 gene and product 332, 520f, 523, 523f, 526
ovary 402, 425
oviducts 103, 426f, 426
oxidative damage 619

P

P elements 76b
P granules 404f, 404–5
p53 tumor-suppressor gene 580, 678
p63 transcription factor 476b
pair-rule gene expression 78–82, 79f
pallium 520
pancreas 389–91, 391f, 394, 394f, 395, 580–1, 581f
pancreatic cancer 579, 580
Panderichthys 696f, 697
Paneth cells 351, 351f, 352, 392b
PAR (partitioning) proteins 64, 536f, 536
parallel evolution 685–6
paralogs 223b
parasegments, *Drosophila*
   boundaries 444–5
   boundary stabilization 84–9
   establishment 78–82, 79f
parathyroid-hormone-related protein (PHRP) 498, 587, 588f
Parazoa 680b
parental conflict theory 415
Parhyale 690, 691f
Parkinson's disease 389
Patched-1 702
patched gene and product 32b, 85f, 85, 87b, 88f, 473b
   butterfly wing markings 706, 706f
pattern formation 18–19, 20
   brain, antero-posterior axis 526
   Drosophila
      imaginal discs 444–5, 453–4, 457
      leg 454–6, 457
      wing 446–51, 453–4
   Hox genes in axial patterning 221–6
   lateral inhibition 35–6, 36f
   mesoderm
      chick 204f, 204
      mouse 204f, 204
      threshold responses 139–40, 139f, 140f
      Xenopus 139–40, 189–90
      zebrafish 147
   nervous system 197–8, 198f
   plants
      flowers 658–63
      leaves 648–50
      roots 653–6, 653f, 654f
   positional information 33–5, 33f, 34f
   somites 216–32
      cell fate determination 229–31

pattern formation *(Continued)*
  vertebrate limb development
    digit separation 487*f*, 487
    Hox gene role 479–82, 480*f*, 481*f*, 482*f*, 483
    limb bud 466–70
    muscle 483–5, 485*f*
    self-organization 482–3, 482*f*
    Sonic hedgehog (Shh) role 471–5, 473*b*, 473*f*, 479, 479*f*
    see also segmentation, *Drosophila*
Pax genes 223*b*, 230, 231, 354*f*, 354, 529*f*, 681
  eye development and 507, 508, 508*f*, 686
  somitogenesis 383*b*
  vertebrate limb development and 485
PCP pathway 275–6
Pdx1 transcription factor 390, 391*f*, 391, 394
pedomorphosis 714, 714*f*
peramorphosis 714, 714*f*, 715
periclinal divisions 635, 635*f*
perivitelline space, *Drosophila* 58*f*, 58
personalized medicine 394, 476*b*
PHABULOSA gene 649
PHAVOLUTA gene 649
phenotype 10*f*, 10
phenotypic suppression 94
photoperiodism 663–4, 667
photoreceptors 504*f*, 504, 506, 552
  evolution 685–6
phyllotaxis 648*f*, 650–1, 651*f*
phylogeny 676
phylotypic stage 104*f*, 106*f*, 106, 683, 683*f*
pie-1 gene and product 404*f*, 404–5
pig
  human organs grown in 394–5, 395*f*
  limbs 701, 702
PIN proteins 651*f*, 652*f*, 652–3, 654–5
PIN1 protein 650–1, 651*f*, 654
PIN7 protein 636, 636*f*
pioneer axons 549
pioneer factors 320
pipe gene and product 59
PISTILLATA gene and products 658, 659, 660*b*, 660*f*, 661
Pitx1 transcription factor 461*f*, 462, 707–8, 709*b*
Pitx2 gene 708
Pitx2 transcription factor 239, 500
placenta 103, 107
  human 173–5, 174*f*
  mouse 159–60, 162
placode 295
  eye 504, 506, 507
  mammary gland development 496, 497, 498
  tooth 489, 490*f*
Placozoa 678*f*, 680*b*, 680, 681*f*

planar cell 273, 273*f*
  polarity 19, 273–6, 273*f*, 465–6, 496, 598*b*
planarians, regeneration 596, 596*f*, 598*b*, 598*f*, 599*f*
plant development 629–31
  embryonic development 632–42
  flower development 657–68
    homeotic gene role 658–62, 659*f*
  leaf arrangement on stem 650–1, 650*f*
  meristems 642–50, 653–7
    cell fate 644–6, 645*f*, 646*f*, 647*f*
    stem cells 643–4, 644*f*
  regeneration capacity 617–18, 618*f*
  root hairs 656*f*, 656
  root tissues 653–6, 653*f*, 654*f*
  secondary shoots 651–3, 651*f*, 652*f*
  transformation and genome editing 640*b*, 640*f*
  see also Arabidopsis thaliana
plasmin 603
plasmodesmata 630
plasticity 564
  mammalian brain 564
Platynereis 519, 686*f*
PLETHORA gene and products 655
Pleurobrachia bachei 680
plexins 546
pluripotent stem cells 381–7, 382*f*, 384*f*, 385*f*, 386*f*
polar body 119, 120*f*, 409, 410*b*, 410*f*
polarity 19
  Hydra 599*f*, 600
  mouse morula 261–2, 262*f*
  planar cell polarity 19, 273–6, 273*f*, 598*b*
  planarians 598*b*
  sea urchin egg and blastula 261
  vertebrate limb development 469–70
  see also body axes, establishment of
pole cells 403
pole plasm 403, 403*f*
polonaise movements 283, 284*f*
Polycells auricularia 686*f*
Polycomb gene family 93, 323*b*, 416, 666
polydactyly 471, 472*b*, 472*f*, 473*b*, 474, 482
  pre-axial 471, 472*b*, 472*f*
polyspermy prevention 420–1, 420*f*, 421*f*
polytene chromosomes 594*f*, 594
pons 520
population asymmetry model 350*f*, 351
Porifera 678*f*, 680*b*, 681*f*
positional identity 221–6
  plant leaves 650
positional information 33–5, 33*f*, 34*f*
  nervous system development 522, 526
  morphogen gradients 449*b*, 449*f*
  root hairs 656*f*, 656
  vertebrate limb development 465–6, 470, 478–82, 480*f*, 481*f*, 482*f*

Sonic hedgehog (Shh) role 471–5, 473*b*, 473*f*, 479, 479*f*
  see also body axes, establishment of; intercalation; positional identity
positive feedback 26, 27*f*, 326
postbithorax mutation 91*f*
posterior dominance 94, 226
posterior marginal zone 155, 156*f*
Prader–Willi syndrome 416
PRC2 protein 666
pre-axial polydactyly 471, 472*b*, 472*f*
prechordal mesendoderm 521
precursor cells 334
preformation 4, 5
preimplantation genetic screening 172*b*, 172*f*, 173*f*
prepattern 483
PRESSED FLOWER (PRS) gene 649, 649*f*
prethalamus 521
Prickle (Pk) protein 274*f*, 274
primary body formation 212
primary cilium 474–5
primitive streak 148, 286–7, 287*f*
  chick 149, 150*f*, 151*f*, 152*f*, 152, 153*f*, 155*f*, 155, 156*f*, 168, 204*f*, 204, 282–6, 283*f*, 284*f*, 287*f*
  formation 282–6, 283*f*, 284*f*
  human 171*f*, 171
  mouse 161, 162*f*, 168*f*, 168, 204*f*, 282–6, 284*f*
    cell differentiation 332, 333
primordial germ cells 372, 402
primordium 642
  lateral-line 301–2, 302*f*
  leaf 647, 648*f*, 648–51, 650*f*
  organ 442
principal components analysis (PCA) 330*b*
Prod1 gene and product 607, 608*f*, 609, 611*b*
progenesis 715
progenitor cells 350–1, 350*f*
  nervous system development 519–21, 523*f*, 524, 527–9, 532, 537, 538*b*, 538–9, 541*b*, 542, 546, 547*b*
progesterone 498
programmed cell death *see* apoptosis
promoters 317
proneural clusters 532–6, 533*f*, 534*f*, 535*f*
proneural genes 533–4, 534*f*
Prospero protein 536*f*, 537
proteins 21–3
  evolutionary changes 675
  gene-regulatory proteins 23–6, 27*f*
  genetic control 21–3, 22*f*
  post-translational modification 22, 22*f*
  tissue-specific 21
prothoracicotropic hormone (PTTH) 591, 591–2, 592*f*
proto-oncogenes 579, 580
protoplasts 639, 640*b*

protostomes 682
Prox1 gene and product 297
proximo-distal axis
   Drosophila
      leg disc 454–6
      wing 453–4
   leaves 648, 649–50
   vertebrate limb 458*f*, 458, 464–5, 466*f*, 466–9, 467*f*
      regeneration 606–9, 606*f*, 607*f*, 608*f*, 609*f*
Prx1 gene and products 713
pseudotime 331
PU.1 transcription factor 342*f*, 342
pupa, *Drosophila* 48
pupation 591
Purdy, Jean 172*b*
python 703, 704*b*, 704*f*
   Hox genes 693, 694*f*

Q
quail 113, 118, 205, 234*f*, 234
quiescent center 653–4, 654*f*, 655*f*

R
rachischisis 293*b*, 293*f*
radial axis 18
radial glial cells 538*b*, 538, 539*f*, 539
radial symmetry 680*b*, 687
Raldh genes 214*b*, 228
Rana temporaria 609, 610*f*
RAR δ2 receptor 609
Ras-MAPK signaling pathway 196
rat 294*f*
ray 697
reaction–diffusion mechanism 207*b*, 483, 484*b*, 484*f*, 601
reactive oxygen radicals 619
recombination 409
redundancy 38
reeler mutation 540
regeneration 572, 596–616
   Hydra 596, 596*f*
   limbs 34, 596, 601–13, 611*b*, 616
      amphibians 601–10, 602*f*, 610*f*
      nerve influence 605–6, 606*f*
   liver 580–1, 581*f*, 596
   mammalian heart 615–16
   newt lens 357–8, 358*f*, 596
   planarians 596, 596*f*, 598*b*, 598*f*, 599*f*
   plants 617–18, 618*f*
   starfish 596, 596*f*
   zebrafish heart 596, 614–15, 614*f*
regenerative medicine 2, 356
   stem cell potential 375, 387–95, 388*f*, 391*f*, 392*b*, 392*f*, 393*f*, 394*f*, 395*f*
regulation 7*f*, 8
   chick 156
   Xenopus 140–1

regulative development 28
   see also regulation
reporter gene 24
repressors 317
reprogramming 356–8, 373, 374*f*
retina 504*f*, 504, 507, 531
   neural connections 562–3
retino-tectal map 562–3
retino-tectal system 552–5, 553*f*, 554*f*, 555*f*
retinoblastoma (*RB*) gene and product 337, 579, 579*f*, 603
retinoic acid 214*b*, 214*f*, 215, 218*f*, 219, 228–9, 328, 431, 467*f*, 468–9, 471, 495, 501, 524, 525, 531
   limb regeneration and 607, 608–9, 608*f*, 609*f*, 610*f*
Rett syndrome 560*b*
reverse genetics 14, 111, 112, 114, 116, 118, 177
REVOLUTA (*REV*) gene 649
Rho-family GTPases 275–6
rhomboid gene and product 69, 69*f*, 70*f*, 89*f*, 456
rhombomeres 235–6, 520, 524*f*, 524–6, 525*f*
ribosomal RNA (rRNA) 38
Ripply2 expression 383*b*, 383*f*
RNA 22*f*
   antisense 23
      morpholino 112
   interference (RNAi) 23, 112, 690, 691*f*, 706
   microRNAs 22, 22*f*, 415, 649
   non-coding RNAs (ncRNAs) 415, 416
      long (lncRNAs) 326
   processing 22
   seq analysis 109
   short interfering (siRNA) 112
   single-cell RNA sequencing (scRNA-seq) 109, 521
      digit tip regeneration 610
   single-cell transcriptional analysis 330*b*
   splicing 22, 22*f*, 315–16
   velocity 330*b*
RNA polymerase II 317, 317*f*
Robinow syndrome 276
Robo proteins 546, 550–1
roof plate 527*f*, 527, 528*f*
root hairs 656*f*, 656
root tissues 653–6, 653*f*, 654*f*
rotundifolia mutation 640, 641*f*
Roux, Wilhelm 7*f*, 7–8
rubella 476*b*

S
salivary glands 498
Sall genes 475, 476*b*
Sasquatch mutation 472*b*
satellite cells 334–5, 353–4, 354*f*, 538*b*, 538*f*, 578

Scalloped transition factor 452, 453, 578
Scaramanga gene 497
SCARECROW (*SCR*) gene 654, 656
Schistocerca 716*f*, 716, 717*b*, 718*f*
Schleiden, Matthias 5
Schmidtea mediterranea 598*f*, 598*b*
Schwann, Theodor 5
Schwann cells 538*b*, 538*f*
Scleraxis transcription factor 230
sclerotome formation 230, 230*f*
SCRAMBLED protein 656
scute gene 535*f*, 535, 543
SDC-2 protein 435
SDF-1 protein 408
sea urchins
   fertilization 420–1, 420*f*, 421*f*, 422
   gastrulation 20*f*, 266–70, 267*f*, 269*f*, 289
   life cycle 715–16
   polarity 261
secondary body formation 212, 212*f*
secondary shoots 651–3, 651*f*, 652*f*
segmentation, *Drosophila* 48, 48*f*
   parasegment boundaries 444–5
   parasegment boundary stabilization 84–9
   parasegment establishment 78–82, 79*f*, 444–5
   segment identity specification 90–5
   segmentation gene expression 82–90, 85*f*
selector gene 84
self-assembly 378, 379*b*, 379*f*
self-organization 379*b*, 379*f*, 394, 482–3, 506, 506*f*
   lateral motor column motor neurons 552
   nervous system 523, 523*f*
semaphorins 297, 300, 523, 523*f*, 541, 545–6, 546*f*
senescence 617
   cell senescence 588, 619–20
   genetic influences 618–19
SEPALLATA (*SEP*) genes and products 661
septate junctions 261
Serrate protein 128*f*, 219*b*, 451*f*, 451, 456
SEX-1 protein 429
Sex combs reduced gene 93
sex determination 424–36
   Caenorhabditis 429–30, 430*f*, 432*f*, 432
   Drosophila 427–9, 428*f*, 429*f*, 432
   genetic constitution and intercellular signals 430–2, 431*f*, 432*f*
   mammals 424–7, 430–1, 431*f*
   Y chromosome 424–5
Sex-lethal (Sxl) protein 428–9, 428*f*, 429*f*, 432
SHANK3 gene and product 560*b*
shark 697
shield in zebrafish 144*f*, 144, 145, 146*f*, 200–2
SHOOT MERISTEMLESS (*STM*) gene 637, 637*f*, 643

Short gastrulation (Sog) protein 72–3, 73f, 190, 694, 694f
short-germ development 716f, 716, 717b, 718f
SHORT HYPOCOTYL 2 (*SHY2*) gene 655, 655f
SHORTROOT (*SHR*) gene and products 654, 656
Shox gene 585
shrimp, body plan 689f
siamois gene 129f, 130, 146f, 190
sickle-cell anemia 344, 345, 346b
signal transduction 30, 31b
signaling center 129–30, 129f, 520f, 521
  see also organizer region
silent heart mutant 502
sine oculis gene 508
single-cell analysis of cell fate decisions 329–34, 330b, 333f
single-cell RNA sequencing (scRNA-seq) 109, 521, 599b
  amphibian limb regeneration 601, 604–5
  mouse digit tip regeneration 610
single-cell transcriptional analysis 109
single-minded gene and product 69, 69f
situs inversus 237
Six3 transcription factor 508
skin 346–8
Slit proteins 546, 550–1
slug gene and product 267
Smad signal proteins 139, 147, 190, 528–9
small interfering RNAs (siRNAs) 23, 649
small open reading frames (smORFs) 44
Smoothened protein 473b
snail gene and product 69, 69f, 70f, 233, 267, 271
snakes 693–4, 694f, 702–3, 704b, 704f
somatic cell nuclear transfer 359–61, 359f, 360f
somatic cells 5, 5f
  plants, regeneration potential 617–18, 618f
somatic transgenesis 111, 118
somatostatin 584f, 584
somites 105
  cell fate determination 229–31
  chick 150, 153f, 153, 154f, 217–21, 217f, 218f, 220f
  formation 216–21, 217f, 218f, 220f
    sequence of 217–19
  identity 221–6, 222f, 224f, 225f
  mouse 162f, 162
  neural crest cells 300f, 300
  patterning within 230
  Xenopus 199
  zebrafish 145
somitogenesis 383b, 383f, 384
Sonic hedgehog (Shh) 32b, 32f, 231, 231f, 239, 479, 697, 701, 702
  brain development 521, 522b

eye development and 507
  non-development in *Astyanax* 708
limb development and 449b, 462f, 462, 463f, 471–5, 473b, 473f, 479, 479f
limblessness 703
lung development and 495, 496
nervous system development 198, 520f, 523f, 527–8, 527f, 528f
primary cilium and 473b, 473f
spinal cord development 528–9, 529f
Sox2 transcription factor 163, 332, 376, 495f, 495
Sox9 transcription factor 475, 697
SoxE transcription factor 532
spalt gene and product 447–8, 447f, 448f
  butterfly wing markings 705, 705f, 706, 707
spatial transcriptomics 109
Spätzle protein 59, 60f, 60b
specification 27, 28f
  germ cells 403–5, 403f, 404f, 405f, 417
  neuronal identity 519–32
Spemann, Hans 8, 9f
Spemann organizer 8, 9f, 21, 29, 30, 120, 127f, 127, 129, 189–200
  body axis patterning 189–93
  chick and mouse equivalents 203–6, 204f
  genes expressed 206
  mesoderm induction and patterning 136–7, 136f, 137f
  neural induction and neural patterning 193–200
  proteins expressed 191–2
  transplantation study 139, 139f
sperm 417–20, 418f
  acrosome reaction 418–19, 418f
  capacitation 418
  see also fertilization
sperm aster 125–6, 125f
spermatogenesis 411f, 412
spider, body plan 689f
spina bifida 293b, 293f
spinal cord 212
  antero-posterior pattern 531, 531f
  cell differentiation 526–8, 527f, 529f
  chick 212
  dorso-ventral pattern 526–8, 527f, 529f
  motor neurons 527, 530–1, 531b
  mouse 212f, 212
  progenitor domains 528–30, 529f
  zebrafish 202, 202f
spinal muscular atrophy 389
spiracles 295, 684f
spleen 581
Splotch mutants 231
spotted gar 699, 699b, 699f, 700
Sprouty gene and product 295f, 296, 496
Squint protein 148
Sry gene 425, 425f

starfish 596, 596f
Staufen RNA-binding protein 66, 66f
stem cell embryo models (SEMs) 384–7, 386f
stem-cell niche 340
stem cells 2
  asymmetric division 36–7, 37f, 350–1, 350f
  blood cell formation 338–41, 339f
  division, models of 350–1, 350f
  embryonic see embryonic stem cells
  epiblast (EpiSCs) 371, 372f, 375, 375f
  epidermal 346–8, 347f, 349b
  extra-embryonic endoderm cells (XEN cells) 370
  induced pluripotent see induced pluripotent stem cells
  intestinal epithelia 351–3, 351f
  medical applications 387–95
  meristem 643–4, 644f
  multipotent 37, 234–5, 338–41, 339f
  muscle 353–5, 353f, 354f
  plant meristem 643–4, 644f
  pluripotent 37, 372–7, 373b
  regenerative medicine potential 375, 387–95, 388f, 391f, 392b, 392f, 393f, 394f, 395f
  trophoblast (TS cells) 370, 378, 380, 381f, 381
Steptoe, Patrick 172b
stickleback 707, 707f, 708, 709b, 709f
strigolactones 652f, 652–3
string gene and product 575–6, 575f
Strongylocentrotus purpuratus 10
  see also sea urchins
substantia nigra 523, 523f
subventricular zone 539
superior colliculus 554, 555f
SUPERMAN gene 662
Survivin protein 561
Sxl gene and product 435
symmetric cell division 260
synapses 518, 519, 556
  formation and refinement 556–65
    reciprocal interactions 556–9
  neural activity 562f, 562–4
synaptic cleft 556, 557f
synaptic plasticity 564
synaptic vesicles 556, 558f
syncytial blastoderm 45–6, 52, 53f, 56
syncytiotrophoblast 173–4, 174f
syncytium 45–6, 52

T

T-cell-specific transcription factors 201–2
T lymphocytes 338
tadpole
  loss of this stage 715, 716f
  Xenopus 594–5, 594f

tadpole shrimp 690
tailbud, *Xenopus* 106*f*, 122
tailless gene and product 73*f*, 75
Takifugu rubripes 688
TALENS (transcription activator-like effector nucleases) 111, 640*b*
tangential migration 540–1, 543*f*
target of rapamycin (TOR) pathway 573–4, 573*f*, 592–3, 618, 619*f*
Tbx genes and products 461–2, 461*f*, 475, 497, 503, 697, 702
Tbx6 expression 383*b*, 383*f*
TbxT gene 139, 139*f*, 140*f*, 140, 201, 280*f*, 280, 287*f*, 375, 376
 cell differentiation 332, 333, 334*f*, 334
 embryonic stem cells 379*b*, 379*f*, 380–1
 semi-dominant mutation, mouse 13*f*, 13, 14*f*
 somitogenesis 383*b*, 383*f*
TCF transcription factors 319
Tcf15 expression 383*b*, 383*f*
tectum 520, 521–2, 523, 552, 553*f*, 553
teeth 489–92, 490*f*, 491*f*
telencephalon 520, 523, 523*f*
telomeres 620
tendons 485–6
teratocarcinoma 371, 388
teratogens 32*b*, 475, 476*b*, 476*f*
teratoma 371, 373*f*, 375
terminal cell differentiation 313
testis 402, 425, 425*f*, 426
testosterone 328, 426, 498
tetraploid chimeric assay 369
TFIID transcription factor 337, 337*f*
thalamus 520, 521, 522, 555
thale-cress *see Arabidopsis thaliana*
thalidomide 475, 476*b*, 476*f*
Thick veins protein 448, 449*b*
thrombin 603
thymus gland 581
thyroid hormones 594–5, 594*f*
thyroid-stimulating hormone 594, 594*f*
tight junctions 250*b*, 261, 264*f*, 264, 265*f*
TILLING (targeting-induced local lesions in genomes) 116, 116*b*, 117*f*
tinman gene 503
tissue growth 573–4, 573*f*
toll gene and product 59, 60*f*
Toll signaling pathway 60*b*, 60*f*
tolloid gene and product 69*f*, 70, 72–3, 72*f*
tooth identity 489–92, 491*f*
topless-1 gene and product 637
topographic maps 552–5
 neural activity 562–4
topologically associating domains (TADs) 320–2, 321*f*, 325*f*, 336
TOR pathway 573, 592, 618, 619
Torpedo protein 65
torso gene and product 54, 54*f*, 58*f*, 58

Torso-like protein 58, 62
totipotency
 egg 37, 413
 plants 630
Townes–Brocks syndrome 475
trachea
 Drosophila 294*f*, 295–6, 295*f*
 vertebrates 294
transcription 21, 22, 22*f*, 315, 316
 control of 316–19
transcription factors 23–5, 26*b*, 26, 27*f*, 316–19, 326, 327*f*, 328*f*, 328
 blood cell differentiation 341–2, 342*f*
 embryonic stem cells 372–4
 eye development 507–8, 508*f*
 general 317–19
 nervous system development 520*f*, 522–3, 523*f*, 524*f*, 525–6, 528*f*, 528–9, 529*f*, 530*b*, 531–2, 533, 535*f*, 536–7, 542–3, 543*f*, 552
 single-cell transcriptional analysis 330–3, 330*b*
 tissue-specific 319
 Xenopus blastula 139, 139*f*
transcription initiation complex 317
transdetermination 357
transdifferentiation 335, 357–9, 358*f*, 389, 390, 391, 603
 replacement cell generation 390–1, 391*f*
transfection 115*b*
transfer RNA (tRNA) 44
transformer (*tra*) genes 428, 428*f*, 430, 430*f*
transforming growth factor-β (TGF-β) family 31*b*, 124*b*, 124*f*
 mesoderm induction 133*b*, 133*f*, 137
 neural development and 532
 receptors 138*b*
 see also bone morphogenetic proteins (BMP); Nodal protein
transgenic techniques 24, 111, 112, 114*f*, 114, 115*b*, 115*f*
 chick 113*f*, 113, 114*f*, 118
 Drosophila 76*b*
 mouse 112, 114*f*, 114, 115*b*, 115*f*
 plants 638–9, 639*f*
 Xenopus 114*f*, 118
 zebrafish 112, 114*f*
transient transgenesis 111, 118
transit-amplifying cells 350, 350*f*
translation 21, 22, 22*f*, 315–16
transplantation studies 111, 139, 140
 Hensen's node 205
 human organs grown in other animals 394–5
 Hydra hypostome region 600*f*, 600
 somatic cell nuclear transfer 359–61, 359*f*, 360*f*
 Spemann organizer 139
tribbles gene and product 576

Tribolium 716*f*, 716, 717*b*, 718*f*
Trichoplax 679–80, 681
trimethadione 476*b*
Triops 690
triploblasts 19*b*, 680*b*
trisomy 410–12
Trithorax gene family 93, 323*b*
trophectoderm 105, 159*f*, 159, 160*f*, 162–4, 163*f*, 164*f*, 173, 174
trophoblast stem cells (TS cells) 370, 378, 380, 381*f*, 381
tropospheres 378, 378*f*, 380
trunk organizer 192
Trunk protein 58
TRYPTYCHON protein 656
Tsix gene 434
Tulerpeton 696*f*, 697
tumor-suppressor genes and products 579–80, 579*f*
Turing mechanisms 484*b*
Turing patterns 379*b*
Turner syndrome 425, 585
Twine protein 576
twinning 10*f*, 10, 175, 176*b*, 176*f*, 176
twist gene and product 69, 69*f*, 70*f*, 271
Twisted gastrulation (Tsg) protein 72–3
type 1 diabetes 389–90

U

ulnar-mammary syndrome 475
Ultrabithorax (*Ubx*) gene and product 91–2, 92*f*, 93–4, 444, 456, 688, 688*f*, 690*f*, 690–1, 691*f*, 692
umbilical cord 174, 174*f*
Uncx4 expression 383*b*
uniform manifold and projection (UMAP) plots 330*b*, 330*f*, 332, 333*f*
UNUSUAL FLORAL ORGANS (UFO) gene and product 661, 661*f*
Urbilateria 679, 680*b*

V

valproic acid 476*b*
Van Gogh (Vang) protein 274*f*, 274, 275*f*
Vanessa cardui 706
vas deferens 426*f*, 426
vasa mRNA 405, 405*f*
vascular epithelial growth factor (VEGF) 268, 297–8, 297*f*
vascular system 296–8, 296*f*, 297*f*
vasculogenesis 296–7, 296*f*
vegetal pole 3*f*, 3*b*, 119, 123
VegT transcription factor 123, 129–30, 133*f*, 134
 mesoderm induction and patterning role 137, 137*f*, 138
veins 297
ventral closure 153, 303–4
ventralized embryos 59, 70, 70*f*

768 Index

ventricular zone 539
vernalization 665-7, 666f
vertebrae 684, 693, 693f
vertebrate limb development 458-88
   apical ectodermal ridge role 463-4, 464f
   axes 458f, 458
      positional information 466-70, 470f, 478
   cartilage, muscles and tendons 485-6
   chick wing as model 458f, 458-61, 460f, 461f
   digits 471-5, 472b
      polydactyly 471, 472b, 472f, 473b, 474, 482
   Hox gene roles 461-2, 461f, 479-82, 480f, 481f, 482f
   joint formation 486, 486f
   long bones 585-9, 587f, 588f
   patterning 466-70, 479-82, 480f, 481f, 482f
      integration 478-9, 479f
   muscle 483-5, 485f
      self-organization 482-3, 482f
   polarizing region 469-75, 470f, 479, 479f
      see also zone of polarizing activity
   position and type of limb 460-3, 461f, 462f, 463f
   programmed cell death role 487f, 487
   progress zone 466, 467f
   signal-progress-zone model 466-9, 467f
   Sonic hedgehog (Shh) role 449b, 462f, 462, 463f, 471-5, 473b, 473f, 479, 479f
   timing model 466-9, 467f
   two-signal model 467f, 468
vertebrates
   axial skeleton 693-4, 694f
   body plan 694-5, 695f
   eye 503-8
      structure 503-4, 503f
   heart development 499-503, 500f, 501f, 502f
   limb development see vertebrate limb development
   lung development 492-6
      endoderm 492-5, 494f
      morphogenesis 495-6
   planar cell polarity 273-4, 273f
vestigial gene 13f, 444, 446, 451-3, 451f, 452f
visual centers 552-5, 553f, 554f, 555f, 562-4, 563f
visual cortex 541b, 555, 563f, 563
vitamin A 476b
vitelline envelope, *Drosophila* 58
vitelline membrane
   Drosophila 58f, 58
   sea urchin 421, 421f
   Xenopus 119

W

Waddington, Conrad 209, 313b
Warts (Wts) protein kinase 576-7, 577b
Weismann, August 5, 6, 6f, 8
WEREWOLF protein 656
Werner syndrome 619
whale 701, 701f, 702, 713
white matter 539
Wieschaus, Eric 49
wing development
   butterfly 691-2
      wing markings 703-7, 703f, 705f, 706f
   chick 701
      see also vertebrate limb development
   Drosophila 442-54, 576-8, 578f, 658-9, 660b
      antero-posterior axis 446-50, 447f, 448f, 449f
      axis establishment 446-51, 453-4
      compartment boundaries 444-50
      metamorphosis 445-6, 445f, 446f
      pattern formation 446-51, 453-4
      proximo-distal axis patterning 453-4
      vestigial gene 444, 451-3, 451f, 452f
   evolutionary comparisons 701
   see also Wnt protein signal
wingless gene and product 84-5, 85f, 86f, 86-9, 87b, 88f, 89f, 124f, 446, 451f, 451, 452, 453, 455, 455f, 704
Wnt-1 520f, 522, 523
Wnt protein signaling 31b, 84, 123, 124b, 190f, 190, 191, 191f, 192f, 192, 201-2, 370, 375, 376, 528-9, 681
   antero-posterior gradient 204
   cancer and 580
   cell differentiation and 319, 352-3, 354
      single-cell transcriptional analysis 332
   convergent extension role 278b, 290
   gastruloids 379b, 379f, 382f, 382
   Hydra 601, 601f
   limb bud outgrowth 465-6
   limb development 478f, 478, 486, 495f, 495, 498
   mammary gland development 497
   mesoderm induction and patterning 191-2, 192f
   planarians 598b
   somite patterning 231
   trophectoderm 380
   see also canonical Wnt/β-catenin pathway; *wingless* gene and product
Wolffian ducts 426f, 426-7
WOX1 gene 649, 649f
WOX2 gene 637
WOX8 gene 637
wunen gene 408
WUSCHEL (*WUS*) gene 643, 644f, 661, 661f

X

X chromosomes 424, 424f, 425f, 425, 427-30, 428f, 429f, 430f
   X inactivation 433-5, 433f
X-linked genes 433-5
Xenopus 2f, 2, 10, 11f, 12, 118-42
   antero-posterior axis 213
   axon navigation 549
   blastula 105, 107
   body axes 122-31
   cleavage 17, 18f
   cytoplasmic determinants 36
   developmental stages 2, 3f, 3b
   dorso-ventral axis 125f, 125-9, 189-90, 190f, 215, 694, 695f, 695
   egg 103, 119f, 122, 123
   egg cylinder 105
   embryo development 104f, 107
      phylotypic stage 106f, 106
   embryonic stem cells 390
   eye 508
   fate map 110b, 110f, 113, 131f, 132
   fertilization 103, 123, 125, 125f, 126, 420, 422-3, 423f
   gastrulation 120-2, 121f, 140, 189, 191-3, 193f, 199, 199f, 276f, 276-80, 277f, 278b, 278f, 279f, 281-2, 282f, 285-6, 286f, 289
   germ cell development 405, 407
   germ layers 132-42
      convergent extension 280
      mesoderm induction and patterning 132-40, 134f, 136f, 137f
   heart 500
   imaging 112, 113
   left-right asymmetry 238, 239-40
   life cycle 119f
   metamorphosis 594-5, 594f
   microsurgical manipulation 112, 112f
   mid-blastula transition 130f, 130
   mutagenesis screen 114
   neural crest migration 299f
   neural induction 193-200
      default model 195-6
   neurulation 106f, 106, 122, 199, 199f
   notochord 281-2, 282f
      directed dilation 282
   organizer region see Spemann organizer
   organogenesis 122
   regulation 140-1
   somatic cell nuclear transfer 359-60, 359f, 360f
   somite formation 199, 217
   Spemann organizer see Spemann organizer
   transgenic technique 114f, 118
   transient transgenesis studies 118
Xenopus Nodal-related (*Xnr*) genes and products 133b, 137, 137f

mesoderm induction and patterning role 137
Xist gene 434
xol-1 gene 429–30, 430*f*, 435

Y

Y chromosome 424–5, 424*f*, 425*f*
YABBY (*YAB*) genes 649, 649*f*
Yamanaka, Shinya 23, 374
Yap/Taz transcriptional effector 263
Yap transcriptional co-activator 163
yolk sac 153, 154*f*, 171*f*, 171, 332
Yorkie (Yki) transcription factor 576, 577*b*, 577, 578*f*, 578

Z

zebrafish 10, 11*f*, 11, 12, 142–8
  antero-posterior axis 213
  body axes 145
  brain development 525
  cleavage 103, 142, 143*f*
  CRISPR-Cas9 system 15*b*, 118
  dorso-ventral axis 215
  egg 103, 142
  embryo development 104*f*, 107
  embryonic vascular system 294*f*
  fate map 113, 146*f*, 146–7, 201*f*
  fin development 458, 697–700, 698*f*, 699*b*, 699*f*, 700*f*, 700
  gastrulation 144*f*, 144, 280, 281*f*, 285–6, 286*f*, 289
  gene regulatory network 203*b*, 203*f*
  germ cell development 405, 405*f*, 407–8, 407*f*
  germ layers 146–8, 146*f*, 147*f*
  heart 499, 500, 502
    regeneration 596, 614–15, 614*f*
  Hox genes 222, 223*b*, 688, 698, 699*b*, 699*f*, 700*f*, 700
  imaging 112
  lateral-line primordium 302*f*
  left–right asymmetry 238, 239
  life cycle 143*f*
  mesoderm induction and patterning 147
  microsurgical manipulation 112*f*, 113
  mid-blastula transition 146*f*, 146
  mutagenesis screen 114, 116, 116*b*, 117*f*, 147
  neural crest migration 299, 299*f*
  neurulation 145
  somite formation 217
  transgenic techniques 112, 114*f*
zerknüllt gene and product 69*f*, 70*f*, 70, 71
Zika virus 476*b*
zinc-finger nucleases (ZFNs) 111
zona limitans intrathalamica (ZLI) 520*f*, 521
zona pellucida 158, 418–19, 419*f*
zone of polarizing activity (ZPA) 462*f*, 462, 469–75, 470*f*, 479, 479*f*
zone of polarizing activity regulatory sequence (ZRS) 472*b*, 703, 704*b*, 704*f*
zygote 5
zygotic genes
  Drosophila 53–4, 53*f*, 68–77
  timing of expression, *Xenopus* 135*b*, 135*f*